PHYSICS

Mechanics II

B.M. Sharma

Andover • Melbourne • Mexico City • Stamford, CT • Toronto • Hong Kong • New Delhi • Seoul • Singapore • Tokyo

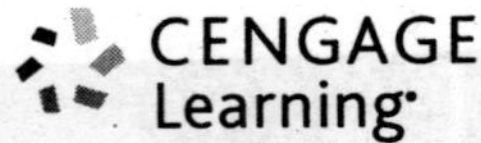

Physics: Mechanics II

B.M. Sharma

For permission to use material from this text or product, email your request to
www.cengage.com/permissions

ISBN-13: 978-81-315-1920-2
ISBN-10: 81-315-1920-1

Cengage Learning India Pvt. Ltd
418, F.I.E., Patparganj
Delhi 110092

Cengage Learning is a leading provider of customized learning solutions with office locations around the globe, including Andover, Melbourne, Mexico City, Stamford (CT), Toronto, Hong Kong, New Delhi, Seoul, Singapore, Tokyo. Locate your local office at: **www.cengage.com/global**

Cengage Learning products are represented in Canada by Nelson Education, Ltd.

For product information, **visit www.cengage.co.in**

Printed in India
Third Impression 2013

Brief Contents

Contents

Chapter 5 Gravitation 5.1

Preface

While the paper-setting pattern and assessment methodology have been revised many times over and newer criteria devised to help develop more aspirant-friendly engineering entrance tests, the need to standardize the selection processes and their outcomes at the national level has always been felt. The Joint Entrance Examination (JEE) to India's prestigious engineering institutions (IITs, IIITs, NITs, ISM, IISERs, and other engineering colleges) aims to serve as a common national-level engineering entrance test, thereby eliminating the need for aspiring engineers to sit through multiple entrance tests.

While the methodology and scope of an engineering entrance test are prone to change, there are two basic objectives that any test needs to serve:

1. The objective to test an aspirant's caliber, aptitude, and attitude for the engineering field and profession.
2. The need to test an aspirant's grasp and understanding of the concepts of the subjects of study and their applicability at the grassroot level.

Students appearing for various engineering entrance examinations cannot bank solely on conventional shortcut measures to crack the entrance examination. Conventional techniques alone are not enough as most of the questions asked in the examination are based on concepts rather than on just formulae. Hence, it is necessary for students appearing for joint entrance examination to not only gain a thorough knowledge and understanding of the concepts but also develop problem-solving skills to be able to relate their understanding of the subject to real-life applications based on these concepts.

This series of books is designed to help students to get an all-round grasp of the subject so as to be able to make its useful application in all its contexts. It uses a right mix of fundamental principles and concepts, illustrations which highlight the application of these concepts, and exercises for practice. The objective of each book in this series is to help students develop their problem-solving skills/accuracy, the ability to reach the crux of the matter, and the speed to get answers in limited time. These books feature all types of problems asked in the examination—be it MCQs (one or more than one correct), assertion-reason type, matching column type, comprehension type, or integer type questions. These problems have skillfully been set to help students develop a sound problem-solving methodology.

Not discounting the need for skilled and guided practice, the material in the books has been enriched with a number of fully solved concept application exercises so that every step in learning is ensured for the understanding and application of the subject. This whole series of books adopts a multi-facetted approach to mastering concepts by including a variety of exercises asked in the examination. A mix of questions helps stimulate and strengthen multi-dimensional problem-solving skills in an aspirant.

It is imperative to note that this book would be as profound and useful as you want it to be. Therefore, in order to get maximum benefit from this book, we recommend the following study plan for each chapter.

Step 1: Go through the entire opening discussion about the fundamentals and concepts.
Step 2: After learning the theory/concept, follow the illustrative examples to get an understanding of the theory/concept.

Overall the whole content of the book is an amalgamation of the theme of physics with ahead-of-time problems, which equips the students with the knowledge of the field and paves a confident path for them to accomplish success in the JEE.

With best wishes!

B.M. Sharma

CHAPTER

1

Centre of Mass, Conservation of Linear Momentum and Collision

THE CENTRE OF MASS

In this section, we will discuss the overall motion of a system of particles in terms of a very special point called **the centre of mass** of the system. This notion gives us confidence in the particle model because we will see that the centre of mass accelerates as if all the system's mass were concentrated at that point and all external forces act there.

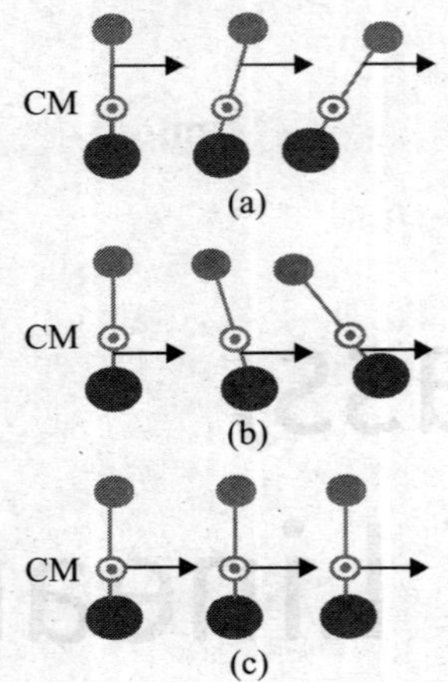

Fig. 1.1

Consider a system consisting of a pair of particles connected by a light rigid rod. The centre of mass as indicated in Fig. 1.1 is located on the rod and is closer to the larger mass in the figure. If a single force is applied at some point on the rod above the centre of mass, the system rotates clockwise as it translates through space [Fig. 1.1(a)]. If the force is applied at a point on the rod below the centre of mass, the system rotates counterclockwise [Fig. 1.1(b)]. If the force is applied exactly at the centre of mass, the system moves in the direction of $\vec{F}$ without rotating [Fig. 1.1(c)] as if the system is behaving as a particle. Therefore, in theory, the centre of mass can be located with this experiment.

If we were to analyse the motion [Fig. 1.1(c)], we would find that the system moves as if all its mass were concentrated at the centre of mass. Furthermore, if the external net force on the system is $\Sigma\vec{F}$ and the total mass of the system is M, the centre of mass moves with an acceleration given by $\vec{a} = \Sigma\vec{F}/M$.

Centre of Mass of a System of 'N' Discrete Particles

Consider a system of N point masses $m_1, m_2, m_3, \ldots, m_n$ whose position vectors from origin O are given by $\vec{r}_1, \vec{r}_2, \vec{r}_3, \ldots, \vec{r}_n$, respectively.

Then, the position vector of the centre of mass C of the system is given by

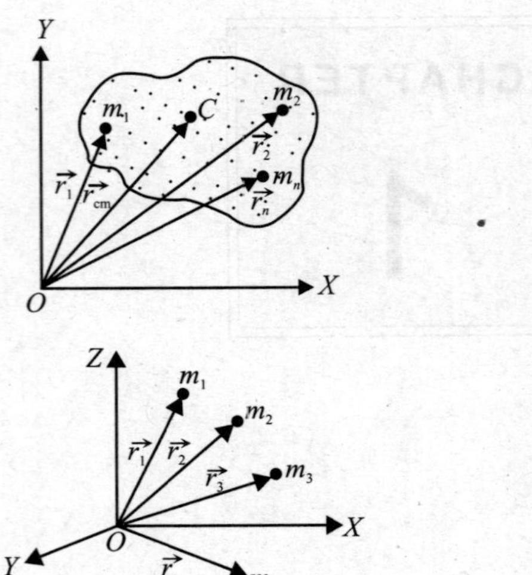

Fig. 1.2

$$\vec{r}_{CM} = \frac{m_1\vec{r}_1 + m_2\vec{r}_2 + \cdots + m_n\vec{r}_n}{m_1 + m_2 + \cdots + m_n}$$

$$= \frac{\Sigma_{i=1}^{n} m_i\vec{r}_i}{\Sigma_{i=1}^{n} m_i}$$

$$= \frac{1}{M}\Sigma_{i=1}^{n} m_i\vec{r}$$

where $M(=\Sigma_{i=1}^{n} m_i)$ is the total mass of the system.

Consider a system of point masses $m_1, m_2, m_3, \ldots$ located at the coordinates $(x_1, y_1, z_1), (x_2, y_2, z_2), \ldots$, respectively. The centre of mass of this system of masses is a point whose coordinates are (x_{cm}, y_{cm}, z_{cm}), which are given by

$$x_{cm} = \frac{m_1x_1 + m_2x_2 + \cdots}{m_1 + m_2 + \cdots}, \quad y_{cm} = \frac{m_1y_1 + m_2y_2 + \cdots}{m_1 + m_2 + \cdots}$$

$$z_{cm} = \frac{m_1z_1 + m_2z_2 + \cdots}{m_1 + m_2 + \cdots}$$

Fig. 1.3

Illustration 1.1 **Four particles of masses 1 kg, 2 kg, 3 kg and 4 kg are placed at the four vertices A, B, C and D, respectively, of a square of side 1 m. Find the position of centre of mass of the particles.**

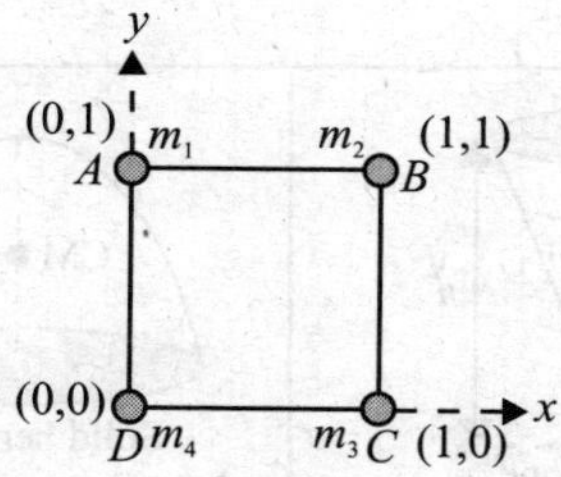

Fig. 1.4

Sol. Assuming D as the origin, DC as x-axis and DA as y-axis, we have

$$m_1 = 1\text{ kg}, (x_1, y_1) = (0, 1\text{ m})$$
$$m_2 = 2\text{ kg}, (x_2, y_2) = (1\text{ m}, 1\text{ m})$$
$$m_3 = 3\text{ kg}, (x_3, y_3) = (1\text{ m}, 0)$$
$$m_4 = 4\text{ kg}, (x_4, y_4) = (0, 0)$$

Coordinates of their CM are calculated as follows.

$$x_{CM} = \frac{m_1x_1 + m_2x_2 + m_3x_3 + m_4x_4}{m_1 + m_2 + m_3 \quad m_4}$$

$$= \frac{(1)(0) + 2(1) + 3(1) + 4(0)}{1 + 2 + 3 + 4} = \frac{5}{10} = \frac{1}{2}\text{ m} = 0.5\text{ m}$$

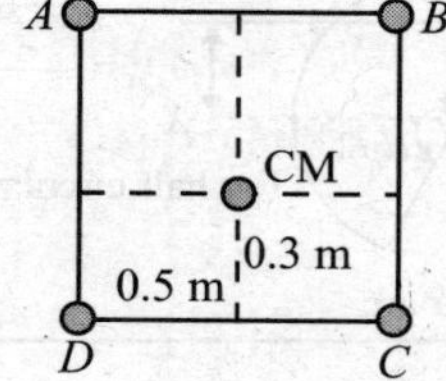

Fig. 1.5

Similarly, $y_{CM} = \dfrac{m_1y_1 + m_2y_2 + m_3y_3 + m_4y_4}{m_1 + m_2 + m_3 + m_4}$

$$= \frac{(1)(1) + 2(1) + 3(0) + 4(0)}{1 + 2 + 3 + 4} = \frac{3}{10}\text{ m}$$

$$= 0.3\text{ m}$$

$\therefore \quad (x_{CM}, y_{CM}) = (0.5\text{ m}, 0.3\text{ m})$

Illustration 1.2 **Consider a two-particle system with the particles having masses m_1 and m_2. If the first particle is pushed towards the centre of mass through a distance d, by what distance should the second particle be moved so as to keep the centre of mass at the same position?**

Sol. Consider Fig. 1.6. Suppose the distance of m_1 from the centre of mass C is x_1 and that of m_2 from C is x_2. Suppose mass m_2 is moved through a distance d' towards C so as to keep the centre of mass at C.

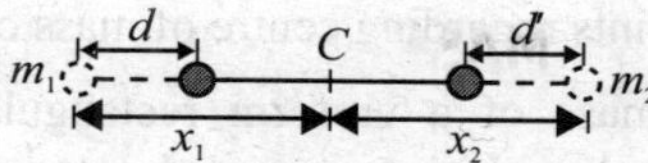

Fig. 1.6

Then, $\quad m_1x_1 = m_2x_2 \quad$ (i)

and $\quad m_1(x_1 - d) = m_2(x_2 - d') \quad$ (ii)

Subtracting Eq. (ii) from Eq. (i), $m_1d = m_2d'$ or $d' = (m_1/m_2)\,d$.

Centre of Mass of a Continuous Mass Distribution

For a continuous mass distribution, the centre of mass can be located by replacing summation sign with an integral sign. Proper limits for the integral are chosen according to the situation.

$$x_{CM} = \frac{\int x\,dm}{\int dm}, y_{CM} = \frac{\int y\,dm}{\int dm}, z_{CM} = \frac{\int z\,dm}{\int dm}$$

$$\int dm = M \text{ (mass of the body)}$$

$$\vec{r}_{CM} = \frac{1}{M}\int \vec{r}\,dm$$

Note: ***If an object has symmetric uniform mass distribution about x-axis, then y-coordinate of CM is zero and vice versa.***

Centre of Mass of a Uniform Rod

Suppose a rod of mass M and length L is lying along the x-axis with its one end at $x = 0$ and the other at $x = L$.

$$\text{Mass per unit length of the rod} = \frac{M}{L}$$

Hence, the mass of the element dx situated at $x = x$ is

$$dm = \frac{M}{L}\,dx$$

→dx←
P Q
X=0 X=X X=L

Fig. 1.7

The coordinates of the element PQ are $(x, 0, 0)$. Therefore, x-coordinate of CM of the rod will be

$$x_{CM} = \frac{\int_0^L x\,dm}{\int dm} = \frac{\int_0^L x\frac{M}{L}dx}{M} = \frac{1}{L}\int_0^L x\,dx = \frac{L}{2}$$

The y-coordinate of CM is $y_{CM} = \dfrac{\int y\,dm}{\int dm} = 0$

Similarly, $\quad z_{CM} = 0$.

i.e., the coordinates of CM of the rod are $(L/2, 0, 0)$, or it lies at the centre of the rod.

The following points regarding centre of mass can be noted.

1. Centre of mass of a uniform rectangular, square or circular plate lies at its geometrical centre.

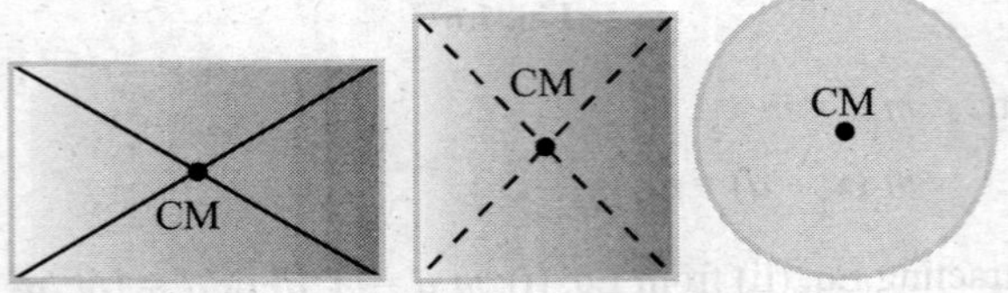

Fig. 1.8

2. For a lamina type (two-dimensional) body with uniform negligible thickness, the formulae for finding the position of centre of mass are as follows:

$$\vec{r}_{CM} = \frac{m_1\vec{r}_1 + m_2\vec{r}_2 + \cdots}{m_1 + m_2 + \cdots}$$

$$= \frac{\rho A_1 t\vec{r}_1 + \rho A_2 t\vec{r}_2 + \cdots}{\rho A_1 t + \rho A_2 t + \cdots}, (m = \rho A t)$$

$$= \frac{A_1\vec{r}_1 + A_2\vec{r}_2 + \cdots}{A_1 + A_2 + \cdots}$$

Here, A stands for area and r for density.

3. If some mass is removed from a rigid body, then the position of centre of mass of the remaining portion is obtained from the following formulae:

(i) $\vec{r}_{CM} = \dfrac{m_1\vec{r}_1 - m_2\vec{r}_2}{m_1 - m_2}$ or $\vec{r}_{CM} = \dfrac{A_1\vec{r}_1 - A_2\vec{r}_2}{A_1 - A_2}$

(ii) $x_{CM} = \dfrac{m_1x_1 - m_2x_2}{m_1 - m_2}$ or $x_{CM} = \dfrac{A_1x_1 - A_2x_2}{A_1 - A_2}$

(iii) $y_{CM} = \dfrac{m_1y_1 - m_2y_2}{m_1 - m_2}$ or $y_{CM} = \dfrac{A_1y_1 - A_2y_2}{A_1 - A_2}$

(iv) $z_{CM} = \dfrac{m_1z_1 - m_2z_2}{m_1 - m_2}$ or $z_{CM} = \dfrac{A_1z_1 - A_2z_2}{A_1 - A_2}$

Here, $m_1, A_1, \vec{r}_1, x_1, y_1$ and z_1 are the values for the whole mass while $m_2, A_2, \vec{r}_2, \vec{x}_2, y_2$ and z_2 are the values for the mass which has been removed.

Note: ***We can imagine a rigid body as a system of masses and hence every rigid body has a centre of mass. In case of a regularly shaped uniform rigid body, centre of mass is simply the geometric centre of the body. The centre of mass of some continuous geometrical figures of uniform mass density is given in Fig. 1.9.***

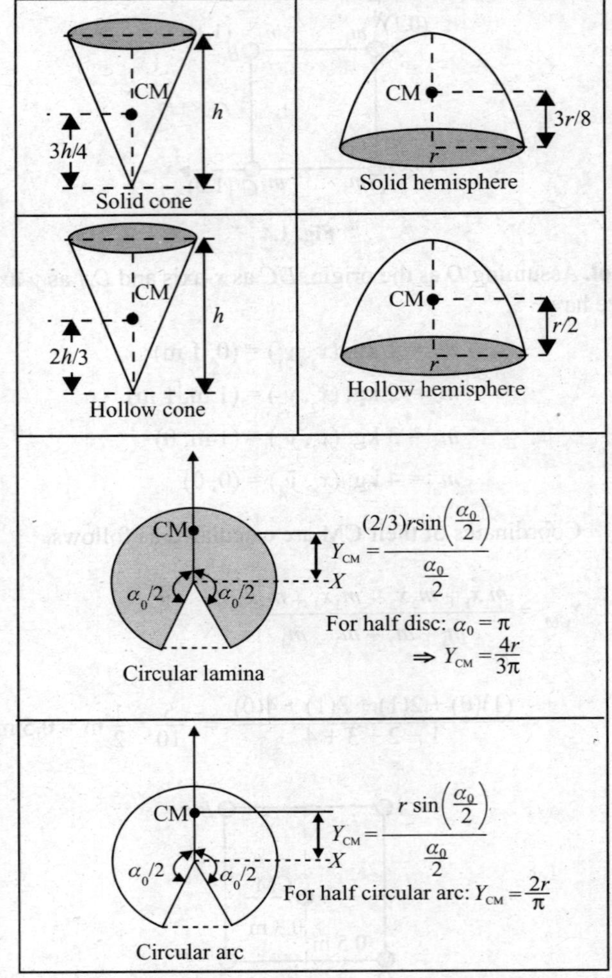

Fig. 1.9

Illustration 1.3 **Find the centre of mass of a uniform L-shaped lamina (a thin flat plate) with dimensions as shown in Fig. 1.10. The mass of lamina is 3 kg.**

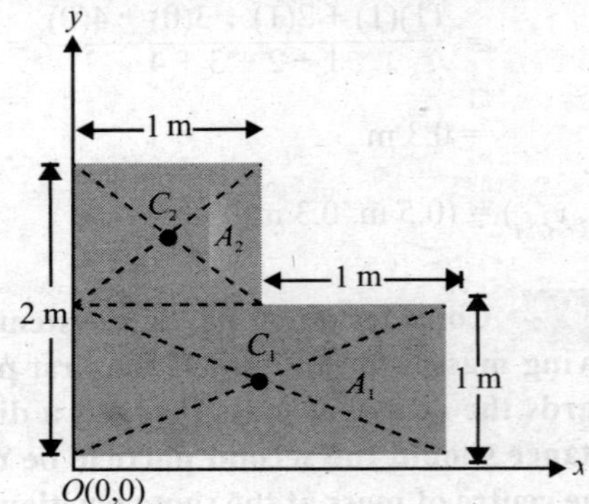

Fig. 1.10

Sol. The plate has uniform density and same thickness everywhere. So its CM will coincide with the centroid.

Divide the given plate into two parts of area A_1 and A_2 as shown in the figure. We have

$A_1 = 2 \times 1\ \text{m}^2$ with its centroid C_1 (1, 1/2) and

$A_2 = 1 \times 1\ \text{m}^2$ with its centroid C_2 (1/2, 3/2)

The centroid of the whole plate can be defined as

$$\bar{x} = \frac{A_1 x_1 + A_2 x_2}{A_1 + A_2} = \frac{2 \times 1 + 1 \times \frac{1}{2}}{2+1} = \frac{5}{6}\ \text{m}$$

$$\bar{y} = \frac{A_1 y_1 + A_2 y_2}{A_1 + A_2} = \frac{2 \times \frac{1}{2} + 1 \times \frac{3}{2}}{2+1} = \frac{5}{6}\ \text{m}$$

Illustration 1.4 **Find the position of centre of mass of the uniform lamina shown in Fig. 1.11.**

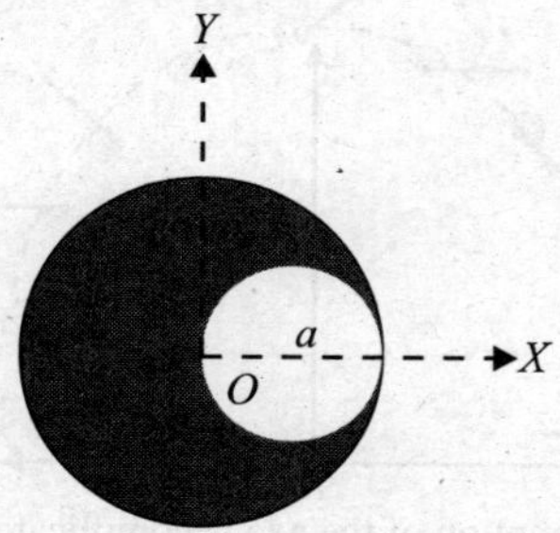

Fig. 1.11

Sol. Here, A_1 = area of complete circle $= \pi a^2$

A_2 = area of small circle $= \pi\left(\frac{a}{2}\right)^2 = \frac{\pi a^2}{4}$

(x_1, y_1) = coordinates of centre of mass of the large circle = (0, 0)

(x_2, y_2) = coordinates of centre of mass of the small circle $= \left(\frac{a}{2}, 0\right)$

Using $x_{\text{CM}} = (A_1 x_1 - A_2 x_2)/(A_1 - A_2)$, we get

$$x_{\text{CM}} = \frac{\pi a^2 \times 0 - \frac{\pi a^2}{4} \times \frac{a}{2}}{\pi a^2 - \frac{\pi a^2}{4}} = -\frac{a}{6}$$

$y_{\text{CM}} = 0$ (as y_1 and y_2 both are zero)

Therefore, coordinates of CM of the lamina shown in Fig. 1.11 are $(-a/6, 0)$.

Illustration 1.5 **Figure 1.12 shows a uniform disc of radius R, from which a hole of radius $R/2$ has been cut out from left of the centre and is placed on the right of the centre of the disc. Find the CM of the resulting disc.**

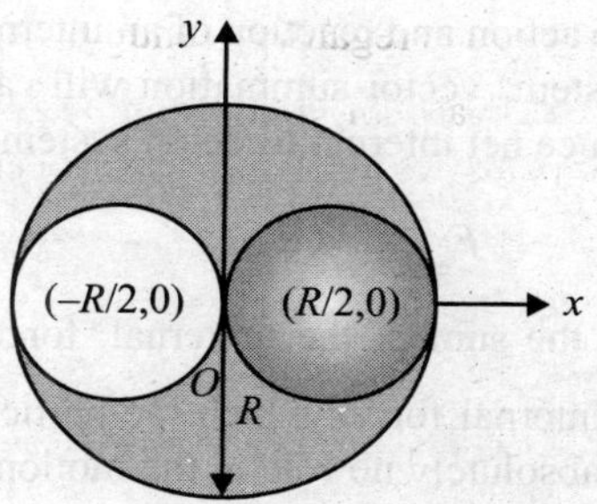

Fig. 1.12

Sol. Mass of the cut-out disc is

$$m = \frac{M}{\pi R^2} \times \pi\left(\frac{R}{2}\right)^2 = \frac{M}{4}$$

Let centre of the disc is at the origin of the coordinates.

Then we can write the CM of the system as

$$x_{\text{CM}} = \frac{M\vec{R} - m\vec{r} + m\vec{r}}{M - m + m} = \frac{M \times 0 - \frac{M}{4}\left(\frac{-R}{2}\right) + \frac{M}{4}\left(\frac{R}{2}\right)}{M - \frac{M}{4} + \frac{M}{4}} = \frac{R}{4}$$

$$y_{\text{CM}} = 0$$

MOTION OF THE CENTRE OF MASS

Motion of Centre of Mass and Conservation of Momentum: Velocity of Centre of Mass of System

$$\vec{v}_{\text{CM}} = \frac{m_1 \frac{d\vec{r}_1}{dt} + m_2 \frac{d\vec{r}_2}{dt} + m_3 \frac{d\vec{r}_3}{dt} + \cdots + m_n \frac{d\vec{r}_n}{dt}}{M}$$

$$= \frac{m_1 \vec{v}_1 + m_2 \vec{v}_2 + \cdots}{m_1 + m_2 + \cdots} \Rightarrow \vec{v}_{\text{CM}} = \frac{\vec{P}_{\text{total}}}{M_{\text{total}}}$$

Here numerator of the right-hand-side term is the total momentum of the system, i.e., summation of momentum of the individual components (particles) of the system.

Hence, velocity of centre of mass of the system is the ratio of momentum of the system per unit mass of the system.

Acceleration of Centre of Mass of System

$$\vec{a}_{\text{CM}} = \frac{m_1 \frac{d\vec{v}_1}{dt} + m_2 \frac{d\vec{v}_2}{dt} + m_3 \frac{d\vec{v}_3}{dt} + \cdots + m_n \frac{d\vec{v}_n}{dt}}{M}$$

$$= \frac{m_1 \vec{a}_1 + m_2 \vec{a}_2 + m_3 \vec{a}_3 + \cdots + m_n \vec{a}_n}{M}$$

$$= \frac{\vec{F}_{\text{ext}}}{M} = \frac{\text{Net external force}}{M}$$

(Both the action and reaction of an internal force must be within the system. Vector summation will cancel all internal forces and hence net internal force on system is zero.)

$$\therefore \qquad \vec{F}_{ext} = M\vec{a}_{CM}$$

where $\vec{F}_{ext}$ is the sum of the 'external' forces acting on the system. The internal forces which the particles exert on one another play absolutely no role in the motion of the centre of mass.

If no external force is acting on a system of particles, the acceleration of centre of mass of the system will be zero. If $a_{CM} = 0$, it implies that v_{CM} must be a constant and if v_{CM} is a constant, it implies that the total momentum of the system must remain constant. It leads to the principle that the total momentum of the system must remain constant. It leads to the principle of conservation of momentum in the absence of external forces.

If $\vec{F}_{ext} = 0$, then $\vec{P}_{total}$ = constant.

If no external force is acting on the system, net momentum of the system must remain constant.

Centre of Mass at Rest

If $F_{ext} = 0$ and $v_{CM} = 0$, then centre of mass remains at rest. Individual components of a system may move and have non-zero momentum due to mutual forces (internal), but the net momentum of the system remains zero.

Some examples:

- All the particles of the system are at rest.
- Particles are moving such that their net momentum is zero. Example: Net momentum $= 2m \times 2 + m(-4) = 0$

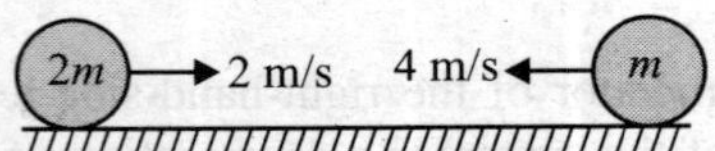

Fig. 1.13

- A bomb at rest suddenly explodes into various smaller fragments, all moving in different directions. Since the explosive forces are internal and there is no external force on the system for explosion, the centre of mass of the bomb will remain at the original position and the fragments fly such that their net momentum remains zero.
- Two men standing on a frictionless platform push each other, then also their net momentum remains zero because the push forces are internal for the two-man system.
- A boat floating in a lake also has the net momentum zero if the people on it change their positions, because the friction force required to move the people is internal of the boat system.
- Objects, initially at rest, if moving under mutual forces (electrostatic or gravitation) also have net momentum zero.
- A light spring of spring constant k is kept compressed between two blocks of masses m_1 and m_2 on a smooth horizontal surface. When released, the blocks acquire velocities in opposite directions, such that the net momentum is zero.
- **Centre of mass in motion:** An axe is thrown at some angle with horizontal. Centre of mass of axe will move in such a way that if all the *mg* force acts on itself. So centre of mass will move in a parabolic path, however the whole motion of the axe will be complicated.

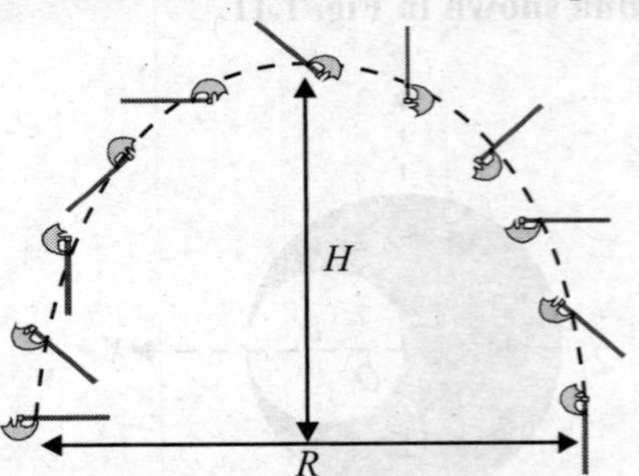

The motion of the axe is complicated but the CM is moving in a parabolic motion.

Fig. 1.14

Illustration 1.6 **A projectile is fired at a speed of 100 m/s at an angle of 37° above the horizontal. At the highest point, the projectile breaks into two parts of mass ratio 1:3, the smaller coming to rest. Find the distance from the launching point to the point where the heavier piece lands.**

Sol. Internal forces do not affect the motion of the centre of mass, the centre of mass hits the ground at a position where the original projectile would have landed. The range of the original projectile is

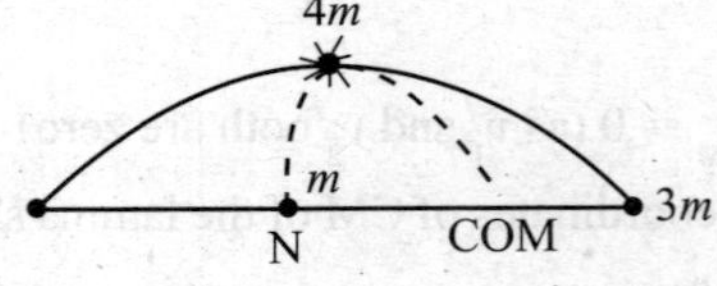

Fig. 1.15

$$x_{CM} = \frac{2u^2 \sin\theta \cos\theta}{g} = \frac{2 \times 10^4 \times \frac{3}{5} \times \frac{4}{5}}{10} \text{ m} = 960 \text{ m}$$

The centre of mass will hit the ground at this position. As the smaller block comes to rest after breaking, it falls down vertically and hits the ground at half of the range, i.e., at $x = 480$ m. If the heavier block hits the ground at x_2, then

$$x_{CM} = \frac{m_1x_1 + m_2x_2}{m_1 + m_2} \Rightarrow 960 = \frac{(m)(480) + (3m)(x_2)}{(m + 3m)}$$

$\therefore\ x_2 = 1120$ m

Illustration 1.7 **Two balls with masses $m_1 = 3$ kg and $m_2 = 5$ kg have initial velocities $v_1 = v_2 = 5$ m/s in the directions shown in Fig. 1.16. They collide at the origin.**

(a) Find the velocity of the CM 3 s before the collision.

(b) Find the position of the CM 2 s after the collision.

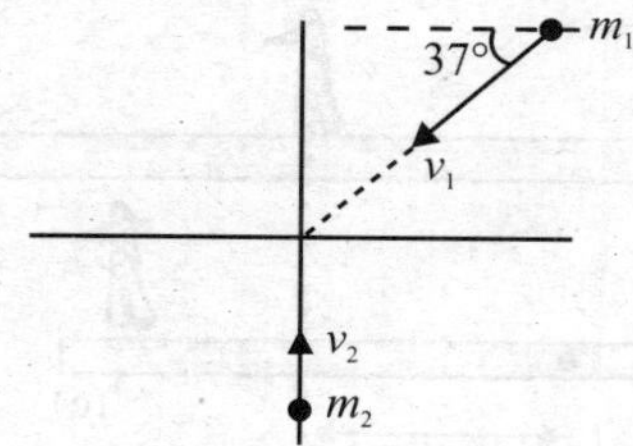

Fig. 1.16

Sol. (a) The given time is of no consequence since v_{CM} is fixed for all times.

$$v_{CM}(x) = \frac{m_1v_{1x} + m_2v_{2x}}{m_1 + m_2}$$

$$= \frac{(3)(-5\cos 37°) + (5)(0)}{8 \text{ kg}} = -1.5 \text{ m/s}$$

$$v_{CM}(y) = \frac{m_1v_{1y} + m_2v_{2y}}{m_1 + m_2}$$

$$= \frac{(3)(-5\sin 37°) + (5 \times 5)}{8 \text{ kg}} = +2 \text{ m/s}$$

$$\therefore \quad \vec{v}_{CM} = -1.5\hat{i} + 2\hat{j} \text{ m/s}$$

(b) Since the collision occurs at the origin ($\vec{r}_1 = \vec{0}$), the position of the CM 2 s later is $\vec{r}_{CM} = \vec{r}_1 + \vec{v}_{CM}t = \vec{v}_{CM}t = -3\hat{i} + 4\hat{j}$ m

Illustration 1.8 **Two particles of masses 2 kg and 4 kg are approaching towards each other with accelerations 1 m/s^2 and 2 m/s^2, respectively, on a smooth horizontal surface. Then find the acceleration of centre of mass of the system and direction of acceleration of CM.**

Sol: The acceleration of CM of the system is given by

$$a_{CM} = \frac{m_1a_1 + m_2a_2}{m_2 + m_2} \Rightarrow a_{CM} = \frac{2\times1 + 4(-2)}{2+4} = -1 \text{ m/s}^2$$

Negative sign indicates that acceleration of CM will be in the direction of acceleration of 4 kg mass.

Illustration 1.9 **A pulley fixed to the ceiling carried a thread with bodies of masses m_1 and m_2 attached to its ends. The masses of the pulley and the thread are negligible and friction is absent. Find the acceleration of the centre of mass of this system.**

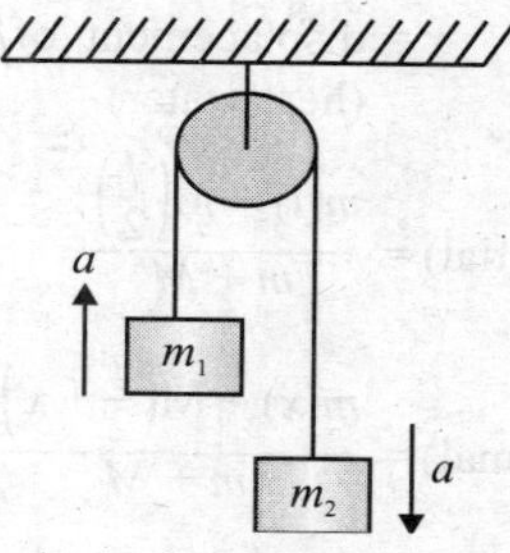

Fig. 1.17

Sol. Let as assume that $m_2 > m_1$. We can see that the masses have equal and opposite acceleration of the same magnitude.

$$a = \frac{m_2 - m_1}{m_2 + m_1} g \text{ and tension in string is } T = \frac{2m_1m_2g}{m_2 + m_1}$$

Taking downward direction as positive, $a_1 = -a$; $a_2 = +a$

$$a_{CM} = \frac{m_1a_1 + m_2a_2}{m_1 + m_2} = \frac{(m_2 - m_1)\,a}{m_2 + m_1}$$

Substituting for the value of a, we have

$$a_{CM} = \left(\frac{m_2 - m_1}{m_2 + m_1}\right)^2 g$$

Alternative method:

$$a_{CM} = \frac{f_{ext}}{m_1 + m_2} = \frac{(m_1g + m_2g) - 2T}{m_1 + m_2} = g - \frac{2T}{m_1 + m_2}$$

$$= g - \frac{4m_1m_2g}{(m_2 + m_1)^2} = \left(\frac{m_2 - m_1}{m_2 + m_1}\right)^2 g \text{ (downwards)}$$

Illustration 1.10 **A log of wood of length l and mass M is floating on the surface of a river perpendicular to the banks. One end of the log touches the banks. A man of mass m standing at the other end walks towards the bank. Calculate the displacement of the log when he reaches the nearer end of the log.**

Sol. Let PQ be the log of wood. As there is no external force, the centre of mass of man and the log system remains at rest. Let the bank of the river be the origin A. Initially, the man is at point Q.

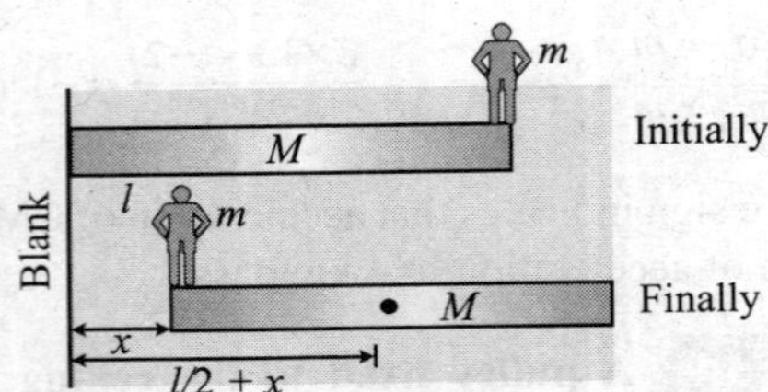

Fig. 1.18

Let m = mass of man,

M = mass of log,

x = displacement of log w.r.t. ground (here water)

$$X_{CM}\text{ (initial)} = \frac{m(l) + M\left(\frac{l}{2}\right)}{m+M}$$

$$X_{CM}\text{ (final)} = \frac{m(x) + M\left(\frac{l}{2}+x\right)}{m+M}$$

Now, X_{CM} (initial) = X_{CM} (final)

$$\Rightarrow \quad ml + \frac{Ml}{2} = mx + \frac{Ml}{2} + Mx$$

$$\Rightarrow \quad x = \frac{ml}{m+M}$$

Hence, the log moves away from the bank through a distance of $\dfrac{m\ell}{(m+M)}$.

Alternative method: Displacement of the log = $\Delta x_1 = x$

Displacement of the man = $\Delta x_2 = \ell - x$

Apply $m_1x_1 = m_2x_2$ $\Rightarrow$ (if centre of mas remains at the same place)

$$\Rightarrow Mx = m(l-x) \quad \Rightarrow \quad x = \frac{m\ell}{(m+M)}$$

Illustration 1.11 **A plank of mass M and length L is at rest on a frictionless floor. The top surface of the plank has friction. At one end of it a man of mass m is standing as shown in Fig. 1.19. If the man walks towards the other end, find the distance, which the plank moves (a) till the man reaches the centre of the plank, (b) till the man reaches the other end of the plank.**

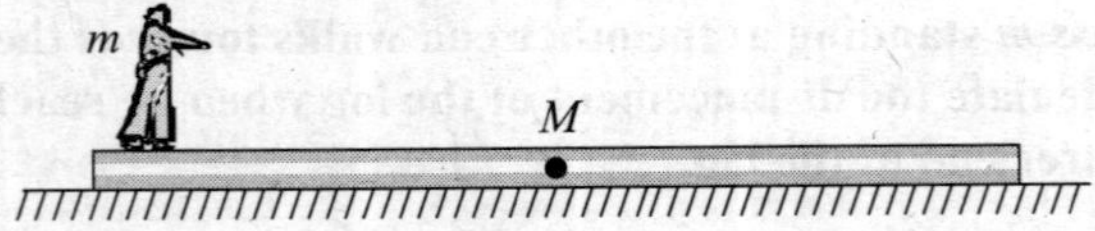

Fig. 1.19

Sol. Method 1: The corresponding situation can be better explained with the help of Fig. 1.19. Consider the man and the plank as a system on which no external force is acting. The centre of mass of the system must remain stationary. The only interaction force between the man and the plank is the friction as shown in Fig. 1.19, due to which the man walks along the plank and the friction on plank would be in opposite direction, due to which plank moves towards left, such that the centre of mass of the plank plus man remains at rest. Just before the motion started, the initial distance of the centre of mass from the centre of the plank is

$$x_c = \frac{m \times \frac{L}{2} + m \times 0}{m+M} = \frac{m \times \frac{L}{2}}{m+M}$$

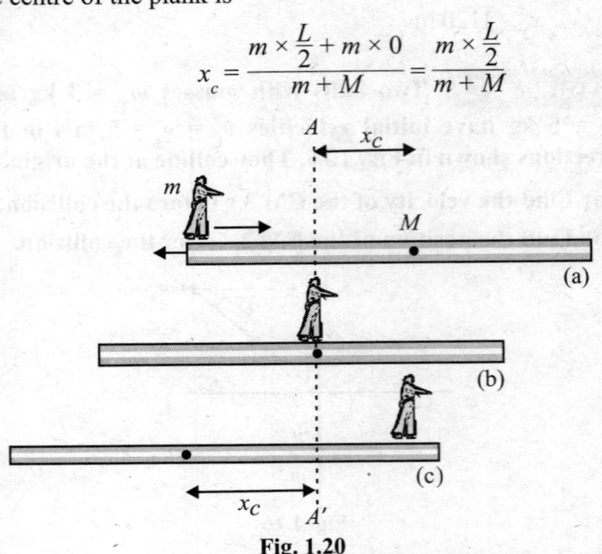

Fig. 1.20

Initially, the centre of mass of the system is on line AA' as shown in Fig. 1.20. During motion of the man, this centre of mass must remain at this line only. As the man moves towards right, the plank will move towards left such that centre of mass remains on AA'. Thus, when the man reaches the centre of the plank, the plank's centre must also reach the same point so that the centre of mass is at the same position. Up to this instant the plank moves by a distance x_c. Similarly, when the man reaches the other end, plank has to move towards left further by x_c to maintain the position of centre of mass.

Method 2: As there is no external force on the system (man + plank) and the system is initially at rest, there should not be movement of centre of mass of the system during the motion of the man.

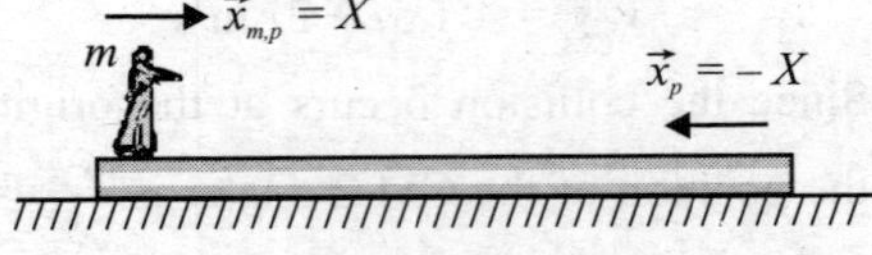

Fig. 1.21

The displacement of centre of mass is Δx_{CM} and is given by

$$\Delta x_{CM} = \frac{m\Delta\vec{x}_m + M\Delta\vec{x}_p}{m+M} \qquad \text{(i)}$$

$\Rightarrow \Delta\vec{x}_m$ = displacement of the man with respect to ground

$\Delta\vec{x}_m = \Delta\vec{x}_{m,p} + \Delta\vec{x}_p = (x - X)$ where $\Delta\vec{x}_p = -X$

(towards left)

As $\Delta\vec{x}_{CM} = 0$, from Eq. (i),

$$0 = \frac{m(x - X) - MX}{m + M} \Rightarrow X = \frac{mx}{m + M}$$

(a) When $x = \frac{L}{2}$, displacement of block $X = \frac{m\frac{L}{2}}{m + M}$

(b) When $x = L$, displacement of block $X = \frac{mL}{m + M}$

Illustration 1.12 **An explosion blows a rock into three parts. Two pieces go off at right angles to each other; 1.0 kg piece with a velocity of 12 m/s and other 2.0 kg piece with a velocity of 8 m/s. If the third piece flies off with a velocity of 40 m/s, compute the mass of the third piece.**

Sol. Let m_1, m_2 and m_3 be the masses of the three pieces.

$$m_1 = 1.0 \text{ kg}, \quad m_2 = 2.0 \text{ kg}.$$

Let $v_1 = 12$ m/s, $v_2 = 8$ m/s, $v_3 = 40$ m/s. Let v_1 and v_2 be directed along x-and y-axes, respectively, and v_3 be directed as shown.

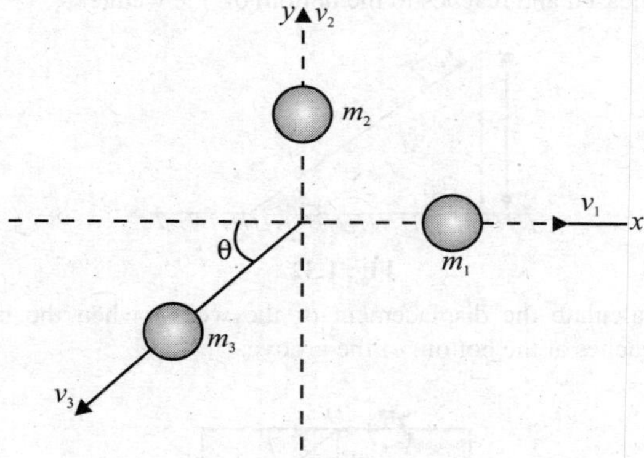

Fig. 1.22

By the principle of conservation of momentum, initial momentum is zero. Hence,

along x-axis: $0 = m_1v_1 - m_3v_3 \cos\theta$

along y-axis: $0 = m_2v_2 - m_3v_3 \sin\theta$

$\Rightarrow$ $m_1v_1 = m_3v_3 \cos\theta$ and $m_2v_2 = m_3v_3 \sin\theta$

By squaring and adding, we get $m_1^2v_1^2 + m_2^2v_2^2 = m_3^2v_3^2$

$$\Rightarrow m_3^2 = \frac{1^2(12)^2 + (2)^2(8)^2}{(40)^2} \Rightarrow m_3 = 0.5 \text{ kg}$$

Concept Application Exercise 1.1

1. Two children A and B of same mass (including their caps) M are sitting on a sea-saw as shown in Fig. 1.23. Initially, the beam is horizontal. At once, child B throws away his cap (mass $M/25$) which falls at point Q, midpoint of the left half of the beam, due to this the balance of beam is disturbed. To balance it again what is the mass m required to be put at point P on the right half of the beam?

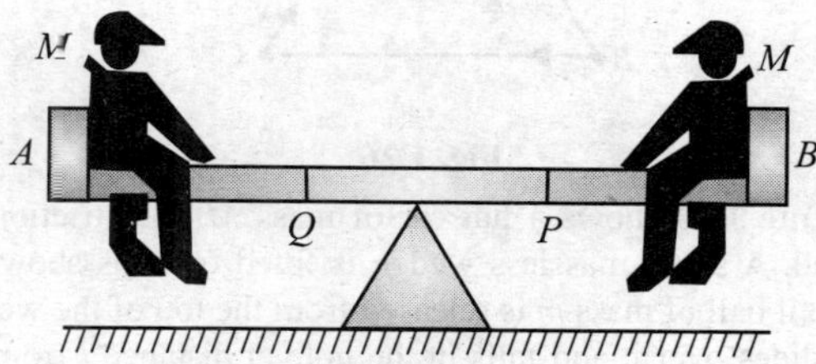

Fig. 1.23

2. Figure 1.24 shows a fixed wedge on which two blocks of masses 2 kg and 3 kg are placed on its smooth inclined surfaces. When the two blocks are released from rest, find the acceleration of centre of mass of the two blocks.

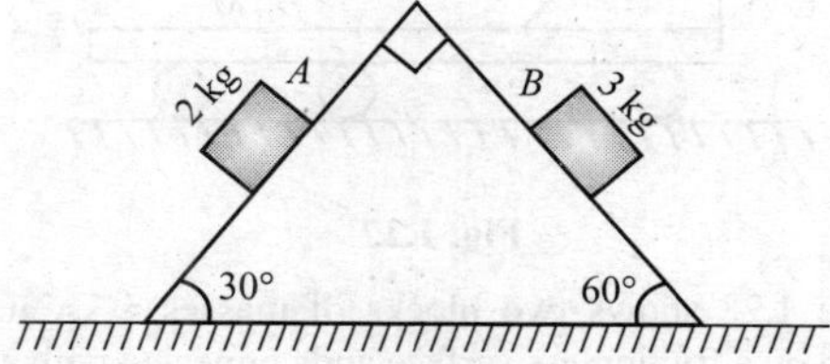

Fig. 1.24

3. Consider a rectangular plate of dimensions $a \times b$. If the plate is considered to be made up of four rectangles of dimensions $\frac{a}{2} \times \frac{b}{2}$ and we now remove one (the lower right) out of the four rectangles, find the position where the centre of mass of the remaining system will be.

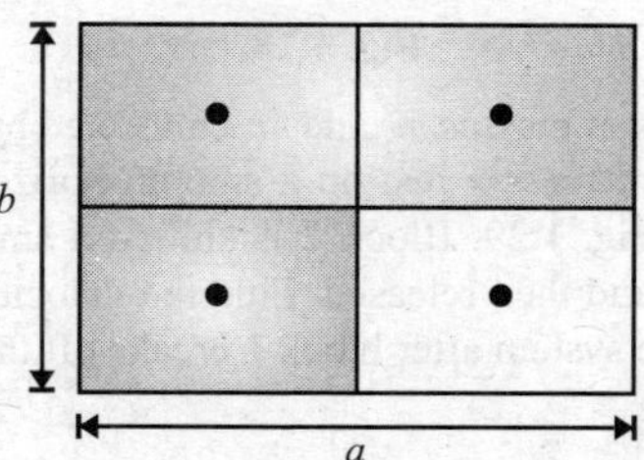

Fig. 1.25

4. There are two masses m_1 and m_2 placed at a distance l apart, let the centre of mass of this system is at a point named C. If m_1 is displaced by l_1 towards C and m_2 is displaced by l_2 away from C, find the distance from C where the new centre of mass will be located.
5. Let there are three equal masses situated at the vertices of an equilateral triangle, as shown in Fig. 1.26. Now particle A starts with a velocity v_1 towards line AB, particle B starts with the velocity v_2 towards line BC and particle C starts with velocity v_3 towards line CA. Find the displacement of the centre of mass of the three particles A, B and C after time t. What would it be if $v_1 = v_2 = v_3$?

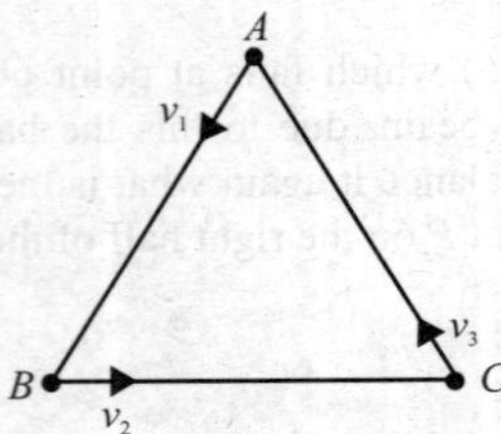

Fig. 1.26

6. Figure 1.27 shows a flat car of mass M on a frictionless road. A small massless wedge is fitted on it as shown. A small ball of mass m is released from the top of the wedge, it slides over it and falls in the hole at distance l from the initial position of the ball. Find the distance the flat car moves till the ball gets into the hole.

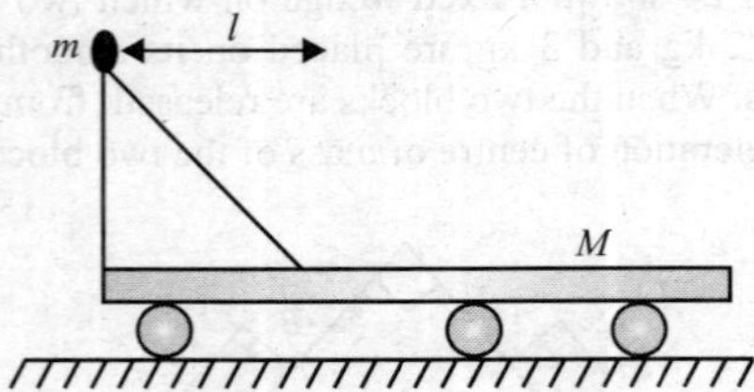

Fig. 1.27

7. Figure 1.28 shows two blocks of masses 5 kg and 2 kg placed on a frictionless surface and connected with a spring. An external kick gives a velocity of 14 m/s to the heavier block in the direction of lighter one. Deduce (a) velocity gained by the centre of mass and (b) the separate velocities of the two blocks in the centre of mass coordinates just after the kick.

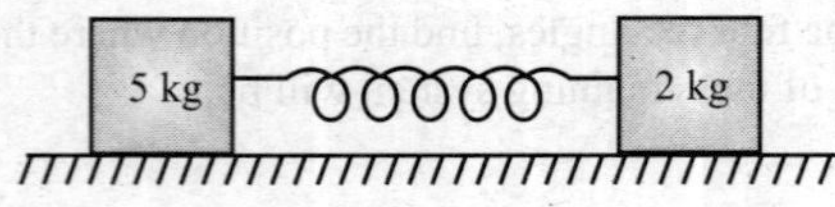

Fig. 1.28

8. Two blocks of masses m_1 and m_2 connected by a weightless spring of stiffness k rest on a smooth horizontal plane as shown in Fig. 1.29. Block 2 is shifted a small distance x to the left and then released. Find the velocity of centre of mass of the system after block 1 breaks off the wall.

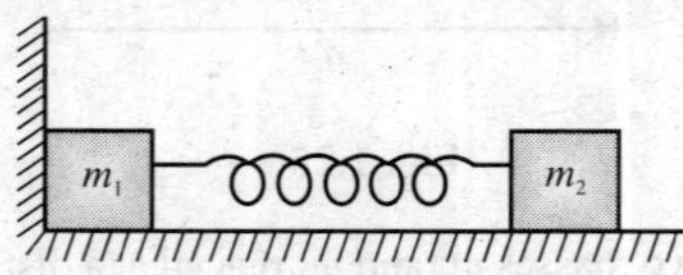

Fig. 1.29

9. Mr. Verma (50 kg) and Mr. Mathur (60 kg) are sitting at the two extremes of a 4 m long boat (40 kg) standing still in water. To discuss a mechanics problem, they come to the middle of the boat. Neglecting friction with water, how far does the boat move in the water during the process?

10. A cart of mass M is at rest on a frictionless horizontal surface and a pendulum bob of mass m hangs from the roof of the cart. The string breaks, the bob falls on the floor, makes several collisions on the floor and finally lands up in a small slot made on the floor. The horizontal distance between the string and the slot is L. Find the displacement of the cart during this process.

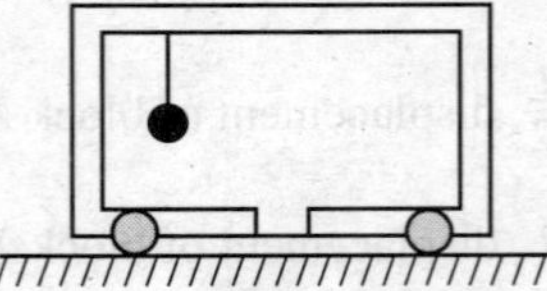

Fig. 1.30

11. Find the displacement of the wedge when 'm' comes out of the wedge. There is no friction anywhere.

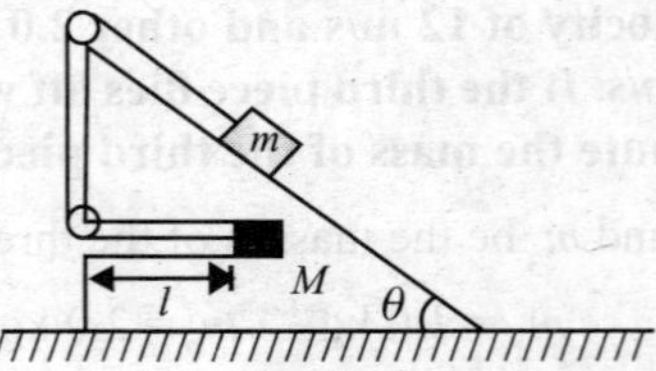

Fig. 1.31

12. A block of mass m is initially lying on a wedge of mass M with an angle of inclination θ, as shown in Fig. 1.32. Calculate the displacement of the wedge when the block is released and reaches to the bottom of the wedge.

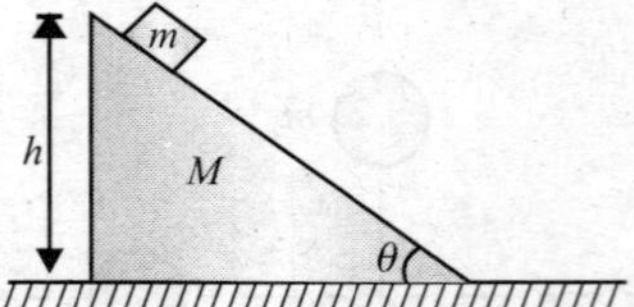

Fig. 1.32

13. Calculate the displacement of the wedge when the ball reaches at the bottom of the groove.

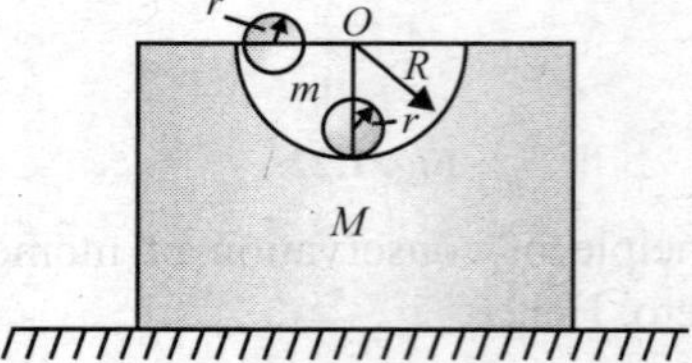

Fig. 1.33

14. A block is released on the convex surface of a hemispherical wedge as shown in Fig. 1.34. Determine the displacement of the wedge when the block reaches the angular position θ.

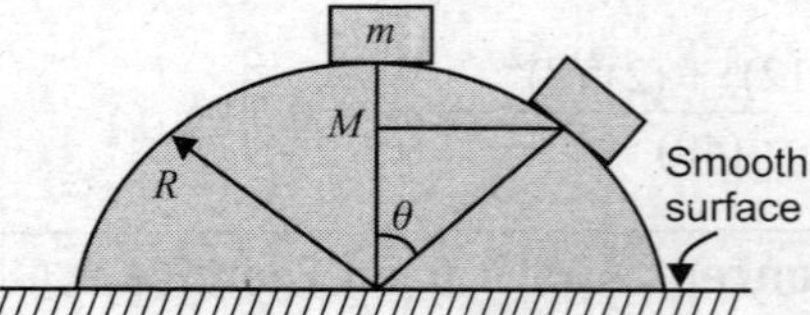

Fig. 1.34

15. Two masses m_1 and m_2 are moving with velocities v_1 and v_2. Find their total kinetic energy in the reference frame of centre of mass.

16. Figure 1.35 the system is at rest initially with $x = 0$. A man and a woman both are initially at the extreme carrier of the platform. The man and the woman start to move towards each other. Obtain an expression for the displacement s of the platform when the two meet in terms of the displacement x_1 of the man relative to the platform.

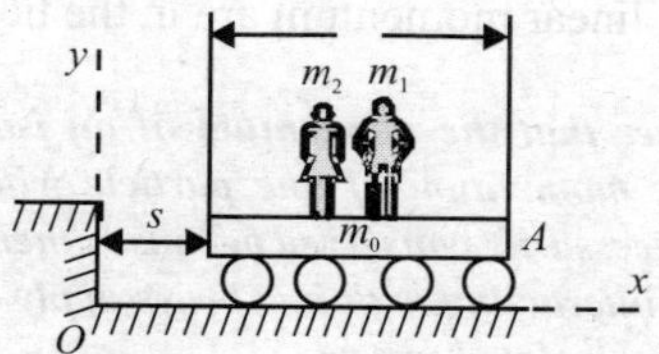

Fig. 1.35

17. A 30 kg projectile moving horizontally with a velocity $\vec{v}_0 = (120 \text{ m/s})\, \hat{i}$ explodes into two fragments A and B of masses 12 kg and 18 kg, respectively. Taking point of explosion as origin and knowing that 3 s later the position of fragment A is (300 m, 24 m, –48 m), determine the position of fragment B at the same instant.

18. Two 20 kg cannon balls are chained together and fired horizontally with a velocity of 200 m/s from the top of a 30 m wall. The chain breaks during the flight of the cannon balls and one of them strikes the ground at $t = 2$ s, at a distance of 250 m from the foot of the wall, and 5 m to the right of the line of fire. Determine the position of the other cannon ball at that instant. Neglect the resistance of air.

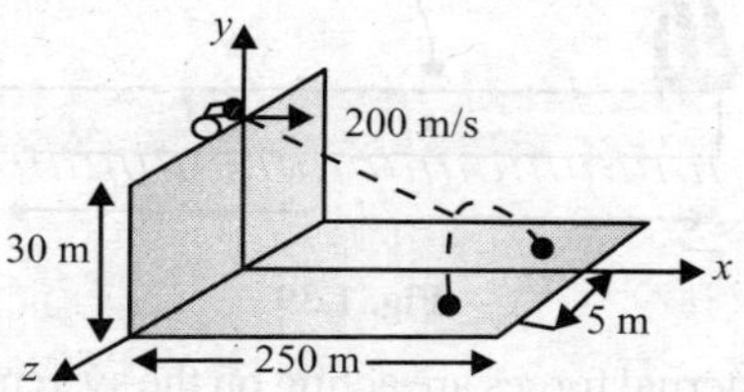

Fig. 1.36

19. A juggler juggles three balls in a continuous cycle. Any one ball is in contact with his hand for one-fifth of the time. Describe the motion of the centre of mass of the three balls. What average force does the juggler exert on one ball while he is touching it?

20. A cannon and a supply of cannon balls are inside a sealed rail road car. The cannon fires to the right, the car recoils to the left. The canon balls remain in the car after hitting the far wall. Show that no matter how the cannon balls are fired, the rail road car cannot travel more than L, assuming it starts from rest.

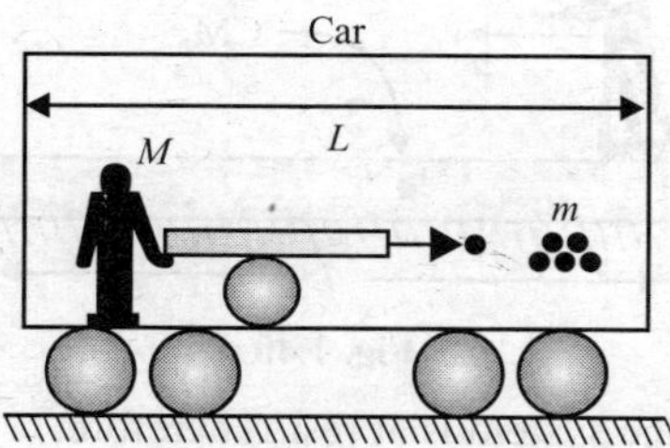

Fig. 1.37

CONSERVATION OF LINEAR MOMENTUM

We were able to solve problems involving these situations by applying a conservation principle, conservation of energy. Consider another situation. A 60 kg archer stands on frictionless ice and fires a 0.50 kg arrow horizontally at 50 m/s. From Newton's third law, we know that the force that the bow exerts on the arrow will be matched by a force in the opposite direction on the bow (and the archer). This force will cause the archer to begin to slide backward on the ice. But with what speed? We cannot answer this question using either Newton's second law or an energy approach because there is not enough information.

Let us conceptualize an isolated system of two particles (Fig. 1.38) with masses m_1 and m_2 and moving with velocities $\vec{v}_1$ and $\vec{v}_2$ at an instant of time. Because the system is isolated, the only force on one particle is that from the other particle, and we can categorize this situation as one in which Newton's laws can be applied. If a force from particle 1 (e.g., a gravitational force) acts on particle 2, there must be a second force, equal in magnitude but opposite in direction, that particle 2 exerts on particle 1. That is, the forces form a Newton's third law action–reaction pair so that $\vec{F}_{12} = -\vec{F}_{21}$. We can express this condition as a statement about the system of two particles as follows:

$$\vec{F}_{12} + \vec{F}_{21} = 0$$

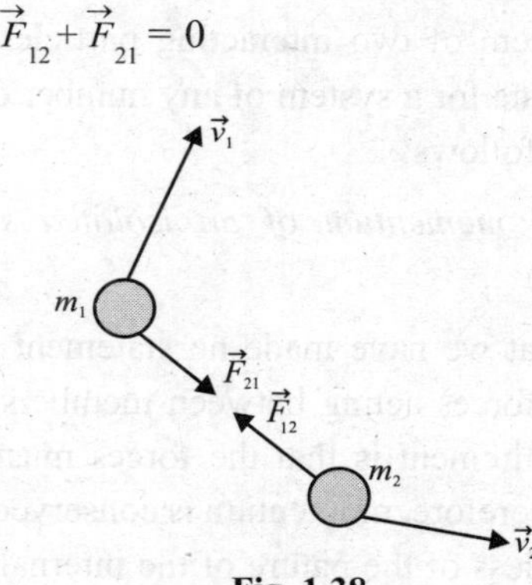

Fig. 1.38

Let us further analyze this situation by incorporating Newton's second law. Over some time interval, the interacting particles in the system will accelerate. Therefore, replacing each force with $m\vec{a}$ gives

$$m_1\vec{a}_1 + m_2\vec{a}_2 = 0 \Rightarrow m_1\frac{d\vec{v}_1}{dt} + m_2\frac{d\vec{v}_2}{dt} = 0$$

If the masses m_1 and m_2 are constant, we can bring them into the derivatives, which gives

$$\frac{d(m_1\vec{v}_1)}{dt} + \frac{d(m_2\vec{v}_2)}{dt} = 0 \text{ or } \frac{d}{dt}(m_1\vec{v}_1 + m_2\vec{v}_2) = 0 \qquad \text{(i)}$$

Using the definition of momentum, Eq. (i) can be written as

$$\frac{d}{dt}(\vec{p}_1+\vec{p}_2)=0$$

Because the time derivative of the total system momentum $\vec{p}_{tot} = \vec{p}_1 + \vec{p}_2$ is zero, we conclude that the total momentum $\vec{p}_{tot}$ must remain constant:

$$\vec{p}_{tot} = \text{constant} \quad \text{(ii)}$$

or equivalently,

$$\vec{p}_{1i} + \vec{p}_{2i} = \vec{p}_{1f} + \vec{p}_{2f} \quad \text{(iii)}$$

where $\vec{p}_{1i}$ and $\vec{p}_{2i}$ are initial values and $\vec{p}_{1f}$ and $\vec{p}_{2f}$ are final values of the momentum during a period over which the particles interact. Equation (iii) in component form states that the momentum components of the isolated system in the x, y and z directions are all independently constant; that is,

$$\sum_{\text{system}} p_{ix} = \sum_{\text{system}} p_{fx},\ \sum_{\text{system}} p_{iy} = \sum_{\text{system}} p_{fy},\ \sum_{\text{system}} p_{iz} = \sum_{\text{system}} p_{fz} \quad \text{(iv)}$$

This result, known as the law of conservation of linear momentum, is the mathematical representation of momentum version of the isolated system model. It is considered one of the most important laws of mechanics. We have generated this law for a system of two interacting particles, but it can be shown to be true for a system of any number of particles. We can state it as follows:

The total momentum of an isolated system remains constant.

Notice that we have made no statement concerning the nature of the forces acting between members of the system. The only requirement is that the forces must be internal to the system. Therefore, momentum is conserved for an isolated system regardless of the nature of the internal forces, even if the forces are non-conservative.

As linear momentum depends on frame of reference, observers in different frames would find different values of linear momentum of a given system but each would agree that his own value of linear momentum does not change with time provided the system is isolated and closed, i.e., law of conservation of linear momentum is independent of the frame of reference though linear momentum depends on the frame of reference.

Conservation of linear momentum is equivalent to Newton's third law of motion.

This law is universal, i.e., it applies to both macroscopic and microscopic systems. It holds good even in atomic and nuclear physics where classical mechanics fails. Further it is more generally applicable than the law of 'conservation of mechanical energy' because 'internal forces' are often non-conservative and so mechanical energy is not conserved but momentum is (if $F_{ext} = 0$). Principal applications of conservation of linear momentum are in the field of collisions.

Note: ***Remember that the momentum of an isolated system is conserved. The momentum of one particle within an isolated system is not necessarily conserved because other particles in the system may be interacting with it. Always apply conservation of momentum to an isolated system.***

Illustration 1.13 **A man of mass m moves on a plank of mass M with a constant velocity u with respect to the plank, as shown in Fig. 1.39.**

(i) If the plank rests on a smooth horizontal surface, determine the velocity of the plank.

(ii) If the man travels a distance L with respect to the plank, find the distance travelled by the plank with respect to the ground.

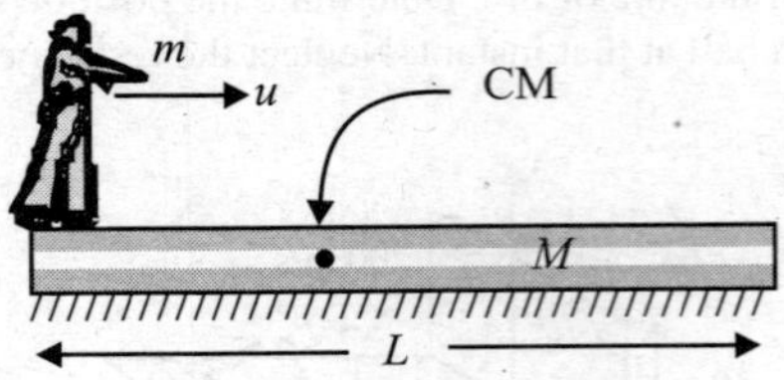

Fig. 1.39

Sol. As no external forces are acting on the system in horizontal direction, so its linear momentum remains constant in that direction. Also there will be no shift in position of centre of mass, i.e., $\Delta x_{CM} = 0$.

(i) Let v_1 and v_2 be the velocities of the man and the plank w.r.t. ground. Then, we have

$$[\vec{v}_{man}]_{plank} = [\vec{v}_{man}]_{ground} - [\vec{v}_{plank}]_{ground}$$

Then, $$[\vec{v}_{man}]_{ground} = [\vec{v}_{man}]_{plank} + [\vec{v}_{plank}]_{ground}$$

or $$v_1 = u - v_2$$

Fig. 1.40

Initially, the system is at rest. Therefore,

$$0 = mv_1 - Mv_2$$

or $$0 = m(u - v_2) - Mv_2$$

or $$v_2 = \frac{mu}{m + M}$$

ii. Let x be the displacement of plank in backward direction, the displacement of mass is $L - x$ in forward dirction.

$$\Delta x_{CM} = \frac{M(-x) + m(L - x)}{M + m}$$

$$\Rightarrow \quad 0 = -Mx + m(L - x) \Rightarrow x = \frac{mL}{M + m}$$

Illustration 1.14 **A shell is fired from a cannon with a speed of 100 m/s at an angle 60° with the horizontal (positive x-direction). At the highest point of its trajectory, the shell explodes into two equal fragments. One of the fragments moves along the negative x-direction with a speed of 50 m/s. What is the speed of the other fragment at the time of explosion.**

Sol. The velocity of the shell at the highest point of trajectory is

$$v_M = u \cos\theta = 100 \cos 60° = 50 \text{ m/s}$$

Let v_1 be the speed of the fragment which moves along the negative x-direction and the other fragment has speed v_2, which must be along the positive x-direction. Now from momentum conservation, we have

$$mv = \frac{-m}{2} v_1 + \frac{m}{2} v_2 \text{ or } 2v = v_2 - v_1$$

or $$v_2 = 2v + v_1 = (2 \times 50) + 50 = 150 \text{ m/s}$$

Illustration 1.15 **A man of mass m is standing at one end of a boat of mass M and length l. The man walks to the other end. What is the displacement of the centre of mass? What is the displacement of boat?**

Sol. In the process of walking, no external force acts on the system (boat + man) in the horizontal direction. So displacement of centre of mass will be zero. The horizontal momentum of the system is conserved.

(Since initially the momentum of the system is zero as it was at rest)

$$m\vec{v}_m + M\vec{v}_b = 0 \Rightarrow m(\vec{v}_{mb} + \vec{v}_b) + M\vec{v}_b = 0$$

$$v_b = \frac{mv_{mb}}{M + m} \Rightarrow \int_0^t v_b dt = \frac{m}{M + m} \int_0^t V_{mb} dt$$

$$X_b = \frac{m}{m + M} X_{mb}$$

X_{mb} = displacement of man relative to boat = l

$$X_b = \frac{ml}{M + m}$$

Illustration 1.16 **Two blocks of masses m_1 and m_2, interconnected with a spring of stiffness K, are kept on a smooth horizontal surface. Find out the ratio of velocity, displacement and acceleration of block with mass m_1 to block with mass m_2.**

Sol. Here x is the displacement, v is the speed, p is the momentum, a is the acceleration, KE is the kinetic energy, F is the force.

$$(F_{ext})_{system} = 0 \Rightarrow \vec{p}_1 + \vec{p}_2 = 0$$

$$m_1 v_1 = m_2 v_2 \Rightarrow \frac{v_1}{v_2} = \frac{m_2}{m_1}$$

$$m_1 \int_0^t v_1 dt = m_2 \int_0^t v_2 dt$$

where t is the time interval of motion of each block.

$$m_1 x_1 = m_2 x_2 \Rightarrow \frac{x_1}{x_2} = \frac{m_2}{m_1}$$

$$\frac{\text{KE}_1}{\text{KE}_2} = \frac{P_1^2/2m_1}{P_2^2/2m_2} = \frac{m_2}{m_1} \quad (\because p_1 = p_2)$$

$$\Rightarrow \quad \frac{F_1}{F_2} = \frac{kx}{kx}$$

where x is the deformation of the spring.

$$\therefore \quad \frac{F_1}{F_2} = 1 \Rightarrow \frac{m_1 a_1}{m_2 a_2} = 1 \Rightarrow \frac{a_1}{a_2} = \frac{m_2}{m_1}$$

Illustration 1.17 **A flat car of mass M is at rest on a frictionless floor with a child of mass m standing at its edge. If the child jumps off from the car towards right with an initial velocity u, with respect to the car, find the velocity of the car after its jump.**

Sol. Let the car attain a velocity v. The net velocity of the child with respect to the earth will be $u - v$, as u is its velocity with respect to the car.

Initially, the system was at rest, thus according to momentum conservation, momentum after jump must be zero. Hence,

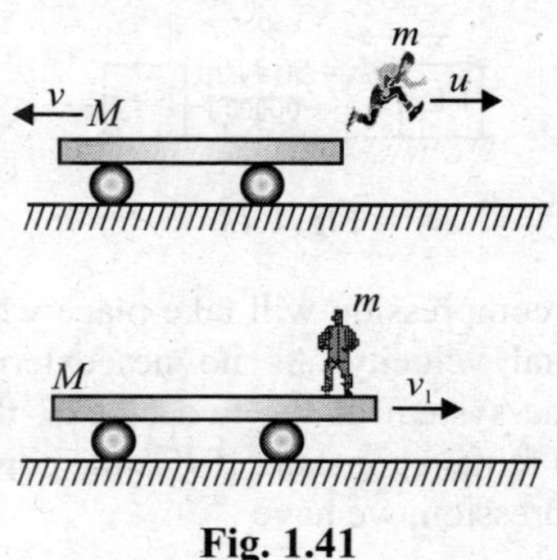

Fig. 1.41

$$m(u - v) = Mv \Rightarrow v = \frac{mu}{m + M}$$

Illustration 1.18 **A flat car of mass M with a child of mass m is moving with a velocity v_1. The child jumps in the direction of motion of car with a velocity u with respect to the car. Find the final velocities of the child and that of the car after jump.**

Sol. This case is similar to the previous example, except now the car is moving before jump. Here, also no external force is acting on the system in horizontal direction, hence momentum remains conserved in this direction. After the jump, car attains a velocity v_2 in the same direction, which is less than v_1, due to backward push of the child for jumping. After the jump, child attains a velocity $u + v_2$ in the direction of motion of car with respect to ground.

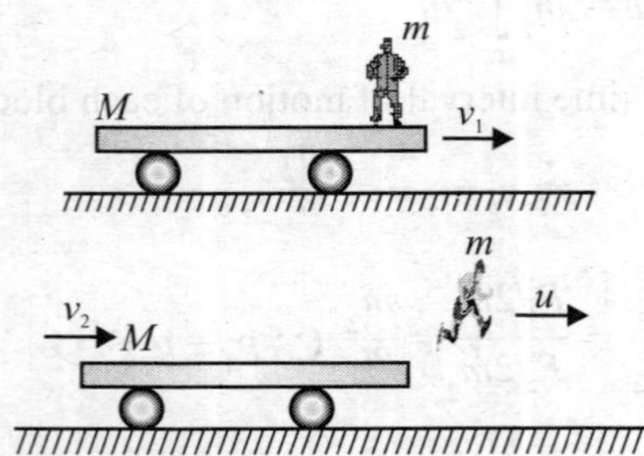

Fig. 1.42

According to momentum conservation,

$$(M + m)v_1 = Mv_2 + m(u + v_2)$$

Velocity of car after jump is $v_2 = \dfrac{(M+m)v_1 - mu}{M+m}$

Velocity of child after jump is $u + v_2 = \dfrac{(M+m)v_1 + Mu}{M+m}$

Illustration 1.19 **Each of the blocks shown in figure has mass 1 kg. The rear block moves with a speed of 2 m/s towards the front block kept at rest. The spring attached to the front block is light and has a spring constant 50 N/m. Find the maximum compression of the spring.**

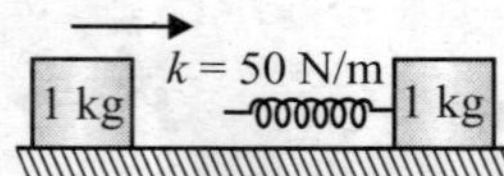

Fig. 1.43

Sol. Maximum compression will take place when the blocks move with equal velocity. As no net external horizontal force acts on the system of the two blocks, the total linear momentum will remain constant. If V is the common speed at maximum compression, we have

$$(1\text{ kg})(2\text{ m/s}) = (1\text{ kg})\,V + (1\text{ kg})\,V \text{ or } V = 1\text{ m/s}.$$

Initial kinetic energy $= \frac{1}{2}(1\text{ kg})(2\text{ m/s})^2 = 2\text{ J}$

Final kinetic energy $= \frac{1}{2}(1\text{ kg})(1\text{ m/s})^2 + \frac{1}{2}(1\text{ kg})(1\text{ m/s})^2 = 1\text{ J}$

The kinetic energy lost is stored as the elastic energy in the spring. Hence,

$$\frac{1}{2}(50\text{ N/m})\,x^2 = 2\text{ J} - 1\text{ J} = 1\text{ J or } x = 0.2\text{ m}$$

Illustration 1.20 **A light spring of spring constant k is kept compressed between two blocks of masses m and M on a smooth horizontal surface. When released, the blocks acquire velocities in opposite directions. The spring loses contact with the blocks when it acquires natural length. If the spring was initially compressed through a distance x, find the final speeds of the two blocks.**

Sol. Consider the two blocks plus the spring to be the system. No external force acts on this system in horizontal direction. Hence, the linear momentum will remain constant. Suppose, the block of mass M moves with a speed V and the other block with a speed v after losing contact with the spring. From conservation of linear momentum in horizontal direction, we have

$$MV - mv = 0 \text{ or } V = \frac{mv}{M} \qquad \text{(i)}$$

Initially, the energy of the system $= \frac{1}{2}kx^2$

Finally, the energy of the system $= \frac{1}{2}mv^2 + \frac{1}{2}MV^2$

As there is no friction, mechanical energy will remain conserved. Therefore,

$$\frac{1}{2}mv^2 + \frac{1}{2}MV^2 = \frac{1}{2}kx^2 \qquad \text{(ii)}$$

Solving Eqs. (i) and (ii), we get $v = \left[\dfrac{kM}{m(M+m)}\right]^{\frac{1}{2}} x$

and $\qquad V = \left[\dfrac{km}{M(M+m)}\right]^{\frac{1}{2}} x$

Illustration 1.21 **A block of mass m is connected to another block of mass M by a massless spring of spring constant k. The blocks are kept on a smooth horizontal plane and are at rest. The spring is unstretched when a constant force F starts acting on the block of mass M to pull it. Find the maximum extension of the spring.**

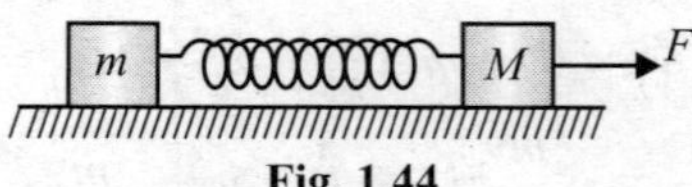

Fig. 1.44

Sol. We solve the situation in the reference frame of centre of mass. As only F is the external force acting on the system, due to this force, the acceleration of the centre of mass is $F/(M + m)$. Thus, with respect to centre of mass, there is a pseudoforce on the two masses in the opposite direction, the free body diagram of m and M with respect to centre of mass (taking centre of mass at rest) is shown in Fig. 1.45.

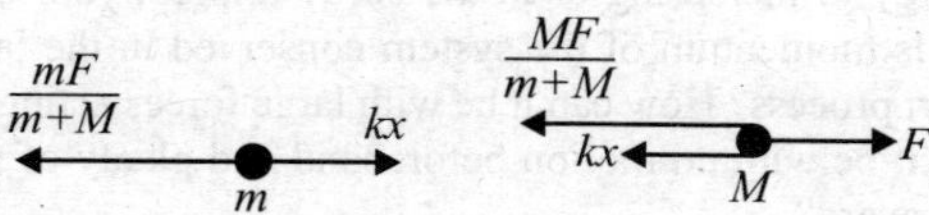

Fig. 1.45

Taking centre of mass at rest, if m moves maximum by a distance x_1 and M moves maximum by a distance x_2, then the work done by external forces (including pseudoforce) will be

$$W = \frac{mF}{m+M}x_1 + \left(F - \frac{MF}{m+M}\right)x_2$$

$$= \frac{mF}{m+M}(x_1 + x_2)$$

This work is stored in the form of potential energy of the spring as $U = (1/2)\,k\,(x_1 + x_2)^2$.

Thus, on equating we get the maximum extension in the spring, as after this instant the spring starts contracting.

$$\frac{1}{2}k(x_1 + x_2)^2 = \frac{mF}{m+M}(x_1 + x_2)$$

$$\Rightarrow \quad x_{max} = x_1 + x_2 = \frac{2mF}{k(m+M)}$$

Concept Application Exercise 1.2

1. A bomb, initially at rest, explodes into several pieces. (a) Is linear momentum of the system conserved? (b) Is kinetic energy of the system conserved? Explain.
2. You are standing perfectly still and then you take a step forward. Before the step your momentum was zero, but afterward you have some momentum. Is the principle of conservation of momentum violated in this case?
3. Does the centre of mass of a rocket in free space accelerate? Explain. Can the speed of a rocket exceed the exhaust speed of the fuel? Explain.
4. Discuss the possibility of conservation of linear momentum of a block moving on a rough inclined plane if $\mu = \tan\theta$.

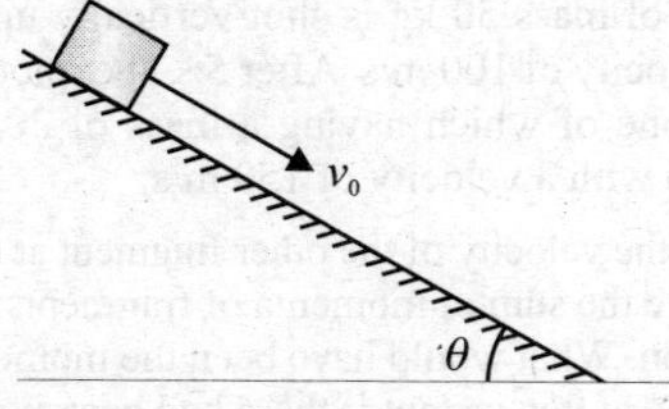

Fig. 1.46

5. A shell is fired from a cannon with a speed of 100 m/s at an angle 30° with the vertical (y-direction). At the highest point of its trajectory, the shell explodes into two fragments of masses in the ratio 1:2. The lighter fragment moves vertically upwards with an initial speed of 200 m/s. What is the speed of the heavier fragment at the time of explosion?
6. Figure 1.47 shows a block A of mass 6 m having a smooth semicircular groove of radius a placed on a smooth horizontal surface. A block B of mass m is released from a position in groove where its radius is horizontal. Find the speed of the bigger block when the smaller block reaches its bottommost position.

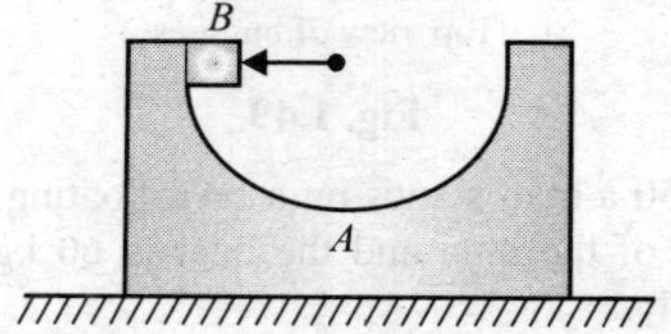

Fig. 1.47

7. Two friends A and B (each weighing 40 kg) are sitting on a frictionless platform some distance d apart. A rolls a ball of mass 4 kg on the platform towards B which B catches. Then B rolls the ball towards A and A catches it. The ball keeps on moving back and forth between A and B. The ball has a fixed speed of 5 m/s on the platform.
   a. Find the speed of A after he rolls the ball for the first time.
   b. Find the speed of A after he catches the ball for the first time.
   c. Find the speed of A and B after the ball has made five round trips and is held by A.
   d. How many times can A roll the ball?
   e. Where is the centre of mass of the system $A + B +$ ball at the end of the nth trip?
8. A smooth wedge of mass M rests on a smooth horizontal surface. A block of mass m is projected from its lowermost point with velocity v_0. What is the maximum height reached by the block?

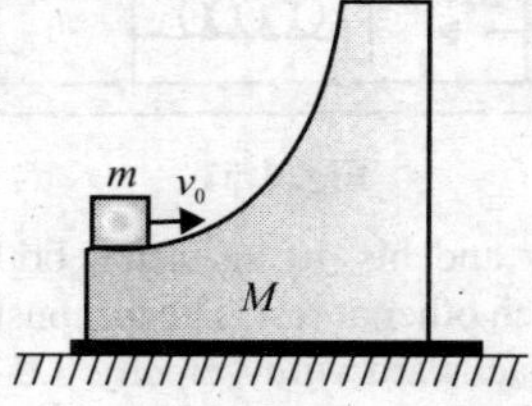

Fig. 1.48

9. Two identical buggies 1 and 2 with one man in each move along parallel rails. When the buggies are opposite to each other, the men jump in a direction perpendicular to the direction of motion of buggies, so as to exchange their places. As a consequence, buggy 1 stops and buggy 2 keeps moving in the same direction with its final velocity v. Find the initial velocities v_1 and v_2 of buggies. Mass of each buggy (without man) equals M, mass of each man is m; ignore frictional effects anywhere and the buggies are constrained to move along the rails only.

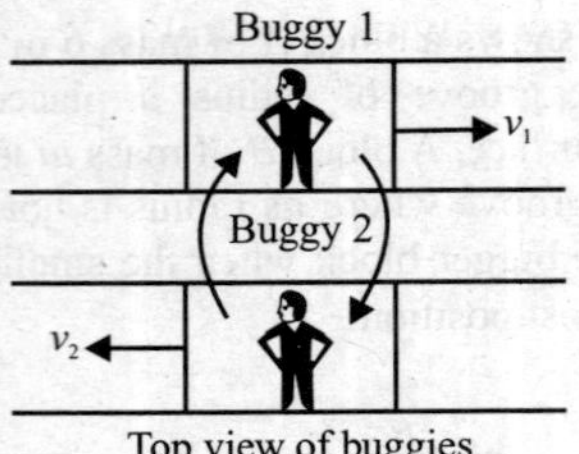

Fig. 1.49

10. In Fig. 1.50 a man stands on a boat floating in still water. The mass of the man and the boat is 60 kg and 120 kg, respectively.

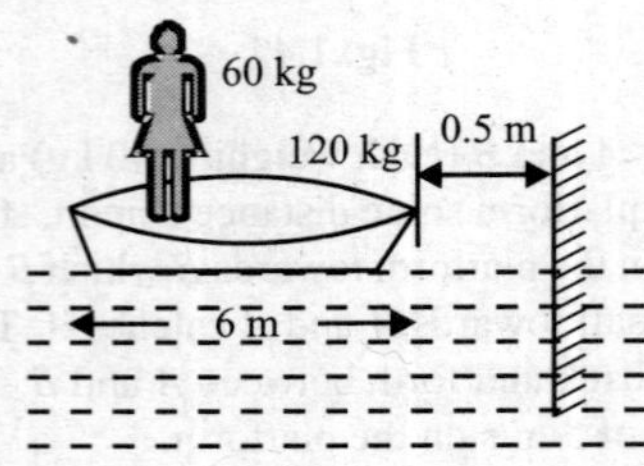

Fig. 1.50

a. If the man walks to the front of the boat and stops, what is the separation between the boat and the pier now?

b. If the man moves at a constant speed of 3 m/s relative to the boat, what is the total kinetic energy of the system (boat + man)? Compare this energy with the kinetic energy of the system if the boat was tied to the pier.

11. Two blocks of masses $m_1 = 2$ kg and $m_2 = 5$ kg are moving in the same direction along a frictionless surface with speeds 10 m/s and 3 m/s, respectively, m_2 being ahead of m_1. An ideal spring with $k = 1120$ N/m is attached to the back side of m_2. Find the maximum compression of the spring when the blocks collide. What are the final velocities of the blocks when they separarate?

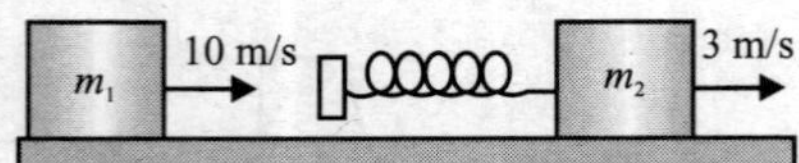

Fig. 1.51

12. An 80 kg boy and his 40 kg sister, both wearing roller blades, face each other at rest. The girl pushes the boy hard, sending him backward with velocity 3.0 m/s towards the west. Ignore friction. (a) Describe the subsequent motion of the girl. (b) How much chemical energy is converted into mechanical energy in the girl's muscles? (c) Is the momentum of the boy–girl system conserved in the pushing apart process? How can it be with no motion beforehand and plenty of motion afterward?

13. Two blocks of masses M and $3M$ are placed on a horizontal, frictionless surface. A light spring is attached to one of them and the blocks are pushed together with the spring between them. A cord initially holding the blocks together is burned; after that, the block of mass $3M$ moves to the right with a speed of 2.00 m/s. (a) What is the velocity of the block of mass M. (b) Find the system's original elastic potential energy, taking $M = 0.350$ kg. (c) Is the original energy in the spring or in the cord? Explain your answer. (d) Is momentum of the system conserved in the bursting apart process? How can it be with large forces acting? How can it be with no motion beforehand and plenty of motion afterward?

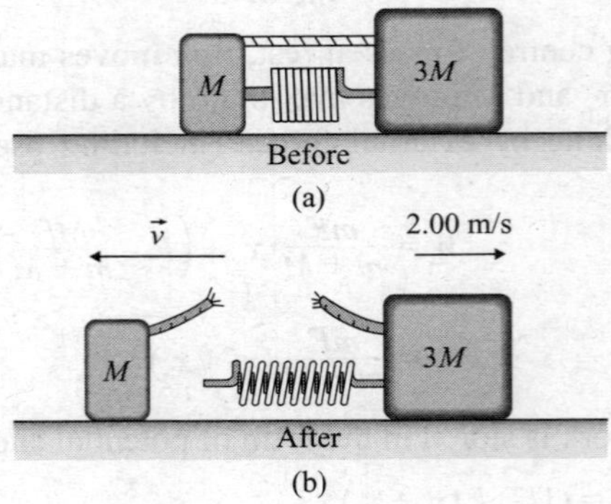

Fig. 1.52

14. A pendulum bob of mass 10^{-2} kg is raised to a height 5×10^{-2} m and then released. At the bottom of its swing, it picks up a mass 10^{-3} kg. To what height will the combined mass rise?

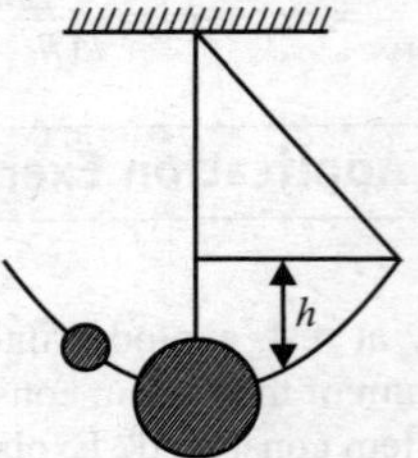

Fig. 1.53

15. A rifle man, who together with his rifle has a mass of 100 kg, stands on a smooth surface and fires 10 shots horizontally. Each bullet has a mass 10 g and a muzzle velocity of 800 m/s.

a. What velocity does the rifle man acquire at the end of 10 shots?

b. If the shots are fired in 10 s, what will be the average force exerted on him?

c. Compare his kinetic energy with that of 10 bullets.

16. A projectile of mass 50 kg is shot vertically upwards with an initial velocity of 100 m/s. After 5 s, it explodes into two fragments, one of which having a mass of 20 kg travels vertically up with a velocity of 150 m/s.

a. What is the velocity of the other fragment at that instant?

b. Calculate the sum of momenta of fragments 3 s after the explosion. What would have been the momentum of the projectile at this instant if there had been no explosion?

17. a. A rail road flat car of mass M can roll without friction along a straight horizontal track (Fig. 1.54). Initially, a man of mass m is standing on the car which is moving to the right with speed v_0. What is the change in velocity of the car if the man runs to the left so that his speed relative to the car is v_{rel} just before he jumps off at the left end?

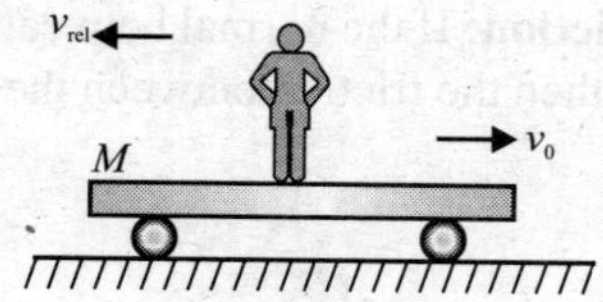

Fig. 1.54

b. If there are n men each of mass m on the car, should they all run and jump off together or should they run and jump one by one in order to give a greater velocity to the car?

18. i. A rail road car of mass M is moving without friction on a straight horizontal track with a velocity u. A man of mass m lands on it normally from a helicopter. What will be the new velocity of the car?

ii. If now the man begins to run on it with speed v_{rel} with respect to car in a direction opposite to motion of the car, what will be the new velocity of the car?

19. A shell of mass 2 kg moving at a rate of 4 m/s suddenly explodes into two equal fragments. The fragments go in directions inclined with the original line of motion with equal velocities. If the explosion imparts 48 J of translational kinetic energy to the fragments, find the velocity and direction of each fragment.

20. A mud ball at rest explodes into three fragments of masses in the ratio 1:2:1. The two equal masses move with velocities $2\hat{i}+5\hat{j}-6\hat{k}$ and $-4\hat{i}+3\hat{j}+2\hat{k}$. Find the velocity of the third mass.

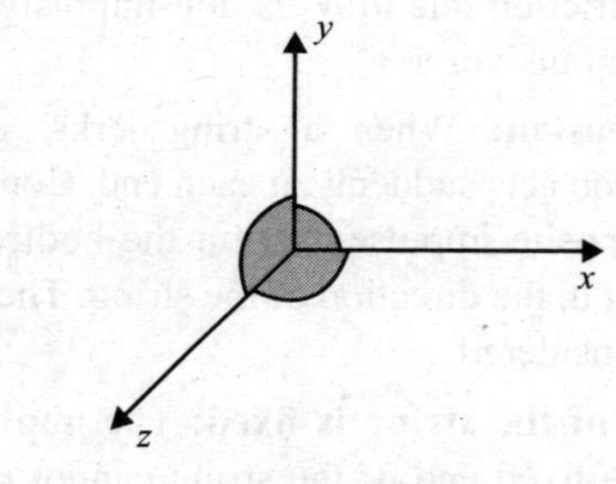

Fig. 1.55

IMPULSE

Impulse of a force F acting on a body is defined as

$$J=\int_{t_i}^{t_f} F dt$$

$$=\int F dt=\int m\frac{dv}{dt}dt=\int m dv$$

It is also defined as change in momentum

$$\vec{J}=\Delta\vec{P} \text{ (impulse–momentum theorem)}$$

Instantaneous Impulse

There are many occasions when a force acts for such a short time that the effect is instantaneous, e.g., a bat striking a ball. In such cases, although the magnitude of the force and the time for which it acts may each be unknown, the value of their product (i.e., impulse) can be known by measuring the initial and final momenta. Thus, we can write

$$\vec{J}=\int \vec{F}dt=\Delta\vec{P}=\vec{P}_f-\vec{P}_i$$

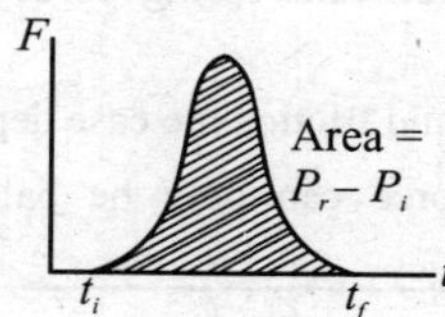

Fig. 1.56

Regarding the impulse, it is important to note that impulse applied to an object in a given time interval can also be calculated from the area under the force–time (F–t) graph in the same time interval.

Points to Remember

1. Impulse is a vector quantity.
2. Dimensions are $[MLT^{-1}]$.
3. SI unit is kg m/s.
4. Direction is along the change in momentum.
5. Magnitude is equal to area under the F–t graph.
6. $J=\int F dt=F_{av}\,\Delta t$.
7. It is not a property of any particle, but it is a measure of the degree to which an external force changes the momentum of the particle.

Illustration 1.22 **The hero of a stunt film fires 50 g bullets from a machine gun, each at a speed of 1.0 km/s. If he fires 20 bullets in 4 s, what average force does he exert against the machine gun during this period?**

Sol. The momentum of each bullet = (0.050 kg) (1000 m/s) = 50 kg m/s.

The gun is imparted this much of momentum by each bullet fired. Thus, the rate of change of momentum of the gun is (50 kg m/s × 20)/4 s = 250 N

In order to hold the gun, the hero must exert a force of 250 N against the gun.

Impulsive Force

A force of relatively higher magnitude and acting for relatively shorter time is called impulsive force.

An impulsive force can change the momentum of a body in a finite magnitude in a very short time interval.

Impulsive force is a relative term. There is no clear boundary between an impulsive and a non-impulsive force.

Note: ***Usually colliding forces are impulsive in nature.***

Since the application time is very small, hence, very little motion of the particle takes place.

Points to Remember

1. Gravitational force and spring force are always non-impulsive.
2. Normal, tension and friction are case dependent.
3. An impulsive force can only be balanced by another impulsive force.

1. **Impulsive normal:** In case of collision, normal forces at the surface of collision are always impulsive.

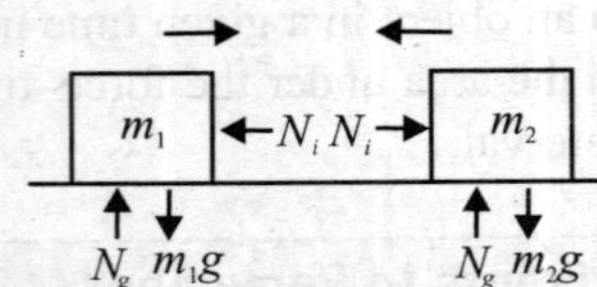

Fig. 1.57

N_i is impulsive; N_g is non-impulsive.

In Fig. 1.58, both normals are impulsive.

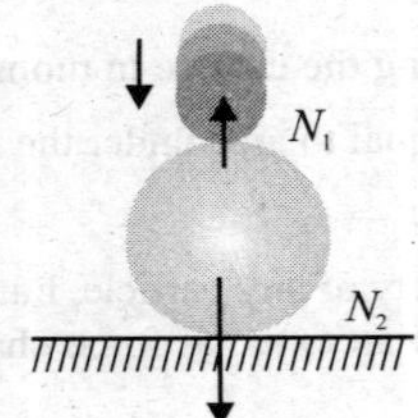

Fig. 1.58

In Fig. 1.59, N_1 and N_3 are impulsive; N_2 is non-impulsive.

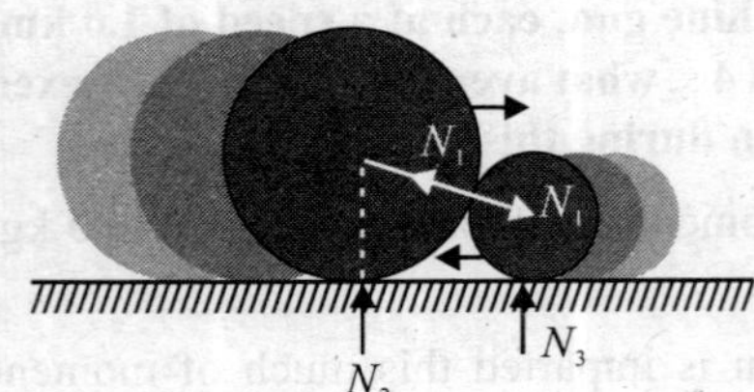

Fig. 1.59

In Fig. 1.60, both normals are impulsive.

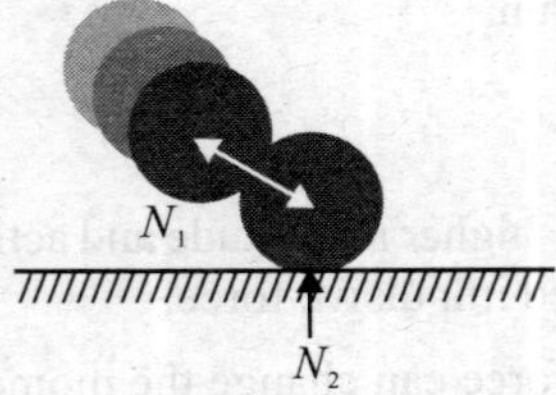

Fig. 1.60

2. **Impulsive friction:** If the normal between the two objects is impulsive, then the friction between the two will also be impulsive.

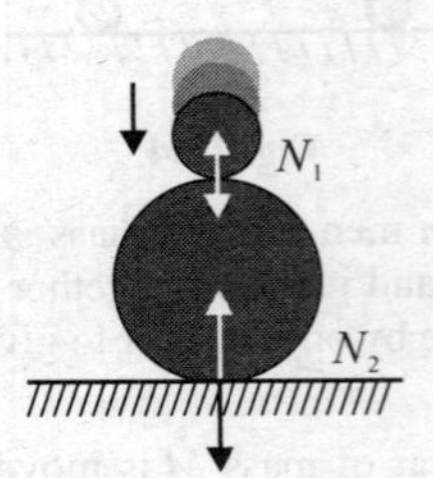

Fig. 1.61

In Fig. 1.61, friction at both surfaces is impulsive.

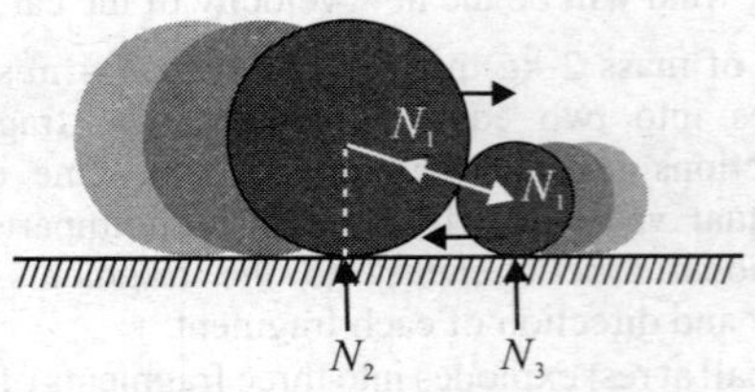

Fig. 1.62

In Fig. 1.62, friction due to N_2 is non-impulsive and that due to N_3 is impulsive.

3. **Impulsive tension:** When a string jerks, equal and opposite tension acts suddenly at each end. Consequently, equal and opposite impulses act on the bodies attached with the string in the direction of the string. There are two cases to be considered.

   (a) One end of the string is fixed: The impulse which acts at the fixed end of the string cannot change the momentum of the fixed object there. The object attached to the free end, however, will undergo a change in momentum in the direction of the string. The momentum remains unchanged in a direction perpendicular to the string where no impulsive forces act.

   (b) Both ends of the string are attached to movable objects: In this case, equal and opposite impulses act on the two objects, producing equal and opposite changes in momentum. The total momentum of the system, therefore, remains constant, although the momentum of each individual object is changed in the direction of the string. However, no impulse acts perpendicular to the string and the momentum of each particle in this direction is unchanged.

Note: ***In case of rod, tension is always impulsive. In case of spring, tension is always non-impulsive.***

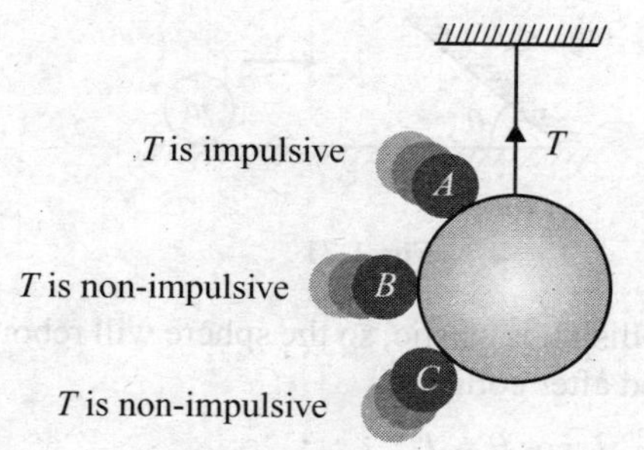

All normal are impulsive but tension
T is impulsive only for ball A.

Fig. 1.63

Illustration 1.23 **A block of mass m and a pan of equal mass are connected by a string going over a smooth light pulley. Initially, the system is at rest; then a particle of mass m falls on the pan and sticks to it. If the particle strikes the pan with a speed v, find the speed with which the system moves just after the collision.**

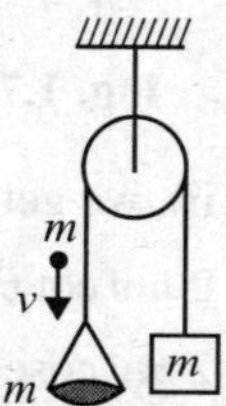

Fig. 1.64

Sol. Let the required speed be V. Further, let J_1 be the impulse between the particle and the pan and J_2 be the impulse imparted to the block and the pan by the string.

Using impulse as the change in momentum, we have the following equations.

For particle,

$$J_1 = mv - mV \quad \text{(i)}$$

For pan,

$$J_1 - J_2 = mV \quad \text{(ii)}$$

For block,

$$J_2 = mV \quad \text{(iii)}$$

Solving Eqs. (i)–(iii), we get $V = v/3$.

Illustration 1.24 **Two identical blocks A and B, connected by a massless string, are placed on a frictionless horizontal plane. A bullet having the same mass, moving with speed u, strikes block B from behind as shown. If the bullet gets embedded into block B, then find**

(a) the velocity of A, B, C after collision;

(b) impulse on A due to tension in the string;

(c) impulse on C due to normal force of collision;

(d) impulse on B due to normal force of collision.

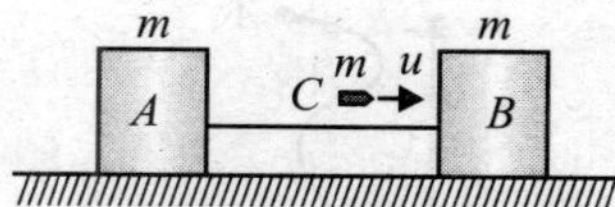

Fig. 1.65

Sol. (a) By conservation of linear momentum, $v = u/3$.

(b) $\int T dt = \dfrac{mu}{3}$

(c) $\int N dt = m\left(\dfrac{u}{3} - u\right) = \dfrac{-2mu}{3}$

(-ve sign indicates left direction)

(d) $\int (N - T)\, dt = \int N dt - \int T dt = \dfrac{mu}{3} \Rightarrow \int N dt = \dfrac{2mu}{3}$

Illustration 1.25 **A ball of mass 1 kg is attached to an inextensible string. The ball is released from the position shown in Fig. 1.66. Find the impulse imparted by the string to the ball immediately after the string becomes taut.**

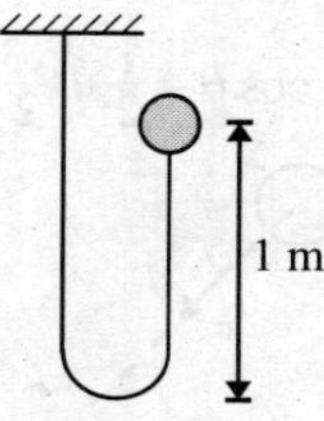

Fig. 1.66

Sol. The string will become taut when the particle will fall through a distance 2 m in downward direction. So the required impulse: $J = mu = 1 \times \sqrt{2 \times 10 \times 2} = \sqrt{40}$ kg m/s.

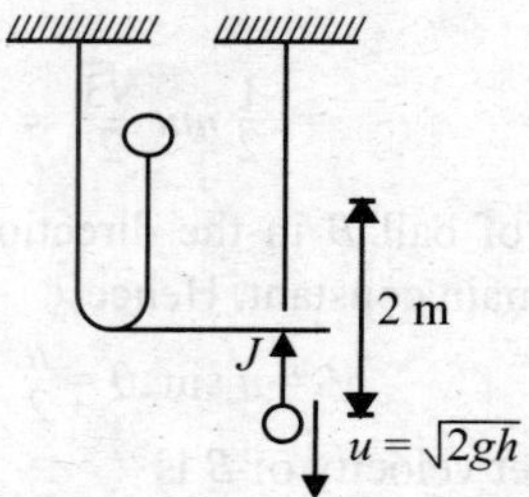

Fig. 1.67

Illustration 1.26 **Two particles A and B of equal mass m are attached by a string of length $2l$ and initially placed over a smooth horizontal table in the position shown in Fig. 1.68. Particle B is projected across the table with speed u perpendicular to AB as shown in the figure. Find the velocities of each particle after the string becomes taut and the magnitude of the impulse tension.**

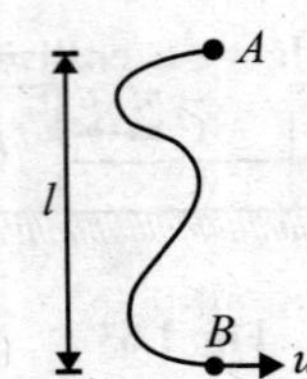

Fig. 1.68

Sol. When the string becomes taut, it can be depicted as in Fig. 1.69. $\theta = 30°$

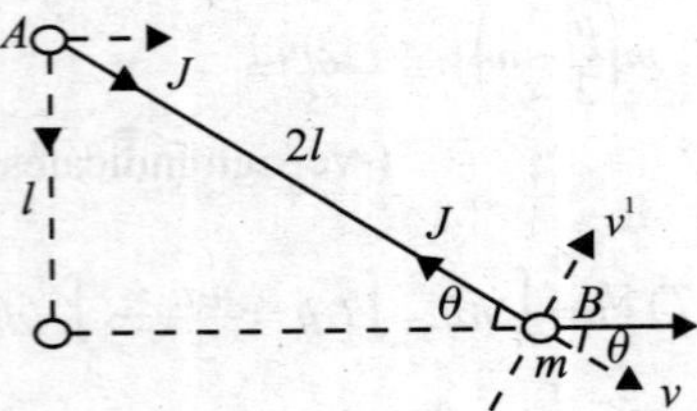

Fig. 1.69

Linear momentum will be conserved in any direction. Let us conserve the linear momentum in the direction of thread. Then,

$$mu \cos \theta = 2\, mv$$

$$\Rightarrow \quad v = \frac{1}{2} u \cos \theta = \frac{1}{2} u \frac{\sqrt{3}}{2} = \left(\frac{\sqrt{3}}{4}\right) u \quad \text{(i)}$$

Fig. 1.70

From the impulse diagram of A (Fig. 1.70),

$$J = mv \Rightarrow J = \frac{1}{2} mu \cos \theta$$

$$= \frac{1}{2} mu \frac{\sqrt{3}}{2} = \frac{\sqrt{3}\, mu}{4}$$

Velocity of ball B in the direction perpendicular to the string will remain constant. Hence,

$$v' = u \sin \theta = \frac{u}{2}$$

Hence, net velocity of B is

$$v_B = \sqrt{\left(\frac{u}{2}\right)^2 + (v)^2} = \frac{\sqrt{7}\, u}{4} \text{ and } v_A = \frac{\sqrt{3}\, u}{4}$$

Illustration 1.27 **A sphere of mass m slides with velocity v on a frictionless surface towards a smooth inclined wall as shown in Fig. 1.71. If the collision with the wall is perfectly elastic, find (a) the impulse imparted by the wall on the sphere, (b) the impulse imparted by the floor on the sphere.**

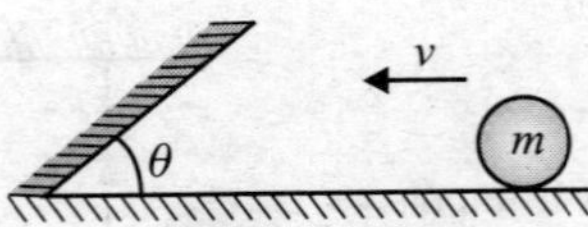

Fig. 1.71

Sol. As the collision is elastic, so the sphere will rebound with the same speed after collision.

$$J_1 \cos \theta = J_2 \quad \text{(i)}$$

$$-mv + J_1 \sin \theta = mv \quad \text{(ii)}$$

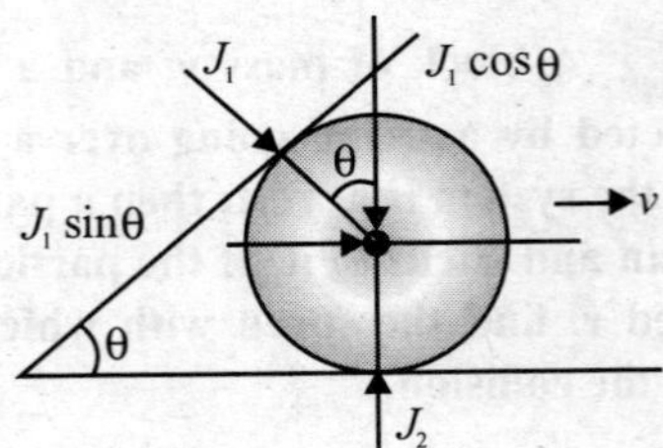

Fig. 1.72

Solving Eqs. (i) and (ii), we get

$$J_2 = 2\, mv \cot \theta$$

$$J_1 = 2\, mv \operatorname{cosec} \theta$$

COLLISION OR IMPACT

Collision is an isolated event in which a strong force acts between two or more bodies for a short time, which results in change in their velocities. In a collision, a relatively large force acts on each colliding particle for a relatively short time. The basic idea of a 'collision' is that the motion of the colliding particles (or of at least one of them) changes rather abruptly and that we can make a relatively clean separation of times that are 'before the collision' and those that are 'after the collision'.

- *It is not necessary that a physical contact takes place in a collision, e.g., when an alpha particle (He^4) 'collides' with the nucleus of gold (Au^{197}), the force acting between them being repulsive—the particles may not touch, even then it may be called a 'collision'.*
- *When a space probe approaches a large planet, swings around it, and then continues its course with increased speed (a slingshot encounter), that too is a collision. The probe and planet do not actually "touch", but a collision does not require contact, and a collision force does not have to be a force of contact; it can just as easily be a gravitational force, as in this case.*

Note:

- ***In a collision, particles may or may not come in physical contact.***
- ***The duration of collision Δt is negligible as compared to the usual time intervals of observation of motion.***
- ***In a collision, the effects of external non-impulsive forces such as gravity are not taken into account as due to small duration of collision (Δt). Average impulsive force responsible for collision is much larger than external forces acting on the system.***

The collision is, in fact, a redistribution of total momentum of the particles. Thus, law of conservation of linear momentum is indispensable in dealing with the phenomenon of collision between particles.

Line of Impact

The line passing through the common normal to the surfaces in contact during the impact is called line of impact. The force during collision acts along this line on both the bodies.

Direction of line of impact can be determined by

(a) geometry of colliding objects such as spheres, discs, wedge, etc;
(b) direction of change of momentum.

If one particle is stationary before collision, then the line of impact will be along its motion after collision.

Classification of Collisions

(a) On the Basis of Line of Impact

(i) Head-on collision: The velocities of the particles are along the same line before and after the collision.

(ii) Oblique collision: The velocities of the particles are along different lines before and after the collision.

(b) On the Basis of Energy

(i) Elastic collision: In an elastic collision, the particles regain their shape and size completely after collision. That is, no fraction of mechanical energy remains stored as deformation potential energy in the bodies. Thus, kinetic energy of a system after collision is equal to kinetic energy of a system before collision. Thus, in addition to the linear momentum, kinetic energy also remains conserved before and after collision.

(ii) Inelastic collision: In an inelastic collision, the particles do not regain their shape and size completely after collision. Some fraction of mechanical energy is retained by the colliding particles in the form of deformation potential energy. Thus, the kinetic energy of the particles no longer remains conserved. However, in the absence of external forces, law of conservation of linear momentum still holds good.

(iii) Perfectly inelastic collision: If velocity of separation just after collision becomes zero, then the collision is perfectly inelastic. Collision is said to be perfectly inelastic if both the particles stick together after collision and move with the same velocity.

Note: ***Actually collision between all real objects is neither perfectly elastic nor perfectly inelastic, its inelastic in nature.***

Illustrations of Line of Impact and Collisions Based on Line of Impact

(i) Two balls A and B are approaching each other such that their centres are moving along line CD.

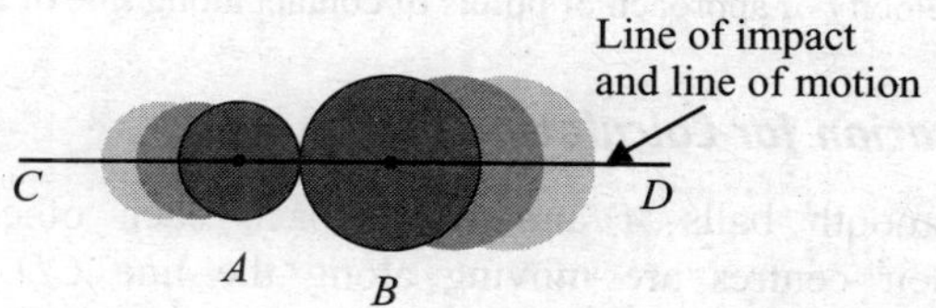

Head-on collision

Fig. 1.73

(ii) Two balls A and B are approaching each other such that their centres are moving along dotted lines as shown in Fig. 1.74.

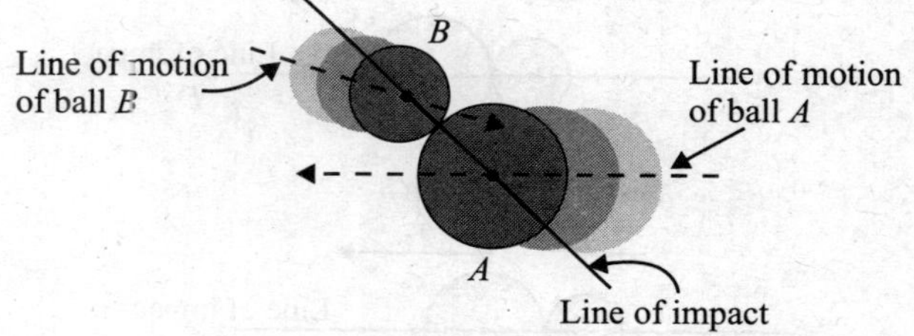

Oblique collision

Fig. 1.74

(iii) Ball is falling on a stationary wedge.

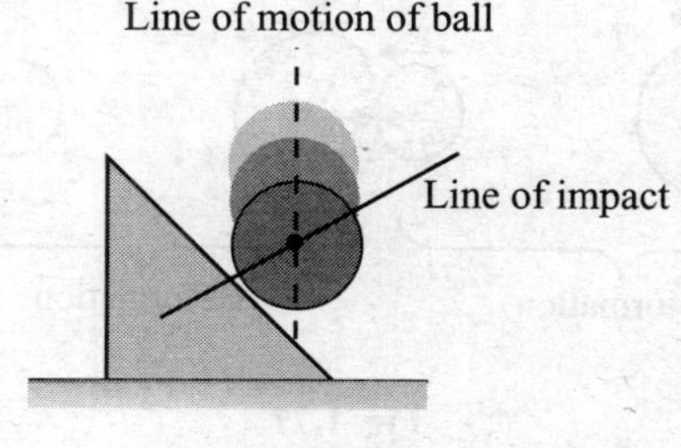

Oblique collision

Fig. 1.75

Coefficient of Restitution (e)

The coefficient of restitution is defined as the ratio of the impulses of recovery and deformation of either body.

$$e = \frac{\text{Impulse of reformation}}{\text{Impulse of deformation}} = \frac{\int F_r\,dt}{\int F_d\,dt}$$

$$= \frac{\text{Velocity of separation along line of impact}}{\text{Velocity of approach along line of impact}}$$

The most general expression for coefficient of restitution is

$$e = \frac{\text{Velocity of separation of points of contact along line of impact}}{\text{Velocity of approach of points of contact along line of impact}}$$

Illustration for Calculation of e

Two smooth balls A and B approach each other such that their centres are moving along the line CD in the absence of external impulsive force. Let the velocities of A and B just before collision be u_1 and u_2, respectively, and the velocities of A and B just after collision be v_1 and v_2, respectively.

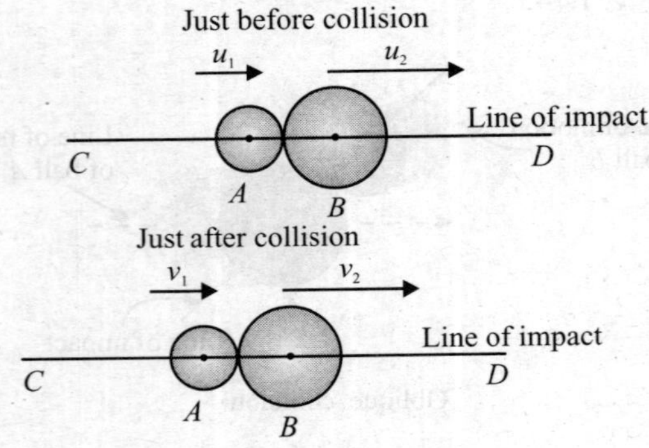

Fig. 1.76

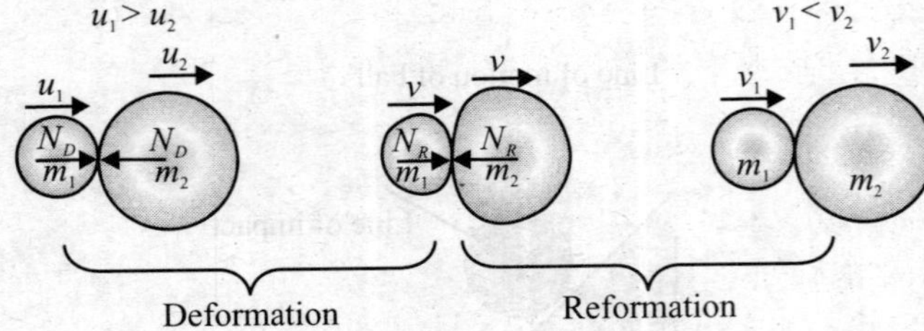

Fig. 1.77

Since, momentum is conserved for the system.

$\therefore \quad F_{ext} = 0$

$$\Rightarrow \quad m_1u_1 + m_2u_2 = (m_1 + m_2)v = m_1v_1 + m_2v_2$$

$$\Rightarrow \quad v = \frac{m_1u_1 + m_2u_2}{m_1 + m_2} = \frac{m_1v_1 + m_2v_2}{m_1 + m_2} \quad \text{(i)}$$

Impulse of Deformation

J_D = change in momentum of any one body during deformation

$= m_2(v - u_2)$ for m_2

$= m_1(-v + u_1)$ for m_1

Impulse of Reformation

J_R = change in momentum of any one body during reformation

$= m_2(v_2 - v)$ for m_2

$= m_1(v - v_1)$ for m_1

$$e = \frac{\text{Impulse of reformation } (J_R)}{\text{Impulse of deformation } (J_D)} = \frac{v_2 - v}{v - u_2}$$

$$= \frac{v_2 - v_1}{u_1 - u_2}$$

[substituting v from Eq. (i)]

$$= \frac{\text{Velocity of separation along line of impact}}{\text{Velocity of approach along line of impact}}$$

This is also known as Newton's experimental law.

Note: ***e is independent of shape and mass of the object but depends on the material.***

The coefficient of restitution is constant for two particular objects.

(a) For $e = 1$

$\Rightarrow$ Impulse of reformation = Impulse of deformation

$\Rightarrow$ Velocity of separation = Velocity of approach

$\Rightarrow$ Kinetic energy is conserved

$\Rightarrow$ Elastic collision

(b) For $e = 0$

$\Rightarrow$ Impulse of reformation = 0

$\Rightarrow$ Velocity of separation = 0

$\Rightarrow$ Kinetic energy is not conserved

$\Rightarrow$ Perfectly inelastic collision

(c) For $0 < e < 1$

$\Rightarrow$ Impulse of reformation < Impulse of deformation

$\Rightarrow$ Velocity of separation < Velocity of approach

$\Rightarrow$ Kinetic energy is not conserved

$\Rightarrow$ Inelastic collision

Note:

- *In case of contact collisions, e is always less than unity.*
- u_1, u_2, v_1 *and* v_2 *can be positive, negative or zero.*

Illustration 1.28 **A ball is projected with a velocity u at an elevation from a point distance d from a smooth vertical wall in a plane perpendicular to it. After rebounding from the wall, it returns to the point of projection, prove that $u^2 \sin 2\alpha = gd(1 + 1/e)$. Hence, find the maximum distance d for which the ball can return to the point of projection.**

Sol. The only vertical force on the ball is mg throughout its motion because during impact it experiences a horizontal force from the wall. We can use

$$u_y t - \frac{1}{2} gt^2 = s_y$$

Let t be the total time of flight.

$$\therefore \quad 0 = u \sin \alpha t - \frac{1}{2} gt^2$$

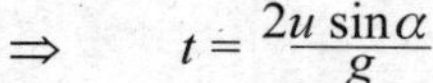

$$\Rightarrow \quad t = \frac{2u \sin\alpha}{g}$$

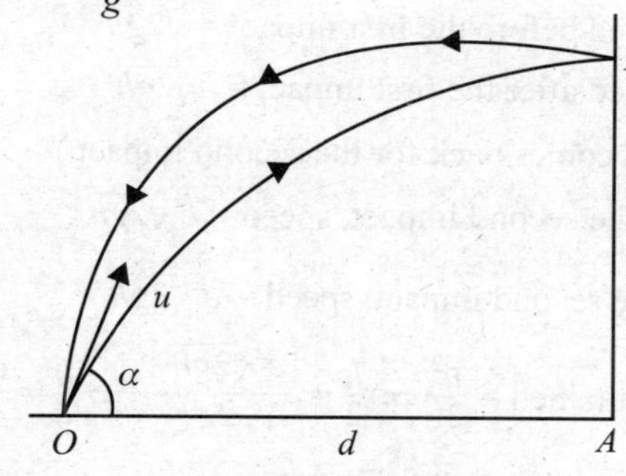

Fig. 1.78

Due to impact with the wall at B, the normal component (i.e., horizontal component) of velocity is reversed and becomes e times.

Horizontal velocity before impact = $u \cos \alpha$

and horizontal velocity after impact = $eu \cos \alpha$

Time taken to reach the wall, $t_1 = d/(u \cos \alpha)$

and time taken to come back to O from B,

$t_2 = d/(eu \cos \alpha)$ we have $t_1 + t_2 = t$

$$\Rightarrow \frac{d}{u\cos\alpha} + \frac{d}{eu\cos\alpha} = \frac{2u\sin\alpha}{g} \Rightarrow u^2 \sin 2\alpha = gd\left[1 + \frac{1}{e}\right]$$

As $\sin 2\alpha \le 1$,

$$\frac{gd}{u^2}\left[1 + \frac{1}{e}\right] \le 1 \Rightarrow d \le \frac{eu^2}{g(1+e)}$$

General Equation for Direct Impact

If u_1, u_2 are the velocities before the impact of the masses m_1, m_2 and v_1, v_2 are the velocities after the impact, then applying conservation of momentum, we get

$$m_1 u_1 + m_2 u_2 = m_1 v_1 + m_2 v_2 \text{ and } v_1 - v_2 = -e(u_1 - u_2)$$

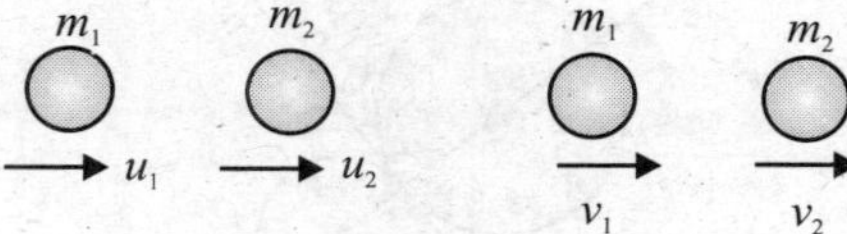

Fig. 1.79

Combining these equations, we get

$$v_1 = \frac{m_1 - em_2}{m_1 + m_2} u_1 + \frac{(1+e)m_2}{m_1 + m_2} u_2; \; v_2 = \frac{(1+e)m_1}{m_2 + m_1} u_1 + \frac{m_2 - em_1}{m_2 + m_1} u_2$$

For a perfectly elastic collision, we can substitute $e = 1$.

Special case: For $e = 1$ and $m_1 = m_2 = m$, we get

$$v_1 = u_2 \text{ and } v_2 = u_1$$

i.e., when two particles of equal mass collide elastically and the collision is head-on, they exchange their velocities, e.g.,

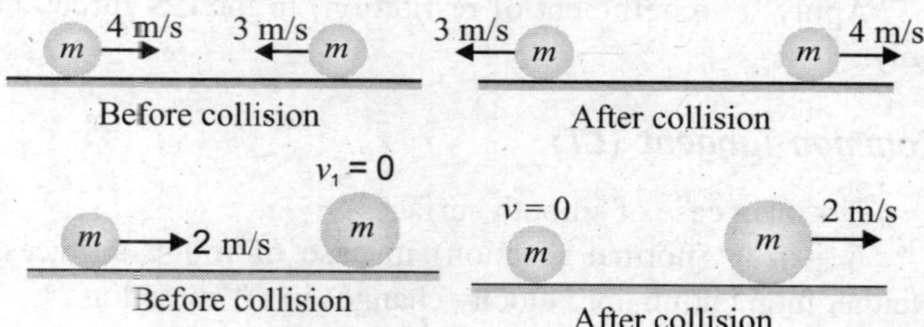

Fig. 1.80

If a body of mass m with initial velocity u strikes on another identical body but at rest and e is the coefficient of restitution, then their velocities v_1 and v_2 after collision will be

$$v_1 = \frac{1-e}{2} u; \; v_2 = \frac{1+e}{2} u$$

If a ball of mass m falls on ground from a vertical height h and rebounds with e as coefficient of restitution between them (Fig. 1.81), then the upward velocity of the ball after nth collision will be $(e^n u)$ and the maximum height attained by the ball after nth collision will be $(e^{2n} h)$.

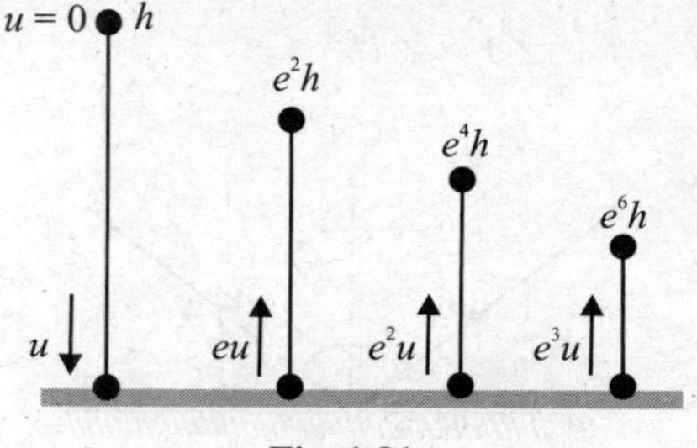

Fig. 1.81

Oblique Collision in Case of Smooth Surfaces

Common Normal (CN)

Force is exerted in CN direction only. Both bodies exert equal and opposite forces (action and reaction) on each other.

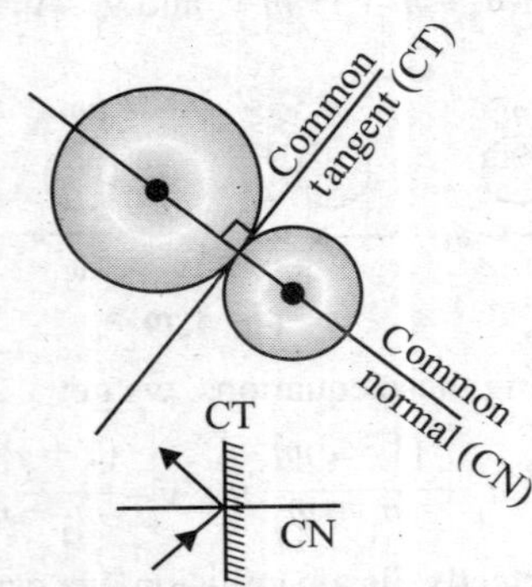

These CN and CT directions have nothing to do with the directions of velocity of the two bodies.

Fig. 1.82

Hence, momentum and velocities change accordingly in CN direction.

Apply 'e' (coefficient of restitution) in the CN direction only.

Common Tangent (CT)

$F = 0$ (in case of smooth surfaces)

$F = \mu \times$ (normal reaction) in case of rough surfaces. Neither momentum nor velocity changes in CT direction.

Illustration 1.29 What will be the angle of reflection in case of an inelastic collision? and v = ?

Sol. $v_1 = eu_1$ (CN by definition of e)

$$v_{\parallel} = u_{\parallel} \text{ (CT)}$$

$$\therefore \quad v\cos r = eu\cos i$$

$$v\sin r = u\sin i$$

Squaring and adding, we get

$$v = u\sqrt{\sin^2 i + e^2\cos^2 i}$$

Dividing, we get $\tan r = \dfrac{\tan i}{e}$

$$\Rightarrow \quad \angle r > \angle i$$

Fig. 1.83

Illustration 1.30 If a ball strikes with a velocity u_1 at the wall which itself is approaching it with a velocity u_2, then find the velocity of the ball after collision with the wall.

Sol. As the wall is heavy so after collision it will continue to move with the same velocity u_2. Relative velocity of separation is equal to relative velocity of approach. Hence,

$$v_1 - v_2 = -e(u_1 - u_2)$$

$$\Rightarrow \quad v_1 - (-u_2) = -e[u_1 - (-u_2)]$$

$$\Rightarrow \quad v_1 = -e(u_1 + u_2) - u_2$$

In case of perfectly elastic collision, $e = 1$.

$$v_1 = -u_1 - u_2 - u_2 = -(u_1 + 2u_2)$$

–ive sign indicates backward direction.

Illustration 1.31 A ball drops from a ceiling of a room and after rebounding twice from the floor reaches a height equal to half that of the ceiling. Show that the coefficient of restitution is $\sqrt[4]{1/2}$.

Sol. Let R = height of ceiling

$\Rightarrow$ Speed before the first impact $= \sqrt{2gh}$

$\Rightarrow$ Speed after the first impact $= e\sqrt{2gh}$

The ball comes back for the second impact.

Before the second impact, speed $= e\sqrt{2gh}$

After the second impact, speed $= e^2\sqrt{2gh}$

Height attained $= \dfrac{u^2}{2g} \Rightarrow \dfrac{h}{2} = \dfrac{[e^2\sqrt{2gh}\,]^2}{2g} \Rightarrow \dfrac{h}{2} = \dfrac{e^4(2gh)}{2g}$

$$\Rightarrow e = \frac{1}{\sqrt[4]{2}}$$

A particle 'B' moving along the dotted line collides with a rod also in state of motion as shown in Fig. 1.84. The particle B comes in contact with point C on the rod.

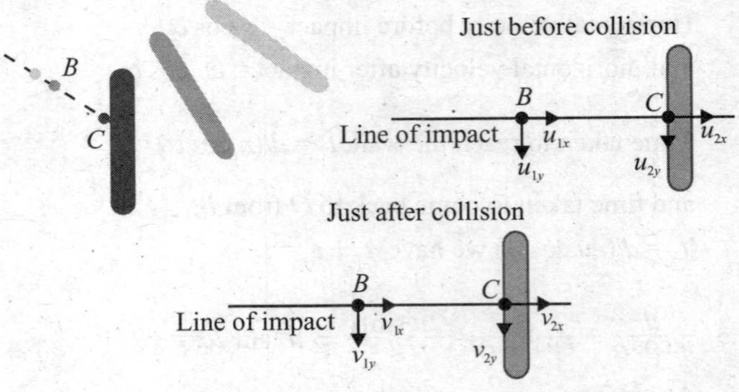

Fig. 1.84

To write down the expression for coefficient of restitution, e, we first draw the line of impact. Then we resolve the components of velocities of points of contact of both the bodies along line of impact just before and just after the collision Then,

$$e = \frac{v_{2x} - v_{1x}}{u_{1x} - u_{2x}}$$

Illustration 1.32 **A ball of mass m hits the floor with a speed v_0 making an angle of incidence α with the normal. The coefficient of restitution is e. Find the speed of the reflected ball and the angle of reflection of the ball.**

Sol. The component of velocity v_0 along common tangent direction $v_0 \sin \alpha$ will remain unchanged. Let v be the component along common normal direction after collision. Applying relative speed of separation = e × (relative speed of approach) along common normal direction, we get

$$v = ev_0 \cos \alpha$$

Thus, after collision, components of velocity v' are $v_0 \sin \alpha$ and $ev_0 \cos \alpha$.

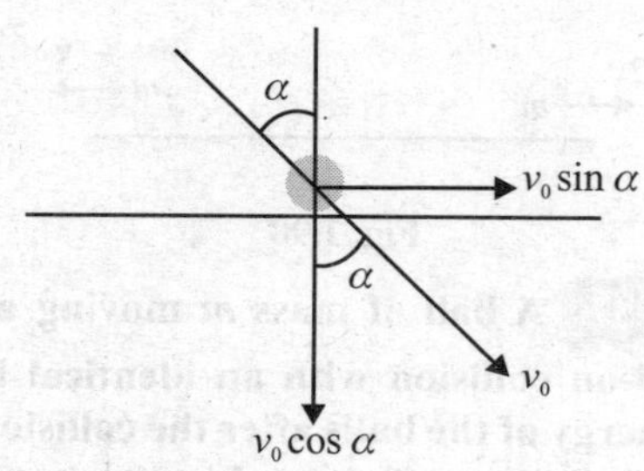

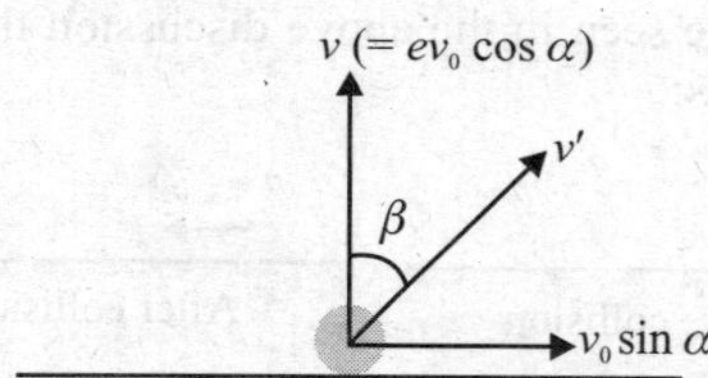

Fig. 1.85

$$\therefore \quad v' = \sqrt{(v_0 \sin \alpha)^2 + (ev_0 \cos \alpha)^2}$$

and

$$\tan \beta = \frac{v_0 \sin \alpha}{ev_0 \cos \alpha} = \frac{\tan \alpha}{e}$$

Note: ***For elastic collision, e* = 1.**

$$\therefore \quad v' = v_0 \text{ and } \beta = \alpha$$

Collision in One Dimension (Head-on)

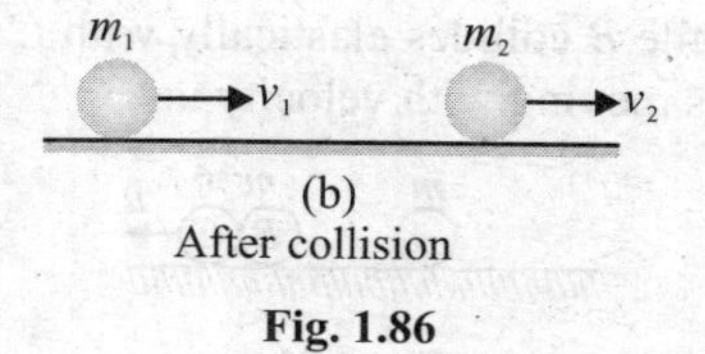

Fig. 1.86

$$u_1 > u_2,\ v_2 > v_1$$

$$e = \frac{v_2 - v_1}{u_1 - u_2} \Rightarrow (u_1 - u_2)e = (v_2 - v_1)$$

By momentum conservation, we get

$$m_1u_1 + m_2u_2 = m_1v_1 + m_2v_2$$

Now,

$$v_2 = v_1 + e(u_1 - u_2)$$

Hence,

$$v_1 = \frac{m_1u_1 + m_2u_2 - m_2e(u_1 - u_2)}{m_1 + m_2}$$

$$v_2 = \frac{m_1u_1 + m_2u_2 + m_1e(u_1 - u_2)}{m_1 + m_2}$$

Illustration 1.33 **Two identical balls are approaching towards each other on a straight line with velocity 2 m/s and 4 m/s, respectively. Find the final velocities, after elastic collision between them.**

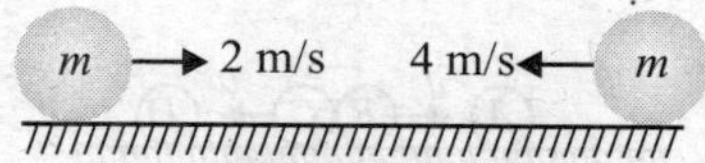

Fig. 1.87

Sol. The two velocities will be exchanged and the final motion is the reverse of the initial motion for both.

Fig. 1.88

Illustration 1.34 **Three balls *A*, *B* and *C* of same mass '*m*' are placed on a frictionless horizontal plane in a straight line as shown. Ball *A* is moved with velocity *u* towards the middle ball *B*. If all the collisions are elastic, find the final velocities of all the balls.**

Fig. 1.89

Sol. *A* collides elastically with *B* and comes to rest but *B* starts moving with velocity *u*.

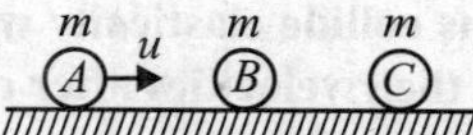

Fig. 1.90

After a while B collides elastically with C and comes to rest but C starts moving with velocity u.

Fig. 1.91

Therefore, final velocities, $V_A = 0$, $V_B = 0$ and $V_C = u$.

Illustration 1.35 **Four identical balls A, B, C and D are placed in a line on a frictionless horizontal surface. A and D are moved with the same speed 'u' towards the middle as shown. Assuming elastic collisions, find the final velocities.**

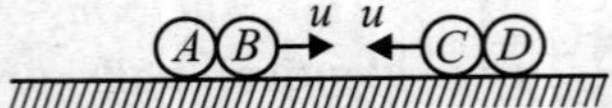

Fig. 1.92

Sol. A and D collide elastically with B and C, respectively, and come to rest but B and C start moving with velocity u towards each other as shown.

Fig. 1.93

B and C collide elastically and exchange their velocities to move in the opposite directions

Fig. 1.94

Now, B and C collide elastically with A and D, respectively, and come to rest but A and D start moving with velocity u away from each other as shown.

Fig. 1.95

Therefore, final velocities, $V_A = u\ (\leftarrow)$, $V_B = 0$, $V_C = 0$ and $V_D = u\ (\rightarrow)$.

Illustration 1.36 **Two particles of masses m and $2m$ moving in opposite directions collide elastically with velocity $2v$ and v, respectively. Find their velocities after collision.**

Fig. 1.96

Sol. Let the final velocities of m and $2m$ be v_1 and v_2, respectively, as shown in Fig. 1.97.

Fig. 1.97

By conservation of momentum, we get

$$m(2v) + 2m(-v) = m(v_1) + 2m(v_2)$$

or $$0 = mv_1 + 2mv_2$$

or $$v_1 + 2v_2 = 0 \quad \text{(i)}$$

and since the collision is elastic,

$$v_2 - v_1 = 2v - (-v)$$

or $$v_2 - v_1 = 3v \quad \text{(ii)}$$

Solving the above two equations, we get

$$v_2 = v \text{ and } v_1 = -2v$$

That is, mass $2m$ returns with velocity v while mass m returns with velocity $2v$ in the direction shown in Fig. 1.98.

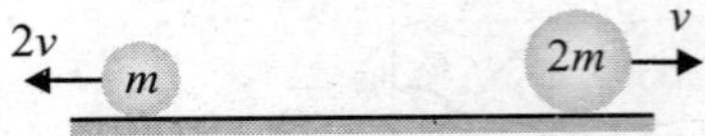

Fig. 1.98

Illustration 1.37 **A ball of mass m moving at a speed v makes a head-on collision with an identical ball at rest. The kinetic energy of the balls after the collision is 3/4th of the original. Find the coefficient of restitution.**

Sol. As we have seen in the above discussion that under the given conditions:

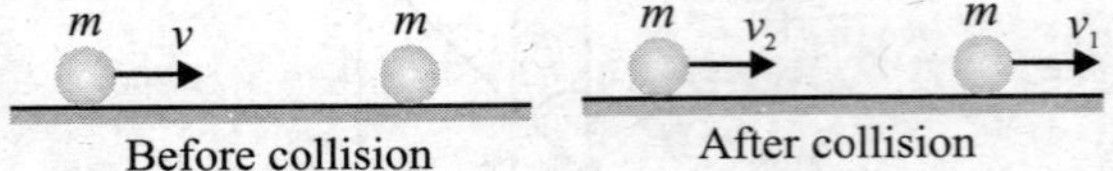

Fig. 1.99

$$v_1 = \left(\frac{1+e}{2}\right) v \text{ and } v_2 = \left(\frac{1-e}{2}\right) v$$

Given that $K_f = \frac{3}{4} K_i$

or $$\frac{1}{2} mv_1^2 + \frac{1}{2} mv_2^2 = \frac{3}{4}\left(\frac{1}{2} mv^2\right)$$

Substituting the values, we get

$$\left(\frac{1+e}{2}\right)^2 + \left(\frac{1-e}{2}\right)^2 = \frac{3}{4}$$

or $(1 + e)^2 + (1 - e)^2 = 3$ or $2 + 2e^2 = 3$

or $$e^2 = \frac{1}{2} \text{ or } e = \frac{1}{\sqrt{2}}$$

Illustration 1.38 **A ball is moving with velocity 2 m/s towards a heavy wall moving towards the ball with speed 1 m/s as shown in Fig. 1.100. Assuming collision to be elastic, find the velocity of the ball immediately after the collision.**

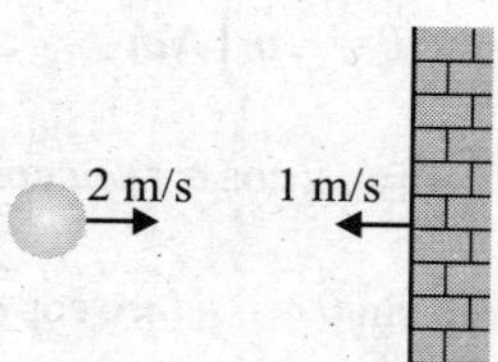

Fig. 1.100

Sol. The speed of the wall will not change after the collision. Let v be the velocity of the ball after collision in the direction shown in Fig. 1.101. Since collision is elastic ($e = 1$),

Speed of separation speed = Speed of approach

or $\quad v - 1 = 2 + 1$ or $v = 4$ m/s

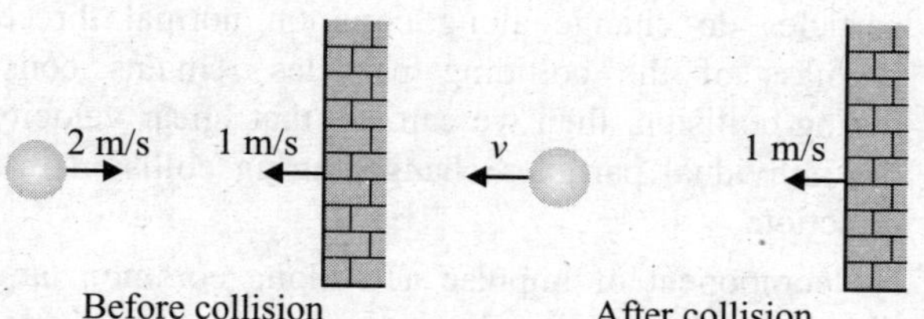

Fig. 1.101

Illustration 1.39 **Two balls of masses 2 kg and 4 kg are moved towards each other with velocities 4 m/s and 2 m/s, respectively, on a frictionless surface. After colliding, the 2 kg ball returns back with velocity 2 m/s. Then find**

(a) velocity of the 4 kg ball after collision

(b) coefficient of restitution *e*;

(c) impulse of deformation J_D;

(d) maximum potential energy of deformation;

(e) impulse of reformation J_R.

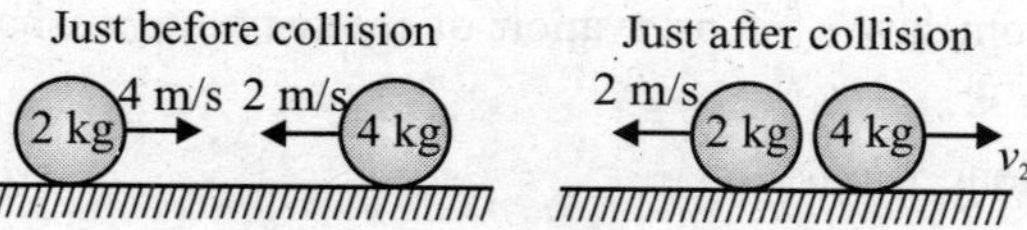

Fig. 1.102

Sol. (a) By momentum conservation,

$$2(4) - 4(2) = 2(-2) + 4(v_2)$$

$$\Rightarrow \quad v_2 = 1 \text{ m/s}$$

(b) $e = \dfrac{\text{Velocity of separation}}{\text{Velocity of approach}} = \dfrac{1-(-2)}{4-(-2)} = \dfrac{3}{6} = 0.5$

(c) At the maximum deformed state, by conservation of momentum, common velocity is $v = 0$.

$$J_D = m_2(v - u_2) = 4[(0 - (-2)] = 8 \text{ N s}$$

(d) Potential energy at the maximum deformed state,

U = loss in kinetic energy during deformation

or $\quad U = \left(\frac{1}{2} m_1 u_1^2 + \frac{1}{2} m_2 u_2^2\right) - \frac{1}{2}(m_1 + m_2) v^2$

$$= \left(\frac{1}{2} 2(4)^2 + \frac{1}{2} 4(2)^2\right) - \frac{1}{2}(2+4)(0)^2$$

$$= 24 \text{ J}$$

(e) $J_R = m_2(v_2 - v) = 4(1 - 0) = 4$ N s

also $J_R = eJ_D = 0.5 \times 8 = 4$ N s

Illustration 1.40 **Two point particles *A* and *B* are placed in line on a frictionless horizontal plane. If particle *A* (mass 1 kg) is moved with velocity 10 m/s towards stationary particle *B* (mass 2 kg) and after collision the two move at an angle of 45° with the initial direction of motion, then find**

Fig. 1.103

(a) velocities of *A* and *B* just after collision.

(b) coefficient of restitution.

Sol. The very first step to solve such problems is to find the line of impact which is along the direction of force applied by *A* on *B*, resulting the stationary particle *B* to move. Thus, by watching the direction of motion of *B*, line of impact can be determined. In this case, the line of impact is along the direction of motion of *B*, i.e., 45° with the initial direction of motion of *A*.

(a) Let us apply the principle of conservation of momentum.

Along *x*-direction:

$$m_A u_A = m_A v_A \cos 45° + m_B v_B \cos 45°$$

or $\quad 1(10) = 1(v_A \cos 45°) + 2(v_B \cos 45°)$

or $\quad v_A + 2v_B = 10\sqrt{2} \qquad$ (i)

Along *y*-direction:

$$0 = m_A v_A \sin 45° - m_B v_B \sin 45°$$

or $\quad 0 = 1(v_A \sin 45°) - 2(v_B \sin 45°)$

or $\quad v_A = 2v_B \qquad$ (ii)

Solving the two equations, $v_A = \dfrac{10}{\sqrt{2}}$ m/s and $v_B = \dfrac{5}{\sqrt{2}}$ m/s

Fig. 1.104

(b) $e = \dfrac{\text{Velocity of separation along line of impact}}{\text{Velocity of approach along line of impact}}$

$$= \frac{v_B - v_A \cos 90°}{u_A \cos 45°} = \frac{\frac{5}{\sqrt{2}} - 0}{\frac{10}{\sqrt{2}}} = \frac{1}{2}$$

Hence, P is constant.

Illustration 1.41 **A bullet of mass 50 g is fired from below into the bob of mass 450 g of a long simple pendulum as shown in Fig. 1.105. The bullet remains inside the bob and the bob rises through a height of 1.8 m. Find the speed of the bullet.**

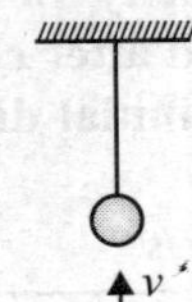

Fig. 1.105

Sol. Let the speed of the bullet be v. Let the common velocity of the bullet and the bob, after the bullet is embedded into the bob, be V. By the principle of conservation of linear momentum,

$$V = \frac{(0.05 \text{ kg})\, v}{0.45 \text{ kg} + 0.05 \text{ kg}} = \frac{v}{10}$$

The string becomes loose and the bob will go up with a deceleration of $g = 10 \text{ m/s}^2$. As it comes to rest at a height of 1.8 m, using the equation $v^2 = u^2 + 2ax$, we get

$$1.8 \text{ m} = \frac{\left(\frac{v}{10}\right)^2}{2 \times 10 \text{ m/s}^2} \quad \text{or} \quad v = 60 \text{ m/s}$$

Illustration 1.42 **A small ball of mass m collides with a rough wall having coefficient of friction μ at an angle θ with the normal to the wall. If after collision the ball moves with angle α with the normal to the wall and the coefficient of restitution is e, then find the reflected velocity v of the ball just after collision.**

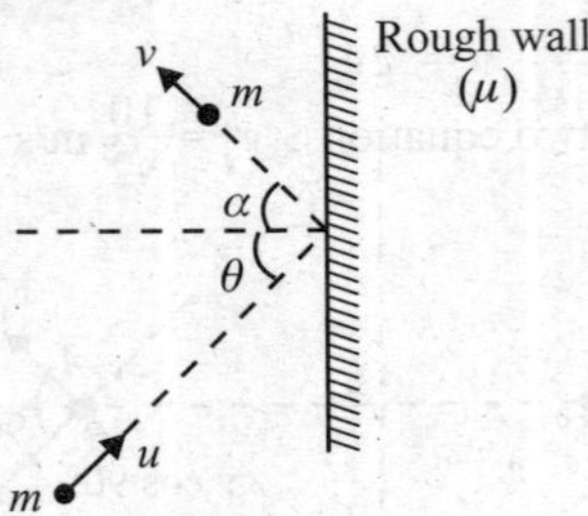

Fig. 1.106

Sol. $mv \cos \alpha - [m(-u \cos \theta)] = \int N dt$

$$mv \sin \alpha - mu \sin \theta = -\mu \int N dt$$

and $e = \dfrac{v \cos \alpha}{u \cos \theta} \Rightarrow v \cos \alpha = eu \cos \theta$ (i)

or $mv \sin \alpha - mu \sin \theta = -\mu\,(mv \cos \alpha + mu \cos \theta)$ (ii)

From Eqs. (i) and (ii), $v = \dfrac{u}{\sin\alpha}\,[\sin \theta - \mu \cos \theta\,(e + 1)]$

Collision in Two Dimensions (Oblique)

1. A pair of equal and opposite impulses act along common normal direction. Hence, linear momentum of individual particles do change along common normal direction. If mass of the colliding particles remains constant during collision, then we can say that linear velocity of the individual particles change during collision in this direction.
2. No component of impulse acts along common tangent direction. Hence, linear momentum or linear velocity of individual particles (if mass is constant) remains unchanged along this direction.
3. Net impulse on both the particles is zero during collision. Hence, net momentum of both the particles remains conserved before and after collision in any direction.
4. Definition of coefficient of restitution can be applied along common normal direction, i.e., along common normal direction we can apply.

Relative speed of separation = e × (relative speed of approach)

If the velocities of colliding masses are not linear, then it is known as oblique collision.

From law of conservation of momentum, we have the following.

Along x-axis:

$$m_1 u_1 \cos \alpha + m_2 u_2 \cos \beta = m_1 v_1 \cos \theta_1 + m_2 v_2 \cos \theta_2$$

Along y-axis:

$$m_1 u_1 \sin \alpha + m_2 u_2 \sin \beta = m_1 v_1 \sin \theta_1 + m_2 v_2 \sin \theta_2$$

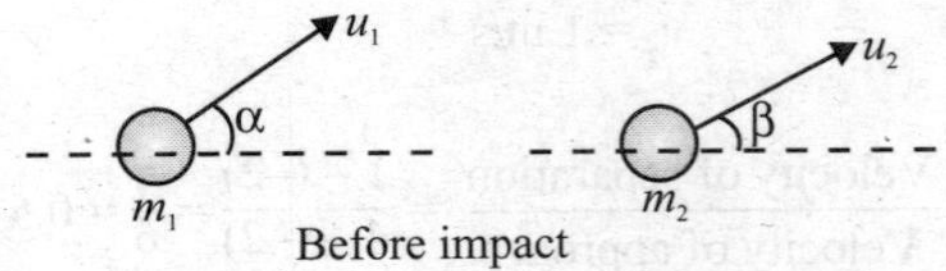

v_1, m_1, θ_1, v_2, θ_2, m_2

After impact

Fig. 1.107

For Oblique Impact

When two bodies collide obliquely, their relative velocity, resolved along their common normal after the impact, is in a constant ratio to their relative velocity before impact (resolved along common normal) and is in the opposite direction.

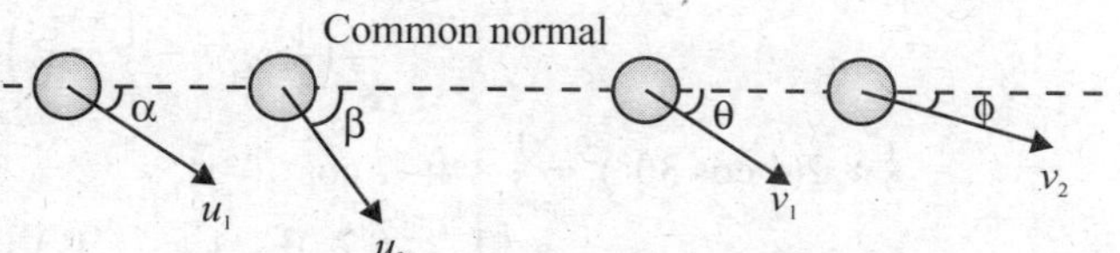

Fig. 1.108

$$\frac{v_1 \cos\theta - v_2 \cos\phi}{u_1 \cos\alpha - u_2 \cos\beta} = -e$$

$$\Rightarrow v_1 \cos\theta - v_2 \cos\phi = -e(u_1 \cos\alpha - u_2 \cos\beta)$$

Illustration 1.43 **A ball of mass *m* makes an elastic collision with another identical ball at rest. Show that if the collision is oblique, the bodies go at right angles to each other after collision.**

Sol. Method 1: In head-on elastic collision between two particles, they exchange their velocities. In this case, the component of ball 1 along common normal direction, $v\cos\theta$, becomes zero after collision, while that of ball 2 becomes $v\cos\theta$. While the components along common tangent direction of both the particles remain unchanged. Thus, the components along common tangent and common normal direction of both the balls in tabular form are given in Table 1.1.

Table 1.1

Ball	Component along common tangent direction		Component along common normal direction	
	Before collision	After collision	Before collision	After collision
1	$v\sin\theta$	$v\sin\theta$	$v\cos\theta$	0
2	0	0	0	$v\cos\theta$

From Table 1.1 and Fig. 1.109, we see that both the balls move at right angle after collision with velocities $v\sin\theta$ and $v\cos\theta$.

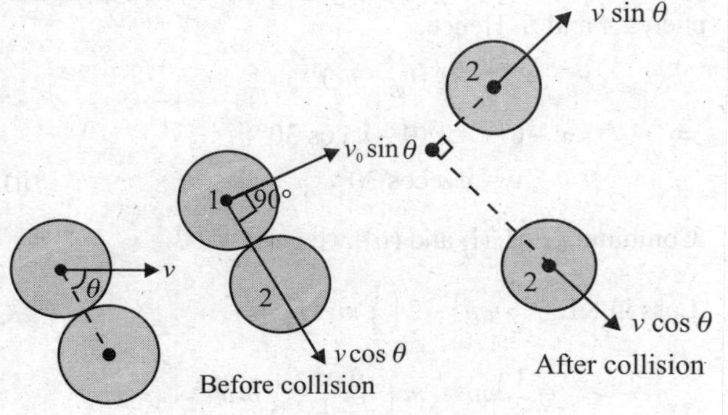

Fig. 1.109

Method 2: $mu = mv_1 \cos\theta_1 + mv_2 \cos\theta_2$ (i)

$0 = mv_1 \sin\theta_1 + mv_2 \sin\theta_2$ (ii)

$\frac{1}{2}mu^2 = \frac{1}{2}mv_1^2 + \frac{1}{2}mv_2^2$ (iii)

Squaring and adding Eqs. (i) and (ii), we get

$$u^2 = v_1^2 + v_2^2 + 2v_1v_2\cos(\theta_1 + \theta_2)$$

Using Eq. (iii), we have $u^2 = v_1^2 + v_2^2$.

Note:

- ***When two identical bodies have an oblique elastic collision, with one particle at rest before collision, then the two particles will go in perpendicular directions.***
- ***The colliding balls of the same mass in an elastic direct impact interchange their velocities.***

Illustration 1.44 **Two spheres are moving towards each other. Both have the same radius but their masses are 2 kg and 4 kg. If the velocities are 4 m/s and 2 m/s, respectively, and coefficient of restitution is $e = 1/3$, find**

(a) the common velocity along the line of impact;

(b) final velocities along line of impact;

(c) impulse of deformation;

(d) impulse of reformation;

(e) maximum potential energy of deformation;

(f) loss in kinetic energy due to collision.

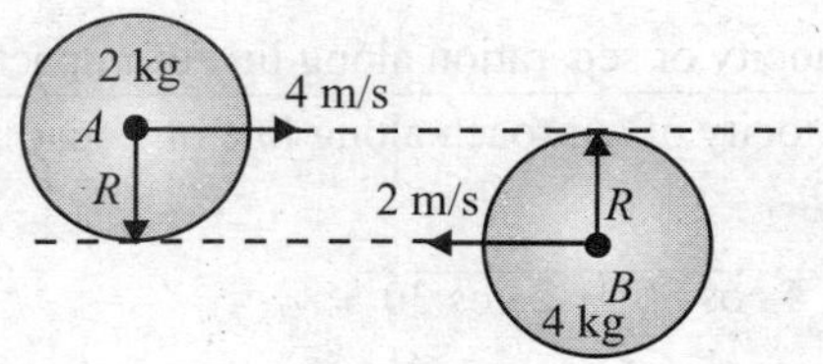

Fig. 1.110

Sol. In ΔABC, $\sin\theta = \frac{BC}{AB} = \frac{R}{2R} = \frac{1}{2} \quad \Rightarrow \theta = 30°$

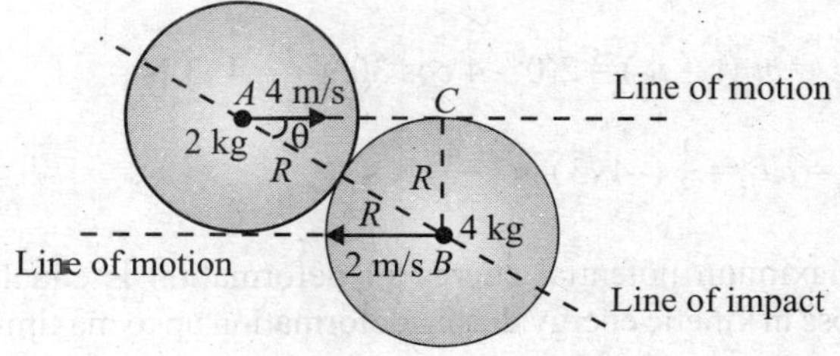

Fig. 1.111

(a) By conservation of momentum along line of impact

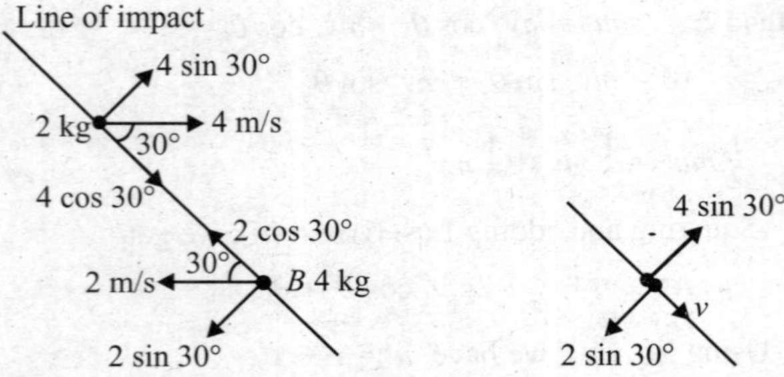

Just before collision along line of impact Maximum deformed state

Fig. 1.112

$$2(4\cos 30^\circ) - 4(2\cos 30^\circ) = (2+4)v$$

or $v = 0$ (common velocity along line of impact)

(b)

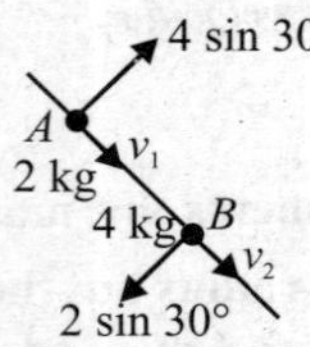

Just after collision along line of impact

Fig. 1.113

Let v_1 and v_2 be the final velocities of A and B, respectively, then by conservation of momentum along the line of impact,

$$2(4\cos 30^\circ) - 4(2\cos 30^\circ) = 2(v_1) + 4(v_2)$$

or $0 = v_1 + 2v_2$ (i)

Coefficient of restitution,

$$e = \frac{\text{Velocity of separation along line of impact}}{\text{Velocity of approach along line of impact}}$$

or $\dfrac{1}{3} = \dfrac{v_2 - v_1}{4\cos 30^\circ + 2\cos 30^\circ}$

or $v_2 - v_1 = \sqrt{3}$ (ii)

From the above two equations,

$$v_1 = \frac{-2}{\sqrt{3}} \text{ m/s and } v_2 = \frac{1}{\sqrt{3}} \text{ m/s}$$

(c) $J_D = m_1(v - u_1) = 2(0 - 4\cos 30^\circ) = -4\sqrt{3}$ N s

(d) $J_R = eJ_D = \frac{1}{3}(-4\sqrt{3}) = -\dfrac{4}{\sqrt{3}}$ N s

(e) Maximum potential energy of deformation is equal to loss in kinetic energy during deformation up to maximum deformed state. Hence,

$$U = \frac{1}{2}m_1(u_1\cos\theta)^2 + \frac{1}{2}m_2(u_2\cos\theta)^2 - \frac{1}{2}(m_1 + m_2)v^2$$

$$= \frac{1}{2}\times 2(4\cos 30^\circ)^2 + \frac{1}{2}\times 4(-2\cos 30^\circ)^2 - \frac{1}{2}\times(2+4)(0)^2$$

or $U = 18$ J

(f) Loss in kinetic energy,

$$\Delta KE = \frac{1}{2}m_1(u_1\cos\theta)^2 + \frac{1}{2}m_2(u_2\cos\theta)^2 - \left(\frac{1}{2}m_1v_1^2 + \frac{1}{2}m_2v_2^2\right)$$

$$= \frac{1}{2}\times 2(4\cos 30^\circ)^2 + \frac{1}{2}\times 4(-2\cos 30^\circ)^2 - \left[\frac{1}{2}\times 2\left(\frac{2}{\sqrt{3}}\right)^2 + \frac{1}{2}\times 4\left(\frac{1}{\sqrt{3}}\right)^2\right]$$

$\Delta KE = 16$ J

Illustration 1.45 **Two equal spheres of mass *m* are in contact on a smooth horizontal table. A third identical sphere impinges symmetrically on them and is reduced to rest. Prove that $e = 2/3$ and find the loss in KE.**

Sol. Let u be the velocity of sphere A before impact. As the spheres are identical, the triangle ABC formed by joining their centres is equilateral. The spheres B and C will move in directions AB and AC after impact making an angle of 30° with the original line of motion of sphere A.

Let v be the speed of the other spheres after impact.

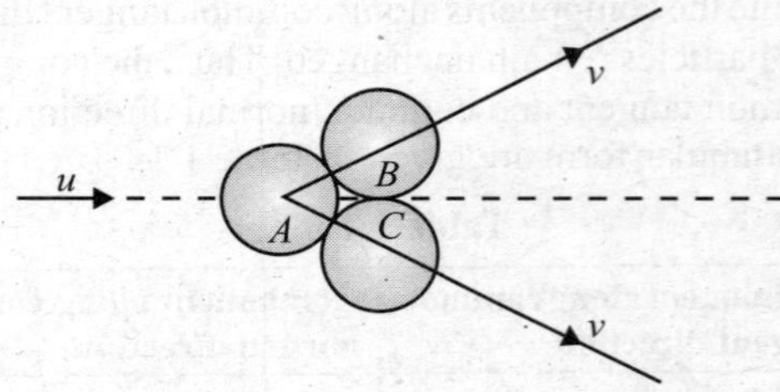

Fig. 1.114

From momentum conservation,

$$mu = mv\cos 30^\circ + mv\cos 30^\circ$$

$$u = v\sqrt{3} \quad \text{(i)}$$

From Newton's experimental law, for an oblique collision, we have to take components along the normal, i.e., along AB for spheres A and B. Hence,

$$v_B - v_A = -e(u_B - u_A)$$

$$\Rightarrow \quad v - 0 = -e(0 - u\cos 30^\circ)$$

$$v = e\,u\cos 30^\circ \quad \text{(ii)}$$

Combining Eqs. (i) and (ii), we get $e = 2/3$.

$$\text{Loss in KE} = \frac{1}{2}mu^2 - 2\left(\frac{1}{2}mv^2\right)$$

$$= \frac{1}{2}mu^2 - m\left(\frac{u}{\sqrt{3}}\right)^2 = \frac{1}{6}mu^2$$

Concept Application Exercise 1.3

1. A ball is dropped on a floor from a height h. If the coefficient of restitution is e, find the height to which the ball will rise after touching the floor and the time it will take to come to rest again.

2. An elevator platform is going up at a speed of 20 m/s and during its upward motion a small ball of 50 g mass, falling in downward direction, strikes the platform at speed 5 m/s. Find the speed with which the ball rebounds.

3. A ball of mass m is distributed from the top of a fixed smooth circular tube in a vertical plane and falls impinging on a ball of mass 2 m at the bottom. The coefficient of restitution is 1/2. Find the heights to which the balls rise after a second impact.

4. A smooth ball is dropped from a height h on a smooth incline, as shown in Fig. 1.115. After collision the velocity of the ball is directed horizontally. Find the coefficient of restitution.

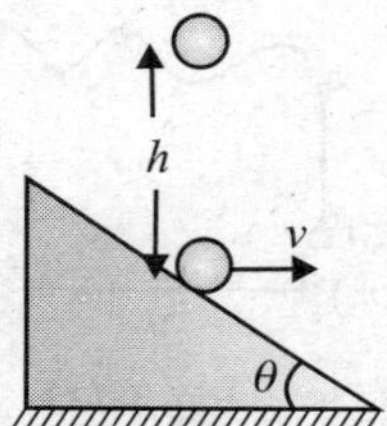

Fig. 1.115

5. A particle of mass 1 kg is attached to a string of length 5 m. The string is attached to a fixed point O. It is released from the position as shown in Fig. 1.116. Calculate

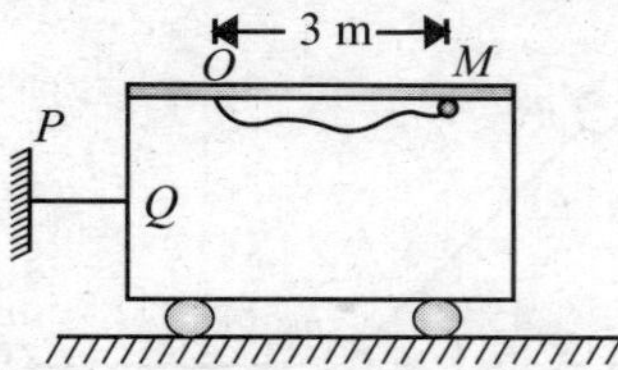

Fig. 1.116

(i) the impulse developed in the string when it becomes taut;
(ii) the velocity of the particle just after the string becomes taut;
(iii) the impulse developed in this string PQ at this instant.

6. A heavy ball of mass $2M$ moving with a velocity v_0 collides elastically head-on with a cradle of three identical balls each of mass M as shown in Fig. 1.117. Determine the velocity of each ball after collision.

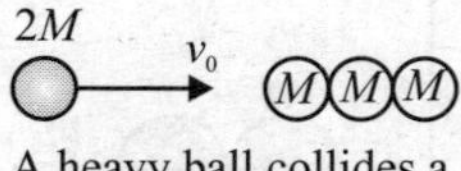

A heavy ball collides a cradle of three identical balls.

Fig. 1.117

7. In Fig. 1.118, there are n identical spheres of mass m suspended with wires of equal length. The spheres are almost in contact with each other. Sphere 1 is pulled aside and released. If sphere 1 strikes sphere 2 with velocity u, find an expression for velocity v_n of the nth sphere immediately after being struck by the one adjacent to it. The coefficient of restitution for all the impacts is e.

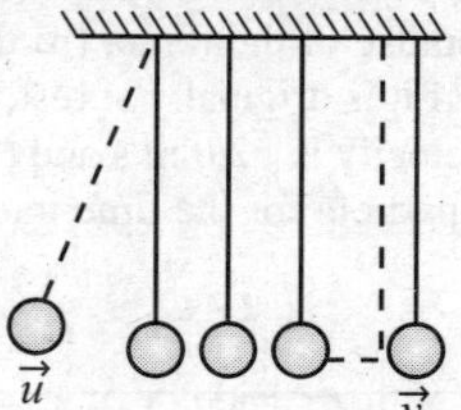

Fig. 1.118

8. A smooth sphere of mass m is moving on a horizontal plane with a velocity $3\hat{i} + \hat{j}$ when it collides with a vertical wall which is parallel to the $\hat{j}$ vector. If the coefficient of restitution between the sphere and the wall is 1/2, find
(a) the velocity of the sphere after impact,
(b) the loss in kinetic energy caused by the impact.
(c) the impulse that acts on the sphere.

9. Two smooth spheres, A and B, having equal radii, lie on a horizontal table. A is of mass m and B is of mass $3m$. The spheres are projected towards each other with velocity vector $5\hat{i} + 2\hat{j}$ and $2\hat{i} - \hat{j}$, respectively, and when they collide the line joining their centres is parallel to the vector $\hat{i}$. If the coefficient of restitution between A and B is 1/3, find the velocities after impact and the loss in kinetic energy caused by the collision. Find also the magnitude of the impulses that act at the instant of impact.

10. Show that the loss of KE in a direct impact of two spheres of masses m_1 moving with initial velocities u_1 and u_2, respectively, is $\frac{m_1 m_2}{2(m_1 + m_2)}(u_1 - u_2)^2 (1 - e^2)$
(e in the coefficient of restitution)

11. A block m_1 strikes a stationary block m_3 inelasticly. Another block m_2 is kept on m_3. Neglecting the friction between all contacting surfaces, calculate the fractional decrease in KE of the system in collision.

12. A mass m_1 moves with a great velocity. It strikes another mass m_2 at rest in head-on collision. It comes back along its path with low speed after collision. Then find out whether $m_1 < m_2$ or $m_1 > m_2$.

13. A ball of mass 4 kg moving with a velocity of 12 m/s impinges directly on another ball of mass 8 kg moving with velocity of 4 m/s in the same direction. Find their velocities after impact and calculate the loss of KE due to impact if $e = 0.5$.

14. A bullet of mass 2 g travelling at a speed of 500 m/s is fired into a ballistic pendulum of mass 1.0 kg suspended from a cord 1.0 m long. The bullet penetrates the pendulum and emerges with a velocity of 100 m/s. Through what vertical height will the pendulum rise?

15. A 3.00 kg steel ball strikes the wall with a speed of 10.0 m/s at an angle of 60° with the surface. It bounces off with the same speed and angle. If the ball is in contact with the wall for 0.200 s, what is the average force exerted by the wall on the ball?

16. The magnitude of the net force exerted in the x-direction on a 2.50 kg particle varies in time as shown in Fig. 1.119. Find (a) the impulse of the force, (b) the final velocity the particle attains if it is originally at rest, (c) its final velocity if its original velocity is −2.0 m/s and (d) the average force exerted on the particle for the time interval between 0 and 5.00 s.

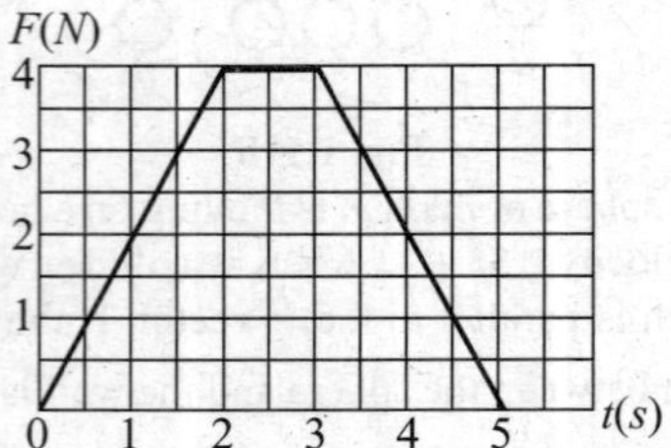

Fig. 1.119

17. A spherical imperfectly elastic ball strikes a plane with velocity 8 m/s at an angle of 30° with the plane. Determine the magnitude and direction of the velocity after impact if $e = 0.5$ (neglect gravity).

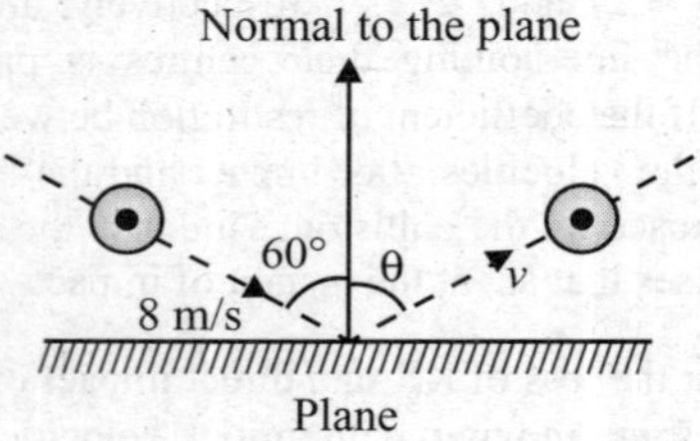

Fig. 1.120

18. Three balls of masses m_1, m_2 and m_3 are lying in a straight line. The first ball is moved with a certain velocity so that it strikes the second ball directly and itself comes to rest. The second ball collides with the third and is itself reduced to rest. If e is the coefficient of restitution for each ball, write down the relation of m_3 in terms of m_1 and m_2.

19. Two identical balls A and B lie on a smooth horizontal surface, which gradually merges into a curve to a height 3.2 m. Ball A is given a velocity of 10 m/s, to collide head-on with ball B, which then takes up the curved path. What is the minimum coefficient of restitution, e, for the collision between A and B in order that ball B reaches the highest point C of the curve.

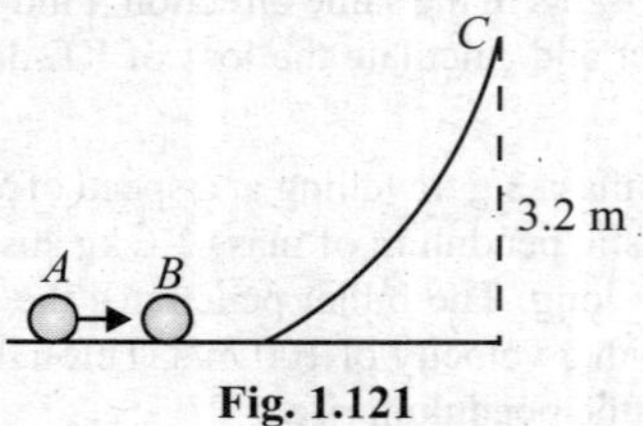

Fig. 1.121

20. A smooth ball of mass m is suspended from a light string of length 1 m. Another ball of mass $2m$ strikes the ball of mass m horizontally with a speed of $u = \sqrt{35}$ m/s. The coefficient of restitution for the collision is e. The string becomes loose, when it makes an angle of 30° with the horizontal; find the value of e.

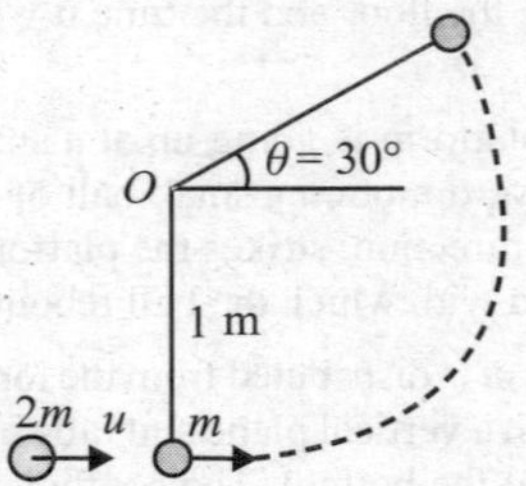

Fig. 1.122

21. Two small discs, A of mass $2m$ and B of mass m, are connected by a light, inextensible string of length $13l$, placed on a smooth horizontal plane, separated by a distance of $5l$. Disc B is given a velocity $26v$ in a direction normal to AB. Find the velocity of A when the string just becomes taut.

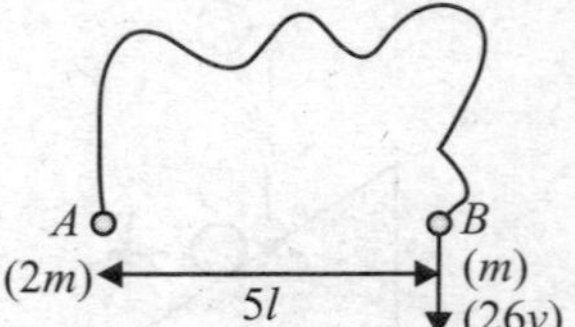

Fig. 1.123

22. Three particles A, B and C of respective masses m_1, m_2 and m_3 lie on a smooth horizontal surface, and are fastened to two light inextensible strings as shown in Fig. 1.124. The particle A is imparted an impulse J along $\overrightarrow{BA}$. Find the initial speed of each particle.

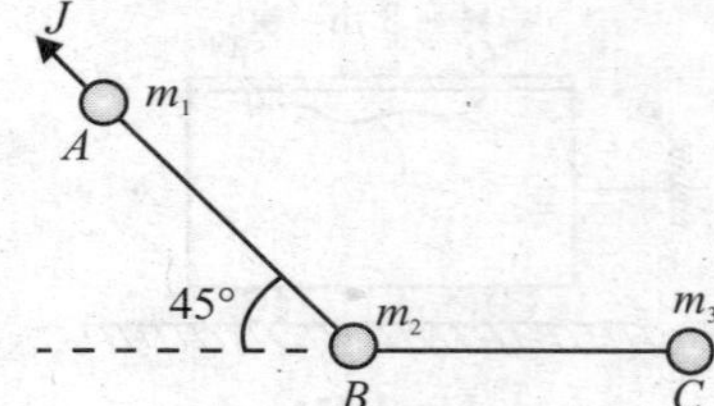

Fig. 1.124

23. A ball moving along a straight line collides elastically with another stationary ball of the same mass. At the moment of collision, the angle between the straight line passing through the centres of the balls and the direction of initial motion of the striking ball is θ, find the fraction of the kinetic energy of the striking ball converted into potential energy at the moment of the maximum deformation.

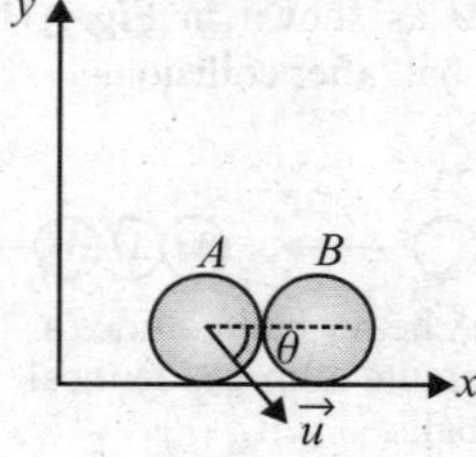

Fig. 1.125

VARIABLE MASS SYSTEM

Consider at any time t, a system of mass m travelling with velocity v as shown. After a small time dt, let its mass becomes $m + dm$ and velocity $v + dv$. Here a small mass dm travelling with velocity u adds into this system of mass m. F_{ext} is the force on m due to an external agent. For the system of $(m + dm)$:

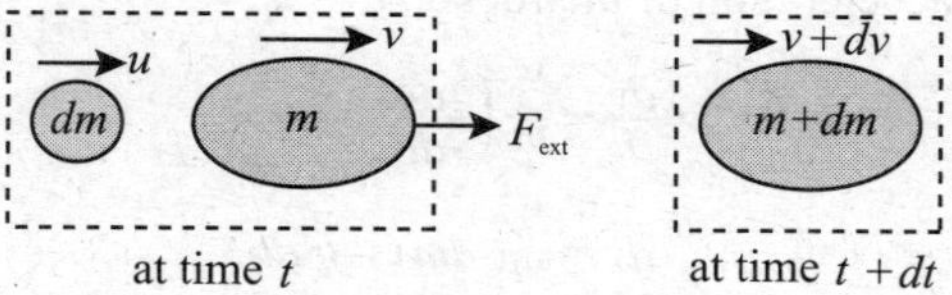

Fig. 1.126

Impulse = change in momentum

$\Rightarrow \quad F_{ext}\, dt = (m + dm)(v + dv) - mv - dm\, u$

$\quad = mdv + vdm + dm.dv - dmu$

but $dm \cdot dv$ is very small, so it is negligible.

$\Rightarrow \quad F_{ext}\, dt = mdv - (u - v)\, dm$

$\Rightarrow \quad F_{ext}\, dt + v_{rel}\, dm = mdv$

where $v_{rel} = u - v$ is the relative velocity of the elementary mass dm w.r.t. the initial mass of system.

$$\Rightarrow \quad F_{ext} + v_{rel}\frac{dm}{dt} = m\frac{dv}{dt}$$

The above equation is also valid in vector form

$$\vec{F}_{ext} + \vec{v}_{rel}\frac{dm}{dt} = m\frac{d\vec{v}}{dt}$$

Note: ***The term $\vec{v}_{rel}$ = (dm/dt) is known as thrust. It is the force applied by mass dm being added to the system.***

Illustration 1.46 **A flat car of mass m_0 starts moving to the right due to a constant horizontal force F. Sand spills on the flat car from a stationary hopper. The rate of loading is constant and equal to μ kg/s. Find the time dependence of the velocity and the acceleration of the flat car in the process of loading. The friction is negligibly small.**

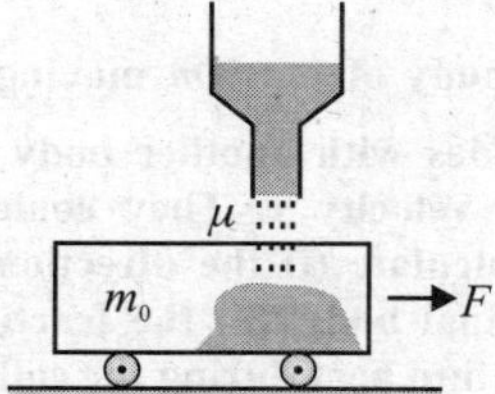

Fig. 1.127

Sol. Initial velocity of the flat car is zero. Let v be its velocity at time t and m be its mass at that instant. Then

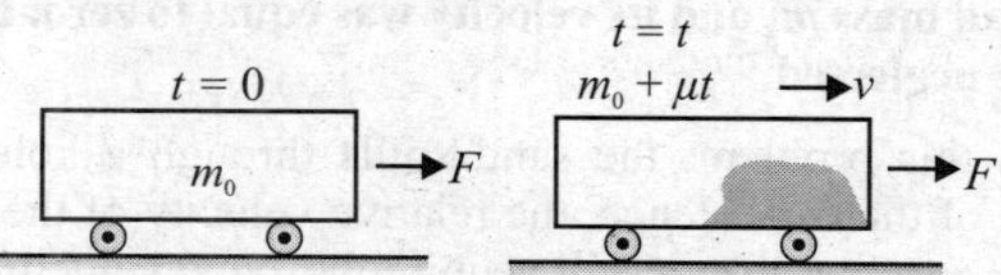

Fig. 1.128

At $\quad t = 0,\ v = 0$ and $m = m_0$

At $\quad t = t,\ v = v$ and $m = m_0 + \mu t$

Here, $v_r = u - v = 0 - v = -v$

Here u is the velocity of mass being added in the horizontal direction which is zero.

$$\frac{dm}{dt} = \mu$$

Apply $F + v_{rel}\dfrac{dm}{dt} = \dfrac{mdv}{dt}$, we get

$$F + (-v)\mu = (m_0 + \mu t)\frac{dv}{dt} \qquad \text{(i)}$$

or $$\int_0^v \frac{dv}{F - \mu v} = \int_0^t \frac{dt}{m_0 + \mu t}$$

$$\Rightarrow \quad -\frac{1}{\mu}[\ln(F - \mu v)]_0^v = \frac{1}{\mu}[\ln(m_0 + \mu t)]_0^t$$

$$\Rightarrow \quad \ln\left(\frac{F}{F - \mu v}\right) = \ln\left(\frac{m_0 + \mu t}{m_0}\right)$$

$$\Rightarrow \quad \frac{F}{F - \mu v} = \frac{m_0 + \mu t}{m_0}$$

or $$v = \frac{Ft}{m_0 + \mu t} \qquad \text{(ii)}$$

From Eq. (i),

$$a = \frac{F - \mu v}{m}$$

$$= \left(\frac{F - \dfrac{F\mu t}{m_0 + \mu t}}{m_0 + \mu t}\right)$$

or $$a = \frac{Fm_0}{(m_0 + \mu t)^2}$$

Note: ***We can also get acceleration by differentiating Eq. (ii) w.r.t. time.***

Illustration 1.47 **A cart loaded with sand moves along a horizontal floor due to a constant force F coinciding in direction with the cart's velocity vector. In the process, the sand spills through a hole in the bottom with a constant rate**

μ kg/s. Find the acceleration and velocity of the cart at the moment t, if at the initial moment $t = 0$ the cart with loaded sand had mass m_0 and its velocity was equal to zero. Friction is to be neglected.

Sol. In this problem, the sand spills through a hole at the bottom of the cart. Hence, the relative velocity of the sand v_r will be zero, because it will acquire the same velocity as that of the cart at the moment of separation.

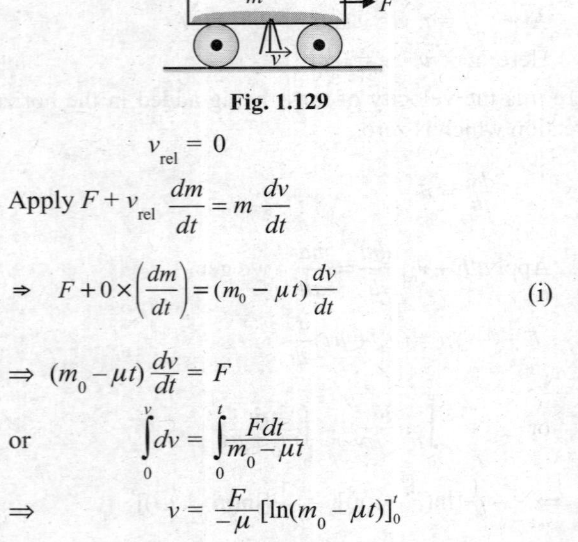

Fig. 1.129

$$v_{rel} = 0$$

Apply $F + v_{rel}\dfrac{dm}{dt} = m\dfrac{dv}{dt}$

$$\Rightarrow \quad F + 0 \times \left(\frac{dm}{dt}\right) = (m_0 - \mu t)\frac{dv}{dt} \qquad \text{(i)}$$

$$\Rightarrow \quad (m_0 - \mu t)\frac{dv}{dt} = F$$

or $$\int_0^v dv = \int_0^t \frac{F dt}{m_0 - \mu t}$$

$$\Rightarrow \quad v = \frac{F}{-\mu}[\ln(m_0 - \mu t)]_0^t$$

or $$v = \frac{F}{\mu}\ln\left(\frac{m_0}{m_0 - \mu t}\right)$$

From Eq. (i), acceleration of the cart,

$$a = \frac{dv}{dt} = \frac{F}{m} \quad \text{or} \quad a = \frac{F}{m_0 - \mu t}$$

Rocket Propulsion

Let m_0 be the mass of the rocket at time $t = 0$, m its mass at any time t and v its velocity at that moment. Initially, let us suppose that the velocity of the rocket is u.

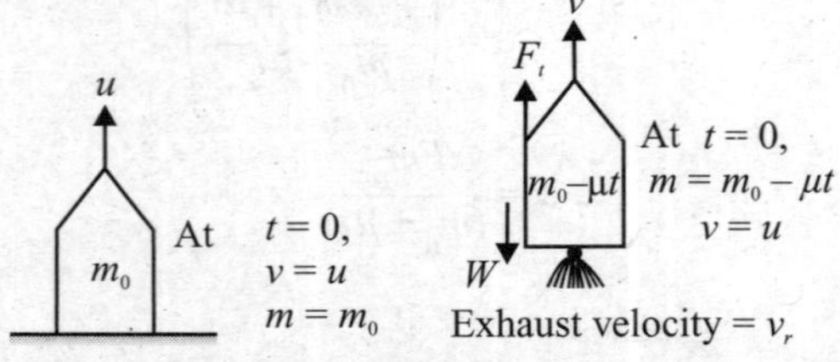

Fig. 1.130

Further, let $-dm/dt$ be the mass of the gas ejected per unit time and v_r the exhaust velocity of the gases with respect to rocket. Usually, $-dm/dt$ and v_r are kept constant throughout the journey of the rocket. Now, let us write a few equations which can be used in the problems of rocket propulsion. At time $t = t$,

1. Thrust force on the rocket, $F_t = v_r\,(dm/dt)$ (upwards)
2. Weight of the rocket, $W = mg$ (downwards)
3. Net force on the rocket,

$$F_{net} = F_t - W \text{ (upwards)}$$
$$= v_r\left(\frac{dm}{dt}\right) - mg$$

4. Net acceleration of the rocket, $a = \dfrac{F}{m}$

or $$\frac{dv}{dt} = \frac{v_r}{m}\left(\frac{dm}{dt}\right) - g$$

or $$dv = \frac{v_r}{m}(dm) - g\,dt$$

or $$\int_u^v dv = v_r\int_{m_0}^m \frac{dm}{m} - g\int_0^t dt$$

Thus, $$v = u - gt - v_r \ln\left(\frac{m_0}{m}\right) \qquad \text{(i)}$$

Note:

- ***$F_t = v_r\,(dm/dt)$ is upwards as v_r is negative (downwards) and dm/dt is negative.***
- ***If gravity is ignored and initial velocity of the rocket $u = 0$, Eq. (i) reduces to $v = v_r \ln (m_0/m)$.***

Illustration 1.48 A rocket, with an initial mass of 1000 kg, is launched vertically upwards from rest under gravity. The rocket burns fuel at the rate of 10 kg per second. The burnt matter is ejected vertically downwards with a speed of 2000 m/s relative to the rocket. Find the velocity of the rocket after 1 min of start.

Sol. The velocity equation is given by

$$v = u - gt + v_r \ln\left(\frac{m_0}{m}\right)$$

Here $u = 0$, $t = 60$ s, $g = 10$ m/s^2, $v_r = 2000$ m/s, $m_0 = 1000$ kg and $m = 1000 - 10 \times 60 = 400$ kg

Putting these values, we get

$$v = 0 - 600 + 2000 \ln\left(\frac{1000}{400}\right)$$

or $$= 2000 \ln 2.5 - 600 = 1232.6 \text{ m/s}$$

Solved Examples

Example 1.1 A body of mass m moving with velocity V in the x-direction collides with another body of mass M moving in y-direction with velocity v. They coalesce into one body during collision. Calculate (i) the direction and magnitude of momentum of the final body, (ii) the fraction of initial kinetic energy transformed into heat during the collision in terms of the two masses.

(IIT-JEE, 1978)

Sol. Let the combined mass travels with velocity V' at angle θ as shown.

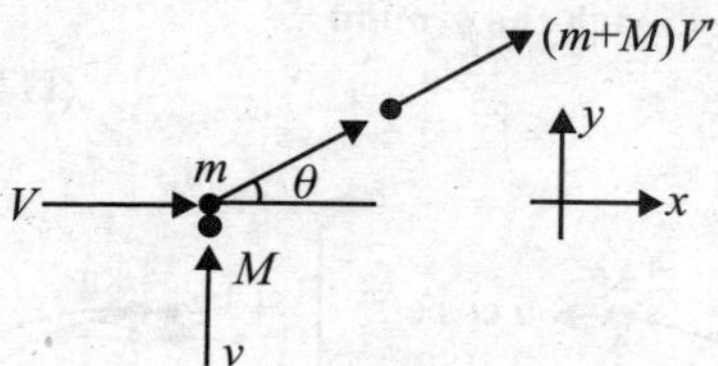

Fig. 1.131

Applying conservation of linear momentum in x-direction, we get

$$(m + M)\, V' \cos \theta = mV \quad \text{(i)}$$

Applying conservation of linear momentum in y-direction, we get

$$(m + M)\, V' \sin \theta = Mv \quad \text{(ii)}$$

Dividing Eqs. (i) and (ii), we get

$$\tan \theta = \frac{Mv}{mV} \Rightarrow \theta = \tan^{-1}\left(\frac{Mv}{mV}\right)$$

This gives the direction of the momentum of the final body.

Squaring and adding Eqs. (i) and (ii), we get

$$(m + M)^2 V'^2 \cos^2 \theta + (m + M)^2\, V'^2 \sin^2 \theta = m^2 V^2 + M^2 v^2$$

$$\therefore \quad V' = \frac{\sqrt{m^2 V^2 + M^2\, v^2}}{m + M}$$

(ii) Fraction of initial KE converted into heat is

$$f = \frac{\mathrm{KE}_i - \mathrm{KE}_f}{\mathrm{KE}_i} = 1 - \frac{\mathrm{KE}_f}{\mathrm{KE}_i} = 1 - \frac{\frac{1}{2}(M+m)V'^2}{\frac{1}{2}mV^2 + \frac{1}{2}Mv^2}$$

Solve to get $f = \dfrac{mM\,(v^2 + V^2)}{(m+M)(mv^2 + Mv^2)}$

Example 1.2 **A body of mass 1 kg, initially at rest, explodes and breaks into three fragments of masses in the ratio 1:1:3. The two pieces of equal mass fly off perpendicular to each other with a speed of 30 m/s each. What is the velocity of the heavier fragment? (IIT-JEE, 1981)**

Sol. The momentum of the system is zero before explosion. Since the momentum is zero initially, it will be zero just after explosion according to conservation of linear momentum (as there is no external force acting on the system). The resultant momentum of the fragments flying off perpendicular to each other should be equal and opposite to the third fragment. Masses of fragments are 1/5 kg, 1/5 kg and 3/5 kg.

Momentum of each of the smaller fragments is $\frac{1}{5} \times 30 = 6$ kg m/s.

Applying the principle of conservation of linear momentum in both the directions, we get

$$6 = \frac{3}{5}\, v \sin \theta \quad \text{(i)}$$

$$6 = \frac{3}{5}\, v \cos \theta \quad \text{(ii)}$$

Squaring and adding, we get $36 + 36 = \frac{9}{25} v^2$

$$\therefore \quad v = 10\sqrt{2} \text{ m/s}$$

Dividing the two equations, we get

$$\tan \theta = 1, \text{ or } \theta = 45°$$

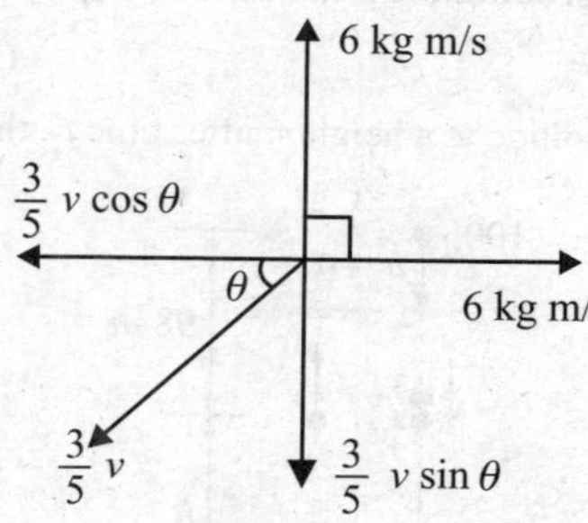

Fig. 1.132

Example 1.3 **Two bodies A and B of masses m and $2m$, respectively, are placed on a smooth floor. They are connected by a spring. A third body C of mass m moves with velocity v_0 along the line joining A and B and collides elastically with A as shown in Fig. 1.133. At a certain instant of time t_0 after collision, it is found that the instantaneous velocities of A and B are the same. Further at this instant, the compression of the spring is found to be x_0. Determine (i) the common velocity of A and B at time t_0 and (ii) the spring constant. (IIT-JEE, 1984)**

Sol. Since the collision between C and A is elastic and their masses are equal and A was initially at rest, therefore the result of collision will be that C will come to rest and A will initially start moving with a velocity v_0. But since A is connected to B with a spring, the spring will get compressed.

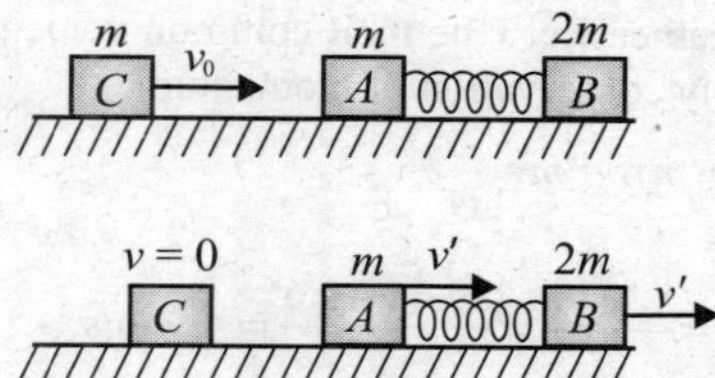

Fig. 1.133

At $t = t_0$, the velocities of A and B become the same.

(i) Applying the momentum conservation, we get

$$mv_0 = mv' + 2mv' \;\therefore\; v' = \frac{v_0}{3} \quad \text{(i)}$$

(ii) Applying energy conservation,

$$\frac{1}{2}\, mv_0^2 = \frac{1}{2} mv'^2 + \frac{1}{2} 2mv'^2 + \frac{1}{2} kx_0^2$$

where x_0 is the compression in the spring at $t = t_0$.

$$\therefore \quad v_0^2 = 3v'^2 + \frac{k}{m} x_0^2 \quad \text{(ii)}$$

From Eqs. (i) and (ii), $v_0^2 - 3 \times \dfrac{v_0^2}{9} = \dfrac{k}{m} x_0^2 \Rightarrow k = \dfrac{2mv_0^2}{3x_0^2}$

Example 1.4 **A ball of mass 100 g is projected vertically upwards from the ground with a velocity of 49 m/s. At the same time, another identical ball is dropped from a height of 98 m to fall freely along the same path as that followed by the first ball. After some time, the two balls collide and stick together and finally fall to the ground. Find the time of flight of the masses.**

(IIT-JEE, 1985)

Sol. Let the balls collide at a height h after time t_1, then for ball 1:

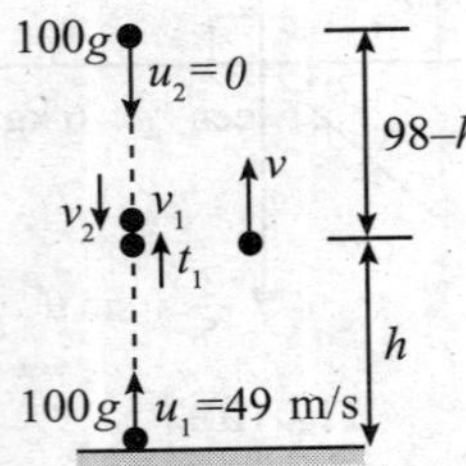

Fig. 1.134

$$h = 49t_1 - \frac{1}{2}gt_1^2 \quad \text{(i)}$$

For ball 2: $98 - h = \frac{1}{2}gt_1^2$ (ii)

$$98 = 49\,t_1 \Rightarrow t_1 = 2 \text{ s}$$

Put the value of t_1 in Eq. (i) to get $h = 78.4$ m

Velocity of ball 1 after time t_1: $v_1 = 49 - 9.8 \times 2 = 29.4$ m/s

Velocity of ball 2 after time t_1: $v_2 = gt_1 = 9.8 \times 2 = 19.6$ m/s

After collision, balls collide and stick together. Now onwards balls will move together. Let v be their common velocity just after collision, then apply conservation of momentum,

$$(m + m)v = mv_1 - mv_2$$

$$\Rightarrow \quad v = \frac{v_1 - v_2}{2} = \frac{29.4 - 19.6}{2} = 4.9 \text{ m/s}$$

Let the balls take time t_2 after collision to come to ground, then

$$-h = vt_2 - \frac{1}{2}gt_2^2$$

$$-78.4 = 4.9t_2 - 4.9t_2^2$$

$$\Rightarrow \quad t_2^2 - t_2 - 16 = 0.$$

$$\Rightarrow \quad t_2 = \frac{1 + \sqrt{1 + 64}}{2} = 4.53 \text{ s}$$

So the time of flight: $T = t_1 + t_2 = 2 + 4.53 = 6.53$ s

Example 1.5 **An object of mass 5 kg is projected with a velocity of 20 m/s at an angle of 60° to the horizontal. At the highest point of its path the projectile explodes and breaks up into two fragments of masses 1 kg and 4 kg. The fragments separate horizontally after the explosion. The explosion releases internal energy such that the kinetic energy of the system at the highest point is doubled. Calculate the separation between the two fragments when they reach the ground.**

(IIT-JEE, 1990)

Sol.

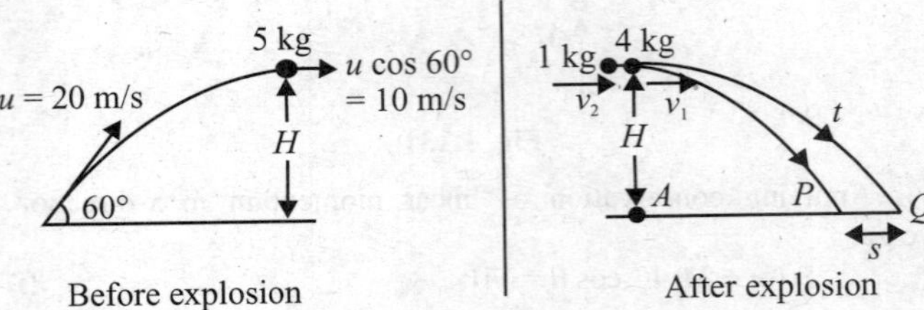

Fig. 1.135

The maximum height reached by mass 5 kg:

$$H = \frac{u^2 \sin^2 60°}{2g} = \frac{20^2 \times 3/4}{2 \times 10} = 15 \text{ m}$$

Velocity of 5 kg mass at the highest point before explosion is

$$u \cos 60° = 20 \times \frac{1}{2} = 10 \text{ m/s}$$

Now here 5 kg explodes into two fragments of masses 4 kg and 1 kg. Let their velocities be v_1 and v_2 as shown. v_1 and v_2 will be horizontal because it is given that masses separate horizontally. Apply conservatio of momentum at the highest point:

$$4v_1 + 1 \times v_2 = 5 \times 10 \quad \text{(i)}$$

It is given that KE of the system becomes doubled due to explosion, so

$$\frac{1}{2}4v_1^2 + \frac{1}{2}1v_2^2 = 2\left[\frac{1}{2} \times 5 \times 10^2\right]$$

$$\Rightarrow \quad 4v_1^2 + v_2^2 = 1000 \quad \text{(ii)}$$

Put the value of v_2 from Eq. (i) into Eq. (ii), $4v_1^2 + (50 - 4v_1)^2 = 1000$

$$\Rightarrow \quad v_1^2 - 20v_1 + 75 = 0 \quad \Rightarrow (v_1 - 15)(v_1 - 5) = 0$$

First solution: $v_1 = 15$ m/s, $v_2 = -10$ m/s

Second solution: $v_1 = 5$ m/s, $v_2 = 30$ m/s

The first solution is possible if 4 kg is ahead of 1 kg and second solution is possible if 1 kg is ahead of 4 kg. Since initial velocities of masses after explosion are in horizontal direction, so they will take same time to return to the ground:

$$t = \sqrt{\frac{2H}{g}} = \sqrt{\frac{2 \times 15}{10}} = \sqrt{3} \text{ s}$$

Now $AQ = v_1 t$, $AP = v_2 t$

Separation between them on reaching the ground:

$$S = AQ - AP = (v_1 - v_2)t = [15-(-10)]\sqrt{3} = 25\sqrt{3}\text{ m}$$

For the second solution:

$$S = (5-30)\sqrt{3} = -25\sqrt{3}\text{ m}$$

-ve sign indicates that 1 kg will be ahead of 4 kg.

Example 1.6 **A block 'A' of mass $2m$ is placed on another block 'B' of mass $4m$ which in turn is placed on a fixed table. The two blocks have the same length $4d$ and they are placed as shown in Fig. 1.136. The coefficient of friction (both static and kinetic) between block 'B' and table is μ. There is no friction between the two blocks. A small object of mass m moving horizontally along a line passing through the centre of mass (CM) of block B and perpendicular to its face with a speed v collides elastically with block B at a height d above the table.**

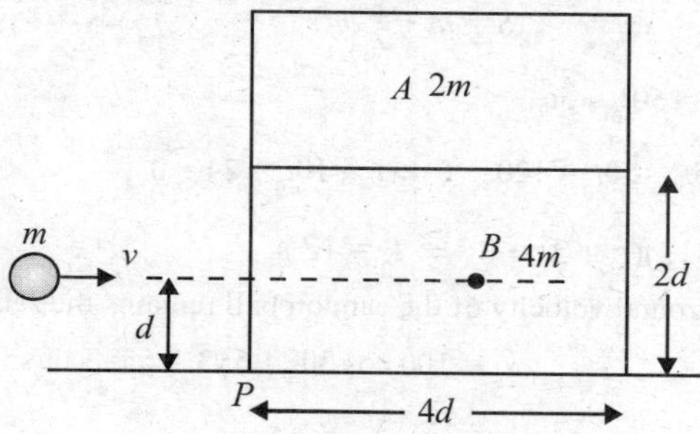

Fig. 1.136

(a) What is the minimum value of v (call it v_0) required to make block A topple?

(b) If $v = 2v_0$, find the distance (from point P in the figure) at which mass m falls on the table after collision. (Ignore the role of friction during the collision.)

(IIT-JEE, 1991)

Sol. (a) Since the collision is elastic in nature, applying conservation of linear momentum and conservation of kinetic energy,

$$mv = (4m)u + mv'$$

where u is the velocity of mass $4m$ after collision and v' is the velocity of mass m.

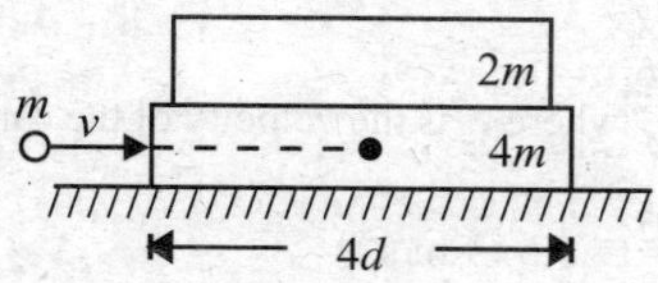

Fig. 1.137

$$\therefore \quad v' = v - 4u \qquad \text{(i)}$$

Also,
$$\frac{1}{2}mv^2 = \frac{1}{2}(4m)u^2 + \frac{1}{2}mv'^2$$

$$\Rightarrow \quad v^2 = 4u^2 + v'^2 \qquad \text{(ii)}$$

From Eqs. (i) and (ii),

$$v^2 = 4u^2 + (v-4u)^2 \Rightarrow u = \frac{2v}{5}$$

For block A to topple, it should move a distance $2d$.

The retardation produced in B due to frictional force between B and the table is

$$f = \mu N = (4m)a \Rightarrow \mu[6mg] = (4m)a$$

$$\Rightarrow \quad a = 1.5\mu g$$

For the motion of B, $u = \frac{2v}{5}$, $v = 0$, $s = 2d$, $a = -1.5g$

Now,
$$v^2 - u^2 = 2as \Rightarrow (0)^2 - \left(\frac{2v}{5}\right)^2 = 2(-1.5\mu g)2d$$

$$\Rightarrow \quad v = \frac{5}{2}\sqrt{6\mu gd} = v_0$$

(b) For elastic collision between two bodies,

$$v_1 = \frac{(m_1 - m_2)u_1}{m_1 + m_2} + \frac{2m_2u_2}{m_1 + m_2}$$

Here
$$m_1 = m,\ m_2 = 4m,\ u_1 = 5\sqrt{6\mu gd},\ u_2 = 0$$

$$v_1 = \frac{(m-4m)5\sqrt{6\mu gd}}{m+4m} + 0$$

$$= -3 \times 5\frac{\sqrt{(6\mu gd)}}{5} = -3\sqrt{6\mu gd}$$

Fig. 1.138

The negative sign shows that mass m rebounds. It then follows a projectile motion. Considering the vertical direction, motion of this projectile is

$$u_y = 0,\ s_y = d,\ a_y = g,\ t_y = ?$$

$$S = ut + \frac{1}{2}at^2 \Rightarrow d = \frac{1}{2}gt^2 \Rightarrow t = \sqrt{\frac{2d}{g}}$$

The horizontal distance travelled by mass m during this time t,

$$x = 3\sqrt{6\mu gd} \times \sqrt{\frac{2d}{g}} = 6\sqrt{3\mu d^2} = 6d\sqrt{3\mu}$$

Example 1.7 **A cart is moving along +ve x-direction with a velocity of 4 m/s. A person on the cart throws a stone with a velocity of 6 m/s relative to himself. In the frame of reference of the cart, the stone is thrown in y-z plane making an angle of 30° with the vertical z-axis. At the highest point of its trajectory, the stone hits an object of equal mass hung vertically from the branch of a tree by means of a string of length L. A completely inelastic collision occurs, in which the stone gets embedded in the object. Determine**

(i) the speed of the combined mass immediately after the collision with respect to an observer on the ground,

(ii) the length L of the string such that the tension in the string becomes zero when the string becomes horizontal during the subsequent motion of the combined mass.

(IIT-JEE, 1997)

Sol.

Fig. 1.139

$u_x = 4$ m/s

$$u_y = v\sin 30° = 6\times\frac{1}{2} = 3\,\text{m/s}$$

$$u_1 = \sqrt{u_x^2 + u_y^2} = 5 \text{ m/s}$$

$$u_z = v\cos 30° = 6\times\frac{\sqrt{3}}{2} = 3\sqrt{3}\,\text{m/s}$$

$u_2 \rightarrow$ net velocity

When the stone reaches point Q, the component of velocity in the +ve Z-direction ($u_z = v \cos \theta$) becomes zero due to the gravitational force in the –ve Z-direction.

The velocity of stone at Q will be $u_1 = 5$ m/s as shown.

(i) Applying conservation of linear momentum at Q for collision with a mass of equal magnitude, $m \times 5 = 2m \times v_1$. (Since the collision is completely inelastic, the two masses will stick together. v_1 is the velocity of the two masses just after collision.)

$\therefore \quad v_1 = 2.5$ m/s

(ii) When the string is undergoing circular motion, at any arbitary position,

$$T - 2mg\cos\theta = \frac{2mv^2}{L}$$

Given that $T = 0$ when $\theta = 90°$, therefore, $0 - 0 = \dfrac{2mv^2}{L} \Rightarrow v = 0$

Velocity is zero when $\theta = 90°$, i.e., in the horizontal position.

Applying energy conservation from Q to M, we get

$$\frac{1}{2}\times 2\,mv_1^{\,2} = 2mgL$$

$$\Rightarrow \quad L = \frac{v_1^2}{2g} = \frac{(2.5)^2}{2\times 10} = 0.3125 \text{ m}$$

Example 1.8 **A car P is moving with a uniform speed of $5\sqrt{3}$ m/s towards a carriage of mass 9 kg at rest kept on the rails at point B as shown in Fig. 1.140. The height AC is 120 m. Cannon balls of 1 kg are fired from the car with an initial velocity 100 m/s at an angle 30° with the horizontal. The first cannon ball hits the stationary carriage after a time t_0 and sticks to it. Determine t_0. At t_0, the second cannon ball is fired. Assume that the resistive force betweeen the rails and the carriage is constant and ignore the vertical motion of the carriage throughout. If the second ball also hits and sticks to the carriage, what will be the horizontal velocity of the carriage just after the second impact?**

(IIT-JEE, 2001)

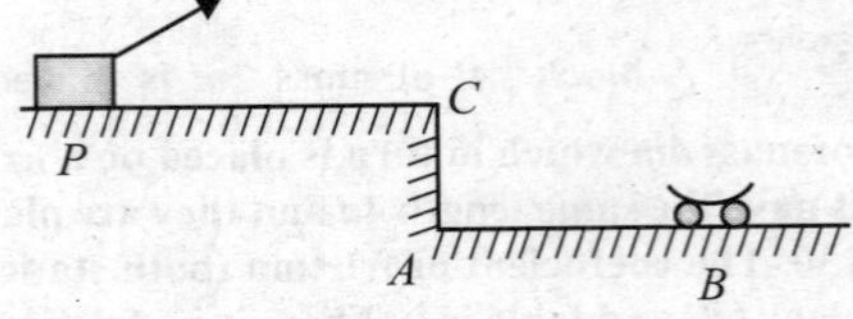

Fig. 1.140

Sol. Consider the vertical motion of the cannon ball, $u_y = +100 \sin 30°$, $S_y = -120$ m, $a_y = -10$ m/s², $t = t_0$.

$$S = ut + \frac{1}{2}at^2,$$

$$\Rightarrow -120 = 50t_0 - 5t_0^2$$

$$\therefore \quad 5t_0^2 - 50t_0 - 120 = 0 \Rightarrow t_0^2 - 10t_0 - 24 = 0$$

$$\Rightarrow (t_0 - 12)(t_0 + 2) = 0 \quad \Rightarrow t_0 = 12 \text{ s}$$

The horizontal velocity of the cannon ball remains the same.

$$\therefore \quad u_x = 100 \cos 30° + 5\sqrt{3} = 55\sqrt{3} \text{ m/s}$$

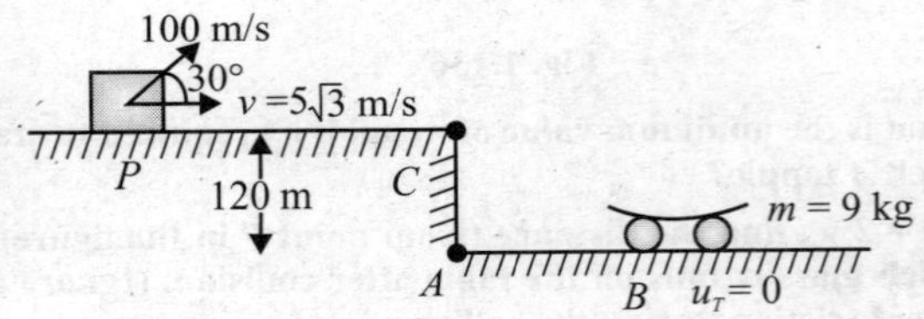

Fig. 1.141

Therefore, applying conservation of linear momentum to the cannon ball–trolley system in horizontal direction. If m is the mass of the cannon ball and M is the mass of the trolley, then

$$mu_x + m \times 0 = (m + M)v_x$$

$\therefore\ v_x = \dfrac{mu_x}{m + M}$ where v_x is the velocity of the cannon ball–trolley system $= \dfrac{1 \times 55\sqrt{3}}{1 + 9} = 5.5\sqrt{3}$ m/s.

The second ball was projected after 12 s. The horizontal distance covered by car $P = 12 \times 5\sqrt{3} = 60\sqrt{3}$ m.

Since the second ball also strikes the trolley, so in time 12 s the trolley covers a distance of $60\sqrt{3}$ m.

Now, $t = 12$ s, $u = 5.5\sqrt{3}$ m/s, $a = ?$, $t = 12$, $S = 60\sqrt{3}$, we have

$$S = ut + \frac{1}{2}at^2$$

$$\Rightarrow\ 60\sqrt{3} = 5.5\sqrt{3} \times 12 + \frac{1}{2} \times a \times 144 \Rightarrow \quad a = -\frac{\sqrt{3}}{12}\,\text{m/s}^2$$

$$\therefore \quad a = -0.144\ \text{m/s}^2$$

$$\Rightarrow \quad V = u + at = 5.5\sqrt{3} - \frac{\sqrt{3}}{12} \times 12 = 4.5\sqrt{3}\ \text{m/s}$$

To find the final velocity of the carriage after the second impact, we again apply conservation of linear momentum in the horizontal direction,

$$mV_x + (M + m)\,4.5\sqrt{3} = (M + 2m)V_f$$

$$\therefore\ 1 \times 55\sqrt{3} + (9 + 1)\,4.5\sqrt{3} = (9 + 2)\,V_f \Rightarrow V_f = 15.75\ \text{m/s}$$

Example 1.9 **A particle of mass m, moving in a circular path or radius R with a constant speed v_2, is located at point $(2R, 0)$ at time $t = 0$ and a man starts moving with a velocity v_1 along the +ve y-axis from origin at time $t = 0$. Calculate the linear momentum of the particle w.r.t. the man as a function of time.**

(IIT-JEE, 2003)

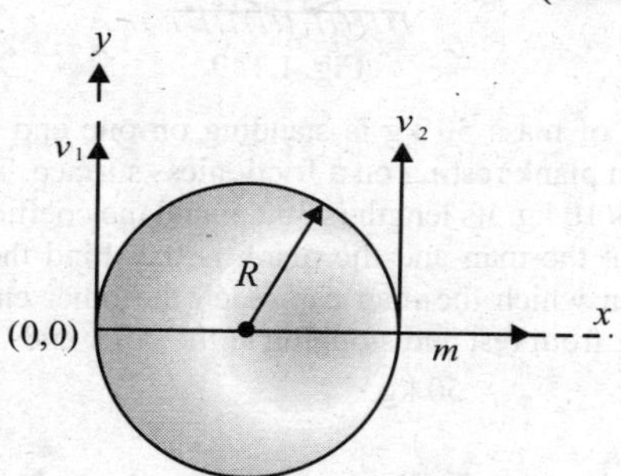

Fig. 1.142

Sol. $\vec{v}_2 = (-v_2 \sin \omega t\hat{i} + v_2 \cos \omega t\hat{j})$;

$$v_1 = v_1\hat{j}$$

$$\vec{v}_{PM} = \vec{v}_2 - \vec{v}_1$$

$$= -v_2 \sin \omega t\hat{i} + (v_2 \cos \omega t\hat{j} - v_1)\hat{j}$$

$$\vec{p}_{PM} = m\vec{v}_{PM}$$

$$= -m\vec{v}_2 \sin \omega t\hat{i} + m(v_2 \cos \omega t - v_1)\hat{j}$$

where $\omega = \dfrac{v_2}{R}$

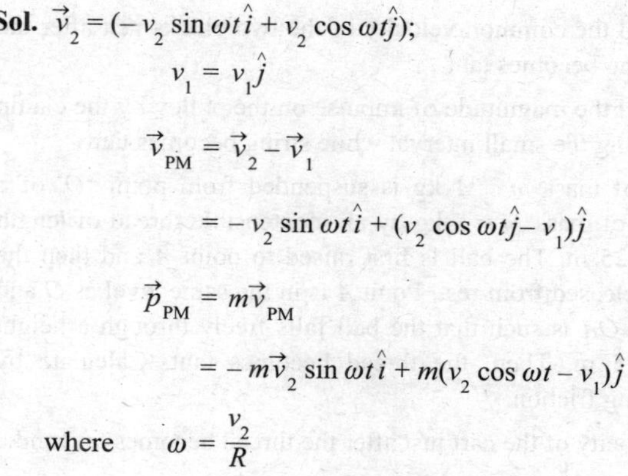

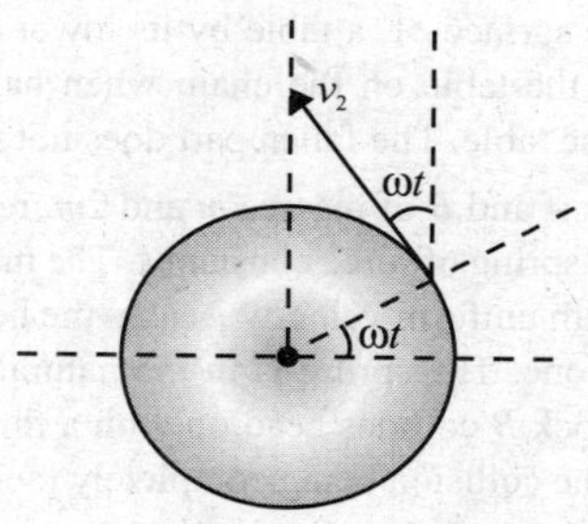

Fig. 1.143

EXERCISES

Subjective Type

Solutions on page 1.77

1. Two identical smooth balls are projected from points O and A on the horizontal ground with same speed of projection. The angle of projection in each case is 30° (see Fig. 1.144). The distance between O and A is 100 m. The balls collide in mid-air and return to their respective points of projection. If the coefficient of restitution is 0.7, find the speed of projection of either ball (in m/s) correct to nearest integer. (Take g = 10 m s^{-2} and $\sqrt{3} = 1.7$).

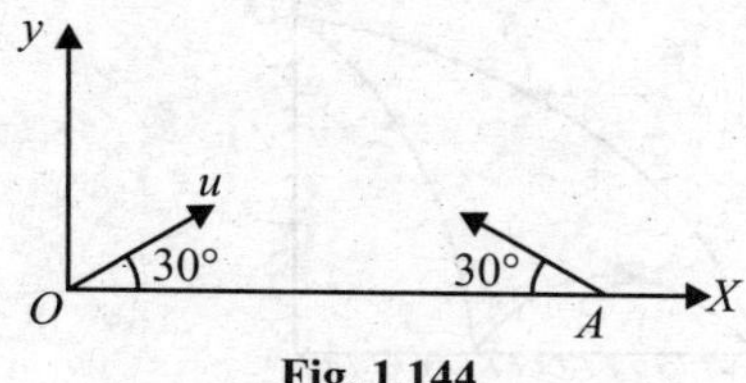

Fig. 1.144

2. A stationary light, smooth pulley can rotate without friction about a fixed horizontal axis. A light rope passes over the pulley. One end of the rope supports a ladder with man and the other end supports a counterweight of mass M. Mass of the man is m. Initially, the centre of mass of the counterweight is at a height h from that of man as shown in Fig. 1.145.
If the man starts to climb up the ladder slowly, calculate work done by him to reach his centre of mass in level with that of the counterweight.

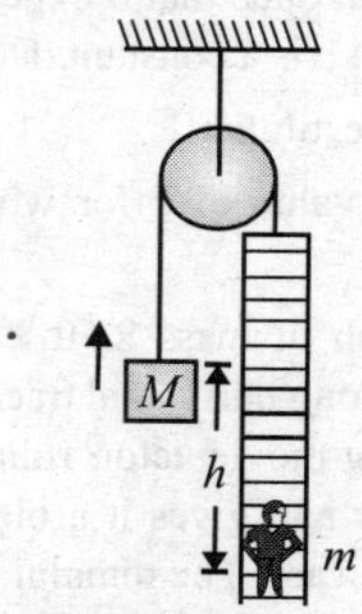

Fig. 1.145

3. Two blocks A and B are joined by means of a slacked string passing over a massless pulley as shown in Fig. 1.146. The system is released from rest and it becomes taut when B falls a distance 0.5 m.

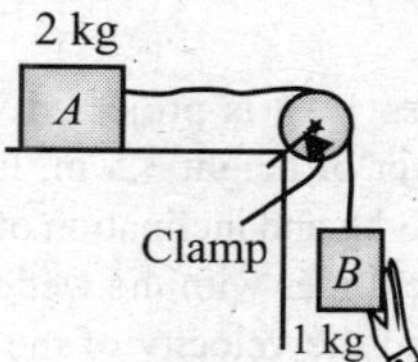

Fig. 1.146

a. Find the common velocity of the two blocks just after the string becomes taut.

b. Find the magnitude of impulse on the pulley by the clamp during the small interval while string becomes taut.

4. A ball of mass $m = 1$ kg is suspended from point 'O' of a toy cart of mass $M = 3$ kg by an inextensible thread of length $l = 1.5625$ m. The ball is first raised to point A and then the ball is released from rest. Point A is in the same level as O and distance OA is such that the ball falls freely through a height $h = 1.25$ m. Then, the thread becomes taut. Calculate by neglecting friction,

i. velocity of the cart just after the thread becomes taut and

ii. loss of energy when thread becomes taut.

5. A uniform chain of mass m and length l hangs on a thread and touches the surface of a table by its lower end. Find the force exerted by the table on the chain when half of its length has fallen off the table. The fallen part does not form a heap.

6. Two blocks A and B of masses m and $2m$, respectively are connected by a spring of force constant k. The masses are moving to the right with uniform velocity v each, the heavier mass leading the lighter one. The spring is in the natural length during this motion. Block B collides head on with a third block C of mass m, at rest, the collision being completely inelastic. Calculate the maximum compression of the spring.

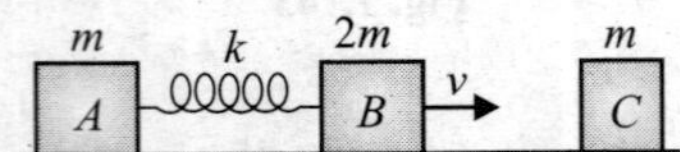

Fig. 1.147

7. Two small spheres A and B of equal radius but different masses of $3m$ and $2m$ are moving towards each other and impinge directly. The speeds of A and B before collision are, respectively, $4u$ and u. The collision is such that B experiences an impulse of magnitude $6mcu$, where c is a constant. Determine

a. the coefficient of restitution,

b. the limits for the value of c for which such collision is possible.

8. Four railroad cars, each of mass 2.50×10^4 kg, are coupled together and coasting along horizontal tracks at speed v_i towards the south. A very strong movie actor, riding on the second car, uncouples the front car and gives it a big push, increasing its speed to 4.00 m/s southward. The remaining three cars continue moving south, now at 2.00 m/s. (a) Find the initial speed of the cars. (b) How much work did the actor do?

9. A particle of mass m is made to move with uniform speed u along the perimeter of a regular polygon of n sides. What is the magnitude of impulse applied by the particle at each corner of the polygon?

10. A smooth ball of mass 1 kg is projected with velocity 7 m/s horizontal from a tower of height 3.5 m. It collides elastically with a wedge of mass 3 kg and inclination of 45° kept on ground (Fig 1.148). The ball collides with the wedge at a height of 1 m above the ground. Find the velocity of the wedge and the ball after collision. (Neglect friction at any contact.)

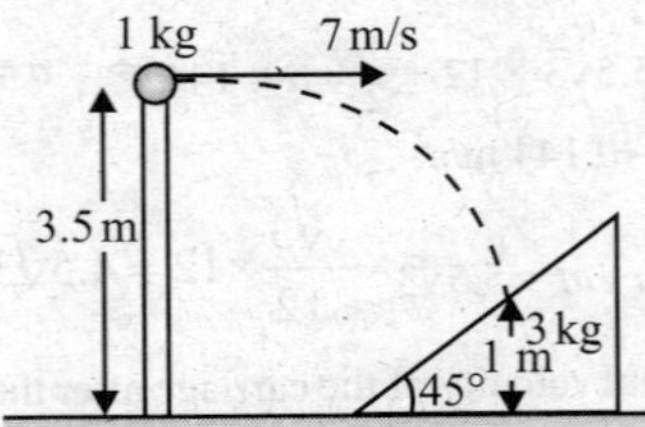

Fig. 1.148

11. Two beads each of mass m are positioned at the top of a frictionless hoop of mass M and radius R, which stands vertically on the ground. The beads are given tiny kicks and they slide down the hoop, one to the right and one to the left, as shown in Fig. 1.149. What is the smallest value of m/M for which the hoop will rise up off the ground at some time during the motion?

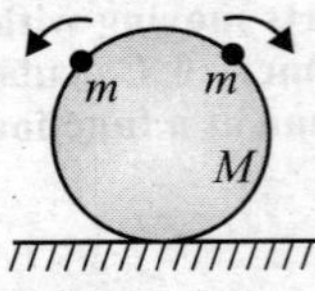

Fig. 1.149

12. A man of mass 50 kg is standing on one end of a stationary wooden plank resting on a frictionless surface. The mass of the plank is 10 kg. Its length is 300 m and the coefficient of friction between the man and the plank is 0.2. Find the shortest time (in s) in which the man can reach the other end of the plank starting from rest and stopping at the other end.

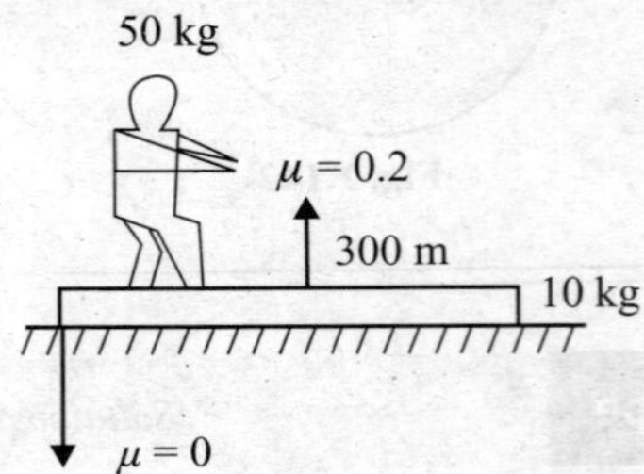

Fig. 1.150

13. A small ball is projected from point P towards a vertical wall as shown in Fig. 1.151. It hits the wall when its velocity is horizontal. Ball reaches point P after one bounce on the floor. The coefficient of restitution assuming it to be same for two collisions is $n/2$. All surfaces are smooth. Find the value of n.

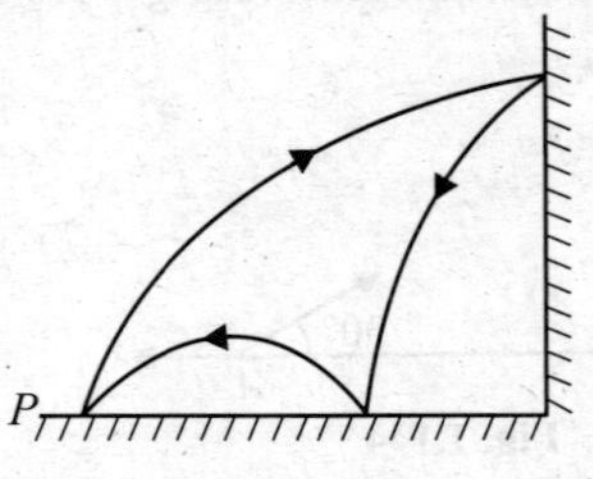

Fig. 1.151

14. A small steel ball A is suspended by an inextensible thread of length $l = 1.5$ m from O. Another identical ball is thrown vertically downwards such that its surface remains just in contact with thread during downward motion and collides elastically with the suspended ball. If the suspended ball just completes vertical circle

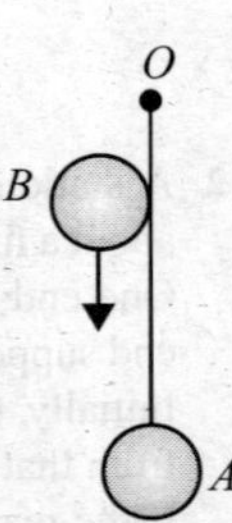

Fig. 1.152

after collision, calculate the velocity (in cm/s) of the falling ball just before collision ($g = 10 \text{ m s}^{-2}$).

15. A tennis ball with (small) mass m_2 rests on the top of a basketball of mass m_1 which is at a height h above the ground, and the bottom of the tennis ball is at height $h + d$ above the ground. The balls are dropped. To what height does the tennis ball bounce with respect to ground? (Assume all collisions to be elastic and $m_1 >> m_2$)

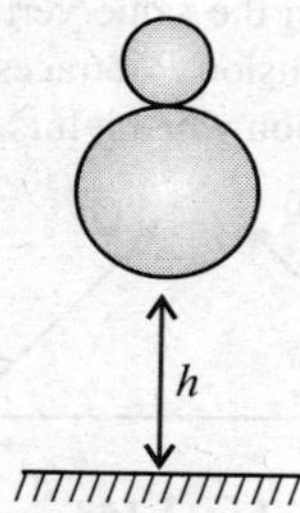

Fig. 1.153

16. A ball of mass m hits a wedge of mass M vertically with speed u, which is placed, on a smooth horizontal surface. Find the maximum compression in the spring, if the collision is perfectly elastic and no friction any where. Spring constant of spring is K.

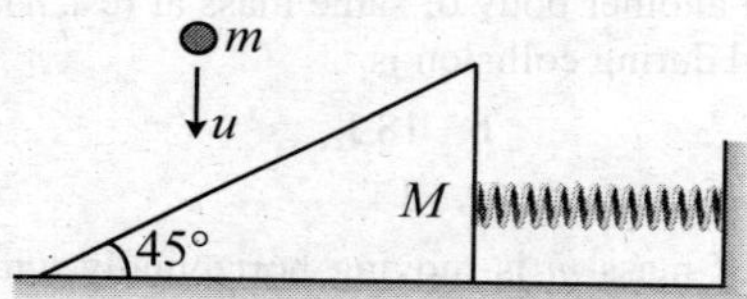

Fig. 1.154

17. A ball of mass m moving with constant horizontal velocity u strikes a stationary wedge of mass M on its inclined surface as shown in the figure. After collision, the ball starts moving up the inclined plane. The wedge is placed on frictionless horizontal surface.

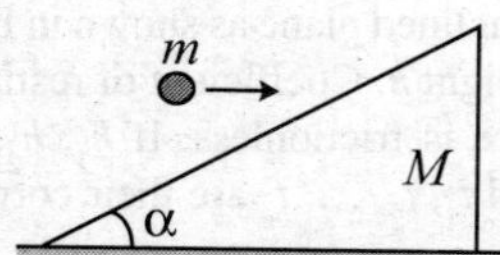

Fig. 1.155

i. Calculate the velocity of wedge immediately after collision.

ii. Calculate the maximum height the ball can ascend on the wedge.

18. A ball of mass m collides with a stationary wedge of mass M, perpendicular to its inclined face, inclined at an angle as shown in the figure. If the coefficient of restitution between the wedge and ball is e, calculate the ratio of modulus of velocity of the ball immediately after and before collision. Also calculate the velocity of wedge just after collision.

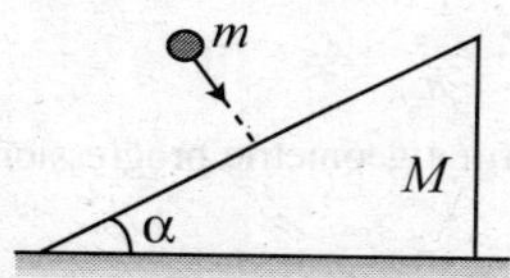

Fig. 1.156

Objective Type

Solutions on page 1.80

1. A shell explodes and many pieces fly off in different directions. Which of the following is conserved?

a. Kinetic energy
b. Momentum
c. Neither momentum nor KE
d. Momentum and KE

2. A ball falls under gravity from a height of 10 m with an initial downward velocity v_0. It collides with the ground, loses 50% of its energy in collision and then rises back to the same height. Find (i) the initial velocity v_0 and (ii) the height to which the ball would rise after collision if the initial velocity v_0 was directed upwards instead of downwards?

a. 14 m/s, 5 m **b.** 14 m/s, 10 m
c. 7 m/s, 5 m **d.** 7 m/s, 10 m

3. A bullet of mass 0.01 kg and travelling at a speed of 500 m/s strikes a block of 2 kg which is suspended by a string of length 5 m. The centre of gravity of the block is found to rise a vertical distance of 0.1 m. What is the speed of the bullet after it emerges from the block?

a. 200 m/s **b.** 220 m/s
c. 204 m/s **d.** 284 m/s

4. A mass of 2.9 kg is suspended from a string of length 50 cm and is at rest. Another body of mass 100 g which is moving horizontally with a velocity of 150 m/s strikes it. After striking the two bodies combine together. The tension in the string, when it makes an angle of 60° with the vertical is (Take g = 9.8 m/s²)

a. 135.3 N **b.** 165.7 N
c. 142 N **d.** 90 N

5. A ball of mass m moving with velocity v strikes the bob of a pendulum at rest. The mass of the bob is also m. If the collision is perfectly inelastic, the height to which the bob will rise is given by

a. $\frac{v^2}{8g}$ **b.** $\frac{v^2}{2g}$
c. $\frac{2v^2}{g}$ **d.** $\frac{v^2}{g}$

6. If in the previous problem the collision is elastic, the height to which the bob will rise will be

a. $\frac{v^2}{8g}$ **b.** $\frac{v^2}{2g}$
c. $\frac{2v^2}{g}$ **d.** $\frac{v^2}{g}$

7. A ball collides elastically with another ball of the same mass. The collision is oblique and initially one of the balls was at rest. After the collision, the two balls move with same speeds. What will be the angle between the velocities of the balls after the collision?

a. 30° **b.** 45°
c. 60° **d.** 90°

8. The first ball of mass m moving with a velocity u collides head on with the second ball of mass m at rest. If the coefficient of restitution is e, then the ratio of the velocities of the first and the second ball after the collision is

a. $\frac{1-e}{1+e}$ b. $\frac{1+e}{1-e}$

c. $\frac{1+e}{2}$ d. $\frac{1-e}{2}$

9. In problem 8, the ratio of the final and initial velocities of the first ball is

a. $\frac{1-e}{1+e}$ b. $\frac{1+e}{1-e}$

c. $\frac{1+e}{2}$ d. $\frac{1-e}{2}$

10. In problem 8, the ratio of the final velocity of the second ball to the initial velocity of the first ball is

a. $\frac{1-e}{1+e}$ b. $\frac{1+e}{1-e}$

c. $\frac{1+e}{2}$ d. $\frac{1-e}{2}$

11. An object initially at rest explodes into three fragments A, B and C. The momentum of A is $p\hat{i}$ and that of B is $\sqrt{3}p\hat{j}$ where p is a +ve number. The momentum of C will be

a. $(1+\sqrt{3})\,p$ in a direction making angle 120° with that of A.

b. $(1+\sqrt{3})\,p$ in a direction making angle 150° with that of B.

c. $2p$ in a direction making angle 150° with that of A.

d. $2p$ in a direction making angle 150° with that of B.

12. A ball impinges directly on another ball at rest. The first ball is brought to rest by the impact. If half of the kinetic energy is lost by the impact, the value of coefficient of restitution is

a. $\frac{1}{2}$ b. $\frac{1}{\sqrt{3}}$

c. $\frac{1}{\sqrt{2}}$ d. $\frac{\sqrt{3}}{2}$

13. A plastic ball is dropped from a height of 1 m and rebounds several times from the floor. If 1.03 s elapse from the moment it is dropped to the second impact with the floor, what is the coefficient of restitution?

a. 0.03 b. 0.64

c. 0.02 d. 0.05

14. A 20 g bullet pierces through a plate of mass $M_1 = 1$ kg and then comes to rest inside a second plate of mass $M_2 = 2.98$ kg as shown in Fig. 1.157. It is found that the two plates, initially at rest, now move with equal velocities. Find the percentage loss in the initial velocity of the bullet when it is between M_1 and M_2. Neglect any loss of material of the plates due to the action of bullet.

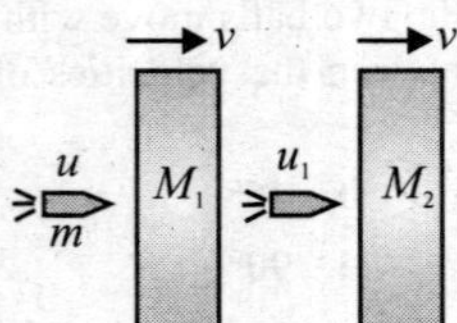

Fig. 1.157

a. 50% b. 25%

c. 100% c. 75%

15. Particles P and Q of masses 20 g and 40 g, respectively, are projected from positions A and B on the ground. The initial velocities of P and Q make angles of 45° and 135°, respectively, with the horizontal as shown in the Fig. 1.158. Each particle has an initial speed of 49 m/s. The separation AB is 245 m. Both particles travel in the same vertical plane and undergo a collision. After the collision P retraces its path. The distance of Q from its initial position when it hits the ground is

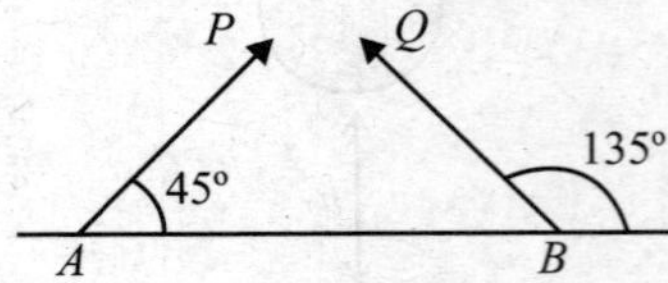

Fig. 1.158

a. 245 m b. $\frac{245}{3}$ m

c. $\frac{245}{2}$ m d. $\frac{245}{\sqrt{2}}$ m

16. A body of mass 2 kg moving with a velocity of 6 m/s strikes inelastically another body of same mass at rest. The amount of heat evolved during collision is

a. 36 J b. 18 J

c. 9 J d. 3 J

17. A particle of mass m is moving horizontally with a constant velocity v towards a rigid wall that is moving in opposite direction with a constant speed u. Assuming elastic impact between the particle and wall, the work done by the wall in reflecting the particle is equal to

a. $(1/2)\,m(u+v)^2$ b. $(1/2)\,m(u+v)$

c. $(1/2)\,muv$ d. None of these

18. A ball falls on an inclined plane as shown in Fig. 1.159. The ball is dropped from height h. Coefficient of restitution for collision is e and the surface is frictionless. If $h_1, h_2, \ldots, h_n$ are heights of n projectiles and $t_1, t_2, \ldots, t_n$ are their corresponding time of flights, then

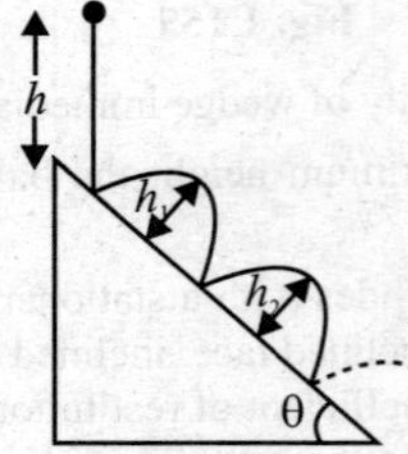

Fig. 1.159

i. $t_1, t_2, \ldots, t_n$ form a geometric progression of common ratio e.

ii. $h_1 > h_2 > h_3 \cdots > h_n$.

iii. $t_1, t_2, \ldots, t_n$ form a geometric progression of common ratio e^2.

iv. $h_1, h_2, \ldots, h_n$ form a geometric progression of common ratio e.

Evaluate the above statements and choose the correct option from the following

a. Statements i, ii are true and iii, iv are false
b. Statements i, ii are false and iii, iv are true
c. All statements are true
d. All statements are false

19. Two identical balls A and B are released from the position shown in Fig. 1.160. They collide elastically with each other on the horizontal portion. The ratio of heights attained by A and B after collision is (neglect friction)

a. 1:4 **b.** 2:1 **c.** 4:13 **d.** 2:5

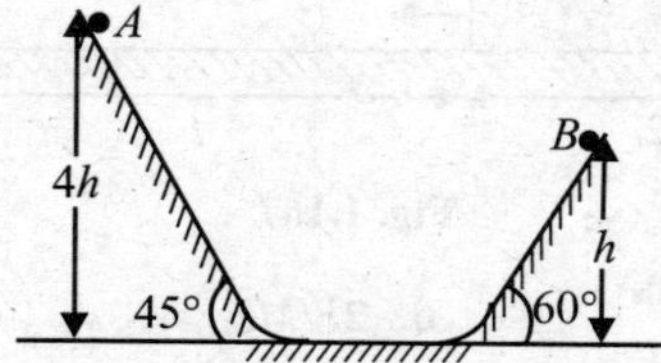

Fig. 1.160

20. Block A of mass M = 2 kg is connected to another block B of mass m = 1 kg with a string and a spring of force constant k = 600 N/m as shown in Fig. 1.161. Initially, spring is compressed to 10 cm and whole system is moving on a smooth surface with a velocity v = 1 m/s. At any time, thread is burnt, the velocity of block A, when B is having maximum velocity w.r.t. ground, is

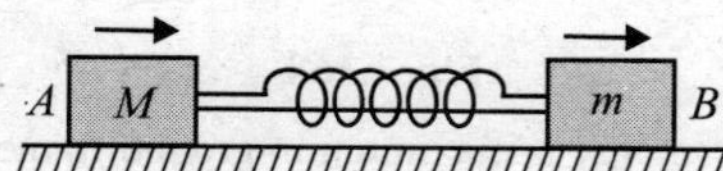

Fig. 1.161

a. zero **b.** 1 m/s **c.** 3 m/s **d.** none of these

21. Two identical blocks, having mass M each, are conjugated and placed on a smooth floor as shown in Fig. 1.162. A small block of mass m is released from position as shown. Velocity of block B is maximum

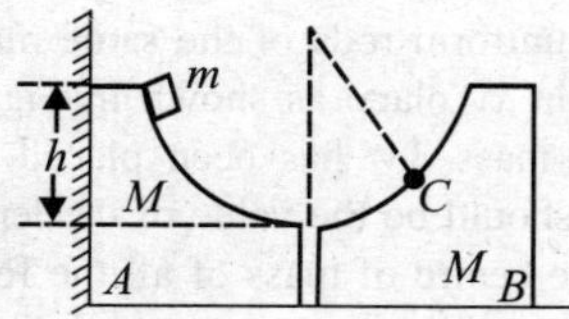

Fig. 1.162

a. when m is at the highest position on B.

b. when m is at the lowest position and moving left.

c. when m is at C.

d. when m is at lowest position and moving right.

22. A bob of mass $10m$ is suspended through an inextensible string of length l. When the bob is at rest in equilibrium position, two particles, each of mass m, strike it as shown in Fig. 1.163. The particles stick after collision. Choose the correct statement from the following:

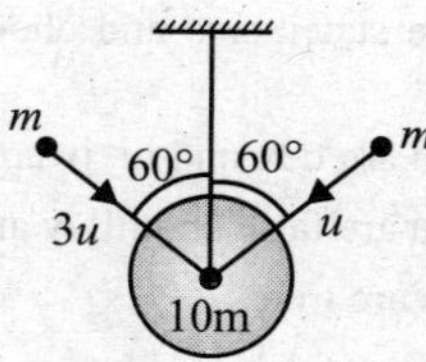

Fig. 1.163

a. Impulse in the string due to tension is $2mu$.

b. Velocity of the system just after collision is $v = \dfrac{u\sqrt{3}}{14}$.

c. Loss of energy is $\dfrac{137}{28} mu^2$.

d. Loss of energy is $\dfrac{137}{56} mu^2$.

23. In Fig. 1.164, a hollow tube of mass M is free in horizontal direction. The system is released from rest. There is no friction present. Region within the box represents the system.

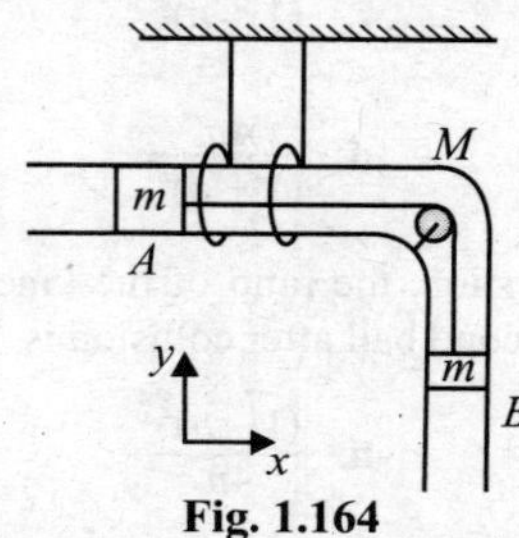

Fig. 1.164

i. Momentum of the system is conserved in x-direction.
ii. Speed of A w.r.t. M = speed of B w.r.t. M.
iii. Trajectory of centre of mass is X-constant.
iv. Centre of mass has finite acceleration.

Evaluate the above statements and choose the correct option from the following:

a. Statements i, ii are true and iii, iv are false.
b. Statements i, ii are false and iii, iv are true.
c. All statements are true.
d. All statements are false.

24. An elastic collision takes place between two smooth, rubber balls of same radius as shown in Fig. 1.165. Initially, one ball is at rest and the other is moving with velocity u. At maximum compression,

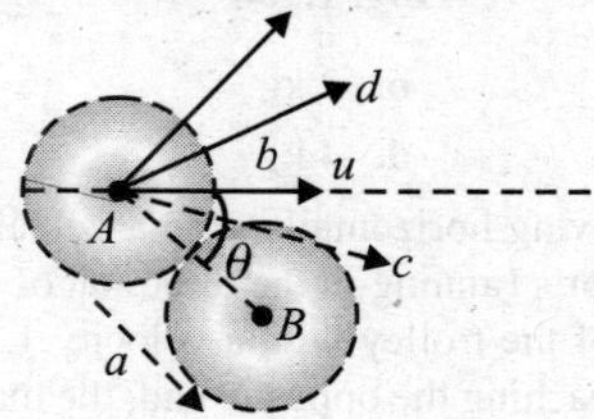

Fig. 1.165

i. ratio of potential to initial KE of the system is $(\cos^2\theta)/2$.
ii. ball B is moving along direction a.
iii. ball A is moving along direction c or d.
iv. value of e cannot be used to finalise maximum compression velocities.

Evaluate the above statements and choose the correct option from the following:

a. Statements i, ii are true and iii, iv are false.

b. Statements i, ii are false and iii, iv are true.

c. All statements are true.

d. All statements are false.

25. A particle of mass m travelling with velocity v and kinetic energy E collides elastically to another particle of mass nm, at rest. What is the fraction of total energy retained by the particle of mass m?

a. $\left(\frac{n+1}{n}\right)^2$ b. $\left(\frac{n+1}{n-1}\right)^2$

c. $\left(\frac{n-1}{n+1}\right)^2$ d. None of these

26. A ball of mass m moving with speed u undergoes a head-on elastic collision with a ball of mass nm initially at rest. The fraction of the incident energy transferred to the second ball is

a. $\frac{n}{1+n}$ b. $\frac{n}{(1+n)^2}$

c. $\frac{2n}{(1+n)^2}$ d. $\frac{4n}{(1+n)^2}$

27. In the above question, the ratio of the kinetic energies of the first ball to the second ball after collision is

a. $\frac{(1-n)^2}{2n}$ b. $\frac{(1-n)^2}{4n}$

c. $\frac{(1+n)^2}{2n}$ d. $\frac{(1+n)^2}{4n}$

28. The graph between kinetic energy and momentum of a particle is plotted as shown in Fig. 1.166. The mass of the moving particle is

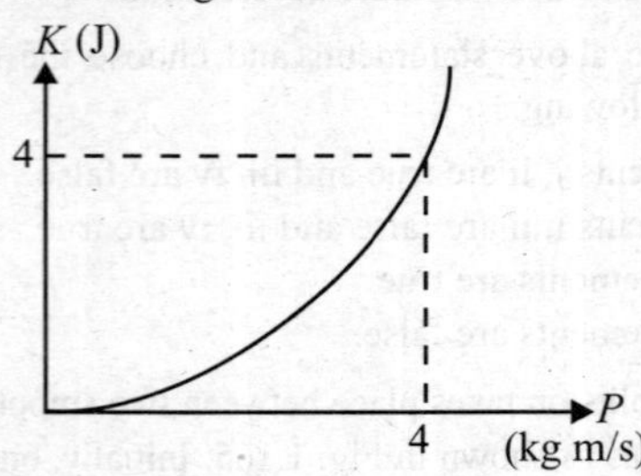

Fig. 1.166

a. 1 kg b. 2 kg

c. 3 kg d. 4 kg

29. A trolley is moving horizontally with a velocity of v m/s w.r.t. earth. A man starts running in the direction of motion of trolley from one end of the trolley with a velocity 1.5 v m/s w.r.t. the trolley. After reaching the opposite end, the man turns back and continues running with a velocity of 1.5 v m/s w.r.t. trolley in the backward direction. If the length of the trolley is L, then the displacement of the man with respect to earth, measured as a function of time, will attain a maximum value of

a. $\frac{4}{3}L$ b. $\frac{2}{3}L$

c. $\frac{5L}{3}$ d. $1.5L$

30. As shown in Fig. 1.167, there are two blocks of the same mass M, one on top of the other, lying on a frictionless horizontal surface. Both the blocks are at rest. The upper block is much smaller than the lower block. A force F is applied on the lower block and both the blocks start moving together without any relative motion. Suddenly, the lower block hits a fixed obstacle and comes to rest. The upper block continues to slide on the lower block. The upper block just manges to reach the opposite end of the lower block. What is the coefficient of friction between the two blocks?

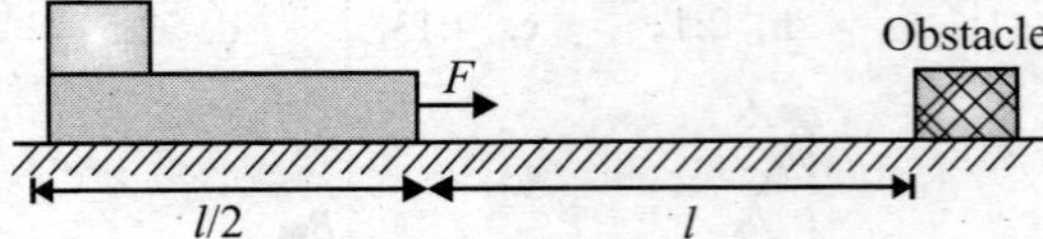

Fig. 1.167

a. F/Mg b. $2F/Mg$

c. $F/2Mg$ d. None

31. Two blocks of masses m and $4m$ lie on a smooth horizontal surface connected with a spring in its natural length. Mass m is given velocity v_0 through an impulse as shown in Fig. 1.168. Which of the following is *not true* about subsequent motion?

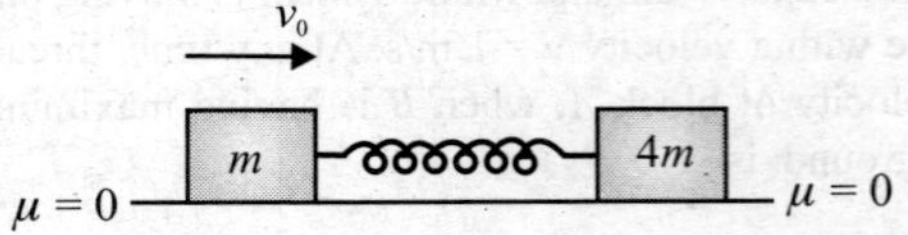

Fig. 1.168

a. Kinetic energy is maximum in ground frame and centre of mass (CM) frame simultaneously.

b. Value of maximum and minimum kinetic energy is same in CM and ground frame.

c. Minimum kinetic energy is zero in CM frame but non-zero in ground frame.

d. Maximum and minimum kinetic energy of m in ground frame is, respectively, $\frac{1}{2}mv_0^2$ and zero.

32. Three identical uniform rods of the same mass M and length L are arranged in xy plane as shown in Fig. 1.169. A fourth uniform rod of mass $3M$ has been placed as shown in the xy plane. What should be the value of the length of the fourth rod such that the centre of mass of all the four rods lie at the origin?

a. $3L$ b. $2L$

c. $L(\sqrt{2}+1)/3$ d. $L(2\sqrt{2}+1)/3$

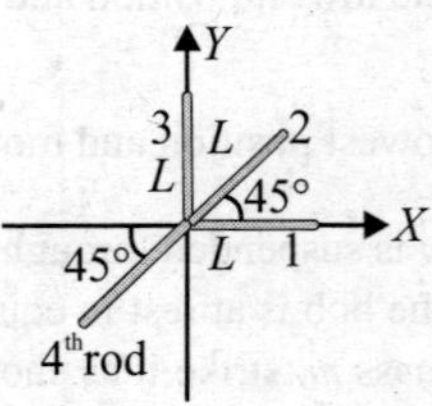

Fig. 1.169

33. A vessel at rest explodes breaking it into three pieces. Two pieces having equal mass fly off perpendicular to one another with the same speed of 30 m/s. The third piece has three times the mass of each of the other two pieces. What is the direction (w.r.t. the pieces having equal masses) and magnitude of its velocity immediately after the explosion?

a. $10\sqrt{2}$, 135° **b.** $10\sqrt{2}$, 90°

c. $10\sqrt{2}$, 60° **d.** $10\sqrt{2}$, 30°

34. A radioactive nucleus initially at rest decays by emitting an electron and neutron at right angles to one another. The momentum of the electron is 3.2×10^{-23} kg m/s and that momentum of the neutron is 6.4×10^{-23} kg m/s. The direction of the recoiling nucleus with that of the electron motion is

a. $\pi - \tan^{-1}(2)$ **b.** $\tan^{-1}(2)$

c. $\tan^{-1}(0.5)$ **d.** $\frac{\pi}{2} + \tan^{-1}(2)$

35. A shell is fired from a cannon with velocity v m/s at an angle θ with the horizontal direction. At the highest point in its path it explodes into two pieces of equal mass. One of the pieces retraces its path to the cannon, then the speed in m/s of the other piece immediately after the explosion is

a. $3v\cos\theta$ **b.** $2v\cos\theta$

c. $\frac{3v}{2}\cos\theta$ **d.** $\frac{\sqrt{3}v\cos\theta}{2}$

36. Three particles A, B and C of equal masses move with equal speeds v along the medians of an equilateral triangle. They collide at the centroid O of the triangle. After collision A comes to rest while B retraces its path with speed v. The velocity of C is then

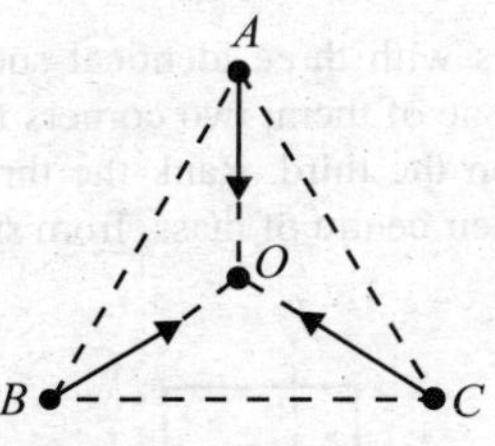

Fig. 1.170

a. v, direction $\overrightarrow{OA}$ **b.** $2v$, direction $\overrightarrow{OA}$

c. $2v$, direction $\overrightarrow{OB}$ **d.** v, direction $\overrightarrow{BO}$

37. In perfectly inelastic collisions, the relative velocity of the bodies

a. before impact is zero

b. before impact is equal to that after impact

c. after impact is zero

d. is characterized by none of the above

38. A body dropped from a tower explodes into two pieces of equal mass in mid-air. Which of the following is not possible?

a. Each part will follow parabolic path

b. Only one part will follow parabolic path

c. Both parts move along a vertical line

d. One part reaches the ground earlier than the other

39. A car of mass 400 kg travelling at 72 km/h crashes into a truck of mass 4000 kg travelling at 9 km/h in the same direction. The car bounces back at a speed of 18 km/h. The speed of the truck after the impact is

a. 9 km/h **b.** 18 km/h

c. 27 km/h **d.** 36 km/h

40. A bomb of mass 12 kg at rest explodes into two pieces of masses 4 kg and 8 kg. The velocity of 8 kg mass is 6 m/s. The kinetic energy of the other mass is

a. 48 J **b.** 32 J

c. 24 J **d.** 288 J

41. A ball moving with a velocity of 6 m/s strikes an identical stationary ball. After collision each ball moves at an angle of 30° with the original line of motion. What are the speeds of the balls after the collision?

a. $\sqrt{3}/2$ m/s **b.** 3 m/s

c. $2\sqrt{3}$ m/s **d.** $\sqrt{3}$ m/s

42. A projectile is moving at 20 m s^{-1} at its highest point where it breaks into equal parts due to an internal explosion. One part moves vertically up at 30 m s^{-1} with respect to the ground. Then the other part will move at

a. 20 m/s **b.** $10\sqrt{13}$ m/s

c. 50 m/s **d.** 30 m/s

43. A stationary body of mass 3 kg explodes into three equal pieces. Two of the pieces fly off at right angles to each other. One with a velocity of $2\hat{i}$ m/s and the other with a velocity of $3\hat{j}$ m/s. If the explosion takes place in 10^{-5} s, the average force acting on the third piece in newtons is

a. $(2\hat{i} + 3\hat{j}) \times 10^{-5}$ **b.** $-(2\hat{i} + 3\hat{j}) \times 10^{5}$

c. $(3\hat{j} + 2\hat{i}) \times 10^{5}$ **d.** $(2\hat{i} + 3\hat{j}) \times 10^{-5}$

44. A ball of mass m collides horizontally with a stationary wedge on a rough horizontal surface, in the two orientations as shown. Neglect friction between the ball and the wedge. The students comment on the system of ball and wedge in these situations

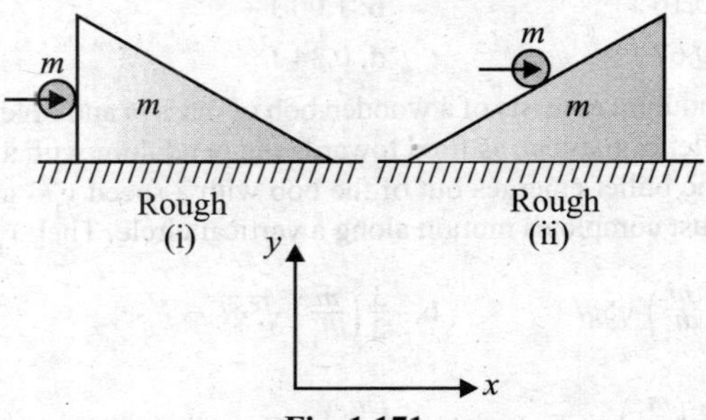

Fig. 1.171

Saurav: Momentum of the system in x-direction will change by significant amount in both the cases.

Rahul: There are no impulsive external forces in x-direction in both cases, hence the total momentum of the system in x-direction can be treated as conserved in both cases.

a. Saurav is incorrect and Rahul is correct

b. Saurav is correct and Rahul is incorrect

c. Both are correct

d. Both are incorrect

45. A body is hanging from a rigid support by an inextensible string of length l. It is struck inelastically by an identical body of mass m moving with horizontal velocity $v = \sqrt{2gl}$, the tension in the string increases just after striking by

a. mg b. $3mg$

c. $2mg$ d. none of these

46. A ball is let fall from a height h_0. There are n collisions with the earth. If the velocity of rebound after n collisions is v_n and the ball rises to a height h_n, then coefficient of restitution e is given by

a. $e^n = \sqrt{\frac{h_n}{h_0}}$ b. $e^n = \sqrt{\frac{h_0}{h_n}}$

c. $ne = \sqrt{\frac{h_n}{h_0}}$ d. $\sqrt{ne} = \sqrt{\frac{h_n}{h_0}}$

47. A body X with a momentum p collides with another identical stationary body Y one dimensionally. During the collision, Y gives an impulse J to body X. Then coefficient of restitution is

a. $\frac{2J}{p} - 1$ b. $\frac{J}{p} + 1$

c. $\frac{J}{p} - 1$ d. $\frac{J}{2p} - 1$

48. A particle of mass m moving towards the east with speed v collides with another particle of the same mass and same speed v moving towards the north. If the two particles stick to each other, the new particle of mass $2m$ will have a speed of

a. v b. $v/2$

c. $\frac{v}{\sqrt{2}}$ d. $v\sqrt{2}$

49. A block of mass 0.50 kg is moving with a speed of 2.00 m/s on a smooth surface. It strikes another stationary mass of 1.00 kg and then they move together as a single body. The energy loss during the collision is

a. 0.16 J b. 1.00 J

c. 0.67 J d. 0.34 J

50. A pendulum consists of a wooden bob of mass m and of length l. A bullet of mass m_1 is fired towards the pendulum with a speed v_1. The bullet emerges out of the bob with a speed $v_1/3$ and the bob just completes motion along a vertical circle. Then v_1 is

a. $\left(\frac{m}{m_1}\right)\sqrt{5gl}$ b. $\frac{3}{2}\left(\frac{m}{m_1}\right)\sqrt{5gl}$

c. $\frac{2}{3}\left(\frac{m_1}{m}\right)\sqrt{5gl}$ d. $\left(\frac{m_1}{m}\right)\sqrt{gl}$

51. Two pendulums each of length l are initially situated as shown in Fig. 1.172. The first pendulum is released and strikes the second. Assume that the collision is completely inelastic and neglect the mass of the string and any frictional effects. How high does the centre of mass rise after the collision?

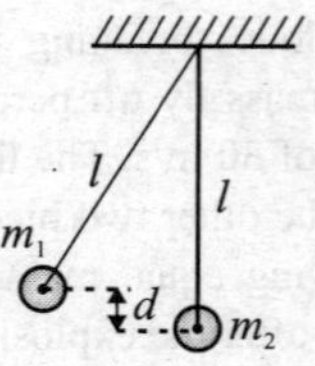

Fig. 1.172

a. $d\left[\frac{m_1}{(m_1+m_2)}\right]^2$ b. $d\left[\frac{m_1}{(m_1+m_2)}\right]$

c. $d\left[\frac{(m_1+m_2)}{m_2}\right]^2$ d. $d\left[\frac{m_2}{(m_1+m_2)}\right]^2$

52. A wooden block of mass 10 g is dropped from the top of a tower 100 m high. Simultaneously, a bullet of mass 10 g is fired from the foot of the tower vertically upwards with a velocity of 100 m/s. If the bullet is embedded in it, how high will the block rise above the top of tower before it starts falling? ($g = 10$ m/s^2)

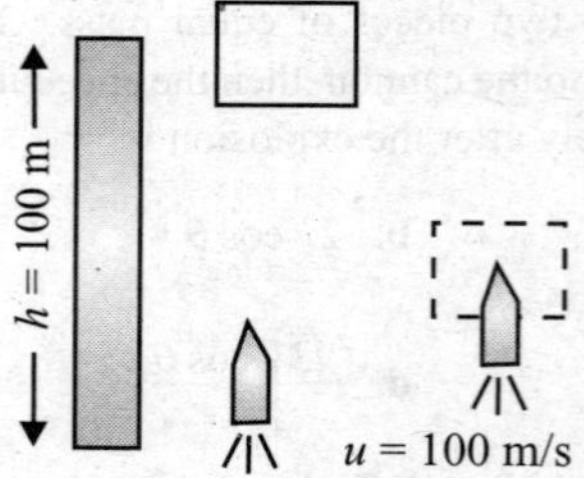

Fig. 1.173

a. 75 m b. 85 m

c. 80 m d. 10 m

53. A machinist starts with three identical square plates but cuts one corner from one of them, two corners from the second and three corners from the third. Rank the three according to the x-coordinate of their centre of mass, from smallest to largest.

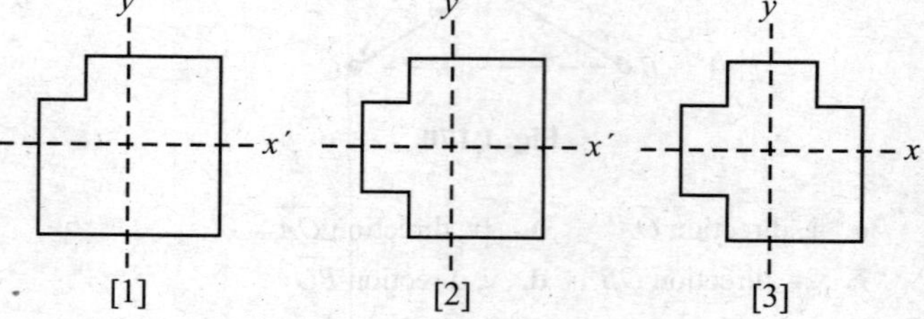

Fig. 1.174

a. 3, 1, 2 b. 1, 3, 2

c. 3, 2, 1 d. 1 and 3 tie, then 2

54. A particle of mass $2m$ is projected at an angle of 45° with the horizontal with a velocity of $20\sqrt{2}$ m/s. After 1 s of explosion, the particle breaks into two equal pieces. As a result of this one part comes to rest. The maximum height from the ground attained by the other part is ($g = 10$ m/s^2)

a. 50 m **b.** 25 m

c. 40 m **d.** 35 m

55. A smooth sphere is moving on a horizontal surface with velocity vector $2\hat{i} + 2\hat{j}$ immediately before it hits a vertical wall. The wall is parallel to $\hat{j}$ vector and the coefficient of restitution between the sphere and the wall is $e = 1/2$. The velocity vector of the sphere after it hits the wall is

a. $\hat{i} - \hat{j}$ **b.** $-\hat{i} + 2\hat{j}$

c. $-\hat{i} - \hat{j}$ **d.** $2\hat{i} - \hat{j}$

56. Two bodies of masses m and $4m$ are attached with a string as shown in Fig. 1.175. The body of mass m hanging from a string of length l is executing oscillations of angular amplitude θ_0 while the other body is at rest. The minimum coefficient of friction between the mass $4m$ and the horizontal surface should is

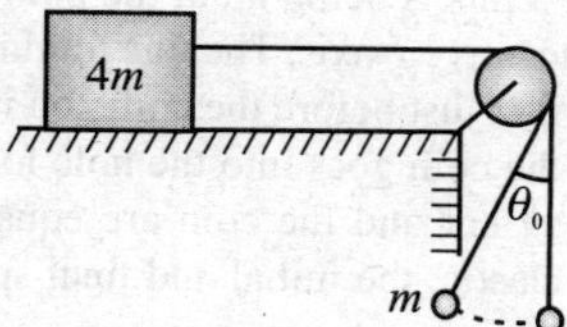

Fig. 1.175

a. $\left(\frac{2 - \cos\theta_0}{3}\right)$ **b.** $2\cos^2\left(\frac{\theta_0}{2}\right)$

c. $\left(\frac{1 - \cos\theta_0}{2}\right)$ **d.** $\left(\frac{3 - 2\cos\theta_0}{4}\right)$

57. Two equal spheres A and B lie on a smooth horizontal circular groove at opposite ends of a diameter. At time $t = 0$, A is projected along the groove and it first impinges on B at time $t = T_1$ and again at time $t = T_2$. If e is the coefficient of restitution, the ratio T_2/T_1 is

a. $\frac{2}{e}$ **b.** $\frac{(2 + e)}{2}$

c. $\frac{2(e + 1)}{e}$ **d.** $\frac{(2 + e)}{e}$

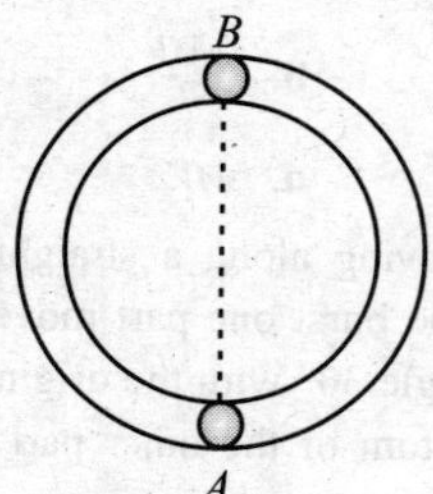

Fig. 1.176

58. A block 'A' of mass m_1 hits horizontally the rear side of a spring (ideal) attached to a block B of mass m_2 resting on a smooth horizontal surface. After hitting, 'A' gets attached to the spring.

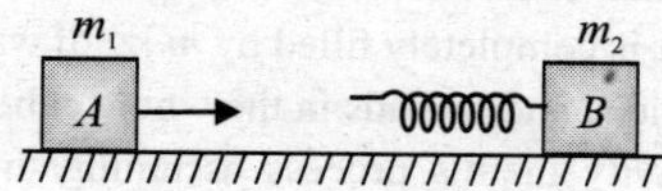

Fig. 1.177

Some statements are given at any moment of time:

i. If velocity of A is greater than B, then kinetic energy of the system will be decreasing.

ii. If velocity of A is greater than B, then kinetic energy of the system will be increasing.

iii. If velocity of A is greater than B, then momentum of the system will be decreasing.

iv. If velocity of A is greater than B, then momentum of the system will be increasing.

Now select correct alternative:

a. only iv **b.** only i

c. ii. and iv **d.** i and iii

59. Imagine a hard surface along xz plane to be fixed. A particle moving along the line $4x + 3y - 12 = 0$, with a speed of 10 m/s, from the positive side of y-axis, approaches towards the plane and collides. If the coefficient of restitution be $e = 0.75$, then the speed of the particle after collision will be

a. 5 m/s **b.** 8 m/s

c. $6\sqrt{2}$ m/s **d.** $5\sqrt{2}$ m/s

60. A particle of mass $4m$ is projected from the ground at some angle with horizontal. Its horizontal range is R. At the highest point of its path it breaks into two pieces of masses m and $3m$, respectively, such that the smaller mass comes to rest. The larger mass finally falls at a distance x from the point of projection, where x is equal to

a. $\frac{2R}{3}$ **b.** $\frac{7R}{6}$

c. $\frac{5R}{4}$ **d.** none of these

61. A block of mass M is tied to one end of a massless rope. The other end of the rope is in the hands of a man of mass $2M$ as shown in Fig. 1.178. The block and the man are resting on a rough wedge of mass M. The whole system is resting on a smooth horizontal surface. The man starts walking towards right while holding the rope in his hands. Pulley is massless and frictionless. Find the displacement of the wedge when the block meets the pulley. Assume wedge is sufficiently long so that man does not fall down.

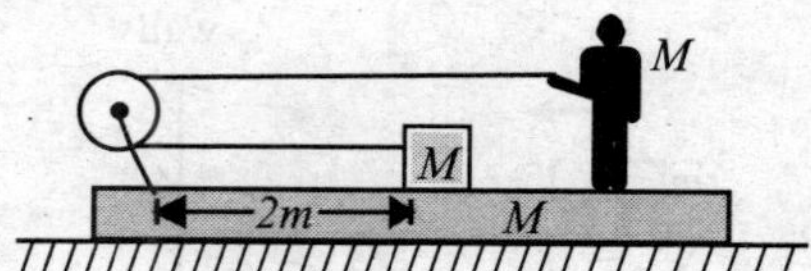

Fig. 1.178

a. 1/2 m towards right

b. 1/2 m towards left

c. The wedge does not move at all

d. 1 m towards left

62. A particle of mass m_1 moving with velocity v in a positive direction collides elastically with a mass m_2 moving in opposite direction also at velocity v. If $m_2 >> m_1$, then

a. the velocity of m_1 immediately after collision is nearly $3v$

b. the change in momentum of m_1 is nearly $4m_1v$

c. the change in kinetic energy of m_1 is nearly $4mv_2$

d. all of the above

63. A strip of wood of mass M and length l is placed on a smooth horizontal surface. An insect of mass m starts at one end of the strip and walks to the other end in time t, moving with a constant speed. The speed of the insect as seen from the ground is

a. $\frac{l}{t}\left(\frac{M}{M+m}\right)$ b. $\frac{l}{t}\left(\frac{m}{M+m}\right)$

c. $\frac{l}{t}\left(\frac{M}{m}\right)$ d. $\frac{l}{t}\left(\frac{m}{M}\right)$

64. The system in Fig. 1.179 is released from rest from the position shown. After blocks have moved distance $H/3$, collar B is removed and block A and C continue to move. The speed of C just before it strikes the ground is

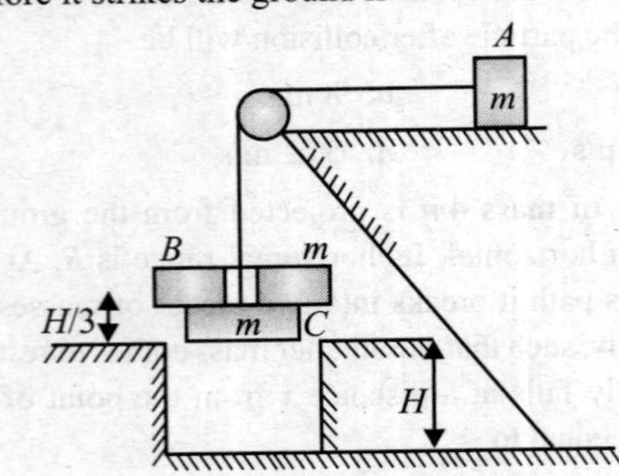

Fig. 1.179

a. $\frac{4}{3}\sqrt{gH}$ b. $2\frac{\sqrt{gH}}{3}$

c. $\frac{\sqrt{13gH}}{3}$ d. $2\sqrt{2gH}$

65. A car of mass m is initially at rest on the boat of mass M tied to the wall of dock through a massless, inextensible string. The car accelerates from rest to velocity v_0 in times t_0. At $t = t_0$ the car applies brake and comes to rest relative to the boat in negligible time. Neglect friction between the boat and water; the time 't' at which boat will strike the wall is

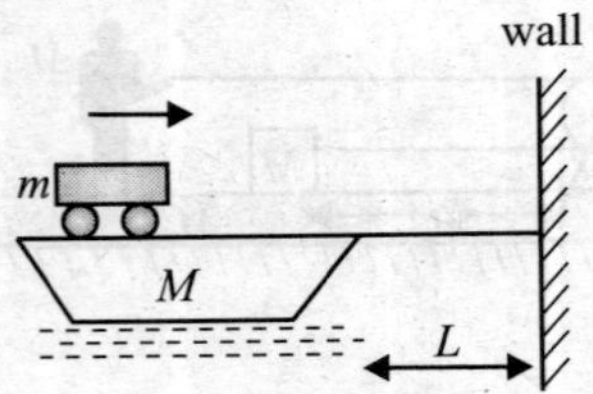

Fig. 1.180

a. $\frac{L(M+m)}{mv_0}$ b. $t_0 + \frac{L(M+m)}{mv_0}$

c. $t_0 + \frac{L(M+m)}{Mv_0}$ d. none of these

66. A particle of mass m moving with velocity 1 m/s collides perfectly elastically with another stationary particle of mass $2m$. If the incident particle is deflected by 90°, the heavy mass will make an angle θ with the initial direction of m equal to

a. 60° b. 45°

c. 15° d. 30°

67. Two particles A and B start moving due to their mutual interaction only. If at any time 't', $\vec{a}_A$ and $\vec{a}_B$ are their respective accelerations, $\vec{v}_A$ and $\vec{v}_B$ are their respective velocities, and up to that time W_A and W_B are the work done on A and B, respectively, by the mutual force, m_A and m_B are their masses, respectively, then which of the following is always correct?

a. $\vec{v}_A + \vec{v}_B = 0$ b. $m_A\vec{v}_A + m_B\vec{v}_B = 0$

c. $W_A + W_B = 0$ d. $\vec{a}_A + \vec{a}_B = 0$

68. On a smooth carom board, a coin moving in negative y-direction with a speed of 3 m/s is being hit at the point (4, 6) by a striker moving along negative x-axis. The line joining the centres of the coin and the striker just before the collision is parallel to x-axis. After collision the coin goes into the hole located at the origin. Masses of the striker and the coin are equal. Considering the collision to be elastic, the initial and final speeds of the striker in m/s will be

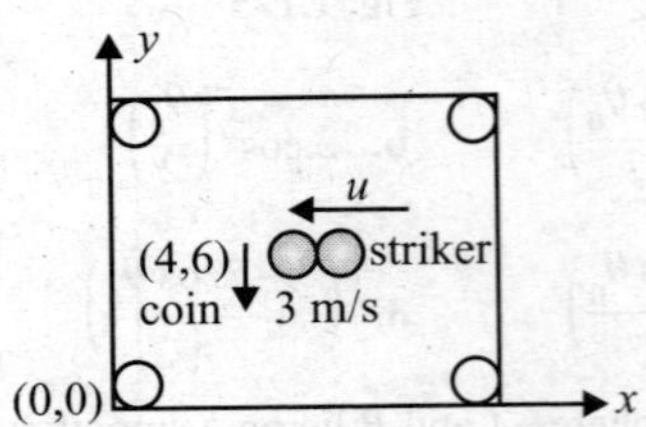

Fig. 1.181

a. (1, 0) b. (2, 0)

c. (3, 0) d. none of these

69. A train of mass M is moving on a circular track of radius 'R' with constant speed V. The length of the train is half of the perimeter of the track. The linear momentum of the train will be

a. 0 b. $\frac{2MV}{\pi}$

c. MVR d. MV

70. A canon shell moving along a straight line bursts into two parts. Just after the burst one part moves with momentum 20 N s making an angle 30° with the original line of motion. The minimum momentum of the other part of shell just after the burst is

a. 0 N s b. 5 N s

c. 10 N s d. 17.32 N s

71. Figure 1.182 shows a hollow cube of side 'a' and volume 'V'. There is a small chamber of volume $V/4$ in the cube as shown. The chamber is completely filled by m kg of water. Water leaks through a hole H and spreads in the whole cube. Then the work done by gravity in this process assuming that the complete water finally lies at the bottom of the cube is

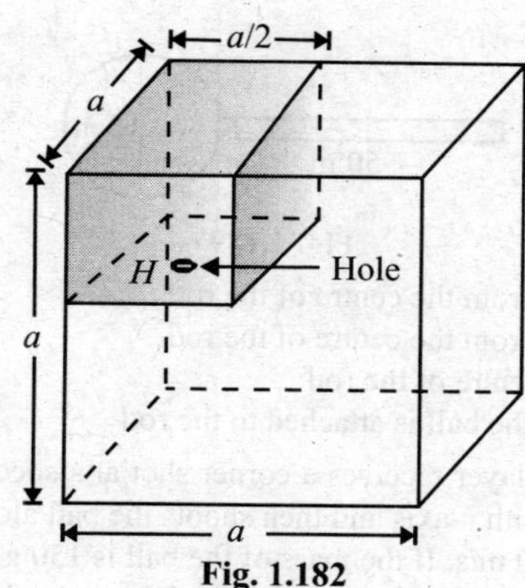

Fig. 1.182

a. $\frac{1}{2}mg\,a$ **b.** $\frac{3}{8}mg\,a$

c. $\frac{5}{8}mg\,a$ **d.** $\frac{1}{8}mg\,a$

72. A balloon having mass 'm' is filled with gas and is held in hands of a boy. Then suddenly it gets released and gas starts coming out of it with a constant rate. The velocity of the ejected gas is also constant (2 m/s) with respect to the balloon. Find out the velocity of the balloon when the mass of gas is reduced to half.

a. ln 2 **b.** 2 ln 4

c. 2 ln 2 **d.** None of these

73. In a vertical plane inside a smooth hollow thin tube, a block of same mass as that of tube is released as shown in Fig. 1.183. When it is slightly disturbed it moves towards right. By the time the block reaches the right end of the tube, the displacement of the tube will be (where 'R' is the mean radius of the tube, assume that the tube remains in vertical plane)

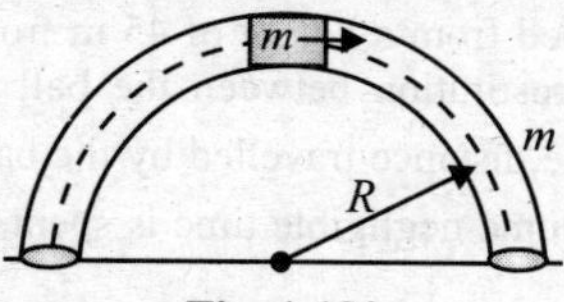

Fig. 1.183

a. $\frac{2R}{\pi}$ **b.** $\frac{4R}{\pi}$

c. $\frac{R}{2}$ **d.** R

74. Two men 'A' and 'B' are standing on a plank. 'B' is at the middle of the plank and 'A' is at the left end of the plank. Lower surface of the plank is smooth. System is initially at rest and masses are as shown in Fig. 1.184. 'A' and 'B' start moving such that the position of 'B' remains fixed with respect to ground, then 'A' meets 'B'. Then the point where A meets B is located at

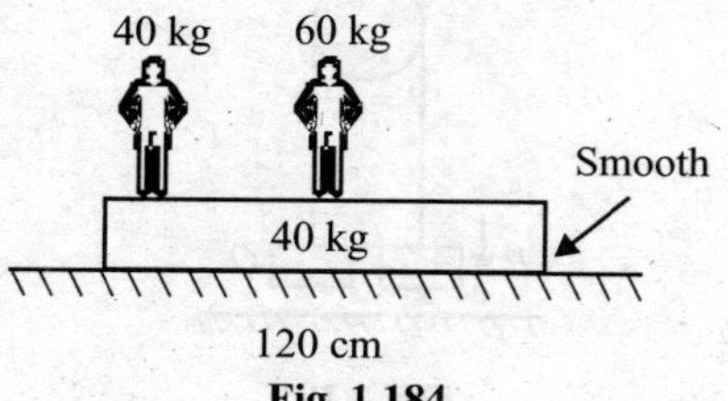

Fig. 1.184

a. the middle of the plank

b. 30 cm from the left end of the plank

c. the right end of the plank

d. none of these

75. A gun which fires small balls of mass 20 g is firing 20 balls per second on the smooth horizontal table surface $ABCD$. If the collision is perfectly elastic and balls are striking at the centre of table with a speed of 5 m/s at an angle of 60° with the vertical just before collision, then force exerted by one of the legs on ground is (assume total weight of the table is 0.2 kg)

a. 0.5 N **b.** 1 N

c. 0.25 N **d.** 0.75 N

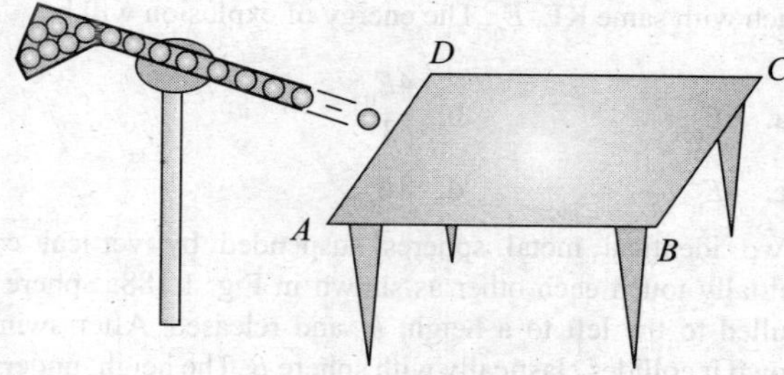

Fig. 1.185

76. Figure 1.186 shows the velocity–time graph for two masses R and S that collided elastically. Which of the following statements is true?

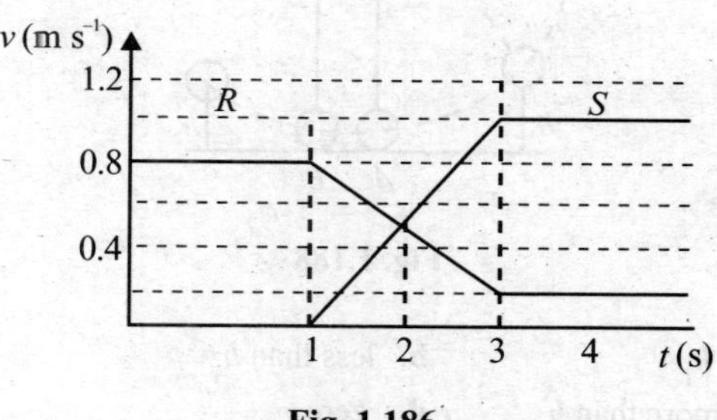

Fig. 1.186

i. R and S moved in the same direction after the collision.

ii. The velocities of R and S were equal at the mid time of the collision.

iii. The mass of R was greater than mass of S.

Which of the following is true?

a. i only **b.** ii only

c. i and ii. only **d.** i, ii. and iii

77. A system of two blocks A and B are connected by an inextensible massless string as shown in Fig. 1.187. The pulley is massless and frictionless. Initially, the system is at rest. A bullet of mass 'm' moving with a velocity 'u' as shown hits block 'B' and gets embedded into it. The impulse imparted by tension force to the block of mass $3m$ is

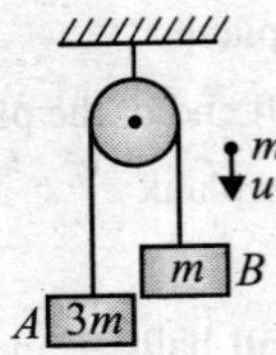

Fig. 1.187

a. $\frac{5\,mu}{4}$ b. $\frac{4\,mu}{5}$

c. $\frac{2\,mu}{5}$ d. $\frac{3\,mu}{5}$

78. A stationary body explodes into four identical fragments such that three of them fly off mutually perpendicular to each other, each with same KE, E_0. The energy of explosion will be

a. $6E_0$ b. $\frac{4E_0}{3}$

c. $4E_0$ d. $8E_0$

79. Two identical metal spheres suspended by vertical cords, initially touch each other as shown in Fig. 1.188. Sphere *A* is pulled to the left to a height h_1 and released. After swinging down it collides elastically with sphere *B*. The height undergone by sphere *B* will be

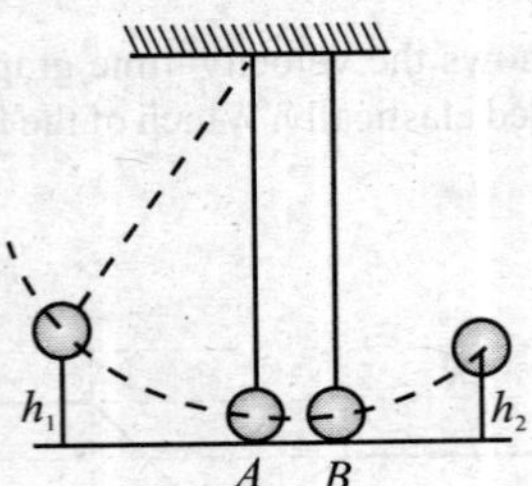

Fig. 1.188

a. h_1 b. less than h_1

c. more than h_1 d. zero

80. A man stands at one end of the open truck which can run on frictionless horizontal rails. Initially, the man and the truck are at rest. Man now walks to the other end and stops. Then which of the following is true?

a. The truck moves opposite to direction of motion of the man even after the man ceases to walk.

b. The centre of mass of the man and the truck remains at the same point throughout the man's walk.

c. The kinetic energy of the man and the truck are exactly equal throughout the man's walk.

d. The truck does not move at all during the man's walk.

81. Figure 1.189 shows a thin uniform rod 50 cm long and has a mass of 100 g. A hollow metal ball is filled with air and has a diameter 10 cm and total mass 50 g is fixed to one end of the rod. At what point along its length will the ball and rod balance horizontally?

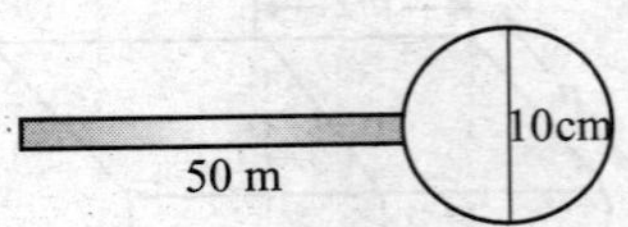

Fig. 1.189

a. 20 cm from the centre of the rod

b. 10 cm from the centre of the rod

c. at the centre of the rod

d. where the ball is attached to the rod

82. A hockey player receives a corner shot at a speed of 15 m/s at angle 30° with *y*-axis and then shoots the ball along *x*-axis with the speed 30 m/s. If the mass of the ball is 150 g and it remains in contact with the hockey stick for 0.01 s, the force exerted on the ball along *x*-axis is

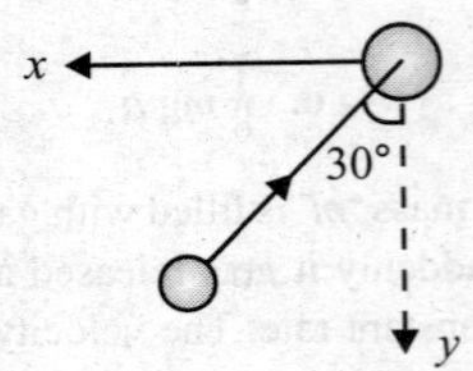

Fig. 1.190

a. 281 N   b. 187.5 N

c. 562.5 N   d. 375N

83. In an inelastic collision

a. momentum is conserved but kinetic energy is not conserved

b. momentum is not conserved but kinetic energy is conserved

c. neither momentum nor kinetic energy is conserved

d. both momentum and kinetic energy are conserved

84. A ball is dropped from a height of 45 m from the ground. The coefficient of restitution between the ball and the ground is 2/3· What is the distance travelled by the ball in 4th second of its motion. Assume negligible time is spent in rebounding. Let $g = 10$ m s^2.

a. 5 m   b. 20 m

c. 15 m   d. 10 m

85. *P* and *Q* are two identical masses at rest suspended by an inextensible string passing over a smooth frictionless pulley. Mass *P* is given a downward push with a speed *v* as shown in Fig. 1.191. It collides elastically with the floor and rebounds immediately. What happens immediately after collision?

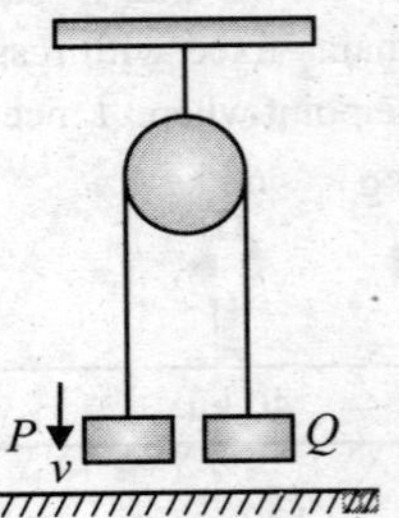

Fig. 1.191

a. P and Q both move upwards with equal speeds.
b. P and Q both move upwards with different speeds.
c. P moves upwards and Q moves downwards with equal speeds.
d. Both P and Q are at rest.

86. A bag of mass M hangs by a long massless rope. A bullet of mass m, moving horizontally with velocity u, is caught in the bag. Then for the combined (bag + bullet) system, just after collision

a. momentum is $muM/(M + m)$
b. kinetic energy is $mu^2/2$
c. momentum is $mu(M + m)/M$
d. kinetic energy is $m^2u^2/2(M + m)$

87. A man stands at one end of a boat which is stationary in water. Neglect water resistance. The man now moves to the other end of the boat and again becomes stationary. The centre of mass of the 'man plus boat' system will remain stationary with respect to water

a. only when the man is stationary initially and finally
b. only if the man moves without acceleration on the boat
c. only if the man and the boat have equal masses
d. in all cases

88. Two particles are shown in Fig. 1.192. At $t = 0$, a constant force $F = 6$ N starts acting on the 3 kg man. Find the velocity of the centre of mass of these particles at $t = 5$ s.

2 kg 3 kg $F = 6$ N

Fig. 1.192

a. 5 m/s b. 4 m/s
c. 6 m/s d. 3 m/s

89. Two blocks of masses 5 kg and 2 kg are placed on a frictionless surface and connected by a spring. An external kick gives a velocity of 14 m/s to the heavier block in the direction of lighter one. The magnitudes of velocities of two blocks in the centre of mass frame after the kick are, respectively,

a. 4 m/s, 4 m/s b. 10 m/s, 4 m/s
c. 4 m/s, 10 m/s d. 10 m/s, 10 m/s

90. In a system of particles 8 kg mass is subjected to a force of 16 N along +ve x-axis and another 8 kg mass is subjected to a force of 8 N along +ve y-axis. The magnitude of acceleration of centre of mass and the angle made by it with x-axis are given, respectively, by

a. $\frac{\sqrt{5}}{2}$ ms^2, $\theta = 45°$ b. $3\sqrt{5}$ ms^2, $\theta = \tan^{-1}\left(\frac{2}{3}\right)$
c. $\frac{\sqrt{5}}{2}$ ms^2, $\theta = \tan^{-1}\left(\frac{1}{2}\right)$ d. 1 ms^2, $\theta = \tan^{-1}\sqrt{3}$

91. Two particles of equal masses have velocities $\vec{v}_1 = 3\hat{i}$ m/s and $\vec{v}_2 = 2\hat{j}$ m/s at any instant. The first particle has a constant acceleration $\vec{a}_1 = (3\hat{i} + 3\hat{j})$ m/s^2 while the acceleration of the other particle is zero. The centre of mass of the two particles moves in a

a. circle b. parabola
c. straight line d. ellipse

92. A circular plate of uniform thickness has a diameter of 28 cm. A circular portion of diameter 21 cm is removed from the plate as shown. O is the centre of mass of complete plate. The position of centre of mass of remaining portion will shift towards left from 'O' by

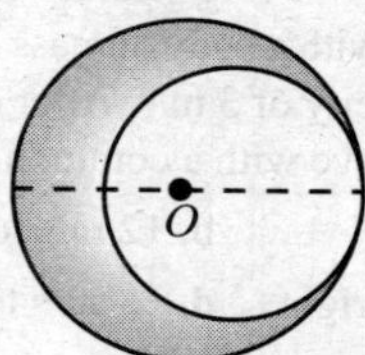

Fig. 1.193

a. 5 cm b. 9 cm
c. 4.5 cm d. 5.5 cm

93. A heavy chain of length 1 m and weight 20 kg hangs vertically with one end attached to a peg and carries a block of mass 10 kg at the other end. Find the work done in winding 50 cm of chain round the peg.

a. 85 J b. 100 J
c. 120 J d. 125 J

94. In the above question, find the height by which the centre of mass of the system of 'chain + block' rises.

a. 0.417 m b. 0.437 m
c. 0.365 m d. 0.405 m

95. A cannon of mass 1000 kg located at the base of an inclined plane fires a shell of mass 50 kg in horizontal direction with velocity 180 km/h. The angle of inclination of the inclined plane with the horizontal is 45°. The coefficient of friction between the cannon and inclined plane is 0.5. The maximum height, in metre, to which the cannon can ascend the inclined plane as a result of recoil is

a. $\frac{5}{6}$ b. $\frac{5}{24}$
c. $\frac{5}{12}$ d. none of these

96. A U-shaped wire has a semicircular bending between A and B as shown in Fig. 1.194. A bead of mass m moving with uniform speed v through a wire enters the semicircular bend at A and leaves at B with velocity $v/2$ after time T. The average force exerted by the bead on the part AB of the wire is

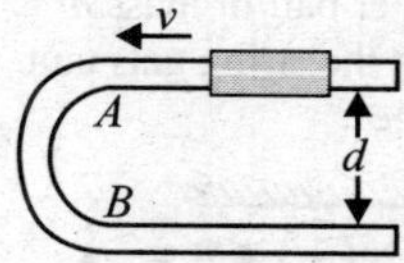

Fig. 1.194

a. 0 b. $\frac{3mv}{2T}$
c. $\frac{3mv}{T}$ d. none of these

97. A particle of mass 2 kg moving with a velocity of 3 m/s is acted upon by a force which changes its direction of motion by an angle of 90° without changing its speed. What is the magnitude of impulse experienced by the particle?

a. 6 N s **b.** 2 N s

c. $3\sqrt{2}$ N s **d.** $6\sqrt{2}$ N s

98. A body of mass 3 kg moving with a velocity of 4 m/s towards left collides head on with a body of mass 4 kg moving in opposite direction with a velocity of 3 m/s. After collision the two bodies stick together and move with a common velocity which is

a. zero **b.** 12 m/s towards left

c. 12 m/s towards right **d.** $\frac{12}{7}$ m/s towards left

99. Two blocks of masses $m_1 = 2$ kg and $m_2 = 4$ kg are moving in the same direction with speeds $v_1 = 6$ m/s and $v_2 = 3$ m/s, respectively on a frictionless surface as shown in figure. An ideal spring with spring constant $k = 30000$ N/m is attached to the back side of m_2. Then the maximum compression of the spring after collision will be

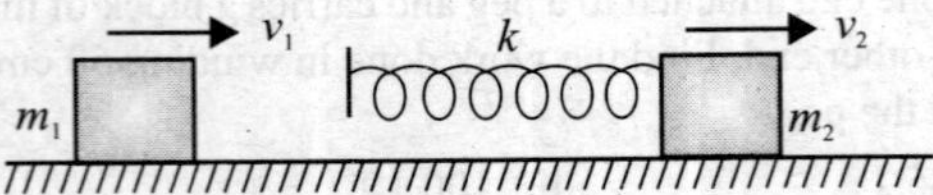

Fig. 1.195

a. 0.06 m **b.** 0.04 m

c. 0.02 m **d.** none of these

100. A cracker is thrown into air with a velocity of 10 m/s at an angle of 45° with the vertical. When it is at a height of 0.5 m from the ground, it explodes into a number of pieces which follow different parabolic paths. What is the velocity of centre of mass, when it is at a height of 1 m from the ground? ($g = 10$ m/s^2)

a. $4\sqrt{5}$ m/s **b.** $2\sqrt{5}$ m/s

c. $5\sqrt{4}$ m/s **d.** 10 m/s

101. A force exerts an impulse I on a particle changing its speed from initial velocity u to final velocity $2u$. The applied force and the initial velocity are oppositely oriented along the same line. The work done by the force is

a. $\frac{3}{2}Iu$ **b.** $\frac{1}{2}Iu$

c. Iu **d.** $2\,Iu$

102. In Fig. 1.196, a heavy ball of mass 2 m rests on the horizontal surface and the lighter ball of mass m is dropped from a height $h > 2l$. At the instant the string gets taut, the upward velocity of the heavy ball will be

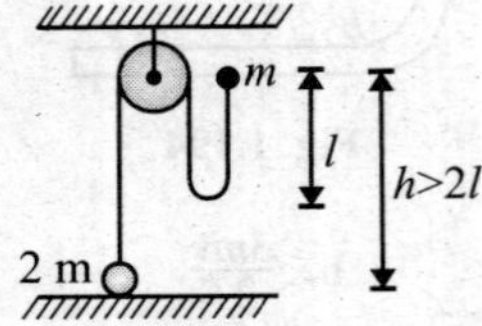

Fig. 1.196

a. $\frac{2}{3}\sqrt{gl}$ **b.** $\frac{4}{3}\sqrt{gl}$

c. $\frac{1}{3}\sqrt{gl}$ **d.** $\frac{1}{2}\sqrt{gl}$

103. A ball falls vertically onto a floor with momentum p, and then bounces repeatedly. If the coefficient of restitution is e, then the total momentum imparted by the ball on the floor till the ball comes to rest is

a. $p(1 + e)$ **b.** $\frac{p}{1 - e}$

c. $p\left(1 + \frac{1}{e}\right)$ **d.** $p\left(\frac{1+e}{1-e}\right)$

104. A ball released from a height h_0 above a horizontal surface rebounds to a height h_1 after one bounce. The graph that relates h_0 to h_1 is shown Fig. 1.197. If the ball (of the mass m) was dropped from an initial height h and made three bounces, the kinetic energy of the ball immediately after the third impact with the surface was

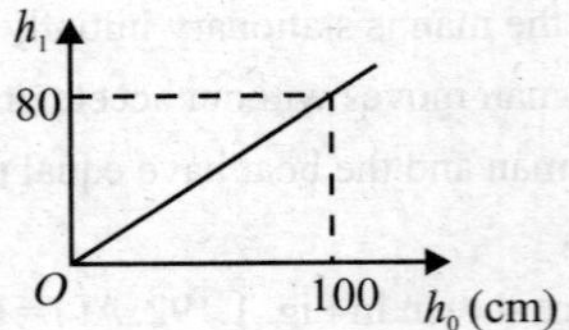

Fig. 1.197

a. $(0.8)^3\,mgh$ **b.** $(0.8)^2\,mgh$

c. $0.8\,mg(h/3)$ **d.** $[1 - (0.8)^3]mgh$

105. Two identical balls, of equal masses A and B, are lying on a smooth surface as shown in Fig. 1.198. Ball A hits ball B (which is at rest) with a velocity $v = 16$ m/s. What should be the minimum value of coefficient of restitution between A and B so that B just reaches the highest point of inclined plane?

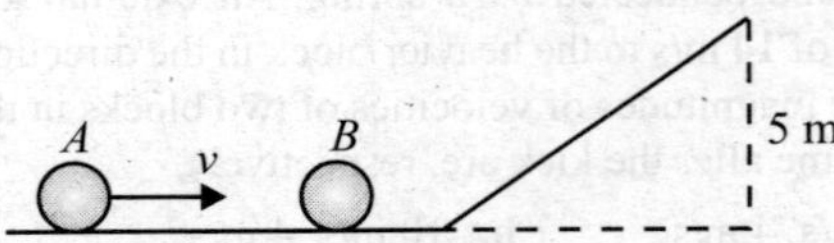

Fig. 1.198

a. $\frac{2}{3}$ **b.** $\frac{1}{4}$

c. $\frac{1}{2}$ **d.** $\frac{1}{3}$

106. A particle of mass m is made to move with uniform speed v_0 along the perimeter of a regular hexagon, inscribed in a circle of radius R. The magnitude of impulse applied at each corner of the hexagon is

a. $2\,mv_0 \sin\frac{\pi}{6}$ **b.** $mv_0 \sin\frac{\pi}{6}$

c. $mv_0 \sin\frac{\pi}{3}$ **d.** $2mv_0 \sin\frac{\pi}{3}$

107. A continuous stream of particles, of mass m and velocity v, is emitted from a source at a rate of n per second. The particles travel along a straight line, collide with a body of mass M and get embedded in the body. If the mass M was originally at rest, its velocity when it has received N particles will be

a. $\frac{mvN}{Nm+n}$ **b.** $\frac{mvN}{Nm+M}$

c. $\frac{mv}{Nm+M}$ **d.** $\frac{Nm+M}{Nm}$

108. A ball kept in a closed box moves in the box making collisions with the walls. The box is kept on a smooth surface. The velocity of the centre of mass

a. of the box remains constant

b. of the (box + ball) system remains constant

c. of the ball remains constant

d. of the ball relative to the box remains constant

109. A particle of mass m moving with a velocity v makes an elastic one dimensional collision with a stationary particle of mass m establishing a contact with it for extremely small time T. Their force of contact increases from zero to F_0 linearly in time $T/4$, remains constant for a further time $T/2$ and decreases linearly from F_0 to zero in further time $T/4$ as shown in Fig. 1.199. The magnitude possessed by F_0 is

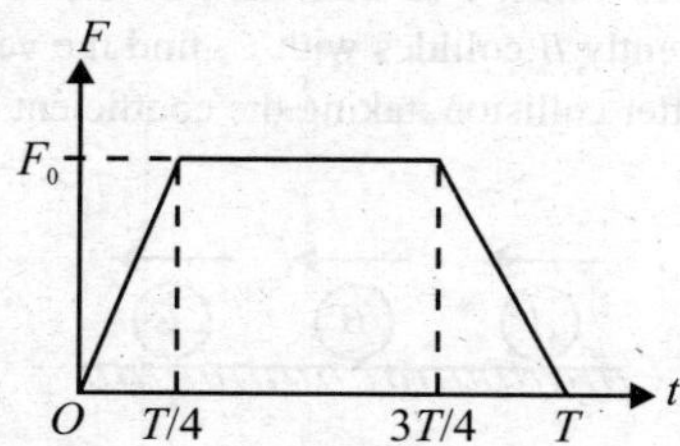

Fig. 1.199

a. $\frac{mu}{T}$ **b.** $\frac{2mu}{T}$

c. $\frac{4mu}{3T}$ **d.** $\frac{3mu}{4T}$

110. A stationary body explodes into two fragments of masses m_1 and m_2. If momentum of one fragment is p, the energy of explosion is

a. $\frac{p^2}{2(m_1+m_2)}$ **b.** $\frac{p^2}{2\sqrt{m_1 m_2}}$

c. $\frac{p^2(m_1+m_2)}{2m_1 m_2}$ **d.** $\frac{p^2}{2(m_1-m_2)}$

111. A ball of mass m is projected with a speed v into the barrel of a spring gun of mass M initially at rest lying on a frictionless surface. The mass sticks in the barrel at the point of maximum compression in the spring. The fraction of kinetic energy of the ball stored in the spring is

a. $\frac{m}{M}$ **b.** $\frac{M}{m+M}$

c. $\frac{m}{M+M}$ **d.** none of the these

112. A railway flat car has an artillery gun installed on it. The combined system has a mass M and moves with a velocity V. The barrel of the gun makes an angle α with the horizontal. A shell of mass m leaves the barrel at a speed v relative to the barrel. The speed of the flat car so that it may stop after the firing is

a. $\frac{mv}{M+m}$ **b.** $\left(\frac{Mv}{M+m}\right)\cos\alpha$

c. $\left(\frac{mv}{M+m}\right)\cos\alpha$ **d.** $(M+m)v\cos\alpha$

113. Figure 1.200 shows three particles A, B and C on the x-axis. They are given velocities of v_1 = 3 m/s, v_2 = 2 m/s and v_3 = 5 m/s, respectively, in the directions shown. The position of centre of mass of A, B and C at time t = 1 s will be

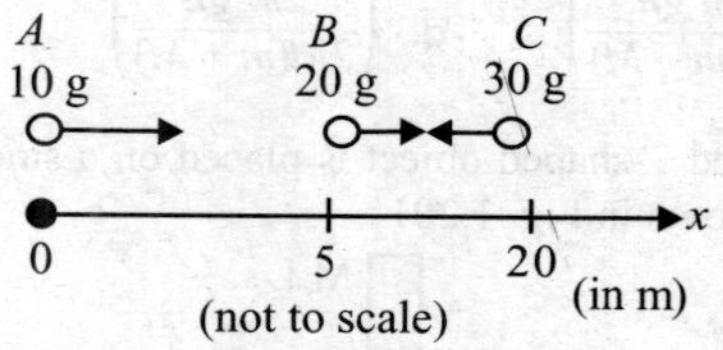

Fig. 1.200

a. $x = 11\frac{2}{3}$ m **b.** $x = 15\frac{1}{3}$ m

c. $x = 10\frac{1}{3}$ m **d.** $x = 10\frac{2}{3}$ m

114. A particle of mass 200 g is dropped from a height of 50 m and another particle of mass 100 g is simultaneously projected up from the ground along the same line with a speed of 100 m/s. The acceleration of the centre of mass after 1 s is

a. 10 m/s^2 **b.** $\frac{10}{3}$ m/s^2

c. 0 **d.** none of these

115. In the above problem, the velocity of the centre of mass after 1 s will be

a. $\frac{20}{3}$ m/s, vertically downwards

b. $\frac{20}{3}$ m/s, vertically upwards

c. $\frac{70}{3}$ m/s, vertically downwards

d. $\frac{70}{3}$ m/s, vertically upwards

116. Two blocks of masses 6 kg and 4 kg are attached to the two ends of a massless string passing over a smooth fixed pulley. If the system is released, the acceleration of the centre of mass of the system will be

a. g, vertically downwards

b. g/5, vertically downwards

c. g/25, vertically downwards

d. zero

117. The momentum of a moving particle is vectorially given as $\vec{p} = p_0 (\cos t\,\hat{i} + \sin t\,\hat{j})$, where t stands for time. Choose the correct option:

a. The applied force is constant.

b. The momentum is constant.

c. The applied force always remains perpendicular to the momentum.

d. The applied force is always parallel to the momentum.

118. A gun of mass M, fires a shell of mass m horizontally and the energy of explosion is such as would be sufficient to project the shell vertically to a height 'h'. The recoil velocity of the gun is

a. $\left(\frac{2m^2gh}{M(m+M)}\right)^{\frac{1}{2}}$ **b.** $\left(\frac{2m^2gh}{M(m-M)}\right)^{\frac{1}{2}}$

c. $\left(\frac{2m^2gh}{2M(m-M)}\right)^{\frac{1}{2}}$ **d.** $\left(\frac{2m^2gh}{2M(m+M)}\right)^{\frac{1}{2}}$

119. An inverted T-shaped object is placed on a smooth horizontal floor as shown in Fig. 1.201.

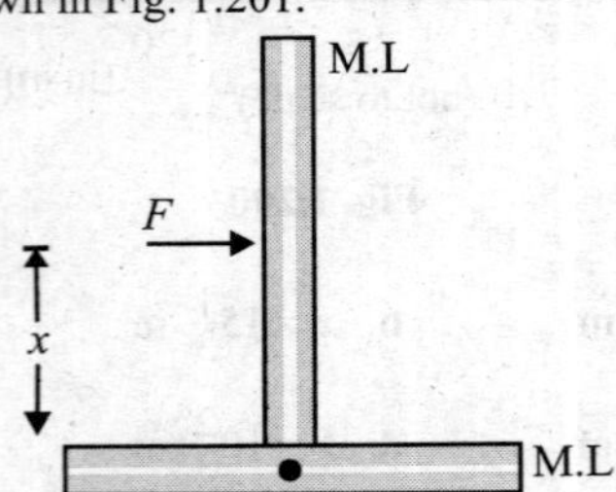

Fig. 1.201

A force F is applied on the system as shown in Fig. 1.201. The value of x so that the system performs pure translational motion is

a. $\frac{L}{4}$ **b.** $\frac{3L}{4}$

c. $\frac{L}{2}$ **d.** $\frac{3L}{2}$

120. Two blocks m_1 and m_2 are pulled on a smooth horizontal surface, and are joined together with a spring of stiffness k as shown in Fig. 1.202. Suddenly, block m_2 receives a horizontal velocity v_0, then the maximum extension x_m in the spring is

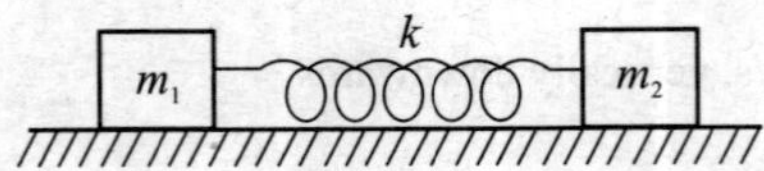

Fig. 1.202

a. $v_0\sqrt{\frac{m_1m_2}{m_1+m_2}}$ **b.** $v_0\sqrt{\frac{2m_1m_2}{(m_1+m_2)k}}$

c. $v_0\sqrt{\frac{m_1m_2}{2(m_1+m_2)k}}$ **d.** $v_0\sqrt{\frac{m_1m_2}{(m_1+m_2)k}}$

121. A particle at rest is constrained to move on a smooth horizontal surface. Another identical particle hits the fractional particle with a velocity v at an angle $\theta = 60°$ with horizontal. If the particles move together, the velocity of the combination just after impact is equal to

a. v **b.** $\frac{v}{2}$

c. $\frac{\sqrt{3}v}{4}$ **d.** $\frac{v}{4}$

122. A particle of mass m comes down on a smooth inclined plane from point B at a height of h from rest. The magnitude of change in momentum of the particle between position A (just before arriving on horizontal surface) and C (assuming the angle of inclination of the plane as θ with respect to the horizontal) is

a. 0 **b.** $2m\sqrt{(2gh)}\sin\theta$

c. $2m\sqrt{(2gh)}\sin\left(\frac{\theta}{2}\right)$ **d.** $2m\sqrt{(2gh)}$

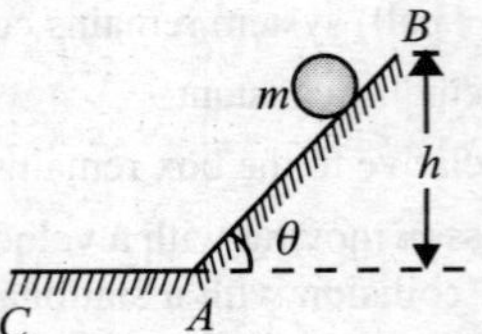

Fig. 1.203

123. Three balls A, B and C of masses 2 kg, 4 kg and 8 kg, respectively, move along the same straight line and in the same direction, with velocities 4 m/s, 1 m/s and 3/4 m/s. If A collides with B and subsequently B collides with C, find the velocity of ball A and ball B after collision, taking the coefficient of restitution as unity.

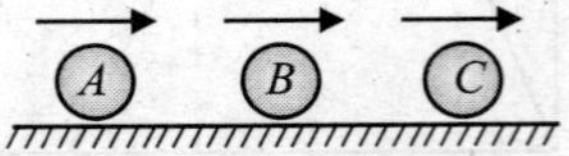

Fig. 1.204

a. $V_A = 3, V_B = 9/4$ **b.** $V_A = 0, V_B = 3$

c. $V_A = 3, V_B = 0$ **d.** $V_A = 0, V_B = 0$

124. A small ball of mass m is attached with upper end of a uniform straight rod of equal mass m and length l. The rod is held vertical over a smooth horizontal surface as shown in Fig. 1.205. When the system is released, the lower end slips freely and the systems falls down. Assuming the initial position of the lower end to be origin and initially rod to be along y-axis as shown in Fig. 1.205, the equation of trajectory of point P of the rod (P is at distance $l/4$ from the lower end) is

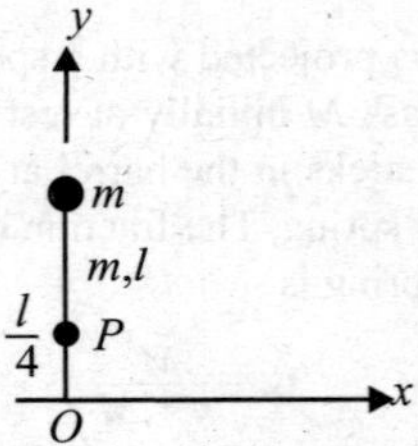

Fig. 1.205

a. $4x^2 + 16y^2 = 4l^2$ **b.** $4x^2 + 16y^2 = l^2$
c. $x^2 + 4y^2 = l^2$ **d.** $x^2 + 4y^2 = 16l^2$

125. A ball of mass m moving with velocity v_0 collides with a wall as shown in Fig. 1.206. After impact it rebounds with a velocity $v_0/2$. The component of impulse acting on the ball along the wall is

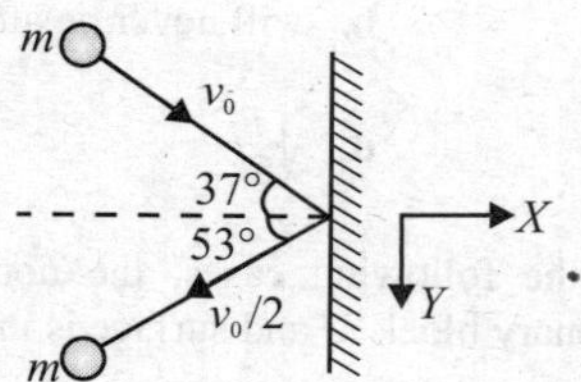

Fig. 1.206

a. $\frac{mv_0}{2}\hat{j}$ **b.** $-\frac{mv_0}{2}\hat{j}$
c. $-\frac{mv_0}{5}\hat{j}$ **d.** none of these

126. Five balls are placed one after the other along a straight line as shown in the figure. Initially, all the balls are at rest. Then the second ball has been projected with speed v_0 towards the third ball. Mark the correct statements. (Assume all collisions to be head-on and elastic.)

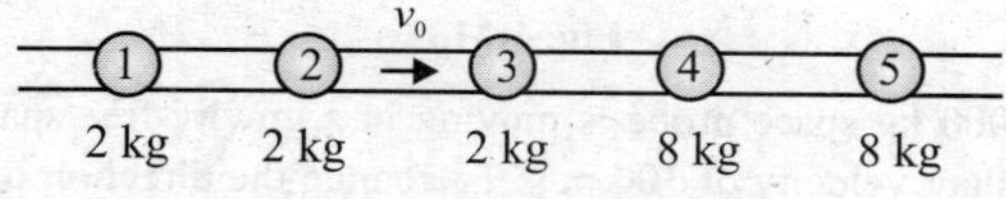

Fig. 1.207

a. Total number of collisions in the process is 5.
b. Velocity of separation between the first and fifth ball after the last possible collision is v_0.
c. Finally, three balls remain stationary.
d. All of the above.

127. Two objects are at rest on a level frictionless surface. The objects are not connected. A force F is applied to one of the objects, which then moves with acceleration a. Mark the correct statement(s).

a. The concept of centre of mass cannot be applied because the external force does not act on both the objects.
b. The centre of mass moves with acceleration that could be greater than a.
c. The centre of mass moves with acceleration that must be equal to a.
d. The centre of mass moves with acceleration that must be less than a.

128. Seven identical eagles are flying south together at constant speeds. A hunter shoots one of them, which immediately dies and falls on the ground. The other six continue flying south at the original speeds. After one eagle has hit the ground, the centre of mass of all the seven eagles

a. continues south at the original speed, but is now located at some distance below the flying eagles
b. continues south but at 6/7 of the original speed and is located at some distance below the flying eagles
c. continues south but at 6/7 of the original speed and is located at some distance behind and below the flying eagles
d. continues south but at 6/7 of the original speed and is located at the original location

129. A particle moves along the x-axis. The x-component of its momentum as a function of time is shown in Fig. 1.208. Rank the numbered regions according to the magnitude of the force acting on the particle from the least to the greatest.

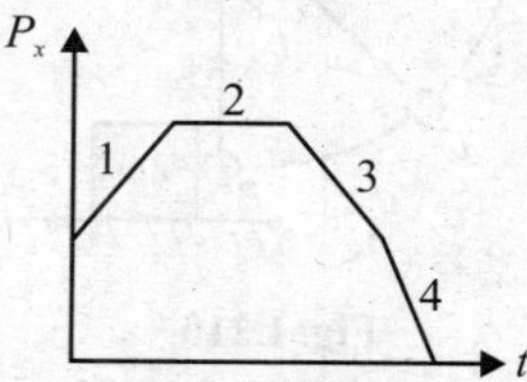

Fig. 1.208

a. 1,2,3,4 **b.** 2,3,1,4
c. 2,3,4,1 **d.** 2,4,3,1

130. A highly elastic ball moving at a speed of 3 m/s approaches a wall moving towards it with a speed of 3 m/s. After the collision, the speed of the ball will be

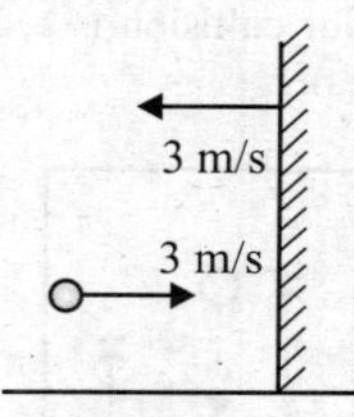

Fig. 1.209

a. 3 m/s **b.** 6 m/s
c. 9 m/s **d.** zero

131. A ball of mass m moving with velocity v_0 makes an oblique elastic collision with a stationary ball of mass $2m$. The angle of divergence between the balls after collision in ground frame, if the ball of mass m turns by an angle of 30° in centre of mass frame is

a. $\frac{\pi}{2}$ **b.** $\frac{\pi}{3}$
c. $75° + \sin^{-1}\left(\frac{1}{1+\sqrt{3}}\right)$ **d.** $75° + \tan^{-1}\left(\frac{1}{1+\sqrt{3}}\right)$

132. Two identical billiard balls undergo an oblique elastic collision. Initially, one of the balls is stationary. If the initially stationary ball after collision moves in a direction which makes an angle of 37° with direction of initial motion of the moving ball, then the angle through which initially moving ball will be deflected is

a. 37° **b.** 60°
c. 53° **d.** > 53°

133. A ball of mass m is attached to a cord of length L, pivoted at point O, as shown in Fig. 1.210. The ball is released from rest at point A, swings down and makes an inelastic collision with a block of mass $2m$ kept on a rough horizontal floor. The coefficient of restitution of collision is $e = 2/3$ and coefficient of friction between block and surface is μ. After collision, the ball comes momentarily to rest at C when cord makes an angle of θ with the vertical and block moves a distance of $3L/2$ on rough horizontal floor before stopping. The values of μ and θ are, respectively,

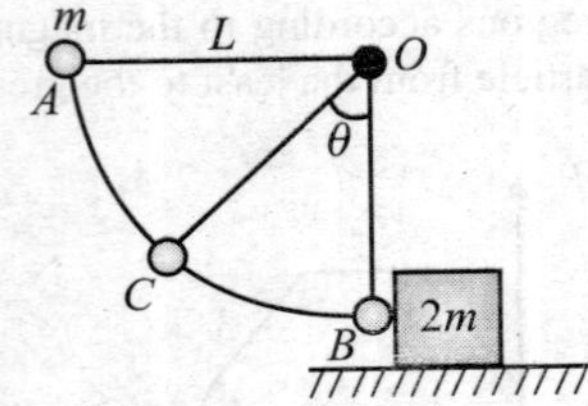

Fig. 1.210

a. $\frac{50}{243}, \cos^{-1}\left(\frac{80}{81}\right)$ b. $\frac{50}{81}, \cos^{-1}\left(\frac{80}{81}\right)$

c. $\frac{2}{81}, \cos^{-1}\left(\frac{80}{243}\right)$ d. $\frac{2}{243}, \cos^{-1}\left(\frac{80}{243}\right)$

134. A ball of mass m is released from rest relative to elevator at a height h_1 above the floor of the elevator. After making collision with the floor of the elevator it rebounces to height h_2. The coefficient of restitution for collision is e. For this situation, mark the correct statement(s).

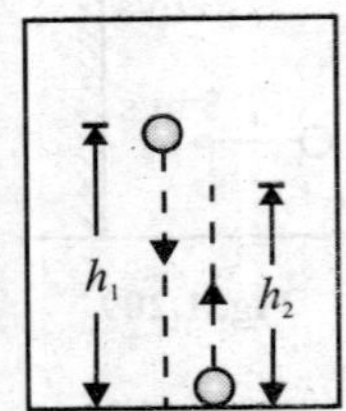

Fig. 1.211

a. If elevator is moving down with constant velocity v_0, then $h_2 = e^2 h_1$.

b. If elevator is moving down with constant velocity v_0, then $h_2 = e^2 h_1 - \frac{v_0^2}{2g}$

c. If elevator is moving down with constant velocity v_0, then impulse imparted by floor of the elevator to the ball is $m(\sqrt{2gh_2} + \sqrt{2gh_1} + 2v_0)$ in the upward direction.

d. If elevator is moving with constant acceleration of $g/4$ in upward direction, then it is not possible to determine a relation between h_1 and h_2 from the given information.

135. Two blocks A and B of masses m and $2m$, respectively, are connected with the help of a spring having spring constant, k as shown in Fig. 1.212. Initially, both the blocks are moving with same velocity v on a smooth horizontal plane with the spring in its natural length. During their course of motion, block B makes an inelastic collision with block C of mass m which is initially at rest. The coefficient of restitution for the collision is 1/2. The maximum compression in the spring is

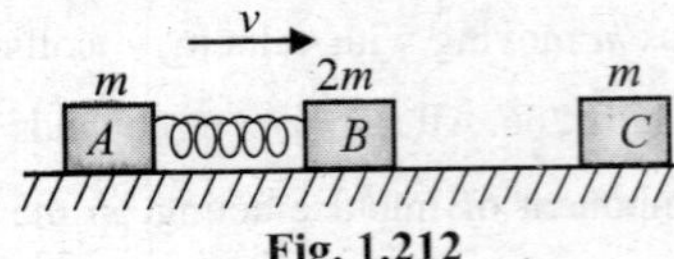

Fig. 1.212

a. $\sqrt{\frac{2m}{k}}$ b. will never be attained

c. $\sqrt{\frac{m}{12k}}\, v$ d. $\sqrt{\frac{m}{6k}}\, v$

136. In which of the following cases, the normal contact force between stationary block M and surface is impulsive?

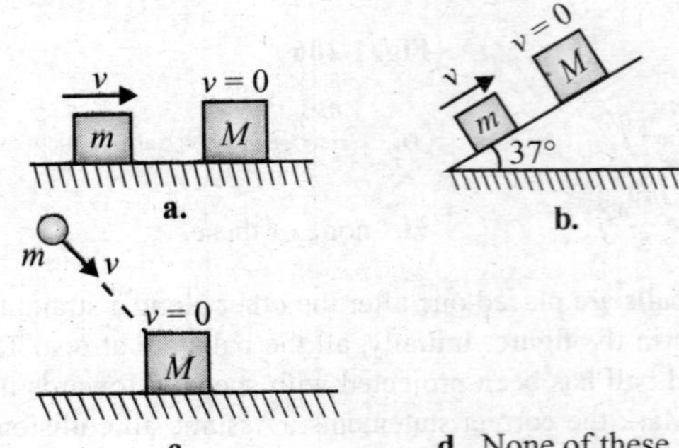

Fig. 1.213

137. A 3000 kg space probe is moving in a gravity free space at a constant velocity of 300 m/s. To change the direction of space probe, rockets have been fired in a direction perpendicular to the direction of initial motion of the space probe, the rocket firing exerts a thrust of 4000 N for 225 s. The space probe will turn by an angle of (neglect the mass of the rockets fired)

a. 30° b. 60°

c. 45° d. 37°

138. After a totally inelastic collision, two objects of the same mass and same initial speeds are found to move together at half of their initial speeds. The angle between the initial velocities of the objects is

a. 120° b. 60°

c. 150° d. 45°

139. An object of mass 10 kg is launched from the ground at $t = 0$, at an angle of 37° above the horizontal with a speed of 30 m/s. At some time after its launch, an explosion splits the projectile into two pieces. One piece of mass 4 kg is observed at (105 m, 43 m) at $t = 2$ s. Find the location of second piece at $t = 2$ s?

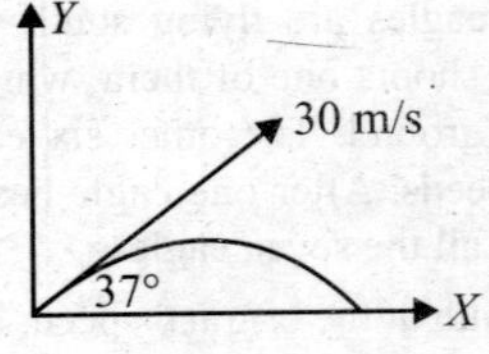

Fig. 1.214

a. (10, 2) b. (48, 16)

c. (10, – 2) d. Information insufficient

140. For the system shown in Fig. 1.215, the string is light and pulley is frictionless. The 4 kg block is given an upward velocity of 1 m/s.

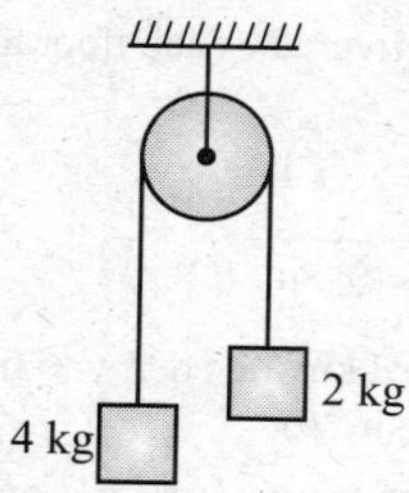

Fig. 1.215

The centre of mass of the two blocks will [neglect the impulse duration]

a. accelerate down with $g/3$.

b. initially accelerate downwards with g and then after some time accelerate down with $g/3$.

c. initially accelerate with $\vec{g}$ and then the acceleration is 0

d. initially accelerate with $\vec{g}$ and then with accelerate with $g/3$.

Multiple Correct Answers Type

Solutions on page 1.96

1. A block of mass m is pushed towards a movable wedge of mass nm and height h with a velocity u. All surfaces are smooth. Choose the correct statement from the following.

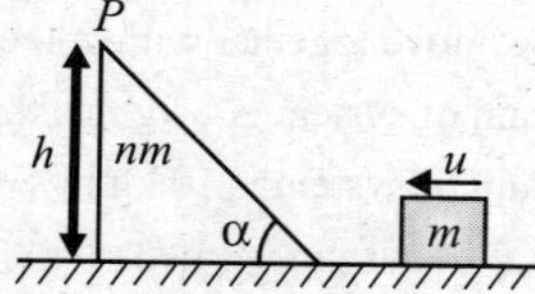

Fig. 1.216

a. Block will reach the top of the wedge if $u = \sqrt{2gh\left(1 - \frac{1}{n}\right)}$.

b. Block will reach the top of the wedge if $u = \sqrt{2gh\left(1 + \frac{1}{n}\right)}$

c. If the block overshoots P, the angle of projectile is α.

d. If the block overshoots P, the angle of projectile is less than α.

2. A ball of mass 1 kg is thrown up with an initial speed of 4 m/s. A second ball of mass 2 kg is released from rest from some height as shown in Fig. 1.217. Choose the correct statement (s).

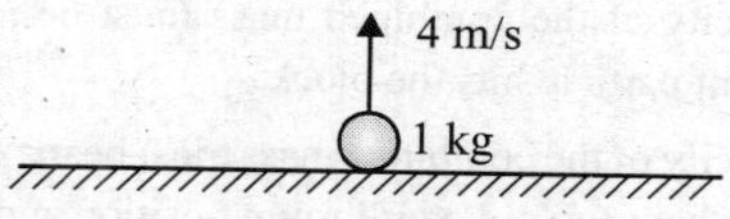

Fig. 1.217

a. The centre of mass of the two balls comes down with acceleration $g/3$.

b. The centre of mass first moves up and then comes down.

c. The acceleration of the centre of mass is g downwards.

d. The centre of mass of the two balls remains stationary.

3. Velocity of a particle of mass 2 kg changes from $\vec{v}_1 = -2\hat{i} - 2\hat{j}$ m/s to $\vec{v}_2 = (\hat{i} - \hat{j})$ m/s after colliding with a plane surface.

a. The angle made by the plane surface with the positive x-axis is $90° + \tan^{-1}\left(\frac{1}{3}\right)$.

b. The angle made by the plane surface with the positive x-axis is $\tan^{-1}\left(\frac{1}{3}\right)$.

c. The direction of change in momentum makes an angle $\tan^{-1}\left(\frac{1}{3}\right)$ with the positive x-axis.

d. The direction of change in momentum makes an angle $90° + \tan^{-1}\left(\frac{1}{3}\right)$ with the plane surface.

4. In an elastic collision between two particles

a. the total kinetic energy of the system is always conserved.

b. the kinetic energy of the system before collision is equal to the kinetic energy of the system after collision.

c. the linear momentum of the system is conserved.

d. the mechanical energy of the system before collision is equal to the mechanical energy of the system after collision.

5. The mathematical statement $\vec{v} = \vec{v}_c + \vec{v}'$, where $\vec{v}_c$ is the velocity of centre of mass, $\vec{v}'$ is the velocity of the point with respect to the centre of mass and $\vec{v}$ is the total velocity of the point with respect to the ground,

a. is true for a rolling sphere

b. is true for a block moving on a frictionless horizontal surface

c. is true for a rolling cylinder

d. none of these

6. A block of mass 'm' is hanging from a massless spring of spring constant K. It is in equilibrium under the influence of gravitational force. Another particle of same mass 'm' moving upwards with velocity u_0 hits the block and sticks to it. For the subsequent motion, choose the *incorrect* statements:

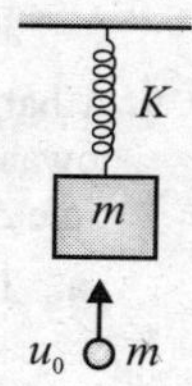

Fig. 1.218

a. Velocity of the combined mass must be maximum at natural length of the spring.

b. Velocity of the combined mass must be maximum at the new equilibrium position.

c. Velocity of the combined mass must be maximum at the instant particle hits the block.

d. Velocity of the combined mass must be maximum at a point lying between old equilibrium position and natural length.

7. Which of the following is/are correct?

a. If centre of mass of three particles is at rest and it is known that two of them are moving along different lines, then the third particle must also be moving.

b. If centre of mass remains at rest, then net work done by the forces acting on the system must be zero.

c. If centre of mass remains at rest, then the net external force must be zero.

d. If speed of centre of mass is changing, then there must be some net work being done on the system from outside.

8. Two masses $2m$ and m are connected by an inextensible light string. The string is passing over a light frictionless pulley. The mass $2m$ is resting on a surface and mass m is hanging in air as shown in Fig. 1.219. A particle of mass m strikes the mass m from below in case (I) with a velocity v_0 and in case (II) strikes mass m with a velocity v_0 from top and sticks to it.

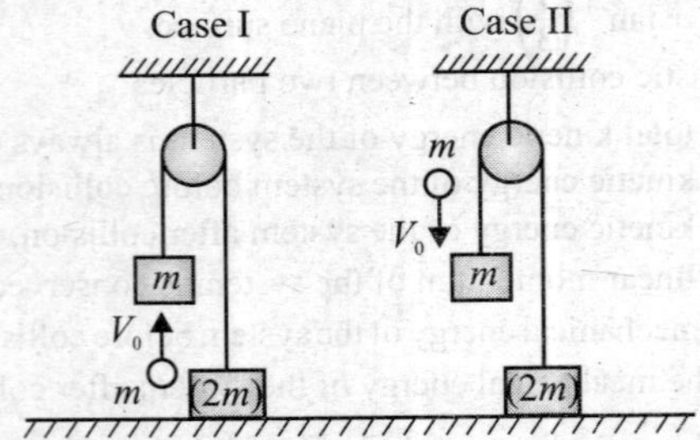

Fig. 1.219

a. The conservation of linear momentum can be applied in both the cases just before and just after collision.

b. The conservation of linear momentum can be applied in case I but cannot be applied in case II just before and just after collision.

c. The ratio of velocities of mass m just after collision in first and second cases is 1/2.

d. The ratio of velocities of mass m just after collision in first and second case is 2.

9. A ball moving with a velocity v hits a massive wall moving towards the ball with a velocity u. An elastic impact lasts for time Δt.

a. The average elastic force acting on the ball is $[m(u + v)]/\Delta t$.

b. The average elastic force acting on the ball is $[2m(u + v)]/\Delta t$.

c. The kinetic energy of the ball increases by $2mu(u + v)$.

d. The kinetic energy of the ball remains the same after the collision.

10. A particle strikes a horizontal smooth floor with a velocity u making an angle θ with the floor and rebounds with velocity v making an angle ϕ with the floor. If the coefficient of restitution between the particle and the floor is e, then

a. the impulse delivered by the floor to the body is $mu(1 + e) \sin\theta$

b. $\tan\phi = e \tan\theta$

c. $v = u\sqrt{1 - (1 - e)^2 \sin^2\theta}$

d. the ratio of final kinetic energy to the initial kinetic energy is $(\cos^2\theta + e^2 \sin^2\theta)$

11. Four bricks, each of length l, are put on the top of one another (Fig. 1.220) in such a way that part of each extends beyond the one beneath. For the largest equilibrium extensions,

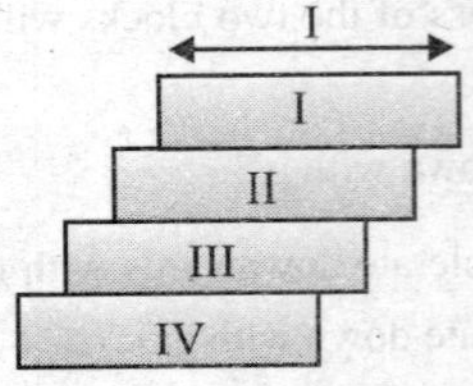

Fig. 1.220

a. the top brick overhanging the one below by $l/4$.

b. the second brick from top over hanging the one below by $l/4$.

c. the third brick from top overhanging the bottom one by $l/6$.

d. the total overhanging length on the edge of the bottom brick is $(11/12)l$.

12. A body of mass 2 kg moving with a velocity 3 m/s collides with a body of mass 1 kg moving with a velocity of 4 m/s in opposite direction. If the collision is head on and completely inelastic, then

a. both particles move together with velocity (2/3) m/s

b. the momentum of system is 2 kg m/s throughout

c. the momentum of system is 10 kg m/s

d. the loss of KE of system is (49/3) J

13. Two small rings, each of mass 'm', are connected to the block of same mass 'm' through an inextensible massless string of length 'l'. Rings are constrained to move over smooth rod AB. Initially, the system is held at rest as shown in Fig. 1.221. Let u and v be the velocities of ring and block, respectively when string makes an angle 60° with the vertical.

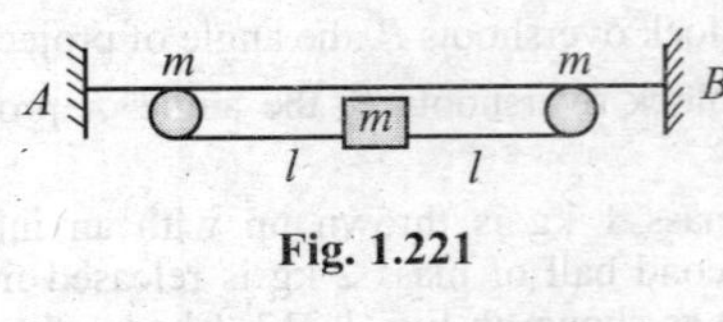

Fig. 1.221

a. $u = \sqrt{\dfrac{gl}{5}}$ **b.** $u = \sqrt{\dfrac{8gl}{5}}$

c. $v = \sqrt{3gl}$ d. $v = \sqrt{\frac{3gl}{5}}$

14. When two particles collide and stick together:
    a. ME is conserved b. energy is conserved
    c. KE is conserved d. work is negative

15. A body moving towards a body of finite mass at rest collides with it. It is possible that
    a. both bodies come to rest
    b. both bodies move after collision
    c. the moving body stops and the body at rest starts moving
    d. the stationary body remains stationary and the moving body rebounds

16. A ball strikes a wall with a velocity $\vec{u}$ at an angle θ with the normal to the wall surface and rebounds from it at an angle β with the surface. Then
    a. $(\theta + \beta) < 90°$, if the wall is smooth
    b. if the wall is rough, coefficient of restitution = $\tan \beta / \cos \theta$
    c. if the wall is rough, coefficient of restitution < $\tan \beta / \cot \theta$
    d. none of the above

17. A ball strikes a smooth horizontal floor obliquely and rebounds inelastically.
    a. The kinetic energy of the ball just after hitting the floor is equal to the potential energy of the ball at its maximum height after rebound.
    b. Total energy of the ball is not conserved.
    c. The angle of rebound with the vertical is greater than the angle of incidence.
    d. None of the above.

18. Choose the correct statements from the following:
    a. The general form of Newton's second law of motion is $\vec{F}_{ext} = m\vec{a}$.
    b. A body can have energy and get no momentum.
    c. A body having momentum must necessarily have kinetic energy.
    d. The relative velocity of two bodies in a head-on elastic collision remains unchanged in magnitude and direction.

19. An ideal spring is permanently connected between two blocks of masses M and m. The blocks–spring system can move over a smooth horizontal table along a straight line along the length of the spring as shown in Fig. 1.222. The blocks are brought nearer to compress the spring and then released. In the subsequent motion,

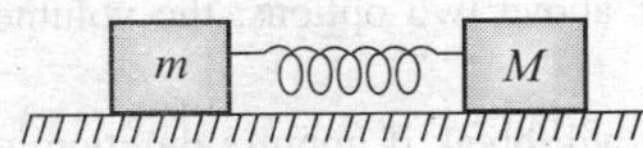

Fig. 1.222

    a. initially they move in opposite directions with velocities inversely proportional to their masses
    b. the ratio of their velocities remains constant
    c. linear momentum and energy of the system remain conserved
    d. the two blocks will oscillate about their centre of mass, which remains stationary

20. A handball falls on the ground and rebounds elastically along the same line of motion. Then
    a. the linear momentum of the universe remains conserved
    b. the linear momentum of the ball is not conserved
    c. during the collision, the whole of the kinetic energy of the ball is converted into potential energy and then completely converted into kinetic energy of the ball
    d. during the collision, the kinetic energy remains constant

21. A flat cart of mass m_0 starts moving to the right due to a constant horizontal force F at $t = 0$. Sand spills on the flat cart from a stationary hopper. The velocity of loading is constant and is equal to μ kg/s.

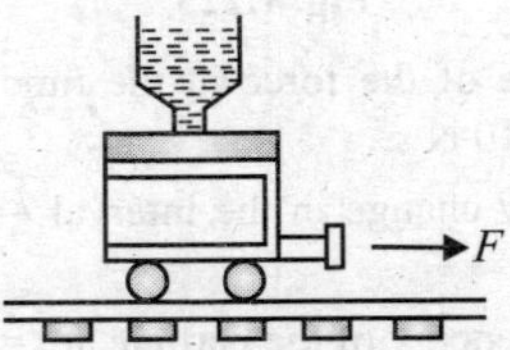

Fig. 1.223

    a. Initial acceleration is equal to F/m_0.
    b. Acceleration at time t is $F/(m_0 + \mu t)$.
    c. Kinetic energy of loaded cart at an instant is equal to work done by force F up to that instant.
    d. Momentum of loaded cart at an instant is equal to impulse of force F up to that instant.

22. A steel ball of mass $2m$ suffers one-dimensional elastic collision with a row of three steel balls, each of mass m. If mass $2m$ has collided with velocity v and the three balls numbered 1, 2, 3 were initially at rest, then after the collision

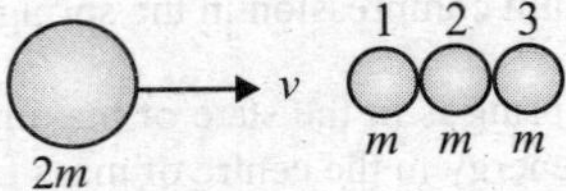

Fig. 1.224

    a. balls 1, 2 and 3 would start moving to the right, each with velocity $v/3$
    b. balls 2 and 3 would start moving to the right, each with velocity $v/2$
    c. balls 2 and 3 would start moving to the right, each with velocity v
    d. ball 1 and ball of mass $2m$ would remain at rest

23. A man standing on the edge of the terrace of a high rise building throws a stone, vertically up with a speed of 20 m/s. Two seconds later, an identical stone is thrown vertically downwards with the same speed of 20 m/s. Then
    a. the relative velocity between the two stones remains constant till one hits the ground
    b. both will have the same kinetic energy, when they hit the ground

c. the time interval between their hitting the ground is 2 s
d. if the collision on the ground is perfectly elastic, both will rise to the same height above the ground

24. Figure 1.225 gives force versus time graph. The force is acting on a particle of mass 2.0 kg at rest at $t = 0$ and particle is moving in one dimension. Which of the following are correct?

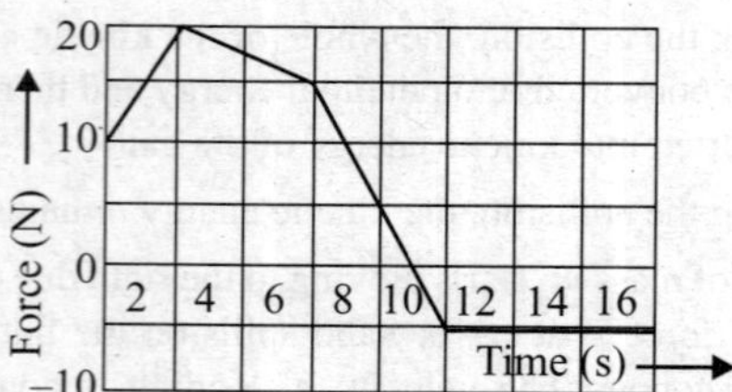

Fig. 1.225

a. The impulse of the force in the time interval $t = 8$ s to $t = 12$ s is –10 N s.
b. The velocity change in the interval $t = 8$ s to $t = 16$ s is –15 m/s.
c. The kinetic energy of the particle at $t = 6$ s is 2500 J.
d. The kinetic energy of the particle at $t = 6$ s is 500 J.

25. A block of mass m moving with a velocity v_0 collides with a stationary block of mass M to which a spring of stiffness k is attached, as shown in Fig. 1.226. Choose the correct alternative(s).

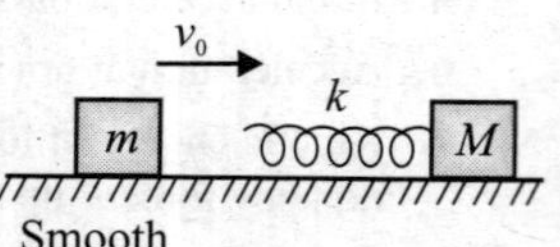

Fig. 1.226

a. The velocity of the centre of mass is v_0.
b. The initial kinetic energy of the system in the centre of mass frame is $\frac{1}{4}\left(\frac{mM}{M+m}\right)v_0^2$.
c. The maximum compression in the spring is $v_0\sqrt{\left(\frac{mM}{m+M}\frac{1}{k}\right)}$.
d. When the spring is in the state of maximum compression, the kinetic energy in the centre of mass frame is zero.

26. In which of the following cases, the centre of mass of a rod is certainly not at its centre?
a. The density continuously increases from left to right.
b. The density continuously decreases from left to right.
c. The density decreases from left to right up to the centre and then increases.
d. The density increases from left to right up to the centre and then decreases.

27. Two particles of masses m_1 and m_2 and velocities u_1 and αu_1 ($\alpha \neq 0$) make an elastic head on collision. If the initial kinetic energies of the two particles are equal and m_1 comes to rest after collision, then

a. $\frac{u_1}{u_2} = \sqrt{2} + 1$ b. $\frac{u_1}{u_2} = \sqrt{2} - 1$

c. $\frac{m_2}{m_1} = 3 + 2\sqrt{2}$ d. $\frac{m_2}{m_1} = 3 - 2\sqrt{2}$

28. A pendulum bob of mass m connected to the end of an ideal string of length l is released from rest from horizontal position as shown in Fig. 1.227.

At the lowest point, the bob makes an elastic collision with a stationary block of mass $5m$, which is kept on a frictionless surface. Mark out the correct statement(s) for the instant just after the impact.

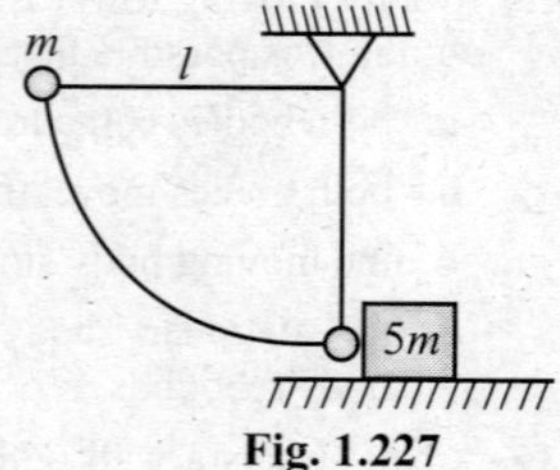

Fig. 1.227

a. Tension in the string is (17/9) mg.
b. Tension in the string is 3 mg.
c. The velocity of the block is $\sqrt{2gl/3}$.
d. The maximum height attained by the pendulum bob after impact is (measured from the lowest position) $4l/9$.

29. A string of length $3l$ is connected to a fixed cylinder whose top view is shown in Fig. 1.228. The string is initially slack. The other end of the string (connected to a marble) is moving at a constant velocity of 10 m/s as shown. The string will get stretched at some instant and impulsive tension occurs in the string. If hinge is exerting a force of 40000 N for 0.25 ms on the cylinder to bear up the impact of impulsive tension, then mark the correct statements. (Take string to be light, breaking tension of the string is 2×10^5 N.)

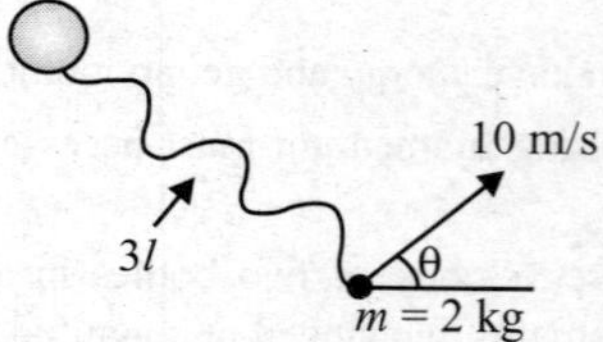

Fig. 1.228

a. The angle made by the velocity of marble with the length of string when it is just stretched is 60°.
b. The marble will move in a circular path of varying radius with constant speed of $5\sqrt{3}$ m/s, after the string is taut.
c. To answer above two options, the volume of θ must be given.
d. The string will break if impulse duration is less than 0.05 min.

30. If the external forces acting on a system have zero resultant, the centre of mass

a. may not move b. must not accelerate
c. may move d. may accelerate

Assertion-Reasoning Type

Solutions on page 1.99

In the following questions, a statement of assertion (Statement I) is given which is followed by a corresponding statement of reason (Statement II). Examine the statements carefully and choose the correct option according to the following options.

a. Both assertion and reason are true and reason is the correct explanation of assertion.
b. Both assertion and reason are true but reason is not the correct explanation of assertion.
c. Assertion is true and reason is false.
d. Assertion is false and reason is true.

1. Statement I: A particle strikes head-on with another stationary particle such that the first particle comes to rest after collision. The collision should necessarily be elastic.

Statement II: In elastic collision, there is no loss of momentum of the system of the particles.

2. Statement I: No external force acts on a system of two spheres which undergo a perfectly elastic head-on collision. The minimum kinetic energy of this system is zero if the net momentum of the system is zero.

Statement II: If any two bodies undergo a perfectly elastic head-on collision, at the instant of maximum deformation, the complete kinetic energy of the system is converted to deformation potential energy of the system.

3. Statement I: If a sphere of mass m moving with speed u undergoes a perfectly elastic head-on collision with another sphere of heavier mass M at rest ($M > m$), then direction of velocity of sphere of mass m is reversed due to collision (no external force acts on system of two spheres).

Statement II: During a collision of spheres of unequal masses, the heavier mass exerts more force on the lighter mass in comparison to the force which lighter mass exerts on the heavier one.

4. Statement I: If a ball projected up obliquely from the ground breaks up into several fragments in its path, the centre of the system of all fragments moves in the same parabolic path compared to initial one till all fragments are in air.

Statement II: In the situation of Statement 1, at the instant of breaking, the fragments may be thrown in different directions with different speeds.

5. Statement I: In a two-body collision, the momenta of the particles are equal and opposite to one another, before as well as after the collision when measured in the centre of mass frame.

Statement II: The momentum of the system is zero from the centre of mass frame.

6. Statement I: If no external force acts on a system of particles, then the centre of mass will not move in any direction.

Statement II: If net external force is zero, then the linear momentum of the system remains constant.

7. Statement I: A block is kept at the top of a smooth wedge which is kept on a smooth horizontal surface. As the block slides down the wedge, centre of the mass of the system will be accelerated.

Statement II: When external force acting on the system is zero, centre of mass is in rest.

8. Statement I: A particle of mass m strikes a smooth wedge of mass M as shown in Fig. 1.229. Linear momentum of the particle along the surface of wedge is conserved during collision.

Statement II: Wedge exerts a force perpendicular to the inclined face of the wedge on particle during collision.

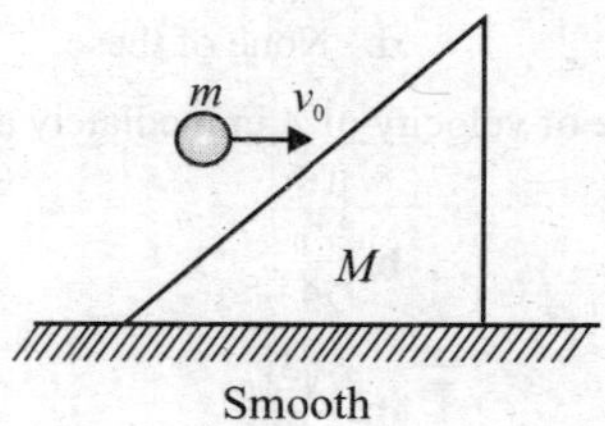

Fig. 1.229

9. Statement I: A body cannot have energy without having momentum but it can have momentum without having energy.

Statement II: Momentum and energy have different dimensions.

10. Statement I: Centre of mass of an isolated system has a constant velocity.

Statement II: If centre of mass of an isolated system is already at rest, it remains at rest.

Comprehension Type

Solutions on page 1.100

For Problems 1–5

Three spheres, each of mass m, can slide freely on a frictionless, horizontal surface. Spheres A and B are attached to an inextensible, inelastic cord of length 'l' and are at rest in the position shown where sphere B is struck by sphere C which is moving to the right with a velocity v_0. Knowing that the cord is taut where sphere B is struck by sphere C and assuming 'head on' inelastic impact between B and C, we cannot conserve kinetic energy of the entire system.

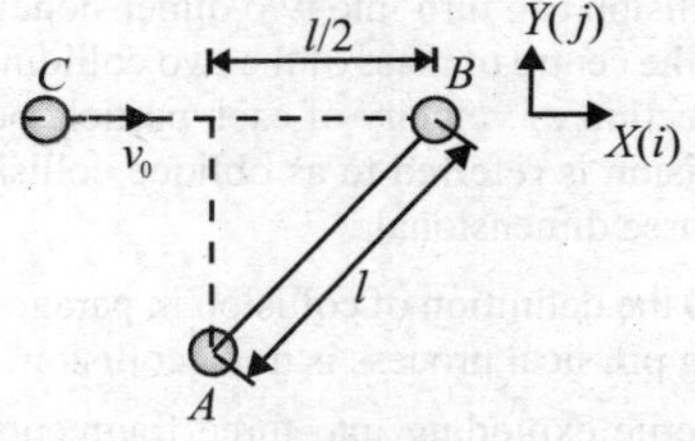

Fig. 1.230

1. The velocity of B immediately after collision is along unit vector

a. $\hat{i}$ **b.** $\frac{\hat{i}+\hat{j}}{\sqrt{2}}$
c. $\frac{\sqrt{3}}{2}\hat{i}+\frac{1}{2}\hat{j}$ **d.** none of these

2. Velocity of A immediately after collision is along unit vector

a. $\hat{i}$ b. $\frac{1}{2}\hat{i}+\frac{\sqrt{3}}{2}\hat{j}$

c. $\hat{j}$ d. None of these

3. If velocity of C immediately after collision becomes $v_0/2$ in the initial direction of motion, the impulse due to string on sphere A is

a. $\frac{mv_0}{8}$ b. $\frac{mv_0}{2}$

c. $\frac{\sqrt{mv_0}}{4}$ d. None of these

4. The magnitude of velocity of A immediately after collision is

a. $\frac{\sqrt{3}v_0}{4}$ b. $\frac{v_0}{8}$

c. $\frac{v_0}{4}$ d. $\frac{\sqrt{3}v_0}{8}$

5. The loss in kinetic energy of the system during collision is

a. $\frac{17mv_0^2}{64}$ b. $\frac{15mv_0^2}{64}$

c. $\frac{12mv_0^2}{25}$ d. None of these

For Problems 6–10

Collision is a physical process in which two or more objects, either particle masses or rigid bodies, experience very high force of interaction for a very small duration. It is not essential for the objects to physically touch each other for collision to occur. Irrespective of the nature of interactive force and the nature of colliding bodies, Newton's second law holds good on the system. Hence, momentum of the system before and after the collision remains conserved if no appreciable external force acts on the system during collision.

The amount of energy loss during collision, if at all, is indeed dependent on the nature of colliding objects. The energy loss is observed to be maximum when objects stick together after collision. The terminology is to define collision as 'elastic' if no energy loss takes place and to define collision as 'plastic' for maximum energy loss. The behaviour of system after collision depends on the position of colliding objects as well. A unidirectional motion of colliding objects before collision can turn into two dimensional after collision if the line joining the centre of mass of the two colliding objects is not parallel to the direction of velocity of each particle before collision. Such type of collision is referred to as oblique collision which may be either two or three dimensional.

6. According to the definition of collision in paragraph 1, which of the following physical process is not a collision?

a. A projectile exploding into three fragments at its highest point.

b. Two soap bubbles coalescing to form a bubble of larger radius.

c. A vertically upward thrown particle changing direction at its highest point.

d. A piece of magnet thrown on a metallic surface.

7. For which of the following collisions, the external force acting on the system during collision is not appreciable as mentioned in paragraph 1.

a. A ball striking a rigid wall (consider ball as system).

b. A 5 kg mass thrown vertically up exploding during its motion (5 kg mass is the system).

c. A particle hitting a rigid bar of length L lying on a frictionless surface (consider rigid bar as the system).

d. Two particles moving towards each other due to gravitational attraction and hitting each other (consider any particle as system).

8. According to the definition of oblique collision in the paragraph, which of the following collisions cannot be oblique?

a. Collision between two point masses.

b. Collision between two rings of same radius.

c. Collision between two rings of different radius.

d. Collision between a particle and a ring of finite radius.

9. Which of the following collisions is one-dimensional?

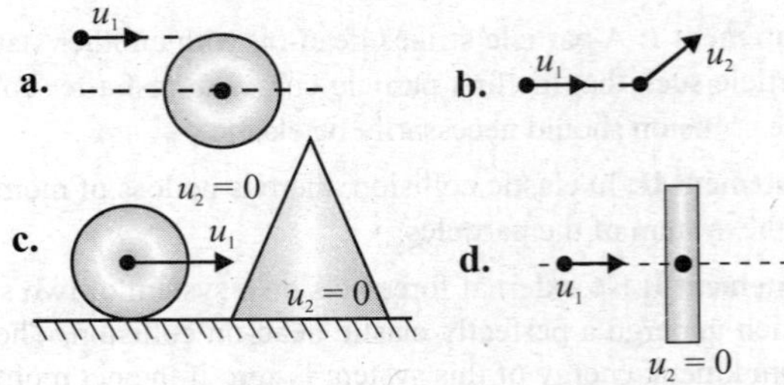

Fig. 1.231

10. A 4 kg sphere moving with a velocity of 4 m/s collides with an identical sphere of 2 kg moving with 2 m/s as shown. Final kinetic energy is less than initial kinetic energy. What type of collision is this?

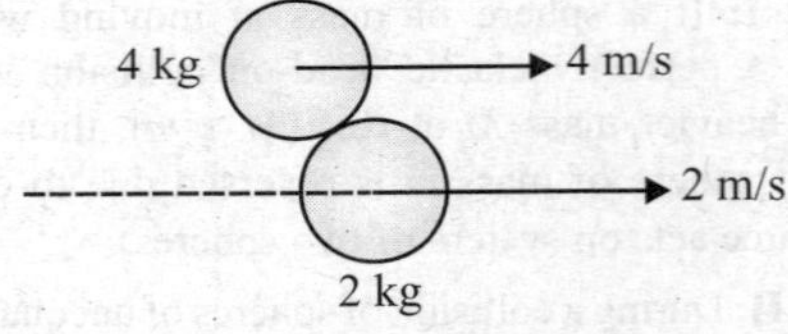

Fig. 1.232

a. Elastic, one-dimensional

b. Inelastic, one-dimensional

c. Elastic, two-dimensional

d. Inelastic, two-dimensional

For Problems 11–13

According to the principle of conservation of linear momentum if the external force acting on the system is zero, the linear momentum of the system will remain conserved. It means if the centre of mass of a system is initially at rest, it will remain at rest in the absence of external force, that is, the displacement of centre of mass will be zero.

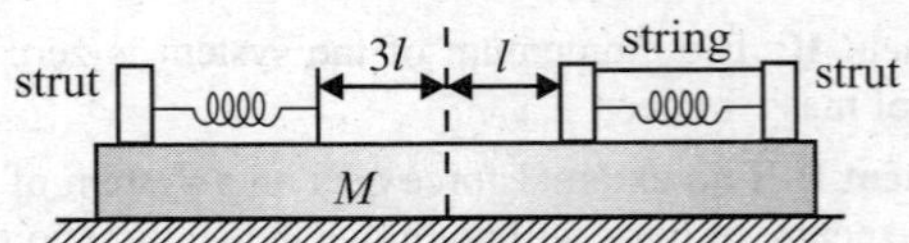

Fig. 1.233

A plank of mass M is placed on a smooth horizontal surface. Two light identical springs, each of stiffness K, are rigidly connected to struts at the end of the plank as shown in Fig. 1.233. When the springs are in their unextended position, the distance between their free ends is $3l$. A block of mass m is placed on the plank and pressed against one of the springs so that it is compressed to l. To keep the block at rest it is connected to the strut by means of a light string. Initially, the system is at rest. Now the string is burnt.

11. The maximum displacement of the plank is

a. $\frac{5ml}{M}$ **b.** $\frac{5ml}{M+m}$

c. $\frac{3ml}{M+m}$ **d.** $\frac{4ml}{M+m}$

12. The maximum velocity of the plank is

a. $\sqrt{\frac{Km}{(M+m)}}\,l$ **b.** $\sqrt{\frac{k}{(M+m)}}\,l$

c. $\sqrt{\frac{Km}{M(M+m)}}\,l$ **d.** $\sqrt{\frac{kM}{m(M+m)}}\,\ell$

13. The maximum kinetic energy of the block m is

a. $\frac{Kml^2}{2M(M+m)}$ **b.** $\frac{Kml^2}{M(M+m)}$

c. $\frac{KMl^2}{2(M+m)}$ **d.** None of these

For Problems 14–16

Two identical balls A and B, each of mass 2 kg and radius R, are suspended vertically from inextensible strings as shown in Fig. 1.234. The third ball C of mass 1 kg and radius $r = (\sqrt{2} - 1)R$ falls and hits A and B symmetrically with 10 m/s. Speed of both A and B just after the collision is 3 m/s.

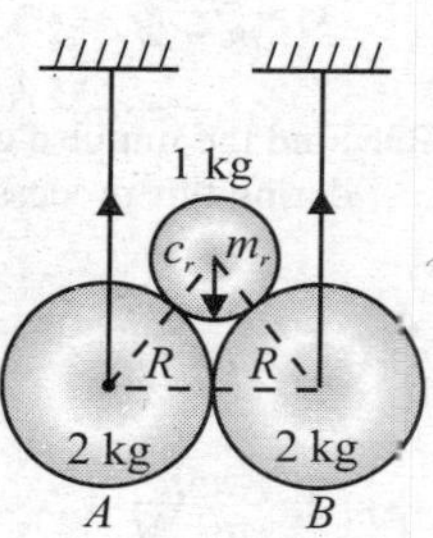

Fig. 1.234

14. Speed of C just after collision is

a. 2 m/s **b.** $2\sqrt{2}$ m/s

c. 5 m/s **d.** $(\sqrt{2} - 1)$ m/s

15. Impulse provided by each string during collision is

a. $6\sqrt{2}$ N s **b.** 12 N s

c. $3\sqrt{2}$ N s **d.** 6 N s

16. The value of coefficient of restitution is

a. $\frac{1}{4}$ **b.** $\frac{1}{\sqrt{2}}$

c. $\sqrt{2} - 1$ **d.** $\frac{1}{2}$

For Problems 17–18

An elastic collision takes place between two masses, m_1 and m_2, moving on a frictionless surface as shown in Fig. 1.235. The spring constant is $k = 600$ N/m.

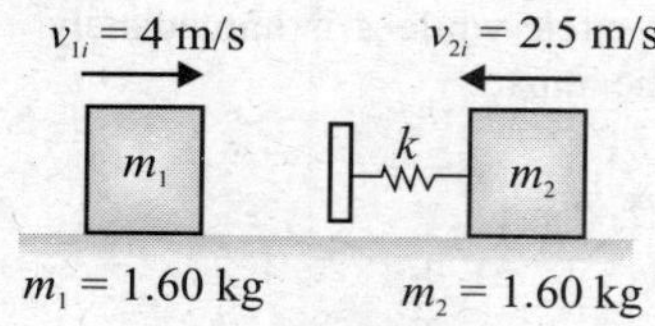

Fig. 1.235

17. What is the velocity of block 2 at the instant when block 1 is moving to the right with a velocity 3.00 m/s?

a. 1.50 m/s **b.** – 2.62 m/s

c. 4.64 m/s **d.** – 1.74 m/s

18. What is the compression of the spring at that instant?

a. 0.52 m **b.** 0.4 m

c. 0.171 m **d.** 0.84 m

For Problems 19–20

A smooth ball is dropped from height h on a smooth incline as shown in Fig. 1.236. After collision, the velocity of the ball is directed horizontally.

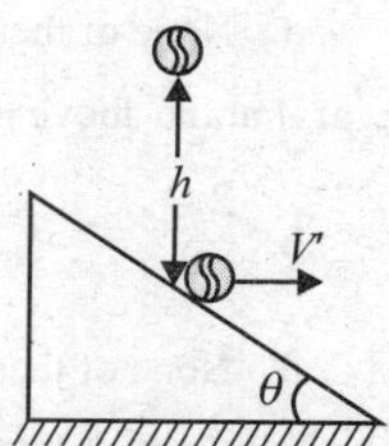

Fig. 1.236

19. Find the coefficient of restitution.

a. $\cot^2\theta$ **b.** $\sin^2\theta$

c. $\tan^2\theta$ **d.** $\cos^2\theta$

20. If the collision were elastic, what is the impulse on the ball?

a. $2m\sqrt{2gh}\tan\theta$ **b.** $2m\sqrt{2gh}\sin\theta$

c. $2m\sqrt{2gh}\cot\theta$ **d.** $2m\sqrt{2gh}\cos\theta$

For Problems 21–22

After falling from rest through a height h, a body of mass m begins to raise a body of mass M $(M > m)$ connected to it through a pulley.

21. Determine the time it will take for the body of mass M to return to its original position.

a. $\frac{2m}{M+m}\sqrt{\frac{2h}{g}}$ **b.** $\frac{2m}{M-m}\sqrt{\frac{2h}{g}}$

c. $\frac{2m}{M-m}\sqrt{\frac{h}{g}}$ **d.** $\frac{m}{M-m}\sqrt{\frac{2h}{g}}$

22. Find the fraction of kinetic energy lost when the body of mass M is jerked into motion.

a. $\frac{M}{M+m}$ **b.** $\frac{M}{M-m}$

c. $\frac{2M}{M+m}$ **d.** $\frac{M}{2(M+m)}$

For Problems 23–25

Three identical balls are connected by light inextensible strings with each other as shown and rest over a smooth horizontal table. Length of each string is l.

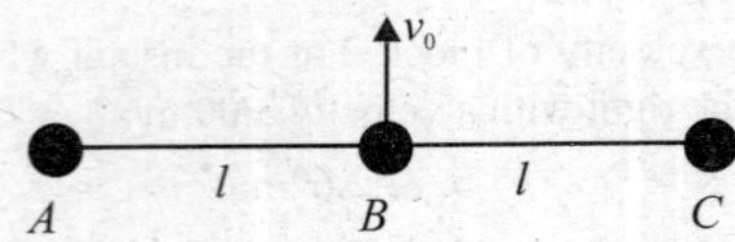

Fig. 1.237

At moment $t = 0$, ball B is imparted a velocity v_0 perpendicular to the strings and then the system is left on its own.

23. Calculate the velocity of B just before A collides with ball C.

a. $\frac{v_0}{3}$ **b.** $\frac{2v_0}{3}$

c. $\frac{2v_0}{5}$ **d.** None of these

24. Calculate the velocity of A at the above given instant.

a. $\frac{v_0}{3}$ **b.** $\frac{2v_0}{3}$

c. $\frac{2v_0}{5}$ **d.** None of these

25. If collision between the balls is completely inelastic, then

a. there is no loss of kinetic energy of the system

b. entire kinetic energy of the system is lost

c. kinetic energy loss in the system is less than 50%

d. kinetic energy loss in the system is more than 50%

For Problems 26–28

In Fig. 1.238, a pulley is shown which is frictionless and a ring of mass m can slide on the string without any friction. One end of the string is attached to point B and to the other end, a block 'P' of mass m is attached. The whole system lies in vertical plane.

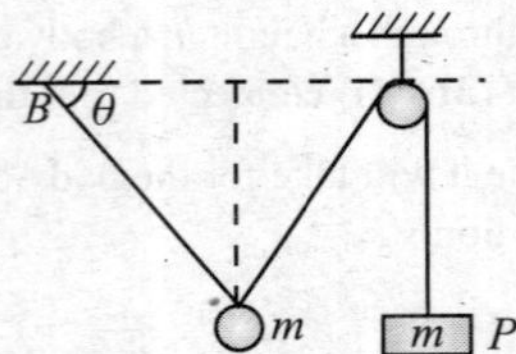

Fig. 1.238

26. If the system is released from rest, it is found that the system remains at rest. What is the value of θ?

a. 30° **b.** 45° **c.** 60° **d.** 75°

27. Now another block 'C' of same mass m is attached to the block 'P' and system is released from rest. If a_1 and a_2 are the magnitudes of initial accelerations of ring and blocks, respectively, then

a. $a_1 + a_2 = g$ **b.** $a_1 + 2a_2 = g$

c. $a_1 = 2a_2$ **d.** $2a_1 = a_2$

28. If block 'C' mentioned above was released from some height and collided with 'P' with some velocity u, then find the velocity of the ring just after the collision. The collision is perfectly inelastic.

a. u **b.** $\frac{u}{2}$ **c.** $\frac{u}{4}$ **d.** $\frac{u}{3}$

For Problems 29–31

A system of men and trolley is shown in Fig. 1.239. To the left end of the string, a trolley of mass M is connected on which a man of mass m is standing. To the right end of the string another trolley of mass m is connected on which a man of mass M is standing. Initially, the system is at rest. All of a sudden both the men leap upwards simultaneously with the same velocity u w.r.t. ground.

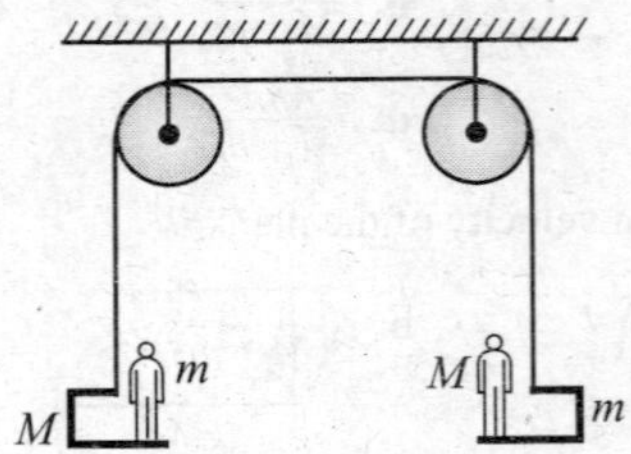

Fig. 1.239

29. Find the relative velocity of left man with respect to his trolley just after he leaps upwards.

a. $\frac{mu}{m+M}$ **b.** $\frac{Mu}{m+M}$

c. $\frac{2mu}{m+M}$ **d.** $\frac{2Mu}{m+M}$

30. Find the impulse generated in the string connecting the trolleys during this process.

a. $\frac{Mmu}{m+M}$ **b.** $\frac{(M^2+m^2)u}{m+M}$

c. $\frac{m^2u}{m+M}$ **d.** $\frac{M^2u}{m+M}$

31. Which is correct?

a. When the men are at the highest point of their motion, the trolleys will also be instantaneously at rest.

b. When the men are at the highest point of their motion, then the left trolley will be moving downwards.

c. Impulse acting on both the men will be same in the given process.

d. None of these.

For Problems 32–34

Two beads A and B of masses m_1 and m_2, respectively, are threaded on a smooth circular wire of radius a fixed in a vertical plane. B is stationary at the lowest point when A is gently dislodged from rest at the highest point. A collided with B at the lowest point. The impulse given to B due to collision is just great enough to carry it to the level of the centre of the circle while A is immediately brought to rest by the impact.

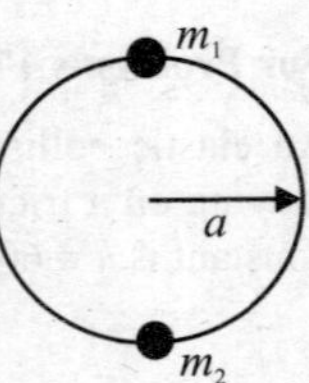

Fig. 1.240

32. Find the ratio $m_1:m_2$.

a. 1 **b.** $\sqrt{2}$

c. $1/\sqrt{2}$ **d.** 2

33. What is the coefficient of restitution between the beads?

a. 1 **b.** 1/2

c. 1/3 **d.** $1/\sqrt{2}$

34. If m_2 again comes down and collides with m_1, then after the collision

a. m_1 will rise the same height as risen by m_2

b. m_1 will rise the less height as risen by m_2

c. m_1 will rise the more height as risen by m_2

d. m_1 and m_2 will move in opposite directions

For Problems 35–37

Two blocks of equal mass m are connected by an unstretched spring and the system is kept at rest on a frictionless horizontal surface. A constant force F is applied on the first block pulling away from the other as shown in Fig. 1.241.

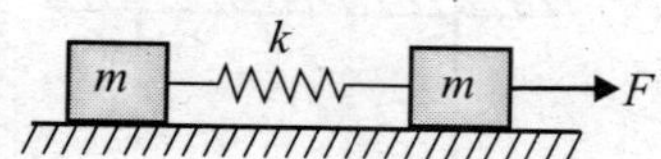

Fig. 1.241

35. Then the displacement of the centre of mass m at time t is

a. $\frac{Ft^2}{2m}$ **b.** $\frac{Ft^2}{3m}$

c. $\frac{Ft^2}{4m}$ **d.** $\frac{Ft^2}{m}$

36. If the extension of the spring is x_0 at time t, then the displacement of the first block at this instant is

a. $\frac{1}{2}\left(\frac{Ft^2}{2m}+x_0\right)$ **b.** $-\frac{1}{2}\left(\frac{Ft^2}{2m}+x_0\right)$

c. $\frac{1}{2}\left(\frac{Ft^2}{2m}-x_0\right)$ **d.** $\frac{Ft^2}{2m}+x_0$

37. If the extension of the spring is x_0 at time t, then the displacement of the second block at this instant is

a. $\frac{Ft^2}{2m}-x_0$ **b.** $\frac{1}{2}\left(\frac{Ft^2}{2m}+x_0\right)$

c. $\frac{1}{2}\left(\frac{2Ft^2}{m}-x_0\right)$ **d.** $\frac{1}{2}\left(\frac{Ft^2}{2m}-x_0\right)$

For Problems 38–39

A ball of mass m makes head-on elastic collision with a ball of mass nm which is initially at rest.

38. The fractional transfer of energy by the first ball to the second ball is

a. $\frac{4n}{(n-1)^2}$ **b.** $\frac{2n}{(n+1)^2}$

c. $\frac{2n}{(n-1)^2}$ **d.** $\frac{4n}{(n+1)^2}$

39. Deduce the value of n for which the transfer is maximum.

a. 0.75 **b.** 1.25

c. 1.0 **d.** 0.50

For Problems 40–41

A ball falls on the ground from height h_1 and rebounds to a height of h_2.

40. The fractional loss in the kinetic energy of ball during the collision with the ground is

a. $\frac{h_1-h_2}{h_1}$ **b.** $\frac{h_1-h_2}{h_2}$

c. $\frac{h_2}{h_1}$ **d.** $\sqrt{h_1-h_2/h_1}$

41. Now if the ball again collides with the ground, to what height will it rise after collision? Is fractional loss in kinetic energy same in both the collisions?

a. $\frac{h_1^2}{h_2}$, yes **b.** $\frac{h_2^2}{h_1}$, no

c. $\frac{h_1^2}{h_2}$, no **d.** $\frac{h_2^2}{h_1}$, yes

For Problems 42–44

Two equal spheres B and C, each of mass m, are in contact on a smooth horizontal table. A third sphere A of same size as that of B or C but mass $m/2$ impinges symmetrically on them with a velocity u and is itself brought to rest.

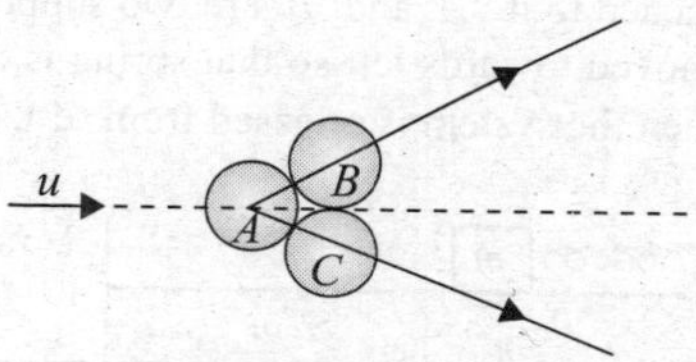

Fig. 1.242

42. The coefficient of restitution between the two spheres A and B (or between A and C) is

a. $\frac{1}{3}$ **b.** $\frac{1}{4}$

c. $\frac{2}{3}$ **d.** $\frac{3}{4}$

43. Find the velocity acquired by each of the spheres B and C after collision.

a. $\frac{u}{\sqrt{3}}$ **b.** $\frac{2u}{\sqrt{3}}$

c. $\frac{u}{2\sqrt{3}}$ **d.** $\frac{u}{2}$

44. The loss of KE during collision is

a. $\frac{mu^2}{8}$ **b.** $\frac{mu^2}{6}$

c. $\frac{mu^2}{3}$ **d.** $\frac{mu^2}{2}$

For Problems 45–47

A pendulum consists of a wooden bob of mass M and length l. A bullet of mass m is fired towards the pendulum with a speed v. The bullet emerges immediately out of the bob from the other side with a speed of $v/2$ and the bob starts rising. Assume no loss of mass of bob takes place due to penetration.

45. What is the momentum transferred to the bob by the bullet?

a. mv b. $\frac{mv}{2}$

c. $\frac{Mv}{2}$ d. Mv

46. If the bob stops where the string becomes horizontal, then v is

a. $\frac{2M}{m}\sqrt{3gl}$ b. $\frac{2M}{m}\sqrt{5gl}$

c. $\frac{2M}{m}\sqrt{gl}$ d. $\frac{2M}{m}\sqrt{2gl}$

47. If the bob is just able to complete the circular motion, then tension at the lowest point just when the bob starts rising will be

a. $6Mg$ b. $6Mg$

c. Mg d. Mg

For Problems 48–50

According to the principle of conservation of linear momentum, if the external force acting on the system is zero, the linear momentum of the system will remain conserved. It means if the centre of mass of a system is initially at rest, it will remain at rest in the absence of external force, that is, the displacement of centre of mass will be zero.

Two blocks of masses 'm' and '$2m$' are placed as shown in Fig. 1.243. There is no friction anywhere. A spring of force constant 'k' is attached to the bigger block. Mass 'm' is kept in touch with the spring but not attached to it. 'A' and 'B' are two supports attached to '$2m$'. Now m is moved towards left so that spring is compressed by distance 'x' and then the system is released from rest.

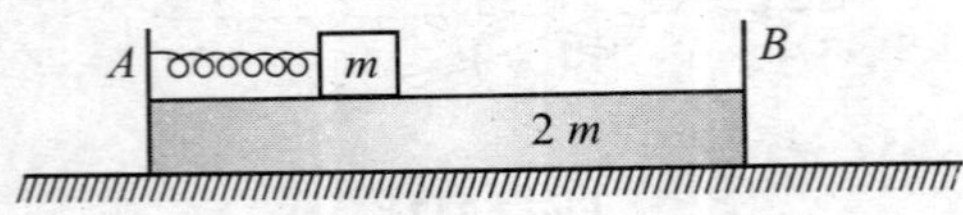

Fig. 1.243

48. Find the relative velocity of the blocks after 'm' leaves contact with the spring.

a. $\sqrt{\frac{2k}{3m}}$ b. $3x\sqrt{\frac{k}{2m}}$

c. $x\sqrt{\frac{3k}{2m}}$ d. None of these

49. Now m arrives at B. Due to its inertia of motion 'm' breaks the support 'B' and due to some resistance offered by 'B', the resulting velocity of 'm' is reduced to half of its previous value. Then what you can say about the velocity of '$2m$'?

a. It is also reduced to half of its previous value.

b. It is reduced to less than half of its previous value.

c. It is reduced to more than half of its previous value.

d. It remains the same.

50. What is the loss in the energy of the system due to breaking of 'B'?

a. $\frac{3}{8}kx^2$ b. $\frac{3}{4}kx^2$

c. $\frac{1}{8}kx^2$ d. $\frac{5}{8}kx^2$

For Problems 51–54

A ballistic pendulum is a device that was used to measure the speeds of bullets before the development of electronic timing devices. The device consists of a large block of wood of mass M, hanging from two long cords. A bullet of mass m is fired into the block, the bullet comes quickly into rest and the block + bullet rises to a vertical distance h before the pendulum comes momentarily to rest as the ends of the arc.

In the process, the linear momentum is conserved. In such a collision, some kinetic energy is dissipated as heat; so mechanical energy is not conserved. When there is a loss in mechanical energy, the collision is said to be inelastic. Further when two bodies coalesce, the collision is said to be perfectly inelastic.

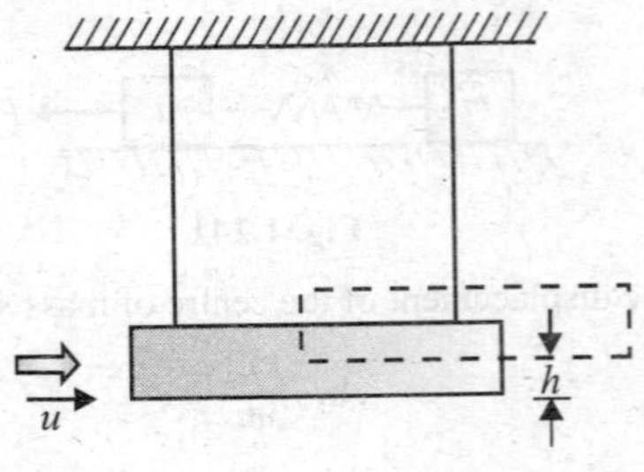

Fig. 1.244

51. After collision what is the combined velocity of the bullet + block system?

a. $\frac{m}{M}u$ b. $\frac{m}{M+m}u$

c. $\frac{M}{M+m}u$ d. $\frac{M+m}{M}u$

52. What is the initial speed of the bullet in terms of height h?

a. $\frac{m}{M}\sqrt{2gh}$ b. $\frac{m}{M+m}\sqrt{2gh}$

c. $\frac{M+m}{m}\sqrt{2gh}$ d. $\sqrt{\frac{2m}{M+m}gh}$

53. The collision of block–bullet system is

a. perfectly elastic b. partially inelastic

c. partially elastic d. perfectly inelastic

54. The energy dissipated as heat in the collision is

a. $\frac{1}{2}mu^2 - mgh$ b. $\frac{1}{2}mu^2 - (M+m)gh$

c. $\frac{1}{2}(M+m)u^2 - (M+m)gh$ d. cannot be estimated

For Problems 55–56

Three particles of equal masses travelling with velocities of 10 m/s, 20 m/s and 30 m/s, respectively, along x-axis, at an angle of 30° to the direction of positive x-axis and y-axis (as shown in Fig. 1.245) collide simultaneously and get sticked to each other.

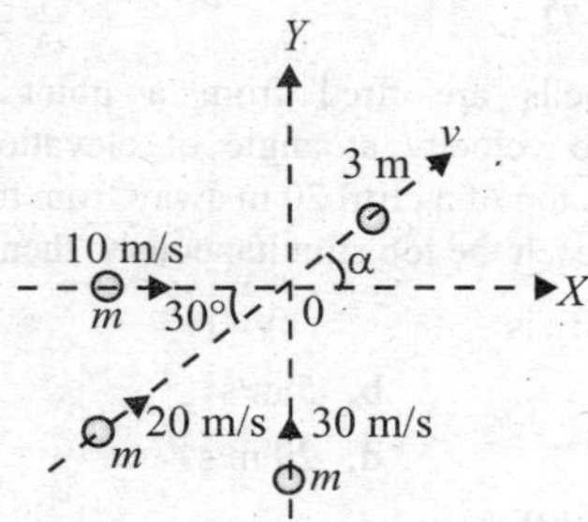

Fig. 1.245

55. The combined particle will move with velocity

a. $\frac{10}{3}\sqrt{20+2\sqrt{3}}$ m/s **b.** $\frac{10}{3}\sqrt{20-2\sqrt{3}}$ m/s

c. $\frac{10}{3}\sqrt{(5-\sqrt{3})}$ m/s **d.** $\frac{10}{3}\sqrt{(5+\sqrt{3})}$ m/s

56. Combined particle will move at angle α with x-axis where

a. $\alpha = \tan^{-1}[2(\sqrt{3}-1)]$ **b.** $\alpha = \tan^{-1}[2(\sqrt{3}+1)]$

c. $\alpha = \tan^{-1}[(\sqrt{3}-2)]$ **d.** $\alpha = \tan^{-1}[2(\sqrt{3}+2)]$

For Problems 57–59

Two identical balls, each of mass m, are tied with a string and kept on a frictionless surface. Initially, the string is slack. They are given velocities $2u$ and u in the same direction. Collision between the balls is perfectly elastic.

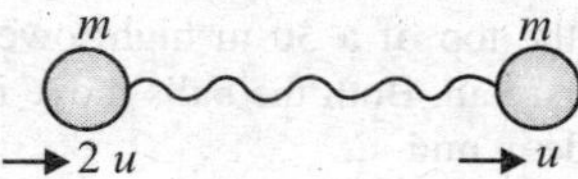

Fig. 1.246

57. After the first collision, what is the total loss in kinetic energy of the balls?

a. $2mu^2$ **b.** mu^2

c. $3mu^2$ **d.** zero

58. What is the final loss in kinetic energy of the balls?

a. $mu^2/4$ **b.** $mu^2/2$

c. $3mu^2/4$ **d.** None of these

59. What is the impulse generated in the string during the second collision?

a. $mu/2$ **b.** $mu/4$

c. $2mu/3$ **d.** None of these

For Problems 60–63

The following problems illustrate the effect of a time-dependent force of a large average magnitude which acts on a moving object only for a short duration. Such forces are called 'impulsive' forces. According to the impulse–momentum theorem, impulse delivered to a body is equal to the change of linear momentum of the body.

A ball of mass 250 g is thrown with a speed of 30 m/s. The ball strikes a bat and it is hit straight back along the same line at a speed of 50 m/s. Variation of the interaction force, as long as the ball remains in contact with the bat, is as shown in Fig. 1.247.

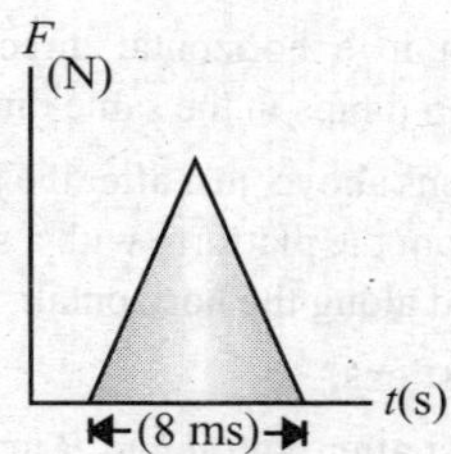

Fig. 1.247

Answer these questions.

60. Maximum force exerted by the bat on the ball is

a. 2500 N **b.** 5000 N

c. 7500 N **d.** 1250 N

61. Average force exerted by the bat on the ball is

a. 5000 N **b.** 1250 N

c. 2500 N **d.** 7500 N

Let us consider another example. The given ball of mass 250 g is dropped from a height 5 m on a hard floor. Force exerted by the floor on the ball, as long as these are in contact, varies with time as shown in Fig. 1.248.

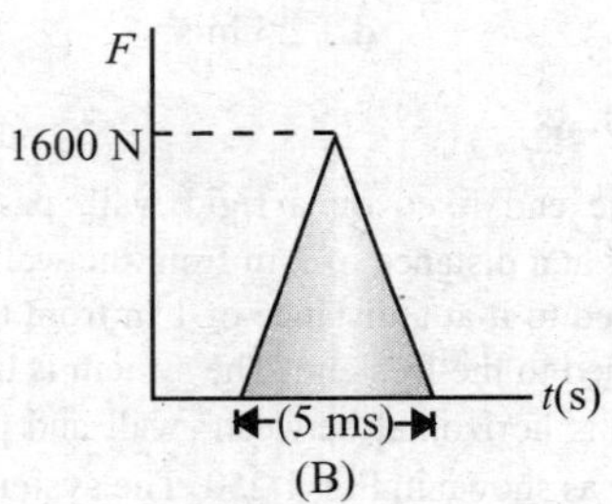

Fig. 1.248

62. Linear momentum of the ball immediately after colliding with the floor will be

a. 1.5 kg m/s **b.** 2.5 kg m/s

c. 4 kg m/s **d.** 0.5 kg m/s

63. After collision with the hard floor, the ball will bounce to a height

a. 0.6 m **b.** 2.4 m

c. 1.2 m **d.** 1.8 m

For Problems 64–66

Two persons, A of mass 60 kg and B of mass 40 kg, are standing on a horizontal platform of mass 50 kg. The platform is supported on wheels on a horizontal frictionless surface and is initially at rest. Consider the following situations.

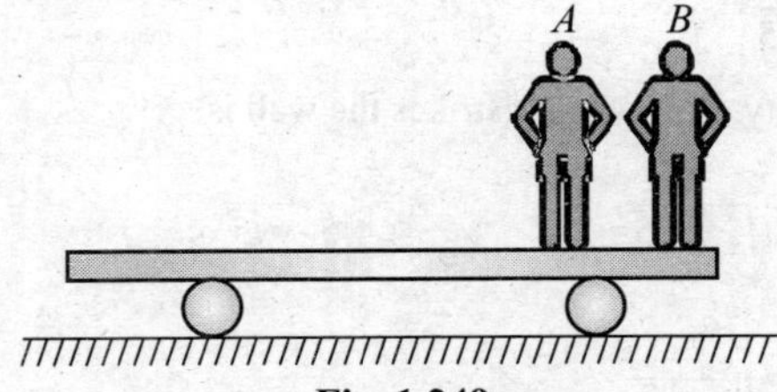

Fig. 1.249

(i) Both A and B jump from the platform simultaneously and in the same horizontal direction.

(ii) A jumps first in a horizontal direction and after a few seconds B also jumps in the same direction.

In both the situations above, just after the jump, the person (A or B) moves away from the platform with a speed of 3 m/s relative to the platform and along the horizontal.

Answer these questions.

64. In situation (i), just after both A and B jump from the platform, velocity of centre of mass of the system (A, B and the platform) is

a. 2 m/s **b.** 6 m/s
c. 5 m/s **d.** none of these

65. Final speed of the platform in situation (i), i.e., just after both A and B have jumped will be

a. 2 m/s **b.** 6 m/s
c. 5 m/s **d.** 3 m/s

66. Final speed of the platform in situation (ii), i.e., just after B has jumped, will be nearly

a. 7.5 m/s **b.** 5.5 m/s
c. 4.5 m/s **d.** 2.5 m/s

For Problems 67–68

A string with one end fixed on a rigid wall, passing over a fixed frictionless pulley at a distance of 2 m from the wall, has a point mass M of 2 kg attached to it at a distance of 1 m from the wall. A mass m of 0.5 kg is attached to the free end. The system is initially held at rest so that the string is horizontal between wall and pulley and vertical beyond the pulley as shown in Fig. 1.250. The system is released from the rest from the position as shown.

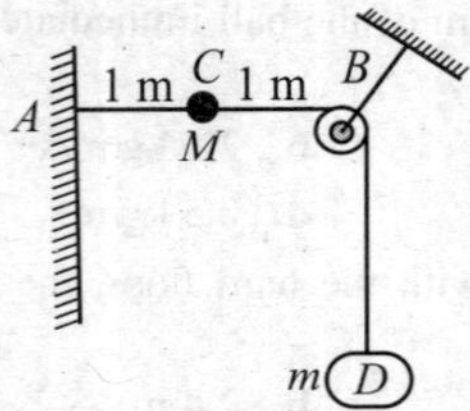

Fig. 1.250

67. The ratio of velocity of M and m when M strikes the wall is

a. $\frac{\sqrt{5}}{2}$ **b.** $\frac{2}{\sqrt{5}}$
c. $\frac{3}{\sqrt{5}}$ **d.** $\frac{\sqrt{5}}{3}$

68. Velocity of M when it strikes the wall is

a. $2\sqrt{\frac{5+\sqrt{5}}{6}}$ m/s **b.** $2\sqrt{\frac{5-\sqrt{5}}{6}}$ m/s
c. $5\sqrt{\frac{5-\sqrt{5}}{6}}$ m/s **d.** $5\sqrt{\frac{5+\sqrt{5}}{6}}$ m/s

For Problems 69–72

Two identical shells are fired from a point on the ground with same muzzle velocity at angle of elevation $\alpha = 45°$ and $\beta = \tan^{-1} 3$ towards top of a cliff, 20 m away from the point of firing. If both the shells reach the top simultaneously, then

69. muzzle velocity is

a. 10 m/s **b.** 5 m/s
c. 15 m/s **d.** 20 m/s

70. height of the cliff is

a. 20 m **b.** 10 m
c. 15 m **d.** 30 m

71. time interval between two firings is

a. $\sqrt{10} - \sqrt{2}$ s **b.** $\sqrt{10} + \sqrt{2}$ s
c. $\sqrt{10} - \sqrt{3}$ s **d.** $\sqrt{10} + \sqrt{3}$ s

72. If just before striking the top of cliff, the two shells get stuck together, considering elastic collision of combined body with the top of cliff, then maximum height reached by the combined body is

a. 20 m **b.** 10 m
c. 24 m **d.** 12 m

For Problems 73–76

A ball of mass m is thrown at an angle of 45° to the horizontal from the top of a 65 m high tower AB as shown in Fig. 1.251 at $t = 0$. Another identical ball is thrown with velocity 20 m/s horizontally towards AB from the top of a 30 m high tower CD 1 s after the projection of the first ball. Both the balls move in the same vertical plane. If they collide in mid-air,

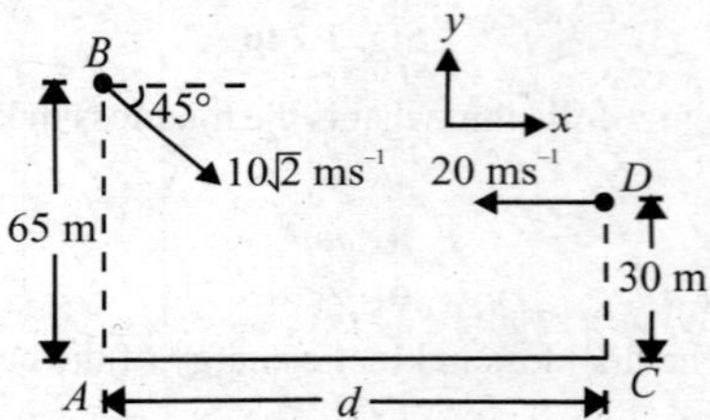

Fig. 1.251

73. the two balls will collide at time t =

a. 2 s **b.** 5 s
c. 10 s **d.** 3 s

74. the height from the ground where the two balls will collide is

a. 15 m **b.** 25 m
c. 10 m **d.** 20 m

75. the distance AC is

a. 20 m **b.** 30 m
c. 40 m **d.** 50 m

76. the velocity of combined ball just after they struck together is

a. $-5\hat{i} - 20\hat{j}$ m/s **b.** $10\hat{i} - 20\hat{j}$ m/s
c. $-10\hat{i} - 20\hat{j}$ m/s **d.** $-5\hat{i} + 10\hat{j}$ m/s

For Problems 77–79

A ball A of mass m is suspended by a thread of length $r = 1.2$ m. Another ball B of mass $2m$ is projected from the ground with velocity $u = 9$ m/s such that at the highest point of its trajectory it collides head-on elastically with ball A. It is observed that during subsequent motion, tension in the thread at the highest point is equal to mg.

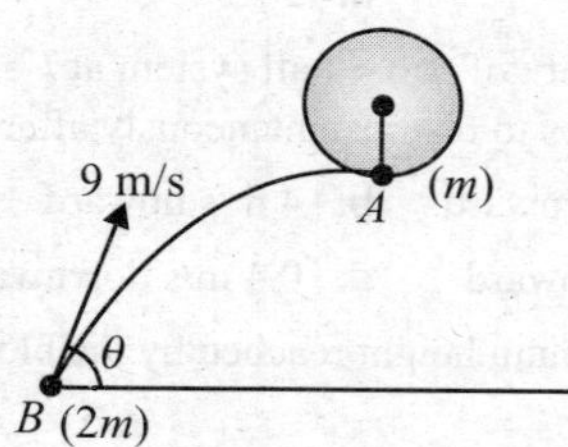

Fig. 1.252

77. At highest point of B, the velocity of ball A is

a. $6\sqrt{2}$ m/s **b.** $2\sqrt{6}$ m/s
c. $3\sqrt{2}$ m/s **d.** $3\sqrt{6}$ m/s

78. The angle of projection (θ) of ball B is

a. 30° **b.** 60°
c. 45° **d.** 75°

79. The height of the point of suspension of ball A from the ground is

a. $\frac{81}{40}$ m **b.** $\frac{129}{40}$ m
c. $\frac{81}{20}$ m **d.** $\frac{101}{40}$ m

For Problems 80–81

Two identical buggies each of mass 150 kg move one after the other without friction with same velocity 4 m/s. A man of mass m rides the rear buggy. At a certain moment, the man jumps into the front buggy with a velocity v relative to his buggy. As a result of this process, rear buggy stops. If the sum of kinetic energies of the man and the front buggy just after collision with the front buggy differs from that just before collision by 2700 J, then

80. the mass m of the man is

a. 60 kg **b.** 75 kg
c. 50 kg **d.** 90 kg

81. the velocity v of the man relative to the buggy is

a. 16 m/s **b.** 8 m/s
c. 10 m/s **d.** 15 m/s

For Problems 82–83

In the arrangement shown in Fig. 1.253, the ball and the block have the same mass $m = 1$ kg each, $\theta = 60°$ and length $l = 2.50$ m. Coefficient of friction between the block and the floor is 0.5. When the ball is released from the position shown in Fig. 1.253, it collides with the block and the block stops after moving a distance 2.50 m.

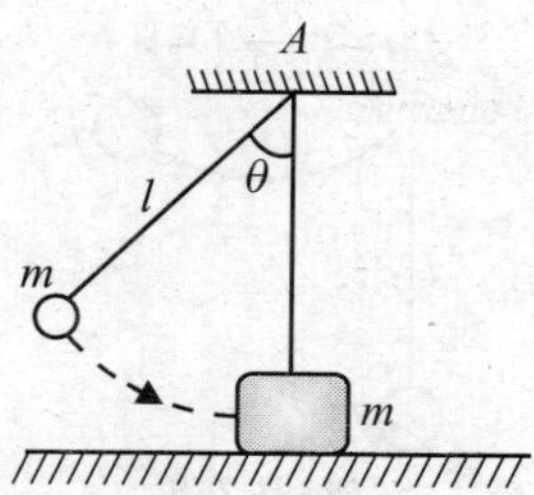

Fig. 1.253

82. The velocity of block just after collision is

a. 10 m/s **b.** 5 m/s
c. 2.5 m/s **d.** 3 m/s

83. The coefficient of restitution for collision between the ball and the block is

a. 0.50 **b.** 0.75
c. 1.0 **d.** 0.30

For Problems 84–86

A ball of mass $m = 1$ kg is hung vertically by a thread of length $l = 1.50$ m. Upper end of the thread is attached to the ceiling of a trolley of mass $M = 4$ kg. Initially, the trolley is stationary and it is free to move along horizontal rails without friction. A shell of mass $m = 1$ kg, moving horizontally with velocity $v_0 = 6$ m/s, collides with the ball and gets stuck with it. As a result, the thread starts to deflect towards right.

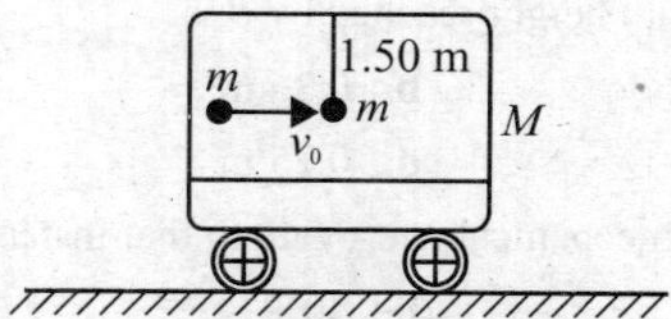

Fig. 1.254

84. The velocity of the combined body just after collision is

a. 2 m/s **b.** 3 m/s
c. 1 m/s **d.** 4 m/s

85. At the time of maximum deflection of the thread with vertical, the trolley will move with velocity

a. 2 m/s **b.** 3 m/s
c. 1 m/s **d.** 4 m/s

86. The maximum deflection of the thread with the vertical is

a. $\cos^{-1}\left(\frac{4}{5}\right)$ **b.** $\cos^{-1}\left(\frac{3}{5}\right)$
c. $\cos^{-1}\left(\frac{2}{3}\right)$ **d.** $\cos^{-1}\left(\frac{3}{4}\right)$

For Problems 87–88

Two balls of masses $m_1 = 100$ g and $m_2 = 300$ g are suspended from point A by two equal inextensible threads, each of length $l = 32/35$ m. Ball of mass m_1 is drawn aside and held at the same level as A but at a distance $(\sqrt{3}/2)l$ from A, as shown in Fig. 1.255. When ball m_1 is released, it collides elastically with the stationary ball of mass m_2.

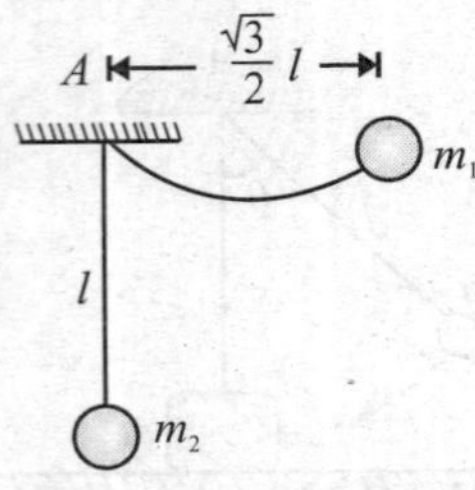

Fig. 1.255

87. Velocity u_1 with which the ball of mass m_1 collides with the other ball is

a. 1 m/s b. 2 m/s c. 3 m/s d. 4 m/s

88. The maximum rise of centre of mass of the ball of mass m_2 is

a. 0.20 m b. 0.50 m c. 0.75 m d. 1 m

For Problems 89–91

Two identical blocks A and B each of mass 2 kg are hanging stationary by a light inextensible flexible string, passing over a light and frictionless pulley, as shown in Fig. 1.256. A shell C of mass 1 kg moving vertically upwards with velocity 9 m/s collides with block B and gets stuck to it.

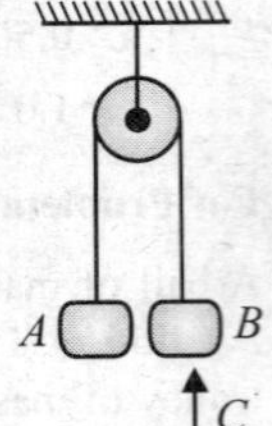

Fig. 1.256

89. Calculate the time after which block B starts moving downwards.

a. 0.90 s b. 1 s
c. 0.60 s d. 0.30 s

90. The maximum height reached by B is

a. 0.81 m b. 0.36 m
c. 0.49 m d. 0.25 m

91. The loss of mechanical energy up to that instant is

a. 32.4 J b. 40 J
c. 16.5 J d. 12.5 J

For Problems 92–95

A pan of mass $m = 1.5$ kg and a block of mass $M = 3$ kg are connected to each other by a light inextensible string, passing over a light pulley as shown in Fig. 1.257. Initially, the block is resting on a horizontal floor. A ball of mass $m_0 = 0.5$ kg collides with the pan at a speed $v_0 = 20$ m/s. Consider this instant of collision as $t = 0$. Assume collision to be perfectly in-elastic. Now, answer the following questions based on the above information.

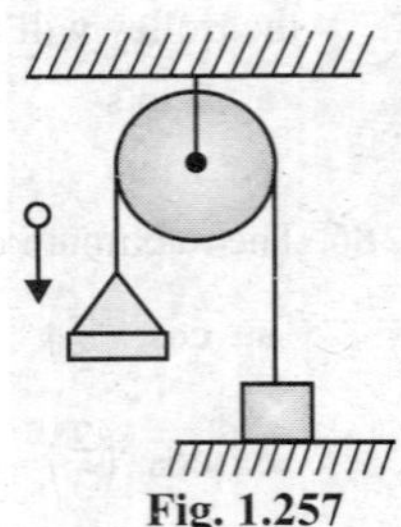

Fig. 1.257

92. Mark the correct statement(s) for this situation.

a. After the collision, the pan + ball system moves downwards with decreasing speed.
b. After the collision, the block is moving upwards with the same speed with which the ball + pan system is moving downwards.
c. The block will jerk for a number of times during its motion.
d. All of these.

93. Find the time t at which the block strikes the floor for the first time.

a. 1 s b. 2 s
c. 4 s d. 5 s

94. Find the velocity of pan + ball system at $t = 2.6$ s. Assume that the block comes to rest instantaneously after striking the floor.

a. 4 m/s downward b. 4 m/s upward
c. 0.6 m/s upward d. 0.4 m/s downward

95. Find the maximum height reached by the block after the second jerk.

a. 0.2 m b. 0.64 m
c. 0.16 m d. No jerk for the second time

Matching Column Type

Solutions on page 1.110

1. A block A of mass $M_A = 1$ kg is kept on a smooth horizontal surface and attached by a light thread to another block B of mass $M_B = 2$ kg. Block B is resting on ground, and thread and pulley are massless and frictionless. A bullet of mass $m = 0.25$ kg moving horizontally with velocity of $u = 200$ m/s penetrates through block A and comes out with a velocity of 100 m/s.

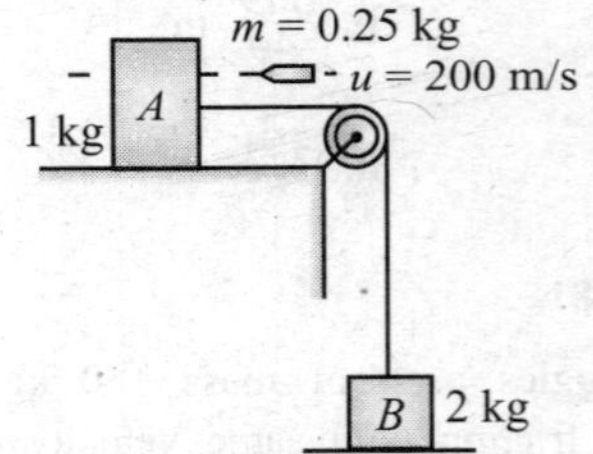

Fig. 1.258

Column I	Column II (Values are in their respective SI units)
i. Velocity of the 2 kg block just after the bullet comes out	a. 50/3
ii. Maximum displacements of 1 kg block in left direction	b. 25
iii. Impulse by the string on block B	c. 25/3
iv. Impulse by the particle on block A	d. 5.2

2. Two blocks A and B of masses m and $2m$, respectively, are connected by a massless spring of spring constant K. This system lies over a smooth horizontal surface. At $t = 0$, block A has velocity u towards right as shown, while the speed of block B is zero, and the length of the spring is equal to its natural length

at that instant. In each situation of column I, certain statements are given and corresponding results are given in column II. Match the statements in column I with the corresponding results in column II.

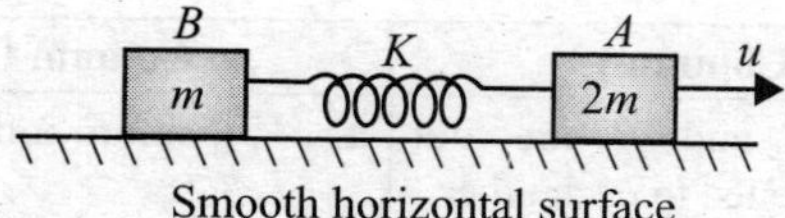

Fig. 1.259

Column I	Column II
i. The velocity of block A	**a.** can never be zero
ii. The velocity of block B	**b.** may be zero at certain instants of time
iii. The kinetic energy of the system of two blocks	**c.** is minimum at maximum compression of spring
iv. The potential energy of spring	**d.** is maximum at maximum extension of spring

3. In each situation of column I, a system involving two bodies is given. All strings and pulleys are light and friction is absent everywhere. Initially, each body of every system is at rest. Consider the system in all situations of column I from rest till any collision occurs. Then match the statements in column I with the corresponding results in column II.

Column I	Column II
i. The block plus wedge system is placed over smooth horizontal surface. After the system is released from rest, the centre of mass of system **Fig. 1.260**	**a.** shifts towards right
ii. The string connecting both the blocks of mass m is horizontal. The left block is placed over smooth horizontal table as shown. After the two blocks system is released from rest, the centre of mass of system **Fig. 1.261**	**b.** shifts downwards
iii. The block and the monkey have the same mass. The monkey starts climbing up the rope. After the monkey starts climbing up, the centre of mass of monkey + block system 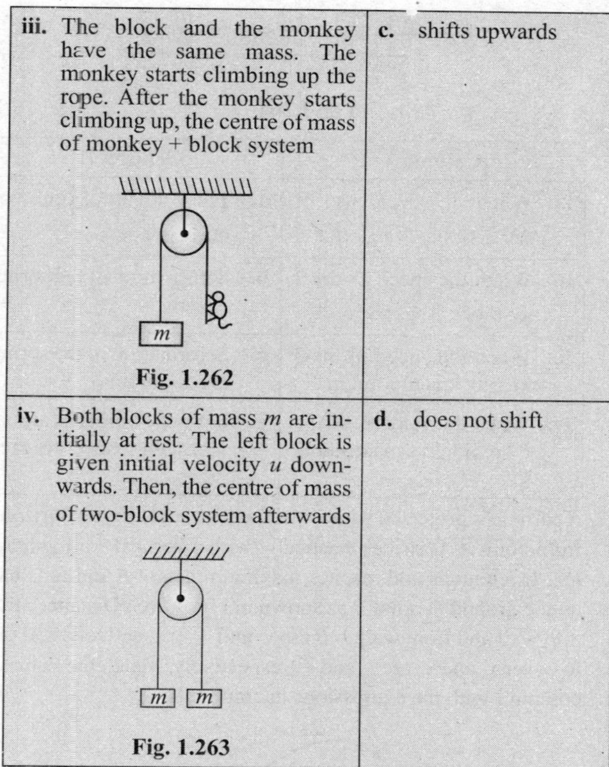 **Fig. 1.262**	**c.** shifts upwards
iv. Both blocks of mass m are initially at rest. The left block is given initial velocity u downwards. Then, the centre of mass of two-block system afterwards **Fig. 1.263**	**d.** does not shift

4. Column I contains physical quantity/process while column II contains formula/principle. Match columns I and II such that the formula/principle is correct corresponding to the quantity in column I.

Column I	Column II
i. Momentum	**a.** $m(v_2 - v_1)$
ii. Impulse	**b.** only momentum is conserved
iii. Elastic collision	**c.** momentum and kinetic energy both are conserved
iv. Inelastic collision	**d.** mv

5. Centre of mass of a system is a unique point which is very helpful in solving the problems related to the motions of rigid bodies. In columns I and II, some statements regarding centre of mass are given. Match the columns:

Column I	Column II
i. In the absence of external forces, the internal forces	**a.** inside the material of the body
ii. Centre of mass of a body can be	**b.** may affect the motion of individuals
iii. The kinetic energy of the system of two blocks	**c.** do not affect the motion of centre of mass
iv. Centre of mass of a solid cylinder is	**d.** outside the material of the body

6. Two blocks of masses 3 kg and 6 kg are connected by an ideal spring and are placed on a frictionless horizontal surface. The 3 kg block is imparted a speed of 2 m/s towards left. (Consider left as positive direction.)

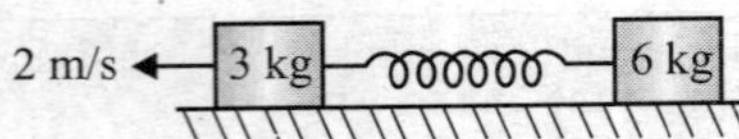

Fig. 1.264

Column I	Column II
i. When the velocity of the 3 kg block is $\frac{2}{3}$ m/s	**a.** The velocity of centre of mass is $\frac{2}{3}$ m/s
ii. When the speed of the 3 kg block is $\frac{2}{3}$ m/s	**b.** deformation of the spring is zero
iii. When the speed of the 3 kg block is minimum	**c.** deformation of the spring is maximum
iv. When the velocity of the 6 kg block is maximum	**d.** both the blocks are at rest with respect to each other

7. A particle is projected with a speed u at an angle θ with horizontal from point A. It strikes elastically with a vertical wall at height $h/2$. It rebounds and reaches maximum height h and falls back on the ground at point B as shown in Fig. 1.265. Distances from A to wall and from wall to B are x_1 and x_2, respectively, and time to cover x_1 and x_2 are t_1 and t_2, respectively. Match the values in column I with the expressions in column II.

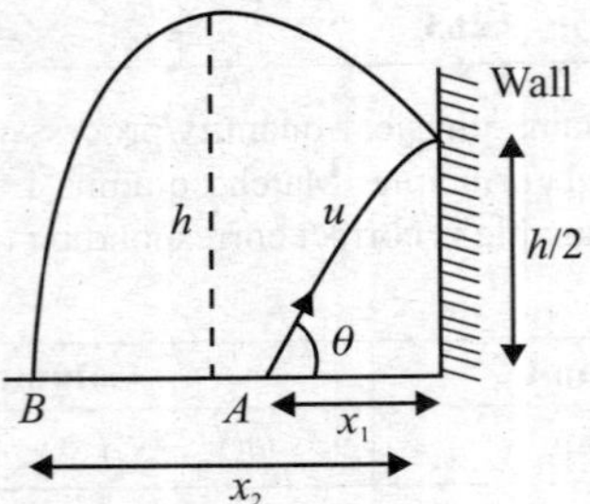

Fig. 1.265

Column I	Column II
i. $\sqrt{2}$	**a.** $\frac{x_2 - x_1}{x_2 + x_1}$ or $\frac{x_2 + x_1}{x_2 - x_1}$
ii. $\frac{1}{\sqrt{2}}$	**b.** $\frac{t_2 - t_1}{t_2 + t_1}$ or $\frac{t_2 + t_1}{t_2 - t_1}$
iii. 1	**c.** $\frac{u \sin\theta}{g(t_1 + t_2)}$
iv. $\frac{1}{2}$	**d.** $\frac{u \cos\theta\,(t_1 + t_2)}{(x_1 + x_2)}$

8. Two spheres A and B move on a smooth horizontal surface with the same velocity V and have some separation between them. A third sphere C is moving in opposite direction on the same surface with same speed. All the spheres are of equal mass. The collisions are elastic. V_{CM} represents the centre of mass velocity of all the three spheres. Column II represents the values after all the possible impacts have occurred.

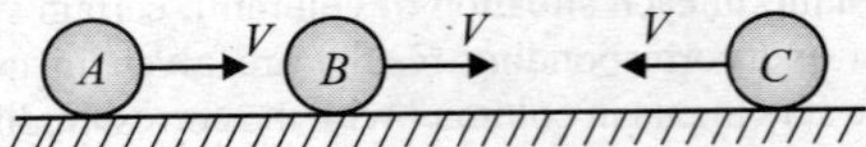

Fig. 1.266

Column I	Column II
i. If A and B are not connected to each other	**a.** V_{CM} before collision $= V/3$
ii. If A and B are connected to each other by a massless rigid rod	**b.** V_{CM} after all the collision $= \frac{V}{3}$
iii. If A and B are connected by an ideal string	**c.** Momentum of system ($B + C$) is not conserved
iv. If A and B are connected by an ideal spring which is initially unstretched	**d.** Momentum of system ($A + B$) is conserved

9. Two balls of masses m_1 and m_2 are moving towards each other with speeds u_1 and u_2, respectively. They collide head-on and their speeds are v_1 and v_2 after collision ($m_1 = 8$ kg, $m_2 = 2$ kg, $u_2 = 3$ m/s).

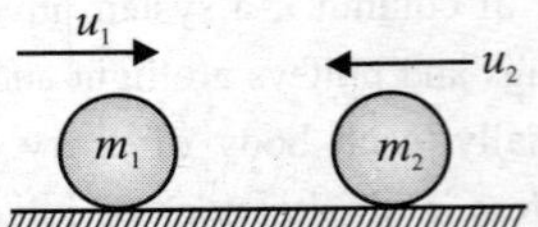

Fig. 1.267

Column I	Column II (SI Units)
i. The speed u_1 (in m/s) so that both balls move in same direction if co-efficient of restitution is $e = 0.5$	**a.** $\frac{1}{14}$
ii. The speed u_1 (in m/s) so that the maximum fraction of energy is transformed to m_2 (assume elastic collision)	**b.** $\frac{1}{8}$
iii. Co-efficient of restitution if m_2 stops after collision and $u_1 = 0.5$ m/s	**c.** 2
iv. If collision is inelastic and $u_1 = 3$ m/s, the loss of kinetic energy (in J) after collision may be	**d.** 4

10. Match the following:

Column I	Column II
i. Inelastic collisions	**a.** kinetic energy of the system may decrease
ii. Elastic collisions	**b.** kinetic energy of the system may increase

iii.	Total work done is zero	**c.**	kinetic energy of the system may remain constant
iv.	Non-conservative forces are not present	**d.**	just before and after collision, momentum remains constant

11. In column I, nature of collision between two bodies is given while in column II some physical quantity that may remain conserved during the collision is given. Match the entries of column I with the entries of column II.

Column I	**Column II**
i. Elastic collision	**a.** kinetic energy is conserved
ii. Inelastic collision	**b.** kinetic energy of the system may increase
iii. Perfectly inelastic collision	**c.** kinetic energy is not conserved
iv. Collision between two cars moving at high speeds	**d.** total energy is conserved

Integer Answer Type

Solutions on page 1.111

1. Three balls A, B and C ($m_A = m_C = 4m_B$) are placed on a smooth horizontal surface. Ball B collides with ball C with an initial velocity v as shown in Fig. 1.268. Find the total number of collisions between the balls (all collisions are elastic).

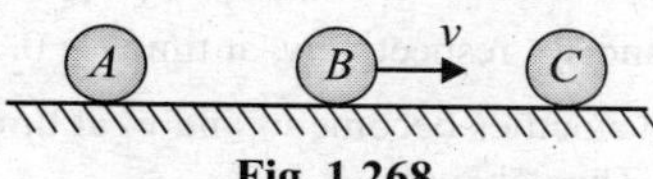

Fig. 1.268

2. A man standing on a trolley pushes another identical trolley (both trolleys are at rest on a rough surface), so that they are set in motion and stop after some time. If the ratio of mass of first trolley with man to mass of second trolley is 3, then find the ratio of the stopping distances of the second trolley to that of the first trolley. (Assume coefficient of friction to be the same for both the trolleys)

3. A particle with a mass of 1 kg is having a velocity of 10 m/s in + ve x-direction at $t = 0$. Forces $\vec{F}_1$ and $\vec{F}_2$ act on the particle whose magnitudes are changing with time according to the variation shown in Fig. 1.269. The magnitude of the velocity of the particle at $t = 3$ s (neglect gravity effect) is found to be $n\sqrt{5}$ m/s. Find the value of n.

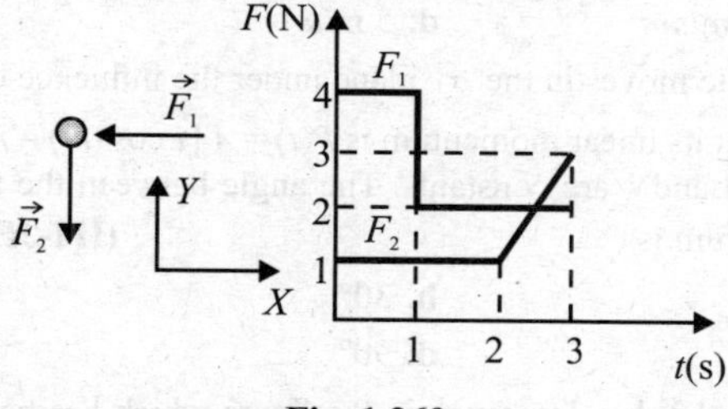

Fig. 1.269

4. A bullet is fired on a fixed target. It penetrates inside the target through distance $d = 3.75$ cm and then stops, mass of the bullet is $m = 1$ kg and of the target is $M = 4$ kg. Now an identical bullet moving with the same velocity is fired on the identical target which is placed at rest on a frictionless horizontal surface. Then find the distance (in cm) to which the bullet will penetrate inside the target?

5. A frog sits on the end of a long board of length $L = 5$ m. The board rests on a frictionless horizontal table. The frog wants to jump to the opposite end of the board. What is the minimum take-off speed (in m/s), i.e., relative to ground 'v' that allows the frog to do the trick? The board and the frog have equal masses.

6. A ball of mass 1 kg moving with a velocity of 5 m/s collides elastically with rough ground at an angle θ with the vertical as shown in Fig. 1.270. What can be the minimum coefficient of friction if ball rebounds vertically after collision? (given $\tan \theta = 2$)

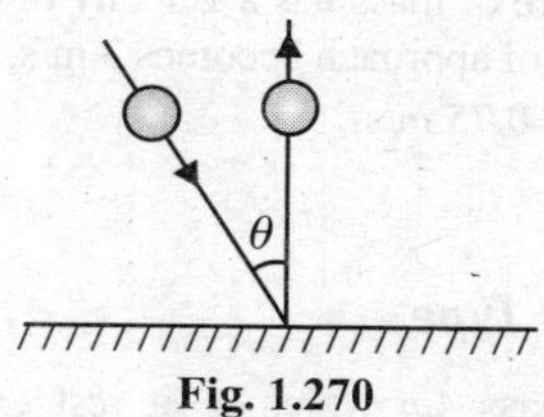

Fig. 1.270

7. A small sphere of mass $m = 1$ kg is moving with a velocity $(4\hat{i} - \hat{j})$ m/s. It hits a fixed smooth wall and rebounds with velocity $(\hat{i} + 3\hat{j})$ m/s. The coefficient of restitution between the sphere and the wall is $n/16$. Find the value of n.

8. A child of mass 4 kg jumps from cart B to cart A and then immediately back to cart B. The mass of each cart is 20 kg and they are initially at rest. In both the cases the child jumps at 6 m/s relative to the cart. If the cart moves along the same line with negligible friction with the final velocities of V_B and V_A, respectively, find the ratio of $6v_B$ and $5v_A$.

Fig. 1.271

9. A man of mass $M = 58$ kg jumps from an aeroplane as shown in Fig. 1.272. He sees the hard ground below him and a lake at a distance $d = 1$ m from the point directly below him. He immediately puts off his jacket (mass $m = 2$ kg) and throws it in a direction directly away from the lake. If he just fails to strike the ground, find the distance (in 10^1 m) he should walk now to pick his jacket. (Neglect air resistance and take the velocity of man at the time of jump with respect to earth zero.)

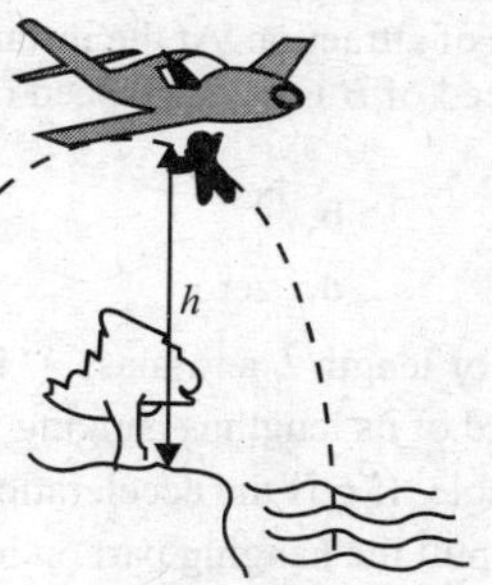

Fig. 1.272

10. A ball of mass m makes head-on elastic collision with a ball of mass nm which is initially at rest. Show that the fractional transfer of energy by the first ball is $4n/(1+n)^2$. Deduce the value of n for which the transfer is maximum.

Archives

Solutions on page 1.113

True or False Type

1. Two particles of masses 1 kg and 3 kg move towards each other under mutual force of attraction. No other force acts on them. When the relative velocity of approach of the two particles is 2 m/s, their centre of mass has a velocity of 0.5 m/s. When the relative velocity of approach becomes 3 m/s, the velocity of the centre of mass is 0.75 m/s. **(IIT-JEE, 1989)**

Fill in the Blanks Type

1. A particle of mass $4m$ which is at rest explodes into three fragments. Two of the fragments, each of mass m, are found to move with speed v each in mutually perpendicular directions. The total energy released in the process of explosion is ______. **(IIT-JEE, 1987)**

2. The magnitude of force (in newtons) acting on a body varies with time t (in microseconds) as shown in Fig. 1.273. AB, BC and CD are straight line segments. The magnitude of total impulse of the force on the body from $t = 4$ s to $t = 16$ s is ___ N s. **(IIT-JEE, 1994)**

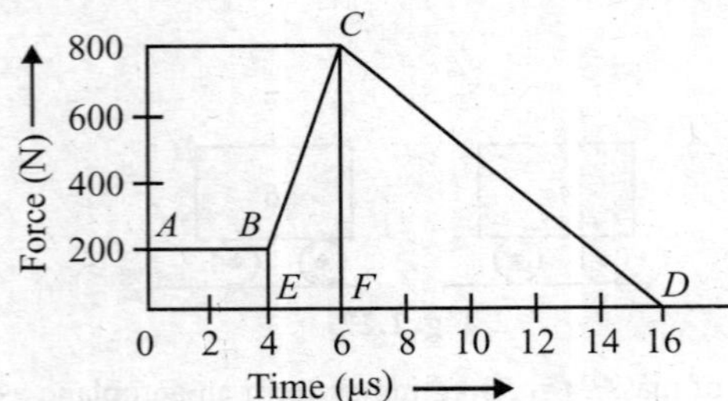

Fig. 1.273

Multiple Choice Questions with One Correct Answer Type

1. Two particles A and B, initially at rest, move towards each other by a mutual force of attraction. At the instant when the speed of A is v and the speed of B is $2v$, the speed of the centre of mass of the system is **(IIT-JEE, 1982)**

 a. $3v$ b. v
 c. $1.5v$ d. zero

2. A uniform chain of length L and mass M is lying on a smooth table and one-third of its length is hanging vertically down over the edge of the table. If g is the acceleration due to gravity, the work required to pull the hanging part on to the table is **(IIT-JEE, 1985)**

 a. MgL b. $MgL/3$
 c. $MgL/9$ d. $MgL/18$

3. A ball hits the floor and rebounds after an inelastic collision. In this case **(IIT-JEE, 1986)**

 a. the momentum of the ball just after collision is same as that just before the collision
 b. the mechanical energy of the ball remains the same in collision
 c. the total momentum of the ball and the earth is conserved
 d. the total mechanical energy of the ball and the earth is conserved

4. A shell is fired from a cannon with a velocity v (m/s) at an angle θ with the horizontal direction. At the highest point in its path it explodes into two pieces of equal mass. One of the pieces retraces its path to the cannon and the speed (in m/s) of the other piece immediately after the explosion is **(IIT-JEE, 1986)**

 a. $3v\cos\theta$ b. $2v\cos\theta$
 c. $\frac{3}{2}v\cos\theta$ d. $\sqrt{\frac{3}{2}}\,v\cos\theta$

5. An isolated particle of mass m is moving in horizontal plane xy along the x-axis, at a certain height above the ground. It suddenly explodes into two fragment of masses $m/4$ and $3m/4$. An instant later, the smaller fragment is at $y = +15$ cm. The larger fragment at this instant is at **(IIT-JEE, 1997)**

 a. $y = -5$ cm b. $y = +20$ cm
 c. $y = +5$ cm d. $y = -20$ cm

6. Two particles of masses m_1 and m_2 in projectile motion have velocities $\vec{v}_1$ and $\vec{v}_2$, respectively, at time $t = 0$. They collide at time t_0. Their velocities become $\vec{v}_1'$ and $\vec{v}_2'$ at time $2t_0$ while still moving in air. The value of $\left|\left(m_1\vec{v}_1' + m_2\vec{v}_2'\right)\right| - \left|\left(m_1\vec{v}_1 + m_2\vec{v}_2\right)\right|$ is **(IIT-JEE, 2001)**

 a. zero b. $(m_1+m_2)gt_0$
 c. $\frac{1}{2}(m_1+m_2)gt_0$ d. $2(m_1+m_2)gt_0$

7. Two blocks of masses 10 kg and 4 kg are connected by a spring of negligible mass and placed on a fractionless horizontal surface. An impulse gives a velocity of 14 m/s to the heavier block in the direction of the lighter block. The velocity of the centre of mass is **(IIT-JEE, 2002)**

 a. 30 m/s b. 20 m/s
 c. 10 m/s d. 5 m/s

8. A particle moves in the xy plane under the influence of a force such that its linear momentum is $\vec{P}(t) = A[\hat{i}\cos(kt) - \hat{j}\sin(kt)]$, where A and k are constants. The angle between the force and momentum is **(IIT-JEE, 2007)**

 a. 0° b. 30°
 c. 45° d. 90°

9. Look at the drawing given in the figure which has been drawn with ink of uniform line-thickness. The mass of ink used to draw each of the two inner circles, and each of the two line segments is m. The mass of the ink used to draw the outer circle is $6m$.

The coordinates of the centres of the different parts are: outer circle (0, 0), left inner circle $(-a, a)$, right inner circle (a, a), vertical line (0, 0) and horizontal line $(0, -a)$. The y-coordinate of the centre of mass of the ink in this drawing is

(IIT-JEE, 2009)

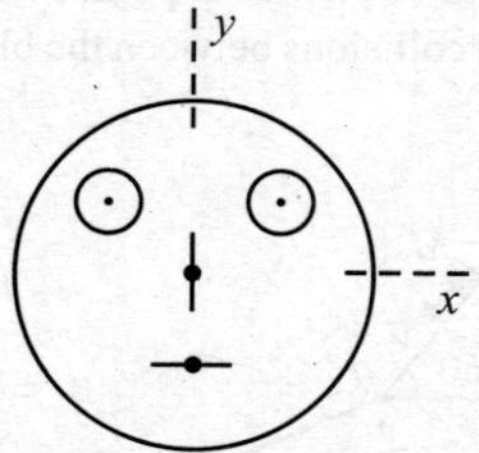

Fig. 1.274

a. $\frac{a}{10}$ **b.** $\frac{a}{8}$

c. $\frac{a}{12}$ **d.** $\frac{a}{3}$

10. Two small particles of equal masses start moving in opposite directions from a point A in a horizontal circular orbit. Their tangential velocities are v and $2v$, respectively, as shown in the figure. Between collisions, the particles move with constant speeds. After making how many elastic collisions, other than that at A, these two particles will again reach the point A?

(IIT-JEE, 2009)

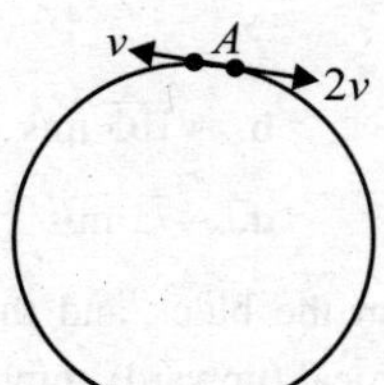

Fig. 1.275

a. 4 **b.** 3

c. 2 **d.** 1

11. A block of mass 2 kg is free to move along the x-axis. It is at rest and from $t = 0$ onwards it is subjected to a time-dependent force $F(t)$ in the x direction. The force $F(t)$ varies with t as shown in the figure. The kinetic energy of the block after 4.5 seconds is

(IIT-JEE, 2010)

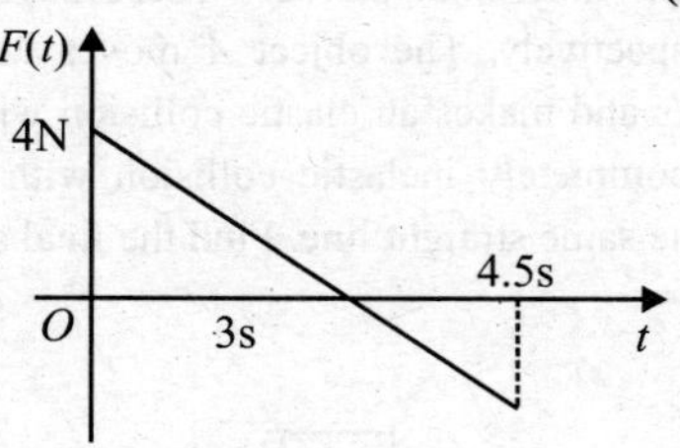

Fig. 1.276

a. 4.50 J **b.** 7.50 J

c. 5.06 J **d.** 14.06 J

12. A ball of mass 0.2 kg rests on a vertical post of height 5 m. A bullet of mass 0.01 kg, travelling with a velocity V m/s in a horizontal direction, hits the centre of the ball.

(IIT-JEE, 2011)

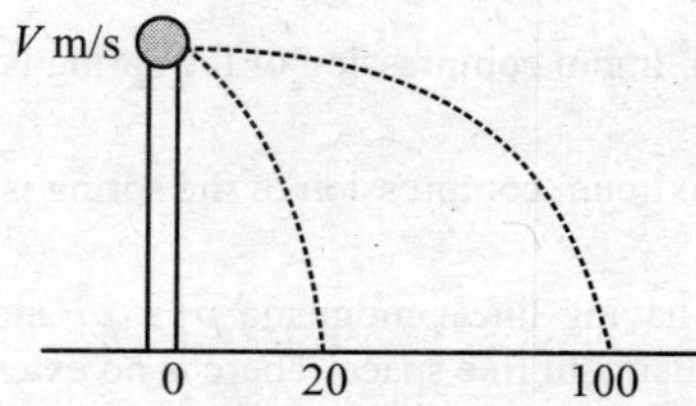

Fig. 1.277

After the collision, the ball and the bullet travel independently. The ball hits the ground at a distance of 20 m and the bullet at a distance of 100 m from the foot of the post. The initial velocity V of the bullet is

a. 250 m/s **b.** $250\sqrt{2}$ m/s

c. 400 m/s **d.** 500 m/s

Multiple Choice Questions with One or More than One Correct Answer Type

1. A uniform bar of length $6a$ and mass $8m$ lies on a smooth horizontal table. Two point masses m and $2m$ moving in the same horizontal plane with speeds $2v$ and v, respectively, strike the bar (as shown in the figure) and stick to the bar after collision. Denoting angular velocity (about the centre of mass), total energy and centre of mass velocity by ω, E and V_c, respectively, we have after collision

(IIT-JEE, 1991)

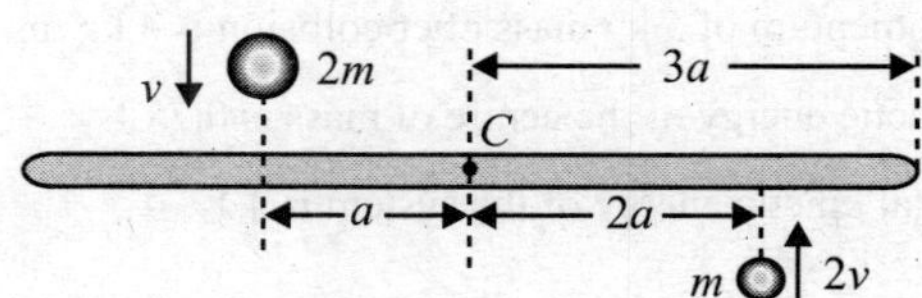

Fig. 1.278

a. $V_c = 0$ **b.** $\omega = \frac{3v}{5a}$

c. $\omega = \frac{v}{5a}$ **d.** $E = \frac{3mv^2}{5}$

2. Two blocks A and B, each of mass m, are connected by a massless spring of natural length L and spring constant K. The blocks are initially resting on a smooth horizontal floor with the spring at its natural length, as shown in Fig. 1.279. A third identical block C, also of mass m, moves on the floor with a speed v along the line joining A and B, and collides elastically with A. Then

(IIT-JEE, 1993)

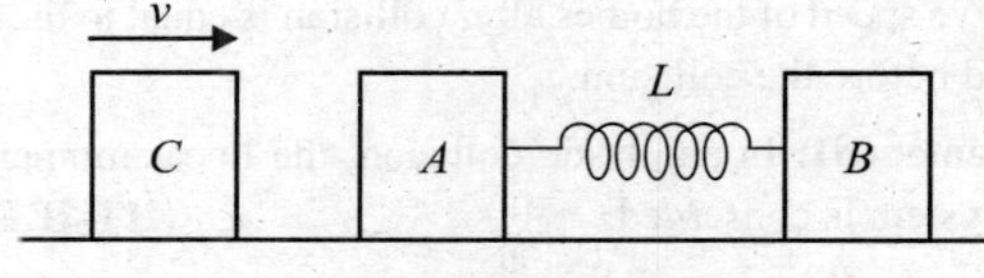

Fig. 1.279

a. the kinetic energy of the A–B system, at maximum compression of the spring, is zero.

b. the kinetic energy of the A–B system, at maximum compression of the spring, is $\frac{mv^2}{4}$.

c. the maximum compression of the spring is $v\sqrt{\left(\frac{m}{K}\right)}$

d. the maximum compression of the spring is $v\sqrt{\left(\frac{m}{2K}\right)}$

3. Two balls, having linear momenta $\vec{p}_1 = p\hat{i}$ and $\vec{p}_2 = -p\hat{i}$, undergo a collision in free space. There is no external force acting on the ball. Let $\vec{p}_1'$ and $\vec{p}_2'$ be their final momenta. Which of the following option (s) is (are) NOT ALLOWED for any non-zero value of p, a_1, a_2, b_1, b_2, c_1 and c_2. **(IIT-JEE, 2008)**

a. $\vec{p}_1' = a_1\hat{i} + b_1\hat{j} + c_1\hat{k}$, $\vec{p}_2' = a_2\hat{i} + b_2\hat{j}$

b. $\vec{p}_1' = c_1\hat{k}$, $\vec{p}_2' = c_2\hat{k}$

c. $\vec{p}_1' = a_1\hat{i} + b_1\hat{j} + c_1\hat{k}$, $\vec{p}_2' = a_2\hat{i} + b_2\hat{j} - c_2\hat{k}$

d. $\vec{p}_1' = a_1\hat{i} + b_1\hat{j}$, $\vec{p}_2' = a_2\hat{i} + b_1\hat{j}$

4. A point mass of 1 kg collides elastically with a stationary point mass of 5 kg. After their collision, the 1 kg mass reverses its direction and moves with a speed of 2 ms^{-1}. Which of the following statement(s) is (are) correct for the system of these two masses? **(IIT-JEE, 2010)**

a. Total momentum of the system is 3 kg ms^{-1}

b. Momentum of 5 kg mass after collision is 4 kg ms^{-1}

c. Kinetic energy of the centre of mass is 0.75 J

d. Total kinetic energy of the system is 4 J

Assertion and Reasoning Type

In the following questions, a statement of Assertion (Statement I) is given which is followed by a corresponding statement of reason (Statement II). Examine the statements carefully choose the correct option according to the following options.

a. Both Statement I and Statement II are true and Statement II is the correct explanation of Statement 1.

b. Both Statement I and Statement II are true but Statement II is not the correct explanation of Statement I.

c. Statement I is true and Statement II is false.

d. Statement I is false and Statement II is true.

1. **Statement I:** In an elastic collision between two bodies, the relative speed of the bodies after collision is equal to the relative speed before the collision.

Statement II: In an elastic collision, the linear momentum of the system is conserved. **(IIT-JEE, 2007)**

Comprehension Type

For Problems 1–3

A small block of mass M moves on a frictionless surface of an inclined plane, as shown in the figure. The angle of the incline suddenly changes from 60° to 30° at point B. The block is initially at rest at A. Assume that collisions between the block and the incline are totally inelastic. **(IIT-JEE, 2008)**

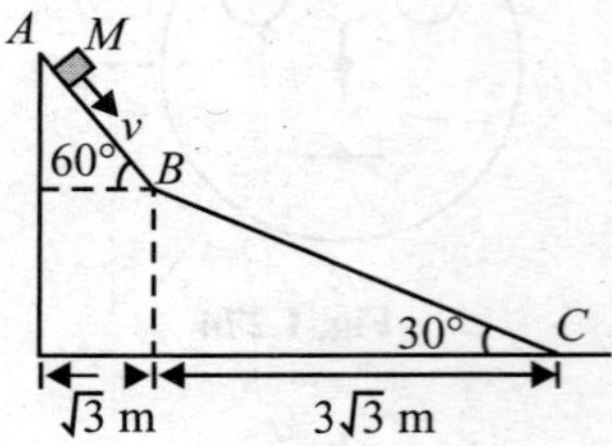

Fig. 1.280

1. The speed of the block at point B immediately after it strikes the second incline is

a. $\sqrt{60}$ m/s b. $\sqrt{45}$ m/s

c. $\sqrt{30}$ m/s d. $\sqrt{15}$ m/s

2. The speed of the block at point C; immediately before it leaves the second incline is

a. $\sqrt{120}$ m/s b. $\sqrt{105}$ m/s

c. $\sqrt{90}$ m/s d. $\sqrt{75}$ m/s

3. If collision between the block and the incline is completely elastic, then the vertical (upward) component of the velocity of the blocks at point B, immediately after it strikes the second incline is

a. $\sqrt{30}$ m/s b. $\sqrt{15}$ m/s

c. 0 d. $-\sqrt{15}$ m/s

Integer Answer Type

1. Three objects A, B and C are kept in a straight line on a frictionless horizontal surface. These have masses m, $2m$ and m, respectively. The object A moves towards B with a speed 9 m/s and makes an elastic collision with it. Thereafter, B makes completely inelastic collision with C. All motions occur on the same straight line. Find the final speed (in m/s) of object C. **(IIT-JEE, 2009)**

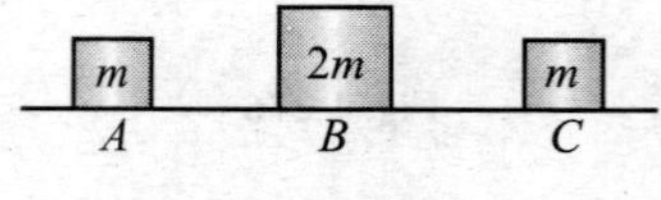

Fig. 1.281

ANSWERS AND SOLUTIONS

Subjective Type

1. The collision does not alter the vertical component of the velocity of either ball, so $t_1 + t_2 = T$

$$\Rightarrow \frac{50}{u \cos 30°} + \frac{50}{0.7u \cos 30°} = \frac{2u \sin 30°}{10} \Rightarrow u \approx 38 \text{ m/s}$$

2. Mass of ladder $= M - m$, because net mass on both sides of string is same. Let in the given process, block M rises by height y, then ladder will come down by distance y and when man comes in the level of M, then height raised by man w.r.t. ground will be $h + y$.

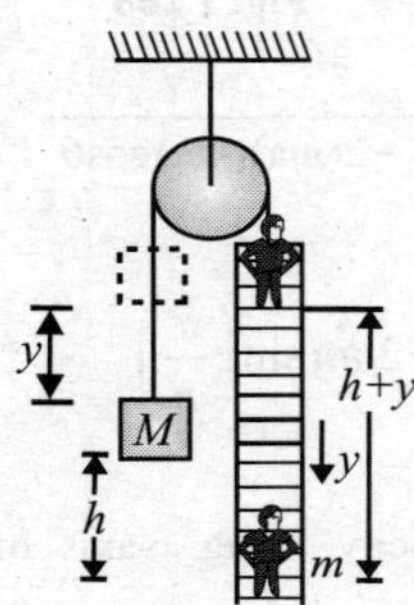

Fig. 1.282

Now centre of mass (man + ladder) system will be raised by y, then

$$y = \frac{m(h+y)+(M-m)(-y)}{M} \qquad \Rightarrow \quad y = \frac{mh}{2(M-m)}$$

Now work done by man:

W = total change in potential energy of system

$\Rightarrow W = Mgy + mg(h+y) - (M-m)gy$

Put the values of y and solve to get $W = \dfrac{Mmgh}{M-m}$

3. Velocity of B just before the string is taut

$v_B = \sqrt{2g\Delta h} = \sqrt{10}$ m/s

a. Common velocity $= v$

$\therefore (m_A + m_B)v = m_B v_B \Rightarrow v = \dfrac{v_B}{3} = \dfrac{\sqrt{10}}{3}$ m/s

b. Magnitude of impulse on A = magnitude of impulse on

$$B = 1\left(\sqrt{10} - \frac{\sqrt{10}}{3}\right) = \frac{2}{3}\sqrt{10} \text{ Ns}$$

Therefore, impulse on pulley $= \sqrt{2}$ impulse on $A = \frac{4}{3}\sqrt{5}$ Ns

4. $\cos\theta = \dfrac{h}{\ell} = \dfrac{1.25}{1.5625} = \dfrac{4}{5}$, $u = \sqrt{2gh} = 5$ m/s

After the string becomes tight:

Let velocity of cart be v and that of ball be v_1 and v_2 in horizontal and vertical direction as shown, all velocities being w.r.t ground.

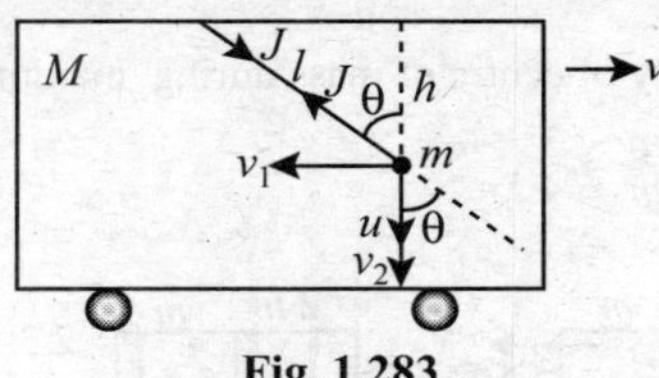

Fig. 1.283

Now

$$J \sin\theta = Mv = mv_1 \qquad \text{(i)}$$

$$J \cos\theta = m(u - v_2) \qquad \text{(ii)}$$

Velocity of both masses along the string should be same:

$$v \sin\theta = v_2 \cos\theta - v_1 \sin\theta \qquad \text{(iii)}$$

From (i), (ii) and (iii)

$$v = \frac{5}{7} \text{ m/s}, v_1 = \frac{15}{7} \text{ m/s}, v_2 = \frac{15}{7} \text{ m/s}$$

Loss of energy $= \text{KE}_i - \text{KE}_f$

$$= \frac{1}{2}mu^2 - \frac{1}{2}Mv^2 - \frac{1}{2}mv_1^2 - \frac{1}{2}mv_2^2$$

$$= \frac{1}{2}\times1\times5^2 - \frac{1}{2}\times3\times\left(\frac{5}{7}\right)^2 - \frac{1}{2}\times1\left(\left(\frac{15}{7}\right)^2 + \left(\frac{15}{7}\right)^2\right)$$

$$= \frac{350}{49} \text{ J}$$

5. Velocity of the falling part when half of part has fallen

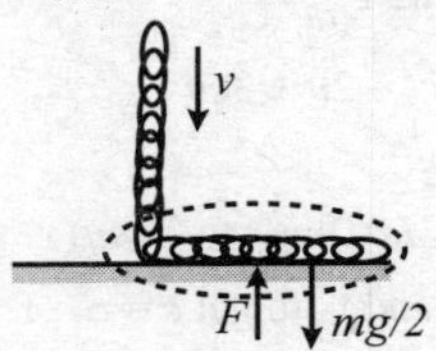

Fig. 1.284

$$v = \sqrt{2g(\ell/2)} = \sqrt{g\ell}$$

Consider the FBD of the part of chain lying on table.

$$\vec{F}_{\text{ex}} + \vec{u}_{\text{rel}}\frac{dm}{dt} = m\frac{d\vec{v}}{dt}$$

$$\Rightarrow \left(F - \frac{mg}{2}\right) + (-v)\frac{dm}{dt} = m \times 0 \quad \Rightarrow F = \frac{mg}{2} + v\frac{dm}{dt}$$

Now, $\dfrac{dm}{dt} = \lambda v = \dfrac{m}{\ell}v$

So, $F = \dfrac{mg}{2} + v\dfrac{m}{\ell}v = \dfrac{mg}{2} + \dfrac{m}{\ell}g\ell = \dfrac{3mg}{2}$

6. First considering an inelastic collision between B and C,

$$2mv = (m + 2m)\, v_B$$

$$v_B = \frac{2}{3} v \text{ velocity of } (B \text{ and } C)$$

Then velocity of centre of mass during maximum compression

$$v_{CM} = \frac{3mv}{4m} = \frac{3}{4} v$$

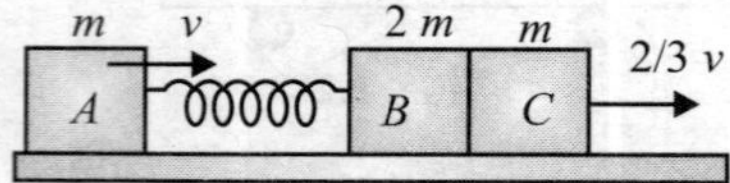

Fig. 1.285

Using conservation of energy just after collision and at the time of maximum compression.

$$\frac{1}{2} mv_A^2 + \frac{1}{2} 3mv_{(B+C)}^2 = \frac{1}{2} 4mv_{CM}^2 + \frac{1}{2} kx^2$$

Then substituting the value $x = \sqrt{\dfrac{mv^2}{12k}}$

7. Using conservation of momentum

$4u$ →	← u	v_1 →	v_2 →
$3m$	$2m$	$3m$	$2m$
Before collision		After collision	

Fig. 1.286

$$m_1u_1 + m_2u_2 = m_1v_1 + m_2v_2$$

Putting the values, we get

$$(3m)(4u) - (2m)u = 3m\, v_1 + 2m\, v_2$$

$$\Rightarrow \quad 3v_1 + 2v_2 = 10\, u \qquad \text{(i)}$$

Using equation of restitution

$$v_2 - v_1 = 5\, ue \qquad \text{(ii)}$$

From (i) and (ii), $v_2 = 2u + 3eu$

Now impulse on m_2:

$$J = 2mv_2 - (-2mu)$$

Put the value of J and v_2 to get $e = c - 1$

As $0 \le e \le 1 \quad \Rightarrow \quad 1 \le c \le 2$

8. a. The internal forces exerted by the actor do not change the total momentum of the system of the four cars and the movie actor $(4m)\, v_1 = (3m)(2.00 \text{ m/s}) + m(4.00 \text{ m/s})$

$$v_1 = \frac{6.00 \text{ m/s} + 4.00 \text{ m/s}}{4} = 2.50 \text{ m/s}$$

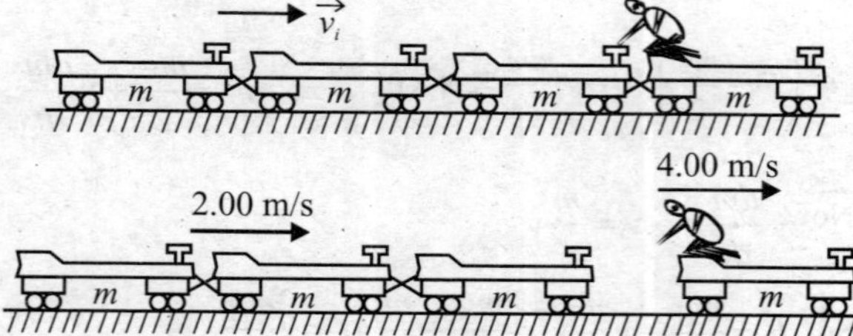

Fig. 1.287

b. $W_{actor} = K_f - K_i = \frac{1}{2}[(3m)(2.00 \text{ m/s})^2 + m(4.00 \text{ m/s})^2]$

$$- \frac{1}{2}(4 \text{ m})(2.50 \text{ m/s})^2$$

$$= \frac{(2.50 \times 10^4 \text{ kg})}{2}(12.0 + 16.0 - 25.0)(\text{m/s})^2 = 37.5 \text{ kJ}$$

9. $\theta = \dfrac{2\pi}{n}$, impulse = change in momentum

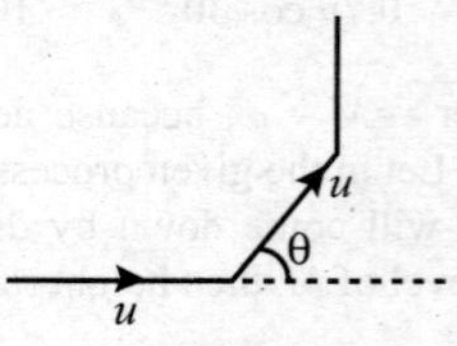

Fig. 1.288

$$= \sqrt{(mu)^2 + (mu)^2 - 2(mu)(mu)\cos\theta}$$

$$= 2mu \sin\left(\frac{\theta}{2}\right) = 2mu \sin\left(\frac{\pi}{n}\right)$$

10. $v_x = 7$ m/s, velocity along y-axis of the ball just before collision.

$$v_y = \sqrt{2 \times 9.8 \times 2.5} = 7 \text{ m/s}$$

As $v_x = v_y$, so it strikes the plane of incline perpendicularly. Let the ball rebound with velocity V and let v_1 be the velocity of the wedge.

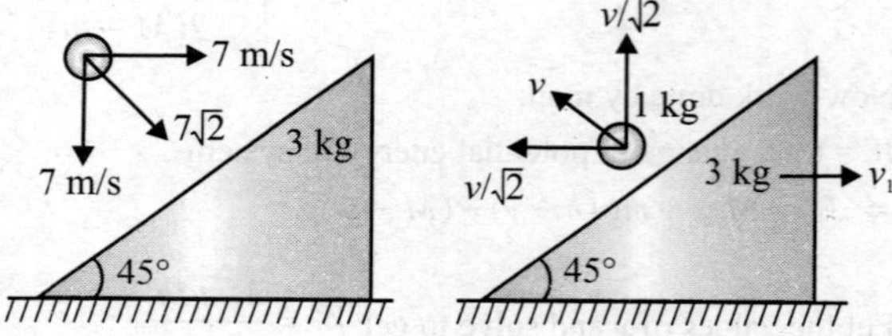

Fig. 1.289

Applying the principle conservation of momentum in horizontal direction, we get

$$1 \times 7 = -1 \times \frac{v}{\sqrt{2}} + 3v_1$$

$$7\sqrt{2} = 3\sqrt{2}\, v_1 - v \qquad \text{(i)}$$

Applying the equation for coefficient of restitution, we get

$$e = 1 = \frac{v_1/\sqrt{2} + v}{7\sqrt{2}},\; 7\sqrt{2} = \frac{v_1}{\sqrt{2}} + v \qquad \text{(ii)}$$

Solving Eqs. (i) and (ii), $v_1 = 4$ m/s and $v = 5\sqrt{2}$ m/s.

11. Let θ be the angle through which the beads have fallen, and let N be the normal force from the hoop on the beads, with inward taken to be positive. Then the radial $F = ma$ equation for the bead is

$$N + mg\cos\theta = \frac{mv^2}{R} \qquad \text{(i)}$$

$$\frac{1}{2}mv^2 = mgR(1-\cos\theta)$$

$$\Rightarrow \quad v^2 = 2gR(1-\cos\theta) \quad \text{(ii)}$$

$$N = \frac{mv^2}{R} - mg\cos\theta = 2mg(1-\cos\theta) - mg\cos\theta$$

$$= mg(2-3\cos\theta) \quad \text{(iii)}$$

For loop to rise:

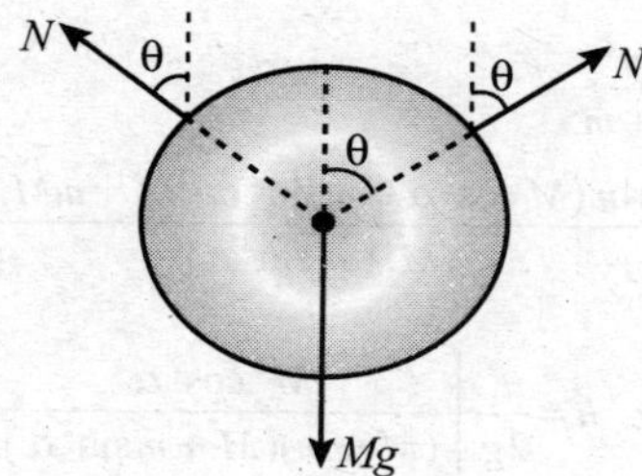

Fig. 1.290

$2N\cos\theta = Mg$

$\Rightarrow 2mg(2-3\cos\theta)\cos\theta = Mg \Rightarrow \dfrac{m}{M} = \dfrac{1}{2(2-3\cos\theta)\cos\theta}$

For m/M to be minimum, $y = 2(2-3\cos\theta)\cos\theta$ is to be maximum, put $dy/d\theta = 0 \Rightarrow \cos\theta = 1/3 \Rightarrow (m/M)_{min} = 3/2$

12. $A_{m/p} = 0.2\times 10 + \dfrac{50\times 0.2\times 10}{10} = 12 \text{ m/s}^2$

$v_{max} = \sqrt{2\times 12\times 150} = 60 \text{ m/s}$

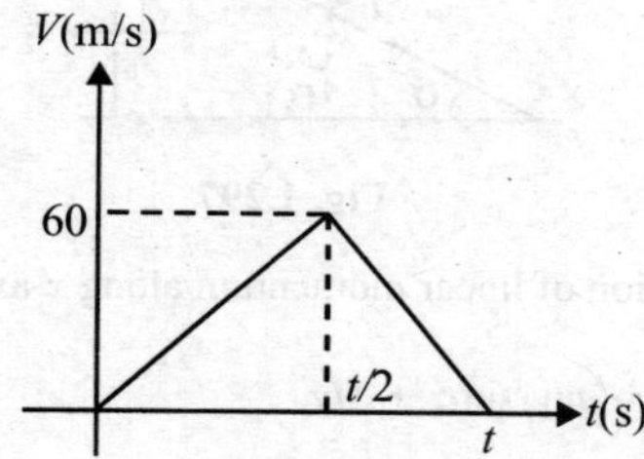

Fig. 1.291

Considering the velocity–time graph

$$60\frac{t}{2} = 300 \quad \Rightarrow \quad t = 10 \text{ s}$$

13. u time (AB) = eu (time BC) + eu (time CA)

$$u\,\frac{v}{g} = eu\left(\frac{v}{g}\right) + eu\times 2\frac{ev}{g} \Rightarrow e = \frac{1}{2}$$

$n = 1$

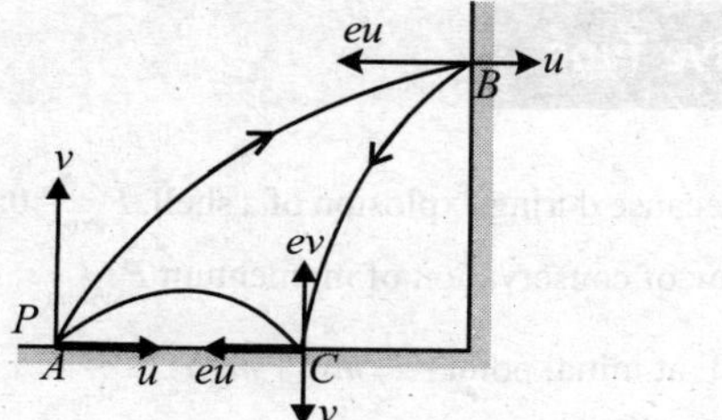

Fig. 1.292

14. $v_2 = \sqrt{5g\ell} = \sqrt{5\times 10\times 1.5} = 5\sqrt{3}$ m/s

For A: $J\sin\theta = mv_2 = 5\sqrt{3}$ m

$$\Rightarrow J\times\frac{1}{2} = 5\sqrt{3}\text{ m} \Rightarrow J = 10\sqrt{3}\text{ m}$$

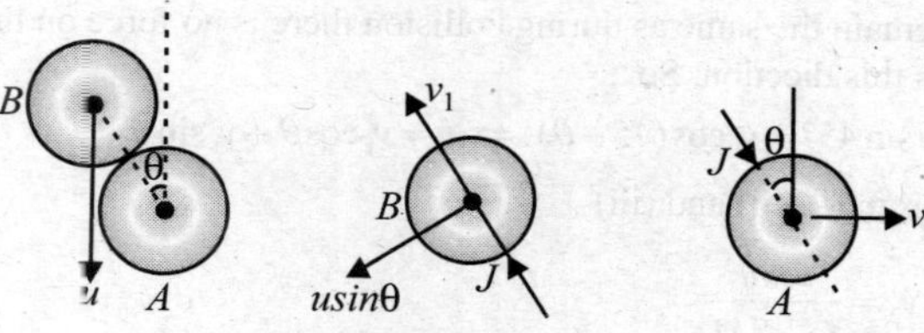

Fig. 1.293

For B: $J = m(v_1 + u\cos\theta)$

$$\Rightarrow 10\sqrt{3}m = m\left(v_1 + \frac{u\sqrt{3}}{2}\right)$$

$$\Rightarrow \quad 20\sqrt{3} = 2v_1 + u\sqrt{3} \quad \text{(i)}$$

By Newton's experimental law:

$$e = 1 = \frac{v_2\sin\theta + v_1}{u\cos\theta} \Rightarrow \frac{u\sqrt{3}}{2} = \frac{v_2}{2} + v_1$$

$$\Rightarrow \quad u\sqrt{3} = 5\sqrt{3} + 2v_1 \quad \text{(ii)}$$

From (i) and (ii), $u = 12.5$ m/s = 1250 cm/s

15. Just before collision, speed of any ball, $v = \sqrt{2gh}$

Just after collision m_1 moves up with v while m_2 moves down with v.

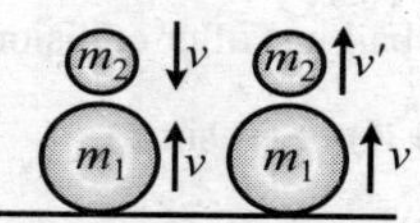

Fig. 1.294

After their collision, $e = \dfrac{v'-v}{v-(-v)} = 1 \Rightarrow v' = 3v$

$$0^2 = (3v)^2 - 2gh'$$

$$\Rightarrow h' = \frac{9v^2}{2g} \Rightarrow \quad H = d + \frac{9v^2}{2g} = d + 9h$$

16. Let after collision, the ball has velocity v_1 and wedge v_2 as shown.

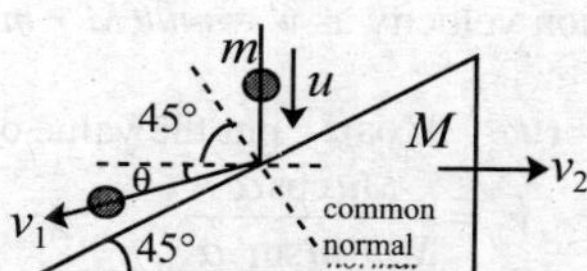

Fig. 1.295

Apply conservation of momentum in horizontal direction:

$$mv_1 \cos\theta = Mv_2 \quad \text{(i)}$$

From Newton's experimental law:

$$e = 1 = \frac{v_2 \sin 45° - [-v_1 \cos(45+\theta)]}{u \cos 45°}$$

$$\Rightarrow u = v_2 + v_1 \cos\theta - v_1 \sin\theta \quad \text{(ii)}$$

Velocity component of the ball along the surface of wedge will remain the same as during collision there is no force on the ball in this diection. So

$$u \sin 45° = v_1 \cos(45-\theta) \Rightarrow u = v_1 \cos\theta + v_1 \sin\theta \quad \text{(iii)}$$

From (i), (ii) and (iii)

$$v_2 = \frac{2mu}{m+2M}$$

To find maximum compression in the spring, after collision apply conservation of energy for the system of wedge and spring:

$$\frac{1}{2}Mv_2^2 = \frac{1}{2}Kx^2 \Rightarrow x = \sqrt{\frac{M}{K}}\, v_2 \Rightarrow x = \left(\frac{2mu}{m+2M}\right)\sqrt{\frac{M}{K}}$$

17. i. Let after collision velocity of wedge be V and that of ball be v_0 as shown. v_0 is w.r.t. wedge. Apply conservation of linear momentum along x-axis, we get

$$mu = Mv + m(V + v_0 \cos\alpha) \quad \text{(i)}$$

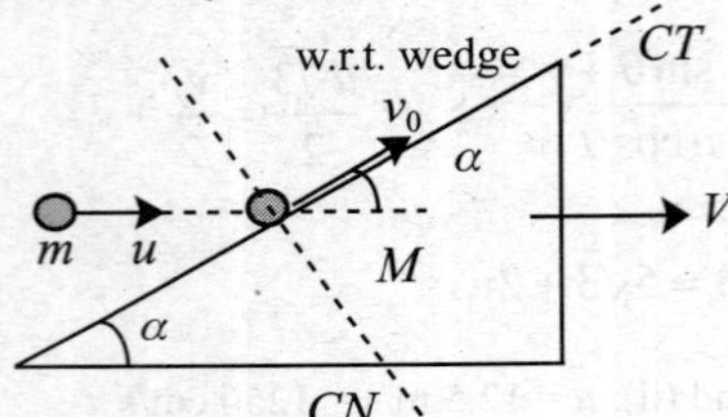

Fig. 1.296

Velocity along CT remains, constant

$$u\cos\alpha = v_0 + V\cos\alpha \quad \text{(ii)}$$

Net velocity of ball just after collision is

$$\vec{v}_b = (v_0 \cos\alpha + V)\hat{i} + (v_0 \sin\alpha)\hat{j}$$

From (i) and (ii),

$$mu = MV + mV + m\cos\alpha(u\cos\alpha - V\cos\alpha)$$

$$= (M+m)V + mu\cos^2\alpha - mV\cos^2\alpha$$

$$\Rightarrow V = \frac{mu\sin^2\alpha}{M + m\sin^2\alpha}$$

ii. When the ball ascends maximum height, then ball will be at rest w.r.t. wedge and their velocity will be same. Let their common velocity be v'. From conservation of momentum we can find this common velocity as $v' = mu/(M+m)$.

From (ii), $v_0 = (u-V)\cos\alpha$, put the value of V in this to get

$$v_0 = \frac{Mu\cos\alpha}{M + m\sin^2\alpha}$$

Also, $$v_0\cos\alpha + V = \left[\frac{M\cos^2\alpha + m\sin^2\alpha}{M + m\sin^2\alpha}\right]u$$

To find maximum height:

From energy conservation:

$$mgh + \frac{1}{2}(m+M)v'^2 = \frac{1}{2}MV^2 + \frac{1}{2}m(v_0\cos\alpha + V)^2 + \frac{1}{2}m(v_0\sin\alpha)^2$$

$$2mgh + \frac{m^2u^2}{M+m} = MV^2 + m(v_0\cos\alpha + V)^2 + m(v_0\sin\alpha)^2$$

$$2mgh + \frac{m^2u^2}{M+m}$$

$$= \frac{Mm^2u^2\sin^4\alpha + m(M\cos^2\alpha + m\sin^2\alpha)^2u^2 + mM^2u^2\cos^2\alpha\sin^2\alpha}{(M + m\sin^2\alpha)^2}$$

Solve to get: $$h = \frac{u^2}{2g}\left[\frac{M^2\cos^2\alpha}{(M+m)(M+m\sin^2\alpha)}\right]$$

18. Let velocity of wedge after collision be v_2 and that of ball be v_1. We can show that $v_1 < u$, because otherwise final KE would be greater than initial KE which is not possible. v_1 will be perpendicular to wedge surface.

From Newton's experimental law:

$$e = \frac{v_2\sin\alpha - (-v_1)}{u} \Rightarrow v_2 = \frac{eu - v_1}{\sin\alpha} \quad \text{(i)}$$

Fig. 1.297

Conservation of linear momentum along x-axis:

$$mu\sin\alpha = -mv_1\sin\alpha + Mv_2 \quad \text{(ii)}$$

From (i) and (ii), $$\frac{v_1}{u} = \frac{Me - m\sin^2\alpha}{M + m\sin^2\alpha}$$

Put the value of v_1 in Eq. (i) to get v_2

$$v_2 = \frac{1}{\sin\alpha}\left[eu - \frac{(Me - m\sin^2\alpha)u}{M + m\sin^2\alpha}\right] = \frac{mu\sin\alpha(e+1)}{M + m\sin^2\alpha}$$

Objective Type

1. b. Because during explosion of a shell, $F_{ext} = 0$; hence according to law of conservation of momentum $\vec{P}_{system}$ = constant.

2. b. TE at initial point $= \frac{1}{2}mv_0^2 + mgh$

After collision, remaining energy

$$= \frac{\frac{1}{2}mv_0^2 + mgh}{2}$$

Hence, $\dfrac{\frac{1}{2}mv_0^2 + mgh}{2} = mgh$

(since after collision, it rebounds to same height h)

$$\frac{1}{2}mv_0^2 + mgh = 2mgh$$

$$v_0 = \sqrt{2gh} = \sqrt{2 \times 9.8 \times 10} = 14 \text{ m/s}$$

The ball would rise to the same height after collision if the initial velocity was directed upwards instead of downwards. Since while returning to the same point, the body would possess the same velocity in the downward direction, hence $h = 10$ m.

3. b. Suppose the velocity of the bullet of mass m is v and it strikes the block of mass M. After collision, the linear velocity of the block is V and that of the bullet is v'.

Applying law of conservation of linear momentum, we get

$$mv = MV + mv'$$

or $500 \times 0.01 = 2V + 0.01v'$

or $5 = 2V + 0.01v'$ (i)

Now, the kinetic energy gained by the block $\frac{1}{2}MV^2$ raises it to a height h where it gains gravitational potential energy Mgh. By conservation of energy, we get

$$\frac{1}{2}MV^2 = Mgh$$

or $V = \sqrt{2gh} = \sqrt{2 \times 9.8 \times 0.1} = 1.4$ m/s

Putting the value of V in Eq. (i), we get

$$5 = 2 \times 1.4 + 0.01v' \text{ or } v' = 220 \text{ m/s}$$

4. a.

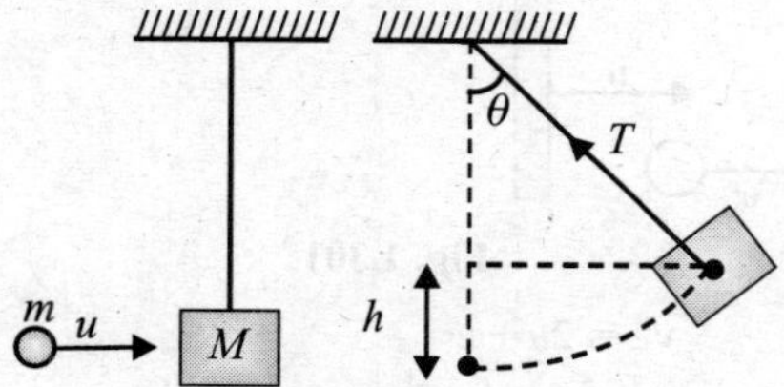

Fig. 1.298

Applying the law of conservation of linear momentum,

$$mu = (M+m)V$$

or $V = \dfrac{mu}{(M+m)} = \dfrac{0.1 \times 150}{(0.1 + 2.9)} = 5$ m/s

If V_1 is the speed of the combined mass at the instant when the string makes an angle 60° with the vertical, then

$$T = (M+m)g\cos 60° + \frac{(M+m)V_1^2}{l}$$

$$= 3 \times 9.8 \times \frac{1}{2} + \left(\frac{3V_1^2}{0.5}\right) = 14.7 + 6V_1^2 \quad \text{(i)}$$

Applying the law of conservation of mechanical energy, we have

$$(\text{KE})_{\text{lowest point}} = (\text{KE} + \text{PE})_{\text{at any other point}}$$

or $\frac{1}{2}(m+M)V^2 = \frac{1}{2}(m+M)V_1^2 + (m+M)gh$

or $\frac{1}{2}(m+M)V^2 = \frac{1}{2}(m+M)V_1^2 + (m+M)\,gl\,(1 - \cos\theta)$

$$\frac{1}{2} \times 3 \times 25 = \frac{3}{2}V_1^2 + 3 \times 9.8 \times 0.5\left(1 - \frac{1}{2}\right) \quad \text{(ii)}$$

Solving for V_1^2, we get $V_1^2 = 20.1\ (\text{m/s})^2$

Putting this value in Eq. (i), we get

$$T = 14.7 + 6 \times 20.1 = 135.3 \text{ N}$$

5. a. Because the collision is perfectly inelastic, hence the two blocks stick together. By conservation of liner momentum,

$$2mV = mv \text{ or } V = \frac{v}{2}$$

By conservation of energy,

$$2mgh = \frac{1}{2}2\,mV^2 \text{ or } h = v^2/8g$$

6. b. If the collision is elastic, whole of the kinetic energy of the ball is transferred to the bob. Hence,

$$\frac{1}{2}mv^2 = mgh \qquad \text{or } h = \frac{v^2}{2g}$$

7. d.

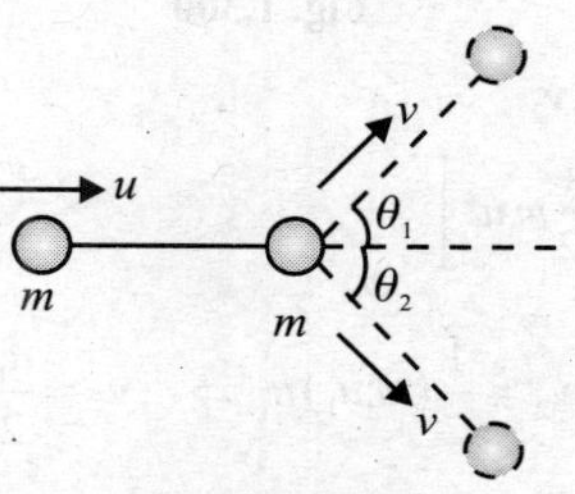

Fig. 1.299

Applying law of conservation of momentum, we get

$$mv\sin\theta_1 - mv\sin\theta_2 = 0$$

$\Rightarrow$ $\theta_1 = \theta_2$ (i)

Also,

$$mu = mv(\cos\theta_1 + \cos\theta_2)$$
$$= 2mv\cos\theta\,[\theta_1 = \theta_2 = \theta \text{ (say)}]$$

$$\cos\theta = \frac{u}{2v} \quad \text{(ii)}$$

According to law of conservation of KE,

$$\frac{1}{2}mu^2 = \frac{1}{2}mv^2 + \frac{1}{2}mv^2$$

$$u^2 = 2v^2 \text{ or } u = \sqrt{2}\,v \quad \text{(iii)}$$

From Eqs. (ii) and (iii), we have $\cos\theta = \dfrac{\sqrt{2}v}{2v}$

or $\cos\theta = \dfrac{1}{\sqrt{2}}$ or $\theta = 45°$

$\therefore$ $\theta_1 + \theta_2 = 90°$

8. a., 9. d., 10. c.

As the collision is head-on,

$$mu + 0 = mv_1 + mv_2 \text{ or } u = v_1 + v_2 \quad \text{(i)}$$

Further, coefficient of restitution is

$$e = \frac{v_2 - v_1}{u_1 - u_2} = \frac{v_2 - v_1}{u} \qquad (\because u_2 = 0)$$

or $eu = v_2 - v_1$ (ii)

Adding Eqs. (i) and (ii), we get

$$v_1 = \frac{u(1-e)}{2},\ v_2 = \frac{u(e+1)}{2} \quad \text{(iii)}$$

Now $\dfrac{v_1}{v_2} = \dfrac{1-e}{1+e}$ (iv)

$$\frac{v_1}{u} = \frac{1-e}{2} \quad \text{(v)}$$

$$\frac{v_2}{u} = \frac{1+e}{2} \quad \text{(vi)}$$

11. d.

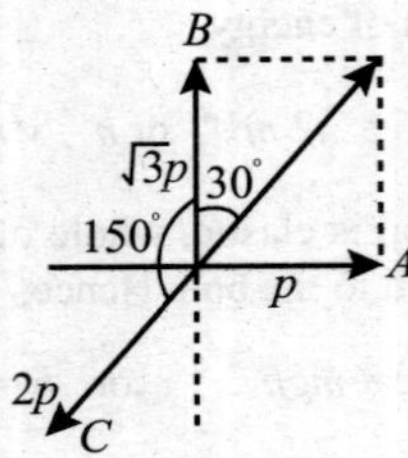

Fig. 1.300

12. a. $m_1u_1 = m_2v_2$

$$\frac{1}{2}m_2v_2^2 = \frac{1}{2}\left[\frac{1}{2}m_1u_1^2\right]$$

$$\Rightarrow \quad (m_2v_2)v_2 = \frac{1}{2}(m_1u_1)u_1 \Rightarrow \quad v_2 = \frac{u_1}{2}$$

$$e = \frac{v_2 - v_1}{u_1 - u_2} = \frac{v_2 - 0}{u_1 - 0} = \frac{v_2}{u_1} \quad \Rightarrow \quad e = \frac{1}{2}$$

13. b. The time elapsed from the moment it is dropped to the second impact with the floor is

$$t = \sqrt{\frac{2h}{g}}\,(1+2e)$$

Align where h is the initial height of the body from the ground

$$1.03 = \sqrt{\frac{2}{9.8}}\,(1+2e)$$

Solving, we get $e = 0.64$.

14. b. Apply conservation of momentum after piercing M_1:

$mu = M_1v + mu_1$ (i)

After it enters into M_2, finally bullet also moves with velocity v, so

$mu_1 = (M_2 + m)v$ (ii)

From (i) and (ii), we get

$mu = (M_1 + M_2 + m)v$ (iii)

Required percentage loss $= \dfrac{u - u_1}{u}\times 100 = \dfrac{m(u-u_1)}{mu}\times 100$

$$= \frac{M_1v}{(M_1+M_2+m)v}\times 100 = \frac{1}{1+2.98+0.02}\times 100 = 25\%$$

15. c. Both the particles will collide at the highest point of their path. Momentum of the particle P just before collision.

$$m_AV_A\cos 45° = 20\times 10^{-3}\times 49\times\frac{1}{\sqrt{2}}$$

$$= \frac{980}{\sqrt{2}}\times 10^{-3}\ \text{kg m/s}$$

Momentum of particle Q just before collision is

$$m_BV_B\cos 135° = -40\times 10^{-3}\times 49\times\frac{1}{\sqrt{2}} = \frac{1960}{\sqrt{2}}\times 10^{-3}\ \text{kg m/s}$$

After collision, let velocity of Q be V and velocity of P be 49 cos 45°. Therefore, momentum after collision is

$$40\times 10^{-3}\,V - \frac{49\times 20\times 10^{-3}}{\sqrt{2}}$$

Applying law of conservation of momentum,

$$\frac{980}{\sqrt{2}}\times 10^{-3} - 1960\times\frac{10^{-3}}{\sqrt{2}} = 40\times 10^{-3}\,V - \frac{980\times 10^{-3}}{\sqrt{2}}$$

or $V = 0$

Thus, particle Q will fall freely after collision. So, the distance travelled by it will be 245/2 m, i.e., 122.5 m.

16. b. $m_1u_1 + m_2u_2 = (m_1 + m_2)V$

$2\times 6 + 2\times 0 = (2+2)V$ or $V = 3$ m/s

Heat evolved = loss in KE $= \dfrac{1}{2}\times 2(6)^2 - \dfrac{1}{2}(2+2)(3)^2 = 18$ J

17. d. $e = \dfrac{v_2 - v_1}{u_1 - u_2} \Rightarrow 1 = \dfrac{-u-(-v')}{v-(-u)}$

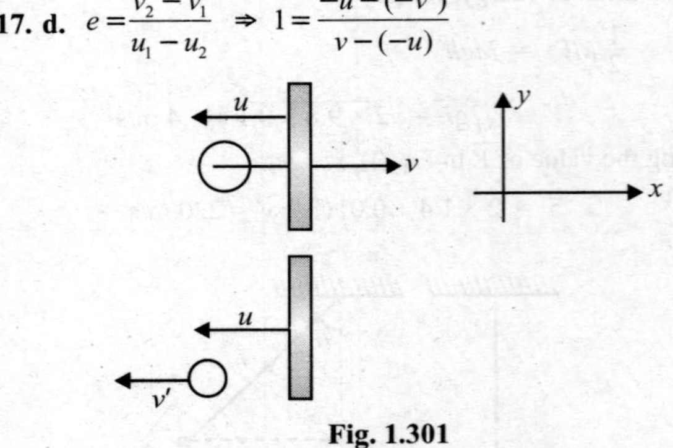

Fig. 1.301

$\Rightarrow \quad v' = 2u + v$

$\therefore \quad W = \Delta\text{KE} = \dfrac{1}{2}mv'^2\ \dfrac{1}{2}mv^2$

$$= \frac{1}{2}m[(2u+v)^2\ v^2] = 2\,mu\,(u+v)$$

18. a. After the first collision, velocity perpendicular to the inclined plane becomes $ev\cos\theta$. Time of flight,

$$t_1 = \frac{2eV}{g},\ t_2 = \frac{2e^2V}{g},\ t_3 = \frac{2e^3V}{g}$$

$$h_1 = \frac{e^2v^2\cos\theta}{2g},\ h_2 = \frac{e^4v^2\cos\theta}{2g}$$

19. c. When the two balls collide with each other, as the mass of the two balls is equal, they exchange their velocities on colliding elastically. Let the speed of the ball B when it reaches back to the initial position be v. Then

$$4\,mgh = \frac{1}{2}mv^2 + mgh \Rightarrow v = \sqrt{6gh}$$

Height reached by particle B (from highest point on the incline) is

$$H_B = \frac{v^2\sin^2 60°}{2g} = \frac{9h}{4};\ \text{total height} = h + \frac{9h}{4} = \frac{13h}{4}$$

After collision the particle A reaches the maximum height = h

$$\text{Ratio} = \frac{H_A}{H_B} = \frac{4}{13}$$

20. c. The velocity of B will be maximum when spring comes to its natural length. Let at that time their velocities are v_1 and v_2 as shown.

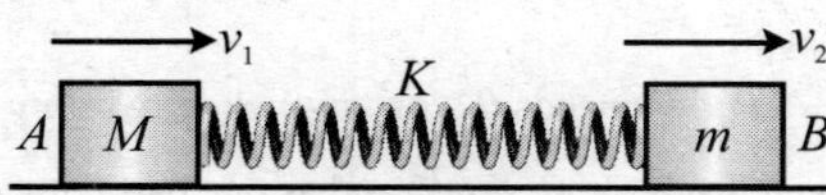

Fig. 1.302

Conservation of momentum:

$$2v_1 + 1v_2 = (2+1)\times 1 \Rightarrow 2v_1 + v_2 = 3 \quad \text{(i)}$$

Conservation of energy:

$$\frac{1}{2}2v_1^2 + \frac{1}{2}1v_2^2 = \frac{1}{2}600\left(\frac{10}{100}\right)^2 + \frac{1}{2}3(1)^2 \Rightarrow 2v_1^2 + v_2^2 = 9 \quad \text{(ii)}$$

From Eqs. (i) and (ii),

$v_1 = 2$ m/s, $v_2 = -1$ m/s or $v_1 = 0$, $v_2 = 3$ m/s

The second solution is possible.

21. b. Till then m remains on B, B will continue to accelerate due to contact force between m and B. Hence, velocity of B will be maximum when finally m leaves B.

22. a. Momentum in vertical direction

$mu\cos 60° + 3mu\cos 60° = 4\,mu\cos 60° = 2mu$

This momentum becomes zero due to impulse in string.

Hence, impulse in string = $2mu$.

Conservation of linear momentum in horizontal direction

$m(3u)\sin 60° - mu\sin 60° = 12\,mv$

$$\Rightarrow \quad v = \frac{\sqrt{3}u}{12}$$

$$\text{Loss of energy} = \frac{1}{2}m(9u^2) + \frac{1}{2}mu^2 - \frac{1}{2}\times 12\,m\times\left(\frac{\sqrt{3}u}{12}\right)^2$$

$$= \frac{39\,mu^2}{8}$$

23. c. Momentum of the system in x-direction is conserved because all the forces on the system in x-direction are balanced. But the force in y-direction is not balanced. So trajectory of motion of centre of the system is X-constant.

24. c. Initial KE $= \frac{1}{2}mu^2$

At the time of maximum compression,

$$\text{KE} = \frac{1}{2}(2m)\left(\frac{u\cos\theta}{2}\right)^2 + \frac{1}{2}mu^2\sin^2\theta$$

$$\text{Change in KE} = \frac{mu^2}{4}\cos^2\theta$$

$$\text{Change in KE} = \text{PE}$$

$$\frac{\text{PE}}{\text{KE}} = \frac{1}{2}\cos^2\theta$$

At the time of max. compression, impulse on B is along direction a and that on A is in opposite direction. But the component of velocity of A in direction perpendicular to A is unchanged.

25. c. $mv = mv_1 + nmv_2$

$$v = v_1 + nv_2$$

$$= v_2\ v_1$$

$$\Rightarrow \quad v_2 = \frac{2v}{(n+1)},\ v_1 = \left(\frac{1-n}{1+n}\right)v$$

$$\frac{\text{KE}_1}{\text{KE}_2} = \frac{\frac{1}{2}mv_1^2}{\frac{1}{2}mv^2} = \left(\frac{n-1}{n+1}\right)^2$$

26. d. As the collision is elastic, we can find

$$v_1 = \frac{1-n}{1+n}u,\ v_2 = \frac{2u}{1+n}$$

Before collision: m, u; nm, $u_2 = 0$. After collision: m, v_1; nm, v_2.

Fig. 1.303

Hence, the required fraction is

$$\frac{\frac{1}{2}nmv_2^2}{\frac{1}{2}mu^2} = n\left(\frac{v_2}{u}\right)^2 = \frac{4n}{(1+n)^2}$$

27. b.

$$\frac{\frac{1}{2}mv_1^2}{\frac{1}{2}nmv_2^2} = \frac{v_1^2}{nv_2^2} = \frac{(1-n)^2}{4n}$$

28. b. $K = \frac{P^2}{2m}$

From the graph, $4 = \frac{4^2}{2m} \Rightarrow m = 2$ kg

29. c. Since velocity of man w.r.t. trolly is greater than velocity of trolly w.r.t. earth, after the man turns back displacement of the man will decrease, so maximum displacement will be at the moment when man turns back.

$$\therefore \quad t = \frac{L}{1.5\,V}$$

$$\text{Displacement} = (v + 1.5\,v)t$$

$$= 2.5\,v\frac{L}{1.5v} = \frac{5}{3}L$$

30. a. Velocity of upper block when lower block hits an obstacle,

$$v = \sqrt{2al} = \sqrt{2\frac{F}{2M}l} = \sqrt{\frac{Fl}{M}}$$

Now after collision retardation of upper block w.r.t. earth,

$$a = \frac{\mu Mg}{M} = \mu g$$

$$\therefore \quad v^2 = u^2 + 2as$$

$$\therefore \quad 0 = \frac{Fl}{M} - 2\mu g\frac{l}{2} \Rightarrow \mu = \frac{F}{Mg}$$

31. b. In both CM and ground frame, K_{max} is there, when x is zero in spring, which occurs simultaneously.

$$v_{CM} = \frac{m(v_0) + 0}{5m} = \frac{v_0}{5}$$

$$K_{max\,CM} = \frac{1}{2}m\left(\frac{4v_0}{5}\right)^2 + \frac{1}{2}(4m)\left(\frac{v_0}{5}\right)^2 = \frac{2}{5}mv_0^2$$

$$K_{max\ ground} = \frac{1}{2} mv_0^2$$

$$K_{mim\ CM} = 0$$

$$K_{min\ ground} = \frac{1}{2}(m + 4m)\, v_{CM}^2 = \frac{mv_0^2}{10}$$

$$K_{max\ m} = \frac{1}{2} mv_0^2 \text{ (ground frame)}$$

$K_{min\ m}$=0 (ground frame when energy is shared by spring and 4*m* only and *m* will reverse direction of motion)

32. c. $$X_{CM} = \frac{3M\left(\frac{-x}{2\sqrt{2}}\right) + M \times \frac{L}{2} + M \times \frac{L}{2\sqrt{2}} + M \times 0}{6M} = 0$$

$$\Rightarrow \quad x = \left(\frac{\sqrt{2}+1}{3}\right) L$$

33. a. Suppose *m* is the mass of each piece flying off perpendicular to one another with same speed *u* (= 30 m/s). Then 3*m* is the mass of the third piece. Let *v* be the velocity of the third piece.

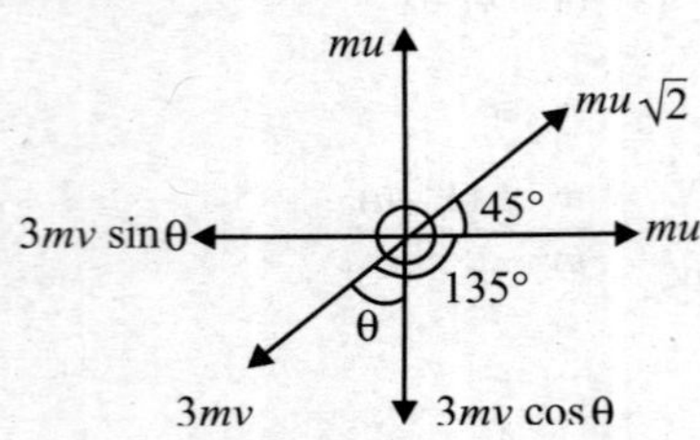

Fig. 1.304

According to Fig. 1.304,

$3\ mv \cos\theta = mu$, $3\ mv \sin\theta = mu$

$$\tan\theta = 1 \text{ or } \theta = 45°$$

Hence, $3\ mv \cos 45° = mu$

or $\frac{3v}{\sqrt{2}} = u = 30$ m/s or $v = 10\sqrt{2}$ m/s

(inclined at 135° w.r.t. direction of each one)

34. a. $P_e = 3.2\times10^{-23}$ kg/s

$P_n = 6.4 \times 10^{-23}$ kg/s

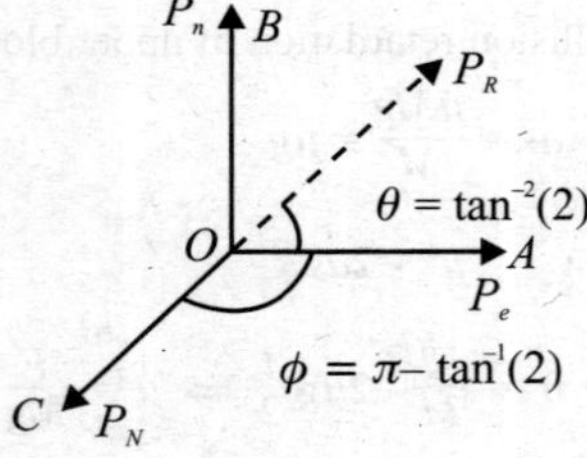

Fig. 1.305

$$\tan\theta = \frac{P_n}{P_e} = \frac{6.4\times10^{-23}}{3.2\times10^{-23}} = 2$$

According to law of conservation of momentum, the residual nucleus must move in a direction just opposite to that of $\vec{P}_R$, as shown in the figure. Hence, direction of the recoiling nucleus with that of the electron motion $\phi = \pi - \theta = \pi - \tan^{-1}(2)$.

35. a. In case of projectile motion as at the highest point

$$(v)_{vertical} = 0 \text{ and } (v)_{horizontal} = v\cos\theta$$

The initial linear momentum of the system will be $mv\cos\theta$. Now as force of blasting is internal and force of gravity is vertical, so linear momentum of the system along horizontal is conserved, i.e.,

$$P_1 + P_2 = mv\cos\theta \quad \text{or} \quad m_1v_1 + m_2v_2 = mv\cos\theta$$

But it is given that $m_1 = m_2 = m/2$ and as one part retraces its path,

$$v_1 = -v\cos\theta$$

$$\frac{1}{2} m(-v\cos\theta) + \frac{1}{2} mv_2 = mv\cos\theta$$

Solving, we get $v_2 = 3v\cos\theta$.

36. d. Initial momentum of the system is zero. So, final momentum should also be zero. Finally, *A* comes to rest after collision so to have final momentum zero, *B* and *C* should move in opposite directions.

37. c. As in a perfectly inelastic collision, two bodies stick together after the collision and move with same velocity, so their relative velocity after the impact is zero.

38. b. If the velocities acquired due to explosions are in vertical direction, then options (c) and (d) are possible. If one of them follows parabolic path (or acquires velocity in horizontal direction), then other also has to follow parabolic path (or acquire velocity in horizontal direction to keep the momentum zero in horizontal direction). Hence, option (b) can never be possible.

39. b. Applying the law of conservation of momentum,

$$400 \times 20 + 4000 \times 2.5 = -400 \times 5 + 4000v_2$$

$$\Rightarrow 8000 + 10000 + 2000 = 4000v_2$$

$$\Rightarrow \quad v_2 = 5 \text{ m/s} = 5 \times \frac{18}{5} \text{ km/h} = 18 \text{ km/h}$$

40. d. By applying conservation of momentum, first find velocity of the second particle and then find its KE.

41. c. Applying the law of conservation of momentum in perpendicular to the initial line of motion,

$$0 = mv_1 \sin 30 - mv_2 \sin 30° \text{ or } v_1 = v_2$$

Hence, speed of both will be same after collision.

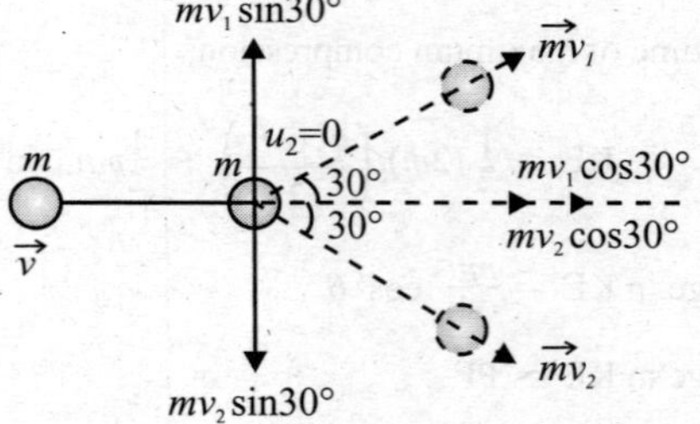

Fig. 1.306

Now along the line of motion,

$$mv = mv_1 \cos 30° + mv_2 \cos 30° \quad \text{(ii)}$$

Putting Eq. (i) in Eq. (ii), $mv = 2mv_1 \cos 30°$

or $v_1 = \frac{v}{\sqrt{3}} = \frac{6}{\sqrt{3}} = 2\sqrt{3}$ m/s

42. c. The downward component of velocity of other part will be 30 m/s. Now applying conservation of momentum along horizontal direction,

$$m \times 20 = \left(\frac{m}{2}\right) v_x \text{ or } v_x = 40 \text{ m/s}$$

So the net velocity of the other part $= \sqrt{30^2 + 40^2} = 50$ m/s.

43. b. Since the body explodes into three equal parts, therefore

$$m_1 = m_2 = m_3 = \frac{m}{3} = 1 \text{ kg}$$

Let the velocity of the third part be $\vec{v}$. According to the principle of conservation of linear momentum,

Momentum of system before explosion = Momentum of system after explosion

or $mv = m_1v_1 + m_2v_2 + m_3v_3$

or $3 \times 0 = 1 \times 2\hat{i} + 3\hat{j} + 1 \times \vec{v}$

or $v = -(2\hat{i} + 3\hat{j})$ m/s

Average force acting on the third particle is

$$\vec{F} = \frac{m\vec{v}}{t} = \frac{-1 \times (2\,\hat{i} + 3\,\hat{j})}{10^{-5}} = -(2\,\hat{i} + 3\,\hat{j}) \times 10^5 \text{ N}$$

44. d. In case I there is no horizontal external impulse on ball and mass system while in case II friction is impulsive as normal with ground is impulsive. Hence, both statements are incorrect.

45. c. $mv = (m + m)v_1$ or $v_1 = \frac{v}{2}$

$$T = \frac{2mv_1^2}{l} + 2\,mg = \frac{2mv^2}{4l} + 2\,mg$$

$$= \frac{m(2gl)}{2l} + 2\,mg = 3mg$$

Initial tension $= mg$, increase in tension $= 2mg$

46. a. In this problem, the velocity of the earth before and after the collision may be assumed zero. Hence, coefficient of restitution will be

$$e^n = \frac{v_1}{v_0} \times \frac{v_2}{v_1} \times \frac{v_3}{v_2} \times \cdots \times \frac{v_n}{v_n - 1}$$

where v_n is the velocity after n^{th} rebounding and v_0 is the velocity with which the ball strikes the earth for the first time. Hence,

$$e^n = \frac{v_n}{v_0} = \frac{\sqrt{2gh_n}}{\sqrt{2gh_0}}$$

where h_n is the height to which the ball rises after n^{th} rebounding. Hence,

$$e^n = \frac{v_n}{v_0} = \frac{\sqrt{h_n}}{\sqrt{h_0}}$$

47. a. $u_1 = \frac{p}{m}, u_2 = 0, v_1 = \frac{(p - J)}{m}, v_2 = \frac{J}{m}$

Now apply $e = \frac{v_2 - v_1}{u_1 - u_2}$

48. c. New momentum should be the resultant of individual moment. Therefore,

$$(2mV)^2 = (mv)^2 + (mv)^2$$

or $V = \frac{v}{\sqrt{2}}$

49. c. $(m_1 + m_2)v = m_1u_1 + m_2u_2$

or $v = \frac{2}{3}$ m/s

Here $u_2 = 0$.

$\therefore$ Energy loss $= \frac{1}{2}(0.5) \times (2)^2 - \frac{1}{2}(1.5) \times \left(\frac{2}{3}\right)^2 = 0.67$ J

50. b. Applying the law of conservation of momentum,

$$m_1v_1 = m_1\frac{v_1}{3} + mv \text{ or } v = \frac{2m_1v_1}{3m}$$

To describe a vertical circle v should be $\sqrt{5gl}$. So,

$$\frac{2m_1v_1}{3m} = \sqrt{5gl} \text{ or } v_1 = \left(\frac{m}{m_1}\right)\frac{3}{2}\sqrt{5gl}$$

51. a. Applying the law of conservation of momentum,

$$m_1v_1 = (m_1 + m_2)V \quad \text{(i)}$$

where $v_1 = \sqrt{2gd}$ is the velocity with which m_1 collides with m_2. Therefore,

$$V = \frac{m_1}{(m_1 + m_2)}\sqrt{2gd}$$

Now, let the centre of mass rise through a height h after collision. In this case, the kinetic energy of $m_1 + m_2$ system is converted into potential energy at maximum height h.

$$\Rightarrow \frac{1}{2}(m_1 + m_2)V^2 = (m_1 + m_2)gh$$

$$\Rightarrow \frac{1}{2}(m_1 + m_2)\left\{\frac{m_1}{m_1 + m_2}\right\}^2 2\,gd = (m_1 + m_2)\,gh$$

$$\Rightarrow \quad h = d\left\{\frac{m_1}{m_1 + m_2}\right\}^2$$

52. a. The bullet and block will meet after time

$$t = \frac{h}{u_{\text{rel}}} = \frac{100}{100} = 1$$

During this time, distance travelled by the block,

$$s_1 = \frac{1}{2}gt^2 = \frac{1}{2} \times 10 \times 1^2 = 5 \text{ m}$$

Distance travelled by the bullet,

$$s_2 = 100 - s_1 = 95 \text{ m}$$

Velocity of the bullet before collision,

$$u_2 = u - gt = 100 - 10 \times 1 = 90 \text{ m/s}$$

Velocity of the block before collision,

$$u_1 = gt = 10 \text{ m/s}$$

Let V be the combined velocity after collision.

According to the law of conservation of momentum,

$$m_1u_1 + m_2u_2 = (m_1 + m_2)\,V$$

or $0.01 \times (-10) + 0.01 \times 90 = 0.02V$

(Velocity in upward direction is considered positive.)

Solving, we get $V = 40$ m/s.
Maximum height risen by the block $= \frac{V^2}{2g} = 80$ m
Height reached above the top of the tower is

$$80 - s_1 = 80 - 5 = 75 \text{ m}$$

53. b.

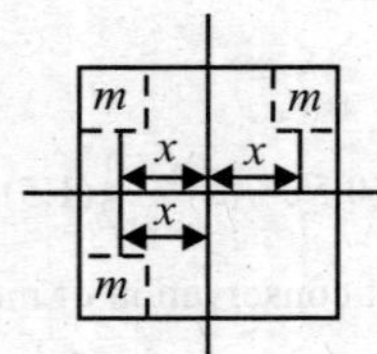

Fig. 1.307

$$x_1 = \left(\frac{mx}{M-m}\right) \Big| x_2 = \left(\frac{2mx}{M-2m}\right) \Big| x_3 = \left(\frac{mx}{M-3m}\right)$$

$$x_1 < x_3 < x_2$$

54. d. $u_x = 20\sqrt{2} \cos 45° = 20$ m/s

$u_y = 20\sqrt{2} \sin 45° = 20$ m/s

After 1 s, horizontal component remains unchanged while vertical component becomes

$$v_y = u_y - gt = 20 - 10 = 10 \text{ m/s}$$

Due to explosion, one part comes to rest. Hence, from conservation of linear momentum, vertical component of second part will become $v_{y1} = 20$ m/s. Therefore, maximum height attained by the second part will be

$$H = h_1 + h_2$$

Here, h_1 = height attained in 1 s
$= (20)(1) - \frac{1}{2} \times 10 \times 1^2 = 15$ m

and h_2 = height attained after 1 s

$$\frac{v_{y1}^2}{2g} = \frac{(20)^2}{2 \times 10} = 20 \text{ m}$$

$\therefore$ $H = 15 + 20 = 35$ m

55. b. $\hat{j}$ component, i.e., component of velocity parallel to wall remains unchanged while $\hat{i}$ component will become (–1/2) $(2\hat{i})$ or $-\hat{i}$. Therefore, velocity vector of the sphere after it hits the wall is $-\hat{i} + 2\hat{j}$.

56. d. Maximum tension in the string is in its lowest position. Speed of mass m in its lowest position is

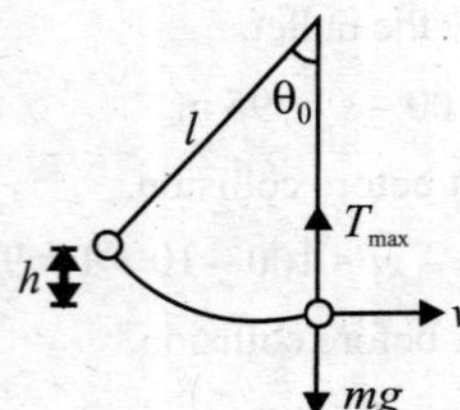

Fig. 1.308

$$v^2 = 2gh = 2gl(1 - \cos\theta_0)$$

$$T_{max} - mg = \frac{mv^2}{l}$$

$$T_{max} = mg + 2mg(1 - \cos\theta_0)$$

$$= mg(3 - 2\cos\theta_0)$$

Block of mass $4m$ does not move. So, $\mu(4mg) \geq T_{max}$

or $4\mu mg \geq mg(3 - 2\cos\theta_0)$

or $\mu \geq \left(\frac{3 - 2\cos\theta_0}{4}\right)$

57. d. $T_1 = \frac{\pi R}{u_1}$ (i)

$$\frac{v_2 - v_1}{u_1} = e \Rightarrow v_2 - v_1 = eu_1$$

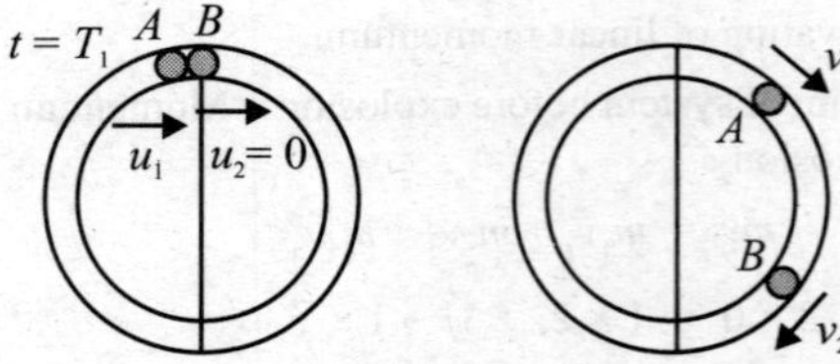

Fig. 1.309

Time taken to collide A and B again is

$$T_2 - T_1 = \frac{2\pi R}{v_2 - v_1} \Rightarrow T_2 - T_1 = \frac{2\pi R}{eu_1} \quad \text{(ii)}$$

Dividing (ii) by (i), we get $\frac{T_2}{T_1} = \frac{2+e}{e}$

58. b. Statement (i) is correct because in this case, spring would be compressing. So PE in the spring will be increasing. It means KE of the system will be decreasing as the total mechanical energy of the system is constant.

Further, momentum of the system will always be conserved. It will neither be increasing nor be decreasing.

59. c. $v \sin\phi = u \sin\theta$, $v \cos\phi = eu \cos\theta$

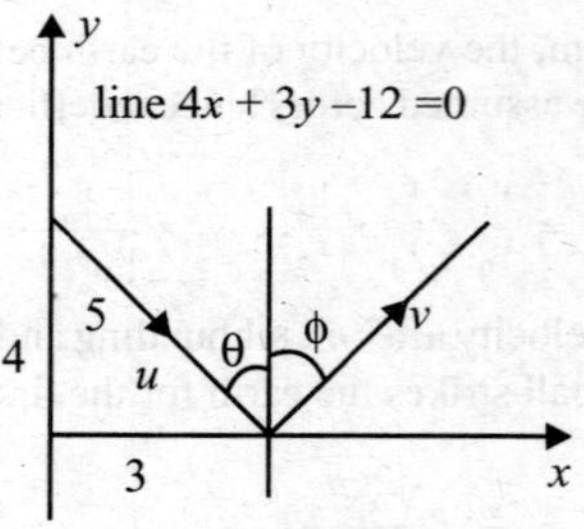

Fig. 1.310

$$\Rightarrow \quad v = u\sqrt{e^2 \cos^2\theta + \sin^2\theta}$$

$$= 10\sqrt{(0.75)^2\left(\frac{4}{5}\right)^2 + \left(\frac{3}{5}\right)^2} = 6\sqrt{2} \text{ m/s}$$

60. b. Apply conservation of momentum at highest point:

$$4mu\cos\theta = 3mv \Rightarrow v = \frac{4}{3}u\cos\theta \quad \text{(i)}$$

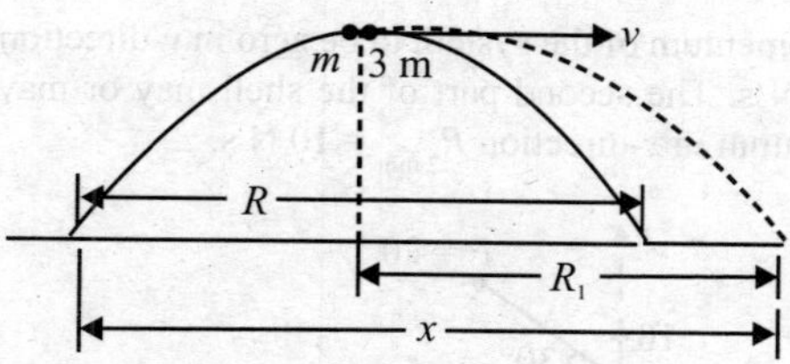

Fig. 1.311

$$R = (u\cos\theta)T$$

$$R_1 = v\frac{T}{2} = \frac{4}{3}\frac{u\cos\theta\, T}{2}$$

$$= \frac{2R}{3} \Rightarrow x = \frac{R}{2} + R_1 = \frac{R}{2} + \frac{2R}{3} = \frac{7R}{6}$$

61. b. Let x be displacement of wedge w.r.t. ground, y be displacement of man w.r.t. ground and z be displacement of block w.r.t. ground. As the length of string will remain same, so

$$(l_1 + z - x) + (l_2 + y - x) = l_1 + l_2$$

$$\Rightarrow \quad z + y = 2x \qquad \text{(i)}$$

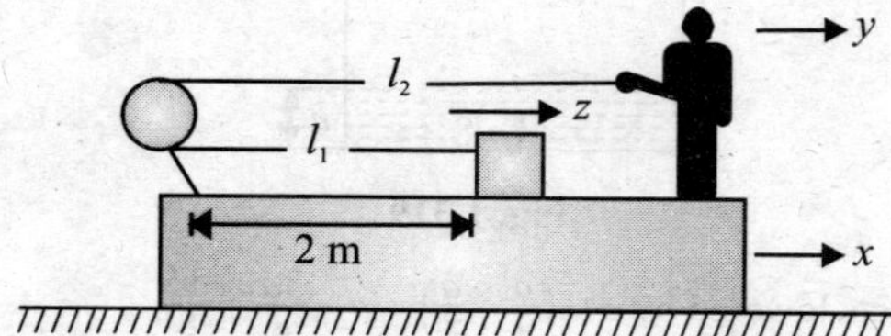

Fig. 1.312

Also, if the pulley and the block meet, then we can write

$$x = z + 2 \qquad \text{(ii)}$$

$$\Delta x_{cm} = 0 \Rightarrow \frac{Mx + 2My + Mz}{4M} = 0$$

$$\Rightarrow \quad x + 2y + z = 0 \qquad \text{(iii)}$$

From Eqs. (i), (ii) and (iii),

$$x = \frac{-1}{2}\text{ m}, y = \frac{3}{2}\text{ m}, z = \frac{-5}{2}\text{ m}$$

62. d.

m_1 O → V m_2 O ← V m_1 O → V_1 m_2 O → V_2

Fig. 1.313

$$v_1 = \left(\frac{m_1 - m_2}{m_1 + m_2}\right)u_1 + \frac{2m_2u_2}{m_1 + m_2}$$

$= -u_1 + 2u_2$ (neglecting m_1 in comparison to m_2)

$$= -v + 2(-v) = -3v$$

Hence, (a) is correct.

Change in momentum of $m_1 = P_f - P_i = mv_1 - mv$

$$= -m3v - mv = -4\,mv$$

Hence, (b) is correct.

Change in KE of $m_1 = K_f - K_i = \frac{1}{2}mv_1^2 - \frac{1}{2}mv^2$

$$= \frac{1}{2}m(3v)^2 - \frac{1}{2}mv^2 = 4mv^2$$

Hence, (c) is correct.

63. a.

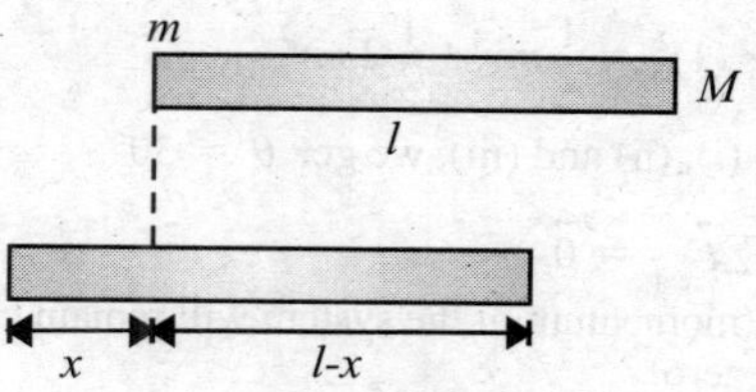

Fig. 1.314

$$m(l - x) = Mx \Rightarrow x = \frac{ml}{m + M}$$

Distance moved by insect w.r.t. ground is

$$l - x = l - \frac{ml}{m + M} = \frac{Ml}{m + M}$$

$$\text{Speed} = \frac{\text{distance}}{\text{time}} = \frac{Ml}{t(M + m)}$$

64. c. Initial acceleration of the system before the collar is removed:

$$a_1 = \frac{2mg}{3m} = \frac{2g}{3}$$

Velocity at the time when the collar is removed:

$$v_1 = \sqrt{2a_1\frac{H}{3}}$$

Acceleration after the collar is removed

$$a_2 = \frac{mg}{2m} = \frac{g}{2}$$

Final velocity: $v_2^2 = v_1^2 + 2a_2H$

$$= \frac{2a_1H}{3} + \frac{2g}{2}H = \frac{2}{3}\frac{2g}{3}H + gH$$

$$\Rightarrow \quad v_2 = \frac{\sqrt{13gH}}{3}$$

65. b. Till time t_0, friction on car will be in forward direction and on boat in backward direction. So the boat will not move till t_0. Let v be the velocity of car + boat system just after applying the brakes. From conservation of momentum,

$$(m + M)v = mv_0 \Rightarrow v = \frac{mv_0}{m + M}$$

Time taken is $t_0 + L/v = t_0 + \dfrac{L(m + M)}{mv_0}$

66. d. From conservation of momentum along incident direction

$$m \times 1 = 2\,mv_2\cos\theta \qquad \text{(i)}$$

From conservation of momentum along perpendicular direction,

$$m \times 0 + 2\,m \times 0 = mv_1 - 2\,mv_2\sin\theta$$

$$\Rightarrow \quad v_1 = 2v_2\sin\theta \qquad \text{(ii)}$$

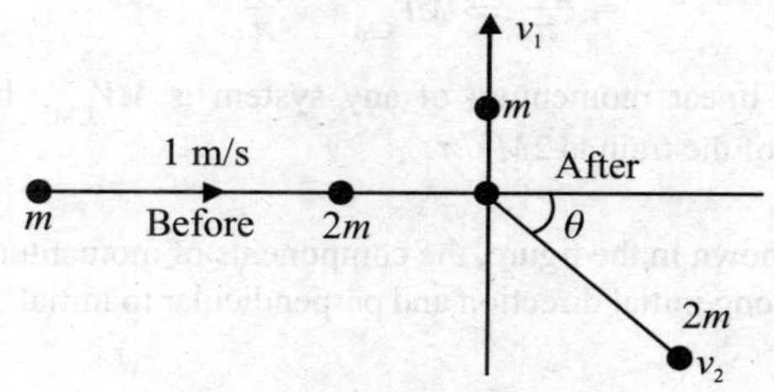

Fig. 1.315

From energy conservation,

$$\frac{1}{2}m \times (1)^2 = \frac{1}{2}mv_1^2 + \frac{1}{2}2\,mv_2^2 \quad \text{(iii)}$$

From Eqs. (i), (ii) and (iii), we get $\theta = 30°$.

67. b Since $\Sigma \vec{F}_{ext} = \vec{0}$

Therefore, momentum of the system will remain conserved and is equal to zero.

68. b. The line of impact for duration of collision is parallel to the x-axis. The situation of striker and coin just before the collision is given as

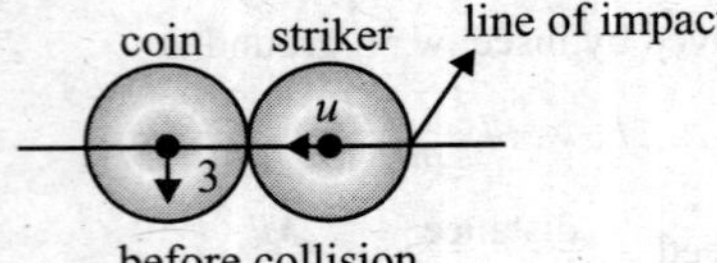

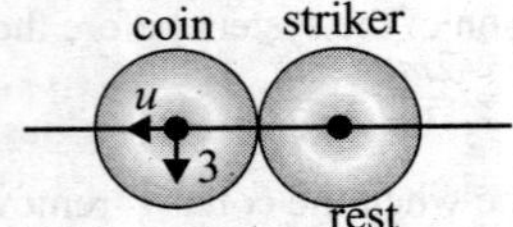

Fig. 1.316

Because masses of coin and striker are same, their components of velocities along the line of impact shall exchange. Hence, the striker comes to rest and the x- and y-components of velocities of coin are u and 3 m/s, respectively, as shown in Fig. 1.317.

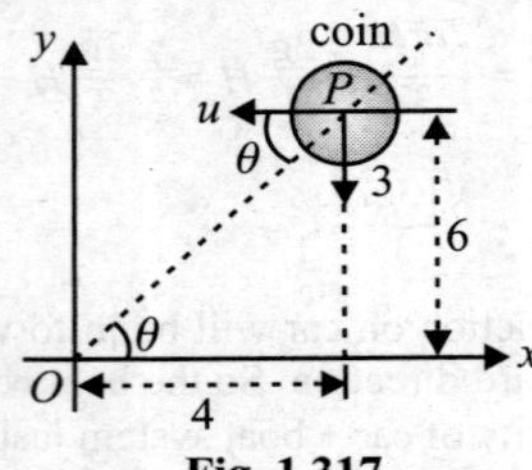

Fig. 1.317

For coin to enter hole, its velocity must be along PO.

$$\tan\theta = \frac{6}{4} = \frac{3}{u} \text{ or } u = 2 \text{ m/s}$$

69. b. If we treat the train as a ring of mass 'M', then its CM will be at a distance $2R/\pi$ from the centre of the circle. Velocity of the centre of mass is

$$V_{CM} = R_{CM}\,\omega = \frac{2R}{\pi}\frac{V}{R} \quad \left(\because \omega = \frac{V}{R}\right)$$

$$= \frac{2V}{\pi} \Rightarrow MV_{CM} = \frac{2MV}{\pi}$$

As the linear momentum of any system is MV_{CM}, the linear momentum of the train is $2MV/\pi$.

70. c. As shown in the figure, the components of momentum of one shell along initial direction and perpendicular to initial direction are

$$P_{1x} = 10\sqrt{3} \text{ N s and } P_{1y} = 10 \text{ N s}$$

For momentum of the system to be zero in y-direction, P_{2y} must be 10 N s. The second part of the shell may or may not have momentum in x-direction $P_{2\,min} = 10$ N s.

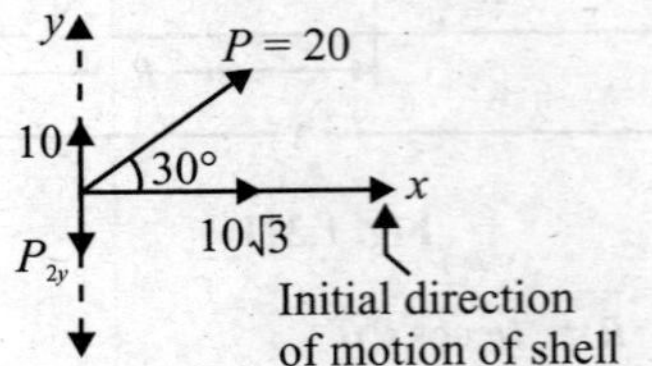

Fig. 1.318

71. c. Let h be the height of water surface. Then,

$$a^2h = a \times \frac{a}{2} \times \frac{a}{2}$$

$$h = \frac{a}{4}$$

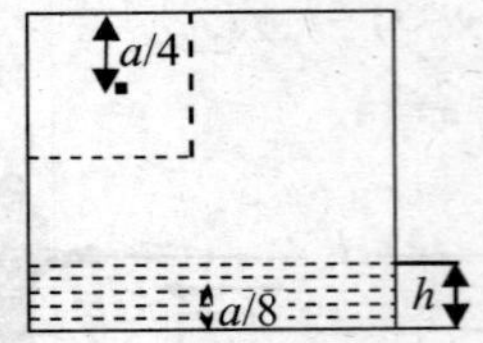

Fig. 1.319

CM gets lowered by $a - \left(\frac{a}{4} + \frac{a}{8}\right) = a - \frac{3a}{8} = \frac{5a}{8}$

Work done by gravity $= mg\frac{5a}{8}$

72. c. Neglecting gravity,

$$v = u\ln\left(\frac{m_0}{m_t}\right)$$

u = ejection velocity w.r.t. balloon

m_0 = initial mass

m_t = mass at any time t

$$v = 2\ln\left(\frac{m_0}{m_0/2}\right) = 2\ln 2 \text{ m/s}$$

73. c. Let the tube be displaced by x towards left. Then

$$mx = m(R - x) \Rightarrow x = \frac{R}{2}$$

74. c. Taking the origin at the centre of the plank,

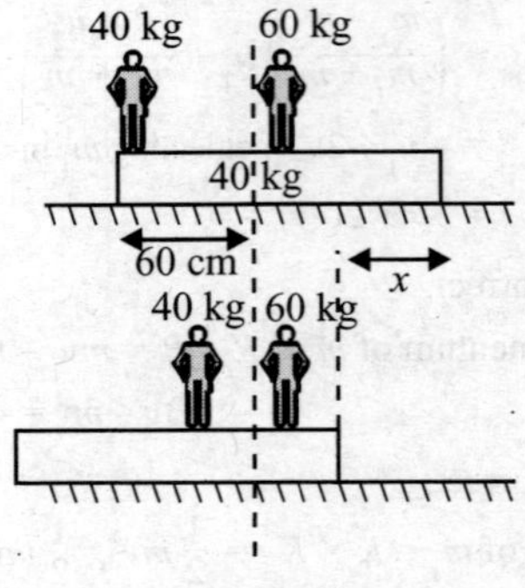

Fig. 1.320

$$m_1\Delta x_1 + m_2\Delta x_2 + m_3\Delta x_3 = 0 \quad (\because \Delta x_{CM} = 0)$$

(Assuming the centres of the two men are exactly at the axis as shown in Fig. 1.320.)

$$60 \times 0 + 40 \times 60 + 40\,(-x) = 0 \;\Rightarrow\; x = 60 \text{ cm}$$

Hence, A and B meet at the right end of the plank.

75. b. Force on table due to collision of balls:

$$F_{\text{dynamic}} = \frac{dp}{dt} = 2 \times 20 \times 20 \times 10^{-3} \times 5 \times 0.5 = 2 \text{ N}$$

Net force on one leg $= \frac{1}{4}(2 + 0.2 \times 10) = 1$ N

76. d. i. Since both have positive final velocities, hence, both moved in the same direction after collision.

ii. At $t = 2$ s, both had equal velocities.

iii. By conservation of linear momentum.

$$m_1(0.8) = m_1(0.2) + m_2(1) \quad \Rightarrow \frac{m_1}{m_2} = \frac{5}{3} \;\Rightarrow m_1 > m_2$$

77.d. By conservation of linear momentum along the string

$$mu = (m + m + 3\,m)\,v \text{ or } v = \frac{u}{5}$$

and impulse on block $A = 3\,m\,(v - 0) = \frac{3\,mu}{5}$

78. a. Let the three mutually perpendicular directions be along xy and z-axis, respectively.

$$\vec{p}_1 = mv_0\hat{i}, \qquad \vec{p}_2 = mv_0\hat{j}$$

where

$$\frac{1}{2}mv_0^2 = E_0$$

$$\vec{p}_3 = mv_0\hat{k} \text{ and } \vec{p}_4 = m\vec{v}$$

By linear momentum conservation,

$$0 = \vec{p}_1 + \vec{p}_2 + \vec{p}_3 + \vec{p}_4$$

or

$$\vec{v} = -v_0\,(\hat{i} + \hat{j} + \hat{k})$$

or

$$v = v_0\sqrt{1^2 + 1^2 + 1^2} = v_0\sqrt{3}$$

$$\text{Total energy} = 3\left(\frac{1}{2}mv_0^2\right) + \frac{1}{2}mv^2 = 3E_0 + 3E_0 = 6E_0$$

79. a. The whole of the energy of A will be transferred to B. So B will rise to the same height.

80. b. In the absence of external force, v_{CM} = constant.

Here initially $v_{\text{CM}} = 0$; so $v_{\text{CM}} = 0$ throughout, i.e., position of centre of mass, remains unshifted.

81. b. The rod will balance about centre of mass; so distance of centre of mass from the left end,

$$x_{\text{CM}} = \frac{m_1x_1 + m_2x_2}{m_1 + m_2} = \frac{100 \times 25 + 50 \times 55}{100 + 50} = 35 \text{ cm}$$

Therefore, distance from centre of rod $= 35 - 25$

$= 10$ cm to the right

82. c.

$$F_x = \frac{\Delta P_x}{\Delta t} = \frac{mv_{2x} - mv_{1x}}{\Delta t} = \frac{m(v_{2x} - v_{1x})}{\Delta t}$$

$$= \frac{(150 \times 10^{-3})[30 - (-15 \sin 30°)]}{0.01}$$

$$= \frac{(150 \times 10^{-3})(30 + 7.5)}{0.01} = 562.5 \text{ N}$$

83. a. From theory.

84. c. Time taken to fall through 45 m $= \sqrt{\frac{2 \times 45}{10}} = 3$ s

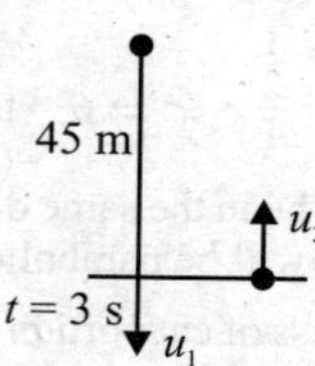

Fig. 1.321

$$u_1 = \sqrt{2gh} = \sqrt{2 \times 10 \times 45} = 30 \text{ m/s}$$

$$u_2 = eu_1 = \frac{2}{3} \times 30 = 20 \text{ m/s}$$

Distance travelled in the fourth second = distance travelled in 1 s after rebounding $= 20 \times 1 - \frac{1}{2} \times 10\,(1)^2 = 15$ m

85. a. Since collision is elastic, P will rebound with same velocity as it strikes with. And at the time of collision, velocities of P and Q will be same in magnitude.

86. d. Let combined velocity of the system be v.

$$(m + M)v = mu \Rightarrow v = \frac{mu}{m + M}$$

$$\text{KE} = \frac{1}{2}(m + M)v^2 = \frac{m^2u^2}{2(M + m)}$$

87. d. As long as no external force acts on the system of 'man + boat', its centre of mass will remain stationary.

88. c. Method-1: Acceleration of 3 kg:

$$a_2 = \frac{6}{3} = 2 \text{ m/s}^2$$

Velocity of 3 kg at $t = 5$ s: $v_2 = u_2 + a_2t = 0 + 2 \times 5 = 10$ m/s

$$v_{\text{CM}(t=5s)} = \frac{2 \times 0 + 3 \times 10}{2 + 3} = 6 \text{ m/s}$$

Method-2:

$$a_{\text{CM}} = \frac{F}{m_1 + m_2} = \frac{6}{2 + 3} = 1.2 \text{ m/s}^2$$

$$v_{\text{CM}} = u_{\text{CM}} + a_{\text{CM}}t = 0 + 1.2 \times 5 = 6 \text{ m/s}$$

89. c.

$$v_{\text{CM}} = \frac{5 \times 14}{5 + 2} = 10 \text{ m/s}$$

Velocity of the 5 kg block w.r.t. CM $= 14 - 10 = 4$ m/s

Velocity of the 2 kg block w.r.t. CM $= 0 - 10 = -10$ m/s

90. c.

$$\vec{a}_{\text{CM}} = \frac{\vec{F}}{m_1 + m_2} = \frac{16\,\hat{i} + 8\,\hat{j}}{8 + 8} = \hat{i} + \frac{1}{2}\hat{j}$$

$$|\vec{a}_{CM}| = \sqrt{1^2 + \left(\frac{1}{2}\right)^2} = \frac{\sqrt{5}}{2}\ \text{m/s}^2$$

$$\tan\theta = \frac{a_y}{a_x} = \frac{\frac{1}{2}}{1} = \frac{1}{2} \Rightarrow \theta = \tan^{-1}\left(\frac{1}{2}\right)$$

91. b. v_{CM} and a_{CM} will not be in the same direction and acceleration is constant, hence path will be parabolic.

92. c. C_1 is the centre of mass of cut portion and C_2 that of remaining portion. We have to find x_2.

$$x_1 = 14 - 10.5 = 3.5\ \text{cm}$$

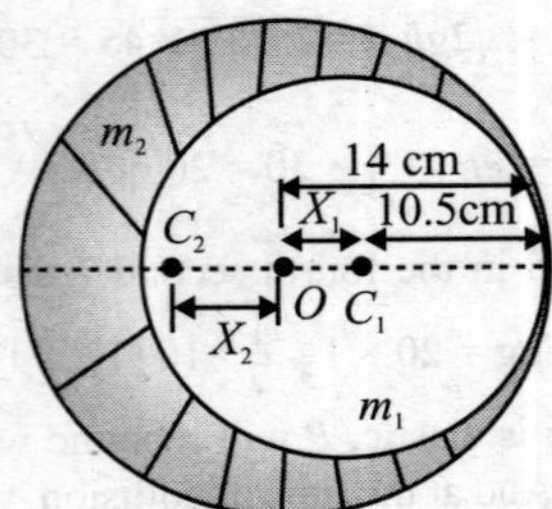

Fig. 1.322

Mass will be proportional to area. So mass of the whole disc is

$$M = k\pi\,(14)^2$$

Mass of cut portion $m_1 = k\pi\,(10.5)^2$

Mass of the remaining portion $m_2 = M - m_1 = k\pi\,(14^2 - 10.5^2)$

$$= k\pi\,(24.5) \times (3.5)$$

Now, $m_1x_1 = m_2x_2$

$$\Rightarrow \quad x_2 = \frac{m_1x_1}{m_2} = \frac{k\pi\,(10.5)^2 \times 3.5}{k\pi\,(24.5) \times 3.5} = 4.5\ \text{cm}$$

93. d. Work done,

$$W = 10 \times g \times \frac{1}{4} + 10 \times g \times \frac{1}{2} + 10 \times g \times \frac{1}{2} = 125\ \text{J}$$

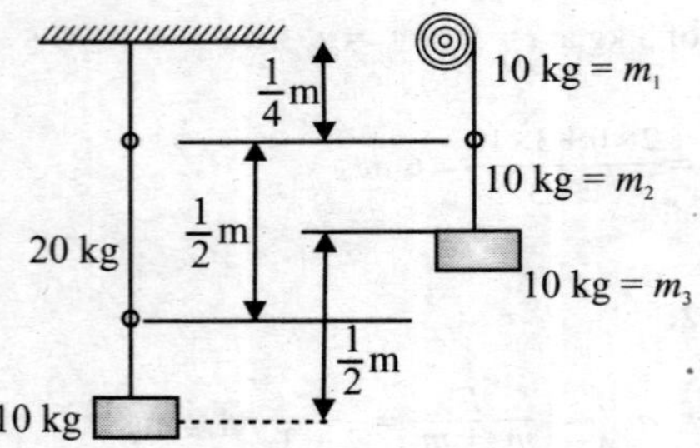

Fig. 1.323

94. a.

$$\Delta y_{CM} = \frac{m_1\,\Delta y_1 + m_2\,\Delta y_2 + m_3\,\Delta y_3}{m_1 + m_2 + m_3}$$

$$= \frac{10 \times \frac{1}{4} + 10 \times \frac{1}{2} + 10 \times \frac{1}{2}}{10 + 10 + 10} = 0.417\ \text{m}$$

95. b. Velocity of shell = 180 km/h = 50 m/s

Let V be the velocity of the cannon after firing, then from the conservation of linear momentum,

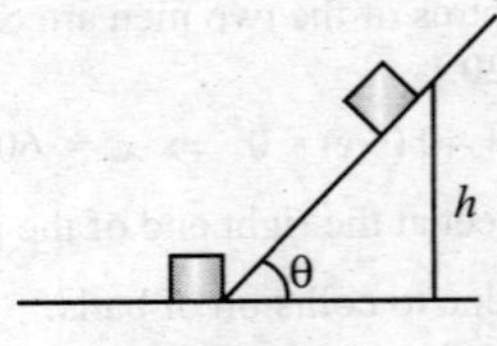

Fig. 1.324

$$0 = 50 \times 50 - 1000\,V \Rightarrow V = 2.5\ \text{m/s}$$

Let the cannon go up by height h.

Net work done by external forces = ΔKE

$$\Rightarrow -Mgh - \mu Mg\cos\theta \times \frac{h}{\sin\theta} = 0 - \frac{1}{2}MV$$

where M is the mass of cannon,

Solving, we get $h = 5/24$ m.

96. b. The momentum of the bead at A is $P_i = -mv\,\hat{i}$

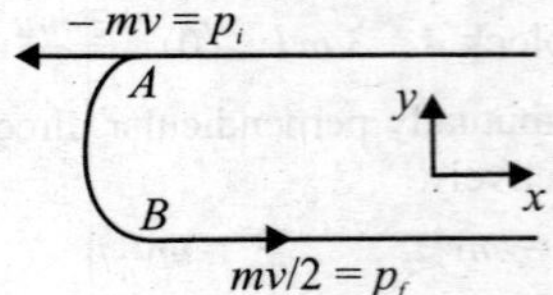

Fig. 1.325

The momentum of the bead at B is $\vec{P}_f = \left(\frac{mv}{2}\right)\hat{i}$

Therefore, the magnitude of the change in momentum between A and B is

$$\Delta P = \left|\vec{P}_f - \vec{P}_i\right| = \left(\frac{3}{2}\right)mv$$

Average force exerted by the bead on the wire is

$$F_{av} = \frac{\Delta p}{\Delta t} = \frac{3mv}{2T}$$

97. d. $\vec{P}_1 = 2 \times 3\hat{i} = 6\hat{i},\ \vec{P}_2 = 6\hat{j}$

$$\vec{I} = \vec{P}_2 - \vec{P}_1 = 6\hat{j} - 6\hat{i}$$

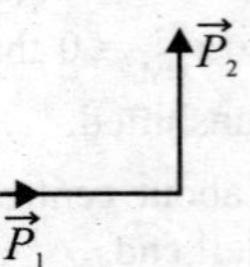

Fig. 1.326

or $\quad I = \sqrt{6^2 + 6^2} = 6\sqrt{2}$ N s

98. a. Initial momentum: $3 \times 4 + 4 \times (-3) = 0$

Final momentum will also be zero. But both stick together, hence both come at rest.

99. c. At the time of maximum compression, the speeds of blocks will be the same. Let that speed be v and maximum compression be x.

Applying conservation of momentum,

$$(m_1 + m_2)\,v = m_1v_1 + m_2v_2$$

$$\Rightarrow \quad v = 4\ \text{m/s}$$

Applying conservation of mechanical energy,

$$\frac{1}{2}kx^2 + \frac{1}{2}(m_1 + m_2)v^2 = \frac{1}{2}m_1v_1^2 + \frac{1}{2}m_2v_2^2$$

Solving , we get $x = 0.02$ m.

100. a. Motion of centre of mass is exactly similar to the motion of a body had it not exploded.

$$u_x = u\cos\theta = \frac{10}{\sqrt{2}} \text{ m/s}, \; u_y = u\sin\theta = \frac{10}{\sqrt{2}} \text{ m/s}$$

$$v_x = u_x = \frac{10}{\sqrt{2}} \text{ m/s}$$

(since there is no change in the horizontal velocity)

$$v_y^2 - u_y^2 = 2(-g)(h)$$

$$\Rightarrow \quad v_y^2 = \frac{100}{2} - 2 \times 10 \times 1 = 30$$

Therefore, net velocity of CM $= \sqrt{v_x^2 + v_y^2}$

$$= \sqrt{\frac{100}{2} + 30} = \sqrt{80} = 4\sqrt{5} \text{ m/s}$$

101. b.

u → • ← I

← • $2u$

Fig. 1.327

$$I = m\,[2u - (-u)] = 3mu$$

$$W = \frac{1}{2}m\,[(2u)^2 - u^2] = \frac{3mu^2}{2} = \frac{Iu}{2}$$

102. a. Velocity of m when string gets tight, $v_1 = \sqrt{2g\,2l} = 2\sqrt{gl}$

Combined velocity, $v_2 = \dfrac{mv_1}{3m}$

$$= \frac{v_1}{3} = \frac{2}{3}\sqrt{gl}$$

103. d.

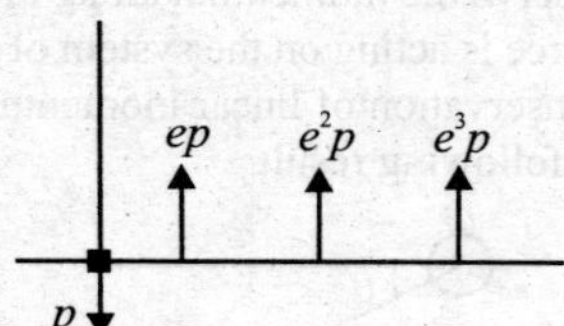

Fig. 1.328

$$\Delta p = (p + ep) + (ep + e^2p) + (e^2p + e^3p) + \cdots$$

$$= p(1 + e)\,[1 + e + e^2 + \cdots] = \frac{p(1 + e)}{1 - e}$$

104. a. $\dfrac{h_1}{h_0} = e^2 \Rightarrow \dfrac{80}{100} = e^2 \Rightarrow e = \sqrt{0.8}$

Velocity before the first impact $u = \sqrt{2gh}$

Velocity after the third impact $v_3 = e^3\sqrt{2gh}$

$$\text{KE} = \frac{1}{2}mv_3^2 = e^6\,mgh = (0.8)^3\,mgh$$

105. b. $v = 16$ m/s

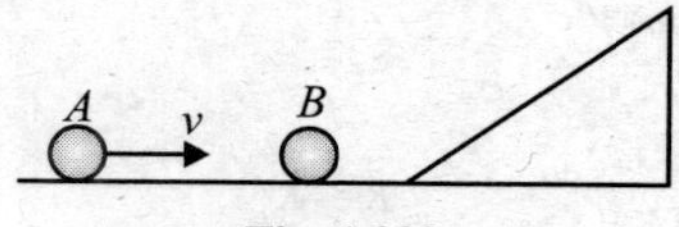

Fig. 1.329

Let velocity of B after collision be v_2 and that of A be v_1. Then

$$v_2 = \sqrt{2gs} = \sqrt{2 \times 10 \times 5} = 10 \text{ m/s}$$

$$mv_1 + mv_2 = mv$$

$$v_1 + 10 = 16 \Rightarrow v_1 = 6 \text{ m/s}$$

$$e = \frac{v_2 - v_1}{v} = \frac{10 - 6}{16} = \frac{1}{4}$$

106. a.

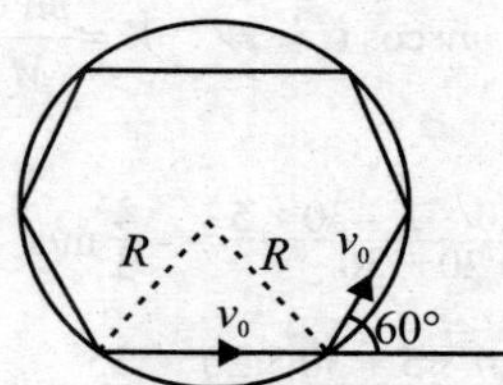

Fig. 1.330

I = change in momentum

$$= \sqrt{(mv_0)^2 + (mv_0)^2 - 2(mv_0)^2\cos 60°}$$

$$= mv_0\sqrt{2 - 2\cos(\pi/3)} = 2\,mv_0 \sin\frac{\pi}{6}$$

107. b.

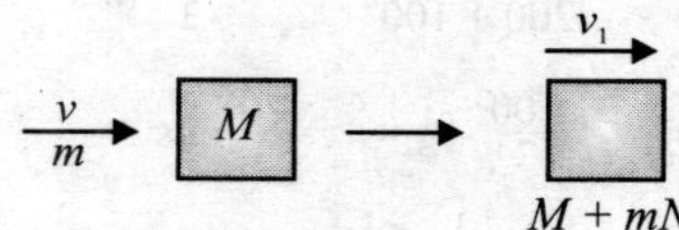

Fig. 1.331

Momentum of all particles colliding = final momentum

$$\Rightarrow \quad mNv = (M + mN)\,v_1$$

$$v_1 = \frac{mvN}{M + mN}$$

108. b. As no external force acts on the ball + box system, hence velocity of the system remains constant.

109. c. Impulse = area of trapezium

$$= \frac{1}{2}\left(T + \frac{T}{2}\right)F_0 = \frac{3TF_0}{4}$$

According to impulse–momentum theorem,

Impulse = Change in momentum

$$\Rightarrow \quad \frac{3TF_0}{4} = mu \Rightarrow F_0 = \frac{4mu}{3T}$$

110. c.

$$u_1 = \frac{p}{m_1},\; u_2 = \frac{p}{m_2}$$

$$E = \frac{1}{2}m_1u_1^2 + \frac{1}{2}m_2u_2^2$$

$$= \frac{1}{2}\left[m_1\frac{p^2}{m_1^2} + m_2\frac{p^2}{m_2^2}\right] = \frac{p^2(m_1 + m_2)}{2m_1m_2}$$

111. b. At maximum compression when m sticks to M,

$mv = (M + m)\, V$

$V = \frac{mv}{M+m}$

Fraction of KE Stored in spring $f = \frac{K_i - K_f}{K_i}$

$\therefore f = 1 - \frac{\frac{1}{2}(M+m)\,V^2}{\frac{1}{2}mv^2} \Rightarrow f = 1 - \left(\frac{m}{M+m}\right) = \frac{M}{M+m}$

112. c. As the car stops, so velocity of bullet after firing is same w.r.t. ground and w.r.t car. Conserving momentum in horizontal direction.

$$(M+m)v = mv\cos\alpha \quad \Rightarrow \quad V = \frac{mv\cos\alpha}{M+m}$$

113. c.

$$v_{CM} = \frac{10\times 3 + 20\times 2 - 30\times 5}{10+20+30} = -\frac{4}{3}\ \text{m/s}$$

$$x_{CM} = \frac{10\times 0 + 20\times 5 + 30\times 20}{10+20+30} = \frac{35}{3}\ \text{m/s}$$

$x = x_{CM} + v_{CM}t$

$$x = \frac{35}{3} + \left(-\frac{4}{3}\right)\times 1 = \frac{31}{3} = 10\frac{1}{3}\ \text{m}$$

114. a. As acceleration in both will be $g = 10\ \text{m/s}^2$, hence acceleration of centre of mass will also be g.

115. d. $v_{CM-i} = \frac{200\times 0 + 100\times 100}{200+100} = \frac{100}{3}\ \text{m/s}$

$$v = v_{CM-i} + a_{CM}t = \frac{100}{3} - g\times 1 = \frac{70}{3}\ \text{m/s}$$

116. c. $a = \frac{6-4}{6+4}g = \frac{g}{5}$, $a_{CM} = \frac{-6\times a + 4a}{6+4} = -\frac{a}{5} = -\frac{g}{25}$

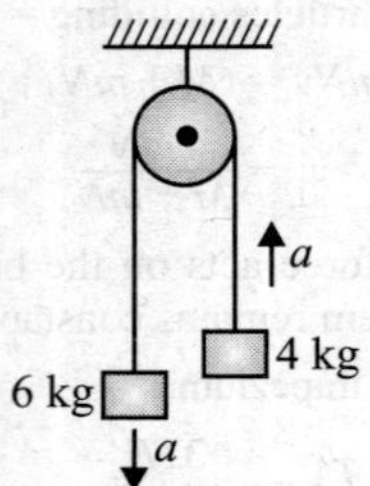

Fig. 1.332

117. c. $p = p_0(\cos t\,\hat{i} + \sin t\,\hat{j})$

$$\vec{F} = \frac{d\vec{p}}{dt} = p_0\,[-\sin t\,\hat{i} + \cos t\,\hat{j}]$$

Both $\vec{p}$ and $\vec{F}$ are functions of time. So none of them is constant.

We see that $\vec{F}\cdot\vec{p} = 0$, So $\vec{F} \perp^r \vec{p}$.

118. a. If E is the energy of explosion, then since it is just sufficient to carry the bullet to a height h, so

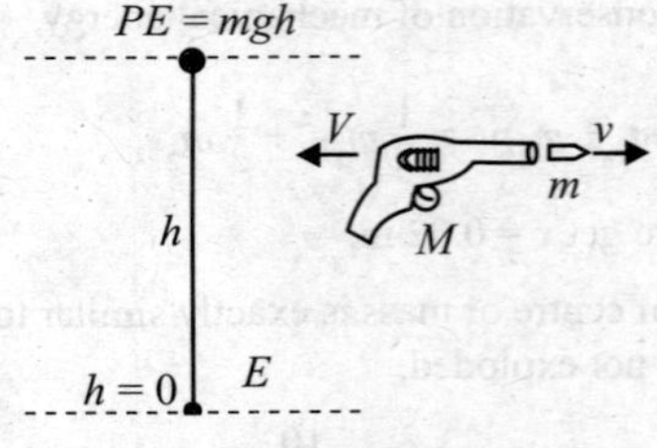

Fig. 1.333

We have $E = mgh$. Now, if v and V are the respective velocities of the bullet and the gun, respectively, then, from the law of conservation of linear momentum,

$$mv + M(-V) = 0 \Rightarrow v = \frac{MV}{m} \quad \text{(i)}$$

Also, $E = \frac{1}{2}mv^2 + \frac{1}{2}MV^2$

$\Rightarrow$ $mgh = \frac{1}{2}m\left(\frac{MV}{m}\right)^2 + \frac{1}{2}MV^2$ [from Eq. (i)]

$\Rightarrow$ $2mgh = \frac{M^2V^2}{m} + MV^2$

$\therefore \quad \Rightarrow \quad V = \left[\frac{2m^2gh}{M(M+m)}\right]^{1/2}$

119. a. The force has to be applied at the centre of mass of the system for pure translation motion.

120. d. Using conservation of linear momentum, we have

$$m_2v_0 = (m_1 + m_2)v \quad \Rightarrow \quad v = \frac{m_2}{m_1+m_2}v_0$$

From work–energy conservation,

$$\frac{1}{2}m_2v_0^2 - \frac{1}{2}(m_1+m_2)\frac{m_2^2v_0^2}{(m_1+m_2)^2} = \frac{1}{2}kx^2$$

$$x = v_0\sqrt{\frac{m_1m_2}{k(m_1+m_2)}}$$

121. d. The impact takes place along the line joining them. Since the particle at rest is constrained to move on the horizontal surface, we cannot conserve the momentum along the line joining them. Since no net force is acting on the system of particles parallel to the surface, conservation of linear momentum along that direction yields the following result.

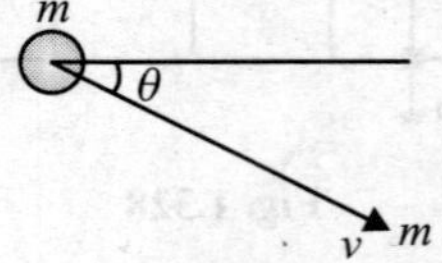

Fig. 1.334

$$mv\cos\theta = (m+m)v' \Rightarrow v' = \frac{v}{2}\cos 60° = \frac{v}{4}$$

122. c. The velocity of the particle when it reaches point A from B is

$$v = \sqrt{(2gh)}\ \text{(directed along } BA)$$

$$(p_i)x = mv\cos\theta \quad \text{(i)}$$

$$(p_i)y = mv\sin\theta \quad \text{(ii)}$$

When the particle reaches point C, the momentum is as shown in Fig. 1.335.

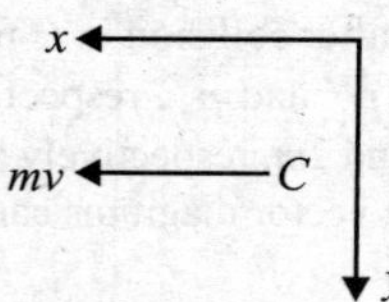

Fig. 1.335

$$(p_f)_x = mv \qquad \text{(iii)}$$

$$(p_f)_y = 0 \qquad \text{(iv)}$$

Therefore, $\quad (\Delta P)_x = (p_f)_x - (p_i)_x$

$$= mv - mv\cos\theta \qquad \text{(v)}$$

$$(\Delta p)_y = (p_f)y - (p_i)_y = -mv\sin\theta$$

Magnitude of change in momentum is

$$\sqrt{[mv(1-\cos\theta)]^2 + (mv\sin\theta)^2} = 2m(\sqrt{2gh})\sin\left(\frac{\theta}{2}\right)$$

123. d. First consider the collision of balls A and B. Let the velocities of these two balls after their collision be v and v'.

Momentum after impact = Momentum before impact

Fig. 1.336

$\therefore \quad 2v + 4v' = 2 \times 4 + 4 \times 1$

$\therefore \quad 2v + 4v' = 12$

$\Rightarrow \quad v + 2v' = 6 \qquad$ (i)

Relative velocity after impact = $- e \times$ relative velocity before impact

$$v - v' = -1\,(4-1)$$

$$v - v' = (-3) \qquad \text{(ii)}$$

Subtracting Eq. (ii) from Eq. (i), we get

$$3v' = 9$$

$$v' = 3 \text{ m/s}$$

Substituting in Eq. (ii), we get

$$v - 3 = -3 \quad \Rightarrow v = 0$$

Fig. 1.337

Hence, after the collision ball A is brought to rest, while ball B will move with a velocity of 3 m/s. Now consider the collision of balls B and C. Let the velocities of these balls after collision be V and V', respectively.

Total momentum after impact = Total momentum before impact

$$4V + 8V' = 4 \times 3 + 8 \times \frac{3}{4} = 18$$

$$V + 2V' = \frac{9}{2} \qquad \text{(iii)}$$

Relative velocity after impact = $- e \times$ relative velocity before impact

$$V - V' = (-1)(3 - 3/4)$$

$$V - V' = -9/4 \qquad \text{(iv)}$$

Subtracting Eq. (iv) from Eq. (iii), we get

$$3V' = 27/4 \Rightarrow V' = 9/4 \text{ m/s}$$

$$V = 0$$

$$V_A = 0,\ V_B = 0$$

124. b. Since the floor is smooth, therefore no horizontal force is acting on the system. Only three forces are acting on it, weight mg of the ball, weight mg of the rod and vertically upward reaction of the floor.

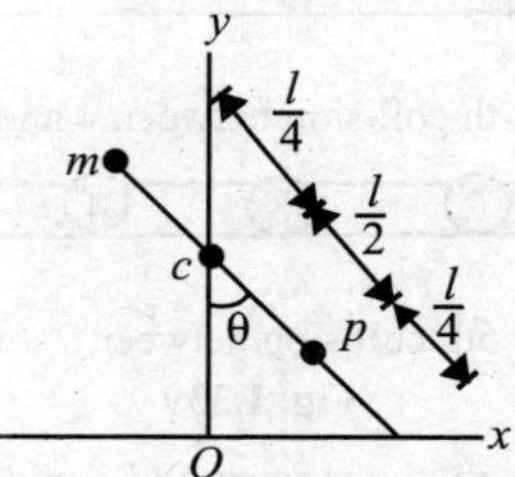

Fig. 1.338

Since the rod is released from rest and no horizontal force is acting on it, therefore centre of mass of the system will not displace horizontally; it means it will fall vertically downward.

Since masses of the rod and ball are equal, therefore centre of mass 'c' is at mid-point of centre of the rod and the ball; it means at distance $l/4$ from the ball or $3l/4$ from the lower end.

If at an instant the rod makes angle θ with the vertical, then point P and the system will be as shown in the figure.

Co-ordinates of the point P are

$$x = \frac{l}{2}\sin\theta$$

$$y = \frac{l}{4}\cos\theta$$

$$\therefore \quad \sin\theta = \frac{2x}{l}$$

$$\cos\theta = \frac{4y}{l}$$

Using $\sin^2\theta + \cos^2\theta = 1$, we have

$$4x^2 + 16y^2 = l^2$$

125. c. Component of initial momentum along the wall is

$$P_{iy} = mv_0 \sin 37°\,\hat{j}$$

Component of final momentum along the wall is

$$P_{fy} = \frac{mv_0}{2}\sin 53°\,\hat{j}$$

So, component of impulse along the wall is $J_y = P_{fy} - P_{iy}$

$$\frac{mv_0}{2} \times \frac{4}{5} - \frac{mv_0}{1} \times \frac{3}{5}\hat{j} = -\frac{mv_0}{5}\hat{j}$$

126. d. The situation after various collisions is as shown in Fig. 1.339.

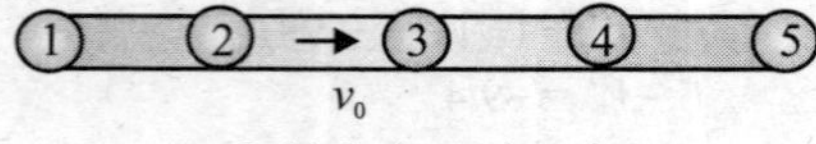

Initial given situation

After 1st collision between 2 and 3

$0.6v_0$ $0.4v_0$

After 2nd collision between 3 and 4

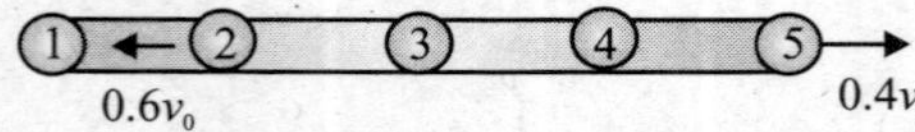

After 3rd and 4th collision between 4 and 5 and 3 and 2

After 5th collision between 2 and 1

Fig. 1.339

After first collision, the momenta of 1 and 2 will be exchanged (property of elastic collision) and hence second ball starts to move towards 3 with velocity v_0.

127. d. Let the masses be m_1 and m_2. Then

$$F = m_1 a \Rightarrow a = \frac{F}{m_1}$$

$$a_{CM} = \frac{m_1 a + m_2 \times 0}{m_1 + m_2} = \frac{F}{m_1 + m_2} < a$$

128. c. $\vec{v}_{CM} = \dfrac{m_1 v_1 + m_2 v_2 + \cdots}{m_1 + m_2 + \cdots} = \dfrac{6mv}{7m} = \dfrac{6v}{7}$

129. b. $\vec{F} = \dfrac{d\vec{p}}{dt}$, find slope and then proceed.

130. c. For elastic collision, $e = 1$. Let speed of the ball be v towards left after collision, then w.r.t wall, incident velocity = reflected velocity. We get $3 + 3 = v - 3 \Rightarrow v = 9$ m/s.

131. d. Let us solve this question in C frame. As the collision is elastic, the magnitude of momenta of all balls remains same after collision in this frame.

Velocity of C frame w.r.t. ground frame is

$$v_C = \frac{mv_0}{3m} = \frac{v_0}{3}$$

The momentum diagram in C frame is drawn as shown in Fig. 1.340.

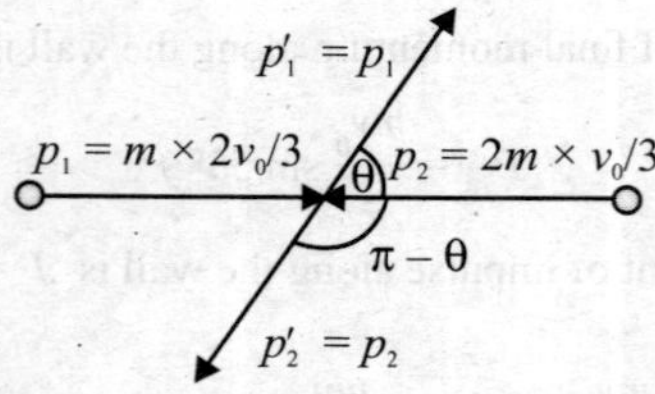

Fig. 1.340

Now, velocity of m after collision in C frame is $2v_0/3$ while that of $2m$ is $v_0/3$ along $\vec{p}_1{}'$ and $\vec{p}_2{}'$, respectively. Let $v_1{}'$ and $v_2{}'$ be the velocities of m and $2m$, respectively after collision in ground frame, then velocity vector diagrams can be drawn as shown.

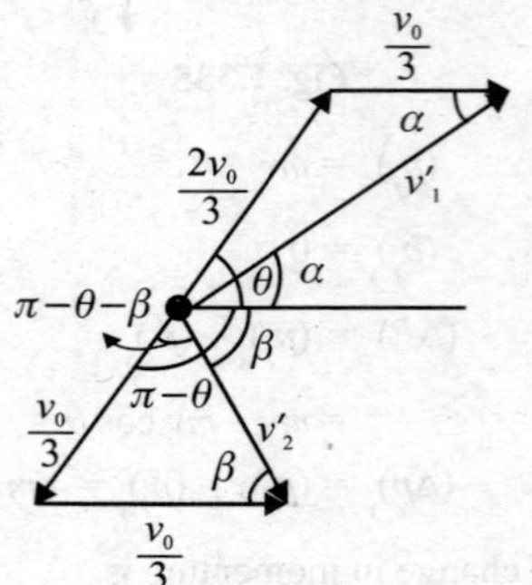

Fig. 1.341

Using trigonometry, from upper triangle

$$\sin\alpha = 2\sin(\theta - \alpha) \Rightarrow \alpha = \tan^{-1}\left[\frac{1}{1+\sqrt{3}}\right]$$

From lower triangle:

$$\pi - \theta - \beta = \beta \Rightarrow \beta = \frac{\pi - \theta}{2} = 75° \ [\because \theta = 30°]$$

So, angle of divergence is $\alpha + \beta = 75° + \tan^{-1}\left(\dfrac{1}{1+\sqrt{3}}\right)$

Alternative method Write down conservation of momentum equations, coefficient of restitution equation, energy conservation and use the given conditions.

132. c. When two identical balls undergo elastic oblique collisions with one of them stationary, then after collision they will move at right angles to each other.

Alternative method I: Just like in previous question, you can proceed using centre of mass frame of reference.

Alternative method II: Using basic equations of conservation of momentum, conservation of energy and coefficient of restitution, you can have the desired result.

133. a. Velocity of the ball just before collision is $v_0 = \sqrt{2gL}$

v_0 $2m$

Before collision

v_1 $2m$ v_2

After collision

Fig. 1.342

Applying momentum conservation along horizontal direction (because momentum is conserved in collision along the line of impact), we get

$$mv_0 = -mv_1 + 2mv_2$$

Applying coefficient of restitution equation, we get

$$e = \frac{2}{3} = \frac{v_1 + v_2}{v_0}$$

Solving the above two equations, we get

$$v_1 = \frac{v_0}{9} \text{ and } v_2 = \frac{5v_0}{9}$$

As the block moves a distance of $3L/2$ before coming to rest, so from work–energy theorem,

$$0 - \frac{1}{2}(2m)v_2^2 = -\mu \times 2\,mg\,\frac{3L}{2} \Rightarrow v_2^2 = 3\mu gL$$

$$\frac{25}{81} \times 2\,gL = 3\,\mu gL \Rightarrow \mu = \frac{50}{243}$$

For ball, KE is converted to gravitational potential energy after collision, so

$$0 - \frac{mv_1^2}{2} = -mg \times L\,(1 - \cos\theta)$$

$$\cos\theta = \frac{80}{81}$$

134. a. As coefficient of restitution equation is valid in all frames of reference, so in the frame of lift

$$e = \frac{\text{Velocity of separation}}{\text{Velocity of approach}} = \frac{v_f}{v_i} = \frac{\sqrt{2gh_2}}{\sqrt{2gh_1}} \qquad \text{(i)}$$

$$h_2 = e^2 h_1$$

Impulse delivered by the elevator to ball is

$$J = m\vec{v}_f - m\vec{v}_i = (m\sqrt{2gh_2} + \sqrt{2gh_1})$$

If the elevator is accelerating, then replacing g with $g + g/4$ we get $h_2 = e^2 h_1$.

135. d. For collision of B and C

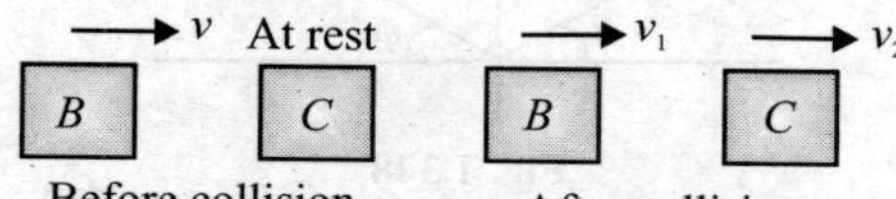

Fig. 1.343

$$2mv = 2mv_1 + mv_2$$

$$\frac{1}{2} = \frac{v_2 - v_1}{v} \Rightarrow 2v_2 - 2v_1 = v$$

Solving above equations, $v_1 = \dfrac{v}{2}$ and $v_2 = v$

Now for blocks A and B plus spring system, using reduced mass concept, and applying work–energy theorem, let maximum compression in spring be x_0 and at the time of maximum compression relative velocity of blocks be zero. Reduced mass is given by

$$\mu = \frac{m \times 2m}{3m} = \frac{2m}{3}$$

$$0 - \frac{\mu \times (v - v/2)^2}{2} = -\frac{kx_0^2}{2} \Rightarrow \quad x_0 = \left(\sqrt{\frac{m}{6k}}\right) v$$

136. c. Because here momentum changes abruptly in vertical direction.

137. c. From impulse–momentum theorem,

$$\vec{J} = \vec{p}_f - \vec{p}_i \Rightarrow \vec{p}_f = \vec{J} + \vec{p}_i$$

Here $p_i = 3000 \times 300 = 9 \times 10^5$ kg m/s

$$J = 4000 \times 225 = 9 \times 10^5 \text{ kg m/s}$$

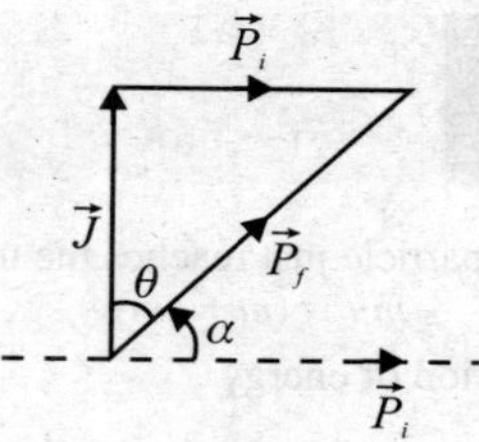

Fig. 1.344

$$\tan\theta = \frac{p_i}{J} = 1 \Rightarrow \theta = 45°$$

So, the angle by which the space probe rotates is

$$\alpha = \frac{\pi}{2} - \theta = 45°$$

138. a. Let angle between initial velocities be $\pi - \theta$ and the situation is as shown in Fig. 1.345. Conserve the momentum along X- and Y-directions.

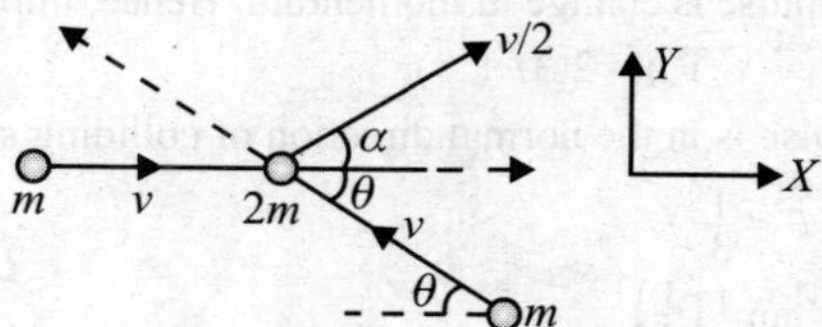

Fig. 1.345

$$mv - mv\cos\theta = 2m\frac{v}{2}\cos\alpha, \text{ for } X\text{-axis}$$

$$mv\sin\theta = 2m\frac{v}{2}\sin\alpha, \text{ for } Y\text{-axis}$$

Solving above equations, $\theta = 60°$

Required angle $= \pi - 60° = 120°$

139. d. As only the gravity force is acting on the system, the centre of mass of the system follows a parabolic path.

At $t = 2$ s,

$$x_{CM} = 30\cos 37° \times 2 = 48 \text{ m}$$

$$y_{CM} = 30\sin 37° \times 2 - \frac{1}{2} \times 10 \times 2^2 = 16 \text{ m}$$

Let coordinates of the second piece, i.e., 6 kg piece be (x, y). Then

$$x_{CM} = 48 = \frac{6x + 4 \times 105}{10} \Rightarrow x = 10 \text{ m}$$

$$y_{CM} = 16 = \frac{6y + 4 \times 43}{10} \Rightarrow y = -2 \text{ m}$$

Negative value of y shows that the second piece collides the ground before $t = 2$ s.

140. d. Initially, both the blocks are under the gravity effect, so

$$a_{CM} = \frac{4g + 2g}{4 + 2} = g$$

But once the jerk in the string occurs and string becomes taut, the 4 kg block moves down with acceleration $g/3$ and 2 kg block moves up with acceleration $g/3$. So,

$$a_{CM} = \frac{4 \times \frac{g}{3} - \frac{2g}{3}}{6} = \frac{g}{9}$$

Multiple Correct Answers Type

1. b., d. When the particle just reaches the top of the wedge

$$mu = (m + nm)v \quad \text{(i)}$$

Apply conservation of energy

$$\frac{1}{2}mu^2 = \frac{1}{2}(M+m)v^2 + mgh$$

Solve to get $u = \sqrt{2gh\left(1+\frac{1}{n}\right)}$

Angle of projection w.r.t. wedge. But angle of projection as observed by the ground will be less than α because horizontal component velocity will be more in this case.

$$\vec{v}_{\text{block/ground}} = \vec{v}_{\text{block/wedge}} + \vec{v}_{\text{wedge/ground}}$$

2. b., c. The initial velocity of CM is upward. The acceleration of the CM is 'g' downward.

3. a., c. Impulse is change in momentum. Hence, impulse

$$= 2(\vec{v}_2 - \vec{v}_1) = 2(3\hat{i} + \hat{j})$$

As impulse is in the normal direction of colliding surface

$$\tan\theta = \frac{1}{3}$$

$$\theta = \tan^{-1}\left(\frac{1}{3}\right)$$

$$\alpha = 90° + \tan^{-1}\left(\frac{1}{3}\right)$$

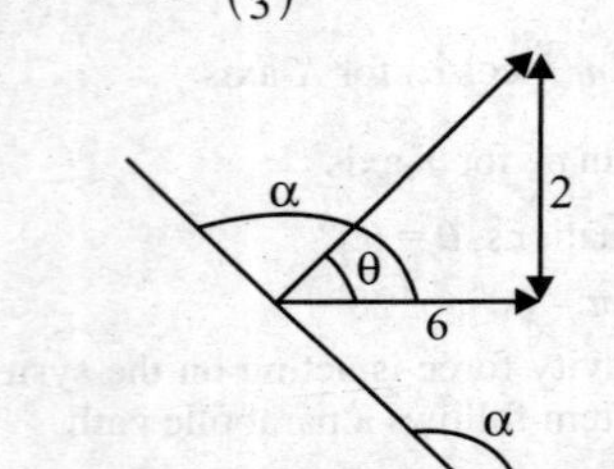

Fig. 1.346

4. b., c., d. In an elastic collision, linear momentum of the system is always conserved i.e., in all the three stages of collision. But the kinetic energy of the system is not always conserved. It is conserved only before and after the collision. During collision it gets converted into potential energy.

5. a., b., c. This relation is applicable to a body undergoing any kind of motion.

6. a., c., d. In an SHM, the maximum velocity is at the equilibrium position.

7. a., c.

a. As $\Sigma\vec{P} = 0$ and $\vec{P}_1 + \vec{P}_2 \neq 0$, $\vec{P}_3$ cannot be zero.

b. For any system of two or more than two particles, KE can change due to work done by internal forces while the centre of mass remains at rest

c. $\vec{a}_{CM} = 0 \Rightarrow \Sigma\vec{F}_{ext} = 0$

d. It is possible to have changing speed of CM without any work done by ext. forces on the system. When a person accelerates himself on a rough horizontal surface without any flipping between his shoes and ground, work done by friction, normal reaction and weight all are zero but speed of centre of mass changes.

8. b., d. In the first case: $(m+m)v_1 = mv_0 \Rightarrow v_1 = \frac{v_0}{2}$

In the second case:

$$-\int T\,dt = (m+m)v - mv_0 \quad \text{(i)}$$

$$\int T\,dt = 2\,mv \quad \text{(ii)}$$

From Eqs. (i) and (ii),

$4\,mv = mv_0 \quad \Rightarrow \quad v = \frac{v_0}{4}$ and required ratio = 2

9. b., c. In an elastic collision: $v_{sep} = v_{app}$

or $\quad v' - u = v + u \quad$ or $\quad v' = v + 2u$

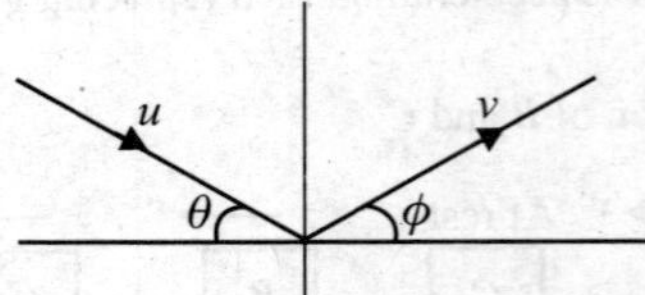

Fig. 1.347

Change in momentum of ball is

$$|p_f - p_i| = |m(-v') - mv| = m(v' + v) = 2\,mu(u+v)$$

Average force is $\frac{\Delta P}{\Delta t} = \frac{2m(u+v)}{\Delta t}$

Change in KE, $K_f - K_i = \frac{1}{2}mv'^2 - \frac{1}{2}mv^2 = 2m\,u\,(u+v)$

10. a., b., d. $v\sin\varphi = eu\sin\theta, \quad v\cos\varphi = u\cos\theta$

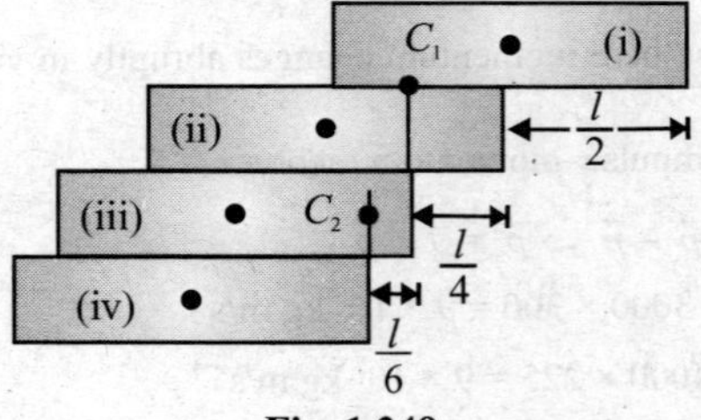

Fig. 1.348

$$v = u\sqrt{\cos^2\theta + e^2\sin^2\theta} = u\sqrt{1 - \sin^2\theta + e^2\sin^2\theta}$$

$$= u\sqrt{1-(1-e^2)\sin^2\theta}$$

$$I = m(v\sin\phi + u\sin\theta) = mu\sin\theta\,(1+e)$$

$$\text{Ratio of KE} = \frac{\frac{1}{2}mv^2}{\frac{1}{2}mu^2} = \cos^2\theta + e^2\sin^2\theta$$

11. a., b., c., d.

Let the weight of each brick be W and length l. As bricks are homogeneous, the centre of gravity of each brick must be at the midpoint. Therefore, the topmost brick will be in equilibrium if its centre of gravity lies at the edge of brick below it, i.e., II brick. Thus the topmost brick can have maximum equilibrium extension of $l/2$.

Fig. 1.349

C_1 is the centre of mass of the top two bricks which lies on the edge of the third brick.

C_2 is the centre of mass of the top three bricks which lies on the edge of the fourth brick.

Thus, the maximum overhanging length to top from the edge of bottom brick is

$$\frac{l}{2}+\frac{l}{4}+\frac{l}{6}=\frac{11}{12}l$$

12. a., b., d. In completely inelastic collision, bodies coalesce

$$m_1u_1 - m_2u_2 = (m_1 + m_2)V$$

$$2 \times 3 - 1 \times 4 = (2 + 1)\ V;\ V = \frac{2}{3}\ \text{m/s}$$

Net momentum of system $P = m_1u_1 - m_2u_2$

$$2 \times 3 - 1 \times 4 = 2\ \text{kg m/s}$$

$$\text{Loss in KE} = \frac{1}{2}m_1u_1^2 + \frac{1}{2}m_2u_2^2 - \frac{1}{2}(m_1 + m_2)V^2$$

$$= \frac{1}{2}\times 2 \times (3)^2 + \frac{1}{2}\times (1) \times (4)^2 - \frac{1}{2}\times (2+1)\times\left(\frac{2}{3}\right)^2$$

$$= 9 + 8 - \frac{2}{3} = 17 - \frac{2}{3} = \frac{49}{3}\ \text{J}$$

13. a., d. $v\cos\theta = u\sin\theta$

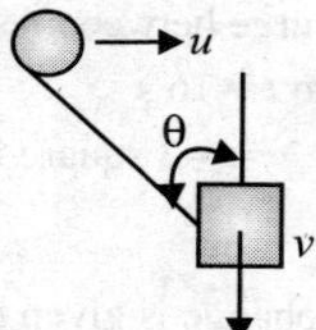

Fig. 1.350

From conservation of energy,

$$mgl\cos\theta = 2\left(\frac{1}{2}mu^2\right) + \frac{1}{2}mv^2$$

$$u = \sqrt{\frac{2gl\cos\theta}{2+\tan^2\theta}} = \sqrt{\frac{gl}{5}},\ \text{using}\ \theta = 60°$$

$$v = \sqrt{\frac{3gl}{5}}$$

14. b., d. When two particles collide and stick together, energy is conserved and work done is negative because KE decreases. So, options (b) and (d) are correct.

15. b., c. When the second body is at rest, velocities after the elastic collision are given by

$$v_1 = \left(\frac{m_1 - m_2}{m_1 + m_2}\right)u_1 \quad \text{and} \quad v_2 = \left(\frac{2m_1}{m_1 + m_2}\right)u_1$$

It is obvious that both bodies move after the collision. If $m_1 = m_2$, then $v_1 = 0$ and $v_2 = u_1$. Further, v_2 cannot be zero. So, options (b) and (c) are correct.

16. a., c. If wall is smooth $v\cos\alpha = eu\cos\theta$

$v\sin\alpha = u\sin\theta$

$\cot\alpha = eu\cot\theta$

$\alpha > \theta$

$\alpha + \beta > \theta + \beta$

$90° > \theta + \beta$

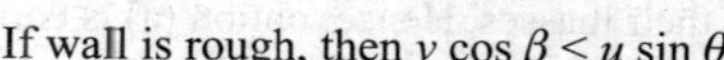

Fig. 1.351

If wall is rough, then $v\cos\beta < u\sin\theta$ (i)

$v\sin\beta = eu\cos\theta$ (ii)

From (i) and (ii), $e < \tan\beta/\cot\theta$

17. b., c. Let the ball strike the floor at angle α with the vertical and velocity at the instant of striking the floor be u and let the ball rebound with velocity v at angle β with the vertical as shown in Fig. 1.352.

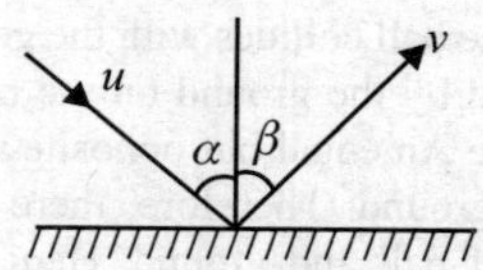

Fig. 1.352

Since the floor is smooth, therefore horizontal component of velocity remains unchanged. Hence

$$v\sin\beta = u\sin\alpha \quad \text{(i)}$$

Since, the floor is inelastic, therefore normal component of velocity just after the collision is less than that just before the collision. Hence,

$$v\cos\beta < u\cos\alpha \quad \text{(ii)}$$

Dividing Eq. (i) by Eq. (ii), $\tan\beta > \tan\alpha$ or $\beta > \alpha$

Hence, option (c) is correct.

Since the ball has a horizontal component of velocity, therefore, at highest point, its kinetic energy is not equal to zero.

It means, at highest point, potential energy of the ball is less than the kinetic energy just after rebound. Hence, option (a) is incorrect.

Since the floor is inelastic, therefore there is some loss of energy during the collision. Hence, the total energy of the ball does not remain conserved. So option (b) is also correct.

18. b., c. The general form of Newton's second law is

$$\vec{F}_{\text{ext}} = \frac{d\vec{p}}{dt} = \frac{d}{dt}(m\vec{v}) = m\frac{d\vec{v}}{dt} + \vec{v}\frac{dm}{dt}$$

The form $\vec{F}_{\text{ext}} = m\vec{a}$ is valid only if $\frac{dm}{dt} = 0$, i.e., mass does not change with time. Hence, option (a) is incorrect.

Option (b) is correct because a body at rest may have potential energy yet no momentum. Option (c) is also correct. A body has momentum if it has mass and velocity and a body having mass and velocity must have kinetic energy. Option (d) is incorrect because the relative velocity remains unchanged in magnitude and gets reversed in direction:

$$v_2 - v_1 = -(u_2 - u_1)$$

Hence, the correct options are (b) and (c).

19. a., b., c., d. Since the system of two blocks and a spring is placed on a smooth horizontal floor, therefore no external horizontal force acts on the system.

Since the system is initially at rest, therefore centre of mass of the system will remain stationary and the block will oscillate with the same frequency. It means option (d) is correct.

Since the system is initially at rest and centre of mass stationary, therefore centre of mass cannot move or the momentum of the system remains equal to zero. It means, momenta of two blocks are numerically equal but they are in opposite directions. Therefore, ratio of magnitude of velocity of the two blocks will be in inverse ratio of their masses. Hence, option (a) is correct.

Since ratio of mass is constant, therefore the ratio of their velocity will remain constant. It means that option (b) is also correct.

Since the system is on smooth horizontal floor, therefore no energy loss takes place. Hence, linear momentum and energy of the system remain conserved. Therefore, option (c) is also correct.

20. a., b., c. When the ball collides with the ground elastically, then a force is exerted by the ground on the ball along the vertical upward direction. An equal but opposite reaction is offered by the ball on the ground. Therefore, there is no resultant force on the system of ball and ground. Hence, the momentum of universe remains constant. Hence, option (a) is correct.

Since during collision, a force is exerted by the ground on the ball, therefore due to this impulsive reaction, momentum of the ball gets changed.

If the ball collides with the ground with velocity v, then change in momentum of the ball is equal to 2 mv and this change takes place along vertically upward direction. Hence, momentum of the ball does not remain conserved. Therefore, option (b) is correct.

During the collision, at an instant, centre of mass of the ball comes to rest. At that instant the ball has the maximum deformation. Due to that deformation, the ball tries to regain its original shape. Hence, it gets rebound. Hence, at an instant, during the collision, the whole of KE gets converted into its deformation energy which ultimately converts into KE. Therefore, option (c) is correct.

21. a., d. Let at an instant 't', velocity of cart be v. Due to the sand falling on the cart, it experiences a retarding force, which is equal to μv. It means net accelerating force on the cart is equal to $(F - \mu v)$. But at this instant, mass of the loaded cart is equal to $(m_0 + \mu t)$.

Therefore, acceleration at time t will be equal to $\dfrac{(F-\mu v)}{(m_0+\mu t)}$. It means option (b) is incorrect.

If in the above equation, t is substituted as zero, then initial acceleration of the cart is

$$a_0 = \frac{(F-\mu v)}{m_0}$$

But at initial moment, speed of flat cart is equal to zero. Hence, the initial acceleration becomes F/m_0. It means, option (a) is correct.

If we consider a system of flat cart and falling sand particles, then sand particles exert a retarding force on the cart which is equal to μv and the cart exerts an accelerating force on sand particles which is also equal to μv. It means the only external force or the only resultant force acting on the system is equal to F.

It means the momentum of the system at any instant t will be equal to impulse of the force F. Hence, option (d) is correct.

The collision of the sand particles with the cart is perfectly inelastic because just after falling on the cart, sand particles start to move horizontally rightward with the cart while just before coming onto the cart, these sand particles have zero horizontal velocity. It means there is a loss of energy during the collision. Hence, kinetic energy of the loaded cart at an instant will be less than the work done by the force F up to that instant. Hence, option (c) is incorrect.

22. c., d. In option (a), neither momentum nor kinetic energy is conserved so collision is not elastic. Hence, option (a) is not correct.

b. Same difficulty lies with option (b). So, only options (c) and (d) are correct.

23. a., b., c., d.

a. $v_1 = 20 - gt$

$v_2 = -20 - g(t-2)$

$v_1 - v_2 = [20 - gt] + 20 + g(t-2)] = 20$ m/s

Clearly, it is time independent, i.e., relative velocity between the two stones remains constant.

b. In both the cases, the total energy remains the same.

c. and **d.** These are simple concepts.

24. a., b., c. Each square has a value $2\text{ s} \times 5\text{ N} = 10\text{ N s}$

a. Area under the curve between 8 s and 12 s = –1 square

b. Area for $t = 8$ s to $t = 16$ s

= – 3 square = – 30 N s

As mass = 2 kg

Hence, velocity change is given by

$$\frac{-30\text{ N s}}{2\text{ kg}} = -15\text{ m/s}$$

c. Area of $t = 0$ to $t = 6$ s is

10 square = 100 N s

$$\text{Kinetic energy} = \frac{p^2}{2m} = \frac{100\times 100}{2\times 2} = 2500\text{ J}$$

25. c., d. Using conservation of linear momentum,

$$mv_0 = (M+m)v$$

$$v = \frac{mv_0}{M+m}$$

KE in centre of mass frame:

$$KE_c = \frac{1}{2}m(v_0 - v)^2 + \frac{1}{2}Mv^2 = \frac{1}{2}\left(\frac{Mm}{M+m}\right)v_0^2$$

From energy conservation,

$$\frac{1}{2}mv_0^2 - \frac{1}{2}(m+M)v^2 = \frac{1}{2}Kx^2$$

$$\Rightarrow \qquad x = v_0\sqrt{\frac{mM}{k(M+m)}}$$

26. a., b. Distribution is uniformly uneven. Heavier part will have the CM closer than the lighter part.

27. a., c. Since both masses have equal kinetic energy, so

$$\frac{1}{2}m_1u_1^2 = \frac{1}{2}m_2(\alpha u_1)^2$$

$$\Rightarrow \quad m_2 = \frac{m_1}{\alpha^2} \quad \text{(i)}$$

By law of conservation of momentum

$$m_1u_1 + m_2u_2 = m_1v_1 + m_2v_2$$

$$m_1u_1 + \frac{m_1}{\alpha^2}(\alpha u_1) = 0 + \left(\frac{m_1}{\alpha^2}\right)v_2$$

$$v_2 = \alpha u_1(1+\alpha) \quad \text{(ii)}$$

Further by law of conservation of energy, we have

$$\frac{1}{2}m_1u_1^2 + \frac{1}{2}\left(\frac{m_1}{\alpha^2}\right)(\alpha u_1)^2 = 0 + \frac{1}{2}\left(\frac{m_1}{\alpha^2}\right)v_2^2$$

$$m_1u_1^2 = \frac{1}{2}\left(\frac{m_1}{\alpha^2}\right)v_2^2$$

$$v_2 = \sqrt{2}(\alpha u_1) \quad \text{(iii)}$$

From Eqs. (ii) and (iii), we get

$$\frac{v_2}{\alpha u_1} = \sqrt{2} = 1 + \alpha \Rightarrow \alpha = \sqrt{2} - 1$$

$$\frac{u_2}{u_1} = \sqrt{2} - 1$$

$$\frac{u_1}{u_2} = \sqrt{2} + 1 \text{ [on rationaliting]}$$

$$\frac{m_2}{m_1} = \frac{1}{\alpha^2} = \left(\frac{1}{\sqrt{2}-1}\right)^2 = (\sqrt{2}+1)^2$$

$$\frac{m_2}{m_1} = 3 + 2\sqrt{2}$$

28. a., d. The velocity of bob just before the impact is $v = \sqrt{2gl}$ along the horizontal direction.

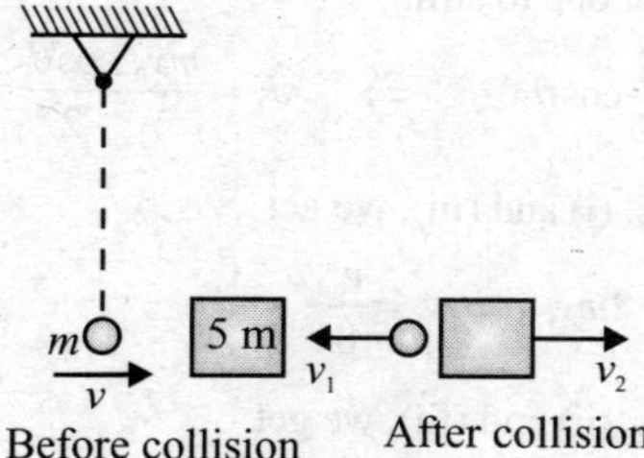

Fig. 1.353

From momentum conservation,

$$mv = -mv_1 + 5\,mv_2$$

From coefficient of restitution equation,

$$1 = \frac{v_1 + v_2}{v} \Rightarrow v_1 + v_2 = v$$

Solving above equations, we get $v_1 = \frac{2v}{3}$

$$v_2 = \frac{v}{3}$$

For tension in string, $T - mg = \frac{mv_1^2}{l}$

$$\Rightarrow \quad T = \frac{17}{9}mg$$

Let the maximum height attained by the bob be h, then $\frac{mv_1^2}{2} = mgh$

$$\Rightarrow \quad h = \frac{4l}{9}$$

29. a., b., d. The impulse exerted by the string on the cylinder is equal and opposite to the impulse exerted by the cylinder on the string.

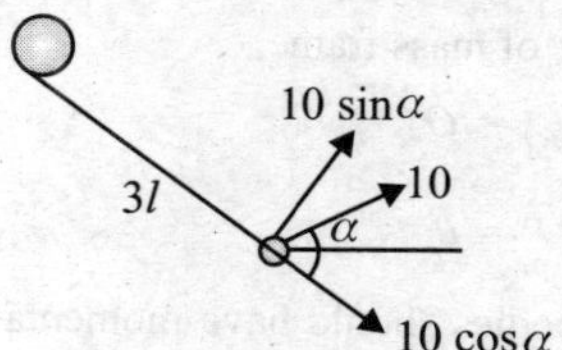

Fig. 1.354

$|\vec{J}| = 40000 \times 0.25 \times 10^{-3} = 10$ N s along the string.

For marble,

$$J = 0 - (-2 \times 10\cos\alpha)$$

As final velocity of marble along the string is zero

Putting, J = 10 Ns we get $\alpha = 60°$.

The string wraps up around the cylinder and marble moves in a circular path of varying radius with speed 10 sin $\alpha = 5\sqrt{3}$ m/s. The speed remains constant as only the tension force is present which is perpendicular to the velocity of the marble.

For string to break, the impulsive tension has to be

$$\geq 2 \times 10^5 \text{ N}$$

$$\therefore \; 40000 \times 0.25 \times 10^{-3} \geq 2 \times 10^5 \times \Delta t$$

$$\Rightarrow \quad \Delta t \leq 0.05 \text{ ms}$$

30. a., b., c. In the absence of external forces: $a_{CM} = 0$

$(\because \Sigma F_{ext} = Ma_{CM})$

So CM will not accelerate. But if the system was moving initially, it will continue to move with same velocity.

Assertion-Reasoning Type

1. d. If the collision is elastic, then masses of both particles should be same. But if masses are different, then collision is inelastic. Further, momentum remains conserved in any kind of collision.

2. c. For a system of two isolated spheres having non-zero initial kinetic energy, the complete kinetic energy can be converted to other forms of energy if the momentum of the system is zero. This is due to the fact that for an isolated system, the net momentum remains conserved. If an isolated system has non-zero momentum, for the momentum to remain constant, complete kinetic energy of the system cannot become zero. Hence, Statement I is true while Statement II is false.

3. **c.** Statement II contradicts Newton's third law and hence it is false.
4. **b.** The centre of mass of the fragments will continue its parabolic path. After the breakage, fragments may move in different directions. Both statements are correct but Statement II is not the explanation of Statement I.
5. **a.** The momentum of a system of particles from any frame is given by

$$\vec{P} = m\vec{v}_{CM}$$

From the centre of mass frame,

$$\vec{v}_{CM} = \vec{O}$$

$$\Rightarrow \quad \vec{P} = 0$$

So individual bodies should have momenta in opposite directions to make net momentum zero.

6. **d.** Net external force is zero. Therefore, no acceleration for CM. So, CM may move with a constant velocity.
7. **c.** Net force acting on the wedge and block system is gravity, therefore it is accelerated in downward direction. If external force acting on the system is zero, CM may be in rest or moving with constant velocity.
8. **a.** During collision, force exerted by the wedge on the particle is perpendicular to the inclined face. So linear momentum of wedge is conserved along the face of wedge.

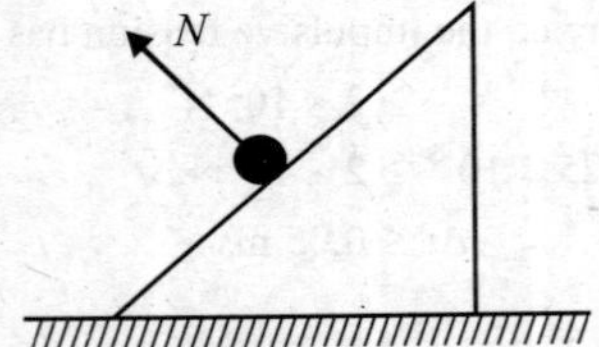

Fig. 1.355

9. **d.** A body may not have momentum but may have potential energy by virtue of its position. But if the body has no energy, then its kinetic energy is zero and therefore its momentum is zero.
Dimension of momentum = $[MLT^{-1}]$
Dimension of energy = $[ML^2T^{-2}]$
10. **b.** For an isolated system, no external force is acting in the absence of external force, linear momentum of centre of mass is unchanged. And so velocity of centre of mass remains constant.

Comprehension Type

For Problems 1–5

1. **d.**, 2. **b.**, 3. **a.**, 4. **b.**, 5. **a.**

Sol. Let after collision, the velocities are as shown in figure.

1. **d.** Here $\theta = 60°$. Let impulse between C and B during collision be J.

For the system of A and B:

$$J\cos\theta = 2mv_1 \quad \text{(i)}$$

$$J\sin\theta = mv_2 \quad \text{(ii)}$$

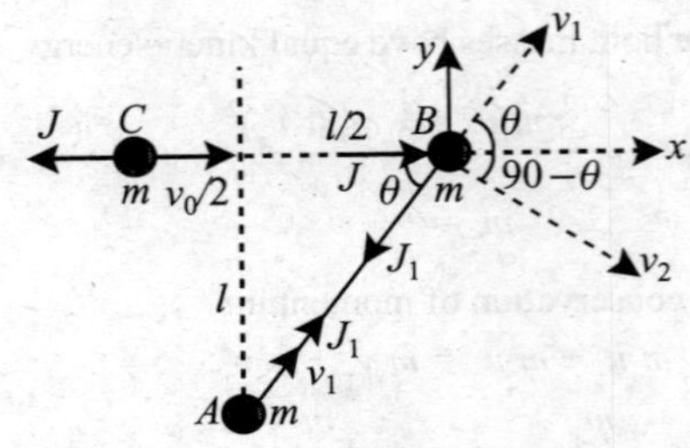

Fig. 1.356

$$\Rightarrow \quad v_1 = \frac{J}{m}\frac{\cos\theta}{2},\ v_2 = \frac{J}{m}\sin\theta$$

Velocity of B:

$$\vec{V}_B = v_1\cos\theta\hat{i} + v_1\sin\theta\hat{j} + v_2\sin\theta\hat{i} - v_2\cos\theta$$

$$= (v_1\cos\theta + v_2\sin\theta)\hat{i} + (v_1\sin\theta - v_2\cos\theta)$$

$$= \frac{J}{m}\left[\left(\frac{\cos^2\theta}{2} + \sin^2\theta\right)\hat{i} + \left(\frac{\cos\theta\sin\theta}{2} - \cos\theta\sin\theta\right)\hat{j}\right]$$

$$= \frac{J}{m}\left[\frac{7}{8}\hat{i} - \frac{\sqrt{3}}{8}\hat{j}\right]$$

From here we can find direction of unit vector in the direction of $\vec{v}_B$.

2. **b.** Velocity of A: $\vec{v}_1 = v_1\cos\theta\,\hat{i} + v_1\sin\theta\,\hat{j}$

$$\Rightarrow \quad \vec{v}_1 = \cos\theta\,\hat{i} + \sin\theta\,\hat{j} = \frac{1}{2}\hat{i} + \frac{\sqrt{3}}{2}\hat{j}$$

3. **a.** For C: $J = mv_0 - \dfrac{mv_0}{2} \Rightarrow J = \dfrac{mv_0}{2}$ (iii)

Impulse on A due to string:

$$J_1 = mv_1 = \frac{J}{2}\cos\theta \quad \Rightarrow \quad J_1 = \frac{mv_0}{2}\frac{\cos\theta}{2} = \frac{mv_0}{8}$$

4. **b.** From Eqs. (i) and (iii), we get

$$\frac{mv_0}{2}\cos\theta = 2mv_1 \quad \Rightarrow v_1 = \frac{v_0}{8}$$

5. **a.** From Eqs. (ii) and (iii), we get

$$\frac{mv_0}{2}\sin\theta = mv_2 \quad \Rightarrow v_2 = \frac{\sqrt{3}v_0}{4}$$

$$\text{Loss in KE} = \frac{1}{2}mv_0^2 - \left[\frac{1}{2}m\left(\frac{v_0}{2}\right)^2 + \frac{1}{2}mv_1^2 + \frac{1}{2}mv_1^2 + \frac{1}{2}mv_2^2\right]$$

Put the values to get loss in KE $= \dfrac{17}{64}mv_0^2$

For Problems 6–10

6. **c.**, 7. **b.**, 8. **a.**, 9. **d.**, 10. **d.**,

Sol.

6. **c.** A vertically upward thrown particle does not feel 'very high force'.

7. b. *mg* is small compared to force of explosion.

8. a. Point masses do not have dimensions.

9. d. See the definition of one-dimensional collision in the paragraph.

10. d. See the definition of two-dimensional collision.

For Problems 11–13

11. b., 12. c., 13. c.

Sol.

11. b. Since the centre of mass will remain at rest.

$$m(5l - x) = Mx$$

$$x = \frac{5ml}{(M+m)}.$$

12. c. $Mv_M - mv_m = 0$ (i)

$$\frac{1}{2}kl^2 = \frac{1}{2}mv_m^2 + \frac{1}{2}Mv_M^2 \quad \text{(ii)}$$

Solving Eqs. (i) and (ii),

$$v_M = \sqrt{\frac{Km}{M(M+m)}}\, l$$

13. c. $v_m = \frac{M}{m}\sqrt{\frac{Km}{M(M+m)}}\, l = \sqrt{\frac{KM}{m(M+m)}}\, l$

$$\text{KE} = \frac{1}{2}mv_m^2 = \frac{KMl^2}{2(M+m)}$$

For Problems 14–16

14. a., 15. d., 16. d.

Sol.

14. a. First calculate, $\theta = 45°$. As the balls A and B are constrained to move horizontally (immediately after collision), if 'I' be the impulse imparted by ball 'C' to each of A and B, the impulse received by ball C from them would be $2I \sin \theta$.

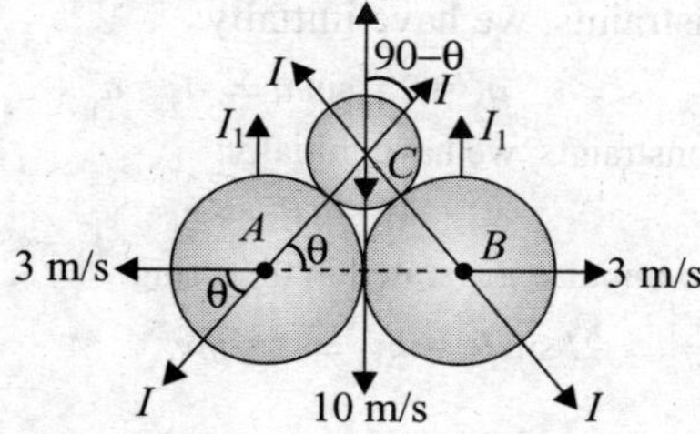

Fig. 1.357

Now, each of ball B and C received impulse 'I' as shown in Fig. 1.357, but moves horizontally as its vertical comp. gets balanced by impulse imparted to ball B and C by the respective strings and hence,

$$I \cos \theta = M_A V_A = M_B V_B$$

$$\Rightarrow \quad I = \frac{M_A V_A}{\cos \theta} \quad (I = \text{magnitude of impulse})$$

Now, for ball C, if its final velocity is V_C' downwards, we have

$$M_C V_C' = M_C V_C - 2I \sin \theta$$

$$\Rightarrow \quad V_C' = V_C - 2\frac{MA}{MC} V_A \quad (\because \theta = 45°)$$

$= -2$ m/s (–ve sign indicates that it is directed upwards)

15. d. Impulse provided by each string

$$I_1 = I \sin \theta = 6 \text{ N s}$$

16. d.

$$e = \frac{\text{Velocity of separation}}{\text{Velocity of approach}}$$

$$\Rightarrow \quad e = \frac{3\cos\theta + V_C' \sin\theta}{10 \sin\theta} \Rightarrow e = \frac{3+2}{10} = \frac{1}{2}$$

For Problems 17–18

17. a., 18. c.

Sol.

17. a. Applying conservation of momentum,

$$m_1 \times 4 + m_2 \times (-2.5) = m_1 \times 3 + m_2 v_2$$

$$\Rightarrow \quad v_2 = -1.5 \text{ m/s}$$

18. c. Let compression of the spring be x, then apply conservation of energy: Loss in KE = gain in PE

$$\Rightarrow \quad \frac{1}{2}m_1(4)^2 + \frac{1}{2}m_2(2.5)^2 - \frac{1}{2}m_1(3)^2 - \frac{1}{2}m_2(1.5)^2 = \frac{1}{2}kx^2$$

Solving, we get $x = 0.171$ m.

For Problems 19–20

19. c., 20. d.

Sol.

19. c.

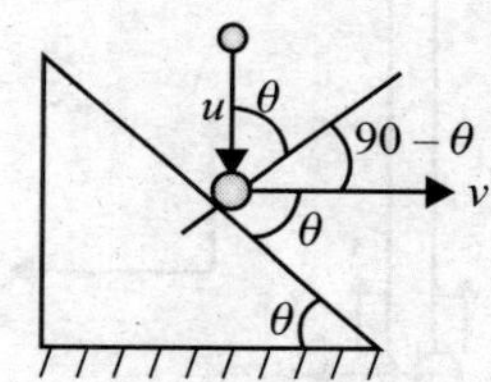

Fig. 1.358

$$e = \frac{v' \cos(90 - \theta)}{u \cos\theta}$$

$$= \frac{v'}{u} \tan\theta \quad \text{(i)}$$

$$u \sin\theta = v' \cos\theta$$

$$\Rightarrow \quad \frac{v'}{u} = \tan\theta \quad \text{(ii)}$$

From Eqs. (i) and (ii), $e = \tan^2\theta$.

20. d.

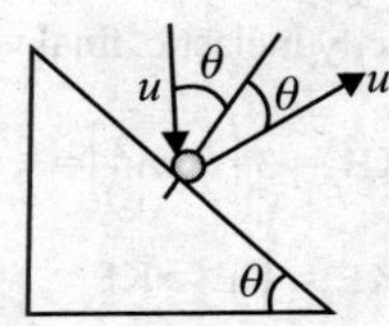

Fig. 1.359

$$\text{Impulse} = 2mu \cos\theta$$

$$= 2m\sqrt{2gh} \cos\theta$$

For Problems 21–22

21. b., 22. a.

Sol.

21. b. $u = \sqrt{2gh}$ is the velocity of m just before collision, and v is the velocity of system $M + m$; just after collision, then $(M + m) v = mu$

$$\Rightarrow \quad v = \frac{m\sqrt{2gh}}{M+m}$$

$$a = \left(\frac{M-m}{M+m}\right) g$$

$$t = \frac{2v}{a} = \left(\frac{2m}{M-m}\right)\sqrt{\frac{2h}{g}}$$

22. a. Fraction of KE lost $= \dfrac{\frac{1}{2}mu^2 - \frac{1}{2}(M+m)v^2}{\frac{1}{2}mu^2}$

$$= 1 - \frac{M+m}{m}\left(\frac{v}{u}\right)^2 = 1 - \frac{M+m}{m}\left(\frac{m}{M+m}\right)^2 = \frac{M}{M+m}$$

For Problems 23–25

23. a., 24. b., 25. d

Sol. 23. a. Just before 'A' collides with 'C', the velocities of all particles will be same and along y-axis. Let it be v. At this time let velocity of 'A' be v_1 along +ve x-axis, and that of 'C' be v_1 along –ve x-axis.

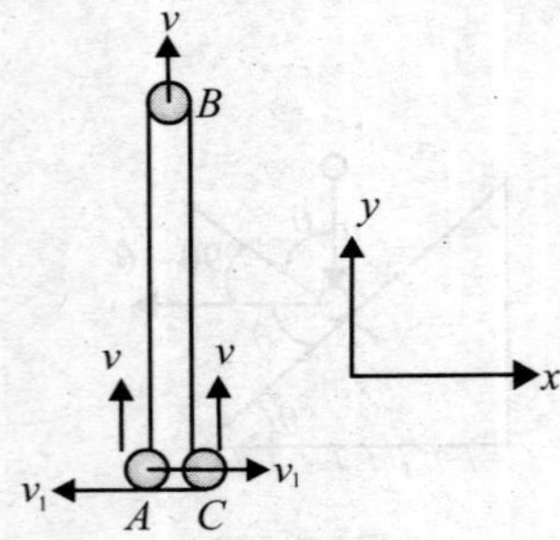

Fig. 1.360

From conservation of momentum along y-axis,

$$mv_0 = 3mv \Rightarrow v = \frac{v_0}{3}$$

24. b. Total kinetic energy will remain same:

$$\frac{1}{2}mv_0^2 = \frac{1}{2}mv^2 + 2\left[\frac{1}{2}m(v^2 + v_1^2)\right]$$

Put the value of v to get: $v_1^2 = \dfrac{v_0^2}{3}$

Velocity of A: $\sqrt{v^2 + v_1^2} = \dfrac{2}{3}v_0$

25. d If collision is perfectly inelastic, final velocity of each particle will be v.

$$\text{KE}_f = 3\left[\frac{1}{2}mv^2\right] = \frac{3}{2}m\left(\frac{v_0}{3}\right)^2 = \frac{mv_0^2}{6}$$

$$\text{Loss in KE} = \text{KE}_i = \text{KE}_f$$

$$= \frac{1}{2}mv_0^2 - \frac{mv_0^2}{6} = \frac{mv_0^2}{3}$$

$$\text{Percentage loss} = \frac{\frac{mv_0^2}{3}}{\frac{1}{2}mv_0^2} \times 100 = 66.67\ \%$$

which is more than 50%.

For Problems 26–28

26. a., 27. b., 28. d.

Sol.

26. a.

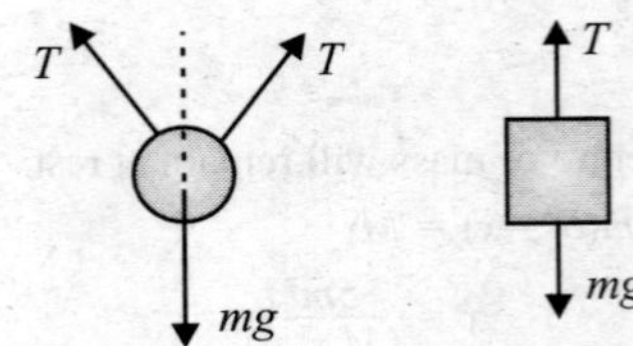

Fig. 1.361

$$2T \sin\theta = mg,\ T = mg$$

$$\sin\theta = \frac{1}{2} \Rightarrow \theta = 30°$$

27. b. $2T\sin\theta - mg = ma_1$, $\theta = 30°$

$$2\,mg - T = 2\,ma_2$$

Simplify to get: $a_1 + 2a_2 = g$

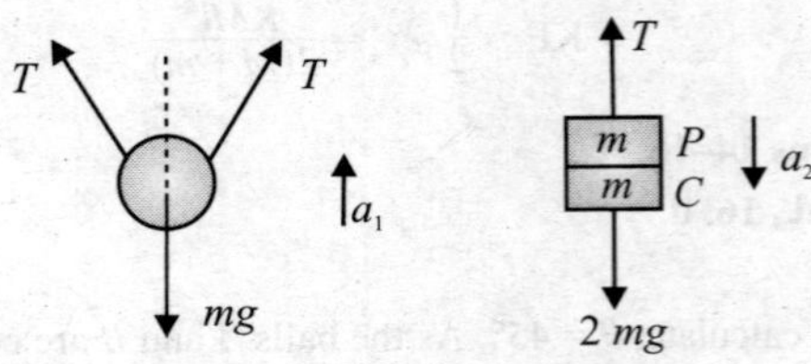

Fig. 1.362

From constraints, we have initially:

$$a_2 = 2a_1 \sin\theta \Rightarrow a_2 = a_1$$

28. d. From constraints, we have initially:

$$v_2 = 2v_1 \sin\theta \Rightarrow v_2 = v_1$$

Let I be the impulse generated in the string,

$$2I\sin\theta = mv_1 \Rightarrow I = mv_1$$

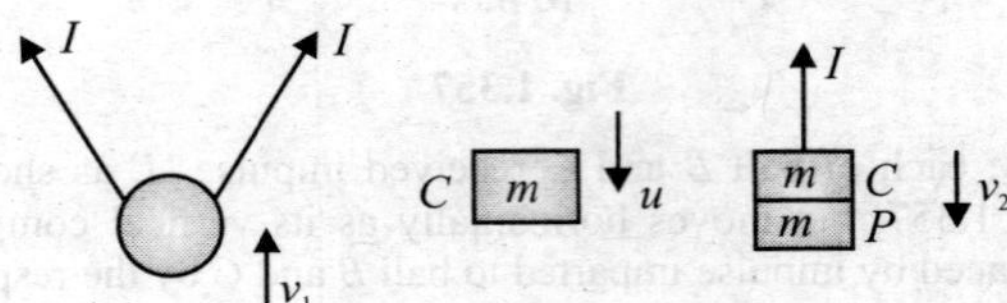

Fig. 1.363

$$I = -2mv_2 - (-mu) \Rightarrow mv_1 + 2mv_2 = mu$$

$$\Rightarrow \quad v_1 + 2v_1 = u \Rightarrow v_1 = \frac{u}{3}$$

For Problems 29–31

29. c., 30. b., 31. a.

Sol.

29. c. $I_1 = mu$, $I_2 = Mu$,

$$I_1 - I = Mu_0,\ I - I_2 = mu_0$$

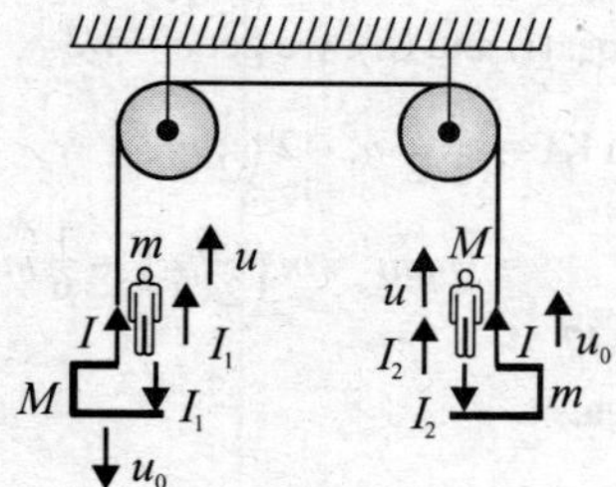

Fig. 1.364

Solve to get: $u_0 = \frac{(m-M)u}{m+M}$

Relative velocity of left man w.r.t. his trolley:

$$u + u_0 = u + \frac{(m-M)u}{m+M} = \frac{2mu}{m+M}$$

30. b. $I = I_1 - Mu_0 = mu - Mu_0 = mu - \frac{M(m-M)}{m+M}u = \frac{(m^2+M^2)u}{M+m}$

31. a. Time taken for men to come to rest:

$$t_1 = \frac{u}{g}$$

Acceleration of trolleys (after the men have jumped):

$$a = \frac{(m-M)}{m+M}g$$

Time taken by trolleys to come to rest:

$$t_1 = \frac{u_0}{a} = \frac{u}{g}$$

For Problems 32–34

32. c., 33. d., 34. a.

Sol.

32. c. Just before collision, velocity of m_1

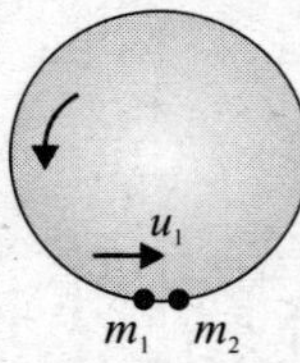

Fig. 1.365

$$u_1 = \sqrt{2g2a} = 2\sqrt{ga}$$

Just after collision, m_1 is brought to rest and let velocity of m_2 be v_2.

From conservation of linear momentum, $m_2 v_2 = m_1 u_1$ (i)

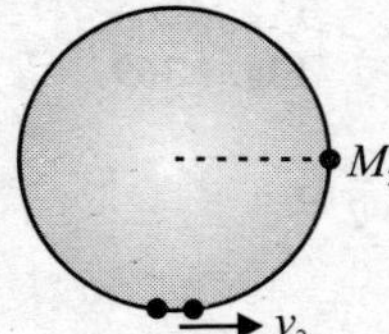

Fig. 1.366

Now $v_2 = \sqrt{2ga}$ so that m_2 can rise up to point M.

Putting the value of u_1 and v_2 in Eq. (i), we get

$$m_2\sqrt{2ga} = m_1 2\sqrt{ga} \Rightarrow \frac{m_1}{m_2} = \frac{1}{\sqrt{2}}$$

33. d. $e = \frac{v_2 - v_1}{u_1 - u_2} = \frac{\sqrt{2ga} - 0}{2\sqrt{ga} - 0} = \frac{1}{\sqrt{2}}$

34. a. $m_1 v_1 + m_2 v_2 = m_2 u_2$

$$\Rightarrow \quad m_1 v_1 + \sqrt{2}m_1 v_2 = \sqrt{2}m_1 u_2$$

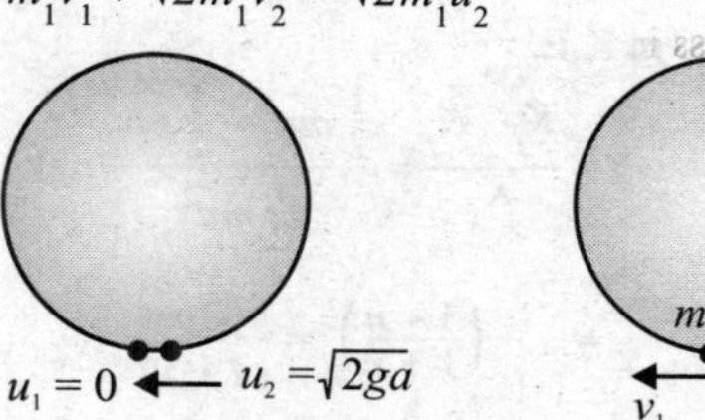

Fig. 1.367

$$\Rightarrow \quad v_1 + \sqrt{2}v_2 = \sqrt{2}u_2 \quad \text{(i)}$$

$$e = \frac{1}{\sqrt{2}} = \frac{v_1 - v_2}{u_2 - u_1} \Rightarrow \frac{1}{\sqrt{2}} = \frac{v_1 - v_2}{u_2}$$

$$\Rightarrow \quad v_1 - v_2 = \frac{u_2}{\sqrt{2}} \quad \text{(ii)}$$

Solving Eqs. (i) and (ii), we get $v_1 = u_2 = \sqrt{2ga}$

$$v_2 = \left(\frac{\sqrt{2}-1}{\sqrt{2}}\right)u_2 = (\sqrt{2}-1)\sqrt{ga}$$

Both are positive, so both move in same direction. Height raised by m_1:

$$\frac{v_1^2}{2g} = a, \text{ same as that of } m_2$$

For Problems 35–37

35. c., 36. a., 37. d.

Sol.

35. c. The acceleration of the centre of mass is $a_{CM} = \frac{F}{2m}$

The displacement of the centre of mass at time t will be

$$x = \frac{1}{2}a_{CM}t^2 = \frac{Ft^2}{4m}$$

36. a.

37. d. Suppose the displacement of the first block is x_1 and that of the second is x_2. Then,

$$x = \frac{mx_1 + mx_2}{2m}, \frac{Ft^2}{4m} = \frac{x_1 + x_2}{2}$$

or $\quad x_1 + x_2 = \frac{Ft^2}{2m}$ (i)

Further, the extension of the spring is $x_1 - x_2$. Therefore,

$$x_1 - x_2 = x_0 \quad \text{(ii)}$$

From Eqs. (i) and (ii),

$$x_1 = \frac{1}{2}\left(\frac{Ft^2}{2m} + x_0\right) \Rightarrow x_2 = \frac{1}{2}\left(\frac{Ft^2}{2m} - x_0\right)$$

For Problems 38–39

38. d., 39. c.

Sol. Let u be initial velocity of the ball of mass m. Then

$$mu = mv_1 + nmv_2 \Rightarrow v_1 + nv_2 = u \quad \text{(i)}$$

For elastic collision, Newton's experimental formula ($u_2 = 0$)

$$v_1 - v_2 = -(u_1 - u_2)$$

$$\Rightarrow \quad v_2 = v_1 + u \quad \text{(ii)}$$

Solving Eqs. (i) and (ii), $v_1 = \frac{1-n}{1+n}u$

Fractional loss in K.E.

$$f = \frac{K_i - K_f}{K_i} = \frac{\frac{1}{2}mu^2 - \frac{1}{2}mv_1^2}{\frac{1}{2}mu^2} = 1 - \left(\frac{v_1}{u}\right)^2$$

$$= 1 - \left(\frac{1-n}{1+n}\right)^2 = \frac{4n}{(n+1)^2}$$

The transfer of energy is maximum (or 100%) when $f = 1$.

$$\therefore \quad \frac{4n}{(n+1)^2} = 1 \Rightarrow n = 1$$

That is the transfer of energy is maximum when the mass ratio is unity.

For Problems 40–41

40. a., 41. d.

Sol.

40. a. Energy loss: $mgh_1 - mgh_2$

$$\text{Fractional loss} = \frac{\text{Energy loss}}{\text{Initial energy}} = \frac{mgh_1 - mgh_2}{mgh_1} = \frac{h_1 - h_2}{h_1}$$

41. d. $h_2 = e^2 h_1 \Rightarrow e^2 = \frac{h_2}{h_1}$.

Height after another collision

$$h_3 = e^2 h_2 = e^4 h_1 = \frac{h_2^2}{h_1}$$

$$\text{Fractional loss} = \frac{mgh_2 - mgh_3}{mgh_2}$$

$$= \frac{h_2 - h_3}{h_2} = \frac{e^2h_1 - e^2h_2}{e^2h_1} = \frac{h_1 - h_2}{h_1} \rightarrow \text{same}$$

For Problems 42–44

42. a., 43. c., 44. b.

Sol. u = velocity of sphere A before impact. As the spheres are identical, the triangle ABC formed by joining their centres is equilateral. The spheres B and C will move in direction AB and AC after impact making an angle of 30° with the original lines of motion of ball A.

Let v be the speed of the balls B and C after impact.

Momentum conservation gives

$$\left(\frac{m}{2}\right)u = mv\cos 30° + mv\cos 30°$$

$$u = 2\sqrt{3}\,v \Rightarrow v = \frac{u}{2\sqrt{3}} \quad \text{(i)}$$

From Newton's experimental law, for an oblique collision, we have to take components along normal, i.e., along AB for balls A and B.

$$v_B - v_A = -e(u_B - u_A)$$

$$v - 0 = -e(0 - u\cos 30°)$$

$$v = eu\cos 30° \quad \text{(ii)}$$

Combining Eqs. (i) and (ii), we get $e = 1/3$.

$$\text{Loss in KE} = \frac{1}{2}\frac{m}{2}u^2 - 2\left(\frac{1}{2}mv^2\right)$$

$$= \frac{1}{4}mu^2 - m\left(\frac{u}{2\sqrt{3}}\right)^2 = \frac{1}{6}mu^2$$

For Problems 45–47

45. b., 46. d., 47. a.

Sol.

45. b.

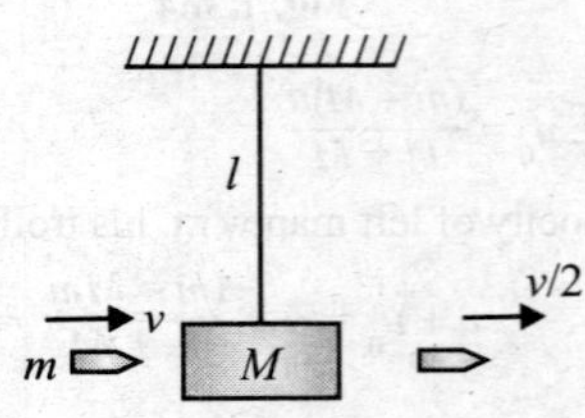

Fig. 1.368

Momentum lost by the bullet $= mv - \frac{mv}{2} = \frac{mv}{2}$

This will be transferred to the bob.

46. d. Let velocity of bob after collision be v_1. Then

$$Mv_1 = \frac{mv}{2} \Rightarrow v_1 = \frac{mv}{2M}$$

Height raised by bob $= l$

$$l = \frac{v_1^2}{2g} \Rightarrow l = \frac{m^2v^2}{8M^2g} \Rightarrow v = \frac{2M}{m}\sqrt{2gl}$$

47. a. Here $v_1 = \sqrt{5gl}$

$$T = Mg + \frac{Mv_1^2}{l} = Mg + \frac{M5gl}{l} = 6\,Mg$$

For Problems 48–50

48. c., 49. a., 50. a.,

Sol.

48. c. $mv_1 = 2mv_2$

$$\Rightarrow \quad v_1 = 2v_2$$

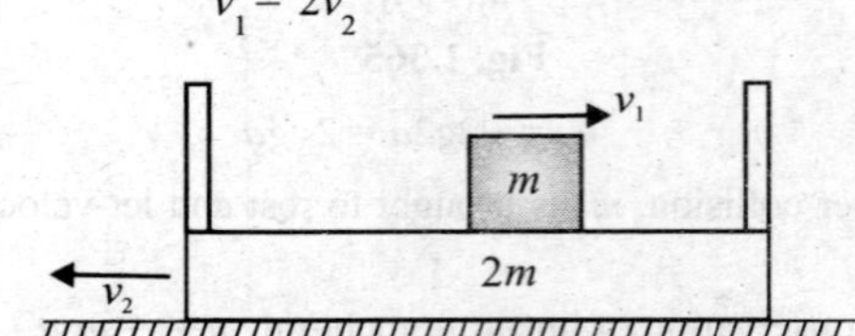

Fig. 1.369

$$\frac{1}{2}mv_1^2 + \frac{1}{2}\times 2mv_2^2 = \frac{1}{2}kx^2$$

Solving, we get

$$v_1 = 2\left(\sqrt{\frac{k}{6m}}\right)x,\; v_2 = \left(\sqrt{\frac{k}{6m}}\right)x$$

Relative velocity:

$$v_1 + v_2 = 3\left(\sqrt{\frac{k}{6m}}\right)x = x\sqrt{\frac{3}{2}\frac{k}{m}}$$

49. a. Momentum of both will be same in magnitude and opposite in direction at any time.

50. a. $KE_i = \frac{1}{2}mv_1^2 + \frac{1}{2}\times 2mv_2^2 = \frac{1}{2}kx^2$

$$KE_f = \frac{1}{2}m\left(\frac{v_1}{2}\right)^2 + \frac{1}{2}\times 2m\left(\frac{v_2}{2}\right)^2 = \frac{1}{4}\left[\frac{1}{2}kx^2\right]$$

$$\text{Loss} = KE_i - KE_f = \frac{3}{8}kx^2$$

For Problems 51–54

51. b., 52. c., 53. d., 54. b.,

Sol.

51. b. Using conservation of linear momentum in horizontal direction just before and just after collision,

$$mu = (m+M)V \Rightarrow V = \frac{mu}{(M+m)}$$

52. c. Using conservation of mechanical energy,

$$\frac{1}{2}(M+m)V^2 = (m+M)gh$$

$$V^2 = 2gh \Rightarrow \left[\frac{mu}{(M+m)}\right]^2 = 2gh$$

$$u^2 = \left(\frac{M+m}{m}\right)^2 2gh$$

$$\Rightarrow \quad u = \frac{[M+m]}{m}\sqrt{2gh}$$

53. d. As bullet and block coalesce after collision, hence collision is perfectly inelastic.

54. b. Energy dissipated in collision

$$= \text{initial energy} - \text{final energy}$$

$$= \frac{1}{2}mu^2 - (m+M)gh$$

For Problems 55–56

55. a., 56. a.

Sol. Let us adopt the vector approach. Let the mass of each particle be 'm' and let they be denoted by A, B and C. Before collision,

$$\text{velocity of } A = 10\hat{i}$$

$$\text{velocity of } B = 20(\cos 30°\,\hat{i} + \sin 30°\,\hat{j})$$

$$= 10(\sqrt{3}\,\hat{i} + \hat{j})$$

$$\text{velocity of } C = 30\hat{j}$$

$$\text{Initial momentum, } \vec{P} = m[10\hat{i} + 10(\sqrt{3}\,\hat{i} + \hat{j}) + 30\hat{j}]$$

$$= 10m[(1+\sqrt{3})\,\hat{i} + 4\hat{j}] \quad \text{(i)}$$

Let the final velocity be $\vec{v}$, then final momentum

$$\vec{p_2} = 3m\vec{v}$$

According to the law of conservation of linear momentum, we have initial momentum = final momentum

i.e., $10m[(1+\sqrt{3})\,\hat{i} + 2\hat{j}] = 3m\vec{v}$

$$\vec{v} = \frac{10}{3}[(1+\sqrt{3})\,\hat{i} + 4\hat{j}]$$

Therefore, magnitude of $\vec{v}$, i.e.,

$$|\vec{v}| = \frac{10}{3}\sqrt{(1+\sqrt{3})^2 + 4^2} = \frac{10}{3}\sqrt{20+2\sqrt{3}} \text{ m/s}$$

If the final velocity $\vec{v}$ makes an angle α with the positive direction of X-axis, then $\tan\alpha = \frac{4}{1+\sqrt{3}} = 2(\sqrt{3}-1)$;

$$\alpha = \tan^{-1}[2(\sqrt{3}-1)]$$

For Problems 57–59

57. d., 58. a., 59. a.

Sol.

57. d. Since the collision is elastic, there is no loss in KE.

58. a. After the first collision, the velocities are as shown in Fig. 1.370:

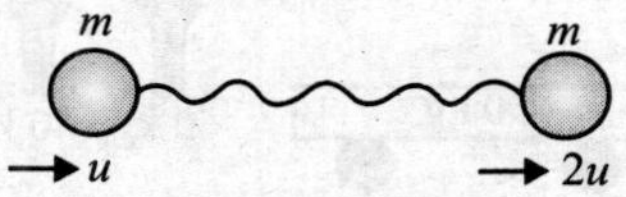

Fig. 1.370

Now after some time, the string will become tight. Let at this point the common velocity of the balls be v.

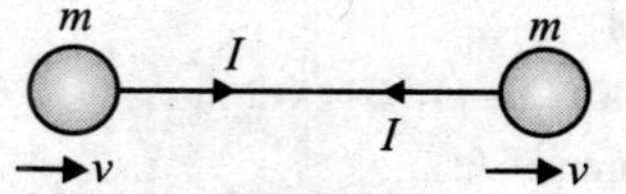

Fig. 1.371

$$2mv = mu + 2mu \Rightarrow v = 3u/2$$

$$KE_i = \frac{1}{2}mu^2 + \frac{1}{2}m(2u)^2 = 5mu^2/2$$

$$KE_f = \frac{1}{2}2mv^2 = \frac{1}{2}2m\left(\frac{3u}{2}\right)^2 = \frac{9mu^2}{4}$$

$$KE_{Loss} = KE_i - KE_f = \frac{5mu^2}{2} - \frac{9mu^2}{4} = mu^2/4$$

59. a. For left ball: $I = mv - mu = m\frac{3u}{2} - mu = mu/2$

For Problems 60–63

60. b., 61. c., 62. a., 63. d.

Sol.

60. b. Using area of Δ = change in momentum

$$\frac{1}{2}F_{max}\Delta t = 250\times 10^{-3}(80)$$

$$\frac{1}{2}F_{max}(8\times 10^{-3}) = 250\times 10^{-3}\times 80$$

$$F_{max} = 5000 \text{ N}$$

61. c.

$$\int F dt = \Delta p$$

$$F_v = \frac{\Delta p}{\int dt} = \frac{250\times 10^{-3}(50+30)}{8\times 10^{-3}} = \frac{250\times 80}{8}$$

$$F_v = 2500 \text{ N}$$

62. a. $\text{Impulse} = \frac{1}{2}\times 600\times 5\times 10^{-3} = 4\text{ N}$

$P_f = \text{Impulse} + P_i \Rightarrow P_f = 4 + \frac{250}{1000}(-4)$

where $u = \sqrt{2gh_1} = \sqrt{2\times 10\times 5} = 10\text{ m/s}$

$\Rightarrow \quad p_f = 1.5 \text{ N}$

63. d. $v = \frac{P_f}{m} = \frac{1.5}{0.250} = 6\text{ m/s}$, $h_2 = \frac{v^2}{2g} = \frac{6^2}{2\times 10} = 1.8\text{ m}$

For Problems 64–66

64. d., 65. a., 66. d.

Sol.

64. d. Centre of mass will remain at rest because there is no external force on the system of A, B and platform.

65. a. When both jump simultaneously.

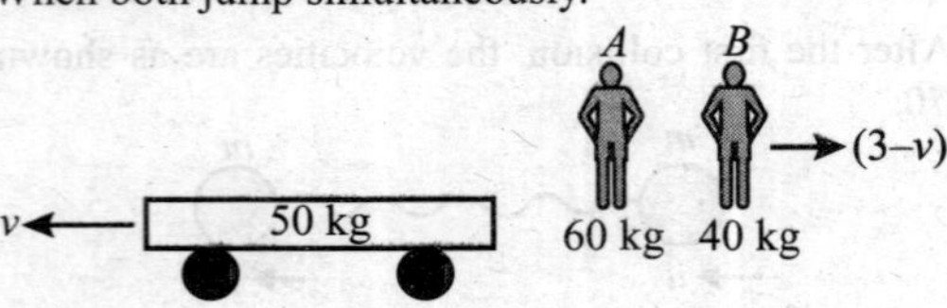

Fig. 1.372

Let speed of the platform be v, then speed of A and B will be $3-v$ as shown because relative velocity is 3 m/s. Conserving momentum is

$50v = (60 + 40)(3 - v) \Rightarrow v = 2$ m/s

66. d. After A jumps:

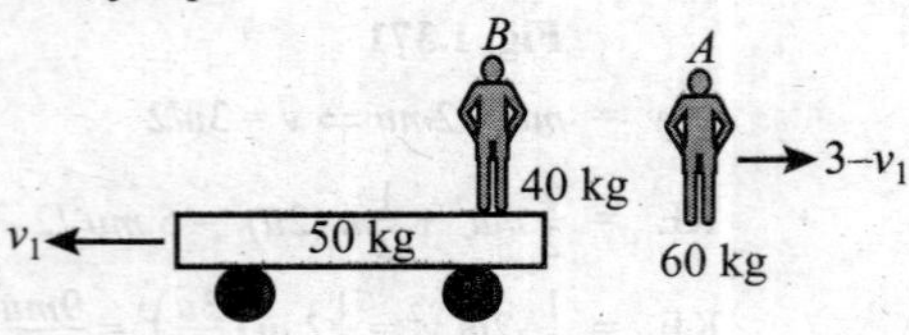

Fig. 1.373

$(50 + 40)\, v_1 = 60\,(3 - v_1) \Rightarrow v_1 = 1.2$ m/s

After B jumps:

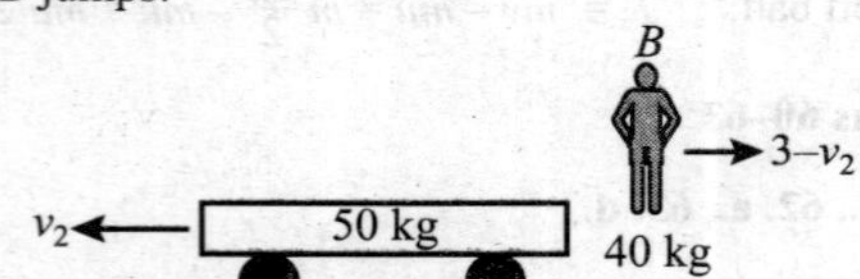

Fig. 1.374

$$(50 + 40)\, v_1 = 50v_2 - 40\,(3 - v_2)$$

$$\Rightarrow 90 \times 1.2 = 50v_2 - 120 + 40v_2 \Rightarrow v_2 = \frac{228}{90} = 2.53 \text{ m/s}$$

For Problems 67–68

67. a., 68. c.

Sol. When M strikes the wall, vertically downward component of its displacement from initial position is 1 m and its distance from pulley B is

$$C'B = \sqrt{1^2 + 2^2} = \sqrt{5} \text{ m}$$

While its initial distance from the pulley was $CB = 1$ m. It means vertically upward displacement of mass m is $(\sqrt{5} - 1)$ m.

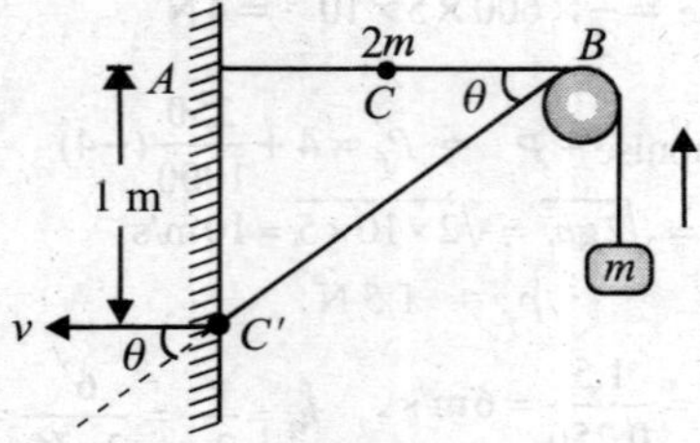

Fig. 1.375

Let M strike the wall with velocity v. Since the string between the two blocks always remains taut, therefore at any instant speed of m is equal to that component of velocity of M which is along the string $C'B$. Hence, velocity of m when M strikes the wall is $v \cos\theta$, where

$$\cos\theta = \frac{2}{\sqrt{5}}$$

$$\therefore \quad \frac{v_M}{v_m} = \frac{v}{v\cos\theta} = \frac{\sqrt{5}}{2}$$

According to law of conservation of energy, loss of potential energy of M = increase in PE of m + KE of M + KE of m

$$Mg \times 1 = mg\,(\sqrt{5} - 1) + \frac{1}{2}Mv^2 + \frac{1}{2}m\,(v\cos\theta)^2$$

$$v = 5\sqrt{\frac{5 - \sqrt{5}}{6}} \text{ m/s}$$

For Problems 69–72

69. d., 70. b., 71. a., 72. d.

Sol. Let muzzle velocity be u_0 and height of cliff be h. Assuming horizontal direction to be positive x-axis and vertically upward direction to be positive y-axis, coordinates of top of cliff become $(20, h)$ as shown in the figure.

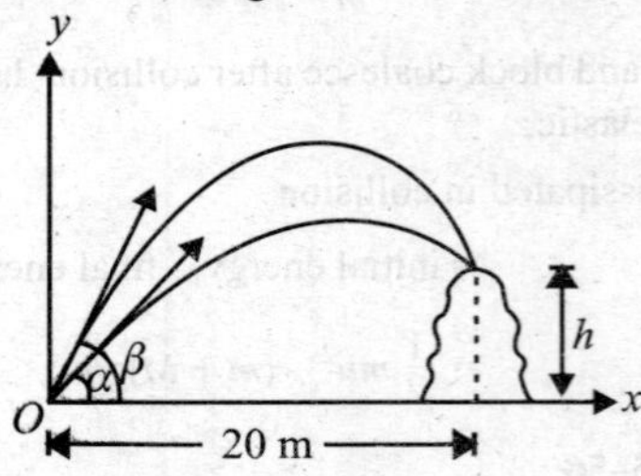

Fig. 1.376

Using equation of trajectory of a projectile for two shells,

$$h = 20\tan\alpha - \frac{g\,(20)^2}{2u_0^2\cos^2\alpha}$$

$$= 20\tan\beta - \frac{g\,(20)^2}{2u_0^2\cos^2\beta}$$

From the above two equations,

$$u_0 = 20 \text{ m/s and } h = 10 \text{ m}$$

Time taken by the shell, fired at angle α, to reach the top of cliff is

$$t_1 = \frac{20}{u_0\cos\alpha} = \sqrt{2} \text{ s}$$

Time taken by the other shell is

$$t_2 = \frac{20}{u_0\cos\beta} = \sqrt{10} \text{ s}$$

Hence, the shell having angle of projection β was fired first and the other shell (having angle of projection α) was fired later.

Time interval between two firings is

$$t_2 - t_1 = (\sqrt{10} - \sqrt{2}) \text{ s} = 1.74 \text{ s}$$

Consider vertical velocities of shells just before striking the top of cliff.

For the first shell,

$$v_{1y} = u_0 \sin\alpha - gt_1 = 20 \times \frac{1}{\sqrt{2}} - 10\sqrt{2} = 0$$

For the second shell,

$$v_{2y} = u_0 \sin\beta - gt_2 = 20 \times \frac{3}{\sqrt{10}} - 10\sqrt{10} = -4\sqrt{10} \text{ m}$$

Let v_y be combined vertical velocity of the shell after sticking together, then from conservation of momentum in vertical direction, we get

$$(m+m)v_y = m \times 0 + m(-4\sqrt{10} \Rightarrow v_y = -2\sqrt{10} \text{ m/s}$$

Collision with the top of cliff is perfectly elastic, so combined shell will rebound with velocity $2\sqrt{10}$ m/s in vertical direction. Maximum height above top of cliff will be $(2\sqrt{10})^2/2g = 2$ m.

Net maximum height = 10 + 2 = 12 m

For Problems 73–76

73. a., 74. b., 75. c., 76. a.

Sol. When two balls collide, their height from ground is same. Let it be h. Then vertical displacement of the first and second balls is $(65 - h)$ m and $(30 - h)$ m, respectively. Let the time of flight of the first ball up to the instant be n seconds. Then the time of flight that of second ball is $(n - 1)$ seconds. Considering vertically downward component of motion of first ball,

$$u = 10\sqrt{2} \sin 45° = 10 \text{ m/s}$$

$$a = g = 10 \text{ ms}^{-1}, s = y_1 = (65 - h), t = n \text{ s}$$

$$s = ut + \frac{1}{2}at^2$$

$$(65 - h) = 10n + 5n^2 \quad \text{(i)}$$

For vertically downward component of motion of second ball,

$$u = 0, a = g = 10 \text{ ms}^{-2}, t = (n - 1)$$

$$s = y_2 = (30 - h)$$

$$(30 - h) = 5(n - 1)^2 \quad \text{(ii)}$$

From Eqs. (i) and (ii), $n = 2$ s, $h = 25$ m

Displacement AC = Horizontal distance moved by first ball up to the instant of collision + that moved by the second ball up to the same instant

$$\therefore \quad AC = (10\sqrt{2} \cos 45°)\, n + 20(n - 1) = 40 \text{ m}$$

Just before collision, vertically downward component of velocity of first ball is

$$v_{y1} = (10\sqrt{2} \sin 45°) + gn = 30 \text{ m/s}$$

And that of second ball is $v_{y2} = g(n - 1) = 10$ m/s

According to the law of conservation of momentum, vertically downward component v_y of velocity of combined body is given by

$$mv_{y1} + mv_{y2} = (m+m)v_y \quad \text{or} \quad v_y = 20 \text{ ms}^{-1}$$

Similarly, horizontally leftward component v_x of velocity of combined body (just after collision) is given by

$$m \times 20 - m \times 10\sqrt{2} \cos 45° = (m+m)v_x$$

$$\Rightarrow \quad v_x = 5 \text{ m/s}$$

For Problems 77–79

77. a., 78. b., 79. b.

Sol. Let velocity of ball A at highest point be v then first considering its FBD at this point

$$mg + T = \frac{mv^2}{r}$$

where $\quad T = mg$

$$\Rightarrow \quad v^2 = 2rg$$

$$\Rightarrow \quad v = \sqrt{2rg} = 2\sqrt{6} \text{ m/s}$$

mg

$\frac{mv^2}{r}$

$T = mg$

Fig. 1.377

If velocity of ball A just after the collision is v_1 then according to law of conservation of energy,

its KE at A = (KE + PE) at highest point

$$\frac{1}{2}mv_1^2 = \frac{1}{2}mv^2 + mg(2r)$$

$$v_1 = \sqrt{6rg} \text{ or } v_1 = 6\sqrt{2} \text{ m/s}$$

If angle of projection of ball B is θ, then horizontal component of its velocity will be $9\cos\theta$ and it remains constant.

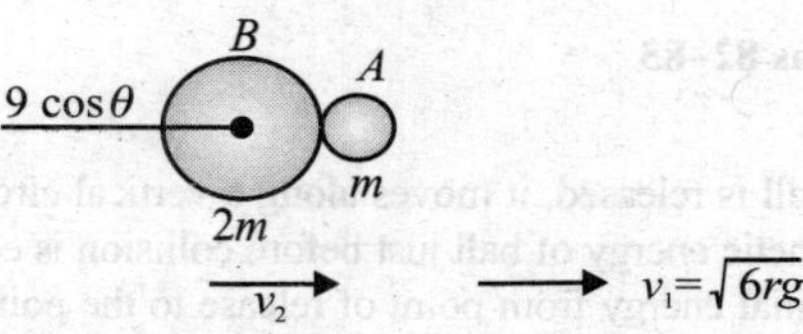

Fig. 1.378

Now considering collision of balls B and A, let velocities of balls A and B, just after the collision, be v_1 and v_2, respectively as shown in Fig. 1.378.

Then according to the law of conservation of momentum,

$$mv_1 + 2mv_2 = 2m(9\cos\theta) \quad \text{(i)}$$

Since collision is elastic, therefore, $e = 1$

$$\frac{v_1 - v_2}{9\cos\theta} = 1 \text{ or } v_1 - v_2 = 9\cos\theta \quad \text{(ii)}$$

Solving Eqs. (i) and (ii), we get

$$\cos\theta = \frac{1}{\sqrt{2}} \text{ or } \theta = 45°$$

$$v_2 = \frac{3}{\sqrt{2}} \text{ ms}^{-1}$$

Height of initial position of ball A (lowest position) is equal to maximum height ascended by the ball B. It is equal to

$$H = \frac{(9\sin\theta)^2}{2g} = \frac{81}{40} \text{ m}$$

Height of point of suspension from the ground is

$$H + r = \frac{129}{40} \text{ m}$$

For Problems 80–81

80. c., 81. a.

Sol. Since rear buggy stops and velocity of man relative to this buggy is v, therefore, absolute velocity of man at the time of jump is v. Applying law of conservation of momentum on the system of rear buggy and man,

$$mv = (150 + m)\, 4 \qquad \text{(i)}$$

Kinetic energy of man and front buggy (just before collision) is

$$E_1 = \frac{1}{2} mv^2 + \frac{1}{2} \times 150 \times 4^2$$

Let velocity of the front buggy, after collision, be v_c.

Applying law of conservation of momentum on the system of front buggy and man, we get

$$mv + 150 \times 4 = (m + 150)v_c \qquad \text{(ii)}$$

Kinetic energy of the system after collision is

$$E_2 = \frac{1}{2}(m + 150)\, v_c^2 \qquad \text{(iii)}$$

But $\quad E_1 - E_2 = 2700$ J

From above three equations, $v_c = 7$ m/s

$m = 50$ kg

$v = 16$ m/s

For Problems 82–83

82. b., 83. c.

Sol. When ball is released, it moves along a vertical circle with centre at A. Kinetic energy of ball just before collision is equal to loss of its potential energy from point of release to the point of collision. Therefore, velocity v_1 of ball, just before collision, is given by

$$\frac{1}{2} mv_1^2 = mg\,(l - l\cos\theta) \Rightarrow v_1 = 5 \text{ m/s}$$

After collision, the block starts to move towards right. But it is retarded by force of friction and ultimately it comes to rest. According to law of conservation of energy,

Kinetic energy of the block just after collision = Work done by it against friction

Therefore, its velocity v_2 just after collision is given by

$$\frac{1}{2} mv_2^2 = \mu mgs$$

where $m = 0.5$ and $s = 2.50$ m. $\quad \therefore v_2 = 5$ m/s

Coefficient of restitution, $e = -\dfrac{v_2 - v_1}{u_2 - u_1}$

where $u_2 = 0$, $u_1 = 5$ m/s and $v_2 = 5$ m/s.

$$\therefore e = -\frac{5 - v_1}{0 - 5} \quad \text{or} \quad v_1 = (5 - 5e) \qquad \text{(i)}$$

Applying law of conservation of momentum,

$$mu_1 + mu_2 = mv_1 + mv_2$$

$$5\,m + (m \times 0) = mv_1 + 5\,m \quad \text{or} \quad v_1 = 0$$

Substituting in Eq. (i), we get $e = 1$.

For Problems 84–86

84. b., 85. c., 86. a.

Sol. When shell strikes the ball and gets stuck with it, combined body of mass $2m$ starts to move to the right. Let velocity of the combined body (just after collision) be v_1. According to law of conservation of momentum,

$$(m + m)\, v_1 = mv_0$$

$$v_1 = \frac{v_0}{2} = 3 \text{ m/s}$$

As soon as the combined body starts to move rightwards, thread becomes inclined to the vertical. Horizontal component of its tension retards the combined body while trolley accelerates rightwards due to the same component of tension.

Inclination of thread with the vertical continues to increase till velocities of both (combined body and trolley) become identical or combined body comes to rest relative to the trolley. Let velocity at that instant of maximum inclination of thread be v. According to law of conservation of momentum,

$$(2m + M)v = 2\, mv_1$$

or $\quad v = 1$ m/s

During collision of ball and shell, a part of energy is lost. But after that there is no loss of energy. Hence, after collision, kinetic energy lost is used up in increasing gravitational potential energy of the combined body.

If maximum inclination of threads with the vertical is θ, then according to law of conservation of energy,

$$\frac{1}{2}(2m)v_1^2 - \frac{1}{2}(2m + M)\, v^2 = 2\, mg\,(l - l\cos\theta)$$

$$\cos\theta = 0.8 \;\text{ or } \theta = 37°$$

For Problems 87–88

87. d., 88. a.

Sol. Ball of mass m_1 falls freely till thread becomes taut. At that instant inclination θ of the thread with the vertical be given by

$$\sin\theta = \frac{\left(\frac{\sqrt{3}}{2}\, l\right)}{l} = \frac{\sqrt{3}}{2} \text{ or } \theta = 60°$$

Ball of mass m_1 falls freely through height.

$$l\cos\theta = \frac{l}{2}$$

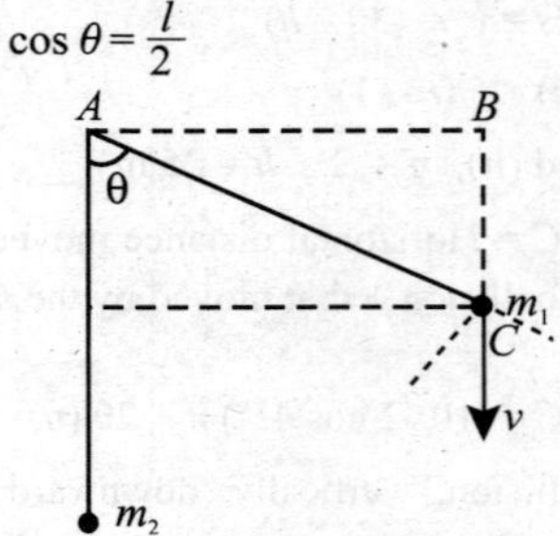

Fig. 1.379

Velocity of this ball at this instant is $v = \sqrt{2g \times \dfrac{l}{2}} = \sqrt{gl}$

It can be resolved into two components.

i. $v\cos\theta$, along the thread. But thread is inextensible, hence this component decreases to zero (due to tension developed in the thread).

ii. $v\sin\theta$, perpendicular to the thread. Due to this component ball starts to move along a circle whose centre is at A.

According to law of conservation of energy,

Kinetic energy of ball m_1 just before collision = Its kinetic energy at AC + Further loss of its potential energy

$$\frac{1}{2} m_1 u_1^2 = \frac{1}{2} m_1 (\sqrt{gl} \sin \theta)^2 + m_1 g (l - l \cos \theta)$$

$$u_1 = 4 \text{ m/s}$$

According to law of conservation of momentum,

$$m_1 v_1 + m_2 v_2 = m_1 u_1 + m_2 u_2$$

And coefficient of restitution, $e = -\frac{v_2 - v_1}{u_2 - u_1} = 1$

Substituting $u_1 = 4$ m/s, $u_2 = 0$, $m_1 = 0.1$ kg and $m_2 = 0.3$ kg, $v_2 = 2$ m/s, the height to which it rises is

$$h = \frac{v_2^2}{2g} = 0.20 \text{ m}$$

For Problems 89–91

89. a., 90. a., 91. a.

Sol. When C strikes with B, the combined body starts to rise vertically upward. According to law of conservation of momentum, velocity v_0 of combined body (just after collision) is given by

$$(2 + 1) v_0 = 1 \times 9 \quad \text{or} \quad v_0 = 3 \text{ m/s}$$

But A is at rest. Therefore, string becomes slack and bodies move under gravity, combined body upwards and A downwards till string again becomes taut. This happens when downward displacement of A becomes equal to upward displacement of combined body. Let it happen at instant t_1. Then,

$$\frac{1}{2} g t_1^2 = 3 t_1 - \frac{1}{2} g t_1^2$$

or $\quad t_1 = 0.3$ s

Displacement of each block at that instant,

$$h_1 = \frac{1}{2} g t_1^2 = 0.45 \text{ m}$$

And velocity of combined body is $3 - g t_1 = 0$ and velocity of A is $g t_1 = 3$ m/s (downward).

Since at this instant, velocities are different, therefore an impulse is developed in the string and magnitude of velocities of bodies becomes equal. Let that velocity magnitude be v' and impulse developed be J. This impulse acts upwards on both the bodies.

Hence for A,

$$2 \times 3 - J = 2v' \qquad \text{(i)}$$

And for combined body,

$$J = 3v' \qquad \text{(ii)}$$

From above equations,

$$v' = 1.2 \text{ m/s}$$

Now combined body starts to move upwards with velocity $v' = 1.2$ m/s and retardation a, where

$$a = \frac{3 - 2}{3 + 2} g = 2 \text{ m/s}^2$$

Combined body comes to an instantaneous rest after time

$$t_2 = \frac{v'}{a} = 0.6 \text{ s}$$

After this instant, combined body starts moving down. Time interval between this instant and instant of collision is given by

$$t_1 + t_2 = 0.3 + 0.6 = 0.9 \text{ s}$$

Further height ascended by combined body during time t_2 is

$$h_2 = v' t_2 - \frac{1}{2} a t_2^2 = 0.36 \text{ m}$$

Maximum height raised by combined body is

$$h = h_1 + h_2 = 0.81 \text{ m}$$

Loss of energy during collision of B and C is

$$E_1 = \frac{1}{2} \times 1 \times 9^2 - \frac{1}{2} 3 v_0^2 = 27 \text{ J}$$

Loss of energy when impulse is developed in string is

$$E_2 = \frac{1}{2} \times 2 \times 3^2 - \frac{1}{2} (2 + 3) (v')^2 = 5.4 \text{ J}$$

Loss of mechanical energy till B reaches highest point is

$$E = E_1 + E_2 = 27 + 5.4 = 32.4 \text{ J}$$

For Problems 92–95

92.d., 93. b., 94. d., 95. c.

Sol. Due to collision, an impulsive tension occurs in the string.

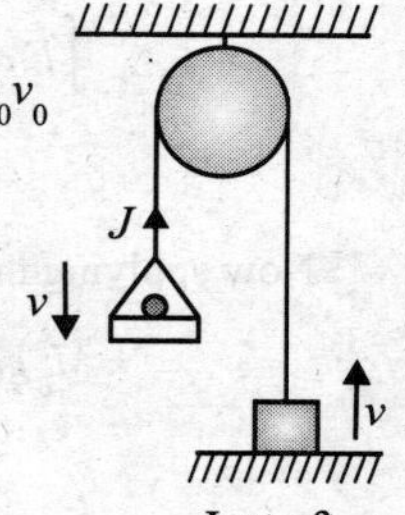

Just after collision

Fig. 1.380

For ball + pan system, $-J = (m + m_0) v - m_0 v_0$

For block, $J = Mv \Rightarrow v = 2$ m/s

As the string is taut and the mass of the block is greater than the mass of ball + pan, so the block and ball + pan are under deceleration, given by

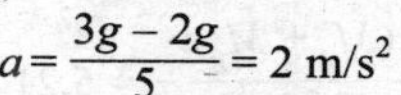

$$a = \frac{3g - 2g}{5} = 2 \text{ m/s}^2$$

So, block will again come to ground after time t, given by

$$0 = 2t - \frac{1}{2} \times 2t^2 \quad \Rightarrow t = 2 \text{s}$$

At this instant, the pan + ball system is moving with $v = 2$ m/s and the string gets slacked, so the system is moving under gravity.

To determine the speed of the system at $t = 2.6$ s, first find the time after which string is again jerked. That will happen when pan + ball system crosses its $t = 2$ s instant during its downward journey, i.e., after its time of flight of motion under gravity.

$$T = \frac{2 \times 2}{10} = 0.4 \text{ s}$$

i.e., string will again jerk at $t = 2.4$ s.

For $t > 2.4$ s, the system is again under a retardation of 2 m/s^2. Before this, find the velocity of various components of system just after jerk. We have

$$-J' = 2v' - 2 \times 2 \text{ and } J = 3v'$$

$$5v = 4 \Rightarrow v = \frac{4}{5} = 0.8 \text{ m/s}$$

With this velocity, the ball + pan system moves down for 0.2 s under a retardation of 2 m/s^2.

$$v = 0.8 - 2 \times 0.2 = 0.4 \text{ m/s (downwards)}$$

The maximum height attained by the block after the second jerk is given by $0 = (0.8)^2 - 2 \times 2 H \Rightarrow H = 0.16$ m.

Matching Column Type

1. i. → c.; ii. → d.; iii. → a.; iv. → b.

$M_A = 1$ kg, $M_B = 2$ kg

For bullet, $\int F\,dt = m(v_f - v_i)$

$$= 0.25[(-100) - (-200)] = 25 \text{ kg m/s} \quad \text{(i)}$$

This is also impulse by particle on A.

For block A, $\int (F - T)\,dt = M_A(v - 0)$ (ii)

For block B, $\int T\,dt = M_B(v - 0)$ (iii)

From Eqs. (ii) and (iii), we get

$$\int F\,dt = (M_A + M_B)v \Rightarrow 25 = 3v$$

$$v = \frac{25}{3} \text{ m/s}, \int T\,dt = M_B v = \frac{2 \times 25}{3} = \frac{50}{3}$$

Now applying the law of conservation of energy,

$$+M_B gh = +\frac{1}{2}(M_A + M_B)v^2$$

$$\Rightarrow \quad h = \frac{1}{2}\frac{(M_A + M_B)v^2}{M_B g} = 5.21 \text{ m}$$

2. i. → a.; ii. → b.; iii. → a., c.; iv. → b., d.

i. If velocity of block A is zero, from conservation of momentum, speed of block B is $2u$. Then KE of block B $= \frac{1}{2}m(2u)^2 = 2mu^2$ is greater than net mechanical energy of the system. Since this is not possible, velocity of A can never be zero.

ii. Since initial velocity of B is zero, it shall be zero for many other instants of time.

iii. Since momentum of system is non-zero, KE of system cannot be zero. Also KE of system is minimum at maximum extension of the spring.

iv. The potential energy of spring shall be zero whenever it comes to natural length. Also PE of spring is maximum at maximum extension of the spring.

3. i. → b.; ii. → a., b.; iii. → c.; iv. → d.

i. Initial velocity of centre of mass of given system is zero and net external force is in vertical direction. Since there is shift of mass downwards, the centre of mass has only downward shift.

ii. Obviously, there is shift of centre of mass of given system downwards. Also the pulley exerts a force on string which has a horizontal component towards right. Hence, centre of mass of system has rightward shift.

iii. Both block and monkey move up, hence centre of mass of the given system shifts vertically upwards.

iv. Net external force on given system is zero. Hence, centre of mass of the given system remains at rest.

4. i. → d.; ii. → a.; iii. → c.; iv. → b.

From theory.

i. Momentum $= mv$

ii. Impulse = change in momentum $= m(v_2 - v_1)$

iii. In elastic collision, momentum and KE both are conserved.

iv. In inelastic collision, momentum is conserved but KE is not conserved.

5. i. → b., c.; ii. → a., d.; iii. → c.; iv. → a.

i. Internal forces may affect the motion of individual particles of a system, but not that of the centre of mass.

ii. Centre of momentum of a body can be inside or outside the body.

iii. Kinetic energy of a system of two blocks may increase, decrease or remain same with time, but it will have no effect on motion of the centre of mass.

iv. Centre of mass of a solid cylinder is inside the cylinder.

6. i. → a., c., d.; ii. → a., b., c., d.; iii. → a.; iv. → a., b.

$$v_{CM} = \frac{3 \times 2 + 6 \times 0}{3 + 6} = \frac{2}{3} \text{ m/s}$$

So velocity of centre of mass is always 2/3 m/s.

i. If velocity of 3 kg block is 2/3 m/s, then velocity of 6 kg block is also 2/3 m/s. In this situation, spring will be maximum elongated. Their relative velocity will be zero.

ii. When the speed is 2/3 m/s, velocity can be 2/3 m/s, if $v_{3kg} = -2/3$ m/s.

$$v_{CM} = \frac{2}{3} = \frac{3 \times \left(-\frac{2}{3}\right) + 6v_2}{3 + 6}$$

$$v_2 = \frac{4}{3} \text{ m/s}$$

Let deformation in spring be x. Then

$$\frac{1}{2} \times 3\,(2)^2 = \frac{1}{2} \times 3\left(\frac{2}{3}\right)^2 + \frac{1}{2} \times 6\left(\frac{4}{3}\right)^2 + \frac{1}{2}kx^2$$

$$\Rightarrow \quad x = 0$$

iii. Minimum speed of 3 kg block is zero. At this speed, spring will have some deformation, but not maximum.

iv. Speed of 6 kg block will be maximum when spring returns to its normal length after passing through elongation.

7. i. → a., b.; ii. → a., b.; iii. → d.; iv. → c.

$$x_1 = u\cos\theta\, t_1;\ x_2 = u\cos\theta\, t_2$$

$$\sqrt{\frac{2h}{g}} = \frac{T}{2} = \frac{t_1 + t_2}{2};\ \sqrt{\frac{2h}{2g}} = \frac{t_2 - t_1}{2}$$

Dividing, $\dfrac{t_2 + t_1}{t_2 - t_1} = \sqrt{2}$

Also, $\dfrac{x_2 - x_1}{x_2 + x_1} = \dfrac{t_2 - t_1}{t_2 + t_1}$

$$T = \frac{2u\sin\theta}{g};\ t_1 + t_2 = \frac{2u\sin\theta}{g}$$

$$\frac{1}{2} = \frac{u \sin\theta}{g(t_1 + t_2)}$$

Also, $x_1 + x_2 = u \cos\theta\,(t_1 + t_2)$

$$\frac{u \cos\theta\,(t_1 + t_2)}{x_1 + x_2} = 1$$

8. i. → a., b.; c.; ii. → a., b., c.; iii. → a., b.; iv. → a., b.

$$v_{CM} = \frac{mv + mv + m(-v)}{3m} = \frac{v}{3}$$

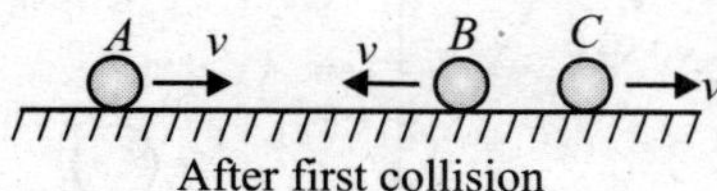

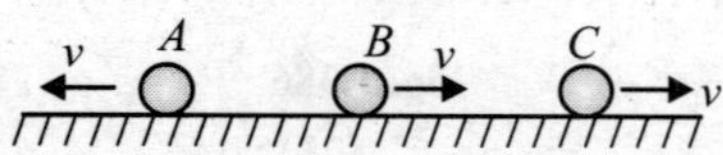

Fig. 1.381

If A and B are connected, then they will act as a mass of $2m$.
If A and B are connected by an ideal string, then finally A and B will be at rest.

9. i. → c., d.; ii. → c.; iii. → a.; iv. → a., b., c., d.

i. $8u_1 + 2(-3) = 8v_1 + 2v_2$

$$\frac{v_2 - v_1}{u_1 - (-3)} = e = 0.5$$

Solving, we get $v_1 = \frac{7u_1 - 9}{10}$, $v_2 = \frac{12u_1 + 6}{10}$

v_2 is always positive.

For $v_1 > 0$ $\quad u_1 > \frac{9}{7}$ m/s

ii. $8u_1 + 2(-3) = 8v_1 + 2v_2$

$$\frac{v_2 - v_1}{u_1 - (-3)} = e = 1$$

For maximum energy to transfer to m_2, $v_1 = 0$

Solving, we get $u_1 = 2$ m/s.

iii. For this $v_2 = 0$

iv. Depends upon e.

10. i. → a., d.; ii. → c., d.; iii. → c.; iv. → a., b., c.

In inelastic collision, kinetic energy is always lost. In elastic collision, kinetic energy remains constant.
If total work done is zero, then kinetic energy will remain constant. If non-conservative forces are not present, then kinetic energy may decrease, increase or remain constant, depending upon other forces.

11. i. → a., d.; ii. → c., d.; iii. → c., d.; iv. → c., d.

In elastic collision, kinetic energy, momentum and total energy all remain conserved. In inelastic collision, kinetic energy is lost but other quantities remain conserved.

Integer Answer Type

1. (2) For the first collision, $e = 1$, $v = v_1 + v_2$

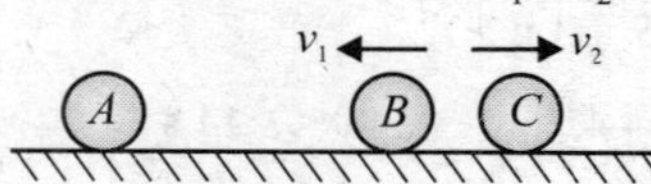

Fig. 1.382

$$\Rightarrow \quad v_2 = v - v_1 \qquad \text{(i)}$$

By momentum conservation

$$m_B v = -m_B v_1 + m_C v_2$$

$$m_B v = -m_B v_1 + 4\, m_B v_2$$

$$v_2 = \frac{v_1 + v}{4} \qquad \text{(ii)}$$

From Eqs. (i) and (ii), $v_1 = \frac{3}{5}v$ and $v_2 = \frac{2}{5}v$

For the second collision, $e = 1$

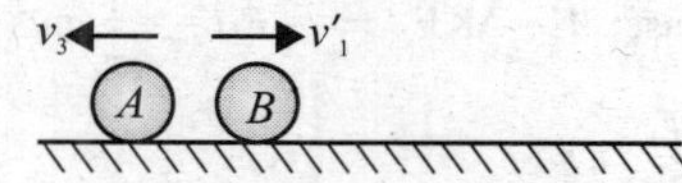

Fig. 1.383

$$v_1 = v_1' + v_3 \Rightarrow v_3 = v_1 - v_1' \qquad \text{(iii)}$$

By momentum conservation, $-m_B v_1 = m_B v_1' - m_A v_3$

or $\quad -m_B v_1 = m_B v_1' - 4\, m_B v_3 \quad (\because m_A = 4\, m_B)$

$$v_3 = \frac{v_1' + v_1}{4} \qquad \text{(iv)}$$

From Eqs. (iii) and (iv), $v_1' = \frac{3}{5}v_1 = \frac{3}{5}\left(\frac{3}{5}v\right) = \frac{9}{25}v$

Clearly, $\frac{9}{25}v < \frac{2}{5}v$

Therefore 'B' cannot collide with 'C' for the second time.

Hence, the total number of collisions is 2.

2. (9) Trolleys gain momentum due to force applied by man which will be internal force for the system of trolleys and man and there is no other external force. Here we assume that man applies force for a very short time, during which effect of friction can be neglected.
Momentum just before pushing = momentum just after pushing.

$$0 = 3\,mv_1 - mv_2 \Rightarrow v_1 = \frac{v_2}{3}$$

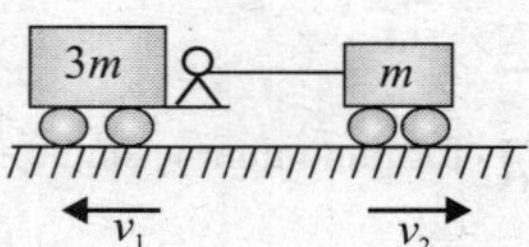

Fig. 1.384

From work–energy theorem for individual trolleys,

$$f_1 S_1 = \frac{1}{2}3mv_1^2, \quad f_2 S_2 = \frac{1}{2}mv_2^2$$

Here $f_1 = \mu 3mg,\ f_2 = \mu mg$

Solve to get $\frac{S_2}{S_1} = \left(\frac{v_2}{v_1}\right)^2 = 9$

3. (2) $F_1 = 4N = -4\hat{i}$ $\quad 0 < t \le 1$ s

$= 2N = -2\hat{i}$ $\quad 1 \le t \le 3$ s

$F_2 = 1N = -\hat{j}$ $\quad 0 < t \le 2$ s

$= 2t - 3N = -(2t-3)\hat{j}$ $\quad 2 \le t \le 3$ s

Initial velocity of the particle is $\vec{u} = 10\hat{i}$

From impulse momentum theorem, $\int d\vec{p} = \int \vec{F}\,dt$

$$m\vec{v} - m\vec{u} = \int^3 (\vec{F}_1 + \vec{F}_2)dt$$

(where $\vec{v}$ is the required velocity)

$1 \times \vec{v} - 1 \times 10\hat{i} = -8\hat{i} - 4\hat{j}$ or $\vec{v} = 2\hat{i} - 4\hat{j}$

or $v = \sqrt{2^2 + 4^2} = 2\sqrt{5}$ m/s

4. (3) For first case: $W = \Delta KE \Rightarrow -Rd = -\frac{1}{2}mu^2$

$$Rd = \frac{1}{2}mu^2 \quad \text{(i)}$$

For second case: $mu = (M+m)v$, v is common velocity

$$\Rightarrow v = \frac{m}{(M+m)}u$$

$$-Rd' = \frac{1}{2}(M+m)v^2 - \frac{1}{2}mu^2 = \frac{1}{2}\frac{m^2u^2}{(M+m)} - \frac{1}{2}mu^2$$

$$\Rightarrow Rd' = \frac{1}{2}\frac{Mm}{(M+m)}u^2 \quad \text{(ii)}$$

Solving Eqs. (i) and (ii), $\frac{Rd}{Rd'} = \frac{M+m}{M} \Rightarrow d' = 3$ cm

5. (2) Apply conservation of momentum in horizontal direction:

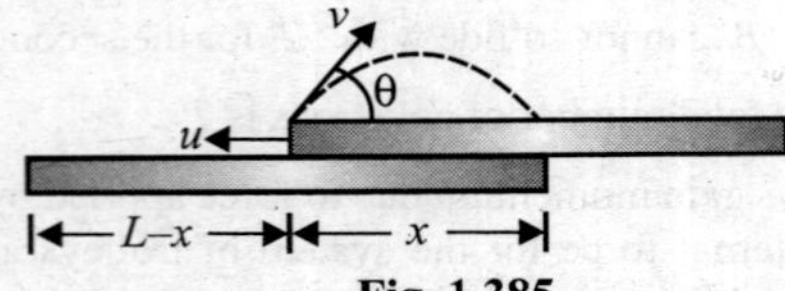

Fig. 1.385

$mv\cos\theta - \text{mu} = 0 \Rightarrow \quad u = v\cos\theta$

$L - x = ut, x = v\cos\theta t$

Solve to get, $x = \frac{L}{2}$

$$x = \frac{v^2\sin 2\theta}{g} \Rightarrow \frac{L}{2} = \frac{v^2\sin 2\theta}{g}$$

$$\Rightarrow v = \sqrt{\frac{gL}{2\sin\theta}} \text{ for minumum } v, \sin 2\theta = 1$$

$$v_{\min} = \sqrt{\frac{gL}{2}} = \sqrt{\frac{10\times 5}{2}} = 2\sqrt{5} \text{ m/s, Hence } n = 2$$

6. (1) From impulse–momentum theorem,

$$\int N\,dt = m(v + 5\cos\theta) \quad \text{(i)}$$

$$\int f\,dt = m5\sin\theta$$

$$\mu\int N\,dt = m5\sin\theta \quad \text{(ii)}$$

$$\Rightarrow \mu m(v + 5\cos\theta) = m5\sin\theta$$

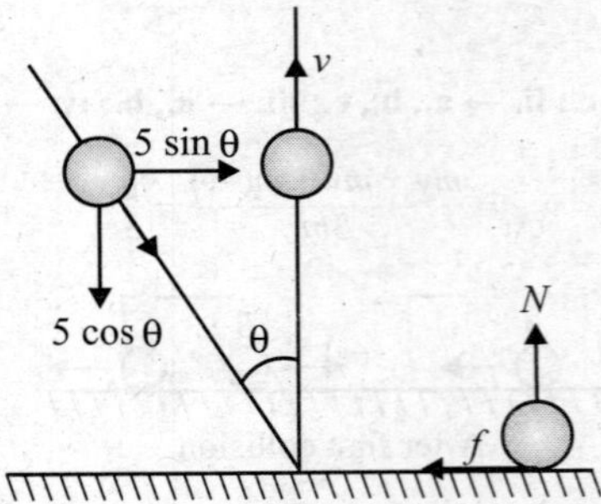

Fig. 1.386

According to Newton's law of restitution,

$$v = e\,5\cos\theta$$

Solve to get $\mu = 1$

7. (9) $\int N dt = 1[(\hat{i} + 3\hat{j}) - (4\hat{i} - \hat{j})] = -3\hat{i} + 4\hat{j}$

Component of $4\hat{i} - \hat{j}$ along $-3\hat{i} + 4\hat{j}$

$$= \frac{-12-4}{25}(-3\hat{i} + 4\hat{j}) = -\frac{16}{25}(-3\hat{i} + 4\hat{j})$$

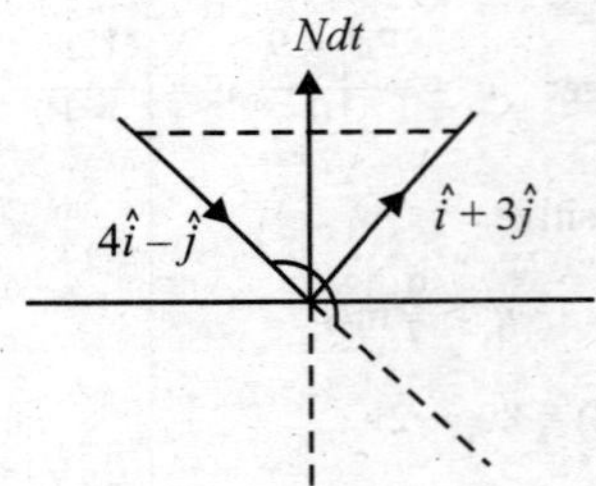

Fig. 1.387

Speed of approach $= \frac{16}{25}\sqrt{25} = \frac{16}{5}$

Component of $\hat{i} + 3\hat{j}$ along $-3\hat{i} + 4\hat{j}$ is $\frac{-3+12}{25}(-3\hat{i} + 4\hat{j})$

Speed of separation = 9/5

Speed of separation = $e \times$ speed of approach

$$e = \frac{9}{16} \qquad n = 9$$

8. (1) All the velocities shown in diagrams are w.r.t. ground.

After first jump:

v_2 ← 4kg

A 20 kg	B 20 kg → v_1

Fig. 1.388

$20v_1 = 4v_2$ and $v_1 + v_2 = 6$ (given)

Solve to get $v_1 = 1$ m/s, $v_2 = 5$ m/s

When child arrives on A:

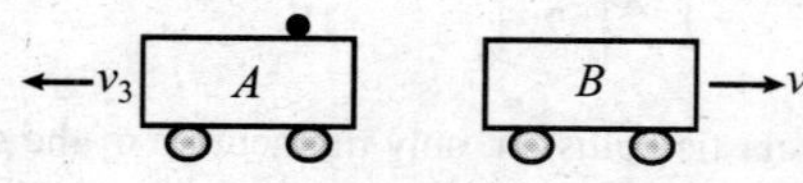

Fig. 1.389

$(20 + 4)\, v_3 = 4v_2 \Rightarrow v_3 = 5/6$ m/s

After the second jump:

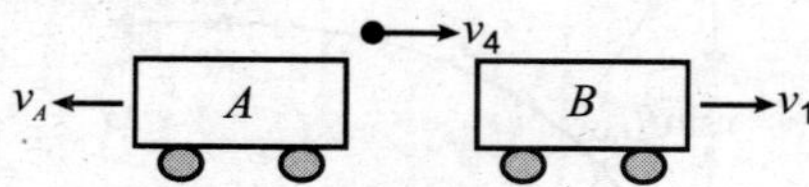

Fig. 1.390

$v_4 + v_A = 6,\ 24v_3 = 20v_A - 4v_4$

Solve to get $v_A = \dfrac{11}{6}$ m/s, $v_4 = \dfrac{25}{6}$ m/s

when child arrives on B:

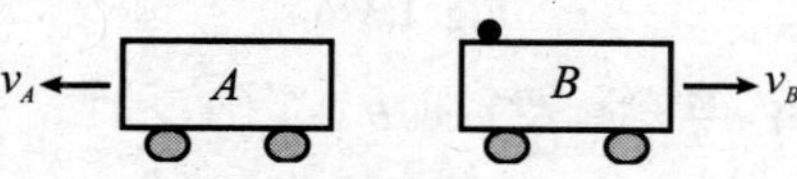

Fig. 1.391

$24v_B = 4v_4 + 20v_1$

$$\Rightarrow \quad 24v_B = 4\left(\frac{25}{6}\right) + 20 \times 1 \Rightarrow v_B = \frac{55}{36} \text{ m/s}$$

$$\text{Now } \frac{6v_B}{5v_A} = \frac{6 \times 55 \times 6}{36 \times 5 \times 11} = 1$$

9. (3) To save himself, the man throws his jacket in opposite direction to the lake. According to momentum conservation, he himself gets a velocity in the direction of the lake. During the motion as gravity is the only external force on the system (man plus jacket), centre of mass will not be displaced horizontally. Thus, centre of mass of the system falls vertically and when the man falls in the lake, jacket falls at a point such that the centre of mass of the man and the jacket will be directly below the point from where the man jumps).

As it is given that man falls at a distance d from this point, it implies that jacket will fall at a distance x in the opposite direction such that

$mx = Md \Rightarrow x = \dfrac{M}{m}d = 29$ m, man has to travel a distance $x + d = 29 + 1 = 30$ m to pick his jacket.

10. (1) Let u be the initial velocity of the ball of mass m. Then

$$mu = mv_1 + nmv_2 \Rightarrow v_1 + nv_2 = u \qquad \text{(i)}$$

For elastic collision, Newton's experimental formula is ($u_2 = 0$)

$$v_1 - v_2 = -(u_1 - u_2) \Rightarrow v_1 - v_2 = -u \qquad \text{(ii)}$$

Solving Eqs. (i) and (ii), $v_1 = \dfrac{1-n}{1+n}u$

Fractional loss in KE (= fractional transfer of KE)

$$f = \frac{K_i - K_f}{K_i} = \frac{\frac{1}{2}mu^2 - \frac{1}{2}mv_1^2}{\frac{1}{2}mu^2} = 1 - \left(\frac{v_1}{u}\right)^2$$

$$f = 1 - \left(\frac{1-n}{1+n}\right)^2 = \frac{4n}{(n+1)^2}$$

The transfer of energy is maximum when $f = 1$ or 100%

$$\frac{4n}{(n+1)^2} = 1 \Rightarrow n = 1$$

That is, the transfer of energy is maximum when the mass ratio is unity.

Archives

True or False Type

1. False: Since net force on the system is zero, velocity of centre of mass will remain constant.

Fill in the Blanks Type

1. $2\,mv' \cos\theta = mv$ (i)

$2\,mv' \sin\theta = mv$

$$\Rightarrow \sin\theta = \cos\theta = \frac{1}{\sqrt{2}}$$

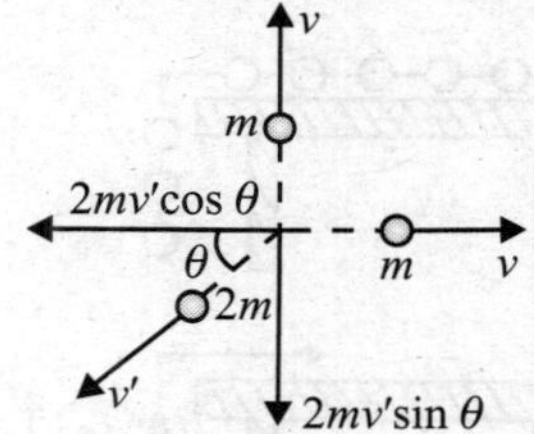

Fig. 1.392

Putting this value in Eq. (i), we get

$$\frac{2mv'}{\sqrt{2}} = mv \qquad \Rightarrow \qquad v' = \frac{v}{\sqrt{2}}$$

$$\text{Total KE} = \frac{1}{2}mv^2 + \frac{1}{2}mv^2 + \frac{1}{2}(2m)\left(\frac{v}{\sqrt{2}}\right)^2$$

$$= \frac{1}{2}mv^2 + \frac{1}{2}mv^2 + \frac{mv^2}{2} = \frac{3}{2}mv^2$$

2. We know that area under the F–t graph gives the impulse imparted to the body.

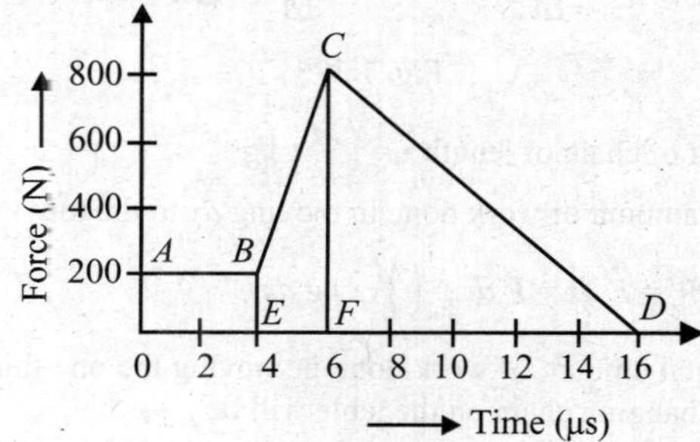

Fig. 1.393

The magnitude of total impulse of force on the body from $t = 4\,\mu$ s to $t = 16\,\mu$ s = area (*BCDFEB*)

= area of *BCFEB* + area *CDFC*

$= \frac{1}{2}(200 + 800) \times 2 \times 10^{-6} + \frac{1}{2} \times 10 \times 800 \times 10^{-6}$

$= 0.001 + 0.004 = 0.005$ N s

Multiple Choice Questions with One Correct Answer Type

1. **d.** Net external force on system is zero. Therefore, centre of mass always remains at rest.

2. **d. i. Mehtod I:**
The hanging part of the chain which is to be pulled can be considered as a point mass situated at the centre of the hanging part. The equivalent diagram is drawn. The work done in bringing the mass up will be equal to the change in potential energy of the mass.

W = Change in potential energy

$$= \frac{M}{3} \times g \times \frac{L}{6} = \frac{MgL}{18}$$

Fig. 1.394

Method II:

The mass per unit length of the chain = M/L. Let us consider a finitesimally small length of the chain dx at a distance x from the bottom. To move dx to the top, a force equal to the weight of chain x will have to be applied upwards.

Fig. 1.395

Weight of chain of length $x = \left(\frac{M}{L}x\right)g$

Small amount of work done in moving dx to the top

$$dW = \vec{F}\cdot\vec{dx} = F\,dx = \left(\frac{M}{L}x\right)g\,dx$$

The total amount of work done in moving the one-third length of the hanging chain on the table will be

$$W = \int_0^{L/3} \frac{M}{L} xg dx = \frac{M}{L} g$$

$$\int_0^{L/3} x\,dx = \frac{M}{L} g \left[\frac{x^2}{2}\right]_0^{L/3} = \frac{MgL}{18}$$

3. **c.** In an inelastic collision, only momentum of the system may remain conserved. Some energy can be lost in the form of heat, sound, etc.

4. **a.** Applying conservation of linear momentum at the highest point

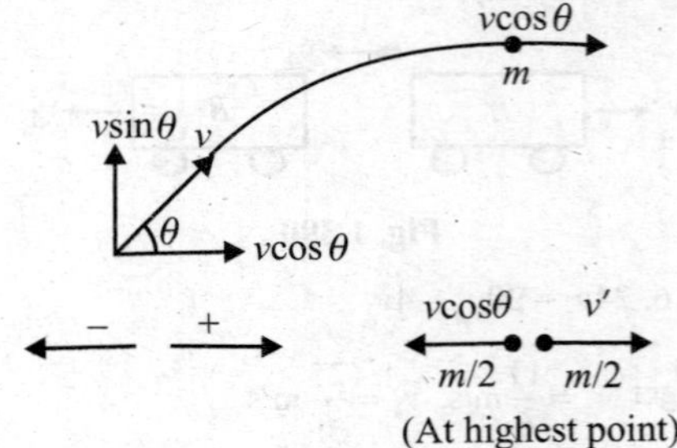

Fig. 1.396

$$m(v\cos\theta) = \frac{m}{2} \times v' - \frac{m}{2} v\cos\theta$$

$$3v\cos\theta = v'$$

5. **a.** Before explosion, particle was moving along x-axis, i.e., it has no y-component of velocity. Therefore, the centre of mass will not move in y-direction or we can say $y_{CM} = 0$.

Now, $$y_{CM} = \frac{m_1 y_1 + m_2 y_2}{m_1 + m_2}$$

Therefore, $$0 = \frac{(m/4)(+15) + (3m/4)(y)}{(m/4 + 3m/4)}$$

or $$y = -5 \text{ cm}$$

6. **d.** $|(m_1\vec{v}'_1 + m_2\vec{v}'_2)| - |(m_1\vec{v}_1 + m_2\vec{v}_2)|$
= |change in momentum of the two particles|
= |external force on the system | × time interval
$= (m_1 + m_2)\,g\,(2t_0) = 2(m_1 + m_2)\,gt_0$

7. **c.** Just after giving the impulse:

$$V_c = \frac{10 \times 14 + 4 \times 10}{10 + 4} = 10 \text{ m/s}$$

Since spring force is internal force, it cannot change the linear momentum of the (two mass + spring) system. Therefore, V_c remains the same.

8. **d.** $\vec{F} = \frac{d\vec{P}}{dt} = -kA\sin(kt)\,\hat{i} - kA\cos(kt)\,\hat{j}$

$$\vec{P} = A\cos(kt)\,\hat{i} - A\sin(kt)\,\hat{j}$$

Now $\vec{F}\cdot\vec{P} = 0$

Therefore, angle between $\vec{F}$ and $\vec{P}$ should be 90°.

9. **a.**

$$Y_{CM} = \frac{6m \times 0 + m \times a + m \times a + m \times 0 + + m \times (-a)}{10m}$$

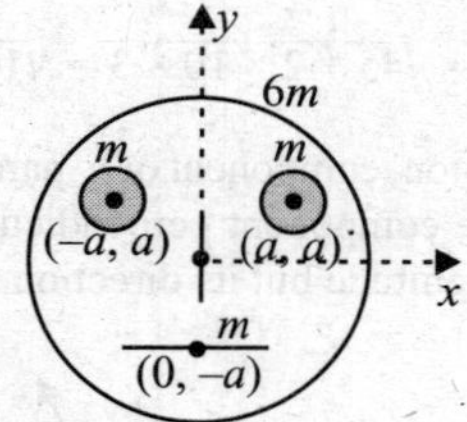

Fig. 1.397

$$Y_{CM} = \frac{a}{10}$$

10. c. Velocity will exchange after each collision

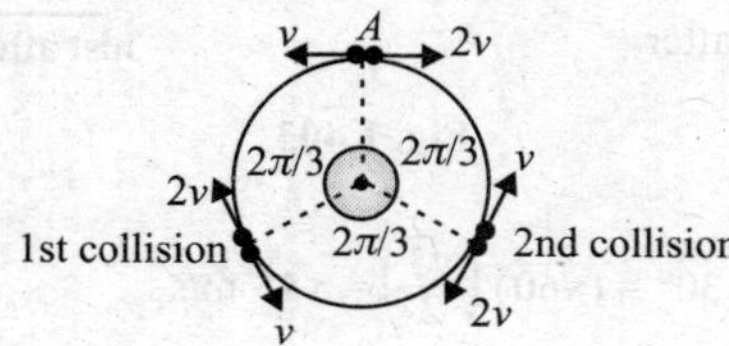

Fig. 1.398

11. c. Area under F-t curve = 4.5 kg m/sec

$$\text{K.E} = \frac{1}{2}(2)\left(\frac{4.5}{2}\right)^2 = 5.06 \text{ J}$$

12. d.

$$R = u\sqrt{\frac{2h}{g}}$$

$$\Rightarrow 20 = V_1\sqrt{\frac{2\times5}{10}} \text{ and } 100 = V_2\sqrt{\frac{2\times5}{10}}$$

$$\Rightarrow V_1 = 20 \text{ m/s}, V_2 = 100 \text{ m/s}$$

Applying momentum conservation just before and just after the collision $(0.01)(V) = (0.2)(20) + (0.01)(100)$

$V = 500$ m/s

Multiple Choice Questions with One or More than One Correct Answer Type

1. a., c., d. Applying conservation of linear momentum

$$2m(-v) + m(2v) + 8m \times 0 = (2m + m + 8m)\,V_c$$

$$\Rightarrow V_c = 0$$

Fig. 1.399

Applying conservation of angular momentum about centre of mass, we get

$$2mv \times a + m(2v) \times 2a = I\omega$$

where

$$I = \frac{1}{12}(8m)(6a)^2 + 2m \times a^2 + m \times 4a^2 = 30\,ma^2$$

$$6\,mva = 30\,ma^2 \times \omega \quad \Rightarrow \quad \frac{v}{5a} = \omega$$

Energy after collision,

$$E = \frac{1}{2}I\omega^2 = \frac{1}{2} \times 30\,ma^2 \times \frac{v^2}{25a^2} = \frac{3mv^2}{5}$$

2. b., d. In situation (i), mass C is moving towards right with velocity v. A and B are at rest. In situation (ii), which is just after the collision of C with A, C stops and A acquires a velocity v. When A starts moving towards right, the spring suffers a compression due to which B also starts moving towards right. The compression of the spring continues till there is a relative velocity between A and B. Once this relative velocity becomes zero, both A and B move with the same velocity v' and the spring is in a state of maximum compression.

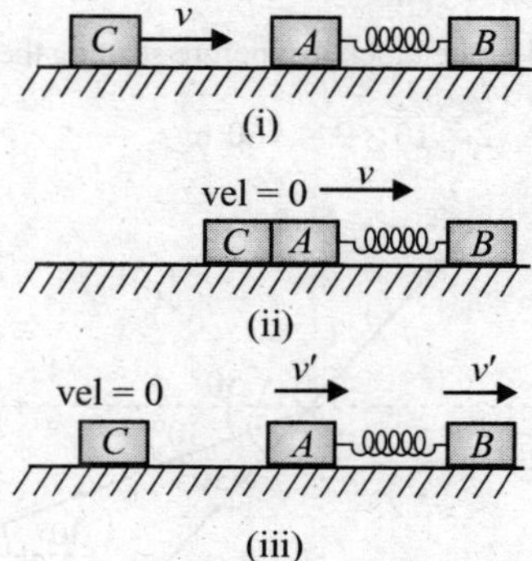

Fig. 1.400

Applying momentum conservation in situations (ii) and (iii),

$$mv = mv' + mv' \Rightarrow v' = \frac{v}{2}$$

Therefore, KE of the system in situation (iii) is

$$\frac{1}{2}mv'^2 + \frac{1}{2}mv'^2 = mv'^2 = \frac{mv^2}{4}$$

Applying energy conservation, we get

$$\frac{1}{2}mv^2 = \frac{1}{2}mv'^2 + \frac{1}{2}mv'^2 + \frac{1}{2}Kx^2$$

Solve to get $x = v\sqrt{\frac{m}{2K}}$

3. a., d. Initial momentum of the system,

$$\vec{p}_1 + \vec{p}_2 = 0$$

Therefore, final momentum $\vec{p}_1' + \vec{p}_2'$ should also be zero.

Option (b) is allowed because if we put $c_1 = -c_2 \neq 0$, $\vec{p}_1' + \vec{p}_2'$ will be zero. Similarly, we can check other options.

Hence correct options are (a) and (d).

4. a., c. By conservation of linear momentum

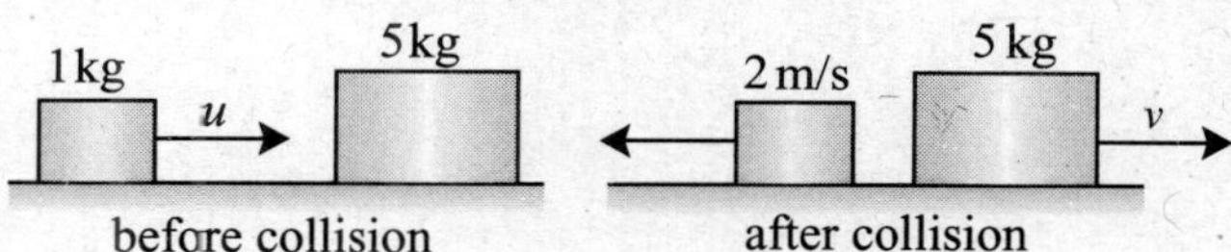

Fig. 1.401

$u = 5v - 2$ (i)

By Newton's experimental law of collision

$u = v + 2$ (ii)

Using Eqs. (i) and (ii) we have $v = 1$ m/s and $u = 3$ m/s

Kinetic energy of the centre of mass $= \frac{1}{2} m_{system} v_{CM}^2 = 0.75$ J

Assertion and Reasoning Type

1. d. In case of elastic collision, coefficient of restitution, $e = 1$. Magnitude of relative velocity of approach = Magnitude of relative velocity of separation. But relative speed of approach ≠ relative speed of separation.

Comprehension Type

1. b. Between A and B, height fallen by the block is

$h_1 = \sqrt{3} \tan 60° = 3$ m

Hence, speed of the block just before striking the second incline is

$v_2 = \sqrt{2gh_1} = \sqrt{2 \times 10 \times 3} = \sqrt{60}$ m/s

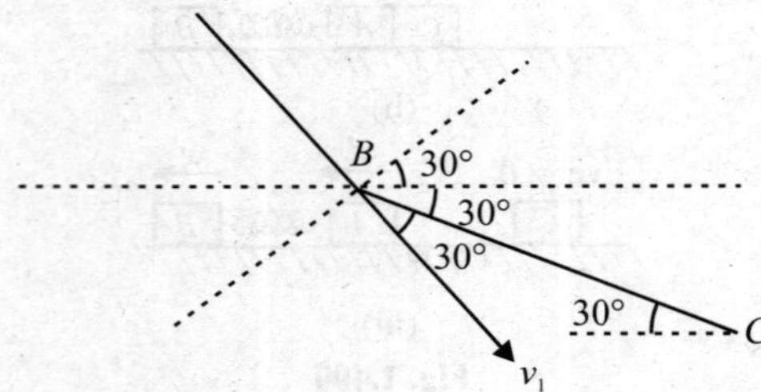

Fig. 1.402

In perfectly inelastic collision, component of v_1 perpendicular to BC will become zero, while component of v_1 parallel to BC will remain unchanged.

v_2 = component of v_1 along BC

$= v_1 \cos 30° = (\sqrt{60})\left(\frac{\sqrt{3}}{2}\right) = (\sqrt{45})$ m/s

2. b. Height fallen by the block from B to C

$h_2 = 3\sqrt{3} \tan 30° = 3$ m

Let v_3 be the speed of the block, at point C, just before it leaves the second incline, then

$v_3 = \sqrt{v_2^2 + 2gh_2} = \sqrt{45 + 2 \times 10 \times 3} = \sqrt{105}$ m/s

3. c. In elastic collision, component of v_1 parallel to BC will remain unchanged, while component perpendicular to BC will remain unchanged in magnitude but its direction will be reversed.

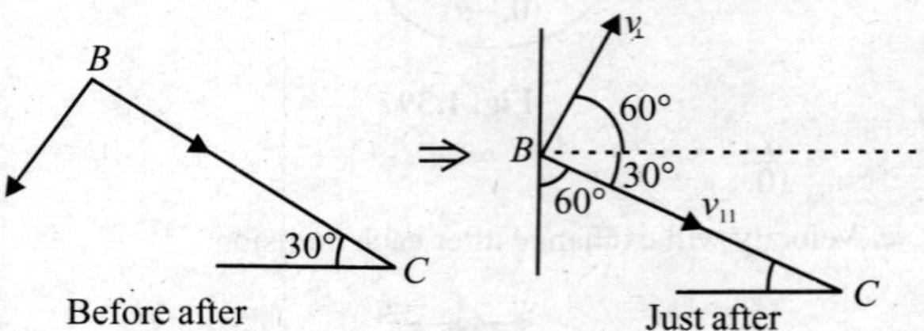

Fig. 1.403

$v_{||} = v_1 \cos 30° = (\sqrt{60})\left(\frac{\sqrt{3}}{2}\right) = \sqrt{45}$ m/s

$v_{\perp} = v_1 \sin 30° = (\sqrt{60})\left(\frac{1}{2}\right) = \sqrt{15}$ m/s

Now the vertical component of velocity of the block:

$v' = v_{\perp} \cos 30° - v_{||} \cos 60°$

$= (\sqrt{15})\left(\frac{\sqrt{3}}{2}\right) - (\sqrt{45})\left(\frac{1}{2}\right) = 0$

Integer Answer Type

1. After the first collision,

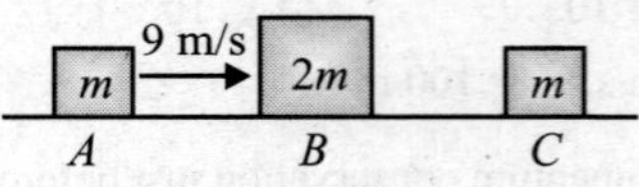

Fig. 1.404

$mv_A = mv'_A + 2mv'_B$

$-1 = \frac{v'_B - v'_A}{0 - v_A} \Rightarrow v'_B = 6$ m/s

After the second collision,

$2mv'_B = (2m + m)v_C \Rightarrow v_C = (2/3)\, v'_B \Rightarrow v_C = 4$ m/s

CHAPTER

2

Rigid Body Dynamics

- Rigid Body
- Moment of Inertia
- Radius of Gyration
- Moment of Inertia of Some Important Cases
- Torque
- Equilibrium of Rigid Bodies
- Analogue of Newton's Second Law of Motion for Pure Rotation
- Application of Torque Equation and Equation of Motion Combined
- Kinetic Energy of Rotation
- Rotational Work and Power
- Work Energy Theorem
- Angular Momentum
- Angular Impulse: Rotational Analog of Impulse
- Laws of Conservation of Angular Momentum
- Angular Impulse–Angular Momentum Theorem
- Angular Momentum of a Rigid Body due to Translation and Rotation Both
- Laws of Conservation of Angular Momentum
- Combination of Translation and Rotation
- Rolling Motion
- Instantaneous Axis of Rotation
- Direction of Friction in Case of Translation and Rotation Combined
- Rolling Bodies in Moving Platform
- Conservation of Angular Momentum in Case of Collision

RIGID BODY

Rigid body is defined as a system of particles in which distance between each pair of particles remains constant (with respect to time), which means the shape and size do not change during the motion.

For every pair of particles in a rigid body, there is no velocity of separation or approach between the particles. In Fig. 2.1, velocities of A and B with respect to the ground are V_A and V_B, respectively.

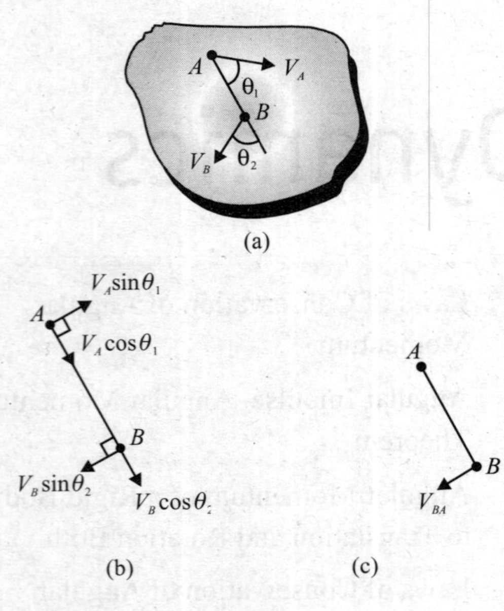

Fig. 2.1

- If the above body is rigid, $V_A \cos \theta_1 = V_B \cos \theta_2$
- With respect to any particle of the rigid body, the motion of any other particle of that rigid body is circular.
  V_{BA} = relative velocity of B with respect to A.

General Rigid Body Motion

Translation

If a body is moving such that a line drawn between any two of its internal points remains parallel to itself, the motion is said to be translation. In translation, all the particles of the body move along parallel paths. If these parallel paths are straight lines, the motion is said to be a rectilinear translation; if the paths are curved lines, the motion is a curvilinear translation (Fig. 2.2). For a rigid body in translation, all the points of the body have the same velocity and the same acceleration at any given instant.

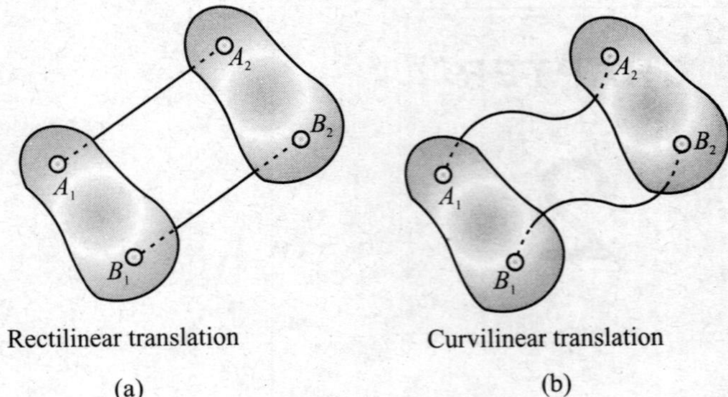

Fig. 2.2

Rotation

A rigid body is said to be in pure rotation if every particle of the body moves in a circle and the centres of all the circles lie on a straight line called the axis of rotation. Ail points in the body which are perpendicular to the axis of rotation turn through the same angle in the same time interval.

If the line joining any two internal points does not remain parallel to itself, the body is in rotation.

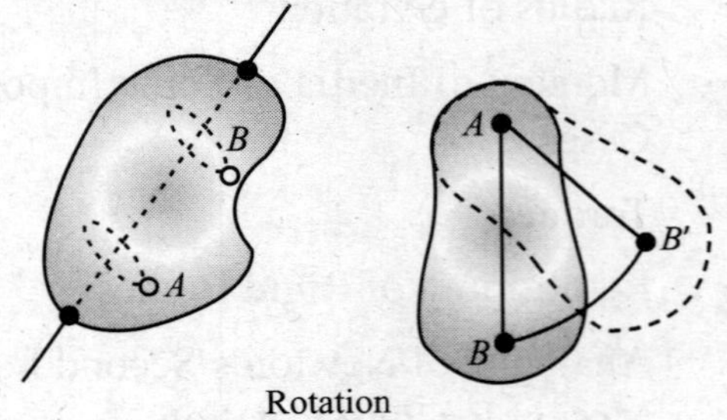

Fig. 2.3

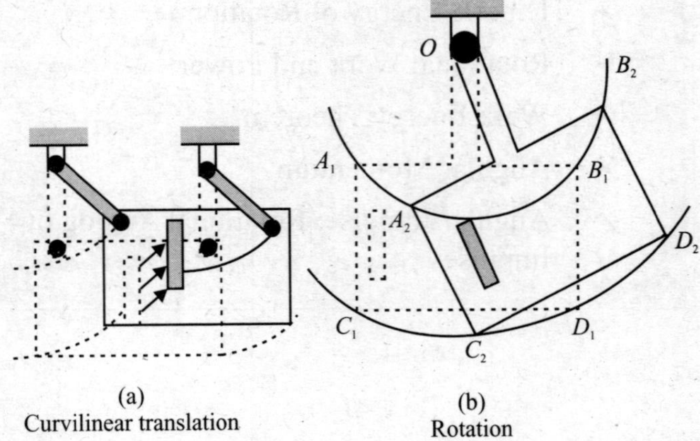

Fig. 2.4

MOMENT OF INERTIA

Inertia is a property of a body to resist the change in linear state of motion. It is measured by the mass of the body. The moment of inertia I of a body is a measure of its ability to resist change in its rotational state of motion.

If we compare for rotational motion with corresponding relations for translatory motion, we find that moment of inertia plays the same role in rotatory motion as is played by mass in translatory motion, i.e., if a body has large moment of inertia, it is difficult to start rotation or to stop it if rotating. Large moment of inertia also helps in keeping the motion uniform. This is why stationary engines are provided with fly wheels having large moment of inertia.

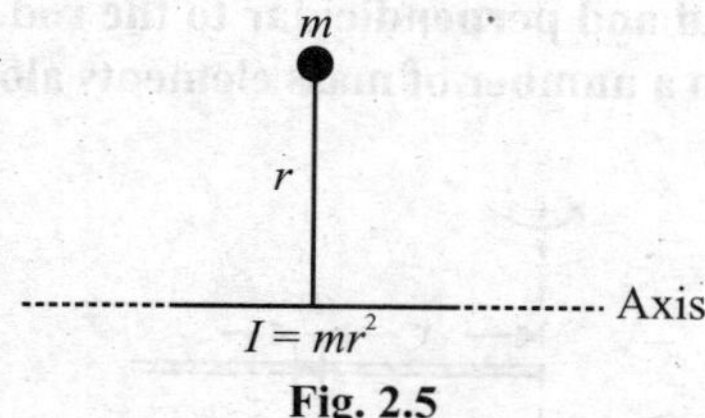

Fig. 2.5

Moment of inertia of a particle about an axis is defined as $I = mr^2$ where r is the distance (perpendicular) of the particle from the axis of rotation (Fig. 2.5).

For a system of particles of masses $m_1, m_2, m_3,..., m_n$ at perpendicular distances $r_1, r_2, r_3,..., r_n$ (Fig. 2.6), respectively, from the axis of rotation, the moment of inertia is given by

$$I = m_1 r_1^2 + m_2 r_2^2 + \cdots + m_n r_n^2 = \Sigma m_i r_i^2$$

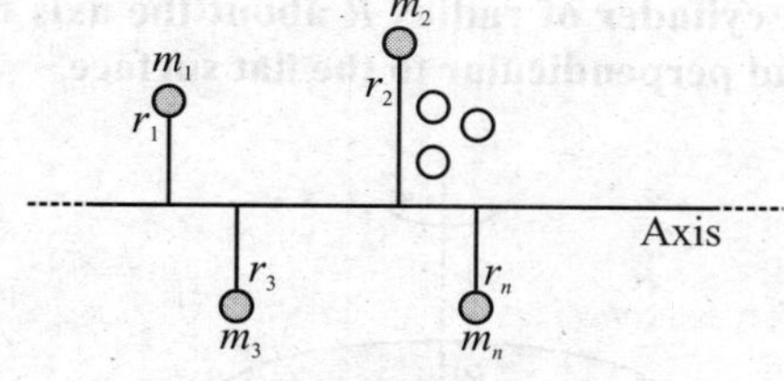

Fig. 2.6

Moment of inertia of a rigid body about an axis is given by $I = \int r^2\, dm$ where r is the perpendicular distance of a particle of mass dm of rigid body from the axis of rotation.

Note:

- ***MI is not constant for a body. It depends on the axis of rotation.***
- ***MI depends on the mass of the body. The higher the mass, the higher is the MI.***
- ***MI depends on the distribution of the mass about an axis. The farther the mass is distributed from the axis, the higher will be the MI.***

Moment of inertia does not change if the mass

i. is shifted parallel to the axis of the rotation.

ii. is rotated with constant radius about axis of rotation.

Illustration 2.1 **Four particles, each of mass m, are kept at the four corners of a square of edge a. Find the moment of inertia of the system about a line perpendicular to the plane of the square and passing through the centre of the square.**

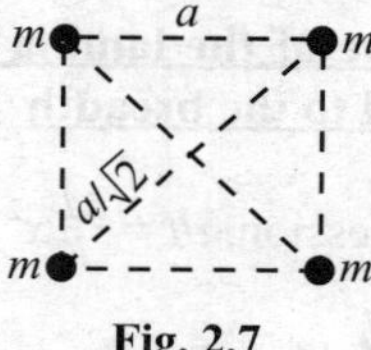

Fig. 2.7

Sol. The perpendicular distance of every particle from the given line is $a/\sqrt{2}$. The moment of inertia of one particle is, therefore, $m(a/\sqrt{2})^2 = \frac{1}{2}ma^2$. The moment of inertia of the system is, therefore,

$$4 \times \frac{1}{2}ma^2 = 2ma^2$$

Moment of Inertia of a Uniform Rectangular Lamina

Length, breadth and mass of the lamina are l, b and M, respectively.

a. <u>Axis in the plane of the lamina passing through its centre and parallel to its breadth</u>

Mass per unit area $= \dfrac{M}{lb}$

Consider a rectangular differential mass element parallel to the breadth at a distance 'x' from the axis of rotation (Fig. 2.8).

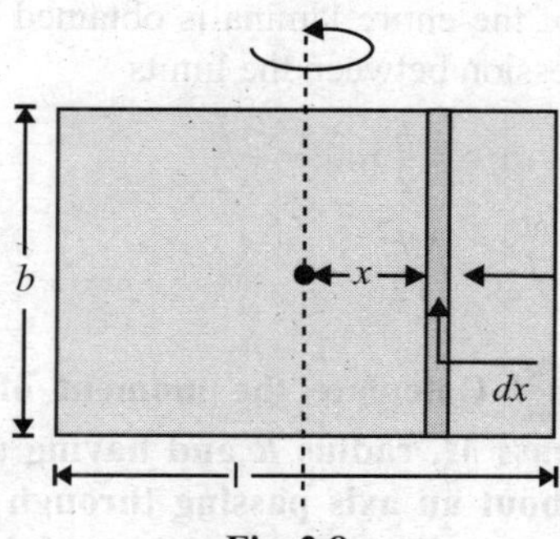

Fig. 2.8

Mass of the strip $dm = \dfrac{M}{lb}(bdx) = \dfrac{M}{l}dx$

Since each particle lying on the differential strip is at the same perpendicular distance 'x' from the axis of rotation, hence the MI of the strip about the given axis is

$$dI = dm\, x^2 = \left(\frac{M}{l}dx\right)x^2$$

Total MI of the entire rectangular lamina can be obtained by integrating the above expression between the limits $-l/2$ to $+l/2$.

$$I=\int_{-l/2}^{+l/2}\left(\frac{M}{l}x^2\,dx\right)=\frac{M}{l}\left[\frac{x^3}{3}\right]_{-l/2}^{+l/2}$$

$$=\frac{Ml^2}{12}$$

b. Axis in the plane of the lamina passing through the end and parallel to the breadth

Integrate the expression, $dI=\frac{M}{l}x^2,\ dx=\frac{Ml^2}{3}$

$$I=\int_0^l \frac{M}{l}x^2\,dx\,\frac{Ml^2}{3};\ I=\frac{Ml^2}{3}$$

c. Axis xx' in the plane of the lamina, passing through its centre and parallel to the length

Consider a rectangular differential strip of width dy parallel to the length and at a distance of 'y' from the axis (Fig. 2.9).

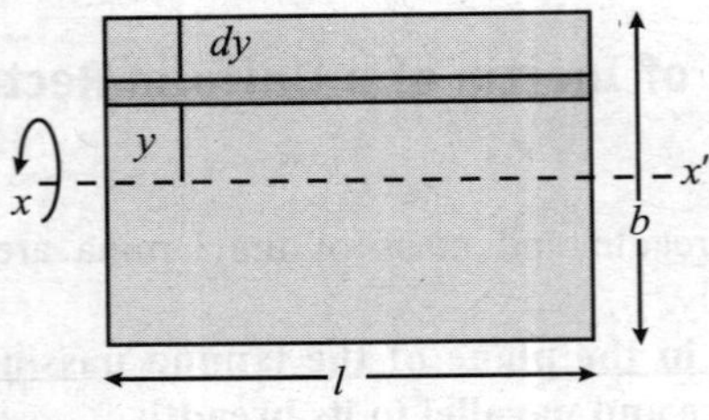

Fig. 2.9

MI of the strip about the required axis is given by

$$dI_{xx'}=\left(\frac{M}{b}dx\right)x^2$$

Total MI of the entire lamina is obtained by integrating the above expression between the limits

$$y=-\frac{b}{2}\text{ to }y=+\frac{b}{2}$$

$$\Rightarrow\quad I_{xx'}=\frac{Mb^2}{12}$$

Illustration 2.2 **Calculate the moment of inertia of a ring having mass M, radius R and having uniform mass distribution about an axis passing through the centre of the ring and perpendicular to the plane of the ring?**

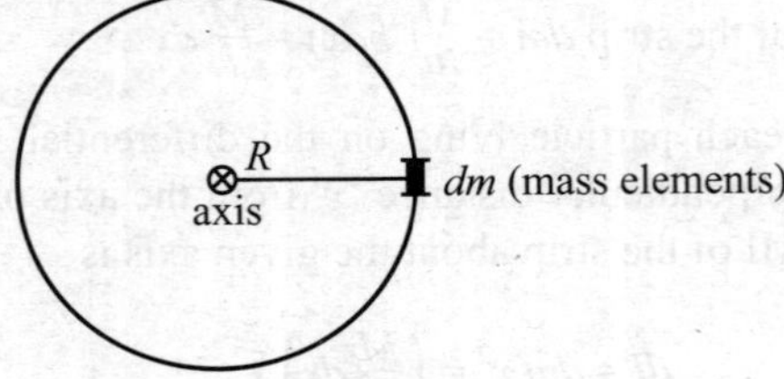

Fig. 2.10

Sol. For a continuous body, moment of inertia is given by

$$I=\int(dm)r^2$$

Because each element is at equal distance from the axis,

so $\quad r=R,\ I=R^2\int dm=MR^2$

Note: ***Answer will remain the same even if the mass is nonuniformly distributed because $\int dm=M$ always.***

The moment of inertia of a hollow cylinder of the same mass and same radius will be equal as the ring.

Illustration 2.3 **Calculate the moment of inertia of a uniform rod of mass M and length l about an axis passing through an end and perpendicular to the rod. The rod can be divided into a number of mass elements along the length of the rod.**

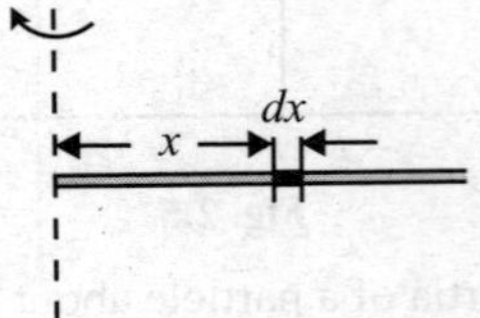

Fig. 2.11

Sol. Taking an element of mass dm at a distance x from the relation axis.

$$I=\int(dm)\,r^2-\int_0^l\left(\frac{M}{l}dx\right)x^2=\frac{Ml^2}{3}$$

Illustration 2.4 **Find the moment of inertia of a circular disk or solid cylinder of radius R about the axis through the centre and perpendicular to the flat surface.**

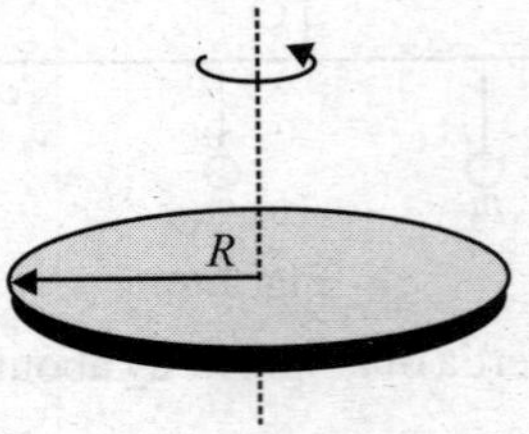

Fig. 2.12

Sol. The figure shows that the appropriate mass element is a circular ring of radius r and width dr.

Its area is $dA=2\pi\,rdr$ and its mass is $dm=\sigma dA$;

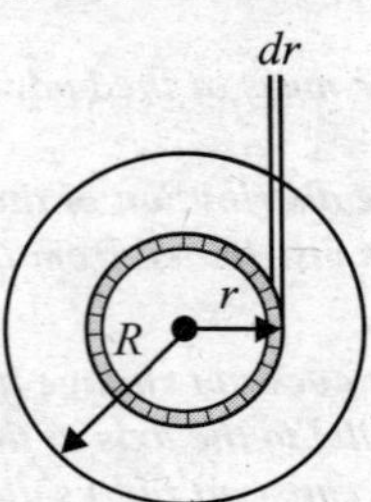

Fig. 2.13

where $\sigma = M/A$ is the real mass density.

The moment of inertia of this element is

$$dI = dm\, r^2 = 2\pi\sigma r^3 dr$$

For the whole body,

$$I = 2\pi\sigma \int_0^R r^3 dr = \frac{1}{2}\pi\sigma R^4$$

The mass of the whole disk or cylinder is

$$M = \sigma A = \sigma\pi R^2, \text{ and so } I = \frac{1}{2} MR^2$$

Note: ***The moment of inertia of a cylinder having the same mass and same radius as disc will also be same as MI is independent of the size of the object parallel to its axis of rotation.***

Perpendicular-Axis Theorem

Consider a particle of mass m (Fig. 2.14) located on the x-y plane at (x, y)

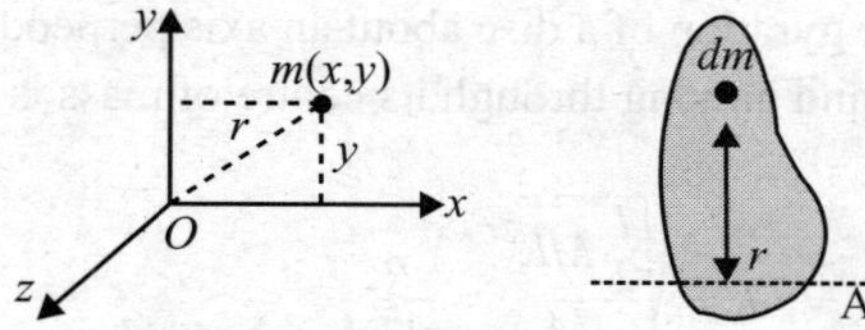

Fig. 2.14

$$I_x = my^2 \text{ and } I_y = mx^2$$

$$\Rightarrow \quad I_x + I_y = my^2 + mx^2 = m(x^2 + y^2) = mr^2 = I_z$$

The distance of the particle from the z-axis $= r$, $I_z = mr^2$.

Now consider a lamina object lying in the x-y plane. Consider a particle of mass dm at (x, y),

$$I_x = y^2\, dm \text{ and } I_y = \int x^2\, dm$$

$$I_x + I_y = \int y^2\, dm + \int x^2\, dm = \int (x^2 + y^2)\, dm = \int r^2\, dm = I_z$$

$I_z = I_x + I_y$ (when object is in x-y plane)

where I_z = moment of inertia of the body about z-axis

I_x = moment of inertia of the body about x-axis

I_y = moment of inertia of the body about y-axis

$I_y = I_x + I_z$ (when object is in x-z plane)

$I_x = I_y + I_z$ (when object is in y-z plane)

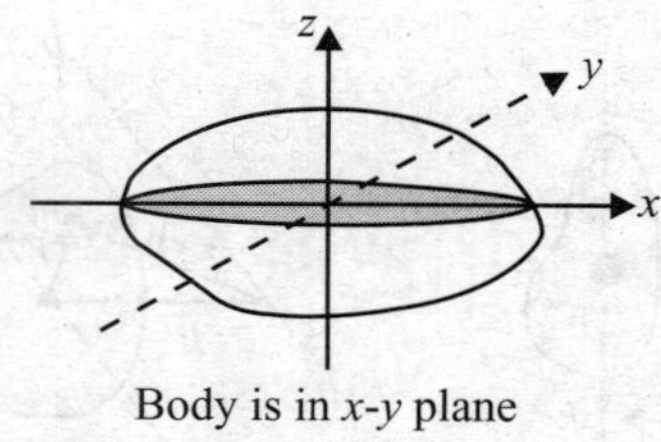

Fig. 2.15

The sum of moments of inertia of a laminar object about two mutually perpendicular axes lying in the plane of lamina is equal to the moment of inertia about an axis normal to the plane of the lamina and passing through the two perpendicular axes.

This theorem is applicable only for a laminar (thin sheet kind of) object.

Illustration 2.5 **Two uniform identical rods each of mass *M* and length *l* are joined to form a cross as shown in Fig. 2.16. Find the moment of inertia of the cross about a bisector shown in dotted line in the figure.**

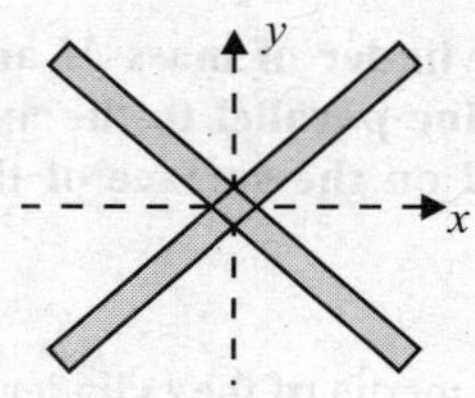

Fig. 2.16

Sol. Consider the line perpendicular to the plane of the figure through the centre of the cross. The moment of inertia of each rod about this line (say about z-axis) $I'_z = Ml^2/12$. Hence the moments of inertia of the cross $I_z = 2I'_z = Ml^2/6$. The moments of inertia of the cross about the two bisector are equal by symmetry $I_X = I_Y$ according to the theorem of perpendicular axis, $I_Z = I_X + I_Y$ the moment of inertia of the cross about the bisector is $Ml^2/12$.

Parallel-Axis Theorem

In Fig. 2.17, bb' is an axis parallel to aa' (an axis passing through centre of mass)

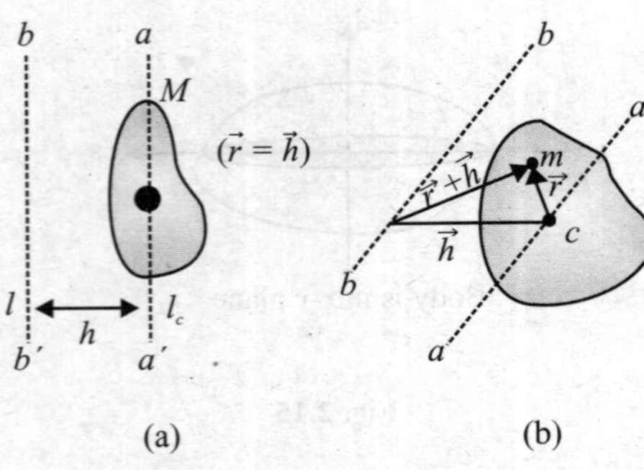

Fig. 2.17

Moment of inertia of the body about centroidal axis (aa'), $I_c = mr^2$

Moment of inertia of the body about an axis parallel to the line bb',

$$I = m\,|\vec{r} + \vec{h}|^2$$

$$= \sum m\,(\vec{r} + \vec{h})(\vec{r} + \vec{h})$$

$$= \sum mr^2 + h^2 \sum m + 2h \sum m\vec{r} = I_c + Mh^2$$

$$\Rightarrow \quad I_{bb} = I_{aa} + mR^2$$

(About centre of mass frame $\sum m_i r_i = 0$)

Illustration 2.6 **Find the moment of inertia of a solid cylinder of mass M and radius R about a line parallel to the axis of the cylinder and on the surface of the cylinder.**

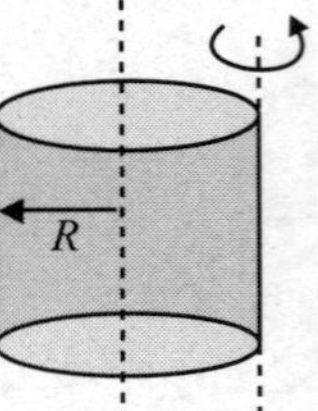

Sol. The moment of inertia of the cylinder about its axis is $\dfrac{MR^2}{2}$

Using parallel axis theorem,

$$I = I_0 + MR^2 = \frac{MR^2}{2} + MR^2 = \frac{3}{2}\,MR^2$$

Similarly, the moment of inertia of a solid sphere about a tangent is

$$\frac{2}{5}\,MR^2 + MR^2 = \frac{7}{5}\,MR^2$$

Note:

- ***Theorem of parallel axes is applicable for any type of rigid body whether it is two dimensional or three dimensional, while the theorem of perpendicular axes is applicable for laminar type or two dimensional bodies only.***
- ***In theorem of perpendicular axes, the point of intersection of the three axes (x, y and z) may be any point on the plane of the body (it may even lie outside the body). This point may or may not be the centre of the mass of the body.***

RADIUS OF GYRATION

Radius of gyration (K) of a body about an axis is the effective distance from the axis where the whole mass can be assumed to be concentrated so that the moment of inertia remains the same (Fig. 2.18). Thus,

$$I = MK^2 \text{ or } K = \sqrt{\frac{I}{M}}$$

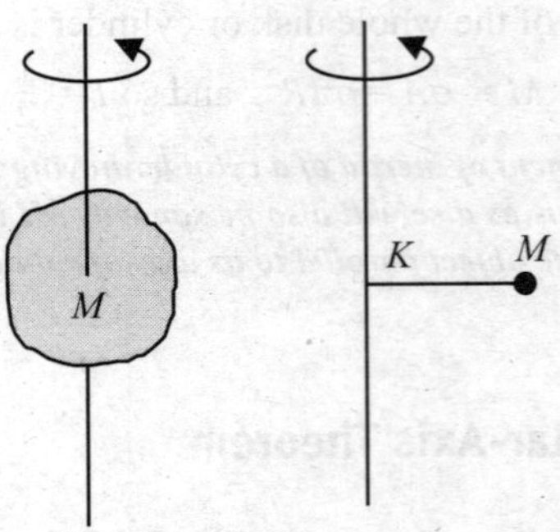

Fig. 2.18

- Radius of gyration of a disc about an axis perpendicular to its plane and passing through its centre of mass is

$$K = \sqrt{\frac{\frac{1}{2}MR^2}{M}} = \frac{R}{\sqrt{2}}$$

- Radius of gyration of a solid sphere, $K = \sqrt{\frac{2}{5}}\,R$
- Radius of gyration of a hollow sphere, $K = \sqrt{\frac{2}{3}}\,R$

MOMENT OF INERTIA OF SOME IMPORTANT CASES

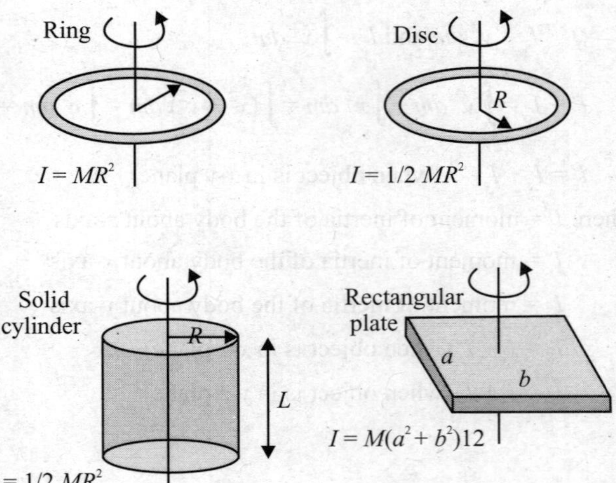

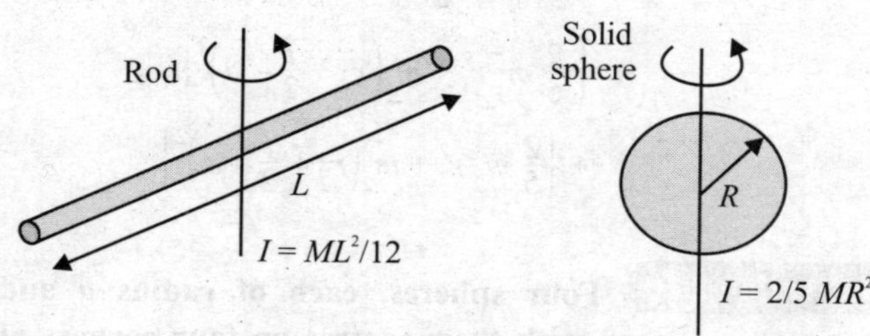

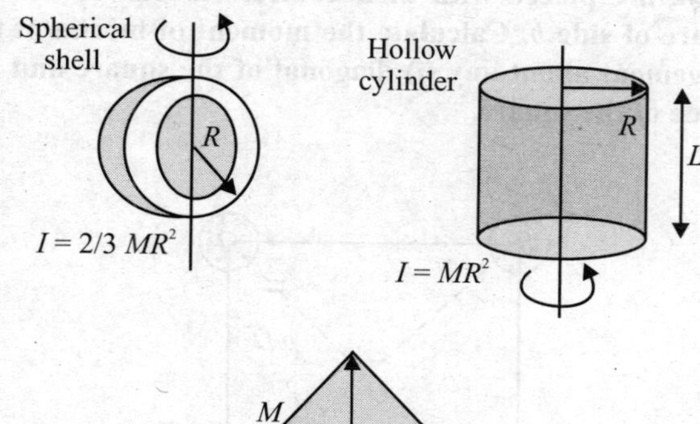

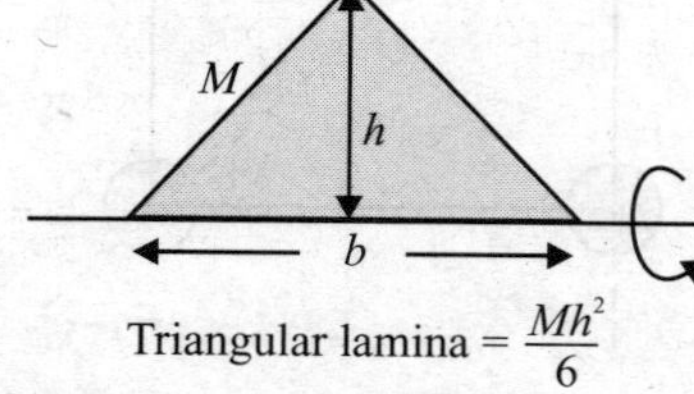

Triangular lamina = $\dfrac{Mh^2}{6}$

Fig. 2.19

Note:

As moment of inertia of a body having uniform mass distribution is independent of size of the body parallel to axis of rotation. Consider this case:

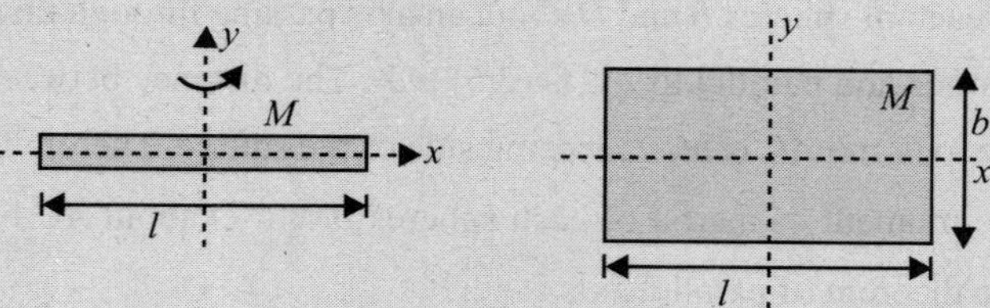

Fig. 2.20

The moment of inertia of a rectangular plate about the y-axis will be same as the moment of inertia of the same mass of the rod about the y-axis. In both the cases $I_y = Ml^2/2$.

We can say directly the moment of inertia of the rectangular plate about x-axis, $I_x = Mb^2/12$.

If we are asked to find moment of inertia of the plate about the x-axis, we can use perpendicular axis theorem, i.e.,

$$I_z = I_x + I_y = \frac{Mb^2}{12} + \frac{Ml^2}{12} = \frac{M}{12}(b^2 + \ell^2)$$

Illustration 2.7 **Calculate the moment of inertia of**

a. a ring of mass M and radius R about an axis coinciding with the diameter of the ring.

b. a thin disc about an axis coinciding with the diameter.

Sol. Let X and Y axes be along two perpendicular diameters of the ring.

By symmetry $I_x = I_y$ and by perpendicular axis theorem $I_z = I_x + I_y$

But we know that $I_z = MR^2$

$$MR^2 = I_x + I_y$$

$$\Rightarrow \quad MR^2 = 2I_x \quad \Rightarrow \quad I_x = I_y = \frac{MR^2}{2}$$

Similarly for a thin disc (i.e., a circular plate), the moment of inertia about a diameter is

$$I = \frac{1}{2}\left(\frac{1}{2}MR^2\right) = \frac{1}{4}MR^2$$

Illustration 2.8 **Find out the moment of inertia of a ring having uniform mass distribution of mass M and radius R about an axis which is tangent to the ring and (i) in the plane of the ring, (ii) perpendicular to the plane of the ring.**

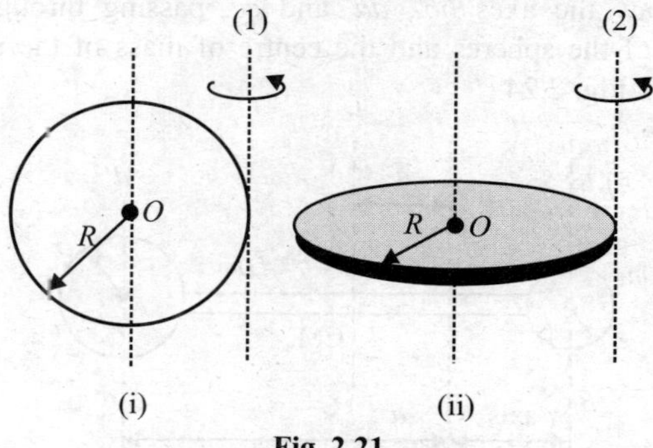

Fig. 2.21

Sol. The moment of inertia of a ring about an axis passing through centre and perpendicular to plain of ring is $I_z = MR^2$ because of symmetry we can say $I_x = I_y$ and using perpendicular axis theorem,

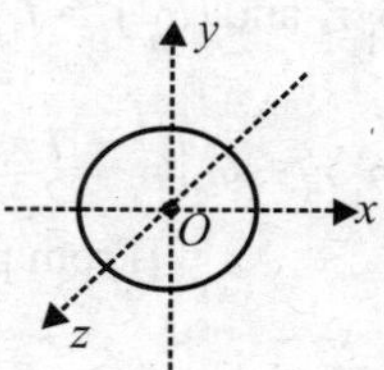

Fig. 2.22

$$I_z = I_x + I_y \Rightarrow MR^2 = 2I_x$$

$$\Rightarrow \quad I_x = \frac{MR^2}{2} = I_0 \text{ [in Fig. 2.21(i)]}$$

For case (i): Using parallel axis theorem,

$$I_1 = I_o + MR^2 = \frac{MR^2}{2} + MR^2 = \frac{3}{2}MR^2$$

For case (ii): Using parallel axis theorem,

$$I_2 = I_z + MR^2 = MR^2 + MR^2 = 2MR^2$$

Illustration 2.9 **Two uniform solid spheres of masses m_1 and m_2 and radii r_1 and r_2, respectively, are connected at the ends of a uniform rod of length l and mass m. Find the moment of inertia of the system about an axis perpendicular to the rod and passing through a point at a distance of a from the centre of mass of the rod as shown in Fig. 2.23.**

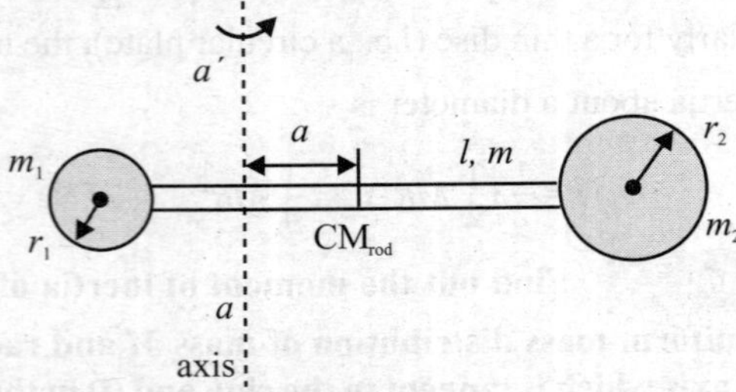

Fig. 2.23

Sol. Draw the axes bb', dd' and cc' passing through the centres of the spheres and the centre of mass of the rod as shown in Fig. 2.24.

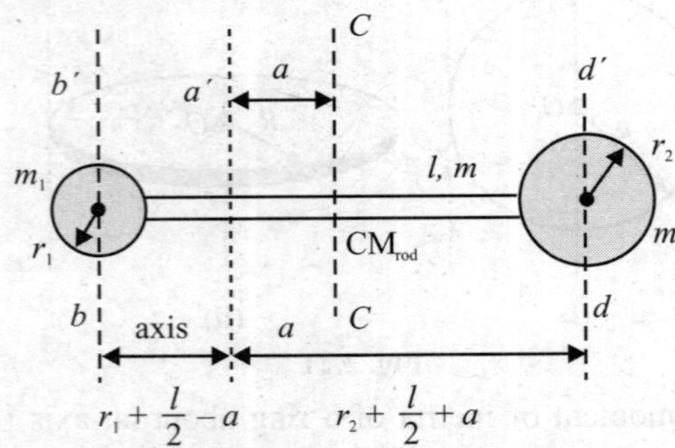

Fig. 2.24

For m_1: $I_{bb'} = \frac{2}{5}m_1 r_1^2$ and $(I_{aa'})_1 = I_{bb'} + m_1\left(r_1 + \frac{l}{2} - a\right)^2$

$$= \frac{2}{5}m_1 r_1^2 + m_1\left(r_1 + \frac{l}{2} - a\right)^2$$

[From parallel axis theorem]

For m: $I_{cc'} = \frac{ml^2}{12}$; $(I_{aa'})_2 = I_{cc'} + ma^2 = \frac{ml^2}{12} + ma^2$

For m_2: $I_{dd'} = \frac{2}{5}m_2 r_2^2 \Rightarrow (I_{aa'})_3 = I_{dd'} + m_2\left(r_2 + \frac{l}{2} + a\right)^2$

$$= \frac{2}{5}m_2 r_2^2 + m_2\left(r_2 + \frac{l}{2} + a\right)^2$$

The moment of inertia of the given system about aa',

$$I_{aa'} = (I_{aa'})_1 + (I_{aa'})_2 + (I_{aa'})_3$$

$$= \left\{\frac{2}{5}m_1 r_1^2 + m_1\left(r_1 + \frac{l}{2} - a\right)^2\right\}$$

$$+ \left\{\frac{2}{5}m_2 r_2^2 + m_2\left(r_2 + \frac{l}{2} - a\right)^2\right\}$$

Illustration 2.10 **Four spheres, each of radius a and mass m, are placed with their centres on four corners of a square of side b. Calculate the moment of inertia of the arrangement about any (i) diagonal of the square and (ii) any side of the square.**

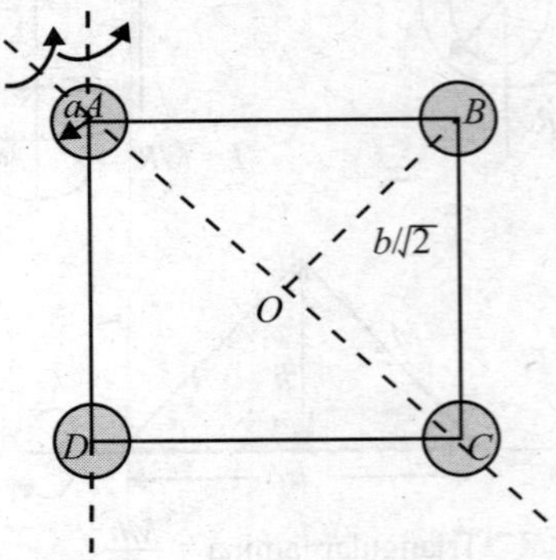

Fig. 2.25

Sol. i. Moment of inertia of the arrangement about the diagonal AC:

The moment of inertia of each of spheres A and C about their common diameter $AC = (2/5)\,ma^2$. The moment of inertia of each of spheres B and D about an axis passing through their centres and parallel to AC is $(2/5)\,ma^2$. The distance between the axis and AC is $b/\sqrt{2}$, b being side of the square. Therefore, the moment of inertia of each spheres B and D about AC by the theorem of parallel axis is

$$\frac{2}{5}ma^2 + m\left(\frac{b}{\sqrt{2}}\right)^2 = \frac{2}{5}ma^2 + \frac{mb^2}{2}$$

Therefore, the moment of inertia of all the four spheres about diagonal AC is

$$I = 2\left(\frac{2}{5}ma^2\right) + 2\left[\frac{2}{5}ma^2 + \frac{mb^2}{2}\right]$$

ii. The moment of inertia of each of spheres A and D about side AD is $(2/5)\,ma^2$.

The moment of inertia of each of spheres B and C about AD is $(2/5)\,ma^2 + mb^2$.

Therefore, the moment of inertia of the whole arrangement about side AD is

$$2\left[\frac{2}{5}ma^2 + \left(\frac{2}{5}ma^2 + mb^2\right)\right] = \frac{2m}{5}[4a^2 + 5b^2]$$

Illustration 2.11 **A circular hole of radius $R/2$ is cut from a circular disc of radius R. The disc lies in the xy-plane and its centre coincides with the origin. If the remaining mass of the disc is M, then**

i. determine the initial mass of the disc, and

ii. determine its moment of inertia about the z-axis.

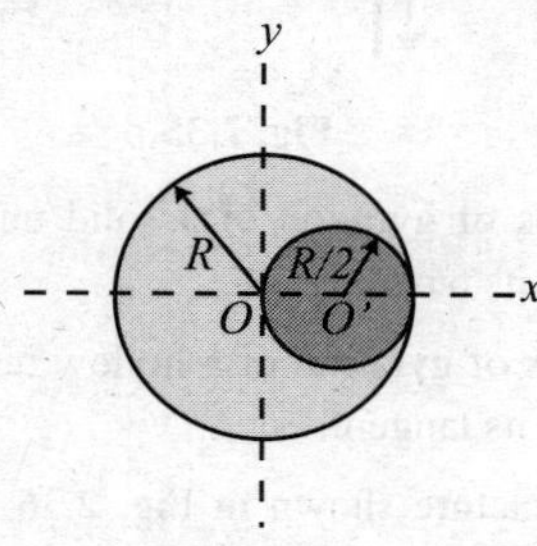

Fig. 2.26

Sol. i. Let σ be the surface mass density of the plate, it is given by

$$\sigma = \frac{M}{\pi R^2 - \pi\left(\frac{R}{2}\right)^2} = \frac{4}{3}\frac{M}{\pi R^2}$$

Now, initial mass of the plate, $m_1 = \sigma(\pi R^2) = \frac{4}{3}M$

Mass of the plate removed, $m_2 = \sigma\left(\frac{\pi R^2}{4}\right) = \frac{M}{3}$

The given arrangement may be considered as a combination of a circular plate of mass (4/3) M and circular plate of negative mass (M/3).

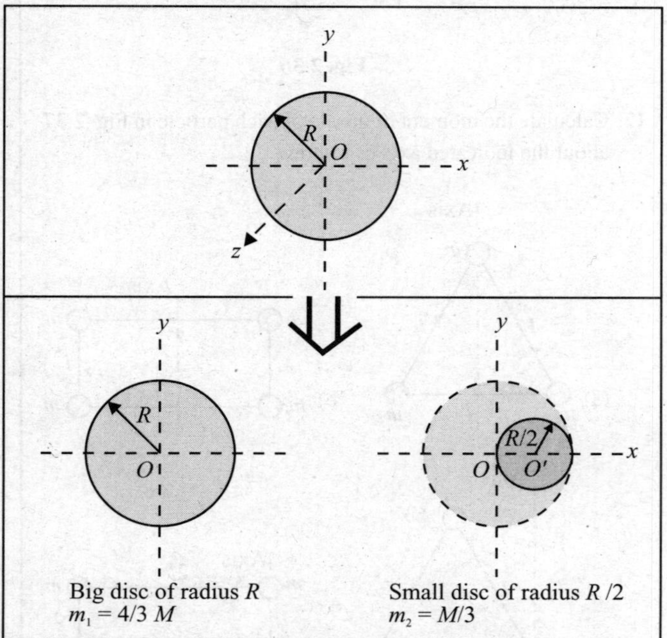

Fig. 2.27

The moment of inertia of the big disc of radius R and mass m_1 about O and parallel to the z-axis is

$$I_{O1} = \frac{m_1 R^2}{2}$$

The moment of inertia of the small disc of mass m_2 and radius $R/2$ about O' and parallel to z-axis

$$I_{O_1} = \frac{m_2 (R/2)^2}{2} = \frac{m_2 R^2}{8}$$

The moment of inertia of the small disc about O (using parallel axis theorem)

$$I_{O_2} = \frac{m_2 R^2}{8} + \frac{m_2 R^2}{4} = \frac{3m_2 R^2}{8}$$

As the part of the disc of mass m_2 is removed from disc m_1, hence the net moment of inertia of structure is $I_O = I_{O_1} - I_{O_2}$.

$$I_o = \frac{m_1 R^2}{2} - \frac{3m_2 R^2}{8}$$

Substituting the values of m_1 and m_2, we get

$$I_0 = \frac{13}{24} MR^2$$

Concept Application Exercise 2.1

1. Two heavy particles having masses m_1 and m_2 are situated in a plane perpendicular to line AB at a distance of r_1 and r_2, respectively.

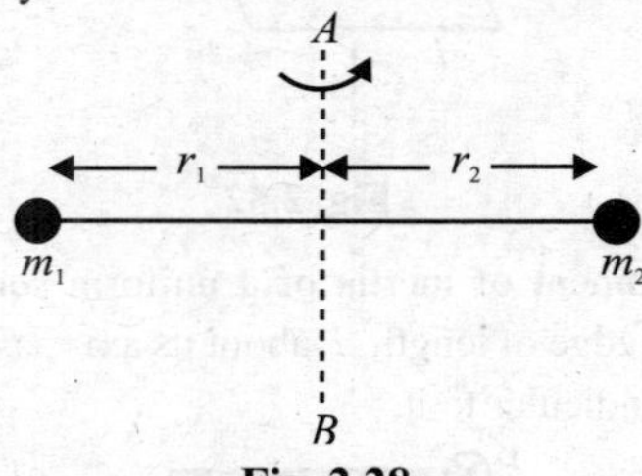

Fig. 2.28

   i. What is the moment of inertia of the system about axis AB?

   ii. What is the moment of inertia of the system about an axis passing through m_1 and perpendicular to the line joining m_1 and m_2?

   iii. What is the moment of inertia of the system about an axis passing through m_1 and m_2?

2. Find out the moment of inertia of the circular arcs shown, each having mass M, radius R and having uniform mass distribution about an axis passing through the centre and perpendicular to the plane ?

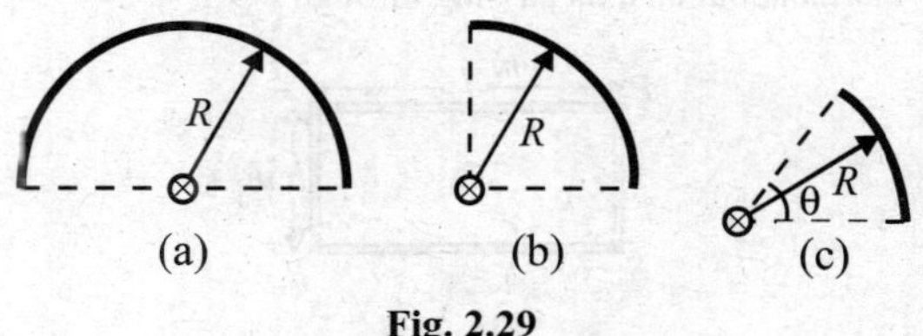

Fig. 2.29

3. Calculate the moments of inertia of the figures shown, each having mass M, radius R and having uniform mass distribution about an axis perpendicular to the plane and passing through the centre?

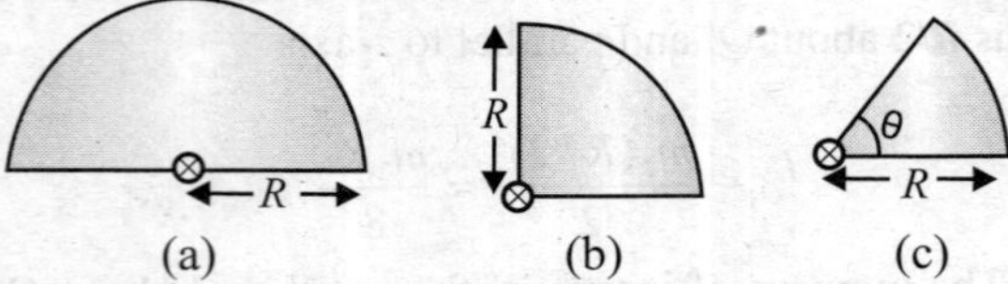

Fig. 2.30

4. In Fig. 2.31, find moment of inertia of a plate having mass M, length l and width b about axes 1, 2, 3 and 4. Assume that C is the centre and mass is uniformly distributed.

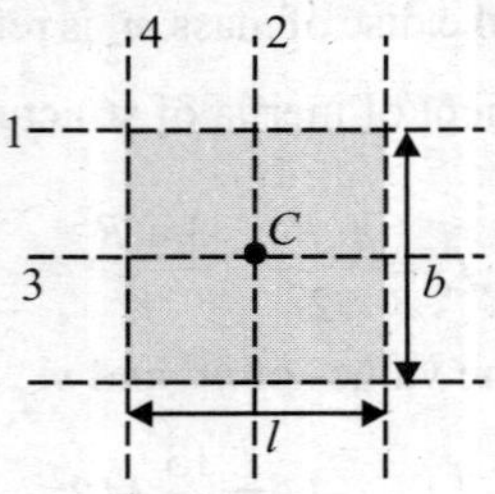

Fig. 2.31

5. Find the moment of inertia of a uniform rectangular plate of mass M and edges of length 'l' and 'b' about its axis passing through the centre and perpendicular to it.

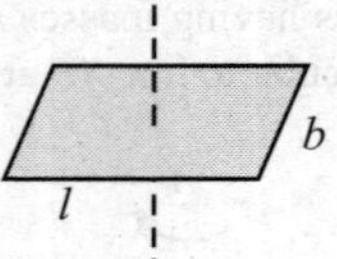

Fig. 2.32

6. Find the moment of inertia of a uniform square plate of mass M and edge of length 'l' about its axis passing through P and perpendicular to it.

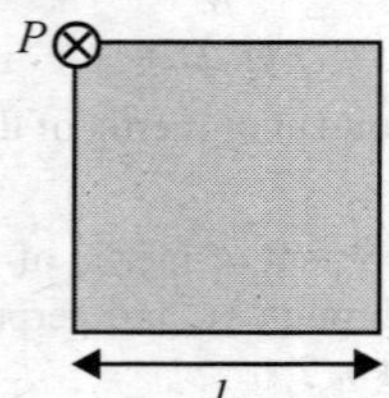

Fig. 2.33

7. Calculate the moment of inertia of a rectangular frame formed by uniform rods having mass m each as shown in Fig. 2.34 about an axis passing through its centre and perpendicular to the plane of frame. Also find moment of inertia about an axis passing through PQ?

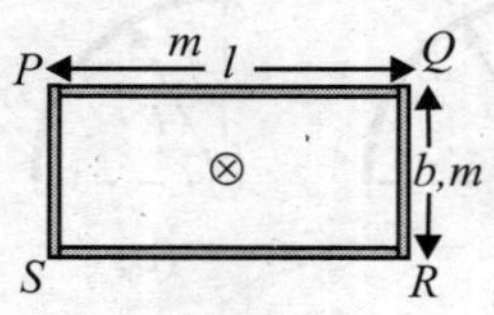

Fig. 2.34

8. Find the moment of inertia of the two uniform joint rods about point P as shown in Fig. 2.35. Use parallel axis theorem. Mass of each rod is M.

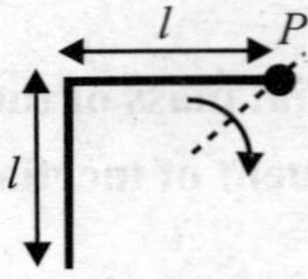

Fig. 2.35

9. Find the radius of gyration of a solid uniform sphere of radius R about its tangent.

10. Find the radius of gyration of a hollow uniform sphere of radius R about its tangent.

11. The square structure shown in Fig. 2.36 consists of four point masses connected by rods of negligible mass. Find the moment of inertia of the structure about the following axes: (a) axis A, passing through the centre of the structure and normal to its plane; (b) axis B, passing through one of the point masses and normal to the plane of the structure; (c) axis CC', passing through two adjacent point masses and (d) axis DD', along the diagonal of the structure.

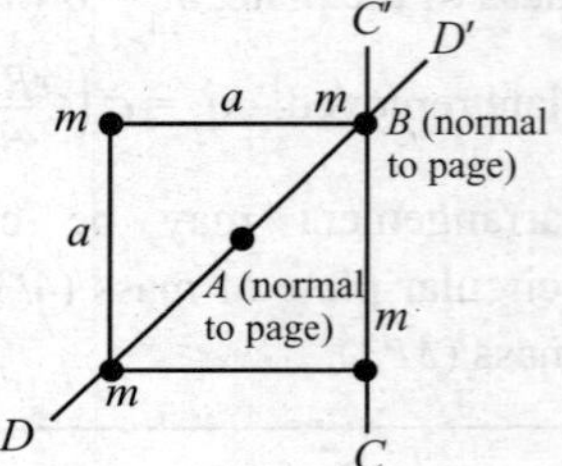

Fig. 2.36

12. Calculate the moment of inertia of each particle in Fig. 2.37 about the indicated axis of rotation.

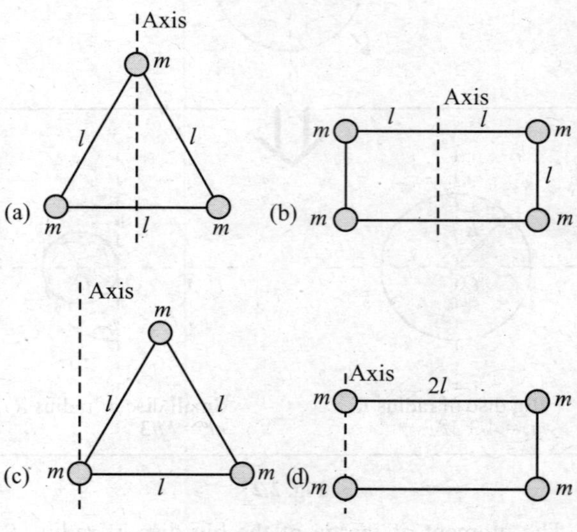

Fig. 2.37

13. A uniform disc of mass m and radius R has an additional rim of mass m as well as four symmetrically placed masses, each of mass $m/4$, tied at positions $R/2$ from the centre as shown in Fig. 2.38. What is the total moment of inertia of the disc about an axis perpendicular to the disc through its centre?

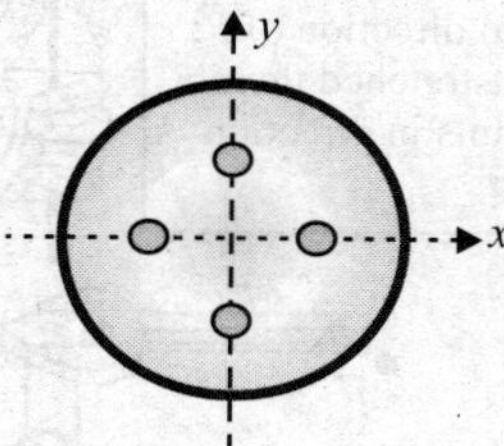

Fig. 2.38

14. Find the moment of inertia of a spherical ball of mass m and radius r attached at the end of a straight rod of mass M and length l, if this system is free to rotate about an axis passing through the end of the rod (end of the rod opposite to sphere).

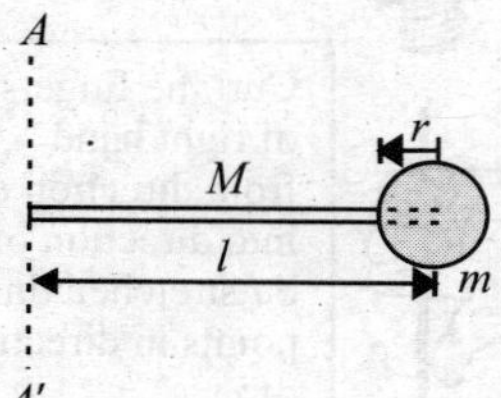

Fig. 2.39

15. Find the moment of inertia of a cylinder of mass M, radius R and length L about an axis passing through its centre and perpendicular to its symmetry axis. Do this by integrating an elemental disc along the length of the cylinder.

16. Find MI of a triangular lamina of mass M about the axis of rotation AB shown in Fig. 2.40.

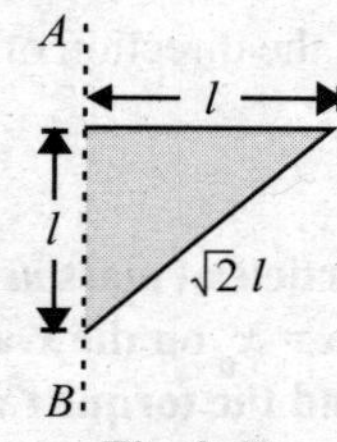

Fig. 2.40

17. Find the moment of inertia of a square lamina of mass M and edge 'a' about its diagonal.

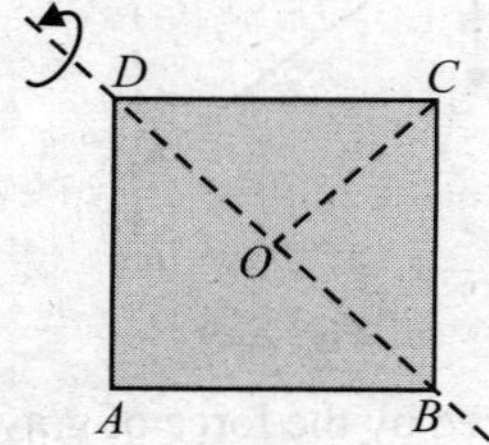

Fig. 2.41

18. Find the moment of inertia of a rectangular lamina of mass M and size $B \times L$ about its diagonal.

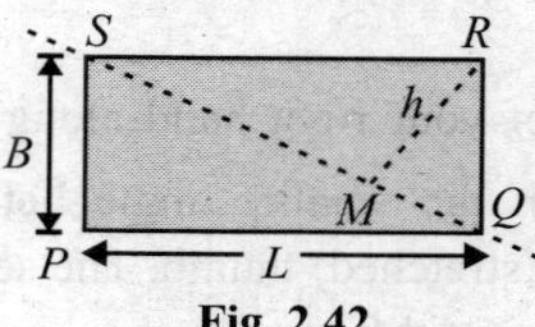

Fig. 2.42

19. Four identical rods, each of mass m and length l, make a square frame in the xy plane as shown in Fig. 2.43.
    i. Calculate its moment of inertia about the x-and y-axes.
    ii. Also, calculate its moment of inertia about the z-axis.

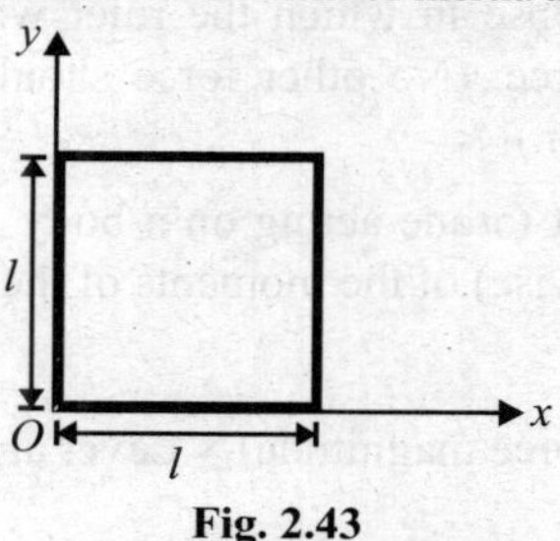

Fig. 2.43

TORQUE

Force causes linear acceleration, whereas torque causes angular acceleration.

It is defined as the moment of force. In other words, the turning ability of a force about an axis is called its torque about that axis.

Consider force $\vec{F}$ which acts through a point whose position vector is r, as shown in Fig. 2.44.

Its torque about the origin is defined as $\vec{\tau} = \vec{r} \times \vec{F}$.

The direction of the torque can be obtained from the right hand rule and its magnitude is given by $\tau = rF \sin \theta$,

where θ is the angle between the vector r and F.

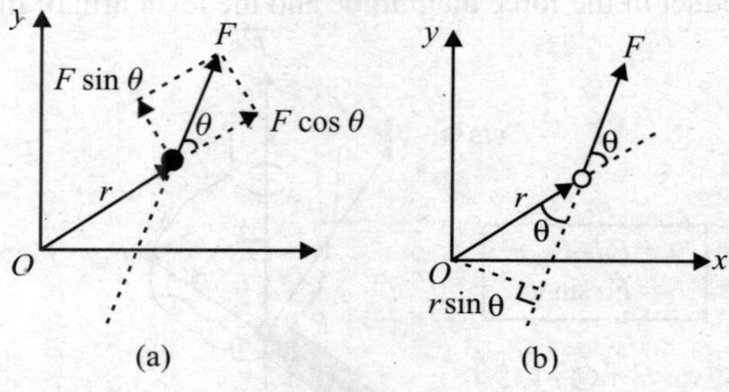

Fig. 2.44

$$\tau = r(F \sin \theta) = rF_{\perp}$$

where $F_{\perp}$ is the component of F perpendicular to r.

$$\tau = (r \sin \theta)\, F = r_{\perp} F$$

where $r_{\perp}$ is the perpendicular distance from the origin to the line of action of the force. It is also called the lever arm.

To Get the Direction of Torque, Use Right Hand Rule

Place the fingers of your right hand along $\vec{r}$ and curl them towards $\vec{F}$ through the smaller angle between them, $\hat{n}$ is directed along the (stretched) thumb. The sense of torque can also be determined in a different way.

Consider an imaginary ruler placed along the perpendicular drawn on the line of action of force from the point about which torque is required, hinge the ruler at this point, find the sense in which the ruler will turn under the action of that force. (No other force should be taken to be acting on the ruler.)

The resultant torque acting on a body is the vector sum (or added sense wise) of the moments of the individual forces acting on it.

Torque = (Force magnitude) × Lever arm

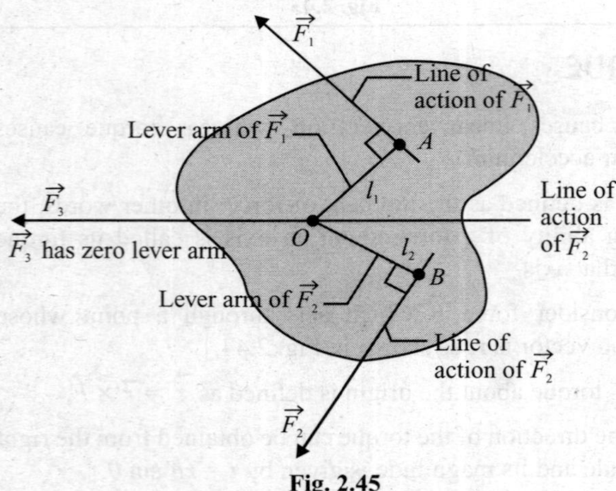

Fig. 2.45

The torque or moment of a force about a point is the product of the force magnitude and the lever arm of the force.

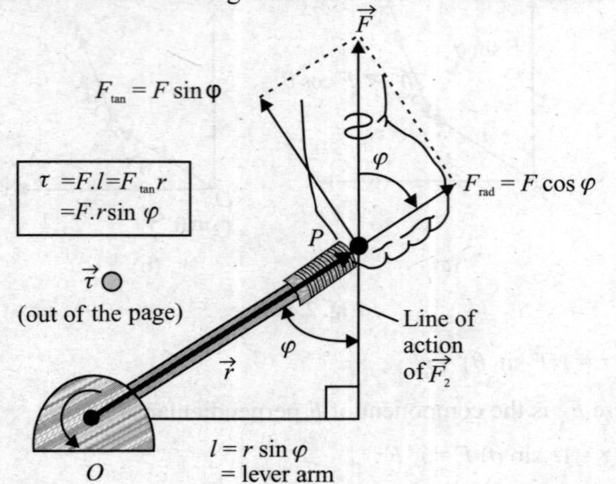

Fig. 2.46

The torque of the force $\vec{F}$ about the point O is defined as $\vec{\tau} = \vec{r} \times \vec{F}$. The magnitude of $\vec{\tau}$ is $Fr \sin\phi$. In Fig. 2.46, $\vec{r}$ and $\vec{F}$ are in the plane of the page and points out of the page towards you.

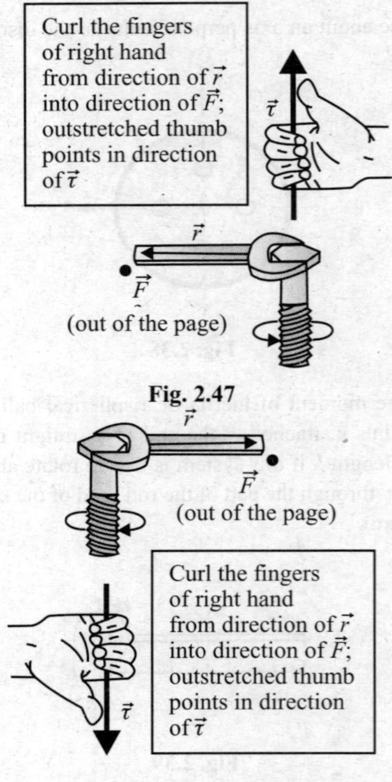

Fig. 2.47

Fig. 2.48

The torque vector $\vec{\tau} = \vec{r} \times \vec{F}$ is directed along the axis of the bolt, perpendicular to both $\vec{r}$ and $\vec{F}$. The direction of $\vec{\tau}$ is given by the right hand rule. Note that the fingers of the right hand curl in the direction of the rotation that the torque tends to cause.

Illustration 2.12 **A particle of mass m is released in vertical plane from a point P at $x = x_0$ on the x-axis. It falls vertically parallel to the y-axis. Find the torque τ acting on the particle at a time t about origin.**

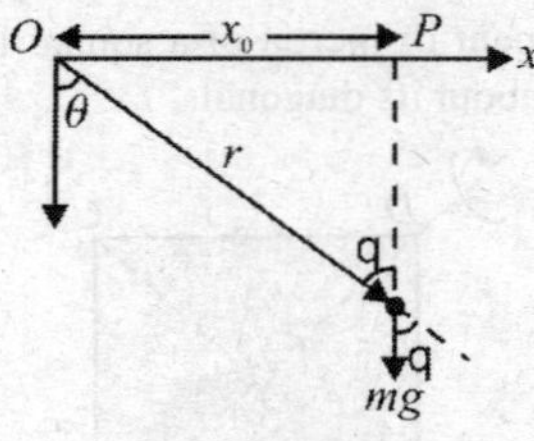

Fig. 2.49

Sol. Torque is produced by the force of gravity.

$$\vec{\tau} = rF\sin\theta \quad \text{or} \quad \tau = r\,mg\,(x_0/r) = mgx_0\,\hat{k}$$

COUPLE

Two equal and opposite forces whose lines of action do not coincide make a couple

Moment of a couple is given by $\tau = r_{\perp}F$, where $r_{\perp}$ is the separation of the lines of action of the forces and F is the magnitude of either of the forces.

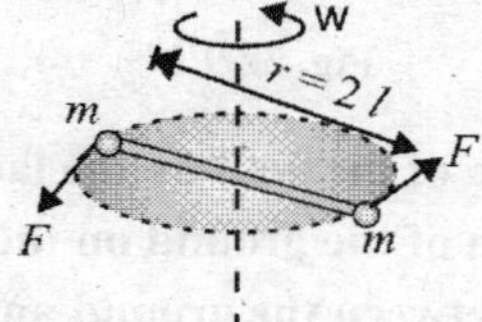

Two balls each of mass m are fixed at the ends of a rigid body of length $2l$.

Fig. 2.50

Point of Application of Force

Point of application of force is the point at which if net force is assumed to be acting then it will produce the same effect, of both translation and rotational nature, as was produced earlier.

OR

If net force is applied at the point of application in the opposite direction, then the body will be in equilibrium (translational and rotational both). Let us understand this through illustration.

Illustration 2.13 **Determine the point of application of force, when forces of 20 N and 30 N are acting on rod as shown in Fig. 2.51.**

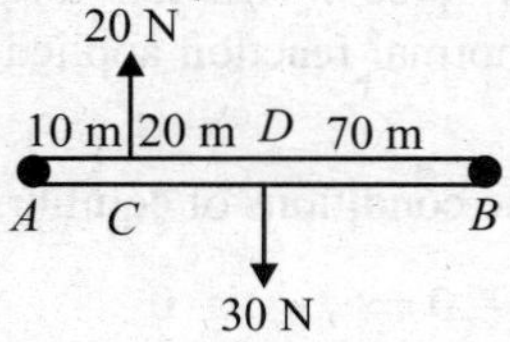

Fig. 2.51

Sol. Net force acting on the rod, $F_{rel} = 10$ N

Net torque acting on the rod about point C is

$$\tau_c = (20 \times 0) + (30 \times 20) = 600 \text{ clockwise}$$

Let the point of application be at a distance x from point C (Fig. 2.52)

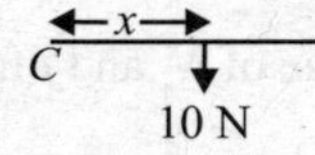

Fig. 2.52

$$600 = 10x \Rightarrow x = 60 \text{ CM}$$

Therefore, 70 CM from A is the point of application of force.

EQUILIBRIUM OF RIGID BODIES

The conditions of equilibrium for a rigid body are different from that of a particle. Unlike particle such as bodies, rigid bodies have a tendency to rotate, therefore, one has to consider the rotational equilibrium also, in addition to the translational equilibrium.

The equations of equilibrium (in two dimensions) for a rigid body are stated as:

Translational equilibrium: $\sum \vec{F} = 0$

Rotational equilibrium: $\sum \vec{\tau}_C = 0$

Note that for rotational equilibrium of a rigid body, the net torque about its centre of mass must be zero.

Action of two forces on a particle and a rigid body is shown in Figs. 2.53 and 2.54.

A particle is in equilibrium under the action of two equal and opposite forces.

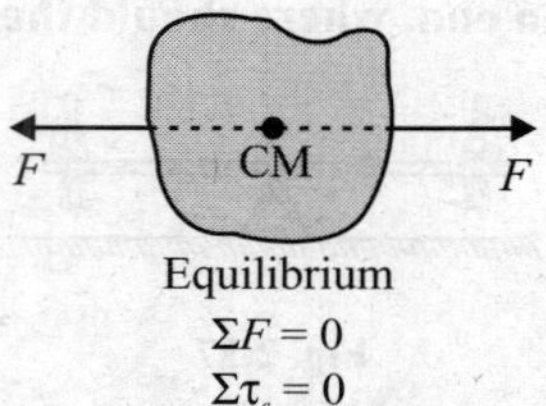

Fig. 2.53

The body have both translational and rotational equilibrium.

A rigid body may or may not be in equilibrium under the action of two equal and opposite forces.

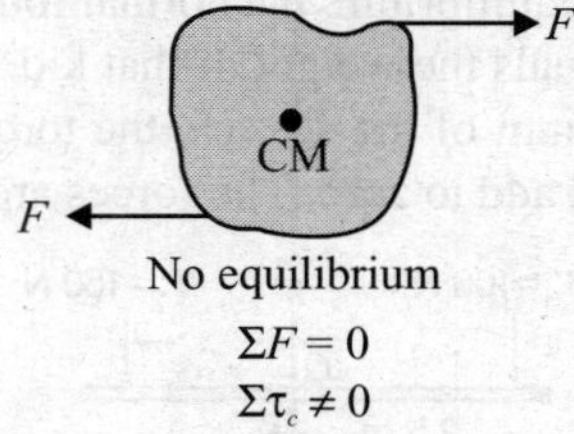

Fig. 2.54

The body is in translational equilibrium but not in rotatinal equilibrium.

Figures 2.55 and 2.56 show equilibrium of a rigid body under the action of three parallel or concurrent forces.

A rigid body may be in equilibrium under the action of three forces if their lines of action are parallel to each other.

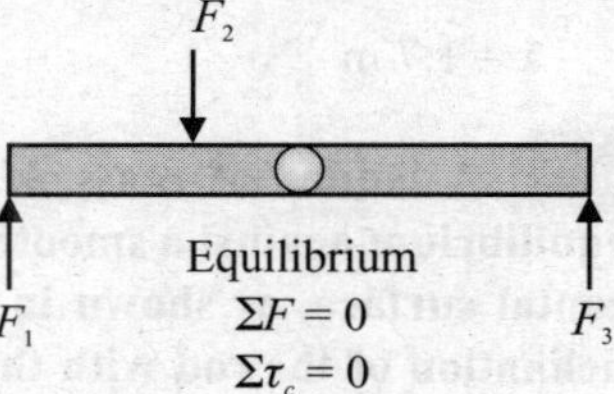

Fig. 2.55

A rigid body may be in equilibrium under the action of three concurrent forces (concurrent means having a common point of application).

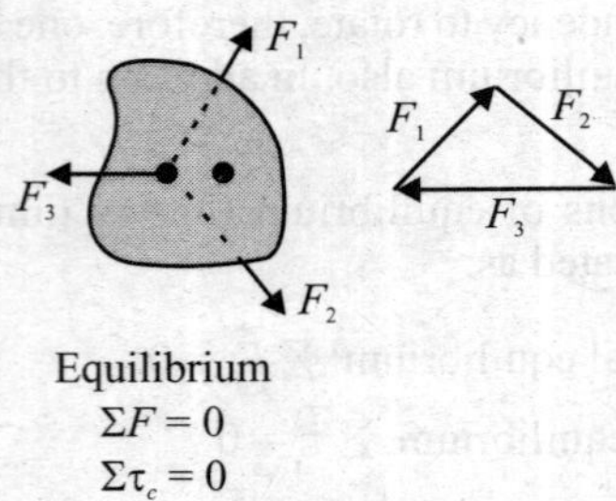

Fig. 2.56

Illustration 2.14 **Two small kids weighing 10 kg and 15 kg, respectively, are trying to balance a seesaw of total length 5.0 m, with the fulcrum at the centre. If one of the kids is sitting at an end, where should the other sit?**

Fig. 2.57

Sol. If the kid of mass 15 kg sit at the end, he will produce maximum torque about fulcrum. The seesaw will not be balance for any position of kid of mass 10 kg. It is clear that the 10 kg kid should sit at the end and the 15 kg kid should sit closer to the centre. Suppose his distance from the centre is x. As the kids are in equilibrium, the normal force between a kid and the seesaw equals the weight of that kid. Considering the rotational equilibrium of the seesaw, the torque of the forces acting on it should add to zero. The forces are as follows:

W_1=100 N W W_2=150 N
O x
2.5 m N

Fig. 2.58

a. (15 kg) g = 150 N downwards by the 15 kg kid,

b. (10 kg) g = 100 N downwards by the 10 kg kid,

c. weight of the seesaw, and

d. the normal force by the fulcrum.

Taking torques about the fulcrum, $150 \times x = 100 \times 2.5$

or $x = 1.7$ m

Illustration 2.15 **A ladder of mass M and length L is supported in equilibrium against a smooth vertical wall and a rough horizontal surface, as shown in Fig. 2.59. If θ be the angle of inclination of the rod with the horizontal, then calculate**

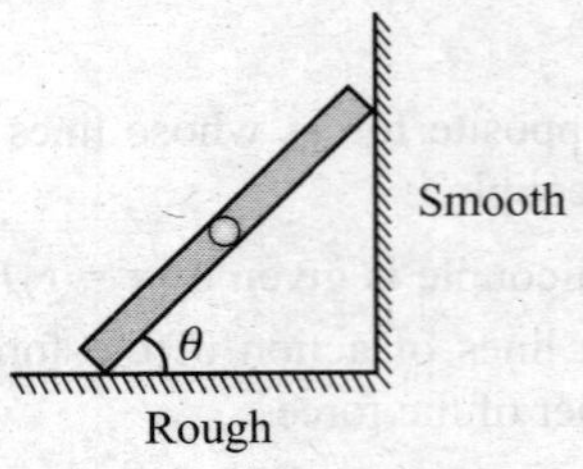

Fig. 2.59

i. normal reaction of the wall on the ladder

ii. normal reaction of the ground on the ladder

iii. friction force between the ground and the ladder

iv. net force applied by the ground on the ladder

Sol. i. The free-body diagram of the ladder is shown in Fig. 2.60.

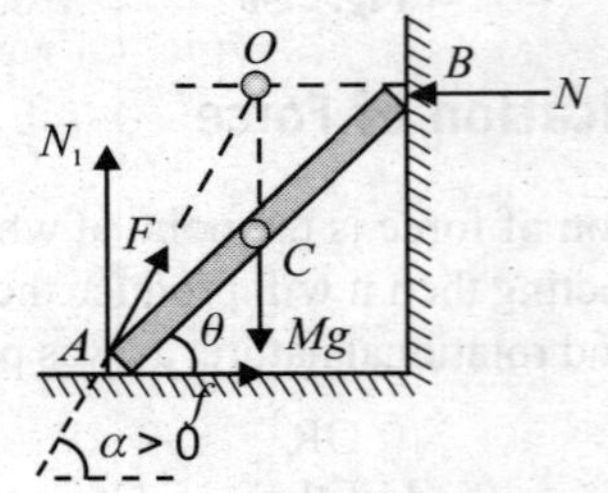

Fig. 2.60

Note that all the three forces acting on the ladder have to pass through a common point O, otherwise it cannot be in equilibrium.

The total reaction force F from the horizontal surface is inclined at an angle α ($> \theta$) to the horizontal. The horizontal component of this force is friction force and its vertical component is the normal reaction applied by the ground on the ladder.

ii. Applying the conditions of equilibrium, we get

$$\Sigma F_x = 0 \Rightarrow f - N = 0$$

$$\Rightarrow \quad f = N \quad \text{(i)}$$

$$\Sigma F_y = 0 \Rightarrow N_1 - Mg = 0$$

$$\Rightarrow \quad N_1 = Mg \quad \text{(ii)}$$

Taking torque about centre of the rod 'C', $\Sigma \tau_C = 0$

$$(N_1)\frac{L}{2}\cos\theta - (f)\frac{L}{2}\sin\theta - N\frac{L}{2}\sin\theta = 0 \quad \text{(iii)}$$

Substituting the value of N_1 and f from Eqs. (i) and (ii) in Eq. (iii), we get

$$Mg\frac{L}{2}\cos\theta - NL\sin\theta = 0$$

$$\Rightarrow \quad N = \frac{Mg}{2\tan\theta} = \frac{1}{2}Mg\cot\theta = f \quad \text{(iv)}$$

Thus, the net force applied by ground on the ladder is

$$F = \sqrt{N_1^2 + f^2}$$

$$F = \frac{1}{2} Mg \sqrt{4 + \cot^2 \theta}$$

Alternative method:

As ladder is in equilibrium, the net torques about A will be zero. $\sum \tau_A = 0$, taking torque about end A of the rod.

$$Mg\left(\frac{L}{2}\cos\theta\right) - N(L\sin\theta) = 0$$

$$N = \frac{Mg}{2\tan\theta} = f$$

It is the same result as we get by taking torque about C.

Illustration 2.16 **At the bottom edge of a smooth wall, an inclined plane is kept at an angle of 45°. A uniform ladder of length l and mass M rests on the inclined plane against the wall such that it is perpendicular to the incline.**

a. If the plane is also smooth, which way will the ladder slide?

b. What is the minimum coefficient of friction necessary so that the ladder does not slip on the incline.

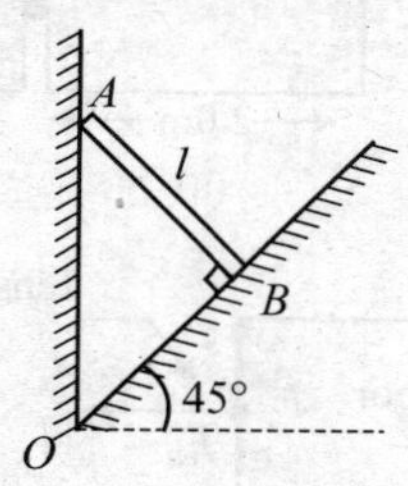

Fig. 2.61

Sol.

i. The ladder has a tendency to slip by rotating clockwise about the point A

ii. The free-body diagram of the ladder is shown in Fig. 2.62.

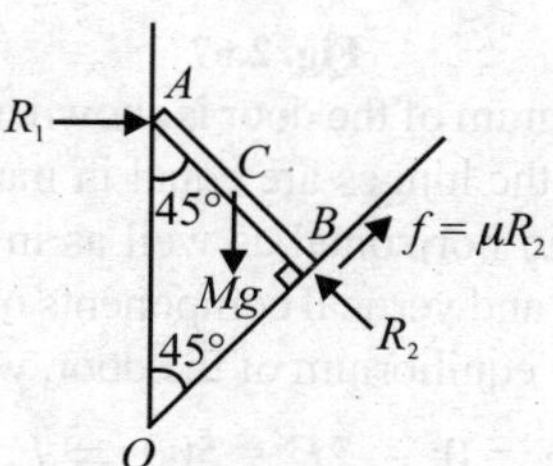

Fig 2.62

Balancing torque about point A, $fl = Mg\left(\frac{l}{2}\sin 45°\right)$

$$\therefore \quad f = \frac{Mg}{2\sqrt{2}} \qquad \text{(i)}$$

Balancing forces in the vertical direction,

$$Mg = R_2 \cos 45° + f \sin 45° \qquad \text{(ii)}$$

From Eqs. (i) and (ii), we get

$$R_2 = \frac{3Mg}{2\sqrt{2}} \qquad \text{(iii)}$$

From Eqs. (i) and (iii), $f = \mu R_2$

or $\frac{Mg}{2\sqrt{2}} = \frac{3Mg}{2\sqrt{2}}\mu$ or $\mu = \frac{1}{3}$

Illustration 2.17 **A heavy block of length b and height h is placed at rest on a rough inclined plane of inclination θ with the horizontal, as shown in Fig. 2.63.**

i. The normal reaction from the surface cannot pass through the centre mass. Why?

ii. If μ is the coefficient of friction between the block and the inclined plane, then discuss the conditions of translational and rotational equilibrium of the block.

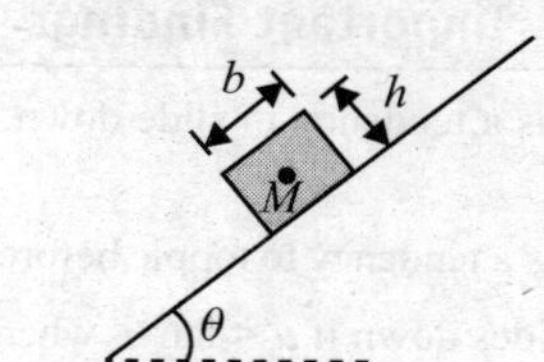

A heavy block is placed at rest on a rough inclined plane.

Fig. 2.63

Sol.

i. The free-body diagram of the block is shown in Fig. 2.64.

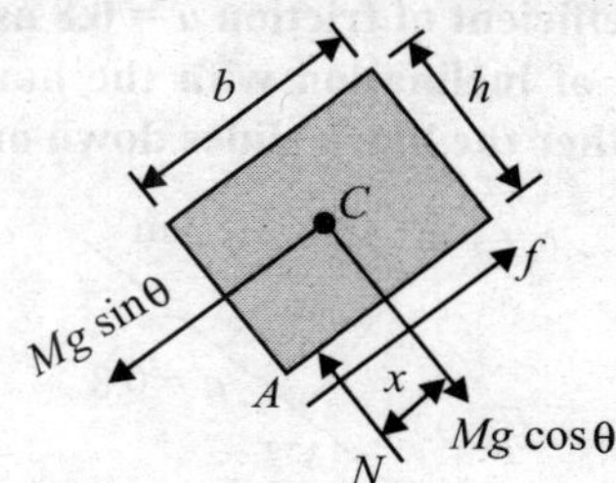

Fig. 2.64

Note that the point of application of the normal reaction N is displaced through x in the downward direction. It has to produce a clockwise torque about the centre of mass that may balance the anticlockwise torque produced by the friction force.

ii. There are two tendencies of the block:

- to slide down
- to rotate (or topple) about the point A

For translational equation $\sum \vec{F} = 0$

$$f = Mg\sin\theta \qquad \text{(i)}$$

$$N = Mg\cos\theta \qquad \text{(ii)}$$

For rotational equilibrium, $\sum \vec{\tau}_C = 0$

$$N \times x = f \times \frac{h}{2} \Rightarrow Mg\cos\theta \times x = \frac{fh}{2}$$

$$\Rightarrow \quad f = \frac{2Mg\cos\theta \times x}{h}$$

If the block topples over about A, $x = \frac{b}{2}$

Hence, $f = \frac{b\,Mg\cos\theta}{h}$ (iii)

From Eq. (i) if block do not slide down, $f < \mu N$

$$Mg\sin\theta < \mu \times Mg\cos\theta \Rightarrow \tan\theta \le \mu$$

or we can say if the block slides $\mu < \tan\theta$

From Eq. (iii) if the block topples before sliding,

$$f \le \mu\,(Mg\cos\theta)$$

$$\frac{b\,Mg\cos\theta}{h} \le \mu\,Mg\cos\theta \Rightarrow \frac{b}{h} \le \mu$$

Hence, if the block topples before sliding, $\mu \ge \frac{b}{h}$

Important Findings

1. The block has a tendency to slide down before toppling if $\mu < b/h$.
2. The block has a tendency to topple before sliding if $\mu \ge b/h$.
3. The block slides down if $\mu < \tan\theta$, where θ is the angle of inclination of the incline with respect to the horizontal.
4. The block topples over if $\mu > \tan\theta$ and $\mu \ge b/h$.

Illustration 2.18 **A tall block of mass $M = 50$ kg and base width $b = 1$ m and height $h = 3$ m is kept on a rough inclined surface with coefficient of friction $\mu = 0.8$ as shown in Fig. 2.65. The angle of inclination with the horizontal is 37°. Determine whether the block slides down or topples over.**

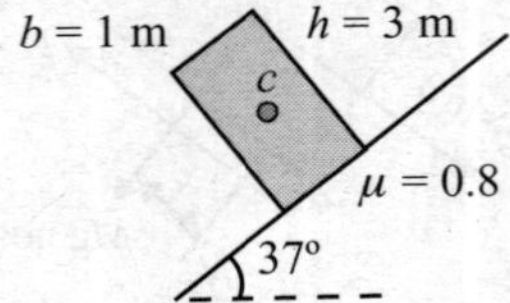

A tall block is resting on a rough inclined plane.

Fig. 2.65

Sol. Here $\mu = 0.8$, $\frac{b}{h} = \frac{1}{3} = 0.33$; $\tan\theta = \frac{3}{4} = 0.75$

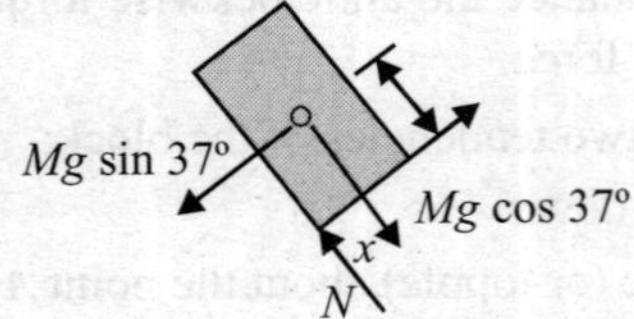

Fig. 2.66

Since $\mu > b/h$ and $\mu > \tan\theta$, therefore the block has a greater tendency to topple over. From the condition of translation equilibrium, we have

$$N = Mg\cos 37° = (50)(10)\left(\frac{4}{5}\right) \text{ or } N = 400 \text{ N}$$

$$f = Mg\cos 37° = (50)(10)\left(\frac{3}{5}\right)$$

$$= 300 \text{ N} < f_{max} = \mu N = 0.8\,(400) = 320 \text{ N}$$

For rotational equilibrium, $\sum \tau_c = 0$

$$N_x = f\frac{h}{2}$$

Thus, $x = \frac{f}{N}\frac{h}{2} = \left(\frac{300}{400}\right)\left(\frac{3}{2}\right) = \frac{9}{8} > \frac{b}{2}$

Since the value of x is more than $b/2$ which is not possible, therefore the block topples over.

Illustration 2.19 **The door of an almirah is 4 m high, 1 m wide and weighs 50 kg. The door is supported by two hinges situated at a distance of 1.0 m from the ends. If the magnitude of each forces exerted by the hinges on the door is equal, find this magnitude.**

Sol.

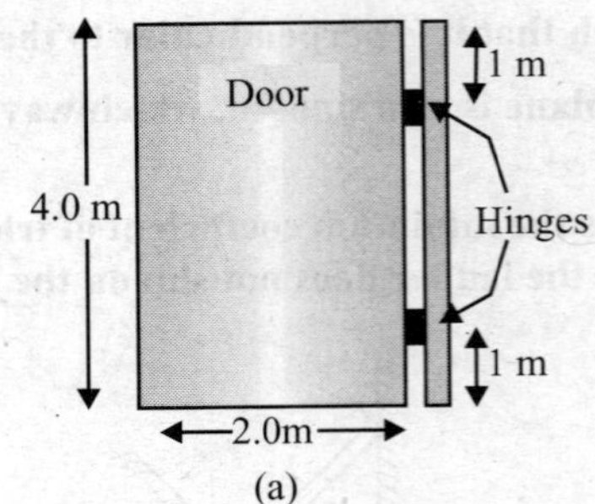

(a)

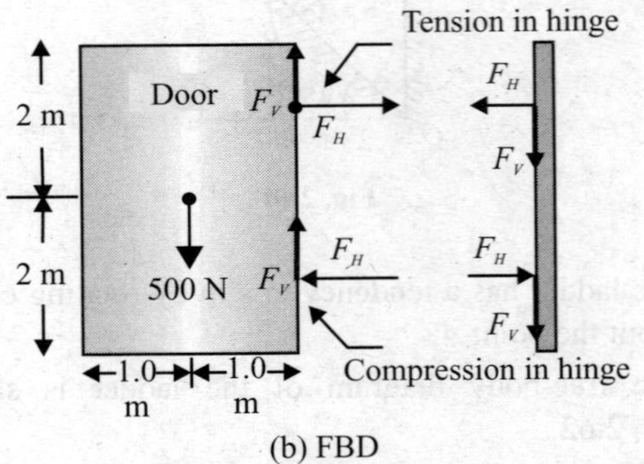

(b) FBD

Fig. 2.67

The free-body diagram of the door is shown in Fig. 2.67(*b*). The forces exerted on the hinges are equal in magnitude. The door is in equilibrium in horizontal as well as in vertical direction, i.e., the horizontal and vertical components of the forces will be equal. For vertical equilibrium of the door, we have

$$\sum F_v = 0; \quad 2F_v = 500 \Rightarrow F_v = 250 \text{ N}$$

For rotational equilibrium of the door, we have $\sum \tau = 0$

Taking moment of all forces acting on the door about lower hinge (may be upper hinge or any other point), we get

$$500 \times 1.0 - F_H \times 2 = 0 \text{ which gives } F_H = 250 \text{ N}$$

The force on the door exerted by either hinge, in magnitude, is

$$F = \sqrt{F_H^2 + F_v^2} = \sqrt{(250)^2 + (250)^2} = 250\sqrt{2} \text{ N}$$

Concept Application Exercise 2.2

1. A uniform cube of side a and mass m rests on a rough horizontal table. A horizontal force F is applied normal to one of the faces at a point directly above the centre of the face, at a height $3a/4$ above the base. What is the minimum value of F for which the cube begins to tip about an edge?
2. A uniform rod is made to lean between a rough vertical wall and the ground. The coefficient of friction between the rod and the ground is μ_1 and between the rod and the wall is μ_2. Find the angle at which the rod can be leaned without slipping.
3. A beam of weight W supports a block of weight W. The length of the beam is L and weight is at a distance $L/4$ from the left end of the beam. The beam rests on two rigid supports at its ends. Find the reactions of the supports.

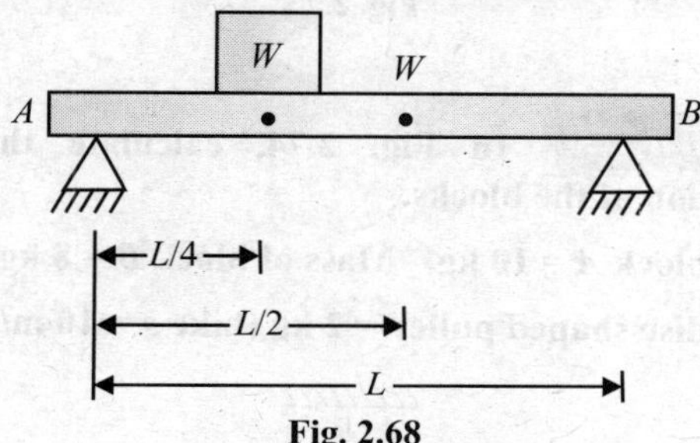

Fig. 2.68

4. A uniform ladder of mass 10 kg leans against a smooth vertical wall making an angle of 53° with it. The other end rests on a rough horizontal floor. Find the normal force and the friction force that the floor exerts on the ladder.
5. A uniform ladder of length 10.0 m and mass 16.0 kg is resting against a vertical wall making an angle of 37° with it. The vertical wall is frictionless but the ground is rough. An electrician weighing 60.0 kg climbs up the ladder. If he stays on the ladder at a point 8.00 m from the lower end, what will be the normal force and the force of friction on the ladder by the ground? What should be the minimum coefficient of friction for the electrician to work safely?
6. A uniform rod of length L rests against a smooth roller as shown in Fig. 2.69. Find the friction coefficient between the ground and the lower end if the minimum angle that the rod can make with the horizontal is θ.

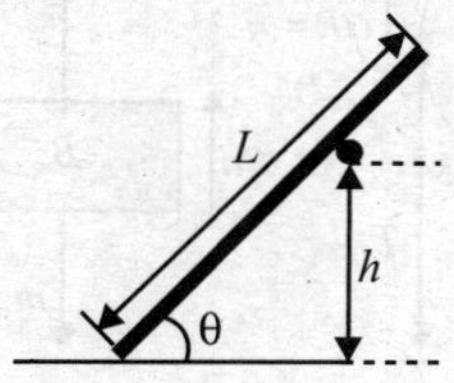

Fig. 2.69

7. The ladder shown in Fig. 2.70 has negligible mass and rests on a frictionless floor. The crossbar connects the two legs of the ladder at the middle. The angle between the two legs is 60°. The fat person sitting on the ladder has a mass of 80 kg. Find the contact forces exerted by the floor on each leg and the tension in the crossbar.

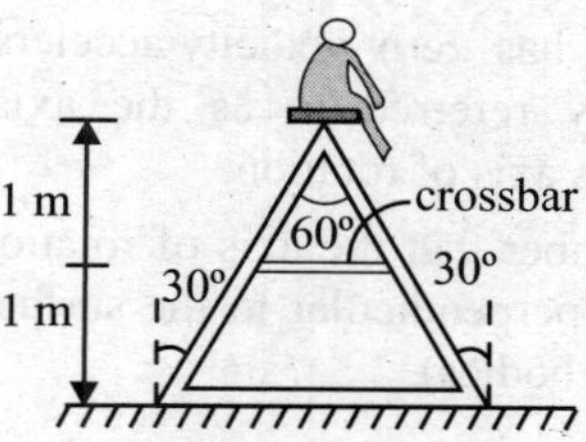

Fig. 2.70

8. A uniform rod of length l and mass m is hung from two strings of equal length from a ceiling as shown in Fig. 2.71. Determine the tensions in the strings?

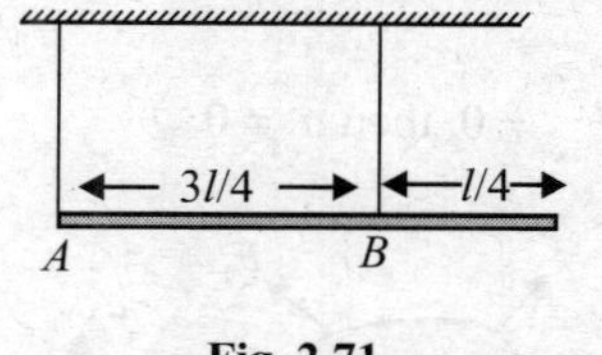

Fig. 2.71

ANALOGUE OF NEWTON'S SECOND LAW OF MOTION FOR PURE ROTATION

Equation analogous to force equation $\sum \vec{F} = M\vec{a}_{CM}$ for pure rotation is $\tau = I\alpha$, where τ is the torque acting on the body and α is its angular acceleration about the axis of rotation provided moment of inertial (I) of the rigid body about the axis of rotation remains constant throughout the motion. As such torque is a three dimensional vector quantity. In the problems we encounter at this level, rotation takes place in two dimensions and τ and α are taken about the axis of rotation in the third dimension (normal to the plane of rotation) and the torque equation is written as

$$\tau_{\text{about the axis of rotation}} = I_{\text{about the axis of rotation}}\ \alpha_{\text{about the axis of rotation in the sense of }\tau} \Rightarrow \tau = I\alpha$$

The above equation is called 'torque equation'.

The torque equation is, however, more complex than the force equation. Why?

- Whereas force is an absolute value, torque depends on the point about which torque is being calculated.
- Mass is an absolute value for a body, whereas MI can have infinite values for a body. MI is defined only for a particular axis of the body.

Remember the following points about the torque equation:

Torque equation can be applied only about two points in a body. These are as follows:

- Centre of mass.
- Point which has zero velocity/acceleration. This point is commonly referred to as the axis of rotation or instantaneous axis of rotation.

Now, remember that the axis of rotation passes through the point and is perpendicular to the surface (valid only for two dimensional bodies).

APPLICATION OF TORQUE EQUATION AND EQUATION OF MOTION COMBINED

If the vector sum of all the forces acting on a rigid body is not equal to zero, then its centre of mass has a linear acceleration. That is

If $\quad \sum F_{net} \neq 0$, then $a_c \neq 0$

Two dimensional motion of a rigid body

Fig. 2.72

If the vector sum of all the torques about the centre of mass is not equal to zero, then the body rotate about its centre of mass. That is,

If $\quad \sum \tau_c \neq 0$ then $\alpha \neq 0$

According to Newton's second law, for two dimensional motion of a rigid body as shown in Fig. 2.73, we can state that

- Translational motion: $\sum F_x = ma_x$, $\sum F_y = ma_y$
- Rotational motion: $\sum \tau_c = I_c \alpha$

Constraints Due to Ropes and Pulleys

While solving the problems of rotational dynamics we encounter situations where pulleys are connected via strings or belts to other objects. We assume that the string turns on the pulley without slipping. Due to this constraint the string's speed v_{string} must be equal to the speed of the rim of the pulley which has magnitude $|\vec{v}_{rim}| = |\vec{\omega}| R$. If the pulley has angular acceleration, the string acceleration a_{string} must be equal to the tangential acceleration of the rim of the pulley $|\vec{a}_t| = |\vec{\alpha}| R$.

In nutshell we can write $\vec{v}_{block} = |\vec{\omega}| R$; $\vec{a}_{block} = |\vec{\alpha}| R$.

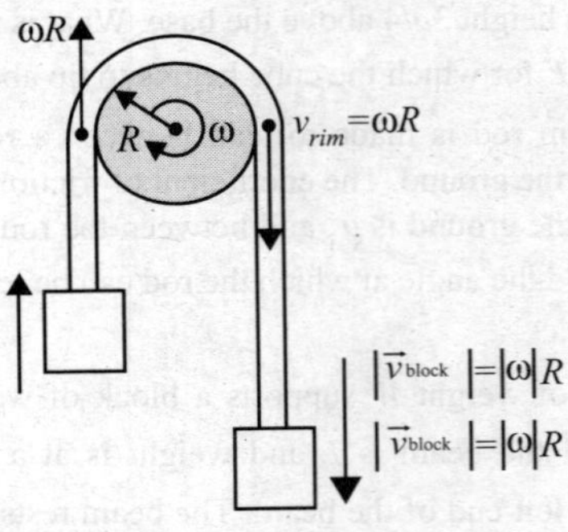

Fig. 2.73

Illustration 2.20 **In Fig. 2.74, calculate the linear acceleration of the blocks.**

Mass of block A = 10 kg, Mass of block B = 8 kg

Mass of disc shaped pulley = 2 kg (take g = 10 m/s²)

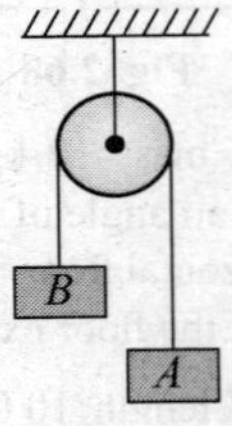

Fig. 2.74

Sol. Let R be the radius of the pulley and T_1 and T_2 be the tensions in the left and right portions of the string. Unlike laws of motion where tension in string is the same on both the sides of the pulley. In this case, the string does not slide and the pulley rotates. The tensions on both the sides of the pulley will be different. This different tensions will provide angular acceleration to pulley.

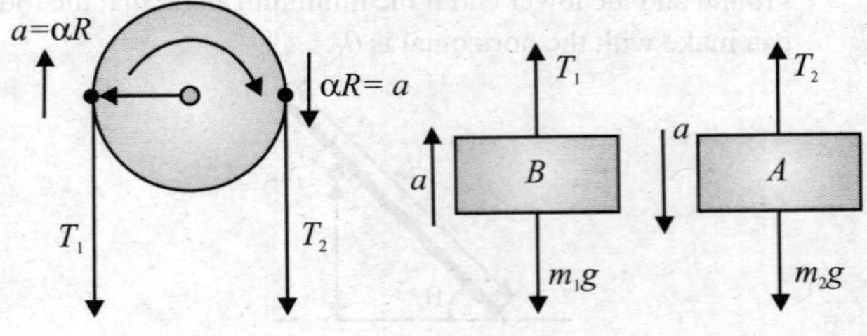

Fig. 2.75

Let m_1 = 8 kg; m_2 = 10 kg; M = 2 kg.

Let a be the acceleration of blocks.

For the blocks (linear motion):

$T_1 - m_1 g = m_1 a$ (i)

$m_2 g - T_2 = m_2 a$ (ii)

For the pulley (rotation):

Net torque $\tau_{net} = I\alpha$

$T_2 R - T_1 R = \frac{1}{2} MR^2 \alpha$ (iii)

The linear acceleration of the blocks is the same as the tangential acceleration of any point on the circumference of the pulley which is $R\alpha$.

$a = R\alpha$

Dividing Eq. (iii) by R and adding to Eqs. (i) and (ii), we get

$$m_2 g - m_1 g = m_2 a + m_1 a + \frac{M}{2} R\alpha \Rightarrow m_2 g - m_1 g$$

$$= \left(m_2 + m_1 + \frac{M}{2}\right) a$$

$$a = \frac{m_2 - m_1}{m_2 + m_1 + \frac{M}{2}} g = \frac{(10-8)g}{10+8+\frac{2}{2}} = \frac{20}{19} \text{ m/s}^2$$

Illustration 2.21 **A block of mass m is attached at the end of an inextensible string which is wound over a rough pulley of mass M and radius R [Fig. 2.76(a)]. Assume the string does not slide over the pulley. Find the acceleration of the block when released.**

Sol. This problem is different from the problems which we have solved in laws of motion where we assumed the string is sliding over the pulley without friction. Here, the pulley is rough and string is not sliding. We will go step by step to analyse the situation.

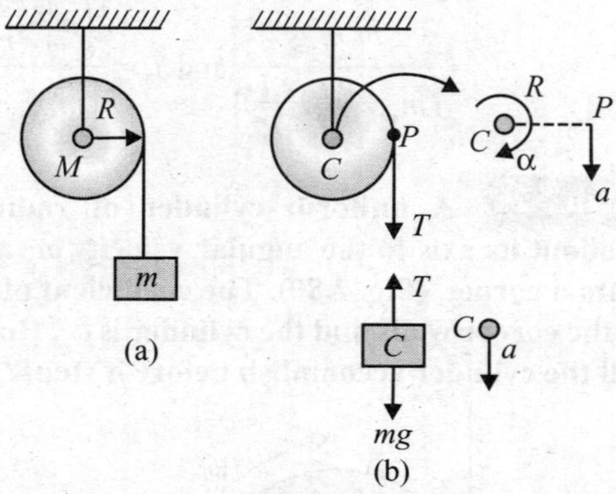

Fig. 2.76

Step I

- We will draw the free-body diagrams of the given situation as in Fig. 2.76 (b).

Step II

- We should apply Newton's second law for each assumed acceleration.

Equation of motion of the block:

$mg - T = ma$ (i)

Step III

- Formulate the constraint relations, if required.

Constraint relation: $a = \alpha R$ (ii)

Step IV

- Solve the equation to get the desired unknown.

Torque equation for the pulley: $\tau_c = I_c \alpha$

$$TR = \left(\frac{MR^2}{2}\right)\alpha \Rightarrow \alpha = \frac{2T}{MR} \quad \text{(iii)}$$

The acceleration of the block is a downwards; hence, acceleration of the point P will also be a downwards.

After solving Eqs. (i), (ii) and (iii), we get

$$a = \frac{2mg}{2m + M}$$

Illustration 2.22 **An extensible string is wound over a rough pulley of mass M_1 and radius R and a cylinder of mass M_2 and radius R such that as the cylinder rolls down, [Fig. 2.77(a)] the string unwounds over the pulley as well the cylinder. Find the acceleration of cylinder M_2.**

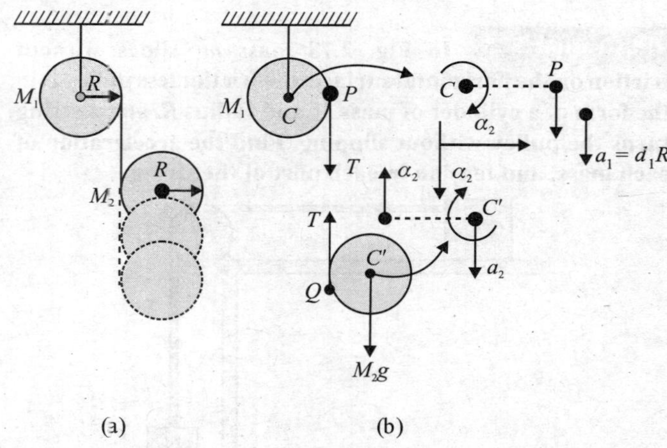

Fig. 2.77

Sol. Through this illustration we will learn the application of torque equation and constraint relation in more complex case here. We have changed the block with cylinder M_2. Pulley M_1 will have only pure rotation while cylinder M_2 will have rotation and translation combined. Let us analyse step by step in the same way as the previous illustration.

Step I: Analyse the motions of the pulley and the cylinder.

Pulley: One rotational acceleration α_1 (clockwise)

cylinder: One rotational acceleration α_2 (clockwise) and a linear acceleration a_2(downward).

Step II: Equation of motion for M_2:

$$M_2 g - T = M_2 a_2 \quad \text{(i)}$$

Step III: Torque equation for pulley: $\tau_c = I_c\alpha$

$$TR = \left(\frac{M_1R^2}{2}\right)\alpha_1 \Rightarrow \alpha_1 = \frac{2T}{M_1R} \quad \text{(ii)}$$

Torque equation for the cylinder (about centre of mass of cylinder): $\tau_c = I_c\alpha$

$$\Rightarrow \quad TR = \left(\frac{M_2R^2}{2}\right)\alpha_2$$

$$\alpha_2 = \frac{2T}{M_2R} \quad \text{(iii)}$$

Step IV: The acceleration of P and Q should be equal as both are connected with the same inextensible string.

Acceleration of P, $a_M = \alpha_1 R$ (downwards) (iv)

Acceleration of Q; $a_N = a_2 - \alpha_2 R$ (downwards) (v)

Hence, constraint relation: $\alpha_1 R = a_2 - \alpha_2 R$ (vi)

Step V: Solving equations:

After solving Eqs. (i), (ii), (iii) and (iv), we get

$$a_2 = \left[\frac{2(M_1 + M_2)}{3M_1 + 2M_2}\right]g$$

Illustration 2.23 **In Fig. 2.78 mass m_1 slides without friction on the horizontal surface, the frictionless pulley is in the form of a cylinder of mass M and radius R, and a string turns the pulley without slipping. Find the acceleration of each mass, and tension in each part of the string.**

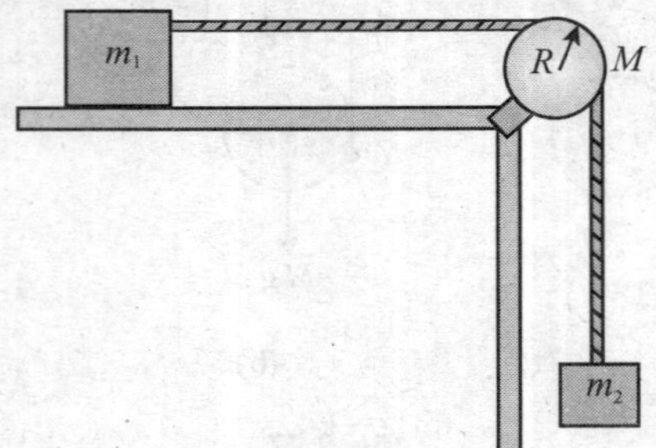

Fig. 2.78

Sol. Step I: Analysis of motion:

In this case the string rotates the pulley, i.e., the string does not slide. The pulley will have angular acceleration. Hence, the tension on the string on both the sides of the pulley will not be equal. The motion of m_1 and m_2 will be translational. Positive direction of angular acceleration α is taken along the corresponding acceleration a_1 which is taken positive in the direction of net force.

Step II: Equation of motion:

Let these be T_1 and T_2. The equation of motion respectively, for masses m_1 and m_2 is

$$T_1 = m_1 a \quad \text{(i)}$$

$$\text{and} \quad m_2 g - T_2 = m_2 a \quad \text{(ii)}$$

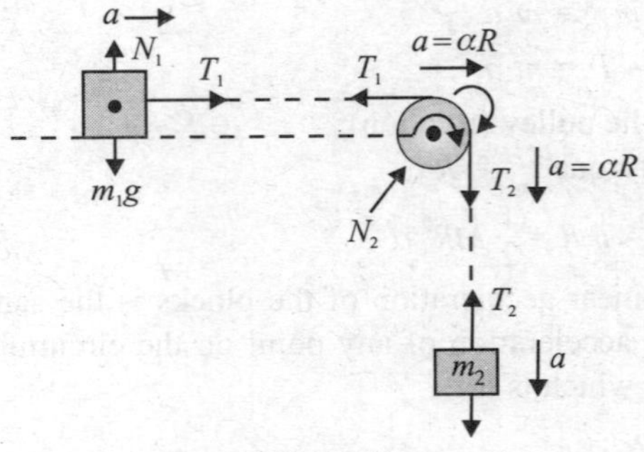

Fig. 2.79

Step III: Application of torque equation:

The torque equation for the pulley,

$$T_2R - T_1R = I\alpha \quad \text{(iii)}$$

(N_2 constitutes no torque about the axis of rotation)

There is no slipping of the string over the pulley.

Step IV: Constraint relation:

$$a = \alpha R \Rightarrow \alpha = \frac{a}{R} \quad \text{(iv)}$$

From Eqs. (iii) and (iv), we get

$$\therefore \quad T_2R - T_1R = I\frac{a}{R}$$

$$\text{or} \quad T_2 - T_1 = \frac{Ia}{R^2} \quad \text{(v)}$$

Step V: Solving equations:

Solving the above equation, we get $a = \dfrac{m_2 g}{\left(m_1 + m_2 + \dfrac{I}{R^2}\right)}$

Here $\quad I = \dfrac{MR^2}{2}$

$$T_1 = \frac{m_1 m_2 g}{\left(m_1 + m_2 + \frac{I}{R^2}\right)} \text{ and } T_2 = \frac{\left(m_1 + \frac{I}{R^2}\right) m_2 g}{\left(m_1 + m_2 + \frac{I}{R^2}\right)}$$

Illustration 2.24 **A uniform cylinder of radius R is spinned about its axis to the angular velocity ω_0 and then placed into a corner, (Fig. 2.80). The coefficient of friction between the corner walls and the cylinder is μ_k. How many turns will the cylinder accomplish before it stops?**

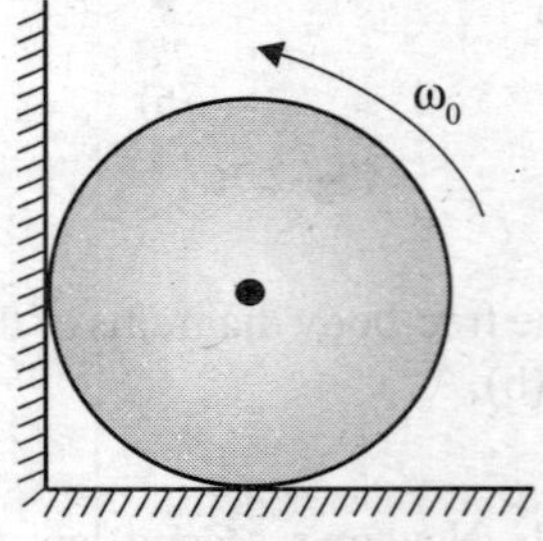

Fig. 2.80

Sol. As the centre of mass of the cylinder does not accelerate, hence $\Sigma F = 0$.

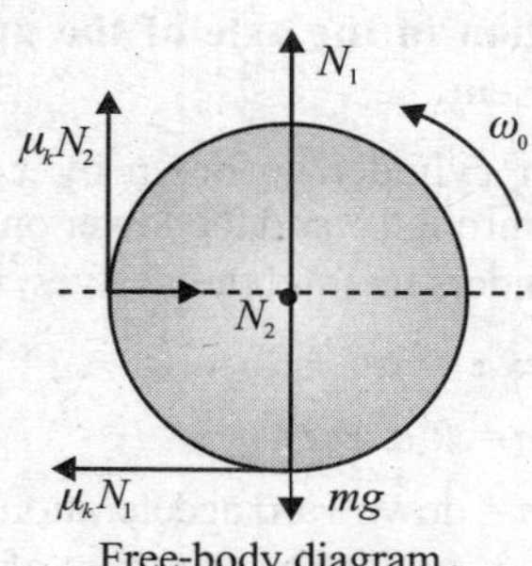

Free-body diagram of the cylinder

Fig. 2.81

$$\Sigma F_x = 0, N_2 - \mu_k N_1 = 0 \quad \text{(i)}$$

$$\Sigma F_y = 0, N_1 + \mu k N_2 - mg = 0 \quad \text{(ii)}$$

Solving these equations: $N_1 = \dfrac{mg}{1+\mu_k^2}, N_2 = \dfrac{\mu_k mg}{1+\mu_k^2}$

The torque on the cylinder about the axis of rotation,

$$\tau_{CM} = mg \times 0 + N_1 \times 0 + N_2 \times 0 + \mu_k N_2 R$$

$$= \frac{\mu_k(1+\mu_k)}{(1+\mu_k^2)} mgR$$

The moment of inertia about axis of rotation, $I_{CM} = \frac{1}{2} mR^2$

The torque equation, $\tau = I\alpha$

$$\frac{\mu_k(1+\mu_k)}{(1+\mu_k^2)} mgR = \left(\frac{1}{2} mR^2\right)\alpha \Rightarrow \alpha = \frac{2\mu_k(1+\mu_k)g}{(1+\mu_k^2)R}$$

Using equation $\omega^2 = \omega_0^2 + 2\alpha\theta$, calculate the angular displacement θ,

$$0^2 = \omega_0^2 - 2\left[\frac{2\mu_k(1+\mu_k)g}{(1+\mu_k^2)R}\right]\theta \text{ (ω and α are in the opposite sense)}$$

$$\Rightarrow \quad \theta = \frac{(1+\mu_k^2)\,\omega_0^2 R}{4\mu_k(1+\mu_k)g}$$

Revolution accomplished,

$$n = \frac{\theta}{2\pi} = \frac{(1+\mu_k^2)\,\omega_0^2 R}{2\pi \times 4\mu_k(1+\mu_k)g} = \frac{(1+\mu_k^2)\,\omega_0^2 R}{8\pi\mu_k(1+\mu_k)g}$$

Illustration 2.25 **A uniform rod of length L and mass M is pivoted freely at one end and placed in vertical position.**

a. **What is the angular acceleration of the rod when it is at an angle θ with the vertical?**
b. **What is the tangential linear acceleration of the free end when the rod is horizontal?**

Sol. Figure 2.82 shows the rod at an angle θ to the vertical. If we take torques about the pivot we need not be concerned with the force due to the pivot.

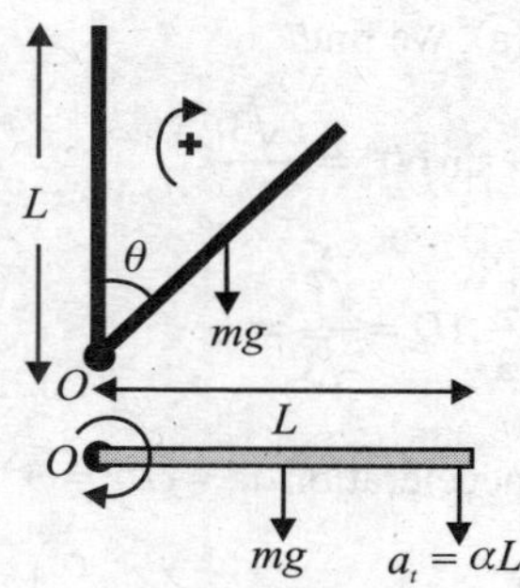

Fig. 2.82

The torque due to the weight about O,

$$\tau_{mg} = \frac{mgL}{2} \sin\theta$$

Using torque equation about O. Here we apply torque equation about fixed axis passing through O. We can apply torque equation about fixed axis or about centre of mass only.

a. $\tau = I\alpha \Rightarrow \dfrac{mgL}{2} \sin\theta = \dfrac{ML^2}{3}\alpha$

Thus, $\alpha = \dfrac{3g \sin\theta}{2L}$

When the rod is horizontal, $\theta = \dfrac{\pi}{2}$ and $\alpha = \dfrac{3g}{2L}$

b. The tangential linear acceleration of the free end is

$$a_t = \alpha L = \frac{3g}{2}$$

Illustration 2.26 **Three particles A, B and C each of mass m are connected to each other by three massless rigid rods to form a rigid, equilateral triangular body of side l. This body is placed on a horizontal frictionless table (x-y plane) and is hinged at point A so that it can move without friction about the vertical axis through A (see Fig. 2.83). The body is set into rotational motion on the table about this axis with a constant angular velocity ω. (a) Find the magnitude of the horizontal force exerted by the hinge on the body. (b) At time T, when side BC is parallel to x-axis, force F is applied on B along BC (as in the figure). Obtain the x-component and the y-component of the force exerted by the hinge on the body, immediately after time T.**

(IIT-JEE, 2002)

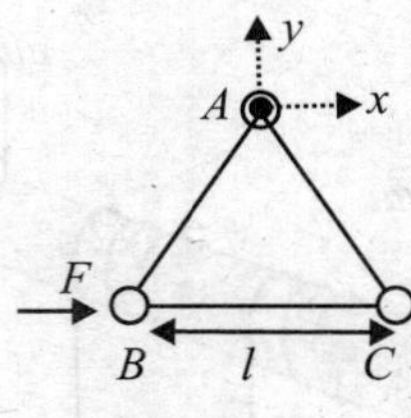

Fig. 2.83

Sol. a. The centre of mass of the system is at the centroid of a triangular assembly. The CM moves along a circular path with constant angular velocity. Therefore, there must be a horizontal centripetal force directed towards the axis at the hinge.

From Fig. 2.84 (a), we find

$$AD = \ell \sin 60° = \frac{\ell\sqrt{3}}{2}$$

$$AO = \frac{2}{3}AD = \frac{\ell^2}{\sqrt{3}} = r$$

The centripetal acceleration $a_c = \omega^2 r = \omega^2\left(\frac{\ell}{\sqrt{3}}\right)$

Tangential acceleration $a_t = \alpha r = \alpha\left(\frac{\ell}{\sqrt{3}}\right)$

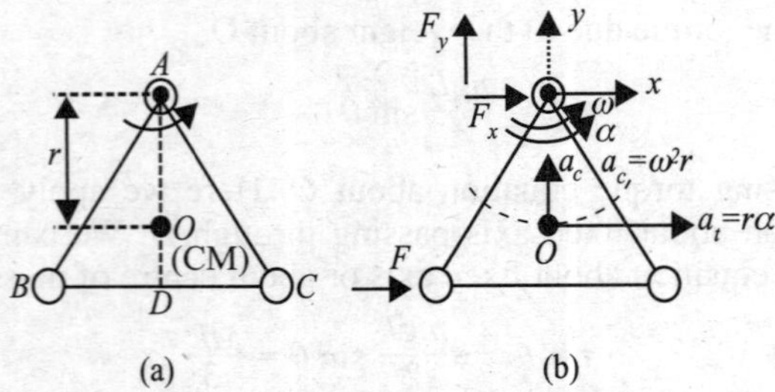

Fig. 2.84

Let F_x and F_y be the forces applied by the hinges along x-axis and y-axis, respectively. The system is in non-centroidal rotation. The three equations of motion are

$$\Sigma F_x = F_x + F = (3m)a_t = 3m\left(\frac{l}{\sqrt{3}}\alpha\right) \quad \text{(i)}$$

$$\Sigma F_y = F_y = 3m\left(\frac{l}{\sqrt{3}}\right)\omega^2 \quad \text{(ii)}$$

$$\Sigma\tau = F \times \left(\frac{\sqrt{3}}{2}l\right) = 2ml^2\alpha \quad \text{(iii)}$$

From Eq. (iii), $\alpha = \frac{\sqrt{3}F}{4ml}$

From Eq. (i), $F_x + F = 3m\left(\frac{1}{\sqrt{3}}\alpha\right) \times \frac{\sqrt{3}F}{4ml} = \frac{3F}{4}$

$\Rightarrow \quad F_x = \frac{F}{4}$

From Eq. (ii), $F_y = \sqrt{3}\, ml\omega^2$

Illustration 2.27 **The arrangement shown in Fig. 2.85 consists of two identical, uniform, solid cylinders, each of mass m, on which two light threads are wound symmetrically.**

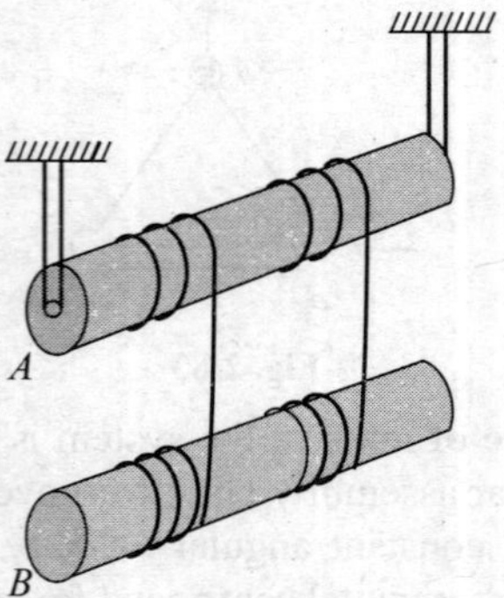

Fig. 2.85

Find the tensions of each thread in the process of motion. The friction in the axle of the upper cylinder is assumed to be absent.

Sol. Let the lower cylinder go down by x when the upper cylinder A rotates through θ and the lower one rotates through θ'. Since both cylinders unwind themselves, therefore

$$x = R(\theta + \theta') \Rightarrow x = R(\theta + \theta')$$

we can write $a = R(\alpha + \alpha')$

where a = downward acceleration of B

α = angular acceleration of A

α' = angular acceleration of B

Fig. 2.86

Considering the rotational motion of A,

$$\tau = 2TR = \left(\frac{1}{2}mR^2\right) \times \alpha \Rightarrow \alpha = \frac{4T}{R}$$

Considering the rotational motion of B,

$$\tau = 2TR = \left(\frac{1}{2}mR^2\right) \times \alpha' \Rightarrow \alpha' = \frac{4T}{R}$$

Considering the downward motion of B,

$$mg - 2T = ma$$

$$mg - 2T = mR(\alpha + \alpha') = mR\left(\frac{4T}{R} + \frac{4T}{R}\right)$$

$$mg - 2T = 8T \Rightarrow T = \frac{1}{10}mg$$

Concept Application Exercise 2.3

1. A uniform rod of mass m and length l can rotate in a vertical plane about a smooth horizontal axis hinged at point H.

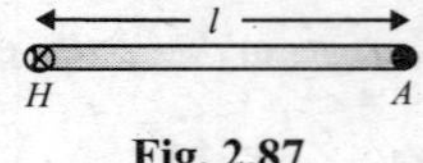

Fig. 2.87

   i. Find angular acceleration α of the rod just after it is released from initial horizontal position from rest ?

   ii. Calculate the acceleration (tangential and radial) of point A at this moment.

2. A uniform rod of mass m and length l can rotate in a vertical plane about a smooth horizontal axis hinged at point H. Find the force exerted by the hinge just after the rod is released from rest, from an initial horizontal position?

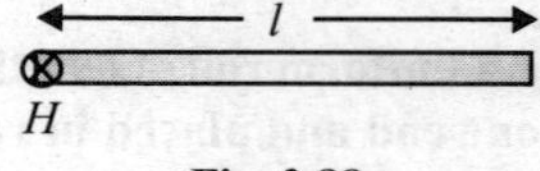

Fig. 2.88

3. A wheel of radius r and moment of inertia I about its axis is fixed at the top of an inclined plane of inclination θ as shown in Fig. 2.89. A string is wrapped around the wheel and its free end supports a block of mass M which can slide on the plane. Initially, the wheel is rotating at a speed ω in a direction such that the block slides up the plane. How far will the block move before stopping?

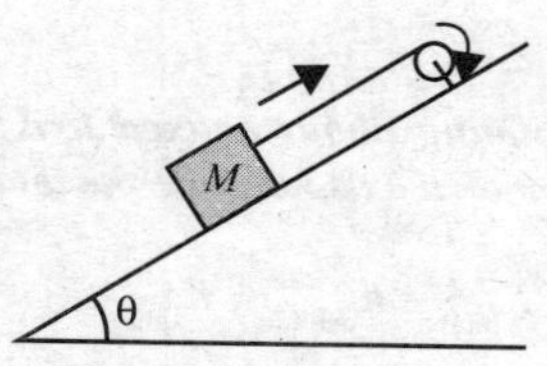

Fig. 2.89

4. A uniform rod AB of mass $m = 2$ kg and length $l = 1.0$ m is placed on a sharp support P such that $a = 0.4$ m and $b = 0.6$ m. A spring of force constant $k = 600$ N/m is attached to end B as shown in Fig. 2.90. To keep the rod horizontal, its end A is tied with a thread such that the spring is elongated by 1 CM. Calculate reaction of support P when the thread is burnt.

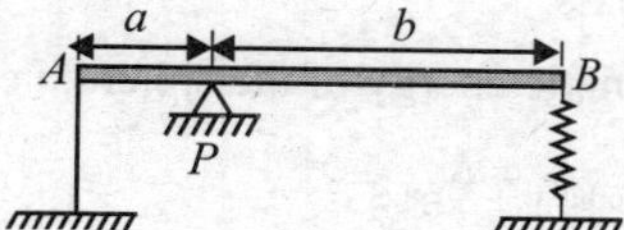

Fig. 2.90

5. A cotton reel of mass m, radius R and moment of inertia I is kept on a smooth horizontal surface. If the string is pulled horizontally by a force F, find the (i) acceleration of CM, (ii) angular acceleration of the cotton reel.

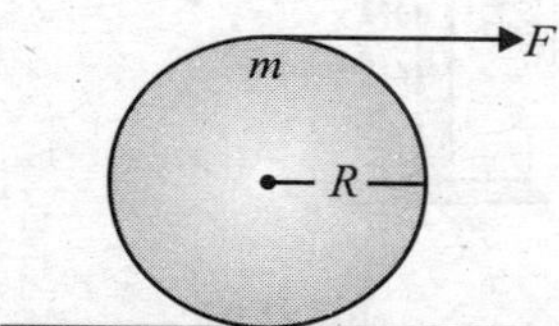

Fig. 2.91

6. Find the acceleration a and angular acceleration α. If $F = 2$ N, $m = 1$ kg and $l = 2$ m.

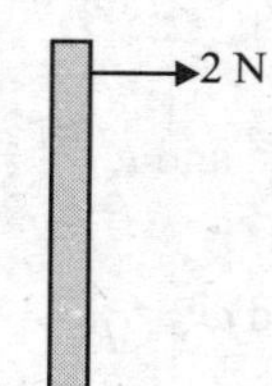

Fig. 2.92

7. Find α, a_Q and the point of zero acceleration when the horizontal force F acts on the smooth rod of mass m and length l which is kept on a horizontal surface.

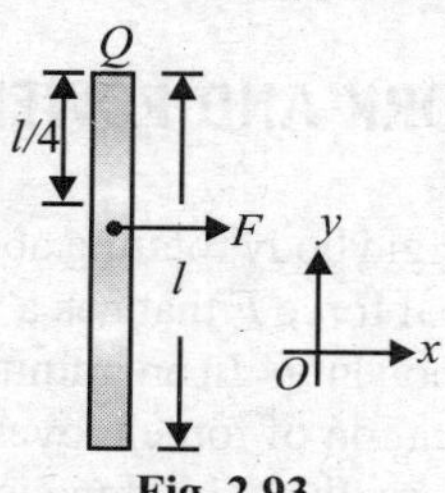

Fig. 2.93

KINETIC ENERGY OF ROTATION

Figure 2.94 shows the cross section of a body with this paper. Let the body rotate about an axis passing through P and normal tc the plane of this page, with an angular velocity ω. The body can be assumed to be consisting of a large number of particles of mass m_1, m_2, m_3,... at distances r_1, r_2, r_3,... from the axis of rotation, respectively.

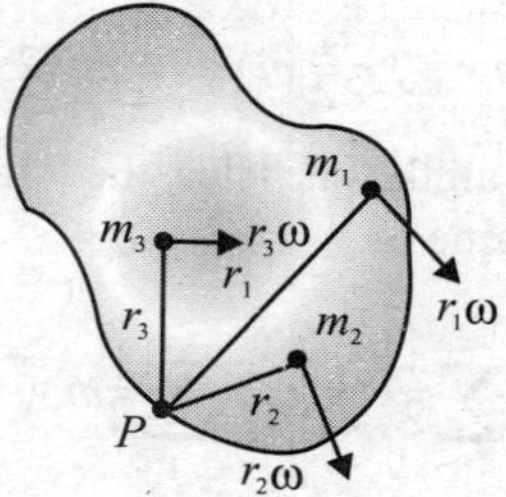

Fig. 2.94

The particles move in circular orbits of radii r_1, r_2, r_3,... with the respective speed $r_1\omega_1$, $r_2\omega_2$, $r_3\omega_3$,... The rotational kinetic energy of the body is equal to the sum of the kinetic energies of the individual particles,

$$K = \frac{1}{2}m_1(r_1\omega)^2 + \frac{1}{2}m_2(r_2\omega)^2 + \frac{1}{2}m_3(r_3\omega)^2 + \cdots$$

$$= \frac{1}{2}(m_1r_1^2 + m_2r_2^2 + m_3r_2^2 + m_2r_3^2 + \cdots)\omega^2 = \frac{1}{2}I\omega^2$$

$$= \frac{1}{2}I\omega^2$$

where $I = \Sigma m_i r_i^2$ is the moment of inertia about the axis of rotation.

Kinetic Energy of a Body in Combined Rotation and Translation

Consider a body in combined translational and rotational motion in the lab frame. Suppose in the frame of the centre of mass, the body is making a pure rotation with an angular velocity ω. The centre of mass itself is moving in the lab frame at a velocity $\vec{v}_0$.

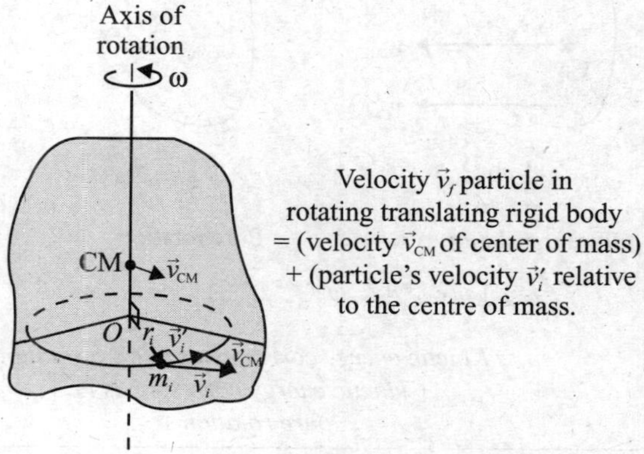

Fig. 2.95

The velocity of a particle of mass m_i is $\vec{v}_{i,\text{CM}}$ with respect to the centre of mass frame and $\vec{v}_i$ with respect to the lab frame. We have $\vec{v}_i = \vec{v}_{i,\text{CM}} + \vec{v}_0$.

The kinetic energy of the particle in the lab frame is

$$\frac{1}{2}m_i v_i^2 = \frac{1}{2}m_i(\vec{v}_{i,\text{CM}} + \vec{v}_0)(\vec{v}_{i,\text{CM}} + \vec{v}_0)$$

$$= \frac{1}{2}m_i v_{i,\text{CM}}^2 + \frac{1}{2}m_i v_0^2 + \frac{1}{2}m_i(2\vec{v}_{i,\text{CM}} \cdot \vec{v}_0)$$

Summing over all the particles, the total kinetic energy of a body in the lab frame is

$$K = \sum_i \frac{1}{2}m_i v_i^2 = \sum_i \frac{1}{2}m_i v_{i,\text{CM}}^2 + \sum_i \frac{1}{2}m_i v_0^2 + \left(\sum_i m_i \vec{v}_{i,\text{CM}}\right)\vec{v}_0$$

Now, $\sum_i \frac{1}{2}m_i v_{i,\text{CM}}^2$ is the kinetic energy of the body in the centre of mass frame. In this frame, the body is making pure rotation with an angular velocity ω. Thus, this term is equal to $\frac{1}{2}I_{\text{CM}}\omega^2$. Also, $\sum\left(m_i\vec{v}_{i,\text{CM}}/M\right)$ is the velocity of the centre of mass in the centre of mass frame; hence, $\left(\sum m_i\vec{v}_{i,\text{CM}} = 0\right)$ obviously is zero.

Thus, $\boxed{K = \frac{1}{2}I_{\text{CM}}\omega^2 + \frac{1}{2}mv_0^2}$

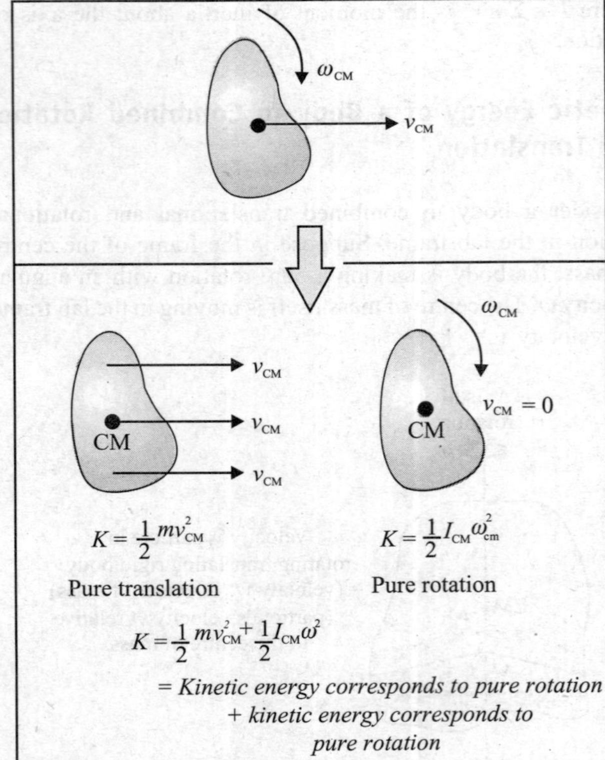

Fig. 2.96

Note:

If a body is rotating about a fixed axis and spinning (or rotating) about it own axis we write the total kinetic energy of the body as

$$K_{\text{total}} = K_{\text{orbital}} + K_{\text{spin}}$$

Illustration 2.28 **A uniform sphere of mass *m* and radius *r* rolls without sliding over a horizontal plane, rotating about a horizontal axle *OA*. In the process, the centre of the sphere moves with a velocity *v* along a circle of radius *R*. Find the kinetic energy of the sphere.**

Sol. Here the motion of the body is orbital and spin combined. The sphere is spining about a horizontal axis and orbiting about a vertical axis.

We write total kinetic energy of the sphere, i.e.,

$$\text{KE} = K_{\text{orbital}} + K_{\text{spin}}$$

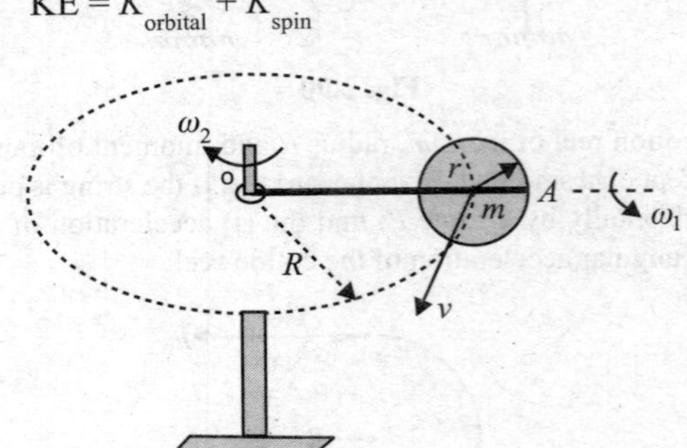

Fig. 2.97

The kinetic energy of the sphere due to its rotation about its own axis and its motion along circular path with velocity *v*.

$$\text{KE} = \frac{1}{2}I_1\omega_1^2 + \frac{1}{2}I_2\omega_2^2$$

where $I_1 = \frac{2}{5}mr^2$ and $I_2 = \left(\frac{2}{5}mr^2 + mR^2\right)$

$$\omega_1 = \frac{v}{r} \text{ and } \omega_2 = \frac{v}{R}$$

Substituting these values in the above equation, we get

$$\text{KE} = \frac{7m}{10}v^2\left[1 + \frac{2r^2}{7R^2}\right]$$

ROTATIONAL WORK AND POWER

Figure 2.98 shows a rigid body rotating about a fixed axis. It is acted on by an external force F that has a radial component F_r and a tangential component F_t. In an infinitesimal time interval dt, the point of application of force moves in a circular arc of length $r = d\theta$. Therefore, the work done is

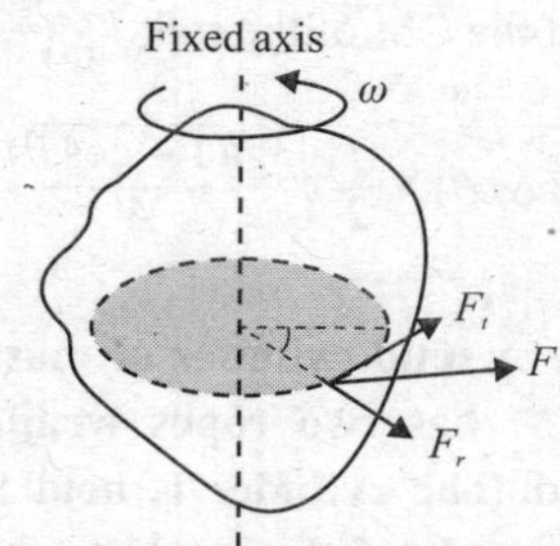

Fig. 2.98

$$dW_{rot} = F_t(rd\theta) = (F_t r)d\theta = \tau d\theta$$

For a finite angular displacement, $W_{rot} = \int_{\theta_i}^{\theta_f} \tau d\theta$

By definition, the instantaneous power is defined as

$$P = \frac{dW}{dt} = \tau \frac{d\theta}{dt} \Rightarrow P = \tau\omega$$

The rotational analogue of work–kinetic energy theorem may be stated as

$$W_{rot} = \Delta K_{rot} = K_f - K_i \quad \text{or} \quad W_{rot} = \frac{1}{2}I\omega_f^2 - \frac{1}{2}I\omega_i^2$$

The work done by a torque on a rigid body rotating about a fixed axis produces a change in its rotational kinetic energy.

$$W_{rot} = \int \tau d\theta;\ W_{trans} = \int Fdx$$

Power imparted by torque in rotation $P_{rot} = \tau\omega$; which is equivalent to expression for power in translation motion $P_{trans} = Fv$.

Illustration 2.29 **Calculate the torque developed by an airplane engine whose output is 2000 HP at an angular velocity of 2400 rev/min.**

Sol. $\omega = 2\pi\ (2400/60) = 80\pi$ rad/s

Work done by torque = (torque) × (angular displacement)

$$\text{Power} = \text{work done per sec} = \tau \frac{\Delta\theta}{\Delta t}$$

$$\text{Power} = \tau\omega \Rightarrow \tau = \frac{2000 \times 746}{80\pi} = 5937 \text{ Nm}$$

CONSERVATION OF MECHANICAL ENERGY

When a rigid body moves, the constraint forces are reactions at the pivot. The normal reaction sometimes does not perform work if the point of application of the forces does not move. For instance, when a rigid body rotates about a fixed axis, the reaction forces R_x and R_y do not perform work.

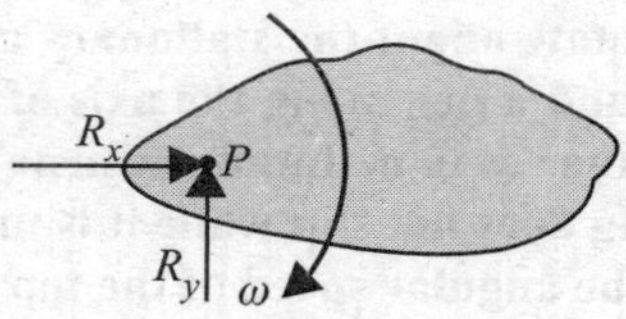

Fig. 2.99

If a rigid body does not experience any non-conservative force or even though a non-conservative force like force of static friction acts, if it does not perform any work (as described in pure rolling etc), the work–energy equation assumes the form $W_{cons} = \Delta K$.

Since conservative forces perform work at the expense of their potential enegy, we can write $W_{cons} = \Delta U$.

Then, we have $-\Delta U = \Delta K$. This gives $\Delta U + \Delta K = 0$.

WORK–ENERGY THEOREM

i. The work done by a force acting on a rigid body may be classified into two parts: translational work and rotational work.

A force can do translational work if the centre of mass of the body is free to move.

Mathematically, it is defined as $W_{trans} = \int F ds_c$

where ds_c is the infinitesimal displacement of the centre of mass.

A force can do rotational work if the body is free to rotate.

Mathematically, it is defined as $W_{rot} = \int \tau_c d\theta$

where $d\theta$ is the infinitesimal angular displacement about the centre of mass and τ_c is the torque produced by the force about the centre of mass of the body.

ii. In an inertial frame, the work–energy theorem may be stated as

$$W_C + W_{NC} + W_{oth} = \Delta K$$
$$W_{NC} + W_{oth} = \Delta K + \Delta U = \Delta E$$

where W_C is the work done by conservative forces, W_{NC} is the work done by non-conservative forces, W_{oth} is the work done by all other forces which are not included in the above two categories.

Note:

- ***Each term of work includes both the translational and rotational work.***
- ***The kinetic energy also includes both the translational and rotational kinetic energies.***

Illustration 2.30 **The top in the figure has a moment of inertia equal to 4.00 × 10^{-4} kgm^2 and is initially at rest. It is free to rotate about the stationary axis AA'. A string wrapped around a peg along the axis of the top is pulled in such a manner as to maintain a constant tension of 5.57 N. If the string does not slip while it is unwound from the peg, what is the angular speed of the top after 80.0 CM of string has been pulled off the peg?**

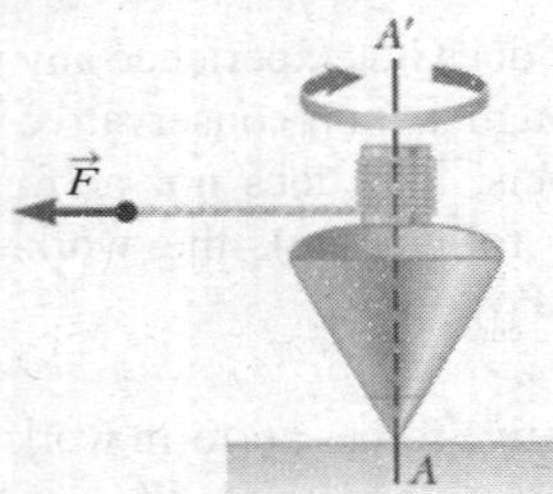

Fig. 2.100

Sol. Work done = $F\Delta r$ = (5.57 N) (0.800 m) = 4.46 J

and work = $\Delta K = \frac{1}{2}I\omega_f^2 - \frac{1}{2}I\omega_i^2$

(The last term is zero because the top starts from rest).

Thus, $4.46 \text{ J} = \frac{1}{2}(4.00 \times 10^{-4} \text{ kg} \times \text{m}^2)\omega_f^2$

and from this, ω_f = 149 rad/s.

Illustration 2.31 **A rod of length l is pivoted about a horizontal, frictionless pin through one end. The rod is released from rest in a vertical position. Find the velocity of the CM of the rod when the rod is inclined at an angle θ from the vertical.**

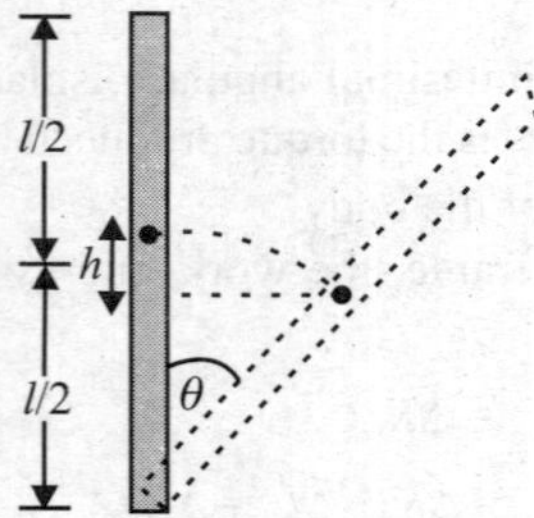

Fig. 2.101

Sol. The fall in position of the CM of the rod, $h = \frac{l}{2}(1 - \cos\theta)$.

In the process, decrease in PE is equal to increase in rotational KE of the rod, so $mgh = \frac{1}{2}I\omega^2$

or $mg\frac{l}{2}(1 - \cos\theta) = \frac{1}{2}\left(\frac{ml^2}{3}\right)\omega^2$

$\therefore \quad \omega = \sqrt{\frac{3g}{l}(1 - \cos\theta)}$

The velocity of the CM of the rod, $V_{CM} = \omega r$ is

$$\sqrt{\frac{3g}{l}(1 - \cos\theta)} \times \frac{l}{2}\sqrt{\frac{3gl(1 - \cos\theta)}{4}}$$

Illustration 2.32 **A solid cylinder of mass m = 4 kg and radius R = 10 CM has two ropes wrapped around it, one near each end. The cylinder is held horizontally by fixing the two free ends of the cords to the hooks on the ceiling such that both the cords are exactly vertical. The cylinder is released to fall under gravity. Find the linear acceleration of the cylinder.**

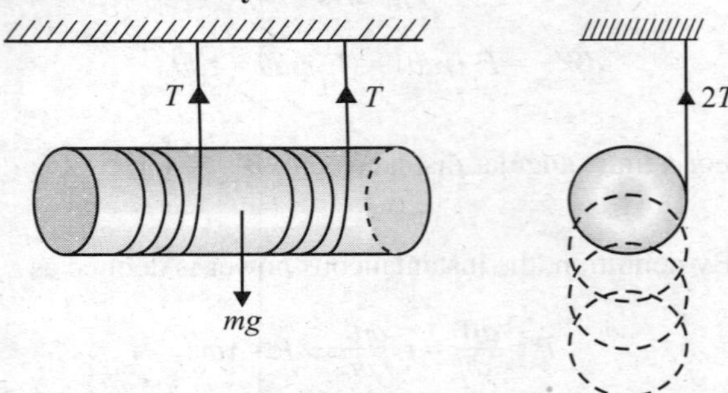

Fig. 2.102

Sol. Method 1: Force/Torque Method:

Let a = linear acceleration and α = angular acceleration of the cylinder.

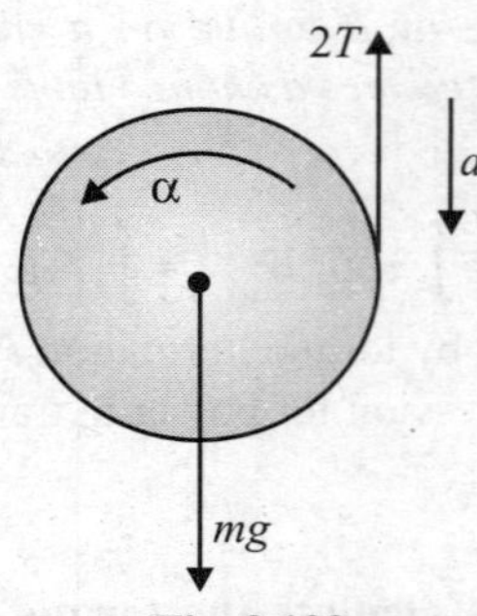

Fig. 2.103

For the linear motion of the cylinder: $mg - 2T = ma$

For the rotational motion: Net torque = $I\alpha$

$$2TR = (1/2\, mR^2)\alpha$$

Also, the linear acceleration of the cylinder is the same as the tangential acceleration of any point on its surface, $a = R\alpha$

Combining the three equations, we get

$$mg = ma + \frac{m}{2}a \Rightarrow a = \frac{2}{3}g$$

Method 2: Energy Method

The motion of the cylinder is rotational and translational combined. Using conservation of mechanical energy.

$\Delta K + \Delta U = 0$ or Loss in PE = gain in KE

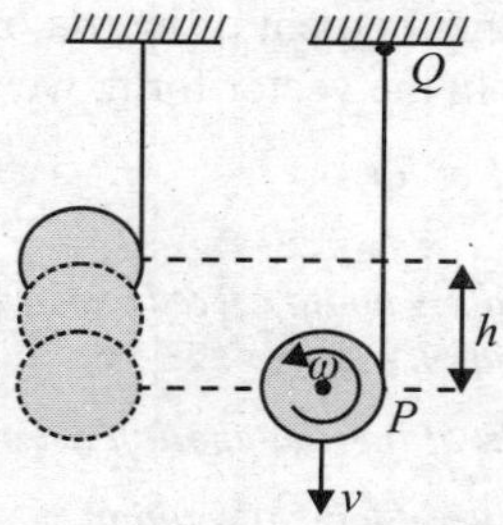

Fig. 2.104

$$mgh = \frac{1}{2}mv^2 + \frac{1}{2}I\omega^2 \quad \text{(i)}$$

Constraint relation:

Velocity of point P, $v_P = 0 = v - \omega R$

$$\omega = \frac{v}{R} \quad \text{(ii)}$$

From Eqs. (i) and (ii), $mgh = \frac{1}{2} mv^2 + \frac{1}{2}\left(\frac{mR}{2}\right)\left(\frac{v}{R}\right)^2$

$$= \frac{m}{2}\left(v^2 + \frac{v^2}{2}\right)$$

$$2gh = \frac{3}{2}v^2 \Rightarrow v^2 = \frac{4}{3}gh \quad \text{(iii)}$$

Differentiating Eq. (iii) with respect to time, we get

$$2v\frac{dv}{dt} = \frac{4}{3}g\frac{dh}{dt}$$

$$2va = \frac{4}{3}gv \Rightarrow a = \frac{2}{3}g$$

Illustration 2.33 **A uniform circular disc has radius R and mass m. A particle, also of mass m, is fixed at point A on the edge of the disc as shown in Fig. 2.105. The disc can rotate freely about a fixed horizontal chord PQ that is at a distance $R/4$ from the centre C of the disc. The line AC is perpendicular to PQ. Initially, the disc is held vertical with point A at its highest position. It is then allowed to fall so that it starts rotating about PQ. Find the linear speed of the particle as it reaches its lowest position.**

(IIT-JEE, 1998)

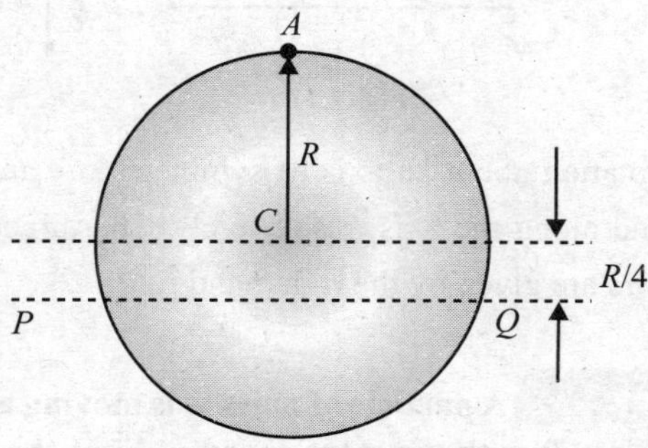

Fig. 2.105

Sol. During the fall, the disc–mass system gains rotational kinetic energy. This is at the expense of potential energy.

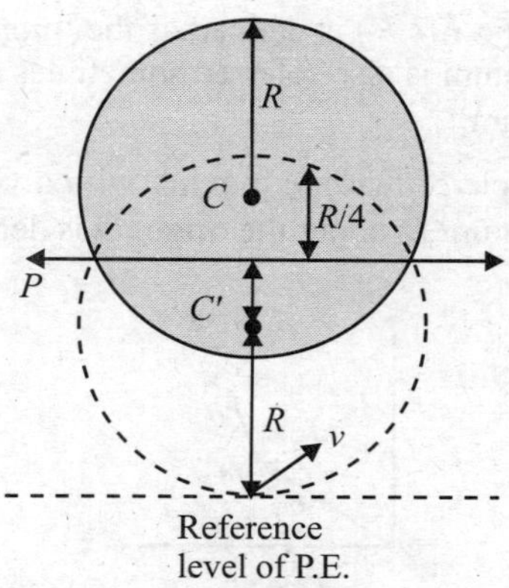

Fig. 2.106

Applying conservation of mechanical energy, we get

Inital total energy = Final total energy

$$mg\left(2R + \frac{2R}{4}\right) + mg\left(R + \frac{2R}{4}\right) = mgR + \frac{1}{2}I\omega^2$$

where I = MI of the disc–mass system about PQ

$$= mg + \frac{10R}{4} + mg\frac{6R}{4} = mgR + \frac{1}{2}\omega^2$$

$$\Rightarrow \quad 3mgR = \frac{1}{2}I\omega^2$$

$$\Rightarrow \quad \omega = \sqrt{\frac{6mgR}{1}} \quad \text{(i)}$$

$$(I)_{PQ} = (I_{\text{disc}})_{PQ} + (I_{\text{mass}})_{PQ}$$

$$= \left[\frac{mR^2}{4} + M\left(\frac{R}{4}\right)^2\right] + m\left(\frac{5R}{4}\right)^2$$

[Therefore, MI of disc about the diameter = (1/4) MR^2]

$$= \frac{mR^2[4 + 1 + 25]}{16} = \frac{15mR^2}{8} \quad \text{(ii)}$$

From Eqs. (i) and (ii), $\omega = \sqrt{\frac{6mgR \times 8}{15mR^2}} \times 8\sqrt{\frac{16g}{5R}}$

Let v be the velocity of mass m at the lowest point of rotation

$$v = \omega\left(R + \frac{R}{4}\right)$$

$$= \sqrt{\frac{16g}{5R}} \times \frac{5R}{4} = \sqrt{5gR}$$

Angular Momentum: Rotational Analogue of Momentum

The angular momentum of a moving particle about a point is defined as

$$\vec{L} = \vec{r} \times \vec{p} \quad \text{(i)}$$

where $\vec{p}$ is the linear momentum of particle and $\vec{r}$ is its position vector from the point. Regarding angular momentum, it is worth nothing that it is an axial vector (i.e., always perpendicular to the plane of motion) having the dimensions $[ML^2T^{-1}]$.

As torque ($=\vec{r}\times\vec{F}$) is defined as the 'moment of force', angular momentum is also referred sometimes as 'moment of linear momentum'.

For a particle of mass m moving with a velocity $\vec{v}$, the angular momentum $\vec{L}$ about the origin O is defined as $\vec{L}=\vec{r}\times m\vec{v}$

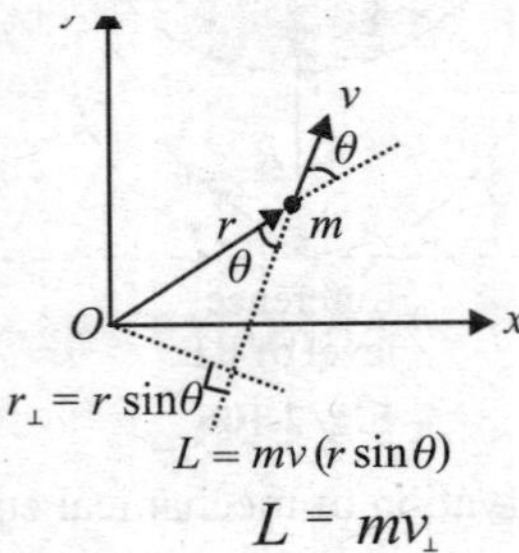

Fig. 2.107

The direction of L vector is given by the right hand rule, and the magnitude is given by

$$L=(mvr\sin\theta)=m(v\sin\theta)r$$

$$=(mv_\perp).\,r \text{ where } v_\perp = v\sin\theta$$

Alternatively, the above expression may be written as

$$L=mv\,(r\sin\theta)$$

$$=mvr_\perp,\text{ where } r_\perp = r\sin\theta$$

For a system of particles, the angular momentum L with respect to a stationary axis (fixed in position) is the sum of angular momentum of the particles.

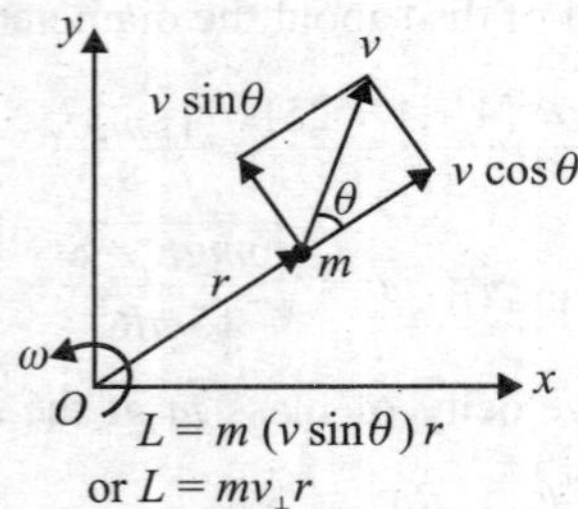

Fig. 2.108

$$\vec{L}_i=(\vec{r}_i\times m\vec{v}_i)$$

If the whole system rotates with an angular velocity ω, the velocity of the ith particle will be $\vec{v}_i=\vec{\omega}\times\vec{r}_i$.

It is obvious from the figure that $\vec{\omega}$, $\vec{r}$ and $\vec{v}$ are perpendicular to each other, therefore, the direction of angular momentum $\vec{L}$ coincides with the direction of $\vec{\omega}$.

And, the magnitude is given by $L=\left|\sum\left(mr_i^2\right)\right|\omega$

or $\qquad L=I\omega$

where $I=\Sigma m_i r_i^2$ is the moment of inertia of the system about the axis of rotation. In the vector form, we may write

$$L=I\omega$$

Note: ***If a body is rotating about a fixed axis, we can write angular momentum of the body L = Iω***

I = Moment of inertia, about fixed axis

ω = angular velocity of rotation

Note that the above expression is analogous to $P=mv$

$I\Rightarrow m \qquad \omega\Rightarrow v \qquad L\Rightarrow p$

Note:

- ***Direction of angular momentum.***

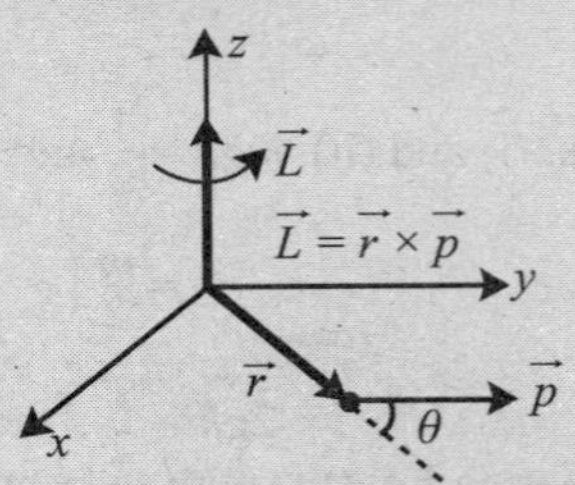

Fig. 2.109

- ***We define angular momentum even if the particle is not moving in a circular path. Even a particle moving in a straight line has angular momentum about any axis displaced from the path of the particle.***

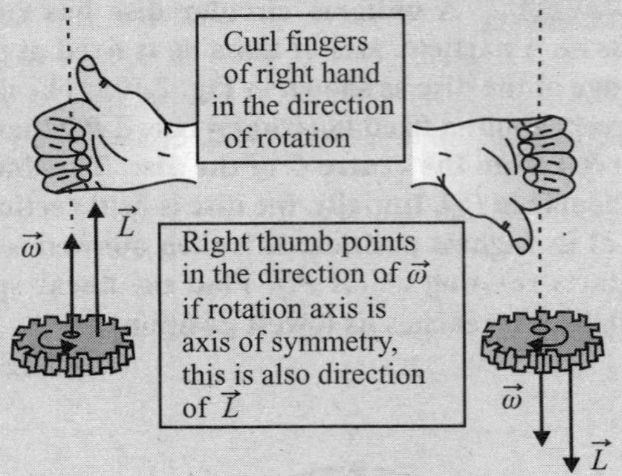

Fig. 2.110

For rotation about an axis of symmetry, $\vec{\omega}$ and $\vec{L}$ are parallel and along the axis, respectively. The directions of both vectors are given by the right hand rule.

Illustration 2.34 **A particle of mass m is moving along the line $y = b$, $z = 0$ with constant speed v. Find the angular momentum of the particle about the origin.**

Sol. $|\vec{L}|=mvr\sin\theta=mvr_\perp\,mvb$

$\therefore |\vec{L}|$ = constant as m, v and b all are constants.

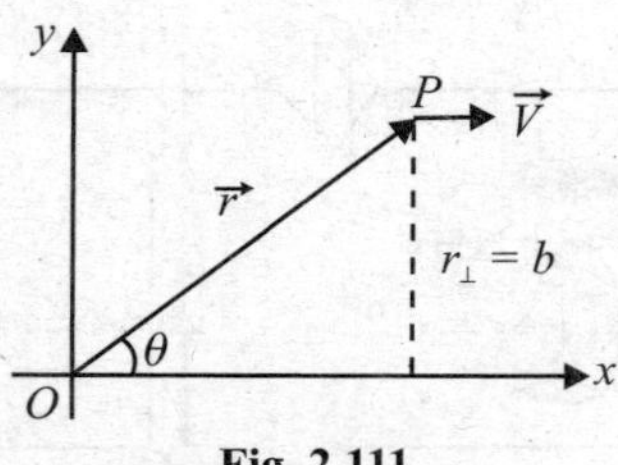

Fig. 2.111

Direction of $\vec{r} \times \vec{v}$ also remains the same. Therefore, angular momentum of the particle about the origin remains constant with due course of time.

Note: ***In this problem,*** $|\vec{r}|$ ***is increasing,*** θ ***is decreasing but*** $r \sin\theta$***, i.e.,*** b ***remains constant. Hence, the angular momentum remains constant.***

Illustration 2.35 **A particle is projected at time $t = 0$ from a point P with a speed v_0 at an angle of 45° to the horizontal. Find the magnitude and the direction of the angular momentum of the particle about the point P at time $t = v_0/g$. (IIT-JEE, 1984)**

Sol. Let us take the origin at P, x-axis along the horizontal and y-axis along the vertically upward direction as shown in Fig. 2.112.

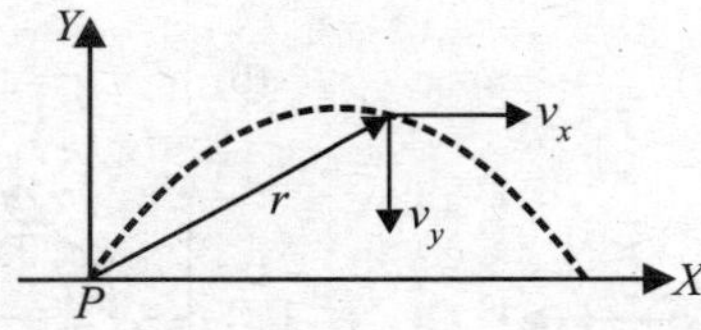

Fig. 2.112

For horizontal motion during the time 0 to t,

$$v_x = v_0 \cos 45° = \frac{v_0}{\sqrt{2}}$$

and $$x = b_x t = \frac{v_0}{\sqrt{2}}\frac{v_0}{g} = \frac{v_0^2}{\sqrt{2}g}$$

For vertical motion, using $v_y = v_{oy} - gt$

$$v_y = v_0 \sin 45° - g\left(\frac{v_0}{g}\right) = \frac{(1-\sqrt{2})}{\sqrt{2}} v_0$$

and $$y = (v_0 \sin 45°)\, t - \frac{1}{2}gt^2$$

$$\Rightarrow\ y = \frac{v_0}{\sqrt{2}}\left(\frac{v_0}{g}\right) - \frac{1}{2}g\left(\frac{v_0}{g}\right)^2 = \frac{v_0^2}{2g}(\sqrt{2}-1)$$

The angular momentum of the particle at time t about the origin is

$$L = \vec{r} \times \vec{p} = m(\vec{r} \times \vec{v})$$

$$= (\hat{i}x + \hat{j}y) \times (v_x\hat{i} + v_y\hat{j}) = m\,(\vec{k}xv_y - \vec{k}yv_x)$$

$$= m\,\vec{k}\left[\left(\frac{v_0^2}{\sqrt{2}g}\right)\frac{v_0}{\sqrt{2}}(1-\sqrt{2}) - \frac{v_0^2}{2g}(\sqrt{2}-1)\frac{v_0}{\sqrt{2}}\right]$$

$$= -\vec{k}\,\frac{mv_0^3}{2\sqrt{2}g}$$

Thus, the angular momentum of the particle is $mv_0^3/(2\sqrt{2}g)$ in the negative Z-direction, i.e., perpendicular to the plane of motion, going into the plane.

ANGULAR IMPULSE: ROTATIONAL ANALOGUE OF IMPULSE

In complete analogue with the linear impulse, the angular impulse $\vec{J}$ for a rigid body rotating about a fixed axis is defined as $\vec{J} = \int \vec{\tau}\, dt$.

The impulse momentum theorem for rotational motion may be stated as

$$\text{Angular impulse} = \Delta\vec{L} = \vec{L}_f - \vec{L}_i$$

$$\Rightarrow\quad \vec{J} = I\vec{\omega}_f - I\vec{\omega}_i = I(\vec{\omega}_f - \vec{\omega}_i)$$

As linear impulse changes the linear momentum, angular impulse changes the angular momentum.

ANGULAR MOMENTUM OF A RIGID BODY DUE TO TRANSLATION AND ROTATION BOTH

Method I:

Consider a body of mass m is rotating with angular velocity ω about CM axis and translating with a linear velocity $\vec{v}$ as shown in Fig. 2.113.

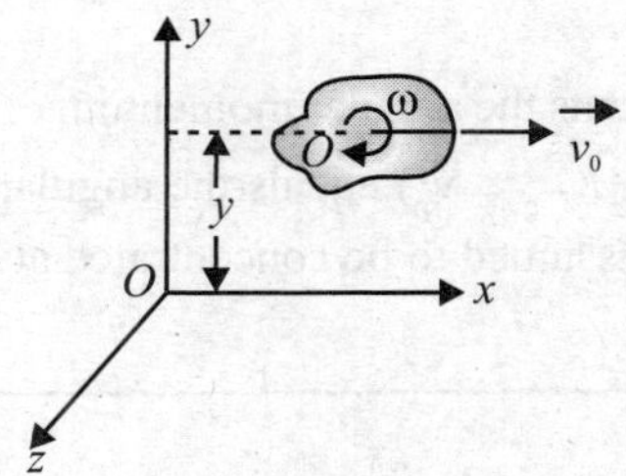

Fig. 2.113

The angular momentum of the body is

$$\vec{L} = \vec{L}_{\text{translation}} + \vec{L}_{\text{rotation}}$$

or $$L = mv_0 y(-\hat{k}) + I_0\omega(-\hat{k})$$

$$= -(mv_0 y + I_0\omega)\hat{k}\ \text{kg m}^2/\text{s}$$

where I is the moment of inertia of the body about the perpendicular axis through O (through which the body is rotating about its own axis).

Method II:

Consider the situation described in method *I*. Choose a particle of mass m at position $\vec{r}$ having velocity $\vec{v}$ in CM frame. Its position and velocity in ground frame at O will be $(\vec{r}+\vec{r}_0)$ and $(\vec{v}+\vec{v}_0)$, respectively.

Angular momentum of the whole body about O,

$$\vec{L}=\Sigma m(\vec{r}+\vec{r}_0)\times(\vec{v}+\vec{v}_0)$$

Here, $\vec{r}_0$ and $\vec{v}_0$ are the position vector and velocity vector of the CM.

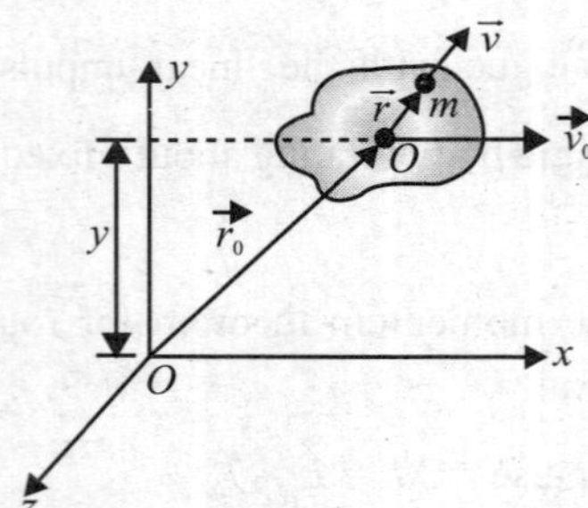

Fig. 2.114

Thus $\quad \vec{L}=\Sigma m(\vec{r}\times\vec{v})+\Sigma m(\vec{r}_0\times\vec{v}_0)+\Sigma m(\vec{r}\times\vec{v}_0)+\Sigma m(\vec{r}_0\times\vec{v}_0)$

$$=\Sigma m(\vec{r}\times\vec{v})+\vec{r}_0\times(\Sigma m\vec{v})\times\vec{v}_0+\Sigma m(\vec{r}_0\times\vec{v}_0)$$

As $\quad \Sigma m\vec{v}=(M\vec{v}_0)_{CM}=0$ and $\Sigma m\vec{r}=(Mr_0)_{CM}=0$

$$\therefore \quad \vec{L}=\Sigma m(\vec{r}\times\vec{v})+\Sigma m(\vec{r}_0\times\vec{v}_0)$$

or $\quad \vec{L}=\vec{L}_{CM}+m\left(\vec{r}_0\times\vec{v}_0\right)$

Here $\vec{L}_{CM}$ represents the angular momentum of the body about CM frame, and $M(\vec{r}_0\times\vec{v}_0)$ equals the angular momentum of the body if it is assumed to be concentrated at CM translating with velocity $\vec{v}_0$.

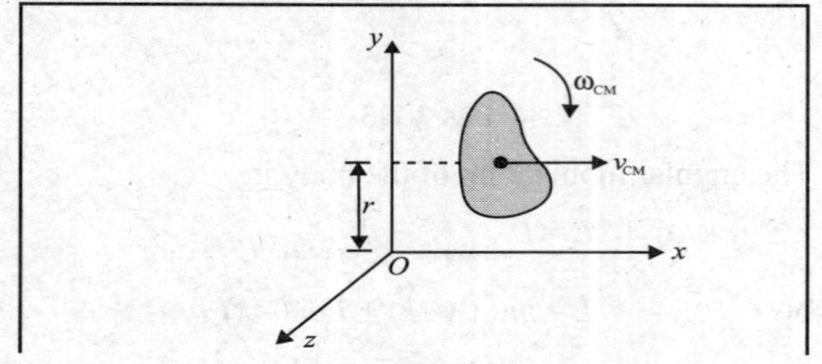

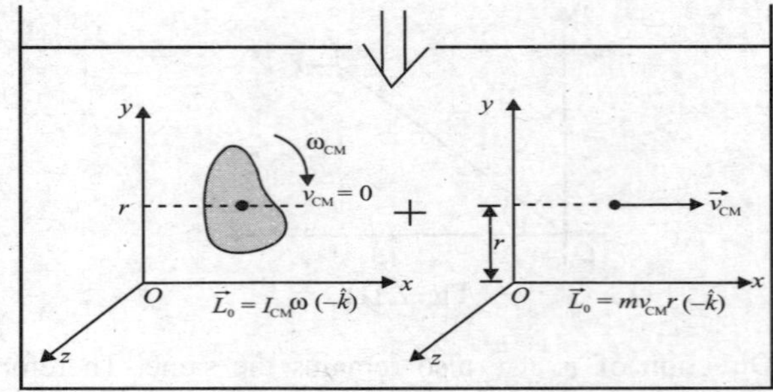

$$\vec{L}=I_{CM}\omega(-k)+mv_{CM}r(-k)$$

Fig. 2.115

Illustration 2.36 **Find the angular momentum of a disc about the axis shown in Fig. 2.116 in the following situations.**

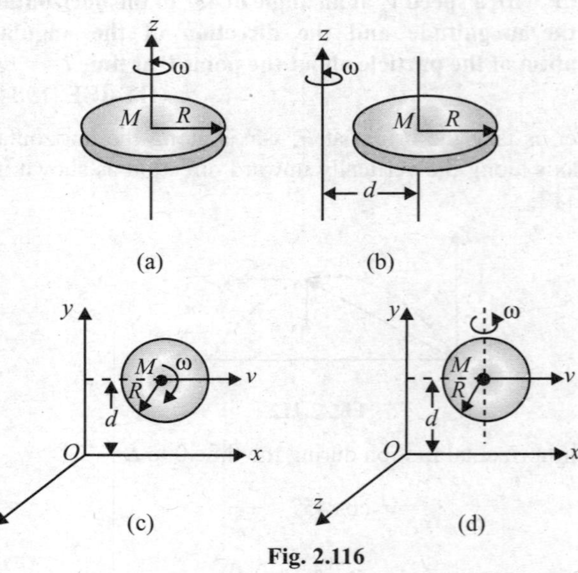

Fig. 2.116

Sol. Case (a): In this case the disc is rotating about fixed axis which is parallel to the z-axis.

Hence, $\vec{L}=I_z(\hat{k})=\left(\frac{MR^2}{2}\right)\omega\,(\hat{k})$

Direction can be determined using right hand rule.

Case (b): In this case also the disc rotates about fixed axis. We can write angular momentum for this situation.

$$\vec{L}=I_z\omega\,(\hat{k})$$

Using parallel axis we can find the value of $I_z=\frac{MR^2}{2}+Md^2$

Hence, $\quad \vec{L}=\left(\frac{MR^2}{2}+Md^2\right)\omega\,(\hat{k})$

Case (c): In this case the disc is rotating about an axis parallel to the z-axis and centre of mass of disc is also translating.

Hence, angular momentum.

$$\vec{L} = I_{cm}\vec{\omega} + \vec{r}\times m\vec{v}$$

$$= \frac{MR^2}{2}\omega(-\hat{k}) + Mvd(-\hat{k}) = \left(\frac{MR^2}{2}\omega + Mvd\right)(-\hat{k})$$

Case (d): In this case also the disc is rotating about an axis passing through it centre and parallel to y-axis and centre of mass of disc is translating with velocity v parallel to x-axis.

$$\vec{L} = I_{CM}\omega(\hat{j}) + Mvd(\hat{k})$$

Here $\quad I_{CM} = \dfrac{MR^2}{4}$

Hence, $\quad \vec{L} = \dfrac{MR^2}{4}\omega(\hat{j}) + Mvd(\hat{k})$

LAW OF CONSERVATION OF ANGULAR MOMENTUM

In case of translation motion, we define force as $\vec{F} = \dfrac{d\vec{p}}{dt}$.

For rotational motion,we define torque.

$$\vec{\tau} = \frac{d\vec{L}}{dt} = \left[\equiv \frac{d\vec{p}}{dt} = \vec{F}\right]$$

So if the net torque on a particle (or system) is zero; $\dfrac{d\vec{L}}{dt} = 0$, i.e., $\vec{L}$ = constant.

Therefore, angular momentum of a system (may be particle or body) remains constant if the resultant torque acting on it is zero. This principle is called 'law of conservation of angular momentum'.

For a system of particle angular momentum,

$$\vec{L} \Rightarrow \vec{L}_1 + \vec{L}_2 + \cdots$$

Hence, when the total angular momentum is conserved,

$$\vec{L} \Rightarrow \vec{L}_1 + \vec{L}_2 + \cdots = \text{constant}$$

This shows that the angular momentum of an individual particle may change, but their sum remains constant in the absence of an external torque.

ANGULAR IMPULSE

The sum of torques of all external forces acting on a system of particles (or rigid body) about any fixed point is equal to the rate of change in angular momentum about the point.

Since $\vec{\tau} = \dfrac{d\vec{L}}{dt} \Rightarrow d\vec{L} = \vec{\tau}\,dt$

The change in angular momentum of a rigid body during a time interval Δt is given by

$$\int d\vec{L} = \int \vec{\tau}\,dt \Rightarrow \Delta\vec{L} = \int_{t_1}^{t_2} \vec{\tau}\,dt$$

Hence, the angular impulse $\Delta\vec{L} = \int_{t_1}^{t_2} \vec{\tau}\,dt$.

Illustration 2.37 **A turntable turns about a fixed vertical axis, making one revolution in 10 s. The moment of inertia of the turntable about the axis is 1200 kgm². A man of 80 kg, initially standing at the centre of the turntable, runs out along the radius. What is the angular velocity of the turntable when the man is 2 m from the centre?**

Sol. Let I_0 be the initial moment of inertia of the system (man + table)

$$I_0 = I_{man} + I_{table}$$
$$= 0 + 1200 = 1200 \text{ kg m}^2$$
$(I_{man} = 0$ as the man is at the axis)

I = final moment of inertia of the system

$$= I_{man} + I_{table}$$
$$= mr^2 + 1200$$
$$= 80(2)^2 + 1200 = 1520 \text{ kg m}^2$$

Fig. 2.117

As no external torque about axis, we can conserve angular momentum.

By conservation of angular momentum: $I_0\omega_0 = I\omega$

Now $\quad \omega_0 = 2\pi/T_0 = 2\pi/10 = \pi/5$ rad/s

$$\Rightarrow \quad \omega = \frac{I_0\omega_0}{I} = \frac{1200\times\pi}{1520\times 5} = 0.51 \text{ rad/s}$$

Illustration 2.38 **A man of mass m stands on a horizontal platform in the shape of a disk of mass M and radius R, pivoted on a vertical axis through its centre about which it can freely rotate. The man starts to move around the centre of the disk in a circle of radius r with a velocity v relative to the disk. Calculate the angular velocity of the disk.**

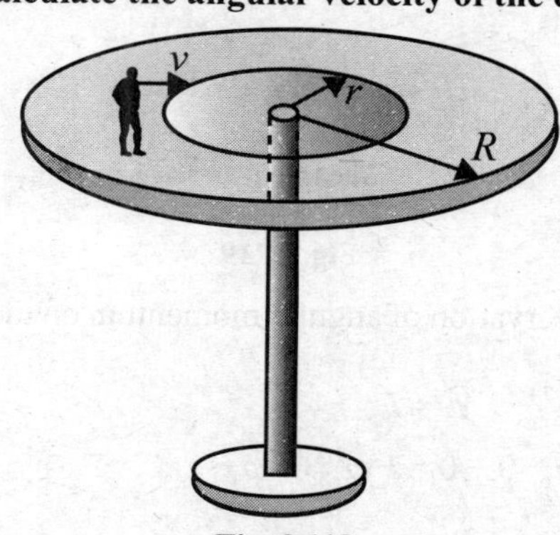

Fig. 2.118

Sol. Since there is no torque acting about the axis of rotation of the disk, so the angular momentum of the system (disk + man) remains constant. Initially it is zero. Suppose ω is the angular velocity of the disk (take positive in the sense of motion of the man). The velocity of the man with respect to the ground observer will be:

$$\vec{v}_{\text{man, disc}} = \vec{v}_{\text{man}} - \vec{v}_{\text{disc}}$$

$$\Rightarrow \vec{v}_{\text{man}} = \vec{v}_{\text{man, disc}} + \vec{v}_{\text{disc}}$$

or $\vec{v}_{\text{man}} = v - \omega r$

Thus, angular momentum of the man $= m(v - \omega r)r$

And angular momentum of the disk is

$$I_{\text{disk}}\,\omega = \frac{MR^2}{2}\omega$$

By conservation of angular momentum, we have

$$m(v-\omega r)r = \frac{MR^2}{2}\omega$$

After solving, we get $\omega = \dfrac{mvr}{\left(mr^2 + \dfrac{MR^2}{2}\right)}$

Illustration 2.39 **A man of mass 100 kg stands at the rim of a turntable of radius 2 m and moment of inertia 4000 kg m^2 mounted on a vertical frictionless shaft at its centre. The whole system is initially at rest. The man now walks along the outer edge of the turnable with a velocity of 1 m/s relative to the earth.**

a. With what angular velocity and in what direction does the turntable rotate?

b. Through what angle will it have rotated when the man reaches his initial position on the turntable?

c. Through what angle will it have rotated when the man reaches his initial position relative to the earth?

Sol. As for the man and table system, initial angular momentum is zero and it will remain zero as no external torque about rotation axis. When the man starts moving in the anticlockwise direction to make angular momentum zero, the table will start rotating in the clockwise direction.

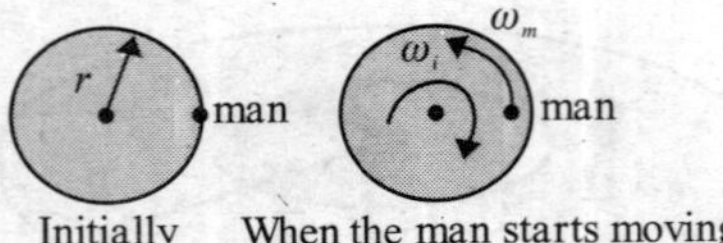

Fig. 2.119

a. By conservation of angular momentum on the man–table system,

$$L_i = L_f$$

$$\Rightarrow \quad 0 + 0 = I_m\omega_m + I\omega_t$$

$$\Rightarrow \quad \omega_t = \frac{-I_m\omega_m}{I_t}, \text{ where } \omega_m = \frac{v}{r} = \frac{1}{2}\text{ rad/s}$$

$$= -\frac{100(2)^2 \times \left(\frac{1}{2}\right)}{4000}$$

(negative sign represents clockwise direction)

$$\Rightarrow \quad \omega_m = v/r = 1/2 \text{ rad/s} \Rightarrow \omega_t = -\frac{1}{20}\text{ rad/s}$$

Thus, the table rotates clockwise (opposite to the man) with angular velocity 0.05 rad/s.

b. If the man completes one revolution relative to the table. Angular displacement of the man w.r.t table is

$$\theta_{mt} = 2\pi \Rightarrow 2\pi = \theta_m - \theta_t$$

$2\pi = \omega_m t - \omega_t t$ (where t is the time taken)

$$t = \frac{2\pi}{\omega_m - \omega_t} = \frac{2\pi}{0.5-(-0.05)} = \frac{2\pi}{0.05}\text{ s}$$

$\Rightarrow$ Angular displacement of table is

$$\theta_t = \omega_t t = 0.05 \times \left(\frac{2\pi}{0.55}\right) = -\left(\frac{2\pi}{11}\right)\text{ rad}$$

The table rotates through $2\pi/11$ rad clockwise.

c. If the man completes one revolution relative to the earth, then

$$\theta_m = 2\pi$$

$$\text{Time} = 2\pi/\omega_m = 2\pi/0.5$$

During this time, angular displacement of the table

$$\theta_t = \omega_t\,(\text{time}) = -0.05 \times \left(\frac{2\pi}{0.5}\right)$$

$$= -\pi/5 \text{ radians}$$

$$= 36° \text{ in clockwise direction}$$

Illustration 2.40 **A small block of mass 4 kg is attached to a cord passing through a hole in a horizontal frictionless surface. The block is originally revolving in a circle of radius 0.5 m about the hole, with a tangential velocity of 4m/s. The cord is then pulled slowly from below, shortening the radius of the circle in which the block revolves. The breaking strength of the cord is 600 N. What will be the radius of the circle when the cord breaks?**

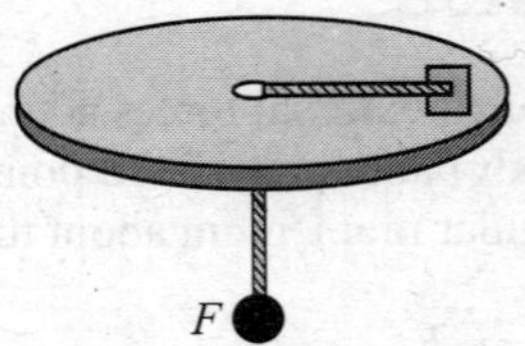

Fig. 2.120

Sol. The tension of the rope is the only net force on the block and it does not exert any torque about the axis of rotation. Hence, the angular momentum of the block about the axis should remain conserved.

$\Rightarrow \quad mvr = \text{constant}$

Let $\quad r_1 = 0.5$ m and $v_1 = 4$ m/s

Let r_2, v_2, T_2 be the radius, velocity and tension, respectively, when the string breaks.

$\Rightarrow \quad T_2 = 600$ N

$$mv_1r_1 = mv_2r_2 \text{ and } T_2 = mv_2^2/r_2$$

$$\Rightarrow \quad mv_1r_1 = m\sqrt{\frac{r_2T_2}{m}}\, r_2$$

$$r_2 = \left(\frac{mv_1^2 r_1^2}{T_2}\right)^{\frac{1}{3}} = \left(\frac{4 \times 16 \times 0.25}{600}\right)^{\frac{1}{3}}$$

$$= \left(\frac{16}{600}\right)^{\frac{1}{3}} = 3.0 \text{ m}$$

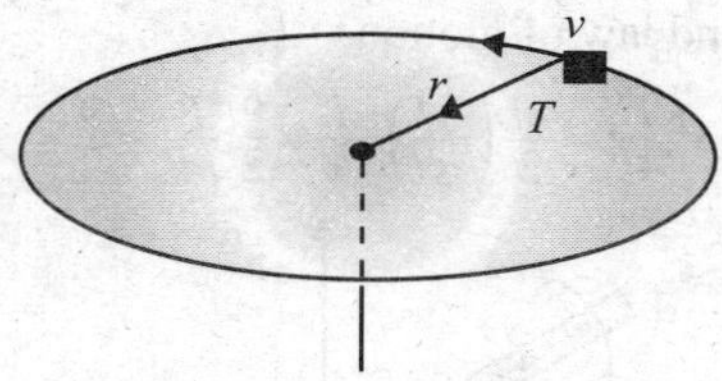

Fig. 2.121

Note: ***The tension in the string is inversely proportional to the cube of the radius.***

Illustration 2.41 **Two skaters, each of mass 50 kg, approach each other along parallel paths separated by 3 m. They have equal and opposite velocities of 10 m/s. The first skater carries a long light pole, 3 m long, and the second skater grabs the end of it as he passes (assume frictionless ice).**

a. Describe quantitatively the motion of the skaters after they are connected by the pole.

b. By pulling on the pole the skaters reduce their distance to 1 m. What is their motion then?

c. Compare the KEs of the system in parts (a) and (b). Where does the change come from?

Sol.

a. As the net linear momentum of the system (skater + pole) is zero, the centre of mass will be at rest before and after the collision.

The skaters and the pole will rotate around the centre of mass (at the midpoint of the pole).

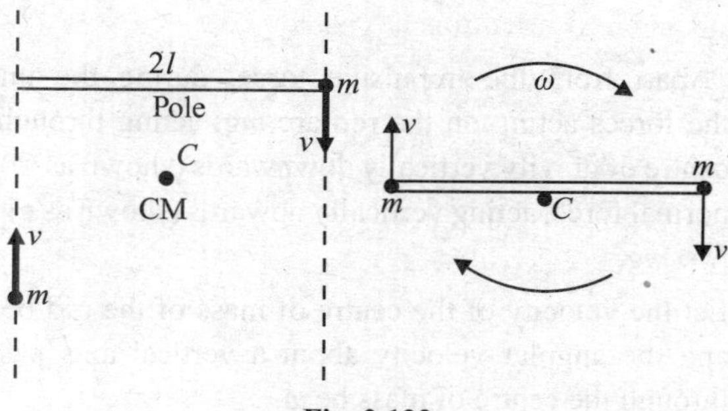

Fig. 2.122

Applying the conservation of angular momentum about an axis through C and perpendicular to the plane of the figure,

$$mvl + mvl = I\omega, \text{ where } I = 2m\,(l)^2$$

$$\omega = mvl/I = v/l$$

$$= 20/3 \text{ rad/s}$$

b. As the separation reduces to $2l' = 1$ m

$$I\omega = I'\omega'$$

Again applying conservation of angular momentum, we get

$$\omega' = \frac{I\omega}{I'} = \frac{2ml'^2\omega}{2ml'^2} = 9\omega = 60 \text{ rad/s}$$

Therefore, angular velocity increases.

c. $$\frac{KE_i}{KE_f} = \frac{\frac{1}{2}I'\omega'^2}{\frac{1}{2}I\omega^2} = \frac{I'\omega'^2}{I^2\omega^2} = 9$$

The kinetic energy increases because the skaters do positive work in pulling themselves towards the centre of pole.

Illustration 2.42 **A uniform rod of mass m = 5.0 kg and length = 90 CM rests on a smooth horizontal surface. One of the ends of the rod is stuck with impulse J = 3.0 Ns in a horizontal direction perpendicular to the rod as shown in Fig. 2.123. As a result the rod obtains momentum p = 3.0 Ns. Find**

a. the angular velocity of the rod about a vertical axis passing through its centre of mass.

b. instantaneous velocity of point A π/8 s later.

c. the force with which one-half of the rod will act on the other in the process of motion.

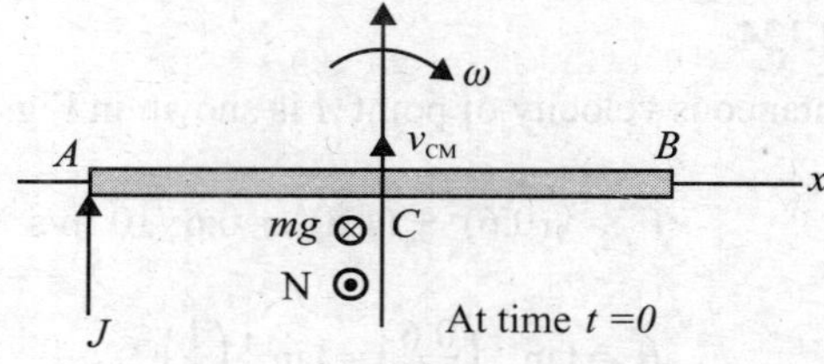

Fig. 2.123

Sol.

a. Apart from the impulsive force, during the impact the forces acting on the rod are mg, acting through the centre of gravity vertically downwards (shown as ⊗) and normal force, acting vertically upwards (shown as •), i.e., $N = mg$.

Let the velocity of the centre of mass of the rod be v_{CM} and the angular velocity about a vertical axis passing through the centre of mass be ω.

Now, impulse imparted to the rod equals change in its linear momentum.

$\Rightarrow \quad \vec{J} = \Delta\vec{p}$

$\Rightarrow \quad J = p - 0$ (already satisfied)

and $\quad J\frac{l}{2} = I\omega$ (angular impulse = change in angular momentum)

$$\Rightarrow \quad J\frac{l}{2} = \frac{ml^2}{12}\omega \Rightarrow \quad \omega = \frac{6J}{ml} = \frac{6 \times 3}{5 \times 0.9} = 4 \text{ rad/s}$$

b. Velocity of the centre of mass of the rod is v_{CM}

$p = mv_{CM}$

$3 = 5v_{CM} \quad \Rightarrow \quad v_{CM} = 0.6$ m/s

Angle through which the rod rotates in $\frac{\pi}{8}$ s $= \frac{\pi}{8} \times 4 = \frac{\pi}{2}$ rad.

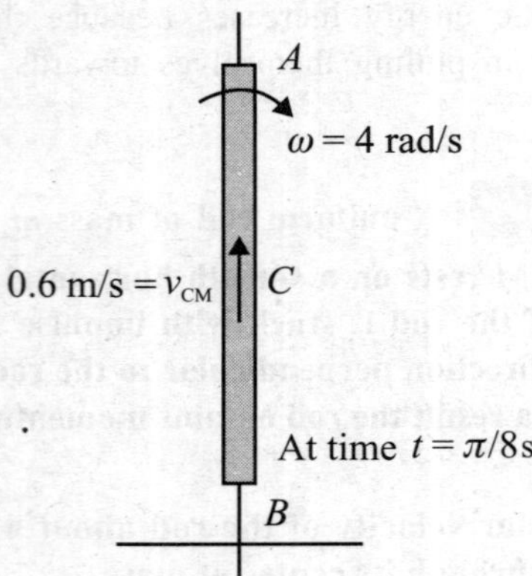

Fig. 2.124

The orientation of the rod at $t = \pi/8$ s is shown in Fig. 2.124.

Instantaneous velocity of point A is shown in Fig. 2.124.

$$V_A = \sqrt{(0.6)^2 + (1.8)^2} = 0.6\sqrt{10} \text{ m/s}$$

$$\theta_x = \tan^{-1}\left(\frac{0.6}{1.8}\right) = \tan^{-1}\left(\frac{1}{3}\right)$$

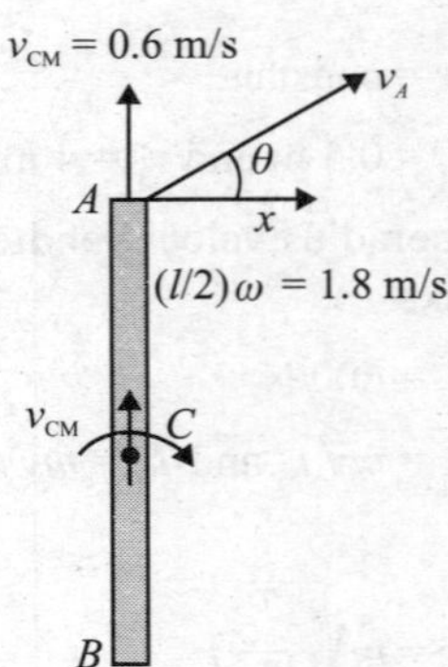

Fig. 2.125

c. The forces acting on one-half of the rod, say AC, are shown in Fig. 2.125. They are (i) the weight $mg/2$, vertically downwards, (ii) the normal force N acting vertically upwards and (iii) the force F on AC by CA.

About the centre of mass of the rod, which is an inertial reference frame (V_{CM} = constant), the centre of mass of AC accelerates with an acceleration $\omega l^2/4$ towards C. From Newton's second law of motion,

$$F = \left(\frac{m}{2}\right)\left(\omega^2 \frac{l}{4}\right) = \frac{m}{2}\left(\frac{6J}{ml}\right)^2 \frac{l}{4} = \frac{9}{2}\frac{J^2}{ml} = \frac{9}{2} \times \frac{(3)^2}{5 \times 0.9} = 9 \text{ N}$$

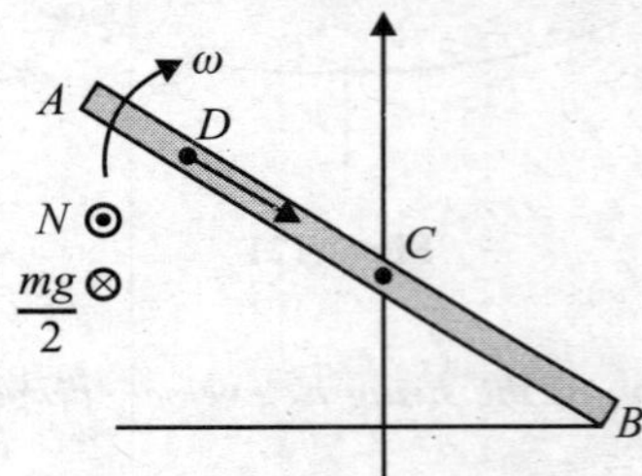

Fig. 2.126

Concept Application Exercise 2.4

1. A circular disc of mass M_1 and radius R, initially moving with a constant angular speed ω_0 is gently placed coaxially on a stationary circular disc of mass M_2 and radius R, as shown in Fig. 2.127. There is a frictional force between the two discs.

i. If disc M_2 is placed on a smooth surface, then determine the final angular speed of each disc.

ii. Determine the work done by friction.

iii. Determine the frictional loss in kinetic energy, i.e., $\Delta K/K_i$.

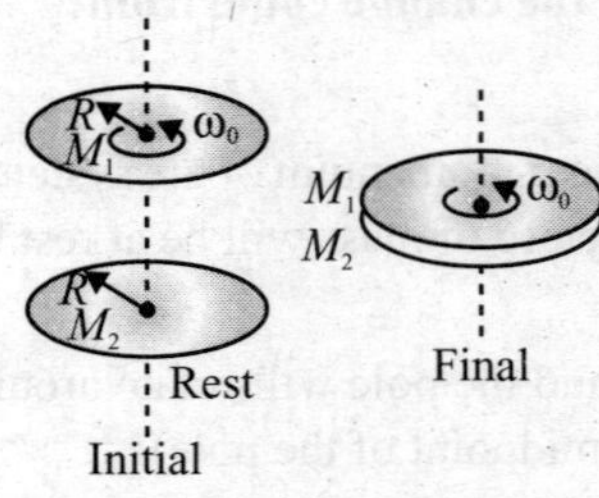

Fig. 2.127

2. A child of mass m is standing on the periphery of a circular platform of radius R, which can rotate about its central axis. The moment of inertia of the platform is I. Child jumps off from the platform with a velocity u tangentially relative to the platform. Find the angular speed of the platform after the child jumps off.
3. Suppose in the previous problem, the child stays at rest on the platform and one of his friend throws a ball of mass m_1 towards him with a velocity of v from a direction tangential to the platform and the child on the platform catches the ball. Find the angular velocity of the platform after he catches the ball.
4. In the previous problem, initially the platform with the child is rotating with an angular speed ω_1. If the child starts walking along its periphery in opposite direction with speed u relative to the platform, what will be the new angular speed of the platform.
5. Consider a cylinder rolling on a horizontal plane. Its linear velocity is v and rotational angular velocity is ω. Find its angular momentum about point P on the ground as shown in Fig. 2.128. What happens to this angular momentum if the cylinder is rotating in opposite direction but moving translationally in the same direction.

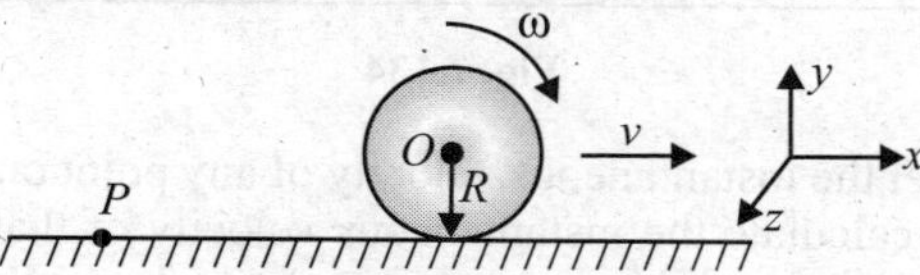

Fig. 2.128

6. A particle of mass m is released from rest at point A in Fig. 2.129, falling parallel to the (vertical) y-axis. Find the angular momentum of the particle at any time t with respect to the same origin O.

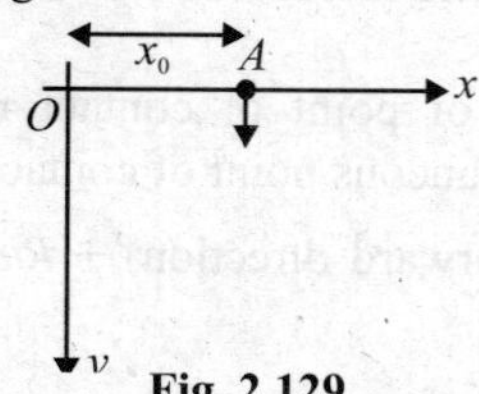

Fig. 2.129

7. A uniform spherical shell of mass M and radius R rotates about a vertical axis on frictionless bearing. A massless cord passes around the equator of the shell, over a pulley of rotational inertia I and radius r and is attached to a small object of mass m that is otherwise free to fall under the influence of gravity. There is no friction of pulley's axle; the cord does not slip on the pulley. What is the speed of the object after it has fallen a distance h from rest? Use work–energy considerations.

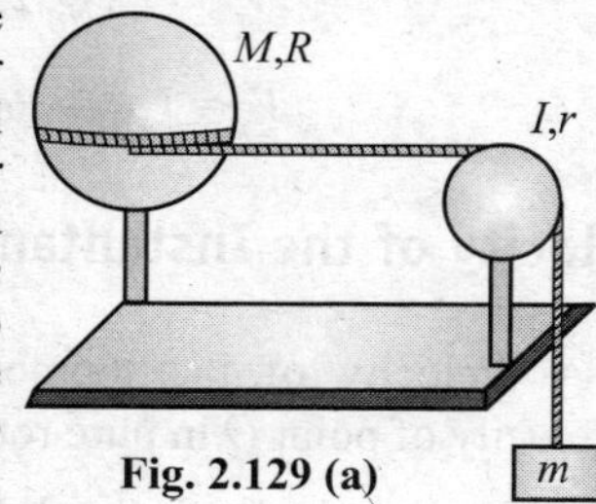

Fig. 2.129 (a)

8. A uniform rod of mass m and length l is kept vertical with the lower end clamped. It is slightly pushed to let it fall down under gravity. Find its angular speed when the rod is passing through its lowest position. Neglect any friction at the clamp. What will be the linear speed of the free end at this instant?
9. A rigid body is made of three identical thin rods, each with length l, fastened together in the form of letter H. The body is free to rotate about a horizontal axis that runs along the length of one of the arms of H. The body is allowed to fall from rest from a position in which the plane of the H is horizontal. What is the angular speed of the body when the plane of H is vertical?

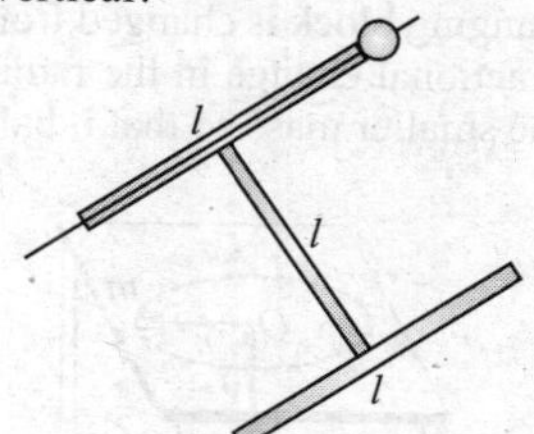

Fig. 2.129 (b)

10. The steel balls A and B have a mass of 500 g each and are rotating about the vertical axis with an angular velocity of 4 rad/s at a distance of 15 CM from the axis. Collar C is now forced down until the balls are at a distance of 5 CM from the axis. How much work must be done to move the collar down?

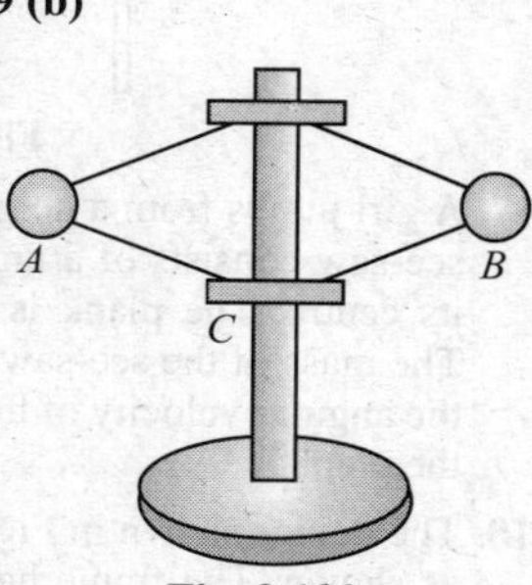

Fig. 2.130

11. Discs A and B are mounted on a shaft SS and may be connected or disconnected by a clutch C. The moment of inertia of disc A is one half that of disc B. With the clutch disconnected, A is brought up to an angular velocity ω_0. The accelerating torque is then removed from A and it is coupled to disc B by the clutch (Bearing friction may be neglected). It is found that 2000 J of heat is developed in the clutch when the connection is made. What is the original kinetic energy of disc A?
12. A circular disc of mass m and radius R is set into motion on a horizontal floor with a linear speed v in the forward direction and an anguler speed $\omega = v / R$ in the clockwise direction as shown in Fig. 2.131. Find the magnitude of the total angular momentum of the disc about the bottommost point O of the disc.

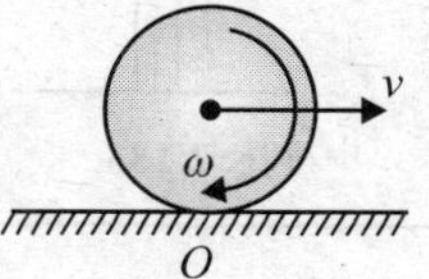

Fig. 2.131

13. A particle of mass m is projected with a speed u at an angle θ to the horizontal at time $t = 0$. Find its angular momentum about the point of projection O at time t, vectorially. Assume the horizontal and vertical lines through O as X and Y axes, respectively.
14. Point masses M_1 and M_2 are placed at the opposite ends of a rigid rod of length L and of negligible mass. The rod is to be set rotating about an axis perpendicular to it. Find the position on this rod through which the axis should pass in order that the work required to set the rod rotating with angular velocity ω_0 should be minimum.
15. A thin circular disc of mass M and radius R is rotating in a horizontal plane about an axis passing through its centre and perpendicular to its plane with an angular velocity ω. Another disc of the same dimension but of mass $M/4$ is placed gently on the first disc coaxially. Find the angular velocity of the system.

16. A thread is passing through a hole at the centre of a frictionless table. At the upper end a block of mass 0.5 kg is tied and a block of mass 8.0 kg is tied at the lower end which is freely hanging. The smaller mass is rotated on the table with a constant angular velocity about the axis passing through the hole so as to balance the heavier mass. If the mass of the hanging block is changed from 8.0 kg to 1.0 kg, what is the fractional change in the radius and the angular velocity of the smaller mass so that it balances the hanging mass again?

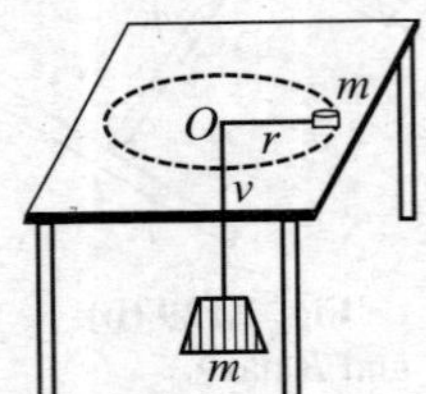

Fig. 2.132

17. A girl jumps from a height h on the end of a see-saw. The see-saw consists of a uniform plank of length l pivoted at its centre. The plank is horizontal before the girl jumps. The mass of the see-saw is twice the mass of the girl. Find the angular velocity of the plank just after the girl jumps on the plank.

18. The device shown in Fig. 2.133 rotates on the vertical axle as shown. The frame has negligible mass as compared to the four masses each of mass m. Initial angular velocity of the system is ω_0. Due to an internal mechanism the spokes in the frame lengthen so that the radii of the masses become $2a$. Initially, it was a. What will be the new angular velocity of the system?

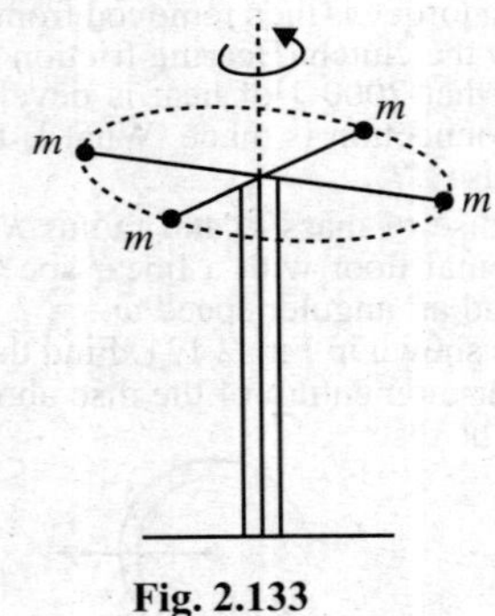

Fig. 2.133

COMBINATION OF TRANSLATION AND ROTATION

When the axis of rotation is not fixed (stationary in space), the motion of a rigid body is considered as combination of motion of centre of mass plus a rotation about centre of mass. The two components of motion are described by

$$\Sigma \vec{F}_{ext} = m\vec{a}_{CM} \text{ and } \Sigma \vec{\tau}_{CM} = I_{CM}\vec{\alpha}$$

Superposition of Translation and Rotation

How the translational motion and rotational motion about the centre of mass are superimposed to get the motion of a rigid body (say a disc of radius R) is illustrated in the following figures.

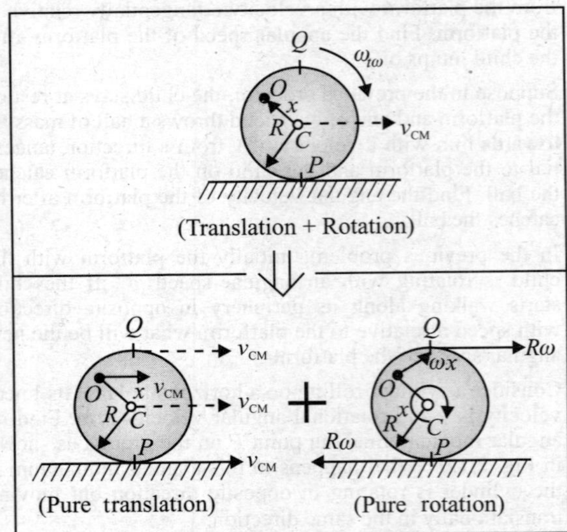

Fig. 2.134

To get the instantaneous velocity of any point on the rigid body we calculate the instantaneous velocity of that point in pure translation and in pure rotation and add them vectorically.

$$\vec{V}_{net} = \vec{V}_{translation} + \vec{V}_{rotation}$$

Velocity of the Instantaneous Point of Contact *P*

V_P = Velocity of point of contact in pure translation + velocity of instantaneous point of contact in pure rotation

$V_P = V_{CM}$ (in the forward direction) + $R\omega$ (in the backward direction)

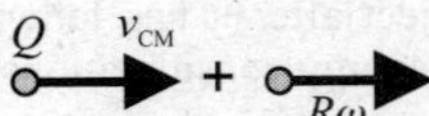

Fig. 2.135

$$V_P = V_{CM} - R\omega, \text{ in the forward direction}$$

Velocity of the Instantaneous Topmost Point *Q*

V_Q = Velocity of the topmost point in pure translation + velocity of point Q in pure rotation

$V_Q = V_{CM}$, in the forward direction + $R\omega$, in the forward direction.

Fig. 2.136

$$V_Q = V_{CM} + R\omega, \text{ in the forward direction}$$

Velocity of Point *O*

V_0 = Velocity of point O in pure translation + instantaneous velocity of point O in pure rotation

$V_0 = V_{CM}$ in the forward direction + $x\omega$, perpendicular to CO as shown in the figure

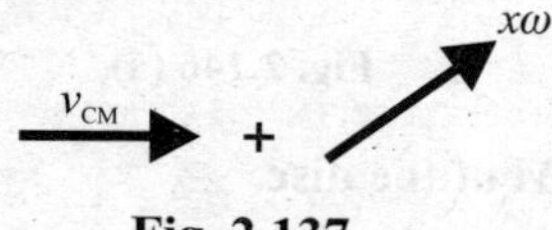

Fig. 2.137

ACCELERATION OF DIFFERENT POINTS

Acceleration of Instantaneous Point of Contact *P*

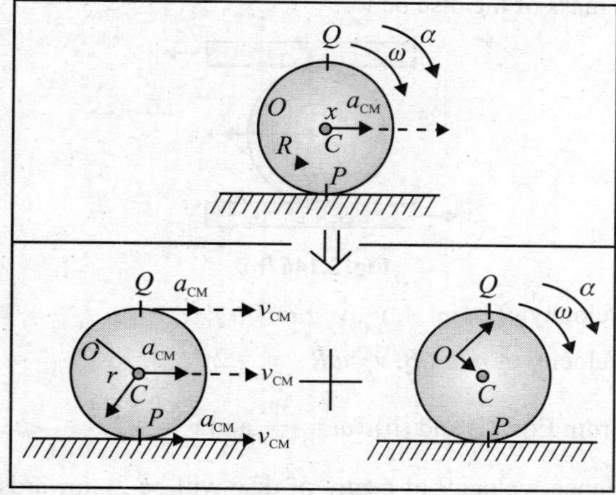

Fig. 2.138

Figure 2.138 illustrates how the instantaneous accelerations are superimposed. Acceleration of the instantaneous point of contact w.r.t. supporting surface, a_P = acceleration of point of contact in pure translation + acceleration of instantaneous point of contact in pure rotation = (a_{CM}, in the forward direction) + ($R\alpha$, in the backward direction + $\omega^2 R$, vertically upward towards the centre of mass)

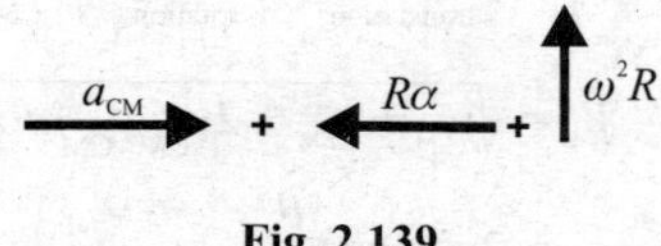

Fig. 2.139

Acceleration of the Instantaneous Topmost Point

a_Q = Acceleration of the topmost point in pure translation + acceleration of the instantaneous topmost point in pure rotation

a_{CM} + $R\alpha$ + $\omega^2 R$

Fig. 2.140

(a_{CM} in the forward direction) + ($R\alpha$, in the forward direction + $\omega^2 R$, vertically downwards towards the centre of mass)

ROLLING MOTION

Consider a wheel moving along a straight track, the centre of the level moves forward in pure translation. A point on the rim of the wheel, however, traces out a complex curve called cycloid. We can analyse the motion of a rolling wheel as a combination of translation and pure rotation.

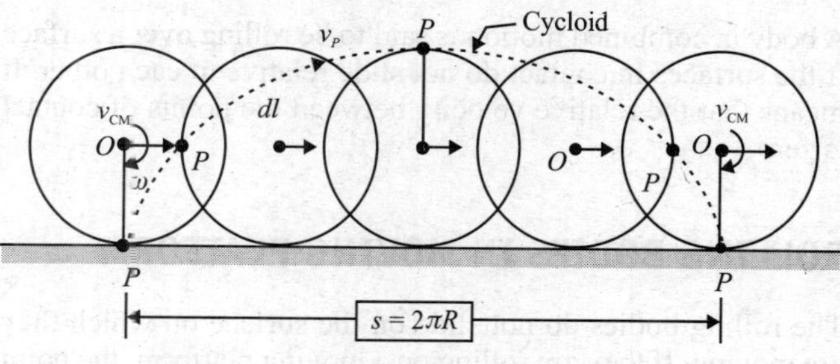

Fig. 2.141

In one complete rotation of wheel, the CM of the wheel moves a distance $s = 2\pi R$. In this motion there is no relative motion between the point of contact of the moving body and the surface. If T is the time to complete a rotation, then we have

$$\frac{s}{T} = \frac{2\pi R}{T} \text{ or } v_{CM} = \omega R$$

Thus in pure rolling, the velocity of the point of the contact is zero and the velocity of centre of mass is $v_{CM} = \omega R$. If the wheel moves through a distance greater than $2\pi R$ in one full rotation, then $v_{CM} > \omega R$ and the wheel slips forward. This type of motion occurs when you apply sudden brakes to a car. The car stops after a long distance but the wheels rotate only a little during the period.

When the wheel moves a distance shorter than $2\pi R$ in full rotation, $v_{CM} < \omega R$ and the wheel slips backward. It happens when you drive the car on a muddy road; the wheel rotates more than the forward motion of the car.

Rolling

Take a uniform disc of radius R, (say), place it vertically on a smooth horizontal surface. Spin the disc and simultaneously push it. In consequence, the disc will rotate with an angular velocity ω and its centre of mass C moves with a velocity v. The velocity of the lowest point P of the disc can be given as

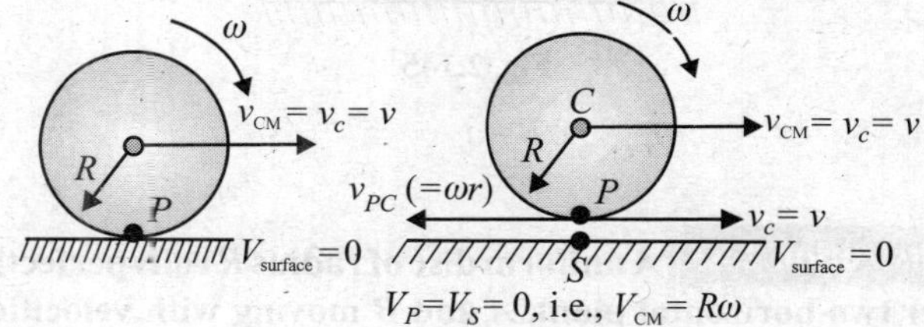

Fig. 2.142

$$\vec{v}_P = \vec{v}_{PC} + \vec{v}_C$$

where $\vec{v}_{PC} = -R\omega\hat{i}$ and $\vec{v}_C = v\hat{i}$

Then, $\vec{v}_P = (v - R\omega)\hat{i}$

The above expression tells us that for different v and ω, the point P will move with different speeds. For suitable velocities v and ω, let us assume that P moves with a velocity which is equal to the velocity of the horizontal surface S, then

$$\vec{v}_P = \vec{v}_S$$

A body in combined motion is said to be rolling over a surface if the surfaces in contact do not slide relative to each other. It means that the relative velocity between the points of contact is zero.

ROLLING BODIES IN MOVING PLATFORM

The rolling bodies do not slide on the surface on which they are moving. If they are rolling on a moving platform, the point of contact of the body with the platform should have the same velocity as the platform.

1. If point of contact of the surface is moving with a velocity u with respect to the ground, then

$$v_{CM} - \omega R = u$$

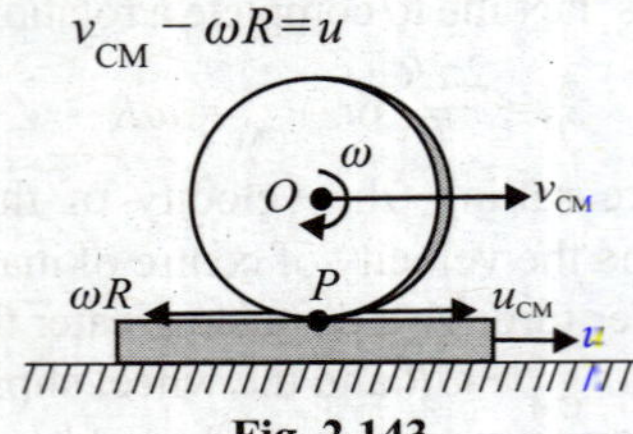

Fig. 2.143

2. For no sliding on the moving platform, $u = \omega R - v_{CM}$

Fig. 2.144

3. For accelerated surface,

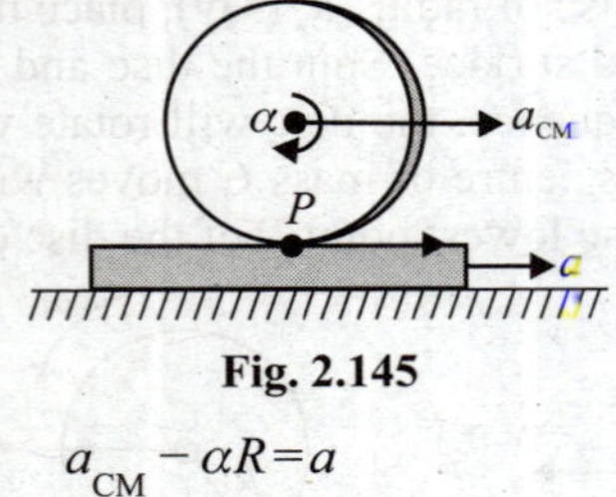

Fig. 2.145

$$a_{CM} - \alpha R = a$$

Illustration 2.43 **A uniform disc of radius R rolls perfectly over two horizontal plank A and B moving with velocities v and $2v$, respectively. Find the**

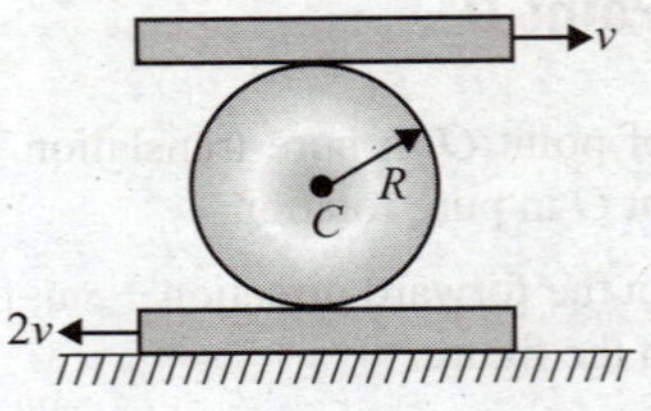

Fig. 2.146 (a)

a. velocity of CM of the disc.

b. angular velocity of the disc.

Sol. As the disc is rolling points A and B which are on the disc and in contact with the planks will not slide. The velocities of points A and B should be the same as the planks. Let the disc rotate with an angular velocity ω and let the velocity of centre of the mass of the disc be v_c.

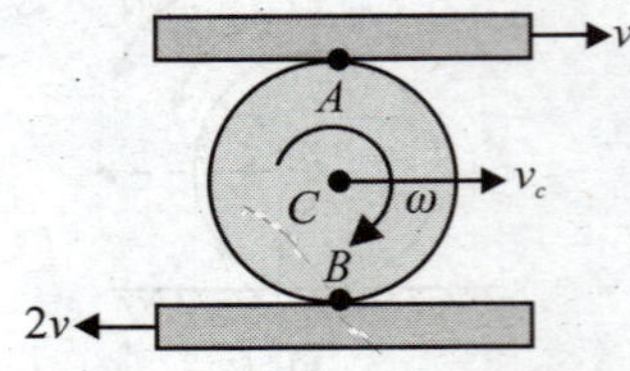

Fig. 2.146 (b)

Velocity of point A, v_A, $v_c + \omega R = v$ (i)

Velocity of point B, v_B, $\omega R - v_c = 2v$ (ii)

From Eqs. (i) and (ii), $\omega = \dfrac{3v}{R}$ and $v_c = -2v$

Hence, velocity of centre of disc will be $2v$ towards the left and the angular velocity will be $3v/R$ the in the clockwise sense.

Velocity of a Point on the Rolling Body

Consider a point P on a rolling body, the velocity of point P is the vector sum of velocity due to translation and due to rotation. Thus,

$$\vec{V}_P = \vec{V}_{translation} + \vec{V}_{rotation} = \vec{V}_{CM} + (\vec{\omega} \times \vec{r})$$

or $$V_P = \sqrt{v_{CM}^2 + v_{CM}^2 + 2v_{CM}v_{CM}\cos\theta}$$

$$= 2v_{CM}\cos\frac{\theta}{2}$$

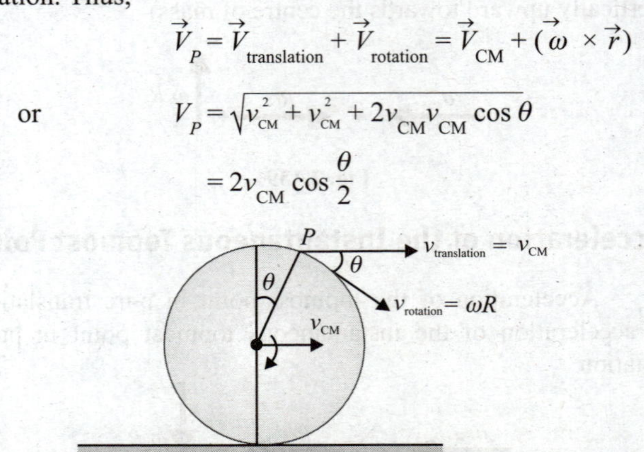

Fig. 2.147

Illustration 2.44 **A wheel rolls purely on ground. Find a point on the periphery of a body which has a velocity equal to the velocity of the centre of mass of the body.**

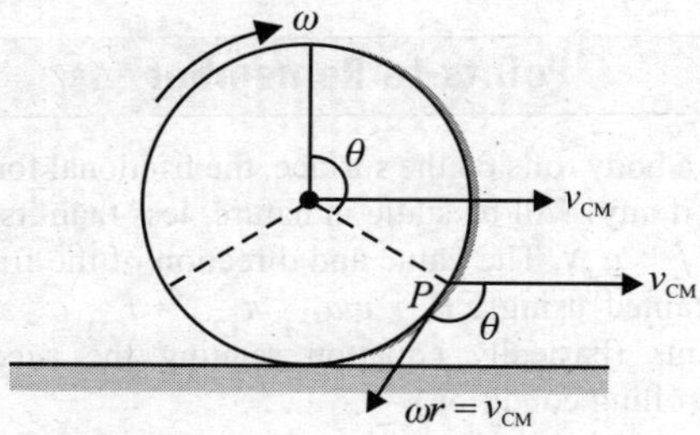

Fig. 2.148

Sol. Velocity of any point on the circumference of a rolling body,

$$v_p = 2v_{CM} \cos\frac{\theta}{2}$$

For $v_p = v_{CM}$, we have

$$v_{CM} = 2v_{CM} \cos\frac{\theta}{2} \text{ or } \cos\frac{\theta}{2} = \frac{1}{2} \text{ or } \theta = 120^\circ$$

Kinetic Energy of Rolling Wheel

Let us calculate the kinetic energy of a rolling wheel as measured by the stationary observer.

KE of a rolling wheel = KE associated with the translational motion + KE associated with the rotational motion

or $\quad K_{rolling} = \frac{1}{2}Mv_{CM}^2 + \frac{1}{2}I_{CM}\omega^2$

For pure rolling, $v_{CM} = \omega R$

$\therefore \quad K_{rolling} = \frac{1}{2}M(\omega R)^2 + \frac{1}{2}I_{CM}\omega^2 = \frac{1}{2}(MR^2 + I_{CM})\omega^2$

Body	I_{CM}	Rolling KE
Ring	MR^2	$MR^2\omega^2$
Disc	$\frac{MR^2}{2}$	$\frac{3}{4}MR^2\omega^2$
Sphere	$\frac{2}{5}MR^2$	$\frac{7}{10}MR^2\omega^2$

As we know $(I_{CM} + MR^2)$ is the moment of inertia of the wheel about the point of contact I, say it is I_p. Thus,

$$K_{rolling} = \frac{1}{2}I_p\omega^2$$

From the above relation, we can say a rolling body performs pure relation about point of contact.

Rolling as Pure Rotation

For a rolling wheel, KE $= \frac{1}{2}I_P\omega^2$. Thus, rolling can be a pure rotation with angular velocity ω, about an axis that always extends through the point of contact P. It is clear from Fig. 2.149 that velocity of point P,

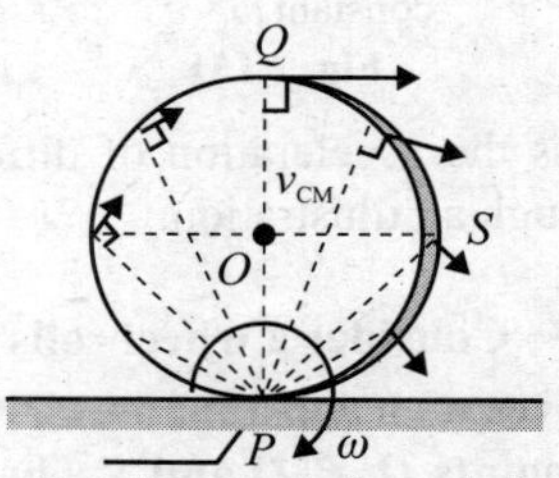

Fig. 2.149

$v_P = 0 \quad$ or $\quad v_O = \omega R$

$v_Q = \omega(2R) = 2\omega R = 2v_{CM}$

$v_s = \omega(\sqrt{2}R) = \sqrt{2}\omega R = \sqrt{2}v_{CM}$

Angular Momentum of Rolling Wheel

Angular momentum of a rolling wheel about an axis passing through the point of contact P and perpendicular to the plane of wheel:

$$\vec{L} = \vec{L}_{translation} + \vec{L}_{rotation} = m(\vec{R} \times \vec{v}_{CM}) + I_{CM}\vec{\omega}$$

or $\quad L = m\omega R^2 + I_{CM}\omega$

or $\quad = (I_{CM} + mR^2)\omega = I_p\omega$

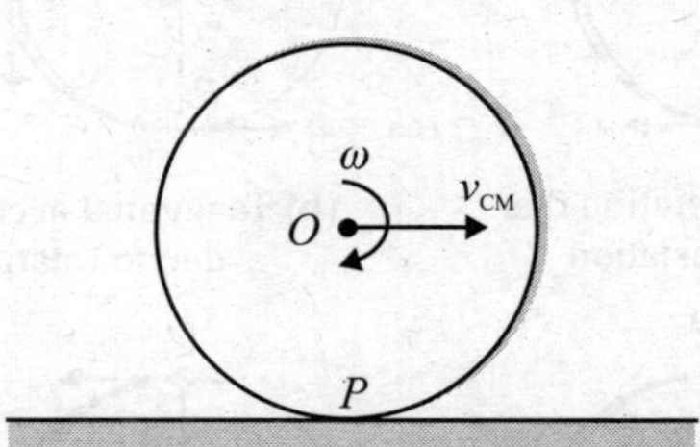

Fig. 2.150

Above relation of the angular momentum is the same as the rolling wheel is having rotation about point of contact.

Acceleration of a Point on a Rolling Wheel

Consider a rolling wheel; each point on the periphery of the wheel is rotating in a circle of radius R, due to which centripetal acceleration of each point on the wheel is ω^2R. As the wheel is moving with constant speed, acceleration due to translation motion of each point is zero.

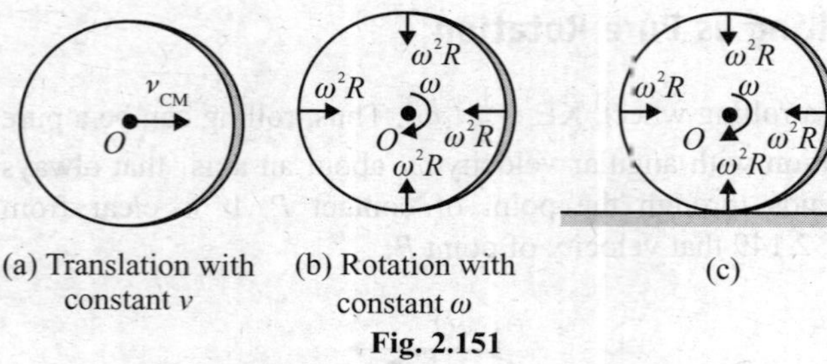

Fig. 2.151

Let us discuss the acceleration of different points on a rolling wheel through an illustration.

Illustration 2.45 **Consider a wheel rolls without slipping and its centre moves with constant acceleration *a*. Find the accelerations of points *O*, *P*, *Q* and *S* when linear velocity of the centre of wheel is *v*.**

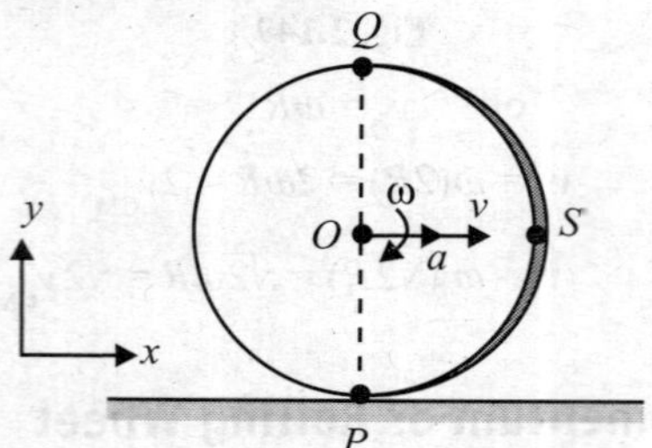

Fig. 2.152

Sol. As body is rolling, we can write the following:

Acceleration of centre of mass $a = \alpha R$

Velocity of centre of mass $v = \omega R$

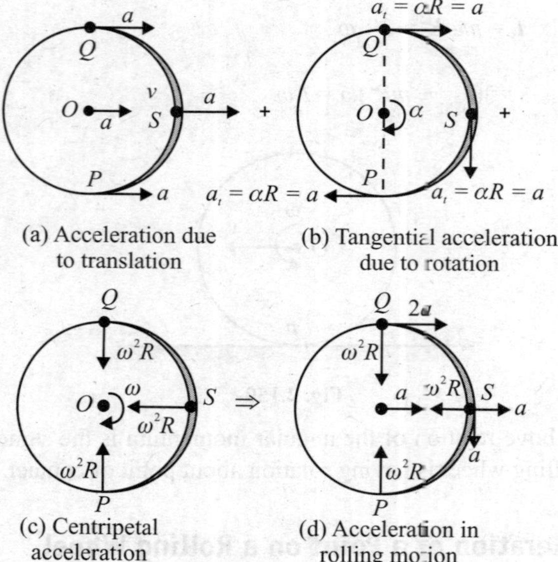

Fig. 2.153

From Fig. 2.153 (d), we get the following:

Resultant acceleration of point O, $a_O = a\hat{i}$

Acceleration of P, $a_p = a\hat{i} - a\hat{i} + \omega^2R\hat{j} = \omega^2R\hat{j}$

Acceleration of Q, $a_Q = a\hat{i} + a\hat{i} - \omega^2R\hat{j} = 2a\hat{i} - \omega^2R\hat{j}$

Acceleration of S, $a_S = a\hat{i} - \omega^2R\hat{j} = (a - \omega^2R)\hat{i} - a\hat{j}$

Points to Remember

1. When a body rolls on the surface, the frictional force on the body (if any) will be static in nature, less than its limiting value, $f_s < \mu_s N$. The value and direction of the friction can be obtained using $\Sigma F = ma_{CM}$, $\tau_{CM} = I_{CM}\,\alpha$, and given constants (basically equation relating the accelerations under rolling condition).
2. When the point of contact moves w.r.t. the supporting surface, $V_{PC'S} \neq 0$, the frictional force (if the surface is rough) is equal to μN, opposite to direction of motion of the point of contact w.r.t. the surface.
3. Rolling motion of a rigid body is also equivalent to pure rotation about the point of contact.

Illustration 2.46 **A force of magnitude *F* is acting on a rolling body of mass *m* and radius *R* as shown in Fig. 2.154. What happens when it enters on a smooth surface?**

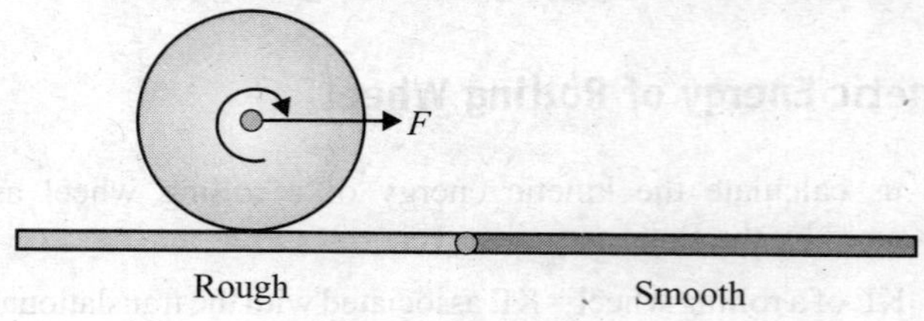

Fig. 2.154

Sol. When the body is rolling on the rough surface, the friction force acting on the surface acts in the backward direction. The force F increases acceleration of centre of mass while friction force provides torque for providing angular acceleration. The velocity of centre of mass is always such that $v_c = \omega R$, i.e., rolling is present. When the body enters on the smooth surface, the pulling force F is still acting, i.e., $F \neq 0$.

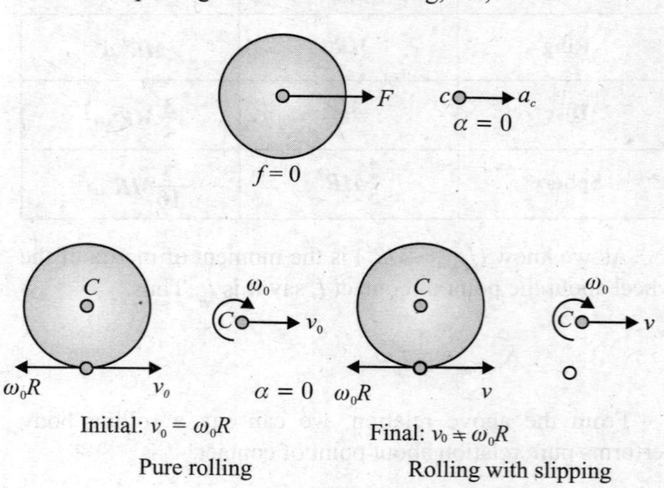

Fig. 2.155

When the rolling body enters on a smooth surface, the force of friction disappears, i.e., $f = 0$. As a result, the angular acceleration of the body about its centre of mass becomes zero.

Whatever is the angular speed attained by the body, it continues to rotate with that. On the contrary, the linear acceleration of the centre of mass suddenly increases to F/m.

As a result, its linear speed starts increasing and the condition $v_c = \omega R$ does not hold true and the body will not roll on the smooth surface. The point of contact slips forward ($v > \omega_0 R$).

Illustration 2.47 **A rolling body of mass m and radius R is rolling on a rough surface without any pulling force as shown in Fig. 2.156. What happens when it enters on a smooth surface?**

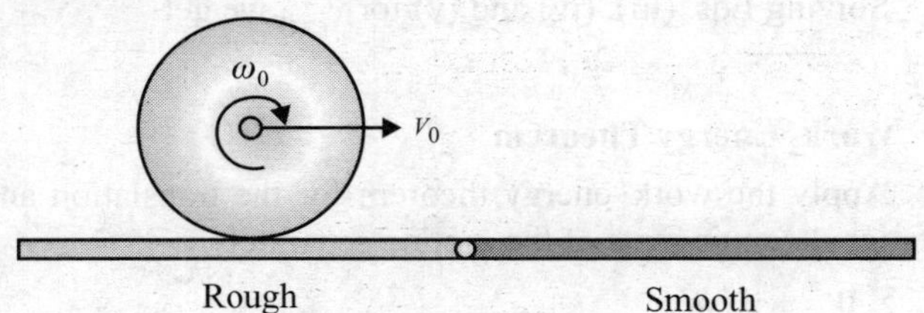

Fig. 2.156

Sol. The pulling force F is not acting, i.e., $F = 0$.

When body is rolling on the rough surface no friction force acts in this case.

When the body enters the smooth surface. In this case, the linear speed v_0 also remains constant. The body keeps on rolling on the smooth horizontal surface and the point of contact is stationary with respect to the surface.

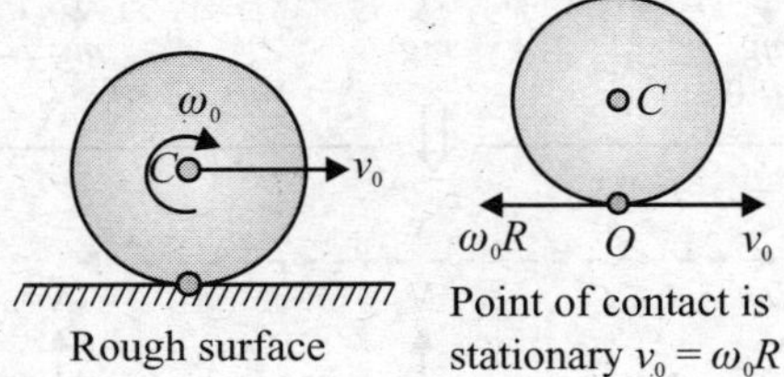

Fig. 2.157

Illustration 2.48 **A sphere of mass M and radius r shown in the figure slips on a rough horizontal plane. At some instant it has translational velocity v_0 and rotational velocity about the centre $v_0/2r$. Find the translational velocity after the sphere starts pure rolling.**

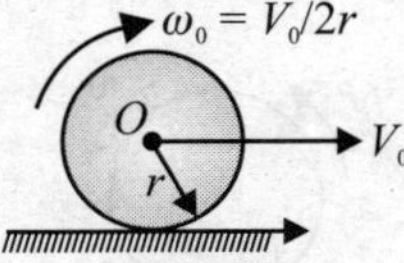

Fig. 2.158

Sol. Velocity of the centre $= v_0$ and the angular velocity about the centre $= v_0/2r$. Thus, $v_0 > \omega_0 r$. The sphere slips forward and thus the friction by the plane on the sphere will act backward. As the friction is kinetic, its value is $\mu N = \mu Mg$ and the sphere will be decelerated by $a_{CM} = f/M$.

Hence,

$$v(t) = v_0 - \frac{f}{M}t \qquad \text{(i)}$$

This friction will also have a torque $= fr$ about the centre. This torque is clockwise and in the direction of ω_0. Hence, the angular acceleration about the centre will be

$$\alpha = f\frac{r}{(2/5)\,Mr^2} = \frac{5f}{2Mr}$$

and the clockwise angular velocity at time t will be

$$\omega(t) = \omega_0 + \frac{5f}{2Mr}$$

$$t = \frac{v_0}{2r} + \frac{5f}{2Mr}t$$

Pure rolling starts when $v(t) = r\omega(t)$,

i.e., $$v(t) = \frac{v_0}{2} + \frac{5f}{2M}t \qquad \text{(ii)}$$

Eliminating t from Eqs. (i) and (ii), we get

$$\frac{5}{2}v(t) + v(t) = \frac{5}{2}v_0 + \frac{v_0}{2}$$

or $$v(t) = \frac{2}{7} \times 3v_0 = \frac{6}{7}v_0$$

Thus, the sphere rolls with translational velocity $6v_0/7$ in the forward direction.

Alternative method:

We can solve this problem by using conservation of angular momentum. We can conserve the angular momentum about any point situated on the ground. (As torque of friction force about any point on the ground is zero.)

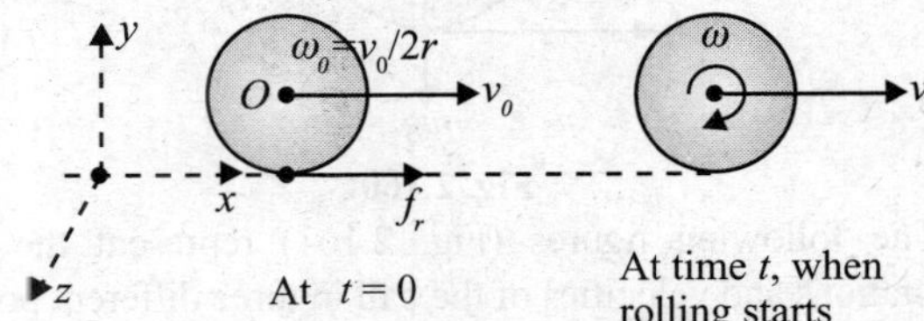

Fig. 2.159

Angular momentum about any point O situated on ground at $t = 0$

$$\vec{L}_{\text{initial}} = \vec{L}_{CM} + \vec{r} \times m\vec{v}$$

$$= I_{CM}\omega_0(-\hat{k}) + Mv_0 r(-\hat{k})$$

$$= \left[\frac{2}{5}Mr^2\left(\frac{v_0}{2r}\right) + Mv_0 r\right](-\hat{k})$$

$$= \frac{6}{5}Mv_0 r(-\hat{k}) \qquad \text{(i)}$$

Angular momentum at time t when rolling starts

$$\vec{L}_{\text{final}} = \vec{L}_{cm} + \vec{r} \times m\vec{v} = \frac{2}{5}mr^2\omega(-\hat{k}) + mvr(-\hat{k})$$

When rolling starts $v = \omega R$ or $\omega = v/r$

Hence, $\vec{L}_{\text{final}} = \left[\frac{2}{5}mr^2\left(\frac{v}{r}\right) + mvr\right](-\hat{k})$

$$= \frac{7}{5}mvr(-\hat{k}) \quad \text{(ii)}$$

Conserving angular momentum, we get

$$\vec{L}_{\text{initial}} = \vec{L}_{\text{final}}$$

$$\frac{6}{5}mv_0 r(-\hat{k}) = \frac{7}{5}mvr(-\hat{k})$$

which gives $v = \frac{6}{7}v_0$

Illustration 2.49 **A ball (solid sphere) is thrown down the alley in such a way that it slides with a speed v_0 initially without rolling. Prove that it will roll without any sliding when its speed falls to $(5/7)v_0$. The transition from pure sliding to pure rolling is gradual, so that both sliding and rolling take place during this interval.**

Sol. 1. Method 1: Force/Torque Method

The forces acting on the ball are as shown in Fig. 2.160. Let the acceleration of the centre of mass be a_{CM} and its angular acceleration about the centre of mass axis be α as shown in the figure.

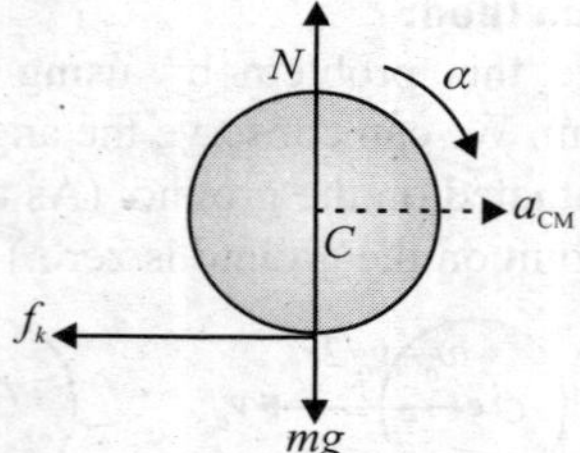

Fig. 2.160

The following figures (Fig. 2.161) represent the forces, accelerations and velocities of the ball in three different positions.

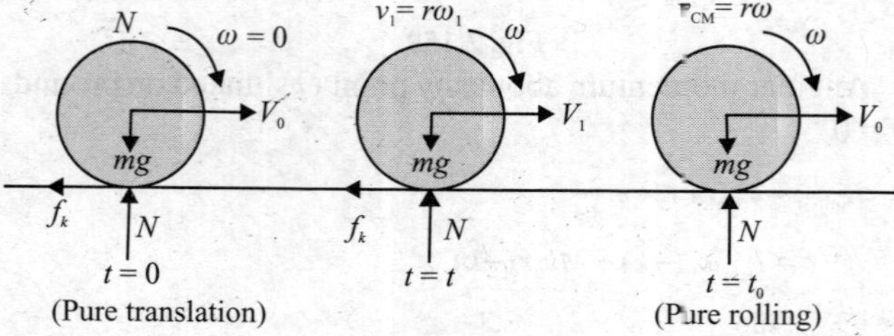

Fig. 2.161

Force equations: $N - mg = 0$ (in the vertical direction)

$$-f_k = ma_{\text{CM}} \text{ (in the horizontal direction)}$$

$$\Rightarrow \quad a_{\text{CM}} = -\frac{f_k}{m} \quad \text{(i)}$$

And from $\tau_{\text{CM}} = I_{\text{CM}}\alpha$

$$\Rightarrow \quad f_k r = \left(\frac{2}{5}mr^2\right)\alpha$$

$$\Rightarrow \quad \alpha = \left(\frac{f_k r}{\frac{2}{5}mr^2}\right) \quad \text{(ii)}$$

Let the velocity of the centre of mass be v_{CM} and angular velocity about the centre of mass axis be ω when rolling starts.

$$v_{\text{CM}} = v_0 + \left(\frac{-f_k}{m}\right)t \quad \text{(using } v = u + at\text{)} \quad \text{(iii)}$$

and $$\omega = 0 + \left(\frac{f_k r}{\frac{2}{5}mr^2}\right)t \quad (\omega = \omega_0 + at) \quad \text{(iv)}$$

Also $$v_{\text{CM}} = r\omega$$

Solving Eqs. (iii), (iv) and (v) for v_{CM}, we get

$$v_{\text{CM}} = \frac{5}{7}v_0$$

2. Work–Energy Theorem

Apply the work–energy theorem for the translation and rotational components of the motion separately.

1. $\sum W_{\text{force}} = \Delta K_{\text{translational}}$
2. $\sum W_{\text{torque}} = \Delta K_{\text{rotational}}$

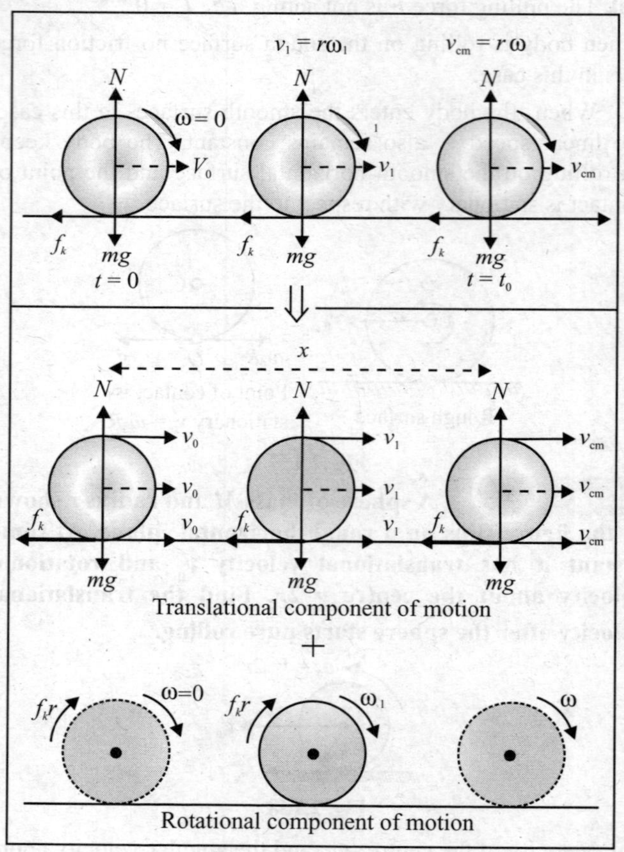

Fig. 2.162

Mathematically, $W_{f_k} + W_N + W_{mg} = (K_2 - K_2)_{\text{translational}}$

$$\Rightarrow \quad -f_k x + 0 + 0 = \frac{1}{2}mv^2 - \frac{1}{2}mv_0^2$$

$$\Rightarrow \quad -f_k(v_{av}t) = \frac{1}{2}(v^2 - v_0^2) \qquad [\because x = v_{av}t]$$

$$\Rightarrow \quad -f_k\left(\frac{v+v_0}{2}\right)t = \frac{1}{2}m(v-v_0)(v+v_0) \qquad \text{(i)}$$

(In case of constant acceleration only, $v_{av} = \frac{v_i + v_f}{2}$)

And for the rotational component of motion,

$$f_k r\theta = \frac{1}{2}I_{CM}\omega^2 - 0$$

$$\Rightarrow \quad f_k r(\omega_{av}t) = \frac{1}{2}I_{CM}\omega^2$$

$$\Rightarrow \quad f_k r\left(\frac{0+\omega}{2}\right)t = \frac{1}{2}\times\frac{2}{5}mr^2\times\omega^2$$

[In case of constant angular acceleration, $\omega_{av} = \frac{\omega_i + \omega_f}{2}$]

$$\Rightarrow \quad f_k r\frac{\omega}{2}t = \frac{1}{2}\times\frac{2}{5}mr^2\omega^2 \qquad \text{(ii)}$$

From Eqs. (i) and (ii), we get

$$\frac{-f_k\frac{v+v_0}{2}t}{f_k r\frac{\omega}{2}t} = \frac{\frac{1}{2}m(v+v_0)(v-v_0)}{\frac{1}{2}\times\frac{2}{5}mr^2\omega^2}$$

$$\Rightarrow \quad -1 = \frac{v-v_0}{\frac{2}{5}\omega r}$$

But $\quad r\omega = v,$

$$-1 = \frac{v-v_0}{\frac{2}{5}\omega r}$$

$$\Rightarrow \quad -\frac{2}{5}v = v - v_0$$

$$\Rightarrow \quad v_0 = v + \frac{2}{5}v = \frac{7}{5}v$$

$$\Rightarrow \quad v = \frac{5}{7}v_0$$

Method 3: By Conservation of Angular Momentum

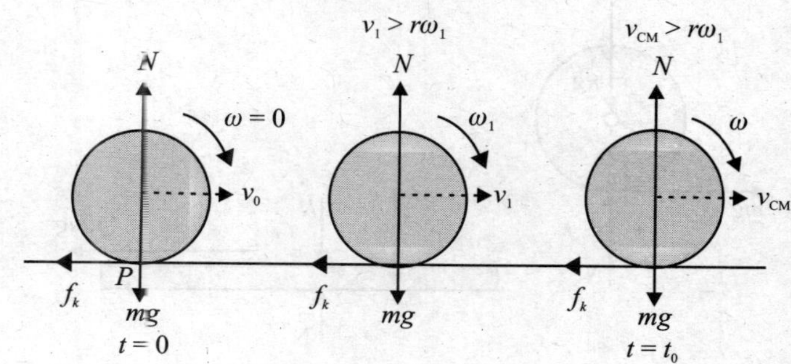

Fig. 2.163

If $\vec{\tau} = 0$, $\frac{d\vec{L}}{dt} = 0$, L = constant

The torque acting on the ball for all its positions, about point P (a point on the ground as shown in Fig. 2.163) is zero. So the angular momentum of the ball about point P will remain constant.

$$\Rightarrow \quad mv_0 r = I_{CM}\,\omega + mvr$$

$$\Rightarrow \quad mv_0 r = \frac{2}{5}mr^2\left(\frac{v}{r}\right) + mvr \qquad \left(\omega = \left(\frac{v}{r}\right)\right)$$

$$\Rightarrow \quad v_0 = \frac{2}{5}v + v$$

$$\Rightarrow \quad v = \frac{5}{7}v_0$$

Illustration 2.50 **A plank of mass M, whose top surface is rough with coefficient of friction μ is placed on a smooth ground (Fig. 2.164). Now a disc of mass $m = M/2$ and radius r is placed on the plank. The disc is now given a velocity V_0 in the horizontal direction, at $t = 0$.**

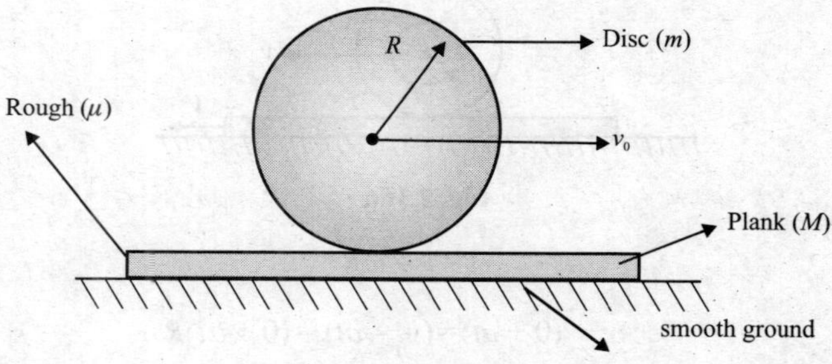

Fig. 2.164

1. **Find the time when the disc starts rolling.**
2. **Find the velocity of the plank and the disc up to that time.**
3. **Find the distance travelled by the plank up to this instant.**
4. **Find the work done by the friction force up to this instant.**

Sol. Step 1: Free-body diagram,

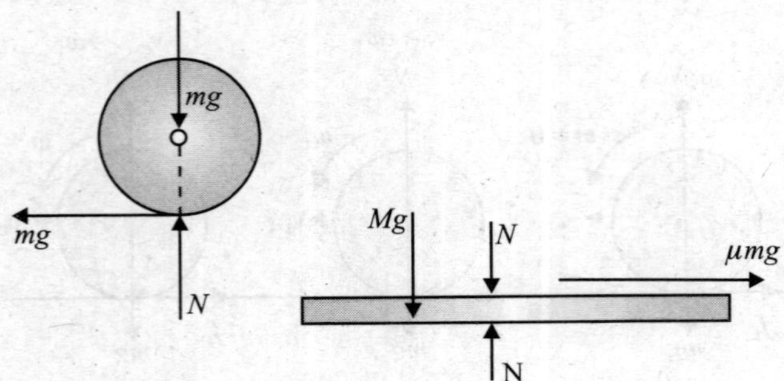

Fig. 2.165

Step 2: Equation of motion

For the plank: $\mu mg = MA$

$\Rightarrow A = \dfrac{\mu mg}{M}$ (towards right) (i)

$\Rightarrow A = \dfrac{\mu g}{2}$ (as $m = M/2$) (will be positive)

For the disc: $\mu mg = ma$

$a = \mu g$ (backwards, towards left)

(will be negative)

Step 3: Torque equation

$$\mu mgR = I\alpha$$

$$\mu mgR = \frac{mR^2}{2}\alpha$$

$$\alpha = \frac{2\mu g}{R} \text{ (clockwise)}$$

(will be positive)

Step 4: Constraint relations

At the time of rolling, point P should not slide with respect to the plank.

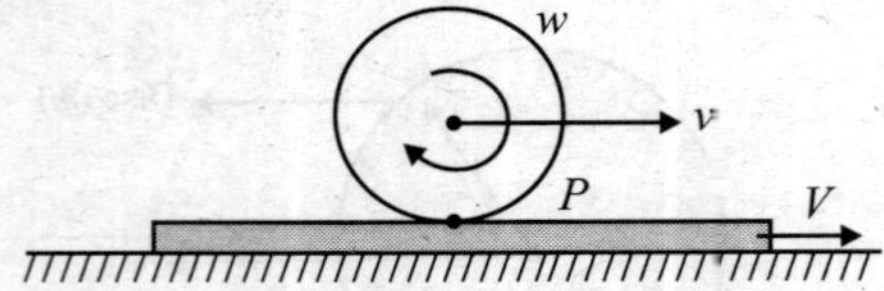

Fig. 2.166

$$V = v - \omega R$$

$$(0 + at) = (v_0 - at) - (0 + \alpha t)R$$

Now substituting the values of A and a, we can calculate

$$\frac{\mu mg}{M}t = (v_0 - \mu gt) - \left(\frac{2\mu g}{R}t\right)R$$

$$\left(\frac{\mu mg}{M} + \mu g + 2\mu g\right)t = v_0$$

which gives $t = \dfrac{v_0}{\mu g\left(\dfrac{m}{M} + 3\right)} = \dfrac{v_0}{5\mu g}$ (as $m = M/2$)

Velocity of the plank when rolling starts,

$$V = At = \left(\frac{\mu g}{2}\right)\left(\frac{v_0}{5\mu g}\right) = \frac{v_0}{10}$$

Velocity of the disc when rolling starts,

$$v = v_0 + at$$

$$= v_0 + (-\mu g)\left(\frac{v_0}{5\mu g}\right) = \frac{4}{5}v_0$$

Angular velocity at the time of rolling,

$$\omega = \alpha t = \left(\frac{2\mu g}{R}\right)\left(\frac{v_0}{5\mu g}\right) = \frac{2}{5}\frac{v_0}{R}$$

Distance travelled by the plank,

$$S_{plank} = \frac{1}{2}At^2 = \frac{1}{2}\left(\frac{\mu g}{2}\right)\left(\frac{v_0}{5\mu g}\right)^2 = \frac{v_0^2}{100\mu g}$$

Work done by friction force,

$$W_{friction} = \Delta K_{translation} + \Delta K_{rotation}$$

$$= [\Delta K_{disc} + \Delta K_{plank}]_{translation} + \Delta K_{rotation}$$

$$= \left[\left(\frac{1}{2}mv^2 - \frac{1}{2}mv_0^2\right) + \left(\frac{1}{2}MV^2 - 0\right) + \left(\frac{1}{2}I\omega^2 - 0\right)\right]$$

Substituting the values of v, V and ω, we get the work done by friction, i.e.,

$$W_{friction} = -\frac{13}{200}Mv_0^2$$

Illustration 2.51 **A uniform cylinder (mass M) of radius R is kept on an accelerating platform (mass M) as shown in Fig. 2.167. If the cylinder rolls without slipping on the platform, determine the magnitude of acceleration of the centre of mass of the cylinder. Assuming the coefficient of friction μ = 0.40, determine the maximum acceleration the platform may have without slip between the cylinder and the platform.**

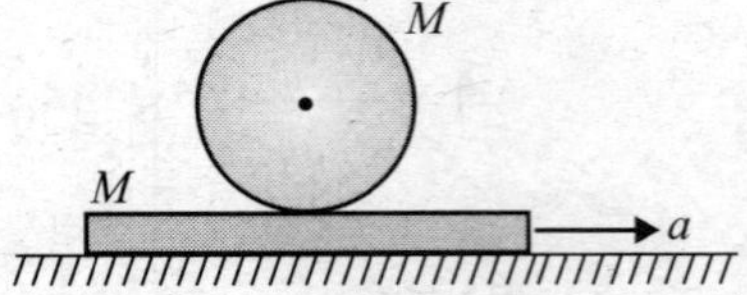

Fig. 2.167

Sol. Figure 2.168 shows the free-body diagram of the cylinder. We have assumed the angular acceleration of the cylinder to be clockwise. In case of pure rolling the contact point of cylinder with surface does not slide. Therefore, the acceleration of the point P must be equal to that of the platform.

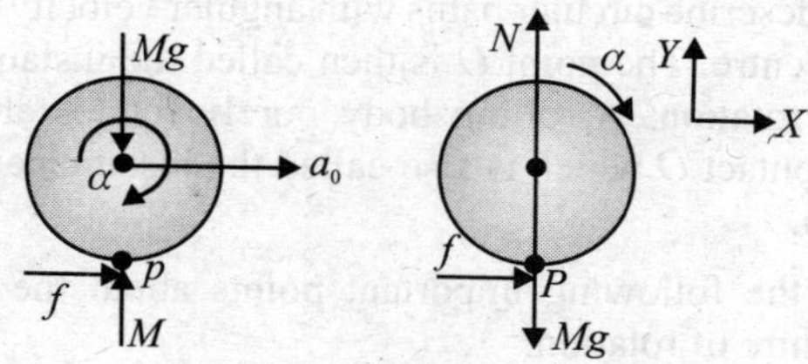

Fig. 2.168

But $\qquad a = a_0 - R\alpha \qquad$ (i)

which yields $a_0 = R\alpha + a$

The equations of motion are

$$\Sigma F_x = f = (M)\, a_0 \qquad \text{(ii)}$$

$$\Sigma \tau = -f \times R = \left(\frac{1}{2}\ MR^2\right)\alpha \qquad \text{(iii)}$$

Negative sign shows that the torque of friction is opposite to the assumed α.

From Eqs. (i) and (ii), we get

$$f = M\,(R\alpha + a) \qquad \text{(iv)}$$

Substituting the expression for f in Eq. (ii), we obtain

$$R\alpha = -\frac{2}{3}\,a$$

From Eq. (i), $\qquad a_0 = -\frac{2}{3}\,a + a = \frac{1}{3}\,a$

The maximum value of frictional force $f_{max} = \mu N$, where $N = Mg$, from Eq. (ii),

$$0.40N = M\left(\frac{1}{3}\,a\right)$$

$$0.40Mg = M\left(\frac{1}{3}\,a\right)$$

$$a = (0.40)\ (3g) = 12\ \text{m/s}^2$$

which is the maximum acceleration the platform may have without slipping between the cylinder and the platform.

INSTANTANEOUS AXIS OF ROTATION

Consider a rod AB which always remains in the same vertical plane with its ends in contact with smooth wall and smooth floor as shown in Fig. 2.169. As the rod slides, its centre of mass translates and the rod rotates about an axis passing through centre of mass and perpendicular to it. The motion of such an object can be analysed by making equations for translation and rotation separately.

Motion of such an object may be looked as pure rotation about a point which has zero velocity to simplify the study. Such a point is called instantaneous centre of rotatsion and the axis passing through this point and perpendicular to the plane of motion is called instantaneous axis of rotation (IAOR).

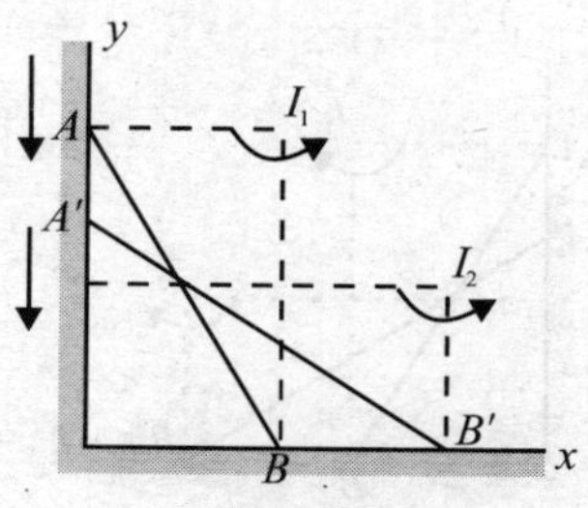

Fig. 2.169

In Fig. 2.169, I_1 and I_2 are the instantaneous centres of rotation of rod at two different instants.

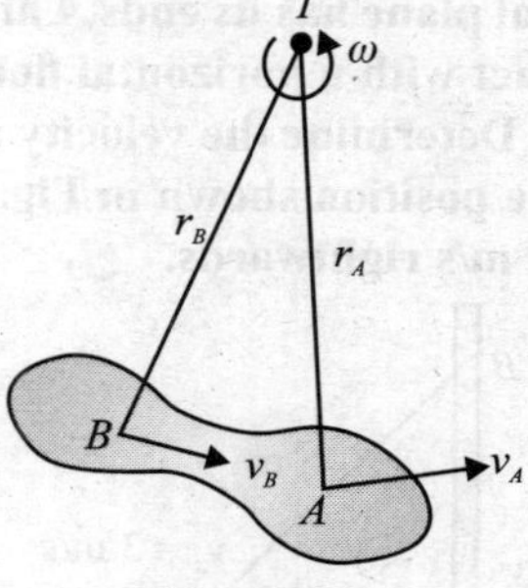

Fig. 2.170

To understand instantaneous centre, consider a rigid body which has plane motion. Let A be a point having velocity v_A at the instant considered. Now locate a point I on perpendicular to the direction of v_A at a distance r_A. The motion of A can be split into translation of I and rotation about I. Thus, we can write

$$v_A = v_I + \omega r_A$$

If we take $r_A = v_A/\omega$, then

$$v_A = v_I + \omega\left(\frac{v_A}{\omega}\right)$$

$$\therefore \qquad v_I = 0$$

Thus, point 'I' is selected at a distance v_A/ω along the perpendicular to the direction of A, the plane motion of point A can be reduced to pure rotation about I. Hence, I is the instantaneous centre of rotation.

If B is any other point on the rigid body, then its velocity will be given by

$$v_B = v_I + \omega r_B$$

or $\qquad v_B = \omega r_B \ (v_I = 0)$

Method of Locating Instantaneous Centre

Instantaneous centre can be located by any of the following two methods:

1. If the angular velocity ω and linear velocity v_A are known, instantaneous centre can be located at a distance (v_A/ω) along the perpendicular to the direction of v_A at A as discussed earlier.
2. If the linear velocities of two points of a rigid body are known, say v_A and v_B, drop a perpendiculars to them at A and B. The intersection point is the instantaneous centre of rotation (see Fig. 2.171)

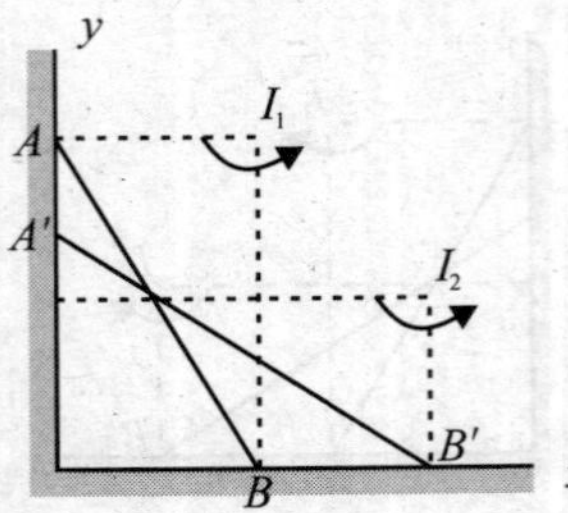

Fig. 2.171

Illustration 2.52 **A rod *AB* of length 5 m which remains in the same vertical plane has its ends *A* and *B* constrained to remain in contact with a horizontal floor and a vertical wall, respectively. Determine the velocity and acceleration of the end *B* at the position shown in Fig. 2.172 if point *A* has a velocity of 3 m/s rightwards.**

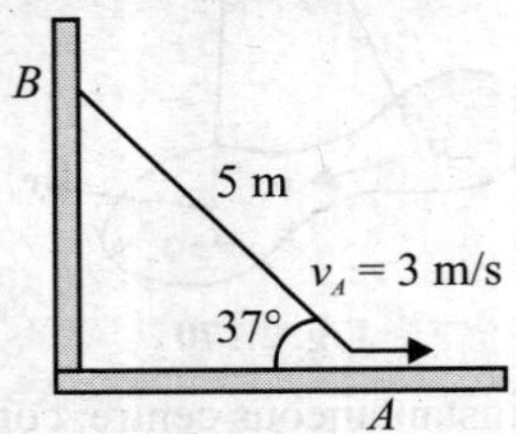

Fig. 2.172

Sol.

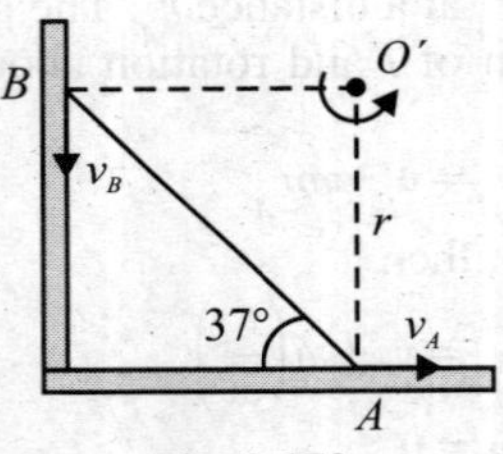

Fig. 2.173

Velocity of end A, $\vec{v}_A$, is horizontal, while velocity of the end B, $\vec{v}_B$, is vertical downwards. For finding the position of instantaneous centre of rotation, drop perpendiculars to the directions of $\vec{v}_A$ and $\vec{v}_B$ at points A and B, respectively. The intersection point O' is the instantaneous centre of rotation.

For end A,

$$V_A = \omega(O'A)$$

$$\therefore \quad \omega = \frac{V_A}{OA} = \frac{2}{5 \sin 37^\circ} = \frac{3}{8} = 1 \text{ rad/s}$$

For end B,

$$V_B = \omega(O'B) \times 1 \times 5 \times \cos 37^\circ = 4 \text{ m/s}$$

Instantaneous Centre of Rotation in Case of Rolling Body

Since a rolling body does not slip, therefore the point of contact O with the surface is instantaneously at rest and the ring momentarily rotates about this point. All the particles appear to describe circular paths with angular velocity ω about O as the centre. The point O is then called the instantaneous centre of rotation. A rolling body purely rotates about the point of contact O which is also called the instantaneous axis of rotation.

Note the following important points about the instantaneous centre of rotation:

- It is stationary with respect to ground.
- It may lie within or outside the body.
- The angular velocity of the body about this point is the same as that about its centre of mass.

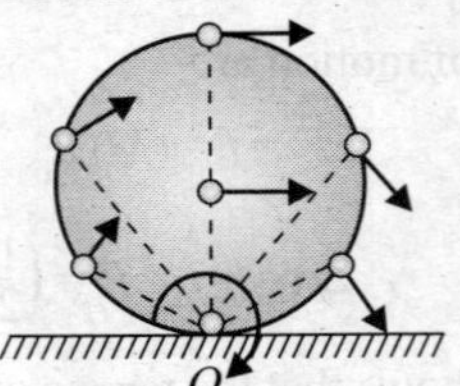

Fig. 2.174

- The magnitude of linear velocity of any point is given by the product of angular velocity ω with the distance of that point from the point of contact. The direction of the velocity vector is perpendicular to the line joining the point and the point of contact.

For an object performing pure rolling motion, the point of contact is at rest (relative to the surface), so in this case instantaneous axis of rotation will pass through the point of contact.

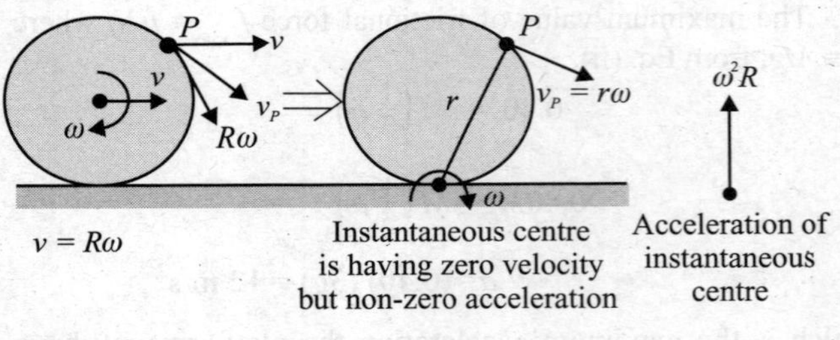

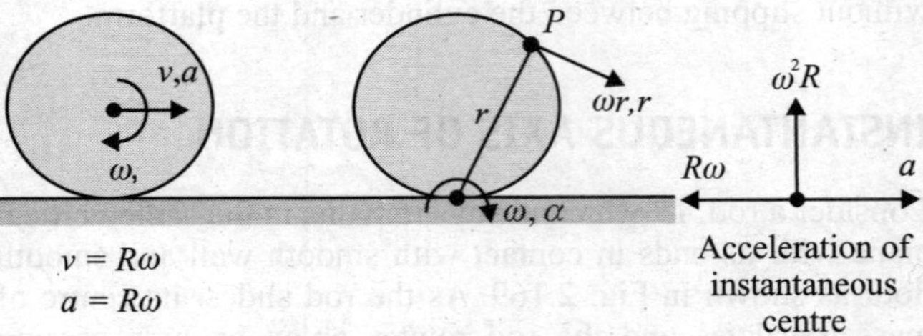

Fig. 2.175

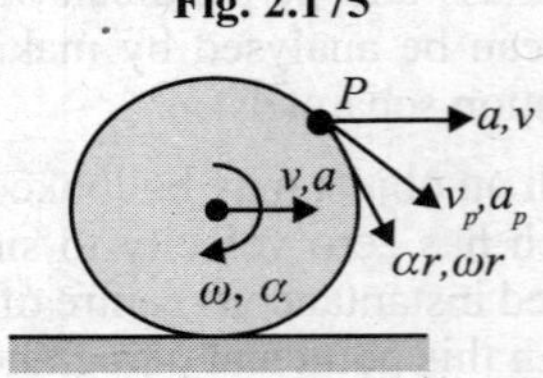

Fig. 2.176

It is not necessary that in every plane motion of a body, the instantaneous axis of rotation is passing through the point of contact. The basic point is that instantaneous axis of rotation passes through the point of zero velocity.

DIRECTION OF FRICTION IN CASE OF TRANSLATION AND ROTATION COMBINED

In case of translation motion we can decide the direction of friction force by direct observation. But the direction of friction cannot be found by direct observation in case of rotational motion, as the body is translating as well as rotating. The direction of friction force is determined after deciding the motion tendency of the point of contact of the body under consideration with the ground.

Let us take an example of analysis of direction of friction in this case.

In the following diagram (Fig. 2.177), a rolling object of mass M and radius R is placed on a rough horizontal surface. A force F is applied as shown.

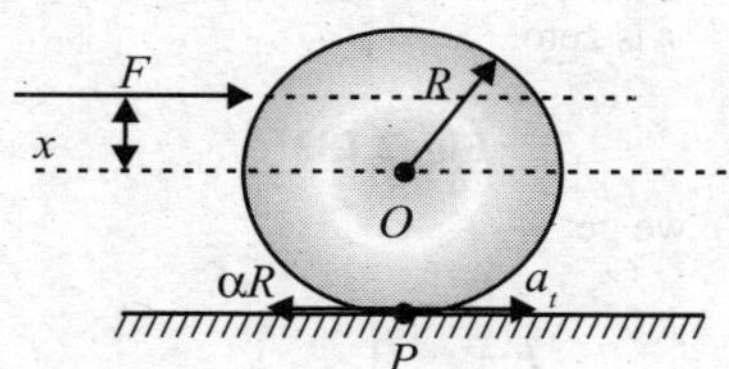

Fig. 2.177

To find the direction of friction at P, let us consider the surface to be frictionless.

Acceleration of point P due to translation only,

$$a_t = \frac{F}{M} (\rightarrow)$$

Acceleration of point P due to rotation only,

$$a_r = \alpha R = \frac{\tau}{I} R \Rightarrow \frac{FR}{I} x (\leftarrow)$$

Net acceleration of point P is $\vec{a}_P = \vec{a}_t + \vec{a}_r$

$$a_p = \frac{F}{M} - \frac{FRx}{I} (\rightarrow) \quad \text{(i)}$$

From Eq. (ii), it is clear that motion tendency at point P depends upon both x and I.

Equation (i) can be written as

$$a_p = \frac{F}{M}\left(1 - \frac{Rx}{K^2}\right) \quad \text{(ii)}$$

The motion tendency of point P will be in forward direction if

$$1 - \frac{Rx}{K^2} > 0 \Rightarrow 1 > \frac{Rx}{K^2} \Rightarrow K^2 > Rx \quad \text{(iii)}$$

If this condition is satisfied, friction will act in backward direction.

We can summerise the situation as

If $K^2 > Rx$: friction will act in backward direction.
If $K^2 = Rx$: no friction will act.

If $K^2 < Rx$: friction will act in forward direction.

If force acts in lower diametric plane, friction will act in backward direction only.

Illustration 2.53 **Draw the direction of friction force in the following cases. If rolling object is**

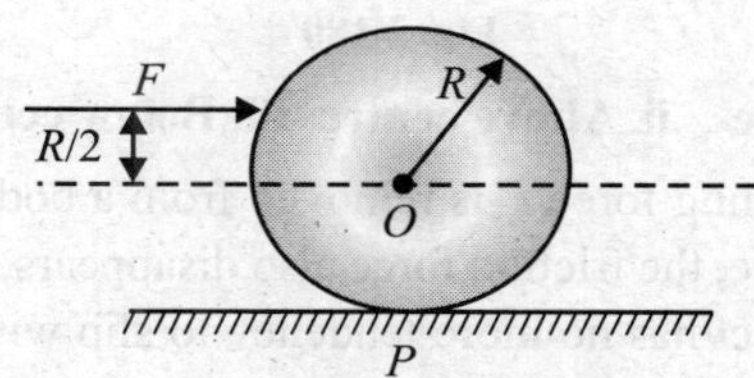

Fig. 2.178

i. a ring **ii. a disc**
iii. a solid sphere **iv. a hollow sphere**

Sol.

Case I: $K^2 = R^2 \Rightarrow Rx = R\frac{R}{2}$

As $K^2 > Rx$,

The friction will act in the backward direction

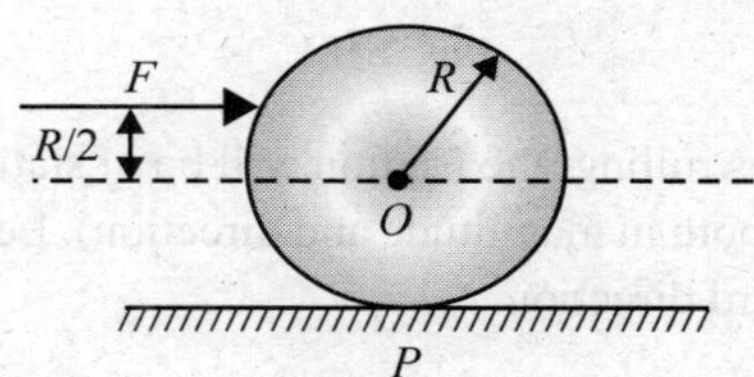

Fig. 2.179

Case ii: $K^2 = \frac{R^2}{2} \Rightarrow Rx = R\frac{R}{2} = \frac{R^2}{2}$

$\Rightarrow$ $K^2 = Rx$

No friction will act.

Case iii: $K^2 = \frac{2}{5}R^2 \Rightarrow Rx = R\frac{R}{2} = \frac{R^2}{2}$

$\Rightarrow$ $K^2 < Rx$

The friction will act in the forward direction.

Case iv: $K^2 = \frac{2}{3}R^2 \Rightarrow Rx = R\frac{R}{2} = \frac{R^2}{2}$

$\Rightarrow$ $K^2 > Rx$

The friction will act in the forward direction.

Illustration 2.54 **A force of magnitude F is acting on a rolling body of mass m and radius R as shown in**

Fig. 2.180. What happens if the pulling force F is removed? Discuss the different possibilities when the pulling force acts on the body.

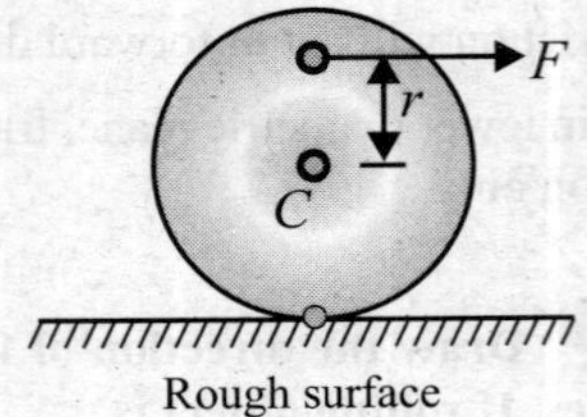

Fig. 2.180

i. At centre ii. Above centre iii. Below centre

Sol. If the pulling force F is removed from a body rolling on a rough surface, the friction force also disappears, because the point of contact has no more tendency to slip with respect to the surface. The linear and angular speeds of the body remain constant and the body keeps on rolling.

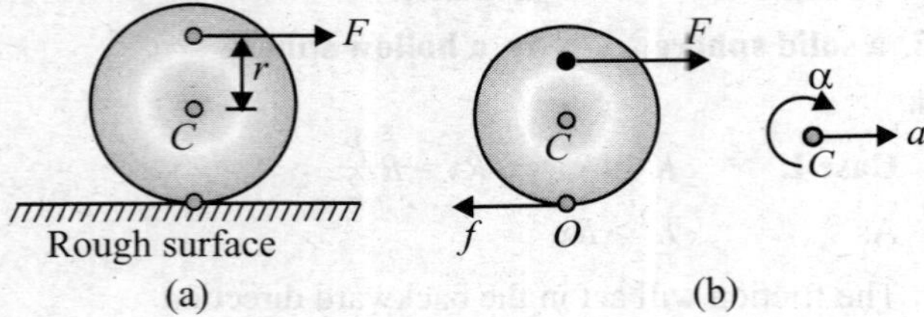

Fig. 2.181

As body is rolling, the friction will be of static nature and Q-unknown (both in magnitude and direction). Let friction act in the backward direction.

Equation of motion: $F - f = ma_c$ (i)

Here we can apply torque equation about centre of mass as well as about point of contact with ground (instantaneous centre of rotation)

About centre of mass: $Fx - fR = I_c\alpha$ (ii)

About point of contact 'O': $F(R + r) = I_c\alpha$

$$\Rightarrow \quad F(R + r) = (I_c + MR^2)\alpha \quad \text{(iii)}$$

Taking torque equation about O, we can directly calculate the value of α.

$$\text{Hence,} \quad \alpha = \frac{F(R+r)}{(I_c + MR^2)} \quad \text{(iv)}$$

$$\text{As the body is rolling, } a_c = \alpha R = \frac{F(R+r)R}{(I_c + MR^2)}$$

$$\text{or} \quad a_c = \frac{F}{M}\left[\frac{\left(1+\frac{r}{R}\right)}{\left(1+\frac{I_C}{MR^2}\right)}\right] \quad \text{(v)}$$

Subsituting the value of a_c in Eq. (i), we get

$$\text{force of friction,} \quad f = F\left[\frac{I_C - MrR}{I_C + MR^2}\right] \quad \text{(vi)}$$

We can find the values of a_c and f by using Eqs. (i) and (ii). But by taking torque equation about point of contact, the calculation steps are reduced.

i. If force acts through C (Fig. 2.182), i.e., $r = 0$:

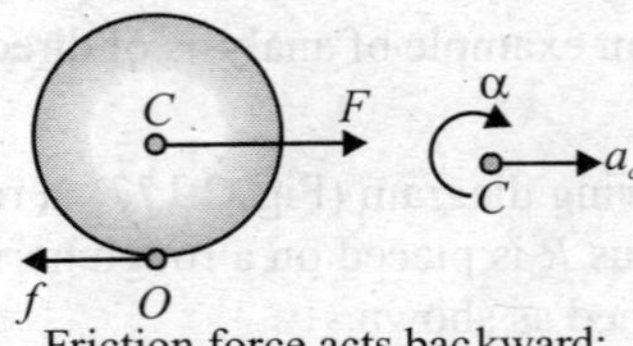

Friction force acts backward: r is zero.

Fig. 2.182

From Eq. (v), we get

$$a_C = \frac{F}{M}\left[\frac{1}{1+\frac{I_C}{MR^2}}\right]$$

From Eq. (vi), we get

$$f = F\left[\frac{1}{1+\frac{MR^2}{I_C}}\right]$$

From the values of a_c and f, it is clear that

- body moves forward
- friction acts backward

ii. Force acts above C (Fig. 2.182), i.e., r is positive:

From Eqs. (v) and (vi), it is clear that

- the body moves forward
- rotation about the centre of mass is clockwise
- the friction force may act forward or backward

Backward, if $r < \frac{I_c}{MR}$

Forward, if $r > \frac{I_c}{MR}$

No friction force acts when $r = \frac{I_c}{MR}$

iii. Force acts below C, i.e., r is negative:

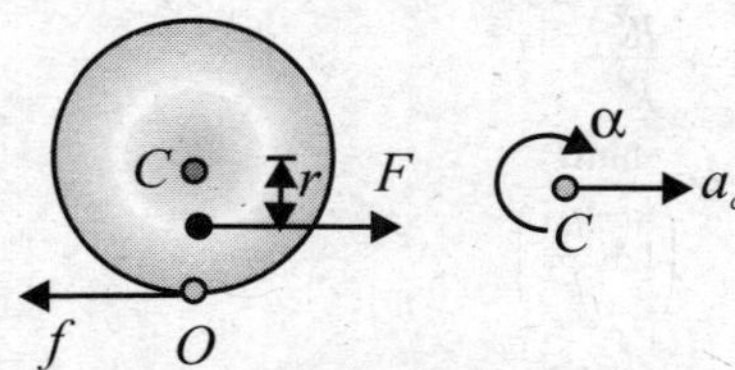

Friction force acts backward:
r is negative.

Fig. 2.183

- The body moves forward.
- Rotation about the centre of mass is clockwise.
- Friction force acts backward ($f < F$).

Illustration 2.55 **A force F acts tangentially at the highest point of a sphere of mass m kept on a rough horizontal plane. If the sphere rolls without slipping, find the acceleration of the centre of the sphere.**

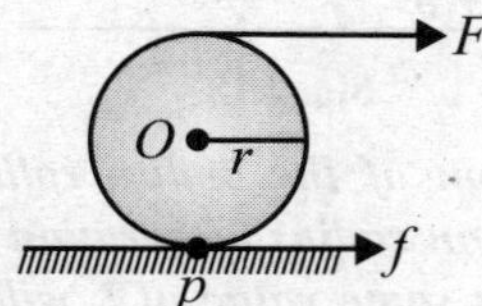

Fig. 2.184

Sol. The situation is shown in Fig. 2.184. As force F rotates the sphere, the point of contact has a tendency to slip towards left so that the static friction on the sphere will act towards right. Let r be the radius of the sphere and a be the linear acceleration of the centre of the sphere. The angular acceleration about the centre of the sphere is $\alpha = a/r$, as there is no slipping.

For the linear motion of the centre,

$$F + f = ma \quad \text{(i)}$$

and for the rotational motion about the centre,

$$Fr - fr = I\alpha = \left(\frac{2}{5}mr^2\right)\frac{a}{r}$$

or

$$F - f = \frac{2}{5}ma \quad \text{(ii)}$$

From Eqs. (i) and (ii), we get

$$2F = \frac{7}{5}ma \quad \text{or} \quad a = \frac{10F}{7m}$$

Alternative method:

As the sphere is rolling, we can apply torque equation about point of contact.

i.e., $F(2r) = I_p\alpha = \left(\frac{2}{5}mr^2 + mr^2\right)\alpha$

$\Rightarrow F(2r) = \left(\frac{7}{5}mr^2\right)\alpha$

Hence, angular acceleration $\alpha = \frac{10F}{7mr}$ (iii)

As sphere is rolling, acceleration of centre of mass is

$$a = \alpha r = \frac{10F}{7m}$$

Illustration 2.56 **A billiard ball (of radius R), initially at rest is given a sharp impulse by a cue. The cue is held horizontally at a distance h above the central line as shown Fig. 2.185. The ball leaves the cue with a speed V_0. It rolls and slides while moving forward and eventually acquires a final speed of 9/7 V_0. Show that h = 4/5 R.**

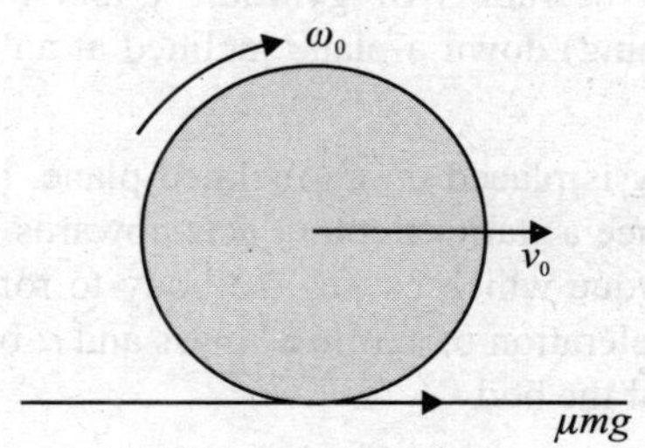

Fig. 2.185

Sol. Let v_0 be the linear velocity and ω_0 be the angular velocity imparted.

$$\text{Linear impulse} = \int F dt = mv_0$$

$$\text{Angular impulse} = \int F h\, dt = I\omega_0$$

$$\Rightarrow \quad h = \frac{\left(\frac{2}{5}mR^2\right)\omega_0}{mv_0}$$

$$\Rightarrow \quad v_0 = \frac{2R^2\omega_0}{5h} \quad \text{(i)}$$

As linear velocity increases to 9/7 v_0, friction must be in forward direction and hence it opposes angular motion.

For linear motion: Friction (μmg) increases the velocity. Hence, $\mu mg = ma$

$$\Rightarrow \quad a = \mu g$$

Let t be the time when rolling starts.

For rotation: Frictional torque (μmgR) opposes rotation and hence decreases ω.

$$\alpha = -\frac{\mu mgR}{I}$$

$$\Rightarrow \quad \omega = \omega_0 - \frac{\mu mgR}{\left(\frac{2}{5}mR^2\right)}t$$

After time t, $V = \frac{9}{7}v_0 = R\omega$

Combining all the equations, we get

$$\frac{9}{7}v_0 - v_0 = -\frac{9v_0}{7R} - \omega_0 \frac{2R}{5} \quad \text{(eliminating } \mu gt\text{)}$$

Substituting for v_0 from Eq. (i), we get

$$\frac{2}{7}\left(\frac{2}{5}\frac{R^2\omega_0}{h}\right) = \left(\omega_0 - \frac{9}{7}\frac{2R\omega_0}{5h}\right)\frac{2R}{5} \Rightarrow h = \frac{4R}{5}$$

ROLLING OF A BODY ON AN INCLINED PLANE

A rigid body of radius of gyration k and radius R rolls (without slipping) down a plane inclined at an angle θ with the horizontal.

When a body is placed on an inclined plane, it tries to slip down and hence a static friction f acts upwards. This friction provides a torque which causes the body to rotate. Let a be the linear acceleration of centre of mass and α be the angular acceleration of the body.

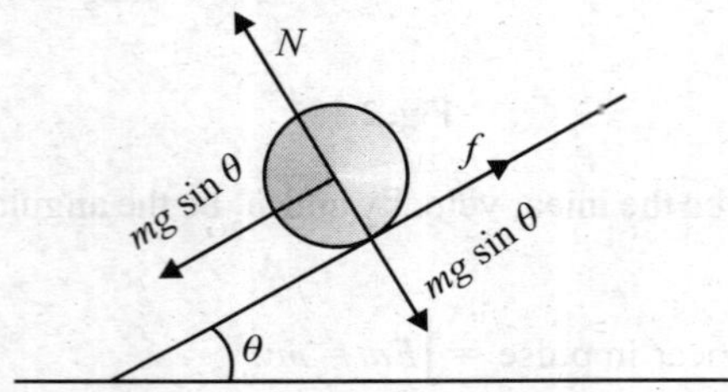

Fig. 2.186

From force diagram: For linear motion parallel to the plane,

$$mg \sin\theta - f = ma$$

For rotation around the axis through centre of mass, net torque = $I\alpha$

$$\Rightarrow \quad fR = (mk^2)\alpha$$

As there is no slipping, the point of contact of the body with the plane is instantaneously at rest.

$$\Rightarrow \quad v = R\omega \text{ and } a = R\alpha$$

Solving the following three equations for a and f,

$$mg \sin\theta - f = ma$$

$$\Rightarrow \quad fR = mk^2\alpha$$

$$a = R\alpha$$

$$a = \frac{g\sin\theta}{\left(1 + \frac{k^2}{R^2}\right)} \quad \text{and} \quad f = \frac{mg\sin\theta}{\left(1 + \frac{R^2}{k^2}\right)}$$

We can also derive the condition for pure rolling (rolling without slipping), to avoid slipping, $f \le \mu_s N$.

$$\frac{g\sin\theta}{1 + \frac{R^2}{k^2}} \le \mu_s mg\cos\theta$$

$$\Rightarrow \quad \mu_s \ge \frac{\tan\theta}{\left(1 + \frac{R^2}{k^2}\right)}$$

This is the condition of μ_s so that the body rolls without slipping.

Note:

- ***If a body rolls up an inclined plane the direction of friction force in this case also in upward direction. As body rolls up velocity of centre of mass decreases, hence torque of friction should decrease angular velocity.***

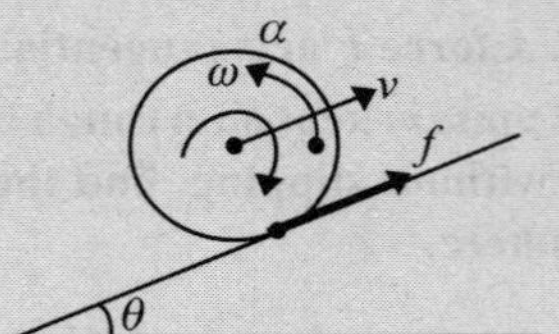

Fig. 2.187

- ***The acceleration of the bodies rolling on inclined plane depends on radius of gyration (k). We can say the bodies with same value of k will have the same acceleration.***
- ***For example, two uniform solid spheres made of different material, unequal mass and unequal radii are released from the same inclined plane if they rolls, they will have same accelerations.***
- ***If a solid sphere and a hollow sphere of the same mass and same radii are released, the hollow sphere will take a longer time and acquires less velocity.***
- ***If a solid sphere, a hollow sphere, a disc and a ring of same mass and radius are released from top of incline and they roll. The solid sphere take the least time while the ring will take maximum time to reach the bottom.***
- ***If all the four objects are released from the top of a smooth incline. All the four bodies will reach the bottom at the same time.***
- ***If all the rolling objects are released on a rough incline, such that friction is not sufficient for pure rolling, still the objects reach the bottom of incline at the same time because acceleration of centre of mass is the same in each case. Each object will have the same velocity of centre of mass but different ω.***

Illustration 2.57 **A solid cylinder rolls down an inclined plane of height h and inclination θ. Calculate its speed at the bottom of the plane using acceleration method and energy method. Also calculate the time taken to reach the bottom.**

Sol. Energy Method: Let v and ω be the velocity of the centre of mass and the angular velocity of cylinder, respectively, at the bottom of the plane.

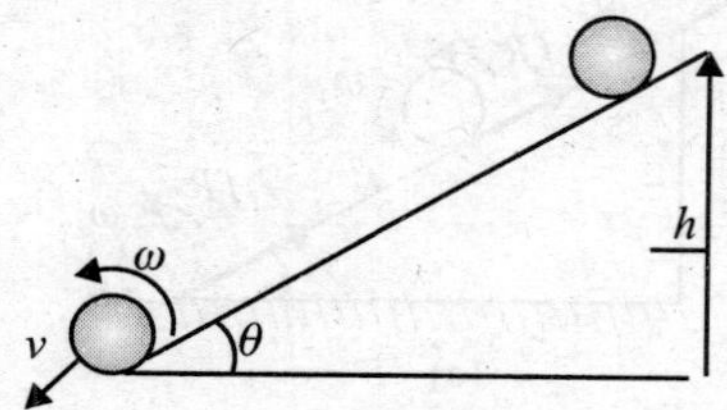

Fig. 2.188

As the cylinder rolls down:

Loss in G PE = Gain in translation KE + gain in rotational KE

$$mgh = \frac{1}{2}mv^2 + \frac{1}{2}\omega^2$$

As the cylinder is rolling without slipping, $v = R\omega$

$$\Rightarrow \quad mgh = \frac{1}{2}mv^2 + \frac{1}{2}\left(\frac{mR^2}{2}\right)\frac{v^2}{R^2}$$

$$\Rightarrow \quad v = \sqrt{\frac{4gh}{3}}$$

Force/Torque Method: From the result of the last example (using $k^2 = R^2/2$),

$$\text{Acceleration of the cylinder is } \frac{g \sin\theta}{1+\frac{1}{2}} = \frac{2}{3} g \sin\theta$$

Using $v^2 = u^2 + 2as$ down the plane (taking downward direction positive)

$$\Rightarrow \quad v = \sqrt{0 + 2a\frac{h}{\sin\theta}}$$

$$= \sqrt{2\left(\frac{2}{3}g\sin\theta\right)\frac{h}{\sin\theta}} = \sqrt{\frac{4gh}{3}}$$

$$\text{Time to reach bottom} = t = \frac{v - v}{a}$$

$$= \frac{\sqrt{\frac{4gh}{3}} - 0}{\frac{2}{3}g\sin\theta} = \sqrt{\frac{3h}{g}}\frac{1}{\sin\theta}$$

In the forthcoming illustration, we will learn the application of various methods, i.e., force/torque, conservation of mechanical energy, work–energy theorem and concept of angular momentum in problem solving.

Illustration 2.58 **Consider a uniform solid cylinder of mass m and radius r rolling (without slipping) down at an angle θ with the horizontal, as shown in Fig. 2.189. Find the speed of its centre of mass when the cylinder reaches the bottom.**

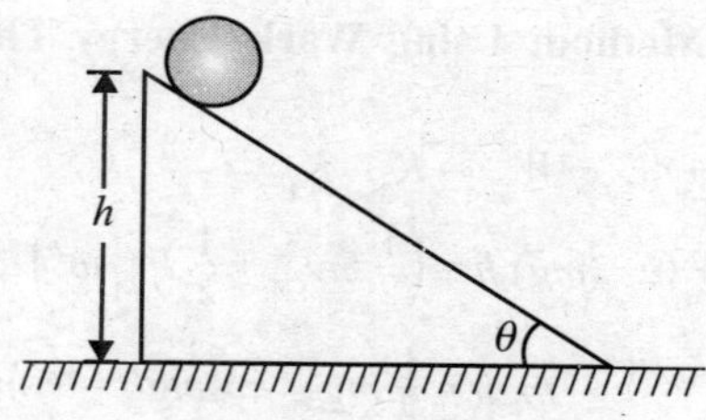

Fig. 2.189

Sol. i. First Method: Force/Torque Method

We can use the formula which we derived

$$a = \frac{g\sin\theta}{1+\frac{k^2}{R^2}} \quad \text{(i)}$$

$$\text{For cylinder, } I = \frac{mR^2}{2} \Rightarrow k^2 = \frac{R^2}{2}$$

Hence, acceleration of the cylinder $a = \frac{2}{3}g\sin\theta$

Let v_{CM} be the velocity of the centre of mass at the bottom.

From equation $\quad v^2 = u^2 + 2as,$

$$v_{CM}^2 = 0^2 + 2\left(\frac{2}{3}g\sin\theta\right)\left(\frac{h}{\sin\theta}\right) = \frac{4}{3}gh$$

$$\Rightarrow \quad v_{CM} = \sqrt{\frac{4}{3}gh}$$

ii. Second Method: By Mechanical Energy Conservation

When the non-conservative forces do not perform any work on a system ($W_{ncf} = 0$), the mechanical energy ($K + U$) of the system remains constant.

Here, $W_f = 0$ (because the velocity of the instantaneous point of contact is always zero)

$W_N = 0$ (because the velocity of the instantaneous point of contact is always zero)

Thus, $\quad W_{ncf} = 0$

The mechanical energy of the cylinder will remain constant.

$$(K+U)_{top} = (K+U)_{bottom}$$

$$\Rightarrow \quad 0 + mgh = \frac{1}{2}mv_{CM}^2 + \frac{1}{2}I_{CM}\omega^2 + 0$$

$$= \frac{1}{2}mv_{CM}^2 + \frac{1}{2}\left(\frac{1}{2}mr^2\right)\left(\frac{v_{CM}}{r}\right)^2$$

$$(v_{CM} = r\omega, \text{ for rolling})$$

$$= \frac{1}{2}mv_{CM}^2 + \frac{1}{4}mv_{CM}^2 = \frac{3}{4}mv_{CM}^2$$

$$\Rightarrow \quad v_{CM} = \sqrt{\frac{4}{3}gh}$$

iii. Third Method: Using Work–Energy Theorem ($\sum W = \Delta K$)

$$W_{f_s} + W_N + W_{mg} = K_2 - K_1$$

$$\Rightarrow \quad 0 + 0 + (mg)\,h = \left(\frac{1}{2} m v_{CM}^2 + \frac{1}{2} I_{CM}\,\omega^2\right)$$

$$\Rightarrow \quad mgh = \frac{1}{2} m v_{CM}^2 \Rightarrow mgh = \frac{3}{4} m v_{CM}^2$$

$$\Rightarrow \quad v_{CM} = \sqrt{\frac{4}{3} gh}$$

iv. Fourth Method: Using Work–Energy Theroem Separately in Translational Motion and Rotational Motion.

The rolling motion of the cylinder is obtained by superimposing the translational and rotational motions: translational motion of the centre of mass and rotational motion about an axis passing through the centre of mass, cylinder's own axis.

You can apply the work–energy theorem for the translational and rotational components of the motion separately as under:

$$\sum W_{force} = \Delta K_{translation}$$

and $$\sum W_{torque} = \Delta K_{rotational}$$

The two forces acting at the point of contact, f_s and N can be translated to the centre of mass after applying an equivalent torque as shown in Fig. 2.190.

For the translational part of the motion,

$$W_{f_s} + W_N + W_{mg} = K_2 - K_1$$

$$\Rightarrow \quad -f_s\left(\frac{h}{\sin\theta}\right) + 0 + (mg)\,h = \frac{1}{2} m v_{CM}^2 - 0$$

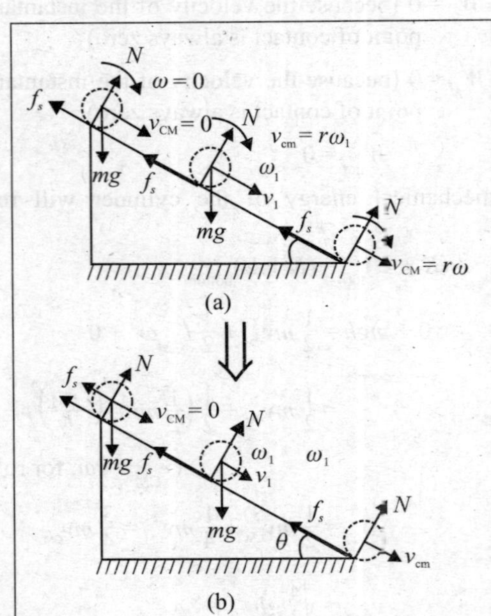

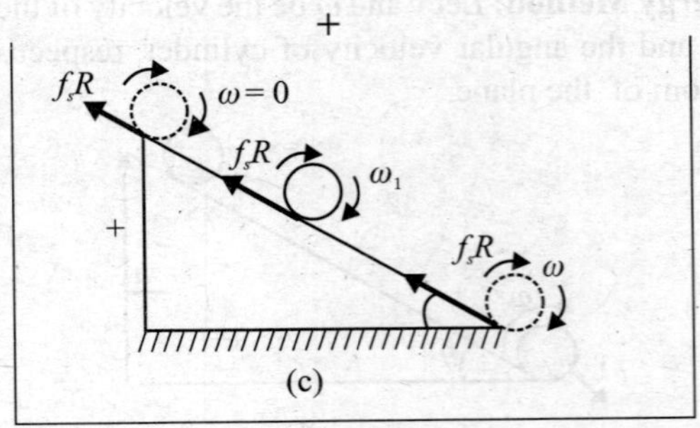

Fig. 2.190

For the rotational part,

$$\text{Torque} \times \text{Angular displacement} = \frac{1}{2} I_{CM}\,\omega^2 - 0$$

$$\Rightarrow \quad (f_s r)\left(\frac{h/\sin\theta}{r}\right) = \frac{1}{2} I_{CM}\,\omega^2$$

($W\tau = \tau\theta$ if τ is constant, and θ_1 and τ are in the same sense of rotation, where θ_1 is the angle through which the cylinder rotates about the centre of mass axis. Here, for rolling, $v_{CM} = r\omega$, $a_{CM} = r\alpha$ and $s = r\theta_1$.)

Thus, $$f_s\left(\frac{h}{\sin\theta}\right) = \frac{1}{2} I_{CM}\,\omega^2$$

From Eqs. (iii) and (iv), we get

$$mgh = \frac{1}{2} m v_{CM}^2 + \frac{1}{2} I_{CM}\,\omega_{CM}^2$$

$$= \frac{3}{4} m v_{CM}^2,\; I_{CM} = \frac{1}{2} mr^2$$

$$\Rightarrow \quad v_{CM} = \sqrt{\frac{4}{3} gh}$$

v. Fifth Method: Using the equation $\vec{\tau} = \frac{d\vec{L}}{dt}$

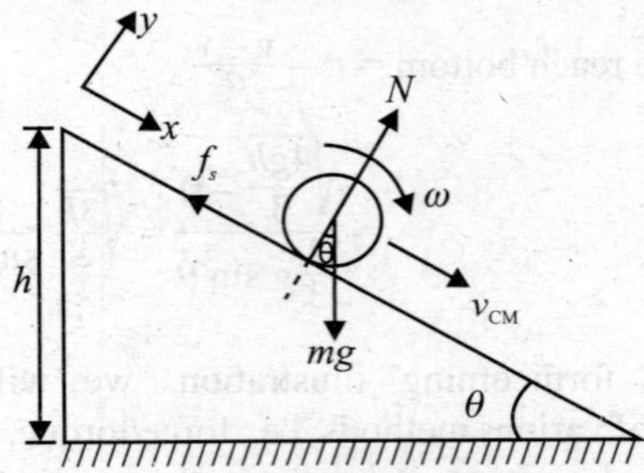

Fig. 2.191

About the instantaneous point of contact, the torque acting on the cylinder is

$$\vec{\tau} = f_s \times 0 + N \times 0 + (mg)\,(r \sin\theta)$$
$$= mg\,r \sin\theta\,(-\hat{k}) \text{ (clockwise)}$$

Angular momentum of the cylinder about the instantaneous point of contact in the clockwise sense,

$$\vec{L} = [I_{CM}\,\omega + mv_{CM}\,r](-\hat{k})$$

$$\left[\left(\frac{1}{2}mr^2\right)\left(\frac{v_{CM}}{r}\right) + mv_{CM}r\right](-\hat{k}) = \frac{3}{2}mrv_{CM}(-\hat{k})$$

Now, from the equation $\vec{\tau} = \frac{d\vec{L}}{dt}$;

$$mgr\sin\theta = \frac{d}{dt}\left(\frac{3}{2}mrv_{CM}\right)$$

(Here $\vec{\tau} = d\vec{L}/dt$ has been applied about the instantaneous point of contact.)

$$\Rightarrow \quad mgr\sin\theta = \frac{3}{2}mr\frac{dv_{CM}}{dt} = \frac{3}{2}mra_{CM} \qquad \text{(i)}$$

$$\Rightarrow \quad a_{CM} = \frac{3}{2}g\sin\theta$$

Now from the equation $v^2 = u^2 + 2as$, we get

$$v_{CM}^2 = 0^2 + 2\left(\frac{2}{3}g\sin\theta\right)\left(\frac{h}{\sin\theta}\right)$$

$$\Rightarrow \quad v_{CM} = \sqrt{\frac{4}{3}gh}$$

Alternative Method: You can also manipulate Eq. (i) as follows:

$$mgr\sin\theta = \frac{d}{dt}\left(\frac{3}{2}mrv_{CM}\right)$$

$$g\sin\theta = \frac{3}{2}\frac{dv_{CM}}{dt} = \frac{3}{2}\frac{dv_{CM}}{dx}\frac{dx}{dt} = \frac{3}{2}v_{CM}\left(\frac{dv_{CM}}{dx}\right)$$

$$\Rightarrow \quad g\sin\theta\,dx = \frac{3}{2}v_{CM}\,dv_{CM}$$

$$\int_0^{h/\sin\theta} g\sin\theta\,dx = \int_0^{v_{CM}} \frac{3}{2}v_{CM}\,dv_{CM}$$

$$\Rightarrow \quad gh = \frac{3}{4}v_{CM}^2 \Rightarrow v_{CM} = \sqrt{\frac{4}{3}gh}$$

Concept Application Exercise 2.5

1. Figure 2.192 (a) and (b) shows two different arrangements of a spool being pulled with a constant force F in each case. Identify the direction of motion of the centre of mass, and the direction of friction force in each case.

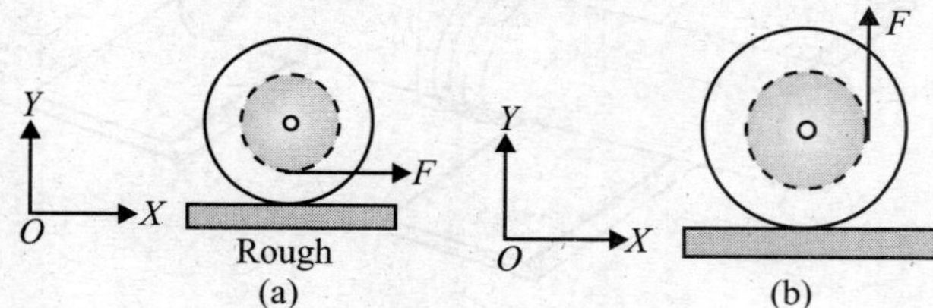

Fig. 2.192

i. The CM moves along.............................x-axis.

ii. The friction force acts along ....x-axis.

iii. The CM moves along..............................x-axis.

iv. The friction force acts along ....x-axis.

2. A sphere moving with a velocity v_0 on a smooth surface suddenly enters on a rough horizontal surface as shown Fig. 2.193. State which of the following statements are true or false

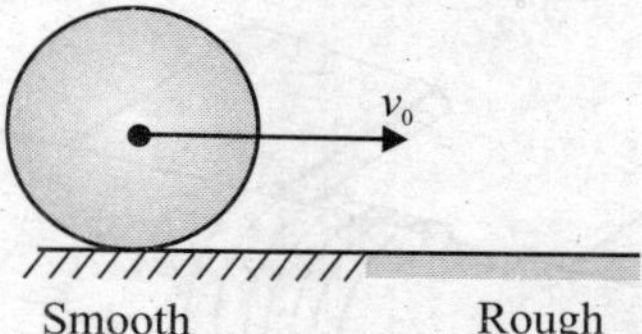

Fig. 2.193

i. The sphere loses translational kinetic energy and gains rotational kinetic energy.

ii. The total energy of the sphere is conserved.

iii. The angular momentum of the sphere about any point on the surface is conserved.

iv. The final velocity attained by the centre of mass is $2v_0/3$.

3. A cylinder of mass m and radius R rolls on a stationary plank of mass M. The lower surface of the plank is smooth and the upper surface is sufficiently rough with a coefficient of friction μ. A man is to hold the plank stationary with respect to the ground, as shown Fig. 2.194.

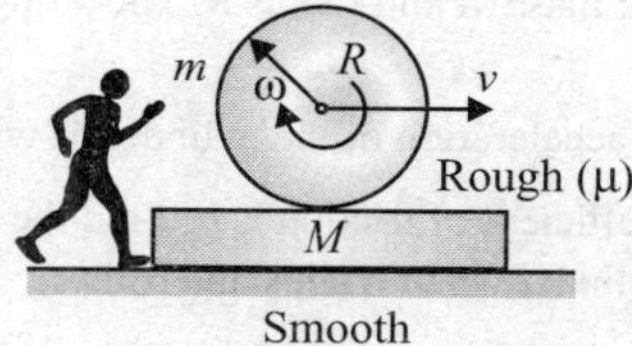

Fig. 2.194

The force exerted by the man to keep the plank stationary is equal to.......

4. Calculate the kinetic energy of a tractor crawler belt of mass m if the tractor moves with a velocity v. There is no slipping. Neglect the size of the wheels.

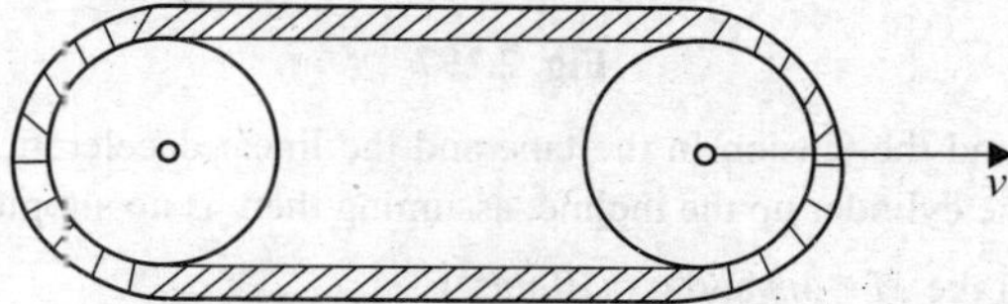

Fig. 2.195

5. A constant horizontal force of 10 N is applied at the centre of a wheel of mass 10 kg and radius 0.30 m. The wheel rolls without slipping on the horizontal surface, and the acceleration of the centre of mass is 0.60 m/s^2.

a. What are the magnitude and direction of the frictional force on the wheel?

b. What is the rotational inertia of the wheel about an axis through its centre of mass and perpendicular to the plane of the wheel?

6. A uniform solid cylinder of mass m rests on two horizontal planks. A thread is wound on the cylinder. The hanging end of the thread is pulled vertically down with a constant force F (Fig. 2.196).

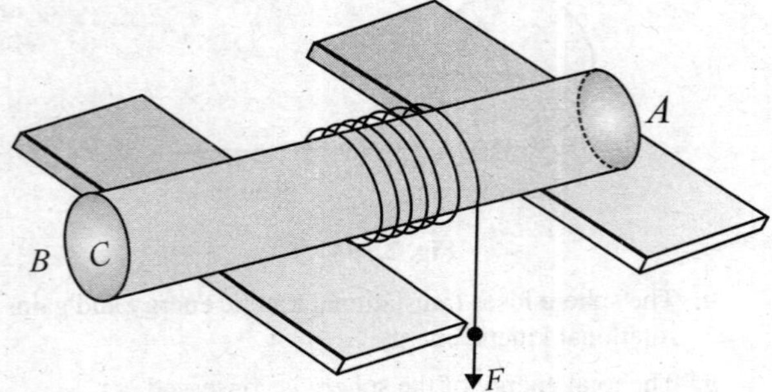

Fig. 2.196

Find the maximum magnitude of the force F which may be applied without bringing about any sliding of the cylinder, if the coefficient of friction between the plank and the cylinder is equal to μ. What is the maximum acceleration of the centre of mass over the planks?

7. A load of mass m is attached to the end of a string wound on a cylinder of mass M and radius R. The string passes round a pulley.

(i) Find the acceleration of the cylinder M when it rolls.

(ii) If the coefficient of friction is μ. Find the acceleration of the cylinder when it slips and rolls.

(iii) Find the minimum value of coefficient of friction for which the cylinder rolls always.

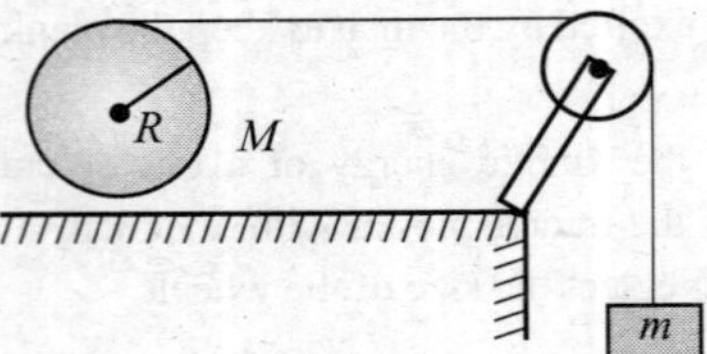

Fig. 2.197

8. Find the tension in the tape and the linear acceleration of the cylinder up the incline, assuming there is no slipping. (Take $M = 4m$ and $g = 10\ \text{m/s}^2$)

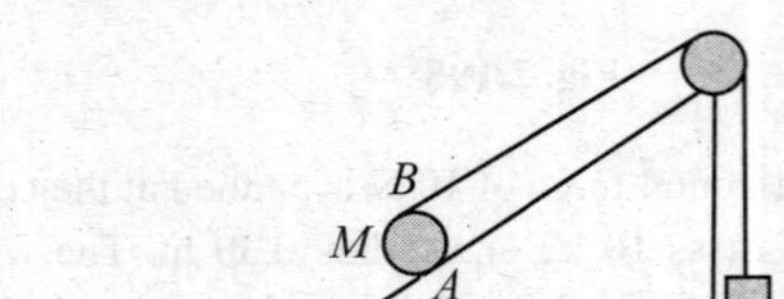

Fig. 2.198

9. A spool (consider it as a double disc system joined by a short tube at their centre) is placed on a horizontal surface as shown Fig. 2.199. A light string wound several times over the short connecting tube leaves it tangentially and passes over a light pulley. A weight of mass m is attached to the end of the string. The radius of the connecting tube is r and mass of the spool is M and radius is R. Find the acceleration of the falling mass m. Neglect the mass of the connecting tube and slipping of the spool.

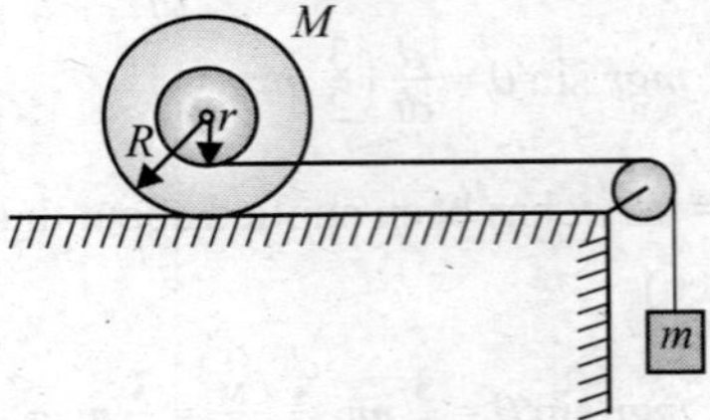

Fig. 2.199

10. A solid cylinder wheel of mass M and radius R is pulled by a force F applied to the centre of the wheel at 37° to the horizontal. If the wheel is to roll without slipping, what is the maximum value of $|F|$? The coefficients of static and kinetic friction are $\mu_S = 0.40$ and $\mu_K = 0.30$. ($\sin 37° = 3/5$).

11. Find the acceleration of a system consisting of a cylinder of mass m and radius R and a plank of mass M placed an a smooth surface if it is pulled with a force F as shown in Fig. 2.200. Given that sufficient friction is present between the cylinder and the plank surface to prevent sliding of the cylinder.

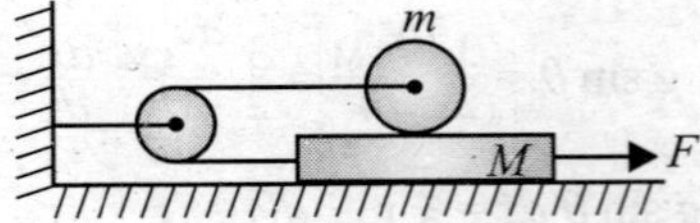

Fig. 2.200

12. A uniform solid cylinder of mass m rests on two horizontal planks as shown in Fig. 2.201. A thread is wound on the cylinder. The hanging end of the thread is pulled vertically up with a force F. What is the maximum magnitude of the force F which still does not bring about any sliding of the cylinder, if the coefficient of friction between the cylinder and the planks is equal to μ. What is the maximum acceleration of the axis of the cylinder?

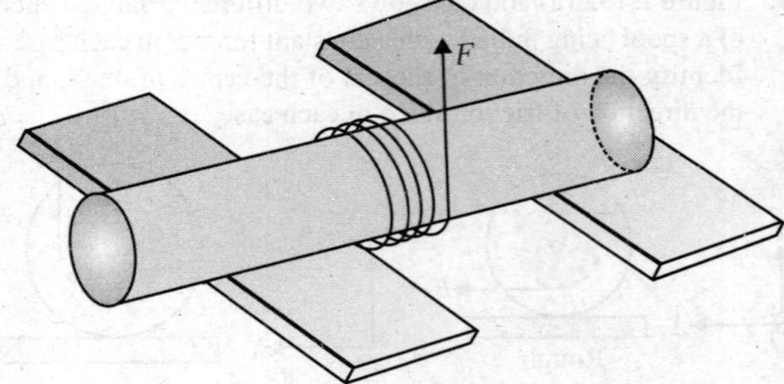

Fig. 2.201

13. A bullet of mass m moving with a velocity of u just grazes the top of a solid cylinder of mass M and radius R resting on a rough horizontal surface as shown and is embedded in the cylinder after impact. Assuming that the cylinder rolls without slipping, find the angular velocity of the cylinder and the final velocity of the bullet.

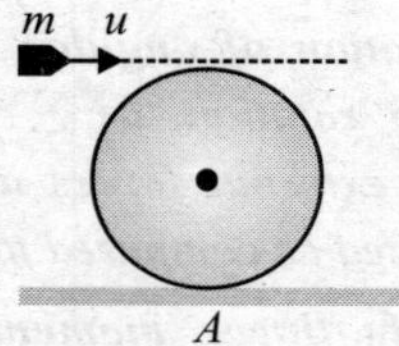

Fig. 2.202

14. A small solid marble of mass M and radius r rolls down along the loop track, without slipping. Find the height h above the base, from where it has to start rolling down the incline such that the sphere just completes the vertical circular loop of radius R.

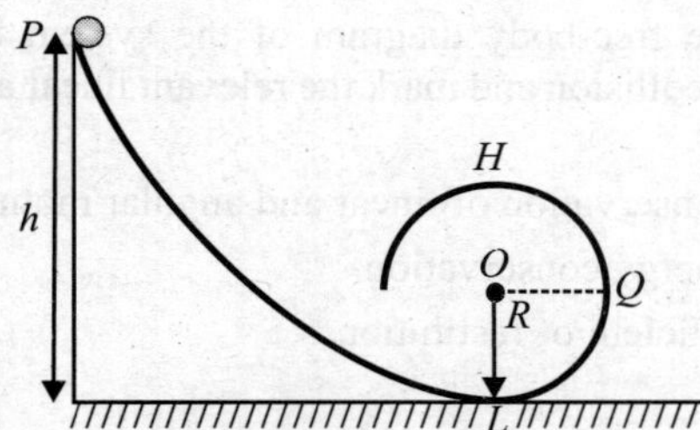

Fig. 2.203

15. A solid sphere of mass 'M' is placed on the top of a plank of the same mass, after giving an angular velocity ω_0 at $t = 0$. Find the velocity of the plank and the sphere when the sphere starts rolling.

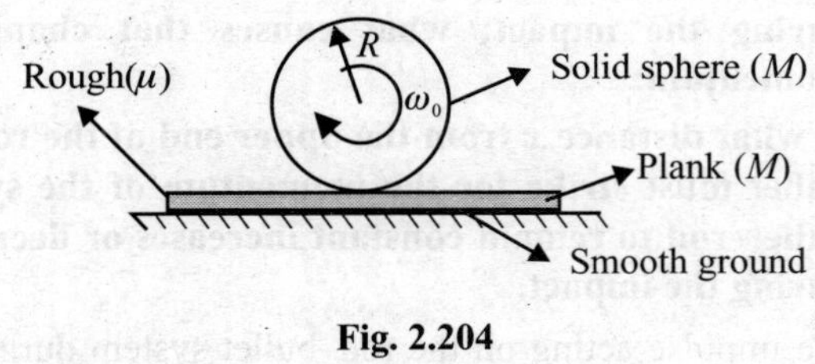

Fig. 2.204

CONSERVATION OF ANGULAR MOMENTUM IN CASE OF COLLISION

The angular momentum of a system is conserved if the net external impulse acting on the system is zero. That is, when $\vec{\tau}_{ext} = 0$, then $\Delta\vec{L} = 0$ or $\vec{L}$ = constant.

In collision problems, the conservation of angular momentum is commonly applied. It has been observed that in situations where linear momentum conservation is not applicable, the conservation of angular momentum is still applicable. Various situations showing the applications of this principle are illustrated below as follows.

Collision in One Dimension

Elastic collision: Consider two particles of masses m_1 and m_2 moving with velocities u_1 and u_2 along a straight line. After collision, the velocities of m_1 and m_2 are v_1 and v_2, respectively. An observer is observing the particles from origin. The following conservative principles can be applied for this situation.

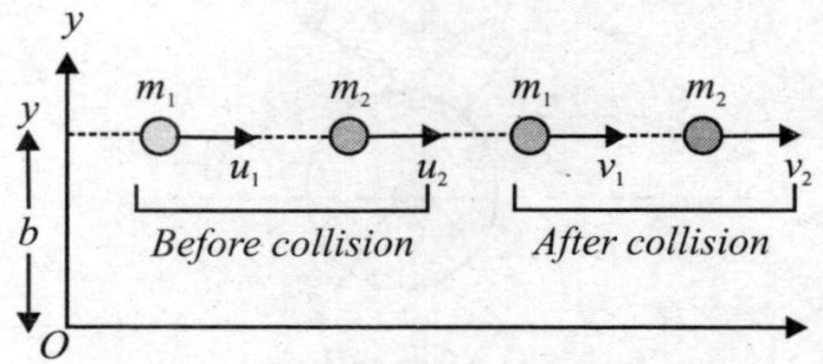

Fig. 2.205

- Kinetic energy is conserved.
- Linear momentum is conserved.
- Angular momentum about any fixed point, say O, is conserved.

Inelastic collision: Consider two particles of masses m_1 and m_2 are moving with velocities u_1 and u_2 along a straight line. After collision m_1 and m_2 are combined and start moving with a common velocity v. An observer standing at O will draw the following conclusions.

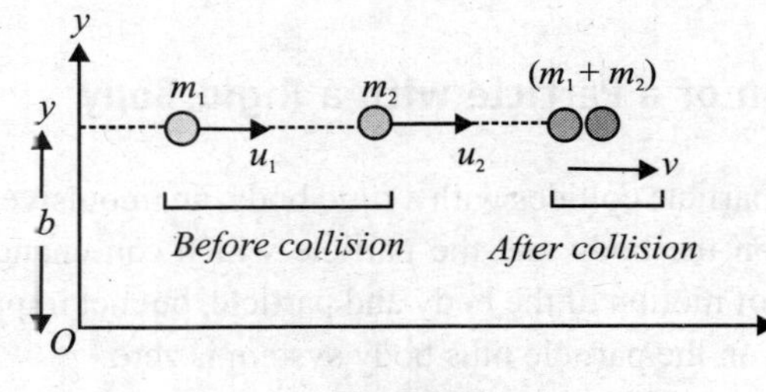

Fig. 2.206

- Kinetic energy is not conserved
- Linear momentum is conserved
- Angular momentum about any point, say O is conserved.

Oblique Collision

A small particle of mass m moving with a velocity v_0 in the horizontal direction collides elastically with a smooth stationary sphere of radius R and mass M as shown in Fig. 2.207 (a).

After collision, the particle moves tangentially with respect to the smooth sphere with a velocity v_1 and the ball moves with velocity v_2 along the normal as shown in Fig. 2.207 (b).

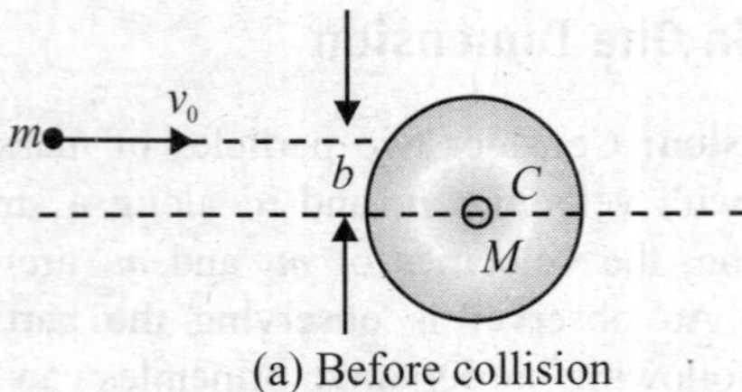

(a) Before collision

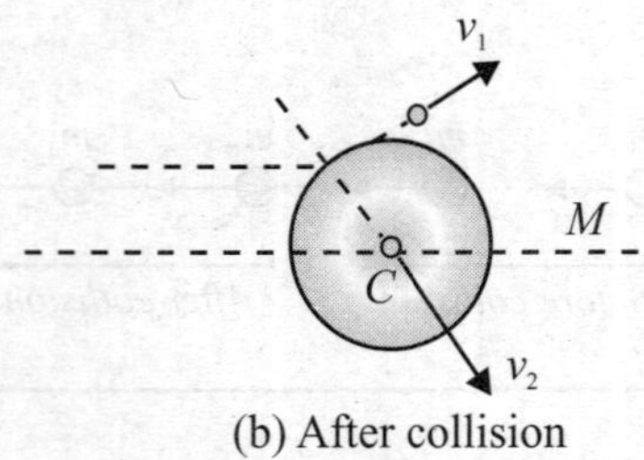

(b) After collision

Fig. 2.207

Note:

- ***Kinetic energy of the particle is not conserved.***
- ***Linear momentum of the particle is not conserved because the sphere exerts a normal impulse on the particle.***
- ***Angular momentum of the particle w.r.t. the centre C of the sphere is conserved, as there is no angular impulse acting on the particle about C.***

Collision of a Particle with a Rigid Body

- As a particle collides with a rigid body, an impulsive force acts on the body and the particle which can change the state of motion of the body and particle, but net impulsive force on the particle plus body system is zero.
- Depending on the fact that whether the body is restricted to rotate about fixed axis of rotation or is free to rotate in any way, we are here classifying this topic into two ways:

  a. when the axis of rotation is fixed

  b. when the body is free to rotate as well as translate

When Axis of Rotation is Fixed

Following steps should be keep in mind in solving the questions:

- ***Mark the forces acting on system (body + particle).***
- ***Mark the velocity (linear and angular) of both the components of the system before and after collision.***
- ***Use the conservation of angular momentum about the fixed axis of rotation, as $\sum \vec{\tau}_{ext} = 0$ under the assumption that external forces are not present or have been neglected as compared to impulsive force.***
- ***We cannot apply linear momentum conservation principle as $\sum F_{ext} \neq 0$, because external force is acting on the body due to fixed axis of rotation.***
- ***Use coefficient of restitution equation.***
- ***Use energy conservation, if required.***

When Body is Free to Rotate and Translate

- Draw the free-body diagram of the system before and after the collision and mark the relevant linear and angular variables.
- Apply conservation of linear and angular momenta.
- Apply energy conservation.
- Use coefficient of restitution.

Illustration 2.59 **A vertically oriented uniform rod of mass M and length l can rotate about its upper end. A horizontally flying bullet of mass m strikes the lower end of the rod and gets stuck in it; as a result, the rod swings through an angle α. Assuming that $m << M$, find**

a. the velocity of the flying bullet;

b. the momentum change in the system bullet–rod during the impact; what causes that change of momentum;

c. at what distance x from the upper end of the rod the bullet must strike for the momentum of the system bullet–rod to remain constant increases or decreases during the impact.

Sol. The impulse acting on the rod–bullet system during the impact is shown in Fig. 2.208. J_1 is the impulse acting on the rod by the hinge. Here the role of J_1 is a little interesting force. J_1 could be greater than zero, $J_1 > 0$; F_1 could be equal to zero, $J_1 = 0$; and F_1 could be less than zero, $J_1 < 0$. And as a consequence the linear momentum of the system may increase, may remain constant or may decrease.

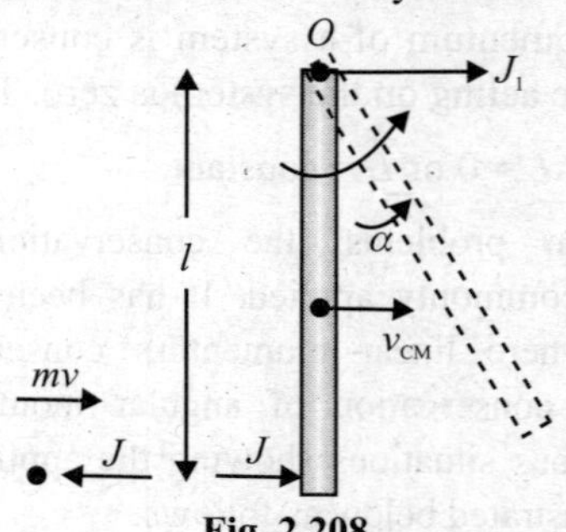

Fig. 2.208

a. During the impact, the moment of impulse acting on the rod–bullet system about the hinge is zero. Angular momentum about the hinge will remain constant. Let ω be the angular velocity of the system about the hinge.

[Angular momentum of the system]$_{\text{initial}}$ = [Angular momentum of the system]$_{\text{final}}$

$$\Rightarrow \quad mv \times l = I_0 \times \omega \qquad \text{(i)}$$

where I_0 is the moment of inertia of the rod–bullet system after collision about O

$$I_0 = I_{\text{rod}} + I_{\text{bullet}} = \frac{1}{3}Ml^2 + ml^2 \qquad \text{(ii)}$$

$$mvl = \left(\frac{1}{3}Ml^2 + ml^2\right)\omega \qquad \text{(iii)}$$

As the system swings about hinge, the mechanical energy of the system will remain constant. $\Delta k + \Delta U = 0$

$$\left[\frac{1}{2}\left(\frac{1}{3}Ml^2 + ml^2\right)\omega^2 - 0\right]$$

$$= mg\frac{l}{2}(1-\cos\alpha) + mg(1-\cos\alpha)$$

From Eq. (iii), $\omega = \dfrac{3mv}{l(M+3m)}$ (iv)

Substituting this value of ω in Eq. (iv), we get

$$\frac{1}{2}\left(\frac{1}{3}Ml^2 + ml^2\right)\left(\frac{3mv}{l(M+3m)}\right)^2$$

$$= gl\frac{(1-\cos\alpha)}{2}(M+2m)$$

$$\Rightarrow \frac{1}{2}\times\frac{1}{2}l^2(M+3m)\frac{9m^2v^2}{l^2(M+3m)^2}$$

$$= \frac{gl(1-\cos\alpha)}{2}(M+2m)$$

$$\Rightarrow \quad v^2 = \frac{2gl}{3}\frac{(M+2m)(M+3m)}{m^2}\sin^2\frac{\alpha}{2}$$

Now for $\quad m << M, (M+3m)(M+3m) \approx M^2$

Thus, $\quad v^2 = \dfrac{2}{3}gl\dfrac{M^2}{m^2}\sin^2\dfrac{\alpha}{2} \Rightarrow v = \dfrac{M}{m}\sqrt{\dfrac{2}{3}gl}\sin\dfrac{\alpha}{2}$

b. Change in linear momentum during the impact,

$$\Rightarrow \quad \Delta p = p_f - p_i$$

$$= \left\{M\left(\frac{l}{2}\omega\right) + m(l\omega)\right\} - mv$$

$$= l\left(\frac{M+2m}{2}\right)\omega - mv \quad \text{(v)}$$

From Eqs. (iv) and (v),

$$= l\left(\frac{M+2m}{2}\right)\frac{3mv}{l(M+3m)} - mv$$

$$= \frac{3}{2}mv - mv = \frac{1}{2}mv$$

$$= \frac{1}{2}m\frac{M}{m}\sqrt{\frac{1}{2}gl}\sin\frac{\alpha}{2}$$

$$= M\sqrt{\frac{1}{6}gl}\sin\frac{\alpha}{2}$$

What causes linear momentum to change? Obviously, impulse J_1 at hinge. Here Δp is positive, linear momentum increases, so J_1 must be towards the right.

c. Angular momentum about the hinge remains conversed irrespective of the distance from the upper end where the bullet strikes the rod; whereas the linear momentum remains conserved for a particular distance. Let it be x from the upper end of the rod.

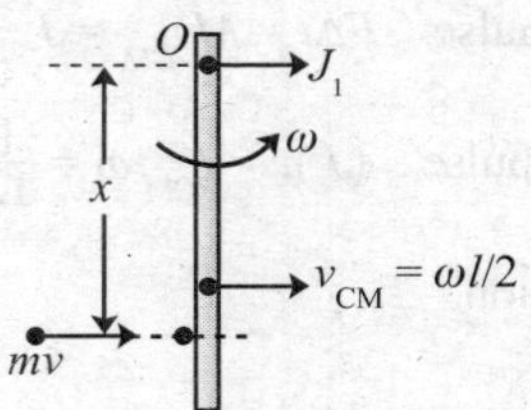

Fig. 2.209

Therefore, $mv\,x = \left(\dfrac{1}{3}Ml^2 + mx^2\right)\omega$ (i)

and $\quad mv = m\omega x + M\dfrac{l}{2}\omega$ (ii)

From these two equations, $\dfrac{mvx}{mv} = \dfrac{\left(\frac{1}{3}Ml^2 + mx^2\right)\omega}{mx\omega + \frac{Ml}{2}\omega}$

$$\Rightarrow \quad x = \frac{\frac{1}{3}Ml^2 + mx^2}{mx + \frac{Ml}{2}}$$

$$\Rightarrow \quad mx^2 + \frac{Mlx}{2} = \frac{1}{3}Ml^2 + mx^2 \Rightarrow x = \frac{2}{3}l$$

Let us understand why does the hinge apply a force of short duration on the rod in the horizontal direction during the impact? Under what conditions such a force will not develop? When is $J_1 > 0$? When is $J_1 = 0$? When is $J_1 < 0$? During the impact the upper end of the rod tends to move either towards the right or towards the left; for a particular point of impact of the bullet, the upper point of the rod will not have any tendency of motion at all. These facts can be understood in a better way by considering a rod hung in vertical position in a gravity free space. (No hinge in needed; on impact the rod will move freely). Let the horizontally flying bullet hit the rod as shown in Fig. 2.210.

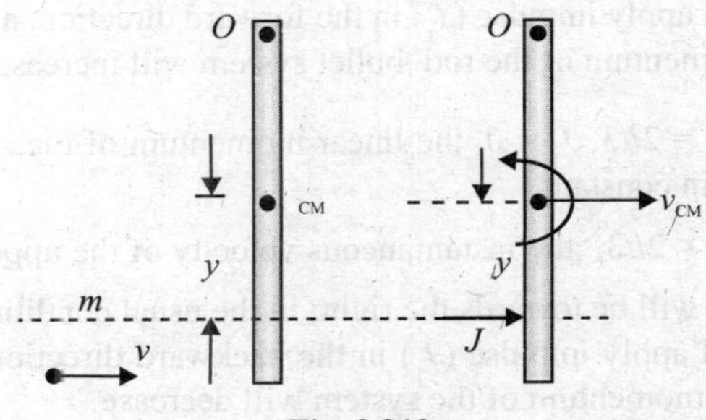

Fig. 2.210

During impact an horizontal impulse acts on the rod. As a result the centre of mass of the rod acquires a velocity (say v_{CM}) and about the centre of mass the rod acquires an angular velocity (say ω). The instantaneous velocity of the upper end will be towards the right, will be zero or will be towards the left depending on whether

$$v_{CM} > \frac{l}{2}\omega, \text{ or } v_{CM} = \frac{l}{2}\omega \text{ or } v_{CM} < \frac{l}{2}\omega$$

Now linear impulse $F\Delta t = Mv_{CM} = J$ (i)

and angular impulse $(J)y = I_{CM}\,\omega = \frac{1}{12}Ml^2\omega$ (ii)

From the equation,

$$\frac{J}{Jy} = \frac{Mv_{CM}}{\frac{1}{12}Ml^2\omega} \Rightarrow \frac{1}{y} = \frac{Mv_{CM}}{\frac{1}{12}Ml^2\omega} \quad \text{(iii)}$$

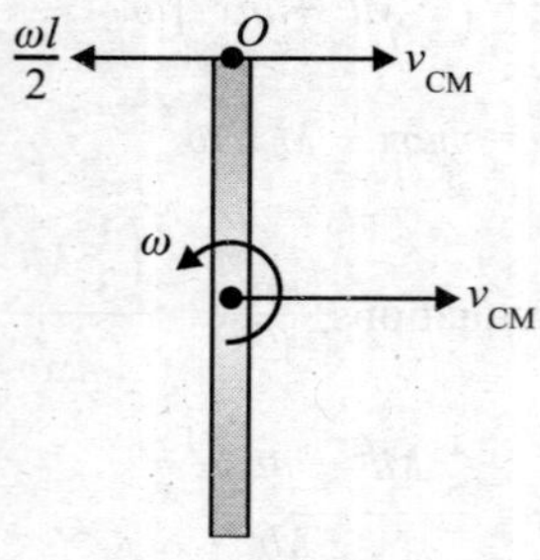

Fig. 2.211

For the instantaneous velocity of the upper point O to be equal to zero, $v_{CM} = \frac{l}{2}\omega$

For this condition, Eq. (iii) will be

$$\frac{1}{y} = \frac{Mv_{CM}}{\frac{1}{12}Ml^2\left(v_{CM}/\frac{l}{2}\right)} \Rightarrow y = \frac{l}{6}$$

Thus, $x = \frac{l}{2} + y = \frac{l}{2} + \frac{l}{6} = \frac{2l}{3}$

For $x > 2l/3$, the instantaneous velocity of the upper end of the rod will be towards the left; in the usual condition the hinge will apply impulse (J_1) in the forward direction, and the linear momentum of the rod–bullet system will increase.

For $x = 2l/3$, $J_1 = 0$, the linear momentum of the system will remain constant.

For $x < 2l/3$, the instantaneous velocity of the upper end of the rod will be towards the right; in the usual condition the hinge will apply impulse (J_1) in the backward direction, and the linear momentum of the system will decrease.

Illustration 2.60 **A rod AB of mass M and length L is lying on a horizontal frictionless surface. A particle of mass m travelling along the surface hits the end A of the rod with a velocity v_0 in a direction perpendicular to AB. The collision is completely elastic. After the collision the particle comes to rest.**

a. Find the ratio m/M.

b. A point P on the rod is at rest immediately after the collision. Find the distance AP.

c. Find the linear speed of a point P at time $\pi L/3v_0$ after the collision.

Sol. a. If we take rod and particle as a system, the impulse between them will be internal. We can conserve the linear momentum of rod + particle's system. Let V_c and ω be the velocity and angular velocity of the CM just after collision.

Using conservation of linear momentum, we have

$$mv_0 = 0 + Mv_c \quad \text{(i)}$$

Here in this case we can conserve angular momentum about any point on the rod.

And by conservation of angular momentum about the entire of mass of the rod, we get

$$mv_0\frac{L}{2} = I\omega \text{ or } mv_0\frac{L}{2} = \left(\frac{ML^2}{12}\right)\omega \quad \text{(ii)}$$

From Eqs. (i) and (ii), $v_c = \frac{mv_0}{M}$ and $\omega = \frac{6mv_0}{ML}$

We can apply conservation of angular momentum about point of collision.

$$\vec{L}_{\text{initial}} = \vec{L}_{\text{final}}$$

$$O = I_0\omega(-\hat{k}) + Mv_0\frac{\ell}{2}(\hat{k})$$

$$\frac{ML^2}{12}\omega = \frac{Mv_cL}{2}$$

which gives $\omega L = 6v_c$ (iii)

Solving Eqs. (i) and (iii), we get the same results.

Since collision is completely elastic, therefore KE before collision is equal to KE after collision

$$\text{or} \quad \frac{1}{2}mv_0^2 + 0 = \frac{1}{2}mv_c^2 + \frac{1}{2}I\omega^2$$

$$\text{or} \quad \frac{1}{2}mv_0^2 = \frac{1}{2}M\left(\frac{mv_0}{M}\right)^2 + \frac{1}{2}\frac{ML^2}{12}\left(\frac{6mv_0}{ML}\right)^2$$

which gives $\frac{m}{M}=\frac{1}{4}$ and $v_c=\frac{v_0}{4}$

Alternative method: We can apply coefficient of restitution equation at the position of point of collision, we have

$$e=\left(\frac{v_2-v_1}{u_1-u_2}\right)$$

Here u_1, u_2, v_1 and v_2 are the velocities of particles and point on the rod at collision point.

$u_1=v_0$, $u_2=0$

$v_1=0$ and $v_2=v_c+\frac{\omega l}{2}$

As collision in elastic $e=1$

Substituting the values in restitution equation, we get

$$1=\frac{\left(v_c+\frac{\omega L}{2}\right)-0}{v_0-0} \Rightarrow v_0=\left(\frac{mv_0}{M}\right)+\left(\frac{6mv_0}{ML}\right)\frac{L}{2}$$

which gives $\frac{m}{M}=\frac{1}{4}$

b. Velocity of point P immediately after collision is zero, let it be at a distance y from CM

$$\therefore \quad v_c-\omega y=0 \text{ or } \frac{mv_0}{M}-\left(\frac{6mv_0}{ML}\right)y=0$$

which gives $y=\frac{L}{6}$

$$\therefore \quad AP=\frac{L}{2}+\frac{L}{6}=\frac{2L}{3}$$

c. Angle rotated by the rod at time $\frac{\pi L}{3v_0}$

$$\theta=\omega t=\left(\frac{6mv_0}{ML}\right)\times\frac{\pi L}{3v_0}=\frac{\pi}{2}$$

The rod turns through $\pi/2$ in this interval of time. The velocity of point P in y-direction will be

$$v_y=\omega y=\left(\frac{6mv_0}{M}\right)\times\frac{L}{6}=\frac{v_0}{4}$$

The resultant velocity of point P,

$$v=\sqrt{v_c^2+v_y^2}=\sqrt{\left(\frac{v_0}{4}\right)^2+\left(\frac{v_0}{4}\right)^2}\Rightarrow\frac{v_0}{2\sqrt{2}}$$

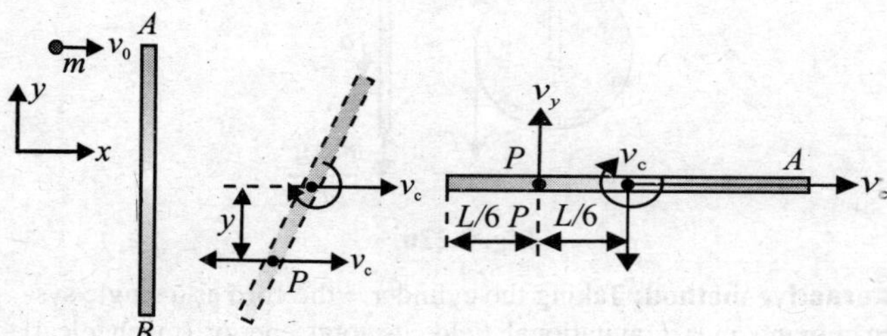

Fig. 2.212

Concept Application Exercise 2.6

1. A sphere rolling on a horizontal rough surface collides elastically with a smooth vertical wall, as shown in Fig. 2.213. State which of the following statements is true or false.

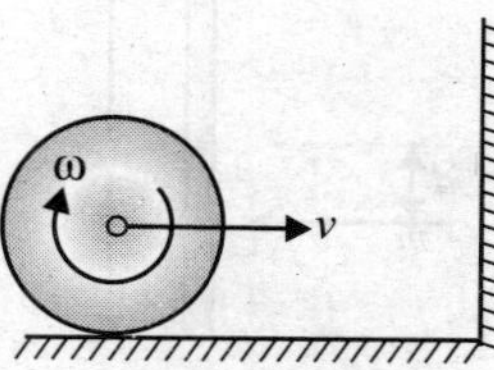

Fig. 2.213

i. After collision, the velocity of the centre of mass gets reversed.
ii. Angular momentum of the sphere about the point of contact with the wall is conserved.
iii. Angular momentum of the sphere about a stationary point on the horizontal surface is conserved.
iv. Just after collision the point of contact with the horizontal surface is moving towards the wall.
v. After collision the friction force acts on the sphere such that it decreases the linear speed and increases the angular speed.
vi. Finally, when the sphere starts rolling, it is moving away from the wall.

2. A uniform bar of length $6a$ and mass $8m$ lies on a smooth horizontal table. Two point masses m and $2m$ moving in opposite direction in the same horizontal plane with speed $2v$ and v, respectively, strike the bar and stick to the bar after collision. Calculate

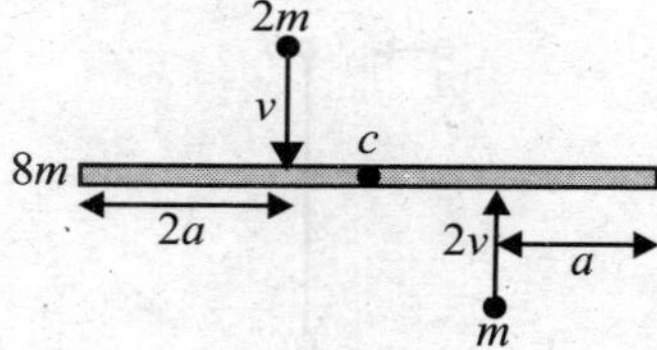

Fig. 2.214

a. the velocity of the centre of mass after collision
b. the angular velocity of the centre of mass after collision
c. total energy

3. A circular wooden hoop of mass m and radius R rests flat on a frictionless surface. A bullet, also of mass m and moving with a velocity v, strikes the hoop and gets embedded in it. The thickness of the hoop is much smaller than R. Find the angular velocity with which the system rotates after the bullet strikes the hoop.

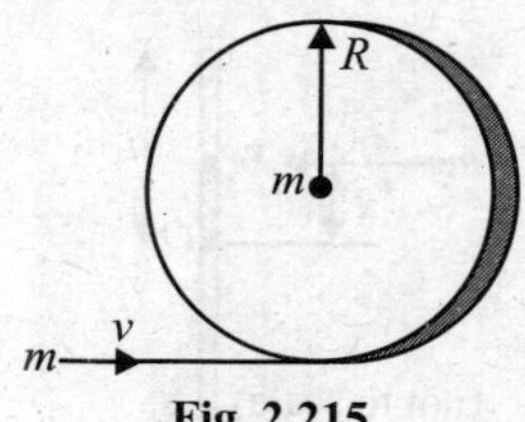

Fig. 2.215

4. A stick of length l and mass m lies on a frictionless horizontal surface on which it is free to move in anyway. A ball of mass m moving with speed v collides elastically with the stick. What must be the mass of the ball so that it remains at rest immediately after collision?

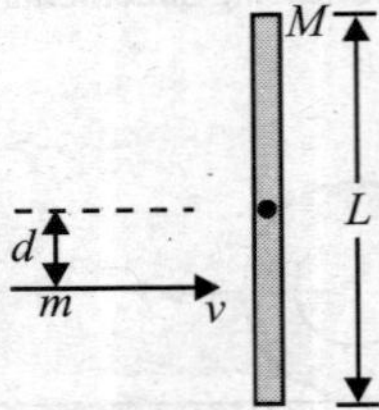

Fig. 2.216

5. A thin uniform rod of length l is initially at rest with respect to an inertial frame of reference. The rod is tapped at one end perpendicular to its length. How far the centre of mass translates while the rod completes one revolution about its centre of mass. Neglect gravitational effect.

6. A thin spherical shell of radius R lying on a rough horizontal surface is hit sharply and horizontally by a cue. Where should it be hit so that the shell does not slip on the surface?

7. A wheel rolling along a rough horizontal surface with an angular velocity ω strikes a rough vertical wall, normally. Find the initial angular velocity, as it tends to roll up the wall (neglect any slippage) due to impulse.

8. Two identical particles B and C each of mass 50 g are connected by a light rod of length 30 CM. Another particle A of same mass moving with a speed u = 60 CM/s strikes B, in a direction perpendicular to AB, and sticks to it. The whole process takes place on a smooth horizontal plane. Find the angular velocity ω of the system about its centre of mass, immediately after the impact.

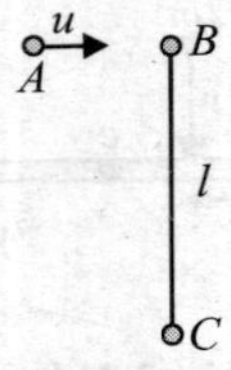

Fig. 2.217

9. In the above problem, find the velocity of C relative to "$A + B$", immediately after the impact.

10. A rod AB of mass M and length $8l$ lies on a smooth horizontal surface. A particle, of mass 'm' and velocity v_0, strikes the rod perpendicular to its length, as shown in Fig. 2.218. As a result of the collision, the centre of mass of the rod attains a speed of $v_0/8$ and the particle rebounds back with a speed of $v_0/4$. Find the following:

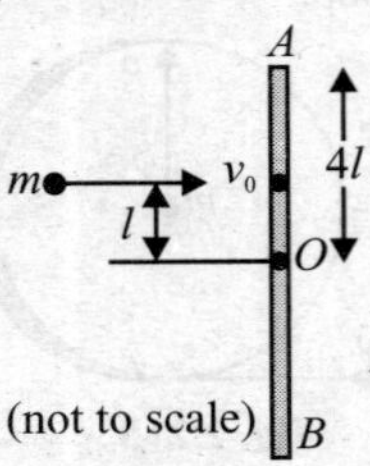

Fig. 2.218

a. The ratio M/m
b. The angular velocity of the rod about O
c. The coefficient of restitution 'e' for the collision
d. The velocities of the ends 'A' and 'B' of the rod, namely, v_A and v_B, respectively.

Solved Examples

Example 2.1 **A uniform cylinder of radius 'r' and mass 'm' can rotate freely about a fixed horizontal axis. A thin cord of length 'l' and mass 'm_0' is wound on the cylinder in a single layer. Find the angular acceleration of the cylinder as a function of the length x of the hanging part of the cord. The wound part of the cord is supposed to have its centre of gravity on the cylinder axis as shown in Fig. 2.219.**

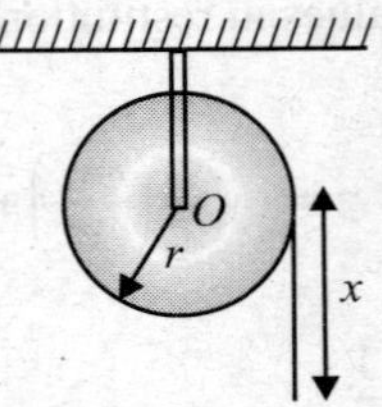

Fig. 2.219

Sol. Figure 2.220 shows the free-body diagram of the two bodies, viz., cylinder and overhanging portion of the cord.

If α is the angular acceleration of the cylinder, then from $\tau = I\alpha$,

we have $$Tr = \left[\frac{mr^2}{2} + \frac{m_0}{l}(l - x)\, r^2\right]\alpha$$

$$T = \left[\frac{m}{2} + m_0 - \frac{m_0 x}{l}\right] r\alpha \quad \text{(i)}$$

If a be the acceleration of the hanging part, then from $F = ma$,

$$\frac{m_0 xg}{l} - T = \frac{m_0 x}{l} a = \frac{m_0 x}{l}(\alpha r) \quad \text{(ii)}$$

[Linear acceleration of the rim of the cylinder is same as that of the cord]

Adding Eqs. (i) and (ii), we get

$$\frac{m_0 x}{l} g = \alpha\left[\frac{m}{2} + m_0\right] r \Rightarrow \alpha = \frac{m_0 xg}{rl\left[\frac{m}{2} + m_0\right]}$$

Fig. 2.220

Alternative method: Taking the cylinder + the cord as a single system moving in a gravitational field, its total energy (mechanical) should remain constant.

Assuming the level of the cylinder's centre as datum or reference level (as shown in Fig. 2.221), we have

$$\frac{dE}{dt} = 0$$

i.e., $\frac{d}{dt}\left[\frac{1}{2}\left(\frac{mr^2}{2}\right)\left(\frac{v}{r}\right)^2 + \frac{1}{2}m_0v^2 - \frac{m_0x}{l}g\left(\frac{x}{2}\right)\right] = 0$

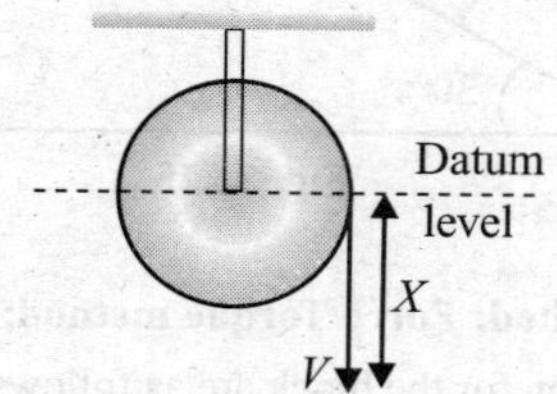

Fig. 2.221

[Here v is the speed of the hanging part]

$$\Rightarrow \left(\frac{m}{2} + m_0\right)2v\left(\frac{dv}{dt}\right) - \frac{m_0 g}{l}\left(2x\frac{dx}{dt}\right) = 0$$

$$\Rightarrow \left(m_0 + \frac{m}{2}\right)\frac{dv}{dt} = \frac{m_0 gx}{l} \Rightarrow \frac{dv}{dt} = \frac{m_0 gx}{l\,(m/2 + m_0)}$$

Therefore, the angular acceleration a of the cylinder

$$\alpha = \frac{a}{r} = \frac{m_0 gx}{rl\left(\frac{m}{2} + m_0\right)}$$

Example 2.2 **A heavy plank of mass 102.5 kg is placed over two cylindrical rollers of radii R = 10 CM and r = 5 CM. Mass of rollers is 40 kg and 20 kg, respectively. Plank is pulled towards right by applying a horizontal force F = 25 N as shown in Fig. 2.222. During the first second of motion, the plank gets displaced by 10 CM. If the plank remains horizontal and slipping does not take place, calculate the magnitude and direction of force of friction acting between**

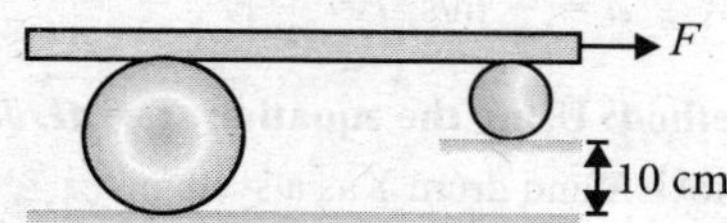

Fig. 2.222

a. plank and bigger roller
b. plank and smaller roller
c. bigger roller and floor
d. smaller roller and floor

Sol. Since force F is constant, therefore the plank moves with a constant acceleration. Let that acceleration be a. For the motion of the plank,

$$u = 0, t = 1 \text{ second}, s = 10 \text{ CM} = 0.10 \text{ m}$$

Using, $s = ut + \frac{1}{2}at^2$; $a = 0.20$ m/s²

Accelerations of the top points of the rollers are also equal to a because these points are in contact with the plank and there is no slipping. Hence, angular accelerations α_1 and α_2 of the bigger and smaller rollers (respectively) are given by $\alpha_1 = a/2R$ and $\alpha_2 = a/2r$

or $\alpha_1 = 1$ rad/s² and $\alpha_2 = 2$ rad/s²

Acceleration of the centre of mass of the bigger and smaller rollers are $a_1 = R\alpha_1$ and $a_2 = r\alpha_2$, respectively.

$$\therefore \quad a_1 = a_2 = 0.10 \text{ m/s}^2$$

Now considering the free-body diagram (as shown in Fig. 2.223).

Since, plank tries to slide (towards right) over the top surfaces of rollers, therefore the rollers exert frictional forces on the plank towards the left.

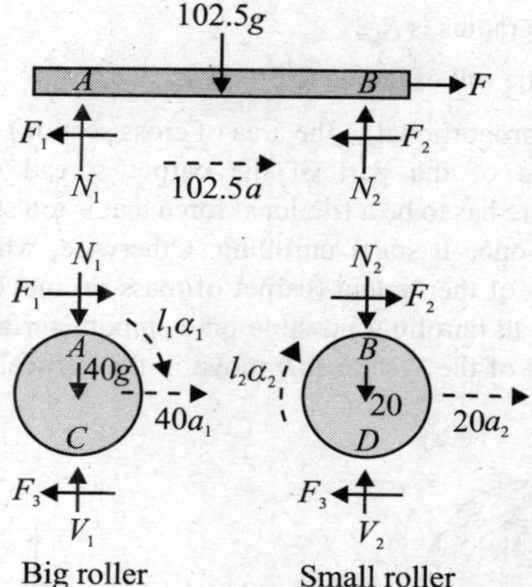

Fig. 2.223

For the vertical forces on the plank,

$$N_1 + N_2 = 102.5g \quad \text{(i)}$$

For the horizontal forces on the plank,

$$F - F_1 - F_2 = 102.5a \quad \text{(ii)}$$

Since instantaneous axes of rotation of the bigger and smaller rollers are C and D, respectively, therefore taking moments of forces acting on these rollers about their respective instantaneous axes of rotation are as follows:

For bigger roller, $F_1(2R) = I_1\alpha$ (iii)

For smaller roller, $F_2(2r) = I_2\alpha_2$ (iv)

where $F_1 = (3/2)(40)R^2 - 0.6$ kg m² (moment of the inertia of the bigger roller about axis C)

Similarly, $I_2 = \frac{3}{2}(20)r^2 = 0.075$ kg m²

From Eqs. (iii) and (iv), $F_1 = 3$ N ; $F_2 = 1.50$ N

For the horizontal forces acting on the bigger roller,

$$F_1 - F_3 = 40a_1 \text{ or } F_3 = F_1 - 40a_1$$

or $F_1 = -1$ N

(–ve sign indicates that friction between bigger roller and floor is opposite to its assumed direction)

$\therefore \quad |F_3| = 1$ N (rightwards)

For the horizontal forces acting on the smaller roller,

$$F_2 - F_4 = 20a_2$$

$\therefore \quad F_4 = -0.50$ N

or $|F_4| = 0.50$ N (Rightwards)

Example 2.3 **A carpet of mass M made of inextensible material is rolled along its length in the form of a cylinder of radius R and is kept on a rough floor. The carpet starts unrolling without sliding on the floor when a negligibly small push is given to it. Calculate the horizontal velocity of the axis of the cylindrical part of the carpet when its radius reduces to $R/2$.**

Sol. Figure 2.224 shows the forces acting on the carpet in two cases:

i. when the radius of the roll is R

ii. when the radius is $R/2$

Mass of the roll of radius $R/2 = M/4$

(Mass is proportional to the area of cross section)

The mass of the part of the carpet spread on the floor $= (3/4) M$. There has to be a frictional force that is too static in nature on the carpet once it starts unrolling. Otherwise, why should the centre of mass of the system (carpet of mass m) move towards the right? Is such an unrolling possible on a smooth surface? Does the centre of mass of the system (M) move in the vertical direction? If yes, why?

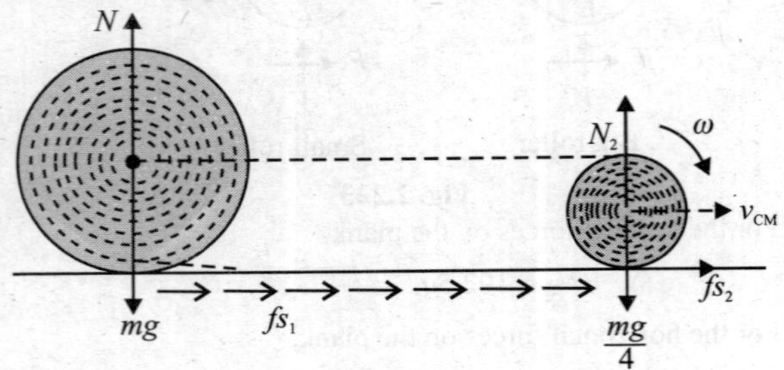

Fig. 2.224

Here, $W_{fs} = 0, W_N = 0$

Therefore, $W_{ncf} = 0$.

The mechanical energy of the system will remain constant. The part of the carpet spread on the ground will have neither kinetic energy nor potential energy.

Therefore, $MgR = \left(\frac{M}{4}\right)g\left(\frac{R}{2}\right)+\frac{1}{2}\left(\frac{M}{4}\right)v_{CM}^2+\frac{1}{2}\left(\frac{1}{2}\frac{M}{4}\left(\frac{R}{2}\right)^2\right)\omega^2$

$$\Rightarrow \quad \frac{7}{8}MgR = \frac{M}{8}v_{CM}^2 + \frac{1}{16}M\left(\frac{R}{2}\right)^2\left(\frac{v_{CM}}{R/2}\right)^2$$

$$\left(v_{CM} = \frac{R}{2}\omega\right)$$

$$\Rightarrow \quad \frac{7}{8}MgR = \frac{3}{8}m\,v_{CM}^2 \Rightarrow v_{CM} = \sqrt{\frac{14}{3}gR}$$

Example 2.4 **A block X of mass 0.5 kg is held by a long massless string on a frictionless inclined plane of inclination 30° to the horizontal. The string is wound on a uniform solid cylindrical drum Y of mass 2 kg and of radius 0.2 m as shown in Fig. 2.225. The drum is given an initial angular velocity such that block X starts moving up the plane.**

a. Find the tension in the string during the motion.

b. At a certain instant of time, the magnitude of the angular velocity of Y is 10 rad/s. Calculate the distance travelled by X from that instant of time until it comes to rest. [$g = 10$ m/s²]

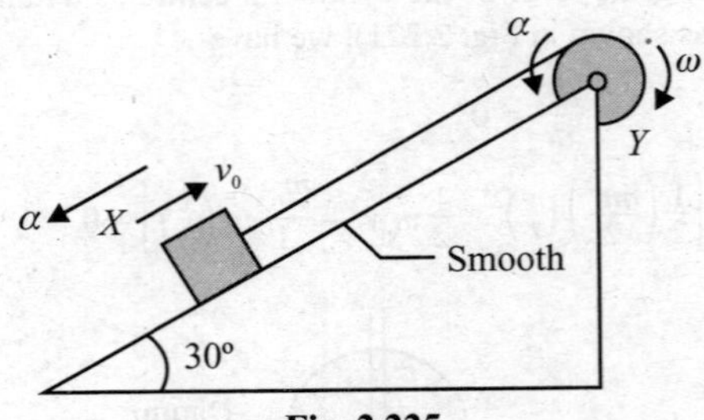

Fig. 2.225

Sol.

a. i. First Method: Force/Torque method:

Force equation for the block are as follows:

$$N - 5\cos 30° = 0 \quad \text{(i)}$$

$$5\sin 30° - T = 0.5 \times a \quad \text{(ii)}$$

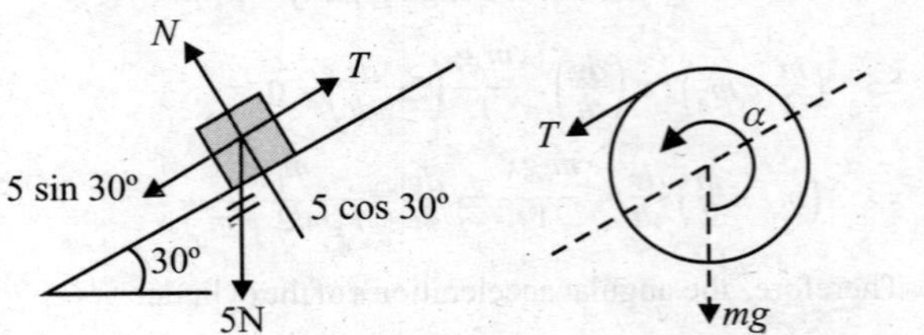

Fig. 2.226

Torque equation for the drum,

$$TR = \left(\frac{1}{2}MR^2\right)\alpha$$

$$\Rightarrow \quad T = \frac{1}{2}MR\alpha = \frac{1}{2}\times 2\times 0.2\alpha \Rightarrow T = 0.2\alpha \quad \text{(iii)}$$

Here, $a = R\alpha$ (The string does not slip on the drum)

Thus $\quad T = (0.2)\left(\frac{a}{R}\right) = 0.2\left(\frac{a}{0.2}\right) = 1 \times a \quad$ (iv)

From Eqs. (ii) and (iv), we get

$$a = \frac{5}{3}\text{ m/s}, T = \frac{5}{3}\text{ N}$$

ii. Second Method: Using the equation $\vec{\tau} = d\vec{L}/dt$

Take both block X and drum Y as a system.

Torque acting on the system about the centre of mass of the drum $= (mg\sin 30°)R$, anticlockwise. Angular momentum of the system about the centre of mass of the drum,

$L = mvR + \frac{1}{2}MR^2\omega = m(R\omega)R + \frac{1}{2}MR^2\omega = \left(mR^2 + \frac{1}{2}MR^2\right)\omega$, clockwise

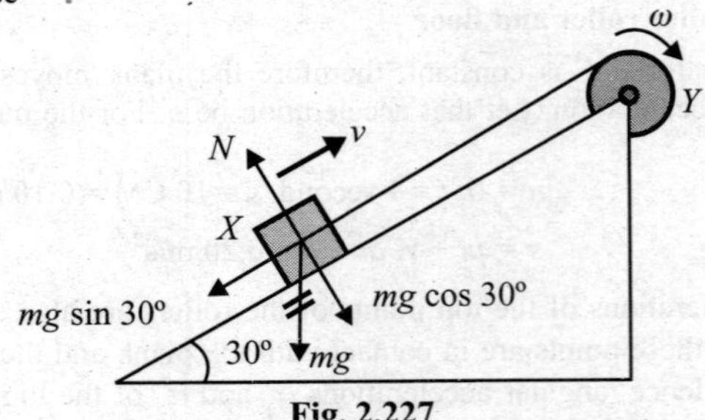

Fig. 2.227

Apply equation $\vec{\tau} = \frac{d\vec{L}}{dt}$

$$-mgR\sin 30° = \frac{d}{dt}\left(mR^2\omega + \frac{1}{2}MR^2\omega\right)$$

$$= \left(mR^2 + \frac{1}{2}MR^2\right)\frac{d\omega}{dt}$$

$$\Rightarrow \quad \frac{d\omega}{dt} = \frac{-mgR\sin 30°}{\left(mR^2 + \frac{MR^2}{2}\right)}$$

$$\Rightarrow \quad \left(\frac{d\omega}{dt}\right)_{\text{anticlockwise}} = \frac{mgR\sin 30°}{\left(mR^2 + \frac{MR^2}{2}\right)}$$ (ω has been taken in the clockwise sense, $d\omega/dt$ is negative. So $d\omega/dt$ is in the anticlockwise sense)

Clearly, $\tau = I\alpha$ $\quad TR = I\left(\frac{d\omega}{dt}\right)_{\text{anticlockwise}}$

$$= \left(\frac{1}{2}MR^2\right)\left(\frac{mgR\sin 30°}{mR^2 + \frac{MR^2}{2}}\right)$$

$$\Rightarrow \quad T = \left(\frac{1}{2}M\right)\left(\frac{mg\sin 30°}{m + \frac{M}{2}}\right) = 5/3\ \text{N}$$

b. i. First Method:

Angular velocity of $Y = 10$ rad/s

Velocity of block $X = R\omega = 0.2 \times 10 = 2$ m/s, along the plane upwards

Acceleration of $X = \frac{5}{3}$ m/s² along the plane downwards

Using equation $v^2 = u^2 - 2as$,

$0^2 = (2) - 2 \times 5/3 \times s \Rightarrow s = \frac{4 \times 3}{2 \times 5} = 1.25$ m

ii. Second Method:

You can also calculate the distance X moves upwards from the energy considerations:

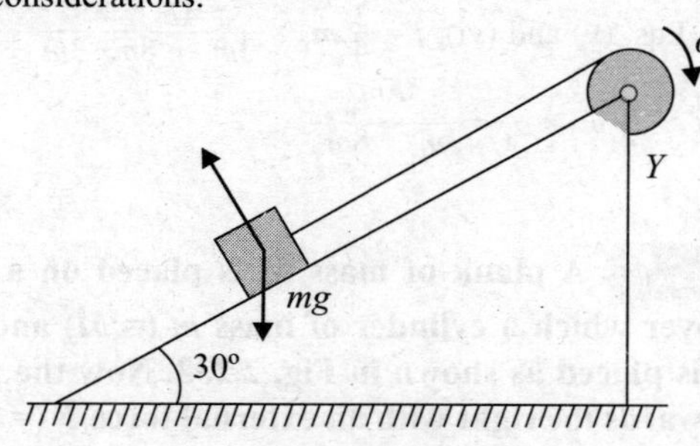

Fig. 2.228

$$W_N + W_{mg} + W_F + W_{Mg} = K_2 - K_1$$

$$0 - (mg\sin\theta)s + 0 + 0 = \left(\frac{1}{2}mv^2 + \frac{1}{2}I\omega^2\right)$$

$$\Rightarrow \quad (mg\sin\theta)s = \frac{1}{2}mv^2 + \frac{1}{2I\omega^2}$$

$$\therefore \quad s = \frac{mv^2 + I\omega^2}{2mg\sin\theta}$$

$$= \frac{(0.5)(2)^2 + \frac{1}{2} \times 2 \times (0.2)^2 \times (10)^2}{2 \times 0.5 \times 10 \times \frac{1}{2}} = 1.2\ \text{m}$$

iii. Third Method: By Mechanical Energy Conservation

Here, the only force which does work on the system mg, the algebraic work done by other forces is zero, is conservative in nature; therefore, the mechanical energy of the system will remain constant. Let the distance moved by the block X along the plane be s.

Gain in gravitational potential energy of $X = mgs\sin\theta$.

Loss in the kinetic energy of the block and the drum is

$$\frac{1}{2}mv_0^2 + \frac{1}{2}I\omega_0^2 (v_0 = \omega_0 r)$$

Thus, $\quad mgs\sin\theta = \frac{1}{2}mv_0^2 + \frac{1}{2}I\omega_0^2$

$$s = \frac{mv_0^2 + I\omega_0^2}{2\,mg\sin\theta}$$

After substituting the given values, $s = 1.02$ m.

Example 2.5 **A small sphere rolls down without slipping from the top of a track in a vertical plane. The track has an elevated section and a horizontal part. The horizontal part is 1.0 m above the ground level and the top of the track is 2.4 m above the ground. Find the distance on the ground with respect to point B (which is vertically below the end of the track as shown in Fig. 2.229) where the sphere lands. During its flight as a projectile, does the sphere continue to rotate about its centre of mass? Explain.**

(IIT-JEE, 1987)

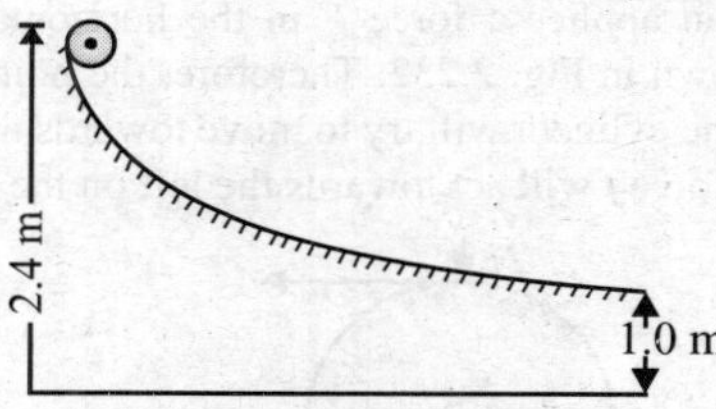

Fig. 2.229

Sol. Applying law of conservation of energy at point D and point A

$$\text{PE at } D = \text{PE at } A + (\text{KE})_T + (\text{KE})_R$$

$$mg(2.4) = mg(1) + \frac{1}{2}mv^2 + \frac{1}{2}I\omega^2$$

Since the case is of rolling without slipping

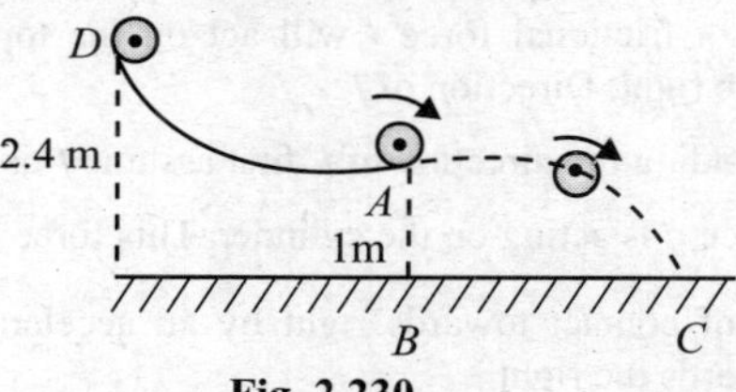

Fig. 2.230

$$\therefore \quad v = r\omega$$

$$\therefore \quad \omega = \frac{v}{r}, \text{ where } r \text{ is the radius of the sphere}$$

Also $\quad I = \frac{2}{5}mr^2$, therefore,

$$mg(2.4) = mg(1) + \frac{1}{2}mv^2 + \frac{1}{2} \times \frac{2}{5}mr^2 \times \frac{v^2}{r^2}$$

$$\Rightarrow \quad v = 4.43\ \text{m/s}$$

After point A, the body takes a parabolic path. The vertical motion parameters of parabolic motion will be

$$u_y = 0, S_y = 1 \text{ m}, a_y - 9.8 \text{ m/s}^2, t_y = ?$$

$$S = ut + \frac{1}{2}at^2, 1 = 4.9\, t_y^2, t_y = \frac{1}{\sqrt{4.9}} = 0.45 \text{ s}$$

Applying this time in horizontal motion of parabolic path, $BC = 4.43 \times 0.45 = 2$ m. During its flight as projectile, the sphere continues to rotate because of conservation of angular momentum.

Example 2.6 **A man pushes a cylinder of mass m_1 with the help of a plank of mass m_2 as shown in Fig. 2.231. There is no slipping at any contact. The horizontal component of the force applied by the man is F.**

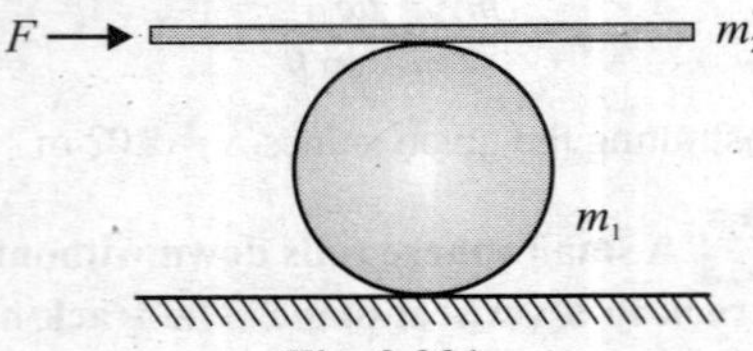

Fig. 2.231

Find:

a. the acceleration of the plank and the centre of mass of the cylinder

b. the magnitudes and directions of frictional forces at contact points **(IIT-JEE, 1999)**

Sol. ***Analysis of the direction of friction force (f)***

The man applies a force F in the horizontal direction on the plank as shown in Fig. 2.232. Therefore, the point of contact of the plank with the cylinder will try to move towards the right. Therefore, the friction force f will act towards the left on the plank.

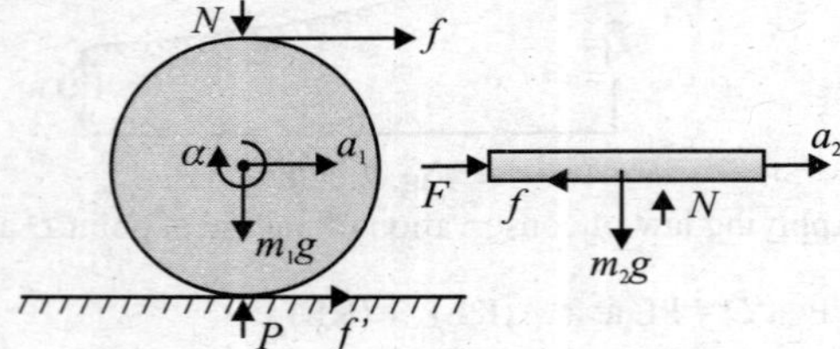

Fig. 2.232

To each and every action, there is an equal and opposite reaction. Therefore, a frictional force f will act on the top of the cylinder towards the right. Direction of f:

For finding the direction of f' first assume f' is not acting.

A force f is acting on the cylinder. This force is trying to move the point of contact towards right by an acceleration $a_{CM} = f/m_1$ acting towards the right.

At the same time, the force f is trying to rotate the cylinder about its center of mass in the clockwise direction.

$$f \times R = I \times \alpha \Rightarrow \alpha = \frac{f \times R}{I} = \frac{f \times R}{\frac{1}{2}m_1R^2} = \frac{2f}{m_1R}$$

Therefore, acceleration of the point of contact,

$$a_p = a_{CM} - \alpha R = \frac{f}{m_1} - \frac{2f}{m_1R} \times R = \frac{-f}{m_1}, \text{ i.e., towards the left.}$$

Therefore, the point of contact of the cylinder with the ground moves towards the left. Therefore, friction force acts towards the right on the cylinder.

Here $k^2 < Rx$ as $k^2 = \frac{R^2}{2}$

and $Rx = R.R = R^2$

Hence, friction should act in forward direction.

Applying Newton's law on the plank, we get $F - f = m_2a_2$ (i)

Also $a_2 = 2a_1$ (ii)

because a_2 is the acceleration of the topmost point of the cylinder and there is no slipping. Applying Newton's law on the cylinder

$$m_1a_1 = f + f' \quad \text{(iii)}$$

The torque equation for the cylinder is

$$f \times R - f' \times R = I\alpha = \frac{1}{2}m_1R^2 \times \left(\frac{a}{R}\right),$$

$\because$ $I = m_1R^2\ R\alpha = a_1$

$\therefore$ $(f - f')R = \frac{1}{2}m_1Ra_1$

$\therefore$ $f + f' = \frac{1}{2}m_1Ra_1$ (iv)

Solving Eqs. (iii) and (iv), we get $f = \frac{3}{4}m_1a_1$ (v)

and $f' = \frac{1}{4}m_1a_1$ (vi)

From Eqs. (i) and (iii), $F - f = 2m_2a_1$

$\Rightarrow$ $F = \frac{3}{4}m_1a_1 = 2m_2a_1$

$\therefore$ $a_1 = \frac{4F}{3m_1 + 8m_2}$

$\therefore$ $a_2 = \frac{8F}{3m_1 + 8m_2}$

From Eqs. (v) and (vi), $f = \frac{3}{4}m_1 \times \frac{4F}{3m_1 + 8m_2} \frac{3Fm_1}{3m_1 + 8m_2}$ and

$$f' = \frac{1}{4}m_1 \times a_1 = \frac{3Fm_1}{3m_1 + 8m_2}$$

Example 2.7 **A plank of mass M is placed on a smooth surface over which a cylinder of mass m (= M) and radius R = 1 m is placed as shown in Fig. 2.233. Now the plank is pulled towards the right with an external force F (= $2Mg$). If the cylinder does not slip over the surface of the plank, find the linear acceleration of the plank and the cylinder and the angular acceleration of the cylinder. (Take $g = 10$ m/s²)**

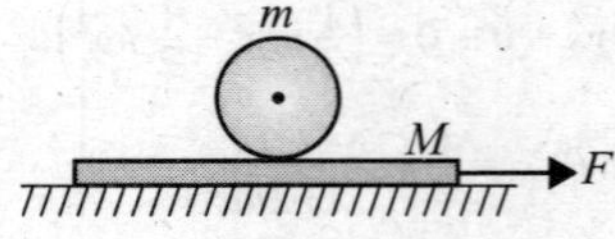

Fig. 2.233

Sol. Here we are given that cylinder does not slip over the plank surface, it is the case of pure rolling, we can use friction on cylinder in any direction. Here we choose towards the right.

As friction is acting on the cylinder towards the right, it must be towards the left on the plank as shown in the force diagram of Fig. 2.234.

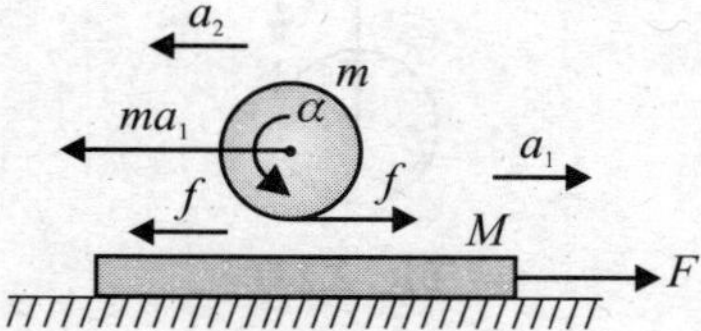

Fig. 2.234

Let the plank move towards the right with an acceleration a_1, the cylinder will experience a pseudoforce ma_1 in the left direction, due to which it will roll towards the left with respect to the plank with an acceleration a_2. As we have used pseudoforce, a_2 must be with respect to the plank. Let its angular acceleration during rolling be α, we have $a_2 = R\alpha$.

For translational motion of the plank, we have

$$F - f = Ma_1 \qquad \text{(i)}$$

For translational motion of the cylinder with respect to the plank, we have

$$ma_1 - f = ma_2 \qquad \text{(ii)}$$

For rotational motion of the cylinder with respect to the plank, we have

$$fR = I\alpha$$

$$fR = \left(\frac{1}{2}mR^2\right)\left(\frac{a_2}{R}\right)$$

or

$$f = \frac{1}{2}ma_2 \qquad \text{(iii)}$$

From Eqs. (i) and (ii), we get

$$ma_1 - \frac{1}{2}ma_2 = ma_2$$

$$a_1 = \frac{3}{2}a_2 \qquad \text{(iv)}$$

Using Eqs. (i), (iii) and (iv), we get

$$F - \frac{1}{2}ma_2 = \frac{3}{2}Ma_2$$

$$a_2 = \frac{2F}{3M + m} = 10 \text{ m/s}^2$$

From Eq. (iv), $\quad a_1 = \dfrac{3F}{3M + m} = 15 \text{ m/s}^2$

As we have already discussed that the value of a_2 is relative to the plank, the net acceleration of the cylinder will be given as $a_1 - a_2$.

Hence, the acceleration of the cylinder is

$$a_{\text{cylinder}} = a_1 - a_2 = 15 - 10 = 5 \text{ m/s}^2$$

The angular acceleration of the cylinder is

$$\alpha = a_2/R = 10 \text{ rad/s}^2$$

EXERCISES

Subjective Type

Solutions on page 2.109

1. A uniform thin circular ring of mass m ($m = 0.4$ kg) and radius R has a small particle of the same mass m fixed on it as shown in Fig. 2.235. The line joining the particle to centre is initially horizontal. The ground is frictionless. Find the contact force (magnitude) exerted by the ground on the ring, when the system is released from rest.

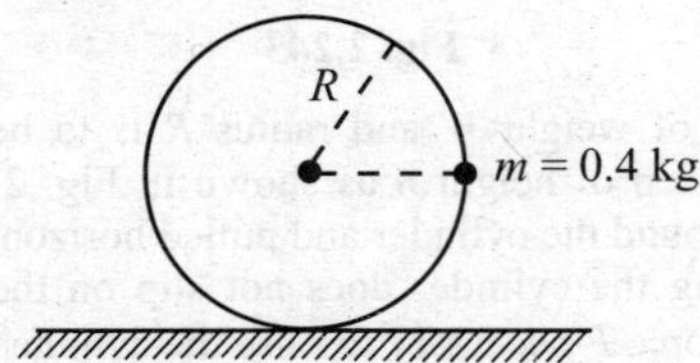

Fig. 2.235

2. Two uniform spheres shown in Fig. 2.236 rotate separately on two parallel horizontal axis passing through their centres. The upper sphere has an angular speed ω_0 and the lower sphere is at rest. Now the two spheres are moved together so that their surfaces touch. After a short time the two spheres, now in contact, are rotating without slipping. Find the final rate of rotation of upper sphere in rad/s [given moment of inertia of the upper sphere $I_1 = 4$ kg m^2, moment of inertia of lower sphere $I_2 = 2$ kg m^2, $a = 2$ m, $b = 1$ m, $\omega_0 = 12$ rad/s].

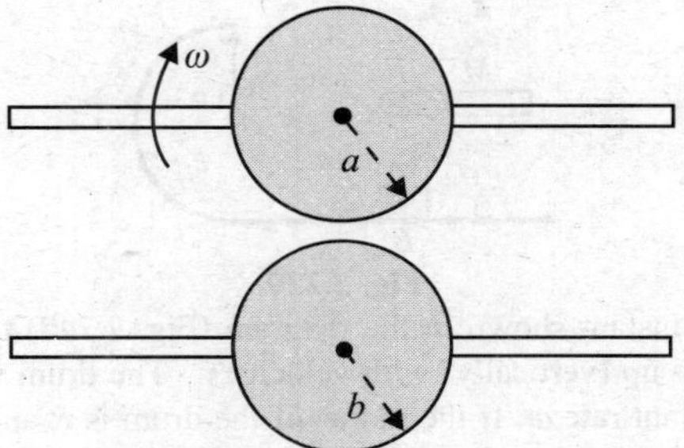

Fig. 2.236

3. One side of a spring of initial, unstretched length $l_0 = 1$ m lying on a frictionless table, is fixed, the other one is fastened to a small puck of mass $m = 0.1$ kg. The puck is given velocity in a direction perpendicular to the spring, at an initial speed $v_0 = 11$ m/s. In the course of the motion, the maximum elongation of the spring is $l = l_0/10$. What is the force constant of the spring (in SI units)?

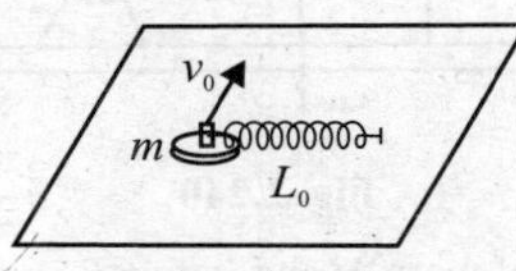

Fig. 2.237

4. A weightless inextensible string which first runs over a fixed weightless pulley D and then coils on a spool B of outer radius R and inner radius $R/2$ tightly. The smaller pulley of spool can roll without sliding along a horizontal fixed rail, as shown. The total mass of the spool is M. The axis O of the spool is perpendicular

to the plane of the drawing and moment of inertia relative to O is 1/2 MR^2. If the end A of the string is pulled downward with constant acceleration $g/2$, then find

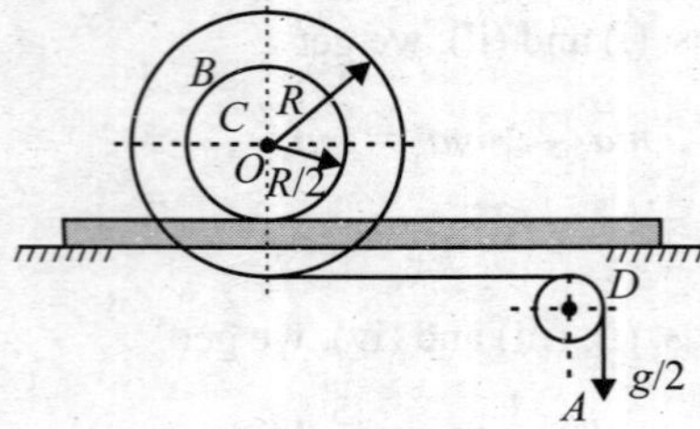

Fig. 2.238

a. direction in which the centre moves.

b. acceleration of O.

c. tension in the string.

d. direction and magnitude of friction force between the shaft and rail.

5. Determine the minimum coefficient of friction between a thin rod and a floor at which a person can slowly lift the rod from the floor, without slipping, to the vertical position applying at its end a force always perpendicular to its length.

6. A rod of length R and mass M is free to rotate about a horizontal axis passing through hinge P as shown in Fig. 2.239. First it is taken aside such that it becomes horizontal and then released. At the lowest point the rod hits block B of mass m and stops. Find the ratio of masses such that block B completes the circle. Neglect any friction.

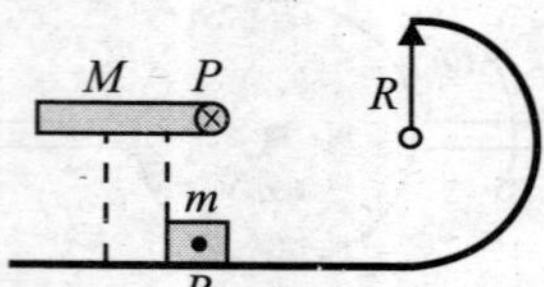

Fig. 2.239

7. In the instant shown in the diagram (Fig. 2.240) the board is moving up (vertically) with velocity v. The drum winds up at a constant rate ω. If the radius of the drum is R and the board always remains horizontal, find the value of velocity v in terms of R, θ, ω.

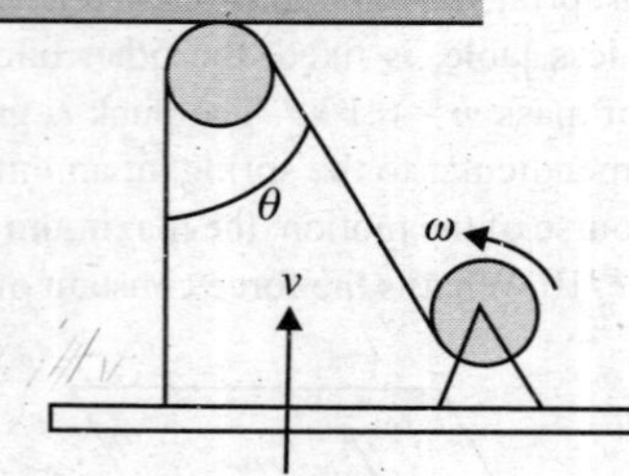

Fig. 2.240

8. A uniform disc of mass M and radius R is hanging from a rigid support and is free to rotate about a horizontal axis passing through its centre in vertical plane as shown in Fig. 2.241. An insect of mass m falls vertically and hits the disc at a point at horizontal diameter with a velocity v_0 so that the disc completes the vertical circular motion. Find the value of minimum value of v.

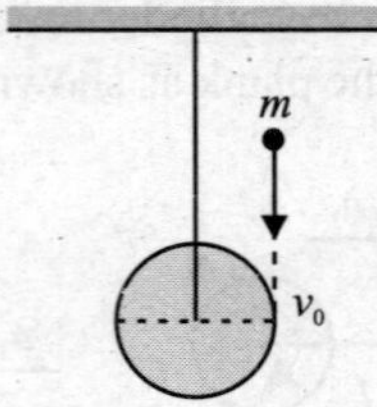

Fig. 2.241

9. A weightless rod of length l with a small load of mass m at the end is hinged at point A as shown and occupies a strictly vertical position, touching a body of mass M. A light jerk sets the system in motion.

a. For what mass ratio M/m will the rod form an angle $\alpha = \pi/6$ with the horizontal at the moment of the separation from the body?

b. What will be the velocity u of the body at this moment? Friction should be neglected.

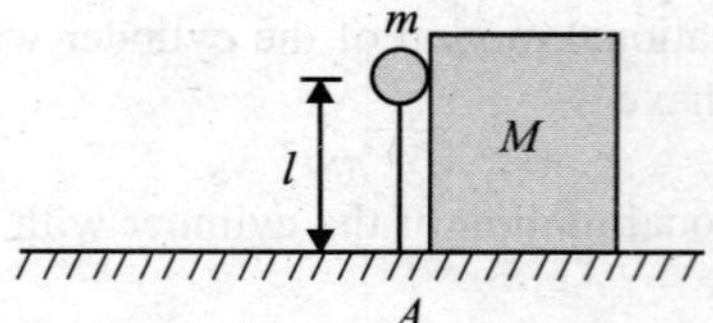

Fig. 2.242

10. A solid cylinder of mass $m = 6$ kg and radius $r = 0.1$ m is kept in balance on a slope of inclination $\alpha = 37°$ with the help of a thread fastened to its jacket. The cylinder does not slip on the slope. Find the minimum required coefficient of friction to keep the cylinder in balance when the thread is held vertically.

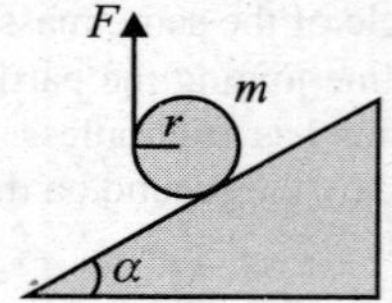

Fig. 2.243

11. A cylinder of weight W and radius R is to be raised onto a horizontal step of height h as shown in Fig. 2.244. A rope is wrapped around the cylinder and pulled horizontally with force F. Assuming the cylinder does not slip on the step, find the minimum force F necessary to raise the cylinder.

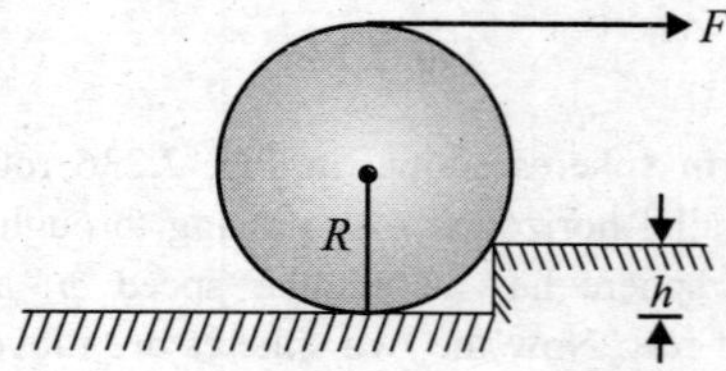

Fig. 2.244

12. A spherical ball of mass M and radius a moving with velocity v strikes elastically with a rigid surface at an angle θ to the normal. Assume that slipping occurs while the sphere is in contact with the surface, the frictional coefficient is μ_k; show that

a. The ball is reflected at an angle ϕ to the normal where $\tan\theta - \tan\phi| = 2\mu$

b. The change in angular velocity is $5\mu v/a \cos\theta$

13. Two identical steel balls are connected by a 40 CM long rigid bar of negligible mass. They fall in a horizontal position from a height 10 CM on two heavy steel and brass plates as shown in Fig. 2.245. The coefficient of restitution between the balls and steel and brass plates are 0.6 and 0.4, respectively. Assuming that the two balls hit the respective plates at the same instant, find the angular velocity ω of the bar immediately after impact

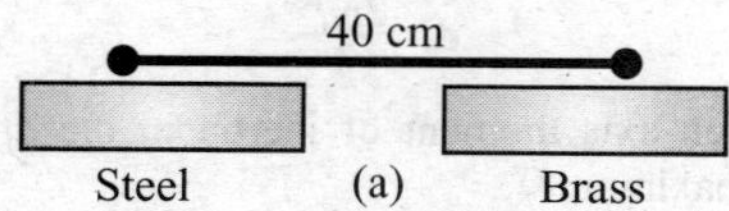

Fig. 2.245

14. A spinning cylinder with angular velocity ω_0 of mass M and radius R is lowered on a rough inclined plane of angle 30° with the horizontal and $\mu = 1/\sqrt{3}$. The cylinder is released at a height of $3R$ from horizontal. Find the total time taken by the cylinder to reach the bottom of the incline.

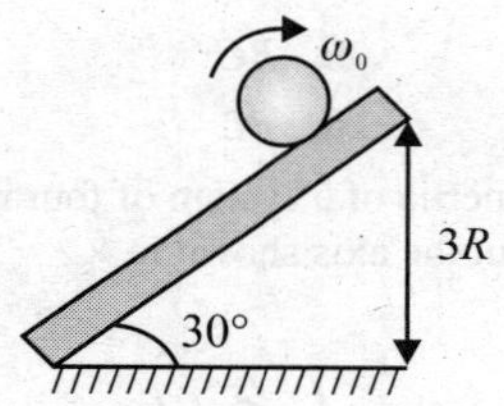

Fig. 2.246

15. A hoop of mass m is projected on a floor with linear velocity v_0 and reverse spin ω_0. The coefficient of friction between the hoop and the ground is μ.

a. Under what condition will the hoop return back?

b. How far will it go?

c. How long will it continue to slip when its centre of mass becomes stationary?

d. What is the velocity of return?

16. A body of mass M and radius r, rolling on a smooth horizontal floor with velocity v, rolls up an irregular inclined plane up to a vertical height $(3v^2/4g)$. Compute the moment of inertia of the body.

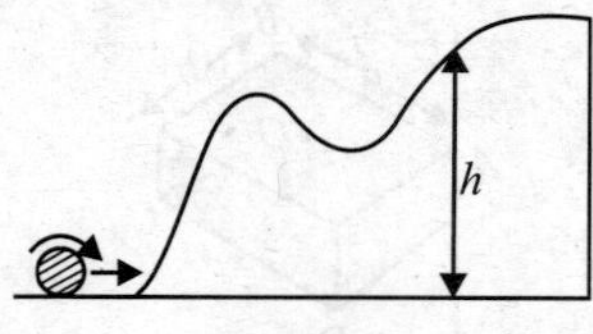

Fig. 2.247

17. A ring of mass 0.3 kg and radius 0.1 m and a solid cylinder of mass 0.4 kg and of the same radius are given the same kinetic energy and released simultaneously on a flat horizontal surface such that they begin to roll as soon as released towards a wall which is at the same distance from the ring and the cylinder. Which body, ring or cylinder, will reach the wall first?

Objective Type

Solutions on page 2.113

1. Two rings of same radius and mass are placed such that their centres are at a common point and their planes are perpendicular to each other. The moment of inertia of the system about an axis passing through the centre and perpendicular to the plane of one of the rings is (mass of the ring = m, radius = r)

a. $\frac{1}{2}mr^2$ **b.** mr^2

c. $\frac{3}{2}mr^2$ **d.** $2mr^2$

2. The moment of inertia of a solid sphere about an axis passing through the centre of gravity is $1/2\ MR^2$; then its radius of gyration about a parallel axis at a distance $2R$ from first axis is

a. $5R$ **b.** $\sqrt{\frac{22}{5}}R$

c. $\frac{5}{2}R$ d. $\sqrt{\frac{12}{5}}R$

3. From a given sample of uniform wire, two circular loops P and Q are made, P of radius r and Q of radius nr. If the M.I. of Q about its axis is four times that of P about its axis (assuming the wire to be diameter much smaller than either radius), the value of n is

a. $(4)^{\frac{2}{3}}$ **b.** $(4)^{\frac{1}{3}}$

c. $(4)^{\frac{1}{2}}$ **d.** $(4)^{\frac{1}{4}}$

4. Two circular discs A and B are of equal masses and thicknesses but made of metal with densities d_A and d_B ($d_A > d_B$). If their moments of inertia about an axis passing through their centres and perpendicular to circular faces are I_A and I_B, then

a. $I_A = I_B$ **b.** $I_A > I_B$

c. $I_A < I_B$ **d.** $I_A \geq I_B$

5. Four identical rods are joined end to end to form a square. The mass of each rod is M. The moment of inertia of the square about the median line is

a. $\frac{Ml^2}{3}$ **b.** $\frac{Ml^2}{4}$

c. $\frac{Ml^2}{6}$ **d.** none of these

6. In problem 5, the moment of inertia of the system about an axis passing through the point of intersection of diagonals and perpendicular to the plane of the square is

a. $\frac{4Ml^2}{3}$ **b.** $\frac{13Ml^2}{3}$

c. $\frac{Ml^2}{6}$ **d.** $\frac{13Ml^2}{6}$

7. In problem 5, the moment of inertia of the system about one of the diagonals is

a. $\frac{2Ml^2}{3}$ **b.** $\frac{13Ml^2}{3}$

c. $\frac{Ml^2}{6}$ **d.** $\frac{13Ml^2}{6}$

8. Two circular discs are of same thickness. The diameter of A is twice that of B. The moment of inertia of A as compared to that of B is

a. twice as large **b.** four times as large
c. eight times as large **d.** 16 times as large

9. Two thin discs, each of mass M and radius r m, are attached as shown in Fig. 2.248, to form a rigid body. The rotational inertia of this body about an axis perpendicular to the plane of disc B passing through its centre is

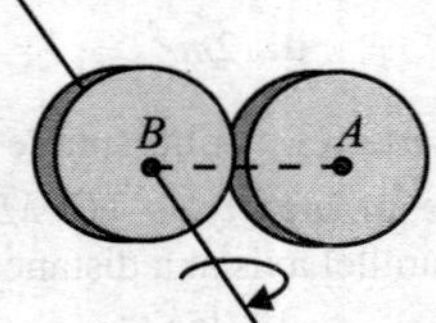

Fig. 2.248

a. $2Mr^2$ **b.** $3Mr^2$
c. $4Mr^2$ **d.** $5Mr^2$

10. An isosceles triangular piece is cut from a square plate of side l. The piece is one-fourth of the square and mass of the remaining plate is M. The moment of inertia of the plate about an axis passing through O and perpendicular to its plane is

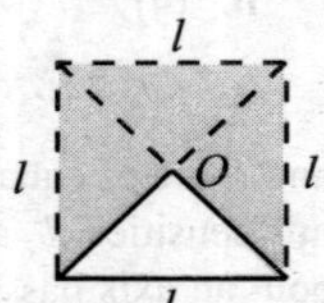

Fig. 2.249

a. $\frac{Ml^2}{6}$ **b.** $\frac{Ml^2}{12}$
c. $\frac{Ml^2}{24}$ **d.** $\frac{Ml^2}{3}$

11. Three rings, each of mass m and radius r, are so placed that they touch each other. Find the moment of inertia about the axis as shown in Fig. 2.250.

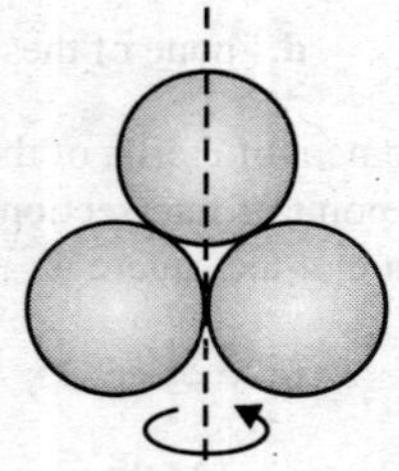

Fig. 2.250

a. $5mr^2$ **b.** $\frac{5}{7}mr^2$
c. $7mr^2$ **d.** $\frac{7}{2}mr^2$

12. Three identical rods, each of mass m and length l, form an equaliteral triangle. Moment of inertia about one of the sides is

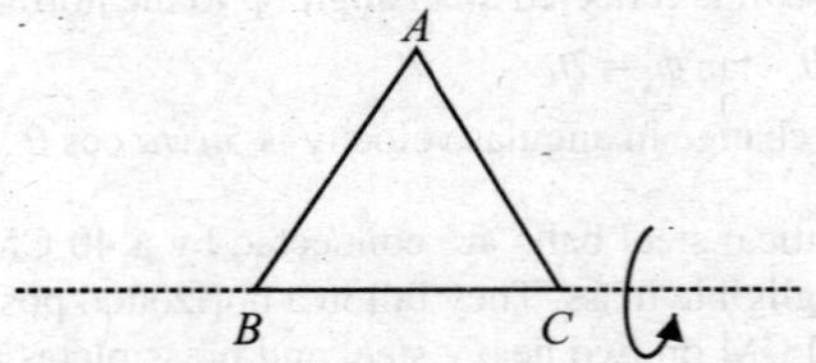

Fig. 2.251

a. $\frac{ml^2}{4}$ **b.** ml^2
c. $\frac{3ml^2}{4}$ **d.** $\frac{2ml^2}{3}$

13. About which axis moment of inertia in the given triangular lamina is maximum?

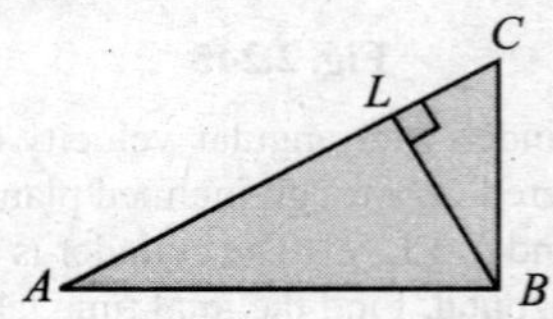

Fig. 2.252

a. AB **b.** BC
c. AC **d.** BL

14. The moment of inertia of a system of four rods, each of length l and mass m, about the axis shown is

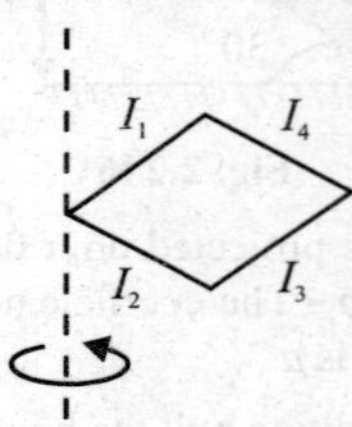

Fig. 2.253

a. $\frac{2}{3}ml^2$ **b.** $2ml^2$
c. $3ml^2$ **d.** $\frac{8}{3}ml^2$

15. Figure 2.254 shows a uniform solid block of mass M and edge lengths a, b and c. Its M.I. about an axis through one edge and perpendicular (as shown) to the large face of the block is

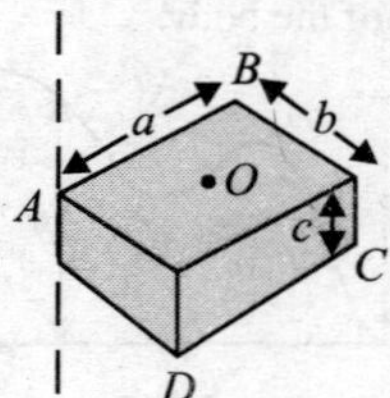

Fig. 2.254

a. $\frac{M}{3}(a^2 + b^2)$ **b.** $\frac{M}{4}(a^2 + b^2)$
c. $\frac{7M}{12}(a^2 + b^2)$ **d.** $\frac{M}{12}(a^2 + b^2)$

16. In rectangle *ABCD*, $AB = 2l$ and $BC = l$. Axes *xx* and *yy* pass through the centre of the rectangle. The moment of inertia is least about

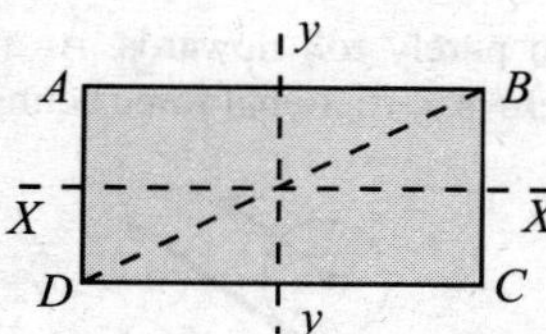

Fig. 2.255

a. *DB* **b.** *BC*

c. *xx* **d.** *yy*

17. Figure 2.256 shows a thin metallic triangular sheet *ABC*. The mass of the sheet is *M*. The moment of inertia of the sheet about side *AC* is

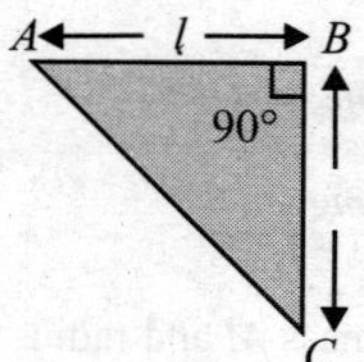

Fig. 2.256

a. $\frac{Ml^2}{18}$ **b.** $\frac{Ml^2}{12}$

c. $\frac{Ml^2}{6}$ **d.** $\frac{Ml^2}{4}$

18. The moment of inertia of a door of mass *m*, length $2l$ and width l about its longer side is

a. $\frac{11ml^2}{24}$ **b.** $\frac{5ml^2}{24}$

c. $\frac{ml^2}{3}$ **d.** none of these

19. A square plate of edge $a/2$ is cut out from a uniform square plate of edge 'a' as shown in Fig. 2.257. The mass of the remaining portion is *M*. The moment of inertia of the shaded portion about an axis passing through '*O*' (centre of the square of side *a*) and perpendicular to the plane of plate is

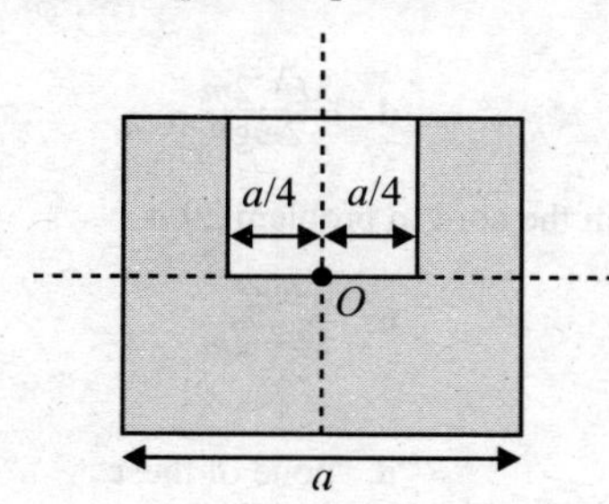

Fig. 2.257

a. $\frac{9}{64}Ma^2$ **b.** $\frac{3}{16}Ma^2$

c. $\frac{5}{12}Ma^2$ **d.** $\frac{Ma^2}{6}$

20. A uniform ball of radius *r* rolls without slipping down from the top of a sphere of radius *R*. The angular velocity of the ball when it breaks from the sphere is

a. $\sqrt{\frac{5g(R+r)}{17r^2}}$ **b.** $\sqrt{\frac{10g(R+r)}{17r^2}}$

c. $\sqrt{\frac{5g(R-r)}{10r^2}}$ **d.** $\sqrt{\frac{10g(R+r)}{7r^2}}$

21. A solid sphere rests on a horizontal surface. A horizontal impulse is applied at height *h* from centre. The sphere starts rolling just after the application of impulse. The ratio *h*/*r* will be

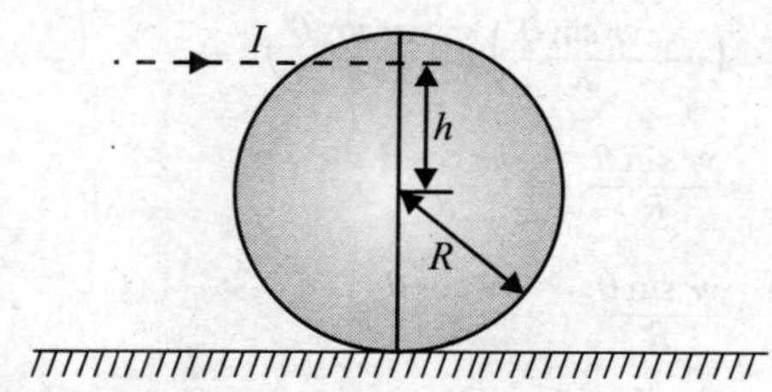

Fig. 2.258

a. $\frac{1}{2}$ **b.** $\frac{2}{5}$

c. $\frac{1}{5}$ **d.** $\frac{2}{3}$

22. In Fig. 2.259, a sphere of radius 2 m rolls on a plank. The accelerations of the sphere and the plank are indicated. The value of α is

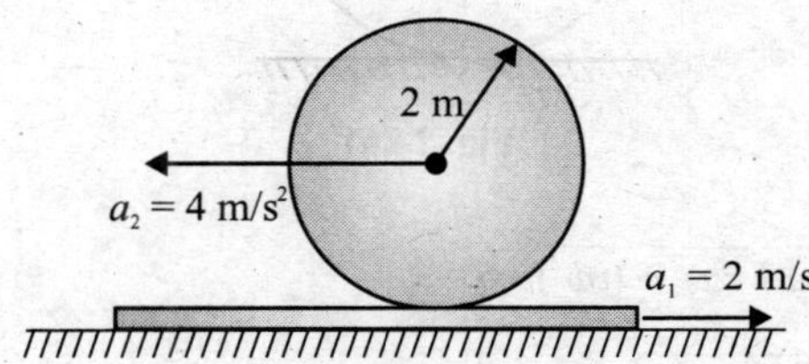

Fig. 2.259

a. 2 rad/s^2 **b.** 4 rad/s^2

c. 3 rad/s^2 **d.** 1 rad/s^2

23. A ring of radius *R* is first rotated with an angular velocity ω_0 and then carefully placed on a rough horizontal surface. The coefficient of friction between the surface and the ring is μ. Time after which its angular speed is reduced to half is

a. $\frac{\omega_0 \mu R}{2g}$ **b.** $\frac{\omega_0 g}{2\mu R}$

c. $\frac{2\omega_0 R}{\mu g}$ **d.** $\frac{\omega_0 R}{2\mu g}$

24. A disc of radius *R* rolls without slipping at speed *v* along positive *x*-axis. Velocity of point *P* at the instant shown in Fig. 2.260 is

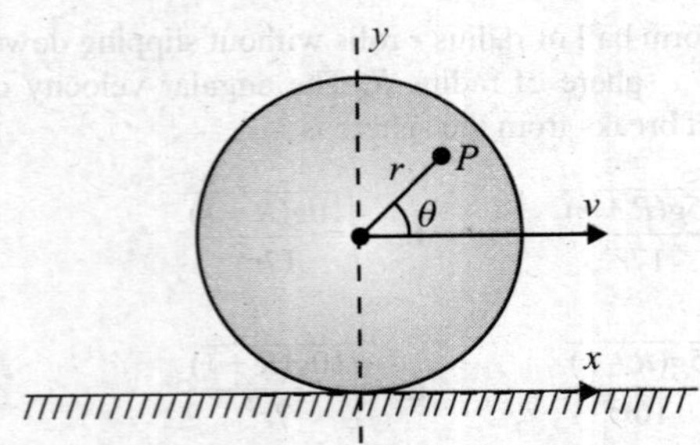

Fig. 2.260

a. $\vec{V}_p = \left(v + \frac{vr \sin\theta}{R}\right)\hat{i} + \frac{vr \cos\theta}{R}\hat{j}$

b. $\vec{V}_p = \left(v + \frac{vr \sin\theta}{R}\right)\hat{i} - \frac{vr \cos\theta}{R}\hat{j}$

c. $\vec{V}_p = \frac{vr \sin\theta}{R}\hat{i} + \frac{vr \cos\theta}{R}\hat{j}$

d. $\vec{V}_p = \frac{vr \sin\theta}{R}\hat{i} - \frac{vr \cos\theta}{R}\hat{j}$

25. A disc of radius R rolls on a horizontal ground with linear acceleration a and angular acceleration α as shown in Fig. 2.261. The magnitude of acceleration of point P as shown in the figure at an instant when its linear velocity is v and angular velocity is ω will be

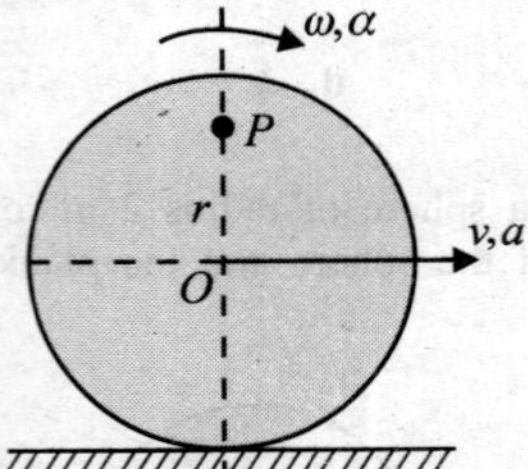

Fig. 2.261

a. $\sqrt{(a + r\alpha)^2 + (r\omega^2)^2}$ b. $\frac{ar}{R}$

c. $\sqrt{r^2\alpha^2 + r^2\omega^4}$ d. $r\alpha$

26. Two bodies with moments of inertia I_1 and I_2 ($I_1 > I_2$) have equal angular momenta. If their kinetic energies of rotation are E_1 and E_2, respectively, then

a. $E_1 = E_2$ b. $E_1 < E_2$

c. $E_1 > E_2$ d. $E_1 \geq E_2$

27. In Fig. 2.262, the velocities are in ground frame and the cylinder is performing pure rolling on the plank, velocity of point 'A' would be

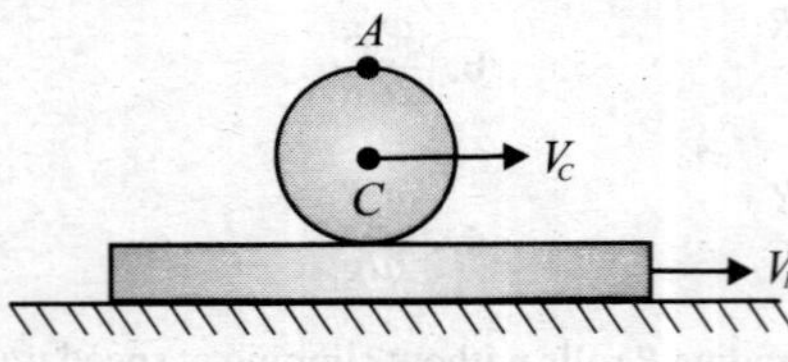

Fig. 2.262

a. $2V_C$ b. $2V_C + V_P$

c. $2V_C - V_P$ d. $2(V_C - V_P)$

28. A sphere has to purely roll upwards. At an instant when the velocity of sphere is v, frictional force acting on it is

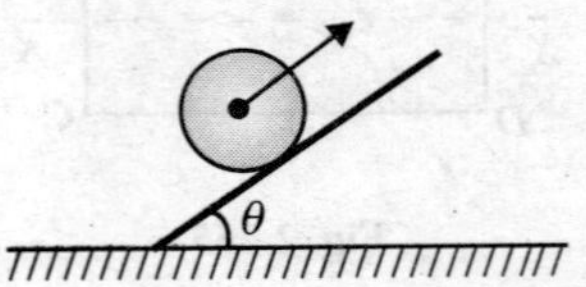

Fig. 2.263

a. downwards and $\mu mg \cos\theta$

b. downwards and $\frac{2mg \sin\theta}{7}$

c. upwards and $\mu mg \cos\theta$

d. upwards and $\frac{2mg \sin\theta}{7}$

29. A uniform disc of mass M and radius R is mounted on an axle supported in frictionless bearings. A light cord is wrapped around the rim of the disc and a steady downward pull T is exerted on the cord. The angular acceleration of the disc is

a. $\frac{T}{MR}$ b. $\frac{MR}{T}$

c. $\frac{2T}{MR}$ d. $\frac{MR}{2T}$

30. In problem 29, the tangential acceleration of a point on the rim is

a. $\frac{T}{M}$ b. $\frac{MR}{T}$

c. $\frac{2T}{MR}$ d. $\frac{MR}{2T}$

31. If we hang a body of mass m with the cord in problem 29, the tangential acceleration of the disc will be

a. $\frac{mg}{M+m}$ b. $\frac{mg}{M+2m}$

c. $\frac{2mg}{M+2m}$ d. $\frac{M+2m}{2mg}$

32. The tension in the cord in problem 29 is

a. $\frac{mg}{M+m}$ b. $\frac{mg}{M+2m}$

c. $\frac{2mg}{Mmg}$ d. none of these

33. A triangular plate of uniform thickness and density is made to rotate about an axis perpendicular to the plane of the paper and (a) passing through A, (b) passing through B, by the application of some force F at C (mid-point of AB) as shown in Fig. 2.264. In which case is angular acceleration more?

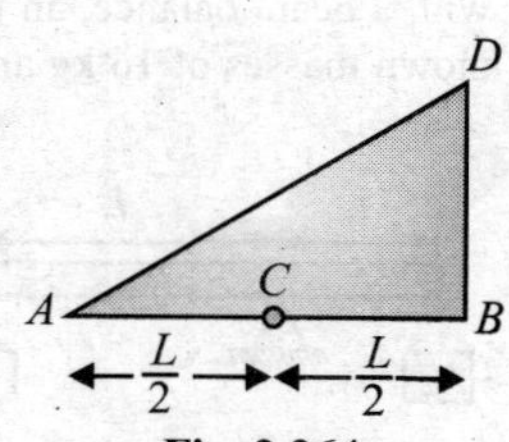

Fig. 2.264

a. in case (a) b. in case (b)

c. both (a) and (b) d. none of these

34. A uniform rod of mass M and length L is pivoted at one end such that it can rotate in a vertical plane. There is negligible friction at the pivot. The free end of the rod is held vertically above the pivot and then released. The angular acceleration of the rod when it makes an angle θ with the vertical is

a. $g \sin\theta$ b. $\frac{g}{L}\sin\theta$

c. $\frac{3g}{2L}\sin\theta$ d. $6gL\sin\theta$

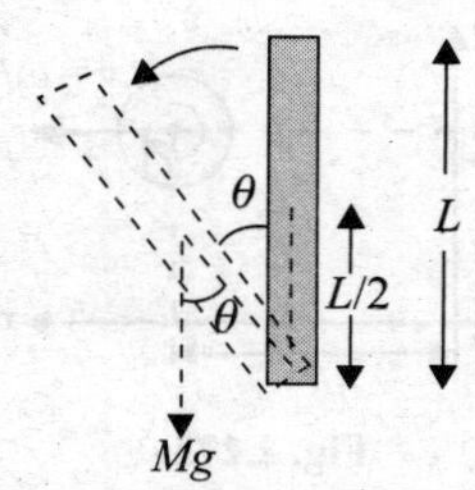

Fig. 2.265

35. A string is wrapped around a cylinder of mass M and radius R. The string is pulled vertically upwards to prevent the centre of mass from falling as the cylinder unwinds the string. The tension in the string is

a. $\frac{Mg}{6}$ b. $\frac{Mg}{3}$

c. $\frac{Mg}{2}$ d. $\frac{2Mg}{3}$

36. In problem 35, work done on the cylinder for reaching an angular speed ω is

a. $\frac{MR^2\omega^2}{4}$ b. $\frac{MR^2\omega^2}{2}$

c. $\frac{MR^2\omega^2}{3}$ d. $\frac{2MR^2\omega^2}{3}$

37. In Fig. 2.266, the bar is uniform and weighing 500 N. How large must W be if T_1 and T_2 are to be equal?

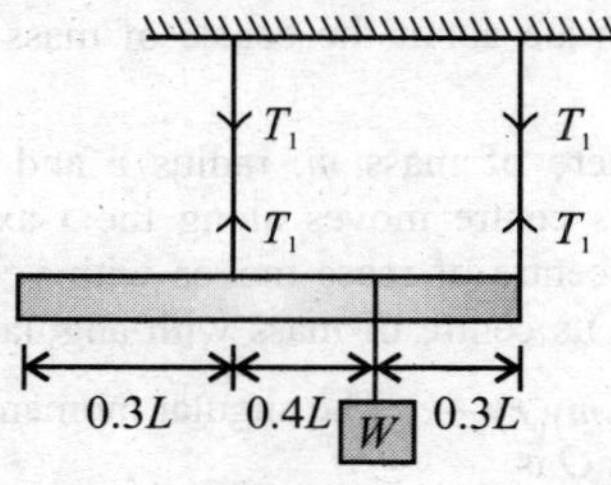

Fig. 2.266

a. 500 N b. 300 N

c. 750 N d. 1500 N

38. A smooth uniform rod of length L and mass M has two identical beads (1 and 2) of negligible size, each of mass m, which can slide freely along the rod. Initially the two beads are at the centre of the rod and the system is rotating with angular velocity ω_0 about an axis perpendicular to the rod and is passing through its midpoint. There are no external forces when the beads reach the ends of the rod, the angular velocity of the system is

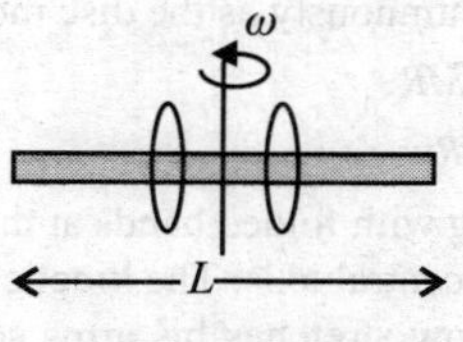

Fig. 2.267

a. $\frac{M\omega_0}{M+6m}$ b. $\frac{M\omega_0}{m}$

c. $\frac{M\omega_0}{M+12m}$ d. ω_0

39. A uniform thin rod of length l and mass m is hinged at a distance $l/4$ from one of the end and released from horizontal position as shown in Fig. 2.268. The angular velocity of the rod as it passes the vertical position is

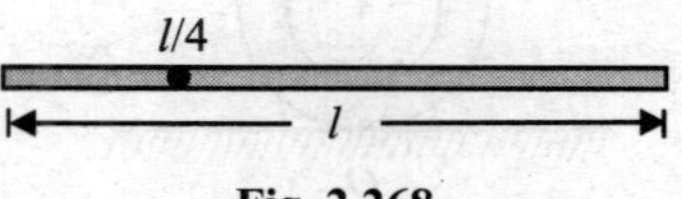

Fig. 2.268

a. $2\sqrt{\frac{5g}{7l}}$ b. $2\sqrt{\frac{6g}{7l}}$

c. $\sqrt{\frac{3g}{7l}}$ d. $2\sqrt{\frac{g}{l}}$

40. A disc is rotated about its axis with a certain angular velocity and lowered gently on a rough inclined plane as shown in Fig. 2.269, then

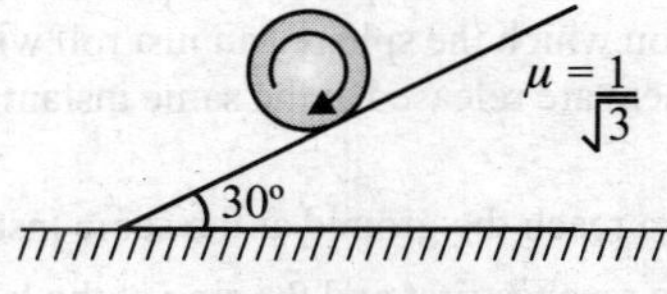

Fig. 2.269

a. It will rotate at the position where it was placed and then will move downwards

b. It will go downwards just after it is lowered

c. It will go downwards first and then climb up

d. It will climb upwards and then move downwards

41. A disc of mass M and radius R rolls without slipping on a horizontal surface. If the velocity of its centre is v_0, then the total angular momentum of the disc about a fixed point P at a height $3/2\ R$ above the centre C

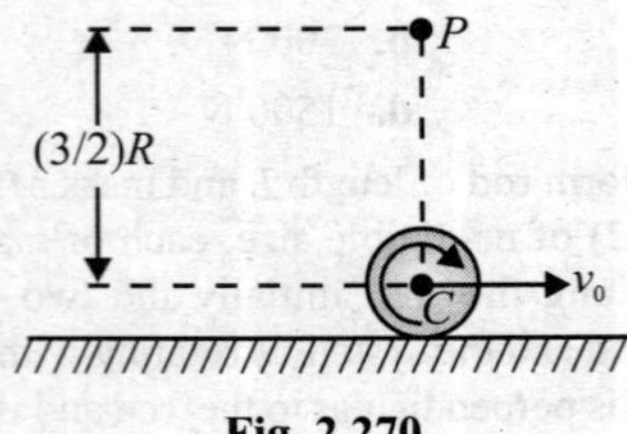

Fig. 2.270

a. increases continuously as the disc moves away

b. decreases continuously as the disc moves away

c. is equal to $2MRv_0$

d. is equal to MRv_0

42. A child is standing with folded hands at the centre of a platform rotating about its central axis. The kinetic energy of the system is K. The child now stretches his arms so that the moment of inertia of the system doubles. The kinetic energy of the system now is

a. $2K$ b. $\frac{K}{2}$

c. $\frac{K}{4}$ d. $4K$

43. A yo-yo is placed on a rough horizontal surface and a constant force F, which is less than its weight, pulls it vertically. Due to this

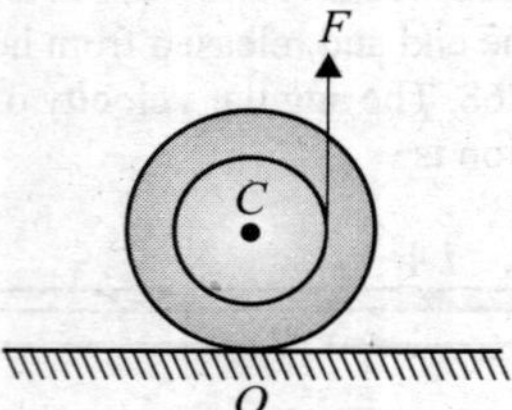

Fig. 2.271

a. friction force acts towards left, so it will move towards left

b. friction force acts towards right, so it will move towards right

c. it will move towards left, so friction force acts towards left

d. it will move towards right so friction force acts towards right

44. A ring, cylinder and solid sphere are placed on the top of a rough incline on which the sphere can just roll without slipping. When all of them are released at the same instant from the same position, then

a. all of them reach the ground at the same instant

b. the sphere reaches first and the ring at the last

c. the sphere reaches first and the cylinder and ring reach together

d. none of the above

45. If a spherical ball rolls on a table without slipping, the fraction of its total energy associated with rotation is

a. $\frac{3}{5}$ b. $\frac{2}{7}$

c. $\frac{2}{5}$ d. $\frac{3}{7}$

46. In an experiment with a beam balance, an unknown mass m is balanced by two known masses of 16 kg and 4 kg as shown in Fig. 2.272.

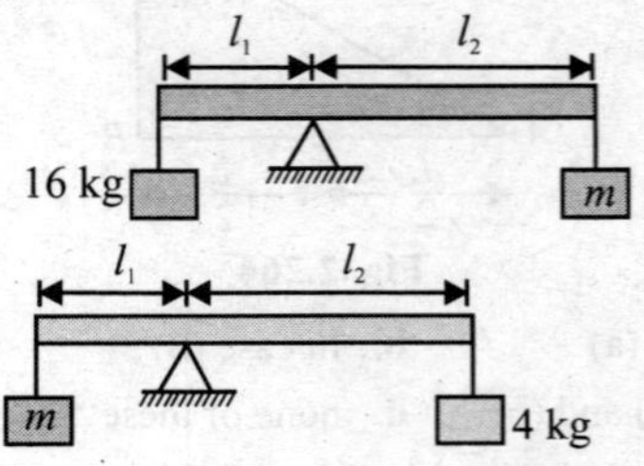

Fig. 2.272

The value of the unknown mass m is

a. 10 kg b. 6 kg

c. 8 kg d. 12 kg

47. A disc of mass m and radius R moves in the x-y plane as shown in Fig. 2.273. The angular momentum of the disc about the origin O at the instant shown is

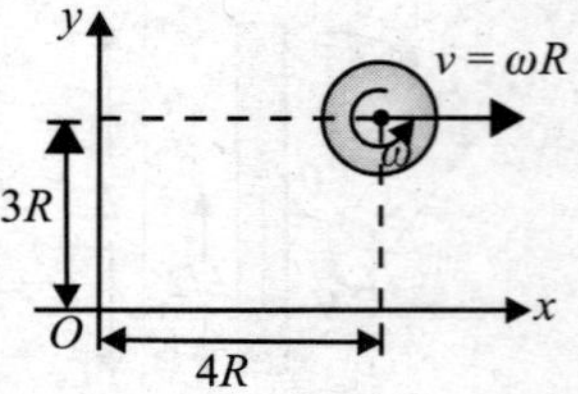

Fig. 2.273

a. $-\frac{5}{2}mR^2\omega\,\hat{k}$ b. $\frac{7}{3}mR^2\omega\,\hat{k}$

c. $-\frac{9}{2}mR^2\omega\,\hat{k}$ d. $\frac{5}{2}mR^2\omega\,\hat{k}$

48. A spool is pulled horizontally by two equal and opposite forces as shown in Fig. 2.274. Which of the following statements are correct?

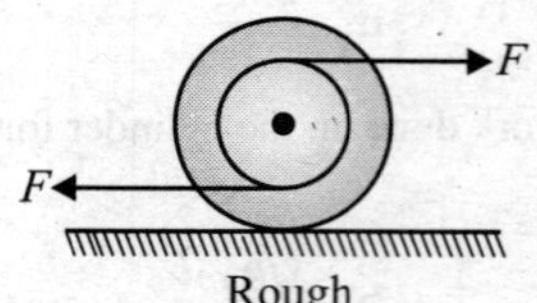

Fig. 2.274

a. The centre of mass moves towards left

b. The centre of mass moves towards right

c. The centre of mass remains stationary

d. The net torque about the centre of mass of the spool is zero

49. A uniform sphere of mass m, radius r and moment of inertia I about its centre moves along the x-axis as shown in Fig. 2.275. Its centre of mass moves with velocity $= v_0$, and it rotates about its centre of mass with angular velocity $= \omega_0$. Let $\vec{L} = (I\omega_0 + mv_0r)(-k)$. The angular momentum of the body about the origin O is

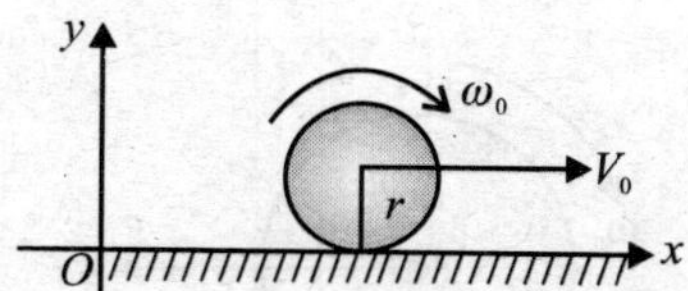

Fig. 2.275

a. $\vec{L}$, only if $v_0 = \omega_0 r$

b. greater than $\vec{L}$, if $v_0 > \omega_0 r$

c. less than $\vec{L}$, if $v_0 > \omega_0 r$

d. $\vec{L}$, for all values of ω_0 and v_0

50. An impulse $J = mv$ at one end of a stationary uniform frictionless rod of mass m and length l which is free to rotate in a gravity-free space. The impact is elastic. Instantaneous axis of rotation of the rod will pass through

a. its centre of mass

b. the centre of mass of the rod plus ball

c. the point of impact of the ball on the rod

d. the point which is at a distance $2l/3$ from the striking end

51. A solid sphere of mass m is lying at rest on a rough horizontal surface. The coefficient of friction between the ground and sphere is μ. The maximum value of F, so that the sphere will not slip, is equal to

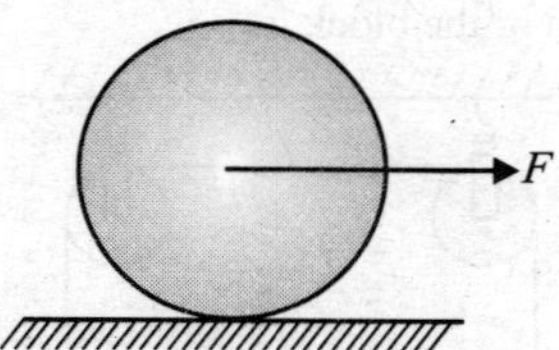

Fig. 2.276

a. $\frac{7}{5}\mu mg$ b. $\frac{4}{7}\mu mg$

c. $\frac{5}{7}\mu mg$ d. $\frac{7}{2}\mu mg$

52. A sphere is moving towards +ve x-axis with a velocity v_c and rotates clockwise with angular speed ω as shown in Fig. 2.277, such that $v_c > \omega R$. The instantaneous axis of rotation will be

a. on point P b. on point P'

c. inside the sphere d. outside the sphere

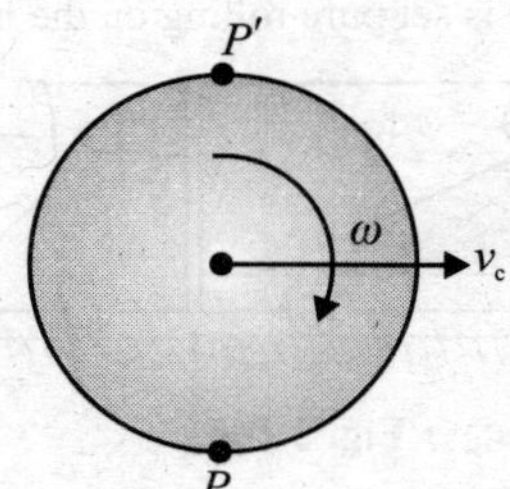

Fig. 2.277

53. A cylinder of height H and diameter $H/4$ is kept on a frictional turntable as shown in Fig. 2.278. The axis of the cylinder is perpendicular to the surface of the table and the distance of axis of the cylinder is $2H$ from the centre of the table. The angular speed of the turntable at which the cylinder will start toppling (assume that friction is sufficient to prevent slipping) is

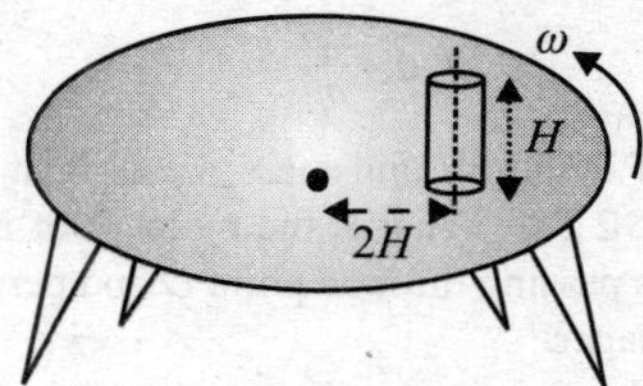

Fig. 2.278

a. $\sqrt{\frac{g}{2}\left(\frac{1}{2} - H\right)}$ b. $\sqrt{g\left(\frac{1}{2} - H\right)}$

c. $\sqrt{\frac{g}{4H}}$ d. $\sqrt{\frac{g}{8H}}$

54. In Fig. 2.279, the rod AB of mass $2m$ and length $4R$ is rigidly attached to a disc of mass m, and radius R in the same plane. The system has prescribed motion in its own plane defined by the velocity v of its centre of mass G and its angular velocity ω. If the end A of the rod is suddenly fixed by a pin, the new angular velocity ω' around point A is

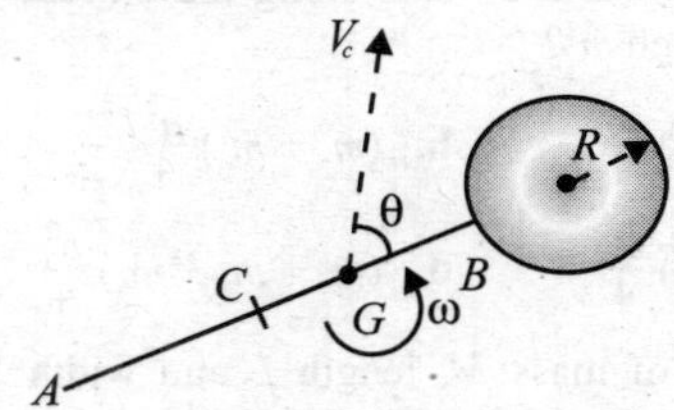

Fig. 2.279

a. $\dfrac{13\, v_c \cos\theta + 55\, R\omega}{73R}$ b. $\dfrac{18\, v_c \sin\theta + 15\, R\omega}{73R}$

c. $\dfrac{54\, v_c \sin\theta + 55\, R\omega}{217\, R}$ d. none of these

55. A uniform rod AB of length l and mass m hangs from point A in a car moving with velocity v_0 on an inclined plane as shown in Fig. 2.280. The rod can rotate in vertical plane about the axis at point A. If the car suddenly stops, the angular speed with which the rod starts rotating is

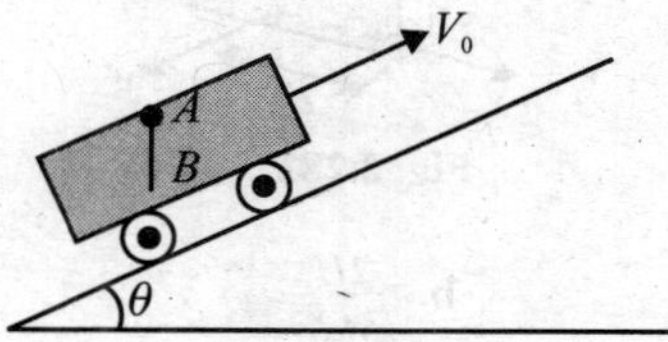

Fig. 2.280

a. $\frac{3}{2}\frac{v_0}{l}\cos\theta$ b. $\frac{v_0 \cos\theta}{2\, l}$

c. $\frac{3}{2}\frac{v_0}{l}\sin\theta$ d. $\frac{5}{2}\frac{v_0}{l}\sin\theta$

56. A uniform rod of mass m, length l rests on a smooth horizontal surface. Rod is given a sharp horizontal impulse p perpendicular to the rod at a distance $l/4$ from the centre. The angular velocity of the rod will be

a. $\frac{3p}{ml}$ **b.** $\frac{p}{ml}$

c. $\frac{p}{2ml}$ **d.** $\frac{2p}{ml}$

57. A thin rod of length $4l$ and mass $4m$ is bent at the points as shown in Fig. 2.281. What is the moment of inertia of the rod about the axis passing through point O and perpendicular to the plane of the paper.

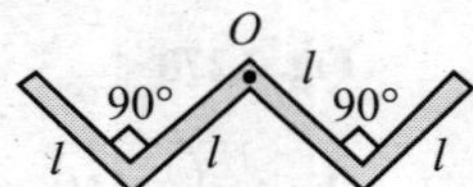

Fig. 2.281

a. $\frac{Ml^2}{3}$ **b.** $\frac{10Ml^2}{3}$

c. $\frac{Ml^2}{12}$ **d.** $\frac{Ml^2}{24}$

58. Three point masses m_1, m_2 and m_3 are located at the vertices of an equilateral triangle of side 'a'. What is the moment of inertia of the system about an axis along the altitude of the triangle passing through m_1?

a. $(m_1 + m_2)\frac{a^2}{4}$ **b.** $(m_2 + m_3)\frac{a^2}{4}$

c. $(m_1 + m_3)\frac{a^2}{4}$ **d.** $(m_1 + m_2 + m_3)\frac{a^2}{4}$

59. A thin plate of mass M, length L and width $2d$ is mounted vertically on a frictionless fixed axle along the z-axis as shown. Initially, the object is at rest. it is then tapped with a hammer to provide a torque τ, which produces an angular impulse H about the z-axis of magnitude $H = \int \tau\, dt$. What is the angular speed ω of the plate about the z-axis after the tap?

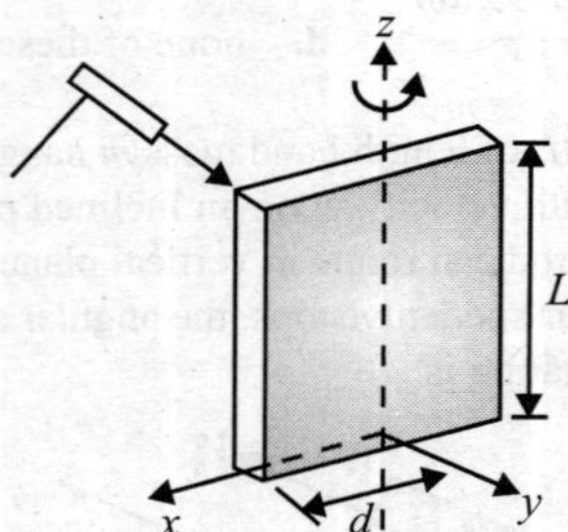

Fig. 2.282

a. $\frac{H}{Md^2}$ **b.** $\frac{2H}{Md^2}$

c. $\frac{3H}{Md^2}$ **d.** $\frac{4H}{Md^2}$

60. A uniform circular disc of radius r is placed on a rough horizontal surface and given a linear velocity v_0 and angular velocity ω_0 as shown. The disc comes to rest after moving some distance to the right. It follows that

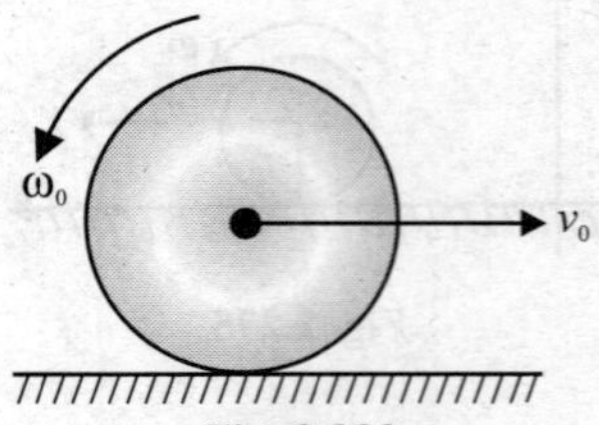

Fig. 2.283

a. $3v_0 = 2\omega_0 r$ **b.** $2v_0 = \omega_0 r$

c. $v_0 = \omega_0 r$ **d.** $2v_0 = 3\omega_0 r$

61. A disc is freely rotating with an angular speed ω on a smooth horizontal plane. If it is hooked at a rigid page P and rotates without bouncing about a point on its circumference. Its angular speed after the impact will be equal to.

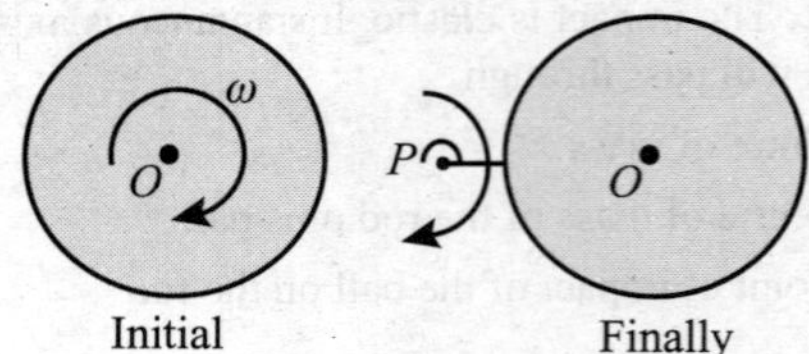

Fig. 2.284

a. $\frac{2\omega}{3}$ **b.** $\frac{\omega}{3}$

c. $\frac{\omega}{2}$ **d.** none of these

62. Each pulley in Fig. 2.285 has radius r and moment of inertia I. The acceleration of the block is

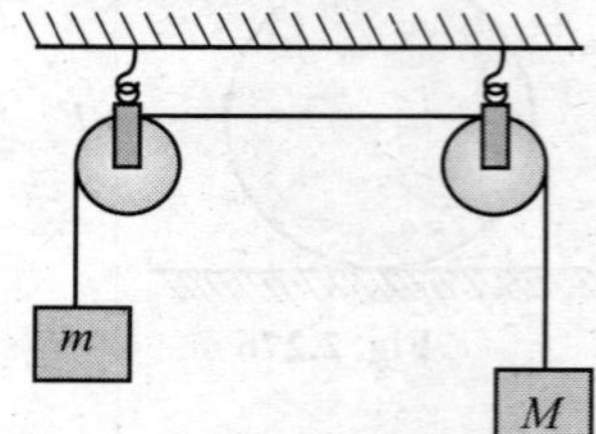

Fig. 2.285

a. $\frac{(M-m)g}{\left(M+m+\frac{2I}{r^2}\right)}$ **b.** $\frac{(M-m)g}{\left(M+m-\frac{2I}{r^2}\right)}$

c. $\frac{(M-m)g}{\left(M+m+\frac{I}{r^2}\right)}$ **d.** $\frac{(M-m)g}{\left(M+m-\frac{I}{r^2}\right)}$

63. A smooth inclined plane is fixed in a car accelerating at $a = g \tan\theta$. If the sphere is set pure rolling on the incline, then

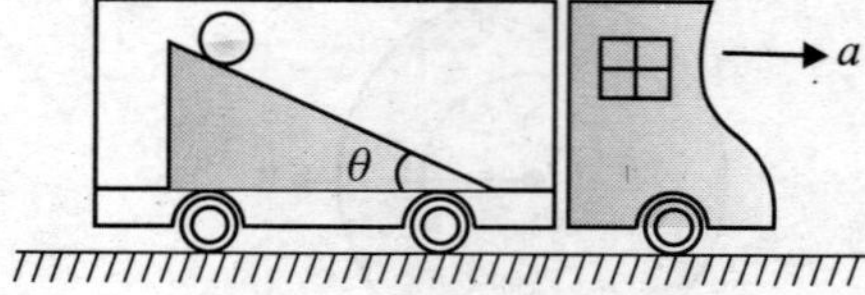

Fig. 2.286

a. it will continue rolling

b. it will slip down

c. its linear velocity will increase

d. its linear velocity will decrease

64. A cube of side a and mass m is to be tilted at point A by applying a force F as shown in Fig. 2.287. The minimum force required is

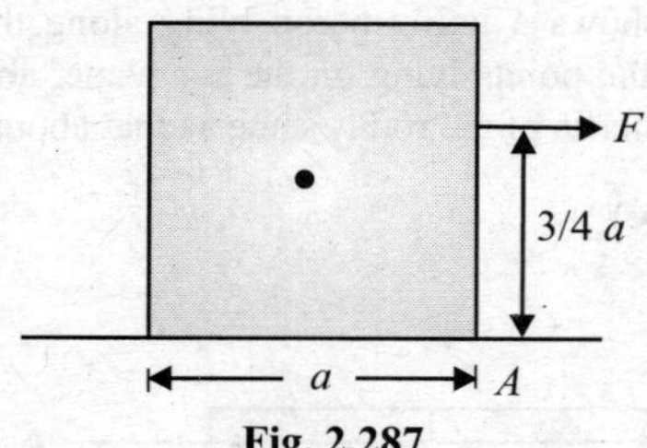

Fig. 2.287

a. mg b. $\frac{2}{3}mg$

c. $\frac{3}{2}mg$ d. $\frac{3}{4}mg$

65. A ladder of length l and mass m is placed against a smooth vertical wall, but the ground is not smooth. Coefficient of friction between the ground and the ladder is μ. The angle θ at which the ladder will stay in equilibrium is

a. $\theta = \tan^{-1}(\mu)$ b. $\theta = \tan^{-1}(2\mu)$

c. $\theta = \tan^{-1}\left(\frac{\mu}{2}\right)$ d. none of these

66. A small solid sphere of radius r rolls down an incline without slipping which ends into a vertical loop of radius R. Find the height above the base so that it just loops the loop

a. $\frac{5}{2}R$ b. $\frac{5}{2}(R-r)$

c. $\frac{25}{10}(R-r)$ d. $\frac{27}{10}R - \frac{17r}{10}$

67. A sphere is rolling down an inclined plane without slipping. The ratio of rotational kinetic energy to total kinetic energy is

a. $\frac{5}{7}$ b. $\frac{2}{5}$

c. $\frac{2}{7}$ d. none of these

68. A cube of side a is placed on an inclined plane of inclination θ. What is the maximum value of θ for which the cube will not topple?

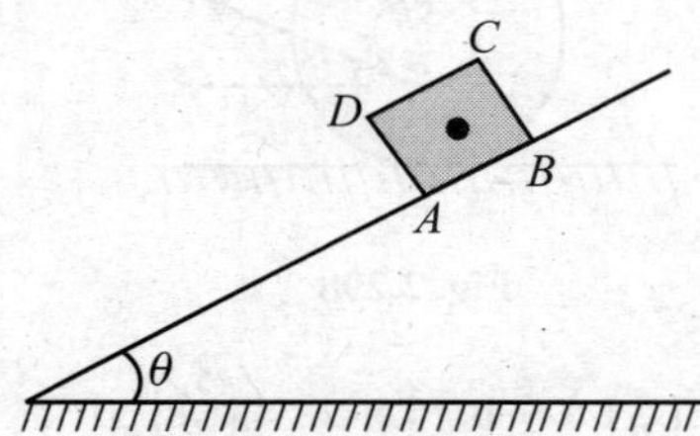

Fig. 2.288

a. 15° b. 30°

c. 45° d. 60°

69. A conical pendulum consists of a mass M suspended from a strong sling of length l. The mass executes a circle of radius R in a horizontal plane with speed v. At time t, the mass is at position $R\hat{i}$ and has $v\hat{j}$ velocity. At time t, the angular momentum vector of mass M about the point from which the string passes on the ceiling is

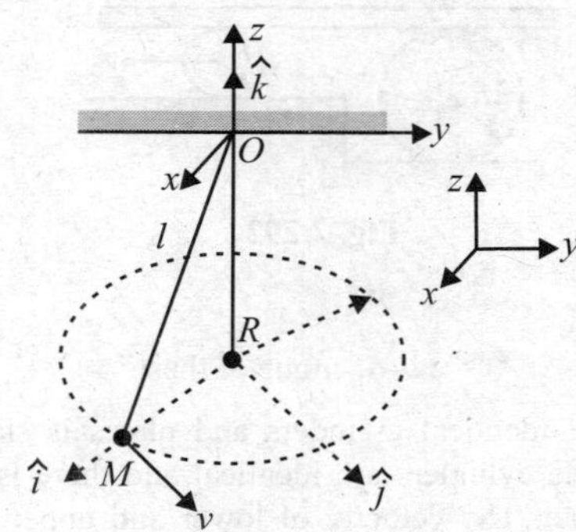

Fig. 2.289

a. $MvR\hat{k}$ b. $Mvl\hat{k}$

c. $Mvl\left[\sqrt{\frac{l^2-R^2}{l}}\,\hat{i} + \frac{R}{l}\hat{k}\right]$ d. $-Mvl\left[\sqrt{\frac{l^2-R^2}{l}}\,\hat{i} + \frac{R}{l}\hat{k}\right]$

70. A thin uniform rod of mass m and length l is kept on a smooth horizontal surface such that it can move freely. At what distance from centre of rod should a particle of mass m strike on the rod such that the point P at a distance $l/3$from the end of the rod is instantaneously at rest just after the elastic collision?

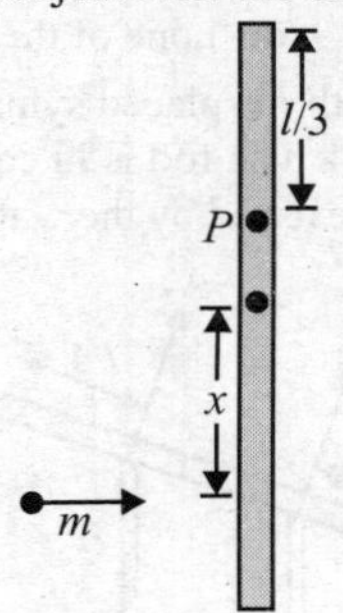

Fig. 2.290

a. $\frac{1}{2}$ b. $\frac{1}{3}$

c. $\frac{1}{6}$ d. $\frac{1}{4}$

71. A yo-yo, arranged as shown, rests on a frictionless surface. When a force F is applied to the string, the yo-yo

a. moves to the left and rotates counterclockwise

b. moves to the right and rotates counterclockwise

c. moves to the left and rotates clockwise

d. moves to the right and rotates clockwise

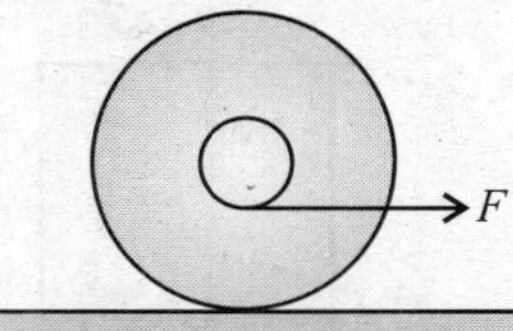

Fig. 2.291

72. Velocity of the centre of a small cylinder is v. There is no slipping anywhere. The velocity of the centre of the larger cylinder is

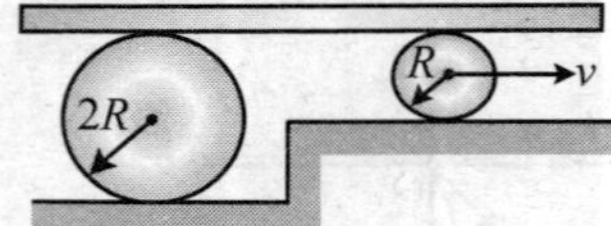

Fig. 2.292

a. $2v$ b. v

c. $\frac{3v}{2}$ d. none of these

73. A system of identical cylinders and plates is shown in Fig. 2.293. All the cylinders are identical and there is no slipping at any contact. The velocity of lower and upper plates are V and $2V$, respectively, as shown in Fig. 2.293. Then the ratio of angular speeds of the upper cylinders to lower cylinders is

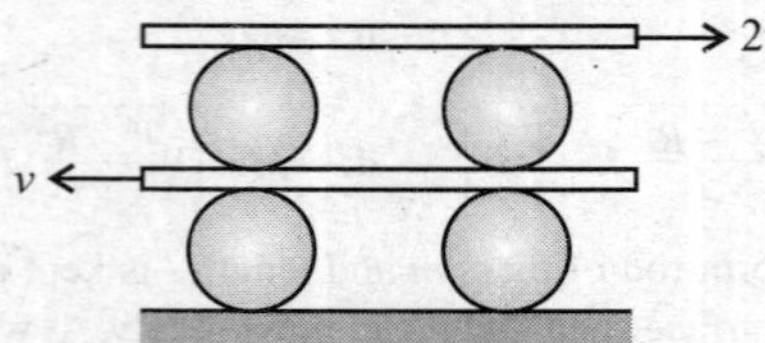

Fig. 2.293

a. $\frac{1}{3}$ b. 3

c. 1 d. none of these

74. A uniform rod of length l is placed symmetrically on two walls as shown in Fig. 2.294. The rod is in equilibrium. If N_1 and N_2 are the normal forces exerted by the walls on the rod, then

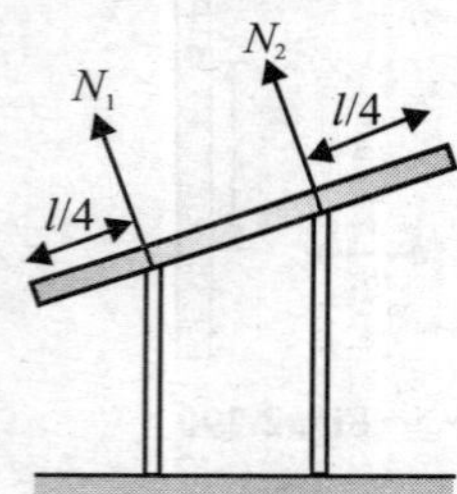

Fig. 2.294

a. $N_1 > N_2$

b. $N_1 < N_2$

c. $N_1 = N_2$

d. N_1 and N_2 would be in the vertical directions

75. A square plate of mass M and edge L is shown in the figure. The moment of inertia of the plate about the axis in the plane of plate and passing through one of its vertex making an angle 15° horizontal is

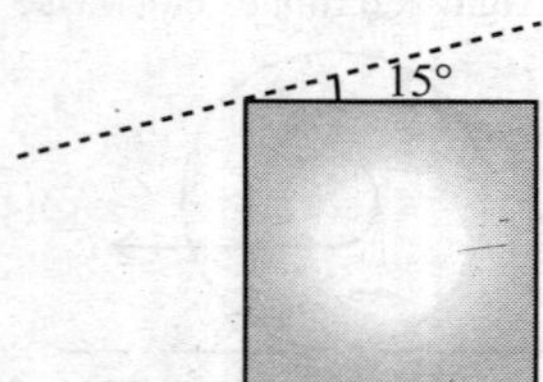

Fig. 2.295

a. $\frac{ML^2}{12}$ b. $\frac{11ML^2}{24}$

c. $\frac{7ML^2}{12}$ d. none of these

76. The figure shows a uniform rod lying along the x-axis. The locus of all the points lying on the x-y plane, about which the moment of inertia of the rod is same as that about O, is

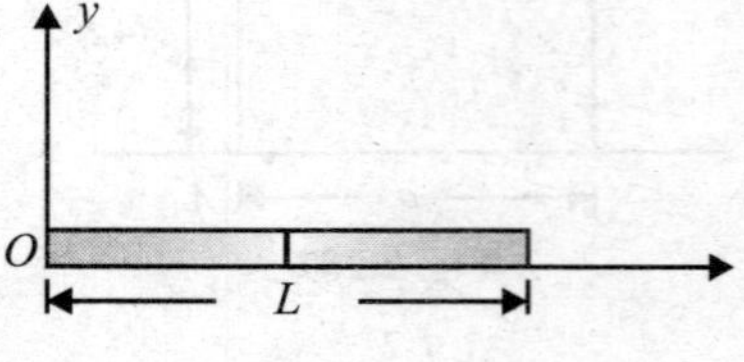

Fig. 2.296

a. an ellipse b. a circle

c. a parabola d. a straight line

77. Find the minimum height of the obstacle so that the sphere can stay in equilibrium.

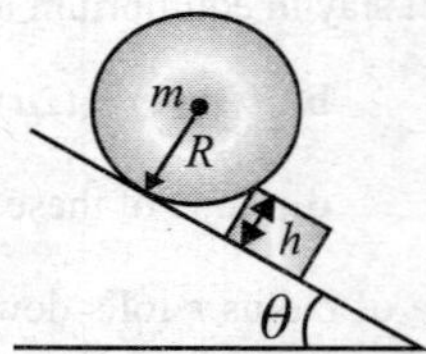

Fig. 2.297

a. $\frac{R}{1+\cos\theta}$ b. $\frac{R}{1+\sin\theta}$

c. $R(1-\sin\theta)$ d. $R(1-\cos\theta)$

78. The spool shown in the figure is placed on a rough horizontal surface and has inner radius r and outer radius R. The angle θ between the applied force and the horizontal can be varied. The critical angle (θ) for which the spool does not roll and remains stationary is given by

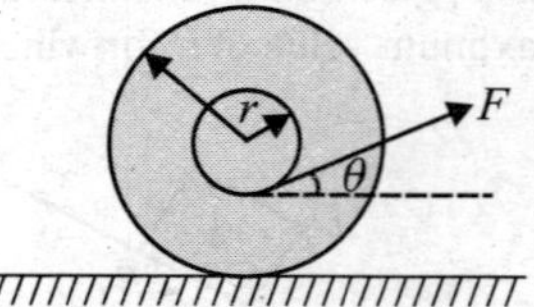

Fig. 2.298

a. $\theta = \cos^{-1}\left(\frac{r}{R}\right)$ b. $\theta = \cos^{-1}\left(\frac{2r}{R}\right)$

c. $\theta = \cos^{-1}\sqrt{\frac{r}{R}}$ d. $\theta = \sin^{-1}\left(\frac{r}{R}\right)$

79. A small bead of mass m moving with velocity v gets threaded on a stationary semicircular ring of mass m and radius R kept on a horizontal table. The ring can freely rotate about its centre. The bead comes to rest relative to the ring. What will be the final angular velocity of the system?

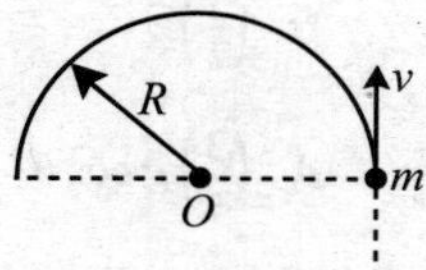

Fig. 2.299

a. $\frac{v}{R}$ b. $\frac{2v}{R}$

c. $\frac{v}{2R}$ d. $\frac{3v}{R}$

80. A sphere is placed rotating with its centre initially at rest in a corner as shown in figures (a) and (b). Coefficient of friction between all surfaces and the sphere is 1/3. Find the ratio of the friction forces f_a/f_b by ground in situations (a) and (b).

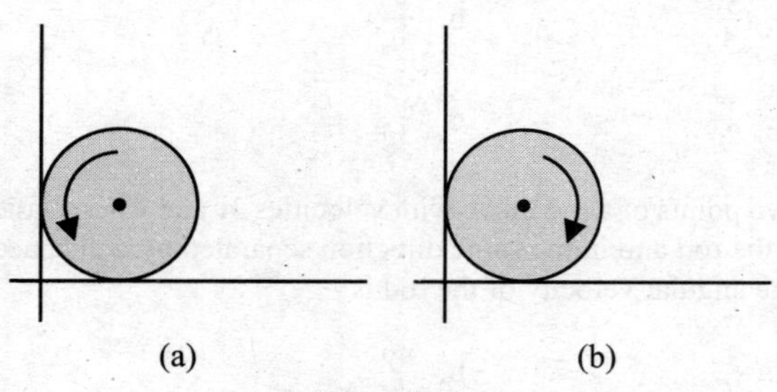

Fig. 2.300

a. 1 b. $\frac{9}{10}$

c. $\frac{10}{9}$ d. none of these

81. A block of mass *m* is attached to a pulley disc of equal mass *m* and radius *r* by means of a slack string as shown. The pulley is hinged about its centre on a horizontal table and the block is projected with an initial velocity of 5 m/s. Its velocity when the string becomes taut will be

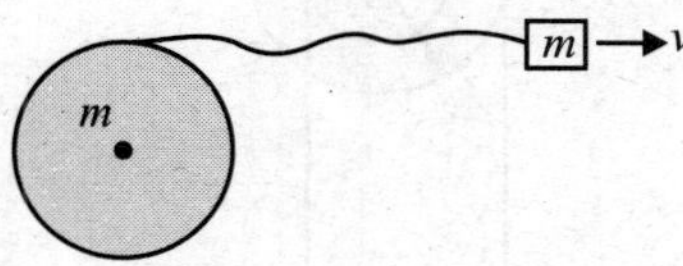

Fig. 2.301

a. 3 m/s b. 2.5 m/s

c. 5/3 m/s d. 10/3 m/s

82. In the figure shown, suppose the compound pulley and the disc have the same angular acceleration in clockwise direction. If a_1 is the upward acceleration of the compound pulley's centre (inner radius *r*, outer radius 3*r*); a_2 is the downward acceleration of the block while a_3 is the upward acceleration of the centre of the disc (radius 2*r*). From kinematics constraints of the thread, the relation between them is

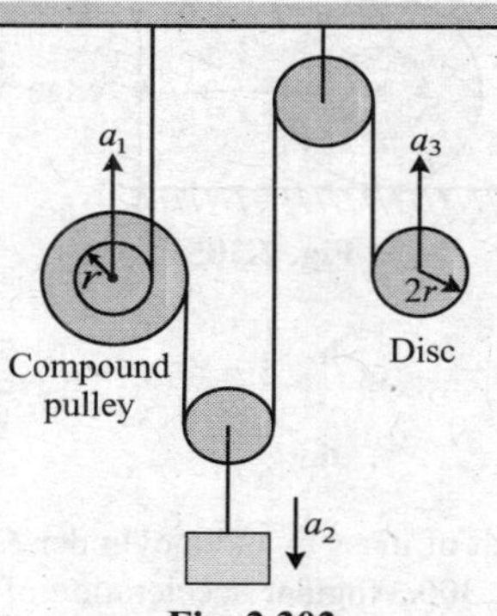

Fig. 2.302

a. $\frac{a_2 - a_3}{2} = 2a_1$ b. $a_2 + a_3 = \frac{a_1}{3}$

c. $2a_2 - a_3 = 4a_1$ d. $a_2 + 2a_3 = a_1$

83. Four rods of side length *l* have been hinged to form a rhombus. Vertex *A* is fixed to a rigid support, vertex *C* is being moved along the *x*-axis with constant velocity *V* as shown in Fig. 2.303. The rate at which vertex *B* is nearing the *x*-axis at the moment the rhombus is in the form of a square is

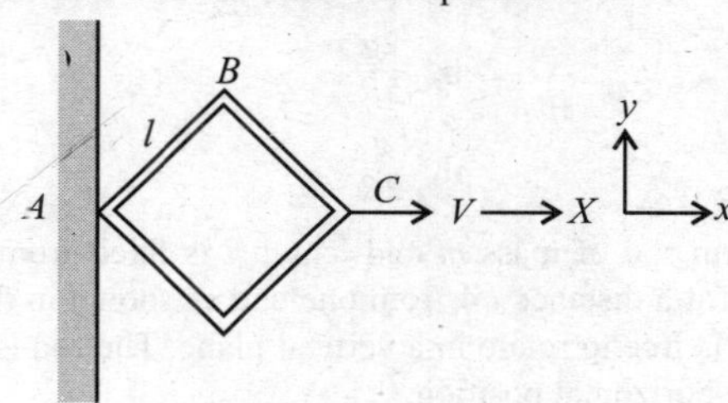

Fig. 2.303

a. $\frac{V}{4}$ b. $\frac{V}{\sqrt{2}}$

c. $\frac{V}{2}$ d. $\frac{V^2}{g}$

84. In the figure shown, the instantaneous speed of end *A* of the rod is *v* to the left. The angular velocity of the rod of length *L* must be

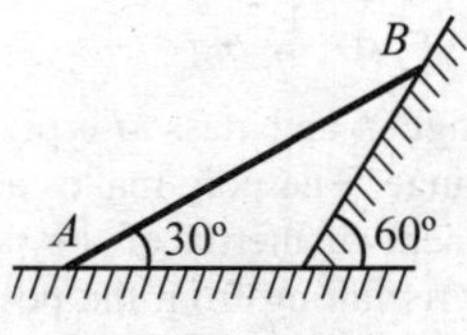

Fig. 2.304

a. $\frac{v}{2L}$ b. $\frac{v}{L}$

c. $\frac{v\sqrt{3}}{2L}$ d. none of these

85. A uniform solid sphere of radius *r* is rolling on a smooth horizontal surface with velocity *V* and angular velocity $\omega = (V = \omega r)$. The sphere collides with a sharp edge on the wall as shown in Fig. 2.305. The coefficient of friction between the sphere and the edge $\mu = 1/5$. Just after the collision the angular velocity of the sphere becomes equal to zero. The linear velocity of the sphere just after the collision is equal to

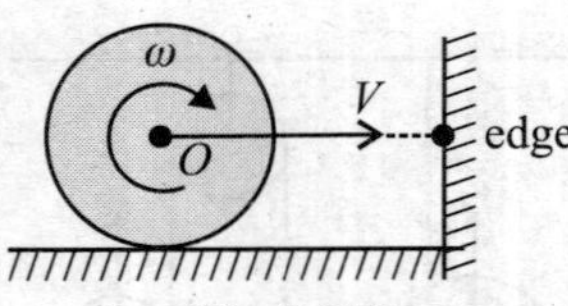

Fig. 2.305

a. V b. $\frac{V}{5}$

c. $\frac{3V}{5}$ d. $\frac{V}{6}$

86. Two blocks each of mass m and a cylinder C are connected as shown in Fig. 2.306. Angular acceleration of the cylinder C of radius R is (all strings and pulley are ideal):

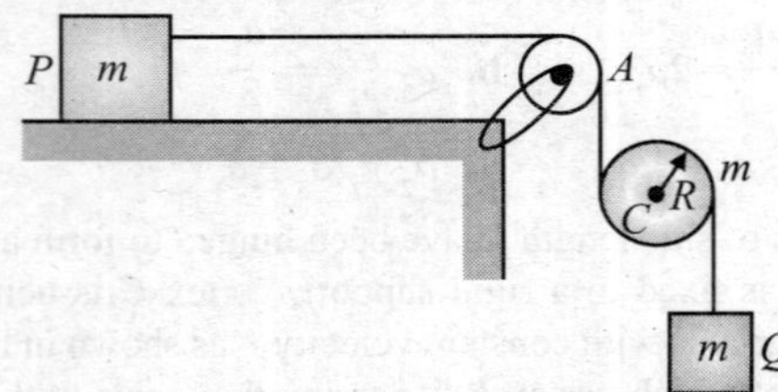

Fig. 2.306

a. $\frac{2g}{3R}$ b. $\frac{2g}{5R}$

c. $\frac{2g}{R}$ d. $\frac{g}{2R}$

87. A uniform rod of mass m and length l is fixed from point A, which is at a distance $l/4$ from one end as shown in the figure. The rod is free to rotate in a vertical plane. The rod is released from the horizontal position.

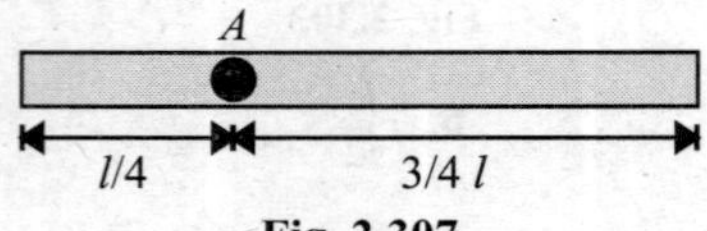

Fig. 2.307

What is the reaction at the hinge, when kinetic energy of the rod is maximum?

a. $\frac{4}{7}$ b. $\frac{5}{7}mg$

c. $\frac{13}{7}mg$ d. $\frac{11}{7}mg$

88. A uniform pole of length L and mass M is pivoted on the ground with a frictionless hinge. The pole makes an angle θ with the horizontal. The moment of inertia of the pole about one end is $(1/3)\ ML^2$. If it starts falling from the position shown in the accompanying figure, the linear acceleration of the free end of the pole immediately after release would be

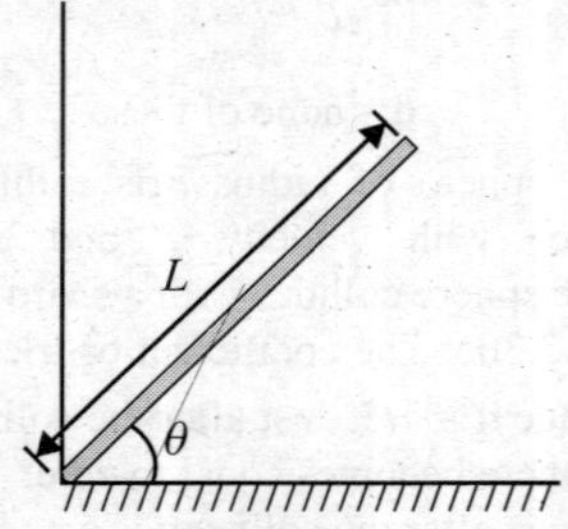

Fig. 2.308

a. $\left(\frac{2}{3}\right) g \cos\theta$ b. $\left(\frac{2}{3}\right) g$

c. g d. $\left(\frac{3}{2}\right) g \cos\theta$

89. In the figure shown, a ring A is initially rolling without sliding with a velocity v on the horizontal surface of the body B (of same mass as A). All surfaces are smooth. B has no initial velocity. What will be the maximum height reached by A on B?

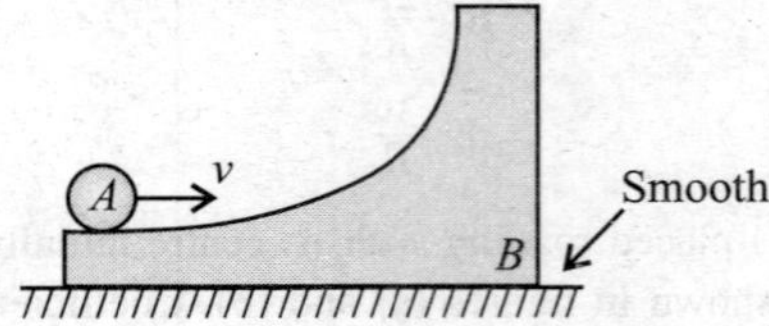

Fig. 2.309

a. $\frac{3v^2}{4g}$ b. $\frac{v^2}{4g}$

c. $\frac{v^2}{2g}$ d. $\frac{v^2}{3g}$

90. Two points of a rod move with velocities $3v$ and v perpendicular to the rod and in the same direction separated by a distance 'r'. The angular velocity of the rod is

a. $\frac{3v}{r}$ b. $\frac{4v}{r}$

c. $\frac{5v}{r}$ d. $\frac{2v}{r}$

91. In the pulley system shown, if radii of the bigger and smaller pulley are 2 m and 1 m, respectively, and the acceleration of block A is 5 m/s^2 in the downward direction, the acceleration of block B will be

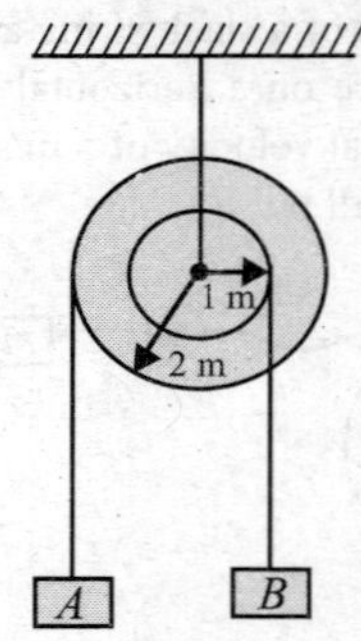

Fig. 2.310

a. 0 m/s^2 b. 5 m/s^2

c. 10 m/s^2 d. $\frac{5}{2}$ m/s^2

92. A sphere is released on a smooth inclined plane from the top. When it moves down, its angular momentum is

a. conserved about every point
b. conserved about the point of contact only
c. conserved about the center of the sphere only
d. conserved about any point on a line parallel to the inclined plane and passing through the centre of the ball

93. A smooth tube of certain mass is rotated in a gravity-free space and released. The two balls shown in Fig. 2.311 move towards the ends of the tube. For the whole system, which of the following quantities is not conserved.

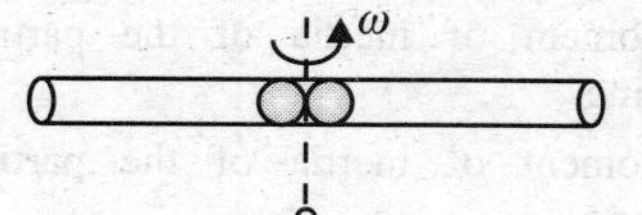

Fig. 2.311

a. Angular momentum **b.** Linear momentum
c. Kinetic energy **d.** Angular speed

94. In the figure shown, a ball without sliding on a horizontal surface. It ascends a curved track up to height h and returns. The value of h is h_1 for sufficiently rough curved track to avoid sliding and is h_2 for smooth curved track, then

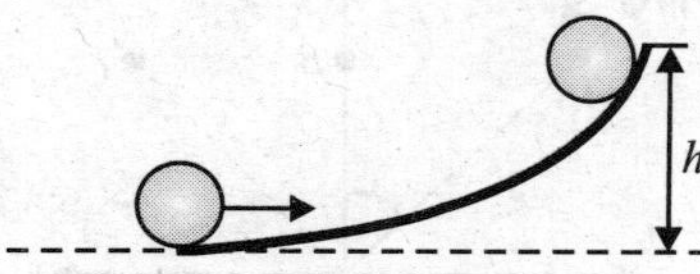

Fig. 2.312

a. $h_1 = h_2$ **b.** $h_1 < h_2$
c. $h_1 > h_2$ **d.** $h_2 = 2h_1$

95. A ring of radius R rolls without sliding with a constant velocity. The radius of curvature of the path followed by any particle of the ring at the highest point of its path will be

a. R **b.** $2R$
c. $4R$ **d.** none of these

96. A uniform ring of radius R is given a back spin of angular velocity $V_0/2R$ and thrown on a horizontal rough surface with velocity of centre to be V_0. The velocity of the centre of the ring when it starts pure rolling will be

a. $\frac{V_0}{2}$ **b.** $\frac{V_0}{4}$
c. $\frac{3V_0}{4}$ **d.** 0

97. A planar object made up of a uniform square plate and four semicircular discs of the same thickness and material is being acted upon by four forces of equal magnitude as shown in Fig. 2.313. The coordinates of point of application of forces is given by

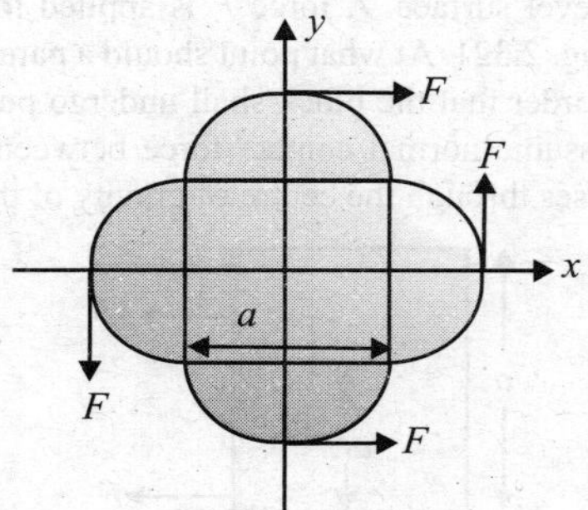

Fig. 2.313

a. $(0, a)$ **b.** $(0, -a)$
c. $(a, 0)$ **d.** $(-a, 0)$

98. A disc is performing pure rolling on a smooth stationary surface with constant angular velocity as shown in Fig. 2.314. At any instant, for the lower most point of the disc,

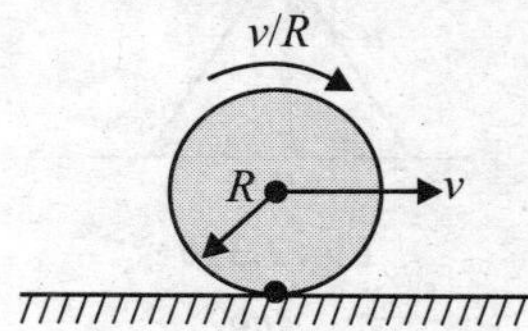

Fig. 2.314

a. velocity is v, acceleration is zero
b. velocity is zero, acceleration is zero
c. velocity is v, acceleration is v^2/R
d. velocity is zero, acceleration is v^2/R

99. A particle of mass 'm' is rigidly attached at 'A' to a ring of mass '$3m$' and radius 'r'. The system is released from rest and rolls without sliding. The angular acceleration of ring just after release is

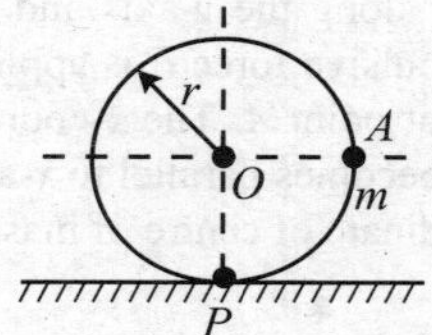

Fig. 2.315

a. $\frac{g}{4r}$ **b.** $\frac{g}{6r}$
c. $\frac{g}{8r}$ **d.** $\frac{g}{2r}$

100. A uniform smooth rod (mass m and length l) placed on a smooth horizontal floor is hit by a particle (mass m) moving on the floor, at a distance $l/4$ from one end elasetically ($e = 1$). The distance travelled by the centre of the rod after the collision when it has completed three revolutions will be

a. $2\pi l$ **b.** cannot be determined
c. πl **d.** none of these

101. A solid sphere, a hollow sphere and a disc, all having the same mass and radius, are placed at the top of an incline and released. The friction coefficients between the objects and the incline are same and not sufficient to allow pure rolling. The least time will be taken in reaching the bottom by

a. the solid sphere **b.** the hollow sphere
c. the disc **d.** all will take the same time

102. In the previous problem, the smallest kinetic energy at the bottom of the incline will be achieved by

a. the solid sphere **b.** the hollow sphere
c. the disc
d. all will achieve the same kinetic energy

103. An equilateral prism of mass m rests on a rough horizontal surface with coefficient of friction μ. A horizontal force F is applied on the prism as shown in Fig. 2.316. If the coefficient of friction is sufficiently high so that the prism does not slide before toppling, the minimum force required to topple the prism is

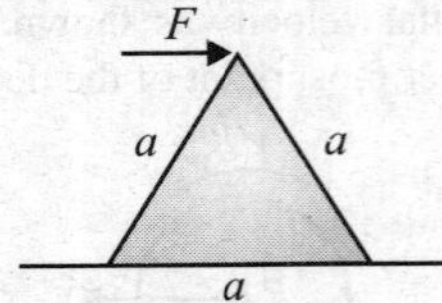

Fig. 2.316

a. $\frac{mg}{\sqrt{3}}$ b. $\frac{mg}{4}$

c. $\frac{\mu mg}{\sqrt{3}}$ d. $\frac{\mu mg}{4}$

104. A uniform disc of radius R lies in the x-y plane, with its centre at origin. Its moment of inertia about z-axis is equal to its moment of inertia about line $y = x + c$. The value of c will be

a. $-\frac{R}{2}$ b. $\pm\frac{R}{\sqrt{2}}$

c. $+\frac{R}{4}$ d. $-R$

105. A uniform rod of mass m and length l is placed over a smooth horizontal surface along the y-axis and is at rest as shown in Fig. 2.317. An impulsive force F is applied for a small time Δt along x-direction at point A. The x-coordinate of end A of the rod when the rod becomes parallel to x-axis for the first time is [initially, the coordinate of centre of mass of the rod is (0, 0)]

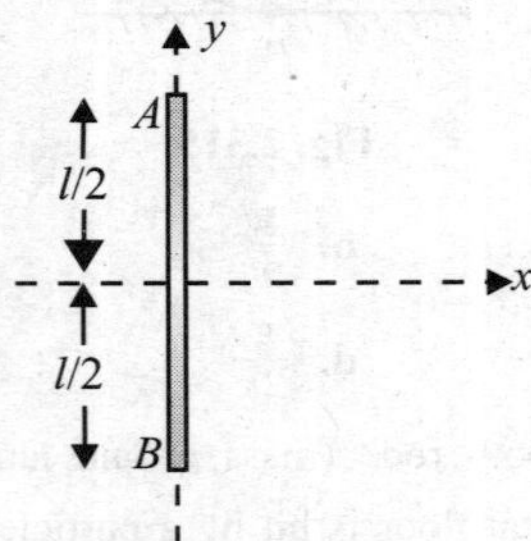

Fig. 2.317

a. $\frac{\pi l}{12}$ b. $\frac{l}{2}\left(1+\frac{\pi}{12}\right)$

c. $\frac{1}{2}\left(1-\frac{\pi}{6}\right)$ d. $\frac{l}{2}\left(1+\frac{\pi}{6}\right)$

106. A mass m is moving at speed v perpendicular to a rod of length d and mass M = 6 m which pivots around a frictionless axle running through its centre. It strikes and sticks to the end of the rod. The moment of inertia of the rod about its centre is $Md^2/12$. Then the angular speed of the system just after the collision is

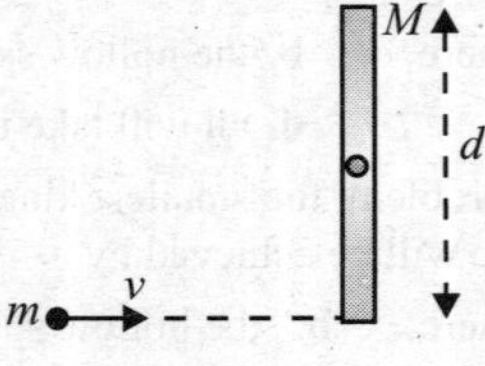

Fig. 2.318

a. $\frac{2v}{3d}$ b. $\frac{2v}{d}$

c. $\frac{v}{d}$ d. $\frac{3v}{2d}$

107. A particle falls freely near the surface of the earth. Consider a fixed point O (not vertically below the particle) on the ground. Then pick up the correct alternative or alternatives.

a. Angular momentum of the particle about O is increasing.

b. The moment of inertia of the particle about O is decreasing.

c. The moment of inertia of the particle about O is decreasing.

d. The angular velocity of the particle about O is increasing.

108. A ball of mass moving with constant velocity u collides with a smooth horizontal surface at O as shown in Fig. 2.319. Neglect gravity and friction. The y-axis is drawn normal to the horizontal surface at the point of impact O and x-axis is horizontal as shown. About which point will the angular momentum of ball be conserved?

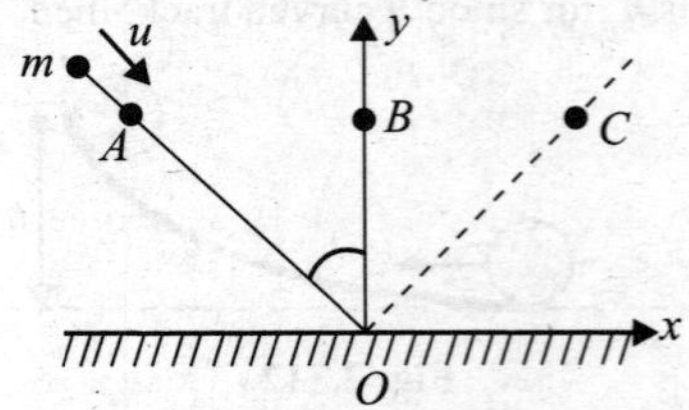

Fig. 2.319

a. Point A b. Point B

c. Point C d. None of these

109. In the figure shown, a small ball of mass 'm' can move without sliding in a fixed semicircular track of radius R in a vertical plane. It is released from the top. The resultant force on the ball at the lowest point of the track is

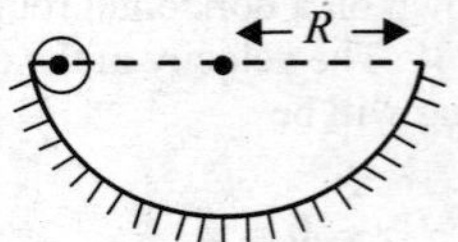

Fig. 2.320

a. $\frac{10mg}{7}$ b. $\frac{17mg}{7}$

c. $\frac{3mg}{7}$ d. zero

110. A rectangular block of mass M and height a is resting on a smooth level surface. A force F is applied to one corner as shown in Fig. 2.321. At what point should a parallel force $3F$ be applied in order that the block shall undergo pure translational motion? Assume normal contact force between the block and surface passes through the centre of gravity of the block.

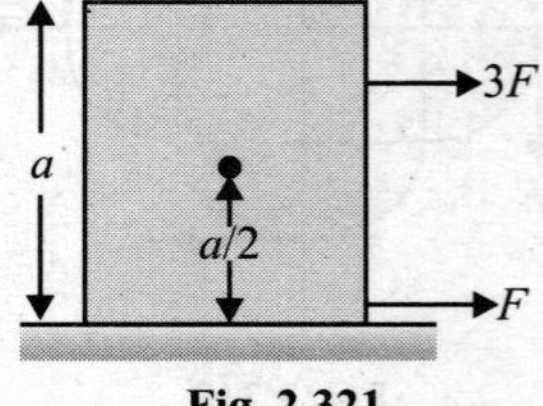

Fig. 2.321

a. $\frac{a}{3}$ vertically above centre of gravity

b. $\frac{a}{6}$ vertically above centre of gravity

c. No such point exists

d. it is not possible

111. A uniform rod of mass 15 kg is held stationary with the help of a light string as shown in Fig. 2.322. The tension in the string is

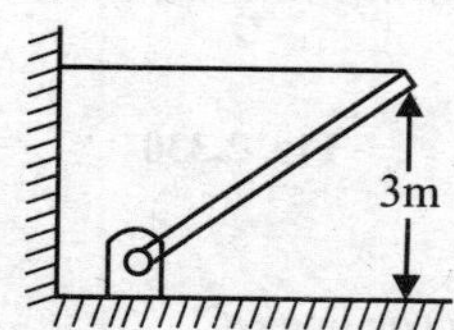

Fig. 2.322

a. 150 N **b.** 225 N

c. 100 N **d.** None of the above

112. A solid sphere of mass M and radius R is placed on a rough horizontal surface. It is struck by a horizontal cuestick at a height h above the surface. The value of h so that the sphere performs pure rolling motion immediately after it has been struck is

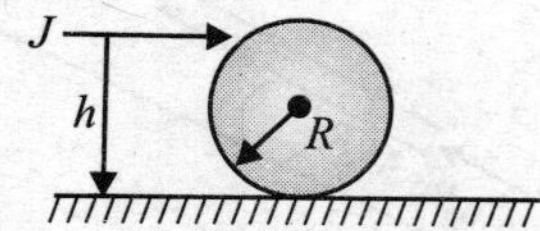

Fig. 2.323

a. $\frac{2R}{5}$ **b.** $\frac{5R}{2}$

c. $\frac{7R}{5}$ **d.** $\frac{9R}{5}$

113. An object of mass M and radius R is performing pure rolling motion on a smooth horizontal surface under the action of a constant force F as shown in Fig. 2.324. The object may be

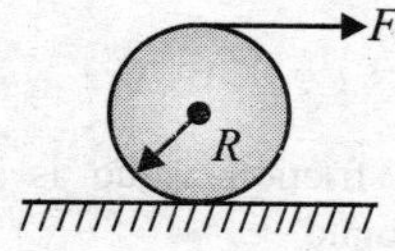

Fig. 2.324

a. disk **b.** ring

c. solid cylinder **d.** hollow sphere

114. A uniform rod of mass M and length L is free to rotate in X-Z plane, i.e., $\vec{F} = (3\hat{i} + 2\hat{j} + 6\hat{k})$ N is acting on the rod at $(L/2, 0, 0)$ in the situation shown in Fig. 2.325. The angular acceleration of the rod is (Take $M = 6$ kg and $L = 4$ m)

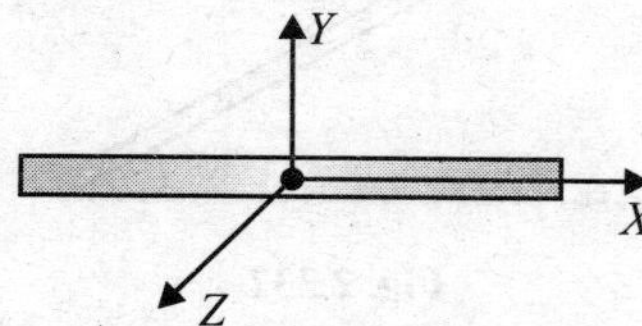

Fig. 2.325

a. $-\frac{3}{2}\hat{j} + 1/2\hat{k}$ **b.** $-\frac{3}{2}\hat{j}$

c. $\frac{1}{2}\hat{k}$ **d.** $4\hat{j}$

115. A cord is wrapped on a pulley (disk) of mass M and radius R as shown in Fig. 2.326. To one end of the cord, a block of mass M is connected as shown and to other end in (a) a force of 2 Mg and in (b) a block of mass $2M$. Let angular acceleration of the disk in A and B is α_A and α_B, respectively, then (cord is not slipping on the pulley).

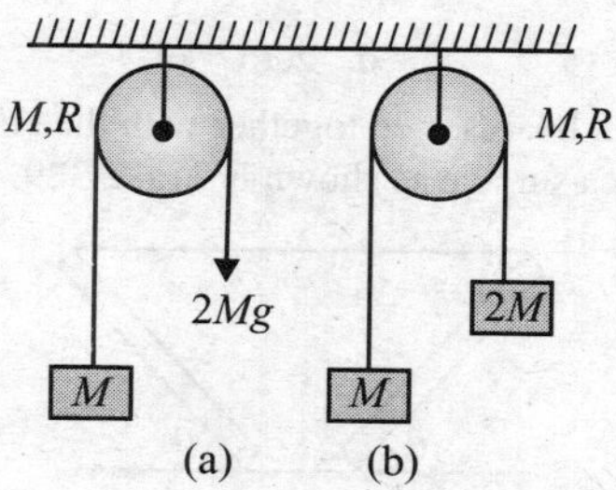

Fig. 2.326

a. $\alpha_A = \alpha_B$ **b.** $\alpha_A > \alpha_B$

c. $\alpha_A < \alpha_B$ **d.** None of these

116. A cylinder having moment of inertia, which is free to rotate about its axis, receives an angular impulse of J kg m^2/s initially, followed by similar impulse after every 4 s. What is the angular speed of the cylinder 30 s after the initial impulse?

a. $\frac{7J}{I}$ **b.** $\frac{8J}{I}$

c. $\frac{J}{I}$ **d.** zero

117. A small object of uniform density rolls up a curved surface with an initial velocity v. It reaches up to a maximum height of $3v^2/4g$ w.r.t. the initial position. The object is

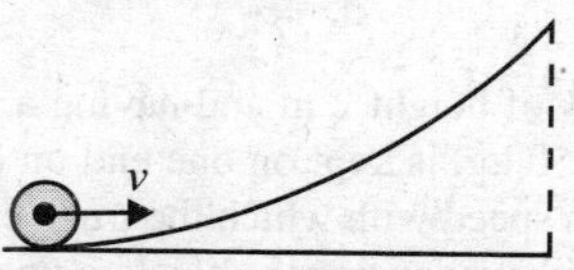

Fig. 2.327

a. ring **b.** solid sphere

c. hollow sphere **d.** disc

118. A solid cylinder of mass M and radius R is resting on a horizontal platform (which is parallel to the x-y plane) with its axis fixed along Y-axis and free to rotate about its axis. The platform is given a motion in the X-direction given by $x = A \cos(\omega t)$. There is no slipping between the cylinder and the platform. The maximum torque acting on the cylinder during its motion is

a. $\frac{M\omega^2 AR}{3}$ **b.** $\frac{M\omega^2 AR}{2}$

c. $\frac{2}{3} \times M\omega^2 AR$ **d.** The situation is not possible

119. A cubical block of side a is moving with velocity v on a horizontal smooth plane as shown in Fig. 2.328. It hits a ridge at point O. The angular speed of the block after it hits O is

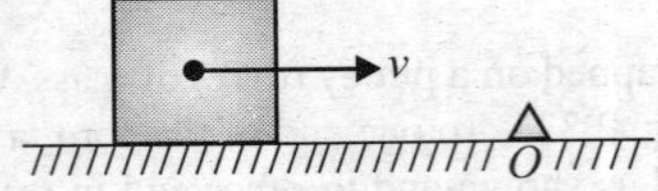

Fig. 2.328

a. $\frac{3v}{4a}$ **b.** $\frac{3v}{2a}$

c. $\sqrt{\frac{3}{2}}\, a$ **d.** zero

120. Two uniform boards, tied together with the help of a string, are balanced on a surface as shown in Fig. 2.329.

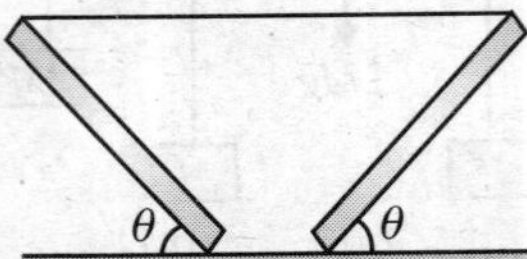

Fig. 2.329

The coefficient of static friction between boards and surface is 0.5. The minimum value of θ, for which this type of arrangement is possible is

a. 30°
b. 45°
c. 37°
d. It is not possible to have this type of balanced arrangement

121. A solid cylinder is placed on the end of an inclined plane. It is found that the plane can be tipped at an angle θ before the cylinder starts to slide. When the cylinder turns on its sides and is allowed to roll, it is found that the steepest angle at which the cylinder performs pure rolling is ϕ. The ratio $\tan\phi/\tan\theta$ is

a. 3 **b.** $\frac{1}{3}$

c. 1 **d.** $\frac{1}{2}$

122. A uniform box of height 2 m and having a square base of side 1 m, weight 150 kg, is kept on one end on the floor of a truck. The maximum speed with which the truck can round a curve of radius 20 m without causing the block to tip over is (assume that friction is sufficient so that there is no sliding).

a. 15 m/s
b. 10 m/s
c. 8 m/s
d. depends on the value of coefficient of friction

123. Angular momentum of a particle about a stationary point O varies with time as $\vec{L} = \vec{a} \times \vec{b}t^2$ where $\vec{a}$ and $\vec{b}$ are constant vectors with $\vec{a}$ perpendicular to $\vec{b}$. The torque $\vec{\tau}$ acting on the particle when angle between $\vec{\tau}$ and $\vec{L}$ is 45° is

a. $\sqrt{\frac{a}{b}} \times \vec{b}$ **b.** $2\sqrt{\frac{b}{a}} \times \vec{b}$

c. $\sqrt{\frac{a}{b}}$ **d.** None of these

124. A disc of radius 0.2 m is rolling with slipping on a flat horizontal surface, as shown in Fig. 2.330. The instantaneous centre of rotation is (the lowest contact point is O and centre of disc is C)

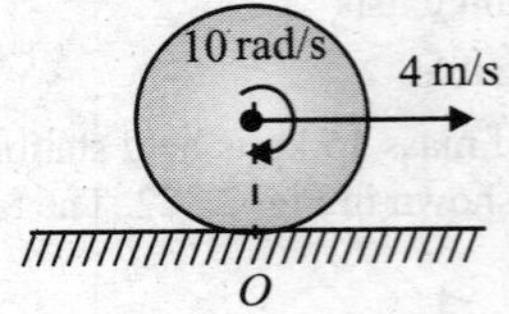

Fig. 2.330

a. zero
b. 0.1 m above O on line OC
c. 0.2 m below O on line OC
d. 02 m above O on line OC

125. A solid cylinder of mass M and radius $2R$ is rolled up on an incline with the help of a plank of mass $2M$ as shown in Fig. 2.331. A force (constant) F is acting on the plank parallel to the incline. There is no slipping at any of the contact. The friction force between the plank and the cylinder is given by

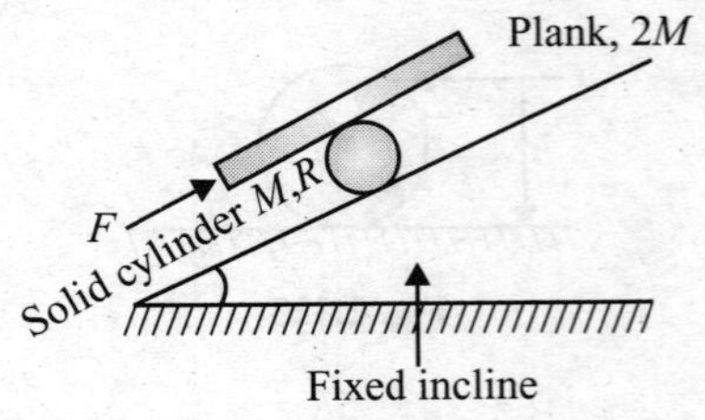

Fig. 2.331

a. $\frac{F}{2}$

b. $\frac{3F + 29Mg\sin\theta}{57}$

c. $\frac{3F + 2Mg\sin\theta}{19}$

d. Coefficient of friction value is not given and hence CANNOT be found

126. A slender rod of mass m and length L is pivoted about a horizontal axis through one end and released from rest at an angle of 30° above the horizontal. The force exerted by the pivot on the rod at the instant when the rod passes through a horizontal position is

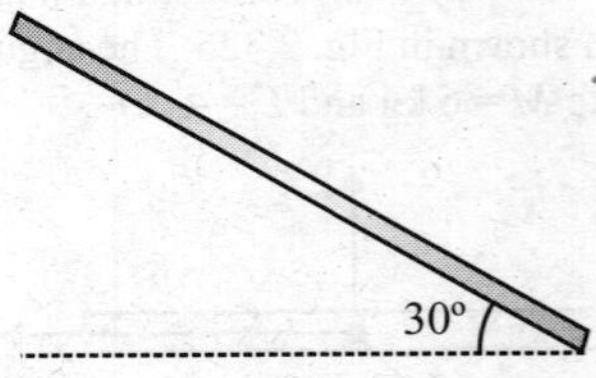

Fig. 2.332

a. $\sqrt{\frac{10}{4}}\, mg$ along horizontal

b. mg along vertical

c. $\frac{\sqrt{10}}{4}\, mg$ along a line making an angle of $\tan^{-1}\left(\frac{1}{3}\right)$ with the horizontal

d. $\frac{\sqrt{10}}{4}\, mg$ along a line making an angle of $\tan^{-1}(3)$ with the horizontal

127. A slender rod of mass M and length L rests on a horizontal frictionless surface. The rod is pivoted about one of its ends. The impulse of the force exerted on the rod by the pivot when the rod is struck by a blow of impulse J perpendicular to the rod at other end is

a. J b. $\frac{J}{2}$

c. $\frac{J}{3}$ d. Information is insufficient

128. Two painters are working from a wooden board 5 m long suspended from the top of a building by two ropes attached to the ends of the plank. Either rope can withstand a maximum tension of 1040 N. Painter A of mass 80 kg is working at a distance of 1 m from one end. Painter B of mass 60 kg is working at a distance of x m from the centre of mass of the board on the other side. Take mass of the board as 20 kg and $g = 10\ \text{m/s}^2$. The range of x so that both the painters can work safely is

a. $\frac{1}{3} < x < \frac{11}{6}$ b. $0 < x < \frac{11}{6}$

c. $0 < x < 2$ d. $\frac{1}{3} < x < 2$

129. In Fig. 2.333, a massive rod AB is held in horizontal position by two massless strings. If the string at B breaks and if the horizontal acceleration of centre of mass, vertical acceleration and angular acceleration of rod about the centre of mass are a_x, a_y and α, respectively, then

Fig. 2.333

a. $2\sqrt{3}a_y = \sqrt{3}\alpha l + 2a_x$ b. $\sqrt{3}a_y = \sqrt{3}\alpha l + a_x$

c. $a_y = \sqrt{3}\alpha l + 2a_x$ d. $2a_y = \alpha l + 2\sqrt{3}a_x$

130. Two identical small balls, each of mass m, are connected by a massless and inextensible string of length l and placed on a smooth horizontal xy plane. An external agent starts pulling the string from its mid-point along y-axis with velocity v_0 as shown in Fig. 2.334. When the separation between the two balls reduces to $l/2$, then the speed of each ball will be

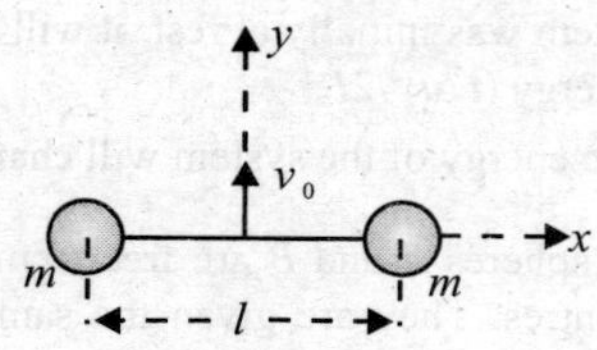

Fig. 2.334

a. $2v_0$ b. v_0

c. $\frac{v_0}{2}$ d. none of the above

Multiple Correct Answers Type

Solutions on page 2.128

1. The disc of radius r is confined to roll without slipping at A and B. If the plates have the velocities shown, then

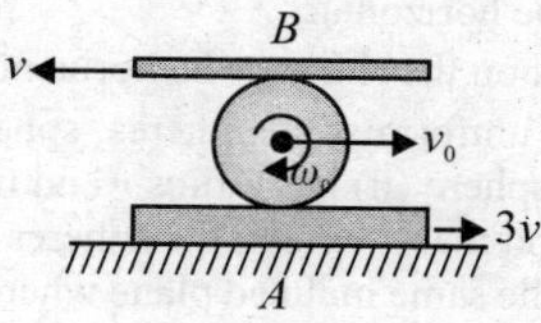

Fig. 2.335

a. Angular velocity of the disc is $2V/r$

b. Linear velocity, $V_0 = V$

c. Angular velocity of the disc is $3V/2r$

d. None of these

2. A solid sphere of mass M and radius R is pulled horizontally on a rough surface as shown in Fig. 2.336. Choose the incorrect alternatives.

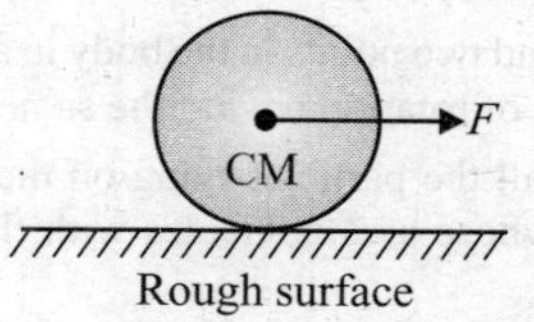

Fig. 2.336

a. The magnitude of the frictional force is $F/3$.

b. The frictional force on the sphere acts forward.

c. The acceleration of the centre of mass is $2F/3M$.

d. The acceleration of the centre of mass is F/M.

3. A disc is given an initial angular velocity ω_0 and placed on a rough horizontal surface as shown Fig. 2.337. The quantities which will not depend on the coefficient of friction is/are

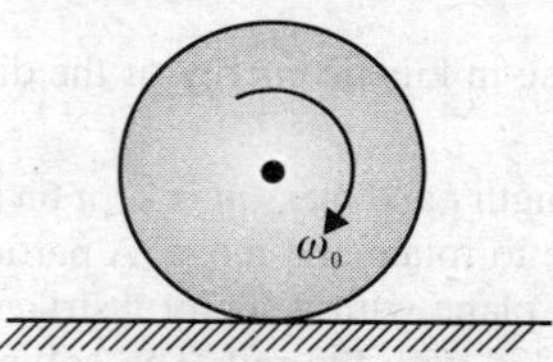

Fig. 2.337

a. the time until rolling begins

b. the displacement of the disc until rolling begins

c. the velocity when rolling begins

d. the work done by the force of friction

4. A wheel is rolling on a horizontal plane. At a certain instant, it has a velocity 'v' and acceleration 'a' of CM as shown in Fig. 2.338. Acceleration of

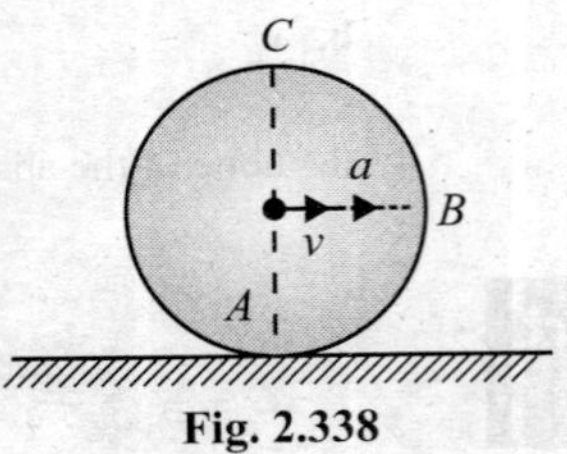

Fig. 2.338

a. A is vertically upwards
b. B may be vertically downwards
c. C cannot be horizontal
d. some point on the rim may be horizontal leftwards

5. Consider three uniform solid spheres, sphere (i) has radius 'r' and mass 'm', sphere (ii) has radius r and mass $3m$, sphere (iii) has radius $3r$ and mass 'm'. All the spheres can be placed at the same point on the same inclined plane where they will roll without slipping to the bottom. If allowed to roll down the incline, then at the bottom of the incline
a. sphere (i) will have the largest speed
b. sphere (ii) will have the largest speed
c. sphere (ii) will have the largest kinetic energy
d. all the spheres will have equal speed

6. A rigid body is in pure rotation, that is, undergoing fixed axis rotation. Then which of the following statement(s) are true?
a. You can find two points in the body in a plane perpendicular to the axis of rotation having the same velocity.
b. You can find two points in the body in a plane perpendicular to the axis of rotation having the same acceleration.
c. Speed of all the particles lying on the curved surface of a cylinder whose axis coincides with the axis of rotation is the same.
d. Angular speed of the body is the same as seen from any point in the body.

7. The uniform speed of a body is the same as seen from any point in the body. A light cord is wrapped around the rim of the disc and mass of 1 kg is tied to the free end. If it is released from rest,
a. the tension in the cord is 5 N
b. in the first 4 s the angular displacement of the disc is 40 rad
c. the work done by the torque on the disc in the first 4 s is 200 J
d. the increase in kinetic energy of the disc in the first 4 s is 200 J

8. A rod AC of length l and mass m is kept on a horizontal smooth plane. It is free to rotate and move. A particle of same mass m moving on the plane with velocity v strikes the rod at point B making angle 37° with the rod. The collision is elastic. After collision,

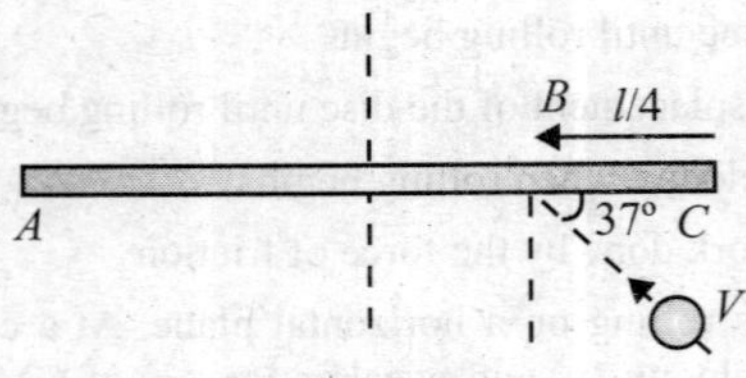

Fig. 2.339

a. the angular velocity of the rod will be $72/55\ v/l$
b. the centre of the rod will travel a distance $\pi l/3$ in the time in which it makes half rotation
c. impulse of the impact force is $24mV/55$
d. none of these

9. The moments of inertia of a thin square plate $ABCD$ of uniform thickness about an axis passing through the centre O and perpendicular to the plate are

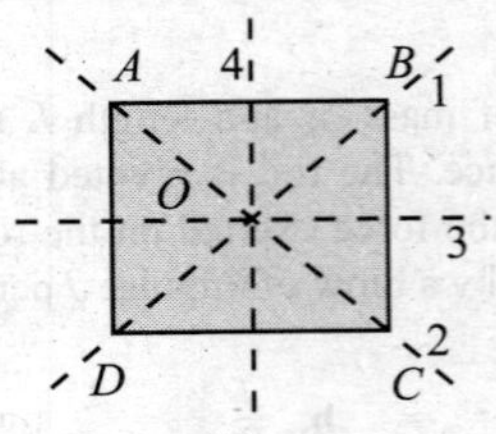

Fig. 2.340

a. $I_1 + I_2$
b. $I_3 + I_4$
c. $I_1 + I_3$
d. $I_1 + I_2 + I_3 + I_4$

(where I_1, I_2, I_3 and I_4 are, respectively, the moments of inertia about axes 1, 2, 3 and 4, where axes are in the plane of the plate)

10. A horizontal disc rotates freely about a vertical axis through its centre. A ring, having the same mass and radius as the disc, is now gently placed on the disc. After some time, the two rotate with a common angular velocity, then
a. some friction exists between the disc and the ring
b. the angular momentum of the "disc plus ring" is conserved
c. the final common angular velocity is (2/3)rd of the initial angular velocity of the disc
d. (2/3)rd of the initial kinetic energy changes to heat

11. Two horizontal discs of different radii are free to rotate about their central vertical axes. One is given some angular velocity, the other is stationary. Their rims are now brought in contact. There is friction between the rims. Then
a. the force of friction between the rims will disappear when the discs rotate with equal angular speeds
b. the force of friction between the rims will disappear when they have equal linear velocities
c. the angular momentum of the system will be conserved
d. the rotational kinetic energy of the system will not be conserved

12. A constant external torque τ acts for a very brief period Δt on a rotating system having moment of inertia I, then
a. the angular momentum of the system will change by $\tau\Delta t$
b. the angular velocity of the system will change by $\tau\Delta t/I$
c. if the system was initially at rest, it will acquire rotational kinetic energy $(\tau\Delta t)^2/2I$
d. the kinetic energy of the system will change by $(\tau\Delta t)^2/2I$

13. Two identical spheres A and B are free to move and to rotate about their centres. They are given the same impulse J. The

lines of action of the impulses pass through the centre of A and away from the centre of B, then

a. A and B will have the same speed

b. B will have greater kinetic energy than A

c. they will have the same kinetic energy, but the linear kinetic energy of B will be less than that of A

d. the kinetic energy of B will depend on the point of impact of the impulse on B

14. A disc of circumference s is at rest at a point A on a horizontal surface when a constant horizontal force begins to act on its centre. Between A and B there is sufficient friction to prevent slipping and the surface is smooth to the right of B, $AB = s$. The disc moves from A to B in time T. To the right of B,

a. the angular acceleration of the disc will disappear, linear acceleration will remain unchanged

b. linear acceleration of the disc will increase

c. the disc will make one rotation in time $T/2$

d. the disc will cover a distance greater than s in further time T

15. A solid sphere starts from rest at the top of an incline of height h and length l and moves down. The force of friction between the sphere and the incline is F. This is sufficient to prevent slipping. The kinetic energy of the sphere at the bottom of the incline is W, then

a. the work done against the force of friction is Fl

b. the heat produced is Fl

c. $W = mgh - Fl$

d. $W > (mgh - Fl)$

16. A ring (R), a disc (D), a solid sphere (S) and a hollow sphere with thin walls (H), all having the same mass but different radii, start together from rest at the top of an inclined plane and roll down without slipping. Then

a. all of them will reach the bottom of the incline together

b. the body with the maximum radius will reach the bottom first

c. they will reach the bottom in the order S, D, H and R

d. all of them will have the same kinetic energy at the bottom of the incline

17. A solid sphere rolls without slipping on a rough horizontal floor, moving with a speed v. It makes an elastic collision with a smooth vertical wall. After impact

a. it will move with a speed v initially

b. its motion will be rolling without slipping

c. its motion will be rolling without slipping initially and its rotational motion will stop momentarily at some instant

d. its motion will be rolling without slipping only after some time.

18. A thin uniform rod of mass m and length l is free to rotate about its upper end. When it is at rest, it receives an impulse J at its lowest point, normal to its length. Immediately after impact,

a. the angular momentum of the rod is Jl

b. the angular velocity of the rod is $3J/ml$

c. the kinetic energy of the rod is $3J^2/2m$

d. the linear velocity of the midpoint of the rod is $3J/2m$

19. Which of the following statements are correct for instantaneous axis of rotation?

a. Acceleration of every point lying on the axis must be equal to zero.

b. Velocity of a point distance r from the axis is equal to $r\omega$.

c. If moment of inertia of a body about the axis is I and angular velocity is ω, then kinetic energy of the body is equal to $I\omega^2/2$.

d. Moment of inertia of a body is least about instantaneous axis of rotation among all the parallel axes.

20. Inner and outer radii of a spool are r and R I, respectively. A thread is wound over its inner surface and spool is placed over a rough horizontal surface. Thread is pulled by a force F as shown in Fig. 2.341. In case of pure rolling, which of the following statements are false?

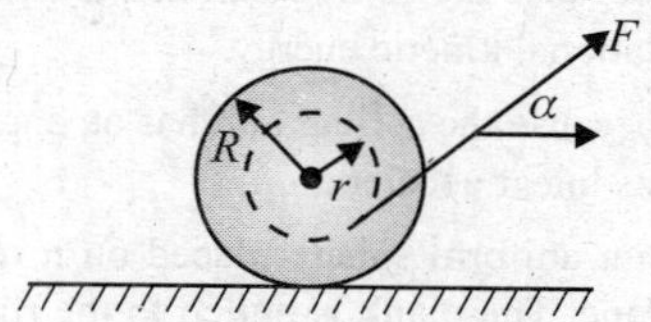

Fig. 2.341

a. Thread unwinds, spool rotates anticlockwise and friction acts leftwards.

b. Thread winds, spool rotates clockwise and friction acts leftwards.

c. Thread winds, spool moves to the right and friction acts rightwards.

d. Thread winds, spool moves to the right and friction does not come into existence.

21. A horizontal plane supports a fixed vertical cylinder of radius R and a particle is attached to the cylinder by a horizontal thread AB as shown in Fig. 2.342. The particle initially rest on a horizontal plane. A horizontal velocity v_0 is imparted to the particle, normal to the threading during subsequent motion. Point out the false statements:

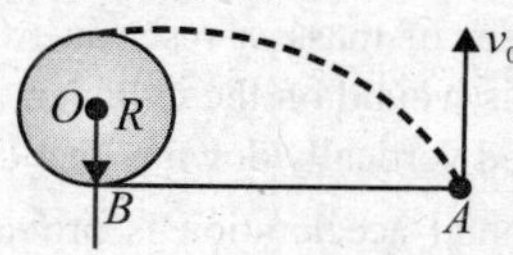

Fig. 2.342

a. Angular momentum of particle about O remains constant.

b. Angular momentum about B remains constant.

c. Momentum and kinetic energy both remain constant.

d. Kinetic energy remains constant.

22. A uniform rod is resting freely over a smooth horizontal plane. A particle moving horizontally strikes at one end of the rod normally and gets stuck. Then

a. the momentum of the particle is shared between the particle and the rod and remains conserved.

b. the angular momentum about the mid-point of the rod before and after the collision is equal.

c. the angular momentum about the centre of mass of the combination before and after the collision is equal.

d. the centre of mass of the rod particle system starts to move translationally with the original momentum of the particle.

23. A ball rolls down an inclined plane and acquires a velocity v_r when it reaches the bottom of the plane. If the same ball slides without friction and acquires rolling from the same height down an equally inclined smooth plane and acquires a velocity v_s (then which of the following statements are not correct?

a. $v_r < v_s$, because a work is done by the rolling ball against the frictional force.

b. $v_r > v_s$, because the angular velocity acquired makes the rolling ball to travel faster.

c. $v_r = v_s$, because kinetic energy of the two balls is same at the bottom of the planes.

d. $v_r < v_s$, because the rolling ball acquires rotational as well as translational kinetic energy.

e. $v_r > v_s$, because the rolling ball has an angular acceleration as well as linear acceleration.

24. A plank with a uniform sphere placed on it rests on a smooth horizontal plane. The plank is pulled to the right by a constant force F. If the sphere does not slip over the plank, then

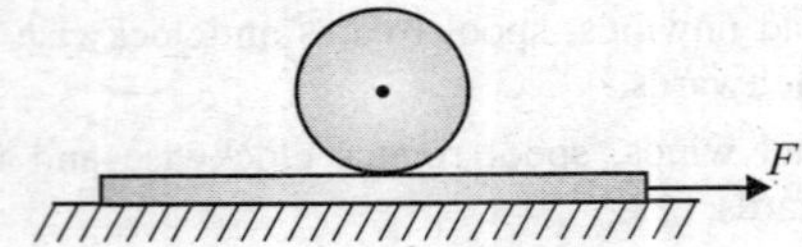

Fig. 2.343

a. both have the same acceleration

b. acceleration of the centre of sphere is less than that of the plank

c. work done by friction acting on the sphere is equal to its total kinetic energy

d. total kinetic energy of the system is equal to work done by the force F

25. A uniform cylinder of mass m rests on two rough horizontal planks. A thread is wound on the cylinder. The hanging end of the thread is pulled vertically down with a constant force F.

a. Since horizontal acceleration is provided by the friction acting on the cylinder, its translation kinetic energy $(1/2\, mv^2)$ is equal to work done by this friction.

b. Since moment about instantaneous axis of rotation is produced by force F, kinetic energy $(1/2\ I\omega^2)$ is equal to work done by F where I is moment of inertia about the instantaneous axis of rotation and ω is the angular velocity.

c. Since cylinder is moving, energy is lost against friction.

d. Work done by F + work done by friction on cylinder = total KE of the cylinder.

26. Consider the rotation of a rod of mass m and length l from position AB to AB'. Which of the following statements are correct?

a. Weight of the rod is lowered by $l/2$.

b. Loss of gravitational potential energy is $\frac{1}{2}\, mgl$.

c. Angular velocity is $\sqrt{3g/l}$.

d. Rotational kinetic energy is $\frac{ml^2\omega^2}{6}$.

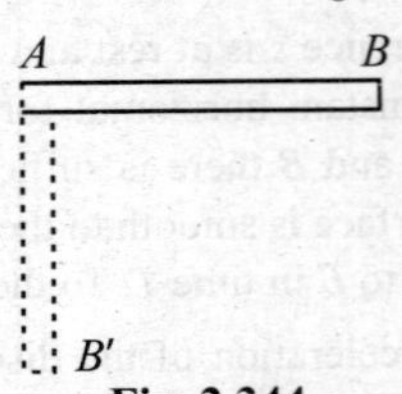

Fig. 2.344

27. A 1 kg solid sphere rolls without slipping on a rough horizontal surface under the influence of a 7 N force. The force acts tangentially at the highest point of the sphere. Which of the following statements are correct? (7 N force acts towards right).

a. The frictional force on the sphere acts towards right.

b. The value of the frictional force is 3 N.

c. The acceleration of the centre of the sphere is 9.8 m/s^2.

d. The acceleration of the centre of the sphere is 10 m/s^2.

28. A bucket of water of mass 21 kg is suspended by a rope wrapped around a solid cylinder 0.2 m in diameter. The mass of the solid cylinder is 21 kg. The bucket is released from rest. Which of the following statements are correct?

a. The tension in the rope is 70 N.

b. The acceleration of the bucket is (20/3) m/s^2.

c. The acceleration of the bucket is independent of the mass of the bucket.

d. All of these

29. A particle, moving horizontally, collides perpendicularly at one end of a rod having equal mass and placed on a smooth horizontal surface.

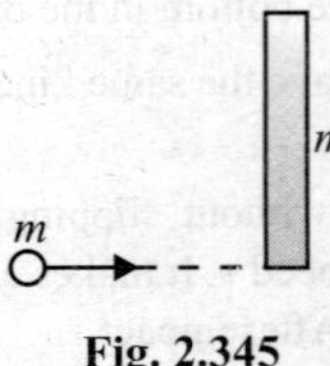

Fig. 2.345

a. Particle comes to rest if collision is perfectly elastic and centre of rod starts to move with the same velocity.

b. Particle continues to move along the same direction, whatever is the value of e.

c. Particle may get rebound back.

d. Velocity of mid-point of the rod will be less than $v_0/2$ if the particle gets stuck.

30. A uniform cylinder of mass M and radius R is placed on a rough horizontal board, which in turn is placed on a smooth surface.

The coefficient of friction between the board and the cylinder is μ. If the board starts accelerating with constant acceleration a, as shown in the figure, then

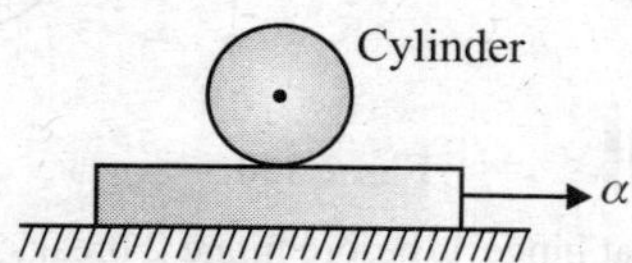

Fig. 2.346

a. for pure rolling motion of the cylinder, direction of frictional force is forward and its magnitude is $Ma/3$

b. the maximum value of a, so that the cylinder performs pure rolling is $3\,\mu g$

c. the acceleration of the centre of mass of the cylinder under pure rolling condition for the given a is $a/3$

31. A sphere A moving with speed u and rotating with an angular velocity ω makes a head-on elastic collision with an identical stationary sphere B. There is no friction between the surfaces of A and B. Choose the correct alternative(s). Disregard gravity.

a. A will stop moving but continue to rotate with an angular velocity ω.

b. A will come to rest and stop rotating.

c. B will move with speed u without rotating.

d. B will move with speed u and rotate with an angular velocity ω.

Assertion-Reasoning Type

Solutions on page 2.133

In the following questions, each question contains Statement I: (Assertion) and Statement II: (Reason). Each question has 4 choices (a), (b), (c), and (d) out of which ONLY ONE is correct.

(a) Statement I is True, Statement II is True; Statement II is a correct explanation for Statement I.

(b) Statement I is True, Statement II is True; Statement II is **NOT a** correct explanation for Statement 1.

(c) Statement I is True, Statement II is False.

(d) Statement I is False, Statement II is True.

1. Statement I: A disc is allowed to roll purely on an inclined plane as shown in Fig. 2.347. A force F parallel to the incline and passing through the centre of the disc acts which remains constant during the motion.

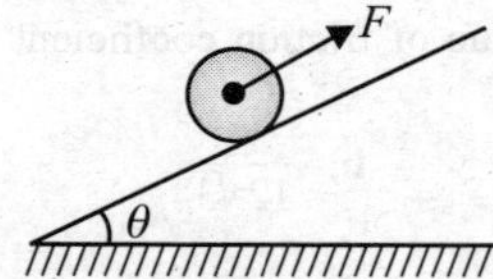

Fig. 2.347

It is possible that for certain values of F, the friction on the disc is acting along downward direction and for certain other values of F, the friction on the disc be acting along upward direction and there is no other possibility.

Statement II: The friction (if acting) will be static and not kinetic in nature.

2. Statement I: Many great rivers flow towards the equator. The sediments that they carry increase the time of rotation of the earth about its own axis.

Statement II: The angular momentum of the earth about its rotation axis is conserved.

3. Statement I: The mass of a body cannot be considered to be concentrated at the centre of mass of the body for the purpose of computing its moment of inertia.

Statement II: Then the moment of inertia of every body about an axis passing through its centre of mass would be zero.

4. Statement I: A ladder is more likely to slip when a person is near the top than when he is near the bottom.

Statement II: The friction between the ladder and the floor decreases as he climbs up.

5. Statement I: A solid sphere is rolling on a rough horizontal surface. Acceleration of contact point is zero.

Statement II: A solid sphere can roll on the smooth surface.

6. Statement I: A disc is rolling on an inclined plane without slipping. The velocity of centre of mass is V. The other points on the disc lie on a circular arc having same speed as that of the centre of mass.

Statement II: When a disc is rolling on an inclined plane, the magnitude of velocities of all the points from the contact point is the same, having distance equal to radius r.

7. Statement I: A sphere is performing pure rolling on a rough horizontal surface with constant angular velocity. Frictional force acting on the sphere is zero.

Statement II: Velocity of contact point is zero.

8. Statement I: If the earth shrinks (without change in mass) to half it's present size, length of the day would become 6 hr.

Statement II: As size of earth changes, its moment of inertia changes.

9. Statement I: A disc is rolling on a rough horizontal surface. The instantaneous speed of the point of contact during perfect rolling is zero with respect to the ground.

Statement II: The force of friction can help in achieving pure rolling condition.

10. Statement I: When a diver dives, the rotational kinetic energy of the diver increases during several somersaults.

Statement II: When diver pulls his limbs, the moment of inertia decreases and on account of conservation of angular momentum his angular speed increases.

11. Statement I: A horizontal force F is applied such that the block remains stationary because N will produce torque.

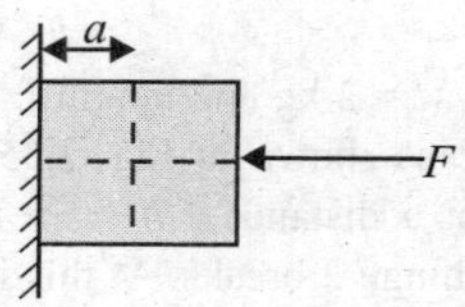

Fig. 2.348

Statement II: The torque produced by friction force is equal and opposite to the torque produced due to normal reaction (N).

12. **Statement I:** The velocity of a body at the bottom of an inclined plane of a given height is more when it slides down the plane compared to when it is rolling down the same plane.

Statement II: In rolling down, a body acquires both kinetic energy of translation and rotation.

13. **Statement I:** A rigid disc rolls without slipping on a fixed rough horizontal surface with uniform angular velocity. Then the acceleration of lowest point on the disc is zero.

Statement II: For a rigid disc rolling without slipping on a fixed rough horizontal surface, the velocity of the lowest point on the disc is always zero.

14. **Statement I:** A uniform cubical block (of side a) undergoes translational motion on a smooth horizontal surface under action of horizontal force F as shown in Fig. 2.349. Under the given condition, the horizontal surface exerts normal reaction non-uniformly on the lower surface of the block.

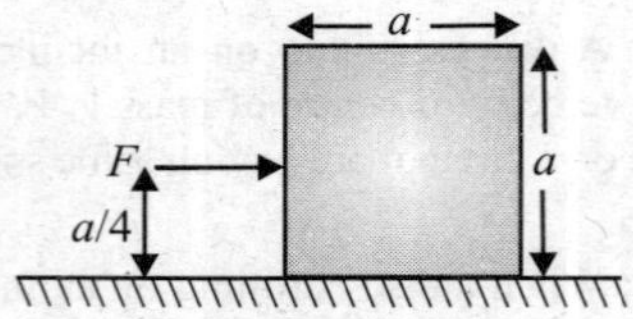

Fig. 2.349

Statement II: For the cubical block given in Statement 1, the horizontal force F has a tendency to rotate the cube about its centre in clockwise sense. Hence, the lower right edge of the cube presses the horizontal surface harder in comparison to the force exerted by the lower left edge of the cube on horizontal surface.

15. **Statement I:** A uniform thin rod of length L is hinged about one of its ends and is free to rotate about the hinge without friction. Neglect the effect of gravity. A force F is applied at a distance x from the hinge on the rod such that the force is always perpendicular to the rod. As the value of x is increased from zero to L, the component of reaction by the hinge on the rod perpendicular to the length of rod increases.

Statement II: Under the conditions given in Statement 1 as x is increased from zero to L, the angular acceleration of the rod increases.

Comprehension Type

Solutions on page 2.134

For Problems 1–3

A uniform rod of mass $M = 2$ kg and length L is suspended by two smooth hinges 1 and 2 as shown in Fig. 2.350. A force $F = 4$ N is applied downward at a distance $L/4$ from hinge 2. Due to the application of force F, hinge 2 breaks. At this instant, applied force F is also removed. The rod starts to rotate downward about hinge 1. ($g = 10$ m/s^2)

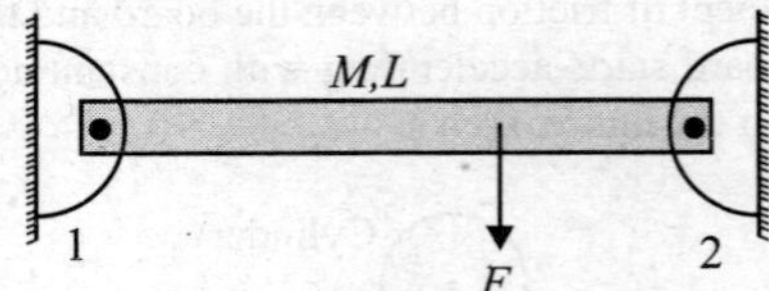

Fig. 2.350

1. The reaction at hinge 1, before hinge 2 breaks, is
   a. 24 N b. 12 N
   c. 11 N d. 10 N
2. The reaction at hinge 1, just after breaking of hinge 2, is
   a. 20 N b. 10 N
   c. 5 N d. 0
3. The acceleration of the end point of the rod of small mass dm at the end point of the rod, when the rod becomes vertical is
   a. 30 m/s^2 b. 20 m/s^2
   c. 10 m/s^2 d. 0

For Problems 4–6

Four identical rods of mass $M = 6$ kg each are welded at their ends to form a square and then welded to a massive ring having mass $m = 4$ kg and radius $R = 1$ m. If the system is allowed to roll down the incline of inclination $\theta = 30°$, determine the minimum value of the coefficient of static friction that will prevent slipping.

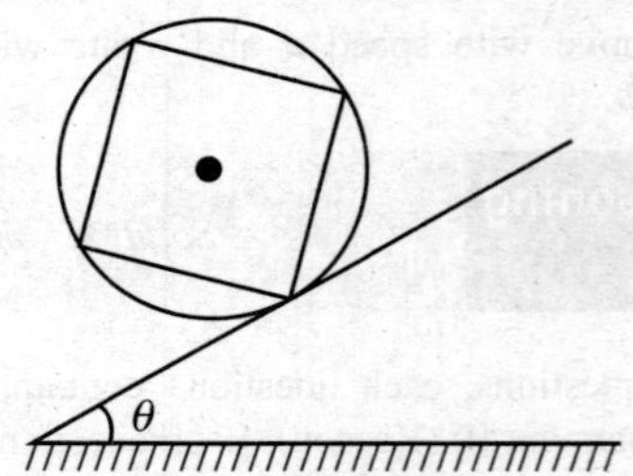

Fig. 2.351

4. The moment of inertia of the system about the centre of ring will be
   a. 20 kg m^2 b. 40 kg m^2
   c. 10 kg m^2 d. 60 kg m^2
5. The acceleration of the system will be
   a. $\frac{g}{2}$ b. $\frac{g}{4}$
   c. $\frac{7g}{24}$ d. $\frac{g}{8}$
6. The minimum value of friction coefficient to prevent slipping is
   a. $\frac{5}{7}$ b. $\frac{5}{12\sqrt{3}}$
   c. $\frac{5\sqrt{3}}{7}$ d. $\frac{7}{5\sqrt{3}}$

For Problems 7–9

Figure 2.352 shows a uniform smooth solid cylinder A of radius 4 m rolling without slipping on the 8 kg plank which, in turn, is supported by a fixed smooth surface. Block B is known to accelerate down with 6 m/s^2 and block C moves down with acceleration 2 m/s^2.

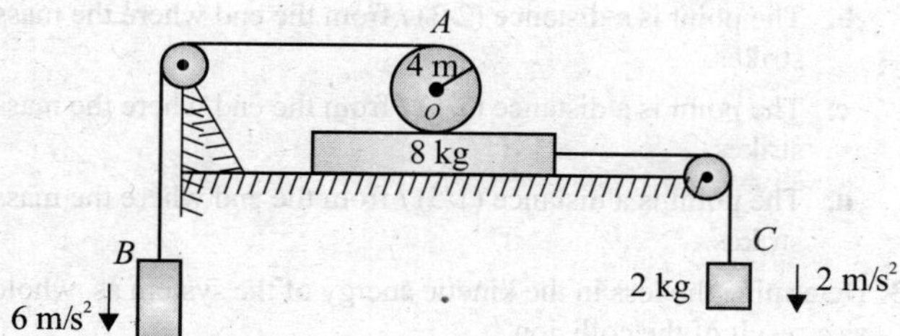

Fig. 2.352

7. What is the angular acceleration of the cylinder?

a. $\frac{4}{5}$ rad s^{-2} **b.** $\frac{6}{5}$ rad s^{-2}

c. 2 rad s^{-2} **d.** 1 rad s^{-2}

8. What is the ratio of the mass of the cylinder to the mass of block *B*?

a. 1 **b.** 2

c. 3 **d.** 4

9. If unwrapped length of the thread between the cylinder and block *B* is 20 m at the beginning, when the system was released from rest, what would it be 2 s later ?

a. 28 m **b.** 30 m

c. 22 m **d.** 32.5 m

For Problems 10–12

A uniform cylinder of radius *R* is spinned about its axis to the angular velocity ω_0 and then placed into a corner, see the figure. The coefficient of friction between the corner walls and the cylinder is μ_k.

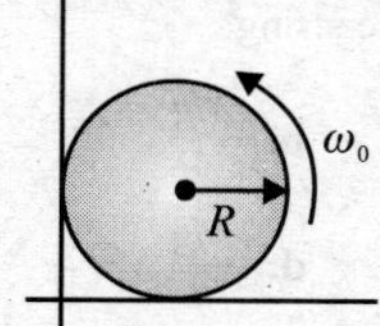

Fig. 2.353

10. The normal reaction imparted by the wall on the cylinder is

a. $\frac{2\mu_k mg}{1+\mu_k^2}$ **b.** $\frac{\mu_k mg}{1+\mu_k^2}$

c. $\frac{\mu_k mg}{1+2\mu_k^2}$ **d.** $\frac{\mu_k mg}{2+\mu_k^2}$

11. Calculate the angular acceleration of the cylinder.

a. $\frac{\mu_k(1+\mu_k)g}{(1+\mu_k^2)R}$ clockwise direction

b. $\frac{\mu_k(1+\mu_k)g}{(1+\mu_k^2)R}$ anticlockwise direction

c. $\frac{2\mu_k(1+\mu_k)g}{(1+\mu_k^2)R}$ clockwise direction

d. $\frac{2\mu_k(1+\mu_k)g}{(1+\mu_k^2)R}$ anticlockwise direction

12. How many turns will the cylinder accomplish before it stops?

a. $\frac{(1+\mu_k^2)\omega_0^2 R}{\pi\mu_k(1+\mu_k)g}$ **b.** $\frac{(1+\mu_k^2)\omega_0^2 R}{8\pi\mu_k(1+\mu_k)g}$

c. $\frac{8(1+\mu_k^2)\omega_0^2 R}{\pi\mu_k(1+\mu_k)g}$ **d.** $\frac{3(1+\mu_k^2)\omega_0^2 R}{\pi\mu_k(1+\mu_k)g}$

For Problems 13–15

A man of mass 100 kg stands at the rim of a turntable of radius 2 m and moment of inertia 4000 kg m^2 mounted on a vertical frictionless shaft at its centre. The whole system is initially at rest. The man now walks along the outer edge of the turntable (anticlockwise) with a velocity of 1 m/s relative to the earth.

13. With what angular velocity and in what direction does the turntable rotate?

a. The table rotates anticlockwise (in the direction of the man motion) with angular velocity 0.05 rad/s.

b. The table rotates clockwise (opposite to the man) with angular velocity 0.1 rad/s.

c. The table rotates clockwise (opposite to the man) with angular velocity 0.05 rad/s.

d. The table rotates anticlockwise (in the direction of the man motion) with angular velocity 0.1 rad/s.

14. Through what angle will it have rotated when the man reaches his initial position on the turntable?

a. The table rotates through $2\pi/11$ radians anticlockwise

b. The table rotates through $4\pi/11$ radians clockwise

c. The table rotates through $4\pi/11$ radians anticlockwise

d. The table rotates through $2\pi/11$ radians clockwise

15. Through what angle will it have rotated when the man reaches his initial position relative to earth ?

a. 36° in clockwise direction

b. 36° in anticlockwise direction

c. 72° in clockwise direction

d. 72° in anticlockwise direction

For Problems 16–17

A small block of mass 4 kg is attached to a cord passing through a hole in a horizontal frictionless surface. The block is originally revolving in a cirlce of radius of 5 m about the hole, with a tangential velocity of 4 m/s. The cord is then pulled slowly from below, shortening the radius of the circle in which the block revolves. The breaking strength of the cord is 200 N.

16. What will be the radius of the circle when the cord breaks?

a. 4.0 m **b.** 1.0 m

c. 3.0 m **d.** 2.0 m

17. Velocity of the block at the time of breaking of the string

a. 10 m/s **b.** 20 m/s

c. 7.5 m/s **d.** 12.5 m/s

For Problems 18–20

End A of a rod AB is being pulled on the floor with a constant velocity v_0 as shown. Taking the length of the rod as l, at an instant when the rod makes an angle 37° with the horizontal, calculate

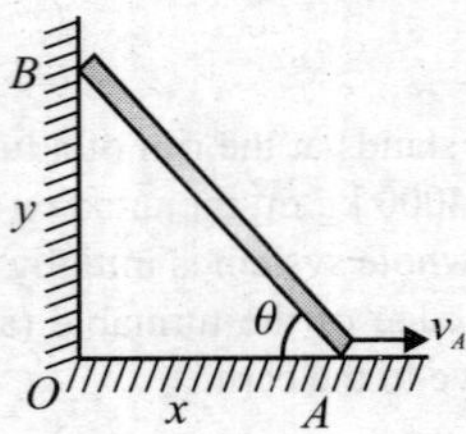

Fig. 2.354

18. the velocity of end B

a. $\frac{3}{5}v_0$ **b.** $\frac{4}{5}v_0$

c. $\frac{5}{3}v_0$ **d.** $\frac{5}{4}v_0$

19. the angular velocity of the end

a. $\frac{5v_0}{3l}$ **b.** $\frac{3v_0}{5l}$

c. $\frac{5v_0}{4l}$ **d.** $\frac{4v_0}{5l}$

20. The velocity of the CM of the rod

a. $\frac{5}{7}v_0$ at $\tan^{-1}\frac{4}{3}$ below horizontal

b. $\frac{5}{7}v_0$ at $\tan^{-1}\frac{3}{4}$ below horizontal

c. $\frac{5}{6}v_0$ at $\tan^{-1}\frac{3}{4}$ below horizontal

d. $\frac{5}{6}v_0$ at $\tan^{-1}\frac{4}{3}$ below horizontal

For Problems 21–23

A uniform rod of length l and mass $2m$ rests on a smooth horizontal table. A point mass m moving horizontally at right angles to the rod with initial velocity v collides with one end of the rod and sticks to it.

21. Determine the angular velocity of the system after the collision.

a. $\frac{v}{l}$ **b.** $\frac{2v}{l}$

c. $\frac{5v}{3l}$ **d.** $\frac{3v}{2l}$

22. Determine the position of the point on the rod which remains stationary immediately after collision.

a. The point is a distance (1/3) l from the end where the mass strikes.

b. The point is a distance (2/3) l from the end where the mass strikes.

c. The point is a distance (2/5) l from the end where the mass strikes.

d. The point is a distance (1/5) l from the end where the mass strikes.

23. Determine the loss in the kinetic energy of the system as whole as a result of the collision.

a. $\frac{1}{6}mv^2$ **b.** $\frac{1}{3}mv^2$

c. mv^2 **d.** $\frac{mv^2}{5}$

For Problems 24–26

An L shaped uniform rod of mass $2M$ and length $2L$ ($AB = BC = L$) is held as shown in Fig. 2.355 with a string fixed between C and wall so that AB is vertical and BC is horizontal. There is no friction between the hinge and the rod at A.

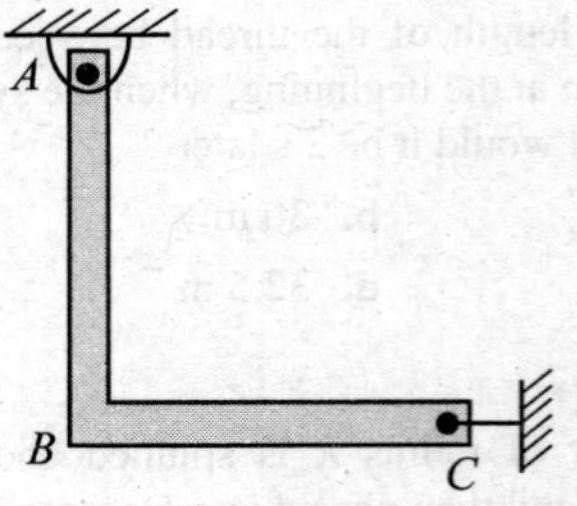

Fig. 2.355

24. Find the tension in the string.

a. $\frac{Mg}{3}$ **b.** $\frac{Mg}{4}$

c. Mg **d.** $\frac{Mg}{2}$

25. What will be the reaction between hinge and rod at point A?

a. $\sqrt{65}\,\frac{Mg}{4}$ **b.** $2\,Mg$

c. $\sqrt{17}\,\frac{Mg}{4}$ **d.** $\sqrt{17}\,\frac{Mg}{2}$

26. If the string is burnt, find the angle between AB and the vertical at equilibrium position.

a. $\tan^{-1}\left(\frac{1}{3}\right)$ **b.** $\tan^{-1}\left(\frac{1}{4}\right)$

c. $\tan^{-1}(3)$ **d.** $\tan^{-1}\left(\frac{1}{2}\right)$

For Problems 27–29

A uniform rod of length L and mass M is lying on a frictionless horizontal plane and is pivoted at one of its ends as shown in Fig. 2.356. There is no friction at the pivot. An inelastic ball of mass m is fixed with the rod at a distance $L/3$ from O. A horizontal impulse J is given to the rod at a distance $2L/3$ from O in a direction perpendicular to the rod. Assume that the ball remains in contact with the rod after the collision and impulse J acts for a small time interval Δt.

Now answer the following questions:

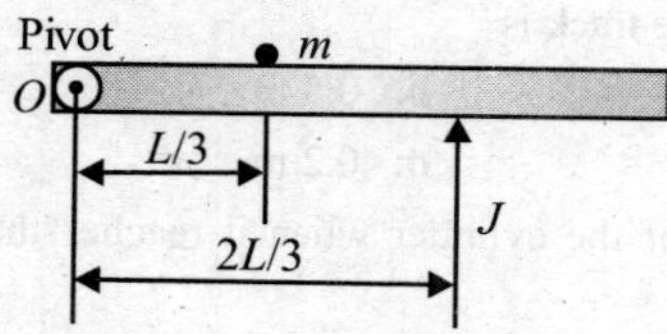

Fig. 2.356

27. Find the resulting instantaneous angular velocity of the rod after the impulse.

a. $\frac{3J}{(m+3M)L}$ b. $\frac{6J}{(m+3M)L}$

c. $\frac{3J}{(3m+M)L}$ d. $\frac{6J}{(3m+M)L}$

28. Find the impulse acted on the ball during the time interval Δt.

a. $\frac{2MJ}{(3m+M)}$ b. $\frac{2MJ}{(m+3M)}$

c. $\frac{2mJ}{(3m+M)}$ d. $\frac{2mJ}{(m+3M)}$

29. Find the magnitude of the impulse applied by the pivot during the time interval Δt.

a. $\frac{mJ}{(m+3M)}$ b. $\frac{mJ}{(3m+M)}$

c. $\frac{MJ}{(m+3M)}$ d. $\frac{MJ}{(3m+M)}$

For Problems 30–32

A long horizontal plank of mass 'm' is lying on a smooth horizontal surface. A sphere of same mass 'm' and radius 'r' is spinned about its own axis with angular velocity ω_0 and gently placed on the plank. The coefficient of friction between the plank and the sphere is μ. After some time the pure rolling of the sphere on the plank will start Answer the following questions.

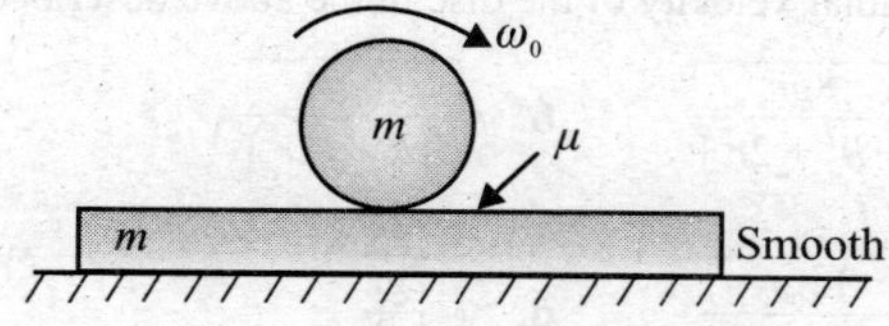

Fig. 2.357

30. Find the time t at which the pure rolling starts.

a. $\frac{\omega_0 r}{9\mu g}$ b. $\frac{2\omega_0 r}{9\mu g}$

c. $\frac{\omega_0 r}{3\mu g}$ d. $\frac{2\omega_0 r}{\mu g}$

31. Find the velocity of the sphere after the pure rolling starts.

a. $\frac{2\omega_0 r}{9}$ b. $\frac{\omega_0 r}{9}$

c. $\frac{\omega_0 r}{3}$ d. $\frac{2\omega_0 r}{3}$

32. Find the displacement of the plank till the sphere starts pure rolling.

a. $\frac{\omega_0^2 r^2}{81\mu g}$ b. $\frac{2\omega_0^2 r^2}{27\mu g}$

c. $\frac{4\omega_0^2 r^2}{81\mu g}$ d. $\frac{2\omega_0^2 r^2}{81\mu g}$

For Problems 33–35

A disc of mass 'm' and radius 'R' is free to rotate in a horizontal plane about a vertical smooth fixed axis passing through its centre. There is a smooth groove along the diameter of the disc and two small balls of mass $m/2$ each are placed in it on either side of the centre of the disc as shown in Fig. 2.358. The disc is given an initial angular velocity ω_0 and released.

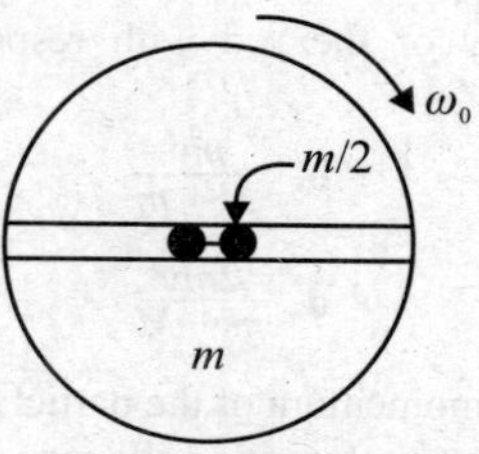

Fig. 2.358

33. The angular speed of the disc when the balls reach the end of the disc is

a. $\frac{\omega_0}{2}$ b. $\frac{\omega_0}{3}$

c. $\frac{2\omega_0}{3}$ d. $\frac{\omega_0}{4}$

34. The speed of each ball relative to the ground just after they leave the disc is

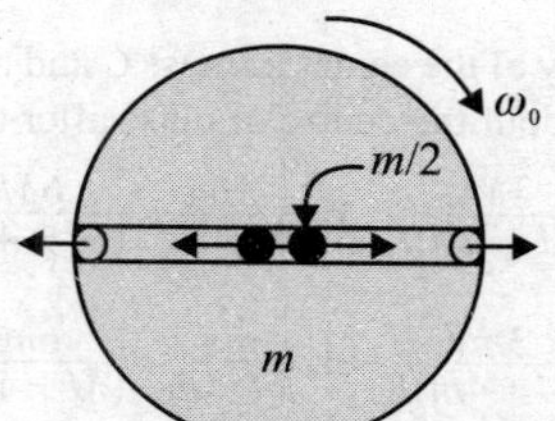

Fig. 2.359

a. $\frac{R\omega_0}{\sqrt{3}}$ b. $\frac{R\omega_0}{\sqrt{2}}$

c. $\frac{2R\omega_0}{3}$ d. none of these

35. The net work done by forces exerted by the disc on the ball for the duration ball remains on the disc is

a. $\frac{2mR^2\omega_0^2}{9}$ b. $\frac{mR^2\omega_0^2}{18}$

c. $\frac{mR^2\omega_0^2}{6}$ d. $\frac{mR^2\omega_0^2}{9}$

For Problems 36–40

A uniform rod of length L lies on a smooth horizontal table. The rod has a mass M. A particle of mass m moving with speed v strikes the rod perpendicularly at one of the ends of the rod and sticks to it after collision.

36. Find the velocity of the centre of mass C of the system constituting "the rod plus the particle".

a. $\dfrac{2Mv}{M-m}$ b. $\dfrac{2mv}{M+m}$

c. $\dfrac{Mv}{M+m}$ d. $\dfrac{mv}{M+m}$

37. Find the velocity of the particle with respect to C before the collision.

a. $\dfrac{Mv}{M+m}$ b. $\dfrac{mv}{M+m}$

c. $\dfrac{2mv}{M+m}$ d. $\dfrac{2Mv}{M+m}$

38. Find the velocity of the rod with respect to C before the collision

a. $\dfrac{Mv}{M+m}$ b. $\dfrac{mv}{M+m}$

c. $\dfrac{2Mv}{m+M}$ d. $\dfrac{2mv}{m+M}$

39. Find the angular momentum of the particle and of the rod about the centre of mass C before the collision.

a. $\dfrac{M^2mv\,l}{4(m+M)^2}, \dfrac{m^2Mv\,l}{4(m+M)^2}$

b. $\dfrac{M^2mv\,l}{2(M+m)^2}, \dfrac{Mm^2vl}{2(M+m)^2}$

c. $\dfrac{M^2mvl}{(m+M)^2}, \dfrac{m^2Mvl}{(m+M)^2}$

d. $\dfrac{2M^2vml}{(m+M)^2}, \dfrac{2m^2Mvl}{(M+m)^2}$

40. Find the velocity of the centre of mass C and the angular velocity of the system about the centre of mass after the collision.

a. $\dfrac{2mv}{M+m}, \dfrac{3Mv}{(M+2m)L}$ b. $\dfrac{Mv}{M+m}, \dfrac{6Mv}{(M+4m)L}$

c. $\dfrac{2Mv}{M+m}, \dfrac{3mv}{(M+4m)L}$ d. $\dfrac{mv}{M+m}, \dfrac{6mv}{(M+4m)L}$

For Problems 41–43

A uniform solid cylinder of mass 2 kg and radius 0.2 m is released from rest at the top of a semicircular track of radius 0.7 m cut in a block of mass $M = 3$ kg as shown in Fig. 2.360. The block is resting on a smooth horizontal surface and the cylinder rolls down without slipping.

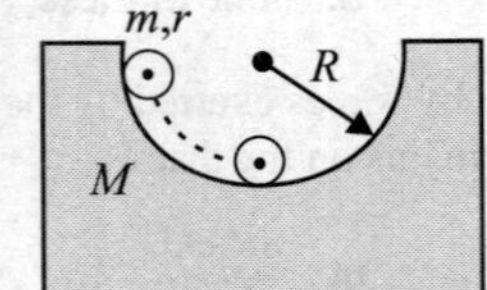

Fig. 2.360

Based on the above information, answer the following questions:

41. The distance moved by the block when the cylinder reaches the bottom of the track is

a. 0.3 m b. 0.5 m

c. 0.7 m d. 0.2 m

42. The speed of the cylinder when it reaches the bottom of the track is

a. $\dfrac{2}{\sqrt{11}}$ m/s b. $\dfrac{8}{\sqrt{11}}$ m/s

c. $\dfrac{8}{\sqrt{11}}$ m/s d. $\dfrac{6}{\sqrt{11}}$ m/s

43. The speed of the point of contact of the cylinder with the block w.r.t. ground, when the cylinder reaches the bottom of the track is

a. $\dfrac{2}{\sqrt{11}}$ m/s b. $\dfrac{8}{\sqrt{11}}$ m/s

c. $\dfrac{4}{\sqrt{11}}$ m/s d. $\dfrac{6}{\sqrt{11}}$ m/s

For Problems 44–46

A disc of mass M and radius R can rotate freely in a vertical plane about a horizontal axis at O distance r from the centre of the disc as shown in Fig. 2.361. The disc is released from rest in the shown position. Answer the following questions based on the above information

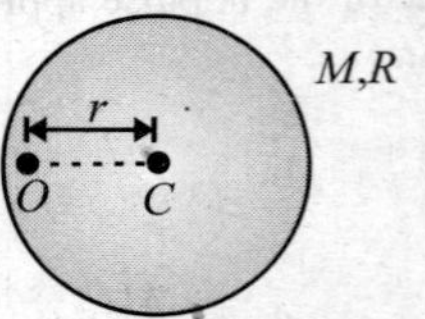

Fig. 2.361

44. The angular acceleration of the disc when OC rotates by an angle of 37° is

a. $\dfrac{8rg}{5[R^2+2r^2]}$ b. $\dfrac{5rg}{4[R^2+2r^2]}$

c. $\dfrac{10rg}{3[R^2+2r^2]}$ d. $\dfrac{8rg}{5R^2}$

45. The angular velocity of the disc in the above described case is

a. $\sqrt{\dfrac{8gr}{5[R^2+2r^2]}}$ b. $\sqrt{\dfrac{6gr}{5[R^2+2r^2]}}$

c. $\sqrt{\dfrac{12gr}{5[R^2+2r^2]}}$ d. $\sqrt{\dfrac{12gr}{5R^2}}$

46. Reaction force exerted by the hinge on the disc at this instant is

a. $\dfrac{Mg}{5(R^2+2r^2)} \times \sqrt{9(R^2+6r^2)^2+(4R^2)^2}$

b. $\dfrac{Mg}{5(R^2+2r^2)} \times 3(R^2+6r^2)$

c. $\dfrac{4Mg}{5(R^2+2r^2)} \times R^2$

d. $\dfrac{Mg}{5(R^2+2r^2)} \times 4R^2$

For Problems 47–48

In Fig. 2.362, $F = 10$ N, $R = 1$ m, mass of the body is 2 kg and moment of inertia of the body about an axis passing through O and perpendicular to the plane of body is 4 kg-m^2. O is the centre of mass of the body.

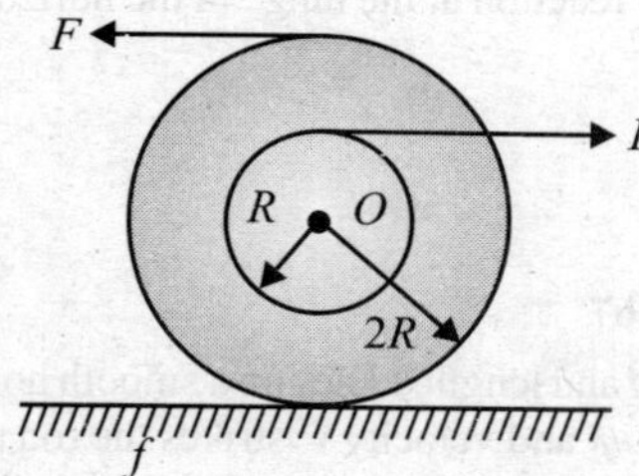

Fig. 2.362

47. If the ground is smooth, what is the total kinetic energy of body at $t = 2$ s. Initially, the body was at rest.

a. 25 J **b.** 16.67 J

c. 50 J **d.** 37.5 J

48. If the ground is sufficiently rough to ensure rolling, what is the kinetic energy of the body now in the given time interval?

a. 18.75 J **b.** 25.67 J

c. 16.67 J **d.** None of these

For Problems 49–50

A wheel of radius R rolls without slipping and its centre O has an acceleration a_0. A point A on the wheel is at a distance r from O. For given values of a_0, R and r, determine the angle θ and velocity v_0 of the wheel for which A has no acceleration in this position:

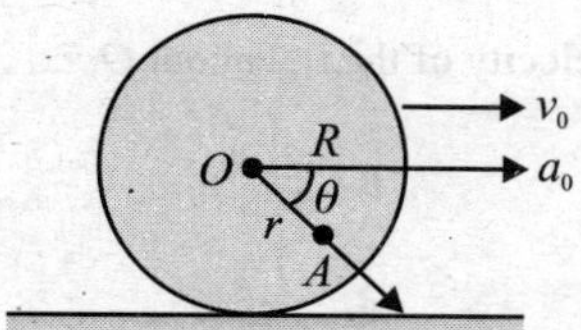

Fig. 2.363

49. Angle θ is given by

a. 30° **b.** $\sin^{-1}\left(\frac{r}{R}\right)$

c. 60° **d.** $\cos^{-1}\left(\frac{r}{R}\right)$

50. Velocity v_0 is given by $\left(\frac{r}{R}\right)$

a. $v_0 = \sqrt{\frac{a_0 R}{r}}(R^2 - r^2)^{\frac{1}{4}}$ **b.** $v_0 = \sqrt{\frac{a_0 R}{r}}(R^2 + r^2)^{\frac{1}{4}}$

c. $v_0 = \sqrt{\frac{a_0 R}{r}}\frac{1}{(R^2 + r^2)^{\frac{1}{4}}}$ **d.** $v_0 = \sqrt{\frac{a_0 R}{r}}\frac{1}{(R^2 - r^2)^{\frac{1}{4}}}$

For Problems 51–53

A disc of radius 20 CM is rolling with slipping on a flat horizontal surface. At a certain instant the velocity of its centre is 4 m/s and its angular velocity is 10 rad/s. The lowest contact point is O.

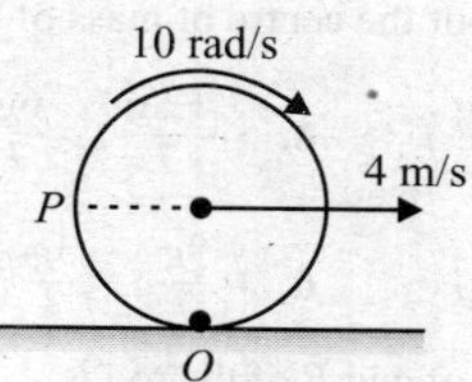

Fig. 2.364

51. Velocity of point O is

a. 2 m/s **b.** 4 m/s

c. 1 m/s **d.** 3 m/s

52. Instantaneous centre of the rotation of disc is located at

a. 0.2 m below O **b.** 0.2 m above O

c. 0.6 m above O **d.** 0.4 m below O

53. Velocity of point P is

a. $\sqrt{10}$ m/s **b.** $2\sqrt{5}$ m/s

c. $\sqrt{5}$ m/s **d.** 5 m/s

For Problems 54–55

A solid sphere of mass M and radius R is initially at rest. Solid sphere is gradually lowered onto a truck moving with constant velocity v_0.

Fig. 2.365

54. What is the final speed of the sphere's centre of mass in ground frame when eventually pure rolling sets in

a. $\frac{5}{7}v_0$ **b.** $\frac{2}{7}v_0$

c. $\frac{7}{5}v_0$ **d.** $\frac{7}{2}v_0$

55. If a sphere with twice the radius and four times the mass had been used; what would have been its final speed?

a. $\frac{5}{7}v_0$ **b.** $\frac{2}{7}v_0$

c. $\frac{7}{5}v_0$ **d.** $\frac{7}{2}v_0$

For Problems 56–58

A uniform rod AB hinged about a fixed point P is initially vertical. A rod is released from vertical position. When rod is in horizontal position:

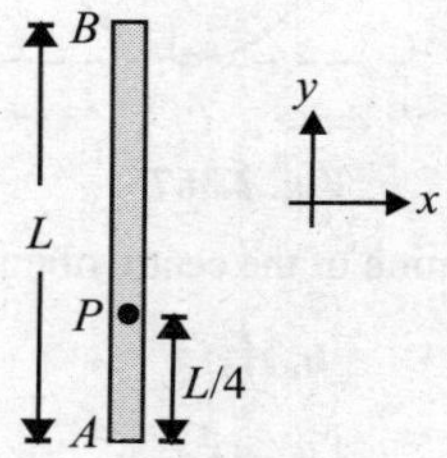

Fig. 2.366

56. The acceleration of the centre of mass of the rod is

a. $-\frac{6g}{7}\hat{i}-\frac{12g}{7}\hat{j}$ b. $-\frac{12g}{7}\hat{i}-\frac{6g}{7}\hat{j}$

c. $-\frac{3g}{7}\hat{i}-\frac{9g}{7}\hat{j}$ d. $-\frac{9g}{7}\hat{i}-\frac{3g}{7}\hat{j}$

57. The acceleration of end B of the rod is

a. $-\frac{6g}{7}\hat{i}-\frac{12g}{7}\hat{j}$ b. $-\frac{5g}{7}\hat{i}-\frac{9g}{7}\hat{j}$

c. $-\frac{18g}{7}\hat{i}-\frac{5g}{7}\hat{j}$ d. $-\frac{18g}{7}\hat{i}-\frac{9g}{7}\hat{j}$

58. The reaction at hinge P is

a. $-\frac{8mg}{7}\hat{i}-\frac{12mg}{7}\hat{j}$ b. $-\frac{3mg}{7}\hat{i}-\frac{9mg}{7}\hat{j}$

c. $-\frac{6mg}{7}\hat{i}+\frac{4mg}{7}\hat{j}$ d. $-\frac{12mg}{7}\hat{i}-\frac{6mg}{7}\hat{j}$

For Problems 59–60

A uniform disc of mass m and radius R rotates about a fixed vertical axis passing through its centre with angular velocity ω. A particle of same mass m and moving horizontally with velocity $2\omega R$ towards centre of the disc collides with the disc and sticks to its rim.

59. The angular velocity of the disc after the particle sticks to it is

a. $\frac{\omega}{5}$ b. $\frac{\omega}{3}$

c. $\frac{\omega}{2}$ d. $\frac{\omega}{4}$

60. The impulse on the particle due to the disc is

a. $\frac{\sqrt{17}}{3}m\omega R$ b. $\frac{\sqrt{35}}{3}m\omega R$

c. $\frac{\sqrt{37}}{3}m\omega R$ d. $\frac{\sqrt{29}}{3}m\omega R$

For Problems 61–63

When a body is hinged at a point and a force is acting on the body in such a way that the line of action of force is at some distance from the hinged point, the body will start rotating about the hinged point. The angular acceleration of the body can be calculated by finding the torque of that force about the hinged point. A disc of mass m and radius R is hinged at point A at its bottom and is free to rotate in the vertical plane. A force of magnitude F is acting on the ring at the top most point.

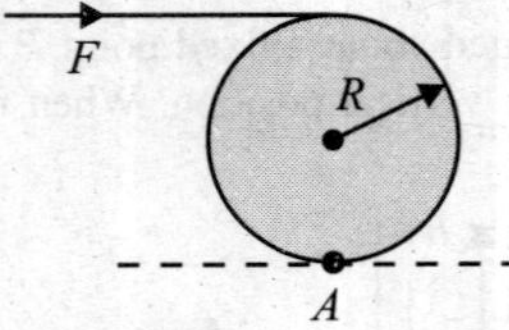

Fig. 2.367

61. Tangential acceleration of the centre of mass is

a. $\frac{3F}{4m}$ b. $\frac{F}{m}$

c. $\frac{2F}{3m}$ d. $\frac{4F}{3m}$

62. Component of reaction at the hinge in the vertical direction is

a. $\frac{4}{3}mg$ b. mg

c. $\frac{mg}{2}$ d. $\frac{2}{3}mg$

63. Component of reaction at the hinge in the horizontal direction is

a. $\frac{F}{4}$ b. F

c. $\frac{F}{3}$ d. $\frac{F}{2}$

For Problems 64–67

A rod AB of mass M and length $8\,l$ lies on a smooth horizontal surface. A particle of mass 'm' and velocity v_0 strikes the rod perpendicular to its length, as shown Fig. 2.368. As a result of the collision, the centre of mass of rod attains a speed of $v_0/8$ and the particle rebounds back with a speed of $v_0/4$. Find the following.

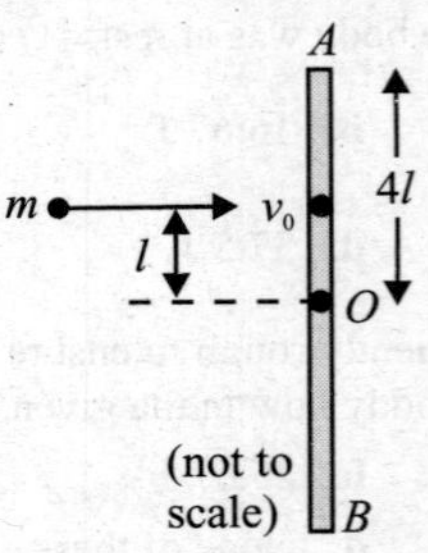

Fig. 2.368

64. The ratio $\frac{M}{m}$,

a. $\frac{M}{m}=10$ b. $\frac{M}{m}=4$

c. $\frac{M}{m}=8$ d. $\frac{M}{m}=5$

65. The angular velocity of the rod about O,

a. $\frac{5v_0}{128\,l}$ b. $\frac{3v_0}{128\,l}$

c. $\frac{v_0}{128\,l}$ d. $\frac{7v_0}{128\,l}$

66. The coefficient of restitution 'e' for the collision,

a. $\frac{51}{128}$ b. $\frac{61}{128}$

c. $\frac{21}{128}$ d. $\frac{31}{128}$

67. The velocities of the ends 'A' and 'B' of the rod, namely v_A and v_B, respectively.

a. $v_A=\frac{5}{32}v_0;\ v_B=\frac{1}{32}v_0$ b. $v_A=\frac{7}{32}v_0;\ v_B=\frac{3}{32}v_0$

c. $v_A=\frac{7}{32}v_0;\ v_B=\frac{1}{32}v_0$ d. $v_A=\frac{5}{32}v_0;\ v_B=\frac{3}{32}v_0$

For Problems 68–69

A block of mass m is kept on a horizontal ruler. The friction coefficient between the ruler and the block is μ. The ruler is fixed at one end and the block is at a distance L from the fixed end. The ruler is rotated about the fixed end in the horizontal plane through the fixed end.

68. What can the maximum angular speed be for which the block does not slip?

a. $\left(\frac{\mu g}{L}\right)^{1/4}$ **b.** $\left(\frac{\mu g}{2L}\right)^{1/2}$

c. $\left(\frac{2\mu g}{L}\right)^{1/2}$ **d.** $\left(\frac{\mu g}{L}\right)^{1/2}$

69. If the angular speed of the ruler is uniformly increased from zero at an angular acceleration α, at what angular speed will the block slip?

a. $\left[\left(\frac{\mu g}{L}\right)^2 - \alpha^2\right]^{1/4}$ **b.** $\left[\left(\frac{\mu g}{2L}\right)^2 + \alpha^2\right]^{1/2}$

c. $\left[\left(\frac{\mu g}{2L}\right)^2 - \alpha^2\right]^{1/4}$ **d.** $\left[\left(\frac{\mu g}{4L}\right)^2 - \alpha^2\right]^{1/2}$

For Problems 70–72

A long slender rod of mass 2 kg and length 4 m is placed on a smooth horizontal table. Two particles of masses 2 kg and 1 kg strike the rod simultaneously and stick to the rod after collision as shown in Fig. 2.369.

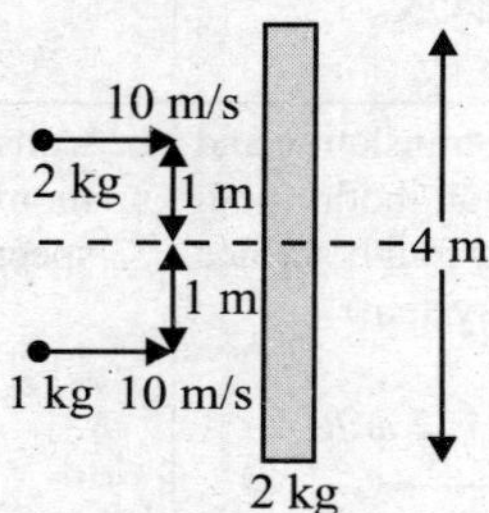

Fig. 2.369

70. Velocity of the centre of mass of the rod after collision is

a. 12 m/s **b.** 9 m/s

c. 6 m/s **d.** 3 m/s

71. Angular velocity of the rod after collision is

a. 11/17 rad/s **b.** rad/s

c. 19/17 rad/s **d.** 30/17 rad/s

72. If the two particles strike the rod in opposite direction, then after collision, as compared to the previous situation the rod will

a. rotate faster but translate at the same rate

b. show no change in linear or angular velocity

c. rotate slower and translate faster

d. rotate faster and translate slower

For Problems 73–75

A spherical ball of mass M moving with initial velocity v collides elastically with another ball of mass M, which is fixed at one end of L shaped rigid massless frame as shown in Fig. 2.370(a). The L shaped frame contains another mass M connected at the other end.

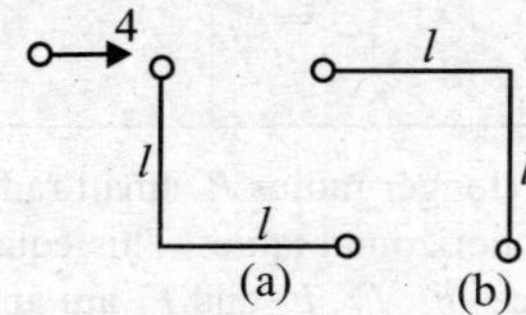

Fig. 2.370

73. The speed of the striking mass after collision is

a. $u/2$ backwards **b.** 0

c. $u/3$ in same direction **d.** $u/7$ backwards

74. The angular speed of L frame immediately after collision is

a. $\frac{4u}{7l}$ **b.** $\frac{u}{3l}$

c. $\frac{4u}{2l}$ **d.** $\frac{u}{7l}$

75. How soon will the frame come to the orientation shown in Fig. 2.370(b) after collision?

a. $\frac{\pi l}{4v}$ **b.** $\frac{7\pi l}{8v}$

c. $\frac{\pi l}{v}$ **d.** $\frac{7\pi l}{2v}$

Matching Column Type

Solutions on page 2.143

1. A cubical block having square cross section of side 2 m and of mass $M = 10$ kg is resting over a platform moving at constant acceleration $a = 1$ m/s^2. Coefficient of friction between the block and the platform is $\mu = 0.1$. A force F acts at the top of the cube as shown in Fig. 2.371. Now match Column I with Column II.

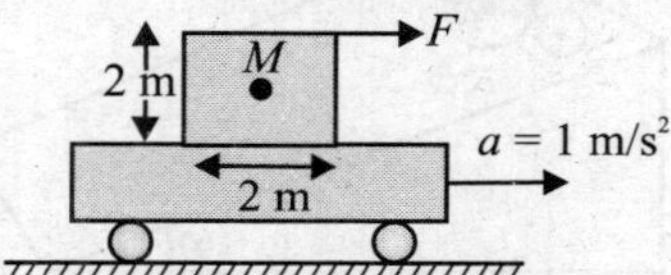

Fig. 2.371

Column I	Column II
i. $F = 0$	**a.** Block neither topples nor slips over the platform
ii. $F = 45$ N	**b.** Block topples but does not slip over the platform
iii. $F = 15$ N	**c.** Block slips but does not topple over the platform
iv. $F = 25$ N	**d.** Block slips as well as topples on the platform

2. Figure 2.372(a) shows two discs A and B, rotating about a common axis with constant angular speeds ω_A and ω_B, respectively. Moment of inertia of disc A is I_A and that of disc B is I_B. The discs are then gently pushed towards each other by forces that act along the axis. Finally, the discs rub against each other and attain a common angular speed ω, as shown in Fig. 2.372 (b). Match Columns I and II.

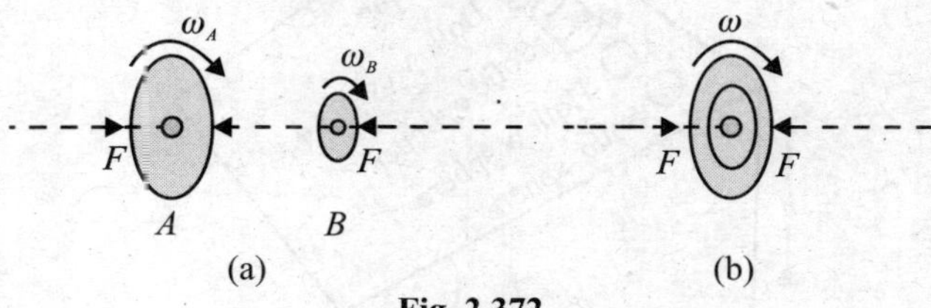

Fig. 2.372

Column I	Column II
i. As the discs stick together, the final angular speed, ω	**a.** will be zero
ii. Torque on each disc, due to the force F applied on the disc	**b.** will be decreasing
iii. When the discs are rubbing against each other, the kinetic energy of their system	**c.** $\dfrac{I_A\omega_A + I_B\omega_B}{I_A + I_B}$
iv. Angular momentum of the system	**d.** there will be loss of energy in doing work against the non-conservative internal forces

3. A solid sphere X of mass $'M'$ and radius R and another sphere Y of mass $'2M'$ and radius $'R/2'$ are simultaneously released at rest from the top of an inclined plane as shown in Fig. 2.373. The spheres roll without slipping. Also we may consider the spheres and the surface on which they roll to be perfectly rigid. Match Columns I and II.

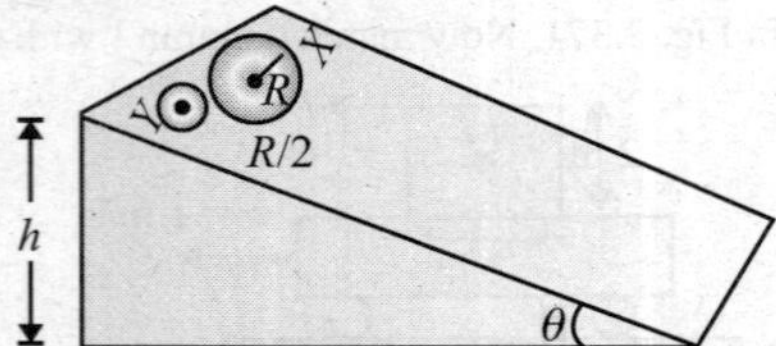

Fig. 2.373

Column I	Column II
i. More for sphere X	**a.** Speeds of the two spheres at the bottom of the incline
ii. More for sphere Y	**b.** Time taken by the spheres to reach the bottom
iii. Same for both spheres	**c.** Angular acceleration
iv. Zero	**d.** Kinetic energy at the bottom

4. A solid sphere, a thin-walled hollow sphere, a solid cylinder, a thin-walled hollow cylinder and a ring, each of mass m and radius R, are simultaneously released at rest from the top of an inclined plane, as shown in Fig. 2.374. The objects roll down the plane without slipping. Also we may consider the objects and the surface on which they roll to be perfectly rigid. Match Columns I and II.

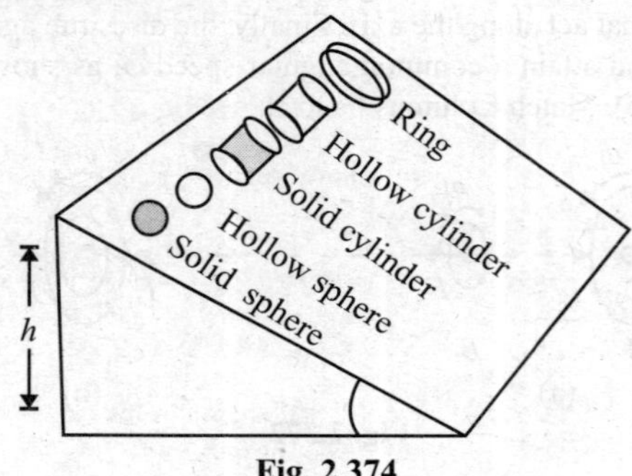

Fig. 2.374

Column I	Column II
i. Time taken to reach the bottom is maximum for	**a.** Solid sphere
ii. Angular acceleration is maximum for	**b.** Hollow cylinder
iii. Kinetic energy at the bottom is the same for	**c.** Hollow sphere
iv. Rotational kinetic energy is maximum for	**d.** Ring

5. Choose the correct option.

Column I	Column II
i. Earth moving in an elliptical orbit (only earth is system) Sun, Earth Fig. 2.375	**a.** Conservation of linear momentum along any direction
ii. A disc having translation and rotation motion both (both slipping) on a rough surface (only disc in system) Disc, ω_0, $v_0 \neq \omega_0 R$, v_0 Fig. 2.376	**b.** Conservation of linear momentum along specific direction
iii. A sphere rolling without slipping on a curved surface (only the sphere in system) ω, $V = \omega R$, v, Rough surface Fig. 2.377	**c.** Conservation of angular momentum about any point in the space
iv. Projection of a particle from the surface of earth (only particle in system). α, v Fig. 2.378	**d.** Conservation of angular momentum about a specific point in the space

6. A uniform yo-yo (longer radius R, small radius r) is resting on a perfectly rough horizontal table. This equation should be attempted after forces F_1, F_2, F_3 and F_4 are applied separately as shown.

Column I	Column II
i. $\cos\theta = r/R$ **Fig. 2.379**	**a.** Centre of mass of the yo-yo accelerates towards the left
ii. $\cos\theta < (r/R)$ **Fig. 2.380**	**b.** Center of mass of the yo-yo accelerates towards the right
iii. $\cos\theta = (r/R)$ **Fig. 2.381**	**c.** Friction acts on the yo-yo towards the left
	d. Friction acts on the yo-yo towards the right

7. From a uniform disc of mass M and radius R, a concentric disc of radius $R/2$ is cut out. For the remaining annular disc: I_1 is the moment of inertia about axis '1', I_2 about '2', I_3 about '3' and I_4 about '4'. Axes '1' and '2' are perpendicular to the disc and '3' and '4' are in the plane of the disc.

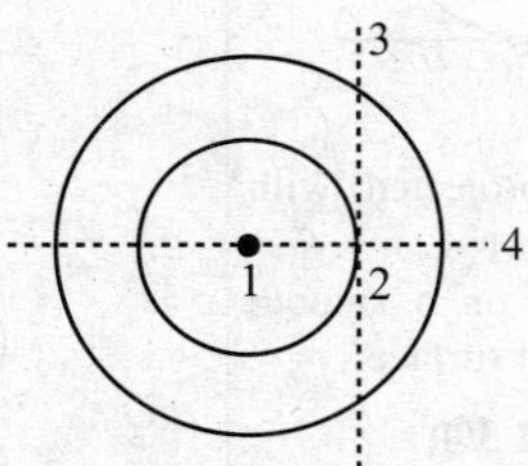

Axes '2', '3' and '4' intersect at a common point.

Match the following

Column I	Column II
i. I_1 is equal to	**a.** $\frac{21}{32}MR^2$
ii. I_2 is equal to	**b.** $I_1/2$
iii. $I_3 + I_4$ is equal to	**c.** $\frac{15}{32}MR^2$
iv. $I_2 - I_3$ is equal to	**d.** none of these

8. A particle P moves with constant speed on a circle in anticlockwise direction as shown in Fig. 2.382

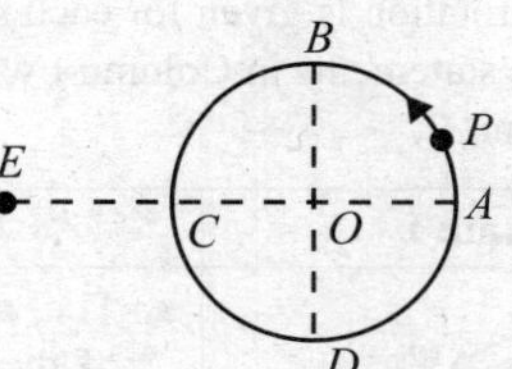

Fig. 2.382

Match Column I with Column II:

Column I	Column II
i. Angular momentum of the particle about O	**a.** is minimum when the particle is at A
ii. Angular momentum of the particle about E	**b.** is maximum when the particle is at A
iii. Angular velocity of the particle about O	**c.** does not remain constant
iv. Angular velocity of the particle about E	**d.** remains constant

9. A uniform disc rolls without slipping on a rough horizontal surface with uniform angular velocity. Point O is the centre of disc and P is a point on disc as shown in Fig. 2.383. In each situation of column I a statement is given and the corresponding results are given in Column II. Match the statements in Column I with the results in Column II.

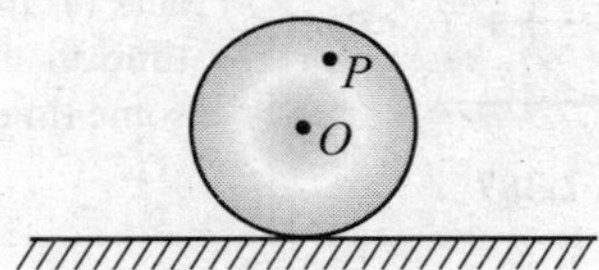

Fig. 2.383

Column I	Column II
i. The velocity of point P on the disc	**a.** changes in magnitude with time
ii. The acceleration of point P on the disc	**b.** always directed from that point (the point on disc given in Column I) towards the centre of disc.
iii. The tangential acceleration of point P on the disc	**c.** is always zero
iv. The acceleration of the point on the disc which is in contact with rough horizontal surface	**d.** is non-zero and remains constant in magnitude

10. In each situation of Column I, a uniform disc of mass m and radius R rolls on a rough fixed horizontal surface as shown. At $t = 0$ (initially) the angular velocity of the disc is ω_0 and velocity of the centre of mass of the disc is v_0 (in horizontal direction). The relation between v_0 and ω_0 for each situation and also initial sense of rotation is given for each situation in column I. Then match the statements in Column I with the corresponding results in Column II.

Column I	Column II
i. 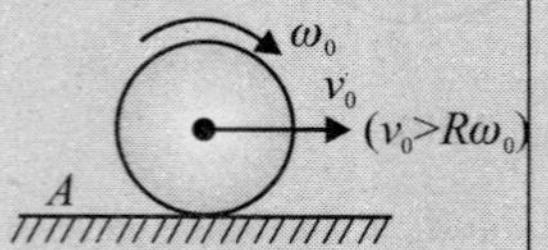 Fig. 2.384	**a.** The angular momentum of the disc about point A (as shown in the figure) remains conserved.
ii. 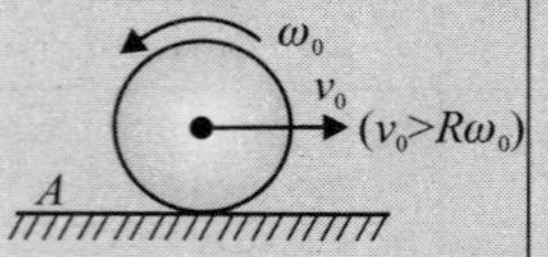 Fig. 2.385	**b.** The kinetic energy of the disc after it starts rolling without slipping is less than its initial kinetic energy.
iii. 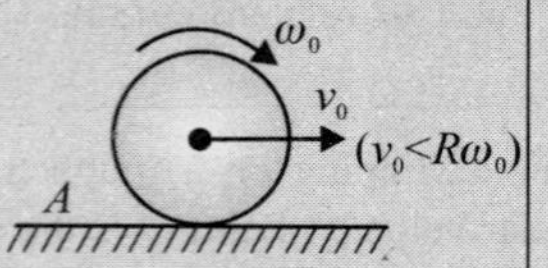 Fig. 2.386	**c.** In the duration disc rolls with slipping, the friction acts on the disc towards the left.
iv. 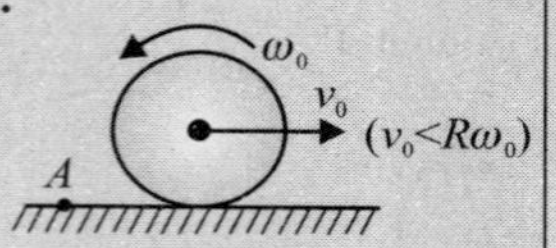 Fig. 2.387	**d.** In the duration disc rolls with slipping, the friction acts on the disc for some time to the right and for some time to the left.

11. Match the statements in Column I with those in Column II. One or more matching is possible.

Column I	Column II
i. Solid sphere rolling with slipping on a rough horizontal surface	**a.** Total kinetic energy is conserved
ii. Solid sphere in pure rolling on a rough horizontal surface and no other force is acting	**b.** Angular momentum about the CM is conserved
iii. Solid sphere in pure rolling on a smooth horizontal surface	**c.** Angular momentum about a contact point on the contact surface is conserved
iv. Solid sphere in pure rolling on a rough incline	**d.** Momentum is conserved
	e. Total mechanical energy is conserved

12.

Column I	Column II
i. 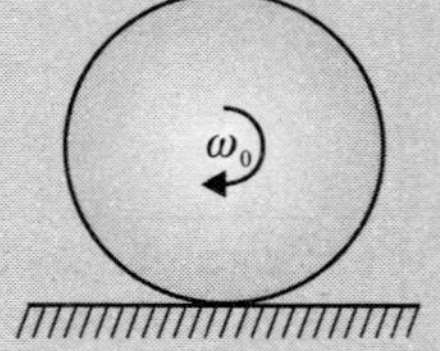 A spinning ball lowered on a rough surface Fig. 2.388	**a.** Net work done by friction force is negative
ii. 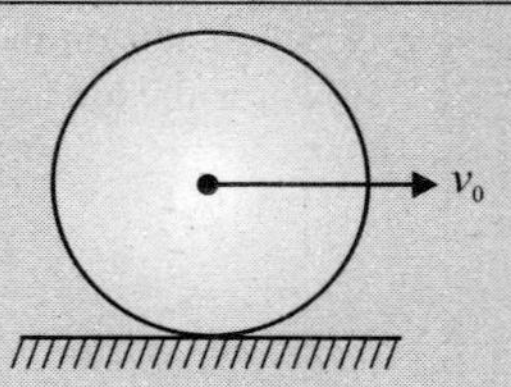 A ball is projected with velocity v_0 on a rough surface Fig. 2.389	**b.** Angular momentum about the CM is conserved
iii. 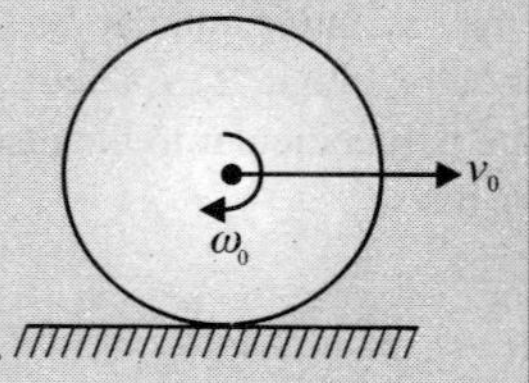 A ball projected with velocity $v_0 = R\omega_0$ projected on a smooth horizontal surface Fig. 2.390	**c.** Translational work done by friction is positive
iv. 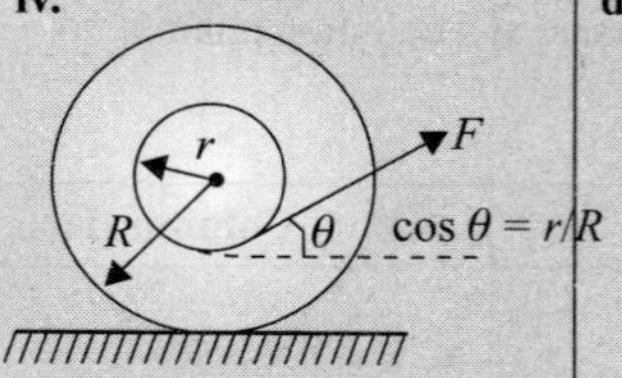 A spool pulled by a force F, rolling without slipping on a rough surface Fig. 2.391	**d.** Net work done by friction is zero

13. Assume sufficient friction to prevent slipping.

Column I	Column II
i. F (applied at top); Ring **Fig. 2.392**	**a.** Body accelerates forward
ii. F (applied at centre); Disc **Fig. 2.393**	**b.** Rotation about the centre of mass is clockwise
iii. F (applied at $R/2$ above centre); Solid Cylinder **Fig. 2.394**	**c.** Friction force acts backward
iv. F (applied below centre); Solid Cylinder **Fig. 2.395**	**d.** No friction acts

14. Match the entries in Column I with those in Column II.

Column I	Column II
i. F; $h = (2/5)R$ An impulsive force acts at a height $h = (2/5)\,R$ on a solid sphere, lying on a rough surface. **Fig. 2.396**	**a.** Momentum is conserved

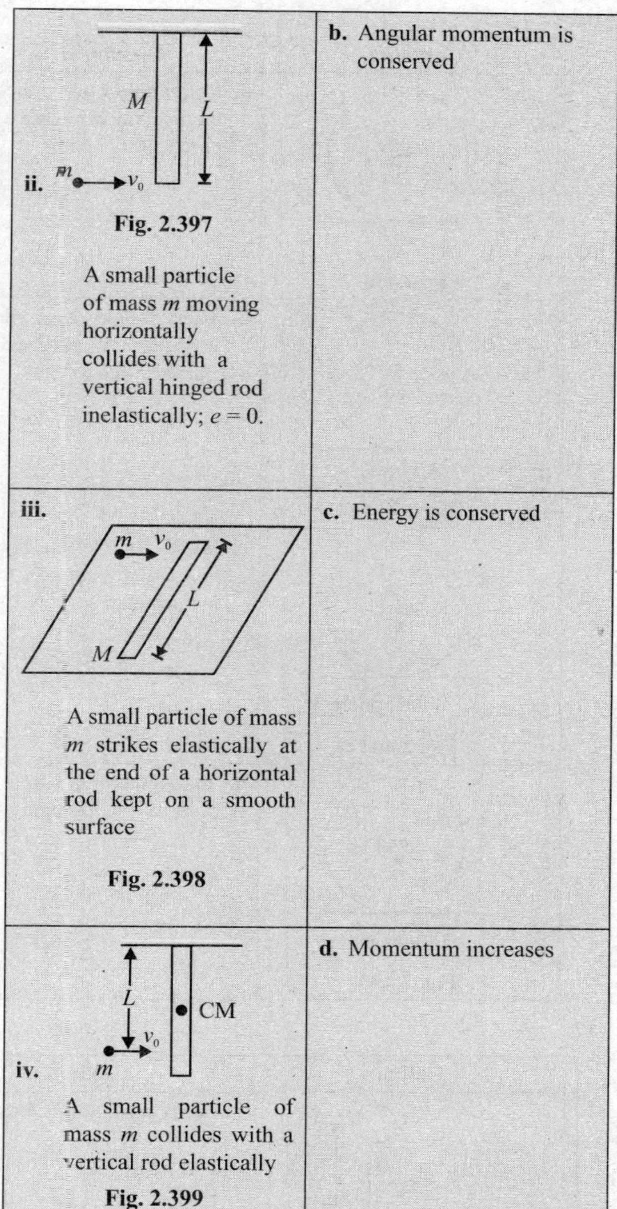

Column I	Column II
ii. M, L, m, v_0 **Fig. 2.397** A small particle of mass m moving horizontally collides with a vertical hinged rod inelastically; $e = 0$.	**b.** Angular momentum is conserved
iii. m, v_0, L, M A small particle of mass m strikes elastically at the end of a horizontal rod kept on a smooth surface **Fig. 2.398**	**c.** Energy is conserved
iv. L, CM, v_0, m A small particle of mass m collides with a vertical rod elastically **Fig. 2.399**	**d.** Momentum increases

15. Column I gives the bodies which are released one by one from the top of an inclined plane. Assume each body rolls purely.

Column I	Column II
i. Wooden solid cylinder	**a.** KE_{rot} is maximum at the bottom
ii. Iron solid cylinder	**b.** KE_{trans} is minimum at the bottom
iii. Thin iron cylindrical shell	**c.** Takes minimum time to reach the bottom
iv. Thin wooden cylindrical shell	**d.** Takes maximum time to reach the bottom

16.

Column I	Column II
i. ω_0 Ring Fig. 2.400	a. Rotational work done by the friction is negative till pure rolling begins
ii. ω_0 Cylinder Fig. 2.401	b. Translational work done is positive till pure rolling starts
iii. ω_0 Solid sphere Fig. 2.402	c. When pure rolling begins velocity of centre of mass is minimum
iv. ω_0 Hollow sphere Fig. 2.403	d. Takes maximum time for pure rolling to begin

17.

Column I	Column II
i. M, M, l, J (impulse) Rod is massless, dumb-bell is placed on a smooth horizontal surface Fig. 2.404	a. Translation occurs
ii. M, M, l, J Fig. 2.405	b. Rotation occurs
iii. J Fig. 2.406	c. Angular momentum increases
iv. Hinge J Fig. 2.407	d. Linear momentum increases

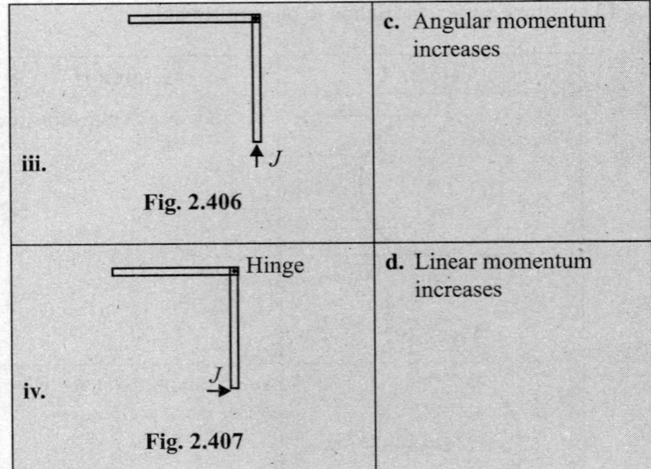

18.

Column I	Column II
i. A small particle of mass m is given an initial velocity in horizontal plane and winds its cord around the fixed vertical shaft of radius a. Fig. 2.408	a. Conservation of angular momentum
ii. A smooth rod rotates with some angular velocity. A small sleeve starts sliding along the rod. Fig. 2.409	b. Conservation of kinetic energy
iii. Two ice-skaters approach each other at equal speeds along the parallel path separated by some distance. They link hands as they pass by and pull each other to reduce their separation.	c. Conservation of total mechanical energy

iv. 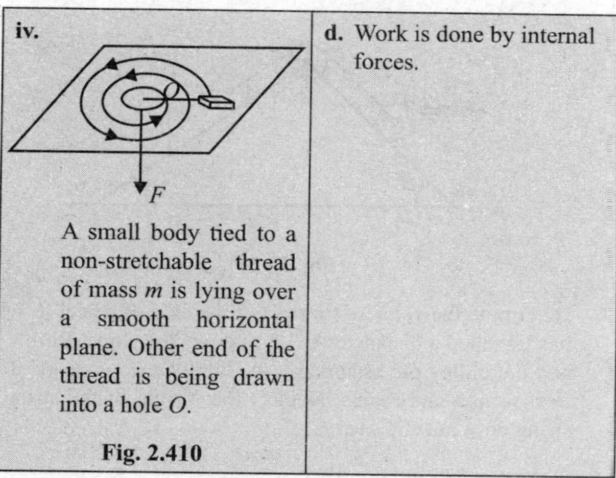 A small body tied to a non-stretchable thread of mass m is lying over a smooth horizontal plane. Other end of the thread is being drawn into a hole O. Fig. 2.410	**d.** Work is done by internal forces.

19. In Column I, information about the force(s) acting on a body is mentioned, while in Column II information about the motion of a body is given. Match the entries in Column II with the entries in Column I.

Column I	Column II
i. A single force through centre of mass	**a.** Rotational motion
ii. Equal and opposite forces separated by non-zero distance	**b.** Translational motion
iii. Equal and opposite forces acting at the same point	**c.** No motion
iv. A single force not through centre of mass	**d.** Centre of mass performs curvilinear motion

Integer Answer Type

Solutions on page 2.146

1. A solid cylinder with $r = 0.1$ m and mass $M = 2$ kg is placed such that it is in contact with the vertical and a horizontal surface as shown in Fig. 2.411. The coefficient of static friction is $\mu = (1/3)$ for both the surfaces. Find the distance d (in CM) from the centre of the cylinder at which a force $F = 40$ N should be applied vertically so that the cylinder just starts rotating in anti-clockwise direction.

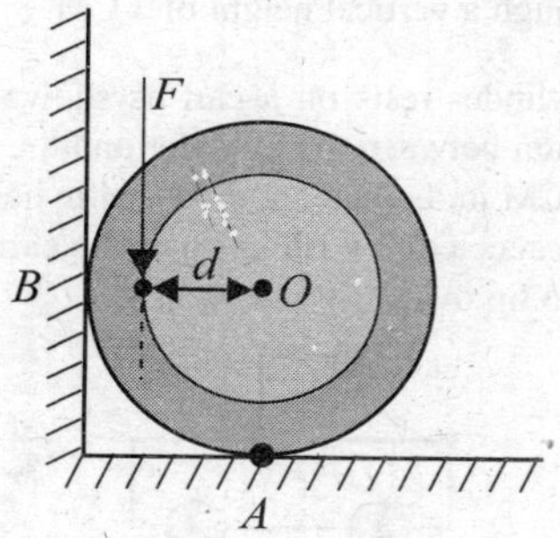

Fig. 2.411

2. A uniform rod of length 1 m and mass 2 kg is suspended. Calculate tension T (in N) in the string at the instant when the right string snaps ($g = 10$ m/s^2).

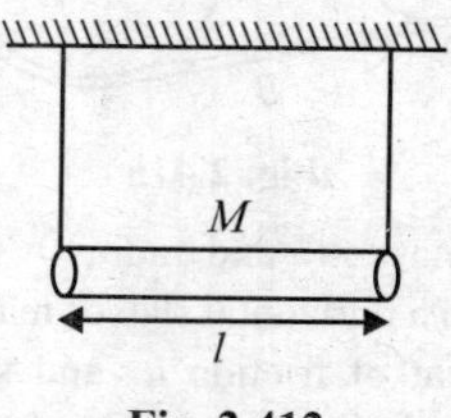

Fig. 2.412

3. A uniform rod AB of mass 2 kg is hinged at one end A. The rod is kept in the horizontal position by a massless string tied to point B. Find the reaction of the hinge (in N) on end A of the rod at the instant when string is cut. ($g = 10$ m/s^2)

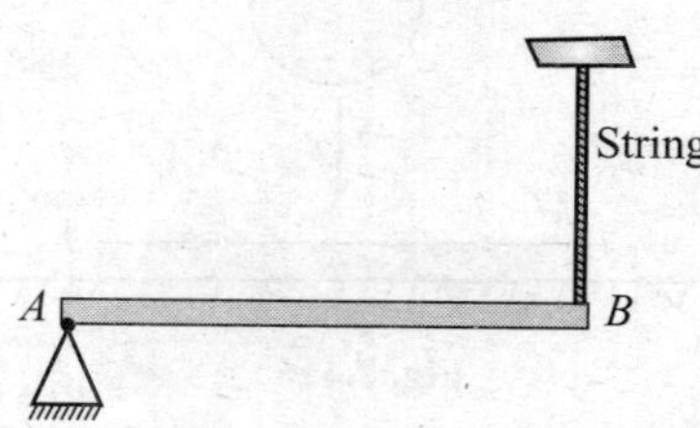

Fig. 2.413

4. A point mass is tied to one end of a cord whose other end passes through a vertical hollow tube, caught in one hand. The point mass is being rotated in a horizontal circle of radius 2 m with a speed of 4 m/s. The cord is then pulled down so that the radius of the circle reduces to 1 m. Compute the ratio of kinetic energies under the final and initial states.

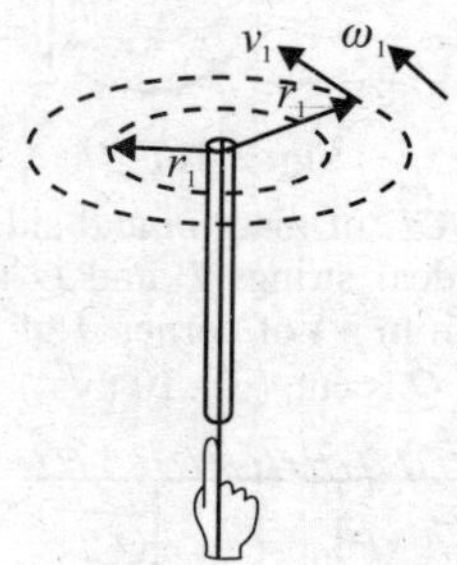

Fig. 2.414

5. A stone of mass m, tied to the end of a string, is whirled around in a horizontal circle (neglect the force due to gravity). The length of the string is reduced gradually keeping the angular momentum of the stone about the centre of the circle constant. Then the tension in the string is given by $T = A/r^n$, where A is a constant, r is the instantaneous radius of the circle and n is____.

6. Two wheels A and C are connected by a belt B as shown in Fig. 2.415. The radius of C is three times the radius of A. What would be the ratio of the rotational inertias (I_A / I_C) if both the wheels have the same rotational kinetic energy?

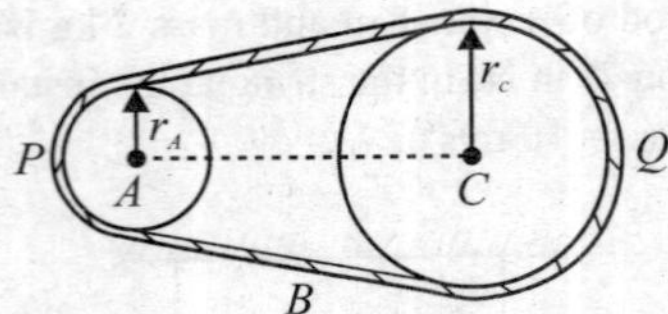

Fig. 2.415

7. A solid ball of mass m and radius r spinning with angular velocity ω falls on a horizontal slab of mass M with rough upper surface (coefficient of friction μ) and smooth lower surface. Immediately after collision the normal component of velocity of the ball remains half of its value just before collision and it stops spinning. Find the velocity of the sphere in horizontal direction immediately after the impact (given: $R\omega = 5$).

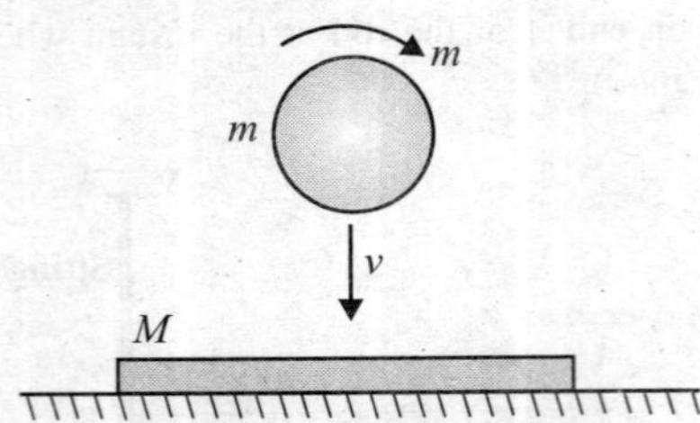

Fig. 2.416

8. An L shaped thin uniform rod of total length $2l$ is free to rotate in a vertical plane about a horizontal axis at P as shown in Fig. 2.417. The bar is released from rest. Neglect air and contact friction. Determine the angular velocity (in rad/s) at the instant when it has rotated through 90° and reached the dotted position as shown. (Take $l = 1/3$ m)

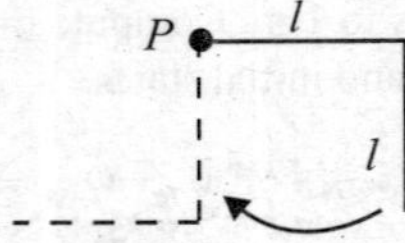

Fig. 2.417

9. A square plate $ABCD$ of mass m and side l is suspended with the help of two ideal strings P and Q as shown. Determine the acceleration (in m/s^2) of corner A of the square just at the moment the string Q is cut. ($g = 10$ m/s^2)

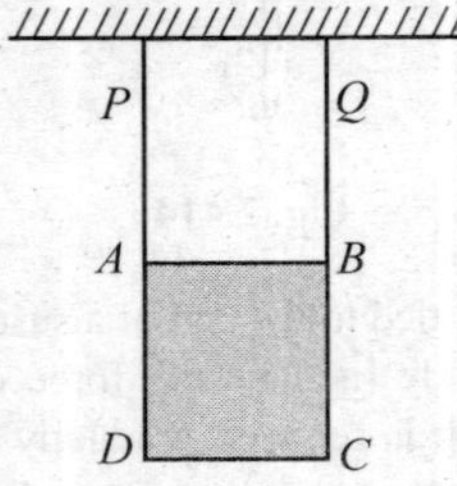

Fig. 2.418

10. A wheel A is connected to a second wheel B by means of inextensible string, passing over a pulley C, which rotates about a fixed horizontal axle O, as shown in the figure. The system is released from rest. The wheel A rolls down the inclined plane OK thus pulling up wheel B which rolls along the inclined plane ON.

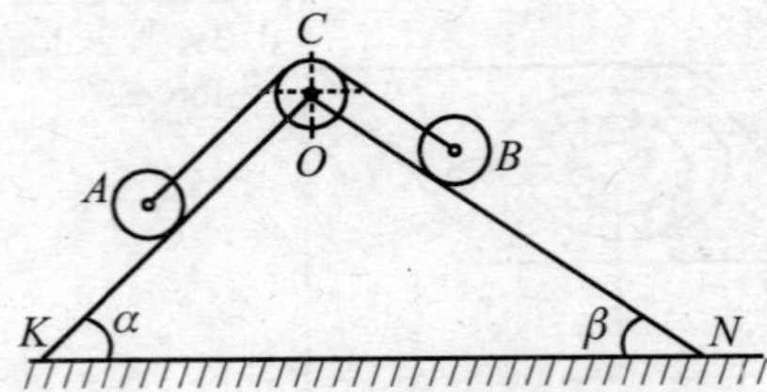

Fig. 2.419

Determine the velocity (in m/s) of the axle of wheel A, when it has travelled a distance $s = 3.5$ m down the slope. Both wheels and the pulley are assumed to be homogeneous disks of identical weight and radius. Neglect the weight of the string. The string does not slip over C.

[Take $\alpha = 53°$ and $\beta = 37°$]

11. Two forces of magnitude F are acting on a uniform disc kept on a horizontal rough surface as shown in the figure. Friction force by the horizontal surface on the disc is nF. Find the value of n.

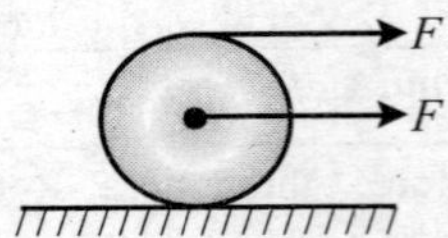

Fig. 2.420

12. A light inextensible thread is wound round a solid cylindrical reel of mass $m = 1.5$ kg and radius 10 CM. The end of the string is held fixed and the reel is allowed to fall so that the thread unwinds. If the axis of the reel remains horizontal, the tension in the thread will be $T =$ '*' N. What is the value of '*'.

13. A disc of radius R is rolling purely on a flat horizontal surface, with a constant angular velocity. The angle between the velocity and acceleration vectors of point P is given by $\sin^{-1}(\sqrt{2}/n)$. What is the value of n?

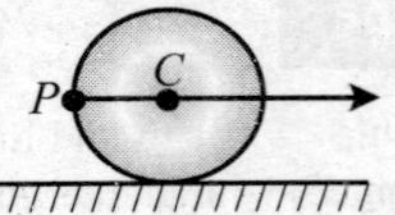

Fig. 2.421

14. A hollow sphere of radius 30 CM is released from rest on a rough inclined plane. The friction is enough to prevent any slipping. Find the angular velocity (in rad/s) of the sphere, when it descends through a vertical height of 3 CM.

15. A uniform cylinder rests on a cart as shown. The coefficient of static friction between the cylinder and the cart is 0.5. If the cylinder is 4 CM in diameter and 10 CM in height, then what is the minimum acceleration (in m/s^2) of the cart needed to cause the cylinder to tip over?

Fig. 2.422

16. A binary star consists of two stars A (mass 2.2 Ms) and B (mass 11 Ms), where Ms is the mass of the sun. They are separated by distance d and are rotating about their centre of mass, which is stationary. The ratio of the total angular momentum of the binary star to the angular momentum of star B about the centre of mass is.

17. A boy is pushing a ring of mass 2 kg and radius 0.5 m with a stick as shown in the figure. The stick applies a force of 2 N on the ring and rolls it without slipping with an acceleration of 0.3 m/s^2. The coefficient of friction between the ground and the ring is large enough that rolling always occurs and the coefficient of friction between the stick and the ring is (P/10). The value of P is.

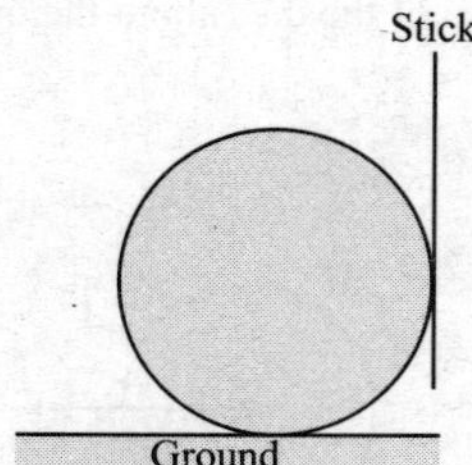

Fig. 2.423

18. Four solid spheres each of diameter $\sqrt{5}$ CM and mass 0.5 kg are placed with their centres at the corners of a square of side 4 cm. The moment of inertia of the system about the diagonal of the square is $N \times 10^{-4}$ kg m^2, then N is.

Archives

Solutions on page 2.149

Fill in the Blanks Type

1. A uniform cube of side a and mass m rests on a rough horizontal table. A horizontal force F is applied normal to one of the faces at a point that is directly above the centre of the face, at a height $3a/4$ above the base. The minimum value of F for which the cube begins to tip about the edge is____(Assume that the cube does not slide) **(IIT-JEE, 1984)**

2. A smooth uniform rod of length L and mass M has two identical beads of negligible size, each of mass m, which can slide freely along the rod. Initially, the two beads are at the centre of the rod and the system is rotating with an angular velocity ω_0 about an exis perpendicular to the rod and passing through the midpoint of the rod (see Fig. 2.424). There are no external forces. When the beads reach the ends of the rod, the angular velocity of the system is____. **(IIT-JEE, 1988)**

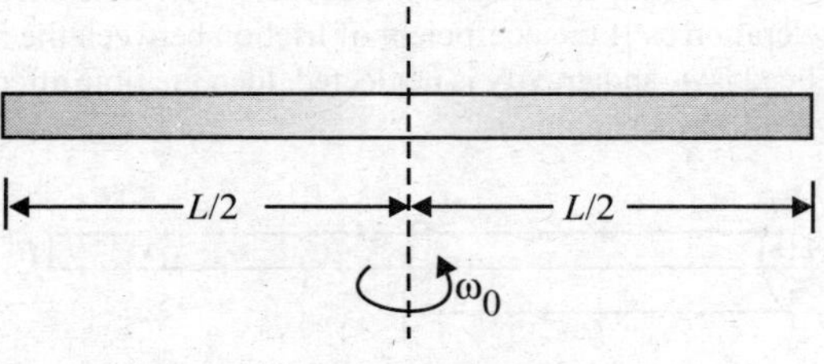

Fig. 2.424

3. A cylinder of mass M and radius R is resting on a horizontal platform (which is parallel to the x-y plane) with its axis fixed along the y-axis and free to rotate about its axis. The platform is given a motion in the x-direction given by $x = A\cos(\omega t)$. There is no slipping between the cylinder and the platform. The maximum torque acting on the cylinder during its motion is ______. **(IIT-JEE, 1988)**

4. A stone of mass m, tied to the end of a string, is whirled around in a horizontal circle (neglect the force due to gravity). The length of the string is reduced gradually keeping the angular momentum of the stone about the centre of the circle constant. Then the tension in the string is given by $T = Ar^n$, where A is a constant, r is the instantaneous radius of the circle and n = ____. **(IIT-JEE, 1993)**

5. A uniform disc of mass m and radius R is rolling up a rough inclined plane which makes an angle of 30° with the horizontal. If the coefficients of static and kinetic friction are each equal to μ and the only force acting are gravitational and frictional, then the magnitude of the frictional force acting on the disc is____and its direction is________(write up or down) the inclined plane. **(IIT-JEE, 1997)**

6. A rod of weight w is supported by two parallel knife-edges A and B and is in equilibrium in a horizontal position. The knives are at a distance d from each other. The centre of mass of the rod is at distance x from A. The normal reaction on A is ... and on B is________. **(IIT-JEE, 1999)**

7. A symmetric lamina of mass M consists of a square shape with a semicircular section over of the edge of the square as shown in Fig. 2.425. The side of the square is $2a$. The moment of inertia of the lamina about an axis through its centre of mass and perpendicular to the plane is $1.6Ma^2$. The moment of inertia of the lamina about the tangent AB in the plane of the lamina is________. **(IIT-JEE, 1997)**

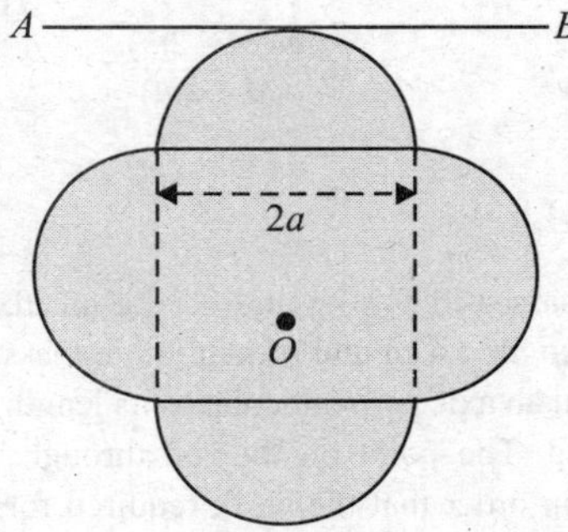

Fig. 2.425

True or False Type

1. A triangular plate of uniform thickness and density is made to rotate about an axis perpendicular to the plane of the paper and (a) passing through A, (b) passing through B, by the application of the same force, F, at C (midpoint of AB) as shown in Fig. 2.426. The angular acceleration in both the cases will be the same. **(IIT-JEE, 1985)**

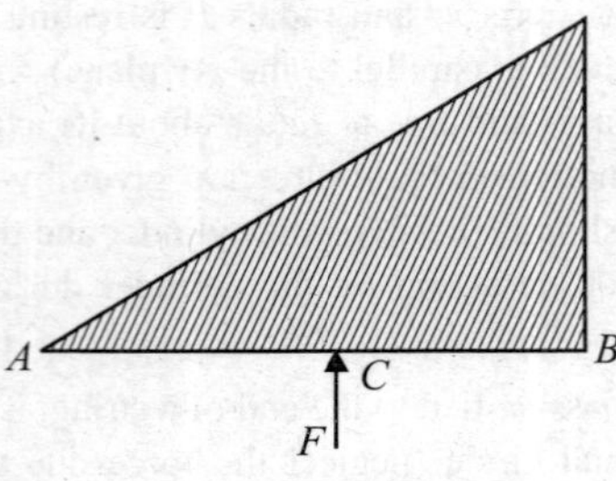

Fig. 2.426

2. A thin uniform circular disc of mass M and radius R is rotating in a horizontal plane about an axis passing through its centre and perpendicular to its plane with an angular velocity ω. Another disc of the same dimensions but of mass $M/4$ is placed gently on the first disc coaxially. The angular velocity of the system now is $2\omega/\sqrt{5}$. **(IIT-JEE, 1986)**
3. A ring of masses 0.3 kg and radius 0.1 m and a solid cylinder of mass 0.4 kg and of the same radius are given the same kinetic energy and released simultaneously on a flat horizontal surface such that they begin to roll as soon as released towards a wall which is at the same distance from the ring and the cylinder. The rolling friction in both the cases is negligible. The cylinder will reach the wall first. **(IIT-JEE, 1989)**
4. Two particles of mass 1 kg and 3 kg move towards each other under their mutual force of attraction. No other force acts on them. When the relative velocity of approach of the two particles is 2 m/s, their centre of mass has a velocity of 0.5 m/s, the velocity of the centre of mass is 0.75 m/s. **(IIT-JEE, 1989)**

Multiple Choice Questions with One Correct Answer Type

1. A thin circular ring of mass M and radius r is rotating about its axis with a constant angular velocity ω. Two objects, each of mass m, are attached gently to the opposite ends of the diameter of the ring. The wheel now rotates with an angular velocity **(IIT-JEE, 1983)**

   a. $\dfrac{\omega M}{(M+m)}$ **b.** $\dfrac{\omega(M-2m)}{(M+2m)}$

   c. $\dfrac{\omega M}{(M+2m)}$ **d.** $\dfrac{\omega(M+2m)}{M}$

2. Two point masses of 0.3 kg and 0.7 kg are fixed at the ends of a rod of length 1.4 m and of negligible mass. The rod is set rotating about an axis perpendicular to its length with a uniform angular speed. The point on the rod through which the axis should pass in order that the work required for rotation of the rod is minimum is located at a distance of **(IIT-JEE, 1995)**

   a. 0.42 m from the mass of 0.3 kg
   b. 0.70 m from the mass of 0.7 kg
   c. 0.98 m from the mass of 0.3 kg
   d. 0.98 m from the mass of 0.7 kg

3. A mass m is moving with a constant velocity along a line parallel to the x-axis, away from the origin. Its angular momentum with respect to the origin **(IIT-JEE, 1997)**

   a. is zero
   b. remains constant
   c. goes on increasing
   d. goes on decreasing

4. A smooth sphere A is moving on a frictionless horizontal plane with angular speed ω_A and ω_B, respectively. Then **(IIT-JEE, 1999)**

   a. $\omega_A < \omega_B$ **b.** $\omega_A = \omega_B$

   c. $\omega_A = \omega$ **d.** $\omega_B = \omega$

5. A disc of mass M and radius R is rolling with angular speed ω on a horizontal plane as shown in Fig. 2.427. The magnitude of angular momentum of the disc about the origin O is **(IIT-JEE, 1999)**

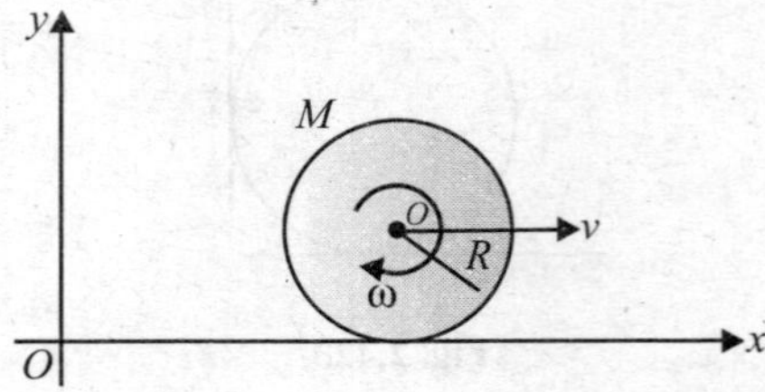

Fig. 2.427

   a. $(1/2)\ MR^2\omega$ **b.** $MR^2\omega$

   c. $(3/2)\ MR^2\omega$ **d.** $2\ Mr^2\omega$

6. A cubical block of side a is moving with velocity V on a horizontal smooth plane as shown in Fig. 2.428. It hits a ridge at point O. The angular speed of the block after it hits O is **(IIT-JEE, 1999)**

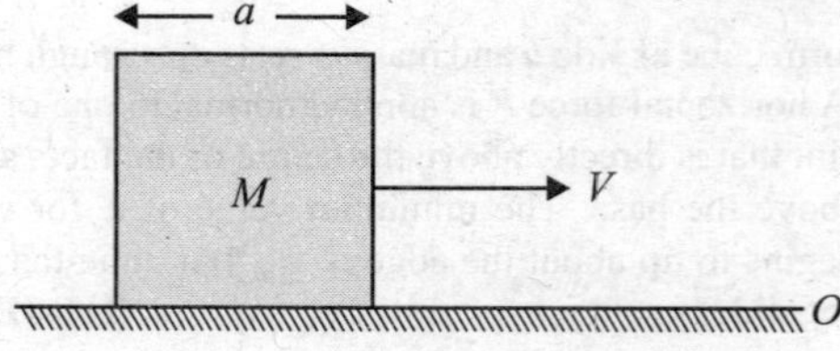

Fig. 2.428

   a. $3V/(4a)$ **b.** $3\ V/(2a)$

   c. $\sqrt{3}V/(\sqrt{2}a)$ **d.** zero

7. A long horizontal rod has a bead which can slide along its length and initially placed at a distance L from one end A of the rod. The rod is set in angular motion about A with constant angular acceleration α. If the coefficient of friction between the rod and the bead is μ, and gravity is neglected, then the time after which the bead starts slipping is

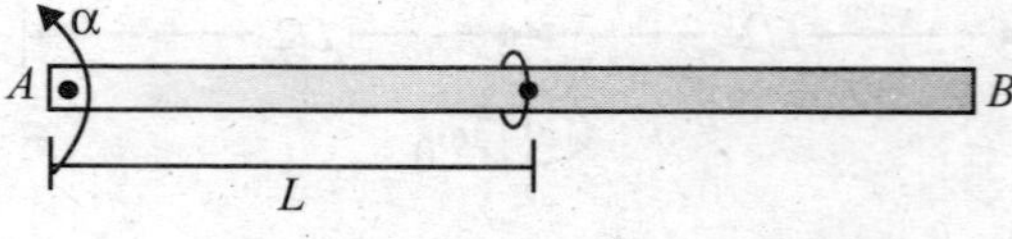

Fig. 2.429

a. $\sqrt{\mu/\alpha}$ b. $\mu/\sqrt{\alpha}$

c. $\frac{1}{\sqrt{\mu\alpha}}$ d. infinitesimal

(IIT-JEE, 2000)

8. A cubical block of side L rests on a rough horizontal surface with coefficient of friction sufficiently high so that the block does not slide before toppling; the minimum force required to topple the block is **(IIT-JEE, 2000)**

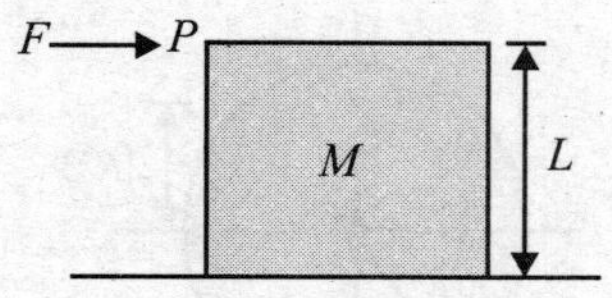

Fig. 2.430

a. infinitesimal b. $mg/4$

c. $mg/2$ d. $mg(1-\mu)$

9. A thin wire of length L and uniform linear mass density ρ is bent into a circular loop with centre at O as shown. The moment of inertia of the loop about the axis XX' is **(IIT-JEE, 2000)**

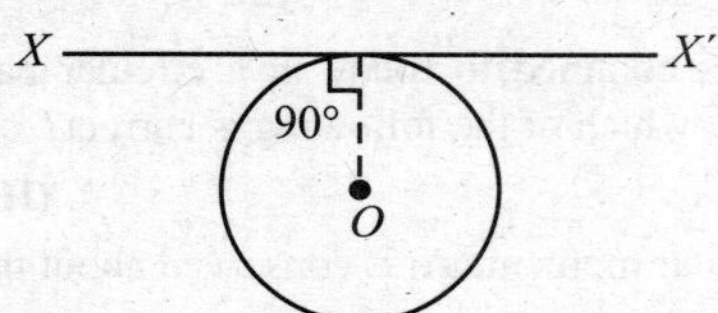

Fig. 2.431

a. $\frac{\pi L^3}{8\pi^2}$ b. $\frac{\pi L^3}{16\pi^2}$

c. $\frac{5\pi L^3}{16\pi^2}$ d. $\frac{3\pi L^3}{8\pi^2}$

10. An equilateral triangle ABC formed from a uniform wire has two small identical beads initially located at A. The triangle is set rotating about the vertical axis AO. Then the beads are released from rest simultaneously and allowed to slide down, one along AB and the other along AC as shown. Neglecting frictional effects, the quantities that are conserved as the beads slide down are **(IIT-JEE, 2000)**

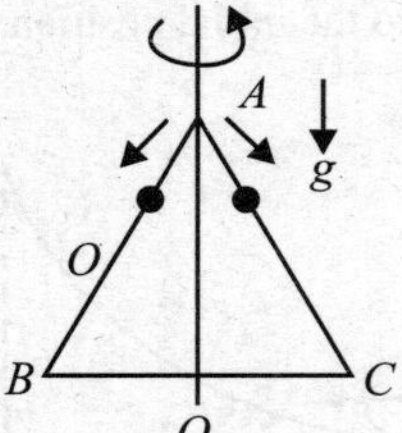

Fig. 2.432

a. angular velocity and total energy (kinetic and potential)

b. total angular momentum and total energy

c. angular velocity and moment of inertia about the axis of rotation

d. total angular momentum and moment of inertia about the axis of rotation

11. One quarter sector is cut from a uniform circular disc of radius R. This sector has mass M. It is made to rotate about a line perpendicular to its plane and passing through the centre of the original disc. Its moment of inertia about the axis of rotation is

(IIT-JEE, 2001)

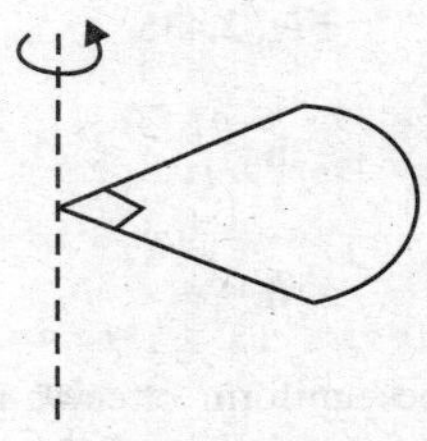

Fig. 2.433

a. $\frac{1}{2}MR^2$ b. $\frac{1}{4}MR^2$

c. $\frac{1}{8}MR^2$ d. $\sqrt{2}\,MR^2$

12. A cylinder rolls up an inclined plane, reaches some height and then rolls down (without slipping throughout these motions). The directions of the frictional force acting on the cylinder are

(IIT-JEE, 2002)

a. up the incline while ascending and down the incline while descending

b. up the incline while ascending as well as descending

c. down the incline while ascending and up the incline while descending

d. down the incline while ascending as well as descending

13. A circular platform is free to rotate in a horizontal plane about a vertical axis passing through its centre. A tortoise is sitting at the edge of the platform. Now, the platform is given an angular velocity ω_0. When the tortoise moves along a chord of the platform with a constant velocity (with respect to the platform), the angular velocity of the platform $\omega(t)$ will vary with time t as

(IIT-JEE, 2002)

a.

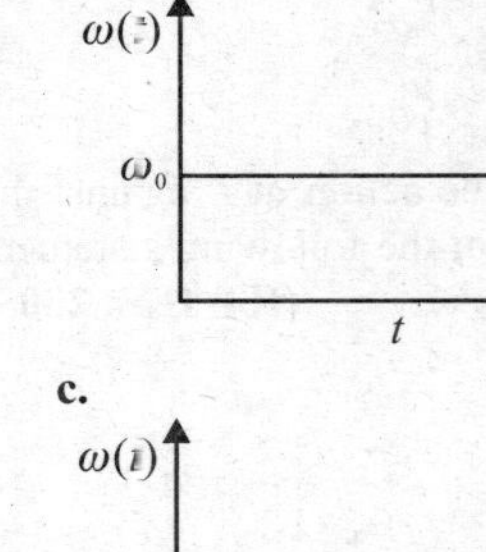

b.

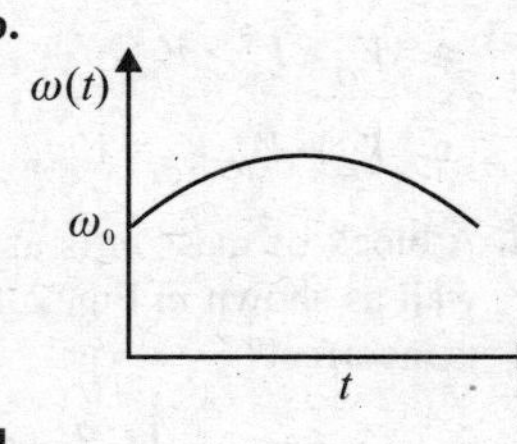

c.

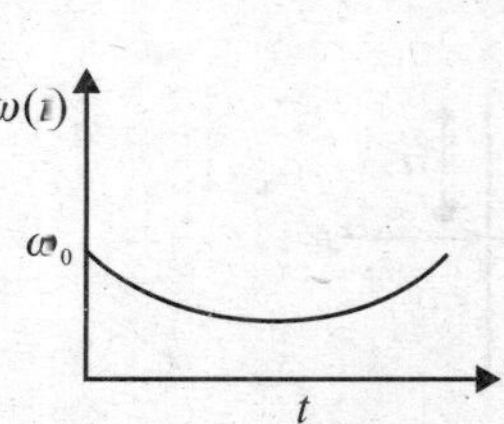

d.

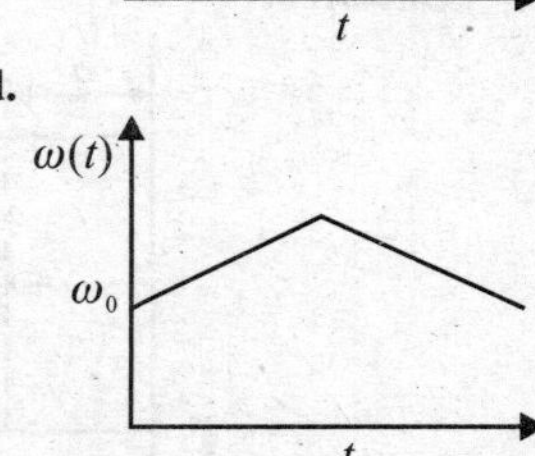

Fig. 2.434

14. Consider a body, shown in Fig. 2.435, consisting of two identical balls, each of mass M connected by a light rigid rod. If an impulse $J = MV$ is imparted to the body at one of its ends, what would be its angular velocity? **(IIT-JEE, 2003)**

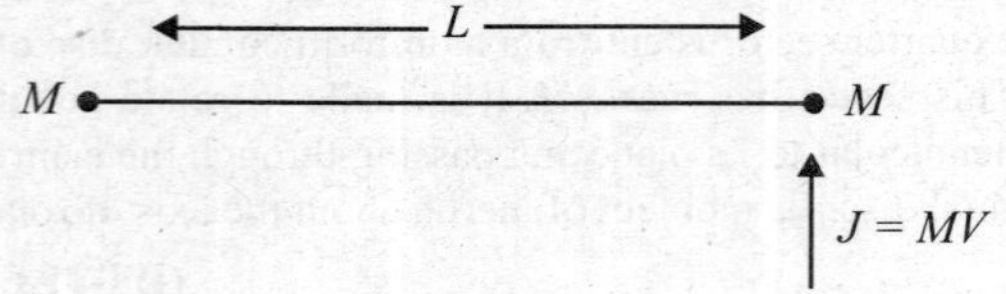

Fig. 2.435

a. $\frac{V}{L}$ **b.** $\frac{2V}{L}$

c. $\frac{V}{3L}$ **d.** $\frac{V}{4L}$

15. A particle undergoes uniform circular motion. About which point on the plane of the circle, will the angular momentum of the particle remain conserved? **(IIT-JEE, 2003)**

a. Centre of the circle

b. On the circumference of the circle

c. Inside the circle

d. Outside the circle

16. A horizontal circular plate is rotating about a vertical axis passing through its centre with an angular velocity ω_o. A man sitting at the centre having two blocks in his hands stretches out his hands so that the moment of inertia of the system doubles. If the kinetic energy of the system is K initially, its final kinetic energy will be **(IIT-JEE, 2004)**

a. $2K$ **b.** $K/2$

c. K **d.** $K/4$

17. A disc is rolling without slipping with angular velocity ω. P and Q are two points equidistant from the centre C. The order of magnitude of velocity is **(IIT-JEE, 2004)**

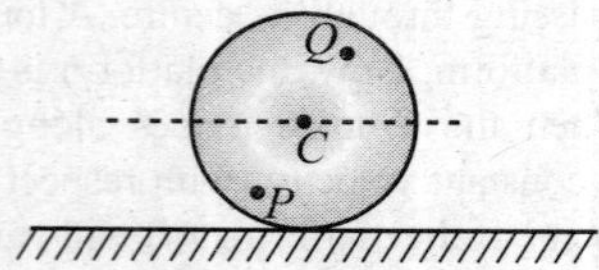

Fig. 2.436

a. $V_Q > V_C > V_P$ **b.** $V_P > V_C > V_Q$

c. $V_Q = V_P$, $V_C = V_P/2$ **d.** $V_P < V_C > V_Q$

18. A block of mass m is at rest under the action of F against the wall as shown in Fig. 2.437. Which of the following statement is incorrect? **(IIT-JEE, 2005)**

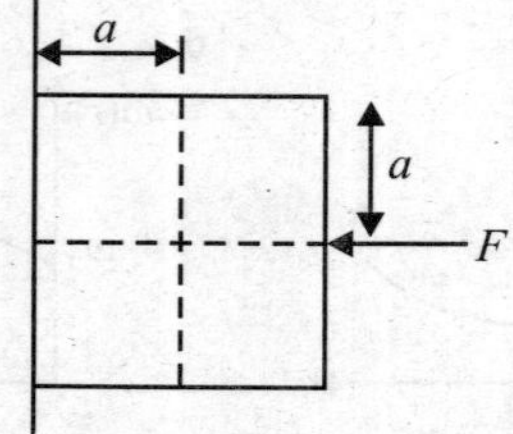

Fig. 2.437

a. $f = mg$ [where f is the friction force]

b. $F = N$ [where N is the normal force]

c. F will not produce torque

d. N will not produce torque

19. From a circular disc of radius R and mass $9M$, a small disc of radius $R/3$ is removed from the disc. The moment of inertia of the remaining disc about an axis perpendicular to the plane of the disc and passing through O is **(IIT-JEE, 2005)**

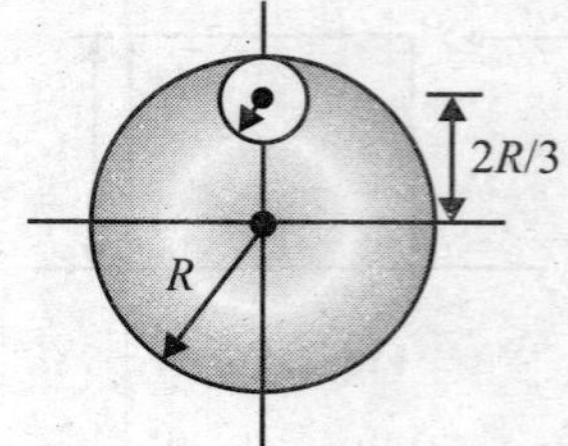

Fig. 2.438

a. $4MR^2$ **b.** $\frac{40}{4}MR^2$

c. $10MR^2$ **d.** $\frac{37}{9}MR^2$

20. A particle is confined to rotate in a circular path decreasing linear speed, which of the following is correct? **(IIT-JEE, 2005)**

a. $\vec{L}$ (angular momentum) is conserved about the centre

b. Only direction of angular momentum $\vec{L}$ is conserved

c. It spirals towards the centre

d. Its acceleration is towards the centre

21. A solid sphere of mass M and radius R having moment of inertia I about its diameter is recast into a solid disc of radius r and thickness t. The moment of inertia of the disc about an axis passing the edge and perpendicular to the plane remains I. Then R and r are related as **(IIT-JEE, 2006)**

a. $r = \sqrt{\frac{2}{15}}R$ **b.** $r = \sqrt{\frac{2}{15}}R$

c. $r = \frac{2}{15}R$ **d.** $r = \frac{2}{\sqrt{5}}R$

22. A small object of uniform density rolls up a curved surface with an initial velocity v. It reaches up to a maximum height of $3v^2/4g$ with respect to the initial position. The object is a **(IIT-JEE, 2007)**

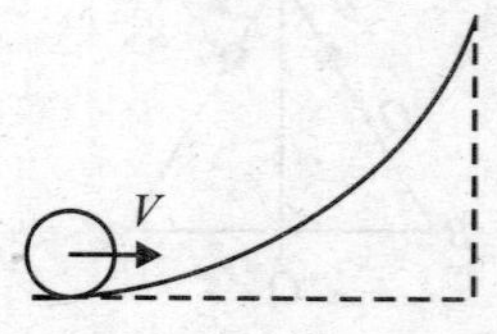

Fig. 2.439

a. ring **b.** solid sphere

c. hollow sphere **d.** disc

23. If the resultant of all the external forces acting on a system of particles is zero, then from an inertial frame, one can surely say that **(IIT-JEE, 2009)**

a. linear momentum of the system does not change in time
b. kinetic energy of the system does not change in time
c. angular momentum of the system does not change in time
d. potential energy of the system does not change in time

24. A block of base 10 cm × 10 cm and height 15 cm is kept on an inclined plane. The coefficient of friction between them is 3. The inclination θ of this inclined plane from the horizontal plane is gradually increased from 0°. Then **(IIT-JEE, 2009)**

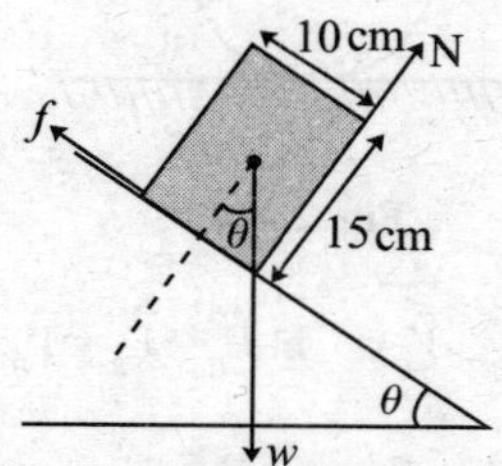

Fig. 2.440

a. at $\theta = 30°$, the block will start sliding down the plane
b. the block will remain at rest on the plane up to certain θ and then it will topple
c. at $\theta = 60°$, the block will start sliding down the plane and continue to do so at higher angles
d. at $\theta = 60°$, the block will start sliding down the plane and on further increasing θ, it will topple at certain θ

25. A thin ring of mass 2 kg and radius 0.5 m is rolling without slipping on a horizontal plane with velocity 1 m/s. A small ball of mass 0.1 kg, moving with velocity 20 m/s in the opposite direction, hits the ring at a height of 0.75 m and goes vertically up with velocity 10 m/s. Immediately after the collision **(IIT-JEE, 2011)**

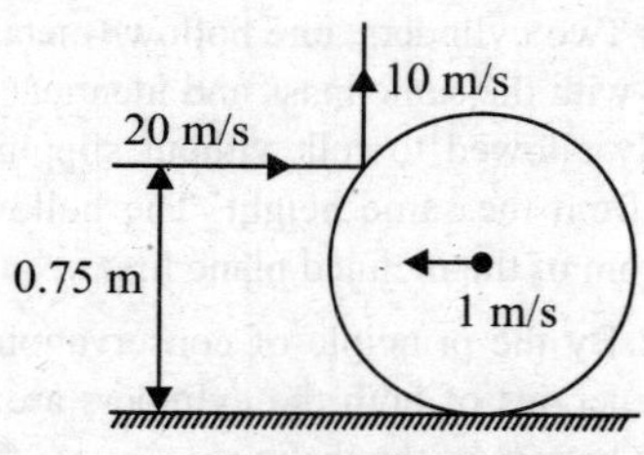

Fig. 2.441

a. the ring has pure rotation about its stationary
b. the ring comes to a complete stop
c. friction between the ring and the ground is to the left
d. there is no friction between the ring and the ground

Multiple Choice Questions with One or More than One Correct Answer Type

1. Two particles A and B, initially at rest, move towards each other under a mutual force of attraction. At the instant when the speed of A is V and the speed of B is $2V$, the speed of the centre of mass of the system is **(IIT-JEE, 1982)**

a. $3V$ **b.** V
c. $1.5V$ **d.** zero

2. A mass M is moving with a constant velocity parallel to the x-axis. Its angular momentum with respect to the origin **(IIT-JEE, 1995)**

a. is zero
b. remains constant
c. goes on increasing
d. goes on decreasing

3. When a bicycle is in motion, the force of friction exerted by the ground on the two wheels is such that it acts **(IIT-JEE, 1990)**

a. in the backward direction on the front wheel and in the forward direction on the rear wheel
b. in the forward direction on the front wheel and in the backward direction on the rear wheel
c. in the backward direction on both the front and the rear wheels
d. in the forward direction on both the front and the rear wheels

4. A particle of mass m is projected with a velocity V making an angle of 45° with the horizontal. The magnitude of angular momentum of the projectile about the point of projection when the particle at its maximum height h is **(IIT-JEE, 1990)**

a. zero **b.** $\dfrac{mv^3}{4\sqrt{2}g}$
c. $\dfrac{mV^3}{\sqrt{2}g}$ **d.** $\dfrac{mV}{\sqrt{2gh^3}}$

5. The moment of inertia of a thin square plate $ABCD$, Fig. 2.442, of uniform thickness about an axis passing through the centre O and perpendicular to the plane is **(IIT-JEE, 1992)**

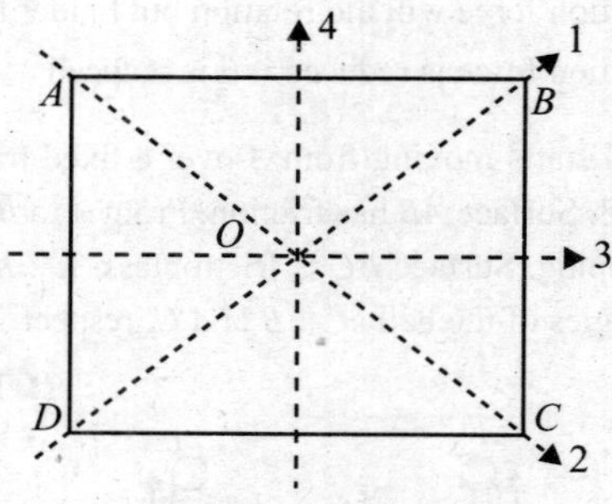

Fig. 2.442

a. $I_1 + I_2$ **b.** $I_3 + I_4$
c. $I_1 + I_3$ **c.** $I_1 + I_2 + I_3 + I_4$

6. A tube of length L is filled completely with an incompressible liquid of mass M and closed at both the ends. The tube is then rotated in a horizontal plane about one of its ends with a uniform angular velocity ω. The force exerted by the liquid at the other end is **(IIT-JEE, 1992)**

 a. $\frac{M\omega^2 L}{2}$ **b.** $M\omega^2 L$

 c. $\frac{M\omega^2 L}{4}$ **d.** $\frac{M\omega^2 L^2}{4}$

7. A car is moving in a circular horizontal track of radius 10 m with a constant speed of 10 m/s. A plumb bob is suspended from the roof of the car by a light rigid rod of length 1.00 m. The angle made by the rod with the track is **(IIT-JEE, 1992)**

 a. zero **b.** 30°

 c. 45° **d.** 60°

8. Let I be the moment of inertia of a uniform square plate about an axis AB that passes through its centre and is parallel to two of its sides. CD is a line in the plane of the plate that passes through the centre of the plate and makes an angle θ with AB. The moment of inertia of the plate about the axis CD is then equal to **(IIT-JEE, 1998)**

 a. I **b.** $I\sin^2\theta$

 c. $I\cos^2\theta$ **d.** $I\cos^2(\theta/2)$

9. The torque τ on a body about a given point is found to be equal to $\boldsymbol{A} \times \boldsymbol{L}$ where $\boldsymbol{A}$ is a constant vector, and $\boldsymbol{L}$ is the angular momentum of the body about that point. From this it follows that **(IIT-JEE, 1998)**

 a. $d\boldsymbol{L}/dt$ is perpendicular to $\boldsymbol{L}$ at all instants of time

 b. the component of $\boldsymbol{L}$ in the direction of $\boldsymbol{A}$ does not change with time

 c. the magnitude of $\boldsymbol{L}$ does not change with time

 d. $\boldsymbol{L}$ does not change with time

10. A solid cylinder is rolling down a rough inclined plane of inclination θ. Then **(IIT-JEE, 2006)**

 a. the friction force is dissipative

 b. the friction force is necessarily changing

 c. the friction force will aid rotation but hinder translation

 d. the friction force is reduced if θ is reduced

11. A small ball starts moving from A over a fixed track as shown in Fig. 2.443. Surface AB has friction. From A to B the ball rolls without slipping. Surface BC is frictionless. K_A, K_B and K_C are kinetic energies of the ball at A, B and C, respectively. Then **(IIT-JEE, 2006)**

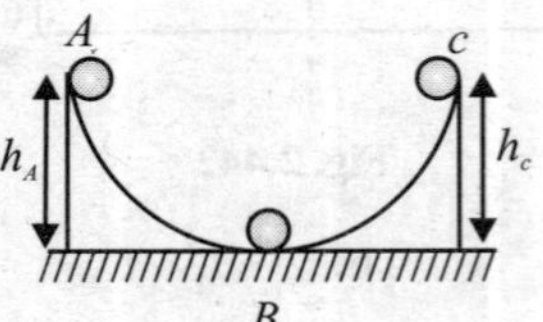

Fig. 2.443

 a. $h_A > h_C$; $K_B > K_C$ **b.** $h_A > h_C$; $K_C > K_A$

 c. $h_A = h_C$; $K_B = K_C$ **d.** $h_A < h_C$; $K_B > K_C$

12. A sphere is rolling without slipping on a fixed horizontal plane surface. In Fig. 2.444, A is the point of contact, B is the centre of the sphere and C is its topmost point. Then, **(IIT-JEE, 2009)**

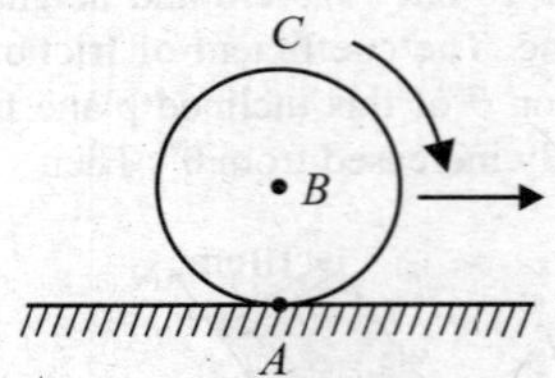

Fig. 2.444

 a. $\vec{V}_C - \vec{V}_A = 2(\vec{V}_B - \vec{V}_C)$ **b.** $\vec{V}_C - \vec{V}_B = \vec{V}_B - \vec{V}_A$

 c. $|\vec{V}_C - \vec{V}_A| = 2|\vec{V}_B - \vec{V}_C|$ **d.** $|\vec{V}_C - \vec{V}_A| = 4|\vec{V}_B|$

Assertion and Reasoning Type

 a. Statement 1 is True, Statement 2 is True; Statement 2 is a correct explanation for Statement 1.

 b. Statement 1 is True, Statement 2 is True; Statement 2 is **NOT** a correct explanation for Statement 1.

 c. Statement 1 is True, Statement 2 is False.

 d. Statement 1 is False, Statement 2 is True.

1. **Statement 1:** If there is no external torque on a body about its center of mass, then the velocity of the centre of mass remains constant.

 Statement 2: The linear momentum of an isolated system remains constant. **(IIT-JEE, 2007)**

2. **Statement 1:** Two cylinders, one hollow (metal) and the other solid (wood), with the same mass and identical dimensions are simultaneously allowed to roll without slipping down an inclined plane from the same height. The hollow cylinder will reach the bottom of the inclined plane first.

 Statement 2: By the principle of conservation of energy, the total kinetic energies of both the cylinders are identical when they reach the bottom of the incline. **(IIT-JEE, 2008)**

Comprehension Type

For Problems 1–3

Two discs A and B are mounted coaxially on a vertical axle. The discs have moments of inertia I and $2I$, respectively, about the common axis. Disc A is imparted an initial angular velocity 2ω using the entire potential energy of a spring compressed by a distance x_1. Disc B is imparted an angular velocity ω by a spring having the same spring constant and compressed by a distance x_2. Both the discs rotate in the clockwise direction. **(IIT-JEE, 2007)**

1. The ratio x_1/x_2 is

a. 2 b. $\frac{1}{2}$

c. $\sqrt{2}$ d. $\frac{1}{\sqrt{2}}$

2. When disc B is brought in contact with disc A, they acquire a common angular velocity in time t. The average frictional torque on one disc by the other during this period is

a. $\frac{2I\omega}{3t}$ b. $\frac{9I\omega}{2t}$

c. $\frac{9I\omega}{4t}$ d. $\frac{3I\omega}{2t}$

3. The loss of kinetic energy during the above process is

a. $\frac{I\omega^3}{2}$ b. $\frac{I\omega^2}{3}$

c. $\frac{I\omega^2}{4}$ d. $\frac{I\omega^2}{6}$

For Problems 4–6

A uniform thin cylindrical disc of mass M and radius R is attached to two identical massless springs of spring constant k which are fixed to the wall as shown in Fig. 2.445. The springs are attached to the axle of the disk symmetrically on either side at a distance d from its centre. The axle is massless and both the springs and the axle are in horizontal plane. The unstretched length of each spring is L. The disc is initially at its equilibrium position with its centre of mass (CM) at a distance L from the wall. The disc rolls without slipping with velocity $V_0 = V_0\hat{i}$. The coefficient of friction is μ.

(IIT-JEE, 2008)

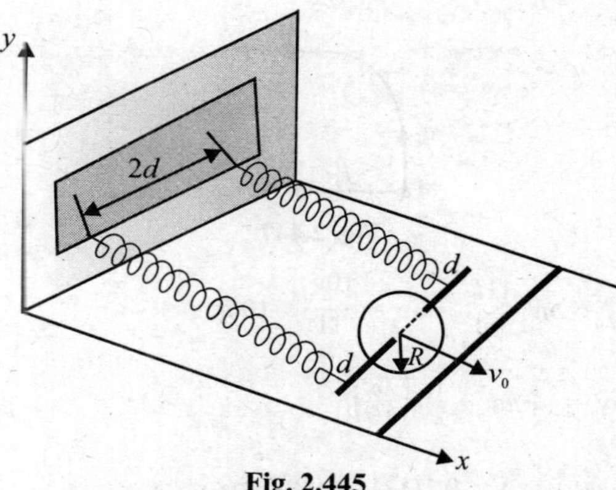

Fig. 2.445

4. The net external force acting on the disc when its centre of mass is at displacement x with respect to its equilibrium position is

a. $-kx$ b. $-2kx$

c. $-\frac{2kx}{3}$ d. $-\frac{4kx}{3}$

5. The centre of mass of the disc undergoes simple harmonic motion with angular frequency ω equal to

a. $\sqrt{\frac{k}{M}}$ b. $\sqrt{\frac{2k}{M}}$ c. $\sqrt{\frac{2k}{3M}}$ d. $\sqrt{\frac{4k}{3M}}$

6. The maximum value of V_0 for which the disc will roll without slipping is

a. $\mu g\sqrt{\frac{M}{k}}$ b. $\mu g\sqrt{\frac{M}{2k}}$

c. $\mu g\sqrt{\frac{3M}{k}}$ d. $\mu g\sqrt{\frac{5M}{2k}}$

ANSWERS AND SOLUTIONS

Subjective Type

1. Applying torque equation about centre of mass

$$\tau_{CM} = I_{CM}\,\alpha \quad \text{(i)}$$

$$\Rightarrow \quad N\frac{R}{2} = (I_{CM})\,\alpha \quad \text{(ii)}$$

$$I_{CM} = mR^2 + m\left(\frac{R}{2}\right)^2 + m\left(\frac{R}{2}\right)^2 = \frac{2}{3}mR^2$$

The CM of system (ring + particle) will be at a distance $R/2$ from the centre of the ring. We can apply torque equation about CM of system.

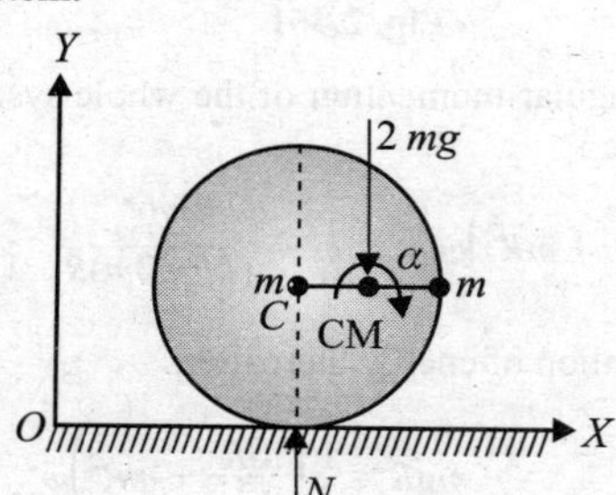

Fig. 2.446

$$2\,mg - N = 2m\,(a_{CM})_y \quad \text{(iii)}$$

Using constraint, acceleration of centre of circle is zero.

$$(a_c)_y = 0$$

$$\Rightarrow \quad \frac{R}{2}\alpha = (a_{CM})_y \quad \text{(iv)}$$

From Eqs. (ii), (iii) and (iv), we get

$$N = \left[\frac{6}{4}mg\right]$$

2. When the two spheres come in contact, they exert equal and opposite impulse.

$$\int Fa\,dt = I_1\,(\omega_0 - \omega_1) \quad \text{(i)}$$

$$\int Fb\,dt = I_2\,\omega_2 \quad \text{(ii)}$$

Finally $\quad \omega_1 a = \omega_2 b$

$$I_1(\omega_0 - \omega_1) = I_2\frac{a^2}{b^2}\omega_1$$

$$\omega_1 = \frac{I_1\omega_0}{I_1 + \frac{a^2}{b^2}I_2} = 4 \text{ rad/s}$$

3. At the maximum elongation, velocity will be perpendicular to the spring conserving angular momentum about O.

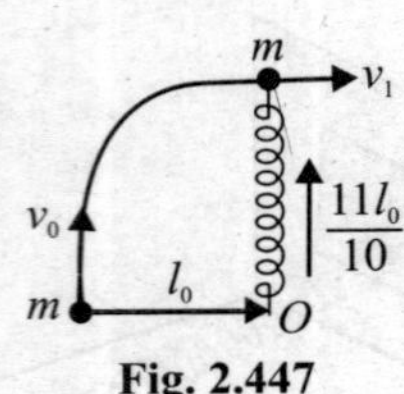

Fig. 2.447

$$mv_0 l_0 = mv_1 \frac{11 l_0}{10} \Rightarrow v_1 = \frac{10 v_0}{11} = 10 \text{ m/s}$$

$$\frac{1}{2} mv_0^2 = \frac{1}{2} mv_1^2 + \frac{1}{2} k\left(\frac{l_0}{10}\right)^2$$

$$k = \frac{m(v_0^2 - v_1^2)}{(0.1)^2} = \frac{0.1(121 - 100)}{(0.1)^2} = \frac{21}{0.1} = 210$$

4.

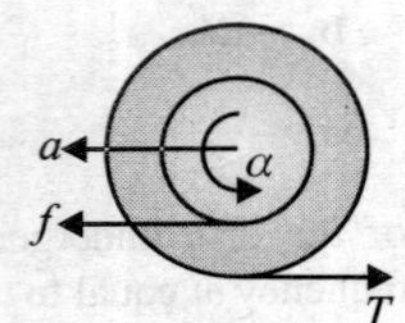

Fig. 2.448

$$a = \frac{\alpha R}{2}, \quad \alpha R - a = \frac{g}{2} \Rightarrow a = \frac{g}{2}$$

$$f - T = Ma \Rightarrow \quad TR - f\frac{R}{2} = \frac{MR^2}{2}\alpha$$

Solving, $\quad T = \frac{3Mg}{2}, f = 2Mg$

5. Consider a general position of the rod as shown in the figure. Let the applied force be 'F'. As the end is moved slowly, some rod has no acceleration. Applying Newton's law in x-direction, $f - F\sin\theta = 0$

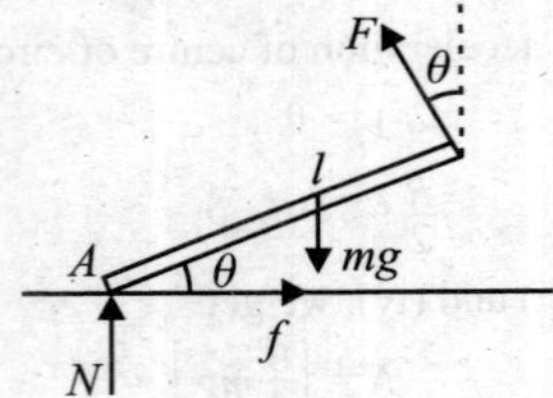

Fig. 2.449

$$f = F\sin\theta \quad \text{(i)}$$

In y-direction, $F\cos\theta + N = mg$ (ii)

Making $\quad \tau_A = 0 \Rightarrow Fl - mg\frac{1}{2}\cos\theta = 0$

$$F = \frac{mg\cos\theta}{2} \quad \text{(iii)}$$

For no slipping, $\quad f \le \mu N \quad \Rightarrow \mu \ge \frac{f}{N}$

$$\mu \ge \frac{F\sin\theta}{mg - F\cos\theta} \Rightarrow \quad \mu \ge \frac{\sin\theta\cos\theta}{2 - \cos^2\theta}$$

$$\mu_{\min} = \left(\frac{\sin\theta\cos\theta}{2 - \cos^2\theta}\right) = \frac{1}{2\sqrt{2}}$$

6. The minimum velocity required by block 'm' to complete the motion is $\sqrt{5gR}$.

Fig. 2.450

Conserving mechanical energy, $\frac{1}{2} I\omega^2 = Mg \times \frac{R}{2}$

$$\Rightarrow \quad \omega = \sqrt{\frac{MgR}{I}}$$

Conservation of angular momentum w.r.t. P before and after collision,

$$I\omega = mR\sqrt{5gR}$$

$$I\sqrt{\frac{MgR}{I}} = mR\sqrt{5gR}$$

$$MgRI = m^2R^2 5gR$$

Putting $\; I = \frac{ML^2}{3} = \frac{MR^2}{3}$ (since $L = R$)

$$\frac{M}{m} = \sqrt{15}$$

7. $\quad l = y + z$ (total length at any time t) $= y + \sqrt{y^2 + l_1^2}$

$$\frac{dl}{dt} = \frac{dy}{dt} + \frac{y}{\sqrt{y^2 + l_1^2}}\frac{dy}{dt}$$

$$\omega R = v(1 + \cos\theta), \; v = \frac{\omega R}{1 + \cos\theta}$$

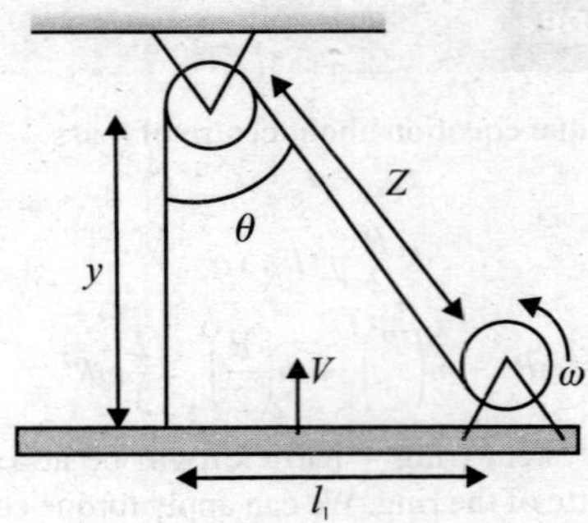

Fig. 2.451

8. Conserving angular momentum of the whole system about the CM of the disc

$$mv_0 R = \left(\frac{MR^2}{2} + mR^2\right)\omega \Rightarrow \omega = \frac{2mv_0}{(M + 2m)R} \quad \text{(i)}$$

From conservation of energy thereafter,

$$mgR = \frac{1}{2}\left(\frac{MR^2}{2} + mR^2\right)\omega^2 \quad \text{(ii)}$$

Putting values of ω from Eq. (i) in Eq. (ii), we get

$$v_0 = \sqrt{\frac{(M+2m)gR}{m}}$$

9. $a = a_t \sin\alpha - a_c \cos\alpha$

At the time of separation,

For M, $\Sigma F_x = 0$ and $a = 0$

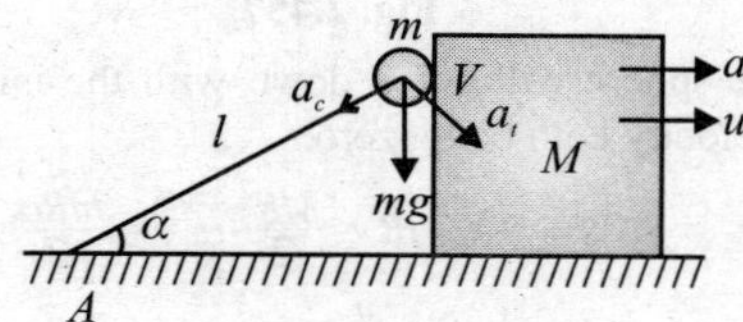

Fig. 2.452

Acceleration of the load in horizontal direction,

$$a = 0 = a_t \sin\alpha - \frac{v^2}{l}\cos\alpha$$

$$a_t \sin\alpha = \frac{v^2}{l}\cos\alpha$$

But $a_t = g\cos\alpha$

Hence, $v = \sqrt{gl\sin\alpha}$

$$u = v\sin\alpha = \sin\alpha\sqrt{gl\sin\alpha}$$

$$mgl = mgl\sin\alpha + \frac{mv^2}{2} + \frac{1}{2}Mv^2\sin\alpha$$

$$\frac{M}{m} = \frac{2 - 3\sin\alpha}{\sin^3\alpha} = 4$$

$$u = v\sin\alpha = \frac{1}{2}\sqrt{\frac{gl}{2}}$$

10. $N = \frac{4mg}{5} - \frac{4F}{5}$ (i)

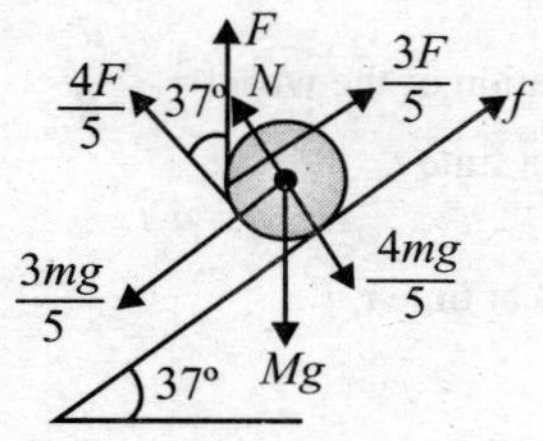

Fig. 2.453

$\therefore$ $$f = \mu\left(\frac{4mg}{5} - \frac{4F}{5}\right) \quad \text{(ii)}$$

$$f + \frac{3F}{5} = \frac{3mg}{5} \Rightarrow F = \frac{3}{8}mg \quad \text{(iii)}$$

$$fr = Fr$$

$\therefore$ $$f = F \quad \text{(iv)}$$

After solving, we get $\mu = \frac{3}{4}$

11. $d = \sqrt{R^2 - (R-h)^2} = \sqrt{2Rh - h^2}$

Taking torque about P, $\tau_F > \tau_W$

For minimum F, $Wd = F(2R - h)$

or $$F = \frac{W\sqrt{2Rh - h^2}}{(2R - h)}$$

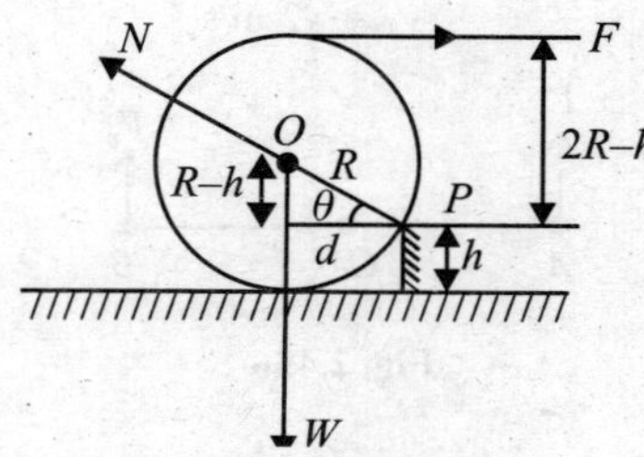

Fig. 2.454

12. a. If the ball were not rotating initially, it could not skid against the surface, in that case angle θ and ϕ would be identical. Let the initial and final angular velocities be ω_1 and ω_2, respectively, as shown in Fig 2.455.

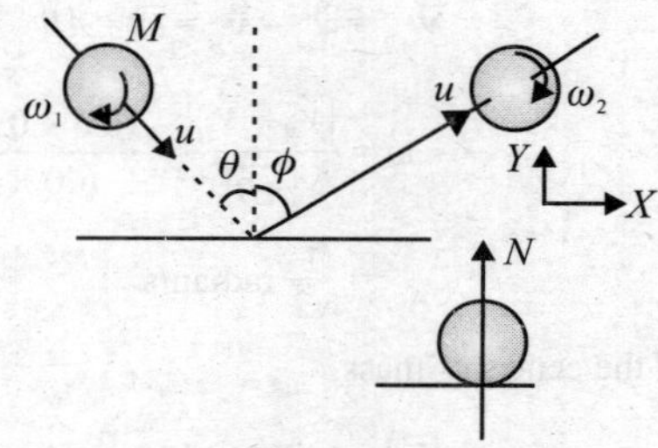

Fig. 2.455

Since the collision is elastic, the normal component of the velocity will reverse in direction. So

$$v\cos\theta = u\cos\phi \quad \text{(i)}$$

From impulse–momentum equation,

$$N\Delta t = p_f - p_i = 2Mv\cos\theta \quad \text{(ii)}$$

Component of impulse parallel to the surface is

$$\mu_k N\Delta t = 2Mv\mu_k\cos\theta$$

If the ball is rotating clockwise as shown in the figure, the impulse will be to the right but it would be to the left if the ball is rotating anticlockwise.

From impulse–momentum equation,

$$2Mv\mu_k\cos\theta = Mu\sin\phi - Mv\sin\theta \quad \text{(iii)}$$

Eqs. (i) and (iii) give

$$\tan\phi - \tan\theta = 2\mu$$

But if the original angular velocity were anticlockwise instead of clockwise, we would obtain

$$\tan\phi - \tan\theta = -2\mu$$

Therefore, $|\tan\theta - \tan\phi| = 2\mu$

b. Angular impulse on the ball = $(2Mv\mu\cos\theta)\,a$

From angular impulse equation,

$$\Delta\omega = \frac{2\mu Mv \cos\theta}{I} = \frac{2\mu Mva \cos\theta}{\frac{2}{5}Ma^2}$$

$$= \frac{5\,mv\cos\theta}{a}$$

13. The velocity of the ball just before the impact

$$v = \sqrt{2gh} = \sqrt{2 \times (20)(10 \times 10^{-2})}$$

$$= \sqrt{2}\text{ m/s}$$

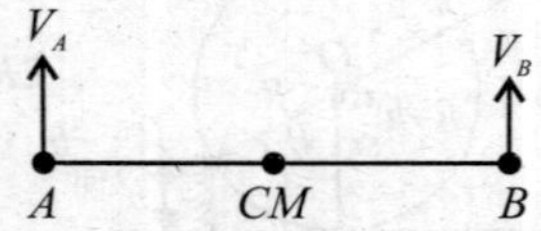

Fig. 2.456

Applying the coefficient of restitution equation,

Velocity of rebound of $A, v_A = 0.6v$

Velocity of rebound of $B, v_B = 0.4v$

After collision the system will rotate about the centre of mass.

$$\vec{v}_{AB} = \vec{v}_A - \vec{v}_B = \vec{\omega} \times \overrightarrow{AB}$$

$$\vec{\omega} = \frac{|\vec{v}_A - \vec{v}_B|}{|\overrightarrow{AB}|} = \frac{(0.6 - 0.4)(\sqrt{2})}{(40 \times 10^{-2})}$$

$$= \frac{1}{\sqrt{2}}\text{ radian/s}$$

Velocity of the centre of mass,

$$V_{CM} = \frac{m_A v_A + M_B v_B}{m_A + M_B}$$

$$= \frac{(0.6 + 0.4)}{2} v = \frac{1}{\sqrt{2}}\text{ m/s}$$

Angular momentum of the system,

$$m(v_A - v_B)r = I_{CM}\,\omega$$

where r is the distance of CM from either balls.

Thus $$\omega = \frac{m(vA - vB)r}{2mr^2} = \frac{v_A - v_B}{2r}$$

$$= \left(\frac{0.6 \times 0.4}{2 \times 0.2}\right)\sqrt{2} = \frac{1}{\sqrt{2}}\text{ radian/s}$$

14. We will check whether the cylinder slips down the incline or not.
As $mg \sin 30° = \mu mg \cos 30°$, initially the cylinder slips at its place till the angular velocity becomes zero.

$$\Sigma\tau = (\mu\, mg \cos\theta)R = \frac{mR^2}{2}\alpha$$

$$\alpha = \frac{2\mu\cos\theta}{R} g$$

$$0 = \omega_0 - \alpha_1 \text{ or } t_1 = \frac{\omega_0}{\alpha}$$

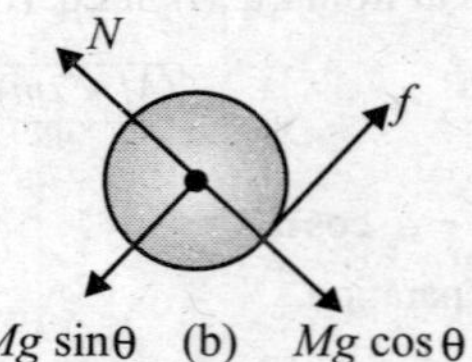

Fig. 2.457

Next the sphere will come down with the initial angular and linear velocity both being zero.

$$fR = \frac{MR^2}{2}\alpha, f = \frac{mR\alpha}{2}$$

and $$mg \sin\theta - f = ma$$

If pure rolling takes place, $R = R\alpha$

$$f - mg\sin\theta - 2f \text{ or } f = \frac{1}{3} mg \sin\theta = \frac{mg}{6}$$

As $$f_{max} = \mu mg \cos\theta$$

$$= \frac{1}{\sqrt{3}} mg \times \frac{\sqrt{3}}{2} = \frac{mg}{2},$$

and $f_{max} < f$, so pure rolling is possible.

$$a = \frac{2F}{m} = \frac{2 - (mg/6)}{m} = \frac{g}{3}$$

Let t_2 be the time taken to reach the foot of the incline.

$$\frac{3R}{\sin 30°} = \frac{1}{2}\frac{g}{3}t_2^2 \text{ or } t_2 = 6\sqrt{\frac{R}{6}}$$

Hence, the total time, $t = t_1 + t_2$

$$= \left(\frac{\omega_0 R}{g} + 6\sqrt{\frac{R}{g}}\right)$$

15. a. The acceleration of the hoop, $a = -\mu_g$

Angular acceleration of the wheel $= -\frac{\mu_g}{R}$

Linear velocity at time t,

$$v = v_0 - \mu_g t \qquad \text{(i)}$$

Angular velocity at time t,

$$\omega = \omega_0 - \frac{\mu_g}{R}t \qquad \text{(ii)}$$

Let linear velocity be zero at t. So,

$$0 = v_0 - \mu_g t_i \text{ or } t_1 = \frac{\mu_0}{\mu_g}$$

Angular velocity of the ring at t is

$$\omega = \omega_0\left(\frac{\mu_g}{R}\right)\left(\frac{v_0}{\mu_g}\right) \qquad \text{(iii)}$$

$$= \omega_0 - \frac{v_0}{R}$$

The hoop will return if it has reverse spin at the instant its linear velocity is zero. So the condition for hoop to return is

$$\omega_0 - \frac{v_0}{R} > 0 \qquad \text{or} \qquad \omega_0 > \frac{v_0}{R}$$

b. Distance moved by the hoop, $S = \frac{v_0^2}{2\mu g}$

c. Let t_2 be the time flow during which the hoop slips on the floor and v be the return velocity.

Then $v = \mu_g t_2$

and $\omega' = \omega - \frac{\mu_g}{R} t_2$

No slip condition is achieved when

$$\frac{\mu_g}{R} t_2 = \omega - \frac{\mu_g}{R} t_2$$

From Eq. (iii), $\frac{2\mu_g}{R} t_2 = \omega = \omega_0 - \frac{v_0}{R}$

or $t_2 = \frac{R}{2\mu_g}\left(\omega_0 - \frac{v_0}{R}\right)$

d. $V = \mu_g t_2 = \frac{\omega_0 R}{2} - \frac{v_0}{2}$

16. The total kinetic energy of the body

$$K = K_T + K_R = \frac{1}{2} Mv^2 + \frac{1}{2} I\omega^2$$

or, $= \frac{1}{2} Mv^2\left[1 + \left(\frac{I}{Mr^2}\right)\right]$ [as $v = r\omega$]

When it rolls up an irregular inclined plane of height ($h = 3v^2/4g$), its KE is converted into PE, so by conservation of mechanical energy,

$$\frac{1}{2} Mv^2\left[1 + \frac{I}{Mr^2}\right] = Mg\left[\frac{3v^2}{4g}\right]$$

which on simplification gives $I = Mr^2/2$. This result clearly indicates that the body is either a disc or a cylinder.

17. In case of rolling

$$K = \frac{1}{2} Mv^2 + \frac{1}{2} I\omega^2 \text{ with } v = r\omega$$

so, $K = \frac{1}{2} Mv^2\left[1 + \left(\frac{I}{Mr^2}\right)\right]$

Now as for the ring

$$I = Mr^2$$

$$K_R = Mv_R^2, \text{ i.e., } v_R = \sqrt{\frac{K_R}{0.3}}$$

and for the cylinder,

$$I = \frac{1}{2} Mr^2$$

$$K_C = \frac{3}{4} Mv_C^2,$$

i.e., $v_C = \sqrt{\frac{4K_C}{3 \times 0.4}} = \sqrt{\frac{K_C}{0.3}}$

Now according to the given problem, $K_R = K_C$, so $v_R = v_C$ and as the motion is uniform, both will reach the wall simultaneously.

Objective Type

1. c. Because the planes of two rings are mutually perpendicular and centres are coincident, hence an axis, which is passing through the centre of one of the rings and to its plane, will be along the diameter of other ring. Hence, moment of inertia of the system

$$= I_{CM} + I_{diameter} = mr^2 + \frac{mr^2}{2}$$

$$= \frac{3}{2} mr^2$$

2. b. According to the theorem of parallel axis,

$$I = I_{CG} + M(2R)^2$$

where I_{CG} = *M.I.* about an axis through the centre of gravity.

$$I = \frac{2}{5} MR^2 + 4MR^2 = \frac{22}{5} MR^2$$

or $MK^2 = \frac{22}{5} MR^2 \Rightarrow K = \sqrt{\frac{22}{5}} R$

3. b. Given that $I_Q = 4I_P$

Here I = mass × (radius)^2. If s is the linear mass density (mass/length) of the wire, then

$$\sigma \times 2\pi (nr) \times n^2r^2 = 4 (\sigma \times 2\pi r \times r^2)$$

$$\Rightarrow \quad n^3 = 4 \quad \Rightarrow n = (4)^{\frac{1}{3}}$$

4. c. Let M be the mass of each disc. Let R_A and R_B be the radii of discs A and B, respectively. Then $M = \pi R_A^2 t d_A = \pi R_B^2 t d_B$.

As $d_A = d_B$, so $R_A^2 < R_B^2$. Now,

$$I_A = \frac{1}{2} MR_A^2, \quad I_B = \frac{1}{2} MR_B^2$$

$$\frac{I_A}{I_B} = \frac{R_A^2}{R_B^2} < 1, \text{ i.e., } I_A < I_B$$

5. d.

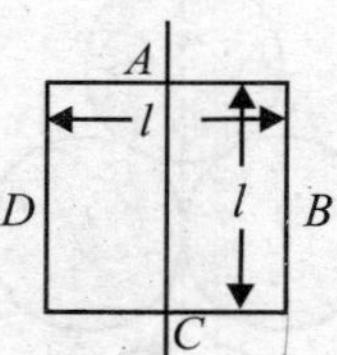

Fig. 2.458

$$I_{median\ line} = I_A + I_B + I_C + I_D$$

$$= 2 \times \frac{Ml^2}{12} + 2M\left(\frac{l}{2}\right)^2 = \frac{Ml^2}{6} + \frac{Ml^2}{2} = \frac{2}{3} Ml^2$$

6. a.

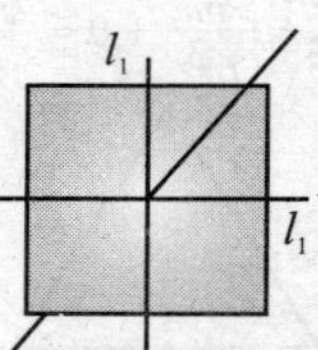

Fig. 2.459

$$I = I_1 + I_1 = 2 \times \frac{2}{3} Ml^2 = \frac{4}{3} Ml^2$$

7. a. The moment of inertia about CM system $= \frac{4}{3} Ml^2$

From perpendicular axis theorem,

$$\frac{4}{3} Ml^2 = Id_1 + Id_2 \qquad (Id_1 = Id_2)$$

$$Id = \frac{2}{3} Ml^2 \quad \text{or} \quad I = 4\frac{Ml^2}{3}(\sin 45°)^2$$

8. d. Mass of disc ∝ area, $M_A = 4M_B$ (as $R_A = 2R_B$)

$$\frac{I_A}{I_B} = \frac{\frac{1}{2} M_A R_A^2}{\frac{1}{2} M_B R_B^2}$$

9. d. Moment of inertia of discs A and B about the axis through their centre of mass and perpendicular to the plane will be

$$I_{AA} = I_{BB} = \frac{1}{2} Mr^2$$

Now, moment of inertia of disc A about an axis through B, by theorem of parallel axis will be

$$I_{AA} = I_{BB} + M(2r)^2 = \frac{1}{2} Mr^2 + 4Mr^2 = \frac{9}{2} Mr^2$$

So $I = I_{BB} + I_{AB} = \frac{1}{2} Mr^2 + \frac{9}{2} Mr^2 = 5Mr^2$

10. a. Mass of each of the four parts $= \frac{M}{3}$

Mass of the plate including the cut piece $= \frac{4M}{3}$

Moment of inertia of the whole plate (including the cut piece) about the said axis $= \left(\frac{4M}{3}\right)\frac{l^2}{6}$

Now moment of inertia of the remaining portion should be 3/4 of the above $= Ml^2/6$.

11. d. $I = I_1 + I_2 + I_3$

$$I_1 = I_2 = \frac{3}{2} mr^2$$

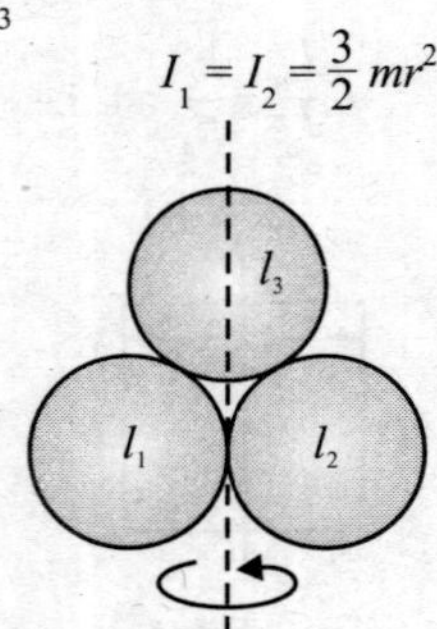

Fig. 2.460

$$I_1 = I_2 = \frac{3}{2} mr^2$$

$$I = I_1 + I_2 + I_3 = \frac{7}{2} mr^2$$

12. d. $I = I_1 + I_2 + I_3 = \frac{ml^2}{3} + \frac{ml^2}{3} + 0 = \frac{2ml^2}{3}$

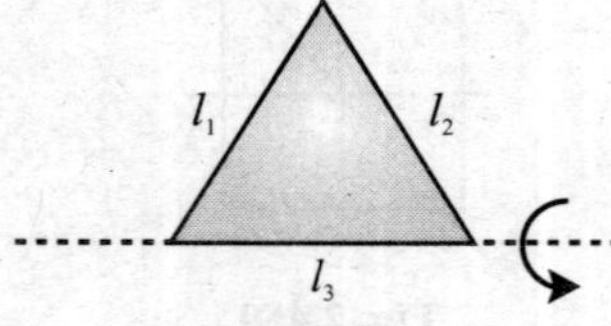

Fig. 2.461

13. b. MOI is $\Sigma m_i r_i^2$. About BC masses are spread far away than about any other axis.

14. d. Moment of inertia about 2:

$$I_2 = 4\left(\frac{ml^2}{3}\sin^2 45°\right) = \frac{2ml^2}{3}$$

Apply perpendicular axis theorem,

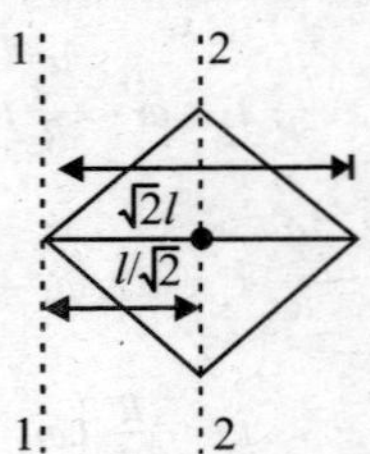

Fig. 2.462

$$I_1 = I_2 + mh^2 = \frac{2ml^2}{3} + 4m\left(\frac{1}{\sqrt{2}}\right)^2 = \frac{8}{3} ml^2$$

15. a. $(\text{MI})_{CG} = M\left(\frac{a^2 + b^2}{12}\right)$

According to the theorem of parallel axis,

$$(\text{MI})_{\text{required axis}} = (\text{MI})^{CG} + M(OA)^2$$

$$= M\left(\frac{a^2 + b^2}{12}\right) + M\left(\frac{a^2 + b^2}{4}\right)$$

$$= M\left(\frac{a^2 + b^2}{3}\right)$$

16. c. The distribution of mass is nearest to axis xx, hence moment of inertia is least about the xx-axis.

17. b. Moment of inertia of ABC about $AC = \frac{1}{2} \times$ moment of inertia of square sheet $ABCD$ about $AC = \frac{1}{2} \times [2M] \times \frac{l^2}{12} = \frac{Ml^2}{12}$

18. c. (About YY') $= \frac{ml^2}{12}$

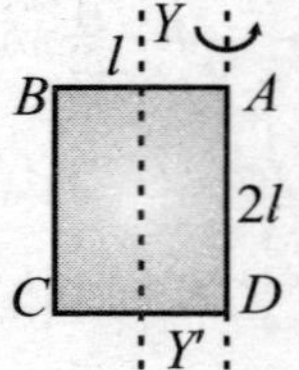

Fig. 2.463

Using parallel axis theorem,

$$I \text{ (about } AD) = \frac{ml^2}{12} + \frac{ml^2}{4} = \frac{ml^2}{3}$$

19. b. Let m_1 be the mass of the square plate of side 'a'. Then

$$m_1 = \sigma\left(\frac{a}{2}\right)^2; \; m_2 = \sigma(a)^2;$$

(σ being the real density)

And $\qquad m_2 - m_1 = M$

$$I = \frac{m_2 a^2}{6} - \left\{ \frac{m_1\left(\frac{a}{2}\right)^2}{6} + m_1\left(\frac{a}{4}\right)^2 \right\}$$

$$= \frac{\sigma a^4}{6} - \left\{ \frac{\sigma\left(\frac{a}{2}\right)^4}{6} + \sigma\left(\frac{a}{4}\right)^2\left(\frac{a}{4}\right)^2 \right\}$$

Also, $\qquad M = \sigma\left(1 - \frac{1}{4}\right)a^2 \Rightarrow \sigma = \frac{4}{3}\frac{M}{a^2}$

After solving, we get $I = \left(\frac{4}{3}\frac{M}{a^2}\right)a^4\left\{\frac{27}{12 \times 16}\right\} \Rightarrow I = \frac{3Ma^2}{16}$

20. b. $\frac{mv^2}{(R+r)} = mg\cos\theta,$

$$mgh = \frac{1}{2}mv^2 + \frac{1}{2}I\omega^2$$

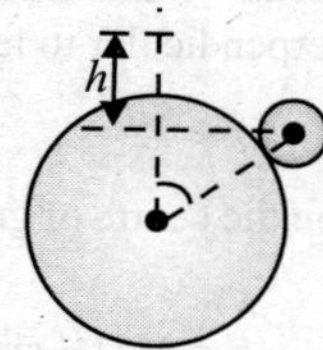

Fig. 2.464

$$mg(R+r)(1-\cos\theta) = \frac{1}{2}mv^2 + \frac{1}{5}mv^2 = \frac{7}{10}mv^2$$

$$\frac{10}{7}mg(R+r)(1-\cos\theta) = mg\cos\theta$$

$$mv^2 = \frac{10}{7}mg(R+r)(1-\cos\theta)$$

$$\frac{10}{7} = \frac{17}{7}\cos\theta \text{ or } \cos\theta = \frac{10}{17}$$

$$v = \sqrt{g(R+r)\cos\theta} = \sqrt{\frac{10}{7}g(R+r)}$$

and $\qquad \omega = \frac{v}{r} = \sqrt{\frac{10g(R+r)}{17r^2}}$

21. b. Rolling is rotation about point of contact. Applying impulse momentum equation about P.

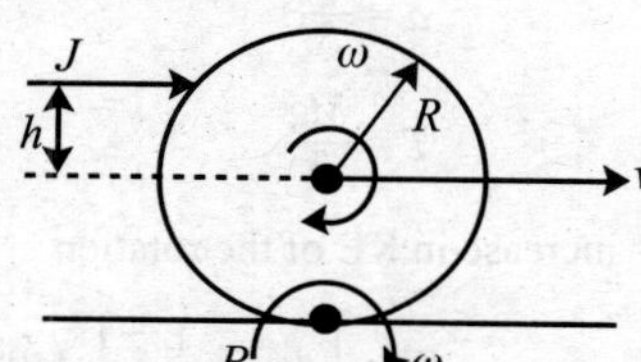

Fig. 2.465

$$J(R+h) = I_P\omega \qquad \text{(i)}$$

and $\qquad J = mv \qquad$ (ii)

As sphere rolls $v = \omega R$, and $I = \frac{2}{5}mR^2 + mR^2 = \frac{7}{5}mR^2$

After solving, we get $\frac{h}{R} = \frac{2}{5}$

22. c. As sphere rolls, the lowest point of the sphere should have the same acceleration as the plank.

Hence, $\qquad a_1 = \alpha R - a_2$

$$2 = 2\alpha - 4 \;\Rightarrow\; \alpha = 3 \text{ rad/s}^2$$

23. d. Taking τ about CM

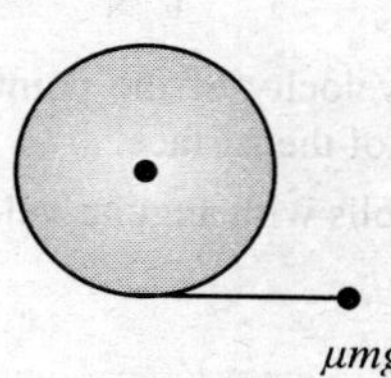

Fig. 2.466

$$\mu mgR = MR^2\alpha$$

$$\mu g = R\alpha$$

$$\alpha = \frac{\mu g}{R}$$

$$\omega = \omega_0 - \frac{\mu g}{R}f = \frac{\omega_0}{2}$$

$$\frac{\omega_0}{2} = \frac{\mu g}{R}f \Rightarrow f = \frac{\omega_0 R}{2\mu g}$$

24. b. Here, $\omega = \frac{v}{R}$

$$V_{P_x} = \left(v + \frac{v}{R}r\sin\theta\right)\hat{i}$$

$$V_{P_y} = -\left(\frac{v}{R}r\cos\theta\right)\hat{j}$$

$$\therefore \qquad \vec{V}_P = \left(v + \frac{vr\sin\theta}{R}\right)\hat{i} - \frac{vr\cos\theta}{R}\hat{j}$$

$$= \left(v + \frac{vr\sin\theta}{R}\right)\hat{i} - \frac{vr\cos\theta}{R}\hat{j}$$

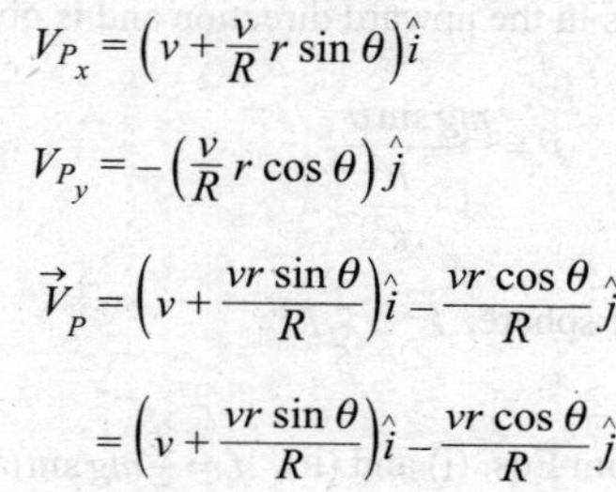

Fig. 2.467

25. a. FBD of point P

In the earth frame, we have

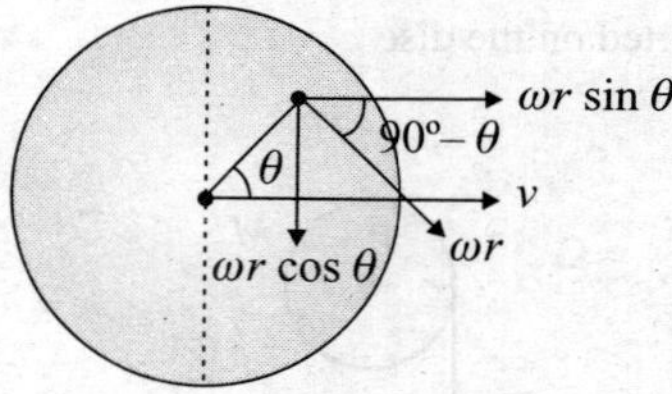

Fig. 2.468

So, resultant $= \sqrt{(a + r\alpha)^2 + (\omega^2 r)^2}$

26. b. $E_1 = \frac{1}{2}I_1\omega_1^2 = \frac{1}{2}(I_1\omega_1)\omega_1 = \frac{1}{2}k\omega_1$

Now $\quad I_1\omega_1 = I_2\omega_2 = k$ (say)

$$E_2 = \frac{1}{2}I_2\omega_2^2 = \frac{1}{2}(I_2\omega_2)\omega_2 = \frac{1}{2}k\omega_2$$

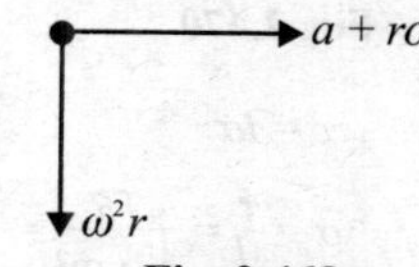

$$I_1 > I_2 \text{ and } \omega_2 > \omega_1$$

so, $$E_2 > E_1$$

27. c. For pure rolling velocity of the point of contact has to be equal to the velocity of the surface.

Let us say cylinder rolls with angular velocity ω.

At point B,

$$V_C - \omega r = V_P \Rightarrow \omega r = V_C - V_P$$

At point A,

$$V_A = V_C + \omega r = 2V_C - V_P$$

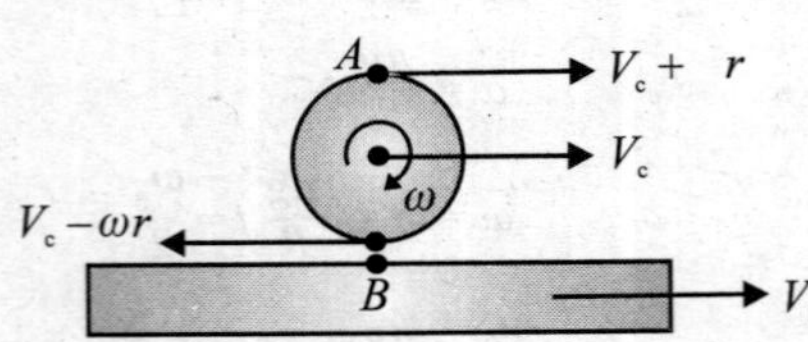

Fig. 2.469

28. d. In case of rolling in the inclined plane, friciton is static and acts in the upward direction and is given by

$$f = \frac{mg\sin\theta}{1+\frac{R^2}{k^2}} \quad \text{(i)}$$

For sphere, $k^2 = \frac{2}{5}R^2$ (ii)

From Eqs. (i) and (ii), $f = \frac{2}{7}mg\sin\theta$ (upwards)

29. c. Torque exerted on the disc

$$\tau = TR$$

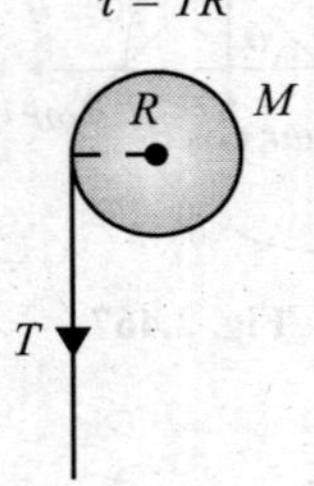

Fig. 2.470

Now $$\tau = I\alpha$$

$$\alpha = \frac{\tau}{I} = \frac{TR}{\frac{1}{2}MR^2} = \frac{2TR}{MR^2} = \frac{2T}{MR}$$

30. c. Tangential acceleration

$$a = r\alpha = R\left(\frac{2T}{MR}\right) = \frac{2T}{M}$$

31. c. From the above problem, $T = \frac{Ma}{2}$

Now

$$mg - T = ma$$

or $$mg - M\frac{a}{2} = ma$$

or $$mg = \left(m + \frac{M}{2}\right)a$$

$$a = \frac{2mg}{M+2m}$$

32. b. According to problems 30 and 31,

$$T = \frac{Ma}{2} = \frac{Mmg}{M+2m}$$

33. b. $\tau = I\alpha$ or $\alpha = \frac{\tau}{I}$

Here, as the force is applied at midpoint,

$$\tau_A = \tau_B = F \times \frac{L}{2}$$

And as moment of inertia depends on the distribution of mass w.r.t. the axis of rotation, hence

$$I_A > I_B \text{ or } \alpha_A > \alpha_B$$

i.e., angular acceleration is more in case (b).

34. c. The moment of inertia of the uniform rod about an axis through one end and perpendicular to length is

$$I = \frac{ML^2}{3}$$

Torque ($t = la$) acting on the centre of gravity of the rod is given by

$$\tau = Mg\left[\frac{L}{2}\sin\theta\right]$$

or $$\frac{ML^2}{3}\alpha = Mg\frac{L}{2}\sin\theta$$

$$\alpha = \frac{3g}{2L}\sin\theta$$

35. b. Equation of motion is

$$Mg - T = Ma \quad \text{(i)}$$

Taking torque about the axis passing through the centre of the spool and perpendicular to it,

$$TR = I\alpha = \frac{1}{2}MR^2\left(\frac{a}{2}\right)$$

$$T = \frac{1}{2}Ma \quad \text{(ii)}$$

From Eqs. (i) and (ii),

$$Ma = Mg - \frac{1}{2}Ma$$

$$a = \frac{2g}{3}$$

$\therefore$ $$T = \frac{Mg}{3}$$

36. a. Work done = increase in KE of the rotation

$$W = \frac{1}{2}I\omega^2 = \frac{1}{2} \times \frac{1}{2}MR^2\omega^2 = \frac{1}{4}MR^2\omega^2$$

37. d. Taking torque about the attachment point for W, we get

$$-T_1(0.4L) + T_2(0.3L) + 500(0.2L) = 0$$

$$T = 1000 \text{ N, where } T_1 = T_2 = T$$

$$\Sigma F_y = 0 \Rightarrow 2T - W - 500 = 0$$

$\Rightarrow$ $$W = 1500 \text{ N}$$

38. a. From conservation of angular momentum,

$$\frac{ML^2}{12}\omega_0 = \left[\frac{ML^2}{12} + 2m\left(\frac{L}{2}\right)^2\right]\omega,$$

$$\omega = \left(\frac{M}{M+6m}\right)\omega_0$$

39. b. $mg\left(\frac{l}{4}\right)=\frac{1}{2}\left[\frac{ml^2}{12}+m\left(\frac{l}{4}\right)^2\right]\omega^2$

$\omega=\sqrt{\frac{24g}{7l}}=2\sqrt{\frac{6g}{7l}}$

40. a. Here kinetic friction force will balance the force of gravity. So it will rotate at its initial position and will not move up or down. As its angular velocity becomes zero (friction also becomes zero), it will move downwards.

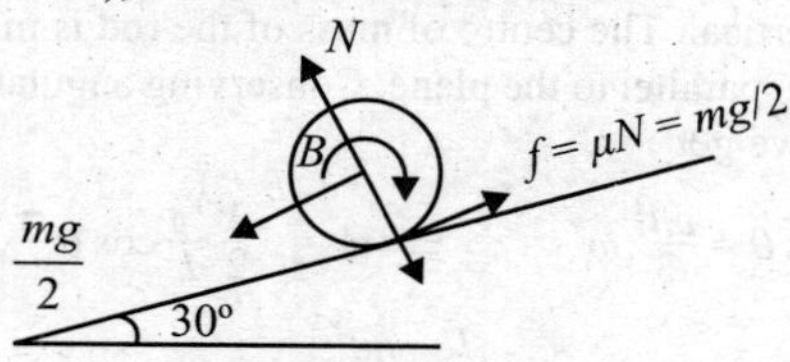

Fig. 2.471

41. d. $\vec{L}=I_{CM}\vec{\omega}+m\vec{r}\times\vec{v}_{CM}$

$=\frac{MR^2}{2}\omega(-\hat{k})+M\frac{3}{2}R(-\hat{j})\times v_0\hat{i}$

$=MRv_0\hat{k}$

42. b. KE $=\frac{L^2}{2I}$, L remains constant. I doubles so KE becomes half.

43. a. F will provide anticlockwise torque about the centre due to which the bottommost point will tend to move towards right, so friction will act towards left. So it will move towards left.

44. a. For a body to roll without slipping on an incline of angle θ:

$\mu\geq\frac{\tan\theta}{1+\frac{mR^2}{I}}$

I is minimum for the sphere, hence μ required is least for the sphere. Here it is given that μ is just sufficient to roll the sphere purely. It means the ring and the cylinder will not roll purely so kinetic friction will act on them. Ultimately, friction acting on each of the three will be $\mu mg\cos\theta$ and each of them will have the same acceleration. Hence, all of them reach at the same time.

45. b.

$$f_r=\frac{\frac{1}{2}I\omega^2}{\frac{1}{2}mv^2+\frac{1}{2}I\omega^2},\text{ where } I=\frac{2}{5}mr^2,$$

$$\omega=\frac{v}{r}$$

Solve to get $f_r=\frac{2}{7}$

46. c. Balancing the torque:

for the first case: $16l_1=ml_2$

for the second case: $ml_1=4l_2$

Divide them to get $m=8$ kg

47. a.

$$\overrightarrow{L_O}=I_{CM}\vec{\omega}+m(\vec{r}\times\vec{v})$$

$$=\frac{1}{2}mR^2\omega\hat{k}+m(4R\hat{i}+3R\hat{j})\times(\omega R\hat{i})$$

$$=\frac{1}{2}mR^2\omega\hat{k}-3mR^2\omega\hat{k}$$

$$=-\frac{5}{2}mR^2\omega\hat{k}$$

48. b. Friction force will act towards right.

49. d.

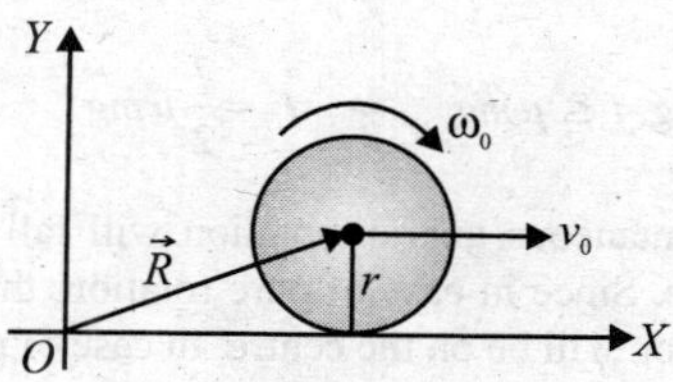

Fig. 2.472

$\vec{L}=\vec{R}\times m\vec{v}_0+I\vec{\omega}_0$

$=mv_0+I\omega_0$ (which is constant)

50. d. $J=mv=mv'$

$\Rightarrow$ Velocity of the CM of rod $=v$

Applying impulse momentum equation about the CM of rod

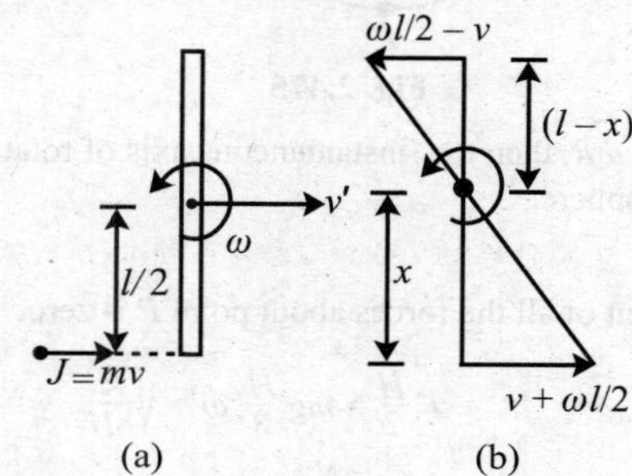

Fig. 2.473

$J\frac{l}{2}=I_{CM}\omega\Rightarrow\frac{mvl}{2}=\left(\frac{ml^2}{12}\right)\omega\quad\Rightarrow\omega=\frac{6V}{\ell}$

About instantaneous axis of rotation the rod is considered to have pure rotation.

Let instantaneous axis of rotation be located at a distance x from the colliding end [fig (b)]

$$\frac{\frac{\omega\ell}{2}-v}{\ell-x}=\frac{v+\frac{\omega\ell}{2}}{x}\qquad\text{(i)}$$

Substituting the value of $\omega=6V/l$ in Eq. (i), we get $x=\frac{2}{3}\ell$.

51. d. As sphere is rolling we can apply torque equation about point of contact (about P).

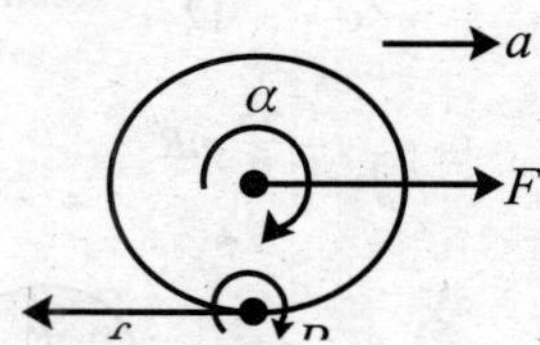

Fig. 2.474

$FR=I_P\alpha\Rightarrow FR=\left(\frac{7}{5}MR^2\right)\alpha$

$\Rightarrow R\alpha=\frac{5F}{7M}=a$ (acceleration of cm)

Equation of translatory motion

$$F - f = Ma \quad \Rightarrow \quad f = \frac{2}{7}F \qquad \text{(i)}$$

For no sliding, $f \le \mu mg \quad \Rightarrow \quad F = \frac{7}{2}\mu mg$

52. d. The instantaneous axis of rotation will fall on the outside of the sphere. Since in case of pure rotation, the instantaneous axis of rotation will be on the centre. In case of pure rolling, the instantaneous axis of rotation will be on P.

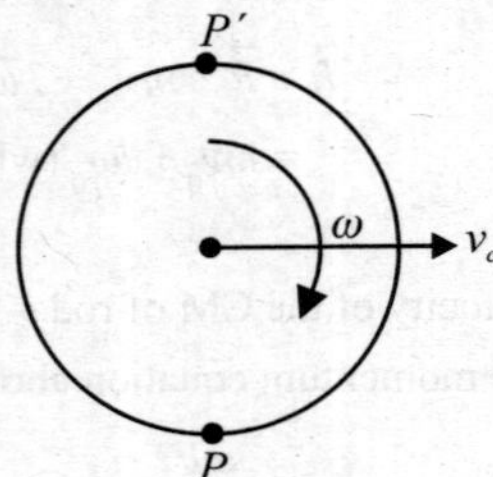

Fig. 2.475

But if $v_C > \omega R$, then the instantaneous axis of rotation will fall out of the sphere.

53. d. $N = mg$ (i)

Net moment of all the forces about point P is zero.

$$F\frac{H}{2} = mg\frac{H}{8},\ \omega = \sqrt{\frac{g}{8H}}$$

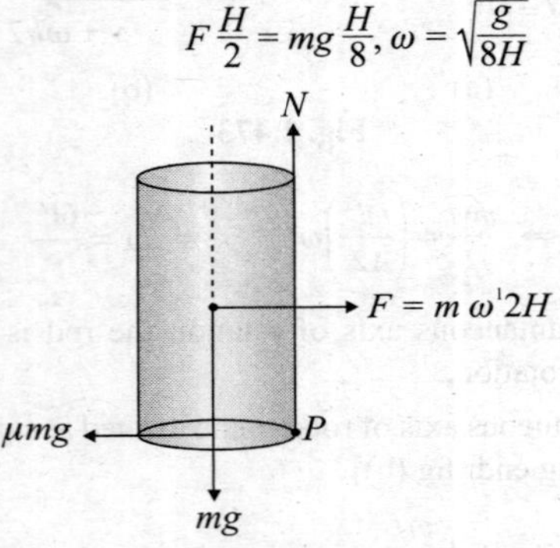

Fig. 2.476

54. c. Let x be the position of CM of the system from CM of the rod.

$$2mx = m(R + 2R - x) \quad \Rightarrow \quad x = R \qquad \text{(i)}$$

$$I_G = \frac{2m(4R)^2}{12} + 2mR^2 + \frac{mR^2}{2} + m(2R)^2$$

$$= \frac{55}{6}mR^2 \qquad \text{(ii)}$$

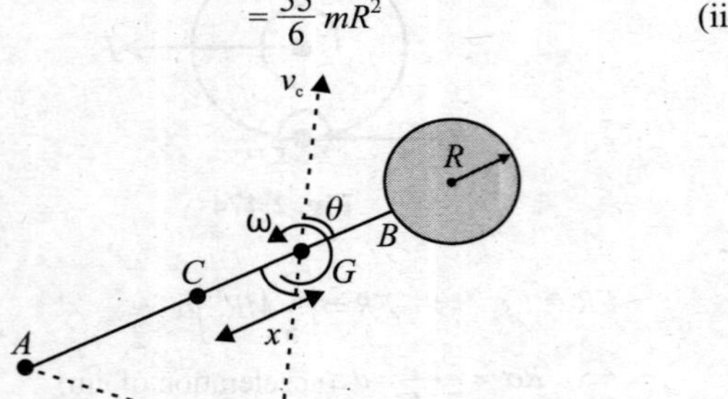

Fig. 2.477

Applying conservation of angular momentum,

$$3mv_c \sin\theta\, 3R + I_G\omega = (I_G + 3m\,9R^2)\omega'$$

$$\omega' = \frac{54mv_cR\sin\theta + 55mR^2\omega}{217mR^2} = \frac{54v_c\sin\theta + 55R\omega}{217R}$$

55. a. When the car is moving with the constant speed, the rod will be vertical. The centre of mass of the rod is moving with a velocity v_0 parallel to the plane. Conserving angular momentum about A, we get

$$mv_0\frac{l}{2}\cos\theta = \frac{ml^2}{3}\omega \qquad \Rightarrow \omega = \frac{3}{2}\frac{v_0}{l}\cos\theta$$

56. a. $$p\frac{l}{4} = \frac{ml^2}{12}\omega$$

57. b. Total MI $= I_1 + I_2 + I_3 + I_4$

$$= 2I_1 + 2I_2$$

$$= 2(I_1 + I_2) \qquad [I_1 = I_3;\ I_1 = I_4]$$

Now,

$$I_2 = I_3 = \frac{Ml^2}{3}$$

Using parallel axis theorem, we have

$$I = I_{CM} + Mx^2 \text{ and } x = \sqrt{l^2 + \left(\frac{l}{2}\right)^2}$$

$$I_1 = I_4 = \frac{Ml^2}{12} + M\left[\sqrt{l^2 + \left(\frac{l}{2}\right)^2}\right]^2$$

Putting all values, we get

$$I = \frac{10\,Ml^2}{3}$$

58. b. $$I = m_1\left(\frac{a}{2}\right)^2 + m_2\left(\frac{a}{2}\right)^2$$

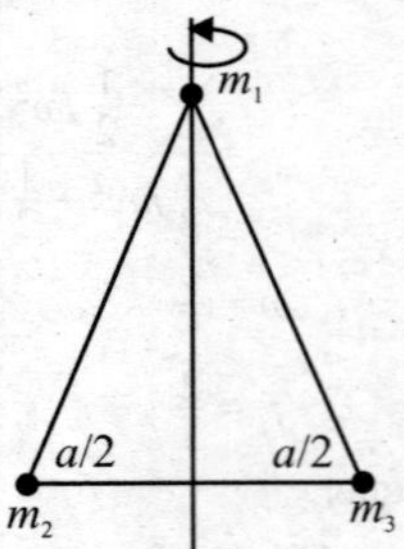

Fig. 2.478

59. c. $H = I\omega$, here $I = \frac{M(2d)^2}{12} = \frac{Md^2}{3}$

$$\omega = \frac{H}{(Md^2/3)} = \frac{3H}{Md^2}$$

60. b. Since the disc comes to rest, it stops rotating and translating simultaneously $v = 0$ and $\omega = 0$. That means, the angular momentum about the instantaneous point of contact just after the time of stopping is zero. We know that the angular momentum of the disc about P remains constant because frictional force f, N and mg pass through point P and thus produce no torque about this point.

$\Rightarrow \quad L_{initial} = L_{final} \Rightarrow mvr - I_0\omega_0 = 0$

$\Rightarrow \quad mvr = \frac{1}{2}mr^2\omega_0 \Rightarrow 2v_0 = \omega_0 r$

61. b. During the impact, the impact forces pass through point P. Therefore, the torque produced by it about P is equal to zero.

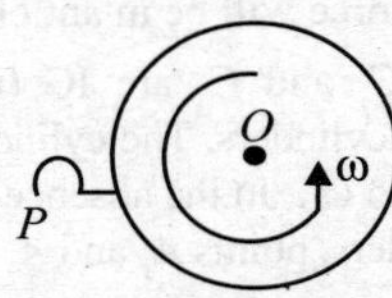

Fig. 2.479

Consequently, the angular momentum of the disc about P, just before and after the impact, remains the same

$\Rightarrow \quad L_2 = L_1$ (i)

where L_1 = angular momentum of the disc about P just before the impact

$$I_0\omega = \frac{1}{2}mr^2\omega$$

L_2 = angular momentum of the disc about P just after the impact

$$I_0\omega = \left(\frac{1}{2}mr^2 + mr^2\right)\omega' = \frac{3}{2}mr^2\omega'$$

Just before the impact, the disc rotates about O. But just after the impact, the disc rotates about P.

$\Rightarrow \quad \frac{1}{2}mr^2\omega = \frac{3}{2}mr^2\omega' \quad \Rightarrow \quad \omega' = \frac{1}{3}\omega$

62. a. $T_1 - mg = ma$ (i)

$r(T_2 - T_2) = I\alpha$ (ii)

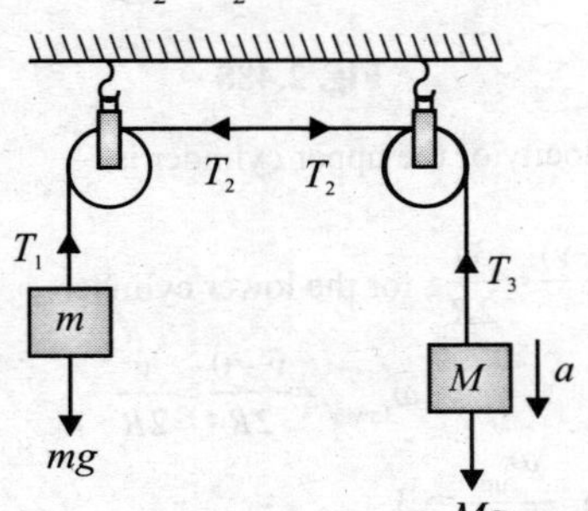

Fig. 2.480

$Mg - T_3 = Ma$ (iii)

$r(T_3 - T_2) = I\alpha$ (iv)

and $a = R\alpha$

From Eqs. (ii) and (iv), we get

$$T_3 - T_1 = \frac{2Ia}{R^2}$$

From Eqs. (i) and (iii), we get

$$(M-m)g = (M+m)a + T_3 - T_1$$

$$(M-m)g = (M-m)a + \frac{2Ia}{r^2}$$

$$\Rightarrow \quad a = \frac{(M-m)g}{\left(M+m+\frac{2I}{r^2}\right)}$$

63. a. The sphere will continue rolling because in the reference frame of incline (or car), the resultant force along the incline becomes zero.

$ma\cos\theta = m(g\tan\theta)\cos\theta = mg\sin\theta$

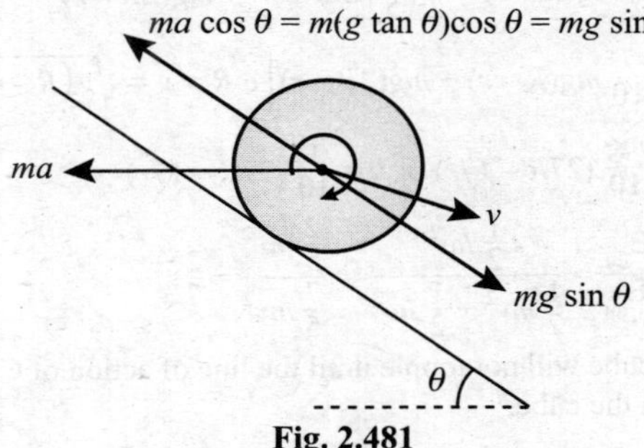

Fig. 2.481

64. b. See the figure. For tilting about A the clockwise torque (due to F) should be greater than the anticlockwise torque about A.

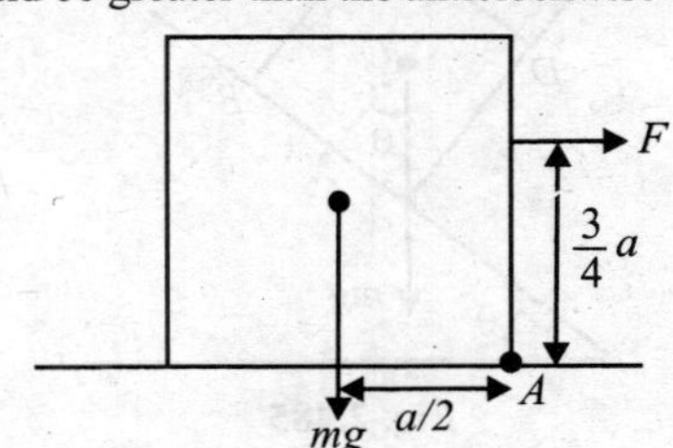

Fig. 2.482

$$mg \times \frac{a}{2} = \frac{F3a}{4};\ F = \frac{2mg}{3}$$

65. d. $mg = N_1$

$$\mu N_1 = N_2 \text{ or } N_2 = \mu mg$$

Taking moment about O, we get

$$\mu mgl\sin\theta = mg\frac{1}{2}\cos\theta$$

$$\tan\theta = \frac{1}{2\mu} \text{ or } \theta = \tan^{-1}\left(\frac{1}{2\mu}\right)$$

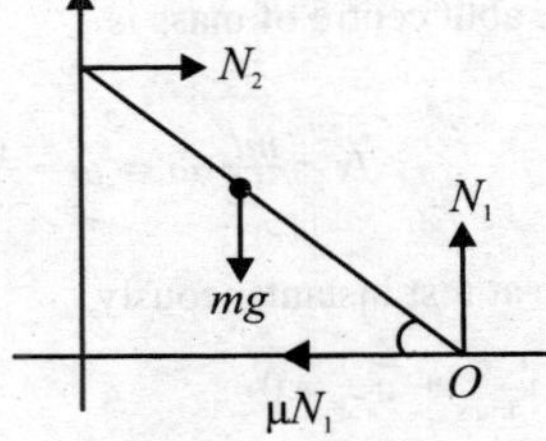

Fig. 2.483

66. d. The minimum velocity at P, top of the loop, should be $v = \sqrt{g(R-r)}$, if the sphere keeps on rolling at top $v = \omega R$

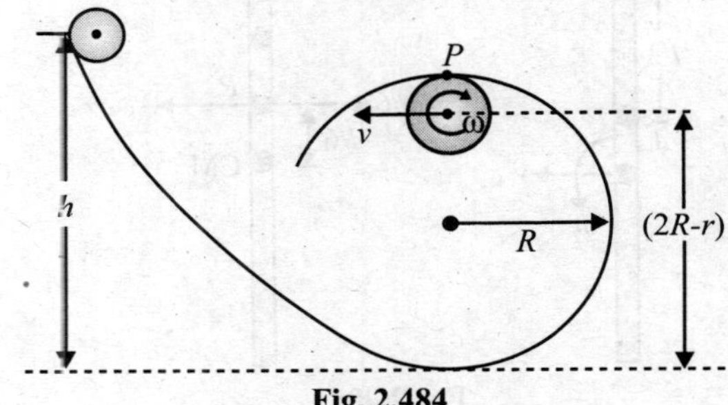

Fig. 2.484

$$mgh = \frac{1}{2}mv^2 + \frac{1}{2}I\omega^2 + mg(2R - r)$$

$$= \frac{1}{2}mg(R - r) + \frac{1}{2}\left(\frac{2}{5}\right)mR^2 2\omega^2 + mg(2R - r)$$

$$= \frac{7}{10}mg(R - r) + mg(2R - r)\left[\omega R = v = \sqrt{g(R - r)}\right]$$

$$= \frac{mg}{10}(27R - 17r) \text{ or } h = \frac{1}{10}(27R - 17r)$$

67. c. $\dfrac{KE_{rot}}{KE_{tot}} = \dfrac{\frac{1}{2}I\omega^2}{\frac{1}{2}mv^2 + \frac{1}{2}I\omega^2} = \dfrac{\frac{2}{5}mv^2}{\frac{7}{5}mv^2} = \dfrac{2}{7}$

68. c. The cube will not topple until the line of action of CM passes through the cube.

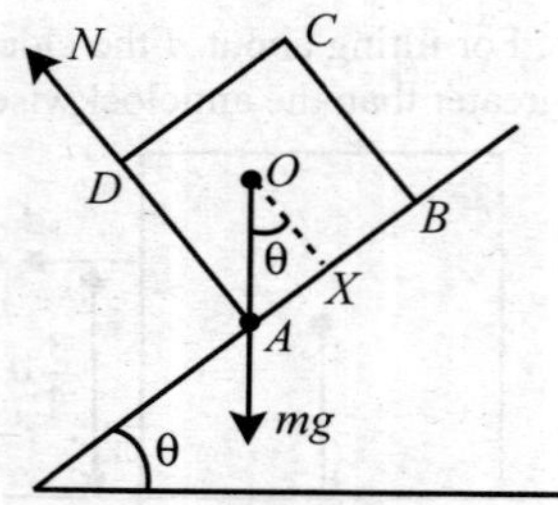

Fig. 2.485

That is, $\tan\theta = \dfrac{AX}{OX} = \dfrac{a/2}{a/2}$ or $\theta = 45°$

69. c. From the figure in question, position vector of mass

$$\vec{r} = R\hat{i} - \sqrt{l^2 - R^2}\hat{k}$$

Linear moment $\vec{p} = Mv\hat{j}$

$$\vec{L} = \vec{r} \times \vec{p}$$

Substituting the values, we get result (c)

70. a. Let the impulse due to the particle be J, $J = mv_{CM} \Rightarrow v_{CM} = \dfrac{J}{m}$

Angular impulse abut centre of mass is

$$Jx = \frac{ml^2}{12}\omega \Rightarrow \omega = \frac{12Jx}{ml^2}$$

For point A to be at rest instantaneously,

$$\vec{v}_{trans} + \vec{v}_{rot} = 0$$

or $$v_{CM} = \frac{\omega \ell}{6}$$

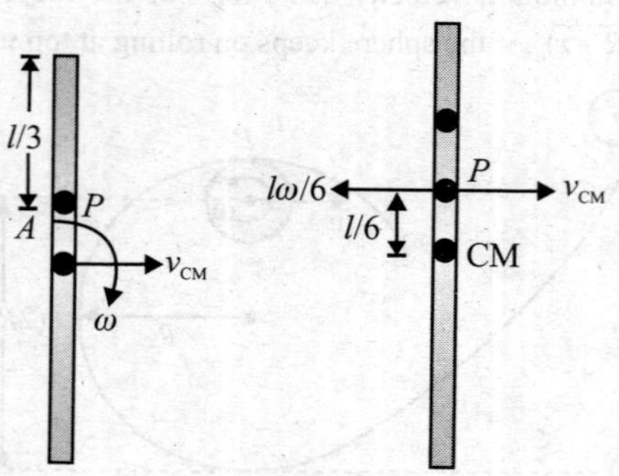

Fig. 2.486

Thus $$\frac{J}{m} = \frac{l}{6} \times \frac{12Jx}{ml^2}$$

which gives, $$x = \frac{l}{2}$$

71. b. As there is no friction, net force is F which is towards right. Torque due to this force will be in anticlockwise direction.

72. b. In Fig. 2.487, C_1 and C_2 are IC (instantaneous centre of rotation) of the two cylinders. The cylinder can be considered as rotating about C_1 and C_2. In the absence of slipping between the plank and the cylinders, points A_1 and A_2 have the same velocity. Angular velocity of the larger cylinder is $\dfrac{2v}{4R} = \dfrac{v}{2R}$

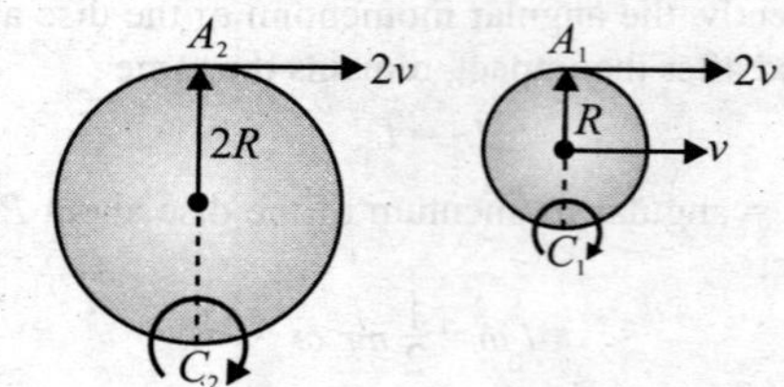

Fig. 2.487

$$v_{CM} = (2R)\left(\frac{v}{2R}\right) = v$$

73. b.

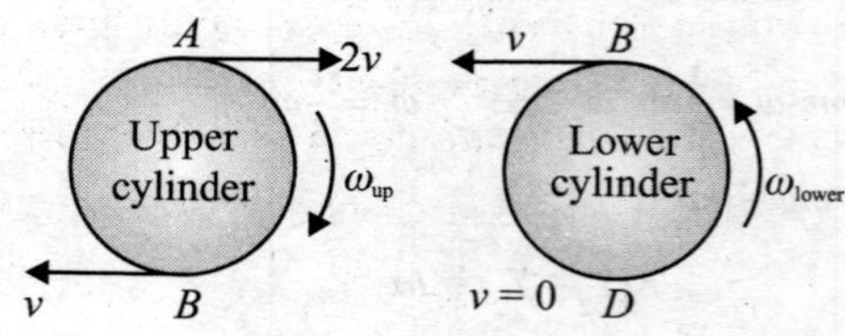

Fig. 2.488

Angular velocity of the upper cylinder is

$$\omega_{up} = \frac{2v - (-v)}{2R} = \frac{3v}{2R}; \text{ for the lower cylinder}$$

$$\omega_{lower} = \frac{v - 0}{2R} = \frac{v}{2R}$$

Hence, ratio, $\dfrac{\omega_{up}}{\omega_{lower}} = 3$

74. c. As rod is in euilibrium. The torque of all the forces about the CM is zero.

$$N_1\frac{l}{4} = N_2\frac{l}{4} \Rightarrow N_1 = N_2$$

Fig. 2.489

75. b. From parallel axis theorem,

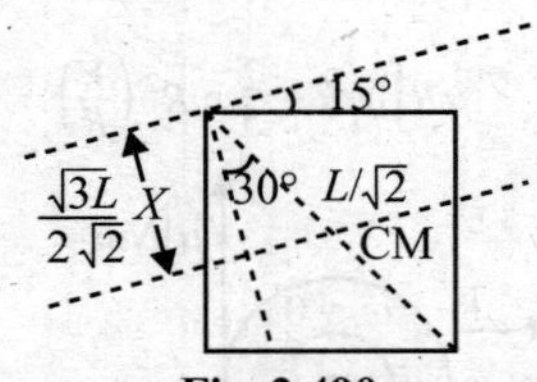

Fig. 2.490

$$I_1 = \frac{ML^2}{12}, I_2 = I_1 + Mx^2 = \frac{ML^2}{12} + M\left(\frac{\sqrt{3}L}{2\sqrt{2}}\right)^2 = \frac{11ML^2}{24}$$

76. b. Let us take any point P on x-y plane. The MI of the rod about O is

$$I_P = I_{CM} + Mr^2$$

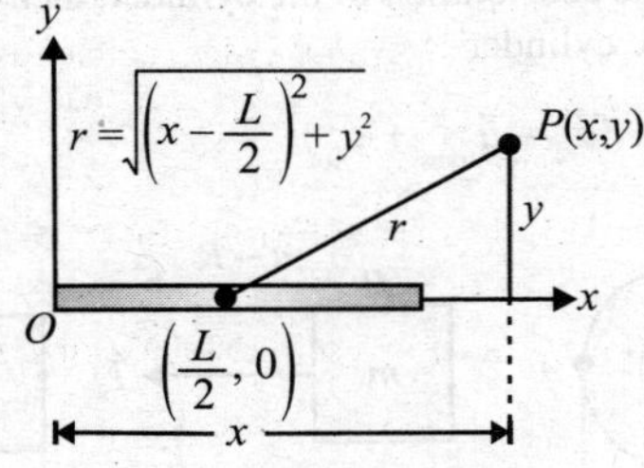

Fig. 2.491

Hence, $$I_P = \frac{ML^2}{12} + M\left[\left(x - \frac{L}{2}\right)^2 + y^2\right]$$

MI of the rod about O, $I_0 = \frac{ML^2}{3}$

Applying $I_P = I_0$, we get an equation of a circle.

77. d. The sphere is on the verge of toppling when a line of action of weight passes through the edge.

$$\cos\theta = \frac{(R-h)}{R} \quad \Rightarrow h = R - R\cos\theta$$

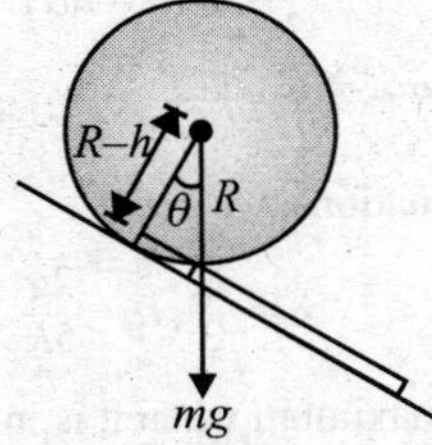

Fig. 2.492

78. a. If the spool is not to translate,

$$F\cos\theta = f \quad \text{(i)}$$

If the spool is not to rotate,

$$Fr = Ffr \quad \text{(ii)}$$

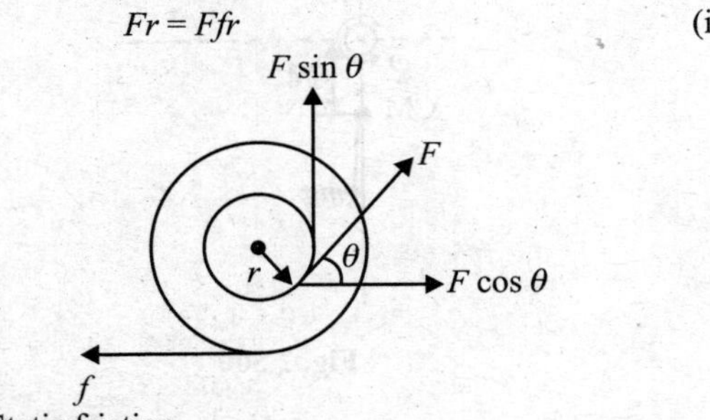

Fig. 2.493

or $$\frac{fR}{r}\cos\theta = f$$

or $$\cos\theta = \frac{r}{R} \quad \Rightarrow \theta = \cos^{-1}\left(\frac{r}{R}\right)$$

Also the line of action of F should pass through the bottom-most point.

79. c. Using conversation of angular momentum about O, we get

$$mvR = (mR^2 + mR^2)\omega;\ \omega = \frac{v}{2R}$$

80. b. In figure, for translational equilibrium in vertical direction,

$$\mu N_1 + N_2 = mg \quad \text{(i)}$$

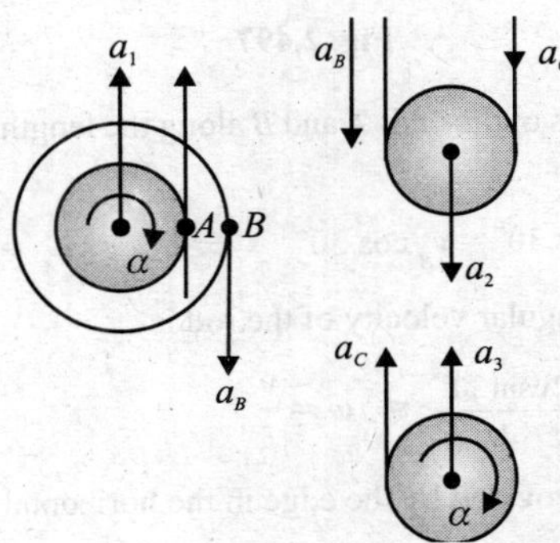

Fig. 2.494

In horizonal direction,

$$N_1 = \mu N_2 \quad \text{(ii)}$$

Solving, Eqs. (i) and (ii), $N_2 = \frac{mg}{1+\mu^2}$

The required friction is

$$\mu N_2 = f_a = \frac{3}{10}mg$$

In figure, the sphere has the tendancy to move towards the right;

$$N_1 = 0;\ N_2 = mg$$

$$f_b = \mu N_2 = \frac{mg}{3}$$

which gives $$\frac{f_a}{f_b} = \frac{9}{10}$$

81. d. Apply conservation of angular momentum about the hinge, we get

$$mvR = \frac{m}{2}R^2\omega + m(R\omega)R$$

$$R\omega = \frac{2v}{3} = \frac{2\times 5}{3} = \frac{10}{3}\text{ m/s}$$

82. c. As the upper end of the string passing through A is connected with the roof (i.e., at rest). Hence, acceleration of A is zero.

For A; $a_1 - \alpha r = 0$ (i)

Acceleration of point B; $a_B = 3\alpha r - a_1$ (ii)

a_1
a_B
a_C
A
B
α
a_2
a_C
a_3
a_B
α

Fig. 2.495

From Eqs. (i) and (ii), $a_B = 2a_1$

For moving pulley,

$$\frac{a_B + a_C}{2} = a_2 \quad \text{(iii)}$$

or $a_C = 2a_2 - a_B = 2a_2 - 2a_1$

For disc,

$$a_C = a_3 + 2r\alpha \quad \text{(iv)}$$

From Eqs. (iii) and (iv), we get $2a_2 - a_3 = 4a_1$

83. c. Let the velocity components of point B along x and y directions are v_x and v_y. As length of the rod is constant, along the length of the rod, velocity of each point must be the same (rigid body constraint).

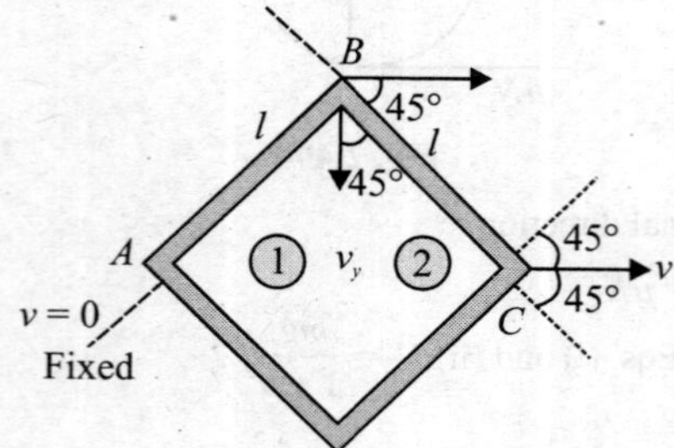

Fig. 2.496

Along rod AC, $\frac{v_x}{\sqrt{2}} + \frac{v_y}{\sqrt{2}} = \frac{V}{\sqrt{2}}$

Along rod AB, $\frac{v_x}{\sqrt{2}} - \frac{v_y}{\sqrt{2}} = 0$

$$v_x = v_y = \frac{V}{2}$$

84. b.

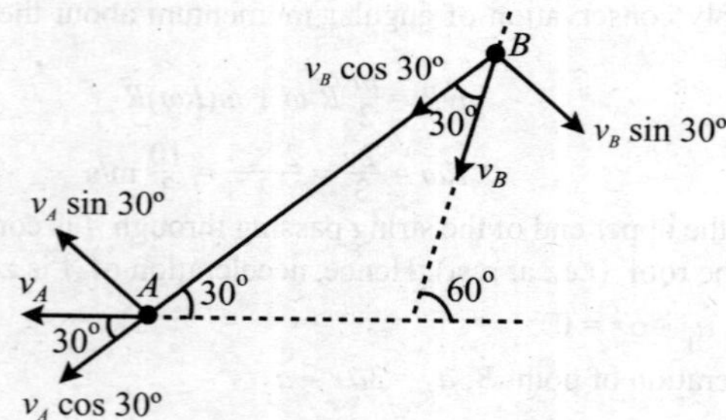

Fig. 2.497

The velocities of the ends A and B along the length of rod should be the same.

Hence, $v_A \cos 30° = v_B \cos 30° \quad \Rightarrow \quad v_A = v_B = v$

Hence, the angular velocity of the rod is

$$\omega = \frac{(v_{AB})_\perp}{l} = \frac{2v \sin 30°}{l} \Rightarrow \omega = \frac{v}{\ell}$$

85. a. Impulse provided by the edge in the horizontal direction:

$$-\int Ndt = -mV' - (mV) \quad \text{(i)}$$

Friction impulse in the vertical direction:

$$\mu R \int Ndt = \frac{2}{5} mR^2 \left(\frac{V}{R}\right) \quad \text{(ii)}$$

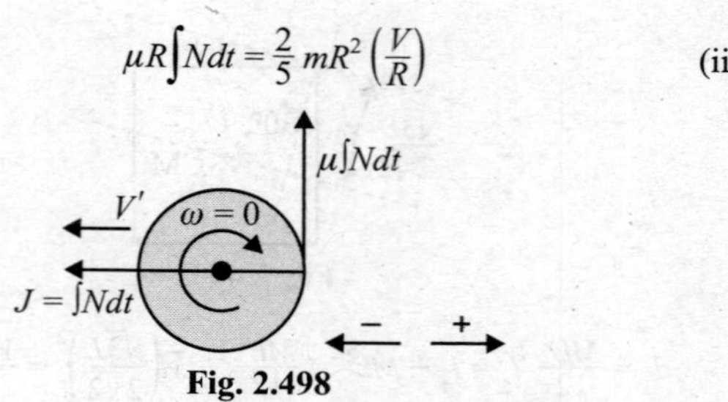

Fig. 2.498

From Eqs. (i) and (ii), we get

$$\int Ndt = 2mV \text{ and } V' = V$$

86. b. Let a be the acceleration of the cylinder, then acceleration of point A on the cylinder

$$\vec{a}_A = \vec{a}_{trans} + \vec{a}_{rot}$$

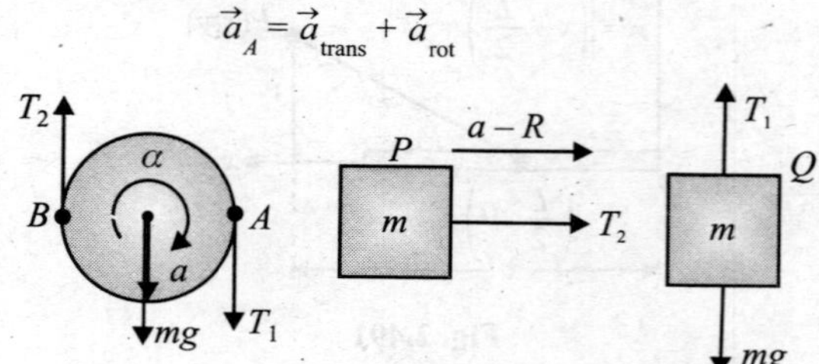

Fig. 2.499

$$|\vec{a}_A| = a + R\alpha = a_Q$$

Similarly, the acceleration of point B on the cylinder is

$$|\vec{a}_B| = a - R\alpha = a_P$$

$$mg - T_1 = m(a + R\alpha) = a_P \quad \text{(i)}$$

$$T_1 + mg - T_2 = ma \quad \text{(ii)}$$

$$T_2 = m(a - R\alpha) \quad \text{(iii)}$$

$$T_1R + T_2R = \frac{mR^2}{2}\alpha \quad \text{(iv)}$$

On solving these equations, we get

$$a = \frac{2g}{3}, \alpha = \frac{2g}{5R}$$

87. c. KE of the rod is maximum when it is in the vertical position. From conservation of energy, we get

$$\frac{mgl}{4} = \frac{1}{2}\left[\frac{ml^2}{12} + \frac{ml^2}{16}\right]\omega^2$$

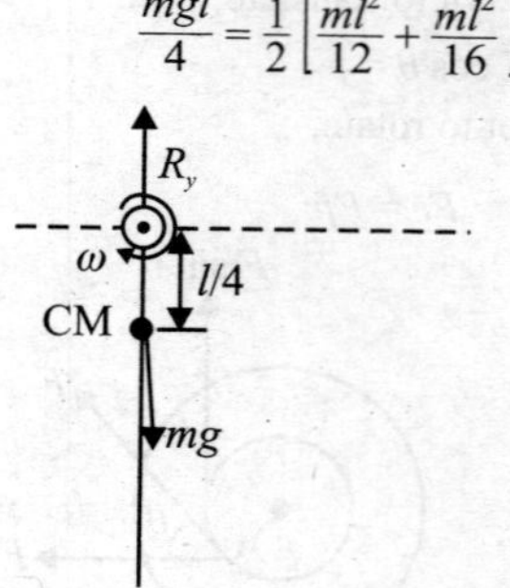

Fig. 2.500

$$\frac{\omega^2 l}{4} = \frac{6g}{7}$$

and $R_y - mg = m(l/4)\,\omega^2$

$$R_y = mg + m\left(\frac{6g}{7}\right)$$

$$= \frac{13mg}{7}$$

88. d. $\frac{MgL}{2}\cos\theta = \frac{ML^2}{3}\alpha$

$$\alpha = \frac{3}{2}\frac{g\cos\theta}{L}$$

$$a = \alpha L = \frac{3}{2}g\cos\theta$$

89. b. When the ring is at the maximum height, the wedge and the ring have the same horizontal component of velocity. As all the surfaces are smooth, in the absence of friction between the ring and the wedge surface, angular velocity of the ring remains constant.

For conservation of mechanical energy, we get

$$\frac{1}{2}mv^2 + \frac{1}{2}I\omega^2 = \frac{1}{2}mv'^2 + \frac{1}{2}I\omega^2 + \frac{1}{2}mv'^2 + mgh$$

where v' is final common velocity;

$v' = \frac{v}{2}$ (from conservation of momentum) and $h = \frac{v^2}{4g}$

90. d. $\omega_{rod} = \frac{(v_{rel})_\perp}{r}$

$(V_{rel})_\perp$ is the velocity of one point w.r.t. other perpendicular to rod.

$\omega_{rod} = \frac{3v - v}{r}$, and '$r$' being the distance between the points.

$$\omega_{rod} = \frac{2v}{r}$$

91. d. Given $\alpha_A = 2\alpha = 5$ m/s^2 $\Rightarrow$ $\alpha = \frac{5}{2}$ m/s^2

Hence, acceleration of B, $a_B = 1(\alpha) = \frac{5}{2}$ m/s^2

92. d. As the inclined plane is smooth, the sphere can never roll, rather it will just slip down. Hence, the angular momentum remains conserved about any point on a line parallel to the inclined plane and passes through the centre of the ball.

93. d. As $\sum\tau = 0$, angular momentum and linear momentum remain conserved. As the two balls will move radially out, I changes. In order to keep the angular momentum ($L = I\omega$) conserved, angular speed (ω) should change.

94. c. If the track is smooth (case A), only translational kinetic energy changes to gravitational potential energy.

But, if the track is rough (case B), both translational and rotational kinetic energies change to potential energy.

Therefore, potential energy (= mgh) will be more in case B than in case A.

Hence, $h_1 > h_2$.

95. c. Radius of curvature = $\frac{(\text{Velocity})^2}{\text{Normal acceleration}}$

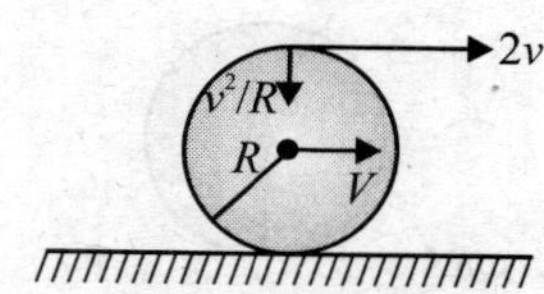

Fig. 2.501

$$\text{Radius of curvature} = \frac{(2v)^2}{v^2/R} = 4R$$

96. b. Here, $u = V_0$, $\omega_0 = -\frac{V_0}{2R}$

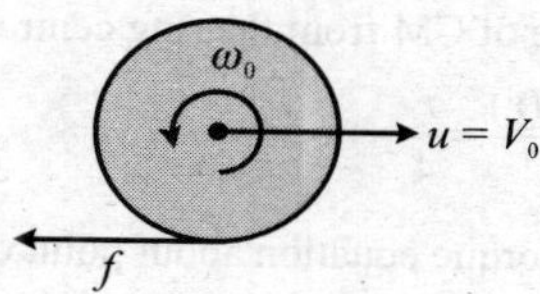

Fig. 2.502

At pure rolling:

$$V = V_0 - \left(\frac{f}{m}\right)t$$

and

$$\frac{V}{R} = -\frac{V_0}{2R} + \left(\frac{f}{mR}\right)t$$

(in pure rolling, $V = R\omega$)

$$\left(\alpha = \frac{\tau}{I} = \frac{fR}{mR^2} = \frac{f}{mR}\right)$$

$$V_0 - V = V + \frac{V_0}{2}$$

$$2V = \frac{V_0}{2} \Rightarrow V = \frac{V_0}{4}$$

97. b. The two forces along the y-direction balance each other. Hence, the resultant force is $2F$ along the x-direction.

Let the point of application of the force be at $(0, y)$.

(By symmetry, x-coordinate will be zero).

For rotational equilibrium:

$F(a) + F(a) + F(a + y) - F(a - y) = 0$

$\Rightarrow$ $y = -a$

Alternative method:

Torque will only be produced by the two forces acting along y-direction in anticlockwise direction. To balance this torque, we should apply a force $2F$ in order to produce a torque in the clockwise direction, which is only possible if we apply a force at a point below the x-axis.

Then $\tau = F(a) + F(a) - 2F \times y = 0$

$\Rightarrow$ $y = a$

98. d. As the disc is in combined rotation and translation, each point has a tangential velocity and a linear velocity in the forward direction.

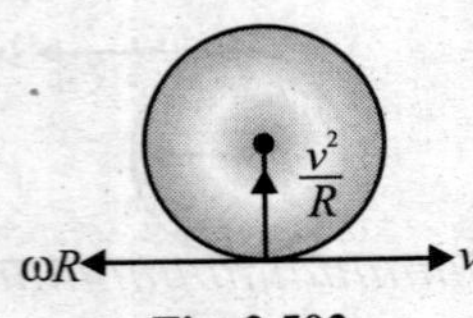

Fig. 2.503

From the figure,

V_{net} (for the lowest point) $= v - R\omega = v - v = 0$

and acceleration $\frac{v^2}{R} + 0 = \frac{v^2}{R}$

(Since the linear speed is constant)

99. b. The distance of CM from the ring centre O

$$x = \frac{3m(0) + m(r)}{3m + m} = \frac{r}{4}$$

We can apply torque equation about point of contact as the ring is rolling.

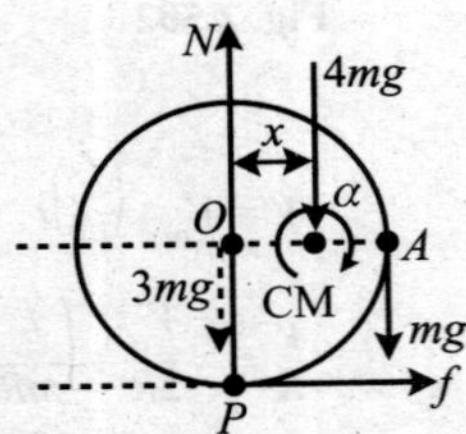

Fig. 2.504

$$\tau_P = I_P\alpha$$

$$4mg\left(\frac{r}{4}\right) = \left[(3mr^2 + mr^2) + m(AP)^2\right]\alpha$$

$$\Rightarrow mg\,r = \left[4mr^2 + m(\sqrt{2}\,r)^2\right]\alpha$$

$$\Rightarrow mg\,r = 6mr^2\alpha \quad \Rightarrow \quad \alpha = \frac{g}{6r}$$

100.a. Applying conservation of linear momentum,

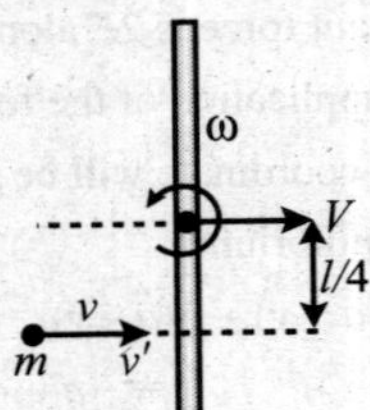

Fig. 2.505

$$mv = mv' = mV \quad \Rightarrow \quad v = v' + V \qquad \text{(i)}$$

Applying conservation of angular momentum about point of collision.

$$0 = \left(\frac{m\ell^2}{12}\right)\omega - mV\left(\frac{\ell}{4}\right) \quad \Rightarrow \quad \ell\omega = 3V \qquad \text{(ii)}$$

Applying restituting equtionn,

$$(u_1 - u_2)_n = (v_2 - v_1)_n \Rightarrow (v - 0) = (V - v' \qquad \text{(iii)}$$

Solving Eqs. (i), (ii), and (iii) we get $V = v$ and $\omega = \frac{3v}{\ell}$

Time taken to complete three revolutions ($\theta = 6\pi$)

$$t = \frac{\theta}{\omega} = \frac{6\pi}{\omega} = \frac{6\pi\,\ell}{3v} = \frac{2\pi}{v}$$

Hence, distance travelled by the centre of the rod is

$$s = Vt = v\left(\frac{2\pi l}{v}\right) = 2\pi l$$

101. d. Since linear acceleration is same for all ($a = Mg\sin\theta - \mu\,Mg\cos\theta$) as they have same mass 'M' and same 'μ'.

Hence, all will reach the bottom simultaneously.

102. b. For all the bodies, torque is the same.

Now, $\quad \text{KE} = \frac{1}{2}mv^2 + \frac{L^2}{2I}$

Linear velocity 'v' is the same for all as the same force acts on them.

Therefore, the more value of moment of inertia implies the lesser kinetic energy.

Among all, the hollow sphere has the maximum moment of inertia, $I = \left(\frac{2}{3}MR^2\right)$

103. a. The tendency of rotating will be about point C. For the minimum force, the torque of F about C has to be equal to the torque of mg about C.

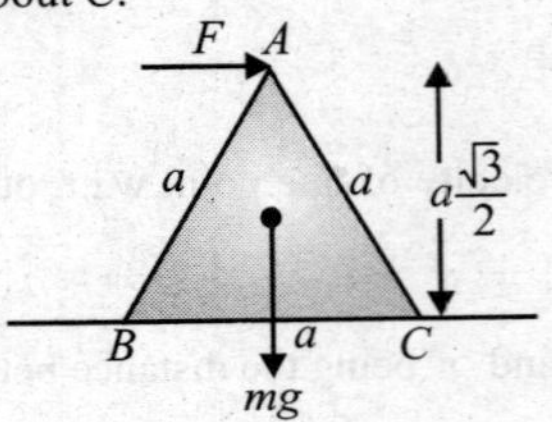

Fig. 2.506

$$F\left(a\frac{\sqrt{3}}{2}\right) = mg\left(\frac{a}{2}\right) \Rightarrow F = \frac{mg}{\sqrt{3}}$$

104. b.

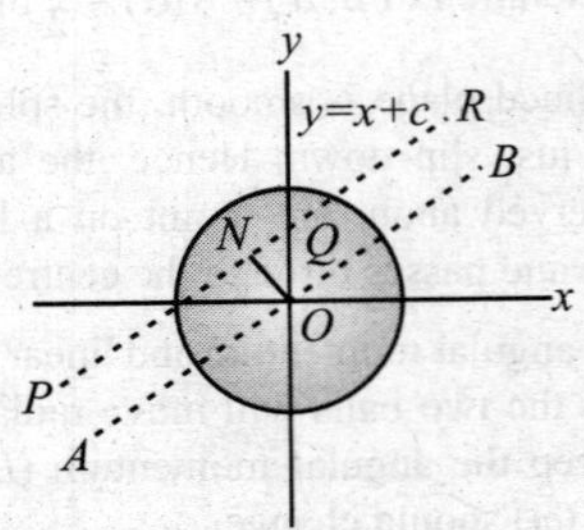

Fig. 2.507

$$I_{PQR} = I_{AOB} + M(ON)^2$$

$$= \frac{1}{2}MR^2 + M\left(\frac{c}{\sqrt{2}}\right)^2$$

But $\quad = \frac{1}{4}MR^2 = I_2$

$$c = \pm\frac{R}{\sqrt{2}}$$

105. d. As torque = change in angular momentum

$$F\Delta t = mv \text{ (linear momentum)} \qquad \text{(i)}$$

and $\left(F\frac{l}{2}\right)\Delta t = \frac{ml^2}{12}\omega$ (angular momentum) (ii)

Dividing Eqs. (i) and (ii), we get

$$2 = \frac{12v}{\omega l} \Rightarrow \omega = \frac{6v}{l}$$

Using $S = ut,$

Displacement of CM is $\frac{\pi}{2} = \omega t = \left(\frac{6v}{l}\right)t$

and $x = vt$

Dividing, we get

$$\frac{2x}{\pi} = \frac{l}{6} \Rightarrow x = \frac{\pi l}{12}$$

Coordinates of A will be $\left[\frac{\pi l}{12} + \frac{l}{2}, 0\right]$.

106. a. By conservation of angular momentum about pivot

$$L = I\omega$$

$$\frac{mvd}{2} = \left[\frac{Md^2}{12} + m\left(\frac{d}{2}\right)^2\right]\omega$$

$$= \left(\frac{md^2}{2} + \frac{md^2}{4}\right)\omega$$

$$= \frac{3}{4}md^2\omega \Rightarrow \frac{2}{3}\frac{v}{d} = \omega$$

107. b. The magnitude of angular momentum of the particle about $O = mvd$.

Since speed v of the particle increases, its angular momentum about O increases.

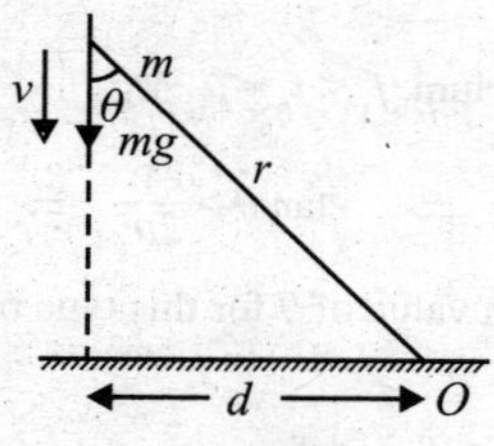

Fig. 2.508

Magnitude of inertia of the particle about $O = mr^2$

Hence, MI of the particle about O decreases.

Angular velocity of the particle about $O = \frac{v\sin\theta}{r}$

Therefore, v and $\sin\theta$ increase and r decreases.

Therefore, angular velocity of the particle about O increases.

108. b. As the normal force exerted by the horizontal surface passes through point B, the external torque on the ball is zero about point B. So the angular momentum of the ball is conserved about point B. ($\because \tau = \frac{dL}{dt}$)

109. a. From conservation of energy, the kinetic energy of the ball at the lowest point is (v_c = speed of centre of ball)

$$\frac{1}{2}mv_c^2 + \frac{1}{2}\times\frac{2}{5}mv_c^2 = mgR$$

or $$\frac{7}{10}mv_c^2 = mgR$$

Since the net tangential force on the sphere at the lowest point is zero, the net force on the sphere at the lowest position is

$$= \frac{mv_c^2}{R} = \frac{10}{7}mg \text{ upwards}$$

110. b. The free-body diagram of the block can be drawn as shown. As the body has to move in pure translational motion, the torque about the centre of gravity must be zero.

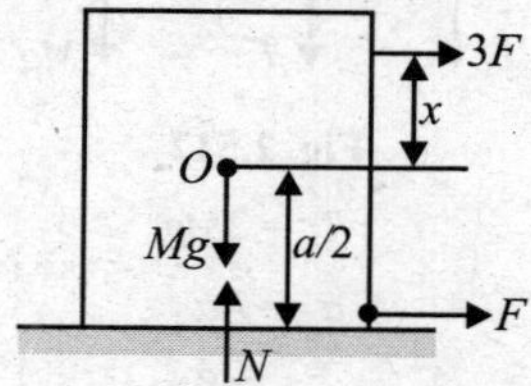

Fig. 2.509

$$3F \times x = F \times \frac{a}{2} \Rightarrow x = \frac{a}{6}$$

111. c. The free-body diagram of the rod is shown in the figure. (Force exerted by the hinge on the rod are not shown.)

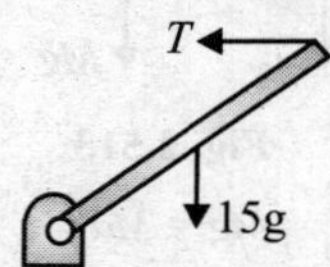

Fig. 2.510

Take torque about the hinge,

$$T \times 3 - 15\,g \times 2 = 0 \Rightarrow T = 100\text{ N}$$

112. c. Let v be the velocity of the centre of mass of the sphere and ω be the angular velocity of the body about an axis passing through the centre of mass.

$$J = Mv$$

$$J(h - R) = \frac{2}{5}MR^2 \times \omega$$

From the above two equations, $v(h - R) = \frac{2}{5}r^2\omega$

From the condition of pure rolling, $v = R\omega$

$$h - R = \frac{2R}{5} \Rightarrow h = \frac{7R}{5}$$

113. b. $F = Ma$

$$FR = I\alpha$$

For pure rolling, $a = R\alpha$

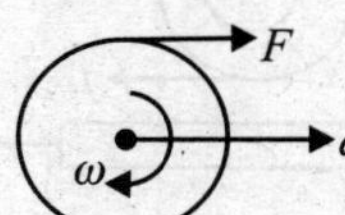

Fig. 2.511

Solving the above equations, we get, $I = MR^2$.

So the object may be a ring or a hollow cylinder.

114. b. $I = \frac{ML^2}{12} = \frac{6\times 4\times 4}{12} = 8\text{ kg m}^2$

From $\vec{\tau} = \vec{r}\times\vec{F} = [2\hat{i}\times(3\hat{i} + 2\hat{j} + 6\hat{k})] = 4\hat{k} - 12\hat{j}$

$$\tau_y = I\alpha \Rightarrow -12j = 8\alpha \Rightarrow \alpha = -\frac{3}{2}j\text{ rad/s}^2$$

115. b. In Fig. 2.512, $(T_2 - T_1)R = \frac{MR^2}{2}\alpha_A$

$$T_1 - Mg = Ma_A$$

Fig. 2.512

$$T_2 = 2Mg$$

$$a_A = R\alpha_A$$

$$\alpha_A = \frac{2g}{3R}$$

From Fig. 2.513,

$$(T_2 - T_1) \times R = \frac{MR^2}{2}\alpha_B$$

Fig. 2.513

$$T_1 - Mg = Ma_B$$

$$2Mg - T_2 = 2Ma_B$$

$$a_B = R\alpha_B$$

$$\alpha_B = \frac{2g}{7R}$$

So, $\alpha_A > \alpha_B$

116. b. There can be a total of eight impulses given to the cylinder in 30 s. Using $\vec{J} = \vec{L}_f - \vec{L}_i$ for every impulse, we get

$$(8 \cdot J) = I \cdot \omega - 0 \Rightarrow \omega = \frac{8J}{I}$$

117. d. $\frac{mv^2}{2} + \frac{I\omega^2}{2} = mg \times \frac{3v^2}{4g} \Rightarrow I = \frac{mR^2}{2}$

For pure rolling condition, $v = R\omega$.

118. b. The cylinder is having a fixed axis, so it can perform rotational motion only.

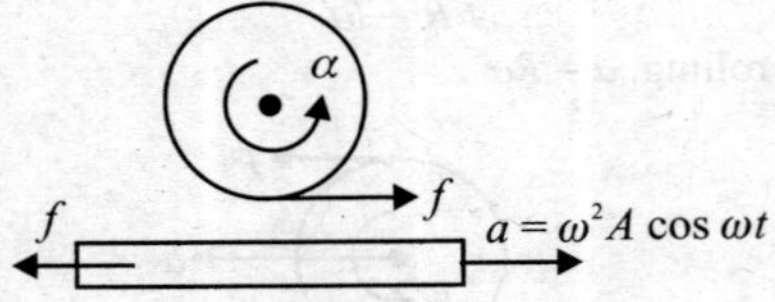

Fig. 2.514

$$fR = \frac{MR^2}{2}\alpha \Rightarrow f = \frac{MR}{2}\alpha$$

For no slipping,

$R\alpha$ = acceleration of the platform

$$\frac{2f}{M} = \omega^2 A \cos \omega t$$

For the maximum torque, f would be maximum and $f_{max} = M\omega^2 A/2$

So the maximum torque,

$$\tau_{max} = f_{max} \times R = \frac{M^2\omega^2 AR}{2}$$

119. a. Conserving angular momentum about O (just before and just after) is

$$mv\frac{a}{2} = \left[\frac{ma^2}{6} + \frac{ma^2}{2}\right]\omega \Rightarrow \omega = \frac{3v}{4a}$$

120. b. For the equilibrium of the entire structure, $N_1 + N_2 = 2Mg$

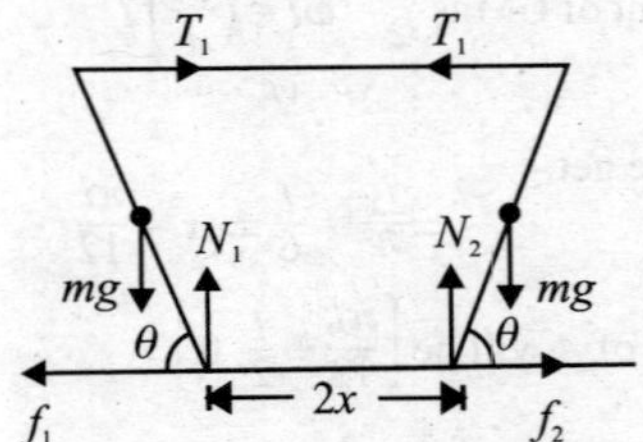

Fig. 2.515

$$Mg\left(\frac{l}{2}\cos\theta\right) + N_2 \times 2x - Mg\left(\frac{l}{2}\cos\theta + 2x\right) = 0$$

Given $N_2 = Mg, N_1 = Mg$

For individual boards, $T_1 = f_1, N_1 = Mg$

and $T_1 \times l \sin\theta = Mg\frac{l}{2}\cos\theta$

$$f_1 = T_1 = \frac{Mg}{2\tan\theta}$$

For safe equilibrium, $f_1 < f_L = \mu_s Mg$

$$\frac{Mg}{2\tan\theta} = \mu_s Mg \Rightarrow \tan\theta > \frac{1}{2\mu_s} \Rightarrow \theta = 45°$$

So the minimum value of θ for this type of arrangement is 45°.

121. a. $f = Mg \sin\theta$

For no sliding, $f < f_L$

$$mg\sin\theta < \mu_s mg\cos\theta$$

$$\tan\theta = \mu_s$$

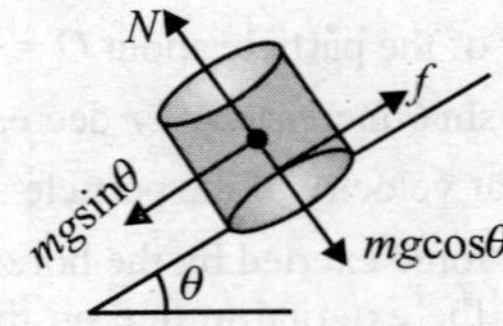

Fig. 2.516

For rolling friction, force $f = \frac{mg\sin\theta}{1 + \frac{R^2}{k^2}}$

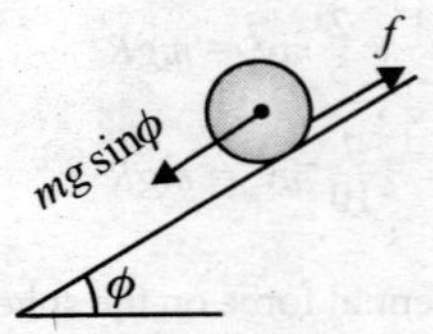

Fig. 2.517

$$\Rightarrow \qquad f = \frac{mg \sin \phi}{3}$$

For pure rolling, $\frac{mg \sin \phi}{3} < \mu_s \, mg \cos \phi$

$\tan \phi < 3\mu_s$, so $\frac{\tan \phi}{\tan \theta} = 3$

122. b. The free-body diagram of the box w.r.t. the truck is as shown in Fig. 2.518.

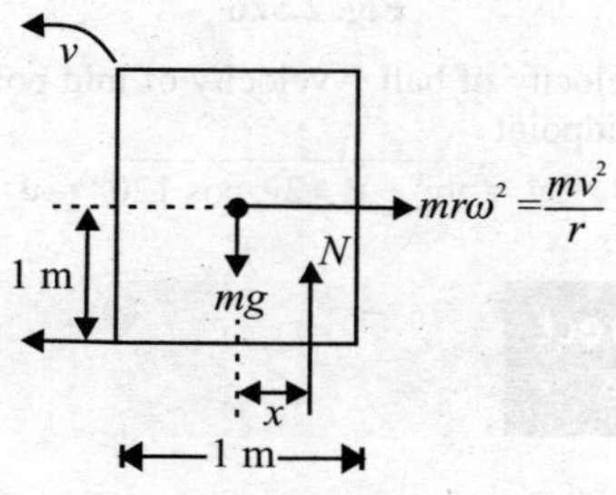

Fig. 2.518

For vertical equilibrium, $mg = N$

For horizontal equilibrium, $f = (mv^2)/r$

For rotational equilibrium, $f \times 1 = N \times x$

For no tipping to take place, $x < (1/2)$ m

So, $mg \times x = \frac{mv^2}{r} \Rightarrow x = \frac{v^2}{rg} < \frac{1}{2} \Rightarrow v < \sqrt{\frac{rg}{2}} = 10$ m/s

123. a. $\vec{b} = \vec{a} + \vec{b}t^2$

$$\frac{dL}{dt} = \vec{\tau} + 2\vec{b}t$$

Let the angle between $\vec{L}$ and $\vec{\tau}$ be 45° at $t = t_0$, then

$$\tan 45^\circ = \frac{|\vec{b}|\, t_0^2}{|\vec{a}|}$$

$$t_0 = \sqrt{\frac{|\vec{a}|}{|\vec{b}|}}, \text{ so } \vec{\tau} \ (\text{at } t = t_0) = \sqrt{\frac{a}{b}} \times \vec{b}$$

124. c. If instantaneous centre of rotation is P at a distance r from the centric line OC, then from the definition of instantaneous centre of rotation, the velocity of point P should be zero.

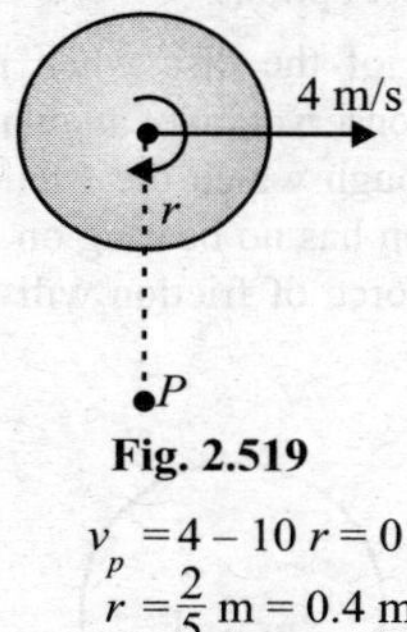

Fig. 2.519

$$v_p = 4 - 10\, r = 0$$

$$r = \frac{2}{5} \text{ m} = 0.4 \text{ m}$$

So, instantaneous centre of rotation is at distance 0.2 m from O (below O) on line OC.

125. c. The free-body diagram of the plank and the cylinder is as shown in Fig. 2.520.

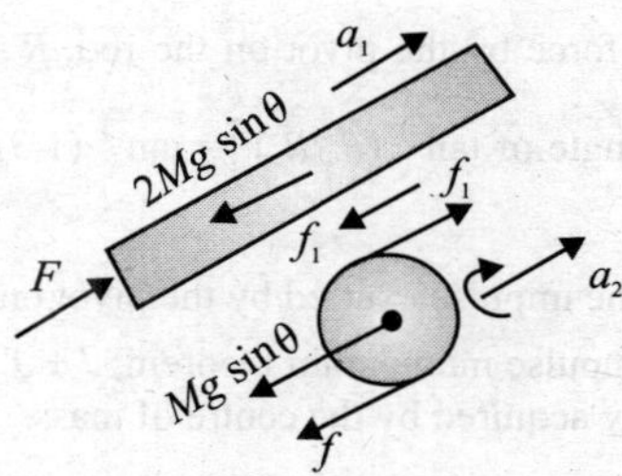

Fig. 2.520

For plank, $F - f_1 - 2\, Mg \sin \theta = 2\, Mg_1$

For cylinder, $f_1 - f_2 - Mg \sin \theta = Ma_2$

$$(f_1 + f_2)R = \frac{MR^2}{2} \alpha$$

For no slipping between the plank and the cylinder, $a_2 + R\alpha = a_1$

For no slipping between the cylinder and the incline, $a_2 = R\alpha$

After solving the above equation, $f_1 = \frac{3F + 2Mg \sin \theta}{19}$

126. c. The angular velocity of the rod about the pivot when it passes through the horizontal position is given by $mg \times \frac{L}{2} \sin 30^\circ = \frac{mL^2}{3} \times \frac{\omega^2}{2}$

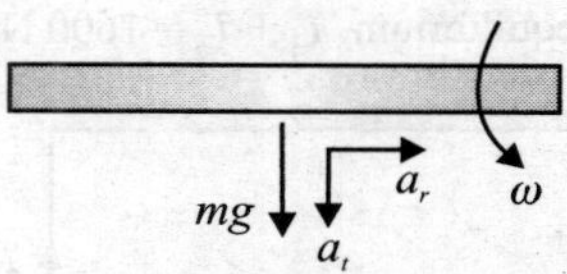

Fig. 2.521

$$\omega = \sqrt{\frac{3g}{2L}}$$

Radial acceleration of the centre of mass (as centre of mass is moving in a circle of radius $L/2$) is given by $a_r = \omega^2 \frac{L}{2} = \frac{3g}{4}$

Torque about pivot, in the horizontal position, is $\tau = mg \frac{L}{2} = I\alpha$

$$\alpha = \frac{mgL/2}{mL^2/3} = \frac{3g}{2L}$$

Tangential acceleration of the centre of mass, $a_t = \frac{L}{2} \alpha = \frac{3g}{4}$

Draw the FBD of the rod at an instant when it passes through the horizontal position. Use Newton's second law of equation.

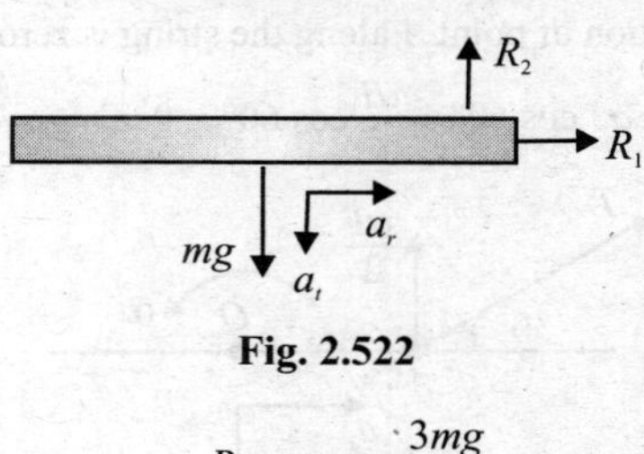

Fig. 2.522

$$R_1 = ma_r = \frac{3mg}{4}$$

$$mg - R_2 = m \times a_t = \frac{3mg}{4}$$

$$R_2 = \frac{mg}{4}$$

So, reaction force by the pivot on the rod, $R = R_1^2 + R_2^2 = \sqrt{10}$ $mg/4$ at an angle of $\tan^{-1}(R_2/R_1)$ $[= \tan^{-1}(1/3)]$ with the horizontal.

127. b. Let J' be the impulse exerted by the pivot on the rod.

Then from impulse momentum theorem, $J + J' = mv_0$, where v_0 is the velocity acquired by the centre of mass.

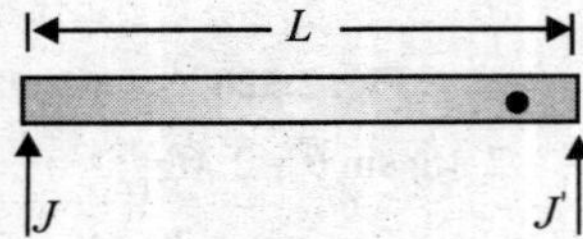

Fig. 2.523

Now apply, angular impulse = change in angular momentum about the pivot.

$$J \times L = \frac{ML^2}{3}\omega$$

$J \times \frac{ML}{3}\omega$, where ω is the angular velocity of the rod.

$$v_0 = \frac{\omega L}{2} \Rightarrow Mv_0 = \frac{M\omega L}{2}$$

$$J + J' = \frac{3J}{2} \Rightarrow J' = \frac{J}{2}$$

128. c. For vertical equilibrium, $T_1 + T_2 = 1600$ N

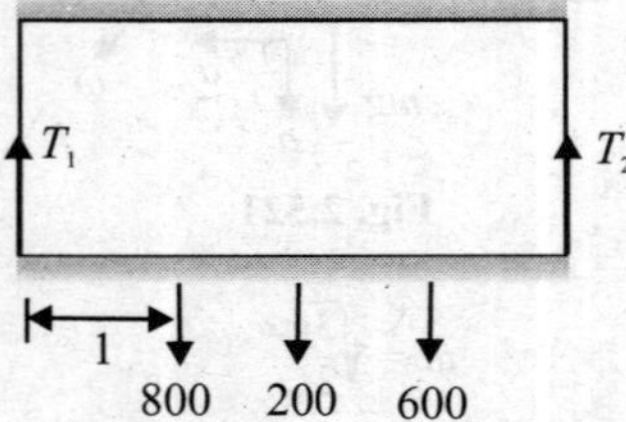

Fig. 2.524

Take moment about O,

$T_1 \times 2.5 + 600x$

$T_2 \times 2.5 + 800 \times 1.5$

$$T_1 = 1040 - 120x$$

and $$T_2 = 560 + 120x$$

For safe working, $T_1 < 1040$ N and $T_2 < 1040$ N

$x > 0$ and $x < 2$, so the range is $0 < x < 2$.

129. d. Acceleration of point A along the string is zero. Therefore,

$$a_x \cos 30° - a_y \cos 60° + \frac{\alpha l}{2}\cos 60° = 0$$

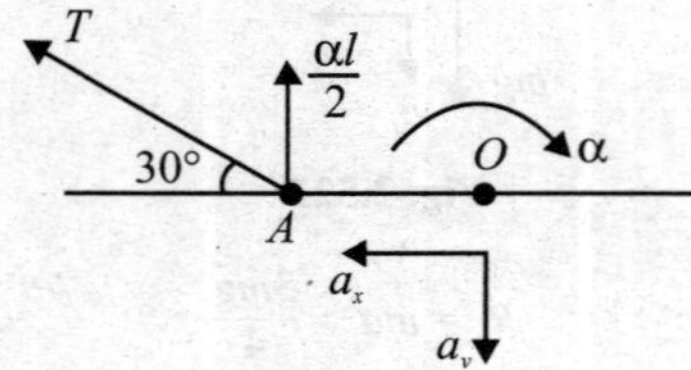

Fig. 2.525

130. b. Using conservation of angular momentum about the midpoint, we can find the velocity of each ball relative to the midpoint, i.e., $v' = v_0$.

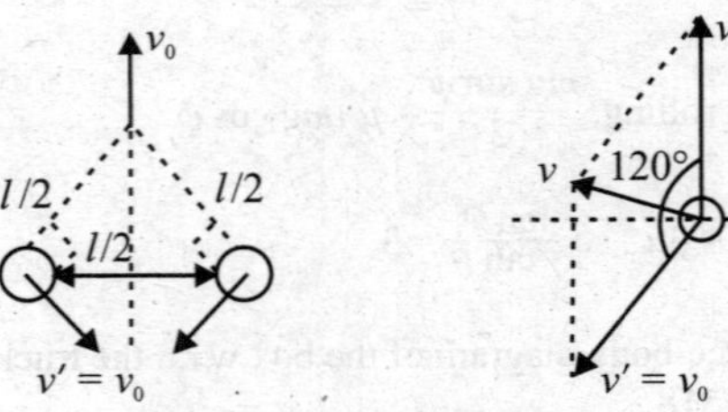

Fig. 2.526

Resultant velocity of ball = velocity of mid point + velocity of ball w.r.t. midpoint

or $$v = \sqrt{v_0^2 + v_0^2 + 2v_0^2 \cos 120°} = v_0$$

Multiple Correct Answers Type

1. a., b.

$$v_0 - r\omega_0 = 3v \quad \text{(i)}$$

$$v_0 + r\omega_0 = -v \quad \text{(ii)}$$

From Eqs. (i) and (ii), we get

$$2v_0 = 2v \Rightarrow v_0 = v$$

$$v_0 + v_0 = -r\omega_0 \Rightarrow \omega_0 = \frac{2v}{r}$$

2. a., b., c., d. Acceleration of centre of mass,

$$a_c = \frac{F}{M}\frac{1}{1+\frac{K^2}{R^2}} = \frac{5F}{7M}\left(\because K = \sqrt{\frac{2}{5}}R\right)$$

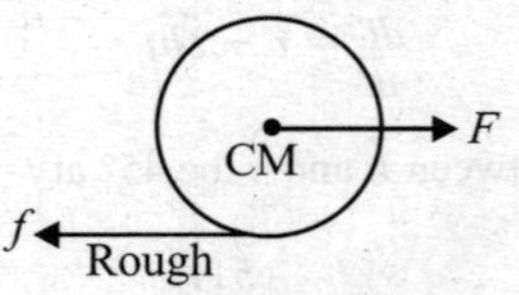

Fig. 2.527

Frictional force, $$f = F\left[\frac{1}{1+\frac{R^2}{K^2}}\right] = \frac{2F}{7}$$

Since $f = +$ ve, friction is directed backward.

Options (a), (b), (c) and (d) are incorrect. So options (a), (b), (c) and (d) are the correct options.

3. c., d. The velocity of the disc when rolling begins can be obtained using the conservation of angular momentum principle about the point through which the friction force acts. So, the coefficient of friction has no bearing on the final velocity. The work done by the force of friction will simply be changed to kinetic energy.

4. a., b., c., d. At A:

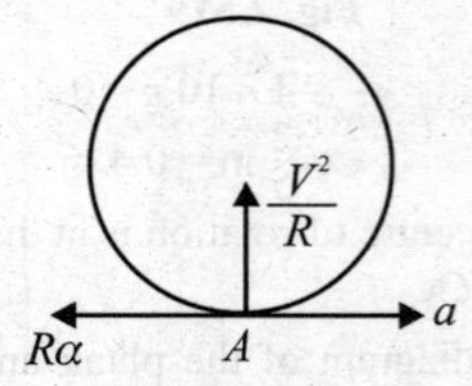

Fig. 2.528

for a rolling wheel, $a = R\alpha$

Therefore, option (a) is correct.

At B:

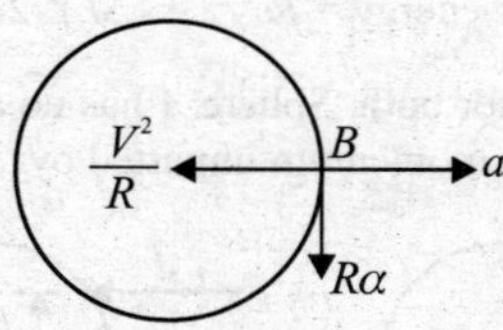

Fig. 2.529

If $V^2/R = a$, then a_B may be vertically downwards

Therefore, option (b) is correct.

At C:

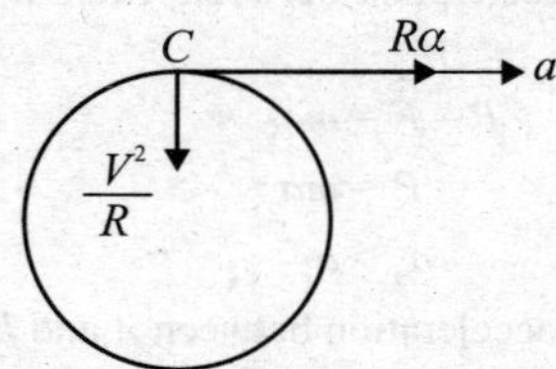

Fig. 2.530

Therefore, option (c) is correct.

Consider this diagram:

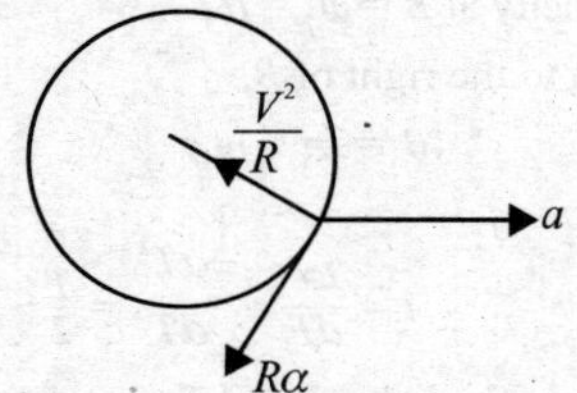

Fig. 2.531

Therefore, option (d) is correct.

5. c., d. $mgh = 1/2\ mv^2\ [1 + (k^2/R^2)\ ]$, k = radius of gyration. For all solid spheres, $k/R = 2v^2/5$ = independent of m and R.

6. c., d. All points in the body, in plane perpendicular to the axis of rotation, revolve in concentric circles. All points lying on the circle of same radius have same speed (and also same magnitude of acceleration) but different directions of velocity

Hence there cannot be two points in the given plane with same velocity or with same acceleration. As mentioned above, points lying on circle of same radius have same speed.

Angular speed of body at any instant w.r.t. any point on the body is same by definition.

7. a., b., c., d. By FBD of particle,

$$mg - T = ma \quad \text{(i)}$$

By FBD of the disc,

$$TR = I\alpha = I\frac{a}{R} \Rightarrow T = \frac{mR^2}{2}\frac{a}{R^2}$$

$$T = \frac{ma}{2} = a \quad \text{(ii)}$$

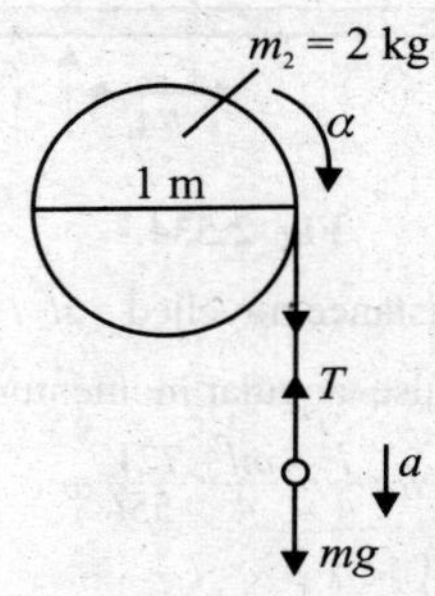

Fig. 2.532

By Eqs. (i) and (ii), we get the following results.

a. $a = 5$ m/s^2 and $T = 5$ N

and $\alpha = \frac{a}{R} = 5$ rad/s^2

b. For angular displacement of disc:

$$\theta = \omega t + \frac{1}{2}\alpha t^2$$

$$= \frac{1}{2} \times 5 \times 4^2 = 40 \text{ rad}$$

c. Work done by torque is

$$\int \tau d\theta = \tau \int d\theta = 5 \times 40 = 200 \text{ J}$$

d. $\Delta KE = \Delta W = 200$ J

$$k_2 - k_1 = 200 \text{ J}$$

8. a., b., c. The ball has V' component of its velocity perpendicular to the length of the rod immediately after the collision. u is the velocity of CM of the rod and ω is angular velocity of the rod just after collision. The ball strikes the rod with a speed of $v \cos 53°$ in the perpendicular direction and its component along the length of the rod after the collision is unchanged.

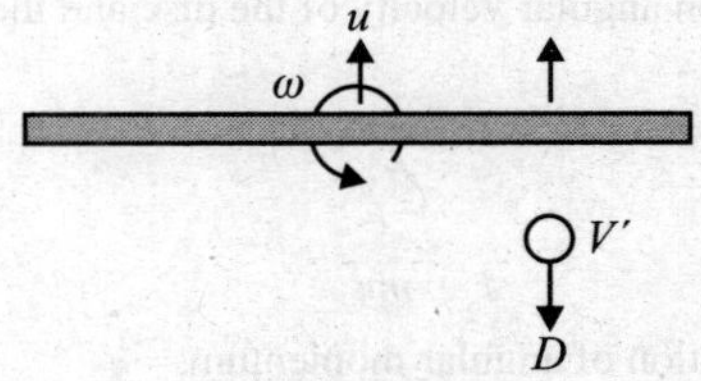

Fig. 2.533

Using for the point of collision.

Velocity of separation = Velocity of approach

$$\frac{3V}{5} = \left(\frac{\omega l}{4} + u\right) + V' \quad \text{(i)}$$

Conserving linear momentum (of rod + particle) in the direction perpendicular to the rod,

$$mV\frac{3}{5} = mu - mV' \quad \text{(ii)}$$

Conserving angular momentum about point 'D' as shown in the figure.

$$0 = 0 + \left[mu\frac{l}{4} - \frac{ml^2}{12}\omega\right] \Rightarrow u = \frac{\omega l}{3} \quad \text{(iii)}$$

$$\Rightarrow \quad u = \frac{24V}{55},\ W = \frac{72V}{55l}$$

Time taken to rotate by π angle, $t = \frac{\pi}{\omega}$

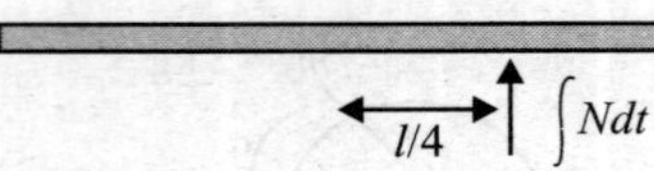

Fig. 2.534

In the same time, distance travelled $= u_2 t = \dfrac{\pi l}{3}$

Using angular impulse–angular momentum equation,

$$\int N dt \frac{l}{4} = \frac{ml^2}{4} \cdot \frac{72V}{55l}\,\omega$$

$$\left\{\because \int N dt \frac{l}{4} = \frac{24mV}{55}\right\}$$

$$\left[\text{Using impulse–momentum equation on the rod} \quad \int N dt = mu = \frac{24mv}{55}\right]$$

9. a., b., c. According to the theorem of perpendicular axis: Moment of inertia of the plate about an axis passing through centre O and perpendicular to the plate is $I_0 = I_1 + I_2 = I_3 + I_4$. (Because diagonals 1 and 2 as well as 3 and 4 are mutually perpendicular)

Now, according to symmetry,

$$I_1 = I_2 = I \text{ (say)}$$

and $$I_3 = I_4 = I' \text{ (say)}$$

For a square plate, $$I_0 = \frac{M(L^2 + L^2)}{12} = \frac{ML^2}{6}$$

Hence, $I = I' = \dfrac{ML^2}{12}$

i.e., $$I_1 = I_2 = I_3 = I_4 = \frac{ML^2}{12}$$

Hence, $I_0 = I_1 + I_2 = I_3 + I_4 = I_1 + I_3$

i.e., options (a), (b) and (c) are correct.

10. a., b., d. Let ω_1 = the initial angular velocity of the disc ω_2 = the final common angular velocity of the disc and the ring.

For the disc, $$I_1 = \frac{1}{2}mr^2$$

For the ring, $$I_2 = mr^2$$

By conservation of angular momentum,

$$L = I_1\omega_1 = (I_1 + I_2)\omega_2$$

$$\omega_2 = \frac{I_1\omega_1}{I_1 + I_2} = \frac{\omega_1}{3}$$

$$\text{Initial kinetic energy} = E_1 = \frac{1}{2} I_1 \omega_1^2$$

$$\text{Final kinetic energy} = E_2 = \frac{1}{2}(I_1 + I_2)\omega_2^2$$

$$\text{Heat produced} = \text{loss in KE}$$

Ratio of heat produced to initial kinetic energy $= (E_1 - E_2)/E_1 = 2/3$

11. b., d. The force of friction between the two surfaces in contact disappears when there is no relative (linear) motion between them. Angular momentum will not be conserved as the discs will have the final angular velocities in the opposite directions.

12. a., b., c. Let L = angular momentum

$$\tau = \frac{dL}{dt} \text{ or } dL = \tau\, dt$$

For constant torque, $\Delta L = \tau \Delta t = I\Delta\omega$

or $\quad I\omega$ (if $\omega_I = 0$)

Rotational kinetic energy $= I\omega^2/2 = (\Delta L)^2/2I$

13. a., b., d. $J = mv$ for both. Sphere A has no angular motion. For sphere B, angular momentum imparted by $J = L = Jh$.

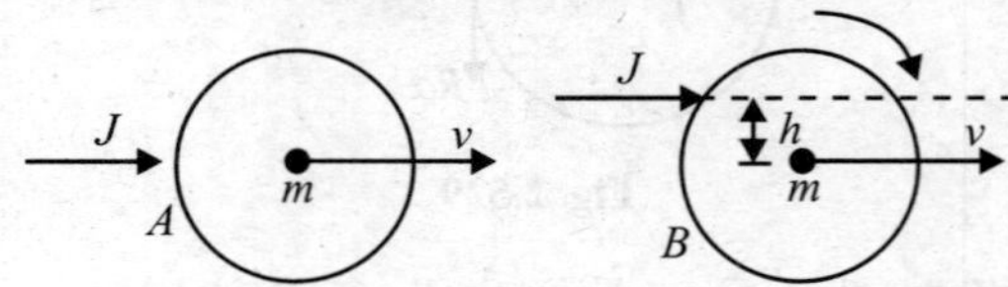

Fig. 2.535

14. b., c., d. Let P = external force, F = force of friction between A and B, a_1 = acceleration between A and B, a_2 = acceleration beyond B

$$P - F = ma_1$$

$$P = ma_2$$

$$a_2 > a_1$$

Let α = angular acceleration between A and B

For one rotation, $$\theta = 2\pi = \frac{1}{2}\alpha T^2$$

$$T = \sqrt{\frac{4\pi}{\alpha}} = \text{time of travel from } A \text{ to } B$$

$$\text{Angular velocity at } B = \omega_B = \alpha T$$

For one rotation to the right of B,

$$\theta = 2\pi = \omega_B t$$

$$t = \frac{2\pi}{\alpha T} = \frac{\frac{1}{2}\alpha T^2}{\alpha T} = \frac{T}{2}$$

15. a., d. In rolling with slipping, the force of friction produces a torque which gives an angular acceleration to the body. Hence, part of the work done against friction is converted into rotational kinetic energy, which adds to the total kinetic energy. Only the remaining part of the work done against friction is converted into heat.

16. c., d. In rolling without slipping, no work is done against friction. Hence, loss in gravitational potential energy of a body is equal to its total kinetic energy, i.e., linear plus rotational kinetic energies.

Also, $v = \omega r$

$$\text{Total kinetic energy} = \frac{1}{2}mv^2 + \frac{1}{2}I\omega^2$$

$$= \frac{1}{2}mv^2 + \frac{1}{2}(mK^2)\left(\frac{v^2}{r^2}\right)$$

$$= \frac{1}{2}mv^2\left[1 + \frac{K^2}{r^2}\right]$$

where K = radius of gyration

As all the bodies have the same final total kinetic energy, their final velocities will depend upon the ratio K/r. Bodies with the smaller value of K/r will have the greater v and hence will reach the bottom earlier.

17. a., c., d. After impact, the force of friction will act in a direction opposite to that of the motion. The body will retain its initial angular motion (clockwise). The force of friction will cause linear retardation, reducing v. It will also cause an anticlockwise angular acceleration, which will reduce ω to zero and then introduce anticlockwise ω till rolling without slipping begins. F will then disappear.

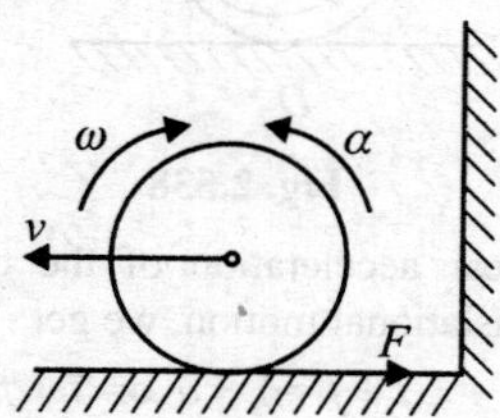

Fig. 2.536

18. a., b., c., d. Angular momentum = linear momentum × perpendicular distance from the point of rotation

or $\quad L = JI$

Also, $I = \dfrac{ml^2}{3}$

Fig. 2.537

$$\omega = \frac{L}{I} = \frac{JI}{ml^2/3} = \frac{3J}{ml}$$

$$\text{Kinetic energy} = \frac{L^2}{2I}$$

$$= \frac{J^2l^2}{2\dfrac{ml^2}{3}} = \frac{3J^2}{2m}$$

$$v_c = \omega\frac{l}{2} = \frac{3J}{2m}$$

19. b., c. In case of a wheel rolling on a horizontal plane, instantaneous axis of rotation of the wheel passes through A (point of contact of wheel with plane). But at A, acceleration is vertically upwards. It means, acceleration at A has non-zero value. Hence, option (a) is incorrect.

If the point is at a distance r from the instantaneous axis of rotation, then its velocity (relative to the axis) will be equal to $r\omega$. But by definition, every point lying on the instantaneous axis of rotation is at rest at that instant. Therefore, the resultant velocity of the particle becomes equal to that relative velocity $r\omega$. Hence, option (b) is correct.

If distance of centre of mass of a body from the instantaneous axis of rotation is equal to r and if its moment of inertia parallel to instantaneous axis of rotation but passing through centre of mass is equal to I_0, then $I = (I_0 + mr^2)$.

Rotational KE of the body will be equal to $I_0\omega^2/2$ and translational KE of the body will be $mv^2/2$, where $v = r\omega$. Therefore, translational KE becomes equal to $mr^2\omega^2/2$. Hence, the total KE becomes equal to $(I_0\omega^2/2) + (mr^2\omega^2/2) = I\omega^2/2$. Hence, option (c) is correct.

In case of a wheel rotating on a plane, instantaneous axis of rotation of the wheel passes through A while the centre of mass is at distance R from A. In that case, the moment of inertia about instantaneous axis of rotation is equal to $I = (I_0 + mR^2)$, where I_0 is moment of inertia about the centroidal axis. Hence, moment of inertia of the wheel about the instantaneous axis of rotation is greater than the minimum possible value. Hence, option (d) is incorrect.

20. a., c., d. Since, the spool rolls over the horizontal surface, instantaneous axis of rotation passes through the point of contact of spool with the horizontal surface.

About the instantaneous axis of rotation, moment produced by F is clockwise. Therefore, the spool rotates clockwise. In that case, acceleration will be rightward and thread will wind.

If rotational motion of the spool is considered about its own, then the resultant moment on it must be clockwise. But moment produced by the force is anticlockwise and its magnitude is equal to F_1/r. Hence, moment produced by the friction (about its own axis) must be clockwise and its magnitude must be greater than Fr. It is possible only when friction acts leftwards. Therefore, option (b) is correct.

21. a., b., c. Weight of the particle is balanced by normal reaction of the floor. Hence, the resultant force acting on the particle is provided by the tension in the thread.

The line of action of this tension passes through 'B' at the initial moment only. After that it does not pass through 'B'. Hence, the tension produces a moment about B. It means, angular momentum about B does not remain conserved. The tension always produces a moment about O. Magnitude of this moment is equal to TR in an anticlockwise direction. Hence, options (a), (b) and (c) are incorrect. Since neither external work is done on the particle nor energy is lost against friction, therefore KE of the ball remains constant. Hence, option (d) is correct.

22. a., b., c., d. When the particle strikes at one end of the rod then an impulse is exerted by the particle on the rod along the direction of its original motion. An equal but opposite impulsive reaction is exerted by the rod on the particle. If system of rod and the particle is considered, then there is no external impulse on the system, hence momentum of the system remains conserved. Hence, option (a) is correct.

Since, both the impulsive reactions act at the same point, one acts on the rod and other on the particle, hence no angular impulse is produced by these impulsive reactions about any point. It means, angular momentum of the system remains conserved about every point. Hence, options (b) and (c) are correct.

23. a., b., c. When the ball moves along a smooth plane, the accelerating force on it is $mg \sin \theta$. Hence, its acceleration is equal to $g \sin \theta$.

When the ball rolls down the rough inclined plane, then $mg \sin \theta$ acts down the plane but a friction comes into existence which acts up the plane. That friction produces angular acceleration on the ball and the net accelerating force on the ball becomes equal to ($mg \sin \theta$ – friction). Therefore, it has a smaller acceleration. Therefore, its velocity at the bottom of the plane is less than that of the ball moving down a smooth plane or $v_s > v_r$.

In fact, at the bottom of the planes both the balls have the same KE because loss of potential energy of both the balls is the same. But the ball rolling down a rough plane has translational as well as rotational kinetic energy at the bottom of the plane while ball sliding down a smooth plane has translational KE only. Therefore, translational KE of the rolling ball will be less than the translational KE of the ball sliding down the smooth plane. Hence, option (d) is correct.

24. b., c., d. Due to force F, the plank has a tendency to slide to the right below the sphere. Hence, friction acts on the plank to the left and that on the sphere acts to the right. The friction not only acts on accelerating force on the sphere but also produces an anticlockwise moment. Therefore, the sphere experiences a rightward translational acceleration and an anticlockwise angular acceleration simultaneously.

If the angular acceleration of the sphere is α, then its centre of mass has a backward acceleration $r\alpha$ relative to the plank. Hence, its net acceleration becomes less than that of the plank. Hence, option (b) is correct. Obviously, option (a) is incorrect. Since there is no sliding between the surfaces of the sphere and the plank, therefore, no energy is lost against friction.

Since friction, acting on the sphere, is the only force which provides motion to it, therefore KE of the sphere at any instant is equal to work done by the friction acting on it. Hence, option (c) is correct.

Since no energy is lost against friction, therefore KE of the whole system is equal to work done by force F. Hence, option (d) is also correct.

25. b., d. When force F acts vertically downwards on the thread, the thread unwinds and the cylinder rotates clockwise. Hence, its surface tries to slip over the planks in a backward direction. Hence, a friction comes into existence and acts on the cylinder in a forward direction (rightward direction). It is the only horizontal force acting on the cylinder. Due to the friction, cylinder experiences a rightward translational acceleration. But the point at which the friction acts remains always at rest, therefore no energy is lost against the friction.

Since the work is done by F and no energy is lost against the friction, therefore total kinetic energy of the cylinder at any instant is equal to work done by F. Hence, options (a) and (c) are incorrect and option (d) is correct.

If I is the moment of inertia of a body about its instantaneous axis of rotation, then the total KE of the rotating body is equal to $I\omega^2/2$. Hence, option (b) is correct.

26. a., b., c., d. Equating loss of gravitational potential energy with the gain of rotational kinetic energy, we get

$$\frac{1}{2}I\omega^2 = \frac{1}{2}mgl$$

$$\frac{1}{2}\times\frac{ml^2}{3}\times\omega^2 = \frac{1}{2}mgl$$

$$\omega = \sqrt{\frac{3g}{l}}$$

$$\text{Rotational kinetic energy} = \frac{1}{2}\frac{ml^2}{3}\omega^2 = \frac{ml^2\omega^2}{6}$$

27. a., b., d. The point of contact between the sphere and the surface slips to the left. So, the frictional force f acts towards the right.

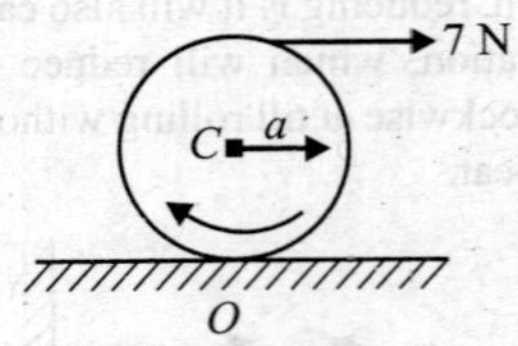

Fig. 2.538

If a is the linear acceleration of the centre of mass, then considering translational motion, we get

$$7 + f = 1\times a \text{ or } 7 + f = a \quad \text{(i)}$$

Considering rotational motion about centre C, we get

$$\text{Net torque} = 7R - fR = I\alpha$$

$$R[7 - f] = \left[\frac{2}{3}\times 1\times R^2\right]\frac{a}{R}$$

or

$$7 - f = \frac{2}{5}a \quad \text{(ii)}$$

Adding Eqs. (i) and (ii), we get

$$14 = \frac{7a}{5} \text{ or } a = 10 \text{ m/s}^2$$

From Eq. (i), $\quad 7 + f = 10$ or $f = 3$ N

28. a., b., c., d. $210 - T = 21a \quad$ (i)

Again, $\tau = R = I\alpha$

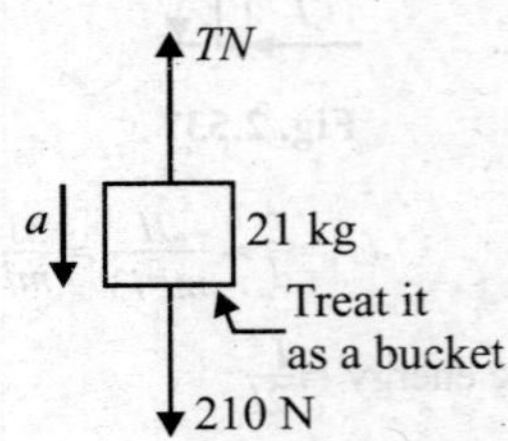

Fig. 2.539

$$TR = \frac{Ia}{R}$$

$$T = \frac{Ia}{R^2} = \frac{\frac{1}{2}\times 2I\times R^2\times a}{R^2} = \frac{1}{2}\times 2Ia$$

$$2Ia = 2T$$

From Eq. (i), $\quad 10 - T = 2T$

or $\quad 3T = 210$

or $\quad T = 0$ N

Again from Eq. (i), $210 - 70 = 21a$

$$a = \frac{140}{21}\text{ ms}^{-2} = \frac{20}{3}\text{ ms}^{-2}$$

29. b., d. Whenever two particles having equal mass collide head-on elastically, their velocities get interchanged. Therefore, if the particle collides at midpoint of the rod, then velocities would get interchanged. In that case option (a) would be correct. But now the particle strikes at an end of the rod, hence particle head-on collision does not take place. Therefore, option (a) is incorrect.

If the particle gets rebound back to its original path, then its final momentum will become negative. Since mass of the particle and rod is equal, therefore law of conservation of momentum will be satisfied only when velocity of centre of mass of the rod is greater than original velocity of the particle. Hence, kinetic energy of the system (just after collision) will become greater than that (just before the collision), which is impossible. Hence, the particle cannot rebound back or it will continue to move along the same direction. Hence, option (b) is correct but option (c) is incorrect. If the particle gets stick to the rod, centre of mass of the system will lie at a distance $l/4$ from the end at which the particle sticks. According to the law of conservation of momentum, centre of mass will start to move with velocity $v/2$. But the particle will simultaneously start to rotate about the centre of mass in an anticlockwise direction. Angular velocity ω of that rotational motion can be calculated by applying law of conservation of angular momentum. Hence, the resultant velocity of the midpoint of the rod will be equal to $[(v_0/2) - (l\omega/4)]$, which is obviously less than $v_0/2$. Therefore, option (d) is correct.

30. a., b., c.

Let the frictional force be in the forward direction, then

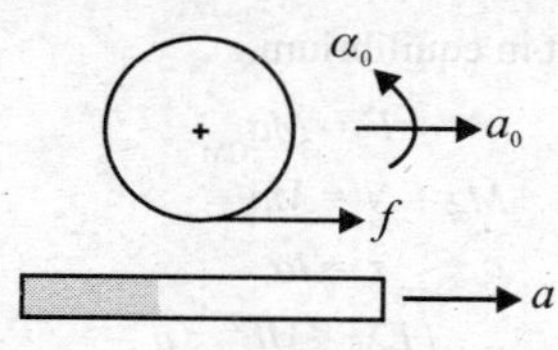

Fig. 2.540

$$f = ma_0$$

and $$fR = \frac{MR^2}{2}\alpha_0$$

$$a_0 = \frac{f}{M} \text{ and } \alpha_0 = \frac{2f}{MR}$$

For pure rolling, $a_0 + R\alpha_0 = a$

$$\frac{f}{M} + \frac{2f}{M} = a \Rightarrow f = \frac{Ma}{3}$$

For pure rolling, $f < f_L = \mu mg$

$$\frac{Ma}{3} < \mu Mg \Rightarrow a < 3\mu g$$

31. a., c. There will be exchange of linear velocities only. However, the two spheres cannot exert torques on each other, as their surfaces are frictionless and the angular velocities of the spheres do not change.

Assertion-Reasoning Type

1. d. If $F > mg \sin\theta$, then friction on disc will be in downward direction and if $F < mg \sin\theta$, then friction on disc will be in upward direction. If $F = mg \sin\theta$, then no friction will act on disc. Hence, statement I is false.

Statement II is true because in pure rolling, friction can be either static or no friction.

2. a. Sediment deposited at the equator (away from the axis of rotation) increases the moment of inertia (not mass) of the Earth. Since $I\omega$ = constant, ω decreases and thus $I = 2\pi/\omega$ increases.

3. a. Moment of inertia is then sum of $m\,r^2$ terms. We cannot change all the r's, keep m's the same and expect $\sum m_i r_i^2$ to remain unchanged.

4. c. As the person climbs up, normal reaction and friction between the ladder and the wall both increase. This decreases normal reaction from the floor, decreasing limiting value of friction there. This increases the possibility of the ladder to ship.

5. d.

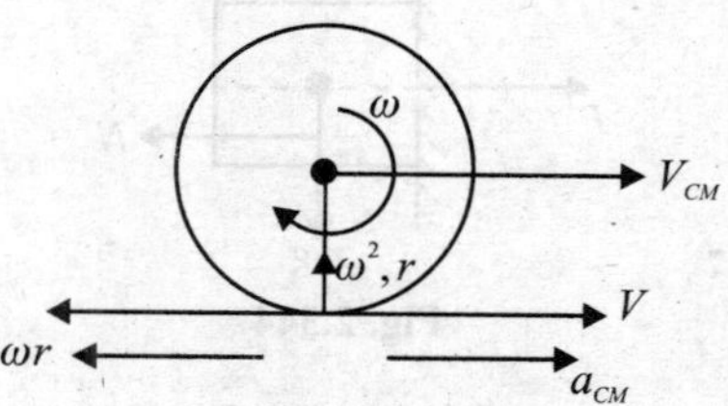

Fig. 2.541

During the motion of rolling, there is radial acceleration towards the centre. Hence, the contact point moves vertically upward.

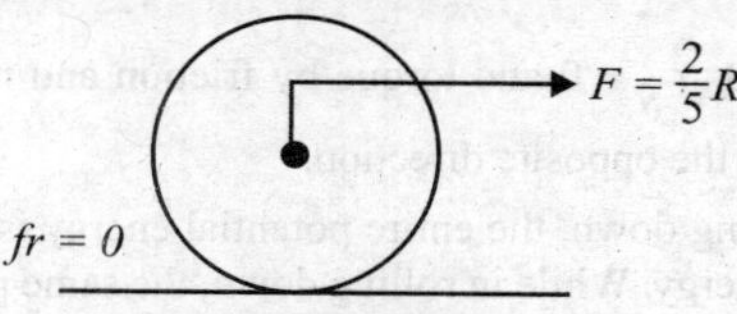

Fig. 2.542

6. a.

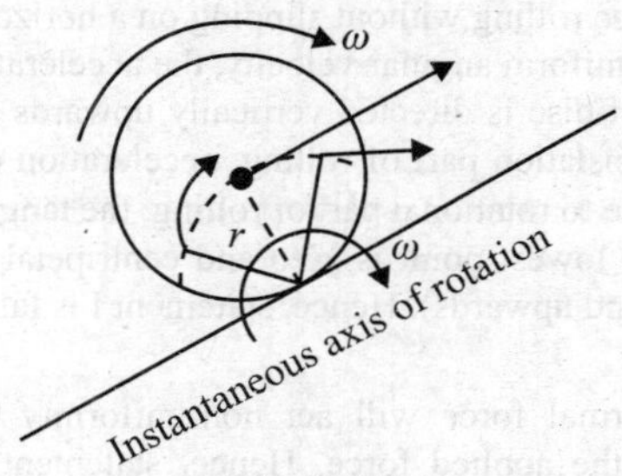

Fig. 2.543

7. b. If angular velocity is constant, then frictional force acting on the sphere is zero. In case of pure rolling, velocity of contact point is zero.

8. a. When the earth shrinks, its angular momentum remains constant, i.e.,

$$L = I\omega = \frac{2}{5}mR^2 \times \frac{2\pi}{T} = \text{constant}$$

$T \propto I \propto R^2$, it means if size of the earth changes, then its moment of inertia also changes.

If radius is half, the time period will become $T/4$. Hence, $24/4 = 6$ h.

9. b. Velocity of point of contact $V = (V_{CM} - \omega R)$

When pure rolling occurs, $V_{CM} = \omega R$.

Hence, $V = 0$.

Also frictional force can provide torque which further helps in achieving the pure rolling condition.

10. a. $I_1\,\omega_1 = I_2\,\omega_2$

$\because$ $I_2 < I_1$

$\therefore \quad \omega_2 > \omega_1$

The diver does work by pulling his limbs and thus, ω increases or rotational kinetic energy increases.

11. a. As the block remains stationary

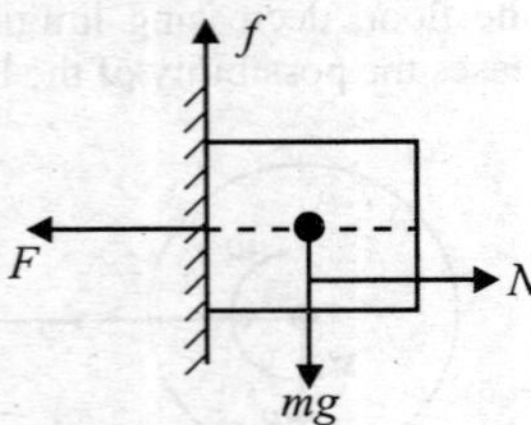

Fig. 2.544

$$\Sigma fx = 0, \text{ i.e., } F = N$$

$$\Sigma fy = 0, \text{ i.e., } f = mg$$

and $\quad \Sigma\tau = 0, \therefore \vec{\tau}_f + \vec{\tau}_N = 0$

As $\vec{\tau}_f \neq 0$, $\vec{\tau}_N \neq 0$ and torque by friction and normal reaction will be in the opposite direction.

12. a. In sliding down, the entire potential energy is converted into kinetic energy. While in rolling down, the same part of potential energy is converted into KE of rotation, therefore linear velocity acquired is less.

13. d. For a disc rolling without slipping on a horizontal rough surface with uniform angular velocity, the acceleration of the lowest point of disc is directed vertically upwards and is not zero (due to translation part of rolling, acceleration of lowest point is zero. Due to rotational part of rolling, the tangential acceleration of the lowest point is zero and centripetal acceleration is non-zero and upwards). Hence, statement I is false. Statement 2 is true.

14. c. The normal force will act non-uniformly to balance the torque of the applied force. Hence, statement I is true. The applied horizontal force F has tendency to rotate the cube in an anticlockwise sense about the centre of the cube. Hence, Statement 2 is false.

15. d. As x increases, the required component of reaction decreases to zero and then increases (with direction reversed). Hence, statement 1 is false.

Comprehension Type

For Problems 1–3

1. c., 2. c., 3. a.

Sol.

1. c.

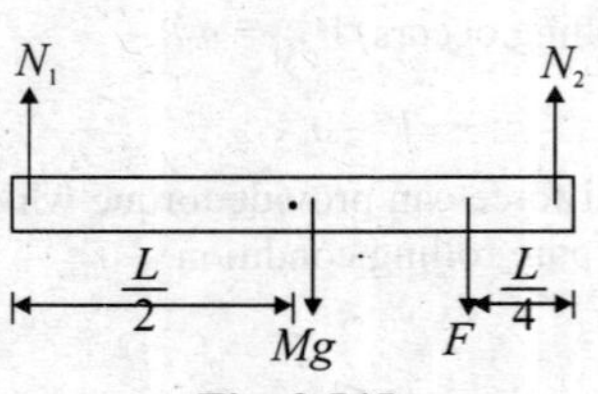

Fig. 2.545

As the rod is in equilibrium,

$$\therefore \quad \sum F_x = 0$$

$$\sum F_y = 0,$$

$$\sum \tau = 0$$

Taking torque about hinge 2, we get

$$N_1 \times L = Mg \times \frac{L}{2} + F \times \frac{L}{4}$$

$$\therefore \quad N_1 = \frac{Mg}{2} + \frac{F}{4} = 11 \text{ N}$$

2. c.

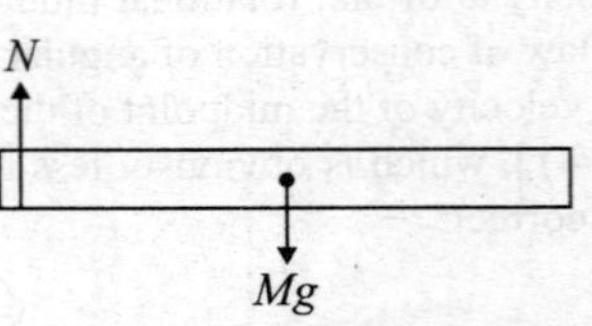

Fig. 2.546

Now, rod is not in equilibrium,

$$\therefore \quad F = Ma_{CM}$$

$$\Rightarrow \quad Mg - N = Ma \quad \text{(i)}$$

and $\quad \tau = I\alpha$

$$\Rightarrow \quad Mg\left(\frac{L}{2}\right) = \frac{ML^2}{3}\frac{a}{(L/2)}$$

$$\left(\because I = \frac{ML^2}{3}, a = \alpha\left(\frac{L}{2}\right)\right)$$

$$\Rightarrow \quad a = \frac{3}{4}g \quad \text{(ii)}$$

From Eqs. (i) and (ii),

$$N = \frac{Mg}{4} = 5 \text{ N}$$

3. a. By conservation of energy,

$$Mg\left(\frac{L}{2}\right) = \frac{1}{2}I\omega^2 = \frac{1}{2}\left(\frac{ML^2}{3}\right)\omega^2$$

$$\therefore \quad \omega = \sqrt{\frac{3g}{L}}$$

$$\text{Acceleration} = \omega^2 L = 3g = 30 \text{ m/s}^2$$

For Problems 4–6

4. a., 5. c., 6. b.

Sol.

$$I = \left\{\frac{M(R\sqrt{2})^2}{12} + M\left(\frac{R}{\sqrt{2}}\right)^2\right\}4 + mR^2$$

$$= \frac{8}{3}MR^2 + mR^2 = 20 \text{ kg m}^2$$

$$(4M + m)\, g \sin\theta - f = (4M + m)a \quad \text{(i)}$$

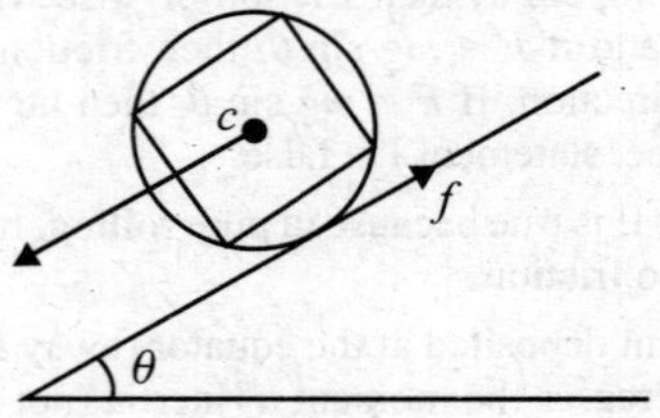

Fig. 2.547

$$fR = I\frac{a}{R} \quad \text{(ii)}$$

Solving Eqs. (i) and (ii),

$$a = \frac{7g}{24}$$

$$f = 20\,a = \mu \times 28g\cos 30°$$

$$\mu = \frac{5}{12\sqrt{3}}$$

For Problems 7–9

7. d., 8. b., 9. a.

Sol.

7. d. Acceleration of the bottom of the cylinder = acceleration of the plank = 2 m/s^2 towards the right

Acceleration of the top of the cylinder = acceleration of B = 6 ms^{-2} towards the left

Let linear acceleration of the cylinder be a towards the left and angular acceleration of it α in an anticlockwise sense. Writing constraint, we get

$$a + R\alpha = 6 \text{ and } R\alpha - a = 2$$

$$\Rightarrow \quad \alpha = 1 \text{ rad s}^{-2},\ a = 2 \text{ m/s}^2.$$

8. b.

For block B; $m_B g - T = m_B 6$ (i)

For cylinder; $T = m_{cyl} 2$ (ii)

Solving Eqs. (i) and (ii), $m_{cyl} : m_B = 2 : 1$

9. a. The acceleration of the top part of the cylinder (thread) with respect to CM of the cylinder is

$$a_{AO} = \alpha R = 1 \times 4 = 4 \text{ m/s}^2$$

Length of the thread $= 20 + \frac{1}{2}(R\alpha)t^2 = 28$ m.

For Problems 10–12

10. b., 11. a., 12. b.

10. b. Construct the free-body diagram of the cylinder:

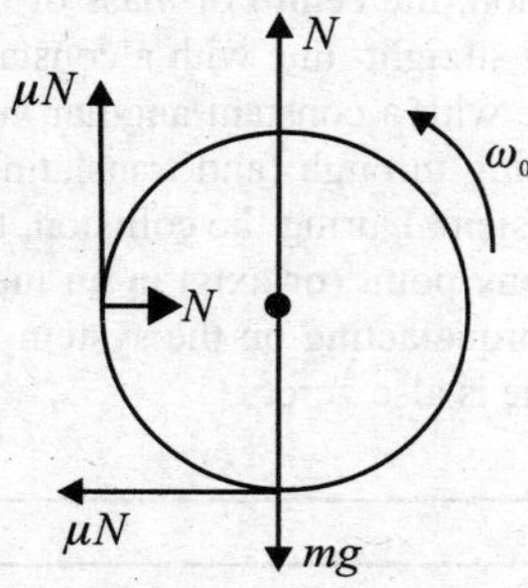

Fig. 2.548

Calculate the force acting on the cylinder: As the centre of mass of the cylinder does not accelerate

$$N_2 - \mu_k N_1 = 0 \quad \text{(i)}$$

$$N_1 + \mu_k N_2 - mg = 0 \quad \text{(ii)}$$

Solving these equations:

$$N_1 = \frac{mg}{1 + m_k^2},\ N_2 = \frac{\mu_k mg}{1 + \mu_k^2}$$

11. a. The torque on the cylinder about the axis of rotation,

$$\tau_{im} = mg \times 0 + N_1 \times 0 + N_2 \times 0 + \mu_k N_1 R + \mu_k N_2 R$$

$$= \frac{\mu_k(1 + \mu_k)}{1 + \mu_k^2} mgR$$

The moment of inertia about the axis of rotation,

$$I_{CM} = \frac{1}{2} mR^2$$

Torque equation, $\tau = I\alpha$,

$$\frac{m_k(1 + m_k)}{1 + m_k^2} mgR = \left(\frac{1}{2} mR^2\right)\alpha$$

$$\alpha = \frac{2\mu_k(1 + \mu_k)g}{1 + \mu_k^2 R}$$

12. b. Using equation $\omega^2 = \omega_0^2 + 2\alpha\theta$, the angular displacement θ,

$$\theta^2 = \omega_0^2 + \left[\frac{2 - 2\mu_k(1 + \mu_k)g}{(1 + \mu_k^2)R}\right]\theta$$

(ω and α are in opposite sense)

$$\theta = \frac{(1 + \mu_k^2)\,\omega_0^2 R}{4\mu_k(1 + \mu_k)g}$$

Revolution accomplished,

$$n = \frac{\theta}{2\pi} = \frac{(1 + \mu_k^2)\,\omega_0^2 R}{2\pi \times 4\mu_k(1 + \mu_k)g}$$

$$= \frac{(1 + \mu_k^2)\omega_0^2 R}{8\pi\mu_k(1 + \mu_k)g}$$

For Problems 13–15

13. c., 14. d., 15. a.

13. c. By conservation of angular momentum on the man–table system,

$$L_i = L_f$$

$$0 + 0 = I_m\omega_m + I_t\omega_t$$

$$\Rightarrow \quad \omega_t = -\frac{I_m\omega_m}{I_t} \text{ where } \omega_m = \frac{v}{r} = \frac{1}{2} \text{ rad/s}$$

$$\Rightarrow \quad \omega_t = -\frac{100\,(2)^2 \times \left(\frac{1}{2}\right)}{4000}$$

$$\Rightarrow \quad \omega_m = \frac{v}{r} = \frac{1}{2} \text{ rad/s}$$

$$\Rightarrow \quad \omega_t = -\frac{1}{20} \text{ rad/s}$$

Thus, the table rotates clockwise (opposite to the man) with angular velocity 0.05 rad/s.

14. d. If the man completes one revolution relative to the table, then

$$\theta_{mt} = 2\pi$$

$$\Rightarrow \quad 2\pi = \theta_m - \theta_t$$

$$2\pi = \theta_m t - \omega_t t \text{ (where } t \text{ is the time taken)}$$

$$t = \frac{2\pi}{\omega_m - \omega_t} = \frac{2\pi}{0.5 - (-0.05)} = \frac{2\pi}{0.55}\text{s}$$

Angular displacement of the table is

$$\theta_t = \omega_t t = -0.05 \times \left(\frac{2\pi}{0.55}\right)$$

$$= -\left(\frac{2\pi}{11}\right) \text{ radians}$$

The table rotates through $2\pi/11$ radians clockwise.

15. a. If the man completes one revolution relative to the earth, then

$$\theta_m = 2\pi$$

$$\text{Time} = \frac{2\pi}{\theta_m} = \frac{2\pi}{0.5}$$

During this time, the angular displacement of the table,

$$\theta_t = \omega_t (\text{time}) = -0.05 \times \left(\frac{2\pi}{0.5}\right)$$

$$= -\frac{\pi}{5} \text{ radians}$$

$$\theta_t \equiv 36° \text{ in clockwise direction}$$

For Problems 16–17

16. d., 17. a.

Sol. The tension of the rope is the only net force on the block and it does not exert any torque about the axis of rotation. Hence, the angular momentum of the block about the axis should remain conserved.

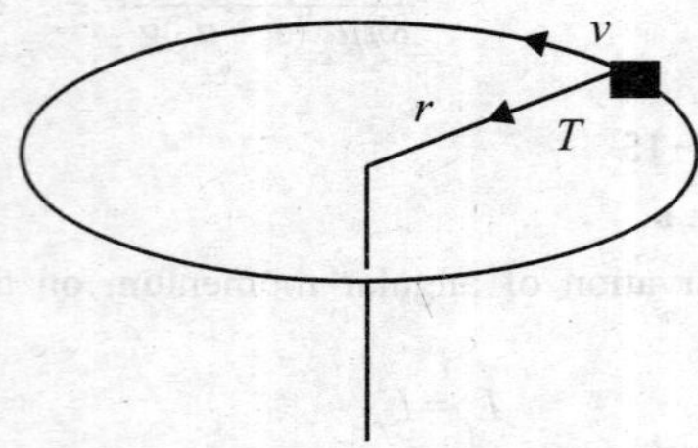

Fig. 2.549

$$\Rightarrow \quad mvr = \text{constant}$$

Let $r_1 = 5$ m and $v_1 = 4$ m/s

Let r_2, v_2, T_2 be the radius, velocity and tension when the string breaks.

$$\Rightarrow \quad T_2 = 200 \text{ N}$$

$$mv_1 r_1 = mv_2 r_2 \text{ and } T_2 = \frac{mv_2^2}{r_2} \quad \text{(i)}$$

$$\Rightarrow \quad mv_1 r_1 = m\sqrt{\frac{r_2 T_2}{m}}\, r_2$$

$$r_2 = \left(\frac{mv_1^2 r_1^2}{T_2}\right)^{\frac{1}{3}} = \left(\frac{4 \times 16 \times 25}{200}\right)^{\frac{1}{3}}$$

$$= 2 \text{ m}$$

Note: ***The tension in the string is inversely proportional to the cube of the radius.***

For Problems 18–20

18. b., 19. a., 20. d.

Sol.

18. b.

Let $OA = x$, $OB = y$

$$x^2 + y^2 = l^2 \text{ and } x = l\cos\theta$$

Differentiate $x^2 + y^2 = l^2$ with respect to time.

$$2x\frac{dx}{dt} + 2y\frac{dy}{dt} = 0$$

$$V_B = \frac{dy}{dt} = -\frac{x}{y}\frac{dx}{dt} = -V_A \cot\theta = \frac{4}{5}V_0 \downarrow$$

19. a. Differentiate $x = l\cos\theta$ with respect to time

$$\frac{dx}{dt} = -l\sin\theta\frac{d\theta}{dt}$$

$$\frac{d\theta}{dt} = -\frac{V_A}{l\sin\theta}$$

(negative sign indicates that θ is decreasing)

$$\omega = \left|\frac{d\theta}{dt}\right| = \frac{V_A}{l\sin 37°} = \frac{5v_0}{3l}$$

20. d.

$$x_{\text{CM}} = \frac{x}{2}$$

$$\Rightarrow \quad V_{\text{CM}_x} = \frac{1}{2}\frac{dx}{dt} = \frac{v_0}{2}(\rightarrow)$$

$$\Rightarrow \quad y_{\text{CM}} = \frac{y}{2}$$

$$\Rightarrow \quad V_{\text{CM}_y} = \frac{1}{2}\frac{dy}{dt} = \frac{v_B}{2} = \frac{2}{3}v_0 \downarrow$$

$$V_{\text{CM}} = \sqrt{\frac{v_0^2}{4} + \frac{4v_0^2}{9}} = \frac{5}{6}v_0 \text{ at } \tan^{-1}\left(\frac{4}{3}\right) \text{ below horizontal}$$

For Problems 21–23

21. a., 22. b., 23. a.

21. a. After collision, the centre of mass of the composite system will move in a straight line with a constant speed and the system will rotate with a constant angular velocity about the vertical axis passing through (and translating with) the centre of mass of the system. During the collision, the external force and torque about any point (or axis) in an inertial reference frame is zero. The torque acting on the system in the centre of mass reference frame is also zero.

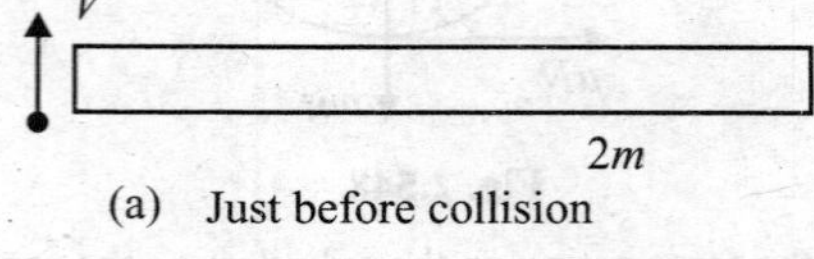

(a) Just before collision

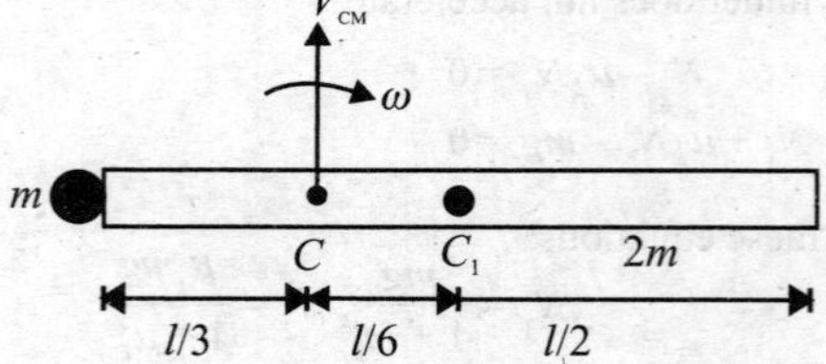

(b) Just after collision

Fig. 2.550

There are many (mathematically infinite) points (axes) about which the angular momentum of the system remains constant during the collision. The centre of mass of the composite system is the most convenient point.

The distance of the centre of mass of the system from the end where m sticks is given by

$$x_{CM} = \frac{m \times 0 + 2m \times \frac{l}{2}}{m + 2m} = \frac{l}{3}$$

Let the velocity of the centre of mass of the system after collision be v_{CM} and the angular velocity of the system about the vertical axis through the centre of mass be ω as shown in the figure.

From the conservation of the angular momentum about the centre of mass axis,

$$\frac{mvl}{3} = \left\{ m\left(\frac{l}{3}\right)^2 + \frac{(2m)l^2}{12} + (2m)\left(\frac{l}{6}\right)^2 \right\}\omega \quad \text{(i)}$$

Moment of inertia of the composite system about the centre of mass axis is

$$I_{CM} = m\left(\frac{l}{3}\right)^2 + \frac{(2m)l^2}{12} + (2m)\left(\frac{l}{6}\right)^2 = \frac{ml^3}{3}a \quad \text{(ii)}$$

From Eqs. (i) and (ii)

$$\Rightarrow \quad mv\frac{l}{3} = \frac{ml^2}{3}\omega \Rightarrow \omega = \frac{v}{l}$$

22. b. From the conservation of the linear momentum.

$$mv = (m + 2m)\, v_{CM}$$

$$\Rightarrow \quad v_{CM} = \frac{v}{3}$$

Fig. 2.551

Let the point whose instantaneous velocity just after collision is zero be at a distance l_0 from the centre of mass of the composite system.

$$v_{CM} - l_0\omega = 0$$

$$\Rightarrow \quad \frac{v}{3} - l_0\left(\frac{v}{l}\right) = 0 \quad \Rightarrow l_0 = \frac{l}{3}$$

The point is a distance (2/3) l from the end where the mass strikes.

23. a. Loss in kinetic energy of the system as a result of collision

$$= K_{\text{Initial}} - K_{\text{Final}} = \frac{1}{2}mv^2 - \left\{\frac{1}{2}(m + 2m)\,v_{CM}^2 + \frac{1}{2}I_{CM}\omega^2\right\}$$

$$= \frac{1}{2}mv^2 - \left\{\frac{1}{2}(3m)\left(\frac{v}{3}\right)^2 + \frac{1}{2}\left(\frac{ml^2}{3}\right)\left(\frac{v}{l}\right)\right\}$$

For the composite $I_{CM} = \frac{ml^2}{3} = \frac{1}{2}mv^2 - \frac{1}{3}mv^2 = \frac{1}{6}mv^2$

For Problems 24–26

24. d., 25. d., 26. a.

Sol.

24. d. Taking torque about A: $Mg\frac{L}{2} = TL$

$$\Rightarrow \quad T = \frac{Mg}{2}$$

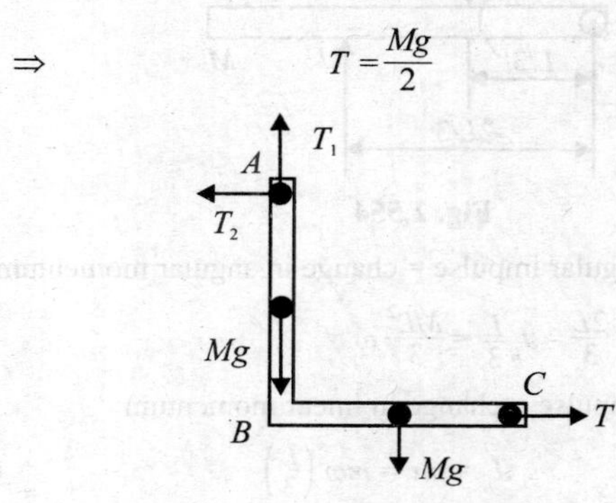

Fig. 2.552

25. d. $T_1 = 2Mg,\ T_2 = T = \frac{Mg}{2}$

$$\text{Net reaction} = \sqrt{T_1^2 + T_2^2}$$

$$= \sqrt{(2Mg)^2 + (Mg/2)^2} = \frac{\sqrt{17}\,Mg}{2}$$

26. a. D is the CM of the rods

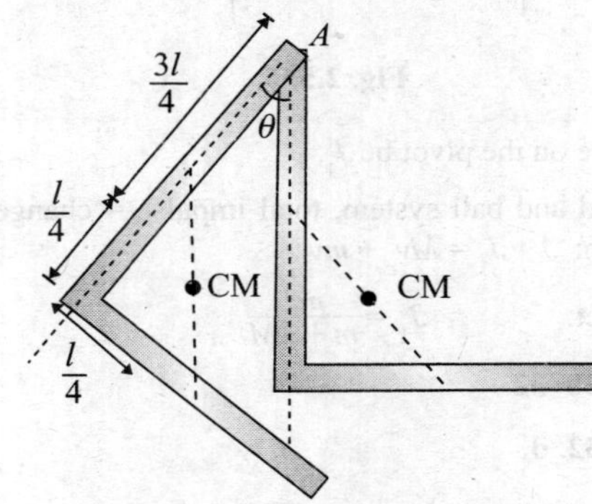

Fig. 2.553

In triangle ADE:

$$\tan\theta = \frac{L/4}{3L/4}$$

$$\Rightarrow \quad \tan\theta = \frac{1}{3} \Rightarrow \theta = \tan^{-1}(1/3)$$

Here D is the centre of mass. In equilibrium, D will be vertically below A. So in equilibrium, AB will make angle θ, as calculated above, with the vertical.

For Problems 27–29

27. b., 28., d., 29. a.

Sol.

Let the system starts with angular velocity ω. Angular velocity of the ball will also be ω as it remains struck to the rod.

Velocity of the ball $\quad v = \omega\frac{L}{3}$

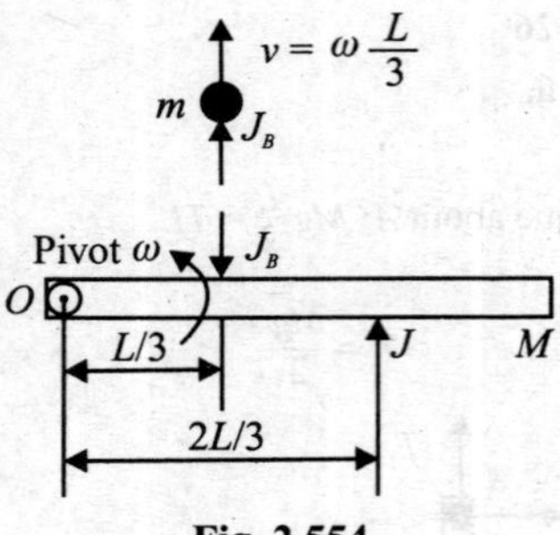

Fig. 2.554

For the rod, angular impulse = change in angular momentum:

$$J\frac{2L}{3} - J_b\frac{L}{3} = \frac{ML^2}{3}\omega \quad \text{(i)}$$

For the ball, impulse = change in linear momentum

$$J_b = mv = m\omega\left(\frac{L}{3}\right) \quad \text{(ii)}$$

From Eqs. (i) and (ii), $\omega = \dfrac{6J}{(m+3M)L}$

and

$$J_b = \frac{2mJ}{(m+3M)}$$

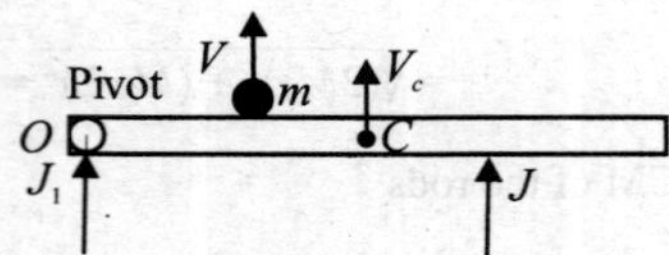

Fig. 2.555

Let impulse on the pivot be J_1.

For the rod and ball system, total impulse = change in linear momentum: $J + J_1 = Mv_c + mv$

Solve to get: $J_1 = \dfrac{mJ}{m+3M}$

For Problems 30–32

30. b., 31. a., 32. d.

Sol.

$$a_1 = \frac{\mu mg}{m} = \mu g,\ a_2 = \frac{\mu mg}{m} = \mu g$$

$$v_1 = a_1 t = \mu gt,\ v_2 = a_2 t = \mu gt$$

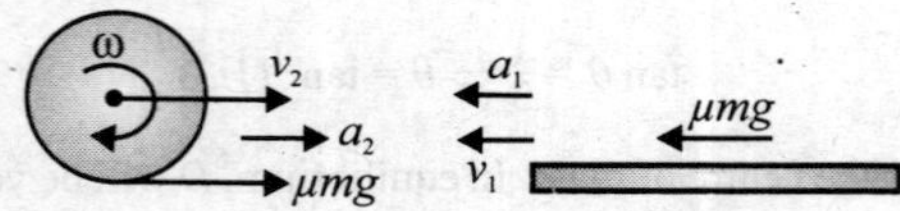

Fig. 2.556

$$\tau = \mu mgr$$

$$\omega = \omega_0 + \alpha t,$$

where $\alpha = -\dfrac{\tau}{I} = -\dfrac{\mu mgr}{\frac{2}{5}mr^2} = -\dfrac{5}{2}\dfrac{\mu g}{r}$

$$\Rightarrow \quad \omega = \omega_0 - \frac{5}{2}\frac{\mu g}{r}t$$

When the pure rolling starts,

$$\omega = \frac{v_1 + v_2}{r}$$

$$\Rightarrow \quad \omega_0 - \frac{5}{2}\frac{\mu gt}{r} = \frac{2\mu gt}{r}$$

$$\Rightarrow \quad t = \frac{2\omega_0 r}{9\mu g}$$

Velocity of the sphere: $v_2 = \mu gt = \dfrac{2\omega_0 r}{9}$

$$s_p = \frac{1}{2}a_1 t^2 = \frac{1}{2}\mu g\left(\frac{2\omega_0 r}{9\mu g}\right)^2 = \frac{2}{81}\frac{\omega_0^2 r^2}{\mu g}$$

For Problems 33–35

33. b., 34. c., 35. d.

Sol.

33. b. Let the angular speed of the disc when the balls reach the end be ω. From conservation of angular momentum,

$$\frac{1}{2}mR^2\omega_0 = \frac{1}{2}mR^2\omega + \frac{m}{2}R^2\omega + \frac{m}{2}R^2\omega$$

or $$\omega = \frac{\omega_0}{3}$$

34. c. The angular speed of the disc just after the balls leave the disc is $\omega = \omega_0/3$.

Let the speed of each ball just after they leave the disc be v.

From conservation of energy,

$$\frac{1}{2}\left(\frac{1}{2}mR^2\right)\omega_0^2 = \frac{1}{2}\left(\frac{1}{2}mR^2\right)\omega^2 + \frac{1}{2}\left(\frac{m}{2}\right)v^2 + \frac{1}{2}\left(\frac{m}{2}\right)v^2$$

Solving, we get $$v = \frac{2R\omega_0}{3}$$

Note: $v = \sqrt{(\omega R)^2 + v_r^2}$

v_r = *radial velocity of the ball*

35. d. Work done by all force is equal

$$K_f - K_i = \frac{1}{2}\left(\frac{m}{2}\right)v^2 = \frac{mR^2\omega_0^2}{9}$$

For Problems 36–40

36. d., 37. a., 38. b., 39. b., 40. d.

Sol.

Velocity of the centre of mass of the 'rod + particle' system,

$$V_{CM} = \frac{mv}{(M+m)}$$

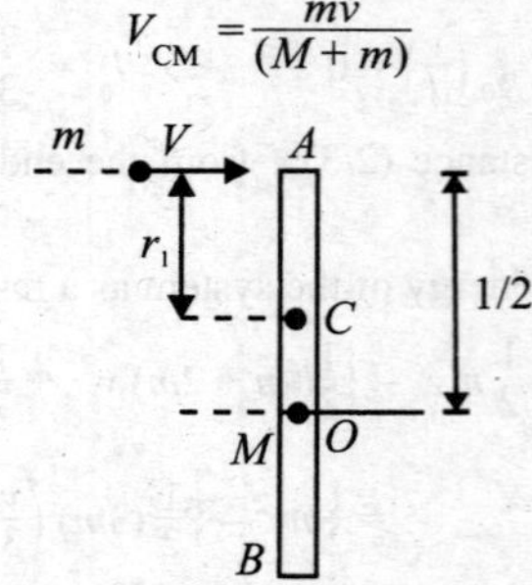

Fig. 2.557

$$\vec{v}_{p, CM} = \vec{v}_p - \vec{v}_c$$

$$v_c = \frac{mv}{M+m} \text{ and } v_p = v$$

$$\vec{v}_{p, CM} = \frac{Mv}{M+m}$$

$$\vec{v}_{rod, CM} = \vec{v}_{rod} - \vec{v}_{CM} = 0 - \frac{mv}{M+m}$$

$$v_{rod, CM} = -\frac{mv}{M+m}$$

Location of the centre of mass of the system 'rod + particle' from A

$$r_1 = \frac{m \times 0 + Ml/2}{(m+M)}$$

$$= \frac{Ml}{2(m+M)}$$

Angular momentum of the particle just before collision about C,

$$L_{p,c} = m\, v_{p,c}\, r_1$$

$$= \frac{M^2 mvl}{2\,(M+m)^2}$$

Angular momentum of the rod about the centre of mass of system C,

$$L_{rod, CM} = Mv_{rod, CM}\left(\frac{l}{2} - r_1\right)$$

$$= M\left(\frac{mv}{M+m}\right)\left(\frac{ml}{2(m+M)}\right)$$

$$= \frac{Mm^2\, vl}{2\,(M+m)^2}$$

Moment of inertia about the vertical axis passing through C,

$$I_{c, rod} = I_0 + M\left(\frac{l}{2} - r_1\right)^2$$

$$I_{c,rod} = \frac{Ml^2}{12} + M\left(\frac{ml}{2(M+m)}\right)^2$$

$$I_{c, particle} = mr_1^2 = m\left(\frac{Ml}{2(M+m)}\right)^2$$

$$I_c = \frac{Ml^2}{12} + M\left(\frac{ml}{2(M+m)}\right)^2 + m\left(\frac{Ml}{2\,(M+m)}\right)^2$$

$$I_c = \frac{M(M+4m)}{12\,(M+m)} L^2$$

As no external forces are acting on the system 'rod + particle', hence the velocity of the centre of mass of the system will remain constant.

$$v_{CM} = \frac{mv}{M+m}$$

For angular velocity about C,

$$mvr_1 = I_c\, \omega$$

$$\omega = \frac{mvr_1}{I_c} = \frac{mv\left(\frac{ml}{2(M+m)}\right)}{\left(\frac{M(M+4m)L^2}{12(M+m)}\right)}$$

$$= \frac{6mv}{(M+4m)L}$$

For Problems 41–43

41.d., 42. d., 43. d.

41. d. As no external force is acting on the system (block + cylinder) in the horizontal direction, the linear momentum of the system remains conserved in the horizontal direction.

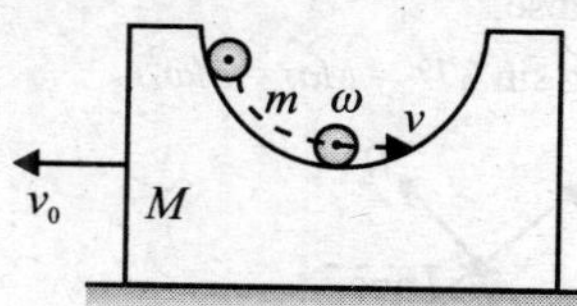

Fig. 2.558

As the cylinder reaches the bottommost point of the track, its centre of mass covers a horizontal distance of 0.25 m ($R - r$) w.r.t. block towards the right. Let the block move by x in backward direction, then

$$Mx = m(0.5 - x)$$

$$x = \frac{2 \times 0.5}{2+3} = 0.2 \text{ m}$$

42. d. v and v_0 are the velocities w.r.t. ground: $\omega = \frac{v + v_0}{r}$

$$mv = Mv_0$$

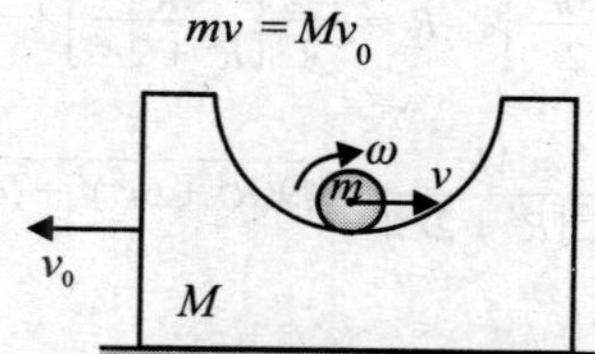

Fig. 2.559

$$mg\,(R - r) = \frac{1}{2}mv^2 + \frac{1}{2}I\omega^2 + \frac{1}{2}Mv_0^2$$

$$\Rightarrow \quad v_0 = \frac{4}{\sqrt{11}} \text{ m/s}$$

$$v = \frac{6}{\sqrt{11}} \text{ m/s}$$

For Problems 44–46

44. a., 45. c., 46. a.

Sol. From $\tau = I\alpha$,

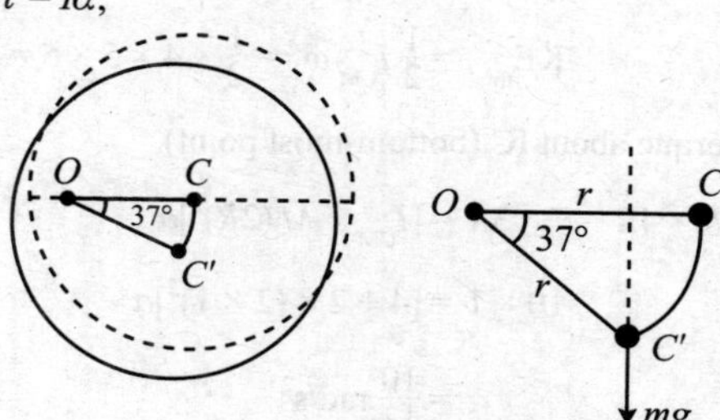

Fig. 2.560

$$Mg \times r \cos 37° = \left[\frac{MR^2}{2} + Mr^2\right]\alpha$$

$$\alpha = \frac{8rg}{5[R^2 + 2r^2]}$$

From energy conservation,

$$\frac{I\omega^2}{2} = Mg \times r \sin 37°$$

$$\left[\frac{MR^2}{2} + Mr^2\right]\frac{\omega^2}{R} = Mgr \times \frac{3}{5}$$

$$\omega = \sqrt{\frac{12gr}{5[R^2 + 2r^2]}}$$

For FBD of the disc,

$$R_x - Mg \sin 37° = Ma_r = M\omega^2 r$$

Fig. 2.561

$$Mg \cos 37° - R_y = Ma_t = Mr\alpha$$

$$R_x = \frac{3Mg}{5}\left[\frac{R^2 + 6r^2}{R^2 + 2r^2}\right], \quad R_y = \frac{Mg}{5}\left[\frac{4R^2}{R^2 + 2r^2}\right]$$

$$R = \sqrt{R_x^2 + R_y^2} = \frac{Mg}{5[R^2 + 2r^2]} \times \sqrt{9(R^2 + 6r^2)^2 + (4R)^2}$$

For Problems 47–48

47. c., 48. c.

Sol. When frictional force is absent, the rigid body does not translate, $\sum \vec{F}_{\text{external}} = 0$, therefore $\vec{a}_{\text{CM}} = 0$

Taking torque about CM,

$$F \times 2R - F \times R = I_{\text{CM}}\,\alpha$$

$$10 \times 1 = 4\alpha$$

$$\alpha = \frac{5}{2} \text{ rad/s}^2$$

$$\omega = \alpha t = 5 \text{ rad/s}$$

$$\text{KE}_{\text{total}} = \frac{1}{2} I_{\text{CM}}\,\omega^2 = \frac{1}{2} \times 4 \times 5 \times 5 = 50 \text{ J}$$

Taking torque about IC (bottom-most point),

$$F \times 4R - F \times 3R = [I_{\text{CM}} + M(2R)^2]\alpha$$

$$10 \times 1 = [4 + 2 \times (2 \times 1)^2]\alpha$$

$$\alpha = \frac{10}{12} \text{ rad/s}^2$$

$$\omega = \alpha t = \frac{10}{12} \times 2 = \frac{5}{3} \text{ rad/s}$$

$$\text{KE}_{\text{total}} = \frac{1}{2}[I_{\text{CM}} + M(2R)^2]\omega^2$$

$$= \frac{1}{2}[4 + 2 \times (2\times1)^2] \times \left(\frac{5}{3}\right)^2$$

$$= 16.67 \text{ J}$$

For Problems 49–50

49. b., 50. b.

Sol. As resultant acceleration is zero; resolving all the three components of acceleration and applying $\sum a_c = 0$, we get

$$a_0 \sin\theta = r\alpha = r\left(\frac{a_0}{R}\right)$$

$$\sin\theta = \frac{r}{R}$$

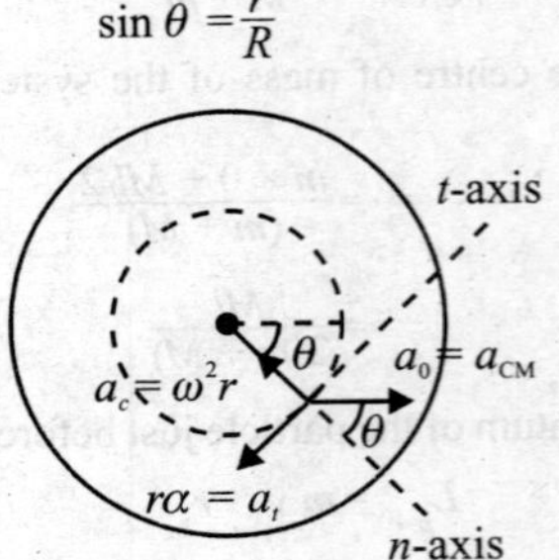

Fig. 2.562

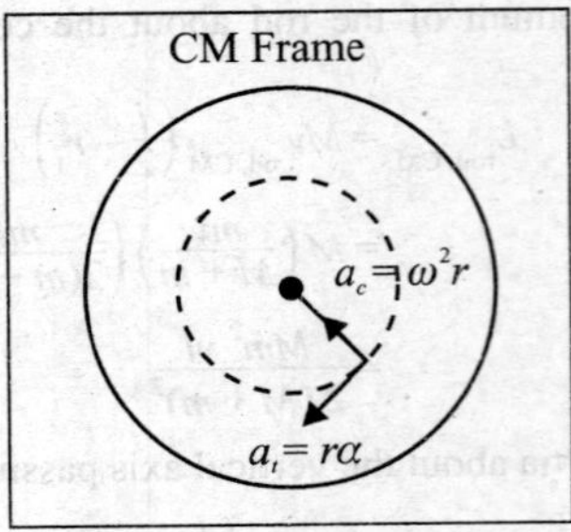

Fig. 2.563

$$\sum a_n = a_0 \cos\theta - \omega^2 r = 0$$

$$\sqrt{1 - \frac{r^2}{R^2}} = \left(\frac{v_0}{R}\right)^2 \times \frac{r}{a_0}$$

On solving for v_0, we get

$$v_0 = \sqrt{\frac{a_0 R}{r}}\,(R^2 - r^2)^{\frac{1}{4}}$$

For Problems 51–53

51. a., 52. a., 53. d.

Sol. As net velocity of contact point is not zero, the disc rolls with slipping, as ω about IC is the same,

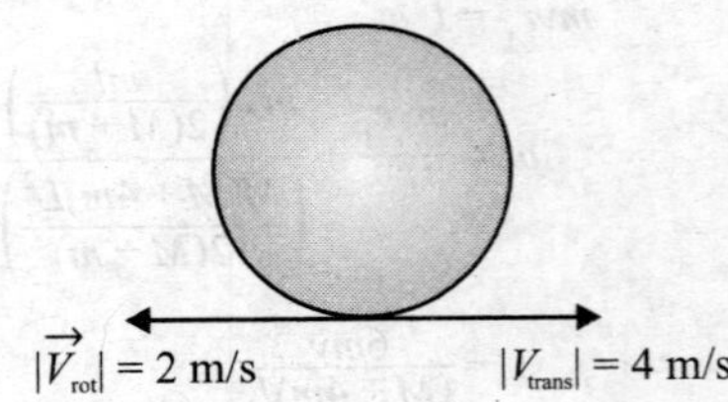

Fig. 2.564

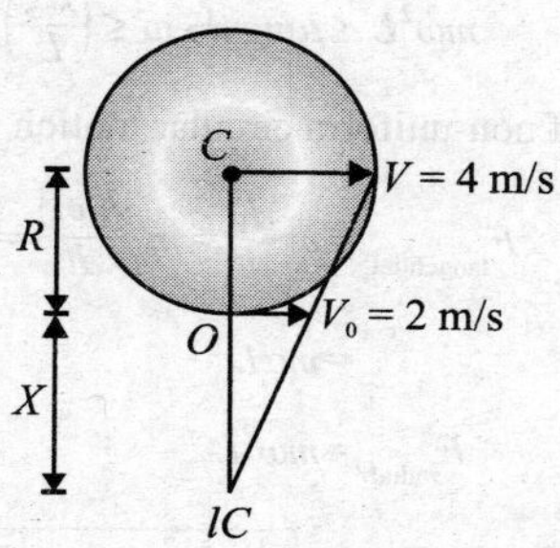

Fig. 2.565

$$\omega = \frac{v_0}{IO} = \frac{v_c}{IC}$$

$$\frac{2}{x} = \frac{4}{x+R} \text{ or } x = 0.2 \text{ m}$$

Velocity of P is given by $v_p = (IP)\omega = \sqrt{5}R \times \frac{2}{x} = 2\sqrt{5}$ m/s

For Problems 54–55

54. b. 55. b.

Sol. b. In reference frame of truck, the angular momentum is conserved about P, i.e.,

$$MvR = Mv_0R + \frac{2}{5}MR^2\omega_0 \quad \text{(i)}$$

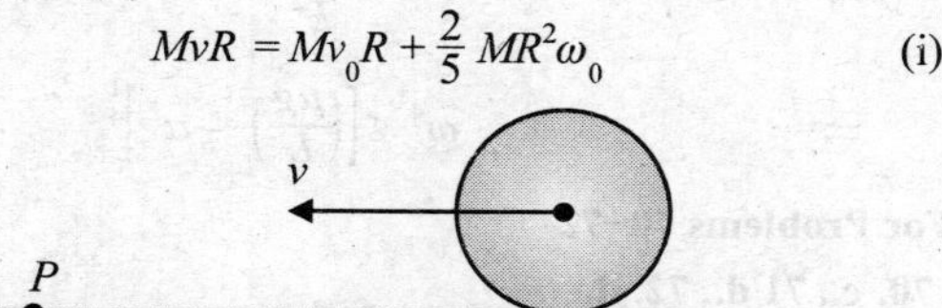

Rolling constraint

Fig. 2.566

$$v = R\omega \quad \text{(ii)}$$

On solving Eqs. (i) and (ii), we get

$$v = \frac{5}{7}v_0$$

Note that v_0 is the speed in truck frame; in ground frame, the velocity is $v_0 - (5v_0/7) = 2v_0/7$ towards the right.

The answer is independent of M or R.

For Problems 56–58

56. a., 57. d., 58. c.

Sol. From Conservation of energy, we get

$$\frac{mgl}{4} = \frac{1}{2}m\left[\frac{l^2}{12} + \frac{l^2}{16}\right]\omega^2; \; \omega^2 = \frac{24g}{7l}$$

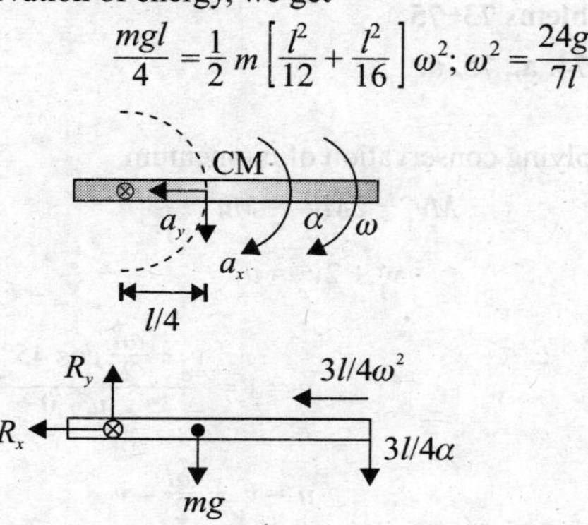

Fig. 2.567

From Newton's law,

$$\frac{mgl}{4} = \left(\frac{ml^2}{12} + \frac{ml^2}{16}\right)\alpha \quad \text{(i)}$$

$$mg - R_y = m\left(\frac{1}{4}\alpha\right) \quad \text{(ii)}$$

$$R_x = \frac{ml}{4}\omega^2 \quad \text{(iii)}$$

$$\alpha = \frac{12g}{7l}; \; \vec{a}_{CM} = -\frac{\omega^2 l}{4}\hat{i} - \frac{1}{4}\alpha\hat{j}$$

$$= -\frac{6g}{7}\hat{i} - \frac{3g}{7}\hat{j}$$

$$\vec{a}_B = -\omega^2\left(\frac{3l}{4}\right)\hat{i} - \left(\frac{3l}{4}\right)\alpha\hat{j}$$

$$\vec{R}_x = -\frac{6mg}{7}\hat{i}; \; \vec{R}_y = +\frac{4mg}{7}\hat{j}$$

For Problems 59–60

59. b., 60. c

Sol. Apply conservation of angular momentum,

$$\frac{mR^2}{2}\omega = \left(\frac{mR^2}{2} + mR^2\right)\omega'$$

$$\omega' = \frac{\omega}{3}$$

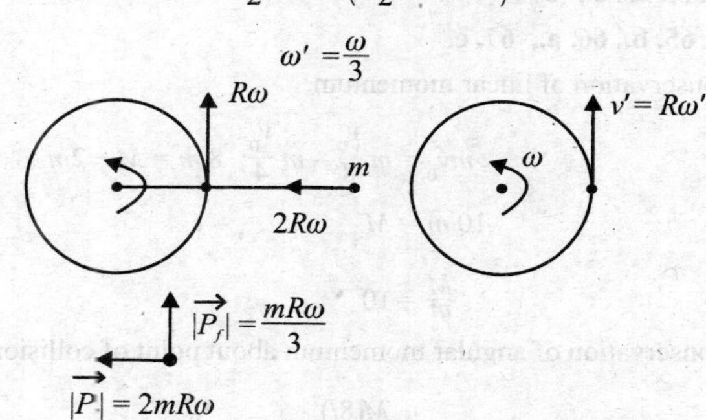

Fig. 2.568

Impulse on particle = change in linear momentum,

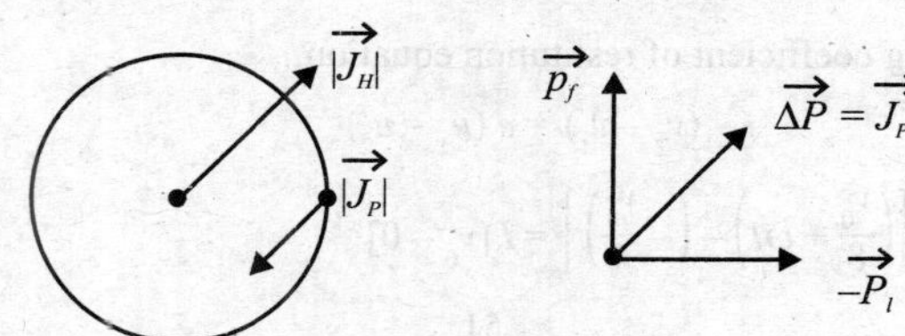

Fig. 2.569

$$\vec{J} = \Delta\vec{p} = \vec{p}_j - \vec{p}_i$$

$$|\Delta\vec{p}| = \sqrt{\left(\frac{mR\omega}{3}\right)^2 + (2mR\omega)^2} = \sqrt{\frac{37}{3}}\, mR\omega$$

Impulse on hinge is negative of impulse on disc.

For Problems 61–63

61. d., 62. b. 63. c.

Sol. 61. d. Using torque equation about point A

$$\Rightarrow \quad FR = I_A\alpha$$

$$\Rightarrow \quad FR = \frac{3}{2}MR^2\left(\frac{a}{R}\right) \Rightarrow \quad a_t = \frac{4}{3}\frac{F}{m}$$

62. b.

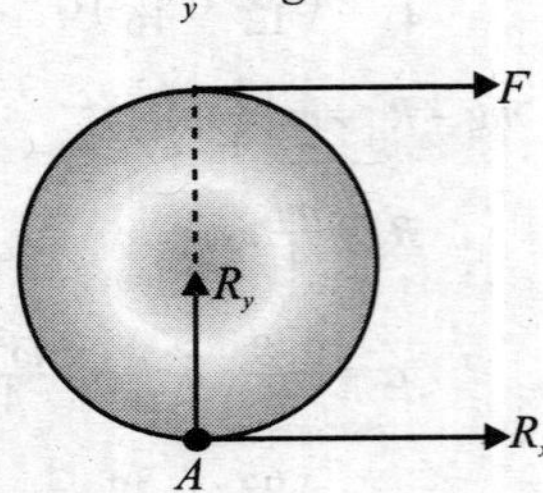

Fig. 2.570

63. c. Torque about the CM,

$$(F - R_x)R = I\alpha$$

$$\Rightarrow \quad (F - R_x)R = \frac{Mr^2}{2}\frac{a_t}{r}$$

$$\Rightarrow \quad (F - R_x)R = \frac{2F}{3}$$

$$\Rightarrow \quad F - \frac{2F}{3} = R_x \Rightarrow R_x = \frac{F}{3}$$

For Problems 64–67

64 a., 65. b., 66. a., 67. c.

Sol. Conservation of linear momentum,

$$mv_0 = m\frac{v_0}{8} - m\frac{v_0}{4}; \; 8\,m = M - 2m$$

$$\Rightarrow \quad 10\,m = M$$

$$\Rightarrow \quad \frac{M}{m} = 10$$

Conservation of angular momentum about point of collision.

$$0 = \frac{M(8l)^2}{12}\omega - M\frac{v_0}{8}l$$

$$\Rightarrow \quad \omega = \frac{3v_0}{128\,l}$$

Using coefficient of restitution equation,

$$(v_2 - v_1) = e\,(u_1 - u_2)$$

$$\left[\left(\frac{v_0}{8} + \omega l\right) - \left(-\frac{v_0}{4}\right)\right] = I\,[v_0 - 0]$$

$$\Rightarrow \quad e = \frac{51}{128}$$

Velocity at end A,

$$v_A = \frac{v_0}{8} + \omega\, 4l = \frac{28}{128}v_0$$

$$\Rightarrow \quad v_A = \frac{7}{32}v_0$$

Velocity at end B,

$$v_B = \frac{v_0}{8} - \omega\, 4l = \frac{4}{128}v_0$$

$$= \frac{1}{32}v_0$$

For Problems 68–69

68. d., 69. a.

Sol. If the block does not slip, $f \le \mu N$

$$m\omega^2 L \le \mu mg \Rightarrow \omega \le \left(\frac{\mu g}{L}\right)^{\frac{1}{2}}$$

It is the case of non-uniform circular motion

$$F_{\text{tangential}} = m\frac{dv}{dt} = m\frac{d(\omega l)}{dt} = mL\frac{d\omega}{dt} = m\alpha L$$

$$= m\alpha L \qquad \text{(i)}$$

$$F_{\text{radial}} = m\omega^2 L \qquad \text{(ii)}$$

Net force, $F_{\text{net}} = \sqrt{(m\alpha L)^2 + (m\omega^2 L)^2}$

Friction force, $f = F_{\text{net}}$

$$= \sqrt{(m\alpha L)^2 + (m\omega^2 L)^2}$$

$$f \le \mu N$$

$$\Rightarrow \quad \sqrt{(m\alpha L)^2 + (m\omega^2 L)^2} \le \mu mg$$

$$\alpha^2 L^2 + \omega^4 L^2 \le \mu^2 g^2$$

$$\Rightarrow \quad \omega^4 \le \frac{\mu^2 g^2}{L^2} - \alpha^2$$

$$\Rightarrow \quad \omega^4 \le \left[\left(\frac{\mu g}{L}\right)^2 - \alpha^2\right]^{\frac{1}{4}}$$

For Problems 70–72

70. c., 71 d., 72. d.

Sol.

70. c. Applying conservation of linear momentum,

$$(2 + 1 + 2)\,v_c = 2 \times 10 + 1 \times 10$$

$$v_c = 6 \text{ m/s}$$

71. d. Applying conservation of angular momentum about the centre of the rod,

$$\left[\frac{2 \times (4)^2}{12} + 2 \times (1)^2 + 1 \times (1)^2\right]\omega = 2 \times 10 \times 1 - 1 \times 10 \times 1$$

$$\omega = \frac{30}{17} \text{ rad/s}$$

72.d. Linear momentum will decrease and angular momentum will increase.

For Problems 73–75

73. d., 74. a., 75. d.

Sol.

Applying conservation of momentum,

$$Mv_1 + 2Mv_c = Mu$$

$$v_1 + 2v_c = u \qquad \text{(i)}$$

$$e = 1 = \frac{v_c + \frac{\omega l}{2}\cos 45° - v_1}{u - 0}$$

$$u = v_c + \frac{\omega l}{2} - v_1 \qquad \text{(ii)}$$

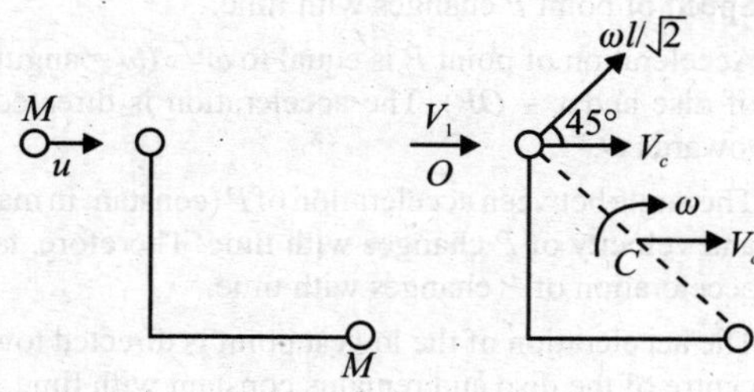

Fig. 2.571

Applying conservation of angular momentum about *C*,

$$Mv_1\frac{l}{2}+2M\left(\frac{l}{\sqrt{2}}\right)^2\omega=Mu\frac{l}{2}$$

$$v_1+2\omega l=u \quad \text{(iii)}$$

Solving Eqs. (i), (ii) and (iii), we get

$$v_1=-\frac{u}{7};\ \omega=\frac{4u}{7l}$$

Angle rotated, $\theta=\pi$

$$t=\frac{\theta}{\omega}=\frac{7\pi l}{4u}$$

Matching Column Type

1. i. → a.; ii.→ d.; iii.→ a.; iv.→ c.

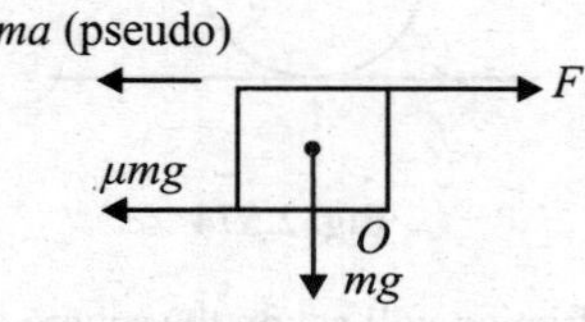

Fig. 2.572

The force required to topple the block about *O*,

$$Fa=\frac{Ma}{g}g+\frac{Ma}{2}\times 1$$

$$F=M+\frac{M}{2}=55\text{ N}$$

and $f_{max}=\mu mg=0.1\times 10\times 10=10\text{ N}$

Force between 0–10 N will do neither sliding or slipping.

Force between 10–55 N will cause slipping.

Force above 55 N will topple the block.

2. i. → c.; ii.→ a.; iii.→ b., d.; iv. → (does not match)

Force *F* passes through the CM of the discs and hence its torque about CM has to be zero.

Also, force of friction will be internal.

After the discs stick together, let the final angular speed be ω,

$$(I_A\omega_A+I_B\omega_B)=(I_A+I_B)\omega$$

$$\omega=\frac{I_A\omega_A+I_B\omega_B}{I_A+I_B}$$

when the discs rub against each other, the kinetic energy will decrease as there will be some loss of energy in doing work against the non-conservative force of friction.

$$\text{Loss of KE}=\frac{1}{2}(I_A+I_B)\omega^2-\frac{1}{2}I_A\omega_A^2-\frac{1}{2}I_B\omega_B^2$$

3. i. → (does not match); ii. → c.; iii → a., b.; iv. → (does not match)

Velocity of two spheres will be the same at the bottom.

$$V_R=\sqrt{\frac{2gh}{\beta}},\ \text{where } \beta=\left[1+\frac{1}{Mr^2}\right]$$

So kinetic energy of smaller mass will be more at the bottom.

Also $\quad a_r=\frac{g\sin\theta}{\beta}$ and $a=\alpha r$

$\Rightarrow \quad \alpha=\frac{a}{r}$

As $r=R/2$, for smaller sphere, $\alpha=2a/R$ and for larger sphere $\alpha=a/R$.

Whereas $t=\frac{1}{\sin\theta}\sqrt{\beta\left(\frac{2h}{g}\right)}$, time will be the same for both the spheres as β will be same.

4. i.→b., d.; ii. →a.; iii → a., b., c., d.; iv. →b., d.

$$Mgh=\frac{1}{2}Mv^2+\frac{1}{2}I\omega^2$$

But as in rolling, $\quad v=r\omega$

$$mgh=\frac{1}{2}mv^2\left[1+\frac{I}{mr^2}\right]$$

$$\left[\text{Putting } \beta=1+\frac{I}{mr^2}\right]$$

$$=\frac{1}{2}\beta mv^2$$

$$\Rightarrow \quad v=\sqrt{\frac{2gh}{\beta}}$$

Solving for acceleration, we have

$$a=\frac{g\sin\theta}{\beta}$$

and time taken

$$t=\frac{1}{\sin\theta}\sqrt{\beta\left(\frac{2h}{g}\right)}$$

5. i.→ c.; ii.→ d.; iii. → (does not match); iv.→ d.

i. For earth moving in an elliptical orbit, centripetal force passes through CM, so its torque will be zero at all points of motion. So angular momentum can be conserved at all points of motion. But as external force is acting, linear momentum will be unconserved.

ii. Angular momentum can be conserved about the point of contact as work done about this point by force of friction is zero. Linear momentum will not be conserved as force of friction is an external force here.

iii. External forces acting on the system are normal reaction, $mg\sin\theta$ and friction. Torque due to normal reaction and $mg\sin\theta$ will be zero about the centre of mass but not of friction. So, angular momentum will not be conserved. Also in the presence of external forces, the linear momentum will not be conserved.

iv. For the projected particle, force of gravity which is external in this case will render linear momentum unconserved. But torque of this force will be zero as it will pass through the CM. So angular momentum is conserved instantaneously.

6. i → b., c.; ii → a., c.; iii → c.

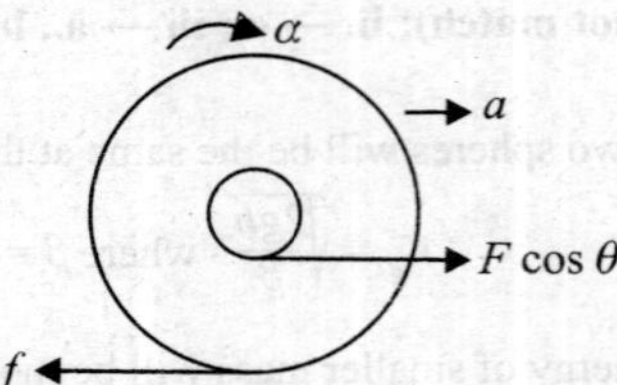

Fig. 2.573

$$F\cos\theta - f = ma$$

$$fR - Fr = \frac{Ia}{R}$$

$$F\cos\theta - F\frac{r}{R} = a(m + \frac{I}{R^2})$$

If $\cos\theta > r/R$, then a is positive

$$a = \frac{F\left(\cos\theta - \frac{r}{R}\right)}{\left(m + \frac{I}{R^2}\right)}$$

If $\cos\theta < r/R$, then a is negative.

In (i), since a = positive, α should be clockwise hence f is towards the left.

Similarly for others.

7. i. → c.; ii. → a.; iii. → a.; iv. → b.

Total mass: M

Mass of cutout portion: $m_1 = \frac{M}{4}$

$$I_1 = \frac{1}{2}MR^2 - \frac{1}{2}m_1\left(\frac{R}{2}\right) = \frac{15}{32}MR^2$$

$$I_2 = I_1 + \frac{3M}{4}\left(\frac{R}{2}\right)^2 = \frac{21}{32}MR^2$$

Also $I_3 + I_4 = I_2$

$$I_2 = I_1 + \frac{3M}{4}\left(\frac{R}{2}\right)^2$$

$$I_3 = \frac{I_1}{2} + \frac{3M}{4}\left(\frac{R}{2}\right)^2$$

$$\Rightarrow \quad I_2 - I_3 = \frac{I_1}{2}$$

8. i. → d.; ii. → b.; c; iii. → d.; iv. → a., c.

We know that angular momentum = linear momentum × perpendicular distance. So about O, angular momentum will remain constant. But about E it will keep on changing. It is maximum about E when the particle is at A, because then the perpendicular distance is maximum.

We know that angular velocity (ω) = v/r. So about O, angular velocity will remain constant. But about E it will keep on changing. It is minimum about E when the particle is at A, because then the perpendicular distance (r) is maximum.

9. i. → a.; ii. → b., d.; iii. → a.; iv. → b., d.

a. Speed of point P changes with time.

b. Acceleration of point P is equal to $\omega^2 x$ (ω = angular speed of disc and $x = OP$). The acceleration is directed from P towards O.

c. The angle between acceleration of P (constant in magnitude) and velocity of P changes with time. Therefore, tangential acceleration of P changes with time.

d. The acceleration of the lowest point is directed towards the centre of the disc and remains constant with time.

10. i → a., b., c.; ii → a., b., c.; iii → a., b.; iv. → a., b., c.

Since all forces on the disc pass through the point of contact with the horizontal surface, the angular momentum of the disc about the point on the ground in contact with disc is conserved. Also the angular momentum of the disc in all cases is conserved about any point on the line passing through the point of contact and parallel to the velocity of the centre of mass.

The KE of the disc is decreased in all cases due to work done by friction. From the calculation of velocity of the lowest point on disc, the direction of friction in case (i), (ii) and (iv) is towards the left and in case r it is towards the right.

The direction of frictional force cannot change in any given case

11. i. → c.; ii. → a., b., c., d., e.; iii. → a., b., c., d., e.; iv. → e.

a. When the sphere rolls on the rough surface, its angular momentum about contact point A remains constant because torque of friction f_k about this point is zero.

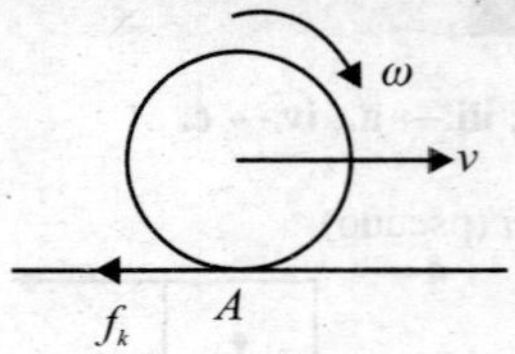

Fig. 2.574

b. Here no friction will act on the sphere. So here everything will remain the same.

c. Here also no friction will act, so everything will remain the same.

d. Here KE will increase. Potential energy will decrease but the total mechanical energy will remain constant. Angular momentum will not be conserved, because $mg\sin\theta$ will provide the torque about contact point.

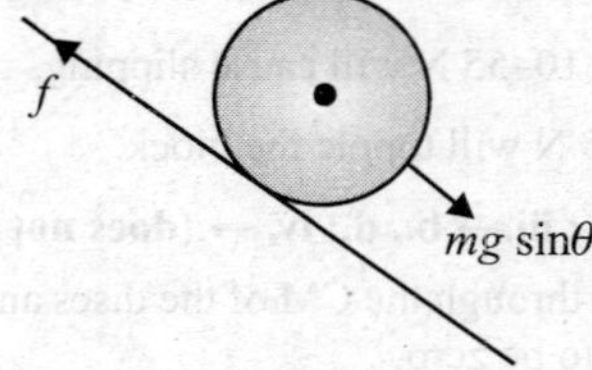

Fig. 2.575

12. i. → a., b., c.; ii. → a.; iii. → d; iv. → d.

a. v increases, ω decreases, hence translational work by friction is positive and rotational work is negative. But overall work will be negative because the friction is kinetic.

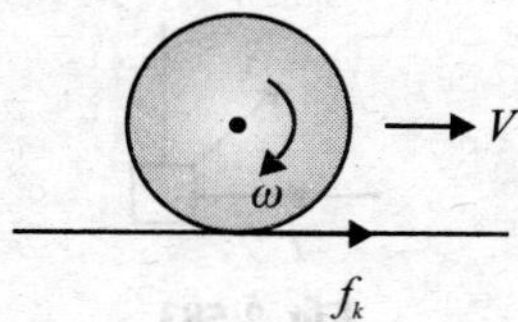

Fig. 2.576

b. v decreases, ω increases, overall work by friction will be negative.

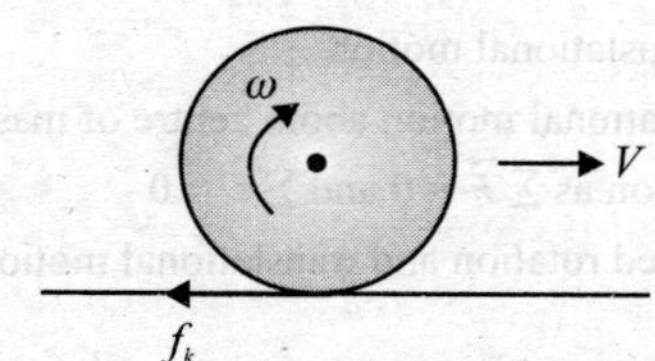

Fig. 2.577

c. This is pure rolling with constant velocity. No friction will act.

d. Here F will pass through the contact point. It will not move. Hence, work by friction is zero.

13. i. → a., b., d.; ii. → a., b., c.; iii. → a., b., d; iv. → a., b., c.

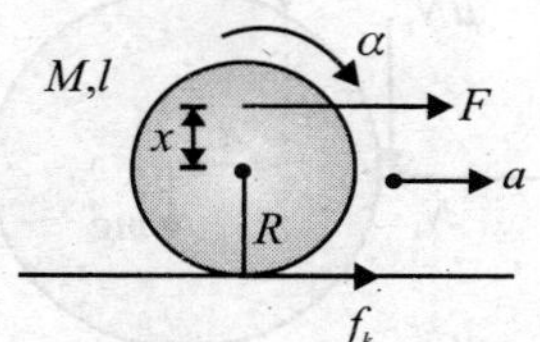

Fig. 2.578

$$F + f = ma$$

$$Fx - fR = I\alpha, \alpha = \frac{a}{R}$$

Solve to get
$$f = \frac{F(MxR - I)}{I + MR^2}$$

$$f = 0, \text{ if } x = \frac{I}{MR}$$

$$f > 0, \text{ if } x > \frac{I}{MR}$$

So friction will act in the forward direction.

$F < 0$ if $x < I/MR$, so friction will act in the backward direction.

Also solved to get

$$a = \frac{FR(x + R)}{I + MR^2}$$

A is positive for all cases, hence body will accelerate in the forward direction in all cases.

It means α will be in the clockwise direction. Hence, bodies will rotate in the clockwise direction.

14. i. → d.; ii. → b., d.; iii. → a., b., c.; iv. → b., c., d.

a. Immediately the sphere will gain velocity, so momentum increases.

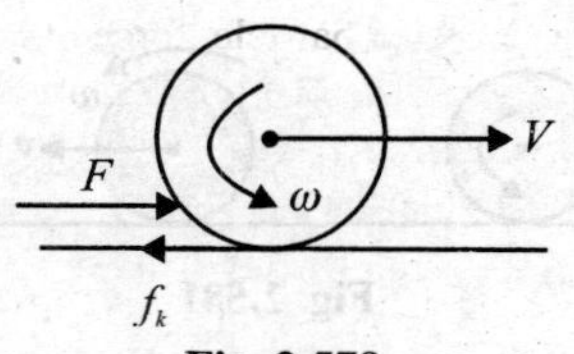

Fig. 2.579

b. Here angular momentum about the hinged point will remain constant.

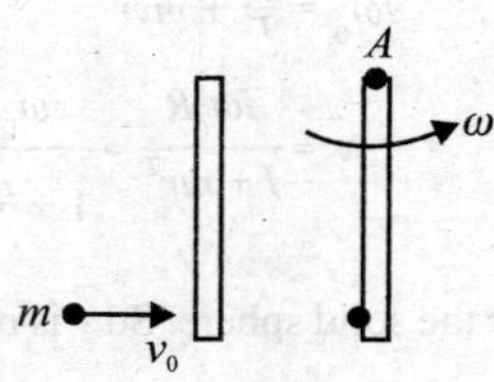

Fig. 2.580

Applying conservation of angular momentum about A,

$$mv_0L = \left(mL^2 + \frac{ML^2}{3}\right)\omega$$

Final momentum, $\omega = \dfrac{mv_0}{\left(m + \dfrac{M}{3}\right)^2}$

$$m\omega L + M\omega\frac{L}{2} = \left(m + \frac{M}{2}\right)\frac{mv_0}{\left(m + \dfrac{M}{3}\right)}$$

This is greater than mv_0.

c. Here no external force or external torque acts. So all remain conserved, as the collision is elastic.

d. Here also momentum increases. As the collision is elastic, so energy is also conserved. Angular momentum is conserved about the hinged point.

15. i. → c.; ii. → c.; iii, → a., b., c.; iv. → a., b., c.

$$a = \frac{g \sin\theta}{1 + \dfrac{I}{MR^2}}$$

For a solid cylinder: $\quad I = \frac{1}{2}MR^2$

So $\quad a_1 = \frac{2}{3} g \sin\theta$

For a thin shell: $\quad I = MR^2$

$$a_2 = \frac{g \sin\theta}{2}$$

So, $a_1 > a_2$, hence the solid cylinder reaches earlier. Now $v = \sqrt{2al}$, acceleration for the solid cylinder is more, so v on reaching the bottom is more for the solid cylinder.

F_{tran} is maximum and KE_{rot} is minimum for the solid cylinder. KE_{tran} is minimum and KE_{rot} maximum for the thin cylinder.

16. i. → a., b., d.; ii. → a., b.; iii. → a., b., c.; iv. → a., b.

In each case, the friction force will act in the forward direction. This friction will increase translational velocity and decrease angular velocity.

Apply conservation of angular momentum about the lowest point,

Fig. 2.581

$$I\omega_0 = I\omega + mvr$$

$$I\omega_0 = \frac{Iv}{r} + mvr$$

$$v = \frac{I\omega_0 R}{I + mr^2} = \frac{\omega_0 r}{1 + \frac{mr^2}{I}}$$

I is minimum for the solid sphere. So v is minimum for the solid sphere.

Similarly, v will be maximum for the ring. From $v = u + at$, time taken by the ring will be maximum, because acceleration is the same for all.

17. i. → a., b., c., d.; ii. → a., d.; iii. → a., b., c., d.; iv. → b., c., d.

a. Since impulse is given to the body, so its momentum will increase and translation will occur. The impulse J does not pass through the centre of mass. Hence, the dumb-bell will rotate and its angular momentum will increase.

b. Here J passes through centre of mass. So rotation does not occur and angular momentum remains zero.

c. This situation is similar to part (a).

d. Here body is hinged. So translation does not occur, rotation will be there so angular momentum will increase.

Linear momentum increases as centre of mass gains some velocity.

18. i. → b., c.; ii. → a., b., c.; iii. → a., d.; iv. → a., d.

a. Angular momentum will not be conserved as T will provide torque about C. Kinetic energy will remain conserved as tension is perpendicular to velocity and it does not do any work. Obviously, the total mechanical energy also remains conserved.

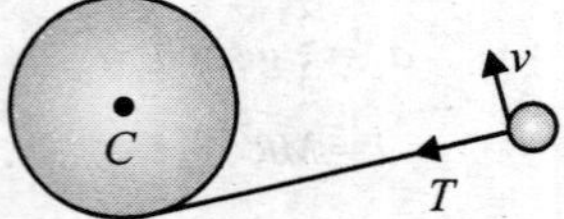

Fig. 2.582

b. Angular momentum remains conserved as no external torque acts. Mechanical energy and kinetic energy of (rod + sleeve) system will remain the same because no work is done by external forces and there is no dissipating force in the system.

c. Angular momentum will remain conserved, because there is no external force (friction is zero because of ice).

Kinetic energy and mechanical energy will decrease because of work done by internal forces (due to muscles) between the icekaters.

d. Angular momentum will remain conserved, because torque of T will be zero.

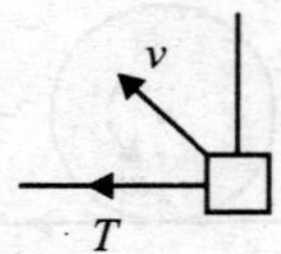

Fig. 2.583

T will do positive work on the block due to which kinetic energy and mechanical energy of the block will increase.

19. i.→ b.; ii. → a.; iii. → c.; iv. → a., b., d.

i. Pure translational motion

ii. Pure rotational motion about centre of mass

iii. No motion as $\sum \vec{F} = 0$ and $\sum \vec{\tau} = 0$

iv. Combined rotation and translational motion

Integer Answer Type

1. **(6)** For equilibrium of the cylinder in the horizontal direction,

$$N_1 = \mu N_2 \quad \text{(i)}$$

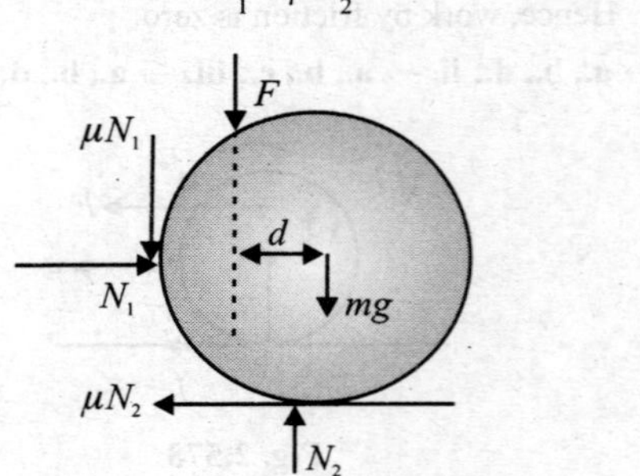

Fig. 2.584

In the vertical direction,

$$N_2 + \mu N_1 = F + Mg \quad \text{(ii)}$$

Solving these two equations with $F = 40$ N, $M = 2$ kg and $\mu = 1/3$, we get

$$N_1 = 18 \text{ N}$$

and $$N_2 = 54 \text{ N}$$

Now, $$Fd = \mu(N_1 + N_1)r$$

$$\therefore \quad d = \frac{M(N_1 + N_2)r}{F}$$

$$= \frac{(1/3)(18 + 54)(0.1)}{40}$$

$$= 0.06 \text{ m}$$

2. **(5)** $$Mg - T = Ma_y \quad \text{(i)}$$

$$T\left(\frac{L}{2}\right) = \frac{ML^2}{12}\alpha \quad \text{(ii)}$$

and $$a_y = \frac{L}{2}\alpha \quad \text{(iii)}$$

on solving, we get

$$T = \frac{Mg}{4} = 5\text{ N}$$

Fig. 2.585

3. **(5)** When the string is cut, the weight of the rod constitutes torque about the hinge, so

$$\tau_A = mg\frac{l}{2} \qquad \text{(i)}$$

According to Newton's second law,

$$\tau_A = I\alpha \qquad \text{(ii)}$$

where α is the angular acceleration of the rod about the end A. From Eqs. (i) and (ii), we get

$$I\alpha = mg\frac{l}{2}$$

or

$$\alpha = \frac{mg\frac{l}{2}}{I}$$

Here $I = ml^2/3$, therefore

$$\therefore \quad \alpha = \frac{mgl/2}{ml^2/3} = \frac{3}{2}\frac{g}{l}$$

Acceleration of the CM of the rod is

$$a_{CM} = \alpha r = \frac{3}{2}\frac{g}{l} \times \frac{l}{2} = \frac{3g}{4}$$

Again by Newton's second law,

$$mg - R_A = ma_{CM}$$

or

$$mg - R_A = m \times \frac{3g}{4}$$

$$\therefore \quad R_A = \frac{Mg}{4} = \frac{2\times 10}{4} = 5\text{ N}$$

4. **(4)** Here the force on the point mass due to the cord is radial and hence the torque about the centre of rotation is zero. Therefore, the angular momentum must remain constant as the cord is shortened.

Let m, v_1 and ω_1 be the mass, linear velocity and angular velocity of the point mass, respectively, in the circle of radius r_1. Further let v_2 and ω_2 be the linear and angular velocities, respectively, of the point mass in a circle of radius r_2. Now, initial angular momentum = final angular momentum

$$I_1\omega_1 = I_2\omega_2$$

$$mr_1^2\frac{v_1}{r_1} = mr_2^2\frac{v_2}{r_2}$$

$$r_1 v_1 = r_2 v_2$$

$$v_2 = \frac{r_1}{r_2}v_1 = \frac{2}{1} \times 4 = 8\text{ m/s}$$

and

$$\omega_2 = \frac{v_2}{r_2} = \frac{8}{1} = 8\text{ rad/s}$$

$$\frac{\text{Final KE}}{\text{Initial KE}} = \frac{\frac{1}{2}I_2\omega_2^2}{\frac{1}{2}I_1\omega_1^2}$$

$$= \frac{mr_2^2(v_2/v_1)^2}{mr_1^2(v_1/r_1)^2} = \frac{v_2^2}{v_1^2} = \frac{(8)^2}{(4)^2} = 4$$

5. **(3)** For circular motion of the stone,

$$\frac{mv^2}{r} = T\,[\text{as } g = 0] \qquad \text{(i)}$$

and as A.M. is constant,

$$mvr = K,$$

i.e.,

$$v = \frac{K}{mr} \qquad \text{(ii)}$$

Eliminating v between Eqs. (i) and (ii), we get

$$\frac{m}{r}\left[\frac{K}{mr}\right]^2 = T,$$

$$T = \frac{K^2}{m}r^{-3}$$

or

$$T = Ar^{-3} \text{ with } A = \left(\frac{K^2}{m}\right) \qquad \text{(iii)}$$

comparing Eqs. (iii) with $T = A/r^n$, we find $n = 3$.

6. **(9)** As the belt does not slip, $v_P = v_Q$

i.e.,

$$r_A\omega_A = r_C\omega_C\,[\text{as } v = r\omega] \qquad \text{(i)}$$

According to the given problem if $r_A = r$, $r_C = 3r$, so Eq. (i) becomes

$$\omega_A = 3\omega_C \qquad \text{(ii)}$$

If both the wheels have the same rotational kinetic energy, then

$$\frac{1}{2}I_A\omega_A^2 = \frac{1}{2}I_C\omega_C^2$$

or

$$\frac{I_A}{I_C} = \left[\frac{\omega_C}{\omega_A}\right]^2 = \left[\frac{1}{3}\right]^2 = \frac{1}{9}$$

7. **(2)** $J = \int N\,dt = m\frac{v}{2} - (-mv) = \frac{3}{2}mv \qquad \text{(i)}$

$$\mu JR = \int \mu\,(N\,dt)\,R = \left(\frac{2}{5}mR^2\omega - 0\right) = \frac{2}{5}mR^2\omega \qquad \text{(ii)}$$

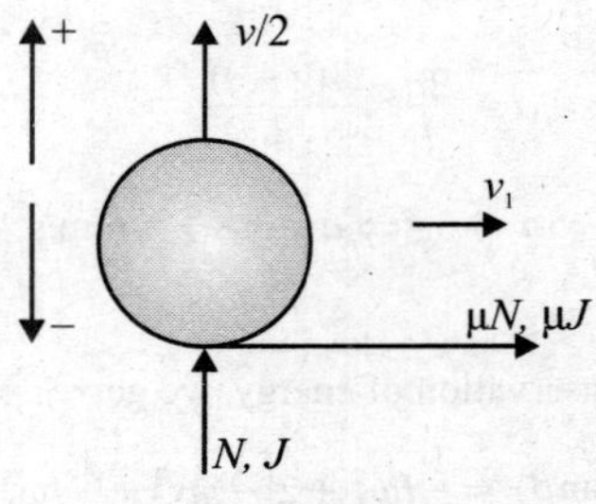

Fig. 2.586

From Eqs. (i) and (ii), we get

$$\frac{3}{2}mv\,R\mu = \frac{2}{5}mR^2\omega \qquad \text{(iii)}$$

Let V and V_1 be the speeds of the plank and the sphere, respectively, in the horizontal direction.

$$\mu J = \int \mu\, N dt = MV = mV_1 \quad \text{(iv)}$$

From Eqs. (i) and (iv), $\mu\left(\frac{3}{2}\right)mv = MV$

$$V = \frac{3}{2}\frac{\mu mv}{M} = \frac{3}{2}\frac{4}{15}\frac{mR\omega}{M} = \frac{2}{5}\frac{m}{M}R\omega$$

and $\quad V_1 = \frac{2}{5}R\omega = 2 \text{ m/s}$

8. (6) Using conservation of mechanical energy,

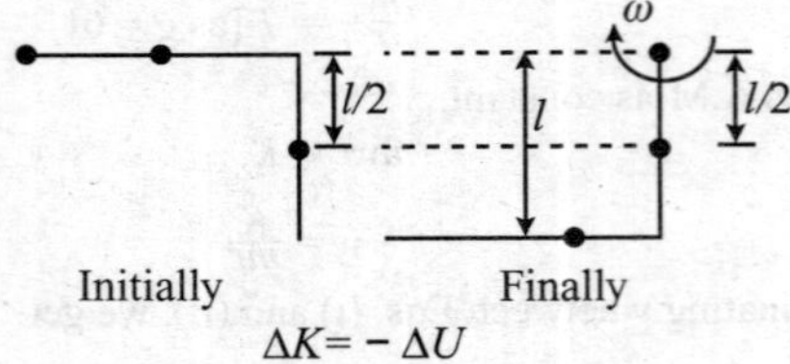

Fig. 2.587

$$mgl = \frac{1}{2}\frac{ml^2}{3}\omega^2 + \frac{1}{2}\left[\frac{ml^2}{12} + m\frac{5l^2}{4}\right]\omega^2$$

$\Rightarrow \quad \omega = 6 \text{ rad/s}$

9. (6) $T\frac{l}{2} = \frac{ml^2}{6} \times \alpha;\ mg - T = ma$

$$\alpha = \frac{3T}{ml} = \frac{3}{ml}(mg - ma)$$

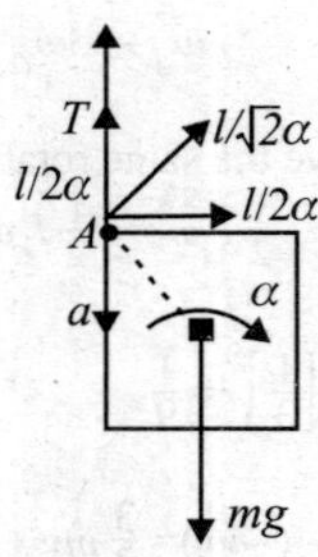

Fig. 2.588

Constraint relation: Acceleration of A in the vertical direction should be zero

$$\frac{\ell}{2}\alpha = a \quad \Rightarrow \quad \alpha = \frac{2a}{l} = \frac{3(g-a)}{l}$$

$$5a = 3g \Rightarrow a = 3 \times \frac{g}{5} = 6 \text{ m/s}^2$$

10. (2)

Applying conservation of energy, we get

$$mgs(\sin\alpha - \sin\beta) = \frac{1}{2}I\omega^2 + 2\left(\frac{1}{2}mv^2 + \frac{1}{2}I\omega^2\right)$$

where $\omega = v/r$ and $I = mr^2/2$

Putting values and solving, we get $v = 2$ m/s.

11. (0)

Let f be the friction force acting on the disc in the backward direction. Then $2F - f = ma$ and $(F + f)r = I\omega$

$$\Rightarrow \quad (F + f)r = \frac{1}{2}mr^2\frac{a}{r}.$$

$\therefore$ solving the above equations, we get $f = 0$.

12. (5) Let T be the tension in the thread and f, the linear acceleration of the reel as it falls.

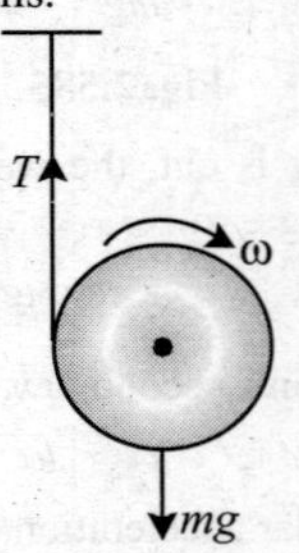

Fig. 2.589

For the downward translation,

$$(mg - T) = mf \quad \text{(i)}$$

For the rotational motion of the reel, angular acceleration is

$$\alpha = \left(\frac{f}{a}\right) \text{ and } T = \frac{mg}{3} \quad \text{(ii)}$$

From Eqs. (i) and (ii), $T = mg - mf = mg - ma\alpha$
$= mg - 2T$

$\Rightarrow 3T = mg$

$\therefore \quad T = \frac{mg}{3} = 1.5 \times 10/3 = 5$ N

13. (2)

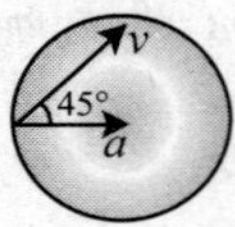

Fig. 2.590

The required angle is 45°, so $\frac{\sqrt{2}}{n} = \frac{1}{\sqrt{2}} \Rightarrow n = 2$

14. (2)

Loss in PE = gain in KE

$$mgh = \frac{1}{2}mv^2 + \frac{1}{2}I\omega^2$$

$$\Rightarrow mgh = \frac{1}{2}m(\omega r)^2 + \frac{1}{2}\left(\frac{2}{3}mr^2\right)\omega^2$$

$$m \times 10\left(\frac{3}{100}\right) = m\frac{5}{6} \times \left(\frac{30}{100}\right)^2 \omega^2 \Rightarrow \omega = 2 \text{ rad/s}$$

15. (4) Maximum acceleration of the cart so that the cylinder does not slip:

$$a_m = \mu g = 0.5 \times 10 = 5 \text{ m/s}^2$$

For tipping over: Let acceleration of the cart be a.

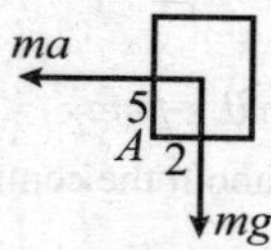

Fig. 2.591

Considering torque about A, we get

$ma \times 5 \geq mg \times 2 \Rightarrow a \geq 2g/5 = 4 \text{ m/s}^2$

16. (6)

$$\frac{L_{\text{total}}}{L_B} = \frac{m_1 r_1^2}{m_2 r_2^2} + 1$$

17. (4)

Note: If net force applied by the rod is considered to be 2 N.

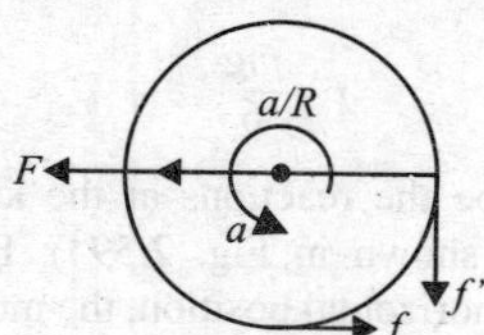

Fig. 2.592

$$\sqrt{f'^2 + F^2} = 2 \qquad \text{(i)}$$

$$FR - f'R = 2mR^2 \frac{a}{R}$$

$$F - f' = 2ma = 1.2 \qquad \text{(ii)}$$

From Eqs. (i) and (ii), $(1.2 + f')^2 + f'^2 = 2^2$

$2f'^2 + 2.4 f' + 1.44 = 4$

$f'^2 + 1.2 f' + 0.72 - 2 = 0$

$f'^2 + 1.2 f' - 1.28 = 0$

$$f' = \frac{-1.2 \pm \sqrt{1.44 + 4 \times (1.28)}}{2}$$

$$= 0.6 \pm \sqrt{0.36 + 1.28} = -0.6 \pm \sqrt{0.64} = 0.68$$

From Eq. (ii), $F = 1.88$.

$$\nu = \frac{0.68}{1.88} = \frac{P}{10} \Rightarrow P = 3.16 \approx 4$$

Note: But if only normal reaction applied by the rod is considered to be 2 N.

II Law $\Rightarrow$ $2 - f = 2[0.3]$ $\Rightarrow$ $f = 2 - 0.6$

$f - 1.4\ Nx$ (i)

$A = R\alpha$

$\Rightarrow$ $0.3 = \alpha[0.5]$

$\Rightarrow$ $\alpha = \frac{3}{5}$ rad/s (ii)

$\tau_c = I_c \alpha \Rightarrow fR - 2\mu R = mR^2\alpha$

$f - 2\mu = mR\alpha$

$$1.4 - 2\mu = \frac{2}{2}\left(\frac{3}{2}\right)$$

$1.4 - 0.6 = 2\mu$

$$0.8 = 2\mu \Rightarrow \mu = 0.4 = \frac{P}{10}$$

$\therefore P = 4$

18. (9)

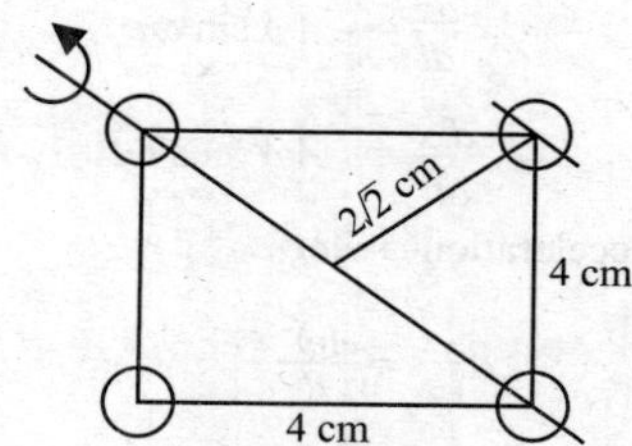

Fig. 2.593

$$I = \left(\frac{2}{5}MR^2\right)2 + \left(\frac{2}{5}MR^2 + Mx^2\right)2$$

$$= \left(\frac{2}{5}MR^2\right)2 + \left(\frac{2}{5}MR^2\right)2 = (Mx^2)2$$

$$= 4\left(\frac{2}{5}MR^2\right) + 2Mx^2 = \frac{8}{5}MR^2 + 2mx^2$$

$$= \left[\frac{8}{5} \times 0.5 \times \left(\frac{\sqrt{5}}{2}\right)^2 + 2 \times (0.5) \times (4 \times 2)\right]10^{-4}$$

$$= \left[\frac{5}{5} + 8\right] \times 10^{-4} = 9 \times 10^{-4} = N \times 10^{-4}$$

So, $N = 9$.

Archives

Fill in the Blanks Type

1. Taking moment of force about D,

$$F = \frac{3a}{4} = mg \times \frac{a}{2} = \frac{2}{3}mg$$

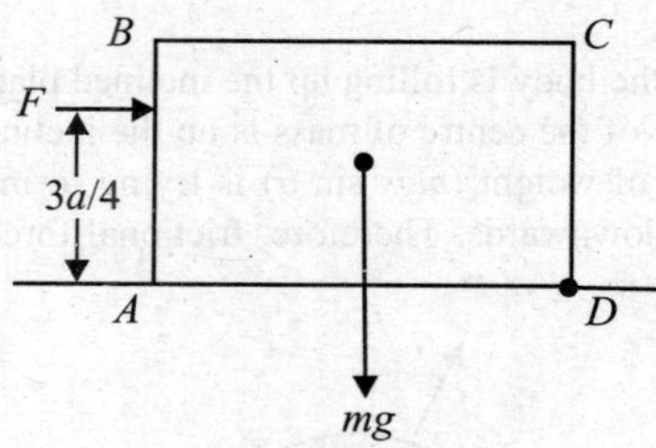

Fig. 2.594

2. Since no force and hence no torque is applied, the angular momentum remains constant

$\therefore$ $I_1\omega_1 = I_2\omega_2$

$\therefore$ $\omega_2 = \frac{I_1\omega_1}{I_2}$

$$= \frac{\frac{ML^2}{12} \times \omega_0}{\frac{ML^2}{12} + 2m \times \left(\frac{L}{2}\right)^2} = \frac{M\omega_0}{M + 6m}$$

3. $x = A \cos \omega t$

$\Rightarrow \quad \frac{dx}{dt} = -A\omega \sin \omega t$

$\Rightarrow \quad \frac{d^2x}{dt^2} = -A\omega^2 \cos \omega t$

$\therefore \quad |\text{Max acceleration}| = A\omega^2$

$\therefore \quad \alpha_{max} = \frac{A\omega^2}{R}$

Max torque $= I\alpha_{max}$

$$= \frac{1}{2}MR^2 \times \frac{A\omega^2}{R} = \frac{1}{2}MRA\omega^2$$

4. Let at any instant of time t, the radius of the horizontal surface be r.

$$T = mr\omega^2 \quad \text{(i)}$$

where m is the mass of the stone and ω is the angular velocity at that instant of time t.

At time t

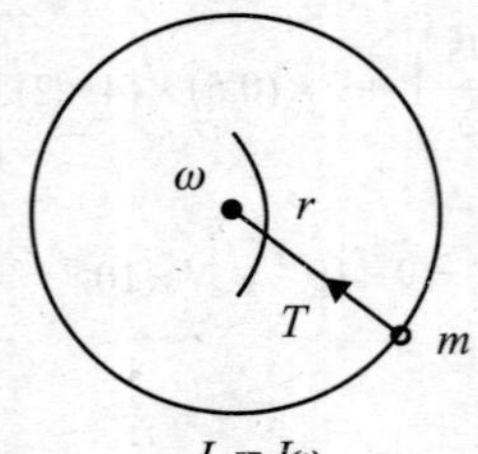

$L = I\omega$

Fig. 2.595

Also, $\quad L = I\omega \quad$ (ii)

From Eqs. (i) and (ii)

$$T = \frac{mrL^2}{I^2} = \frac{mL^2}{(mr^2)^2} \times r$$

$\Rightarrow \quad T \propto r^{-3}$ ($\because$ m, L are constant)

5. Given that the body is rolling up the inclined plane. Therefore, the velocity of the centre of mass is up the inclined plane. The component of weight ($mgs \sin \theta$) is trying to move the point of contact downwards. Therefore, frictional force will act upwards.

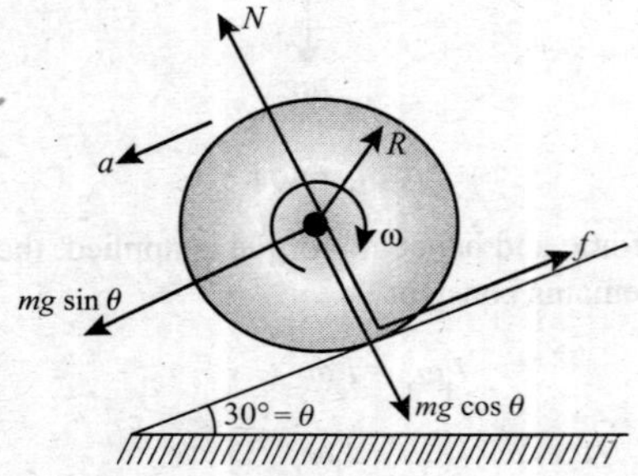

Fig. 2.596

From force diagram,

$$mg \sin \theta = f\, ma_c \quad \text{(i)}$$

For rotational motion about the centre of mass of disc O,

$$f \times R = I\alpha$$

$\Rightarrow \quad$ But $a_c = \alpha R$

$\Rightarrow \quad \alpha = \frac{a_c}{R} f \times R = \frac{Ia_c}{R}$

$\Rightarrow \quad a_0 = \frac{fR^2}{I} = \frac{fR^2}{\frac{1}{2}mR^2} = \frac{2f}{m} \quad$ (ii)

From Eqs. (i) and (ii),

$\Rightarrow \quad 3f = mg \sin \theta = mg \sin 30°$

$\Rightarrow \quad f = \frac{mg}{6}$

6. Let R_A and R_B be the reactions at the knife edges A and B, respectively (as shown in Fig. 2.591). For the rod to be in equilibrium in a horizontal position, the moment of forces about the knife edges must be equal to zero.

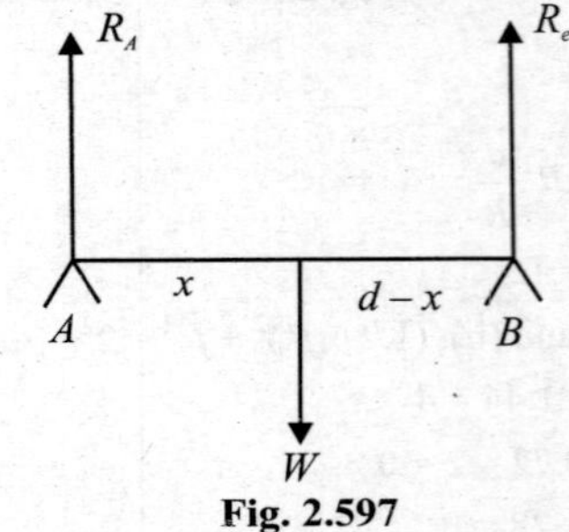

Fig. 2.597

Moment of forces about A,

$$R_A \times 0 + x\, W - R_B d = 0$$

i.e., $\quad R_B = \frac{xW}{d}$

Moment of forces about B,

$$R_A\, d - (d - x)\, W - R_B\, 0 = 0$$

i.e., $\quad R_A = \left(\frac{d - x}{d}\right) W$

7. Assuming symmetric lamina to be in xy plane, we will have

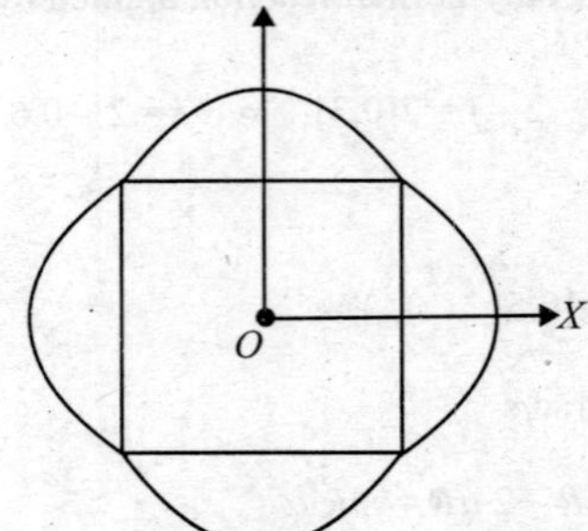

Fig. 2.598

$I_x = I_y$ (Since the mass distribution is the same about the x-axis and y-axis)

$I_x = I_y = I_z$ (perpendicular-axis theorem)

It is given that

$$I_z = 1.6\ Ma^2$$

Hence, $I_x = I_y = \frac{I_z}{2} = 0.8Ma^2$

Now, according to the parallel-axis theorem, we get

$$I_{AB} = I_{XT}\ M(2a)^2$$

$$= 0.8Ma^2 + 4Ma^2 = 4.8Ma^2$$

True or False Type

1. False: $\tau = I_a$, therefore $\alpha = \frac{\tau}{L}$

τ = Force × Perpendicular distance

Torque is the same in both the cases. But since I will be different due to different mass distribution about the axis, therefore α will be different.

2. False: $\vec{\tau} = \frac{\vec{dt}}{dt}$

Since $\vec{\tau} = 0$, $\vec{L}$ = const

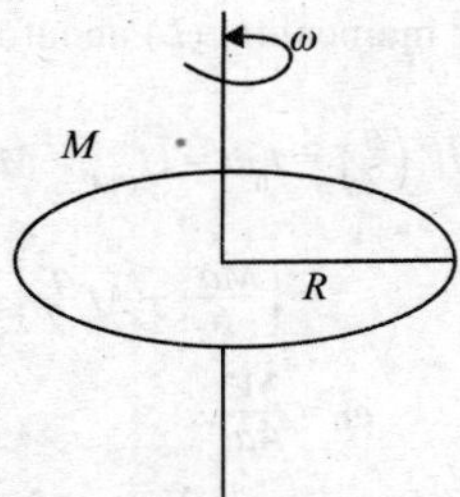

Fig. 2.599

$\therefore \quad I_1\omega_1 = I_2\omega_2$

$$I_1 = \frac{1}{2}MR^2, \omega_1 = \omega$$

$$I_2 = \frac{1}{2}MR^2 + \frac{1}{2}\frac{M}{4}R^2$$

$$= \left(\frac{4+1}{8}\right)MR^2 = \frac{5}{8}MR^2$$

$$\omega_2 = ?$$

$$\omega_2 = \frac{I_1\omega_1}{I_2} = \frac{\frac{1}{2}MR^2 \times \omega}{\frac{5}{8}MR^2} = \frac{8}{2 \times 5}\omega = \frac{4}{5}\omega$$

3. False: Total energy of the ring = $(KE)_{Rotation} + (KE)_{Translational}$

$$= \frac{1}{2}I\omega^2 + \frac{1}{2}mV_c^2$$

$$= \frac{1}{2} \times mr^2\omega^2 + \frac{1}{2}m(r\omega)^2$$

$\therefore \quad I = mr^2, V_c = r\omega$

Total KE of the cylinder

$$= (KE)_{Rotation} + (KE)_{Translational}$$

$$= \frac{1}{2} = I'\omega'^2 + \frac{1}{2}MV_c'^2$$

$$= \frac{1}{2}\left(\frac{1}{2}Mr^2\right)\omega'^2 + \frac{1}{2}M(r\omega')^2$$

$$= \frac{3}{4}Mr^2\omega'^2 \quad \text{(i)}$$

Equating Eqs. (i) and (ii), $mr^2\omega^2 = \frac{3}{4}Mr^2\omega'^2$

$$\Rightarrow \quad \frac{\omega'^2}{\omega^2} = \frac{4m}{3M} = \frac{4}{3} \times \frac{0.3}{0.4} = 1$$

$$\Rightarrow \quad \omega' = \omega$$

Both will reach at the same time.

4. False: Since no external force is acting on the two particle system,

$\therefore \quad a_{CM} = 0$

$\Rightarrow \quad V_{CM}$ = constant

Multiple Choice Questions with One Correct Answer Type

1. c. Since the objects are placed gently, therefore no external torque is acting on the system. Therefore, angular momentum is constant.

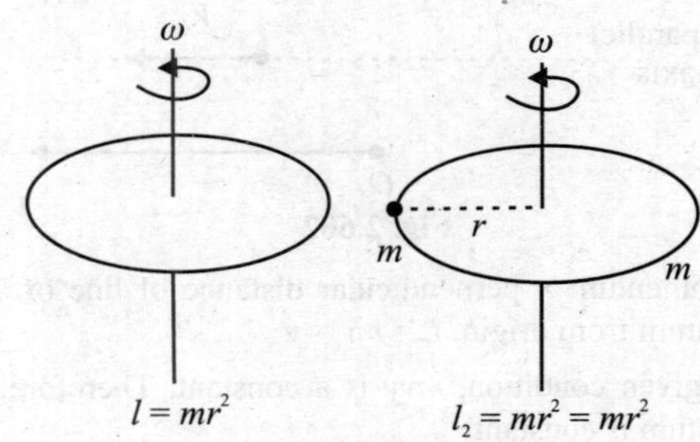

Fig. 2.600

i.e., $\quad I_1\omega_1 = I_2\omega_2$

$$Mr^2 \times \omega_1 = (Mr^2 + 2mr^2)\omega_2$$

$\therefore \quad \omega_2 = \frac{M\omega}{M + 2m}$

2. c. The moment of inertia of the system about axis of rotation O is

$$I = I_1 + I_2$$

$$= 0.3x^2 + 0.7(1.4 - x)^2$$

$$= 0.3x^2 + 0.7(1.96 + x^2 - 2.8\pi)$$

$$= x^2 + 1.372 - 1.96x$$

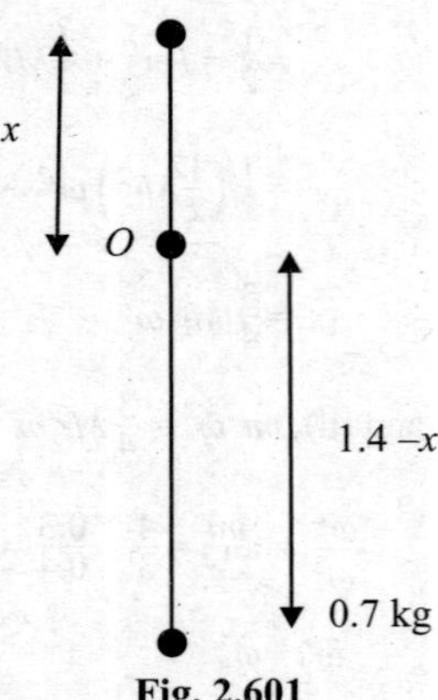

Fig. 2.601

The work done in rotating the rod is converted into its rotational kinetic energy.

$$\therefore \quad W = \frac{1}{2}I\omega^2 = \frac{1}{2}[x^2 + 1.372 - 1.96x]\omega^2$$

For work done to be minimum

$$\frac{d\omega}{dx} = 0$$

$$\Rightarrow \quad 2x - 1.96 = 0$$

$$\Rightarrow \quad x = \frac{1.96}{2} = 0.98 \text{ m}$$

3. b. Angular momentum of mass *m* moving with a constant velocity about origin is

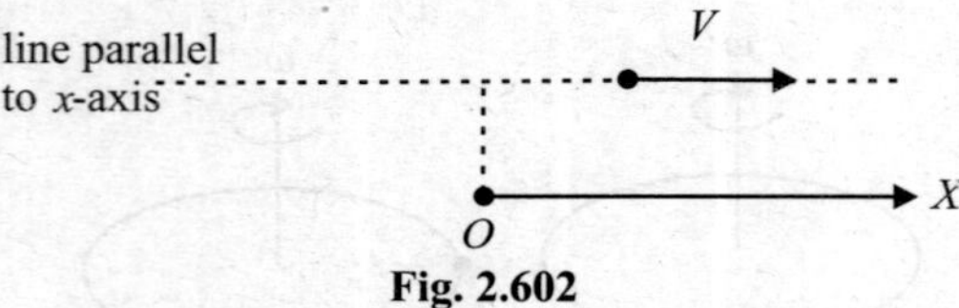

Fig. 2.602

L = momentum × perpendicular distance of line of action of momentum from origin, $L = mv \times y$

In the given condition, *mvy* is a constant. Therefore, angular momentum is constant.

4. c. As the spheres are smooth, there will be no transfer of angular momentum. Thus, *A*, after collision, will remain with its initial angular momentum.

5. c. The disc has two types of motion, namely, translational and rotational. Therefore, there are two types of angular momentum and the total angular momentum is the sum of these two.

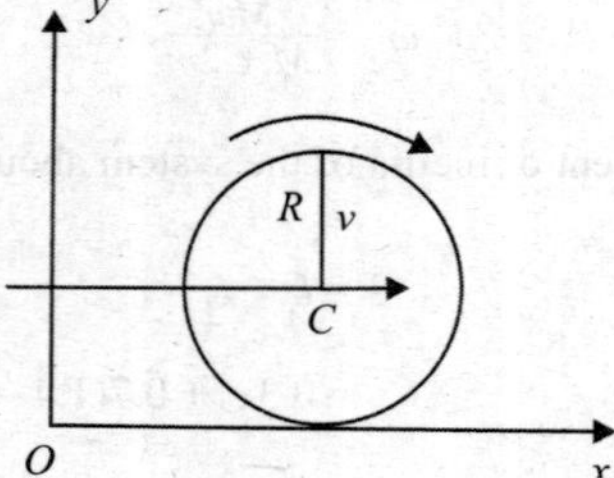

Fig. 2.603

$L = L_T + L_R$, L_T = angular momentum due to translational motion

L_R = angular momentum due to rotational motion about CM

$L = MV \times R + I_{CM}\, \omega I_{CM}$

= MI about center of mass *C*

$= M(R\omega)R + \frac{1}{2}MR^2\omega$ ($V = R\omega$ in case of rolling motion and surface at rest)

$= \frac{3}{2}MR^2\omega$

6. a.

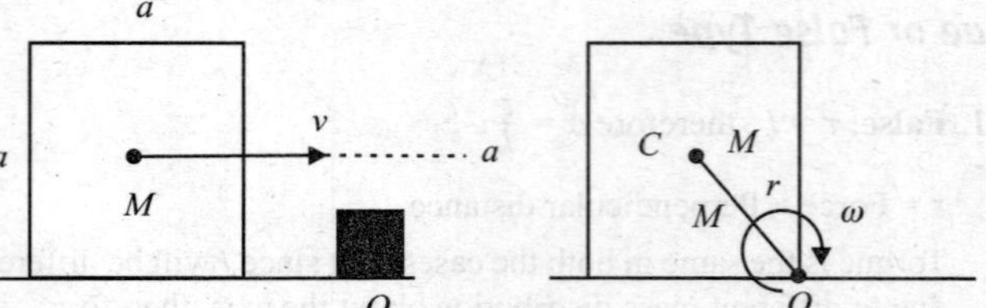

Fig. 2.604

$$r = \sqrt{2}\frac{a}{2} \text{ or } r^2 = \sqrt{2}\frac{a^2}{2}$$

Net torque about *O* is zero.

Therefore, angular momentum (*L*) about *O* will be conserved, or $L_i = L_f$

$$MV\left(\frac{a}{2}\right) = I_0\omega = (I_{CM} + Mr^2)\omega$$

$$= \left\{\frac{Ma^2}{6} + M\frac{a^2}{2}\right\}\omega$$

$$\omega = \frac{3V}{4a}$$

7. a. When we are giving an angular acceleration to the rod, the bead is also having an instantaneous acceleration $a = L\alpha$. This will happen when a force is exerted on the bead by the rod. The bead has a tendency to move away from the centre. But due to friction between the bead and the rod, this does not happen to the extent to which frictional force is capable of holding the bead. The frictional force here provides the necessary centripetal force. If the instantaneous angular velocity is ω, then

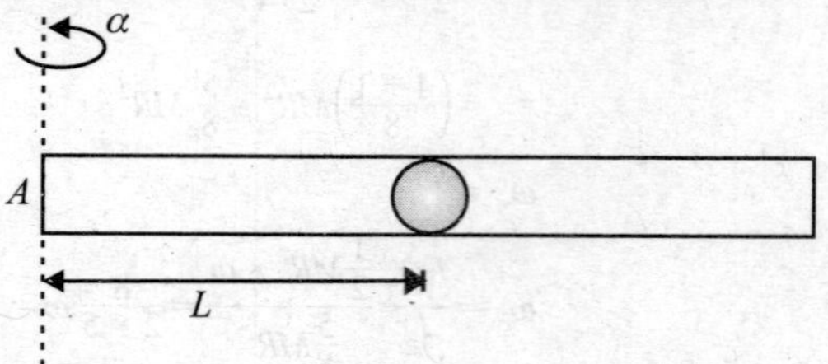

Fig. 2.605

$$mL\omega^2 = \mu(ma) = \mu mL\alpha$$

$$\Rightarrow \quad \omega^2 = \mu\alpha$$

By applying $\omega = \omega_0 + \alpha t$, we get $\omega = \alpha t$

$$\therefore \quad \alpha^2 t^2 = \mu\alpha \Rightarrow t = \sqrt{\frac{\mu}{\alpha}}$$

8. c. The applied force shifts the normal reaction to one corner as shown in Fig. 2.606. At this situation, the cubical block starts topping about *O*. Taking torque about *O*, we get

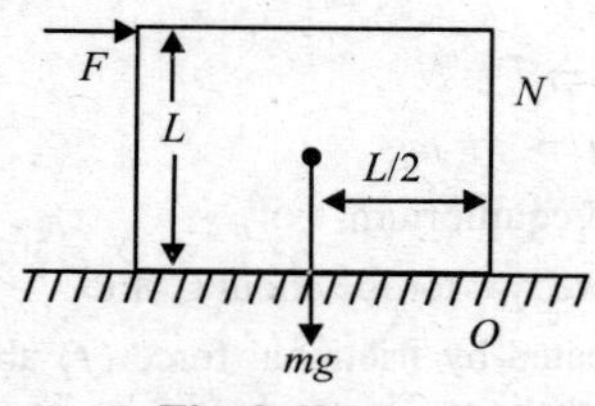

Fig. 2.606

$$F \times L = mg \times \frac{L}{2}$$

$$\Rightarrow \quad F = \frac{mg}{2}$$

9. d. The moment of inertia of the loop about XX' axis is

$$I_{XX} = \frac{mR^2}{2} + mR^2 + \frac{3}{2} mR^2$$

where m = mass of the loop and R = radius of the loop

Here $m = L\rho$ and $R = \frac{L}{2\pi}$; therefore,

$$I_{XX'} = \frac{3}{2}(L\rho)\left(\frac{L}{2\pi}\right)^2 = \frac{3L^3\rho}{8\pi^2}$$

10. b. The MI about the axis of rotation is not constant as the perpendicular distance of the bead with the axis of rotation increases.

Also since no external torque is acting, therefore

$$\tau_{ext} = \frac{dL}{dt}$$

$$\Rightarrow \quad L = \text{constt.}$$

$$I\omega = \text{constt.}$$

Since I increases, ω decreases.

11. a. We cannot consider the quadrant as a single mass as the distance of different particles is different from the axis of rotation. So we take the help of calculus. Let us consider a segment as shown in Fig. 2.607. All masses lying in this segment are at a distance r from the axis and hence considered as a small differential mass dm. Let the thickness of the segment be dr.

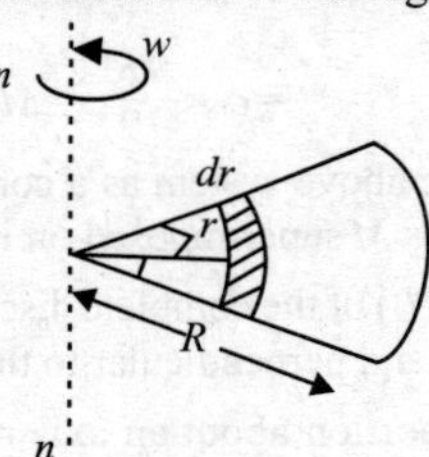

Fig. 2.607

The mass per unit area of the quadrant $= \frac{M}{\pi R^2/4} = \frac{4M}{\pi R^2}$

Area of the segment $= \frac{1}{4}[\pi(r+dr)^2 - \pi r^2] = \frac{1}{4} \times 2\pi r dr = \frac{\pi r dr}{2}$

Mass of the segment $dm = \frac{\pi r dr}{2} \times \frac{4M}{\pi R^2} = \frac{2M}{R^2} r^2 dr$

MI of this segment about $nn' = \frac{2M}{R^2} r dr \times r^2 \frac{2M}{R^2} r^3 dr$

MI of the quadrant about $nn' = \int_0^R \frac{2M}{R^2} r^3 dr$

$$= \frac{2M}{R^2} \times \frac{R^4}{4} \times \frac{MR^2}{2}$$

12. b. Imagine the cylinder to be moving on a frictionless surface. In both the cases the acceleration of the centre of mass of the cylinder is $g \sin\theta$. This is also the acceleration of the point of contact of the cylinder with the inclined surface. Also no torque (about the center of cylinder) is acting on the cylinder since we assumed the surface to be frictionless and the forces acting on the cylinder is mg and N which pass through the centre of the cylinder. Therefore, the net movement of the point of contact in both the cases is in the downward direction as shown in Fig. 2.608. Therefore, the frictional force will act in the upward direction in both the cases.

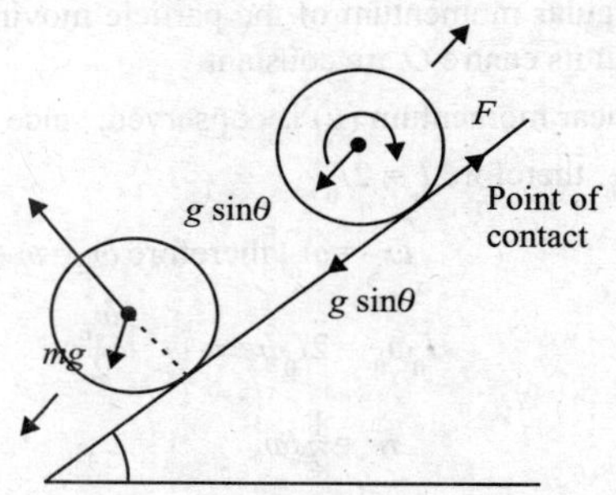

Fig. 2.608

Please note that, in géneral, we find the acceleration of the point of contact due to translational and rotational motion and then find the net acceleration of the point of contact. The frictional force acts in the opposite direction to that of net acceleration of point of contact.

13. b. Since there is no external torque, angular momentum remains conserved. As moment of inertia initially decreases and then increases, so ω will increase initially and then decrease.

Moment of inertia when tortoise is at B,

$$I = mr^2 + \frac{MR^2}{2};$$

where $r^2 = a^2 + [\sqrt{\sqrt{2} - a^2} - vt]^2$; m = mass of tortoise ; M = mass of disc

Therefore, the variation of ω is non-linear.

14. a. Change in angular momentum of the system = angular impulse given to the system (angular momentum)$_f$ – (angular momentum)$_i$

$$= MV \times \frac{L}{2} \qquad \text{(i)}$$

Let the system starts rotating with the angular velocity ω.

Moment of inertia of the system about its axis of rotation (centre of mass of the system) $= M\left(\frac{L}{2}\right)^2 + M\left(\frac{L}{2}\right)^2 = \frac{2ML^2}{4}$

$$= \frac{ML^2}{2} \text{ (From (i))}$$

$$I\omega - 0 = MV \times \frac{L}{2}$$

$$\Rightarrow \omega = \frac{MV}{I} \times \frac{L}{2} = \frac{MV}{ML^2/2} \times \frac{L}{2} = \frac{V}{L}$$

15. a. $|\vec{L}| = mvr$

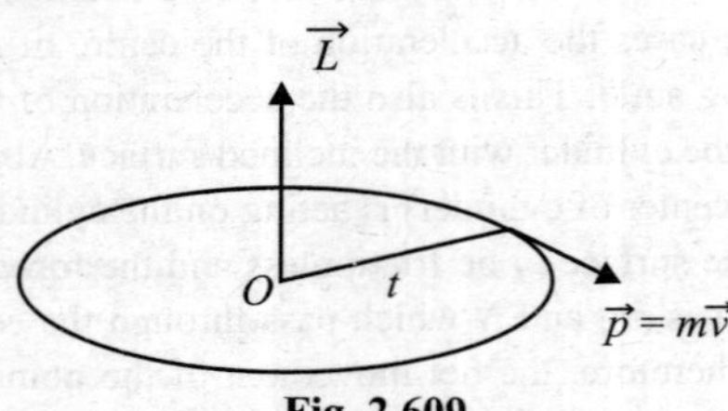

Fig. 2.609

The direction of $\vec{L}$ (about the centre) is perpendicular to the plane containing the circular path. Both magnitude and direction of the angular momentum of the particle moving in a circular path about its centre O are constant.

16. b. The linear momentum (L) is conserved, since τ_{ext} is zero.

Let $I_1 = I_0$, therefore $I_f = 2I_0$

$$\omega_1 = \omega_0, \text{ therefore } \omega_f = \omega \text{ (say)}$$

$$\therefore \quad I_0\omega_0 = 2I_0\omega \Rightarrow \omega = \frac{\omega_0}{2}$$

$$K = \frac{1}{2}I\omega_0^2$$

$$K_f = \frac{1}{2}(2I_0) \times \left(\frac{\omega_0}{2}\right)^2$$

$$= \frac{1}{2} \times 2I_0 \times \frac{\omega_0^2}{4}$$

17. c. Here $V_P = \sqrt{V_c^2 + \omega^2 r^2}$

$$V_Q = \sqrt{V_c^2 + \omega^2 r^2 + 2V_c\omega r\cos\theta}$$

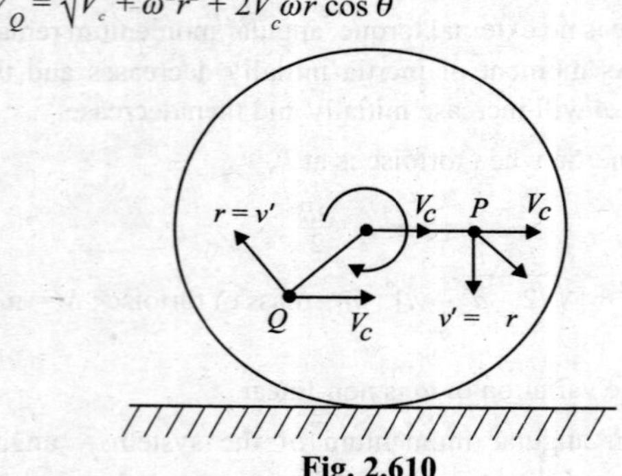

Fig. 2.610

18. d. The cubical block is in equilibrium. For translational equilibrium,

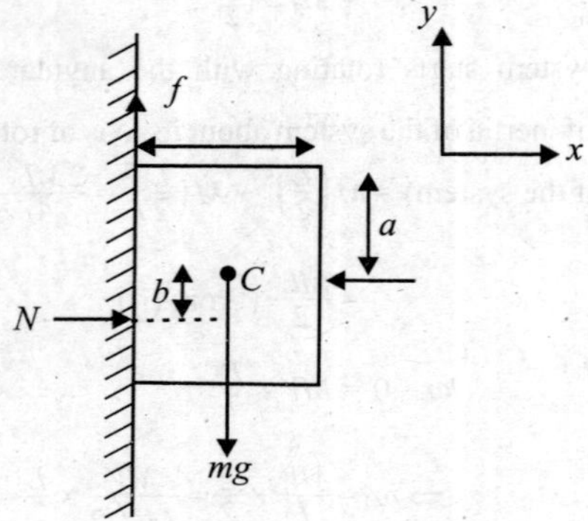

Fig. 2.611

a. $\Sigma F_x = 0 \Rightarrow F = N$

b. $\Sigma F_v = 0 \Rightarrow f = mg$

For rotational equilibrium,

$\Sigma\tau_c = 0$, where τ_c = torque about CM C

Torque is created by frictional force (f) about C = $f \times a$ in a clockwise direction. There should be another torque which should counter this torque. The normal reaction N on the block acts as shown in Fig. 2.611. This will create a torque $N \times b$ in the anticlockwise direction. Such that $f \times a = N \times b$

Note: ***The normal force does not act through the centre of the body always. The point of application of the normal force depends on all the forces acting on the body.***

19. a. Let σ be the mass per unit area. The total mass of the disc = $\sigma \times \pi R^2 = 9M$.

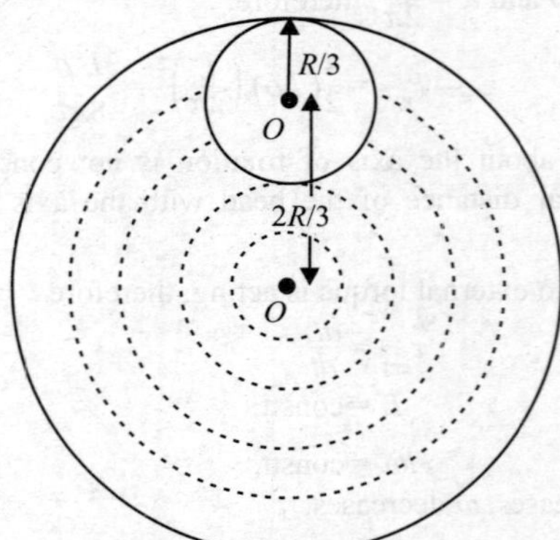

Fig. 2.612

M = Mass of the circular disc cut

$$= \sigma \times \pi\left(\frac{R}{3}\right)^2$$

$$= \sigma \times \frac{\pi R^2}{9} = M$$

Let us consider the above system as a complete disc of mass $9M$ and a negative mass M superimposed on it.

Moment of inertia (I_1) of the complete disc = $9MR^2/2$ about an axis passing through O and perpendicular to the plane of the disc.

MI of the cutout portion about an axis passing through O' and perpendicular to the plane of disc is

$$\frac{1}{2} \times M \times \left(\frac{R}{3}\right)^2$$

Therefore, MI (I_2) of the cutout portion about an axis passing through O and perpendicular to the plane of disc is

$$\left[\frac{1}{2} \times M \times \left(\frac{R}{3}\right)^2 + M \times \left(\frac{2R}{3}\right)^2\right]$$

[Using perpendicular axis theorem]

Therefore, the total MI of the system about an axis passing through O and perpendicular to the plane of the disc is

$$I = I_1 + I_2$$

$$= \frac{1}{2}9\,MR^2 - \left[\frac{1}{2} \times M \times \left(\frac{R}{3}\right)^2 + M \times \left(\frac{2R}{3}\right)^2\right]^2$$

$$=\frac{1}{2}\ 9MR^2 - MR^2\left[\frac{1}{18}+\frac{4}{9}\right]$$

20. b. $|\vec{L}| = mvr$. Since v is changing (decreasing), L is not conserved in magnitude. Since it is given that a particle is confined to rotate in a circular path, it cannot have spiral path. Since the particle has two accelerations a_c and a_t, therefore the net acceleration is not towards the centre.

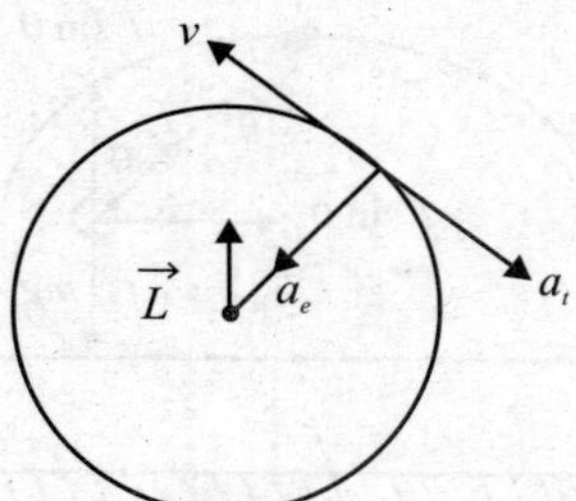

Fig. 2.613

The direction of $\vec{L}$ remains the same even when the speed decreases.

21. b. $I_{AB} = \frac{2}{5}MR^2 = I$ (given) (i)

$$I_{A'B'} = I_{YY'} + Mr^2 = I\frac{1}{2}Mr^2 + Mr^2$$

$$= \frac{3}{2}Mr^2 = 1$$

(given) (ii)

From Eqs. (i) and (ii), $\frac{2}{5}MR^2 = \frac{3}{2}Mr^2 \Rightarrow r = \frac{2}{\sqrt{15}}R$

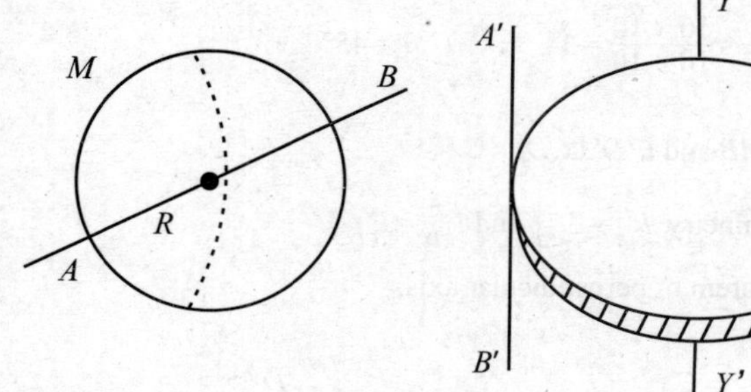

Fig. 2.614

22. d. By the concept of energy conservation,

$$\frac{1}{2}mV^2 + \frac{1}{2}I\frac{V^2}{R^2} = \frac{3}{4}mV^2$$

$$\therefore \quad \frac{1}{2}I\frac{V^2}{R^2} = \frac{3}{4}mV^2 - \frac{1}{2}mV^2 = \frac{1}{4}mV^2$$

$$\Rightarrow \quad I = \frac{1}{2}mR^2$$

23. a. Linear momentum remains constant if net external force on the system of particles is zero.

24. b. For sliding, $\tan\theta \geq \sqrt{3}$ $(= 1.732)$

For toppling, $\tan\theta \geq \frac{2}{3}$ $(= 0.67)$

25. c. Friction force on the ring.

Multiple Choice Questions with One or More than One Correct Answer Type

1. d. We know that $F_{ext} = Ma_{CM}$ (i)

We consider the two particles in a system. Mutual force of attraction is an internal force. There are no external forces acting on the system. From Eq. (i), we get

$$a_{CM} = 0$$

Since initial $v_{CM} = 0$, therefore, final $v_{CM} = 0$

2. b. Angular momentum

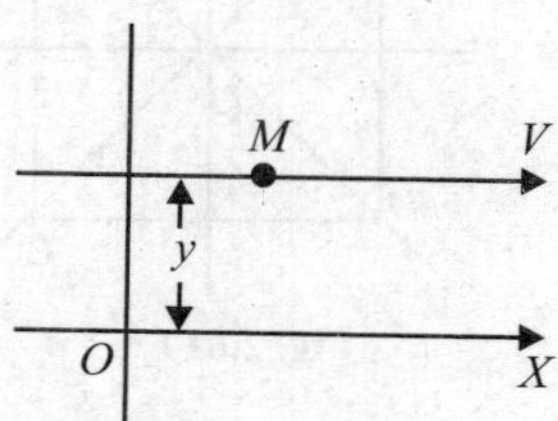

Fig. 2.615

$$\vec{L} = \vec{r} \times \vec{p}$$

L = Momentum × perpendicular distance of line of action of momentum w.r.t. point of rotation

$$L = MV \times y$$

The quantities on the right side of the equation are not changing. Thus, magnitude is constant.

3. c. When the cycle is not pedaled but it is in motion (due to previous effort) the wheels move in the direction such that the centre of mass of the wheel moves forward. Rolling friction will act in the opposite direction to the relative motion of the centre of mass of the body with respect to the ground. Therefore the rolling friction will act in the backward direction in both the wheels The sliding friction will act in the forward direction of the rear wheel during pedaling.

4. b, d. Angular momentum = (momentum) × (perpendicular distance of the line of action of momentum from the axis of rotation)

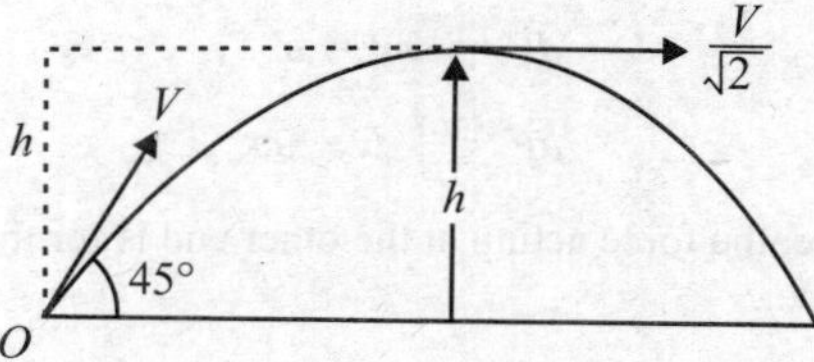

Fig. 2.616

Angular momentum about O, $L = \frac{mv}{\sqrt{2}} \times h$ (i)

Now, $$h = \frac{V^2 \sin^2\theta}{2g} = \frac{V^2}{4g}$$

$[\because \theta = 45°]$ (ii)

From Eqs. (i) and (ii), we get

$$L = \frac{mv}{\sqrt{2}}(2\sqrt{gh})h = m\sqrt{2gh^3}$$

Also from Eqs. (i) and (ii), we get

$$L = \frac{mV}{\sqrt{2}} \times \frac{V^2}{4g} = \frac{mV^3}{4\sqrt{2}\,g}$$

5. a., b., c. To find the moment of inertia of *ABCD* about an axis passing through centre *O* and perpendicular to the plane of the plate, we use perpendicular axis theorem. If we consider *ABCD* to be in the *X-Y* plane, then we know that

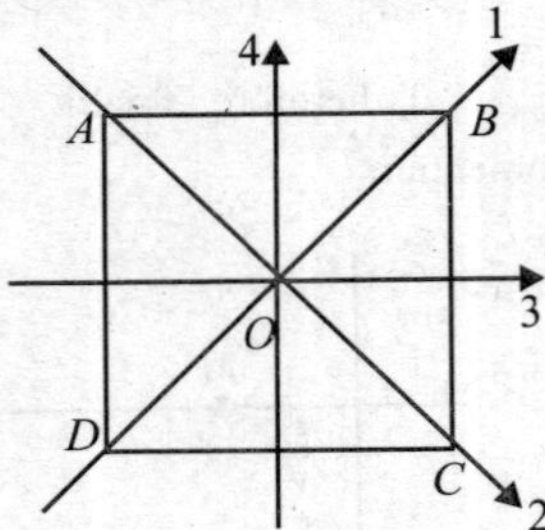

Fig. 2.617

$$I_{zz'} = I_{xx'} + I_{yy'}$$

$\therefore$ $I_{zz'} = I_1 + I_2$ (i)

Also $I_{zz'} = I_3 + I_4$ (ii)

Adding Eqs. (i) and (ii), we get

$$2I_{zz'} = I_1 + I_2 + I_3 + I_4$$

But $I_1 = I_2$ and $I_3 = I_4$ (by symmetry)

$\therefore$ $2I_{zz'} = I_1 + I_1 + I_3 + I_3$

$$= 2I_1 + 2I_3$$

6. a. The force acting on the mass of liquid of length *dx* at a distance *x* from the axis of rotation *O* is as follows:

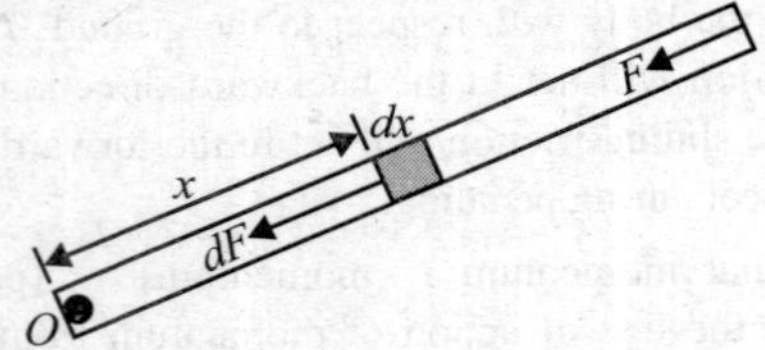

Fig. 2.618

$$dF = (dm) \times \omega^2$$

$\therefore$ $dF = \frac{M}{L} dx \times x\omega^2$

Therefore, the force acting at the other end is for the whole liquid in tube.

$$F = \int_0^L \frac{M}{L}\omega^2 x dx = \frac{M}{L}\omega^2 \int_0^L x dx$$

$$= \frac{M}{L}\omega^2 \left[\frac{x^2}{2}\right]_0^L = \frac{M}{L}\omega^2 \left[\frac{L^2}{2} - 0\right]$$

$$= \frac{ML\omega^2}{2}$$

7. c. When the car is moving in a circular horizontal track of radius 10 m with a constant speed, then the bob is also undergoing a circular motion. The bob is under the influence of two forces.

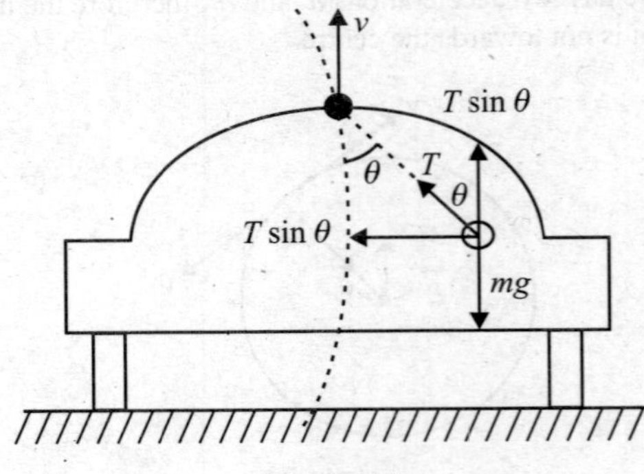

Fig. 2.619

i. *T* (tension in the rod)

ii. *mg* (weight of the bob)

Resolving tension, we get

$$T\cos\theta = mg \quad \text{(i)}$$

and $T\sin\theta = \frac{mv^2}{r}$ (ii)

(Here $T\sin\theta$ is producing the necessary centripetal force for the circular motion)

Dividing Eqs. (i) and (ii), we get

$$\tan\theta = \frac{v^2}{rg} = \frac{10 \times 10}{10 \times 10} = 1 \Rightarrow \theta = 45°$$

8. a. $A'B' \perp AB$ and $C'D' \perp CD$

From symmetry, $I_{AB} = I_{A'B'}$ and $I_{CD} = I_{C'D'}$

From theorem of perpendicular axis,

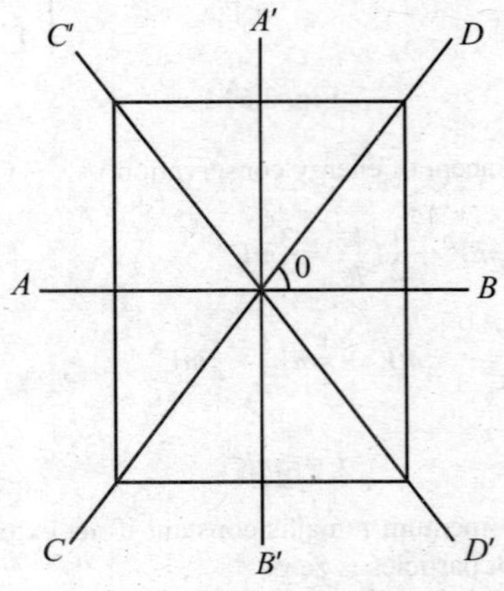

Fig. 2.620

$$I_{ZZ} = I_{AB}\, I_{A'B'} = I_{CD} + I_{CD'}$$

$\Rightarrow$ $$2I_{AB} = 2I_{CD}$$

$\therefore$ $$I_{AB} = I_{CD}$$

Alternative method:

The relation between I_{AB} and I_{CD} should be true for all values of θ.

At $\theta = 0, I_{CD} = I_{AB}$

Similarly, at $\theta = \dfrac{\pi}{2}$

$$I_{CD} = I_{AB} \text{ (By symmetry)}$$

Keeping these things in mind, only option (a) is correct.

9. a., b., c. $$\vec{\tau} = \frac{d\vec{L}}{dt}$$

Given that $$\vec{\tau} = \vec{A} \times \vec{L}$$

$\Rightarrow$ $$\frac{d\vec{L}}{dt} = \vec{A} \times \vec{L}$$

From the cross-product rule, $\dfrac{d\vec{L}}{dt}$ is always perpendicular to the plane containing $\vec{A}$ and $\vec{L}$. By the dot product definition, $\vec{L}\cdot\vec{L} = \vec{L}^2$

Differentiating with respect to time, we get

$$\vec{L}\cdot\frac{d\vec{L}}{dt} + \vec{L}\cdot\frac{d\vec{L}}{dt} = 2L\frac{dL}{dt}$$

$$2\vec{L}\cdot\frac{d\vec{L}}{dt} = 2L\frac{dL}{dt}$$

Since $d\vec{L}/dt$ is perpendicular to $\vec{L}$

$\Rightarrow$ $$\frac{dL}{dt} = 0 \Rightarrow L = \text{constant}$$

10. c., d. As shown in the figure, the component of weight $Mg \sin\theta$ tends to slide the point of contact (of the cylinder with inclined plane) along its direction. The sliding friction acts in the opposite direction to this relative motion. Because of frictional force the cylinder rolls. Thus frictional force aids rotation but hinders translational motion.

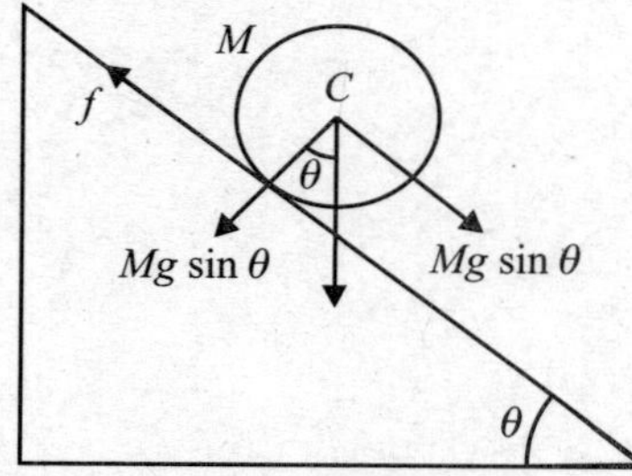

Fig. 2.621

Applying $F_{net} = ma$ along the direction of inclined plane, we get,

$$Mg \sin\theta - f = Ma_c,$$

where a_c = acceleration of centre of mass of the cylinder

$\therefore$ $$f = Mg \sin\theta - Ma_c \quad \text{(i)}$$

where a_c = acceleration of centre of mass cylinder.

But $$a_c = \frac{g \sin\theta}{I + \frac{I_c}{MR^2}} = \frac{g \sin\theta}{I + \frac{MR^2/2}{MR^2}}$$

$$= \frac{2}{3} g \sin\theta \quad \text{(ii)}$$

From Eqs. (i) and (ii), $$f = \frac{Mg \sin\theta}{3}$$

If θ is reduced, then frictional force is reduced.

11. a.,d. Since no work is being done in the process, the sum of kinetic energy, translational kinetic energy, rotational kinetic energy and potential energy will be constant.

$$KE_A + mgh_A = KE_B = KE_C + mgh_C$$

$$KE_B > KE_A \text{ and } KE_B > KE_C$$

Now if $h_C > h_A$, then $KE_A < KE_C$.

12. b., c.

$$V_A = 0,\ V_B = \omega r,\ V_C = 2\omega r$$

$\vec{V}_C - \vec{V}_A$ is towards the right and $\vec{V}_B - \vec{V}_C$ is towards left,

Hence, option (a) is incorrect.

But in magnitude, option (c) is correct.

$$\vec{V}_C - \vec{V}_A = 2\omega r = 2\vec{V}_B$$

Assertion and Reasoning Type

1. d. For statement 1: Force may be acting through centre of mass as shown. For this, torque about CM will be zero. The force F will produce an acceleration in the CM.

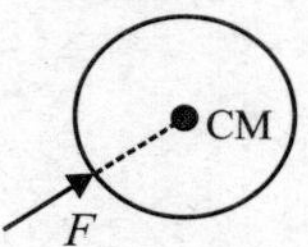

Fig. 2.622

2. d. $a = \dfrac{g \sin\theta}{1 + \dfrac{I_{CM}}{mR^2}}$, I_{CM} of hollow cylinder is less, so it will have more acceleration and will take less time to reach the bottom.

Comprehension Type

1. c. $\frac{1}{2} I(2\omega)^2 = \frac{1}{2} kx_1^2$ and $\frac{1}{2} 2\, I(\omega)^2 = \frac{1}{2} kx_2^2$

Dividing them, we get

$$\frac{x_1}{x_2} = \sqrt{2}$$

2. a. Let final angular velocity be ω_1. Applying conservation of angular momentum:

$$(I+2I)\omega_1 = I(2\omega) + 2I\omega \Rightarrow \omega_1 = \frac{4\omega}{3}$$

Now angular impulse = change in angular momentum:

$$\tau t = 2I(\omega_1 - \omega) \Rightarrow \tau t = 2I\frac{\omega}{3} \Rightarrow t = \frac{2I\omega}{3t}$$

3. b. $KE_i - KE_f$

$$\Rightarrow \frac{1}{2}I(2\omega)^2 + \frac{1}{2}2\,I\omega^2 - \frac{1}{2}3\,I\omega_1^2 = \frac{1}{3}\,I\omega^2$$

4. d. $f - 2kx = Ma, fR = -I\alpha \Rightarrow fR = -(1/2)MR^2(a/R)$

Solve to get net force $= Ma = -\dfrac{4kx}{3}$

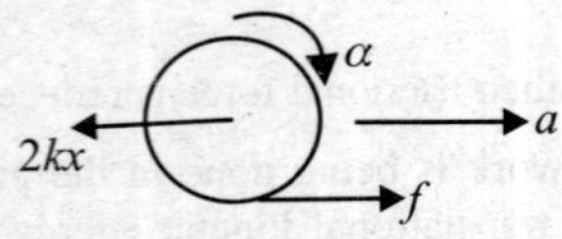

Fig. 2.623

Negative sign is because this force will act in the backward direction.

5. d. From the previous question, $a = -\dfrac{4kx}{3M}$

Comparing with $a = -\omega^2 x$, we get

$$\omega = \sqrt{\frac{4k}{3M}}$$

6. c. From the previous question, we can get

$$f = \frac{2kx}{3}$$

$$f_{max} = \mu Mg$$

For disc not to slip, $\quad f_{max} \geq f \Rightarrow x \leq \dfrac{3}{2}\dfrac{\mu Mg}{k}$ (i)

From conservation of energy:

$$\frac{1}{2}(2k)\,x^2 = \frac{1}{2}I_{CM}\,\omega^2 + \frac{1}{2}MV_0^2$$

$$\Rightarrow \quad x = \sqrt{\frac{3}{4}\frac{MV_0^2}{k}}$$

Putting the value of x in Eq. (i), we get

$$\Rightarrow \quad V_0 \leq \mu g\sqrt{\frac{3M}{k}}$$

CHAPTER

3

Fluid Mechanics

HYDROSTATICS

A fluid is a substance, such as a liquid or a gas, that has no rigidity like solids. Liquids are distinguished from the gases by the presence of a surface. As the fluids have no rigidity, they fail to support a shearing stress. When a fluid is subjected to a shearing stress, the fluid layers slide relative to each other. This characteristic gives fluid the ability to flow or change shapes. However, in this chapter, we confine ourselves to the study of liquids only.

Pressure

While considering the equilibrium of solids we usually think in terms of forces exerted by one body on another body. But in case of liquids, we usually speak in terms of pressure instead of forces. If a uniform force F is exerted normal to an area A, then the pressure P is defined as

$$P=\frac{F}{A}$$

Note: ***The area A in the above expression is the actual area of contact. In case of solids, the actual area of contact between two surfaces is less than the apparent area of contact. Therefore, the concept of pressure is of no use in case of solids. On the contrary, the contact between a liquid and a solid surface or between two liquid surfaces is rather a smoother one, that is, the actual area of contact is equal to the apparent area of contact. Therefore, one can easily use the concept of pressure.***

If the applied force F is not uniform, then the pressure at a point is defined as the intensity of force at that point. That is, 1

$$P=\lim_{\Delta A\to 0}\frac{\Delta F}{\Delta A}$$

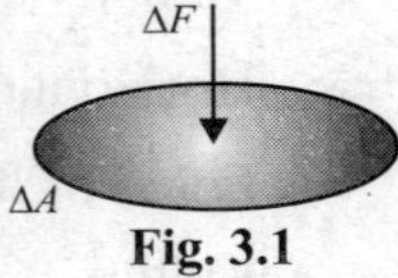

Fig. 3.1

Pressure is a scalar quantity. Pressure acts normal to a surface and it is always compressive in nature. Therefore, only its magnitude is required for its complete description. The SI unit of pressure is N/m^2 and is also called pascal (Pa). The other common units of pressure are atmosphere and bar.

$$1\text{ atm}=1.013\times 10^5\text{ Pa}\Rightarrow 1\text{ bar}=10^5\text{ Pa}$$

Measuring Pressure

Figure 3.2 shows a mercury barometer which was invented by Torricelli. It is a device used for measuring pressure. A long glass tube (may be 1 m long) is filled with mercury and inverted with its open end in a dish of mercury. The mercury level in the tube falls and comes to rest at a vertical height of 76 cm above the mercury level in the dish.

The space above mercury in the tube is almost a vacuum. So $P_A = 0$. At point C, there is only atmospheric pressure. So $P_C = P_B$ which is equal to the pressure at B, because points B and C are at the same level.

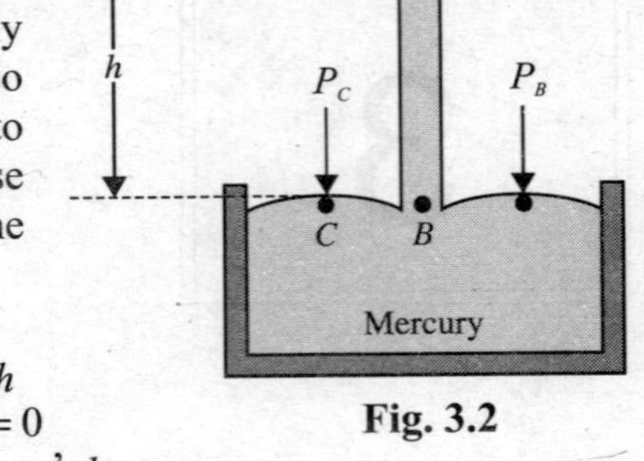

Fig. 3.2

Thus, $P_C = P_B = P_A + \rho gh$

$\therefore \quad P_A = 0$

$\because P_C = \rho gh$

For mercury barometer, $h = 76$ cm $= 0.76$ m,

$\rho = 13.6\times 10^3$ kg/m^3

$P_C = P_B = 0.76\times 13.6\times 10^3\times 9.81 = 1.013\times 10^5$ N/m^2

Note: ***The space above the free surface of mercury in the tube is vacuum. In fact, there is a little vapour of mercury present in that free space due to the evaporation of mercury, but it does not give a considerable pressure. In this way, the vapour pressure of mercury in the free space (or physical vacuum) can be neglected. Hence, P = 0 in the free space.***

Some Properties of Pressure

1. Pressure is isotropic

The pressure exerted by a liquid at a point is the same in all directions. Figure 3.3(a) shows a small cubic element of liquid with a force acting on each face of this element. Since the element is in equilibrium, the force acting on each lateral face of the cube must be equal.

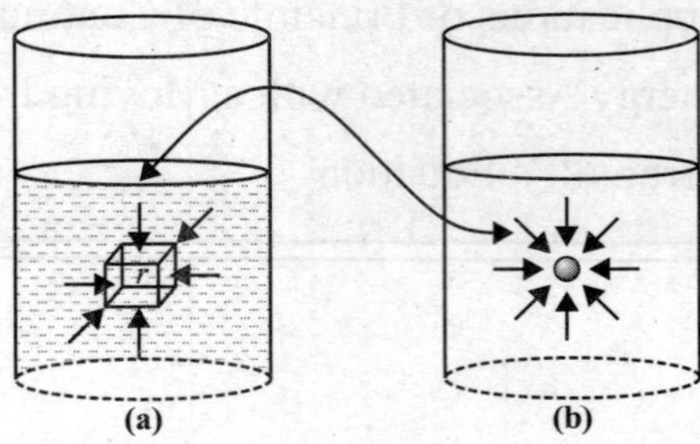

Fig. 3.3

As the element reduces to a point, the top and bottom forces also become equal. Thus, every point within the liquid is equally compressed from all directions, as shown in Fig. 3.3(b).

2. Pressure is uniform on a horizontal plane

For a liquid at rest (or moving with uniform velocity or vertical acceleration), the pressure at different points on the same horizontal level within the liquid remains the same. Consider a cylindrical volume of liquid within the liquid contained in a vessel. Let 'ds' be the elemental cross-sectional area of the horizontal cylinder. The force F_1 on face A is $P_1\,ds$ and force F_2 on face B is $P_2 ds$. Here, P_1 and P_2 are the respective pressures on the faces A and B.

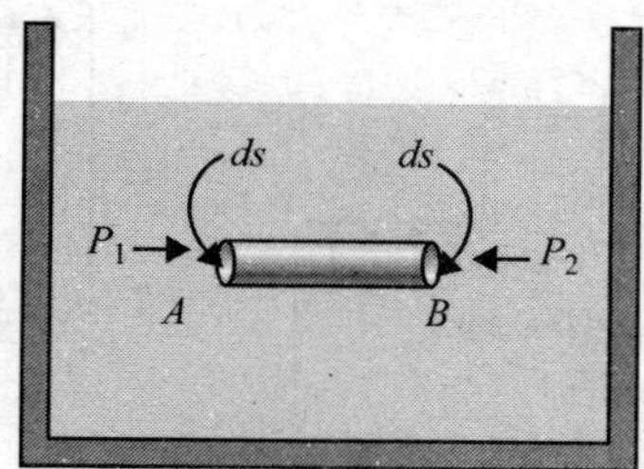

Fig. 3.4

For equilibrium, $F_1 = F_2 \Rightarrow P_1 ds = P_2 ds$

$\Rightarrow \quad P_1 = P_2$

On any horizontal plane in a stationary liquid, pressure is the same at all the points. In Fig. 3.5 pressures at A, B, C and D are equal, i.e., $P_A = P_B = P_C = P_D$.

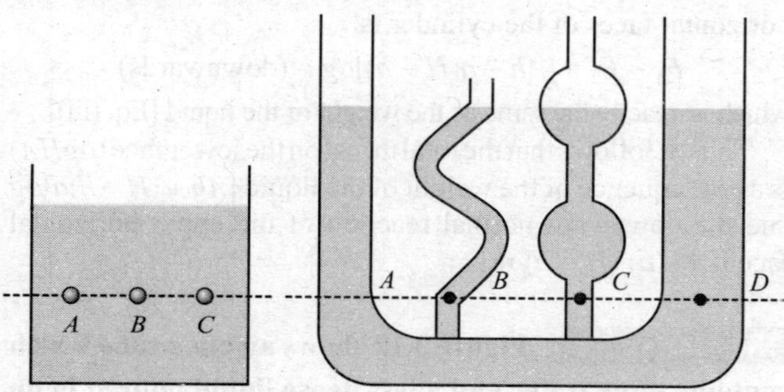

Fig. 3.5

3. Fluids in equilibrium exert pressure at right angles to the surface of contact

Let us consider some liquid contained in a beaker as shown in Fig. 3.6. Consider the force exerted by the liquid at any point A of the container. Let it act along some direction PA.

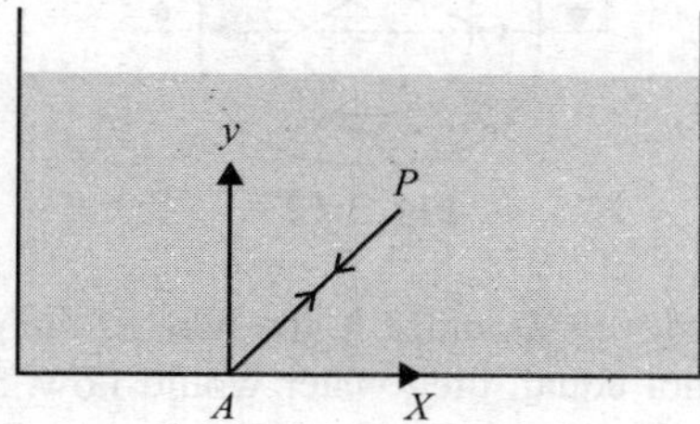

Fig. 3.6 Fluid in equilibrium

Since action and reaction are equal in magnitude and opposite in direction, the surface of the container will give an equal reaction to water along AP. Let us resolve the reaction force into two rectangular components—along AX and AY. As liquid cannot resist any tangential force, the liquid near A should begin to flow along AX parallel to the surface. But the liquid is at rest. So, there cannot be any component along AX. Thus, the following points can be noted.

(i) The force acting on a liquid in equilibrium has to be perpendicular to the surface of the container.

(ii) The force exerted by a liquid on any surface in contact with it is called thrust of the liquid.

(iii) The thrust exerted by a liquid (at rest) per unit area of the surface in contact with the liquid is called pressure.

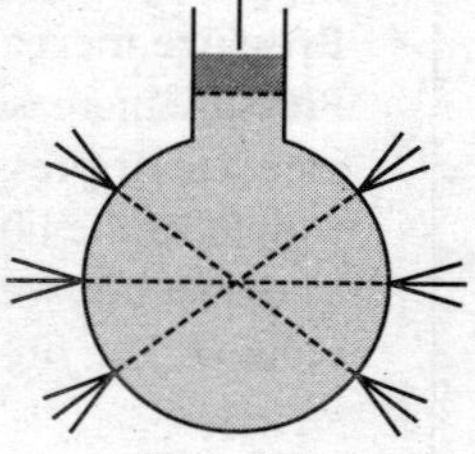

Fig. 3.7

The fact that a liquid exerts pressure at right angles to the surface of the container at every point can be demonstrated in a simple mechanical manner. Take a globe which has a number of holes in its surface and is fitted with a piston (Fig. 3.7). Fill the globe and a part of the barrel with water. Now push the piston inwards. Water will be seen to flow out radially from all the holes. This demonstrates that water exerts pressure at right angles to the walls of the container.

4. Pressure varies with depth and height

Let us look at Fig. 3.8 and consider a small volume element in between the liquid at a depth y.

Let A be the area of cross section of this volume element, dy be its thickness, P_0 be the atmospheric (i.e., outside) Pressure acting on the free surface of liquid, F be the total force acting downward on the top face of the volume element $= P \times A$, F' be the total upward force acting on its bottom face $= (P + dP) \times A$, dW be the weight of this volume element, acting downwards.

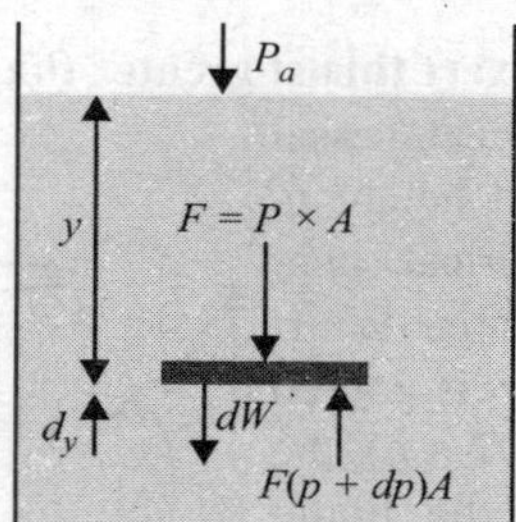

Fig. 3.8

As this volume element is in equilibrium,

$$(P + dP) - PA = dW$$

$$dP \times A = (A \times dy)\rho g$$

$$\text{or } dP = \rho g dy$$

$$\therefore \quad \int_{P_a}^{P} dP = \int_{0}^{h} \rho g dy$$

$$[P]_{P_a}^{P} = \rho g [y]_0^h \Rightarrow P = P_a + \rho g h$$

Thus, we can note the following points:

- The liquid pressure is the same on all points at the same depth.

- The total (or absolute) pressure at a depth h below the free liquid surface is more than the outside atmospheric pressure by an amount ρgh.
- The pressure at the bottom of the container will not depend upon the shape or size of the container.
- **Pressure increases with depth linearly**
  Pressure increases with depth linearly, that is, the rate of increase of pressure with depth is equal to the specific weight of the liquid at that point.
  The specific weight is defined as the weight per unit volume, i.e., ρg. Thus,
$$\frac{dp}{dy} = \rho g$$
- **Pressure decreases with height linearly**
  The rate of decrease of pressure with height is constant and is given by

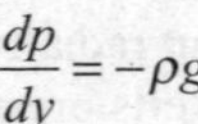

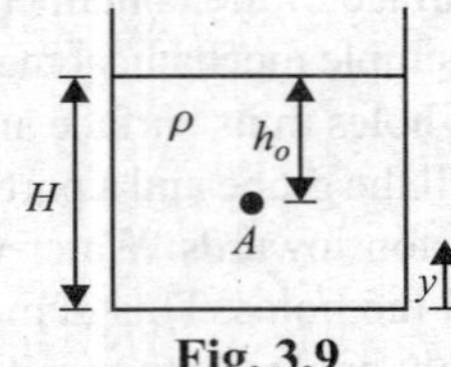

Fig. 3.9

The pressure at the point A is obtained as follows.
$$dp = -\rho g\, dy$$
$$\Rightarrow \int dp = -\rho g \int dy \Rightarrow p = -\rho gy + c$$
Now, at $y = H, p = p_0$. Therefore, $c = p_0 + \rho gH$

Thus, $p = \rho g(H - y) + p_0$

From Fig. 3.9, since $H - y = h_o$ $p = \rho gH + p_o$

5. A liquid may exert thrust greater than its weight

Consider a cylindrical vessel of base area A and height h, with a thin cylindrical stem of cross–sectional area 'a'. If a liquid of density ρ is poured in it to a height $H (H > h)$, the thrust exerted by the liquid on the base of the vessel = pressure at the base due to liquid column × base area.

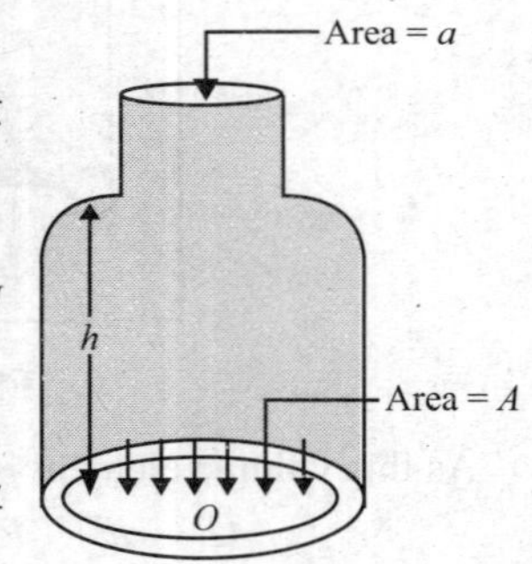

Fig. 3.10

$$F_{\text{base}} = (\rho gH)A \quad \text{(i)}$$
The weight of the liquid contained in the vessel is given by
$$\text{Volume} \times \text{density} \times g = [AH - (A - a)(H - h)]\rho g \quad \text{(ii)}$$
$$W_{\text{liquid}} = [Ah - a(H - h)]\rho g \quad \text{(iii)}$$
Comparing Eqs. (i) and (ii) it is evident that the force exerted by the liquid on the base of the vessel exceeds the weight of the liquid.

To understand the situation, let us draw the pressure diagram on the horizontal faces of the vessel.

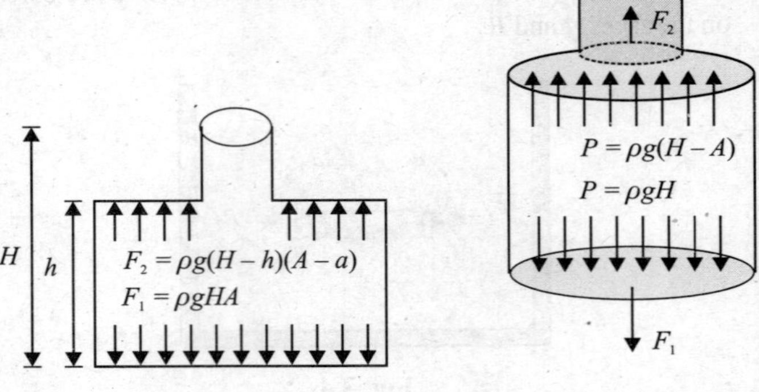

Fig. 3.11

The thrust exerted by the liquid on the bottom face is
$$F_1 = \rho gHA \quad \text{(downwards)}$$
and that on the top face is
$$F_2 = \rho g(H - h)(A - a) \quad \text{(upwards)}$$
Therefore, the net thrust exerted by the liquid on the horizontal faces of the cylinder is
$$F_1 - F_2 = [Ah - a(H - h)]\rho g \quad \text{(downwards)}$$
which is exactly the same as the weight of the liquid [Eq. (iii)].

Thus, it follows that the total thrust on the lower face (ρgHA) is a consequence of the weight of the liquid $[Ah + (H - h)a]\rho g$ and the downward normal reaction of the upper horizontal face $[(A - a)(H - h)\rho g]$.

Illustration 3.1 **Figure 3.12 shows an open tube which contains some water and a less dense liquid poured in on the right-hand side. If the density of the unknown liquid is ρ_x, show that $\rho_x = \frac{\rho_w h_w}{h_x}$.**

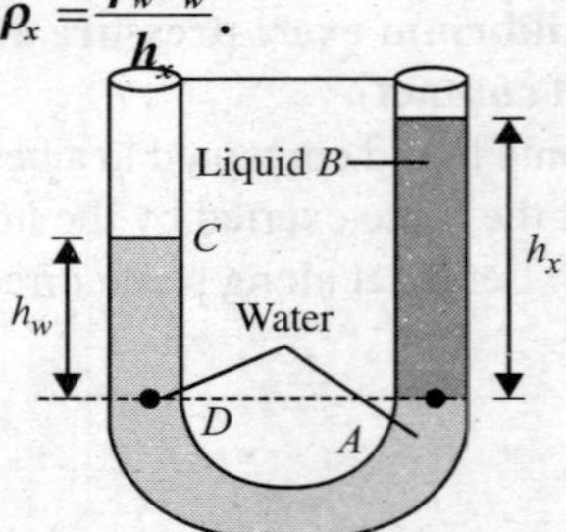

Fig. 3.12

Sol. Consider levels D and A in the tube. If the pressures at D and A are not equal, then water would flow. Since water does not flow, the pressures at D and A are equal.
$$P_A = P_{\text{atm}} + \rho_w gh_w \text{ and } P_D = P_{\text{atm}} + \rho_x gh_x$$
Equating P_A and P_D, we obtain
$$\rho_x = \frac{\rho_w h_w}{h_x}$$

Illustration 3.2 **A vertical U tube of uniform cross section contains mercury in both of its arms. A glycerine**

($d = 1.3$ g/cm³) column of length 10 cm is introduced into one of the arms. Oil of density 0.8g/cm³ is poured in the other arm until the upper surfaces of the oil and glycerine are in the same horizontal level. Find the length of oil column. Density of mercury is 13.6 g/cm³.

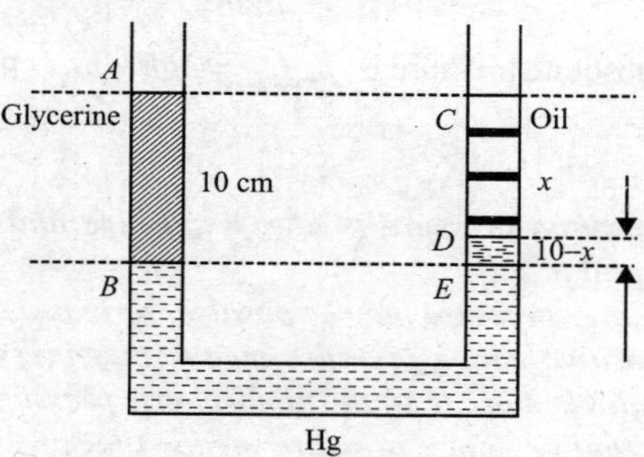

Fig. 3.13

Sol. Let the length of oil column = $CD = x$ cm.
Length of glycerine = AB = 10 cm

$\therefore \quad DE = (10 - x)$ cm

Equating pressure at horizontal level BE in both arms, we have

$$P_{\text{oil}} + P_{\text{Hg}} = P_{\text{glycerine}}$$

$$xd_{\text{oil}}g + (10 - x)\rho_{\text{Hg}}g = 10d_{\text{glycerine}}g$$

or $\quad x(0.8) + (10 - x)13.6 = 10(1.3) \Rightarrow x = \dfrac{136 - 13}{12.8} = 9.6$ cm

Illustration 3.3 **Two communicating cylindrical tubes contain mercury. The diameter of one vessel is four times larger than the diameter of the other. A column of water of height 70 cm is poured into the narrow vessel. How much will the mercury level rise in the other vessel and how much will it sink in the narrow one? How much will the mercury level rise in the narrow vessel, if a column of water of the same height is poured into the broad vessel?**

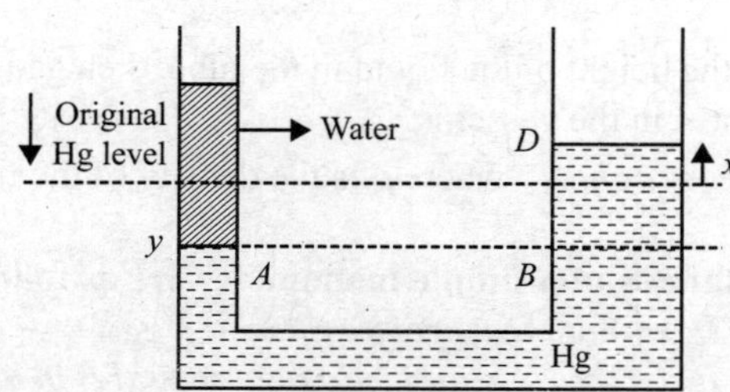

Fig. 3.14

Sol. Let x = rise of Hg level in the broad vessel, y = drop of Hg level in the narrow vessel. Then,

$$xA_1 = yA_2 \Rightarrow 16x = y$$

(because volume displaced will be the same)
Equating pressures at A and B on the same horizontal plane, we get

$$70 \times 1 \times g = (x + y) \times 136 \times g \Rightarrow x + y = \frac{70}{13.6}$$

Solving the equations, we get $x = 0.3$ cm, $y = 4.8$ cm.
If 70 cm water column is poured into the broad vessel, then following the same pattern, we get the equations

$$x' = 16y' \text{ and } (x' + y') \times 13.6 = 70 \times 1$$

$$\Rightarrow \quad x' = 4.8 \text{ cm and } y' = 0.3 \text{ cm}$$

PASCAL'S LAW

Pressure applied to a confined liquid is transmitted equally undiminished to every part of the liquid and to all the walls of the containing vessel, and is always exerted at right angles to the walls.

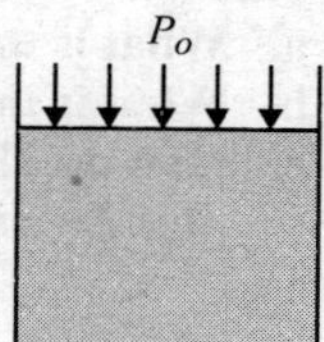

Fig. 3.15

Figure 3.15 shows a liquid confined by solid boundaries from the four sides and the base. At the free surface the atmosphere is exerting a pressure p_0 on this confined liquid which is transmitted to every part of the liquid.

This principle is used in a hydraulic jack or a lift where a heavy load can be lifted up by a small force. In Fig. 3.16, the pressure due to a small force F_1 applied to a piston of area A_1 is transmitted to the larger piston of area A_2. The pressure at the two pistons is the same because they are at the same level.

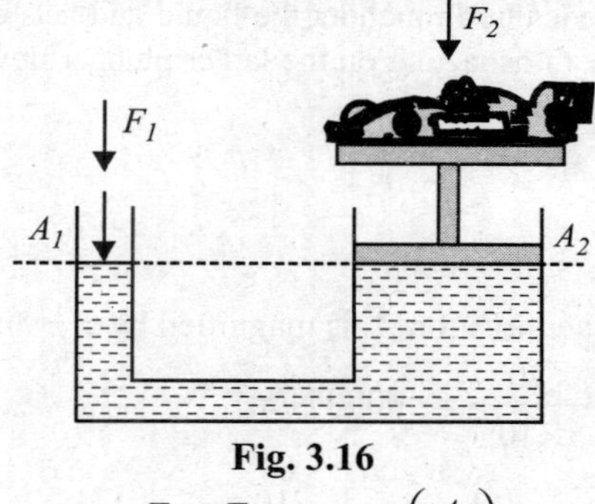

Fig. 3.16

$$p = \frac{F_1}{A_1} = \frac{F_2}{A_2} \Rightarrow F_2 = \left(\frac{A_2}{A_1}\right)F_1$$

Thus, a small force F_1 acting on a small area A_1 results in a larger force F_2 acting on a larger area A_2.

Illustration 3.4 **The area of cross section of the two arms of a hydraulic press is 1 cm² and 20 cm², respectively (Fig. 3.17). A force of 10 N is applied on the water in the thinner arm. What force should be applied on the water in the thicker arms so that the water may remain in equilibrium?**

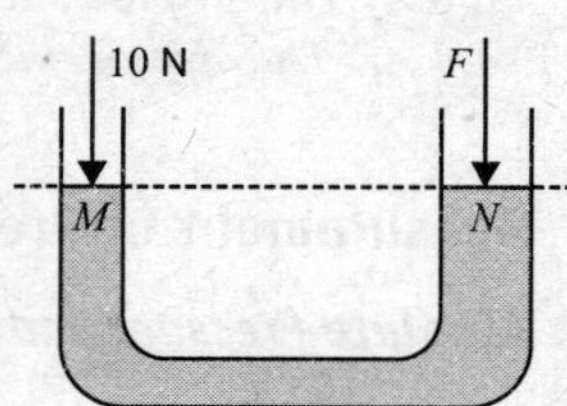

Fig. 3.17

Sol. In equilibrium, the pressures at the two surfaces should be equal as they lie in the same horizontal level. If the atmospheric pressure is P_0 and a force F is applied to maintain the equilibrium, the pressure just below the pistons are

$$P_M = P_0 + \frac{10\text{ N}}{1\text{ cm}^2} \text{ and } P_N = P_0 + \frac{F}{10\text{ cm}^2}$$

These pressures should be the same, i.e., $P_M = P_N$ which gives $F = 200$ N.

Illustration 3.5 **In a simple hydraulic press, the cross-sectional area of the two cylinders is 5×10^{-4} m² and 10^{-2} m², respectively. A force of 20 N is applied at the small plunger.**

a. What is the pressure produced in the cylinders?
b. What is the thrust exerted on the large plunger?
c. How much work is done by the operator, if the smaller plunger moves down 0.1 m?

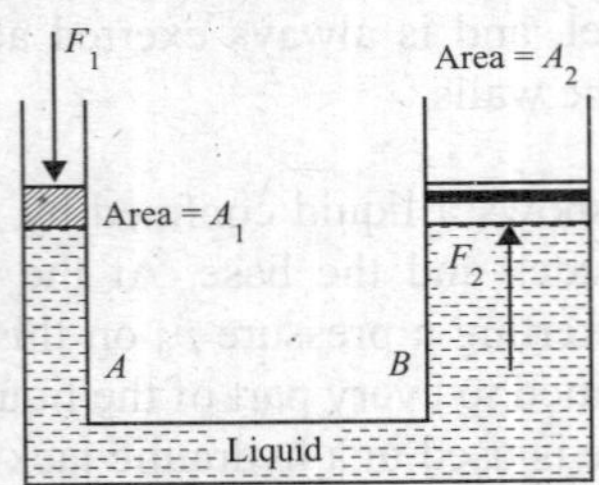

Fig. 3.18

Sol. In a hydraulic press, a force F_1 applied to the smaller plunger creates a pressure F_1/A_1 in the liquid and this pressure is transmitted equally throughout the liquid and acts on the larger plunger. The thrust acting on the larger plunger upwards due to this pressure is

$$F_2 = A_2\left(\frac{F_1}{A_1}\right)$$

Hence, the thrust on A_2 is magnified by $\frac{A_2}{A_1}$ times

a. $\frac{F_1}{A_1} = \frac{20}{5 \times 10^{-4}} = 40000 \text{ N/m}^2$

b. $F_2 = F_1 \times \left(\frac{A_2}{A_1}\right) = 20 \times \frac{10^{-2}}{5 \times 10^{-4}} = 400 \text{ N}$

c. Work done by force $F_1 = F_1 d_1 = 20 \times 0.1 = 2 \text{ J}$

Note: ***The displacement d_2 of the larger plunger = d_1 A_2 / A_1. The displacement of the larger plunger is less.***

Measurement of Pressure

Absolute Pressure and Gauge Pressure

Absolute pressure is the total pressure at a point including the contribution of the liquid as well as that of the atmosphere. Gauge pressure is the pressure measured at a point relative to the local atmospheric pressure. In other words, it does not include the contribution of atmospheric pressure. Absolute pressure and gauge pressure are related as

$$p_{\text{absolute}} = p_{\text{gauge}} + p_{\text{atm}}$$

Consider a point A at a depth h from the free surface of a liquid of density ρ, as shown in Fig. 3.19. The gauge pressure at the point A is

$$p_{\text{gauge}} = \rho g h$$

and the absolute pressure is $p_{\text{absolute}} = \rho g h + p_0$

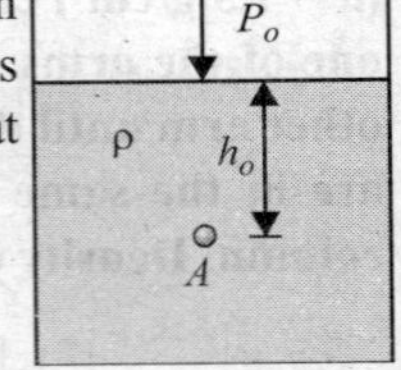

Fig. 3.19

Note:

- ***Absolute pressure is always positive and is never equal to zero.***
- ***Gauge pressure may be positive, zero or negative.***
- ***Positive gauge pressure means that pressure at a point is more than the atmospheric pressure.***
- ***Negative gauge pressure means that pressure at a point is less than the atmospheric pressure.***
- ***Practically, the exact value of absolute pressure cannot be measured due to the uncertainty in the value of atmospheric pressure. The atmospheric pressure varies from place to place and at a place from time to time.***

Manometers: Measurement of Gauge Pressure

The manometers are simple devices used for measuring gauge pressure of fluids. There are two types of manometers as discussed in the following.

1. Simple manometer

It consists of a vertical glass tube, the lower end of which is attached to the point where the pressure is to be measured and the upper end is left open to the atmosphere. It works on a very simple principle. The height of the liquid in the tube balances the pressure in the container to which it is connected.

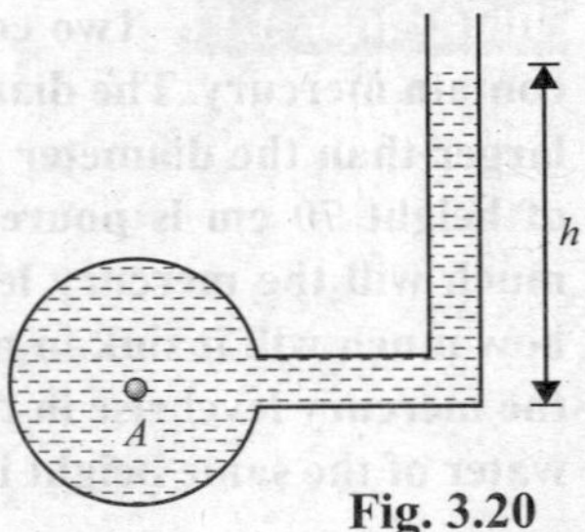

Fig. 3.20

If h is the height of the liquid in the tube, then gauge pressure at point A in the vessel is

$P_A = \rho g h$ where ρ is the density of the liquid.

Limitations of a simple manometer are as follows.

- ***It cannot measure pressure in a gas.***
- ***It can measure positive gauge pressures in liquids only.***
- ***It cannot measure very large pressures as it may require very long tube.***

2. U-tube manometer

An U-tube manometer consists of two tubes connected at the bottom to form a U-shaped tube. It can measure positive and negative gauge pressures in liquids as well as in gases. It works on Pascal's law. The pressure at a point in a vessel is obtained by the manometric equation.

Writing Manometric Equation

We start from a point where pressure is to be obtained and move towards the free end through the tube. As we move down in the tube pressure increases, as we move up in the tube pressure decreases and as we move horizontally pressure does not change in accordance with Pascal's law.

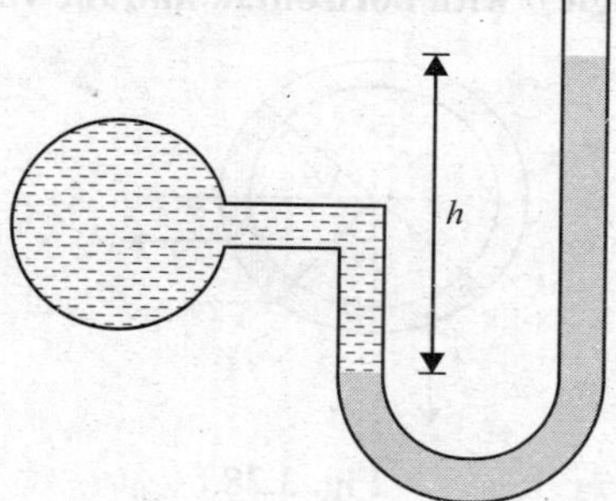

Fig. 3.21

The increase in pressure is taken as positive, the decrease in pressure is taken as negative, and finally we reach the free end of the tube where we equate it with the pressure at the free surface. If we put pressure at the free surface equal to zero, we obtain gauge pressure at that point in the vessel; and if we put pressure at free surface equal to atmospheric pressure, we obtain the absolute pressure at that point in the vessel.

Illustration 3.6 **A container 'A' is having some fluid of density θ_0. It is connected with a U-tube manometer shown in Fig. 3.22. Find the pressure of the container in both the cases given in Fig. 3.22.**

Sol.

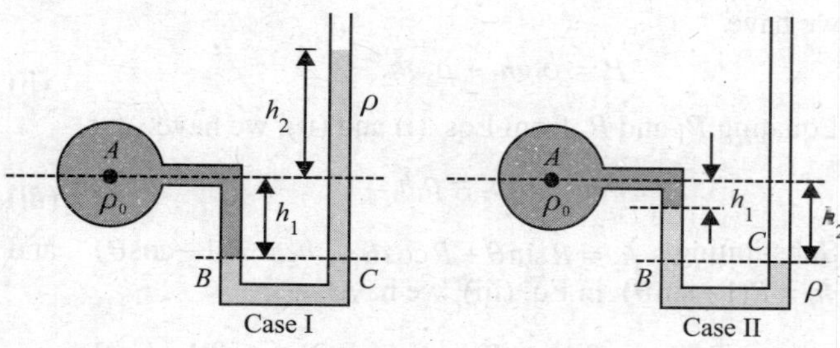

Fig. 3.22

Sol.

Case I

Let the gauge pressure at A be p_A. Starting from container move along the manometer by a height h_1 downwards. The change in pressure will be $+\rho_0 g h_1$. The pressure at B and C will be the same. To move to the open end, we have to move by a distance $(h_1 + h_2)$ upward. Hence, change in pressure will be $-\rho g(h_1 + h_2)$. Finally, we reach the top, i.e., open to atmosphere. Hence, the manometric equation of the gauge pressure will be

$$p_A + \rho_0 g h_1 - \rho g (h_1 + h_2) = 0 \Rightarrow p_A = \rho g (h_1 + h_2) - \rho_0 g h_1$$

In terms of absolute pressure at A, the manometric equation will be

$$p'_A + \rho_0 g h_1 - \rho g(h_1 + h_2) = p_0$$

$$p'_A = (h_1 + h_2) - \rho_0 g h_1 + p_0$$

Case II

Gauge pressure at, A $\quad p_A + \rho_0 g h_1 + \rho g(h_2 - h_1) = 0$

Absolute pressure at, A $\quad p'_A + \rho_0 g h_1 + \rho g(h_2 - h_1) = p_0$

Note:

- ***Pressures at B and C are equal.***
- ***Absolute pressure is always greater than or equal to zero, while gauge pressure can be negative also.***
- ***In the same liquid, pressure will be same at all points at the same level. For example, in Fig. 3.23*** $P \neq P_2$, $P_3 = P_4$, $P_5 = P_6$.

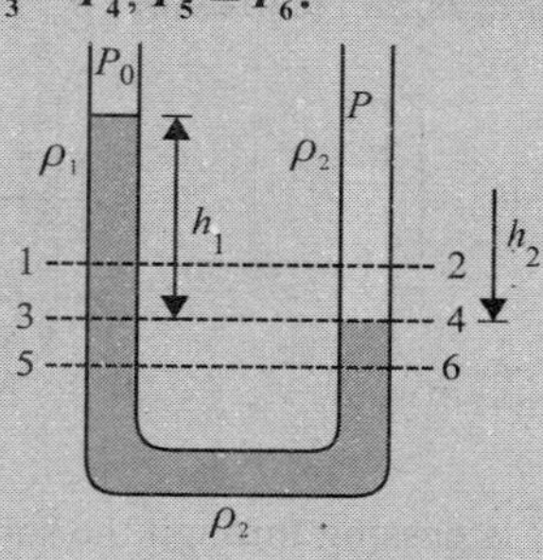

Fig. 3.23

Illustration 3.7 **An inverted U-tube manometer shown in Fig. 3.24 is used to measure the difference in water level between the two tanks. Calculate this difference for the conditions indicated.**

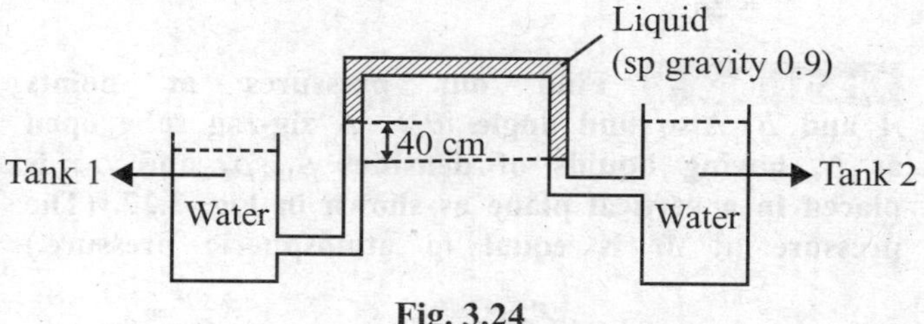

Fig. 3.24

Sol. The inverted U-tube manometer is used for finding the difference of water level in two tanks. Draw a horizontal straight line passing through the lower level and upper level liquid in the manometer as shown in Fig. 3.25. Starting from top of tank 2 and moving through the tube and reaching the top of tank 1, the manometric equation can be written as

$$p_a + h_1 \rho g - 40\rho_1 g + 40\rho g = p_a + h_2 \rho g$$

$$h_2 \rho g - h_1 \rho g = 40 \rho g - 40 \rho_1 g$$

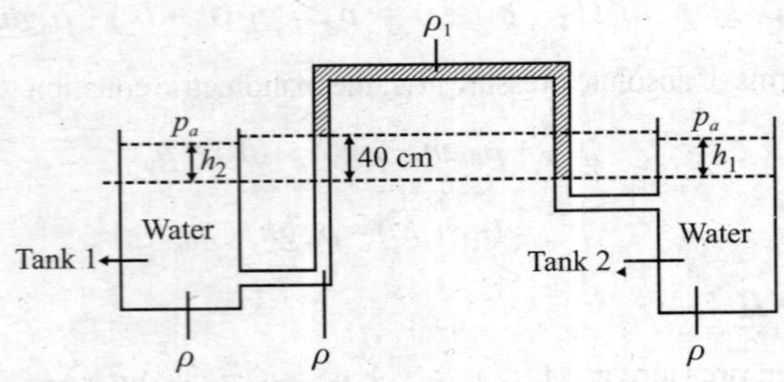

Fig. 3.25

But we are given

$$\rho_1 = 0.9\rho$$
$$(h_2 - h_1)\rho g = 40\rho g - 36\rho g$$
$$h_2 - h_1 = 4 \text{ cm}$$

Illustration 3.8 **In a given U-tube open at left end and closed at right end, if the pressure above the liquid in the right arm is p, find the value of p. Given $d_2 = 2 \times 13.6$ g/cm³, $d_1 = 13.6$ g/cm³. Atmospheric pressure $p_a = 76$ cm of Hg.**

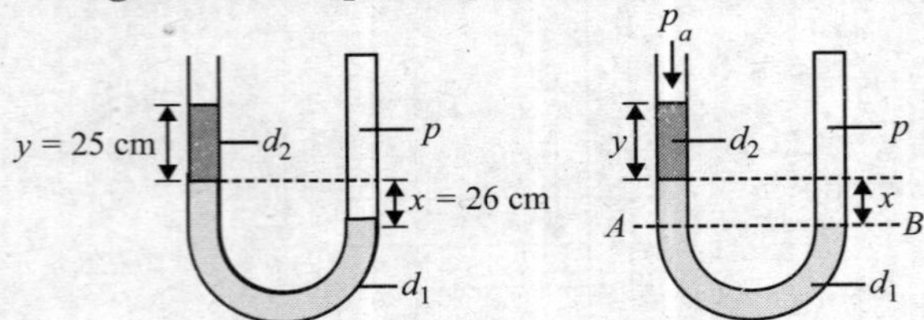

Fig. 3.26

Sol. The line AB is passing through the same liquid; hence, the pressure in the liquid at the same level is the same, i.e., at A and B, the pressures will be equal. The manometric equation is

$$p_a + d_2 yg + d_1 xg = p$$

$$\Rightarrow \quad p_a + 13.6 \times 2 \times 0.25 \times g + 13.6 \times 26 \times g = p$$

$$p_a + 13.6 \times g/100[50 + 26] = p$$

$$\Rightarrow \quad p = 2 \times 13.6 \times g \times \frac{76}{100} \quad [\because p_a = 13.6 \times g \times 76/100]$$

$$= 2p_a$$

Illustration 3.9 **Find out pressures at points A and B. Also find angle 'θ'. A zig-zag tube open at N, having liquids of densities ρ_1, ρ_2 and ρ_3, is placed in a vertical plane as shown in Fig. 3.27. (The pressure at M is equal to atmospheric pressure.)**

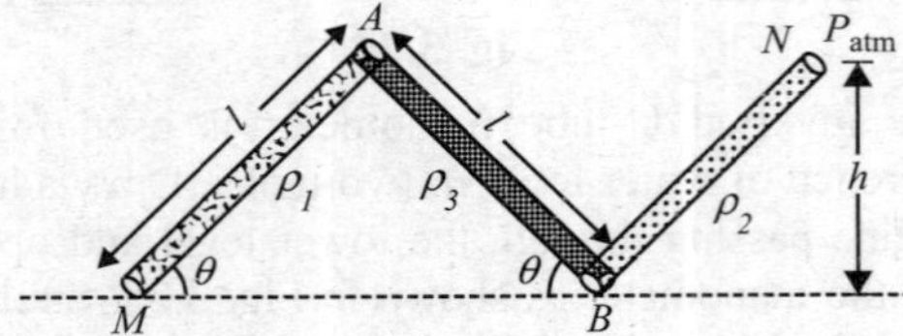

Fig. 3.27

Sol. Pressure at A is $P_A = P_{atm} - \rho_1 gl \sin\theta$

Pressure at B is $P_B = P_{atm} + \rho_2 gh$

But P_B is also given by $P_B = P_A + \rho_3 gl \sin\theta$

Hence, $P_{atm} + \rho_2 gh = P_A + \rho_3 gl \sin\theta$

$$\Rightarrow \quad P_{atm} + \rho_2 gh = P_{atm} - \rho_1 gl \sin\theta + \rho_3 gl \sin\theta$$

$$\Rightarrow \quad \sin\theta = \frac{\rho_2 h}{(\rho_3 - \rho_1)l}$$

Illustration 3.10 **A circular tube of uniform cross section is filled with two liquids of densities ρ_1 and ρ_2 such that half of each liquid occupies a quarter of volume of the tube. If the line joining the free surfaces of the liquid makes an angle θ with horizontal, find the value of θ.**

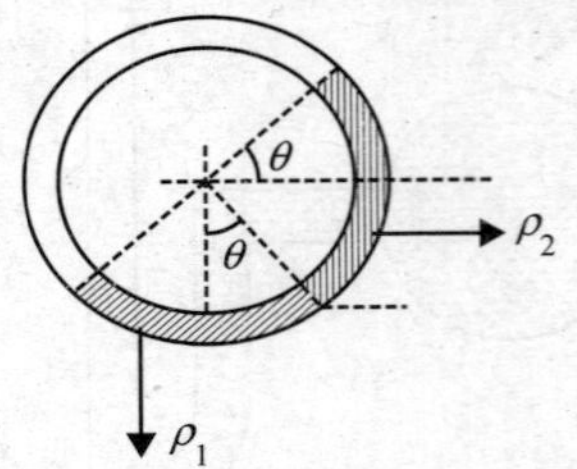

Fig. 3.28

Sol. Let us find the pressure at the lowest point 1. Since the liquid has density ρ_2 and height of liquid column is h_2' on the right-hand side of point 1, we have

$$P_1 = \rho_1 gh_1 \quad \text{(i)}$$

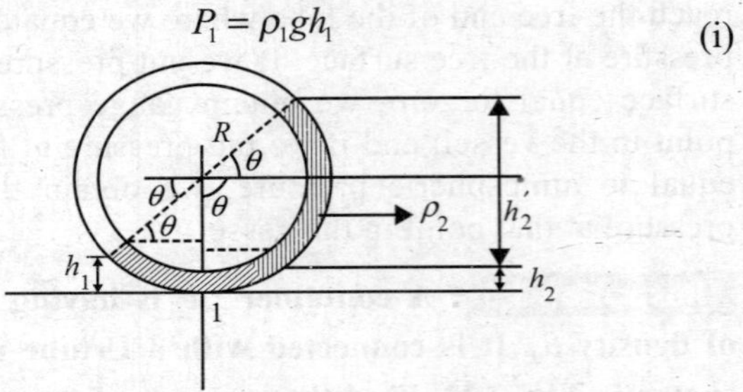

Fig. 3.29

Since two liquid columns of heights h_1 and h_2 and densities ρ_1 and ρ_2 are situated above point 1, on the left-hand side, we have

$$P_2 = \rho_1 gh_2 + \rho_2 gh_2' \quad \text{(ii)}$$

Equating P_1 and P_2 from Eqs. (i) and (ii), we have

$$\rho_1 h_2 + \rho_2 h_2' = \rho_1 h_1 \quad \text{(iii)}$$

Substituting $h_2' = R\sin\theta + R\cos\theta$, $h_2 = R(1-\cos\theta)$ and $h_1 = R(1-\sin\theta)$ in Eq. (iii), we have

$$\rho_1 R(1-\cos\theta) + \rho_2 R(\sin\theta + \cos\theta) = \rho_1 R(1-\sin\theta)$$

This gives

$$\frac{\cos\theta + \sin\theta}{\cos\theta - \sin\theta} = \frac{\rho_1}{\rho_2} \Rightarrow \tan\theta = \frac{\rho_1 - \rho_2}{\rho_1 + \rho_2}$$

$$\text{Then } \theta = \tan^{-1}\left(\frac{\rho_1 - \rho_2}{\rho_1 - \rho_3}\right)$$

Pressure Diagrams and Force on Boundaries

A pressure diagram is a graphical representation of the variation of the pressure intensity over a surface. Such a diagram may be prepared by plotting to some convenient scale the pressure intensities at various points on the surface. Net force (total pressure) as well as point of application of force (centre of pressure) for a plane surface wholly submerged in a static liquid, either vertically or inclined, may also be determined by drawing a pressure diagram.

Since force at any point acts in the direction normal to the surface, the pressure intensities at various points on the surface are plotted normal to the surface. A diagram representing the magnitude of pressure at various points, to some suitable scale, is known as pressure diagram. The areas of pressure diagrams yield the thrust exerted by a liquid on a surface per unit width. A common application of pressure diagram is in the field of construction of dams. Figure 3.30 shows typical pressure diagrams for horizontal, vertical and inclined plane surfaces.

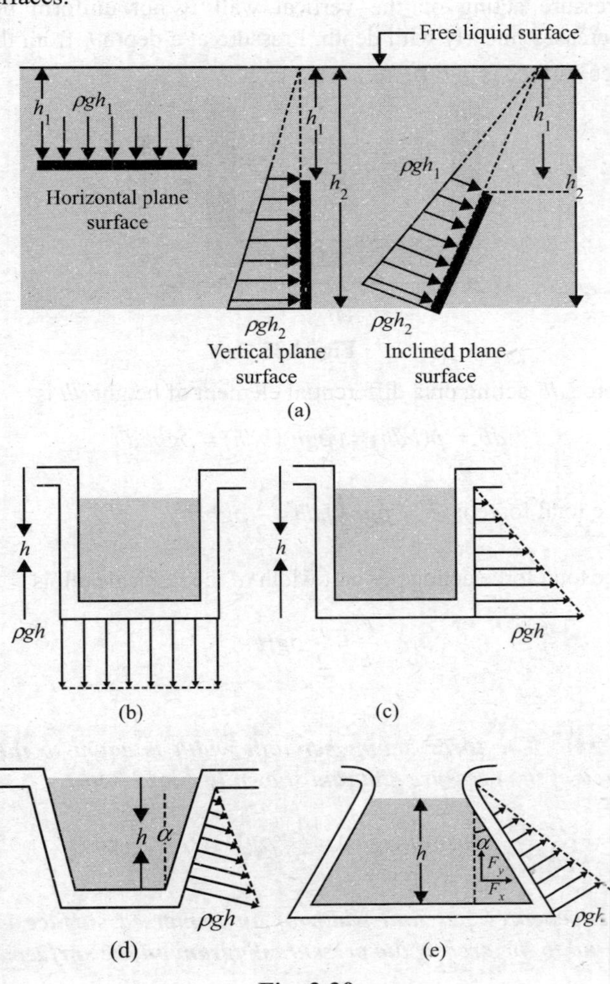

Fig. 3.30

Force on Surface Immersed in a Liquid and Centre of Force

Consider a rectangular plane surface of area 'A' held vertically in a static liquid of density ρ as shown in Fig. 3.31. Let top and bottom edges of the plane surface be at vertical depths of h_1 and h_2, respectively, below the free surface of the liquid. The pressure intensity at the top edge is $P_1 = \rho g h_1$. The pressure intensity at the bottom edge is $P_2 = \rho g h_2$. As the pressure intensity increases linearly from $\rho g h_1$ to $\rho g h_2$, so the average pressure intensity over the entire surface is

$$P_{av} = \frac{P_1 + P_2}{2} = \rho g\left(\frac{h_1 + h_2}{2}\right)$$

Thus, the net force on the vertical surface is

$$F = P_{av} \times \text{Area of the surface} \quad \text{or} \quad F = \rho g\left(\frac{h_1 + h_2}{2}\right) \times A$$

This result can be generated for a surface of any shape. Let us analyse the same case by another approach. It is easy to see that the thrust due to the liquid on any surface can be obtained as

Thrust = (Area) × (Hydrostatic pressure at the centroid of the area)

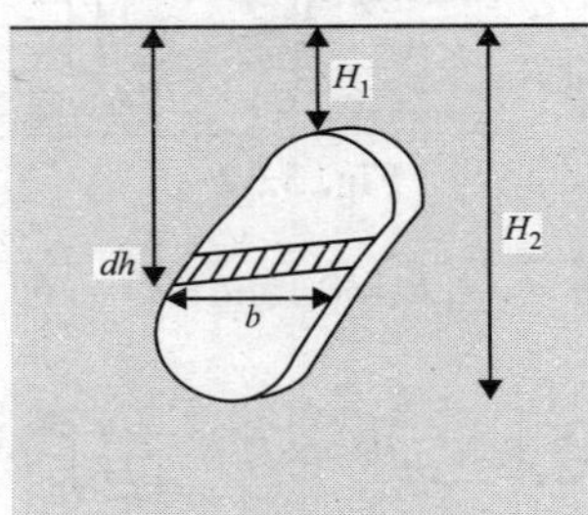

Fig. 3.31

Consider a plane surface of any arbitrary shape, immersed in a liquid of density ρ. Let us subdivide the area into infinitesimally thin horizontal strips. Consider one such strip of width 'b' and thickness dh, at a depth h below the free liquid surface. Normal thrust on this strip is dF = Pressure × Area = $\rho g h\,(bdh)$. Total thrust exerted by the liquid on the area is

$$F = \Sigma dF = \Sigma \rho g h b dh = \rho g \sum_{h=h_1}^{h_2} bhdh$$

Now, if $\bar{h}$ be the depth of the centroid of the area, below the free surface and A is the total area, then

$$\bar{h} = \sum_{h=h_1}^{h_2} \frac{(bdh)h}{A} \Rightarrow \sum_{h=h_1}^{h_2} (bdh)h = A\bar{h}$$

$$\therefore \quad F = (\rho g\,\bar{h})A \qquad \text{(i)}$$

Hence force on a vertical immersed surface is the product of the pressure on the centroid of the surface and the area of the surface.

Centre of Force

The point where the entire force can be thought of as acting, producing the same rotational effect as the individual thrusts on each elemental strip, is known as the centre of force.

Taking moments of forces due to liquid pressure about a horizontal axis passing through the surface, we have

$$M = \sum_{h=h_1}^{h_2} (bdh)(h\rho g)h \quad \text{(a)}$$

Now if H is the depth of the centre of force below the free surface, then

$$M = (\rho g \bar{h} A)H \quad \text{(b)}$$

From Eqs. (a) and (b), we get

$$H = \frac{1}{A}\bar{h}\left[\sum_{h=h_1}^{h_2} (bh^2)dh\right] \quad \text{(ii)}$$

As a special case, consider a rectangular plate of dimension $b \times h$, kept vertical, with its upper edge coinciding with the free horizontal surface.

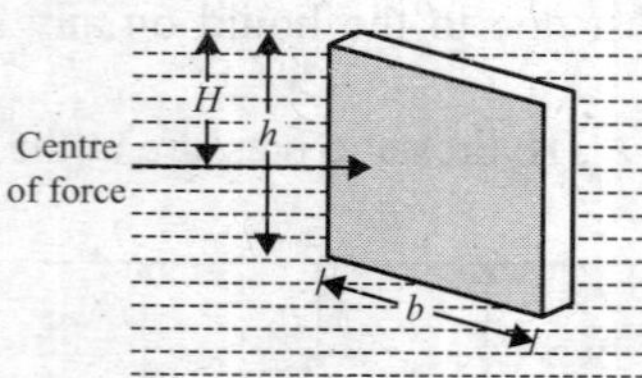

Fig. 3.32

Then $A = bh$, $\bar{h} = \frac{h}{2}$, $h_1 = 0$ and $h_2 = h$. From Eq. (ii), we have

$$H = \frac{1}{(bh)}\left(\frac{h}{2}\right)\sum_{h=0}^{h}(bh^2)dh \Rightarrow H = \frac{2b}{bh^2}\int_0^h h^2 dh = \frac{2}{3}h$$

Illustration 3.11 **Consider a regular tank of size ($l \times b$) filled with a liquid of density ρ to a height H as shown in Fig. 3.33. Find the force at the base and on the walls of the tank.**

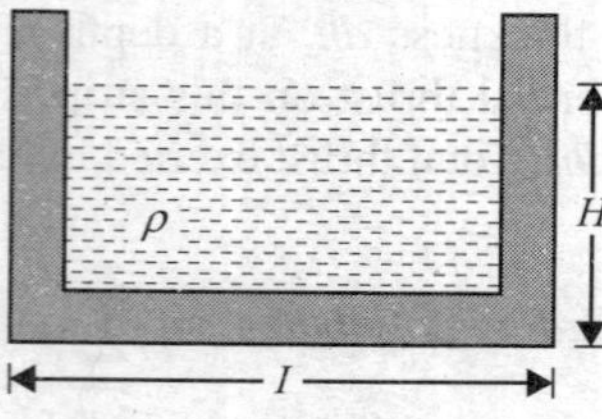

Fig. 3.33

Sol. Whenever a liquid comes in contact with solid boundaries it exerts a force on it. The force on the boundary may be obtained by integrating the pressure over the entire area of the boundary. The variations of liquid pressure acting at the base and at the wall are shown in Figs. 3.34(a) and (b), respectively.

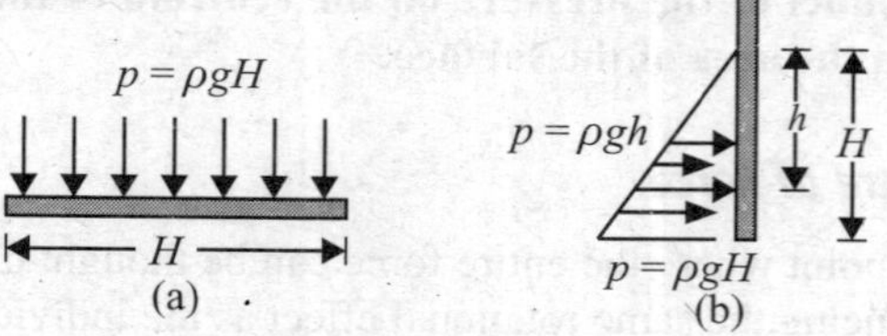

Fig. 3.34

(i) Force at the base

Since the pressure is uniform at the base, force acting at the base is given by

$$F = p \times \text{(area of the base)}$$

Since $p = \rho g H$, therefore

$$F = \rho g H(lb) = \rho g(lbH)$$

Since $lbH = V$ (volume of the liquid), so

$$F = \rho g V = \text{weight of the liquid inside the tank}$$

Note: ***The force acting at the base per unit width (F/b) is equal to the area of the pressure diagram. That is,***

$$\frac{F}{b} = pl = \rho g H l$$

(ii) Force acting on the vertical wall

Pressure acting on the vertical wall is not uniform but increases linearly with depth. Pressure at a depth h from the free surface is $p = \rho g h$.

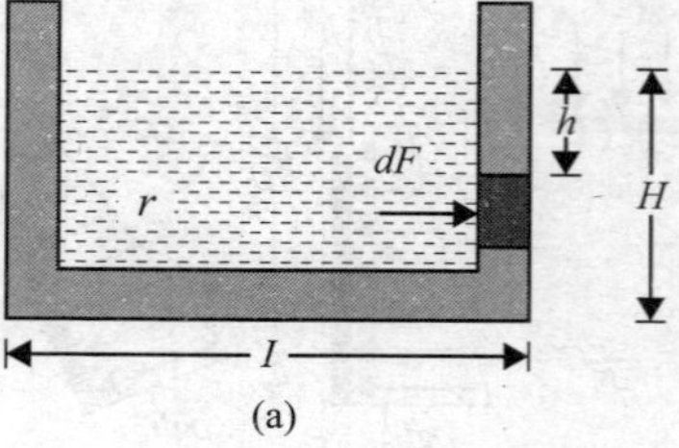

Fig. 3.35

Force dF acting on a differential element of height dh is

$$dF = p(bdh) = (\rho g h)(bdh) = \rho g b h dh$$

The total force is $F = \rho g b \int_0^H h dh = \frac{1}{2}\rho g b H^2$

The total force acting per unit width of the vertical wall is

$$\frac{F}{b} = \frac{1}{2}\rho g H^2$$

Note: ***The force acting per unit width is equal to the area of the pressure diagram shown in Fig. 3.34(b).***

$$\frac{F}{b} = \frac{1}{2}(\textit{base})(\textit{height}) = \frac{1}{2}(\rho g H)(H) = \frac{1}{2}\rho g H^2$$

Hence, force per unit width of an immersed surface is equal to the area of the pressure diagram on the surface.

The point of application (or the centre of force) of the total force from the free surface is given by $h_C = \frac{1}{F}\int_0^H hF$.

Here, $\int_0^H h\, dF = \int_0^H h(\rho g b h\, dh) = \rho g b \int_0^H h^2\, dh = \frac{1}{3}\rho g b H^3$

$$F = \frac{1}{2}\rho g b H^2 \Rightarrow h_C = \frac{2}{3}H$$

The total force acts at a depth 2/3 H from the force diagram.

Note: ***The total force acts through the centroid of the pressure diagram, at a depth $\frac{2}{3}$ H from the top of the liquid level.***

Illustration 3.12 **A tank of square cross section (2 m × 2 m) is filled with water up to a height of 2.5 m. Find the thrust experienced by the vertical and bottom face of the tank (g = 10 m/s²).**

Sol. Thrust at the bottom surface:

The pressure is uniformly distributed over the bottom surface.

Thrust = Pressure × Area

$$= (2.5\rho g)A = 2.5 \times (1 \times 10^3)[10(2 \times 2)]$$

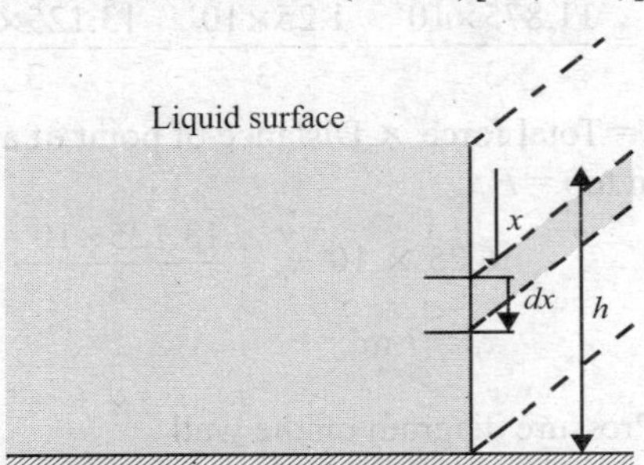

Fig. 3.36

Thrust at the vertical surface:

On a vertical surface in contact with the liquid, the pressure is not the same at all points. Points at greater depth experience the larger pressures and hence the greater thrust. Consider a vertical surface $ABCD$ ($l \times h$) in contact with the liquid of height h.

Divide the vertical surface $ABCD$ into rectangular strips of height dx and area ldx. Then,

$$F = \int PdA = \int_0^h (x\rho g)ldx = \frac{\rho g l h^2}{2}$$

Expressing F as $F = \left[\frac{h}{2}\rho g\right]hl$, we get

F = [Pressure at CG of vertical surface] × [Area of surface]

$$= \left(\frac{2.5}{2} \times \rho \times g\right)(2 \times 2.5) = \frac{2.5}{2} \times 10^3 \times 10 \times 5 = 6.25 \times 10^4$$

Illustration 3.13 **Figure 3.37 shows a cross-sectional view of a masonry dam whose length perpendicular to the diagram is 30 m. The depth of water behind the dam is 10 m. The masonry of which the dam is constructed has a density of 3000 kg/m³.**

a. Find the dimensions x and $2x$ if the weight of the dam is to be 10 times the horizontal force exerted on it by water.

b. Check the stability of the dam with respect to its overturning about the edge through point O.

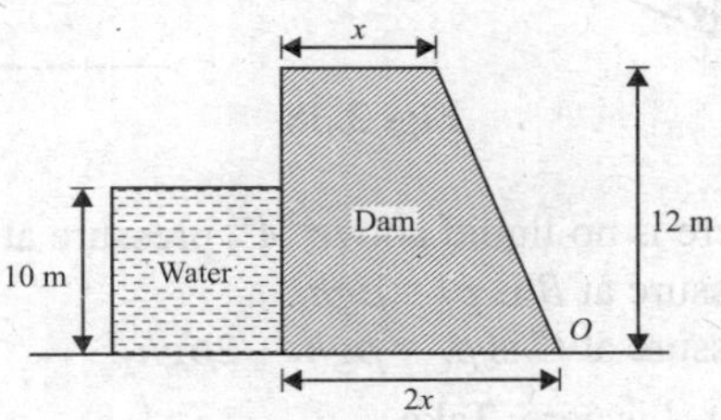

Fig. 3.37

Sol. a. Weight of dam = volume × density × g

$$= \left[\frac{1}{2}(x + 2x)12 \times 30\right]3000 \times 10$$

$$= 162x \times 10^5 \text{ N}$$

Horizontal thrust due to water(on a vertical surface) = pressure at G × Area = 5 × 10³ × 10 × (30 × 10) = 15 × 10⁶ N

According to the problem, $162x \times 10^6 = 10(15 \times 10^6)$

$$\Rightarrow \quad x = 9.26 \text{ m} \quad \text{and} \quad 2x = 18.52 \text{ m}$$

b. Let G be the centre of gravity of dam. Taking O as origin and the horizontal axis towards left-hand side of O as $+X$-axis, let r be the distance of centre of gravity from O

$$r = \frac{(12x)\frac{3x}{2} + (6x)\frac{2x}{3}}{12x + 6x} = \frac{11}{9}x$$

Anticlockwise torque exerted by the weight of the dam is

$$W\left(\frac{11}{9}x\right) = 162 \times 10^6 \times \frac{11}{9}x^2 = 169.78 \times 10^9 \text{ Nm}$$

Clockwise torque due to liquid thrust,

$$\int_0^{10} (x\rho g)(30dx)(10 - x) = 30\rho g \left[5x^2 - \frac{x^3}{3}\right]_0^{10} = 5000\rho g$$

$$= 5 \times 10^9 \text{ Nm}$$

As the torque due to the liquid thrust is much less than that due to the weight of the dam, the dam is stable.

Illustration 3.14 **Water and liquid are filled up behind a square wall of side l. Find out**

a. pressures at A, B and C;

b. forces in part AB and BC;

c. total force and point of application of force. (neglect atmospheric pressure in every calculation)

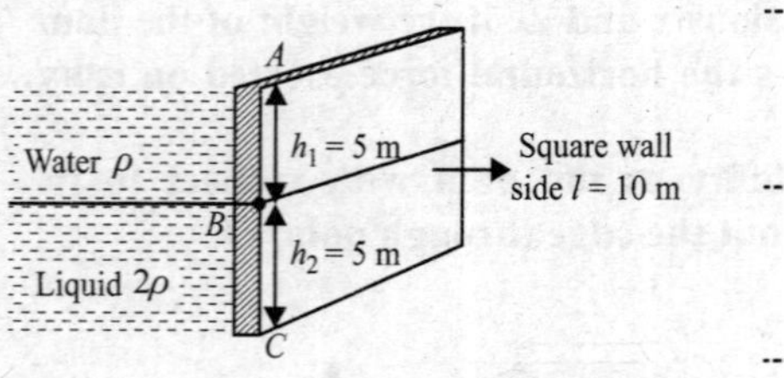

Fig. 3.38

Sol. a. As there is no liquid above 'A', pressure at A is $p_A = 0$

Pressure at B is $p_B = \rho g h_1$

Pressure at C is $p_C = \rho g h_1 + 2\rho g h_2$

b. Force at A is zero. Take a strip of width 'dx' at a depth 'x' in part AB. Pressure is equal to $\rho g x$.

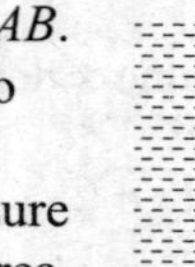

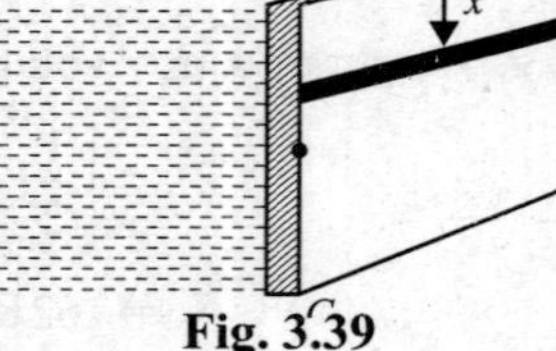

Fig. 3.39

Force on strip = pressure × area

$$dF = \rho g\, l dx$$

Total force up to B,

$$F_B = \int_0^{h_1} \rho g\, l dx = \frac{\rho g\, l h_1^2}{2} = \frac{1000 \times 10 \times 10 \times 5 \times 5}{2}$$

$= 1.25 \times 10^6$ N

In part BC for force take an elementary strip of width dx in portion BC. Pressure is given by

$$\rho g h_1 + 2\rho g(x - h_1)$$

Force on the elementary strip = pressure × area

$$\therefore \quad dF = [\rho g h_1 + 2\rho g(x - h_1)]\, l dx$$

The limit of x will be $x = h_1$ to $x = h_1 + h_2 = l$

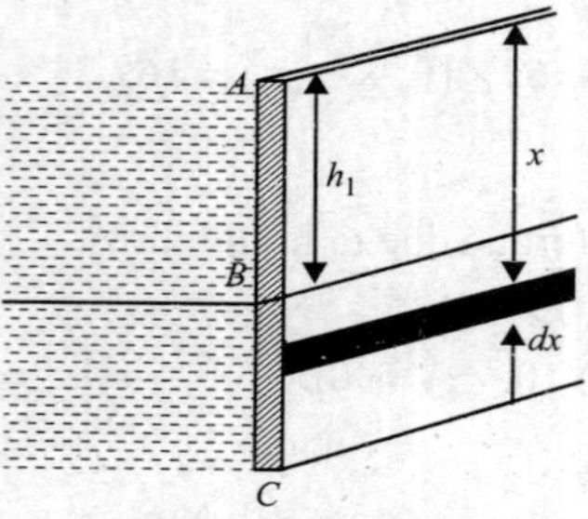

Fig. 3.40

Total force on part BC is $F = \int_{h_1}^{h_1+h_2} [\rho g h_1 + 2\rho g(x - h_1)] l$

But $h_1 + h_2 = l = \left[\rho g h_1 x + 2\rho g\left[\frac{x^2}{2} - h_1 x\right]\right]_{h_1}^{l} l$

$= \rho g h_1 h_2 l + \rho g l\, (l - h_1)^2$

$= \rho g h_2 l\, [h_1 + h_2] = \rho g h_2 l^2$

$= 1000 \times 10 \times 5 \times 10 \times 10$

$= 5 \times 10^6$ N

c. Method 1:

Total force $= 5 \times 10^6 + 1.25 \times 10^6 = 6.25 \times 10^6$ N

Taking torque about A, the total torque of force in AB is

$$\int dFx = \int_0^{h_1} \rho g x l dx x$$

$$\left[\frac{\rho g l x^3}{3}\right]_0^{h_1} = \frac{\rho g l h_1^3}{3} = \frac{1000 \times 10 \times 10 \times 125}{3} = \frac{1.25 \times 10^7}{3} \text{ Nm}$$

The total torque of force in BC about A is given by

$$\int dFx = \int_B^l \left[\rho g h_1 + 2\rho g(x - h_1)\right] l dx . x$$

$$\rho g h_1 h_2 l \left[h_1 + \frac{h_2}{2}\right] + \rho g h_2^2 l \left[h_1 + \frac{2h_2}{3}\right]$$

$$= 2.5 \times 7.5 \times 10^6 + \frac{62.5}{3} \times 10^6 = \frac{118.75}{3} \times 10^6 \text{ Nm}$$

$$\text{Total torque} = \frac{11.875 \times 10^7}{3} + \frac{1.25 \times 10^7}{3} = \frac{13.125 \times 10^7}{3} \text{ Nm}$$

Total torque = Total force × Distance of point of application of force from top = $F x_p$

$$\Rightarrow \quad 6.25 \times 10^6 x_p = \frac{13.125 \times 10^7}{3} \text{ Nm}$$

$$\Rightarrow \quad x_p = 7 \text{ m}$$

Method 2: Pressure diagram on the wall

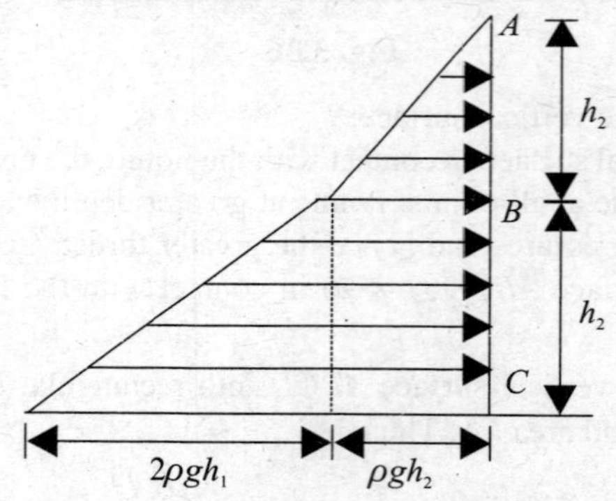

Fig. 3.41

Total force on the wall per unit width of the wall = Area of pressure diagram.

$$\frac{F}{b} = \frac{1}{2}(\rho g h_1) h_1 + \rho g h_1 h_2 + \frac{1}{2}(2\rho g h_2) h_2$$

$$= \rho g\left[\frac{h_1^2}{2} + h_2^2\right] + \rho g h_1 h_2$$

$$F_{\text{del}} = F = \frac{1}{2}\rho g l h_1^2 + \rho g l h_2\, [h_1 + h_2]$$

$$= 6.25 \times 10^6 \text{ N}$$

Total force on AB part, $F_{AB} = \frac{1}{2}\rho g \ell h_1 = 1.25\times10^6\ N$

Total force on BC part, $F_{BC} = \rho g l h_2[h_1 + h_2] = 5\times10^6$ N
The total force will act at the centroid of the pressure diagram. Let the distance of centroid of pressure diagram be 'r' from A.

$$r = \frac{\frac{\rho g h_1^2}{2}\times\frac{2}{3}h_1 + \frac{2\rho g h_2^2}{2}\left(h_1+\frac{2}{3}h_2\right)}{}$$

$$+\frac{\rho g h_1 h_2\left(h_1+\frac{h_2}{2}\right)}{\frac{\rho g h_1^2}{2}+\rho g h_2^2+\rho g h_1 h_2}$$

$$= \frac{\frac{h_1^3}{3}+h_2^2\left(h_1+\frac{2}{3}h_2\right)+h_1h_2\left(h_1+\frac{h_2}{2}\right)}{\frac{h_1^2}{2}+h_2^2+h_1h_2}$$

$$r = \frac{\left[\frac{1}{3}+\left(1+\frac{2}{3}\right)+\left(1+\frac{1}{2}\right)\right]5}{\frac{1}{2}+1+1} = \frac{\frac{7}{2}}{\frac{5}{2}}\times5 = 7\,\text{m}$$

Hydrostatic Forces on Inclined Surfaces Immersed in Liquid

Method 1: Consider a plane surface of length 'a' and width 'b' which is inside a liquid of density ρ as shown in Fig. 3.42.

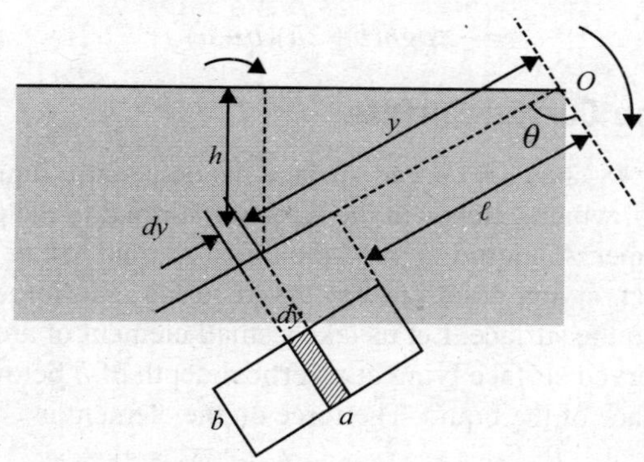

Fig. 3.42

Choose an element of length (dy) at a distance y from O. The depth of the element $h = y\sin\theta$. The intensity of pressure at the position of the element is $P = \rho g h = \rho g\,(y\sin\theta)$.
The force on the element is $dF = P(bdy)$

$$= \rho g(y\sin\theta)\,bdy$$
$$= \rho g b\sin\theta\ y\,(dy)$$

The force on the entire surface is

$$F = \int_l^{(l+a)} dF = \rho g b\sin\theta \int_l^{(l+a)} y dy = \rho g\frac{b\sin\theta}{2}\left|y^2\right|_l^{(l+a)}$$

$$= \rho g b\frac{\sin\theta}{2}\left[(l+a)^2 - l^2\right]$$

$$= \rho g b\frac{\sin\theta}{2}\left[(l+a)^2 - l^2\right]$$

$$= \frac{\rho g b\sin\theta}{2}\left[a^2+2al\right] = \frac{\rho g a b}{2}[a+2l]$$

Method 2: Force per unit width on the surface can be found out by taking area of pressure diagram on the surface. Figure 3.43 shows the pressure diagram on the surface.

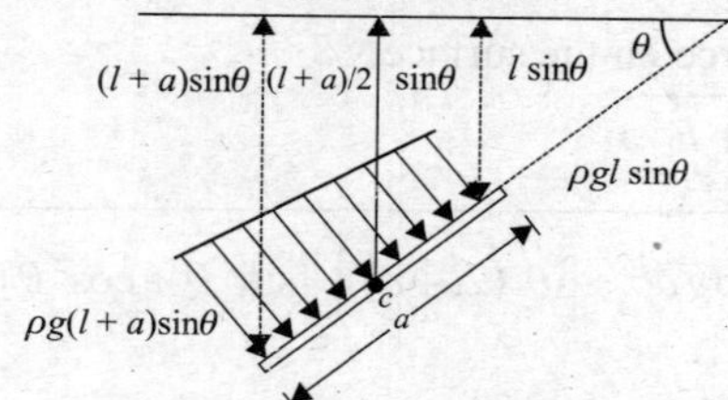

Fig 3.43

Area of pressure diagram = Force per unit width of surface

$$\frac{F}{b} = \frac{1}{2}[\rho g(l+a)\sin\theta + \rho g l\sin\theta]a$$

$$= \frac{\rho g a\sin\theta}{2}[l+a+l]$$

$$F = \frac{\rho g a b\sin\theta}{2}[a+2l]$$

Method 3: Force on inclined surface is given by

$$F = P_{av} \times \text{area of surface}$$

$$= \frac{1}{2}[\rho g l\sin\theta + \rho g(l+a)\sin\theta]\ \ ab$$

$$= \frac{1}{2}\rho g ab\sin\theta\ [a+2l]$$

Method 4: Force on inclined surface is given by
F = Pressure at the centroid of the surface × Area of surface

$$= \left[\rho g\left(l+\frac{a}{2}\right)\sin\theta\right]ab$$

$$= \frac{\rho g ab\sin\theta}{2}[a+2l]$$

Method 5:

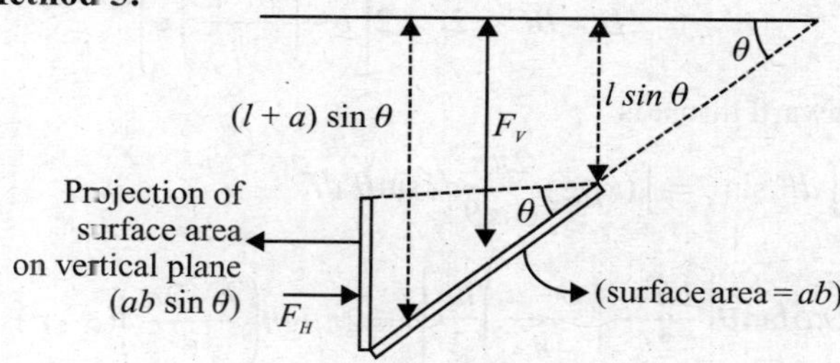

Fig. 3.44

Horizontal component of total force,

F_H = Force on the area projected on the vertical plane

$$= \frac{1}{2}[\rho g l \sin\theta + \rho g(l+a)\sin\theta](ab\sin\theta)$$

$$= \frac{1}{2}\rho g ab \sin^2\theta\,(2l+a) \qquad \text{(i)}$$

Vertical component of the force,

F_V = Weight of water prism above the inclined surface

$$= \frac{1}{2}[l\sin\theta + (l+a)\sin\theta]a\cos\theta\, b\rho g$$

$$= \frac{1}{2}\rho g ab \sin\theta\cos\theta(2l+a) \qquad \text{(ii)}$$

Hence, net force on the surface,

$$F = \sqrt{F_H^2 + F_V^2}$$

$$= \sqrt{\left(\frac{1}{2}\rho g ab \sin\theta\,(2l+a)\right)^2 [\sin^2\theta + \cos^2\theta]}$$

$$= \frac{1}{2}\rho g ab \sin\theta\,(2l+a)$$

Illustration 3.15 **Find the net downward thrust on the inclined surface of a bucket full of water of height *h* and radii *a* and *b***

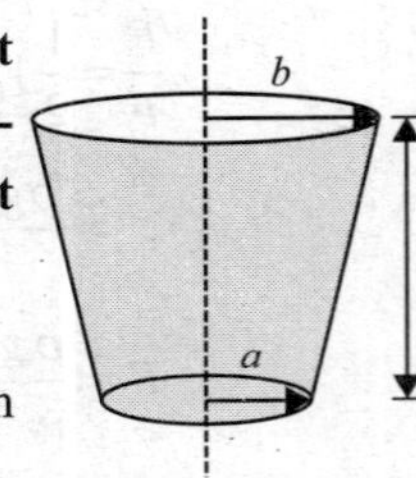

Fig. 3.45

Sol. Method 1

Consider an element of height dx at depth x below the surface.

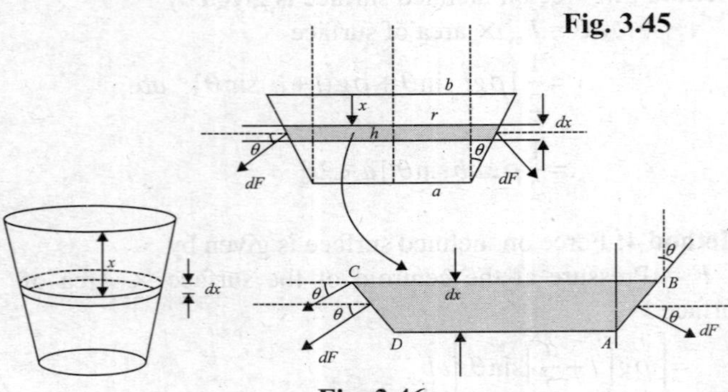

Fig. 3.46

$$\text{Area} = 2\pi r(AB) = 2\pi r\frac{dx}{\cos\theta}$$

$$AD \approx BC = 2r = 2\left[b - \left(\frac{b-a}{h}\right)x\right]$$

Downward thrust is

$$F = \int dF\sin\theta = \int_0^h (x\rho g)\frac{2\pi r}{\cos\theta}dx\sin\theta\, dF$$

$$= 2\pi\rho\tan\theta\left[\frac{bh^2}{2} - \left(\frac{b-a}{h}\right)\frac{h^3}{3}\right] = 2\pi\rho g h\left(\frac{b+2a}{6}\right)(b-a)$$

Method 2

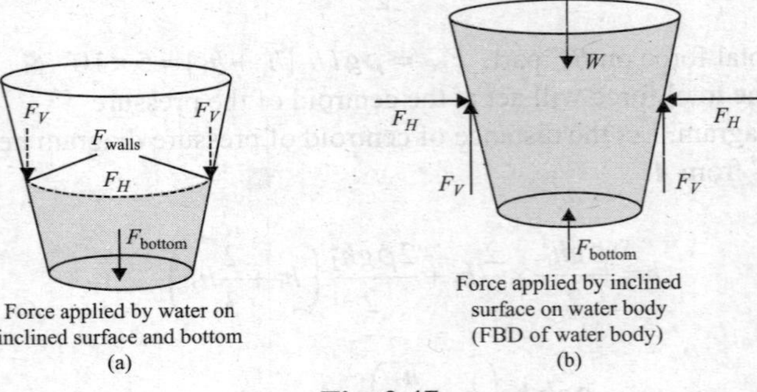

Fig. 3.47

The water will apply normal force on the inclined wall and at the bottom of the container as shown in Fig. 3.47(a).

Force at the bottom of the container is

$$F_{bottom} = \text{Pressure at bottom} \times \text{Area}$$

$$= (\rho g h)(\pi a^2)$$

$$= \rho g h \pi a^2 \qquad \text{(i)}$$

Weight of water in container is

$$W = \frac{1}{3}\pi h(a^2 + ab + b^2)\rho g \qquad \text{(ii)}$$

From the free-body diagram of the water body,

$$W = F_V + F_{bottom}$$

$$F_V = W - F_{bottom}$$

$$= \frac{1}{3}\pi h(a^2 + ab + b^2)\rho g - \rho g h \pi a^2$$

$$= \frac{1}{3}\pi h \rho g\left[a^2 + ab + b^2 - 3a^2\right]$$

$$= \frac{1}{3}\pi\rho g h\left[ab + b^2 - 2a^2\right]$$

$$= \frac{1}{3}\pi\rho g h(b+2a)(b-a)$$

Force on Curved Surface

Figure 3.48 shows a curved surface inside a static liquid of density ρ, which extends in the direction normal to the plane of the paper. As liquid imparts the thrust normal to the point of contact, at any point on the curved surface the force acts normal to the surface. Let us take a small element of area dA of the curved surface lying at a vertical depth of h below the free surface of the liquid. The force on the element is

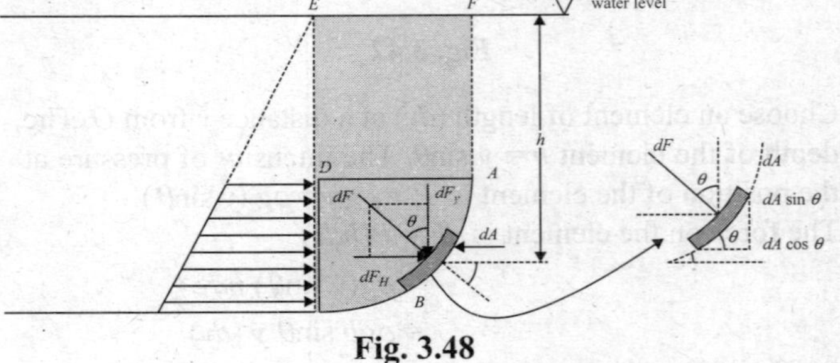

Fig. 3.48

$$dF = P(dA) = \rho gh(dA) \quad \text{(i)}$$

The force dF acting on the element can be resolved into horizontal and vertical components. Let dF is inclined at an angle θ with the vertical.

$$dF_H = dF \sin\theta = \rho gh(dA) \sin\theta \quad \text{(ii)}$$

$$dF_V = dF \cos\theta = \rho gh(dA) \cos\theta \quad \text{(iii)}$$

The total horizontal and vertical components of the force can be obtained by integrating Eqs. (ii) and (iii). So

$$F_H = \int dF_H = \rho g \int h(dA) \sin\theta \quad \text{(iv)}$$

$$F_V = \int dF_v = \rho g \int h(dA) \cos\theta \quad \text{(v)}$$

In Eq. (iv), $dA \sin\theta$ is the vertical projection of the area dA and in Eq. (v), $dA \cos\theta$ is the horizontal projection of the elementary area dA. Thus, $\rho g \int h(dA)\sin\theta$ represents the total force on the vertical projection of the curved surface, i.e., the horizontal component of the net force is equal to the total force on the projected area of the curved surface on the vertical plane, the trace of which is represented by *CD*.

Further, $\rho g \int h(dA)\sin\theta$ represents the total force on the horizontal projection of the curved surface, and it is equal to the weight of the liquid contained in the portion extending above the curved surface in the portion *ABCDEFA*, i.e., vertical component of net force is equal to the weight of the liquid contained in the portion extending vertically above the curved surface up to the free surface of the liquid.

The resultant force on the curved surface is $F = \sqrt{F_H^2 + F_V^2}$.

Illustration 3.16 **Water is retained by a curved gate *AB*. Compute the horizontal and vertical components of the resultant of the hydrostatic pressure distribution on the gate *AB*, which is a quarter of a cylinder. Assume that the gate is 4 m wide.**

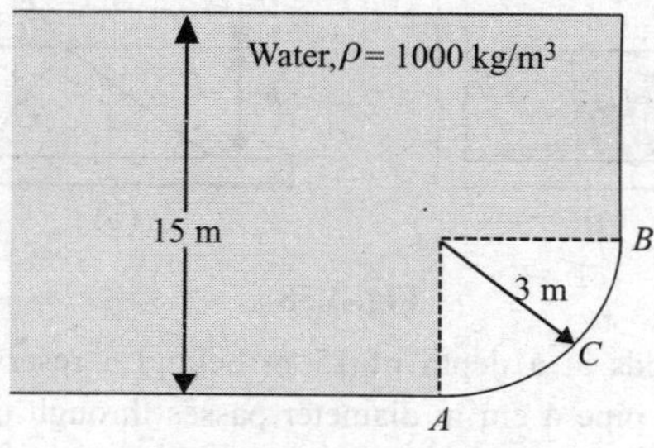

Fig. 3.49

Sol. (a) Horizontal component: The horizontal component of the resultant force on the gate is equal to the force on the vertical projection BA' of the gate. The pressure at a depth y from the free surface of the water is $P = \rho gy$. The force on the strip of area $4dy$ of the gate is

$$dF_H = P(4dy) = \rho gy(4dy) = 4\rho gydy$$

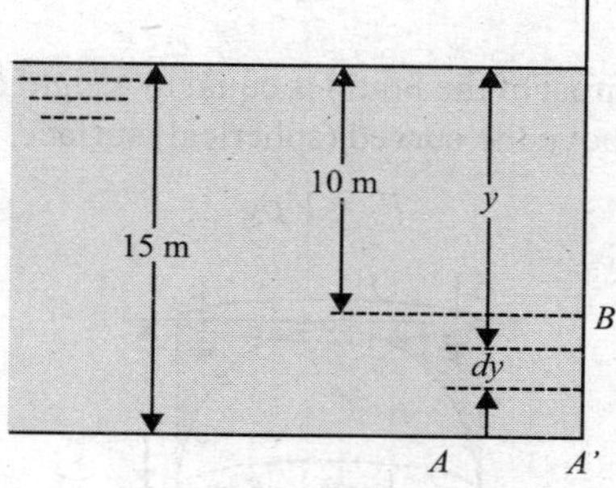

Fig. 3.50

Total horizontal force is

$$F_H = 4\rho g \int_{10}^{15} ydy = 4\rho g \left|\frac{y^2}{2}\right|_{10}^{15}$$

$$= 4 \times 1000 \times 10 \left(\frac{15^2 - 10^2}{2}\right) = 250 \times 10^4 \text{ N}$$

(b) Vertical component: It is equal to the weight of the water above the curved surface. Dividing the volume above the curved surface into a rectangular parallelepiped and a quarter cylinder, we thus have

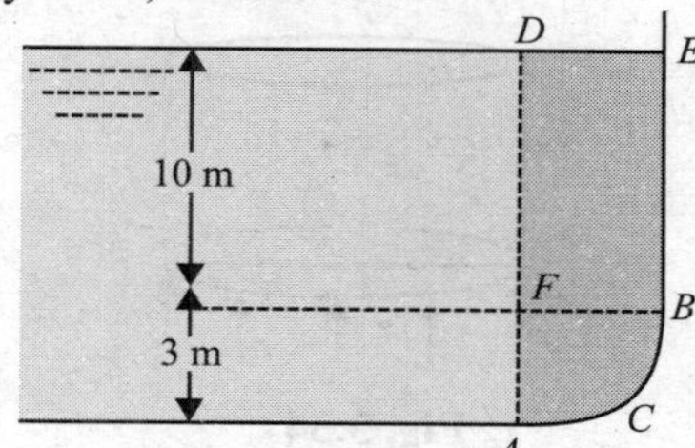

Fig. 3.51

$$F_V = \text{Weight of water in } DEFB + \text{Weight of water in } ACBF$$

$$= \rho[V_1 + V_2]g = 1000\left[10 \times 4 \times 4 + \frac{\pi(5)^2 \times 4}{4}\right] \times 10$$

$$= [8 \times \pi] \times 10^4 \text{ N}$$

After calculating the values of F_V and F_H we can calculate the resultant force on the gate $F = \sqrt{F_H^{\,2} + F_V^{\,2}}$.

Illustration 3.17 **A solid hemisphere of radius *R* is made to just sink in a liquid of density ρ. Find (a) the vertical thrust on the curved surface, (b) the vertical thrust on the flat surface, (c) the side thrust on the hemisphere, (d) the total hydrostatic force acting on the hemisphere.**

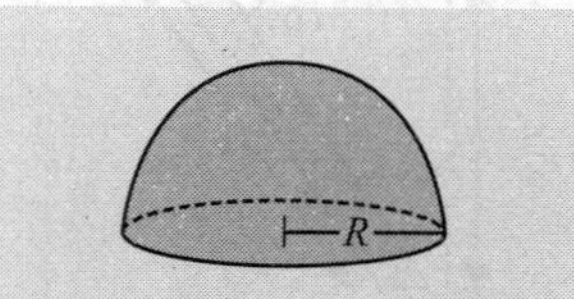

Fig. 3.52

Sol.

(a) Vertical thrust of the liquid is equal to weight of the liquid column above the curved (spherical) surface, i.e.,

$$F_V = V\rho g$$

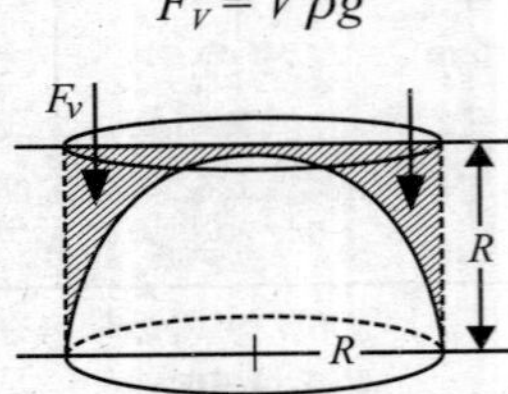

Fig. 3.53

where V = Volume of the shaded portion
= Volume of the cylinder − Volume of the hemisphere

$$= (\pi R^2)R - \frac{2}{3}\pi R^3 = \frac{\pi R^3}{3}$$

Substituting V in the equation $F_V = V\rho g$, we have

$$F_V = \frac{\pi R^3}{3}\rho g \text{ (down)}$$

(b) The upward thrust on the base of the hemisphere is

$$F'_V = V'\rho g$$

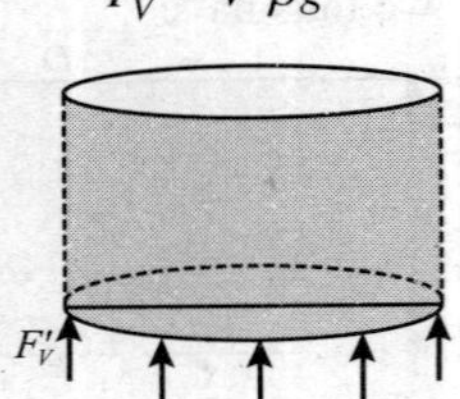

Fig. 3.54

where V' = Volume of the liquid column above the base = $(\pi R^2)R = \pi R^3$. Then, we have $F'_V = \pi R^3 \rho g$ (upward)

(c) Side thrust $F_h = 0$ (as discussed earlier)

(d) The total hydrostatic force is

$$F = F'_V - F_V = \pi R^3 \rho g - \frac{\pi R^3}{3}\rho g = \frac{2}{3}\pi R^3 \rho g \text{ (upward)}$$

Concept Application Exercise 3.1

1. What is the absolute pressure on a swimmer 10 m below the surface of a lake? Take atmospheric pressure 1×10^5 N/m².
2. What will be the length of mercury column in a barometer tube when the atmospheric pressure is 75 cm of mercury and the tube is inclined at an angle of 60° with the horizontal direction?

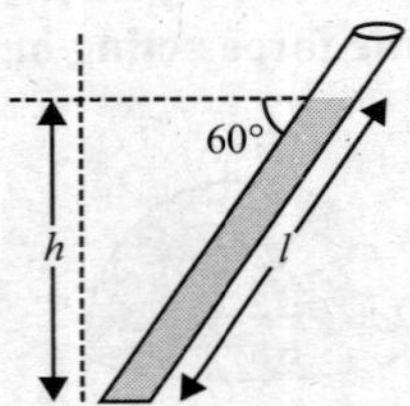

Fig. 3.55

3. A U-tube contains water and methylated spirit separated by mercury. The mercury columns in the two arms are in the same level with 10 cm of water in one arm and 12.5 cm of spirit in the other. What is the specific gravity of spirit?

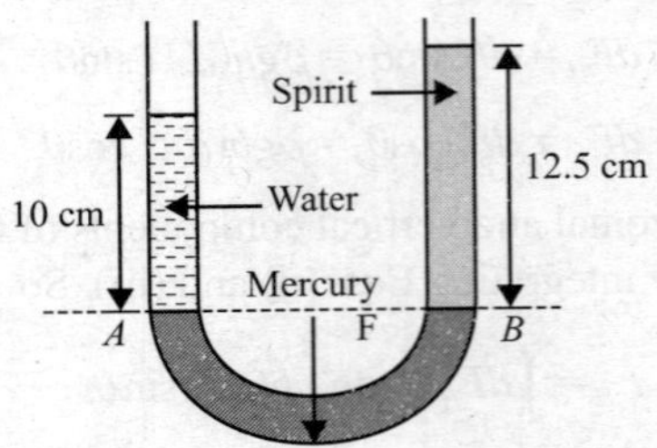

Fig. 3.56

4. A U-tube in which the cross-sectional area of the limb on the left is one-third of the limb on the right contains mercury (density 13.6 g/cm³). The level of mercury in the narrow limb extends to a distance of 30 cm from the upper end of the tube. What will be the rise in the level of mercury in the right limb, if the left limb is filled to the top with water (neglect surface tension effects)?

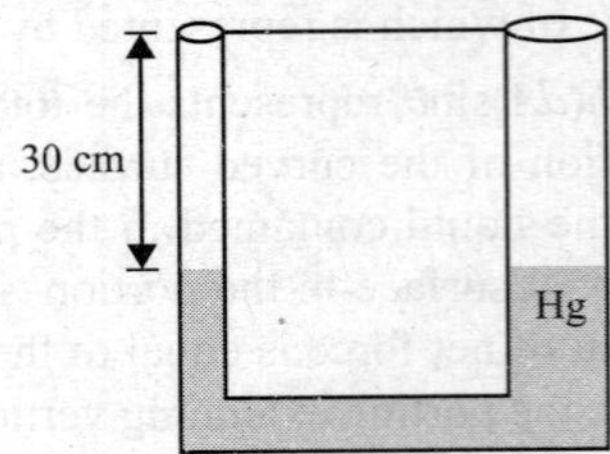

Fig. 3.57

5. Figure 3.58(b) shows the variation of pressure (p) with depth (h) along a vertical wall supporting two liquids of different densities as shown in Fig. 3.58(a). Identify the correct variation of pressure in the liquid of density ρ_2.

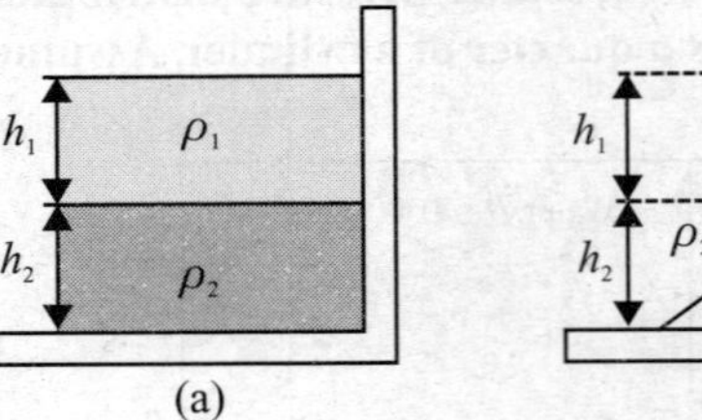

Fig. 3.58

6. Water stands at a depth of 15 m behind a reservoir dam. A horizontal pipe 4 cm in diameter passes through the dam 6 m below the surface of water as shown. There is a plug which secures the pipe opening. Then find the friction between the plug and pipe wall.

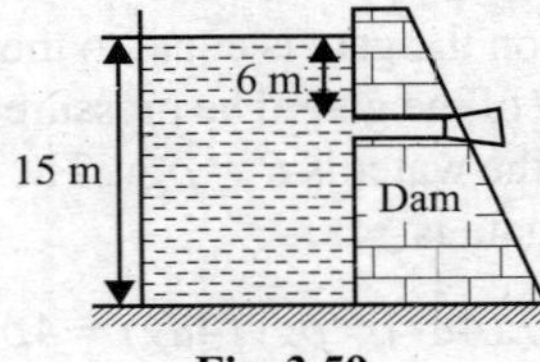

Fig. 3.59

7. A glass full of water has a bottom of area 20 cm^2, top of area 20 cm^2, height 20 cm and volume half a litre.

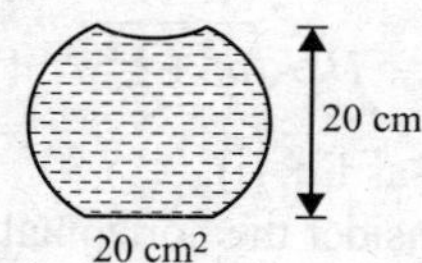

Fig. 3.60

(a) Find the force exerted by the water on the bottom.

(b) Considering the equilibrium of the water, find the resultant force exerted by the sides of the glass on the water. Atmospheric pressure = 1.0 × 10^5 N/m^2. Density of water = 1000 kg/m^3 and g = 10 m/s^2.

8. An isosceles triangular plate of base 3 m and altitude 3 m is immersed in oil vertically with its base coinciding with the free surface of the oil of relative density 0.8. Determine the total thrust.

9. Two lock gates of 7.5 m height are provided in a canal of 16 m width meeting at an angle of 120°. Calculate the force acting on each gate when the depth of water in upstream side is 5 m.

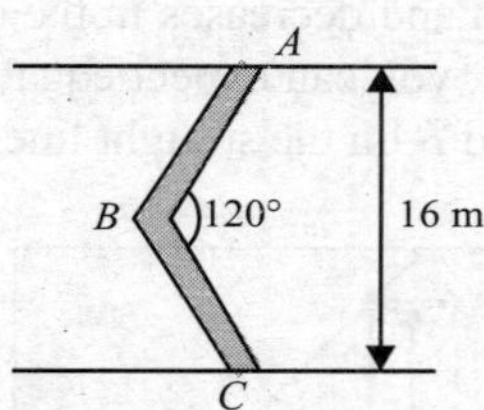

Fig. 3.61

10. Figure 3.61 shows a semi-cylindrical massless gate pivoted at the point O holding a stationary liquid of density ρ. The length of the cylinder is l. Calculate, the horizontal force exerted by the liquid on the gate.

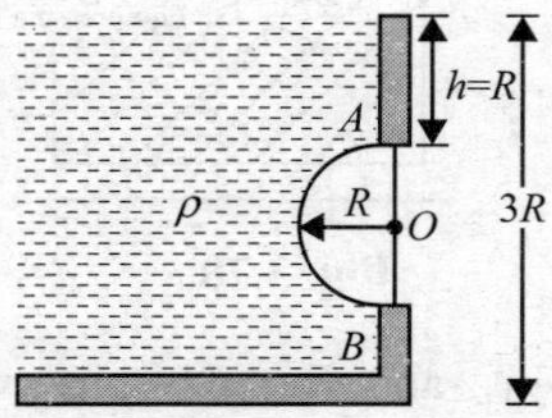

Fig. 3.62

11. Consider a container having an inclined wall, inclined an angle θ with horizontal as shown in Fig. 3.63. Calculate the force per unit width on the wall.

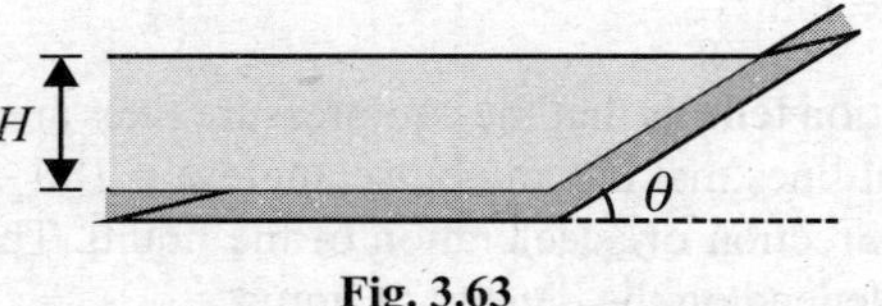

Fig. 3.63

Fluid in Uniformly Accelerating Motion

Liquid Subjected to Constant Horizontal Acceleration

Consider a liquid in a tank which is moving on a horizontal surface with constant acceleration a. The free surface of the liquid takes the shape as shown in Fig. 3.64. Consider a cylinder of liquid of length l and area of cross section A. The force on the left face of the cylinder is $F_1 = P_1A$ and force on the right face of the cylinder is $F_2 = P_2A$. Here $P_1 = \rho g y_1$ and $P_2 = \rho g y_2$. Mass of the liquid cylinder is $m = A\,l\rho$. Using Newton's second law for the liquid cylinder,

$$F_1 - F_2 = ma$$

or $P_1A - P_2A = ma$ or $(\rho g y_1 - \rho g y_2)A = Al\rho a$

or
$$\frac{y_1 - y_2}{l} = \frac{a}{g}$$

From Fig. 3.64, $\frac{y_1 - y_1}{l} = \tan\theta$

$$\therefore \quad \tan\theta = \frac{a}{g}$$

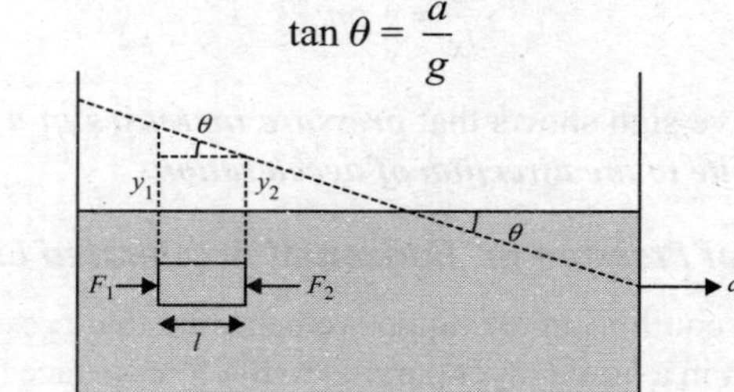

Fig. 3.64

The behaviour of a liquid contained in a horizontally accelerated container can be compared to a pendulum suspended from the ceiling of a horizontally accelerated trolley.

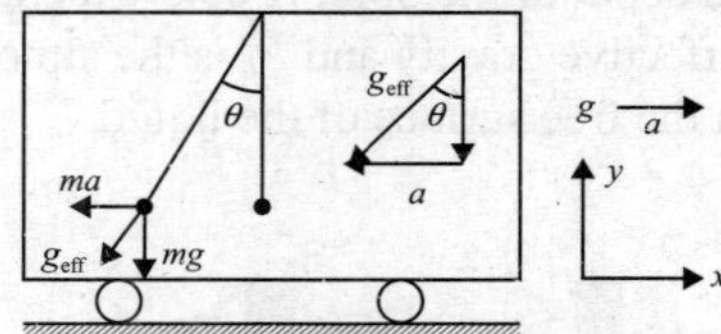

Fig. 3.65

The weight (mg) acts downward and the pseudo-force (ma) on the pendulum acts horizontally. Hence, it attains a new equilibrium position with respect to the trolley. The pendulum orients itself in the direction of net effective gravity (g_{eff}).

$$\tan\theta = \frac{a}{g}$$

Every fluid element attains an equilibrium position under the action of gravity and pseudo-force. The free surface of the liquid orients itself perpendicular to the direction of net effective gravity.

$$\tan\theta = \frac{a}{g}$$

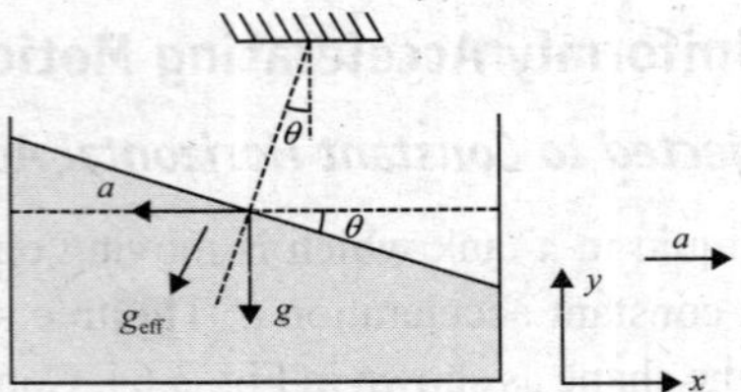

Fig.3.66

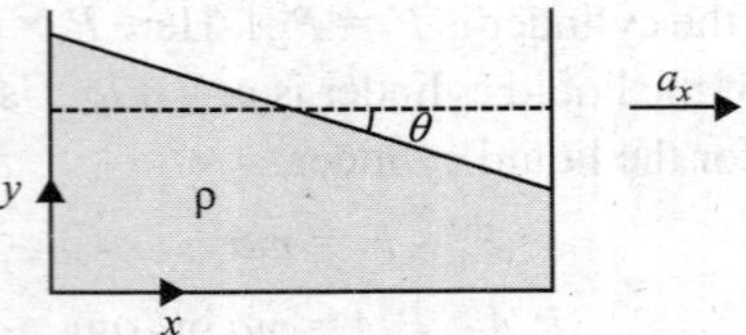

Fig. 3.67

In a horizontally accelerated fluid, pressure varies in the horizontal direction. In the arrangement shown in Fig. 3.67, rate of change of pressure in x direction is

$$\frac{dp}{dx} = -\rho a_x$$

The negative sign shows that ***pressure increases in a direction opposite to the direction of acceleration.***

Analysis of Pressure in Horizontal Accelerated Liquid

Now from equilibrium of liquid we can state that pressure at every point in a liquid layer parallel to the free surface (which is not horizontal) remains the same. For example, pressure at a point A in the accelerated container as shown in Fig. 3.68(a) is given as

$$P_A = P_0 + H\rho\sqrt{a^2+g^2} \tag{i}$$

where H is the depth of the point A below the free surface of liquid along effective gravity and P_0 is the atmospheric pressure acting on the free surface of the liquid.

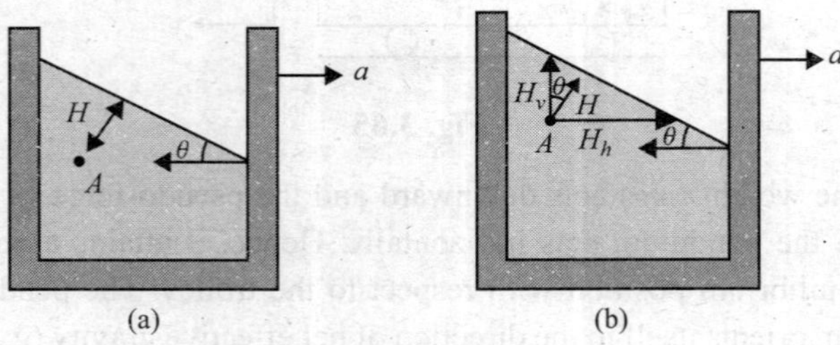

Fig. 3.68

The pressure at point A can also be obtained in another way as shown in Fig. 3.68(b). If H_V and H_h are the vertical and horizontal distances of point A from the free surface of liquid, then pressure at point A can also be given.

According to Pascal's law, the pressure at point A due to the vertical height of liquid above A is given as

$$P_a = P_0 + H_V\rho g \tag{ii}$$

Here we can write H_V as $H_V = H\sec\theta = \dfrac{H\sqrt{a^2+g^2}}{g}$

From Eq. (ii), $P_A = P_0 + H\rho\sqrt{a^2+g^2}$ (iii)

This equation is same as Eq. (i).

Similarly, if we consider the horizontal distance of point A from the free surface of liquid, which is H_h, then due to pseudo-acceleration of container the pressure at point A is given as

$$P_A = P_0 + H_h\rho a \tag{iv}$$

Here, H_h is given as $H_h = H\operatorname{cosec}\theta = \dfrac{H\sqrt{g^2+a^2}}{a}$

From Eq. (iv), we have

$$P_A = P_0 + H\rho\sqrt{g^2+a^2} \tag{v}$$

This equation is also same as Eq. (i).

Equipressure Lines

We have seen that pressure at different points at same horizontal line will be different. Since the pressure increases vertically downward and decreases in the direction of acceleration of the vessel, you can expect equal pressure between any two points A and B on the straight line AB.

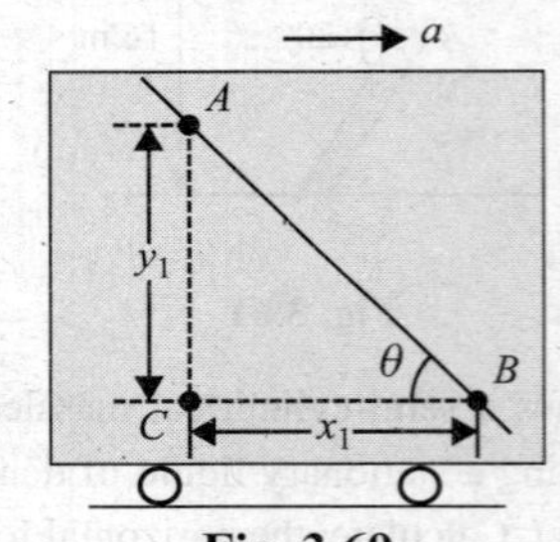

Fig. 3.69

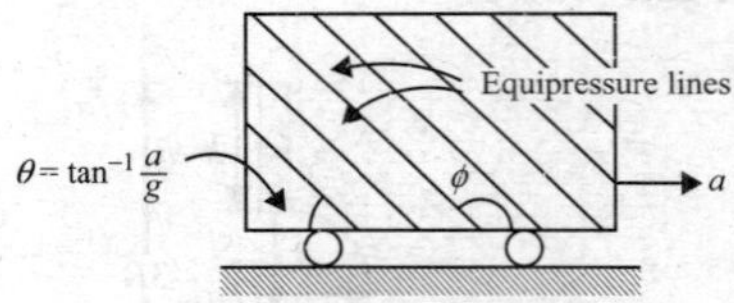

Fig. 3.70

Since $P_C = P_A + \rho g y_1$ and $P_B = P_C - \rho a x_1$, we have

$$P_B - P_A = \rho(gy_1 - ax_1)$$

Substituting $P_B = P_A$,

we have $y_1 = \dfrac{a}{g}x_1 \Rightarrow \dfrac{y_1}{x_1} = \tan\theta$

Hence, $\theta = \tan^{-1}\dfrac{a}{g}$

This equation tells us that the equipressure lines are the parallel straight lines making an obtuse angle $\phi = 180 - \tan^{-1} a/g$ with the direction of acceleration of the liquid. The angle ϕ does not depend on the density of liquid.

Fluid Subjected to Constant Vertical Acceleration

Pressure difference when liquid is accelerating in vertical direction

Consider a cylindrical element of height h and area A. The force on the top face of the element is P_1A and force on the bottom face is P_2A. If a is the acceleration of the liquid, then

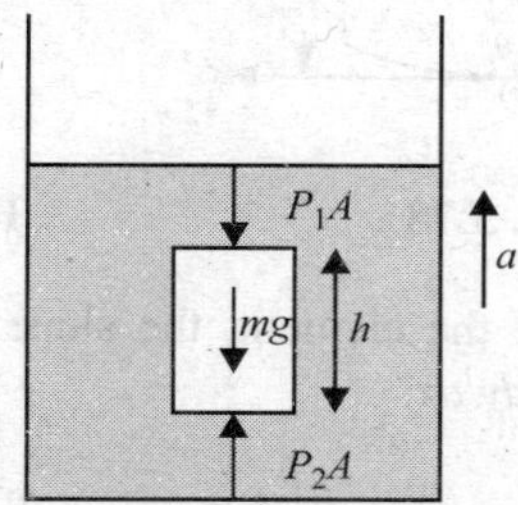

Fig. 3.71

$$P_2A - (mg + P_1A) = ma$$

Here m is the mass of the element of the liquid which is equal to $hA\rho$. Thus, we have $P_2A - (hA\rho g + P_1A) = (hA\rho)\,a$

After simplification, we get $P_2 - P_1 = \rho(g + a)h = \rho g_{\text{eff}} h$.

Comparison of an Accelerated Fluid with an Accelerated Pendulum

Constant downward acceleration ($a < g$)

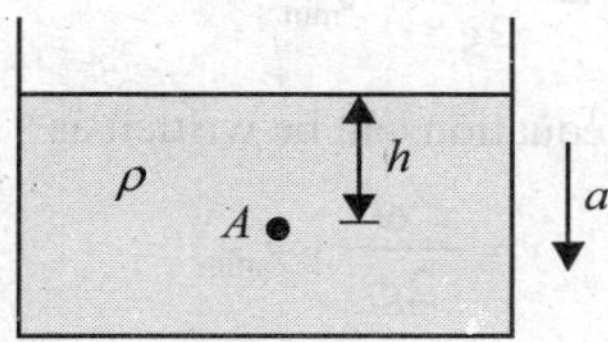

Fig. 3.72

The free surface remains horizontal (Fig. 3.72). Pressure decreases at every point below the surface. At the point A,

$$P = \rho(g - a)h\, \rho g_{eff} h$$

Pressure becomes zero everywhere as $a = g$.

Constant downward acceleration ($a > g$)

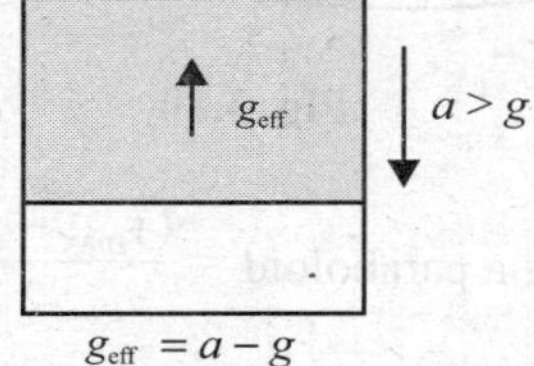

Fig. 3.73

In this case, the fluid occupies the upper part of the container as shown in Fig. 3.73.

Fluid Subjected to Combined Horizontal and Vertical Acceleration

In general, when the fluid has both horizontal and vertical acceleration, the inclination of the free surface with the horizontal may be obtained as

$$\tan\theta = \frac{dy}{dx} = \frac{dp/dx}{dp/dy} = \frac{a_x}{g + a_y}$$

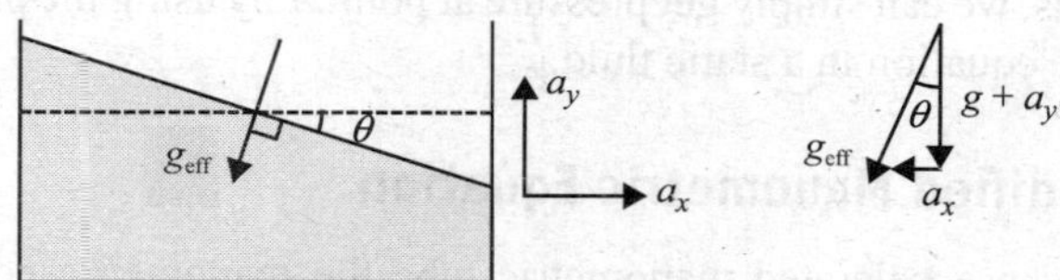

Fig. 3.74

Note: ***In a non-accelerating fluid, pressure increases with depth and decreases with height:***

$$\frac{dp}{dh} = \rho g \qquad \Rightarrow \qquad \frac{dp}{dy} = -\rho g$$

But when a fluid is subjected to a vertical acceleration, the above two equations may be modified as

Variation with depth: $\dfrac{dp}{dh} = \rho(g + a_y)$

Variation with height: $\dfrac{dp}{dy} = -\rho(g + a_y)$

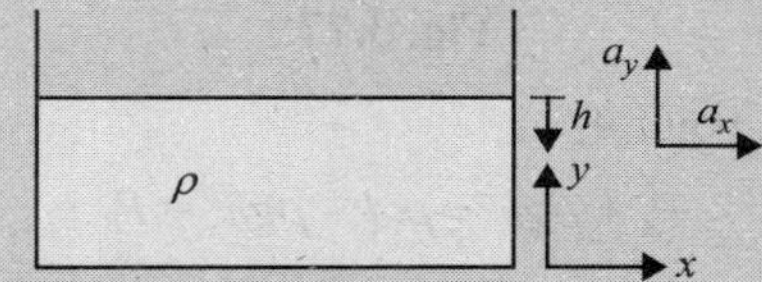

A container accelerated in the vertical direction.

Fig. 3.75

a_y is positive for upward acceleration and a_y is negative for downward acceleration.

Pressure distribution in a closed accelerated container

Consider the situation shown in Fig. 3.76. A closed tank car filled with water is accelerating on a horizontal track with an acceleration a. In the body of the whole liquid, the least-pressure point is P. If container is completely closed, then P can be taken as a zero-pressure point. If we wish to find pressure at a general point A at a depth h below the top surface of container and at a distance ℓ from the right wall of container, then this can again be obtained directly by writing the pressure equation for the liquid with respect to the tank car from point P to A through point M. It can be written as

$$0 + h\rho g + \ell\rho a = P_A$$

or

$$P_A = h\rho g + \ell\rho a \tag{i}$$

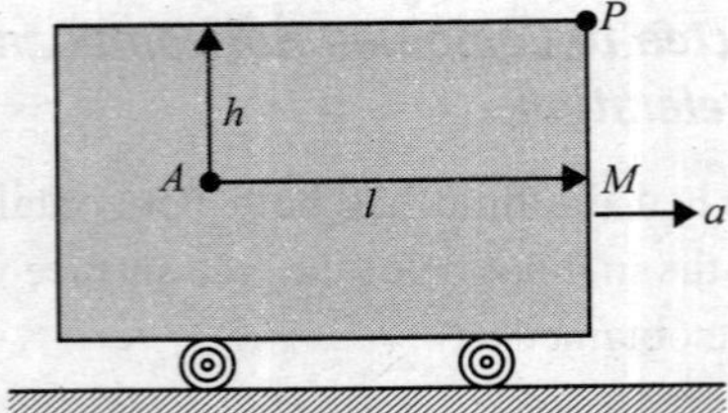

Fig. 3.76

Thus, we can simply get pressure at point A by using the pressure equation in a static fluid.

Modified Manometric Equation

For an accelerated manometric tube the manometric equation is modified at a particular point, that is, when we move horizontally pressure does not remain constant; it increases in the opposite direction of acceleration and decreases in the direction of acceleration. For example, for an arrangement shown in Fig. 3.77 the manometric equation may be written as follows:

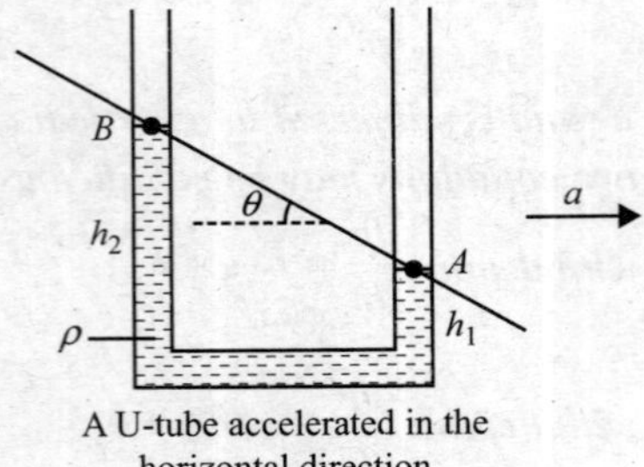

A U-tube accelerated in the horizontal direction

Fig. 3.77

Starting from point A,

$$P_A + \rho g h_1 + \rho a l - \rho g h^2 = P_B$$

Starting from point B,

$$P_B + \rho g h_2 + \rho a l - \rho g h_1 = P_A$$

Note down the positive and negative signs of the term $\rho a l$.

Liquid in a Vessel Rotating with Constant Angular Velocity

When the vessel is rotated there is no relative motion between fluid elements. The liquid surface orients itself in permanent position relative to the vessel. We will analyse a differential element dm in the reference frame of vessel. The forces acting on it are shown in Fig. 3.78 and 3.79. R is the force exerted on the element due to neighbouring elements of fluid.

$$R\cos\theta = dmg \quad \text{(i)}$$

$$R\sin\theta = dmxr\omega^2 \quad \text{(ii)}$$

Dividing Eq. (ii) by Eq. (i), we have

$$\tan\theta = \frac{\omega^2 x}{g} \quad \text{(iii)}$$

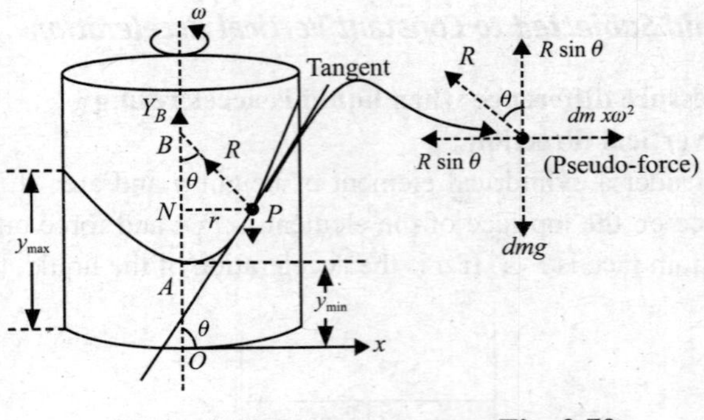

Fig. 3.78 Fig. 3.79

At the position of the element, the slope of the tangent to curve is given by dy/dx.

$$\tan\theta = \frac{dy}{dx} = \frac{\omega^2 x}{g} \Rightarrow dy = \frac{\omega^2 x}{g}dx$$

which on integration yields $y = \dfrac{\omega^2 r^2}{2g} + c$

which represents a parabola. Thus, liquid surface is a paraboloid of revolution.

At $x = R$ and $y = y_{max} = \dfrac{R^2\omega^2}{2g} + c$

At $x = 0$, $y = y_{min} = C$

Therefore, $y_{max} = \dfrac{R^2\omega^2}{2g} + y_{min}$

and the general equation can be written as

$$y = \frac{x^2\omega^2}{2g} + y_{min} \quad \text{(i)}$$

Volume of a paraboloid of revolution is half that of the circumscribing cylinder.

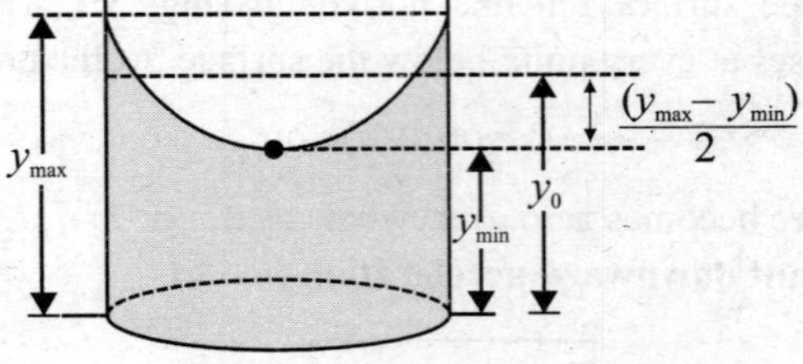

Fig. 3.80

Hence, volume of a paraboloid $= \dfrac{(y_{max} - y_{min})}{2}A$

If originally liquid filled up to height y_0, then

$$y_0 = y_{min} + \frac{y_{max} - y_{min}}{2} = \frac{y_{max} + y_{min}}{2}$$

$$y_{max} = y_0 + \frac{\omega^2 R^2}{4g}$$

Illustration 3.18 **A cylindrical tank of radius 20 cm and height 50 cm has water up to 30 cm of height. What will be the rise in level of liquid at the periphery if the cylinder be given an angular velocity of 10 rad s^{-1}? Also determine the frequency of rotation when water just starts spilling over the sides of the vessel.**

Sol. At an angular speed of 10 rad s^{-1}, let us assume that the water does not spill.

We have

$$h_2 = H + \frac{\omega^2 r^2}{4g} \Rightarrow h_2 - H = \frac{\omega^2 r^2}{4g}$$

Therefore, rise in the liquid level at the periphery is 10 cm.

For water to just spill over the sides, the maximum height (h_{max}) is 50 cm.

Again, $h_{max} = y_0 + \frac{\omega^2 R^2}{4g}$

Here, h_{max} = 50 cm, y_0 = 30 cm, R = 20 cm, g = 10 m s^{-2}

$$\therefore \quad 50 \times 10^{-2} = 30 \times 10^{-2} + \frac{\omega^2 (20 \times 10^{-2})^2}{4 \times 10}$$

$$\Rightarrow \quad \omega^2 \times 10^{-3} = 20 \times 10^{-2}$$

$$\omega^2 = 200 \Rightarrow \omega = 10\sqrt{2} \text{ rad s}^{-1}$$

Hence, frequency of rotation is $f = \frac{\omega}{2\pi} = \frac{5\sqrt{2}}{\pi} \text{s}^{-1}$.

Illustration 3.19 **A cylindrical vessel of diameter 0.3 m and height 0.6 m is filled two-thirds with a liquid of specific gravity 0.8. The vessel is rotated about its axis. (a) Determine the speed of rotation when the liquid just starts spilling. (b) Find the speed of rotation when the base is just visible. (c) What is the percentage of liquid left in the vessel?**

Sol. Initial height of liquid in the vessel $y = 0.6 \times \frac{2}{3} = 0.4$ m

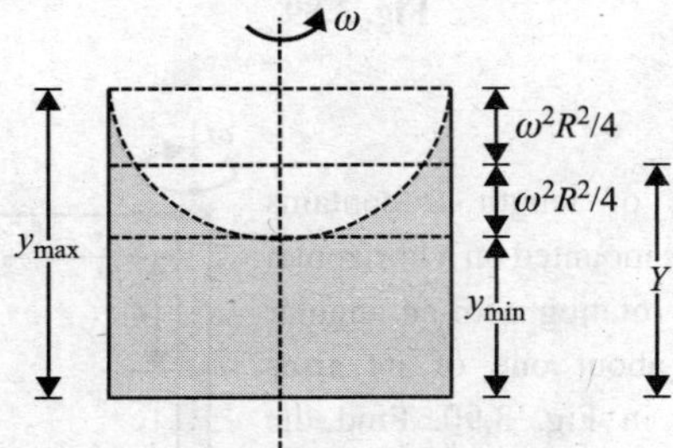

Fig. 3.81

Height of the container = 0.6 m

$$y_{max} - y_{min} = \frac{\omega^2 R^2}{2g} \quad \text{(i)}$$

$$y_{max} = y + \frac{\omega^2 R^2}{4g} \quad \text{(ii)}$$

$$y_{min} = y - \frac{\omega^2 R^2}{4g} \quad \text{(iii)}$$

(a) $y_{max} = 0.6$ m

From Eq. (ii), $0.6 = 0.4 + \frac{\omega^2 \times \left(\frac{0.3}{2}\right)^2}{4 \times 9.81}$

(b) $y_{max} = 0.6$ m, $y_{min} = 0 \Rightarrow y_{max} = \frac{R^2\omega^2}{2g}$

$$0.6 = \frac{(0.3)^2 \times \omega^2}{8 \times 9.81} \Rightarrow \omega = 22.9 \text{ rad/s}$$

(c) The volume of paraboloid of revolution is half the volume of the enclosing cylinder. Volume of the liquid left in the vesse = (1/2) × 0.6 × A, where A is cross-sectional area. Initial volume was 0.4A m^3. Percentage of the liquid left $= \frac{0.3A}{0.4A} \times 100 = 75\%$.

Illustration 3.20 **A U-tube rotates with angular velocity ω about the vertical axis AB. What is the difference in fluid level 'h' in terms of ω, the radii r_1 and r_2 and the fluid density ρ?**

Sol. Let y_{min} be the lowest point on the dotted parabola shown in Fig. 3.82.

$$y_1 = \frac{r_1^2\omega^2}{2g} + y_{min} \text{ and } y_2 = \frac{r_2^2\omega^2}{2g} + y_{min}$$

$$\therefore \quad h = y_2 - y_1 = \frac{(r_2^2 - r_1^2)\omega^2}{2g} = \frac{\omega^2}{2g}(r_2^2 - r_1^2)$$

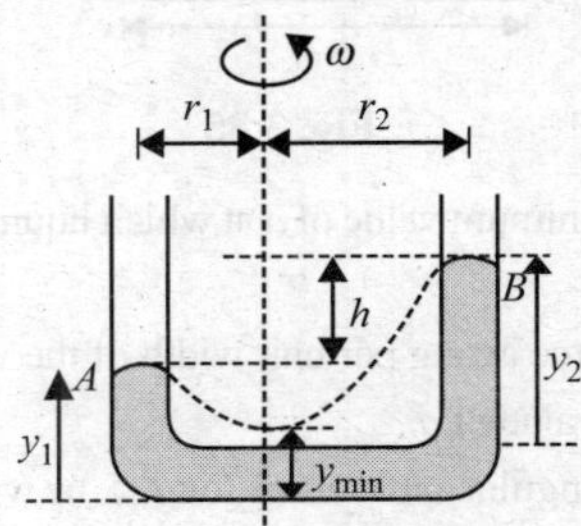

Fig. 3.82

Alternatively, we may apply Bernoulli's equation between points A and B.

$$P_{atm} + \frac{1}{2}\rho(r_1\omega)^2 = P_{atm} + \frac{1}{2}\rho(r_2\omega)^2 + \rho g h \Rightarrow h = \left(\frac{r_2^2 - r_1^2}{2g}\right)\omega^2$$

Concept Application Exercise 3.2

1. A rectangular box is completely filled with a liquid of density ρ, as shown in Fig. 3.83. The box is accelerated horizontally with a constant acceleration a. Determine the gauge pressures at the four points A, B, C and D.

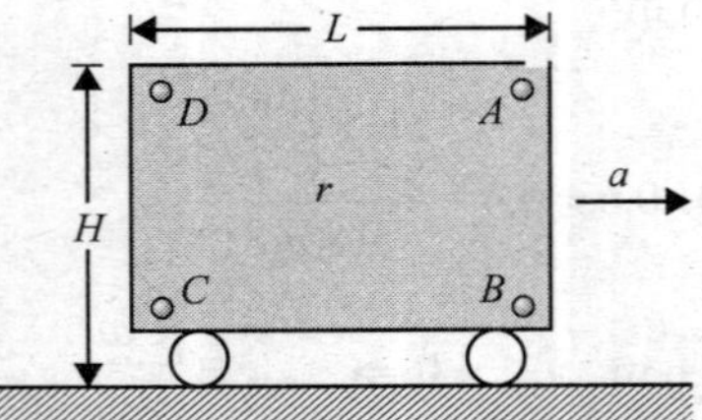

Fig. 3.83

2. Figure 3.84 shows an L-shaped tube filled with a liquid to a height H. What should be the horizontal acceleration a of the tube so that the pressure at the point A becomes atmospheric.

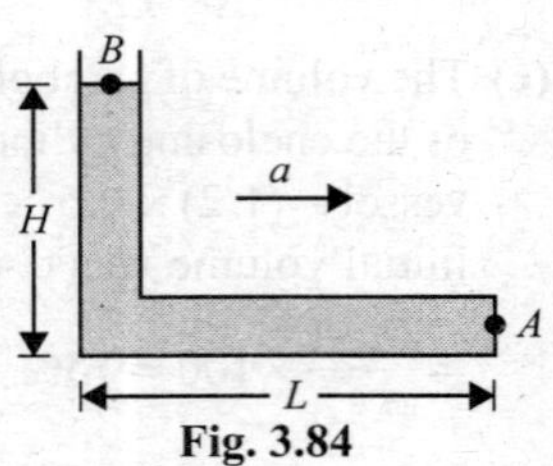

Fig. 3.84

3. A barometer kept in an elevator accelerating upwards reads 76 cm of Hg. What will be the possible air pressure inside the elevator?
4. A barometer kept in an elevator reads 76 cm when it is at rest. What will be the barometric reading when the elevator accelerates upwards?
5. A container shown in Fig. 3.85 is accelerated horizontally with acceleration a.

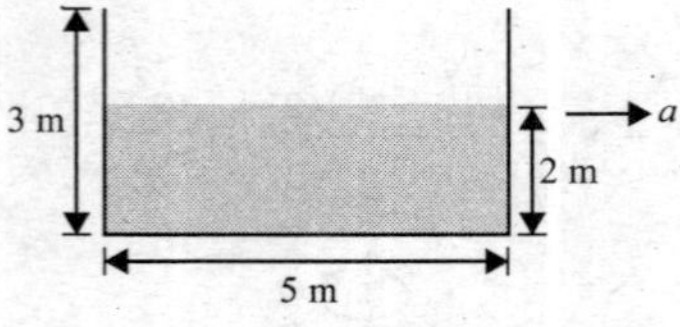

Fig. 3.85

(i) Find the minimum value of a at which liquid just starts spilling out.

(ii) Find the force acting per unit width of the vertical wall at this minimum value of a.

6. A closed rectangular tank 10 m long, 5 m wide and 3 m deep is completely filled with an oil of specific gravity 0.92. Find the pressure difference between the rear and front corners of the tank, if it is moving with an acceleration of 3 m/s² in the horizontal direction.
7. A trolley containing a liquid slides down a smooth inclined plane of angle α with the horizontal. Find the angle of inclination θ of the free surface with the horizontal.

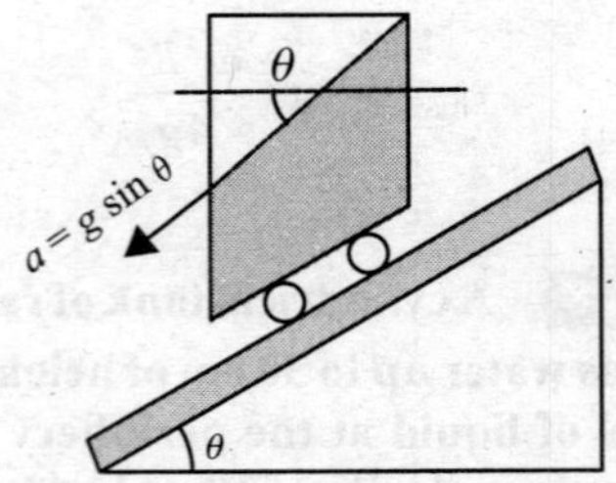

Fig. 3.86

8. A cylinder of radius R and height H is filled with a liquid to an unknown height h. When it is rotated at an unknown constant angular velocity ω, the base of the cylinder gets exposed when the liquid just starts spilling out, as shown in Fig. 3.87.

(i) Find the height h of the liquid.

(ii) Find the angular speed ω of the cylinder.

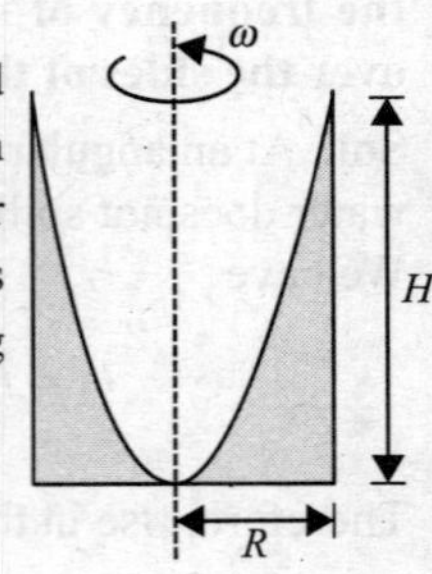

Fig. 3.87

9. A cylinder of radius $R = 1$ m and height $H = 3$ m, two-thirds filled with water, is rotated about its vertical axis, as shown in Fig. 3.88. Determine the speed of rotation when

(i) the water just starts spilling over the rim.

(ii) the point at the centre of the base is just exposed.

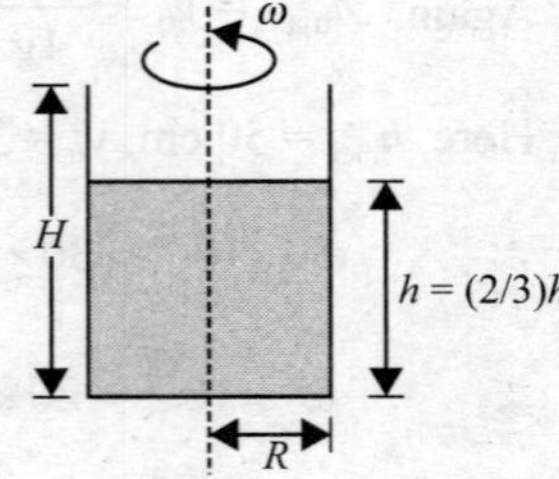

Fig. 3.88

10. A U-tube shown in Fig. 3.89 containing two liquids of specific gravities ρ_1 and ρ_2 is given an acceleration a in the horizontal direction. What is the ratio $\rho_1 : \rho_2$?

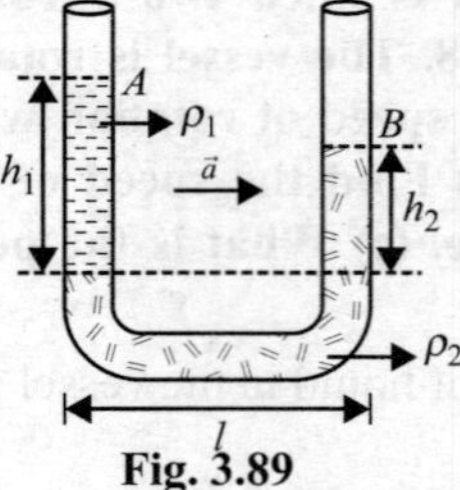

Fig. 3.89

11. A U-tube of length L contains liquid. It is mounted on a horizontal turn table rotating with an angular speed ω about one of its arms as shown in Fig. 3.90. Find the difference in heights (h) between the liquid columns in two vertical arms.

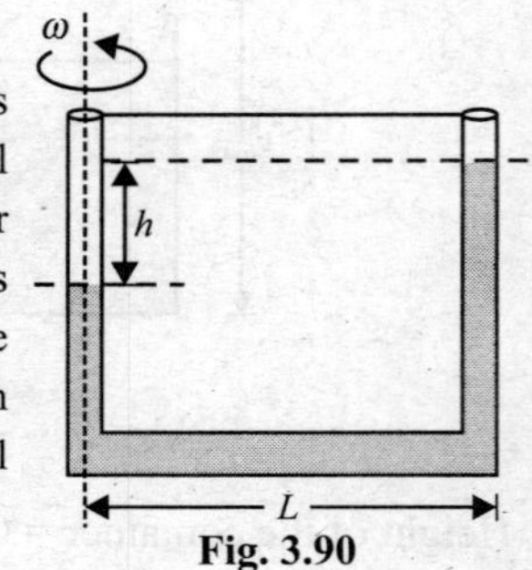

Fig. 3.90

Archimedes' Principle

Consider an irregularly shaped object at rest within a fluid. Liquid exerts a thrust on sides of the solid surface in contact with it due to its pressure. Because the liquid pressure increases with depth, the force exerted by the liquid on the surface of the object is greater for those portions which are more deeply immersed. In Fig. 3.91, liquid thrust at point B is greater than that at A. The net effect on the entire body is an upward or lifting force which is called the buoyant force.

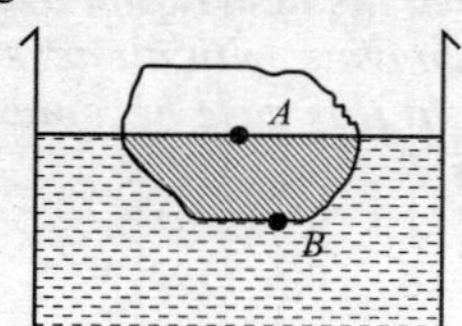

Fig. 3.91

The magnitude of this upward buoyant force (B) is given by Archimedes' principle. We have $B = v\rho g$, where v is the volume occupied by the solid inside the liquid and ρ is the density of the liquid.

A body that is partially or entirely submerged in a fluid feels an upward force equal in magnitude to the weight of the displaced fluid. In Fig. 3.91, a body is partially dipped in the liquid.

Let v = volume of the liquid displaced = space occupied by the body in the liquid = volume of the shaded portion. Then $B = v\rho g$, where ρ is the density of the liquid.

Calculation of Buoyant Force

The force on the bottom of a submerged object is greater than that on its top because pressure increases with depth. Therefore, the fluid exerts a net upward buoyant force on the object.

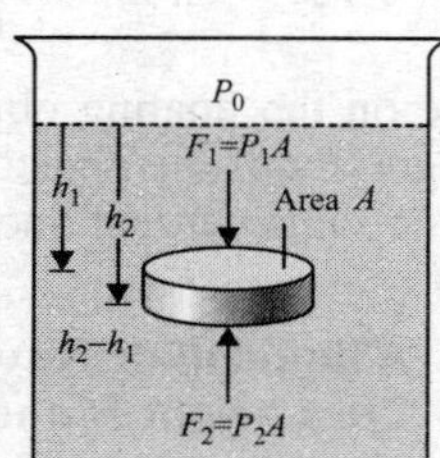

Fig. 3.92

Consider a submerged cylinder as shown in Fig. 3.92. The top of the cylinder is at depth h_1 and its bottom is at depth h_2. The horizontal forces on the cylinder acting at any depth between h_1 and h_2 cancel out, because for each force on one side there is a force of equal magnitude but opposite direction on the other.

$$F_1 = P_1 A$$

where $P_1 = P_0 + \rho g h_1$

From the bottom, $F_2 = P_2 A$

where $P_2 = P_0 + \rho g h_2$

Net buoyant force which acts upward on the cylinder is

$$F_b = F_2 - F_1 = P_2 A - P_1 A$$

$$= (P_0 + \rho g h_2)A - (P_0 + \rho g h_1)A = \rho g h(h_2 - h_1)$$

$A(h_2 - h_1)$ is the volume of the cylinder or in other words it is the volume of the fluid displaced by the cylinder V.

Therefore, $F_b = (\rho V)g$

where ρV is the mass of the fluid displaced and we have

$F_b = mg$ (the weight of the fluid displaced)

which is Archimedes' principle. It states that 'the buoyant force acting on a submerged volume is equal to the weight of an equivalent volume of the fluid displaced'.

Illustration 3.21 **A beaker containing water is placed on the pan of a balance which shows a reading of M. A lump of sugar of mass m and volume v is now suspended by a thread (from an independent support) in such a way that it is completely immersed in water without touching the beaker and without any overflow of water. How will the reading change as time passes on?**

Sol. Initially, the balance will show the weight of M plus thrust. But we know that

$$\text{Thrust} = v\sigma g = \frac{m}{\rho} \times 1 \times g \left[\text{as } v = \frac{m}{\rho} \text{ and } \sigma = 1\right]$$

$$\therefore \quad W_1 = \left(M + \frac{m}{\rho}\right)g = \left(M + \frac{m}{2\rho} + \frac{m}{2\rho}\right)g$$

Now as the sugar dissolves, the thrust and hence the reading of the balance should decrease. However, as dissolved sugar comes in solution, the reading of the balance should increase. To find which of these effects predominates, consider the situation when half the sugar has dissolved. In this situation the thrust reduces to ($mg/2\rho$) from (mg/ρ) while weight of the solution increases by (m/2)g (due to sugar dissolved). So reading of the balance will become

$$W_2 = \left(M + \frac{m}{2\rho} + \frac{m}{\rho}\right)g$$

Finally, when all the sugar is dissolved, thrust will become zero and weight of the solution will increase by mg. So the reading will become

$$W_3 = (M + m)g = \left(M + \frac{m}{2} + \frac{m}{2}\right)g$$

Now as $\rho > 1$, $m/2 > m/2\rho$. So comparing Eqs. (i), (ii) and (iii), we find that the reading of the balance will gradually increase till all the sugar dissolves in water and finally will become constant equal to $(M + m)g$.

Illustration 3.22 **A piece of ice is floating in water. What will happen to the level of water when all ice melts? What will happen if the beaker is filled not with water but with liquid (a) denser than water, (b) lighter than water?**

Sol. If M grams of ice is floating in a liquid of density σ_L, then for its equilibrium, weight of the ice = thrust

or $$Mg = V_D \sigma_L g$$

So the volume of the liquid displaced by the floating ice is

$$V_D = (M / \sigma_L) \quad \text{(i)}$$

Now if M grams of ice melts completely, water formed will have mass M grams (as mass is conserved). Now if σ_w is the density of water, the volume of water formed will be

$$V_F = (M / \sigma_W) \quad \text{(ii)}$$

Here the liquid is water, i.e., $\sigma_L = \sigma_W$; so water displaced by floating ice is equal to the water formed by melting of the whole ice, and hence the level of water will remain unchanged. The following two points may be noted:

(a) If $\sigma_L > \sigma_W$, then $M/\sigma_L < M/\sigma_W$, i.e., $V_D < V_F$

i.e., water displaced by the floating ice will be lesser than water formed and so the level of the liquid in the beaker will rise.

(b) If $\sigma_L < \sigma_W$, $M/\sigma_L > M/\sigma_W$, i.e., $V_D > V_F$

i.e., water displaced by the floating ice will be more than water formed and so the level of the liquid in the beaker will fall.

Floating Body

Whenever the buoyant force on a body equals its weight, the net force acting on the body becomes zero. Hence, the body floats in equilibrium. Let a body of total volume V and density d be floating on a liquid of density ρ. Let v be the volume of the body inside the liquid.

Then $B = mg$ (for a floating body)

or $v\rho g = Vdg$

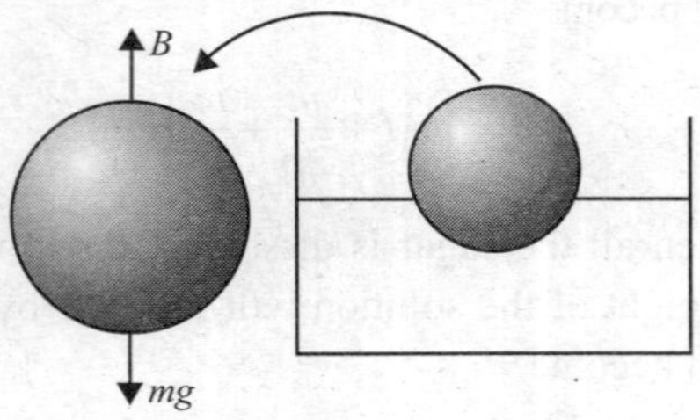

Fig. 3.93

Note:

- *When a body is in air, the net downward force on it is due to earth's gravity only. This net force = mg = Vdg. This force is known as weight of the body.*
- *When the same body is immersed in a liquid of density σ, the net downward force on it is = mg – B.*
- *Buoyant force = Vdg – $v\rho g$.*
- *This downward force is also known as apparent weight in the liquid.*
- *When a body is immersed in a liquid, it feels lighter due to buoyant force. Hence, the buoyant force is also known as loss in weight (= $v\rho g$).*
- *Buoyant force in accelerating fluids: Suppose a body is dipped inside a liquid of density ρ_L placed in an elevator moving with an acceleration. The buoyant force F in this case becomes $F = V\rho_L g_{eff}$. Here, $g_{eff} = |\vec{g} - \vec{a}|$.*

Centre of Buoyancy

Centre of gravity of the displaced fluid is called buoyant centre or centre of buoyancy. Buoyant force is applied on buoyant centre.

Some important points regarding floating bodies

- The net force on the floating object should be zero, i.e., weight of the object is same as the buoyant force on it.

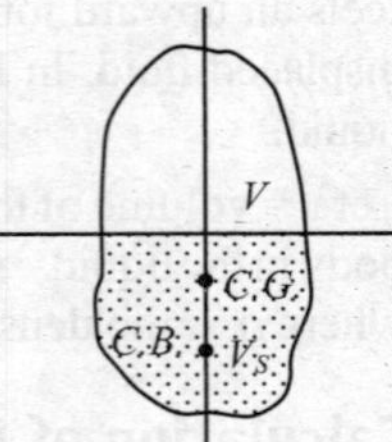

Fig. 3.94

- Let density of the solid be ρ_s, density of the fluid be ρ_l, submersed volume of the floating object be V_s and total volume of the object be V. Then

$$W = B \Rightarrow V\rho_s g = V_s \rho_l g \Rightarrow \frac{V_s}{V} = \frac{\rho_s}{\rho_l}$$

- The net torque on the floating object should be zero, i.e., centre of gravity of the object and buoyant centre should lie on the same vertical line.

Illustration 3.23 **A liquid of density ρ is filled in a beaker of cross section S to a height H and then a cylinder of mass m and cross section s is made to float in it as shown in Fig. 3.95. If the atmospheric pressure is p_0, find the pressure (a) at the top face A of the cylinder (b) at the bottom face C of the cylinder and (c) at the base B of the beaker. Can ever these three pressures be equal?**

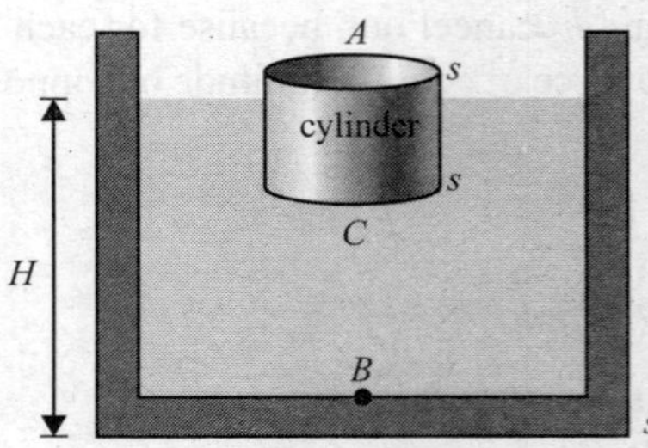

Fig. 3.95

Sol.

(a) Above the cross section A, there is external pressure due to atmosphere only. So, p_A = atmospheric pressure = p_0.

(b) At point C the pressure will be due to atmosphere and also due to the weight of the cylinder, i.e.,

$$p_C = p_0 + \frac{mg}{s}$$

If the system is in a state of free fall (as in a satellite), i.e., $g \to 0$, then $p_A = p_C = p_B = p_0$ [as weight = 0]

Illustration 3.24 **What fraction of an iceberg lies beneath the surface of the sea? Density of sea water (ρ) = 1.028 × 10³ kg/m³, density of ice (d) = 0.917 × 10³ kg/m³.**

Sol. For floating conditions, we know

Weight of the iceberg = Buoyant force

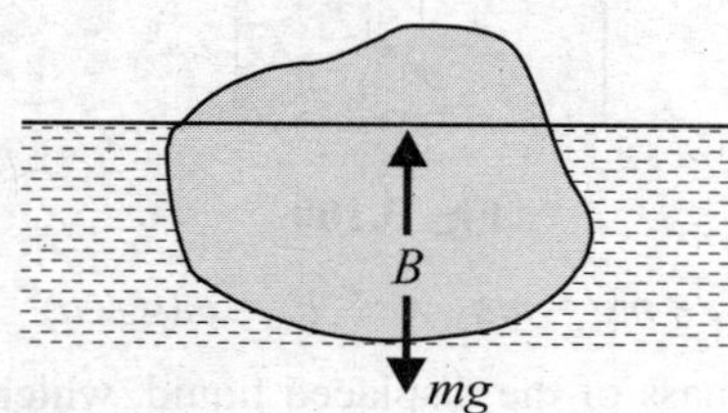

Fig. 3.96

Let V be the total volume of the iceberg and v be the volume inside the sea water. Then

$$Vdg = v\rho g$$

$$\frac{v}{V} = \frac{d}{\rho} \Rightarrow \frac{v}{V} = \frac{0.917 \times 10^3}{1.028 \times 10^3} \Rightarrow \frac{v}{V} = 0.892$$

About 89.2% of the iceberg remains inside.

Illustration 3.25 **A cubical block of wood 10 cm along each side floats at the interface between an oil and water with its lowest surface 2 cm below the interface. If the heights of oil and water columns are 10 cm each and ρ_{oil} = 0.8 g/cc, find (a) the mass of the block, (b) the liquid pressure at the lower surface of the block.**

Sol.

(a) For the floating cube,

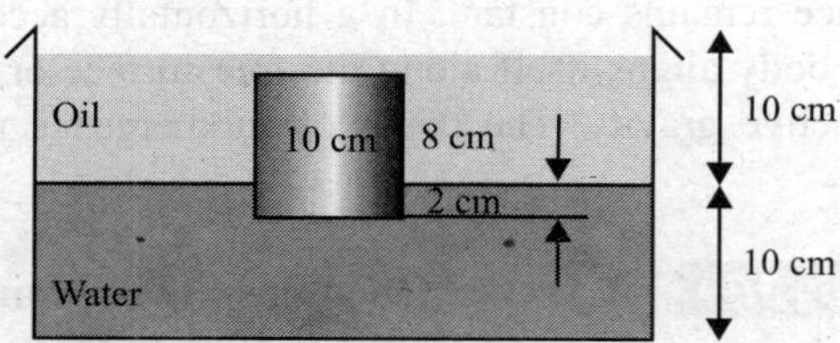

Fig. 3.97

Weight of the cube = Buoyant force

$$\Rightarrow \quad mg = B_{oil} + B_{water}$$

$$= 8 \times 10^2 \times \rho_{oil} g + 2 \times 10^2 \rho_w \times g$$

$$\Rightarrow \quad m = 800 \times 0.8 + 200 \times 1$$

$$= 640 + 200 = 840 \text{ g}$$

(b) The pressure at the lower surface of the cube is

$$P_{oil} + P_w = \frac{10}{100} \times (0.8 \times 10^3) 9.8 + \frac{2}{100} \times (1 + 10^3) 9.8 = 980 \text{ N/m}^2$$

(Here we have used h_{oil} = 10 cm and h_w = 2 cm above the lower surface.)

Illustration 3.26 **A cube of wood supporting a 200 g mass just floats in the water. When the mass is removed, the cube rises by 2 cm. What is the size of the cube?**

Sol. The cube 'just floats' means that it is fully under water while floating.

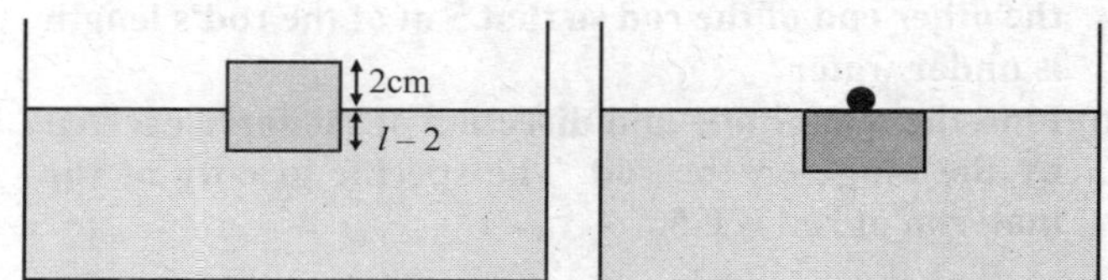

Fig. 3.98

Let m be the mass of the cube. Then

$$(m + 200)g = \ell^3 \rho g \quad \text{(i)}$$

When the mass is removed,

$$mg = (\ell - 2)\ell^2 \rho g \quad \text{(ii)}$$

Subtracting Eq. (ii) from Eq. (i), we have
$200 = \rho\,[l^3 - (l-2)\,l^2] \Rightarrow l = 10$ cm (using ρ = 1 g/cc of water)

Illustration 3.27 **A glass sphere has a cavity inside it. When placed in water, it floats with 20% of its volume above the water surface. Find the volume of the cavity as a fraction of total volume (density of glass = 2.5 g/cc).**

Sol. Let the total volume of the sphere be V.
Weight of the sphere = Buoyant force

$$\Rightarrow \quad \text{mass of glass} \times g = \frac{80}{100} V \times 1 \times g$$

$$\Rightarrow \quad \text{volume of glass} \times d = \frac{80V}{100}$$

If the volume of the cavity is v, then the volume of the glass is $(V - v)$

$$(V - v)d = 0.8V \Rightarrow \frac{V - v}{V} = \frac{0.8}{2.5}$$

$$\Rightarrow \quad \frac{V-v}{V}=0.32 \Rightarrow \frac{v}{V}=0.68$$

Hence, volume of the cavity is 68% of total volume.

$$\Rightarrow \quad \text{volume of glass} \times d = \frac{80V}{100}$$

If the volume of the cavity is v, then the volume of the glass is $(V - v)$,

$$(V-v)d = 0.8V \Rightarrow \frac{V-v}{V} = \frac{0.8}{2.5}$$

$$\Rightarrow \quad \frac{V-v}{V}=0.32 \Rightarrow \frac{v}{V}=0.68$$

Hence, volume of the cavity is 68% of the total volume.

Illustration 3.28 **A rod of length 6 m has a mass of 12 kg. It is hinged at one end of the rod at a distance of 3 m below the water surface.**

(a) What must be the weight of a block that is attached to the other end of the rod so that 5 m of the rod's length is under water?

(b) Find the magnitude and direction of the force exerted by the hinge on the rod. The specific gravity of the material of rod is 0.5.

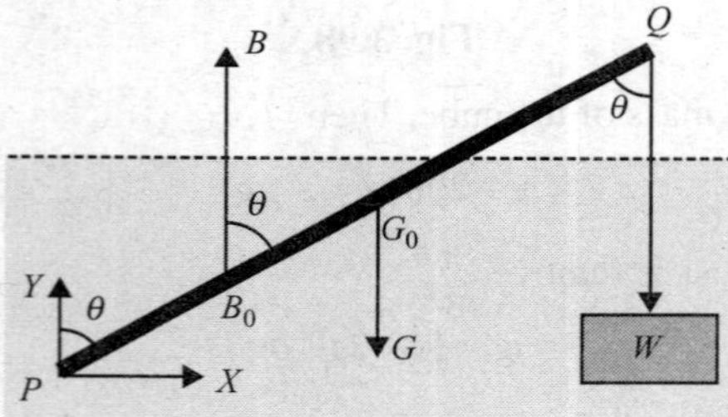

Fig. 3.99

Sol. Let w be the weight attached at the end Q; X, Y be the components of force exerted by hinge P on the rod; B be the buoyant force acting at the centre of buoyancy B_0 (mid-point of submerged portion); G be the weight of rod and θ be the angle of the rod with vertical. Now the balancing force is given by $B + Y = G + W$ and $X = 0$.

Balancing torque about P is given by

$$B\sin\theta(PB_0) = G\sin\theta(PG_0) + W\sin\theta(PQ)$$

$$\Rightarrow \quad B\sin\theta(2.5) = G\sin\theta(3) + W\sin\theta(6)$$

If area of cross section of the rod is A, then buoyant force $B = 5A\rho g$ then

$$\Rightarrow \quad (5A)\rho g(2.5) = (6A)dg(3) + W(6)$$

$$\Rightarrow \quad 10^3(5\times1\times2.5) = 10^3(6\times0.5\times3) + \frac{6W}{Ag}$$

Mass of the rod $= 12 \text{ kg} = 6Ad = (6A)\, 0.5 \times 10^3 \Rightarrow A = \frac{4\ m^2}{1000}$

$$\therefore \quad W = \frac{3.5\times4\times9.8}{6} = 2.33\times9.8 = 22.83 \text{ N (or 2.33 kg)}$$

$$Y = G + W - B = (6A\,dg) + 2.33\,g - (5A\,\rho g) = -5.66\,g$$

Therefore, the horizontal component of force by hinge is $X = 0$ and the vertical component is 5.66 kg acting downwards.

Buoyant Force in Accelerating Fluid

Suppose the body is submerged into a liquid of density ρ, which is accelerating upwards. If a is the acceleration of the liquid, then

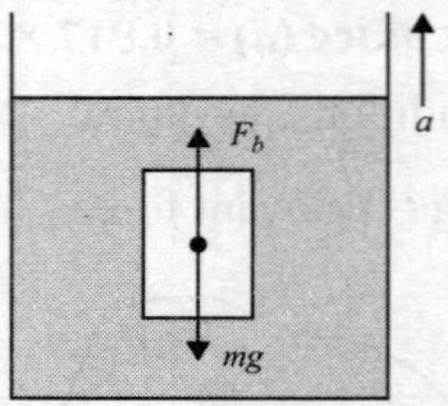

Fig. 3.100

$$F_b - mg = ma \quad \text{or} \quad F_b = m(g + a)$$

Here m is the mass of the displaced liquid, which is equal to $V\rho$. Therefore,

$$F_b = \rho V(g + a)$$

Expression for the Buoyant Force for an Accelerated Fluid

Instead of taking gravity 'g' we take the effective gravity g_{eff}. Then

$$F_B = \rho V g_{\text{eff}}$$

where ρ is the density of the fluid, V is the volume of the displaced fluid and g_{eff} is the effective gravity.

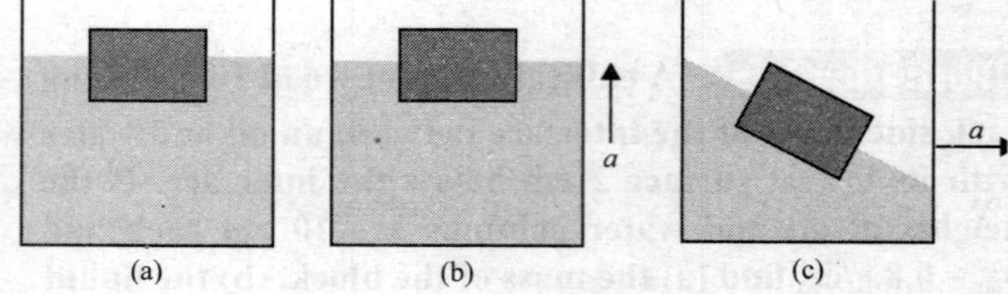

Fig. 3.101

Figure 3.101(a) shows that a body floats on the surface of a stationary liquid. In a vertically accelerated liquid, the depth of submergence remains constant. In a horizontally accelerated liquid, the body aligns itself along the free surface or normal to the effective gravity. The depth of submergence remains constant.

Illustration 3.29 **A vessel contains two immiscible liquids of density ρ_1 = 1000 kg/m³ and ρ_2 = 1500 kg/m³. A solid block of volume $V = 10^{-3}$ m³ and density d = 800 kg/m³**

is tied to one end of a string and the other end is tied to the bottom of the vessel as shown in Fig. 3.102.

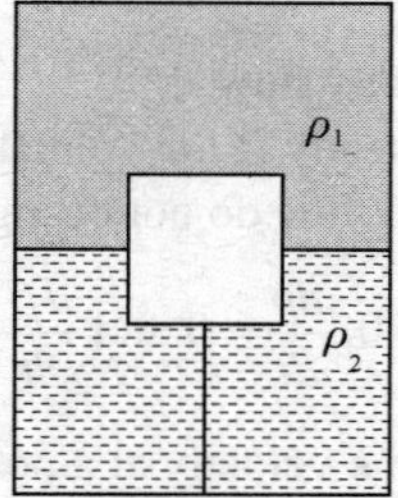

Fig. 3.102

The block is immersed with two-fifths of its volume in the liquid of lower density. The entire system is kept in an elevator which is moving upwards with an acceleration of $a = g/2$. Find the tension in the string.

Sol. We will analyse this problem from the reference frame of elevator. Total buoyant force on the block is

$$F_B = \left(\frac{2}{5}V\rho_2 + \frac{3}{5}V\rho_1\right)(g+a)$$

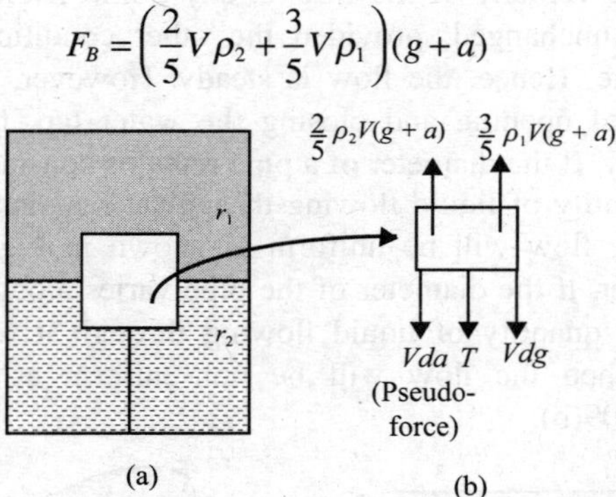

Fig. 3.103

From the condition of equilibrium,

$$F_B = T + Vd(g+a)$$

$$\Rightarrow \qquad T = F_B - Vd(g+a) = (g+a)V\left[\frac{2}{5}\rho_2 + \frac{3}{5}\rho_1 - d\right]$$

$$= 15 \times 10^{-3}\left[\frac{2}{5} \times 1500 + \frac{3}{5} \times 1000 - 800\right] = 6 \text{ N}$$

Illustration 3.30 **A solid sphere of mass m = 2 kg and specific gravity s = 0.5 is held stationary relative to a tank filled with water as shown. The tank is accelerating vertically upward with acceleration a = 2 ms^{-2}.**

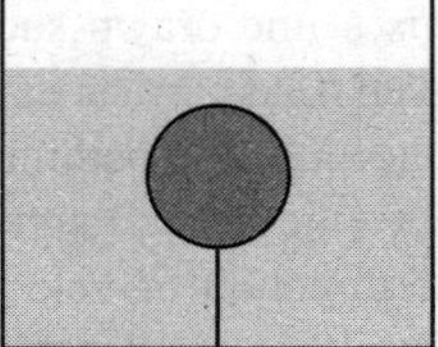

Fig. 3.104

(i) Calculate tension in the thread connected between the sphere and the bottom of the tank.

(ii) If the thread snaps, calculate acceleration of the sphere with respect to the tank (density of water is ρ = 1000 kg m^{-3}) (g = 10 ms^{-2}).

Sol. Density of the spherical material = specific gravity × density of water = $s\rho$

Its volume is $V = \dfrac{m}{s\rho}$

Hence, mass of water displaced by the sphere is

$$V_\rho = \frac{m}{s\rho}\rho = \frac{m}{s} = \frac{2}{0.5} = 4 \text{ kg}$$

Since, the tank is accelerating upwards with acceleration a; therefore apparent value of gravitational acceleration is

$$g' = g + a = 12 \text{ ms}^{-2}$$

Hence, upthrust exerted by water on the sphere is

$$F = V_\rho(g+a) = 48 \text{ N}$$

Now considering the free-body diagram of the sphere accelerating with the tank,

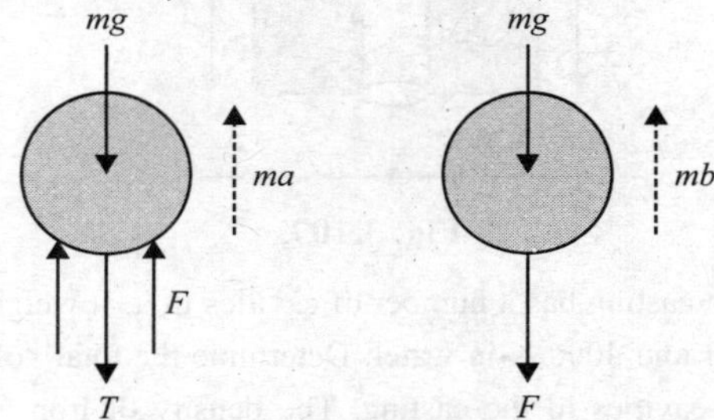

Fig. 3.105

$$F - mg - T = ma \;\Rightarrow T = F - mg - ma = 24 \text{ N}$$

When thread snaps, tension T disappears. Let the sphere now starts accelerating upwards with acceleration b. Considering the free-body diagram,

$$F - mg = mb$$

$$\therefore \qquad b = 14 \text{ ms}^{-2}$$

This is the absolute acceleration of the sphere. But the tank itself is accelerating upwards with acceleration a. Therefore, upward acceleration of the sphere relative to the tank is $b - a$ = 12 ms^{-2}.

Concept Application Exercise 3.3

1. A beaker exactly full of water has an ice piece floating on it. As the cube melts what happens to the water level if (a) the cube contains an air bubble, (b) the cube contains (i) a lead piece and (ii) a cork piece.
2. A boat containing some piece of material is floating in a pond. What will happen to the level of water in the pond if on unloading the pieces in the pond, the piece (a) floats, (b) sinks?
3. A body of density ρ floats with volume V_1 of its total volume V immersed in a liquid of density ρ_1 and with the remainder of volume V_2 immersed in another liquid of density ρ_2, where $\rho_1 > \rho_2$. Find the volume immersed in two liquids. (V_1 and V_2).

4. A rock is found to weigh w in air when suspended from a spring scale. When completely submerged in water while attached to the spring scale, it weighs $w_{submerg}$ as shown in Fig. 3.106. Find the density of the rock ρ_{rock} in terms of the scale reading and the density of water.

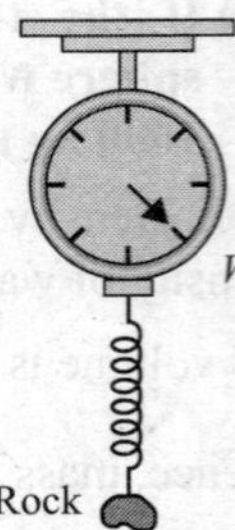

Fig. 3.106

5. A large block of ice 5 m thick has a vertical hole drilled through it and is floating in the middle of a lake. What is the minimum length of a rope required to scoop up a bucket full of water through the hole? Relative density of ice = 0.9.

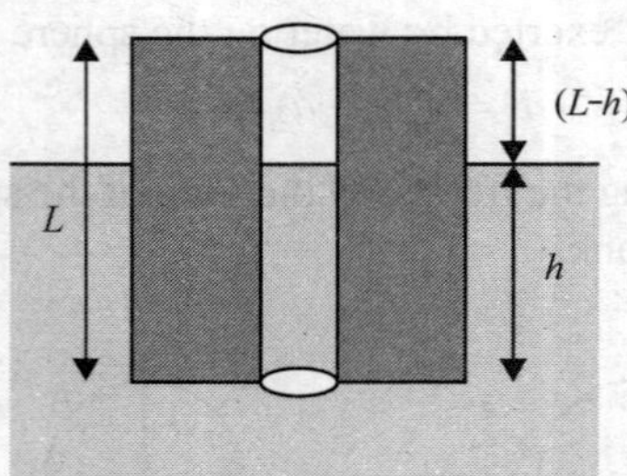

Fig. 3.107

6. An iron casting has a number of cavities in it. It weighs 6000 N in air and 4000 N in water. Determine the total volume of all the cavities in the casting. The density of iron (without cavities) is 8.0 g/cm³, $\rho_w = 1$ g/cm³.

7. A necklace weighs 50 g in air, but it weighs 46 *g* in water. Assume that copper is mixed with gold to prepare the necklace. Find how much copper is present in it. (Specific gravity of gold is 20 and that of copper is 10.)

8. A block of ice of area A and thickness 0.5 m is floating in fresh water. In order to just support a man of 100 kg, find the area A. (The specific gravity of ice is 0.917 and density of water is 1000 kg/m³.)

9. As Fig. 3.108 shows, S_1 and S_2 are spring balances. Block A is hanging from spring balance S_1 and immersed in liquid L which is contained in beaker B. The mass of beaker B is 1 kg and mass of liquid L is 1.5 kg. Balances S_1 and S_2 reads 2.5 kg and 7.5 kg, respectively. What will be the readings of S_1 and S_2 when block A is pulled up out of the liquid. Find the reading of S_1 and S_2?

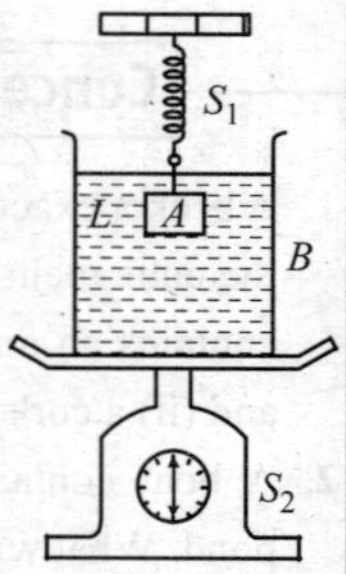

Fig. 3.108

10. A ball floats on the surface of water in a container exposed to the atmosphere. Will the ball remain immersed at its initial depth or will it sink or rise somewhat if the container is shifted to the moon?

HYDRODYNAMICS

Types of Flow

(i) Steady and unsteady flow

A flow is said to be steady if the velocity, pressure and density at any point in the flow do not change with time, i.e.,

$$\frac{dv}{dt} = 0, \frac{dp}{dt} = 0 \text{ and } \frac{d\rho}{dt} = 0$$

In an unsteady flow, the velocity at a point in the flow varies with time, i.e., $\frac{dv}{dt} \neq 0$.

(ii) Uniform and non-uniform flow

A flow is said to be *uniform* at any instant of time if the velocity (both in magnitude and in direction) does not vary along the direction of flow. In a non-uniform flow, velocity varies in the direction of flow. For example, if a water tap is kept fully open and left untouched, then after some time the velocity of the flow at any point in the pipe will remain unchanged, provided the other conditions remain the same. Hence, the flow is steady. However, during the period of opening and closing the water tap, the flow is unsteady. If the diameter of a pipe remains constant and also the quantity of liquid flowing through it remains the same, then the flow will be uniform as shown in Fig. 3.109(a). However, if the diameter of the pipe varies along its length and the quantity of liquid flowing through it remains the same, then the flow will be non-uniform as shown in Fig. 3.109(b).

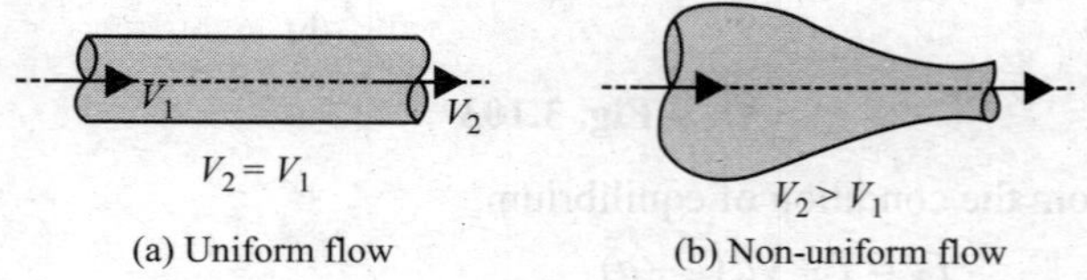

Fig. 3.109

Visualization of Flow

The fluid flow pattern may be visualized in terms of pathlines and streamlines.

(i) Pathline

It is a line drawn such that it describes the trajectory of a given fluid particle as it moves with the passage of time. The concept of the pathline in fluids is identical to the trajectory of a solid body. Tangent drawn at a point on a pathline gives the direction of velocity at that point at the time when the particle passes that point.

Important properties of a pathline

- A pathline can intersect itself at different times.
- Pathlines are the history lines of individual fluid particles over a period of time.

(ii) Streamline

It is the most common graphical concept of flow visualization. It is a line drawn such that a tangent at every point of it is in the direction of velocity at that instant.

A collection of streamlines drawn in this manner represents the complete flow pattern at an instant. Thus, for an unsteady flow, the streamline pattern may or may not remain the same at the next instant. Note that if the unsteadiness is due to a change in the magnitude of the velocity, then the streamline pattern still remains the same at different instants of time. On the other hand, if the unsteadiness is due to change in the direction of the velocity, the pattern will change from time to time.

Unlike pathlines, a streamline cannot intersect itself; furthermore, two streamlines cannot meet or intersect at a point. This is because instantaneously fluid can have a unique velocity at a point.

Streamline Flow

If every point of a steadily flowing liquid follows exactly the same path that has been followed by the particles preceding it, the flow is said to be streamlined. The path is known as 'streamline'. In Fig. 3.110, the paths 1, 2, 3 are streamlines. If a liquid follows the path *ABC*, particles following it move along the same path.

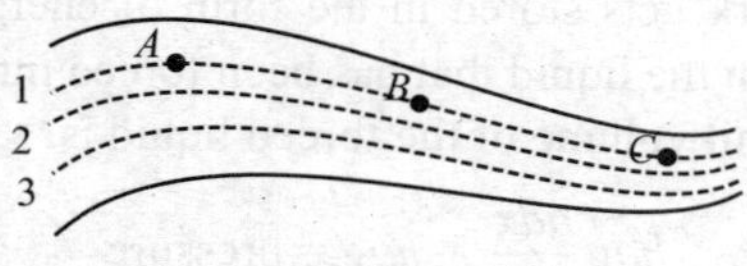

Fig. 3.110

Turbulent Flow

When the velocity exceeds a certain critical value, the nature of flow becomes complicated. Random, irregular, local currents (called vortices) develop throughout the fluids. The resistance to the flow increases tremendously. This type of flow is called 'turbulent flow'.

PRINCIPLE OF CONTINUITY: THE CONTINUITY EQUATION

The continuity equation is a mathematical expression of the law of conservation of mass in fluid dynamics.

It states that 'in a steady flow, the amount (mass) of liquid entering a section is the same as that leaving another section in a given time interval'. If a_1, a_2, a_3, etc., are the cross-sectional areas of a pipe (carrying a liquid in a steady flow) at various sections, and v_1, v_2, v_3, etc., are the corresponding velocities (average), with ρ_1, ρ_2, ρ_3, etc. as the densities, then the mass of liquid flowing past these sections in time Δt would be $\rho_1 v_1 a_1 \Delta t, \rho_2 v_2 a_2\, \Delta t, \rho_3 v_3 a_3\, \Delta t$, etc. Since there is no accumulation of the liquid within the pipe, so

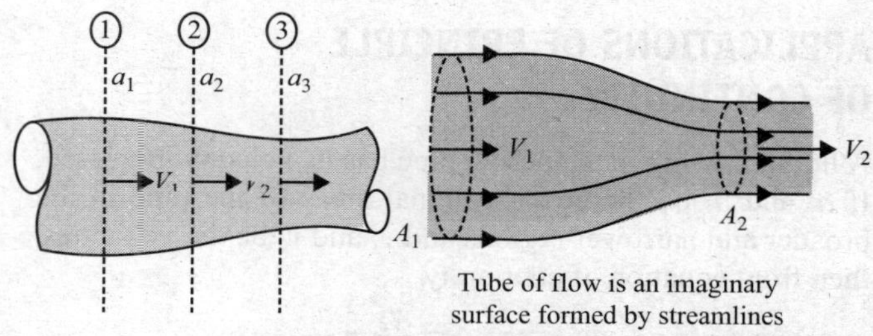

Fig. 3.111

$$\rho_1 v_1 a_1 \Delta t = \rho_2 v_2 a_2 \Delta t = \rho_3 v_3 a_3 \Delta t = \cdots$$

$$\Rightarrow \quad \rho_1 v_1 a_1 = \rho_2 v_2 a_2 = \rho_3 v_3 a_3 = \cdots$$

In case of incompressible liquids, $\rho_1 = \rho_2 = \rho_3 = \cdots$

$$\therefore \quad v_1 a_1 = v_2 a_2 = v_3 a_3 = \cdots$$

The above equation is known as the 'equation of continuity',

$$A_1 v_1 = A_2 v_2$$

Illustration 3.31 **A pipe of non-uniform cross section has two distinct sections 1 and 2 with areas 2 cm² and 4 cm², respectively. If the velocity of flowing liquid at section 1 is 5 cm/s, determine the velocity at section 2.**

Sol. From the equation of continuity, $a_1 v_1 = a_2 v_2$

Given, $a_1 = 2$ cm², $a_2 = 4$ cm², $v_1 = 5$ cm/s, $v_2 = ?$

$\therefore \quad 2(5) = 4(v_2) \Rightarrow v_2 = 2.5$ cm/s.

Illustration 3.32 **A broad pipe having a radius 10 cm branches into two pipes of radii 5 cm and 3 cm. If the velocity of flowing water in the pipe of radius 3 cm is 5 cm/s, determine the velocities of water in the remaining two pipes. Given that the rate of discharge through the main branch is 600π cm³/s.**

Sol. Consider any three sections (1), (2) and (3) in the three pipes of different radii as shown in Fig. 3.112. If v_1 and v_2 be the velocities of water at sections (1) and (2), respectively; then

$$a_1 = \pi(10)^2 \text{ cm}^2, \quad a_2 = \pi(5)^2 \text{ cm}^2$$

and $\quad a_3 = \pi(3^2)$ cm²

$$v_1 = ?, \quad v_2 = ?, \quad v_3 = 5 \text{ cm/s}$$

The rates of discharge through the three pipes are

$$a_1 v_1 = 100\pi v_1 \text{ cm}^3/\text{s}$$

$$a_2 v_2 = 25\pi v_2 \text{ cm}^3/\text{s}$$

$$a_3 v_3 = 9\pi \times 5 = 45\pi \text{ cm}^3/\text{s}$$

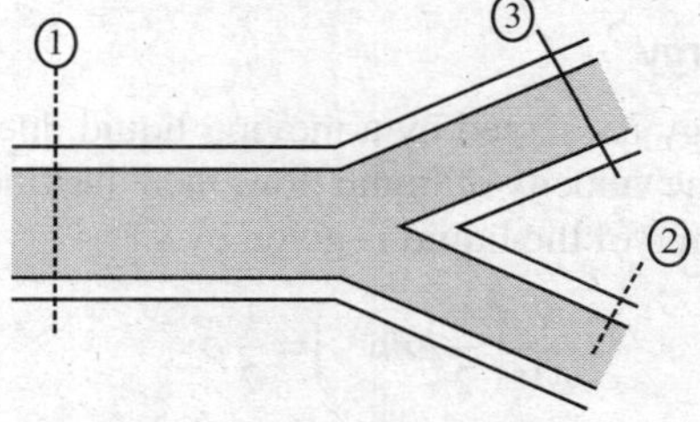

Fig. 3.112

Now, $Q = a_1 v_1 = a_2 v_2 + a_3 v_3 \quad \Rightarrow \quad 600\pi = 100\pi v_1 = 25\pi v_2 + 45\pi$

Solving we get $v_1 = 6$ cm/s and $v_2 = 22.2$ cm/s

APPLICATIONS OF PRINCIPLE OF CONTINUITY

A liquid flowing in a tapering pipe has its velocity increased. If a_1 and a_2 be the cross-sectional areas of the pipe in the broader and narrower regions and v_1 and v_2 be the velocities, then from equation of continuity,

$$a_1v_1 = a_2v_2 \Rightarrow \frac{v_2}{v_1} = \frac{a_1}{a_2}$$

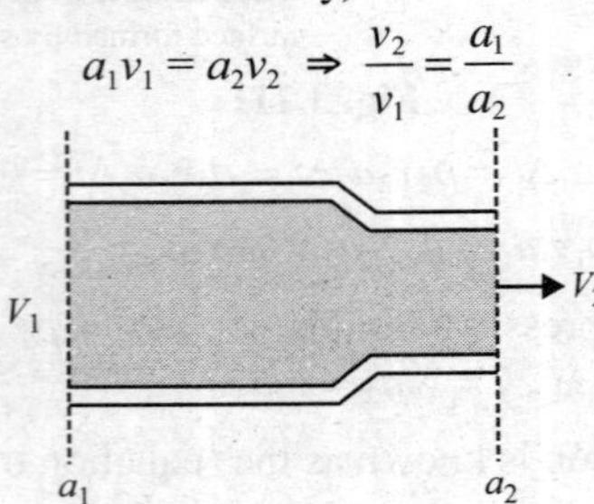

Fig. 3.113

Now, $\frac{a_1}{a_2} > 1 \Rightarrow \frac{v_2}{v_1} > 1 \Rightarrow v_2 > v_1$

Thus, in case of a steady flow of liquid, the greater the cross-sectional area of the pipe, the lesser is the velocity and vice versa.

ENERGY ASSOCIATED WITH A MOVING LIQUID

Potential Energy

It is the energy possessed by a flowing liquid, by virtue of its position, i.e., height above a fixed datum level (see Fig. 3.114).

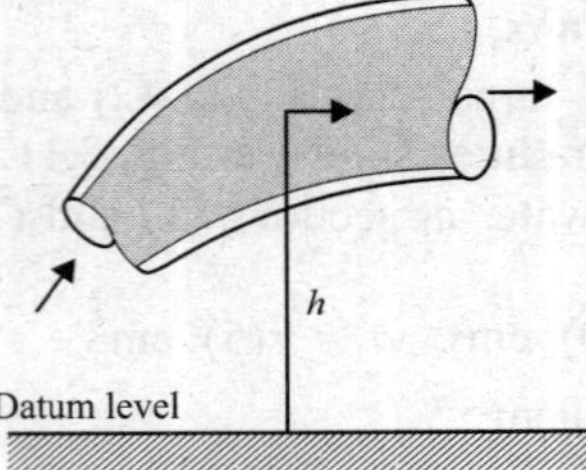

Fig. 3.114

The potential energy possessed by a liquid at a height h above the datum level per unit volume will be

$$\frac{\Delta m}{\Delta V}gh = \rho gh$$

Kinetic Energy

It is the energy possessed by a moving liquid due to its motion. If v be the velocity of liquid flow, then the kinetic energy per unit volume of the liquid is given by

$$\frac{1}{\Delta V}\left(\frac{1}{2}\Delta mv^2\right) = \frac{1}{2}\rho v^2$$

Pressure Energy

As a direct consequence of Pascal's law of transmission of pressure, it is easy to see that a liquid can be made to possess energy (in the form of pressure energy) by applying external pressure. To understand the concept of pressure energy, let us consider a broad vessel, fille1 with water (or other liquid) up to a height 'h', and fitted with a thin, side horizontal tube at its bottom. Let the water within the side tube be stopped by a frictionless cork.

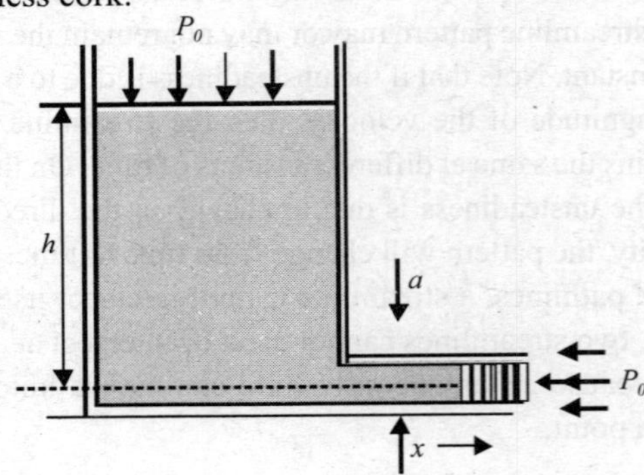

Fig. 3.115

Let the cork be applied an external force, pushing the water in the side tube into the vessel by a distance dx. Work done by the external agency against the hydrostatic pressure is given by

$$\Delta W = \text{Force} \times \text{Displacement} = (\text{Pressure} \times \text{Area}) \times \text{Displacement} = h\rho g\, adx$$

Now this work gets stored in the form of energy (pressure energy) within the liquid that has been forced into the vessel. Energy per unit volume of the forced liquid is

$$h\rho g\frac{adx}{adx} = h\rho g = \text{pressure}$$

Thus, the pressure energy per unit volume of a liquid is equal to the hydrostatic pressure.

BERNOULLI'S EQUATION

According to Bernoulli's theorem, the sum total of these three kinds of energies, kinetic energy, potential energy and pressure energy remains constant along a streamline in a steady flow of an ideal fluid. That is,

$$\frac{1}{2}\rho v^2 + \rho gy + p = \text{constant}$$

An ideal fluid is inviscous and incompressible. In other words, there is no dissipation of energy due to internal friction between adjacent layers of the fluid and density of this fluid remains constant. In other words, the sum of total energy per unit volume (pressure + kinetic + potential) is constant for an ideal fluid.

Other Forms of Bernoulli's Equation

If the kinetic, potential and pressure energies are expressed in terms of per unit mass or per unit weight, then the Bernoulli's equation may be written as

$$\frac{p}{\rho}+\frac{v^2}{2}+gy=\text{constant} \quad \text{(per unit mass)}$$

$$\frac{p}{\rho g}+\frac{v^2}{2g}+y=\text{constant} \quad \text{(per unit weight)}$$

The Bernoulli's equation in the per unit weight form is the most commonly used expression. Each term in this equation has the dimensions of length. The term $v^2/2g$ is known as the velocity head, $p/\rho g$ is known as the pressure head and y is known as the potential head.

Applications and Limitations

- The flow is assumed to be steady, i.e., there is no change in pressure, velocity and density of the fluid at any point with respect to time.
- However, in the problems of unsteady flow with gradually changing conditions, Bernoulli's equation can be applied without appreciable error. For example, a problem of emptying a large tank can be solved by applying Bernoulli's equation.
- The flow is assumed to be incompressible. Since liquids are incompressible, Bernoulli's equation can be applied to all liquids. However, it can be applied to the problems of gas flow when there is little variation in pressure, velocity and temperature so that density of gas can be assumed to be constant.
- The flow is assumed to be irrotational. The irrotationality means the net angular momentum at any point in the fluid flow is zero.
- The fluid is assumed to be ideal, i.e., the energy loss due to friction is assumed to be absent.

PROOF OF BERNOULLI'S EQUATION

Figure 3.116 shows flow of an ideal fluid through a tube of varying cross section and height. The fluid at the left end exerts a force P_1A_1, which moves the fluid through a distance Δx_1 and therefore does work $W_1 = F_1\Delta x_1 = P_1A_1\Delta x_1$ on the fluid. At the same time an equal volume of fluid at the upper right end is displaced through Δx_2. So the work done on this fluid volume is

$$W_1 = F_1\Delta x_1 = P_1A_1\Delta x_1$$

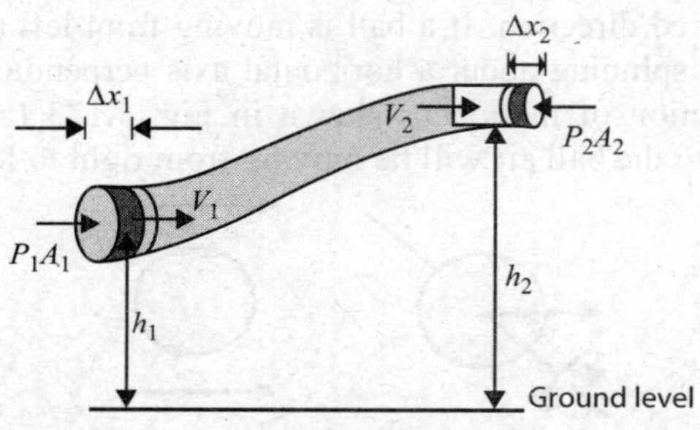

Fig. 3.116

At right end of tube the work done on this fluid element is negative because the force on it is in the opposite direction to the displacement. The total work done on the entire fluid element shown is

$$W = W_1 + W_2 = P_1A_1\Delta x_1 - P_2A_2\Delta x_2$$

Using equation of continuity,

$$A_1\Delta x_1 = A_2\Delta x_2 = \Delta V$$

$$\Rightarrow \quad W = (P_1 - P_2)\Delta V$$

The change in kinetic energy between the right and left end is

$$\Delta K = \frac{1}{2}\Delta m v_2^2 - \frac{1}{2}\Delta m v_1^2$$

And the change in potential energy is

$$\Delta U = \Delta mgh_2 - \Delta mgh_1$$

Using energy conservation,

$$\Delta W = \Delta K + \Delta U$$

$$\Rightarrow \quad (P_1 - P_2)\Delta V = \left(\frac{1}{2}\Delta m v_2^2 - \frac{1}{2}\Delta m v_1^2\right) + (\Delta mgh_2 - \Delta mgh_1)$$

Dividing this equation by ΔV and substituting $\rho = \Delta m/\Delta V$, we find

$$P_1 + \frac{1}{2}\rho v_1^2 + \rho gh_1 = P_2 + \frac{1}{2}\rho v_2^2 + \rho gh_2$$

$$\therefore \quad P + \frac{1}{2}\rho v^2 + \rho gh = \text{constant} \quad \text{(i)}$$

Equation (i) is called Bernoulli's equation. This equation is, in fact, the conservation of energy.

Illustration 3.33 **A liquid moves in a smooth tube of uniform cross section. Find the pressure P_2 at point 2 of the tube if the pressure at point 1 of the extended tube is P_1 and the vertical distance between points 1 and 2 is h.**

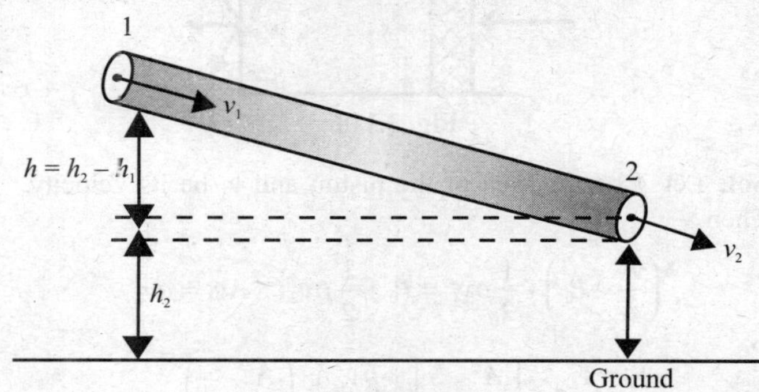

Fig. 3.117

Sol. Since the cross section of the tube is uniform,

$A_1 = A_2 = A.$

Then, applying equation of continuity at points 1 and 2, we have $v_1 = v_2$ (i)

Substituting $v_1 = v_2$ (= v) in Bernoulli's equation, we have

$$P_1 + \frac{1}{2}\rho v^2 + \rho g h_1 = P_2 + \frac{1}{2}\rho v^2 + \rho g h_2$$

This gives $P_1 + \rho g h_1 = P_2 + \rho g h_2$

Since $h_1 - h_2 = h$, we have $P_2 = P_1 + \rho g h$

Illustration 3.34 **A bent tube is lowered into a water stream as shown in Fig. 3.118. The velocity of the stream is V_1. The closed upper end is located at a height h_0 and has a small orifice. To what height h will the liquid get spurt?**

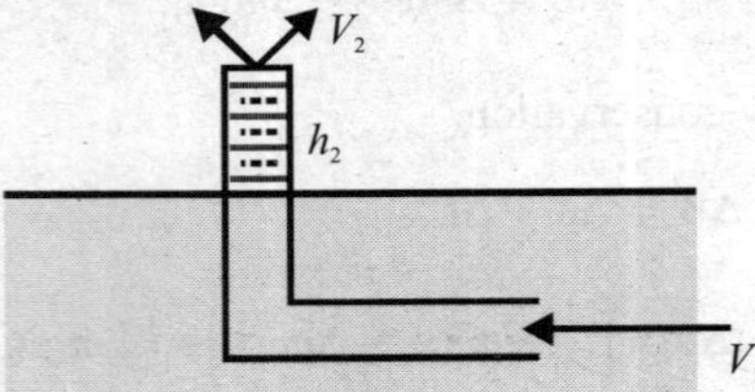

Fig. 3.118

Sol. $\frac{1}{2}\rho v_1^2 + P_0 + 0 = \frac{1}{2}\rho v_2^2 + P_0 + \rho g h_0 \Rightarrow V_2^2 = V_1^2 - 2gh_0$

If the liquid reaches a height h, then $V_2^2 = 2gh$,

$$\Rightarrow \quad 2gh = V_1^2 - 2gh_0 \Rightarrow h = \frac{V_1^2}{2g} - h_0$$

Illustration 3.35 **A horizontal cylinder of length l contains water. On the left a constant force F is applied on the piston and on the right there is a small hole of area a. What is the velocity of the piston? Hence, calculate the work done in emptying the vessel in time t.**

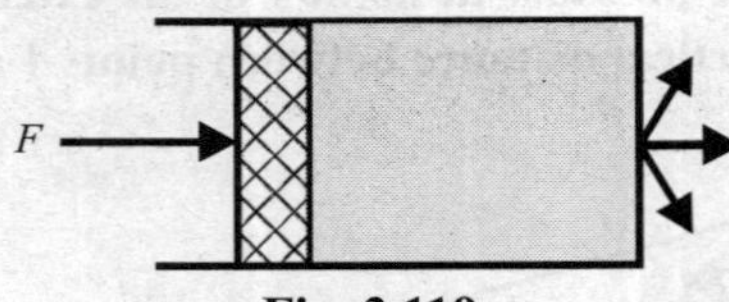

Fig. 3.119

Sol. Let A be the area of the piston and v_1 be its velocity. Then

$$\left(\frac{F}{A} + P_0\right) + \frac{1}{2}\rho v_1^2 = P_0 + \frac{1}{2}\rho v_2^2; \quad Av_1 = av_2$$

$$\Rightarrow \quad \frac{F}{A} = \frac{1}{2}\rho v_1^2\left[\frac{A^2}{a^2} - 1\right] = \frac{\rho v_1^2 A^2}{a^2}\left(\frac{A^2}{a^2} - 1\right)$$

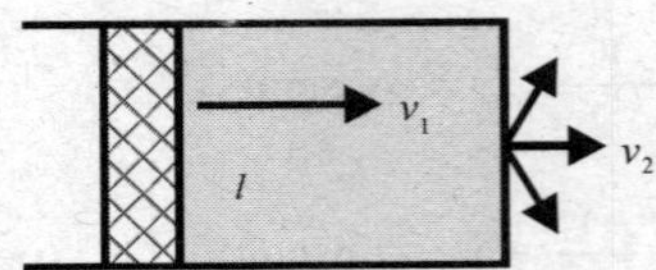

Fig. 3.120

Net work done is $F \times \text{distance} = Fl = \frac{2v_1^2 A^2 l}{2a^2}$

Blowing off roofs by wind storms

During tornado or hurricane, when a high-speed wind blows over a straw or tin roof, it creates a low pressure (p) in accordance will Bernoulli's principle. However, the pressure below the roof (i.e., inside the room) is still atmospheric (= p_0). So due to this difference of pressure the roof is lifted up and is then blown off by the wind.

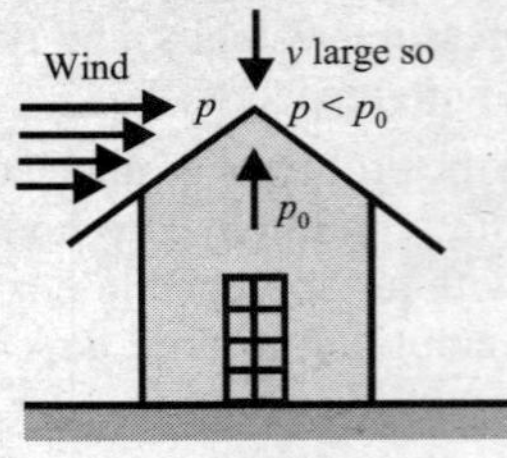

Fig. 3.121

Attraction between two closely parallel moving boats (or buses)

When two boats or buses move side by side in the same direction, the water (or air) in the region between them moves faster than that on the remote sides. Consequently in accordance with Bernoulli's principle the pressure between them is reduced and hence due to pressure difference they are pulled towards each other creating the so-called attraction.

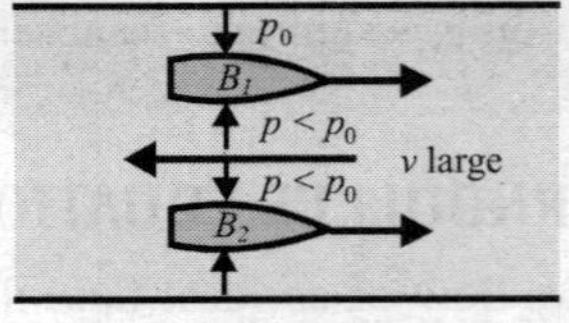

Fig. 3.122

Magnus effect

When a spinning ball is thrown, it deviates from its usual path in flight. This effect is called Magnus effect and plays an important role in tennis, cricket, soccer, etc., as by applying appropriate spin the moving ball can be made to curve in any desired direction. If a ball is moving from left to right and also spinning about a horizontal axis perpendicular to the direction of motion as shown in Fig. 3.123 (a), then relative to the ball air will be moving from right to left.

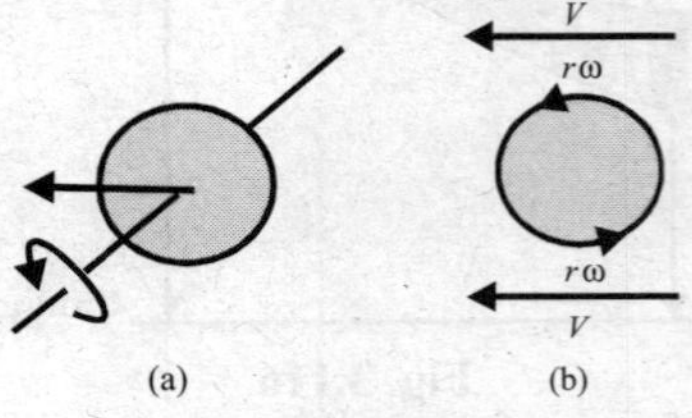

Fig. 3.123

The resultant velocity of air above the ball will be $V + \rho\omega$ while below the ball it is $V - \rho\omega$. So in accordance with Bernoulli's principle, pressure above the ball will be less than that below it. Due to this difference of pressure, an upward force will act on the ball and hence the ball will deviate from its usual path OA_0 and will hit the ground at A_1 following the path PA_1. If a ball is thrown with back spin, the pitch will curve less sharply prolonging the flight.

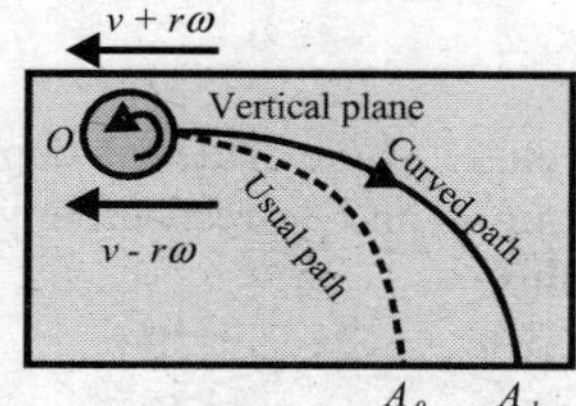

Fig. 3.124

Action of atomiser

The action of aspirator, carburettor, paint-gun, scent-spray or insect-sprayer is based on Bernoulli's principle. In all these, by means of motion of piston P in cylinder C, high speed air is passed over tube T dipped in liquid L to be sprayed. High speed air creates low pressure over the tube due to which liquid (paint, scent, insecticide or petrol) rises in it and is then blown off in very small droplets with expelled air.

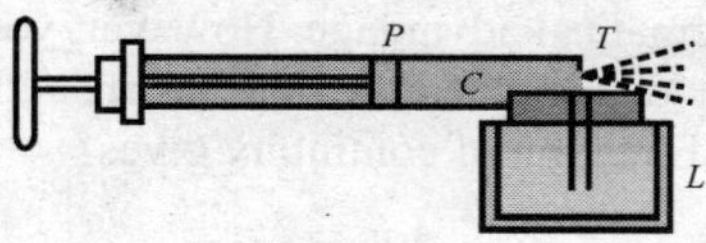

Fig. 3.125

Working of aeroplane

This is also based on Bernoulli's principle. The wings of the aeroplane are having tapering as shown. Due to this specific shape of wings when the aeroplane runs, air passes at higher speed over it as compared at its lower surface.

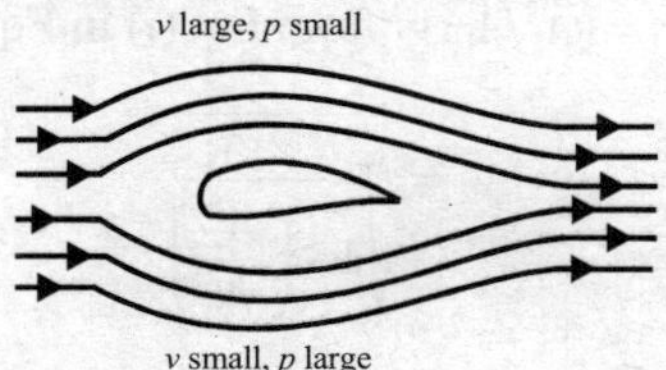

Fig. 3.126

VENTURI METER

This is an instrument for measuring the rate of flow of liquids and gases (fluids). The instrument is connected horizontally in the tube in which the rate of flow is to be measured. We have

$$P_A - P_B = h\rho g$$

where h is the difference of heights of liquids of density r in vertical tubes. The velocities of fluid at A and B are v_1 and v_2, respectively.

For liquids

$$P_A + \rho\frac{v_1^2}{2} = P_B + \rho\frac{v_2^2}{2}$$

$$\Rightarrow v_2^2 - v_1^2 = \frac{2}{\rho}(P_A - P_B) = \frac{2}{\rho}h\rho g \quad \Rightarrow v_2^2 - v_1^2 = 2hg$$

$$\Rightarrow \frac{Q^2}{A_2^2} - \frac{Q^2}{A_1^2} = 2hg \quad (\because Q = Av) \quad \Rightarrow Q = A_1A_2\sqrt{\frac{2hg}{A_1^2 - A_2^2}}$$

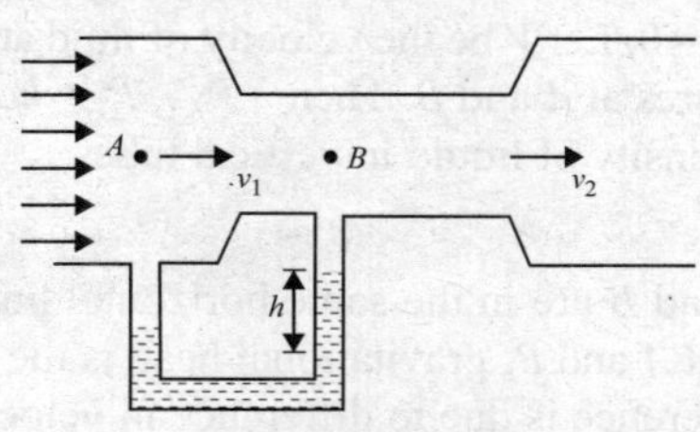

Fig. 3.127

For gases

Let d be the density of the gas.

Fig. 3.128

$$P_A - P_B = h\rho g \Rightarrow P_A + d\frac{V_1^2}{2} = P_B + d\frac{V_2^2}{2}$$

$$\Rightarrow \qquad V_2^2 - V_1^2 = \frac{2h\rho g}{d}$$

$$\Rightarrow \qquad Q = A_1A_2\sqrt{\frac{2h\rho g}{d\left(A_1^2 - A_2^2\right)}}$$

STATIC PRESSURE AND DYNAMIC PRESSURE

The static pressure at a point in a fluid flow may be obtained by connecting a single-tube manometer perpendicular to the direction of the fluid flow as shown by tube A in Fig. 3.129. The height h_s to which the liquid rises in it with respect to the reference level (datum) is a measure of the static pressure in the fluid at a point on the axis. That is,

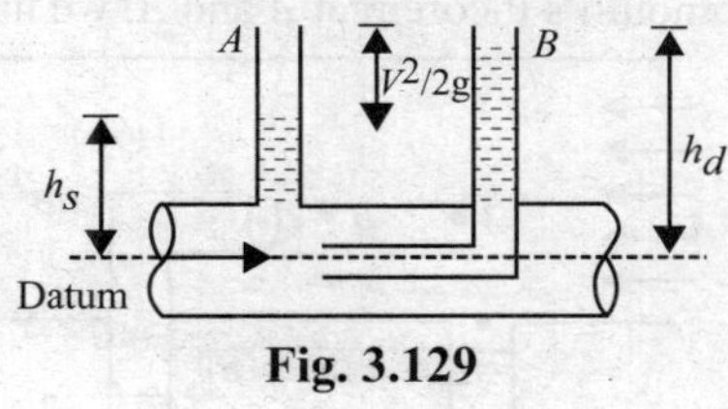

Fig. 3.129

$$p_s = \rho g h_s \quad \text{or} \quad \frac{p_s}{\rho g} = h_s$$

The height h_s is also called the static pressure head. When a tube bent at right angle is inserted in the fluid flow such that the open end faces against the direction of fluid flow as shown by tube B in Fig. 3.129. At the open end of the tube the velocity of the fluid suddenly reduces to zero and the kinetic energy gets converted into pressure energy. The liquid in tube B rises to a higher level than that in tube A. The difference in the levels of liquid is a measure of velocity of the fluid. Tube B measures the dynamic pressure and the height h_d of the liquid in the tube is called the dynamic head, which is equal to the sum total of the static pressure head and velocity head.

That is, $h_d = h_s + \frac{v^2}{2g}$.

PITOT TUBE

Pitot tube is used to measure the speed of flow of a fluid. The cross section of tube at B is perpendicular to the direction of flow and tube is at rest, so that at B, velocity of fluid becomes zero, i.e., $V_B = 0$. Let V be the velocity of fluid at A and P_A and P_B are pressures at A and B. Then $P_B - P_A = h\rho g$, where ρ is density of liquid in vertical tubes.

For liquids

If points A and B are in the same horizontal line and area of is the same at A and B, gravitational head is the same and the pressure difference is due to difference in velocity. Applying Bernoulli's theorem at A and B, we have

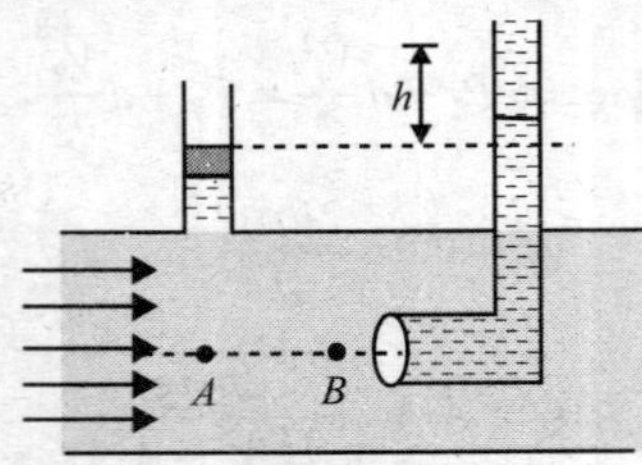

Fig. 3.130

$$P_A + \frac{1}{2}\rho V^2 = P_B + 0 \qquad \Rightarrow \frac{2(P_B - P_A)}{\rho} = V^2$$

$$\Rightarrow V = \sqrt{\frac{2(P_B - P_A)}{\rho}} \qquad \Rightarrow V = \sqrt{2gh}$$

For gases

Let r be the density of the liquid in the vertical tube.

$$P_B - P_A = h\rho g$$

Applying Bernoulli's theorem at A and B, we have

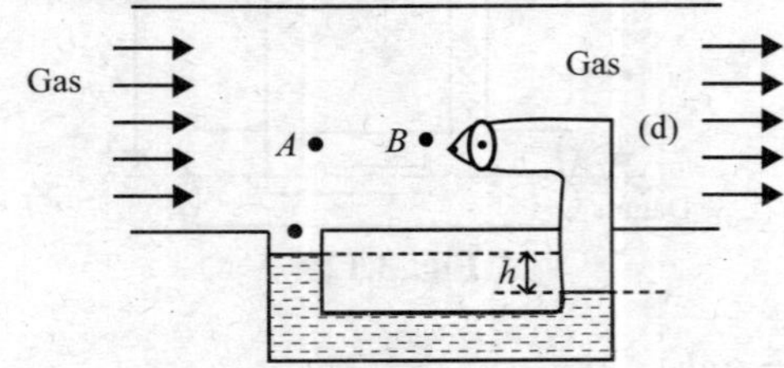

Fig. 3.131

$$P_A + \frac{1}{2}dV^2 = P_B \qquad \Rightarrow V^2 = \frac{2(P_B - P_A)}{d}$$

$$\Rightarrow V^2 = \frac{2h\rho g}{d} \qquad \Rightarrow V = \sqrt{\frac{2h\rho g}{d}}$$

VELOCITY OF EFFLUX: TORRICELLI'S THEOREM

The liquid comes out of a hole made at a depth y. The areas of the base and hole are A_1 and A_2, respectively. Let us calculate the velocity of efflux.

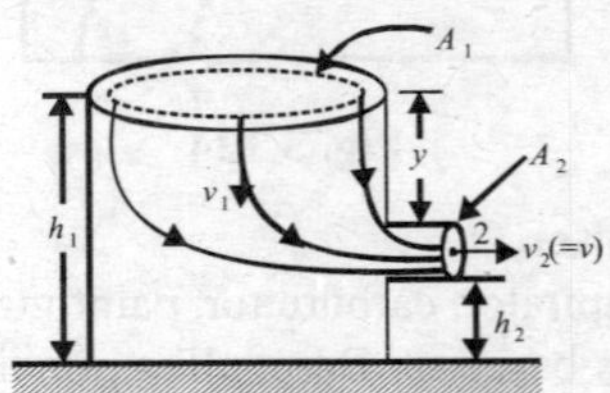

Fig. 3.132

Choose two points 1 and 2 at the free surface of liquid and at just outside the hole, respectively. Since the points 1 and 2 are exposed to atmosphere, $P_1 = P_2 = P_{atm}$. We have done this to get a mathematical advantage. However, you can take any other point inside the tube instead of taking at the free surface of the liquid. Equation of continuity gives

$$A_1 v_1 = A_2 v_2 \qquad \text{(i)}$$

$$P_{atm} + \frac{\rho v_1^2}{2} + \rho g h_1 = P_{atm} + \frac{\rho v_2^2}{2} + \rho g h_2$$

Substituting $h_1 - h_2 = y$, we have

$$v_2 = \sqrt{v_1^2 + 2gy} \qquad \text{(ii)}$$

Substituting $v_1 = (A_2/A_1)v_2$ from Eq. (i) in Eq. (ii), we have

$$v_2 = \sqrt{\frac{2gy}{1 - \left(\frac{A_2}{A_1}\right)^2}}$$

Note:

- ***If area of opening is much lesser than area of cross section of tank ($A_2 << A_1$), then velocity of efflux*** $v = \sqrt{2gh}$ ***which is the same speed that an object would acquire in falling from rest through a distance h under gravtiy.***
- ***The velocity of efflux is the velocity of escaping liquid relative to the container (but not necessarily relative to ground when the container moves).***

HORIZONTAL RANGE OF THE ESCAPING LIQUID

Water stands at a depth H in a tank whose sides are vertical. A hole is made in one of the walls at a depth h below the water surface. Let container remains at rest.

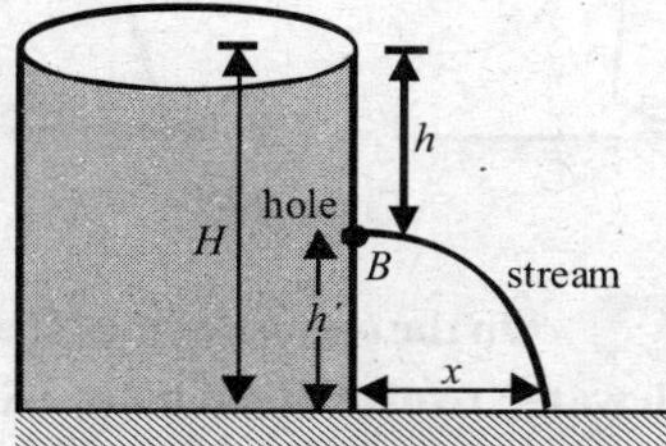

Fig. 3.133

When the liquid stream emerges out of the hole, it goes along a parabolic path. So time taken by water to fall through a height of $H - h$ is given as

$$H - h = 0 \times t + 1/2\, gt^2 \quad \Rightarrow \quad t = \sqrt{\frac{2(H-h)}{g}}$$

Horizontal range is given by

$$V \times t = \sqrt{2gh}\sqrt{\frac{2(H-h)}{g}} = \sqrt{4h(H-h)} = 2\sqrt{hh'}$$

Horizontal range is given by $x = \sqrt{4h(H-h)}$

$$= \sqrt{4hH - 4h^2 + H^2 - H^2}$$

$$= \sqrt{H^2 - (H^2 - 4hH + 4h^2)}$$

$$= \sqrt{H^2 - (2h - H)^2}$$

For the range to be maximum, $(2h - H)^2 = 0$

$\Rightarrow \quad 2h - H = 0 \Rightarrow h = H/2$

So the maximum range $x = 2\sqrt{hh'}$.

- ***The formula** $= 2\sqrt{hh'}$ **tells us if we make two holes at equal vertical distances from top and bottom, both liquids jets will strike the same spot (but not simultaneously).***

Illustration 3.36 **A cylindrical vessel of area of cross section A is filled with a liquid up to a height H. A very small hole of area of cross section a is made at the bottom of the vessel. Find the time taken by the vessel to become empty.**

Sol. The liquid escapes with a velocity $(= \sqrt{2gh})$ which varies with the level of liquid. Hence, we have to use integral. Let y be the height of the liquid at an instant. This height changes by dy in time dt.

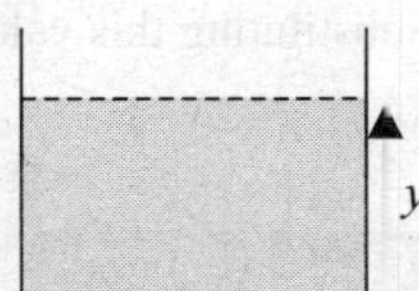

Fig. 3.134

Volume of water leaving out per second $= -A\, dy/dt$.

At the hole, volume escaping per second is $av = a\sqrt{2gy}$

$$\therefore \quad a\sqrt{2gy} = -A\frac{dy}{dt} \quad \Rightarrow \int_H^0 \frac{-dy}{\sqrt{y}} = \frac{a\sqrt{2g}}{A}\int_0^t dt$$

$$\Rightarrow \quad t = \frac{A}{a\sqrt{2g}} 2\sqrt{H} \quad \text{(assuming } a \ll A\text{)}$$

Illustration 3.37 **A wide vessel with a small hole in the bottom is filled with water and kerosene. Find the velocity of water flow if the thickness of water layer is $h_1 = 30$ cm and that of kerosene is $h_2 = 20$ cm.**

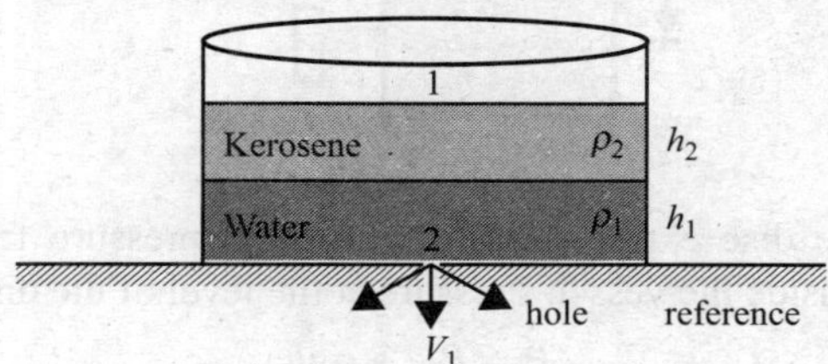

Fig. 3.135

Sol. Method 1: Let the densities of water and kerosene be ρ_1 and ρ_2, respectively. Let the area of the vessel be A and area of the hole be a.

Applying Bernoulli's theorem at cross sections 1 and 2, we have

$$\frac{1}{2}\rho_1 V_1^2 + P_{atm} + 0 = \frac{1}{2}\rho_1 V_1^2 + (P_{atm} + h_2\rho_2 g) + h_1\rho_1 g$$

$$av_1 = Av_2$$

$$\Rightarrow \frac{1}{2}\rho_1 V_1^2\left[1 - \frac{a^2}{A^2}\right] = \left[h_1 - \frac{h_2 P_2}{\rho_1}\right]\rho_1 g$$

$$\Rightarrow v_1 = \sqrt{2g\left(h_1 + \frac{h_2\rho_2}{\rho_1}\right)}$$

Method 2:

We can replace kerosene with water column which produces the same pressure as kerosene column at interface of kerosene and water.

Pressure produced by kerosene $P_{interface} = \rho_2 gh_2$.

Let the equivalent height of water column be h'_1.

Hence, $P_{interface} = \rho_1 gh'_1 = \rho_2 gh_2$

$$\Rightarrow \quad h'_1 = \frac{\rho_2 h_2}{\rho_1} \qquad \text{(i)}$$

Total height of water above orifice, which produces the same pressure as before, is

$$H = h_1 + h'_1 = h_1 + \frac{\rho_2}{\rho_1}h_2$$

Using Torricelli's formula for velocity of efflux,

$$v = \sqrt{2gH} = \sqrt{2g\left(h_1 + \frac{\rho_2}{\rho_1}h_2\right)}$$

FORCE OF REACTION DUE TO EJECTION OF LIQUID

A cylindrical vessel has an opening of cross-sectional area '*a*' near the bottom. A disc is held against the opening to prevent liquid of density ρ from coming out. If the height of the liquid above opening is h. Let us analyse the force on the disc in this situation.

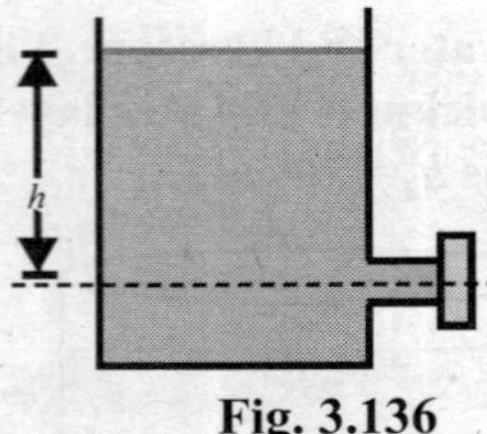

Fig. 3.136

The disc experiences a hydrostatic pressure from the liquid inside the vessel. Pressure at the level of the disc is

$$P_1 = P_{atm} + \rho gh$$

The air pressure on the outside of disc is $P_2 = P_{atm}$.
The net outward force $= (p_2 - p_1)a = \rho gha$.

Now the disc is moved a short distance away in horizontal direction. The liquid comes out, strikes the disc inelastically and drops vertically downward. The water in this case will impart impulsive (impact) force on the disc.

When the disc is moved away, the liquid moves out with speed $v = \sqrt{2gh}$.

The (mass per second), i.e, the rate of mass coming out of the opening is given by $dm/dt = \rho av = \rho a\sqrt{2gh}$.

Momentum per second imparted by water rightward is given by

$$\frac{dm}{dt}v = (\rho av)v = \rho av^2 = 2\rho gha$$

The change in momentum per second after striking the disc is

$$\Delta\vec{P} = \vec{P}_f - \vec{P}_i \Rightarrow 0 - (\rho av^2) = -\rho av^2 = -2\rho gha$$

Taking the rightward direction as positive, the force on the liquid is towards the left and its reaction with disc is towards the right. The force obtained is twice the hydrostatic force. The force due to atmospheric pressure is cancelled out on the both sides.

Note:

- ***If the velocity of a liquid (or gas) of density ρ coming out through an opening of area of cross section a is v, then there will be a thrust force $F = \rho av^2$ due to liquid coming out the opening. The direction of the thrust force will be just opposite to the velocity direction.***

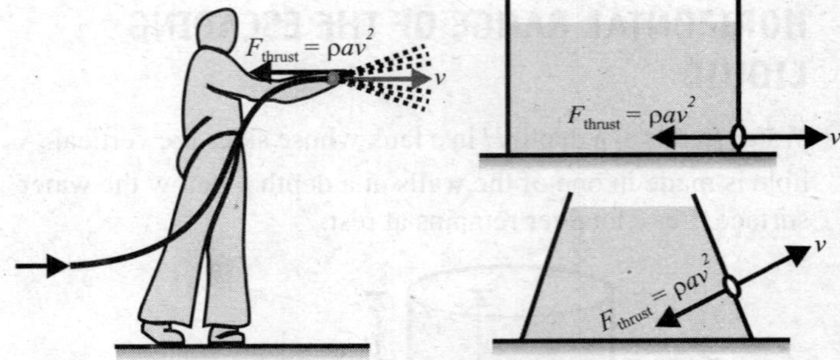

Fig. 3.137

Illustration 3.38 **On the opposite sides of a wide vertical vessel filled with water, two identical holes are opened, each having cross-sectional area '*a*'. The height difference between them is equal to *h*. Find the resultant force of reaction of water flowing out of vessel.**

Sol. Let A and B be two openings, each of area '*a*' the velocity of water coming out of opening are V_A and V_B, respectively.
Volume of water emerging per second through A is aV_A.
Volume of water emerging per second through B is aV_B.
Force of reaction due to water coming out at A is given by
F_A = Rate of change of momentum per second at A

$$(aVA\rho)V_A = a\rho V_A^2$$

Similarly, force of reaction at B is given by $F_B = a\rho V_B^2$

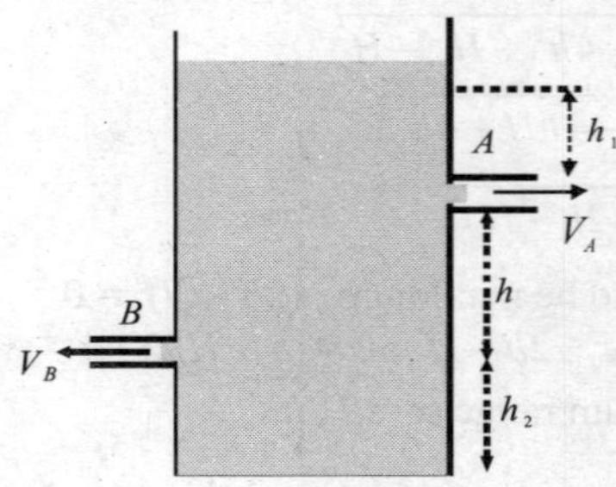

Fig. 3.138

Therefore, net force on vessel is

$$F = F_B - F_A = a\rho\left(V_B^2 - V_A^2\right) \quad \text{(i)}$$

Now, applying Bernoulli's theorem at A and B, we get

$$p_a + \frac{1}{2}\rho V_A^2 + (h_2 + h)\rho g = p_0 + \frac{1}{2}\rho V_B^2 + \rho gh_2 \quad \text{(ii)}$$

where p_a is atmospheric pressure.

$$\therefore \quad V_B^2 - V_A^2 = 2gh$$

Substituting this value in Eq. (i), we get $F = a\rho\, 2gh$

or $$F = 2\rho g\, ah$$

Illustration 3.39 **The side wall of a wide vertical vessel of height *h* = 75 cm has a narrow slit (vertical) running all the way down to the bottom of the vessel.**

The length of the slit is $l = 50$ cm and the width is $b = 1$ mm. With the slit closed, water is filled to the top. Find the resultant reaction force of water coming out as the slit is opened.

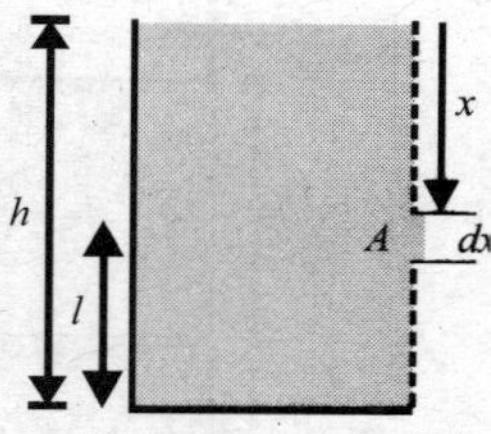

Fig. 3.139

Sol. Let us consider an infinitesimal portion of length dx of the slit at a depth x below water level. Reaction force due to this portion is given by

$$dF = Av^2\rho = (bdx)2gx\rho$$

Total reaction force is

$$F = \int df = \int_{h-\ell}^{h} 2g\rho b(xdx)$$

$$= \frac{2g\rho b}{2}\left[h^2 - (l-h)^2\right] = g\rho\, bl(2h-l)$$

Concept Application Exercise 3.4

1. A non-viscous liquid of constant density 500 kg/m³ flows in a variable cross-sectional tube. The area of cross section of the tube at two points P and Q at heights of 3 m and 6 m are 2×10^{-3} m³ and 4×10^{-3} m³, respectively. Find the work done per unit volume by the forces of gravity as the fluid flows from point P to Q.

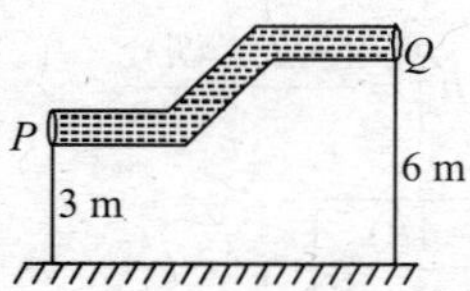

Fig. 3.140

2. A horizontal pipe of varying cross-sectional area carries water flowing at a constant rate. The pressure at sections (1) and (2) are 50 cm and 20 cm of Hg. If the velocity of section (1) is 5 m/s, what is the velocity of section (2). Also determine the ratio of areas at sections (1) and (2) (as shown in Fig. 3.141).

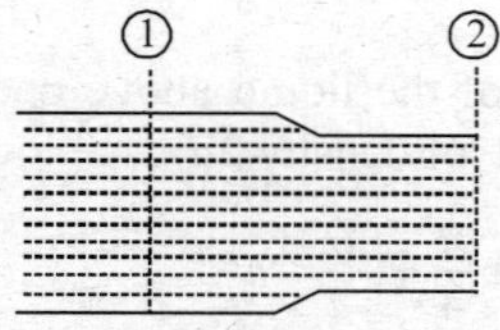

Fig. 3.141

3. The speed of flow past the lower surface of a wing of an aeroplane is 50 ms^{-1}. What speed of flow over the upper surface will give a dynamic lift of 1000 Pa? Density of air = 1.3 kg m^{-3}.

4. A gardener holds a pipe of inside cross-sectional area 3.60 cm² at a height of 1.50 m from ground. The opening at the nozzle of pipe has cross-sectional area 0.250 cm². Water velocity in the segment of the hose that lies on the ground is 50.0 cm/s.
   (a) What is the velocity of water leaving the nozzle?
   (b) What is the thrust force experienced by the gardener due to flow of water through nozzle.
   (c) What is the water pressure in the hose on the ground?

5. Figure 3.142 shows a stream of fluid emerging from a tube in the base of an open tank. Obtain an expression for 'y' in terms of θ and 'h'. Can 'y' exceed 'h'?

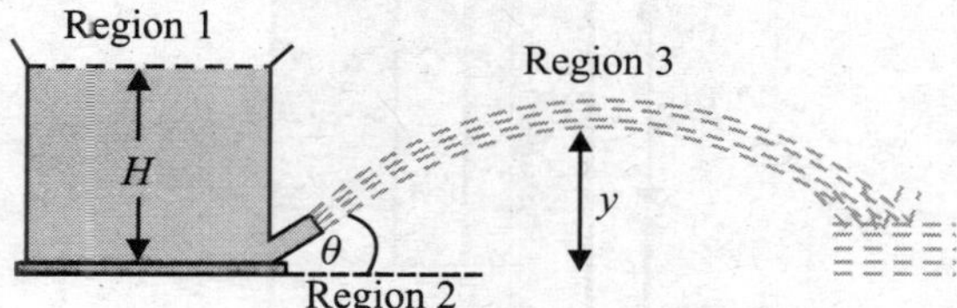

Fig. 3.142

6. If water flows horizontally through a pipe of varying cross section and the pressure of water equals 10 cm of Hg at a point where the velocity of flow is 40 cm/s, what is the pressure at another point where the velocity of flow is 50 cm/s?

7. A pipe is running full of water. At a certain point A it tapers from 60 cm diameter to 2.0 cm diameter at B. The pressure difference between A and B is 100 cm of water column. Find the rate of flow of water through the pipe.

8. A pitot tube is fixed in a water pipe of diameter 14 cm and difference of pressure by the gauge is 20 cm of water column. Calculate the rate of flow of water.

9. Water is used as the manometric liquid in a pitot tube mounted in an aircraft to measure air speed. If the maximum difference in height between the liquid columns is 0.1 m, what is the maximum air speed that can be measured? Density of air $d_{air} = 1.3$ kg/m³.

10. Water flows out of a big tank along a tube bent at right angles; the inside of the tube is of radius r. The length of the horizontal portion is l. The rate of water flow is Q. Find the moment of reaction forces of flowing water, acting on the tube walls relative to point O.

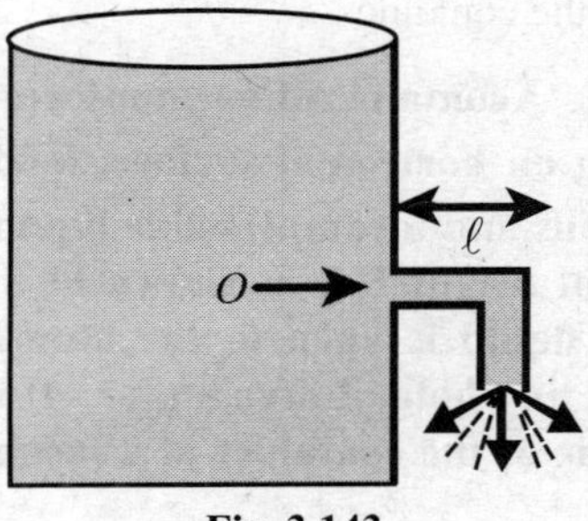

Fig. 3.143

Example 3.1 **A container of a large uniform cross-sectional area A resting on horizontal surface holds two immiscible, non-viscous and incompressible liquids of densities d and $2d$, each of height $H/2$ as shown in the figure. The lower density liquid is open to atmosphere. A homogeneous solid cylinder of length L ($L < H/2$), cross-sectional area $A/5$ is immersed such that it floats with its axis vertical of the liquid–liquid interface with length $L/4$ in the denser liquid. Determine (i) density D of the solid and (ii) the total pressure at the bottom of the container (atmospheric pressure = P_0).** **(IIT-JEE, 1995)**

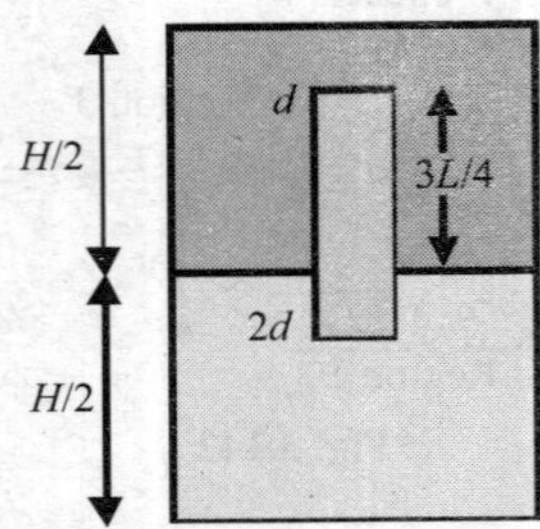

Fig. 3.144

Sol.

(i) For equilibrium of the cylinder,

Weight of the cylinder = Weight of the liquid displaced

$$\Rightarrow \quad L\left(\frac{A}{5}\right)Dg = \frac{3L}{4}\left(\frac{A}{5}\right)dg + \frac{L}{4}\left(\frac{A}{5}\right)2dg$$

This gives $D = \frac{3}{4}d + \frac{d}{2}$ or $D = \frac{5}{4}d$

(ii) Total pressure at the bottom of cylinder is

$$P = \frac{\text{weight of liquids} + \text{weight of cylinder}}{A} + P_0$$

$$= \frac{\left[Ad\left(\frac{H}{2}\right)g + A2d\left(\frac{H}{2}\right)g\right] + \frac{A}{5}\left(\frac{5}{4}d\right)Lg}{A} + P_0$$

$$= \left(\frac{6H + L}{4}\right)dg + P_0$$

Here we are asked to find the total pressure at the bottom of the container and not only the hydrostatic pressure; hence, we need to consider the total weight of liquid and cylinder in the container.

Example 3.2 **Acontaineroflargeuniformcrosssectional area A, resting on horizontal surface, holds two immiscible non-viscous and incompressible liquids of density d and $2d$, each of height $H/2$ as shown in the figure. The lower density liquid is open to the atmosphere having pressure P_0. A tiny hole of area s ($s \ll A$) is punched on the vertical side of the container at a height h ($h < H/2$). Determine**

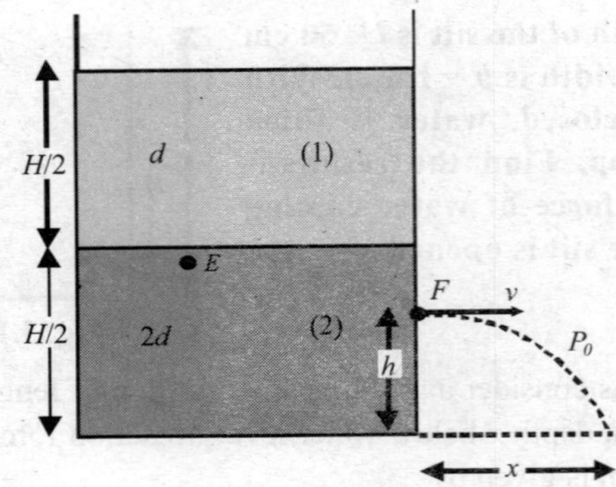

Fig. 3.145

(i) the initial speed of efflux of the liquid at the hole

(ii) the horizontal distance x travelled by the liquid initially

(iii) the height h_m at which the hole should be punched so that the liquid travels the maximum distance x_m initially. Also calculate x_m (neglect air resistance in the calculations) **(IIT-JEE, 1995)**

Sol.

(i) Applying Bernoulli's theorem at points E (a point on the interface of the liquids and on the top surface of the liquid of density $2d$) and F, i.e., pressure energy + kinetic energy + potential energy = constant

$$\Rightarrow P_0 + \frac{H}{2}dg + \frac{1}{2}(2d) \times 0 + \frac{H}{2}2dg = P_0 + \frac{1}{2}(2d)v^2 + 2hdg$$

$$\Rightarrow \frac{3}{2}Hdg = dv^2 + 2hdg$$

$$\Rightarrow v^2 = \frac{3}{2}Hg - 2hg = \frac{(3H - 4h)g}{2}$$

$$\therefore v = \sqrt{\left\{\frac{(3H - 4h)}{2}g\right\}}$$

Method 2: Let the equivalent depth of liquid (1) of density d in terms of density $2d$ which gives the same pressure as liquid (1) at interface, i.e.,

$$\frac{H}{2}dg = H'(2d)g$$

which gives $H' = \frac{H}{4}$

Hence, depth of the liquid above opening (in terms of depth of liquid of density $2d$),

$$H' + \left(\frac{H}{2} - h\right) = \frac{H}{4} + \left(\frac{H}{2} - h\right)$$

$$H_{net} = \left(\frac{3H - 4h}{4}\right)$$

And velocity of efflux is given $v = \sqrt{2g\,H_{net}}$

$$v = \sqrt{\left(\frac{3H-4h}{2}\right)g}$$

Time taken by the liquid to fall through a vertical height h is given by

$$H = \frac{1}{2}gt^2 \Rightarrow t = \sqrt{\frac{2h}{g}}$$

Therefore, horizontal distance traversed by liquid is

$$x = vt = \sqrt{\frac{(3H-4h)g}{2}} \times \sqrt{\frac{2h}{g}} = \sqrt{(3H-4h)h}$$

(iii) For the maximum horizontal distance x,

$$\frac{dx}{dh} = 0 \Rightarrow \frac{d}{dh}(3Hh - 4h^2)^{1/2} = 0$$

or $\frac{1}{2}(3Hh-4h^2)^{-1/2} \times (3H-8h) = 0$

This gives $3H - 8h = 0 \Rightarrow h = \frac{3H}{8}$

This is the length of the liquid for which the range is maximum, i.e.,

$$h \to h_m = \frac{3H}{8}$$

The maximum horizontal distance is given by

$$x_m = \sqrt{(3H-4h_m)h_m} = \sqrt{\left\{\left(3H - 4\frac{3H}{8}\right)\frac{3H}{8}\right\}} = \frac{3}{4}H$$

Example 3.3 **A large open top container of negligible mass and uniform cross-sectional area A has a small hole of cross-sectional area $A/100$ in its side wall near the bottom. The container is kept on a smooth horizontal floor and contains a liquid of density ρ and mass M_0. Assuming that the liquid starts flowing out horizontally through the hole at $t = 0$, calculate (i) the acceleration of the container and (ii) its velocity when 75% of the liquid has drained out. (IIT-JEE, 1997)**

Sol. Liquid is escaping from the container. Let y be the height of the liquid at any instant t after start.

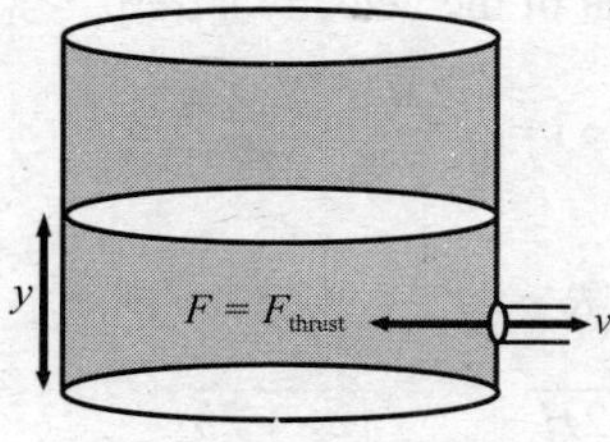

Fig. 3.146

Then velocity of escape is given by

$$v = \sqrt{2gy}$$

Mass of the liquid flowing in time $dt = av\rho\, dt$, where a = area of the hole = $A/100$.

Force is the rate of change of momentum. Hence,

$$F = \frac{dp}{dt} = \left(\frac{av\rho dt}{dt}\right)v = av^2\rho = a\,(2gy)\,\rho$$

As F = mass of container × acceleration = Ma'

$$\therefore \quad a' = \frac{F}{M}$$

But M = mass of container = $Ay\rho$

$$\therefore \quad a' = \frac{a(2gy)\rho}{M} = \frac{\left(\frac{A}{100}\right)2gy\rho}{Ay\rho} = \frac{g}{50}$$

When 75% of the liquid has drained out, the height of the liquid y is given by

$$Ay\rho = \frac{25}{100}M_0 \Rightarrow y = \frac{M_0}{4A\rho}$$

But $v = \sqrt{2gy} = \sqrt{2g\left(\frac{M_0}{4A\rho}\right)} = \sqrt{\frac{M_0 g}{2A\rho}}$

Example 3.4 **A wooden stick of length L, radius R and density ρ has a small metal piece of mass m (of negligible volume) attached to its one end. Find the minimum value for the mass m (in terms of given parameters) that would make the stick float vertically in equilibrium in a liquid of density ρ ($>\rho$). (IIT-JEE, 1999)**

Sol. The mass m cannot be at the top, because for stable equilibrium, the centre of mass should be the lowest possible. Therefore, m is at the lower end of the stick. Let x be the length of the stick immersed in liquid. For the vertical equilibrium,

$$Mg + mg = B$$

$$\Rightarrow \quad \pi R^2 L\rho g + mg = (\pi R^2 x\sigma)g$$

$$\therefore \quad x = \frac{m + \pi R^2 L\rho}{\pi R^2 \sigma} \qquad \text{(i)}$$

The distance of the centre of mass from m is

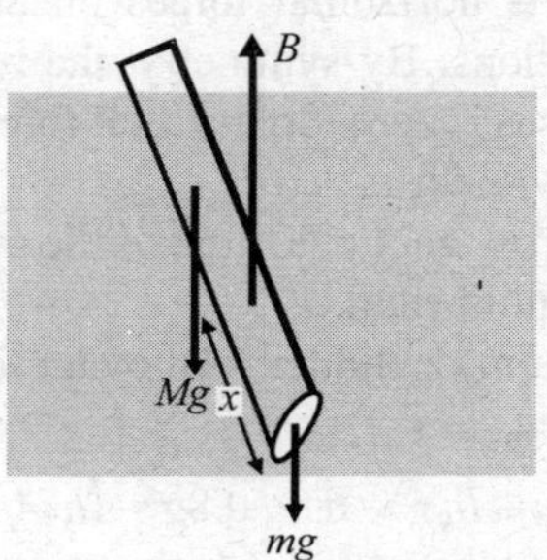

Fig. 3.147

$$y_{CM} = \frac{m_1y_1 + m_2y_2}{m_1 + m_2} = \frac{m \times 0 + \pi R^2 L\rho\left(\frac{L}{2}\right)}{m + \pi R^2 L\rho}$$

For rotational equilibrium of the stick, the point of application of upthrust of the liquid must pass above the centre of mass of the floating system, i.e., $y_{CM} \le \frac{x}{2}$

$$\frac{\pi R^2 L\rho\left(\frac{L}{2}\right)}{m+\pi R^2 L\rho} \le \frac{1}{2}\left(\frac{m+\pi R^2 L\rho}{\pi R^2 \sigma}\right)$$

$\Rightarrow \quad (\pi R^2 L)^2 \rho\sigma \le (m+\pi R^2 L\rho)^2$

$\Rightarrow \quad \pi R^2 L\sqrt{\rho\sigma} \le m+\pi R^2 L\rho$

$\Rightarrow \quad m \ge \pi R^2 L\left\{\sqrt{\rho\sigma}-\rho\right\}$

$\Rightarrow \quad m_{\min} = \pi R^2 L\left\{\sqrt{\rho\sigma}-\rho\right\} = \pi R^2 L\rho\left\{\frac{\sqrt{\sigma}}{\rho}-1\right\}$

Example 3.5 **A uniform solid cylinder of density 0.8 g/cm³ floats in equilibrium in a combination of two non-mixing liquids A and B with its axis vertical. The densities of liquids A and B are 0.7 g/cm³ and 1.2 g/cm³, respectively. The height of liquid A is h_A = 1.2 cm. The length of the part of the cylinder with liquid B is h_B = 0.8 cm.**

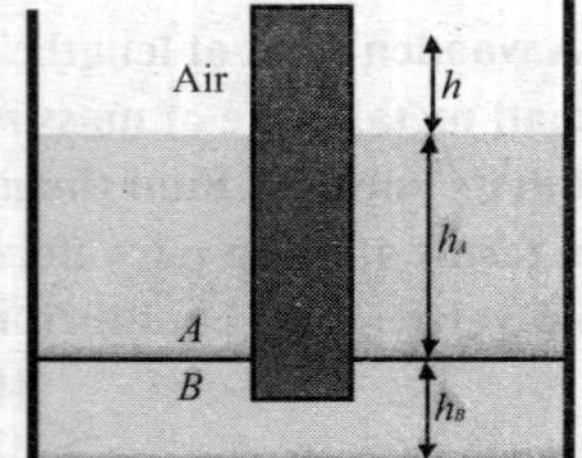

Fig. 3.148

(a) Find the total force exerted by liquid A on the cylinder.
(b) Find h, the length of part of the cylinder in air.
(c) The cylinder is depressed in such a way that its top surface is just below the upper surface of liquid A and is then released. Find the acceleration of the cylinder immediately just after it is released. (IIT-JEE, 2002)

Sol.
(a) Liquid A exerts horizontal forces on all area elements from all directions. By symmetry, the resultant of these horizontal forces is zero. So the total force exerted by liquid A on the cylinder is zero.
(b) Let A be the cross-sectional area of the cylinder. For floating equilibrium,
Weight of floating cylinder = Buoyant force = weight of liquid displaced.

$(h+h_A+h_B)\times A\times 0.8g = (h_A A\rho_A g + h_B A\rho_B g)$

or $\quad (h+h_A+h_B)\times 0.8 = h_A\rho_A + h_B\rho_B$

$\Rightarrow \quad (h+1.2+0.8)\times 0.8 = 1.2\times 0.7 + 0.8\times 1.2$

$\Rightarrow \quad 0.8h + 1.6 = 1.8$ or $0.8h = 0.2 \Rightarrow h = 0.25$ cm

(c) Mass of cylinder $= (h+h_A+h_B)A\rho$
$= (0.2+1.2+0.8)A\times 0.8 = 1.8A$

When the cylinder is depressed just completely, its height h goes in liquid B, so extra buoyant force due to liquid B acts on it. Due to this force, cylinder is accelerated upward.

Extra buoyant force $F = (h_A)\rho_B g$
$= 0.25A\times 1.2g = 0.3Ag$

Acceleration is given by $a = \frac{F}{m} = \frac{0.3Ag}{1.8A} = \frac{g}{6}$

$= \frac{10}{6} = \frac{5}{3}\ \text{m/s}^2$

Example 3.6 **Consider a horizontally oriented syringe containing water located at a height of 1.25 m above the ground. The diameter of the plunger is 8 mm and the diameter of the nozzle is 2 mm. The plunger is pushed with a constant speed of 0.25 m/s. Find the horizontal range of water stream on the ground. Take g = 10 m/s². (IIT-JEE, 2004)**

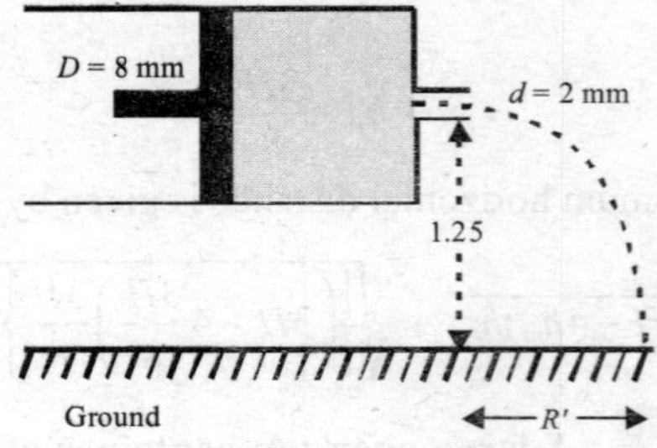

Fig. 3.149

Sol. From equation of continuity,

$$A_1 v_1 = A_2 v_2 \Rightarrow \frac{A_1}{A_2} v_1 = \left(\frac{\pi r_1^2}{\pi r_2^2}\right) v_1$$

or $v_2 = \left(\frac{D}{d}\right)^2 v_1 = \left(\frac{8\times 10^{-3}}{2\times 10^{-3}}\right)^2 \times 0.25\ \text{m/s} = 4\ \text{m/s (horizontal)}$

Vertical component of the velocity is zero.

Now, $H = \frac{1}{2}gt^2 \Rightarrow t = \sqrt{\frac{2H}{g}}$

Range is given by $R = v_2 t = v_2 \sqrt{\frac{2H}{g}} = 4\times\sqrt{\frac{2\times 1.25}{10}} = 2$ m

Example 3.7 **A non-viscous liquid of constant density 1000 kg/m³ flows in a streamline motion along a tube of variable cross section. The tube is kept inclined in a vertical plane as shown in the figure.**

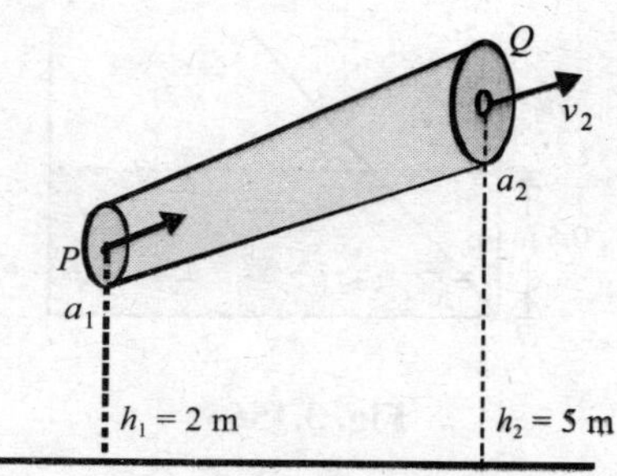

Fig. 3.150

The areas of cross section of the tube at two points P and Q at heights of 2 m and 5 m, respectively, are 4×10^{-3} m^2 and 8×10^{-3} m^2. The velocity of the liquid at point P is 1 m/s. Find the work done per unit volume by the pressure and the gravity forces as the fluid flows from points P and Q. (IIT-JEE, 2004)

Sol. From equation of continuity at points P and Q, we have

$$a_1 v_1 = a_2 v_2$$

$$v_2 = \frac{a_1}{a_2} v_1 = \frac{4\times10^{-3}}{8\times10^{-3}} \times 10 \text{ m/s} = 0.5 \text{ m/s}$$

From Bernoulli's theorem at points P and Q,

$$p_1 + \frac{1}{2}\rho v_1^2 + \rho g h_1 = p_2 + \frac{1}{2}\rho v_2^2 + \rho g h_2$$

Each term in the equation represents energy per unit volume.

Work done by pressure per unit volume is

$$p_1 - p_2 = \frac{1}{2}\rho(v_2^2 - v_1^2) + \rho g(h_2 - h_1)$$

$$\frac{1}{2}\times(1000)(0.5^2 - 1^2) + 1000\times10(5-2)$$

$$= \frac{1}{2}\times(1000)\times0.75 + 1000\times10\times3$$

$$= -0.375 \times 10^3 + 30 \times 10^3 = 30.375 \times 10^3$$

$$= 3.0375 \times 10^4 \text{ J/m}^3$$

Work done by gravity $= \rho g h_1 - \rho g h_2 = \rho g(h_1 - h_2)$

$$= -1000 \times 10 \times 2.5$$

$$= -2.5 \times 10^4 \text{ J/m}^3$$

EXERCISES

Subjective Type

Solutions on page 3.69

1. Two immiscible liquids of densities ρ and 2ρ and thickness (heights) h and $2h$, respectively, push a vertical plate. Find the total side thrust given by the liquids on the vertical plate.

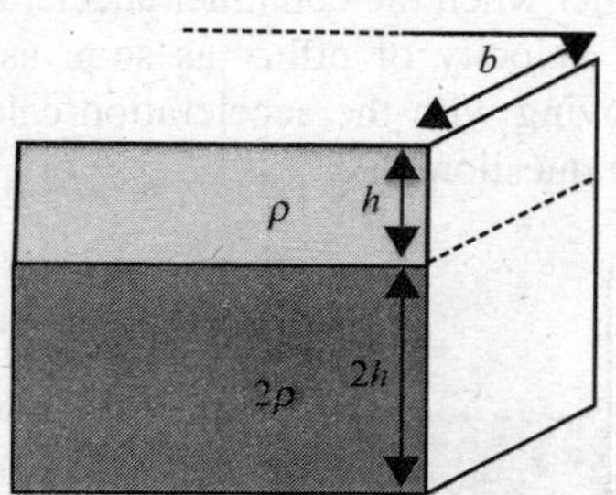

Fig. 3.151

2. A composite plate comprises a triangular area (1) and rectangular area (2) as shown in the figure. The triangular plate experiences forces due to the liquids of densities ρ_1 and the rectangular plate experiences the force due to liquid of density ρ_2. Find the hydrostatic forces on **(a)** area 1, **(b)** area 2, **(c)** total composite area of the plate. Assume $\rho_1 = \rho$ and $\rho_2 = 2\rho$.

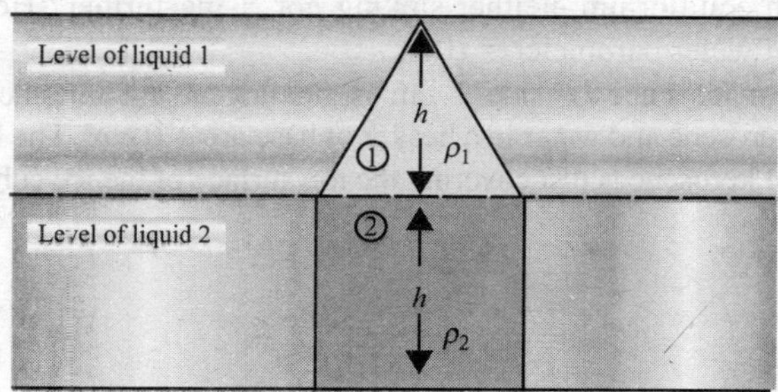

Fig. 3.152

3. A rectangular tube of uniform cross section has three liquids of densities ρ_1, ρ_2 and ρ_3. Each liquid column has length l equal to length of sides of the equilateral triangle. Find the length x of the liquid of density ρ_1 in the horizontal limb of the tube, if the triangular tube is kept in the vertical plane.

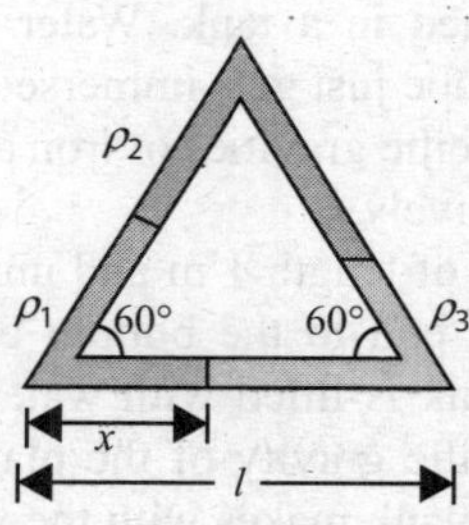

Fig. 3.153

4. A simple accelerometer (an instrument for measuring acceleration) can be made in the form of a tube filled with a liquid and bent as shown. During motion, the level of the liquid in the left arm will be at a height h_1 and in the right arm at a height h_2. Determine the acceleration a of a carriage in which the accelerometer is installed, assuming that the diameter of the tube is much smaller than h_1 and h_2.

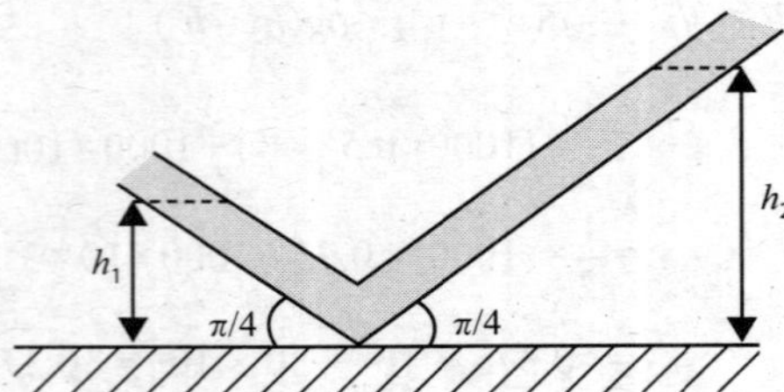

Fig. 3.154

5. A spherical tank of 1.2 m radius is half filled with oil of relative density 0.8. If the tank is given a horizontal acceleration of 10 m/s^2, the maximum pressure on the tank is $\sqrt{2}P$ pascal. Find the value of P.
6. An expansible balloon filled with air is floating on the top surface of a lake, with 2/3 of its volume is submerged in water. How deep should it be sunk in water without changing the temperature of air in it, in order that it is just in equilibrium, neither sinking nor rising further? Height of water barometer = 10 m.
7. An ice cube of side 1 cm is floating at the interface of kerosene and water in a beaker of base area 10 cm^2. The level of kerosene is just covering the top surface of the ice cube.

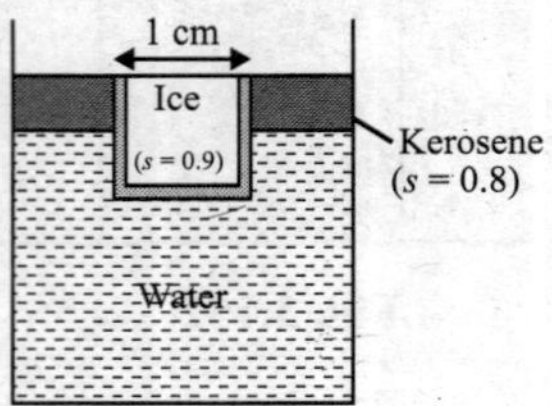

Fig. 3.155

a. Find the depth of submergence in the kerosene and that in the water.
b. Find the change in the total level of the liquid when the whole ice melts into water.
8. A cube of iron of edge 5 cm floats on the surface of mercury, contained in a tank. Water is then poured on top, so that the cube just gets immersed. Find the depth of water layer. (Specific gravities of iron and mercury are 7.8 and 13.6, respectively.)
9. A wooden plank of length 1 m and uniform cross section is hinged at one end to the bottom of a tank as shown in figure. The tank is filled with water up to a height of 0.5 m. The specific gravity of the plank is 0.5. Find the angle θ that the plank makes with the vertical in the equilibrium position (exclude the case $\theta = 0$).

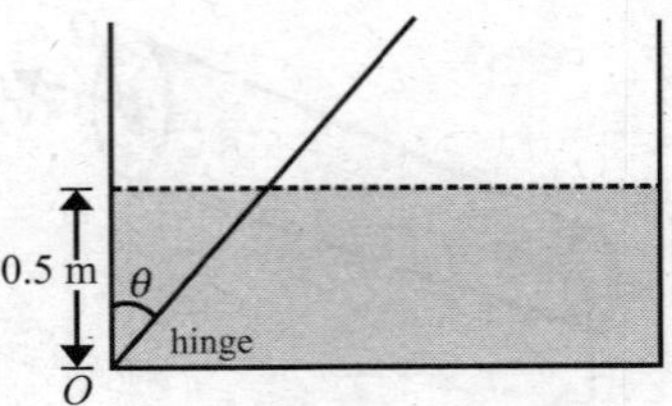

Fig. 3.156

10. Oil enters the bend of a pipe in the horizontal plane with velocity 4 ms^{-1} and pressure 280 × 10^3 Nm^{-2} as shown in the figure. Find the pressure of oil at the point Q in (kNm^{-2}). (Take specific gravity of oil as 0.9 and sin 37° = 0.6.)

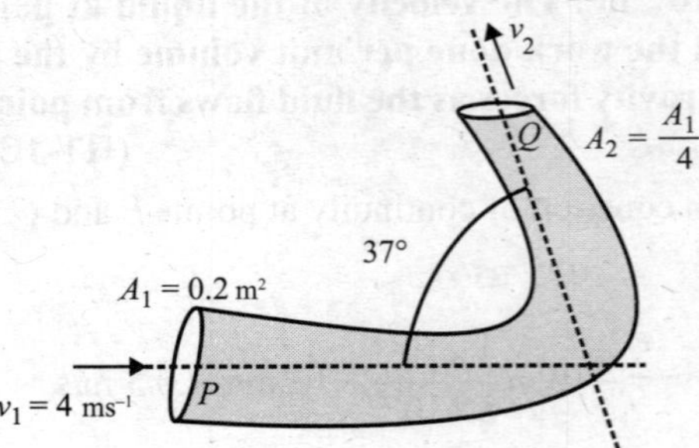

Fig. 3.157

11. Water is filled to a height of 2.5 m in a container lying at rest on a horizontal surface as shown in the figure. The bottom of the container is square shaped of side $\sqrt{3}$ m and its top is open. There is a very small hole in the vertical wall of the container near the bottom. A small cap closes the hole. The hole gets opened when pressure near the bottom becomes more than 4 × 10^4 pascal.
a. What minimum acceleration can be given to the container in the positive x-direction so that the cap will come out of the hole? Assume water does not spill out of container when the container accelerates.
b. Find the velocity of efflux as soon as the container starts moving with the acceleration calculated in part (a) of the question.

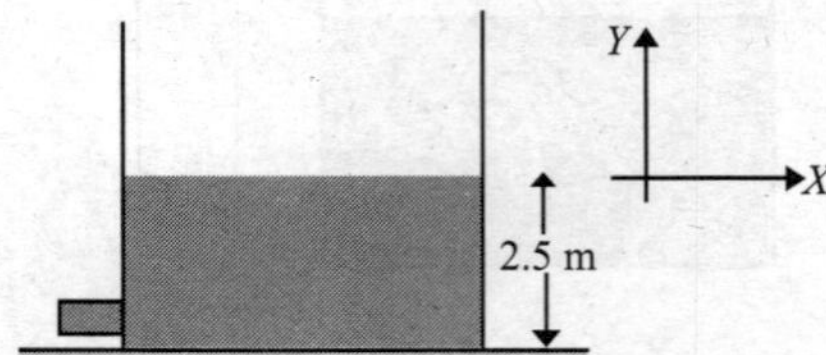

Fig. 3.158

12. A tank with a small orifice contains oil on top of water. It is immersed in a large tank of the same oil. Water flows through the hole. Assuming the level of oil outside the tank above orifice does not change. **(i)** What is the velocity of this flow initially? **(ii)** When the flow stops, what would be the position of the oil–water interface in the tank? **(iii)** Determine the time at which the flow stops. Density of oil = 800 kg/m^3.

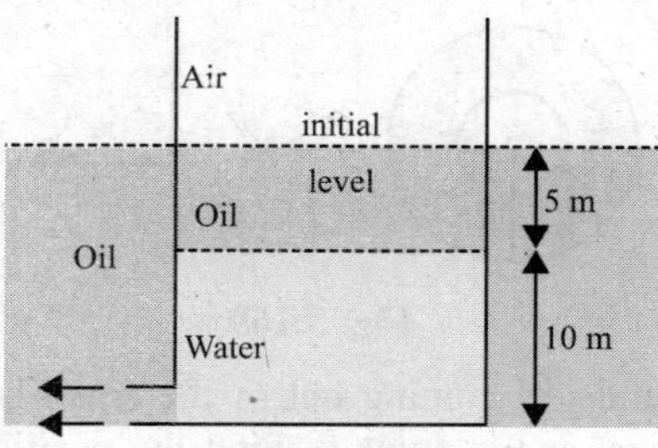

Fig. 3.159

13. A ball of weight W is supported on a vertical jet of water. If the stream of water flowing from the nozzle has a diameter D and velocity u, determine the value of H. Assume that no loss of energy takes place.

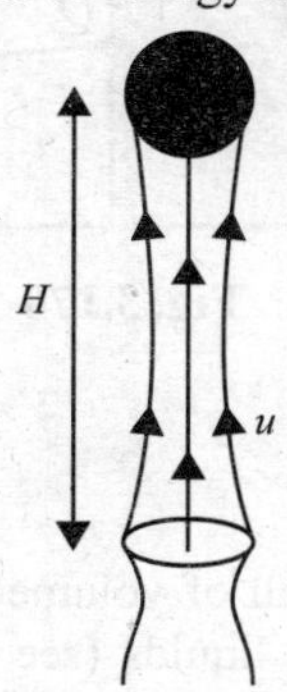

Fig. 3.160

14. A rocket ejects the fuel (hot gases) of density ρ from the chamber which is maintained at a pressure P_1, say. The atmospheric pressure is P_0. If the base area of the cylindrical chamber is A_1 and the area of the hole through which hot gases escape from the rocket is A_2, find (a) velocity of efflux of the fuel, (b) thrust exerted on the rocket given by the escaping fuel.

15. A vertical U-tube has two liquids 1 and 2. The heights of liquids columns in both the limbs are h and $2h$, as shown in the figure. If the density of the liquid 1 is 2ρ.

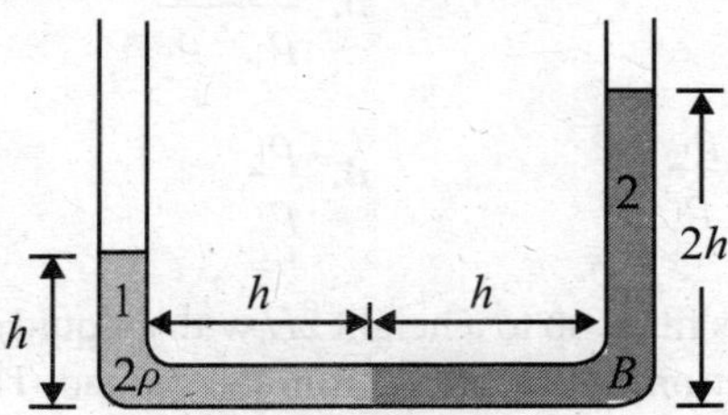

Fig. 3.161

(a) Find the density of liquid 2.

(b) If we accelerate the tube towards right till the heights of liquid columns will be the same, find the acceleration of the tube.

16. A cup ABC is kept in a liquid of density ρ as shown in the figure. Find the vertical thrust on the cup by the liquid present in region 2, if the maximum volume of the liquid that can be kept in the cup is equal to V.

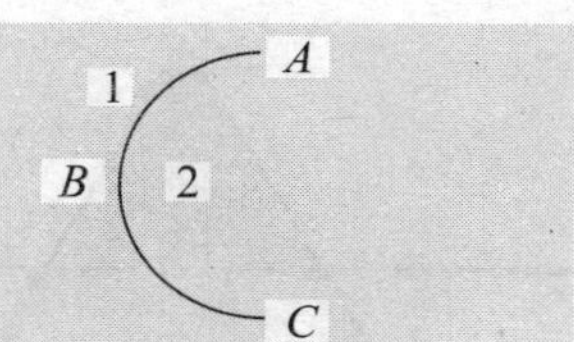

Fig. 3.162

17. A vertical U-tube is spinned about the vertical axis passing through the left hand limb, with an angular speed ω such that the liquid columns in the limbs have heights h and $3h$, as shown in the figure. Find the value of ω.

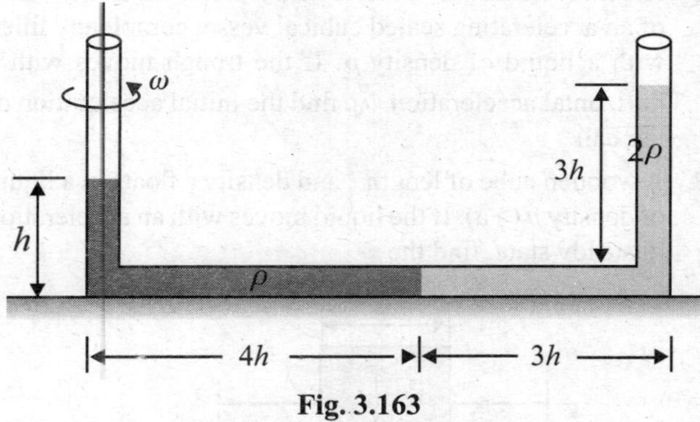

Fig. 3.163

18. A vertical U-tube has a liquid up to a height h. If the tube is slowly rotated to an angular speed $\omega = \sqrt{g/h}$, find the (i) heights of the liquid column, (ii) pressures at the points 1, 2 and 3 in the limbs in steady state.

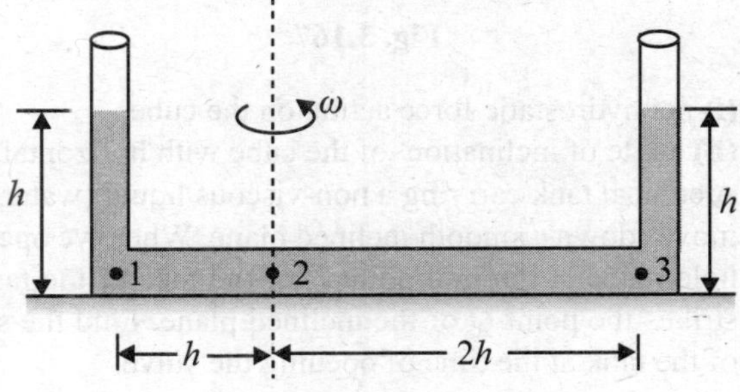

Fig. 3.164

19. A sphere is just immersed in a liquid. Find the ratio of hydrostatic force acting on top and bottom half of the sphere.

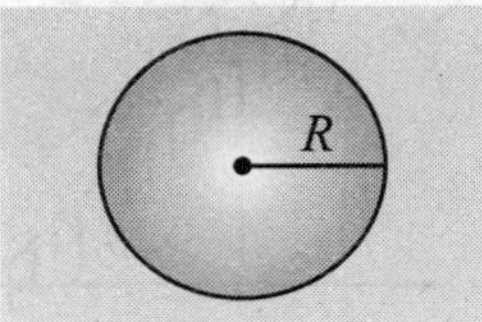

Fig. 3.165

20. The light cone is in equilibrium under the action of hydrostatic forces of two liquids of densities ρ_1 and ρ_2. Find ρ_1/ρ_2.

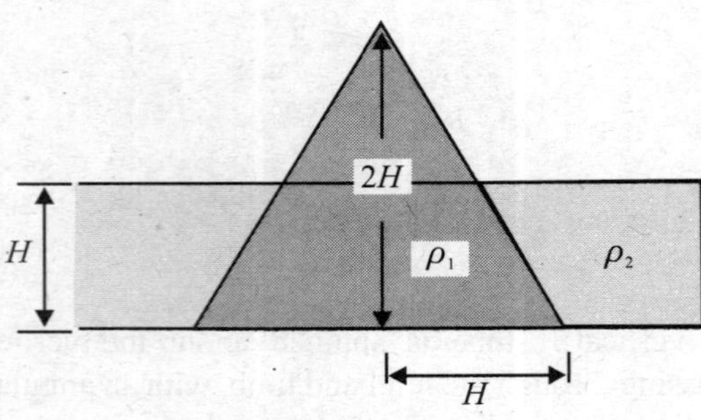

Fig. 3.166

21. A ball of density σ is released from rest from the centre of an accelerating sealed cubical vessel completely filled with a liquid of density ρ. If the trough moves with a horizontal acceleration $\vec{a}_0$, find the initial acceleration of the ball.

22. A wooden cube of length ℓ and density a floats in a liquid of density p (> a). If the liquid moves with an acceleration in steady state, find the

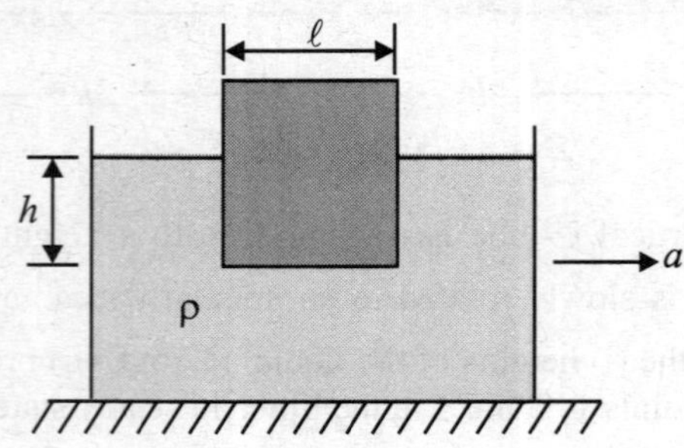

Fig. 3.167

(i) net hydrostatic force acting on the cube
(ii) angle of inclination of the cube with horizontal.

23. A cubical tank carrying a non-viscous liquid (water, say) moves down a smooth inclined plane. When we open the hole made at the mid-point P of the face of the tank, it strikes the point Q of the inclined plane. Find the speed of the tank at the time of opening the valve.

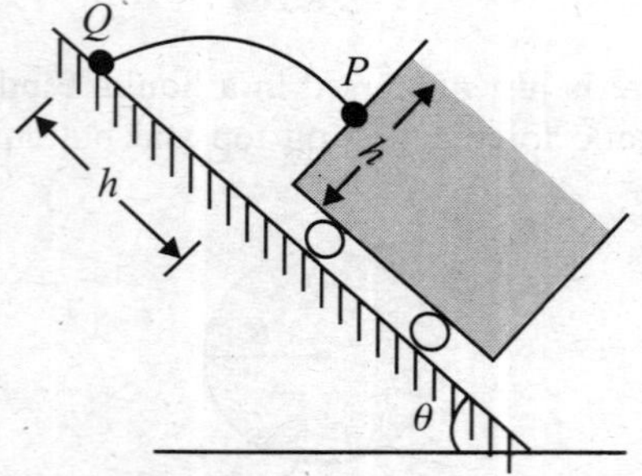

Fig. 3.168

24. A basket ball of radius R is spun with an angular speed $\vec{\omega} = \omega\hat{k}$ and its CM moves with a speed v. Find the aerodynamic lift experienced by the ball. Assume ρ = density of air.

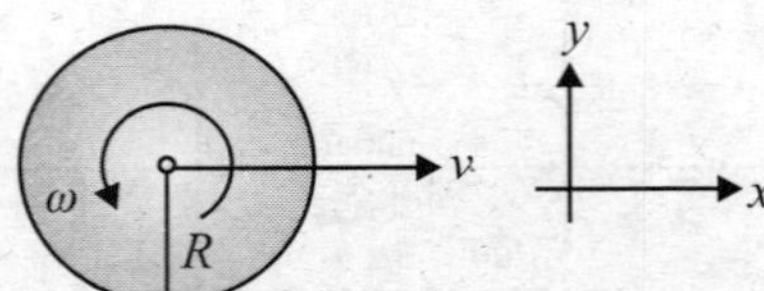

Fig. 3.169

25. Two liquid jets coming out of the small holes at P and Q intersect at the point R. Find the position of R if we maintain the liquid level constant.

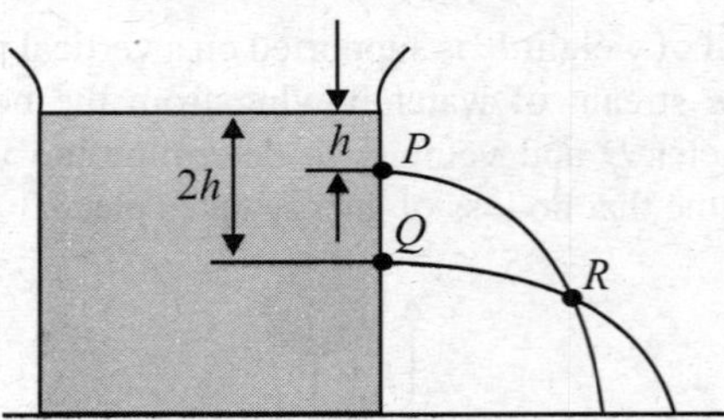

Fig. 3.170

Objective Type

Solutions on page 3.75

1. A solid uniform ball of volume V floats on the interface of two immiscible liquids (see the figure). The specific gravity of the upper liquid is ρ_1 and that of lower one is ρ_2 and the specific gravity of ball is ρ $(\rho_1 < \rho < \rho_2)$. The fraction of the volume of the ball in the upper liquid is

Fig. 3.171

a. $\dfrac{\rho_2}{\rho_1}$ b. $\dfrac{\rho_2 - \rho}{\rho_2 - \rho_1}$

c. $\dfrac{\rho - \rho_1}{\rho_2 - \rho_1}$ d. $\dfrac{\rho_1}{\rho_2}$

2. A tank is filled up to a height $2H$ with a liquid and is placed on a platform of height H from the ground. The distance x from the ground where a small hole is punched to get the maximum range R is

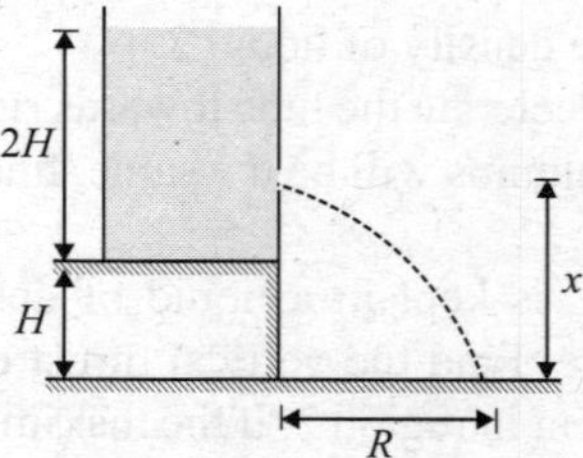

Fig. 3.172

a. H **b.** $1.25H$
c. $1.5H$ **d.** $2H$

3. The minimum horizontal acceleration of the container so that the pressure at point A of the container becomes atmospheric is (the tank is of sufficient height)

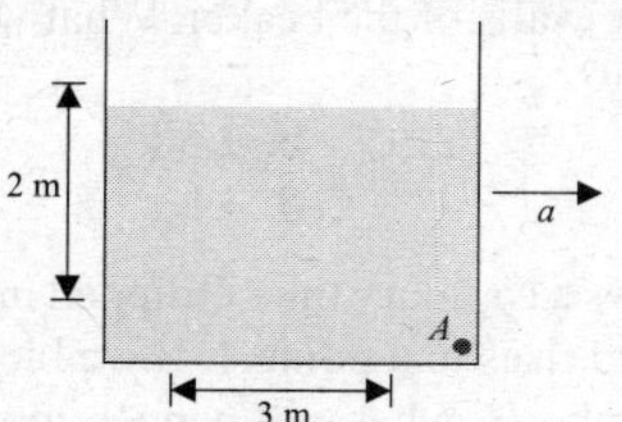

Fig. 3.173

a. $\frac{3}{2}g$ **b.** $\frac{4}{3}g$ **c.** $\frac{4}{2}g$ **d.** $\frac{3}{4}g$

4. A solid cone of height H and base radius $H/2$ floats in a liquid of density ρ. It is hanging from the ceiling with the help of a string. The force by the fluid on the curved surface of the cone is (P_0 = atmospheric pressure)

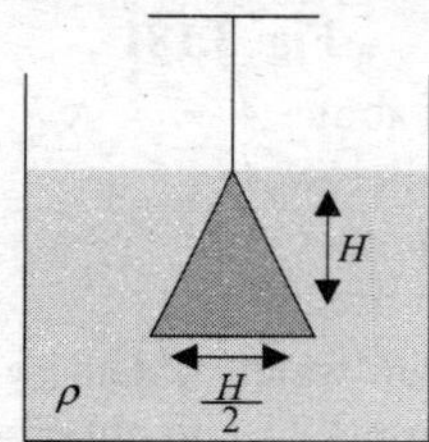

Fig. 3.174

a. $\pi H^2\left(\frac{P_0}{4}+\frac{\rho gH}{3}\right)$ **b.** $\pi H^2\left(\frac{P_0}{4}+\frac{\rho gH}{6}\right)$

c. $\frac{\pi H^2}{4}\left(\frac{P_0}{4}+\rho gH\right)$ **d.** $\frac{\pi H^2}{4}(P_0+\rho gH)$

5. Three identical vessels A, B and C contain same quantity of liquid. In each vessel balls of different densities but same masses are placed. In vessel A, the ball is partly submerged; in vessel B, the ball is completely submerged but floating and in vessel C, the ball has sunk to the base. If F_A, F_B and F_C are the total forces acting on the base of vessels A, B and C, respectively, then

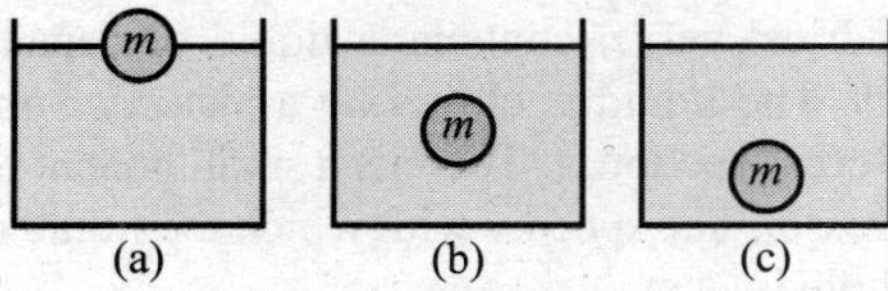

Fig. 3.175

a. $F_A = F_B = F_C$ **b.** $F_A < F_B < F_C$
c. $F_A = F_B < F_C$ **d.** $F_A < F_B < F_C$

6. In the figure, the cross-sectional area of the smaller tube is a and that of the larger tube is $2a$. A block of mass m is kept in the smaller tube having the same base area a, as that of the tube. The difference between water levels of the two tubes is

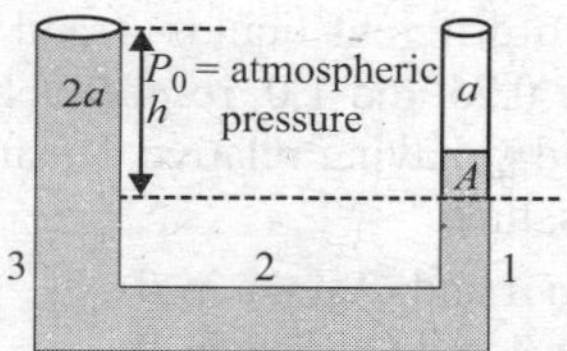

Fig. 3.176

a. $\frac{P_0}{\rho g}+\frac{m}{a\rho}$ **b.** $\frac{P_0}{\rho g}+\frac{m}{2a\rho}$

c. $\frac{m}{a\rho}$ **d.** $\frac{m}{2\rho a}$

7. An ornament weighing 36 g in air weighs only 34 g in water. Assuming that some copper is mixed with gold to prepare the ornament, find the amount of copper in it. Specific gravity of gold is 19.3 and that of copper is 8.9.

a. 2.2 g **b.** 4.4 g
c. 1.1 g **d.** 3.6 g

8. Water flows through a horizontal tube as shown in the figure. If the difference of heights of water column in the vertical tubes is $h = 0.02$ m, and the areas of cross section at A and B are 4×10^{-4} m^2 and 2×10^{-4} m^2, respectively, then the rate of flow of water across any section is

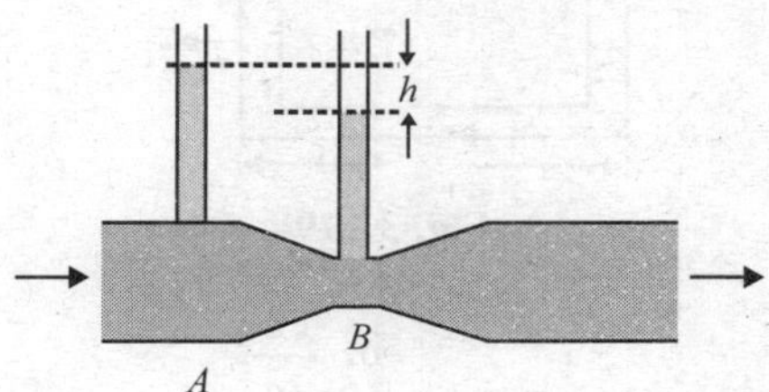

Fig. 3.177

a. 130×10^{-6} m^3/s **b.** 146×10^{-6} m^3/s
c. 160×10^{-6} m^3/s **d.** 170×10^{-6} m^3/s

9. In a U-tube in which the cross-sectional area of the limb on the left is one-quarter, the limb on the right contains mercury (density 13.6 g/cm^3), and the level of mercury in the narrow limb is at a distance of 36 cm from the upper end of the tube. What will be the rise in the level of mercury in the right limb if the left limb is filled to the top with water?

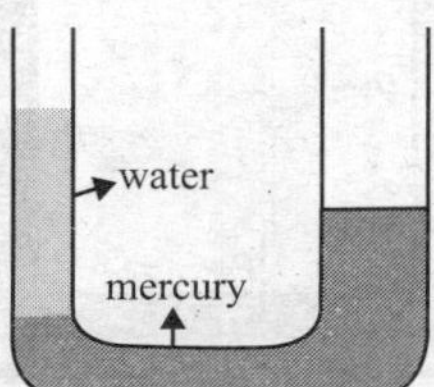

Fig. 3.178

a. 1.2 cm **b.** 2.35 cm
c. 0.56 cm **d.** 0.8 cm

10. We have two different liquids A and B whose relative densities are 0.75 and 1.0, respectively. If we dip solid objects P and Q having relative densities 0.6 and 0.9 in these liquids, then

a. P floats in A and Q sinks in B
b. P sinks in A and Q floats in B
c. P floats in B and Q sinks in A
d. P sinks in B and Q floats in A

11. A cubical block of wood of side a and density ρ floats in water of density 2ρ. The lower surface of the cube just touches the free end of a massless spring of force constant K fixed at the bottom of the vessel. The weight W put over the block so that it is completely immersed in water without wetting the weight is

a. $a(a^2\rho g + K)$ **b.** $a(a\rho g + 2K)$
c. $a\left(\frac{a\rho g}{2} + 2K\right)$ **d.** $a\left(a^2\rho g + \frac{K}{2}\right)$

12. When at rest, a liquid stands at the same level in the tubes as shown in the figure. But as indicated, a height difference h occurs when the system is given an acceleration a towards the right. Then h is equal to

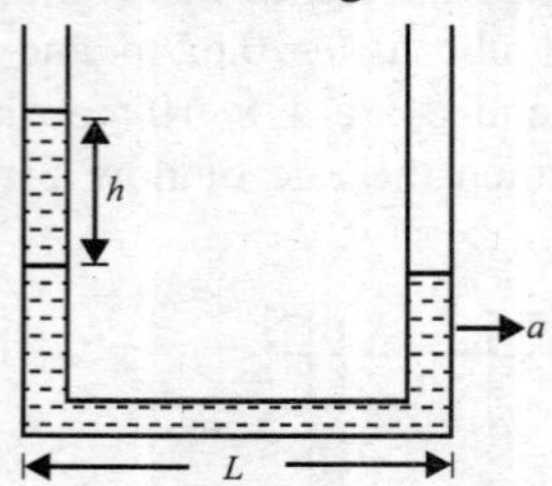

Fig. 3.179

a. $\frac{aL}{2g}$ **b.** $\frac{gL}{2a}$
c. $\frac{gL}{a}$ **d.** $\frac{aL}{g}$

13. Three liquids having densities ρ_1, ρ_2 and ρ_3 are filled in a U-tube. Length of each liquid column is equal to I, $\rho_1 > \rho_2 > \rho_3$ and liquids remain at rest (relative to the tube) in the position shown in the figure. It is possible that

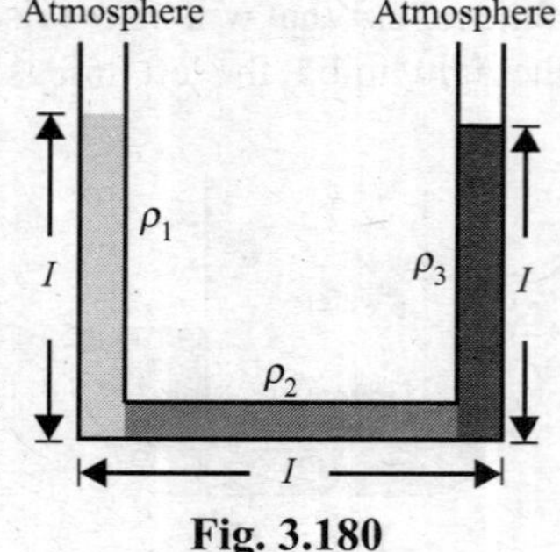

Fig. 3.180

a. U-tube is accelerating leftwards
b. U-tube is accelerating upwards with acceleration g
c. U-tube is moving with a constant velocity
d. none of the above

14. A beaker containing water is placed on the platform of a spring balance. The balance reads 1.5 kg. A stone of mass 0.5 kg and density 10^4 kg/m^3 is immersed in water without touching the walls of the beaker. What will be the balance reading now?

a. 2 kg **b.** 2.5 kg
c. 1 kg **d.** 3 kg

15. Figure shows a capillary tube C dipped in a liquid that wets it. The liquid rises to a point A. If we blow air through the horizontal tube H, what will happen to the liquid column in the capillary tube?

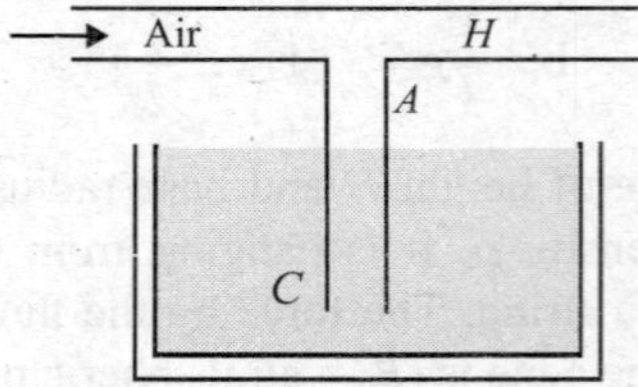

Fig. 3.181

a. Level will rise above A
b. Level will fall below A
c. Level will remain at A
d. It is difficult to predict

16. A tank is filled with water of density 10^3 kg/m^3 and oil of density 9×10^3 kg/m^3. The height of water layer is 1 m and that of the oil layer is 4 m. The velocity of efflux from an opening in the bottom of the tank is

a. $\sqrt{85}$ m/s **b.** $\sqrt{88}$ m/s
c. $\sqrt{92}$ m/s **d.** $\sqrt{98}$ m/s

17. A hole is made at the bottom of a tank filled with water (density = 10^3 kg/m^3). If the total pressure at the bottom of the tank is 3 atm (1 atm = 10^5 N/m^2), then the velocity of efflux is

a. $\sqrt{400}$ m/s **b.** $\sqrt{200}$ m/s
c. $\sqrt{600}$ m/s **d.** $\sqrt{500}$ m/s

18. A cylindrical vessel contains a liquid of density ρ up to height h. The liquid is closed by a piston of mass m and area of cross section A. There is a small hole at the bottom of the vessel. The speed v with which the liquid comes out of the hole is

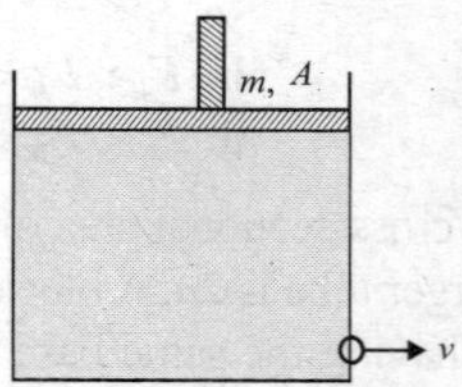

Fig. 3.182

a. $\sqrt{2gh}$ **b.** $\sqrt{2\left(gh+\frac{mg}{\rho A}\right)}$

c. $\sqrt{2\left(gh+\frac{mg}{A}\right)}$ **d.** $\sqrt{2gh+\frac{mg}{A}}$

19. The opening near the bottom of the vessel shown in the figure has an area A. A disc is held against the opening to keep the liquid from running out. Let F_1 be the net force on the disc applied by liquid and air in this case. Now the disc is moved away from the opening a short distance. The liquid comes out and strikes the disc inelastically. Let F_2 be the force exerted by the liquid in this condition. Then F_1/F_2 is

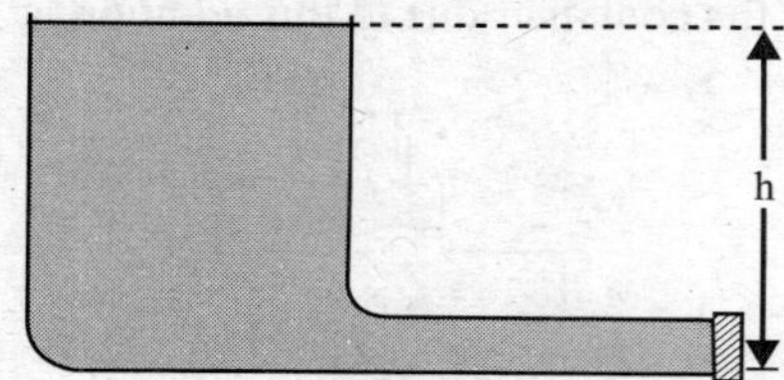

Fig. 3.183

a. $\frac{1}{2}$ **b.** 1

c. $\frac{2}{1}$ **d.** $\frac{1}{4}$

20. There are two identical small holes of area of cross section a on the opposite sides of a tank containing a liquid of density ρ. The difference in height between the holes is h. The tank is resting on a smooth horizontal surface. The horizontal force which will have to be applied on the tank to keep it in equilibrium is

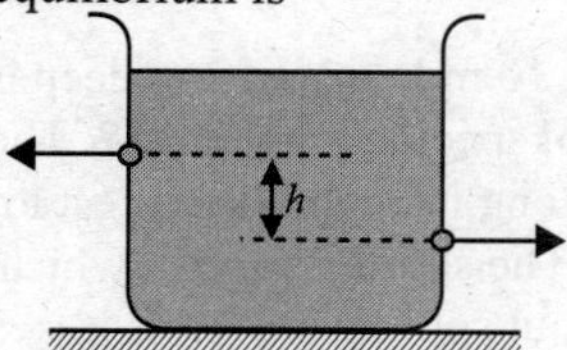

Fig. 3.184

a. $gh\rho a$ **b.** $\frac{2gh}{\rho a}$

c. $2\rho agh$ **d.** $\frac{\rho gh}{a}$

21. Equal volume of two immiscible liquids of densities ρ and 2ρ are filled in a vessel as shown in the figure. Two small holes are punched at depths $h/2$ and $3h/2$ from the surface of lighter liquid. If v_1 and v_2 are the velocities of efflux at these two holes, then v_1/v_2 is

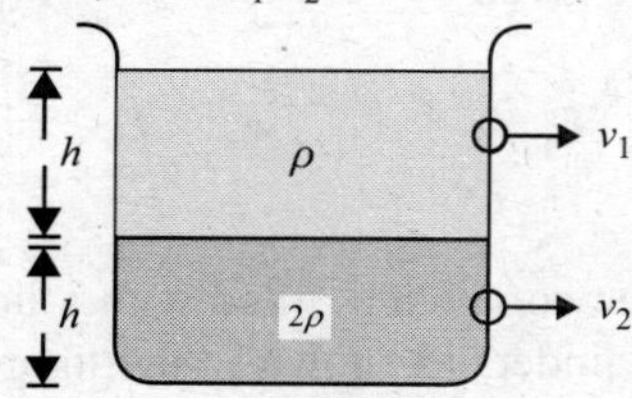

Fig. 3.185

a. $\frac{1}{2\sqrt{2}}$ **b.** $\frac{1}{2}$

c. $\frac{1}{4}$ **d.** $\frac{1}{\sqrt{2}}$

22. Figure 3.186 shows a liquid flowing through a tube at the rate of 0.1 m³/s. The tube is branched into two semicircular tubes of cross-sectional area $A/3$ and $2A/3$. The velocity of liquid at Q is (the cross section of the main tube is $A = 10^{-2}$ m² and $v_P = 20$ m/s)

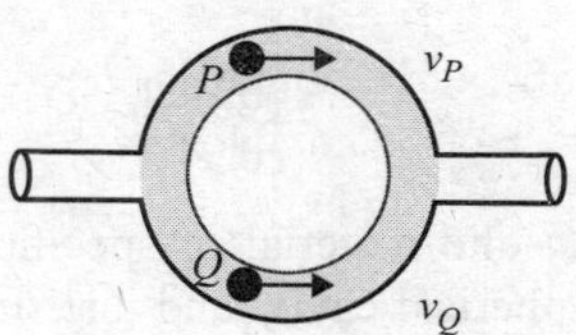

Fig. 3.186

a. 5 m/s **b.** 30 m/s

c. 35 m/s **d.** none of these

23. In a cylindrical water tank there are two small holes Q and P on the wall at a depth of h_1 from the upper level of water and at a height of h_2 from the lower end of the tank, respectively, as shown in the figure. Water coming out from both the holes strike the ground at the same point. The ratio of h_1 and h_2 is

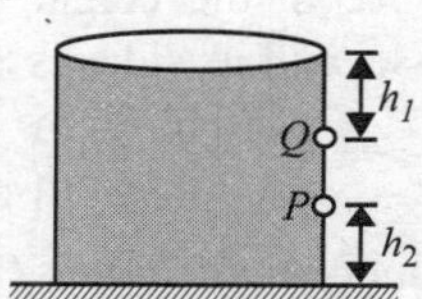

Fig. 3.187

a. 1 **b.** 2

c. >1 **d.** <1

24. The tension in a string holding a solid block below the surface of a liquid (where $\rho_{\text{liquid}} > \rho_{\text{block}}$) as in shown in the figure is T when the system is at rest.

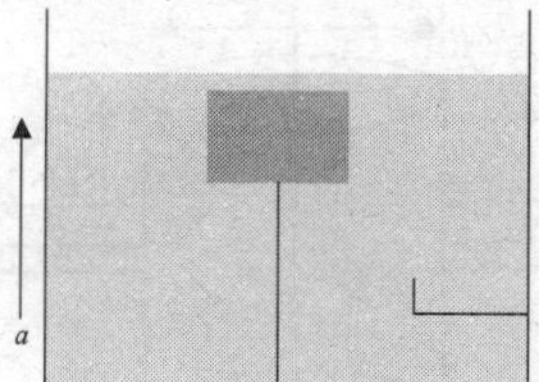

Fig. 3.188

Then what will be the tension in the string if the system has upward acceleration a?

a. $T\left(1-\frac{a}{g}\right)$ **b.** $T\left(1+\frac{a}{g}\right)$

c. $T\left(\frac{a}{g}-1\right)$ **d.** $\frac{a}{g}T$

25. An ideal fluid flows in the pipe as shown in the figure. The pressure in the fluid at the bottom P_2 is the same as it is at the top P_1. If the velocity of the top $v_1 = 2$ m/s. Then the ratio of areas A_1/A_2 is

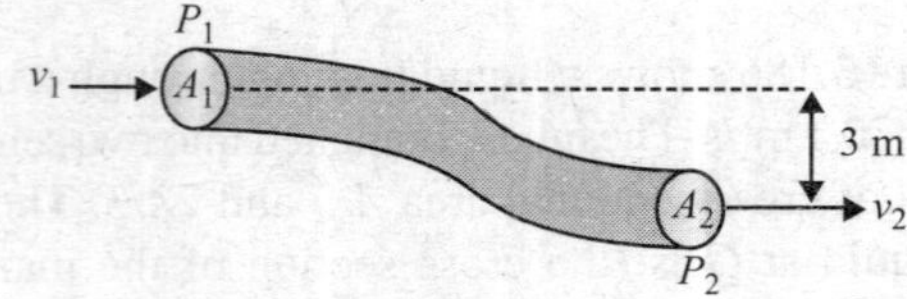

Fig. 3.189

a. 2:1 **b.** 4:1
c. 8:1 **d.** 4:3

26. A sphere of solid material of specific gravity 8 has a concentric spherical cavity and just sinks in water. The ratio of radius of cavity to that of outer radius of the sphere must be

a. $\frac{7^{1/3}}{2}$ **b.** $\frac{5^{1/3}}{2}$
c. $\frac{9^{1/3}}{2}$ **d.** $\frac{3^{1/3}}{2}$

27. A liquid is kept in a cylindrical vessel which is rotated along its axis. The liquid rises at the side. If the radius of the vessel is 5 cm and the speed of rotation is 4 rev/s, then the difference in the height of the liquid at the centre of the vessel and its sides is

a. 8 cm **b.** 2 cm
c. 40 cm **d.** 4 cm

28. A cylindrical container of radius 'R' and height 'h' is completely filled with a liquid. Two horizontal L-shaped pipes of small cross-sectional area 'a' are connected to the cylinder as shown in the figure. Now the two pipes are opened and fluid starts coming out of the pipes horizontally in opposite directions. Then the torque due to ejected liquid on the system is

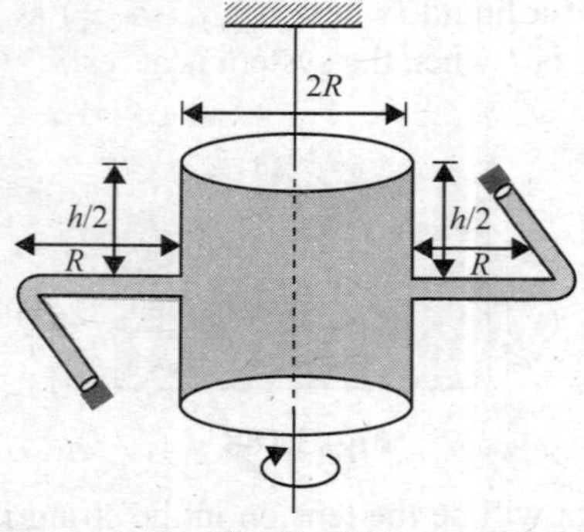

Fig. 3.190

a. $4aghρR$ **b.** $8aghρR$
c. $2aghρR$ **d.** none of these

29. A block of silver of mass 4 kg hanging from a string is immersed in a liquid of relative density 0.72. If relative density of silver is 10, then tension in the string will be

a. 37.12 N **b.** 42 N
c. 73 N **d.** 21 N

30. A vessel contains oil (density = 0.8 g/cm³) over mercury (density = 13.6 g/cm³). A uniform sphere floats with half its volume immersed in mercury and the other half in oil. The density of the material of sphere in g/cm³ is

a. 3.3 **b.** 6.4
c. 7.2 **d.** 12.8

31. Figure 3.191 shows water filled in a symmetrical container. Four pistons of equal area A are used at the four openings to keep the water in equilibrium. Now an additional force F is applied at each piston. The increase in the pressure at the centre of the container due to this addition is

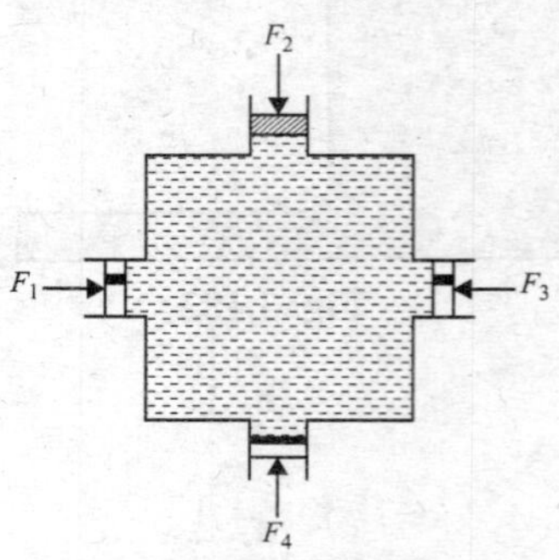

Fig. 3.191

a. $\frac{F}{A}$ **b.** $\frac{2F}{A}$
c. $\frac{4F}{A}$ **d.** 0

32. An open tank 10 m long and 2 m deep is filled up to 1.5 m height of oil of specific gravity 0.82. The tank is uniformly accelerated along its length from rest to a speed of 20 m/s horizontally. The shortest time in which the speed may be attained without spilling any oil is

a. 20 s **b.** 18 s
c. 10 s **d.** 5 s

33. A U-tube of base length 'l' filled with the same volume of two liquids of densities $ρ$ and $2ρ$ is moving with an acceleration 'a' on the horizontal plane. If the height difference between the two surfaces (open to atmosphere) becomes zero, then the height h is given by

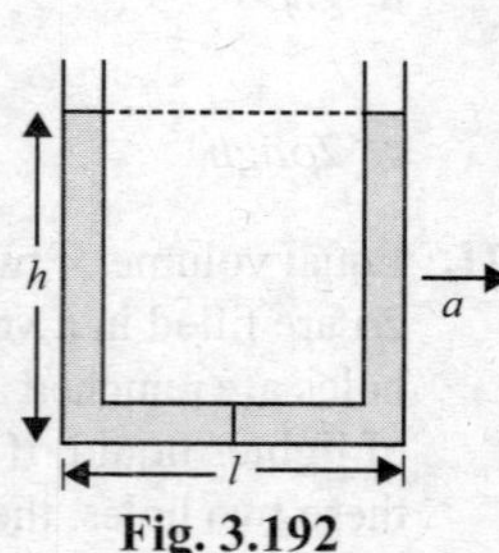

Fig. 3.192

a. $\frac{a}{2g}l$ **b.** $\frac{3a}{2g}l$ **c.** $\frac{a}{g}l$ **d.** $\frac{2a}{3g}l$

34. A narrow tube completely filled with a liquid is lying on a series of cylinders as shown in the figure. Assuming no sliding between any surfaces, the value of acceleration of

the cylinders for which liquid will not come out of the tube from anywhere is given by

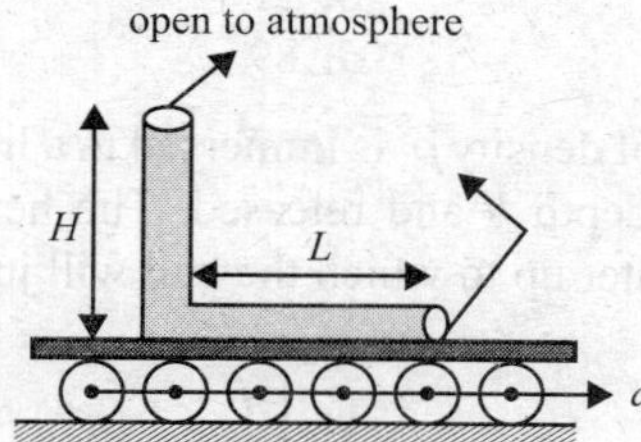

Fig. 3.193

a. $\frac{gH}{2L}$ b. $\frac{gH}{L}$ c. $\frac{2gH}{L}$ d. $\frac{gH}{\sqrt{2}L}$

35. The velocity of the liquid coming out of a small hole of a vessel containing two different liquids of densities 2ρ and ρ as shown in the figure is

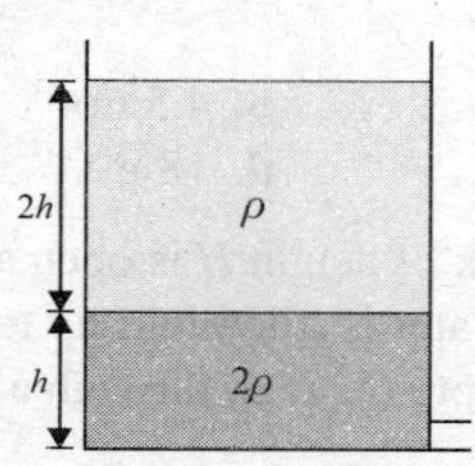

Fig. 3.194

a. $\sqrt{6gh}$ b. $2\sqrt{gh}$ c. $2\sqrt{2gh}$ d. $\sqrt{gh}$

36. A non-uniform cylinder of mass m, length l and radius r is having its centre of mass at a distance $l/4$ from the centre and lying on the axis of the cylinder. The cylinder is kept in a liquid of uniform density ρ. The moment of inertia of the rod about the centre of mass is I. The angular acceleration of point A relative to point B just after the rod is released from the position as shown in the figure is

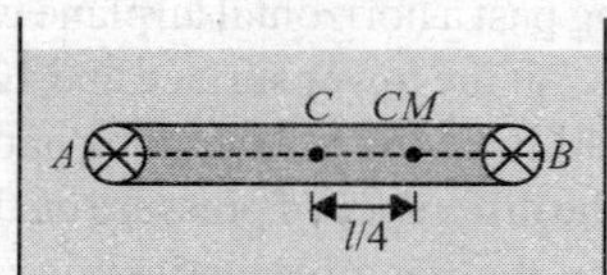

Fig. 3.195

a. $\frac{\pi\rho g l^2 r^2}{I}$ b. $\frac{\pi\rho g l^2 r^2}{4I}$

c. $\frac{\pi\rho g l^2 r^2}{2I}$ d. $\frac{3\pi\rho g l^2 r^2}{4I}$

37. A square box of water has a small hole located in one of he bottom corners. When the box is full and sitting on a level surface, complete opening of the hole results in a flow of water with a speed v_0, as shown in Fig. 3.196(a). When the box is still half empty, it is tilted by 45° so that the hole is at the lowest point. Now the water will flow out with a speed of

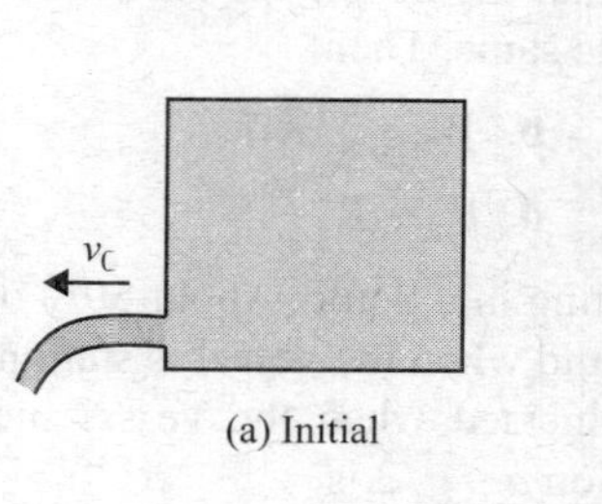

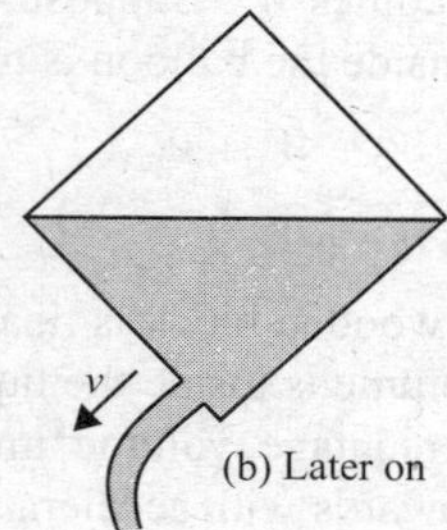

Fig. 3.196

a. v_0 b. $v_0/2$ c. $\frac{v_0}{\sqrt{2}}$ d. $\frac{v_0}{\sqrt[4]{2}}$

38. A cylindrical block is floating (partially submerged) in a vessel containing water. Initially, the platform on which the vessel is mounted is at rest. Now the platform along with the vessel is allowed to fall freely under gravity. As a result, the buoyancy force

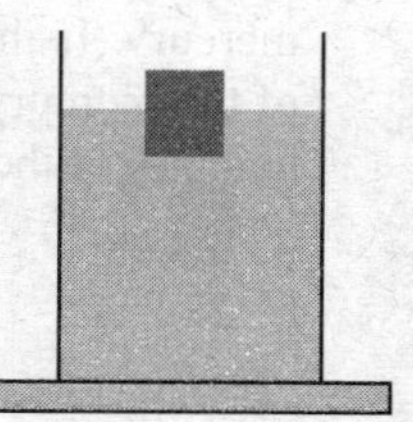

Fig. 3.197

a. becomes zero b. decreases

c. increases d. information is insufficient

39. For the L-shaped vessel shown in the figure, determine the value of acceleration a so that pressure at point A becomes equal to $p_0/2$? (p_0 is the atmospheric pressure)

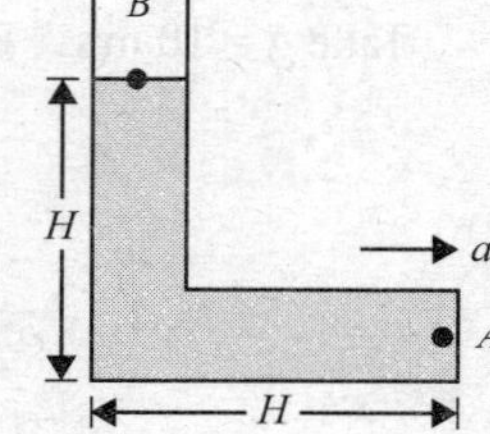

Fig. 3.198

a. g b. $\frac{g}{2}+\frac{p_0}{2\rho H}$

c. $\frac{p_0}{2\rho H}+g$ d. $\frac{3p_0}{2\rho H}+g$

40. A block of wood is floating in water in a closed vessel as shown in the figure. The vessel is connected to an air pump. When more air is pushed into the vessel, the block of wood floats with (neglect compressibility of water)

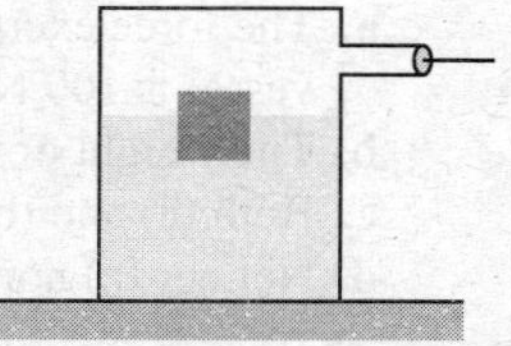

Fig. 3.199

a. larger part in the water b. smaller part in the water

c. some part in the water d. at some instant it will sink

41. A U-tube is partially filled with water. Oil which does not mix with water is next poured into one side until water rises by 25 cm on the other side. If the density of oil is 0.8, the oil level will stand higher than the water level by

a. 6.25 cm b. 12.50 cm

c. 31.75 cm d. 62.50 cm

42. An empty balloon weighs W_1. If air equal in weight to W is pumped into the balloon, the weight of the balloon becomes W_2. Suppose that the density of air inside and outside the balloon is the same. Then

a. $W_2 = W_1 + W$ b. $W_2 = \sqrt{W_1 W}$
c. $W_2 = W_1$ d. $W_2 = W_1 - W$

43. A wooden block is floating in a liquid. About 50% of its volume is inside the liquid when the vessel is stationary. Percentage volume immersed when the vessel moves upwards with acceleration $a = g/2$ is

a. 75% b. 25%
c. 50% d. 33.33%

44. The liquids shown in the U-tube of uniform cross section are water and mercury. If the difference of the heights of the mercury columns is 2 cm, then the height h of the water column is

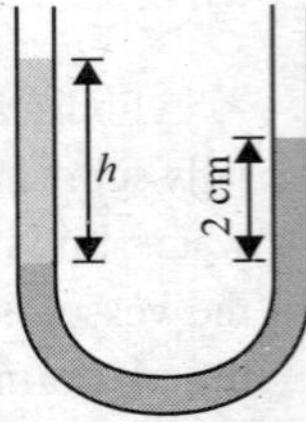

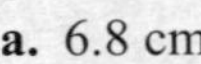
a. 6.8 cm b. 13.6 cm

c. 27.2 cm d. 54 cm

Fig. 3.200

45. A tube 1 cm² in cross section is attached to the top of a vessel 1 cm high and of cross section 100 cm². Water is poured into the system filling it to a depth of 100 cm above the bottom of the vessel as shown in the figure. Take $g = 10\ \text{m s}^{-2}$. Find the correct statement.

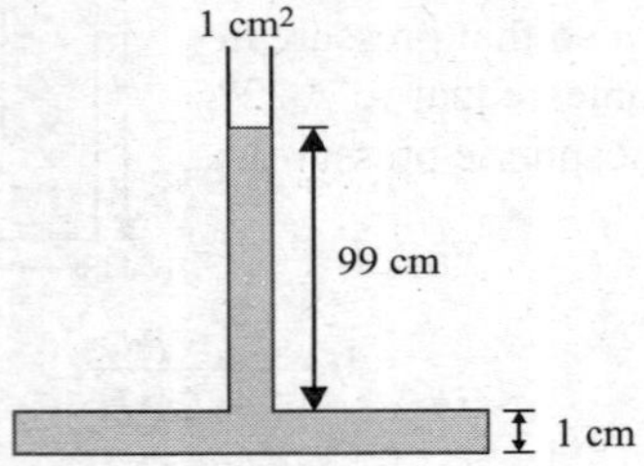

Fig. 3.201

a. The force exerted by water against the bottom of the vessel is 100 N.
b. The weight of water in the system is 1.99 N.
c. Both (a) and (b) are correct.
d. Neither (a) nor (b) is correct.

46. A wooden rod of a uniform cross section and of length 120 cm is hinged at the bottom of the tank which is filled with water to a height of 40 cm. In the equilibrium position, the rod makes an angle of 60° with the vertical. The centre of buoyancy is located on the rod at a distance (from the hinge) of

a. 20 cm b. 40 cm
c. 60 cm d. 75 cm

47. An iceberg is floating partially immersed in sea water. The density of sea water is 1.03 g cm⁻³ and that of ice is 0.92 g cm⁻³. The approximate percentage of total volume of iceberg above the level of sea water is

a. 8 b. 11
c. 34 d. 89

48. A small ball of density ρ is immersed in a liquid of density σ $(>\rho)$ to a depth h and released. The height above the surface of water up to which the ball will jump is

a. $\left(\frac{\sigma}{\rho}-1\right)h$ b. $\left(\frac{\rho}{\sigma}-1\right)h$

c. $\left(\frac{\rho}{\sigma}+1\right)h$ d. $\left(\frac{\sigma}{\rho}+1\right)h$

49. A lawn sprinkler has 20 holes, each of cross-sectional area $2.0 \times 10^{-2}\ \text{cm}^2$, and is connected to a hose pipe of cross sectional area 2.4 cm². If the speed of water in the hose pipe is $1.5\ \text{m s}^{-1}$, the speed of water as it emerges from the holes is

a. $2.25\ \text{m s}^{-1}$ b. $4.5\ \text{m s}^{-1}$
c. $9\ \text{m s}^{-1}$ d. $18\ \text{m s}^{-1}$

50. A cylindrical tank of height H is open at the top end and it has a radius r. Water is filled in it up to a height of h. The time taken to empty the tank through a hole of radius r' in its bottom is

a. $\sqrt{\frac{2h}{g}}\frac{r^2}{r'^2}$ b. $\sqrt{\frac{2H}{g}}\frac{r^2}{r'^2}$

c. $\sqrt{hH}$ d. none of these

51. A pressure meter attached to a closed water tap reads 1.5×10^5 Pa. When the tap is opened, the velocity of flow of water is $10\ \text{m s}^{-1}$ and the reading of the pressure meter is

a. 1.5×10^5 Pa b. 3×10^5 Pa
c. 0.5×10^5 Pa d. 10^5 Pa

52. Air is streaming past a horizontal airplane wing such that its speed is $90\ \text{m s}^{-1}$ at the lower surface and $120\ \text{m s}^{-1}$ over the upper surface. If the wing is 10 m long and has an average width of 2 m, the difference of pressure on the two sides and the gross lift on the wing respectively, are (density of air = 1.3 kg m⁻³)

a. 5 Pa, 900 N b. 95 Pa, 900 N
c. 4095 Pa, 900 N d. 4095 Pa, 81900 N

53. A hemispherical bowl just floats without sinking in a liquid of density 1.2×10^3 kg/m³. If the outer diameter and the density of the bowl are 1 m and 2×10^4 kg/m³, respectively, then the inner diameter of the bowl will be

a. 0.94 m b. 0.97 m
c. 0.98 m d. 0.99 m

54. A container filled with liquid up to height h is placed on a smooth horizontal surface. The container is having a

small hole at the bottom. As the liquid comes out from the hole, the container moves in a backward direction with acceleration a and finally, when all the liquid is drained out, it acquires a velocity v. Neglect mass of the container. In this case

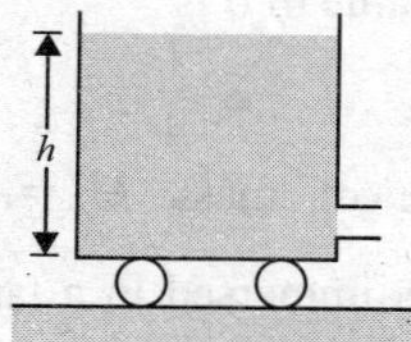

Fig. 3.202

a. both a and v depend on h
b. only a depends on h
c. only v depends on h
d. neither a nor v depends on h

55. A square of size 1 m × 1 m is hinged at its midpoint. A fluid of density ρ fills the space to the left of the gate. The force F required to hold the gas stationary is

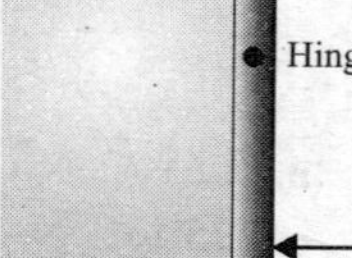

Fig. 3.203

a. 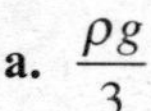 $\frac{\rho g}{3}$ **b.** 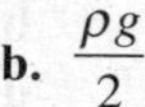 $\frac{\rho g}{2}$

c. $\frac{\rho g}{6}$ **d.** $\frac{\rho g}{8}$

56. Equal masses of three liquids are kept in three identical cylindrical vessels A, B and C. Their densities are ρ_A, ρ_B and ρ_C with $\rho_A < \rho_B < \rho_C$. The force on the base will be

a. maximum in vessel A
b. maximum in vessel B
c. maximum in vessel C
d. same in all the three vessels

57. A metal piece of mass 160 g lies in equilibrium inside a glass of water. The piece touches the glass at small number of points. If the density of the metal is 8000 kg/m³, then the normal force exerted by the bottom of the glass on the metal piece is

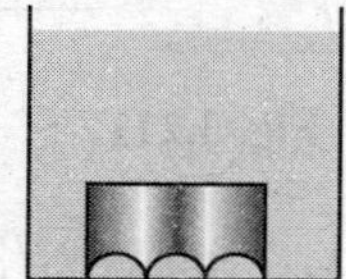

Fig. 3.204

a. 2 N **b.** 8 N **c.** 0.16 N **d.** 1.4 N

58. A body B is capable of remaining stationary inside a liquid at the position shown in Fig. 3.205 (a). If the whole system is gently placed on smooth inclined plane (Fig. 3.205(b)) and is allowed to slide down, then ($0 < \theta < 90°$)

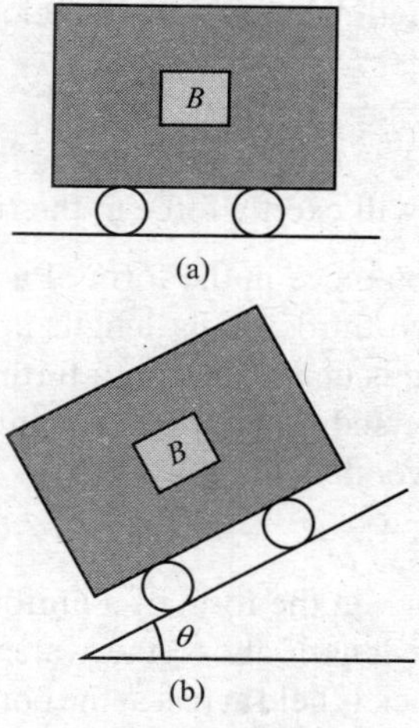

Fig. 3.205

a. The body will move up (relative to liquid)
b. The body will move down (relative to liquid)
c. The body will remain stationary (relative to liquid)
d. The body will move up for some inclination θ and will move down for another inclination θ

59. A sealed tank containing a liquid of density ρ moves with horizontal acceleration a as shown in the figure. The difference in pressure between two points A and B will be

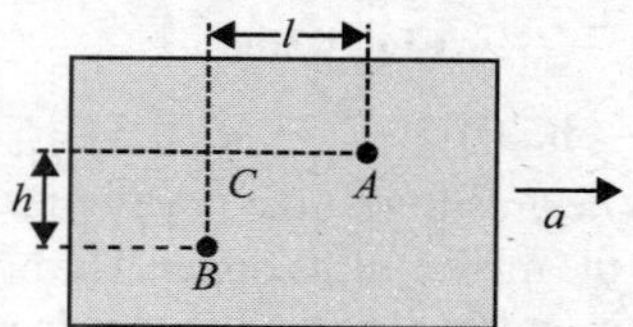

Fig. 3.206

a. $h\rho g$ **b.** $l\rho g$

c. $h\rho g - l\rho a$ **d.** $h\rho g + l\rho a$

60. The U-tube shown in the figure has a uniform cross section. A liquid is filled up to heights h_1 and h_2 in the two arms and then they are allowed to move. Neglecting viscosity and surface tension, find the state of the liquid when the levels equalize in the two arms. Find the correct statement.

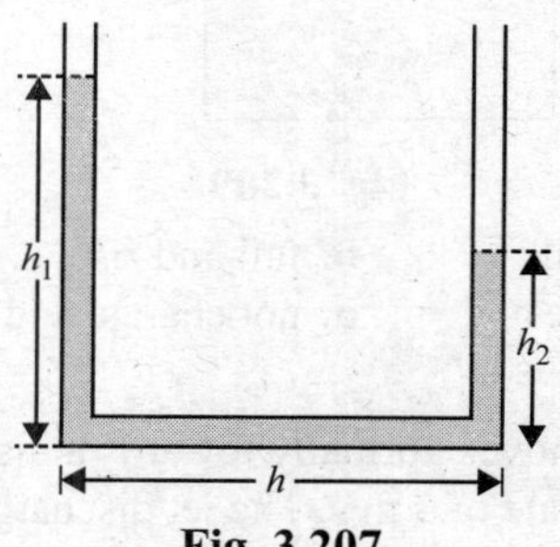

Fig. 3.207

a. The liquid will be at rest.
b. The liquid will be moving with acceleration.

c. The liquid will be moving with velocity

$$\sqrt{\frac{g}{2(h_1 + h_2 + h)}}.$$

d. The liquid will exert a force to the right on the tube.

61. A large wooden piece in the form of a cylinder floats on water with two-thirds of its length immersed. When a man stands on its upper surface, a further one-sixth of its length is immersed. The ratio between the masses of the man and the wooden piece is

a. 1:2 b. 1:3 c. 1:4 d. 1:5

62. A wooden block in the form of a uniform cylinder floats with one-third length above the water surface. A small chip of this block is held at rest at the bottom of a container containing water to a height of 1 m and is then released. The time in which it will rise to the surface of water is ($g = 10$ m/s^2)

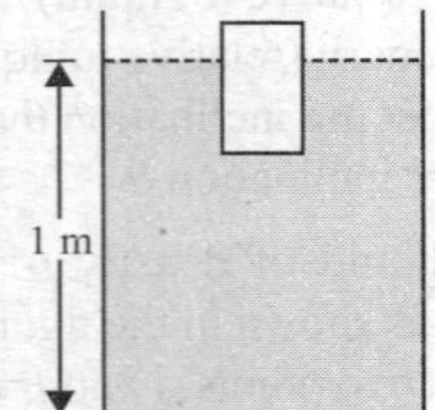

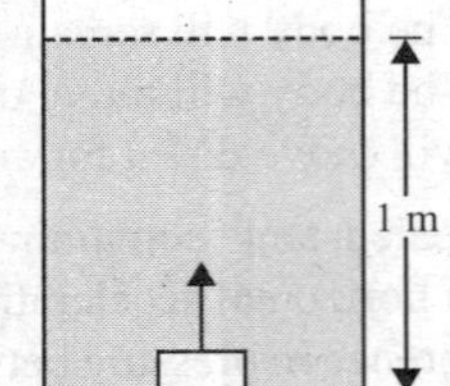

Fig. 3.208

a. 0.84 s b. 0.63 s c. 0.5 s d. 1 s

63. A goldsmith desires to test the purity of a gold ornament suspected to the mixed with copper. The ornament weighs 0.25 kg in air and is observed to displace 0.015 litre of water when immersed in it. Densities of gold and copper with respect to water are, respectively, 19.3 and 8.9. The approximate percentage of copper in the ornament is

a. 5% b. 10% c. 15% d. 25%

64. An ice cube is floating in water above which a layer of a lighter oil is poured. As the ice melts completely, the level of water surface (or interface) and upper level of oil will

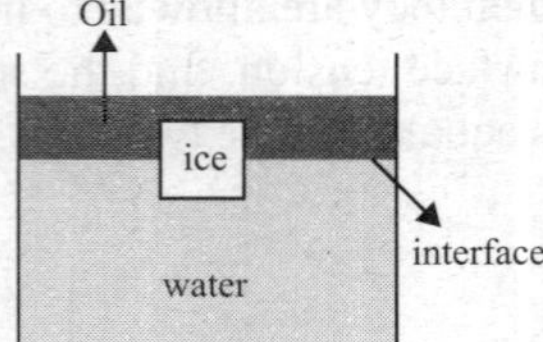

Fig. 3.209

a. rise and fall b. fall and rise
c. not change and change d. not change and fall not

65. A flat plate moves normally towards a discharging jet of water at the rate of 3 m/s. The jet discharges the water at the rate of 0.1 m^3/s and at the speed of 18 m/s. The force exerted on the plate due to the jet is

a. 1800 N b. 2100 N
c. 2450 N d. 1560 N

66. A hollow cylindrical container floats in water with half its length immersed. A liquid of specific gravity σ is slowly poured into the container until the levels of the liquid inside the container and of water outside are the same. It is observed that the container is submerged to two-thirds its length. Then the value of σ is

a. $\frac{1}{2}$ b. $\frac{1}{3}$ c. $\frac{1}{4}$ d. $\frac{1}{3}$

67. A hollow sphere of mass $M = 50$ kg and radius $r = \left(\frac{3}{40\pi}\right)^{1/3}$ m is immersed in a tank of water (density $\rho_w = 10^3$ kg/m^3). The sphere is tied to the bottom of a tank by two wires A and B as shown. Tension in wire A is ($g = 10$ m/s^2)

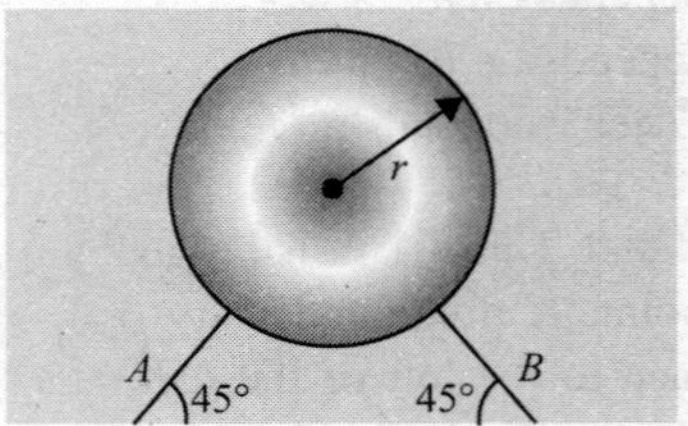

Fig. 3.210

a. $125\sqrt{2}$ N b. 125 N
c. $250\sqrt{2}$ N d. 250 N

68. In a horizontal pipeline of uniform cross section the pressure falls by 8 N/m^2 between two points separated by 1 km. If oil of density 800 kg/m^3 flows through the pipe, find the change in KE per kg of oil at these points.

a. 10^{-2} J/kg b. 10^{-3} J/kg
c. 10^{-4} J/kg d. 10^{-1} J/kg

69. In the figure shown the velocity and pressure of the liquid at the cross section (2) are given by (If P_0 is the atmospheric pressure).

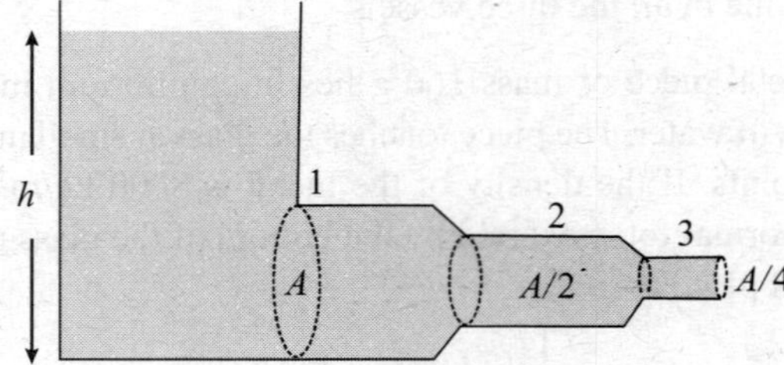

Fig. 3.211

a. $\sqrt{2hg}, P_0 + \frac{\rho hg}{2}$ b. $\sqrt{hg}, P_0 + \frac{\rho hg}{2}$

c. $\sqrt{\frac{hg}{2}}, P_0 + \frac{3\rho hg}{4}$ d. $\frac{\sqrt{hg}}{2}, P_0 + \frac{3\rho hg}{4}$

70. A liquid is kept in a cylindrical vessel which is rotated along its axis. The liquid rises at the sides. If the radius of the vessel is r and the speed of revolution is n rotations/second, find the difference in height of the liquid at the centre of vessel and its sides.

a. $h=\frac{2\pi^2n^2r^2}{g}$ b. $h=\frac{4\pi^2n^2r^2}{g}$

c. $h=\frac{\pi^2r^2n^2}{g}$ d. $h=\frac{\pi^2r^2n^2}{g}$

71. A large cylindrical tank has a hole of area A at its bottom and water is poured into the tank through a tube of cross-sectional area A ejecting water at the speed v. Which of the following is true?

a. Water level in tank keeps on rising.
b. No water can be stored in the tank.
c. Water level will rise to a height $v^2/2g$ and then stop.
d. The water level will be oscillating.

72. A tube filled with water and closed at both ends uniformly rotates in a horizontal plane about the OO' axis. The manometers A and B fixed in the tube at distances r_1 and r_2 from rotational axis indicate pressures p_1 and p_2, respectively. Determine the angular velocity ω of rotation of the tube.

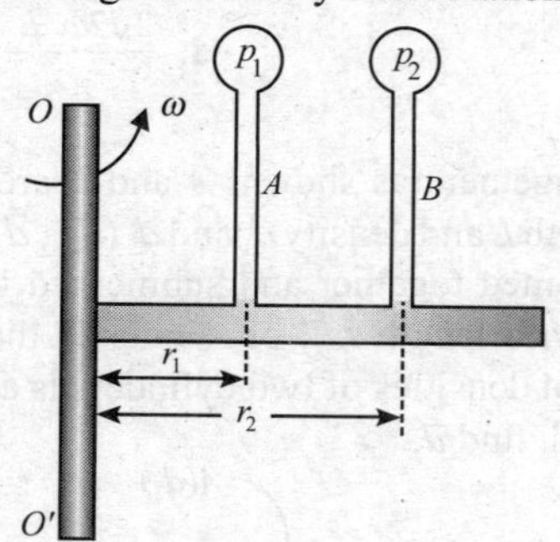

Fig. 3.212

a. $\omega=\sqrt{\frac{2(p_2-p_1)}{\rho(r_2^2-r_1^2)}}$ b. $\omega=\sqrt{\frac{2(p_2-p_1)}{\rho(r_2^2+r_1^2)}}$

c. $\omega=\sqrt{\frac{2(r_2^2-r_1^2)}{p_1-p_2}}$ d. $\omega=\frac{\sqrt{2(p_2-p_1)}}{\rho r_1 r_2}$

73. A horizontal tube has different cross sections at points A and B. The areas of cross section are a_1 and a_2, respectively and pressures at these points are $p_1=\rho g h_1$ and $p_2=\rho g h_2$, where ρ is the density of liquid flowing in the tube and h_1 and h_2 are heights of liquid columns in vertical tubes connected at A and B. If $h_1-h_2=h$, then the flow rate of the liquid in the horizontal tube is

a. $a_1a_2\sqrt{\frac{2gh}{a_1^2-a_2^2}}$ b. $a_1a_2\sqrt{\frac{2g}{h(a_1^2-a_2^2)}}$

c. $a_1a_2\sqrt{\frac{(a_1^2+a_2^2)h}{2g(a_1^2-a_2^2)}}$ d. $\frac{2a_1a_2gh}{\sqrt{a_1^2-a_2^2}}$

74. Figure shows a crude type of atomizer. When bulb A is compressed, air flows swiftly through tube BC causing a reduced pressure in the particles of the vertical tube. Liquid rises in the tube, enters BC and is sprayed out. If the pressure in the bulb is P_a+P, where P is the gauge pressure and P_a is the atmospheric pressure, v is the speed of air in BC, find how large would v need to be to cause the liquid to rise to BC. Density of air = 1.3 kg/m³.

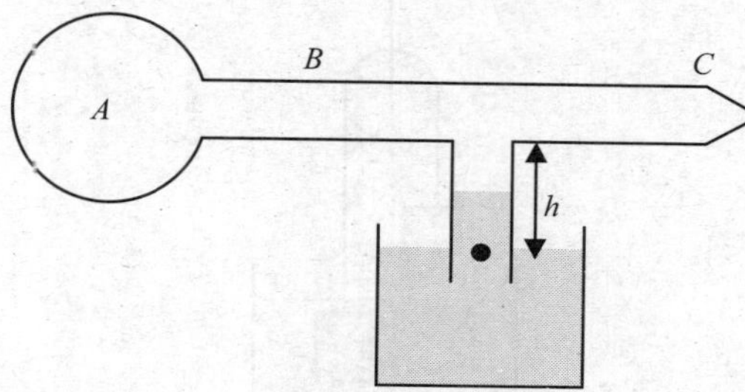

Fig. 3.213

a. $\frac{P+\rho gh}{\sqrt{0.65}}$ b. $\frac{Ph\rho g}{\sqrt{0.65}}$

c. $\frac{P\sqrt{h\rho g}}{0.65}$ d. $\frac{P}{h\rho g\sqrt{0.65}}$

75. A cubic block of side a is connected with two similar vertical springs as shown. Initially, bottom surface of the block of density σ touches the surface of the fluid of density 2σ while floating. A weight is placed on the block so that it is immersed half in the fluid, find the weight.

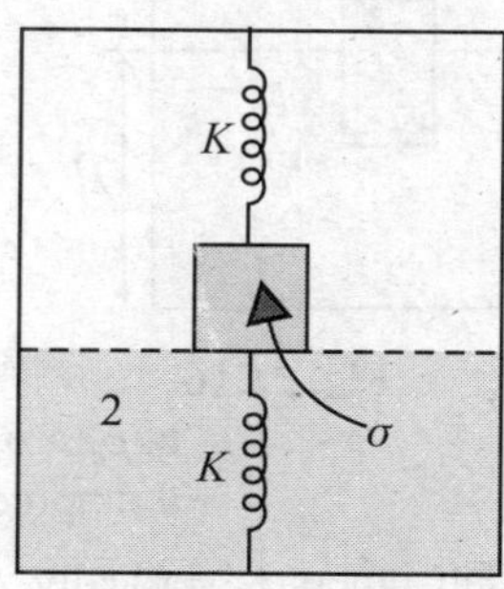

Fig. 3.214

a. $a\left(\frac{K}{2}+a^2\sigma g\right)$ b. $a(K+a^2\sigma g)$

c. $a\left(K+\frac{a^2}{2}\sigma g\right)$ d. $\frac{a}{2}(K+a^2\sigma g)$

76. Two unequal blocks placed over each other of different densities σ_1 and σ_2 are immersed in a fluid of density of σ. The block of density σ_1 is fully submerged and the block of density σ_2 is partly submerged so that ratio of their masses is 1/2 and $\sigma/\sigma_1=2$ and $\sigma/\sigma_2=0.5$. Find the degree of submergence of the upper block of density σ_2.

a. 50% submerged **b.** 25% submerged
c. 75% submerged **d.** Fully submerged

77. In the arrangement as shown, $m_B = 3m$, density of liquid is ρ and density of block B is 2ρ. The system is released from rest so that block B moves up when in liquid and moves down when out of liquid with the same acceleration. Find the mass of block A.

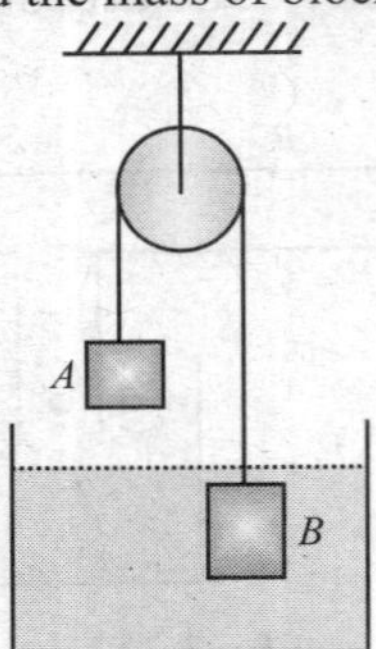

Fig. 3.215

a. $\frac{7}{4}m$ **b.** $2m$

c. $\frac{9}{2}m$ **d.** $\frac{9}{4}m$

78. Two wooden blocks A and B float in a liquid of density ρ_L as shown. The distance L and H are shown. After some time, block B falls into the liquid, so that L decreases and H increases. If density of block B is ρ_B, find the correct option.

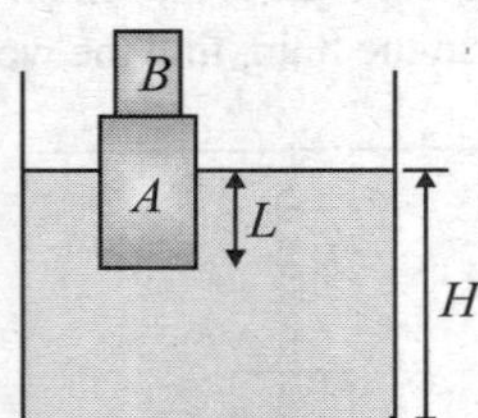

Fig. 3.216

a. $\rho_L = \rho_B$ **b.** $\rho_L > \rho_B$

c. $\rho_L < \rho_B$ **d.** unpredictable

79. Water from a tap emerges vertically downwards with an initial velocity V_0. Assuming pressure is constant throughout the stream of water and the flow is steady, find the distance from the tap at which cross-sectional area of stream is half of the cross-sectional area of stream at the tap:

a. $V_0^2/2g$ **b.** $3V_0^2/2g$

c. $2V_0^2/g$ **d.** $5V_0^2/2g$

80. A large open tank has two holes in the wall. One is a square hole of side L at a depth y from the top and the other is a circular hole of radius R at a depth $4y$ from the top. When the tank is completely filled with water, the quantities of water flowing out per second from both the holes are the same. Then, R is equal to

a. $L/\sqrt{2\pi}$ **b.** $2\pi/L$

c. L **d.** $L/2\pi$

81. A U-shaped tube contains a liquid of density ρ and it is rotated about the left dotted line as shown in the figure. Find the difference in the levels of the liquid column.

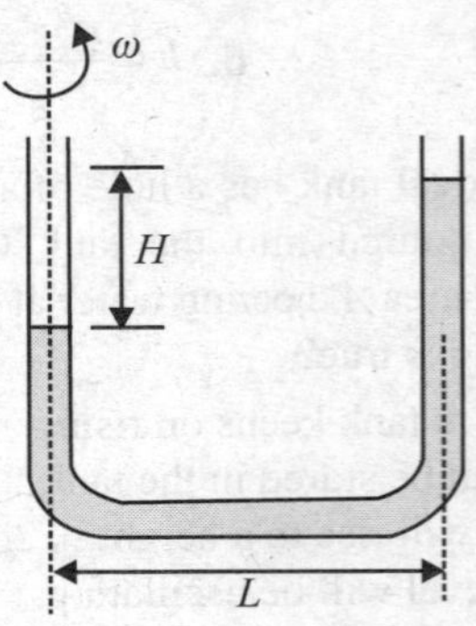

Fig. 3.217

a. $\frac{\omega^2 L^2}{2\sqrt{2}g}$ **b.** $\frac{\omega^2 L^2}{2g}$

c. $\frac{2\omega^2 L^2}{g}$ **d.** $\frac{2\sqrt{2}\omega^2 L^2}{g}$

82. In the arrangement as shown, A and B are two cylinders each of length L and density d_1 and d_2 ($d_1 < d_2$), respectively. They are joined together and submerged in the liquid of density d with length $L/2$ of A outside the liquid. If the difference in densities of two cylinders is equal to density of the liquid, find d_1.

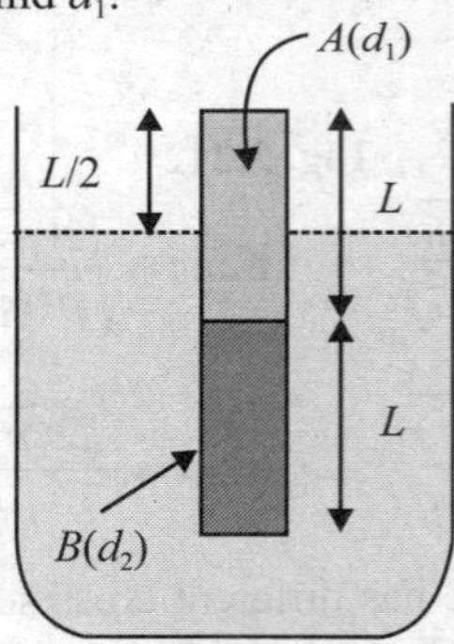

Fig. 3.218

a. $\frac{5}{4}d$ b. $\frac{d}{4}$

c $\frac{3}{4}d$ d. $\frac{d}{2}$

Multiple Correct Answers Type

Solutions on page 3.86

1. A rectangular vessel of dimension ($l \times b \times h$) and mass M contains a liquid of density ρ. The vessel has an orifice at its bottom at a distance c from the rear wall as shown in the figure.

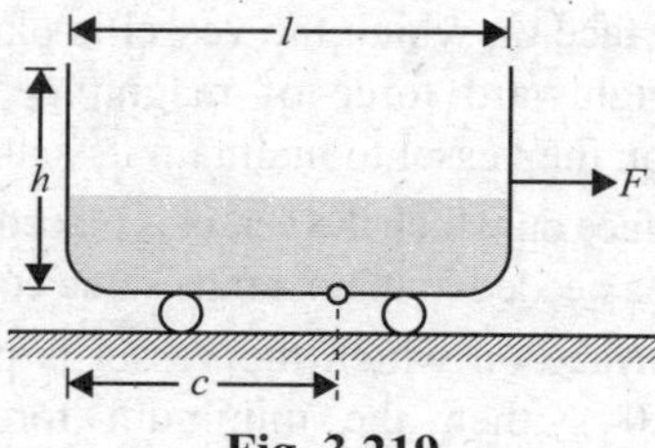

Fig. 3.219

a. The maximum volume of the water that can be stored when the vessel is accelerated is $hcb/2$.
b. The maximum volume of the water that can be stored when the vessel is accelerated is $hlb/2$.
c. Force F that must be applied when maximum water stored is $\left[M + \dfrac{hcb\rho}{2}\right]\dfrac{hg}{c}$.
d. Force F that must be applied when maximum water stored is $\left[M + \dfrac{hcb\rho}{2}\right]\dfrac{lg}{c}$.

2. Figure 3.220 shows a container filled with a liquid of density ρ. Four points A, B, C and D lie on the vertices of a vertical square. Points A and C lie on a vertical line and points B and D lies on a horizontal line. Choose the correct statement(s) about the pressure at the four points.

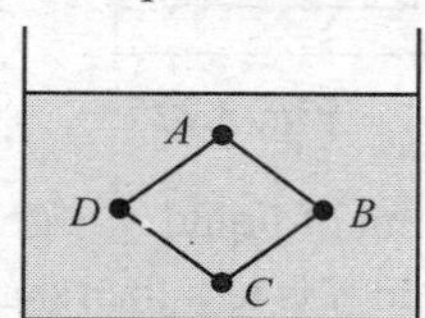

Fig. 3.220

a. $P_D = P_B$
b. $P_A < P_B = P_D < P_C$
c. $P_D = P_B = \dfrac{P_C - P_A}{2}$
d. $P_D = P_B = \dfrac{P_C + P_A}{2}$

3. A circular cylinder of radius R and height H is filled with water to a height $(2/3)H$. It starts rotating about its axis with constantly increasing angular speed. Choose the correct alternatives.

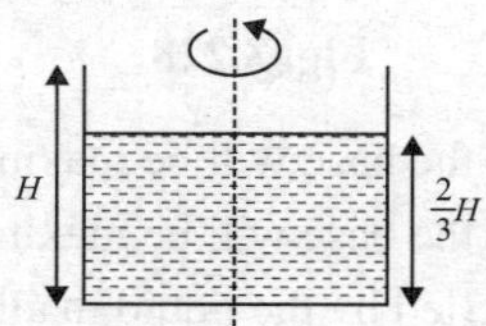

Fig. 3.221

a. At all speeds, shape of the free surface is parabolloid.
b. The free surface touches first the brim of the cylinder and then the base of the cylinder.
c. The free surface cannot touch the base without spilling water.
d. The free surface touches the brim as well as the base at the same instant.

4. A small body of density ρ' is dropped from rest at a height h into a lake of density ρ, where $\rho > \rho'$. Which of the following statements is/are correct if all dissipative effects are neglected (neglect viscosity)?
a. The speed of the body just entering the lake is $\sqrt{2gh}$.
b. The body in the lake experiences upward acceleration equal to $\{(\rho/\rho') - 1\}\, g$.
c. The maximum depth to which the body sinks in the lake is $h\rho'/(\rho - \rho')$.
d. The body does not come back to the surface of the lake.

5. Three identical blocks each of mass m = 1 kg and volume 3 × 10^{-4} m³ are suspended by massless strings from a support as shown. Underneath are three identical containers containing the same amount of water that are placed over the scales. In Fig. 3.222(a), the block is completely out of the water; in Fig. 3.222(b), the block is completely submerged but not touching the beaker and in Fig. 3.222(c), the block rests on the bottom of the beaker. The scale in Fig. 3.222(a) reads 14 N.

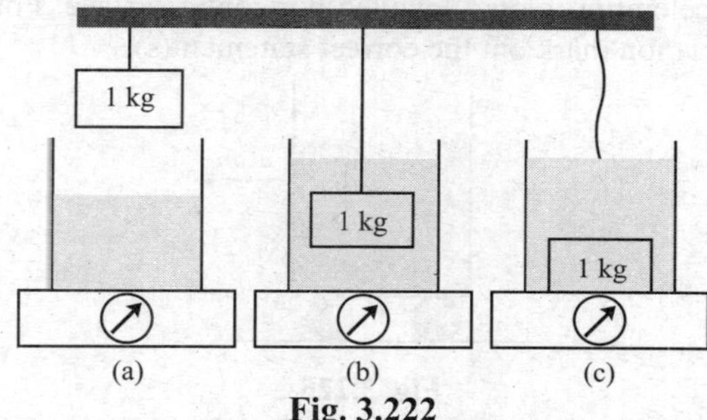

Fig. 3.222

a. The tension in the string in (b) is 10 N.
b. The tension in the string in (b) is 7 N.
c. The reading of the scale in (b) is 17 N.
d. The reading of the scale in (c) is 24 N.

6. A cylindrical vessel of 90 cm height is kept filled up to the brim. It has four holes 1, 2, 3 and 4 which are, respectively, at heights of 20 cm, 30 cm, 40 cm and 50 cm from the horizontal floor PQ. The water falling at the maximum horizontal distance from the vessel comes from

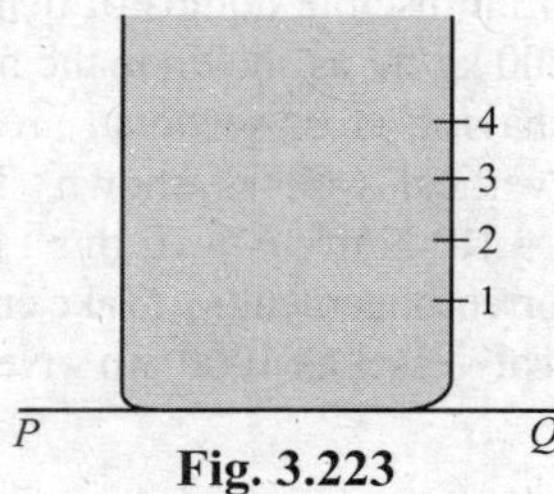

Fig. 3.223

a. hole number 4
b. hole number 3
c. hole number 2
d. hole number 1

7. In the following figure shown, a liquid is filled into the vessel up to a height h and the forces exerted against the liquid by the walls of the vessel are shown by arrows, the force being everywhere perpendicular to the walls of the vessel. For this situation, mark out the correct statement(s).

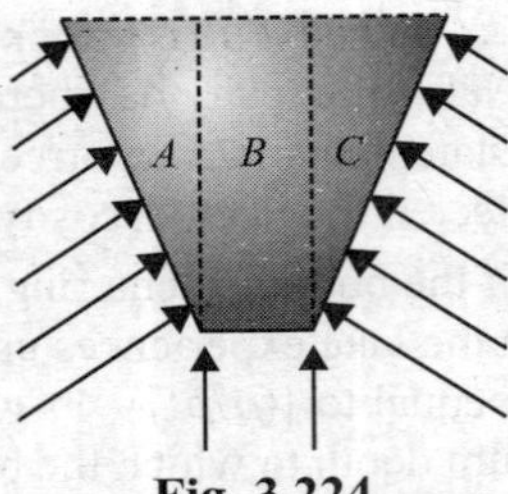

Fig. 3.224

a. The horizontal component of the forces on the left and right side inclined faces will cancel each other.

b. The weight of the section lettered A is supported by vertical component of forces acting on the inclined faces.

c. Force exerted by the liquid on the base of the vessel is greater than $(p_0 + \rho gh)A$ where ρ is the density of the liquid and A is base area of the vessel.

d. Net force exerted by liquid on the base of the vessel is in the vertical downward direction.

8. An open vessel containing liquid is moving with constant acceleration a on a levelled horizontal surface. For this situation mark out the correct statement(s).

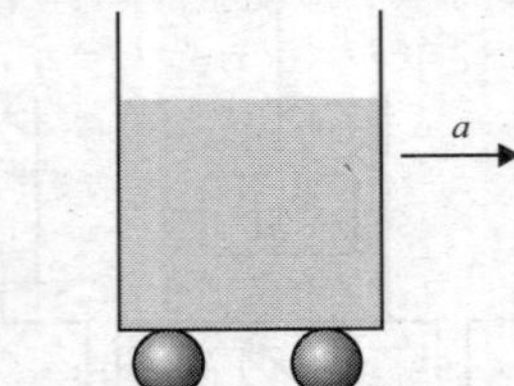

Fig. 3.225

a. The maximum pressure is at the leftmost bottom corner.

b. Along a horizontal line within the liquid as we move from left to right the pressure decreases.

c. The pressure at all points on a line making an angle of $\tan^{-1}\left(\frac{a}{g}\right)$ with horizontal would be same.

d. Along a horizontal line within the liquid as we move from left to right, the pressure remains same.

9. A cylindrical vessel of a very large cross-sectional area is containing two immiscible liquids of density $\rho_1 = 600$ kg/m^3 and $\rho_2 = 1200$ kg/m^3 as shown in the figure.
A small hole having cross-sectional area 5 cm^2 is made in right side vertical wall as shown. Take atmospheric pressure as $p_0 = 10^5$ N/m^2, $g = 10$ m/s^2. For this situation, mark out the correct statement(s). (Take cross-sectional area of the cylindrical vessel as 1000 cm^2. Neglect the mass of the vessel.)

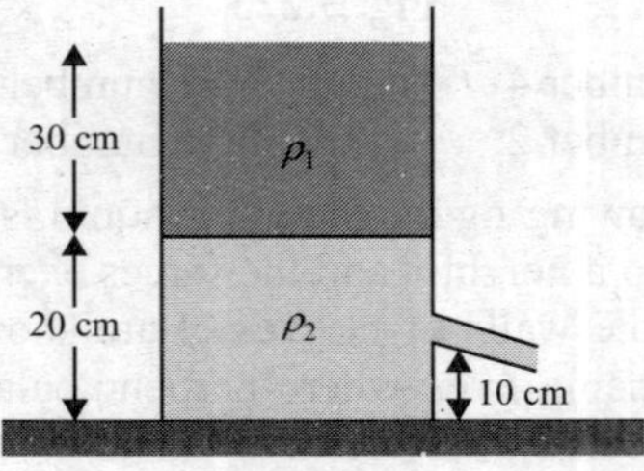

Fig. 3.226

a. If the surface on which the vessel is placed is smooth, then a rightward force of magnitude 3 N is to be applied on the vessel to maintain its static equilibrium.

b. If the surface on which the vessel is placed is smooth, then no force is needed to maintain its static equilibrium.

c. If the surface on which the vessel is placed is rough ($\mu = 0.04$), then the minimum force (horizontal) needed to be applied on the vessel to maintain its static equilibrium is zero.

d. If the surface on which the vessel is placed is rough (= 0.04), then the maximum force (horizontal) needed to be applied on the vessel to maintain its static equilibrium is 19.8 N.

10. A U-tube is held in a vertical plane such that its two limbs are vertical with middle portion horizontal as shown in the figure. Three liquids A, B and C are poured into it and in steady state their interfaces are as shown in the figure. Which of the following statements are correct?

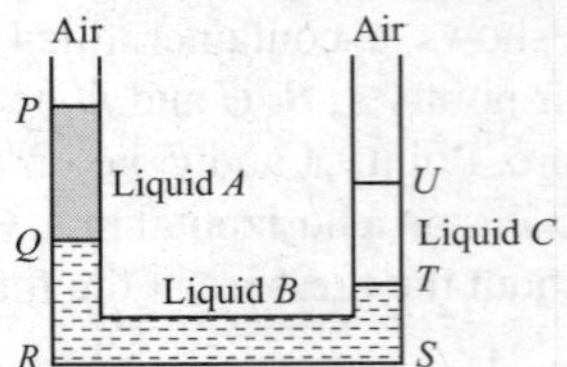

Fig. 3.227

a. Height of column of liquid A is greater than that of C.

b. Density of liquid A is less than that of liquid C.

c. Density of liquid B is maximum among these three liquids.

d. Pressure at R is greater than that at S.

11. Equal volumes of a liquid are poured in the three vessels A, B and C ($h_1 < h_2 < h_3$). All the vessels have the same base area. Select the correct alternatives.

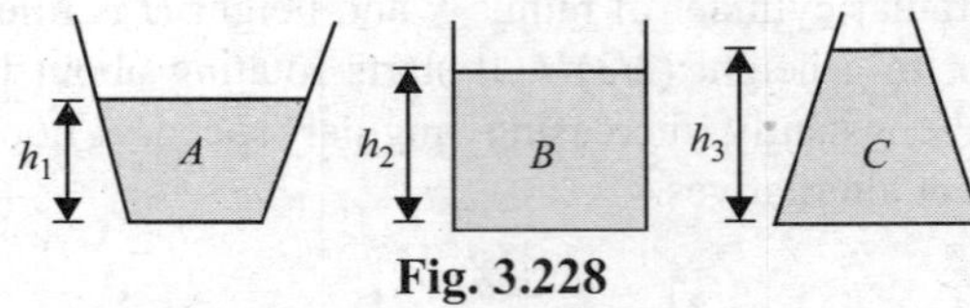

Fig. 3.228

a. The force on the base will be maximum in vessel A.

b. The force on the base will be maximum in vessel C.

c. Net force exerted by the liquid in all the three vessels is equal.

d. Net force exerted by the liquid in vessel A is maximum.

12. If the system is not in free fall, which of the following statements are true about hydrostatic pressure?

a. In a liquid, points at different depths can never be at the same pressure.

b. In a liquid, points at different depths may be at the same pressure.

c. In different liquids, points at different depths can be at the same pressure.

d. In different liquids, points at the same depth can never be at same pressure.

13. In the figure, an ideal liquid flows through the tube, which is of uniform cross section. The liquid has velocities v_A and v_B, and pressures P_A and P_B at the points A and B, respectively. Then

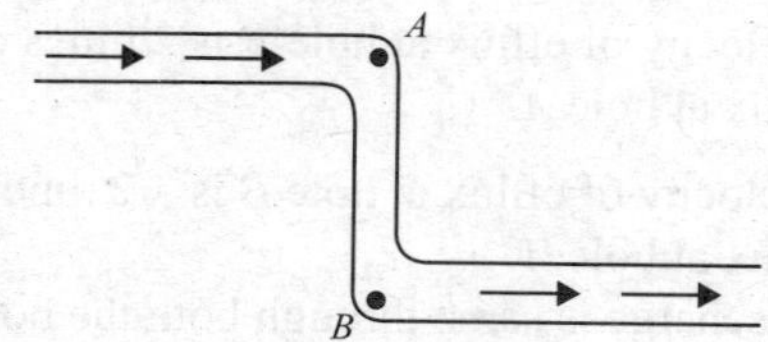

Fig. 3.229

a. $v_A = v_B$ **b.** $v_A > v_B$

c. $P_A = P_B$ **d.** $P_B > P_A$

14. A liquid flows through a horizontal tube. The velocities of the liquid in the two sections, which have areas of cross section A_1 and A_2, are v_1 and v_2, respectively. The difference in the levels of the liquid in the two vertical tubes is h. Then

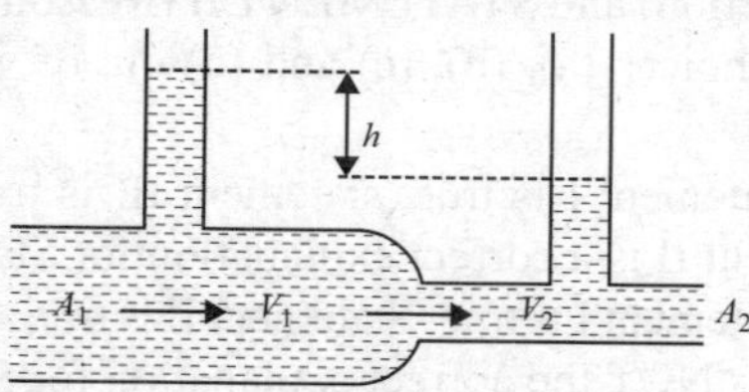

Fig. 3.230

a. the volume of the liquid flowing through the tube in unit time is A_1v_1

b. $v_2 - v_1 = \sqrt{2gh}$

c. $v_2^2 - v_1^2 = 2gh$

d. the energy per unit mass of the liquid is the same in both sections of the tube

15. Figure shows a cylindrical vessel of height 90 cm filled up to the brim. There are four holes in the vessel as shown. The liquid falling at maximum horizontal distance from the vessel would come from:

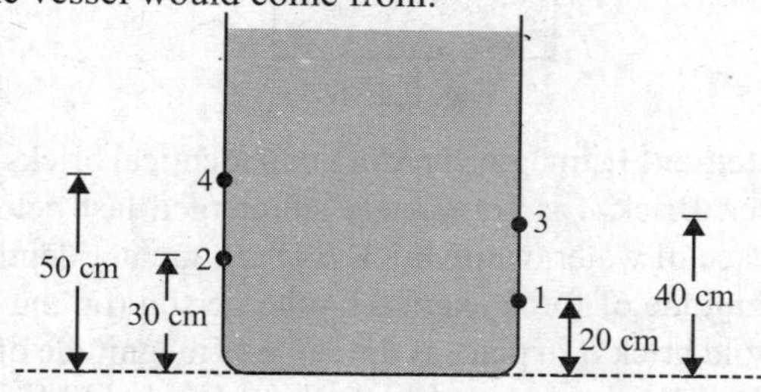

Fig. 3.231

a. hole 1 **b.** hole 2

c. hole 3 **d.** hole 4

16. A non-viscous liquid is flowing through a non-uniform pipe from section A to section B as shown in the figure. Cross sectional area of the pipe at A is less than that at B. Which of the following statements are correct?

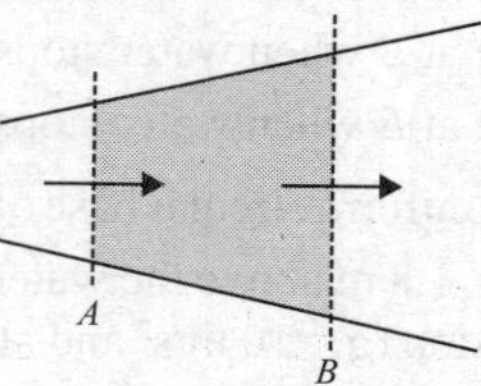

Fig. 3.232

a. The pressure at A is greater than that at B.

b. Velocity at A is greater than that at B.

c. Total energy per unit volume of the liquid is greater at A than at B.

d. Axis of the pipe cannot be horizontal.

17. A spring balance reads W_1 when a ball is suspended from it. A weighing machine reads W_2 when a tank of liquid is kept on it. When the ball is immersed in the liquid, the spring balance reads W_3 and the weighing machine reads W_4. Then, which of the following are not correct?

a. $W_1 < W_3$ **b.** $W_1 < W_3$

c. $W_2 < W_4$ **d.** $W_2 > W_4$

18. A wooden block is floating in a water tank. The block is pressed to the bottom. During the process, a work is done. Which of the following statements are false?

a. Work done is equal to work done against upthrust exerted by the water.

b. Work done is equal to work done against upthrust plus loss of gravitational potential energy of the block.

c. Work done is equal to work done against upthrust minus loss of gravitational potential energy of the block.

d. Work done is equal to loss of gravitational potential energy of the block.

19. Choose the correct statements from the following:

a. A body will sink in a liquid if its weight is equal to or greater than the weight of the liquid displaced by it.

b. A body will float in a liquid if its weight is equal to or less than the weight of liquid displaced by it.

c. When a body floats in a liquid, the portion of the body above the surface of the liquid is independent of the density of the body relative to that of the liquid.

d. In still air, a hydrogen filled balloon rises up to a certain height and then stops rising.

20. A body is floating in a liquid in a beaker. If the whole system falls freely under gravity, the upthrust on the body due to the liquid is

a. zero

b. equal to the weight of the liquid displaced

c. equal to the weight of the body in air

d. equal to the weight of the immersed portion of the body

21. Water is floating smoothly through a closed-pipe system. At one point A, the speed of the water is 3.0 m/s while at another point B, 1.0 m higher, the speed is 4.0 m/s. The pressure at A is 20 kPa when the water is flowing and 18 kPa when the water flow stops. Then

a. the pressure at B when water is flowing is 6.5kPa
b. the pressure at B when water is flowing is 8.0 kPa
c. the pressure at B when water stops flowing is 10 kPa
d. the pressure at B when water stops flowing is 8.0kPa

22. A siphon has a uniform circular base of diameter $8/\sqrt{\pi}$ cm with its crest A, 1.8 m above the water level vessel B is of large cross section ($g = 10$ m/s^2 and atmospheric pressure $P_0 = 0^5$ N/m^2).

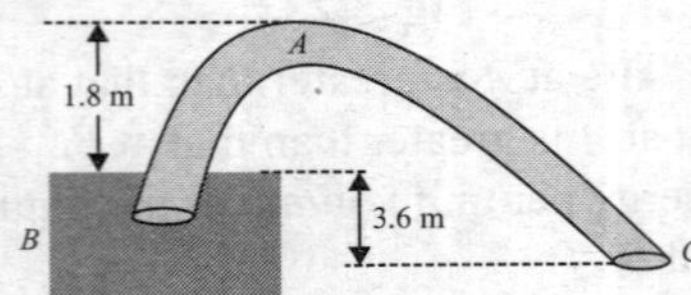

Fig. 3.233

a. Velocity of flow through pipe is $6\sqrt{2}$ m/s .
b. Discharge rate of flow through pipe is $96\sqrt{2}\times10^{-4}$ m^3/s.
c. Velocity of flow through pipe is 6 m/s.
d. Pressure of A is 0.46×10^{-5} N/m^2.

23. The difference of pressure between two points along a horizontal pipe through which water is flowing is 1.4 cm of *Hg*. If due to non-uniform cross section the speed of flow at a point of greater cross section is 60 cm/s, the speed of flow at the other point is

a. 2 m/s
b. less than 60 cm/s
c. not affected by non-uniform cross section
d. greater than 60 cm/s

24. A liquid is filled upto height h in a vessel, as shown. Find correct option (s):

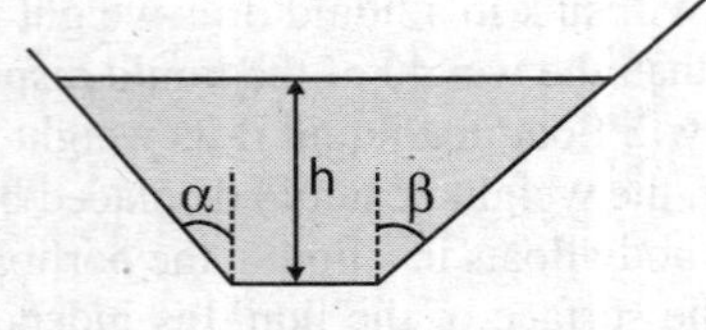

Fig. 3.234

a. If $\alpha = \beta$, horizontal component of forces on left and right side of inclined faces will be equal and opposite.
b. If $\alpha \neq \beta$, horizontal component of forces on left and right side of inclined faces will be equal and opposite.
c. If A is the area of the base of the vessel, then force exerted by liquid on walls of the vessel is greater than $(P_{atm} + \rho gh)A$.
d. As above, the force exerted by liquid on walls is equal to $(P_{atm} + \rho gh)A$.

25. The area of two holes A and B are $2a$ and a, respectively. The holes are at height $(H/3)$ and $(2H/3)$ from the surface of water. Find the correct option (s):

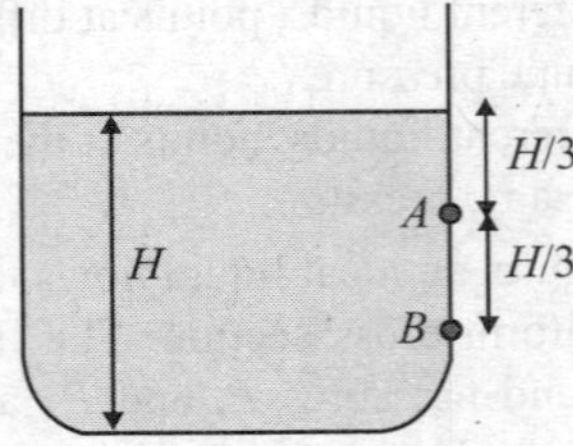

Fig. 3.235

a. The velocity of efflux at hole B is 2 times the velocity of efflux at hole A.
b. The velocity of efflux at hole B is $\sqrt{2}$ time the velocity of efflux at hole A.
c. The discharge is same through both the holes.
d. The discharge through hole A is $\sqrt{2}$ time the discharge through hole B.

Assertion-Reasoning Type

Solutions on page 3.89

In the following questions, each question contains STATEMENT I (Assertion) and STATEMENT II (Reason). Each question has four choices (a), (b), (c) and (d) out of which **ONLY ONE** is correct.

a. Statement I is true, statement II is true and Statement II is a correct explanation for Statement I.
b. Statement I is true, Statement II is true and Statement II is NOT the correct explanation for Statement I.
c. Statement I is true, Statement II is false.
d. Statement I is false, Statement II is true.

1. Statement I: When a body floats such that its parts are immersed into two immiscible liquids, then force exerted by liquid 1 is of magnitude $\rho_1 v_1 g$.
Statement II: Total buoyant force $= \rho_1 v_1 g + \rho_2 v_2 g$.

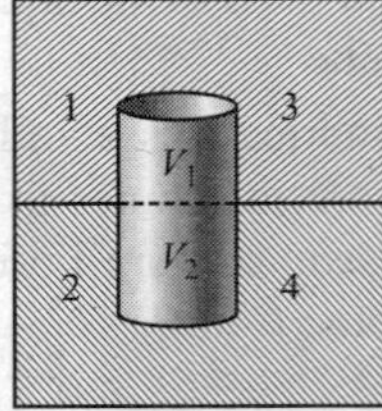

Fig. 3.236

2. Statement I: Imagine holding two identical bricks under water. Brick A is completely submerged just below the surface of water, while brick B is at a greater depth. The magnitude of force exerted by the person (on the brick) to hold brick B in place is the same as magnitude of force exerted by the person (on the brick) to hold brick A in place.

Statement II: The magnitude of buoyant force on a brick completely submerged in water is equal to the magnitude of weight of water it displaced and does not depend on the depth of the brick in water.

3. **Statement I:** A hydrogen-filled balloon stops rising after it has attained a certain height in the sky.
Statement II: The atmospheric pressure decreases with height and becomes zero when the maximum height is attained.

4. **Statement I:** Pascal's law is the working principle of hydraulic lift.
Statement II: Pressure $= \dfrac{\text{Thrust}}{\text{Area}}$.

5. **Statement I:** In taking into account the fact that any object, which floats, must have an average density less than that of water, during World War I, a number of cargo vessels were made of concrete.
Statement II: Concrete cargo vessels were filled with air.

6. **Statement I:** A dam for water reservoir is built thicker at bottom than at the top.
Statement II: Pressure of water is very large at the bottom.

7. **Statement I:** In the three cases shown in the figure, force exerted by liquid on three vessels is the same.
Statement II: Pressure at the bottom in each case is same.

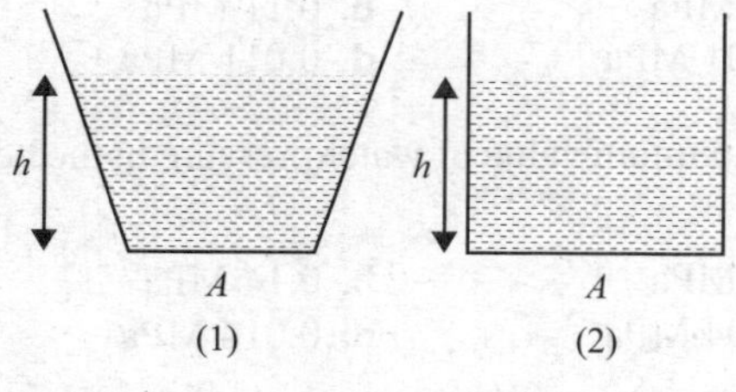

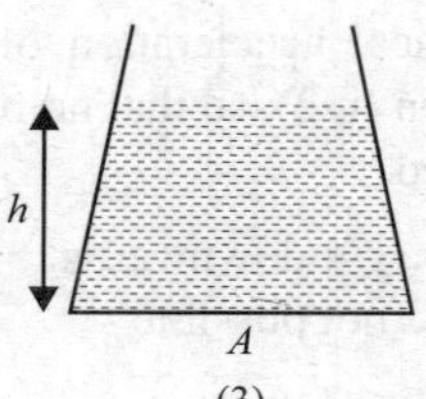

Fig. 3.237

8. **Statement I:** The velocity increases, when water flowing in broader pipe enters a narrow pipe.
Statement II: According to equation of continuity, product of area and velocity is constant.

9. **Statement I:** When spinning ball is thrown it deviates from its usual path in flight.
Statement II: Time of flight will remain same if axis of rotation is vertical.

10. **Statement I:** A liquid will flow faster and more smoothly from a sealed can when two holes are punched in the can than when one hole is punched
Statement II: The flow becomes streamlined with two holes rather than with one hole.

Comprehension Type

Solutions on page 3.89

For Problems 1–3

A small spherical ball of radius r is released from its completely submerged position (as shown in the figure) in a liquid whose density varies with height h (measured from the bottom) as $\rho_L = \rho_0[4 - (3h/h_0)]$. The density of the ball is $(5/2)\rho_0$. The height of the vessel is $h_0 = 12/\pi^2$. Consider $r << h_0$ and $g = 10$ m/s².

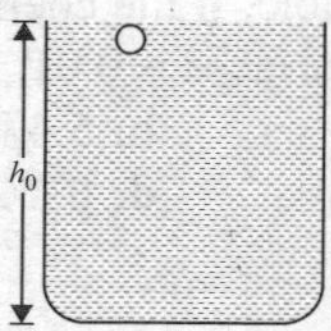

Fig. 3.238

1. Where will the ball come in the equilibrium?
   a. At a depth $h_0/2$ from top
   b. At the bottom of the vessel
   c. At a depth $3h_0/4$ from top
   d. The ball will never be at equilibrium

2. The motion of the ball in the vessel is
   a. oscillatory but not simple harmonic motion
   b. simple harmonic motion with time period 2 s
   c. simple harmonic motion with time period 1 s
   d. the motion is not oscillatory

3. The ball will reach at the bottom of the vessel
   a. in 1 s
   b. in 2 s
   c. in 0.5 s
   d. the ball will not reach at the bottom of the vessel

For Problems 4–6

A cylindrical tank having cross-sectional area $A = 0.5$ m² is filled with two liquids of density $\rho_1 = 900$ kg m^{-3} and $\rho_2 = 600$ kg m^{-3}, to a height $h = 60$ cm each as shown in the figure. A small hole having area $a = 5$ cm² is made in right vertical wall at a height $y = 20$ cm from the bottom. A horizontal force F is applied on the tank to keep it in static equilibrium. The tank is lying on a horizontal surface. Neglect mass of the cylindrical tank in comparison to mass of the liquids (take $g = 10$ m s^{-2}).

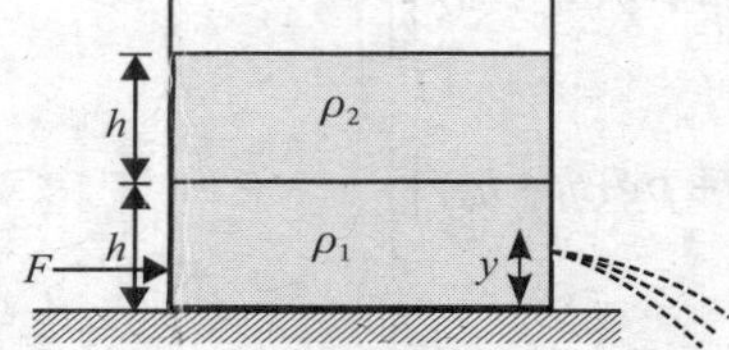

Fig. 3.239

4. The velocity of efflux is
   a. 10 m s^{-1} b. 20 m s^{-1}
   c. 4 m s^{-1} d. 35 m s^{-1}

5. Horizontal force F to keep the cylinder in static equilibrium, if it is placed on a smooth horizontal plane, is
   a. 7.2 N b. 10 N
   c. 15.5 N a. 20.4 N

6. The minimum and maximum values of F to keep the cylinder in static equilibrium just after the water starts to spill through the hole, if the coefficient of static friction between contact surfaces is 0.01, are
   a. 0, 40 N b. 5.4 N, 52.2 N
   c. 0, 70 N d. 0, 52.2 N

For Problems 7–9

Figure 3.240 shows a large closed cylindrical tank containing water. Initially, the air trapped above the water surface has a height h_0 and pressure $2p_0$ where p_0 is the atmospheric pressure. There is a hole in the wall of the tank at a depth h_1 below the top from which water comes out. A long vertical tube is connected as shown.

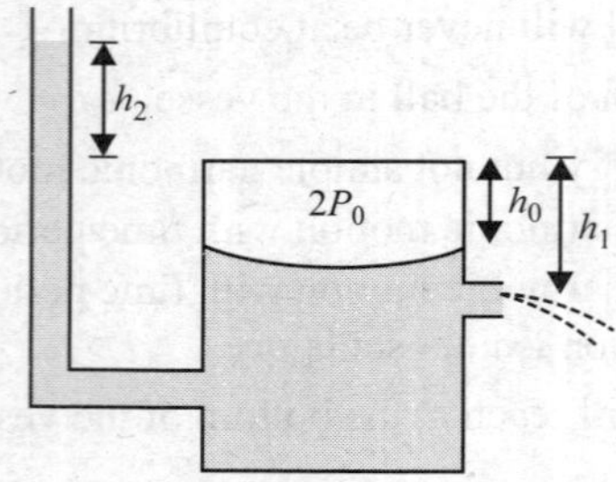

Fig. 3.240

7. Find the height h_2 of the water in the long tube above the top initially.
   a. $\dfrac{3p_0}{\rho g}-\dfrac{h_0}{3}$ b. $\dfrac{2p_0}{\rho g}-\dfrac{h_0}{2}$
   c. $\dfrac{p_0}{\rho g}-h_0$ d. $\dfrac{p_0}{2\rho g}-2h_0$

8. Find the speed with which water comes out of the hole.
   a. $\dfrac{1}{\rho}[p_0-\rho g(h_1-2h_0)]^{1/2}$
   b. $\left[\dfrac{2}{\rho}[p_0+\rho g(h_1-h_0)]\right]^{1/2}$
   c. $\left[\dfrac{3}{\rho}[p_0+\rho g(h_1+h_0)]\right]^{1/2}$
   d. $\left[\dfrac{4}{\rho}[p_0-\rho g(h_1-h_0)]\right]^{1/2}$

9. Find the height of the water in the long tube above the top when the water stops coming out of the hole.
   a. $-2h_0$ b. h_0
   c. h_2 d. $-h_1$

For Problems 10–14

If the container filled with liquid gets accelerated horizontally or vertically, pressure in liquid gets changed. In liquid (a_y) for calculation of pressure, effective g is used. A closed box with horizontal base 6 m by 6 m and a height 2 m is half filled with liquid. It is given a constant horizontal acceleration $g/2$ and vertical downward acceleration $g/2$.

10. The angle of the free surface with the horizontal is equal to
    a. 30° b. $\tan^{-1}\dfrac{2}{3}$ c. $\tan^{-1}\dfrac{1}{3}$ d. 45°

11. Length of exposed portion of top of box is equal to
    a. 2 m b. 3 m c. 4 m d. 2.5 m

12. Water pressure at the bottom of centre of the box is equal to (atmospheric pressure = 10^5 N/m^2, density of water = 1000 kg/m^3, g = 10 m/s^2)
    a. 1.1 MPa b. 0.11 MPa
    c. 0.101 MPa d. 0.011 MPa

13. The maximum value of water pressure in the box is equal to
    a. 1.4 MPa b. 0.14 MPa
    c. 0.104 MPa d. 0.014 MPa

14. What is the value of vertical acceleration of box for given horizontal acceleration ($g/2$), so that no part of the bottom of the box is exposed?
    a. $g/2$ upwards b. $g/4$ downwards
    c. $g/4$ upwards d. not possible

For Problems 15–19

When a jet of liquid strikes a fixed or moving surface, it exerts thrust on it due to rate of change of momentum.

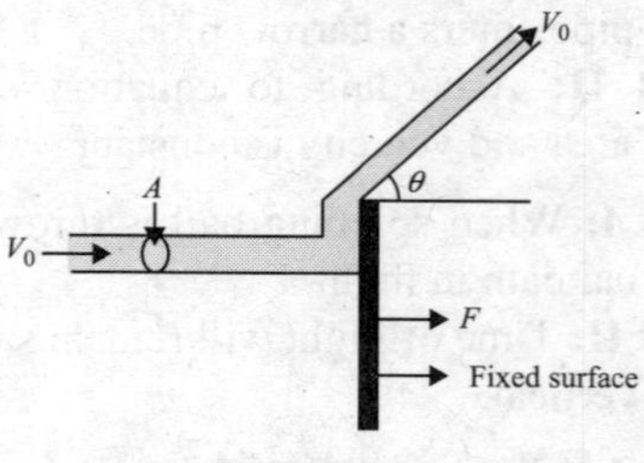

Fig. 3.241

$$F=(\rho AV_0)V_0-(\rho AV_0)V_0\cos\theta$$
$$=\rho AV_0^2\,[1-\cos\theta]$$

If surface is free and starts moving due to thrust of the liquid, then at any instant, the above equation gets modified based on relative change of momentum with respect to surface. Let, at any instant, the velocity of surface be u. Then the above equation becomes

$$F = \rho A(V_0 - u)^2[1 - \cos\theta]$$

Based on the above concept, as shown in the following figure, if the cart is frictionless and free to move in the horizontal direction, then answer the following.

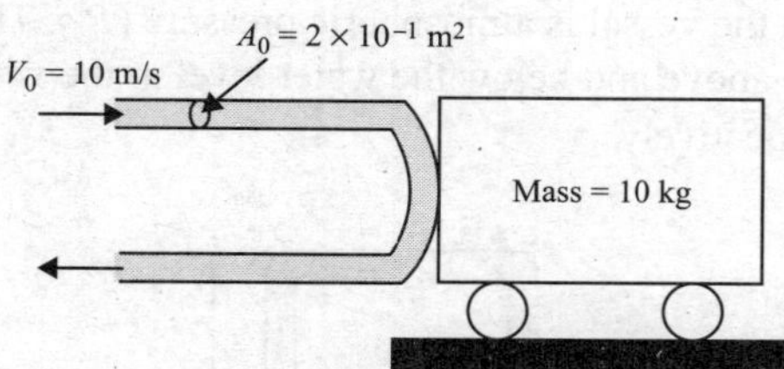

Fig. 3.242

Given that cross-sectional area of jet = 2×10^{-4} m², velocity of jet $V_0 = 10$ m/s, density of liquid = 1000 kg/m³, mass of cart $M = 10$ kg.

15. Initially ($l = 0$), the force on the cart is equal to

a. 20 N **b.** 40 N **c.** 80 N **d.** zero

16. Velocity of cart at $t = 10$ s is equal to

a. 4 m/s **b.** 6 m/s
c. 8 m/s **d.** 5 m/s

17. In the above problem, what is the acceleration of the cart at this instant?

a. 1.6 m/s² **b.** 1 m/s²
c. 0.64 m/s² **d.** 0.16 m/s²

18. The time at which velocity of cart becomes 2 m/s is equal to

a. 1.6 s **b.** 2 s
c. 3.2 s **d.** 4 s

19. The power supplied to the cart when its velocity becomes 5 m/s is equal to

a. 100 W **b.** 25 W **c.** 50 W **d.** 200 W

For Problems 20–21

A uniform rod of length L pivoted at the bottom of a pond of depth $L/2$ stays in stable equilibrium as shown in the figure.

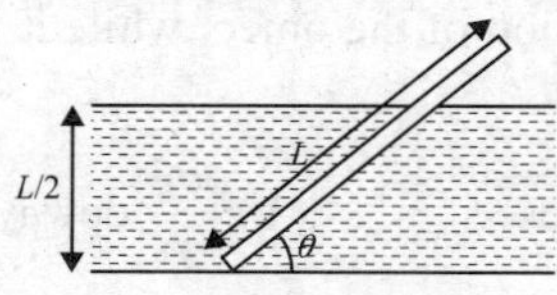

Fig. 3.243

20. Find the angle θ if the density of the material of the rod is half of the density of water.

a. 60° **b.** 30° **c.** 45° **d.** 37°

21. Find the force acting at the pivoted end of the rod in terms of mass m of the rod.

a. $(\sqrt{2} - 1)\,mg$ **b.** $(\sqrt{2} + 1)\,mg$
c. $(\sqrt{3} + 1)\,mg$ **d.** $(\sqrt{2} + 3)\,mg$

For Problems 22–23

A cylindrical object of cork of mass 15 g and cross-sectional area $A_1 = 10$ cm² floats in a pan of water as shown in the figure. An aluminium cylinder of mass 25 g and cross-sectional area $A_2 = 2$ cm² is attached 4 cm below the cork and slides through a watertight frictionless hole in the bottom of the pan. Take density of the cork, $\rho = 0.2$ g/cm³, $\rho_{\text{aluminium}} = 2.7$ g/cm³, $g = 10$ m/s². Based on the above information, answer the following questions.

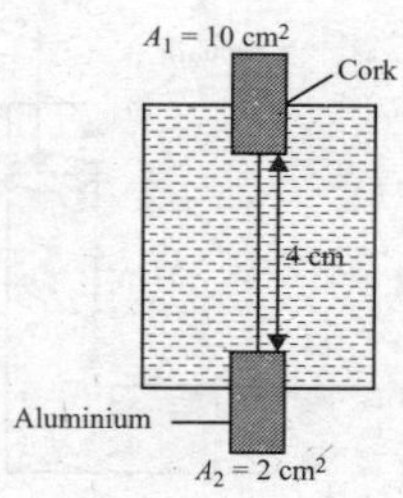

Fig. 3.244

22. The length of the cork cylinder inside the water in equilibrium is

a. 6 cm **b.** 4 cm **c.** 8 cm **d.** 3 cm

23. The tension in the string joining the cork and aluminium cylinder is

a. 2 N **b.** 1 N **c.** 0.45 N **d.** 0.6 N

For Problems 24–26

A liquid having density 6000 kg/m³ stands to a height of 4 m in a sealed tank as shown in the figure. The tank contains compressed air at a gauge pressure of 3 atm. The horizontal outlet pipe has a cross-sectional area of 6 cm² and 3 cm² at larger and smaller sections. Atmospheric pressure = 1 atm, $g = 10$ m/s², 1 atm = 10^5 N/m².

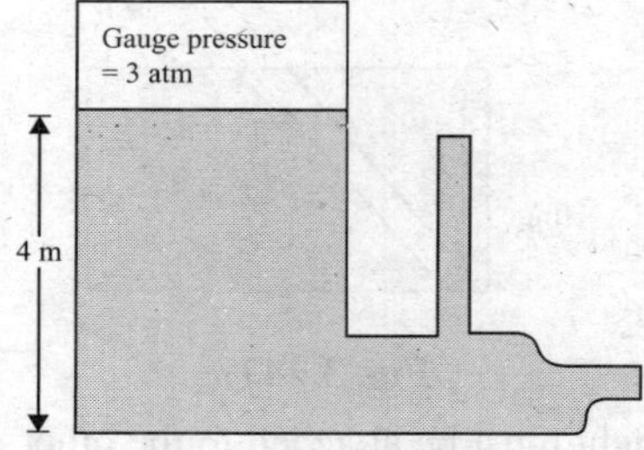

Fig. 3.245

Assume that depth of water in the tank remains constant due to its very large base and air pressure above it remains constant. Based on the above information, answer the following questions.

24. The discharge rate from the outlet is

a. 4.02×10^{-3} m³/s b. 6×10^{-5} m³/s
c. 7.56×10^{-3} m³/s d. 4.02×10^{-5} m³/s

25. The height at which liquid will stand in the open end of the pipe is

a. 4 m b. 3.5 m c. 4.5 m d. 5.6 m

26. If a hole is created in the tank at its top end, then the height of liquid level in the open end of the pipe is

a. 4 m b. 3 m c. 3.5 m d. 3.8 m

For Problems 27–28

The vessel shown in the figure contains water, (take ρ_{water} = 1000 kg/m³, g = 10 m/s²). For this arrangement, answer the following questions:

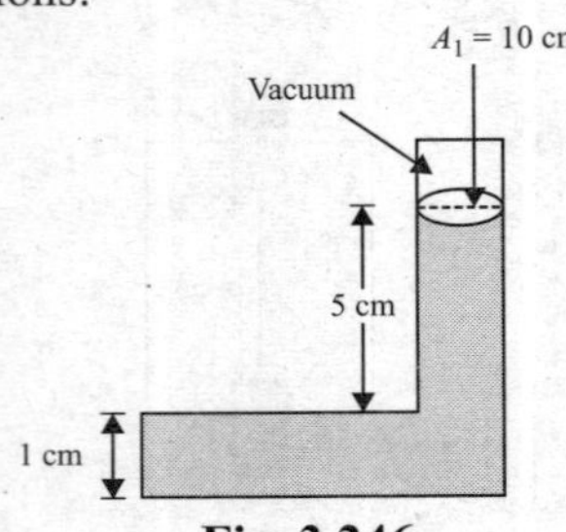

Fig. 3.246

27. The net force acting on the base of the vessel is

a. 1 N b. 2 N c. 3 N d. 5 N

28. Mark out the correct statement(s).

a. Net force acting on the base of the vessel > weight of the liquid inside the vessel.
b. Net force acting on the base of the vessel = weight of the liquid inside the vessel.
c. Net pressure force acting on the liquid = weight of the vessel
d. Both (a) and (c) are correct.

For Problems 29–30

A rod of length 6 m has a mass 12 kg. It is hinged at one end A at a distance of 3 m below the water surface. The specific gravity of the material of rod is 0.5.

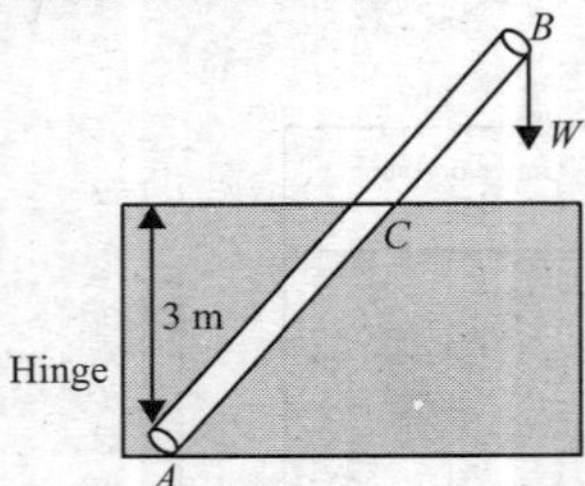

Fig. 3.247

29. What weight must be attached to the other end B so that 5 m of the rod is immersed in water?

a. 7 kgf b. $\frac{17}{3}$ kgf
c. $\frac{7}{5}$ kgf d. $\frac{7}{2}$ kgf

30. Find the magnitude and direction of the force exerted by the hinge on the rod.

a. 17/3 kgf in the downward direction
b. 8 kgf in the downward direction
c. 4 kgf in the downward direction
d. 5 kgf in the downward direction

For Problems 31–33

A tube of uniform cross section is used to siphon water from a vessel V as shown in the figure. The pressure over the open end of water in the vessel is atmospheric pressure (P_0). The heights of the tube above and below the water level in the vessel are h_1 and h_2, respectively.

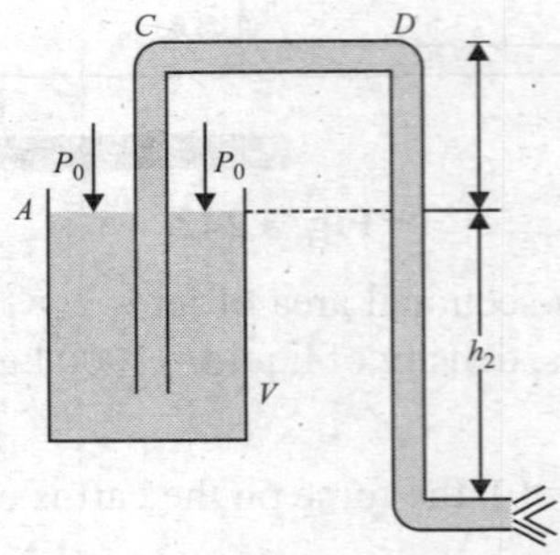

Fig. 3.248

31. Determine the velocity v_B of the water issuing out at B.

a. $v_B = \sqrt{2gh_2}$ b. $v_B = \sqrt{2g(h_1 + h_2)}$
c. $v_B = \sqrt{2g(h_2 - h_1)}$ d. $v_B = \sqrt{gh_2}$

32. If h_2 = 3.0 m, the maximum value of h_1, for which the siphon will work will be

a. 3.0 m b. 6.0 m c. 7.2 m d. 4.8 m

33. Given $h_1 = h_2$ = 3.0 m, the gauge pressure of water in the highest level CD of the tube will be

a. 3.0×10^4 N/m² b. 5.9×10^4 N/m²
c. 5.9×10^4 N/m² d. 1.5×10^4 N/m²

For Problems 34–36

An object of density σ ($<\rho$) is dropped from rest at a height h into a liquid of density ρ kept in a tall vertical cylindrical container. Neglect all dissipative effects and assume that there is no spilling of the liquid.

34. The acceleration of the object while it is inside the liquid will be

a. $\frac{\sigma g}{\rho}$ upwards b. $\frac{\sigma g}{\rho}$ downwards
c. $g\left(\frac{\sigma}{\rho}-1\right)$ upwards d. $g\left(\frac{\rho}{\sigma}-1\right)$ upwards

35. The maximum depth to which the body sinks before it returns to surface is

a. $\left(\dfrac{h\sigma}{\rho-\sigma}\right)$ b. $h\dfrac{\sigma}{\rho}$

c. $h\left(\dfrac{\rho}{\rho-\sigma}\right)$ d. $h\left(\dfrac{\rho}{\sigma}-1\right)$

36. The time taken by the object to reach the starting point (from which it is dropped) is ($\rho = 2\sigma$)

a. $2\sqrt{\dfrac{2h}{g}}$ b. $4\sqrt{\dfrac{2h}{g}}$

c. $3\sqrt{\dfrac{2h}{g}}$ d. it will never reach starting point

Matching Column Type

Solutions on page 3.94

Bernoulli's equation can be written in the following different forms (Column I). Column II lists certain units each of which pertains to one of the possible forms of the equation. Match the unit associated with each of the equations.

1.

Column I	Column II
i. $\dfrac{v^2}{2g}+\dfrac{p}{\rho g}+z$ = constant	**a.** Total energy per unit mass
ii. $\dfrac{2V^2}{2}+P+\rho gz$ = constant	**b.** Total energy per unit weight
iii. $\dfrac{V^2}{2}+\dfrac{P}{\rho}+gz$ = constant	**c.** Total energy per unit volume

2.

Column I	Column II
i. Bernoulli's theorem	**a.** Narrower pipes have less pressure
ii. Ball moving with spin	**b.** Paint gun
iii. Artificial high pressure	**c.** Non-viscous fluids
iv. Streamlined flow	**d.** Conservation of energy
	e. Uplift due to pressure difference

3.

Column I	Column II
i. Magnus energy	**a.** Pascal's law
ii. Loss of energy	**b.** Archimedes' principle
iii. Pressure is same at the same level in a liquid	**c.** Viscous force
iv. Hydraulic machines	**d.** Lifting of asbestos roofs

4.

Column I	Column II
i. A cube of mass m and density σ is pulled by a force $\vec{F}_x = mg\hat{i}$ in an accelerating liquid of density ρ. The value of a_x/a_y is 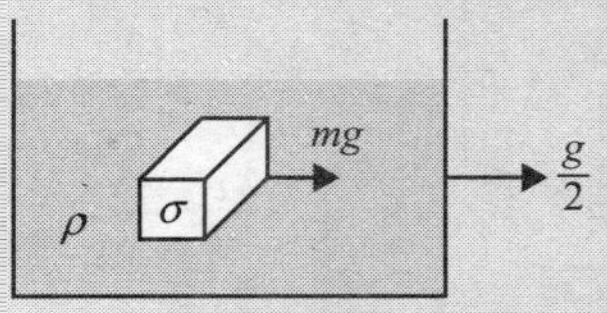 **Fig. 3.249**	**a.** $\dfrac{\rho}{\sigma}$
ii. The sphere of mass m is pulled horizontally by a force $F_x = mg$ in a non-accelerating liquid. The value of a_x/a_y is 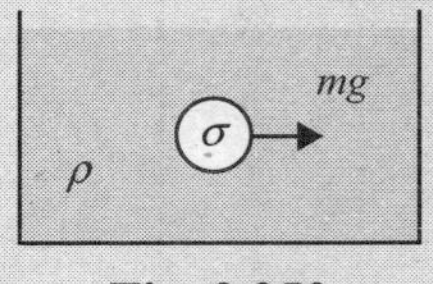 **Fig. 3.250**	**b.** 1:1
iii. A closed tube of length x is placed in a rotating liquid. The value of a_x/g, where a_x is the horizontal acceleration of the cube, is 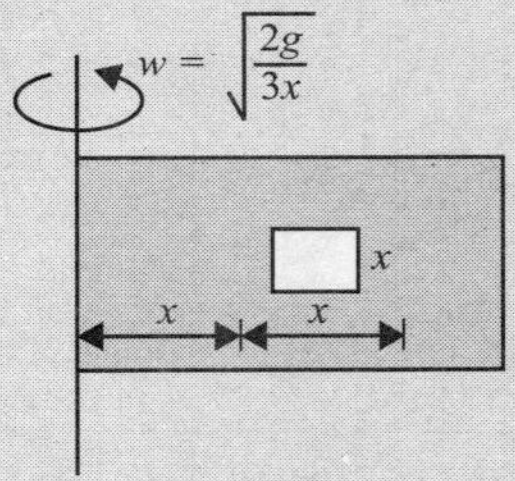 **Fig. 3.251**	**c.** $\dfrac{\sigma}{\rho-\sigma}$
iv. The string of a pendulum makes an angle θ with vertical when the cart moves with an acceleration a. If the cart is filled with liquid of density $\rho < \sigma$ (density of the bob), the value of the angle of inclination with the vertical becomes θ'. Then $\frac{\tan\theta}{\tan\theta'}$ is 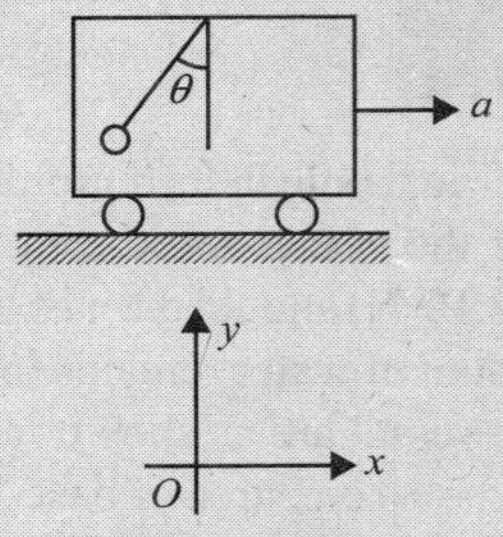 **Fig. 3.252**	**d.** $\dfrac{2\sigma+\rho}{2(\sigma-\rho)}$

5.

Column I	Column II
i. Hydrostatic force on the side wall of the cubical vessel, ρ = density of liquid **Fig. 3.253**	**a.** $A\rho v^2$
ii. Buoyant force on the cube, ρ = density of liquid **Fig. 3.254**	**b.** $\frac{1}{2}A\rho v^2$
iii. Impact (reaction) force on the vessel by the liquid coming out of the vessel, ρ = density of liquid **Fig. 3.255**	**c.** ρghA
iv. Aerodynamic force acting on the flat roof surface of area A, ρ = density of air **Fig. 3.256**	**d.** $\frac{1}{2}\rho ghA$

Integer Answer Type

Solutions on page 3.95

1. A vessel with a symmetrical hole in its bottom is fastened on a cart. The mass of the vessel and the cart is 1.5 kg. With what force F (in $\times\ 10^2$ N) should the cart be pulled so that the maximum amount of water remains in the vessel. The dimensions of the vessel are as shown in the figure. Given that $b = 50$ cm, $c = 10$ cm, area of base $A = 40$ cm^2, $L = 20$ cm, $g = 10$ m/s^2.

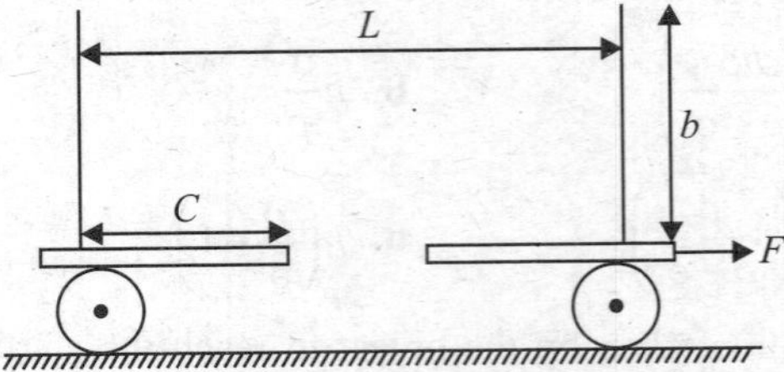

Fig. 3.257

2. A solid cylinder of height h and mass m is floating in a liquid of density ρ as shown in the figure. Find the acceleration of the vessel (in m/s^2) containing liquid for which the relative downward acceleration of the completely immersed cylinder w.r.t. vessel becomes equal to one-third of that of the vessel. (Take $g = 10$ m/s^2)

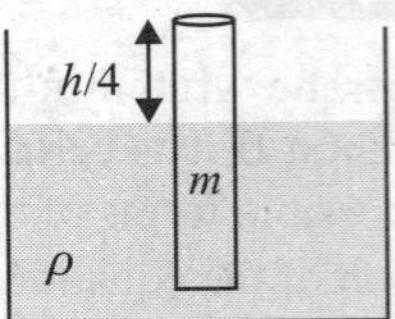

Fig. 3.258

3. Length of horizontal arm of a uniform cross section U-tube is $l = 21$ cm and ends of both of the vertical arms are open to surrounding of pressure 10×500 N/m^2. A liquid of density $\rho = 10^3$ kg/m^3 is poured into the tube such that liquid just fills the horizontal part of the tube. Now, one of the open ends is sealed and the tube is then rotated about a vertical axis passing through the other vertical arm with angular velocity $\omega_0 = 10$ rad/s. If length of each vertical arm is $a = 6$ cm, calculate the length of air column in the sealed arm. (in cm)

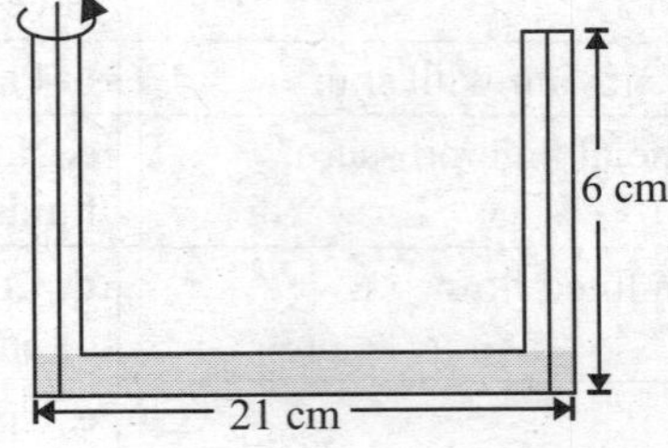

Fig. 3.259

4. A wooden plank of length 1 m and uniform cross section is hinged at one end to the bottom of a tank as shown in the figure. The tank is filled with water up to a height of 0.5 m. The specific gravity of the plank is 0.5. If the angle θ by the inclinaiton of that the plank makes with the vertical in the equilibrium position (exclude the case $\theta = 0$). Find the value of $\frac{1}{\cos^2\theta}$.

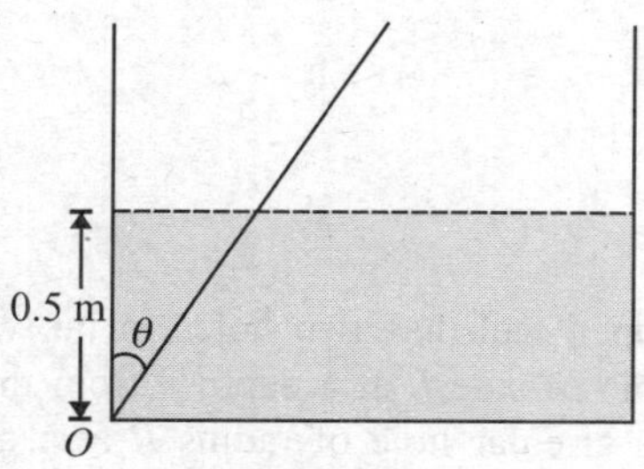

Fig. 3.260

5. A rod of length 6 m has specific gravity ρ (= 25/36). One end of the rod is tied to a 5 m long rope, which in turn is tied to the floor of a pool 10 m deep, as shown. Find the length (in m) of the part of rod which is out of water.

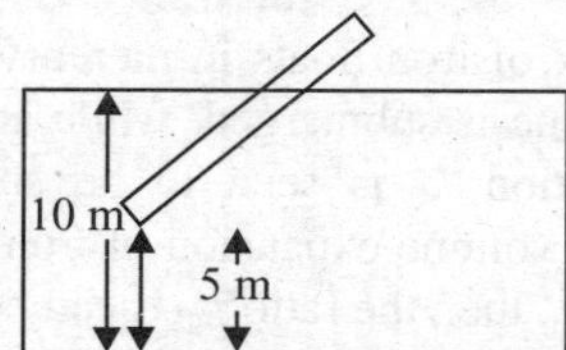

Fig. 3.261

6. Length of a horizontal arm of a U-tube is $L = 1$ m and ends of both the vertical arms are open to atmospheric pressure $P_0 = 10^5$ N/m². A liquid of density $\rho = 10^3$ kg /m³ is poured in the tube such that liquid just fills the horizontal part of the tube as shown in Fig. 3.262. Now one end of the open ends is sealed and the tube is then rotated about a vertical axis passing through the other vertical arm with angular speed $\omega_0 = \sqrt{40/3}$ rad/sec. If length of each vertical arm is a = 1 m and in the sealed end liquid rises to a height $y = 1/2$ m, find pressure in the sealed tube during rotation. (in $\times 10^5$ N/m²)

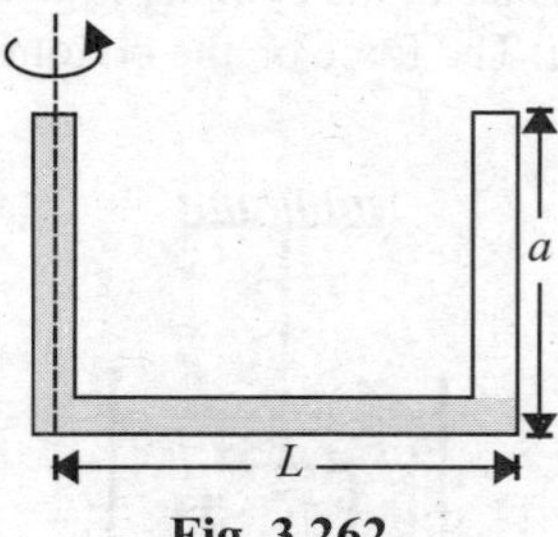

Fig. 3.262

Archives

Solutions on page 3.97

Fill in the Blanks Type

1. A piece of metal floats on mercury. The coefficient of volume expansion of the metal and mercury are γ_1 and γ_2, respectively. If the temperature of both mercury and the metal is increased by an amount ΔT, the fraction of the volume of the metal submerged in mercury changes by the factor ——. **(IIT-JEE, 1991)**
(This problem is mixed concept of heat and fluid mechanics)

2. A horizontal pipeline carries water in a streamline flow. At a point along the pipe, where the cross-sectional area is 10 cm², the water velocity is 1 ms⁻¹ and the pressure is 2000 Pa. The pressure of water at another point where the cross-sectional area is 5 cm² is —— Pa (density of water = 10^3 kg m^{-3}). **(IIT-JEE, 1994)**

True or False Type

1. A man is sitting in a boat which is floating in a pond. If the man drinks some water from the pond, water level in the pond decreases. **(IIT-JEE, 1980)**
2. A barometer made of a very narrow tube (see Fig. 3.263) is placed at normal temperature and pressure. The coefficient of volume expansion of mercury is 0.00018 per degree centigrade and that of the tube is negligible. The temperature of mercury in the barometer is now raised by 1°C, but the temperature of the atmosphere does not change. Then, the mercury height in the tube remains unchanged. **IIT-JEE, 1983)**

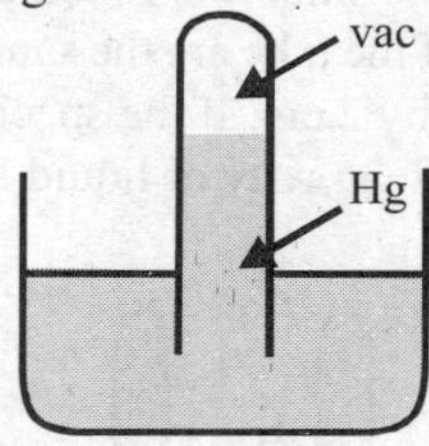

Fig. 3.263

Multiple Choice Questions with One Correct Answer Type

1. A metal ball immersed in alcohol weighs W_1 at 0°C and W_2 at 50°C. The expansion of cubical expansion of the metal is less than that of the alcohol. Assuming that the density of the metal is large compared to that of alcohol, it can be shown that **(IIT-JEE, 1980)**
(This problem is mixed concept of heat and fluid mechanics)
a. $W_1 > W_2$ **b.** $W_1 = W_2$
c. $W_1 > W_2$ **d.** none of these
2. A vessel containing water is given a constant acceleration 'a' towards the right along a straight horizontal path. Which of the following diagrams in the figure represents the surface of the liquid? **(IIT-JEE, 1981)**

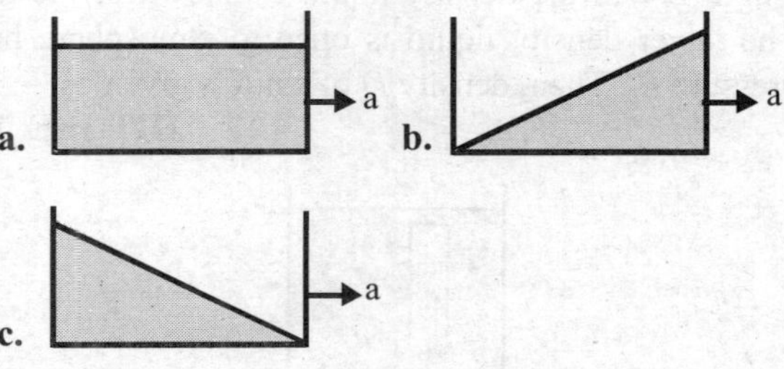

Fig. 3.264

(This problem is mixed concept of heat and fluid mechanics)

3. A body floats in a liquid contained in a beaker. The whole system as shown in the figure falls freely under gravity. The upthrust on the body is **(IIT-JEE, 1982)**

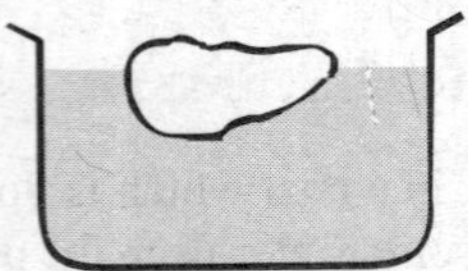

Fig. 3.265

a. zero
b. equal to the weight of the liquid displaced
c. equal to the weight of the body in air
d. equal to the weight of the immersed portion of the body

4. A U-tube of uniform cross section is partially filled with a liquid I. Another liquid II which does not mix with liquid I is poured into one side. It is found that the liquid levels of the two sides of the tube are the same, while the level of liquid I has risen by 2 cm. If the specific gravity of liquid I is 1.1, the specific gravity of liquid II must be **(IIT-JEE, 1983)**

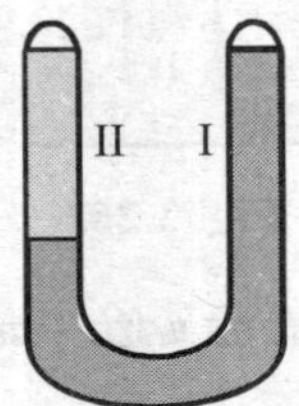

Fig. 3.266

a. 1.12 **b.** 1.1
c. 1.05 **d.** 1.0

5. A vessel contains oil (density = 0.8 g/cm^2) over mercury (density = 13.6 g/cm^3). A homogenous sphere floats with half its volume immersed in mercury and the other half in oil. The density of the material of the sphere in g/cm^3 is **(IIT-JEE, 1988)**

a. 3.3 **b.** 6.4
c. 7.2 **d.** 12.8

6. A homogeneous solid cylinder of length $L(L < H/2)$, cross-sectional area $A/5$ is immersed such that it floats with its axis vertical at the liquid–liquid interface with length $L/4$ in the density liquid as shown in the figure. The lower density liquid is open to atmosphere having pressure P_0. Then, density D of solid is given by **(IIT-JEE, 1995)**

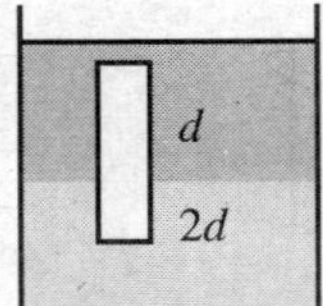

Fig. 3.267

a. $\frac{5}{4}d$ **b.** $\frac{4}{5}d$
c. $4d$ **d.** $\frac{d}{5}$

7. A large open tank has two holes in the wall. One is a square hole of side L at a depth y from the top and the other is a circular hole of radius R at a depth $4y$ from the top. When the tank is completely filled with water, the quantities of water flowing out per second from both holes are the same. Then, R is equal to **(IIT-JEE, 2000)**

a. $L/\sqrt{2\pi}$ **b.** $2\pi L$
c. L **d.** $1/2\pi$

8. When a block of iron floats in mercury at 0°C, fraction k_1 of its volume is submerged, while at the temperature 60°C, a fraction k_2 is seen to be submerged. If the coefficient of volume expansion of iron is γ_{Fe} and that of mercury is γ_{Hg}, then the ratio k_1/k_2 can be expressed as **(IIT-JEE, 2001)**

a. $\frac{1+60\gamma_{Fe}}{1+60\gamma_{Hg}}$ **b.** $\frac{1-60\gamma_{Fe}}{1+60\gamma_{Hg}}$

c. $\frac{1+60\gamma_{Fe}}{1-60\gamma_{Hg}}$ **d.** $\frac{1+60\gamma_{Hg}}{1+60\gamma_{Fe}}$

(This problem is mixed concept of heat and fluid mechanics)

9. A hemispherical portion of radius R is removed from the bottom of a cylinder of radius R. The volume of the remaining cylinder is V and its mass M. It is suspended by a string in a liquid of density ρ where it stays vertical. The upper surface of the cylinder is at a depth h below the liquid surface. The force on the bottom of the cylinder by the liquid is **(IIT-JEE, 2001)**

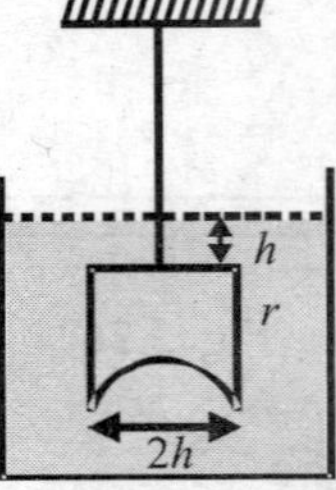

Fig. 3.268

a. Mg **b.** $Mg - V\rho g$
c. $Mg + \pi R^2 h\rho g$ **d.** $\rho g(V + \pi R^2 h)$

10. A wooden block, with a coin placed on its top, floats in water as shown in the figure. The distance l and h are shown here. After some time, the coin falls into the water. Then **(IIT-JEE, 2002)**

a. l decreases and h increases
b. l increases and h decreases
c. both l and h increases
d. both l and h decrease

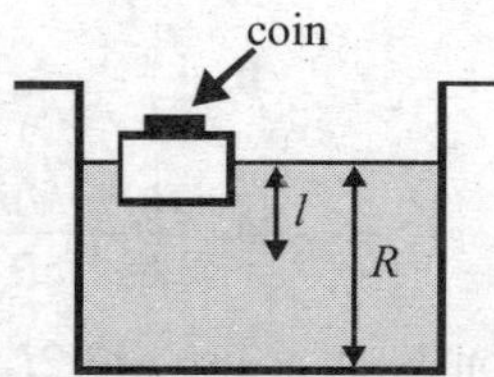

Fig. 3.269

11. Water is filled in a container up to height 3 m. A small hole of area 'a' is punched in the wall of the container at a height 52.5 cm from the bottom. The cross sectional area of the container is A. If $a/A = 0$; then v^2 is (where v is the velocity of water coming out of the hole) **(IIT-JEE, 2005)**
    a. 50 **b.** 51
    c. 48 **d.** 51.5

12. Water is filled up to a height h in a beaker of radius R as shown in the figure. The density of water is ρ, the surface tension of water is T and the atmospheric pressure is P_0. Consider a vertical section $ABCD$ of the water column through a diameter of the beaker. The force on water on one side of the section by water on the other side of this section has magnitude **(IIT-JEE, 2007)**

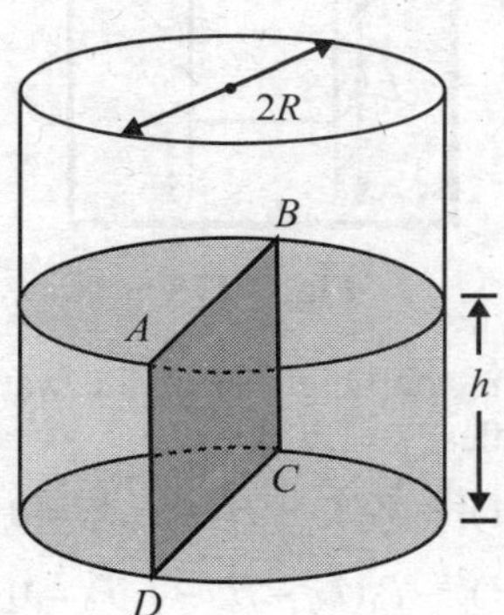

Fig. 3.270

a. $|2P_0Rh + \pi R^2\rho gh - 2RT|$
b. $|2P_0Rh + R\rho gh^2 - 2RT|$
c. $|P_0\pi R^2 + R\rho gh^2 - 2RT|$
d. $|P_0\pi R^2 + R\rho gh^2 + 2RT|$

Assertion and Reasoning Type

In each of the questions, assertion (A) is given by corresponding statement of reason (R) of the statements, mark the correct answer.

a. If both assertion and reason are true and the reason is correct explanation of the assertion.
b. If both assertion and reason are true but reason is not the correct explanation of assertion.
c. If assertion is true, but the reason is false.
d. If assertion is false, but the reason is true.

1. **Assertion:** The stream of water flowing at high speed from a garden hose pipe tends to spread like a fountain when held vertically up, but tends to narrow down when held vertically down.
    Reason: In any steady flow of an incompressible fluid, the volume flow rate of the fluid remains constant. **(IIT-JEE, 2008)**

Multiple Choice Questions With One or More than One Correct Answer Type

1. The spring balance A reads 2 kg with a block m suspended from it. A balance B reads 5 kg when a beaker with liquid is put on the pan of the balance. The two balances are now so arranged that the hinging mass is inside the liquid in the beaker as shown in the figure. In this situation, **(IIT-JEE, 1985)**
    a. the balance A will read more than 2 kg
    b. the balance B will read more than 5 kg
    c. the balance A will read less than 2 kg and B will read more than 5 kg
    d. the balance A and B will read 2 kg and 5 kg, respectively

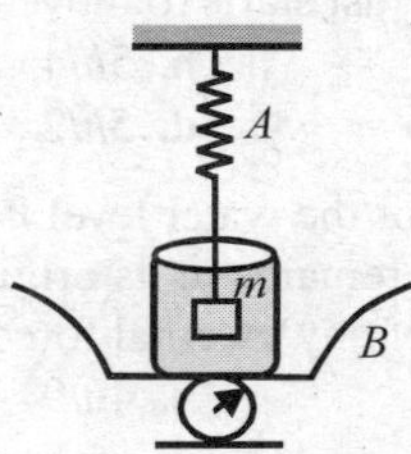

Fig. 3.271

2. Two solid spheres and of equal volumes but of different densities and are connected by a string. They are fully immersed in a fluid of density. They get arranged into an equilibrium state as shown in the figure with a tension in the string. The arrangement is possible only if

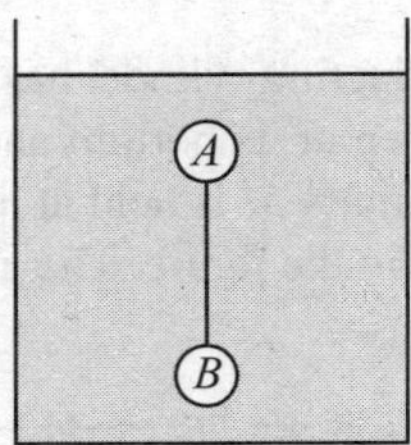

Fig. 3.272

a. $d_A < d_F$ **b.** $d_B > d_F$
c. $d_A > d_F$ **d.** $d_A + d_B = 2d_F$

Comprehension Type

For Problems 1–3

A cylindrical tank has a hole of diameter $2r$ in its bottom. The hole is covered wooden cylindrical block of diameter $4r$, height h and density $\rho/3$.

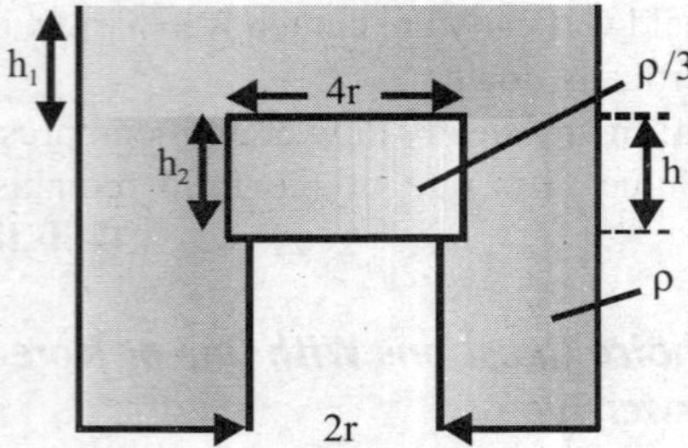

Fig. 3.273

Situation 1: Initially, the tank is filled with water of density ρ to a height such that the height of water above the top of the block is h_1 (measured from the top of the block).

Situation 2: The water is removed from the tank to a height h_2 (measured from the bottom of the block), as shown in the figure. The height h_2 is smaller than h (height of the block) and thus the block is exposed to the atmosphere. **(IIT-JEE, 2006)**

1. Find the minimum value of height h_1 (in situation 1) for which the block just starts to move up?
   a. $2h/3$ **b.** $5h/4$
   c. $5h/3$ **d.** $5h/2$

2. Find the height of the water level h_2 (in situation 2), for which the block remains in its original position without the application of any external force.
   a. $h/3$ **b.** $4h/9$
   c. $2h/3$ **d.** h

3. In situation 2, if h_2 is further decreased, then
   a. cylinder will not move up and remains at its original position
   b. for $h_2 = h/3$, cylinder again starts moving up
   c. for $h_2 = h/4$, cylinder again starts moving up
   d. for $h_2 = h/5$, cylinder again starts moving up

For Problems 4–6

A fixed thermally conducting cylinder has a radius R and height L_0. The cylinder is open at its bottom and has a smaller hole at its top. A piston of mass M is held at a distance L from the top surface, as shown in the figure. The atmospheric pressure is P_0. **(IIT-JEE, 2007)**

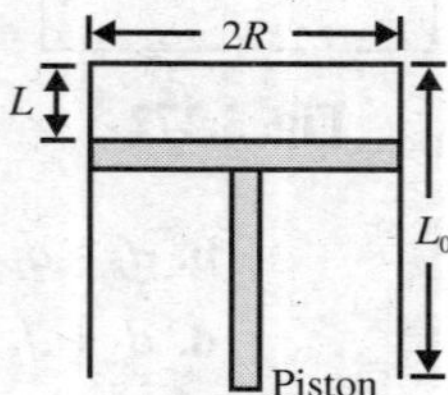

Fig. 3.274

4. The piston is now pulled out slowly and held at a distance $2L$ from the top. The pressure in the cylinder between its top and the piston will then be
   a. P_0 **b.** $\frac{P_0}{2}$
   c. $\frac{P_0}{2} - \frac{Mg}{\pi R^2}$ **d.** $\frac{P_0}{2} - \frac{Mg}{\pi R^2}$

5. While the piston is at a distance $2L$ from the top, the hole at the top is sealed. The piston is then released to a position where it can stay in equilibrium. In this condition, the distance of the piston from the top is
   a. $\left(\frac{2P_0\pi R^2}{\pi R^2 P_0 + Mg}\right)(2L)$ **b.** $\left(\frac{P_0\pi R^2 - Mg}{\pi R^2 P_0}\right)(2L)$
   c. $\left(\frac{P_0\pi R^2 + Mg}{\pi R^2 P_0}\right)(2L)$ **d.** $\left(\frac{P_0\pi R^2}{\pi R^2 P_0 - Mg}\right)(2L)$

6. The piston is taken completely out of the cylinder. The hole at the top is sealed. A water tank is brought below the cylinder and put in a position, so that the water surface in the tank is at the same level as the top of the cylinder as shown in the figure. The density of the water is ρ.

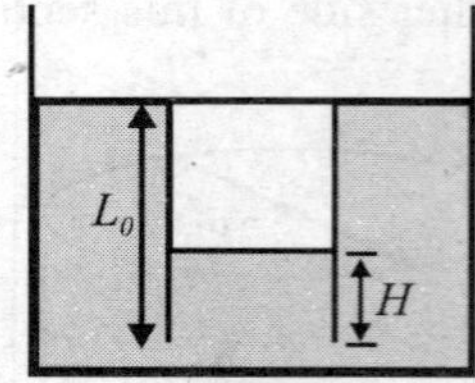

Fig. 3.275

In equilibrium, the height H of the water column in the cylinder satisfies
   a. $\rho g(L_0 - H)^2 + P_0(L_0 - H) + L_0P_0 = 0$
   b. $\rho g(L_0 - H)^2 - P_0(L_0 - H) - L_0P_0 = 0$
   c. $\rho g(L_0 - H)^2 + P_0(L_0 - H) - L_0P_0 = 0$
   d. $\rho g(L_0 - H)^2 - P_0(L_0 - H) + L_0P_0 = 0$

For Problems 7–9

A small spherical monoatomic ideal gas bubble ($\gamma = 5/3$) is trapped inside a liquid of density ρ_1 (see figure). Assume that the bubble does not exchange any heat with the liquid. The bubble contains n moles of gas. The temperature of the gas when the bubble is at the bottom is T_0, the height of the liquid is H and the atmospheric pressure is P_0 (neglect surface tension). **(IIT-JEE, 2008)**

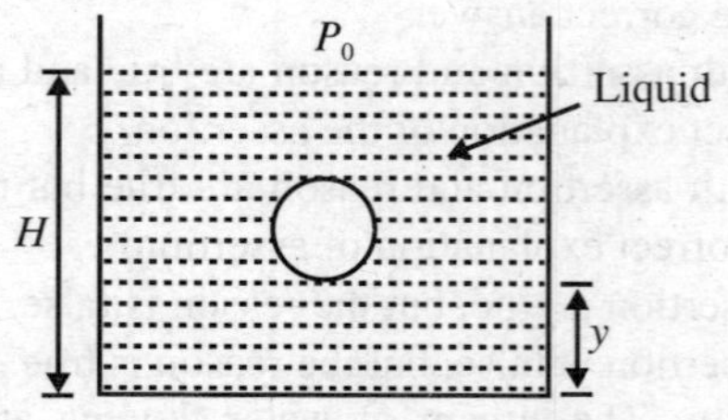

Fig. 3.276

7. As the bubble moves upwards, beside the buoyancy force, the following forces are acting on it.
   a. Only the force of gravity
   b. The force due to gravity and the force due to the pressure of the liquid
   c. The force due to gravity, the force due to the pressure of the liquid and the force due to viscosity of the liquid
   d. The force due to gravity and the force due to viscosity of the liquid

8. When the gas bubble is at height y from the bottom, its temperature is
   a. $T_0\left(\dfrac{P_0+\rho_l gH}{P_0+\rho_l gy}\right)^{2/5}$ **b.** $T_0\left(\dfrac{P_0+\rho_l g(H-y)}{P_0+\rho_l gy}\right)^{2/5}$
   c. $T_0\left(\dfrac{P_0+\rho_l gH}{P_0+\rho_l gy}\right)^{3/5}$ **d.** $T_0\left(\dfrac{P_0+\rho_l g(H-y)}{P_0+\rho_l gy}\right)^{3/5}$

9. The buoyancy force acting on the gas bubble is (assume R is the universal gas constant)
   a. $\rho_l nRgT_0\dfrac{(P_0+\rho_l gH)^{2/5}}{(P_0+\rho_l gy)^{7/5}}$
   b. $\dfrac{\rho_l nRgT_0}{(P_0+\rho_l gH)^{2/5}[(P_0+\rho_l g(H-y))]^{3/5}}$
   c. $\rho_l nRgT_0\dfrac{(P_0+\rho_l gH)^{3/5}}{(P_0+\rho_l gy)^{8/5}}$
   d. $\dfrac{\rho_l nRgT_0}{(P_0+\rho_l gH)^{3/5}[(P_0+\rho_l g(H-y))]^{2/5}}$

Integer Answer Type

This section contains two questions. The answer to each of the questions is a single-digit integer, ranging from 0 to 9. The appropriate bubbles below the respective question numbers in the ORS have to be darkened. For example, if the correct answers to question numbers X, Y, Z and W (say) are 6, 0, 9 and 2, respectively, then the correct darkening of bubbles will look like the following:

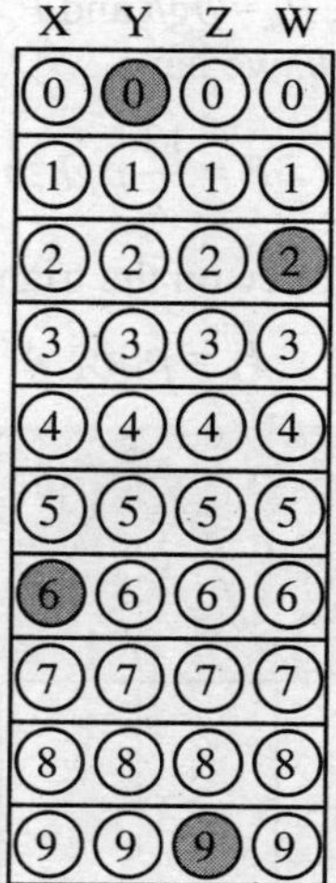

Fig. 3.277

1. A cylindrical vessel of height 500 mm has an orifice (small hole) at its bottom. The orifice is initially closed and water is filled in it up to height H. Now the top is completely sealed with a cap and the orifice at the bottom is opened. Some water comes out from the orifice and the water level in the vessel becomes steady with height of water column being 200 mm. Find the fall in height (in mm) of water level due to opening of the orifice. (Take atmospheric pressure = 1.0×10^5 N/m², density of water = 1000 kg/m³ and g = 10 m/s². Neglect any effect of surface tension.) **(IIT-JEE, 2009)**

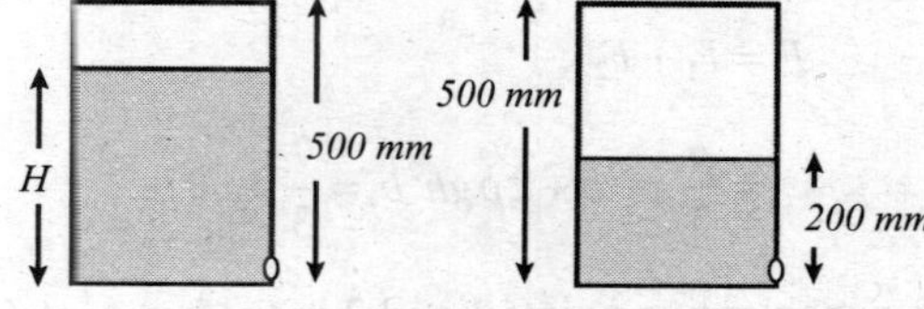

Fig. 3.278

ANSWERS AND SOLUTIONS

Subjective Type

1. Let the hydrostatic forces acting on the rectangular patches of areas A_1 and A_2 of the gate be F_1 and F_2. Then, the total hydrostatic force on the gate is

$F = F_1 + F_2$

where $F_1 = P_{av_1}A_1$ and $F_2 = P_{av_2}A_2$. P_{av_1} and P_{av_2} are the average pressures on the areas A_1 and A_2, respectively.

$$F = P_{av_1}A_1 + P_{av_2}A_2$$

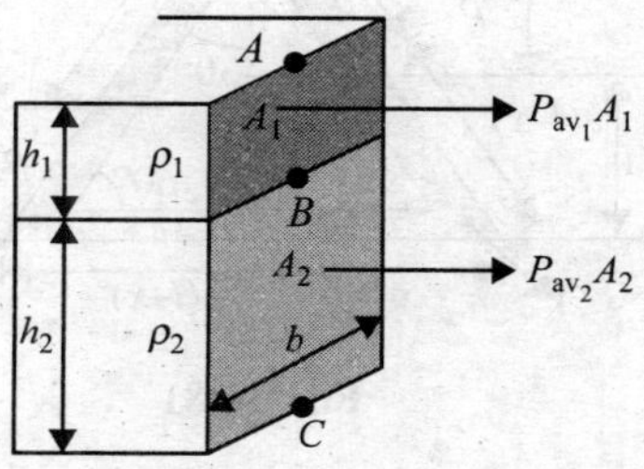

Fig. 3.279

Substituting $P_{avl} = (P_A + P_B)/2$, $A_1 = bh_1 = bh$, $P_{av2} = (P_B + P_C)/2$, $A_2 = bh_2 = b(2h)$, we have

$$F = \frac{(P_A + P_B)}{2}bh + \frac{(P_B + P_C)}{2}2bh$$

Substituting $P_A = 0$, $P_B = \rho gh$ and $P_C = (\rho_1 h_1 + \rho_2 h_2)g = [\rho h + (2\rho)(2h)]g = 5\rho gh$, we have

$$F = \frac{13}{2}\rho gh^2 b$$

2. **a.** The hydrostatic force on the triangular area is

$$F_1 = \rho_1 g y_{C_1} A_1$$

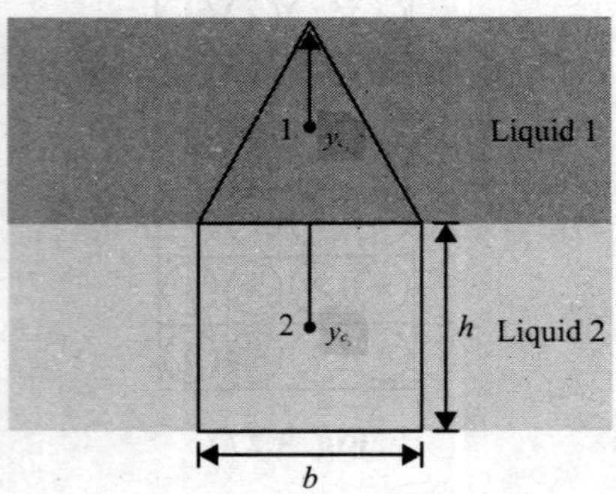

Fig. 3.280

where $\rho_1 = \rho$, $y_{C_1} = 2h/3$ and $A_1 = bh/2$. This gives

$$F_1 = \frac{\rho gh^2 b}{3}$$

b. The hydrostatic force on the rectangular area is

$$F_2 = (\rho_1 gh_1 + \rho_2 ghy_{C_2})A_2$$

Substituting $\rho_1 = \rho$, $h_1 = h$, $\rho_2 = 2\rho$, $y_{C_2} = h/2$ and $A_2 = bh$, we have $F_2 = \rho gh^2 b$.

c. The total hydrostatic force on the composite patch is

$$F = F_1 + F_2$$

$$= \frac{\rho}{3}gh^2 b + 2\rho gh^2 b = \frac{7}{3}\rho gh^2 b$$

3. Let us consider two points 1 and 2 in the horizontal limb.

Pressure at 1 is $P_1 = \rho_1 gh_1 + \rho_2 gh_2$

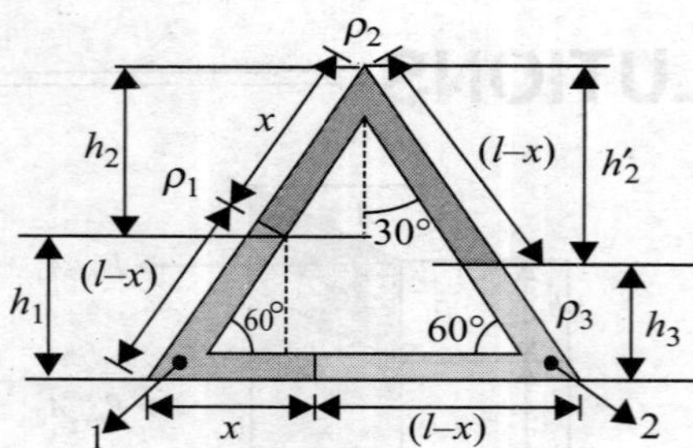

Fig. 3.281

Pressure at 2 is $P_2 = \rho_2 gh'_2 + \rho_3 gh_3$

Since $P_1 = P_2$ for non-accelerating liquid, we have

$$\rho_1 h_1 + \rho_2 h_2 = \rho_2 h'_2 + \rho_3 h_3$$

Substituting $h_1 = h'_2 = (l - x)\sin 60°$ and $h_2 = h_3 = x \sin 60°$, we have

$$x = \frac{(\rho_2 - \rho_1)l}{2\rho_2 - (\rho_1 + \rho_3)}$$

4. $\tan\phi = \frac{a}{g} = \frac{h_2 - h_1}{h_2 + h_1} \Rightarrow a = \frac{g(h_2 - h_1)}{h_2 + h_1}$

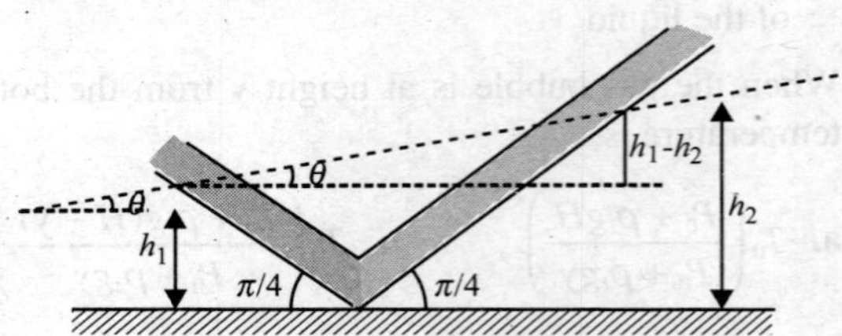

Fig. 3.282

5. $\tan\theta = \frac{a}{g} = 1 \Rightarrow \theta = 45°$

$$h_{max} = \sqrt{2}R$$

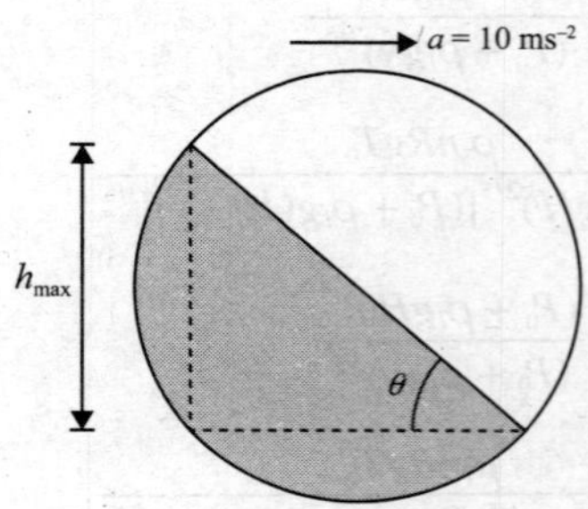

Fig. 3.283

$p_{max} = \rho gh_{max} = 0.8 \times 10^3 \times 10 \times \sqrt{2} \times 1.2 = 9600\sqrt{2}$ Pa above atmosphere

$\therefore \quad P = 9600$

6. Let V be the total volume of the balloon and M be its mass. When it is floating on the surface of lake.

Weight of water displaced = weight of balloon

$$\Rightarrow \quad \frac{2}{3}V\rho g = Mg \quad \text{or} \quad M = \frac{2}{3}V\rho$$

Now, let the depth of immersion be h, so that the new pressure is $(10 + h)$ of water. Applying Boyle's law,

$$P_1V_1 = P_2V_2$$

$$\Rightarrow \quad 10V = (10 + h)V' \Rightarrow \quad V' = \frac{10V}{(10 + h)}$$

Let the balloon be at equilibrium at a depth h from the surface of the lake, then at equilibrium

$$Mg = V'\rho g$$

$$\Rightarrow \quad M = V'\rho$$

$$\therefore \quad h = 5 \text{ m}$$

7. a. According to the condition of floating,

$$0.8\rho_w gh_k + \rho_w gh_w = 0.9\rho_w gh \quad \text{(i)}$$

where h_k and h_w are the submerged depth of the ice in the kerosene and water, respectively.

$$h_k + h_w = h \quad \text{(ii)}$$

Solving Eqs. (i) and (ii), we get $h_k = 0.5$ cm, $h_w = 0.5$ cm.

b. 1 cm³ ice $\xrightarrow{\text{melts}}$ 0.9 cm³ water

After melting the ice, 0.5 cm³ of volume will be occupied by the kerosene. Therefore, the fall in the level of kerosene is $\Delta h_k = 0.5/A$.

Out of 0.9 cm³ of water formed, 0.5 cm³ of water will be filled in the vacant part of ice and remaining (0.9 – 0.5) = 0.4 cm³ of water will be available to increase the level of water.

Rise in the level of water is $\Delta h_w = \dfrac{0.9-0.5}{A} = \dfrac{0.4}{A}$

Hence, the net fall in the overall level is

$$\Delta h = \frac{0.1}{A} = \frac{0.1}{10} = 0.01 \text{ cm} = 0.1 \text{ mm}$$

8. Let the depth of water layer be h_1 cm. So, the depth to which the cube is immersed in mercury will be as follows

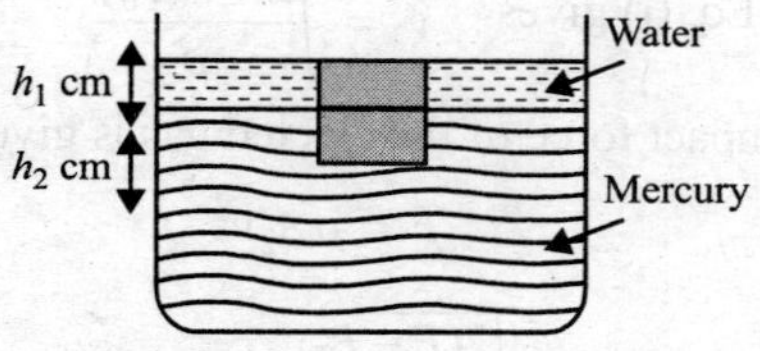

Fig. 3.284

$$h_2 = (5 - h_1) \text{ cm}$$

Weight of water displaced = $(5^2 h_1)g$ dyn and weight of mercury displaced = $5^2(5 - h_1)$ 13.6 g dyn.

Therefore, upthrust offered by water and mercury together is given by $5^2 h_1 g + 5^2 (5 - h_1)\, 13.6\, g$

$$= 5^2 (68 - 12.6\, h_1)\, g \text{ dyn}$$

Weight of cube = $5^3 \times 8.56\, g$ dyn

From the principle of floatation, we have

Upthrust = weight of the floating body

$\therefore \quad 5^2 (68 - 12.6\, h_1)\, g = 5^3 \times 8.56\, g$

$\Rightarrow \quad 68 - 12.6\, h_1 = 41.8$

$\Rightarrow \quad h_1 = 25.2/(12.6) = 2$ cm

9. Let y be the length of the plank inside water.

$$y = \frac{0.5}{\cos\theta}$$

Let A be the cross sectional area of the plank. Then buoyant force on it is $F_b = V\rho_w g = (Ay)\,\rho_w g$

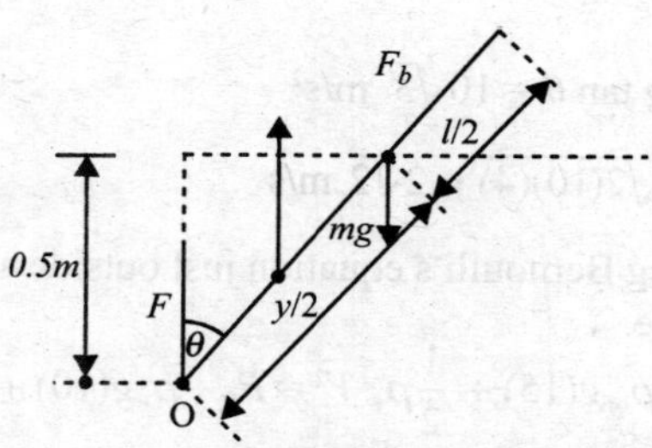

Fig. 3.285

Since plank is in rotational equilibrium, so $\sum \vec{\tau}_0 = 0$

or $\quad mg \times \dfrac{l}{2}\sin\theta - F_b \times \dfrac{y}{2}\sin\theta = 0$ or $\quad mgl - F_b \times y = 0$

or $\quad (Al \times 0.5)gl - (Ay)d_w y = 0$ or $\quad 0.5l^2 = y^2$

or $\quad 0.5 \times (1)^2 = \left(\dfrac{0.5}{\cos\theta}\right)^2$ or $\quad \cos^2\theta = \dfrac{1}{2}$

$\therefore \quad \theta = 45°$

10. Let A_1 and A_2 be the cross-sectional areas of the pipe at points P and Q, respectively.
Let v_1 and v_2 be the velocities of oil at the points P and Q, respectively. By conservation of mass,

$$Q = A_1 v_1 = A_2 v_2 \Rightarrow v_2 = \left(\frac{A_1}{A_2}\right)(v_1) = 4v_1$$

Applying Bernoulli's equation between points P and Q, we have

$$P_1 + \frac{1}{2}\rho v_1^2 = P_2 + \frac{1}{2}\rho v_2^2$$

$$P_2 = P_1 + \frac{1}{2}\rho(v_1^2 - v_2^2) = P_1 + \frac{1}{2} \times 0.9[16 - 256] \times 10^3$$

$$= 280 \times 10^3 + (-108) \times 10^3$$

$P_2 = 172 \times 10^3$ Nm^{-2}

$P_2 = 172$ (kNm^{-2})

11. When the pressure on cap is 4×10^4 Pa,

$$h = 4 \text{ m}$$

From the figure, $\tan\theta = \dfrac{3}{\sqrt{3}} \Rightarrow \theta = 60°$

As water does not spill out, the volume of water will be constant.

Hence, $\quad 2.5 \times \sqrt{3} \times \sqrt{3} = \left(\dfrac{4+y}{2}\right) \times \sqrt{3} \times \sqrt{3}$

$\Rightarrow \quad y = 1$ m

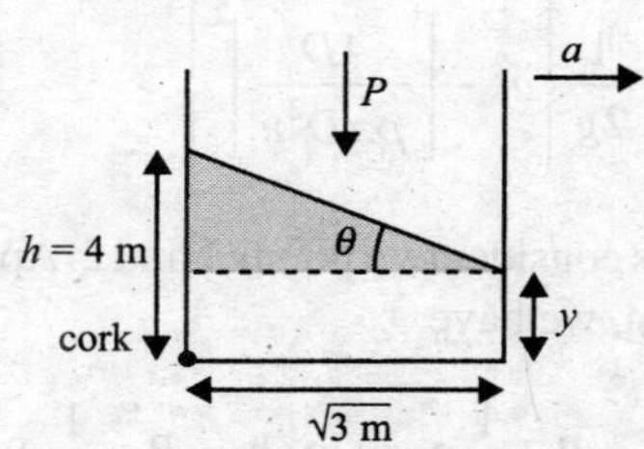

Fig. 3.286

(a) $a = g\tan\theta = 10\sqrt{3}\ \text{m/s}^2$

(b) $v = \sqrt{2(10)(4)} = 2\sqrt{2}\ \text{m/s}$

12. Applying Bernoulli's equation just outside and just inside th orifice.

(i) $P_0 + \rho_{\text{oil}}g(15) + \dfrac{1}{2}\rho_w V^2 = P_0 + \rho_w g(10) + \rho_{\text{oil}}g(5)$

$\therefore\ V = \sqrt{40}\ \ \text{m/s} = 6.3\ \text{m/s}$

(ii) Let h be the height when flow through orifice slopes. Again applying Bernoulli's equation,

$$P_0 + \rho_{\text{oil}}g(5) + \rho_w gh = P_0 + \rho_{\text{oil}}g(15)$$

$$h = \frac{\rho_{\text{oil}}}{\rho_w}10 = 8\ \text{m}$$

(iii) $P_0 + \rho_{\text{oil}}g(5)\,\rho_\omega gh = P_0 + \rho_{\text{oil}}g(15) + \dfrac{1}{2}\rho_\omega V^2$

$$\therefore\ V = \sqrt{20(h-8)}$$

$$\Rightarrow -\frac{dh}{dt} = \sqrt{20(h-8)} \quad \Rightarrow -\int_{10}^{8}\frac{dh}{\sqrt{20(h-8)}} = \int_0^{t} dt$$

$$\therefore\ t = \sqrt{\frac{2}{5}}\ \text{s}$$

13. Applying continuity equation at nozzle end and at top position.

$$a_1 u = a_2 v \quad \Rightarrow \quad v = \frac{a_1 u}{a_2} \qquad \text{(i)}$$

$$\Rightarrow v^2 - u^2 = -2gH$$

$$\Rightarrow H = \frac{1}{2g}\left\{u^2 - \left[\frac{W}{\rho a_1^2 u}\right]^2\right\} \qquad \text{(ii)}$$

$$\Rightarrow \rho a_2 v^2 = W \Rightarrow v = \frac{W}{\rho a_1 u}$$

$$\Rightarrow \frac{\rho a_1 u}{W} = \frac{1}{v} \quad \Rightarrow v^2 = \frac{W}{\rho a_1 u}$$

$$\Rightarrow H = \frac{1}{2g}\left\{u^2 - \left[\frac{W}{\rho a_1^2 u}\right]^2\right\}$$

$$= \frac{1}{2g}\left\{u^2 - \left[\frac{4D}{\rho\pi D^2 u}\right]^2\right\}$$

Fig. 3.287

14. **a.** Let us consider two points 1 and 2. Applying Bernoulli's equation, we have

$$P_1 + \frac{1}{2}\rho v_1^2 + \rho g h_1 = P_2 + \frac{1}{2}\rho v_2^2 + \rho g h_2$$

Since $\rho g(h_1 - h_2) = \rho gh$ is negligible compared to $P_1 = P$ (pressure inside the chamber) and $P_2 = P$ (atmospheric pressure), we have

$$P + \frac{1}{2}\rho v_1^2 = P_0 + \frac{1}{2}\rho v_2^2 \qquad \text{(i)}$$

Equation of continuity gives

$$A_1 v_1 = A_2 v_2 \qquad \text{(ii)}$$

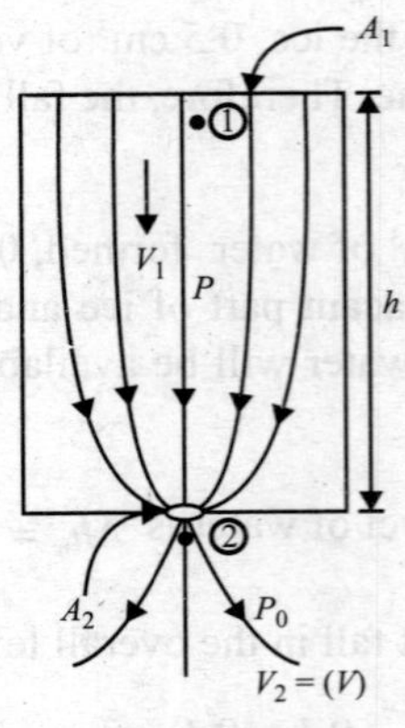

Fig. 3.288

Since $v_1 = (A_2/A_1)v_2$ and $A_2 \ll A_1$, we can ignore v_1 when compared to v.

Then Eq. (i) gives $\quad v_2 = \sqrt{\dfrac{2(P-P_0)}{\rho}}$

b. The impact force of the ejected gas is given as

$$F = \rho A_2 v_2^2$$

where $\quad v_{\text{rel}} = v_2 = \sqrt{\dfrac{2(P-P_0)}{\rho}}$

Then, we have $F = 2A_2(P - P_0)$.

15. (a) First of all, we take two points 1 and 2 at the bottom of the left and right vertical limbs. The pressure at 1 is

$$P_1 = P_0 + (2\rho)gh$$

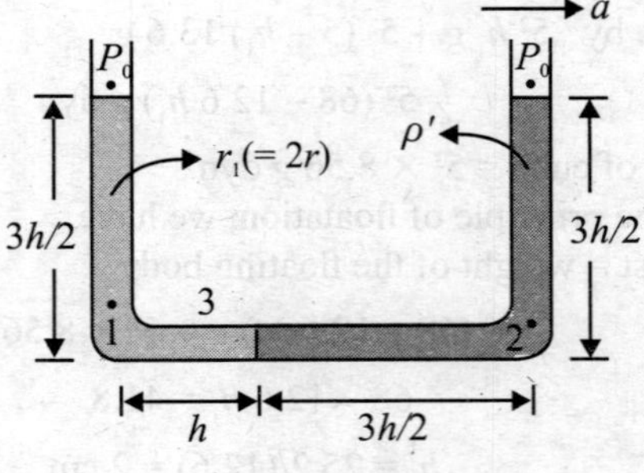

Fig. 3.289

The pressure at 2 is $P_2 = P_0 + \rho' g(2h)$

where ρ' = density of the other liquid

Since $P_1 = P_2$ for stationary liquid, we have

$$P_0 + 2\rho gh = P_0 + \rho' g(2h)$$

This gives $\rho' = \rho$.

(b) If the levels in both the limbs are equal, conserving the volume of the total (two) liquids, the height of each liquid column will be equal to $3h/2$. The lengths of liquids in the horizontal limbs are $h/2$ and $3h/2$, respectively.

The pressure at interface 3 of the liquid is

$$P_3 = P_1 - (2\rho)a\left(\frac{h}{2}\right)$$

Where $P_2 = P_0 + (2\rho)a\left(\frac{3h}{2}\right)$

This gives $P_3 = P_0 + (2\rho)g\left(\frac{3h}{2}\right) - (2\rho)a\left(\frac{h}{2}\right)$

$$= P_0 + 3\rho gh - \rho ah \quad \text{(i)}$$

P_3 can also be given as $P_3 = P_2 + \rho a\left(\frac{3h}{2}\right)$ where $P_2 = P_0 + \rho a\left(\frac{3h}{2}\right)$. This gives

$$P_3 = P_0 + \frac{3}{2}\rho gh + \frac{3}{2}\rho ah \quad \text{(ii)}$$

Eliminating P_3 from Eqs. (i) and (ii), we have

$$a = \frac{3}{5}g$$

16. Let us divide the cup into two parts, AB and BC. Liquid pushes the curved surface AB upwards and BC downwards.

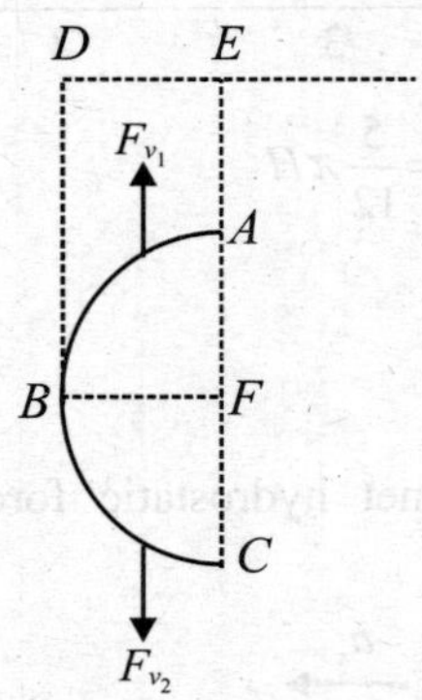

Fig. 3.290

Vertical thrust of the liquid on AB is $F_{V_1} = V_1\rho g$ where V_1 = volume of the portion $ABDE$ above AB

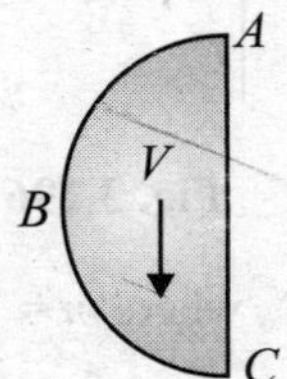

Fig. 3.291

Similarly, the vertical thrust of liquid on BC is $F_{V_2} = V_2\rho g$ where V_2 = volume of the portion $ACBDE$ above BC. Hence the net vertical force on the curved surface ABC of the cup is $F_V = F_{V_2} - F_{V_1} = V_2\rho g - V_1\rho g = (V_2 - V_1)\rho g = V\rho g$,

where V = volume of liquid in the cup shown as shaded portion.

17. Let pressure at P be p^1

Now at a distance x

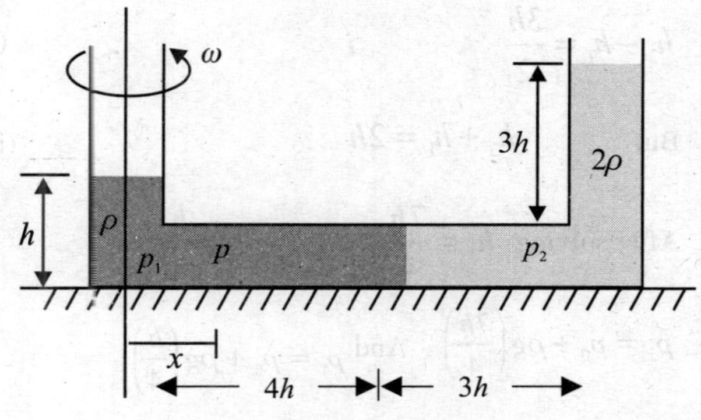

Fig. 3.292

$$p - p_1 = \frac{\rho\omega^2}{2}[x^2]_0^{4h}$$

$$p - p_1 = \frac{\rho\omega^2}{2} \times 16h^2 \quad \text{(i)}$$

$$p_2 - p = \frac{2\rho\omega^2}{2}[x^2]_{4h}^{7h}$$

$p^2 - p = \rho\omega^2 \times (33h^2)$ (ii)

Adding Eqs. (i) and (ii) $p_2 - p_1 = 41\rho\omega^2 h^2$

$$5\rho gh = 41\rho h^2\omega^2; \Rightarrow \sqrt{\frac{5g}{41h}} = \omega$$

18. $\frac{dp}{dx} = \rho(\omega^2 x) \Rightarrow dp = \rho\omega^2 x\, dx$

Integrating both sides, we get

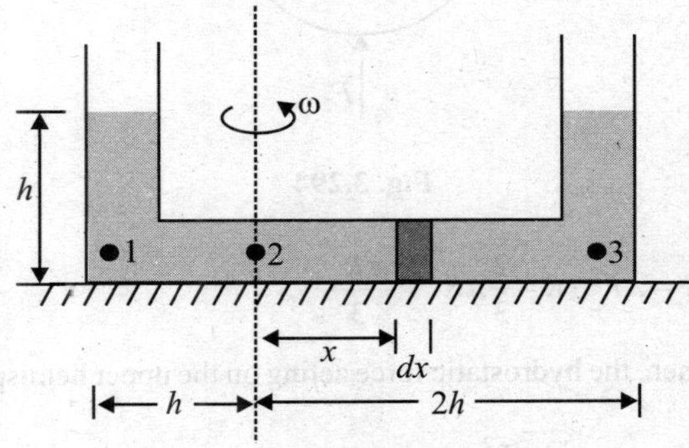

Fig. 3.293

$$\int_{p_1}^{p_3} dp = \rho\omega^2 \int_{-h}^{2h} x dx$$

$$p_3 - p_1 = \rho\omega^2\left[\frac{x^2}{2}\right] = \frac{\rho\omega^2}{2}[4h^2 - h^2]$$

$$\rho g(h_3 - h_1) = \frac{3\rho\omega^2 h^2}{2} = \frac{3\rho}{2}h^2\left(\frac{g}{h}\right)$$

$$h_3 - h_1 = \frac{3h}{2} \qquad i \qquad \text{(i)}$$

But $\quad h_3 + h_1 = 2h \qquad$ (ii)

After solving $h_3 = \dfrac{7h}{4}$ *and* $h_1 = \dfrac{h}{4}$

$$p_3 = p_0 + \rho g\left(\frac{7h}{4}\right) \quad \text{And} \quad p_1 = p_0 + \rho g\left(\frac{h}{4}\right)$$

$$p_3 - p_2 = \frac{\rho\omega^2}{2}[x^2]_0^{2h} = \frac{\rho\omega^2}{2}(4h)^2$$

$$p_3 - p_2 = \frac{\rho}{2}\left(\frac{g}{h}\right)\cdot 4h^2 = 2\rho g h$$

$$p_0 + \rho g\left(\frac{7h}{4}\right) - p_2 = 2\rho g h \Rightarrow p_2 = p_0 - \frac{\rho g h}{4}$$

19. Volume V_1 of the super-incumbent liquid of the upper hemisphere.

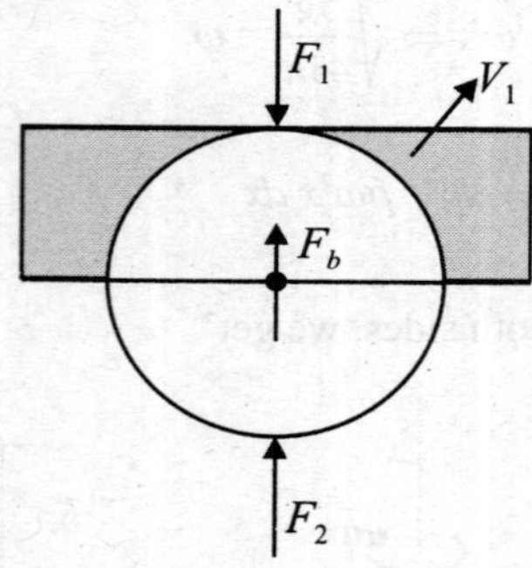

Fig. 3.294

$$V_1 = \pi R^2 \cdot R = \frac{2}{3}\pi R^3 = \frac{\pi R^3}{3}$$

Then, the hydrostatic force acting on the upper hemispher

$$F_1 = \rho V_1 g = \frac{\pi R^2 \rho g}{3}$$

Then, $F_2 - F_1 = F_b$ or $F_2 = F_b + F_1 = \dfrac{5}{3}\pi R^3 \rho g$

Then, $\dfrac{F_1}{F_2} = \dfrac{1}{5}$

20. The upward thrust due to the liquid 1 = downward thrust due to liquid 2.

$F_1 = F_2$

Fig. 3.295

$$F_1 = \left[\rho_1 \pi R^2(2H)g - \rho_1\left(\pi R^2\left(\frac{2H}{3}\right)\right)\right]g = \frac{4\pi R^2 H\rho_1 g}{3}$$

$$= \frac{4\pi\rho_1 H^3 g}{3}$$

$F_2 = V\rho_2 g$, where V = volume of the double shaded position

$$= \pi R^2 H - \left[\frac{\pi R^2(2H)}{3} - \frac{\pi(R/2)^2 H}{3}\right]$$

$$= \frac{5}{12}\pi R^2 H = \frac{5}{12}\pi H^3$$

Then, $\dfrac{\rho_1}{\rho_2} = \dfrac{5}{16}$

21. There are two net hydrostatic forces along x and y directions.

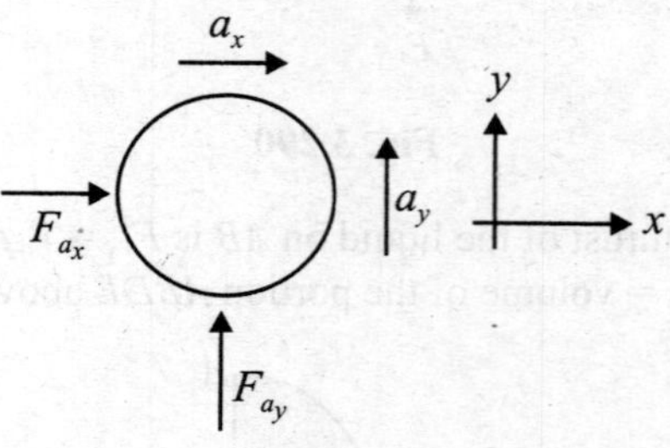

Fig. 3.296

$F_{bx} = V\rho a_0 = ma_x$, where $m = V\sigma$

Or $a_x = \dfrac{\rho}{\sigma}a_0$

$$F_y = V\rho g - mg = ma_y$$

Where $m = V\sigma$

$$a_y = \left(\frac{\rho}{\sigma} - 1\right)g$$

Then, $\vec{a} = \frac{\rho}{\sigma} a_0 \hat{i} + \left(\frac{\rho}{\sigma} - 1\right) g\,\hat{j}$

22. (i) $F = N = \text{mg} \cos \theta$

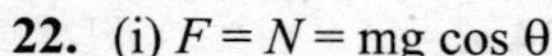

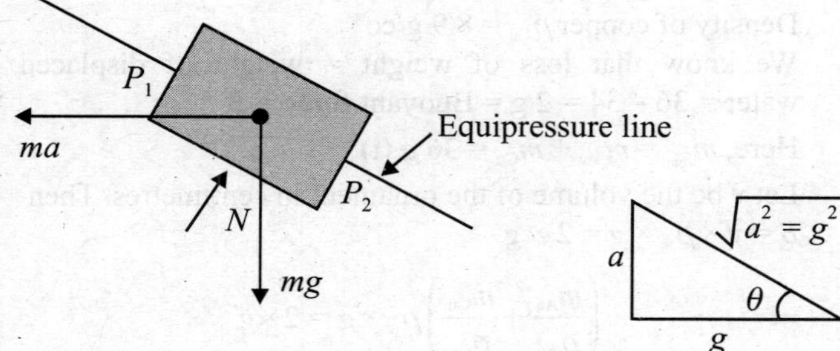

Fig. 3.297

where $\cos\theta = \frac{g}{\sqrt{g^2 + a^2}}$ or $T = \frac{mg^2}{\sqrt{g^2 + a^2}}$

(ii) $\tan\theta = \frac{a}{g}$ or $\theta = \tan^{-1}\frac{a}{g}$

23. $v_{\text{rel}} = \sqrt{2g_{\text{eff}}\frac{h}{2}} = -\sqrt{g_{\text{eff}}h} = \sqrt{gh\cos\theta}$

The actual velocity of the liquid, $v = (v_r - v_0)$

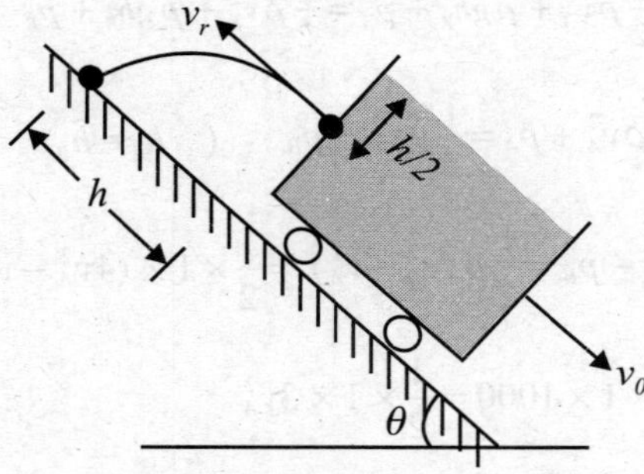

Fig. 3.298

Then, the time of flight, $T = \sqrt{\frac{2(h/2)}{g\cos\theta}} = \sqrt{\frac{h}{g\cos\theta}}$

$h = vt - \frac{1}{2}(g\cos\theta)T^2$, putting T and v, we have

$$v_0 = h\left(1 + \frac{\tan\theta}{2}\right)\sqrt{\frac{g\cos\theta}{h}} + \sqrt{gh\cos\theta}$$

24. $F = (\Delta P)A_{\text{eff}}$, where $A_{\text{eff}} = \pi R^2$ and $\Delta P = |P_1 - P_2|$

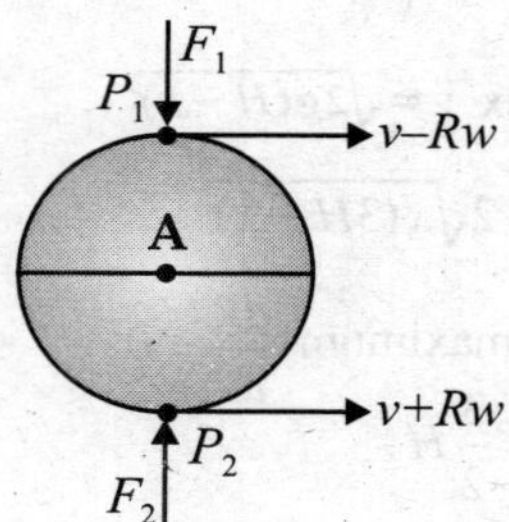

Fig. 3.299

$$\Delta P = \frac{1}{2}\rho\left[(v + R\omega)^2 - (v - R\omega)^2\right] = 2R\rho v\omega$$

Then $\quad F = 2\pi\rho R^3 v\omega$

25. Since $y = gx^2/2v^2$ for a horizontal projection for the same x, we have y_1, y_2 and v_1, v_2. Putting $y_1 - y_2 = h$ and other values, find x and y_s.

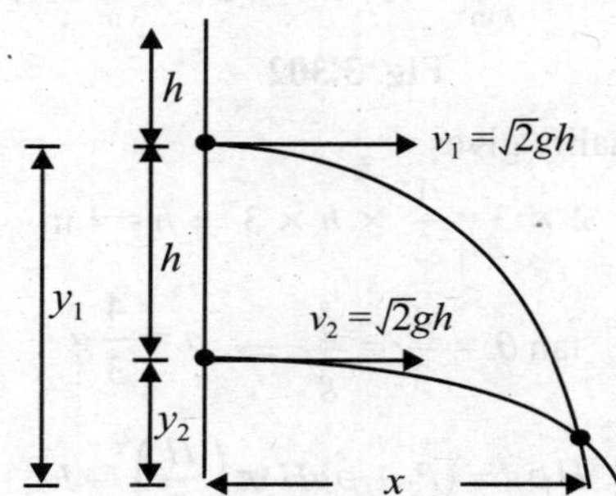

Fig. 3.300

Objective Type

1. b. Let V be the total volume of the ball and v be the volume of the ball in the upper liquid. Then $V - v$ is the volume of the lower liquid displaced.

Using the law of floatation, we have

$$V\rho g = v\rho_1 g + (V - v)\rho_2 g$$

$$V\rho = v\rho_1 + V\rho_2 - v\rho_2$$

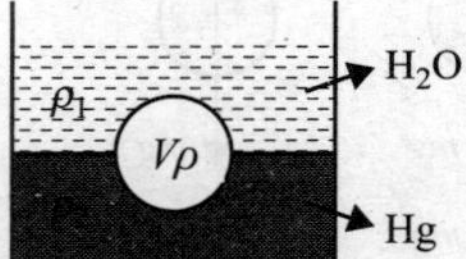

Fig. 3.301

or $\quad V(\rho - \rho_2) = v(\rho_1 - \rho_2)$

$$\frac{v}{V} = \frac{\rho - \rho_2}{\rho_1 - \rho_2} = \frac{\rho_2 - \rho}{\rho_2 - \rho_1}$$

2. c. As $x=\frac{1}{2}gt^2 \Rightarrow t=\sqrt{\frac{2x}{r}}$

Velocity of efflux $v=\sqrt{2g(H-x)}$

Hence $R=vt=2\sqrt{x(3H-x)}$

For range to be maximum $\frac{dR}{dx}=0$

which gives $x=\frac{3}{2}H$

Alternatively:
Maximum range is also equal to average of two heights

$$x=\frac{H+2H}{2}=\frac{3H}{2}$$

3. b.

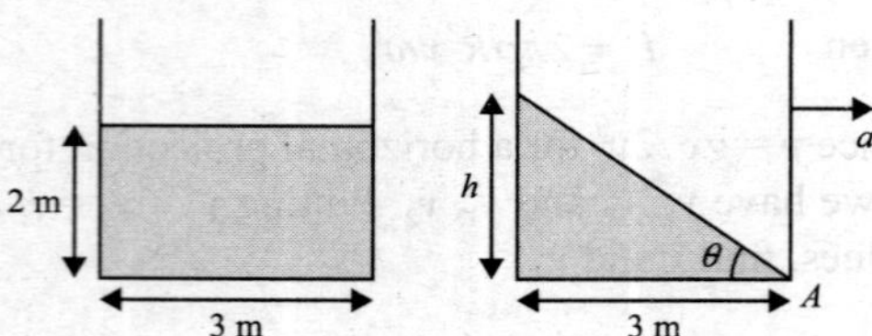

Fig. 3.302

Volume equality gives

$$2\times 3=\frac{1}{2}\times h\times 3 \Rightarrow h=4 \text{ m}$$

$$\therefore \quad \tan\theta=\frac{4}{3}=\frac{a}{g} \Rightarrow a=\frac{4}{3}g$$

4. b. $\frac{1}{3}\pi\left(\frac{H}{2}\right)^2 H\rho g=(P_0+\rho gH)\pi\left(\frac{H}{2}\right)^2 - F_V$

$$F_v=P_0\pi\frac{H^2}{4}+\rho g\frac{H^3}{4}\pi-\frac{\pi H^3\rho g}{12}$$

$$=P_0\frac{\pi H^2}{4}+\frac{\rho gH^3\pi}{6}$$

5. a. W = weight of liquid.
f_B = buoyant force on the ball
mg = weight of the ball
N = normal reaction between the ball and the surface
The free-body diagrams of the balls in each vessel are as follows.

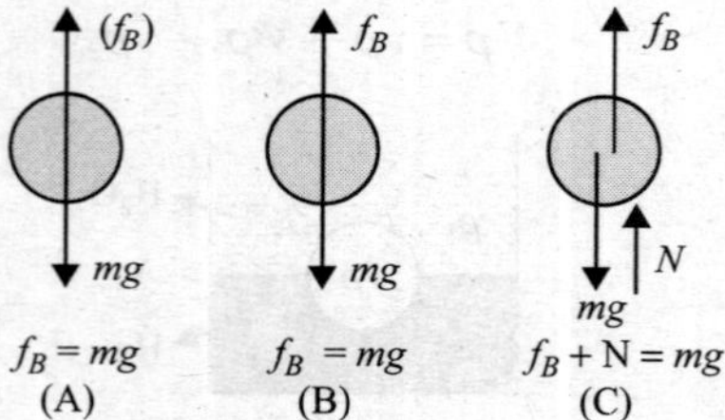

Fig. 3.303

At base, reaction force of buoyant force will act in downward direction.
The forces acting at the base of each tank are

$$F_A=W+f_B=W+mg$$
$$F_B=W+f_B=W+mg$$
$$F_C=W+f_B+N=W+mg$$

Thus, $F_A=F_B=F_C$

6. c. Pressure at the bottom level of the block

$$P=\frac{mg}{a}+P_0 \quad \text{(i)}$$

The pressure at the same level remains the same.

Hence, $\frac{mg}{a}+P_0=h\rho g+P_0 \Rightarrow h=\frac{m}{a\rho}$

7. a. Given that $m_{real}=36$ g, $m_{app}=34$ g.
Density of gold $\rho_{Au}=19.3$ g/cc
Density of copper $\rho_{Cu}=8.9$ g/cc
We know that loss of weight = weight of displaced water = 36 – 34 = 2 g = Buoyant force = B
Here, $m_{real}=m_{Aw}+m_{Cu}=36$ g (i)
Let v be the volume of the ornament in centimetres. Then $B=v\times\rho_w\times g=2\times g$

$$\Rightarrow \left(\frac{m_{Au}}{\rho_{Au}}+\frac{m_{Cu}}{\rho_{Cu}}\right)\rho_w\times g=2\times g$$

$$m_{Au}\rho_{Cu}+m_{Cu}\rho_{Au}=2\rho_{Acc}\rho_{Cu}$$

$$8.9m_{Au}+19.3m_{Cu}=2\times 19.3\times 8.9=343.54 \quad \text{(ii)}$$

From Eqs. (i) and (ii), $8.9m_{Au}+19.3m_{Cu}=343.54$

$$\Rightarrow 8.9(m_{Au}+m_{Cu})+10.4m_{Cu}=343.54$$
$$\Rightarrow 8.9\times 36+10.4m_{Cu}=343.54$$
$$m_{Cu}=2.225 \text{ g}$$

So the amount of copper in the ornament is 2.2 g.

8. b.
$$v_Aa_A=v_B\times a_B=v_A\times 4=v_B\times 2$$
$$v_B=2v_A \quad \text{(i)}$$

Again, $\frac{1}{2}\rho v_A^2+\rho gh_A+p_A=\frac{1}{2}\rho v_B^2+\rho gh_B+p_B$

$$\Rightarrow \frac{1}{2}\rho v_A^2+p_A=\frac{1}{2}\rho v_B^2+p_B \quad (\because h_A=h_B)$$

$$\Rightarrow p_A-p_B=\frac{1}{2}\rho(v_B^2-v_A^2)=\frac{1}{2}\times 1\times(4v_A^2-v_A^2)$$

$$\Rightarrow 2\times 1\times 1000=\frac{1}{2}\times 1\times 3v_A^2$$

($p_A=p_B=2$ cm of water column $=2\times 1\times 1000$ dyn/cm^2)

$$\therefore \quad v_A=\sqrt{\frac{4000}{3}}=36.51 \text{ cm/s}$$

So, rate of flow $=V_aa_A=36.51\times 4=146$ cm^3/s.

9. c. Let the rise of level in the right limb be x cm, the fall of the level of mercury in the left limb be $4x$.
Therefore, the level of water in the left limb is $(36+4x)$ cm.
Equating pressure at the interface of mercury and water, we have (at A_1 and B_1)

$$(36 + 4x) \times 1\, g = 5x \times 13.6 \times g$$

Solving, we get $x = 0.56$ cm

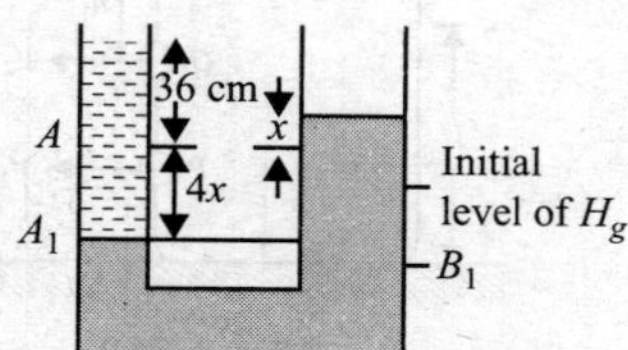

Fig. 3.304

10. c. $\rho_A = 0.75$, $\rho_P = 0.6$
$\rho_B = 1.0$, $\rho_Q = 0.9$

As relative density of P is lesser than B, so it will float in liquid B and as relative density of Q is greater than liquid A so it will sink, because if density of the object is greater than that of the liquid in which it is immersed, then its weight is more than the upthrust and vice versa.

11. d. Since density of block $= \frac{1}{2}$(density of water), 50% of its volume is immersed in water.

In the second case, half of the volume of the block is further immersed in water.

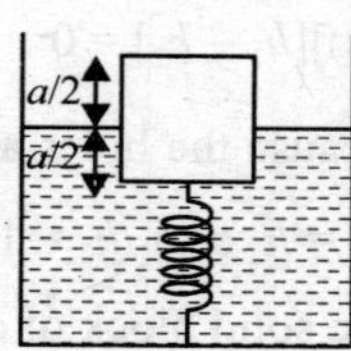

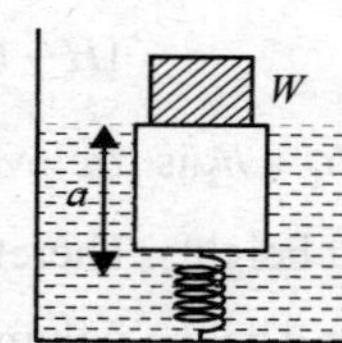

Fig. 3.305

Therefore, W = Extra upthrust + Spring force

$$= a \times a \times \frac{a}{2} \times 2\rho \times g + K\left(\frac{a}{2}\right)$$

$$= a\left(a^2 \rho g + \frac{K}{2}\right)$$

12. d. Let P_1 and P_2 be the pressures at the bottom of the left and right ends of the tube, respectively.

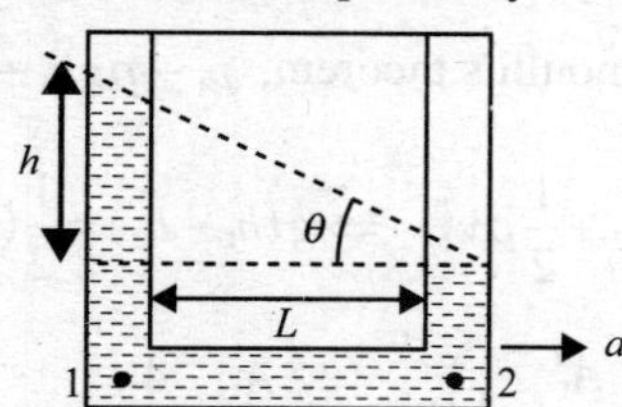

Fig. 3.306

Then $F = (P_1 - P_2)A = \rho g h A$

where A is the cross section of the tube.

The mass of the liquid in the horizontal portion is

$$m = \rho L A$$

Now, $F = ma \Rightarrow \rho g h A = \rho L A a$

$$\therefore \quad h = \frac{aL}{g}$$

13. d. Since both the vertical limbs of U-tube are open to atmosphere, at surface of liquids, the pressure is equal to the atmospheric pressure. Since heights of the liquids in the two vertical limbs are equal and the liquids have different densities, pressure exerted by them will be different from each other. The liquids can remain in static equilibrium relative to the tube only, when the system is accelerating down with acceleration g. In that case, liquids experience weightlessness. Hence, at every point of the tube, the pressure will become equal to atmospheric pressure. But if the tube is not accelerating down under gravity, then the heavier liquid will exert more pressure at the bottom. Hence, at the bottom of the left vertical limb, pressure will be greater than that at the bottom of the right vertical limb. It means on horizontal limb there will be a resultant horizontal force which will be towards the right as shown in the figure. Therefore, the liquid in the horizontal limb will have a rightward acceleration.

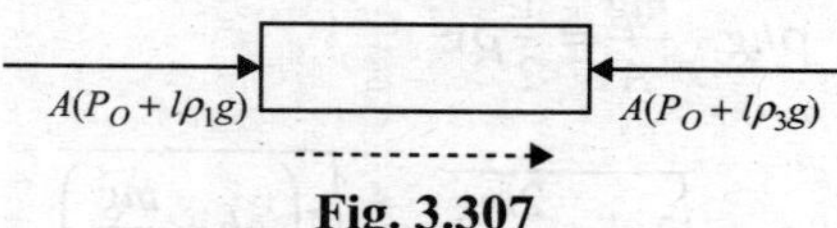

Fig. 3.307

In fact, this whole system must have a horizontally rightward acceleration. Obviously, only option (d) is correct.

14. a. As the water exerts upthrust on the stone, the stone also exerts the same force on water, according to Newton's third law. Upthrust force acting on the stone is

$$U = V_{stone}\, \rho_\omega g$$

$$= \left(\frac{m_{stone}}{\rho_{stone}}\right) = \frac{0.5}{10^4} \times 1000 \times 10 = 0.5 \text{ N}$$

Also, the weight is exerted on the beaker. Therefore the reading will be 1.5 + .5 = 2 kg.

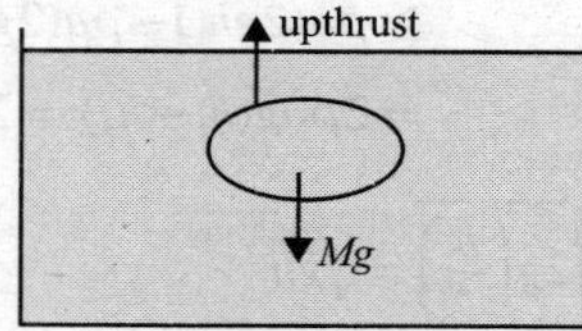

Fig. 3.308

15. a. The air through the horizontal tube will decrease the pressure and more liquid will be pushed into the capillary tube.

16. c. Let d_w and d_0 be the densities of water and oil, respectively. Then the pressure at the bottom of the tank is $h_w d_w g + h_0 d_0 g$.

Let this pressure be equivalent to pressure due to water of height h. Then, $h d_w g = h_w d_w g + h_0 d_0 g$

$$\therefore \quad h = h_w + \frac{h_0 d_0}{d_w} = 100 + \frac{400 \times 0.9}{1}$$

$$= 100 + 360 = 460 = 4.6 \text{ m}$$

According to Toricelli's theorem,

$$v = \sqrt{2gh} = \sqrt{2 \times 10 \times 4.6} \text{ m/s} = \sqrt{92} \text{ m/s}$$

17. a. Total pressure at the bottom = 3 atm
Pressure due to water in the tank = 3 atm − 1 atm = 2 atm = 20 m of water column
Height of water in the tank is h = 20 m
So, velocity of efflux $= \sqrt{2gh} = \sqrt{2 \times 10 \times 20} = \sqrt{400}$ m/s

18. b. Applying Bernoulli's theorem at points 1 and 2, difference in pressure energy between 1 and 2 = difference in kinetic energy between 1 and 2.

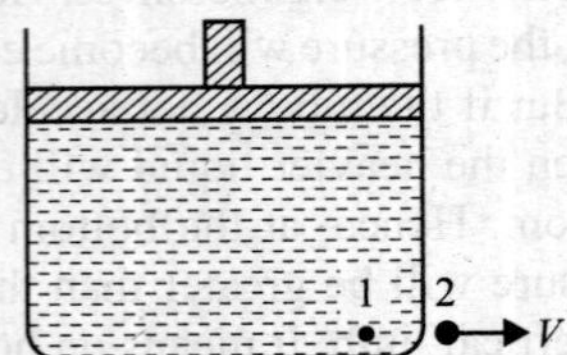

Fig. 3.309

Hence,

$$\rho hg + \frac{mg}{A} = \frac{1}{2}\rho v^2$$

or $$v = \sqrt{2gh + \frac{2mg}{\rho A}} = \sqrt{2\left(gh + \frac{mg}{\rho A}\right)}$$

19. a. Let ρ be the density of the liquid. Then

$$F_1 = (\Delta P)\,A = \rho gh\,A \quad \text{(i)}$$

In the second case, F_2 = rate of change of momentum

$$= \rho A u^2 = \rho A(\sqrt{2gh})^2 = 2\rho ghA \quad \text{(ii)}$$

$$\therefore \quad \frac{F_1}{F_2} = \frac{1}{2}$$

20. c. Thrust force, $F = F_1 - F_2 = \rho a v_1^2 - \rho a v_2^2$

$$= \rho a(2gh_1) - \rho a(2gh_2)$$
$$= 2\rho a g(h_1 - h_2) = 2\rho a g h$$

21. d. $$v_1 = \sqrt{2g\left(\frac{h}{2}\right)} = \sqrt{gh} \quad \text{(i)}$$

From Bernoulli's theorem, $\rho gh + 2\rho g\left(\frac{h}{2}\right) = \frac{1}{2}(2\rho)v_2^2$

$$\Rightarrow \quad v_2 = \sqrt{2gh} \quad \text{(ii)}$$

$$\therefore \quad \frac{v_1}{v_2} = \frac{1}{\sqrt{2}}$$

22. a. From the equation of continuity,

$$\text{Rate} = r = \frac{A}{3}V_p + \left(A - \frac{A}{3}\right)V_Q$$

$$\therefore \quad V_P + 2V_Q = \frac{3r}{A} = \frac{3 \times 0.1}{10^{-2}} = 30 \text{ m/s}$$

As V_P = 20 m/s, so V_Q = 5 m/s.

23. a. The two streams strike at the same point on the ground.

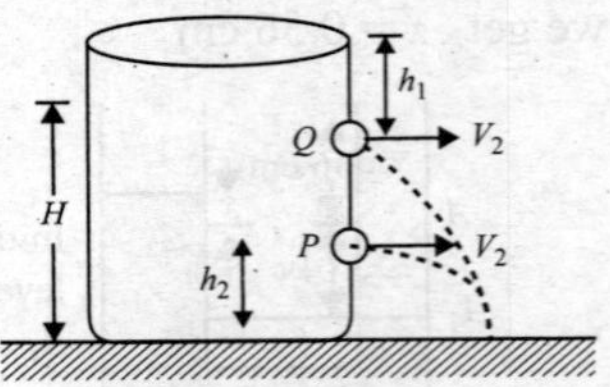

Fig. 3.310

$R_1 = R_2 = R$

$$u_1 t_1 = u_2 t_2 \quad \text{(i)}$$

where u_1 = velocity of efflux at $Q = \sqrt{(2gh_1)}$ and u_2 = velocity of efflux at $P = \sqrt{[2g(H - h_2)]}$

t_1 = time of fall of water stream through Q is

$$= \sqrt{\frac{2(H - h_1)}{g}}$$

t_2 = time of fall of the water stream through $P = \sqrt{\frac{2h_2}{g}}$

Putting these values is Eq. (i), we get

$$(H - h_1)h_1 = (H - h_2)h_2$$

or $$[H - (h_1 + h_2)][h_1 - h_2] = 0$$

$H = h_1 + h_2$ is irrelevant because the holes are at two different heights. Therefore, $h_1 = h_2$ or $h_1/h_2 = 1$

24. b. Let v be the volume of the solid block of density ρ. Let ρ_1 be the density of water. Weight of body = $v\rho g$. When the body is immersed in water,
Tension in the string = Upward thrust − Weight of the body

$$\Rightarrow \quad T = v\rho_1 g - v\rho g = vg(\rho_1 - \rho_2)$$

When the lift is moving upwards with acceleration a, the tension in the string is $T = v(\rho_1 - \rho)(g + a)$
From Eqs. (i) and (ii), $T = T_0(1 + a/g)$.

25. b. Using equation of continuity, we have $v_2 = \frac{A_1}{A_2}v_1$

From Bernoulli's theorem, $p_1 + \rho g h_1 + \frac{1}{2}\rho v_1^2$

$$p_2 + \rho g h_2 + \frac{1}{2}\rho v_2^2 \quad \Rightarrow g(h_1 - h_2) = \frac{1}{2}(v_2^2 - v_1^2)$$

$$\Rightarrow 60 = \left(\frac{A_1^2}{A_2^2} - 1\right)v_1^2 \quad \Rightarrow \frac{A_1}{A_2} = \frac{4}{1}$$

26. a. Let ρ be the density of the material, ρ_0 be the density of water. When the sphere has just started sinking, the weight of the sphere = weight of water displaced (approx.)

$$\Rightarrow \frac{4}{3}\pi(R^3 - r^3)\rho g = \frac{4}{3}\pi R^3 \rho_0 g$$

$$\Rightarrow (R^3 - r^3)\rho = R^3\rho_0 \quad \Rightarrow \frac{(R^3 - r^3)}{R^3} = \frac{\rho_0}{\rho}$$

$$\Rightarrow \frac{r}{R} = \frac{(7)^{1/3}}{2}$$

27. a. Given $R = 5$ cm, $\omega = 2\pi\,(4)$ rad/s

Using $y = \frac{\omega^2 x^2}{2g}$,

the required level difference is y. When $x = R$,

$$y = \frac{(8\pi)^2(5\times10^{-2})}{2\times9.8} = 8 \text{ cm}$$

28. a.

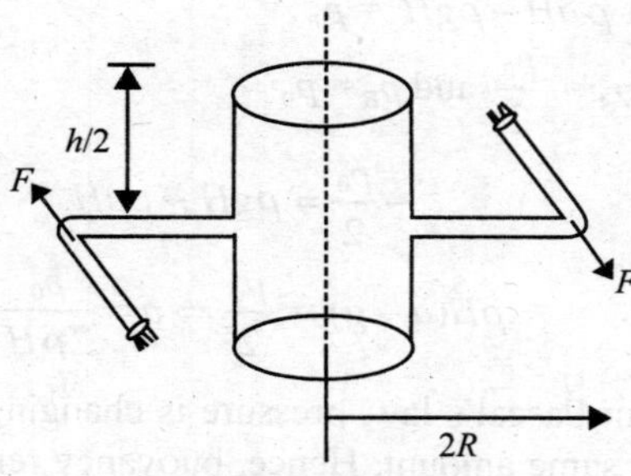

Fig. 3.311

Velocity of efflux of water $v = \sqrt{2g\left(\frac{h}{2}\right)} = \sqrt{gh}$

Force on ejected water = Rate of change of momentum of ejected water

$$F = \rho\, av^2$$

Torque of these forces about central line = $(\rho av^2) \times (4R)$

$$= 4\rho av^2 R = 4\rho\, agh\, R$$

29. a. Let ρ_S and ρ_L be the densities of silver and liquid, respectively, and m and V be the mass and volume, respectively, of the silver block. Therefore,

Tension in the string = mg – buoyant force

$$\Rightarrow \quad T = \rho_S Vg - \rho_L Vg = (\rho_S - \rho_L)\, Vg$$

Also, $V = \frac{m}{\rho_S}$

$$\therefore \quad T = \left(\frac{\rho_S - \rho_L}{\rho_S}\right) mg$$

$$= \frac{(10-0.72)\times10^3}{10\times10^3} \times 4\times10 = 37.12 \text{ N}$$

Fig. 3.312

30. c.

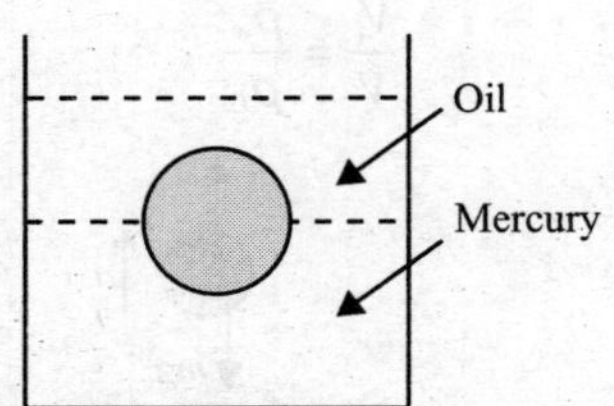

Fig. 3.313

Weight = buoyant force $\frac{V}{2}\rho_{Hg} g + \frac{V}{2}\rho_{oil} g$

$$\therefore \; \rho = \frac{\rho_{Hg} + \rho_{oil}}{2} = \frac{13.6+0.8}{2} = \frac{14.4}{2} = 7.2$$

31. a. The four pistons are initially in equilibrium. As additional force F is applied to each piston, the pressure in the fluid at each point must be increased by F/A so that each piston retains state of equilibrium.

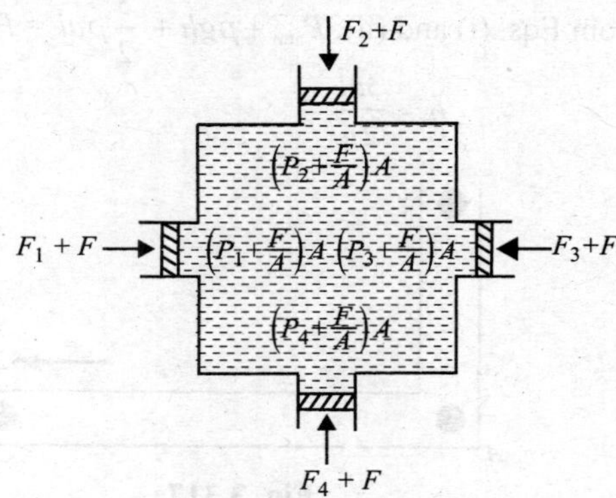

Fig. 3.314

Thus, the increment in the pressure at each point is

$$\Delta P = \frac{F}{A} \quad \text{(by Pascal's law)}$$

32. a.

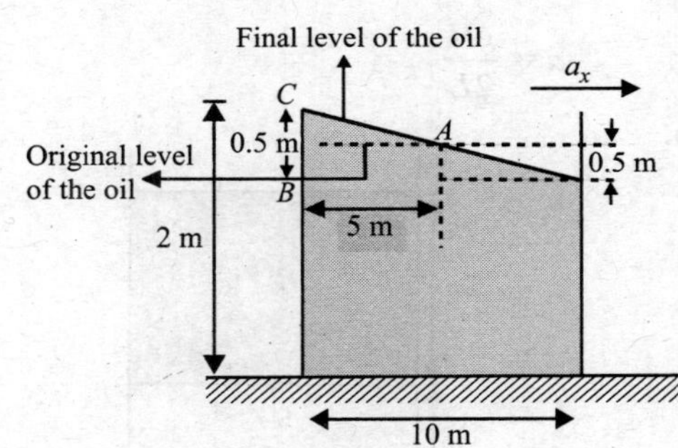

Fig. 3.315

$$v = u + a_x t,\; a_x = v/t$$

$$\tan\theta = \frac{a_x}{g} = \frac{v}{tg} = \frac{0.5}{5} \quad \text{(in triangle } ABC)$$

$$\Rightarrow \quad t = \frac{10\times20}{10} = 20 \text{ s}$$

33. b.

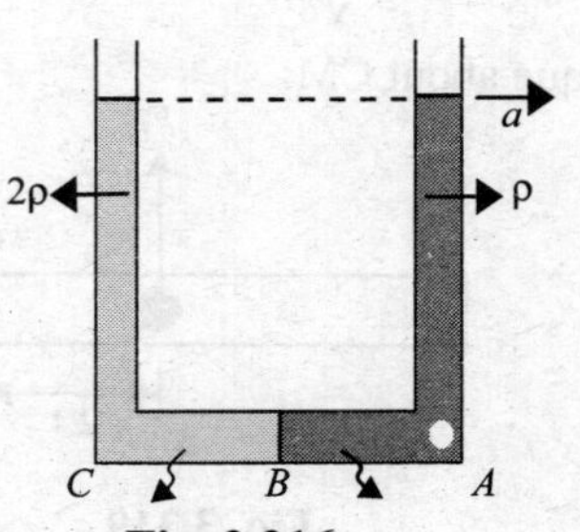

Fig. 3.316

For the given situation, liquid of density 2ρ should be behind that of ρ.
From the right limb, $P_A = P_{atm} + \rho gh$

$$P_B = P_A + \rho a\frac{\ell}{2} = P_{min} + \rho gh + \rho a\frac{\ell}{2}$$

$$P_C = P_B + (2\rho)\, a\frac{l}{2} = P_{atm} + \rho gh + \frac{3}{2}\rho al \quad \text{(i)}$$

But from the left limb, $P_C = P_{atm} + (2\rho)\, gh$ (ii)

From Eqs. (i) and (ii), $P_{atm} + \rho gh + \frac{3}{2}\rho al = P_{atm} + 2\rho gh$

$$\Rightarrow \quad h = \frac{3al}{2g}$$

34. a.

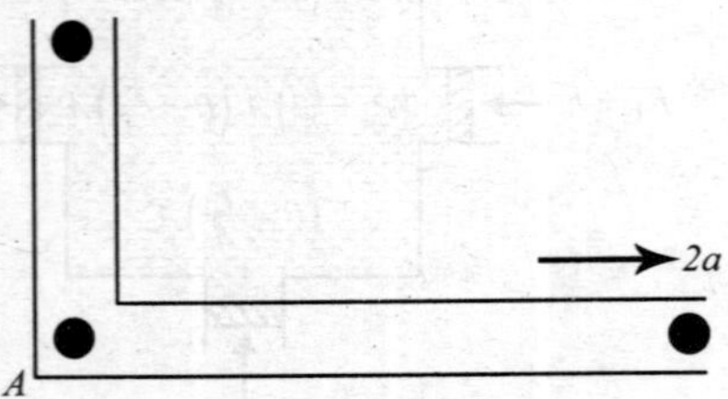

Fig. 3.317

No sliding ⇒ pure rolling

Therefore, acceleration of the tube = $2a$ (since COM of the cylinders are moving at 'a')

$P_A = P_{atm} + \rho(2a)L$ (from the horizontal limb)

Also, $P_A = P_{atm} + \rho gH$ (from the vertical limb)

$$\therefore \quad a = \frac{gH}{2L}$$

35. b.

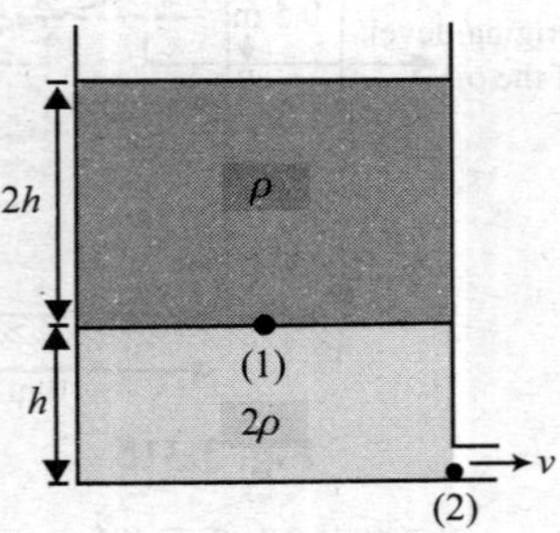

Fig. 3.318

Pressure at (1), $P_1 = P_{atm} + \rho g(2h)$

Applying Bernoulli's theorem between points (1) and (2),

$$(P_{atm} + 2\rho gh) + \rho g(2h) = P_{atm} + \frac{1}{2}(2\rho)\, v^2$$

$$\Rightarrow \quad v = 2\sqrt{gh}$$

36. b. Torque about CM:

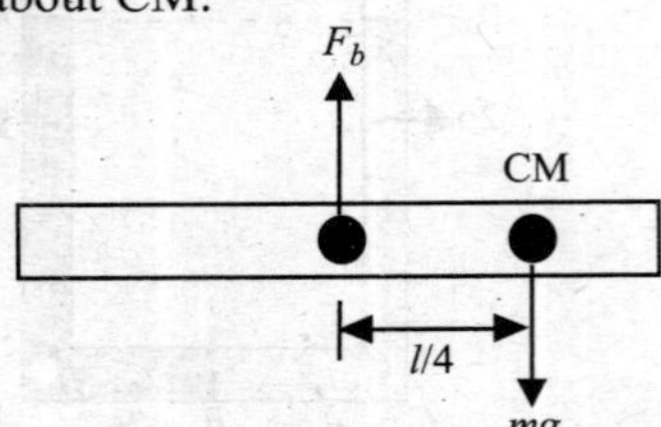

Fig. 3.319

$$F_b\frac{l}{4} = I\alpha \Rightarrow \alpha = \frac{1}{I}(\pi r^2)(l)(\rho)(g)\frac{l}{4}$$

$$\Rightarrow \quad \alpha = \frac{\pi r^2 l^2 g\rho}{4I}$$

37. d. $V_0 = \sqrt{2gh} \Rightarrow V_2 = \sqrt{2g\frac{h}{\sqrt{2}}} = \frac{V_0}{\sqrt[4]{2}}$

38. a. As the vessel is falling freely, the pressure at all the points in the liquids is same and equal to the atmospheric pressure and hence buoyancy becomes zero.

39. c. $p_A + \rho aH - \rho gH = p_B$

Here, $p_A = \frac{p_0}{2}$ and $p_B = p_0$

$$\therefore \quad -\frac{p_0}{2} = \rho gH - \rho aH$$

$$\Rightarrow \quad \rho H(a-g) = \frac{p_0}{2} \Rightarrow a = \frac{p_0}{2\rho H} + g$$

40. c. From Pascal's law, pressure is changing at every point by the same amount. Hence, buoyancy remains the same. So, the part of the block inside water remains the same.

41. b. On pouring oil on the left side water rises by 25 cm from its previous level in the right limb of U-tube, creating a difference of levels of water by 50 cm. Let h cm be the height of oil above level A in the left limb of U-tube. Equating pressures at A and B, we get $P_A = P_B$.

$$\Rightarrow \quad h \times \rho_{oil} \times g = 50 \times \rho_{water} \times g$$

$$\Rightarrow \quad h = \frac{50 \times \rho_{water}}{\rho_{oil}} = \frac{50 \times 1}{0.8} = 62.5 \text{ cm}$$

Difference of levels of oil (C) and water (D) in the two limbs = 62.5 – 50 = 12.5 cm.

42. c. Apparent weight of the balloon = W_2

Apparent weight = real weight – upthrust

or $\quad W_2$ = real weight – W

Now, $\quad$ real weight = $W_1 + W$

$$\therefore \quad W_2 = W_1 + W - W = W_1$$

43. c. When the vessel is stationary,

Weight = upthrust

$$\Rightarrow \quad V\rho_w g = V_i\rho_L g$$

ρ_w = density of wood

ρ_L = density of liquid

$$\text{or} \quad \frac{V_i}{V} = \frac{\rho_w}{\rho_L} \quad \text{(i)}$$

Fig. 3.320

When the vessel moves upwards.

$$\text{Upthrust} - \text{weight} = \text{mass} \times \text{acceleration}$$

$$B - mg = ma$$

or $$V_i' \rho_L \left(g + \frac{g}{2}\right) - V\rho_W g = \frac{V\rho_w g}{2}$$

or $$\frac{V_i'}{V} = \frac{\rho_w}{\rho_L} \quad \text{(ii)}$$

From Eqs. (i) and (ii), $\dfrac{V_i'}{V} = \dfrac{V_i}{V}$

Hence, 50% will be immsersed.

44. c. $h \times 1000 \times g = \dfrac{2}{100} \times 13600 \times g$

or $$h = \frac{2 \times 13.6}{100} \text{ m} = 27.2 \text{ cm}$$

45. c. $P = 100 \text{ cm} \times 1 \text{ g cm}^{-3} \times 1000 \text{ cm s}^{-2} = 10^5 \text{ dyn cm}^{-2}$

$F = 10^5 \times 100 \text{ dyn} = 100 \text{ N}$

Again, $V = 199 \text{ cm}^3$

Weight $= 199 \times 1 \times 1000 \text{ dyn} = 1.99 \text{ N}$

46. b. *OP* is the portion of the rod immersed in water.

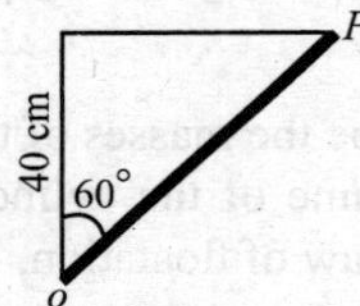

Fig. 3.321

$$\cos 60° = \frac{40}{OP}$$

or $$OP = \frac{40}{\cos 60°} \text{cm} = \frac{40}{1/2} \text{cm} = 80 \text{cm}$$

The centre of buoyancy is at the centre of the immersed part of the rod. So, the required distance is 40 cm.

47. b. $(V - \Delta V) \times 1.03 \times g = V \times 0.92 \times g$

or $\dfrac{V - \Delta V}{V} = \dfrac{0.92}{1.03}$

or $\dfrac{\Delta V}{V} = 1 - \dfrac{0.92}{1.03} = \dfrac{0.11}{1.03}$ or $\dfrac{\Delta V}{V} \times 100 = \dfrac{11}{1.03} \approx 11$

48. a. Let V be the volume of the ball.

$$\text{Net upward force} = V\sigma g - V\rho g$$

Net upward acceleration,

$$a = \frac{V\sigma g - V\rho g}{V\rho} = \frac{(\sigma - \rho)g}{\rho}$$

Velocity at the surface $= \sqrt{\dfrac{2(\sigma - \rho)gh}{\rho}}$

If h_a is the height in air to which the ball rises, then

$$0 - \frac{2(\sigma - \rho)gh}{\rho} = 2(-g)h_a$$

$$\therefore \quad h_a = \frac{(\sigma - r)gh}{g\rho} = \left(\frac{\sigma}{\rho} - 1\right)h$$

49. c. $(2.4 \times 10^{-4}) \times 1.5 = 20 \times 2 \times 10^{-2} \times 10^{-4} \times v$

or $v = \dfrac{2.4 \times 1.5 \times 10^2}{40} \text{m s}^{-1} = \dfrac{24 \times 15}{40} \text{m s}^{-1} = 9 \text{m s}^{-1}$

50. a.

$$-\pi r^2 \frac{dh}{dt} = \pi r'^2 \sqrt{2gh} \text{ or } \frac{dh}{dt} = -\frac{r'^2}{r^2}\sqrt{2gh}$$

or $$dt = -\frac{r^2}{r'^2} \frac{1}{\sqrt{2gh}} dh$$

$$\therefore \quad \int_0^t dt = -\frac{r^2}{r'^2\sqrt{2g}} \int_h^0 h^{-1/2} dh$$

or $$t = -\frac{r^2}{r'^2\sqrt{2g}} \times 2\left[\sqrt{h}\right]_h^0 \text{ or } t = \sqrt{\frac{2h}{g}}, \frac{r^2}{r'^2}$$

51. b. $1.5 \times 10^5 = p + \dfrac{1}{2} \times 10^3 \times 10^2$

or $p = (1.5 - 0.5) \times 10^5 \text{ Pa} = 10^5 \text{ Pa}$

52. d. $P_2 - P_1 = \dfrac{1}{2} \times 1.3\left[120^2 - 90^2\right] = 4095 \text{ Pa}$

$$\text{Lift} = 4.95 \times 2 \times 10 \text{ N} = 81900 \text{ N}$$

53. c. Let d be the inner diameter of the hemispherical bowl. Then, $B = mg$

$$\frac{4}{3}\pi \times \left(\frac{1}{2}\right)^3 \times (1.2 \times 10^3)g = \frac{4}{3}\pi\left[\left(\frac{1}{2}\right)^3 - \left(\frac{d}{2}\right)^3\right] \times (2 \times 10^4)g$$

Solving, we get $d = 0.98$ m.

54. c. The velocity with which the liquid comes out is

$v_0 = \sqrt{2gh}.$

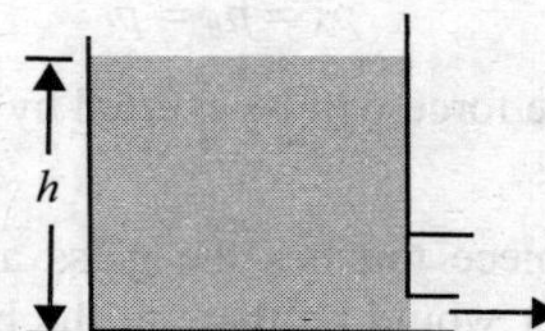

Fig. 3.322

Let cross-sectional area of the hole be A. Then the force exerted by the ejecting fluid due to change in momentum on the container is $F = \rho A v^2$.

$F = \rho A v^2 = ma$, where m is the mass of the liquid inside the container.

$$\rho A v^2 = \rho A h \times a$$

$$\Rightarrow \quad a = 2g$$

Differentiating the above equation, we can have $v = 2gt$ where t is the time in which the liquid comes out completely which is dependent on h.

55. c. At a depth y from the surface of the fluid, the net force acting on the gate element of width dy is

$$dF = (p_0 + \rho g y - p_0) \times 1 dy = \rho g y dy$$

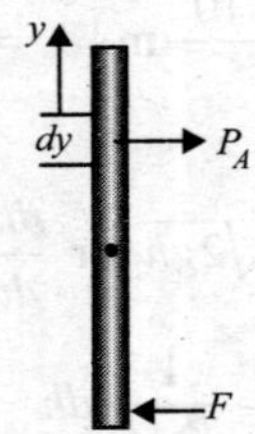

Fig. 3.323

Torque of this force about the hinge is

$$d\tau = \rho g y dy \times \left(\frac{1}{2} - y\right)$$

Net torque experienced by the gate is

$$\tau_{net} = \int d\tau + F \times \frac{1}{2}$$

$$= \int_0^1 \rho g y dy \left(\frac{1}{2} - y\right) + F \times \frac{1}{2} = 0$$

$$\Rightarrow \quad F = \frac{\rho g}{6}$$

56. d. Let m be mass of the liquid. Let h_A, h_B and h_C be the heights of liquid level in three vessels. Then

$$m = \rho_A A \times h_A = \rho_B A \times h_B = \rho_C \times A \times h_C$$

$$\rho_A h_A = \rho_B h_B = \rho_C h_C = \frac{m}{A}$$

Pressure at the bottom in vessel A, $p_A = p_0 + \rho_A g h_A$

Similarly, at B and C, $p_B = p_0 + \rho_B g h_B$

$$p_C = p_0 + \rho_C g h_C$$

i.e., $\quad p_A = p_B = p_C$

So, the same force will be exerted by the liquid in all the three vessels.

57. d. As the piece touches the glass at small number of points, water would be there at the bottom of the piece also and hence buoyancy acts on it.

From equilibrium condition, $mg = F_B + N$

$$F_B = V \rho_{water} g = \frac{m}{\rho_{metal}} \times \rho_{water} \times g$$

$$= \frac{0.16}{8} \times 10 = 0.2 \text{ N}$$

So, $\quad N = mg - F_B = \frac{160}{1000} \times 10 - 0.2 = 1.4 \text{ N}$

58. d. Due to acceleration down the plane, both body B and liquid will received the same component of acceleration down the plane.

59. d. The points A and C are in same horizontal level

hence $\quad P_C - P_A = \rho a l$

(refer to figure in question to identify A, B, C)

Now, $\quad P_B - P_C = \rho g h$

$\Rightarrow P_B - (P_1 + \rho a l) = \rho g h \Rightarrow P_B - P_A = h\rho g + l\rho a$

60. c. When the levels equalize, the height of liquid in each arm is $(h_1 + h_2)/2$.

We may then visualize that a length $h_1 - (h_1 + h_2)/2 = (h_1 - h_2)/2$ of the liquid has been transferred from the left arm to the right arm.

Mass of liquid transferred $= (h_1 - h_2)/2 A\rho$, where A is area of hole and ρ is density.

Vertical distance through which it moves $= (h_1 - h_2)/2$.

Loss in gravitational potential energy $= [(h_1 - h_2)/2]^2 A\rho$.

Mass of the entire liquid $= (h_1 + h_2 + h) A\rho$

If this moves with velocity v,

kinetic energy $= \frac{1}{2}(h_1 + h_2 + h) A\rho v^2$

Equating energies, we get $v = (h_1 - h_2)\sqrt{\frac{g}{2(h_1 + h_2 + h)}}$

61. c. Let M and m be the masses of the wooden piece and the man, V the volume of the cylinder and ρ the density of water. Then by law of floatation,

$$Mg = \frac{2V\rho}{3} g \Rightarrow M = \frac{2V\rho}{3} \quad \text{(i)}$$

and $\quad (M + m)g = \left(\frac{2}{3} + \frac{1}{6}\right) V\rho g = \frac{5V\rho}{6} g$

$$\Rightarrow \quad (M + m) = \frac{5V\rho}{6} \quad \text{(ii)}$$

From Eqs. (i) and (ii), we get $\frac{m}{M} = \frac{1}{4}$.

62. b. Since two-thirds of the wooden block is immersed in water, its density $= 2\rho/3$, where ρ is the density of water. Let the volume of the chip be V. Then its weight is $(2/3)V\rho g$. The upthrust acting on it inside the water is $V\rho g$. Hence, the unbalanced upward force is $V\rho g(1 - 2/3) = V\rho g/3$.

Hence, its acceleration a in the upward direction is

$$a = \frac{V\rho g/3}{(2/3)V\rho} = \frac{g}{2}$$

If t is the required time, then $1 = 0 + \frac{1}{2}at^2 = \frac{1}{2}\frac{g}{2}t^2$

Hence, $t^2 = \frac{4}{g} = \frac{4}{10} = 0.4 \Rightarrow 0.63 \text{ s}$

63. d. Weight of ornament in air = 0.25 kg

Weight of displaced water = 0.015 × 10^{-3} × ρ kg, where ρ is the density of water.

Hence, the relative density of the ornament is 0.25/(0.015 × 10^{-3} × ρ).

Let the percentage of copper in the ornament be n%. Then the relative density of ornament is

$$\frac{1}{100}[(100-n)19.3+8.9n] = \frac{1}{100}[1930-10.4n]$$

Hence,

$$\frac{1}{100}[1930 - 10.4n] = \frac{0.25}{0.015\times10^{-9}\times\rho} = \frac{0.25}{0.015} = \frac{50}{3}$$

⇒ 3[1930 – (10.4 n)] = 5000

$$\therefore n = \frac{(5790-5000)}{3\times10.4} = \frac{790}{31.2} = 25$$

64. a. When the oil is poured, the fraction of ice in the water decreases, i.e., volume of ice melted into water is greater than volume of water displaced by ice. So water level rises. Overall volume of ice will decrease as it melts. So the upper level of oil falls.

65. c. Relative velocity of the water jet with respect to the moving plate = (18 + 3) = 21 m/s. The mass of water hitting the plate per second = $a\rho(21)$, where a is the cross-sectional area of the jet.

F = Force normal to the plate

= Rate of change of momentum along the direction of the water jet

= (Mass) × (Velocity) = $[a\rho\,(21)]\,21 = 441a\rho$

$$\therefore = 441\times\rho\times\left(\frac{0.1}{18}\right)$$

$$= \frac{441\times0.1\times1000}{18}\text{ N} = 2450\text{ N}$$

66. c. Since the container alone floats to half its length in water, its relative density = 0.5. Since the levels of the liquid inside and water outside are the same, the volumes of the liquid and of the displaced water are the same. If this volume is V and the volume of the material of container is v, then the total mass of floating body is $M = [V\sigma + v(0.5)\rho]$, where ρ is the density of water. By law of floatation, $M = [V\sigma + v(0.5)\rho] = V\rho$ giving $V(1-\sigma) = v \times 0.5$.

Let the cross section of the container be A. Then the volume V' of water displaced V by the liquid is

$$LA\left[\frac{2}{3}-\frac{1}{2}\right] = LA\left[\frac{1}{6}\right]$$

But $\quad LA = \dfrac{3V}{2}$

Hence $V' = \dfrac{3V}{2}\times\dfrac{1}{6} = \dfrac{V}{4}$

Hence $V\sigma\rho = \left(\dfrac{V}{4}\right)\rho \quad \Rightarrow \therefore \sigma = \dfrac{1}{4}$

67. c. $T_1 = T_2$, $Mg + \sqrt{2}\,T_1 = \frac{4}{3}\pi r^3\rho_w g$

$T_1 = 250\sqrt{2}$ N

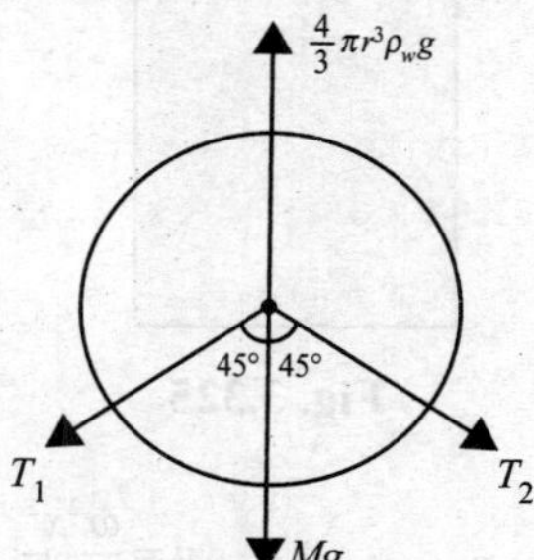

Fig. 3.324

68. a. According to Bernoulli's theorem,

$$P_1 + \frac{1}{2}\rho v_1^2 = P_2 + \frac{1}{2}\rho v_2^2$$

$$\Rightarrow \quad P_1 - P_2 = \frac{1}{2}\rho(v_2^2 - v_1^2)$$

$$\Rightarrow \quad \frac{P_1 - P_2}{\rho} = \frac{1}{2}(v_2^2 - v_1^2)$$

Change in kinetic energy per kilogram mass of oil is $\dfrac{P_1 - P_2}{\rho}$

Substituting the numerical values, change in kinetic energy = $\dfrac{8}{800}$ = 10^{-2} J/kg

69. c. Velocity of the liquid just after coming out of section (3) is

$$v_3 = \sqrt{2gh} \quad \text{(i)}$$

Using continuity equation at section '2' and section '3'

$$\frac{A}{2}v_2 = \frac{A}{4}v_3$$

$$v_2 = \frac{1}{2}v_3 = \sqrt{\frac{hg}{2}}$$

Now using Bernoullie's theorem at section 2 and at the opening end of pipe '3'.

$$P_0 + \frac{1}{2}\rho v_3^2 = P_2 + \frac{1}{2}\rho v_2^2$$

$$\Rightarrow P_2 = P_0 + \frac{1}{2}\rho(v_3^2 - v_2^2) = P_0 + \frac{1}{2}\rho\left(2gh - \frac{gh}{2}\right)$$

$$P_2 = P_0 + \frac{3\rho gh}{4}$$

70. a. When the cylindrical vessel rotates, the velocity at the sides is higher. Hence pressure at the sides is lower.

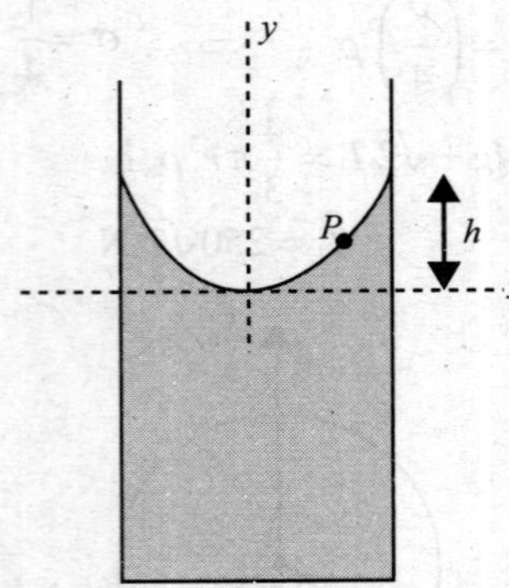

Fig. 3.325

Equation of surface liquid is $y=\frac{\omega^2 x^2}{2g}$

Here $\omega = 2\pi n$

$x = R$

$y = h$

$$h=\frac{(2\pi n)^2(R^2)}{2g} = \frac{2\pi^2 n^2 r^2}{g}$$

71. c. Initially, the water flowing out will be less than that flowing into it. Hence, the water level will go on rising. When the water level is h, the velocity of efflux $=\sqrt{2gh}$. When this becomes equal to the velocity of inflow, the level will become steady as the area of cross section of the filling tube and area of cross section of the hole are equal. This height is given by Torricelli's theorem,

$$v=\sqrt{2gh} \quad \text{or} \quad \frac{v^2}{2g}=h$$

Thereafter the water level will not rise.

72. a. Consider the condition of equilibrium for the mass of water contained between cross section separated by x and $x + dx$ from rotation axis.

$$dp = \rho\omega^2 x dx\,; \quad \int_{p_1}^{p_2} dp = \int_{r_1}^{r_2} \rho\omega^2 x dx$$

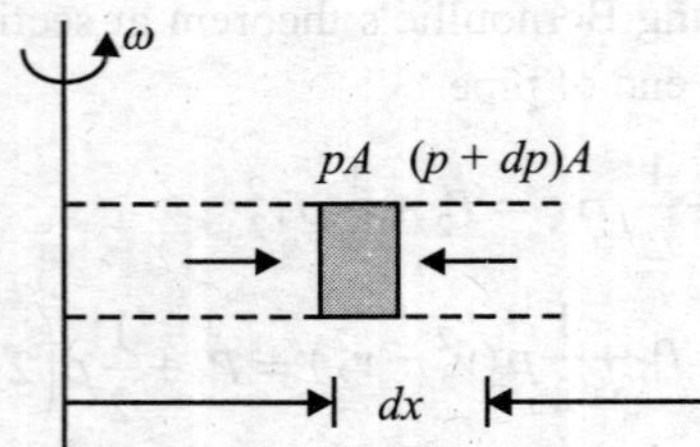

Fig. 3.326

$$p_2 - p_1 = \rho\omega^2\left[\frac{x^2}{2}\right]_{r_1}^{r_2} = \rho\omega^2\left(\frac{r_2^2 - r_1^2}{2}\right)$$

$$\omega = \sqrt{\frac{2\,(p_2 - p_1)}{\rho\,(r_2^2 - r_1^2)}}$$

73. a. Let a_1 and a_2 be cross sectional areas and v_1 and v_2 the velocities of liquid flow at A and B. If p_1 and p_2 are pressures of liquid recorded by manometer, then

$$p_1 + \frac{1}{2}\rho v_1^2 = p_2 + \frac{1}{2}\rho v_2^2 \Rightarrow p_1 - p_2 = \frac{1}{2}\rho v_1^2\left(\frac{v_2^2}{v_1^2} - 1\right)$$

By equation of continuity, $a_1 v_1 = a_2 v_2$

Also, $p_1 - p_2 = h\rho g$

Substituting these values, we have $v_1 = \sqrt{\dfrac{2gh}{\dfrac{a_1^2}{a_2^2} - 1}}$

Rate of flow of liquid is $a_1 v_1 = a_1 a_2\sqrt{\dfrac{2gh}{a_1^2 - a_2^2}}$

74. a. Absolute pressure in bulb = Gauge pressure + Atmospheric pressure $= P + P_a$

Using Bernoulli's equation, $P_a + P = P_{BC} + \dfrac{\rho_a v^2}{2}$

where ρ_a is the density of air.

$$\therefore \quad P_{BC} = P_a + P - \left(\frac{1.3}{2}\right)v^2$$

$$P_{BC} = P_a - \rho g h$$

So equating these two values for P_{BC}, we get

$$V = \frac{P + \rho g h}{\sqrt{0.65}}$$

75. b. Initially, $W - 2\,Kx = 0$ (i)

Finally, $W' - 2K\left(x + \dfrac{a}{2}\right) - a.a.\dfrac{a}{2}2\sigma.g = 0$ (ii)

$W' = W + W_0$

$W + W_0 - 2Kx - Ka - a_3\rho g = 0$ [From Eq. (ii)]

From Eq. (i),

$W_0 = Ka + a^3\,\sigma g = a(K + a^2\sigma g).$

76. d. $\left(m_1 + \dfrac{m_2}{K}\right)g = \left[\dfrac{m_1}{\sigma_1} + \dfrac{m_2}{K\sigma_2}\right]\sigma\, g$

Divide by $m_2 g$, $\dfrac{m_1}{m_2} + \dfrac{1}{K} = \dfrac{m_1\sigma}{m_2\sigma_1} + \dfrac{\sigma}{K\sigma_2}$

$$\frac{m_1}{m_2}.2+\frac{1}{K}.0.5= \text{(values are given)}$$

$$\frac{m_1}{m_2}=\frac{0.5}{K};\ K=1.0, \text{ fully submerged.}$$

77. d. Acceleration when block B is in the liquid,

$$a_1=\frac{m_A g-(m_B g-\text{upthrust})}{(m_A+3m)}$$

$$=\frac{m_A g-\left(3mg-\frac{3m}{2\rho}\rho g\right)}{(m_A+3m)}(\uparrow)$$

Acceleration when block B is outside of the liquid,

$$a_2=\frac{3mg-m_A g}{(m_A+3m)}(\downarrow)$$

Given $a_1=a_2$, we get

$$m_A g-\frac{3}{2}mg=3mg-m_A g$$

$$\Rightarrow 2m_A g-\frac{9}{2}mg \quad \Rightarrow m_A=\left(\frac{9}{4}\right)m$$

78. b. As block B falls, L will decrease.
Block B will displace volume of liquid (V_1)
Equal to its own volume when it is in the liquid.

$$V_1=V_B=\frac{M_B}{\rho_B} \qquad \text{(i)}$$

When block B is on block A, it will displace the volume of liquid (V_2) whose weight is equal to the weight of the block B.

$V_2\rho_L g=M_B g=(V_1\rho_B)g$ or $V_2\rho_L=V_1\rho_B$
Since H increase, $V_1>V_2$
So, $\rho_L>\rho_B$.

79. b. $V_2^2=V_0^2+2gh$ and $A_1V_0=A_2V_2$ (i)

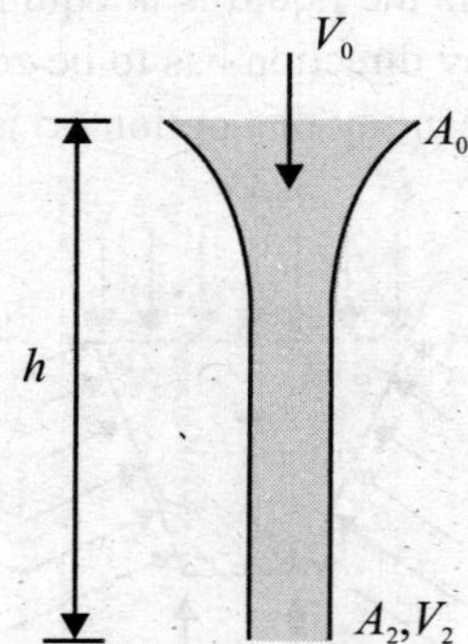

Fig. 3.327

Solving, we get $\frac{A_2}{A_1}=\frac{V_0}{\sqrt{V_0^2+2gh}}$

$$A_2/A_1=\frac{1}{2}=\frac{V_0}{\sqrt{V_0^2+2gh}}$$

$$4V_0^2=V_0^2+2gh$$

$$h=\frac{3V_0^2}{2g}$$

80. a. Velocity of efflux at a depth h is given by $V=\sqrt{2gh}$

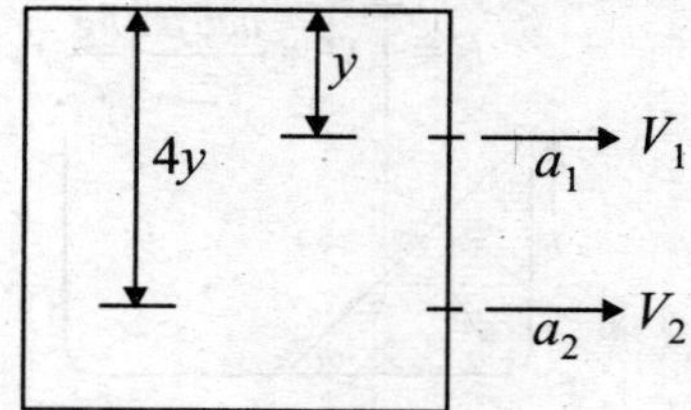

Fig. 3.328

Given, $a_1V_1=a_2V_2$

$L^2\sqrt{2gy}=\pi R^2\sqrt{2g4y}$ or $R=L/\sqrt{2\pi}$

81. b. $dF=dmx\omega^2$
and $(dp)A=(\rho Adx)x\omega^2$

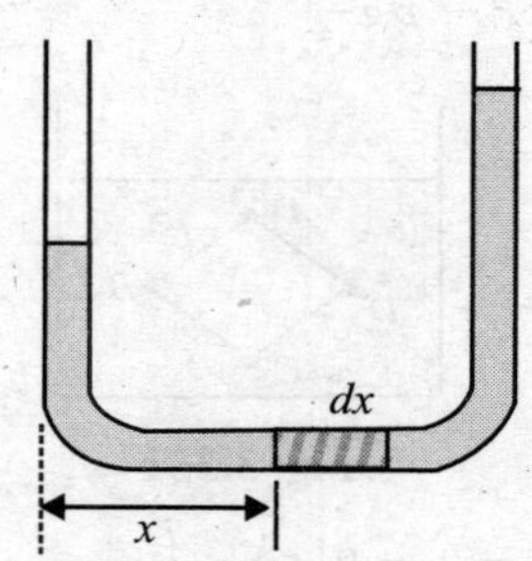

Fig. 3.329

$$\int_{P_1}^{P_2} dp=\rho\omega^2\int_0^L \omega\, dx$$

$$\text{or } P_2-P_1=\frac{\rho\omega^2L^2}{2}$$

$$\Rightarrow \rho gH=\frac{\rho\omega^2L^2}{2}$$

$$H=\frac{\omega^2L^2}{2g}$$

82. b. Upthrust = Net Weight

$$\frac{3}{2}LAdg=(d_1+d_2)LAg$$

$$\Rightarrow d_1+d_2=\frac{3}{2}d$$

and $d_2-d_1=d$

$$\Rightarrow d_2=\frac{5}{4}d \text{ and } d_1=d/4$$

Multiple Correct Answers Type

1. a., c. The maximum amount of water that can be retained is shown in the figure. If θ is the angle made by the water surface with the horizontal, then

$$\tan\theta = \frac{h}{c} = \frac{a}{g} \Rightarrow a = \frac{hg}{c}$$

So the maximum volume that can be retained is (1/2) $\times\, h \times c \times b$ and

$$F = \left[M + \frac{hcb\rho}{2}\right]\frac{hg}{c}$$

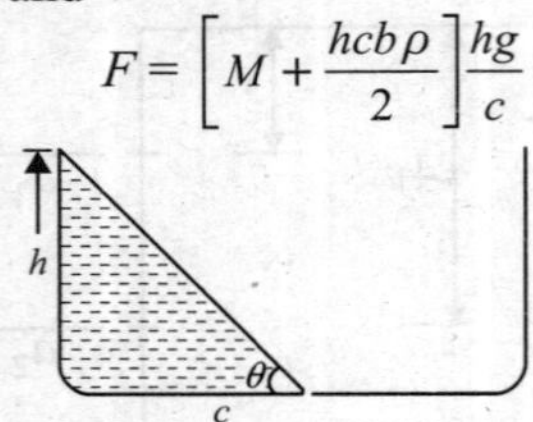

Fig. 3.330

2. a., b., d. According to Pascal's law,

$$P_D = P_B$$

because B and D lie on the same horizontal plane.

Let l be the side of the square. Then the vertical distance between A and C is $l\sqrt{2}$.

$$P_B = P_D = P_C - \rho g\frac{L}{\sqrt{2}} \quad \text{(i)}$$

Fig. 3.331

Similarly, $P_B = P_D = P_A + \rho g\dfrac{L}{\sqrt{2}}$ (ii)

$$P_B = P_D = \frac{P_A + P_C}{2}$$

3. a., b., c. The equation of the free surface is

$$y = \frac{\omega^2 r^2}{2g} + y_0$$

Fig. 3.332

When the free surface touches the brim of the cylinder, the lowest point of the paraboloid is at a height $H/3$ above the base, as shown in Fig. 3.332 (b). When the lowest point of the free surface touches the base of the cylinder, the height of water in the tank must be $H/2$. It means that water must spill out.

4. a., b., c. By the principle of conservation of energy, we have

$$mgh = \left(\frac{1}{2}\right)mv^2$$

or $$v = \sqrt{2gh}$$

Since $\rho > \rho'$, the buoyant force of the lake is greater than the weight of the body. If a is the upward acceleration, then

$$ma = \rho gV - \rho' gV \quad \text{or} \quad (\rho' V)a = \rho gV - \rho' gV$$

Cancelling V on both the sides, we get

$$a = g\left(\frac{\rho - \rho'}{\rho}\right) = g\left(\frac{\rho}{\rho'} - 1\right)$$

Now, $$v^2 = v_0^2 + 2ay$$

where v_0 is the velocity at $y = 0$. $v = 0$ is the velocity at the maximum depth and y is the negative displacement from the surface to the maximum depth. With these, we get $v_0^2 = -2ay$.

Hence, the depth is given by

$$-y = \frac{v_0^2}{2a} = \frac{2gh}{2g\dfrac{(\rho - \rho')}{\rho'}} = \frac{h\rho'}{\rho - \rho'}$$

5. b., c., d.

Density of the block $= \dfrac{1}{3\times10^{-4}}$ kg/m^3 $= \dfrac{10}{3} \times 10^3$ kg/m^3

Buoyant force $B = 3 \times 10^{-4} \times 10^3 \times g = 3$ N

$\therefore$ $$T_B = 10 - 3 = 7 \text{ N}$$

Reading of $B = 14 + 3 = 17$ N

Reading of $C = [14 + (10 - 3) + 3]$N $= 24$ N

6. a., b. The maximum horizontal distance from the vessel comes from hole numbers 3 and 4. Now, $v = \sqrt{2gh}$, where h is the height of the hole from the top.

Horizontal distance,

$$x = vt = \sqrt{2gh}\sqrt{\frac{2(H-h)}{g}}$$

$$= 2\sqrt{h(H-h)}$$

7. a., b., c., d. The free-body diagram of the entire liquid is as shown in the figure, m is the mass of the entire liquid in the vessel. As the liquid is in equilibrium, net force on the liquid in any direction has to be zero. From the above statement, we can say that option (a) is correct.

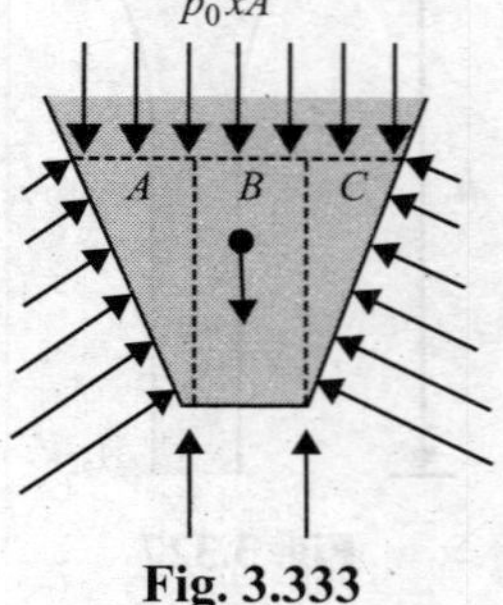

Fig. 3.333

For liquid contained in section lettered A, vertical component of forces acting on the inclined face is equal to atmospheric pressure force and weight of liquid. Force on the base is atmospheric force plus weight in section B. From the above statement, we can say (directly or indirectly) that options (b), (c) and (d) are also correct.

8. a., b., c. Pressure decreases in the direction of acceleration. If we take two points 1 and 2, two points in the horizontal line, than we can write $p_1 - p_2 = \rho a l$.

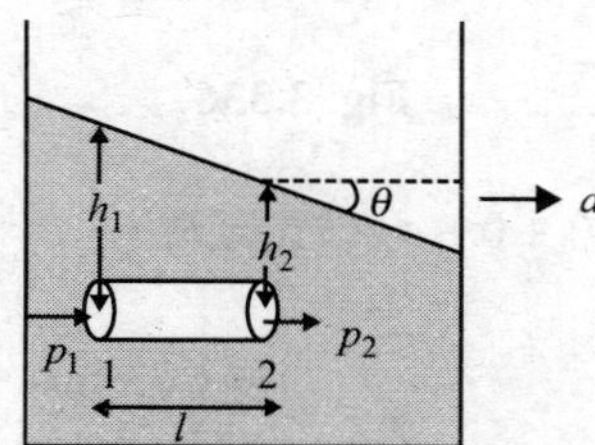

Fig. 3.334

As the inclination of liquid surface with $\tan\theta = a/g$. The equipressure lines will be parallel to this line.

9. a., c., d. The velocity with which the liquid comes out is given by the expression

$$P_0 + \rho_1 g \times 30 \text{ cm} + \rho_2 g \times 10 \text{ cm} = p_0 + \frac{\rho_2 v^2}{2}$$

$$v = 2.24 \text{ m/s}$$

Due to transfer of liquid, change in momentum takes place and hence a force has been exerted on the liquid in backward direction, this force is given by $F = \rho_2 a v^2$, where a is the cross-sectional area of hole $= 1200 \times 5 \times 10^{-4} \times 5 = 3$ N.

Normal contact force between the vessel and the surface is $N = mg$.

$N = (\rho_1 A \times 0.3)g + (\rho_2 \times A \times 0.2)g = 420$

where A is the cross-sectional area of the vessel. Limiting friction force, $f_L = \mu N = 0.04 \times 420 = 16.8$

As $f_L > F$, therefore the minimum force necessary to maintain the static equilibrium is 0. For the maximum value of the applied force, $F_{max} = f_L + F = 16.8 + 3 = 19.8$ N.

10. b., c. Since the liquids are in static equilibrium, pressure at R = pressure at S. Since liquid A is above liquid B, density of liquid A is less than that of B. Similarly, density of liquid C is less than that of B or liquid B has maximum density among these three liquids. Hence, option (c) is correct.

Since T is at lower level than Q, the pressure exerted by C is greater than that exerted by A. It is possible only when density of liquid A is less than that of C. Hence, option (b) is correct.

11. b., c. Height of liquid in vessel C is maximum. Therefore, force on the base of vessel C is maximum [$= (P_0 + h\rho g) A$]. Net force on all the three vessels = weight of the liquid, which is the same for all the three vessels.

12. a., c., d. If the system is not in the state of free fall, pressure at any point in a liquid depends on its depth h w.r.t. its free surface.

$$P = P_0 + hdg$$

i.e., pressure is different at points lying in different depths in the liquid. In different liquids (i.e., density d is different) points at the same depth h will be at different pressures. But points lying at different depths in different liquids can be at the same pressure if

$$h_1 d_1 = h_2 d_2 = \cdots$$

So, options (a), (c) and (d) are correct.

13. a., d. As per the continuity equation, Av = constant. $v_A = v_B$. As point B is at the lower level, hence $p_B > p_A$.

14. a., c., d.

$$\frac{P_1}{\rho} + \frac{v_1^2}{2} = \frac{P_2}{\rho} + \frac{v_2^2}{2}$$

$$P_1 - P_2 = \frac{\rho}{2}(v_2^2 - v_1^2)$$

But

$$P_1 - P_2 = \rho g h = \frac{\rho}{2}(v_2^2 - v_1^2)$$

or $\quad v_2^2 - v_1^2 = 2gh$

15. c., d. $R = 2\sqrt{hh'}$

R would be maximum if hh' is maximum.

a. $hh' = 70 \times 20 = 1400$
b. $hh' = 60 \times 30 = 1800$
c. $hh' = 50 \times 40 = 2000$
d. $hh' = 40 \times 50 = 2000$

Clearly, the liquid coming out from holes 3 and 4 would fall at maximum horizontal distance.

16. a., b., d. Here, liquid is flowing from A to B. Hence, $P_A > P_B$

According to the equation of continuity,

$$a_1 v_1 = a_2 v_2$$

As $a_2 > a_1$, hence $v_1 > v_2$. As the liquid is non-viscous, there is no loss of energy, i.e., energy per unit volume of the liquid is the same at every point.

So option (c) is incorrect.

Here, $P_A > P_B$, so $v_A > v_B$ and energy/volume of liquid is same at every point. So, $(PE)_A < (PE)_B$.

Hence, section A must be at the lower level than section B, i.e., the pipe cannot be horizontal.

17. a., b., d. W_1, W_3 are the forces exerted by the ball and the spring balance on each other before and after immersion. W_2, W_4 are the forces exerted by the tank and the weighing machine on each other before and after immersion.

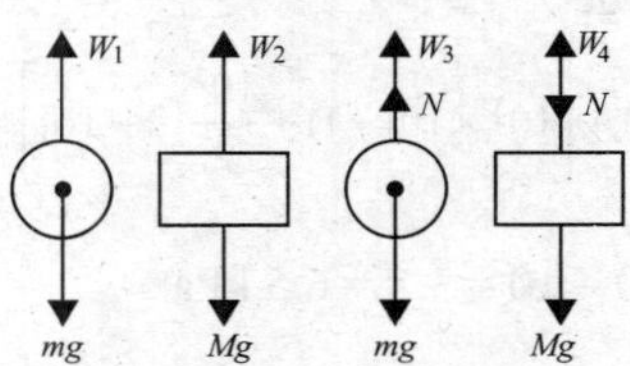

Fig. 3.335

Let m and M be the masses of the ball and the tank, respectively, and let N be the force of interaction between the ball and the liquid in the tank.

$$W_1 = mg$$
$$W_2 = Mg$$
$$W_3 = N = mg$$
$$W_4 = N + Mg$$

18. a., b., d. Initially, the wooden block floats partially immersed in water. Initially, upthrust exerted by water is exactly equal to weight of the block. But when it is exactly pressed down, then the more water is displaced. Hence, upthrust exerted by water increases. The force required to press down the block is F = upthrust – mg. Hence, work done by the force F will be equal to work against (upthrust – mg). It means, work done by the external force = work done against upthrust – loss of gravitational potential energy of the block.

19. b., d. The option (a) is incorrect. A body will sink if its weight is greater than the upthrust. Option (b) is correct. Option (c) is incorrect. Option (d) is correct. The density of hydrogen is less than that of air. Therefore, initially, the balloon rises because the upthrust due to air is greater than the weight of the hydrogen-filled balloon. The density of air decreases, as we go up. Therefore, the upthrust on the balloon also decreases. It will stop rising when it attains a height at which the upthrust becomes equal to its own weight.

20. a., b., c., d. The system consisting of the body floating in water in the beaker is in a state of weightlessness when it is falling freely under gravity. If the body is initially floating (while starting to fall, say, from rest) it will continue to float during the free fall. The body falls along with the displaced liquid and the immersed portion of the body remains unchanged. However, the weight of the immersed portion of the body and the weight of the liquid displaced both becomes zero during the free fall. The upthrust on the body, therefore, becomes zero. The weight of the body in air also becomes zero because it becomes weightless. Hence, all the four options (a), (b), (c) and (d) are correct.

21. a., c. Let p_1, h_1, v_1 and p_2, h_2, v_2 be the pressures, heights and velocities of flow at the two points, respectively. Then Bernoulli's theorem gives

$$p_1 + \rho g h_1 + \frac{1}{2}\rho v_1^2 = p_2 + \rho g h_2 + \frac{1}{2}\rho v_2^2 \qquad \text{(i)}$$

Putting v_1 = 3.0 m/s, v_2 = 4.0 m/s, $(h_2 - h_1)$ = 1 m, p_1 = 20 kPa, we get

$$p_2 = 20 + \left[10^3 \times 10(-1) + \frac{10^3}{2}[9-16]\right] \times 10^{-3}$$

$$= 20 - 10 - 3.5 = 6.5 \text{ kPa}$$

Also when the flow stops, $v_1 = v_2 = 0$ and then from Eq. (i), we have $P_2 = 20 - 10 = 10$ kPa

22. a., b., d. Applying Bernoulli's equation at C and D, we have

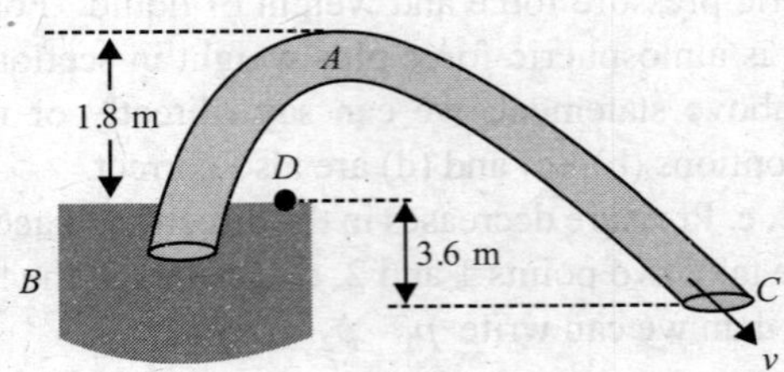

Fig. 3.336

$$P_0 + 0 + \rho g(3.6) = P_0 + \frac{1}{2}\rho v^2 + 0$$

$$\Rightarrow \quad v = 6\sqrt{2} \text{ m/s}$$

Volume blown per unit time = $av = \pi r^2 v = 96\sqrt{2} \times 10^{-4}$ m^3/s

Similarly, at A and C,

$$P_A + \frac{1}{2}\rho v^2 + \rho g(3.6+1.8) = P_0 + \frac{1}{2}\rho v^2 + 0$$

$$\Rightarrow \quad P_A = 0.46 \times 10^5 \text{ N/m}^2$$

23. a., d. According to Bernoulli's equation,

$$P_1 - P_2 = \frac{1}{2}\rho' v_1^2\left[\left(\frac{a_1}{a_2}\right)^2 - 1\right]$$

If h is the difference in level of mercury,

$$h\rho g = \frac{1}{2}\rho' v_1^2\left[\left(\frac{a_1}{a_2}\right)^2 - 1\right]$$

where ρ is density of mercury and $\rho\otimes$ is density of water. Hence,

$$1.4 \times 13.6 \times 980 = \frac{1}{2} \times v_1^2\left[\frac{v_2^2}{v_1^2} - 1\right]$$

$$\because \quad a_1 v_1 = a_2 v_2, \rho' = 1$$

$$\therefore \quad 2.8 \times 13.6 \times 980 = v_2^2 - v_1^2$$

$$\Rightarrow \quad v_2^2 = 2.8 \times 13.6 \times 980 + v_1^2$$

$$= 2.8 \times 13.6 \times 980 + 60^2$$

Solving, we get v_2 = 202 cm/s ≃ 2 m/s

$$\therefore \quad v_2 > 60 \text{ cm/s}$$

24. a.,b.,c. The forces on the walls of the vessel are shown. The liquid is in equilibrium, so net force on the liquid in any direction has to be zero (either $\alpha = \beta$ or $\alpha \neq \beta$).

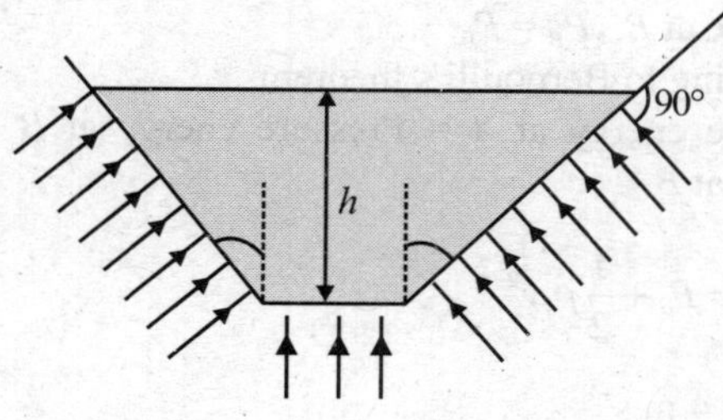

Fig. 3.337

25. b., d. $V_A = \left(2g\dfrac{H}{3}\right)^{1/2}$

$$V_B = \left(\frac{4gH}{3}\right)^{1/2} = \sqrt{2}\left(\frac{2gH}{3}\right)^{1/2} = \sqrt{2}V_A$$

$$Q_A = V_A(2a)$$

$$\text{and } Q_B = V_B(a) = V_A\sqrt{2}(a)$$

$$\frac{Q_A}{Q_B} = \sqrt{2}$$

Assertion-Reasoning Type

1. d.

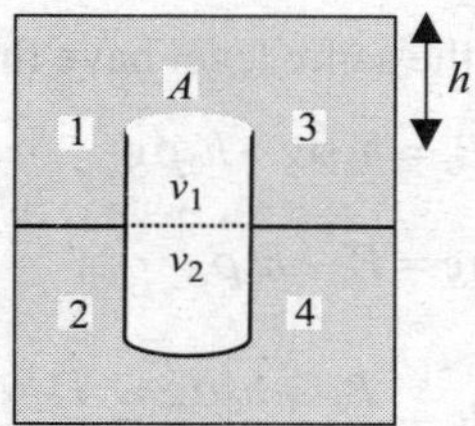

Fig. 3.338

$\rho_1 v_1 g$ is not the force applied by liquid 1 on body, it is $\rho_1 gh \times A$, although net force (buoyant) comes out to be $\rho_1 v_1 g + \rho_2 v_2 g$.

2. a. Since the net buoyant force on the brick completely submerged in water is independent of its depth below the water surface, the man will have to exert the same force on both the bricks. Hence, Statement I is true, Statement II is true; Statement II is a correct explanation for Statement I.

3. c. When air displaced by the balloon is greater than its weight, then a upthrust is applied on it upwards and it rises. But at a certain height it stops rising because at that height the atmospheric pressure becomes equal to the upthrust applied on it.

4. b. A hydraulic lift is an arrangement used to multiply the force. When a force is applied hydraulic pressure is transmitted in all directions. This is the working principle of Pascal's law.

5. b. The density of concrete, of course, is more than that of water and block of concrete will sink like a stone if dropped into water. Concrete cargo were filled with air and as such, average density of cargo vessels is given by

$$\frac{\text{Mass of concrete + mass of air}}{\text{Volume of concrete + volume of air}}$$

It follows that the average density of cargo vessels must be less than that of water. As a result the concrete cargo vessels do not sink.

6. a. With the increase in depth the pressure increases according to the formula $P \propto h$. Therefore, the force perpendicular to the wall of dam increases. Hence, the dam must have greater strength at base than at top. Due to this, dams are made thicker at the base than at the top.

7. a. According to Pascal's law, the pressure at bottom of each vessels containing liquid in equilibrium of rest is the same.

$$P_1 = P_2 = P_3$$

$$\text{Hence } F_1 = F_2 = F_3$$

8. a. Continuity equation av = const. If area of pipe is less velocity of water will be more.

9. a. Due to pressure difference across two sides of the ball, it deviates from its usual path. Force on vertical direction will remain the same.

10. b. When two holes are made, the velocity of flow through each hole is much less than the velocity of flow when there is only one hole. Hence, the flow becomes streamline with two holes. This is the reason why a liquid flows more smoothly and faster with two holes.

Comprehension Type

For Problems 1–3

1. a., 2. b., 3. a.

Sol. Let the net force on this particle be zero at height h. Then weight = upthrust (at h)

$$\Rightarrow \quad \frac{4}{3}\pi r^3 g \frac{5}{2}\rho_0 = \frac{4}{3}\pi r^3 g \rho_0 \left[4 - \frac{3h}{h_0}\right]$$

$$\Rightarrow \quad \frac{3h}{h_0} = 4 - \frac{5}{2} = \frac{3}{2} \quad \Rightarrow h = \frac{h_0}{2} \qquad \text{(i)}$$

Therefore, at $h = h_0/2$, the equilibrium position, the particle has a speed in downward direction due to which it will move downward. Let at a further depth x from $h_0/2$,

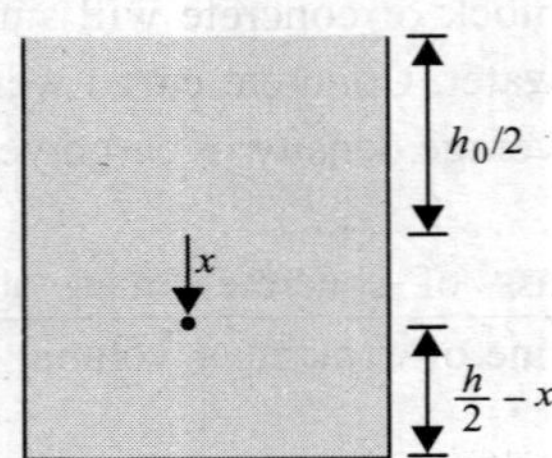

Fig. 3.339

F_{net} = upthrust – weight

$$= \frac{4}{3}\pi r^3 g\rho_0 \left[4 - \frac{3\left(\frac{h_0}{2} - x\right)}{h_0}\right] - \frac{4}{3}\pi r^3 g \frac{5}{2}\rho_0$$

$$= \frac{4}{3}\pi r^3 \rho_0 g \left[4 - \frac{3}{2} + \frac{3x}{h_0} - \frac{5}{2}\right] = \frac{4\pi r^3 \rho_0 g}{h_0} x$$

$$\therefore \quad F = \frac{4\pi r^3 \rho_0 g}{h_0} x$$

Since the net force on the particle and displacement of ball from its mean position are oppositely directed, the particle executes SHM about its mean position, i.e., $h = h_0/2$.

$$\therefore \quad F = ma = -\frac{4\pi r^3 \rho_0 g}{h_0} x$$

$$\Rightarrow \quad \frac{4}{3}\pi r^3 \frac{5}{2}\rho_0 a = -\frac{4\pi r^3 \rho_0 g}{h_0} x$$

$$\Rightarrow \quad a = -\frac{6g}{5h_0} x = -(12/h_0)x = -\omega^2 x$$

$$\text{where } \omega = \sqrt{\frac{12}{h_0}} = \sqrt{\frac{12}{\frac{12}{\pi^2}}} = \pi$$

Hence, time taken by the particle to reach the bottom is

$$t = \frac{T}{2} = \frac{1}{2}\frac{2\pi}{\omega} = \frac{\pi}{\omega} = \frac{\pi}{\pi} = 1 \text{ s}$$

For Problems 4–6

4. c., 5. a., 6. d.

Sol. 4. Since area of a hole is very small in comparison to base area A of the cylinder, velocity of liquid inside the cylinder is negligible. Let velocity of efflux be v and atmospheric pressure be P_0. Consider two points A (inside the cylinder) and B (just outside the hole) in the same horizontal line as shown in the figure in question.

Pressure at A, $P_A = P_0 + h\rho_2 g + (h - y)\rho_1 g$

Pressure at B, $P_B = P_0$

According to Bernoulli's theorem,

Pressure energy at A = Pressure energy at B + Kinetic energy at B

$$\therefore \quad P_A = P_B + \frac{1}{2}\rho_1 v^2$$

$$\therefore \quad v = 4 \text{ m s}^{-1}$$

5. When the cylinder is on the smooth horizontal plane, force F required to keep the cylinder stationary equals horizontal thrust exerted by fluid jet. But that is equal to mass flowing per second × change in velocity of this mass.

$$\therefore \quad F = (av\rho)(v - 0) = a\rho v^2 \quad \text{or} \quad F = 7.2 \text{ N}$$

6. Total mass of the liquid in the cylinder is

$$m = Ah\rho_1 + Ah\rho_2 = 450 \text{ kg}$$

Limiting friction = μmg = 45 N

Since $F <$ limiting friction, therefore minimum force required is zero. Considering the free-body diagram for maximum vertical forces,

$$N = mg$$

Now considering the horizontal forces,

$$F_{max} = F + \mu N = 52.2 \text{ N}$$

For Problems 7–9

7. c., 8. b., 9. d.

Sol. $2P_0 = (h_2 + h_0)\rho g + p_0$

(since liquids at the same level have the same pressure)

$$P_0 = h_2\rho g + h_0\rho g$$

$$h_2\rho g = P_0 - h_0\rho g$$

$$h_2 = \frac{P_0}{\rho g} - \frac{h_0\rho g}{\rho g} = \frac{P_0}{\rho g} - h_0$$

KE of the water = Pressure energy of the water at that layer

$$\frac{1}{2}mV^2 = m \times \frac{P}{\rho}$$

$$V^2 = \frac{2P}{\rho} = \frac{2}{\rho}[P_0 + \rho g(h_1 - h_0)]$$

$$V = \left[\frac{2}{\rho}\{P_0 + \rho g(h_1 - h_0)\}\right]^{1/2}$$

We know $\quad 2P_0 + \rho g(h_1 - h_0) = P_0 + \rho g X$

$$\Rightarrow \quad X = \frac{P_0}{\rho g} + (h_1 - h_0) = h_2 + h_1$$

i.e, X is h_1 metre below the top or X is $-h_1$ above the top.

For Problems 10–14

10. d., 11 c., 12. b., 13. b., 14. a.

Sol. 10. $\dfrac{dy}{dx}=\dfrac{a_x}{a_y+g}=\dfrac{g/2}{-g/2+g}=1$

(Effective acceleration due to gravity will be $g-a=g/2$)

$\therefore \theta = 45°$

11. The slope of the free surface is 45°. Thus, the free surface passes through the centre of the box having coordinates (2,2) at the top of the box.

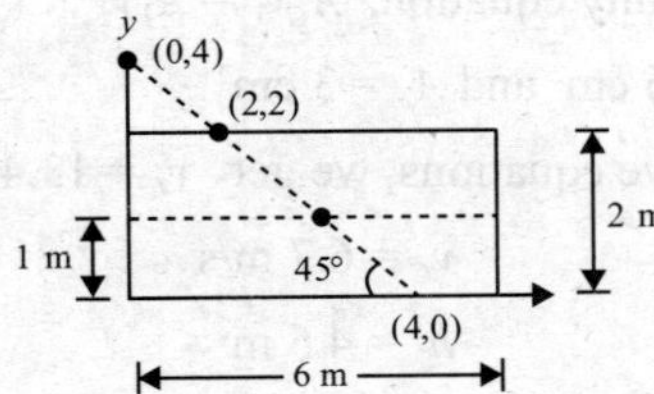

Fig. 3.340

Required length = 6 – 2 = 4 m

12. $P = P_a + \rho gh = 10^5 + 1000 \times 10 \times 1$

$= (10^5 + 10^4)\ \text{N/m}^2 = 0.11\ \text{MPa}$

13. $P = (10^5 + 10^3 \times 10 \times 4)\ \text{N/m}^2$

$= [0.1 + 0.04]\ \text{MPa} = 0.14\ \text{MPa}$

14. As the maximum slope of the free surface is 1/3 for the condition of non-exposure of the bottom of the box, so

$$\frac{a_x}{a_y+g}=\frac{1}{3}$$

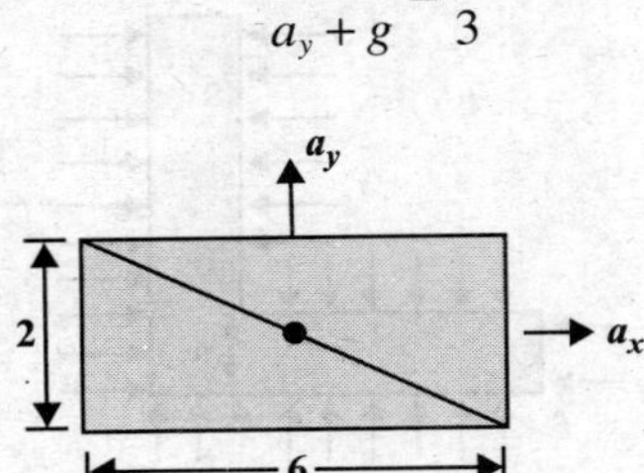

Fig. 3.341

As $a_x = g/2$

$\therefore\ 3a_x = a_y + g$

$\Rightarrow\ ay = g/2$, upwards

For Problems 15–19

15. b., 16. c., 17. d., 18. a., 19. c.

Sol. 15. $F = \rho_A (V_0 - 0)^2\ [1 - \cos 180°]$

$= 2\rho A v^2 = 2 \times 1000 \times 10^{-4} \times 10 \times 10 = 40\ \text{N}$

16. $F = 2\rho A\ (V_0 - u)^2$

u = speed of the cart

$$m\ \frac{du}{dt} = 2\rho A (V_0 - u)^2$$

$$\int_0^u \frac{du}{(V_0-u)^2}=\frac{2\rho A}{m}\int_0^t dt$$

$$\left[\frac{1}{V_0-u}\right]_0^u=\frac{2\rho At}{m}$$

$$\frac{1}{V_0-u}-\frac{1}{V_0}=\frac{2\rho At}{m}=\frac{4t}{100} \qquad \text{(i)}$$

$$\left(\frac{2\rho A}{m}=\frac{2\times10^3\times2\times10^{-4}}{10}=\frac{4}{100}\right)$$

At $\quad t = 10\ \text{s}$

$$\frac{1}{V_0-u}=\frac{4}{10}+\frac{1}{10}=\frac{1}{2}$$

$\Rightarrow\quad V_0 - u = 2$

$u = 8\ \text{m/s}$

17. $F = 2\rho A (V_0 - u)^2 = 2 \times 10^3 \times 10^{-4}\ (10 - 8)^2$

$= 2 \times 10^3 \times 2 \times 10^{-4} \times 4 = 1.6\ \text{N}$

$\therefore\ a = \dfrac{F}{M} = 0.16\ \text{m/s}^2$

18. From Eq. (i), $\dfrac{1}{V_0-u}-\dfrac{1}{V_0}=\dfrac{4t}{100}$

$$\Rightarrow\quad \frac{1}{8}-\frac{1}{10}=\frac{4t}{100} \Rightarrow \frac{2}{80}=\frac{4t}{100}$$

$t = 1.6\ \text{s}$

19. $F = 2\rho A (V_0 - u)^2 = 2 \times 10^3 \times 2 \times 10^{-4} \times 25 = 10\ \text{N}$

$P = Fu = 10 \times 5 = 50\ \text{W}$

For Problems 20–21

20. c., 21. a.

Sol. Weight of the rod $w = LA\rho g$

(A = area of cross section, ρ = density of material)

Submerged length of the rod is given by

$$OT = \frac{L}{2\sin\theta}$$

Buoyant force $\quad B = \left(\dfrac{L}{2\sin\theta}\right)A\rho_w g$

from Archimedes' principle (ρ_w = density of water).

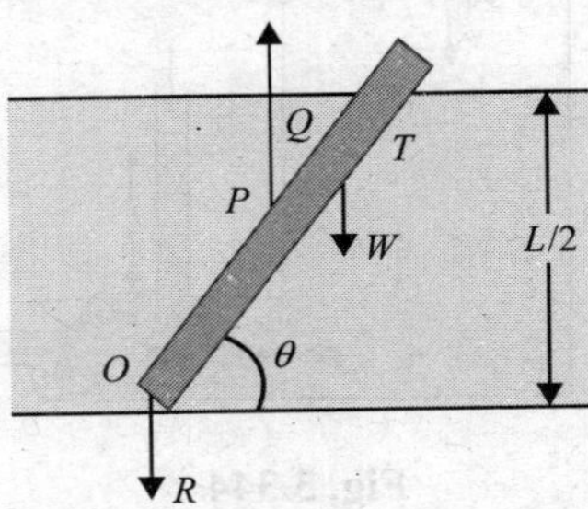

Fig. 3.342

This force acts at P, where $OP = \frac{OT}{2} = \frac{l}{4\sin\theta}$
Balancing torque about O is

$$WOQ \cos\theta = BOP \cos\theta$$

$$(LA\rho g)\, L/2 = \left(\frac{L}{2\sin\theta}\right) A\rho_w g \left(\frac{L}{4\sin\theta}\right)$$

$$\sin^2\theta = \frac{\rho_w}{4\rho} \Rightarrow \sin^2\theta = 1/2 \Rightarrow \theta = 45°$$

Also, balancing forces $R + W = B$

$$R = B - W = \frac{\rho A \rho_w g}{2\sin\theta} - LA\rho g$$

$$= LA\rho g\left[\frac{1}{\sin\theta} - 1\right] \quad (\because \rho_w = 2\rho) = \left(\sqrt{2} - 1\right) mg$$

For Problems 22–23

22. a., 23. c.

Sol. Let x cm be the length of cork cylinder inside the water and T be the tension in the string.

The free-body diagrams of the two cylinders are as shown in the following figures.

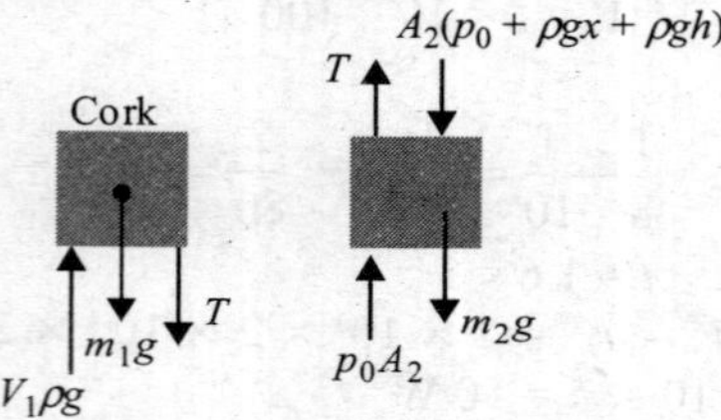

Fig. 3.343

Here, $m_1 = 15$ g, $m_2 = 25$ g, $h = 4$ cm, $v_{in} = A_1 x$ cm^3
For cork cylinder, $T + m_1 g = V_{in}\rho g$
For aluminium cylinder,

$$T + p_0 A_2 = m_2 g + (p_0 + \rho g x + \rho g h) A_2$$

$$\Rightarrow \quad x = 6 \text{ cm}$$

$$T = 0.45 \text{ N}$$

For Problems 24–26

24. a., 25. c., 26. b.

Sol. Let atmospheric pressure be p_0. Then absolute pressure of air is $p + p_0 = 4$ atm

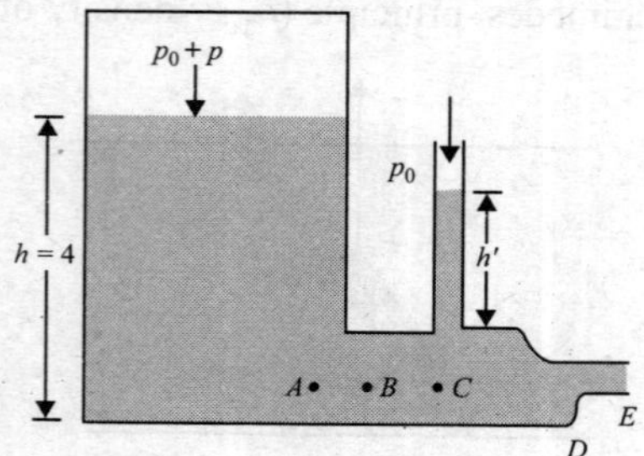

Fig. 3.344

Pressure at A, $p_A = p + p_0 + \rho gh$

From equation of continuity and Bernoulli's theorem,

$$p_B = p_C$$

Applying Bernoulli's theorem at A and E, we have

$$p + p_0 + \rho gh = p_C + \frac{\rho v_C^2}{2} = p_0 + \frac{\rho v_E^2}{2}$$

where v_C and v_E are the velocities of flow of liquid at C and E, respectively.

From hydrostatics, $p_C = p_0 + \rho g h'$

From continuity equation, $A_C v_C = A_E v_E$,

where $A_C = 6$ cm^2 and $A_E = 3$ cm^2

Solving above equations, we get $v_E = 13.4$ m/s

$$v_C = 6.7 \text{ m/s}$$

$$h' = 4.5 \text{ m}$$

Discharge rate

$$Q = A_E v_E = 3 \times 10^{-4} \times 13.4 \text{ m}^3/\text{s}$$

When a hole has been created at the top of the vessel, gauge pressure drops to zero, i.e., in above equation, put $p = 0$ and get h'.

$$\therefore \quad h' = 3 \text{ m}$$

For Problems 27–28

27. c., 28. d.

Sol. Pressure at bottom of vessel,

$$P = \rho g \times 6 \text{ cm} = 1000 \times 10 \times \frac{6}{100} = 600 \text{ N/m}^2$$

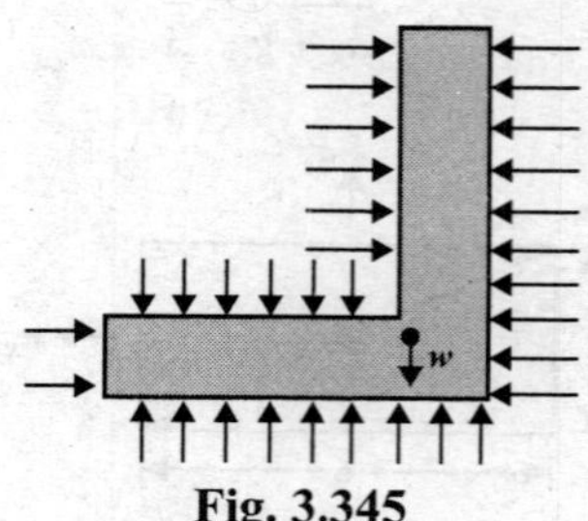

Fig. 3.345

Force acting at the base, $F = pA_2 = 600 \times 50 \times 10^{-4} = 3$ N
Weight of liquid inside the vessel,

$$W = \rho[A_1 \times 5 \text{ cm} + A_2 \times 1 \text{ cm}]g$$

$$= 1000[10 \times 5 + 50 \times 1] \times 10^{-6} \times 10 = 1 \text{ N}$$

So, $F > W$

The free-body diagram of the liquid is as shown. The net force on the liquid is zero.

For Problems 29–30

29. b., 30. a.

Sol. Mass per unit length of rod is 2 kg.

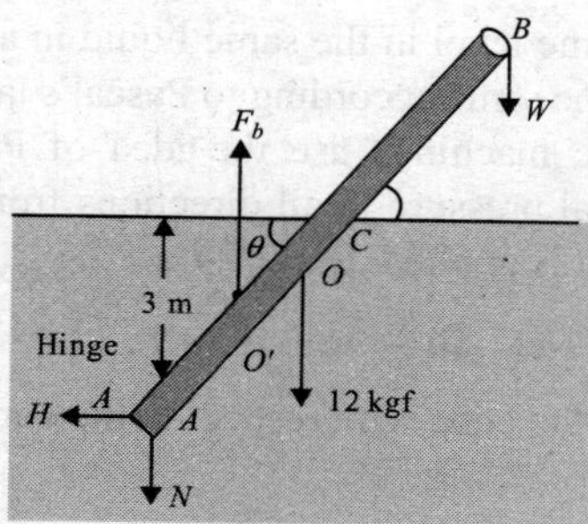

Fig. 3.346

Mass of the submerged portion of the rod is 10 kg and its volume is $\frac{10}{500}$ m^3.

Buoyant force $F_B = \frac{10}{500} \times 1000 = 20$ kgf

Let N and H be the vertical and horizontal reactions of the hinge on the rod, respectively. Considering the horizontal and vertical translational equilibrium of the rod,

$$N + 12 + W = 20$$

where W is the weight to be attached at the end.

$$H = 0$$

Considering the rotational equilibrium of the rod about A,

$$-20 \times g \times 2.5 \cos\theta = 12 \times g \times 3 \cos\theta + W \times 6 \cos\theta = 0$$

$$\Rightarrow \quad 6W = 50g - 36g = 14g$$

$$\Rightarrow \quad W = \frac{7}{3}g = \frac{7}{3}\text{kgf} \Rightarrow N = 8 - \frac{7}{3} = \frac{17}{3}\text{kgf}$$

For Problems 31–33

31. a., 32. c., 33 b.

Sol. 31. As long as water fills the tube (as shown in the figure) and points A and B are open to the atmosphere, the velocity at E will be given by Torricelli's theorem.

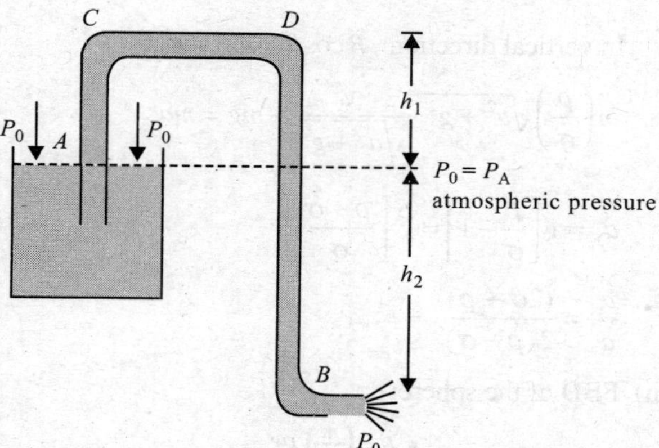

Fig. 3.347

Hence, $v_B = \sqrt{2gh_2}$ where h_2 is the difference in the levels A and B.

32. We apply Bernoulli's theorem for level A and for the highest level, CD (labelled by subscript 1), to get

$$P_A = P_1 + \frac{1}{2}\rho v_1^2 + \rho g h_1 \quad \text{(i)}$$

Since the tube has uniform cross section and water is incompressible,

$$v_1 = v_B = \sqrt{2gh_2} \quad \text{(ii)}$$

From Eqs. (i) and (ii), $P_1 = P_A - \frac{1}{2}\rho\left[\sqrt{2gh_2}\right]^2 - \rho g h_1$

$$= P_{atm} - \rho g(h_1 + h_2) \quad \text{(iii)}$$

The minimum value of $P_1 = 0$ (P_1 cannot be negative because then no water will reach the level). Hence, putting $P_1 = 0$,

$$(h_1)_{max} = \frac{P_{atm}}{\rho g} - h_2 = \frac{1.0 \times 10^5}{1 \times 10^3 \times 9.8} - 3.0$$

$$= 10.2 \text{ m} - 3.0 \text{ m} = 7.2 \text{ m}$$

33. Putting $h_1 = h_2 = 3.0$ m in Eq. (iii) above,

P_1 = pressure at level CD

$$= P_{atm} - (\rho g)(3+3) = 1.0 \times 10^5 - 6\rho g$$

The gauge pressure at level $CD = 6\rho g = 6 \times 10^3 \times 9.8$ N/m^2 $= 5.9 \times 10^4$ N/m^2

For Problems 34–36

34. d., 35. a., 36. b.

Sol. 34. The upward buoyant force $B = \rho g V$, where V is the volume of the body.

Hence, the equation of motion is

$B - mg = ma$, where a is the upward acceleration, i.e.,

$$\rho g V - \sigma V g = \sigma V a$$

Hence, $a = \frac{g(\rho - \sigma)}{\sigma} = g\left(\frac{\rho}{\sigma} - 1\right)$ upward

35. If $v_0 = \sqrt{2gh}$ is the velocity downwards just when the body enters the liquid, the final velocity at the maximum depth it will sink is zero. Its acceleration is a.

Hence, $(-)\,y$ = maximum depth

$$= \frac{v_0^2}{2a} = \frac{2gh}{2g\left(\frac{\rho}{\sigma} - 1\right)}$$

$$\therefore \quad y = \frac{h\sigma}{\rho - \sigma}$$

36. Time taken to reach the surface of liquid, $t_1 = \sqrt{\frac{2h}{g}}$

Time taken by object to travel from surface of liquid to maximum depth,

$$t_2 = \sqrt{\frac{2h\left(\frac{\sigma}{\rho - \sigma}\right)}{g\left(\frac{\rho - \sigma}{\sigma}\right)}}$$

As $\rho = 2\sigma$,

$$t_2 = \sqrt{\frac{2h}{g}}$$

Total time taken to reach starting point is

$$t = 2t_1 + 2t_2 = 4\sqrt{\frac{2h}{g}}$$

Matching Column Type

1. **c.** Kinetic energy is 1/2 mv^2. Hence,

$$\frac{v^2}{2g} = \frac{\left(\frac{1}{2}mv^2\right)}{mg}$$

is energy per unit weight.

$$\frac{\rho v^2}{2} = \frac{\left(\frac{1}{2}mv^2\right)}{m\rho}$$

is energy per unit volume

and

$$\frac{v^2}{2} = \frac{\frac{1}{2}mv^2}{m}$$

is energy per unit mass.

2. **i → c, d.; ii → e; iii → b; iv → c, a.**
Bernoulli's theorem is applicable for incompressible, non-viscous liquids with streamlined flow.
So, **i→c.** It is based on the principal of conservation of energy. So, **i → d.**
A ball moving with a spin will have higher velocity of air above the ball than below. So the pressure below will be more than that above, causing a lift. So, **ii → e.**
Based on the principle of Bernoulli's theorem, artificial high pressure created pushes the paint through the gun. So, **iii → b.**
Streamlined flow is possible in only non-viscous medium. So, **iv → c.**
As the stream passes through smaller cross section or narrower pipes, the velocity increases causing the pressure to fall. So, **iv → a.**

3. **i → d; ii → c; iii → a; iv → a.**
Magnus effect talks of difference in pressure caused due to difference in velocity to lift a structure. Lifting of asbestos roofs during cyclonic storms is based on this phenomenon. So, **i → d.**
Viscous force is wet friction and brings loss in internal energy. So, **ii → c.**
At the same level in the same liquid in a container, the pressure is the same according to Pascal's law. So, **iii → a.**
Hydraulic machines use the idea of Pascal's law to transfer equal pressure in all directions from the free surface where it is applied. So, **iv → a.**

4. **i → d; ii → c; iii → a; iv → b.**
(i) FBD of the cube with respect to the container

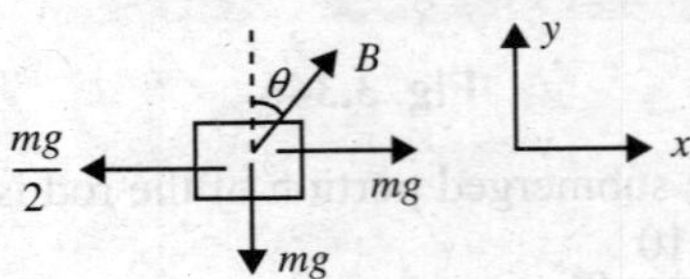

Fig. 3.348

$$B = \left(\frac{m}{\sigma}\right)\rho g_{\text{eff}} = m\left(\frac{\rho}{\sigma}\right)\sqrt{a^2+g^2}$$

$$\tan\theta = \frac{a}{g}$$

$$\sin\theta = \frac{a}{\sqrt{a^2+g^2}}; \cos\theta = \frac{g}{\sqrt{a^2+g^2}}$$

Let acceleration of the cube w.r.t. container in horizontal direction in a_x'

$$mg + B\sin\theta - \frac{mg}{2} = m a_x'$$

$$a_x' = \frac{g}{2} + \frac{\rho}{\sigma}a = a_x \Rightarrow \frac{g}{2} + \frac{\rho}{\sigma}\cdot\frac{g}{2} = a_x'$$

$$\text{Hence, } a_x = a_x' + \frac{g}{2} = g + \frac{g}{2}\frac{\rho}{\sigma}$$

$$a_x = g\left[\frac{2\sigma+\rho}{2\sigma}\right] \quad \text{(i)}$$

In vertical direction: $B\cos\theta - mg = ma_y$

$$m\left(\frac{\rho}{\sigma}\right)\sqrt{a^2+g^2}\cdot\frac{g}{\sqrt{a^2+g^2}} - mg = ma_y$$

$$a_y = g\left[\frac{\rho}{\sigma} - 1\right] = g\left[\frac{\rho-\sigma}{\sigma}\right] \quad \text{(ii)}$$

$$\frac{a_x}{a_y} = \frac{(2\sigma+\rho)}{2(\rho-\sigma)}$$

(ii) FBD of the sphere:

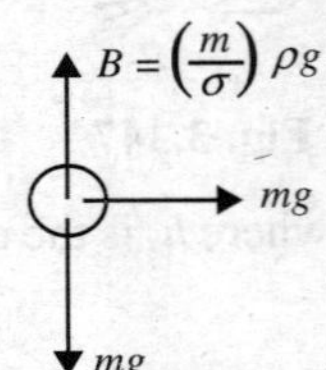

Fig. 3.349

$$a_x = g$$

$$a_y = \frac{-mg + \frac{m\rho}{\sigma}g}{m} = \left(\frac{\rho - \sigma}{\sigma}\right)g$$

$\Rightarrow$ Hence, $\dfrac{a_x}{a_y} = \dfrac{\sigma}{\rho - \sigma}$

(iii) FBD of the cylinder

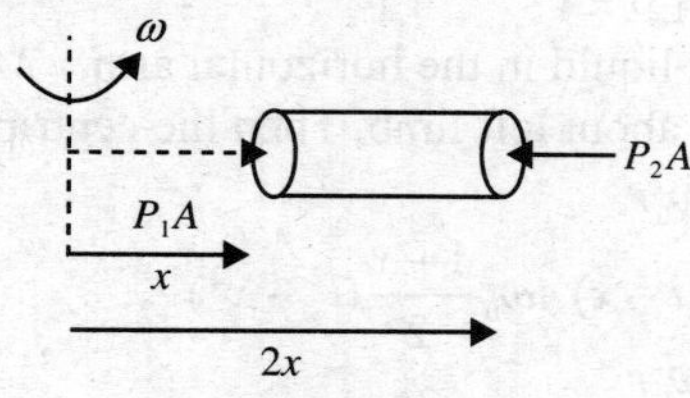

Fig. 3.350

Hence, $(P_2 - P_1)A = ma_x$

$$\frac{\rho\omega^2}{2}\left[(2x)^2 - x^2\right]A = ma_x$$

$$\frac{\rho}{2}\left(\frac{2g}{3x}\right)3x^2 A = (A - x)\sigma a_x$$

This gives $\dfrac{a_x}{g} = \dfrac{\rho}{\sigma}$

(iv) When the cart is not filled with liquid

$$\tan\theta = \frac{a}{g} \quad \text{(i)}$$

When the car is filled with liquid

FBD of the pendulum,

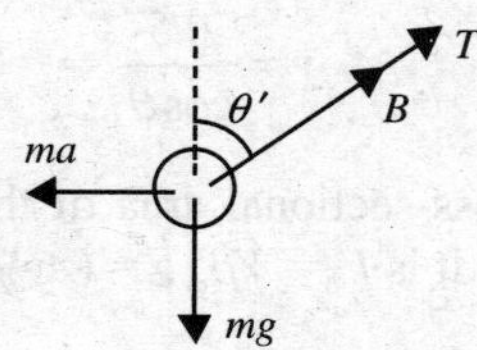

Fig. 3.351

Cart: $(T + B)\sin\theta' = mg$ (i)

$(T + B)\cos\theta' = mg$ (ii)

$$\tan\theta' = \frac{a}{g} \quad \text{(iii)}$$

Hence, $\dfrac{\tan\theta'}{\tan\theta} = \dfrac{1}{1} = \dfrac{\theta'}{\theta} = 1:1$

5. i → d; ii → c; iii → a; iv → b.

(i) Hydrostatic force on horizontal direction

$$\frac{F_H}{h} = \frac{1}{2}\rho g h^2 = \frac{1}{2}\rho g A$$

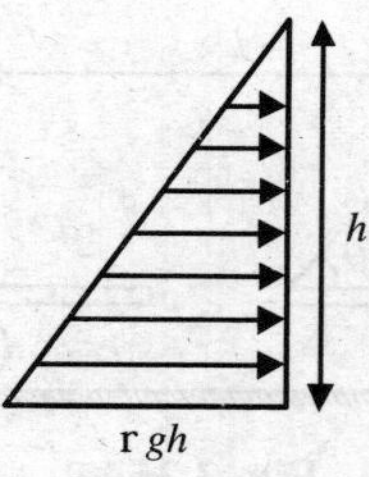

Fig. 3.352

$$F_H = \frac{1}{2}\rho g A h$$

(ii) Buoyant force on the cube

$$F_B = (Ah)\rho g = \rho g h A$$

(iii) The thrust force (impact) due to liquid coming out of nozzle = $= \rho A v^2$

(iv) Applying Bernoulli's equation just above and just below the roof,

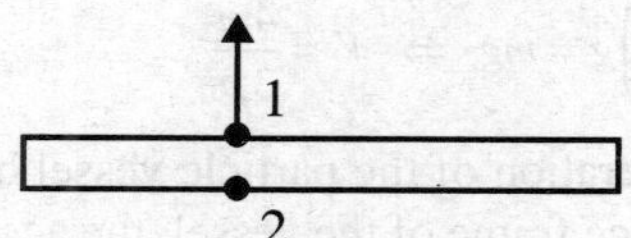

Fig. 3.353

$$P_1 + \frac{1}{2}\rho v_1^2 = P_2 + \frac{1}{2}\rho v_2^2 \quad (v_2 = 0)$$

Hence, $(P_2 - P_1) = \dfrac{1}{2}\rho v^2 = \Delta P$

Hence, lift force $F_{\text{lift}} = \Delta P A$

$$F_{\text{lift}} = \frac{1}{2}\rho A v^2$$

Integer Answer Type

1. **(1)** As the cart is drawn by a force F, the water in the vessel takes up a slant position rising upward at the back wall of the vessel. To prevent water flowing out of the hole H, the acceleration of the vessel should have such a value that it occupies a face area DBH and a width of vessel given by A/L.

Area of $\Delta DBH = \dfrac{1}{2}bc$

Volume of liquid retained $= \dfrac{1}{2}bc \times \dfrac{A}{L}$

Mass of cart and water $= M + \dfrac{bcA\rho}{2L}$

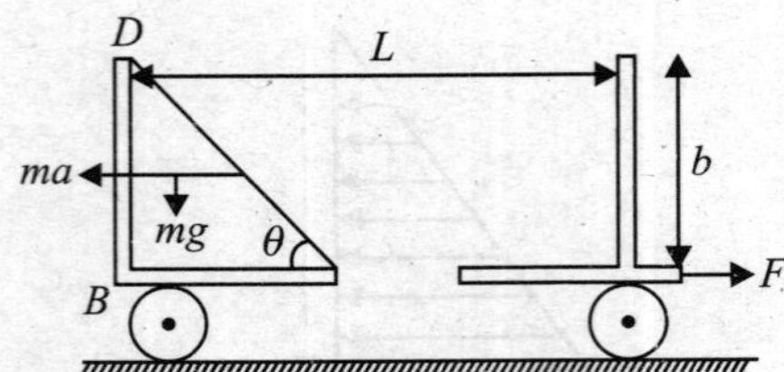

Fig. 3.354

$$\tan\theta = \frac{ma}{mg}$$

$$a = g\tan\theta = g\times\frac{b}{c}$$

$$\text{Required force } \left(M + \frac{bcAr}{2L}\right)\frac{gb}{c} = [1.5 + 0.5]\times 50$$
$$= 2.0\times 50 = 100\,\text{N}$$

2. **(5)** Let the volume of the cylinder be V. When the cylinder is floating, upthrust = weight. Hence,

$$\rho = \left(\frac{3}{4}V\right)g = mg \Rightarrow V = \frac{4m}{3\rho}$$

Let the acceleration of the particle vessel be A (upwards). In the reference frame of the vessel, the acceleration of the cylinder is $A/3$.

$$\therefore mg + mA - \text{uptrust} = m\left(\frac{A}{3}\right)$$

$$mg + mA - \rho Vg' = m\left(\frac{A}{3}\right)$$

where $g' = g + A$ = effective value of g for upthrust.

$$\therefore mg + mA - \rho V(g + A) = m\left(\frac{A}{3}\right)$$

$$\Rightarrow mg - \frac{4}{3}m(g + A) = -m\frac{2}{3}A$$

$$A = \left(-\frac{g}{2}\right)\text{upwards}$$

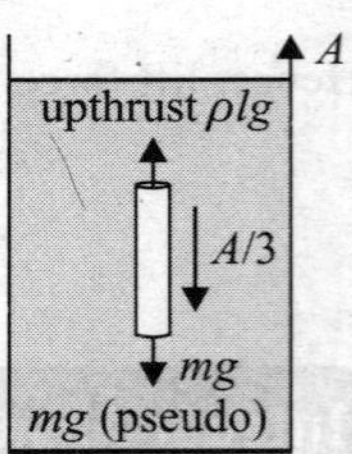

Fig. 3.355

The acceleration of the vessel should be $\frac{g}{2} = \frac{10}{2} = 5\ \text{ms}^{-2}$ (downwards).

3. **(5)** Due to rotation, let the shift of the liquid be x cm. Let the cross-sectional area of the tube be A.

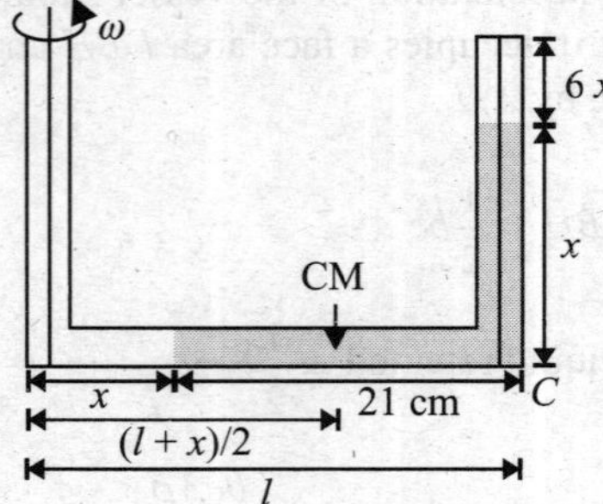

Fig. 3.356

In the right limb for compressed air,

$$p_1 v_1 = p_2 v_2$$
$$p_0 A\times 6 = p_2 A\,(6 - x)$$
$$p_2 = \frac{6p_1}{6 - x} \qquad \text{(i)}$$

Force at the corner 'C' of right limb climb due to liquid above, $p_1 = \left[\frac{6p_1}{6 - x} + x\rho g\right]A$.

Mass of the liquid in the horizontal arm, $M = \rho(l - x)A$

It is rotated about left limb. Then the centripetal force is

$$F_2 = m\omega_0^2 r$$
$$= \rho(l - x)A\omega_0^2\frac{1 + x}{2}(l^2 - x^2)$$
$$F_2 = m\omega_0^2 r$$
$$= \rho(1 - x)A\omega_0^2\frac{1 + x}{2} = \frac{\rho A\omega_0^2}{2}(1^2 - x^2)$$

But $F_1 = F_2$

$$\Rightarrow \frac{\rho A\omega_0^2(1^2 - x^2)}{2} = \left[\frac{\delta p_0}{\delta - x} + x\rho g\right]A$$
$$= \frac{10^5\times 100\times(21^2 - x^2)\times 10^{-4}}{2}$$
$$= \left[\frac{\delta\times 10500}{(\delta - x)} + x\times 10^5\times 10\times 10^{-2}\right]$$

On solving, we get $x = 0.01$ m = 1 cm

The length of the air column = 6 – 1 = 5 cm.

4. **(2)** Let y be the length of the plank inside water.

$$y = \frac{0.5}{\cos\theta}$$

Let A be the cross-sectional area of the plank. Then the buoyant force on it is $F_b = V\rho_\omega g = (Ay)\rho_\omega g$

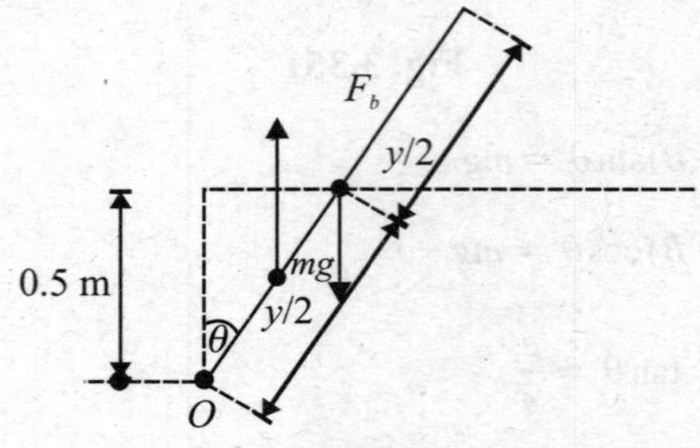

Fig. 3.357

Since the plank is in rotational equilibrium, so $\sum\vec{\tau}_0 = 0$

$$\text{or } mg\times\frac{l}{2}\sin\theta - F_b\times\frac{y}{2}\sin\theta = 0 \text{ or } mgl - F_b\times y = 0$$

$$\text{or } (Al\times 0.5)gl - (Ay)d_v y = 0 \text{ or } 0.5l^2 = y^2$$

$$\text{or } 0.5\times(1)^2 = \left(\frac{0.5}{\cos\theta}\right)^2 \text{ or } \cos^2\theta = \frac{1}{2}$$

5. (1) $(6 - x)^2 A\rho_\omega g/2 = 18A\rho_r g$

Solving we get $x = 1$ m.

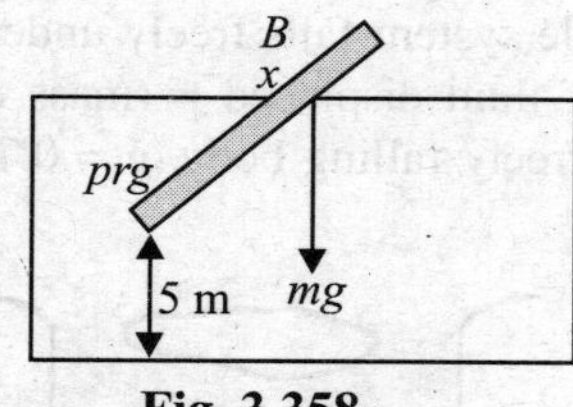

Fig. 3.358

6. (1) When the tube is rotated, the liquid starts to flow radially outwards and air in sealed arm is compressed. Let the shift of the liquid be y as shown in Fig. 3.359.

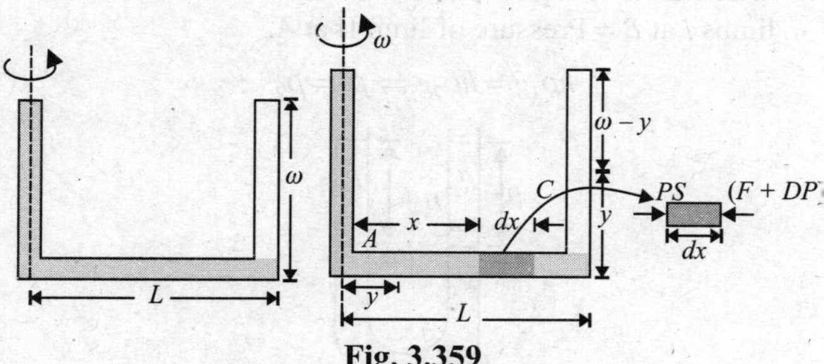

Fig. 3.359

Let the cross-sectional area of the tube be S. Here element of length dx is rotating with the tube. From the free-body diagram of the element, we can write

$$(P + dP)S - PS = dm\,\omega^2 x = (\rho S dx)\,\omega^2 x$$

$$dPS = \rho\omega^2\,Sxdx$$

Here the pressure difference between points A and B can be given by integrating the pressure difference across an element of width dx, which is given as

$$dP = \rho\omega^2\,xdx$$

Now integrating from A to B, we get

$$P_B - P_A = \int_y^L \rho\,\omega^2 x dx = \frac{\rho\,\omega^2}{2}(L^2 - y^2)$$

$$\Rightarrow\ p_B = p_0 + \frac{\rho\,\omega^2}{2}(L^2 - y^2)$$

The pressure at point C can be given as

$$P_C = P_a - y_\rho g$$

At point A, pressure is atmospheric. Thus, we have

$$P_C = \frac{\rho\omega^2}{2}(L^2 - y^2) + P_0 - y_\rho g = 10^5\ \text{N/m}$$

Archives

Fill in the Blanks Type

1. $Vd_{\text{Hg}}g = Vd_{\text{m}}g$

Fraction of volume of metal submerged in mercury

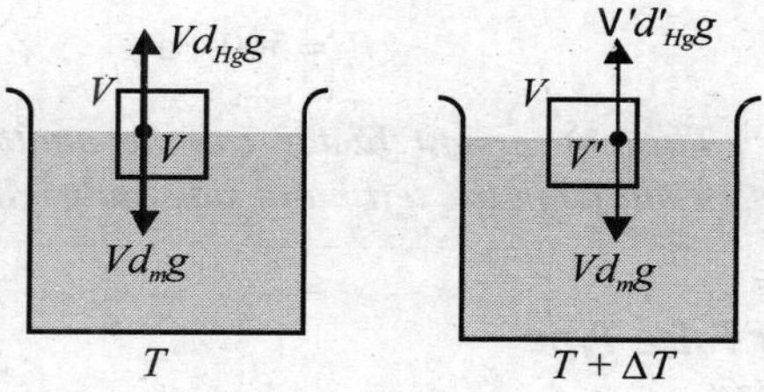

Fig. 3.360

$$\frac{v}{V} = \frac{d_{\text{m}}}{d_{\text{Hg}}} = K_1\ \text{(say)}$$

In second case $(T + \Delta T)$

$$v'd'H_g g = V'd'_{\text{m}}g$$

$$\Rightarrow\quad \frac{v'}{V'} = \frac{d'_{\text{m}}}{d'_{\text{Hg}}}$$

Fraction of volume of metal submerged in mercury $= K_2$ (say)

$$\therefore\quad \frac{K_2}{K_1} = \frac{d'_{\text{m}} \times d_{\text{Hg}}}{d'_{\text{Hg}} \times d_{\text{m}}} = \frac{d'_{\text{m}} \times d'_{\text{Hg}}(1+\gamma_2\Delta T)}{d'_{\text{Hg}} \times d'_{\text{m}}(1+\gamma_1\Delta T)}$$

$$= \frac{1+\gamma_2\Delta T}{1+\gamma_1\Delta T}$$

$$\therefore\quad 1 - \frac{K_2}{K_1} = 1 - \frac{1+\gamma_2\Delta T}{1+\gamma_1\Delta T}$$

$$\Rightarrow\quad \frac{K_1 - K_2}{K_1} = \frac{1-\gamma_1\Delta T - 1 - \gamma_2\Delta T}{1-\gamma_1\Delta T}$$

$$\frac{K_1 - K_2}{K_1} = -\frac{(\gamma_1 + \gamma_2)\Delta T}{1-\gamma_1\Delta T}$$

2. Applying the equation of continuity at section 1 and section 2, we get

$$A_1V_1 = A_2V_2$$

$$\Rightarrow\quad 10 \times 1 = 5 \times V_2$$

$$\Rightarrow\quad V_2 = 2\ \text{m/s}$$

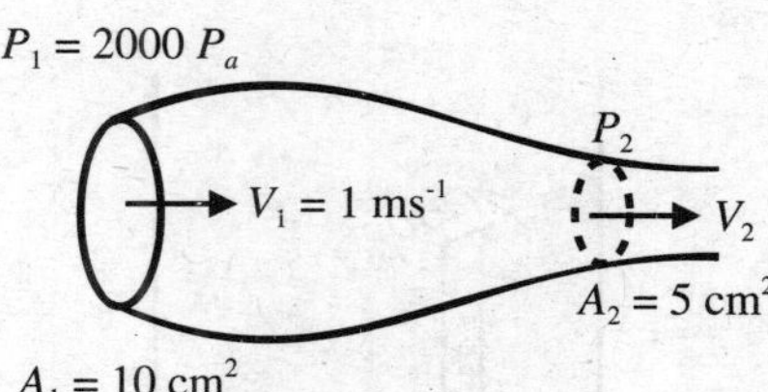

Fig. 3.361

Applying Bernoulli's theorem, we get

$$P_1 + \frac{1}{2}\rho V_1^2 = P_2 + \frac{1}{2}\rho V_2^2$$

$$\Rightarrow \quad 2000+\frac{1}{2}\times 1000\times 1^2 = P_2+\frac{1}{2}\times 1000\times 2^2$$

$$\Rightarrow \quad P_2 = 500 \text{ Pa}$$

Note: ***The conversion factor can be omitted if it gets cancelled out from the left hand side and right hand side.***

True or False Type

1. **False.** When the man drinks some water from the pond, his weight increases and therefore the boat will sink further. The further sinking of the boat will displace the same volume of water in pond as drunk by man. Therefore, there will be no change in the le vel of water in the pond.
2. **False.** Pressure $P_1 = P_2 = 1$ atm $= h\rho g$. On changing the temperature, g will not change and atmosphere pressure will not change.

$\therefore \quad h\times\rho = \text{constt},$

When temperature is increased, the density of Hg decreases and hence h increases.

Multiple Choice Questions With One Correct Answer Type

1. **c.** $W_1 = mg - Vd_a g$

$$W_2 = mg - V'd_a'g$$

$$= mg - V(1+50\gamma_b)\frac{d_a g}{(1+50\gamma_a)}$$

$$= mg - Vd_a g\left[\frac{1+50\gamma_b}{1+50\gamma_a}\right]$$

V = volume of ball at 0°C
d_a = density of alcohol at 0°C
V' = volume of ball at 50°C
d'_a = density of alcohol at 0°C

Given $\quad \gamma_b < \gamma_a$

$\therefore \quad 1+50\,\gamma_b < 1+50\gamma_a$

$\therefore \quad \dfrac{1+50\gamma_b}{1+50\gamma_a} < 1$

$\therefore \quad W_2 > W_1 \quad$ or $\quad W_1 > W_2$

2. **c**

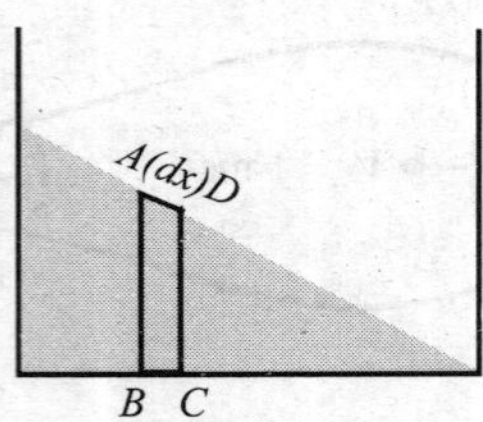

Fig. 3.362

Let us consider a small dotted segment of thickness dx for observation. Since this segment is accelerated towards right, a net force is acting in this segment towards right from the liquid towards the left of $ABCD$. According to Newton's third law, the segment $ABCD$ will also apply a force on the previous section creating a pressure on it which makes the liquid rise.

3. **a.** The whole system falls freely under gravity. Upthrust = weight of fluid displaced = (mass of fluid displaced) × g. For a freely falling body, g = 0.Therefore, upthrust = 0.

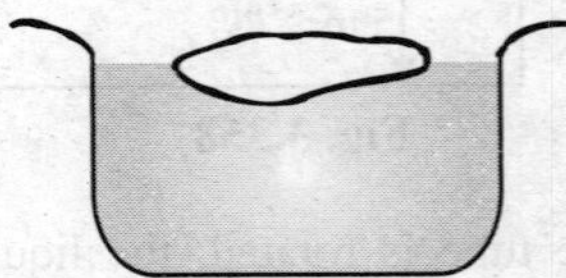

Fig. 3.363

4. **b.** The level of liquid is same in both the limbs. Pressure in limbs I at B = Pressure of limb II at A.

$$h\rho_1 g = h\rho_2 g \Rightarrow \rho_1 = \rho_2$$

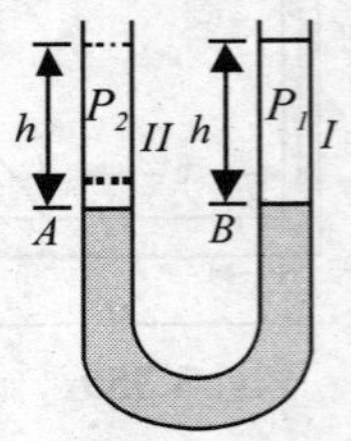

Fig. 3.364

5. **c.** Weight of sphere = Upthrust due to Hg + Upthrust due to oil, i.e., $Vdg = \dfrac{V}{2}d_{\text{Hg}}g + \dfrac{V}{2}d_{\text{oil}}\times g$

$$\Rightarrow \quad d = \frac{d_{\text{Hg}}+d_{\text{oil}}}{2} = \frac{13.6+0.8}{2} = 7.2 \text{ g/cm}^3$$

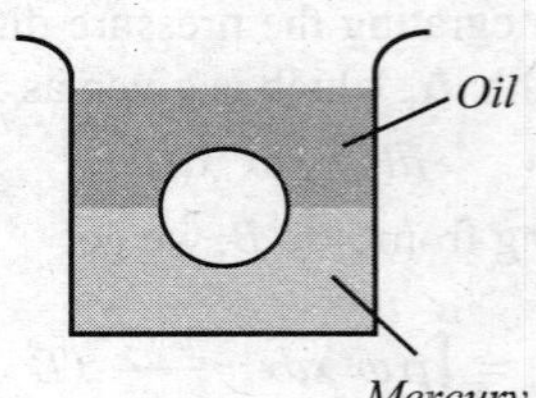

Fig. 3.365

6. **a.** Weight of cylinder = Uptrust due to upper liquid + Uptrust due to lower liquid

$$\Rightarrow D = \frac{A}{5}\times L\times g = d\left(\frac{A}{5}\right)\left[\frac{3}{4}L\right]g + 2d\left[\frac{A}{5}\right]\left[\frac{L}{4}\right]\times g$$

$\therefore \; D = 5d/4$

7. **a.** Equating the rate of flow, we have

$$\sqrt{(2gy)}\times L^2 = \sqrt{(2g\times 4y)}\pi R^2$$

$$\Rightarrow \quad L^2 = 2\pi R^2 \quad \Rightarrow R = \frac{L}{\sqrt{2\pi}}$$

Flow = (area) × (velocity). Here, Vel = $\sqrt{2gx}$,

where $x = ht$ from top

8. a. For equilibrium is case 1 at 0ºC: Upthrust = Weight of the body

$$\Rightarrow \quad K_1 V d_2 g = V d_1 g$$

$$\Rightarrow \quad K_1 = \frac{d_1}{d_2} \quad \text{(i)}$$

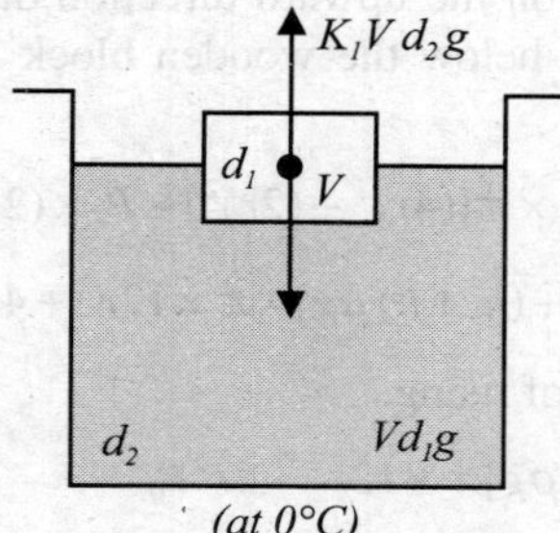

Fig. 3.366

For equilibrium in case 2 at 60ºC: When the temperature is increased, the density will decrease.

$\therefore \quad d_1' = d_1(1 + \gamma_{Fe} \times 60)$ and $d_2' = d_2(1 + \gamma_{Hg} \times 60)$

Again, upthrust = Weight of the body

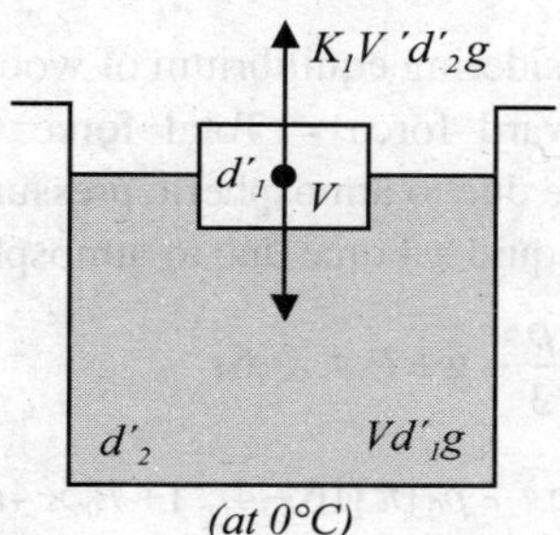

Fig. 3.367

$$\therefore \quad K_2 V' d_2' g = V' d_1' g$$

$$\therefore \quad K_2 \left[\frac{d_2}{1 + \gamma_{Hg} \times 60}\right] = \frac{d_1}{1 + \gamma_{Fe} \times 60}$$

$$\therefore \quad K_2 \left[\frac{1 + \gamma_{Fe} \times 60}{1 + \gamma_{Hg} \times 60}\right] = \frac{d_1}{d_2}$$

$$\Rightarrow \quad \frac{K_1}{K_2} = \frac{1 + \gamma_{Fe} \times 60}{1 + \gamma_{Hg} \times 60}$$

9. d.

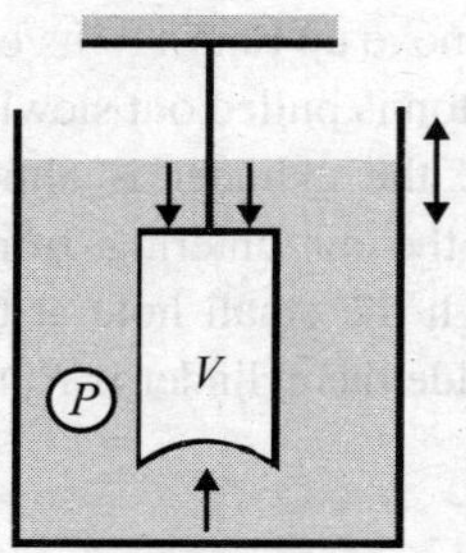

Fig. 3.368

According to Archimedes principle, upthrust = weight of the fluid displaced $F_{bottom} - F_{top} = V\rho g$

$$\therefore \quad F_{bottom} = F_{top} + V\rho g$$
$$= P_1 \times A + V\rho g$$
$$= (h\rho g) \times (\pi R^2) + V\rho g$$
$$= \rho g[\pi R^2 h + V]$$

10. d. l decreases as the block moves up. H will also decrease because when the coin is in the water it will displace equal volume of water, whereas when it is on the block an equal weight of water is displaced.

11. a. The square of the velocity of flux

$$v^2 = \frac{2gh}{\sqrt{1 - \left(\frac{A_0}{A}\right)^2}}$$

$$= \frac{2 \times 10 \times 2.475}{\sqrt{1 - (0.1)^2}} = 50 \text{ m}^2/\text{s}^2$$

A
3 m
2475 = h m
A_0
0.525 m

Fig. 3.369

12. b. Force from right hand side of the liquid on left hand side liquid.

(i) Due to surface tension force = $2RT$ (towards right)

(ii) Due to liquid pressure force

$$= \int_{x=0}^{x=h} (p_0 + \rho g h)(2Rx)\,dx$$

$$= (2p_0 Rh + R\rho g h^2) \text{ (towards left)}$$

$\therefore$ Net force is $| 2p_0 Rh + R\rho g h^2 - 2RT |$

Assertion and Reasoning Type

1. d. From continuity equation, Av = constant

or $$A \propto \frac{1}{v}$$

At lower heights, speed will be more. Therefore, area of cross section will be less.

Multiple Choice Questions With One or More than One Correct Answer Type

1. b., c. When the block of mass m is arranged as shown in the figure, an upthrust F_y will act on the mass which will decrease the reading on A. But according to Newton's second law, to each and every action, there is equal and opposite reaction. So F_y will act on the liquid of the beaker which will increase the reading in B.

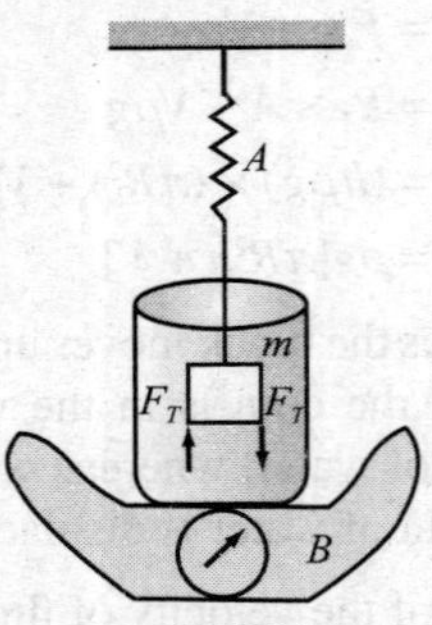

Fig. 3.370(a)

2. **b., d.** For equilibrium

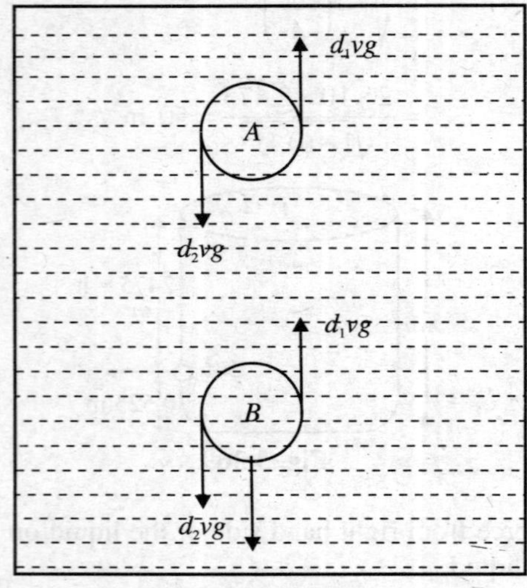

Fig. 3.370(b)

$$d_A vg + d_B vg = d_F vg + d_F vg \Rightarrow d_F = \frac{d_A + d_B}{2}$$

Option (d) is correct to keep the string tight

$$d_B > d_F \text{ and } d_A < d_F$$

Comprehension Type

1. **c.** Consider the equilibrium of wooden block. Forces acting in the downward direction are as follows.

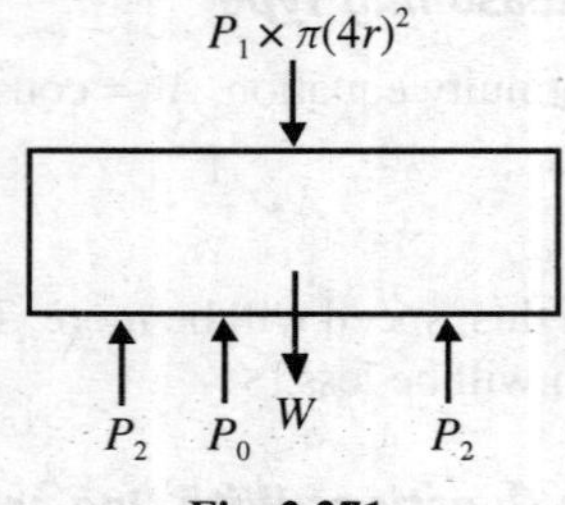

Fig. 3.371

(a) Weight of wooden cylinder

$$= \pi(4r)^2 \times h \times \frac{\rho}{3} \times g = \pi \times 16r^2 \frac{h\rho}{3} g$$

(b) Force due to pressure (P_1) created by liquid of height h_1 above the wooden block is

$$= P_1 \times \pi(4r)^2 = [P_0 + h_1 \rho g] \times \pi(4r)^2$$

$$[\rho_0 + h_1 \rho g]\pi \times 16r^2$$

Force acting on the upward direction due to pressure P_0 exerted from below the wooden block and atmospheric pressure is

$$= p_2 \times \pi[(4r)^2 - (2r)^2] + P_0 \times (2r)^2$$

$$= [P_0 + (h_1 + h)\rho g] \times \pi \times 12r^2 + 4r^2 P_0$$

At the verge of rising,

$$[P_0 + (h_1 + h)\rho g]\pi \times 12r^2 + 4r^2 P_0$$

$$= \pi \times 10r^2 h \times \frac{\rho}{3} g + [P_0 + h_1 \rho g] \times \pi \times 16r^2$$

$$12h_1 + 12h = \frac{16h}{3} + 16h$$

$$12h_1 - \frac{16h}{3} = 4h_1 \Rightarrow \frac{5h}{3} = h_1$$

2. **b.** Again considering equilibrium of wooden block. Total downward force = Total force upwards Wt. of block + Force due to atmospheric pressure = Force due to pressure of liquid + Force due to atmospheric pressure

$$\pi(16r^2)\frac{\rho}{3} + g + P_0 \pi \times 16r^2$$

$$= [h_2 \rho g + p_0]\pi[16 - 4r^2] + P_0 \times 4r^2$$

$$\pi(16r^2) h \frac{\rho}{g} g = h_2 \rho g \times \pi \times 12r^2$$

$$16\frac{h}{3} = 12h_2$$

$$\Rightarrow \quad h_2 = \frac{4}{9}h$$

3. **a.** When the height h_2 of water level is further decreased, then the upward force acting on the wooden block decreases. The total force downward remains the same. This difference will be compensated by the normal reaction by the tank wall on the wooden block. Thus, the block does not move up return to its original position.
4. **a.** When the piston is pulled out slowly, the pressure drop produced inside the cylinder is almost instantaneously neutralized by the air entering from outside into the cylinder (through the small hole at the top). Therefore, the pressure inside the cylinder is P_0 throughout the slow pulling process.

5. **d.** The condition for equilibrium of the piston is

$$Mg = (P_0 - p)\pi R^2$$

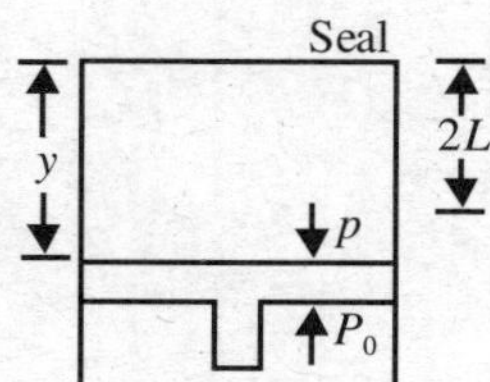

Fig. 3.372

$$p = \frac{-Mg}{\pi R^2} + P_0$$

Since the cylinder is thermally conducting, the temperature remains the same.
Therefore, $P_0 \times (2L \times \pi R^2) = py \times \pi R^2$

$$y = \frac{P_0}{p} \times 2L = \frac{P_0}{\left[P_0 - \dfrac{Mg}{\pi r^2}\right]} \times 2L$$

$$= \frac{P_0 \times \pi R^2}{P_0 \pi R^2 - Mg} \times 2L$$

6. **c.** At equilibrium, $p = P$

$$p = P_0 + (L_0 - H)\rho g \qquad \text{(i)}$$

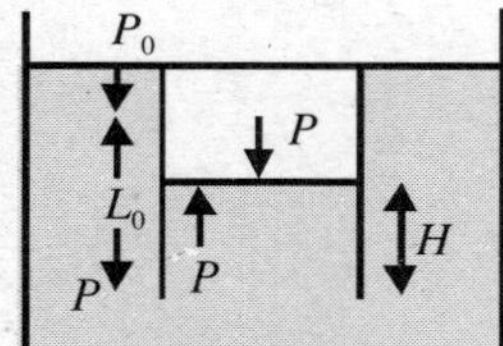

Fig. 3.373

Also $\qquad P_0 \times (\pi R^2 L_0) = p[\pi R^2 (L_0 - H)]$

$$p = \frac{L_0 p_0}{L_0 - H} \qquad \text{(ii)}$$

From Eqs. (i) and (ii), we get

$$\frac{L_0 P_0}{L_0 - H} = P_0 + (L_0 - H)\rho g$$

$$L_0 P_0 = P_0(L_0 - H) + (L_0 - H)^2 \rho g$$

$$\rho g (L_0 - H)^2 + P_0 (L_0 - H) - L_0 P_0 = 0$$

7. **d.** As the bubble moves upwards, besides the buoyancy force (cause of which is pressure difference), only force of gravity and force of viscosity will act.

8. **b.** As there is no exchange of heat, process is adiabatic. Applying,

$$Tp^{\frac{1-\gamma}{\gamma}} = \text{constant}$$

$$T_2 p_2^{\frac{1-\gamma}{\gamma}} = T_1 p_1^{\frac{1-\gamma}{\gamma}}$$

$$T_2 = T_1 \left(\frac{p_1}{p_2}\right)^{\frac{1-\gamma}{\gamma}} = T_1 \left(\frac{p_2}{p_1}\right)^{\frac{1-\gamma}{\gamma}}$$

Substituting the values, we have

$$T_2 = T_0 \left[\frac{p_0 + \rho_1 g (H - y)}{p_0 + \rho_1 g H}\right]^{\frac{\frac{5}{3} - 1}{5/3}}$$

$$T_0 = \left[\frac{p_0 + \rho_l g (H - y)}{p_0 + \rho_l g H}\right]^{2/5}$$

9. **b.** Buoyancy force F = (volume of bubble) $(\rho_1)g$

$$= \left(\frac{nRT_2}{p_2}\right)\rho_l g$$

Here, $\qquad T_2 = T_0 \left[\frac{p_0 + \rho_e g (H - y)}{p_0 + \rho_l g h}\right]^{2/5}$

and $\qquad p_2 = p_0 + \rho_l g\,(H - y)$

Substituting the values, we get

$$F = \frac{\rho_l n R_g T_0}{(p_0 + \rho_l g H)^{2/5} [p_0 + \rho_l g (H - y)]^{3/5}}$$

Integer Answer Type

1. **(6)** $\qquad P = P_0 - \rho g h = 98 \times 10^3$ N/m^2

$$P_0 V_0 = PV$$

$$10^5 [A(500 - H)] = 98 \times 10^3 [A(500 - 200)]$$

$$H = 206 \text{ mm}$$

$$\text{Level fall} = 206 - 200 = 6 \text{ mm}$$

CHAPTER

4

Properties of Solids and Fluids

- Elasticity
- Elastic Behaviour of Solids
- Elasticity and Plasticity
- Stress
- Strain
- Elastic Limit
- Analogy of Rod as a Spring
- Bars of Composite Section
- Stress–Strain Diagram
- Elasticity and Plasticity
- Energy Stored in a Deformed Body
- Interatomic Force Constant
- Viscosity
- Flow of Liquid in a Tube: Critical Velocity
- Viscosity in Fluids
- Similarities and Differences Between Viscosity and Sliding Friction
- Some Applications of Viscosity
- Units of Coefficient of Viscosity
- Effect of Temperature on Viscosity
- Stokes' Law
- Terminal Velocity
- Surface Tension
- Surface Energy
- Angle of Contact
- Excess Pressure Inside a Curved Surface
- Excess Pressure Inside a Liquid Drop and a Bubble
- Capillarity
- Determination of Capillary Rise

ELASTICITY

The elastic behaviour of material plays an important role in engineering design and also constructing bridges, automobiles, ropeways, etc. In this section, we will study the elastic behaviour and the mechanical properties of the solids.

ELASTIC BEHAVIOUR OF SOLIDS

The elastic behaviour of solids can be understood by taking the microscopic nature of a solid body. Accordingly, a solid body is composed of atoms or molecules, where each atom or molecule is surrounded by other atoms or molecules. These atoms or molecules are bonded together by interatomic or intermolecular forces and stay in a stable equilibrium position. This can be visualized by taking a mode of spring ball system as shown in Fig. 4.1. Here the balls represent the atoms or molecules and the springs represent the interatomic or intermolecular forces. In this system, if any ball is displaced a little from its equilibrium position, the springs attached to that ball will either be stretched or compressed. Because of this, the restoring forces will come into play in the springs and they will bring the ball back to its original position. This accounts for the elastic behaviour of the solid body.

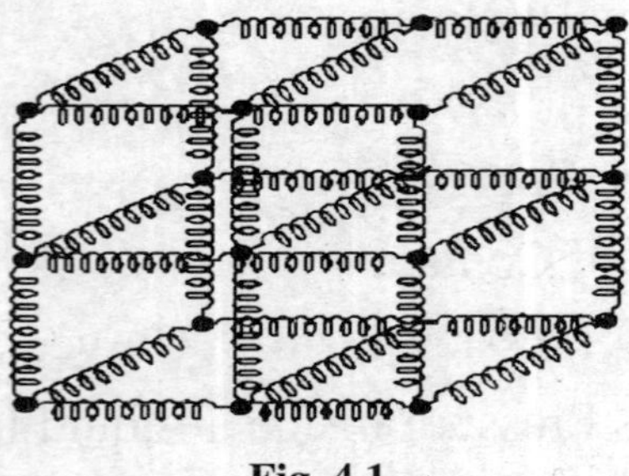

Fig. 4.1

ELASTICITY AND PLASTICITY

The property of a material body by virtue of which it regains its original configuration (i.e., shape and size) when the external deforming force is removed is called elasticity. The property of the material body by virtue of which it does not regain its original configuration when the external force is removed is called plasticity.

Some Definitions

Deforming force

If a force applied on a body produces a change in the normal positions of the molecules of the body, which results in a change in the configuration of the body either in lengths, volume or shape, then the force applied is called deforming force. Thus a deforming force is the one which when applied changes the configuration of the body.

Elasticity

The property of the body to regain its original configuration (length, volume or shape) when the deforming forces are removed is called elasticity.

Perfectly elastic body

A body which regains its original configuration immediately and completely after the removal of deforming force from it is called the perfectly elastic body. Quartz and phosphor bronze are the examples of nearly perfectly elastic bodies.

Perfectly plastic body

A body which does not regain its original configuration at all on the removal of the deforming force, howsoever small the deforming force may be, is called perfectly plastic body. Putty, mud and paraffin wax are the examples of nearly perfectly plastic bodies. In fact, nothing is perfectly elastic or perfectly plastic. Only the degree of elasticity or plasticity differs from body to body.

Cause of Elasticity

In a solid, atoms and molecules are arranged in such a way that each molecule is acted upon by the forces due to the neighbouring molecules. These forces are known as intermolecular forces. When no deforming force is applied on the body, each molecule of the solid (i.e., body) is in its equilibrium position and the intermolecular forces between the molecules of the solid are maximum.

On applying a deforming force on the body, the molecules either come closer or go far apart from each other. As a result of this, the molecules are displaced from their equilibrium position. In other words, intermolecular forces get changed and restoring forces are developed on the molecules. When the deforming force is removed, these restoring forces bring the molecules of the solid to their respective equilibrium positions and hence the solid (or the body) regains its original form.

STRESS

When a deforming force is applied on a body, it changes the configuration of the body by changing the normal positions of the molecules or atoms of the body. As a result, an internal restoring force comes into play which tends to bring the body back to its initial configuration. This internal restoring force acting per unit area of a deformed body is called stress, i.e.,

$$\text{Stress} = \frac{\text{Restoring force}}{\text{Area}} \qquad \text{(i)}$$

If there is no permanent change in the configuration of the body (i.e., in the absence of plastic behaviour of the body), the restoring force is equal and opposite to the external deforming force applied. Thus quantitatively, stress can be given as

$$\text{Stress} = \frac{\text{External deforming force}}{\text{Area}} \qquad \text{(ii)}$$

The SI unit of stress in is N/m^2 or pascal (denoted by Pa) and in CGS system it is dyn/cm^2. The dimensional formula of stress is $(ML^{-1}T^{-2})$.

We know that the internal forces acting on the sectional area of a cut may be of varying magnitudes and directions; and these are generated to maintain equilibrium under the action of externally applied forces. The capacity of a material to resist external forces depends on the intensity of internal forces, which is also a measure of strength of the material. The intensity of internal forces at a point is defined as the stress at that point. Mathematically,

$$\text{Stress} = \lim_{\Delta A \to 0} \frac{\Delta F}{\Delta A} \quad \text{(i)}$$

where ΔF is the internal force, which acts on an infinitesimally small area ΔA. Furthermore, ΔF has two components: ΔF_n (normal to the surface) and ΔF_t (tangential to the surface).

Types of Stress

There are mainly three types of stresses as discussed in the following.

Normal Stress

The intensity of the internal force perpendicular to or normal to the section is called the normal stress or axial stress at a point. That is,

$$\sigma = \lim_{\Delta A \to 0} \frac{\Delta F_n}{\Delta A} \quad \text{(ii)}$$

If restoring force developed is perpendicular to the cross-sectional area chosen, then the corresponding stress is known as normal stress. It is also known as longitudinal stress. It can be of two types.

(a) **Tensile stress:** If there is an increase in the length or extension of the body in the direction of force applied, the stress set up is called tensile stress.

When a solid cylinder is stretched by two equal forces applied normal to its cross-sectional area, Fig. 4.2(a), the restoring force per unit area in the body is called tensile stress.

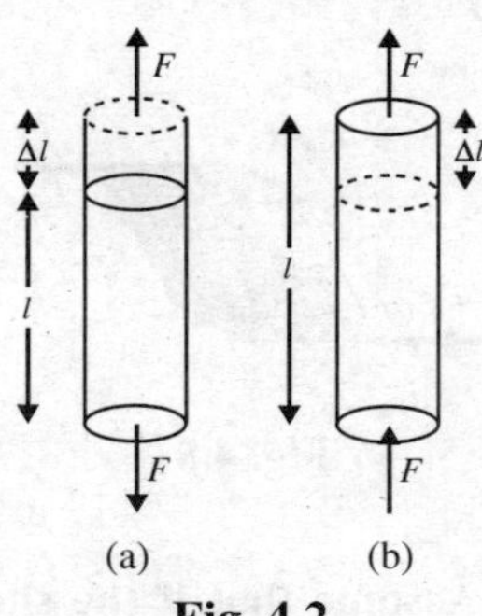

Fig. 4.2

(b) **Compressive stress:** If there is a decrease in the length of the wire or compression of the body due to force applied, the stress set up is called compressive stress. When a solid cylinder is compressed under the action of two equal forces applied normal to its cross-sectional area as shown in Fig. 4.2(b), then the restoring force per unit area in the body is called compressive stress.

Note:

- ***Tensile or compressive stress can also be termed as longitudinal stress.***
- ***If the two parts of the body on the two sides of cross-sectional area pull each other, then the normal stress is known as tensile stress.***

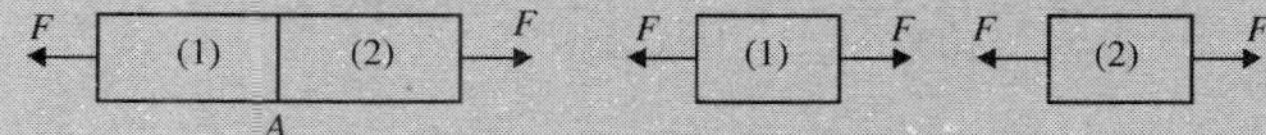

Fig. 4.3

- ***If the two parts of the body on the two sides of cross-sectional area push each other, then the normal stress is known as compressive stress.***

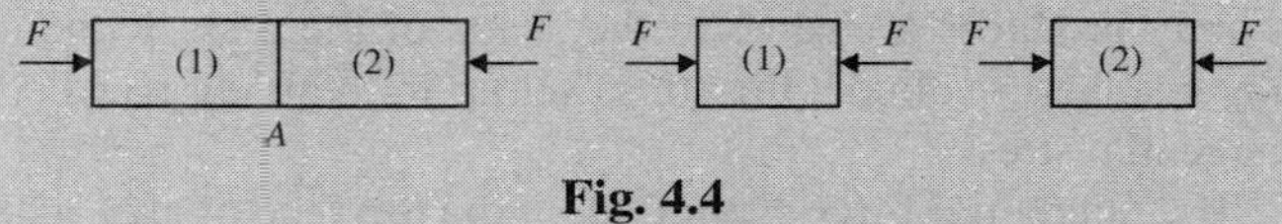

Fig. 4.4

Examples of normal stress

(a) A load is suspended by a wire. Tensile stress is produced in the wire.

(b) A column is supporting a vertical load. Compressive stress is produced in the wire.

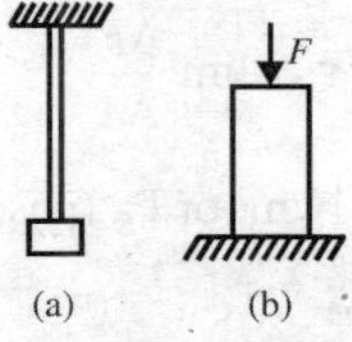

Fig. 4.5

Illustration 4.1 **Two rods A and B, each of equal length but different materials are suspended from a common support as shown in the figure. The rods A and B can support a maximum load of $W_1 = 600$ N and $W_2 = 6000$ N, respectively. If their cross-sectional areas are $A_1 = 10$ mm^2 and $A_2 = 1000$ mm^2, respectively, then identify the stronger material.**

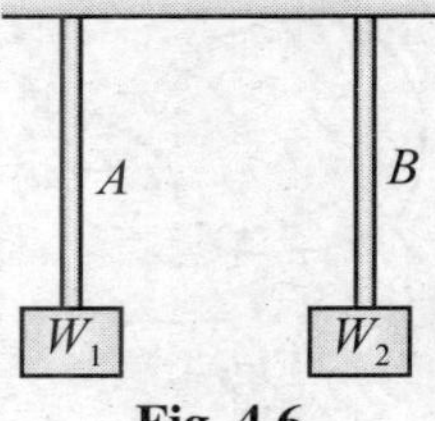

Fig. 4.6

Sol. At first sight, it looks that rod B is stronger as it supports a larger load. But we cannot compare strengths without having a common basis of comparison. Since the rods have different cross-sectional area, therefore, the strengths can be compared by determinig the load capacity per unit area.

For rod A

$$\sigma_1 = \frac{W_1}{A_1} = \frac{600\,\text{N}}{10\,\text{mm}^2} = \frac{600}{10\times10^{-6}} = 60\times10^6\ \text{Nm}^{-2}$$

For rod B

$$\sigma_2 = \frac{W_2}{A_2} = \frac{6000\,\text{N}}{1000\,\text{mm}^2} = \frac{6000}{1000\times10^{-6}} = 6\times10^6\ \text{Nm}^{-2}$$

Since $\sigma_1 > \sigma_2$, therefore, the material of rod A is stronger than that of rod B.

Shearing Stress or Tangential Stress

When a deforming force, acting tangentially to the surface of a body produces a change in the shape of the body without any change in volume, the stress set up in the body is called tangential stress.

If a tangential force F applied on the top face of area a of a cubical body whose bottom face is rigidly fixed changes the shape of the cubical body without changing its volume, then

$$\text{Tangential stress} = F/a$$

The intensity of the internal force parallel or tangential to the section is called the shear stress or tangential stress at a point. That is,

$$\tau = \lim_{\Delta A\to 0} \frac{\Delta F}{\Delta A} \qquad \text{(iii)}$$

The SI unit of stress is N/m² or Pa (pascal).

Note:

- ***Sometimes, stress is simply determined by dividing the normal or tangential force with the cross-sectional area.***

$$\textit{Normal stress} = \frac{F_n}{A} \qquad \textit{(i)}$$

$$\text{Shearing stress} = \frac{F_t}{A} \qquad \textit{(ii)}$$

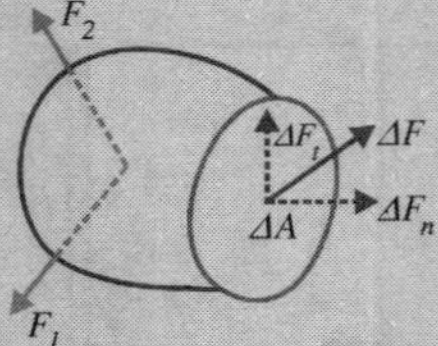

Fig. 4.7

As shown in Fig. 4.7, ΔF_n is the normal component and ΔF_t is the tangential component of internal force ΔF on an infinitesimally small area ΔA of the section of a member loaded in a plane.

- **This method does not give stress at all points in the cross-sectional area but it merely gives the average stress.**
- **When the stress is constant or uniform over the cross section, Eqs. (i) and (ii) define the stress at all points of the cross section.**
- **A uniform distribution can exist only if the resultant of the applied loads passes through the centroid of the cross section.**
- **Stresses multiplied by the respective areas on which they act give forces, and it is the sum of these forces at an imaginary cut that keeps a body in equilibrium. That is,**

$$F_n = \int \sigma\, dA$$

$$F_t = \int \tau\, dA$$

Examples of shear stress

Shearing stress is produced whenever the applied loads cause (or tend) one section of a body to slide past its adjacent section.

(a) A river resists shear stress across its cross-sectional area.
(b) The fibres of a shaft resists shear stress when it is twisted by an external torque.

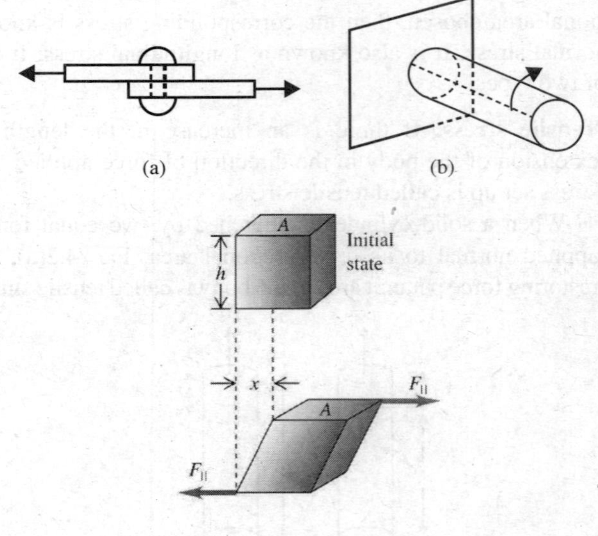

Fig. 4.8

Illustration 4.2 **Assume that if the shear stress in steel exceeds about 4.00 × 10⁸ N/m², the steel reptures. Determine the shearing force necessary to (a) shear a steel bolt 1.00 cm in diameter and (b) punch a 1.00-cm-diameter hole in a steel plate 0.500-cm thick.**

Sol. (a) Force applied $F = (A)$ (stress)

$\pi (5.00 \times 10^{-3}\ \text{m})^2 (4.00 \times 10^8\ \text{N/m}^2) = 3.14 \times 10^4\ \text{N}$

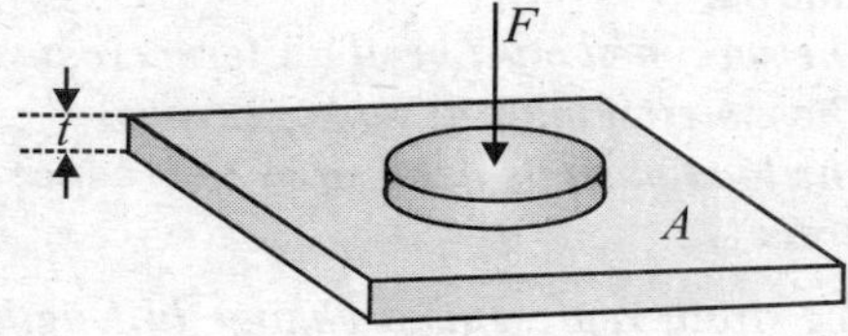

Fig. 4.9

(b) The area over which the shear stress occurs is equal to the circumference of the hole times its thickness. Thus,

$$A = (2\pi r)\, t = 2\pi\, (5.00 \times 10^{-3}\ \text{m})\, (5.00 \times 10^{-3}\ \text{m})$$

$$= 1.57 \times 10^{-4}\ \text{m}^2$$

So, $F = (A)\ \text{Stress} = (1.57 \times 10^{-4}\ \text{m}^2)(4.00 \times 10^8\ \text{N/m}^2)$

$$= 6.28 \times 10^4\ \text{N}$$

Illustration 4.3 **A bar of cross section A is subjected to equal and opposite tensile forces at its ends. Consider a plane section of the bar whose normal makes an angle θ with the axis of the bar.**

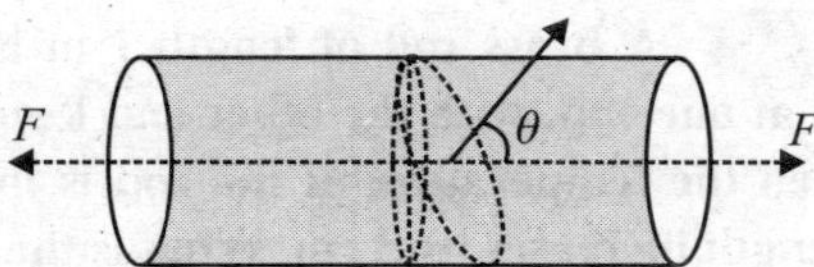

Fig. 4.10

(a) What is the tensile stress on the plane?
(b) What is the shearing stress on this plane?
(c) For what value of θ is the tensile stress maximum?
(d) For what value of θ is the shearing stress maximum?

Sol.

(a) The resolved part of F along the normal is the tensile force on this plane and the resolved part parallel to the plane is the shearing force on this plane.

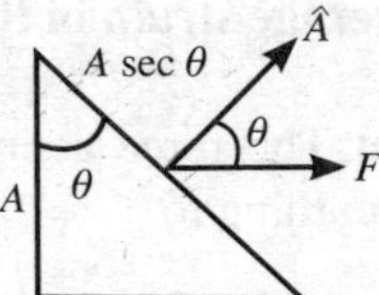

Fig. 4.11

Therefore, tensile stress $= \dfrac{\text{Force}}{\text{Area}} = \dfrac{F\cos\theta}{A\sec\theta} = \dfrac{F}{A}\cos^2\theta$

($\because$ area of section $= A \sec\theta$)

(b) Shearing stress $= \dfrac{\text{Force}}{\text{Area}} = \dfrac{F\sin\theta}{A\sec\theta} = \dfrac{F}{2A}\sin 2\theta$

(c) Obviously, tensile stress on the plane is maximum when $\cos^2\theta$ is maximum, that is, $\cos\theta = 1$ or $\theta = 0°$.

(d) Obviously, shearing stress is maximum when $\sin 2\theta$ is maximum, that is

$$\sin 2\theta = 1 \quad \text{or} \quad 2\theta = 90°$$
$$\text{or} \quad \theta = 45°$$

Illustration 4.4 **(a) Estimate the force with which a karate master strikes a board, assuming the hand's speed at the moment of impact is 10.0 m/s, decreasing to 1.00 m/s during a 0.002 00 s time internal of contact between the hand and the board. The mass of his hand and arm is 1.00 kg. (b) Estimate the shear, assuming this force it exerted on a 1.00 cm-thick pine board that is 10.0 m wide. (c) If the maximum shear stress a pine board can support before breaking is 3.60×10^6 N/m², will the board break?**

Sol. We will use the impulse momentum theorem to find the force on the hand, which is equal in magnitude to the force on the board. Then the definition of shear stress will let us compute that quantity to see if it is large enough to break the board.

The impulse momentum theorem describes the force of the board on his hand: $Ft = mv_f - mv_i$

$F\,(0.002\ 00\text{s}) = (1.00\ \text{kg})\ (1.00\ \text{m/s} - 10.0\ \text{m/s})$

and $\quad F = -4500\ \text{N}$

(a) Therefore, the force of his hand on board is 4500 N.

(b) That force produces a shear stress on the area that is exposed when the board snaps:

$$\text{Stress} = \frac{F}{A} = \frac{4500\,\text{N}}{(10^{-1}\,\text{m})(10^{-2}\,\text{m})} = 4.50 \times 10^6\ \text{N/sm}^2$$

(c) Yes, this is larger than 3.60 MN/m² and suffices to break the board.

We do not have the right combination of information to use the work-kinetic energy theorem to find the force. We could compute the acceleration of the hand and use Newton's second law to find the same 4.50 kN force, but writing the impulse-momentum theorem is more direct. The area in stress = force/area is always the area of the layers of molecules between which the stress produces distortion (strain).

Volumetric Stress

When a solid body undergoes a change in volume without any change in its geometrical shape on applying the force perpendicular to every point on the surface of body, Fig. 4.12, the restoring force per unit area in the body is called volumetric stress.

If the deforming forces act on the body from all sides, then corresponding stress developed is known as volume stress. For example, if a block is kept completely immersed in water, hydrostatic pressure acts on all sides of the block. As stress is the ratio of some force and area, it is a kind of pressure.

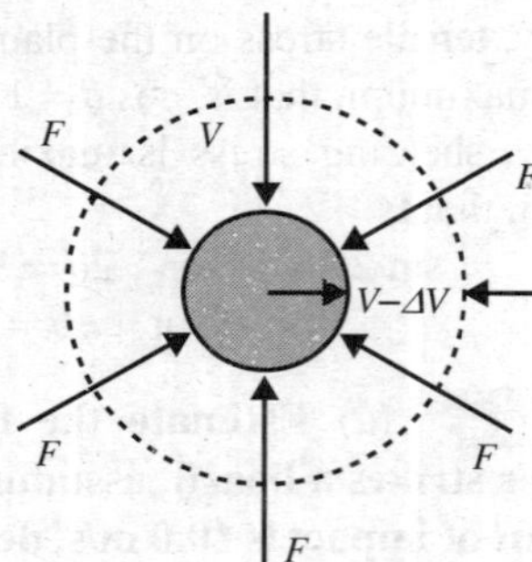

Fig. 4.12

STRAIN

When a deforming force is applied on a body, there is a change in the configuration of the body. The body is said to be strained or deformed. The ratio of the change in configuration (i.e., shape, length or volume) to the original configuration of the body is called strain. Mathematically, strain is given as

$$\in = \frac{\text{Change in configuration}}{\text{Original configuration}}$$

Strain does not have any unit.

Types of Strain

The change in configuration can be a change either in length, volume or shape of the body. There are three types of strain.

(i) **Longitudinal strain:** This type of strain is produced when the deforming force causes a change in length of the body. It is defined as the ratio of the change in length to the original length of the body. Consider a wire of length L. The wire is stretched by a force F. Let the change in length of the wire be ΔL. Then longitudinal strain,

$$\in = \frac{\text{Change in length}}{\text{Original length}} = \frac{\Delta L}{L} \quad \text{(i)}$$

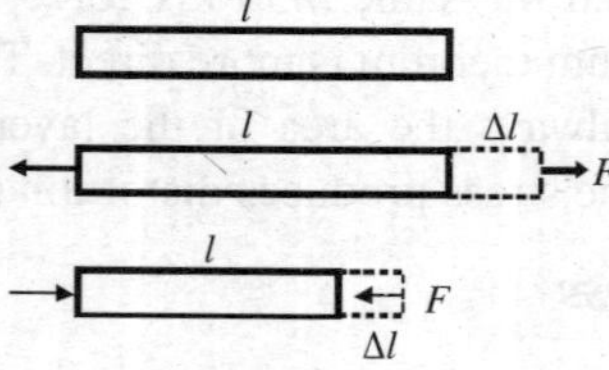

Fig. 4.13

- **The above expression measures only the average value of strain. The correct expression for strain at any position is**

$$\in = \frac{d\delta}{dL}$$

where dχ is the differential elongation of the differential length dL.

- **Under certain conditions the strain produced in a rod is assumed to be uniform and it can be computed from Eq. (i). These conditions are as follows:**
  - ***The specimen must be of uniform cross section.***
  - ***The material must be homogeneous.***
  - ***The load must be axial so as to produce uniform stress.***

Since strain represents a change in length divided by the original length, strain is a dimensionless quantity. Sometimes, its units are used as metre per metre (m/m). However, commonly it is used in percentage.

Illustration 4.5 **A steel wire 2 m long is suspended from the ceiling. When a mass is hung at its lower end, the increase in length recorded is 1 cm. Determine the strain in the wire.**

Sol. Original length = 2 m

Increase in length = 1 cm = 1×10^{-2} m

$$\text{Longitudinal strain} = \frac{\text{Increase in length}}{\text{Original length}}$$

$$= \frac{1 \times 10^{-2}\ \text{m}}{2\ \text{m}} = 5 \times 10^{-3}$$

Illustration 4.6 **A brass rod of length 1 m is fixed to a vertical wall at one end, with the other end keeping free to expand. When the temperature of the rod is increased by 120°C, the length increases by 3 cm. What is the strain?**

Sol. After the rod expands to the new length, there are no elastic forces developed internally in it.

So, strain = 0.

Illustration 4.7 **A 30.0 kg hammer, moving with speed 20.0 m/s, strikes a steel spike 2.30 cm in diameter. The hammer rebounds with speed 10.0 m/s after 0.110 s. What is the average strain in the spike during the impact?**

Sol. The force acting on the hammer changes its momentum according to

$$mv_i + \vec{F}(\Delta t) = mv_f, \text{ so } |\vec{F}| = \frac{m|v_f - v_i|}{\Delta t}$$

$$\text{Hence, } |\vec{F}| = \frac{30.0|-10.0-20.0|}{0.110 s} = 8.18 \times 10^3\ \text{N}$$

By Newton's third law, that is also the magnitude of the average force exerted on the spike by the hammer during the blow. Thus, the stress in the spike is

$$\text{Stress} = \frac{F}{A} = \frac{8.18\times10^3\,N}{\pi(0.0230\,m)^2/4} = 1.97\times10^7\ \text{N/m}^2$$

and the strain is

$$\text{strain} = \frac{\text{Stress}}{Y} = \frac{1.97\times10^7\ \text{N/m}^2}{20.0\times10^{10}\ \text{N/m}^2} = 9.85\times10^{-5}$$

(ii) Volume strain: This type of strain is produced when the deforming force produces a change in volume of the body as shown in Fig. 4.14. It is defined as the ratio of the change in volume to the original volume of the body. If ΔV is change in volume and V is the original volume, then

$$\in_v = \text{Volume strain} = \frac{\Delta V}{V}$$

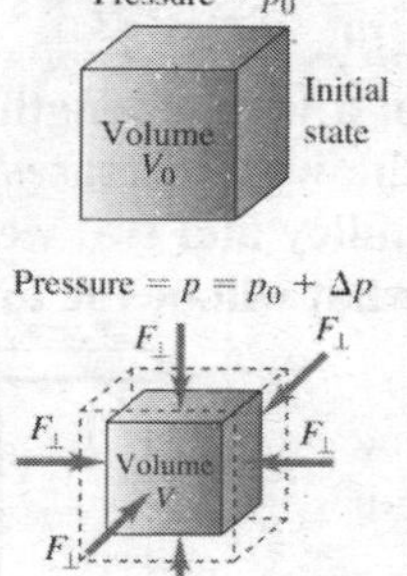

Fig. 4.14

(iii) Shear strain: This type of strain is produced when the deforming force causes a change in the shape of the body without changing its volume. It is defined as the angle (θ) through which a face originally perpendicular to the fixed face is turned as shown in the figure.

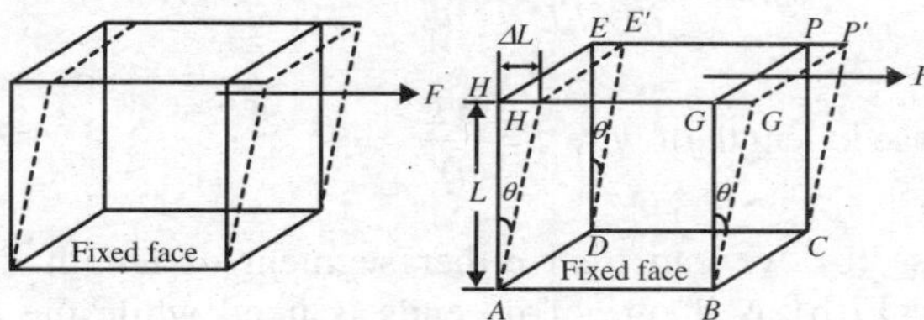

Fig. 4.15

Consider a cubical body as shown in Fig. 4.15. Its lower surface *ABCD* is fixed and a tangential force *F* is applied on the top surface *EPGH*. Let the vertical planes *AHED* and *BCPG* be laterally shifted to positions *AH'E'D* and *BCP'G'*, respectively, through an angle θ, where $\angle\, H\,AH' = 0$. Let $HA = L$ and $HH' = \Delta L$.

$$\text{Shearing strain} = \theta = \frac{\Delta L}{L}$$

Thus the shearing strain is also defined as the ratio of displacement of a surface under a tangential force to the perpendicular distance of the displaced surface from the fixed surface.

- **More precisely, the shearing strain is defined as the angular change between two perpendicular faces of a differential element.**
- **Strain in a rectangular shaped element is determined from the following parameters:**
  (a) ***longitudinal deformation d under the action of normal forces F_n;***
  (b) ***shear deformation χ_s under the action of tangential forces F_t.***

Illustration 4.8 **The shear modulus for a metal is 50000 MPa. Suppose that a shear force of 200 N is applied to the upper surface of a cube of this metal that is 3.0 cm on each edge. How far will the top surface be displaced?**

Sol. Shearing strain = $\Delta L/L = (F/A)/S$

$$= 200/[L^2(5\times10^{10})] = (4\times10^{-9})/L^2$$

Solving for ΔL with $L = 0.030$ m, we have

$$\Delta L = 1.33\times10^{-7}\,\text{m} = 0.133\ \mu\text{m}$$

Note:

(i) A shearing strain is never accompanied by a change in volume, i.e., in a shear strain, the body gets distorted, resulting only in an alternation in its shape but not in size.

(ii) A change in shape/size, i.e., dimensions need not necessarily imply a strain. For example, if a body is heated to expand, its volume changes. It acquires a new size due to expansion. However, the strain remains zero. Unless and until internal elastic forces operate to bring the body to the original state, no strain exists. When a body is heated, the total energy of molecules increases owing to an increase in the kinetic energy of the molecules. This results in a shift (increment) of the 'equilibrium distance' of molecules and the body acquires a new shape and size in the expanded from, whereby the molecules are in 'zero force' state. Hence, there is no strain. However, if the body is restricted to expand during the process of heating, then the molecules become 'strained', and even if there is no apparent change in dimensions of the body, there is strain. In such a case, strain is measured as the ratio of the change in dimension that would have occurred, had the

ELASTIC LIMIT

Elastic limit is the upper limit of deforming force up to which, if deforming force is removed, the body regains its original form completely and beyond which if deforming force is increased, the body loses its property of elasticity and gets permanently deformed.

Hooke's Law and Elastic Moduli

By direct experiment with the extension of uniform rods of different materials, it has been established for many materials that within certain limits the strain produced in the rod is directly proportional to the tensile stress. The simple linear relationship between stress and strain was first formulated by English scientist Robert Hooke. If we consider the stress–strain diagram for the mid-steel then we find that up to a point (called proportional limit), the stress is proportional to strain. So, Hooke's law is applicable only in this region.

Later, Thomas Young introduced a constant of proportionality that came to be known as Young's modulus or modulus of elasticity Y. Thus, at present Hooke's law is stated as follows.

Within certain limits (called proportional limit), stress produced in a material is proportional to the strain and the constant of proportionality is the modulus of elasticity. That is,

$$\sigma \propto \in \quad \text{or} \quad \sigma = Y \in \qquad \text{(i)}$$

Types of Modulus of Elasticity

Corresponding to the three types of strain, there are three types of modulus of elasticity:

1. Young's modulus of elasticity (Y)
2. Bulk modulus of elasticity (K)
3. Modulus of rigidity (η):

Young's Modulus of Elasticity

It is defined as the ratio of the normal stress to the longitudinal strain. That is,

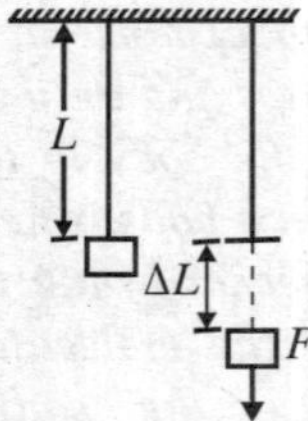

Fig. 4.16

$$\text{Young's modulus } (Y) = \frac{\text{Longitudinal stress}}{\text{Longitudinal strain}}$$

$$\text{Normal stress} = F/A$$
$$\text{Longitudinal strain} = \Delta L/L$$

$$Y = \frac{F/A}{\Delta L/L} = \frac{FL}{A\Delta L} \qquad \text{(ii)}$$

Note:

- ***Extension is proportional to the deforming force and length of the rod, i.e., greater the force larger the deformation; longer the rod larger the extension.***
- ***Extension is inversely proportional to the cross-sectional area A and the modulus of elasticity, i.e., a thicker rod produces less deformation and a stiffer material produces less deformation.***
- ***Eq. (ii) may be rearranged as***

$$F = \left(\frac{YA}{L}\right)\delta \qquad \text{(iii)}$$

It indicates that deforming force is proportional to the elongation and the term in the bracket indicates another constant of proportionality called the stiffness constant k. Thus, Eq. (iii) may be written as

$$F = k\chi \qquad \text{(iv)}$$

where** $k = (YA/L) = $ **constant

- **By using equation (iv), a rod of uniform cross section may be considered as an elastic spring of stiffness $k = (YA/L)$.**
- ***Longer the rod, lesser the stiffness and thicker the rod, greater the stiffness.***

Illustration 4.9 **When a weight W is hung from one end of a wire of length L (other end being fixed), the length of the wire increases by l. If the same wire is passed over a pulley and two weights W each are hung at the two ends, what will be the total elongation in the wire?**

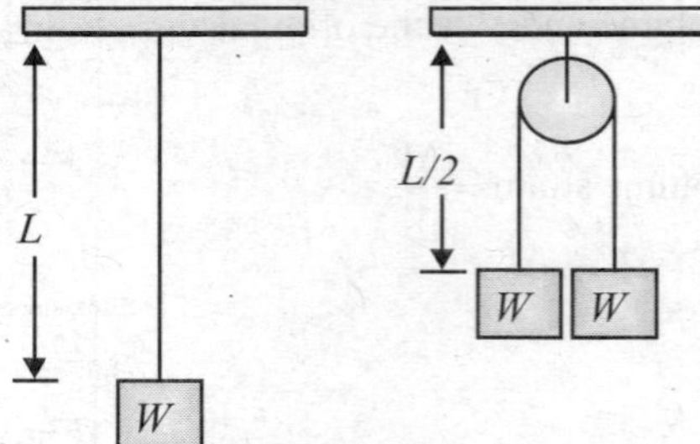

Fig. 4.17

Sol. In case (a), let Y be Young's modulus of the material of the wire. If a is its area of cross section, then we can write

$$Y = \frac{F/a}{l/L}\frac{F/a}{l/L} = \frac{WL}{al} \qquad (\because F = W)$$

Increase in length of wire $l = \dfrac{WL}{aY}$

In case (b), we can treat either segment of length $L/2$ in Fig. 4.17(b) as if one of its ends is fixed while the other end is attached to the load W because the point M remains stationary, i.e., fixed all the time due to symmetrical load. Thus, when the wire is passed over the pulley, let l' be the increase in the length of each segment. Since each segment is of length $L/2$, we have

$$Y = \frac{W(L/2)}{al'}$$

Increase in length of wire of one side of pulley $l' = \dfrac{1}{2}\dfrac{WL}{aY} = \dfrac{l}{2}$

Therefore, increase in length of both the segments of the wire $= l' + l' = l/2 + l/2 = l$.

So the increase in length remains the same.

Illustration 4.10 **A load of 10 kN is supported from a pulley which in turn is supported by a rope of sectional area 1×10^3 mm² and modulus of elasticity 10^3 N/mm², as shown in the figure. Neglecting the friction at the pulley, determine the deflection of the load.**

Sol. Let T be the tension in the rope. Then

$$2T = 10 \text{ kN} \quad \Rightarrow T = 5 \text{ kN}$$

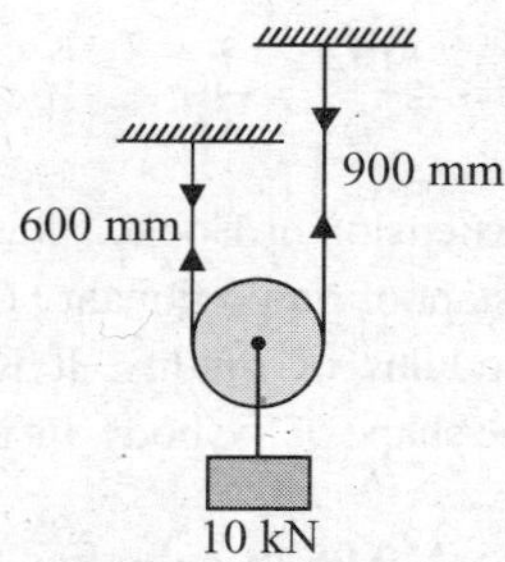

Fig. 4.18

Hence longitudinal stress in the rope is

$$\sigma = \frac{T}{A} = \frac{5\text{ kN}}{10^3\text{ mm}^2} = 5\text{ N/mm}^2$$

$$\text{Extension in the rope} = \frac{\text{Stress}}{Y} \times L$$

$$= \frac{5\text{ N/mm}^2}{10^3\text{ N/mm}^2} \times 1500\text{ mm} = 7.5\text{ mm}$$

$$\text{Therefore, deflection of the load } \delta = \frac{7.5}{2} = 3.75\text{ mm}$$

Illustration 4.11 **A rod of uniform cross-sectional area A and length L has a weight W. It is suspended vertically from a fixed support. If the material of the rod is homogeneous and its modulus of elasticity is Y, then determine the total elongation produced in the rod due to its own weight.**

Sol. In this case, different parts of the rod does not elongate to the same extent. The element closer to the support elongates more as the stress is higher with respect to the elements closer to the free end of the rod.

To determine total elongation of the rod, let us consider a small element of differential length dx at a distance x from the free end of the rod, as shown in the figure.

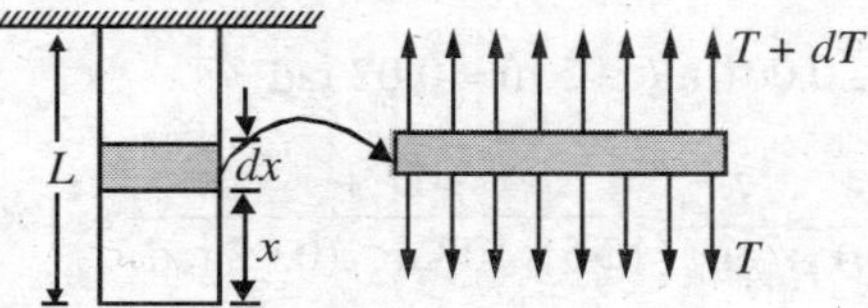

Fig. 4.19

The stress at the position of this element is produced by the weight of the rod of length x lying below it, i.e., $(W/L)x$.

Therefore, stress at this section

$$\sigma = \frac{(W/L)x}{A} = \frac{Wx}{AL}$$

The elongation $d\delta$ produced in the differential element dx is

$$d\delta = \frac{\sigma}{Y}dx = \frac{W}{YAL}x\,dx$$

Thus, total elongation produced in the rod can be calculated as

$$\delta = \int d\delta$$

$$\delta = \frac{W}{YAL}\int_0^L x\,dx = \frac{W}{YAL}\left[\frac{x^2}{2}\right]_0^L$$

or $$\sigma = \frac{1}{2}\frac{WL}{YA} \qquad \text{(i)}$$

Note:
Equation (i) may be visualized as

$$\delta = \frac{(W/2)L}{YA}$$

which suggests that a rod of uniform mass is equivalent to a massless rod with a concentrated load equal to half weight (W/2) of the rod acting at its lower end, as shown in the following figure.

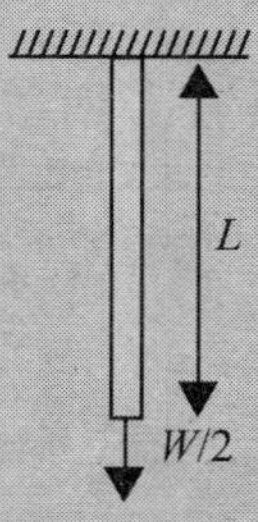

Fig. 4.20

If σ is the density of the material of the rod, then Eq. (i) may be written as

$$\delta = \frac{1}{2}\frac{\rho g L^2}{Y}$$

It indicates that the deflection of the rod is proportional to the square of the length of the rod.

Illustration 4.12 **A uniform rod of length L and mass M is pulled horizontally on a smooth surface with a force F. Determine the elongation of rod if Young's modulus of the material is Y.**

Sol. Under the action of the external force, the rod accelerates with $a = F/m$. The magnitude of stress decreases as one moves away from the point of application of the force. Therefore, the strain also decreases as one moves towards the free end.

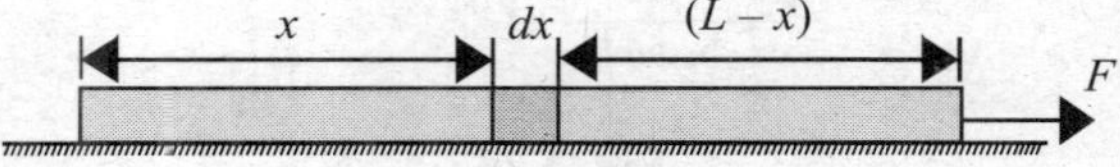

Fig. 4.21

To determine the total elongation of the rod, let us consider a small element of length dx at a distance x from the free end of the rod. The magnitude of force at this section is $F' = Fx/L$

Therefore, the stress at this section is $\sigma = \frac{F'}{A} = \frac{F}{A}\frac{x}{L}$

and elongation $d\delta$ produced in this differential element is

$$d\delta = \frac{F}{YAL}xdx$$

Thus, total elongation is $\delta = \frac{F}{YAL}\int_0^1 x\,dx = \frac{F}{YAL}\left[\frac{x^2}{2}\right]_0^1$

or

$$\delta = \frac{1}{2}\frac{FL}{YA}.$$

Note: ***The elongation produced in the rod is equivalent to the case when a rod of length L is being pulled with two equal and opposite forces, each of magnitude F/2, as shown in the following figure.***

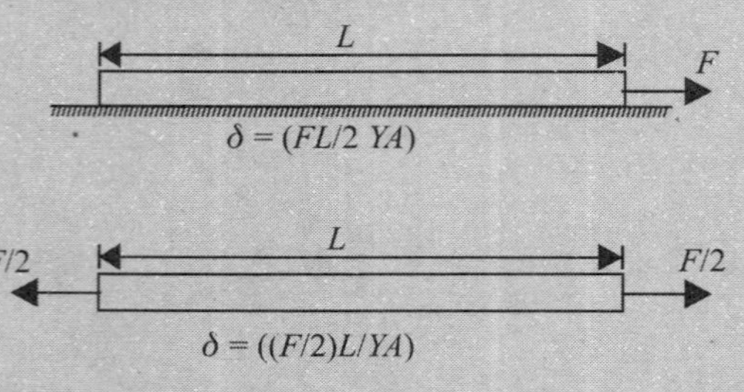

Fig. 4.22

If the rod is pulled with two equal and opposite forces each of magnitude F, then the elongation in the rod is FL/YA. Here, since the rod is being pulled only by a single force, therefore, deformation in the rod reduces to one-half.

Modulus of Rigidity

It is defined as the ratio of the tangential stress to the shear strain. Let us consider a cube whose lower face is fixed and a tangential force F acts on the upper face whose area is A.

$\therefore$ Tangential stress = F/A

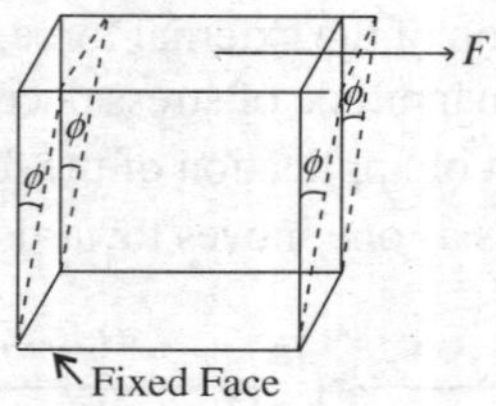

Fig. 4.23

If the vertical sides of the cube shifts through an angle ϕ called shear strain, then the modulus of rigidity is given by

$$\eta = \frac{\text{Tangential stress}}{\text{Shear strain}}$$

or

$$\eta = \frac{F/A}{\phi} = \frac{F}{A\phi} \qquad \text{(v)}$$

Equation (v) is an extension of Hooke's law for shearing stress and strain. The constant of proportionality G is called the ***shear modulus*** or the modulus of rigidity. It is a measure of the resistance to alter the shape of the body. Its units are N/m^2 or Pa.

Illustration 4.13 **A 0.05 m cube has its upper face displaced by 0.2 cm by a tangential force of 8 N. Calculate the shearing strain, shearing stress and modulus of rigidity of the material of the cube.**

Sol. $l = 5\times10^{-2}$ m, $\Delta l = 0.2$ cm $= 0.2 \times 10^{-2}$ m, $F = 8$ N

$$\text{Shearing strain} = \frac{\Delta l}{l} = \frac{0.2}{5} = 0.04$$

$$\text{Shearing stress} = \frac{F}{l\times l} = \frac{8}{(5\times10^{-2})^2} = 3200 \text{ N/m}^2$$

Modulus of rigidity,

$$\eta = \frac{\text{Shearing stress}}{\text{Shearing strain}} = \frac{3200}{0.04} = 80000 \text{ N/m}^2 = 8 \times 10^4 \text{ N/m}^2$$

Illustration 4.14 **A cube of sponge rubber with edge length 5 cm has a force of 2 N applied horizontally to the top face (parallel to an edge) while the bottom face is held fixed. If the top face is displaced horizontally through a distance of 1 mm, find the shear modulus for the sponge rubber.**

Sol. We have $G = \frac{\text{Stress}}{\text{Strain}} = \frac{F/A}{\tan\theta}$

$\tan\theta \cong \theta \cong 0.001$ m/0.05 m $= 0.02$ rad

$$S = \frac{F/A}{\tan\theta} \cong \frac{F}{A\theta} \cong \frac{2 \text{ N}}{(2.5\times10^{-3}\text{ m}^2)(0.02 \text{ rad})} = 4 \times 10^4 \text{ N/m}^2$$

Illustration 4.15 **Two parallel and opposite forces, each of magnitude 4000 N, are applied tangentially to the upper and lower faces of a cubical metal block 25 cm on a side. Find the angle of shear and the displacement of the upper surface relative to the lower surface. The shear modulus for the metal is 80 GPa.**

Sol. We use the approximate from $S = F/(A\phi)$

Putting $S = 8 \times 10^{10}$ N/m^2, $F = 4000$ N and $A = (0.25 \text{ m})^2 = 6.25 \times 10^{-2}$ m^2 in the above equation and solving for ϕ, we get

$$\phi = \frac{(4000 \text{ N})}{(6.25\times10^{-2}\text{ m}^2)(8\times10^{10}\text{ N/m}^2)} = 8.0 \times 10^{-7} \text{ rad}$$

The displacement of the upper surface is given by $d = L\phi$, where L is an edge of the cube.

$$\therefore \quad d = (8.0 \times 10^{-7})(25 \text{ cm}) = 2.0 \times 10^{-5} \text{ cm}$$

Illustration 4.16 **A horizontal aluminium rod of diameter 4.8 cm projects 5.3 cm from a wall. A 1200 kg object is suspended from the end of the rod. The shear modulus of aluminium is 3.0×10^{10} N/m². Neglecting the mass of the rod, find (a) the shearing stress on the rod and (b) the vertical deflection of the end of the rod.**

Sol.

(a) Shearing stress $= \dfrac{F}{A} = \dfrac{Mg}{\pi r^2} = \dfrac{1200 \times 9.8}{3.14 \times (0.024)^2}$

$$= 6.5 \times 10^6 \text{ N/m}^2$$

(b) Let Δx be the vertical deflection of the rod. Then shear modulus,

$$G = \frac{F/A}{\Delta x / l}$$

Thus, $\Delta x = \dfrac{(F/A)l}{G} = \dfrac{6.5 \times 10^6 \times 0.053}{3.0 \times 10^{10}} = 1.1 \times 10^{-5}$ M

Illustration 4.17 **A metal cube of side 10 cm is subjected to a shearing stress of 10^6 N/m². Calculate the modulus of rigidity if the top of the cube is displaced by 0.05 cm with respect to its bottom.**

Sol. Shearing stress, $F/A = 10^6$ N/m²

Length of side, $L = 10$ cm $= 0.1$ m

Shearing displacement, $\Delta x = 0.05$ cm $= 0.005$ cm

The modulus of rigidity is

$$\eta = \frac{F}{A\theta} = \frac{FL}{A\Delta x}\left(\because \theta \cong \tan\theta = \frac{\Delta x}{L}\right) = \frac{10^6 \times 0.1}{0.005} = 2 \times 10^8 \text{ N/m}^2$$

Bulk Modulus

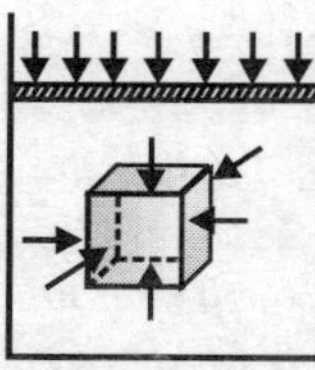

Fig. 4.24

If a body is uniformly compressed from all sides then it undergoes a decrease in volume. Figure 4.24 shows a small metal cube which is uniformly compressed from all sides by placing it inside a liquid column which is compressed externally. If p is the uniform compression on the cube and ΔV is the change in volume V of the cube, then it has been obtained experimentally that $p \propto \Delta V / V$.

A metal placed inside a liquid column is equally compressed from all directions.

Volumetric strain is defined as

$$\epsilon_V = \frac{\Delta V}{V}$$

$$p = -B\left(\frac{\Delta V}{V}\right) \Rightarrow \quad B = -\frac{P}{\Delta V / V} \qquad \text{(vi)}$$

The constant of proportionality B is called the bulk modulus of elasticity. Negative sign shows that increase in pressure (p) causes decrease in volume (ΔV).

Compressibility

The reciprocal of bulk modulus of elasticity is called compressibility. Unit of compressibility in SI is N^{-1} m² or pascal^{-1} (Pa^{-1}).

Bulk modulus of solids is about 50 times that of liquids, and for gases it is 10^{-8} times that of solids.

$$K_{\text{solids}} > K_{\text{liquids}} > K_{\text{gases}}$$

Isothermal modulus of elasticity of gas $K = P$ (pressure of gas).

Adiabatic modulus of elasticity of gas $K = \gamma \times P$ where $\gamma = \dfrac{C_p}{C_v}$.

Note:

- ***Bulk modulus is different from modulus of elasticity. A material with large bulk modulus is difficult to compress but it does not need to be stiff. For example, water.***
- ***The reciprocal of bulk modulus is defined as compressibility (K)***

$$K = (1/B)$$

Illustration 4.18 **Calculate the density of water near the bottom of the ocean where the pressure is about 500 atm and density of water is 1025 kg/m³ approximately. Take bulk modulus of water as for water as 2.0×10^9 N/m².**

Sol. The density of water at the ocean's surface where the pressure is 1 atm is about 1025 kg/m³. Thus, 1 m³ of water at the surface has a mass of 1025 kg. Let us calculate the change in volume ΔV of this water when it is at a pressure of 500 atm.

$$\Delta V = \frac{(\Delta p)V}{K}$$

The pressure change Δp is 500 atm − 1 atm = 499 atm, which is equivalent to 499×10^5 N/m². Thus, the volume change ΔV is

$$\therefore \quad \Delta V = \frac{(499 \times 10^5 \text{ N/m}^2)(1 \text{ m}^3)}{2.0 \times 10^9 \text{ N/m}^2} = 0.025 \text{ m}^3$$

The density of the water is then

$$\rho = \frac{m}{V - \Delta V} = \frac{1025 \text{ kg}}{1.00 \text{ m}^3 - 0.025 \text{ m}^3} = \frac{1025 \text{ kg}}{0.9775 \text{ m}^3} = 1051 \text{ kg/m}^3$$

The water is compressed only 2% to 3% at the bottom of the ocean where the pressure is several hundred atmosphere. Liquids and solids are not easily compressed.

Illustration 4.19 **A hydraulic press contains 0.25 m³ (250 L) of oil. Find the decrease in volume of the oil when it is subjected to a pressure increase $\Delta p = 1.6 \times 10^7$ Pa. The bulk modulus of the oil is $B = 5.0 \times 10^9$ Pa.**

Sol.

$$B = -\frac{\Delta p}{\Delta V / V_0} = -\frac{(0.25 \text{ m}^3)(1.6\times 10^7 \text{ Pa})}{5.0\times 10^9 \text{ Pa}}$$

$$= -8.0 \times 10^{-4} \text{ m}^3 = -0.80 \text{ L}$$

Even through the pressure increase is very large, the fractional change in volume is very small.

$$\therefore \Delta V/V_0 = (-8.0 \times 10^{-4} \text{ m}^3)/(0.25 \text{ m}^3) = -0.0032 \quad \text{or} \quad -0.32\%$$

Illustration 4.20 **Compressibility of water is 5×10^{-10} m²/N. Find the decrease in volume of 100 mL of water when subjected to a pressure of 15 MPa.**

Sol. The compressibility K is the reciprocal of the bulk modulus.

$$\therefore \Delta V = -VK\,\Delta p$$

$$= -(100 \text{ mL})(5 \times 10^{-10} \text{ m}^2/\text{N})(15 \times 10^6 \text{ N/m}^2) = -0.75 \text{ mL}$$

The decrease in volume is thus 0.75 mL.

Illustration 4.21 **Bulk modulus for rubber is 9.8×10^8 N/m². To what depth should a rubber ball be taken in a lake so that its volume is decreased by 0.1%?**

Sol. $K = \dfrac{(\Delta P)V}{\Delta V}$ or $\Delta P = \dfrac{K\Delta V}{V}$

Let h be the required depth and the density of water be ρ. Then

$$h\rho g = \frac{K\Delta V}{V}$$

or $$h = \frac{K\Delta V}{\rho g V} = \frac{9.8\times 10^8 \times 0.1}{10^3 \times 9.8 \times 100} = 100 \text{ m}$$

ANALOGY OF ROD AS A SPRING

A thin wire can be imagined as a parallel combination of arrays of molecular springs. When we pull a wire, we really pull the springs. Let us take a elastic rod of length l, area of cross-section A and modulus of elasticity y and apply a force F as show. If the rod elongate under the action of force Δl then we can write.

Young's modulus,

$$Y = \frac{\text{Stress}}{\text{Strain}} \Rightarrow Y = \frac{Fl}{A\Delta l} \quad \text{or} \quad F = \frac{AY}{l}\Delta l$$

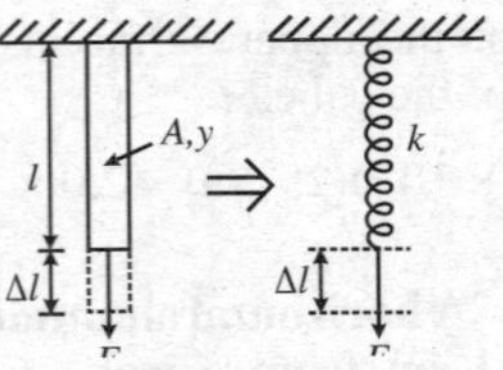

Fig. 4.25

$AY/l = k$ (constant) depends on type of material and geometry of rod.

Now, $F = k\,\Delta l \Rightarrow F \propto \Delta l$

If we compare the relation in Eq. (i) with spring force ($F_{\text{spring}} = kx$) we can say k in Eq. (i) is equivalent to force constant of the spring. Hence we can write,

$$k = \frac{Ay}{l} = \text{equivalent spring constant.}$$

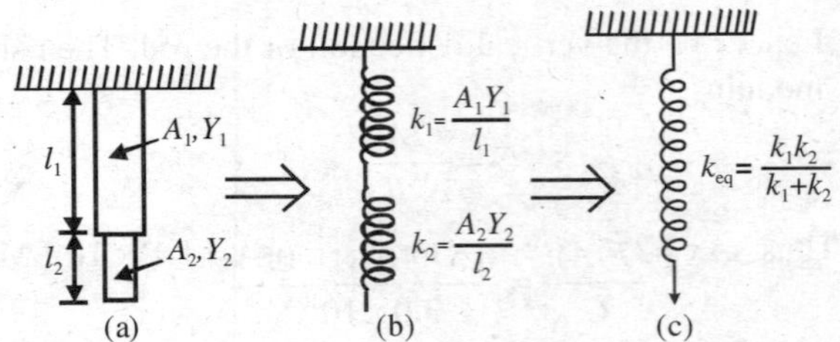

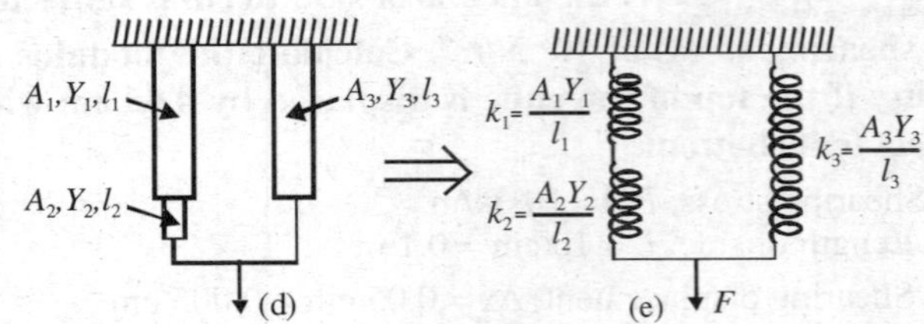

Fig. 4.26

For the system of rods shown in Fig. 4.26(a), the replaced spring system is shown in Fig. 4.26(b) (two springs in series). Figure 4.26(c) represents the equivalent spring system. Figure 4.26(d) represents another combination of rods and their replaced spring system.

Note: ***Since $k_{eq} = YA/l$, we have $k_{eq}l = YA$. It means that the product of stiffness of an elastic wire and its length is always constant. If we cut the wire into two equal parts, the stiffness of each part will be double that of the original wire. Following the same logic we can say that if a spring is cut into N equal parts, each part will have stiffness N times the stiffness of the original spring. Hence, kl = constant, where k = stiffness and l = natural length of the body. It is also equally valid for the springs.***

If F is the stretching force applied on the bar, then the same force is transmitted longitudinally to each bar; however, stresses induced will be different and hence elongations of each bar will also be different. Figure 4.26(e) shows the free-body diagram of the bar. Stress in the bar is given by

$$\sigma_1 = F/A_1$$

Illustration 4.22 **A steel bar $ABCD$ 40 cm long is made up of three parts AB, BC and CD, as shown in Fig. 4.27. The rod is subjected to a pull of 25 kN. Determine the stress in the three parts and the total extension of the rod. Young's modulus for steel = 2×10^{11} Nm^{-2}.**

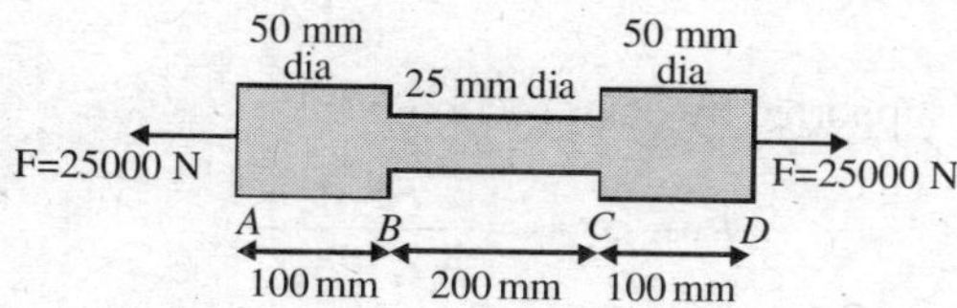

Fig. 4.27

Sol. Method 1: The axial force 25 kN is transmitted to each of the three bars.

Stress in part AB is

$$\frac{25000 \text{ N}}{\frac{\pi}{4}(50)^2 \text{ mm}^2} = \frac{40}{\pi} = 12.73 \text{ N/mm}^2$$

$$\text{Stress in part } BC = \frac{25000}{\frac{\pi}{4}(25)^2} = 50.93 \text{ N/mm}^2$$

Stress in part CD = 12.73 N/mm^2

Therefore, total extension of the rod = extension in the parts $AB + BC + CA$.

$$= \left(\frac{12.73 \text{ N/mm}^2}{2\times10^5 \text{ N/mm}^2}\times 10 \text{ mm}\right)\times 2 + \frac{50.93 \text{ N/mm}^2}{2\times10^5 \text{ N/mm}^2}\times 200 \text{ mm}$$

$$= \frac{12732}{2\times10^5} \text{ mm} = 0.0637 \text{ mm}$$

Method 2: The steel bar $ABCD$ can be replaced with series combination of three springs.

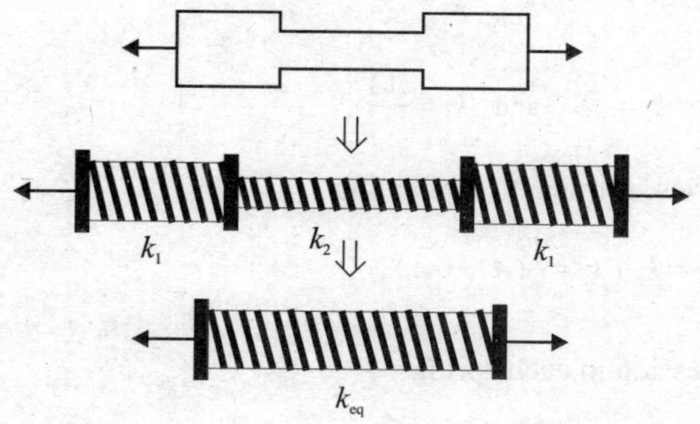

Fig. 4.28

Equivalent spring contact can be written as $\frac{1}{k_{eq}} = \frac{1}{k_1} + \frac{1}{k_2} + \frac{1}{k_1}$

Here $k_1 = \frac{YA_1}{\ell_1}$ and $k_2 = \frac{YA_2}{\ell_2}$

which gives $k_{eq} = \frac{Y(A_1A_2)}{(2A_2\ell_1 + A_1\ell_2)}$

Extension of composite rod $x = \frac{F}{k_{eq}}$

This gives $x = \frac{F(2A_2\ell_1 + A_1\ell_2)}{YA_1A_2} = \frac{1}{5\pi}\text{mm} = 0.0637 \text{ mm}$

In series combination of springs the force in each spring should be equal. Hence, each rod will experience same force F.

Hence, stress in rods AB and CD is

$$\sigma_{AB} = \sigma_{CD} = \frac{F}{A_1} = \frac{25\times10^3}{\left(\frac{\pi}{4}\times50^2\right)} = \frac{40}{\pi} \text{ N/mm}^2$$

Stress in rod BC is

$$\frac{25\times10^3}{\left(\frac{\pi}{4}\times25^2\right)} = \frac{160}{\pi} \text{ N/mm}^2$$

Illustration 4.23 **Three elastic wires, PQ, PR and PS support a body P of mass M, as shown in Fig. 4.29. The wires are of the same material and cross-sectional area, the middle one being vertical. Find the loads carried by each wire.**

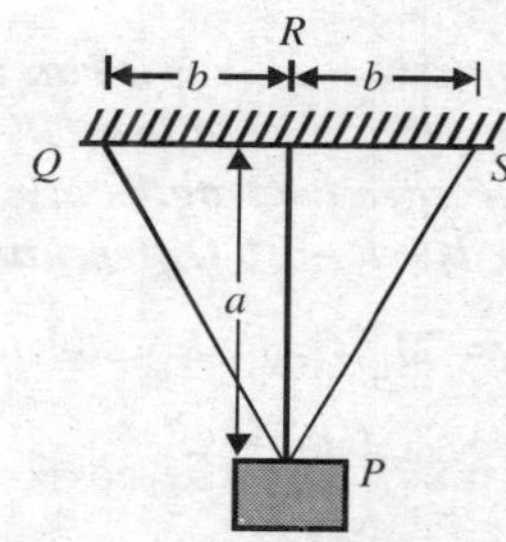

Fig. 4.29

Sol. Figure 4.30 shows the situation when the load (body P) descends due to the stretch in the wires. The new positions of wires PQ and PS are shown by the dotted lines. Let T_1, T_2 and T_3 be the loads (tensions) carried by the three wires PQ, PR and PS, respectively, with the wires PQ and PS making angle θ with the vertical.

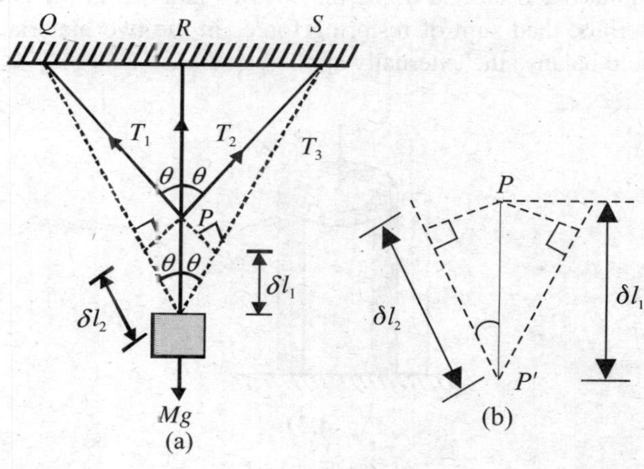

Fig. 4.30

Considering the horizontal equilibrium of point P,

$$T_1 \sin\theta = T_3 \sin\theta \quad \Rightarrow T_1 = T_3 = T \quad \text{(say)}$$

Considering the vertical equilibrium of point P,

$$T_2 + 2T\cos\theta = Mg \qquad \text{(i)}$$

If δl_1 and δl_2 be the elongations in the wires PR and PQ (or PS), respectively, then

$$\delta l_2 = \delta l_1 \cos\theta \quad \text{(from geometry)}$$

If A and Y be the cross-sectional area and Young's modulus of each of the three wires, then

$$\frac{T}{AY}(A\sec\theta) = \frac{T_2}{AY}A\cos\theta \quad \Rightarrow T = T_2\cos^2\theta \qquad \text{(ii)}$$

Solving Eqs. (i) and (ii), we get $T_2 = \dfrac{Mg}{1+2\cos^3\theta}$
and

$$T_1 = T_3 = T = \frac{Mg\cos^2\theta}{1+2\cos^3\theta}$$

Note: *The relation $\delta l_2 = \delta l_1 \cos\theta$ can also be obtained from the constraint relation as follows:*
If l_1 and l_2 are the respective lengths of wires PR and PQ at any instant, then $l_2^2 - l_1^2 - b^2$. Differentiating, we get

$$2l_2\delta l_2 = 2l_1\delta_1 \quad (\because b = \text{constant})$$

$$\Rightarrow \quad \delta l_2 = \delta l_1\left(\frac{l_1}{l_2}\right) = \delta l_1\cos\theta$$

Here the change in the value of π being small is ignored.

BARS OF COMPOSITE SECTION

Consider a column (or strut) made up of two different materials. Figure 4.31 shows the inner section of area A_1 having Young's modulus Y_1 and the outer section of area A_2 having Young's modulus Y_2. If σ_1 and σ_2 be the stresses induced in the two materials, then sum of restoring forces in the two materials should balance the externally applied load F.

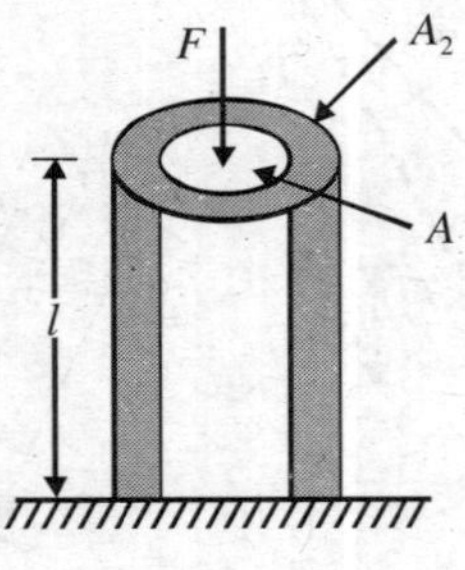

Fig. 4.31

$$\therefore \quad \sigma_1 A_1 + \sigma_2 A_2 = F$$

Also, the compresion in the two portions should be the same. But since their original lengths are same, so strains will be same.

$$\therefore \quad \frac{\sigma_1}{Y_1} = \frac{\sigma_2}{Y_2}$$

Solving equations (i) and (ii), $\sigma_1 = \dfrac{FY_1}{A_1Y_1 + A_2Y_2}$

and load supported by outer section

$$F_2 = \sigma_2 A_2 = \frac{FY_2A_2}{A_1Y_1 + A_2Y_2}$$

Generalizing the above formulae, the loads borne by different portions will be $\dfrac{FY_1A_1}{\Sigma A_iY_i}, \dfrac{FY_2A_2}{\Sigma A_iY_i}$, etc.

Alternative method: We can replace the composite section as combination of two springs in parallel.

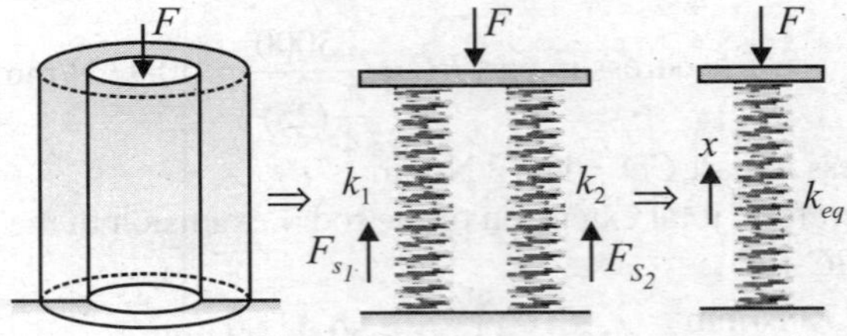

Fig. 4.32

As springs are connected in series, the compressions in each spring will be same

$$x_1 = x_2 = \frac{F_{s_1}}{k_1} = \frac{F_{s_2}}{k_2}$$

Here $\quad k_1 = \dfrac{A_1Y_1}{\ell_1}$ and $k_2 = \dfrac{A_2Y_2}{\ell_2}$

The series combination of spring

$$k_{eq} = k_1 + k_2 = \frac{1}{\ell}(A_1Y_1 + A_2Y_2)$$

Compression in each spring

$$x_1 = x_2 = x = \frac{F}{k_{eq}} = \frac{F\ell}{(A_1Y_1 + A_2Y_2)}$$

Load supported by inner section,

Hence $F_{s_1} = k_1x_1 = \dfrac{FY_1A_1}{(A_1Y_1 + A_2Y_2)}$

Load supported by outer section $F_{s_2} = k_2x_2 = \dfrac{FY_2A_2}{(A_1Y_1 + A_2Y_2)}$

Illustration 4.24 **A composite tube is made by striking a thin steel tube on a brass tube. If A_S and A_B are the respective sectional areas of the steel and brass tubes and Y_S and Y_B**

their Young's moduli, then find the Young's modulus of single tube of the same length and total sectional area, which would behave in the same fashion as that of the composite tube.

Sol. Method 1: Let the composite tube be subjected to an axial (tensile) force F and δl be the corresponding elongation. As discussed in theory, the stress borne by steel tube

$$\rho_{\text{steel}} = \frac{FY_S}{A_S Y_S + A_B Y_B}$$

But we know $Y_S = \dfrac{\sigma_S}{\varepsilon_S} = \dfrac{\sigma_S}{(\delta\ell / L)}$

$$\delta\ell = \frac{\sigma_S L}{Y_S} = \frac{FL}{(A_S Y_S + A_B Y_B)} \quad \text{(i)}$$

If Y be the required Young's modulus of the tube which behaves in the same fashion as that of the composite tube, then, corresponding to the same external force F, the deflection δl should be the same, i.e.,

$$\delta l = \frac{FL}{(A_S + A_B)Y} \quad \text{(ii)}$$

Comparing Eqs. (i) and (ii), we have

$$Y = \frac{A_S Y_S + A_B Y_B}{A_S + A_B}$$

Method 2:

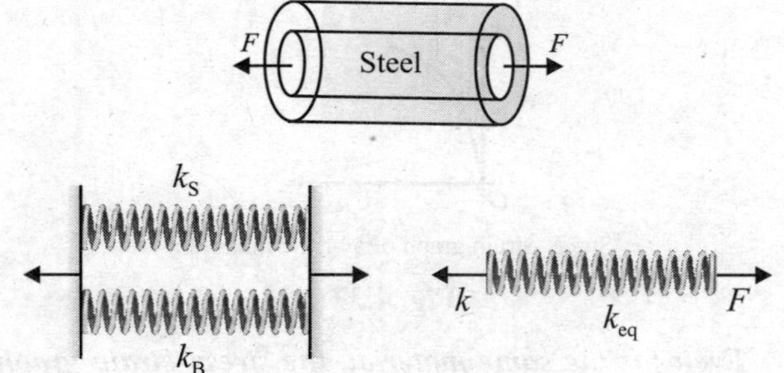

Fig. 4.33

As discussed in the theory section, we can compare composite rod system with spring combination. We can write equivalent force contact of spring as

$$k_{\text{eq}} = \frac{1}{L}(A_S Y_S + A_B Y_B) \quad \text{(iii)}$$

If replace the composite rods into a single rod of length L and area $(A_S + A_B)$

Here $\quad k_{\text{eq}} = \dfrac{(A_S + A_B)Y}{L} \quad$ (iv)

Comparing Eqs. (iii) and (iv) we have equivalent Young's modulus as

$$Y = \frac{A_S Y_S + A_B Y_B}{(A_S + A_B)}$$

Illustration 4.25 **Two vertical rods of equal lengths, one of steel and the other of copper, are suspended from the ceiling, at a distance l apart and are connected rigidly to a rigid horizontal bar at their lower ends. If A_S and A_C be their respective cross-sectional areas, and Y_S and Y_C, their respective Young's moduli of elasticities, where should a vertical force F be applied to the horizontal bar in order that the bar remains horizontal?**

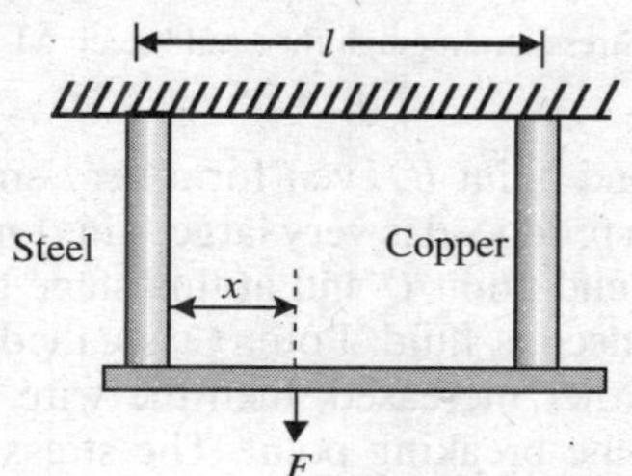

Fig. 4.34

Sol. Let the force F be applied at a distance x from the steel bar, measured along the horizontal bar. Let F_S and F_C be the loads on steel and copper rods, respectively. Then

$$F_S + F_C = F \quad \text{(i)}$$

Since the rigid horizontal bar remains horizontal, the extensions produced in the two rods and hence strains remain the same. That is

$$\frac{F_s}{A_s Y_s} = \frac{F_C}{A_C Y_C} \quad \text{(ii)}$$

Solving Eqs (i) and (ii), we get $F_s = \dfrac{FA_S Y_S}{A_S Y_S + A_C Y_C}$

and $F_C = \dfrac{FA_C Y_C}{A_S Y_S + A_C Y_C}$

Now, taking moments about the steel bar,

$$F_c l = F.x = \frac{F_C}{F} l = \frac{A_C Y_C l}{A_s Y_S + A_c Y_c}$$

or $\quad x = \dfrac{l}{1 + (A_S / A_C)(Y_S / Y_C)}$

STRESS–STRAIN DIAGRAM

When a wire is stretched by a load, it is seen that for small value of load, the extension produced in the wire is proportional to the load. On removing the load, the wire returns to its original length. The wire regains its original dimensions only when load applied is less or equal to a certain limit. This limit is called elastic limit. Thus, elastic limit is the maximum stress on whose removal, the bodies regain their original dimensions. As shown in Fig. 4.35, this type of behaviour is represented by OA portion of the graph. As we increase the stress further, a point B is reached where the increase in strain is proportional to the stress. On reaching point B, we remove the stress. Then the wire does not regain its original dimension.

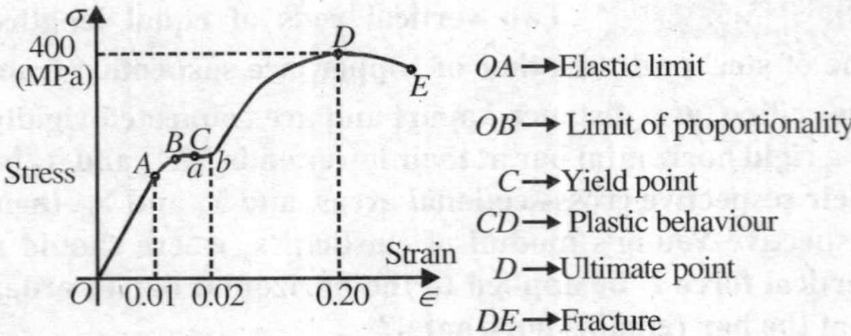

Stress-strain graph for a mild steel rod.

Fig. 4.35

As we go beyond point B, even for a very small increase in stress, the strain produced is very large. This type of behaviour is observed around point C and at this stage the wire begins to flow like a viscous fluid. Point C is called yield point. If the stress is further increased, then the wire breaks off at a point D called the breaking point. The stress corresponding to this point is called breaking stress or tensile strength of the material of the wire. A material for which the plastic range CD is relatively high is called ductile material. These materials get permanently deformed before breaking. The materials for which plastic range is relatively small are called brittle materials. These materials break as soon as the elastic limit is crossed.

The stress–strain diagram is used to study the properties of materials under the action of external forces. In these stress-strain diagrams, conventionally the ordinate scale is used for stresses (σ) and the abscissa for strains ($\in$).

Important Points

- ***Breaking stress = breaking force/area of cross section.***
- ***Breaking stress is constant for a material.***
- ***Breaking force depends upon the area of the section of the wire of a given material.***
- ***The working stress is always kept lower than that of a breaking stress so that 'safety factor = breaking stress/working stress' may have a large value.***
- ***Breaking strain = elongation or compression/original dimension.***
- ***Breaking strain is constant for material.***

Some Important Points in Stress–Strain Diagram

- ***In the yield region, strain is 15–20 times those that takes place up to the proportional limit occurring during yielding.***
- ***A study of stress–strain diagrams shows that the yield point is so close to the elastic limit that for most purposes the two may be taken as one.***
- ***As the deformation is increased further beyond the yield region, the curve moves upward and attains its maximum at a point U (called the ultimate stress) after which the curve moves downward to point B at which the material fails or breaks down.***
- ***The breaking stress is lower than the ultimate stress because the breaking stress is computed by dividing the breaking load by the original cross-sectional area which, although convenient, is incorrect. This error is caused by a phenomenon known as necking. After the point U, the material stretches very rapidly and simultaneously narrows down, as shown in Fig. 4.36, so that the breaking load is distributed over a smaller area. If the breaking area is measured after failure occurs, and divided into the breaking load, the result is a truer value of the actual failure stress, although this is commonly taken as the maximum stress of the material.***

Necking of the mild-steel bar beyond the ultimate point.

Fig. 4.36

- ***The stress–strain diagrams differ considerably for different materials. The steeper curve indicates a stiffer material; a taller graph indicates a stronger material; a long graph parallel to the $\in$-axis indicates a ductile material and the absence of yield point or plastic zone refers to a brittle material.***

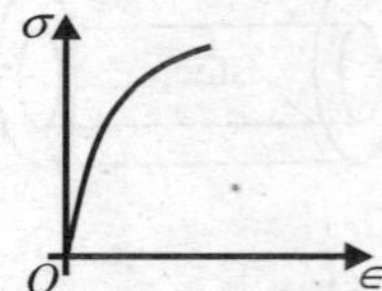

Stress–strain graph of a brittle material.

Fig. 4.37

- ***Even for the same material, the stress–strain graphs differ, depending on the temperature at which the test is conducted, the speed of the test, etc.***
- ***Breaking strain is constant for material.***

ELASTICITY AND PLASTICITY

The property by which a body returns to its original shape after removal of the deforming force (or load) is called elasticity. It has been observed experimentally that some bodies return to their original states immediately after the removal of external load while some other bodies take longer time to return back.

The delay in returning to the original state by an elastic body after the removal of external load is called elastic aftereffect. Quartz and phosphor bronze are free from elastic after-effects.

Plasticity is the property of a body by which it shows a permanent deformation which is not recovered after the removal of external load. In the stress–strain graph as shown in Fig. 4.38, the region of plastic deformation lies on the right of the elastic limit. If the body is loaded to a stress in the plastic region and then unloaded as shown in the figure, the unloading curve does not retrace the loading curve but it returns back along a different path. The area enclosed by the loop *OEAA'* gives the energy lost per unit volume of the body. The loop is also called the hysteresis loop.

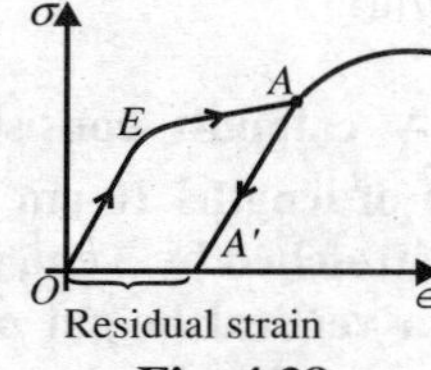

Fig. 4.38

A specimen is loaded to a point *A* in the plastic region and then unloaded. The specimen shows permanent or residual strain.

On the contrary, if the body is loaded in the elastic region and then unloaded, it returns back by retracing the same path. It indicates that the internal forces between the particles of the body act as conservative forces.

Elastic Aftereffect

We know that some material bodies take some time to regain their original configuration when the deforming force is removed. The delay in regaining the original configuration by the bodies on the removal of deforming force is called elastic aftereffect. The elastic aftereffect is negligibly small for quartz fibre and phosphor bronze. For this reason, the suspensions made from quartz and phosphor bronze are used in galvanometers and electrometers. For glass fibre elastic aftereffect is very large. It takes hours for glass fibre to return to its original state on the removal of deforming force.

Elastic Fatigue

The loss of strength of the material due to repeated strains on the material is called elastic fatigue. That is why bridges are declared unsafe after a long time of their use.

ENERGY STORED IN A DEFORMED BODY

We know that when a metal rod is elongated under a gradually increasing load, work is done by the external forces on the rod which is transferred either partially or completely into potential energy of strain. If the strain remains within the elastic limit, the work done is completely transformed into potential energy and can be recovered during a gradual unloading of the strained rod.

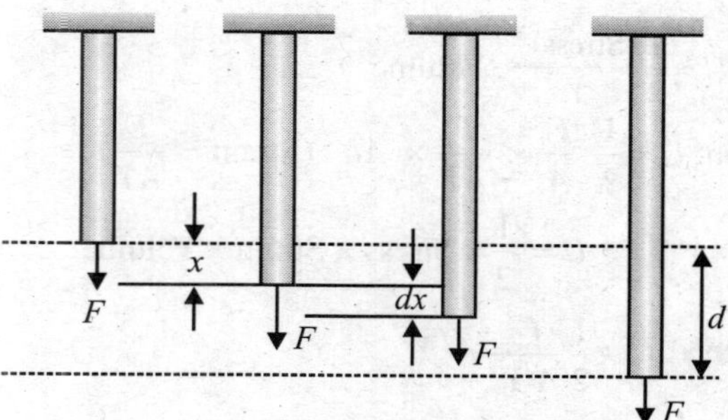

Fig. 4.39

To produce *x* elongation in a rod of length *L* and cross-sectional area *A*, force required is given by

$$F = \frac{YA}{L}x$$

The work done produces an additional deformation *dx* is given by

$$dW = \frac{YA}{L}x\,dx$$

If total deformation of rod is δ. The total work done is

$$W = \frac{YA}{L}\int_0^{\delta} x\,dx \quad \Rightarrow W = \frac{1}{2}\frac{YA}{L}\delta^2$$

According tc work–energy theorem, $W = \Delta U$. Thus, the potential energy stored in the rod is

$$U = \frac{1}{2}\frac{YA}{L}\delta^2$$

Elastic Potential Energy Stored in a Stretched Wire or in a Rod

We can replace a rod with a springs as shown in the figure. Strain energy stored in equivalent spring is

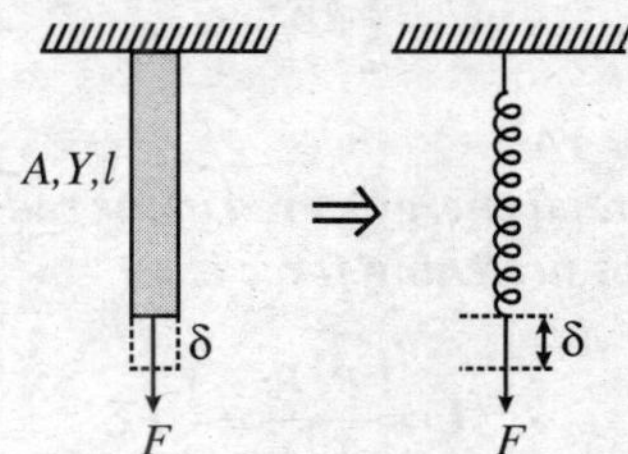

Fig. 4.40

$$U = \frac{1}{2}k\delta^2 = \frac{1}{2}\left(\frac{AY}{L}\right)\delta^2$$

where $\delta = \dfrac{Fl}{AY}$

$$\therefore \text{ Strain energy } U = \frac{1}{2}\frac{AY}{l}\frac{F^2l^2}{A^2Y^2}$$

$$= \frac{1}{2}\frac{F^2\ell}{AY} \quad \text{or} \quad U = \frac{1}{2}F\delta$$

This equation can be rearranged as

$$U = \frac{1}{2}\frac{F^2}{A^2}\cdot\frac{\ell A}{Y} \qquad (lA = \text{volume of the rod, } F/A = \text{stress})$$

$$U = \frac{1}{2}\frac{(\text{Stress})^2}{Y}.\text{Volume}$$

Again $U = \frac{1}{2}\frac{F}{A} \times \frac{F}{AY} \times Al$ (Strain $= \frac{F}{AY}$)

$$U = \frac{1}{2} \times \text{Stress} \times \text{Strain} \times \text{Volume}$$

Again $U = \frac{1}{2}\frac{F^2}{A^2Y^2} AlY$

$$U = \frac{1}{2} Y(\text{Strain})^2 \times \text{Volume}$$

$\therefore$ Strain energy density $u = \frac{\text{Strain energy}}{\text{Volume}} = \frac{1}{2}\frac{(\text{Stress})^2}{Y}$

$$u = \frac{1}{2}Y(\text{Strain})^2 = \frac{1}{2}\text{Stress} \times \text{Strain}$$

$$u = \frac{U}{AL} = \frac{\rho^2}{2Y} = \in \frac{1}{2}Y^2 = \frac{1}{2} \in Y^2$$

where $\in = \delta/L$ and $\rho = F/A$.

- ***The area under the stress–strain diagram gives the potential energy (or strain energy) stored per unit volume of the specimen.***
- ***If stress applied is not constant we can write energy stored in any elastic rod***

$$U = \frac{1}{2Y}\int \rho^2 dV$$

where ρ = stress and dV = change in volume of rod

Note: ***The potential energy stored in the rod in terms of stiffness constant k is given by***

$$U = \frac{1}{2}k\delta^2$$

where k = YA/L.
The potential energy stored in the rod in terms of the maximum stretching force is

$$U = \frac{F^2L}{2YA} = \frac{1}{2}F\delta$$

- ***The poetential energy stored per unit volume of the rod is given by***

Illustration 4.26 **A metallic wire is stretched by suspending a weight to it. If α^2 is the longitudinal strain and Y is its Young's modulus of elasticity, then show that the elastic potential energy per unit volume is given by 1/2 Y^2.**

Sol. Potential energy per unit volume = (1/2) Stress × Strain

= (1/2) (Y × Strain) × Strain

= (1/2) $Y\alpha^2$

Illustration 4.27 **Calculate the elastic potential energy per unit volume of water at a depth of 1 km. Compressibility (α) of water = 5×10^{-10} SI units. Density of water = 10^3 kg/m^3.**

Sol. Energy per unit volume

= (1/2) Stress × Strain = (1/2) Stress × $\frac{\text{Stress}}{K}$

= (1/2) β (stress)^2 = (1/2) β $(h\rho g)^2$

= (1/2) × (5×10^{-10}) × $(1000 \times 1000 \times 9.8)^2$

= 2.4×10^4 J/m^3

Illustration 4.28 **A catapult consists of two parallel rubber strings, each of lengths 10 cm and cross-sectional area 10 mm^2. When stretched by 5 cm, it can throw a stone of mass 100 gm to a vertical height of 25 m. Determine Young's modulus of elasticity of rubber.**

Sol. A stretched catapult has elastic potential energy stored in it (Strain energy stored in both the rubber strings)

$$U = \left(\frac{1}{2}\frac{YAl^2}{L}\right) \times 2$$

This energy, when imparted to the stone, it flies off a height 20 m. Energy possessed by the stone = mgh. Now,

$$U = mgh \Rightarrow \frac{YAl^2}{L} = mgh$$

$$Y = \frac{mghL}{Al^2} = \frac{1 \times 10^{-1} \times 10 \times 25 \times 10^{-1}}{10 \times 10^{-6} \times (5 \times 10^{-2})^2}$$

$\therefore$ $Y = 10 \times 10^8$ N/m^2

INTERATOMIC FORCE CONSTANT

The interatomic force F' developed between the atoms is directly proportional to the change in distance Δr between them.

$$F' = k\Delta r, \quad \text{(i)}$$

where k is interatomic force constant. Let L be the length of a wire and r_0 be the interatomic distance between the atoms of the wire. Suppose, when a force F is applied, the length of the wire increases by l, and the distance between its atoms increases from r_0 to $r_0 + \Delta r$. Then longitudinal strain,

$$\frac{l}{L} = \frac{\Delta r}{r_0} \quad \text{(ii)}$$

Fig. 4.41

If A is the area of cross section of the wire, it will have A/r_0^2 chains of atoms. Therefore, the interatomic force (force acting on chain) is

$$F' = \frac{\text{External force}}{\text{Number of chains}} = \frac{F}{A/r_0^2} = \frac{Fr_0^2}{A}$$

Interatomic force constant is

$$k = \frac{F'}{\Delta r} = \frac{Fr_0^2/A}{\Delta r} = \frac{F}{A}\frac{r_0^2}{\Delta r} \qquad \text{[using Eq. (i)]}$$

But $r_0/\Delta r = L/l$. Hence,

$$k = \frac{F}{A}\frac{L}{l}r_0 = \frac{F/A}{l/L}r_0 \qquad \text{[using Eq. (ii)]}$$

But by definition,

$$\frac{F/A}{l/L} = Y \text{ (Young's modulus of the material of the wire)}$$

$\Rightarrow k = Yr_0$

Thus, interatomic force constant k is equal to the product of Young's modulus of the material of the wire and the normal distance r_0 between the atoms of the wire.

Some Important Points About Elasticity

- ***Breaking stress depends on the material of the wire. It does not depend upon the diameter or length of the wire.***
- ***Breaking load depends on the area of cross section of the wire.***
- ***The strain persists even when the stress is removed and thus lags behind the stress.***
- ***Steel is highly elastic material.***
- ***Steel is more elastic than rubber.***
- ***Young's modulus decreases with rise in temperature.***
- ***A liquid has zero modulus of rigidity.***
- ***Rubber remains elastic even for a strain of the order of thousand.***
- ***A rise in temperature results, in general, in weakening of elastic properties and a fall in temperature results in strengthening them.***
- ***Young's modulus is independent of length or radius of wire.***
- ***Solids are the most elastic and gases are the least elastic.***
- ***Under tensile stress, the restoring forces are caused by interatomic attraction.***
- ***Under compressional stress, the restoring force is due to the interatomic repulsion.***
- ***Force of repulsion, not the force of attraction, is responsible in the interatomic or intermolecular potential for the formation of a solid.***
- ***Solids have all the three types of elasticity whereas liquids and gases have only volume elasticity, i.e., only bulk modulus is applicable to them. Gases, however, have two types of bulk moduli, viz. isothermal and adiabatic.***
- ***Young's modulus is defined for solids but not for liquids and gases. The value of Young's modulus increases on mixing the impurity in the solid and decreases on increasing the temperature of the body.***
- ***The area under the stress-strain graph represents the work done per unit volume.***

Concept Application Exercise 4.1

1. Stress and pressure are both forces per unit area. Then in what respect does stress differ from pressure?
2. The stress and strain graphs for two materials, A and B, are shown here. Answer the following:

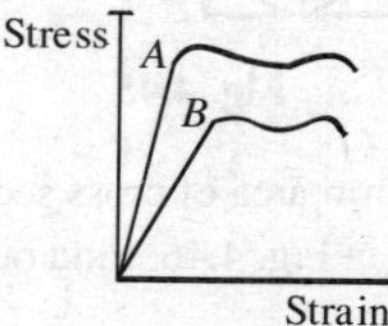

Fig. 4.42

   (a) Which material has greater Young's modulus?
   (b) Which material is more ductile?
   (c) Which material is more brittle?
   (d) Which material has greater tensile strength?
   Give reasons.
3. State whether the following statements are true or false with reasons.
   (a) Elastic forces are always conservative.
   (b) Elastic forces are strictly conservative only when Hooke's law is obeyed.
   (c) When a wire is loaded beyond the elastic limit and then deloaded, the work done disappears completely as heat.

4. A horizontal force of magnitude F acts at the end P of a uniform rigid rod which is welded at point Q. In each case 1 and 2, as shown in Fig. 4.43, find the reaction force acting at a point C at a distance x from the fixed end Q of the rod.

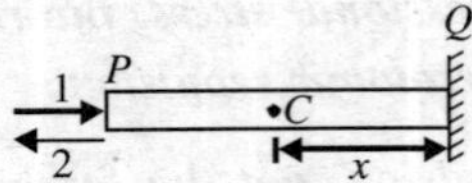

Fig. 4.43

5. A rubber ball of bulk modulus B is taken to a depth h of a liquid of density ρ. Find the fractional change in the radius of the ball.

6. Find out longitudinal stress and tangential stress on a fixed block.

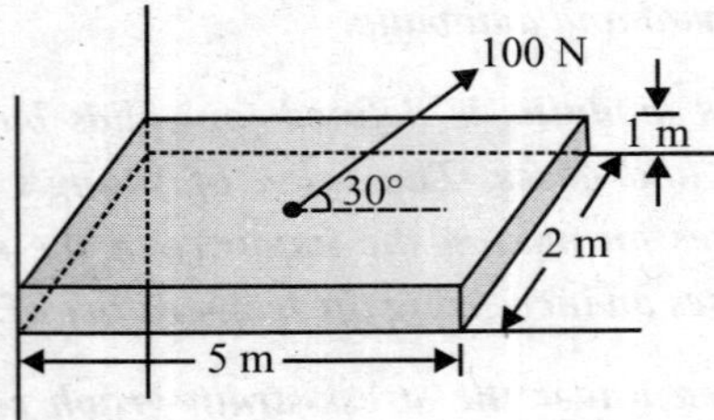

Fig. 4.44

7. Find out bulk stress on the spherical object of radius $10/\pi$ cm if area and mass of piston are 50 cm² and 50 kg, respectively, for a cylinder filled with gas as shown in Fig. 4.45.

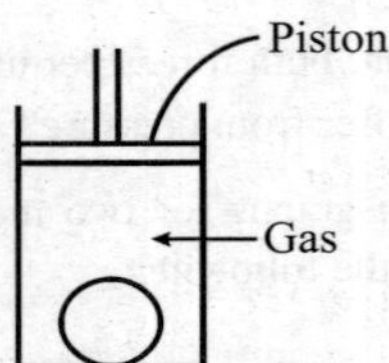

Fig. 4.45

8. Three rods of uniform area of cross section. A = 10^{-7} m² are arragned as shown in Fig. 4.46. Find out the shift in point B, C and D

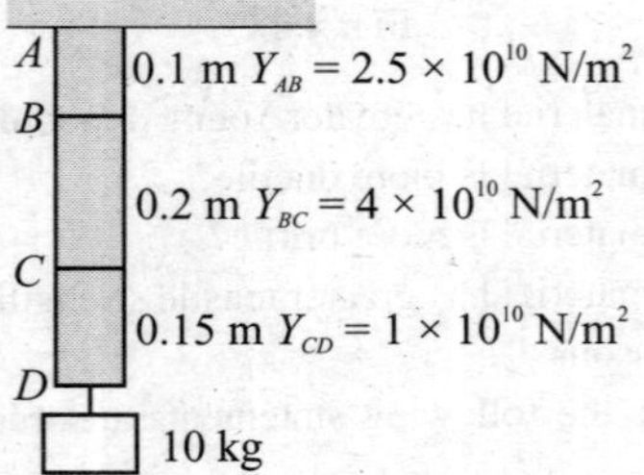

Fig. 4.46

9. Find the depth of lake at which density of water is 1% greater than that at the surface. Given compressibility $k = 50 \times 10^{-6}$ atm^{-1}.

10. A rubber cube of side 5 cm has one side fixed while a tangential force equal to 1800 N is applied to opposite face. Find the shearing strain and the lateral displacement of the strained face. Modulus of rigidity for rubber is 2.4×10^6 N/m².

11. A mass 'm' is attached with rods as shown in Fig. 4.47. This mass is slightly stretched and released. Find out whether the motion of mass is S.H.M. If yes then find out the time period.

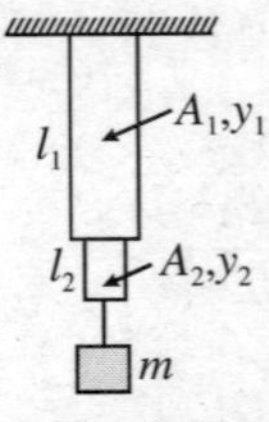

Fig. 4.47

12. A ball of mass 'm' drops from a height 'h', which sticks to a massless hanger after striking it. Neglect overturning. Find out the maximum extension in rod, assuming that the rod is massless.

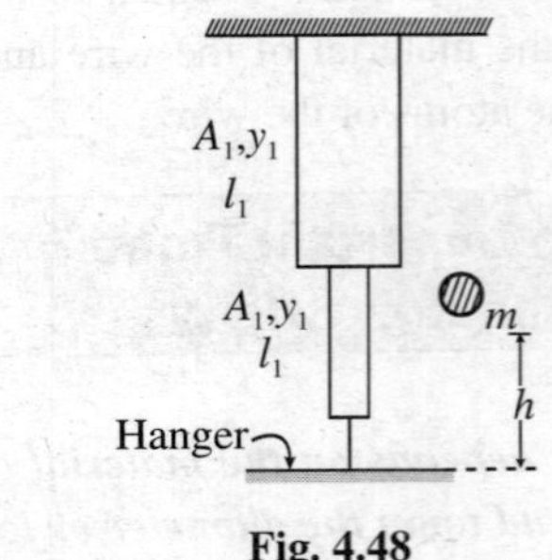

Fig. 4.48

VISCOSITY

When a solid body slides over another solid body, a frictional force begins to act between them. This force opposes the relative motion of the bodies. Similarly, when a layer of a liquid slides over another layer of the same liquid, a frictional force acts between them which opposes the relative motion between the layers. This force is called 'internal frictional force'.

Suppose a liquid is flowing in a streamlined motion on a fixed horizontal surface AB (Fig. 4.49). The layer of the liquid which is in contact with the surface is at rest, while the velocity of other layers increases with a distance from the fixed surface. In Fig. 4.49, the lengths of the arrows represent the increasing velocity of the layers. Thus, there is a relative motion between adjacent layers of the liquid. Let us consider three parallel layers—a, b and c. Their velocities are in an increasing order. The layer a tends to retard the layer b, while b tends to retard c. Thus, each layer tends to decrease the velocity of the layer above it. Similarly, each layer tends to increase the velocity of the layer below it. This means that in between any two layers of the liquid, internal tangential forces act which try to destroy the relative motion between the layers. These forces are called 'viscous forces'. If the flow of the liquid is to be maintained, an external force must be applied to overcome the dragging viscous forces. In the absence of the external force, the viscous forces would soon bring the liquid to rest. *The property of the liquid by virtue of which it opposes the relative motion between its adjacent layers is known as 'viscosity'.*

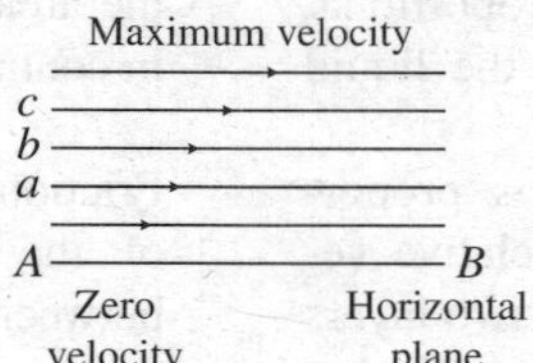

Fig. 4.49

The property of viscosity is seen in the following examples:

(i) A stirred liquid, when left, comes to rest on account of viscosity. Thicker liquids such as honey, coal tar, glycerine, etc. have a larger viscosity than thinner ones such as water. If we pour coal tar and water on a table, the coal tar will stop soon while the water will flow up to quite a large distance.

(ii) If we pour water and honey in separate funnels, water comes out readily from the hole in the funnel while honey takes enough time to do so. This is because honey is much more viscous than water. As honey tends to flow down due to gravity, the relative motion between its layers is opposed strongly.

(iii) We can walk fast in air, but not in water. The reason is again viscosity which is very small for air but comparatively much larger for water.

(iv) The cloud particles fall down very slowly because of the viscosity of air and hence appear floating in the sky.

Viscosity comes into play only when there is a relative motion between the layers of the same material. This is why it does not act in solids.

FLOW OF LIQUID IN A TUBE: CRITICAL VELOCITY

When a liquid flows 'in a tube', the viscous forces oppose the flow of the liquid. Hence a pressure difference is applied between the ends of the tube which maintains the flow of the liquid. If all particles of the liquid, passing through a particular point in the tube, move along the same path, the flow of the liquid is called 'streamlined flow'. This occurs only when the velocity of flow of the liquid is below a certain limiting value called 'critical velocity'. When the velocity of flow exceeds the critical velocity, the flow is no longer streamlined but becomes turbulent. In this type of flow, the motion of the liquid becomes zigzag and eddy currents are developed in it.

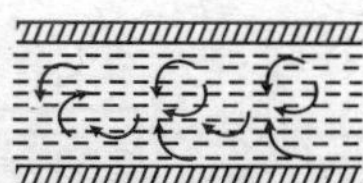

Fig. 4.50

Reynolds proved that the critical velocity for a liquid flowing in a tube is $v_C = k\eta/\rho\, a$, where ρ is density and η is viscosity of the liquid, a is radius of the tube and k is 'Reynolds number' (whose value for a narrow tube and for water is about 1000). When the velocity of flow of the liquid is less than the critical velocity, then the flow of the liquid is controlled by the viscosity, the density having no effect on it. But when the velocity of flow is greater than the critical velocity, the flow is mainly governed by the density, the effect of viscosity becoming less important. It is because of this reason that when a volcano erupts, the lava coming out of it flows speedily in spite of being very thick (i.e., of large viscosity).

VISCOSITY IN FLUIDS

Viscosity is a measure of frictional forces between the adjacent layers of a fluid with which it opposes the relative motion between them. The frictional resistance of fluids is also observed when a solid body moves within the body of a fluid. Viscous forces are present both in liquids and gases. Obviously the liquids are more viscous than gases.

The viscosity of two liquids can be compared in the following two ways:

- ***By allowing them to flow out of identical vessels. For example, a liquid of low viscosity such as water flows rapidly out of it, and on the contrary, a highly viscous liquid such as glue or glycerine makes a slow movement out of it.***
- ***By dropping steel ball bearings through each liquid. The balls fall more slowly in a liquid of high viscosity.***

Newton's Formula

The basic formula for the frictional (or viscous force) F in a liquid was first suggested by Newton. He observed that larger the area A of the surface of liquid considered, larger was the frictional force F; larger the gradient of velocity dv/dy across the layers of liquid, larger was the frictional force F. That is,

$$F \propto A\frac{dv}{dy}$$

Consider a fluid confined between two parallel plates, with the lower plate stationary and the upper plate moving with a velocity v_0, as shown in Fig. 4.51. It has been proved experimentally that for all real fluids possessing any viscosity, the fluid particles in immediate contact with any solid surface move with the velocity of the surface itself. Thus, the topmost layer of the liquid moves with velocity v_0 and the bottommost layer remains stationary. And all the intermediate layers acquire velocities $0 < v < v_0$ as indicated by the lengths of the arrows in the figure.

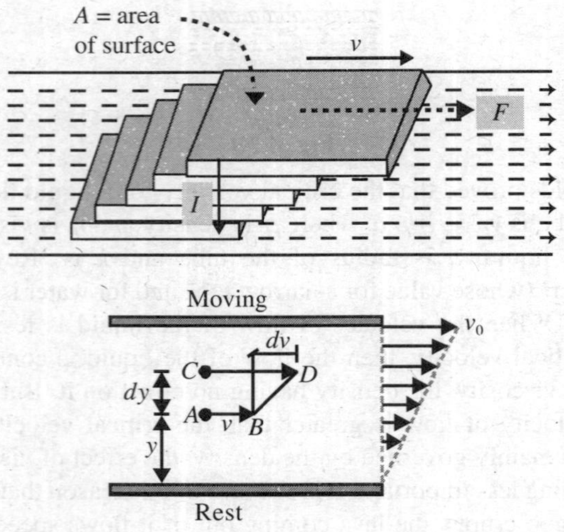

Fig. 4.51

Figure 4.51 shows a layer *AB* of the liquid at a distance *y* from stationary plate which is moving with a velocity *v* parallel to the plate. An adjacent layer *CD* at a distance $y + dy$ moves with a velocity $v + dv$. The velocity gradient between the two layers is dv/dy. Due to relative motion between the adjacent layers of the liquid there exists a frictional resistance, which according to Newton is given by

$$F \propto A\frac{dv}{dy} \quad \text{or} \quad F = -\eta A\frac{dv}{dy} \qquad \text{(i)}$$

η is called coefficient of viscosity. Negative sign shows that the direction of viscous drag (F) is just opposite to the direction of the motion of the liquid.

It is important to note that solids resist relative motion between its adjacent layers to maintain their shapes. The resistance offered by solids in maintaining their shapes is measured in terms of their modulus of rigidity *G*.

Note: ***The coefficient of viscosity η is analogous to the modulus of rigidity of solids. The shear stress of fluids is proportional to the rate of shear strain and that in case of solids is proportional to the shear strain.***

SIMILARITIES AND DIFFERENCES BETWEEN VISCOSITY AND SLIDING FRICTION

Similarities:

1. Both viscosity and sliding friction oppose relative motion. Whereas viscosity opposes the relative motion between two adjacent liquid layers, sliding friction opposes the relative motion between two solid layers.
2. Both come into play whenever there is a relative motion between layers of liquid or solid surfaces as the case may be.
3. Both are due to molecular attractions.

Dissimilarities:

Viscosity	Sliding friction
(i) Viscosity (or viscous drag) between layers of liquid is directly proportional to the area of the liquid layers.	Friction between two solids is independent of the area of solid surfaces in contact.
(ii) Viscous drag is proportional to the relative velocity between two layers of liquid.	Friction is independent of the relative velocity between two surfaces.
(iii) Viscous drag is independent of normal reaction between two layers of liquid.	Friction is directly proportional to the normal reaction between two surfaces in contact.

SOME APPLICATIONS OF VISCOSITY

Knowledge of viscosity of various liquids and gases have been put to use in daily life. Some applications of its knowledge are discussed as follows:

1. As the viscosity of liquids vary with temperature, proper choice of lubricant is made depending upon season.
2. Liquids of high viscosity are used in shock absorbers and buffers at railway stations.
3. The phenomenon of viscosity of air and liquid is used to damp the motion of some instruments.
4. The knowledge of the coefficient of viscosity of organic liquids is used in determining the molecular weight and shape of the organic molecules.
5. It finds an important use in the circulation of blood through the arteries and veins of human body.

UNITS OF COEFFICIENT OF VISCOSITY

From the definition, we have $\eta = \dfrac{F}{A(\Delta v_x/\Delta y)}$

$$\therefore \text{ Dimensions of } \eta = \frac{[MLT^{-2}]}{[L^2][LT^{-1}/L]} = \frac{[MLT^{-2}]}{[L^2T^{-1}]} = [ML^{-1}T^{-1}]$$

Its unit is kg/(m s)

In CGS system, the unit of coefficient of viscosity is dyn s cm^{-2} and is called poise. In SI the unit of coefficient of viscosity is N s m^{-2} and is called decapoise.

$$1 \text{ decapoise} = 1 \text{ N s m}^{-2} = (10^5 \text{ dyn}) \times \text{s} \times (10^2 \text{ cm})^{-2}$$

$$= 10 \text{ dyn s cm}^{-2} = 10 \text{ poise}$$

EFFECT OF TEMPERATURE ON VISCOSITY

Experiments show that η of a liquid decreases sharply with increase in its temperature and becomes zero at its boiling point. On the other hand, η of gases increase with temperature.

The viscosity of liquids decreases with increase in temperature and increases with the decrease in temperature. That is, $\eta \propto 1/\sqrt{T}$. On the other hand, the value of viscosity of gases increases with the increase in temperature and vice versa. That is, $\eta \propto \sqrt{T}$.

Illustration 4.29 **A plate of area 100 cm² is placed on the upper surface of castor oil, 2 mm thick. Taking the coefficient of viscosity to be 15.5 poise, calculate the horizontal force necessary to move the plate with a velocity 3 cm/s.**

Sol. The (horizontal tangential) viscous force is given by

$$F = -\eta A \frac{dv}{dy}$$

Given η = 15.5 poise, A = 100 cm²

$$\frac{dv}{dx} = \frac{v_2 - v_1}{x_2 - x_1} = \frac{(0-3)\,\text{cm/s}}{(2-0)\,\text{mm}} = -\frac{3}{0.2}\,\text{s}^{-1} = -15\,\text{s}^{-1}$$

$$F = -15.5 \times 100 \times (-15) = 2.235 \times 10^4 \text{ dyn} = 0.2325 \text{ N}$$

Illustration 4.30 **A metal plate of area 0.10 m² is connected to a 0.01 kg mass via a string that passes over an ideal pulley (considered to be frictionless), as shown in the figure. A liquid with a film thickness of 3.0 mm is placed between the plate and the table. When released, the plate moves to the right with a constant speed of 0.085 m/s. Find the coefficient of viscosity of the liquid.**

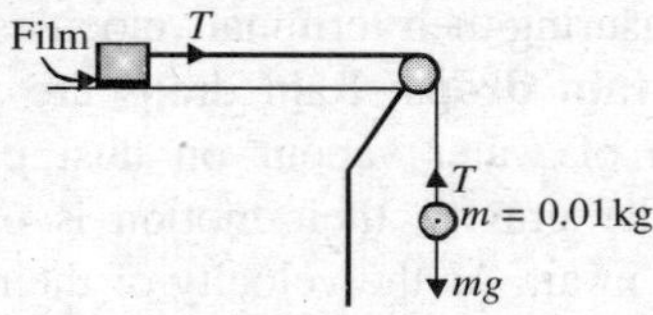

Fig. 4.52

Sol. As the metal plate moves with constant velocity,

$$mg - T = 0 \quad \Rightarrow T = mg$$

If F is the viscous force on the plate, then

$$F = T = mg = 0.01 \times 9.8 = 9.8 \times 10^{-2} \text{ N}$$

By Newton's law of viscosity,

$$F = \eta A \frac{dv}{dy}$$

where $\frac{dv}{dy} = \frac{0.085}{0.3 \times 10^{-3}} = 28.3$ s

$\Rightarrow$ $9.8 \times 10^{-2} = \eta \times 0.10 \times 28.3$

$\Rightarrow$ $\eta = 3.4 \times 10^{-3}$ N s/m²

Illustration 4.31 **A sliding fit cylindrical body of mass of 1 kg drops vertically down at a constant velocity of 5 cm/s. Find the viscosity of the oil.**

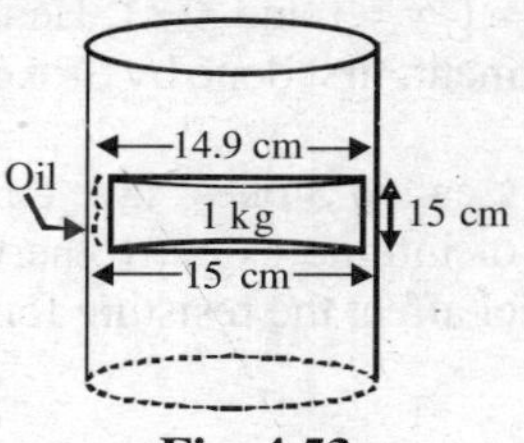

Fig. 4.53

Sol. As the body moves with constant velocity,

$$mg = F_v \quad \Rightarrow \quad 1g = \eta A \frac{dv}{dy}$$

$$\Rightarrow 1g = \eta (2\pi l) \frac{dv}{dy}$$

$$\Rightarrow 1g = \eta (2\pi \times 7.5 \times 10^{-2} \times 15 \times 10^{-2}) \left(\frac{0.05 - 0}{0.05 \times 10^{-2}} \right)$$

After solving, we get $\eta = 1.4$ N s/m².

STOKES' LAW

The streamlines for a fluid flowing slowly past a stationary solid sphere are shown in the following figure.

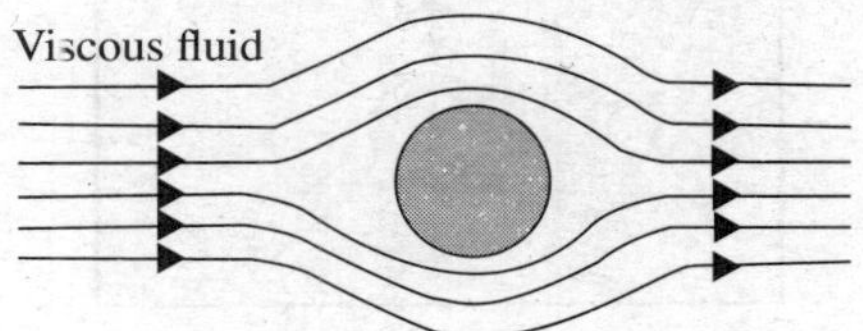

Fig. 4.54

When the sphere moves slowly as compared to the fluid, the pattern is similar but the streamlines then flow in such a way that the apparent motion of the fluid particles is as seen by someone on the moving sphere. In this latter case, it is known that the layer of fluid in contact with the sphere moves with it, thus creating a velocity gradient between this layer and other layers of fluid. Viscous forces are thereby brought into play and constitute the resistance experienced by the moving sphere. If we make the plausible assumption that the viscous retarding force F depends on the size of the body, the velocity with which it moves, the viscosity of the fluid and mass density of the fluid, then an expression can be derived for F by the method of dimensions. Thus,

$$F = k \eta^x v^y r^z$$

where x, y and z are the indices to be found and k is a dimensionless constant. The dimensional equation is

$$[MLT^{-2}] = [ML^{-1}T^{-1}]^x [LT^{-1}]^y [L]^z$$

Equating indices of M, L and T on both sides, we have

$$1 = x$$
$$1 = -x + y + z$$
$$-2 = -x - y$$

Solving, we get $x = 1$, $y = 1$ and $z = 1$. Hence $F = k\eta vr$

A detailed treatment, first done by Stokes, gives $k = 6\pi$ and so $F = 6\pi\eta vr$.

This expression, called Stokes' law, only holds for steady motion in a fluid of infinite extent (otherwise the walls and bottom of the vessel affect the resisting force).

TERMINAL VELOCITY

When a body is dropped in a viscous fluid, it is first accelerated and then its acceleration becomes zero and it attains a constant velocity called terminal velocity. Consider a small ball of radius r, which is gently dropped into the liquid of infinite extent. As the steel ball falls, it experiences three forces: the force of gravity W; the buoyant force F_B; the drag force F_D.

The free-body diagram of the ball is shown in Fig. 4.55. If ρ_0 be the density of the body and ρ be the density of the liquid, then

$$W = \frac{4}{3}\pi r^3 \rho_0 g$$

and $$F_B = \frac{4}{3}\pi r^3 \rho g$$

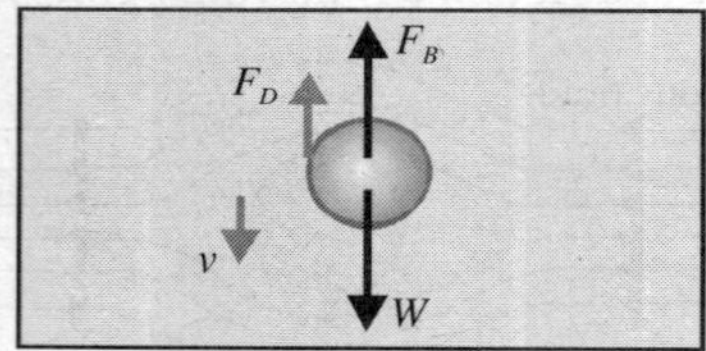

Free-body diagram of a ball falling under gravity in a viscous liquid.

Fig. 4.55

The magnitude of the drag force is $F_D = 6\pi\eta rv$.

As the ball falls under gravity, its net weight $W - F_B$ is opposed by liquid resistance F_D. Initially, the drag force is small and as the body gains velocity the drag force also increases and at a particular velocity, called the terminal velocity, the magnitude of drag force becomes equal to the effective weight of the body. At this moment, the net force on the ball becomes zero and it moves with a constant velocity. That is,

$$F_D = W - F_B$$

or $$6\pi\eta rv = \frac{4}{3}\pi r^3(\rho_0 - \rho)g$$

$$\Rightarrow \quad v = \frac{2}{9}\frac{r^2}{\eta}(\rho_0 - \rho)g \qquad \text{(i)}$$

If we plot the variation of velocity of the falling sphere with time, we obtain a graph as shown in Fig. 4.56. Following points can be noted:

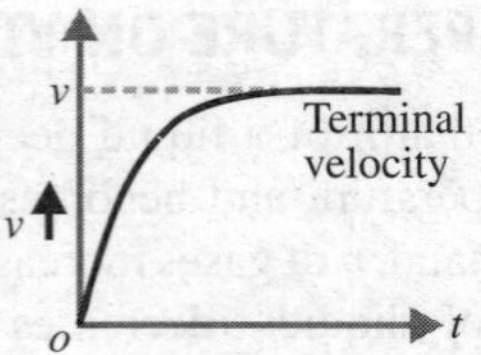

Variation of velocity of sphere with time

Fig. 4.56

1. ***Initially the velocity of the ball increases at a fast rate.***
2. ***The rate of increase of velocity decreases with time.***
3. ***Finally, the rate of increase of velocity becomes zero and the sphere acquires a terminal velocity.***

Important Point

Air bubble in water always goes up. It is because density of air (ρ) is less than the density of water (σ). So the terminal velocity for air bubble is negative, which implies that the air bubble will go up. Positive terminal velocity means the body will fall down.

Examples of Terminal Velocity

Some examples of terminal velocity are discussed as follows:

(i) **Determining the electronic charge by Millikan's experiment**: Stokes' formula is used in Millikan's method for determining the electronic charge. In this method, the formula is applied for finding out the radii of small oil drops by measuring their terminal velocities in air.

(ii) **Velocity of rain drops**: Rain drops are formed by the condensation of water vapour on dust particles. When they fall under gravity, their motion is opposed by the viscous drag in air. As the velocity of their fall increases, the viscous drag also increases and finally becomes equal to the effective force of gravity. The drops then attain a (constant) terminal velocity which is directly proportional to the square of the radius of the drops. In the beginning, the rain drops are very small in size and so they fall with such a small velocity that they appear floating in the sky as cloud. As they grow in size by further condensation, they reach the earth with appreciable velocity.

(iii) **Falling down with the help of a parachute**: When a person jumps down from a flying aeroplane, his parachute is closed. Hence, initially his velocity increases very rapidly while the viscosity of air tries to reduce his velocity. When the parachute opens, the viscosity of air exerts greater viscous force in upward direction (since viscous force is directly proportional to surface area), due to which the velocity of the person begins to decrease and finally he attains the terminal velocity and reaches safely on ground.

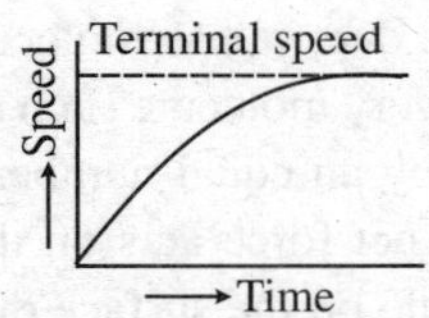

Fig. 4.57

(iv) **Formation of clouds**: When water vapours present in the atmosphere condense, small droplets are formed. The weight of these droplets in air is very small. Therefore, they attain the terminal velocity very soon due to viscosity of air. Because the value of their terminal velocity is very small, they appear to float in the sky as clouds.

Illustration 4.32 **A powder comprising particles of various sizes is stirred up in a vessel filled to a height of 10 cm with water. Assuming the particles to be spherical, find the size of the largest particle that will remain in suspension after 1 h (density of powder = 4 g/cm³, viscosity of water = 0.01 poise).**

Sol. Terminal velocity of the largest particle which is just about to settle at the bottom of the vessel is

$$v_t = \frac{10 \times 10^{-2}}{3600} \text{ m/s}$$

Let r be the radius of that particle. Then

$$v_t = \frac{2}{9} r^2 \frac{(\sigma - \rho)}{\eta} g$$

where $\sigma = 4 \times 10^3$ kg/m³ and $\eta = \frac{0.01}{10}$ Ns/m³

After solving, we get $r = 2.0 \times 10^{-6}$ m.

Illustration 4.33 **Spherical particles of pollen are shaken up in water and allowed to settle. The depth of water is 2 × 10⁻² m. What is the diameter of the largest particles remaining in suspension one hour later? Density of pollen = 1.8 × 10³ kg m⁻³, viscosity of water = 1 × 10⁻² poise and density of water = 1 × 10⁻⁵ kg m⁻³.**

Sol. Terminal velocity, $v = \frac{2r^2}{9} \frac{(\rho - \sigma) g}{\eta}$ (i)

But we know $v = \frac{s}{t}$

$$\therefore \quad \frac{s}{t} = \frac{2}{9} \frac{r^2 (\rho - \sigma) g}{\eta} \Rightarrow r^2 = \frac{9s}{2t} \frac{\eta}{(\rho - \sigma) g}$$

Given $s = 2 \times 10^{-2}$ m, $t = 1$ h $= 3600$ s

$$\therefore \quad \eta = 1 \times 10^{-2} \text{ poise} = 1 \times 10^{-3} \text{ kg m}^{-1} \text{ s}^{-1}$$

Substituting given values, we get

$$r^2 = \frac{9}{2} \times \frac{2 \times 10^{-2}}{3600} \times \frac{1 \times 10^{-3}}{(1.8 \times 10^3 - 1 \times 10^3) \times 10} = \frac{9}{36} \times \frac{1}{8} \times 10^{-10}$$

$$= \frac{1}{32} \times 10^{-10}$$

$$\therefore r = \sqrt{\frac{100}{32}} \times 10^{-6} \text{ m} = 1.77 \times 10^{-6} \text{ m}$$

Diameter $D = 2r = 2 \times 1.77$ μm $= 3.54$ μm

Illustration 4.34 **A small sphere falls from rest in a viscous liquid. Due to friction, heat is produced. Find the relation between the rate of production of heat and the radius of the sphere at terminal velocity. (IIT-JEE, 2004)**

Sol. Viscous force on a falling sphere in a liquid $F = \pi \eta r v_T$ where $v_T = (2/9) r^2 (\rho - \sigma) g/\eta$ is terminal velocity, ρ = density of sphere, σ = density of liquid.

Rate of production of heat (= power) $= F\, v_T = 6\pi\eta r v_T^2$

$$= 6\pi\eta r \left[\frac{2}{9} \frac{r^2 (\rho - \sigma) g}{\eta}\right]^2 = \frac{8}{27} \frac{\pi g^2 (\rho - \sigma)^2}{\eta} r^5$$

Clearly,

$$\frac{dQ}{dt} \propto r^5$$

Concept Application Exercise 4.2

1. A man is rowing a boat with a constant velocity 'v_0' in a river. The contact area of boat is 'A' and coefficient of viscosity is η. The depth of river is 'D'. Find the force required to row the boat.

2. A cubical block (of side 2 m) of mass 20 kg slides on inclined plane lubricated with the oil of viscosity $\eta = 10^{-1}$ poise with constant velocity of 10 m/s. Find out the thickness of the layer of liquid (take $g = 10$ m/s²).

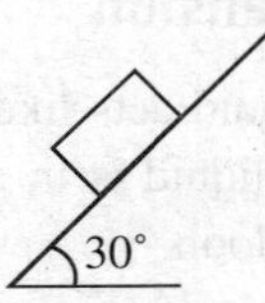

Fig. 4.58

3. A drop of water of radius 0.0015 mm is falling in air. If the coefficient of viscosity of air is 1.8 × 10⁻⁵ kg/(m s), what will be the terminal velocity of the drop? Density of water = 1.0 × 10³ kg/m² and g = 9.8 N/kg. Density of air can be neglected.

4. A metallic sphere of radius 1.0 × 10⁻³ m and density 1.0 × 10⁴ kg/m³ enters a tank of water, after a free fall through a distance of h in the earth's gravitational field. If its velocity

remains unchanged after entering water, determine the value of h. Given: coefficient of viscosity of water = 1.0×10^{-3} N s/m², g = 10 m/s² and density of water = 1.0×10^3 kg/m³.

5. Find the minimum force required to drag a hard polythene plate of area 2 m² on a thin film of oil of thickness 0.25 cm and η = 15 poise. Assume the speed of the plate is 10 cm/s.

6. A force of 3.14 N is required to drag a sphere of radius 4 cm with a speed of 5 m/s in a medium in gravity free space. Find the coefficient of the viscosity of the medium.

SURFACE TENSION

Surface tension is a property of liquid at rest by virtue of which a liquid surface gets contracted to a minimum area and behaves like a stretched membrane.

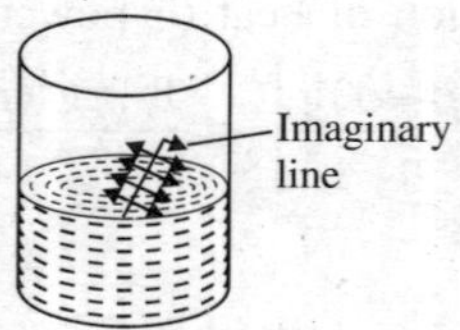

Fig. 4.59

Surface tension of a liquid is measured by force per unit length on either side of any imaginary line drawn tangentially over the liquid surface, force being normal to the imaginary line as shown in Fig. 4.59. Mathematically, surface tension is defined as

$$T = \frac{\text{Total force on either side of the imaginary line } (F)}{\text{Length of the line } (l)}$$

In CGS system the unit of surface tension is dyn/cm (dyn cm^{-1}) and in SI system its units is Nm^{-1}.

Causes of Surface Tension

The free surface of a liquid acts like a stretched membrane, i.e., is, the surface of a liquid is in a state of tension as the surface of an inflated balloon.

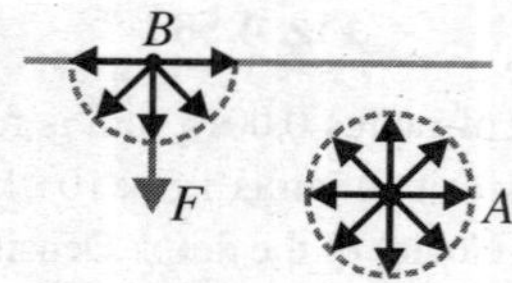

Intermolecular forces in liquids

Fig. 4.60

Surface tension is due to intermolecular attraction. In the volume of a liquid, every molecule such as A, as shown in Fig. 4.60, is surrounded by an equal number of molecules on all sides. Therefore, no net force acts on this molecule. On the other hand, a molecule on the surface of the liquid such as B is surrounded with a very few molecules on the vapour side as compared with the liquid below. Thus, it experiences a net force F in the downward direction, as shown in the figure. If one pulls the molecule B upwards, then one has to work against the intermolecular attraction. Therefore, molecules on the free surface of a liquid have potential energy. If the molecule A is shifted to the position of molecule B, then external work has to be done in breaking the bonds because at position A the molecule forms more bonds with the neighbours as compared to the position of molecule B. Thus, molecules on the surface of a liquid have more potential energy as compared to the molecules in the body of the liquid. Under the action of surface tension forces, the free surface of a liquid tends to have the least area for a given volume of liquid in order to minimize its potential energy. Since a sphere has minimum area for a given volume, therefore, rain drops and small droplets of mercury are approximately spherical in shape.

Note:

- ***The magnitude of T depends on the temperature of the liquid and on the medium on the other side of the free surface. The surface tension of a liquid decreases by increasing temperature.***
- ***The surface tension of a liquid decreases by adding impurities.***
- ***At 20°C, the surface tension of water in contact with air is 7.26×10^{-2} Nm^{-1}. At 20°C, the surface tension of mercury in contact with air is 46.5×10^{-2} Nm^{-1}. The surface tension of a water–olive oil boundary is 2.06×10^{-2} Nm^{-1}; and for mercury–water boundary it is 42.7×10^{-2} Nm^{-1}.***

Force of Cohesion

The force of attraction between the molecules of the same substance is called cohesion. In case of solids, the force of cohesion is very large and due to this solids have definite shape and size. On the other hand, the force of cohesion in case of liquids is weaker than that of solids. Hence, liquids do not have definite shape but have definite volume. The force of cohesion is negligible in case of gases. Because of this, gases have neither fixed shape nor fixed volume.

Examples:

(i) Two drops of a liquid coalesce into one when brought in mutual contact because of the cohesive force.

(ii) It is difficult to separate two sticky plates of glass wetted with water because a large force has to be applied against the cohesive force between the molecules of water.

(iii) It is very difficult to break a drop of mercury into small droplets because of large cohesive force between mercury molecules.

Force of Adhesion

The force of attraction between molecules of different substances is called adhesion.

Examples:

(i) Adhesive force enables us to write on the black board with a chalk.

(ii) Adhesive force helps us to write on the paper with ink.

(iii) Large force of adhesion between cement and bricks helps us in construction work.

(iv) Due to force of adhesion, water wets the glass plate.

(v) Fevicol and gum are used in gluing two surfaces together because of adhesive force.

Measuring of Surface Tension

The molecular theory tells that the surface molecules (molecules of the skin of the liquid surface) of a liquid attract each other more strongly than the molecules in the bulk of liquid. Now, let us draw an imaginary line AB on the free surface of the liquid. It will divide the surface of liquid into two parts. In consequence, we can imagine two rows 1 and 2 of surface molecules separated by the imaginary line AB. Since, each molecule in row 1 attracts the neighbouring molecules in row 2, the surface tension (cohesive) force is uniformly distributed over all molecules of the rows.

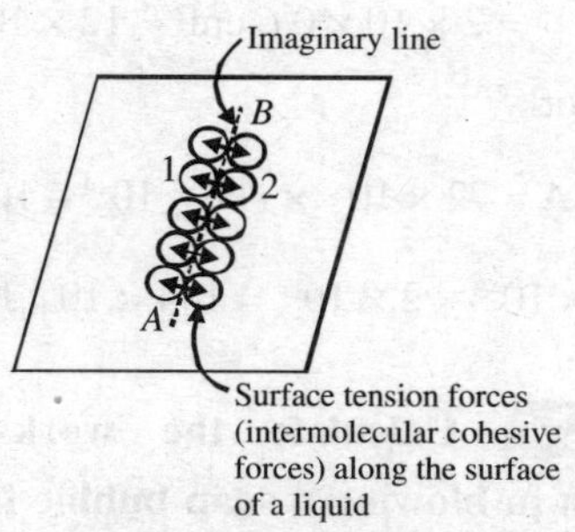

Fig. 4.61

If the total surface tension force acting on the rows of length l is F, for uniform distribution of the surface tension force, dividing F by l, we have 'F/l' which is constant quantity, defined as a measure of surface tension of liquid. We call it surface tension denoted by the letter T. Experimentally it is verified that surface tension of a liquid is constant at any temperature. If temperature increases, cohesive force decreases. Hence, surface tension decreases with temperature.

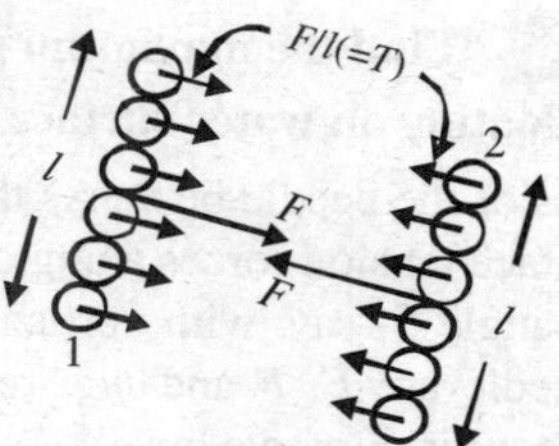

Fig. 4.62

The surface tension of a liquid is given as the linear density of surface tension force (cohesive force along the liquid surface) which is given as

$$T = \frac{F}{l}\left(\text{or } \frac{dF}{dl}\right)$$

T is coherent at a given temperature.

Since the direction of surface tension is perpendicular to any imaginary line, surface tension has no preferred direction. Hence, we can treat the surface tension T as a scalar, whereas force is a vector. Its unit is N/m or dyn/cm (same as the unit of stiffness of a spring).

Some Observed Phenomena Based on Surface Tension

1. Lead balls are spherical in shape.
2. Rain drops and a globule of mercury placed on glass plate are spherical.
3. Hair of a shaving brush/painting brush, when dipped in water spread out, but as soon as it is taken out, its hair stick together.
4. A greased needle placed gently on the free surface of water in a beaker does not sink.
5. Similarly, insects can walk on the free surface of water without drowning.
6. Bits of camphor gum move irregularly when placed on water surface.

Illustration 4.35 **A cube of mass $m = 800$ g floats on the surface of water. Water wets it completely. The cube is 10 cm on each edge. By what additional distance is it buoyed up or down by surface tension? Surface tension of water $= 0.07$ N m^{-1}.**

Sol. If surface tension is neglected the condition for floating gives

$$800 \times 10^{-3}\text{g} = (0.1^2 x\rho)g \quad \text{or} \quad x = 0.08 \text{ m}$$

$$(\rho = 1000 \text{ kg m}^{-3} \text{ for water})$$

Since water wets the cube, the angle of contact is zero and force of surface tension acts vertically downwards. So it is buoyed down by surface tension.

$$\therefore \quad 800 \times 10^{-3}\, g + 4 \times 0.1 \times 0.07 = (0.1^2 x'\rho)g$$

$$\text{or} \quad x' = 0.08 + \frac{0.028}{98} = 0.08 + 2.8 \times 10^{-4}$$

Therefore, the additional distance $= 2.8 \times 10^{-4}$ m

Illustration 4.36 **Find the maximum possible mass of a greased needle floating on water surface.**

Sol. Let the mass of the needle be m. As the liquid surface is distorted, the surface tension forces acing on both sides of the needle make an angle θ, say, with vertical. Since the forces acting on the needle are F, F and mg, resolving the forces vertically for its equilibrium, we have

$$\sum F_y = F\cos\theta + F\cos\theta - mg = 0$$

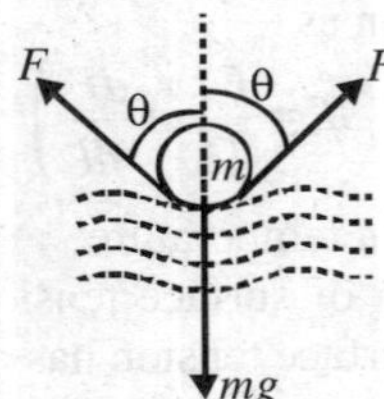

Fig. 4.63

This gives $m = \dfrac{2F\cos\theta}{g}$

where $F = Tl$

Then $m = \dfrac{2Tl\cos\theta}{g}$

For m to be maximum, $\cos\theta = 1$

Hence, $m_{max} = \dfrac{2Tl}{g}$

SURFACE ENERGY

We know that the molecules on the liquid surface experience net downward force. So to bring a molecule from the interior of the liquid to the free surface, some work is required to be done against the intermolecular force of attraction, which will be stored as potential energy of the molecule on the surface. The potential energy of surface molecules per unit area of the surface is called surface energy.

Unit of surface energy is erg cm^{-2} in CGS system and J m^{-2} in SI system. Dimensional formula of surface energy is $[ML^0T^{-2}]$. Surface energy depends on number of surfaces, e.g., a liquid drop is having one liquid air surface while bubble is having two liquid air surfaces.

Relation Between Surface Tension and Surface Energy

Consider a rectangular frame $PQSR$ of wire, whose arm RS can slide on the arms PR and QS. If this frame is dipped in a soap solution, then a soap film is produced in the frame $PQSR$ in Fig. 4.64. Due to surface tension (T), the film exerts a force on the frame (towards the interior of the film). Let l be the length of the arm RS. Then the force acting on the arm RS towards the film is $F = T \times 2l$ (since soap film has two surfaces, the length is taken twice).

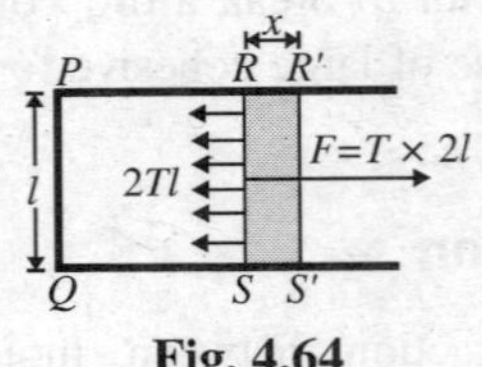

Fig. 4.64

Let the arm RS be displaced to a new position $R'S'$ through a distance x. Then work done, $W = Fx = 2Tlx$

Increase in potential energy of the soap film is give by

$$EA = 2Elx = \text{work done in increasing the area } (\Delta W)$$

where E = surface energy of the soap film per unit area.

According to the law of conservation of energy, the work done must be equal to the increase in the potential energy, i.e., $2Tlx = 2Elx$

or $$T = E = \frac{\Delta W}{A}$$

Thus, surface tension is numerically equal to surface energy or work done per unit increase of surface area.

Illustration 4.37 **A film of water is formed between two straight parallel wires each 10 cm long and at separation 0.5 cm. Calculate the work required to increase 1 mm distance between the wires. Surface tension of water = 72 × 10^{-3} N/m.**

Sol. Initial surface area = 2 × length × separation

$= 2 \times 10\text{ cm} \times 0.5\text{ cm} = 10\text{ cm}^2 = 10 \times 10^{-4}\text{ m}^2$

Final surface area $= 2 \times 10\text{ cm} \times (0.5 + 0.1)\text{ cm}$

$= 2 \times 10 \times 0.6\text{ cm}^2 = 12 \times 10^{-4}\text{ m}^2$

The required work,

$$W = T\Delta A = 72 \times 10^{-3} \times (12 \times 10^{-4} - 10 \times 10^{-4})\text{ J}$$

$$= 72 \times 10^{-3} \times 2 \times 10^{-4} = 144 \times 10^{-7}\text{ J}$$

Illustration 4.38 **Calculate the work done against surface tension in blowing a soap bubble from a radius of 10 cm to 20 cm, if the surface tension of soap solution in 25 × 10^{-3} N/m.**

Sol. Original total surface area $= 2 \times 4\pi r_1^2 = 2 \times 4\pi \times (0.1)^2\text{m}^2$ (as bubble has two surfaces).

Final total surface area $= 2 \times 4\pi r_2^2 = 2 \times 4\pi \times (0.2)^2\text{ m}^2$

Therefore, extension in area $= 2 \times 4\pi[(0.2)^2 - (0.1)^2] = 0.24\pi\text{ m}^2$

Now, work done W = surface tension × extension in area

$= 25 \times 10^{-3} \times 0.14\pi = 6\pi \times 10^{-3}\text{ J}$

ANGLE OF CONTACT

The angle which the tangent to the liquid surface at the point of contact makes with the solid surface inside the liquid is called angle of contact. Those liquids, which wet the walls of the container (say in case of water and glass), have meniscus concave upwards and their value of angle of contact is less than 90° (also called acute angle). However, those liquids, which do not wet the walls of the container (say in case of mercury and glass), have meniscus convex upwards and their value of angle of contact is greater than 90° (also called obtuse angle). The angle of contact of mercury with glass is about 140°, whereas the angle of contact of water with glass is about 8°. But, for pure water, the angle of contact θ with glass is taken as 0°.

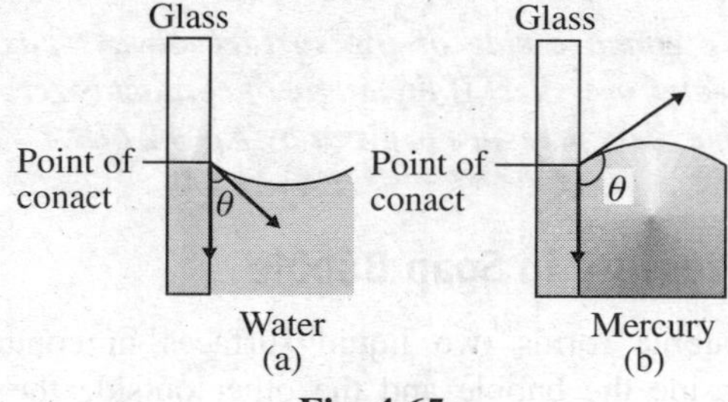

Fig. 4.65

Shape of Liquid Meniscus

When a capillary tube or a tube is dipped in a liquid, the liquid surface becomes curved near the point of contact. This curved surface is due to two forces, viz., (i) the force of cohesion and (ii) the force of adhesion. The curved surface of the liquid is called *meniscus of the liquid*. Various forces acting on molecule A (Fig. 4.66) are as follows:

(i) Force F_1 due to force of adhesion which acts outwards at right angle to the wall of the tube. This force is represented by AB.

(ii) Force F_2 due to force of cohesion which acts at an angle of 45° to the vertical. This force is represented by AD.

(iii) The weight of the molecule A which acts vertically downward along the wall of the tube.

Since the weight of the molecule is negligible as compared to F_1 and F_2, so it can be neglected. Thus, there are only two forces (F_1 and F_2) acting on the liquid molecules. These forces are inclined at an angle of 135°.

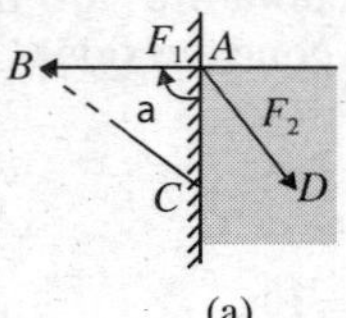

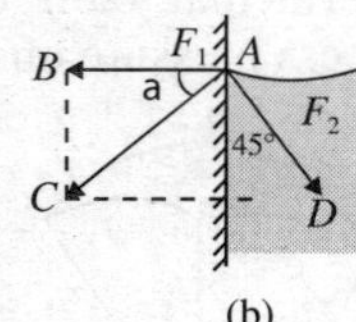

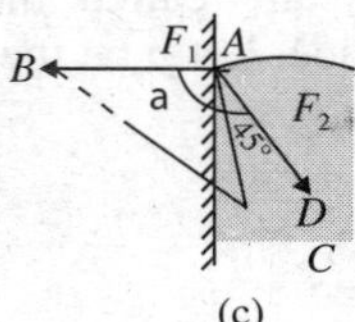

Fig. 4.66

The resultant force represented by AC will depend upon the values of F_1 and F_2. Let the resultant force make an angle α with F_1.

Special Cases

i If $F_2 = \sqrt{2}\ F_1$, then $\tan \alpha = \infty$, $\therefore \alpha = 90°$.
Then the resultant force will act vertically downward and hence the meniscus will be plane or horizontal as shown in Fig. 4.66(a). For example, pure water contained in silver capillary tube.

ii If $F_2 < \sqrt{2}\ F_1$, then $\tan \alpha$ is positive. Hence, α is the acute angle. Thus, the resultant will be directed outside the liquid and hence the meniscus will be concave upward as shown in Fig. 4.66(b). This is possible in case of liquids which wet the walls of the capillary tube. For example, water in glass capillary tube.

iii If $F_2 > \sqrt{2}\ F_1$, then $\tan \alpha$ is negative. Therefore, α is obtuse angle. Thus, the resultant will be directed inside the liquid and hence the meniscus will be convex upward as shown in Fig. 4.66(c). This is possible in case of liquids which do not wet the walls of the capillary tube. For example, mercury in glass capillary tube.

EXCESS PRESSURE INSIDE A CURVED SURFACE

Some examples of plane and curved surfaces (cavities, drops and bubbles) are shown in Fig. 4.67.

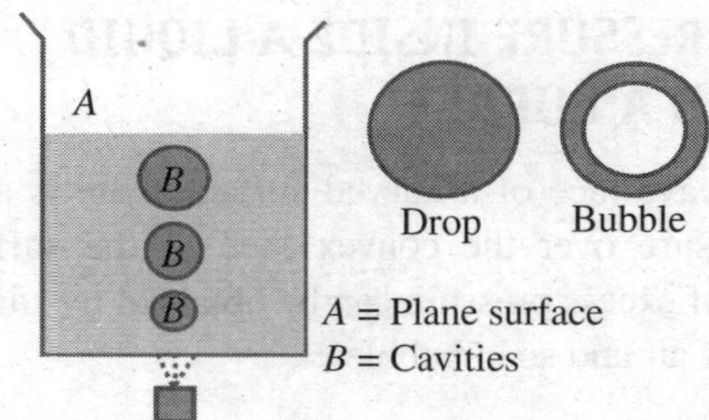

Fig. 4.67

The bubbles, such as soap bubbles, are like blown-up balloons, air inside and air outside with a thin liquid film in between them. This thin film naturally has two free surfaces, one inside and the other outside. Cavities generally have air inside and liquid outside. They have one interface or free surface. Drops generally have water or some other liquid inside and air or some other gas outside them. They too have one exposed surface.

Plane Surface

If the surface of the liquid is plane [as shown in Fig. 4.68(a)], the molecule on the liquid surface is attracted equally in all directions. The resultant force due to surface tension is zero. The pressure, therefore, on the liquid surface is normal.

Concave Surface

If the surface is concave upwards [as shown in Fig. 4.68(b)], there will be upward resultant force due to surface tension acting on the molecule. Since the molecule on the surface is in equilibrium, there must be an excess of pressure on the concave side

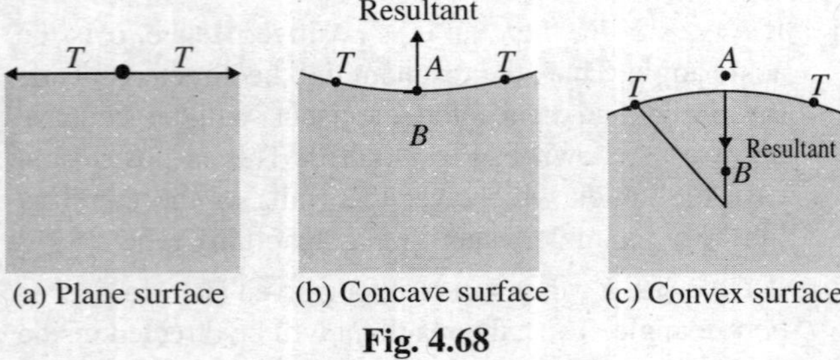

(a) Plane surface (b) Concave surface (c) Convex surface

Fig. 4.68

Convex Surface

If the surface is convex [as shown in Fig. 4.68(c)], the resultant force due to surface tension acts in the downward direction. Since the molecules on the surface are in equilibrium, there must be an excess of pressure on the concave side of the surface acting in the upward direction to balance the downward resultant force of surface tension. Hence, there is always an excess of pressure on concave side of a curved surface over that on the convex side.

EXCESS PRESSURE INSIDE A LIQUID DROP AND A BUBBLE

On the concave face of a curved surface there is always an excess pressure over the convex face of the surface. The magnitude of excess pressure can be obtained by studying the formation of air and soap bubbles.

Excess Pressure in an Air Bubble in Liquid

Figure 4.69 shows one-half cross section of an air bubble formed inside liquid. It is in equilibrium under the action of three forces:

- ***due to external pressure p_1***
- ***due to internal pressure p_2***
- ***due to surface tension of the liquid σ***

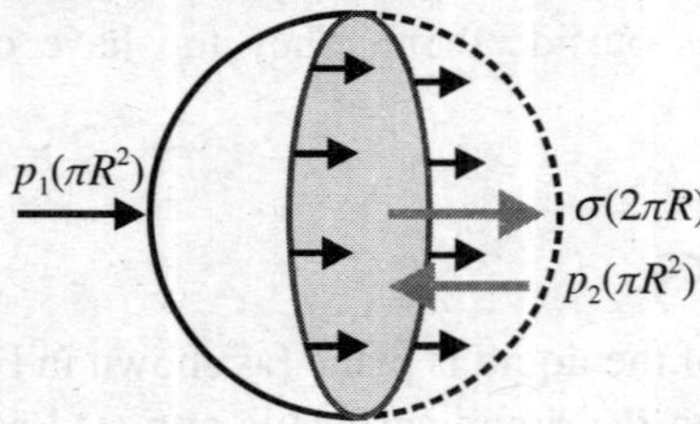

The cross section of an air bubble of radius *R*.

Fig. 4.69

If *R* is the radius of the air bubble, then the forces due to external and internal pressures are $p_1(\pi R^2)$ and $p_2(\pi R^2)$, respectively. Since the surface tension acts around the circumference of the bubble, therefore, the force of surface tension is $\sigma(2\pi R)$.

Thus, from the condition of equilibrium,

$$p_2(\pi R^2) = p_1(\pi R^2) + T(2\pi R)$$

or
$$p_2 - p_1 = \frac{2T}{R}$$

> **Note:**
>
> - ***A curved surface always maintains a pressure difference, which depends on the radius of the curved surface and surface tension of the liquid.***
> - ***The concave side of the surface always possesses greater pressure. If the angle of contact is zero, then the excess pressure is given by $\Delta p = 2T/R$.***

Excess Pressure in Soap Bubble

A soap bubble forms two liquid surfaces in contact with air, one inside the bubble and the other outside the bubble. Figure 4.70 shows one-half cross section of the soap bubble. By considering its equilibrium, we get

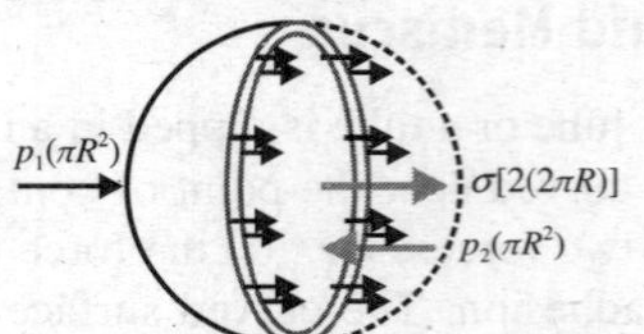

The cross section of an air bubble of radius *R*.

Fig. 4.70

$$p_2(\pi R^2) = p_1(\pi R^2) + T[2(2\pi R)]$$

$\Rightarrow$
$$p_2 - p_1 = \frac{4T}{R}$$

Relation Between Surface Tension, Radii of Curvature and Excess Pressure on a Curved Surface

Let us consider a small element *ABCD* (Fig. 4.71) of a curved liquid surface which is convex on the upper side. R_1 and R_2 are the maximum and minimum radii of curvature, respectively. They are called the 'principal radii of curvature' of the surface. Let *p* be the excess pressure on the concave side.

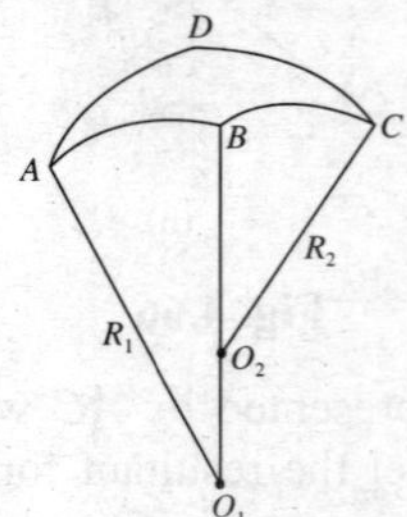

Fig. 4.71

Then $$p = T\left(\frac{1}{R_1}+\frac{1}{R_2}\right)$$

If instead of a liquid surface, we have a liquid film, the above expression will be

$$p = 2T\left(\frac{1}{R_1}+\frac{1}{R_2}\right)$$

because a film has two surfaces.

Excess Pressure on Curved Surfaces in General

The pressure on the concave side (whatever be on its left or right) is always greater than the pressure on the convex side, such that

$$p_{\text{concave}} - p_{\text{convex}} = T\left(\frac{1}{R_1}+\frac{1}{R_2}\right)$$

where σ = surface tension.

This formula is same as what we derived earlier, i.e.,

$$p_i - p_0 = \frac{2T}{R} = T\left(\frac{1}{R}+\frac{1}{R}\right)$$

(i) For a single spherical surface, e.g. drops, cavities, etc.,

$$R_1 = R_2 = R(\text{say})$$

Then $$p_i - p_0 = T\left(\frac{1}{R}+\frac{1}{R}\right) = \frac{2T}{R}$$

(ii) For spherical film, e.g., soap bubble,

$$p_i - p_0 = (2T)\left(\frac{1}{R}+\frac{1}{R}\right) = \frac{4T}{R}$$

because it has two free surfaces.

(iii) For a single cylindrical surface, $R_1 = R, R_2 = \infty$

$$\therefore \quad p_i - p_0 = (T)\left(\frac{1}{R}+\frac{1}{\infty}\right) = \frac{T}{R}$$

Illustration 4.39 **Calculate the difference (h) in water levels in two communicating capillary tubes of radius 1 mm and 1.5 mm. Surface tension of water = 0.07 Nm⁻¹.**

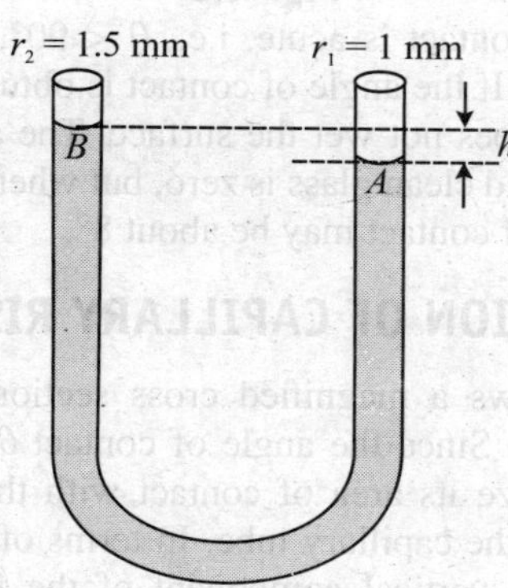

Fig. 4.72

Sol. Pressure at $A = p_0 + \frac{2T}{r_2}$

(Since, pressure inside a curved surface is greater than that outside)

Pressure at $B = p_0 + \frac{2T}{r_1}$

Therefore, pressure difference $= 2T\left(\frac{1}{r_1}-\frac{1}{r_2}\right)$

Let this pressure difference correspond to h units of the liquid. Then,

$$2T\left(\frac{1}{r_1}-\frac{1}{r_2}\right) = \rho g h \Rightarrow h = \frac{2T}{\rho g}\left(\frac{1}{r_1}-\frac{1}{r_2}\right)$$

$$\therefore \quad h = \frac{2\times 0.07}{1000\times 9.8}\left(\frac{1}{1\times 10^{-3}}-\frac{1}{1.5\times 10^{-3}}\right) = 4.76\,\text{mm}$$

Illustration 4.40 **A vessel filled with air under pressure p_0 contains a soap bubble of diameter d. The air pressure have been reduced n-fold, and the bubble diameter increased r-fold isothermally. Find the surface tension of the soap–water solution.**

Sol. For the air enclosed in the bubble we have

$$pV = nRT$$

$$\Rightarrow \quad \left(p_0 + \frac{8T}{d}\right)\frac{\pi}{6}d^3 = \left(\frac{p_0}{n}+\frac{8T}{rd}\right)\frac{\pi}{6}r^3 d^3$$

$$\Rightarrow \quad p_0 + \frac{8T}{d} = \frac{p_0 r^3}{n} + \frac{8Tr^2}{d}$$

$$\Rightarrow \quad p_0\left(1-\frac{r^3}{n}\right) = \frac{8T}{d}(r^2-1) \Rightarrow T = \frac{1}{8}p_0 d \times \frac{1-\frac{r^3}{n}}{r^2-1}$$

$$\Rightarrow \quad p_0\left(1-\frac{r^3}{n}\right) = \frac{8T}{d}(r^2-1) \Rightarrow T = \frac{1}{8}p_0 d \times \frac{1-\frac{r^3}{n}}{r^2-1}$$

Illustration 4.41 **What should be the pressure inside a small air bubble of 0.1 mm radius situated just below the surface of water? Surface tension of water = 72 × 10⁻³ N/m and atmospheric pressure = 1.013 × 10⁵ N/m².**

Sol. The excess pressure for air bubble is

$$p = \frac{2T}{r}$$

or $$p = \frac{2\times 72\times 10^{-3}}{0.1\times 10^{-3}} = 1440\ \text{N/m}^3$$

Since the bubble is just below the water surface, the external pressure on it is equal to the atmospheric pressure P. Hence the pressure inside the bubble is

$$P + p = 1.013 \times 10^5 + 1440 = 1.013 + 10^5 + 0.0144 \times 10^5$$
$$= 1.274 \times 10^5 \text{ N/m}^3$$

Illustration 4.42 **Two separate air bubbles (radii 0.002 cm and 0.004 m) formed of the same liquid (surface tension 0.07 N/m) come together to form a double bubble. Find the radius and the sense of curvature of the internal film surface common to both the bubbles.**

Sol. Let R be the radius of curvature of common surface when bubbles A and B of radii R_A and R_B coalesce. The excess pressure in A and B are $4T/R_A$ and $4T/R_B$, respectively.

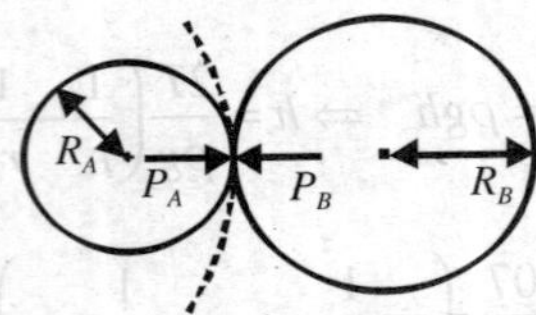

Fig. 4.73

$$\therefore \quad p_A = \frac{4T}{R_A} \text{ and } p_B = \frac{4T}{R_B}$$

If R is radius of common interface, then we must have

$$p_A - p_B = \frac{4T}{R}$$

or
$$\frac{4T}{R_A} - \frac{4T}{R_B} = \frac{4T}{R}$$

This gives

$$R = \frac{R_A R_B}{R_B - R_A} = \frac{0.002 \times 0.004}{0.004 - 0.002} = 0.004 \text{ m}$$

As the excess pressure is always towards concave surface and pressure in smaller bubble is greater than that in larger bubble, the common surface is concave towards the centre of the smaller bubble.

CAPILLARITY

If a narrow glass tube open at both ends is pushed in water as shown in Fig. 4.74, water rises in the tube to a height above its surface. The narrower the tube, the greater is the height to which water rises. This phenomenon is known as capillarity. When the same capillary tubes are placed in mercury, the liquid is depressed below the outside level. The depression increases as the diameter of the capillary tube decreases. A close observation of the capillary phenomenon reveals the following facts.

The free surface (meniscus) of the liquid which rises in the capillary tube is concave upward. The meniscus of liquid which falls in the capillary tube is convex upward.

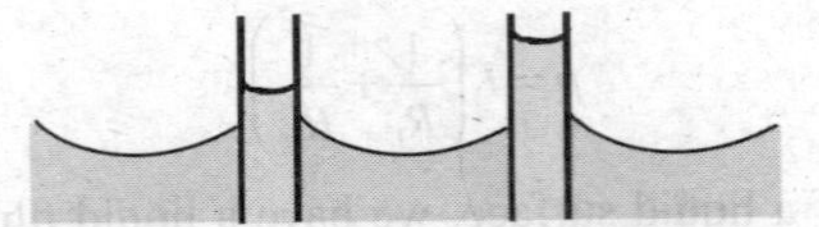
Capillary action with water. Narrower the tube, greater the height.

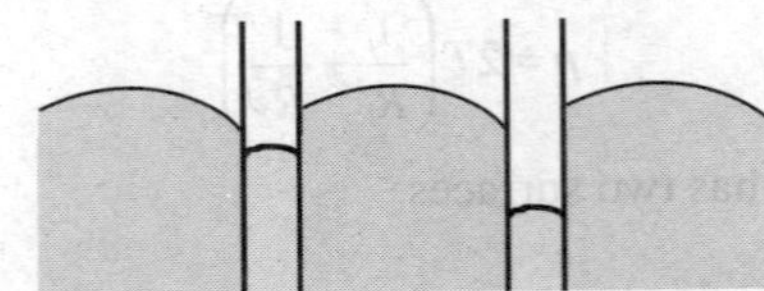
Capillary action with mercury. Narrower the tube, greater the depression in the tube.

Fig. 4.74

Wetting and Non-Wetting Liquids

The liquids which rise in a capillary are called wetting liquids; and the liquids which fall in the capillary are called non-wetting liquids. The wetting and non-wetting action of a liquid can be explained in terms of the cohesive and adhesive forces.

Adhesion is the attractive force between the molecules of solids and liquids or between the molecules of two different liquids.

Cohesion is the attractive force among the molecules of the same liquid.

If the adhesive forces are stronger than the cohesive forces then the liquids wet the solid surface, as water wets the surface of glass. If the cohesive forces are stronger than the adhesive forces, then the liquids do not wet the solid surface, as mercury does not wet the surface of glass.

The wetting and non-wetting action of a liquid can also be explained in terms of the angle of contact. If a tangent is drawn on the meniscus of the liquid at the line of contact between the liquid and the surface, as shown in Fig. 4.75, then the angle of contact is defined as the angle between the tangent and the solid surface measured through the liquid.

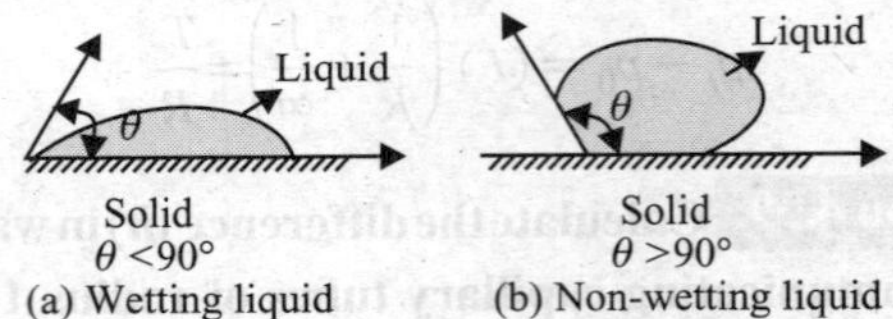

Fig. 4.75

If the angle of contact is acute, i.e., $\theta < 90°$, then the liquid wets the surface. If the angle of contact is obtuse, i.e., $\theta > 90°$, then the liquid does not wet the surface. The angle of contact between water and clean glass is zero, but when the glass is not clean the angle of contact may be about 8°.

DETERMINATION OF CAPILLARY RISE

Figure 4.76 shows a magnified cross section of a capillary tube of radius R. Since the angle of contact θ is acute, water tends to maximize its area of contact with the glass surface. Thus, it rises in the capillary tube. In terms of forces, one can imagine that the vertical component of the force of surface tension pulls the liquid up in the tube to a height such that this force is able to balance the weight of the liquid in the tube.

Since the force of surface tension acts along the circumference of the free surface of the liquid, therefore, it is given by $T = \sigma(2\pi R)$. The weight of water inside the tube is given by

$$W = (V + V')\rho g$$

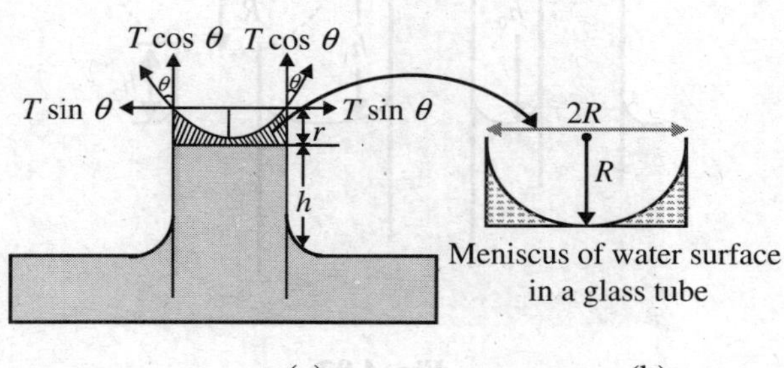

Fig. 4.76

where V = volume of the cylinder of radius R and height h, and V' is the volume of water lying below the meniscus and above the cylinder of height h.

$$\therefore \quad V = \pi R^2 h \Rightarrow V' = (\pi R^2)R - \frac{2}{3}\pi R^3 \quad \Rightarrow V' = \frac{1}{3}\pi R^3$$

Since V' is negligible with respect to V,

$$W = (\pi R^2 h)\rho g$$

By using the condition of equilibrium, we get

$$T\cos\theta = W \quad \Rightarrow (2\pi R\sigma)\cos\theta = (\pi R^2 h)\rho g \quad \Rightarrow \sigma = \frac{\rho g R h}{2\cos\theta}$$

Since the angle of contact between glass and water is 8°, $\cos\theta \approx 1$. Thus

$$\sigma = \frac{1}{2}\rho g R h$$

Alternative method:

Let us take four points A, B, C and D as shown in Fig. 4.77. A and B are the points just above and just below the meniscus, respectively.

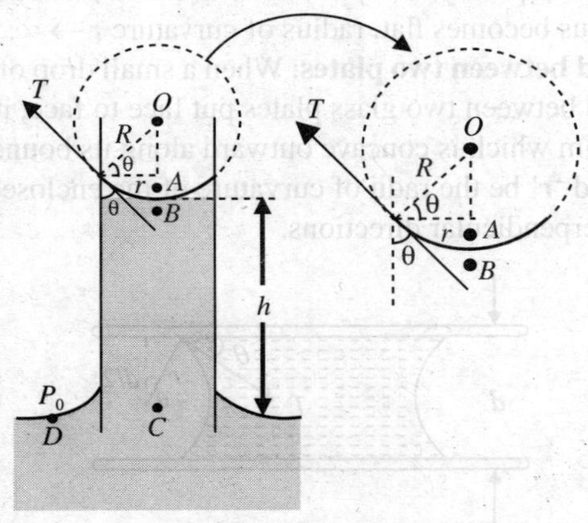

Fig. 4.77

Now we have, $P_A = P_B + \dfrac{2T}{R}$

where R = radius of curvature of liquid meniscus. Substituting $P_A = P_0$, we have

$$P_B = P_0 - \frac{2T}{R} \qquad \text{(i)}$$

If C is the point at the bottom of the excess liquid column, we have $P_C = P_B + \rho g h$

where h = height of the excess liquid column.

Substituting $P_C = P_D = P_0$, we have

$$P_B = P_0 - \rho g h \qquad \text{(ii)}$$

Using Eqs. (i) and (ii), we have

$$P_0 - \rho g h = P_0 - \frac{2T}{R}$$

This gives $h = \dfrac{2T}{\rho g R}$ where $R = \dfrac{r}{\cos\theta}$

Then $h = \dfrac{2T\cos\theta}{\rho g r}$

Important Points

- h depends on surface tension T, angle of contact θ, radius of the tube, density ρ of liquid and acceleration due to gravity (or effective g). Hence the capillary action depends on the nature of liquid and solid (capillary tube) in contact, temperature of the surroundings.
- If $\theta < 90°$, h is positive (wetting) which signifies capillary rise and concave meniscus; if $\theta = 90°$, $h = 0$ (neither wetting nor non-wetting): plane meniscus; if $\theta > 90°$, h = negative (non-wetting) which signifies capillary fall and convex meniscus.

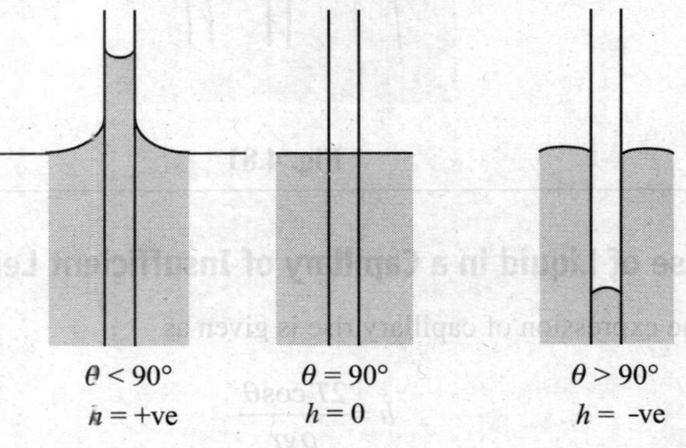

Fig. 4.78

- Since T, ρ and θ are constant at a given temperature, for any given liquid and capillary tube, $hr = c$; this tells us that capillary rise (or fall) will be greater in thinner tubes and vice versa.

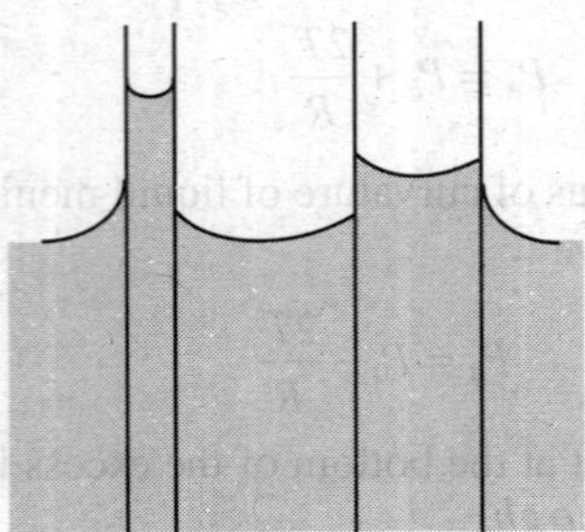
Fig. 4.79

- Since T, ρ, θ, g and R are constant for a given set of capillary and liquid, we have h = constant. It means that if we change the angle β of orientation from vertical, the capillary rise h (or capillary fall) remains constant even though the length of the excess liquid in the tube increases from h to $h \sec \beta$.

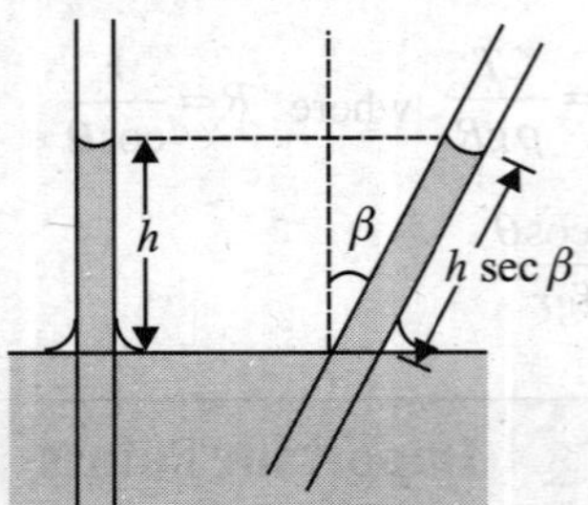

Fig. 4.80

- If the radius of the liquid meniscus remains constant, the weight (but not length) of the liquid column remains the same in capillary tubes of different shapes and sizes.

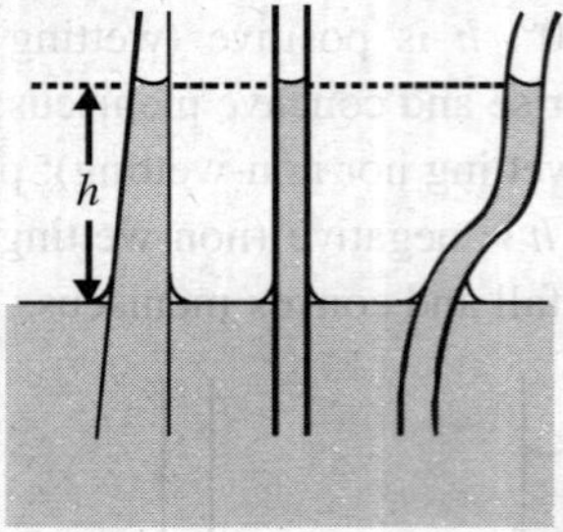

Fig. 4.81

Rise of Liquid in a Capillary of Insufficient Length

The expression of capillary rise is given as

$$h = \frac{2T\cos\theta}{\rho g r}$$

Substituting $r/\cos\theta = R$ (radius of curvature of the liquid meniscus), we have

$$hR = \frac{2T}{\rho g} = \text{Constant}$$

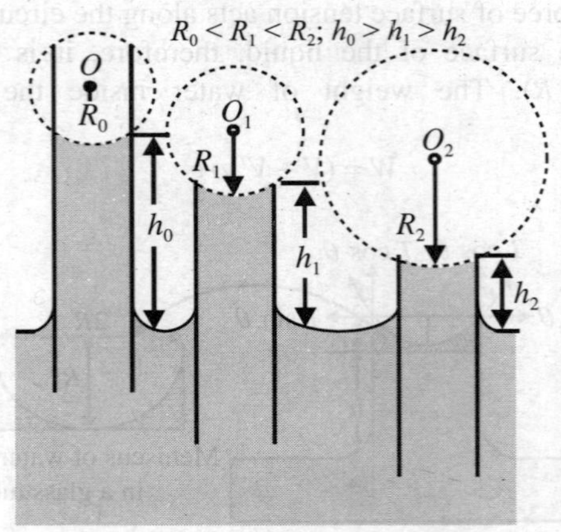

Fig. 4.82

Since T, ρ and g are constant, we have $hR = C$ (constant) $= h_0R_0$ (say), where h = free height of the liquid and R = radius of curvature of free meniscus of liquid.

If we slowly push the capillary tube vertically down (into) the liquid, the height of the liquid column remains the same till the length of the tube above the liquid surface is equal to $2T \cos \theta/(\rho g r)(= h)$. Therefore, the liquid meniscus gets flattened gradually by increasing its radius of curvature to obey the law hR = constant.

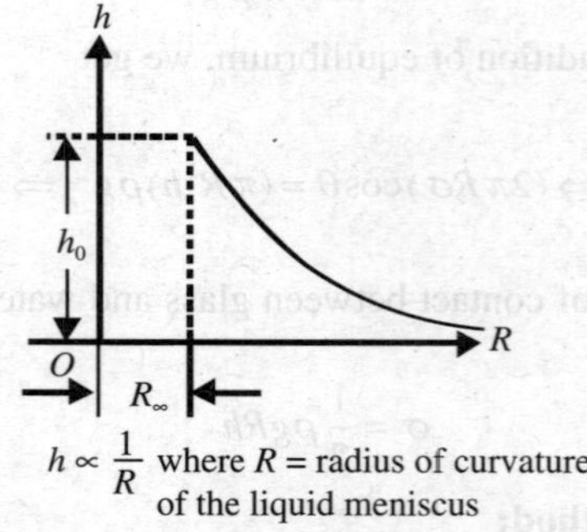

Fig. 4.83

It means that $hr = h_1r_1 = h_2r_2 = \cdots$

When the capillary tube just sinks, i.e., $h = 0$, we can see that the meniscus becomes flat, radius of curvature $r \to \infty$.

Liquid between two plates: When a small drop of water is placed between two glass plates put face to face, it forms a thin film which is concave outward along its boundary. Let 'R' and 'r' be the radii of curvature of the enclosed film in two perpendicular directions.

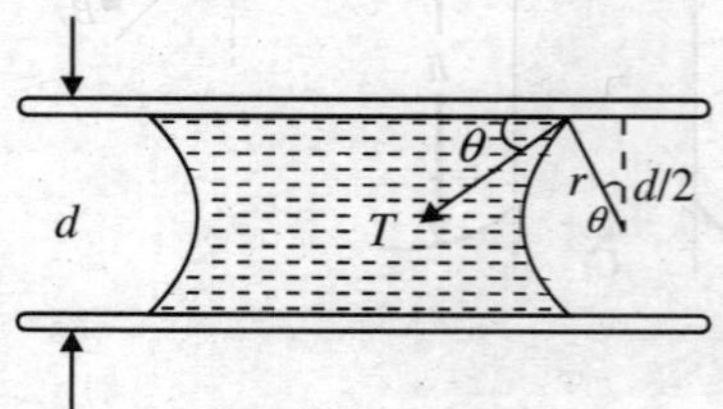

Fig. 4.84

Hence, the pressure inside the film is less than the atmospheric pressure outside it by an amount p given by

$$p = T\left(\frac{1}{r} + \frac{1}{R = \infty}\right)$$

and we have $\quad p = \dfrac{T}{r}$ (i)

If d be the distance between the two plates and θ be the angle of contact for water and glass, then, from the figure,

$$\cos\theta = \frac{\frac{1}{2}d}{r} \quad \text{or} \quad \frac{1}{r} = \frac{2\cos\theta}{d}$$

Substituting for $1/r$ in (i), we get $p = \dfrac{2T}{d}\cos\theta$

θ can be taken zero for water and glass, i.e., $\cos\theta = 1$. Thus, the upper plate is pressed downward by the atmospheric pressure minus $2T/d$. Hence, the resultant downward pressure acting on the upper plate is $2T/d$. If A be the area of the plate wetted by the film, the resultant force F pressing the upper plate downward is given by F = resultant pressure × area = $2TA/d$. For very nearly plane surface, d will be very small and hence the pressing force F is very large. Therefore, it will be difficult to separate the two plates normally.

Illustration 4.43 **A vertical capillary with inside diameter 0.50 mm is submerged into water so that the length of its part emerging outside the water surface is equal to 25 mm. Find the radius of curvature of the meniscus. Surface tension of water is 73 × 10⁻³ N/m.**

Sol. In the capillary tube, the water should rise to a height

$$h = \frac{2T}{r\rho g}$$

Here

$$T = 73 \times 10^{-3}\ \text{N/m}$$

$$r = \frac{0.50\ \text{mm}}{2} = 0.25\times10^{-3}\ \text{m}$$

$$\rho = 10^3\ \text{kg/m}^3$$

$$\therefore h = \frac{2\times73\times10^{-3}}{0.25\times10^{-3}\times10^3\times9.8} = 59\times10^{-3}\ \text{m} = 59\ \text{mm}$$

Now $h > h'$

i.e., length is outside water surface.

Therefore, radius of meniscus > radius of capillary r.

If R is the radius of meniscus, then we have

$$\frac{2T}{R} = h'\rho g$$

or $$R = \frac{2T}{h'\rho g}$$

Here $h' = 25\ \text{mm} = 25\times10^{-3}\ \text{m}$

$$\therefore \quad R = \frac{2\times73\times10^{-3}}{25\times10^{-3}\times10^3\times9.8} = 0.6\times10^{-3}\ \text{m} = 0.6\ \text{mm}$$

Illustration 4.44 **A glass rod of diameter d_1 = 1.5 mm is inserted symmetrically into a glass capillary with inside diameter d_2 = 2.0 mm. Then the whole arrangement is vertically oriented and brought in contact with the surface of water. To what height will the liquid rise in the capillary? Surface tension of water = 73 × 10⁻³ N/m.**

Sol. If R is radius of meniscus, then $\dfrac{2T}{R} = h\rho g$

Here $\quad R = \dfrac{r_2 - r_1}{\cos\theta}$

θ being angle of contact, r_1 = radius of glass rod, r_2 = radius of capillary.

$$\frac{2T\cos\theta}{r_2 - r_1} = h\rho g$$

or $$h = \frac{2T\cos\theta}{(r_1 - r_1)\rho g}$$

Here $$r_1 = d_1/2,\ r_2 = \frac{d_2}{2}.$$

$$\therefore \quad h = \frac{4T\cos\theta}{(d_2 - d_1)\rho g}$$

Substituting given values and $\theta \cong 0°$ for water–glass interface, we have

$$h = \frac{4\times73\times10^{-3}\cos0°}{(2.0-1.5)\times10^{-3}\times10^3\times9.8} = 60\times10^{-3}\ \text{m} = 6\ \text{cm}$$

Illustration 4.45 **The end of a capillary tube with a radius r is immersed in water. Is mechanical energy conserved when the water rises in the tube? The tube is sufficiently long. If not, calculate the energy change.**

Sol. In the equilibrium position (θ = 0° for water and glass)

$$2\pi rT\cos0° = \pi r^2 h\rho g$$

or $$h = \frac{2T}{\rho g r}$$

Work done by surface tension = $(2\pi rT)\times h = \dfrac{4\pi T^2}{\rho g}$

The potential energy of water in the tube, $U = (\pi r^2 h\rho)gh/2$; it is multiplied by $h/2$ because the centre of gravity of the water and the capillary tube is at a height $h/2$

$$\therefore \quad U = \frac{2\pi T^2}{\rho g}$$

Thus, it is seen that the mechanical energy is not conserved.

Therefore, mechanical energy loss $= \dfrac{4\pi T^2}{\rho g} - \dfrac{2\pi T^2}{\rho g} = \dfrac{4\pi T^2}{\rho g}$

This energy is converted into heat.

Illustration 4.46 **A glass capillary tube of internal radius $r = 0.25$ mm is immersed in water. The top end of the tube projects by 2 cm above the surface of water. At what angle does the liquid meet the tube? Surface tension of water $= 0.7\ \text{N m}^{-1}$.**

Sol. Water wets glass and so the angle of contact is zero. Neglecting the small mass in the meniscus, for full rise,

$$2\pi rT = \pi r^2 h\rho g$$

or $\quad h = \dfrac{2T}{r\rho g} = \dfrac{2\times 0.07}{0.25\times 10^{-3}\times 1000\times 9.8} = 0.057\ \text{m} = 5.7\ \text{cm}$

But here the tube is only 2 cm above the water and so water will rise by 2 cm and meet the tube at an angle such that

$$2\pi rT\cos 0^0 = \pi r^2 h'\rho g$$

$$\Rightarrow\ 2T\cos\theta = h'r\rho g \quad \Rightarrow \cos\theta = \frac{h'r\rho g}{2T}$$

$$\therefore\quad \cos\theta = \frac{2\times 10^{-2}\times 0.25\times 10^{-3}\times 1000\times 9.8}{2\times 0.07} \quad \Rightarrow \theta = 70^\circ$$

Illustration 4.47 **Derive an expression for the height of capillary rise between two parallel plates dipping in a liquid of density σ separated by a distance d. The surface tension of the liquid is T.**

Sol.

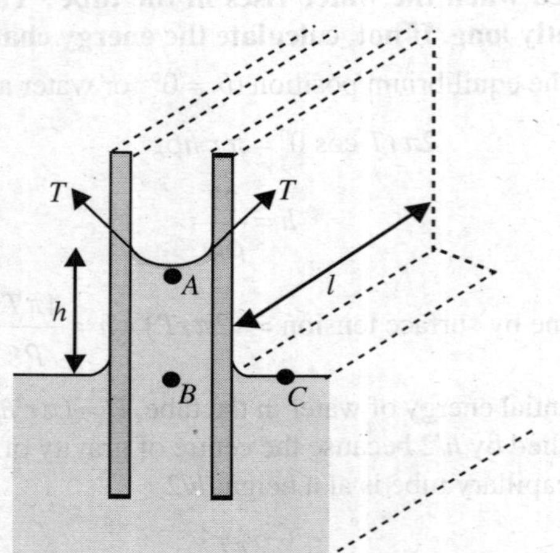

Fig. 4.85

The meniscus between two plates is cylindrical in shape. Pressure at A (the lowest point of the meniscus)

$$p_A = p_0 - T/r$$

Pressure at B = pressure at $C = p_0$ = pressure at $A + \rho gh$

$$\therefore\quad p_0 = p_0 - \frac{T}{r} + \rho gh \quad \times h = \frac{T}{\rho gr} = \frac{2T}{\rho gd}$$

Alternative method:

Force upward $= 2lT\cos\theta = 2lT \quad (\because \theta = 0^0)$

Gravitational pull = (Volume × Density)$g = lhd\rho g$

$$\therefore\quad 2lT = lhd\rho g \quad \Rightarrow h = \frac{2T}{d}$$

Applications of Surface Tension

Certain applications of surface tension are discussed as follows:

(i) The wetting property is made use of in detergents and waterproofing. When the detergent materials are added to liquids, the angle of contact decreases and hence the wettability increases. On the other hand, when waterproofing material is added to a fabric, it increases the angle of contact, making the fabric water-repellant.

(ii) The antiseptics have very low value of surface tension. The low value of surface tension prevents the formation of drops that may otherwise block the entrance to skin or a wound. Due to low surface tension the antiseptic spreads properly over the wound. The lubricating oils and paints also have low surface tension. So they can spread properly.

(iii) Surface tension of all lubricating oils and paints is kept low so that they spread over a large area.

(iv) Oil spreads over the surface of water because the surface tension of oil is less than the surface tension of cold water.

(v) A rough sea can be calmed by pouring oil on its surface.

Practical Applications of Capillarity

1. The oil in a lamp rises in the wick by capillary action.
2. The tip of nib of a pen is split up to make a narrow capillary so that the ink rises up to the tin or nib continuously.
3. Sap and water rise up to the top of the leaves of the tree by capillary action.
4. If one end of the towel dips into a bucket of water and the other end hangs over the bucket the towel soon becomes wet throughout due to capillary action.
5. Ink is absorbed by the blotter due to capillary action.
6. Sandy soil is more drier than clay. It is because the capillaries between sand particles are not so fine as to draw the water up by capillaries.
7. The moisture rises in the capillaries of soil to the surface, where it evaporates. To preserve the moisture in the soil, capillaries must be broken up. This is done by ploughing and levelling the fields.
8. Bricks are porous and behave like capillaries.

Concept Application Exercise 4.3

1. Why is moisture retained longer in the soil if it is harrowed?
2. A capillary tube is dipped in water vertically. It is long enough for the water to rise to the maximum height h in the tube. The length of portion immersed in water is $I < h$. The lower end of the tube is closed and then the tube is taken out and opened again. Will all the water flow out of tube? Explain.
3. Two soap bubbles A and B of different diameters are blown at the two ends of a bent tube. By opening the stopcock S, the two bubbles are put in communication. What will happen?

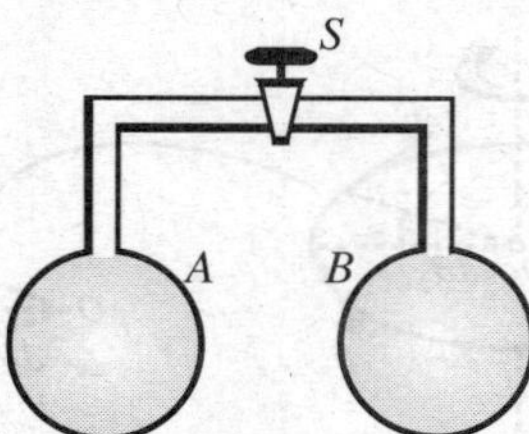

Fig. 4.86

4. A mercury drop of radius R is sprayed into n droplets of equal size. Calculate the energy expanded if surface tension of mercury is T.
5. If a number of little droplets of water, each of radius r, coalesce to form a single drop of radius R, show that the rise in temperature will be given by

$$\frac{3T}{J}\left(\frac{1}{r}-\frac{1}{R}\right)$$

where T is the surface tension of water and J is the mechanical equivalent of heat.
6. A liquid of specific gravity 1.5 is observed to rise 3.0 cm in a capillary tube of diameter 0.50 mm and the liquid wets the surface of the tube. Calculate the excess pressure inside a spherical bubble of 1.0 cm diameter blown from the same liquid.
7. A drop of water of volume 0.05 cm^3 is pressed between two glass plates, as a consequence of which it spreads and occupies an area of 40 cm^2. If the surface tension of water is 70 dyn/cm, find the normal force required to separate out the two glass plates in newton.
8. A glass tube of circular cross section is closed at one end. This end is weighted and the tube floats vertically in water, heavy end down. How far below the water surface is the end of the tube? Given: outer radius of the tube is 0.14 cm, mass of weighted tube is 0.2 g, surface tension of water 73 dyn/cm and $g = 980$ cm/s^2.
9. If a 5 cm long capillary tube with 0.1 mm internal diameter, open at both ends, is slightly dipped in water having surface tension 75 dyn/cm, state whether: (i) water will rise halfway in the capillary, (ii) water will rise up to the upper end of capillary, (iii) water will overflow out of the upper end of capillary. Explain your answer.

Example 4.1 **A boy has a catapult made of a rubber cord of length 42 cm and diameter 6.0 mm. The boy stretches the cord by 20 cm to catapult a stone of mass 20 g The stone flies off with a speed of 20 m/s. Find Young's modulus for rubber. Ignore the change in the cross section of the cord in stretching.**

Sol. The potential energy stored in the cord due to stretching is converted into the kinetic energy of the stone:

$$1/2F\Delta l = \frac{1}{2}mv^2 \quad \Rightarrow \quad \text{force } F = \frac{mv^2}{\Delta l}$$

$$\text{Strain} = \frac{\Delta l}{l}$$

$$\text{Young's modulus} = \frac{\text{Stress}}{\text{Strain}} = \frac{(F/A)}{(\Delta \ell/\ell)} = \frac{(mv^2/\Delta \ell)}{\pi r^2\left(\frac{\Delta \ell}{\ell}\right)}$$

$$= 2.97 \times 10^6 \text{ N/m}^2$$

Example 4.2 **A steel wire of cross-sectional area 0.5 mm^2 is held between two fixed support. If tension in the wire is negligible and it is just taut at 20°C, determine tension when temperature falls to 0°C. Young's modulus of elasticity is 21 $\times$ 10^{11} dyn/cm^{-2} and coefficient of linear expansion is 12 $\times$ 10^{-6} per degree centigrade. Assume that the distance between the supports remains the same.**

Sol. If L be the original length, change in length Δl is given by

$$\Delta l = L\alpha t$$

$$\text{Threfore, longitudinal strain } \frac{\Delta l}{L} = \alpha t$$

$$= 12 \times 10^{-6} \times 20 = 24 \times 10^{-5}$$

Tensile stress in the wire = $Y \times$ Strain

$$= 21 \times 10^{11} \times 24 \times 10^{-5}$$

$$= 5.04 \times 10^8 \text{ dyn/cm}^2$$

Therefore, Tension = Tensile stress × Area of cross section

$$= 5.04 \times 10^8 \times 0.005 \text{ dyn}$$

$$= 25.2 \times 10^5 \text{ dyn} = 25.2 \text{ N}$$

Example 4.3 **A sphere of radius 10 cm and mass 25 kg is attached to the lower end of a steel wire which is suspended from the ceiling of a room. The point of support is 521 cm above the floor. When the sphere is set swinging as a simple pendulum, its lowest point just grazes the floor. Calculate the velocity of the ball at its lowest position.**

Sol. Young's modulus of steel = 20 × 10^{10} N/m^2

Unstretched length of the wire = 500 cm

Radius of the steel wire = 0.05 cm

A wire can be considered as a spring of suitable force constant. For a wire,

$$Y = \frac{F/A}{\Delta l/l} = \frac{FL}{A\Delta l} \Rightarrow F = Y\frac{Al}{l}\Delta l$$

Now $\quad F = k.\Delta l$

Therefore, k (force constant) $= \dfrac{F}{\Delta l} = \dfrac{YA}{l}$

$$\Rightarrow \quad k = \frac{20\times10^{10}\times\pi\left(0.05\times10^{-2}\right)^2}{5} = \pi\times 10\ \text{N/m}^2$$

Now, we may treat the problem as if the mass were suspended by a spring of force constant $k = 10^4\,\pi$ N/m². Considering the dynamics of circular motion at the lowest point

$$T - mg = \frac{mv^2}{r}$$

Obviously, at the lowest point, the wire elongates by 1 cm

$$\therefore \quad T = k\Delta l = \pi \times 10^4 \times 0.01 = \pi \times 10^2\ \text{N}$$

$$\therefore \quad \pi\times10^2 - 25\times9.8 = \frac{25\times v^2}{5.21}$$

$$v^2 = \frac{(100\pi - 245)\times5.21}{25} = 14.42$$

$$\Rightarrow \quad v = 3.8\ \text{m/s}$$

Example 4.4 **A steel bolt is inserted into a copper tube as shown in the figure. Find the forces induced in the bolt and in the tube when the nut is turned through one revolution. Assume that the length of the tube is l, the pitch of the bolt thread is h and the cross-sectional areas of the steel bolt and the copper tube are A_s and A_c, respectively.**

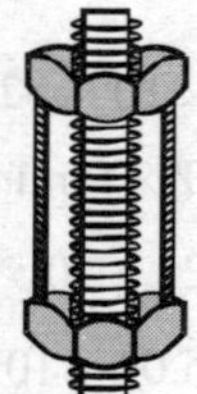

Fig. 4.87

Sol. When a nut is turned, the tube is compressed and so its length decreases. The bolt is subjected to tensile force, so its length increases. The sum of these two lengths is the distance through which the nut moves.

Now $\quad Y_s = \dfrac{F/A_s}{\dfrac{\Delta l}{l}}$ or $\quad \Delta l = \dfrac{F\times l}{Y_sA_s}$

Similarly, for the copper tube

$$\Delta l = \frac{Fl}{Y_cA_c}$$

$$\therefore\ h = \frac{Fl}{Y_sA_s} + \frac{Fl}{Y_cA_c} \Rightarrow F = \frac{h}{l}\frac{Y_sY_cA_sA_c}{(Y_sA_s + Y_cA_c)}$$

Example 4.5 **A circular ring of radius R and mass m made of a uniform wire of cross-sectional area A is rotated about a stationary vertical axis passing through its centre and perpendicular to the plane of the ring. If the breaking stress of the material of the ring is σ_b, then determine the maximum angular speed ω_{max} at which the ring may be rotated without failure.**

Sol. Every element of the ring rotates in a circular path of radius R about the axis of rotation. The radial component of tension in the wire provides the centripetal forces.

The figure shows a free-body diagram of a small element of mass $dm = (m/2\pi)\, d\theta$ of the ring.

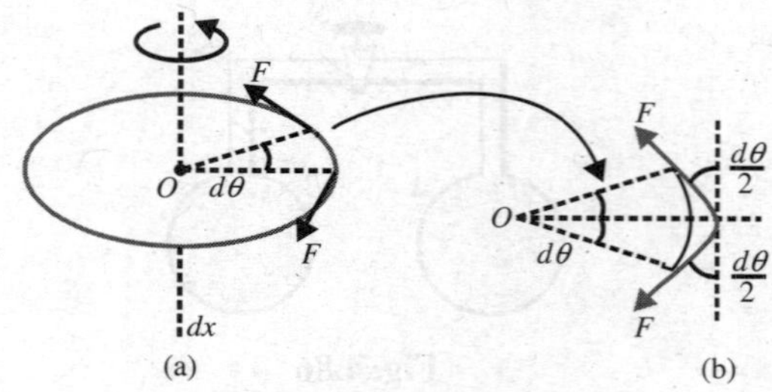

Fig. 4.88

The radial component of tension is

$$F_r = 2F\sin\frac{d\theta}{2} \approx 2F\left(\frac{d\theta}{2}\right) \approx F\,d\theta$$

Applying Newton's second law, we get $F d\theta = (dm)\omega^2 R$

$$F\,d\theta = \left(\frac{m\,d\theta}{2\pi}\right)\omega^2 R \quad \text{or} \quad F = \frac{m\omega^2 R}{2\pi}$$

If the breaking stress is σ_b, then the maximum value of F can be

$$F_{max} = \sigma_b A$$

$$\frac{m\omega_{max}^2 R}{2\pi} = \sigma_b A$$

$$\omega_{max} = \sqrt{\frac{2\pi\sigma_0 A}{mR}}$$

Note: ***If the mass of the ring is expressed in terms of density r, then the maximum speed of rotation is given by*** $\omega_{max} = \dfrac{1}{R}\sqrt{\dfrac{\sigma_0}{\rho}}$ ***a smaller ring can be rotated with a greater angular speed.***

Example 4.6 **In the previous example, if Young's modulus of the material of ring is Y, then determine the increment in the radius ΔR of the ring.**

Sol. We know that the tensile force in the ring is

$$F = \frac{m\omega^2 R}{2\pi}$$

If ΔR is the increment in the radius of the ring, then the elongation in the length of the ring is

$$\delta = 2\pi(R+\Delta R) - 2\pi R = 2\pi(\Delta R)$$

$$\text{Strain } \in = \frac{\delta}{2\pi R} = \frac{\Delta R}{R}$$

Since $\quad Y = \frac{\sigma}{\in} \Rightarrow \in = \frac{\sigma}{Y}$

$$\Delta R = \frac{\sigma}{Y}R = \left(\frac{m\omega^2 R}{2\pi A}\right)\frac{R}{Y} \quad \text{or} \quad \Delta R = \frac{m\omega^2 R^2}{2\pi AY}$$

Note: *If mass of the ring is expressed in terms of density ρ, then* $\Delta R = \frac{\rho\omega^2 R^3}{Y}$

Example 4.7 **A glass plate of length 10 cm, breadth 1.54 cm and thickness 0.20 cm weight 8.2 g in air. It is held vertically with the long side horizontal and the lower half under water. Find the apparent weight of the plate. Surface tension of water = 7.3 × 10^{-2} N/m and g = 9.8 m/s^2.**

Sol. The forces acting on the plate are

(i) weight of plate (mg) acting vertically downward.
(ii) force of surface tension (TL) acting vertically downward.
(iii) upthrust.

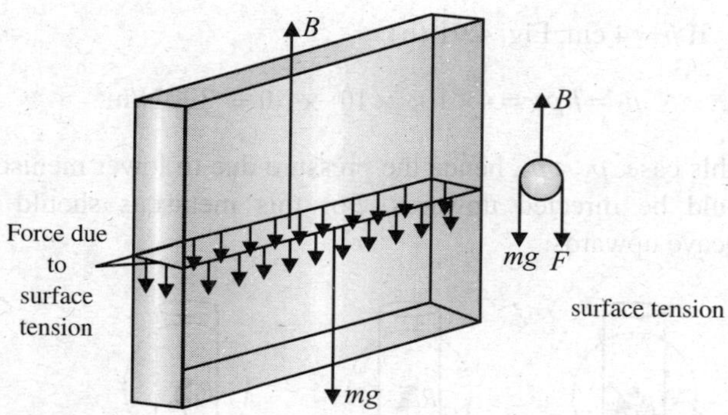

Fig. 4.89

Weight of glass plate = 8.2 g = 8.2 ×10^{-3} kg

Total length of plate in contact in with water surface is

$$l = 2(10+0.2)\text{cm} = 20.4\times10^{-2}\text{ m}$$

Therefore, downward pull upon the plate due to surface tension

$$TL = 7.\times10^{-2}\times20.4\times10^{-2} = 14.892\times10^{-3}\text{N}$$

$$= \frac{14.892\times10^{-3}}{9.8} = 1.5196\times10^{-3}\text{kg wt}$$

Volume of plate immersed in water is

$$v = 10\times\frac{1}{2}(1.54)\times0.2 = 1.54.10^{-6}\text{m}^3$$

Therefore, upthrust = $v\sigma g$ = 1.54 × 10^{-6} × 10^3 × 9.8 N

$$= \frac{1.54\times10^{-3}\times9.8}{9.8}\text{ kg wt} = 1.54\times10^{-3}\text{kg wt}$$

Therefore, apparent weight of plate in water = Weight of plate in air + Force of surface tension – Upthrust

$$= 8.2\times10^{-3} + 1.5196\times10^{-3} - 1.54\times10^{-3}$$
$$= 8.1796\times10^{-6}\text{ kg wt} = 8.1796\text{ kg wt}$$

Example 4.8 **A barometer contains two uniform capillaries of radii 1.44 × 10^{-3} m and 7.2 × 10^{-4} m. If the height of liquid in narrow tube is 0.2 m more than that in wide tube, calculate the true pressure difference. Density of liquid = 10^3 kg/m^3, surface tension = 72 × 10^{-3} N/m and g = 9.8 m/s^2.**

Sol. Let the pressures in wide and narrow limbs be P_1 and P_2, respectively. If R_1 and R_2 be the radii of meniscus in wide and narrow limb pressure just below the meniscus of wide tube $= P_1 - \frac{2T}{R_1}$ and pressure just below the meniscus of narrow limb $= P_2 - \frac{2T}{R_2}$.

Therefore, difference of these pressures

$$\left(P_1 - \frac{2T}{R_1}\right) - \left(P_2 - \frac{2T}{R_2}\right) = h\rho g$$

Therefore, true pressure difference,

$$P_1 - P_2 = h\rho g - 2T\left(\frac{1}{R_2} - \frac{1}{R_1}\right)$$

For the water and glass surface, taking the angle of contact θ to be zero, we have $R_1 = \frac{r_1}{\cos\theta} \approx r_1$ and $R_2 = \frac{r_2}{\cos\theta} \approx r_2$ where r_1 and r_2 are radii of wide and narrow limbs, respectively.

$$\therefore\ P_1 - P_2 = h\rho g - 2T\left(\frac{1}{r_2} - \frac{1}{r_1}\right) = 0.2\times10^3\times9.8 - 2\times72\times$$
$$\times\,10^{-3}\times\left(\frac{1}{7.2\times10^{-4}} - \frac{1}{1.44\times10^{-3}}\right)$$
$$= 1.96\times10^3 - 0.10\times10^3 = 1.86\times10^3\text{ N/m}^2 = 1860\text{ N/m}^2$$

Example 4.9 **A glass capillary sealed at the upper end is of length 0.11 m and internal diameter 2 × 10^{-5} m.**

The tube is immersed vertically into a liquid of surface tension 5.06 × 10⁻² N/m. To what length the capillary has to be immersed so that the liquid level inside and outside the capillary becomes the same. What will happen to water level inside the capillary if the seal is now broken?

Sol. Let P_0 be atmospheric pressure and P_1 the pressure of air within the sealed tube. Then for equilibrium pressure just below the meniscus should be equal to atmospheric pressure because levels of water inside and outside the tube is same.

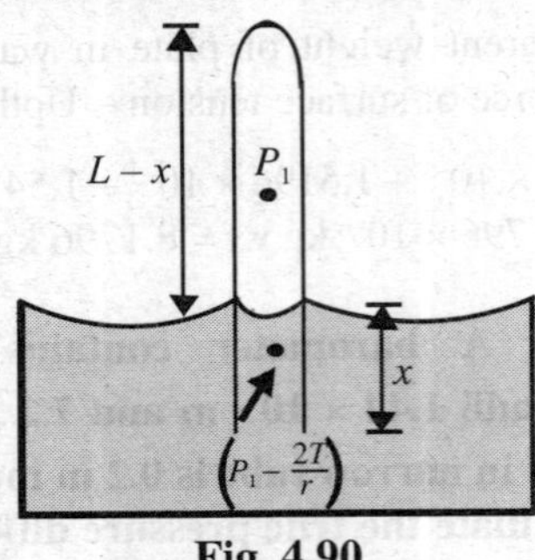

Fig. 4.90

i.e., $\left(P_1 - \frac{2T}{r}\right) = P_0$ or $P_1 = P_0 + \frac{2T}{r}$

If $L = 0.11$ m is the length of tube and x the length of immersed part, then from Boyle's law

$$p_1V_1 = p_2V_2$$

$$P_0La = \left(P_0 + \frac{2T}{r}\right)(L-x)a$$

where a is the cross-sectional area of tube,

i.e., $$P_0 \times L = \left(P_0 + \frac{2T}{r}\right)(L-x)$$

$$P_0 \times L = P_0(L-x) + \frac{2T}{r}(L-x)$$

i.e., $$P_0x = \frac{2T}{r}(L-X)$$

$$1.012\times10^5 \times x = \frac{2\times5.06\times10^{-2}}{1\times10^{-5}}(0.11-x)$$

Solving for x, we get $x = 0.01$ m

Length of capillary to be immersed = 0.01 m

If seal is broken the water level in the capillary rises.

Example 4.10 **An open capillary tube contains a drop of water. When the tube is in its vertical position, the drop forms a column with a length of (a) 2 cm, (b) 4 cm and (c) 2.98 cm. The internal diameter of the capillary tube is 1 mm. Determine the radii of curvature of the upper and lower meniscuses in each case. Consider the wetting to be complete. Surface tension of water = 0.0075 N/m.**

Sol. When capillary is in vertical position, the upper meniscus is concave and pressure due to surface tension

$$P_1 = \frac{2T}{R_1}$$

is directed vertically upward, R_1 being radius of curvature of upper meniscus. When wetting is complete

$$P_1 = \frac{2T}{r},$$

r being radius of capillary tube $= \frac{D}{2} = \frac{1.0}{2}$ mm

$R_1 = r = 0.5$ mm in each case.

We have $$P_1 = \frac{2T}{r} = \frac{2\times75\times10^{-3}}{0.5\times10^{-3}} = 300 \text{ N/m}^2$$

The hydrostatic pressure $p_2 = h\rho g$ is always directed downwards.

(i) $h = 2$ cm, Fig. 4.91 (i)

$$p_2 = h\rho g = 2\times10^{-2}\times10\times10^3 = 200 \text{ N/m}^2$$

Clearly $p_1 > p_2$. That is resulting pressure is directed upward. For equilibrium, the pressure due to lower meniscus should be downwards. This makes the lower meniscus concave downwards. Then

$$P_1 - P_2 = \frac{2T}{R_2}$$

or $$300 - 200 = \frac{2\times75\times10^{-3}}{R_2}$$

$$R_2 = \frac{2\times75\times10^{-3}}{100} = 1.50\times10^{-3} \text{ m} = 1.50 \text{ mm}$$

(ii) If $h = 4$ cm, Fig. 4.91 (ii)

$$p_2 = h\rho g = 4\times10^{-2}\times10^{-3}\times10 = 392 \text{ N/m}^2$$

In this case, $p_1 < p_2$, hence the pressure due to lower meniscus should be directed upwards. For this meniscus should be concave upwards.

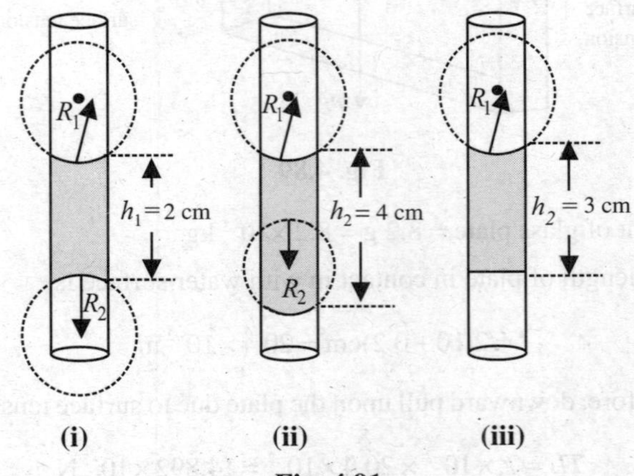

Fig. 4.91

Then $p_2 - p_1 = \dfrac{2T}{R_2}$

$$400 - 300 = \frac{2\times 75\times 10^{-3}}{R_2}$$

$$R_2 = \frac{2\times 75\times 10^{-3}}{100} = 1.50\times 10^{-3}\,\text{m} = 1.50\text{ mm}$$

(iii) If $h = 3.00$ cm [Fig. 4.91(iii)]

$$p_2 = h\rho g = 3.00\times 10^{-2}\times 10^3\times 10 = 300\text{ N/m}^2$$

In this case $p_1 = p_2$

$$p_1 - p_2 = \frac{2T}{R_2}$$

or $\dfrac{2T}{R_2} = 0$ or $R_2 = \infty$

i.e., lower surface will be flat.

Example 4.11 **Two spherical soap bubble coalesce. If V is the consequent change in volume of the contained air and S the change in total surface area, show that**

$$3PV + 4ST = 0$$

where T is the surface tension of soap bubble and P is atmospheric pressure.

Sol. Let a and b be the radii of the two soap bubbles and p_a, p_b the pressure inside them before they coalesce. Let c be the radius and P_c the pressure inside the combined bubbles. Then

$$P_a = P + \frac{4T}{a},\ P_b = P + \frac{4T}{b} \text{ and } P_c = P + \frac{4T}{c}$$

If V_a, V_b and vector are the volumes of two soap bubbles and that of combined soap bubbles, respectively, then by Boyle's law under isothermal conditions

$$P_aV_a + P_bV_b = P_cV_c$$

$$\left(P+\frac{4T}{a}\right)\frac{4}{3}\pi a^3 + \left(P+\frac{4T}{b}\right)\frac{4}{3}\pi b^3 = \left(P+\frac{4T}{c}\right)\frac{4}{3}\pi c^3$$

$$P\left(\frac{4}{3}\pi a^2 + \frac{4}{3}\pi b^2 - \frac{4}{3}\pi c^2\right) + \frac{4T}{3}(4\pi a^2 + 4\pi b^2 + 4\pi c^2) = 0$$

$$\left(\frac{4}{3}\pi a^3 + \frac{4}{3}\pi b^3 - \frac{4}{3}\pi c^3\right) = V$$

= change in volume

And $(4\pi a^2 + 4\pi b^2 - 4\pi c^2) = S$

= change in surface area

or $PV + \dfrac{4T}{3}S = 0$

or $3PV + 4ST = 0$

Example 4.12 **A container of width $2a$ is filled with a liquid. A thin wire of weight per unit length κ is gently placed over the liquid surface in the middle of surface as shown. As a result the liquid surface is depressed by a distance y ($y << a$). Determine the surface tension of the liquid.**

(IIT-JEE, 2004)

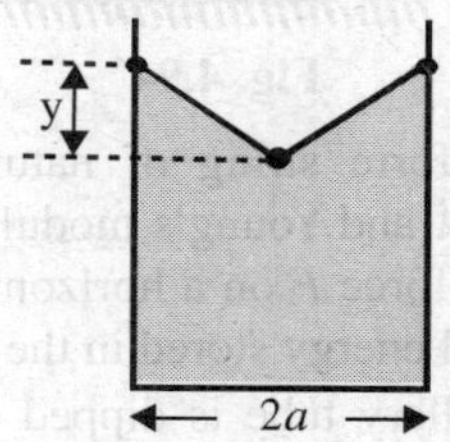

Fig. 4.92

Sol. The free-body diagram of wire is shown in the figure. Let l be the length of wire. If λ is mass per unit length of wire, the weight of wire = $(lx)g$ (acting vertically downward).
Vertical force due to surface tension = $2Tl\cos\theta$ (upward)

Therefore, for vertical equilibrium $2Tl\cos\theta = (l\lambda)g$

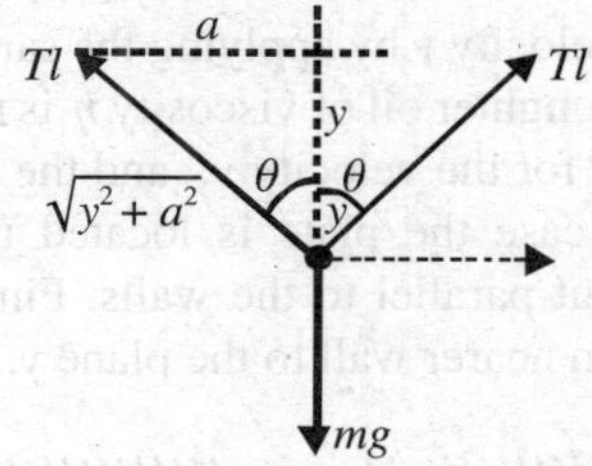

Fig. 4.93

$$T = \frac{\lambda g}{2\cos\theta}$$

From the figure, $\cos\theta = \dfrac{y}{\sqrt{y^2+a^2}} = \dfrac{y}{a}$ (for $y << a$ given)

$$T = \frac{\lambda g}{2\left(\frac{y}{a}\right)} = \frac{\lambda a g}{2y}$$

EXERCISES

Subjective Type

Solutions on page 4.61

1. A bob of mass m hangs from the ceiling of a smooth trolley car which is moving with a constant acceleration a. If Young's modulus, radius and length of the string are Y, r and l, respectively, find the (a) stress in the string and (b) extension of the string when it makes a constant angle relative to vertical.

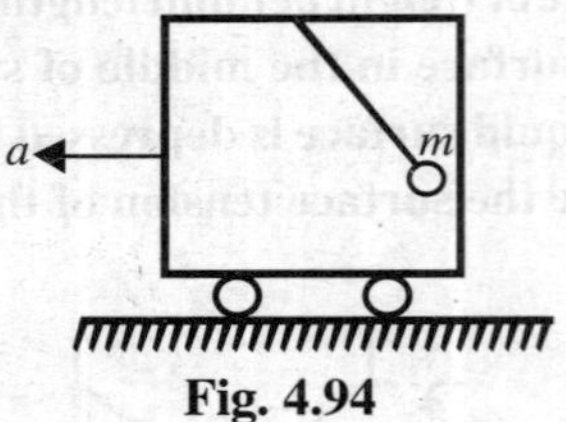

Fig. 4.94

2. A smooth uniform string of natural length l, cross-sectional area A and Young's modulus Y is pulled along its length by a force F on a horizontal surface. Find the elastic potential energy stored in the string.

3. A narrow capillary tube is dipped 10 cm below water surface and a liquid bubble of radius 2 mm formed at the lower end by blowing air through the tube.
   a. Calculate the excess pressure due to surface tension.
   b. What is the pressure required in the tube in order to blow a hemispherical bubble at its end in water?
   The surface tension of water at temperature of the experiment is 7.30×10^{-2} N/m. 1 atmospheric pressure = 10^5 Pa, density of water = 1000 kg/m^3.

4. A thin plate of large area is placed midway in a gap of height h filled with oil of viscosity and the plate is pilled at constant velocity v by applying the same drag force on the plate. If a lighter oil of viscosity η is then in the gap it is found that for the velocity v, and the same drag force as previous case the plate is located unsymmetrically in the gap but parallel to the walls. Find η in terms of distance from nearer wall to the plane y.

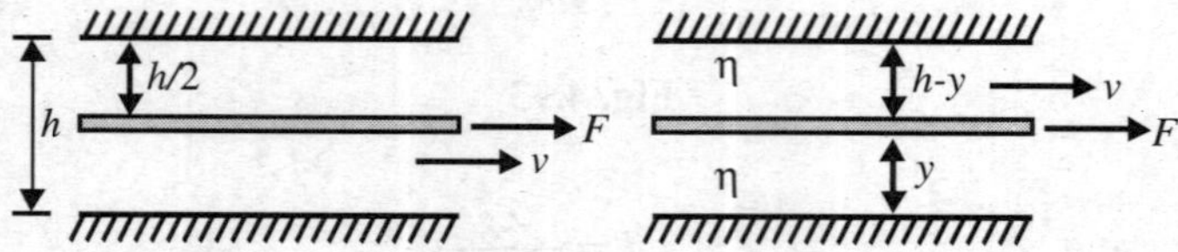

Fig. 4.95

5. When a vertical capillary of length l with the sealed upper end was brought in contact with the surface of a liquid, the level of this liquid rose to the height h. The liquid density is ρ, the inside diameter of the capillary is d, the contact angle is θ, the atmospheric pressure is ρ_0. Find the surface tension of the liquid. (Temperature in this process remains constant.)

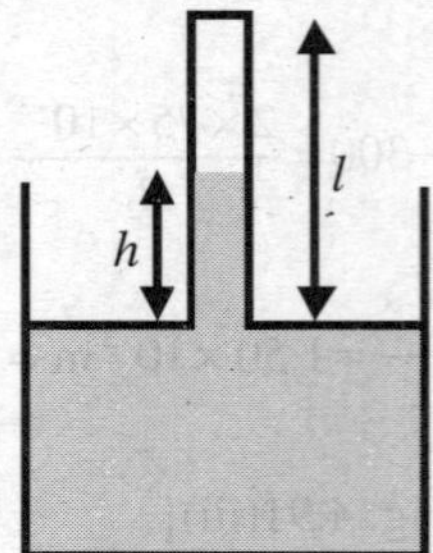

Fig. 4.96

6. The rubber cord of catapult has a cross-sectional area 1 mm^2 and total unstretched length 10.0 cm. It is stretched to 12.0 cm and then released to project a missile of mass 5.0 g. Taking Young's modulus Y for rubber as 5.0×10^8 Nm^2, total elastic energy of catapult is converted into kinetic energy of missile without any heart loss. calculate the velocity of projection (in m/s).

7. Two wires of diameter 0.25 cm, one made of steel and other made of brass, are loaded as shown in the figure. The unloaded length of the steel wire is 1.5 m and that of brass is 1.0 m. Young's modulus of steel is 2.0×10^{11} Pa and that of brass is 1.0×10^{11} Pa. Compute the ratio of elongations of steel and brass wires. $\left(\dfrac{\ell_{steel}}{\ell_{brass}} = ?\right)$

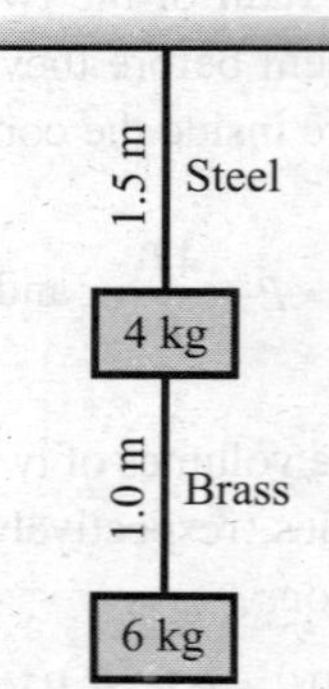

Fig. 4.97

8. A steel rod of length $l_1 = 30$ cm and two identical brass rod of length $l_2 = 20$ cm each support a light horizontal platform as shown in the figure. Cross-sectional area of each of the three rods is $A = 1$ cm^2.
   A vertically downward force $F = 5000$ N is applied on the platform.
   Young's modulus of elasticity for steel $Y_s = 2 \times 10^{11}$ Nm^{-2} and brass $Y_b = 1 \times 10^{11}$ Nm^{-2}.
   Find stress (in MPa) developed in
   (i) Steel rod (ii) Brass rod

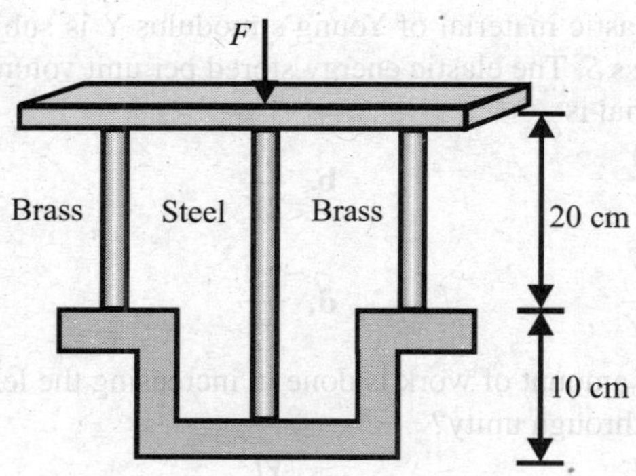

Fig. 4.98

9. The two wires shown in the figure are made of the same material which has a breaking stress of 8×10^8 N/m^2. The area of the cross section of the upper wire is 0.006 cm^2 and that of the lower wire is 0.003 cm^2. The mass m_1 = 10 kg, m_2 = 20 kg and the hanger is light.
(i) Find the maximum load (in kg) that can be put on the hanger without breaking a wire.
(ii) Repeat the above part if m_1 = 10 kg and m_2 = 36 kg.

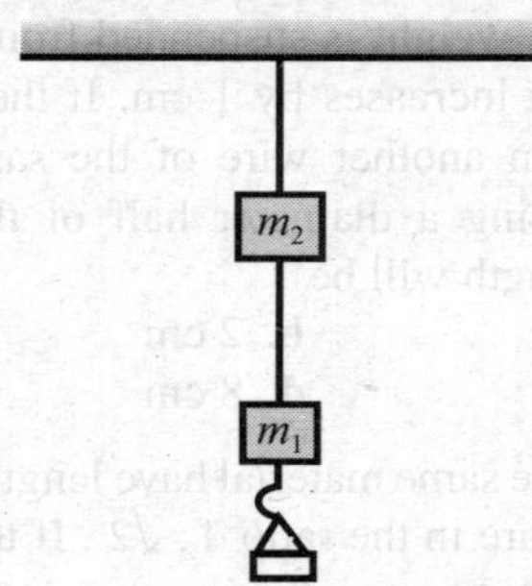

Fig. 4.99

10. A copper wire of negligible mass, 1 m length and cross -sectional area 10^{-6} m^2 is kept on a smooth horizontal table with one end fixed. A ball of mass 1 kg is attached to the other end. The wire and the ball are rotating with an angular velocity of 20 rad/s. If the elongation in the wire is 10^{-3} m.
(i) Find the Young's modulus of the wire (in terms of $\times 10^{11}$ N/m^2).
(ii) If for the same wire as stated above, the angular velocity is increased to 100 rad/s and the wire breaks down, find the breaking stress (in terms of $\times 10^{10}$ N/m^2).

11. An amusement park ride consists of airplane shaped cars attached to steel rods. Each rod has a length of 20.0 m and a cross-sectional area of 8.00 cm^2. Young's modulus for steel is 2×10^{11} N/m^2.
(a) How much is the rod stretched (in mm) when the ride is at rest ? (Assume that each car plus two people seated in it has a total weight of 2000 N.)
(b) When operating, the ride has a maximum angular speed of $\sqrt{19/5}$ rad/s. How much is the rod stretched (in mm) then?

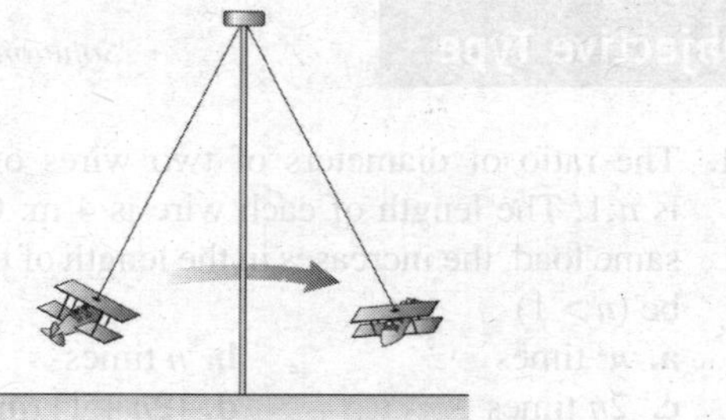

Fig. 4.100

12. A capillary tube of length (i) l = 60 cm, (ii) l = 50 cm and radius r = 1/4 mm is immersed vertically into water. Find the capillary rise in both cases. Angle of contact = 0°. Take coefficient of surface tension as 72 dyne/cm, g = 1000 cm/s^2.

13. An annular disc of radius r_1 = 10 cm and r_2 = 5 cm is placed on a water surface. Find the surface tension force on the disc if we want to pull it from water surface. Take coefficient of surface tension as $\sigma = 7 \times 10^{-3}$ N/m, g = 10 m/s^2.

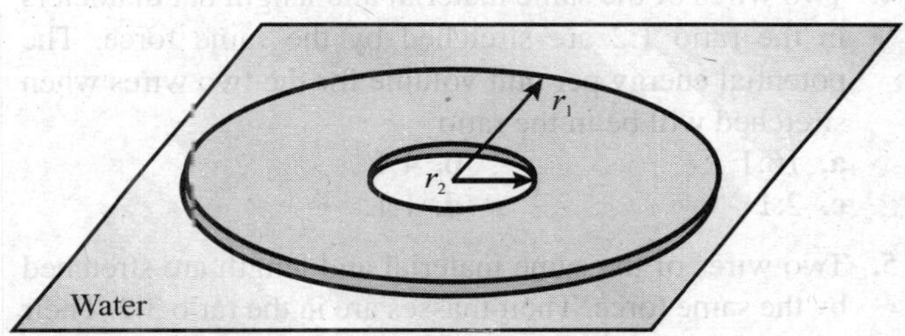

Fig. 4.101

14. Two soap bubbles of radii a and b coalesce to form a single bubble of radius c. If the external pressure is P, find the surface tension of the soap solution.

15. Calculate the pressure inside a small air bubble of radius 0.01 mm situated at a depth of h = 20 m below the free surface of liquid of density $\rho_1 = 10^3$ kg/m^3, ρ_2 = 800 km/m^3 and surface tension $T_2 = 7.5 \times 10^{-2}$ N/m. The thickness of the first liquid is h_1 = 15 m and h_2 = 25 m.

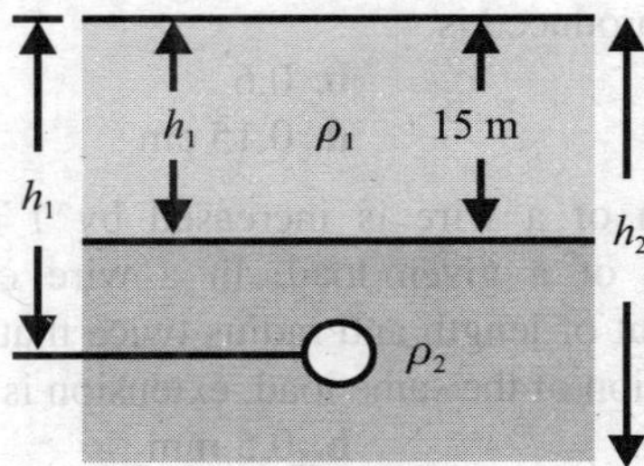

Fig. 4.102

16. Calculate the energy spent in spraying a drop of mercury of r = 1 cm radius into $N = 10^6$ droplets all of same size. If the surface tension of mercury is $T = 35 \times 10^{-3}$ N/m.

Objective Type

Solutions on page 4.64

1. The ratio of diameters of two wires of same material is n:1. The length of each wire is 4 m. On applying the same load, the increases in the length of the thin wire will be ($n > 1$)
   a. n^2 times b. n times
   c. $2n$ times d. $(2n + 1)$ times
2. A nylon rope 2 cm in diameter has a breaking strength of 1.5×10^5 N. The breaking strength of a similar rope 1 cm in diameter is
   a. 0.375×10^5 N b. 2×10^5 N
   c. 6×10^5 N d. 9×10^4 N
3. The dimensions of four wires of the same material are given below. In which wire the increase in the length will be maximum?
   a. Length 100 cm, diameter 1 mm
   b. Length 200 cm, diameter 2 mm
   c. Length 300 cm, diameter 3 mm
   d. Length 50 cm, diameter 0.5 mm
4. Two wires of the same material and length but diameters in the ratio 1:2 are stretched by the same force. The potential energy per unit volume for the two wires when stretched will be in the ratio
   a. 16:1 b. 4:1
   c. 2:1 d. 1:1.
5. Two wires of the same material and length are stretched by the same force. Their masses are in the ratio 3:2. Their elongations are in the ratio
   a. 3:2 b. 9:4
   c. 2:3 d. 4:9
6. Two wires of the same length and same material but radii in the ratio of 1:2 are stretched by unequal forces to produce equal elongation. The ratio of the two forces is
   a. 1:1 b. 1:2
   c. 1:3 d. 1:4
7. When a weight of 5 kg is suspended from a copper wire of length 30 m and diameter 0.5 mm, the length of the wire increases by 2.4 cm. If the diameter is doubled, the extension produced is
   a. 1.2 cm b. 0.6
   c. 0.3 cm d. 0.15 cm
8. The length of a wire is increased by 1 mm on the application of a given load. In a wire of the same material, but of length and radius twice that of the first, on application of the same load, extension is
   a. 0.25 mm b. 0.5 mm
   c. 2 mm d. 4 mm
9. A cube is shifted to a depth of 100 m is a lake. The change in volume is 0.1%. The bulk modulus of the material is nearly
   a. 10 Pa b. 10^4 Pa
   c. 10^7 Pa d. 10^9 Pa
10. An elastic material of Young's modulus Y is subjected to a stress S. The elastic energy stored per unit volume of the material is
    a. $\frac{SY}{2}$ b. $\frac{S^2}{2Y}$
    c. $\frac{S}{2Y}$ d. $\frac{2S}{Y}$
11. What amount of work is done in increasing the length of a wire through unity?
    a. $\frac{YL}{2A}$ b. $\frac{YL^2}{2A}$
    c. $\frac{YA}{2L}$ d. $\frac{YL}{A}$
12. Young's modulus of rubber is 10^4 N/m² and area of cross section is 2 cm⁻². If force of 2×10^5 dyn is applied along its length, then its initial l becomes
    a. $3l$ b. $4l$
    c. $2l$ d. none of these
13. When a certain weight is suspended from a long uniform wire, its length increases by 1 cm. If the same weight is suspended from another wire of the same material and length but having a diameter half of the first one, the increases in length will be
    a. 0.5 cm b. 2 cm
    c. 4 cm d. 8 cm
14. Two wires of the same material have lengths in the ratio 1:2 and their radii are in the ratio $1:\sqrt{2}$. If they are stretched by applying equal forces, the increase in their lengths will be in the ratio
    a. $\sqrt{2}:2$ b. $2:\sqrt{2}$
    c. 1:1 d. 1:2
15. A piece of copper wire has twice the radius of a piece of steel wire. Young's modulus for steel is twice that of the copper. One end of the copper wire is joined to one end of the steel wire so that both can be subjected to the same longitudinal force. By what fraction of its length will the steel have stretched when the length of the copper has increased by 1%?
    a. 1% b. 2%
    c. 2.5% d. 3%
16. The breaking stress for a substance is 10^6 N/m². What length of the wire of this substance should be suspended vertically so that the wire breaks under its own weight? (Given: density of material of the wire = 4×10^3 kg/m³ and $g = 10$ m/s²)
    a. 10 m b. 15 m
    c. 25 m d. 34 m
17. Water rises to a height of 2 cm in a capillary tube. If the tube is tilted 60° from the vertical, water will rise in the tube to a length of

a. 4.0 cm b. 2.0 cm
c. 1.0 cm d. water will not rise at all

18. A spherical liquid drop of radius R is divided into eight equal droplets. If the surface tension is T, then the work done in this process will be
a. $2\pi R^2T$ b. $3\pi R^2T$
c. $4\pi R^2T$ d. $2\pi RT^2$

19. If T is surface tension of soap solution, the amount of work done in blowing a soap bubble from a diameter D to a diameter $2D$ is
a. $2\pi D^2T$ b. $4\pi D^2T$
c. $6\pi D^2T$ d. $8\pi D^2T$

20. A water drop is divided into eight equal droplets. The pressure difference between inner and outer sides of the big drop
a. will be the same as for smaller droplet
b. will be half of that for smaller droplet
c. will be one-fourth of that for smaller droplet
d. will be twice of that for smaller droplet

21. A vessel, whose bottom has round holes with diameter 0.1 mm, is filled with water. The maximum height up to which water can be filled without leakage is
a. 100 cm b. 75 cm
c. 50 cm d. 30 cm

22. Water rises to a height of 10 cm in a capillary tube and mercury falls to a depth of 3.42 cm in the same capillary tube. If the density of mercury is 13.6 g/c.c. and the angles of contact for mercury and water are 135° and 0°, respectively, the ratio of surface tension for water and mercury is
a. 1:0.15 b. 1:3
c. 1:6.5 d. 1.5:1

23. The velocity of small ball of mass M and density d_1 = when dropped a container filled with glycerine becomes constant after some time. If the density glycerine is d_2, the viscous force acting on ball is
a. $\dfrac{Md_1g}{d_2}$ b. $Mg\left(1-\dfrac{d_2}{d_1}\right)$
c. $\dfrac{M(d_1+d_2)}{g}$ d. $M\,d_1d_2$

24. Two soap bubbles, one of radius 50 mm and the other of radius 80 mm, are brought in contact so that they have a common interface. The radius of the curvature of the common interface is
a. 0.003 m b. 0.133 m
c. 1.2 m d. 8.9 m

25. A glass rod of radius r_1 is inserted symmetrically into a vertical capillary tube of radius r_2 such that their lower ends are at the same level. The arrangement is now dipped in water. The height to which water will rise into the tube will be (σ = surface tension of water, ρ = density of water)
a. $\dfrac{2\sigma}{(r_2-r_1)\rho g}$ b. $\dfrac{\sigma}{(r_2-r_1)\rho g}$
c. $\dfrac{2\sigma}{(r_2+r_1)\rho g}$ d. $\dfrac{2\sigma}{(r_2^2+r_1^2)\rho g}$

26. A large number of droplets, each of radius a, coalesce to form a bigger drop of radius b. Assume that the energy released in the process is converted into the kinetic energy of the drop. The velocity of the drop is (σ = surface tension, ρ = density)
a. $\left[\dfrac{\sigma}{\rho}\left(\dfrac{1}{a}-\dfrac{1}{b}\right)\right]^{1/2}$ b. $\left[\dfrac{2\sigma}{\rho}\left(\dfrac{1}{a}-\dfrac{1}{b}\right)\right]^{1/2}$
c. $\left[\dfrac{3\sigma}{\rho}\left(\dfrac{1}{a}-\dfrac{1}{b}\right)\right]^{1/2}$ d. $\left[\dfrac{6\sigma}{\rho}\left(\dfrac{1}{a}-\dfrac{1}{b}\right)\right]^{1/2}$

27. A thick rope of density ρ and length L is hung from a rigid support. The increase in length of the rope due to its own weight is (Y is the Young's modulus)
a. $\dfrac{0.1}{4Y}\rho L^2g$ b. $\dfrac{1}{2Y}\rho L^2g$
c. $\dfrac{\rho L^2g}{Y}$ d. $\dfrac{\rho Lg}{Y}$

28. When the load on a wire is slowly increased from 3 to 5 kg wt, the elongation increases from 0.61 to 1.02 mm. The work done during the extension of wire is
a. 0.16 J b. 0.016 J
c. 1.6 J d. 16 J

29. Two identical wires of iron and copper with their Young's modulus in the ratio 3:1 are suspended at same level. They are to be loaded so as to have the same extension and hence level. Ratio of the weight is
a. 1:3 b. 2:1
c. 3:1 d. 4:1

30. A wire of cross section A is stretched horizontally between two clamps located $2l$ m apart. A weight W kg is suspended from the mid-point of the wire. If the mid-point sags vertically through a distance $x < 1$ the strain produced is
a. $\dfrac{2x^2}{l^2}$ b. $\dfrac{x^2}{l^2}$
c. $\dfrac{x^2}{2l^2}$ d. None of these

31. If in the above question, the Young's modulus of the material is Y, the value of extension x is
a. $\left(\dfrac{Wl}{YA}\right)^{1/3}$ b. $\left(\dfrac{YA}{Wl}\right)^{1/3}$
c. $\dfrac{1}{l}\left[\dfrac{WA}{Y}\right]^{1/3}$ d. $l\left[\dfrac{W}{YA}\right]^{1/3}$

32. A long wire hangs vertically with its upper end clamped. A torque of 8 Nm applied to the free end twists it through 45°. The potential energy of the twisted wire is

a. π J **b.** $\frac{\pi}{2}$ J

c. $\frac{\pi}{4}$ J **d.** $\frac{\pi}{8}$ J

33. Bulk modulus of water is 2×10^9 N/m². The change in pressure required to increase the density of water by 0.1% is

a. 2×10^9 N/m² **b.** 2×10^8 N/m²

c. 2×10^6 N/m² **d.** 2×10^4 N/m²

34. Two rods of different materials having coefficients of linear expansion α_1 and α_2 and Young's moduli, Y_1 and Y_2, respectively, are fixed between two rigid massive walls. The rods are heated such that they undergo the same increase in temperature. There is no bending of rods. If $\alpha_1/\alpha_2 = 2/3$, then the thermal stresses developed in the two rods are equal, provided Y_1/Y_2 is equal to

a. 2:3 **b.** 1:1

c. 3:2 **d.** 4:9

35. One end of uniform wire of length L and of weight W is attached rigidly to a point in the roof and a weight W_1 is suspended from its lower end. If s is the area of cross section of the wire, the stress in the wire at a height $(3L/4)$ from its lower end is

a. $\frac{W_1}{s}$ **b.** $\left[W_1 + \frac{W}{4}\right]s$

c. $\left[W_1 + \frac{3W}{4}\right]/s$ **d.** $\frac{W_1 + W}{s}$

36. A wire is stretched 1 mm by a force of 1 kN. How far would a wire of the same material and length but of four times that diameter be stretched by the same force?

a. $\frac{1}{2}$ mm **b.** $\frac{1}{4}$ mm

c. $\frac{1}{8}$ mm **d.** $\frac{1}{16}$ mm

37. Young's modulis of brass and steel are 10×10^{10} N/m and 2×10^{11} N/m², respectively. A brass wire and a steel wire of the same length are extended by 1 mm under the same force. The radii of the brass and steel wires are R_B and R_S, respectively. Then

a. $R_S = \sqrt{2}\, R_B$ **b.** $R_S = \frac{R_B}{\sqrt{2}}$

c. $R_S = 4R_B$ **d.** $R_S = \frac{R_B}{4}$

38. When the tension in a metal wire is T_1, its length is l_1. When the tension is T_2, its length is l_2. The natural length of wire is

a. $\frac{T_2}{T_1}(l_1 + l_2)$ **b.** $T_1 l_1 + T_2 l_2$

c. $\frac{l_1 T_2 - l_2 T_1}{T_2 - T_1}$ **d.** $\frac{l_1 T_2 + l_2 T_1}{T_2 + T_1}$

39. Two blocks of masses 1 kg and 2 kg are connected by a metal wire going over a smooth pulley as shown in the figure. The breaking stress of the metal is $(40/3\pi) \times 10^6$ N/m². If $g = 10$ m/s², then what should be the minimum radius of the wire used if it is not to break?

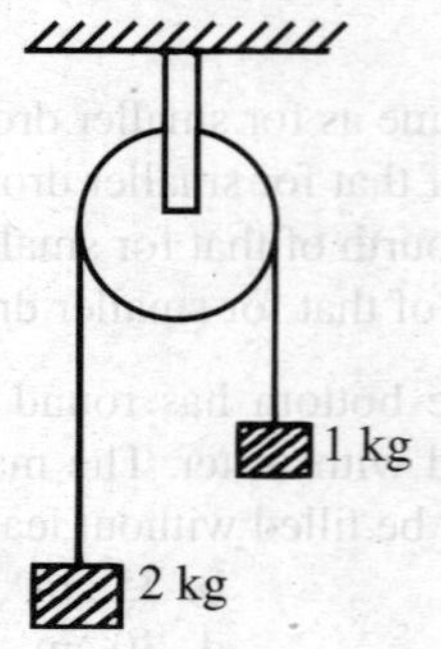

Fig. 4.103

a. 0.5 mm **b.** 1 mm

c. 1.5 mm **d.** 2 mm

40. A long elastic spring is stretched by 2 cm and its potential energy is U. If the spring is stretched by 10 cm, the P.E., will be

a. 5 U **b.** 25 U

c. U/5 **d.** U/20

41. A copper bar of length L and area of cross section A is placed in a chamber at atmospheric pressure. If the chamber is evacuated, the percentage change in its volume will be (compressibility of copper is 8×10^{-12} m²/N and 1 atm = 10^5 N/m)

a. 8×10^{-7} **b.** 8×10^{-5}

c. 1.25×10^{-4} **d.** 1.25×10^{-5}

42. A small but heavy block of mass 10 kg is attached to a wire 0.3 m long. Its breaking stress is 4.8×10^7 N/m². The area of the cross section of the wire is 10^{-6} m². The maximum angular velocity with which the block can be rotated in the horizontal circle is

a. 4 rad/s **b.** 8 rad/s

c. 10 rad/s **d.** 32 rad/s

43. A ball falling in a lake of depth 200 m shows a decrease of 0.1% in its volume at the bottom. The bulk modulus of the elasticity of the material of the ball is (take $g = 10$ m/s²)

a. 10^9 N/m^2 b. 2×10^9 N/m^2
c. 3×10^9 N/m^2 d. 4×10^9 N/m^2

44. A massive stone pillar 20 m high and of uniform cross section rests on a rigid base and supports a vertical load of 5.0×10^5 N at its upper end. If the compressive stress in the pillar is not exceed 16×10^6 N/m, what is the minimum cross-sectional area of the pillar? (Density of the stone = 2.5×10^3 kg/m^3 take g = 10 N/kg.)

a. 0.15 m^2 b. 0.25 m^2
c. 0.35 m^2 d. 0.45 m^2

45. If the work done by stretching a wire by 1 mm is 2 J, the work necessary for stretching another wire of the same material but with the radius of cross section and half the length by 1 mm is

a. $\frac{1}{4}$ J b. 4 J
c. 8 J d. 16 J

46. Two wires of the same material and same mass are stretched by the same force. Their lengths are in the ratio 2:3. Their elongations are in the ratio

a. 3:2 b. 2:3
c. 4:9 d. 9:4

47. A wire is suspended vertically from a rigid support. When loaded with a steel weight in air, the wire extends by 16 cm. When the weight is completely immersed in water, the extension is reduced to 14 cm. The relative density of the material of the weight is

a. 2 g/cm^3 b. 6 g/cm^3
c. 8 g/cm^3 d. 16 g/cm^3

48. Two bars *A* and *B* of circular cross section, same volume and made of the same material, are subjected to tension. If the diameter of *A* is half that of *B* and if the force applied to both the rod is the same and it is in the elastic limit, the ratio of extension of *A* to that of *B* will be

a. 16 b. 8
c. 4 d. 2

49. A uniform cylindrical wire is subjected to a longitudinal tensile stress of 5×10^7 N/m^2. Young's modulus of the material of the wire is 2×10^{11} N/m^2. The volume change in the wire is 0.02%. The factional change in the radius is

a. 0.25×10^{-4} b. 0.5×10^{-4}
c. 1.0×10^{-4} d. 1.5×10^{-4}

50. A material has normal density ρ and bulk modulus *K*. The increase in the density of the material when it is subjected to an external pressure *P* from all sides is

a. $P/\rho K$ b. $K/\rho P$
c. $\rho P/K$ d. $\rho K/P$

51. A rubber rope of length 8 m is hung from the ceiling of a room. What is the increase in length of rope due to its own weight? (Given: Young's modulus of elasticity of rubber = 5 × 106 N/m and density of rubber = 1.5×10^3 kg/m^3. Take g = 10 m/s^2.)

a. 1.5 mm b. 6 mm
c. 24 mm d. 96 mm

52. A straw 6 cm long floats on water. The water film on one side has surface tension of 50 dyn/cm. On the other slide, camphor reduces the surface tension to 40 dyn/cm. The resultant force acting on the straw is

a. (50 × 6 – 40 × 6) dyn b. 10 dyn
c. $\left(\frac{50}{6}-\frac{40}{6}\right)$dyn d. 90 dyn

53. Two glass plates are separated by water. If surface tension of water is 75 dyn/cm and the area of each plate wetted by water is 8 cm^2 and the distance between the plates is 0.12 mm, then the force applied to separate the two plates is

a. 10^2 dyn b. 10^4 dyn
c. 10^5 dyn d. 10^6 dyn

54. A ring is cut from a platinum tube 8.5 cm internal and 8.7 cm external diameter. It is supported horizontally from the pan of a balance, so that it comes in contact with the water in a glass vessel. If an extra 3.103 g.f. is required to pull it away from water, the surface tension of water is

a. 72 dyn/cm b. 70.80 dyn/cm
c. 63.35 dyn/cm d. 60 dyn/cm

55. A soap film of surface tension 3×10^{-2} formed in a rectangular frame can support a straw as shown in the figure. If g = 10 m/s^2, the mass of the straw is

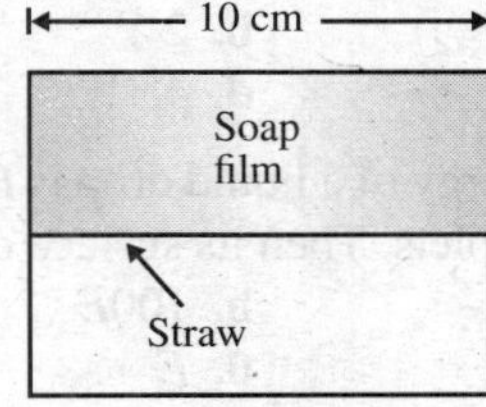

Fig. 4.104

a. 0.006 g b. 0.06 g
c. 0.6 g d. 6 g

56. The lower end of a capillary tube is at a depth of 12 cm and water rises 3 cm in it. The mouth pressure required to blow an air bubble at the lower end will be *x* cm of water column, where *x* is

a. 12 b. 15
c. 3 d. 9

57. A film of soap solution is trapped between a vertical frame and a light wire *ab* of length 0.1 m. If g = 10 m/s^2, then the load *W* that should be suspended from the wire to keep it in equilibrium is

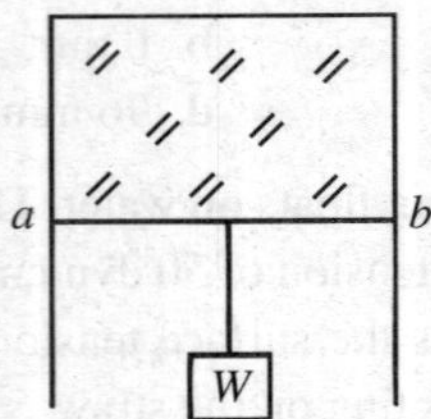

Fig. 4.105

a. 0.2 g **b.** 0.3 g
c. 0.4 g **d.** 0.5 g

58. The angle of contact between glass and water is 0° and water (surface tension 70 dyn/cm) rises in a glass capillary up to 6 cm. Another liquid of surface tension 140 dyn/cm, angle of contact 60° and relative density 2 will rise in the same capillary up to

a. 12 cm **b.** 24 cm
c. 3 cm **d.** 6 cm

59. A hollow sphere has a small hole in it. On lowering the sphere in a tank of water, it is observed that water enters into the hollow sphere at a depth of 40 cm below the surface. Surface tension of water is 7×10^{-2} N/m. The diameter of the hole is

a. $\frac{1}{28}$ mm **b.** $\frac{1}{21}$ mm
c. $\frac{1}{14}$ mm **d.** $\frac{1}{7}$ mm

60. Work W is required to form a bubble of volume V from a given solution. What amount of work is required to be done to form a bubble of volume $2V$?

a. W **b.** $2W$
c. $2^{1/3}W$ **d.** $4^{1/3}W$

61. The surface energy of a liquid drop is E. It is sprayed into 1000 equal droplets. Then its surface energy becomes

a. $1000E$ **b.** $100E$
c. $10E$ **d.** E

62. In the previous question, the work done in spraying is

a. $999E$ **b.** $99E$
c. $9E$ **d.** E

63. A cube with a mass = 20 g wettable water floats on the surface of water. Each face of the cube is $\alpha = 3$ cm long. Surface tension of water is 70 dyn/cm. The distance of the lower face of the cube from the surface of water is ($g = 980$ cm/s²)

a. 2.3 cm **b.** 4.6 cm
c. 9.7 cm **d.** 12.7 cm

64. Consider vertical parallel of semi-circular cross section dipped in a liquid. Assume that the wetting of the tube is complete. The force of surface tension on the flat part and on curved part of the tube are in the ratio

a. $2:\pi$ **b.** $1:\pi$
c. $3:\pi$ **d.** $2.7:\pi$

65. Two vertical parallel glass plates are partially submerged in water. The distance between the plates is d and the length is l. Assume that the water between the plates does not reach the upper edges of the plates and that the wetting is complete. The water will rise to height (ρ = density of water and σ = surface tension of water)

a. $\frac{2\sigma}{\rho g d}$ **b.** $\frac{\sigma}{2\rho g d}$
c. $\frac{4\sigma}{\rho g d}$ **d.** $\frac{5\sigma}{\rho g d}$

66. In the previous question, the force of attraction between the plates is

a. $\frac{l\rho g}{2}h^2$ **b.** $\frac{l\rho g}{2}h$
c. $\frac{l\rho g}{2}$ **d.** $\frac{l\rho g}{2h}$

(Here h represents the height to which the liquid rises.)

67. In Q.65, if water is replaced by another liquid of surface tension 2σ, then the force F of attraction between two plates would become

a. $2F$ **b.** $4F$
c. $8F$ **d.** $16F$

68. A number of droplets, each of radius r, combine to form a drop of radius R. If T is the surface tension, the rise in temperature will be

a. $\frac{2T}{r}$ **b.** $\frac{3T}{R}$
c. $2T\left[\frac{1}{r}-\frac{1}{R}\right]$ **d.** $3T\left[\frac{1}{r}-\frac{1}{R}\right]$

69. A drop of liquid of density ρ is floating half-immersed in a liquid of density d. If ρ is the surface tension the diameter of the drop of the liquid is

a. $\sqrt{\frac{\sigma}{g(2\rho-d)}}$ **b.** $\sqrt{\frac{2\sigma}{g(2\rho-d)}}$
c. $\sqrt{\frac{6\sigma}{g(2\rho-d)}}$ **d.** $\sqrt{\frac{12\sigma}{g(2\rho-d)}}$

70. A drop of water of volume V is pressed between the two glass placed so as to spread to an area A. If T is the surface tension, the normal force required to separate the glass plates is

a. $\frac{TA^2}{V}$ **b.** $\frac{2TA^2}{V}$
c. $\frac{4TA^2}{V}$ **d.** $\frac{TA^2}{2V}$

71. Two soap bubbles of radii a and b combine to form a single bubble of radius c. If P is the external pressure, then the surface tension of the soap solution is

a. $\frac{P(c^3+a^3+b^3)}{4(a^2+b^2-c^2)}$ **b.** $\frac{P(c^3-a^3-b^3)}{4(a^2+b^2-c^2)}$

c. $Pc^3-4a^2-4b^2$ **d.** $Pc^2-2a^2-3b^2$

72. A thin square plate of side 5 cm is suspended vertically from a balance so that lower side just dips into water with side to surface. When the plate is clean ($\theta = 0°$), it appears to weigh 0.044 N. But when the plate is greasy ($\theta = 180°$), it appears to weigh 0.03 N. The surface tension of water is

a. 3.5×10^{-2} N/m **b.** 7.0×10^{-2} N/m

c. 14.0×10^{-2} N/m **d.** 1.08 N/m

73. A loop of 6.28 cm long thread is put gently on a soap film in a wire loop. The film is pricked with a needle inside the soap film enclosed by the thread. If the surface tension of soap solution is 0.030 N/m, then the tension in the thread is

a. 1×10^{-4} N **b.** 2×10^{-4} N

c. 3×10^{-4} N **d.** 4×10^{-4} N

74. A 20-cm-long capillary tube is dipped in water. The water rises up to 8 cm. If the entire arrangement is put in a freely falling elevator, the length of water column in the capillary tube will be

a. 20 cm **b.** 4 cm

c. 10 cm **d.** 8 cm

75. A marble of mass x and diameter $2r$ is gently released in a tall cylinder containing honey. If the marble displaces mass y ($< x$) of the liquid, then the terminal velocity is proportional to

a. $x+y$ **b.** $x-y$

c. $\frac{x+y}{r}$ **d.** $\frac{x-y}{r}$

76. A small metal ball of diameter 4 mm and density 10.5 g/cm³ in dropped in glycerine of density 1.5 g/cm³. The ball attains a terminal velocity of 8/cm/sec. The coefficient of viscosity of glycerine is

a. 4.9 poise **b.** 9.8 poise

c. 98 poise **d.** 980 poise

77. A capillary tube is attached horizontally to a constant pressure head arrangement. If the radius of the capillary tube is increased by 10%, then the rate of flow of the liquid shall change nearly by

a. +10% **b.** 46%

c. –10% **d.** –40%

78. A sphere of brass released in a long liquid column attains a terminal speed v_o. If the terminal speed is attained by a sphere of marble of the same radius and released in the same liquid is $n\, v_o$, then the value of n will be (Given: The specific gravities of brass, marble and liquid are 8.5, 2.5 and 0.8, respectively)

a. $\frac{5}{17}$ **b.** $\frac{17}{77}$

c. $\frac{11}{31}$ **d.** $\frac{17}{5}$

79. Between a plate of area 100 cm² and another plate of area 100 m² there is a 1 mm, thick layer of water, if the coefficient of viscosity of water is 0.01 poise, then the force required to move the smaller plate with a velocity 10 cm/s with reference to large plate is

a. 100 dyn **b.** 10^4 dyn

c. 10^6 dyn **d.** 10^9 dyn

80. A river 10 m deep is flowing at 5 m/s. The shearing stress between horizontal layers of the river is ($\eta = 10^{-3}$ SI units)

a. 10^{-3} N/m² **b.** 0.8×10^{-3} N/m²

c. 0.5×10^{-3} N/m² **d.** 1 N/m²

81. A ball rises to the surface of a liquid with constant velocity. The density of the liquid is four time the density of the material of the ball. The frictional force of the liquid on the rising ball is greater than the weight of the ball by a factor of

a. 2 **b.** 3

c. 4 **d.** 6

82. A spherical ball falls through viscous medium with terminal velocity v. If this ball is replaced by another ball of the same mass but half the radius, then the terminal velocity will be (neglect the effect of buoyancy.)

a. v **b.** $2v$

c. $4v$ **d.** $8v$

83. Each of the pictures shows four objects tied together with rubber bands being pulled to the right across a horizontal frictionless surface by a horizontal force F. All the objects have the same mass; all the rubber bands obey Hooke's law and have the same equilibrium length and the same force constant. Which of these pictures is drawn most correctly?

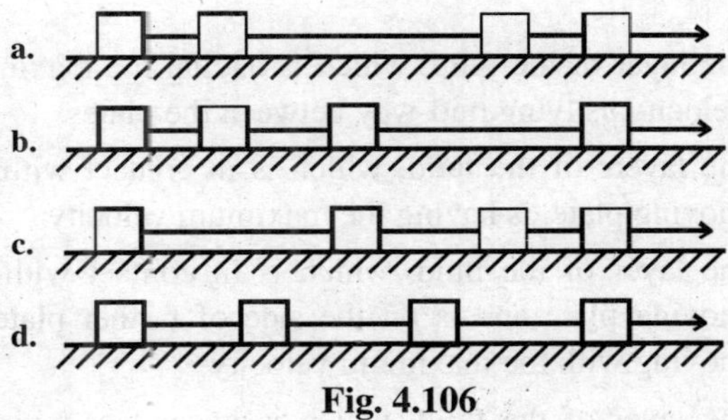

Fig. 4.106

84. A glass rod of radius 1 mm is inserted symmetrically into a glass capillary tube with inside radius 2 mm. Then the whole arrangement is brought in contact of the surface of water. Surface tension of water is 7×10^{-2} N/m. To what height will the water rise in the capillary? ($\theta = 0°$)

a. 1.4 cm b. 4.2 cm
c. 2.1 cm d. 6.8 cm

85. The value V in the bent tube is initially kept closed. Two soap bubbles A (smaller) and B (larger) are formed at the two open ends of the tube. V is now opened and air can flow freely between the bubbles.

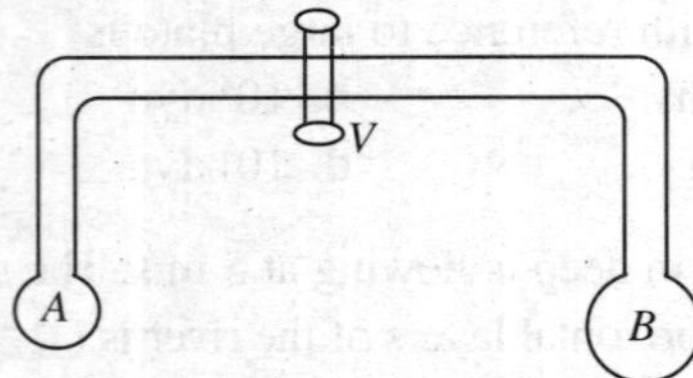

Fig. 4.107

a. There will be change in the size of the bubbles
b. The bubbles will become of equal size
c. A will become smaller and B will become larger
d. The sizes of A and B will be interchanged

86. In the figure shown, forces of equal magnitude are applied to the two ends of a uniform rod. Consider A as the cross-sectional area of the rod. For this situation, mark out the incorrect statements.

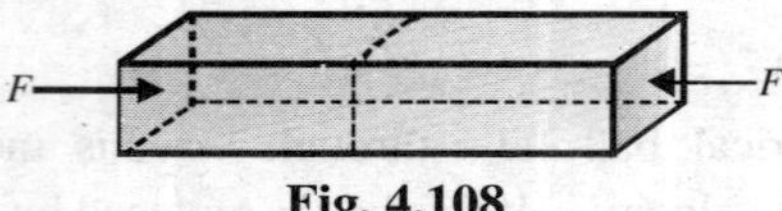

Fig. 4.108

a. The rod is in compressive stress.
b. The numerical value of stress developed in the rod is equal to F/A.
c. The stress is defined as internal force developed at any cross section per unit area.
d. None of the above.

87. The space between two large horizontal metal plates, 6 cm apart, is filled with a liquid of viscosity 0.8 N/m². A thin plate of surface area 0.01 m² is moved parallel to the length of the plate such that the plate is at a distance of 2 m from one of the plates and 4 cm from the other. If the plate moves with a constant speed of 1 m/s, then

a. the layer of the fluid, which is having the maximum velocity, is lying mid-way between the plates
b. the layers of the fluid, which is in contact with the moving plate, is having the maximum velocity
c. the layer of the fluid, which is in contact with the moving plate and is on the side of farther plate, is moving with the maximum velocity
d. the layer of the fluid, which is in contact with the moving plant and is on the side of nearer plate, is moving with the maximum velocity

88. In the previous question, the net force exerted by fluid on moving plate is

a. 0.6 N backwards b. 0.6 N forwards
c. 0.2 N backwards d. 0.2 N backwards

89. The length of a steel cylinder is kept constant by applying pressure at its two ends. When the temperature of rod is increased by 100°C from its initial temperature, the increase in pressure to be applied at its ends is ($Y_{steel} = 2 \times 10^{11}$ N/m², $\alpha_{steel} = 11 \times 10^{-6}$/°C, 1 atm $= 10^5$ N/m².)

a. 22×10^7 atm b. 2.2×10^3 atm
c. zero d. 4.3×10^3 atm

90. Maximum excess pressure inside a thin-walled steel tube of radius r and thickness Δr ($<< r$), so that the tube would not rupture would be (breaking stress of steel is σ_{max})

a. $\sigma_{max} \times \dfrac{r}{\Delta r}$ b. $\sigma_{max} \times \dfrac{\Delta r}{r}$

c. σ_{max} d. $\sigma_{max} \times \dfrac{\Delta 2r}{r}$

91. Two equal and opposite point forces are applied at mid-points of the ends of a rod of square cross section, as shown. Consider the dotted section $ABCD$. If the rod is cut across this cross section, the force exerted by the right part of the rod on left part across this cross section is

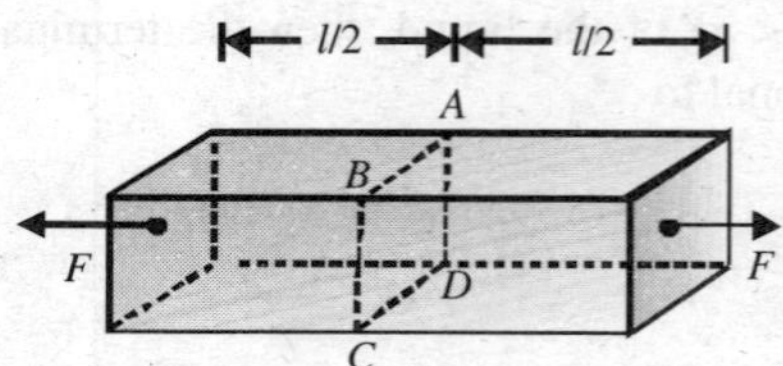

Fig. 4.109

a. acting at point passing through cross section $ABCD$
b. acting at a point but not passing through the centre of cross section $ABCD$
c. uniformly distributed across the cross section $ABCD$
d. non-uniformly distributed across the cross section $ABCD$

92. A 5 kg rod of square cross section 5 cm on a side and 1 m long is pulled along a smooth horizontal surface by a force applied at one end. The rod has a constant acceleration of 2 m/s². Determine the elongation in the rod. (Young's modulus of the material of the rod is 5×10^3 N/m⁹).

a. Zero, as for elongation to be there, equal and opposite forces must act on the rod
b. Non-zero but cannot be determine from the given situation
c. 0.4 μm
d. 16 μm

93. Two soap bubbles of different radii are in communication with each other. Then
a. air follows from the larger bubble into smaller bubble till both bubbles acquire same size
b. air follows from the smaller bubble into larger bubble and the larger bubble grows in size with decrease in size of the smaller bubble
c. air does not flow but the sizes of the bubbles changes
d. sizes of the bubbles remain unchanged

94. A paper disc of radius R from which a hole of radius r is cut out is floating in a liquid of the surface tension S. The force on the disc due to the surface tension is
a. $S \times 2\pi R$ b. $S \times 2\pi r$
c. $S \times 2\pi (R - r)$ d. $S \times 2\pi (R + r)$

95. A glass rod of diameter $d = 2$ mm is inserted symmetrically into a glass capillary tube of radius $r = 2$ mm. Then the whole arrangement is vertically dipped into liquid having surface tension 0.072 N/m. The height to which liquid will rise on capillary is (Take $g = 10$ m/s^2, $\rho_{\text{liq}} = 1000$ kg/m^3. Assume contact angle to be zero of capillary tube to be long enough)
a. 1.44 cm   b. 6 cm
c. 4.86 cm   d. None of these

96. The elastic limit of an elevator cable is 2×10^9 N/m^2. The maximum upward acceleration that an elevator of mass 2×10^3 kg can have when supported by a cable whose cross-sectional area is 10^{-4} m^2, provided the stress in cable would not exceed half of the elastic limit would be
a. 10 m/s^2 b. 50 m/s^2
c. 40 m/s^2 d. Not possible to move up

97. A wire can sustain the weight of 20 kg before breaking. If the wire is cut into two equal parts, each part can sustain a weight of
a. 10 kg   b. 20 kg   c. 40 kg   d. 35 kg

98. A wire of length L and radius r is fixed at one end. When a stretching force F is applied at free end, the elongation in the wire is l. When another wire of same material but of length $2L$ and radius $2r$, also fixed at one end is stretched by a force $2F$ applied at free end, then elongation in the second wire will be
a. $\ell/2$ b. ℓ c. 2ℓ d. $\ell/4$

99. On applying a stress of x N/m^2, the length of wire of some material gets doubled. Value of Young's modulus for the material of and wire in N/m^2, is (assume Hooke's law to be valid and go for approx. results)
a. x b. $2x$ c. $x/2$ d. Insufficient information

100. A copper wire and a steel wire of the same cross-sectional area and length are joined end-to-end (at one end). Equal and opposite longitudinal forces are applied to the free ends giving a total elongation of l. Then the two wires will have
a. same stress and same strain
b. same stress and different strains
c. different stresses and same strain
d. different stresses and different strains

101. A steel wire of length 4.7 m and cross-sectional area 3×10^{-6} m^2 stretches by the same amount as a copper wire of length 3.5 m and cross-sectional area of 4×10^{-6} m^2 under a given load. The ratio of Young's modulus of steel to that of copper is
a. 1.8   b. 3.6   c. 0.6   d. 8.7

102. The edges of an aluminium cube are 10 cm long. One face of the cube is firmly fixed to a vertical wall. A mass of 100 kg is then attached to the opposite face of the cube. Shear modulus of aluminium is 25×10^9 Pa, the vertical deflection in the face to which mass is attached is
a. 4×10^{-4} m   b. 4×10^7 m
c. 25×10^{-6} m   d. 6×10^{-7} m

103. A solid sphere of radius R, made up of a material of bulk modulus K is surrounded by a liquid in a cylindrical container. A massless piston of area A floats on the surface of the liquid. When a mass M is placed on the piston to compress the liquid, the fractional change in the radius of the sphere is
a. $\dfrac{Mg}{AK}$ b. $\dfrac{Mg}{3AK}$
c. $\dfrac{3Mg}{AK}$ d. $\dfrac{Mg}{2AK}$

104. Two parallel wires each of length 10 cm are 0.5 cm apart. A film of water is formed between them. If surface tension of water is 0.072 N/m, then the work done in increasing the distance between the wires by 1 mm is
a. 1.44×10^{-5} J   b. 1.72×10^{-5} J
c. 1.44×10^{-4} J   d. 1.72×10^{-4} J

105. The length of a needle floating on water is 2.5 cm. The minimum force in addition to its weight needed to lift the needle above the surface of water will be (surface tension of water is 0.072 N/m)
a. 3.6×10^{-3} N   b. 10^{-2} N
c. 9×10^{-4} N   d. 6×10^{-4} N

106. A steel wire is stretched by 1 kg wt. If the radius of the wire is doubled, its Young's modulus will
a. remain unchanged   b. become half
c. become double   d. become four times

107. Two long metallic strips are joined together by two rivets each of radius 2 mm. Each rivet can withstand a maximum shearing stress of 1.5×10^9 N/m^2. Assuming that each rivet shares the stretching load equally, the maximum tensile force the strip can exert without rupture is

a. 1.88×10^4 N
b. 3.8×10^4 N
c. 6×10^7 N
d. 3×10^4 N

108. A solid sphere falls with a terminal velocity of 20 m/s in air. If it is allowed to fall in vacuum,

a. terminal velocity will be 20 m/s
b. terminal velocity will be less than 20 m/s
c. terminal velocity will be greater than 20 m/s
d. no terminal velocity will be attained

109. The density of water at the surface of ocean is ρ. If the bulk modulus of water is B, then the density of ocean water at depth, when the pressure is αp_0 and p_0 is the atmospheric pressure, is

a. $\dfrac{\rho B}{B-(\alpha-1)p_0}$
b. $\dfrac{\rho B}{B+(\alpha-1)p_0}$
c. $\dfrac{\rho B}{B-\alpha p_0}$
d. $\dfrac{\rho B}{B+\alpha p_0}$

110. Water rises to a height h in a capillary tube of cross-sectional area A. the height to which water will rise in a capillary tube of cross-sectional area $4A$ will be

a. h
b. $h/2$
c. $h/4$
d. $4h$

111. Neglecting the density of air, the terminal velocity obtained by a raindrop of radius 0.3 mm falling through the air of viscosity 1.8×10^{-5} N/m^2 will be

a. 10.9 m/s
b. 8.3 m/s
c. 9.2 m/s
d. 7.6 m/s

Multiple Correct Answers Type

Solutions on page 4.73

1. A composite rod consists of a steel rod of length 25 cm and area $2A$ and a copper rod of length 50 cm and area A. The composite rod is subjected to an axial load F. If the Young's moduli of steel and copper are in the ratio 2:1, then

a. the extension produced in copper rod will be more
b. the extension in copper and steel parts will be in the ratio 1:2
c. the stress applied to copper rod will be more
d. no extension will be produced in the steel rod

2. Four rods A, B, C and D of the same length and material but of different radii $r, r\sqrt{2}, r\sqrt{3}$ and $2r$, respectively, are held between two rigid walls. The temperature of all rods is increased through the same range. If the rods do not bend, then

a. the stress in the rods A, B, C and D is in the ratio 1:2:3:4
b. the forces on them exerted by the wall are in the ratio 1:2:3:4
c. the energy stored in the rods due to elasticity is in the ratio 1:2:3:4
d. it is independent of area like surface tension while friction depends

3. Viscous force is somewhat like friction as it opposes, the motion and is non-conservative but not exactly so, because

a. it is velocity dependent while friction is not
b. it is velocity independent while friction is
c. it is temperature dependent while friction is not
d. it is independent of area is like surface tension while friction is dependent

4. Excess pressure can be $(2T/R)$ for

a. spherical drop
b. spherical meniscus
c. cylindrical bubble in air
d. spherical bubble in water

5. If a liquid rises to the same height in two capillaries of the same material at the same temperature, then

a. the weight of liquid in both capillaries must be equal
b. the radius of meniscus must be equal
c. the capillaries must be cylindrical and vertical
d. the hydrostatic pressure at the base of capillaries must be same

6. The wires A and B shown in the figure are made of the same material and have radii r_A and r_B, respectively. The block between them has a mass m. When the force F is mg/3, one of the wires breaks. Then

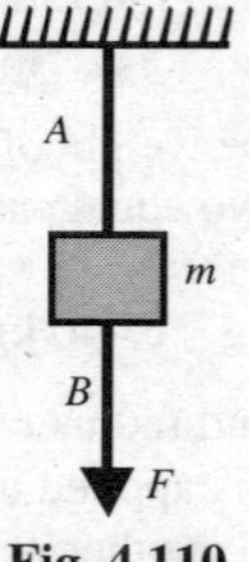

Fig. 4.110

a. A will break before B if $r_A = r_B$
b. A will break before B if $r_A < 2r_B$
c. either A or B may break if $r_A = 2r_B$
d. the lengths of A and B must be known to predict which wire will break.

7. If n drops of a liquid, each with surface energy E, join to form a single drop, then

a. some energy will be released in the process
b. some energy will be released in the process
c. the energy released or absorbed will be $E(n - n^{2/3})$
d. the energy released or absorbed will be $nE(2^{2/3} - 1)$

8. When a capillary tube is dipped in a liquid, the liquid rises to a height h in the tube. The free liquid surface inside the tube is hemispherical in shape. The tube is now pushed down so that the height of the tube outside the liquid is less than h. Then

a. the liquid will come out of the tube like in a small fountain
b. the liquid will ooze out of the tube slowly
c. the liquid will fill the tube but not come out of its upper end
d. the free liquid surface inside the tube will not be hemispherical

9. A vertical glass capillary tube, open at both ends, contains some water. Which of the following shapes may not be taken by the water in the tube?

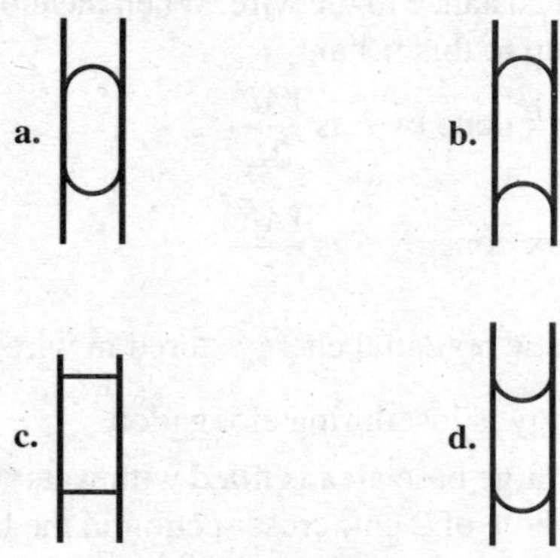

Fig. 4.111

10. A uniform plank is resting over a smooth horizontal floor and is pulled by applying a horizontal force at its one end. Which of the following statements are not correct?
a. Stress developed in plank material is maximum at the end at which force is applied and decrease linearly to zero at the other end.
b. A uniform tensile stress is developed in the plank material.
c. Since plank is pulled at one end only, plank starts to accelerate along direction of the force. Hence, no stress is developed in the plank material.
d. None of the above.

11. A rod is made of uniform material and has non-uniform cross section. It is fixed at both the ends as shown and heated at mid-section. Which of the following statements are not correct?

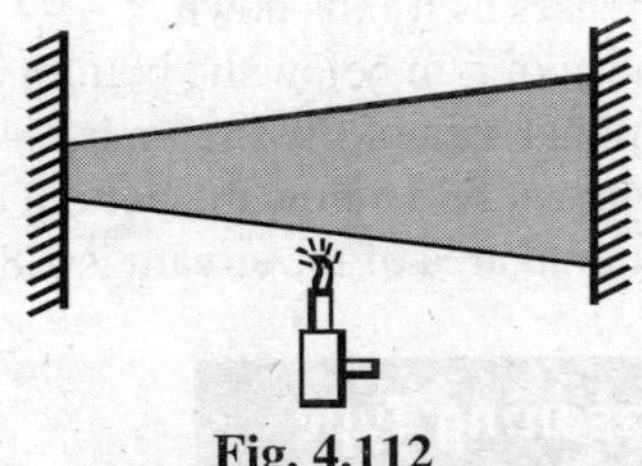

Fig. 4.112

a. Force of compression in the rod will be maximum at mid-section.
b. Compressive stress in the rod will be maximum at left end.
c. Since rod in fixed at both the ends, its length will remain unchanged. Hence, no strain will be induced in it.
d. None of the above.

12. The figure shows the stress–strain graphs for materials A and B. From the graph it follows that

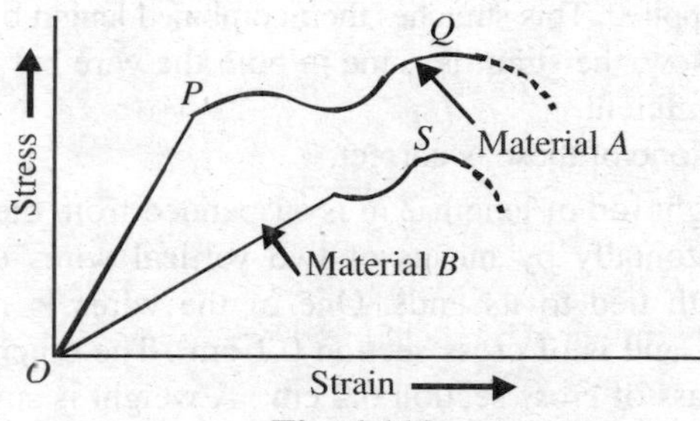

Fig. 4.113

a. material A has a higher Young's modulus
b. material B is more ductile
c. material A can withstand greater stress
d. material A can withstand greater stress

13. Two wires A and B have the same cross section and are made of the same material, but the length of wire A is twice that of B. Then, for a given load
a. the extension of A will be twice that of B
b. the extensions of A and B will be equal
c. the strain in A will be half that in B
d. the strains in A and B will be equal

14. Two wire A and B have equal lengths and are made of the same material, but the diameter of A is twice that of wire B. Then, for a given load
a. The extension of B will be four times that of A
b. the extension of A and B will be equal
c. the strain in B is four times that in A
d. the strains in A and B will be equal

15. In the previous question
a. Young's modulus of A is twice that of B
b. Young's modulus is the same for both A and B
c. A can withstand a greater load before breaking than B
d. A and B will withstand the same load before breaking

16. Choose the correct statements from the following:
a. Steel is more elastic than rubber.
b. The stretching of a coil spring is determined by the Young's modulus of the wire of the spring.
c. The frequency of a tuning fork is determined by the shear modulus of the material of the fork.
d. When a material is subjected to a tensile (stretching) stress the restoring forces are caused by interatomic attraction.

17. Which of the following are correct?
a. For a small deformation of a material, the ratio (stress/strain) decreases.
b. For a large deformation of a material, the ratio (stress/strain) decreases.
c. Two wires made of different materials, having the same diameter and length are connected end to end. A force is

applied. This stretches their combined length by 2 mm. Now, the strain is same in both the wire but stress is different.

d. None of these is correct.

18. A light rod of length 2 m is suspended from the ceiling horizontally by means of two vertical wires of equal length tied to its ends. One of the wires is made of steel and is of cross section 0.1 cm^2. The other wire is a brass of cross section 0.2 cm^2. A weight is suspended from a certain point of the rod such that equal stress are produced in both the wires. Which of the following are correct?

a. The ratio of tension in the steel and brass wires is 0.5.

b. The load is suspended at a distance of 400/3 cm from the steel wire.

c. Both a and b are correct.

d. Neither a nor b is correct.

19. Which of the following are correct?

a. The product of bulk modulus of elasticity and compressibility is 1.

b. A rope 1 cm in diameter breaks if the tension in it exceeds 500 N. The maximum tension that may be given to a similar rope of diameter 2 cm is 2000 N.

c. Both a and b are correct.

d. Neither a nor b is correct.

20. Which of the following are correct?

a. The shear modulus of a liquid is infinite.

b. Bulk modulus of a perfectly rigid body is infinity.

c. According to Hooke's law, the ratio of the stress and strain remains constant.

d. None of the above.

21. A heavy block of mass 150 kg hangs with the help of three vertical wires of equal length and equal cross-sectional area as shown in the figure.

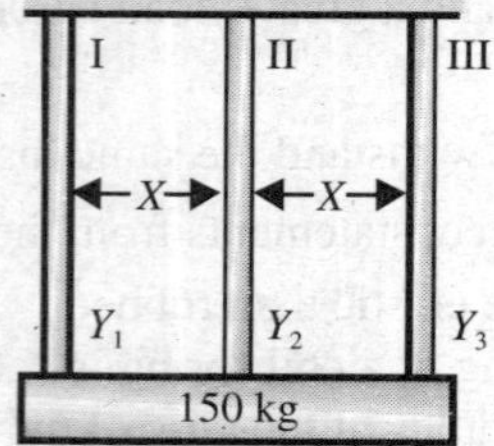

Fig. 4.114

Wire is attached to the mid-point (centre of mass) of block. Take $Y_2 = 2Y_1$. For this arrangement mark out the correct statement(s).

a. The wire I and III should have same Young's modulus.

b. Tension in I and III would be always equal.

c. Tension in I and III would be different.

d. Tension in II is $75g$.

22. When a body of mass M is attached to lower end of a wire (of length L) whose upper end is fixed, then the elongation of the wire is l. In this situation, mark out the correct statement(s).

a. Loss in gravitational potential energy of M is Mgl.

b. Elastic potential energy stored in the wire is $\frac{Mgl}{2}$.

c. Elastic potential energy stored in the wire is Mgl.

d. Elastic potential energy stored in the wire is $\frac{Mgl}{3}$.

23. A metal wire length L, cross-sectional area A and Young's modulus Y is stretched by a variable force F. F is varying in such a way that F is always slightly greater than the elastic forces of resistance in the wire. When the elongation in the wire is l, up to this instant

a. the work done by F is $\frac{YAl^2}{2L}$

b. the work done by F is $\frac{YAl^2}{L}$

c. the elastic potential energy stored in wire is $\frac{YAl^2}{2L}$

d. no energy is lost during elongation

24. A tank of large base area is filled with water up to a height of 5 m. A hole of 2 cm^2 cross section in the bottom allows the water to drain out in continuous streams. For this situation, mark out the correct statement(s) (take ρ_{water} = 1000 kg/m^3, g = 10 m/s^2).

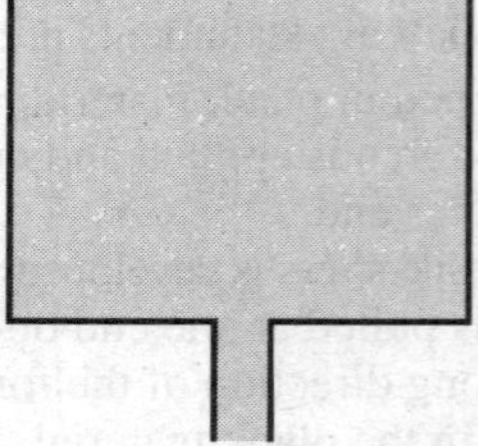

Fig. 4.115

a. The cross-sectional area of the emerging stream of water decreases as it falls down.

b. The cross-sectional area of the emerging stream of water increases as it falls down.

c. At a distance of 5 m below the bottom of the tank, the cross-sectional area of the stream is 1.414 cm^2.

d. At a distance of 5 m below the bottom of the tank, the cross-sectional area of the stream is 2.86 cm^2.

Assertion-Reasoning Type

Solutions on page 4.76

a. Statement I is true, Statement II is true and Statement II is correct explanation for Statement I.

b. Statement I is true, Statement II is true and Statement II is NOT the correct explanation for Statement I.

c. Statement I is true, Statement II is false.

d. Statement I is false, Statement II is true.

1. **Statement I:** Surface tension has the same units as force gradient.
   Statement II: Surface tension is the force gradient along the surface of liquid.
2. **Statement I:** Small liquid drops assume spherical shape.
   Statement II: Due to surface tension liquid drops tend to have minimum surface area.
3. **Statement I:** A small drop of mercury is spherical but bigger drops are oval in shape.
   Statement II: Surface tension of liquid decreases with increase in temperature.
4. **Statement I:** Droplets of liquid are usually more spherical in shape than large drops of the same liquid.
   Statement II: Force of surface tension predominates the force of gravity in case of small drops.
5. **Statement I:** Finer the capillary, greater is the height to which the liquid rises in the tube.
   Statement II: This is in accordance with the ascent formula.
6. **Statement I:** A raindrop after failing through some height attains a constant velocity.
   Statement II: At constant velocity, the viscous drag is just equal to its weight.
7. **Statement I:** A needle placed carefully on the surface of water may float, whereas the ball of the same material will always sink.
   Statement II: The buoyancy of an object depends both on the material and shape of the object.
8. **Statement I:** Dust particles generally settle down in a closed room.
   Statement II: The terminal velocity is inversely proportional to the square of their radii.
9. **Statement I:** Smaller drops of liquid resist deforming forces better than the larger drops.
   Statement II: Excess pressure inside a drop is directly proportional to its surface area.
10. **Statement I:** An object from a greater height reaches a steady terminal velocity.
    Statement II: The viscous forces on a body depends upon its velocity. The greater the velocity the greater is the viscous force.
11. **Statement I:** Spraying of water causes cooling.
    Statement II: For an isolated system, surface energy increase on the expense of internal energy.
12. **Statement I:** While blowing a soap bubble, to increase the size of soap babble, we have to increase the air pressure within the soap bubble.
    Statement II: To increase the size of soap bubble, more air has to be pushed into the bubble.
13. **Statement I:** More is the cohesive force, more is the surface tension.
    Statement II: More cohesive force leads to more shrinking of liquid surface.

Comprehension Type

Solutions on page 4.77

For Problems 1–2

A light rod of length $L = 2$ m is suspended horizontally from the ceiling by two wires A and B of equal lengths. The wire A is made of steel with the area of cross section $A_S = 1 \times 10^{-5}$ m², while the wire B is made of brass of cross-sectional area $A_b = 2 \times 10^{-5}$ m². A weight W is suspended at a distance x from the wire A as shown in the figure.

Take, Young's modulus of steel and brass as $Y_s = 2 \times 10^{11}$ Nm^{-2} and $Y_b = 1 \times 10^{11}$ Nm^{-2}.

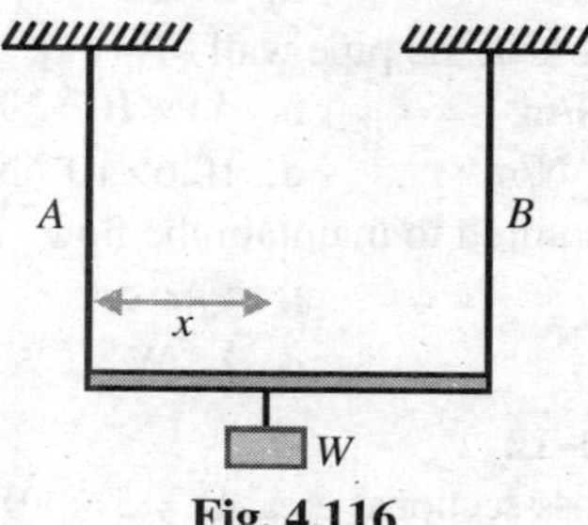

Fig. 4.116

1. Determine the value of x so that equal stresses are produced in each wire.
   a. 1.33 m **b.** 2.5 m
   c. 3.6 m **d.** 2.1 m
2. Determine the value of x so that equal strains are produced in each wire.
   a. 1 m **b.** 2 m
   c. 3 m **d.** 2.2 m

For Problems 3–4

A lead sphere of 1.0 mm diameter and relative density 11.20 attains a terminal velocity of 0.7 cm/s in a liquid of relative density 1.26.

3. Determine the coefficient of dynamic viscosity of the liquid.
   a. 0.45 N/m^2 **b.** 0.85 N/m^2
   c. 0.56 N/m^2 **d.** 0.77 N/m^2
4. What is the value of the Reynolds number?
   a. 0.01 **b.** 0.03
   c. 0.15 **d.** 0.26

For Problems 5–6

A long capillary tube of radius 0.2 mm is placed vertically inside a beaker of water.

5. If the surface tension of water is 7.2×10^{-2} N/m and the angle of contact between glass and water is zero, then determine the height of the water column in the tube.
   a. 3 cm **b.** 9 cm
   c. 7 cm **d.** 5 cm
6. If the tube is now pushed into water so that only 5.0 cm of its length is above the surface, then determine the angle of contact between the liquid and glass surface.

a. $\cos^{-1}\left(\frac{4}{5}\right)$ b. $\cos^{-1}\left(\frac{5}{7}\right)$

c. $\cos^{-1}\left(\frac{3}{5}\right)$ d. $\cos^{-1}\left(\frac{5}{4}\right)$

For Problems 7–9

An oil of relative density 0.9 and viscosity 0.12 kg/ms flows through a 2.5 cm diameter pipe with a pressure drop of 38.4 kN/m² in a length of 30 m. Determine

7. the discharge

a. $2.16\times10^{-4}\ \text{m}^3/\text{s}$ b. $2.9\times10^{-3}\ \text{m}^3/\text{s}$

c. $1\times10^{-4}\ \text{m}^3/\text{s}$ d. $2\times10^{-4}\ \text{m}^3/\text{s}$

8. the shear stress at the pipe wall

a. $8\times10^{-6}\ \text{N/m}^2$ b. $3.9\times10^{-6}\ \text{N/m}^2$

c. $2.3\times10^{-6}\ \text{N/m}^2$ d. $10.6\times10^{-6}\ \text{N/m}^2$

9. the power required to maintain the flow

a. 2.2 W b. 3.84 W

c. 5.6 W d. 9.3 W

For Problems 10–12

A steel bolt of cross-sectional area $A_b = 5\times10^{-5}$ m² is passed through a cylindrical tube made of aluminium. Cross-sectional area of the tube material is $A_t = 10^{-4}$ m² and its length is $l = 50$ cm. The bolt is just taut so that there is no stress in the bolt and temperature of the assembly increases through $\Delta\theta = 10$°C. Given, coefficient of linear thermal expansion of steel, $\alpha_b = 10^{-5}$/°C.

Fig. 4.117

Young's modulus of steel $Y_b = 2\times10^{11}$ N/m².
Young's modulus of Al, $Y_t = 10^{11}$ N/m², coefficient of linear thermal expansion of *Al* $\alpha_t = 2\times10^{-5}$/°C.

10. The compressive strain in tube is

a. 10^{-4} b. 5×10^{-5}

c. 2×10^{-3} d. 10^{-6}

11 The compressive stress in tube is

a. 5×10^{6} N/m² b. 10^{5} N/m²

c. 10^{8} N/m² d. 10^{3} N/m²

12. The tensile stress in bolt is

a. 10^{4} N/m² b. 10^{7} N/m²

c. 2×10^{8} N/m² d. 10^{10} N/m²

For Problems 13–15

According to Hooke's law, within the elastic limit stress/strain = constant. This constant depends on the type of strain or the type of force acting. Tensile stress might result in compressional or elongative strain; however, a tangential stress can only cause a shearing strain. After crossing the elastic limit, the material undergoes elongation and beyond a stage beaks. All modulus of elasticity are basically constants for the materials under stress.

13. Two wires of the same material have length and radius (l, r) and $\left(2l, \frac{r}{2}\right)$. The ratio of their Young's modulus is

a. 1:2 b. 2:3

c. 2:1 d. 1:1

14. Just on crossing the yield region, the material will have

a. reduced stress b. increased stress

c. breaking stress d. constant stress

15. If Stress/Strain is x in elastic region and y in the region of yield, then

a. $x = y$ b. $x > y$

c. $x < y$ d. $x = 2y$

For Problems 16–18

Molecular forces exist between the molecules of a liquid in a container. The molecules on the surface have unequal force leading to a tension on the surface. If this is not compensated by a force, the equilibrium of the liquid will be a difficult task. This leads to an excess pressure on the surface. The nature of the meniscus can inform us of the direction of the excess pressure. The angle of contact of the liquid decided by the forces between the molecules, air and container can make the angle of contact.

16. The direction of the excess pressure in the meniscus of a liquid of angle of contact $2\pi/3$ is

a. upward b. downward

c. horizontal d. cannot be determined

17. If the excess pressure in a soap bubble is p, the excess pressure in an air bubble is

a. $\frac{p}{2}$ b. p c. $2p$ d. $4p$

18. In a meniscus of radius r, with excess pressure p in atmospheric pressure p_0, the force experienced is

a. $(p-p_0)\pi r^2$ b. $(p-p_0)2\pi r$

c. $p\pi r^2$ d. $p_0 2\pi r$

For Problems 19–21

Materials get deformed when force is applied. Some of them regain their status when the applied force is removed. They are termed as elastic. Those of which not regaining are called plastic. There may be delay in the regaining in some materials. They are said to have got elastic aftereffect, since they have gone beyond the elastic limit. Repeated application and removal of force leads to fatigueness in the material. Fatigued materials may break at any point time and so are avoided.

The stress strain graph for two materials *A* and *B* is shown in the following figure:

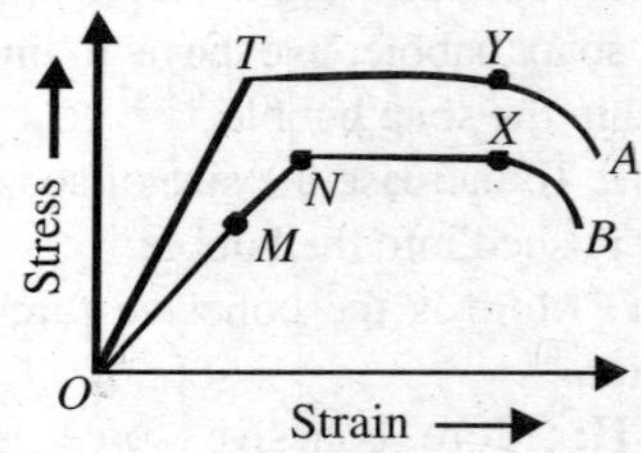

Fig. 4.118

19. If the intensity of A and B is E_A and E_B, respectively, then

a. $E_A = E_B$ b. $E_A > E_B$
c. $E_A < E_B$ d. $E_A \ll E_B$

20. The strength of the material A and B is S_A and S_B, respectively, while the longevity of plastic behaviour is L_A and L_B. Then

a. $S_A > S_B, L_A < L_B$ b. $S_A = S_B, L_A = L_B$
c. $S_A > S_B, L_A > L_B$ d. $S_A < S_B, L_A < L_B$

21. The time in which the two materials regain their original status is t_A and t_B related as $t_B = 2t_B$. Then the material under elastic aftereffect (relatively) is

a. B b. A
c. both A and B d. neither A nor B

For Problems 22–24

The figure shows a capillary tube of radius r dipped into water. The atmospheric pressure is P_0 and the capillary rise of water is h. s is the surface tension for water–glass.

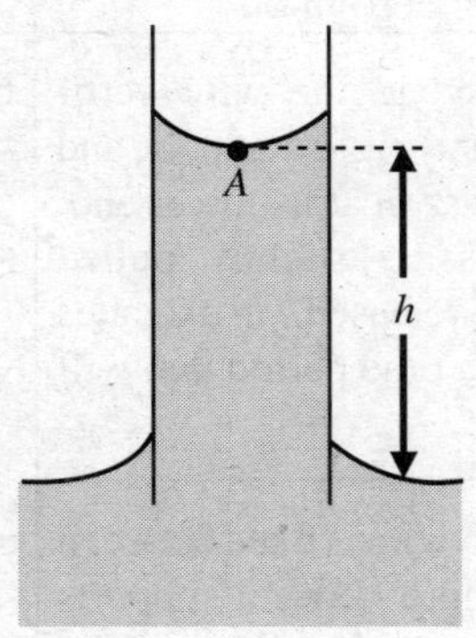

Fig. 4.119

22. The pressure inside water at the point A (lowest point of the meniscus) is

a. P_0 b. $P_0 + \frac{2s}{r}$
c. $P_0 - \frac{2s}{r}$ d. $P_0 - \frac{4s}{r}$

23. Initially, $h = 10$ cm. If the capillary tube is now inclined at 45°, the length of water rising in the tube will be

a. 10 cm b. $10\sqrt{2}$ cm
c. $\frac{10}{\sqrt{2}}$ cm d. None of these

24. Which of the following graphs may represent the relation between the capillary rise h and the radius r of the capillary?

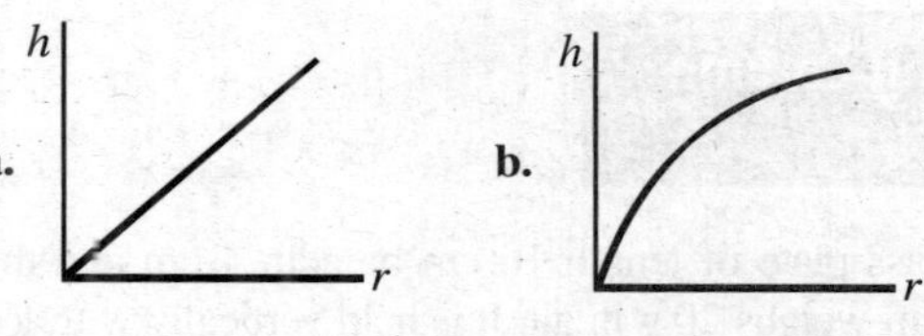

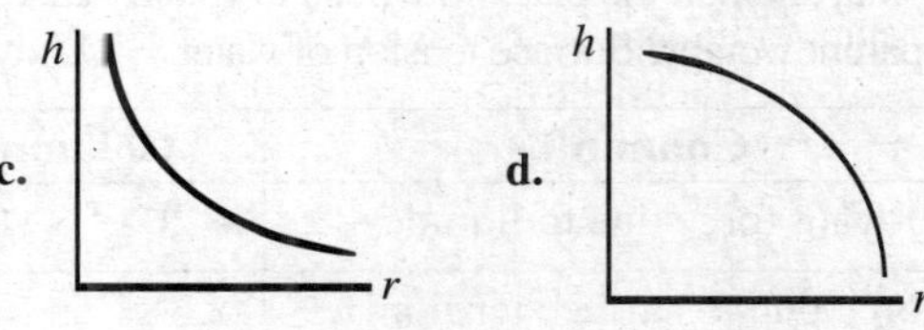

Fig. 4.120

For Problems 25–27

In the figure shown, A and B are two short steel rods each of cross-sectional area 5 cm². The lower ends of A and B are welded to a fixed plate CD. The upper end of A is welded to the L-shaped piece EFG, which can slide without friction on upper end of B. A horizontal pull of 1200 N is exerted at G as shown. Neglect the weight of EFG.

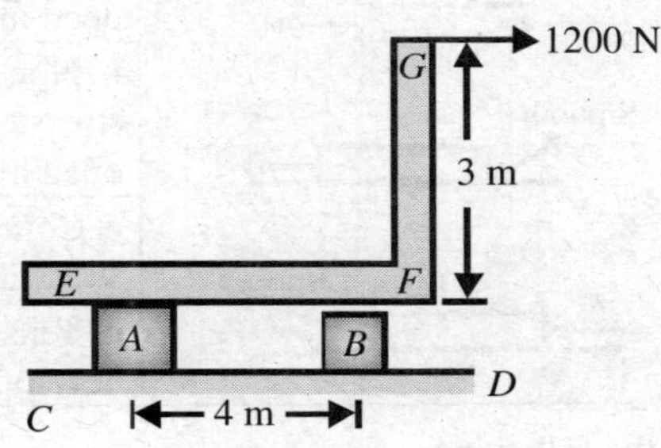

Fig. 4.121

25. Mark out the correct statement(s).
a. Shearing stress in A is zero.
b. Shearing stress in B is zero.
c. Shearing stress in both A and B is zero.
d. Shearing stress in both A and B is non-zero.

26. Longitudinal stress in A is
a. tensile in nature and having magnitude 180 N/m²
b. tensile in nature and having magnitude 240 N/m²
c. compressive in nature and having magnitude 180 N/m²
d. compressive in nature and having magnitude 240 N/m²

27. Longitudinal stress in B is
a. tensile in nature and having magnitude 180 N/cm²
b. tensile in nature and having magnitude 240 N/cm²
c. compressive in nature and having magnitude 180 N/cm²
d. compressive in nature and having magnitude 240 N/cm²

Matching Column Type

Solutions on page 4.80

1. A glass plate of length 10 cm breadth 4 cm and thickness 0.4 cm weighs 20 g in air. It is held vertically with long side horizontal and half the plate immersed in water. What will be its apparent weight? Surface tension of water = 70 J dyn/cm.

Column I	Column II
i. Buoyant force due to liquid	**a.** 20 gf
ii. Force due to surface tension	**b.** 8 gf
ii. Apparent weight of the plate	**c.** 1.5 gf
	d. 13.5 gf

2. In Column I, a uniform bar of uniform cross-sectional area under the application of forces is shown in the figure and in Column II, some effects/phenomena are given. Match the entries of Column I with the entries of Column II.

Column I	Column II
i. $F \rightarrow$ [bar] $\leftarrow F$	**a.** Uniform stresses developed in the rod
ii. $F \leftarrow$ [bar] $\rightarrow F$	
iii. Smooth; [bar] $\rightarrow F$	**b.** Non-uniform stresses developed in the rod
	c. Compressive stress developed
iv. $F \rightarrow$ [bar]; Smooth	**d.** Tensile stresses developed

3. Match the following:

Column I	Column II
i. longitudinal stress	**a.** Volume changes
ii. Shear stress	**b.** Shape changes
iii. Volume changes	**c.** Volume does not change
iv. Tensile stress	**d.** Shape does not change

4. Match the following:

Column I	Column II
i. Splitting of bigger drop into small droplets	**a.** Temperature increases
ii. Formation of bigger drop from small droplets	**b.** Temperature decreases
iii. Spraying of liquid	**c.** Surface energy increases
	d. Surface energy decrease

5. With regard to dependence of quantities given in Columns I and II, match the following:

Column I	Column II
i. Young's modulus of a substance	**a.** Depends on temperature
ii. Bulk modulus of a substance	**b.** Depends on length
iii. Modulus of rigidity of a substance	**c.** Depends on area of cross section
iv. Volume of a substance	**d.** Depends on the nature of material

6. Consider a wire of length l, cross-sectional area A and Young's modulus Y and match Column I with Column II:

Column I	Column II
i. If the wire is pulled at its ends by equal and opposite forces of magnitude F so that it undergoes an elongation x, according to Hooke's law, F kx, where (k) of the wire will depend on	**a.** Young's modulus Y
ii. Let us suspend the wire vertically from a rigid support and attach a mass m at its lower end. If the mass is slightly pulled down and released, it executes S.H.M. of a time period that will depend on	**b.** Elongation (x)
iii. If the given wire is fixed between two r. Length (l) rigid supports and its temperature is increased thermal stress that develops in the rod will depend on	**c.** Length (l)
iv. Work done in stretching the wire to a section length $l + x$ will depend on	**d.** Area of cross-section (A)

Integer Answer Type

Solutions on page 4.80

1. Two opposite forces $F_1 = 120$ N and $F_2 = 80$ N act on an elastic plank of modulus of elasticity $Y = 2 \times 10^{11}$ N/m^2 and length $l = 1$ m placed over a smooth horizontal surface. The cross-sectional area of the plank is $S = 0.5$ m^2. The change in length of the plank is $x \times 10^{-11}$ m. Find the value of x.

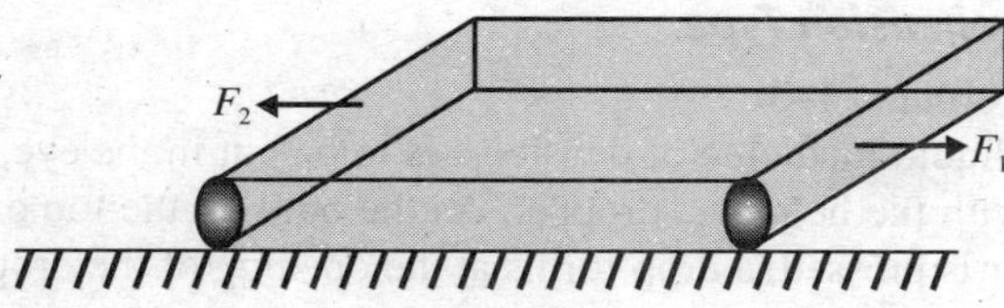

Fig. 4.122

2. A ring of radius r made of wire of density ρ is rotated about a stationary vertical axis passing through its centre and perpendicular to the plane of the ring as shown in the figure. Determine the angular velocity (in rad/s) of ring at which the ring breaks. The wire breaks at tensile stress σ. Ignore gravity. Take $\sigma/\rho = 4$ and $r = 1$ m.

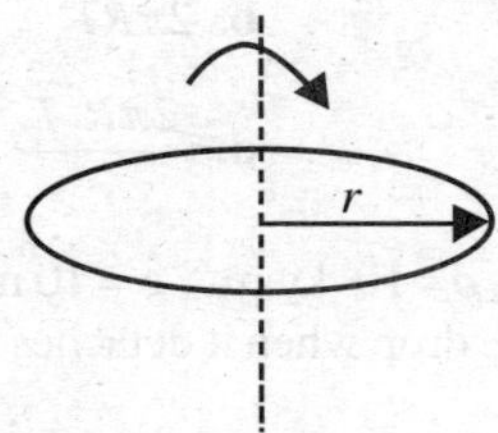

Fig. 4.123

3. A cube of side a and mass m just floats on the surface of water as shown in the figure. The surface tension and density of water are T and ρ_w, respectively. If angle of contact between cube and water surface is zero, find the distance h (in metres) between the lower face of cube and surface of the water.

(Take $m = 1$ kg, $g = 10$ m/s^2, $aT = \dfrac{10}{4}$ unit and $\rho_w a^2 g = 10$ unit)

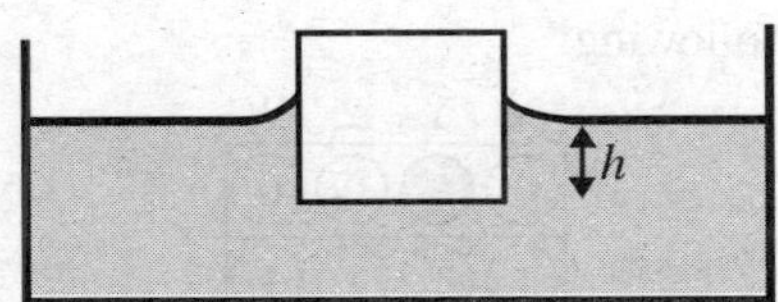

Fig. 4.124

4. A thin plate AB of large area A is placed symmetrically in a small gap of height h filled with water of viscosity η_0 and the plate has a constant velocity v by applying a force F as shown in the figure. If the gap is filled with some other liquid of viscosity 0.75 η_0, at what minimum distance (in cm) from top wall should the plate be placed in the gap, so that the plate can again be pulled at the same constant velocity V, by applying the same force F? (Take $h = 20$ cm)

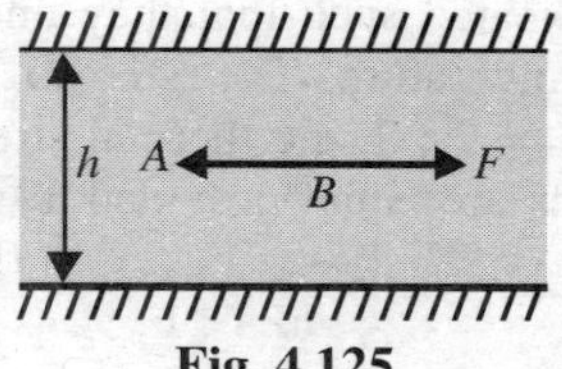

Fig. 4.125

5. The diameter of a gas bubble formed at the bottom of a pond is $d = 4$ cm. When the bubble rises to the surface, its diameter tension of water = $T = 0.07$ N m^{-1}.
6. n drops of water, each of radius 2 mm, fall through air at a terminal velocity of 8 cm/s. If they coalesce to form a single drop, then the terminal velocity of the combined drop is 32 cm/s. The value of n is
7. A substance breaks down under a stress of 10^5 Pa. If the density of the wire is 2×10^3 kg/m^3, find the minimum length of the wire which will break under its own weight ($g = 10$ m/s^2).

Archives

Solutions on page 4.81

Fill in the Blanks Type

1. A wire of length L and cross-sectional area A is made of a material of Young's modulus. If the wire is stretched by an amount x, the work done is ______. **(IIT-JEE, 1987)**
2. A solid sphere of radius R made of a material of bulk modulus K is surrounded by a liquid in cylindrical container. A massless piston of area A floats on the surface of the liquid. When a mass M is placed on the piston to compress the liquid, the fractional change in the radius of the sphere, $\delta R/R$, is ______. **(IIT-JEE, 1988)**
3. A uniform rod of length L and density ρ is being pulled along a smooth floor with a horizontal acceleration a (see figure). The magnitude of the stress at the transverse cross section through the mid-point of the rod is______. **(IIT-JEE, 1993)**

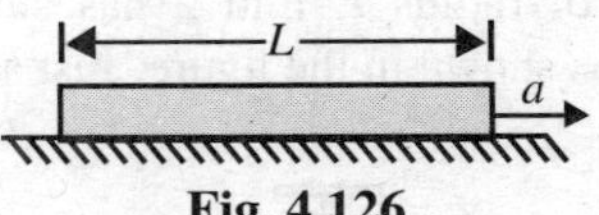

Fig. 4.126

Multiple Choice Questions with One Correct Answer Type

1. The following four wires are made of the same material. Which of these will have the largest extension when the same tension is applied? **(IIT-JEE, 1981)**
   a. Length = 50 cm, diameter = 0.5 mm
   b. Length = 100 cm, diameter = 1 mm
   c. Length = 200 cm, diameter = 2 mm
   d. Length = 300 cm, diameter = 3 mm
2. Two rods of different materials having coefficients of thermal expansionand Young's moduli Y_1, Y_2, respectively are fixed between two rigid massive walls.

The rods are heated such that they undergo the same increase in temperature. There is no bending of the rods. If $\alpha_1 : \alpha_2 = 2 : 3$, the thermal stresses developed in the two rods are equal provided $Y_1 : Y_2$ is equal to **(IIT-JEE, 1989)**

a. 2 : 3 **b.** 1 : 1
c. 3 : 2 **d.** 4 : 9

3. The adjacent graph shows the extension (Δl) of a wire of length 1 m suspended from the top of a roof at one end and with a load *W* connected to the other end. If the cross-sectional area of the wire is 10^{-6} m², Young's modulus of the material of the wire is **(IIT-JEE, 2003)**

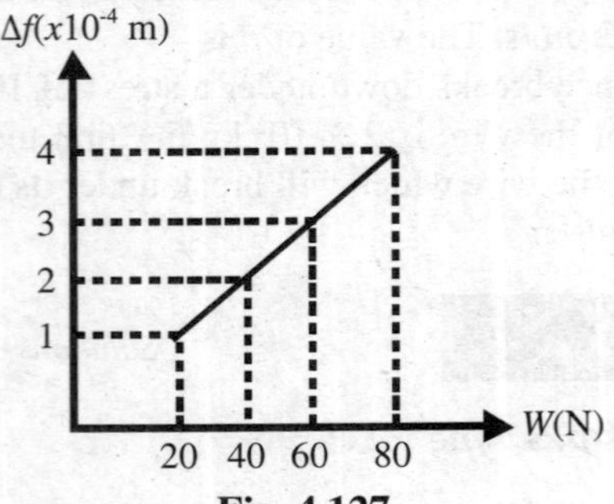

Fig. 4.127

a. 2×10^{11} N/m **b.** 2×10^{-11} N/m
c. 3×10^{-12} N/m **d.** 2×10^{-13} N/m

4. When temperature of a gas is 20°C and pressure is changed from $p_1 = 1.01 \times 10^5$ Pa to $p_2 = 1.165 \times 10^5$ Pa, the volume changes by 10%. The bulk modulus is **(IIT-JEE, 2005)**

a. 1.55×10^5 Pa **b.** 0.115×10^5 Pa
c. 1.4×10^5 Pa **d.** 1.01×10^5 Pa

5. A glass tube of uniform internal radius (*r*) has a valve separating the two identical ends. Initially, the valve is in tightly closed position. End 1 has a hemispherical soap bubble of radius *r*. End 2 has sub-hemispherical soap bubble as shown in the figure. Just after opening the valve **(IIT-JEE, 2008)**

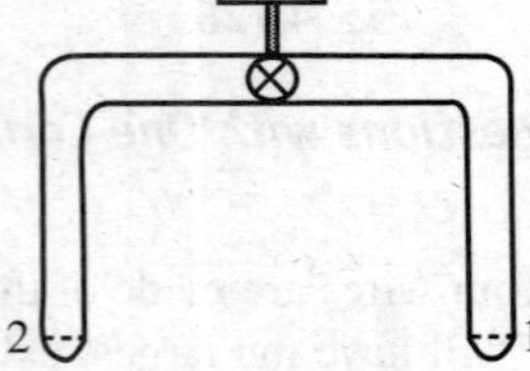

Fig. 4.128

a. Air from end 1 flows towards end 2. There is no change in the volume of the soap bubble.
b. Air from end 1 flows towards end 2. Volume of the soap bubble at end 1 decreases.
c. No change occurs.
d. Air from end 2 flows towards end 1. Volume of the soap bubble at end 1 increases.

Comprehension Type

For Problems 1–3

When liquid medicine of density ρ is to be put in the eye, it is done with the help of a dropper. As the bulb on the top of the dropper is pressed, a drop forms at the opening of the dropper. We wish to estimate the size of the drop. We first assume that the drop formed at the opening is spherical because that requires a minimum increase in its surface energy. To determine the size, we calculate the net vertical force due to the surface tension *T* when the radius of the drop is *R*. When the force becomes smaller than the weight of the drop, the drop gets detached from the dropper.

1. If the radius of the opening of the dropper is *r*, the vertical force due to the surface tension on the drop of radius *R* (assuming $r << R$) is

a. $2\pi rT$ **b.** $2\pi RT$
c. $\frac{2\pi r^2 T}{R}$ **d.** $\frac{2\pi R^2 T}{r}$

2. If $r = 5 \times 10^{-4}$ m, $\rho = 10^3$ kg m^{-3}, $g = 10$ m/s², $T = 0.11$ Nm^{-1}, the radius of the drop when it detaches from the dropper is approximately

a. 1.4×10^{-3} m **b.** 3.3×10^{-3} m
c. 2.0×10^{-3} m **d.** 4.1×10^{-3} m

3. After the drop detaches, its surface energy is

a. 1.4×10^{-6} J **b.** 2.7×10^{-6} J
c. 5.4×10^{-6} J **d.** 8.1×10^{-6} J

Integer Answer Type

This section contains two questions. The answer to each of the questions is a single-digit integer, ranging from 0 to 9. The appropriate bubbles below the respective question numbers in the ORS have to be darkened. For example, if the correct answers to question numbers X, Y, Z and W (say) are 6, 0, 9 and 2, respectively, then the correct darkening of bubbles will look like the following:

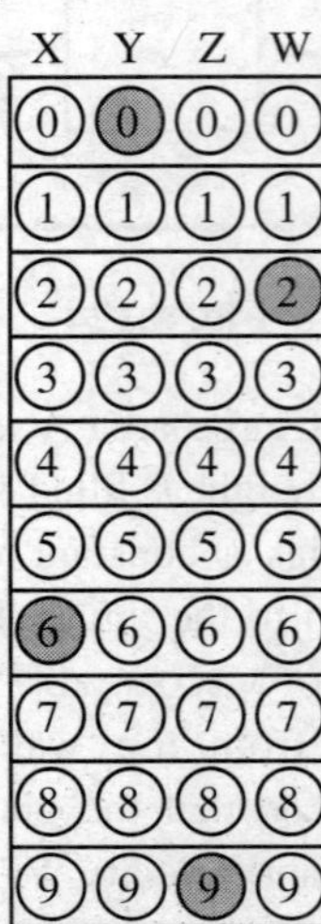

Fig. 4.129

1. Two soap bubbles A and B are kept in a closed chamber where the air is maintained at pressure 8 N/m². The radii of bubbles A and B are 2 cm and 4 cm, respectively. Surface tension of the soap water used to make bubbles is 0.04 N/m. Find the ratio nB/nA, where nA and nB are the number of moles of air in bubbles A and B, respectively. (Neglect the effect of gravity.) **(IIT-JEE, 2009)**

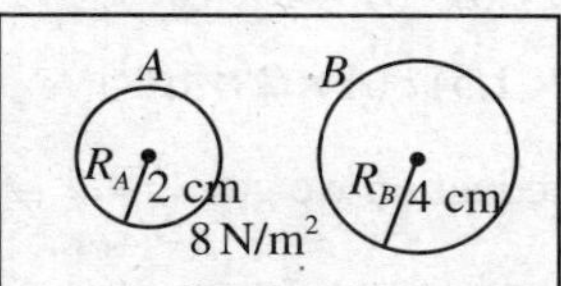

Fig. 4.130

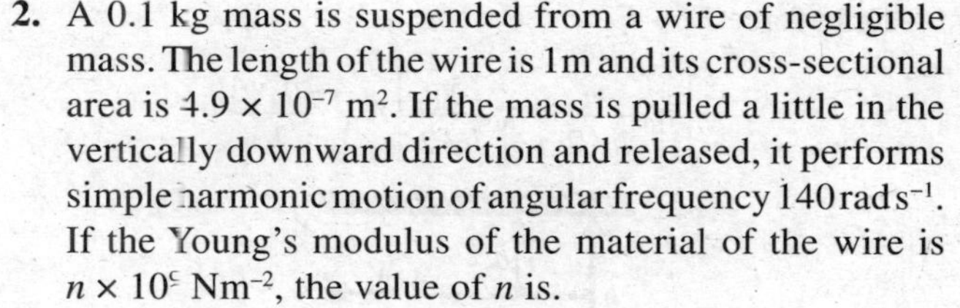

2. A 0.1 kg mass is suspended from a wire of negligible mass. The length of the wire is 1m and its cross-sectional area is 4.9×10^{-7} m². If the mass is pulled a little in the vertically downward direction and released, it performs simple harmonic motion of angular frequency 140 rad s⁻¹. If the Young's modulus of the material of the wire is $n \times 10^{9}$ Nm⁻², the value of n is.

ANSWERS AND SOLUTIONS

Subjective Type

1. a. Relative to the accelerating frame (trolley car), the forces acting on the bob are as shown in the figure.

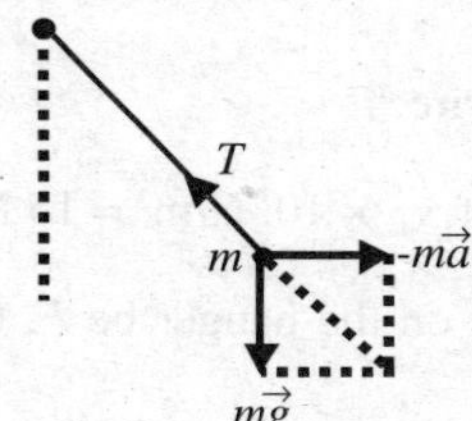

Fig. 4.131

Then $$T = m\sqrt{a^2 + g^2}$$

Since the string is light, same tension is felt at all points of the string.

Putting the area of cross section of the string $A = \pi r^2$ in the formula

$$\sigma = \frac{F'}{A}$$

where $$F' = T = m\sqrt{a^2 + g^2}$$

$$\sigma = \frac{m\sqrt{g^2 + d^2}}{\pi r^2}$$

b. As $$Y = \frac{F/A}{\Delta l/l} = \frac{\sigma l}{\Delta l}$$

which gives $$\Delta l = \frac{\sigma l}{Y}$$

Hence, $$\Delta l = \frac{m|\vec{a} - \vec{g}|l}{\pi r^2 Y} = \frac{m\left(\sqrt{g^2 + a^2}\right)l}{\pi r^2 Y}$$

2. As discussed in *Mechanics*, the tension T in the string at a distance x from its free end is given as

$$T = \frac{F}{l}x$$

Hence, stress, $$p = \frac{T}{A} = \frac{F}{Al}x$$

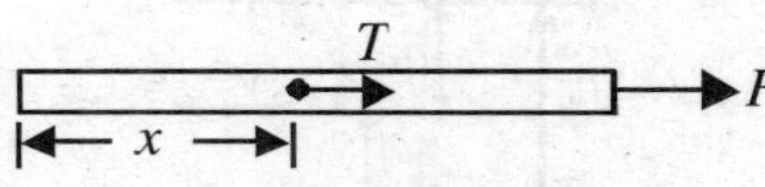

Fig. 4.132

Substituting (p) in the formula $U = \frac{1}{2Y}\int p^2 dV$

We have $U = \frac{1}{2Y}\int_0^l \frac{F^2}{A^2 l^2}x^2 dV$, where $dV = A dx$

This gives $U = \frac{F^2 l}{6AY}$

3. Excess pressure $= 2T/r = 146$ Pa

Pressure h below the surface $= \rho_0 + \rho gh$

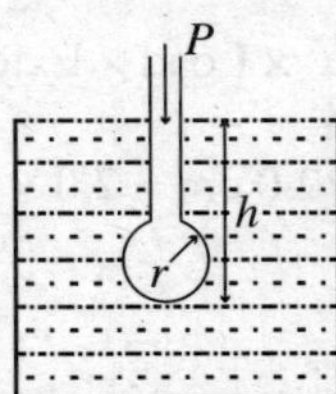

Fig. 4.133

$$\therefore P = (P_0 + \rho gh) + 2T/r$$

$$= 1\times10^5 + 10^3 \times 10 \times 0.1 + 2 \times \frac{7.3\times10^{-2}}{10^3}$$

$$= 1 \times 10^5 + 10^3 + 146$$

$$= 1 \times 10^5 + 1146$$

$$= 1.01146 \times 10^5$$

4. Case I:

Drag force $F_1 + F_2$

$$= \left[\eta_0 \frac{v}{h/2} + \eta_0 \frac{v}{h/2}\right] A = \frac{4\eta_0 vA}{h}$$

Fig. 4.134

Case II:
Drag force on the plate

$$\left[\eta \frac{v}{h-y} + \eta \frac{v}{y}\right] A = \frac{\eta vha}{y(h-y)}$$

In both case, drag force are equal

$$\Rightarrow \quad \frac{4\eta_0 vA}{h} = \frac{\eta vha}{y(h-y)}$$

$$\Rightarrow \quad \eta = \frac{4\eta_0 y}{h}\left(1 - \frac{y}{h}\right)$$

Fig. 4.135

5. Applying Boyle's law

$$P_0 Al = \left[(P_0 - h\rho g) + \frac{4T}{d}\cos\theta\right] A(L-h)$$

$$T = \frac{P_0}{4\cos\theta}\left[\left(\frac{h}{L-h}\right) + h\rho g\right] d$$

6. Assumption: Total elastic energy of catapult is converted into kinetic energy of missile without any heat loss.

Elastic energy = $\frac{1}{2}$ × Load × Extension
Here extension,
$\Delta L = 12.0 - 10.0 = 2.0$ cm $= 2.0 \times 10^{-2}$ m
From relation

$$Y = \frac{FL}{A\Delta L}$$

$$\text{Load, } F = \frac{YA\Delta L}{L}$$

$$= \frac{5.0\times10^8 \times 1\times10^{-6} \times 2.0\times10^{-2}}{10.0\times10^{-2}} = 100 \text{ N}$$

If v is velocity of projection, we have
Elastic energy of catapult = kinetic energy of missile

i.e. $\frac{1}{2}$ × load × extension = $\frac{1}{2}$ mv^2

$$\frac{1}{2} \times 100 \times 2.0 \times 10^{-2} = \frac{1}{2} \times (5.0 \times 10^{-3}) v^2$$

This gives $v^2 = \frac{100 \times 10^{-2} \times 2}{5 \times 10^{-3}} = 400$

$\therefore v^2 = \sqrt{400} = 20$ m/s.

7. $\Delta l_{brass} = (6\text{ g} \times 1.0) / (A \times 1 \times 10^{11})$

$\Delta l_{steel} = (10\text{ g} \times 1.5) / (A \times 2 \times 10^{11})$

After solving equation we get $\frac{\Delta \ell_{steel}}{\Delta \ell_{brass}} = 1.25$

8. Let tension in steel rod be T_1 and in brass rod be T_2, then
$T_1 + 2T_2 = 5000$ (i)
Extension in both steel and brass rods will be same:
$\Delta l_s = \Delta l_b$

$$\frac{T_1 \ell_1}{AY_1} = \frac{T_2 \ell_2}{AY_2} \quad \Rightarrow T_1 = \frac{4T_2}{3} \qquad \text{(ii)}$$

From Eqs. (i) and (ii) $T_1 = 2000$ N
$T_2 = 1500$ N
Stress in steel wire:

$$S_1 = \frac{T_1}{A} = \frac{2000}{10^{-4}} = 2 \times 10^7 \text{ N/m}^2 = 20 \text{ MPa}$$

Stress in brass wire:

$$S_2 = \frac{T_2}{A} = \frac{1500}{10^{-4}} = 1.5 \times 10^7 \text{ N/m}^2 = 15 \text{ MPa}$$

9. **(i)** Let load put on the hanger be F, then stress in lower wire
$S_1 = (m_1 g + F) / (0.003 \times 10^{-4})$
Let $S_1 = 8 \times 10^8$ N/m², then
$\Rightarrow F = 140$ N
Let stress developed in upper wire be S_2, then
$S_2 = [(m_1 + m_2) g + F] / (0.006 \times 10^{-4})$
Let $S_2 = 8 \times 10^8$ N/m²
$\Rightarrow F = 180$ N
We have to select the lower value of F. Hence, maximum load that can be suspended is $140/g = 14$ kg

(ii) $S_1 = (m_1 g + F) / (0.003 \times 10^{-4}) = 8 \times 10^8$
$\Rightarrow 10 \times 10 + F = 240 \Rightarrow F = 140$ N
$S_2 = [(m_1 + m_2) g + F] / (0.006 \times 10^{-4}) = 8 \times 10^8$
$\Rightarrow 46g + F = -480 \Rightarrow F = 20$ N
Hence, the maximum load that can be suspended is $20/g$ = 2 kg

10. The restoring force (F) produced in the wire provides the necessary centripetal force i.e.
$F = ML\omega^2$ (i)
(Since radius of circular path = length of wire = L)
We have, Young's modulus

$$Y = \frac{F/A}{\Delta L / L} \text{ or } F = YA.\frac{\Delta L}{L} \qquad \text{(ii)}$$

Comparing Eqs. (i) and (ii), we get

$YA\frac{\Delta L}{L} = mL\omega^2$ or $Y = \frac{mL^2w^2}{A\Delta L}$

Substituting given values, we get

$Y = 400 \times 10^9 = 4 \times 10^{11}$ N/m^2

The breaking force,

$F = mL\,\omega^2_{max}$

$= (1\text{kg})\,(1\text{m})\,(100\text{ rad/s})^2 = 10^4$ newton

Therefore, breaking stress = $\frac{\text{Breaking force}}{\text{Area}}$

$= \frac{10^4}{10^{-6}} = 10^{10}\,\text{N/m}^2$

11. (i) On substituting the values, we get

$\Delta l = \frac{Fl_0}{YA} = 2.5 \times 10^{-4}$ m.

$= 0.25$ mm

(ii) The force that each car exerts on the cable is

$F = m\omega^2 l_0 = \frac{W}{g}\omega^2 l_0$, so

$$\Delta l = \frac{Fl_0}{YA} = \frac{W\omega^2 l_0^2}{gYA}$$

On substituting the value we get 3.8×10^{-4} m = 0.38 mm

12. $h = h_0 = \frac{2T\cos\theta}{\rho g r}$

After substituting the value = 57.6 cm

(i) Since l (= 60 cm) > h_0.

$h = 57.6$ cm

(ii) Since l (= 50cm) < h_0.

$h = 50$ cm

13. $F = 2\pi r_1 T + 2\pi r_2 T$

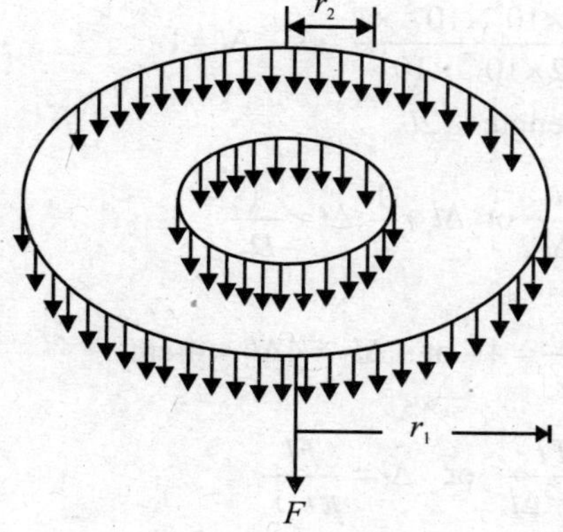

Fig. 4.136

$= 2\pi(r_1 + r_2)\,T$

$= 2 \times 3.14(10 + 5)(72) = 6782.4$ dyne

14. For mass conservation,

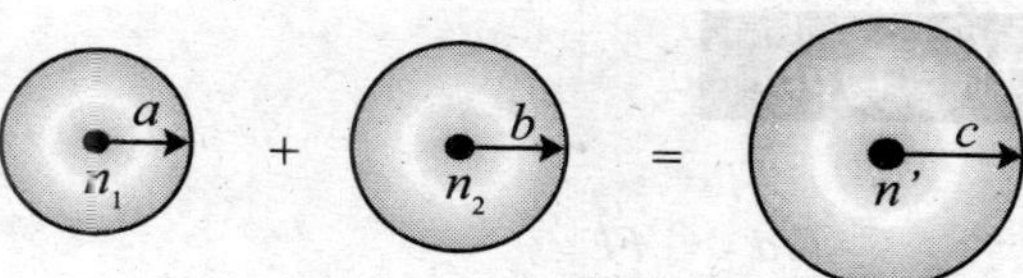

Fig. 4.137

$n_1 + n_2 = n'$

$$\frac{P_1V_1}{RT} + \frac{P_2V_2}{RT} = \frac{PV}{RT}$$

$P_1V_1 + P_2V_2 = PV$ (i)

$P_1 = \frac{4T}{c} + P_0$, $P_2 = \frac{4T}{b} + P_0$

and $P = \frac{4T}{C} + P_0$ in Eq. (i) to get the given answer.

15. $P_C = P_B + \frac{2T}{r}$

$P_B = P_A = \rho_1 g h_1 + \rho_2 g(h - h_1)$

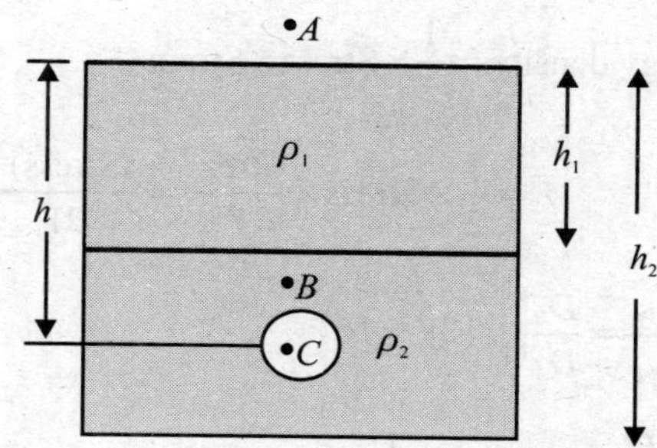

Fig. 4.138

where $P_A = P_0$

$$P_C = P_0 + \rho_1 g h_1 + \rho_2 g(h - h_1) + \frac{2T}{r}$$

Then evaluate by putting the numerical values.

16. $\Delta U = N(4\pi r_1^2)T - 4\pi r^2 T$

$= 4\pi T\,(N r_1^2 - r^2)$ (i)

By mass conservation,

$$N\left(\frac{2}{3}\pi r_1^3 \rho\right) = \frac{4}{3}\pi R^3 \rho$$

$r_1 = \frac{R}{(N)^{1/3}}$ (ii)

Substitute r_1 from Eq. (ii) in Eq. (i).

Objective Type

1. a.

$$Y=\frac{F/a}{\Delta l/l}=\frac{Fl}{a\,\Delta l}$$

or $Y=\dfrac{Fl\times 4}{\pi D^2\times \Delta l}$ or $\Delta l \propto \dfrac{1}{D^2}$ or $\dfrac{\Delta l_2}{\Delta l_1}=\dfrac{D_1^2}{D_2^2}=\dfrac{n^2}{1}$

2. a. Breaking strength = Breaking stress × $\dfrac{\pi D^2}{4}$

Breaking stress is unchanged.

D is halved. So, breaking strength becomes one-fourth, i.e.,

$$\frac{1}{4}\times 1.5\times 10^5 \text{ N} \quad\text{or}\quad 0.375\times 10^5 \text{ N}.$$

3. d. $Y=\dfrac{F/a}{\Delta l/l}=\dfrac{Fl}{a\,\Delta l}$ or $\Delta l \propto \dfrac{1}{D^2}$

a. $\dfrac{100}{l^2}=100$ **b.** $\dfrac{200}{4}=50$

c. $\dfrac{300}{9}=33.33$ **d.** $\dfrac{50}{(1/2)^2}=200$

4. a. Energy density $=\dfrac{1}{2}\times\text{Stress}\times\text{Strain}$

$$=\frac{1}{2}\times\text{Stress}\times\frac{\text{Stress}}{Y}=\frac{(\text{Stress})^2}{2Y}\propto\frac{1}{D^4}$$

Now, $\dfrac{u_A}{u_B}=\dfrac{D_B{}^4}{D_A{}^4}=[2]^4=16$

5. c. $Y=\dfrac{Fl}{a\Delta l}$ or $\Delta l \propto \dfrac{1}{a}$

Again, $m = al\rho$ or $m \propto a$

$\therefore \quad \Delta l \propto \dfrac{1}{m}$

$$\frac{\Delta l_1}{\Delta l_2}=\frac{m_2}{m_1}=\frac{2}{3}$$

6. d. $Y=\dfrac{Fl}{a\,\Delta l}$

In the given problem, Y, l and Δl are constants.

$\therefore \quad F \propto a$

or $\quad F \propto \pi r^2$ or $F \propto r^2$ or $\dfrac{F_1}{F_2}=\dfrac{r_1^2}{r_2^2}=\dfrac{1}{4}$

7. b.

$$Y=\frac{Mg\times 4\times l}{\pi D^2\times \Delta l} \quad\text{or}\quad \Delta l\propto\frac{1}{D^2}$$

When D is doubled, Δl becomes one-fourth, i.e., $\dfrac{1}{4}\times 2.4$ cm, i.e., 0.6 cm.

8. b.

$$Y=\frac{F/a}{\Delta l/l} \quad\text{or}\quad Y=\frac{Fl}{a\,\Delta l}$$

or $\quad \Delta l=\dfrac{Fl}{aY}=\dfrac{Fl}{\pi r^2 Y}$

In the given problem, $\Delta l \propto \dfrac{l}{r^2}$

When both l and r are doubled, Δl is halved.

9. d. 10 m column of water exerts nearly 1 atmosphere pressure. So, 100 m column of water exerts nearly 10 atmosphere pressure, i.e., 10×10^{-5} Pa or 10^6 Pa.

Now, $\quad K=\dfrac{(\Delta P)V}{\Delta V}=\dfrac{10^6\times 100}{0.1}\text{Pa}=10^9\text{ Pa}$

10. b. Energy per unit volume $=\dfrac{1}{2}\times\text{Stress}\times\text{Strain}$

$$=\frac{1}{2}\times\text{Stress}\times\frac{\text{Stress}}{Y} \quad\Big|\quad Y=\frac{\text{Stress}}{\text{Strain}}=\frac{S^2}{2Y}$$

11. c. Work done $=\dfrac{1}{2}F\times\text{Extension}$

$$=\frac{1}{2}\times\frac{YA}{L}\times 1 \quad\Big|\quad Y=\frac{F\times L}{A\times 1}$$

$$=\frac{YA}{2L} \quad\Big|\quad F=\frac{YA}{L}$$

12. c. $\Delta l=\dfrac{Fl}{aY}$

$$\Delta l=\frac{2\times 10^5\times 10^{-5}\times 1}{2\times 10^{-4}\times 10^4} \quad\text{or}\quad \Delta l=1$$

So, new length is $2l$.

13. c. $Y=\dfrac{Fl}{a\,\Delta l}$ or $\Delta l\propto\dfrac{1}{a};\Delta l\propto\dfrac{1}{D^2}$

$$\frac{\Delta l_2}{\Delta l_1}=\frac{D_1{}^2}{D_2{}^2}=4 \quad\text{or}\quad \Delta l_2=4\Delta l_1=4\text{ cm.}$$

14. c. $Y=\dfrac{Fl}{\pi r^2\Delta l}$ or $\Delta l=\dfrac{Fl}{\pi r^2 Y}$

$$\Delta l\propto\frac{1}{r^2},\quad \Delta l'\propto\frac{2l}{(\sqrt{2}r)^2} \quad\text{or}\quad \Delta l'\propto\frac{l}{r^2}$$

$\therefore \quad \dfrac{\Delta l}{\Delta l'}=1$

15. b.

$$\Delta l = \frac{Fl}{\pi r^2 Y} \quad \text{or} \quad \frac{\Delta l}{l} \propto \frac{1}{r^2 Y}$$

$$\frac{\left(\frac{\Delta l}{l}\right)_s}{\left(\frac{\Delta l}{l}\right)_{Cu}} = \frac{r_{Cu}^{\,2}\,_{Cu}}{r_s^{\,2} Y_3} = \frac{4r_s^{\,2}\,\frac{Y_s}{2}}{r_s^{\,2} Y_s} = 2$$

$$\therefore \quad \left(\frac{\Delta l}{l}\right)_s = 2\left(\frac{\Delta l}{l}\right)_{Cu}$$

16. c.

$$\text{Breaking stress} = \frac{\text{Maximum weight}}{\text{Area of cross section}}$$

$$10^6 = \frac{al\rho g}{a} = l\rho g$$

or $$l = \frac{10^6}{\rho g} = \frac{10^6}{4\times 10^3 \times 10}\,\text{m} = 25\text{ m}$$

17. a. Since water rises to height of 2 cm in a capillary

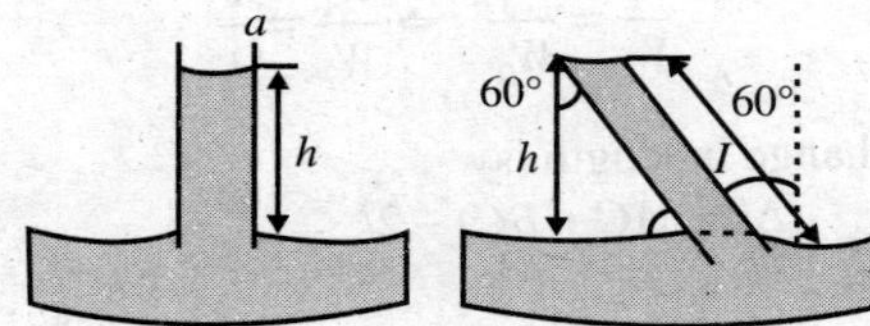

Fig. 4.139

If tube is at 60°, in this case height must be equal to

$$h = 2\text{ cm} \Rightarrow \cos 60° = \frac{h}{l}$$

$$\therefore \quad l = \frac{h}{\cos 60°} = \frac{2}{1/2} = 4\text{ cm}$$

18. c. Radius of the larger drop = R
Suppose radius of the droplets = r
Since volume will be remain constant,

$$\frac{4}{3}\pi R^3 = 8\times\frac{4}{3}\pi r^2 \quad \text{(as no. of droplets = 8)}$$

$$\therefore \quad r = \left(\frac{R^3}{8}\right)^{\frac{1}{3}} = \frac{R}{2}$$

Therefore, work done = Increase in surface area × Surface tension $= \left[84\pi\left(\frac{R}{2}\right)^2 - 4\pi R^2\right]\times T$

$$= (8\pi R^2 - 4\pi R^2)\times T = 4\pi R^2 T$$

19. c. Work done = (increase in surface area) × surface tension

$$= 2\times\left[4\pi\left(\frac{2D}{2}\right)^2 - 4\pi\left(\frac{D}{2}\right)^2\right]\times T$$

(Since for sap bubble, there are two free surfaces)

$$= 2\times(4\pi D^2 - \pi D^2)\,T = 6\,\pi D^2 T$$

20. b. Suppose, R = radius of water drop
and r = radius of droplets

$$\therefore \quad \frac{4}{3}\pi R^3 = 8\times\frac{4}{3}\pi r^3$$

(Since volume remains constant)

$$\therefore \quad r = \frac{R}{2}$$

Since excess pressure inside drop $= \frac{2T}{R}$

(T—surface tension, R—radius)
Therefore, pressure difference between inner and outer surface of big drop will be half of that for smaller droplet.

21. d. For equilibrium,
Total upward force by surface tension
= Weight of the water in tube
$\Rightarrow$ $\pi \times D \times$ surface tension (circumference)
$= \pi\ (D/2)^2 \times h \times$ density $\times g$ (cross section)
where D (diameter) = 0.1 mm = 0.01 cm
Density of water = 1×10^{-3} g/cm^3
$\Rightarrow$ $\pi \times (0.01)\times 75\times 10^{-3}$

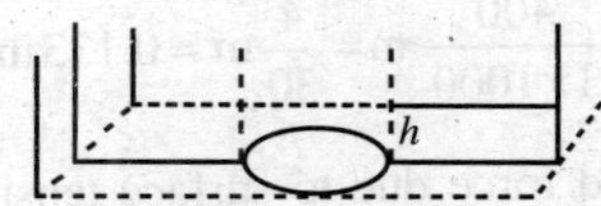

Fig. 4.140

$$= \pi \times\left(\frac{0.01}{2}\right)^2 \times h\times 1\times 10^{-3}\times 1000$$

$$\therefore \quad h = \frac{0.75\times 0.01\times 4}{0.01\times 0.01} = 0.3\,\text{m} = 30\,\text{cm}$$

22. c. Height, $h = \dfrac{2T\cos\theta}{r\rho g}$

$$\therefore \quad \text{For water, } h_w = \frac{2\times T_w\times\cos 0°}{r\times 1\times g}$$

And, for mercury,

$$h_m = -\frac{2\times T_m\times\cos 135°}{r\times 13.6\times g}$$

$$\therefore \quad \frac{h_w}{h_m} = \frac{2\times T_w\times 1}{r\times 1\times g}\times\frac{r\times 13.6\times g\times\sqrt{2}}{2\times T_m\times 1}$$

$[\because \cos 135° = -1/\sqrt{2}]$

$$\Rightarrow \quad \frac{10}{3.42} = \frac{T_w}{T_m} \times 13.6 \times \sqrt{2}$$

$$\therefore \quad \frac{T_w}{T_m} = \frac{10}{3.42 \times 13.6 \times 1.414} = \frac{1}{6.5}$$

$\therefore \quad T_w : T_m = 1:6.5.$

23. b. Since, ball gains the constant velocity. Therefore, net force on the ball = zero.

$\Rightarrow$ Weight of the ball = Buoyancy force + Viscous force (F_v)

$$\Rightarrow Mg = \frac{M}{d_1} \times d_2 \times g + F_v$$

$$\therefore F_v = Mg\left(1 - \frac{d_2}{d_1}\right)$$

24. b. $\left[P_0 + \frac{4\sigma}{R_2}\right] - \left[P_0 + \frac{4\sigma}{R_1}\right] = \frac{4\sigma}{R}$

or $\quad \frac{1}{R} = \frac{1}{R_2} - \frac{1}{R_1}$

or $\quad R = \frac{R_1 R_2}{R_1 - R_2} = \frac{50 \times 80}{30} \text{mm} = \frac{400}{3} \text{mm}$

$$= \frac{400}{3 \times 1000} \text{m} = \frac{4}{30} \text{m} = 0.133 \text{m}$$

25. a. Total upward force due to surface tension $= \sigma(2r_1 + 2\pi_2)$. This supports the weight of the liquid column of height h. Weight of liquid column $= h\,[\pi r_2^2 - \pi r_1^2]\,\rho g$

Equating, we get $h\pi\,(r_2 + r_1)(r_2 - r_1)\rho g = 2\pi\sigma\,(r_1 + r_2)$

or $\quad h\,(r_2 - r_1)\rho g = 2\sigma \quad$ or $\quad h = \frac{2\sigma}{(r_2 - r_1)\rho g}$

26. d. Energy released $= [n \times 4\pi a^2 - 4\pi b^2]\,\sigma$

Now, $\quad n \times \frac{4}{3}\pi r^3 = \frac{4}{3}\pi b^3 \quad$ or $\quad n = \frac{b^3}{a^3}$

Therefore, energy released is

$$\left[\frac{b^3}{a^3} \times 4\pi a^2 - 4\pi b^2\right]\sigma = 4\pi b^2\left[\frac{b}{a} - 1\right]\sigma$$

Now, $\quad \frac{1}{2}\left(\frac{4}{3}\pi b^3\right)\rho v^2 = 4\pi b^2\left[\frac{b}{a} - 1\right]\sigma$

or $\quad v = \left[\frac{6\sigma}{\rho}\left(\frac{1}{a} - \frac{1}{b}\right)\right]^{1/2}$

27. b. As the weight of wire acts at centre of gravity. Therefore, only half the length of wire gets extended.

Now $\quad Y = \frac{F}{A} \cdot \frac{(L/2)}{\Delta l} = \frac{Mg(L/2)}{A\,\Delta l}$

$$\Rightarrow \quad \Delta l \frac{MgL}{2AY} \Rightarrow \Delta l \frac{AL\rho gL}{2AY}$$

$$\therefore \quad \Delta l = \frac{\rho L^2 g}{2Y}$$

28. b. Work done $= \frac{1}{2} \times$ stretching force $\times$ extension.

Therefore, net work done in increasing the length from 0.164 mm to 1.02 mm is

$$W_2 - W_1 = \frac{1}{2} \times 5 \times 9.8 \times 1.02 \times 10^{-3} - \frac{1}{2} \times 3 \times 9.8 \times 0.61 \times 10^{-3} = 0.016 \text{ J}$$

29. c. Y µ weight applied

$$\therefore \quad \frac{Y_1}{Y_2} = \frac{W_1}{W_2} \Rightarrow \frac{W_1}{W_2} = \frac{3}{1}$$

30. c. Here change in length is

$$\Delta l = (AC + BC) - 2l$$

$$= 2(l^2 + x^2)^{1/2} - 2l = 2l\left(1 + \frac{x^2}{l^2}\right)^{1/2}$$

$$= 2l\left(1 + \frac{1}{2}\frac{x^2}{l^2}\right) - 2l = \frac{x^2}{l}$$

Fig. 4.141

$$\therefore \quad \text{Strain} = \frac{\Delta l}{2l} = \frac{x^2}{2l^2}$$

31. d. We know $Y = \frac{\text{Stress}}{\text{Strain}} = \frac{Wl}{2Ax} \times \frac{2l^2}{x^2}$

$$= \frac{Wl^3}{Ax^3} \text{ or } x = \left(\frac{W}{AY}\right)^{1/3} \times l$$

32. a. $W = \frac{1}{2}\tau\phi = \frac{1}{2} \times 8 \times \frac{\pi}{4} = \pi$ J

33. c. The density would increase by 0.1% if the volume decreases by 0.1%.

$$K = \frac{\Delta P}{\Delta V / V}$$

$$\Delta P = K\frac{\Delta V}{V} = 2\times10^9 \times \frac{0.1}{100} = 2\times10^6 \text{ N/m}^2$$

34. c. Thermal stress = $Y\alpha t$

In the given problem,

$$Y\alpha = \text{constant}$$

$$\frac{Y_1}{Y_2} = \frac{\alpha_2}{\alpha_1} = \frac{3}{2}$$

35. c. Force = Weight suspended + Weight of 3*L*/4 of wire

$$= W_1 + \frac{3W}{4}$$

$$\text{Stress} = \frac{\text{Force}}{\text{Area}}$$

36. d.

$$Y = \frac{Fl}{a\Delta l}$$

Y, *l* and *F* are constants.

$$\therefore \quad \Delta l \propto \frac{1}{D^2}$$

$$\frac{\Delta l_2}{\Delta l_1} = \frac{D_1^2}{D_2^2} = \frac{1}{16}$$

$$\therefore \quad \Delta l_2 = \frac{1}{16}\text{ mm}$$

37. b. $Y = \frac{F/A}{\Delta l/l} = \frac{Fl}{\Delta l A} = \frac{Fl}{\Delta l(\pi R^2)}$

Hence $\quad Y \propto \frac{1}{R^2}$

$$\frac{Y_B}{Y_S} = \frac{R_S^2}{R_B^2}$$

$$\frac{10\times10^{10}}{20\times10^{10}} = \frac{R_S^2}{R_B^2}$$

$$R_B^2 = 2R_S^2$$

$$R_B = \sqrt{2}R_S$$

$$\Rightarrow \quad R_S = \frac{R_B}{\sqrt{2}}$$

38. c. $Y = \frac{Fl}{a\Delta l}$

Y, *l* and *a* are constants

$$\therefore \quad \frac{Fl}{\Delta l} = \text{constant or } \Delta l \propto F$$

Now, $\quad l_1 - l = T_1$ and $l_2 - l = T_2$

Dividing, $\quad \frac{l_1 - l}{l_2 - l} = \frac{T_1}{T_2}$

or $\quad l_1T_2 - lT_2 = l_2T_1 - lT_1 \quad$ or $\quad T(T_1 - T_2) = l_2T_1 - l_1T_2$

or $\quad l = \frac{l_2T_1 - l_1T_2}{T_1 - T_2} \quad$ or $\quad l = \frac{l_1T_2 - l_2T_1}{T_3 - T_1}$

39. b. $T = \frac{2m_1m_2}{m_1 + m_2}g = \frac{2\times1\times2}{1+2}\times10 \text{ N} = \frac{40}{3}\text{ N}$

If *r* is the minimum radius, then

$$\text{Breaking stress} = \frac{\frac{40}{3}}{\pi r^2} \quad \text{or} \quad \frac{40}{3\pi}\times10^6 = \frac{40}{3\pi r^2}$$

or $\quad r^2 = \frac{1}{10^6} \quad$ or $\quad r = \frac{1}{10^3}\text{ m}$

or $\quad r = \frac{1}{10^3}\times10^3 \text{ m} = 1 \text{ mm}$

40. b. The elastic potential is

$$\text{energy} = \frac{1}{2}\text{ Stress} \times \text{Strain}$$

$$= \frac{1}{2}Y(\text{strain})^2 = \frac{1}{2}Y\left(\frac{\Delta l}{L}\right)^2$$

$$\therefore \quad \frac{U_2}{U_1} \propto \left(\frac{\Delta l_2}{\Delta l_1}\right)^2 = \left(\frac{10}{2}\right)^2$$

$$\frac{U_2}{U_1} = 25 \Rightarrow U_2 = 25\,U_1$$

41. b. $\frac{1}{K} = \frac{\Delta V/V}{\Delta P} \quad$ or $\quad \frac{\Delta V}{V} = \Delta P\left(\frac{1}{K}\right)$

or $\quad \frac{\Delta V}{V}\times100 = 10^5\times8\times10^{-12}\times100 = 8\times10^{-5}$

42. a. $mr\omega^2$ = Breaking stress × Cross-sectional area

$$10 \times 0.3\omega^2 = 4.8 \times 10^7 \times 10^{-6} = 48$$

or $\quad \omega^2 = \frac{48}{3} = 16 \quad$ or $\quad \omega = 4 \text{ rad/s}$

43. b. $\Delta P = h\rho g = 200 \times 10^3 \times 10 \text{ N/m}^2 = 2 \times 10^{-6} \text{ N/m}^2$

$$K = \frac{\Delta P}{\frac{\Delta V}{V}} = \frac{2\times10^6}{\frac{0.1}{100}}$$

$$=\frac{2\times10^8}{0.1}\text{N/m}^2=2\times10^9\ \text{N/m}^2$$

44. d. $\frac{20a\times2.5\times10^3\times10+5\times10^5}{a}=1.6\times10^6$

or $500a+500=1600a$

or $1100a=500$ or $a=\frac{5}{11}\text{m}^2=0.45\text{m}^2$

45. d. $W=\frac{1}{2}F\,\Delta l$

$$W=\frac{1}{2}\times\frac{Y\pi r^2\Delta l}{l}\Delta l \quad \text{and } Y=\frac{Fl}{\pi r^2\Delta l}$$

or $W=\frac{Y\pi r^2\Delta l^2}{2l}$ and $F=\frac{Y\pi r^2\Delta l}{l}$

or $W\propto\frac{r^2}{l},\ W'\propto\frac{(2r)^2 2}{l}$

$$\frac{W'}{W}=8 \quad \text{or} \quad W'=8\times2\,\text{J}=16\,\text{J}$$

46. c. $m=alp$, m and ρ are constants.

$$\therefore \quad \frac{a_1}{a_2}=\frac{l_2}{l_1}=\frac{3}{2}$$

Now, $Y=\frac{Fl}{a\,\Delta l}$ or $\Delta l\propto\frac{l}{a}$

or $\frac{\Delta l_1}{\Delta l_2}=\frac{l_1}{l_2}\times\frac{a_2}{a_1}=\frac{2}{3}\times\frac{2}{3}$ or $\frac{\Delta l_1}{\Delta l_2}=\frac{4}{9}$

47. c. $Y=\frac{Fl}{a\,\Delta l}$; Y, l and a are constants

$\therefore F=\Delta l$

In the first case, $V\rho g=16$

In the second case, $(V\rho g-V\times1\times g)=14$

Dividing, $\frac{Vg(\rho-1)}{Vg\rho}=\frac{14}{16}=\frac{7}{8}$

or $\frac{\rho-1}{\rho}=\frac{7}{8}$ or $8\rho-7\rho$ or $\rho=8\ \text{g/cm}^3$

48. a. $Y=\frac{Fl}{a^2\Delta l}$

$$\Delta l\propto\frac{1}{a^2} \quad \text{or} \quad \Delta l\propto\frac{1}{D^4}$$

$$\frac{\Delta l_A}{\Delta l_B}=\frac{D_B{}^4}{D_A{}^4}=\frac{1^4}{\left(\frac{1}{2}\right)^4}=16$$

☞ $$Y=\frac{F}{a}\times\frac{l}{\Delta l}$$

Now, $V=al$ or $l=\frac{Y}{a}$ $\therefore Y=\frac{FV}{a^2\Delta l}$

49. a. $Y=\frac{\text{stress}}{\Delta L/L}$ or $\frac{\Delta L}{L}=\frac{\text{stress}}{Y}=\frac{5\times10^7}{2\times10^{11}}=2.5\times10^{-4}$

Now, $V=\pi r^2L$

$$\frac{\Delta V}{V}=\frac{\pi\,\Delta(r^2L)}{\pi r^2L}$$

or $\frac{\Delta V}{V}=\frac{r^2\Delta L+L\times2r\Delta r}{r^2L}$ or $\frac{\Delta V}{V}=\frac{\Delta L}{L}+2\frac{\Delta r}{r}$

or $2\frac{\Delta r}{r}=\frac{\Delta V}{V}-\frac{\Delta L}{L}$ or $2\frac{\Delta r}{r}=\frac{0.02}{100}-2.5\times10^{-4}$

or $\frac{\Delta r}{r}=1\times10^{-4}-\frac{2.5}{2}\times100^{-4}=-0.25\times10^{-4}$

50. c. $\frac{\rho'}{\rho}=\frac{\frac{M}{V-\Delta V}}{M/V}=\frac{V}{V-\Delta V}$

or $\frac{\rho'}{\rho}=V(V-\Delta V)^{-1}$ or $\frac{\rho'}{\rho}=1+\frac{\Delta V}{V}$

(using binomial theorem)

or $\frac{\Delta V}{V}=\frac{\rho'}{\rho}-1$ or $\frac{\Delta V}{V}=\frac{\rho'-\rho}{\rho}$

Again, $k=k=\frac{\Delta P}{\frac{\Delta V}{V}}=\frac{\rho\Delta P}{\rho'-\rho}$ or $\rho'-\rho=\frac{\rho\Delta R}{K}$

But, $\Delta P=P$ (given)

$$\therefore\ \rho'-\rho=\frac{\rho P}{K}$$

51. d. $Y=\frac{Mg}{A}\times\frac{L/2}{\Delta L}$

(Length is taken as $L/2$ because weight acts at C.G.)

Now, $M=AL\rho$

(For the purpose of calculation of mass, the whole of geometrical length L is to be considered.)

$$\therefore \quad Y=\frac{AL\rho gL}{2A\Delta L}$$

or $\Delta L=\frac{\rho gL^2}{2Y}=\frac{1.5\times10^3\times10\times8\times8}{2\times5\times10^6}\text{m}$

$=9.6\times10^{-2}$ m $=9.6\times10^{-2}\times10^3$ mm

$=96$ mm

52. a. $F=(\sigma_1-\sigma_2)\,l$

53. **c.** $F=\dfrac{2A\sigma}{t}=\dfrac{2\times8\times75\times10}{0.12}\text{dyn}=10^5\text{ dyn}$

54. **a.** $\left[2\pi\times\dfrac{8.7}{2}+2\pi\times\dfrac{8.5}{2}\right]\sigma=3.97\times980$

or $\sigma=\dfrac{3.97\times980\times7}{22\times17.2}\text{dyn/cm}^1=72\text{dyn/cm}^1$

55. **c.** $m\times10=2\times3\times10^{-2}\times\dfrac{10}{100}$

or $m=6\times10^{-4}\text{ kg}=6\times10^{-4}\times10^3\text{ g}=0.6\text{ g}$

56. **b.** Pressure due to 15 cm long liquid column needs to be balanced.

57. **d.** $25\times10^{-3}\times2\times0.1=m\times10$

or $m=\dfrac{5\times10^{-3}}{10}\text{kg}=0.5\text{g}$

58. **c.** $\dfrac{h_2}{h_1}=\dfrac{\sigma_2\cos\theta_2}{\rho_2}\times\dfrac{\rho_1}{\sigma_1\cos\theta_1}$

or $\dfrac{h_1}{h_2}=\dfrac{140\times\frac{1}{2}}{2}\times\dfrac{1}{70\times1}$ or $h_2=\dfrac{h_1}{2}=\dfrac{6}{2}\text{cm}=3\text{cm}$

59. **c.** $\dfrac{40}{100}\times1000\times9.8=\dfrac{2\times7\times10^{-2}}{R}$

or $R=\dfrac{14\times10^{-2}\times100}{40\times1000\times9.8}\text{m}=\dfrac{14\times1000}{40\times1000\times9.8}\text{mm}$

$=\dfrac{1}{28}\text{mm}$

Diameter $=2R=\dfrac{1}{14}\text{mm}$

60. **d.** $V\times\frac{4}{3}\pi R^3$; $2V=\frac{4}{3}\pi R'^3$; $2\times\frac{4}{3}\pi R^3=\frac{4}{3}\pi R'^3$;

$R'=2^{1/3}R$

$W'2\times4\pi\,[2^{1/3}R]^2\sigma=2^{2/3}\times2\times4\pi R^2\sigma=4^{1/3}W$

61. **c.** Final surface energy $=1000\times4\pi r^2\sigma$
Initial surface energy, $E=4\pi R^2\sigma$

Again, $\frac{4}{3}\pi R^3=1000\times\frac{4}{3}\pi r^2$ or $R=10r$

Now, final energy

$1000\times4\pi r^2\sigma=10\times4\pi R^2\sigma=10E$

62. **c.** Work done $=10E-E=9E$.

63. **a.** If y is the required distance, then

$a^2y\rho g=mg+4a\sigma$

or $y=\dfrac{mg\times4a\sigma}{a^2\rho g}=\dfrac{20\times980+4\times3\times70}{3\times3\times1\times980}\text{cm}=2.3\text{ cm}$

64. **a.** Effective length of flat part $=2r$
Effective length of curved part is πr.

$F=\sigma l$

So, required ratio is $2:\pi$.

65. **a.** Total upward force due to surface tension $=2\sigma l$
Weight of lifted liquid $=(hld)\rho g$

Equating, we get $h=\dfrac{2\sigma}{\rho gd}$

66. **a.** Increase in P.E. $=hld\,\rho g\left(\dfrac{h}{2}\right)=\dfrac{ld\rho g}{2}h^2$

If F is the force of attraction between the plates, then

$Fd=\dfrac{ld\rho g}{2}h^2$ or $F=\dfrac{l\rho g}{2}h^2$

67. **b.**

$F\propto h^2$

But $h^2\propto\sigma^2$

$\therefore$ $F\propto\sigma^2$

If σ is doubled, then F will become $4F$.

68. **d.** $\frac{4}{3}\pi R^3=n\times\frac{4}{3}\pi r^3$ or $R^3=nr^3$

Energy evolved $=(n\times4\pi r^2-4\pi r^2)\,T$

Now, $\theta=\dfrac{Q}{V\rho S}=\dfrac{Q}{V}$ (In C.G.S. system, $\rho=1$ and $S=1$)

$\therefore\theta=\dfrac{[n\times4\pi r^2-4\pi R^2]T}{\frac{4}{3}\pi R^3}$

$=\dfrac{3T[nr^2-R^2]}{R^3}$

$=3T\left[\dfrac{nr^2}{nr^3}-\dfrac{R^2}{R^3}\right]=3T\left[\dfrac{1}{r}-\dfrac{1}{R}\right]$

69. **d.** $2\pi r\sigma+\frac{1}{2}\times\frac{4}{3}\pi r^3dg=\frac{4}{3}\pi r^3\rho g$

or $2\pi r\sigma=\dfrac{\pi r^3g}{3}[4\rho-2d]$ or $r^2=\dfrac{3\times2\pi\sigma}{\pi g(2\rho-2d)}$

or $r^2=\dfrac{3\sigma}{g(2\rho-d)}$ or $r=\sqrt{\dfrac{3\sigma}{g(2\rho-d)}}$

Diameter $=2r=\sqrt{\dfrac{12\sigma}{g(2\rho-d)}}$

70. b.

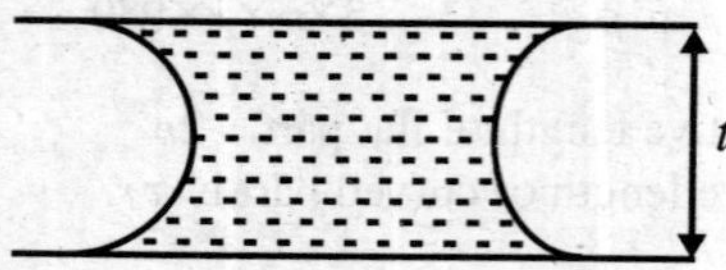

Fig. 4.142

Pressure difference $= \dfrac{T}{R} = \dfrac{T}{t/2} = \dfrac{2T}{t}$

$$\text{Force} = \frac{2T}{t} A$$

But $\quad V = At$ or $t = \dfrac{V}{A}$

$\therefore \quad \text{Force} = \dfrac{2TA^2}{V}$

71. b. Assuming isothermal conditions,

$$\left(P+\frac{4\sigma}{a}\right)\left(\frac{4}{3}\pi a^3\right)+\left(P+\frac{4\sigma}{b}\right)\left(\frac{4}{3}\pi b^3\right)$$

$$=\left(P+\frac{4\sigma}{c}\right)\left(\frac{4}{3}\pi c^3\right)$$

or $\quad P[a^3 + b^3 - c^3] = 4\sigma\,[c^2 - a^2 - b^2]$

or $\quad \sigma = \dfrac{P(c^3 - a^3 - b^3)}{4(a^2 + b^2 - c^2)}$

72. b. Difference in apparent weights is due to differences in forces of surface tension. Due to 180°, the force surface tension in one case is opposite to the force of surface tension in the other case.

$\therefore \quad 2\times\sigma_w\times\dfrac{10}{100} = 0.004 - 0.03$

or $\quad \sigma_w \times \dfrac{0.014}{1} \times 5\ \text{N/m} = 0.07\ \text{N/m}$

73. c. When the soap film enclosed by the thread is pricked, the thread would take up a circular shape. Consider a small element of the thread. While horizontal components of tension get balanced, the vertical components get added up.

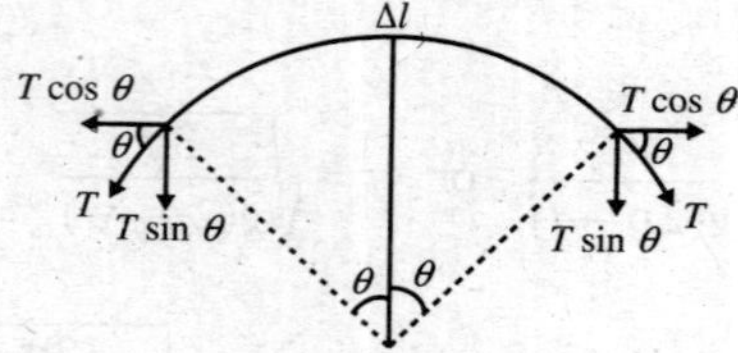

Fig. 4.143

or $\quad 2T\sin\theta = \sigma\,\Delta l$

or $\quad 2T\theta = \sigma\,(2\theta \times r)$

or $\quad T = \sigma r = \sigma \times \dfrac{2\pi r}{2\pi}$

$$= 0.030 \times \frac{6.28\times10^{-2}}{2\pi}\ \text{N}$$

$$= 3\times10^{-4}\ \text{N}$$

74. a. In the condition of weightlessness, water rises to the whole of the available length.

75. d.

$$v_0 = \frac{2}{9}\frac{r^2(\rho-\rho')g}{\eta}$$

Now, $\quad x = \dfrac{4}{3}\pi r^3 \rho$ or $\rho \propto \dfrac{x}{r^3}$

Similarly, $\quad \rho' \propto \dfrac{y}{r^3}$

$\therefore \quad v_0 \propto \dfrac{x-y}{r}$

76. b. $v_0 = \dfrac{2}{9}\dfrac{r^2(\rho-\rho')g}{\eta}$ or $\quad \eta = \dfrac{2r^2(\rho-\rho')g}{9v_0}$

$$= \frac{2\times0.2\times0.2\times9\times980}{9\times8}\ \text{poise} = 9.8\ \text{poise}$$

77. b. $V \propto r^4, V' \propto \left(r + \dfrac{1}{10}r\right)^4$

$$\frac{V'}{V} = \frac{11\times11\times11\times11}{10\times10\times10\times10} = 1.4641$$

or $\quad \left(\dfrac{V'}{V} - 1\right)\times100 = 46.4\%$

78. b. For the same radius, terminal velocity is proportional to the density difference.

79. a. $F = \dfrac{0.01\times100\times10}{0.1}\ \text{dyn} = 100\ \text{dyn}$

80. c. Shearing stress $= \dfrac{|\text{viscous force}|}{\text{area}}$

$$= \frac{\eta A \dfrac{dv}{dx}}{A} = \eta\frac{dv}{dx}$$

$$= 10^{-3}\times\frac{5}{10}\ \text{N/m}^2 = 0.5\times10^{-3}\ \text{N/m}^2$$

81. b. $V\rho g + F = V(4\rho)g$

or $\quad F = 3V\rho g = 3mg$ or $\dfrac{F}{mg} = 3$

82. b. $m = \frac{4}{3}\pi r^3 \rho m$

Keeping m constant, if r is halved, ρ will increased by a factor of 8.

Now, $v_0 \propto r^2 \rho$

$$v_0' \propto \frac{r^2}{4}(8\rho) \quad \text{or} \quad v_0' \propto 2r^2 \rho$$

Dividing, $\frac{v_0'}{v_0} = 2$ or $v_0' = 2v_0$

or $v_0' = 2v$

83. b.

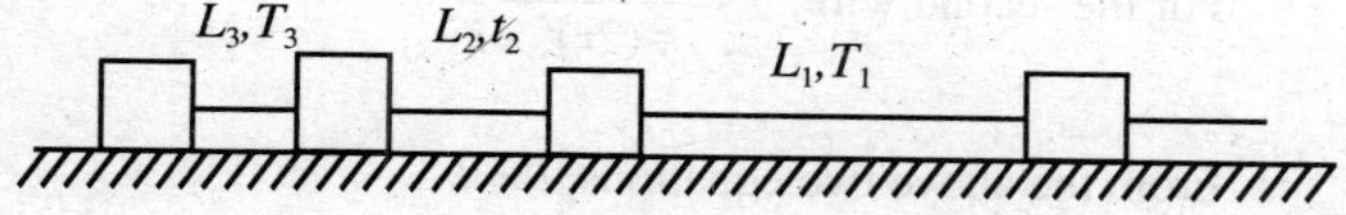

Fig. 4.144

$$T_1 > T_2 > T_3 \Rightarrow \text{lengths } L_1 > L_2 > L_3$$

84. a. Let h be the height of water inside the capillary. Total upward force tending to pull water supports the weight of the water.

$$T(2\pi r_1 + 2\pi r_2) = h(\pi r_2^2 - \pi r_1^2)\rho g$$

$$\therefore \quad h = \frac{2T}{(r_2 - r_1)\rho g} = \frac{2 \times 7 \times 10^{-2}}{10^{-3}(10^3)(10)}$$

$$= 14 \times 10^{-3} \text{ m} = 1.4 \text{ cm}$$

85. c. The air pressure is greater inside the smaller bubble ($4T/r$). Hence, air flows from the smaller to the larger bubble.

86. d. All of the statements are factual statements. Stress is defined as internal force developed in the system per unit cross-sectional area and is not defined as force applied per unit cross-section area, although in equilibrium, stress is numerically equal to applied force per unit area.

87. b. The profile of the velocity of fluid is as shown in the figure. The velocity of the layer of fluid, which is in contact with metal plates (fixed), is zero. As we move towards the centre from either plate, the velocity of the layer of fluid increases and it becomes maximum at the location of moving plate. This maximum value is same as that of the velocity of plate.

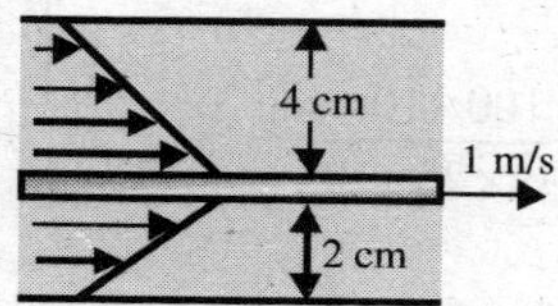

Fig. 4.145

88. a. The force experienced by the plate due to fluid is shown in the figure.

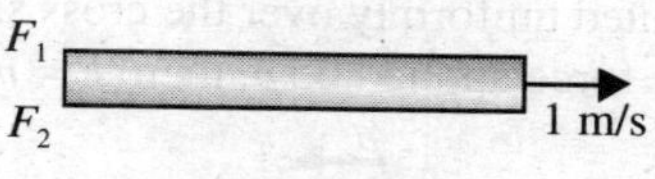

Fig. 4.146

F_1 is the force exerted by the layer in contact with plate on upper side and F_2 is the force exerted by the layer in contact with plate on lower side.

$$F_1 = \eta A \times \frac{v}{4\text{ cm}} = 0.8 \times 0.01 \times \frac{1}{0.04} = 0.2 \text{ N}$$

where v is the velocity of layer in contact with plate.

$$F_2 = \eta A \times \frac{v}{2\text{ cm}} = 0.8 \times 0.01 \times \frac{1}{0.02} = 0.2 \text{ N} = 0.4 \text{ N}$$

Total force, $F = F_1 + F_2 = 0.6$ N in backward direction.

89. b. If the rod is allowed to expand, then it will expand by $\Delta l = l\ \alpha\ \Delta T$ due to increase in temperature by ΔT. But as the length of the cylinder is kept constant by applying pressure, a stress is developed in the cylinder. The decrease in length of cylinder due to elasticity is $\Delta l = l\ \alpha\ \Delta T$ and a compressive stress will develop in it.

Stress = Excess pressure applied at ends

$$= Y \times \frac{\Delta l}{l} Y \alpha \Delta T$$

$$= 2 \times 10^{11} \times 11 \times 10^{-6} \times 100 = 2.2 \times 10^3 \text{ atm}$$

90. b. Consider a small element of the tube.

$$2T \sin\theta = \Delta p \times A$$

where $\Delta p = p_i - p_0$ and A is the area of element. As θ is very small, $\sin\theta \approx \theta$ so, $2T \times \theta = \Delta p \times l \times (2r\theta)$

$$\Rightarrow \Delta p = \frac{T}{lr} \quad \sigma \text{ (stress developed in tube)} = \frac{\Delta T}{\Delta r \times l}$$

where $\Delta r \times i$ is the cross-sectional area.

$$\sigma = \frac{\Delta p \times lr}{\Delta r \times l} = \Delta p \times \frac{r}{\Delta r}$$

For no rupturing, $\sigma \le \sigma_{max}$

So, $\Delta p \times \frac{r}{\Delta r} \le \sigma_{max}$

$$\Delta p(\text{max, value}) = \sigma_{max} \times \frac{\Delta r}{r}$$

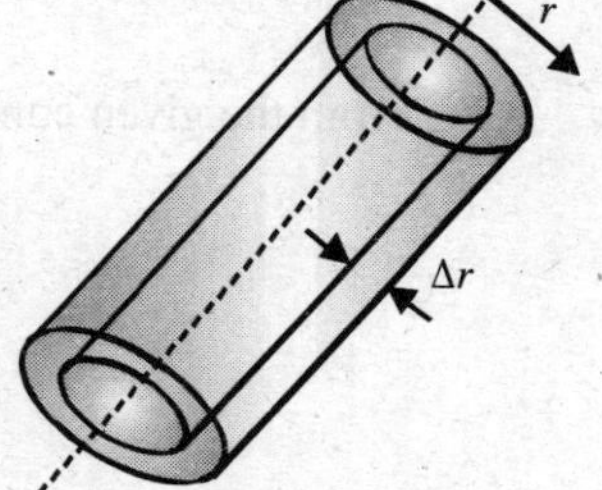

Fig. 4.147

91. c. As already discussed in key concepts, if the cut is not very near to end of the bar, the internal developed force is distributed uniformly over the cross section.

92. c. Let the force applied be p, then $P = ma$

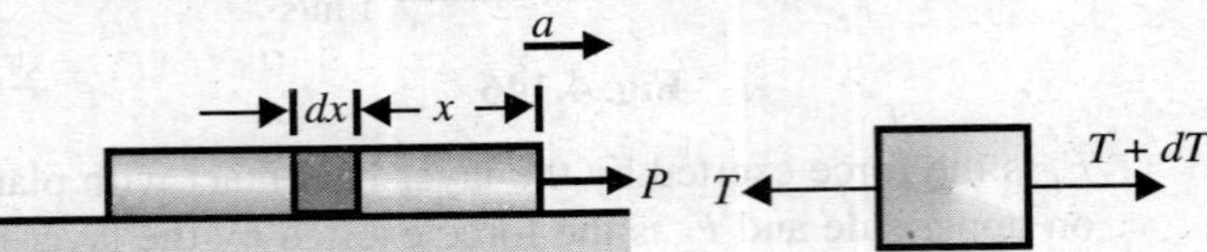

Fig. 4.148

$\Rightarrow \qquad p = 5 \times 2 = 10\text{ N}$

For the element shown in the figure,

$$T = \frac{m}{l}(l - x) \times a$$

$$\Rightarrow \qquad T = \frac{5}{1}(1 - x) \times 2 = 10(1 - x)$$

Elongation in dx is $\dfrac{\Delta(dx)}{dx} = \dfrac{T/A}{A}$

$$\Delta(dx) = \frac{10(1-x)}{(5 \times 10^{-2})^2} \times \frac{1}{5 \times 10^9} dx$$

Total elongation, $\Delta l = \int_0^1 \dfrac{10(1-x)}{25 \times 10^{-4}} \times \dfrac{1}{5 \times 10^9} dx = 0.4 \times 10^{-6}\text{ m}$

93. b. When two bubbles come into contact, they try to minimize the surface area for which air flows from bubble having higher pressure (smaller radius) to bubble having lower pressure.

94. d. Surface tension force, $F = S \times$ length $F = S \times [2\pi R + 2\pi r]$ as liquid surface is on the both sides.

95. a. Here, surface tension force support the weight of liquid.

$$S \times 2\pi\left[r + \frac{d}{2}\right] = \left(r^2 - \frac{d^2}{4}\right) h\rho g$$

$$\Rightarrow \qquad h = 0.0144\text{ m} = 1.44\text{ cm}$$

96. c. From the free-body diagram of the elevator,

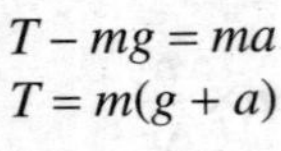

$$T - mg = ma$$

$$T = m(g + a)$$

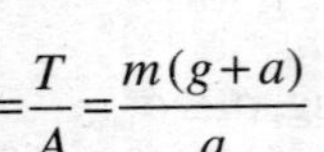

Stress in cable is $\alpha = \dfrac{T}{A} = \dfrac{m(g+a)}{a}$

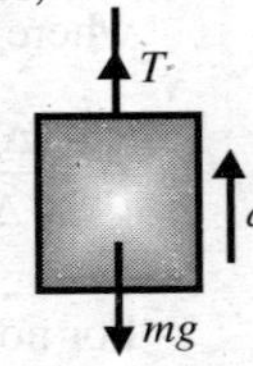

Fig. 4.149

From the given condition, $\alpha \le \dfrac{\alpha_{max}}{2}$

$$\frac{m(g+a)}{A} \le \frac{\sigma_{max}}{2}$$

$$g + a \le \frac{2 \times 10^9}{2} \times \frac{10^{-4}}{2 \times 10^3} = 50$$

$$g + a \le \frac{2 \times 10^9}{2} \times \frac{10^{-4}}{2 \times 10^3} = 50$$

97. b. The maximum load which a wire can sustain is a property of the material of wire and it depends on the cross-sectional area. $\sigma_{breaking}$ is a characteristic property of material of wire and is independent of length or breadth (dimensions) of the wire.

98. b. $Y = \dfrac{F/A}{\Delta l/l}$

For the first wire, $Y = \dfrac{F \times L}{\pi r^2 \times l}$

For the second wire, $Y = \dfrac{2F \times 2L}{\pi(2r)^2 \times l'}$

From the above two equations, $l' = l$

99. a. $Y = \dfrac{\text{Stress}}{\text{Strain}} = \dfrac{x}{\dfrac{2l - l}{l}} = \dfrac{x}{1} = x$

Actually, the above expression is not exact for this much elongation.

100. b. Tension in any of the wire is same at all points as cross-sectional area are the same and tension also. So stress $(= F/A)$ in both the wires would be the same but from Strain = Stress/Strain as Y is different for both, so strain is different in both.

101. a. $Y = \dfrac{F/A}{\Delta l/l}$

So $\qquad \dfrac{Y_s}{Y_c} = \dfrac{F \times l_s}{A_s \times \Delta l} \div \dfrac{F}{A_c} \times \dfrac{l_c}{\Delta l}$

$$\frac{A_c l_s}{A_s l_c} = \frac{4 \times 10^{-6} \times 4.7}{3 \times 10^{-6} \times 3.5} = 1.8$$

102. b. The situation is as shown in the figure. A force of 100g acts on farther face of cube as shown, due to which shear will develop.

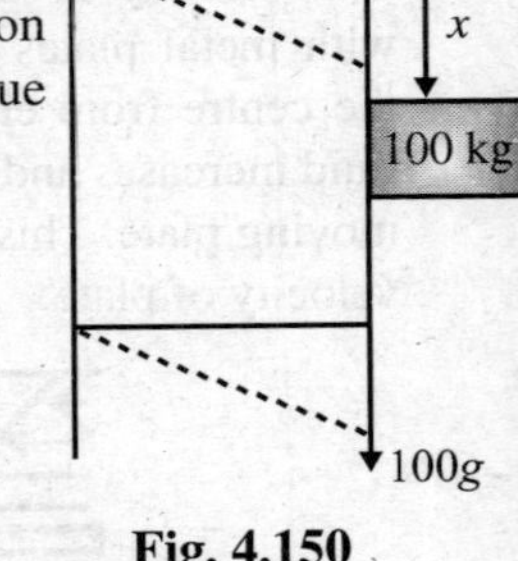

Fig. 4.150

$$\eta = \frac{F/A}{x/l}$$

$$x = \frac{Fl}{A\eta} = \frac{100 \times 10 \times l}{l^2 \times 25 \times 10^9}$$

$$= 4 \times 10^{-7}\text{ m}$$

103. b. Change in pressure due to placing of mass on prism is

$$\Delta p = \frac{Mg}{A}$$

From bulk modulus definition $K = \frac{-dp}{dV/V}$

$$\left|\frac{dV}{V}\right| = \frac{\Delta p}{K} = \frac{Mg}{AK}$$

From $V = \frac{4}{3}\pi r^3$

$$\frac{dV}{V} = \frac{3dR}{R} \Rightarrow \frac{dR}{R} = \frac{1}{3}\frac{dV}{V} = \frac{Mg}{3AK}$$

104. a. As we increase the separation between the wires by 1 mm, the increase in surface area of the film is

$\Delta A = 2\times 0.1\times 10^{-3} = 2\times 10^{-4}\ \text{m}^2$

(as there are two surfaces of film)

So, increase in surface energy, $\Delta U = S\Delta A = 1.44\times 10^{-5}$ J

Work done $= \Delta U$.

105. a. Surface tension force,

$$F = S\times 2l = 0.072\times 2\times \frac{2.5}{100} = 3.6\times 10^{-3}\ \text{N}$$

106. a. Young's modulus is the property of material; it does not depend on shape and size of body.

107. b. Let the maximum tensile force that each rivet can experience be F, then maximum shearing stress on each rivet is $\sigma = \frac{F}{A} = 1.5\times 10^9$

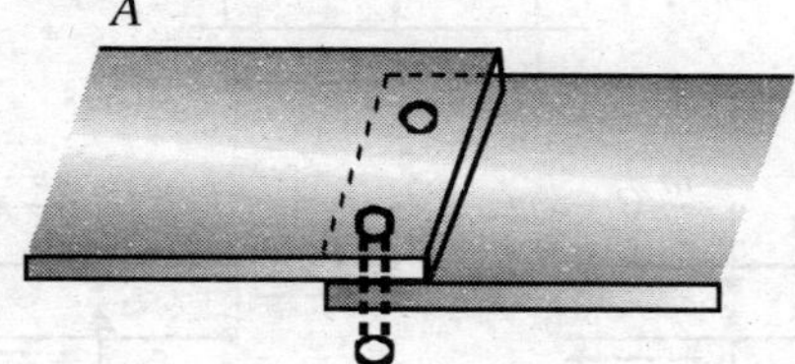

Fig. 4.151

$A = \pi r^2$, where $r = 2\times 10^{-3}$ m.

Required force is $2F$.

108. d. As in vacuum, there is no viscosity, no resistive force and hence no terminal velocity can be acquired.

109. a. From the definition of bulk modulus $B = -\frac{dp}{dV/V}$

As we move from surface to placed where pressure changes to αp_0, let us assume volume changes by ΔV, then

$$B = \frac{V\Delta p}{\Delta V} = \frac{V(\alpha - 1)p_0}{\Delta V}$$

New volume, $V' = V - \Delta V = V\left[1 - \frac{(\alpha-1)p_0}{B}\right]$

Density at the given depth, $\rho' = \rho V/V'$, where ρ is density at surface

$$\rho' = \frac{\rho\times B}{B - (\alpha-1)p_0}$$

110. b. $h = \frac{2S\cos\theta}{r\rho g}$

Cross-sectional area increases four times, which means radius gets doubled.

So, $h' = \frac{2S}{2(r\rho g)} = \frac{h}{2}$

111. a. The terminal velocity of the spherical raindrop of radius r is given by

$v_t = \frac{2r^2\rho g}{9\eta}$ where ρ is the density of water and η the viscosity of air.

Substituting $r = 0.3$ mm $= 0.3\times 10^{-3}$m, $\rho = 10^3$ kg/m^3

$g = 9.8$ m/s^2 and $\eta = 1.8\times 10^{-5}$ Ns/m^2

We get $v_t = \frac{2\times(0.3)^2\times 10^{-3}\times 9.8}{9\times 1.8\times 10^{-5}} = 10.9$ m/s

Multiple Correct Answers Type

1. a., b., c. $(\text{stress})_S = \frac{F}{2A}$, $(\text{stress})_{Cu} = \frac{F}{A}$

Given that $\frac{Y_S}{Y_{Cu}} = \frac{2}{1}$

$$\frac{(\text{strain})}{(\text{strain})_{Cu}} = \frac{(\text{stress})_S/Y_S}{(\text{stress})_{Cu}/Y_{Cu}}$$

$$= \frac{(\text{stress})_S}{(\text{stress})_{Cu}}\times\frac{Y_{Cu}}{Y_S} = \frac{(F/2A)}{(F/A)}\times\frac{1}{2} = \frac{1}{4}$$

or $\frac{L_S/\Delta L_S}{L_{Cu}/\Delta L_{Cu}} = \frac{1}{4}$ or $\frac{L_S/\Delta L_{Cu}}{L_{Cu}/\Delta L_S} = \frac{1}{4}$

or $\frac{25}{50}\times\frac{\Delta L_{Cu}}{\Delta L_S} = \frac{1}{4}$ or $\frac{\Delta L_{Cu}}{\Delta L_S} = \frac{1}{2}$

So, options (a), (b) and (c) are correct.

2. b., c. Thermal force $= YA\alpha\, d\theta = Y\pi r^2\alpha\, d\theta$

$$r_1 = r,\quad r_2 = r\sqrt{2},\quad r_3 = r\sqrt{3},\quad r_4 = 2r$$

$$F_1{:}F_2{:}F_3{:}F_4 = 1{:}2{:}3{:}4$$

Thermal stress = $Y\alpha\, d\theta$

As Y and α are same for all the rods, hence stress developed in each rod will be the same. As strain = $\alpha\, d\theta$, so strain will also be the same.

$$E = \text{Energy stored}$$

$$= \frac{1}{2} Y(\text{strain})^2 \times A \times L$$

$$\therefore \quad E_1{:}E_2{:}E_3{:}E_4 = 1{:}2{:}4$$

So, options (b) and (c) are correct.

3. a., c. Viscous force is somewhat like a frictional force but not exactly the same because it differs from it due to the two main following statements:

i. Viscous force is velocity dependent while frictional force is not.

ii. Viscous force is temperature dependent while frictional force is not.

So, options (a) and (c) are correct.

4. a., b., c., d.

i. Excess pressure inside a spherical liquid drop = $2T/R$, because there is only one free surface here

ii. In case of spherical meniscus of radius of curvature R, also excess pressure = $2T/R$ because again there is only one free surface.

iii. Excess pressure inside a cylindrical drop of radius $R = T/R$. Hence, for a cylindrical bubble in air, excess pressure is $2T/R$, because there exist two free surfaces in this case.

iv. For a spherical bubble in water, excess pressure = T/R, as there is only one free surface.

Hence, all the four options are correct.

5. a., b. $h = \dfrac{T}{Rdg}$

Given that h, T, d and g are fixed, hence R must be same. As weight of the liquid is balanced by force due to surface tension ($T\cos\theta$ and R are fixed for a given liquid, given material of capillary at a constant temperature) hence, weights of liquids in both the capillaries must be equal

So, options (a) and (b) are correct.

6. a., b., c. Tension in $B = T_B = \dfrac{mg}{3}$

$$\text{Tension in } A = T_A = T_B + mg = \frac{4mg}{3}$$

$$\therefore \quad T_A = 4T_B$$

$$\text{Stress } \frac{T}{\pi r^2} = S$$

Wire breaks when S is equal to breaking stress

$$\text{For } r_A = r_B, \quad S_A = 4S_B$$

Therefore, A breaks before B

For $\quad r_A = 2r_B$

$$S_B = \frac{T_B}{\pi r_A^2} - \frac{T_B}{\pi r_A^2} = \frac{4T_B}{\pi(2r_B)^2} = \frac{T_B}{\pi r_B^2} = S_B$$

As the stresses are equal, either of the wires may break.

7. a., c. Let S = surface tension

= surface energy per unit area

r = radius of each small drop

R = radius of a single drop

$$n \times \frac{4}{3}\pi r^3 = \frac{4}{3}\pi R^3 \quad \text{or} \quad R = rn^{1/3}$$

Initial surface energy,

$$E_i = n \times 4\pi r^2 \times S = nE$$

Final surface energy,

$$E_f = 4\pi R^2 S = 4\pi r^2 n^{2/3} S = n^{2/3} E$$

Therefore, energy released = $E_i - E_f = E(n - n^{2/3})$

8. c., d. The angle of contact at the free liquid surface inside the capillary tube will change such that the vertical component of the surface tension forces just balances the weight of the liquid column.

9. a., b., c. The two free liquid surfaces must provide a net upward force due to surface tension to balances the weight of liquid column.

10. b., c. Since, force is applied at one end only, the plank starts to accelerate along the direction of this force and a stress is developed in its material. Hence, option (c) is incorrect.

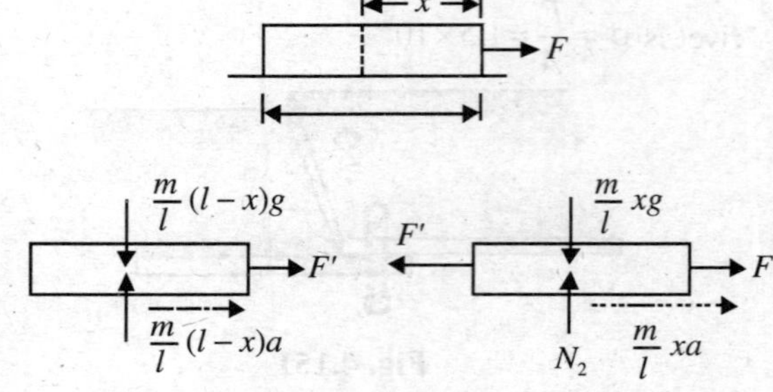

Fig. 4.152

To calculate stress in the material of the plank, at a distance x from the end at which force is applied, the free-body diagrams are considered as shown in the figure.

$$\therefore \quad F' = \frac{m}{l}(l - x)a \qquad \text{(i)}$$

$$F - F' = \frac{m}{l}xa$$

From these two equations, $\quad F' = \dfrac{F(l - x)}{l}$

$$\therefore \quad \text{Stress} = \frac{F'}{A} = \frac{F(l-x)}{Al}$$

where A is cross-sectional area of the plank.

It shows that stress varies linearly with x. It is maximum at $x = 0$ and zero at $x = l$.

Hence, option (a) is correct.

11. **a., c.** The compressive force in the rod will be same at left end therefore, compressive stress will be maximum at this end. Hence, option (b) alone is correct.

12. **a., d.** The slope of the linear portion of the curve gives the Young's modulus of the material. The slope of the linear portion OP for material A is greater than that of the linear portion OR for material B. Hence, statement (a) is correct. The plastic region for material A (from P to Q) is greater than that (from R to S) for material B, which indicates that material A is more ductile. Hence, statement (b) is correct. The breaking stress for material B (i.e., stress corresponding to point S) is less than that for material A (i.e., stress corresponding to point Q), which implies that material B can break more easily than material A. Thus, material B is more brittle. Hence, choice (c) is also incorrect. Material A is stronger than material B because it can withstand a greater stress before it breaks. The breaking stress is the stress corresponding to point Q for material A and to point S for material B. Hence, the correct choices are (a) and (d).

13. **a., d.** We know that $Y = \dfrac{FL}{Al}$

$$\therefore \quad l = \frac{FL}{AY}$$

Since, the two wires are made of the same material, Young's modulus Y is same for both. Since, load F and the cross-sectional area A are the same,

$$\therefore \quad l \propto L$$

i.e., extension is proportional to the length of the wire. Hence, option (a) is correct. The strain in a wire is given by

$$\frac{l}{L} = \frac{4F}{AY}$$

Since F, A and Y are the same, the strains in wires A and B will be equal. Thus, the correct options are (a) and (d)

14. **a., c.** Area of cross section,

$$A = \frac{\pi d^2}{4}$$

where d is the diameter of the wire. Therefore,

$$l = \frac{4FL}{\pi d^2 Y}$$

Since F, L and Y are the same for wires A and B

$$\therefore \quad l \propto \frac{l}{d^2}$$

i.e., the extension is inversely proportional to the square of the diameter. Hence, choice (a) is correct. The strain is

$$\frac{l}{L} = \frac{4F}{\pi d^2 Y}$$

Thus, strain $\propto \dfrac{1}{d^2}$

i..e, the extension is inversely proportional to the square of the diameter. Hence, choice (a) is correct. The strain is

$$\frac{l}{L} = \frac{4F}{\pi d^2 Y}$$

Thus, strain $\propto \dfrac{1}{d^2}$

Hence, the correct choices are (a) and (c)

15 **b., c.** Since, wires A and B are made of the same material they have the same Young's modulus. Now, breaking load = breaking stress × cross-sectional area.

The wire having a greater area of cross section can withstand a greater load. Hence, the correct options are (b) and (c)

16. **a., d.** Statement (a) is correct because the Young's modulus of steel is greater than that of rubber.

Statement (b) is incorrect. If a spring is stretched, both the total length of the wire in the coil and the volume of the wire do not change. Only the shape of the coils of the wire undergoes a change. Hence, it is the shear modulus that determines the stretching of the coil.

Statement (c) is also incorrect. The bending moment of the prongs of a tuning fork is determined by the Young's modulus of the material. Hence, the restoring force on the prongs depends on Young's modulus which determines the frequency of the fork. Statement (d) is correct. When the material is not subjected to any stress, its atoms are in their normal (equilibrium) positions. When a tensile stress is applied, the separation R between the atoms becomes greater than the equilibrium separation R_0. For $R > R_0$, the interatomic forces are attractive.

17. **a., b.**

a. Within elastic limit,

$$\text{Stress} = \text{strain}$$

b. Beyond elastic limit, a small increase in stress produces large deformation, i.e., large strain. So,

$\dfrac{\text{Stress}}{\text{Strain}}$ decreases.

c. Both the wires will have the same stress

Again $\quad Y = \dfrac{\text{Stress}}{\text{Strain}} \quad$ or $\quad \text{Strain} = \dfrac{\text{Stress}}{Y}$

Since Y is different, strain will be different. So, option (c) is not correct.

18. a., b., c. $\frac{T_1}{0.1}=\frac{T_2}{0.2}$ or $\frac{T_1}{T_2}=\frac{1}{2}=0.5$

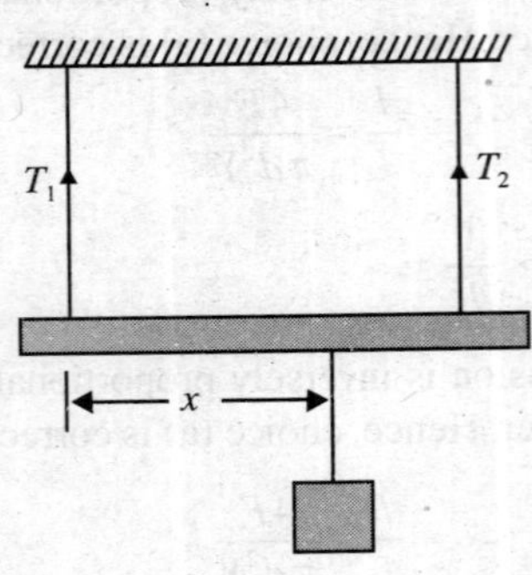

Fig. 4.153

Further, $T_1x=T_2(2-x)$

or $\frac{T_1}{T_2}=\frac{2-x}{x}$

Hence, $\frac{2-x}{x}=0.5=\frac{1}{2}$ or $4-2x=x$

or, $x=\frac{4}{3}\text{ m}=\frac{400}{3}\text{ cm}$

19. a., b., c.

a. Bulk modulus is reciprocal of compressibility.

b. Breaking force D^2.

20. a., b., c. Even a small stress causes large strain. In the case of a perfectly rigid body, strain is zero.

21. a., b., d. Here, for all three wires, length l, elongation Δl and cross-sectional area A are the same. From

$$Y=\frac{\text{Stress}}{\text{Strain}}=\frac{T/A}{\Delta l/l}$$

$$\Rightarrow \quad T\propto Y$$

Let T_1, T_2 and T_3 be tension in three wires I, II and III, respectively.

For vertical equilibrium,

$$T_1\times T_2+T_3=150\text{ g}$$

For rotation equilibrium,

$$T_1\times x-T_3x=0$$

$$\Rightarrow \quad T_1=T_3 \quad \text{so} \quad Y_1=Y_3$$

$$\Rightarrow \quad \frac{T_2}{T_1}=\frac{T_2}{Y_1}=2$$

$$\Rightarrow \quad T_2=2T_1$$

So, $2T_1+T_2=150\text{ g}$

$$\Rightarrow \quad T_2=75\text{ g}$$

22. a., b. Loss in gravitational potential energy of M is Mgl as M falls down by l.

Elastic potential energy stored in wire is

$$U=\frac{1}{2}\times\text{Stress}\times\text{Strain}\times\text{Volume}$$

$$=\frac{1}{2}\frac{1}{2}\times\frac{Mg}{A}\times\frac{l}{L}\times AL=\frac{Mgl}{2}$$

Now, work done by gravity force is not equal to elastic potential energy stored in wire. This is due to the fact that some work has been done against air friction etc., which increases the internal energy of wire.

23. a., c., d. In this question, the applied force is not constant. Let us consider at any instant the wire elongates by x. Then work done by F to elongate it more by dx is

$$dW=Fdx=\frac{YAx}{L}dx\,[x\ll L]$$

$$W=\int_0^l\frac{YAxdx}{L}=\frac{YAl^2}{2L}$$

As internal restoring force and applied force are almost same (equal and opposite), numerical value of work done by these two forces would be equal.

24. a., c. Velocity of water as it comes out through the hole is

$$v=\sqrt{2gh}=\sqrt{2\times10\times5}=10\text{ m}$$

From equation of continuity, Av = constant as water falls down, after coming through hole its speed increases and hence A decreases. From Bernoulli's theorem

$$P_0+\frac{\rho v^2}{2}+0=p_0+\frac{\rho v_1^2}{2}-\rho gh$$

(Just outside the hole and 5 m below the bottom of tank)

$$\Rightarrow \quad \frac{\rho v_1^2}{2}=\frac{\rho v^2}{2}+\rho gh \Rightarrow v_1=14.14\text{ m/s}$$

From $\quad Av=\text{constant.}$

$$\Rightarrow \quad 2\times10=A_1\times14.14\Rightarrow A_1=1.414\text{ cm}^2$$

Assertion-Reasoning Type

1. a. The surface tension of liquid can be measured as the force per unit length (force gradient) on an imaginary line drawn on the liquid surface, which acts perpendicularly to the line on its either side at every point and tangentially to the liquid surface.

Therefore, surface tension and force gradient have the same dimensions.

2. a. The free surface of a liquid drop tries to acquire minimum surface area due to surface tension. Since for a given volume, the surface area of sphere is minimum, the liquid drops acquire sphere is minimum, the liquid drops acquire spherical shape.

3. c. In a small drop, the force due to the surface tension is very large as compared with its weight and hence it is spherical in shape. A big drop becomes oval in shape due to its large weight. The surface tension of liquid decreases with increase of temperature.

4. a. The shape of the liquid is actually controlled by two forces viz. surface tension and gravitational force. When size of drop is small, surface tension and force > gravitational force, which means net force is due to surface tension. In this case, PE is minimum when the surface area is minimum. So, the drop is spherical.

When the size of the drop is larger, gravitational force > surface tension. It means net force is due to gravitational force. In this case, gravitational PE is minimum when centre of gravity of the drop at the minimum. So, the drop is no more spherical.

5. a. According to the ascent formula

$$h = \frac{2T\cos\theta}{r\rho g} \Rightarrow h \propto \frac{1}{r}$$

From the relation, radius is less, and the height to which the liquid rises will be greater.

6. a. When a raindrop falls in air (viscous medium), after falling through the same height, the viscous drag balances the weight of the drop. Through the rest of its height, velocity is constant or it attains a terminal velocity.

7. c. When a needle is placed carefully on the surface of water, it floats on the surface of water due to the surface tension of water, which does not allow the needle to sink. In case of a steel ball, the surface tension of the water is not sufficient to keep it floating, so it sinks down.

8. c. Since the particles of dust are like spheres of very small radii and when it acquires the terminal velocity they begin through air. As from the definition, the terminal velocity of dust particles is directly proportional to the square of its radii. Thus, the terminal velocity of dust particle is very small and so they settle down in a closed room after some time.

9. c. From the formula,

$$\text{Express pressure} = \frac{2T}{r}$$

Here, T is surface tension; r is the radius of liquid drop. Hence, excess pressure is inversely proportional to radius and hence the surface, area. Therefore, the excess pressure inside a smaller drop is larger as compared to the larger drop due to which smaller crop of liquid resists deforming forces better than a larger drop.

10. b. The viscous force on a body depends on its velocity. The greater the velocity, the greater its viscous force. When a body falls from a sufficient height it acquires enough velocity to produce a viscous force that balances its height. The resultant force on the body being zero, the body moves with uniform velocity called terminal velocity.

11. a. On spraying, surface area increases and hence surface energy; this increase in surface energy is on the expense of decrease in internal energy and hence temperature decreases.

12. d. Excess pressure inside a soap bubble is given by $p_i - p_0 = 4S/r$ as $p_i - p_0 \propto 1/r$, so as r increases, p_1 decreases (p_0 constant); therefore Statement I is wrong.

Statement II is correct as more air has to be forced in to cause the bubble to grow but contrary to the process of blowing up a rubber balloon, the excess pressure required to force air into the bubble decrease with the bubble size.

13. a. Surface tension can be understood as a property of a surface due to which it tries to acquire the minimum possible area. If cohesive force is more, the liquid is having the capability to shrink its surface more, i.e., surface area of the liquid is less and hence more is the surface tension.

Comprehension Type

For Problems 1–2

1. a., 2. a.

Sol. 1. a. Let F_s and F_b be the forces in the wires A and B, respectively. The free-body diagram of the rod is shown in the figure.

Applying the condition of rotational equilibrium about the point of application of load, we get,

$$F_s X = F_b(L - x)$$

Free body diagram of the rod

Fig. 4.154

$$\frac{F_s}{F_b} = \frac{L-x}{x} \quad \text{(i)}$$

If the stresses are equal in two wires, we have

$$\frac{F_s}{F_b} = \frac{A_s}{A_b}$$

Here $A_s = 1 \times 10^{-5}$ m^2 and $A_b = 2 \times 10^{-5}$ m^2

$$\frac{F_s}{F_b} = \frac{1}{2} \quad \text{(ii)}$$

From Eqs. (i) and (ii), we get

$$\frac{L-x}{x}=\frac{1}{2} \quad \text{or} \quad x=\frac{2L}{3}$$

Since $L = 2$ m, $x = (4/3) = 1.33$ m

2. a. For equal strains,

$$\in=\frac{\sigma}{Y}=\frac{F}{AY};\frac{F_s}{F_b}=\frac{1}{2}$$

$$\frac{F_s}{F_b}=\frac{A_s Y_s}{A_b Y_b}=\frac{(1\times10^{-5})\,(2\times10^{11})}{(2\times10^{-5})\,(1\times10^{11})}=\frac{1}{1} \quad \text{(iii)}$$

Therefore, from Eqs. (i) and (iii), we get

$$x=\frac{L}{2}=\frac{2}{2}=1\text{ m}$$

For Problems 3–4

3. d., 4. a.

Sol.

3. d. If ρ_w is the density of water, then the density of the lead sphere and the density of the liquid are given by

$$\rho_0 = 11.20\,\rho_w \quad \text{and} \quad \rho = 1.26\,\rho_w$$

Thus, the equation $T=2\pi\sqrt{\frac{l}{2g}}$ may be written as $\eta=\frac{2}{9}\frac{r^2}{v}(11.20-1.26)\rho_w g$

Here, $r = 0.5 \times 10^{-3}$ m; $v = 0.7 \times 10^{-2}$ m/s; $g = 9.8$ m/s²; $\rho_w = 10^3$ kg/m³.

$$\eta=\frac{2}{9}\frac{(0.5\times10^{-3})^2}{0.7\times10^{-2}}(11.20-1.26)(10^3)(9.8)$$

$$= 0.77 \text{ N s/m}^2$$

4. a. Reynolds number $\text{Re}=\frac{\rho v d}{\eta}$

$$\text{Re}=\frac{(1.26\times10^3)(0.7\times10^{-2})(1\times10^{-3})}{0.77}$$

Re = 0.01 < 0.20 (maximum limit of Stokes' law).

For Problems 5–6

Sol. 5. c., 6. b.

Here $s = 7.0 \times 10^{-2}$ N/m; $r = 0.2 \times 10^{-3}$ m; $r = 10^3$ kg/m³ we know that the height of liquid rise in the tube is $h = 2\sigma\cos\theta/\rho Rg$ where R is the radius of the meniscus, and θ is the angle of contact.

Since $\theta = 0$ (given), radius of the meniscus is equal to the radius of the capillary tube, i.e., $R = r$.

$$\therefore \quad h=\frac{2(7.0\times10^{-2})}{(10^3)(0.2\times10^{-3})(10)}=0.07\,\text{m}$$

or $\quad h = 7.0$ cm

When the length of the capillary tube above the free surface of the liquid is less than the height of liquid that rise in the tube, radius R of the free surface is not equal to the radius of the tube. It is greater than r as the surface tends to be flatter.

According to the equation, $p_1 - p_2 = (4\sigma/R)$, the pressure difference across the surface is given by

$$\Delta p=\frac{2\sigma\cos\theta}{R}$$

If p_1 and p_2 are the pressure just above and below the meniscus, respectively, then $p_1 - p_2 = \rho g h_0$

$$\therefore \quad \rho g h_0=\frac{2\sigma}{r} \quad \text{(i)}$$

In part (a) we have seen that when $h_0 = h$; $\theta = 0$, $R = r$.

$$\text{and} \quad \rho g h=\frac{2\sigma}{r} \quad \text{(ii)}$$

Therefore, dividing Eq. (i) by Eq. (ii), we have $\frac{r}{R}=\frac{h_0}{h}$

From the figure, it is clear that $\cos\theta=\frac{r}{R}$

Therefore, the angle of contact is

$$\theta=\cos^{-1}\frac{h_0}{h}=\cos^{-1}\left(\frac{5}{7}\right)\approx 44°$$

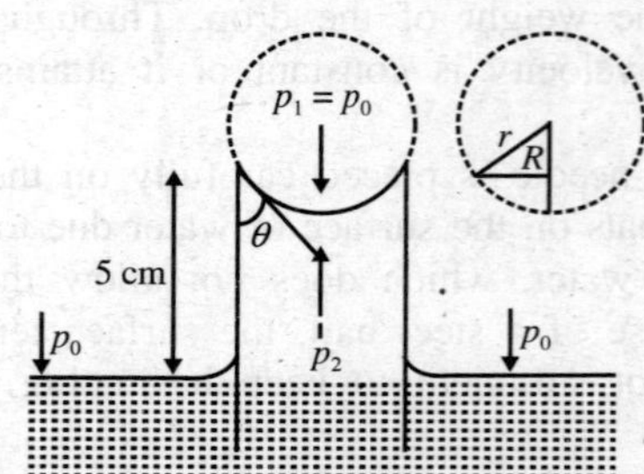

Fig. 4.155

For Problems 7–9

7. c., 8. a., 9. b.

Sol. a. Using Poiseuille's equation, we get

$$Q=\frac{(p_1-p_2)\pi D^4}{128\eta L}$$

Here, $\quad p_1 - p_2 = 38.4 \times 10^3$ N/m²

$D = 2.5$ cm $= 2.5 \times 10^{-2}$ m

$L = 30$ m; $h = 0.12$ N/m²

$$\therefore Q=\frac{(38.4\times10^3)(3.14)(2.5\times10^{-2})^4}{128(0.12)(30)}=1.0\times10^{-4}\,\text{m}^3/\text{s}$$

b. We know that $\tau = -\frac{dp}{dx}\frac{r}{2}$

At the wall of the pipe, i.e., $r = R$

$$\tau_{\max} = -\frac{dp}{dx}\frac{R}{2} = \left(\frac{p_1 - p_2}{L}\right)\frac{D}{4}$$

$$= \left(\frac{38.4\times10^{-3}}{30}\right)\frac{2.5\times10^{-2}}{4} = 8\times10^{-6}\ \text{N/m}^2$$

c. We know that $P = \frac{128\eta Q^2 L}{\pi D^4} = (p_1 - p_2)Q$

$\therefore \quad P = (38.4\times10^3)(1\times10^{-4}) \quad \text{or} \quad P = 3.84W$

For Problems 10–12

10.b., 11.a., 12.b.

Sol.

10. b. Since $\alpha_t > \alpha_b$, the tube tries to expand more than the bolt when assembly is heated. But the tube is tightened by the bolt therefore its expansion cannot be more than that of the bolt. Hence, a compressive stress is developed in the tube and a tensile stress in bolt.

Let the initial length of the assembly be l and let its elongation be Δl. Elongation of the tube if it would be free to expand $= l\alpha_t\Delta\theta$. But its actual elongation is Δl, therefore elongation prevented in it is equal to $(l\alpha_t\Delta\theta - \Delta l)$. Therefore, compressive strain in tube

$$E_t = \frac{(l\alpha_t\Delta\theta - \Delta l)}{l} = 5\times10^{-5}$$

11. a. Compressive stress in tube

$$\sigma_t = E_t Y_t = 5\times10^6\ \text{N/m}^2$$

12. b. Tensile stress in bolt

$$\sigma_b = \frac{Y_b(\Delta l - l\alpha_b\Delta\theta)}{l} = 10^7\ \text{N/m}^2$$

For Problems 13–15

Sol.

13. d. Young's modulus is a material constant. So, length and radius do not affect its value. So, ratio has to be 1:1.

14. a., c. On crossing the yield region, the material will experience the breaking stress and further elongation causes reduction in stress and breaking of the wire. So, options (a) and (c) are correct.

15. b. Stress–strain curve flattens on crossing the elastic limiting region. So, (b) is correct.

For Problems 16–18

16.a., 17.a., 18.c.

Sol.

16. a. Excess pressure is always on the concave side of the surface. For angle of contact of $2\pi/3$, the liquid should have a convex surface. So, the excess pressure should be in the upward direction.

17. a. In soap bubble, there will be two free surfaces. So, excess pressure will be double than that in an air bubble. So, air bubble is $p/2$.

18. c. Difference in pressure between the inner and outer surface is $(p_i - p_0)$.

$$\therefore \quad \text{Force} = (p_i - p_0)\pi r^2$$

Since excess pressure is given, $p = (p_i - p_0)$

$$\therefore \quad F = p\pi r^2$$

So (c) is correct and the rest are incorrect.

For Problems 19–21

19.b., 20.c., 21.a.

Sol.

19. b. Bernoulli's theorem is used in this arrangement. Plant gun uses this principle. Spherical shape of bullets and spray are also working under this pressure concept.

20. c. *TY* and *NX* are showing the plastic regions in *A* and *B*.

So $\quad L_A > L_B$

Since *A* withstands more stress before yielding than *B*, $S_A > S_B$

So (c) is correct and the rest are incorrect.

21. a. Time to regain original status in *B* is greater than the time to regain original status in *A*. So relatively *B* is more fatigued than *A*. So (a) is correct and the rest are incorrect. However, this answer can only be relative and never absolute.

For Problems 22–24

22.c., 23.b., 24.b.

Sol.

22. c. $P_A = P_0 - \frac{2S}{r}$

The pressure inside a concave meniscus is less than the pressure outside (atmospheric). Assuming the meniscus to be spherical (as for thin capillaries), excess pressure is $2S/r$ where r is the radius of the hemispherical surface.

23. b. When the capillary tube is tilted by an angle θ with the vertical, the capillary rise to distance l in the tube will be such that the meniscus will remain at the same vertical height above the level of water in the container.

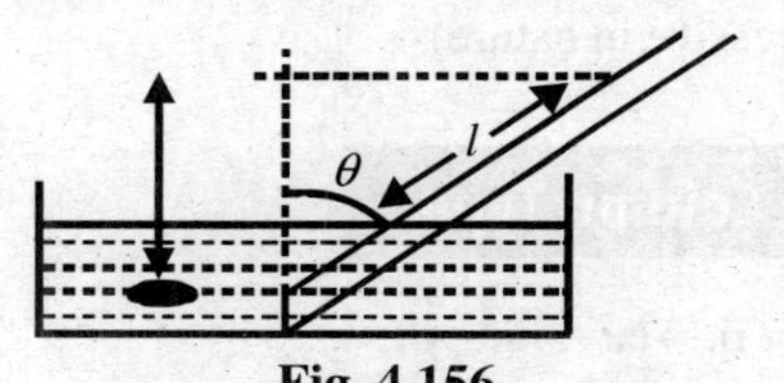

Fig. 4.156

Hence, $h = l\cos\theta$

$$l = \frac{h}{\cos\theta} = \frac{10}{\cos 45^\circ} = 10\sqrt{2}\text{ cm}$$

24. c. The capillary rise h is given by, $h = \frac{2S\cos\theta}{\rho g r}$

Hr = constant for a given glass and liquid. This is shown correctly in (c).

For Problems 25–27
25. b., 26. a., 27. c.

Sol. Here, first of all, draw the free-body diagram of the steel rods and the L-shaped structure as shown in the figure.

The direction of N_1 and N_2 can be found by using the equilibrium condition.

For L-shaped structure,

For vertical equilibrium, $N_1 = N_2$,

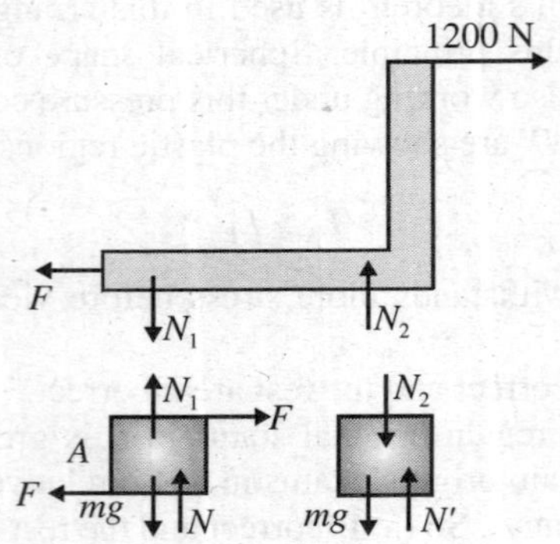

Fig. 4.157

For horizontal equilibrium, $F = 1200$ N
For rotational equilibrium, torque about E (or any other point) = 0

$$\Rightarrow \quad N_2 \times 4 - 1200 \times 3 = 0$$

$$\Rightarrow \quad N_2 = 900 \text{ N}$$

Shearing stress in $A = \frac{F}{A} = \frac{1200}{5}$ N/m² = 240 N/cm²

Shearing stress in $B = 0$

Longitudinal stress in $A = \frac{N_1}{A} = 180$ N/cm²

Longitudinal stress in $B = \frac{N_2}{A} = 180$ N/cm²

(Compressive in nature)

Matching Column Type

1. i.→b.; ii.→c.; iii.→d.

The forces acting on the plate are weight of the plate vertically downwards

$$W = mg = 20 \text{ gf}$$

i. Buoyant force due to liquid,

$$F_b = V\rho_l g$$

$$= \left(\frac{lbt}{2}\right)\rho_l g = \left(\frac{10\times 4\times 0.4}{2}\right)\times 1\text{gf}$$

$$= 8 \text{ gf}$$

ii. Force due to surface tension, vertically downward
$F = T \times$ perimeter of plate in contact with water

$$= \frac{70}{980}\times 2(10+0.4)\text{ gf} = 1.5\text{ gf}$$

iii. Apparent weight of the plate = $(W + F) - F$
= (20 + 1.5) − 8 = 13.5 gf

2. i.→a.,c.; ii.→a.,d.; iii.→b.,d.; iv.→b.,c.

For A and B, the rod is in equilibrium and hence internal restoring force developed per unit area across any cross-section is same and stress developed in the rod is uniform, while in C and D, the case is reverse.

3. i.→a.,d.; ii.→b.,c.; iii.→a.,d.; iv.→a.,d.

Directly from the theory. (Here, we assumed that in longitudinal and tensile stress that cubicle and parallelopiped are of the same shape, i.e., both are parallel lepipeds).

4. i.→b., c.; ii.→a.,d.; iii.→b.,c.

For A and C: In this case, the final surface area is greater than the initial surface area, and hence surface energy increases the expanse of internal energy; hence it causes cooling.

For B, reverse of the above reasoning.

5. Based on theory

i.→a.,d.; ii.→a.,d.; iii.→a.,d.; iv.→a.,b.,c.

6. Based on theory

i.→a.,c,d.; ii.→a.,c.,d.; iii.→a.; iv.→a.,b.,c.,d.

Integer Answer Type

1. (1) Consider an element of thickness dx. Change in the length of the element is $dl = \frac{T\,dx}{S\,Y}$ and $T = F_1 - (F_1 - F_2)\frac{x}{l}$

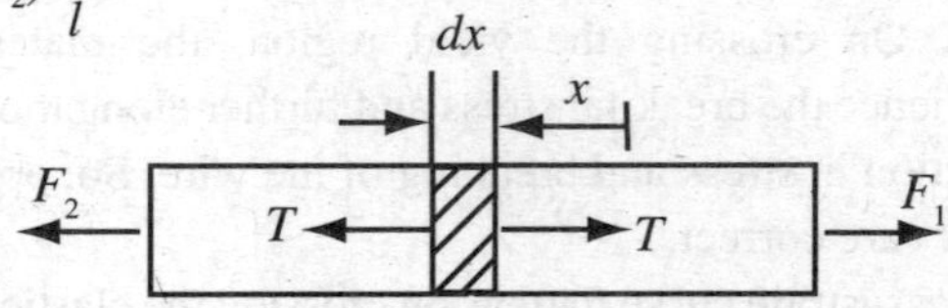

Fig. 4.158

$$\int_0^{\Delta l} dl = \int_0^{l} \frac{F_1 - \frac{(F_1 - F_2)x}{l}}{SY}dx$$

$$\Delta l = \frac{(F_1 + F_2)}{2SY} l = \frac{200\times 1}{2\times 0.5\times 2\times 10^{11}}$$

$$= 1\times 10^{-9} \text{ m}$$

$$x = 1$$

2. (2) $2T\sin\frac{\Delta\theta}{2} = dm\omega^2 r$

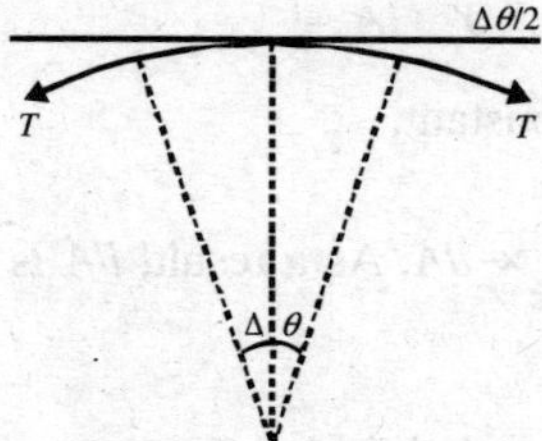

Fig. 4.159

$$2T\left(\frac{\Delta\theta}{2}\right) = \rho\times A\times r^2\omega^2$$

$$\sigma = \frac{T}{A} = \rho r^2\omega^2$$

$$\therefore \quad \omega = \frac{1}{r}\sqrt{\frac{\sigma}{\rho}} = 2 \text{ rad/s}$$

3. (2) $Mg + F_{ST} = B$

$mg + 4aT = a^2 h\rho_w g$

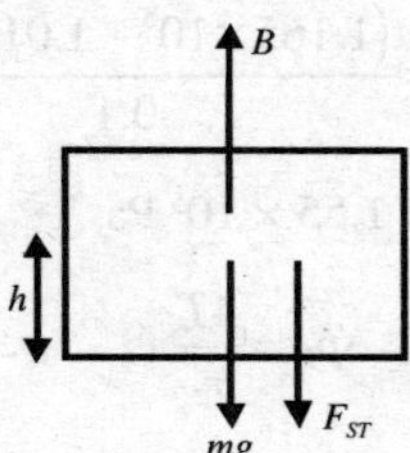

Fig. 4.160

$$10 + 4\times\frac{10}{4} = 10h$$

$$h = 2 \text{ m}$$

4. (5) Case I: $y_1 = y_2 = h/2$

$$\frac{\partial u}{\partial y} = \frac{v}{h/2} = \frac{2v}{h}$$

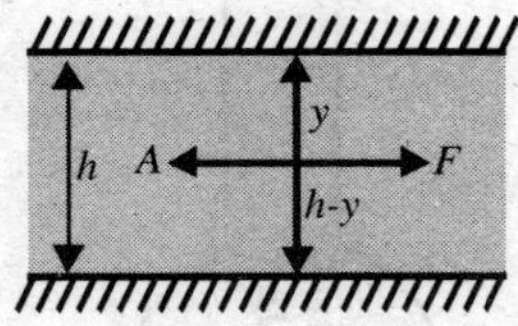

Fig. 4.161

$$\text{Viscous force} = \frac{\eta_0 2V}{h}A$$

The plate has two sides.

$$\text{Total force } F = \frac{4\eta_c va}{h}$$

Case II:

$$F = \frac{\eta vA}{y} + \frac{\eta vA}{y-h} = \frac{\eta vAh}{y(h-y)}$$

$$\frac{\eta vAh}{y(h-y)} = \frac{4v^2 Av}{h}$$

$$\frac{3}{16}h^2 = y(h-y)$$

Solving, we get $y = \frac{h}{4}, \frac{3h}{4}$

Hence, for minimum distance from top $y = \frac{20}{4} = 5$ cm

5. (5) We have

$$pV = mRT$$

$$\left(p_0 + \rho_w gh - \frac{4T}{d}\right)\frac{\pi}{6}d^3 = \left(p_0 + \frac{4T}{d}d\right)\frac{\pi}{6}n^3 d^3$$

(the bubble has only one surface)

$$\therefore \quad h = \frac{d(n^3-1)p_0 + 4T(n^2-1)}{\rho_w gd}$$

or

$$h = \frac{4\times 10^{-6}(1\times 1^3 - 1)\times 10^5 + 4\times 0.07(1\times 1^2 - 1)}{1000\times 9.8\times 4\times 10^{-6}} = 5 \text{ m}$$

6. (8)

$$\frac{4}{3}\pi R^3 = n\times\frac{4}{3}\pi r^3$$

or $\quad R^3 = nr^3$ or $R = n^{1/3}r$

or $\quad R = 2\pi^{1/3}$ mm

$$v_0 \propto r^2, v_0' \propto R^2$$

Now, $\quad \frac{v_0'}{v_0} = \frac{R^2}{r^2} = \frac{4n^{2/3}}{4}f$

or $\quad \frac{32}{8} = n^{2/3}$ or $n^{2/3} = 4$

or $\quad n = 4^{3/2} = \sqrt{64}$

or $\quad n = 8$

7. (5)

$Mg = PA$; A is sectional area

$$Alqg = 10^5 A$$

$$l = \frac{10^5}{2\times 10^3\times 10} = 5 \text{ m}$$

Archives

Fill in the Blanks Type

1. Let during the stretching of the wire, its extension at an instant of time is x and it is further stretched by dx. The restoring force acting at that instant is

$$Y=\frac{F/A}{x/L} \Rightarrow F=\frac{Yx}{1\times L}\times A$$

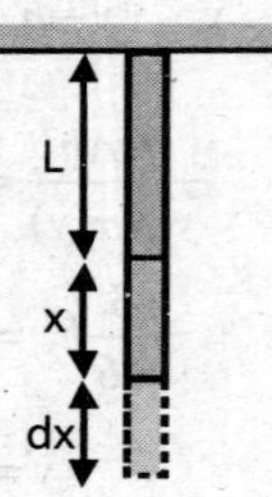

Fig. 4.162

The small amount of work done during the stretching by dx by the applied force

$$dW = Fdx = \frac{YxA}{L}dx$$

Total work done for stretching for distance x can be found by integrating above equation.

$$W=\int_0^x \frac{YxAdx}{L}=\frac{YA}{L}\int_0^x xdx=\frac{YAx^2}{2L}$$

2. $k=\frac{-\Delta p}{\Delta V/V}$; where $\Delta p=\frac{Mg}{A}$

$$\therefore \quad -\frac{\Delta V}{V}=\frac{Mg}{Ak}$$

$$\Rightarrow \quad -\frac{(V_f-V_i)}{V_i}=\frac{Mg}{Ak}$$

$$\Rightarrow \quad -\frac{(V_i-V_f)}{V_i}=\frac{Mg}{Ak}$$

$$\Rightarrow \quad \frac{\frac{4}{3}\pi R^3-\frac{4}{3}\pi(R-\delta R)^3}{\frac{4}{3}\pi R^3}=\frac{Mg}{Ak}$$

$$\Rightarrow \quad \frac{R^3-[R^3-3R^2\delta R]}{R^3}=\frac{Mg}{Ak}$$

$$\Rightarrow \quad \frac{\delta R}{R}=\frac{Mg}{3Ak}$$

3. Let A be the area of cross section of the rod.

Mass m = volume × density $=\left(\frac{L}{2}A\right)\rho$

$$T=ma=(\frac{L}{2}A\rho a)$$

Therefore, stress $=\frac{T}{A}=\frac{1}{2}\rho a\ L$

Multiple Choice Questions with One Correct Answer Type

1. a. $Y=\frac{T/A}{\Delta l/l}$

$$\therefore \Delta l=\frac{T\times l}{A\times Y}=\frac{T}{Y}\times\frac{l}{A}$$

Here, $\frac{T}{Y}$ is constant.

Therefore, $\Delta l \propto l/A$. As a result, l/A is largest in the first case.

2. c. $Y=\frac{\text{Stress}}{\text{Strain}}$ and $\Delta l = l\alpha\Delta T$

$$\therefore \text{ Strain}=\frac{\Delta l}{l}=\alpha\Delta T \quad \Rightarrow \text{Stress} = Y\alpha\Delta T$$

For first rod, stress $=\gamma_1\alpha_1\Delta T$. For the second rod, stress $=\gamma_2\alpha_2\Delta T$. Since stresses are equal, Therefore,

$$\gamma_1\alpha_1\Delta T \quad \Rightarrow \frac{Y_1}{Y_2}=\frac{\alpha_2}{\alpha_1}=\frac{3}{2}$$

3. a. $Y=\frac{F}{A}\Big/\frac{\Delta l}{l}=\frac{20\times 1}{10^{-6}\times 10^{-4}}=2\times 10^{11}\ \text{N/m}^2$

4. a. $B=-\frac{\Delta p}{\Delta v/V}=\frac{(1.165\times 10^5-1.01\times 10^5)}{0.1}$

$$= 1.55 \times 10^5 \text{ Pa}$$

5. b. $\Delta p_1=\frac{4T}{r_1}$ and $\Delta p_2=\frac{4T}{r_2}$

$$r_1 < r_2 \quad \therefore \Delta p_1 > \Delta p_2$$

Therfore, air will flow from 1 to 2 and volume of bubble at end 1 will increase.

Comprehension Type

1. c. Surface tension force $=2\pi rT\frac{r}{R}=\frac{2\pi r^2T}{R}$

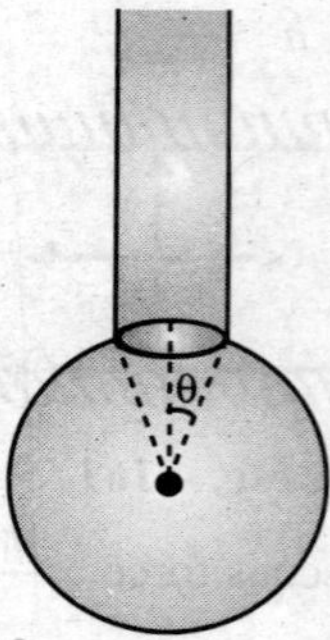

Fig. 4.163

2. a. $\frac{2\pi r^2 T}{R} = mg = \frac{4}{3}\pi R^3 \rho g$

3. b. Surface energy = $T(4\pi R^2) = 2.7 \times 10^{-6}$ J

Integer Answer Type

1. (6)

$$P_A = P_0 + \frac{4T}{R_A} = 16 \text{ N/m}^2$$

$$P_B = P_0 + \frac{4T}{R_B} = 12 \text{ N/m}^2$$

$$\frac{n_B}{n_A} = \frac{P_B}{P_A}\left(\frac{R_B}{R_A}\right)^3 = 6$$

2. (4)

$$\omega = \sqrt{\frac{YA}{mL}}$$

CHAPTER

5

Gravitation

- Gravitation
- Kepler's Laws
- Newton's Law of Gravitation
- Deduction of Newton's Law of Gravitation From Kepler's Laws
- Principle of Superposition of Gravitation
- Inertial and Gravitational Masses
- Gravitational Field
- Intensity of Gravitational Field
- The Forces of Attraction Between a Hollow Spherical Shell of Uniform Density and a Point Mass Body
- Gravity
- Gravitational Potential
- Gravitational Potential Energy
- Gravitational Self-Energy
- Principle of Launching a Satellite
- Geostationary or Geosynchronous Satellites

GRAVITATION

Gravitation is the name given to the force of attraction between any two bodies of the universe. It was discovered by Newton in 1665, when he saw an apple falling down from the tree.

Gravitational force is the weakest force among the four fundamental forces of nature, but is the most important force because it played an important role in initiating the birth of stars, and controlling the entire structure of the universe. The gravitational force has significant applications in the present advancement of physics.

KEPLER'S LAWS

A heavenly body revolving around a planet in an orbit is called a natural satellite. For example, Moon revolves around the planet earth, so Moon is the satellite of earth. Their motions can be studied with the help of Kepler's laws, as stated in the following:

I. Law of orbit: Each planet moves around the Sun in an elliptical orbit with the Sun at one of the foci as shown in Fig. 5.1. The eccentricity of an ellipse is defined as the ratio of the distance SO and AO, i.e., $e = SO/AO$.

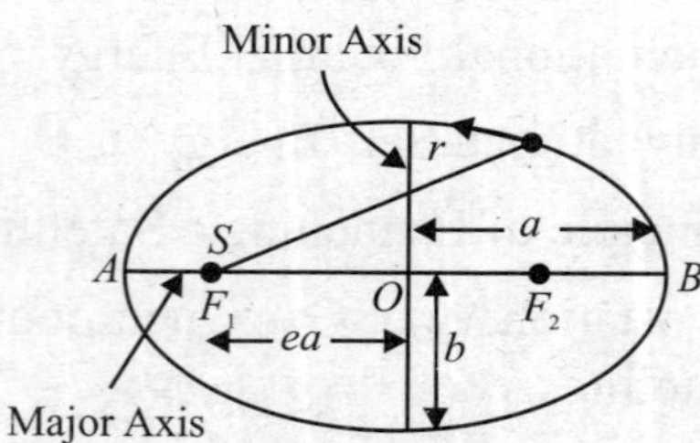

Fig. 5.1

$$\therefore \quad e = \frac{SO}{a}$$

$$SO = ea$$

The distance of closest approach with the Sun at F_1 is AS. This distance is called perigee. The greatest distance (BS) of the planet from the Sun is called apogee.

Perigee $\quad (AS) = AO - OS = a - ea = a(1 - e)$

Apogee $\quad (BS) = OB + OS = a + ea = a(1 + e)$

II. Law of areas: The line joining the Sun and a planet sweeps out equal areas in equal intervals of time. A planet takes the same time to travel from A to B as from C to D as shown in Fig. 5.2 (the shaded areas are equal.) Naturally, the planet has to move faster from C to D. In other words, areal velocity of a planet is constant at $\frac{dA}{dt}$ = const. The law of areas is identical with the law of conservation of angular momentum.

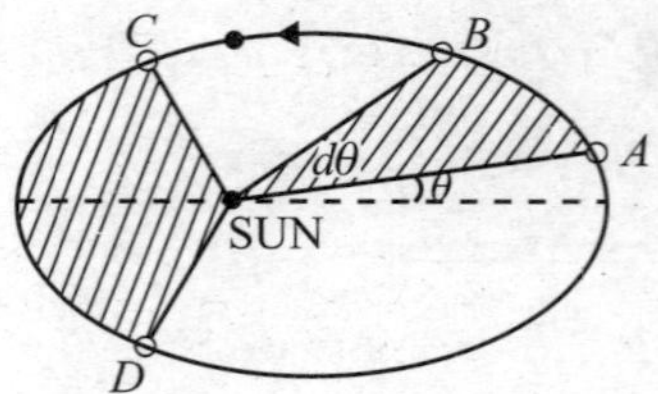

Fig. 5.2

$$\text{Areal velocity} = \frac{\text{Area swept}}{\text{Time}} = \frac{\frac{1}{2}r\,(r\,d\theta)}{dt}$$

$$= \frac{1}{2}r^2\frac{d\theta}{dt} = \text{constant}$$

Hence, $\quad \frac{1}{2}r^2\,\omega = \text{constant}$

or $mr^2\omega = \text{constant}$

where $mr^2 = MI$ of the planet relative to Sun $= I$

$I\omega = \text{constant} \quad$ or $\quad \vec{L} = \text{constant}$

III. Law of periods: The square of the time for a planet to complete a revolution about the Sun is proportional to the cube of semimajor axis of the elliptical orbit.

$$T^2 \propto a^3 \qquad \text{(i)}$$

T = Time period of revolution of the planet

a = semi-major axis of the ellipse. The orbits of all the planets except Mercury and Pluto are very close to being circular. Hence, we can take the orbits of planets as sufficiently circular since the semi-major axis becomes the radius of the circle. Hence, equation (i) can be written as

$$T^2 = Kr^3$$

Kepler's laws are applicable not only to the solar system but also equally well for satellites, either natural or artificial, orbiting around a planet (earth or any massive natural central body).

Kepler's laws which are empirical laws (i.e., laws based on observation, not theory), sum up neatly how planets of the solar system behave without indicating why they do so.

Later, Sir Issac Newton arrived at the universal law of gravitation with the help of Kepler's laws.

Illustration 5.1 **The time period of Jupiter is 11.6 years. How far is Jupiter from the Sun? Distance of the Earth from the Sun is 1.5 × 10¹¹ m.**

Sol. Here $T_J = 11.6$ years; $r_J = ?$

$$T_e = 1 \text{ yr} ; r_e = 1.5 \times 10^{11} \text{ m}$$

As $$\frac{T_J^2}{T_e^2} = \frac{r_J^3}{r_e^3} \text{ or } r_J = r_e \left(\frac{T_J}{T_e}\right)^{2/3}$$

$$r_J = 1.5 \times 10^{11} \left(\frac{11.6}{1}\right)^{2/3} = 7.68 \times 10^{11} \text{ m}$$

Illustration 5.2 **The mean distance of Mars from the Sun is 1.524 times that of the Earth from the Sun. Find the number of years required for Mars to make one revolution about the Sun.**

Sol. For planets revolving around the Sun, $T^2 \propto r^3$

$\Rightarrow \frac{T_1^2}{T_2^2} = \frac{r_1^3}{r_2^3}$ (T_1 = time period of Mars, T_2 = time period of the Earth)

$$\Rightarrow \frac{T_1}{T_2} = \left(\frac{r_1}{r_2}\right)^{3/2} \Rightarrow T_1 = T_2\left(\frac{r_1}{r_2}\right)^{3/2} = (1 \text{ year})\,(1.524)^{3/2}$$

$$= 1.88 \text{ years}$$

Illustration 5.3 **Let the speed of the planet at the perihelion P in Fig. 5.3 be v_P and the Sun–planet distance SP be r_P. Relate r_P, v_P to the corresponding quantities at the aphelion (r_A, v_A). Will the planet take equal times to traverse BAC and CPB?**

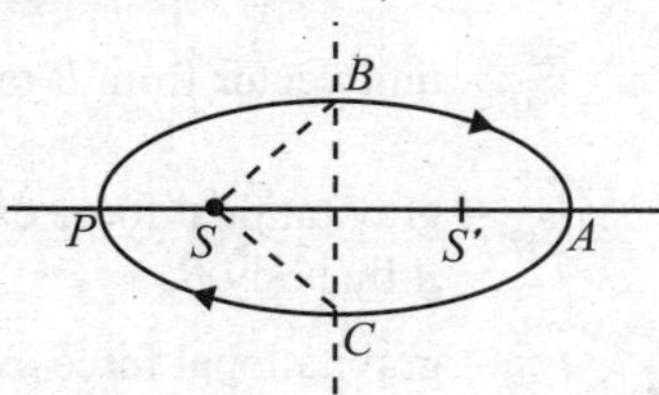

Fig. 5.3

Sol. Refering Fig. 5.4, we note that $\vec{r}_P$ and $\vec{v}_P$ are perpendicular to each other. Similarly, $\vec{r}_A$ and $\vec{v}_A$ are perpendicular to each other.

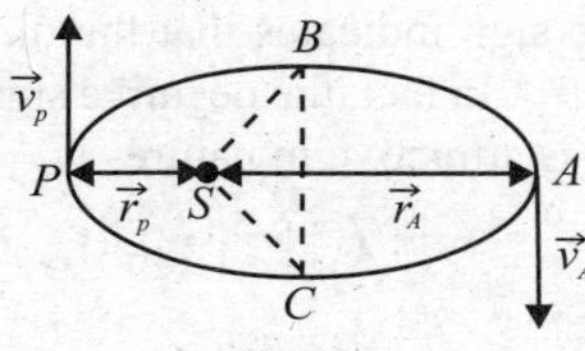

Fig. 5.4

Using the law of conservation of angular momentum,

Angular momentum of the planet at P = Angular momentum of the planet at A

$$\Rightarrow \quad m_P v_P r_P = m_A v_A r_A \text{ or } \frac{v_P}{v_A} = \frac{r_A}{r_P}$$

Since $\quad r_A > r_P,\ v_P > v_A$

Here area $SBAC$ is greater than area $SCPB$. As the areal velocity of a planet is constant around the Sun, i.e., equal areas are swept in equal times. Hence the planet will take longer time to traverse BAC than CPB.

NEWTON'S LAW OF GRAVITATION

It states that every body in this universe attracts every other body with a force which is directly proportional to the product of their masses and is inversely proportional to the square of the distance between them.

Consider two point mass bodies A and B of masses m_1 and m_2. Let r be the distance between their centres and F be the force of attraction between them.

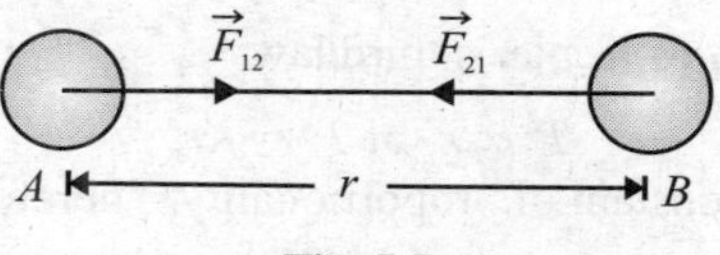

Fig. 5.5

According to Newton's law of gravitation,

$$F \propto \frac{m_1 m_2}{r^2} \text{ or } F = \frac{G m_1 m_2}{r^2} \quad \text{(ii)}$$

Definition of G

Let $\quad m_1 = m_2 = 1$ and $r = 1$.

From Eq. (iv), $\quad F = G\frac{1 \times 1}{1^2} = G$ or $G = F$

Thus universal gravitational constant is equal to the force of attraction acting between two bodies each of unit mass, whose centres are placed unit distance apart.

Gravitational constant is a scalar quantity. Its value is same throughout the universe and is independent of the nature and size of the bodies as well as the nature of the medium between the bodies. The value of G in SI unit is 6.67×10^{11} Nm²kg⁻² and in CGS system its value is 6.67×10^{-8} dyne cm² g⁻².

The dimensional formula for G is

$$\frac{Fr^2}{m_1 m_2} = \frac{(MLT^{-2})}{M \times M}$$
$$= [M^{-1}L^3T^{-2}]$$

***Note:* Newton's law of gravitation holds good for objects lying at very large distances and also at very short distances. It fails when the distance between the objects is less than 10^{-9} m (i.e., of the order of intermolecular distances).**

DEDUCTION OF NEWTON'S LAW OF GRAVITATION FROM KEPLER'S LAWS

Newton's law of gravitation can be easily obtained from Kepler's laws of planetary motion.

Suppose a planet of mass m is revolving around the Sun of mass M in a nearly circular orbit of radius r, with a constant angular velocity ω. Let T be the time period of revolution of the planet around the Sun.

$$\omega = \frac{2\pi}{T} \quad \text{(i)}$$

The centripetal force acting on the planet for its circular motion is

$$F = mr\omega^2 = mr\left(\frac{2\pi}{T}\right)^2$$

$$= \frac{4\pi^2 mr}{T^2} \quad \text{[from Eq. (i)]}$$

According to Kepler's third law,

$$T^2 \propto r^3 \text{ or } T^2 = Kr^3$$

where K is a constant of proportionality. Therefore,

$$F = \frac{4\pi^2 mr}{Kr^3} = \frac{4\pi^2}{K} \times \frac{m}{r^2} \quad \text{(ii)}$$

$$\Rightarrow \quad F \propto \frac{m}{r^2} \left(\text{Since } \frac{4\pi^2}{K} \text{ is a constant}\right) \quad \text{(iii)}$$

This centripetal force is provided by the gravitational attraction exerted by the Sun on the planet. According to Newton, the gravitational attraction between the Sun and the planet is mutual. If force F is directly proportional to the mass of the planet, m [from Eq. (iii)], it should also be directly proportional to the mass of the Sun, M. Hence, the factor $4\pi^2/K \propto M$ or $4\pi^2/K = GM$. Putting this value in Eq. (ii), we get $F = GMm/r^2$, which is Newton's law of gravitation.

On the basis of Kepler's laws, Newton concluded the following:

i. A force acting on the planet due to Sun is the centripetal force which is directed towards the Sun.
ii. The force acting on the planet must be inversely proportional to the square of the distance from the Sun.
iii. The force acting on the planet is directly proportional to the product of the masses of the planet and the Sun.

Illustration 5.4 A sphere of mass 40 kg is attracted by a second sphere of mass 15 kg, when their centres are 20 cm apart, with a force of 0.1 milligram weight. Calculate the value of gravitational constant.

Sol. Here, $m_1 = 40$ kg, $m_2 = 15$ kg

$$r = 0.20 \text{ m}$$

$$F = 0.1 \text{ milligram wt.} = 0.1 \times 10^{-6} \text{ kg wt.} = 0.1 \times 10^{-6} \times 9.8 \text{ N}$$

Now, $$G = \frac{Fr^2}{m_1 m_2} = \frac{(0.1 \times 10^{-6} \times 9.8) \times 0.20)^2}{40 \times 15}$$

$$= 6.53 \times 10^{-11} \text{ Nm}^2\text{kg}^{-2}$$

Evidence in Support of Newton's Law of Gravitation

1. The rotation of earth around the Sun or the rotation of Moon around the earth is explained on the basis of this law.
2. The prediction about solar and lunar eclipses, made on the basis of this law, always came out to be true.
3. The formation of tides in the ocean is due to force of attraction between the Moon and ocean water.
4. The prediction about the orbits and time period of the modern artificial satellites made on the basis of this law are found to be very accurate.

Vector Form of Newton's Law of Gravitation

Consider two point mass bodies A and B of masses m_1 and m_2 placed distance r apart.

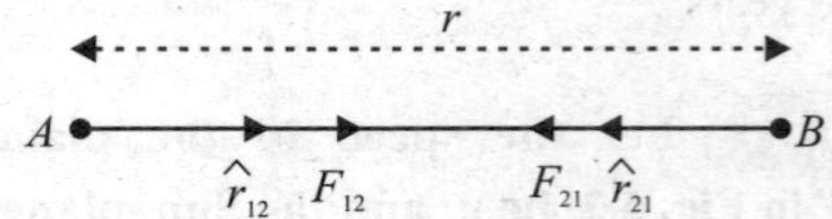

Fig. 5.6

Let $\vec{r}_{12}$ = unit vector from A to B,

$\vec{r}_{21}$ = unit vector from B to A,

$\vec{F}_{12}$ = gravitational force exerted on body A by body B,

$\vec{F}_{21}$ = gravitational force exerted on body B by body A.

According to Newton's law of gravitation,

$$\vec{F}_{12} = -G\frac{m_1 m_2}{r^2}\hat{r}_{21} \quad \text{(iv)}$$

Here, negative sign indicates that the direction of $\vec{F}_{12}$ is opposite to that of $\vec{F}_{21}$. In fact, the negative sign shows that the gravitational force is attractive in nature.

As $$r = |\vec{r}_{12}| = |\vec{r}_{21}|$$

so $$\vec{F}_{12} = -\frac{Gm_1 m_2}{|\vec{r}_{21}|^2}\hat{r}_{21} \quad \text{(v)}$$

Similarly, $\vec{F}_{21} = -\frac{Gm_1m_2}{|\vec{r}_{12}|^2}\hat{r}_{12} = \frac{Gm_1m_2}{|\vec{r}_{12}|^2}\hat{r}_{21}$ (vi)

$[\because \hat{r}_{12} = -\hat{r}_{21}]$

From Eqs. (vi) and (vii), we have $\vec{F}_{21} = -\vec{F}_{12}$

If $\vec{r}_1$ and $\vec{r}_2$ are the position vectors of point masses m_1 and m_2 at points A and B, as shown in Fig. 5.7, then $\vec{r}_{12} = \vec{r}_2 - \vec{r}_1$.

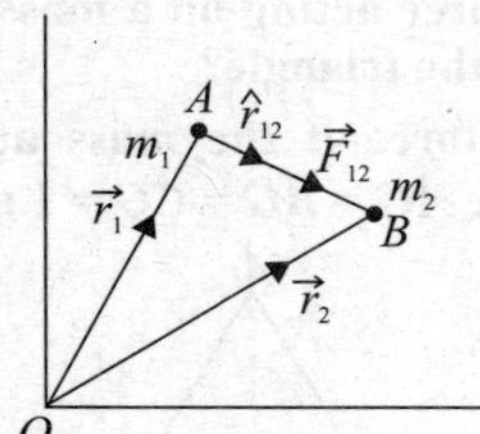

Fig. 5.7

Now, $\vec{F}_{21} = -\frac{Gm_1m_2}{r^3}r\hat{r}_{12} = -\frac{Gm_1m_2}{|\vec{r}_{12}|^3}\vec{r}_{12}$

$$= -\frac{Gm_1m_2}{|\vec{r}_2 - \vec{r}_1|^3}(\vec{r}_2 - \vec{r}_1)$$

Important Characteristics of Gravitational Force

According to Newton's law of gravitation, the gravitational force of attraction between two given bodies

i. is independent of nature of intervening medium.

ii. is independent of the presence or absence of other bodies.

iii. is independent of nature and size of the bodies, till their masses remain the same and the distance between their centres is fixed.

iv. forms action and reaction pair, i.e., the gravitational forces are equal in magnitude and opposite in direction and hence obey Newton's third law of motion.

v. is a central force as it acts along the line joining the centres of the two bodies.

vi. is conservative in nature.

Illustration 5.5 **The mass of planet Jupiter is 1.9 × 10^{27} kg and that of the Sun is 1.99 × 10^{30} kg. The mean distance of Jupiter from the Sun is 7.8 × 10^{11} m. Calculate the gravitational force which Sun exerts on Jupiter. Assuming that Jupiter moves in circular orbit around the Sun, also calculate the speed of Jupiter. G = 6.67 × 10^{-11} Nm^2 kg^{-2}.**

Sol. Here, $M_j = 1.9 \times 10^{27}$ kg

$M_s = 1.99 \times 10^{30}$ kg ; $r = 17.8 \times 10^{11}$ m

$G = 6.67 \times 10^{-11}$ Nm^2 kg^{-2} ; $F = ?$

Now $F = \frac{GM_JM_s}{r^2} = \frac{6.67 \times 10^{-11} \times 1.9 \times 10^{27} \times 1.99 \times 10^{30}}{(7.8 \times 10^{11})^2}$

$= 4.15 \times 10^{23}$ N

Since the gravitational pull of the Sun on Jupiter provides the required centripetal force to Jupiter, therefore the velocity of Jupiter v can be given by the relation,

$$F = \frac{M_Jv^2}{r} \text{ or } v = \sqrt{\frac{F_r}{M_J}}$$

$$= \sqrt{\frac{4.15 \times 10^{23} \times 7.8 \times 10^{11}}{1.9 \times 10^{27}}} = 1.3 \times 10^4 \text{ ms}^{-1}$$

Illustration 5.6 **Gravitational force between two point masses m and M separated by a distance r is F. Now if a point mass $3m$ is placed next to m, what will be the (a) force on M due to m, (b) total force on M?**

Sol. a. If r is the distance between two point masses m and M, then the gravitational force on m due to mass M is $F = GMm/r^2$.

Since the gravitational force between two point masses is independent of the presence of other masses, so if a point mass $3m$ is placed next to m, the force on M due to m is $F = \frac{GMm}{r^2}$.

b. Total force on the body of mass M is F, i.e.,

$$F = \frac{GM \times (m + 3m)}{r^2} = \frac{4GMm}{r^2} = 4F$$

Illustration 5.7 **Two particles of masses m_1 and m_2, initially at rest at infinite distance from each other, move under the action of mutual gravitational pull. Show that at any instant their relative velocity of approach is $\sqrt{2G(m_1 + m_2)/R}$, where R is their separation at that instant.**

Sol. The gravitational force of attraction on m_1 due to m_2 at a separation r is

$$F_1 = \frac{Gm_1m_2}{r^2}$$

Therefore, the acceleration of m_1 is

$$a_1 = \frac{F_1}{m_1} = \frac{Gm_2}{r^2}$$

Similarly, the acceleration of m_2 due to m_1 is

$$a_2 = -\frac{Gm_1}{r^2}$$

The negative sign is put as a_2 is directed opposite to a_1. The relative acceleration of approach is

$$a = a_1 - a_2 = \frac{G(m_1 + m_2)}{r^2} \quad \text{(i)}$$

If v is the relative velocity, then

$$a = \frac{dv}{dt} = \frac{dv}{dr}\frac{dr}{dt}$$

But $-dr/dt = v$ (negative sign shows that r decreases with increasing t).

$$\therefore \quad a = -\frac{dv}{dr}v \qquad \text{(ii)}$$

From Eqs. (i) and (ii), we have

$$v\,dv = -\frac{G(m_1 + m_2)}{r^2}dr$$

Integrating, we get

$$\frac{v^2}{2} = \frac{G(m_1 + m_2)}{r} + C$$

At $r = \infty$, $v = 0$ (given), and so $C = 0$. Therefore,

$$v^2 = \frac{2G(m_1 + m_2)}{r}$$

Let $v = v_R$ when $r = R$. Then

$$v_R = \sqrt{\frac{2(Gm_1 + m_2)}{R}}$$

PRINCIPLE OF SUPERPOSITION OF GRAVITATION

It states that the resultant gravitational force $\vec{F}$ acting on a particle due to a number of point masses is equal to the vector sum of the forces exerted by the individual masses on the given particle, i.e.,

$$\vec{F} = \vec{F}_{01} + \vec{F}_{02} + \cdots + \vec{F}_{0n} = \sum_{i=1}^{n}\vec{F}_{0i}$$

where $\vec{F}_{01}, \vec{F}_{02}, \ldots, \vec{F}_{0n}$ are the gravitational forces on a particle of mass m_0 due to particles of masses $m_1, m_2, \ldots, m_n$, respectively.

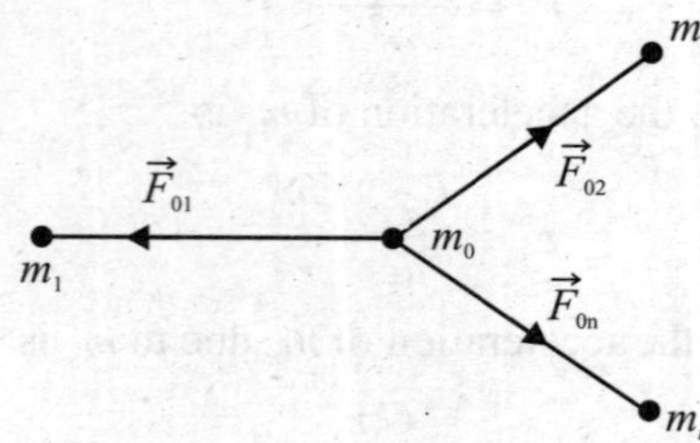

Fig. 5.8

$$\vec{F}_{01} = -\frac{Gm_0M_1}{r^2}\hat{r}_{10} = \frac{Gm_0m_1}{|\vec{r}_{01}|^2}\hat{r}_{01}$$

$$\vec{F}_{02} = \frac{Gm_0m_2}{|\vec{r}_{02}|^2}\hat{r}_{02} \text{ and } \vec{F}_{0n} = \frac{Gm_0m_n}{|\vec{r}_{0n}|^2}\hat{r}_{0n}$$

$$\vec{F}_0 = Gm_0\left[\frac{m_1\hat{r}_{01}}{|\hat{r}_{01}|^2} + \frac{m_2\hat{r}_{02}}{|\hat{r}_{02}|^2} + \cdots + \frac{m_n\hat{r}_{0n}}{|\vec{r}_{0n}|^2}\right]$$

Illustration 5.8 **Three equal masses of m kg each are fixed at the vertices of an equilateral triangle ABC.**

a. What is the force acting on a mass $2m$ placed at the centroid G of the triangle?

b. What is the force if the mass at the vertex A is doubled? Take $AG = BG = CG = 1$ m.

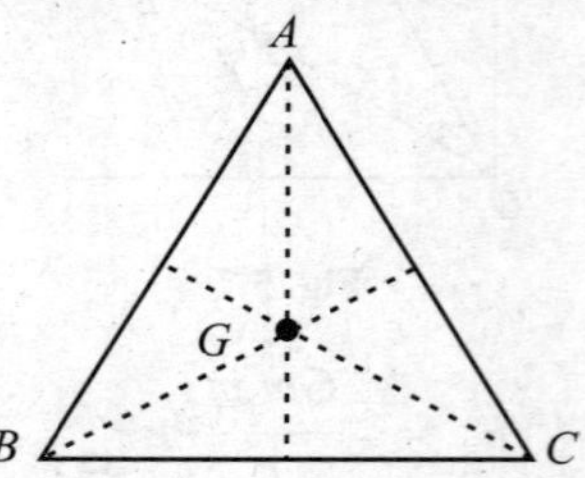

Fig. 5.9

Sol. a. From Fig. 5.10, the gravitational force on mass $2m$ at G due to mass at A is

$$F_1 = G\frac{m \times 2m}{1^2} = 2Gm^2 \text{ along } GA$$

Gravitational force on mass $2m$ at G due to mass at B is

$$F_2 = G\frac{m \times 2m}{1^2} = 2Gm^2 \text{ along } BG$$

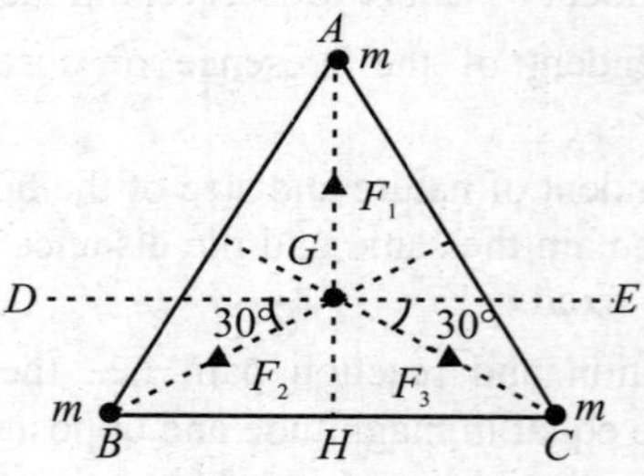

Fig. 5.10

Gravitational force on mass $2m$ at G due to mass at C is

$$F_3 = G\frac{m \times 2m}{1^2} = 2Gm^2 \text{ along } GC$$

Draw DE parallel to BC passing through point G. Then $\angle EGC = 30° = \angle DGB$.

Resolving $\vec{F}_2$ and $\vec{F}_3$ into two rectangular components, we have

$F_2 \cos 30°$ along GD and $F_2 \sin 30°$ along GH

$F_3 \cos 30°$ along GE and $F_3 \sin 30°$ along GH

Here, $F_2 \cos 30°$ and $F_3 \sin 30°$ are equal in magnitude and acting in opposite directions, and cancel out each other. The resultant force on mass $2m$ at G is

$$F_1 - (F_2 \sin 30° + F_3 \sin 30°)$$
$$= 2Gm^2 - \left(2Gm^2 \times \frac{1}{2} + 2Gm^2 \times \frac{1}{2}\right)$$
$$= 0$$

b. When mass at A is $2m$, then gravitational force on mass $2m$ at G due to mass $2m$ at A is

$$F_1' = G\frac{2m \times 2m}{1^2} = 4Gm^2 \text{ along } GA$$

The resultant force on mass $2m$ at G due to masses at A, B and C is

$$F_1' - (F_2 \sin 30° + F_3 \sin 30°)$$
$$= 4Gm^2 - \left(2Gm^2 \times \frac{1}{2} + 2Gm^2 \times \frac{1}{2}\right)$$
$$= 2Gm^2 \text{ along } GA$$

Illustration 5.9 **Three identical bodies, each of mass *m*, are located at the vertices of an equilateral triangle with side *r*. At what speed must they move if they all revolve under the influence of one another's gravitation in a circular orbit circumscribing the triangle while still preserving the equilateral triangle?**

Sol. Refer to Fig. 5.11, force of attraction on body at C due to body at A is

$$F_1 = \frac{Gmm}{r^2} = \frac{Gm^2}{r^2} \text{ along } CA$$

Force of attraction on body at C due to body at B is

$$F_2 = \frac{Gmm}{r^2} = \frac{Gm^2}{r^2} \text{ along } CB$$

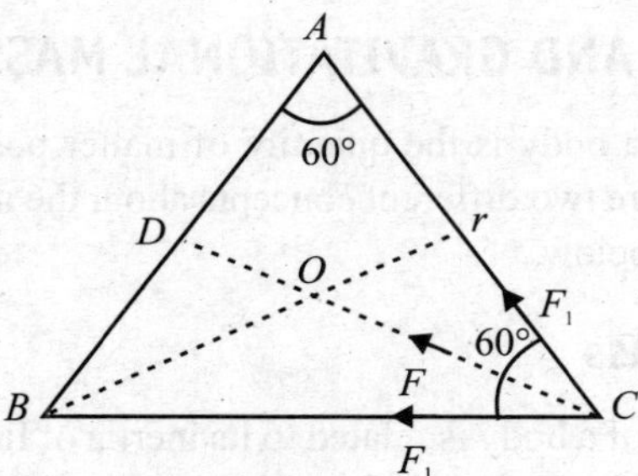

Fig. 5.11

These forces, $\vec{F}_1$ and $\vec{F}_2$, are inclined at an angle 60°. The resultant force on the body at C is

$$F = \sqrt{F_1^2 + F_2^2 + 2F_1F_2 \cos 60°}$$
$$= \sqrt{F_1^2 + F_2^2 + 2F_1F_2 (1/2)}$$
$$= \sqrt{3}F_1 = \sqrt{3}\frac{Gm^2}{r^2} \text{ acting along } CD$$

$$\left[\because F_1 = F_2 = \frac{Gm^2}{r^2}\right]$$

Here, $$OC = \frac{2}{3}CD = \frac{2}{3}AC \sin 60°$$
$$= \frac{2}{3}r \times \frac{\sqrt{3}}{2} = \frac{r}{\sqrt{3}}$$

When each body is describing a circular orbit with centre of orbit at O, the force F provides the required centripetal force. The radius of the circular orbit is $OC = r/\sqrt{3}$. If v is the speed of the body in circular orbit, then

$$\frac{Mv^2}{r/\sqrt{3}} = \frac{\sqrt{3}Gm^2}{r^2} \text{ or } v = \sqrt{\frac{Gm}{r}}$$

Illustration 5.10 **A solid sphere of lead has mass *M* and radius *R*. A spherical hollow is dug out from it (see Fig. 5.12). Its boundary passing through the centre also touches the boundary of the solid sphere. Deduce the gravitational force on a mass *m* placed at *P*, which is at a distance *r* from *O* along the line of centres.**

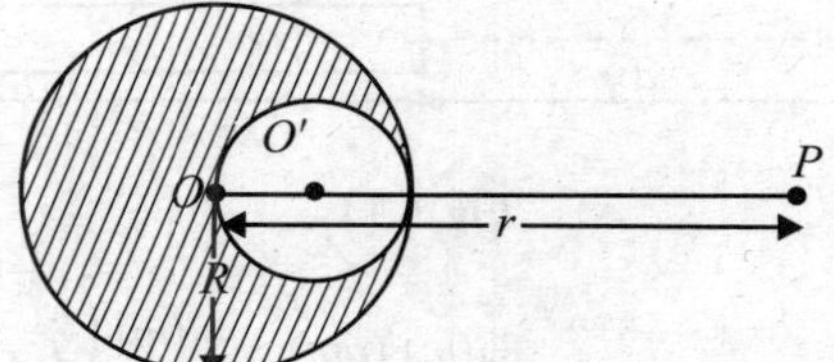

Fig. 5.12

Sol. Let O be the centre of the solid sphere and O' that of the hollow (Fig. 5.13). For an external point the sphere behaves as if its entire mass is concentrated at its centre. Therefore, the gravitational force on a mass 'm' at P due to the original sphere (of mass M) is

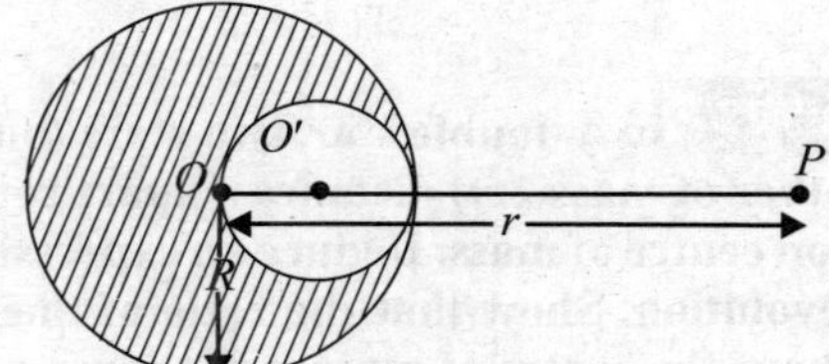

Fig. 5.13

$$F = G\frac{Mm}{r^2}, \text{ along } PO$$

The diameter of the smaller sphere (which would be cut off) is R, so that its radius OO' is $R/2$. The force on m at P due to this sphere of mass M' (say) would be

$$F' = G\frac{M'm}{(r - \frac{R}{2})^2} \text{ along } PO'$$

($\because$ distance $PO' = r - R/2$)

As the radius of this sphere is half of that of the original sphere, we have

$$M' = \frac{m}{8}$$

Therefore, $$F' = G\frac{Mm}{8(r - R/2)^2} \text{ along } PO'$$

As both F and F' point along the same direction, the force due to the hollowed sphere is

$$F - F' = \frac{GMm}{r^2} - \frac{GMm}{8r^2(1 - \frac{R}{2r})^2}$$

$$= \frac{GMm}{r^2}\left\{1 - \frac{1}{8(1 - \frac{R}{2r})^2}\right\}$$

Illustration 5.11 **What is the magnitude of the gravitational force F_1 on the particle due to the rod?**

Sol. As m is a point mass, consider an element on the rod at distance r from m

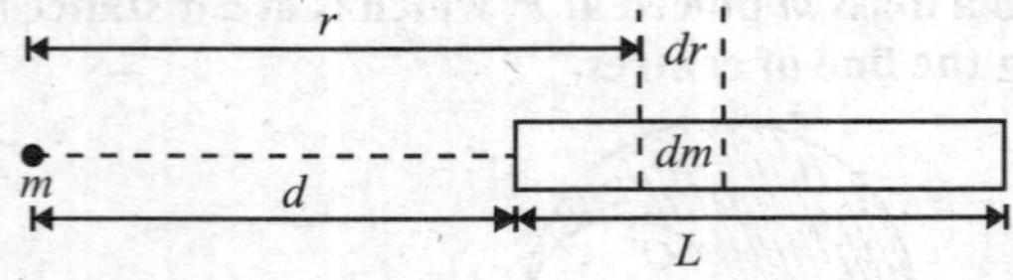

Fig. 5.14

$$dF = \frac{G(m)\,(dm)}{r^2} = \frac{G(m)\left(\frac{M}{L}\right)dr}{r^2}$$

or $$F = \int_{d}^{L+d} \frac{GMm}{L}\frac{1}{r^2}\,dr$$

$$= \frac{GMm}{L}\int_{d}^{L+d}\frac{1}{r^2}\,dr = \frac{GMm}{d(L+d)}$$

Illustration 5.12 **In a double star, two stars (one of mass m and the other of mass $2m$) distance d apart rotate about their common centre of mass. Deduce an expression for the period of revolution. Show that the ratio of their angular momenta about the centre of mass is the same as the ratio of their kinetic energies.**

Sol. The centre of mass C will be at distances $d/3$ and $2d/3$ from masses $2m$ and m, respectively. Both the stars rotate round C in their respective orbits with the same angular velocity ω. The gravitational force acting on each star due to the other supplies the necessary centripetal force.

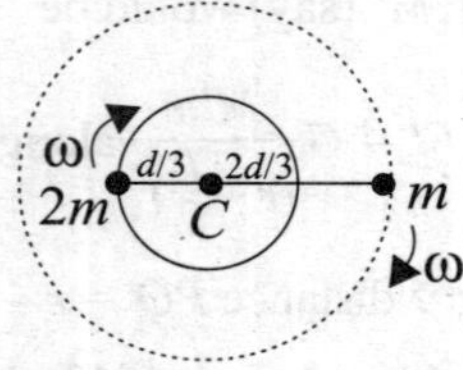

Fig. 5.15

The gravitational force on either star is $G(2m)m/d^2$. If we consider the rotation of the smaller star, the centripetal force ($mr\omega^2$) is $[m\,(2d/3)\,\omega^2]$ and $[(2md\omega^2)/3]$ for the bigger star, i.e., the same.

$$\therefore \quad \frac{G\,(2m)m}{d^2} = m\left(\frac{2d}{3}\right)\omega^2 \text{ or } \omega = \sqrt{\left(\frac{3Gm}{d^3}\right)}$$

Therefore, the period of revolution is given by

$$T = \frac{2\pi}{\omega} = 2\pi\sqrt{\left(\frac{d^3}{3Gm}\right)}$$

The ratio of the angular momenta is

$$\frac{(I\omega)_{\text{big}}}{(I\omega)_{\text{small}}} = \frac{I_{\text{big}}}{I_{\text{small}}}$$

$$= \frac{(2m)\left(\frac{d}{3}\right)^2}{m\left(\frac{2d}{3}\right)^2} = \frac{1}{2}$$

since ω is the same for both.

The ratio of their kinetic energies is

$$\frac{(\frac{1}{2}I\omega^2)_{\text{big}}}{(\frac{1}{2}I\omega^2)_{\text{small}}} = \frac{I_{\text{big}}}{I_{\text{small}}} = \frac{1}{2}$$

which is the same as the ratio of their angular momenta.

INERTIAL AND GRAVITATIONAL MASSES

The mass of a body is the quantity of matter possessed by a body. There are two different concepts about the mass of body as discussed below.

Inertial Mass

Inertial mass of a body is related to its inertia of linear motion, and is defined by Newton's second law of motion.

Let a body of mass m_i move with acceleration a under the action of an external force F. According to Newton's second law of motion,

$$F = m_i a \text{ or } m_i = \frac{F}{a} \qquad \text{(vii)}$$

The mass m_i of the body in this sense is the inertial mass of the body. If $a = 1$, then from Eq. (vii) $m_i = F$.

Thus, inertial mass of a body is equal to the magnitude of external force required to produce unit acceleration in the body.

In fact, inertial mass of a body is the measure of the ability of the body to oppose the production of acceleration in its motion by an external force.

Properties of Inertial Mass

1. Inertial mass of a body is proportional to the quantity of matter contained in the body.
2. Inertial mass of a body is independent of size, shape and state of the body.
3. Inertial mass of a body does not depend upon the temperature of the body.
4. Inertial mass of a body is not affected by the presence or absence of other bodies near it.
5. Inertial mass is conserved when the two bodies combine physically or chemically.
6. Inertial mass can be added by simple laws of algebra, irrespective of the materials of the bodies.
7. Inertial mass of a body increases with the speed of the body. When a body moves with a velocity v, its inertial mass m is given by

$$m = \frac{m_0}{\sqrt{1 - v^2/c^2}} \quad \text{(viii)}$$

where m_0 is the rest mass of the body, c is the velocity of light in vacuum. The mass m is affected only when the velocity of the body is comparable with the velocity of light.

Gravitational Mass

Gravitational mass of a body is related to gravitational pull on the body and is defined by Newton's law of gravitation.

If a body of mass m_G is placed on the surface of earth of radius R and mass M, then gravitational pull on the body is given by

$$F = \frac{GMm_G}{R^2}$$

$$m_G = \frac{F}{(GM/R^2)} = \frac{F}{I} \quad \text{(ix)}$$

The mass m_G of the body in this sense is the gravitational mass of the body. The inertia of the body has no effect on the gravitational mass of the body. If $I = 1$ then from Eq. (ix), $m_G = F$. Thus gravitational mass of a body is defined as the magnitude of gravitational pull experienced by the body in a gravitational field of unit intensity.

The properties of gravitational mass are the same as those of inertial mass except those mentioned below in the comparison of inertial mass and gravitational mass.

Comparison of Inertial Mass and Gravitational Mass

1. Both inertial and gravitational masses are measured in the same units.
2. Both are scalar quantities.
3. Both do not depend on the shape and state of the body.
4. Gravitational mass of a body is affected by the presence of other bodies near it whereas the inertial mass of a body remains unaffected by the presence of other bodies near it.
5. The inertial mass of a body is measured by finding its acceleration, produced by the applied force and gravity plays no part in the measurement. On the other hand, the gravitational mass of a body is measured by finding the gravitational pull on it and inertia plays no part in the measurement.
6. The gravitational mass is measured by spring balance whereas inertial mass is measured by inertial balance.
7. Both are equivalent to each other.

Equivalence of Inertial and Gravitational Masses

Consider two bodies A and B of gravitational masses m_G and m'_G, respectively. Let these be placed at equal distance R from a third body C (say earth). Let M be the gravitational mass of the body C. According to Newton's law of gravitation, the gravitational force exerted by C on A is given by

$$F = \frac{GMm_G}{R^2}$$

and the gravitational force exerted by C on B is

$$F' = \frac{GMm'_G}{R^2}$$

Therefore, $$\frac{F}{F'} = \frac{GMm_G/R^2}{GMm'_G/R^2} = \frac{m_G}{m'_G} \quad \text{(x)}$$

Let m_i and m'_i be the inertial masses of bodies A and B, respectively. If both the bodies are allowed to fall in vacuum from the same place under the action of gravity, where acceleration due to gravity is g, then $F = m_i g$ and $F' = m'_i g$.

Therefore, $$\frac{F}{F'} = \frac{m_i g}{m'_i g} = \frac{m_i}{m'_i} \quad \text{(xi)}$$

From Eqs. (x) and (xi), we have $\frac{m_G}{m'_G} = \frac{m_i}{m'_i}$

It means that the gravitational mass of a body is proportional to its inertial mass. In fact, inertial and gravitational masses of a body are identical.

Importance of the Equivalence of Two Masses

This idea of equivalence of inertial mass and gravitational mass of a body led Einstein to predict the general theory of relativity which is the modern theory of gravitation.

Concept Application Exercise 5.1

1. Gravitational force is weak force but still it is considered the most important force. Why?
2. The earth is continuously pulling the Moon towards its centre. Why the moon does not fall on to the earth?
3. Which has longer period of revolution, a satellite revolving close or away from the surface of earth?
4. What would happen if gravity suddenly disappears?
5. Two identical copper spheres of radius R are in contact with each other. If the gravitational attraction between them is F, find the relation between F and R.
6. A mass M is broken into two parts of masses m_1 and m_2. How are m_1 and m_2 related so that force of gravitational attraction between the two parts is maximum?
7. Imagine a light planet revolving around a very massive star in a circular orbit of radius R with a period of revolution T. If the gravitational force of attraction between the planet and the star is proportional to $R^{-5/2}$, then find the relation between T and R.
8. A planet of mass m moves around the Sun of mass M in an elliptical orbit. The maximum and minimum distance of the planet from the Sun are r_1 and r_2, respectively. Find the relation between the time period of the planet in terms of r_1 and r_2.
9. Suppose the gravitational force varies inversely as the nth power of the distance. Show that the time period of a planet in circular orbit of radius R around the Sun will be proportional to $R^{(n+1)/2}$.
10. The distance of planet Jupiter from the Sun is 5.2 times that of the earth. Find the period of revolution of Jupiter around the Sun.
11. The distance of the two planets from the Sun are 10^{13} m and 10^{12} m, respectively. Find the ratio of time periods and speeds of the two planets.
12. Given that $T^2 = kR^3$, express the constant k of the above relation in days and kilometres. Given, $k = 10^{-13} s^2 m^{-3}$. The Moon is at a distance of 3.84×10^5 km from the earth. Obtain its time period of revolution in days.
13. A sphere of mass 40 kg is attracted by a second sphere of mass 60 kg with a force equal to 4 mg. If $G = 6 \times 10^{-11}$ N m^2 kg^{-2}, calculate the distance between them. Acceleration due to gravity = 10 ms^{-2}.

GRAVITATIONAL FIELD

Consider a scenario where a person fires from a machine gun. Also assume that the man swings the machine gun around at an angle of 180° so fast that bullets fly all about him in his front. Now let us look at somebody who stands close to the gun. He is likely to get more bullets than a man who stands far off from the gun. The gun is the source or the mass, the bullet is the field and the effect is felt by a person through the field. The higher the mass of the sources, or the more powerful the gun is, the higher will be the number of bullets, i.e., the field will be higher. As you go farther away from the gun, the effect of the bullets will be felt less by you. The field decreases as you go farther from the source.

We know that earth attracts every body towards its centre. According to Newton's law of gravitation, the force of attraction varies inversely as the square of the distance of the body from the centre of earth. This shows that in the space all around the earth, its gravitational pull can be experienced by other material bodies. The same is true for material body. Hence, the space around a material body in which its gravitational pull can be experienced is called its gravitational field.

According to the field concept, a material particle placed in space modifies the space around it. This modified space is called gravitational field. When another material particle is brought in this field, it experiences a gravitational force of attraction due to interaction of gravitational field on the mass of the particle.

INTENSITY OF GRAVITATIONAL FIELD

The intensity of the gravitational field of a body at a point in the field is defined as the force experienced by a body of unit mass placed at that point provided the presence of unit mass does not disturb the original gravitational field. It is always directed towards the centre of gravity of the body whose gravitational field is considered. Intensity of the gravitational field at a point is a vector quantity and is denoted by $\vec{E}$.

Let M be the mass of a body with centre of gravity at O. Let F be the gravitational force of attraction experienced by a test mass m_0 when placed at P in the gravitational field of the body, such that $OP = x$. According to Newton's law of gravitation,

$$F = MG\,m_0/x^2$$

Intensity of gravitational field at P will be

$$E = F/m_0 = \frac{GMm_0/x^2}{m_0} = \frac{GM}{x^2} \quad \text{(i)}$$

In vector form, $$\vec{E} = -\frac{Gm}{x^2}\hat{x} \quad \text{(ii)}$$

Here the negative sign shows that the gravitational intensity is of attractive force. From Eq. (ii), we note that as x increases, E decreases and when $x = \infty$, $E = 0$. It means that the intensity of gravitational field is zero only at infinite distance from the body.

If the test mass is free to move at P, it will be accelerated due to force of attraction F, by the body of mass m_0. Let a be the acceleration produced in the test mass due to force F, then

$$F = m_0\, a$$

or $$\frac{F}{m_0} = a \qquad \text{(iii)}$$

From Eqs. (ii) and (iii), $E = a$.

It means that the intensity of gravitational field at a point in the gravitational field is equal to the acceleration of test mass placed at that point.

If the gravitational field is due to earth and point P lies on the surface of the earth then $x = R$. Now

$$E = \frac{GM}{R^2} = g$$

where g is the acceleration due to gravity at the surface of earth.

Unit of intensity of gravitational field in SI system is N kg^{-1} or m s^{-2} and in CGS system, it is dyne g^{-1} or cm s^{-2}.

Dimensional formula of gravitational intensity

$$E = F/m_0 = \frac{MLT^{-2}}{[M]} = [M^0LT^{-2}]$$

Illustration 5.13 **Find the relation between the gravitational field on the surface of two planets A and B of masses m_A, m_B and radii R_A and R_B, respectively, if**

i. they have equal mass,

ii. they have equal (uniform) density.

Sol. Let E_A and E_B be the gravitational field intensities on the surface of planets A and B.

Then, $$E_A = \frac{GM_A}{R_A^2} = \frac{G\frac{4}{3}\pi R_A^3 \rho_A}{R_A^2} = \frac{4G\pi}{3}\rho_A R_A$$

Similarly, $$E_B = \frac{Gm_B}{R_B^2} = \frac{4G}{3}\pi \rho_B R_B$$

i. For $m_A = m_B$, $\dfrac{E_A}{E_B} = \dfrac{R_B^2}{R_A^2}$

ii. For $\rho_A = \rho_B$, $\dfrac{E_A}{E_B} = \dfrac{R_A}{R_B}$

Illustration 5.14 **Calculate the gravitational field intensity at the centre of the base of a hollow hemisphere of mass M and radius R. (Assume the base of hemisphere to be open)**

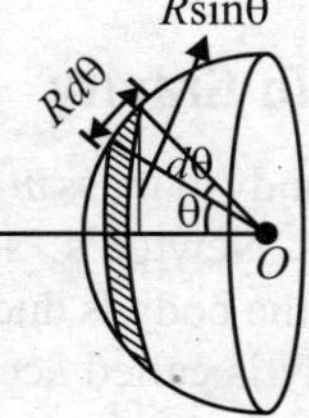

Fig. 5.16

Sol. We consider the shaded elemental ring of mass

$$dm = \frac{M}{(2\pi R^2)}\, 2\,\pi R \sin\theta\, (Rd\theta)$$

Field due to this ring at O,

$$dE = \frac{Gdm\, R\cos\theta}{R^3}$$

(see formulae for field due to a ring)

or $$dE = \frac{GM}{R^2}\sin\theta \cos\theta\, d\theta$$

Hence, $$E = \int_0^{\pi/2} dE = \int_{R^2}^{\pi/2} \frac{GM}{R^2}\sin\theta\cos\theta\, d\theta$$

or $$E = \frac{GM}{2R^2}$$

Illustration 5.15 **Two masses 90 kg and 160 kg are at a distance 5 m apart. Compute the magnitude of intensity of the gravitational field at a point distance 3 m from the 90 kg and 4 m from the 160 kg mass. $G = 6.67 \times 10^{-11}$ kg^{-2}.**

Sol. Let A and B be the positions of the masses and P be the point at which the gravitational intensity is to be computed.

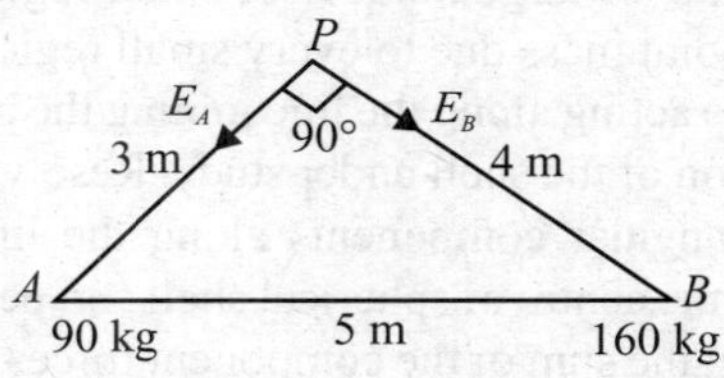

Fig. 5.17

Gravitational intensity at P due to mass at A will be $E_A = G\dfrac{90}{3^2} = 10G$, along PA.

Gravitational intensity at P due to mass at B will be $E_B = G\dfrac{160}{4^2} = 10G$, along PB.

In $\Delta\, APB$, $AB^2 = AP^2 + PB^2$ $\quad (\because \angle APB = 90°)$

Therefore, the magnitude of resultant gravitational intensity at P will be

$$E = \sqrt{E_A^2 + E_B^2} = \sqrt{(10\,G)^2 + (10\,G)^2} = 10\sqrt{2}G$$

$$= 10\sqrt{2} \times 6.67 \times 10^{-11}\ \text{N/kg} = 9.43 \times 10^{-10}\ \text{N kg}^{-1}$$

Illustration 5.16 **Two bodies of masses 100 kg and 10,000 kg are at a distance 1 m apart. At which point on the line joining them will the resultant gravitational field intensity be zero?**

Sol. Let x be the distance of point P from the 100 kg body where the resultant gravitational intensity is zero.

Therefore, gravitational intensity at P due to body of mass 100 kg is equal and opposite to that due to body of mass 10,000 kg. Hence,

$$\frac{G \times 100}{x^2} = \frac{G \times 10{,}000}{(1 - x^2)}$$

$$\Rightarrow \quad 100\,x^2 = (1 - x^2)$$

$$\Rightarrow \quad 10\,x = 1 - x$$

$$\Rightarrow \quad 11\,x = 1 \text{ or } x = \frac{1}{11} \text{ m}$$

THE FORCES OF ATTRACTION BETWEEN A HOLLOW SPHERICAL SHELL OF UNIFORM DENSITY AND A POINT MASS BODY

Let m be the mass of the point mass body and m be the mass of a hollow spherical shell of uniform density with centre at O and radius R.

Case I

Point mass is situated outside the shell. Consider the spherical shell to be made of large number of small regions. Calculate the force on point mass due to every small region of spherical shell. It will be acting along the line joining the point mass and the small region of the shell under study. Resolving each force into two rectangular components along the line joining the point mass to the centre of spherical shell and perpendicular to it, we note that the sum of the component forces perpendicular to the line joining the centre of the shell to point mass cancel out whereas the sum of the component force along the line joining the point mass to centre of the shell will have a resultant force whose magnitude is

$$F = \frac{GMm}{r^2}$$

where r is the distance between the centre of spherical shell and the point mass body.

The above force F is the same as if the entire mass of the spherical shell is concentrated at the centre of the spherical shell.

Case II

The point mass is situated inside the hollow spherical shell of uniform density. The resultant gravitational force on this point mass is zero. It is so because the gravitational forces on the point mass due to various small regions of the spherical shell will be acting in various directions which cancel out each other. Thus, we conclude the following:

i. The gravitational force on a point mass body situated outside the hollow spherical shell of uniform density will be the same as if the entire mass of the spherical shell is concentrated at its centre.

ii. The gravitational force on a point mass body situated inside the hollow spherical shell of uniform density will be zero.

GRAVITY

Gravity is the force of attraction exerted by earth towards its centre on a body lying on or near the surface of earth. Gravity is merely a special case of gravitation and is also called earth's gravitational pull.

For illustration, the force of attraction between two bodies on the surface of earth is the force of gravitation, but the force of attraction between each body and the earth is the force of gravity.

Force of gravity acting on a body is the measure of weight of the body. Thus, weight of a body is defined as the force of attraction exerted by the earth on the body towards its centre.

If m is the mass of the body placed on the surface of earth, where acceleration due to gravity is g, then

$$\text{Gravity pull} = \text{Weight of body} = mg.$$

The unit and dimensions of gravity pull or weight are the same as those of force.

Difference Between Gravitation and Gravity

1. Gravitation is the force of attraction acting between any two bodies of the universe. Gravity is the earth's gravitational pull on the body lying on or near the surface of earth.
2. The gravitational force on a body A of mass m_1 due to a body B of mass m_2 placed a distance r apart is $F = Gm_1m_2/r^2$, where G is a universal gravitational constant. The force of gravity on a body of mass m is $F = mg$, where g is the acceleration due to gravity.
3. The direction of gravitational force on body A due to body B is along AB. The direction of force of gravity on a body is along the line joining the body and the centre of the earth and is directed towards the centre of the earth.
4. The force of gravitation between the two bodies can be zero if the separation between two bodies becomes infinity. The force of gravity on a body is zero at the centre of the earth.

Acceleration Due To Gravity

If a force F acting on a body of mass m produces an acceleration a then according to Newton's second law of motion $F = ma$. If the force on the body is due to gravity of earth, then acceleration in the body is called acceleration due to gravity, which is denoted by g, i.e., $a = g$.

Then, $F = mg$ or $g = \frac{F}{m}$

If $m = 1$, then $g = \frac{F}{1} = F$. Thus, acceleration due to gravity is defined as the force of gravity acting on a unit mass of the body placed on or near the surface of the earth.

If the body is falling freely under the effect of gravity, then the acceleration in the body is also called acceleration due to gravity. Thus, acceleration due to gravity is the acceleration set up in the body when it falls freely under the effect of gravity alone. It is denoted by g. The SI unit of acceleration due to gravity is ms^{-2} or N/kg^{-1}.

g is a vector. Its direction is towards the centre of the earth. At a given place, the value of g is constant. However, it differs from place to place on the surface of the earth. It also varies with altitude, depth and rotation of the earth. On the surface of the earth, the value of g is taken to be 9.8 ms^{-2} for all practical purposes. The dimensional formula for g is $[M^0 LT^{-2}]$.

The value of g is independent of the shape, size, mass, etc. of the body but depends upon mass and radius of the earth or planet due to which there is a gravity pull. When two bodies of different masses are allowed to fall from same height in vacuum, they reach the earth at the same time. If the two bodies of same mass but of different volumes are allowed to fall from same height in air medium, they will not be reaching simultaneously. The body of smaller volume will reach the earth earlier than the other body, since the upward thrust due to air on the body of smaller volume will be less than that on the body of larger volume.

Relation Between *g* and *G*

Consider earth to be a spherical body of mass M, radius R with centre O. Suppose a body of mass m is placed on the surface of earth, where acceleration due to gravity is g.

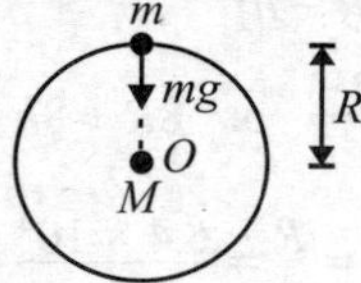

Fig. 5.18

The force on the body of mass m, outside the surface of earth is due to earth whose mass M is concentrated at the centre O. Let F be the force of attraction between the body and the earth.

According to Newton's law of gravitation

$$F = \frac{GMm}{R^2}$$

From gravity pull, $F = mg$

$$\therefore \quad mg = \frac{GMm}{R^2} \Rightarrow g = \frac{GM}{R^2} \quad \text{(i)}$$

From Eq. (i), we note that the value of acceleration due to gravity is independent of mass, shape and size of the body but depends upon mass and radius of the earth.

Illustration 5.17 **Assuming the earth to be a uniform sphere of radius 6400 km and density 5.5 g/c.c., find the value of *g* on its surface. $G = 6.66 \times 10^{-11}$ Nm^2 kg^{-2}.**

Sol. Here, $R = 6400 \times 10^3$ m

$= 6.4 \times 10^6$ m

$\rho = 5.5$ g/c.c.

$= 5.5 \times 10^3$ kg/m^3

Now, $$g = \frac{GM}{R^2} = \frac{G}{R^2} \times \frac{4}{3}\pi R^3 \times \rho$$

$$= \frac{4}{3}\pi GR\rho$$

$$= \frac{4}{3} \times \frac{22}{7} \times 6.66 \times 10^{-11} \times 6.4 \times 10^6 \times 5.5 \times 10^3$$

$$= 9.82 \text{ ms}^{-2}$$

Mass and Density of earth

Mass of Earth

From Eq. (i), we find

$$M = \frac{gR^2}{G} \quad \text{(i)}$$

We know that $R = 6.4 \times 10^6$ m

$G = 6.66 \times 10^{-11}$ Nm^2 kg^{-2}

and $g = 9.8$ ms^{-2}

Then, $$M = \frac{9.8 \times (6.4 \times 10^6)^2}{6.66 \times 10^{-11}}$$

$$= 6.018 \times 10^{24} \text{ kg}$$

Density of Earth

Consider earth to be a spherical body of radius R. Let ρ be the uniform density of the material of the earth.

Now, density $$\rho = \frac{\text{Mass}}{\text{Volume}}$$

$$= \frac{M}{\frac{4}{3}\pi R^3} = \frac{3M}{4\pi R^3} \quad \text{(ii)}$$

From Eqs. (i) and (ii), we get

$$\rho = \frac{\frac{3gR^2}{G}}{4\pi R^3} = \frac{3g}{4\pi RG} \quad \text{(iii)}$$

We know that $g = 9.8\ ms^{-2}$

$$R = 6.4 \times 10^6\ m$$
$$G = 6.66 \times 10^{-11}\ Nm^2\ kg^{-2}$$

Hence, $$\rho = \frac{3 \times 9.8}{4 \times \frac{22}{7} \times 6.4 \times 10^6 \times (6.67 \times 10^{-11})}$$

$$= 5.4783 \times 10^3\ kg\ m^{-3} \approx 5.5 \times 10^3\ kg\ m^{-3}$$

Variation of Acceleration Due to Gravity

The value of acceleration due to gravity changes with height (i.e., altitude), depth, shape of the earth and rotation of earth about its own axis. The effect of each of the above factors on the value of g has been discussed below.

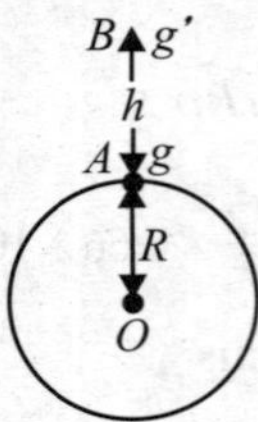

Fig. 5.19

Effect of Altitude

Consider earth to be a sphere of mass *M*, radius *R* with centre at *O*. Let *g* be the value of acceleration due to gravity at a point A on the surface of earth.

$$g = \frac{GM}{R^2} \quad \text{(iv)}$$

If g' is the acceleration due to gravity at point *B*, at a height *h* above the surface of earth, then the force on the body at *B* is due to earth whose mass *M* is concentrated at the centre *O* of earth.

$$g' = \frac{GM}{(R+h)^2} \quad \text{(v)}$$

Dividing Eq. (v) by Eq. (iv), we get

$$\frac{g'}{g} = \frac{GM}{(R+h)^2} \times \frac{R^2}{GM} = \frac{R^2}{(R+h)^2} \quad \text{(vi)}$$

$$= \frac{R^2}{R^2(1+\frac{h}{R})^2} = \left(1+\frac{h}{R}\right)^{-2}$$

If $h \ll R$ then $\frac{h}{R}$ is very small compared to 1. Expanding the R.H.S. of the above equation by Binomial theorem and neglecting the square and higher powers of h/R, we get

$$\frac{g'}{g} = 1 - \frac{2h}{R}$$

$$g' = g\left(1 - \frac{2h}{R}\right) \quad \text{(vii)}$$

From Eqs. (vi) and (vii), we note that the value of acceleration due to gravity decreases with height.

It is due to this reason that the value of acceleration due to gravity is lesser at mountains that in plains. At a height equal to the radius of the earth (i.e., $h = R = 6400$ km), from Eq. (vi),

$$g' = \frac{gR^2}{(R+R)^2} = \frac{g}{4}$$

Note:

- ***Relation (vi) is used to find the value of acceleration due to gravity at a height h, when h is comparable to the radius of earth R and relation (vii) is used to find g′ when h is very small as compared to R.***
- ***With height h, the decrease in the value of g is***

$$= g - g' = \frac{2hg}{R}$$

Therefore, fractional decrease in the value of

$$g = \frac{g'-g}{g} = \frac{2h}{R}$$

Therefore, % decrease in the value of

$$g = \left(\frac{g-g'}{g}\right) \times 100 = \frac{2h}{R} \times 100$$

Illustration 5.18 How much above the surface of earth does the acceleration due to gravity reduce by 36% of its value on the surface of earth? Radius of earth = 6400 km.

Sol. Since the acceleration due to gravity reduces by 36%, the value of acceleration due to gravity there is 100 – 36 = 64%. It means, $g' = \frac{64}{100} g$. If *h* is the height of location above the surface of earth, then

$$g' = g\frac{R^2}{(R+h)^2}$$

or $$\frac{64}{100} g = g\frac{R^2}{(R+h)^2}$$

$$\frac{8}{10} = \frac{R}{R+h} \quad \text{or} \quad 8R + 8h = 10R$$

$$\Rightarrow \quad h = \frac{2R}{8} = \frac{R}{4} = \frac{6.4 \times 10^6}{4} = 1.6 \times 10^6\ m$$

Effect of Depth

Consider earth to be a homogeneous sphere of radius *R* and mass *M* with centre at *O*. Let *g* be the value of acceleration due to gravity at a point *A* on the surface of the earth. Then

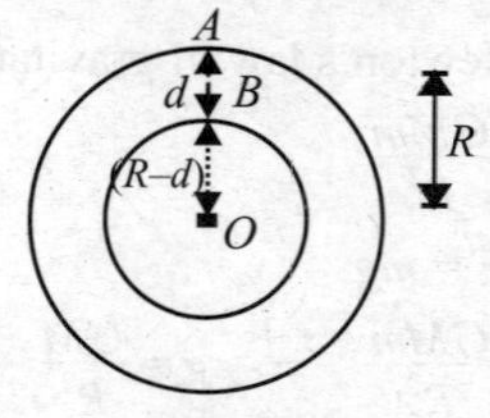

Fig. 5.20

$$g = \frac{GM}{R^2}$$

If ρ is uniform density of the material of earth, then

$$M = \frac{4}{3}\pi R^3 \rho$$

$$g = \frac{G \times \frac{4}{3}\pi R^3 \rho}{R^2} = \frac{4}{3}\pi GR\rho \qquad \text{(viii)}$$

Let g' be the acceleration due to gravity at the point B at a depth d below the surface of earth. The distance of the point B from the centre of the earth is $(R - d)$. The earth can be supposed to be made of

i. a smaller sphere of radius $(R - d)$ and

ii. a spherical shell of thickness d.

The body at B is inside the spherical shell of thickness d. The force on the body of mass m at B due to the spherical shell is zero, as explained. The body at B is outside the surface of the smaller sphere of radius $(R - d)$. The force on the body of mass m at B is due to smaller sphere of earth of radius $(R - d)$ is just as if the entire mass M' of the smaller sphere of earth is concentrated at the centre O.

$$g' = \frac{GM'}{(R-d)^2}$$

and $\quad M' = \frac{4}{3}\pi (R-d)^3 \rho$

$$g' = \frac{G \times \frac{4}{3}\pi (R-d)^3 \rho}{(R-d)^2} = \frac{4}{3}\pi G(R-d)\rho \qquad \text{(ix)}$$

Dividing Eq. (ix) by Eq. (viii), we get

$$\frac{g'}{g} = \frac{\frac{4}{3}\pi G(R-d)\rho}{\frac{4}{3}\pi GR\pi} = \frac{R-d}{R} = \frac{R}{R} - \frac{d}{R}$$

$$\Rightarrow \quad g' = g\left(1 - \frac{d}{R}\right) \qquad \text{(x)}$$

From Eq. (x), we note that the value of acceleration due to gravity decreases with depth.

At the centre of the earth, $d = R$, $g' = g_0$ (say)

From Eq. (x) we get, $g' = g\left(1 - \frac{d}{R}\right)$ (xi)

It means the acceleration due to gravity is zero at the centre of the earth. Therefore, the weight of the body of mass m at the centre of the earth $= mg_0 = 0$, but the mass of the body will not be zero. Thus, the value of acceleration due to gravity is maximum at the earth's surface and becomes zero at the centre of the earth.

Note: *Decrease in the value of g with depth d is* $g - g' = \frac{dg}{R}$

Therefore, fractional decrease in the value of g with depth

$$= \frac{g - g'}{g} = \frac{d}{R} \qquad \text{(xii)}$$

Therefore, % decrease in the value of g

$$= \frac{g - g'}{g} \times 100 = \frac{d}{R} \times 100 \qquad \text{(xiii)}$$

Illustration 5.19 **Find the percentage decrease in the weight of the body when taken to a depth of 32 km below the surface of earth. Radius of the earth is 6400 km.**

Sol. Here, $d = 32$ km; $R = 6400$ km

Weight of body at depth d is $= mg' = mg\left(1 - \frac{d}{R}\right)$

$$\% \text{ decrease in weight} = \frac{mg - mg'}{mg}$$

$$= \frac{d}{R} \times 100 = \frac{32}{6400} \times 100$$

$$= 0.5\%$$

Effect of Shape of Earth

earth is not a perfect sphere. It is flattened at the poles and bulges out at the equator. Equatorial radius R_e of the earth is about 21 km greater than the polar radius R_p.

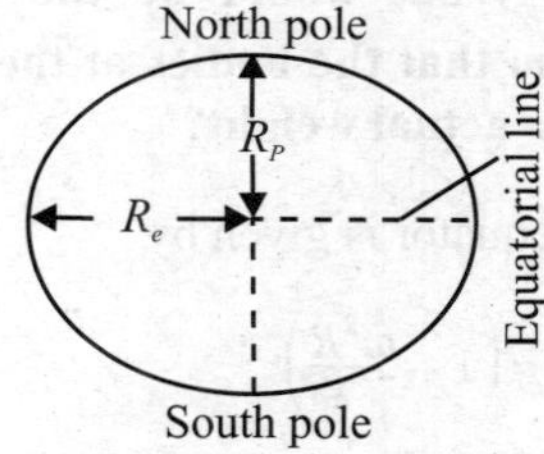

Fig. 5.21

Now,

$$g = \frac{GM}{R^2}$$

Since G and M are constants. Therefore,

$$g \propto \frac{1}{R^2}$$

Thus, we conclude that the value of g is least at the equator and maximum at the pole. It means, the value of acceleration due to gravity increases as we go from equator to the pole.

At sea level, the value of g at pole is greater than its value at equator by 1.80 cm s^{-2}.

Illustration 5.20 **What is the fractional decrease in the value of free-fall acceleration g for a particle when it is lifted from the surface to an elevation h? ($h << R$).**

Sol. $g = \frac{GM}{R^2} \Rightarrow \frac{dg}{dR} = -\frac{2GM}{R^3}$

$\Rightarrow \quad \frac{dg}{g} = -\frac{2GM}{R^2}\frac{1}{R}$

$\Rightarrow \quad \frac{dg}{g} = -2\left(\frac{h}{R}\right)$

Illustration 5.21 **(a) Find the height from the earth's surface where *g* will be 25% of its value on the surface of earth (*R* = 6400 km).**

(b) Find the percentage increase in the value of '*g*' at a depth '*h*' from the surface of earth.

Sol. (a) At a height *h* above the earth's surface, we have

$$g' = g\left(\frac{R}{R+h}\right)^2 \Rightarrow \frac{g}{4} = g\left(\frac{R}{R+h}\right)^2 \Rightarrow \frac{R}{R+h} = \frac{1}{2}$$

$\Rightarrow \quad h = R = 6400$ km

(b) At a depth *h* below earth's surface, we have

$$g' = g\left(1 - \frac{h}{R}\right) \Rightarrow \frac{g'}{g} = 1 - \frac{4}{6400}$$

$$\frac{g-g'}{g} = \frac{4}{6400} \Rightarrow \frac{\Delta g}{g} = \frac{1}{16}$$

$\therefore$ %age $= \frac{\Delta g}{g} \times 100 = \frac{1}{16}$ % increase

Illustration 5.22 **What would be the time period of rotation of earth so that the bodies at the equator would weigh 40% of their actual weight?**

Sol. Gravity at the equator is given by

$$g' = g\left(1 - \frac{\omega^2 R}{g}\right)$$

Apparent weight at the equator is

$$W' = mg\left(1 - \frac{\omega^2 R}{g}\right)$$

$$\Rightarrow \quad \frac{40}{100} mg = mg\left(1 - \frac{\omega^2 R}{g}\right)$$

$$\Rightarrow \quad \omega = \sqrt{\frac{0.6}{R}} \Rightarrow \frac{2\pi}{\omega} = 2\pi\sqrt{\frac{R}{0.6\,g}} = 2\pi\sqrt{\frac{6400 \times 10^3}{0.6 \times 10}}$$

$= 6489.245$

Illustration 5.23 **On a planet whose size is the same and mass four times as that of our earth, find the amount of work done to lift 3 kg mass vertically upwards through 3 m distance on the planet. The value of *g* on the surface of earth is 10 m s^{-2}.**

Sol. Here, $R_P = R$; $M_P = 4\,M_e$

$M = 3$ kg; $h = 3$ m; $g_e = 10$ m/s^{-2}

On the surface of the earth, $g_e = \frac{GM_e}{R_e^2}$

On the surface of the planet,

$$g_P = \frac{GM_P}{R_P^2} = \frac{G4M_e}{R_e^2} = 4\frac{GM_e}{R_e^2} = 4\,g_e$$

$$g_P = 4 \times 10 = 40 \text{ m/s}^2$$

Work done $= mg_p h = 3 \times 40 \times 3 = 360$ J

Concept Application Exercise 5.2

1. The radii of two planets are R_1 and R_2, respectively, and their densities are, ρ_1 and ρ_2 respectively. What is the ratio of the acceleration due to gravity at their surfaces?
2. Draw graphs showing the variation of acceleration due to gravity with (a) height above the earth's surface, (b) depth below the earth's surface.
3. The mass and the diameter of a planet are twice those of earth. What will be the time period of that pendulum on this planet, which is a seconds pendulum on the earth.
4. Will 1 kg sugar be more at poles or at the equator?
5. The magnitude of gravitational field at distances r_1 and r_2 from the centre of a uniform sphere of radius *R* and mass *M*, respectively. Find the ratio of (I_1/I_2) if $r_1 > R$ and $r_2 > R$.
6. A person sitting in an artificial satellite of earth feels weightlessness but a person standing on Moon has weight though Moon is also a satellite of earth. Why?
7. Why does a body lose weight at the centre of the earth?
8. Does the concentration of the earth's mass near its centre change the variation of *g* (acceleration due to gravity) with height from its surface?
9. The Sun's tide-raising power is only half as great as that of the Moon. The direct pull of the Sun on the earth, however, is about 175 times that of the Moon. Why is it then that the Moon causes larger tides?
10. If a planet of given density were made larger, its force of attraction for an object on its surface would increase because of the greater distance from the object to the centre of the planet. Which effect predominates?
11. A stone is dropped along the centre of a deep vertical mine shaft. Assume no air resistance but consider the earth's rotation. Will the stone continue along the centre of the shaft? If not, describe the motion.
12. Show that if the earth were not rotating about its axis the value of *g* at the equator would exceed its present value by 3.36 cm/s^{-2}. Given the radius of earth = 6.37×10^6 m and angular speed of earth = 7.27×10^{-5} rad/s^{-1}.

13. How far away from the earth does the acceleration due to gravity become 10% of its value on earth's surface? Radius of earth = 6.37×10^6 m.

14. (a) Assuming the earth to be a sphere of uniform density, calculate the value of acceleration due to gravity at a point (i) 1600 km above the earth, (ii) 1600 km below the earth, (b) Also find the rate of variation of acceleration due to gravity above and below the earth's surface. Radius of earth = 6400 km, $g = 9.8$ m/s^2.

15. How much faster than its present speed should the earth (radius 6.37×10^6 m) rotate so that the bodies lying on the equator may fly off into space? (At equator, $g = 9.78$ m/s^{-2}.)

GRAVITATIONAL POTENTIAL

Gravitational potential at a point in a gravitational field of a body is defined as the amount of work done in bringing a body of unit mass from infinity to that point without acceleration.

If W is the amount of work done in bringing a body of mass m_0 from infinity to a point P in the gravitational field without acceleration, then gravitational potential at P is $V = W/m_0$

Gravitational potential is a scalar quantity, since work done is a scalar. Unit of gravitational potential at a point in SI system is J kg^{-1} and in cgs system it is erg g^{-1}. Dimensional formula for gravitational potential is given by

$$= \frac{W}{m_0} = \frac{[ML^2T^{-2}]}{[M]} = [M^0L^2T^{-2}]$$

Expression for Gravitational Potential at a Point

Let earth to be a perfect sphere of radius R and mass M. The mass of the earth can be supposed to be concentrated at its centre O. Let us calculate the gravitational potential at any point P, where $OP = r$ and $r > R$. Join OP and produce it onwards. Take two points A and B on this line such that $OA = x$ and $AB = dx$.

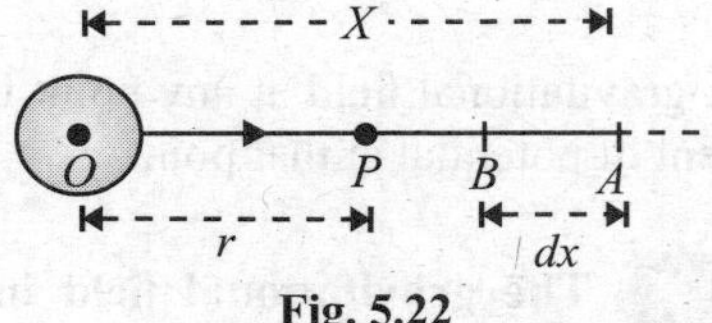

Fig. 5.22

The gravitational force of attraction on a body of unit mass at point A will be

$$F = \frac{GM \times 1}{x^2} = \frac{GM}{x^2}$$

Small amount of work done in bringing the unit mass of the body without acceleration through a small distance AB (= dx) is

$$dW = Fdx = \frac{GM}{x^2}dx$$

Total work done in bringing the body of unit mass from infinity to point P can be obtained by intergrating the above expression within the limits $x = \infty$ to $x = r$ we get

$$\therefore \quad W = \int_{\infty}^{r} \frac{GM}{x^2}dx = -\left(\frac{GM}{x}\right)_{\infty}^{r} = -GM\left[\frac{1}{r} - \frac{1}{\infty}\right]$$

$$= -\frac{GM}{r}$$

This work done is the measure of gravitational potential at P (i.e., V_P).

$$V_p = W = -\frac{GM}{r} \qquad \text{(i)}$$

From Eq. (i), we note that

i. the gravitational potential at a point is always negative

ii. when $r = \infty$, from (1), $V_P = 0$, hence gravitational potential is maximum (= zero) at infinity

iii. at the surface of earth, $r = R$, therefore $V_P = -GM/R$ at the surface of earth

The unit of gravitational potential is J kg^{-1}. Dimensional formula of gravitational potential

$$= \frac{\text{Work}}{\text{Mass}} = \frac{[ML^2T^{-2}]}{[M]} = [M^0L^2T^{-2}]$$

Gravitational potential is a scalar field because at any point in space it has only magnitude and no direction.

Gravitational Potential at a Point Due to the earth

Outside the earth:

The work done in bringing a body of mass m from infinity to a point at a distance r from the centre of earth is $W = -GMm/r$ for $r > R$.

Hence, gravitational potential, $V = W/m$

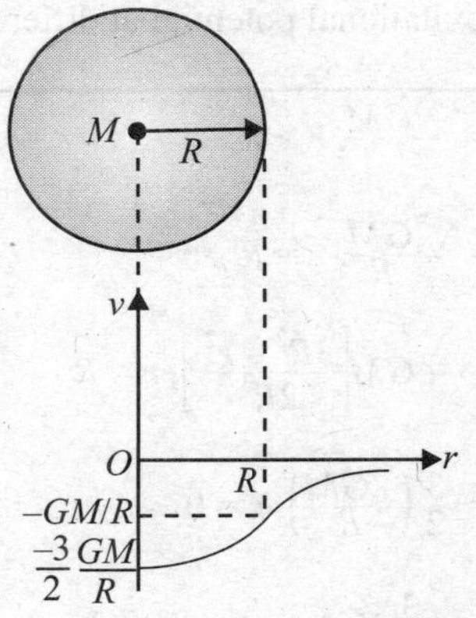

Fig. 5.23

$$= \frac{-GMm/r}{m}$$

$$= \frac{-GM}{r}, r > R$$

Inside the earth:

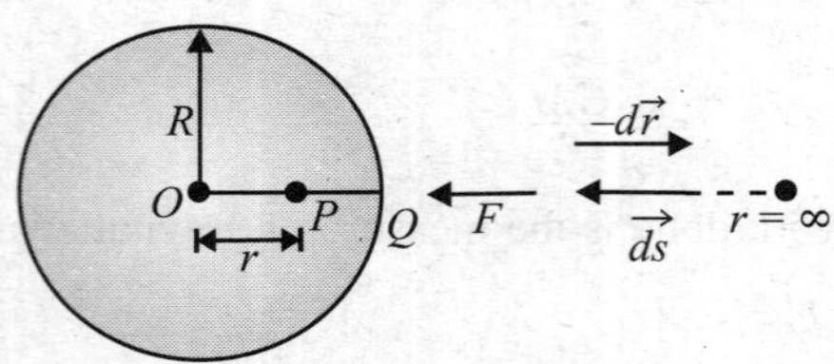

Fig. 5.24

Potential at point P is

$$V_P = \frac{[W_{\infty \to P}]}{m}$$

$$= -\left[\frac{W_{\infty \to Q} + W_{Q \to P}}{m}\right] = -\frac{1}{m}\left[\int_{\infty}^{R} \vec{F}.\vec{ds} + \int_{R}^{r} \vec{F}.\vec{ds}\right]$$

$$= -\frac{1}{m}\left[\int_{\infty}^{R} Fds + \int_{R}^{r} Fds\right]$$

$$[\vec{F}.\vec{ds} = Fds\cos\theta = Fds]$$

$$= -\frac{1}{m}\left[\int_{\infty}^{R} \frac{GMm}{r^2}(-dr) + \int_{R}^{r} \frac{GMmr}{r^2}(-dr)\right]$$

After solving, we get

$$V_P = -GM\left[\frac{3R^2 - r^2}{2R^2}\right]$$

We can write gravitational potential at different points

- $V = \frac{-GM}{r}, r > R$
- $V = -\frac{GM}{R}, r = R$
- $V = -GM\left[\frac{3R^2 - r^2}{2R^3}\right], r < R$
- $V = \frac{3}{2}\left(\frac{-GM}{R}\right), r = 0$

or $V_{\text{centre}} = \frac{3}{2} V_{\text{surface}}$

Note:

Spherical Shell of Mass *M* and Radius *R*

Intensity of Gravitational Field and Potential

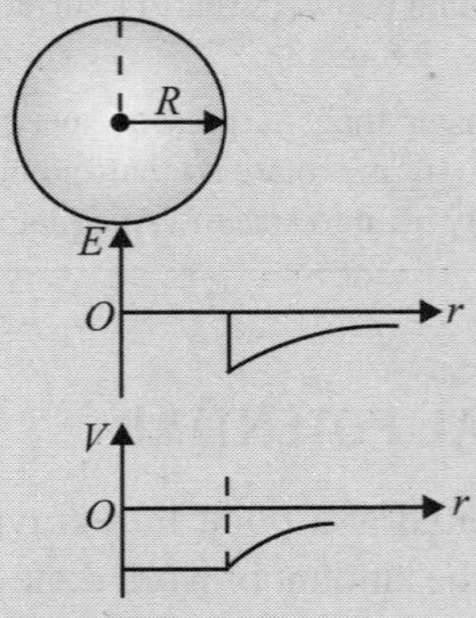

Fig. 5.25

Intensity of Gravitational Field and Potential	Gravitational Potential
i. $E = -\frac{GM}{r^2}, r > R$	i. $V = -\frac{GM}{r}, r > R$
ii. $E = -\frac{GM}{r^2}, r = R$	ii. $V = -\frac{GM}{R}, r = R$
iii. $E = 0, r < R$	iii. $V = -\frac{GM}{R}, r < R$

Relation Between Gravitational Field and Potential

The work done by an external agent to move unit mass from a point to another point in the direction of the field E, slowly through an infinitesimal distance dr = Force by external agent × distance moved = $-Edr$.

Thus, $dV = -Edr \Rightarrow E = \frac{-dV}{dr}$

Therefore, gravitational field at any point is equal to the negative gradient of potential at that point.

Illustration 5.24 **The gravitational field in a region is given by $\vec{E} = -(20 \text{ N kg}^{-1})(\hat{i} + \hat{j})$. Find the gravitational potential at the origin (0, 0) in J kg^{-1}.**

Sol. We know the relation between field and potential

$$V = -\int E.dr = \left[\int E_x.dx + \int E_y.dy\right]$$

$$= 20x + 20y$$

$$\Rightarrow \quad V = 0 \text{ at the origin } (0, 0)$$

Illustration 5.25 **Calculate the gravitational field intensity and potential at the centre of the base of a solid hemisphere of mass *m*, radius *R*.**

Sol. We consider the shaded elemental disc of radius $R \sin\theta$ and thickness $Rd\theta$

Its mass,

$$dM = \frac{M}{\frac{2}{3}\pi R^3} \pi (R\sin\theta)^2 (Rd\theta \sin\theta)$$

or $$dM = \frac{3M}{2} \sin^3\theta \, d\theta$$

Field due to this plate at O,

$$dE = \frac{2GdM(1-\cos\theta)}{(R\sin\theta)^2} \quad \text{(see field due to a uniform disc)}$$

or $$dE = \frac{3GM\sin\theta\,(1-\cos\theta)d\theta}{R^2}$$

Therefore, $$E = \int_0^{\frac{\pi}{2}} dE = \int_0^{\frac{\pi}{2}} \frac{3GM\sin\theta(1-\cos\theta)}{R^2} d\theta$$

$$= \frac{3GM}{R^2}\left[-\cos\theta + \frac{\cos^2\theta}{2}\right]_0^{\frac{\pi}{2}}$$

or $$E = \frac{3GM}{2R^2}$$

Now, potential due to the element under consideration at the centre of the base of the hemisphere,

$dV = -2GdM/r\,(\operatorname{cosec}\theta - \cot\theta)$ (see potential due to a circular plate)

or $$dV = \frac{-3GM\sin^3\theta\,(\operatorname{cosec}\theta - \cot\theta)\,d\theta}{(R\sin\theta)}$$

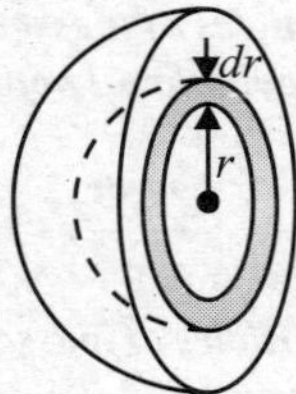

Fig. 5.26

Therefore, $$V = -\frac{3GM}{R}\int_0^{\frac{\pi}{2}} (\sin\theta - \cos\theta\sin\theta)\,d\theta$$

$$= -\frac{3GM}{R}\left[-\cos\theta + \frac{\cos^2\theta}{2}\right]_0^{\frac{\pi}{2}}$$

or $$V = \frac{-3GM}{2R}$$

Alternative method: Consider a hemispherical shell of radius r and thickness dr

Its mass,

$$dm = \frac{M}{\frac{2}{3}\pi R^3}(2\pi r^2 dr) \text{ or } dm = \frac{3Mr^2 dr}{R^3}$$

Since all points of this hemispherical shell are at the same distance r from O, hence potential at O is

$$dV = \frac{-Gdm}{r} = \frac{-3GMrdr}{R^3}$$

$$\therefore \quad V = \int_0^R dV = \frac{-3GM}{2R}$$

GRAVITATIONAL POTENTIAL ENERGY

We know that potential energy of a body at a given position is defined as the energy stored in the body at that position. If the position of a body changes on account of the forces acting on it, then change in its potential energy is equal to the amount of work done on the body by the forces acting on it, for the given change in position. As this work done is independent of the path followed, so the gravitational forces are conservative forces.

When the body is at infinity with respect to another body, the gravitational attraction on the body is zero. Therefore, its potential energy is zero, which is called zero level of potential energy.

If W_0 is the zero level of potential energy of the body and W is the work done in raising the body from the position of zero level potential energy to present location, then PE of the body $= W + W_0$.

If initially the body is at infinity, then the zero level of PE is zero, i.e., $W_0 = 0$.

Now, PE of the body $= W$

Gravitational potential energy of a body at a point in a gravitational field of another body is defined as the amount of work done in bringing the given body from infinity to that point without acceleration.

Expression for Gravitational Potential Energy

Consider a gravitational field due to earth of mass M and radius R. The mass of the earth can be supposed to be concentrated at its centre O. Let us calculate the gravitational potential energy of the body of mass m placed at point P in the gravitational field, where $OP = r$ and $r > R$. Join OP and produce it onwards. Take two points A and B on this line such that $OA = x$ and $AB = dx$.

The gravitational force of attraction on the body at A will be $F = GMm/x^2$.

Small amount of work done in bringing the body without acceleration through a very small distance AB (= dx) is given by

$$dW = Fdx = \frac{GMm}{x^2}\,dx$$

Total work done in bringing the body from infinity to point P is given by

$$W = \int_{\infty}^{r} \frac{GMm}{x^2}\,dx = -GMm\int_{\infty}^{r} x^{-2}\,dx$$

$$= -GMm\left[\frac{1}{x}\right]_{\infty}^{r} = -GMm\left[\frac{1}{r} - \frac{1}{\infty}\right]$$

$$= \frac{-GMm}{r} \qquad \text{(ii)}$$

Since this work done is stored in the body as its gravitational potential energy U, therefore gravitational potential energy of the body of mass m placed at point P in the gravitational field of earth of mass M at distance r will be

$$U = \frac{-GMm}{r} \qquad \text{(iii)}$$

From Eqs. (i) and (iii), we note that $U = V_p \times m$.

That is, gravitational potential energy = gravitational potential × mass of the body.

Discussion of Result

1. In Eq. (iii), negative sign shows that the potential energy is due to attractive gravitational force exerted by earth on the body. It means to bring a body from infinity to point P, the work has been performed by the gravitational field of earth, i.e., by attractive force.
2. From Eq. (iii), we note that if the value of r increases, the value of gravitational potential energy increases as it becomes less negative. When $r = \infty$, the value of gravitational potential energy becomes zero (i.e., maximum).
3. If a body of mass m is moved from a point at a distance r_1 to a point at a distance r_2, then change in potential energy of the body is given by

$$2\Delta U = \int_{r_1}^{r_2} \frac{GMm}{x^2}\,dx = GMm\left[\frac{1}{r_1} - \frac{1}{r_2}\right] \qquad \text{(iv)}$$

If $r_1 > r_2$, then ΔU is negative. It means as the body is brought closer to the earth, its potential energy decreases (i.e., becomes more negative).

4. If a body is moved from the surface of earth to a point at a height h above the surface of earth, then $r_1 = R$ and $r = R + h$. The change in potential energy will be

$$\Delta U = GMm\left[\frac{1}{R} - \frac{1}{R+h}\right]$$

$$= \frac{GMm}{R}\left[1 - \frac{R}{R+h}\right]$$

$$= \frac{GMm}{R}\left[1 - \left(1 + \frac{h}{R}\right)^{-1}\right]$$

$$= \frac{GMm}{R}\left[1 - \left(1 - \frac{h}{R} + \ldots\right)\right]$$

(Expanding by binomial theorem)

$$\frac{GMmh}{R^2} = \left(\frac{GM}{R^2}\right)mh = gmh \qquad \left(\because g = \frac{GM}{R^2}\right)$$

$$\therefore \qquad \Delta U = mgh$$

Note:

- ***The gravitational potential energy of a point mass m at a distance r (>R) from the centre of earth is***

$$U = mV = -\frac{GMm}{r}$$

- ***Gravitational potential energy of a point mass m relative to infinity at the surface of earth will be***

$$U = -\frac{GMm}{R} \qquad (\because r = R)$$

- ***The potential energy of a point mass m at the centre of earth relative to infinity will be given by***

$$U = mV_C = -\frac{3}{2}\frac{GMm}{R}$$

$$\left[\because V_C = \frac{3}{2}V_S = -\frac{3GM}{2R}\right]$$

The potential energy of a body at the centre of earth is minimum but not zero.

- ***If there is a gravitational field due to point masses m_1, m_2, m_3,... then gravitational potential energy of a point mass body m_0 at a point in the gravitational field is***

$$U = -\left[\frac{Gm_0m_1}{r_{01}} + \frac{Gm_0m_2}{r_{02}} + \frac{Gm_0m_3}{r_{03}} + \ldots\right]$$

- ***If there are two particles of masses m_1 and m_2 in an isolated system then total gravitational potential energy of the system is***

$$U = -\frac{Gm_1m_2}{r_{12}}$$

- ***If there are three particles of masses m_1, m_2, m_3 in an isolated system, then total gravitational potential energy of the system is***

$$U = -\frac{Gm_1m_2}{r_{12}} + \frac{Gm_2m_3}{r_{23}} + \frac{Gm_1m_3}{r_{13}}$$

Thus, the total gravitational potential energy of an isolated system of particles will be equal to the sum of the gravitational potential energy for all possible pairs of its constituent particles. This is an example of the application of the superposition principle.

Illustration 5.26 **If g is the acceleration due to gravity on the surface of earth, find the gain in potential energy of an object of mass m raised from the surface of earth to a height equal to the radius R of the earth.**

Sol. Let M, R be the mass and radius of the earth. Then

$$g = GM/R^2$$

or $$GM = gR^2 \quad \text{(i)}$$

Potential energy of the object on the surface of earth

$$U_1 = \frac{-GMm}{R}$$

The potential energy of the object at a height equal to radius of the earth is

$$U_2 = \frac{-GMm}{2R}$$

Gain in PE is

$$U_2 - U_1 = -\frac{GMm}{2R} + \frac{GMm}{R} = \frac{GMm}{2R}$$

$$= \frac{(gR^2)m}{2R} = \frac{1}{2} mgR$$

Illustration 5.27 **The gravitational potential difference between the surface of a planet and a point 20 m above it is 16 J/kg. Calculate the work done in moving a 4 kg body by 8 m on a slope of 60° from the horizontal.**

Sol. Here, $h = 20$ m.

Gravitational potential difference = 16 J kg^{-1}.

The vertical distance through which the body has to be raised is $8 \sin 60° = 8 \times (\sqrt{3}/2) = 4\sqrt{3}$ m.

Since gravitational potential difference for a distance of 20 m is 16 J/kg, hence the potential difference for a distance of $4\sqrt{3}$ m for a body of mass 4 kg is $\left(\frac{16}{20} \times 4\sqrt{3}\right)$ J kg^{-1}.

Work done in lifting a 4 kg body through a vertical height of $4\sqrt{3}$ m is

$$\left(\frac{16}{20} \times 4\sqrt{3}\right) \times 4 = \frac{64 \times 1.732}{5} = 22.16 \text{ J}$$

Illustration 5.28 **A projectile is fired vertically from earth's surface with an initial speed of 10 km s^{-1}. Neglecting atmospheric retardation, how far above the surface of earth would the projectile go? (R_{earth} = 6400 km)**

Sol. $$\frac{1}{2} mv^2 - \frac{GMm}{\text{R}} = -\frac{GMm}{R+h}$$

(as final KE = 0)

Substituting $h = 2.5 \times 10^7$ m, we get the desired result.

Illustration 5.29 **A satellite of mass 1000 kg is rotating around the earth in a circular orbit of radius 3R. What extra energy should be given to this satellite if it is to be lifted into an orbit of radius 4R?**

Sol. Energy required = $(\text{TE})_f - (\text{TE})_i$

$$= \left(\frac{-GMm}{2(4R)}\right) - \left(-\frac{GMm}{2(3R)}\right) = -\frac{GMm}{8R} + \frac{GMm}{6R} = \frac{GMm}{24R}(4-3)$$

$$= \frac{GMm}{24R}$$

$$= \left(\frac{9.8 \times (6400 \times 10^3)^2 \times 1000}{24(6400 \times 10^3)}\right) = 2.614 \times 10^9 \text{ J}$$

Illustration 5.30 **Two particles of masses m and M are initially at rest at an infinite distance apart. They move towards each other and gain speeds due to gravitational attraction. Find their speeds when the separation between the masses becomes equal to d.**

Sol. Let v_1 and v_2 be the speeds of two masses m and M, respectively, when they are at a separation d.

As they approach each other, the kinetic energy increases and GPE decreases. Hence, for the system,

Loss in GPE = Gain in KE

$\Rightarrow (\text{GPE})_i - (\text{GPE})_f = \text{KE}_f - \text{KE}_i$

$$\Rightarrow \quad 0 - \left(-\frac{GMm}{d}\right) = \left(\frac{1}{2}mv_1^2 + \frac{1}{2}Mv_2^2\right) - 0$$

$$\frac{GMm}{d} = \frac{1}{2}mv_1^2 + \frac{1}{2}Mv_2^2$$

As there is no external force on this system, its total momentum remains conserved.

$$P_i = P_f$$

$$0 = mv_1 - Mv_2$$

Combining the two equations, we have

$$v_1 = \sqrt{\frac{2GM^2}{d(m+M)}}$$

and $$v_2 = \sqrt{\frac{2GM^2}{d(m+M)}}$$

Illustration 5.31 **A body is released at a distance far away from the surface of the earth. Calculate its speed when it is near the surface of earth. Given g = 9.8 m s^{-2}, radius of earth R = 6.37 × 10^6 m.**

Sol. Conservation of energy implies

$$0 + 0 = \frac{1}{2}mv^2 - \frac{GMm}{R}$$

$$v = \sqrt{\frac{2GM}{R}} = \sqrt{\frac{2(R^2g)}{R}}$$

$$= \sqrt{2Rg} \approx 11.2 \times 10^3 \text{ m s}^{-1}$$

Illustration 5.32 **Calculate the velocity with which a body must be thrown vertically upward from the surface of the earth so that it may reach a height of 10*R*, where *R* is the radius of the earth and is equal to 6.4×10^8 m. (earth's mass = 6×10^{24} kg, gravitational constant $G = 6.7 \times 10^{-11}$ Nm^2 kg^{-2})**

Sol. The gravitational potential energy of a body of mass m on earth's surface is

$$U(R) = -\frac{GMm}{R}$$

where M is the mass of the earth (supposed to be concentrated at its centre) and R is the radius of the earth (distance of the particle from the centre of the earth). The gravitational energy of the same body at a height $10R$ from earth's surface, i.e., at a distance $11R$ from earth's centre is

$$U(11R) = -\frac{GMm}{R}$$

Therefore, change in potential energy,

$$U(11R) - U(R) = -\frac{GMm}{11R} - \left(-\frac{GMm}{R}\right) = \frac{10}{11}\frac{GMm}{R}$$

This difference must come from the initial kinetic energy given to the body in sending it to that height. Now, suppose the body is thrown up with a vertical speed v, so that its initial kinetic energy is $1/2\ mv^2$.

Then $$\frac{1}{2}mv^2 = \frac{10}{11}\frac{GMm}{R} \text{ or } v = \sqrt{\frac{20}{11}\frac{Gmm}{R}}$$

Putting the given values:

$$v = \sqrt{\left(\frac{20 \times (6.7 \times 10^{-11}\ \text{Nm}^2\ \text{kg}^{-2}) \times (6 \times 10^{24}\ \text{kg})}{11(6.4 \times 10^6\ \text{m})}\right)}$$

$$= 1.07 \times 10^4\ \text{m s}^{-1}$$

Illustration 5.33 **Distance between centres of two stars is 10*a*. The masses of these stars are *M* and 16*M* and their radii are *a* and 2*a*, respectively. A body is fired straight from the surface of the larger star towards the smaller star. What should be its minimum initial speed to reach the surface of the smaller star?**

Sol. Let P be the point on the line joining the centres of the two planets such that the net field at that point is zero.

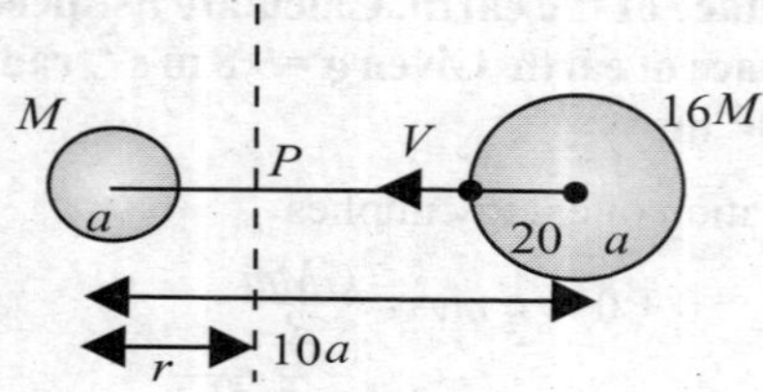

Fig. 5.27

Then, $$\frac{GM}{r^2} - \frac{G16M}{(10a - r)^2} = 0$$

$$\Rightarrow (10a - r)^2 = 16r^2$$

$$\Rightarrow 10a - r = 4r$$

$$\Rightarrow r = 2a$$

Potential at point P,

$$v_P = \frac{-GM}{r} - \frac{G16M}{(10a - r)} = \frac{-GM}{2a} - \frac{2GM}{a}$$

$$= \frac{-5GM}{2a}$$

Now if the particle projected from the larger planet has enough energy to cross this point, it will reach the smaller planet.

For this, the KE imparted to the body must be just enough to raise its total mechanical energy to a value which is equal to PE at point P, i.e.

$$\frac{1}{2}mv^2 - \frac{G(16M)m}{2a} - \frac{GMm}{8a} = mv_P$$

or $$\frac{v^2}{2} - \frac{8GM}{a} - \frac{GM}{8a} = -\frac{5GM}{2a}$$

or $$v^2 = \frac{45GM}{4a}$$

or $$v_{min} = \frac{3}{2}\sqrt{\frac{5GM}{a}}$$

Illustration 5.34 **Estimate whether it takes more energy to get a satellite upto 1600 km above the earth than to put in orbit there. earth's radius is 6400 km. Does your answer remain same for height 3200 km or for height 4800 km?**

Sol. Energy required to raise a satellite at height h,

$$E_1 = -\frac{GMm}{(R+h)} - \left(-\frac{GMm}{R}\right)$$

$$= GMm\left(\frac{1}{R} - \frac{1}{R+h}\right) = \frac{GMmh}{R(R+h)}$$

Energy required to put a satellite in orbit at that height = KE of satellite

$$E_2 = \frac{1}{2}mv^2 = \frac{1}{2}m\left(\frac{GM}{r}\right) = \frac{GMm}{2r}$$

$$= \frac{GMm}{2(R+h)}$$

$$\frac{E_1}{E_2} = \frac{2(R+h)h}{R(R+h)} = \frac{2h}{R} = \frac{2 \times (1600\ \text{km})}{(6400\ \text{km})}$$

$$= \frac{1}{2} = 0.5$$

Clearly, the answer depends on height, if $h = 3200$ km

$$\frac{E_1}{E_2} = \frac{2h}{R} = \frac{2 \times (3200 \text{ km})}{(6400 \text{ km})} = 1.10$$

For height $h = 4800$ km

$$\frac{E_1}{E_2} = \frac{2 \times (4800 \text{ km})}{(6400 \text{ km})} = \frac{3}{2} = 1.5$$

GRAVITATIONAL SELF-ENERGY

The gravitational self-energy of a body (or a system of particles) is defined as the work done by an external agent in assembling the body (or system of particles) from infinitesimal elements (or particles) that are initially at infinite distance apart.

Gravitational Self-Energy of a System of n Particles

Potential energy of n particles at an average distance 'r' due to their mutual gravitational attraction is equal to the sum of the potential energy of all pairs of particle, i.e.

$$U_s = -G \sum_{\text{all pairs } j \neq i} \frac{m_i m_j}{r_{ij}}$$

This expression can be written as U_s

$$= -\frac{1}{2} G \sum_{i=1}^{i=n} \sum_{j=1, j \neq 1}^{n} \left(\frac{m^2}{r}\right)_{ij}$$

If we consider a system of 'n' particles, each of same mass 'm' and separated from each other by the same average distance 'r', then self energy

or $$U_s = -\frac{1}{2} G \sum_{i=1}^{n} \sum_{j=1, j \neq 1}^{n} \left(\frac{m^2}{r}\right)_{ij}$$

Thus on the right-hand side 'i' comes 'n' times while 'j' comes $(n-1)$ times. Thus

$$U_s = -\frac{1}{2} Gn(n-1) \frac{m^2}{r}$$

Illustration 5.35 **Calculate the self-gravitational potential energy of matter forming (a) a thin uniform shell of mass M and radius R, and (b) a uniform sphere of mass m and radius R.**

Sol. In the process of formation of a body, work has to be done by an external agent bit by bit to build up the body. This energy of the external agent is stored as gravitational energy. This is called the self-gravitation potential energy or gravitational energy of mutual gravitational interaction.

(a) Consider a sphere of any radius x.

Mass of the sphere $= 4\pi/3\ x^3 \rho$ where ρ = density of mass

Gravitation potential of the surface $= -4\pi/3\ G\rho x^2$

This is also the work done in adding unit mass to the sphere by the external agent. When the thickness is increased by dx, mass added by the agent is $4\pi x^2\ dx\,\rho$.

Therefore, work done by the agent in increasing the surface from x to x is dx

$$\left(-\frac{4\pi}{3} G\rho x^2\right)\left(4\pi x^2 dx\rho\right) = \frac{16\pi^2}{3} G\rho^2 x^4 dx$$

Therefore, total work done by the agent

$$= \frac{16\pi^2 G\rho^2}{3} \int_0^R x^4\, dx = \frac{16\pi^2 G\rho^2 R^5}{15}$$

Now, $$\rho = \frac{m}{\frac{4\pi}{3}R^3} = \frac{3m}{4\pi R^3}$$

Therefore, $$U \text{ (self-energy)} = -\frac{16\pi^2 R^5 G}{15} \times \frac{9m^2}{16\pi 2R^6}$$

$$= -\frac{3}{5}\frac{GM^2}{R}$$

(b) Consider the shell when mass m has already been piled up by the agent.

Then, potential of the shell $= -GM/R$

This is also the work done by the agent in adding another unit mass to the shell. Therefore, the elementary work done in adding an elementary mass dm is $-Gm/R\ dm$.

Therefore, net work done by the agent in piling the entire mass

$$= -\int_0^m \frac{GMdm}{R} = -\frac{Gm^2}{2R}$$

Therefore, U (self potential energy) $= -\frac{1}{2}\ Gm^2/R$

Escape Speed

The minimum speed required to project a body from the surface of the earth so that it never returns to the surface of the earth is called escape speed. If a velocity greater than the escape velocity is imparted, the body will escape and leave the surface. If a velocity lesser than the escape velocity is given, it will fall back to the surface or be in an orbit. A body thrown with escape speed goes out of the gravitational pull of the earth. Work done to displace the body from the surface of the earth $(r = R_e)$ to infinity $(r = \infty)$ is given by

$$\int dW = \int_{R_e}^{\infty} \frac{GM_e m}{r^2}\, dr$$

or $$W = GM_e m \int_{R_0}^{\infty} \frac{1}{r^2} dr = -GM_e m \left[\frac{1}{r}\right]_{R_e}^{\infty}$$

$$= -GM_e m \left[\frac{1}{\infty} - \frac{1}{R_e}\right] \Rightarrow W = \frac{GM_e m}{R_e}$$

Let v_e be the escape speed of the body of mass m, then kinetic energy of the body is given by

$$\frac{1}{2} mv_e^2 = \frac{GM_e m}{R_e} \Rightarrow v_e = \sqrt{2gR_e} = 11.2 \text{ km s}^{-1}.$$

Important Points

1. Escape speed depends on the mass and size of the planet. That is why escape velocity on the Jupiter is more than that on the earth.
2. Escape speed is independent of the mass of the body.
3. Any body thrown upward with escape speed starts moving around the Sun.
   i. The value of escape speed of a body does not depend upon the mass (m) of the body and its angle of projection from the surface of earth or any other planet.
   ii. The value of escape speed depends upon the mass and radius of the planet from the surface of which the body is to be projected. Clearly, the value of escape speed of a body will be different for different planets.
   iii. For earth, $g = 9.8 \text{ ms}^{-2}$, $R = 6.4 \times 10^6$ m.

   $$\therefore \quad v_e = \sqrt{2gR_e} = \sqrt{2 \times 9.8 \times 6.4 \times 10^6}$$
   $$= 11.2 \times 10^3 \text{ ms}^{-1} = 11.2 \text{ km s}^{-1}$$

   iv. If a body is projected from a planet with a speed v which is smaller than the escape speed v_e (i.e., $v < v_e$) then the body will reach a certain height, may either move in an orbit around the planet or may fall back to the planet.
   v. If speed of projection (v) of the body from the surface of a planet is greater than the escape speed (v_e) of that planet, the body will escape out from the gravitational field of that planet and will move in the interstaller field of that planet with speed v' which can be obtained using law of conservation of energy, according to which

   $$\frac{1}{2} mv^2 + \left(-\frac{GMm}{R}\right) = \frac{1}{2} mv'^2 + 0$$
   $$v'^2 = v^2 - \frac{2GM}{R} = v^2 - v_e^2$$
   $$\left[\because v_e = \sqrt{\frac{2GM}{R}}\right]$$
   $$v' = \sqrt{v^2 - v_e^2}$$

Note:

- *In deriving the expression for escape speed, we have neglected the resistance to the body by earth's atmosphere. Therefore in reality, the value of escape speed is slightly greater than the value calculated from these relations.*
- *Escape speed of Sun is 618 km/s.*

Illustration 5.36 **Calculate the escape speed on the surface of a planet of mass 7.5×10^{25} g and radius 1.6×10^6 m. $G = 6.67 \times 10^{-8}$ dyne cm^2 g^{-2}.**

Sol. Here, $G = 7.5 \times 10^{25}$ g, $R = 1.6 \times 10^8$ cm

$$\text{Escape speed, } v_e = \sqrt{\frac{2GM}{R}} = \sqrt{\frac{2 \times 6.67 \times 10^{-8} \times 7.5 \times 10^{25}}{1.6 \times 10^8}}$$
$$= 2.5006 \times 10^5 \text{ cm s}^{-1}$$

Illustration 5.37 **The escape speed from earth's surface is 11 k ms^{-1}. A certain planet has a radius twice that of earth but its mean density is the same as that of the earth. Find the value of the escape speed from the planet.**

Sol. Let M, R, ρ be the mass, radius and density of earth and M, R, ρ be the corresponding values for the planet. Then escape speed from earth's surface is

$$v_e = \sqrt{\frac{2GM}{R}} = \sqrt{\frac{2G}{R} \times \frac{4}{3}\pi R^3 \rho}$$
$$= \sqrt{\frac{8\pi G\rho R^2}{3}}$$

Escape speed from planet's surface is

$$v_e = \sqrt{\frac{8\pi G\rho(2R)^2}{3}} = 2\sqrt{\frac{8\pi G\rho R^2}{3}} = 2v_e$$
$$= 2 \times 11 = 22 \text{ km/s}$$

Illustration 5.38 **A spaceship goes into a circular orbit close to the earth's surface. What additional velocity must be imparted to the ship so that it is able to escape the gravitational pull of the earth? (R = 6400 km, $g = 9.8$ m s^{-2})**

Sol. The orbital velocity in a circular orbit close to the earth is $v = \sqrt{gR}$.

The velocity required to escape $v_e = \sqrt{2gR}$.

Hence additional velocity required is $v_e - v = (\sqrt{2} - 1)\sqrt{gR}$. Therefore,

$$v_e - v = 0.414 \times \sqrt{9.8 \times 6400 \times 10^3}$$
$$= 3278.71 \text{ m/s} = 3.278 \text{ km s}^{-1}$$

Illustration 5.39 **A rocket starts vertically upward with speed v_0. Show that its speed v at height h is given by**

$$v_0^2 - v^2 = \frac{2hg}{1+\frac{h}{R}}$$

where R is the radius of the earth and g is acceleration due to gravity at earth's surface. Deduce an expression for maximum height reached by a rocket fired with speed 0.9 times the escape velocity.

Sol. The gravitational potential energy of a mass m on earth's surface and at a height h is given by

$$U(R) = -\frac{GMm}{R}$$

and $$U(R+h) = -\frac{GMm}{R+h}$$

$$\therefore \quad U(R+h) - U(R) = -GMm\left(\frac{1}{R+h} - \frac{1}{R}\right)$$

$$= \frac{GMmh}{(R+h)R} = \frac{mhg}{1+\frac{h}{R}} \quad [\because GM = gR^2]$$

This increase in potential energy occurs at the cost of kinetic energy which correspondingly decreases. If v is the velocity of the rocket at height h, then the decrease in kinetic energy is $1/2\, mv_0^2 - 1/2\, mv^2$. Thus,

$$\frac{1}{2}mv_0^2 - \frac{1}{2}mv^2 = \frac{mhg}{1+\frac{h}{R}} \text{ or } v_0^2 - v^2 = \frac{2gh}{1+\frac{h}{R}}$$

Let h_{max} be the maximum height reached by the rocket, at which its velocity has been reduced to zero. Thus, substituting $v = 0$ and $h = h_{max}$ in the last expression, we have

$$v_0^2 \frac{2gh_{max}}{1+h_{max}/R}$$

$$\text{or} \quad v_0^2\left(1+\frac{h_{max}}{R}\right) = 2gh_{max}$$

$$\text{or} \quad v_0^2 = h_{max}\left[2g - \frac{v_0^2}{R}\right]$$

$$\text{or} \quad h_{max} = \frac{v_0^2}{2g - \frac{v_0^2}{R}}$$

Now, it is given that $v_0 = 0.9 \times$ escape velocity

$$= 0.9 \times \sqrt{(2gR)}$$

$$\therefore \quad h_{max} = \frac{(09 \times 0.9)2gR}{2g - \frac{(09 \times 0.9)2gR}{R}}$$

$$= \frac{1.62gR}{2g - 1.62R}$$

$$= \frac{1.62R}{0.38} = 4.26R$$

Illustration 5.40 **For a particle projected in a transverse direction from a height h above earth's surface, find the minimum initial velocity so that it just grazes the surface of earth such that path of this particle would be an ellipse with centre of earth as the farther focus, point of projection as the apogee and a diametrically opposite point on earth's surface as perigee.**

Sol. Suppose velocity of projection at point A is v_A and at point B, the velocity of the particle is v_B.

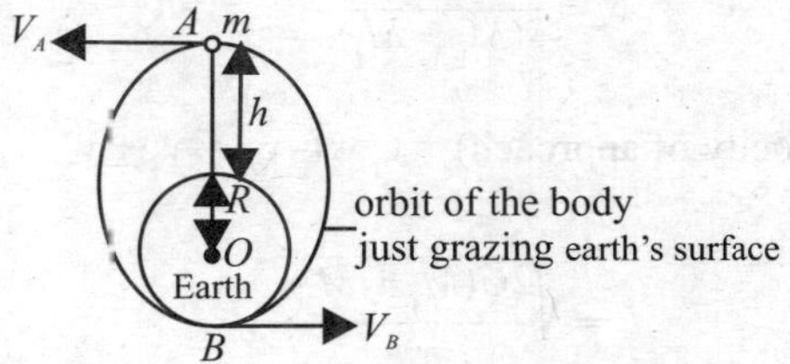

Fig. 5.28

Then applying Newton's second law at points A and B, we get

$$\frac{mv_A^2}{\rho_A} = \frac{GM_e m}{(R+h)^2} \text{ and } \frac{mv_B^2}{\rho_B} = \frac{GM_e m}{R^2}$$

where ρ_A and ρ_B are radii of curvature of the orbit at points A and B of the ellipse, but $\rho_A = \rho_B = \rho$(say).

Now applying conservation of energy at points A and B,

$$\frac{-GM_e m}{R+h} + \frac{1}{2}mv_A^2 = -\frac{GM_e m}{R} + \frac{1}{2}mv_B^2$$

$$\Rightarrow GM_e m\left(\frac{1}{R} - \frac{1}{(R+h)}\right) = \frac{1}{2}(mv_B^2 - mv_A^2)$$

$$= \left(\frac{1}{2}\rho GM_e m\left(\frac{1}{R^2} - \frac{1}{(R+h)^2}\right)\right)$$

$$\text{or} \quad \rho = \frac{2R(R+h)}{2R+h} = \frac{2Rr}{R+r}$$

Thus we get

$$V_A^2 = \frac{\rho GM_e}{(R+h)^2} = 2GM_e \frac{R}{r(r+R)}$$

where r = distance of point of projection from earth's centre = $R + h$.

Illustration 5.41 **Two masses M_1 and M_2 at an infinite distance from each other and initially at rest, start interacting gravitationally. Find their velocity of approach when they are distances apart.**

Sol. Since they move under mutual attraction and no external force acts on them, their momentum and energy are conserved. Therefore,

$$\therefore \quad 0 = \frac{1}{2}M_1 v_1^2 + \frac{1}{2}M_2 v_2^2 - \frac{GM_1M_2}{s}$$

It is zero because in the beginning, both kinetic energy and potential energy are zero.

$$0 = M_1 v_1 + M_2 v_2$$

Solving the equations,

$$v_1^2 = \frac{2GM_2^2}{s(M_1 + M_2)}$$

and

$$v_2^2 = \frac{2GM_1^2}{s(M_1 + M_2)}$$

V (velocity of approach) $= v_1 - (-v_2) = v_1 + v_2$

$$= \sqrt{\frac{2G(M_1 + M_2)}{s}}$$

Concept Application Exercise 5.3

1. Why do different planets have different escape speeds?
2. What are the two factors which determine why some bodies in solar system have atmosphere and others do not?
3. Why does Moon have no atmosphere?
4. What is binding energy of a satellite?
5. If a body is projected with speed v greater than escape speed v_e from the surface of earth, find its speed in intersteller space.
6. The radius of a planet is twice that of the earth but its average density is the same. If the escape speed at the planet and at the earth are v_p and v_e, respectively, then prove that $v_p = 2v_e$.
7. A particle is projected vertically upwards from the surface of the earth (radius R) with a kinetic energy equal to half of the minimum value needed for it to escape. Find the height to which it rises above the surface of earth.
8. What is the escape velocity of the object, if the magnitude of the potential energy per unit mass of the object at the surface of earth is E?
9. Would you expect the total energy of the solar system to be constant? What about the total angular momentum? Explain.
10. The total energy of the earth + Sun system is negative. How do you interpret the negative energy of a system?
11. An artificial satellite in the presence of frictional forces will move into an orbit closer to the earth and may have increased kinetic energy. Explain this.
12. Is the velocity of escape of a particle at rest on the surface of the earth the same as that of a particle just orbiting the earth? Explain.
13. A body is released at a distance far away from the surface of the earth. Calculate its speed when it is near the surface of earth. Given $g = 9.8$ m s^{-2}, radius of earth $R = 6.37 \times 10^6$ m.
14. A rocket is launched from the surface of earth with an initial speed of 10 km s^{-1}. How far above the surface of earth would it go? Radius of earth = 6400 km. Neglect atmospheric resistance.
15. Two earth satellites A and B, each of mass m are to be launched into circular orbits about earth's centre at altitudes 6400 km and 19200 km, respectively. The radius of earth is 6400 km. Find (i) the ratio of potential energy of B to that of A, (ii) ratio of kinetic energy of B to that of A and (iii) which one has greater total energy?
16. Calculate the radius of an isolated sphere of density 3.0 g cm^{-3} from the surface of which the escape velocity be 40 m s^{-1}.

PRINCIPLE OF LAUNCHING A SATELLITE

Consider a very high tower on the surface of earth whose height is more than the height of the earth's atmosphere. Let a body be projected from top of this tower with some horizontal speed. The body will describe a parabolic path under the effect of gravity and hit the surface of earth at point A_1. If the initial horizontal speed of projection of the body is increased, the body will hit the surface of the earth at point A_2, farther from the foot of the tower. Clearly, the horizontal range in second case is more than that in the first case. If we go on increasing the initial horizontal speed of projection of the body, its horizontal range will also increase and finally a stage will come when the body will not hit the earth, but will always be in a state of free fall under gravity in an attempt to fall on earth. Then the body will describe a stable circular path round the earth and becomes a satellite of earth. The minimum speed required to put the satellite into its orbit around earth is called orbital speed of satellite.

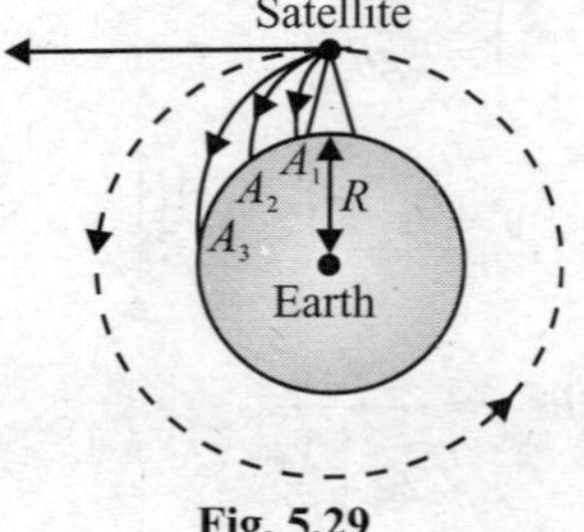

Fig. 5.29

Orbital Speed, Time Period and Height of Satellite

Orbital Speed

Orbital speed of a satellite is the minimum speed required to put the satellite into a given orbit around earth. The value of orbital speed is different for different orbits around earth and is independent of the mass of the satellite. Let

M = mass of earth
R = radius of earth
m = mass of the satellite
v = orbital speed of the satellite
h = height of the satellite above the surface of earth
r = radius of the orbit of the satellite = $R + h$

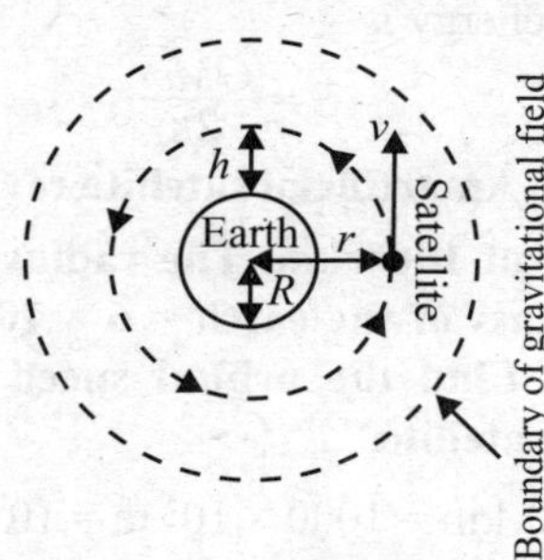

Fig. 5.30

The centripetal force required to keep the satellite in its orbit, $F = mv^2/r$.

According to Newton's law of gravitation, gravitational pull acting on the satellite $= GM\,m/r^2$.

In equilibrium, the gravitational pull provides the required centripetal force to the satellite. Therefore,

$$\frac{mv^2}{r} = \frac{GMm}{r^2}$$

$$\Rightarrow \qquad v = \sqrt{\frac{GM}{r}} \qquad \text{(i)}$$

If g is the value of acceleration due to gravity on the surface of earth, then

$$g = \frac{GM}{R^2} \text{ or } GM = gR^2$$

Putting this value in Eq. (i), we get

$$v = \sqrt{\frac{gR^2}{r}} = R\sqrt{\frac{g}{R+h}} \qquad \text{(ii)}$$

It is clear from Eq. (ii) that orbital speed of a satellite

- is independent of the mass of the satellite.
- decreases with an increase in the radius of orbit or increase in the height of the satellite.
- depends upon the mass and radius of the earth/planet around which the revolution of the satellite is taking place.
- The direction of orbital speed of a satellite at an instant is along the tangent to the orbital path of the satellite at that instant.
- When a satellite is orbiting very close to the surface of earth $h \ll R$, then

$$r = R + h \approx R \text{ and } v = v_0 \text{ (say)}$$

From Eq. (ii), $\quad v_0 = R\sqrt{g/R} = \sqrt{gR}$

Substituting $\quad g = 9.8 \text{ ms}^{-2}$

and $\quad R = 6.4 \times 10^6$ m

We get $\quad v_0 = \sqrt{9.8 \times 6.4 \times 10^6} = 7.92 \times 10^3 \text{ m s}^{-1}$

$$= 7.92 \text{ km s}^{-1}$$

Time Period of Satellite

It is the time taken by a satellite to complete one revolution around the earth and is denoted by T. Thus

$$T = \frac{\text{Distance travelled in one revolution}}{\text{Orbital velocity}} = \frac{2\pi r}{v}$$

$$T = \frac{2\pi r}{R}\sqrt{\frac{r}{g}} = \frac{2\pi}{R}\sqrt{\frac{r^3}{g}} = \frac{2\pi r}{R}\sqrt{\frac{(R+h)^3}{g}} \qquad \text{(iii)}$$

If the earth is supposed to be a sphere of mean density ρ, then the mass of the earth is

$$M = \frac{4}{3}\pi R^3\rho \text{ and } g = \frac{GM}{R^2} = \frac{G}{R^2}\left(\frac{4}{3}\pi R^3\rho\right) = \frac{4\pi aR\pi}{3}$$

Substituting this value of g in Eq. (iii), we get

$$T = \frac{2\pi}{R}\sqrt{\frac{3(R+h)^3}{4\pi GR\rho}} = \sqrt{\frac{4\pi^2}{R^2} \times \frac{3(R+h)^3}{4\pi GR\rho}}$$

$$= \sqrt{\frac{3\pi(R+h)^3}{G\rho R^3}} \qquad \text{(iv)}$$

For a satellite orbiting close to the surface of earth, $h \ll R$.

$$\therefore \qquad h + R \approx R$$

From Eq. (iii), $\quad T = \sqrt{\frac{3\pi}{G\rho}}$

From Eq. (iv), $\quad T = \frac{2\pi}{R}\sqrt{\frac{R^3}{g}} = 2\pi\sqrt{\frac{R}{g}} \qquad \text{(v)}$

By substituting $g = 9.8 \text{ ms}^{-2}$ and $R = 6.4 \times 10^6$ m in Eq. (v), we get the value of $T = 5.08 \times 10^3$ s = 84.6 min.

It means a satellite orbiting close to the surface of the earth has a time period of revolution about 84.6 min.

It is clear from Eq. (iv) that the period of revolution of a satellite depends only upon its height above the earth's surface. The larger is the height of a satellite above the surface of earth, greater will be its period of revolution.

Altitude or Height of Satellite Above the earth's Surface

Squaring both sides of Eq. (iii), we get

$$T^2 = \frac{4\pi^2(R+h)^3}{R^2 g} \Rightarrow (R+h)^3 = \frac{T^2R^2g}{4\pi^2}$$

$$R+h=\left(\frac{T^2R^2g}{4\pi^2}\right)^{1/3} \Rightarrow h=\left(\frac{T^2R^2g}{4\pi^2}\right)^{1/3}-R \quad \text{(vi)}$$

Knowing the values of T, R and g, the height h of the satellite above the surface of earth can be calculated from Eq. (vi).

Angular Momentum, L

When a satellite of mass m is orbiting with linear speed v on the orbital path of radius r around the earth, its angular momentum is given by

$$L = mvr = mr\sqrt{GM/r} = [m^2GMr]^{1/2}$$

From the above equation, it is clear that angular momentum of a satellite depends on both the mass of the satellite (m) and the mass of the planet (M). It also depends upon the radius of the orbit (r) of the satellite.

Energy of an Orbiting Satellite

The total mechanical energy of a satellite revolving around the earth is the sum of its potential energy (U) and kinetic energy (K).

The potential energy of a satellite is due to its position with respect to earth. It appears because of the gravitational pull acting on the satellite due to earth. If a satellite of mass m is revolving around the earth of mass M, radius R, with orbital velocity v, in an orbit of radius r, then potential energy of the satellite is

$$U = -GMm/r$$

The kinetic energy of a satellite,

$$K = \frac{1}{2}mv^2 = \frac{1}{2}m\left(\frac{GM}{r}\right) \quad [\because v = \sqrt{GM/r}]$$

Thus, total mechanical energy of a satellite is $E = U + K$

$$= -\frac{GMm}{r} + \frac{1}{2}\frac{GMm}{r} = -\frac{GMm}{2r} \Rightarrow E = -\frac{GMm}{2r} = -\frac{GMm}{2(R+h)}$$

If the satellite is orbiting close to earth, then $r \approx R$. Now, total energy of the satellite is $E = -GMm/2R$.

The above expression has been obtained considering the circular orbit of the satellite. When the orbit of satellite becomes elliptical, both the PE and KE of the satellite vary from point to point but the total energy remains constant and is negative, as in the case of circular orbit.

In the total energy of a circularly orbiting satellite which is negative, the potential energy is negative and magnitude of potential energy is twice the magnitude of the positive kinetic energy. We know that the total energy of a body is positive or zero, when it is at infinity from the earth. Therefore, the satellite will escape to infinity if its total energy becomes zero or positive. Since a satellite is always at finite distance from the earth, its total energy can never be positive or zero.

Binding Energy of a Satellite

The energy required to remove the satellite from its orbit around the earth to infinity is called binding energy of the satellite. Binding energy is equal to negative value of total mechanical energy of a satellite in its orbit.

Thus binding energy is

$$-E = \frac{GMm}{2r}$$

Illustration 5.42 **An artificial satellite revolves round the earth at a height of 1000 km. The radius of the earth is 6.38×10^3 km. Mass of the earth = 6×10^{24} kg; $G = 6.67 \times 10^{-11}$ Nm^{2}kg^{-2}. Find the orbital speed and period of revolution of the satellite.**

Sol. Here, $h = 1000$ km $= 1000 \times 10^3$ m $= 10^6$ m

$$r = R + h = 6.38 \times 10^6 + 10^6 = 7.38 \times 10^6 \text{ m}$$

Orbital speed,

$$v_0 = \sqrt{\frac{GM}{R+h}} = \sqrt{\frac{6.67 \times 10^{-11} \times 6 \times 10^{24}}{7.38 \times 10^6}} = 7364 \text{ m s}^{-1}$$

Time period,

$$T = \frac{2\pi r}{v_0} = \frac{2 \times (22/7) \times (7.38 \times 10^6)}{7.364 \times 10^3} = 6297 \text{ s}$$

Illustration 5.43 **A satellite orbits the earth at a height of 3.6×10^6 m from its surface. Compute its (a) kinetic energy, (b) potential energy, (c) total energy. Mass of the satellite = 500 kg, mass of the earth = 6×10^{24} kg, radius of the earth = 6.4×10^6, $G = 6.67 \times 10^{-11}$ N m^2 kg^{-2}.**

Sol. Here, $r = R + h = 6.4 \times 10^6 + 3.6 \times 10^6 = 10^7$ m.

Orbital velocity of the satellite around the earth is given by

$$v_0 = \sqrt{\frac{GM}{(R+h)}} = \sqrt{\frac{6.67 \times 10^{-11} \times 6 \times 10^{24}}{10^7}} = \sqrt{6.67 \times 6 \times 10^6} \text{ m s}^{-1}$$

a. KE of the satellite = $1/2\, mv_0^2$

$$= \frac{1}{2} \times 500 \times 6.67 \times 6 \times 10^6 = 10^{10}\, J$$

b. PE of the satellite $= -\dfrac{GMm}{R+h}$

$$= -\frac{6.67 \times 10^{-11} \times (6 \times 10^{24}) \times 500}{10^7} = -2 \times 10^{10}\, J$$

c. Total energy = KE + PE = $10^{10} - 2 \times 10^{10} = -10^{10}\, J$

Illustration 5.44 **You are given the following data: $g = 9.81$ m s^{-2}, radius of earth = 6.37×10^6 m, the distance the Moon from the earth = 3.84×10^8 m and the time period**

of the Moon's revolution = 27.3 days. Obtain the mass of the earth in two different ways. $G = 6.67 \times 10^{-11}$ N m^2 kg^2.

Sol. Here, $g = 9.81$ m s^{-2}

$$R = 6.37\times 10^6 \text{ m}, r = 3.84 \times 10^8 \text{ m}$$

$$T = 27.3 \text{ days} = 27.3 \times 24 \times 60 \times 60 \text{ s}$$

$$\therefore \quad M_e = \frac{gR_e^2}{G} = \frac{9.81 \times (6.37 \times 10^6)^2}{6.67 \times 10^{-11}}$$

$$= 5.97 \times 10^{24} \text{ kg}$$

Alternative method: Since the gravitational pull provides the required centripetal force, so

$$mr(2\pi/T^2) = GM_e m/r^2 \text{ or } M_e = \frac{4\pi^2 r^3}{GT^2}$$

$$= \frac{4 \times (22/7)^2 \times (3.84 \times 10^8)^3}{6.67 \times 10^{-11} \times (27.3 \times 24 \times 60 \times 60)^2} = 6.02 \times 10^{24} \text{ kg}$$

Illustration 5.45 **Assume the radius of the earth to be 6.4×10^6 m.**

a. Calculate the time period T of a satellite on equatorial orbit at 1.4×10^3 m above the surface of the earth.

b. What is the speed of the satellite in this orbit?

c. If the satellite is travelling in the same direction as the rotation of the earth, i.e., west to east, what is the interval between two successive times at which it will appear vertically overhead to an observer at a fixed point on the equator?

Sol. a. $r = R_{\text{earth}} + h = 6.4 \times 10^6 + 1.4 \times 10^6 = 7.8 \times 10^6$ m

$$\Rightarrow r = 7.8 \times 10^6 \text{ m} \Rightarrow T = \left(\frac{4\pi^2 r^3}{GM_{\text{earth}}}\right) = 6831 \text{ s}$$

b. Speed of satellite,

$$V = \sqrt{\frac{GM_{\text{earth}}}{r}} = 7174 \text{ m s}^{-1}$$

c. Relative angular velocity $= \omega_{\text{satellite}} - \omega_{\text{earth}}$

Thus $$T = \frac{2\pi}{(\omega_s - \omega_e)} = \frac{2\pi}{\left[\frac{2\pi}{T_s} - \frac{2\pi}{T_e}\right]}$$

$$\Rightarrow \quad T = \frac{T_s T_e}{T_e - T_s} = 7417 \text{ s}$$

Illustration 5.46 **Two satellites A and B of equal mass move in the equatorial plane of the earth, close to earth's surface. Satellite A moves in the same direction as that of the rotation of the earth while satellite B moves in the opposite direction. Calculate the ratio of the kinetic energy of B to that of A in the reference frame fixed to the earth. ($g = 9.8$ m s^{-2} and radius of the earth = 6.37×10^6 km)**

Sol. Let ω_A and ω_B be the absolute angular speeds of A and B. Since they are in the same orbit, their time periods must be the same, i.e., $\omega_A = \omega_B$. Considering the dynamics of circular motion in the cases,

$$m\omega_A^2 R = \frac{GMm}{R^2} \Rightarrow \omega_A = \sqrt{\frac{g}{R}} \quad (\because GM = gR^2)$$

Similarly $\omega_B = \sqrt{\frac{g}{R}}$

And $$\omega_A = \omega_B = \sqrt{\frac{9.8}{6.37 \times 10^6}} = 124 \times 10^{-5} \text{ rad s}^{-1}$$

Now

$$\omega_{AE} = \omega_A - \omega_E = 124 \times 10^{-5} - 7.3 \times 10^{-5}$$

$$= 116.7 \times 10^{-5} \text{ rad s}^{-1}$$

ω_{BE} (velcicty of B relative to E)

$$= \omega_B - (-\omega_E) = \omega_B + \omega_E$$

$$= 131.1 \times 10^{-5} \text{ rad s}^{-1}$$

Therefore, $$\frac{K_B}{K_A} = \frac{\frac{1}{2} m\,\omega_{BE}^2 r^2}{\frac{1}{2} m\,\omega_{AE}^2 r^2} = \frac{131.3^2}{116.7^2} = 1.27$$

Illustration 5.47 **A satellite is revolving in a circular equatorial orbit of radius $R = 2 \times 10^4$ km from east to west. Calculate the interval after which it will appear at the same equatorial town. Given that the radius of the earth = 6400 km and g (acceleration due to gravity) = 10 m s^{-2}.**

Sol. Let ω be the actual angular velocity of the satellite from east to west and ω_e be the angular speed of the earth (west to east).

Then $\omega_{\text{reletaive}} = \omega - (-\omega_e) = \omega + \omega_e$

$$\Rightarrow \quad \omega = \omega_{\text{rel}} - \omega_e$$

By the dynamics of circular motion

$$\frac{GMm}{R^2} = m\omega^2 R$$

or $$\omega^2 = \frac{gR_e^2}{R^3} \quad (\because GM = gR_e^2)$$

$$\Rightarrow \quad \omega = \sqrt{\frac{gR_e^2}{R^3}} \quad \therefore \omega_{\text{rel}} = \sqrt{\frac{gR_e^2}{R^3}} + \omega_e$$

$$\Rightarrow \quad \omega_{\text{rel}} = \sqrt{\frac{10 \times 6.4^2 \times 10^{12}}{2^3 \times 10^{21}}} + 7.27 \times 10^{-5}$$

$$\left(\because \omega_e = \frac{2\pi}{86400} = 7.27 \times 10^{-5}\right)$$

$$\Rightarrow \quad \omega_{rel} = 22.6 \times 10^{-5} + 7.27 \times 10^{-5}$$

$$= 30 \times 10^{-5} \text{ rad s}^{-1}$$

$$\therefore \quad \tau = \frac{2\pi}{\omega_{rel}} = \frac{2\pi}{30 \times 10^{-5}}$$

$$= 2.09 \times 10^4 \text{ s} = 5 \text{ h } 48 \text{ min}$$

GEOSTATIONARY OR GEOSYNCHRONOUS SATELLITES

A satellite which appears to be at fixed position at a definite height (i.e., stationary) to an observer on earth is called geostationary satellite. This satellite is also called the geosynchronous satellite as its angular speed is synchronized with the angular speed of the earth about its axis, i.e., this satellite revolves around the earth with the same angular speed in the same direction as is done by earth around its axis.

Obviously, the speed of such a satellite relative to earth is zero. That is why it appears to be stationary w.r.t. any point on the surface of earth. Clearly, $T = 24$ h for a geostationary satellite.

To calculate the height of geostationary satellite, we use the relation

$$h = \left(\frac{T^2R^2g}{4\pi^2}\right)^{\frac{1}{3}} - R$$

Putting, $R = 6.4 \times 10^6$ m, $g = 9.8$ m^{-2}, $T = 24$ h

$= 24 \times 60 \times 60$ s, we get $h = 3.6 \times 10^7$ m

$= 36000$ km

The orbital speed of such a satellite, using the relation $v = R\, Rg/R + h$ is found to be 3.1 km s^{-1}.

Geostationary satellites are used for the purposes of communication as they act as reflectors of suitable waves carrying the messages. Therefore, they are also called communication satellites. Telstar was the first communication satellite launched by U.S.A. in 1962. Since then, a large number of communication satellites have been put in use by various countries. India has also joined this race. INSAT 2B and INSAT 2C are the communication satellites of India.

Essential Conditions for Geostationary Satellite

1. A geostationary satellite should be at a height nearly 36000 km above the equator of earth.
2. Its time period of revolution around the earth should be the same as that of the earth about its axis, i.e., exactly 24 h.
3. It should revolve in an orbit concentric and coplanar with the equatorial plane, so the plane of the orbit of the satellite is normal to the axis of rotation of the earth.
4. Its sense of rotation should be the same as that of the earth about its own axis, i.e., from west to east. Its orbital speed is nearly 3.1 km s^{-1}.

Illustration 5.48 **A satellite is launched into a circular orbit 1600 km above the surface of the earth. Find the period of revolution if the radius of the earth is R = 6400 km and the acceleration due to gravity is 9.8 m s^{-2}. At what height from the ground should it be launched so that it may appear stationary over a point on the earth's equator?**

Sol. The orbiting period of a satellite at a height h from earth's surface is

$$T = \frac{2\pi r^{\frac{3}{2}}}{gR^2} \text{ where } r = R + h.$$

Then, $$T = \frac{2\pi(R+h)}{R}\sqrt{\left(\frac{R+h}{g}\right)}$$

Here, $R = 6400$ km, $h = 1600$ km $= R/4$

$$T = \frac{2\pi\left(R + \frac{R}{4}\right)}{R}\sqrt{\left(\frac{R + \frac{R}{4}}{g}\right)} = 2\pi\,(1.25)^{\frac{3}{2}}\sqrt{\frac{R}{g}}$$

Putting the given values,

$$T = 2 \times 3.14 \times \sqrt{\left(\frac{6.4 \times 10^6 \text{ m}}{9.8 \text{ m s}^{-2}}\right)}\,(1.25)^{\frac{3}{2}}$$

$$= 7092 \text{ s} = 1.97 \text{ h}$$

Now, a satellite will appear stationary in the sky over a point on the earth's equator if its period of revolution around the earth is equal to the period of revolution of the earth up around its own axis which is 24 h. Let us find the height h of such a satellite above the earth's surface in terms of the earth's radius. Let it be nR. Then

$$T = \frac{2\pi(R+nR)}{R}\sqrt{\left(\frac{R+nR}{g}\right)}$$

$$= 2\pi\sqrt{\left(\frac{R}{g}\right)}\,(1+n)^{\frac{3}{2}}$$

$$= 2 \times 3.14\sqrt{\left(\frac{6.4 \times 10^6 \text{ m/s}}{9.8 \text{ m/s}^2}\right)}\,(1+n)^{\frac{3}{2}}$$

$$= (5075 \text{ s})\,(1+n)^{\frac{3}{2}} = (1.41\ h)\,(1+n)^{\frac{3}{2}}$$

For $T = 24\ h$, we have $(24\ h) = (1.41\ h)\,(1+n)^{\frac{3}{2}}$

or $$(1+n)^{\frac{3}{2}} = \frac{24}{1.41} = 17$$

or $\quad 1 + n = (17)^{\frac{2}{3}} = 6.61$

or $\quad n = 5.61$

The height of the geostationary satellite above the earth's surface is $nR = 5.61 \times 6400$ km $= 3.59 \times 10^4$ km.

Concept Application Exercise 5.4

1. What are the conditions under which a rocket, fired from the earth, launches an artificial satellite of the earth?
2. Two artificial satellites, one close to the surface and the other away, are revolving around the earth. Which one has larger speed?
3. Should the speed of two artificial satellites of the earth having the different masses but the same orbital radius be the same?
4. Can a pendulum vibrate in an artificial satellite?
5. Air friction increases the velocity of the satellite. Explain.
6. Why is it that we can learn more about the shape of the earth by studying the motion of an artificial satellite than by studying the motion of the Moon?
7. Objects at rest on the earth's surface move in circular paths with a period of 24 h. Are they in 'orbit' in the sense that an earth satellite is in orbit? Explain.
8. Can a satellite move in a stable orbit in a plane not passing through the earth's centre? Explain.
9. As measured by an observer on the earth, would there be any difference in the periods of two satellites, each in a circular orbit near the earth in an equatorial plane, but one moving eastward and the other westward?
10. Two satellites move along a circular orbit in the same direction at a small distance from each other. A container has to be thrown from the first satellite on the second one. When will the container reach the second satellite earlier, if it is thrown in the direction of motion of the first satellite or in the opposite direction? The velocity of the container is small in comparison to that of the satellite.
11. Why is there 'weightlessness' in a satellite?
12. A satellite moves in a circular orbit around the earth at height $R_e/2$ from earth's surface where R_e is the radius of the earth. Calculate its period of revolution. Given $R_e = 6.38 \times 10^6$ m.
13. A small satellite revolves around a planet in an orbit just above planet's surface. Taking the mean density of the planet 8000 kg m^{-3} and $G = 6.67 \times 10^{-11}$ N m^{-2} /kg^{-2}, find the time period of the satellite.
14. Two satellites of the same mass are launched in the same orbit around the earth so as to rotate opposite to each other. They collide inelastically and stick together as wreckage. Obtain the total energy of the system before and just after the collision. Describe the subsequent motion of the wreckage.
15. Mars and earth have masses in the ratio 1:11 and radii in the ratio 42:79. Compare
    - **i.** their densities, assuming them to be spheres of uniform density;
    - **ii.** gravitational field strengths at their surfaces;
    - **iii.** escape velocities from their surfaces;
    - **iv.** periods of their satellites near their surfaces.
16. If a satellite is revolving around a planet of mass M in an elliptical orbit of semi-major axis a. Show that the orbital speed of the satellite when it is at a distance r from the planet will be given by

$$v^2 = GM\left[\frac{2}{r} - \frac{1}{a}\right]$$

EXERCISES

Subjective Type

Solutions on page 5.52

1. Calculate the velocity with which a body must be thrown vertically upward from the surface of the earth so that it may reach a height of $10R$, where R is the radius of the earth and is equal to 6.4×10^6 m.

   (Given: Mass of the earth $= 6 \times 10^{24}$ kg, gravitational constant $G = 6.7 \times 10^{-11}$ N m^2 kg^{-2})
2. A sky laboratory of mass 2×10^3 kg has to be lifted from one circular orbit of radius $2R$ into another circular orbit of radius $3R$. Calculate the minimum energy required if the radius of earth is $R = 6.37 \times 10^6$ m and $g = 9.8$ m s^{-2}.
3. Estimate whether it takes more energy to get a satellite upto 1600 km above the earth than to put in the orbit there. The earth's radius is 6400 km. Does your answer remain the same for height 3200 km or for height 4800 km?
4. The gravitational field in a region is given by $\vec{E} = (3\hat{i} - 4\hat{j})$ N kg^{-1}. Find out the work done (in joule) in displacing a particle by 1 m along the line $4y = 3x + 9$.
5. A body suspended on a spring balance in a ship weighs W_0 when the ship is at rest. When the ship begins to move along the equator with a speed v, show that the scale reading is very close to $W_0(1 \pm 2\omega V/g)$, where ω is the angular speed of the earth.
6. A spaceship is sent to investigate a planet of mass M and radius R. While hanging motionless in space at a distance $5R$ from the centre of the planet, the spaceship fires an instrument package of mass m, which is much smaller than the mass of the spaceship. For what angle θ will the package just graze the surface of the planet?

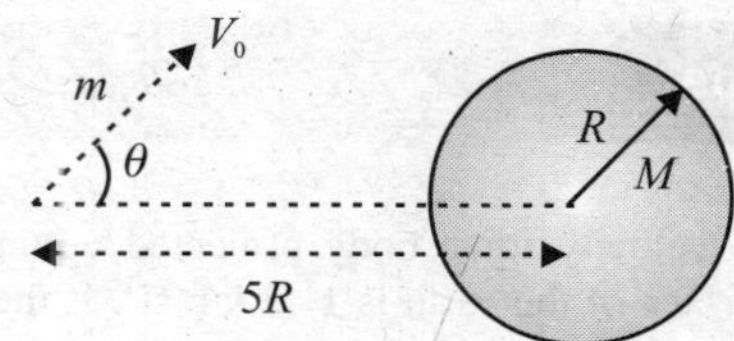

Fig. 5.31

7. Ravi can throw a ball at a speed on the earth which can cross a river of width 10 m. Ravi reaches on an imaginary planet whose mean density is twice that of the earth. Find out the maximum possible radius of the planet so that if Ravi throws the ball at the same speed it may escape from the planet. Given radius of the earth = 6.4×10^6 m.

8. A spaceship approaches the Moon (mass = M and radius = R) along a parabolic path which is almost tangential to its surface. At the moment of the maximum approach, the brake rocket is fired to convert the spaceship into a satellite of the Moon. Find the change in speed.

9. Taking the earth to be a uniform sphere of radius 6400 km and the value of g at the surface to be 10 m s^{-2}, calculate the energy needed to raise a satellite of mass 2000 kg to a height of 800 km above the earth's surface and to set it into circular orbit at that altitude.

10. A satellite is to be put into an orbit 600 km above the surface of the earth. If its vertical velocity after launching is 2400 m/s at this height, calculate the magnitude and direction of the impulse required to put the satellite directly into orbit. The mass of the satellite is 60 kg and the radius of the earth is 6400 km. Take g = 10 m s^{-2}.

11. A pair of stars rotates about a common centre of mass. One of the stars has a mass M and the other m. Their centres are a distance d apart, d being large compared to the size of either star. Derive an expression for the period of revolution of the stars about their common centre of mass. Compare their angular momenta and kinetic energies.

12. The small dense stars rotate about their common centre of mass as a binary system, each with a period of 1 year. One star has mass double than that of the other, while mass of the lighter star is one-third the mass of the Sun. The distance between the two stars is r and the distance of the earth from the Sun is R; find the relation between r and R.

13. A satellite revolving in a circular equatorial orbit of radius $R = 2.0 \times 10^4$ km from west to east appears over a certain point at the equator every 11.6 h. From these data, calculate the mass of the earth. ($G = 6.67 \times 10^{-11}$ N m^2)

14. An artificial satellite of mass m of a planet of mass M, revolves in a circular orbit whose radius is n times the radius R of the planet. In the process of motion, the satellite experiences a slight resistance due to cosmic dust. Assuming resistance force on the satellite depends on velocity as $F = av^2$ where a is constant, calculate the time the satellite will stay in orbit before it falls onto the planet's surface.

15. A satellite of mass m is orbiting the earth in a circular orbit of radius r. It starts losing energy due to small air resistance at the rate of C J s^{-1}. Find the time taken for the satellite to reach the earth.

Objective Type

Solutions on page 5.55

1. The escape velocity for a body projected vertically upwards from the surface of the earth is 11.2 km s^{-1}. If the body is projected in a direction making an angle 45° with the vertical, the escape velocity will be

a. $\frac{11.2}{\sqrt{2}}$ km s^{-1} **b.** $11.2 \times \sqrt{2}$ km s^{-1}

c. 11.2×2 km s^{-1} **d.** 11.2 km s^{-1}

2. A projectile is fired vertically upwards from the surface of the earth with a velocity Kv_e, where v_e is the escape velocity and $k < 1$. If R is the radius of the earth, the maximum height to which it will rise, measured from the centre of the earth, will be (neglect air resistance)

a. $\frac{1-k^2}{R}$ **b.** $\frac{R}{1-k^2}$

c. $R(1-k)^2$ **d.** $\frac{R}{1+k^2}$

3. The radius of a planet is R. A satellite revolves around it in a circle of radius r with angular velocity ω_0. The acceleration due to the gravity on planet's surface is

a. $\frac{r^3\omega_0}{R}$ **b.** $\frac{r^3\omega_0^3}{R^2}$

c. $\frac{r^3\omega_0^2}{R}$ **d.** $\frac{r^3\omega_0^2}{R^2}$

4. A space vehicle approaching a planet has a speed v, when it is very far from the planet. At that moment tangent of its trajectory would miss the centre of the planet by distance R. If the planet has mass M and radius r, what is the smallest value of R in order that the resulting orbit of the space vehicle will just miss the surface of the planet?

a. $\frac{r}{v}\left[v^2 + \frac{2GM}{r}\right]^{\frac{1}{2}}$ **b.** $vr\left[1 + \frac{2GM}{r}\right]$

c. $\frac{r}{v}\left[v^2 + \frac{2GM}{r}\right]$ **d.** $\frac{2GMv}{r}$

5. The two planets with radii R_1, R_2 have densities ρ_1, ρ_2, and atmospheric pressures p_1 and p_2, respectively. Therefore, the ratio of masses of their atmospheres, neglecting variation of g and ρ within the limits of atmosphere, is

a. $\frac{p_1R_2\rho_1}{p_2R_1\rho_2}$ **b.** $\frac{p_1R_2\rho_2}{p_2R_1\rho_1}$

c. $\frac{p_1R_1\rho_1}{p_2R_2\rho_2}$ **d.** $\frac{p_1R_1\rho_2}{p_2R_2\rho_1}$

6. If g is the acceleration due to gravity on the earth's surface, the change in the potential energy of an object of mass m raised from the surface of the earth to a height equal to the radius R of the earth is

a. $\frac{mgR}{2}$ **b.** $2mgR$

c. mgR **d.** $-mgR$

7. A space station is set up in space at a distance equal to the earth's radius from the surface of the earth. Suppose a satellite can be launched from the space station. Let v_1 and v_2 be the escape velocities of the satellite on the earth's surface and space station, respectively. Then

a. $v_2 = v_1$

b. $v_2 < v_1$

c. $v_2 > v_1$

d. (a), (b) and (c) are valid depending on the mass of satellite.

8. The orbital velocity of an artificial satellite in a circular orbit just above the earth's surface is v. For a satellite orbiting at an altitude of half of the earth's radius, the orbital velocity is

a. $\left(\frac{3}{2}\right)v$ b. $\sqrt{\left(\frac{3}{2}\right)}v$

c. $\sqrt{\left(\frac{2}{3}\right)}v$ d. $\left(\frac{2}{3}\right)v$

9. The radius of the earth is about 6400 km and that of Mars is about 3200 km. The mass of the earth is about 10 times the mass of Mars. An object weighs 200 N on the surface of the earth. Its weight on the surface of mars would be

a. 6 N b. 20 N

c. 40 N d. 80 N

10. The distance of two planets from the Sun are 10^{13} and 10^{12} m, respectively. The ratio of time periods of these two planets is

a. $\frac{1}{\sqrt{10}}$ b. 100

c. $\frac{10}{\sqrt{10}}$ d. $\sqrt{10}$

11. The gravitational force between two objects is proportional to $1/R$ (and not as $1/R^2$) where R is separation between them, then a particle in circular orbit under such a force would have its orbital speed v proportional to

a. $\frac{1}{R^2}$ b. R^0 c. R^1 d. $\frac{1}{R}$

12. Two particles of equal mass go around a circle of radius R under the action of their mutual gravitational attraction. The speed of each particle is

a. $v = \frac{1}{2R}\sqrt{\left(\frac{1}{Gm}\right)}$ b. $v = \sqrt{\left(\frac{Gm}{2R}\right)}$

c. $v = \frac{1}{2}\sqrt{\left(\frac{Gm}{R}\right)}$ d. $v = \sqrt{\left(\frac{4Gm}{R}\right)}$

13. A satellite of mass m is revolving around the earth at height R (radius of the earth) from the earth's surface. Its potential energy will be

a. mgR b. $0.67\ mgR$

c. $-\frac{mgR}{2}$ d. $0.33\ mgR$

14. If g is same at a height h and at a depth d, then

a. $R = 2d$ b. $d = 2h$

c. $h = d$ d. none

15. A solid sphere of radius $R/2$ is cut out of a solid sphere of radius R such that the spherical cavity so formed touches the surface on one side and the centre of the sphere on the other side, as shown. The initial mass of the solid sphere was M. If a particle of mass m is placed at a distance $2.5R$ from the centre of the cavity, then what is the gravitational attraction on the mass m?

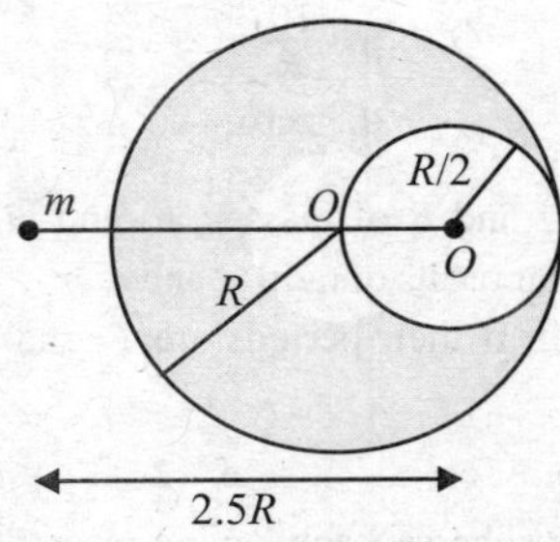

Fig. 5.32

a. $\frac{GMm}{R^2}$ b. $\frac{GMm}{2R^2}$

c. $\frac{GMm}{8R^2}$ d. $\frac{23}{100}\frac{GMm}{R^2}$

16. A geostationary satellite orbits around the earth in a circular orbit of radius 36000 km. Then, the time period of a spy satellite orbiting a few 100 km above the earth's surface ($R_{earth} = 6400$ km) will approximately be

a. $\frac{1}{2}h$ b. $1\ h$

c. $2\ h$ d. $4\ h$

17. If W_1, W_2 and W_3 represent the work done in moving a particle from A to B along three different paths 1, 2 and 3, respectively, (as shown in the figure) in the gravitational field of a point mass m, find the correct relation between W_1, W_2 and W_3.

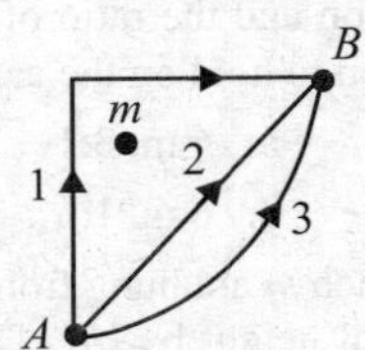

Fig. 5.33

a. $W_1 > W_2 > W_3$ b. $W_1 = W_2 = W_3$

b. $W_1 < W_2 < W_3$ d. $W_2 > W_1 > W_3$

18. If R is the radius of the earth and g the acceleration due to gravity on the earth's surface, the mean density of the earth is

a. $\frac{4\pi G}{3gR}$ b. $\frac{3\pi R}{4gG}$

c. $\frac{3g}{4\pi RG}$ d. $\frac{\pi R}{12G}$

19. A satellite moves around the earth in a circular orbit with speed v. If m is the mass of the satellite, its total energy is

a. $-\frac{1}{2}mv^2$ b. $\frac{1}{2}mv^2$

c. $\frac{3}{2}mv^2$ d. $\frac{1}{4}mv^2$

20. The value of g (acceleration due to gravity) at earth's surface is 10 m s^{-2}. Its value in m s^{-2} at the centre of the earth which is assumed to be a sphere of radius R metre and uniform mass density is

a. 5
b. $\frac{10}{R}$
c. $\frac{10}{2R}$
d. zero

21. Two satellites A and B of masses m_1 and m_2 ($m_1 = 2\,m_2$) are moving in circular orbits of radii r_1 and r_2 ($r_1 = 4r_2$), respectively, around the earth. If their periods are T_A and T_B, then the ratio T_A/T_B is

a. 4 b. 16 c. 2 d. 8

22. If three uniform spheres, each having mass M and radius R, are kept in such a way that each touches the other two, the magnitude of the gravitational force on any sphere due to the other two is

a. $\frac{GM^2}{4r^2}$
b. $\frac{2GM^2}{r^2}$
c. $\frac{2GM^2}{4r^2}$
d. $\frac{\sqrt{3}GM^2}{4r^2}$

23. If the radius of the earth decreases by 10%, the mass remaining unchanged, what will happen to the acceleration due to gravity?

a. Decreases by 19%
b. Increases by 19%
c. Decreases by more than 19%
d. Increases by more than 19%

24. The maximum vertical distance through which a fully dressed astronaut can jump on the earth is 0.5 m. If mean density of the Moon is two-third that of the earth and radius is one quarter that of the earth, the maximum vertical distance through which he can jump on the Moon and the ratio of the time of duration of the jump on the Moon to hold on the earth are

a. 3 m, 6:1
b. 6 m, 3:1
c. 3 m, 1:6
d. 6 m, 1:6

25. Two equal masses each m are hung from a balance whose scale pans differ in vertical height by 'h'. The error in weighing in terms of density of the earth ρ is

a. $\pi G\rho mh$
b. $\frac{1}{3}\pi G\rho mh$
c. $\frac{8}{3}\pi G\rho mh$
d. $\frac{4}{3}\pi G\rho mh$

26. The distances from the centre of the earth, where the weight of a body is zero and one-fourth that of the weight of the body on the surface of the earth are (assume R is the radius of the earth)

a. $0, \frac{R}{4}$
b. $0, \frac{3R}{4}$
c. $\frac{R}{4}, 0$
d. $\frac{3R}{4}, 0$

27. If a man at the equator would weigh (3/5)th of his weight, the angular speed of the earth is

a. $\sqrt{\frac{2}{5}\frac{g}{R}}$
b. $\sqrt{\frac{g}{R}}$
c. $\sqrt{\frac{R}{g}}$
d. $\sqrt{\frac{2}{5}\frac{R}{g}}$

28. The distance between the centres of the Moon and the earth is D. The mass of the earth is 81 times the mass of the Moon. At what distance from the centre of the earth, the gravitational force will be zero?

a. $\frac{D}{2}$ b. $\frac{2D}{3}$ c. $\frac{4D}{3}$ d. $\frac{9D}{10}$

29. Two bodies with masses M_1 and M_2 are initially at rest and a distance R apart. Then they move directly towards one another under the influence of their mutual gravitational attraction. What is the ratio of the distances travelled by M_1 to the distance travelled by M_2?

a. $\frac{M_1}{M_2}$ b. $\frac{M_2}{M_1}$ c. 1 d. $\frac{1}{2}$

30. If g is acceleration due to gravity on the earth's surface, the gain in the potential energy of an object of mass m raised from the surface of earth to a height equal to the radius R of the earth is

a. $\frac{1}{2}mgR$
b. $2mgR$
c. mgR
d. $\frac{1}{4}mgR$

31. Imagine a light planet revolving around a very massive star in a circular orbit of radius r with a period of revolution T. If the gravitational force of attraction between the planet and the star is proportional to $r^{5/2}$, then the square of the time period will be proportional to

a. r^3 b. r^2 c. $r^{2.5}$ d. $r^{3.5}$

32. The masses and radii of the earth and the Moon are M_1, R_1 and M_2, R_2, respectively. Their centres are at distance d apart. The minimum speed with which a particle of mass m should be projected from a point midway the two centres so as to escape to infinity is

a. $\sqrt{\frac{2G(M_1+M_2)}{d}}$
b. $\sqrt{\frac{4G(M_1+M_2)}{d}}$
c. $\sqrt{\frac{4GM_1M_2}{d}}$
d. $\sqrt{\frac{G(M_1+M_2)}{d}}$

33. A space ship is launched into a circular orbit close to the surface of the earth. The additional velocity now imparted to the spaceship in the orbit to overcome the gravitational pull is

a. 11.2 km s^{-1}
b. 8 km s^{-1}
c. 3.2 km s^{-1}
d. 1.414 × 8 km s^{-1}

34. A skylab of mass m kg is first launched from the surface of the earth in a circular orbit of radius $2R$ (from the centre of the earth) and then it is shifted from this circular orbit to another circular orbit of radius $3R$. The minimum energy required to place the lab in the first orbit and to shift the lab from first orbit to the second orbit are

a. $\frac{3}{4}mgR, \frac{mgR}{6}$
b. $\frac{3}{4}mgR, \frac{mgR}{12}$
c. mgR, mgR
d. $2mgR, mgR$

35. Two satellites of the same mass are launched in the same orbit around the earth so as to rotate opposite to each other. If they collide inelastically and stick together as wreckage, the total energy of the system just after collision is

a. $-\frac{2GMm}{r}$
b. $-\frac{GMm}{r}$
c. $\frac{GMm}{2r}$
d. $\frac{GMm}{4r}$

36. A spherical shell is cut into two pieces along a chord as shown in the figure. P is a point on the plane of the chord. The gravitational field at P due to the upper part is I_1 and that due to the lower part is I_2. What is the relation between them?

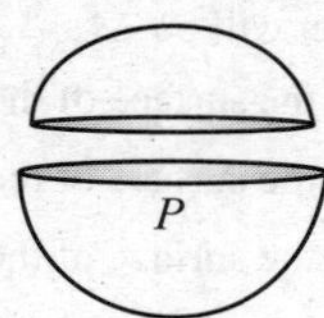

Fig. 5.34

a. $I_1 > I_2$ b. $I_1 < I_2$
c. $I_1 = I_2$ d. no definite relation

37. Two particles of equal mass go around a circle of radius R under the action of their mutual gravitational attraction. The speed of each particle is

a. $v = \frac{1}{2R}\sqrt{\left(\frac{1}{GM}\right)}$ b. $v = \sqrt{\left(\frac{GM}{2R}\right)}$

c. $v = \frac{1}{2}\sqrt{\left(\frac{GM}{R}\right)}$ d. $v = \sqrt{\left(\frac{4GM}{R}\right)}$

38. A rocket is launched vertically from the surface of earth with an initial velocity v. How far above the surface of earth it will go? Neglect the air resistance.

a. $R\left(\frac{2gR}{v^2} - 1\right)^{-1/2}$ b. $R\left(\frac{2gR}{v^2} - 1\right)$

c. $R\left(\frac{2gR}{v^2} - 1\right)^{-1}$ d. $R\left(\frac{2gR}{v^2} - 1\right)^{2}$

39. The gravitational potential due to earth at infinite distance from it is zero. Let the gravitational potential at a point P be -5 J kg^{-1}. Suppose, we arbitrarily assume the gravitational potential at infinity to be $+10$ J kg^{-1}, then the gravitational potential at P will be

a. -5 J kg^{-1} b. $+5$ J kg^{-1}
c. -15 J kg^{-1} d. $+15$ J kg^{-1}

40. A ball of mass m is fired vertically upwards from the surface of the earth with velocity nv_e, where v_e is the escape velocity and $n > 1$. To what height will the ball rise? Neglecting air resistance, take radius of the earth as R

a. $\frac{R}{n^2}$ b. $\frac{R}{(1-n^2)}$

c. $\frac{Rn^2}{(1-n^2)}$ d. Rn^2

41. How many hours would make a day if the earth were rotating at such a high speed that the weight of a body on the equator were zero

a. 6.2 h b. 1.4 h c. 28 h d. 5.6 h

42. Two bodies of masses M_1 and M_2 are placed at a distance R apart. Then at the position where the gravitational field due to them is zero, the gravitational potential is

a. $-G\frac{\sqrt{M_1}}{R}$ b. $-G\frac{\sqrt{M_2}}{R}$

c. $-(\sqrt{M_1} + \sqrt{M_2})^2 \frac{G}{R}$ d. $-(\sqrt{M_1} - \sqrt{M_2})^2 \frac{G}{R}$

43. In the solar system, the Sun is in the focus of the system for Sun-earth binding system. Then the binding energy for the system will be [given that radius of the earth's orbit round the Sun is 1.5×10^{11} m and mass of the earth $= 6 \times 10^{24}$ kg]

a. 2.7×10^{33} J b. 5.4×10^{33} J
c. 2.7×10^{30} J d. 5.4×10^{30} J

44. A body is released from a point of distance R' from the centre of earth. Its velocity at the time of striking the earth will be $(R' > R_e)$

a. $\sqrt{2gR_e}$ b. $\sqrt{R_e g}$

c. $\sqrt{2g(R' - R_e)}$ d. $\sqrt{2gR_e\left(1 - \frac{R_e}{R'}\right)}$

45. A solid sphere of uniform density and radius R applies a gravitational force of attraction equal to F_1 on a particle placed at a distance $2R$ from the centre of the sphere. A spherical cavity of radius $R/2$ is now made in the sphere as shown in the figure. The sphere with the cavity now applies a gravitational force F_2 on the same particle. The ratio F_1/F_2 is

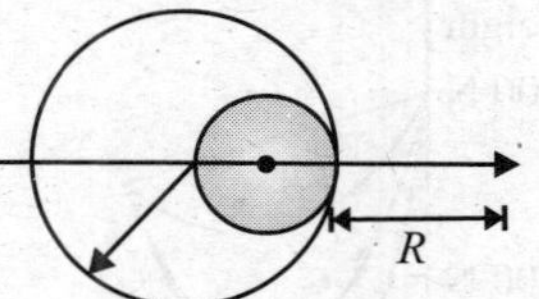

Fig. 5.35

a. $\frac{1}{2}$ b. $\frac{3}{4}$ c. $\frac{7}{8}$ d. $\frac{9}{7}$

46. The value of g at a particular point is 10 m s^{-2}. Suppose the earth shrinks uniformly to half of its present size without losing any mass. The value of g at the same point (assuming that the distance of the point from the centre of the earth does not change) will now be

a. 5 m s^{-2} b. 10 m s^{-2}
c. 3 m s^{-2} d. 20 m s^{-2}

47. Two satellites of masses m_1 and m_2 $(m_1 > m_2)$ are revolving around the earth in a circular orbit of radii r_1 and r_2 $(r_1 > r_2)$, respectively. Which of the following statements is true regarding their speeds v_1 and v_2?

a. $v_1 = v_2$ b. $v_1 > v_2$

c. $v_1 < v_2$ d. $\frac{v_1}{r_1} = \frac{v_2}{v_2}$

48. The earth moves around the Sun in an elliptical orbit as shown in Fig. 5.36. The ratio $OA/OB = x$. The ratio of the speed of the earth at B to that at A is nearly

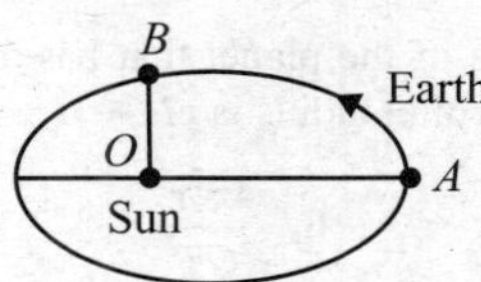

Fig. 5.36

a. $\sqrt{x}$ b. x c. $x\sqrt{x}$ d. x^2

49. The following figure shows two shells of masses m_1 and m_2. The shells are concentric. At which point, a particle of mass m shall experience zero force?

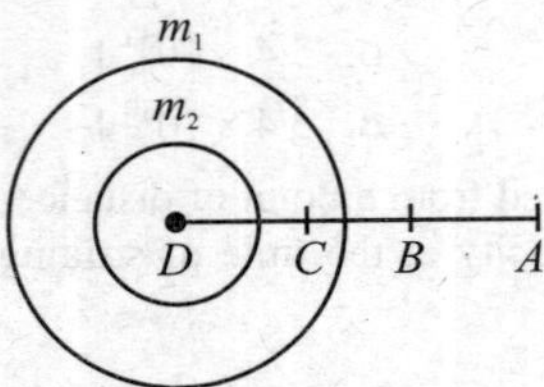

Fig. 5.37

a. A **b.** B **c.** C **d.** D

50. Suppose, the acceleration due to gravity at the earth's surface is $10\ \text{m s}^{-2}$ and at the surface of Mars is $4.0\ \text{m s}^{-2}$. A 60 kg passenger goes from the earth to Mars in a spaceship moving with a constant velocity. Neglect all other objects in the sky. Which part of the figure best represents the weight (net gravitational force) of the passenger as a function of time?

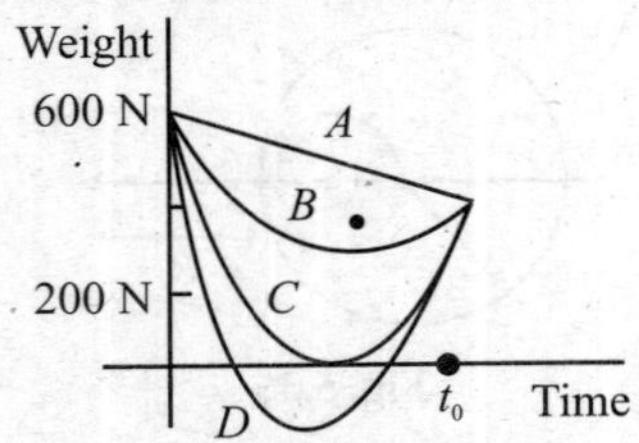

Fig. 5.38

a. A **b.** B

c. C **d.** D

51. Two satellites A and B of the same mass are revolving around the earth in the concentric circular orbits such that the distance of satellite B from the centre of the earth is thrice as compared to the distance of the satellite A from the centre of the earth. The ratio of the centripetal force acting on B as compared to that on A is

a. $\frac{1}{3}$ **b.** 3

c. $\frac{1}{9}$ **d.** $\frac{1}{\sqrt{3}}$

52. The minimum energy required to launch a m kg satellite from the earth's surface in a circular orbit at an altitude $2R$, where R is the radius of earth is

a. $\frac{5}{3}mgR$ **b.** $\frac{4}{3}mgR$

c. $\frac{5}{6}mgR$ **d.** $\frac{5}{4}mgR$

53. What is the mass of the planet that has a satellite whose time period is T and orbital radius is r?

a. $\frac{4\pi^2 r^3}{GT^2}$ **b.** $\frac{4\pi^2 r^3}{GT^2}$

c. $\frac{4\pi^2 r^3}{GT^3}$ **d.** $\frac{4\pi^2 T}{GT^2}$

54. If the mass of a planet is 10% less than that of the earth and the radius is 20% greater than that of the earth, the acceleration due to gravity on the planet will be

a. 5/8 times that on the surface of the earth

b. 3/4 times that on the surface of the earth

c. 1/2 times that on the surface of the earth

d. 9/10 times that on the surface of the earth

55. What should be the angular velocity of rotation of the earth about its own axis so that the weight of a body at the equator reduces to 3/5 or its present value? (Take R as the radius of the earth)

a. $\sqrt{\frac{g}{3R}}$ **b.** $\sqrt{\frac{2g}{3R}}$

c. $\sqrt{\frac{2g}{5R}}$ **d.** $\sqrt{\frac{2g}{7R}}$

56. In order to shift a body of mass m from a circular orbit of radius $3R$ to a higher orbit of radius $5R$ around the earth, the work done is

a. $\frac{3GMm}{5R}$ **b.** $\frac{GMm}{2R}$

c. $\frac{2}{15}\frac{GMm}{R}$ **d.** $\frac{GMm}{5R}$

57. A uniform ring of mass m and radius r is placed directly above a uniform sphere of mass M and of equal radius. The centre of the ring is directly above the centre of the sphere at a distance $r\sqrt{3}$ as shown in the figure. The gravitational force exerted by the sphere on the ring will be

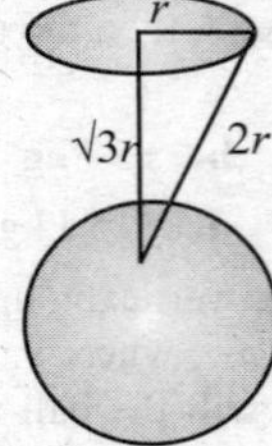

Fig. 5.39

a. $\frac{GMm}{8r^2}$ **b.** $\frac{GMm}{4r^2}$

c. $\sqrt{3}\frac{GMm}{8r^2}$ **d.** $\frac{GMm}{8r^3\sqrt{3}}$

58. A tunnel is dug along a diameter of the earth. If M_e and R_e are the mass and radius, respectively, of the earth, then the force on a particle of mass m placed in the tunnel at a distance r from the centre is

a. $\frac{GM_e m}{R_e^3}r$ **b.** $\frac{GM_e m}{R_e^3 r}$

c. $\frac{GM_e mR_e^3}{r}$ **d.** $\frac{GM_e m}{R_e^2}r$

59. The value of 'g' at a certain height h above the free surface of the earth is $x/4$ where x is the value of 'g' at the surface of the earth. The height h is

a. R b. $2R$ c. $3R$ d. $4R$

60. A planet is revolving in an elliptical orbit around the Sun. Its closest distance from the Sun is r_{min} and the farthest distance is r_{max}. If the velocity of the planet at the distance of the closest approach is v_1 and that at the farthest distance from the Sun is v_2, then $\frac{v_1}{v_2}$

a. $\frac{r_{max}}{r_{min}}$ b. $\frac{r_{min}}{r_{max}}$

c. $\frac{r_{min}+r_{max}}{r_{max}-r_{min}}$ d. none

61. Masses of 1 kg each are placed 1 m, 2 m, 4 m, 8 m, ... from a point P. The gravitational field intensity at P due to these masses is

a. G b. G c. $4G$ d. $4G/3$.

62. Suppose that the acceleration of a free fall at the surface of a distant planet was found to be equal to that at the surface of the earth. If the diameter of the planet were twice the diameter of the earth, then the ratio of mean density of the planet to that of the earth would be

a. 4:1 b. 2:1 c. 1:1 d. 1:2

63. Three uniform spheres, each having mass m and radius r, are kept in such a way that each touches the other two. The magnitude of the gravitational force on any sphere due to the other two is

a. $\frac{Gm^2}{r^2}$ b. $\frac{Gm^2}{4r^2}$ c. $\sqrt{2}\,\frac{Gm^2}{4r^2}$ d. $\sqrt{3}\,\frac{Gm^2}{4r^2}$

64. A body of mass m rises to a height $h = R/5$ from the earth's surface where R is earth's radius. If g is acceleration due to gravity at the earth's surface, the increase in potential energy is

a. mgh b. $\frac{4}{5}mgh$

c. $\frac{5}{6}mgh$ d. $\frac{6}{7}mgh$

65. A man weighs 80 kg on the surface of earth of radius R. At what height above the surface of earth his weight will be 40 kg?

a. $\frac{R}{2}$ b. $\sqrt{2}\,R$

c. $(\sqrt{2}-1)\,R$ d. $(\sqrt{2}+1)\,R$

66. The gravitational potential energy of an isolated system of three particles, each of mass m, at the three corners of an equilateral triangle of side l is

a. $-\frac{Gm^2}{l}$ b. $-\frac{Gm^2}{2l}$

c. $-\frac{2Gm^2}{l}$ d. $-\frac{3Gm^2}{l}$

67. A diametrical tunnel is dug across the earth. A ball is dropped into the tunnel from one side. The velocity of the ball when it reaches the centre of the earth is

[Given: gravitational potential at the centre of earth $= -3/(2\,GM/R)$]

a. $\sqrt{R}$ b. $\sqrt{gR}$

c. $\sqrt{2.5\,gR}$ d. $\sqrt{7.1\,gR}$

68. Consider two solid uniform spherical objects of the same density ρ. One has radius R and the other has radius $2R$. They are in outer space where the gravitational fields from other objects are negligible. If they are arranged with their surface touching, what is the contact force between the objects due to their traditional attraction?

a. $G\pi^2R^4$ b. $\frac{128}{81}G\pi^2R^4\rho^2$

c. $\frac{128}{81}G\pi^2$ d. $\frac{128}{87}\pi^2R^2G$

69. A body starts from rest from a point distant r_0 from the centre of the earth. It reaches the surface of the earth whose radius is R. The velocity acquired by the body is

a. $2GM\sqrt{\frac{1}{R}-\frac{1}{r_0}}$ b. $\sqrt{2GM\left(\frac{1}{R}-\frac{1}{r_0}\right)}$

c. $GM\sqrt{\frac{1}{R}-\frac{1}{r_0}}$ d. $\sqrt{GM\left(\frac{1}{R}-\frac{1}{R_0}\right)}$

70. Four particles, each of mass M, move along a circle of radius R under the action of their mutual gravitational attraction. The speed of each particle is

a. $\frac{GM}{R}$ b. $\sqrt{2\sqrt{2}\,\frac{GM}{R}}$

c. $\sqrt{\frac{GM}{R}(2\sqrt{2}+1)}$ d. $\sqrt{\frac{GM}{R}\left(\frac{2\sqrt{2}+1}{4}\right)}$

71. The mass of the earth is 81 times the mass of the Moon and the distance between the earth and the Moon is 60 times the radius of the earth. If R is the radius of the earth, then the distance between the Moon and the point on the line joining the Moon and the earth where the gravitational force becomes zero is

a. $30R$ b. $15R$ c. $6R$ d. $5R$

72. If three particles, each of mass M, are placed at the three corners of an equilateral triangle of side a, the forces exerted by this system on another particle of mass M placed (i) at the midpoint of a side and (ii) at the centre of the triangle are, respectively,

a. $0, \frac{4GM^2}{3a^2}$ b. $\frac{4GM^2}{3a^2}, 0$

c. $\frac{3GM^2}{a^2}, \frac{GM^2}{a^2}$ d. $0, 0$

73. Suppose the gravitational force varies inversely as the nth power of the distance. Then, the time period of a planet in a circular orbit of radius R around the Sun will be proportional to

a. R^n b. $R^{\frac{(n+1)}{2}}$ c. $R^{\frac{(n-1)}{2}}$ d. R^{-n}

74. Two astronauts have deserted their spaceship in a region of space far from the gravitational attraction of any other body. Each has a mass of 100 kg and they are 100 m apart. They are initially

at rest relative to one another. How long will it be before the gravitational attraction brings them 1 cm closer together?

a. 2.52 days **b.** 1.41 days
c. 0.70 days **d.** 1.41 s

75. A satellite of mass m revolves around the earth of radius R at a height x from its surface. If g is the acceleration due to gravity on the surface of the earth, the orbital speed of the satellite is

a. gx **b.** $\frac{gR}{R-x}$
c. $\frac{gR^2}{R+x}$ **d.** $\left(\frac{gR^2}{R+x}\right)^{\frac{1}{2}}$

76. A projectile is fired vertically upwards from the surface of the earth with a velocity kv_e where v_e is the escape velocity and $k < 1$. If R is the radius of the earth, the maximum height to which it will rise measured from the centre of earth will be (neglect air resistance)

a. $\frac{1-k^2}{R}$ **b.** $\frac{R}{1-k^2}$
c. $R(1-k^2)$ **d.** $\frac{R}{1+k^2}$

77. If an artificial satellite is moving in a circular orbit around the earth with a speed equal to half the magnitude of the escape velocity from the earth, the height of the satellite above the surface of the earth is

a. $2R$ **b.** $\frac{R}{2}$ **c.** R **d.** $\frac{R}{4}$

78. Two concentric shells of masses M_1 and M_2 are having radii r_1 and r_2. Which of the following is the correct expression for the gravitational field on a mass m?

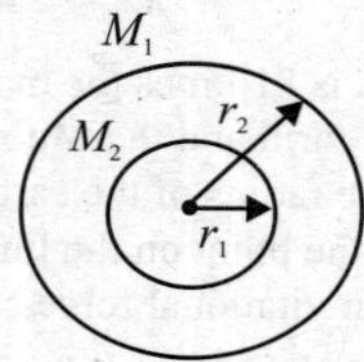

Fig. 5.40

a. $F = \frac{G(M_1+M_2)}{r^2}$, for $r < r_1$

b. $F = \frac{G(M_1+M_2)}{r^2}$, for $r < r_2$

c. $F = \frac{GM_2}{r^2}$, for $r_1 < r < r_2$

d. $F = \frac{GM_1}{r^2}$, for $r_1 < r < r_2$

79. A satellite of mass m is orbiting around the earth at a height h above the surface of the earth. Mass of the earth is M and its radius is R. The angular momentum of the satellite is independent of

a. m **b.** M
c. h **d.** none of these

80. Two concentric shells have masses M and m and their radii are R and r, respectively, where $R > r$. What is the gravitational potential at their common centre?

a. $-\frac{GM}{R}$ **b.** $-\frac{GM}{r}$
c. $-G\left[\frac{M}{R}-\frac{m}{r}\right]$ **d.** $-G\left[\frac{M}{R}+\frac{m}{r}\right]$

81. In problem 80, if x is the distance from the common centre, then what is the gravitational potential at a point for which $r < x < R$?

a. $-G\left[\frac{M}{x}+\frac{m}{r}\right]$ **b.** $-G\left[\frac{M}{x}-\frac{m}{r}\right]$
c. $-G\left[\frac{M}{R}+\frac{m}{x}\right]$ **d.** $-G\left[\frac{M}{R}-\frac{m}{x}\right]$

82. In problem 81, what is the gravitational intensity at a point for which $x < r$?

a. $\frac{Gm}{r^2}$ **b.** $\frac{Gm}{x^2}$
c. $\frac{Gm}{R^2}$ **d.** zero

83. A solid sphere of uniform density and mass M has radius 4 m. Its centre is at the origin of the coordinate system. Two spheres of radii 1 m are taken out so that their centres are at P (0, −2, 0) and Q (0, 2, 0), respectively. This leaves two spherical cavities. What is the gravitational field at the origin of the coordinate axes?

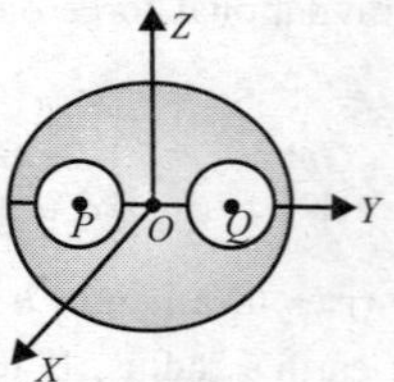

Fig. 5.41

a. $\frac{31GM}{1024}$ **b.** $\frac{Gm}{1024}$
c. $31GM$ **d.** zero

84. In problem 83, what is the gravitational field at the centre of the cavities?

a. $\frac{31GM}{1024}$ **b.** $\frac{Gm}{1024}$
c. $31GM$ **d.** GM

85. In problem 83, what is the gravitational potential at any point on the circle $x^2 + z^2 = 6$?

a. $-\frac{GM}{6}$ **b.** $-\frac{GM}{64\sqrt{10}}$
c. $-\frac{GM}{2}\left[\frac{1}{3}-\frac{1}{32\sqrt{10}}\right]$ **d.** $-\frac{GM}{2}\left[\frac{1}{3}+\frac{1}{32\sqrt{10}}\right]$

86. Three particles, each of mass M, are placed at the three corners of an equilateral triangle of side l. What is the force due to this system of particles on another particle of mass m placed at the midpoint of any side?

a. $\frac{3GMm}{4l^2}$ b. $\frac{4GMm}{3l^2}$

c. $\frac{GMm}{4l^2}$ d. $\frac{4GMm}{l^2}$

87. In problem 86, what is the gravitational potential energy of the mass m?

a. $-\frac{2}{\sqrt{3}}\frac{GMm}{l}(1-2\sqrt{3})$ b. $-\frac{2}{\sqrt{3}}\frac{GMm}{l}(1+2\sqrt{3})$

c. $-\frac{\sqrt{3}}{2}\frac{GMm}{l}(1-2\sqrt{3})$ d. $-\frac{\sqrt{3}}{2}\frac{GMm}{l}(1+2\sqrt{3})$

88. In problem 86, what is the gravitational field strength at the location of m?

a. $\frac{GM}{l^2}$ b. $\frac{4GM}{l^2}$

c. $\frac{4GM}{3l^2}$ d. $\frac{GM}{3l^2}$

89. Figure 5.42 shows a planet in an elliptical orbit around the Sun S. Where is the kinetic energy of the planet maximum?

a. P_1 b. P_2 c. P_3 d. P_4

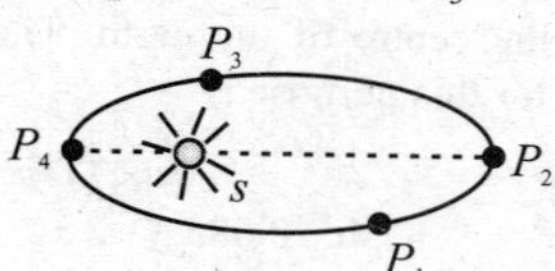

Fig. 5.42

90. The escape velocity corresponding to a planet of mass M and radius R is 50 km s^{-1}. If the planet's mass and radius were $4M$ and R, respectively, then the corresponding escape velocity would be

a. 100 km s^{-1} b. 50 km s^{-1}

c. 200 km s^{-1} d. 25 km s^{-1}

91. A planet is revolving around the Sun in an elliptical orbit. Its closest distance from the Sun is r and farthest distance is R. If the orbital velocity of the planet closest to the Sun is v, then what is the velocity at the farthest point?

a. $\frac{vr}{R}$ b. $\frac{vR}{r}$ c. $v\sqrt{\frac{r}{R}}$ d. $v\sqrt{\frac{R}{r}}$

92. Two planets have radii r_1 and r_2 and their densities are ρ_1 and ρ_2, respectively. The ratio of acceleration due to gravity on them will be

a. $r_1\rho_1 : r_2\rho_2$ b. $r_1\rho_1^2 : r_2\rho_2^2$

c. $r_1^2\rho_1 : r_2^2\rho_2$ d. $r_1\rho_2 : r_2\rho_1$

93. The given figure shows the motion of a planet around the Sun S in an elliptical orbit with the Sun at the focus. The shaded areas A and B are also shown in the figure which can be assumed to be equal. If t_1 and t_2 represent the time taken for the planet to move from a to b and c to d, respectively, then

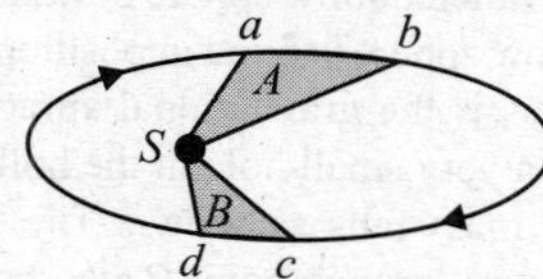

Fig. 5.43

a. $t_1 < t_2$

b. $t_1 > t_2$

c. $t_1 = t_2$

d. from the given information the relation between t_1 and t_2 cannot be determined

94. A body is fired with a velocity of magnitude $\sqrt{gR} < V < \sqrt{2gR}$ at an angle of 30° with the radius vector of the earth. If at the highest point, the speed of the body is $V/4$, the maximum height attained by the body is equal to

a. $\frac{V^2}{8g}$ b. R

c. $\sqrt{2}R$ d. none of these

95. A system of binary stars of masses m_A and m_B are moving in circular orbits of radii r_A and r_B, respectively. If T_A and T_B are the time periods of masses m_A and m_B, respectively, then

a. $\frac{T_A}{T_B} = \left(\frac{r_A}{r_B}\right)^{\frac{3}{2}}$ b. $T_A > T_B$ (if $r_A > r_B$)

c. $T_A > T_B$ (if $m_A > m_B$) d. $T_A = T_B$

96. Two spherical bodies of masses m and $5M$ and radii R and $2R$, respectively, are released in free space with initial separation between their centres equal to $12R$. If they attract each other due to gravitational force only, then the distance covered by the smaller body just before collision is

a. 2.5R b. 4.5R

c. 7.5R d. 1.5R

97. The change in the value of g at a height 'h' above the surface of earth is the same as at a depth 'd' below the earth. When both d and h are much smaller than the radius of earth, then which one of the following is correct?

a. $d = \frac{h}{2}$ b. $d = \frac{3h}{2}$

c. $d = 2h$ d. $d = h$

98. A particle of mass 10 g is kept on the surface of a uniform sphere of mass 100 kg and radius 10 cm. Find the work done against the gravitational force between them to take the particle far away from the sphere.

a. 13.34×10^{-10} J b. 3.33×10^{-10} J

c. 6.67×10^{-9} J d. 6.67×10^{-10} J

99. A tunnel is dug along the diameter of the earth (radius R and mass M). There is a particle of mass 'm' at the centre of the tunnel. The minimum velocity given to the particle so that it just reaches to the surface of the earth is

a. $\sqrt{\frac{GM}{R}}$ b. $\sqrt{\frac{GM}{2R}}$

c. $\sqrt{\frac{2GM}{R}}$

d. it will reach with the help of a negligible velocity.

100. A cavity of radius $R/2$ is made inside a solid sphere of radius R. The centre of the cavity is located at a distance $R/2$ from the centre of the sphere. The gravitational force on a particle of mass 'm' at a distance $R/2$ from the centre of the sphere on the line joining both the centres of the sphere and the cavity is (opposite to the centre of the cavity)
[Here $g = (GM)/R^2$, where M is the mass of the sphere]

a. $\frac{mg}{2}$ **b.** $\frac{3mg}{8}$
c. $\frac{mg}{16}$ **d.** none of these

101. A satellite is seen after each 8 h over the equator at a place on the earth when its sense of rotation is opposite to the earth. The time interval after which it can be seen at the same place when the sense of rotation of the earth and the satellite is the same will be

a. 8 h **b.** 12 h **c.** 24 h **d.** 6 h

102. Four similar particles of mass m are orbiting in a circle of radius r in the same angular direction because of their mutual gravitational attractive force. Velocity of a particle is given by

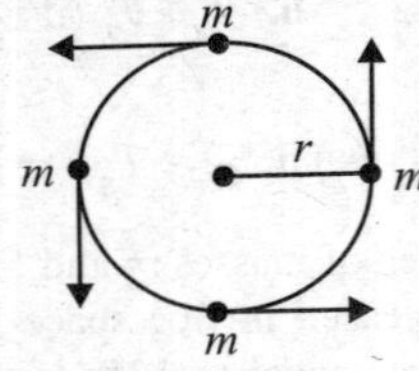

Fig. 5.44

a. $\left[\frac{Gm}{r}\left(\frac{1+2\sqrt{2}}{4}\right)\right]^{\frac{1}{2}}$ **b.** $\sqrt{\frac{Gm}{r}}$
c. $\sqrt{\frac{Gm}{r}}(1+2\sqrt{2})$ **d.** $\left[\frac{1}{2}\frac{Gm}{r}\left(\frac{1+2\sqrt{2}}{2}\right)\right]^{\frac{1}{2}}$

103. The gravitational potential of two homogenous spherical shells A and B of same surface density at their respective centres are in the ratio 3:4. If the two shells coalesce into a single one such that surface charge density remains the same, then the ratio of potential at an internal point of the new shell to shell A is equal to

a. 3:2 **b.** 4:3 **c.** 5:3 **d.** 5:6

104. A point P is on the axis of a fixed ring of mass M and radius R, at a distance $2R$ from the centre O. A small particle starts from P and reaches O under the gravitational attraction only. Its speed at O will be

a. zero **b.** $\sqrt{\frac{2GM}{R}}$
c. $\sqrt{\frac{2GM}{R}(\sqrt{5}-1)}$ **d.** $\sqrt{\frac{2GM}{R}\left(1-\frac{1}{\sqrt{5}}\right)}$

105. The percentage change in the acceleration of the earth towards the Sun from a total eclipse of the Sun to the point where the Moon is on a side of earth directly opposite to the Sun is

a. $\frac{M_s}{M_m}\frac{r_2}{r_1} \times 100$ **b.** $\frac{M_s}{M_m}\left(\frac{r_2}{r_1}\right)^2 \times 100$
c. $2\left(\frac{r_1}{r_2}\right)^2 \frac{M_m}{M_s} \times 100$ **d.** $\left(\frac{r_1}{r_2}\right)^2 \frac{M_m}{M_s} \times 100$

106. A comet is in highly elliptical orbit around the Sun. The period of the comet's orbit is 90 days. Some statements are given regarding the collision between the comet and the earth. Mark the correct statement. [Mass of the Sun = 2×10^{30} kg, mean distance between the earth and the Sun = 1.5×10^{11} m.]

a. Collision is there.
b. Collision is not possible.
c. Collision may or may not be there.
d. Enough information is not given.

107. Imagine that you are in a spacecraft orbiting around the earth in a circle of radius 7000 km (from the centre of the earth). If you decrease the magnitude of mechanical energy of the spacecraft – earth system by 10% by firing the rockets, then what is the greatest height you can take your spacecraft above the surface of the earth? [R_e = 6400 km]

a. 6400 km **b.** 540 km **c.** 2140 km **d.** 3000 km

108. A satellite of mass m is in an elliptical orbit around the earth. The speed of the satellite at its nearest position is $(6GM_e)/(5r)$ where r is the perigee (nearest point) distance from the centre of the earth. It is desired to transfer the satellite to the circular orbit of radius equal to its apogee (farthest point) distance from the centre of the earth. The change in orbital speed required for this purpose is

a. $0.35\sqrt{\frac{GM_e}{r}}$ **b.** $0.085\sqrt{\frac{GM_e}{r}}$
c. $\sqrt{\frac{2GM_e}{r}}$ **d.** zero

109. Two rings having masses M and $2M$, respectively, having the same radius are placed coaxially as shown in the figure.

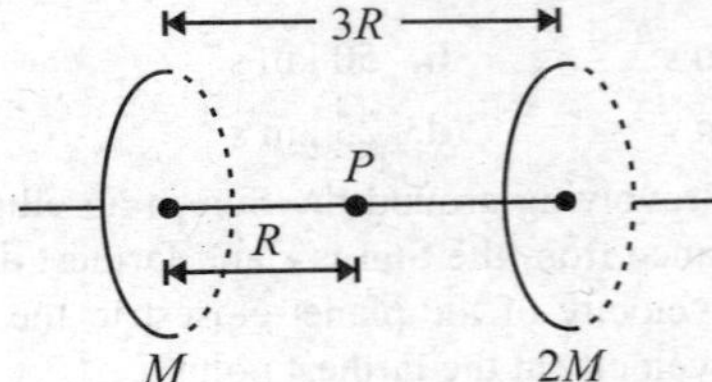

Fig. 5.45

If the mass distribution on both the rings is non-uniform, then the gravitational potential at point P is

a. $-\frac{GM}{R}\left[\frac{1}{\sqrt{2}}+\frac{2}{\sqrt{5}}\right]$
b. $-\frac{GM}{R}\left[1+\frac{2}{2}\right]$
c. zero
d. cannot be determined from the given information

110. A point mass m is released from rest at a distance of $3R$ from the centre of a thin-walled hollow sphere of radius R and mass M as shown. The hollow sphere is fixed in position and the only force on the point mass is the gravitational attraction of the hollow sphere. There is a very small hole in the hollow sphere through which the point mass falls as shown. The velocity of a point mass when it passes through point P at a distance $R/2$ from the centre of the sphere is

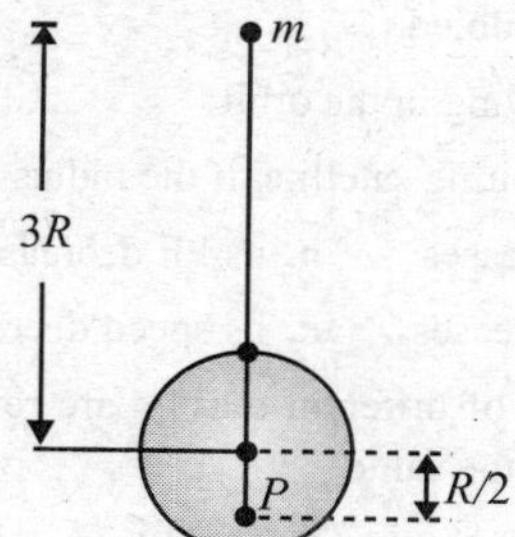

Fig. 5.46

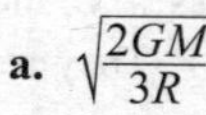

a. $\sqrt{\dfrac{2GM}{3R}}$ **a.** $\sqrt{\dfrac{5GM}{3R}}$

c. $\sqrt{\dfrac{25GM}{24R}}$ **d.** none of these

111. A body is thrown from the surface of the earth with velocity $(gR_e)/2$, where R_e is the radius of the earth at some angle from the vertical. If the maximum height reached by the body is $R_e/4$, then the angle of projection with the vertical is

a. $\sin^{-1}\left(\dfrac{\sqrt{5}}{4}\right)$ **b.** $\cos^{-1}\left(\dfrac{\sqrt{5}}{4}\right)$

c. $\sin^{-1}\left(\dfrac{\sqrt{3}}{2}\right)$ **d.** none of these

112. The gravitational force exerted by the Sun on the Moon is about twice as great as the gravitational force exerted by the earth on the Moon, but still Moon is not escaping from the gravitational influence of the earth. Mark the option which correctly explains the above system.

a. Escape speed is independent of the direction in which it is projected.

b. The rotational effect of the earth plays a role in computation of escape speed, however small it may be.

c. A body thrown in the eastward direction has less escape speed.

d. None of the above.

113. A tunnel has been dug into a solid sphere of non-uniform mass density as shown in the figure.

As one moves from A to B, the magnitude of gravitational field intensity

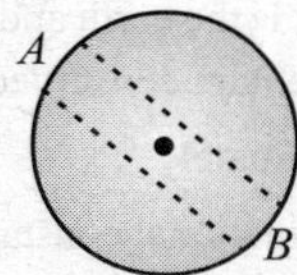

Fig. 5.47

a. will continuously decrease

b. will decrease up to the centre of the sphere and then increase

c. may increase or decrease

d. will continuously increase

114. A ring having non-uniform distribution of mass M and radius R is being considered. A point mass m_0 is taken slowly towards the ring. In doing so, work done by the external force against the gravitational force exerted by ring is

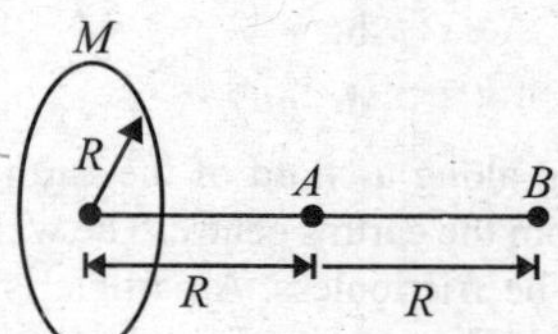

Fig. 5.48

a. $\dfrac{GMm_0}{\sqrt{2}R}$

b. $\dfrac{GMm_0}{R}\left[\dfrac{1}{\sqrt{2}}-\dfrac{1}{\sqrt{5}}\right]$

c. $\dfrac{GMm_0}{R}\left[\dfrac{1}{\sqrt{5}}-\dfrac{1}{\sqrt{2}}\right]$

d. It is not possible to find the required work as the nature of distribution of mass is not known.

115. An artificial satellite of the earth is launched in circular orbit in the equatorial plane of the earth and the satellite is moving from west to east. With respect to a person on the equator, the satellite is completing one round trip in 24 h. Mass of the earth is $M = 6 \times 10^{24}$ kg. For this situation, the orbital radius of the satellite is

a. 2.66×10^4 km **b.** 6400 km

c. 36,000 km **d.** 29,600 km

116. A satellite is orbiting around the earth in a circular orbit of radius r. A particle of mass m is projected from the satellite in a forward direction with a velocity $v = 2/3$ times the orbital velocity (this velocity is given w.r.t. earth). During subsequent motion of the particle, its minimum distance from the centre of earth is

a. $\dfrac{r}{2}$ **b.** r **c.** $\dfrac{2r}{3}$ **d.** $\dfrac{4r}{5}$

Multiple Correct Answers Type

Solutions on page 5.67

1. Figure 5.49 shows the kinetic energy (E_k) and potential energy (E_p) curves for a two-particle system. Name the point at which the system is a bound system.

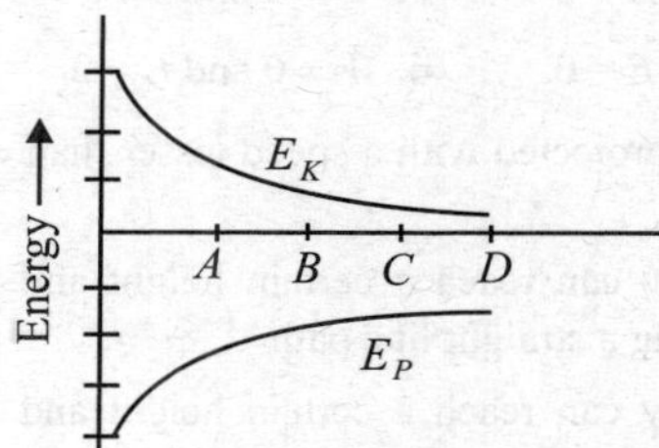

Fig. 5.49

a. A **b.** B **c.** C **d.** D

2. A double star is a system of two stars of masses m and $2m$, rotating about their centre of mass only under their mutual gravitational attraction. If r is the separation between these two stars then their time period of rotation about their centre of mass will be proportional to

a. $r^{\frac{3}{2}}$ b. r

c. $m^{\frac{1}{2}}$ d. $m^{-\frac{1}{2}}$

3. A tunnel is dug along a chord of the earth at a perpendicular distance $R/2$ from the earth's centre. The wall of the tunnel may be assumed to be frictionless. A particle is released from one end of the tunnel. The pressing force by the particle on the wall, and the acceleration of the particle vary with x (distance of the particle from the centre) according to

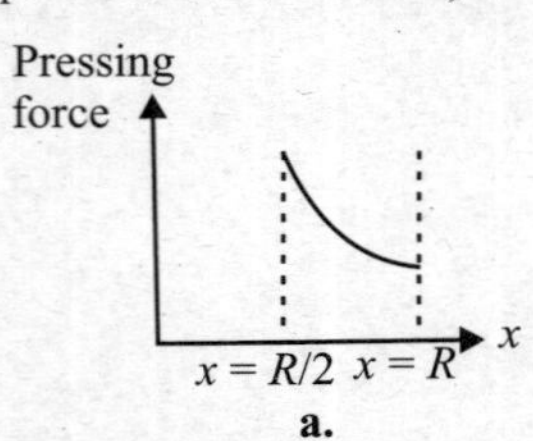

a.

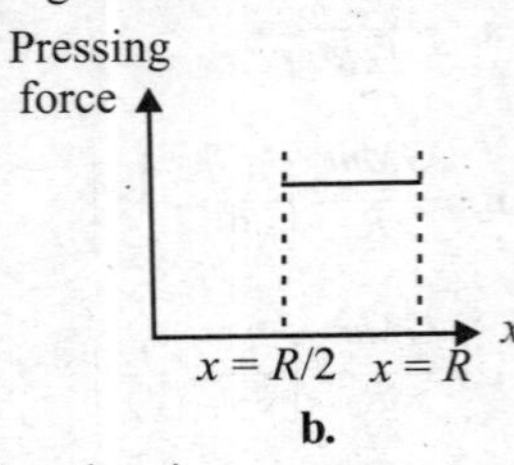

b.

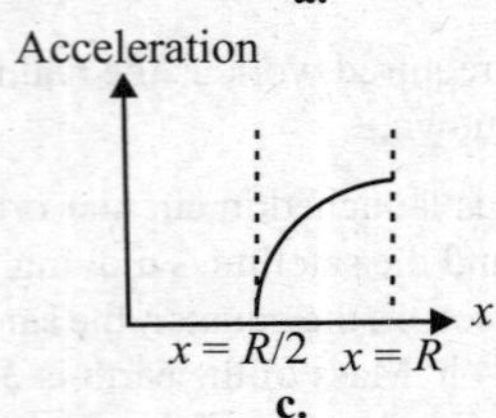

c.

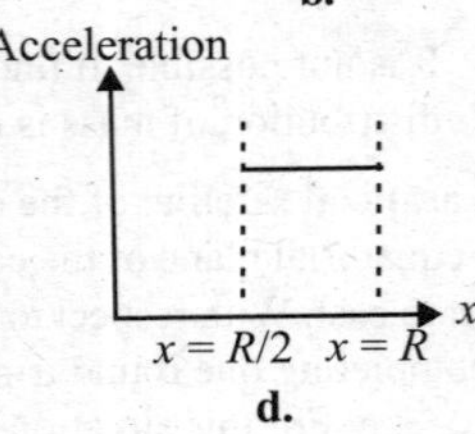

d.

4. Consider a planet moving in an elliptical orbit around the Sun. The work done on the planet by the gravitational force of the Sun
 a. is zero in any small part of the orbit
 b. is zero in some parts of the orbit
 c. is zero in complete revolution
 d. is zero in no part of the motion

5. Which of the following statements are true about acceleration due to gravity?
 a. g decreases in moving away from the centre if $r > R$.
 b. g decreases in moving away from the centre if $r < R$.
 c. g is zero at the centre of earth.
 d. g decreases if earth stops rotating on its axis.

6. Let V and E denote the gravitational potential and gravitational field at a point, respectively. It is possible to have
 a. $V =$ and $E = 0$ b. $V = 0$ and $E \neq 0$
 c. $V \neq$ and $E = 0$ d. $V \neq 0$ and $E \neq 0$

7. If a body is projected with a speed lesser than escape velocity, then
 a. the body can reach a certain height and may fall down following a straight line path
 b. the body can reach a certain height and may fall down following a parabolic path
 c. the body may orbit the earth in a circular orbit
 d. the body may orbit the earth in an elliptical orbit

8. An orbiting satellite will escape if
 a. its speed is increased by 41%
 b. its speed in the orbit is made $\sqrt{(1.5)}$ times of its initial value
 c. its KE is doubled
 d. it stops moving in the orbit

9. In case of an orbiting satellite, if the radius of orbit is decreased
 a. its KE decreases b. its PE decreases
 c. its ME decreases d. its speed decreases

10. If two satellites of different masses are revolving in the same orbit, they have the same
 a. angular momentum b. energy
 c. time period d. speed

11. A double star consists of two stars having masses M and $2M$. The distance between their centres is equal to r. They revolve under their mutual gravitational interaction. Then, which of the following statements are not correct?
 a. Heavier star revolves in orbit of radius $2r/3$.
 b. Both the stars revolve with the same speed, period of which is equal to $(2\pi/r^3)(2GM^2/3)$.
 c. Kinetic energy of the heavier star is twice that of the other star.
 d. None of the above

12. Mark the correct statements.
 a. Gravitational potential at the centre of curvature of a thin hemispherical shell of radius R and mass M is equal to GM/R.
 b. Gravitational field strength at a point lying on the axis of a thin, uniform circular ring of radius R and mass M is equal to $GMx/[(R^2 + x^2)^{3/2}]$, where x is distance of that point from the centre of the ring.
 c. Newton's law of gravitation for gravitational force between two bodies is applicable only when bodies have spherical symmetric distribution of mass.
 d. None of these

13. Suppose an earth satellite, revolving in a circular orbit experiences a resistance due to cosmic dust. Then
 a. its kinetic energy will increase
 b. its potential energy will decrease
 c. it will spiral towards the earth and in the process its angular momentum will remain conserved
 d. it will burn off ultimately

14. Suppose universal gravitational constant starts to decrease, then
 a. length of the day on the earth will increase
 b. length of the year will increase
 c. the earth will follow a spiral path of decreasing radius
 d. kinetic energy of the earth will decrease

15. An astronaut, inside an earth satellite, experiences weightlessness because
 a. no external force is acting on him
 b. he is falling freely

c. no reaction is exerted by the floor of the satellite

d. he is far away from the earth's surface

16. Suppose a smooth tunnel is dug along a straight line joining two points on the surface of the earth and a particle is dropped from rest at its one end. Assume that mass of the earth is uniformly distributed over its volume. Then, which of the following statements are not correct?

a. The particle will emerge from the other end with velocity$(GM_e)/(2R_e)$, where M_e and R_e are earth's mass and radius, respectively.

b. The particle will come to rest at the centre of the tunnel because at this position, the particle is closest to the earth's centre.

c. Potential energy of the particle will be equal to zero at centre of the tunnel if it is along a diameter.

d. Acceleration of the particle will be proportional to its distance from the mid-point of the tunnel.

17. If the radius of the earth suddenly decreases to 80% of its present value, the mass of the earth remaining the same, the value of the acceleration due to gravity will

a. remain unchanged **b.** become 9.8/0.64 m s^{-2}

c. increase by 36% **d.** increase by about 56%

18. Choose the correct statements from the following:

a. The gravitational forces between two particles are an action and reaction pair.

b. Gravitational constant (G) is scalar but acceleration due to gravity (g) is a vector.

c. The values of G and g are to be determined experimentally.

d. G and g are constant everywhere.

19. Choose the correct statements from the following:

a. The magnitude of the gravitational force between two bodies of mass 1 kg each and separated by a distance of 1 m is 9.8 N.

b. The higher the value of the escape velocity for a planet, the higher is the abundance of the lighter gases in its atmosphere.

c. The gravitational force of attraction between two bodies of ordinary mass is not noticeable because the value of the gravitational constant is extremely small.

d. Force of friction arises due to gravitational attraction.

20. Choose the incorrect statements from the following:

a. It is possible to shield a body from the gravitational field of another body by using a thick shielding material between them.

b. The escape velocity of a body is independent of the mass of the body and the angle of projection.

c. The acceleration due to gravity increases due to the rotation of the earth.

d. The gravitational force exerted by the earth on a body is greater than that exerted by the body on the earth.

21. A satellite is orbiting the earth, if its distance from the earth is increased, its

a. angular velocity would increase

b. linear velocity would increase

c. angular velocity would decrease

d. time period would increase

22. Which of the following statements are true? For a particle on the surface of the earth:

a. the linear speed is minimum at the equator

b. the angular speed is maximum at the equator

c. the linear speed is minimum at the poles

d. the angular speed is 7.3×10^{-5} rad s^{-1} at the equator

23. For two satellites at distances R and $7R$ above the earth's surface, the ratio of their

a. total energies is 4 and potential and kinetic energies is 2

b. potential energies is 4

c. kinetic energies is 4

d. total energies is 4

24. If both the mass and radius of the earth decrease by 1%, the value of

a. acceleration due to gravity would decrease by nearly 1%

b. acceleration due to gravity would increase by 1%

c. escape velocity from the earth's surface would decrease by 1%

d. the gravitational potential energy of a body on earth's surface will remain unchanged

25. An object is taken from a point P to another point Q in a gravitational field:

a. assuming the earth to be spherical, if both P and Q lie on the earth's surface, the work done is zero

b. if P is on the earth's surface and Q above it, the work done is minimum when it is taken along the straight line PQ

c. the work done depends only on the position of P and Q and is independent of the path along which the particle is taken

d. there is no work done if the object is taken from P to Q and then brought back to P along any path

26. A small mass m is moved slowly from the surface of the earth to a height h above the surface. The work done (by an external agent) in doing this is

a. mgh, for all values of h

b. mgh, for $h \ll R$

c. $1/2\ mgR$, for $h = R$

d. $-1/2\ mgR$, for $h = R$

27. Which of the following are correct?

a. If R is the radius of a planet, g is the acceleration due to gravity, the mean density of the planet is $3g/4\pi GR$.

b. Acceleration due to gravity is a universal constant.

c. The escape velocity of a body from earth is 11.2 km s^{-1}. The escape velocity from a planet which has double the mass of earth and half its radius is 22.4 km s^{-1}.

d. The ratio of gravitational mass and inertial mass of a body at the surface of earth is 1.

28. Which of the following are correct?

a. An astronaut going from the earth to the Moon will experience weightlessness once.

b. When a thin uniform spherical shell gradually shrinks maintaining its shape, the gravitational potential at its centre decreases.

c. In the case of a spherical shell, the plot of V versus r is continuous.

d. In the case of a spherical shell, the plot of gravitational field intensity I versus r is continuous.

29. Which of the following are not correct?

a, The escape velocity for the Moon is 6 km s^{-1}.

b. The escape velocity from the surface of Moon is v. The orbital velocity for a satellite to orbit very close to the surface of Moon is $v/2$.

c. When an earth satellite is moved from one stable orbit to a further stable orbit, the gravitational potential energy increases.

d. The orbital velocity of a satellite revolving in a circular path close to the planet is independent of the density of the planet.

30. Which of the following are correct?

a. Out of electrostatic, electromagnetic, nuclear and gravitational interactions, the gravitational interaction is the weakest.

b. If the earth were to rotate faster than its present speed, the weight of an object would decrease at the equator but remain unchanged at the poles.

c. The mass of the earth in terms of g, R and G is (gR^2/G).

d. If the earth stops rotating in its orbit around the Sun there will be no variation in the weight of a body on the surface of earth.

31. Two satellites S_1 and S_2 are revolving around the earth in coplanar concentric orbits in the opposite sense. At $t = 0$, the positions of satellites are shown in the diagram. The periods of S_1 and S_2 are 4 h and 24 h, respectively. The radius of orbit of S_1 is 1.28×10^4 km. For this situation, mark the correct statement(s).

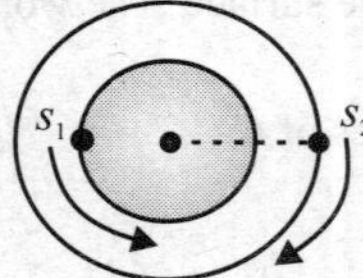

Fig. 5.50

a. The angular velocity of S_2 as observed by S_1 at $t = 12\ h$ is 0.468π rad s^{-1}.

b. The two satellites are closest to each other for the first time at $t = 12$ h and then after every 24 h they are closest to each other.

c. The orbital velocity of S_1 is $0.64\pi \times 10^4$ km.

d. The velocity of S_1 relative to S_2 is continuously changing in magnetic and direction both.

32. Consider two satellites A and B of equal mass m, moving in the same circular orbit about the earth, but in the opposite sense as shown in Fig. 5.51 The orbital radius is r. The satellites undergo a collision which is perfectly inelastic. For this situation, mark out the correct statement(s). [Take mass of earth as M]

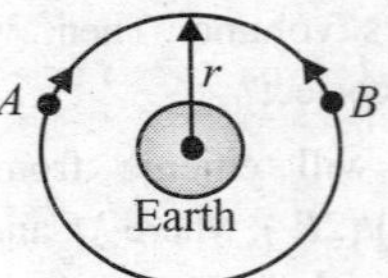

Fig. 5.51

a. The total energy of the two satellites plus earth system just before collision is $-(GMm)/r$.

b. The total energy of the two satellites plus earth system just after collision is $-(2GMm)/r$.

c. The total energy of two satellites plus earth system just after collision is $-(GMm)/2r$.

d. The combined mass (two satellites) will fall towards the earth just after collision.

Assertion-Reasoning Type

Solutions on page 5.70

In the following questions, each question contains Statement I (Assertion) and Statement II (Reason). Each question has four choices (a), (b), (c) and (d) out of which *only one* is correct.

(a) Statement I is True, Statement II is True; Statement II is a correct explanation for Statement I.

(b) Statement I is True, Statement II is True; Statement II is **NOT** a correct explanation for Statement I.

(c) Statement I is True, Statement II is False.

(d) Statement I is False, Statement II is True.

1. Statement I: For the planets orbiting around the Sun, angular speed, linear speed and KE change with time, but angular momentum remains constant.

Statement II: No torque is acting on the rotating planet. So its angular momentum is constant

2. Statement I: For a satellite revolving very near to the earth's surface the time period of revolution is given by 1 h 24 min.

Statement II: The period of revolution of a satellite depends only upon its height above the earth's surface.

3. Statement I: Kepler's second law can be understood by conservation of angular momentum principle.

Statement II: Kepler's second law is related with areal velocity which can further be proved to be based on conservation of angular momentum as $(dA/dt) = (r^2\omega)/2$.

4. Statement I: The force of gravitation between a sphere and a rod of mass M_2 is $= (GM_1M_2)/r$.

Statement II: Newton's law of gravitation holds correct for point masses.

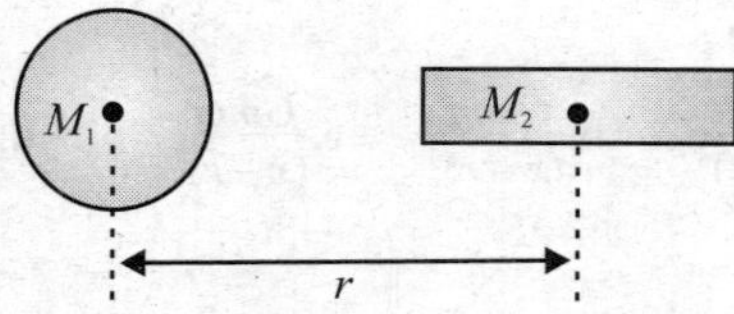

Fig. 5.52

5. **Statement I:** The value of escape velocity from the surface of earth at 30º and 60º is $v_1 = 2v_e$, $v_2 = 2/3\ v_e$.

Statement II: The value of escape velocity is independent of angle of projection.

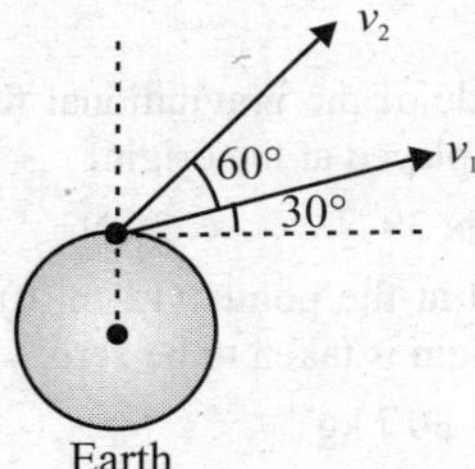

Fig. 5.53

6. **Statement I:** If the earth suddenly stops rotating about its axis, then the acceleration due to gravity will become the same at all the places.

Statement II: The value of acceleration due to gravity is independent of rotation of the earth.

7. **Statement I:** The magnitude of the gravitational potential at the surface of solid sphere is less than that of the centre of sphere.

Statement II: Due to the solid sphere, the gravitational potential is the same within the sphere.

8. **Statement I:** The smaller the orbit of a planet around the Sun, the shorter is the time it takes to complete.

Statement II: According to Kepler's third law of planetary motion, square of time period is proportional to cube of mean distance from Sun.

9. **Statement I:** The value of acceleration due to gravity does not depend upon the mass of the body.

Statement II: Acceleration due to gravity is a constant quantity.

10. **Statement I:** In free space a uniform spherical planet of mass M has a smooth narrow tunnel along its diameter. This planet and another superdense small particle of mass M start approaching towards each other from rest under action of their gravitational forces. When the particle passes through the centre of the planet, sum of kinetic energies of both the bodies is maximum.

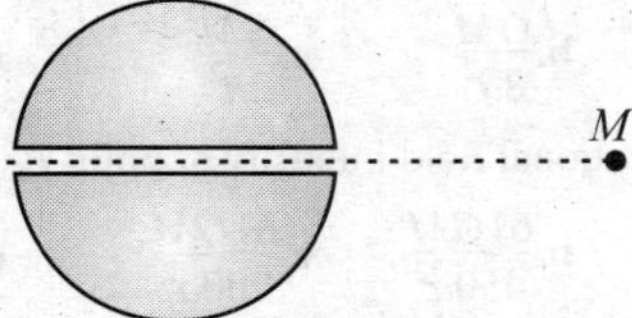

Fig. 5.54

Statement II: When the resultant of all forces acting on a particle or a particle like object (initially at rest) is constant in direction, the kinetic energy of the particle keeps on increasing.

11. **Statement I:** The earth does not retain hydrogen molecules and helium atoms in its atmosphere, but does retain much heavier molecules, such as oxygen and nitrogen.

Statement II: Lighter molecules in the atmosphere have translational speed that is greater or closer to escape speed of earth.

12. **Statement I:** It takes more fuel for a spacecraft to travel from the earth to the Moon than for the return trip.

Statement II: The point of zero gravitational field intensity due to the earth and the Moon is lying nearer to the Moon, i.e., in the diagram shown, for $r < r_0$, E_g is towards the earth's centre and for $r > r_0$, E_g is towards the Moon's centre, and at $r = r_0$, E_g is zero.

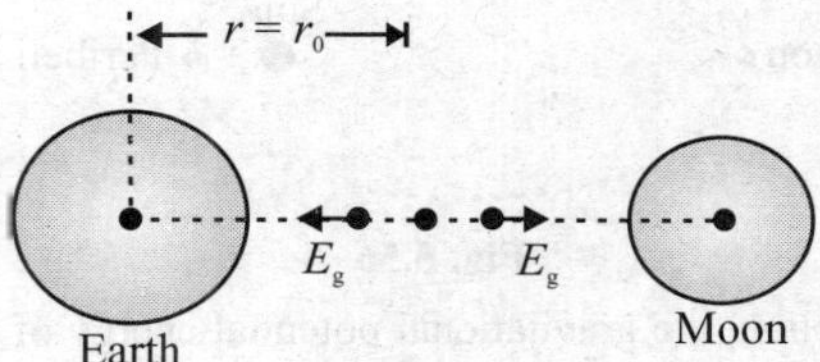

Fig. 5.55

13. **Statement I:** Consider a satellite moving in an elliptical orbit around the earth. As the satellite moves, the work done by the gravitational force of the earth on the satellite for any small part of the orbit is zero.

Statement II: KE of the satellite in the above described case is not constant as it moves around the earth.

14. **Statement I:** If a particle projected horizontally just above the surface of the earth with a speed greater than escape speed, then it will escape from gravitational influence of the earth. Assume that particle has a clear path.

Statement II: Escape velocity is independent of its direction.

15. **Statement I:** If time period of a satellite revolving in circular orbit in equatorial plane is 24 h, then it must be a geostationary satellite.

Statement II: Time period of a geostationary satellite is 24 h.

16. **Statement I:** Two satellites are following one another in the same circular orbit. If one satellite tries to catch another (leading one) satellite, then it can be done by increasing its speed without changing the orbit.

Statement II: The energy of earth–satellite system in circular orbit is given by $E = (-Gms)/(2a)$, where r is the radius of the circular orbit.

Comprehension Type

Solutions on page 5.71

For Problems 1–2

The minimum and maximum distances of a satellite from the centre of the earth are $2R$ and $4R$, respectively, where R is the radius of the earth and M is the mass of the earth.

1. The minimum and maximum speeds are

a. $\sqrt{\frac{GM}{9R}}, \sqrt{\frac{2GM}{R}}$ **b.** $\sqrt{\frac{GM}{5R}}, \sqrt{\frac{3GM}{2R}}$

c. $\sqrt{\frac{GM}{6R}}, \sqrt{\frac{2GM}{3R}}$ d. $\sqrt{\frac{GM}{3R}}, \sqrt{\frac{5GM}{2R}}$

2. The radius of curvature at the point of minimum distance is

a. $\sqrt{\frac{8R}{3}}$ b. $\sqrt{\frac{5R}{3}}$ c. $\sqrt{\frac{4R}{3}}$ d. $\sqrt{\frac{7R}{3}}$

For Problems 3–5

The orbit of Pluto is much more eccentric than the orbits of the other planets. That is, instead of being nearly circular, the orbit is noticeably elliptical. The point in the orbit nearest to the Sun is called the perihelion and the point farthest from the Sun is called the aphelion.

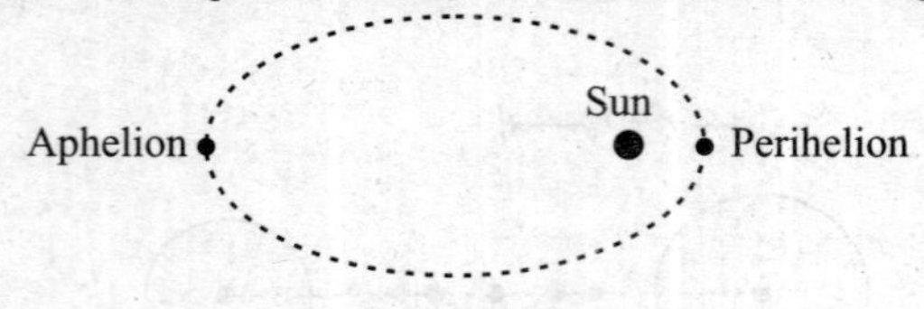

Fig. 5.56

3. At perihelion, the gravitational potential energy of Pluto in its orbit has

a. its maximum value
b. its minimum value
c. the same value as at every other point in the orbit
d. the value which depends on the sense of rotation

4. At perihelion, the mechanical energy of Pluto's orbit has

a. its maximum value
b. its minimum value
c. the same value as at every other point in the orbit
d. the value which depends on the sense of rotation

5. As Pluto moves from the perihelion to the aphelion, the work done by gravitational pull of Sun on Pluto is

a. is zero b. is positive
c. is negative d. depends on sense of rotation

For Problems 6–8

A solid sphere of mass m and radius r is placed inside a hollow thin spherical shell of mass M and radius R as shown in the figure. A particle of masss m' is placed on the line joining the two centres at a distance x from the point of contact of the sphere and the shell. Find the magnitude of the resultant gravitational force on this particle due to the sphere and the shell if

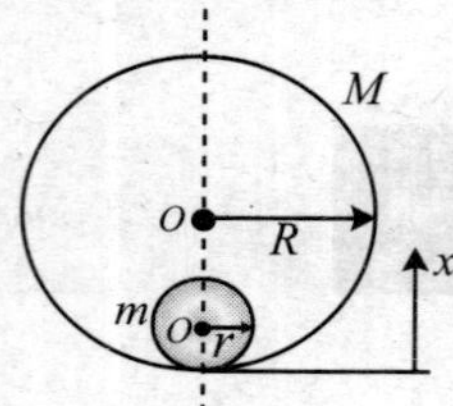

Fig. 5.57

6. $r < x < 2r$

a. $\frac{Gmm'(2r-x)}{2r^3}$ b. $\frac{Gmm'(x-r)}{2r^3}$

c. $\frac{Gmm'(x-r)}{r^3}$ d. $\frac{Gmm'(2x-r)}{r^3}$

7. $2r < x < 2R$

a. $\frac{Gmm'}{4(x-x)^2}$ b. $\frac{Gmm'}{(x-r)^2}$ c. $\frac{Gmm'}{(x-r)^3}$ d. $\frac{2Gmm'}{(x-r)^2}$

8. $x > 2R$

a. $\frac{2\,GMm'}{(x-r)^2} + \frac{Gmm'}{(x+r)^2}$ b. $\frac{GMm'}{2(x-R)^2} + \frac{2\,Gmm'}{(x-r)^2}$

c. $\frac{GMm'}{(x+R)^2} + \frac{Gmm'}{(x+r)^2}$ d. $\frac{GMm'}{(x-R)^2} + \frac{Gmm'}{(x-r)^2}$

For Problems 9–12

The gravitational field in a region is given by $\vec{E} = (5 \text{ N kg}^{-1})\,\vec{i} + (12 \text{ N kg}^{-1})\,\vec{j}$.

9. Find the magnitude of the gravitational force acting on a particle of mass 2 kg placed at the origin.

a. 26 N b. 30 N c. 20 N d. 35 N

10. Find the potential at the points (12 m, 0) and (0, 5 m) if the potential at the origin is taken to be zero.

a. -30 J kg^{-1}, -30 J kg^{-1}
b. -40 J kg^{-1}, -30 J kg^{-1}
c. -60 K kg^{-1}, -60 J kg^{-1}
d. -40 J kg^{-1}, -50 J kg^{-1}

11. Find the change in gravitational potential energy if a particle of mass 2 kg is taken from the origin to the point (12 m, 5 m).

a. -225 J b. -240 J c. -245 J d. -250 J

12. Find the change in potential energy if the particle is taken from (12 m, 0) to (0, 5 m).

a. -10 J b. -50 J c. zero d. -60 J

For Problems 13–14

A uniform metal sphere of radius a and mass M is surrounded by a thin uniform spherical shell of equal mass and radius $4a$. The centre of the shell falls on the surface of the inner sphere.

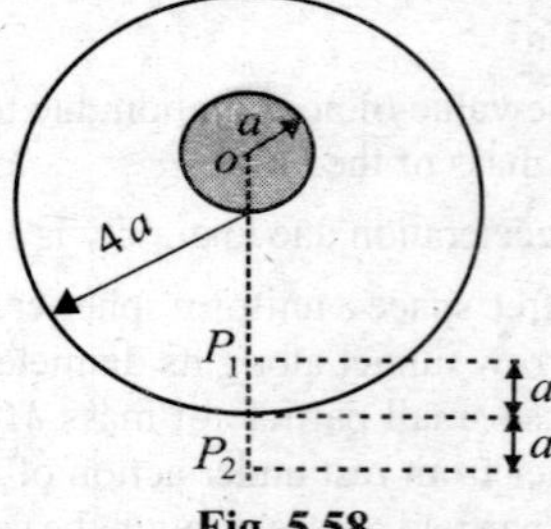

Fig. 5.58

13. Find the gravitational field intensity at point P_1.

a. $\frac{GM}{16a^2}$ b. $\frac{GM}{8a^2}$ c. $\frac{GM}{2a^2}$ d. $\frac{GM}{4a^2}$

14. Find the gravitational field intensity at point P_2.

a. $\frac{21GM}{900a^2}$ b. $\frac{61GM}{450a^2}$ c. $\frac{61GM}{900a^2}$ d. $\frac{61GM}{1800a^2}$

For Problems 15–17

In the graph shown, the PE of earth–satellite system is shown by a solid line as a function of distance r (the separation between earth's centre and satellite). The total energy of the two objects which may or may not be bounded to the earth are shown in the figure by dotted lines.

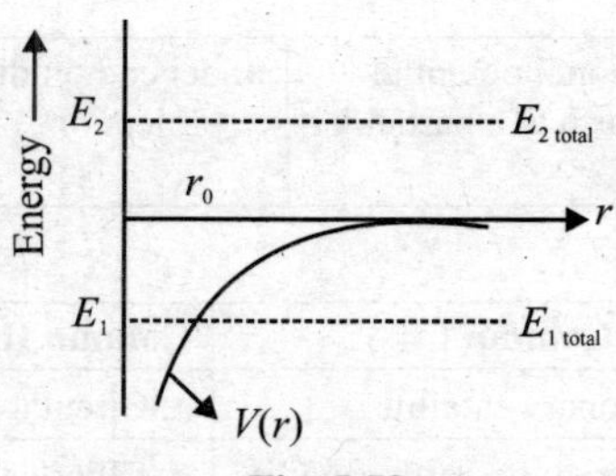

Fig. 5.59

Based on the above information, answer the following questions.

15. Mark the correct statement(s).

- **a.** The object having the total energy E_1 is bounded one.
- **b.** The object having the total energy E_2 is bounded one.
- **c.** Both the objects are bounded.
- **d.** Both the objects are unbounded.

16. If object having total energy E_1 is having the same PE curve as shown in the figure, then

- **a.** r_0 is the maximum distance of the object from the earth's centre
- **b.** this object and the earth system is bounded one
- **c.** the KE of the object is zero when $r = r_0$
- **d.** all the above

17. If both the objects have the same PE curve as shown in the figure, then

- **a.** for objects having total energy E_2, all values of r are possible
- **b.** for the object having total energy E_2, values of $r < r_0$ are only possible
- **c.** for the object having total energy E_1, all values of r are possible
- **d.** none of the above

For Problems 18–22

An unmanned satellite A and a spacecraft B are orbiting around the earth in the same circular orbit as shown.

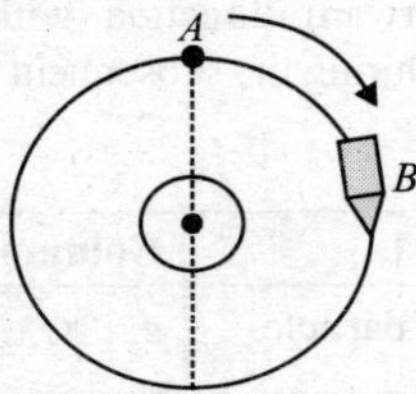

Fig. 5.60

The spacecraft is ahead of the satellite by some time. Let us consider that some technical problem has arisen in the satellite and the astronaut from B has made it correct. For this to be done docking of two (A and B) is required (in layman terms connecting A and B). To achieve this, the rockets of A have been fired in forward direction and docking takes place as shown in the figure below:

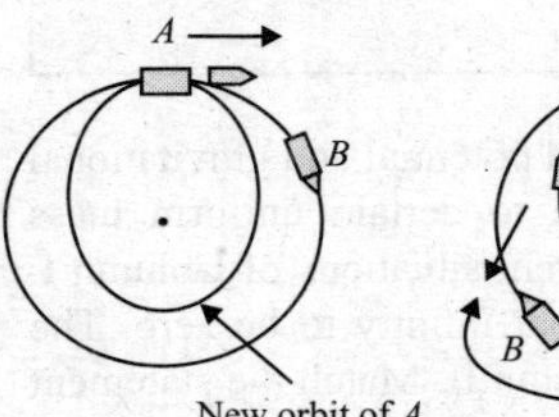

Fig. 5.61

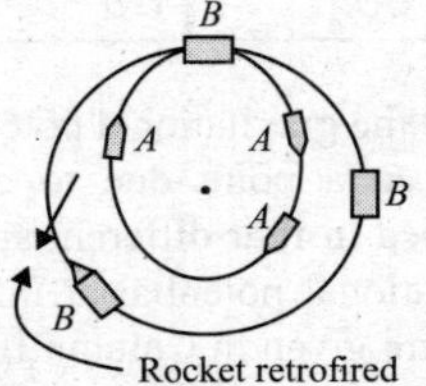

Fig. 5.62

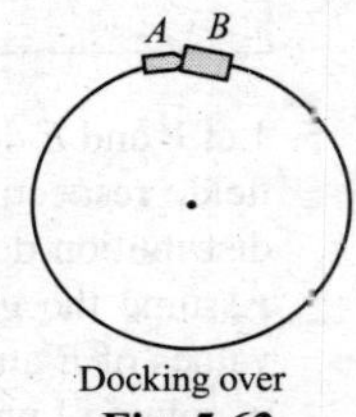

Fig. 5.63

Take mass of the earth = 5.98×10^{24} kg

Radius of the earth = 6400 km

Orbital radius = 9600 km

Mass of satellite A = 320 kg

Mass of spacecraft = 3200 kg

Assume that initially spacecraft B leads satellite A by 100 s, i.e., A arrives at any particular position after 100 s of B's arrival. Based on the above information answer the following questions.

18. To dock A and B in the above-described situation, one can use the rocket system of either one, i.e., either of A or of B. To accomplish docking in the minimum possible time which is the best way?

- **a.** To use rocket system of A.
- **b.** To use rocket system of B.
- **c.** Either (a) or (b).
- **d.** Information insufficient.

19. The initial total energy and time period of satellite are, respectively,

- **a.** -6.65×10^{10} J, 9358 s **b.** -6.65×10^{9} J, 9358 s
- **c.** -6.65×10^{10} J, 9140 s **d.** -6.65×10^{9} J, 9140 s

20. If from the base station, the rocket system of A has been operated, so that its rocket has been fired in forward direction thus reducing the speed of A by 0.5%, then which of the following will happen?

- **a.** Its orbit becomes elliptical with semi-major axis 9505.3 km.
- **b.** Its total energy becomes -6.714×10^{9} J.
- **c.** Its new time period becomes 9219.67 s.
- **d.** All of the above.

21. By doing the above operation, now by how much time is the satellite leading the spacecraft?

a. 38 s **b.** 138 s **c.** Lags by 38 s **d.** Lags by 138 s

22. After once returing to the original point, i.e., the place from where the rockets have been fired, in which direction and with what extent the rockets have to be fired from the satellite to again come back in the original orbit?

- **a.** Forward direction with the same extent.
- **b,** Backward direction with the same extent.
- **c.** Forward direction with the higher extent.
- **d.** Backward direction with the higher extent.

For Problems 23–27

The satellites when launched from the earth are not given the orbital velocity initially, a multistage rocket propeller carries the spacecraft up to its orbit and during each stage rocket has been fired to increase the velocity to acquire the desired velocity for a particular orbit. The last stage of the rocket brings the satellite in circular/elliptical (desired) orbit.

Consider a satellite of mass 150 kg in a low circular orbit. In this orbit, we cannot neglect the effect of air drag. This air opposes the motion of satellite and hence the total mechanical energy of earth–satellite system decreases. That means the total energy becomes more negative and hence the orbital radius decreases which causes the increase in KE When the satellite comes in the low enough orbit, excessive thermal energy generation due to air friction may cause the satellite to burn up. Based on the above information, answer the following questions.

23. It has been mentioned in the passage that as r decreases, E decreases but K increases. The increase in K is [E = total mechanical energy, r= orbital radius, K= kinetic energy]

a. due to increase in gravitational PE
b. due to decrease in gravitational PE
c. due to work done by air friction force
d. both (b) and (c)

24. If due to air drag, the orbital radius of the earth decreases from R to $R - \Delta R$, $\Delta R \ll R$, then the expression for increase in the orbital velocity Δv is

a. $\frac{\Delta r}{2}\sqrt{\frac{GM}{R^3}}$ **b.** $-\frac{\Delta R}{2}\sqrt{\frac{GM}{R^3}}$

c. $\Delta R\sqrt{\frac{GM}{R^3}}$ **d.** $-\Delta R\sqrt{\frac{GM}{R^3}}$

25. For information given in problem 24, the change in KE, ΔK is

a. $-\frac{GMm}{R^2}\times\Delta R$ **b.** $-\frac{GMm}{R^2}\times\Delta R$

c. $\frac{GMm}{2R^2}\times\Delta R$ **d.** $-\frac{GMm}{2R^2}\times\Delta R$

26. For information given in problem 24, the change in PE, ΔU is

a. $-\frac{GMm}{R^2}\times\Delta R$ **b.** $-\frac{GMm}{R^2}\times\Delta R$

c. $\frac{GMm}{2R^2}\times\Delta R$ **d.** $-\frac{GMm}{2R^2}\times\Delta R$

27. For information given in problem 24, the work done by air friction force W is

a. $-\frac{GMm}{R^2}\times\Delta R$ **b.** $-\frac{GMm}{R^2}\times\Delta R$

c. $\frac{GMm}{2R^2}\times\Delta R$ **d.** $-\frac{GMm}{2R^2}\times\Delta R$

Matching Column Type

Solutions on page 5.73

1. Considering the earth to be a homogeneous sphere but keeping in mind its spin, match the following:

Column I	Column II
i. Acceleration due to gravity	**a.** may change from point to point
ii. Orbital angular momentum of the earth as seen from a distant star	**b.** does not depend on direction of projection
iii. Escape velocity from the earth	**c.** remains constant
iv. Gravitational potential due to earth at a particular point	**d.** depend on direction of projection

2.

Column I	Column II
i. Geostationary satellite	**a.** Circular orbit
ii. Total energy of earth–satellite system is constant in	**b.** Elliptical orbit
iii. Angular momentum of satellite about centre of earth is constant in	**c.** Equatorial plane orbit
iv. Orbital speed of satellite is/may be constant	**d.** Non-equatorial plane orbit

3. An artificial satellite is in circular orbit around the earth. One of the rockets of the satellite is momentarily fired, the direction of firing of rocket is mentioned in Column I and corresponding change(s) are given in Column II. Match the entries of Column I with the entries of Column II

Column I	Column II
i. Towards the earth's centre	**a.** Orbit changes and becomes elliptical
ii. Away from the earth's centre	**b.** Orbit plane changes
iii. At right angle to the plane or orbit	**c.** Semi-major axis of orbit increases
iv. In forward direction	**d.** Energy of earth–satellite system increases

4. A satellite is revolving around the earth in a circular orbit of radius 'a' with velocity v_0. A particle is projected from the satellite in forward direction with relative velocity $V = [(\sqrt{5/4}) - 1]\, v_0$. During the subsequent motion of the particle, match the following:

Column I	Column II
i. Total energy of particle	**a.** $(3GM_e m)/a$
ii. Minimum distance of particle from the earth	**b.** $(5GM_e m)/a$
iii. Maximum distance of particle from the earth	**c.** $5a/3$
	d. $2a$
	e. $2a/3$
	f. a

5. Let V and E denote the gravitational potential and gravitational field, respectively, at a point due to certain uniform mass distribution described in four different situations of Column I. Assume the gravitational potential at infinity to be zero. The values of E and V are given in Column II. Match the statement in column I with the results in Column II.

Column I	Column II
i. At the centre of thin spherical shell	a. $E = 0$
ii. At the centre of solid sphere	b. $E \neq 0$
iii. A solid sphere has a non-concentric spherical cavity. At the centre of the spherical cavity	c. $V \neq 0$
iv. At the centre of the joining two point masses of equal magnitude	d. $V = 0$

6. For a planet orbiting about the Sun in a elliptical orbit, some incomplete statements regarding physical quantities are given in Column I, which can be completed by using the entries of Column II. Match the entries of Column I with the entries of Column II.

Column I	Column II
i. Maximum PE of the Sun–planet system	a. is at perihelioin
ii. Maximum speed of the planet	b. is at aphelion
iii. Minimum PE of Sun–planet system	c. is independent of mass of planet
iv. Minimum kinetic energy of planet	d. is independent of semi-major axis of orbit

7.

Column I	Column II
i. Gravitational potential	a. On the surface of planets with density ratio 1 : 2
ii. Escape velocity	b. Conservation of angular momentum
iii. Ratio of the acceleration due to gravity 1 : 3	c. Varies with the reference point
iv. Orbiting satellite	b. Does not depend on the angle
	e. Similar to an atom

Integer Answer Type

Solutions on page 5.74

1. A particle of mass m is subjected to an attractive central force of magnitude k/r^2, k being a constant. If at the instant when the particle is at an extreme position in its closed orbit, at a distance a from the centre of force, its speed is $(k/2ma)$, if the distance of other extreme position is 'b'. Find a/b.

2. The earth (mass = 10^{24} kg) revolves round the Sun with an angular velocity 2×10^{-7} rad s^{-1} in a circular orbit of radius 1.5×10^8 km. Find the force exerted by the Sun on the earth (in $\times 10^{21}$ N).

3. The density of a newly discovered planet is twice that of the earth. The acceleration due to gravity at the surface of the planet is equal to that at the surface of the earth. If the radius of the earth is R and the radius of the planet is R'. Then what is the value of R/R'?

4. Imagine a new planet having the same density as that of the earth but it is 3 times bigger than the earth in size. If the acceleration due to gravity on the surface of the earth is g and that on the new planet is g', then what is the value of g'/g?

Archives

Solutions on page 5.75

Fill in the Blanks Type

1. The numerical value of the angular velocity of rotation of the earth should be ________ rad s^{-1} in order to make the effective acceleration due to gravity equal to zero. **(IIT-JEE, 1984)**

2. According to Kepler's second law, the radius vector to a planet from the Sun sweeps out equal areas in equal intervals of time. This law is a consequence of the conservation of ________. **(IIT-JEE, 1985)**

3. A geostationary satellite is orbiting the earth at a height of $6R$ above the surface of the earth, where R is the radius of the earth. The time period of another satellite at a height of $2.5R$ from the surface of the earth is ________ hours. **(IIT-JEE, 1987)**

4. The masses and radii of the earth and the Moon are M_1, R_1 and M_2, R_2, respectively. Their centres are a distance d apart. The minimum speed with which a particle of mass m should be projected from a point midway between the two centres so as to escape to infinity is ________. **(IIT-JEE, 1988)**

5. A particle is projected vertically upwards from the surface of the earth (radius R_e) with a kinetic energy equal to half of the minimum value needed for it to escape. The height to which it rises above the surface of the earth is ________. **(IIT-JEE, 1987)**

6. The ratio of the earth's orbital angular momentum (about the Sun) to its mass is 4.4×10^{15} m^2 s^{-1}. The area enclosed by the earth's orbit is approximately ________ m^2. **(IIT-JEE, 1987)**

True or False Type

1. It is possible to put an artificial satellite into orbit in such a way that it will always remain directly over New Delhi. **(IIT-JEE, 1984)**

Multiple Choice Questions with One Correct Answer Type

1. If the radius of the earth were to shrink by one per cent, its mass remaining the same, acceleration due to gravity on the earth's surface would **(IIT-JEE, 1981)**
   a. decrease b. remain unchanged
   c. increase d. be zero

2. If g is the acceleration due to gravity on the earth's surface, the gain in the potential energy of an object of mass m raised from the surface of the earth to a height equal to the radius R of the earth is **(IIT-JEE, 1983)**
   a. $\frac{1}{2}mgR$ b. $2\,mgR$ c. mgR d. $\frac{1}{4}mgR$

3. Imagine a light planet revolving around a very massive star in a circular orbit of radius R with a speed of revolution T. If the gravitational force of attraction between the planet and the star is proportional to $R^{-5/2}$, then **(IIT-JEE, 1989)**
   a. T^2 is proportional to R^2
   b. T^2 is proportional to $R^{7/2}$
   c. T^2 is proportional to $R^{3/2}$
   d. T^2 is proportional to $R^{3.75}$

4. If the distance between the earth and the Sun were half its present value, the number of days in a year would have been
   a. 64.5 **b.** 129 **c.** 182.5 **d.** 730
   (IIT-JEE, 1996)

5. A satellite S is moving in an elliptical orbit around the earth. The mass of the satellite is very small compared to the mass of the earth. **(IIT-JEE, 1998)**
   a. The acceleration of S is always directed towards the centre of the earth.
   b. The angular momentum of S about the centre of the earth changes in direction, but its magnitude remains constant.
   c. The total mechanical energy of S varies periodically with time.
   d. The linear momentum of S remains constant in magnitude.

6. A simple pendulum has a time period T_1 when on the earth's surface and T_2 when taken to a height R above the earth's surface, where R is the radius of the earth. The value of T_2/T_1 is
   a. 1 **b.** $\sqrt{2}$ **c.** 4 **d.** 2
   (IIT-JEE, 2001)

7. A geostationary satellite orbits around the earth in a circular orbit of radius 36,000 km. Then, the time period of a spy satellite orbiting a few hundred km above the earth's surface (R_{earth} = 6,400 km) will approximately be **(IIT-JEE, 2002)**
   a. $\frac{1}{2}$ h **b.** 1 h **c.** 2 h **d.** 4 h

8. A double star system consists of two stars A and B which have time periods T_A and T_B. Radius R_A and R_B and mass M_A and M_B. Choose the correct option. **(IIT-JEE, 2006)**
   a. If $T_A > T_B$, then $R_A > R_B$
   b. If $T_A > T_B$, then $M_A > M_B$
   c. $\left(\frac{T_A}{T_B}\right)^2 = \left(\frac{R_A}{R_B}\right)^3$
   d. $T_A = T_B$

9. A spherical symmetric gravitational system of a particle has a mass density $\rho = \begin{cases} \rho_0 & \text{for } r \le R \\ 0 & \text{for } r > R \end{cases}$, where ρ_0 is a constant. A test mass can undergo circular motion under the influence of the gravitational field of particles. Its speed V as a function of distance r $(0 < r < \infty)$ from the centre of the system is represented by

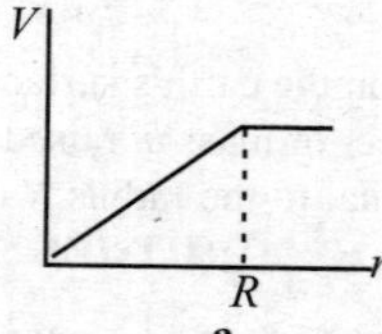

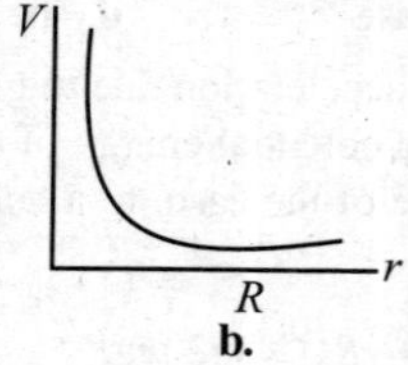

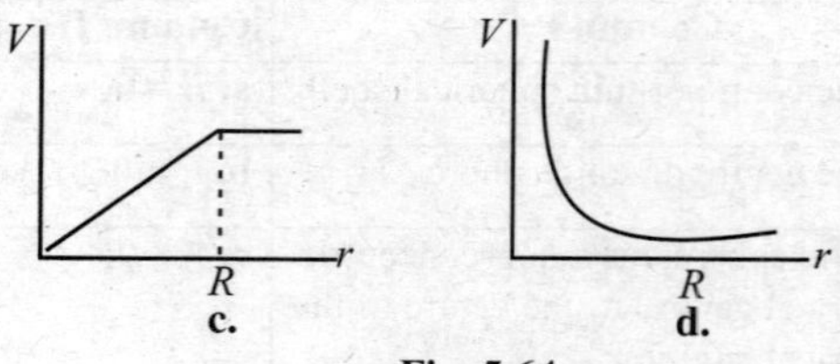

Fig. 5.64

(IIT-JEE, 2008)

10. A thin uniform annular disc (see the figure) of mass M has an outer radius $4R$ and an inner radius $3R$. The work required to take a unit mass from point P on its axis to infinity is **(IIT-JEE, 2010)**

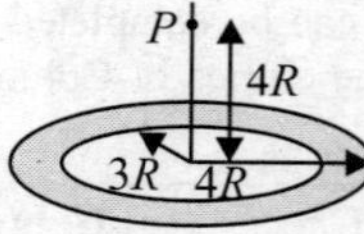

Fig. 5.65

   a. $\frac{2GM}{7R}(4\sqrt{2}-5)$ **b.** $-\frac{2GM}{7R}(4\sqrt{2}-5)$
   c. $\frac{GM}{4R}$ **d.** $\frac{2GM}{5R}(\sqrt{2}-1)$

11. A satellite is moving with a constant speed 'V' in a circular orbit about the earth. An object of mass 'm' is ejected from the satellite such that it just escapes from the gravitational pull of the earth. At the time of its ejection, the kinetic energy of the object is **(IIT-JEE, 2010)**
   a. $\frac{1}{2}mV^2$ **b.** mV^2
   c. $\frac{3}{2}mV^2$ **d.** $2\,mV^2$

Multiple Choice Questions with One or More than One Correct Answer Type

1. A solid sphere of uniform density and radius 4 units is located with its centre at the origin O of coordinates. Two spheres of equal radii 1 unit, with their centres at A (–2, 0,0) and B (2, 0, 0), respectively, are taken out of the solid leaving behind spherical cavities as shown in the figure.

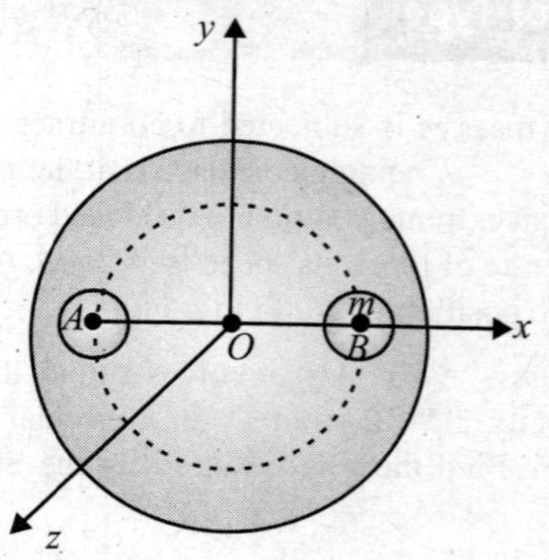

Fig. 5.66

Then

a. the gravitational force due to this object at the origin is zero

b. the gravitational force at point B (2, 0, 0) is zero

c. the gravitational potential is the same at all points of circle $y^2 + z^2 = 36$

d. the gravitational potential is the same at all points on the circle $y^2 + z^2 = 4$

(IIT-JEE, 1993)

2. The magnitudes of the gravitational field at distances r_1 and r_2 from the centre of a uniform sphere of radius R and mass m are F_1 and F_2, respectively. Then

a. $\frac{F_1}{F_2} = \frac{r_1}{r_2}$ if $r_1 < R$ and $r_2 < R$

b. $\frac{r_2^2}{r_2}$ if $r_1 > R$ and $r_2 > R$

c. $\frac{F_1}{F_2} = \frac{r_1}{r_2}$ if $r_1 > R$ and $r_2 > R$

d. $\frac{F_1}{F_2} = \frac{r_1^2}{r_2^2}$ if $r_1 < R$ and $r_2 < R$ **(IIT-JEE, 1994)**

3. A satellite S is moving in an elliptical orbit around the earth. The mass of the satellite is very small compared to the mass of the earth.

a. The acceleration of S is always directed towards the centre of the earth.

b. The angular momentum of S about the centre of the earth changes in direction, but its magnitude remains constant.

c. The total mechanical energy of S varies periodically with time.

d. The linear momentum of S remains constant in magnitude.

(IIT-JEE, 1998)

Assertion and Reasoning Type

In each of the questions, assertion is given by corresponding statement of reason, mark the correct answer.

a. If both assertion and reason are true and the reason is a correct explanation of the assertion.

b. If both assertion and reason are true but reason is not a correct explanation of assertion.

c. If assertion is true, but the reason is false.

d. If assertion is false but the reason is true.

1. Assertion: An astronaut in an orbiting space station above the earth experiences weightlessness.

Reason: An object moving around the earth under the influence of earth's gravitational force is in a state of 'free fall'.

(IIT-JEE, 2008)

Integer Answer Type

1. Gravitational acceleration on the surface of a planet is $\frac{\sqrt{6}}{11}g$, where g is the gravitational acceleration on the surface of the earth. The average mass density of the planet is 2/3 times that of the earth. If the escape speed on the surface of the earth is taken to be 11 kms^{-1}, then find the escape speed on the surface of the planet in kms^{-1}

(IIT-JEE, 2010)

Matching Column Type

1.

Column I	Column II
i. The earth moving in an elliptical orbit (only earth in system) Sun, Earth Fig. 5.67	**a.** Conservation of linear momentum along any direction
ii. A disc having translation and rotation motion both (both slipping) on a rough surface (only disc in system) Disc ω_0, $V_0 \neq \omega_0 R$, V_0 Fig. 5.68	**b.** Conservation of linear momentum along specific direction
iii. A sphere rolling without slipping on a curved surface (only the sphere in system) ω, $V = \omega R$, v, Rough surface Fig. 5.69	**c.** Conservation of angular momentum about any point in the space
iv. Projection of a particle from the surface of the earth (only particle in system) α, v Fig. 5.70	**d.** Conservation of angular momentum about a specific point in the space

ANSWERS AND SOLUTIONS

Subjective Type

1. Conservation of energy gives

$$\frac{1}{2}mv^2 - \frac{GMm}{R} = 0 - \frac{GMm}{(R+10R)}$$

$$v = \sqrt{\frac{20}{11}\frac{GM}{R}} = 1.07 \times 10^4 \text{ m s}^{-1}$$

2. Total energy of sky laboratory, $E = -\frac{GMm}{2r}$

$$E_1 = -\frac{GMm}{2(2R)} \text{ and } E_2 = -\frac{GMm}{2(3R)}$$

Therefore, the energy required,

$$\Delta E = E_2 - E_1 = GMm\left(\frac{1}{4R} - \frac{1}{6R}\right) = \frac{GMm}{12R}$$

$$= \frac{(R^2g)m}{12R} = \frac{mgR}{12} = 1.04 \times 10^{10} \text{ J}$$

3. Energy required to raise a satellite to height h,

$$E_1 = -\frac{GMm}{(R+h)} - \left(-\frac{GMm}{R}\right) = GMm\left(\frac{1}{R} - \frac{1}{R+h}\right) = \frac{GMmh}{R(R+h)}$$

Energy required to put the satellite in the orbit at that height = KE of the satellite

$$E_2 = \frac{1}{2}mv^2 = \frac{1}{2}m\left(\frac{GM}{r}\right) = \frac{GMm}{2r} = \frac{GMm}{2(R+h)}$$

$$\frac{E_1}{E_2} = \frac{2(R+h)h}{R(R+h)} = \frac{2h}{R} = \frac{2 \times (1600 \text{ km})}{(6400 \text{ km})} = \frac{1}{2} = 0.5$$

Clearly, the answer depends on height, if $h = 3200$ km

$$\frac{E_1}{E_2} = \frac{2h}{R} = \frac{2 \times (3200 \text{ km})}{(6400 \text{ km})} = 1.10$$

For height, $h = 4800$ km

$$\frac{E_1}{E_2} = \frac{2 \times (4800 \text{ km})}{(6400 \text{ km})} = \frac{3}{2} = 1.5$$

4. Slope of the displacement vector, $m_1 = \frac{3}{4}$

Slope of the force vector, $m_2 = -\frac{4}{3}$

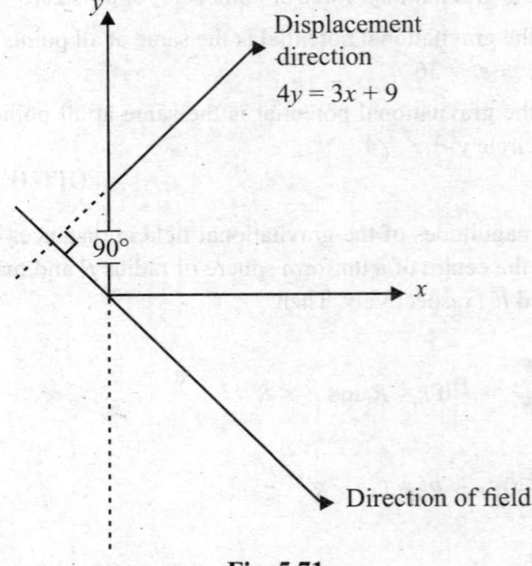

Fig. 5.71

$$m_1 \times m_2 = \frac{3}{4} \times -\frac{4}{3} = -1$$

i.e., force and displacement directions are perpendicular. The work done is zero.

5. Let W_0 be weight of the body when ship is at rest then

$$W_0 = F - \frac{mv_0^2}{R}$$

where F = gravitational pull of the earth when it is at rest and v_e = velocity of the earth at the equator when both move due east, the net speed of the ship

$$v' = v_e + v$$

Therefore, the apparent weight $\quad W = F - \frac{m(v_e + v)^2}{R}$

$$W - W_0 = \frac{mv_e^2}{R} - \frac{m(v_e + v)^2}{R} = \frac{mv_e^2}{R}\left[1 - \left(1 + \frac{v}{v_e}\right)^2\right]$$

As $\quad v << v_e$,

$$W - W_0 = mv_e^2\left[1 - \left(1 + \frac{v}{2v_0}\right)\right] = -\frac{mv_e v}{2R}$$

As $m = \frac{W_0}{g}$ and $\omega = \frac{v_e}{R} \Rightarrow W = W_0\left(1 - \frac{2\omega v}{g}\right)$

Similarly, if the ship moves due west, then the net velocity of the ship is

$$v' = v_e - v$$

$$W = W_0 - \frac{m(v_e - v)^2}{R} \Rightarrow W = W_0\left(1 + \frac{2\omega v}{g}\right)$$

6. Let the speed of the instrument package be v when it grazes the surface of the planet.

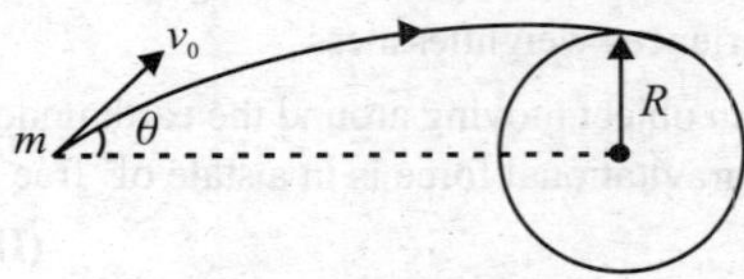

Fig. 5.72

Conserving angular momentum of the package about the centre of the planet

$$mv_0 \times 5R \sin\theta = mvR$$

$$\Rightarrow \quad v = 5v_0 \sin\theta \quad \text{(i)}$$

Conserving mechanical energy, we get

$$-\frac{GMm}{5R} + \frac{1}{2}mv_0^2 = -\frac{GMm}{R} + \frac{1}{2}mv^2$$

$$\Rightarrow \quad \frac{1}{2}m(v^2 - v_0^2) = \frac{4GMm}{5R}$$

$$\Rightarrow \quad v^2 - v_0^2 = \frac{8GM}{5R} \quad \text{(ii)}$$

Substituting the value of v from Eq. (i) in Eq. (ii), we have

$$25\, v_0^2 \sin\theta - v_0^2 = \frac{8GM}{5R}$$

$$\Rightarrow \quad \theta = \sin^{-1}\left(\frac{1}{5}\sqrt{1 + \frac{8GM}{5v_0^2 R}}\right)$$

7. Let the speed of the ball which can cross 10 m wide river be V.

Then $\quad R = \dfrac{V^2 \sin(2 \times 45°)}{g} = 10,\ v = \sqrt{10g}$

Let the radius of planet be 'R_p', then mass of the planet

$$M_p = \frac{4}{3}\pi R_p^3 \times 2\sigma = \frac{4}{3}\pi R_p^3 \times \frac{2 \times M_e}{4/3\pi R_e^3} = \frac{2M_e R_p^3}{R_e^3}$$

Escape velocity of the planet

$$V_e = \sqrt{\frac{2GM_p}{R_p}} = \sqrt{\frac{2G \times 2 \times M_e R_p^3}{R_p \times R_e^3}} = \sqrt{\frac{10GM_e}{R_e^2}}$$

$$\left[\therefore g = \frac{GMe}{R_e^2}\right]$$

$$2R_p = \sqrt{10R_e} \Rightarrow 2R_p = \sqrt{10 \times 6.4 \times 10^6}$$

$$\Rightarrow \quad R_p = \frac{8 \times 10^3}{2} = 4 \times 10^3 = 4 \text{ km}$$

8.

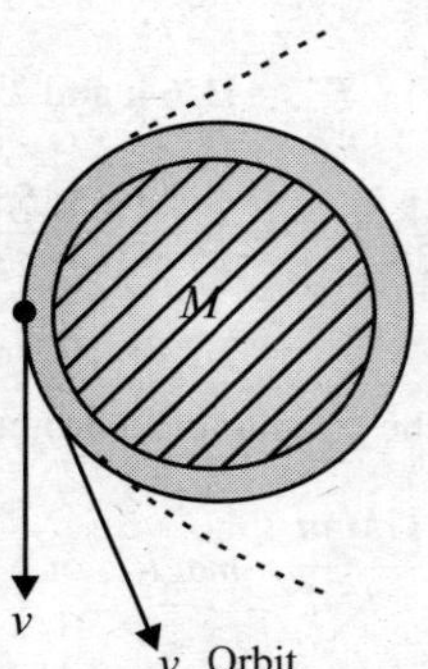

Fig. 5.73

If v is the velocity at the vertex of the parabola, then v is also the escape velocity. It will follow the parabolic path never to return to the Moon. Now $v_{escape} = \sqrt{\dfrac{2GM}{R}}$

$$\Delta v = v_{final} - v_{initial} = v_{orbit} - v_{escape}$$

$$\Rightarrow \quad \Delta v = \sqrt{\frac{GM}{R}} - \sqrt{\frac{2GM}{R}}$$

$$= -\sqrt{\frac{GM}{R}}(\sqrt{2} - 1)$$

The negative sign means the speed has to be decreased. Therefore, the required change in the speed is

$$\sqrt{\frac{GM}{R}}(\sqrt{2} - 1)$$

9. Initial energy of the satellite,

$$E_1 = -\frac{GMm}{R} \text{ (since initial KE = 0)}$$

Final energy of the satellite

Energy needed, $\quad \Delta E = E_2 - E_1$

$$= -\frac{2GMm}{2(R+h)} - \left(-\frac{GMm}{R}\right) = GMm\left(\frac{1}{R} - \frac{1}{2(R+h)}\right)$$

But, $\quad GM = R^2 g$

$$\Delta E = R^2 gM\left(\frac{1}{R} - \frac{1}{2(R+h)}\right) = mgR\left[1 - \frac{1}{2\left(1 + \frac{h}{R}\right)}\right]$$

$$= 7.1 \times 10^{10} \text{ J}$$

10. Orbital velocity,

$$v_0 = \sqrt{\frac{GM}{r}} = \sqrt{\frac{R^2 g}{r}}$$

$$v_0 = \sqrt{\left\{\frac{(6400 \times 10^3)^2 \times 9.8}{7000 \times 10^3}\right\}} = 7572.6 \text{ m s}^{-1}$$

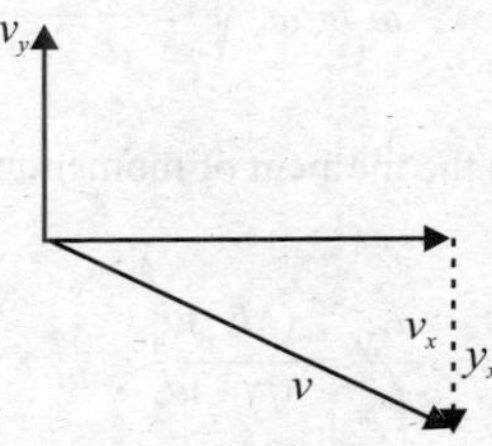

Fig. 5.74

This velocity is horizontal so $v_0 = v_x = 7572.6$ m s^{-1}.

For launching the satellite in the circular orbit, the vertical component of velocity should be zero; so the velocity required is v_x =7572.26 m s^{-1} and v_y (downward) = 2400 m s^{-1}

Therefore, the required impulse is

$$\sqrt{(mv_x)^2+(mv_y)^2} = m\sqrt{v_x^2+v_x^2} = 60\sqrt{(7572.6)^2+(2400)^2}$$

$$= 4.77\times 10^5 \text{ kg m}^{-1}$$

$$\tan\theta = \frac{v_y}{v_x} = \frac{2400}{7572.6} - 0.317$$

$$\theta = \tan^{-1}(0.317) = 17.6°$$

Hence, impulse = 4.77×10^5 m s^{-1} at an angle 17.6° from horizontal (downward).

11. The centre of mass of the system divides the distance between the stars in the inverse ratio of their masses. If d_1 and d_2 are the distance force for their circular motion,

$$d_1 = \frac{d}{M+m}\times m$$

and $$d_2 = \frac{d}{M+m}\times M \Rightarrow \frac{d_1}{d_2} = \frac{m}{M}$$

the stars will rotate in circles of radii d_1 and d_2 about their centre of mass. The same force of attraction provides the necessary centripetal force for their circular motion.

$$\therefore \quad \frac{GmM}{d^2} = M\omega_1^2 d_1 = m\omega_2^2 d_2$$

or $$\omega_1^2 = \frac{Gm}{d^2 d_1} = \frac{Gm}{d^2}\times\frac{M+m}{d\times m}$$

$$= \frac{G(M+m)}{d^3}$$

and $$\omega_2^2 = \frac{GM}{d^2 d_2} = \frac{GM}{d^2}\times\frac{M+m}{d\times M}$$

$$= \frac{G(M+m)}{d^3}$$

$$\therefore \quad \omega_1 = \omega_2\sqrt{\frac{G(M+m)}{d^3}}$$

From the fact that the moment of momentum is also the angular momentum,

$$\frac{L_M}{L_m} = \frac{(Mv_1)d_1}{(mv_2)d_2} = \frac{M}{m}\times\frac{d_1^2}{d_2^2} = \frac{L_M}{L_m} = \frac{m}{M}$$

$$\Rightarrow \frac{K_M}{K_m} = \frac{\frac{1}{2}mv_1^2}{\frac{1}{2}mv_1^2} = \frac{M}{m}\frac{\omega_1^2 d_1^2}{\omega_2^2 d_2^2} = \frac{M}{m}\frac{d_1^2}{d_2^2} \quad (\because \omega_1 = \omega_2)$$

Therefore, $$\frac{K_M}{K_m} = \frac{M}{m}\left(\frac{m}{M}\right)^2 = \frac{m}{M}$$

12. If r_1 and r_2 are distances of stars A and B from the centre of mass, then we have

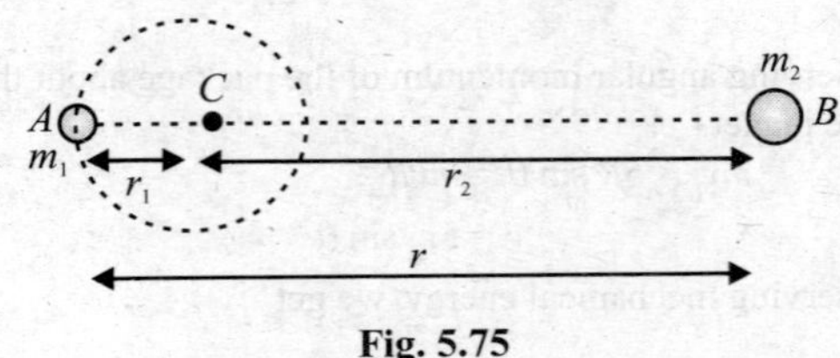

Fig. 5.75

$$m_1 r_1 = m_2 r_2 \quad \text{and} \quad r_1 + r_2 = r$$

This gives $$r_1 = \frac{m_2 r}{m_1 + m_2}$$

Given $$m_1 = 2m_2$$

Therefore, $$r_1 = \frac{m_2 r}{2m_2 + m_2} = \frac{r}{3}$$

Now from the condition of the circular path, if T_1 is period of revolution of star A about the centre of mass, then

$$m_1 r_1\left(\frac{2\pi}{T_i}\right)^2 = \frac{Gm_1m_2}{r^2} = m_1\left(\frac{r}{3}\right)\left(\frac{2\pi}{T_1}\right)^2 = \frac{Gm_1m_2}{r^2}$$

$$\left(\frac{2\pi}{T_1}\right)^2 = \frac{3Gm_2}{r^3}$$

But $m_2 = \frac{M_s}{3}$; where M_s = mass of Sun

$$\left(\frac{2\pi}{T_1}\right)^2 = \frac{GM_s}{r^3} \quad \text{(i)}$$

Again period of revolution of earth (T_e) about the Sun is

$$m_e R\omega_e^2 = \frac{GM_s m_e}{R^2} \Rightarrow \left(\frac{2\pi}{T_e}\right)^2 = \frac{GM_s}{R^3}\left(\because \omega_e = \frac{2\pi}{T_e}\right) \quad \text{(ii)}$$

As $$T_1 = T_e = 1 \text{ year}$$

$$r^3 = R^3 \Rightarrow r = R$$

13. Angular speed of the satellite relative to the earth

$$\omega_{SE} = \omega_S - \omega_E$$

(because they are moving in same direction west to east.)

$$\omega_S = \omega_{SE} - \omega_E = \frac{2\pi}{T_{SE}} + \frac{2\pi}{T_E}$$

Given $$T_{SE} = 11.6 \text{ h and } T_E = 24 \text{ h}$$

$$\omega_S = 2\pi\left[\frac{1}{11.6}+\frac{1}{24}\right]\text{h}^{-1} = \frac{2\pi\times(35.6)}{24\times 11.6\times 3600}\text{ rad s}^{-1}$$

$$= 2.23\times 10^{-4}\text{ rad s}^{-1}$$

The condition of the circular orbit around the earth is

$$\frac{GM_e m}{R^2} = m\omega_s^2 R \quad \text{or} \quad GM_e = \omega_s^2 R^3$$

$$R = 2.0\times 10^4 \text{ km} = 2.0\times 10^7 \text{ m}$$

$$M_e = (2.23\times 10^{-4})^2\times(2.0\times 10^7)6.67\times 10^{-11} \approx 6\times 10^{24}\text{ kg}$$

14. $F = m\frac{dv}{dt} = av^2$ (i)

At $t = 0$, when the satellite is at a distance nR from the centre of the earth. Velocity of the satellite is

$$v_i = \sqrt{\frac{GM}{nR}}$$

At time t when the satellite falls on the earth surface, the velocity of the satellite is

$$v_f = \sqrt{\frac{GM}{R}}$$

from Eq. (i) $m\frac{dv}{v^2} = a\,dt$

Integrating both the sides, we get

$$m\int_{v_i}^{v_f}\frac{dv}{v^2} = a\int_0^t dt \Rightarrow m\left[\frac{v^{-2+1}}{-2+1}\right]_{v_i}^{v_f} = at$$

On solving, we get $t = \frac{m\sqrt{R}(\sqrt{n}-1)}{a\sqrt{(GM)}}$

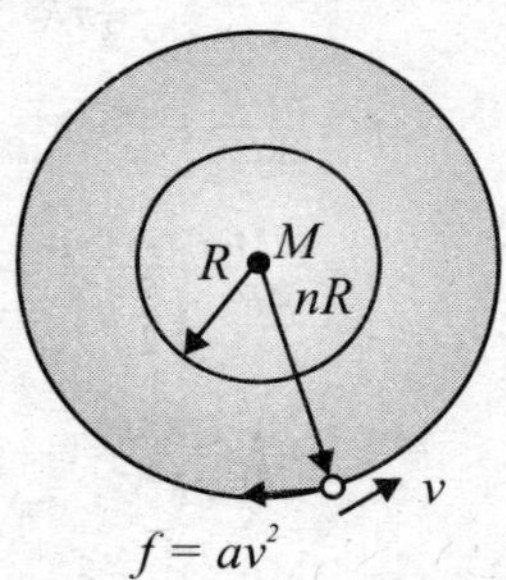

Fig. 5.76

15. $$E = -\frac{GMm}{2r}$$

$$-\frac{dE}{dt} = \frac{GMm}{2}\frac{1}{r^2}\frac{dr}{dt}$$

$$\int_0^t dt = \frac{GMm}{2C}\int_r^R \frac{dr}{r^2}$$

$$t = \frac{GMm}{2C}\left[\frac{1}{R} - \frac{1}{r}\right] = \frac{GMm}{2C}\left[\frac{1}{R} - \frac{1}{r}\right]$$

Objective Type

1. d. Change in energy = $1/(2\,(GMm)/R) = 1/2Mv_e^2$

So it is independent of angle as gravitational field is conservative in nature.

2. b. If a body is projected from the surface of the earth with a velocity v and reaches a height h, then according to law of conservation of energy,

$$\frac{1}{2}mv^2 = \frac{mgh}{1+h/R}$$

Here $v = kv_e = k\sqrt{2gR}$

Therefore, $$\frac{1}{2}mk^2\,3gR = \frac{mg(r-R)}{1+\frac{(r-R)}{R}}$$

$$k^2R\left[1+\frac{r-R}{R}\right] = r - R \text{ or } k^2r = r - R$$

or $$r = \frac{R}{1-k^2}$$

3. d.

$$\frac{GMm}{r^2} = m\omega_0^2 r$$

$$\Rightarrow \quad GM = \omega_0^2 r^3$$

$$\Rightarrow \quad gR^2 = \omega_0^2 r^3$$

$$\Rightarrow \quad g = \frac{\omega_0^2 r^3}{R^2}$$

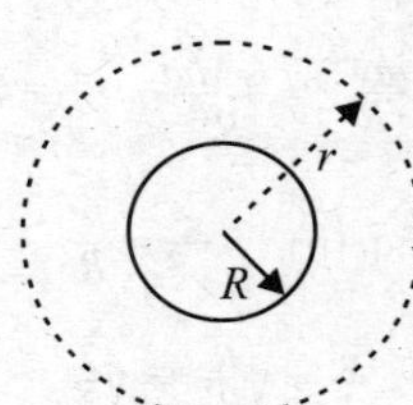

Fig. 5.77

4. a. From the principle of conserving angular momentum, we have

$$mvR = mv'r \quad \text{(i)}$$

[v' = speed when spaceship is just touching the planet]

From conserving of energy, we have

$$\frac{1}{2}mv^2 = \frac{1}{2}mv'^2 - \frac{GMm}{r} \quad \text{(ii)}$$

Solving Eqs. (i) and (ii), we get

$$R = \frac{r}{v}\left[v^2 + \frac{2GM}{r}\right]^{1/2}$$

5. d.

$$p = \rho_{atm}gh = \frac{m}{\frac{4}{3}\pi[(R+h)^3 - R^3]}\frac{GM}{R^2}h$$

$$\Rightarrow p = \frac{m}{\frac{4}{3}\pi R^3\left[\left(1+\frac{h}{R}\right)^3 - 1\right]} \times \frac{G\frac{4}{3}\pi R^3\ell}{R^2} \times h$$

$$= \frac{m}{\frac{4}{3}\pi R^3\left[1+\frac{3h}{R} - 1\right]} \times G\frac{4}{3}\pi R\ell h$$

$$\left(\text{Here } M = \frac{4}{3}\pi R^3\ell\right)$$

Hence, $$p = \frac{mG\rho}{3R}$$

$\therefore \quad m \propto \dfrac{pR}{\rho}$

6. a.

$$U_g = -\frac{GM_1M_2}{R}$$

$$\Rightarrow \quad U_f - U_i = -\frac{GMm}{2R} - \frac{GMm}{R}$$

$$\Delta u = \frac{GMm}{2R} = \frac{mgR}{2} \quad \left(\because g = \frac{GM}{R^2}\right)$$

7. b. From the surface of the earth, the escape velocity is $\sqrt{(2GM)/R}$. From the satellite, the escape velocity is calculated as follows. By conservation of energy,

$$\frac{1}{2}mv_2^2 - \frac{GMm}{2R} = 0$$

$$\Rightarrow \quad v_2^2 = \sqrt{\frac{GM}{R}} \Rightarrow v_2 < v_1$$

8. c.

$$\frac{mv^2}{R} = \frac{GMm}{R^2} \quad \text{(i)}$$

and

$$\frac{mv'^2}{R'} = \frac{GMm}{R'^2} \quad \text{(ii)}$$

Dividing Eq. (i) by Eq. (ii), $\dfrac{v^2}{v'^2} = \dfrac{R'}{R}$

$$\Rightarrow \quad \frac{v'}{v} = \sqrt{\frac{R}{R'}} = \sqrt{\frac{2}{3}} \quad \left[\because R' = \frac{3R}{2}\right]$$

$$\Rightarrow \quad v' = \sqrt{\frac{2}{3}}v$$

9. d. $g_E = (GM)/R^2$

$$g_M = \frac{GM^1}{R^{12}} = \frac{GM/10}{R^2/4}$$

$$= \frac{2}{5}\frac{GM}{R^2}$$

$$= \frac{2}{5}g_E$$

$$W_M = W_E\frac{2}{5}$$

$$= 200 \times \frac{2}{5} = 80 \text{ N}$$

10. c. $T^2 \propto a^3$

$$T_1^2 \propto (10^{13})^3$$

$$(T^2)^2 \propto (10^{12})^3$$

$$\therefore \quad \frac{T^1}{T^2} = \left(\frac{10^{13}}{10^{12}}\right)^{\frac{3}{2}} \Rightarrow \frac{T^1}{T^2} = 10/\sqrt{10}$$

11. b. $F_G = \dfrac{GMm}{R}$ (as $F_G \propto \dfrac{1}{R}$ given)

So $\quad \dfrac{mv^2}{R} = \dfrac{GMm}{R} \Rightarrow v^2 \propto R^0$

12. c. $F_G = \dfrac{Gm^2}{4R^2} \Rightarrow \dfrac{Mv^2}{R} = \dfrac{Gm^2}{4R^2}$

$$\therefore \quad v = \frac{1}{2}\sqrt{\frac{Gm}{R}}$$

13. c.

$$\text{PE} = \frac{-GM}{2R} \text{ (at height } R)$$

$$= -\frac{GMR_m}{2R^2} = -\frac{1}{2}mgR$$

14. b. $g' = g\left[1 - \frac{2h}{R}\right] = g\left[1 - \frac{d}{R}\right]$

$$\frac{2h}{R} = \frac{d}{R} \Rightarrow d = 2h$$

15. d. Let mass of the cavity = M'

Density of the sphere = $M/(4/3\,\pi R^3)$

Mass of the cavity cut out = M'

$$= \frac{4}{3}\pi\frac{R^3}{8} \times \frac{M}{\frac{4}{3}\pi R^3}$$

$$\therefore \quad M' = \frac{M}{8} \Rightarrow F_{net} = F_{Mm} - F_{M'm}$$

$$= \frac{GMm}{4R^2} - \frac{GM'm}{\left(\frac{5}{2}R\right)^2} = \frac{GMm}{4R^2} - \frac{GMm}{50R^2}$$

$$F_{net} = \frac{23}{100}\frac{GMm}{R^2}$$

16. c. $T^2 \propto R^3$

$$T^2 = KR^3 \Rightarrow (24)^2 = K(36000)^3$$

$$K = \frac{1}{9 \times 10^4\sqrt{10}}$$

$$T' = \frac{1}{9 \times 10^4\sqrt{10}}(6400)^{3/2} = 2h$$

17. b. As gravitational force is a conservative force, work done is independent of path.

$$\therefore \quad W_1 = W_2 = W_3$$

18. c.

$$g = \frac{Gm}{R^2}$$

$$M = \frac{4}{3}\omega R^2\rho$$

So $\quad \rho = \dfrac{3g}{4\pi RG}$

19. a. Total mechanical energy is given by

$$E = K + u = -\frac{GMm}{2a} - \frac{GMm}{a}$$

$$= \frac{-GMm}{2a}$$

$$\frac{GM}{a} = v^2 \Rightarrow E = -\frac{1}{2}mv^2$$

20. d.

Inside the earth $\quad g' = g\left[1 - \frac{h}{R}\right]$

At centre, $\quad g' = g\left[1 - \frac{R}{R}\right] = 0$

21. d.

$$m_1 = 2m_2 \Rightarrow \frac{m_1}{m_2} = 2$$

$$r_1 = 4r_2 \Rightarrow \frac{r_1}{r_2} = 4$$

$$T_A^2 \propto r_1^2 \text{ and } T_B^2 \propto r_2^3$$

$\therefore$
$$\frac{T_A}{T_B} = \left(\frac{r_1}{r_2}\right)^{3/2} = (4^3)^{1/2}$$

$\Rightarrow$
$$\frac{T_A}{T_B} = 8$$

22. d.

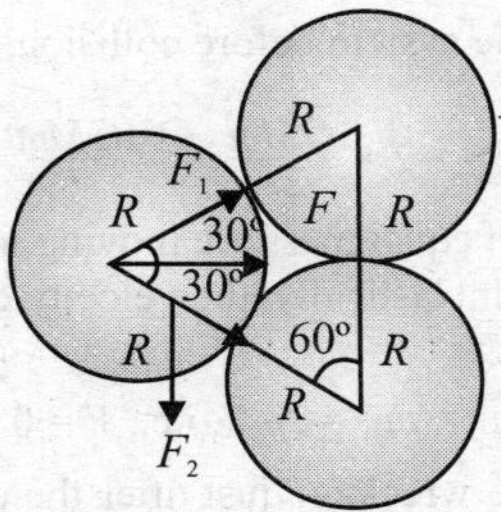

Fig. 5.78

$$\vec{F} = \vec{F}_1 + \vec{F}_2$$

As $\quad |\vec{F}_1| = |\vec{F}_2|$

$\therefore \quad |F| = 2F_1 \cos 30°$

$$= 2\frac{GM^2}{(2r)^2}\frac{\sqrt{3}}{2} = \frac{\sqrt{3}}{4}\frac{GM^2}{r^2}$$

23. d.

$$g = \frac{GM}{R^2} \quad \text{(i)}$$

$$g' = \frac{Gm}{\left(\frac{90}{100}R\right)^2} = \frac{100}{81}\frac{GM}{R^2} \quad \text{(ii)}$$

From Eqs. (i) and (ii),

$$g' = \frac{100}{81}g \Rightarrow \frac{g'}{g} = \frac{100}{81}$$

$$\frac{g'}{g} - 1 = \frac{100}{81} - 1$$

$\therefore \quad \Delta g = \frac{19}{81}g = 23\%$ of g

So increase is more than 19% of g.

24. a.

i. g' = acceleration due to gravity on the Moon

$$= G\frac{4}{3}\pi R'\rho' = G\frac{4}{3}\pi \times \frac{R}{4} \times \frac{2}{3}\rho = \frac{g}{6}$$

Now, $\quad h_m = \frac{u^2}{2g}$ or $h_m \propto \frac{1}{8}$

Hence $\quad h'_m g' = h_m g$

or $\quad h'_m = 6h_m = 6 \times 0.5 = 3$ m

ii. $t = \frac{2u}{g}$ or $t \propto \frac{1}{g}$

$\therefore \quad t'g' = tg \quad$ or $\quad \frac{t'}{t} = \frac{g}{g'} = \frac{6}{1}$

25. c.

$$g_1 = \frac{g}{\left[1 + \frac{h}{r}\right]^2} = g\left[1 - \frac{2h}{R}\right]$$

$W_2 - W_1$ = error in weighing

$$= 2mg\left[\frac{h_1}{R} - \frac{h_2}{R}\right] = 2m\frac{GM}{R^2}\frac{h}{R}$$

$$\left[\because g = \frac{GM}{R^2} \text{ and } h_1 - h_2 = h\right]$$

Therefore, $\quad W_2 - W_1$ = error in weighing

$$= 2mG\frac{4}{3}\pi R^3\rho\frac{h}{R^3} = \frac{8\pi}{3}Gm\rho h$$

26. a. (i) The weight of the body at the centre of the earth is equal to zero because

$$g_{\text{centre}} = g\left(1 - \frac{d}{R}\right) = g\left(1 - \frac{R}{R}\right) = 0$$

$$\frac{g_1}{g} = \left(1 - \frac{d}{R}\right) = \frac{1}{4} \Rightarrow d = \frac{3R}{4}$$

So from the centre, $\quad d' = \frac{R}{4}$

27. a.

$$3/5\ mg = mg - mR\omega^2$$

$$\omega^2 = g - \frac{3}{5}g \Rightarrow \omega = \sqrt{\frac{2}{5}\frac{g}{R}}$$

28. d.

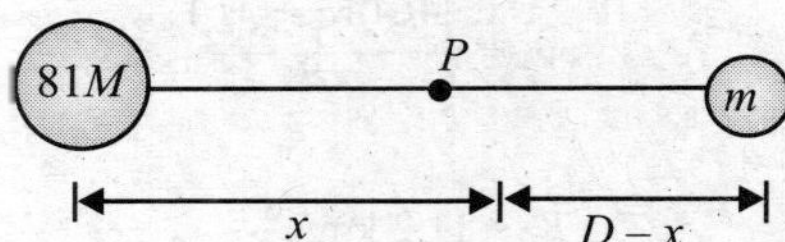

Fig. 5.79

$$\frac{81GM}{x^2} = \frac{GM}{(D-x)^2}$$

$$\frac{9}{x} = \frac{1}{(D-x)}$$

$$D = \frac{9}{10}x$$

$$x = \frac{9}{10}D$$

29. b. Because the gravitational force is the mutual force, hence the position of centre of mass remains unaffected.

$$\therefore \quad M_1R_1 = M_2R_2 \text{ or } \frac{R_1}{R_2} = \frac{M_2}{M_1}$$

30. a. At a distance x from the centre of earth, the gravitational force

$$F = \frac{GMm}{x^2}$$

$$dW = Fdx = \frac{GMm}{x^2}dx$$

We get

$$W = \int_R^{R+h} \frac{GMm}{x^2}dx = GMm\left[\frac{1}{R} - \frac{1}{R+h}\right]$$

$$= gR^2m\left[\frac{1}{R} - \frac{1}{R+h}\right]$$

Therefore, the gain in potential energy is

$$gR^2m\left[\frac{1}{R} - \frac{1}{2R}\right] = \frac{mgR}{2}$$

31. d. $\frac{mv^2}{r} = \frac{GMm}{r^{5/2}}\left[\because F \propto \frac{1}{r^{5/2}}\right]$

or $$r\omega^2 = \frac{GM}{r^{5/2}}$$

or $$r\frac{4\pi^2}{T^2} = \frac{GM}{r^{5/2}}$$

or $$T^2 \propto r^{5/2} \text{ or } T^2 \propto r^{3.5}$$

32. b. Potential energy of mass m when it is midway between masses M_1 and M_2 is

$$U = -\frac{GM_1m}{d/2} - \frac{GM_2m}{d/2}$$

$$= -\frac{2Gm}{d}(M_1 + M_2)$$

According to law of conservation of energy,

$$\frac{1}{2}mv_e^2 = \frac{2Gm}{d}(M_1 + M_2)$$

Therefore, escape velocity,

$$v_e = \sqrt{\frac{4G(M_1 + M_2)}{d}}$$

33. c.

$$V_e = 11.2 \text{ km/s}$$

$$\frac{mv'^2}{R} = \frac{GMm}{R^2}$$

$$v'^2 = \frac{GM}{R^2}R$$

$$v' = \sqrt{10 \times 6400000} = 8 \text{ km/s}$$

Therefore, the additional velocity = (11.2 – 8) = 3.2 km/s

34. b. Energy of the skylab in the first orbit is

$$-\frac{GMm}{2(2R)} = -\frac{GMm}{4R}$$

Total energy required to place the skylab into the orbit of radius $2R$ from the surface of earth is

$$-\frac{GMm}{4R} - \left(-\frac{GMm}{R}\right) = \frac{3GMm}{4R}$$

$$= \frac{3gR^2m}{4R} = \frac{3}{4}mgR$$

Energy of the skylab in the second orbit = – $(GMm)/6R$.

Energy needed to shift the skylab from the first orbit to the second orbit is

$$\frac{GMm}{4R} - \frac{GMm}{6R} = \frac{GMm}{R} \times \frac{2}{24} = \frac{mgR}{12}$$

35. a. Energy of each satellite in the orbit $= \frac{-GMm}{2r}$.

Total energy of the system before collision,

$$E_i = E_1 = E_2 = 2E = -2 \times \frac{GMm}{2r} = -\frac{GMm}{r}$$

As the satellites of equal mass are moving in the opposite directions and collide inelastically, the velocity of the wreckage just after the collision is

$$mv - mv = 2mV, \text{ i.e., } V = 0$$

The energy of the wreckage just after the collision will be totally potential and will be

$$E_f = \frac{GM(2m)}{r} = -\frac{2GMm}{r}$$

As after collision the wreckage comes to standstill in the orbit, it will move along the radius towards the earth under gravity.

36. c. At point P, we have $I_1 - I_2 = 0$ (because the gravitational field inside a shell is zero). Hence, $I_1 = I_2$.

37. c. $F_G = \frac{GM^2}{4R^2} \Rightarrow \frac{mv^2}{R} = \frac{GM^2}{4R^2}$

$$v = \frac{1}{2}\sqrt{\frac{GM}{R}}$$

38. c. On the surface of the earth,

Total energy = Kinetic energy + Potential energy

$$= \frac{1}{2}mv^2 \frac{GmM}{R}$$

At the highest point, $v = 0$,

potential energy = – $(GmM)/(R + h)$

where h is the maximum height.

$$\frac{1}{2}mv^2 - \frac{GmM}{R} = -\frac{GmM}{R+h}$$

$$\frac{1}{2}v^2 = \frac{gRh}{R} + h \text{ or } \frac{R+h}{h} = \frac{2gR}{v^2}$$

$$h = R\left(\frac{2gR}{v^2} - 1\right)^{-1}$$

39. b. According to the problem, as the potential at ∞ increases by $+10$ J kg^{-1}, hence potential will increase by the same amount everywhere (potential gardient will remain constant). Hence, potential at point $P = 10 - 5 = +5$ J kg^{-1}

40. c.

$$\frac{1}{2}mv^2 = \frac{GMm}{R} - \frac{GMm}{R+h}$$

$$= \frac{GMmh}{R(R+h)}$$

$$= \frac{mghR}{R+h}$$

Since $v = nv_e$ and $v_e = \sqrt{2gR}$, hence $= (Rn^2)/(1 + n^2)\cdot$

41. b. $mg = mR\omega^2$

R = radius of earth

$$\omega = \sqrt{\frac{g}{R}}$$

$$T = 2\pi\sqrt{\frac{R}{g}} = 2\pi\sqrt{64000}$$

$$= 2\pi \times 800 \text{ s} = \frac{2\pi \times 800}{3600} \text{ h} = 1.36$$

$$= 0.14 \text{ h}$$

42. c.

M_1 ⟵ (x) ⟶ $(R - x)$ M_2

Fig. 5.80

$$\frac{GM_1}{x^2} = \frac{GM_2}{(R-x)^2}$$

$$\frac{M_2}{M_1}x^2 = R^2 + x^2 - 2Rx$$

Let $\frac{M_2}{M} = k$

$$x^2(k-1) + 2Rx - R^2 = 0$$

$$x = -\frac{2R + \sqrt{4R^2 + 4(k-1)R^2}}{2(k-1)}$$

$$= \frac{R\sqrt{M_1}}{\sqrt{M_1} + \sqrt{M_2}}$$

$$R - x = \frac{R\sqrt{M_2}}{\sqrt{M_1} + \sqrt{M_2}}$$

Gravitational potential at point P is

$$-\left(\frac{GM_1}{x} + \frac{GM_2}{R-x}\right)$$

$$= -\left[\frac{GM_1(\sqrt{M_1} + \sqrt{M_2})}{R\sqrt{M_1}} + \frac{GM_2(\sqrt{M_1} + \sqrt{M_2})}{R\sqrt{M_2}}\right]$$

$$= -\left[\frac{G(\sqrt{M_2} + \sqrt{M_1})}{R}(\sqrt{M_1} + \sqrt{M_2})\right]$$

$$= -\frac{G(\sqrt{M_1} + \sqrt{M_2})^2}{R}$$

43. a. Binding energy $= GMm/R$

M = mass of the Sun $= 10^{30}$ kg

m = mass of the earth $= 6 \times 10^{24}$ kg

$R = 1.5 \times 10^{11}$ m

Binding energy of the system is

$$\frac{6.67 \times 10^{-11} \times 10^{30} \times 6 \times 10^{24}}{1.5 \times 10^{11}} = 2.7 \times 10^{33} \text{ J}$$

44. d. Let the mass of the particle be m

PE at a distance of $R' = (GMm)/R'$

PE at a distance of $R_e = -(GMm)/R_e$

Decrease in PE = Increase in KE

$$\Rightarrow \quad -\frac{GMm}{R'} + \frac{GMm}{R_e} = \frac{1}{2}mv^2$$

$$v^2 = 2GM\left[\frac{1}{R_e} - \frac{1}{R'}\right] \Rightarrow v^2 = \frac{2GM}{R_e}\left[1 - \frac{R_e}{R'}\right]$$

$$\Rightarrow v^2 = \frac{2GM}{R_e}\left(1 - \frac{R_e}{R'}\right) \Rightarrow v = \sqrt{\frac{2GMR_e}{R_e^2}}\left(1 - \frac{R_e}{R'}\right)$$

$$\therefore \quad v = \sqrt{2gR_e}\left(1 - \frac{R_e}{R'}\right)$$

45. d. Mass of the cavity $= M/8$ if mass of the sphere $= M$ as volume of the cavity is 1/8th of the sphere.

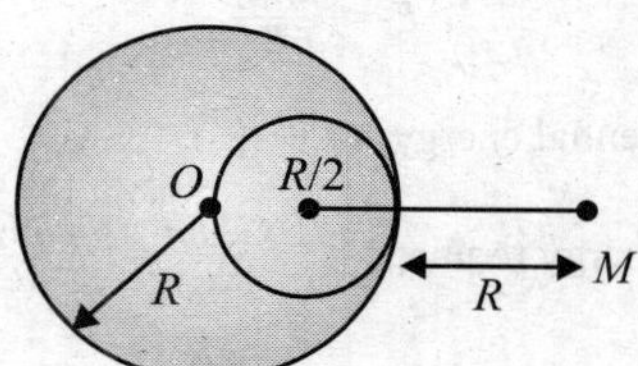

Fig. 5.81

$$F_2 = \frac{GMm}{4R^2} - \frac{GMm}{8\left(\frac{3}{2}R\right)^2} = \frac{GMm}{R^2}\left[\frac{1}{4} - \frac{1}{18}\right] = \frac{GMm}{R^2}\left[\frac{9-2}{36}\right]$$

$$= \frac{GMm}{R^2}\frac{7}{36}$$

$$\therefore \quad \frac{F_1}{F_2} = \frac{36}{4 \times 7} = \frac{9}{7}$$

46. b. $F = (GMm)/R^2 = mg$

As the mass remains the same and radius, i.e., distance from the centre is also the same, therefore g will remain the same, i.e., 10 m s^{-2}.

47. c. $\frac{m_1 v_1^2}{r_1} = m_1 \frac{GM}{r_1^2}$, $\frac{m_2 v_2^2}{r_2} = m_2 \frac{GM}{r_2^2}$

$$\frac{v_1^2}{v_2^2} = \frac{r_2}{r_1};\ r_2 < r_1 \Rightarrow v_1^2 < v_2^2.$$

Hence, $v_1 < v_2$.

48. b. Applying conservation of angular momentum at position A and B

$mv_A \times OA = mv_B \times OB$

Hence, $\frac{v_B}{v_A} = \frac{OA}{OB} = x$

49. d. The gravitational field intensity at a point inside the spherical shell is zero.

Note: ***There is NO gravitational field inside a spherical shell.***

50. c. Note that there must be some point where the gravitational field of the earth is balanced by the gravitational field of Mars.

51. c. $V_A = \sqrt{\frac{GM}{r_A}}$ and $V_B = \sqrt{\frac{GM}{r_B}}$

Given $r_B = 3r_A$

Now $F_A = \frac{mv_A^2}{r_A} = \frac{M}{r_A}\frac{GM}{r_A} = \frac{GMm}{r_A^2}$

$$F_B = \frac{GMm}{r_B^2}$$

$$\frac{F_B}{F_A} = \frac{R_A^2}{r_B^2} = \frac{1}{9}$$

52. c. Total energy, $E = \frac{1}{2}mv^2 - \frac{GmM}{r}$

$$= \frac{GmM}{2r} - \frac{GMm}{r} = -G\frac{mM}{2r}$$

$$r = 2R = R = 3R$$

$$E = \frac{GmM}{6R}$$

Potential energy $= -\frac{GMm}{R}$

Minimum energy required $= \frac{1}{6}\frac{GMm}{R} - \left(\frac{-GMm}{R}\right) = \frac{5}{6}\frac{GMm}{R}$

$$= \frac{5}{6}mgR$$

53. a. Suppose that a satellite of mass m describes a circular orbit around a planet of mass M.

$$F = \frac{GmM}{r^2}$$

This force must be mass times the centripetal acceleration.

$$\therefore \quad F = \frac{mv^2}{r} = m\omega^2 r = m\frac{4\pi^2}{T^2}r$$

$$\therefore \quad M = \frac{4\pi^2 r^3}{GT^2}$$

54. a.

$$g_p = \frac{G\left[M - \frac{10}{100}M\right]}{\left[R + \frac{20}{100}R\right]^2} = \frac{G \times 9M}{10} \times \frac{25}{36R^2} = \frac{5}{8}g$$

55. c. $mg' = mg - mR\omega^2 \cos^2\phi$

Now, $\frac{3}{5}mg = mg - mR\omega^2$

or $mR\omega^2 = mg - \frac{3}{5}mg$

$$= \frac{3}{5}mg \text{ or } \omega = \sqrt{\frac{2g}{5R}}$$

56. c. $W = \left[-\frac{GMm}{5R}\right] - \left[\frac{GMm}{3R}\right]$

$$= \frac{GMm}{3R} - \frac{GMm}{5R}$$

$$= \frac{GMm}{R}\left[\frac{1}{3} - \frac{1}{5}\right] = \frac{2}{15}\frac{GMm}{R}$$

57. c. $dF = G\frac{Mdm}{4r^2}$

$$F = \Sigma dF \cos\theta$$

$$= \Sigma \frac{GMdm}{4r^2}\cos\theta$$

$$= \frac{GM}{4r^2} \times \frac{\sqrt{3}r}{2r}\Sigma dm$$

$$= \frac{\sqrt{3}GMm}{8r^2}$$

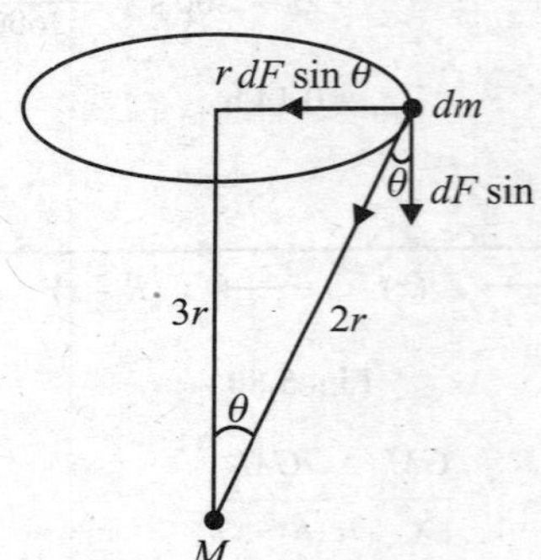

Fig. 5.82

Alternative solution:

The gravitational field due to the ring at a distance $\sqrt{3}r$ is given by

$$E = \frac{Gm(\sqrt{3}r)}{[r^2 + (\sqrt{3}r)^2]^{3/2}} \text{ or } E = \frac{\sqrt{3}Gm}{8r^2}$$

The required force is EM, i.e., $(\sqrt{3}Gm)M/8r^2$.

Note: ***Follow the central argument in this solution: The gravitational field at any point on the ring due to a sphere is equal to the field due to a single particle of mass M located at the centre of the sphere.***

58. a. Let us first calculate the mass of the inner solid sphere of radius r.

Mass of the inner solid sphere is

$$M' = \frac{M_e}{\frac{4}{3}\pi R_e^3} \times \frac{4}{3} = \pi r^2 = \frac{M_e}{R_e^3}r^3$$

Now, $g = \frac{GM_e r^3}{R_e^3} \times \frac{1}{r^2}$ or $g = \frac{GM_e r}{R_e^3}$

Force on the particle of mass $m = mg$

$$= \frac{GM_e mr}{R_e^3}$$

Note: ***Those layers of the sphere which have radii larger than r do not have contribution to the gravitational force.***

59. a. $x = \frac{GM}{R^2}$

Again,

$$\frac{x}{4} = \frac{GM}{(R+h)^2} \text{ or } x = GM\left(\frac{2}{R+h}\right)^2$$

$$\therefore \quad \frac{1}{R^2} = \left(\frac{2}{R+h}\right)^2$$

$$= \frac{2}{R+h} \text{ or } R + h = 2R \text{ or } h = R$$

60. a. Conserving angular momentum,

$$mv_1 r_{min} = mv_2 r_{max} \text{ or } \frac{v_1}{v_2} = \frac{r_{max}}{r_{min}}$$

61. d. $E = \frac{G \times 1}{1^2} + \frac{G \times 1}{2^2} + \frac{G \times 1}{4^2} + \cdots +$

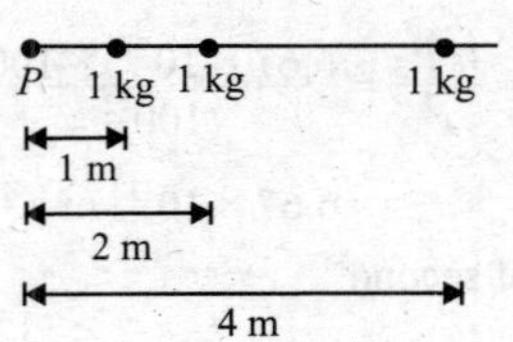

Fig. 5.83

$$E = G\left[1 + \frac{1}{4} + \frac{1}{16} + \cdots\right]$$

$$= G\frac{1}{1 - \frac{1}{4}} = \frac{4G}{3}$$

62. d. $g_p = g_e$; $\frac{GM_p}{R_p^2} = \frac{GM_e}{R_e^2}$; $\frac{4}{3}\pi R_p d_p = \frac{4}{3}\pi R_e d_e$

or $2R_e d_p = R_e d_e \Rightarrow \frac{d_p}{d_e} = \frac{1}{2}$

63. d. The given system may be regarded as a system of three particles located at the three vertices of an equilateral triangle of side $2r$.

Now, $F_A = F_B$

$$= \frac{Gm^2}{(2r)^2} = \frac{Gm^2}{4r^2}$$

F_A and F_B are inclined to each other at an angle of 60°. If F is the resultant of F_A and F_B, then

$$F = \sqrt{3} \times \frac{GM^2}{4r^2}$$

64. c. Increase in gravitational potential energy is

$$\left[\frac{GMm}{R - \frac{R}{5}}\right]\left[-\frac{GMm}{R}\right] = \frac{GMm}{R} - \frac{GMm \times 5}{6R}$$

$$= \frac{GMm}{R}\left[1 - \frac{5}{6}\right] = \frac{GMm}{R} \times \frac{1}{6}$$

$$= \frac{mgR^2}{6R} = \frac{1}{6}mgR = \frac{5}{6}mgh \ [\because R = 5h]$$

65. c. $\frac{40}{80} = \frac{R^2}{(R+h)^2} = \sqrt{2}\,R = R + h$

$$h = (\sqrt{2} - 1)R$$

66. d. Gravitational potential energy of a pair of particles = – $(Gm^2)/l$.

Since we have three pairs, the total gravitational potential energy is – $3Gm^2/l$.

67. b. Gravitational potential at a point on the surface of earth is $-\frac{GM}{R}$.

If the earth is assumed to be a solid sphere, then the gravitational potential at the centre of the earth is – $(3/(2GM)/R)$.

Decrease in gravitational potential is

$$\frac{R}{2} \times \frac{GM}{R^2} = \frac{Rg}{2}$$

Loss in potential energy is

$$\frac{R}{2} \times \frac{GM}{R^2} \times m$$

Now, gain in kinetic energy = loss in potential energy.

Therefore,

$$\frac{1}{2}mv^2 = \frac{1}{2}mgR \text{ or } v = \sqrt{gR}$$

68. b. $m_1 = 4/3\pi R^2 \rho, m_2 = 4/2\pi(2R)^3\rho$, distance between the centres of the two spherical objects, $r = 3R$.

$$F = \frac{GM_1 m_2}{r^2} = G\left(\frac{4}{3}\pi R^3 \rho\right)\left(8 \times \frac{4}{3}\pi R^3 \rho\right) \times \frac{1}{(3R)^2}$$

$$= \frac{128}{81} G\pi^2 R^4 \rho^2$$

69. b. $\frac{GMm}{r_0} = \frac{1}{2}mv^2 - \frac{GMm}{R}$

or $\frac{1}{2}mv^2 = \frac{GMm}{R} - \frac{GMm}{r_0}$

or $\frac{v^2}{v} = \frac{GM}{R} - \frac{GM}{r_0}$

or $v^2 = 2GM\left[\frac{1}{R} - \frac{1}{r_0}\right]$

or $$v = \sqrt{2GM\left(\frac{1}{R} - \frac{1}{r_0}\right)}$$

70. d.
$$F_{12} = \frac{GM^2}{2R^2}$$
$$F_{14} = \frac{GM^2}{2R^2}$$

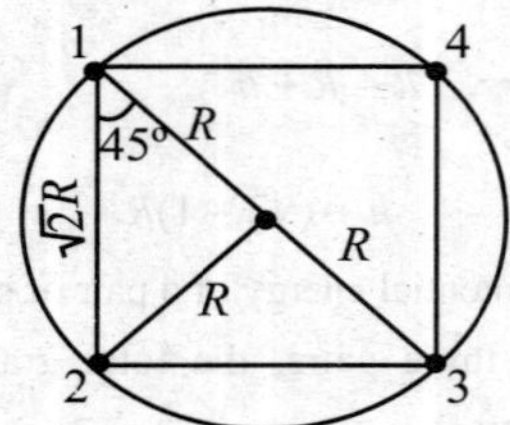

Fig. 5.84

The resultant of these two forces is $(\sqrt{2}GM^2)/2R^2$. Now, $F_{12} = (GM^2)/(4R^2)$.

The combined resultant of all the forces is

$$\frac{\sqrt{2}GM^2}{2R^2} + \frac{GM^2}{4R^2} \text{ or } \frac{GM^2}{R^2}\left[\frac{\sqrt{2}}{2} + \frac{1}{4}\right]$$

Equating this with centripetal force, we get

$$\frac{Mv^2}{R} = \frac{GM^2}{R^2}\left[\frac{2\sqrt{2}+1}{4}\right]$$

or $v^2 = \frac{GM}{R}\left[\frac{2\sqrt{2}+1}{4}\right]$ or $v = \sqrt{\frac{GM}{R}\left(\frac{2\sqrt{2}+1}{4}\right)}$

71. c. The required distance is

$$x \times \frac{d}{\sqrt{\frac{m_2}{m_2}} + 1} = \frac{60R}{\sqrt{\frac{81}{1}} + 1} = 6R$$

72. b. i. Gravitational force on the particle placed at the midpoint D of side AB of length a is $= \vec{F}_1 + \vec{F}_2 + \vec{F}_3$. But $\vec{F}_1$ and $\vec{F}_3$ are equal in magnitude and opposite in direction.

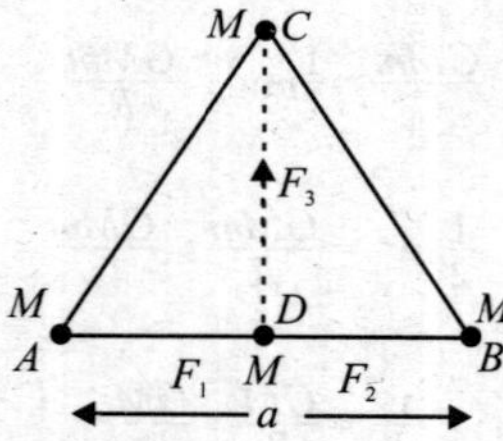

Fig. 5.85

$$F = F_3 = \frac{4GMm}{3a^2} = \frac{4GM^2}{3a^2} \qquad \left[\because CD^2 = \frac{3a^2}{4}\right]$$

ii. Gravitational force on the particle placed at the point of intersection of three medians $= \vec{F}_1 + \vec{F}_2 + \vec{F}_3 = 0$, since the resultant of $\vec{F}_1$ and $\vec{F}_2$ is equal and opposite to $\vec{F}_3$.

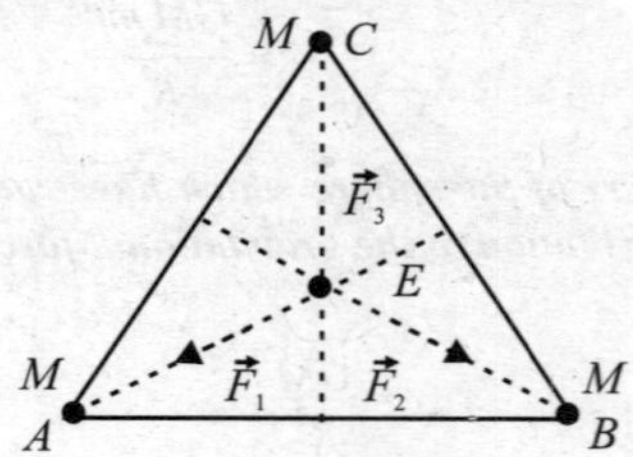

Fig. 5.86

73. b.
$$T = \frac{2\pi R}{v}$$
$$E = \frac{1}{2}mv^2 = \frac{GMm}{R^{n-1}}$$
or $$v = \left[\frac{2GM}{R^{n-1}}\right]^{1/2}$$
$\therefore$ $$T = \frac{2\pi R}{\sqrt{2GM/R^{n-1}}} = \frac{2\pi}{\sqrt{2GM}} \times R^{(n+1)/2}$$
$\Rightarrow$ $$T \propto R^{(n+1)/2}$$

74. b.
a_1 = acceleration of first
$$= \frac{Gm_1m_2}{r_2} \times \frac{1}{m_1} = \frac{Gm_2}{r^2} = \frac{6.67 \times 10^{-11} \times 100}{(100)^2}$$
$$= 6.67 \times 10^{-13} \text{ m s}^{-1}$$
a_2 = acceleration of second
$$= \frac{Gm_1}{r^2} = \frac{6.67 \times 10^{-11} \times 100}{(100)^2} = 6.67 \times 10^{-13} \text{ m s}^{-1}$$
Net acceleration of approach
$$a = a_1 + a_2 = 2 \times 6.67 \times 10^{-13} \text{ m s}^{-1}$$
Now, $$s = \frac{1}{2}at^2$$
$$1 \times 10^{-2} = \frac{1}{2} \times 6.67 \times 10^{-13} \times 2 \times t^2$$
Solving, we get $t = 1.41$ days.

75. d. For the satellite, the gravitational force provides the necessary centripetal force, i.e.,
$$\frac{GM_em}{(R+x)^2} = \frac{mv_0^2}{(R+x)} \text{ or } \frac{GM_e}{R^2} = g$$
$$v_0 = \sqrt{\frac{gR^2}{R+x}}$$

76. b. Using conservation of energy:
$$\frac{1}{2}m(Kv_e)^2 - \frac{GMm}{R} = -\frac{GMm}{R+h}$$
Use $v_e = \sqrt{2gR}$ and $GM = gR^2$

and to solve we get $h + R = R/(1 - K^2)$.

77. c. $\sqrt{\frac{GM}{R+h}} = \frac{1}{2}\sqrt{\frac{2GM}{R}} \Rightarrow h = R$

78. c. Gravitational field inside a shell is zero, but for points outside it, the shell behaves as if whole of its mass is concentrated at its centre. Hence, for a point lying in between the shells there will be a field only due to the inner shell ($= GM_2/r^2$) because the point will be an external point for the inner shell but internal for the outer shell.

79. d.

$$L = mv_0(R+h)$$

$$= m\sqrt{\frac{GM}{(R+h)}}(R+h) = m\sqrt{GM(R+h)}$$

i.e., L depends on m, M as well as h.

80. d. $V = \frac{GM}{R} + \left(-\frac{Gm}{r}\right) = -G\left[\frac{M}{R} + \frac{m}{r}\right]$

81. c. $V = -\frac{GM}{R} + \left(-\frac{Gm}{x}\right) = -G\left[\frac{M}{R} + \frac{m}{x}\right]$

82. d. The point lies inside both the shells, hence gravitational field due to both is zero.

83. d. Let ρ = density of sphere, R = radius of sphere, r = radius of the spherical cavities.

Mass of the complete sphere $= \frac{4}{3}\pi R^3\rho = M$

Mass of the removed sphere $= \frac{4}{3}\pi r^3\rho = m$

Here $\quad m = \frac{Mr^3}{R^3} = \frac{M(1)^3}{(4)^3} = \frac{M}{64}$

Now $\quad \vec{I}_R = \vec{I} + \vec{I}_P + \vec{I}_Q$

Here $\quad I = 0$, also $\vec{I}_P = -\vec{I}_Q \Rightarrow I_R = 0$

84. a. At the centre of the cavities, if the smaller spheres are not removed, then

$$I_R = \frac{GM}{R^3} \times \frac{GM}{64} \times 2 = \frac{GM}{32}$$

Also, $\quad \vec{I}_R = \vec{I} + \vec{I}_P + \vec{I}_Q$

$$\vec{I} = \vec{I}_R + \vec{I}_P + \vec{I}_Q$$

At P, we have $\quad I_P = 0$ and

$$I_Q = \frac{Gm}{2} = G\left(\frac{M}{64}\right) \times \frac{1}{4^2}$$

Hence, $\quad I = \frac{31GM}{1024}$

85. c. The circle lies outside the bigger sphere.

Hence, $\quad V_R = V + V_P + V_Q$

or $\quad V = V_R + V_P + V_Q$

$$= -\frac{Gm}{6} - \left(-\frac{Gm}{r}\right) \times 2$$

where $\quad r = \sqrt{6^2 + 2^2} = \sqrt{40}$

$\therefore \quad V = -\frac{GM}{2}\left[\frac{1}{3} - \frac{1}{32\sqrt{10}}\right]$

86. b. Force on m due to masses at Q and R is zero. So, the net force is due to the mass at P.

Hence, $\quad F = (GMm)/(PL)^2$

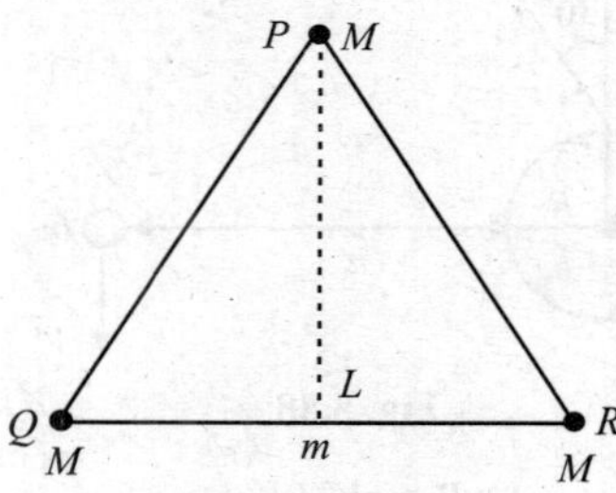

Fig. 5.87

Now, $\quad PL = l\sin 60° = l\frac{\sqrt{3}}{2}$

$$F = \frac{4GMm}{3l^2}$$

87. b. $\quad U_P = -\sum\frac{GMm}{r}$

$$= -GMm\left[\frac{1}{l\sin 60°} + \frac{1}{l/2} + \frac{1}{l2}\right]$$

$$= -\frac{2}{\sqrt{3}}\frac{GMm}{l}[1 + 2\sqrt{3}]$$

88. c. $I = \frac{F}{m} = \frac{4GMm/(3l^2)}{m} = \frac{4GM}{3l^2}$

89. d. Here, angular momentum is conserved. According to it,

$$I_1\omega_1 = I_2\omega_2$$

or $\quad MR_1^2\omega_1 = MR_1^2\omega_2$ or $R_1v_1 = R_2v_2$

At point P_e, the value of R is minimum and hence the velocity is maximum or KE is maximum.

90. a. In the first case, $v_e = \sqrt{\frac{2GM}{R}}$ or $50 = \sqrt{\frac{2GM}{R}}$

In the second case, $v_e = \left[\frac{2G(4M)^2}{R}\right]^{\frac{1}{2}} = 2\sqrt{\frac{2GM}{R}}$

$$= 2 \times 50 = 100 \text{ km s}^{-1}$$

91. a. As angular momentum is conserved, hence

$$I_1\omega_1 = I_2\omega_2$$

or $\quad MR_1^2\omega_1 = MR_1^2\omega_2$ or $MR_1v_1 = MR_2v_2$

$$\left(\because \omega = \frac{v}{R}\right)$$

or $\quad R_1v_1 = R_2v_2$ or $VR = vr$

$\therefore \quad V = \frac{vr}{R}$

92. a.

$$g = \frac{GM}{r^2} = G\frac{\frac{4}{3}\pi r^3\rho}{r^2} = G\frac{4}{3}\pi r\rho$$

$$\frac{g_1}{g_2} = \frac{r_1\rho_1}{r_2\rho_2}$$

93. c. Equal time is taken to cover equal area.

94. b. Conservation of angular momentum of the body about O yield the following:

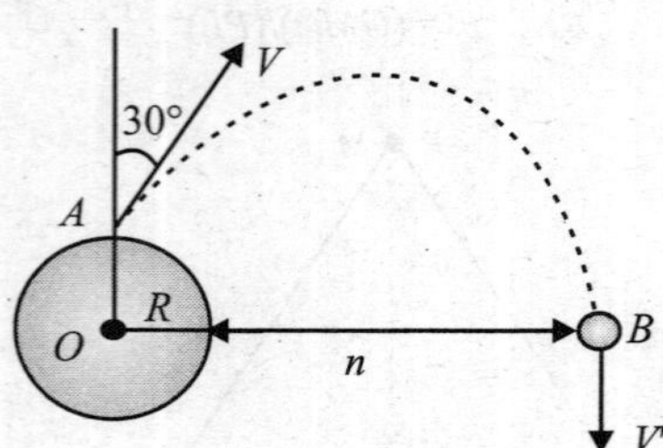

Fig. 5.88

$(mv \sin 30°)$ $\quad R = nV'(R+h)$

$$\frac{V}{2}R = \frac{V}{4}(R+h) \qquad \left[\therefore V' = \frac{V}{4}\right]$$

Therefore, $\quad h = R$

95. d. If the binary stars are rotating about their common centre of mass, then they have to be on the same line all the time, otherwise the centre of mass will be changing. Their angular velocities have to be the same although in the same time, the smaller mass will describe a bigger circle.

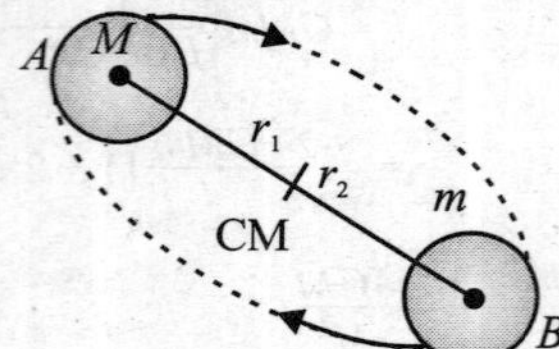

Fig. 5.89

But as $\quad \omega = \frac{2\pi}{T}$, so $T_A = T_B$

96. c. If x_1 and x_2 are the distances covered by the two bodies, then $x_1 + x_2 = 9R$.

Also, $\quad Mx_1 = 5Mx_2 \Rightarrow x_2 = \frac{x_1}{5}$

$$x_1 + \frac{x_1}{5} = 9R \Rightarrow x_1 = 7.5R$$

97. c. At height h above the surface of the earth, therefore

$$g' = g\left(1 - \frac{2h}{R}\right) \Rightarrow \Delta g_1 = \frac{2h}{R}g$$

At depth d below the surface of the earth

$$g' = g\left(1 - \frac{d}{R}\right) \Rightarrow \Delta g_2 = \frac{d}{R}g$$

$$\Delta g_1 = \Delta g_2 \Rightarrow d = 2h$$

98. d. Work done = change in GPE = $U_\infty - U_R$

$$W = 0 - \left(\frac{GMm}{R}\right) = \frac{GMm}{R}$$

$$= \frac{6.67 \times 10^{-11} \times 100 \times 10^{-2}}{10^{-1}} = 6.67 \times 10^{-10}\ \text{J}$$

99. a. Let the minimum speed imparted to the particle of mass m so that it just reaches the surface of the earth is v. Applying the principle of conservation of energy,

$$\frac{1}{2}mv^2 + \left(-\frac{3}{2}\frac{GM}{R}m\right) = \frac{GM}{R}m + 0$$

Solving, we get $v = \sqrt{\frac{GM}{R}}$.

100. b. Mass of sphere $M \propto R^3$. If mass of complete sphere is M, then, the mass of cavity will be $M/8$.

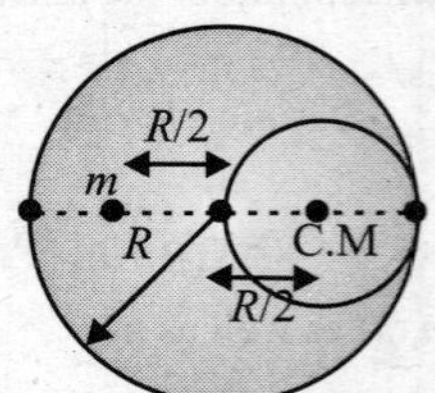

Fig. 5.90

Using the figure, F = force due to whole sphere – force due to cavity.

$$F = \frac{GMm}{R^3}\left(\frac{R}{2}\right) - \frac{G(M/8)m}{(R)^2}$$

$$\frac{GMm}{R^2}\frac{R^2}{2} - \frac{G(M/8)m}{R^2}$$

$$\frac{GMm}{R^2}\left[\frac{4}{8} - \frac{1}{8}\right] = \frac{3}{8}\frac{GMm}{R^2} = \frac{3}{8}mg$$

101. c. Given $8 = \dfrac{2\pi}{\omega_1 + \omega_2} = \dfrac{2\pi}{\frac{2\pi}{T_1} + \frac{2\pi}{T_2}}$

$\Rightarrow \quad T_1 = 24$ h for the earth

$\Rightarrow \quad T_2 = 12$ h (T_2 being the time period of the satellite, it will remain the same as the distance from the centre of the earth remains constant)

$$\Rightarrow \qquad T = \frac{2\pi}{\omega_2 + \omega_1} = \frac{2\pi}{\frac{2\pi}{T_1} - \frac{2\pi}{T_2}} = 24\ \text{h}$$

102. a.

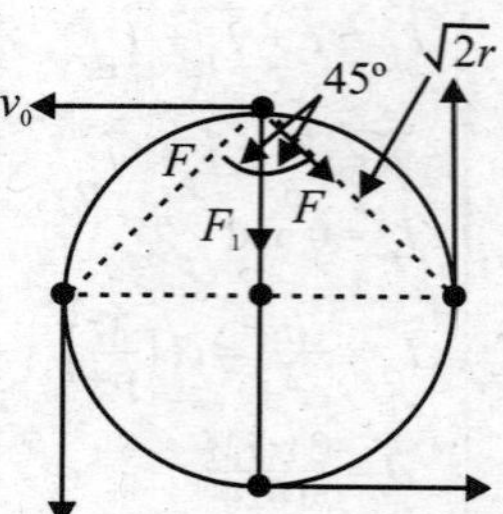

Fig. 5.91

Centripetal force = net gravitational force

$$\Rightarrow \qquad mv_0^2 = 2F\cos 45° + F_1 = \frac{2Gm^2}{(\sqrt{2})^2}\frac{1}{\sqrt{2}} + \frac{Gm^2}{4r^2}$$

$$\frac{mv_0^2}{r} = \frac{GM^2}{4r^2}[2\sqrt{2} + 1]$$

$$\Rightarrow \qquad v_0 = \left[\frac{Gm(2\sqrt{2}+1)}{4r}\right]^{\frac{1}{2}}$$

103. c. $M_A = \sigma 4\pi R_A^2, M_B = \sigma 4\pi R_B^2$
where σ is surface density.

$$\Rightarrow V_A = \frac{-GM_A}{R_A}, V_B = \frac{-GM_B}{R_B}$$

$$\frac{V_A}{V_B} = \frac{M_A R_B}{M_B R_A} = \frac{\sigma 4\pi R_A^2}{\sigma 4\pi R_B^2}\frac{R_B}{R_A} = \frac{R_A}{R_B}$$

Given $\frac{V_A}{V_B} = \frac{M_A}{M_B} = \frac{3}{4}$

Then $R_B = \frac{4}{3} R_A$

For new shell of mass M and radius R,

$$M = M_A + M_B = \sigma 4\pi R_A^2 + \sigma 4\pi R_B^2$$

$$\sigma 4\pi R^2 = 4\pi(R_A^2 + R_B^2)$$

Then $\frac{V}{V_A} = \frac{MR_A}{RM_A} = \frac{\sigma 4\pi[R_A^2 + R_B^2]}{(R_A^2 + R_B^2)^{\frac{1}{2}}}\frac{R_A}{\sigma 4\pi R_A^2} = \frac{\sqrt{R_A^2 + R_B^2}}{R_A} = \frac{5}{3}$

104. d.

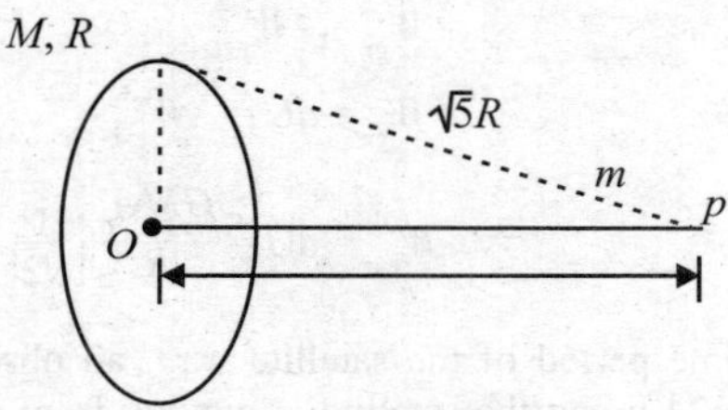

Fig. 5.92

Gravitational potential at 'P',

$$v_P = \frac{-GM}{\sqrt{5}R}$$

Gravitational potential at 'O',

$$v_O = -\frac{GM}{R}$$

Using work energy theorem,

$$W = \Delta K \Rightarrow m[v_P - v_O] = \frac{1}{2} mv^2$$

$$m = \left[\frac{GM}{R} - \frac{GM}{\sqrt{5}R}\right] = \frac{1}{2} mv^2$$

$$\sqrt{\frac{2GM}{R}\left[1 - \frac{1}{\sqrt{5}}\right]} = V$$

105. c. During total eclipse, total attraction due to the Sun and the Moon,

$$F_1 = \frac{GM_sM_e}{r_1^2} + \frac{GM_mM_e}{R_2^2}$$

When the Moon goes on the opposite side of the earth, the effective force of attraction is

$$F_2 = \frac{GM_sM_e}{r_1^2} - \frac{GM_mM_e}{r_2^2}$$

Change in force, $\Delta F = F_1 - F_2 = \frac{2GM_mM_e}{r_2^2}$

Change in acceleration of the earth,

$$\Delta a = \frac{\Delta F}{M_e} = \frac{2GMm}{r_2^2}$$

Average force on the earth,

$$F_{av} = \frac{F_1 + F_2}{2} = \frac{GM_sM_e}{r_1^2}$$

Average acceleration of the earth,

$$a_{av} = \frac{F_{av}}{M_e} = \frac{GM_s}{r_1^2}$$

Percentage change in acceleration is

$$\frac{\Delta a}{a_{av}} \times 100 = \frac{2GM_m}{r_2^2} \times \frac{r_1^2}{GM_s} \times 100 = 2\left(\frac{r_1}{r_2}\right)^2 \frac{M_m}{M_s} \times 100$$

106. b. Using Kepler's third law, the semi-major axis of the comet's orbit is given by

$$T^4 = \frac{4\pi^2}{GM_s} \times a^3$$

$$a = \left[\frac{T^2 \times GM_s}{4\pi^2}\right]^{\frac{1}{3}}$$

$$= \left[\frac{(90 \times 24 \times 3600) \times 6.67 \times 10^{-11} \times 2 \times 10^{30}}{4\pi^2}\right]^{\frac{1}{3}}$$

$$= 5.89 \times 10^{10} \text{ m}$$

So, the major axis of the comet's orbit would be $2a = 1.178 \times 10^{11}$ m

As $2a$ < mean distance between the earth and the Sun, so the collision is not possible.

107. c. As we decrease the magnitude of mechanical energy of the spacecraft–earth system, it means we are increasing the energy of the spacecraft–earth system as the total energy of the bounded system is negative.

As we change the energy, the circular orbit of the spacecraft will become elliptical. Let a be the semi-major axis of this new elliptical orbit.

$$E_{final} = -\frac{GM_m}{2a}$$

$$E_{initial} = \frac{GM_m}{2r}, \text{ where } r = 7000 \text{ km}$$

$$= \frac{GM_m}{2r} = -\frac{GM_m}{2a}$$

$$a = \frac{r}{0.9} = 1.11\, r$$

$r_{max} = 2a$ [where r_{max} is the distance corresponding to aphelion]

$= 2.22r - r = 1.22r$

Required greatest height, $h = r_{max} - R_e = 2140$ km

108. b.

$$E = \frac{m}{2} \times \frac{6GM_e}{5r} - \frac{GM_e m}{2r} = -\frac{2}{3}\frac{GM_e m}{r}$$

which is the total energy of the earth–satellite system.

So, semi-major axis of the elliptical orbit is $a = \frac{5r}{4}$.

Speed of the satellite at the apogee position is $v_A = \frac{v_P \times r}{2a - r}$

$$= \frac{2}{3}\sqrt{\frac{6GM_e}{5r}}$$

For orbit to change to a circle of radius $3r/2 = (2a - r)$, the rocket has to be fired when the satellite is at the apogee position.

New orbital speed is $v_0 = \sqrt{\frac{GM_e}{3r/2}} = \sqrt{\frac{2GM_e}{3r}}$.

Required change in the orbital speed is

$$\Delta v = v_A - v_0 = 0.085\sqrt{\frac{GM_e}{R}}$$

109. a. As all the points on the periphery of either ring are at the same distance from point *P*, the potential at point *P* due to the whole ring can be calculated as $V = -(GM)/(\sqrt{R^2 + x^2})$ where x is the axial distance from the centre of the ring. This expression is independent of the fact whether the distribution of mass is uniform or non-uniform.

So, at *P*, $V = -\frac{GM}{\sqrt{2}R} - \frac{G \times 2M}{\sqrt{5}R} = -\frac{GM}{R}\left[\frac{1}{\sqrt{2}} + \frac{2}{\sqrt{5}}\right]$.

110. d. Inside the spherical shell, *V* is constant, so from energy conservation,

$$\frac{-GMm}{3R} = \frac{mv^2}{2} - \frac{GMm}{R}$$

$$\frac{v^2}{2} = \frac{GM}{R}\left[1 - \frac{1}{3}\right] = \frac{GM}{R} \times \frac{2}{3} \quad \text{or} \quad v = \sqrt{\frac{4GM}{3R}}$$

111. a. Let the speed at the maximum height be v_1, then from energy be conservation,

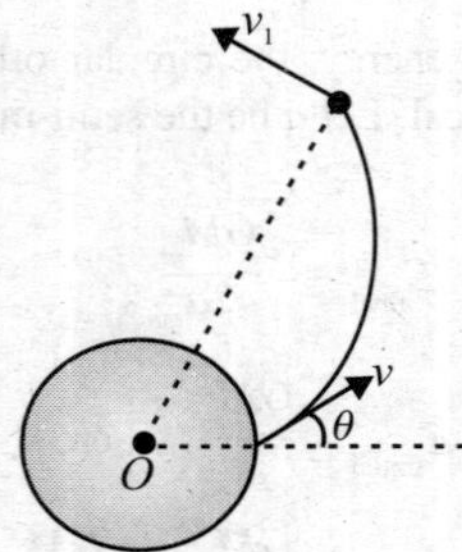

Fig. 5.93

$$\frac{mv^2}{2} - \frac{GM_m}{R_e} = \frac{mv_1^2}{2} - \frac{GMm}{5R_e/4}$$

From angular momentum conservation,

$$mv_{R_e} \sin\theta = mv_1 \times \frac{5R_e}{4}$$

Using $v = \sqrt{\frac{gR_e}{2}}$ and solving the equation, we get $\theta = \sin^{-1}\left(\frac{\sqrt{5}}{4}\right)$.

112. c. A minimum amount of energy equal to total energy of the moon-earth system has to be given to break (unbound) the system. The Sun is exerting force on the Moon but not providing any energy.

113. c. As the sphere is having non-uniform mass density, so nothing can be predicted about the variation of gravitational field intensity.

114. b. Even though the distribution of the mass is unknown, we can find the potential due to the ring on any axial point because from any axial point the entire mass is at the same distance (whatever would be the nature of distribution).

Potential at *A* due to the ring is $V_A = \frac{GM}{\sqrt{2}R}$

Potential at *B* due to the ring is $V_B = \frac{GM}{\sqrt{5}R}$

$$d_U = U_f - U_f - U_i = U_B - U_A = m_0(V_B - V_A)$$

$$= \frac{GMm_0}{R}\left[-\frac{1}{\sqrt{5}} + \frac{1}{\sqrt{2}}\right]$$

$$W_{gr} = -W_{ext}$$

$$W_{gr} = dU = -W_{ext}$$

$$W_{ext} = \text{dU} = \frac{GMm_0}{R}\left[\frac{1}{\sqrt{2}} - \frac{1}{\sqrt{5}}\right]$$

115. a. Here time period of the satellite w.r.t. an observer on the equator is 24 h and the satellite is moving from west to east, so angular velocity of the satellite w.r.t. earth's axis of rotation (considered as fixed) is $\omega = \frac{2\pi}{T_s} + \frac{2\pi}{T_e}$, where T_s and T_e are time periods of satellite and earth, respectively.

$$\omega = \frac{\pi}{6}(\text{h}^{-1}) = 1.45 \times 10^{-4} \text{ rad s}^{-1}$$

From, $$v = \sqrt{\frac{GM}{r}}$$

$$r\omega = \sqrt{\frac{GM}{r}}$$

$$r^{\frac{3}{2}} = \sqrt{\frac{GM}{r}}$$

$$r = 2.66 \times 10^7 \text{ m} = 2.66 \times 10^4 \text{ km}$$

116. a. As the velocity of the particle is less than the orbital velocity of the satellite, the particle goes in the elliptical orbit of the semi-major axis less than *r*.

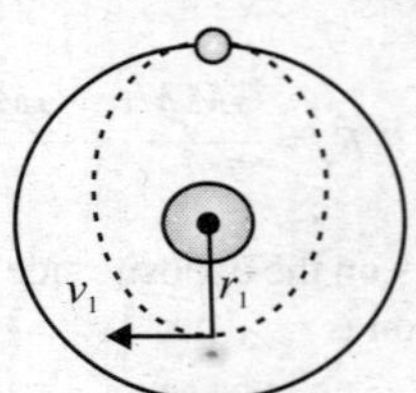

Fig. 5.94

Let r_1 be the minimum distance and v_1 be the velocity of the particle at this position, then

$m_0 \times \sqrt{\frac{2}{3}} \times v_0 r = m_0 v_1 r_1$, where m_0 is the mass of the particle and v_0 is the orbital speed equal to $\sqrt{GM/r}$.

$$v_1 r_1 = \sqrt{\frac{2}{3}}\, v_0 r$$

From energy conservation,

$$\frac{m_0 \times \frac{2}{3} v_0^2}{2} - \frac{Gm_0}{r} = \frac{m_0 v_1^2}{2} - \frac{GMm_0}{r_1}$$

Solving the equations, we get $r_1 = r/2$.

Multiple Correct Answers Type

1. a, b, c, d.

For all the points, $E_P > E_K$ (numerically). So, the total energy is negative. Thus, the system is a bound system corresponding to all the points.

2. a, d.

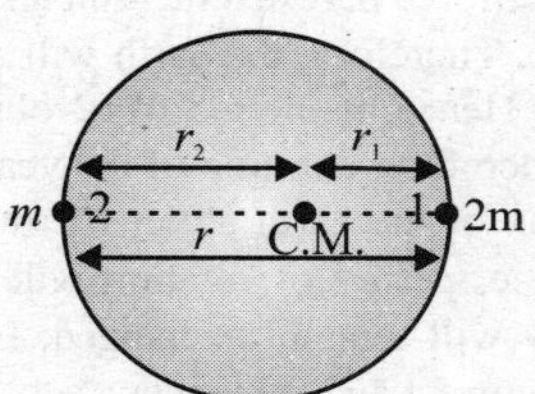

Fig. 5.95

$$r_2 = \frac{2mr}{m + 2m} = \frac{2r}{3}$$

$$T_2^2 = \frac{4\pi^2 r_2^2}{Gm} \Rightarrow T_2^2 = \frac{32\pi^2 r^3}{27Gm}$$

$$T_2 \propto r^{\frac{3}{2}} \quad \text{and} \quad T_2 \propto m^{\frac{1}{2}}$$

3. b., c.

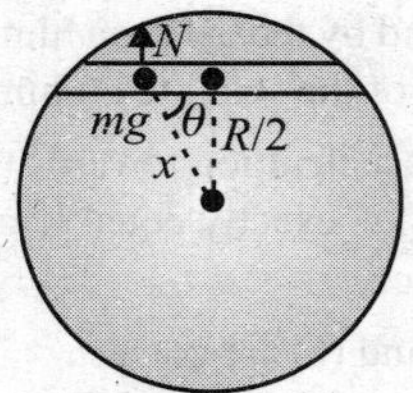

Fig. 5.96

Net force towards the centre of the earth $= mg' = \frac{mGMx}{R^3} = \frac{mgR^2x}{R^3}$ $= (mgx)/R$

Normal force, $N = mg' \sin\theta$

Thus, pressing force, $N = \frac{mgx}{R}\frac{R}{2x}$

$N = \frac{mg}{2}$ is constant and independent of x.

Hence, option (b) is correct.

Tangential force, $F = ma = mg' \cos\theta$

$$a = g' \cos\theta = \frac{gx}{R}\frac{\sqrt{\frac{R^2}{4} - x^2}}{x} \Rightarrow a = \frac{gx}{R}\sqrt{R^2 - 4x^2}$$

Curve is parabolic and at $x = \frac{R}{2}$, $a = 0$.

Hence, option (c) is correct.

4. b., c.

When a planet is on the major axis of the orbit, gravitational force on the planet is normal to its motion. So, no work is done. As that energy remains the same, no work is done in complete revolution

5. a., c.

If $r > R$, $g' = g\frac{R^2}{(R+h)^2} = g\frac{R^2}{r^2}$

If $r < R$, $g' = g\left[\frac{R-d}{R}\right]$

where d is the depth below the surface of earth.

If $d = R, g = 0$.

Further, due to rotation: $g' = g - R\omega^2 \cos^2\theta$

If $\omega = 0$, then g increases.

6. a., c., d.

From relation $|E| = \frac{dV}{dr}$

when $V = 0$, $E = 0$; when $V =$ constant, $E = 0$; when $V \neq 0, E \neq 0$.

So, options (a), (c) and (d) are correct.

7. a., b., c., d.

i. If $v < v_e$ and the body is projected vertically upwards, the body will rise up to that height where its velocity becomes zero. After that it will fall freely due to gravity following a straight line path.

ii. If $v < v_e$ but the body is projected at some angle w.r.t. vertical direction, the body will reach up to a certain height (where vertical component of the velocity becomes zero) and then fall down following a parabolic path (this will be a case of projectile motion).

iii. If $v = v_0 = \sqrt{GM/R}$, then the orbit will be circular.

But if $v_0 < v < v_e$ (escape velocity), then the orbit will be an ellipse.

So, all the options are correct.

8. a., c.

If the speed of the orbiting satellite is made $\sqrt{2}$ times, then the satellite will escape because its velocity becomes equal to escape velocity.

This implies that the speed is increased by 41% and the KE is doubled.

9. b., c. If radius of an orbiting satellite is decreased, then its PE gets decreased. As ME = KE + PE, so ME also decreases.

$$\left[\text{PE} = -\frac{GMm}{r}, \text{KE} = +\frac{GMm}{2r} \text{ and ME} = -\frac{GMm}{2r}\right]$$

10. c., d. If two satellites of different masses are revolving in the same orbit, then they have the same time period and speed because

$$v_0 = \sqrt{\frac{GM}{r}}$$

and $T = 2\pi\sqrt{\frac{r^3}{GM}}$

Angular momentum (= mvr) and energy (= – $(GMm)/2r$) both depend on the mass of the satellites. So, options (c) and (d) are correct.

11. a., c. The centre of mass of the double star system remains stationary and both the stars revolve around in circular orbits which are concentric with the centre of mass.

The distance of centre of mass from the heavier star is equal to $\left(\frac{Mr + 2M.0}{M + 2M}\right) = \frac{r}{3}$.

Hence, the heavier star revolves in a circle of radius $r/3$ while the lighter star in a circle of radius $2r/3$. Hence, option (a) is incorrect.

To calculate period of revolution of a double star system, concept of reduced mass can be used.

The reduced mass of the system is equal to

$$\frac{(M)(2M)}{(M + 2M)} = \frac{2M}{3}.$$

Hence, the period of revolution will be equal to $\dfrac{2\pi}{\sqrt{2\frac{GM}{3}}} r^{\frac{3}{2}}$

where r is distance between two stars. Hence, option (b) is correct.

Kinetic energy of a star will be equal to $1/2\, mv^2$ where v is speed of the star which is equal to (radius of its circular orbit) × ω.

Hence, KE of heavier star is

$$E_1 = \frac{1}{2}(2M)\left(\frac{r}{3}\omega\right)^2$$

And that of lighter star, $E_2 = \frac{1}{2}M\left(\frac{2r}{3}\omega\right)^2$.

It means, KE of the lighter star is twice that of the heavier star. Hence, option (c) is incorrect.

12. b., c. Since every element of the hemispherical shell is at a distance R from the centre of curvature, therefore gravitational potential at its centre = – GM/R. It is not equal to GM/R. Hence, option (a) is incorrect.

Gravitational field strength at a point, lying on the axis of a thin uniform circular ring of radius R, can be calculated by considering two equal elemental lengths of the ring such that these are lengths lying on the ends of a diameter and thus the expression given in option (b) is obtained. Hence, option (b) is correct.

Newton's law of gravitation is applicable to only those bodies which have spherically symmetric distribution of mass. For example, if we consider two solid hemispheres of equal radii; one made of wood and the other made of iron. These two hemispheres are joined together to form a shape of complete sphere, we cannot apply Newton's gravitational law to this sphere. So, option (c) is correct.

13. a., b., d. If a satellite is revolving in a circular orbit then the magnitude of its KE is equal to half of the magnitude of gravitational potential energy. But the potential energy is negative. That is why the total energy of a revolving satellite is negative. If the satellite experiences resistance due to cosmic dust, then it follows a spiral path of decreasing radius. During the process of motion, its potential energy decreases. It means PE becomes more negative or the magnitude of PE increases. Hence, magnitude of KE also increases. Therefore, options (a) and (b) are correct.

Since cosmic dust exerts a tangential force on the satellite, therefore it experiences a retarding moment, hence, its angular momentum does not remain conserved. Hence, option (c) is incorrect. Since KE of the satellite increases, therefore speed of the satellite increases. Power acting against the resisting force is equal to force × speed, therefore it increases. Hence, thermal power generated increases with time. Hence, ultimately it will turn off. Therefore, option (d) is correct.

14. b., d. If universal gravitational constant G starts to decrease, then gravitational force between the Sun and the earth will also start to decrease. Therefore, the earth will try to follow a path of larger radius. Hence, its period of revolution round the Sun will increase. Therefore, duration of the year will increase. But rotational motion of the earth about its own axis will remain unchanged; hence, period of rotation will remain unchanged or length of day will remain unchanged. Hence, option (a) is incorrect, (b) is correct and (c) is incorrect.

Since the radius of the circular path of the earth will increase or the earth will follow a spiral path of increasing radius, therefore its PE will increase but PE is always negative, so the magnitude of PE will decrease. Hence, KE will also decrease. Therefore, option (d) is correct.

15. b., c. For an earth satellite moving in a circular orbit, the centripetal force required for its circular motion is provided by the gravitational force exerted by the earth on it.

It means the resultant force on the astronaut is equal to the gravitational force exerted by the earth on him. Hence, no reaction is exerted by the floor of the satellite on him.

In other words, his acceleration towards the earth's centre (centripetal acceleration) is exactly equal to acceleration caused by the gravitational force.

Hence, options (b) and (c) are correct.

16. a., b., c. When the particle is dropped in the tunnel at its one end, it starts to execute simple harmonic motion with mean position as the centre of the tunnel. Hence, acceleration of the particle is directly proportional to its distance from the centre of the tunnel. It means, option (d) is correct.

The particle has zero velocity at one end because it is dropped from rest at that end, therefore, ends of the tunnel are its extreme positions. Hence, at the other end, the velocity will become equal to zero. It means option (a) is incorrect.

Velocity of a particle executing S.H.M. is maximum at its mean position. Hence, at the centre of the tunnel, its velocity will be maximum possible. Hence, option (b) is incorrect.

When the particle moves from extreme position to centre of the tunnel, its velocity increases. It means KE increases or PE decreases. But initial potential energy of the particle is negative. Hence, potential energy becomes more negative and it is least at the mid-point of the tunnel because KE is maximum there. It means the gravitational PE can never be equal to zero. Hence, option (c) is also incorrect.

17. b., d. Now, $g = GM/R^2$

If R reduces to $R' = 0.8R$, the value of g becomes

$$g' = \frac{GM}{R^2} = \frac{GM}{0.64\,R^2} = \frac{g}{0.64} = \frac{9.8\text{ m s}^{-2}}{0.64}$$

which is choice (b).

$$\text{Increase in the value of } g, = \frac{g}{0.64} = \frac{0.36g}{0.64}$$

$$\text{Therefore, the percentage increase} = \frac{0.36\,g}{0.64\,g} \times 100 = 56.25\%$$

Hence, choices (b) and (d) are correct.

18. a., b., c. Choices (a), (b) and (c) are correct. When the particle exerts a force on another particle; the second particle exerts an equal and opposite force on the first particle. Choice (d) is incorrect because although G is constant everywhere, the value of g varies with height and depth and also from place to place.

19. b., c. Statement (a) is incorrect; the magnitude of the force is 6.67×10^{-11} N. Statement (b) is correct. If the escape velocity for a planet is high, gases are not able to escape. Statement (c) is also correct. Statement (d) is incorrect; force of friction arises due to electrical forces. Hence, correct choices are (b) and (c).

20. a., c., d. Statement (a) is incorrect; there is no method by which a body can be shielded from the gravitational field of another body. Statement (b) is correct. Statement (c) is incorrect. As the earth rotates about its axis, a body on the surface of the earth also rotates with it. Since the body is in a rotating (non-inertial) frame, it experiences an outward centrifugal force against the inward force of gravity. As a result, the acceleration due to gravity decreases due to rotation. Statement (d) is incorrect, the forces are equal in magnitude but opposite in direction. Hence, choices (a), (c) and (d) are wrong.

21. c., d. Linear velocity or orbital velocity is $v = \sqrt{GM/r}$ where r is the distance of the satellite from the centre of the earth. Therefore, v decreases as r is increased.

Also $\quad v = r\omega$, where ω is the angular velocity.

Therefore, $$\omega = \frac{v}{r} = \frac{1}{r}\sqrt{\frac{GM}{r}} = \sqrt{\frac{GM}{r^3}}.$$

Thus, ω decreases with increase in r. The time period of the satellite is given by

$$T = 2\pi\sqrt{\frac{r^3}{GM}}$$

Thus, T increases as r is increased. Hence, the correct choices are (c) and (d).

22. c., d. Linear speed is $v = r\omega$. At the equator, the radius of the earth r is maximum. Therefore, $\omega = \frac{v}{r}$ is minimum. Also,

$$\omega = \frac{2\pi}{T} = \frac{2 \times 3.14}{24 \times 60 \times 60}$$

$$= 7.3 \times 10^{-5}\text{ rad s}^{-1}$$

Hence, the correct choices are (c) and (d).

23. b., c., d. Distances of the two satellites from the centre of the earth are $r_1 = 2R$ and $r_2 = 8R$, respectively (R = earth's radius. Their potential energies are

$$V_1 = -\frac{GmM}{r_1} \text{ and } V_2 = -\frac{GmM}{r_2}$$

Their ratio is $$\frac{V_1}{V_2} = \frac{r_2}{r_1} = \frac{8R}{2R} = 4$$

The kinetic energy of a satellite can be obtained from the relation

$$\frac{mv^2}{r} = \frac{GmM}{r^2}$$

or $$K = \frac{1}{2}mv^2 = \frac{GmM}{2r}$$

$$K_1 = \frac{GmM}{2r_1} \text{ and } K_2 = \frac{GmM}{2r_2}$$

The ratio of their kinetic energies is

$$\frac{K_1}{K_2} = \frac{r_2}{r_1} = \frac{8R}{2R} = 4$$

Their total energies are

$$E_1 = -\frac{GmM}{r_1} + \frac{GmM}{2r_1} = -\frac{GmM}{2r_1}$$

and $$E_2 = -\frac{GmM}{r_2} + \frac{GmM}{2r_2} = -\frac{GmM}{2r_2}$$

Their ratio is $$\frac{E_1}{E_2} = \frac{r_2}{r_1} = \frac{8R}{2R} = 4$$

Hence, the correct choices are (b), (c) and (d).

24. b., d. The acceleration due to gravity is $g = \frac{GM}{R^2}$.

The new value of g would be

$$g' = \frac{G(0.99M)}{(0.99R)^2} = 1.01\frac{GM}{R^2}$$

Thus, g would increase by about 1%. The new escape velocity would be

$$v_e' = \sqrt{\frac{2 \times 0.99M \times G}{0.99R}} = \sqrt{\frac{2MG}{R}} = v_e$$

Thus, the escape velocity will remain unchanged. The potential energy of a body of mass m on the earth's surface would be

$$-\frac{Gm\,(0.99M)}{(0.99R)} = -\frac{GmM}{R}$$

Thus, the potential energy will also remain unchanged. Hence, the correct choices are (b) and (d).

25. a., c., d. Work done is independent of the path chosen and depends only on the initial and final positions of the object. Also the work done on any closed path in a gravitational field will be zero. Since, every point on the surface of the earth is at the same potential, no work is done for points on the surface of the earth. Hence, the correct choices are (a), (c) and (d).

26. b., c.

$$W = U_2 - U_1 = \frac{GMm}{R+h} - \left(-\frac{GMm}{R}\right) = -GMm\left[\frac{1}{R} - \frac{1}{R+h}\right]$$

$$= gR^2m\left[\frac{h}{R(R+h)}\right] = \frac{mgRh}{R+h}$$

For $h << R,\ W = mgh$

For $h = R,\ W = \dfrac{mgR}{2}$

27. a., c., d.

a. $g = \dfrac{GM}{R^2} = \dfrac{G}{R^2} \times \dfrac{4}{3}\pi R^3\rho$

or $\rho = \dfrac{3g}{4\pi GR}$

b. Acceleration due to gravity varies from place to place, hence, not correct.

c. $v_e' = \sqrt{\dfrac{2G(2M)}{\frac{R}{2}}} = 2\sqrt{\dfrac{2GM}{R}} = 2 \times 11.2 \text{ km s}^{-2} = 22.4 \text{ km s}^{-1}$

d. Factual statement.

28. a., b., c.

a. For a proper explanation, the earth–Moon system will be regarded as an isolated system. At a particular point in space, the gravitational force of attraction of the earth on astronaut will be balanced by the gravitational force of attraction of the Moon.

b. $V = -GM/R$. As R decreases, GM/R increases or $-GM/R$ decreases.

c. In case of the spherical shell, plot of I versus r is discontinuous.

29. a., d.

a. For Moon, escape velocity is 2.4 km s^{-1}.

d. $v_0 = \sqrt{\dfrac{2GM}{R}} = \sqrt{\dfrac{2G}{M} \times \dfrac{4}{3}\pi R^3\rho}$

or, $v_0 \propto \sqrt{\rho}$

30. a., b., c., d.

a. Factual statement.

b. $g = \dfrac{GM}{R^2}$ or $M = \dfrac{gR^2}{G}$

d. The weight of a body on the surface of earth is not determined by the orbital motion of earth around Sun.

31. a., b., c., d. From Kepler's law, $T^2 \propto R^3$

So, $\dfrac{T_1}{T_2} = \left(\dfrac{R_1}{R_2}\right)^{\frac{3}{2}}$

$$R_2 = \left(\frac{T_2}{T_1}\right)^{\frac{2}{3}} \times R_1 = \left(\frac{24}{6}\right)^{\frac{2}{3}} \times 1.28 \times 10^4 \text{ km}$$

$$= 3.22 \times 10^4 \text{ km}$$

Orbital velocity of S_1 is

$$v_1 = \frac{2\pi R_1}{T_1} = \frac{2\pi \times 1.28 \times 10^4}{4}$$

$$= 0.64\,\pi \times 10^4 \text{ km}$$

Orbital velocity of S_2 is

$$v_2 = \frac{2\pi R_2}{T_2} = \frac{2\pi \times 1.28 \times 10^4}{4}$$

$$= 0.64\pi \times 10^4 \text{ km}$$

At $t = 12$ h the two satellites are closest to each other and after every 24 h they come at the same position relative to each other. It is clear that direction of v_2 w.r.t. v_1 is changing continuously in both magnitude and direction.

Angular velocity of S_2 w.r.t. S_1 at $t = 12$ h is

$$\omega = v_1 + v_2/R_2 - R_1 = 0.468\pi \text{ rad s}^{-1}$$

32. a., b., d. Just before collision, the total energy of two satellites is

$$E = -\frac{GMm}{2r} - \frac{GMm}{2r} = -\frac{GMm}{r}$$

Let the orbital velocity be v, then from momentum conservation,

$$mv - mv = 2m \times v_1$$

$$v_1 = 0$$

As velocity of combined mass just after collision is zero, the combined mass will fall towards the earth. At this instant, the total energy of the system only consists of the gravitational potential energy given by

$$U = \frac{GM \times 2m}{2r}$$

Assertion-Reasoning Type

1. a. The torque on a body is given by $\vec{\tau} = d\vec{L}/dt$. In case of a planet orbiting around the Sun, no torque is acting on it.

$$\frac{d\vec{L}}{dt} = 0 \Rightarrow \vec{L} = \text{constant}$$

2. a. The time period of the satellite which is very near to the earth is given by $T = 2\pi\sqrt{\dfrac{R}{g}} \simeq 84 \text{ min} = 1 \text{ h } 24 \text{ min}$.

3. a. $\dfrac{dA}{dt} = \dfrac{1}{2}r^2 \quad \Rightarrow \quad \dfrac{d\theta}{dt} = \dfrac{1}{2}r^2\omega$

$$\frac{mr^2\omega}{2m} = \frac{2}{m} = \text{constant}$$

$$\therefore \quad L = \text{constant}$$

4. d. We can take the sphere as a point mass lying at its centre, but the rod will not be taken as point mass lying at its centre of mass.

5. d. The value of escape velocity is derived from the method of conservation of total mechanical energy and energy is independent of direction.

6. c. The value of g at any place is given by

$$g' = g - \omega^2 R_e \cos^2\lambda$$

If $\omega = 0$, then $g' = g$. The value of g will be same at all places.

7. c. $V_{in} = \frac{GM}{2R^3}[3R^2 - r^2]$.

At surface, $V_s = \frac{GM}{R}$ [at $r = R$]

$$V_{in} = \frac{3}{2} V_s$$

$$V_{in} > V_s$$

V is not the same everywhere as indicated by V_{in}.

8. a. According to Kepler's third law of motion, the square of the time period of a planet about the Sun is proportional to the cube of the semi-major axis of the ellipse or mean distance of the planet from the Sun, i.e., $T^2 \propto a^3$. When a is smaller, the time period is shorter.

9. c. Acceleration due to gravity is given by $g = GM/l^2$. Thus, it does not depend on the mass of the body on which it is acting. Also, it is not a constant quantity. It changes with changes in value of both M and R (distance between two bodies).

10. a. Till the particle reaches the centre of the planet, force on both the bodies are in direction of their respective velocities, hence kinetic energies of both keep on increasing. After the particle crosses the centre of the planet, forces on both are retarding in nature. Hence, as the particle passes through the centre of the planet, sum of kinetic energies of both the bodies is maximum. Therefore Statement 1 is true, Statement 2 is true; Statement 2 is a correct explanation for Statement 1.

11. a. From kinetic theory of gas, $v = \sqrt{\frac{3RT}{M}}$

So, for lighter gas molecules, v is greater which is enough to take these molecules away from the earth's atmosphere.

12. a. Here, both the statements are correct and Statement 2 correctly explains Statement 1.

This can be easily proved by writing the basic equations.

13. d. For Statement 1: Gravitational force is not perpendicular to velocity of the satellite. So for any small part of the orbit work done is not zero, although when satellite is at perihelion or aphelion position then work done by gravitational force for small part would be zero.

14. a. Both the statements are true and Statement 2 correctly explains Statement 1.

15. d. For a satellite to be geostationary, the necessary requirements are as follows:

a. its orbit must be in equatorial plane and circular;

b. its time period must be 24 h;

c. its sense of rotation must be the same as that of the earth about its own axis.

16. d. Here, Statement 1 is incorrect because as speed of one satellite increases, its kinetic energy and hence total energy increases, i.e., total energy becomes less negative and hence r increases, i.e., orbit changes.

Comprehension Type

For Problems 1–2

1. c., 2. a.

Sol. Applying the principle of conservation of angular momentum,

$$mv_1(2R) = mv_2(4R)$$

$$v_1 = 2v_2 \quad \text{(i)}$$

From conservation of energy,

$$\frac{1}{2}mv_1^2 - \frac{GMm}{2R} = \frac{1}{2}mv_2^2 - \frac{GMm}{4R} \quad \text{(ii)}$$

Solving Eqs. (i) and (ii), we get

$$v_2 = \sqrt{\frac{GM}{6R}},\ v_1 = \sqrt{\frac{2GM}{3R}}$$

If r is the radius of curvature at point A,

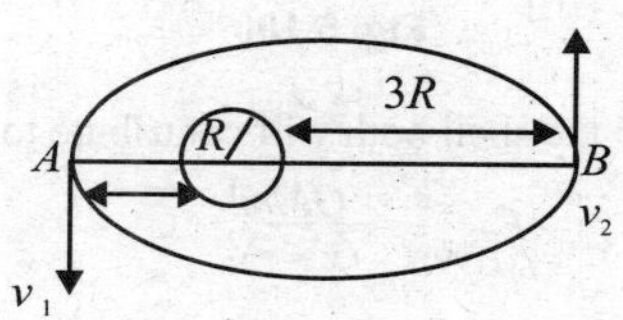

Fig. 5.97

$$\frac{mv_1^2}{r} = \frac{GMm}{(2R)^2}$$

$$r = \frac{4v_1^2}{GM} = \frac{8R}{3} \text{ (putting value of } v_1\text{)}$$

For Problems 3–5

3. b., 4. c., 5. c.

Sol.

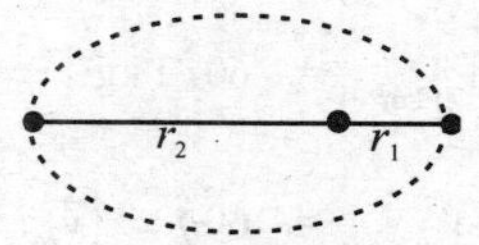

Fig. 5.98

Gravitational PE at perihelion $= -GMm/r_1$ as r_1 is minimum

Therefore, PE is minimum.

4. c. Total energy = constant.

5. c. As Pluto moves away, the displacement has component opposite to the air force, hence work done is –negative.

For Problems 6–8

6. c., 7. b., 8. d.

Sol. When $2r < x < 2R$

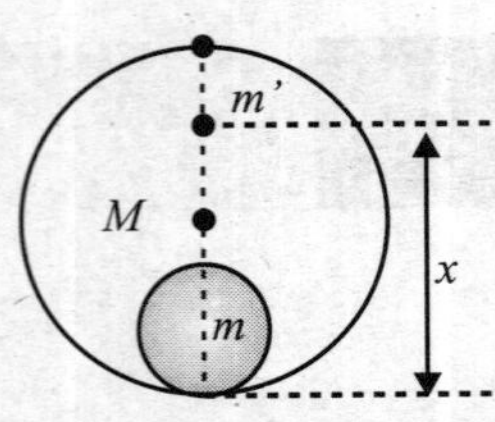

Fig. 5.99

then the force will be due to sphere only.

$$F = \frac{Gmm'}{(x-r)^2}$$

When $x > 2R$

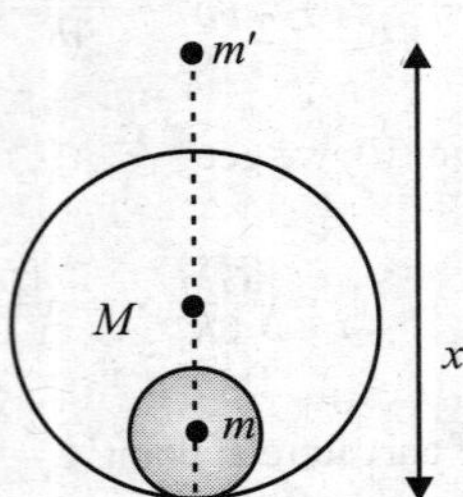

Fig. 5.100

the sphere and the shell both will contribute to the

$$F_{\text{sphere}} = \frac{GMm'}{(x-r)^2}$$

$$F_{\text{shell}} = \frac{GMm'}{(x-R)^2}$$

$$F = \frac{GMm'}{(x-R)^2} + \frac{GMm'}{(x-r)^2}$$

For Problems 9–12

9. a., 10. c., 11. b., 12. c.

Sol.

9. a. $\vec{F} = \vec{E}m = (10\,\hat{i} + 24\,\hat{j})$ N

$= \sqrt{(100) + 576} = 26$ N

10. c. $\vec{v} = \vec{E}r$

$$\vec{v}_{(12\text{ m},\,0)} = -60\hat{i}\text{ J kg}^{-1} \Rightarrow \left|\vec{v}_{(12\text{ m},\,0)}\right| = 60\text{ J kg}^{-1}$$

$$\vec{v}_{(0,\,5\text{ m})} = -60\,\hat{j} \Rightarrow \left|\vec{v}_{(0,\,5\text{ m})}\right| = -60\text{ J kg}^{-1}$$

11. b. $\Delta v = +\int_{(12\text{ m},\,5\text{ m})}^{(0,\,0)} Emdr = +\int_{(12\text{ m},\,5\text{ m})}^{(0,\,0)} (5\,\hat{i} + 12\,\hat{j})\,2dr$

$= +(10\,\hat{i} + 24\,\hat{j})\,|r|_{(12\text{ m},\,5\text{ m})}^{(0,\,0)}$

$= -(10\,\hat{i} + 24\,\hat{j})\,(12\,\hat{i} + 5\,\hat{j}) = -240$ J

12. c. $\Delta v = +\int_{(0,\,5\text{ m})}^{(12\text{ m},\,0)} Emdr$

$= Em\int_{5\hat{j}}^{5\hat{i}} dr = (10\,\hat{i} + 24\,\hat{j})\,[12\,\hat{i} - 5\,\hat{j}] = 0$

For Problems 13–14

13. a., 14. c.

Sol. $E_P = \frac{GM}{(4a)^2} = \frac{GM}{16a^2}$ and $E_{P_2} = \frac{GM}{(6a)^2} + \frac{GM}{5a^2} = \frac{61GM}{900a^2}$

For Problems 15–17

15. a., 16. d., 17. a.

Sol.

15. a. For the system to be bounded one, total energy of the system must be negative. So, object having total energy E_1 is bounded and other is unbounded.

16. d. Total energy = KE + PE = negative, so the system is bounded.

At $r = r_0$, total energy = PE and KE = 0.

For $r > r_0$, total energy = PE and KE = negative, which is not possible, so r_0 is the maximum distance of the object from the earth's centre.

17. a. The object having total energy E_2 is not bounded as its total energy is positive and it is never an intersecting PE curve, so all the values of r are possible for it.

For Problems 18–22

18. c., 19. b., 20. d., 21. a., 22. b.

Sol.

18. c. As it is not known that by how much the velocity has changed, which will cause the change in orbital radius and hence time speed, so we can use either rocket system to carry out the docking in the minimum possible time, though the extent of firing rockets may differ.

19. b. Time period of A and B in a given circular orbit is $T = (2\pi r^{3/2})/\sqrt{GM}$,

where $r = 9600$ km, $G = 6.67 \times 10^{-11}$ N m^2 kg^{-2}, $M = 5.98 \times 10^{24}$ kg, which gives

$$T = 9357.79\text{ s} = 9358\text{ s}$$

Initially, total energy of A is $E = -\frac{GMm}{2r} = -6.65 \times 10^9$

20. d. Speed of A before firing the rocket,

$$v_i = \sqrt{GM/r} = 6445.82\text{ m s}^{-1}$$

After firing the rocket speed of A is

$$V_f = 0.995,\ v_i = 6413.6\text{ m s}^{-1}$$

Energy of the satellite after firing,

$$E_f = 2E + \frac{mv_i^2}{2} = -6.714 \times 10^9\text{ J}$$

as due to firing potential energy does not change.

From $E = -\frac{GMm}{2a}$

the new semi-major axis is given by

$$a = -\frac{GMm}{2E_f} \Rightarrow a = 9505.3\text{ km}$$

New time period, $T' = \frac{2\pi a^{32/2}}{\sqrt{GM}} = 9219.67$ s

21. a. Initially, the time periods of the satellite and the spacecraft were same, equal to 9358 s and after firing rockets of A, it acquires slower and faster orbit and A will come to initial position in time period of 9220 s while B will take 9358 s to come to its initial position or (9358 – 100) s to come to the original position of A as 100 s is the initial time gap between the two.

It means now A leads B by (9358 – 100 – 9220) = 38 s.

22. b. To bring back the satellite in the same orbit, we have to make the speed of the satellite as initial one for which we have to increase v_f by such amount so that it becomes equal to v_i for which we have to fire the rocket in backward direction with the same extent.

For Problems 23–27

23. b., 24. a., 25. c., 26. a., 27. d.

Sol.

23. b. From work–energy theorem, $dK = -dU + W_{\text{air friction}}$

$W_{\text{air friction}}$ is negative, so

$$dK = -dU + \text{(a negative quantity)}$$

As K increases, it means U decreases by an amount greater than magnitude of $W_{\text{air friction}}$.

24. a. We have $v = \sqrt{\frac{GM}{R}}$

Taking log on both the sides and then differentiating,

$$\frac{dv}{v} = \frac{-1 \times dR}{2R}$$

$$\frac{\Delta v}{v} = \frac{\Delta R}{2R} \qquad [\because \Delta R = -dR]$$

$$\Delta v = \frac{\Delta R}{2}\sqrt{\frac{GM}{R^3}}$$

25. c. $g' = g - \omega^2 R_e \cos^2 \lambda$

$$\frac{dK}{K} = \frac{2dv}{v}$$

$$= \frac{2dv}{v} \quad [\therefore dv = \Delta v \text{ as } v \text{ is increasing}]$$

$$= \frac{2\Delta R}{2R} = \frac{\Delta R}{R}$$

$$dK = K \times \frac{\Delta R}{R} = \frac{GMm}{2R} \times \frac{\Delta R}{R} = \frac{GMm}{2R^2} \times \Delta r$$

$\Delta K = dK$ [$\therefore$ KE is increasing]

26. a. $\Delta U = U_f - U_i = -\left[\frac{GMm}{R - \Delta R}\right] - \left[-\frac{GMm}{R}\right]$

$$= -\left[GMm\left(\frac{1}{R - \Delta R} - \frac{1}{R}\right)\right] = -\frac{GMm}{R^2}\Delta R$$

27. d. From the work–energy theorem,

$$dK = -dU + W_{\text{air friction}}$$

$$\frac{GMm}{2R^2} \times \Delta R = -\left[-\frac{GMm}{R^2}\Delta R\right] + W_{\text{air friction}}$$

$$W_{\text{air friction}} = -\frac{GMm}{2R^2} \times \Delta R$$

Matching Column Type

1. i. → a., b.; ii. → c.; iii. → a. b.; iv. → c.

i. Acceleration due gravity is different at different points and it does not depend upon direction of projection.

ii. Angular momentum of the earth remains constant about the Sun.

iii. Escape velocity depends upon g and independent of direction of projection.

iv. Gravitational potential will remain constant.

2. i. → a., c.; ii. → a., b., c., d.; iii. → a., b., c., d; iv. → a., c., d.

i. A geostationary satellite has circular orbit over equator.

ii. Total energy will remain constant in any orbit.

iii. Angular momentum will remain constant in any orbit.

iv. In circular orbit, speed remains the same, whereas in elliptical orbit speed changes.

3. i.→ a.; ii. → a., c., d.; iii. → a., b., c., d.; vi. → a. i, ii, iii.

As rocket has been fired the speed of the satellite increases in all these three cases which leads to the increase in kinetic energy and hence total energy becomes less negative or we can say the semi-major axis increases.

Due to thrust exerted by firing of the rocket in a direction perpendicular to the plane of the orbit, the plane of the orbit changes.

iv. Here speed decreases and hence the kinetic energy, the total energy and the semi-major axis decrease.

4. i. → a.; ii. → c.; iii. → b.

Angular momentum of the particle = $m(v_0 + v)a$

$$= \sqrt{\frac{5}{4}}\, m v_0 a \qquad \left[v_0 = \sqrt{\frac{GM_e}{a}}\right]$$

Total energy of the particle = $\frac{1}{2}m(v_0 + v)^2 - \frac{GM_e m}{a}$

$$= \frac{1}{2} \times \frac{5}{4} m v_0^2 - \frac{GM_e m}{a}$$

$$= \frac{5}{8}\frac{GM_e m}{a} - \frac{GM_e m}{a} = -\frac{3GM_e m}{a}$$

At any distance 'r', T.E. = $\frac{1}{2}mu^2 - \frac{GM_e m}{a}$

But angular momentum conservation gives

$$mur = \sqrt{\frac{5Gm_e}{4a}}$$

$$\Rightarrow \qquad u = \sqrt{\frac{5}{4}\frac{GM_e a}{r^2}}\, a$$

T.E. at any distance 'r' $= \frac{1}{2} m \frac{5}{4} \frac{GM_e a}{r^2} - \frac{GM_e m}{r}$

But through conservation of total energy, we have

$$\frac{1}{2} m \frac{5}{4} \frac{GM_e a}{r^2} - \frac{GM_e m}{r} = -\frac{3GM_e m}{a}$$

On solving, we get

$$3r^2 - 8ar + 5a^2 = 0$$
$$(r - a)(3r - 5a) = 0$$
$$R = a,\ r = \frac{5a}{3}$$

Minimum distance $= a$

Maximum distance $= \frac{5a}{3}$

5. i. → a., c; ii. → a., c.; iii. → a., c.; iv. → a., c.

i. At centre of thin spherical shell, $V \neq 0$, $E = 0$.
ii. At centre of solid sphere, $V \neq 0$, $E = 0$.
iii. At centre of spherical cavity inside solid sphere, $V \neq 0$, $E \neq 0$.
iv. At centre of two point masses, $V \neq 0$, $E = 0$.

6. i. → b., d.; ii. → a., c., d.; iii. → a., d.; iv. → b., c.

Speed of the planet is maximum at perihelion and minimum at aphelion. So, KE is maximum at perihelion and minimum at aphelion.

where KE is maximum, PE is minimum.

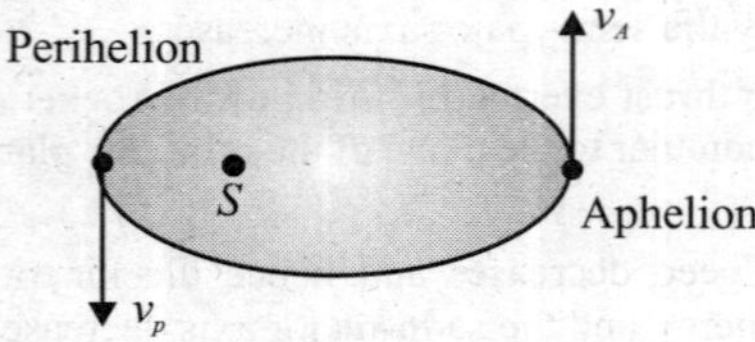

Fig. 5.101

PE and KE of the planet are dependent on both mass of planet and on semi-major axis.

7. i. → c., d.; ii. → d.; iii. → a.; iv. → b., e.

Gravitational potential is defined as the work done in carrying a unit mass from a point outside the field (zero potential) to a point inside the field. So when reference point is altered the potential will vary and the potential difference will remain the same. Also the energy is not dependent on the angle by any means and depends only on the two points (initial and final) in any conservative field like gravitational field.

So i. → c., d.

Escape velocity is given by $\sqrt{GM/R}$ and is dependent on the mass of the planet and does not depend on the angle of projection.

So ii. → d.

Acceleration due to gravity is given by $g = GM/R^2 = 4/3\ \pi\ R\rho G$ and so its ratio in two planets depends on their densities when their mass is not the same.

So iii. → a.

As the Kepler's law, the angular momentum will remain conserved for keeping the area swept in equal time to remain the same. The same concept was proposed by Bohr's model of atom.

So iv. → b., c.

Integer Answer Type

1. (3) $F = -K/r^2$ (negative sign is for attractive force)

Potential energy $\quad U = -\int F\, dr = \int \frac{K}{r^2}\, dr = -\frac{K}{r}$

Conservation of energy gives (let at other extreme position $r = b$)

$$K_1 + U_1 = K_2 + U_2$$
$$\frac{1}{2} m v_1^2 - \frac{K}{a} = \frac{1}{2} m v_2^2 - \frac{K}{b} \qquad \text{(i)}$$

where $\quad v_1 = \sqrt{\frac{K}{2ma}}$

Conservation of angular momentum gives

$$m v_1 a = m v_2 b$$
$$v_2 = \frac{a}{b} v_1 = \frac{a}{b}\sqrt{\frac{K}{2ma}}$$

Therefore, from Eq. (i) $\Rightarrow \frac{1}{2} m \frac{K}{2ma} - \frac{K}{a} = \frac{1}{2} m \left(\frac{a}{b}\right)^2 \frac{K}{2ma} - \frac{K}{b}$

$$-\frac{3K}{4a} = \frac{aK}{4b^2} - \frac{K}{b} \Rightarrow b^2 - \frac{4a}{3} b + \frac{a^3}{3} = 0$$

Hence, $\quad b = a/3 \Rightarrow a/b = 3$

2. (6) $F = mr\omega^2 = 10^{24} \times 1.5 \times 10^8 \times 10^3 \times (2 \times 10^{-7})^2$

$= 6 \times 10^{21}\ N$

3. (2)

$$g = \frac{GM}{R^2} = G \frac{4}{3} \pi R^3 \rho$$
$$g = \frac{4}{3} G\pi R\rho \Rightarrow g \propto R\rho$$
$$g' \propto R'\rho' \Rightarrow \rho' = 2\rho$$

Given, $\quad \frac{g}{g'} = 1$

$$\frac{R}{R'} = 2$$

4. (3) Given $R' = 3R$, $M = \frac{4}{3}\pi R^3 \rho$

$$g' = GM'R'^2$$

Putting $\quad M' = \frac{4}{3}\pi R'^3 \rho,$

$$g' = \frac{G}{R'^2} \frac{4}{3} \pi R'^3 \rho = \frac{27}{9} \frac{G}{R^2} \frac{4}{3} \pi R^3 \rho$$
$$= 3\frac{GM}{R^2} = 3g$$

Archives

Fill in the Blanks Type

1. $g' = g - R\omega^2$

Here $g' = 0$

$\therefore \omega = \sqrt{\frac{g}{R}} = 1.23 \times 10^{-3}$ rad/s

2. $\frac{dA}{dt} = \frac{L}{2m}$ = constant, because angular momentum of planet (L) about the centre of the Sun is constant. Thus, this law comes from law of conservation of angular momentum.

3. According to Kepler's law,

$$\frac{T_1^2}{T_2^2} = \frac{R_1^3}{R_2^3}$$

$$\Rightarrow \quad \frac{24 \times 24}{T_2^2} = \frac{6 \times 6 \times 6 \times R^3}{2.5 \times 2.5 \times 2.5 \times R^3}$$

$$\Rightarrow \quad T_2^2 = \frac{24 \times 24 \times 2.5 \times 2.5 \times 2.5}{6 \times 6 \times 6} = 6.45 \text{ h}$$

4. When the particle of mass m reaches infinity, then the potential energy of the system increases.

Increase in PE of the system $= -\{(\text{PE})_i - (\text{PE})_f\}$

$$= -\left\{\left[-\frac{GM_1M_2}{d} - \frac{GM_1M}{d/2} - \frac{GM_1m}{d/2}\right] - \left[-\frac{GM_1M_2}{d}\right]\right\}$$

$$dK = -dU + W_{\text{air friction}}$$

Increase in PE of the system $= +\frac{GM}{d/2} - (M_1 + M_2)$

This PE is produced as KE to mass m

$$\therefore \quad \frac{1}{2}mv^2 = \frac{Gm}{d/2} - (M_1 + M_2)$$

where v is the velocity with which mass m is projected.

$$\Rightarrow \quad v = \sqrt{\frac{4G}{d}(M_1 + M_2)}$$

5. Applying energy conservation,

Total energy at A = Total energy at B

$$\frac{1}{2}mv^2 = \left(-\frac{GMm}{R}\right) = -\frac{GMm}{R+h} \quad \text{(i)}$$

$$\text{Escape velocity} = V_e = \sqrt{\frac{2GM}{R}}$$

Therefore, kinetic energy required for escape velocity is

$$\frac{1}{2}mv_e^2 = \frac{GMm}{R}$$

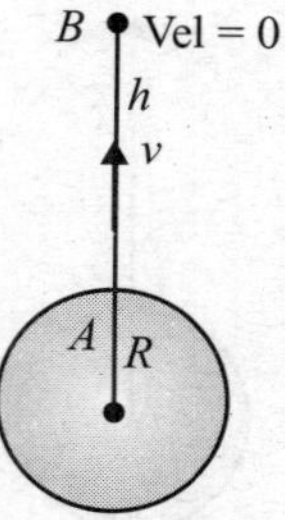

Fig. 5.102

The kinetic energy gives half of this value. Therefore,

$$\frac{1}{2}mv_e^2 = \frac{GMm}{2R} \quad \text{(ii)}$$

From Eqs. (i) and (ii),

$$\frac{GMm}{2R} - \frac{GMm}{R} = -\frac{GMm}{R+h}$$

$$\Rightarrow \quad -\frac{1}{2R} = -\frac{1}{R+h} \Rightarrow R + h = 2R$$

6. Areal velocity of a planet around the Sun is constant and is given by

$$\frac{dA}{dt} = \frac{L}{2m} \Rightarrow dA = \frac{L}{2m}dt$$

Integrating both sides $\int dA = \frac{L}{2m}\int dt \Rightarrow A = \frac{L}{2m}T$

where L = angular momentum of the planet (earth) about the Sun and m = mass of planet (earth).

Hence, $A = \frac{L}{2m}T = \frac{1}{2} \times 4.4 \times 10^{15} \times 365 \times 24 \times 3600 \text{ m}^2$

Area $= 6.94 \times 10^{22}$ m^2

True or False Type

1. Since New Delhi is not on the equatorial plane and geostationary satellite is launched on the equatorial plane, the given statement is false.

Multiple Choice Questions with One Correct Answer Type

1. c. $g = \frac{GM}{R^2}$; $g' = \frac{GM}{(0.99R)^2}$

$$\therefore \frac{g'}{g} = \left(\frac{R^2}{0\,99R}\right)^2 \quad \Rightarrow g' > R$$

2. a. $U_i = -GMm/R$ = Initial potential energy of the system

$U_f = -GMm/2R$ = Final PE of the system

$$\therefore \Delta U = U_f - U_i = -GMm\left[\frac{1}{2R} - \frac{1}{R}\right] = \frac{GMm}{2R} \quad \text{(i)}$$

But $g = GM/R^2$

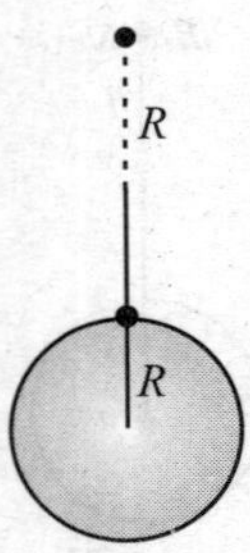

Fig. 5.103

$$dK = -dU + W_{\text{air friction}}$$

$\therefore$ $$GM = gR^2 \quad \text{(ii)}$$

From Eqs. (i) and (ii), $\Delta U = \dfrac{gR^2 m}{2R} = \dfrac{gRm}{2}$

3. b. $\dfrac{mv^2}{R} \propto R^{-5/2}$, $\therefore v \propto R^{-3/4}$

Now, $T = \dfrac{2\pi R}{v}$ or $T^2 \propto \left(\dfrac{R}{v}\right)^2$

$T^2 \propto \left(\dfrac{R}{R^{-3/4}}\right)^2$ or $T^2 \propto R^{7/2}$

4. b. According to Kepler's law,

$\dfrac{T_1^2}{T_2^2} = \dfrac{R_1^3}{R_2^3}$, $T_1 = 365$ days, $T_2 = ?$, $R_1 = R$, $R_2 = \dfrac{R}{2}$

$\Rightarrow T_2 = T_1\left(\dfrac{R_2}{R_1}\right)^{3/2} = 365\left[\dfrac{R/2}{R}\right]^{3/2} = 129$ days

5. a. Force on the satellite is always towards the earth, therefore, acceleration of satellite S is always directed towards the centre of the earth. Net torque of this gravitational force F about the centre of the earth is zero. Therefore, angular momentum (both in magnitude and direction) of S about the centre of the earth is constant throughout. Since force F is conservative in nature, therefore mechanical energy of the satellite remains constant. Speed of S is maximum when it is nearest to the earth and minimum when it is farthest.

6. d. We know that $T_1 = 2\pi\sqrt{l/g}$ and $T_2 = 2\pi\sqrt{l/g'}$

$\therefore$ $$\frac{T_2}{T_1} = \sqrt{\frac{g}{g'}} \quad \text{(i)}$$

Also $$g = \frac{GM}{R^2}$$

$\therefore$ $$g' = \frac{GM}{(2R)^2} = \frac{GM}{4RL}$$

$\therefore$ $$\frac{g}{g'} = 4 \Rightarrow \frac{T_2}{T_1} = 2$$

7. c. $T^2 \propto R^3$

Now, $R_C = 6400$ km, $\dfrac{T^2}{(24)^2} = \left(\dfrac{6400}{36000}\right)^3$

$\Rightarrow$ $T = 1.7$ h

For the spy satellite, R is slightly greater than R_C

$\Rightarrow$ $T_S > T$, $\therefore T_s = 2$ h

8. d. In case of binary star system, angular velocity and the time period of both the stars are equal.

9. c. For $r \le R$: $\dfrac{mv^2}{r} = \dfrac{GmM}{r^2}$

Here $M = \left(\dfrac{4}{3}\pi r^3\right)\rho_0$

Substituting in Eq. (i), we get

$$v \propto r$$

i.e., v–r graph is a straight line passing through the origin.

For $r > R$:

$$\frac{mv^2}{r} = \frac{GM\left(\frac{4}{3}\pi R^3\right)\rho_0}{r^2}$$

$\Rightarrow$ $$v \propto \frac{1}{\sqrt{r}}$$

The corresponding v–r graph will be as shown in option (c).

10. a. $V = -\displaystyle\int_{3R}^{4R} \frac{\sigma 2\pi r dr G}{\sqrt{r^2 + 16R^2}}$

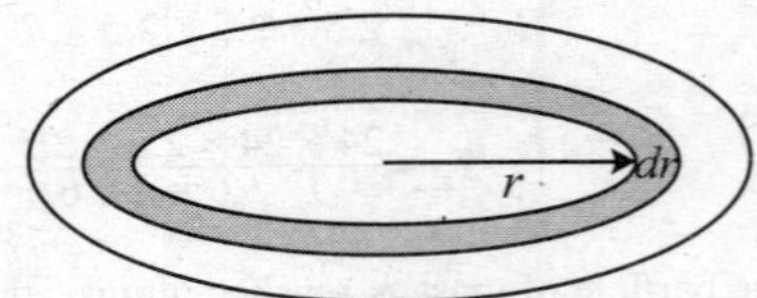

Fig. 5.104

11. b. $V_c = \sqrt{2}v_0$

$$\text{KE} = \frac{1}{2}mv_e^2 = \frac{1}{2}m\left(\sqrt{2}v_0\right)^2 = mv_0^2$$

Multiple Choice Questions with One or More than One Correct Answer Type

1. a., c., d. The gravitational field intensity at point O is zero (as the cavities are symmetrical with respect to O).

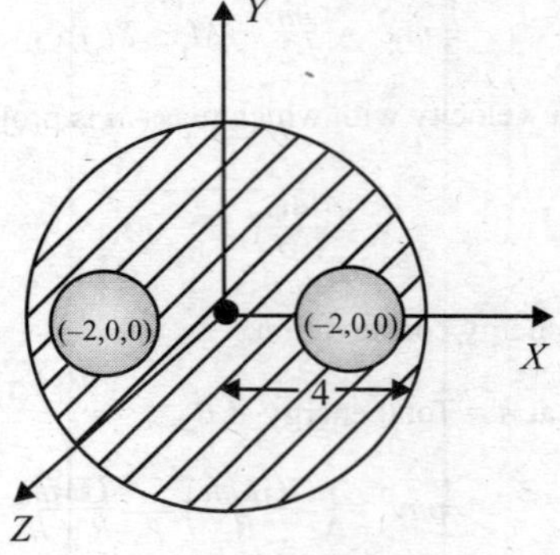

Fig. 5.105

Now the force acting on a test mass m_0 placed at O is given by

$$F = m_0 E = m_0 \times 0 = 0$$

Now, $y^2 + z^2 = 36$ represents the equation of a circle with centre (0, 0, 0) and radius 6 units. The plane of the circle is perpendicular to the x-axis. Since the spherical mass distribution behaves as if the whole mass is at its centre (for a point outside the sphere) and since all the points on the circle are equidistant from the centre of the sphere, the circle is a gravitational equipotential.

2. a., b. For $r > R$, the gravitational field is $F = GM/r^2$

$$\therefore \quad F_1 = \frac{GM}{r_1^2} \text{ and } F_2 = \frac{GM}{r_2^2} \Rightarrow \frac{F_1}{F_2} = \frac{r_2^2}{R_1^2}$$

For $\quad r > R,$

the gravitational field is $F = \dfrac{GM}{R^3} \times r$

$$\therefore \quad F_1 = \frac{GM}{R^3} \times r_1 \text{ and } F_2 = \frac{GM}{R^3} \times r_2$$

$$\Rightarrow \quad \frac{F_1}{F_2} = \frac{r_1}{r_2}$$

3. a., c. Force on the satellite is always towards the earth; therefore, acceleration of satellite S is always directed towards the centre of the earth. Net torque of this gravitational force F about the centre of the earth is zero. Therefore, angular momentum (both in magnitude and direction) of S about the centre of the earth is constant throughout. Since force F is conservative in nature, therefore mechanical energy of the satellite remains constant. Speed of S is maximum when it is nearest to the earth and minimum when it is farthest.

Assertion and Reasoning Type

1. a. Force acting on astronaut is utilized in providing necessary centripetal force, thus he feels weightlessness, as he is in a state of free fall.

Integer Answer Type

1. (3)

$$\frac{g'}{g} = \frac{\sqrt{6}}{11}; \frac{\rho'}{\rho} = \frac{2}{3} \quad \Rightarrow \quad \frac{R'}{R} = \frac{3\sqrt{6}}{22}$$

$$\frac{v'_{esc}}{v_{esc}} \propto \sqrt{\frac{R'^2 \rho'}{R^2 \rho}} = \frac{3}{11} \quad \Rightarrow \quad v'_{esc} = 3 \text{ km/s}$$

Matching Column Type

1. i. → c.; ii. → d.; iii. → No match; iv. → d.

i. For earth moving in an elliptical orbit.

Centripetal force passes through centre of mass. So its torque will be zero at all points of motion. Here, angular momentum can be conserved at all points of motion. But as external force is acting, so linear momentum will be unconserved.

ii. Angular momentum can be conserved about the point of contact as work done about this point by force of friction is zero. Linear momentum will not be conserved as force of friction is an external force here.

iii. External forces acting on the system are normal reaction, $mg \sin\theta$ and friction. Torque due to normal reaction and $mg \sin\theta$ will be zero about centre of mass but not due to friction. So, angular momentum will not be conserved. Also, in the presence of external forces, the linear momentum will not be conserved.

iv. For the projected particle, force of gravity which is external in this case will render linear momentum unconserved. But torque of this force will be zero as it will pass through centre of mass. So angular momentum is conserved instantaneously.

CHAPTER

6

Miscellaneous Assignments

EXERCISES

Objective Type

Solutions on page 6.22

1. Figure 6.1 shows a path followed by a particle and position of a particle at any instant. Four different students have represented the velocity vectors and acceleration vectors at the given instant. Which vector diagram cannot be true in any situation? (In each figure, velocity is tangential to the trajectory).

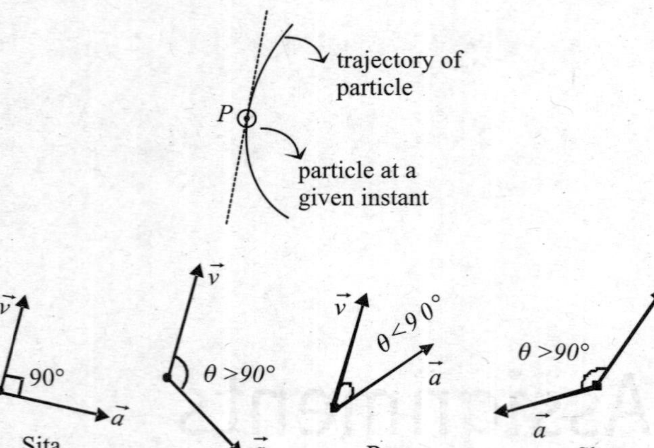

Fig. 6.1

a. 15 m/s **b.** 30 m/s
c. 10 m/s **d.** 25 m/s

2. Two students analysed two different problems of mechanics involving constraint motion. Symbols have their usual meaning.

Student *A*

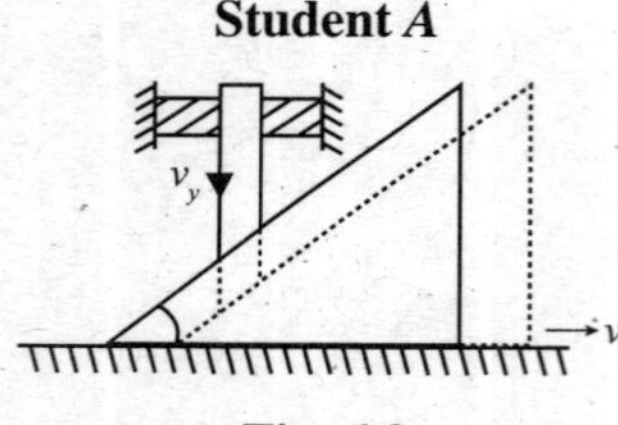

Fig. 6.2

The vertical rod can move only vertically and the wedge can move only horizontally.

$$\frac{y}{x} = \tan\theta;\ y = x\tan\theta,\ v_y = v_x \tan\theta$$

Student *B*

The ends of the rods are slipping on the ground and the wall, respectively.

$$\frac{y}{x} = \tan\theta;\ y = x\tan\theta,\ vy = vx\tan\theta$$

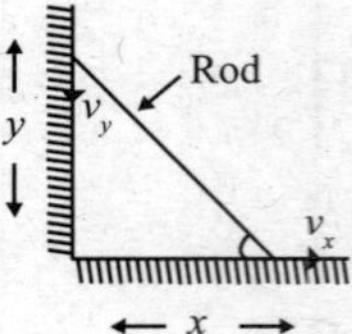

Fig. 6.3

a. Student *A* is correct, *B* is wrong.
b. Student *A* is wrong, *B* is correct.
c. Both are correct.
d. Both are wrong.

3. The system of two weights with masses m_1 and m_2 is connected with a weightless spring as shown. The system is resting on support *S*. The support *S* is quickly removed. The acceleration of each of the weights right after the support *S* is removed is (a_1 is acceleration of m_1 and a_2 that of m_2)

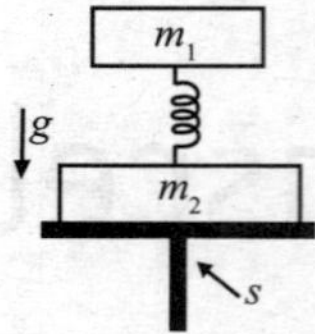

Fig. 6.4

a. $a_1 = 0,\ a_2 = \dfrac{(m_1 + m_2)g}{m_2}$

b. $a_1 = 0,\ a_2 = \dfrac{(m_1 + m_2)g}{m_1}$

c. $a_1 = \dfrac{(m_1 + m_2)g}{m_1},\ a_2 = 0$

d. $a_1 = 0,\ a_2 = 0$

4. The plot shows the position (x) as a function of time (t) for two trains that run on a parallel track. Train *A* is next to train *B* at $t = 0$ and at $t = T_0$.

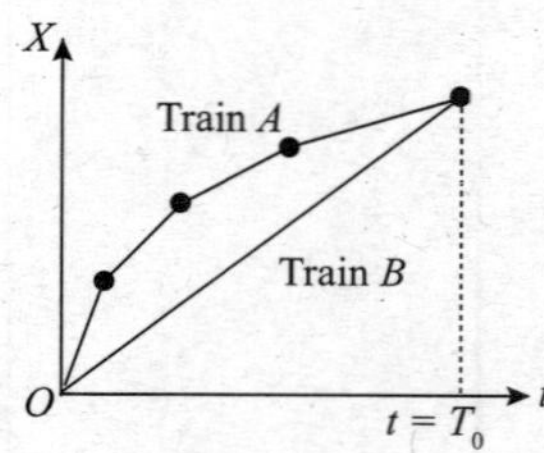

Fig. 6.5

a. At T_0, both the trains have the same velocity.
b. Both the trains speed up all the times.
c. Both the trains have the same velocity at some time before T_0.
d. At T_0, the trains have covered different distances.

5. Two identical balls are set into motion simultaneously from an equal height h. While ball A is thrown horizontally with velocity v, ball B is just released to fall by itself. Choose the alternative that best represents the motion of A and B with respect to an observer who moves with velocity $v/2$ with respect to the ground as shown in the figure.

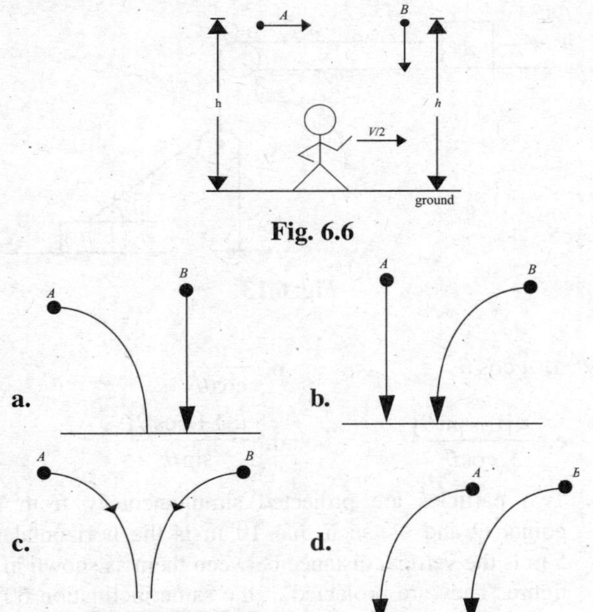

Fig. 6.6

a. b. c. d.

6. The figure shows a method for measuring the acceleration due to gravity. The ball is projected upward by a "gun". The ball passes the electronic "gates" 1 and 2 as it rises and again as it falls. Each gate is connected to a separate timer. The first passage of the ball through each gate starts the corresponding timer, and the second passage through the same gate stops the timer. The time intervals Δt_1 and Δt_2 are thus measured. The vertical distance between the two gates is d. If $d = 5$ m, $\Delta t_1 = 3$ s, $\Delta t_2 = 2$ s, then find the measured value of acceleration due to gravity (in m/s^2).

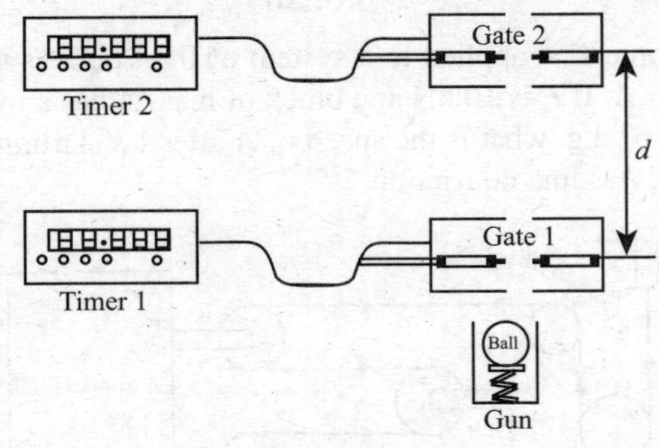

Fig. 6.7

a. 8 m/s^2 b. 4 m/s^2
c. 2 m/s^2 d. 1 m/s^2

7. A particle is projected with a velocity 30° above on an inclined plane, inclination of which from the horizontal is 37°. Choose the appropriate path (air resistance is negligible).

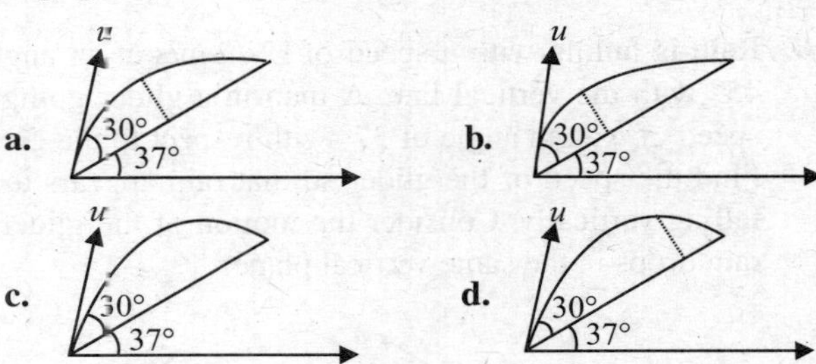

8. A helicopter is moving to the right at a constant horizontal velocity. It experiences three forces $\vec{F}_{gravitational}$, $\vec{F}_{drag}$ and force on it caused by rotor $\vec{F}_{rotor}$. Which of the following diagrams can be a correct free-body diagram representing forces on the helicopter?

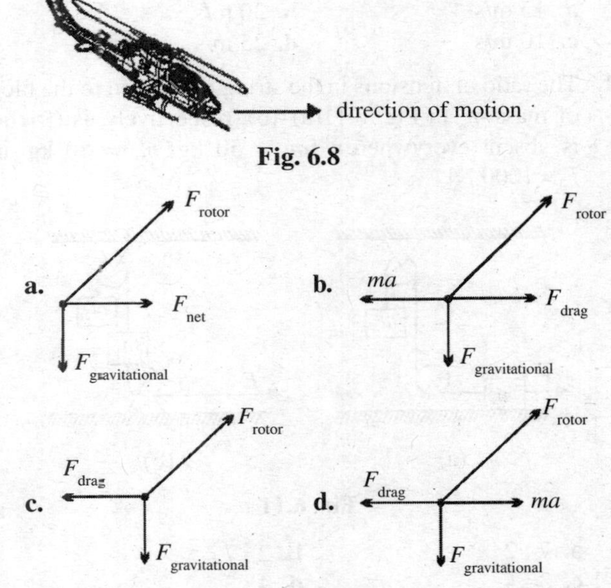

9. Figure 6.9 shows the direction of the total acceleration and velocity of a particle moving clockwise in a circle of radius m at an instant of time. Tangential acceleration at this instant is 5 m/s^2. Which of the following statements is not correct?

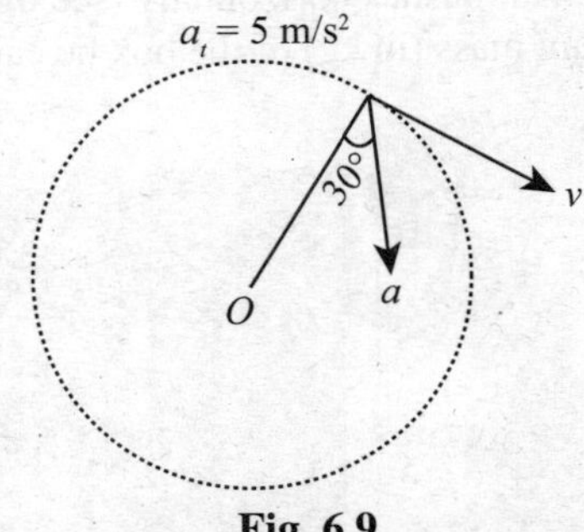

Fig. 6.9

a. The centripetal acceleration is $5\sqrt{3}$ m/s^2.
b. Particle is speeding up.

c. The net acceleration is 10 m/s^2.
d. The particle is slowing down.

10. Rain is falling with a speed of $12\sqrt{2}$ m/s at an angle of 45° with the vertical line. A man in a glider going at a speed of u at an angle of 37° with respect to the ground. Find the speed of the glider so that rain appears to him falling vertically. Consider the motion of the glider and rain drops in the same vertical plane.

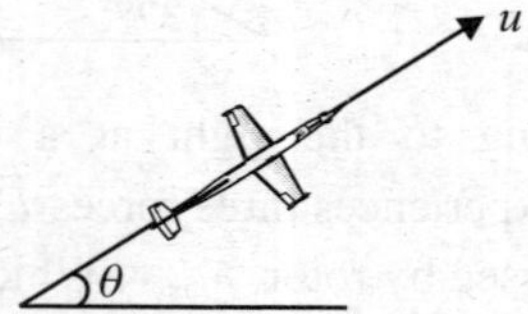

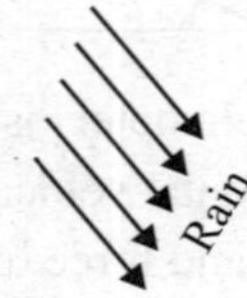

Fig. 6.10

a. 15 m/s **b.** 30 m/s
c. 10 m/s **d.** 25 m/s

11. The ratio of tensions in the string connected to the block of mass m_2 in Fig. 6.11(a)–(b), respectively, is (friction is absent everywhere): [m_1 = 50 kg, m_2 = 80 kg and F = 1000 N]

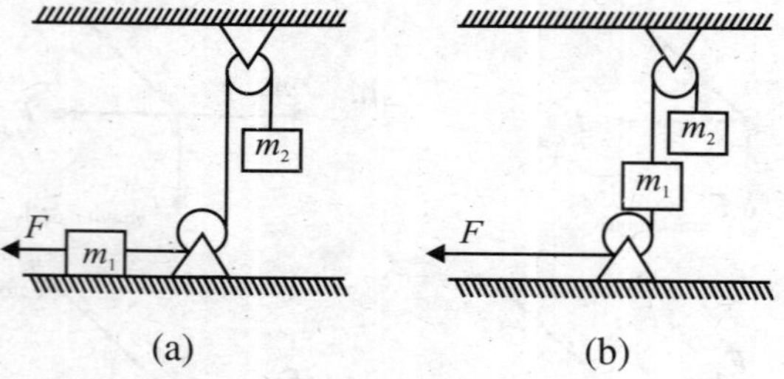

Fig. 6.11

a. 7 : 2 **b.** 2 : 7
c. 3 : 4 **d.** 4 : 3

12. A man of mass 75 kg is pushing a heavy box on a flat floor. The coefficient of kinetic and static friction between the floor and the box is 0.20, and the coefficient of static friction between the man's shoes and the floor is 0.80. If the man pushes horizontally (see figure), what is the maximum mass (in kg) of the box he can move?

Fig. 6.12

a. 300 kg **b.** 600 kg
c. 900 kg **d.** none of these

13. In the figure shown, the lower pulley is free to move in a vertical direction only. Block A is given a uniform velocity u as shown, what is the velocity of block B as a function of angle θ?

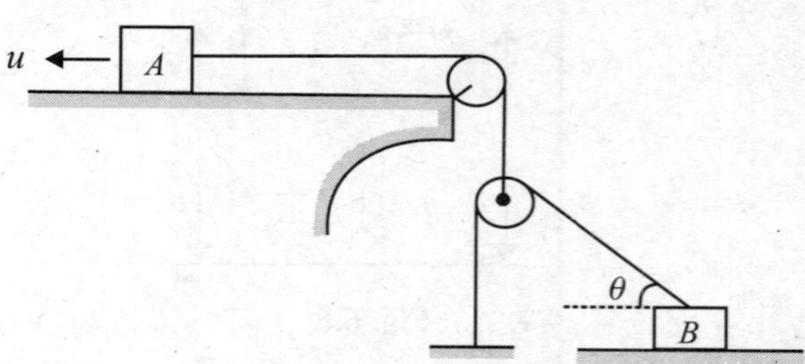

Fig. 6.13

a. $u \cos \theta$ **b.** $\dfrac{u}{\cos\theta}$

c. $\dfrac{u[1+\sin\theta]}{\cos\theta}$ **d.** $\dfrac{u[1+\cos\theta]}{\sin\theta}$

14. Two particles are projected simultaneously from two points O and O' such that 10 m is the horizontal and 5 m is the vertical distance between them as shown in the figure. They are projected at the same inclination 60° to the horizontal with the same velocity 10 ms^{-1}. The time after which their separation becomes minimum is

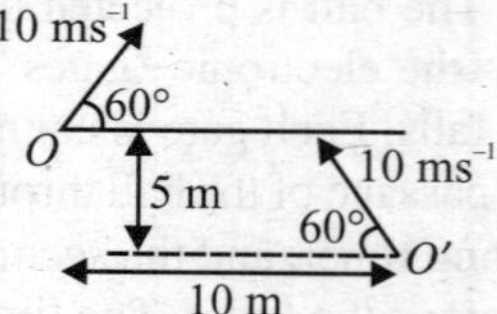

Fig. 6.14

a. 2.5 s **b.** 1 s
c. 5 s **d.** 10 s

15. A force F is applied to a system of light pulleys to pull body A. If F is 10 kN and block of mass A has a mass of 5×10^3 kg, what is the speed of A after 1 s starting from rest? Assume no friction.

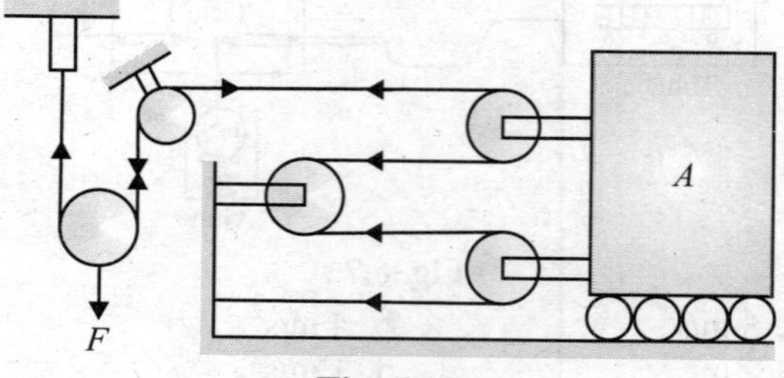

Fig. 6.15

a. 4 m/s **b.** 8 m/s
c. 6 m/s **d.** none of these

16. A block of mass 2 kg slides down the face of a smooth 45° wedge of mass 9 kg as shown in the figure. The wedge is placed on a frictionless horizontal surface. Determine the acceleration of the wedge.

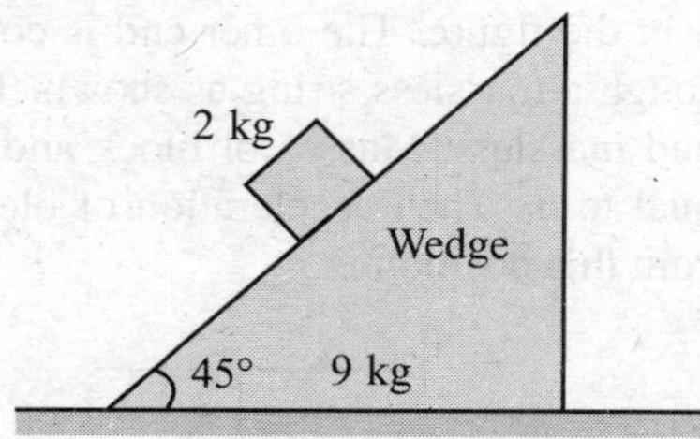

Fig. 6.16

a. 2 m/s^2 **b.** $\frac{11}{\sqrt{2}}$ m/s^2

c. 1 m/s^2 **d.** none of these

17. Four blocks are arranged on a smooth horizontal surface as shown. The masses of the blocks are given (see the figure). The coefficient of static friction between the top and the bottom blocks is μ_s. What is the maximum value of the horizontal force F, applied to one of the bottom blocks as shown, that makes all the four blocks move with the same acceleration?

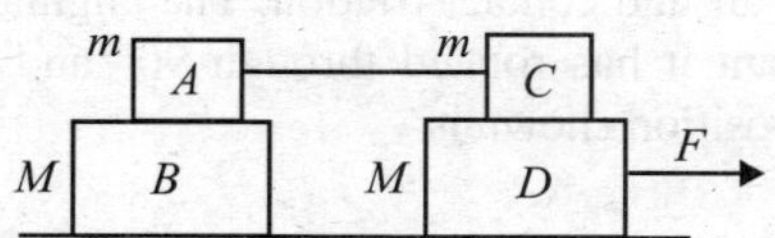

Fig. 6.17

a. $F_{max} = 2\mu_s mg\left(\frac{2m+M}{m+M}\right)$ **b.** $F_{max} = \mu_s mg\left(\frac{m+M}{2m+M}\right)$

c. $F_{max} = 2\mu_s mg\left(\frac{m+M}{2m+M}\right)$ **d.** $F_{max} = \mu_s mg\left(\frac{2m+M}{m+M}\right)$

18. A bomb of mass $3m$ is kept inside a closed box of mass $3m$ and length 4 L at its centre. It explodes in two parts of mass m and $2m$. The two parts move in the opposite directions and stick to the opposite sides of the walls of the box. The box is kept on a smooth horizontal surface.

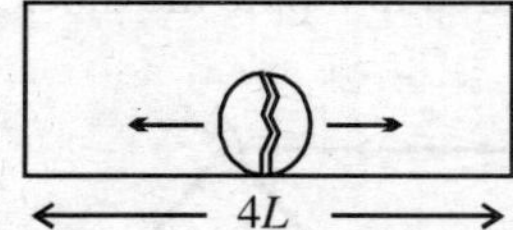

Fig. 6.18

What is the distance moved by the box during this time interval.

a. 0 **b.** $\frac{L}{6}$

c. $\frac{L}{12}$ **d.** $\frac{L}{3}$

19. In the situation as shown in the figure if acceleration of B is a, then find the acceleration of A (B always remains horizontal).

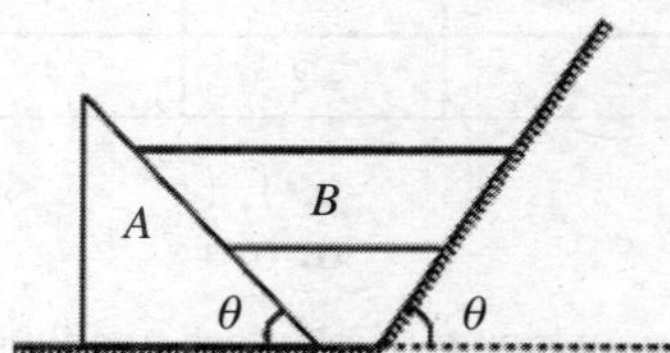

Fig. 6.19

a. $a \sin\theta$ **b.** $a \cot\theta$

c. $2a \tan\theta$ **d.** $2a \cos\theta$

20. Three particles of masses 1 kg, 2 kg and 3 kg are situated at the corners of an equilateral triangle move at speed 6 ms^{-1}, 3 ms^{-1} and 2 ms^{-1} respectively. Each particle maintains a direction towards the particle at the next corner symmetrically. Find velocity of CM of the system at this instant

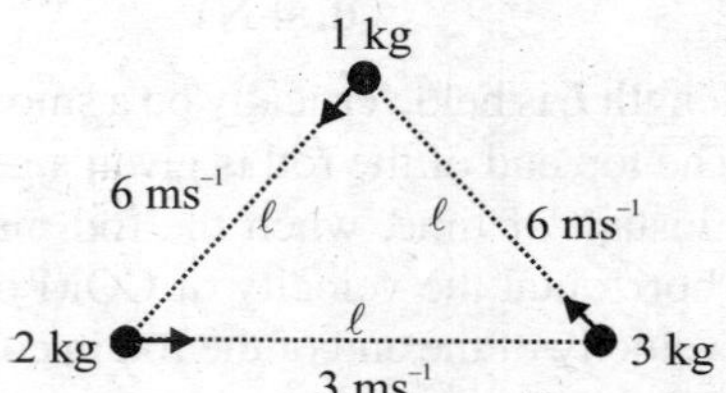

Fig. 6.20

a. 3 ms^{-1} **b.** 5 ms^{-1}

c. 6 ms^{-1} **d.** zero

21. A uniform solid right circular cone of base radius R is joined to a uniform solid hemisphere of radius R and of the same density, as shown. The centre of mass of the composite solid lies at the centre of base of the cone. The height of the cone is

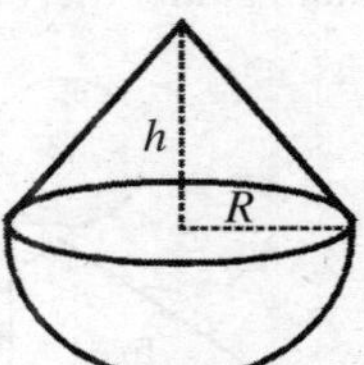

Fig. 6.21

a. 1.5R **b.** $\sqrt{3}R$

c. 3R **d.** $2\sqrt{3}R$

22. Two objects A and B are moving along an xy-plane on a frictionless floor collide. Assume that they form a closed, isolated system. The following table gives some of the momentum components (in kilogram metres per second) before and after the collision. What are the missing values (a, b)?

	Before collision		After collision	
Object	P_X	P_Y	P_X	P_Y
A	–4	5	3	a
B	b	–2	4	2

a. 10, 11
b. 1, 11
c. 5, 7
d. 6, 4

23. Three carts move on a frictionless track with masses and velocities as shown. The carts collide and stick together after successive collisions. Find the total magnitude of the impulse experienced by A.

$m_1 = 3$ kg, $v_1 = 1$ m/s; $m_2 = 1$ kg, $v_2 = 1$ m/s; $m_3 = 2$ kg, $v_3 = 2$ m/s; → +ve

A → B → ← C

Fig. 6.22

a. 1 Ns
b. 2 Ns
c. 3 Ns
d. 4 Ns

24. A rod of length L is held vertically on a smooth horizontal surface. The top end of the rod is given a gentle push. At a certain instant of time, when the rod makes an angle 37° with horizontal the velocity of COM of the rod is 2 m/s. The velocity of the end of the rod in contact with the surface at that instant is:

a. 2 m/s
b. 1 m/s
c. 4 m/s
d. 1.5 m/s

25. A uniform cylinder of mass M lies on a fixed plane inclined at an angle θ with the horizontal. A light string is tied to the cylinder at the rightmost point, and a mass m hangs from the string as shown. Assume that the coefficient of friction between the cylinder and the incline plane is sufficiently large to prevent slipping. For the cylinder to remain static, the value of m is

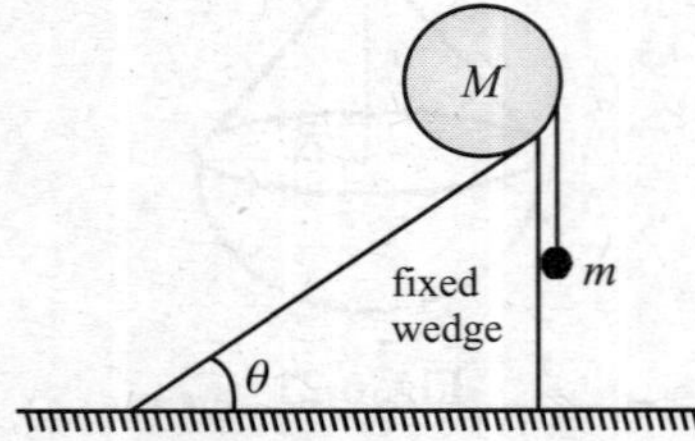

Fig. 6.23

a. $\dfrac{M\sin\theta}{1-\sin\theta}$
b. $\dfrac{M\cos\theta}{1+\sin\theta}$
c. $\dfrac{M\sin\theta}{1+\sin\theta}$
d. $\dfrac{M\cos\theta}{1-\sin\theta}$

26. Uniform rod AB is hinged at end A in horizontal position as shown in the figure. The other end is connected to a block through a massless string as shown. The pulley is smooth and massless. Masses of block and rod is same and is equal to m. Then acceleration of block just after release from this position is

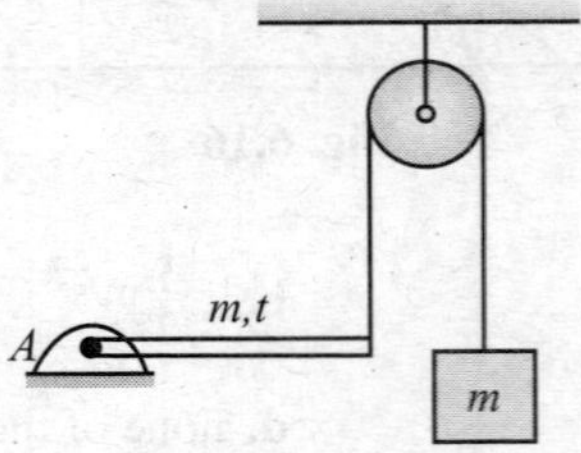

Fig. 6.24

a. $6g/13$
b. $g/4$
c. $3g/8$
d. None

27. An L shaped thin uniform rod of total length $2l$ is free to rotate in a vertical plane about a horizontal axis at P as shown in the figure. The bar is released from rest. Neglect air and contact friction. The angular velocity at the instant it has rotated through 90° and reached the dotted position shown is

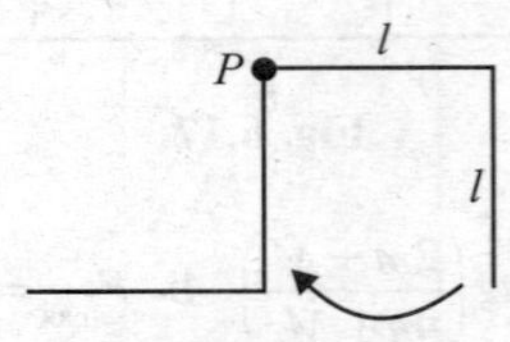

Fig. 6.25

a. zero
b. $\sqrt{\dfrac{6g}{5l}}$
c. $\sqrt{\dfrac{3g}{5l}}$
d. none

28. A horizontal force F is applied at the top of an equilateral triangular block having mass m. The minimum coefficient of friction required to topple the block before translation will be

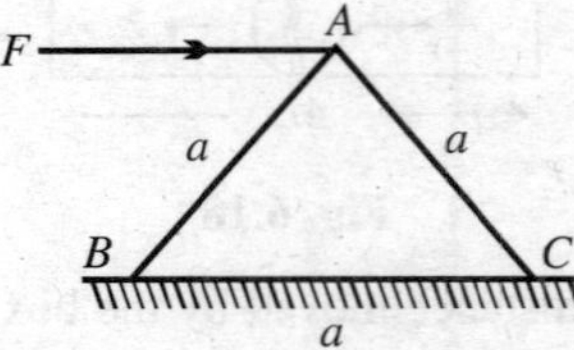

Fig. 6.26

a. $\frac{2}{\sqrt{3}}$ b. $\frac{1}{3}$
c. $\frac{1}{\sqrt{3}}$ d. $\frac{1}{2}$

29. A closed rectangular tank is completely filled with water and is accelerated horizontally with an acceleration towards the right. Pressure is (i) maximum and (ii) minimum at

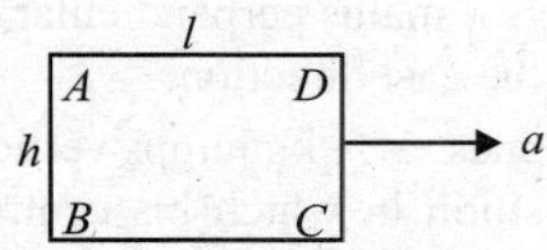

Fig. 6.27

a. (i) B (ii) D b. (i) C (ii) D
c. (i) B (ii) C d. (i) B (ii) A

30. We have a vessel in shape of a cuboid partially filled with water. Its base is square with an area of 4.5 dm^2 (1 dm = 10 cm) and a vessel contains water up to 2 cm height. Then we place a wooden cube inside water. The wood has mass of 4 kg and specific gravity 0.5. The base of the wooden cube is horizontal. Find the height of water level above the base of the wooden block.
a. 10 cm b. 5 cm
c. 15 cm d. 7 cm

31. A square gate of size 1 m × 1 m is hinged at its midpoint. A fluid of density ρ fills the space to the left of the gate. The force F required to hold the gate stationary is

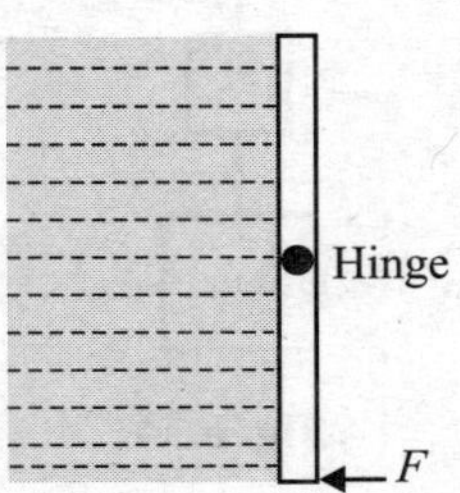

Fig. 6.28

a. $\rho g/3$ b. (1/2) ρg
c. $\rho g/6$ d. none of these

32. A rectangular bar of soap has density 800 kg/m^3 floats in water of density 1000 kg/m^3. Oil of density 300 kg/m^3 is slowly added, forming a layer that does not mix with water. When the top surface of the oil is at the some level as the top surface of the soap. What is the ratio of the oil layer thickness to the soap's thickness, x/L?

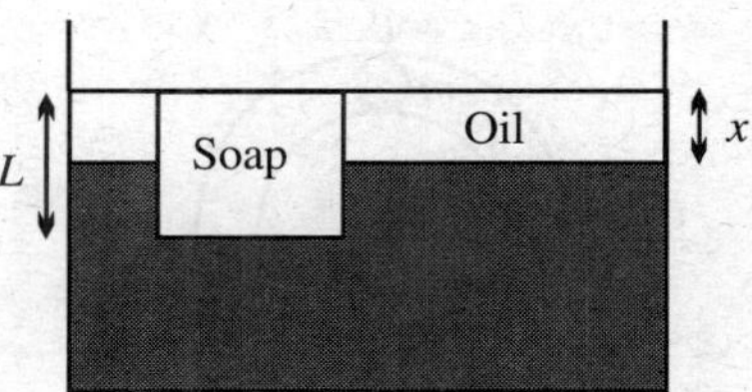

Fig. 6.29

a. $\frac{3}{7}$ b. $\frac{2}{7}$
c. $\frac{3}{10}$ d. $\frac{3}{8}$

33. A soap bubble of radius R is surrounded by another soap bubble of radius $2R$, as shown. Take surface tension $= S$. Then, the pressure inside the smaller soap bubble, in excess of the atmospheric pressure, will be

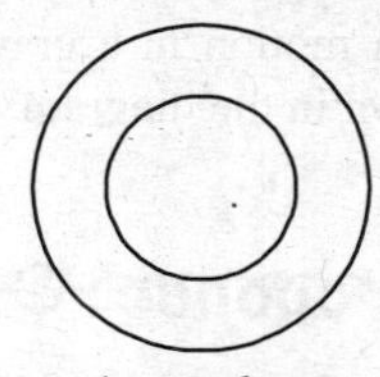

Fig. 6.30

a. $4S/R$ b. $3S/R$
c. $6S/R$ d. None of these

34. A metal ball B_1 (density 3.2 g/cc) is dropped in water, while another metal ball B_2 (density 6.0 g/cc) is dropped in a liquid of density 1.6 g/cc. If both the balls have the same diameter and attain the same terminal velocity, the ratio of viscosity of water to that of the liquid is
a. 2.0
b. 0.5
c. 4.0
d. indeterminate due to insufficient data.

35. The vernier of a circular scale is divided into 30 divisions, which coincides with 29 main scale divisions. If each main scale division is (1/2)°, the least count of the instrument is (1° = 60′)
a. 0.1′ b. 1′
c. 10′ d. 30′

36. In the figure shown, there is a smooth tube of radius 'R', fixed in the vertical plane. A ball 'B' of mass 'm' is released from the top of the tube. B slides down due to gravity and compresses the spring. End 'C' of the spring is fixed and end A is free. Initially, line OA makes an angle of 60° with OC and finally it makes an angle of 30° after compression. Find the spring constant of the spring.

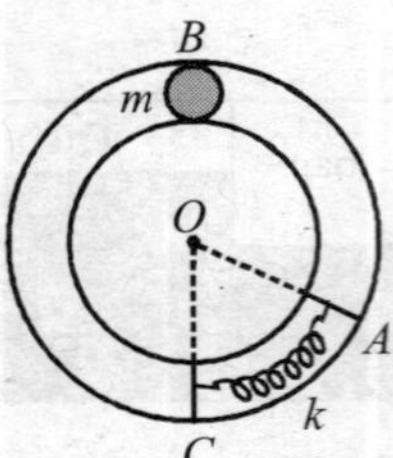

Fix smooth tube

Fig. 6.31

a. $\frac{12mg(2+\sqrt{3})}{\pi^2 R}$ **b.** $\frac{36mg(2+\sqrt{3})}{\pi^2 R}$

c. $\frac{18mg}{\pi^2 R}$ **d.** None of these

37. A system comprises of two small spheres with the same masses '*m*'. Initially, the spring is non-deformed. The spheres are set in motion in a gravity-free space at the velocities as shown in the diagram.

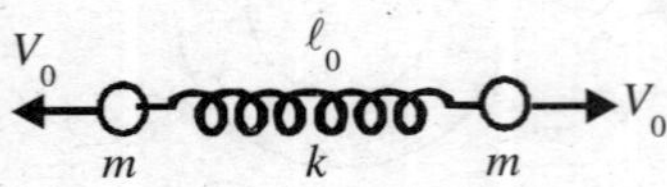

Fig. 6.32

The maximum elastic potential energy stored in the system is

a. $\frac{mv_0^2}{2\sqrt{2}}$ **b.** mv_0^2

c. $\frac{1}{2}mv_0^2$ **d.** $2mv_0^2$

38. In the figure shown, the minimum force F to be applied perpendicular to the incline so that the block does not slide is

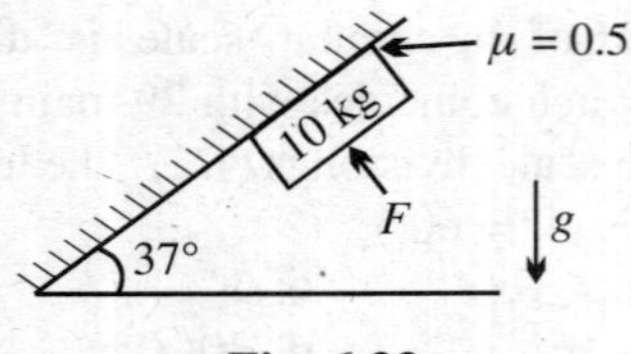

Fig. 6.33

a. 200 N **b.** 40 N
c. 120 N **d.** none

39. A vernier caliper with a least count of 0.01 cm was used to measure diameter of a cylinder as 4 cm and a scale (0–15 cm) with the least count of 1 mm was used to measure a length of 5 cm. The % error in the measurement of volume of the cylinder is

a. 3.0 **b.** 4.0
c. 5.0 **d.** 2.5

40. The angular momentum vector for a spinning wheel lies along its axle and is pointed north. To make this vector point east without changing its magnitude, it is necessary to exert a force of constant magnitude on the north end of the axle in which direction?

a. Always up
b. Always down
c. At the initial moment in the east direction, but the force always remains perpendicular to the axle
d. Always in the east direction

41. A man whose mass is m kg jumps vertically into air from the sitting position in which his centre of mass is at a height h_1 from the ground. When his feet are just about to leave the ground his centre of mass is at height h_2 from the ground and finally centre of mass rises to h_3 above the ground when he is at the top of the jump. What is the average upward force exerted by the ground on him?

a. $\frac{mg(h_3-h_1)}{(h_3-h_2)}$ **b.** $\frac{mg(h_3-h_1)}{h_3}$

c. $\frac{mg(h_3-h_1)}{(h_2-h_1)}$ **d.** none of these

42. If the ratio of lengths, radii and Young's moduli of steel and brass wires in the figure are a, b and c, respectively, then the corresponding ratio of increase in their lengths is

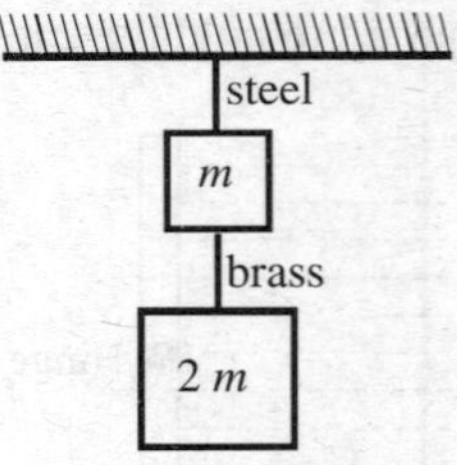

Fig. 6.34

a. $\frac{2a^2c}{b}$ **b.** $\frac{3a}{2b^2c}$

c. $\frac{2ac}{b^2}$ **d.** $\frac{3c}{2ab^2}$

43. An open capillary tube is lowered in a vessel with mercury. The difference between the levels of the mercury in the vessel and in the capillary tube $\Delta h = 4.6$ mm. What is the radius of curvature of the mercury meniscus in the capillary tube? Surface tension of mercury is 0.46 N/m, density of mercury is 13.6 gm/cc.

a. $\frac{1}{340}$ m **b.** $\frac{1}{680}$ m

c. $\frac{1}{1020}$ m **d.** Information insufficient

44. A rain drop starts falling from a height of 2 km. It falls with a continuously decreasing acceleration and attains its terminal velocity at a height of 1 km. The ratio of the work done by the gravitational force in the first half to that in the second half of the drops journey is

a. 1 : 1 and the time of fall of the drop in the two halves is a : 1 (where $a > 1$)

b. 1 : 1 and the time of fall of the drop in the two halves is a : 1 (where $a < 1$)

c. a : 1 (where $a > 1$) and the time of fall of the drop in the two halves is 1 : 1

d. a : 1 (where $a < 1$) and the time of fall of the drop in the two halves is 1 : 1

45. Two wires AC and BC are tied at C of a small sphere of mass 5 kg, which revolves at a constant speed v in the horizontal speed v in the horizontal circle of radius 1.6 m. Find the minimum value of v.

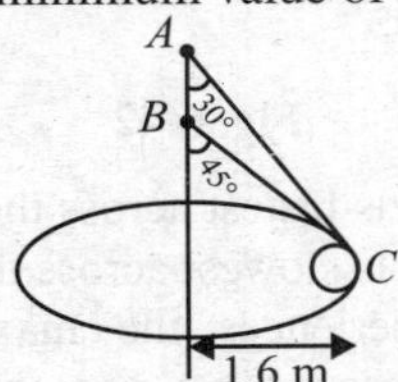

Fig. 6.35

a. 4 m/s **b.** 2 m/s
c. 2.5 m/s **d.** none of these

46. At $t = 0$, the positions and velocities of two particles are as shown in the figure. They are kept on a smooth surface and being mutually attracted by gravitational force. Find the position of centre of mass at $t = 2$ s.

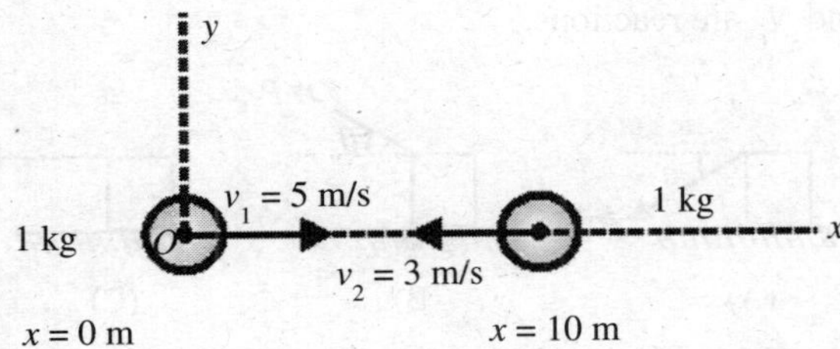

Fig. 6.36

a. $X = 5$ m **b.** $X = 7$ m
c. $X = 3$ m **d.** $X = 2$ m

47. A truck is moving on the ground with a velocity $v = 12$ m/s and a box is moving on the truck with respect to it with a velocity $u = 5$ m/s as shown in the figure. What is the velocity of the box with respect to the ground?

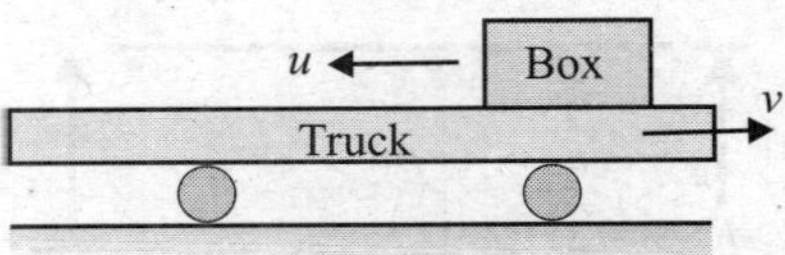

Fig. 6.37

a. 5 m/s towards right
b. 7 m/s towards left
c. 7 m/s towards right
d. 5 m/s towards left

48. A wedge is placed on a smooth horizontal plain and a rat runs on its sloping side. The velocity of the wedge is $v = 4$ m/s towards the right. What should be the velocity of the rat with respect to the wedge (u), so that the rat appears to move in the vertical direction to an observer standing on the ground?

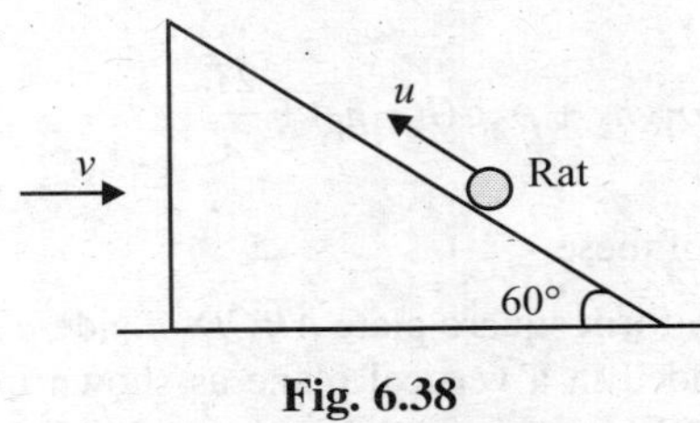

Fig. 6.38

a. 2 m/s **b.** 4 m/s
c. 8 m/s **d.** $4\sqrt{2}$ m/s

49. An annular disc of radii r_1 = 10 cm and r_2 = 5 cm is placed on a water surface. Find the surface tension force on the disc if we want to pull it from the water surface. Take the coefficient of surface tension = g = 10 m/s^2.

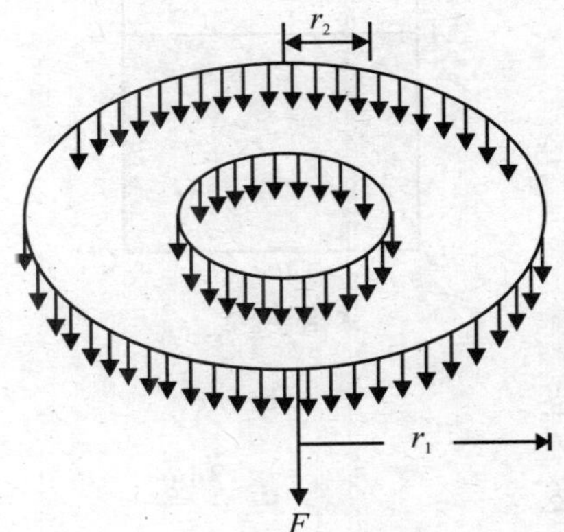

Fig. 6.39

a. 6782.4 dyne **b.** 67.82 dyne
c. 678.24 dyne **d.** none of these

50. Calculate the pressure inside a small air bubble of radius r situated at a depth h below the free surface of liquids of densities ρ_1 and ρ_2 and surface tensions T_1 and T_2. The thickness of the first and second liquids are h_1 and h_2, respectively. Take atmosphere pressure = P_0.

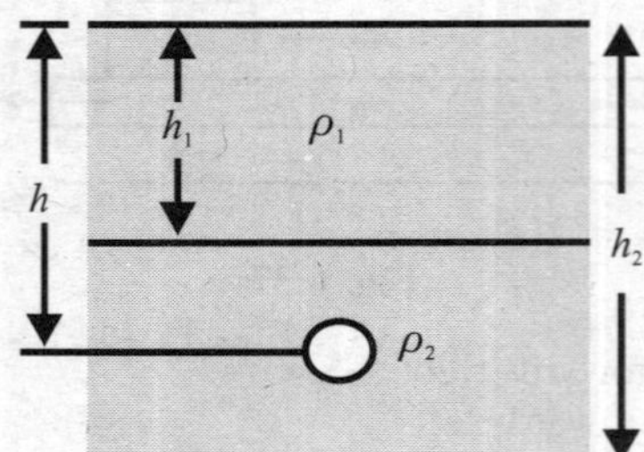

Fig. 6.40

a. $P_0 + \rho_1 g h_1 + \rho_2 g (h - h_1) - \dfrac{2T_2}{r}$

b. $P_0 + \rho_1 g h_1 + \rho_2 g (h - h_1) + \dfrac{2T_2}{r}$

c. $P_0 - \rho_1 g h_2 + \rho_2 g (h - h_1) + \dfrac{2T_2}{r}$

d. none of these

51. A thin uniform square plate *ABCD* of side *a* and mass *m* is suspended in a vertical plane as shown in the figure. *AE* and *BF* are two massless inextensible strings. The line *AB* is horizontal. The tension in *AE* just after *BF* is cut will be

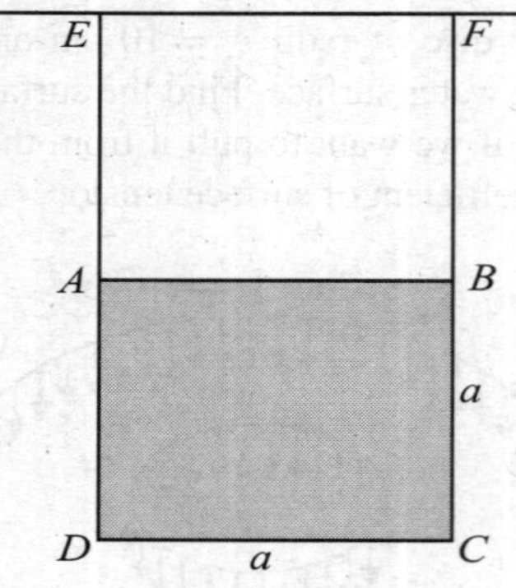

Fig. 6.41

a. $\dfrac{2mg}{5}$ b. mg

c. $\dfrac{2mg}{7}$ d. $\dfrac{3mg}{5}$

Multiple Correct Answers Type

Solutions on page 6.29

1. Figure 6.42 shows top view of an airplane blown off course by wind in various directions. Assume the magnitude of the velocity of the airplane relative to the wind and the magnitude of the velocity of the wind to be the same in each case. $\vec{v}_{A/w}$ = velocity of the airplane relative to the wind, $\vec{v}_{w/g}$ = velocity of the wind in ground frame.

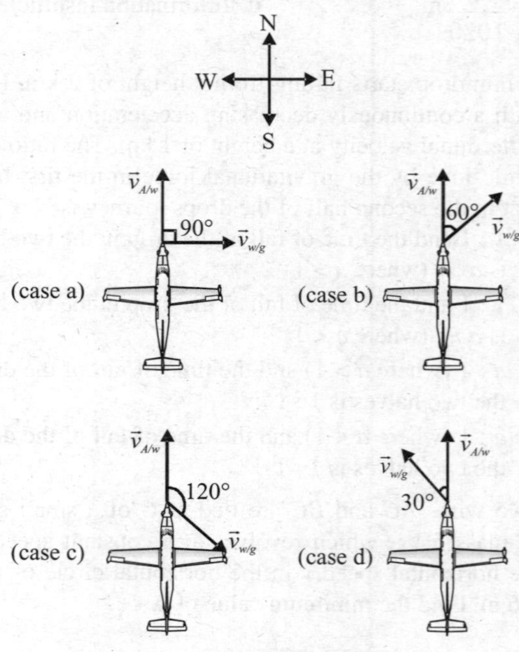

Fig. 6.42

a. Airplane travels fastest across the ground in case (d)
b. Airplane travels slowest across the ground in case (c)
c. Airplane experiences the maximum lateral displacement in case (a) in a given time
d. In none of the cases, the velocity of the wind with respect to the airplanes can be directed along south-west

2. Figure 6.42 shows three blocks on a rough surface under the influence of a force *P* of the same magnitude in all the three cases. Coefficient of friction is the same between each block and the ground. What possible relation holds between the magnitudes of normal reaction and friction forces? (Assume that the blocks do not overturn about edge.) Here, f_A, f_B and f_C are frictional forces and N_A, N_B and N_C are reactions.

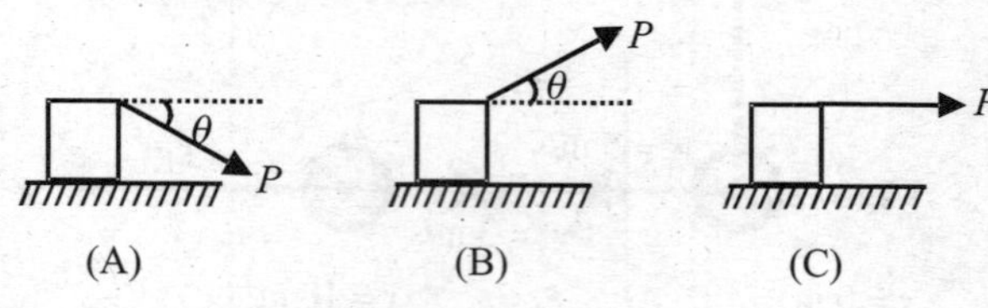

Fig. 6.43

a. $N_A > N_C > N_B$ b. $f_A > f_C > f_B$

c. $f_C > f_A = f_B$ d. $N_C > N_A = N_B$

3. Consider a cart being pulled by a horse with a constant velocity. The horse exerts force $\vec{F}_{C/h}$ on the cart. (The subscript indicates the force on the cart due to the horse.) The first subscript denotes the body on which the force acts and the second due to which it acts.

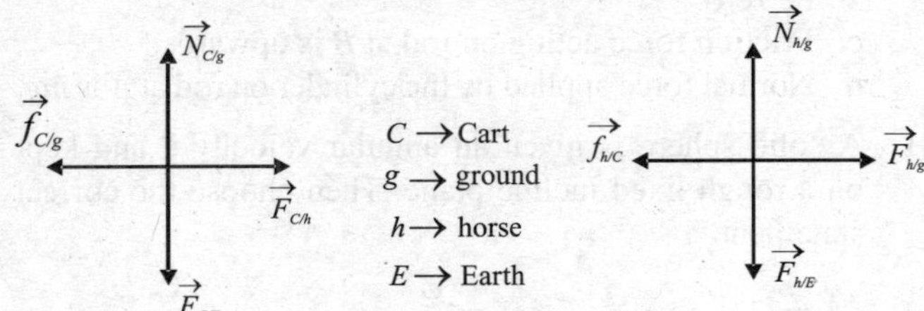

Fig. 6.44

Choose the correct statement(s):

a. $\vec{f}_{C/g}$, $\vec{N}_{C/g}$, $\vec{N}_{h/g}$ are external forces on a system consisting of a horse and a cart.

b. $\vec{f}_{C/g}\vec{N}_{C/g}\vec{N}_{h/g}$

c. $\vec{N}_{C/g}$ and $\vec{F}_{C/E}$ are action–reaction pairs.

d. $\vec{F}_{C/h}\vec{F}_{h/C}$ are action–reaction pairs.

4. A small sphere of mass m is connected by a string to a nail at O and moves in a circle of radius r on the smooth plane inclined at an angle θ with the horizontal. If the sphere has a velocity u at the top position A. Mark the correct options.

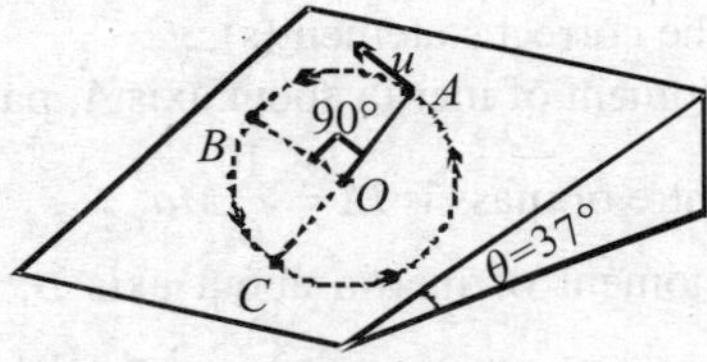

Fig. 6.45

a. Minimum velocity at A so that the string does not get slack instantaneously is $\sqrt{\frac{3}{5}gr}$.

b. Tension at B is $9mg/5$ if the sphere has the required velocity as in option (a)

c. Tension at C is $18mg/5$ in situation of option (b)

d. None of these

5. A man is standing on a plank which is placed on a smooth horizontal surface. There is sufficient friction between the feet of man and plank. Now man starts running over plank, correct statement is/are

Fig. 6.46

a. Work done by friction on the man with respect to the ground is negative.

b. Work done by friction on the man with respect to the ground is positive.

c. Work done by friction on the plank with respect to the ground is positive.

d. Work done by friction on the man with respect to the plank is zero.

6. Consider a block of mass 10 kg which rests on a smooth surface and is subjected to a horizontal force of 6 N. If observer A is in a fixed frame x.

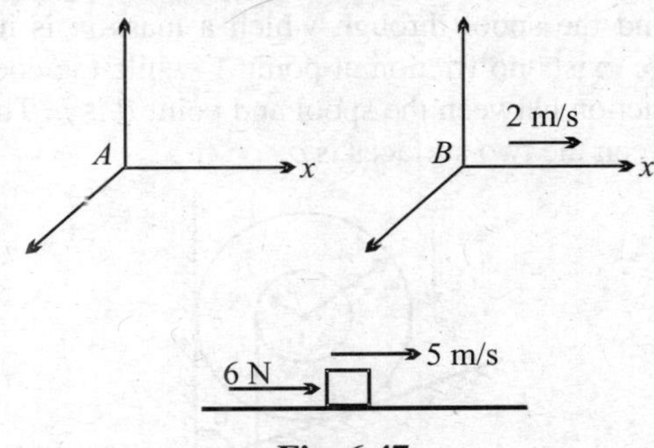

Fig. 6.47

a. The final speed of the block in 4 s is 7.4 m/s, if it has an initial speed of 5 m/s measured from the fixed frame.

b. The same speed will be observed by observer B, attached to the x' axis that moves at a constant velocity of 2 m/s relative to A.

c. Principle of impulse and momentum is valid for observers in any inertial reference frame.

d. Momentum of a body is reference frame dependent.

7. A particle of mass m is released from a height H on a smooth curved surface which ends into a vertical loop of radius R, as shown. Choose the correct alternative(s) if $H = 2R$.

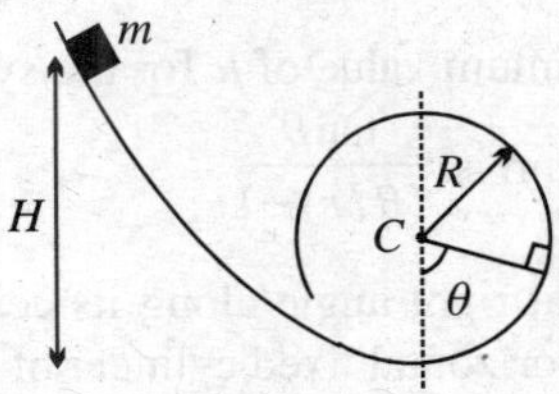

Fig. 6.48

a. The particles reaches the top of the loop with zero velocity.

b. The particle cannot reach the top of the loop.

c. The particle breaks off at a height $h = R$ from the base of the loop.

d. The particle breaks off at a height h from the base of the loop such that $R < h < 2R$

8. In the above question, if θ is instantaneous angle which the line joining the particle and the centre of the loop makes with the downward vertical, then identify the correct statement(s) related to the normal reaction N between the block and the surface.

a. The maximum value of N occurs at $\theta = 0$.
b. The minimum value of N occurs at $\theta = \pi$ for $H > 5R/2$.
c. The value of N becomes negative for $\pi/2 < \theta < 3\theta/2$.
d. The value of N can become zero only for $\theta \geq \pi/2$.

9. A massless spool of inner radius r, outer radius R is placed against a vertical wall and a tilted split floor as shown. A light inextensible thread is tightly wound around the spool through which a mass m is hanging. There exists no friction at point A, while the coefficient of friction between the spool and point B is μ. The angle between the two surfaces is ρ.

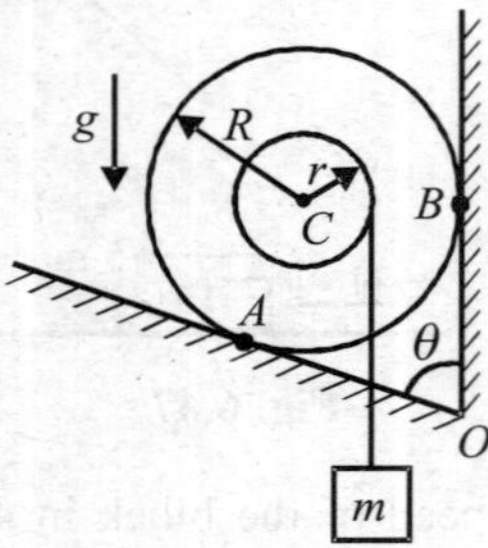

Fig. 6.49

a. The magnitude of the force on the spool at B in order to maintain equilibrium is $mg\sqrt{\left(\frac{r}{R}\right)^2 + \left(1-\frac{r}{R}\right)^2 \frac{1}{\tan^2\theta}}$.

b. The magnitude of the force on the spool at B in order to maintain equilibrium is $mg\left(1-\frac{r}{R}\right)\frac{1}{\tan\theta}$.

c. The minimum value of μ for the system to remain in equilibrium is $\frac{\cot\theta}{(R/r)-1}$.

d. The minimum value of μ for the system to remain in equilibrium is $\frac{\tan\theta}{(R/r)-1}$.

10. A rod bent at right angle along its centre line is placed on a rough horizontal fixed cylinder of radius R as shown in the figure. Mass of the rod is $2m$ and the rod is in equilibrium. Assume that friction force on rod at A and B is equal in magnitude.

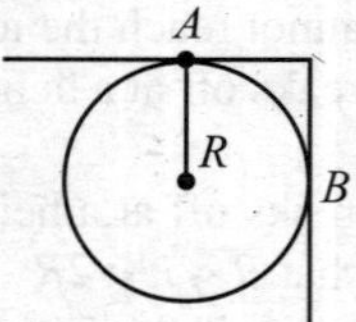

Fig. 6.50

a. Normal force applied by the cylinder on the rod at A is $3\ mg/2$.
b. Normal force applied by the cylinder on rod at B must be zero.
c. Friction force acting on rod at B is upwards.
d. Normal force applied by the cylinder on rod at A is mg.

11. A solid sphere is given an angular velocity ζ and kept on a rough fixed incline plane. Then choose the correct statement.

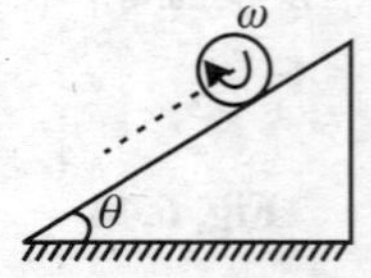

Fig. 6.51

a. If $\mu = \tan\rho$, then the sphere will be in linear equilibrium for some time and after that pure rolling down the plane will start.
b. If $\mu = \tan\rho$, then sphere will move up the plane and frictional force acting all the time will be $2\ mg\sin\rho$.
c. If $\mu = \tan\theta/2$ there will never be pure rolling (consider inclined plane to be long enough.)
d. If the incline plane is not fixed and it is on the smooth horizontal surface, then the linear momentum of the system (wedge and sphere) can be conserved in the horizontal direction.

12. Illustrated is a uniform cubical block of mass M and side a. Mark the correct statement(s).

a. The moment of inertia about axis A, passing through the centre of mass is $IA = \frac{1}{6}Ma^2$.
b. The moment of inertia about axis B, which bisects one of the cube faces is $IB = \frac{5}{12}Ma^2$.
c. The moment of inertia about axis C, along one of the cube edges is $IC = \frac{2}{3}Ma^2$.
d. The moment of inertia about axis D, which bisects one of the horizontal cube faces is $\frac{7}{12}Ma^2$.

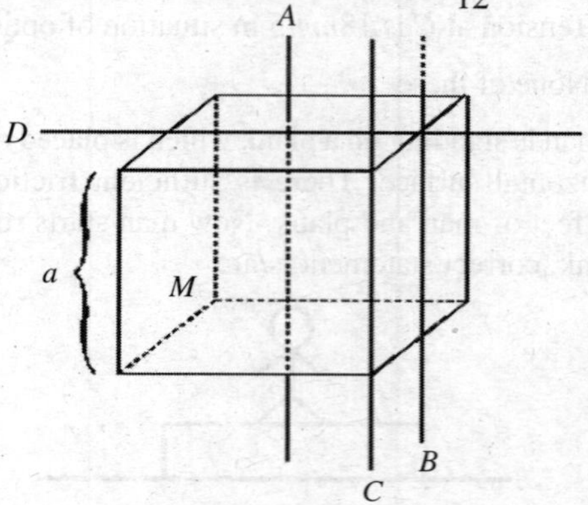

Fig. 6.52

13. A body floats on water and then also on an oil of density 1.25. Which of the following is/are true?
a. The body loses more weight in oil than in water.
b. The volume of water displaced is 1.25 times that of oil displaced.
c. The body experiences equal upthrust from water and oil.
d. Apparent weight is zero in both cases.

14. A body of density d_2 is dropped from rest at a height x into a beaker having a liquid of density d_1 $(d_2 > d_1)$. Which of the following is/are true?

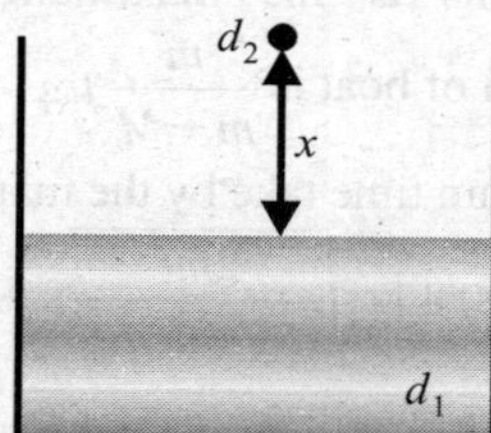

Fig. 6.53

a. The speed with which the body just enters the liquid is $\sqrt{2gx}$.
b. The body has an acceleration of $g\left(1-\dfrac{d_1}{d_2}\right)$ downwards.
c. The body does not come back to the surface of the liquid (observation made after a long time).
d. None

15. The limbs of a U-tube are lowered into beakers A and B. A contains water and B some other liquid. Density of water is 1 g/cm^3. Some air is pumped out from C and then this end is closed, as a result of this liquids rise by 10 cm and 12 cm, respectively, on the left and right side. Which of the following is/are correct?

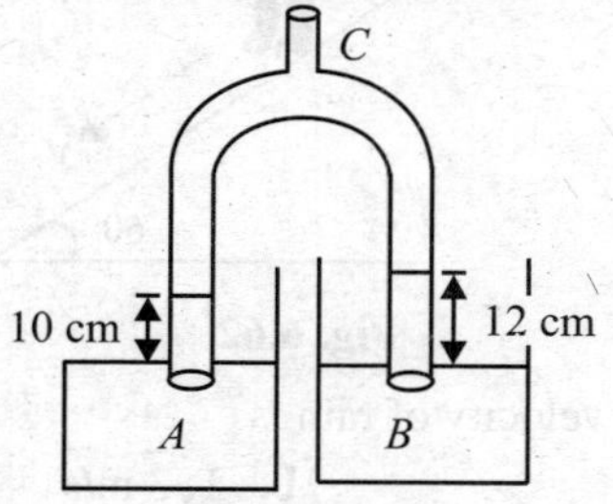

Fig. 6.54

a. Density of the liquid in B is 0.83 g/cc.
b. The heights raised by water and the other liquid in both the sides will change if beakers are kept at different heights.
c. The heights raised by water and the other liquid in both the sides will be independent of the positions of the beakers.
d. Liquid in B is more denser than water in A.

16. A cone of radius R and height H is hanging inside a liquid of density by means of a string as shown in the figure.

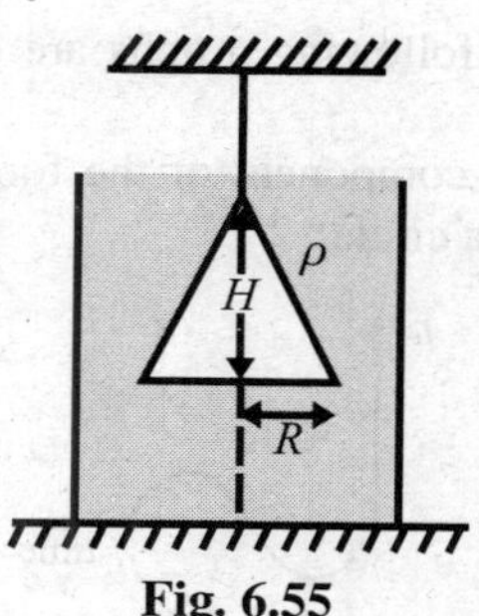

Fig. 6.55

a. Net force on the cone by the liquid is equal to buoyancy force on the cone.
b. Force on the bottom surface of the cone is more than buoyancy force on the cone.
c. Net force on the slant surface of the cone is more than buoyancy force on the cone.
d. Net force on the slant surface of the cone is less than buoyancy force on the cone.

17. Two particles move on a circular path (one just inside and the other just outside) with the angular velocities ω and 5ω starting from the same point. Then
a. they cross each other at regular intervals of time $2\pi/4\omega$ when their angular velocities are oppositely directed
b. they cross each other at points on the path subtending an angle of 60° at the centre if their angular velocities are oppositely directed
c. they cross at intervals of time $\pi/3\omega$ if their angular velocities are oppositely directed
d. they cross each other at points on the path subtending 90° at the centre if their angular velocities are in the same sense

18. A particle starts moving along a straight line path with a velocity 10 m/s. After 5 s, the distance of the particle from the starting point is 50 m. Which of the following statements about the nature of motion of the particle are correct?
a. The body may be speeding up with a constant positive acceleration.
b. The body may be moving with a constant velocity.
c. The body may have a constant negative acceleration.
d. The body may be first accelerated and then retarded.

19. A boy is sitting on a seat of merry-go-round moving with a constant angular velocity. At $t = 0$, the boy is at position A as shown in the figure.

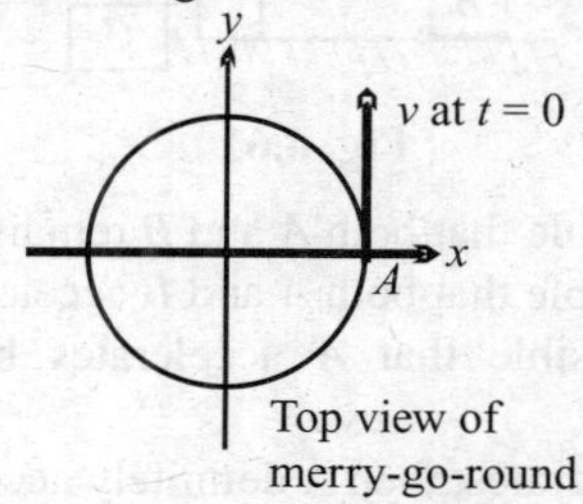

Fig. 6.56

Which of the following graphs are correct? All graphs are sinusoidal.

a. F_y is the y-component of the force keeping the boy moving in a circle.

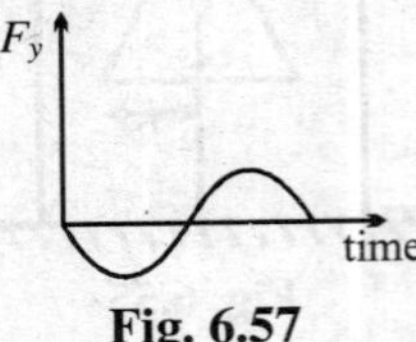

Fig. 6.57

b. x is the x-component of the boy's position.

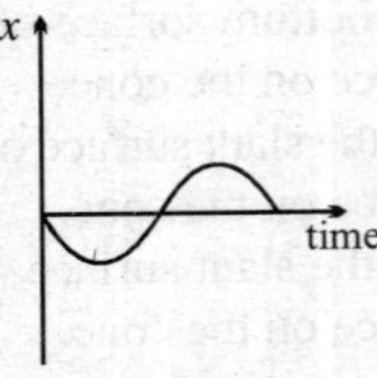

Fig. 6.58

c. θ is the angle that the position vector of the boy makes with the positive x-axis.

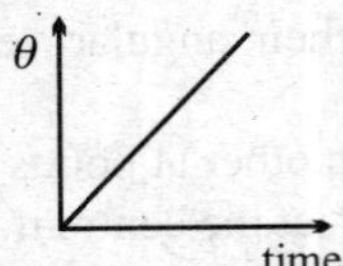

Fig. 6.59

d. V_x is the x-component of the boy's velocity.

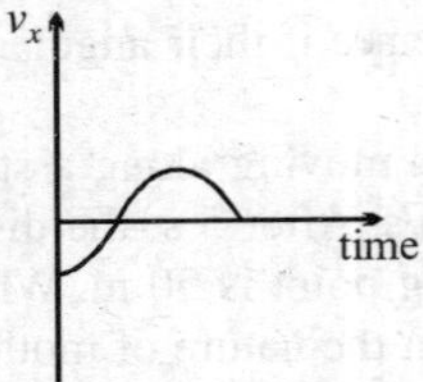

Fig. 6.60

20. Consider three masses A, B and C as shown in the figure. Friction coefficient between all surfaces is 0.5. Pulleys are smooth. (Given $m_A = 1$ kg, $m_B = 1$ kg.) Mass of C may have any value.

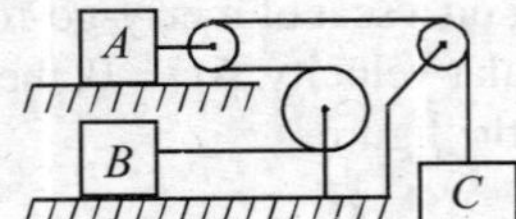

Fig. 6.61

a. It is possible that both A and B remain at rest.
b. It is possible that both A and B accelerate.
c. It is possible that A accelerates but B does not accelerate.
d. If B accelerates, then A definitely accelerates.

21. A man of mass m walks from end A to the other end B of a boat of mass M and length ℓ. The coefficient of friction between the man and the boat is μ and neglect any resistive force between the boat and the water.

a. If the man runs at his maximum acceleration, the acceleration of the boat is $(m/M)g$.

b. The minimum time take by the man to reach the other end of the boat is $\sqrt{\dfrac{2M\ell}{(M+m)\mu g}}$

c. If man runs at his maximum acceleration the acceleration of boat is $\dfrac{m}{m+M}\mu g$.

d. The minimum time take by the man to reach the other end of the boat is $\sqrt{\dfrac{2ml}{(M+m)\mu g}}$.

Comprehension Type

Solutions on page 6.32

For Problems 1–2

A man standing on an inclined plain observes rain is falling vertically. When he starts moving down the inclined plain with velocity $v = 6$ m/s observes rain hitting him horizontally.

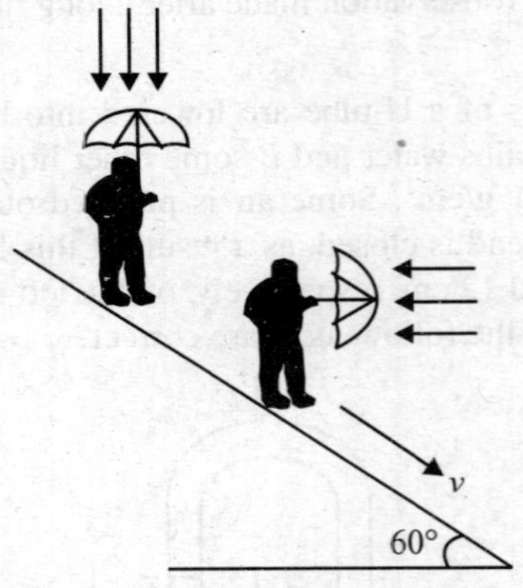

Fig. 6.62

1. The actual velocity of rain is

a. 3 m/s **b.** $3\sqrt{3}$ m/s
c. 4 m/s **d.** none of these

2. The velocity of rain with respect to the man when he is moving down is

a. 3 m/s **b.** $3\sqrt{3}$ m/s
c. 4 m/s **d.** none of these

For Problems 3–5

A straight rigid rod of mass M and length L lies on a smooth horizontal table. Taking the co-ordinate axes as shown in the figure and assuming an average force $\langle F\rangle$ acting for a small time interval (Δt), at the end A, making an angle 30° with the

X-axis, Based on the above passage, answer the following questions:

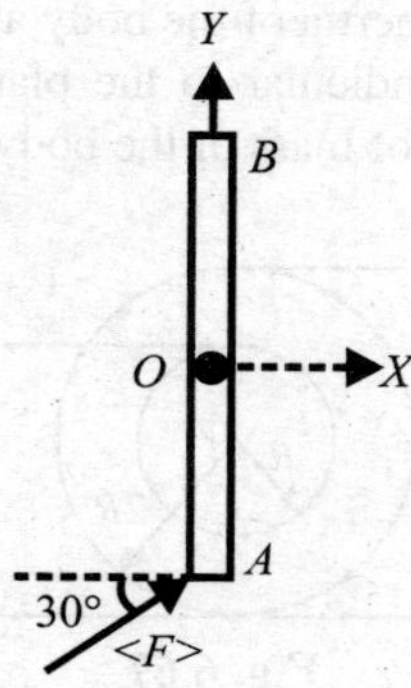

Fig. 6.63

3. The angular impulse, due to the force about *O*, is

a. $\frac{\sqrt{5}}{4} < F > \Delta tL\,\hat{k}$ **b.** $\frac{\sqrt{5}}{2} < F > \Delta\, tL\,\hat{k}$

c. $\frac{\sqrt{3}}{4} < F > \Delta\, tL\,\hat{k}$ **d.** $\frac{\sqrt{3}}{2} < F > \Delta\, tL\,\hat{k}$

4. Angular velocity of rod about *O* is

a. $\frac{3\sqrt{5} < F > (\Delta t)}{ML}$ **b.** $\frac{2\sqrt{5} < F > (\Delta t)}{ML}$

c. $\frac{2\sqrt{3} < F > (\Delta t)}{ML}$ **d.** $\frac{3\sqrt{3} < F > (\Delta t)}{ML}$

5. Linear velocity of the CM of rod is

a. $\frac{< F > (\Delta t)}{2m}(\sqrt{3}\,\hat{i} + \hat{j})$ **b.** $\frac{< F > (\Delta t)}{2m}(\sqrt{5}\,\hat{i} + 2\hat{j})$

c. $\frac{< F > (\Delta t)}{m}(\sqrt{3}\,\hat{i} + \hat{j})$ **d.** $\frac{< F > (\Delta t)}{2m}(\sqrt{5}\,\hat{i} + \hat{j})$

For Problems 6–8

A planet is revolving around the sun in an elliptical orbit. Its KE is different for different points and the total energy is negative. Its linear momentum is not conserved. The eccentricity decides the shape of the orbit.

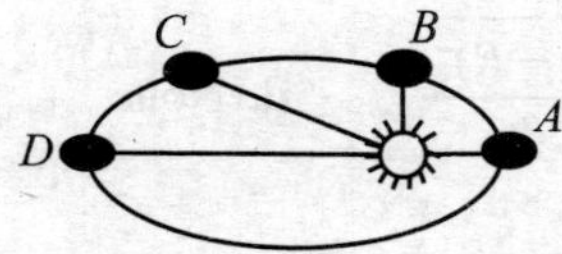

Fig. 6.64

6. Velocity of the planet is minimum at

a. *C* **b.** *D*

c. *A* **d.** *B*

7. Net torque on the planet is

a. constant at all points

b. zero at all point

c. maximum at *A*

d. minimum at *D*

8. Linear momentum of the planet is

a. different for different points of the orbit

b. conserved

c. not conserved

d. none of these

For Problems 9–11

A small toy train is moving with a constant speed of 10 m/s in a circle of radius $16/\pi$ m. The circle lies in a horizontal plane. At time $t = 0$ the train is at point *P* and moving in the anti-clockwise direction. At this instant, a stone is projected from the train with velocity $\left(-6\hat{i} + 8\hat{k}\right)$ m/s relative to the train. Answer following questions based on the above situation.

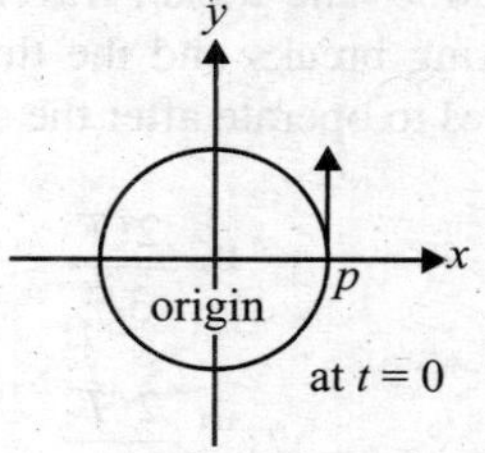

Fig. 6.65

9. What is the velocity vector of the stone relative to the train when the stone is the at highest point of its trajectory?

a. $\left(-4\hat{i} + 10\hat{j}\right)$ **b.** $\left(4\hat{i} - 10\hat{j}\right)$

c. $\left(4\hat{i} + 10\hat{j}\right)$ **d.** $\left(-4\hat{i} - 10\hat{j}\right)$

10. What is the angular displacement of the train during this time interval?

a. $\frac{\pi}{2}$ **b.** $\frac{\pi}{4}$

c. π **d.** $\frac{3\pi}{4}$

11. What is the displacement of the stone till it finally lands on the ground?

a. $\left(-9.2\hat{i} - 8\hat{j}\right)$ **b.** $\left(9.2\hat{i} + 8\hat{j}\right)$

c. $\left(9.6\hat{i} - 16\hat{j}\right)$ **d.** $\left(-9.6\hat{i} + 16\hat{j}\right)$

For Problems 12–14

A model rocket rests on a frictionless horizontal surface and is joined by a string of length ℓ to a fixed point so that the rocket moves in a horizontal circular path of radius *l*. The string will break if its tension exceeds a value *T*. The rocket engine provides a thrust *F* of constant magnitude along the rocket's direction of motion. The rocket has a mass *m* that does not change with time. Answer the following questions based on the above passage.

12. Starting from rest at $t = 0$ at what later time t_1 is the rocket travelling so fast that the string breaks? Ignore any air resistance.

a. $\left(\frac{2m\ell T}{F^2}\right)^{1/2}$ **b.** $\left(\frac{m\ell T}{F^2}\right)^{1/2}$

c. $\left(\frac{m\ell T}{2F^2}\right)^{1/2}$ **d.** $\left(\frac{m\ell F}{T^2}\right)^{1/2}$

13. What was the magnitude of instantaneous net acceleration at time $t_1/2$? Obtain answer in terms of F, T and m.

a. $\frac{[T^2+8F^2]^{1/2}}{m}$ **b.** $\frac{[T^2+4F^2]^{1/2}}{2m}$

c. $\frac{[T^2+16F^2]^{1/2}}{4m}$ **d.** None of these

14. What distance does the rocket travel between the time t_1 when the string breaks and the time $2t_1$? The rocket engine continues to operate after the string breaks.

a. $\frac{3\ell T}{2F}$ **b.** $\frac{2\ell T}{3F}$

c. $\frac{\ell T}{2F}$ **d.** $\frac{2\ell T}{F}$

For Problems 15–17

Two ropes are pulling a large ship at rest of mass 1 × 106 kg into harbour. Rope A exerts a force of 40,000 N and rope B exerts a force of 30,000 N.

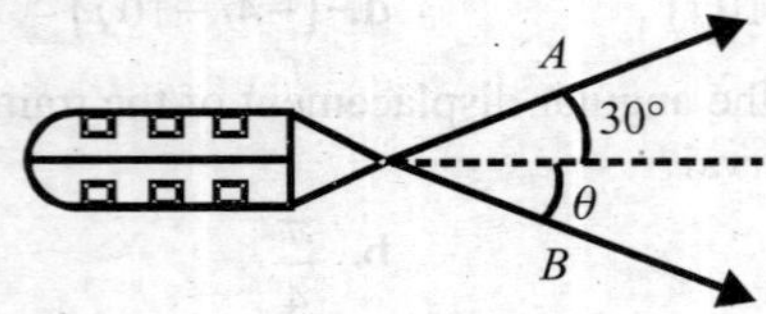

Fig. 6.66

15. The angle θ if the ship is to move directly forward is

a. $\sin^{-1}\left(\frac{4}{9}\right)$ **b.** $\sin^{-1}\left(\frac{2}{3}\right)$

c. $\sin^{-1}\left(\frac{\sqrt{3}}{2}\right)$ **d.** none of these

16. If the acceleration of the ship is 0.03 ms^{-2}, the magnitude of the resistive forces on the ship is

a. 40,000 × 9.8 N **b.** 30,000 × 9.8 N

c. 27,000 N **d.** none of these

17. The time for the ship to travel 6 m to the dock is

a. 1 min 30 s **b.** 1 min

c. 40 s **d.** 20 s

For Problems 18–19

In the given figure, $F = 10$ N, $R = 1$ m, mass of the body is 2 kg and moment of inertia of the body about an axis passing through O and perpendicular to the plane of the body is 4 kg m^2. O is the centre of mass of the body.

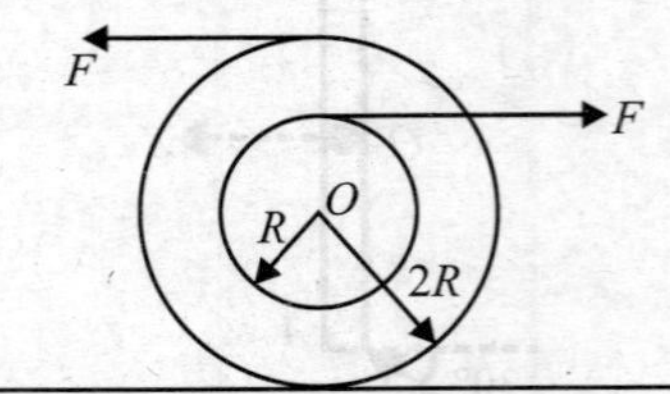

Fig. 6.67

18. If the ground is smooth, what is the total kinetic energy of the body after 2 s?

a. 100 J **b.** 75 J

c. 50 J **d.** 25 J

19. If the ground is sufficiently rough to ensure rolling, what is the kinetic energy of the body now in the given time interval of 2 s?

a. 18.75 J **b.** 16.67 J

c. 5.55 J **d.** Cannot be found

For Problems 20–21

A disc having radius R is rolling without slipping on a horizontal (x-z) plane. Centre of the disc has a velocity v and acceleration a as shown.

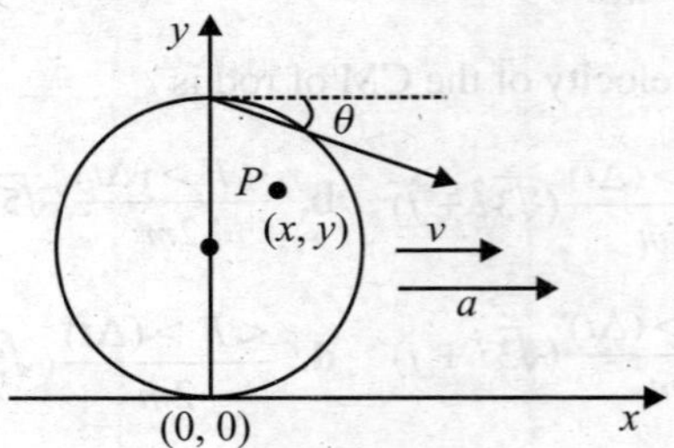

Fig. 6.68

20. Speed of point P having coordinates (x,y) is

a. $\frac{v\sqrt{x^2+y^2}}{R}$ **b.** $\frac{v\sqrt{x^2+(y+R)^2}}{R}$

c. $\frac{v\sqrt{x^2+(y-R)^2}}{R}$ **d.** None

21. If $v = \sqrt{2aR}$ the angle ρ between acceleration of the topmost point and the horizontal is

a. 0 **b.** 45°

c. $\tan^{-1} 2$ **d.** $\tan^{-1}\left(\frac{1}{2}\right)$

For Problems 22–25

A cylindrical container of length L is full to the brim with a liquid which has mass density ρ. It is placed on a weigh-scale; the scale reading is W. A light ball of volume 'V' and mass 'm' which would float on the liquid, if allowed to do so, is pushed gently down and held beneath the surface of the liquid with a rigid rod of negligible volume as shown on the left.

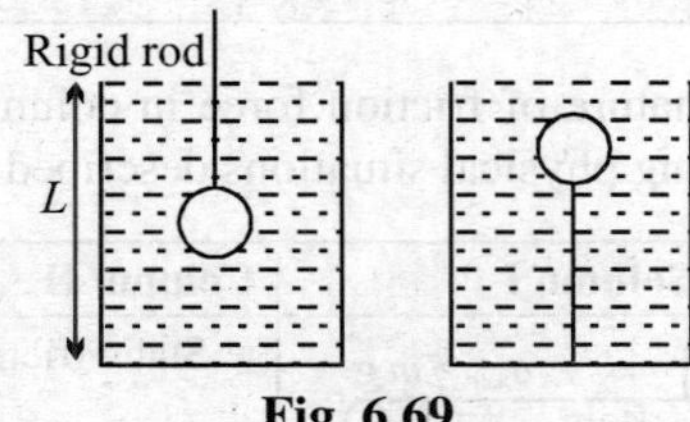

Fig. 6.69

22. What is the mass M of the liquid which overflowed while the ball was being pushed into the liquid?

a. ρV **b.** m
c. $m - \rho V$ **d.** None

23. What is the reading of the scale when the ball is fully immersed?

a. $W - \rho Vg$ **b.** W
c. $W + mg - \rho Vg$ **d.** None

If instead of being pushed down by the rod, the ball is held in place by a thin string attached to the bottom of the container as shown on the right.

24. What is the tension T in the string?

a. $(\rho V - m)g$ **b.** ρVg
c. mg **d.** None

25. What is the reading on the scale?

a. $W - \rho Vg$ **b.** W
c. $W + mg - \rho Vg$ **d.** None

For Problems 26–31

Fluids at rest exert a normal force to the walls of the container or to the surface of the body immersed in the fluid. The pressure exerted by this force at a point inside the liquid is the sum of atmospheric pressure and a factor which depends on the density of the liquid, the acceleration due to gravity and the height of the liquid above that point. The upthrust acting on a body immersed in a stationary liquid is the net force acting on the body in the upward direction. A number of phenomenon of liquids in motion can be explain by Bernoulli's theorem which relates the pressure, flow speed and height for flow of an ideal incompressible fluid.

A container of large uniform cross-sectional area A resting on a horizontal surface holds two immiscible, non-viscous and incompressible liquids of densities d and $2d$, each of height $H/2$ as shown in the figure. The lower density liquid is open to the atmosphere having pressure P_0.

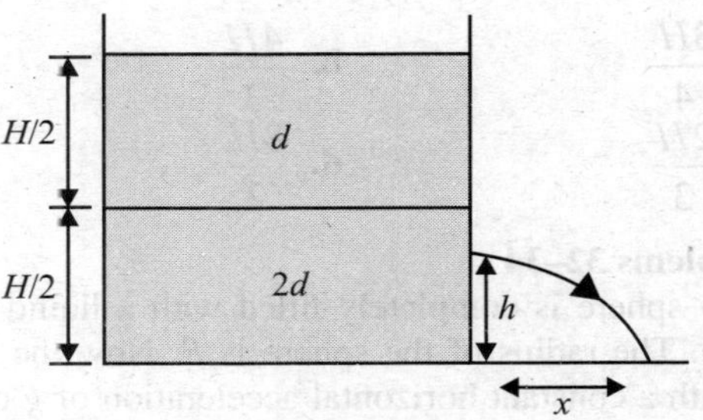

Fig. 6.70

Situation I:

A homogeneous solid cylinder of length L ($L < H/2$), cross-sectional area $A/5$ is immersed such that it floats with its axis vertical at liquid–liquid interface with length $L/4$ in the denser liquid.

26. The density of the solid is

a. $\frac{5d}{4}$ **b.** $\frac{d}{4}$
c. $\frac{2d}{4}$ **d.** $\frac{3d}{4}$

27. The total pressure at the bottom of the container is

a. $\left(\frac{H}{2}+\frac{L}{4}\right)dg$ **b.** $P_0+\left(\frac{3H}{2}+\frac{L}{2}\right)dg$
c. $P_0+\left(H+\frac{L}{4}\right)dg$ **d.** $P_0+\left(\frac{3H}{2}+\frac{L}{4}\right)dg$

Situation II:

The cylinder is removed and the original arrangement is restored. A tiny hole of area s ($s << A$) is punched on the vertical side of the container at a height h ($h < H/2$).

28. The initial speed of efflux of the liquid at the hole is

a. $\frac{\sqrt{(4H-3h)g}}{2}$ **b.** $\frac{\sqrt{(3H-4h)g}}{2}$
c. $\sqrt{\frac{(4H-3h)g}{2}}$ **d.** $\sqrt{(3H-4h)\frac{g}{2}}$

29. The horizontal distance x travelled by the liquid initially is

a. $\sqrt{(H-3h)h}$ **b.** $\sqrt{3Hg}$
c. $\sqrt{(3H-4h)h}$ **d.** $2h$

30. The height h_m at which the hole should be punched so that the liquid travels the maximum distance is

a. $\frac{2H}{3}$ **b.** $\frac{3H}{8}$
c. $\frac{4H}{3}$ **d.** $\frac{4H}{3}$

31. The maximum distance travelled x_m is

a. $\dfrac{3H}{4}$ b. $\dfrac{4H}{3}$

c. $\dfrac{2H}{3}$ d. $\dfrac{8H}{3}$

For Problems 32–34

A hollow sphere is completely filled with a liquid having a density ρ. The radius of the sphere is R. Now the sphere is pulled with a constant horizontal acceleration of g on a horizontal surface. Take centre of sphere as origin of co-ordinate system as shown in the figure.

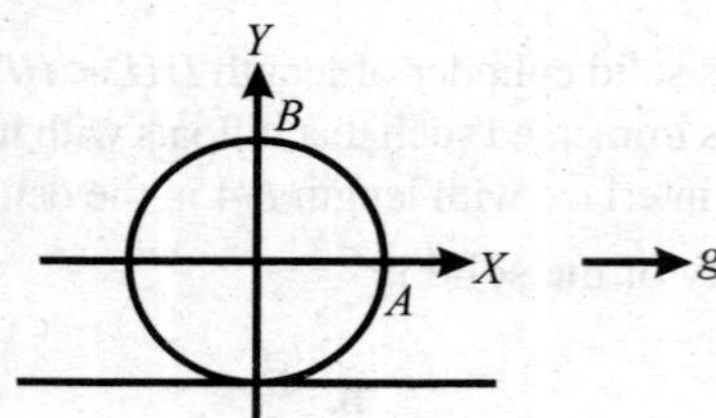

Fig. 6.71

32. Co-ordinate of the point having the minimum pressure is

a. $\dfrac{R}{\sqrt{2}}, \dfrac{R}{\sqrt{2}}$ b. $\dfrac{R}{2}, \dfrac{R}{2}$

c. $\dfrac{-R}{\sqrt{2}}, \dfrac{-R}{\sqrt{2}}$ d. $\dfrac{-R}{2}, \dfrac{-R}{2}$

33. Co-ordinate of the point having the maximum pressure is

a. $\dfrac{R}{\sqrt{2}}, \dfrac{R}{\sqrt{2}}$ b. $\dfrac{R}{2}, \dfrac{R}{2}$

c. $\dfrac{-R}{\sqrt{2}}, \dfrac{-R}{\sqrt{2}}$ d. $\dfrac{-R}{2}, \dfrac{-R}{2}$

34. Consider points A and B as shown in the figure.

a. $P_A = P_B$ b. $P_A > P_B$

c. $P_B > P_A$ d. Cannot say

Matching Column Type

Solutions on page 6.36

1. The arrangement is kept in a vertical plane and all masses are released from rest with strings taut. In column I different values of masses are given. Match with the corresponding parameters in column II. Symbols have their usual meanings. Pulleys and strings are ideal.

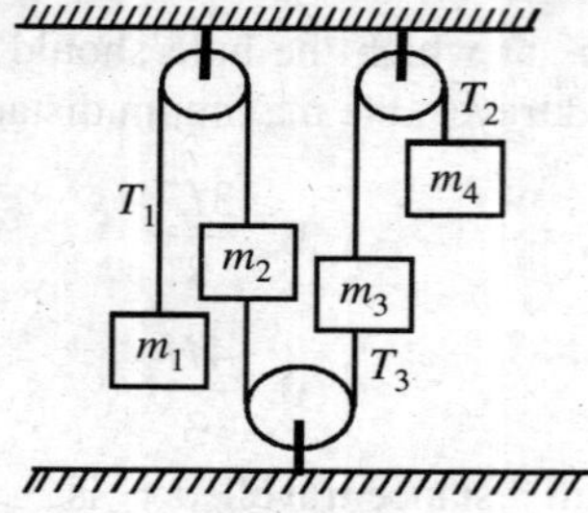

Fig. 6.72

	Column I				Column II
	m_1	m_2	m_3	m_4	
i.	$2m$	m	$3m$	$4m$	**a.** $\lvert a_1 \rvert = \lvert a_3 \rvert \neq 0$
ii.	m	$2m$	$4m$	$3m$	**b.** $T_2 > T_1$
iii.	m	$2m$	$3m$	$4m$	**c.** $\lvert a_1 \rvert > \lvert a_3 \rvert$
iv.	$4m$	$3m$	$2m$	m	**d.** $T_3 = 0$

2. Match the nature of friction force in column II with the corresponding physical situations described in column I.

Column I	Column II
i. A; $a_{\text{belt}} > m_s g$; Belt. A block is placed on a very long conveyor belt. Initially, the block and the belt are at rest. The belt starts to move with an acceleration a as shown. Consider the situation after very long time.	**a.** Static friction
ii. V_{Block}; A; $a_{\text{belt}} > m_k g$; Belt. The belt is moving with an acceleration a as shown. A block is projected on the belt with a velocity v. The block is opposite to the direction of motion of the belt. Consider the situation after very long time.	**b.** Kinetic friction
iii. A car is moving on a banked road without a lateral skidding. Speed of the car $v = \sqrt{gR\tan\theta}$ is constant.	**c.** No friction
iv. rough; F; M; A; smooth; θ. Block A of mass m placed on a wedge, $F = (M + m)g\tan\theta$.	**d.** Cannot say anything about friction.

3. v-t graph for a rectilinear motion is given as shown in the figure. Match the column on the basis of the v-t graph.

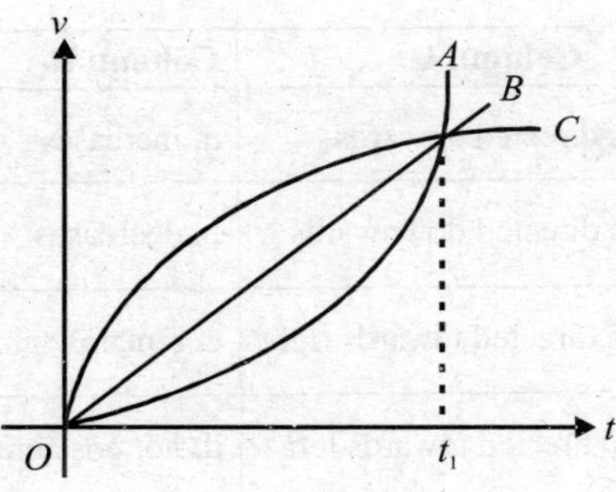

Fig. 6.73

Column I	Column II
i. The magnitude of average velocity from $t = 0$ to $t = t_1$	**a.** Same for A, B, C
ii. Magnitude of acceleration at $t = t_1$	**b.** Maximum for A
iii. Magnitude of acceleration at $t = 0$	**c.** Minimum for C
iv. Speed at t_1	**d.** Maximum for C

4. A small ball is projected along the surface of a smooth inclined plane with speed 10 m/s along the direction shown at $t = 0$. The point of projection is origin, z-axis is along the vertical. The acceleration due to gravity is 10 m/s^2.

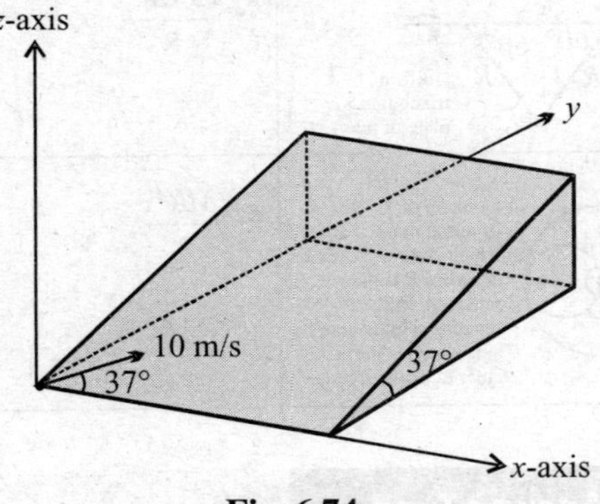

Fig. 6.74

Column I lists the values of certain parameters related to motion of ball and column II lists different time instants. Match appropriately.

Column I	Column II
i. Distance from the x-axis is 2.25 m	**a.** 0.5 s
ii. Speed is minimum	**b.** 1.0 s
iii. Velocity makes an angle 37° with the x-axis	**c.** 1.5 s
iv. The ball is maximum away from the x-axis	**d.** 2.0 s

5. A heavy block C of mass M kept on a frictionless surface and being pulled by two ropes A and B of equal mass m. At $t = 0$, force on the left rope $(\vec{F}_1)$ is withdrawn but the force on the right end $(\vec{F}_2)$ continues to act. Match the following based on the above information. It is given $|\vec{F}_2| > |\vec{F}_1|$.

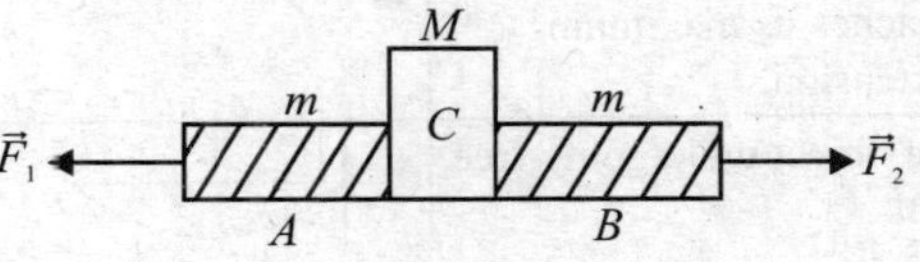

Fig. 6.75

Column I	Column II
i. Magnitude of net force acting on the rope B at $t < 0$ is	**a.** $\dfrac{mF_2}{2m + M}$
ii. Magnitude of net force acting on the rope A at $t < 0$ is	**b.** $\dfrac{m(F_2 - F_1)}{M + 2m}$
iii. Tension force applied by rope B on block C is (at $t > 0$)	**c.** $\dfrac{(m + M)(F_2 - F_1)}{M + 2m}$
iv. Tension force applied by rope A on block C is (at $t > 0$)	**d.** $\dfrac{(M + m)F_2}{(2m + M)}$

6. Column I gives various situations in which an iron sphere weighing 10 N rests in a V shaped smooth trough whose sides form an angle of 60°. Column II gives the reactions forces at points A and B. Match the columns.

Column I	Column II
i. (60°, A)	**a.** $R_A = 10$ N
ii. (A, B, 60°, 60°)	**b.** $R_B = 0$
iii. (A, B, 60°)	**c.** $R_B = 10$ N
iv.	**d.** $R_B = 10/\sqrt{3}$ N

7. Match the physical situation on the left with the graph on the right. The graphs depict the variation of total energy (E), potential energy (U) and kinetic energy (KE) with time.

Column I	Column II
i. A mass attached to an unstretched ideal spring. Released in a vertical plane from rest until it reaches its maximum extension.	**a.** E, KE, U
ii. An object undergoing free fall.	**b.** E, U, KE
iii. An object being pulled on a level, frictionless surface by a constant force in the horizontal direction.	**c.** E, U, KE
iv. A block is coming down on a rough inclined plane with constant velocity.	**d.** E, KE, U, t

8. In the table shown, column II shows the possible outcomes to the water level of a swimming pool when a person standing on a boat in it does any one of the actions shown in column I. Match the possible outcomes.

Column I	Column II
i. He throws a 20 kg iron anchor from the boat into the water, which then settles at the bottom ($\rho_{iron} > \rho_{water}$).	**a.** It becomes lower
ii. He throws out a 20 kg log of wood from the boat. The log floats on water ($\rho_{wood} < \rho_{water}$).	**b.** It becomes higher.
iii. He empties 20 kg of water from the boat into the pool.	**c.** It stays the same.
iv. He drinks some water from the pool.	**d.** Cannot be predicted from the information given.

9. A container is carrying some liquid as shown in the diagram. Column I shows the direction of a small acceleration $\vec{a}$ given to the container and column II indicates the effect on the value of $P_A - P_B$. Match the columns:

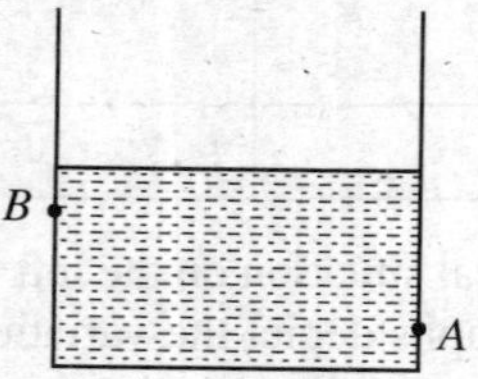

Fig. 6.76

Column I	Column II
i. If $\vec{a}$ is directed upwards	**a.** increases
ii. If $\vec{a}$ is directed downwards	**b.** decreases
iii. If $\vec{a}$ is directed towards right	**c.** remains same
iv. If $\vec{a}$ is directed towards left	**d.** not possible to predict

10. Match the columns. Column I gives some systems whose moment of inertia is listed in column II about the shown axis.

Column I	Column II
i. Uniform rod, M, $l = R$, 30°	**a.** $\dfrac{8MR^2}{11}$
ii. Uniform semicircular ring axis is perpendicular to plane of ring, M ($\pi = 22/7$)	**b.** $\dfrac{MR^2}{12}$
iii. Uniform triangular plate of mass M, R, 60°, 60°	**c.** $\dfrac{13MR^2}{8}$
iv. Uniform disk of initial mass M from which circular portion of radius R is then removed. M.I. of remaining mass about axis which is perpendicular to plane of plate, $2R$, R	**d.** $\dfrac{MR^2}{8}$

Integer Answer Type

Solutions on page 6.38

1. Two blocks are arranged as shown in the figure. Find the ratio of a_1/a_2. (a_1 is acceleration of m_1 and a_2 that of m_2)

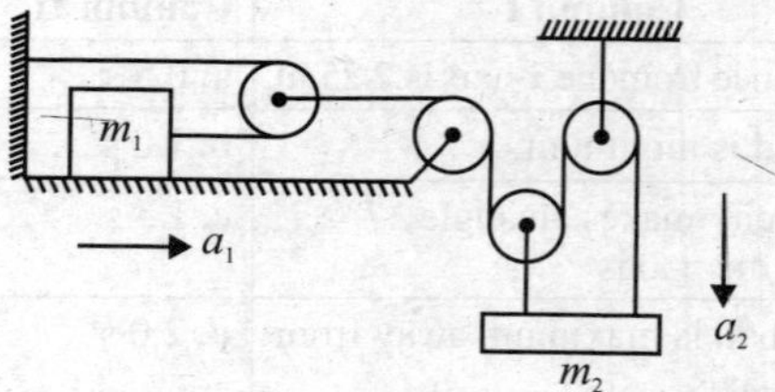

Fig. 6.77

2. A single wire ACB passes through a smooth ring at C which revolves at a constant speed in the horizontal circle of radius $r = 6.4$ m as shown in the figure. Find the speed (in m/s) of revolution of the ring.

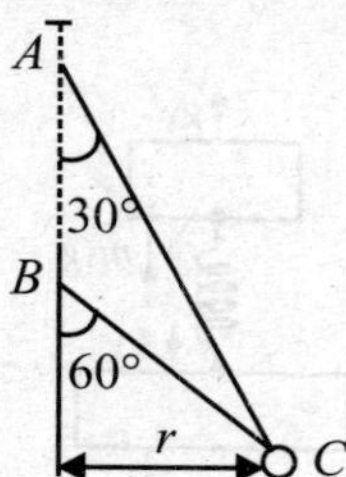

Fig. 6.78

3. We apply a force of 10 N on a cord wrapped around a cylinder of mass 2 kg. The cylinder rolls without slipping on the floor. What is its kinetic energy (in joule) when cylinder has moved by a distance of 3/5 m?

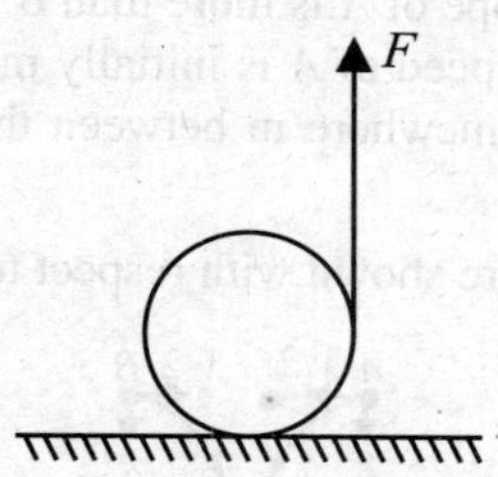

Fig. 6.79

4. The bar shown in the figure is made of a single piece of material. It is fixed at one end. It consists of two segments of equal length $L/2$ each but different cross-sectional area A and $2A$. Find the ratio of total elongation in the bar to the elongation produced in thicker segment under the action of an axial force F. Consider the shape of the joint to remain circular. (Given: Y is Young's modulus).

5. A block of wood of density 500 kg/m^3 has mass m kg in air. A lead block which has apparent weight of 28 kg in water is attached to the block of wood, and both of them are submerged in water. If their combined apparent weight in water is 20 kg, find the value of m. Take density of water = 1000 kg/m^3.

6. The range of water flowing out of a small hole made at a depth 10 m below water surface in a large tank is R. Find the extra pressure (in atm) applied on the water surface so that range becomes $2R$. Take 1 atm = 10^5 Pa.

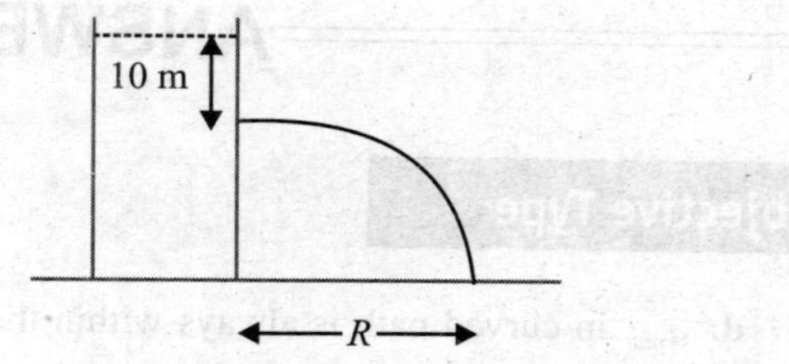

Fig. 6.80

7. An artificial satellite is moving in a circular orbit around the earth with a speed equal to half the magnitude of the escape velocity from the earth. If the satellite is stopped suddenly in its orbit and allowed to fall freely onto the earth, find the speed (in km/sec) with which it hits the surface of the earth ($g = 10$ m/s^2 and $R = 6400$ km).

8. Blocks A and B each of mass 1 kg are moving with 4 m/s and 2 m/s, respectively, as shown. The coefficient of friction for all surfaces is 0.10. Find the distance (in m) by which centre of mass will travel before coming to rest. Assume large distance between the blocks.

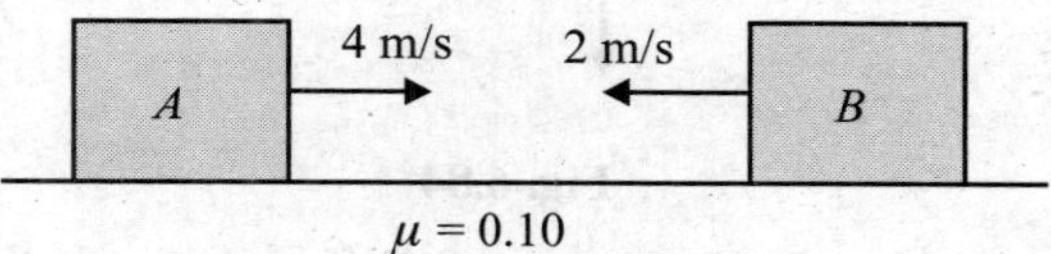

Fig. 6.81

9. Two wheels, each marked with a dot on its rim, are mounted side by side. Initially, the dots are aligned and wheels are at rest. One of the wheels is given a constant angular acceleration of ($\pi/2$) rad/sec^2 and the other wheel is given a constant angular acceleration of ($\pi/4$) rad/sec^2. Both accelerations are in the same direction. Find the time (in s) after which the two dots will become aligned again for the first time.

10. Two steel wires of the same length but radii r and $2r$ are connected together end to end and tied to a wall as shown.

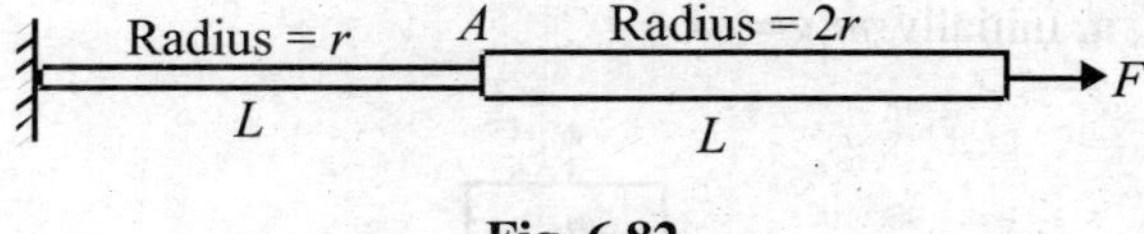

Fig. 6.82

The force stretches the combination by 10 mm. How far does the midpoint A move? (in mm)

ANSWERS AND SOLUTIONS

Objective Type

1. **d.** $\vec{a}_{net}$ in curved path is always within the curvature of trajectory.

2. **a.**

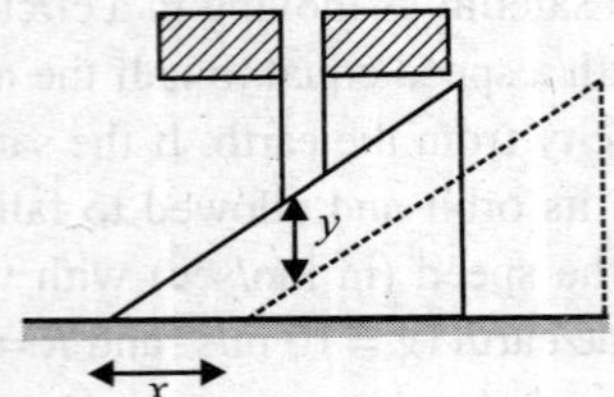

Fig. 6.83

Displacement along the common normal will be the same.

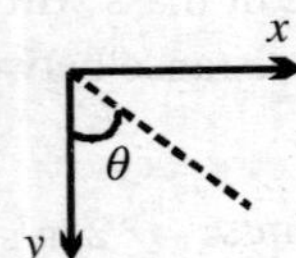

Fig. 6.84

$y \cos\theta = x \sin\theta \Rightarrow y = x \tan\theta$. Differentiating both sides, $vy = vx \tan\theta$.

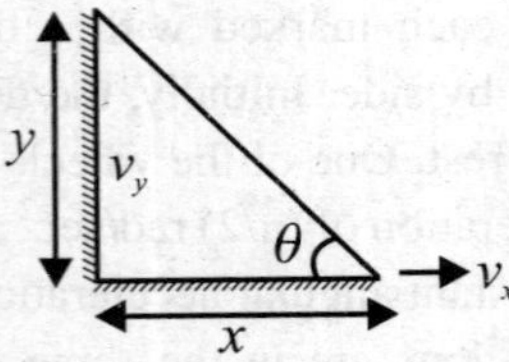

Fig. 6.85

Velocity of the two ends of the rod will be the same along the rod.

$vy \sin\theta = vx \cos\theta \Rightarrow vy = vx \cot\theta$

Student *A* is correct and *B* is wrong.

3. **a.** Initially, $m_1 g = kx$

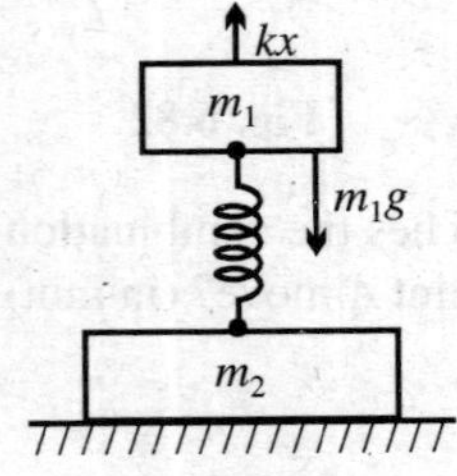

Fig. 6.86

When the support is removed, spring force does not change.

New FBD,

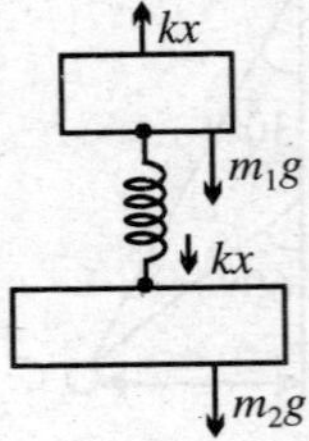

Fig. 6.87

For m_1: $m_1 g - kx = m_1 a_1 \Rightarrow a_1 = 0$

For m_2: $m_2 g + kx = m_2 a_2 \Rightarrow a_2 = \frac{(m_1 + m_2)g}{m_2}$

4. **c.** Initially, slope of *A* is more than *B* and finally slope of *A* is less. So speed of *A* is initially more and finally less than *B*. So somewhere in between their speeds become the same.

5. **c.** Velocities are shown with respect to the man

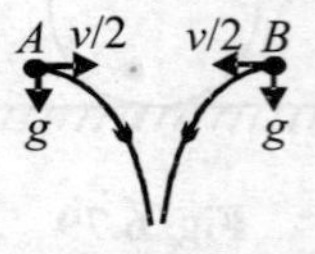

Fig. 6.88

6. **a.** $\frac{v_1}{a} = \frac{\Delta t_1}{2} = 1.5$

$\frac{v_2}{a} = \frac{\Delta t_2}{2} = 1$

$d = \frac{v_1^2 - v_2^2}{2a} \Rightarrow a = 8 \text{ m/s}^2$

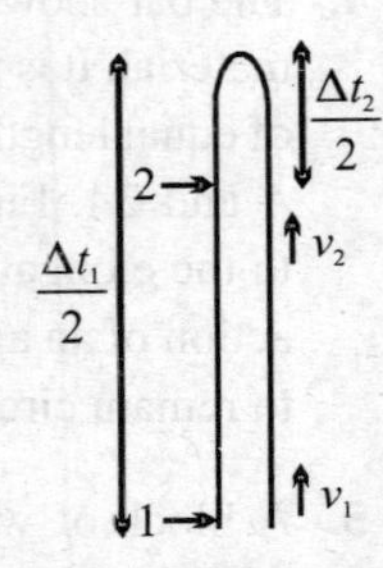

Fig. 6.89

7. **d.** $\beta > \alpha$

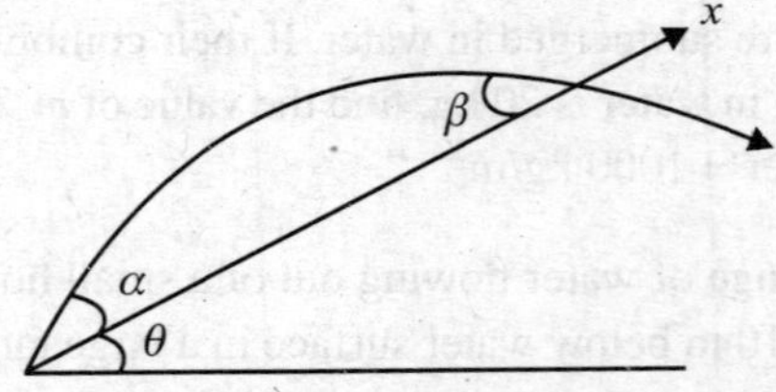

Fig. 6.90

8. **c.** $\vec{F}_{\text{gravitational}}$ acts downwards.

$\vec{F}_{\text{drag}}$ acts opposite to the direction of the motion.

Now let us see how $\vec{F}_{\text{rotor}}$ acts.

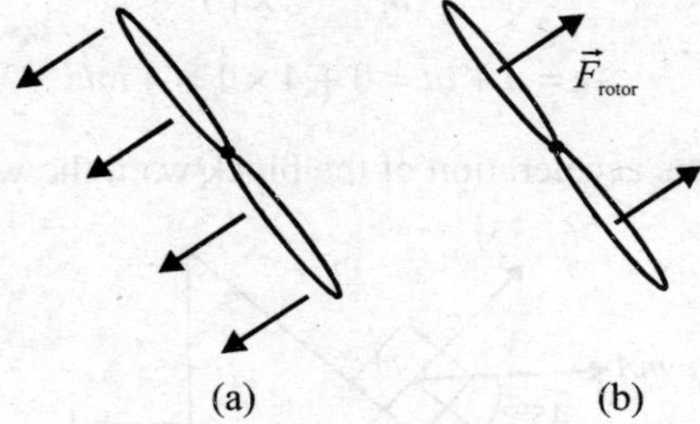

Fig. 6.91

Wings of the helicopter will push the air perpendicular to their plane of rotation as shown in Fig. 6.91(a). From the third law, air will apply force $\vec{F}_{\text{rotor}}$ on the wings as shown in Fig. 6.91(b).

There is no force *ma*. Basically, *ma* is the resultant of all the forces acting on a body. Moreover, here *a* is zero because velocity of the helicopter is constant.

9. d. Since a_t and v are in the same direction, so the particle should be speeding up.

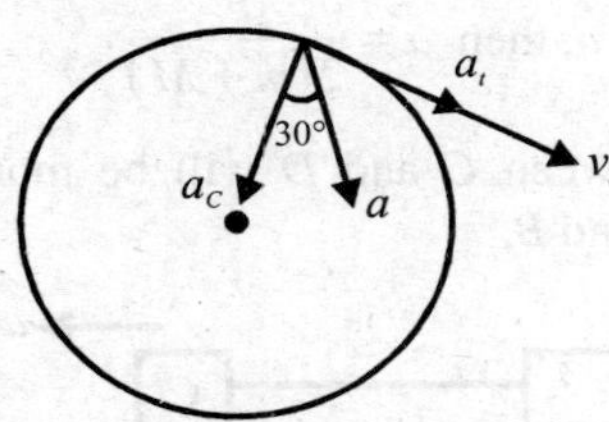

Fig. 6.92

$$\tan 30° = \frac{a_t}{a_c}$$

$$\Rightarrow \frac{1}{\sqrt{3}} = \frac{5}{a_c} \Rightarrow a_c = 5\sqrt{3} \text{ m/s}^2$$

Net acceleration is

$$a = \sqrt{a_c^2 + a_t^2} = \sqrt{(5\sqrt{3})^2 + 5^2} = 10 \text{ m/s}^2$$

10. a. Horizontal component of both should be the same.

$$u \cos \theta = v_{\text{rain}} \cos 45°$$

$$\Rightarrow u \cos 37° = 12\sqrt{2} \times \frac{1}{\sqrt{2}} \Rightarrow u = 15 \text{ m/s}$$

11. c. In Fig. (a) in question, let tension in the string be T_1.

$$F - T_1 = m_1 a$$

$$\Rightarrow 1000 - T_1 = 50a \quad \text{(i)}$$

$$T_1 - m_2 g = m_2 a$$

$$\Rightarrow T_1 - 80\text{g} = 80a \quad \text{(ii)}$$

From Eqs. (i) and (ii), $T_1 = \frac{12,000}{13}$ N

In Fig. (b): Let tension in the string be T_2.

$$F + m_1 g - T_2 = m_1 a$$

$$\Rightarrow 1000 + 50g - T_2 = 50a \quad \text{(iii)}$$

$$T_2 - 80g = 80a \quad \text{(iv)}$$

From Eqs. (iii) and (iv), we get

$$T_2 = \frac{16000}{13} \text{ N} \Rightarrow \frac{T_1}{T_2} = \frac{12000}{16000} = \frac{3}{4}$$

12. a. From FBD of the box, the box will move if

$$F \geq f_b$$

$$\Rightarrow F \geq 0.2mg \quad \text{(i)}$$

where *m* is the mass of box.

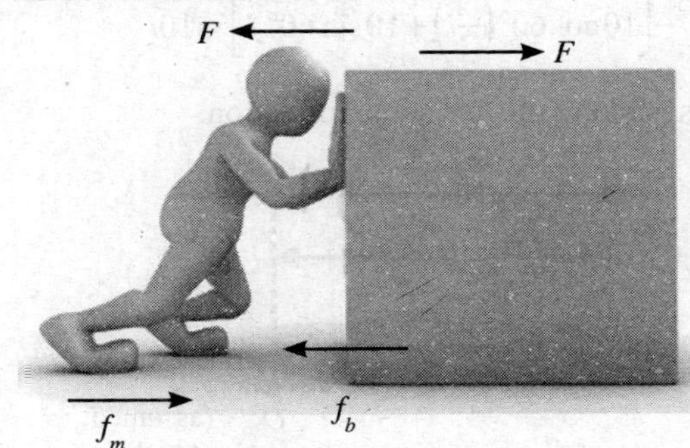

Fig. 6.93

If the man is moving the box, then there is no slipping of the shoes of the man on the floor. So friction on the man f_m will be static.

From FBD of the man $f_m = F$,

but $f_m \leq \mu_s \, 75\, g$

$$\Rightarrow F \leq \mu s \, 75\, g \quad \text{(ii)}$$

From Eqs. (i) and (ii), we get

$$0.2mg \leq F \leq \mu_s 75g \Rightarrow 0.2mg \leq \mu_s 75g$$

$$\Rightarrow \quad m \leq \frac{0.8}{0.2} \times 75 \Rightarrow m \leq 300 \text{ kg}$$

13. c.

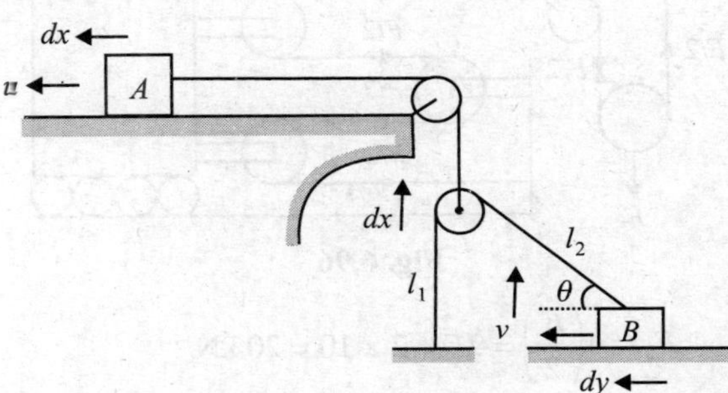

Fig. 6.94

Let in a small time dt, displacement of A is dx and displacement of B is dy. Increase in length $l_1 = dx$.

Increase in the length $\ell_2 = dx\sin\theta - dy\cos\theta$

But net increase in the length should be zero, so

$$dx + dx\sin\theta - dy\cos\theta = 0$$

$$\Rightarrow dy = \frac{(1+\sin\theta)dx}{\cos\theta}$$

$$\Rightarrow \frac{dy}{dt} = \left(\frac{1+\sin\theta}{\cos\theta}\right)\frac{dx}{dt}$$

$$\Rightarrow v = \left(\frac{1+\sin\theta}{\cos\theta}\right)u$$

14. b. Let us find the relative velocity of O. w.r.t. O'.

$$\vec{v}_{O/O'} = \vec{v}_O - \vec{v}_{O'} = \left(10\cos 60°\hat{i} + 10\sin 60°\hat{j}\right)$$
$$-\left[10\cos 60°\left(-\hat{i}\right) + 10\sin 60°\hat{j}\right] = 10\hat{i}$$

This is along the horizontal direction.

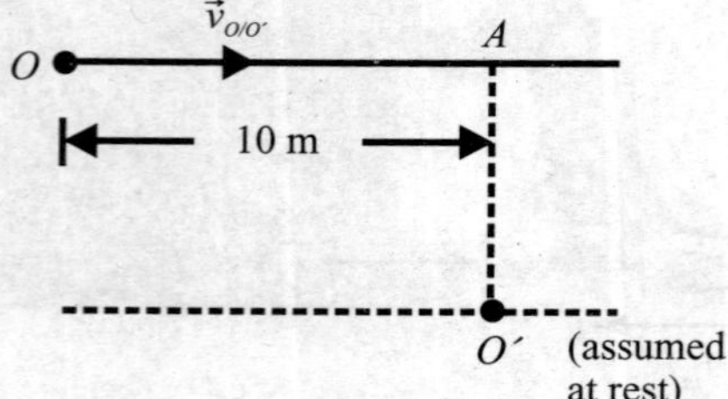

Fig. 6.95

If we assume O' to be at rest, then O will move along OA w.r.t. O' and the minimum separation will be when particle O is at A.

For this, the relative displacement travelled = OA = 10 m

$$\text{Time taken} = \frac{10}{v_{O/O'}} = \frac{10}{10} = 1 \text{ s}$$

15. a. Net force on A:

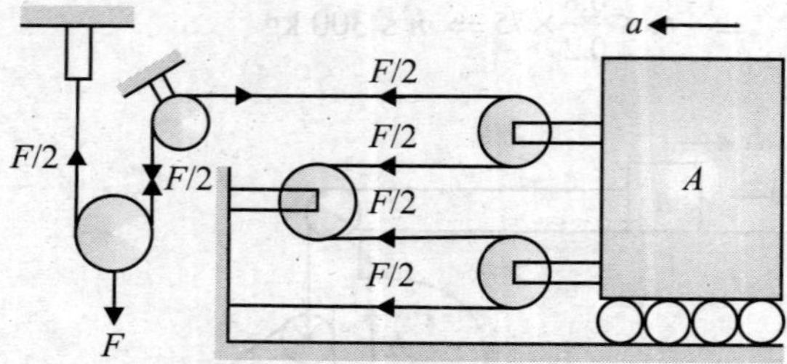

Fig. 6.96

$$F_A = 4\left(\frac{F}{2}\right) = 2F = 2 \times 10 = 20 \text{ kN}$$

$$\text{Acceleration: } a = \frac{F_A}{m} = \frac{20\times10^3}{5\times10^3} = 4 \text{ m/s}^2$$

$$v = u + at = 0 + 4 \times 1 = 4 \text{ m/s}$$

16. c. a is the acceleration of the block w.r.t. the wedge.

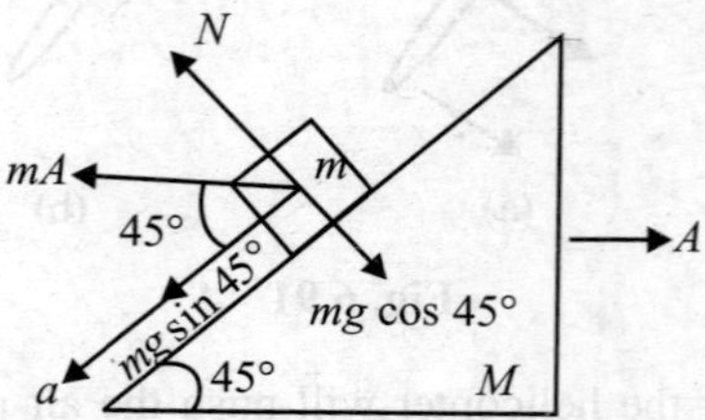

Fig. 6.97

For block:

$N + mA \sin 45° = mg \cos 45°$ (i)

For wedge: $N \sin 45° = MA$ (ii)

From Eqs. (i) and (ii), $A = 1$ m/s^2

17. c. Let all the four blocks move with a common acceleration a, then $a = \dfrac{F}{2(m+M)}$.

Friction between C and D will be more than friction between A and B.

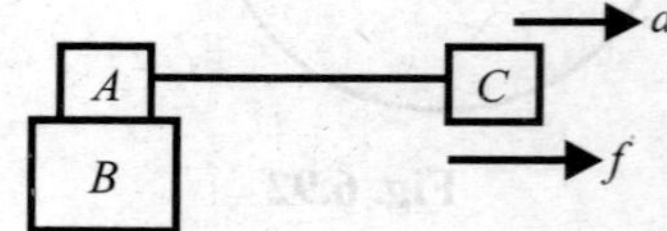

Fig. 6.98

Let this friction be f, then

$$f = (m+m+M)a = \frac{(2m+M)F}{2(m+M)}$$

For no slipping: $f \le f_1$

$$\frac{(2m+M)F}{2(m+M)} \le \mu_s mg \Rightarrow F \le \frac{2\mu_s mg(m+M)}{2m+M}$$

18. d. Let the box move towards the right by a distance x, then the displacement of $2m$ is $x + 2L$ and the displacement of m is $x - 2L$ towards the right.

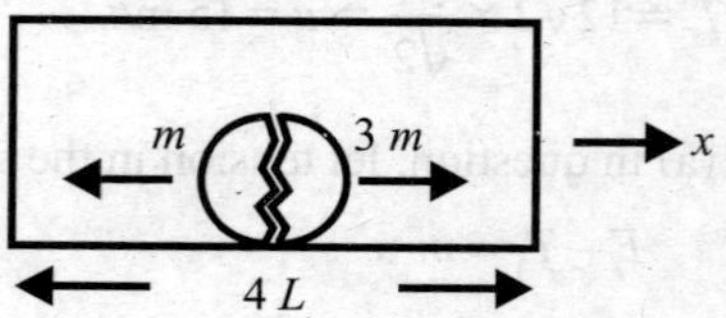

Fig. 6.99

$$\Delta x_{CM} = \frac{m_1\Delta x_1 + m_2\Delta x_2 + m_3\Delta x_3}{m_1 + m_2 + m_3}$$

$\Rightarrow \ 0 = 3mx + 2m(x + 2L) + m(x - 2L)$

$\Rightarrow \ x = -\dfrac{L}{3}$

19. d. Components of accelerations of A and B along normal $(\hat{n})$ to the incline of wedge should be the same.

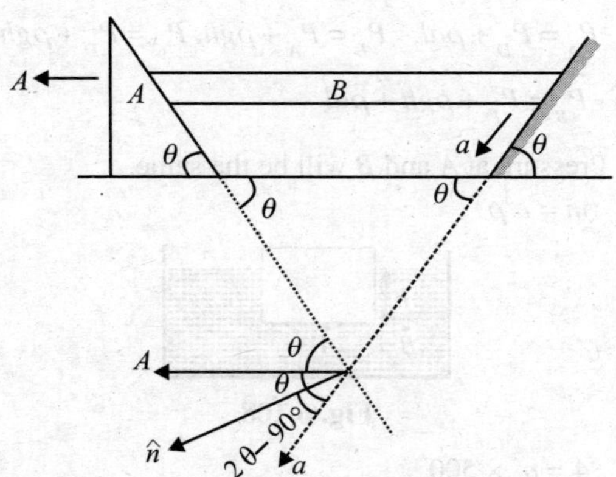

Fig. 6.100

$\Rightarrow \ A\sin\theta = a\cos(2\theta - 90°)$

$\Rightarrow \ A \sin\theta = a \sin 2\theta$

$\Rightarrow \ A = 2a\cos\theta$

20. d.

$$\vec{v}_{CM} = \frac{m_1\vec{v}_1 + m_2\vec{v}_2 + m_3\vec{v}_3}{m_1 + m_2 + m_3}$$

$$= \frac{\text{Total momentum}}{\text{Total mass}}$$

Here total momentum of system is zero, because momentum of each particle is same in magnitude and they are symmetrically oriented as shown.

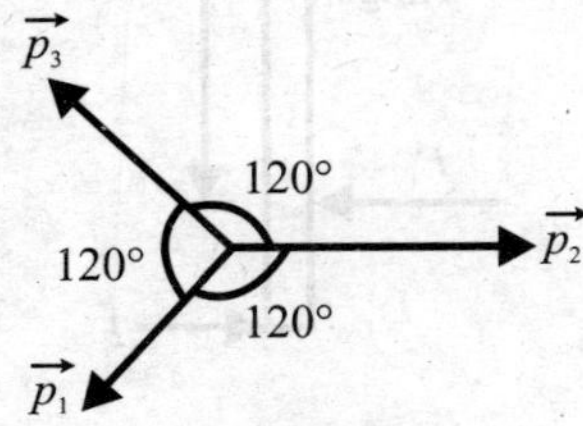

Fig. 6.101

So $\quad \vec{p}_1 + \vec{p}_2 + \vec{p}_3 = 0$

21. b. $y_{CM} = \dfrac{m_1y_1 + m_2y_2}{m_1 + m_2}$

$$= \frac{\rho \times \frac{1}{3}\pi R^2 h \times \frac{h}{4} + \rho \times \frac{2}{3}\pi R^3 \times \frac{3R}{8}}{\rho \times \frac{1}{3}\pi R^2 h + \rho \times \frac{2}{3}\pi R^3}$$

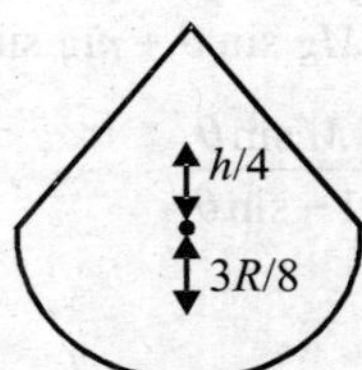

Fig. 6.102

$\Rightarrow \quad h = \sqrt{3}R$

22. b. P_x (before collision) = P_x (after collision)

$\Rightarrow \ -4 + b = 3 + 4 \Rightarrow b = 11$

P_y (before collision) = P_y (after collision)

$5 - 2 = a + 2 \quad \Rightarrow a = 1$

23. c. Let the final velocity of the combined system be V, then applying conservation of linear momentum, we get

$3 \times 1 + 1 \times 1 - 2 \times 2 = 6V$

$V = 0$

Impulse experienced by A = change in momentum of A

$\Rightarrow J = 3[0 - 1] = -3$ Ns

24. d. C is instantaneous centre of rotation

$$\omega\frac{L}{2}\cos\theta = v_0$$

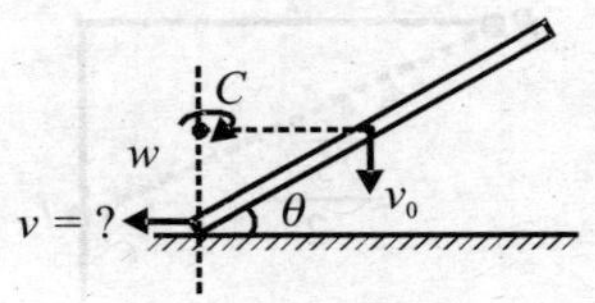

Fig. 6.103

$$\omega\frac{L}{2}\sin\theta = v \Rightarrow v = v_0\tan\theta$$

25. a. $T = mg$

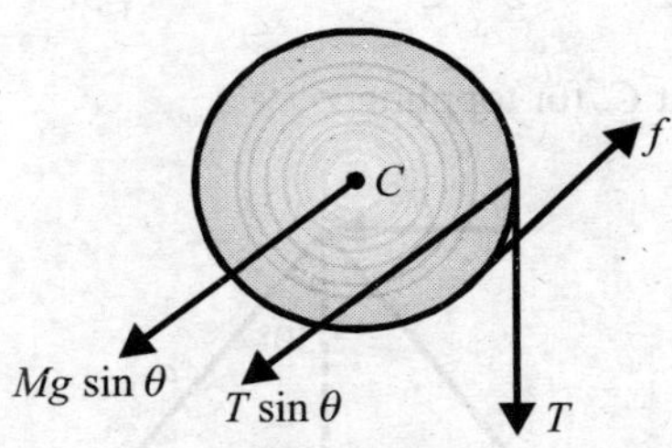

Fig. 6.104

$f = Mg\sin\theta + T\sin\theta$

$= Mg\sin\theta + mg\sin\theta$

Balancing torque about C:

$Tr = fr$

$$\Rightarrow \quad mg = Mg \sin\theta + mg \sin\theta$$

$$\Rightarrow \quad m = \frac{M\sin\theta}{1-\sin\theta}$$

26. c.

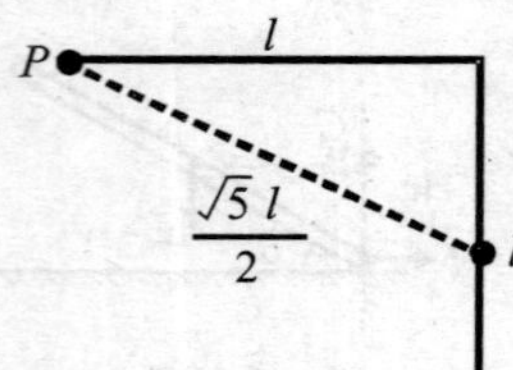

Fig. 6.105

$$\alpha = \frac{a}{\ell} \quad \text{(i)}$$

$$mg - T = ma \quad \text{(ii)}$$

$$T\ell - mg\frac{\ell}{2} = \frac{m\ell^2}{3}\alpha \quad \text{(iii)}$$

Solving the equations, we get $a = \frac{3g}{8}$.

27. b. Moment of inertia of the system of rods about P:

$$I = \left(\frac{m\ell^2}{3}\right) + \left[\frac{m\ell^2}{12} + \frac{m5\ell^2}{4}\right] = \frac{5}{3}m\ell^2$$

Fig. 6.106

Loss in PE = Gain in KE

$$mg\frac{\ell}{2} + mg\frac{\ell}{2} = \frac{1}{2}I\omega^2 \Rightarrow \quad \omega = \sqrt{\frac{6g}{5\ell}}$$

28. c. About C, for toppling $\tau_F > \tau_{mg}$

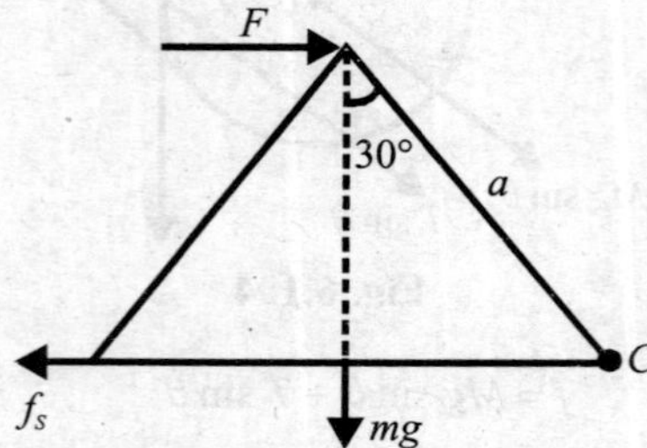

Fig. 6.107

$$Fa\cos 30° > mg\frac{a}{2} \quad \Rightarrow \quad F > \frac{mg}{\sqrt{3}}$$

If no translation, then $f_s = F > \frac{mg}{\sqrt{3}}$

but $\quad f_s \le \mu\, mg$

$$\Rightarrow \frac{mg}{\sqrt{3}} \le \mu\, mg \quad \Rightarrow \mu \ge \frac{1}{\sqrt{3}}$$

29. a. $P_A = P_D + \rho a l,\ \ P_B = P_A + \rho g h,\ P_C = P_D + \rho g h$

$$P_B = P_D + \rho g h + \rho a l$$

30. a. Pressure at A and B will be the same.

$$m = a^3\rho$$

Fig. 6.108

$$4 = a^3 \times 500$$

$$a = \frac{1}{5}$$

$$P_0 + \rho g h = P_0 + \frac{mg}{A}$$

$$h = 10 \text{ cm}$$

31. c. $F_1 = \left(\frac{\rho g h}{2}\right)bh,\ b = h = 1\text{ m}$

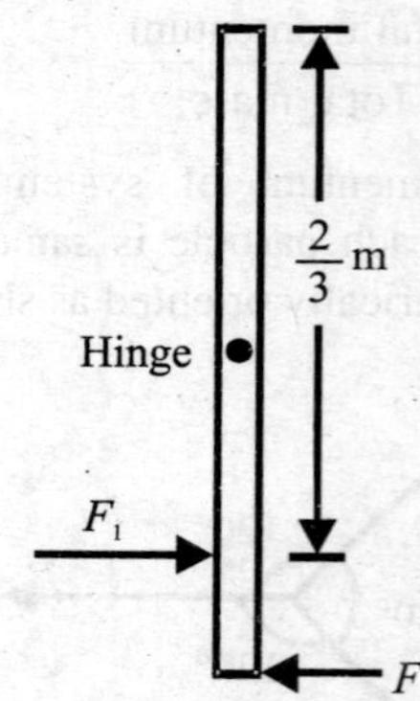

Fig. 6.109

$$F_1 = \frac{\rho g}{2}bh^2 = \frac{\rho g}{2}$$

Torque about hinge is zero.

$$F_1\left[\frac{2}{3} - \frac{1}{2}\right] = F \times \frac{1}{2} \Rightarrow \quad F = \frac{\rho g}{6}$$

32. b. Let A be the area of the soap bar.

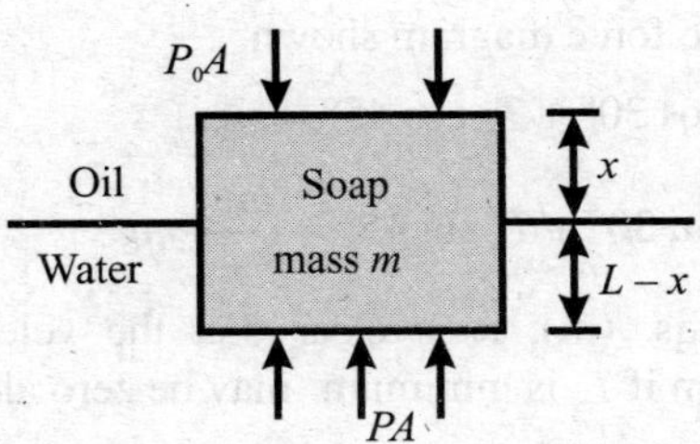

Fig. 6.110

$PA - P_0A = mg \Rightarrow (P - P_0)A = mg$

$\Rightarrow [300gx + 1000g(L - x)A] = AL\ 800g$

$\Rightarrow \dfrac{x}{L} = \dfrac{2}{7}$

33. c. $P_1 - P_0 = \dfrac{4S}{2R}$ (i)

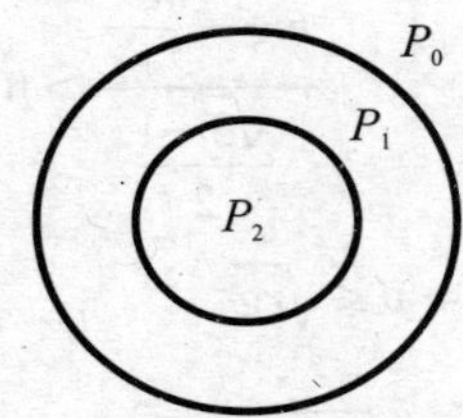

Fig. 6.111

$P_2 - P_1 = \dfrac{4S}{R}$ (ii)

From Eqs. (i) and (ii), $P_2 - P_0 = \dfrac{6S}{R}$

34. b. $v_{t_1} = v_{t_2}$

$$\frac{2}{9}g\frac{r^2}{n_w}[\sigma_1 - \rho_w] = \frac{2}{9}g\frac{r^2}{n_l}(\sigma_2 - \rho_l)$$

$$\Rightarrow \frac{n_w}{n_l} = \frac{\sigma_1 - \rho_w}{\sigma_2 - \rho_l} = \frac{3.2 - 1}{6 - 1.6} = \frac{1}{2}$$

35. b. 30 VS = 29 MS and 1 MS = (1/2)°

Least count = 1 MS – 1VS = 1MS – $\dfrac{29}{30}$ MS

$$= \frac{1}{30}\text{ MS} = \frac{1}{30} \times \frac{1}{2} = \frac{1}{60°} = 1'$$

36. b. By energy conservation,

$$mgR(1 + \cos 30°) = \frac{1}{2}k\left(\frac{\pi R}{6}\right)^2$$

$$\Rightarrow k = \frac{36mg\left(2+\sqrt{3}\right)}{\pi^2 R}$$

37. b. Both the spheres will come at rest simultaneously because net momentum of system is zero. And when they comes to rest, their KE converts into elastic potential energy of spring.

$$U = 2\left[\frac{1}{2}mv_0^2\right] = mv_0^2$$

38. a.

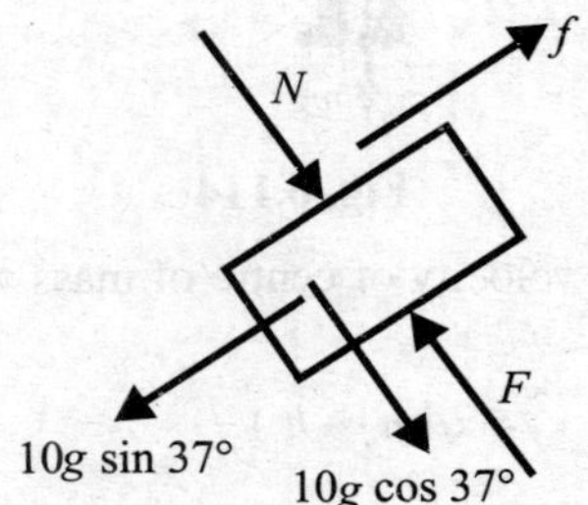

Fig. 6.112

$f_l = \mu N = 0.5[F - 10g \cos 37°]$

$f = 10g \sin 37°$

For the block to not to slide, $f_l > f$.

$\Rightarrow 0.5[F - 10g \cos 37°] > 10g \sin 37°$

$\Rightarrow F > 200$ N

39. d. $V = \dfrac{\pi d^2}{4}\ell$

$$\Rightarrow \frac{\Delta V}{V} = \frac{2\Delta d}{d} + \frac{\Delta \ell}{\ell} = \frac{2 \times 0.01}{4} + \frac{0.1}{5} = \frac{1}{40}$$

% error in volume = $\dfrac{\Delta V}{V} \times 100 = 2.5\%$

40. a.

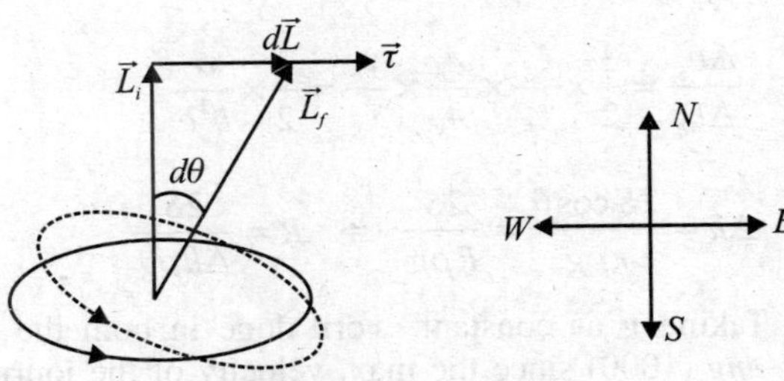

Fig. 6.113

Let the wheel rotate by a small angle $d\theta$. Then $\overline{dL}$ is along east, because magnitude of $\vec{L}$ remains the same. But the torque is in the direction of change in angular momentum. So the torque should be along east. For this, the force should be in the upward direction.

41. c. From h_1 to h_2, let the average force be F. Acceleration a will be upwards.

$$F - mg = ma \Rightarrow a = \frac{F - mg}{m}$$

Fig. 6.114

Let v be the velocity of centre of mass when the contact is left:

$$v^2 = 2a(h_2 - h_1)$$

$$\Rightarrow v^2 = 2\left(\frac{F-mg}{m}\right)(h_2 - h_1)$$

Now he is in free fall. Further the height raised $= h_3 - h_2$.

So $$h_3 - h_2 = \frac{v^2}{2g}$$

$$\Rightarrow h_3 - h_2 = \left(\frac{F-mg}{mg}\right)(h_2 - h_1)$$

$$\Rightarrow F = \frac{mg(h_3 - h_1)}{(h_2 - h_1)}$$

42. b. $TS = 3\,mg$

$$\frac{3mg}{A_s} = y_S \times \frac{\Delta \ell_S}{\ell_S}$$

$$\frac{2mg}{A_b} = y_b \times \frac{\Delta \ell_b}{\ell_b}$$

$$\frac{\Delta \ell_S}{\Delta \ell_b} = \frac{3}{2} \times \frac{\ell_S}{\ell_b} \times \frac{A_b}{A_S} \times \frac{y_b}{y_S} = \frac{3}{2} \times \frac{a}{b^2 c}$$

43. b. $\Delta h = \dfrac{2S\cos\theta}{r\rho g} = \dfrac{2S}{R\rho g} \Rightarrow R = \dfrac{2S}{\Delta h \rho g}$

44. a. Taking g as constant, work done in both the cases is $-mg$ (1000) since the max. velocity of the journey is attained at a height of 1 km, and the second half journey is travelled by max. velocity, the time taken for the second half journey will be less.

45. a.

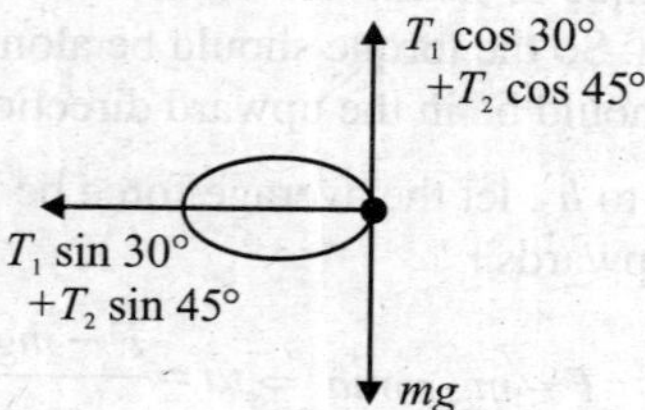

Fig. 6.115

From the force diagram shown

$$T_1 \cos 30° + T_2 \cos 45° = mg \quad \text{(i)}$$

$$T_1 \sin 30° + T_2 \sin 45° = \frac{mv^2}{r} - mg \quad \text{(ii)}$$

From Eqs. (iii), it is clear that the velcoity will be minimum if T_2 is minimum (may be zero also)

$$\frac{T_1}{2} + \frac{T_2}{\sqrt{2}} = \frac{mv^2}{r} \quad \text{(iii)}$$

After solving Eqs. (i) and (ii), we get

$$T_1 = \frac{mg - \dfrac{mv^2}{r}}{\dfrac{(\sqrt{3}-1)}{2}}$$

But $T_1 > 0$, $$\frac{mg - \dfrac{mv^2}{r}}{\dfrac{\sqrt{3}-1}{2}} > 0$$

$$mg > \frac{mv^2}{r} \Rightarrow v < \sqrt{rg}$$

$$v_{max} = \sqrt{rg} = \sqrt{1.6 \times 10} = 4 \text{ m/s}$$

46. b. As no external force is acting on the system. The velocity of centre of the mass will be constant.

$$v_{CM} = \frac{1 \times 5 - 1 \times 3}{1+1} = 1 \text{ m/s}$$

Displacement of the centre of mass $\Delta x_{CM} = 1 \times 2 = 2$ m

Hence, position of the centre of mass $= 5 + 2 = 7$ m

47. c. $\vec{v}_b = \vec{v}_{b,t} + \vec{v}_t = -4\hat{i} + v\hat{i} = (v-u)\hat{i}$

$= (12 - 5)\hat{i} = 7\hat{i} = 7$ m/s towards right

48. c. $\vec{v}_r = \vec{v}_{rw} + \vec{v}_w$

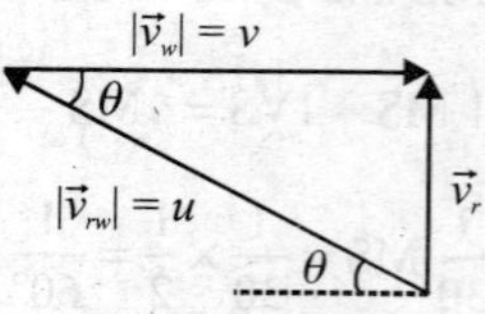

Fig. 6.116

$$\frac{v}{u} = \cos\theta$$

$$\Rightarrow u = \frac{v}{\cos\theta} = \frac{4}{\cos 60°} = 8 \text{ m/s}$$

49. a. $F = 2\pi r_1 T + 2\pi r_2 T$

$= 2\pi(r_1 + r_2)T$

$= 2 \times 3.14(10 + 5)(72) = 6782.4$ dyne

50. b. $P_C = P_B + \frac{2T}{r}$

$P_B = P_A + \rho_1 g h_1 + \rho_2 g (h - h_1)$

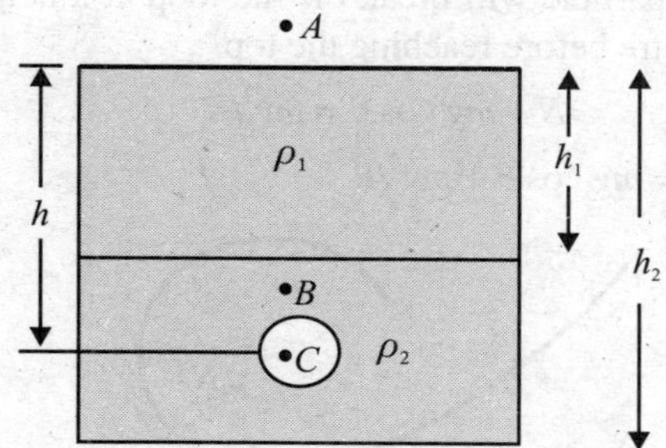

Fig. 6.117

where $P_A = P_0$

$P_C = P_0 + \rho_1 g h_1 + \rho_2 g (h - h_1) + \frac{2T_2}{r}$

Then evaluate by putting the numerical values.

51. a. Let a_0 be the acceleration of centre G and α be the angular acceleration of the plate just after cutting the string BF.

$mg - T = ma_0$ (i)

$\tau_G = I_G \alpha \quad \Rightarrow T\frac{a}{2} = \frac{ma^2}{6}\alpha$

$\Rightarrow T = \frac{ma\alpha}{3}$ (ii)

Just after cutting BF, net acceleration of A is resultant of two accelerations a_0 and $\alpha a/\sqrt{2}$ as shown. But the net acceleration of A in the vertical direction should be zero.

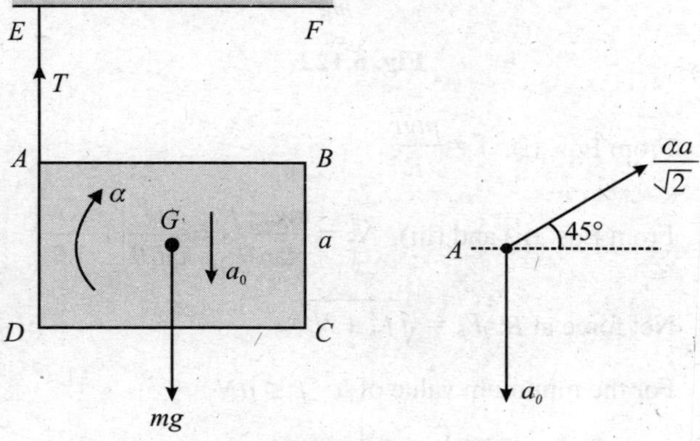

Fig. 6.118

Hence, $a_0 = \frac{\alpha a}{\sqrt{2}} \sin 45°$

$\Rightarrow a_0 = \frac{\alpha a}{2}$ (iii)

From Eqs. (i), (ii) and (iii), we get

$$T = \frac{2mg}{5}$$

Multiple Correct Answers Type

1. a, b, c, d. $\vec{v}_{A/g} = \vec{v}_{A/w} + \vec{v}_{w/g}$

Net velocity of the airplane is the resultant of two given velocities.

The resultant is maximum in case d and minimum in case c.

Lateral velocity (velocity perpendicular to $\vec{v}_{A/w}$) is maximum in case a, hence lateral displacement is maximum in case a.

In each case, $\vec{v}_{A/w}$ is towards north, so $\vec{v}_{w/A}$ will be towards south.

2. a, b, c.

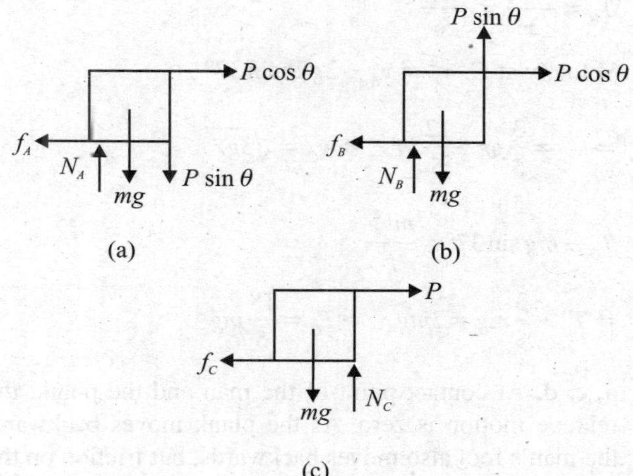

Fig. 6.119

$N_A = mg + P\sin\theta$

$N_B = mg - P\sin\theta$

$N_C = mg$

Clearly, $N_A > N_C > N_B$

Now to find relation betwen frictions:

(i) Let sliding does not occur:

$f_A = P\cos\theta,\ f_B = P\cos\theta,\ f_C = P$

$\Rightarrow f_C > f_A = f_B$

(ii) Let sliding occur at all contact surfaces:

$f_A = \mu N_A, f_B = \mu N_B, f_C = \mu N_C$

$\Rightarrow f_A > f_C > f_B$

3. a, b, d. (a) For the system of (horse + cart), only $\vec{F}_{c/h}$ and $\vec{F}_{h/c}$ will be internal forces. All other forces will be external.

(b) As (horse + cart) system is moving with constant velocity, so net horizontal force on it should be zero.

For this $\vec{F}_{h/g} + \vec{f}_{c/g} = 0$

(c) $\vec{N}_{C/g}$ and $\vec{F}_{C/E}$ cannot be action–reaction pairs, as they act on the same body.

(d) $\vec{F}_{C/h}$ and $\vec{f}_{h/C}$ are action–reaction pairs.

4. a, b, c. Minimum velocity at A should be

$$v_A = \sqrt{g \sin 37° r} = \sqrt{\frac{3}{5} gr}$$

Velocity at B: $v_B^2 = v_A^2 + 2g\, r \sin 37°$

$$\Rightarrow v_B^2 = \frac{3}{5} gr + \frac{6}{5} gr \quad \Rightarrow v_B = \sqrt{\frac{9gr}{5}}$$

$$T_B = \frac{mv_B^2}{r} = \frac{9mg}{5}$$

Velocity at C: $v_C^2 = v_A^2 + 2g2r \sin 37°$

$$\Rightarrow v_C^2 = \frac{3}{5} gr + \frac{12}{5} gr \quad \Rightarrow v_C = \sqrt{3gr}$$

$$T_C - mg \sin 37° = \frac{mv_C^2}{r}$$

$$\Rightarrow T_C - \frac{3}{5} mg = 3mg \quad \Rightarrow T_C = \frac{18}{5} mg$$

5. a, c, d. At contact point of the man and the plank, the relative motion is zero. As the plank moves backward, the man's foot also moves backwards, but friction on the man acts forward. Hence, the work done by friction on the man is negative. But with respect to the plank, the foot will be at rest, hence no work done by friction on the man w.r.t. the plank.

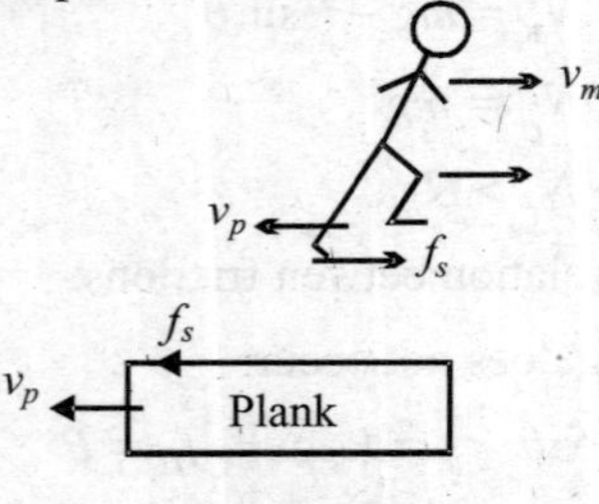

Fig. 6.120

6. a, c, d. Impulse = Change in momentum

$\Rightarrow \quad 6 \times 4 = 10(v - 5) \quad \Rightarrow \quad v = 7.4$ m/s

With respect to observer B, speed will be $v - 2 = 7.4 - 2 = 5.4$ m/s

7. b, d. Velocity on reaching the base of circle is

$$u = \sqrt{2gH} = \sqrt{2g2R} = \sqrt{4gR}$$

This is less than $\sqrt{5gR}$. Hence, the particle cannot complete the circle. But this velocity is more than $\sqrt{2gR}$, so the particle will break off the loop at a height above the centre before reaching the top.

8. a, b, d. $\quad N - mg \cos \theta = mv^2/R$

$\Rightarrow N = mg \cos \theta + mv^2/R$

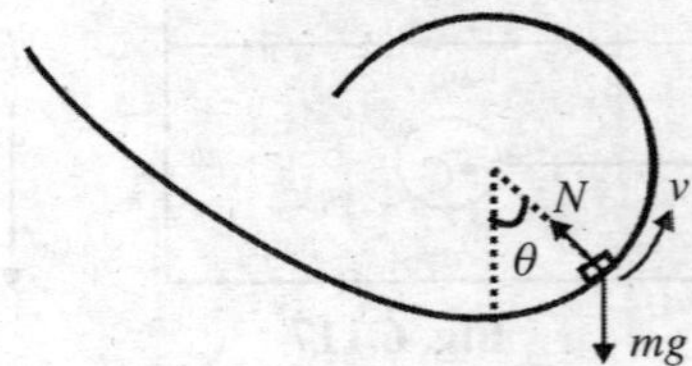

Fig. 6.121

At $\theta = 0$, v is maximum so N is maximum

If $H > 5R/2$, then the particle completes the circle and minimum N occurs at top ($\theta = \pi$).

If $\theta < \pi/2$, then N cannot become zero even if the velocity becomes zero. N can never become negative.

9. a, d. $\quad mgr = fR \quad$ (i)

$N_1 \sin \theta + f = mg \quad$ (ii)

$N_1 \cos \theta = N_2 \quad$ (iii)

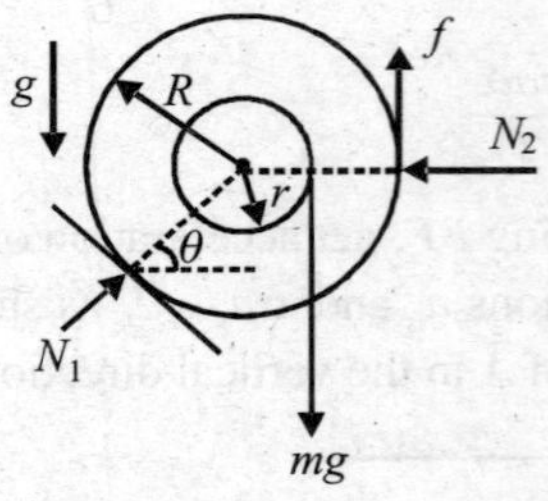

Fig. 6.122

From Eqs. (i), $f = \dfrac{mgr}{R}$

From Eqs. (ii) and (iii), $N_2 = \dfrac{mg - f}{\tan \theta} = \dfrac{mg}{\tan \theta}\left[1 - \dfrac{r}{R}\right]$

Net force at B: $F_B = \sqrt{f^2 + N_2^2}$

For the minimum value of μ: $f \le \mu N_2$

$$\Rightarrow \quad \frac{mgr}{R} \le \frac{\mu mg}{\tan \theta}\left[1 - \frac{r}{R}\right]$$

$$\Rightarrow \quad \mu \ge \frac{\tan \theta}{R/r - 1}$$

10. a, c. $N_A + f_B = 2\,mg,\ N_B = f_A$

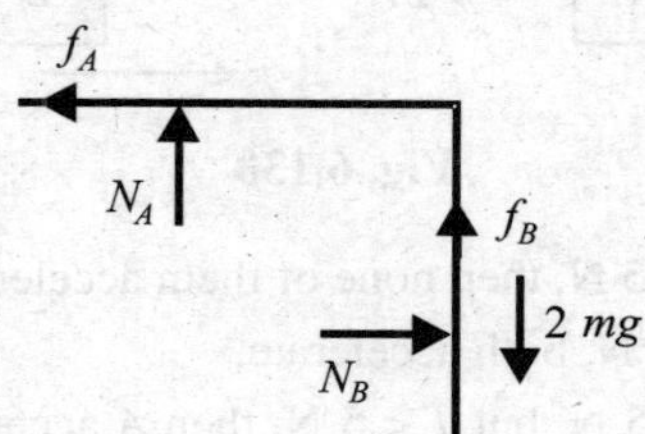

Fig. 6.123

$mgR - f_A R - f_B R = 0$

$mg = f_A + f_B$ and given $f_A = f_B$.

11. a, d. Due to the torque of friction about CM, ζ eventually decreases to zero, initially there is no translation. Friction is sufficient for pure rolling, therefore after sometime pure rolling begins. There is no external force in x direction, therefore momentum is conserved along x direction.

12. a, b, c. Using the parallel axis theorem, check the distance carefully. $I_D = I_B$ (symmetric)

$$I_A = I_B + M\left(\frac{a}{2}\right)^2 = \frac{5}{12}Ma^2$$

$$I_C = I_A + M\left(\frac{a}{\sqrt{2}}\right)^2 = \frac{2}{3}Ma^2$$

13. b, c, d. $V_1 \times d_w g = V_2 \times d_{oil} g = W$

$\Rightarrow V_1 \times 1 = V_2 \times 1.25 \Rightarrow V_1 = 1.25\, V_2$

V_1 = Volume of water displaced, V_2 = volume of the oil displaced

Since the body floats in both the liquids, so in both the cases

Buoyancy = Weight

Entire weight is lost, the apparent weight is zero in the both cases.

Also, Upthrust = Buoyancy

14. a, b, c. $v^2 - u^2 = 2as \Rightarrow v^2 = 2gx \Rightarrow v = \sqrt{2gx}$

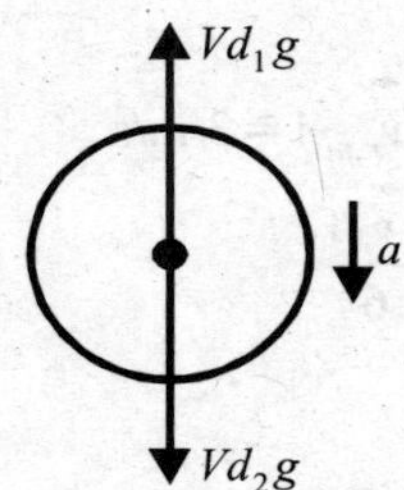

Fig. 6.124

$$Vd_2 g - Vd_1 g = Vd_2 a \quad \Rightarrow \quad a = g\left[1 - \frac{d_1}{d_2}\right]$$

Since $d_2 > d_1$, the body sinks.

15. a, c. $P_5 = P_6 = P_{atm},\ \rho_w = 1$ g/cc

As the pressure is the same at the same horizontal level, so $P_3 = P_6 = P_{atm}$ and $P_4 = P_5 = P_{atm}$

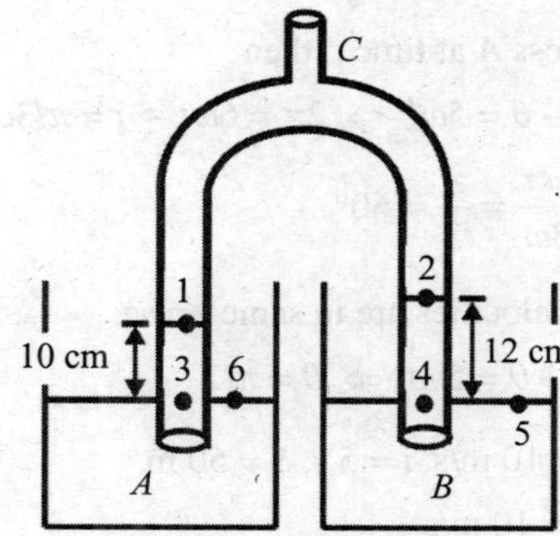

Fig. 6.125

Pressure at 1 and 2 should be the same.

For A: $P_3 = P_1 + \rho_w g h_1$, where $h_1 = 10$ cm

For B: $P_4 = P_2 + \rho_1 g h_2$, where $h_2 = 12$ cm

Solve to get: $\rho_w g h_1 = \rho_1 g h_2$

or $\rho_1 = 0.83$ g/cc

So the liquid is less denser than water.

Pressures at 5 and 6 will be atmospheric, which will be independent of the levels of beakers.

16. a, b, c. $B = F_B - F_S$, where F_S is the net vertical downward force on the slant surface and F_B is net force on the bottom surface.

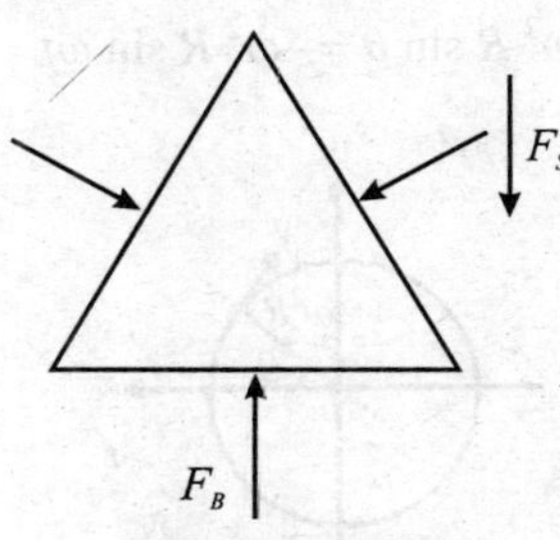

Fig. 6.126

$$B = V\rho g = \frac{1}{3}\pi R^2 \rho g, \quad F_B = \rho g H \pi R^2$$

Now $F_S = F_B - B$

$$\Rightarrow F_S = \rho g H \pi R^2 - \frac{H\rho g \pi R^2}{3} = \frac{2}{3}\rho g H \pi R^2$$

Clearly, $F_B > B,\ F_S > B$

17. b, c, d. If the velocities are oppositely directed:

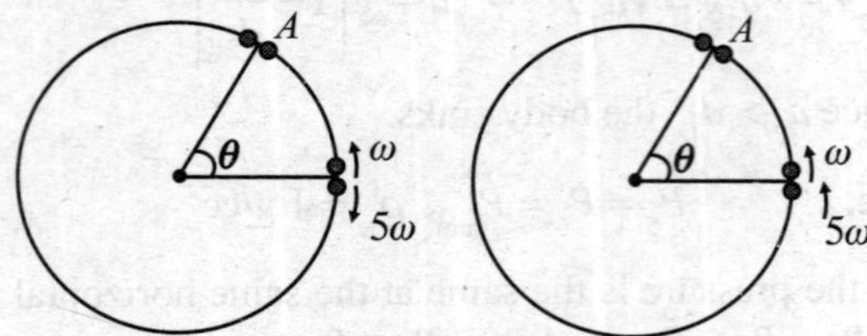

Fig. 6.127

Let they cross A at time t, then

$\theta = \omega t,\ 2\pi - \theta = 5\omega t \Rightarrow 2\pi = 6\omega t \Rightarrow t = \pi/3\omega$

$$\theta = \omega t = \frac{\omega\pi}{3\omega} = \frac{\pi}{3} = 60^\circ$$

If angular velocities are in same sense:

$\theta = \omega t,\ 2\pi + \theta = 5\omega t \Rightarrow \theta = \pi/2$

18. b, c, d. $u = 10$ m/s, $t = 5$ s, $S = 50$ m

$V_{av} = 50/5 = 10$ m/s

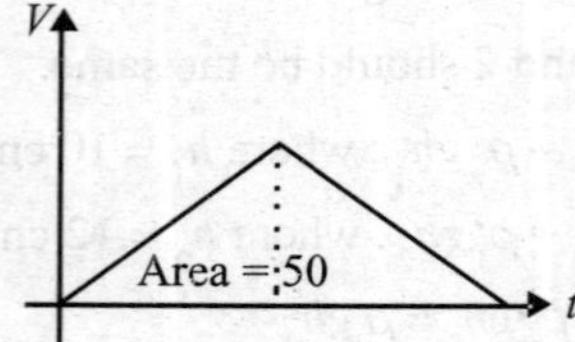

Fig. 6.128

The motion may be with constant velocity. The body may be first accelerated and then retarded. If the body has constant negative acceleration, then it is possible that for some acceleration that body is at a distance of 50 m from the starting point on the opposite side of the starting motion.

19. a, c. (a) $f_y = -\omega^2 R \sin\theta = -\omega^2 R \sin\omega t$

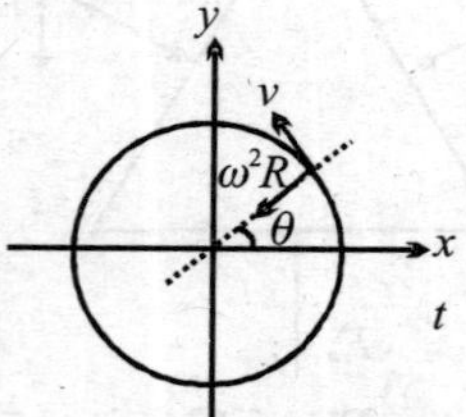

Fig. 6.129

(b) $x = R\cos\theta = R\cos\omega t$

(c) $\theta = \omega t$

(d) $v_x = -v\sin\theta = -v\sin\omega t$

20. a, b, c, d. $f_{l_1} = f_{l_2} = 0.5 \times 1 \times g = 5$ N, T can have any value depending on mass of C.

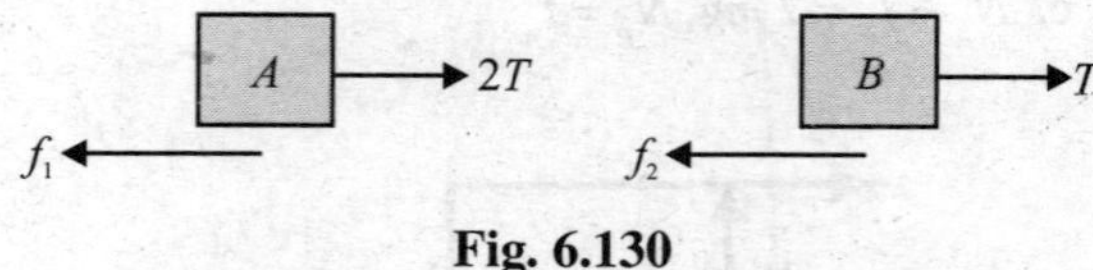

Fig. 6.130

a. If $2T < 5$ N, then none of them accelerates.

b. If $T > 5$ N, both accelerate.

c. If $2T > 5$ N, but $T < 5$ N, then A accelerates, B does not accelerate.

d. If B accelerates, then $T > 5$ N, so $2T > 5$ N, hence A also accelerates.

21. a, b. Maximum acceleration of the man $= \mu mg/m = \mu g$

Maximum acceleration of the boat $= \mu mg/M$

Maximum relative acceleration:

$$a_{rel} = \mu g + \mu g\frac{m}{M} = \frac{(M+m)}{M}\mu g$$

$$S_{rel} = \frac{1}{2}a_{rel}t^2 \Rightarrow t = \sqrt{\frac{2lM}{\mu g(M+m)}}$$

Comprehension Type

For Problems 1–2

1. b., 2. a.

Sol. $\vec{v}_{r,m} = \vec{v}_r - \vec{v}_m = \vec{v}_r + (-\vec{v}_m)$

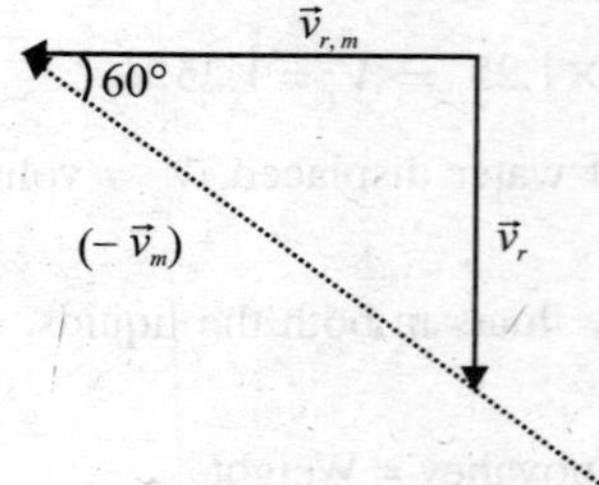

Fig. 6.131

$$\cos 60^\circ = \frac{|\vec{v}_{r,m}|}{|\vec{v}_m|} = \frac{|\vec{v}_r|}{6}$$

$$\frac{1}{2} = \frac{|\vec{v}_{r,m}|}{6} \Rightarrow |\vec{v}_{r,m}| = 3 \text{ m/s}$$

$$\sin 60^\circ = \frac{|\vec{v}_r|}{|\vec{v}_m|} = \frac{|\vec{v}_r|}{6}$$

$$\Rightarrow \frac{\sqrt{3}}{2} = \frac{|\vec{v}_r|}{6} \Rightarrow |\vec{v}_r| = 3\sqrt{3} \text{ m/s}$$

For Problems 3–5

3. c. Impulse applied $J = \langle F \rangle \Delta t$

Angular impulse about O is

$$J_x \times \frac{L}{2} = J\cos 30^\circ \times \frac{L}{2} = \langle F \rangle \times \Delta t \times \frac{\sqrt{3}}{2} \times \frac{L}{2}$$
$$= \frac{\sqrt{3}}{4}\langle F \rangle \Delta t \times L$$

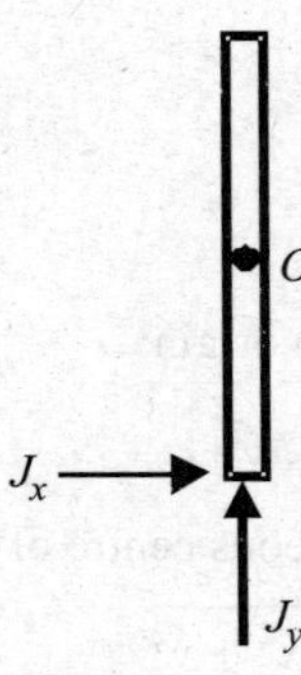

Fig. 6.132

And direction is anticlockwise about O. Hence, $\hat{k}$ direction.

4. d. Angular impulse = $\Delta I\omega$

$$\frac{\sqrt{3}}{4}\langle F \rangle \Delta t \times L = \frac{ML^2}{12} \times \omega \Rightarrow \omega = \frac{3\sqrt{3}\langle F \rangle \Delta L}{ML}$$

5. a. $J_x = MV_x \Rightarrow V_x = \dfrac{J\cos 30^\circ}{M}$

$$J_y = MVy \Rightarrow V_y = \frac{J\cos 30^\circ}{M}$$

Hence, $\vec{v} = v_x\hat{i} + v_y\hat{j}$.

For Problems 6–8

6. b. Angular momentum is conserved, i.e.,

mvr = constant

7. b. Gravitational force is a central force.

For central force, net torque is zero at all points.

$$\vec{\tau} = \frac{d\vec{L}}{dt}$$

Since angular momentum is conserved. So $\vec{\tau} = 0$. then L = constant and conserved.

8. a, c. At net force = 0

$$F = \frac{d\vec{P}}{dt}$$

P = constant or conserved

For Problems 9–11

9. c., 10. a., 11. d.

Sol. At $t = 0$

$$\vec{v}_T = 10\hat{j} \text{ m/s}$$

$$\vec{v}_{s/T} = -6\hat{i} + 8\hat{k} \text{ m/s}$$

$$\vec{v}_S = \vec{v}_{ST} + \vec{v}_T = (-6\hat{i} + 10\hat{j} + 8\hat{k}) \text{ m/s}$$

Time of flight of the stone = $T = \dfrac{2v_g}{g} = \dfrac{8}{5}$ s

For the highest point $t_a = T/2 = 4/5$ s

Angular displacement of train in time $t_a = \omega t_a$

$$= \left(\frac{v}{r}\right)t_a = \left(\frac{10}{16/\pi}\right)(4/5) = \frac{\pi}{2}$$

At the highest point, $\vec{v}_S = -6\hat{i} + 10\hat{j}$

$$\vec{v}_S = -10\hat{i},\ \vec{v}_{ST} = \vec{v}_S - \vec{v}_T = 4\hat{i} + 10\hat{j}$$

$$\vec{S} = \vec{v}_{\text{horizontal}}\, T = [-6\hat{i} + 10\hat{j}]1.6 = -9.6\hat{i} + 16\hat{j}$$

For Problems 12–14

12. b.

Sol. $a_t = \dfrac{F}{m}$

$$v = u + a_t t_1 = 0 + \frac{F}{m}t_1$$

$$T = \frac{mv^2}{l} \Rightarrow T = \frac{m}{l}\frac{F^2}{m^2} \Rightarrow t_1 = \sqrt{\frac{mlT}{F^2}}$$

13. c. At $t = \dfrac{t_1}{v}$

$$v_1 = a_1\frac{t_1}{2} = \frac{F}{m}\frac{t_1}{2}$$

$$a_C = \frac{v_1^2}{l} = \frac{F^2}{m^2}\frac{t_1^2}{4m^2 l}\frac{mlT}{F^2} = \frac{T}{4m}$$

Net acceleration $a = \sqrt{a_C^2 + a_t^2}$

$$= \sqrt{\left(\frac{T}{4m}\right)^2 + \left(\frac{F}{m}\right)^2} = \frac{\sqrt{T^2 + 16F^2}}{4m}$$

14. a. $S = v(2t_1 - t_1) + \dfrac{1}{2}a_1(2t_1 - t_1)^2$

$$= \frac{F}{m}t_1(t_1) + \frac{1}{2}\frac{F}{m}t_1^2 = \frac{3}{2}\frac{F}{m}t_1^2 = \frac{3}{2}\frac{F}{m}\frac{mlT}{F^2} = \frac{3lT}{2F}$$

For Problems 15–17

15. b. $40000 \sin 30 = 30000 \sin\theta \Rightarrow \sin\theta = \frac{2}{3}$

16. c. $40000 \times \frac{\sqrt{3}}{2} + 30000 \times \sqrt{\left(1 - \frac{4}{9}\right)} - F = 10^6 \times 0.03$

$\Rightarrow F = 20000\sqrt{3} + 10000\sqrt{5} - 30000$

= 27,000 N

17. d. $s = \frac{1}{2}at^2 \Rightarrow t^2 = \frac{2s}{a} = \frac{2\times 6}{0.03} = 2\times 200$

$t = 20$ s

For Problems 18–19

18. c. Net torque: $\tau = F2R - FR = FR$

$$\alpha = \frac{\tau}{I} = \frac{FR}{I} = \frac{10\times 1}{4} = 2.5 \text{ rad/s}^2$$

After 2 s: $\omega = \alpha t = 2.5 \times 2 = 5$ rad/s

$$\text{KE} = \frac{1}{2}I\omega^2 = \frac{1}{2}\times 4\times (5)^2 = 50 \text{ J}$$

There will be no translation, as acceleration of centre of mass will be zero.

19. b. $\tau = \frac{\tau_A}{I_A} = \frac{F4R - F3R}{I_0 + m(2R)^2}$

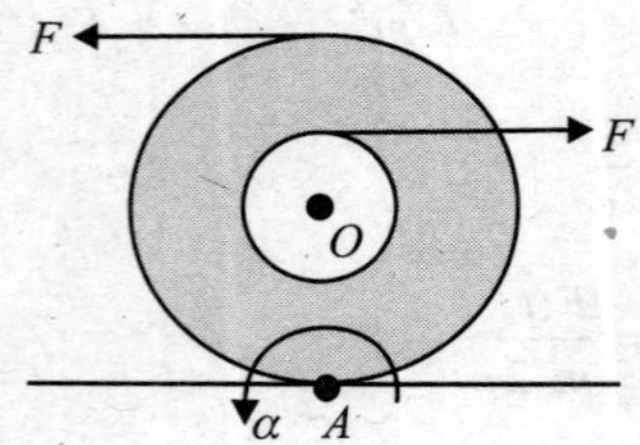

Fig. 6.133

$$\tau = \frac{FR}{4 + 2\times(2\times 1)^2} = \frac{10\times 1}{12} = \frac{5}{6} \text{ rad/s}^2$$

After 2 s: $\omega = \alpha t = \frac{5}{6}\times 2 = \frac{5}{3}$ rad/s

$$\text{KE} = \frac{1}{2}mv^2 + \frac{1}{2}I_0\left(\omega^2\right)$$

$$= \frac{1}{2}\times 2(\omega 2R)^2 + \frac{1}{2}4(\omega)^2$$

$$= \frac{1}{2}\times 2\times\left(\frac{5}{3}\times 2\times 1\right)^2 + \frac{1}{2}\times 4\left(\frac{5}{3}\right)^2$$

$$= \frac{150}{9} = 16.67 \text{ J}$$

For Problems 20–21

20. a. $\omega = \frac{v}{R}$

Distance of P from origin:

$$r = \sqrt{x^2 + y^2}$$

Origin is instantaneous centre of rotation. So,

$$v_P = \omega r = \frac{v\sqrt{x^2 + y^2}}{R}$$

21. b. $a = \alpha R$

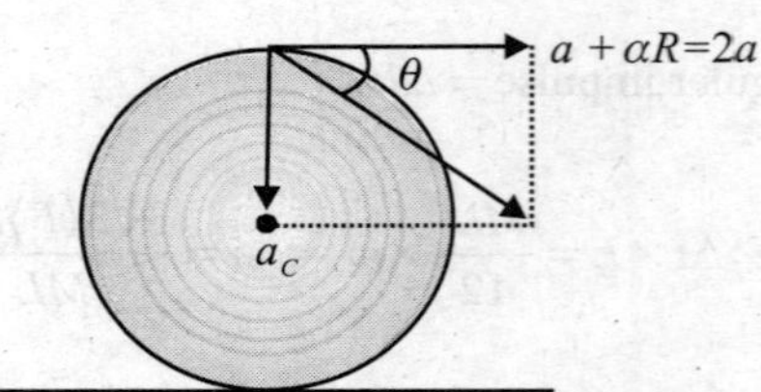

Fig. 6.134

$$a_C = \omega^2 R = \left(\frac{v}{R}\right)^2 R = \frac{v^2}{R} = \frac{2aR}{R} = 2a$$

$$\tan\theta = \frac{2a}{2a} = 1 \Rightarrow \theta = 45°$$

For Problems 22–25

22. a. Volume of the liquid that will flow out is equal to volume of the ball. So mass of the liquid that will overflow is

M = volume overflow × density of liquid

$\Rightarrow M = \rho V$

23. b. Let force F is to be applied on the rod to hold it.

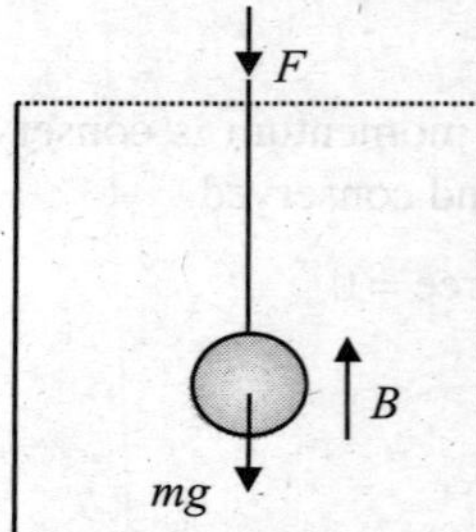

Fig. 6.135

$F + mg = B$

$\Rightarrow \quad F = B - mg \quad \Rightarrow \quad F = V\rho g - mg$

Now reading of the scale:

$W' = F +$ weight of ball + weight of the remaining liquid

$= (V\rho g - mg) + mg + (W - Mg)$

$= V\rho g + W - V\rho g = W$

Alternatively: Reading of the scale will be net force applied by liquid on the container, which will be the same in both cases.

24. a. $T + mg = B$

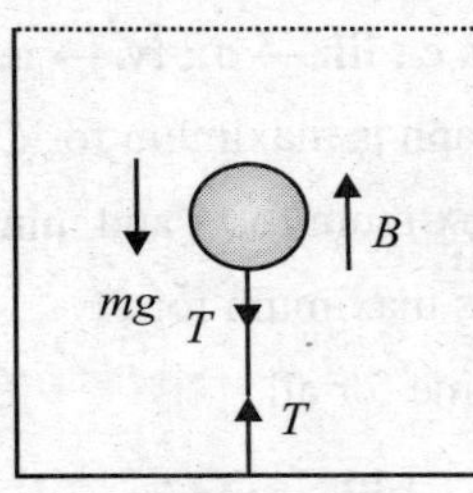

Fig. 6.136

$\Rightarrow T = B - mg = V\rho g - mg = (V\rho - m)g$

25. c. Reading of scale:

$W' =$ weight of ball + weight of the remaining liquid

$= mg + W - Mg$

$= W + mg - V\rho g$

For Problems 26–31

26. a. Since the body is in equilibrium, the net force acting should be zero, i.e., the vector sum of forces should be zero.

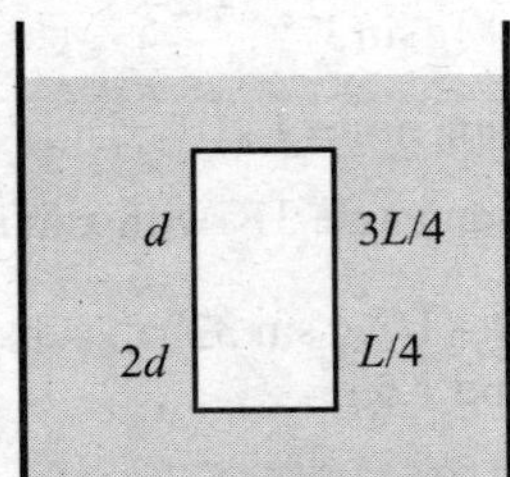

Fig. 6.137

The forces on the body are mg downwards, buoyant force due to denser liquid upwards and buoyant force due to lighter liquid in upward direction.

$$F_{mg} = (\text{Density} \times \text{volume})g = D \times \frac{AL}{5} g \quad \text{(i)}$$

$$F_{\text{buoyant}} = d \times \frac{A}{5} \times \frac{3L}{4} g + 2d \times \frac{A}{5} \times \frac{L}{4} g$$

$$= \left(\frac{5d}{4} \times \frac{A}{5} \times L\right) g = \frac{AdL}{4} g \quad \text{(ii)}$$

From Eqs. (i) and (ii), we have $D \times \frac{AL}{5} g = \frac{AdL}{4} g$

$\Rightarrow \quad D = \frac{5d}{4}$

27. d. Total pressure at the bottom is

$$P = P_0 + \frac{[\text{weight of(liquid + object)}]}{A}$$

$$\Rightarrow P = P_0 + \frac{\left[\frac{AH}{2} dg + \frac{AH}{2} 2dg + \frac{A}{5} L \frac{5d}{4} g\right]}{A}$$

$$= P_0 + \left(\frac{3H}{2} + \frac{L}{4}\right) dg$$

28. d. Pressure at hole, before opening:

$$P = P_0 + dg\frac{H}{2} + 2dg\left(\frac{H}{2} - h\right)$$

After opening the hole, pressure decreases and pressure energy converts into KE. So,

$$P - P_0 = \frac{1}{2}(2d)\, v^2$$

$$\Rightarrow dg\frac{H}{2} + 2dg\left(\frac{H}{2} - h\right) = dv^2$$

$$\Rightarrow v = \sqrt{\left(\frac{3H}{2} - 2h\right) g}$$

$$= \sqrt{\frac{(3H - 4h)g}{2}}$$

29. c. $x = v\sqrt{\frac{2h}{g}}$

$$= \sqrt{\frac{(3H - 4h)g}{2} \times \frac{2h}{g}} = \sqrt{(3H - 4h)h}$$

30. b. $x^2 = 3Hh - 4h^2$

x is maximum if x^2 is maximum.

$$\frac{dx^2}{dh} = 3H - 8h = 0 \quad \Rightarrow h = \frac{3H}{8}$$

31. a. $x_m = \sqrt{\left(3H - 4 \times \frac{3H}{8}\right)\frac{3H}{8}} = \frac{3H}{4}$

For Problems 32–34

32. a., 33. c., 34. a.,

Sol. Effective gravity $(\sqrt{2}g)$ will be along line OC.

Pressure will be minimum at D and maximum at C.

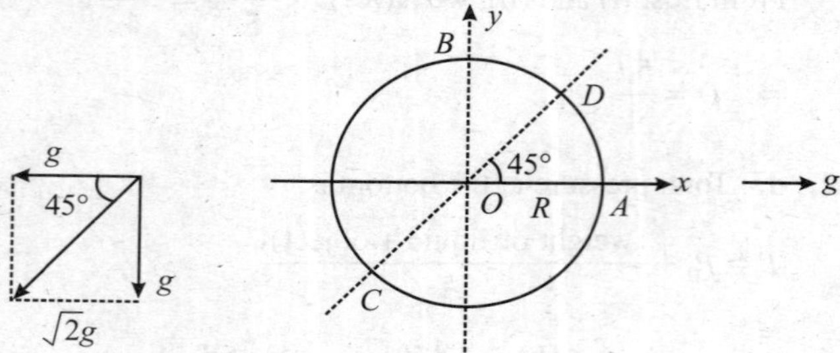

Fig. 6.138

Pressure will be the same at all points on a surface which is perpendicular to line CD. Hence, $P_A = P_B$.

Matching Column Type

1 **i. → b.; ii. → b., c., d.; iii. → b., c., d.; iv. → d.**

Sol. i. m_4 will dominate and move down and m_1 will move up. Let the whole system move with acceleration a,

$m_4g - m_3g + m_2g - m_1g = (m_1 + m_2 + m_3 + m_4)a$

$\Rightarrow$ $(4m - 3m + m - 2m)g = 10a$

$\Rightarrow$ $a = 0$ $\Rightarrow$ $|a_1| = |a_1| = 0$

$T_2 = m_4g = 4mg,\ T_1 = m_1g = 2mg$

$\Rightarrow$ $T_2 > T_1,\ T_3 + m_3g = T_2$

$\Rightarrow$ $T_3 = T_2 - m_3g = 4mg - 3mg = mg$

ii. In pair of m_1 and m_2, m_2 will come down and in pair of m_3 and m_4, m_3 will come down. Hence, $T_3 = 0$, and like at wood machine $T_1 = \dfrac{2m_1m_2}{m_1+m_2}g = \dfrac{4mg}{3}$ and $T_2 = \dfrac{2m_2m_3}{m_3+m_4}g = \dfrac{27}{7}mg$

$$a_1 = \frac{2m-m}{2m+m}g = \frac{g}{3},\quad a_3 = \frac{4m-3m}{4m+3m}g = \frac{g}{7}$$

$\Rightarrow$ $a_1 > a_3$

iii. In pair m_1 and m_2, m_2 will come down with acceleration $g/3$ and in pair of m_3 and m_4, m_3 will go up with acceleration $g/7$. Hence, $T_3 = 0$ and like at wood machine

$$T_1 = \frac{2m_1m_2}{m_1+m_2}g \text{ and } T_2 = \frac{2m_3m_4}{m_3+m_4}g$$

iv. In pair of m_1 and m_2, m_1 will come down with acceleration $g/7$ and in pair of m_3 and m_4, m_3 will come down with acceleration $g/3$. Hence, $T_3 = 0$ and like at wood machine

$$T_1 = \frac{2m_1m_2}{m_1+m_2}g \text{ and } T_2 = g$$

2. **i. → b.; ii. → a.; iii. → c.; iv. → c.**

Sol. i. Acceleration of the block cannot be greater than $\mu_s g$, but acceleration of the belt is $> \mu_s g$. Hence, relative slipping will be there and friction will be kinetic.

ii. After long time, the block will come to rest w.r.t. the belt because acceleration of the belt is less than $\mu_K g$. Now both will move with common acceleration and friction between them will be static.

iii. For the given speed, no friction is required.

iv. No friction will develop between the block and the wedge for the given value of force.

3. **i. → d.; ii. → b. c.; iii. → d.; iv. → a.**

Sol. i. Area of v-t graph is maximum for C.

ii. Slope at t_1 is maximum for A and minimum for C

iii. At $t = 0$, slope is maximum for C

iv. Speed at t_1 is same for all.

4. **i. → a. c.; ii. → b.; iii. → d.; iv. → b.**

Sol. Considering y'-axis along the inclined plane

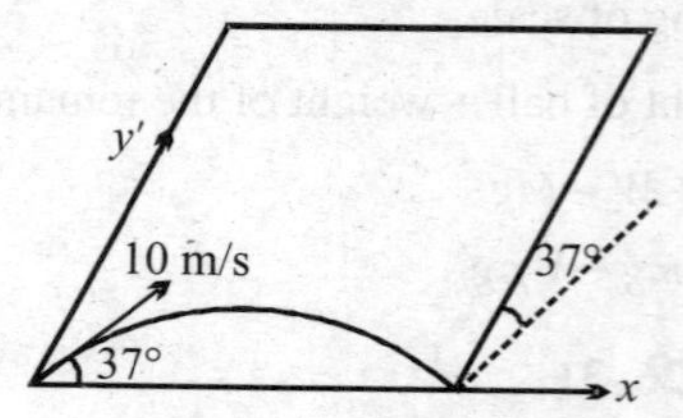

Fig. 6.139

$$\text{Time of flight} = \frac{2u\sin 37°}{g\sin 37°} = 2 \text{ s}$$

Speed is minimum at $t = 1$ s

Velocity makes an angle 37° with x-axis at $t = 2$ s

$y' = 10 \sin 37°\, t - 1/2\, g \sin 37°\, t^2$, substituting $y' = 2.25$, we get $t = 0.5$ and 1.5 s.

5. **i. → b.; ii. → b.; iii. → d.; iv. → a.**

Sol. For $t < 0$, acceleration of system is

$$a = \frac{F_2 - F_1}{M + 2m}$$

Net force on B is

mass × acceleration $= ma$

$$= \frac{m(F_2 - F_1)}{M + 2m}$$

As mass of A is the same as mass of B. So net force on A is also $\dfrac{m(F_2 - F_1)}{M + 2m}$

For $t > 0$; $F_1 = 0$

$$\Rightarrow \quad a = \frac{F_2}{M + 2m}$$

Let T_1 be the tension by B on C

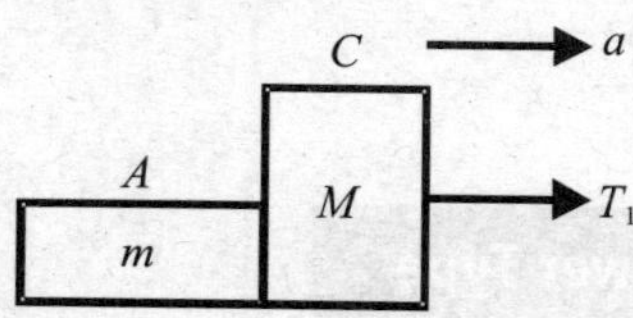

Fig. 6.140

$$T_1 = (M + m)a \qquad \Rightarrow T_1 = \frac{(M + m)F_2}{M + 2m}$$

Let T_2 is tension by A on C. Then C also applies tension T_2 on A.

Fig. 6.141

$$T_2 = ma = \frac{mF_2}{M + 2m}$$

6. i. → a., b.; ii. → a., c.; iii. → d.

Sol. i. $R_B \sin\theta = 0 \Rightarrow R_B = 0$

$R_A = mg + R_B \cos\theta \Rightarrow R_A = mg = 10$ N

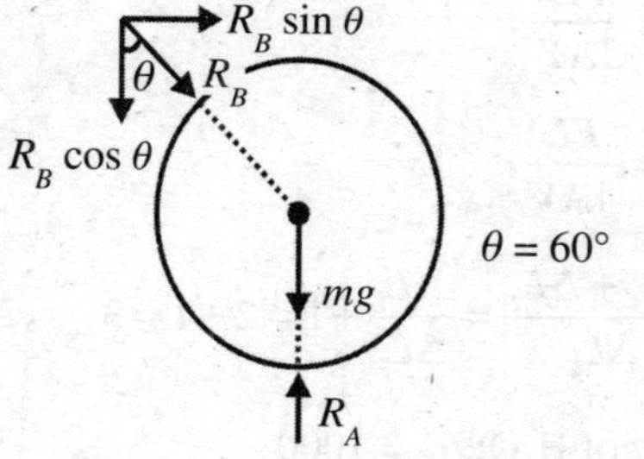

θ = 60°

θ = 60°

R_A mg R_B 30°   30°

Fig. 6.142

ii. $R_A \cos 30° = R_B \cos 30°$

$\Rightarrow \quad R_A = R_B$

$R_A \sin 30° + R_B \sin 30° = mg$

$\Rightarrow \quad R_A = R_B = 10$ N

iii.

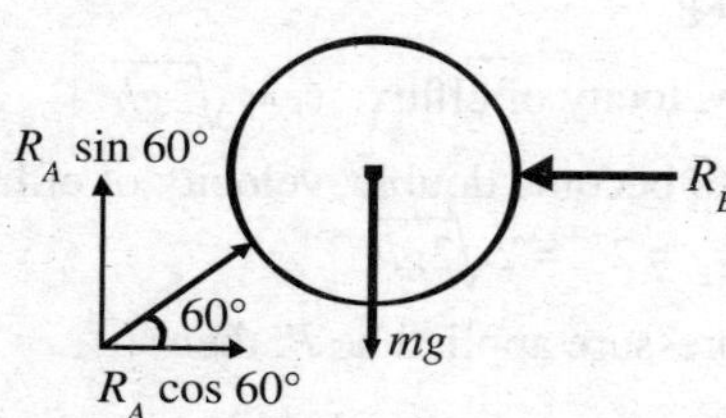

Fig. 6.143

$R_A \sin 60° = mg \qquad \Rightarrow \quad R_A = 20/\sqrt{3}$ N

$$R_B = R_A 60° = \frac{20}{\sqrt{3}} \times \frac{1}{2} = \frac{10}{\sqrt{3}} \text{N}$$

7. i. → b.; ii. → c.; iii. → a.; iv. → d.

Sol. i. Forces are conservative, so E is constant. KE will increase up to equilibrium and then decrease. So U will first decrease and then increase.

ii. Gravitation force is conservative, so E is constant, U decreases and KE increases.

iii.

$$U = mgh = \text{constant}$$

$$E = U + \text{KE}$$

$$\text{KE} = \frac{1}{2}mv^2 = \frac{1}{2}m(0 + at)^2$$

$$\text{or} \quad \text{KE} = \frac{1}{2}m = \left(\frac{F}{m} \times t\right)^2$$

iv. Here velocity is constant, so KE is constant. As the block comes down, U decreases so E decreases.

9. i. → a.; ii. → c.; iii. → c.; iv. → c.

Sol. When the iron is in the boat, it will displace more volume of water than when it is directly in water.

A log of wood will float in both the cases, hence it will displace same amount of water in both cases. No effect on the level of water, if water is transferred either from boat into pool or from pool to boat.

9. i. → a.; ii. → b.; iii. → b.; iv. → a.

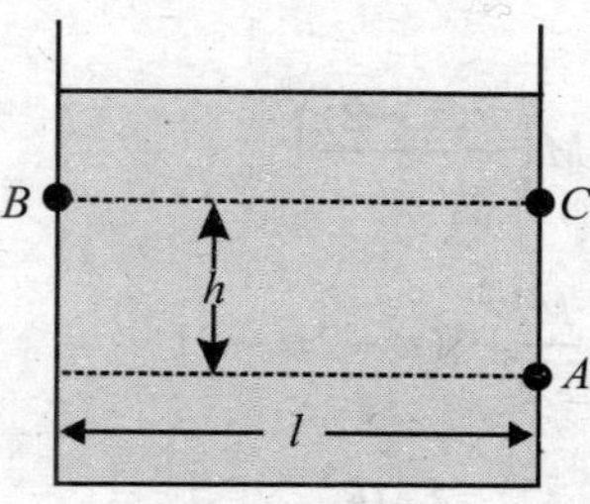

Fig. 6.144

Sol. i. Before giving upward acceleration:

$$P_A - P_B = \rho gh$$

After giving upward acceleration a,

$$P_A - P_B = \rho(g + a)h$$

So $P_A - P_B$ increase.

ii. If $\vec{a}$ is downwards

$$P_A - P_B = \rho(g - a)h$$

Hence, $P_A - P_B$ decreases.

iii. If $\vec{a}$ is towards right:

$$P_A - P_C = \rho gh \quad \text{(i)}$$

$$P_B - P_C = \rho al \quad \text{(ii)}$$

From Eqs. (i) and (ii), $P_A - P_B = \rho(gh - al)$

Hence, $P_A - P_B$ decreases.

iv. If $\vec{a}$ is towards left:

Then replace a with $-a$ in result for (iv):

$$P_A - P_B = \rho(gh + al)$$

So $P_A - P_B$ increases

10. i. → b.; ii. → a.; iii. → d.; iv. → c.

Sol. i. $I_1 = \dfrac{Ml^2}{3}\sin 30° = \dfrac{MR^2}{12}$

ii. $I_1 = MR^2$

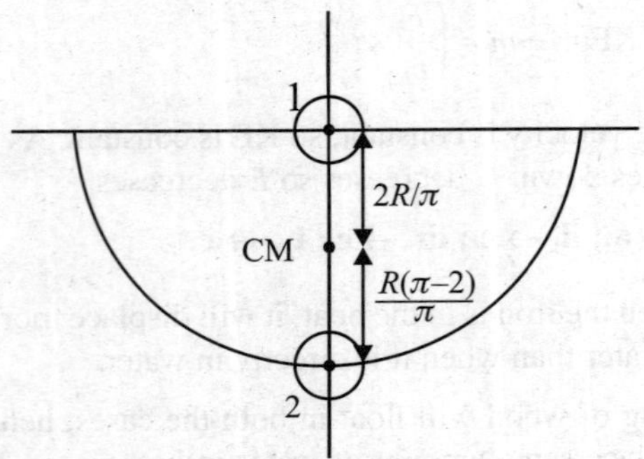

Fig. 6.145

$$I_1 = I_{CM} + M\left(\frac{2R}{\pi}\right)^2$$

$$I_2 = I_{CM} + M\left(\frac{R(\pi-2)}{\pi}\right)^2$$

$$\Rightarrow I_2 - I_1 = \frac{MR^2}{\pi^2}[(\pi-2)^2 - 2]$$

Solve to get: $I_2 = \dfrac{8MR^2}{11}$

iii. $I_1 = \dfrac{1}{6}Mh^2 = \dfrac{1}{6}(R\cos 30°)$

$$= \frac{MR^2}{8}$$

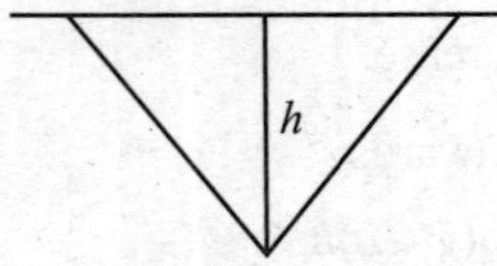

Fig. 6.146

iv. $I = I_{\text{whole}} - I_{\text{cut}}$

$$I_1 = \frac{1}{2}M(2R)^2 = -\left[\frac{1}{2}\times\frac{M}{4}R^2 + \frac{M}{4}R^2\right]$$

$$= \frac{13MR^2}{8}$$

Integer Answer Type

1. **(6)** Applying constraints: $a_1 = 6a_2$.

2. **(8)** Tension in both parts of the string will be the same.

$$T\cos 30° + T\cos 60° = mg \quad \text{(i)}$$

$$T\cos 30° + T\cos 60° = \frac{mv^2}{r} \quad \text{(ii)}$$

From Eqs. (i) and (ii), $v = \sqrt{gr} = \sqrt{10\times 6.4} = 8$ m/s

3. **(6)** Since $10 < mg$, the cylinder rolls on the ground without lifting up. Therefore,

$$\text{KE} = \text{workdone} = F \times \text{displacement}$$

$$= 10\times\frac{3}{5} = 6\text{ J}$$

Note: ***Point of application of F will move upward by same distance as the cylinder moves on the ground.***

4. **(3)** $\Delta L = \dfrac{F(L/2)}{AY} = \dfrac{FL}{2AY}$

$$\Delta L_2 = \frac{F(L/2)}{2AY} = \frac{FL}{4AY}$$

Required ratio $= \dfrac{\Delta L_1 + \Delta L_2}{\Delta L_2} = \dfrac{\Delta L_1}{\Delta L_2} + 1 = 2 + 1 = 3$

5. **(8)** $\rho_w V_w = m$, density of H_2O: $\rho_H = 1000$

Given $\rho_L V_L - \rho_H V_L = 28$ and

$$\rho_L V_L + \rho_w V_w - \rho H(V_L + V_w) = 20$$

$$\Rightarrow 28 + m - \rho_H V_w = 20$$

$$\Rightarrow 28 + m - \rho_H m/\rho_w = 20$$

$$\Rightarrow m = 8\text{ kg}$$

6. **(3)** Initial velocity of efflux: $v_1 = \sqrt{2gh}$

For range to become double, velocity of efflux should be doubled: $v_2 = 2_{v_1} = 2\sqrt{2gh}$

Let extra pressure applied be P, then

$$(p_0 + p_0 + pgh) - p_0 = \frac{1}{2}pv^2{}_2$$

$$\Rightarrow p + pgh = \frac{1}{2} p8gh$$

$$\Rightarrow p + 3\rho gh = 3 \times 10^3 \times 10 \times 10$$

$$= 3 \times 10^5 \text{ Pa} = 3 \text{ atm}$$

7. **(8)** Given $v_0 = v_e/2$

$$\left(\frac{GM}{R+h}\right)^{1/2} = \frac{1}{2}\left(\frac{2GM}{R}\right)^{1/2}$$

On solving, $h = R$.

From law of conservation of energy,

$$-\frac{GMm}{(R+h)} = \frac{1}{2}mv^2 - \frac{GMm}{R}$$

$$\text{or } \frac{1}{2}mv^2 = \frac{GMm}{R} - \frac{GMm}{2R} = \frac{GMm}{2R}$$

$$\text{or } v = \sqrt{\frac{GM}{R}} = \sqrt{gR}$$

$$= [(10)(6.4 \times 10^6)]^{1/2} = 8 \text{ km/s}$$

8. **(3)** Retardation of each block:

$$a = \mu g = 0.10 \times 10 = 1 \text{ m/s}^2$$

Distance travelled by A before it stops:

$$x_1 = \frac{u_1^2}{2a} = \frac{4^2}{2 \times 1} = 8 \text{ m}$$

Distance travelled by B before it stops:

$$x_2 = \frac{u_2^2}{2a} = \frac{2^2}{2 \times 1} = 2 \text{ m}$$

$$\Delta x_{CM} = \frac{m_1 x_1 - m_2 x_2}{m_1 + m_2}$$

$$= \frac{1 \times 8 - 1 \times 2}{1+1} = 3 \text{ m}$$

9. **(4)** Dots will be aligned again after time t, if the slower disc makes one rotation less than the other disc.

$$2\pi n\left(\frac{1}{2}\right)\left(\frac{\pi}{2}\right)t^2 \quad \text{(i)}$$

$$\text{and } 2\pi(n-1) = \left(\frac{1}{2}\right)\left(\frac{\pi}{4}\right)t^2 \quad \text{(ii)}$$

By dividing, $\frac{n}{(n-1)} = \frac{2}{1}$ or $n = 2$

From Eq. (i), $t = 4$ s.

10. **(8)**

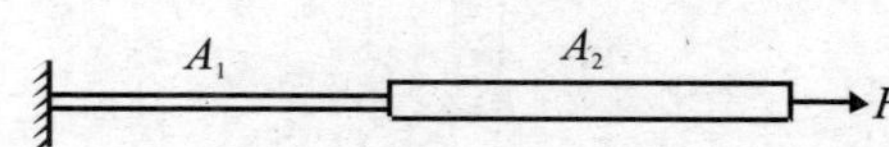

Fig. 6.147

$$\frac{F}{A} = Y\frac{\Delta\ell_1}{\ell} \Rightarrow \Delta\ell_1 = \frac{F\ell}{A_1 Y}$$

$$\frac{F}{A_2} = Y\frac{\Delta\ell_2}{\ell} \Rightarrow \Delta\ell_2 = \frac{F\ell}{A_2 Y} = \frac{F\ell}{4A_1 Y} = 8 \text{ mm}$$

$$\Delta\ell_1 + \Delta\ell_2 = 10 \text{ mm}$$

$$\frac{F\ell}{A_1 Y} + \frac{F\ell}{4A_1 Y} = 10 \text{ mm} \Rightarrow \frac{F\ell}{A_1 Y} = 8 \text{ mm}$$

Appendix

Solutions to Concept Application Exercises

Chapter 1

Exercise 1.1

1. Initially, the beam was horizontal because the centre of mass of the system was at the centre (pivot of seasaw) of the system. When child B threw his cap to point Q, centre of mass of the system shifted to the left of the pivot and that's why the balance got disturbed and beam started rotating in anticlockwise sense. To balance it again, we must put some mass to the right of it so as to displace the centre of mass of the system again at the centre of the system.

If a mass m is put at point P, to bring the centre of mass of the system at the centre, we have

$$M\frac{L}{2}+\frac{M}{25}\frac{L}{4}=\frac{24ML}{25}\frac{}{2}+m\frac{L}{4}$$

or $m = 0.12\,M$

2. Block A slides with acceleration $g/2$ and block B slides with acceleration $\sqrt{3}g/2$. Now the acceleration of centre of mass of the system of blocks A and B can be given both in x and in y directions as

$$a_x=\frac{3\times\frac{\sqrt{3}\,g}{2}\cos 60^\circ-2\times\frac{g}{2}\cos 30^\circ}{5}=\frac{\sqrt{3}g}{20}$$

and $$a_y=\frac{3\times\frac{\sqrt{3}g}{2}\sin 60^\circ+2\times\frac{g}{2}\sin 30^\circ}{5}=\frac{11g}{20}$$

Thus acceleration of the centre of mass of the system is given as

$$a_{CM}=\sqrt{a_x^2+a_y^2}=\sqrt{\left(\frac{\sqrt{3}g}{20}\right)^2+\left(\frac{11g}{20}\right)^2}=\frac{\sqrt{31}}{10}g$$

3. The rectangular plate is shown in the figure of which one part is removed. We can find the x and y coordinates of the centre of mass of this system, taking origin at the centre of the plate. The coordinates of the three remaining rectangles are $(a/4, b/4)$, $(-a/4, +b/4)$ and $(-a/4, -b/4)$. By geometry, masses of these rectangles can be taken as $M/4$. Now x-coordinate of the centre of mass:

$$x_{CM}=\frac{\frac{M}{4}\frac{a}{4}-\frac{M}{4}\frac{a}{4}-\frac{M}{4}\frac{a}{4}}{3\frac{M}{4}}=\frac{a}{12}$$

and y-coordinate of the centre of mass:

$$y_{CM}=\frac{\frac{M}{4}\frac{b}{4}+\frac{M}{4}\frac{b}{4}-\frac{M}{4}\frac{b}{4}}{3\frac{M}{4}}=\frac{b}{12}$$

4. If m_1 and m_2 are placed at a distance l apart, their centre of mass will be located at a distance x from m_1, where

$$x=\frac{m_2 l}{m_1+m_2}$$

If m_1 is displaced by l_1 towards C and m_2 is displaced by l_2 away from C, the new centre of mass C' now will be located at a distance x' from m_1, where

$$x'=\frac{m_2(l-l_1+l_2)}{m_1+m_2}$$

Displacement of the centre of mass is

$$\Delta x=x'+l_1-x=\frac{m_1l_1+m_2l_2}{m_1+m_2}$$

5. First we write the three velocities in vectorial form, taking right direction as positive x-axis and upwards as positive y-axis.

$$\vec{v}_1=-\frac{1}{2}v_1\hat{\imath}-\frac{\sqrt{3}}{2}v_1\hat{\jmath}$$

$$\vec{v}_2=v_2\,\hat{\imath},\ \vec{v}_3=-\frac{1}{2}v_3\hat{\imath}+\frac{\sqrt{3}}{2}v_3\hat{\jmath}$$

Thus the velocity of centre of mass of the system is

$$\vec{v}_{CM}=\frac{\vec{v}_1+\vec{v}_2+\vec{v}_3}{3}$$

$$=\left(v_2-\frac{1}{2}v_1-\frac{1}{2}v_3\right)\hat{\imath}+\frac{\sqrt{3}}{2}(v_3-v_1)\hat{\jmath}$$

which can be written as $\vec{v}_{CM}=v_x\hat{\imath}+v_y\hat{\jmath}$

Thus displacement of the centre of mass in time t is

$$\Delta\vec{r}=v_xt\,\hat{\imath}+v_yt\,\hat{\jmath}$$

If $v_1=v_2=v_3=v$, we have

$$\vec{v}_{CM}=0.$$

Therefore, there is no displacement of centre of mass of the system.

6. When the ball falls into the hole, with respect to the flat car, the ball travels a horizontal distance l. During this motion, to conserve momentum and to maintain the positon of centre of mass, the car moves towards left, say by a distance x. Thus, the total distance travelled by the ball towards right is $(l-x)$. As centre of mass remains at rest, the change in mass moments of the two (ball and car) about any point must be equal to zero. Hence

$$m\,(l-x)=Mx \Rightarrow m=\frac{ml}{M+m}$$

7. a. Velocity of centre of mass is

$$v_{CM}=\frac{5\times 14\times 2\times 0}{5+2}=10\text{ m/s}$$

b. Due to the kick on the 5 kg block, it starts moving with a velocity 14 m/s immediately, but due to inertia, 2 kg block remains at rest, at that moment. Thus, velocity of 5 kg block w.r.t. to centre of mass is $v_1=14-10=4$ m/s.and velocity of 2 kg block w.r.t. to centre of mass is $v_2=0-10=-10$ m/s.

8. If m_2 is shifted by a distance x and released, the mass m_1 will break off from the wall when the spring restores its natural length and m_2 will start going towards right. At the time of breaking m_1, m_2 will be going towards right with a velocity v, which is given as

$$\frac{1}{2}kx^2 = \frac{1}{2}m_2v^2 \Rightarrow v = \left(\sqrt{\frac{k}{m_2}}\right)x$$

and the velocity of the centre of mass at this instant is

$$v_{CM} = \frac{m_1 \times 0 + m_2 \times v}{m_1 + m_2} = \frac{\left(\sqrt{m_2 k}\right)x}{m_1 + m_2}$$

9.

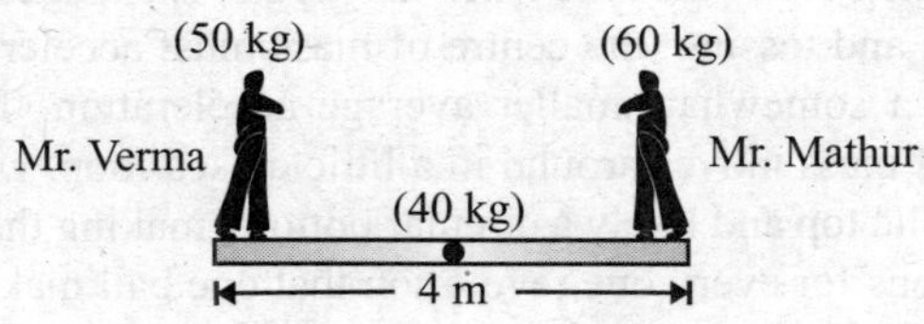

Fig. S-1.1

Let displacement of the boat is $X_b = x$ towards right

The displacement of Mr. Verma: $X_v = (x + 2)$ m

The displacement of Mr. Mathur: $X_m = (x - 2)$ m

As no external forces are acting on the system, the displacement of the centre of mass of the system should be zero.

$$\Delta x_{CM} = \frac{M_b x_b + M_v x_v + M_M x_M}{(M_b + M_v + M_m)}$$

$$\therefore 0 = \frac{40x + 50(x + 2) + 60(x - 2)}{150} \Rightarrow x = 2/15 \text{ m}$$

10. Let displacement of the cart be $x_c = -x$

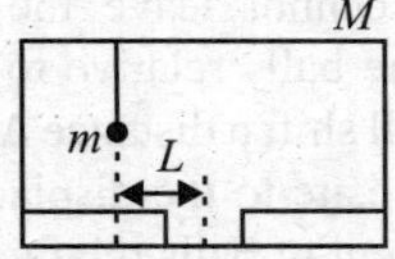

Fig. S-1.2

Displacement of the bob is

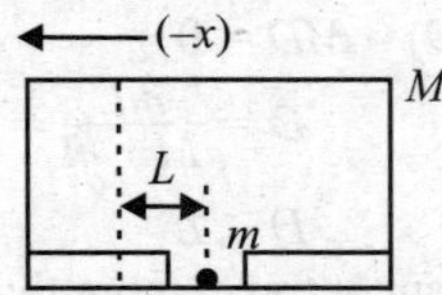

Fig. S-1.3

$$x_b = (L - x)$$

$$(x_{\text{CM system}}) = 0$$

$$M_{\text{cart}}\, x_{\text{cart}} + M_{\text{bob}}\, x_{\text{bob}} = 0$$

$$M(-x) + m(L - x) = 0$$

$$x = \frac{mL}{m + M}$$

11. Let wedge move towards right by distance x; then

$$\Delta x_{CM} = \frac{m(l\cos\theta + x) + Mx + m(x - l)}{m + M + m}$$

$$0 = ml\cos\theta - ml + (m + M + m)x$$

$$x = \frac{ml - ml\cos\theta}{2m + M}$$

12. $l = h\cot\theta$

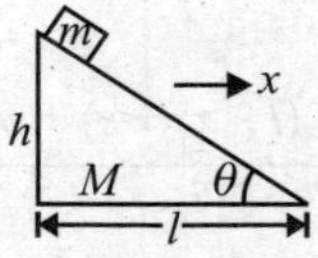

Fig. S-1.4

Let displacement of wedge be x towards right:

$$\Delta x_{CM} = \frac{m(l + x) + Mx}{M + m}$$

$$0 = ml + (M + m)x$$

$$x = \frac{-ml}{M + m} = -\frac{mh\cot\theta}{M + m}$$

13. $$\Delta x_{CM} = \frac{Mx + m(R - r + x)}{M + m}$$

$$0 = Mx + mx + m(R - r)$$

$$x = \frac{-m(R - r)}{M + m}$$

14. $$m_1\Delta\vec{x}_1 + m_2\Delta\vec{x}_2 = 0$$

$$m(R\sin\theta - x) - Mx = 0$$

$$x = \frac{mR\sin\theta}{m + M}$$

15. $K = \frac{1}{2}m_1v_{1c}^2 + \frac{1}{2}m_2v_{2c}^2$ (i)

where v_{1c} and v_{2c} are velocities relative to the CM.

$$v_{1c} = v_1 - v_{cm} = v_1 - \left(\frac{m_1v_1 + m_2v_2}{m_1 + m_2}\right) = \frac{m_2(v_1 - v_2)}{m_1 + m_2}$$

$$v_{2c} = v_2 - v_{CM} = v_2 - \left(\frac{m_1v_1 + m_2v_2}{m_1 + m_2}\right) = \frac{m_1(v_2 - v_1)}{m_1 + m_2}$$

Putting these in Eq. (i), $K = \frac{1}{2}\frac{m_1m_2}{(m_1 + m_2)}(v_1 - v_2)^2$

Note: ***For a system of two particles of masses m_1 and m_2, the total kinetic energy is $k = (1/2)\,\mu v_r^2 + (1/2)\,(m_1 + m_2)\bar{v}_{cm}^2$ where $m = m_1m_2/(m_1 + m_2)$ (known as reduced mass of the system) and $\vec{v}_r = \vec{v}_1 - \vec{v}_2$ (relative velocity).***

16. As no external force is acting on the system, the displacement of the centre of mass should be zero.
Let the displacement of the platform toward right be s,

$$\Delta\vec{x}_1 = \Delta\vec{x}_{1,p} + \Delta\vec{x}_p = (-x_1 + s)$$

$$\Delta\vec{x}_2 = \Delta\vec{x}_{2,p} + \Delta\vec{x}_p = (l - x_1 + s)$$

Hence,

$$0 = \frac{m_1(-x_1 + s) + m_2(l - x_1 + s) + m_0 s}{m_1 + m_2 + m_0}$$

$$s(m_1 + m_2 + m_0) - m_1 x_1 + m_2(l - x_1) = 0$$

$$s = \frac{m_1 x_1 - m_2(l - x_1)}{m_1 + m_2 + m_0}$$

17. Initial co-ordinates of CM = (0, 0, 0)
x co-ordinates of CM after 3 s,

$$x_{CM} = 120 \times 3 = 360 \text{ m}$$

y co-ordinates of CM after 3s,

$$y_{CM} = -\frac{1}{2}gt^2 = -45 \text{ m}$$

After 3 s,

$$(m_1 + m_2)x_{CM} = m_1 x_1 + m_2 x_2$$

$$x_2 = \frac{30 \times 360 - 12 \times 300}{18} = 400 \text{ m}$$

Similarly, $(m_1 + m_2)\, y_{CM} = m_1 y_1 + m_2 y_2$

$$-30 \times 45 = 12 \times 24 + 18 \times y_2$$

$$y_2 = -91 \text{ m}$$

and in z co-ordinate, $m_1 z_1 + m_2 z_2 = 0 \Rightarrow z_2 = 32$ m

18. As no external force acts in z direction, hence z-coordinate of the centre of mass of the ball should be zero. To make z-coordinate zero other ball should fall symmetrically with respect to z-axis.
Hence, z-coordinate of other ball = –5 m.
The balls do not have any external force in x-direction. Hence in x-direction the centre of mass should move with constant velocity. x-coordinate of centre of mass at $t = 1.5$ s = 200 × 2 = 400 m

Hence,

$$x_{CM} = \frac{m_1 x_1 + m_2 x_2}{m_1 + m_2}$$

$$400 = \frac{20 \times 250 + 20\, x_2}{20 + 20}$$

$$x_2 = 800 - 250 = 550 \text{ m}$$

Height fallen by centre of mass at $t = 2$ s,

$$h = \frac{1}{2} \times 10 \times (2)^2 = 20 \text{ m}$$

Hence, y-coordinate of centre of mass = 30 – 20 = 10 m

Hence,

$$y_{CM} = \frac{m_1 y_1 + m_2 y_2}{m_1 + m_2}$$

$$10 = \frac{20 \times 0 + 20 \times y_2}{20 + 20}$$

$$\Rightarrow \quad y_2 = 20 \text{ m}$$

19. The centre of mass of the balls is in free fall, moving up and then down with the acceleration due to gravity, during the 40% of the time when the juggler's hands are empty. During the 60% of the time when the juggler is engaged in catching and tossing, the centre of mass must accelerate up with a somewhat smaller average acceleration. The centre of mass moves around in a little closed loop, with a parabolic top and likely a circular bottom, making three revolutions for every one revolution that one ball makes. Let T represent the time for one cycle and F_g the weight of one ball, we have $F_J 0.6T = 3F_g T$ and $F_J = 5F_g$. The average force exerted by the juggler is five times the weight of one ball.

20. Initially, the whole system is at rest, so $v_{CM} = 0$. As there is no external force acting on the system, $\vec{v}_{CM}$ = constant = 0. So position of centre of mass of system remains fixed.

$$x_{CM} = \frac{mx_1 + Mx_2}{m + M} \quad \text{(i)}$$

where m is mass of the cannon balls and M that of the (car + cannon) system.
As $\Delta x_{CM} = 0$, therefore

$$m\,\Delta x_1 + M\Delta x_2 = 0 \quad \text{(ii)}$$

As cannon balls cannot leave the car, so maximum displacement of the balls relative to the car is L and in doing so the car will shift a distance $\Delta x_2 = D$ (say) relative to the ground, opposite to the displacement of the balls; then the displacement of balls relative to ground will be

$$\Delta x_1 = L - D \quad \text{(iii)}$$

Substituting the value of Δx_1 from Eq. (iii) in Eq. (ii), we get

$$m(L - D) - MD = 0$$

$$\Rightarrow \quad D = \frac{mL}{M + m} = \frac{L}{1 + \frac{M}{m}}$$

$$\Rightarrow \quad D < L$$

i.e., rail road car cannot travel more than L.

Exercise 1.2

1. a. Linear momentum is conserved since there are no external forces acting on the system. The fragments go off in different directions and their vector momenta add to zero.

b. Kinetic energy is not conserved because the chemical potential energy initially in the explosive is converted into kinetic energy of the pieces of the bomb.

2. Momentum conservation is not violated if we choose as our system the planet along with you. When you receive an impulse forward, the Earth receives the same size impulse backwards. The resulting acceleration of the Earth due to this impulse is much smaller than your acceleration forward, but the planet's backward momentum is equal in magnitude to your forward momentum.

3. In empty space, the centre of mass of a rocket plus fuel system does not accelerate during a burn, because no outside force acts on this system. The rocket body does accelerate as it blows exhaust containing momentum out of its back.

   According to the text's basic expression for rocket propulsion, the change in speed of the rocket body will be larger than the speed of the exhaust relative to the rocket, if the final mass is less than 37% of the original mass.

4. Here net force on the block will be zero, hence its momentum or speed will remain conserved.

5. The velocity of shell at the highest point is

$$v = u \sin\theta = 100 \times \sin 30°$$
$$= 50 \text{ m/s}$$

Let m be the mass of the shell. Then the mass of the lighter fragment is $m/3$ and that of heavier fragment is $2m/3$.

Initial momentum of the shell before explosion is

$$mv = 50m$$

As no external force is acting on the shell, we can conserve momentum of the shell before and after its explosion.

Note: ***An external force of gravity is present here during explosion but as explosion is an instantaneous phenomenon, so due to very short duration impulse of gravity is negligible and cannot cause any change in momentum. Thus in all very short duration explosions or collisions, we always ignore the presence of gravity just before and after the occurrence.***

In x-direction,

$$mv = \frac{2m}{3} v_2 \cos\theta \Rightarrow v_2 \cos\theta = \frac{3}{2} v \quad \text{(i)}$$

in y-direction,

$$0 = \frac{m}{3} v_1 - \frac{2m}{3} v_2 \sin\theta \Rightarrow v_2 \sin\theta = \frac{v_1}{2} \quad \text{(ii)}$$

Squaring and adding Eqs. (i) and (ii), we get

$$v_2^2 = \frac{9}{4} v^2 + \frac{1}{4} v_1^2$$

$$\Rightarrow v_2 = \frac{1}{2}\sqrt{9v^2 + v_1^2} = \frac{1}{2}\sqrt{9 \times 2500 + 40000} = 125 \text{ m/s}$$

6. Let the smaller block is moving with speed v_1 relative to the bigger block when it reaches the bottommost position and at this instant bigger block is moving at v_2 (say). Then using conservation of momentum in horizontal direction, we have

$$6mv_2 = m(v_1 - v_2) \quad \text{(i)}$$

Now using energy conservation

$$mga = \frac{1}{2} m(v_1 - v_2)^2 + \frac{1}{2}(6m)v_2^2 \quad \text{(ii)}$$

Solving Eqs. (i) and (ii), we get

$$2ga = 36v_2^2 + 6v_2^2$$

$$v_2 = \sqrt{\frac{ga}{21}}$$

7. After every throw 'A' or 'B' gains a momentum of 20 kgm/s in the direction opposite to the motion of the ball.

First round

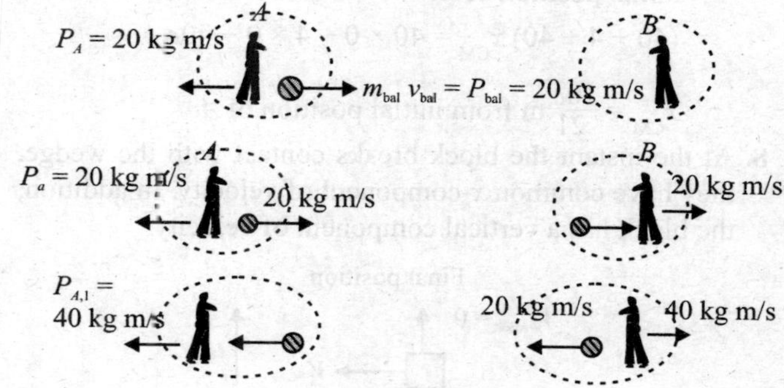

Fig. S-1.5

Second round

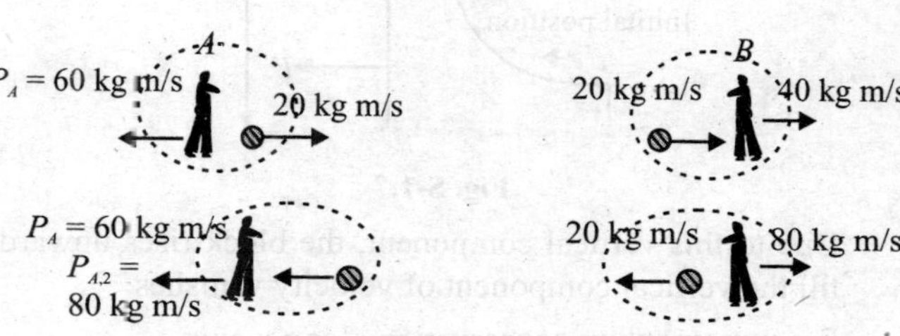

Fig. S-1.6

At the end of each round, both men receive a momentum 40 kg m/s.

a. $p_A = 20 = 40\, v_A$

$$v_A = \frac{1}{2} \text{ m/s}$$

b. $p_{A,1} = 40 = (40 + 4)\, v_{A,1}$

$$v_{A,1} = \frac{40}{44} = \frac{10}{11} \text{ m/s}$$

c. At the end of five round trips, the linear momentum of (ball + A) system,

$$p_{A,5} = 40 \times 5 = 200 \text{ kg m/s}$$

$$200 = (40 + 4)\, v_{A,5}$$

$$v_{A,5} = \frac{50}{11} \text{ m/s}$$

Momentum transferred to B in five trips

$$= 40 \times 5 = 200 \text{ kg m/s}$$

So $\quad 200 = 40\, v_{B,5} \Rightarrow v_{B,5} = 5 \text{ m/s}$

d. 'A' can catch the ball until the velocity of 'A' is less than 5 m/s. After five rounds when 'A' throws the ball, i.e., 6th time, the velocity of 'A' becomes greater than 5 m/s; hence A can throw the ball only six times.

e. As no external force is acting on 'A', 'B' and 'ball' system hence the position of centre of mass of the system should remain constant always.

Let the centre of mass be at a distance x cm from the initial position of A.

$$(40+4+40)\,x_{CM} = 40 \times 0 + 4 \times 0 + 40d$$

$$x_{CM} = \frac{10d}{21} \text{ m from initial position of } A.$$

8. At the instant the block breaks contact with the wedge, they have common x-component of velocity. In addition, the block has a vertical component of velocity.

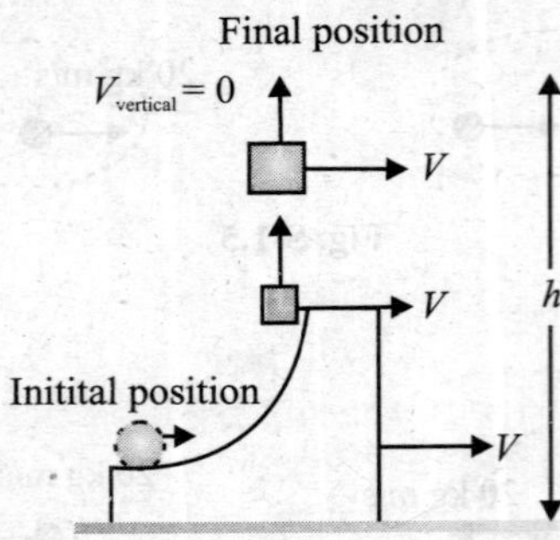

Fig. S-1.7

Due to this vertical component, the block rises upwards till the vertical component of velocity vanishes.

From momentum conservation along x-axis,

$$mv_0 = (m+M)\,V \qquad \text{(i)}$$

$$V = \frac{mv_0}{(m+M)} \qquad \text{(ii)}$$

From energy conservation between initial and final positions of the block,

$$\frac{1}{2}mv_0^2 = \frac{1}{2}(m+M)V^2 + mgh \qquad \text{(iii)}$$

$$\frac{1}{2}mv_0^2 = \frac{1}{2}\left(\frac{m^2}{m+M}\right)v_0^2 + mgh$$

$$h = \frac{v_0^2}{2g}\left[\frac{M}{m+M}\right]$$

9. In the problems involving mass ejection and mass addition, the ejected and added mass both are part of a system. When the men jump they push on the floor of the buggy (action); the same force acts on them as well (reaction).

This action–reaction pair is internal force for the system of man and buggy. Similarly, when men jump in, there is an action–reaction force between them and the buggy.

Let buggy "1" moves to the right and buggy "2" to the left, when the men jump. Man in "1" brings momentum, mv_1 directed towards right in buggy "2".

Similarly, the man in "2" brings momentum mv_2 in "1" directed towards left.

For system 1:

$$P_i = (M+m)v_1 - mv_1 - mv_2 = Mv_1 - mv_2$$

$$P_f = 0$$

From law of conservation of momentum,

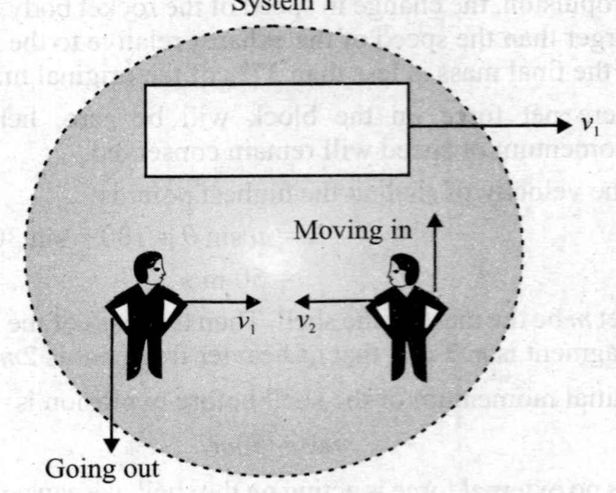

Fig. S-1.8

$$P_i = P_f$$

$$Mv_1 - mv_2 = 0 \qquad \text{(i)}$$

For system 2:

$$P_i = (M+m)\,v_2 - mv_1 - mv_2$$

$$P_f = (M+m)\,v$$

From law of conservation of momentum,

$$P_i = P_f$$

$$(M+m)v_2 - mv_1 - mv_2 = (M+m)v \qquad \text{(ii)}$$

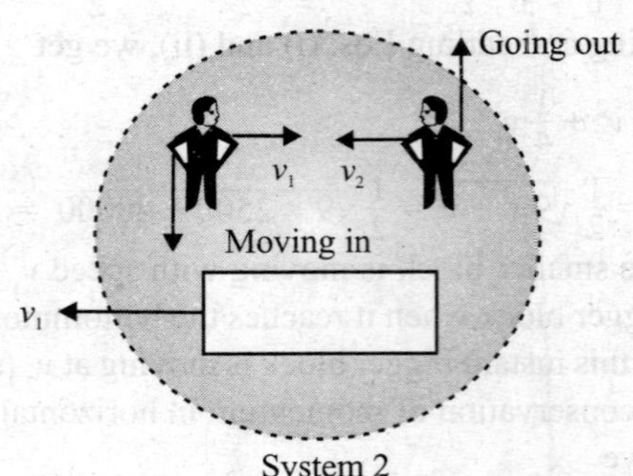

Fig. S-1.9

On solving Eqs. (i) and (ii) simultaneously, we obtain

$$v_1 = \frac{mv}{(M-m)};\ v_2 = \frac{Mv}{(M-m)}$$

10.

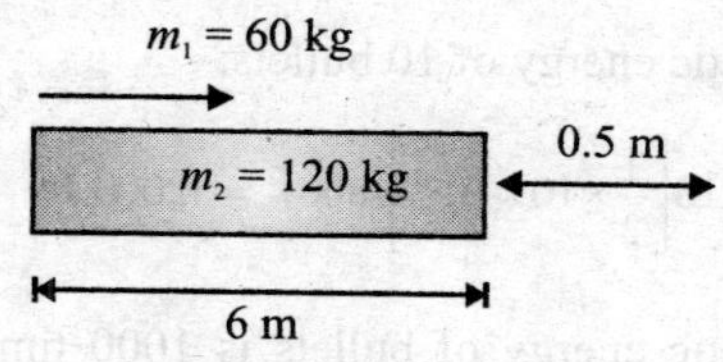

Fig. S-1.10

a. Displacement of man w.r.t. boat, $x_{m,b} = 6$ m

$$x_b = x_{m,b}\left(\frac{m_1}{m_1 + m_2}\right) = -6 \times \frac{60}{60+20} = -2 \text{ m}$$

Therefore, boat displaces by 2 m away from the pier.

$\therefore$ Separation = 2.5 m

b. $\vec{v}_{mb} = \vec{v}_m - \vec{v}_b$

$$m_1\vec{v}_m + m_2\vec{v}_b = 0$$

$$m_1(\vec{v}_{mb} + \vec{v}_b) + m_2\vec{v}_b = 0$$

$$m_1 v_{mb} = (m_1 + m_2)v_b$$

$$v_b = \left(\frac{m_1}{m_1 + m_2}\right)v_{mb}$$

$$v_b = \frac{60}{180} \times 3 = 1 \text{ m/s}$$

KE of the man,

$$= \frac{1}{2}m_m v_m^2 = \frac{1}{2} \times 60 \times (3-1)^2 = 120 \text{ J}$$

$$\text{KE of the boat} = \frac{1}{2} \times 120 \times (1)^2 = 60 \text{ J}$$

Total KE energy = 120 + 60 = 180 J

KE of the system when the boat is tied

$$= \frac{1}{2} \times 60 \times 3^2 = 270 \text{ J}$$

11. From conservation of momentum

$$v = \frac{m_1 v_1 + m_2 v_2}{m_1 + m_2} = 5 \text{ m/s}$$

Conservation of energy.

$$\frac{1}{2}m_1 v_1^2 + \frac{1}{2}m_2 v_2^2 = \frac{1}{2}(m_1 + m_2)v^2 + \frac{1}{2}kx^2$$

$$\Rightarrow x^2 = \frac{m_1 v_1^2 + m_2 v_2^2 - (m_1 + m_2)v^2}{k}$$

$$\Rightarrow x = \sqrt{\frac{70}{1120}} = 0.25 \text{ m}$$

When the blocks have equal speeds, spring has maximum compression. After this instant, the spring again expands and after sometime m_1 loses contact with the spring. Let v_1 and v_2 be the velocities of the blocks after they lose contact.

$$m_1 u_1 + m_2 u_2 = m_1 v_1 + m_2 v_2$$

and $\quad v_1 - v_2 = -1\,(u_1 - u_2)$

[$e = 1$, because there is no loss in KE]

Solving for v_1 and v_2, we get

$$v_1 = 0 \text{ m/s and } v_2 = 7 \text{ m/s}$$

12. a. Whomsoever we consider the aggressor, brother and sister exert equal in magnitude oppositely directed forces on each other, to give each other equal magnitudes of momentum. We take the eastward component of the equation

Total original momentum = total final momentum for the two-sibling system

$$0 = 80 \text{ kg } (-3 \text{ m/s}) + (40 \text{ kg})v \Rightarrow v = 6.0 \text{ m/s},$$

i.e., she moves at 6.0 m/s east .

b. Original chemical energy in the girl's body = total final kinetic energy

$$U_{\text{chemical}} = (1/2)(80 \text{ kg})(3 \text{ m/s})^2 + (1/2)(40 \text{ kg})$$

$$(6.0 \text{ m/s})^2 = 1080\ J$$

c. System's momentum is conserved with the value zero. The net forces on the two siblings are of equal magnitude and opposite directions. Their impulses add to zero. Their final momenta are of equal magnitude in opposite directions, to add as vectors to zero.

13. a. For the system of two blocks $\Delta p = 0$

or $\quad p_i = p_f$

Therefore,

$$0 = Mv_M + (3M)(2.00 \text{ m/s})$$

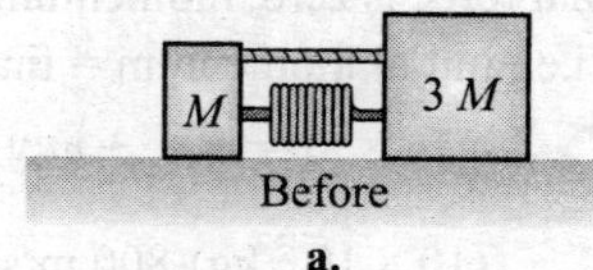

a.

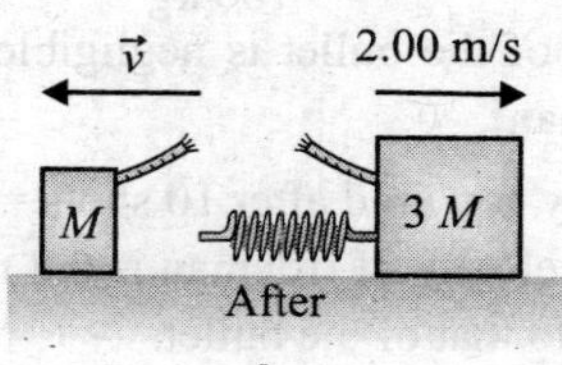

b.

Fig. S-1.11

Solving, $v_M = -6.00$ m/s (motion toward the left)

b. $\frac{1}{2}kx^2 = \frac{1}{2}Mv_M^2 + \frac{1}{2}(3M)v_{3M}^2 = 8.40$ J

c. The original energy is in the spring. A force had to be exerted over a distance to compress the spring, transferring energy into it by work.

d. Momentum of the system is conserved with the

value zero. The forces on the two blocks are of equal magnitude and opposite directions. Their impulses add to zero. The final momenta of the two blocks are of equal magnitude and opposite directions. Plenty of motion afterward is due to the fact that energy stored in the spring converts into the kinetic energy of blocks.

14. Velocity of pendulum bob in mean position,

$$v_1 = \sqrt{2gh} = \sqrt{2 \times 10 \times 5 \times 10^{-2}} = 1 \text{ m/s}$$

When the bob picks up a mass 10^{-3} kg at the bottom, then by conservation of linear momentum, the velocity of the coalesced mass is given by

$$m_1 v_1 + m_2 v_2 = (m_1 + m_2) V$$

$$10^{-2} \times 1 + 10^{-3} \times 0 = (10^{-2} + 10^{-3}) V$$

or $\quad V = \frac{10}{11}$ m/s

If h' is the height risen by the combined mass, we have

$$\frac{1}{2}(m_1 + m_2) V^2 = (m_1 + m_2) gh'$$

$$\therefore h' = \frac{V^2}{2g} = \frac{\left(\frac{10}{11}\right)^2}{2 \times 10} \text{ m} = \frac{5}{121} \text{ m} = 0.0413 \text{ m}$$

15. a. Let m_1 and m_2 be the masses of bullet and the rifleman; v_1 and v_2 their respective velocities after the first shot.

Initially, the rifleman and bullet are at rest, therefore initial momentum of the system = 0.

As external force is zero, momentum of the system is constant i.e., initial momentum = final momentum

$$0 = m_1 v_1 + m_2 v_2$$

or $\quad v_2 = -\frac{m_1 v_1}{m_2} = -\frac{(10 \times 10^{-3} \text{ kg})(800 \text{ m/s})}{100 \text{ kg}} = -0.08 \text{ m/s}$

As mass of the bullet is negligible as compared to (rifle + man),

$\therefore$ Velocity acquired after 10 shots $= 10v_2 = 0.8$ m/s i.e., the velocity of rifleman is 0.8 m/s in a direction opposite to that of the bullet.

b. The momentum acquired by the rifleman is

$$P = m_2 v_2 = 100 \times 0.8 \text{ kg m/s} = 80 \text{ kg m/s}$$

This momentum is acquired in 10 s, therefore the average force exerted $= \frac{\Delta p}{\Delta t} = \frac{80}{10} = 8$ N

c. Kinetic energy of the rifleman,

$$E_1 = \frac{1}{2} m_2 v_2^2 = \frac{1}{2} \times 100 \times (0.8)^2 = 32 \text{ J}$$

Kinetic energy of 10 bullets:

$$E_2 = 10\left[\frac{1}{2} \times 10 \times 10^{-3}\right](800)^2 = 32000 \text{ J}$$

Kinetic energy of bullets is 1000 times the kinetic energy of rifleman.

16. a. Velocity of the projectile after 5 s,

$$v = u - gt = 100 - 9.8 \times 5 = 51 \text{ m/s}$$

$\therefore$ Momentum of projectile before explosion

$$= Mv = 50 \times 51 = 2550 \text{ kg m/s}$$

Momentum of the fragments after explosion

$$= m_1 v_1 + m_2 v_2 = 20 \times 150 + 30v = 3000 + 30\, v_2$$

$\therefore$ Conservation of momentum principle yields

$$2550 = 3000 + 30v_2$$

or $\quad v_2 = -\frac{450}{30} = -15$ m/s

So velocity of the second fragment will be 15 m/s downward.

b. Three seconds after explosion, velocity of first fragment upward

$$v_2' = u' - gt' = 150 - 9.8 \times 3 = 120.6 \text{ m/s}$$

Velocity of the second fragment downward

$$v_2'' = u'' + gt' = 15 + 9.8 \times 3 = 44.4 \text{ m/s}$$

$\therefore$ Sum of momenta,

$$= m_1 \vec{v}_1 + m_2 \vec{v}_2 = 20 \times 120.6 - 30 \times 44.4 = 1080 \text{ kg m/s}$$

If the projectile had not exploded, its velocity after 8 s from firing would have been,

$$v' = u - gt' = 100 - 9.8 \times 8 = 21.6 \text{ m/s}$$

$\therefore$ Momentum of the projectile would have been

$$Mv' = 50 \times 21.6 = 1080 \text{ kg m/s}$$

So net momentum at 8 s is independent of explosion.

17. a. Let v be the velocity of the car relative to the earth, when the man runs to the left, then conservation of linear momentum gives

$$(M + m)v_0 = Mv + m(v - v_{rel})$$

$$\therefore \quad v = v_0 + \frac{m}{M + m} v_{rel} \quad \text{(i)}$$

Change in velocity, $\Delta v = v - v_0 = \frac{mv_{rel}}{M + m}$

b. When there are n men on the car and they jump off together, then velocity of the car after jumping off, from Eq. (i), will be

$$v_1 = v_0 + \frac{mn}{M + mn} v_{rel} \quad \text{(i)}$$

If the men jump one by one, the mass of the system will go on changing till the last man jumps.

Let v' be the velocity when one man jumps off the car, then from Eq. (i)

$$v' = v_0 + \frac{m}{M + (n)\, m} v_{rel}$$

When the second man also jumps off, then

$$v'' = v' + \frac{m}{M + (n-1)\, m} v_{rel}$$

$$= v_0 + \frac{mv_{rel}}{M + (n)\, m} + \frac{mv_{rel}}{M + (n-1)\, m}$$

When the last man jumps, then final velocity

$$v_2 = v_0 + \frac{mv_{rel}}{M + (n)\, m} + \frac{mv_{rel}}{M + (n-1)\, m} + \ldots + \frac{mv_{rel}}{M + m} \quad \text{(ii)}$$

Comparing Eqs. (i) and (ii), we note that

$$v_2 > v_1$$

i.e., the men would impart greater velocity to the car by running and jumping off one by one.

18. i. Let v_1 be the new velocity of the car. Then according to the principle of conservation of linear momentum,

$$Mu = (M + m) v_1$$

$$\Rightarrow \quad v_1 = \frac{m}{M + m} u \quad \text{(i)}$$

ii. Let v_2 be the new velocity of the car relative to the earth; then velocity of the man relative to the earth $= (v_2 - v_{rel})$

$$\therefore \quad (M + m) v_1 = Mv_2 + m(v_2 - v_{rel})$$

$$\therefore \quad v_2 = v_1 + \frac{mv_{rel}}{M + m}$$

19. Conservation of linear momentum along X and Y axes gives

$$mu = m_1 v_1 \cos\theta_1 + m_2 v_2 \cos\theta_2$$

i.e., $\quad 2u = 1 \times v_1 \cos\theta_1 + 1 \times v_2 \cos\theta_2$

$\Rightarrow \quad v_1 \cos\theta_1 + v_2 \cos\theta_2 = 2u \quad \text{(i)}$

and $\quad 0 = m_1 v_1 \sin\theta_1 + m_2 v_2 \sin\theta_2$

As $m_1 = m_2 = 1$ kg

$\Rightarrow \quad v_1 \sin\theta_1 = v_2 \sin\theta_2$

As $v_1 = v_2 = v$ (say)

$\Rightarrow \quad \theta_1 = \theta_2 = \theta$ (say)

From Eq. (i),

$\Rightarrow \quad v\cos\theta + v\cos\theta = 2u$

$\Rightarrow \quad v\cos\theta = u = 4 \quad \text{(ii)}$

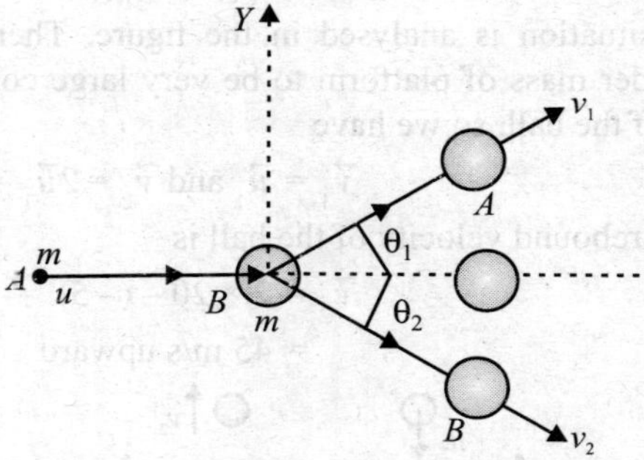

Fig. S-1.12

Also conservation of energy gives

$$\frac{1}{2} mu^2 + 48 = \frac{1}{2}(m_1) v^2 + \frac{1}{2}(m_2) v^2$$

$$\Rightarrow \quad \frac{1}{2} \times 2 \times (4)^2 + 48 = \frac{1}{2} \times 1 v^2 + \frac{1}{2} \times 1 v^2$$

or $\quad v^2 = 64$

$\Rightarrow \quad v = 8$ m/s

From Eq. (ii),

$$\cos\theta = \frac{u}{v} = \frac{4}{8} = \frac{1}{2}$$

$$\Rightarrow \quad \theta = 60°$$

20. Apply conservation of momentum:

$$\Rightarrow \quad m_1 \vec{v}_1 + m_2 \vec{v}_2 + m_3 \vec{v}_3 = 0$$

$$m(2\hat{i} + 5\hat{j} - 6\hat{k}) + 2m\vec{v}_2 + m(-4\hat{i} + 3\hat{j} + 2\hat{k}) = 0$$

$$\vec{v}_2 = \hat{i} - 4\hat{j} + 2\hat{k}$$

Exercise 1.3

1. The ball reaches the floor with velocity $u = \sqrt{2gh}$.

If collision on the floor is elastic, the ball will rebound with the same velocity u and it will reach the same height. But if the coefficient of restitution between the ball and the floor is e, the ball will rebound with a lesser velocity v, given as

$$v = eu \quad \text{(i)}$$

where v is the velocity of separation and u is the velocity of approach.

In this case the ball will rebound to a height h_1; from equations, it is given that

$$h_1 = \frac{v^2}{2g} = e^2 h \quad \text{(ii)}$$

In downward and upward motion the acceleration of the particle is g only, thus total time of motion is the time of downward motion plus the time of upward motion, as

$$t = \sqrt{\frac{2h}{g}} + \sqrt{\frac{2e^2 h}{g}}$$

$$= \sqrt{\frac{2h}{g}} + e\sqrt{\frac{2h}{g}}$$

2. The situation is analysed in the figure. There we can consider mass of platform to be very large compared to that of the ball, so we have

$$\vec{v}_1 = \vec{u}_1 \text{ and } \vec{v}_2 = 2\vec{u}_1 - \vec{u}_2$$

Thus rebound velocity of the ball is

$$\vec{v}_2 = 2 \times 20 - (-5)$$
$$= 45 \text{ m/s upward}$$

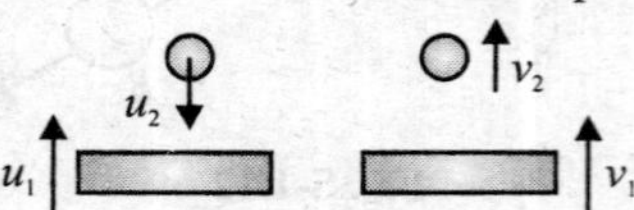

Fig. S-1.13

3. The situation is shown in the figure. If u be the velocity of ball A just before the impact. Then

$$u = \sqrt{2g(2a)}$$

Consider the first impact between A and B. If velocities of the balls become v_1 and v_2, respectively, according to linear momentum conservation

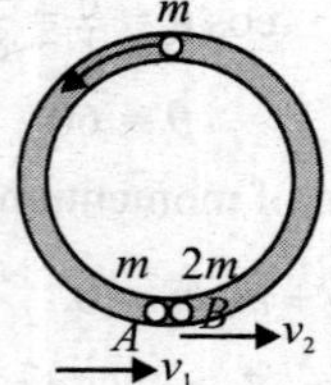

Fig. S-1.14

$$mu + 0 = mv_1 + 2mv_2 \quad \text{(i)}$$

Coefficient of restitution is given as

$$e = \frac{v_2 - v_1}{u} = \frac{1}{2} \quad \text{(ii)}$$

On solving Eqs. (i) and (ii), we get

$$v_1 = 0 \text{ and } v_2 = \frac{u}{2}$$

Ball A comes to rest, so second impact between A and B occurs, when ball B returns to its initial position with the same speed $u/2$. If after the second impact, the velocities of the two balls become v_3 and v_4, again from linear momentum conservation and coefficient of restitution

$$2m\left(\frac{u}{2}\right) + 0 = mv_3 + 2mv_4 \quad \text{(iii)}$$

and
$$e = \frac{v_3 - v_4}{\frac{u}{2}} = \frac{1}{2} \quad \text{(iv)}$$

On solving Eqs. (iii) and (iv), we get

$$v_3 = \frac{u}{2} \text{ and } v_4 = \frac{u}{4}$$

Both the velocities are positive, it implies that both masses will move in the same direction after the second impact. If h_1 and h_2 are the heights to which the masses m and $2m$ will rise after the second impact, according to energy conservation, we have

For ball of mass m,

$$\frac{1}{2}m\left(\frac{u}{2}\right)^2 = mgh_1 \text{ or } h_1 = \frac{u^2}{8g} = \frac{a}{2}$$

For ball of mass $2m$,

$$\frac{1}{2}(2m)\left(\frac{u}{4}\right)^2 = 2mgh_2 \quad \text{or} \quad h_2 = \frac{u^2}{32\,g} = \frac{a}{8}$$

4. Normal axis and tangential axis are shown in Fig. S-1.15. Reaction of the incline is along n-axis and in the absence of friction there is no force along t-axis; therefore velocity along t-axis remains unchanged,

$$v \cos\theta = u \sin\theta \quad \text{(i)}$$

From the definition of coefficient of restitution,

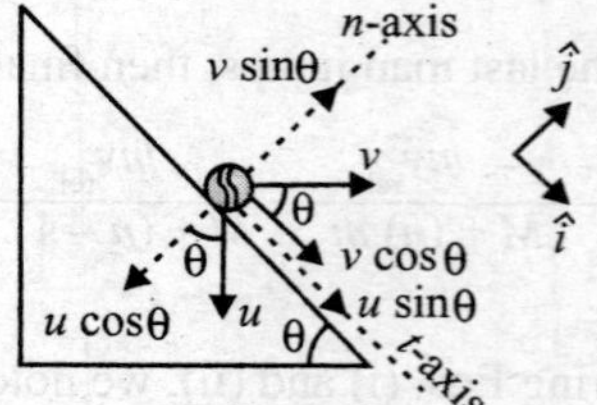

Fig. S-1.15

$$e = \frac{v\sin\theta}{u\cos\theta} \quad \text{(ii)}$$

$$v = eu\cot\theta \quad \text{(iii)}$$

From Eqs. (i) and (iii),

$$(eu\cot\theta)\cos\theta = u\sin\theta$$

$$e = \tan^2\theta$$

5. From M to N the ball will fall a distance 4 m, then impulse will be developed in the string ON.

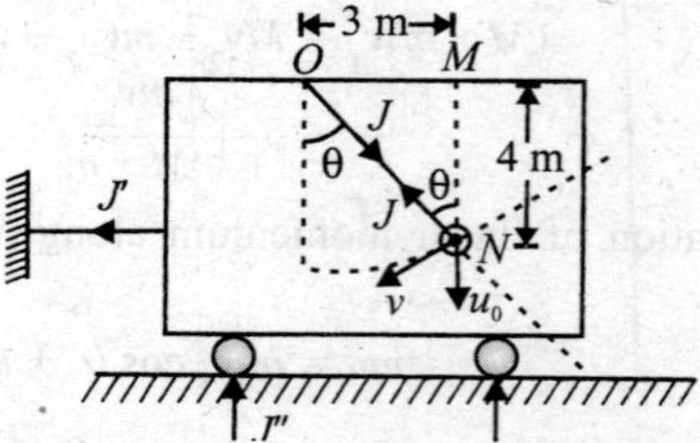

Fig. S-1.16

Velocity of the ball just before the string becomes tight

$$u_0 = \sqrt{2 \times 10 \times 4} = \sqrt{80} \text{ m/s}$$

Velocity of the ball in the direction perpendicular to the impulse will remain same (just after and just before the impulse developed).

$$v = u_0 \sin\theta = \sqrt{80} \times \frac{3}{5} = \frac{12}{\sqrt{5}} \text{ m/s}$$

$$J = mu_0 \cos\theta = 1\sqrt{80} \times \frac{4}{5} = \frac{16}{\sqrt{5}} \text{ kg m/s}$$

$$J' = J\sin\theta = \frac{48}{5\sqrt{5}}\text{ kg m/s}$$

and impulse given by earth,

$$J'' = J\cos\theta = \frac{64}{5\sqrt{5}}\text{ kg m/s}$$

6. Assume some spacing between the balls

$\xrightarrow{v_0}$ (2M) 1 (M) 2 (M) 3 (M)

After collision of 2M with ball '1',

$\xrightarrow{u}$ (2M) 1 (M) $\xrightarrow{v_1}$ 2 (M) 3 (M)

$$u = \left(\frac{2M - M}{2M + M}\right)v_0 = \frac{v_0}{3}, \qquad v_1 = \frac{2(2M)v_0}{2M + M} = \frac{4}{3}v_0$$

Now ball '1' collides with '2' and itself comes to rest, then '2' will collide with '3' and itself comes to rest, so we have

$\xrightarrow{v_0/3}$ (2M) 1 (M) 2 (M) 3 (M) $\xrightarrow{4v_0/3}$

Now again $2M$ will collide with ball '1', after this collision we have

$\xrightarrow{v_0/9}$ (2M) 1 (M) $\xrightarrow{4v_0/9}$ 2 (M) 3 (M) $\xrightarrow{4v_0/3}$

Now '1' will collide with '2' and itself comes to rest, so we have

$\xrightarrow{v_0/9}$ (2M) 1 (M) 2 (M) $\xrightarrow{4v_0/9}$ 3 (M) $\xrightarrow{4v_0/3}$

Now $2M$ will again collide with '1', after this collision we have

$\xrightarrow{v_0/27}$ (2M) 1 (M) $\xrightarrow{4v_0/27}$ 2 (M) $\xrightarrow{4v_0/9}$ 3 (M) $\xrightarrow{4v_0/3}$

So final velocities are as shown in above figure.

7. Let velocities of sphere 1 and 2 after collision be v_1 and v_2, respectively, then

$$\frac{v_2 - v_1}{u} = e \quad and \quad mv_2 + mv_1 = mu$$

From above equations: $v_2 = \dfrac{(1+e)u}{2}$

Now sphere 2 will collide with sphere 3 and after collision velocity of 3 can be formed as

$$v_3 = \left(\frac{1+e}{2}\right)^2 u$$

Hence, similarly velocity of nth sphere can be formed as

$$v_n = \left(\frac{1+e}{2}\right)^{n-1} u$$

8. Let v be the velocity of the sphere after the impact.

Now we separate the velocity components parallel and perpendicular to the wall. Using the law of restitution, the component of velocity parallel to the wall remains unchanged while the component perpendicular to the wall becomes e times in opposite direction.

Thus, $$\vec{v} = -\frac{3}{2}\hat{i} + \hat{j}$$

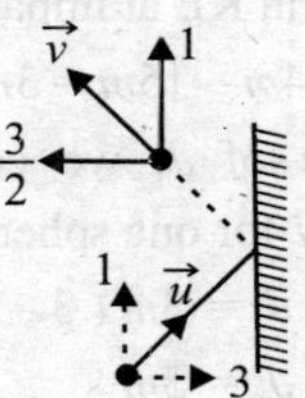

Fig. S-1.17

a. Therefore, the velocity of the sphere after the impact

$$= -\frac{3}{2}\hat{i} + \hat{j}$$

b. The loss in KE $= \frac{1}{2}mu^2 - \frac{1}{2}mv^2$

$$= \frac{1}{2}m(3^2 + 1^2) - \frac{1}{2}m\left(\left\{\frac{3}{2}\right\}^2 + 1^2\right) = \frac{27}{8}\text{ m}$$

c. $\vec{J} = \Delta\vec{P} = \vec{P}_f - \vec{P}_i = m(\vec{v}) - m(\vec{u})$

$$= m\left(-\frac{3}{2}\hat{i} + \hat{j}\right) - m(3\hat{i} + \hat{j}) = -\frac{9}{2}m\hat{i}$$

9. The velocity components of A and B perpendicular to the line connecting centres are unchanged by the impact. It is because no force acts on A and B in a direction perpendicular to the line connecting the centres

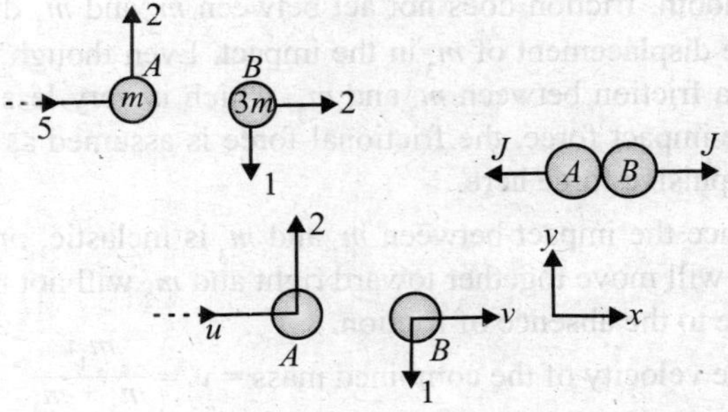

Fig. S-1.18

Applying conservation of linear momentum and the law of restitution, we have in x direction

$$5m + (3m)(2) = mu + 3mv \quad \text{(i)}$$

and $$\frac{1}{3}(5-2) = v - u \quad \text{(ii)}$$

Solving these equations, we have $u = 2$ and $v = 3$

The velocities of A and B after the impact are $2\hat{i} + 2\hat{j}$ and $3\hat{i} - \hat{j}$, respectively

Before impact, the kinetic energy of A is

$$\frac{1}{2}m(5^2 + 2^2) = 29m/2$$

and that of B is

$$\frac{1}{2}(3m)(2^2 + 1^2) = 15m/2$$

After the impact, the kinetic energy of A is

$$\frac{1}{2}m(2^2 + 2^2) = 4m$$

and that of B is

$$\frac{1}{2}(3m)(3^2 + 1^2) = 15m$$

Therefore, the loss in KE at impact is

$$\frac{29}{2}m + \frac{15}{2}m - 4m - 15m = 3m$$

To find the value of J, we consider the change in momentum along $\hat{i}$ for one sphere only.

For sphere B, $\quad J = 3m(3-2)$

or $\quad J = 3m$

10. $K_i = \frac{1}{2}\mu(u_1 - u_2)^2 + \frac{1}{2}(m_1 + m_2)v_{CM}^2$

$\Rightarrow \quad K_f = \frac{1}{2}\mu(v_1 - v_2)^2 + \frac{1}{2}(m_1 + m_2)v_{CM}^2$

Loss in KE

$$= \frac{1}{2}\mu\left[(u_1 - u_2)^2 - (v_1 - v_2)^2\right]$$

we also have

$$v_1 - v_2 = -e(u_1 - u_2)$$

Hence, loss in KE $= \frac{1}{2}\mu(1 - e^2)(u_1 - u_2)^2$

$$= \frac{1}{2}\frac{m_1 m_2}{m_1 + m_2}(1 - e^2)(u_1 - u_2)^2$$

11. Impact takes place between m_1 and m_3 horizontally. Since m_2 is kept on m_3 and all the surfaces in contact are smooth, friction does not act between m_2 and m_3 during the displacement of m_3 in the impact. Even though there is a friction between m_2 and m_3, which is very less than the impact force, the frictional force is assumed as non-impulsive force here.

Since the impact between m_1 and m_3 is inelastic, m_1 and m_3 will move together toward right and m_2 will not move due to the absence of friction.

The velocity of the combined mass $= v' = \frac{m_1 v}{m_1 + m_3}$

$$\frac{|\Delta KE|}{KE} = \frac{\frac{1}{2}m_1 v^2 - \frac{1}{2}(m_1 + m_3)v'^2}{\frac{1}{2}m_1 v^2} = 1 - \left(\frac{m_1 + m_3}{m_1}\right)\left(\frac{v'}{v}\right)^2$$

$$= 1 - \left(\frac{m_1 + m_3}{m_1}\right)\left(\frac{m_1}{m_1 + m_3}\right)^2 = 1 - \frac{m_1}{m_1 + m_3} = \frac{m_3}{m_1 + m_3}$$

If m_2 and m_3 are rigidly attached, both together behave as a single mass $(m_2 + m_3)$ and the answer would have been $\frac{m_2 + m_3}{m_1 + m_3 + m_3}$.

12. Since m_2 is at rest $(u_2 = 0)$, velocity of m_1 after collision:

$$v_1 = \left(\frac{m_1 - em_2}{m_1 + m_2}\right)u_1$$

given, $\quad v_1 < 0 \quad \Rightarrow \quad m_1 - em_2 < 0$

$\Rightarrow \quad m_1 < em_2 \quad \Rightarrow \quad m_1 < m_2$

13. $u_1 = 12$ m/s $\quad m_1 = 4$ kg;

$u_2 = 4$ m/s $\quad m_2 = 8$ kg

Let v_1 and v_2 be the velocities after impact.

Conservation of momentum:

$$m_1 v_1 + m_2 v_2 = m_1 u_1 + m_2 u_2$$

$$\Rightarrow \quad 4v_1 + 8v_2 = 80 \quad \text{(i)}$$

Newton's experimental law: $v_1 - v_2 = -e(u_1 - u_2)$

$$v_1 - v_2 = -0.5(12 - 4) = -4 \quad \text{(ii)}$$

Solving Eqs. (i) and (ii), we get

$$v_1 = 4 \text{ m/s}, v_2 = 8 \text{ m/s}$$

Loss in KE $= \frac{1}{2}m_1 u_1^2 + \frac{1}{2}m_2 u_2^2 - \left(\frac{1}{2}m_1 v_1^2 + \frac{1}{2}m_2 v_2^2\right)$

$$= \frac{1}{2}[4(144) + 8(16)] - \frac{1}{2}[4(16) + 8(64)] = 64 \text{ J}$$

14. Let $m = 2 \times 10^{-3}$ kg; $M = 1.0$ kg

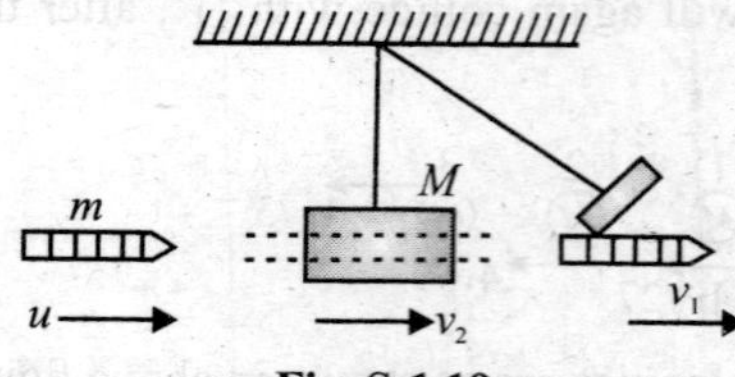

Fig. S-1.19

$u = 500$ m/s $v_1 = 100$ m/s

v_2 = speed of the pendulum after the impact

$mu = mv_1 + Mv_2$ (conservation of momentum)

$$v_2 = \frac{m(u - v_1)}{M} = \frac{2}{1000}(500 - 100) = 0.8 \text{ m/s}$$

The block swings and its kinetic energy gets converted into potential energy.

$$\frac{1}{2}Mv_2^2 = Mgh$$

$$\Rightarrow h = \frac{v_2^2}{2g} = \frac{0.8 \times 0.8}{2 \times 10} = 0.032 \text{ m}$$

15. $\Delta\vec{p} = \vec{F}\,\Delta t$

$\Delta p_y = m(v_{fy} - v_{iy}) = m(v \cos 60°) - mv \cos 60° = 0$

$\Delta p_x = m(-v \sin 60° - v \sin 60°) = -2mv \sin 60°$

$= -2(3.00 \text{ kg})(10.0 \text{ m/s})(0.866)$

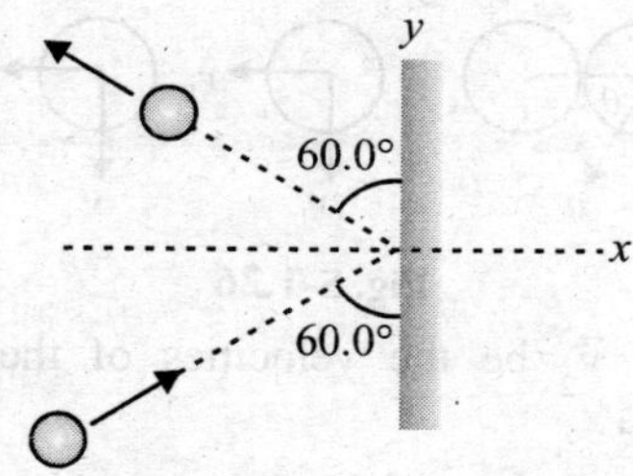

Fig. S-1.20

$= -52.0$ kg m/s

$$F_{\text{avg}} = \frac{\Delta p_x}{\Delta t} = \frac{-52.0 \text{ kg m/s}}{0.200 \text{ s}} = -260 \text{ N}$$

16. a. The impulse is to the right and equal to the area under the $F - t$ graph

$$I = \frac{1}{2} \times 2 \times 4 + 1 \times 4 + \frac{1}{2} \times 2 \times 4 = 12 \text{ Ns}$$

b. $m\vec{v}_i + \vec{F}t = m\vec{v}_f$

$(2.5 \text{ kg})(0) + 12\,\hat{i} \text{ Ns} = (2.5 \text{ kg})\,\vec{v}_f$

$\vec{v}_f = 4.80\,\hat{i}$ m/s

c. From the same equation,

$(2.5 \text{ kg})(-2\,\hat{i} \text{ m/s}) + 12\hat{i} \text{ Ns} = (2.5 \text{ kg})\,\vec{v}_f$

$v_f = 2.80\,\hat{i}$ m/s

d. $\vec{F}_{\text{avg}}\,\Delta t = 12.0\,\hat{i} \text{ Ns} \Rightarrow \vec{F}_{\text{avg}}\,(5s) = 12.0\,\hat{i}$ Ns

$\vec{F}_{\text{avg}} = 2.40\,\hat{i}$ N

17. Resolving the velocity before impact, velocity component along the plane = 8 sin 60° = $4\sqrt{3}$ m/s and that normal to the plane = 8 cos 60° = 4 m/s.

If 'v' m/s is the velocity after the impact and θ the angle made by the velocity with the normal to the plane, then velocity component along the plane = $v \sin \theta$ and that normal to the plane = $v \cos \theta$.

As the velocity along the plane remains unchanged,

$\therefore \quad v \sin \theta = 4\sqrt{3}$ (i)

From Newton's experimental law,

$v \cos \theta = e(4)$ (in magnitude)

$= 2$ (ii)

Squaring and adding Eqs. (i) and (ii),

$v^2 = 52 \Rightarrow v = 2\sqrt{13}$ m/s

Dividing Eq. (i) by Eq. (ii), we get

$\tan\theta = 2\sqrt{3} \Rightarrow \theta = \tan^{-1}(2\sqrt{3})$

18.

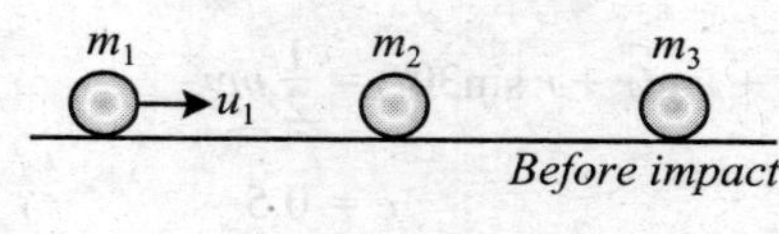

Fig. S-1.21

After first impact: $v_1 = 0$

$$\Rightarrow \left(\frac{m_1 - em_2}{m_1 + m_2}\right)u_1 = 0 \Rightarrow \quad m_1 = em_2 \quad \text{(i)}$$

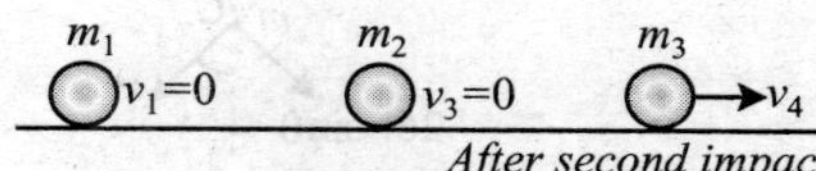

Fig. S-1.22

After second impact: $v_3 = 0$

$$\Rightarrow \left(\frac{m_1 - em_3}{m_1 + m_3}\right)v_2 = 0 \Rightarrow \quad m_2 = em_3 \quad \text{(ii)}$$

From (i) and (ii) $m_2^2 = m_1 m_3$

19. Velocity of B after collision:

$$v_2 = \frac{(1+e)m_1 u_1}{m_1 + m_2} = \frac{(1+e)m \times 10}{m + m} \Rightarrow v_2 = 5\,(1 + e)$$

Height raised by $B = \dfrac{v_2^2}{2g}$

Now, $\dfrac{v_2^2}{2g} \geq 3.2 \Rightarrow \dfrac{(5(1+e))^2}{2 \times 10} \geq 3.2 \Rightarrow e \geq \dfrac{3}{5}$

20. v_1 and v_2 are after collision.

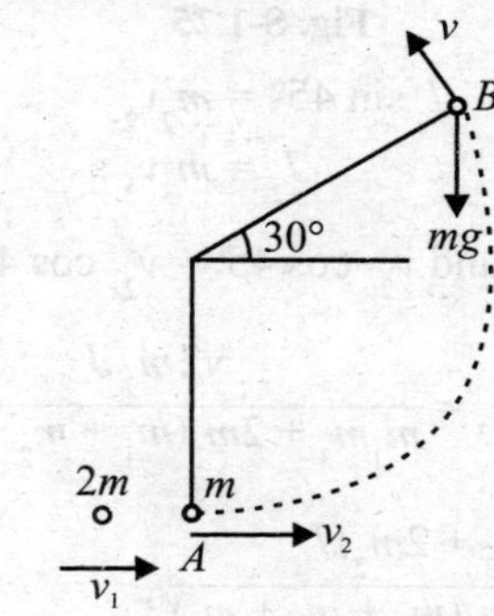

Fig. S-1.23

$$v_2 = \frac{(1+e)2mu}{2m+m} = \frac{2u(1+e)}{3}$$

Since tension becomes zero at B, so

$$mg \sin 30^\circ = \frac{mv^2}{r} \Rightarrow v = \sqrt{\frac{gr}{2}}$$

Apply conservation of energy between A and B

$$\frac{1}{2}mv^2 + mg(r + r\sin 30^\circ) = \frac{1}{2}mv_2^2$$

$$\Rightarrow \qquad e = 0.5$$

21. Apply conservation of momentum along AC:

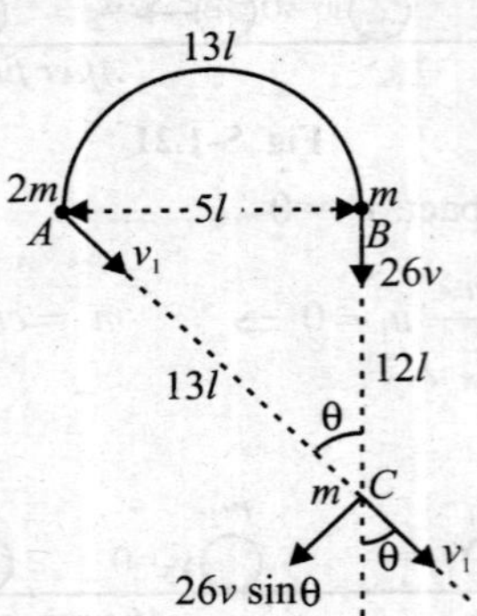

Fig. S-1.24

$$m\,26\,v\cos\theta = (2m + m)v_1$$

$$v_1 = 8v$$

22. $J - J_1 \qquad = m_1v_1 \qquad$ (i)

$$J_1 - J_2(\cos 45^\circ) = m_2v_{2y} \qquad \text{(ii)}$$

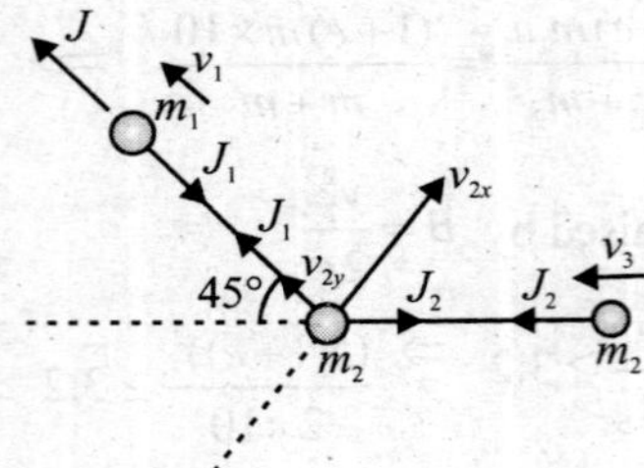

Fig. S-1.25

$$J_2\sin 45^\circ = m_2v_{2x} \qquad \text{(iii)}$$

$$J_2 = m_3v_3 \qquad \text{(iv)}$$

Now $v_{2y} = v_1$ and $v_{2y}\cos 45^\circ - v_{2x}\cos 45^\circ = v_3$

Solve to get: $v_3 = \dfrac{\sqrt{2}m_2J}{m_1m_3 + 2m_2(m_1 + m_2 + m_3)}$,

$$v_1 = \frac{(m_3 + 2m_2)J}{m_1m_3 + 2m_2(m_1 + m_2 + m_3)},$$

$$v_{2x} = \frac{m_3v_3\cos 45^\circ}{m_2} = \frac{m_3J}{m_1m_3 + 2m_2(m_1 + m_2 + m_3)}$$

Net velocity of m_2,

$$= \sqrt{v_1^2 + v_{2x}^2} = \frac{\sqrt{2}(m_3^2 + (m_2 + m_3)^2)^{1/2}}{m_1m_3 + 2m_2(m_1 + m_2 + m_3)}$$

23. Let a ball A of mass m moving with velocity $\vec{v}$ collide with a stationary ball B as shown in figure. As there is no external force, the linear momentum is conserved. As collision is two dimensional, the x and y components of the linear momentum are separately conserved.

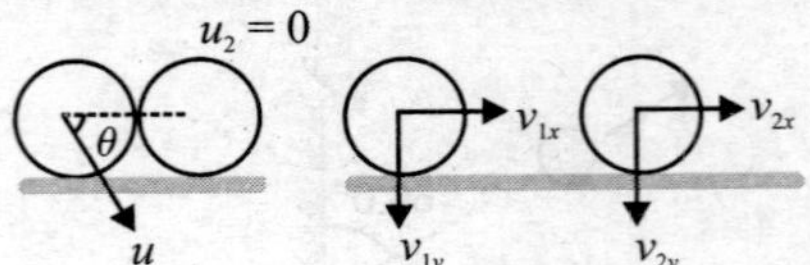

Fig. S-1.26

Let $\vec{v}_1$ and $\vec{v}_2$ be the velocities of the particle after collision.

Then conservation of linear momentum along x-axis gives

$$mu\cos\theta = mv_{1x} + mv_{2x}$$

$$\Rightarrow \qquad v_{1x} + v_{2x} = u\cos\theta \qquad \text{(i)}$$

and conservation of linear momentum along y-axis gives

$$mu\sin\theta = mv_{1y} + mv_{2y}$$

$$\Rightarrow \qquad v_{1y} + v_{2y} = u\sin\theta \qquad \text{(ii)}$$

At the moment of maximum deformation

$$\Rightarrow \qquad v_{1xy} = v_{2x} = \frac{u\cos\theta}{2} \qquad \text{[using Eq. (i)]}$$

and the tangential velocities of balls before and after collision should be equal

$$u\sin\theta = v_{1y} \text{ and } v_{2y} = 0$$

Initial KE $\qquad E_i = \frac{1}{2}mu^2$

Final KE

$$E_f = \frac{1}{2}mv_{1x}^2 + \frac{1}{2}mv_{1y}^2 + \frac{1}{2}mv_{2x}^2 + \frac{1}{2}mv_{2y}^2$$

$$= \frac{1}{2}m(v_{1x}^2 + v_{2x}^2) + \frac{1}{2}mv_{1y}^2 + \frac{1}{2}mv_{2y}^2$$

$$= mv_{1x}^2 + \frac{1}{2}mv_{1y}^2 + 0 \text{ (Since } v_{1x} = v_{2x})$$

$$= m\left(\frac{u\cos\theta}{2}\right)^2 + \frac{1}{2}m(u\sin\theta)^2$$

$$= \frac{1}{4}mu^2\cos^2\theta + \frac{1}{2}mu^2\sin^2\theta$$

$$= \frac{1}{2}mu^2\left\{\frac{1}{2}\cos^2\theta + \sin^2\theta\right\}$$

$$= \frac{1}{2}mu^2\left\{\cos^2\theta + \sin^2\theta - \frac{1}{2}\cos^2\theta\right\}$$

$$= \frac{1}{2}mu^2\left\{1 - \frac{1}{2}\cos^2\theta\right\}$$

Fraction of KE converted into PE at the moment of maximum deformation

$$f = \frac{E_i - E_f}{E_i} = \frac{\frac{1}{2}mu^2 - \frac{1}{2}mu^2\left(1 - \frac{\cos^2\theta}{2}\right)}{\frac{1}{2}mu^2} = \frac{\cos^2\theta}{2}$$

Here $\theta = 45°$,

$$\therefore \quad f = \frac{\cos^2 45°}{2} = \frac{1}{4} = 0.25$$

Chapter 2

Exercise 2.1

1. i. Moment of inertia of the particle on the left is $I_1 = m_1 r_1^2$.

Moment of inertia of the particle on the right is $I_2 = m_2 r_1^2$.

Moment of inertia of the system about AB is $I = I_1 + I_2 = m_1 r_2^2 + m_2 r_2^2$.

ii. Moment of inertia of the particle on the left is $I_1 = 0$.

Moment of inertia of the particle on the right is $I_2 = m_2(r_1 + r_2)^2$.

Moment of inertia of the system about the given axis is $I = I_1 + I_2 = 0 + m_2(r_1 + r_2)^2$.

iii. Moment of inertia of the particle on the left is $I_1 = 0$.

Moment of inertia of the particle on the right is $I_2 = 0$.

Moment of inertia of the system about the given axis is $I = I_1 + I_2 = 0 + 0$.

2. MR^2 (infact M.I. of any part of mass M of a ring of radius R about an axis passing through the geometrical centre and perpendicular to the plane of the ring is MR^2)

3. $\frac{MR^2}{2}$. These cases are same as circular disc. The distribution's with respect to axis of rotation is same pattern as circular disc.

4. Moment of inertia of the plate about axis 1 (by taking rods perpendicular to axis 1)

$$I_1 = \frac{Mb^2}{3}$$

Moment of inertia of the plate about axis 2 (by taking rods perpendicular to axis 2)

$$I_2 = \frac{Ml^2}{12}$$

Moment of inertia of the plate about axis 3 (by taking rods perpendicular to axis 3)

$$I_3 = \frac{Mb^2}{12}$$

Moment of inertia of the plate about axis 4 (by taking rods perpendicular to axis 4)

$$I_4 = \frac{Mb^2}{3}$$

5. $I_3 = I_1 + I_2$

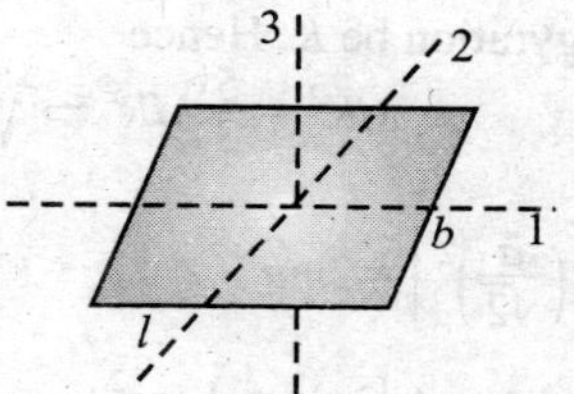

Fig. S-2.1

$$= \frac{Mb^2}{12} + \frac{Ml^2}{12}$$

6. $I_3 = I_1 + I_2 = \frac{Ml^2}{3} + \frac{Ml^2}{3} = \frac{2Ml^2}{3}$

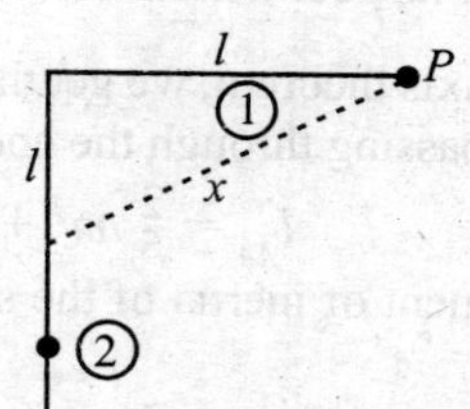

Fig. S-2.2

$$\Rightarrow \quad n = 1$$

7. i. For calculation. Moment of inertia about centre, use parallel axis theorem

$$I = 2\left[\frac{Ml^2}{12} + m\left(\frac{b}{2}\right)^2\right] + 2\left[\frac{mb^2}{12} + m\left(\frac{l}{2}\right)^2\right]$$

$$I = \frac{2}{3}m(l^2 + b^2)$$

ii. Hence, we can also use parallel axis theorem.

$$I_{PQ} = 2\left(\frac{mb}{3}\right)^2 + mb^2 = \frac{5}{3}mb^2$$

8. $I_P = I_{P_1} + I_{P_2}$

Fig. S-2.3

$$I_P = \frac{Ml^2}{3} + \left(\frac{Ml^2}{12} + Mx^2\right)$$

$$x^2 = \ell^2 + \left(\frac{\ell}{2}\right)^2 = \frac{5\ell^2}{4}$$

$$\Rightarrow I_P = \frac{M\ell^2}{3} + \left[\frac{M\ell^2}{12} + M\left(\frac{5\ell^2}{4}\right)\right] = \frac{5}{3}M\ell^2$$

9. $I = \frac{2}{5}mR^2 + mR^2 = \frac{7}{5}mR^2 = mK^2$

$$\Rightarrow \quad K = \sqrt{\frac{7}{5}}R$$

10. Moment of inertia of hollow sphere about its tangent

$$I = \frac{2}{3}MR^2 = \frac{5}{3}MR^2$$

Let radius of gyration be K. Hence

$$MK^2 = \frac{5}{3}MR^2 \Rightarrow \sqrt{\frac{5}{3}}R$$

11.

a. $I_A = 4\left[m\left(\frac{a}{\sqrt{2}}\right)^2\right] = 2ma^2$

b. $I_B = 2[ma^2] + m(\sqrt{2}a)^2 = 4ma^2$

c. $I_C = 2[ma^2] = 2ma^2$

d. $I_D = 2\left[m\left(\frac{a}{\sqrt{2}}\right)^2\right] = ma^2$

12. a. $I = m\left(\frac{l}{2}\right)^2 + m\left(\frac{l}{2}\right)^2 = \frac{ml^2}{2}$

b. $I = 4ml^2$

c. $I = ml^2 + m\left(\frac{l}{2}\right)^2 = \frac{5}{4}ml^2$

d. $I = 2m(2l)^2 = 8ml^2$

13. $I = I_{disc} + I_{ring} + I_{point\ masses}$

$$= \frac{mR^2}{2} + mR^2 + 4\left[\frac{m}{4}\left(\frac{R}{2}\right)^2\right] = \frac{7mR^2}{4}$$

14. In the system shown in figure, the total moment of inertia can be given as the sum of moment of inertia of the rod and that of the spherical ball.

Moment of inertia of the rod about the axis passing through an end is $I_{rod} = Ml^2/3$

Moment of inertia of the spherical ball about the axis passing through its geometrical centre is $I = 2mr^2/5$.

Using parallel axis theorem, we get the moment of inertia about the axis passing through the end of rod as

$$I_{AA'} = \frac{2}{5}mr^2 + ml^2$$

Thus total moment of inertia of the system can be given as

$$I_{total} = I_{rod} + I_{AA'}$$

or $$= \frac{Ml^2}{3} + \left(\frac{2}{5}mr^2 + ml^2\right)$$

15. Moment of inertia of an elemental disc at a distance x from axis (1)

$$dm = \frac{Mdx}{L}$$

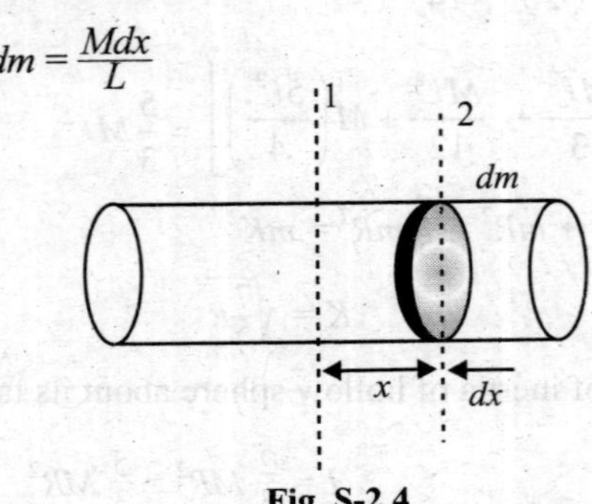

Fig. S-2.4

MI of the elemental disc about axis (1).

$$I_1 = \int\left(\frac{dm}{2}R^2 + dm\,x^2\right)$$

$$= \int_{-L/2}^{L/2}\frac{M}{2L}R^2dx + \int_{-L/2}^{L/2}\frac{M}{L}x^2\,dx$$

$$= \frac{MR^2}{2} + \frac{ML^2}{12}$$

16. We know moment of inertia of the triangular lamina about one of the side is

$$I = \frac{Mh^2}{6}$$

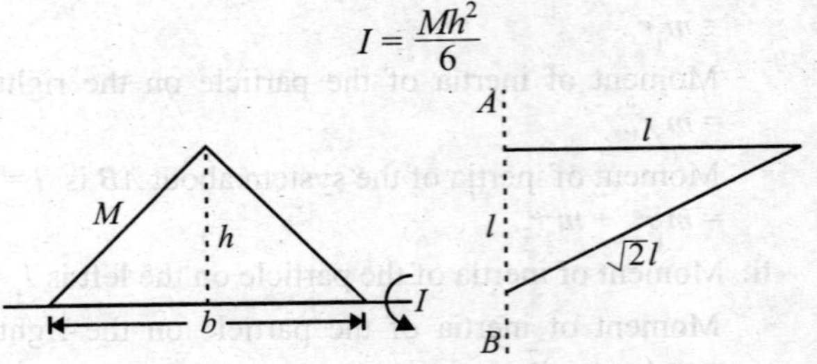

Fig. S-2.5

Here in the problem $\quad I = \dfrac{Ml^2}{6}$

17. The M.I. of a square lamina about its diagonal

Let the mass of the lamina be M and each of its edges be 'a'. The diagonal divides the square lamina into two triangular laminas each of equal mass $M/2$.

Consider the triangular lamina BCD, with base BD and altitude OC

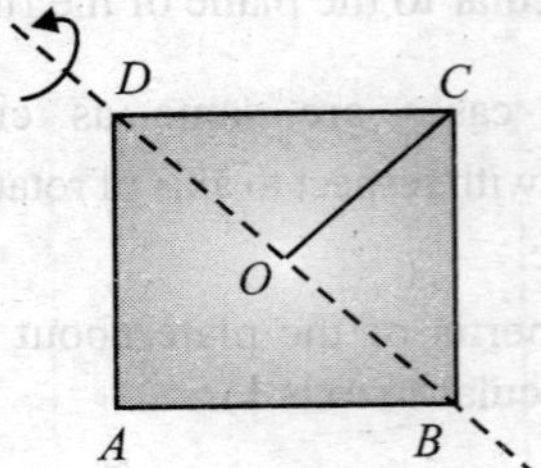

Fig. S-2.6

$$= \frac{1}{2}(\text{diagonal}) = \frac{1}{2}(a\sqrt{2}) = \frac{a}{\sqrt{2}}$$

M.I. of the lamina BCD about BD

$$= \left(\frac{M}{2}\right)\left(\frac{a}{\sqrt{2}}\right)^2 \times \frac{1}{6} = \frac{Ma^2}{24}$$

M.I. of the square lamina about the diagonal BD = algebraic sum of M.I. of laminas ABD and BCD each about their respective bases BD

$$= 2\left(\frac{Ma^2}{24}\right) = \frac{Ma^2}{12}$$

18. M.I. of a rectangular lamina about its diagonal

Let the rectangular lamina $PQRS$ have a mass M and dimensions $L \times B$.

By symmetry, the MI of the rectangular lamina about the diagonal $QS = 2$ (M.I. of the triangular lamina about base QS).

Let 'h' be the altitude of the triangle QRS. Then from area of rectangle $PQRS = 2$(area of ΔQRS),

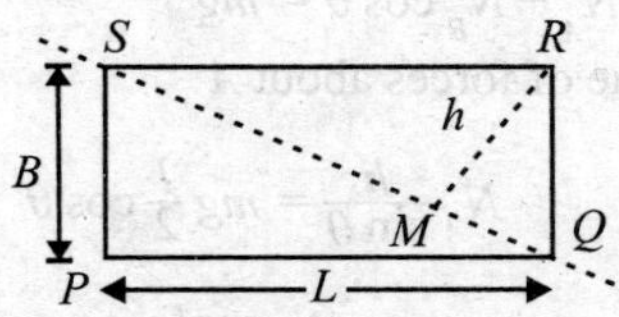

Fig. S-2.7

$$L \times B = 2\left[\frac{1}{2}\sqrt{L^2+B^2} \times h\right]$$

or $$h = \frac{LB}{\sqrt{L^2+B^2}}$$

Required M.I., $$I_{QS} = 2\left[\frac{M}{2} \times \left(\frac{LB}{\sqrt{L^2+B^2}}\right)^2 \times \frac{1}{6}\right]$$

$$I_{QS} = \frac{M}{6}\frac{L^2B^2}{(L^2+B^2)}$$

19. i. Due to symmetry the moment of inertia of the square frame about the x- and y-axes are equal, i.e., $I_x = I_y$.

Now $$I_x = 0 + 2\left[\frac{ml^2}{3}\right] + ml^2$$

$$= \frac{5}{3}ml^2$$

$$I_y = I_x = \frac{5}{3}ml^2$$

ii. Due to perpendicular axis theorem.

$$I_z = I_x + I_y = \frac{5}{3}ml^2 + \frac{5}{3}ml^2$$

$$= \frac{10}{3}ml^2$$

Note that the z-axis does not pass through its centre of mass.

Exercise 2.2

1. In the limiting case normal reaction will pass through O. The cube will tip about O if torque of F exceeds the torque of mg.

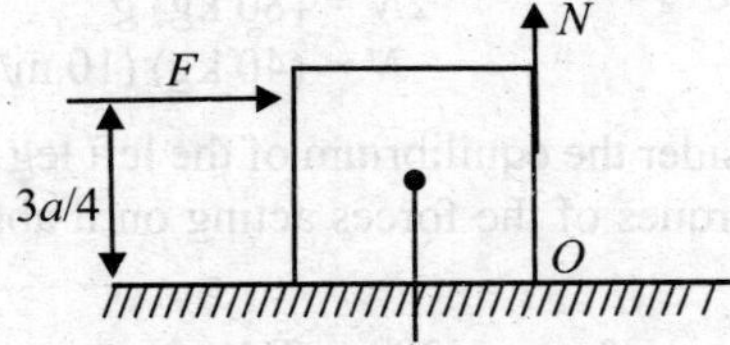

Fig. S-2.8

Hence, $$F\left(\frac{3a}{4}\right) > mg\left(\frac{a}{2}\right) \text{ or } F > \frac{2}{3}mg$$

Therefore, minimum value of F is $2mg/3$.

2.

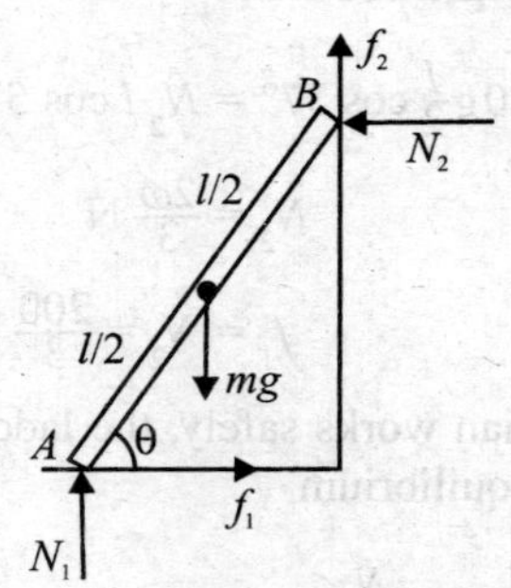

Fig. S-2.9

$$f_1 = \mu_1 N_1, f_2 = \mu_2 N_2$$

$$N_2 = f_1 = \mu_1 N_1 \qquad \text{(i)}$$

$$N_1 + f_2 = mg \qquad \text{(ii)}$$

In limiting case $f_1 = \mu_1 N_1$ and $f_2 = \mu_2 N_2$

Subsituting the values of f_1 and f_2 in (i) and (ii)

After placing the values of f_1 and f_2 solving Eqs. (i) and (ii),

$$N_2 = \frac{\mu_1 mg}{1+\mu_1\mu_2}$$

Taking torque about A and putting it to be zero,

$$mg\frac{l}{2}\cos\theta = f_2 l\cos\theta + N_2 l\sin\theta$$

$$mg\cos\theta = 2\mu_2 N_2\cos\theta + 2N_2\sin\theta$$

Put the value of N_2 and simplify to get

$$\tan\theta = \frac{1-\mu_1\mu_2}{2\mu_1}$$

3. $R_1 + R_2 = 2W \qquad \text{(i)}$

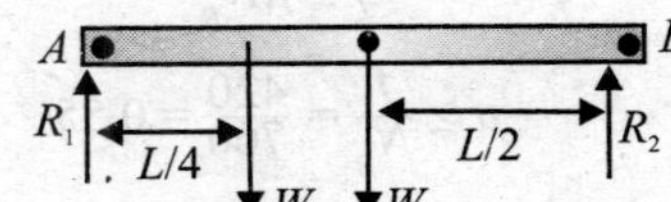

Fig. S-2.10

Balancing torque about A:

$$R_2 L = W\frac{L}{4} + W\frac{L}{2}$$

$$R_2 = \frac{3W}{4} \text{ and } R_1 = \frac{5W}{4}$$

4. $N_1 = 10g = 100\text{N}$

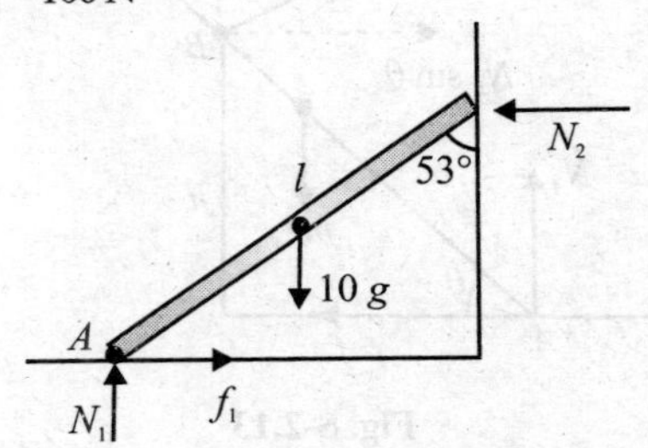

Fig. S-2.11

Balancing torque about A:

$$10g\frac{l}{2}\cos 37^\circ = N_2\, l\cos 53^\circ$$

$$N_2 = \frac{2\omega}{3}\text{ N}$$

Now $$f_1 = N_2 = \frac{200}{3}\text{ N}$$

5. If the electrician works safely, the ladder + man system should be in equilibrium

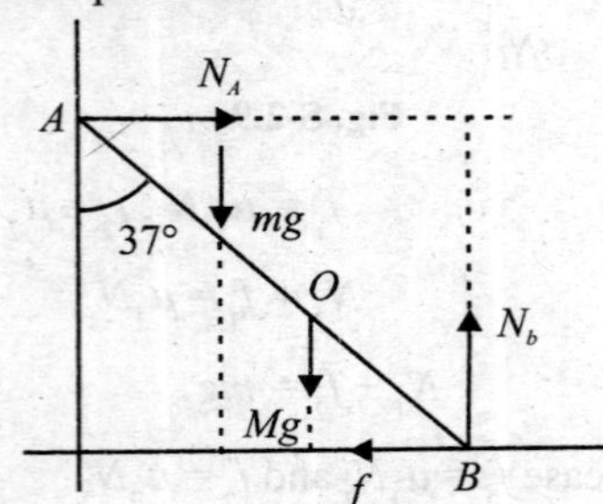

Fig. S-2.12

For equilibrium, $\Sigma F = 0$ and $\Sigma\tau = 0$

for $\Sigma F_x = 0$, $N_A = f$

for $\Sigma F_y = 0$, $N_B = mg + Mg = 60\times 10 + 16\times 10 = 760$ N

Taking torque about B,

$$N_A\, 10\cos 37^\circ = mg\, 8\sin 37^\circ + Mg\, 5\sin 37^\circ$$

$$N_A = \frac{\tan 37^\circ}{10}(65\times 10\times 8 + 16\times 10\times 5)$$

$$f = N_A = 420\text{ N}$$

For maximum value of coefficient of friction,

$$f \le \mu N_B$$

$$\mu \ge \frac{f}{N_B} = \frac{420}{760} = 0.55$$

6. For equilibrium

$$\Sigma F_x = 0, \Sigma F_y = 0$$

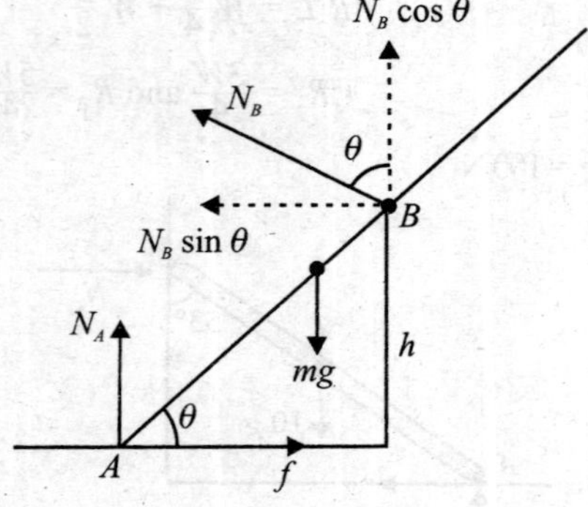

Fig. S-2.13

and $$\Sigma\tau = 0$$

From $\Sigma F_x = 0$

$$f = N_B\sin\theta \qquad \text{(i)}$$

For $$\Sigma F_y = 0$$

$$N_A + N_B\cos\theta = mg \qquad \text{(ii)}$$

Taking torque of forces about A

$$N_A\frac{h}{\sin\theta} = mg\frac{l}{2}\cos\theta$$

That gives

$$N_A = \frac{mgl}{2h}\sin\theta\cos\theta \qquad \text{(iii)}$$

From Eqs. (ii) and (iii),

$$\frac{mgl}{2h}\sin\theta\cos\theta + N_B\cos\theta = mg$$

$$N_B\cos\theta = mg\left(1 - \frac{1}{2h}\sin\theta\cos\theta\right)$$

$$N_B = \frac{mg}{\cos\theta}\left(1 - \frac{1}{2h}\sin\theta.\cos\theta\right) \qquad \text{(iv)}$$

From Eqs. (i) and (iv),

$$f = mg\tan\theta\left(1 - \frac{1}{2h}\sin\theta\cos\theta\right) \qquad \text{(v)}$$

But $$f \le \mu N_A$$

$$\mu \ge \frac{L\cos\theta\sin^2\theta}{2h - L\cos^2\theta\sin\theta}$$

7. The forces acting on different parts are shown in Fig. S-2.14. Consider the vertical equilibrium of "the ladder plus the person" system. The forces acting on this system are its weight (80 kg)g and the contact force $N + N = 2N$ due to the floor.

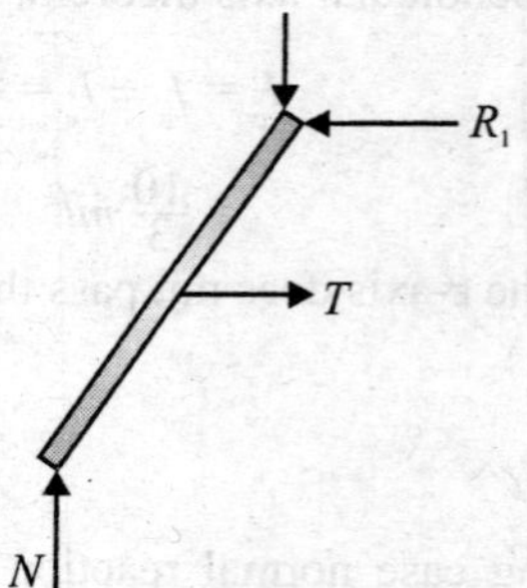

Fig. S-2.14

Thus, $$2N = (80\text{ kg})\,g$$

or $$N = (40\text{ kg})(10\text{ m/s}^2) = 400\text{ N}$$

Next consider the equilibrium of the left leg of the ladder. Taking torques of the forces acting on it about the upper end,

$$N(2m)\tan 30^\circ = T(1\text{ m})$$

or $$T = N\frac{2}{\sqrt{3}} = (400\text{ N})\times\frac{2}{\sqrt{3}}$$

$$= \frac{80}{\sqrt{3}}\text{ N}$$

8.

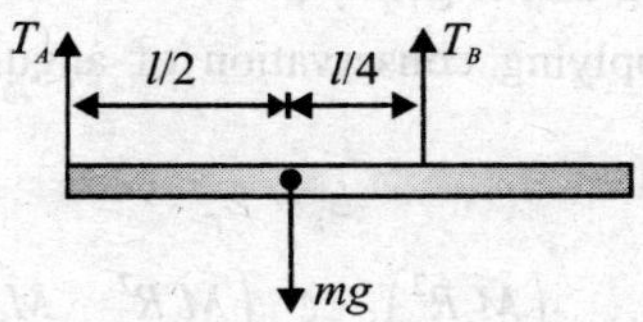

Fig. S-2.15

$$T_A + T_B = mg \qquad \text{(i)}$$

$$T_A \frac{l}{2} = T_B \frac{l}{4} \Rightarrow 2T_A = T_B \qquad \text{(ii)}$$

From Eqs. (i) and (ii)

$$T_A = \frac{mg}{3},\; T_B = \frac{2mg}{3}$$

Exercise 2.3

1. i. $\tau_H = I_H \alpha$

$$mg\frac{l}{2} = \frac{ml^2}{3}\alpha \Rightarrow \alpha = \frac{3g}{2l}$$

ii. $a_{tA} = \alpha l = \frac{3g}{2l} l$

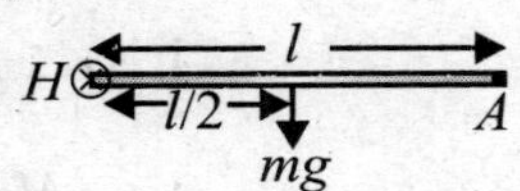

Fig. S-2.16

$= a_{CA} = \omega^2 r = 0 \times l = 0$ ($\omega = 0$ just after release)

2. Free-body diagram of the rod

Fig. S-2.17

In the previous problem

$$a = \frac{3g}{2\ell}$$

Hence, $a_{cm} = a_t = \frac{\alpha\ell}{2} = \frac{3g}{4}$

From the free-body diagram in the vertical direction

$$mg - N_1 = ma_{cm} = m\left(\frac{3g}{4}\right)$$

$$\Rightarrow \qquad N_1 = \frac{mg}{4}$$

In horizontal direction

$$F_{\text{ext}} = ma_{\text{CM}} \Rightarrow N_2 = 0$$

($\because a_{\text{CM}}$ in horizontal $= 0$ as $\omega = 0$ just after release).

3. Suppose the deceleration of the block is a. The linear deceleration of the rim of the wheel is also a. The angular deceleration of the wheel is $\alpha = a/r$. If the tension in the string is T, the equations of motion are as follows:

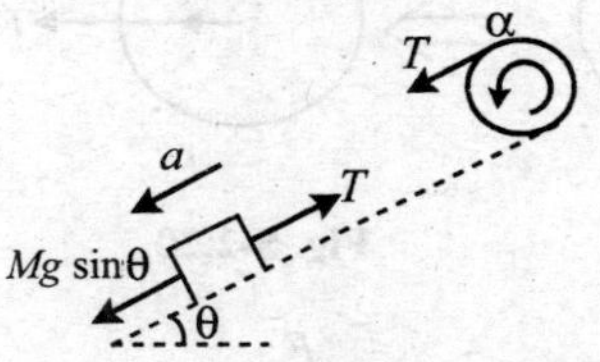

Fig. S-2.18

$$Mg \sin\theta - T = Ma \text{ and } Tr = I\alpha = \frac{Ia}{r}.$$

Eliminating T from these equations,

$$Mg \sin\theta - I\frac{a}{r^2} = Ma$$

$$\Rightarrow \qquad a = \frac{Mgr^2 \sin\theta}{(I + Mr^2)}$$

The initial velocity of the block up the incline is $v = \omega r$. Thus, the distance moved by the block before stopping is

$$x = \frac{v^2}{2a} = \frac{\omega^2 r^2 (I + Mr^2)}{2Mgr^2 \sin\theta} = \frac{(I + Mr^2)\,\omega^2}{2Mg \sin\theta}$$

4. Just after burning of a thread, torque about point P.

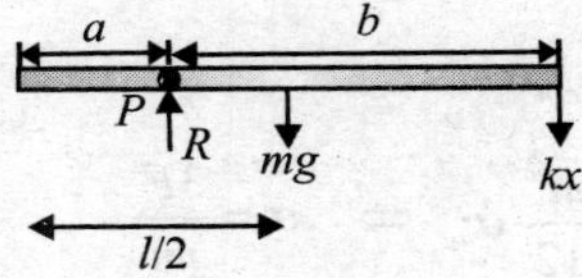

Fig. S-2.19

$$\tau_p = mg\left(\frac{l}{2} - a\right) + kxb$$

$$I_p\alpha = mg\left(\frac{l}{2} - a\right) + kxb$$

$$I_p = \frac{ml^2}{12} + m\left(\frac{l}{2} - a\right)^2$$

$$= \frac{56}{300} \text{ kg m}^2$$

$$\frac{56}{300}\alpha = \frac{28}{5} \Rightarrow \alpha = 30 \text{ rad/s}$$

Equation of motion at this instant

$$mg + kx - R = ma_6 = m\left(\frac{l}{2} - a\right)$$

$$20 + 6 - R = 2 \times 30 \times \frac{1}{10}$$

$$\Rightarrow \qquad R = 20 \text{ N}$$

5. (i) The net force acting on the cotton reel is

$$F = ma_C$$

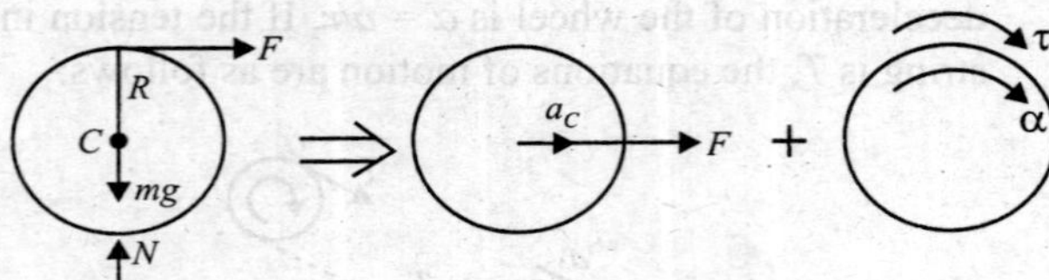

Fig. S-2.20

or $$a_C = \frac{F}{m}$$

(ii) The net torque about the CM is $\tau_C (= FR) = I\alpha$

(as torques produced by mg and N about C are zero as they pass through the CM).

Then, $$\alpha = \frac{FR}{I}$$

6. $$a_{CM} = \frac{F}{m} = \frac{2}{1} = 2\ \text{m/s}^2$$

$$\tau_{cm} = I_{cm}\,\alpha$$

$$\Rightarrow F\frac{l}{2} = \frac{ml^2}{12}\alpha$$

$$\Rightarrow \alpha = \frac{6F}{ml} = \frac{6\times 2}{1\times 2} = 6\ \text{rad/s}^2$$

7. $$\tau_{cm} = I_{cm}\alpha$$

$$F\frac{l}{4} = \frac{ml^2}{12}\alpha \quad \Rightarrow \quad \alpha = \frac{3F}{ml}$$

$$a_{cm} = \frac{F}{m}$$

$$a_Q = a_{cm} + \alpha\frac{l}{2} = \frac{F}{m} + \frac{3F}{ml}\frac{l}{2} = \frac{5F}{2m}$$

Let acceleration of a point P (which is distance x below centre of mass) is zero, then

$$a_P = a_{cm} - \alpha x = 0$$

$$\Rightarrow \quad \frac{F}{m} - \frac{3F}{ml}x = 0 \quad \Rightarrow \quad x = \frac{l}{3}$$

Exercise 2.4

1. i. The frictional force between the two discs exerts a retarding torque on M_1 and an accelerating torque on M_2. Till both the discs start moving together with a common angular speed ω, the angular momentum of each disc changes but the angular momentum of the whole system $(M_1 + M_2)$ remains conserved as there is no external torque acting on the system.

A rotating disc is gently placed on a stationary disc. Thus, applying conservation of angular momentum, we get

$$L_i = L_f$$

$$\left(\frac{M_1R^2}{2}\right)\omega_0 = \left(\frac{M_1R^2}{2} + \frac{M_2R^2}{2}\right)\omega$$

$$\omega = \frac{M_1\omega_0}{M_1 + M_2}$$

ii. The work done by friction is equal to the change in kinetic energy of the system.

$$W_f = K_f - K_i$$

$$K_i = \frac{1}{2}I_1\omega_0^2 = \frac{1}{2}\left(\frac{M_1R^2}{2}\right)\omega_0^2$$

$$= \frac{1}{4}M_1R^2\omega_0^2$$

$$K_f = \frac{1}{2}(I_1 + I_2)\omega^2$$

$$= \frac{1}{2}(M_1 + M_2)\frac{R^2}{2}\left[\frac{M_1\omega_0}{M_1 + M_2}\right]^2$$

$$K_f = \frac{1}{4}\frac{M_1^2}{(M_1 + M_2)}R^2\omega_0^2$$

$$W_f = K_f - K_i = \frac{-1}{4}\left(\frac{M_1M_2}{M_1 + M_2}\right)R^2\omega_0^2$$

Note that the negative sign shows that work done by friction reduces the kinetic energy of the system.

iii. The fractional loss in kinetic energy is

$$\frac{\Delta K}{K_i} = \frac{M_2}{M_1 + M_2}$$

2. Initially the system was at rest, thus the initial angular momentum was zero. As child jumps off from the platform, it gains an angular momentum in the opposite direction. This implies that the platform must also gain the same amount of angular momentum in opposite direction.

When the child jumps off, the platform gains an angular velocity ω. Then the net velocity of child with respect to earth is $u - R\omega$. As no external torque is present, the net angular momentum of the system must be finally zero, thus

$$m(u - R\omega) = I\omega$$

$$\omega = \frac{muR}{mR^2 + I}$$

3. Again in this case no external torque is acting on the system so we can equate the angular momentum before catch and after catch. If after catching the ball, platform starts rotating with an angular speed ω, we have

Angular momentum of the ball about centre of platform before catch = Angular momentum of the platform plus child plus ball after catch

$$m_1uR = (I + mR^2 + m_1R^2)\omega$$

or
$$\omega = \frac{m_1uR}{I + mR^2 + m_1R^2}$$

Here we can conserve angular momentum with respect to the centre pivot of the platform. This is because when the child catches the ball, it has a linear momentum and due to it the platform tends to gain a linear momentum, which develops a normal reaction on the pivot in opposite direction and it will not allow the platform to move translationally. If the platform was resting on a smooth plane without pivot at the centre, after catching the ball, the platform would also move translationally with its rotational motion and in that case linear momentum of system will also remain conserved.

4. As no external torque is present, we can conserve angular momentum before and after starting the walk by the child. As the child walks in opposite direction to the rotation of platform, its angular momentum will be in opposite direction which will tend to decrease the total angular momentum, hence the angular speed of the platform must increase to maintain the angular momentum conservation we have

$$(I + mR^2)\omega_1 = I\omega_2 - m(u - R\omega_2)R$$

The term on left side of the above expression is the angular momentum of platform plus child system when the child was at rest and the first term on right side of the expression is the angular momentum of the platform with increased angular speed (ω_2), when the child starts walking. The second term on right side is the final angular momentum of the child. This term is negative due to his walking in opposite direction and we have taken the velocity of the child as $(u - R\omega_2)$ because u is the relative speed of the child on platform which is rotating with angular speed ω_2 in opposite direction.

Again as we have used in the above problem, be careful, that angular momentum conservation should always be used with respect to earth or some inertial reference frame.

$$\omega_2 = \frac{(I + mR^2)\,\omega_1 + muR}{I + mR^2}$$

5. In the given situation, cylinder has both translational and rotational motion. We find its angular momentum about P by using the equation, $\vec{L} = \vec{L}_{cm} + \vec{r} \times m\vec{v}$

The angular momentum of the cylinder about its axis of rotation passing through its centre of mass is given by

$$\vec{L}_C = I\omega\ (-\hat{k}) = \left(\tfrac{1}{2}mr^2\right)\omega\,(-\hat{k})$$

And the angular momentum of this cylinder's translational motion about P is

$$\vec{r} \times m\vec{v} = mvR(-\hat{k})$$

Total angular momentum of the cylinder about point P is given as

$$\vec{L} = \tfrac{1}{2}mr^2\omega + mvR\,(-\hat{k})$$

If the cylinder rotates in anticlockwise direction and moves in the same direction, its total angular momentum will now be written as

$$\vec{L} = \tfrac{1}{2}mr^2\,\omega\,(\hat{k}) + mvR\,(-\hat{k})$$
$$= \left(\frac{1}{2}mr^2\omega - mvR\right)(-\hat{k})$$

6. $\vec{L} = \vec{r} \times \vec{p}$

The magnitude of $\vec{L}$ is given by, $L = rp\sin\theta\,(\hat{k})$

In this example, velocity of the particle at any time

$v = gt$, $r\sin\theta = x_0$ and $p = mv = mgt$

$$\therefore \qquad \vec{L} = mgx_0t\,(\hat{k}) \qquad \text{(i)}$$

The right hand rule shows that $\vec{L}$ is parallel to torque, i.e., directed perpendicular to paper inwards.

7. Using conservation of energy principle, we have,

Fall in PE of the block = Gain in KE of the block + rotational KE of the pulley + rotational KE of the shell.

or $\quad mgh = \tfrac{1}{2}mv^2 + \tfrac{1}{2}I\omega^2 + \tfrac{1}{2}\left(\frac{2MR^2}{3}\right)\omega'^2$

where $\omega = \frac{v}{r}$ and $\omega' = \frac{v}{R}$

After substituting these values in the above equation, and solving, we get

$$v^2 = \frac{mgh}{\left(\frac{m}{2} + \frac{I}{2r^2} + \frac{M}{3}\right)}$$

8. As the rod reaches its lowest position, the centre of mass is lowered by a distance l. Its gravitational potential energy is decreased by mg. As no energy is lost against friction, this should be equal to the increase in the kinetic energy. As the rotation occurs about the horizontal axis through the clamped end, the moment of inertia is $I = m^2/3$.

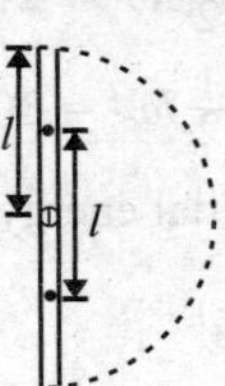

Fig. S-2.21

Thus, $\frac{1}{2} I\omega^2 = mg\frac{1}{2}\left(\frac{ml^2}{3}\right)\omega^2 = mg$

or $\omega = \sqrt{\frac{6g}{l}}$

The linear speed of the free end is $v = l\omega = \sqrt{6gl}$

9. Using principle of conservation of mechanical energy, we have

$$\frac{mgl}{2} + mgl = \frac{1}{2} I\omega^2$$

where ω is the angular speed of the rod when it becomes vertical, and $I = (ml^2/3) + ml^2$, assuming mass of each rod is m.

Substituting value of I in the above equation, we get

$$\omega = \sqrt{\left(\frac{9g}{4l}\right)}$$

10. In the process, angular momentum remains constant

$$I_1\omega_1 = I_2\omega_2$$

where $I_1 = 2m\,(0.15)^2$, $\omega_1 = 4$ rad/s

$$I_2 = 2m\,(0.05)^2$$

Substituting these values in above equation, we get

$$\omega_2 = \frac{I_1\omega_1}{I_2} = 36 \text{ rad/s}$$

Work done on the collar

$$W = \frac{1}{2} I_2\,\omega_2^2 - \frac{1}{2} I_1\omega_1^2$$

or $W = \frac{1}{2}(0.05)^2 \times 36^2 - \frac{1}{2} 2m\,(0.015)^2 \times 4^2$

where $m = \frac{500}{1000} = 0.05$ kg $= 1.44$ J

11. Let $I_A = I$ and $I_B = 2I$

In the connection of disk A with disk B, the angular momentum remains constant

or $I\omega_0 + 0 = (I + 2I)\omega$

which gives $\omega = \frac{\omega_0}{3}$

The change in KE evolved as heat, i.e.,

$$\frac{1}{2} I\omega_0^2 - \frac{1}{2}(3I)\omega^2 = 2000$$

Substituting the value of ω in the above equation, we get

$$\frac{1}{2} I\omega_0^2 - \frac{1}{2}(3I)\left(\frac{\omega_0}{3}\right)^2 = 2000$$

or $\frac{1}{2} I\omega^2 = 2000$

$\therefore$ $\frac{1}{2} I\omega^2 = 3000$ J

where $I\omega^2/2$ is the initial energy of the disk A.

12. $\vec{L} = \vec{L}_{CM} + m\,(\vec{r}_0 \times \vec{v}_0)$ (i)

Here,

$\vec{L}_{CM} = I\omega = \left(\frac{1}{2} mR^2\right)\left(\frac{v}{R}\right) = \frac{1}{2} mvR$ and $m(\vec{r}_0 \times \vec{v}_0)$ (perpendicular to the paper inwards)

Since, both the terms of right hand side of Eq. (i) are in the same direction,

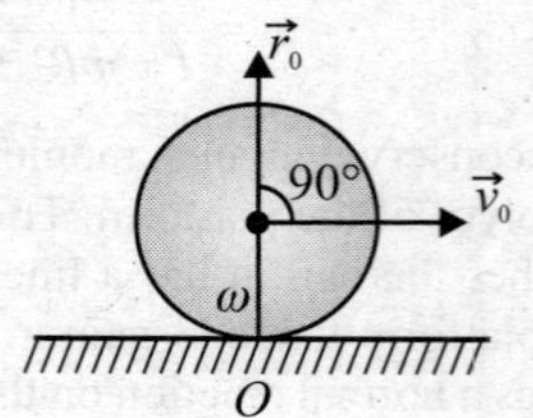

Fig. S-2.22

$\therefore$ $|\vec{L}| = \frac{1}{2} mvR + mvR$

or $|\vec{L}| = \frac{3}{2} mvR$

13. Initial velocity is $\vec{u} = u(\cos\theta\hat{i} + \sin\theta\hat{j})$

$\therefore$ Velocity after time t will be $v = \vec{u} + \vec{a}t$

$\therefore$ $\vec{v} = u\,(\cos\theta\hat{i} + \sin\theta\hat{j}) - gt\,\hat{j}$
$= u\cos\theta\,\hat{i} + (u\sin\theta - gt)\,\hat{j}$

If $\vec{r}$ be the radius vector of the particle P at time t (as shown in figure), then from $\vec{s} = \vec{u}t + \vec{a}t^2/2$

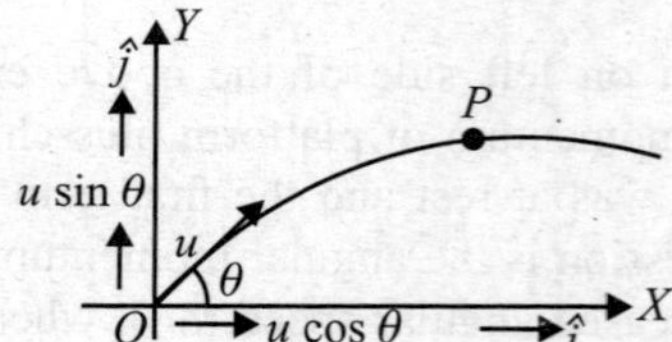

Fig. S-2.23

$\therefore$ $\vec{r} = ut\,(\cos\theta\hat{i} + \sin\theta\hat{j}) - \frac{gt^2}{2}\hat{j}$
$= ut\cos\theta\,\hat{i} + \left(ut\sin\theta - \frac{gt^2}{2}\right)\hat{j}$

$\therefore$ Angular momentum $\vec{L}$ of the particle at time t will be

$$\vec{L} = m\,(\vec{r} \times \vec{v})$$

i.e., $\vec{L} = m\left[ut\cos\theta\, i + \left(ut\sin\theta - \frac{gt^2}{2}\right)\hat{j}\right] \times [ut\cos\theta\hat{i} + (u\sin\theta - gt)\hat{j}]$

$$= \frac{-\,mut^2\,g\cos\theta\,\hat{k}}{2}$$

14. $(KE)_I = 0$, $(KE)_F = \frac{1}{2}(I_1 + I_2)\,\omega_0^2$

So by work–energy theorem,

$$W = (KE)_F - (KE)_I$$
$$= \frac{1}{2}(I_1 + I_2)\,\omega_0^2$$

or, $W = \frac{1}{2}[M_1x^2 + M_2(L - x)^2]\omega_0^2$

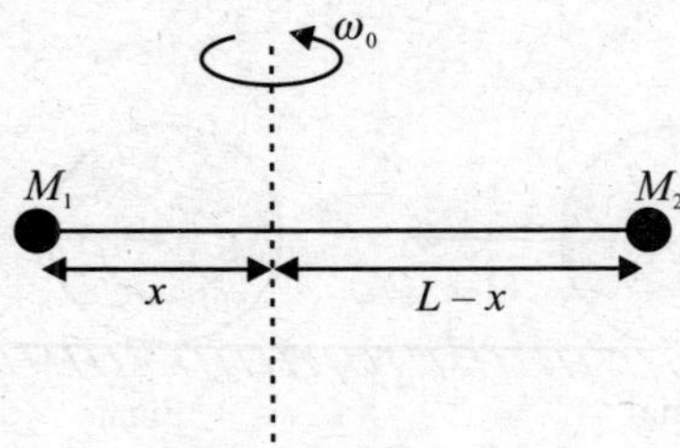

Fig. S-2.24

For W to be minimum,

$$\frac{dW}{dx} = 0, \text{ i.e., } 2M_1x - 2M_2(L-x) = 0$$

so,
$$x = \frac{M_2L}{(M_1+M_2)}$$

This is the position of centre of mass of the rod from M_1. So the required work is minimum when the rod is rotating about an axis passing through its centre of mass and perpendicular to the length of the rod.

15. As moment of inertia of a disc through its centre of mass and perpendicular to its plane is $MR^2/2$, its angular momentum about the given axis when it is rotating with angular velocity ω will be

$$L = I\omega = \frac{1}{2}MR^2\omega \quad \text{(i)}$$

When another disc of same radius but mass $M/4$ is placed on the first, the moment of inertia of the system

$$I' = \frac{1}{2}MR^2 + \frac{1}{2}\left[\frac{M}{4}\right]R^2 = \frac{5}{8}MR^2$$

And so if ω' is the new angular velocity of the system

$$L' = I'\omega' = \frac{5}{8}MR^2\omega' \quad \text{(ii)}$$

Now as no external torque has been applied, by conservation of angular momentum we have

$$L' = L, \text{ i.e., } \frac{1}{2}MR^2\omega' = \frac{5}{8}MR^2\omega'$$

or,
$$\omega' = \left(\frac{4}{5}\right)\omega$$

16. For circular motion of a body tied to a string on a horizontal plane

$$\left(\frac{mv^2}{r}\right) = T$$

Here as tension is provided by the hanging mass M, i.e., $T = g$, therefore

so,
$$\left(\frac{mv^2}{r}\right) = Mg$$

According to the given problem,

$$\frac{(mv_1^2/r_1)}{(mv_2^2/r_2)} = \frac{M_1g}{M_2g} = \frac{8}{1}$$

or,
$$\frac{v_1^2 r_2}{v_2^2 r_1} = \frac{8}{1} \quad \text{(i)}$$

Now as force T is central, so angular momentum is also conserved, i.e.,

$$mv_1r_1 = mv_2r_2 \quad \text{(ii)}$$

So substituting the value of v_1/v_2 from Eq. (ii) in Eq. (i),

$$\left[\frac{r_2}{r_1}\right]^2 \times \frac{r_2}{r_1} = \frac{8}{1}$$

$$\text{i.e., } \frac{r_2}{r_1} = 2 \quad \text{(iii)}$$

so that
$$\frac{\Delta r}{r} = \frac{r_2 - r_1}{r_1} = \frac{r_2}{r_1} - 1 = 2 - 1 = 1$$

Furthermore as in circular motion, $v = r\omega$

so,
$$\frac{\omega_2}{\omega_1} = \frac{v_2}{v_1} \times \frac{r_1}{r_2} = \left[\frac{r_1}{r_2}\right]^2$$

$$\left[\text{as from Eq. (ii), } \frac{v_2}{v_1} = \frac{r_1}{r_2}\right]$$

$$\Rightarrow \quad \frac{\omega_2}{\omega_1} = \left[\frac{1}{2}\right]^2 = \frac{1}{4}$$

$$\left[\text{as from Eq. (iii), } \frac{r_2}{r} = 2\right]$$

so,
$$\frac{\Delta\omega}{\omega} = \frac{\omega_2 - \omega_1}{\omega_1} = \frac{\omega_2}{\omega_1} - 1 = \frac{1}{4} - 1 = -\frac{3}{4}$$

17. When the girl lands on the plank, she exerts a force on it, similarly the plank exerts an equal and opposite force on the girl. For the system consisting of girl and plank these forces are internal; therefore angular momentum is conserved.

$$\vec{L}_i = \frac{1}{2}ml\sqrt{2gh}$$

$$\vec{L}_f = \frac{1}{6}ml^2\omega + \frac{1}{4}ml^2\omega$$

$$\frac{1}{2}ml\sqrt{2gh} = \frac{5}{12}ml^2\omega$$

$$\vec{L}_f = \vec{L}_f \Rightarrow \omega = \frac{6}{5}\sqrt{\frac{2gh}{l}}$$

18. There will be no torque on the system if we ignore frictional torque at the axle. Therefore, the angular momentum of the system is conserved.

$$\vec{L}_{\text{initital}} = \vec{L}_{\text{final}}$$

$$I_{\text{initial}}\,\omega_0 = I_{\text{final}}\,\omega$$

$$4ma^2\omega_0 = 4m(2a)^2\omega$$

from which we have, $\omega = \frac{\omega_0}{4}$

Exercise 2.5

1.

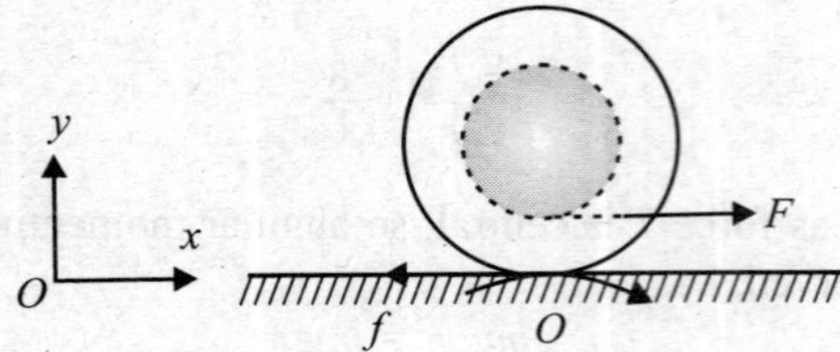

Fig. S-2.25

i. The CM moves along the positive x-axis because the net torque about the point of contact is clockwise.

ii. The friction force acts along the negative x-axis. In the absence of friction force, point O has a tendency to move along the positive x-axis.

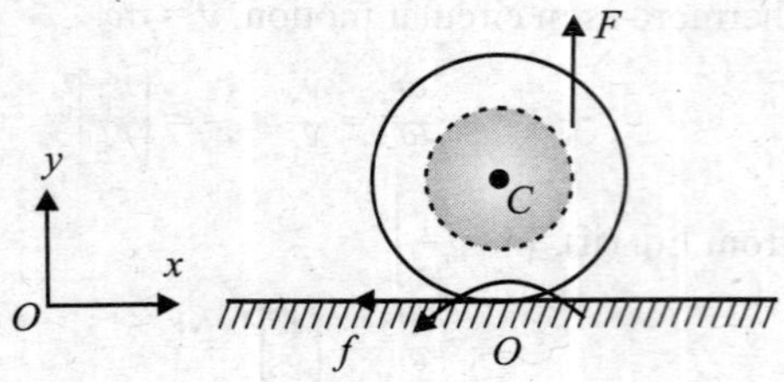

Fig. S-2.26

iii. The CM moves along the negative x-axis because the net torque about the point of contact is anticlockwise.

iv. The friction force acts along the negative x-axis. No other force is acting along the x-axis that can accelerate the centre of mass along the negative x-axis.

2. i. **True:** The friction force retards the motion of centre of mass and accelerates the rotational motion with respect to centre of mass.

ii. **False:** Energy is not conserved because friction force does negative work.

iii. **True:** The frictional force produces no torque about any point on the horizontal surface.

iv. **False:** Conserving angular momentum about a stationary point on the horizontal surface, we get

$$mv_0R = mvR + \left(\frac{2}{5}mR^2\right)\omega$$

Using the condition of rolling motion, we get

$$v = \omega R$$

$$mv_0R = mvR + \frac{2}{5}mvR = \frac{7}{5}mvR$$

$$v = \frac{5}{7}v_0$$

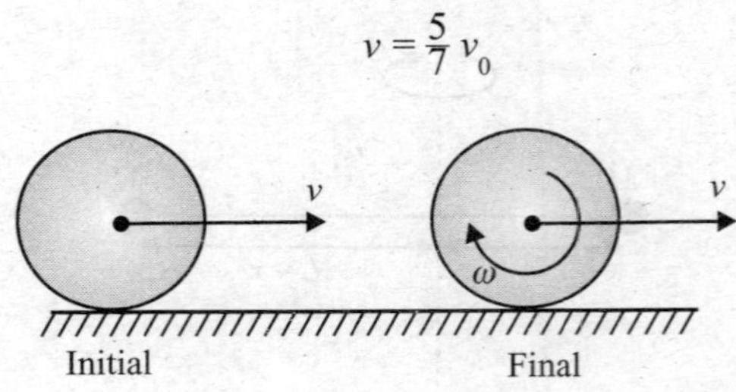

Fig. S-2.27

3. Zero. Since lowest point of the sphere is stationary with respect to the plank, therefore friction force acting on the plank is zero. Thus, no force is required to keep the plank stationary.

4. The velocity of the centre of mass of the tractor wheel is v.

Therefore, velocity of the lower half of the belt at any instant

$$= v - v = 0$$

Velocity of the upper half $= v + v = 2v$

$$\therefore \quad T = \frac{1}{2}\left(\frac{m}{2}\right)(2v)^2 = mv^2$$

5.

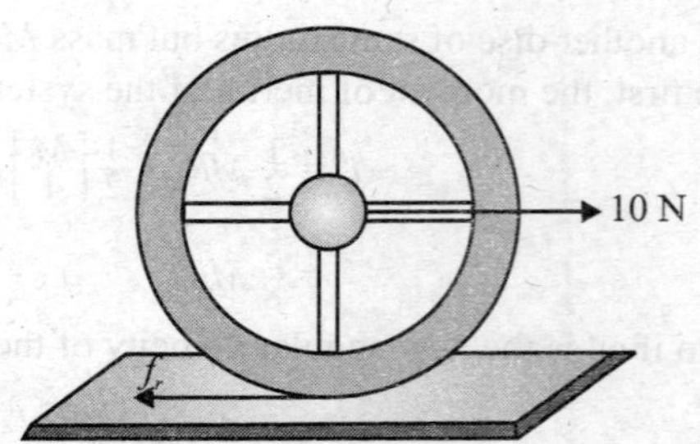

Fig. S-2.28

a. The direction of the frictional force will be in backward direction of motion of the wheel.

$$10 - f = ma$$

Substituting $m = 10$ kg and $a = 0.60$ m/s^2 in the above equation, we get

$$F_r = 4 \text{ N}$$

b. Equation of motion for rotation of wheel

$$f_r \times R = I\alpha$$

which gives

$$I = f_r \times \frac{R}{\alpha} = \frac{4 \times 0.30}{(0.60/0.30)}$$

or $$I = 0.6 \text{ kg m}^2$$

6. The force acting on the cylinder are (i) mg, downwards, (ii) F, applied force downwards, (iii) N, reaction of each plank upwards, (iv) f_r frictional force (not limiting)

between each plank and the cylinder to the right.

Considering vertical motion

$$F + mg = 2N \quad \text{(i)}$$

Considering horizontal motion

$$2f_r = ma \quad \text{(ii)}$$

where a is the horizontal acceleration of the centre of mass of the cylinder to the right. Considering rotational dynamics of the cylinder about its axis

$$\tau = FR - 2f_r R = \left(\tfrac{1}{2}mR^2\right) \times \alpha$$

$$(F - 2f_r)R = \tfrac{1}{2} mR^2 \times \tfrac{a}{R} \quad (\because a = \alpha R)$$

$$F - 2f_r = \tfrac{1}{2} ma$$

$$F = \tfrac{1}{2} ma + 2f_r = \tfrac{3}{2} ma \quad \text{(iii)}$$

$[(\because 2f_r = ma$ by Eq. (ii)$]$

$$F = 3f_r \quad \text{(iv)}$$

Now $f_r \le f_{\lim}$ or $f_r \le \mu N$ $(\because f_{\lim} = \mu N)$

$$\tfrac{1}{3}F \le \mu \frac{F + mg}{2}$$

$$2F \le 3\mu F + 3\mu mg$$

$$F \le \frac{3\mu mg}{2 - 3\mu}$$

$$F_{\max} = \frac{3\mu mg}{2 - 3\mu}$$

When the force is maximum, acceleration is automatically maximum.

$$a_{\max} = \frac{2F_{\max}}{3m} = \frac{2\mu g}{2 - 3\mu}$$

7. The cylinder will roll without slipping so long as the friction force is sufficiently large and it never reaches the limiting value. The forces acting on the cylinder are: (i) its weight Mg, (ii) normal reaction N, (iii) tension T to the right, (iv) f_{fr} to the right.

Let a be the acceleration of the falling mass m.

Then $\qquad a_{CM} = \frac{a}{2}$

For the load,

$$mg - T = ma \Rightarrow T = mg - ma$$

For the cylinder

$$T + f_{fr} = \frac{Ma}{2}$$

and $\qquad \tau = TR - f_{fr} R = \left(\tfrac{1}{2}MR^2\right)\alpha$

$$T - f_{fr} = \tfrac{1}{2}MR\alpha = \tfrac{1}{2}M, R\frac{a_{CM}}{R} = \tfrac{1}{2} Ma_{CM} = \tfrac{1}{2}Ma$$

Solving $\qquad f_{fr} = \tfrac{1}{8}Ma$

$$T = \tfrac{3}{8}Ma \text{ and } a = \frac{mg}{m + \frac{3}{8}M}$$

There will be no slip if

$$f_{fr} \le \mu mg \Rightarrow \mu \ge \frac{Ma}{8mg}$$

$$\mu \ge \frac{M}{8m + 3M}$$

Rolling + slipping: Now $a_{CM} \ne \alpha R$, rather acceleration of the highest point of the cylinder $= a_{CM} + \alpha R =$ acceleration of m.

and $\qquad \tau = TR - f_{\lim} R = \left(\tfrac{1}{2}MR^2\right)\alpha$

$$T - \mu Mg = \tfrac{1}{2}MR\alpha$$

For load $\qquad mg - T = ma$

Solving for α

$$a = \frac{3mg - \mu Mg}{M + 3m}$$

8. Since the cylinder does not slip, the displacement, velocity and acceleration of the hanging mass are, respectively, equal to the displacement, velocity and acceleration of the upper point B of the cylinder as the string leaves the upper point tangentially.

Now v (velocity of the upper point) $= v_m + \omega r = 2v_{CM}$

$$(\because \omega r = v_m)$$

From conservation of energy, when m lowers through h

$$mgh = \tfrac{1}{2}mv^2 + \left(\tfrac{1}{2}mv_{CM}^2 + \tfrac{1}{2}I\omega^2\right)$$

$(\because$ loss in PE = gain in KE$)$

$$= \tfrac{1}{2}mv^2 + \tfrac{1}{2}\frac{Mv^2}{4} + \tfrac{1}{2}\left(\tfrac{1}{2}Mr^2\right)\frac{v_{CM}^2}{r^2}$$

$\left(\because I = \tfrac{1}{2}Mr^2\right)$

$$= \tfrac{1}{2}mv^2 + \tfrac{1}{8}Mv^2 + \tfrac{1}{4}M\frac{v^2}{4}$$

$$= \tfrac{1}{2}mv^2 + \tfrac{3}{16}Mv^2$$

$$16mgh = 8mv^2 + 3Mv^2$$

$$\Rightarrow \qquad v^2 = \frac{16mgh}{8m + 3M}$$

From $\qquad v^2 = 2ah$

$$\Rightarrow \quad a = \frac{8mg}{8m + 3M} = \frac{g}{1 + 3M/8m} = 4 \text{ m/s}^2$$

Considering downward motion of m,

$$mg - T = ma$$

or $$T = mg - ma = 2(10 - 4) = 12 \text{ N}$$

9. Since there is no slipping, the velocity of the point of contact of the string with the pulley is the vector sum of the velocity of the point of contact due to rotation plus the velocity of the centre of mass of the spool. That is

$v_B = v_{CM} - \omega r$, ω is in clockwise direction

But $v_{CM} = \omega R$

$$v_B = v_{CM} - v_{CM}\frac{r}{R} = v_{CM}\left(\frac{R-r}{R}\right)$$

This is also the velocity (v) of the rolling mass.

$$v_B = v_{CM}\left(\frac{R-r}{R}\right) = v\left(\frac{R-r}{R}\right)$$

Considering the conservation of energy when m lowers by h, we get

$$mgh = \frac{1}{2}mv^2 + \left(\frac{1}{2}Mv_{CM}^2 + \frac{1}{2}I\omega^2\right)$$

$$= \frac{1}{2}mv^2 + \left(\frac{1}{2}Mv_{CM}^2 + \frac{1}{2} \times \frac{1}{2}MR^2\frac{v_{CM}^2}{R^2}\right)$$

$$= \frac{1}{2}mv^2 + \frac{3}{4}Mv_{CM}^2$$

$$= \frac{1}{2}mv^2 + \frac{3}{4}Mv^2\left(\frac{R}{R-r}\right)$$

$$v^2 = \frac{4mgh}{2m + 3M[R/(R-r)]^2}$$

comparing with $v^2 = 2ah$

$$a = \frac{v^2}{2h} = \frac{2mg}{2m + 3M[R/(R-r)]^2}$$

10. F.B.D. is shown in Fig. S-2.29

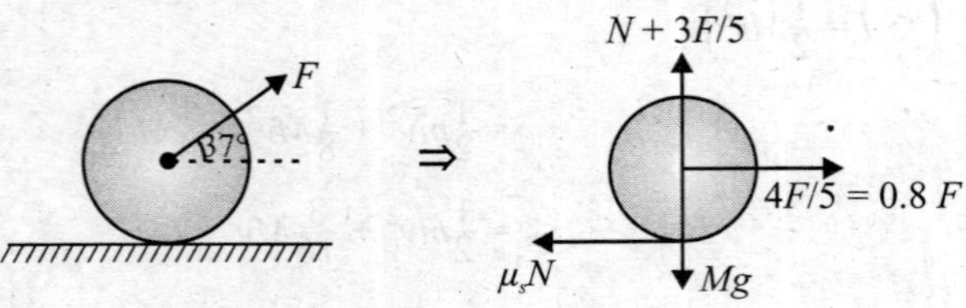

Fig. S-2.29

$$N = Mg - \frac{3F}{5} = Mg - 0.6F$$

For pure rolling, $a = R\alpha$

or $$\frac{0.8F - \mu_s N}{M} = \frac{(R)(\mu_s N)(R)}{\frac{1}{2}MR^2} = \frac{2\mu_s N}{M}$$

$\therefore$ $$0.8F = 3(0.4)(Mg - 0.6F)$$

or $$F = 0.79\ Mg$$

Therefore, maximum value of F is 0.79 Mg.

11. $F - T - f = Ma$ (i)

$$T - f = Ma \quad \text{(ii)}$$

$$\alpha = \frac{fR}{\frac{1}{2}mR^2} = \frac{2f}{mR} \quad \text{(iii)}$$

Constraint relation $a_{plank} = a_p$

$$R\alpha - a = a$$

or, $$R\alpha = 2a \quad \text{(iv)}$$

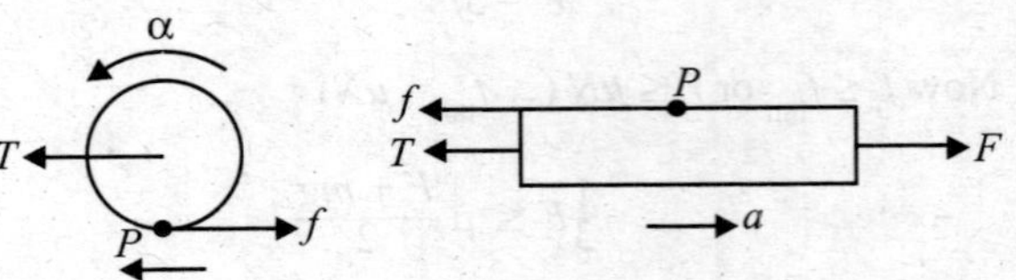

Fig. S-2.30

Solving these equations, we get

$$a = \frac{F}{3m + M}$$

12. As the force F is applied, the cylinder tends to rotate along anticlockwise direction with the point of contact 'C' tending to move to the right. Hence, friction acts along the left (as shown in Fig. S-2.31). If α is the angular acceleration of the cylinder, then from $\tau = I\alpha$, we have

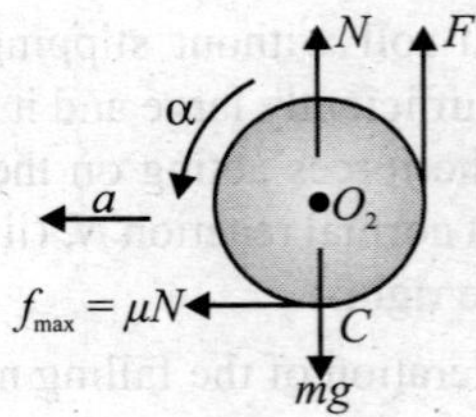

Fig. S-2.31

$$r(F - f) = (mr^2/2)\alpha \quad \text{(i)}$$

If a is the linear acceleration of the cylinder's CM, then from $F = ma$, we have

$$f = ma \quad \text{(ii)}$$

For pure rolling,

$$a = \alpha r \quad \text{(iii)}$$

Dividing Eq. (i) by Eq. (ii), we get

$$\frac{(F-f)}{f}=\frac{r\alpha}{2a}=\frac{1}{2}$$

$$\Rightarrow F=(3f/2)\Rightarrow \quad F_{max}=(3/2)f_{max}$$

$$[f_{max}=\mu N \Rightarrow \mu(mg-F)]$$

$$\Rightarrow \quad F_{max}=\frac{3\mu}{2}[mg-F_{max}]=\frac{3\mu mg}{2+3\mu}$$

From Eq. (ii),

$$a=\frac{f}{m}$$

$$\Rightarrow \quad a_{max}=\frac{f_{max}}{m}=\frac{2}{3m}F_{max}$$

$$\Rightarrow \quad a_{max}=\frac{2}{3m}\left(\frac{3\mu mg}{2+3\mu}\right)=\frac{2\mu g}{2+3\mu}$$

13. Let v and ω be the velocity of the bullet and angular velocity of the cylinder, respectively. Applying conservation of angular momentum about the point of contact of the cylinder with the floor, we get

$$mu\,2R=mv\,2R+(I_{CM}+MR^2)\,\omega$$

Here, $v=2R\omega$ and $I_{CM}=½MR^2$

Substituting and solving, we get

$$v=\frac{8mu}{8m+3M}$$

and $$\omega=\frac{4mu}{(8m+3M)R}$$

14. Here the centre of mass of the marble will move in a circle of radius $(R-r)$ so for just looping the loop, at H

$$v=\sqrt{g(R-r)} \quad \text{(i)}$$

Now as in rolling

$$K=K_T+K_R=½\,Mv^2+½\,I\omega^2$$

and here

$$I=\left(\frac{2}{5}\right)Mr^2 \text{ with } v=r\omega$$

so, $$K=\frac{1}{2}Mv^2+\frac{1}{2}\left[\frac{2}{5}Mr^2\right]\left[\frac{v}{r}\right]^2$$

i.e., $$K=\frac{1}{2}Mv^2+\frac{1}{5}\,Mv^2=\frac{7}{10}Mv^2 \quad \text{(ii)}$$

So in the light of Eq. (i), Eq. (ii) becomes

$$K=\frac{7}{10}Mg\,(R-r) \quad \text{(iii)}$$

As this kinetic energy is provided by loss in PE, applying conservation of mechanical energy between P and H

$$0+Mgh=\frac{7}{10}Mg\,(R-r)+Mg\,(2R-r)$$

$$h=\frac{1}{10}\,(27R-17r)$$

15. Step 1: Free-body diagram,

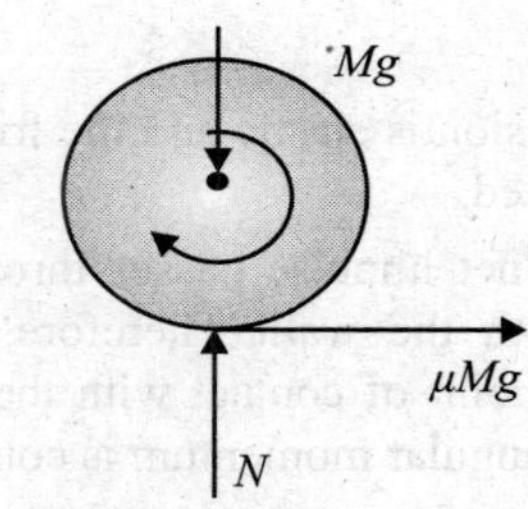

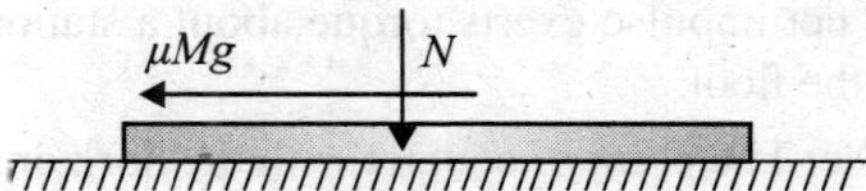

Fig. S-2.32

Step 2: Equation of motion,

Plank:

$$\mu Mg=MA\Rightarrow A=\mu g \quad \text{(backward)}$$

(A will be positive)

Sphere:

$$\mu Mg=MA\Rightarrow a=\mu g \quad \text{(forward)}$$

(a will be positive)

Step 3: Torque equation,

$$\mu Mg\,R=\left(\frac{2}{5}MR^2\right)\alpha$$

$$\alpha=\frac{5}{2}\frac{\mu g}{R} \quad \text{(anticlockwise)}$$

(it will be negative)

Step 4: Constraint relation

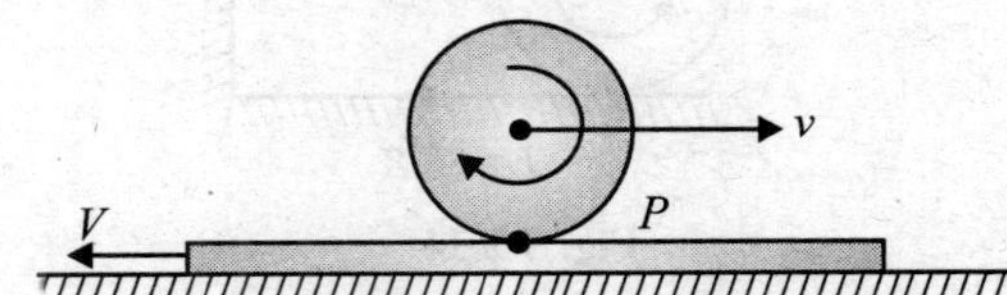

Fig. S-2.33

$$V_P=V=\omega R-v$$

$$(0+\mu g\,t)=\left(\omega_0-\frac{5}{2}\frac{\mu g}{R}t\right)R-(0+\mu g\,t)$$

$$\left(2\mu g\,t+\frac{5}{2}\mu g\,t\right)=\omega_0 R\Rightarrow\frac{9\mu gt}{2}=\omega_0 R$$

$$\Rightarrow \quad t=\frac{2\omega_0 R}{9\mu g}$$

Velocity of the plank when rolling starts,

$$V=0+\mu g\,t$$

$$V_{plank}=\mu g\times\frac{2\omega_0 R}{9\mu g}=\frac{2}{9}\,\omega_0 R$$

Velocity of the sphere when rolling starts,

$$V_{sphere}=0+\mu g\,t=\frac{2}{9}\,\omega_0 R$$

Exercise 2.6

1. **i.** True: Collision is elastic and the frictional impulse is not generated.

   ii. True: The net impulse passes through the point of contact with the wall. Therefore, angular impulse about the point of contact with the wall is zero and hence the angular momentum is conserved.

   iii. False: Angular momentum is not conserved because the net impulse exerts torque about a stationary point on the floor.

   iv. False: The point of contact with the horizontal surface moves away from the wall.

   v. False: Friction force decreases both the speeds.

   vi. True: When the sphere starts rolling, it is moving away from the wall because the angular velocity becomes zero before the linear velocity. The linear acceleration of the CM of the sphere is

$$a = -\mu g$$

The angular acceleration about the CM is

$$\alpha = \frac{-5}{2}\frac{\mu g}{R}$$

The negative sign shows that both the accelerations have the retarding effect. The time after which the linear velocity will become zero is

$$t_1 = \frac{v}{\mu g}$$

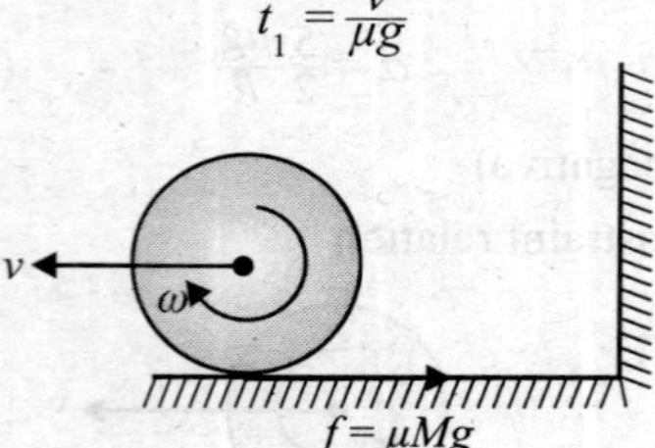

Fig. S-2.34

and the time after which the angular velocity will become zero is

$$t_2 = \frac{2}{5}\frac{\omega R}{\mu g} = \frac{2}{5}\frac{v}{\mu gh} \qquad (\therefore v = \omega R)$$

Note that $t_2 < t_1$.

2. **a.** In the process of collision, the linear momentum remains constant

$$2mv - m^2(2v) + 0 = (2m + m + 8m)\, v_{CM}$$

or

$$v_{CM} = 0$$

   b. Angular momentum of the system also remains constant. Therefore,

$$2mva + m(2v)(2a) + 0 = I\omega \qquad \text{(i)}$$

where I is the moment of inertia of the system after collision.

I = M.I. of the rod + M.I. of the mass $2m$ + M.I. of the mass m

or

$$I = \frac{(8m)(6a)^2}{12} + 2ma^2 + m\,(2a)^2$$

Substituting this value in Eq. (i), we get $\omega = \dfrac{v}{5a}$

   c. Total energy after collision

$$E = \frac{1}{2}I\omega^2 = \frac{1}{2}30ma^2 \times \left(\frac{v}{5a}\right)^2$$

or

$$E = \frac{3mv^2}{5}$$

3. Let the velocity of the CM of the system after the strike be v_c. By conservation of linear momentum,

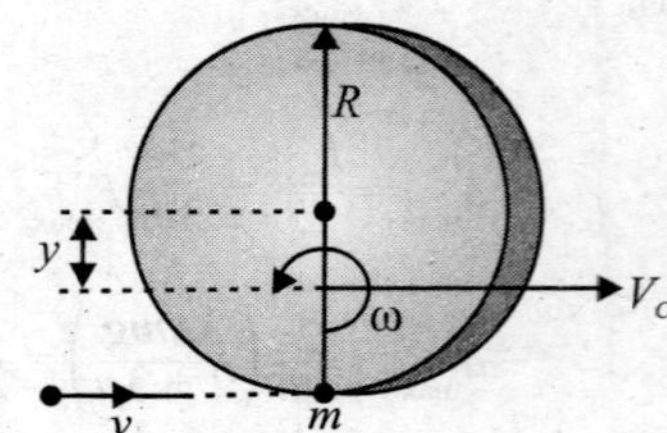

Fig. S-2.35

$$mv = 2mv_c \Rightarrow v_c = \frac{v}{2}$$

Position of CM, $y = \dfrac{m.O + m.R}{(m+m)} + m = \dfrac{R}{2}$

Using conservation of angular momentum about CM of the system (hoop + bullet),

$$mv\frac{R}{2} = I_c\,\omega$$

or

$$\frac{mvR}{2} = (I_{bullet} + I_{hoop})_c\,\omega$$

or

$$\frac{mvR}{2} = \left[m\left(\frac{R}{2}\right)^2 + \left\{mR^2 + m\left(\frac{R}{2}\right)^2\right\}\omega\right]$$

After solving, we get $\omega = \dfrac{v}{R}$

4.

Fig. S-2.36

By conservation of linear momentum and angular momentum, we have

$$mv + 0 = m.O + MV \Rightarrow = V - \frac{mv}{M}$$

and

$$mvd = I\omega \Rightarrow \omega = \frac{mvd}{I} = \frac{mvd}{\left(\dfrac{ML^2}{12}\right)}$$

Since collision is elastic, therefore

$$\frac{1}{2}mv^2 = \frac{1}{2}M\left(\frac{mv}{M}\right)^2 + \frac{1}{2}\left(\frac{ML^2}{12}\right)\left(\frac{mvd}{\frac{mL^2}{12}}\right)^2$$

$$\Rightarrow \quad m = \frac{ML^2}{L^2 + 12d^2}$$

5.

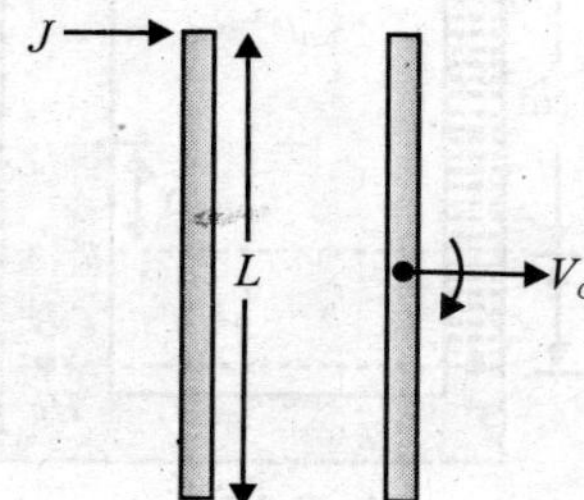

Fig. S-2.37

Let Δt be the duration of impact, then

$$J = \Delta\vec{P}$$

or $\quad J = (Mv_c - 0) \quad$ (i)

Also angular impulse is equal to change in angular momentum

or $$\frac{Jl}{2} = I\Delta\omega$$

$$= I(\omega - 0) = \frac{m\ell^2}{12}\omega \quad \text{(ii)}$$

Dividing Eq. (i) by Eq. (ii), we get

$$\omega = \frac{6v_0}{\ell}$$

Let t be the time in which the rod rotates at an angle 2π, then

$$\omega t = 2\pi \Rightarrow t = \frac{\pi L}{3v_0}$$

Hence, distance travelled by centre of mass $\Rightarrow x_c = \frac{\ell\pi}{3}$

6. Let vz be the velocity attained by its centre of mass and ω be the angular velocity about the centre of mass by the impact of the cue.

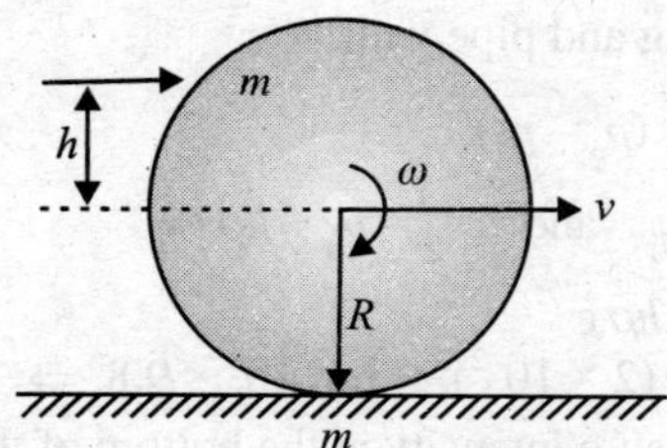

Fig. S-2.38

Suppose J is the impulse provided by the cue

$$J = m(v - o) \quad \text{(i)}$$

and $$J.h = I(\omega - 0) \quad \text{(ii)}$$

From Eqs. (i) and (ii), $Mv \times h = I\omega$

For pure rolling, $\quad \omega = \frac{v}{R}$

$$\therefore \quad mv \times h = \frac{2}{3}mR^2 \times \frac{v}{R}$$

$$\Rightarrow \quad h = \frac{2R}{3}$$

7. Figure S-2.39 shows the two situations of a rolling wheel before and after its striking the wall.

If C_2 is the point on the wall where the wheel strikes, then the impulsive force F passes through it. Moreover, the weight of the wheel mg and the normal reaction offered by the horizontal surface on the wheel to balance each other. So, the angular momentum of the wheel remains conserved about point C_2.

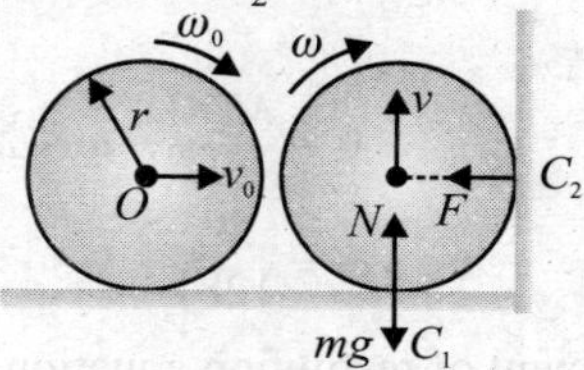

Fig. S-2.39

$\therefore$ Initial angular momentum = final angular momentum

$$\therefore \quad I\omega_0 = I\omega + mvr$$

[where ω and v are the angular and linear speeds of the wheel, just after its striking]

Since slippage is prevented, so $v = \omega r$

$$\therefore \quad I\omega_0 = I\omega + m\omega r^2$$

$$\Rightarrow \quad \frac{mr^2\omega_0}{2} = \frac{mr^2\omega}{2} + m\omega r^2$$

$$\Rightarrow \quad \omega = \frac{\omega_0}{3}$$

8. Clearly, the centre of mass of the system $(A + B + C)$ is at a distance of 10 cm from B along the rod.

Initial angular momentum

$$L_i = m_A v_A r_A$$

$$= 50 \times 10^{-3} \times 60 \times 10^{-2} \times 10 \times 10^{-2}$$

$$= 3 \times 10^{-3} \text{ S.I. unit}$$

Moment of inertia about new centre of mass

$$I = (2 \times 50 \times 10^{-3}) \times (10 \times 10^{-2})^2 + (50 + 10^{-3}) \times (20 \times 10^{-2})^2$$

$$= 3 \times 10^{-3} \text{ S.I. unit}$$

From $L_i = L_f \Rightarrow L_i = I.\omega$

$$\Rightarrow \quad \omega = \frac{L}{I}$$

$$\Rightarrow \quad \omega = 1 \text{ rad/s}$$

9. If 'O' is the centre of mass, then velocity of $A + B$

$$= v_1 = v_{1,0} + v_0 = \omega(10) + v_0$$

$\Rightarrow (10 + v_0)$ (towards right)

Velocity of $C = v_2 = v_{2,0} + v_0 = \omega(-20) + v_0$
$= (-20 + v_0)$ cm/s

(towards right)

$\therefore \quad v_{2,1} = v_2 - v_1 = -30$ cm/s

(towards right)

10. Conservation of linear momentum

$$mv_0 = M\frac{v_0}{8} - m\frac{v_0}{4}$$

$$8\,m = M - 2m \Rightarrow 10\,m = M$$

$$\Rightarrow \quad M/m = 10$$

Conservation of angular momentum about the point of collision

$$0 = \frac{M(81)^2}{12}\omega - M\frac{v_0}{81}$$

$$\Rightarrow \quad \omega = \frac{3v_0}{1281}$$

Using coefficient of restitution equation

$$\left[\left(\frac{v_0}{8} + \omega l\right) - \left(-\frac{v_0}{4}\right)\right] = I(v_0 - 0) \Rightarrow e = \frac{51}{1281}$$

Velocity at end A

$$v_A = \frac{v_0}{8} + \omega.4l = \frac{28}{128}v_0$$

$$\Rightarrow \quad v_A = \frac{7}{32}v_0$$

Velocity at end B

$$v_B = \frac{v_0}{8} + \omega.4l = \frac{4}{128}v_0$$

$$\Rightarrow \quad v_B = \frac{1}{32}v_0$$

Chapter 3

Exercise 3.1

1. Pressure at any depth h from the free surface of the water is given by $P = \rho g h + P_a$

$$= (1000) \times 10 \times 10 + 1.0 \times 10^5 = 2.00 \text{ N/m}^2$$

2. The barometric height, $h = 75$ cm

If l is the length of the mercury column in the tube, then $h/l = \sin 60°$

or $\quad l = \dfrac{h}{\sin 60°} = \dfrac{75}{(\sqrt{3}/2)} = 86.6$ cm

3. As the mercury column in the two arms of the U-tube is the same, so

$$P_A = P_B \Rightarrow \rho_w g h_w = \rho_s g h_s$$

$$\frac{\rho_s}{\rho_w} = \frac{h_w}{h_s} = \frac{10}{12.5} = 0.8$$

Thus, specific gravity of the spirit is 0.8.

4. Suppose area of cross section of narrow limb be a, then area of cross section of right limb will be $3(a)$

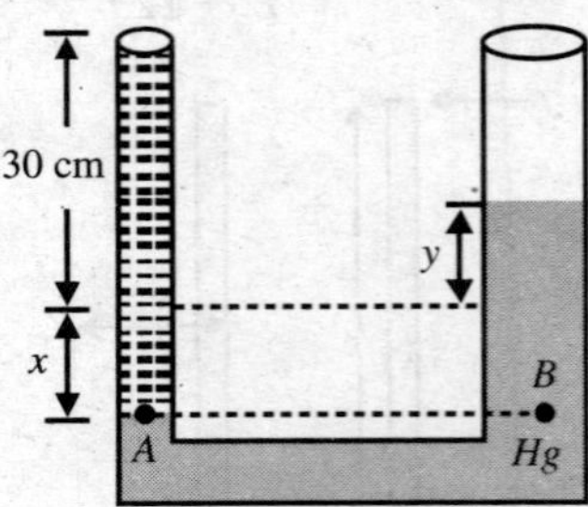

Fig. S-3.1

Let the level of mercury in left limb fall be x and the rise of level in right limb be y, then

$$ax = (3a)y$$
$$x = 3y$$

According to Pascal's law, $P_A = P_B$

$$(30 \times x)\rho_w g = (x + y)\rho_{Hg} g$$

$$(30 \times 3y) \times 1 \times g = (3y + y) \times 13.6 \times g$$

$$y = 0.58 \text{ cm}$$

5. The inclination of the pressure diagram with the vertical wall is proportional with the density of the liquid. Since $\rho_2 > \rho_1$, therefore inclination of the lower diagram is always more than that of the upper one.

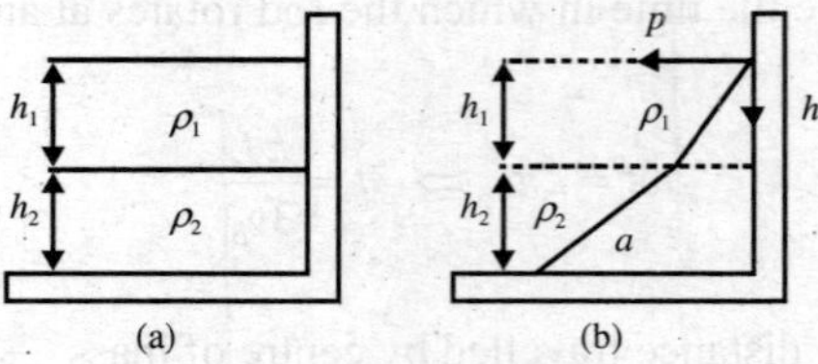

Fig. S-3.2

The diagram (a) is correct.

The diagrams (b) and (c) are not possible.

6. As plug secures the pipe opening the force of friction between plug and pipe wall,

$$F = A\,(p_2 - p_1)$$

But $p_1 = p_0$ and $p_2 = p_0 + h\rho g$

$$F = Ah\rho g$$

$$F = \pi(2 \times 10^{-2})^2 \times 6 \times 10^3 \times 9.8 \Rightarrow F = 74 \text{ N}$$

7. **(a)** The pressure intensity at the bottom of the container is

$$P = h\rho g = 0.20 \times 1000 \times 10 = 2 \times 10^3 \text{ N/m}^2$$

Force at the bottom of container

$= P \times \text{base area} = 2\times10^3 \times (20\times10^{-4})$
$= 4$ N (Force exerted by water only)

(b) Weight of the water in the container

$= \text{Volume} \times \text{density of water} \times g$

$= 0.5\times10^3 \times 1000 \times 10 = 5$ N

$P_2 > P_1$

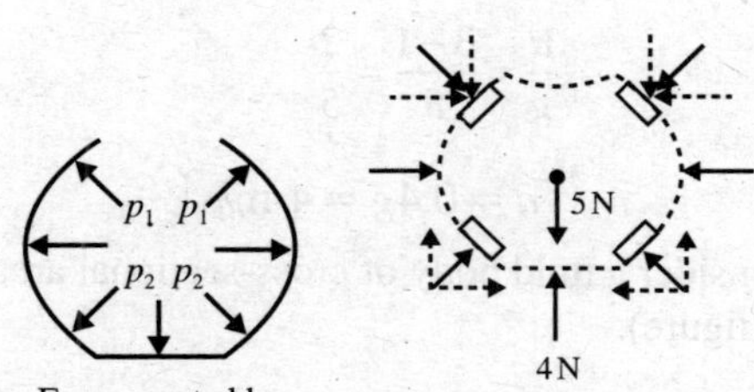

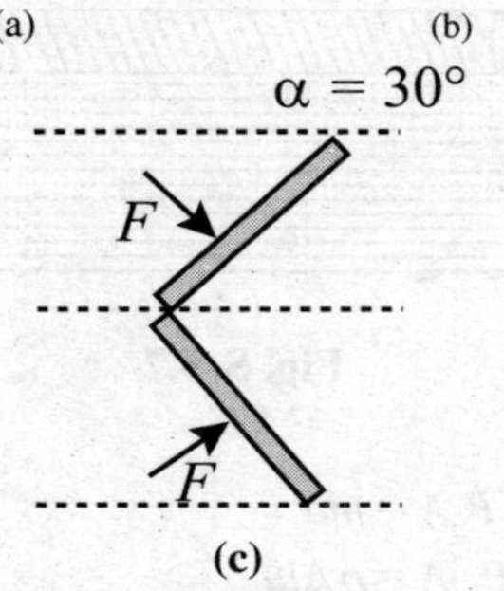

Fig. S-3.3

From the diagram, the force exerted by the container is in upward direction because ($P_2 > P_1$). Let force exerted by sides of container be F, then

$$4 + F = 5 \text{ N} \Rightarrow F = 1 \text{ N}$$

8. Area of the plate = 1/2 × base × altitude

= 1/2 × 3 × 3 = 4.5 m²

Density of oil = 0.8 × 1000 = 800 kg m^{-3}

Depth of CG. of the plate from the free surface

= (1/3) 3 = 1 m

Now, total thrust on the plate

= (plate area) × (pressure at the CG. of the plate)

= (4.5) (800 × 10 × 1) = 36 kN

9. Height of lock gates = 7.5 m
Width of canal = 16 m
Inclination of gates = 120° and H = 5 m
From the geometry of the lock gates, we find that inclination of the lock gates with the walls

$$a = (180° - 120°)/2 = 30°$$

and width of each gate $= \dfrac{16/2}{\cos\alpha} = \dfrac{8}{\cos 30°} = \dfrac{8}{0.866} = 9.24$ m

Therefore, wetted area of each gate,

$$A = 5 \times 9.24 = 46.2 \text{ m}^2$$

and force acting on each gate is

$$F = gA \times (H/2) = 9.81 \times 46.2 \times (5/2) = 1133 \text{ kN}$$

10. The force exerted by liquid on the gate is

$$F_H = P_{av} \times \text{vertical projection area of the gate}$$

$$= \left(\frac{P_A + P_B}{2}\right) \times (2Rl)$$

$$= \left(\frac{R\rho g + 3R\rho g}{2}\right) \times 2Rl \quad = 4\rho g R^2 l$$

11. Since pressure always gets normal to the surface, therefore the pressure diagram for an inclined wall is shown in Fig. S-3.4(a). The base of the triangular pressure diagram is equal to ρgH because pressure at a point depends on the height of liquid column above it.

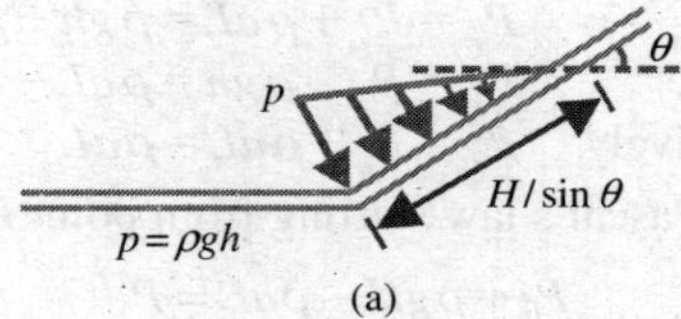

Fig. S-3.4(a)

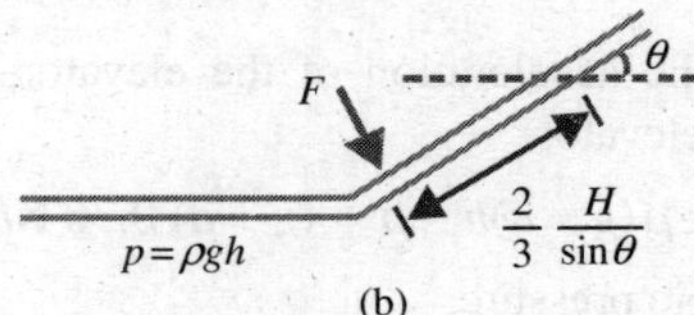

Fig. S-3.4(b)

The total force acting on the inclined wall is

$$\frac{F}{b} = \frac{1}{2}(\rho g H)\left(\frac{H}{\sin\theta}\right)^2 = \frac{1}{2}\rho g\frac{H^2}{\sin\theta}$$

The point of application of the total force is through the centroid of the pressure diagram as shown in Fig. S-3.4(b). Alternatively, the force on the inclined wall may be obtained in two parts: horizontal and vertical.
The horizontal force F_x acts on the vertical projection of the inclined wall.

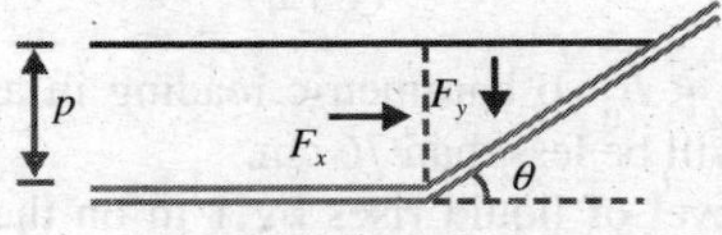

Fig. S-3.5

$$F_x = \frac{1}{2}\rho gbH^2$$

The vertical force F_v acts due to the weight of the liquid supported by the wall.

$$F_y = \frac{1}{2}\rho gb(H)(H\cot\theta)$$

$$= \frac{1}{2}\rho gbH^2\cot^2\theta$$

The magnitude of the resultant force is

$$F = \sqrt{F_x^2 + F_y^2} = \frac{1}{2}\rho gbH^2\text{cosec}\theta$$

$$\Rightarrow \quad \frac{F}{b} = \frac{1}{2}\rho g\frac{H^2}{\sin\theta}$$

Exercise 3.2

1. At point A, $P_A = 0$
   At point B, $P_B = P_A + \rho gh = \rho gh$
   At point C, $P_C = P_B + \rho aL = \rho gh + \rho gL$
   At point D, $P_D = P_C - \rho gh = \rho aL$
   Or alternatively, $P_D = P_A + \rho aL = \rho aL$
2. Applying Pascal's law starting from point B, we get
   $$P_B + \rho gH - \rho aL = P_A$$
   Since $P_B = P_A = P_{atm}$, therefore $a = g\left(\frac{H}{L}\right)$
3. Let a be the acceleration of the elevator, then pressure inside the elevator
   $$P = \rho(g+a)h = \rho \times (g+a)\,0.76 \text{ N/m}^2$$
   Atmospheric pressure,
   $$P_a = \rho g \times 0.76 \text{ N/m}^2$$
   Clearly, the air pressure inside the elevator will be greater than P_a, i.e., 76 cm of Hg.
4. Let a be the acceleration of the elevator and h be the barometer height, then
   $$P_a = \rho(g+a)h \quad \text{(i)}$$
   For the static elevator with barometric height h_0
   $$P_a = \rho gh_0 \quad \text{(ii)}$$
   From Eqs. (i) and (ii), we get
   $$h = \frac{gh_0}{(g+a)}$$
   Clearly, $h < h_0$ so barometric reading in an accelerating elevator will be less than 76 cm.
5. (i) The level of liquid rises by 1 m on the left side and falls by 1 m on the right side.

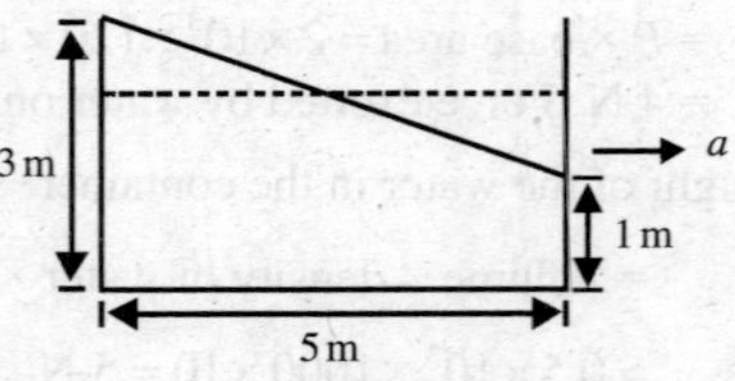

Fig. S-3.6

$$\frac{a}{g} = \frac{3-1}{5} = \frac{2}{5}$$

or $$a = 0.4g = 4 \text{ m/s}^2$$

6. We consider a fluid body of cross-sectional area A (shown in the figure).

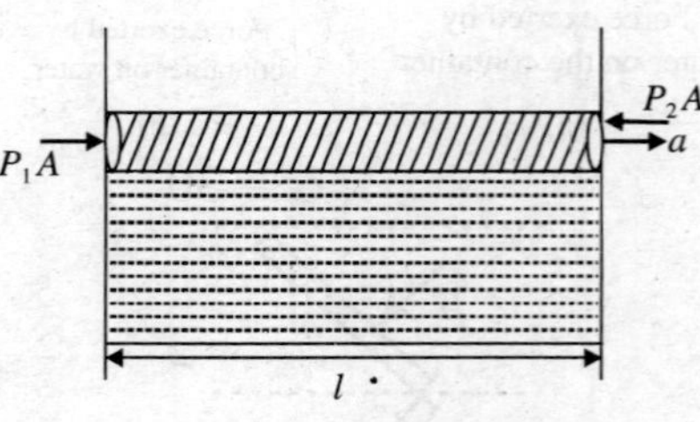

Fig. S-3.7

$$P_1A - P_2A = ma$$

or, $$(P_1 - P_2)A = \rho Ala$$

or, $$P_1 - P_2 = \rho la = 0.92 \times 10 \times 3 = 27.6 \text{ kPa}$$

7. The acceleration of the trolley is $a = g\sin\theta$ down the inclined plane
   $$a_x = a\cos\alpha = (g\sin\alpha)\cos\alpha$$
   $$a_y = a\sin\alpha = (g\sin\alpha)\sin\alpha = g\sin^2\alpha$$
   Let θ be the angle made by free surface of liquid, then
   $$\tan\theta = \frac{a_x}{g - a_y}$$
   $$= \frac{g\sin\alpha\cos\alpha}{g - g\sin^2\alpha} = \tan\alpha$$
   $$\theta = \alpha$$
8. (i) Since the original level of water before rotation is equal to the average of the minimum and maximum water level after rotation, therefore
   $$h = \frac{+H}{2}$$
   (ii) The equation of the parabolic surface is
   $$y = \frac{\omega^2 r^2}{2g} \quad (\because y_0 = 0)$$
   We know that for $r = R$, $y = H$, thus

$$H = \frac{\omega^2 R^2}{2g} \quad \text{or} \quad \omega = \frac{\sqrt{2gH}}{R}$$

9. (i) We know that $y_{max} = H = 3$ m,

$$h = 2/3\ H = 2 \text{ m},$$
$$R = 1 \text{ m},\ g = 10 \text{ m/s}^2$$

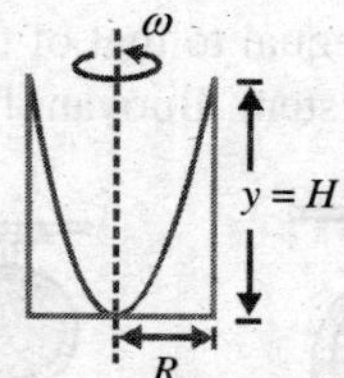

Fig. S-3.8

$$\text{Thus, } \omega = \sqrt{\frac{4g(H-h)}{R^2}} = \sqrt{\frac{4(10)(3-2)}{1^2}} = \sqrt{40} \text{ rad/s}$$

(ii) We know that

$$y = \frac{\omega r^2}{2g} = y_0$$

Here, $y = H = 3$ m, $y_0 = 0$, $r = R = 1$ m, $g = 10$ m/s^2

$$\omega = \sqrt{\frac{2gH}{R^2}} = \sqrt{\frac{2(10)(3)}{1^2}} = \sqrt{60} \text{ rad/s}$$

10. Let A and B be the two free liquid levels in the two limbs. If P_0 be the atmosphere pressure, then starting from point A and reaching point B through the tube, we have

$$P_0 + h_1\rho_1 g - l\rho_2 a - h_2\rho_2 g = P_0$$

$$\Rightarrow \quad h_1\rho_1 g = l\rho_2 a + h_2\rho_2 g = \rho_2(la + h_2 g)$$

$$\Rightarrow \quad \frac{\rho_1}{\rho_2} = \frac{la + h_2 g}{h_1 g}$$

11.

$$\text{Equation of parabola } y = \frac{\omega^2 x^2}{2g} \quad \text{(i)}$$

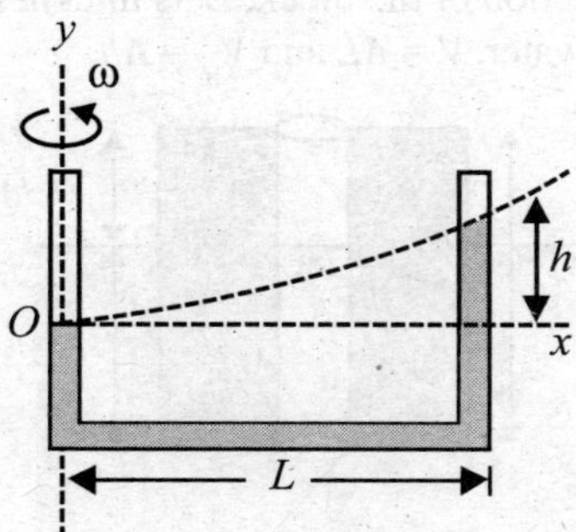

Fig. S-3.9

At A, coordinate will be (L, h)

Substituting in (i) $h = h = \dfrac{\omega^2 L^2}{2g}$

Exercise 3.3

1. (a) If Mg of ice containing an air bubble is floating in water,

$$Mg = V_D \sigma_W g$$

i.e., $\quad V_D = (M/\sigma_W)$

[as air bubble encloses air which is weightless]

Now when Mg ice melts, water formed will be M g and it will occupy volume $V_F = (M/\sigma_W)$. As the volume of water formed by melting of ice is equal to the volume of water displaced by the floating ice, the level of water in the beaker will remain unchange(d)

(b) Let the mass of ice be Mg and that of embedded piece m, then for floating

$$\text{Weight} = \text{upthrust, i.e., } \quad (M + m)g = V_D \sigma_W g$$

So the volume of water displaced by the system when floating

$$V_D = \frac{M}{\sigma_W} + \frac{m}{\sigma_W} \quad \text{(i)}$$

Now when ice melts, water formed will be Mg and so the volume of water formed by melting of ice

$$V_1 = (M/\sigma_W)$$

(i) If the ice cube initially contains the lead piece, then as $\rho_{pb} > \sigma_W$, it will sink when ice melts and so the water displaced by it will be equal to its own volume.

$$V_2 = (m/\rho_{pb}) \quad [\text{as } \rho = m/V]$$

So finally total volume occupied by water formed by melting of ice and lead piece

$$V_1 + V_2 = \frac{M}{\sigma_W} + \frac{m}{\rho_{pb}} \quad \text{(ii)}$$

Now as $\rho_{pb} > \sigma_W$, this volume will be lesser than initial water displaced by the floating system (Eq. i); so the level of water in beaker will go down as the ice melts if the ice contains a sinking impurity.

(ii) If the ice initially contains a cork piece, then as $\rho_{rock} < \sigma_W$, the cork will float on water when ice melts. So the water displaced by the floating cork piece will be

$$mg = V_2' \sigma_W g$$

$$\Rightarrow \quad V_2' = m/\sigma_W$$

[and not its own volume (m/ρ_{rock})]

So in this case, water formed by melting of Mg of ice and water displaced by floating cork,

$$V_1 + V_2' = \frac{M}{\sigma_W} + \frac{m}{\sigma_W} \quad \text{(iii)}$$

and as this is same as initial volume displaced by the floating system (Eq. i), the level of water in the beaker will remain unchange(d)

2. If M is the mass of boat and m of pieces in it, then initially as the system is floating,

$$(M+m)g = V_D \sigma_W g$$

i.e., the system displaces water

$$V_D = \frac{M}{\sigma_W} + \frac{m}{\sigma_W}$$

When the pieces are dropped in the pond, the boat will still float, so it displaces water $M = V_1 \sigma_W$, i.e., $V_1 = (M/\sigma_W)$

(a) Now if the unloaded pieces float in the pond, the water displaced by them

$$m = V_2 \sigma_W$$

$$\Rightarrow \quad V_2 = (m/\sigma_W) \quad \text{(i)}$$

So the total water displaced by the boat and the floating pieces,

$$V_1 + V_2 = \frac{M}{\sigma_W} + \frac{m}{\sigma_W} \quad \text{(ii)}$$

which is same as the water displaced by the floating system initially (Eq. i); so the level of water in the pond will remain unchange(d)

(b) Now if the unloaded pieces sink, the water displaced by them will be equal to their own volume, i.e.,

$$V_2' = \frac{m}{\rho} \quad \left[\text{as } \rho = \frac{m}{V}\right]$$

and so in this situation the total volume of water displaced by the boat and sinking pieces will be

$$V_1 + V_2' = \left(\frac{M}{\sigma_W} + \frac{m}{\rho}\right) \quad \text{(iii)}$$

Now as the pieces are sinking $\rho > \sigma_W$, so this volume will be lesser than initial water displaced by the floating system (Eq. i); so the level of water in the pond will go down (or fall).

3. For a floating body in two liquids,
Weight of the body = Buoyant force of liquid I + buoyant force of liquid II

$$V\rho g = V_1 \rho_1 g + V_2 \rho_2 g$$

$$V\rho = V_1 \rho_1 + V_2 \rho_2 \quad \text{(i)}$$

Also, $$V = V_1 + V_2 \quad \text{(ii)}$$

After solving Eqs. (i) and (ii), we get

$$V_1 = V\left(\frac{\rho - \rho_2}{\rho_1 - \rho_2}\right) \text{ and } V_2 = V\left(\frac{\rho_1 - \rho}{\rho_1 - \rho_2}\right)$$

4. Let the volume of the rock be V, then magnitude of its weight.

$$w = \rho_{\text{rock}} V g$$

In water, rock experiences an additional buoyant force with a magnitude equal to that of the weight of the water displaced by the system. Buoyant force has magnitude.

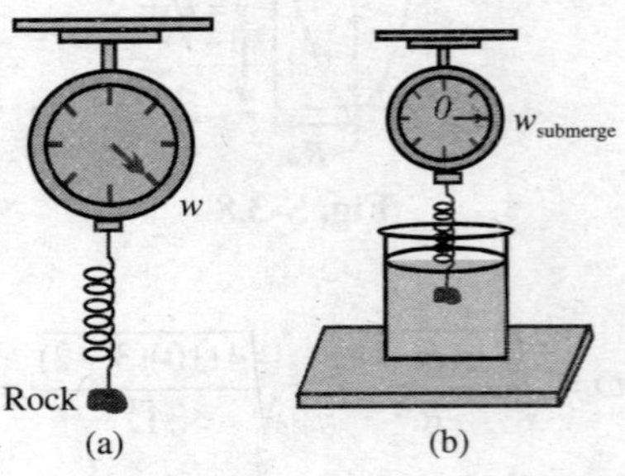

Fig. S-3.10

$$F_{\text{buoy}} = \rho_{\text{water}} V g$$

The new scale reading is the total downward force on the system when it is in the water.

$$w_{\text{submerge}} = w - F_{\text{buoy}}$$
$$= \rho_{\text{rock}} V g - \rho_{\text{water}} V g$$

From the ratio $$\frac{w_{\text{submerge}}}{w} = \frac{\rho_{\text{rock}} V g - \rho_{\text{water}} V g}{\rho_{\text{rock}} V g} = \frac{\rho_{\text{rock}} - \rho_{\text{water}}}{\rho_{\text{rock}}}$$

Solving for the density of the system (i.e., the rock)

$$\rho_{\text{rock}} = \frac{w}{w - w_{\text{submerge}}} \rho_{\text{water}}$$

5. As ice is floating, $W = Th$, i.e., $V\rho g = V_{in} \sigma_W$. Now if A is the cross section of the block, L its thickness and out of L, h is inside water, $V = AL$ and $V_{in} = Ah$.

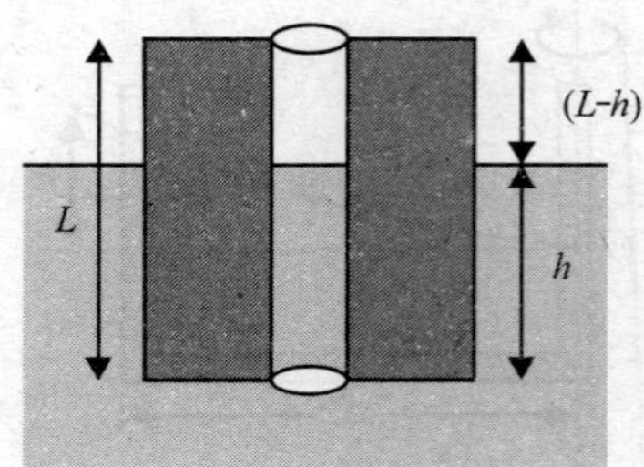

Fig. S-3.11

$$AL\rho_{\text{ice}} = hA\rho_{\text{water}}$$

$$h = \frac{\rho_{ice}}{\rho_{water}} L = 0.9 \times 5 = 4.5 \text{ m}$$

i.e., 4.5 m of ice will be submerged in water. So the level of water in the hole will be 5 – 4.5 = 0.5 m below the top of ice and hence the length of rope required to scoop up water is 0.5 m.

6. Volume of the cavities V_{cav} can be determined by taking difference between the volume V_{cast} of the casting as a whole and the volume of the iron in the casting.

$$V_{cav} = V_{cast} - V_{iron}$$

$$V_{iron} = \frac{W}{\rho_{iron} g}$$

W is the weight of casting; $W_{eff} = W - \rho_w g V_{cast}$

$$V_{cast} = \frac{W - W_{eff}}{\rho_w g} \Rightarrow V_{cav} = \frac{W - W_{eff}}{\rho_w g} - \frac{W}{g\rho_{iron}}$$

$$= \frac{6000 - 4000}{(10^3)(10)} - \frac{6000}{(10)(8 \times 10^3)} = 0.125 \text{ m}^3$$

7. Let m be the mass of copper in neckless.

Mass of gold = $(50 - m)$

Volume of copper = $V_1 = m/10$

Volume of gold = $V_2 = \dfrac{50 - m}{20}$

When immersed in water, $\rho_w = 1$ g/cm³

Decrease in weight = upthrust

$$(50 - 46)\, g = (V_1 + V_2)\rho_w\, g$$

$$4 = \frac{m}{10} + \frac{50 - m}{20}$$

$$m = 30 \text{ g}$$

8. For equilibrium, $(m_1 + m_2)\, g = \rho V g$

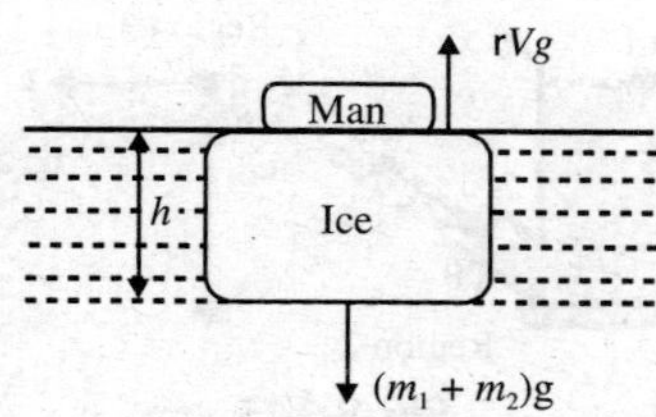

Fig. S-3.12

Here, m_1 = mass of man = 100 kg

m_2 = mass of ice = 0.917 × 1000 V = 917 V

ρ = 1000 kg/m³, h = 0.5 m

$\therefore$ 100 g + 917 $Vg = \rho Vg$ = 1000 Vg

$$V = \frac{100g}{1000g - 917g} \Rightarrow Ah = \frac{100}{83}$$

$$\therefore \quad A = \frac{100}{83 \times 0.5} = 2.41 \text{ m}^2$$

9. When a body is immersed in a liquid, it experiences an upthrust u, which makes spring balance S_1 to read

Apparent weight = true weight – upthrust

$= W_A - u = 2.5$ kg

S_2 reads: Weight of B + weight of L + reaction = upthrust of A

$W_B + W_L + u = 7.5$ kg weight

$W_B = 1$ kg weight, $W_L = 1.5$ kg weight

$1 + 1.5 + u = 7.5$ kg weight

$u = 5$ kg weight

$W_A = u + 2.5 = 7.5$ kg weight.

10. The gravity on moon is about one-sixth of that on the earth, i.e., $g_m = g/6$

On earth, water and air both exert buoyant force on the ball, but on moon as there is no air so weight of the ball is balanced only by buoyant force of the water. Thus

On earth, $\quad mg = V_w \rho_w g + V_a \rho_a g$

or $\quad m = V_w \rho_w + V_a \rho_a \quad$ (i)

On moon,

$$m\left(\frac{g}{6}\right) = V'_w \rho_w \left(\frac{g}{6}\right) \quad \text{(ii)}$$

or $\quad m = V'_w \rho_w$

From Eqs. (i) and (ii), we have $V'_w = V_w + \dfrac{V_a \rho_a}{\rho_w}$

Clearly $V'_w > V_w$. That is, the volume of the ball immersed in water on the moon will be greater than that on the earth. Hence ball will sink slightly more in water on the moon.

Exercise 3.4

1. Work done by per unit volume by force of gravity

= –change in potential energy per unit volume

$\therefore W = -\Delta U = -\rho g\,(h_2 - h_1) = -\rho g\,(6 - 3)$

$= -3 \times 500 \times 10 = -15 \times 10^4$ J/m³

2. Since the flow is horizontal, so the reduced form of Bernoulli's equation can be applied as

$$P_1 + \frac{1}{2}\rho v_1^2 = P_2 + \frac{1}{2}\rho v_2^2$$

Given $P_1 = 50 \times 13.6 \times 1000 = 6.8 \times 10^5$ dyne/cm²

and $P_2 = 20 \times 13.6 \times 1000 = 2.72 \times 10^5$ dyne/cm²

and $\rho = 1$ g/cc and $v_1 = 500$ cm/s

$$\therefore 6.8 \times 10^5 + 1/2 \times 1 \times (500)^2$$
$$= 2.72 \times 10^5 + 1/2 \times 1 \times v_2^2$$

$$\Rightarrow \frac{1}{2}(v_2^2 - 250000) = 4.08 \times 10^5 \Rightarrow v_2 \approx 10.25\ m/s$$

From the principle of continuity, $a_1 v_1 = a_2 v_2$

$$\therefore \frac{a_1}{a_2} = \frac{v_2}{v_1} = \frac{10.25}{5} = \frac{41}{20}$$

3. Applying Bernoulli's theorem to the lower and upper surface of a wing

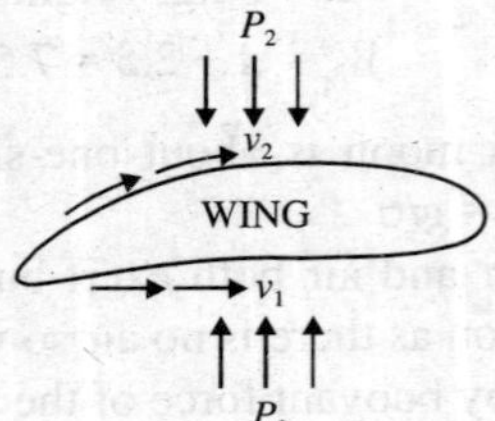

Fig. S-3.13

$$P_1 + \frac{1}{2}\rho v_1^2 = P_2 + \frac{1}{2}\rho v_2^2$$

or, $$P_1 - P_2 = \frac{1}{2}\rho(v_2^2 - v_1^2)$$

$$\Rightarrow \quad 1000 = \frac{1}{2} \times 1.30(v_2^2 - 50^2)$$

$$[\because \text{lift} = P_1 - P_2]$$

or $v_2^2 - 2500 = 2000/1.3$

or, $v_2 = 63.55$ m/s

4. (a) Figure S-3.14 illustrates the situation. From continuity equation, we can find the velocity of the fluid at the nozzle.

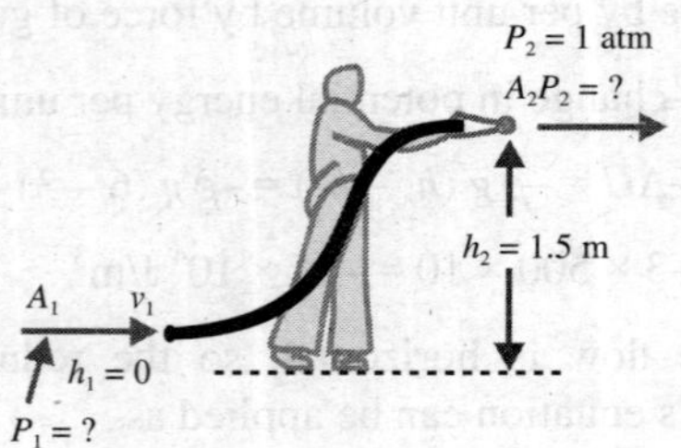

Fig. S-3.14

$$v_2 = \frac{A_1}{A_2} v_1 = \left(\frac{3.60}{0.250}\right)(50.0) = 720 \text{ cm/s}$$
$$= 7.20 \text{ m/s}$$

(b) Thrust force experienced by gardener at hand due to flow of water

$$F_{\text{thrust}} = \rho A_2 v_2^2$$
$$= 1000 \times 0.25 \times 10^{-4} \times (7.2)^2 = 1.3 \text{ N}$$

(c) We will apply Bernoulli's equation to determine pressure P_1.

$h_1 = 0.00$ m, $h_2 = 1.50$ m, $P_2 = 1.01 \times 10^5$ Pa,
$v_1 = 0.50$ m/s, $v_2 = 7.20$ m/s, $\rho = 1.00 \times 10^3$ kg/m³

$$P_1 + \frac{1}{2}\rho v_1^2 + \rho g h_1 = P_2 + \frac{1}{2}\rho v_2^2 + \rho g h_2$$

$$P_1 = P_2 + \frac{1}{2}\rho(v_2^2 - v_1^2) + \rho g(h_2 - h_1)$$

$$= (1.01 \times 10^5) + \frac{1}{2}(1.00 \times 10^3)[(7.20)^2 - (0.50)^2]$$

$$+ (1.00 \times 10^3)(9.80)(1.50 - 0.00)$$

$$= 1.41 \times 10^5 \text{ Pa}$$

5. We apply Bernoulli's equation between points 2 and 3.

$$P_2 + \frac{1}{2}\rho v_2^2 + \rho g y_2 = P_3 + \frac{1}{2}\rho v_3^2 + \rho g y_3$$

$$P_{\text{atm}} + \frac{1}{2}\rho v_2^2 + 0 = P_{\text{atm}} + \frac{1}{2}\rho v_3^2 + \rho g y \qquad \text{(i)}$$

As both the points are open to atmosphere, $P_2 = P_3 = P_{\text{atm}}$. Taking ground level as reference level, potential energy at point 2 is zero.

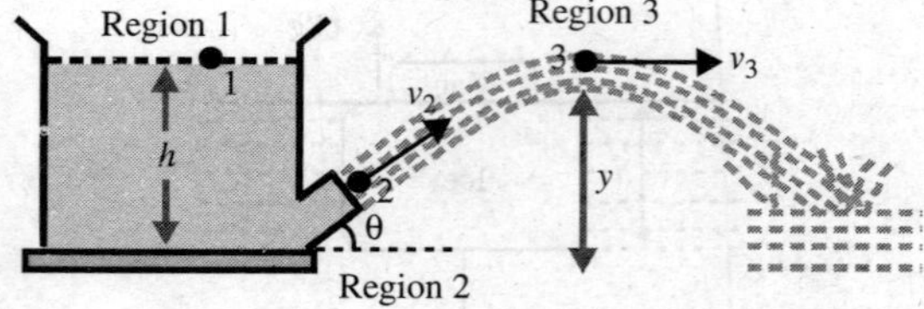

Fig. S-3.15

Since there is no horizontal acceleration, $v_3 = v_2 \cos\theta$. From Eq. (i) we obtain

$$y = \frac{1}{2}\frac{(v_2^2 - v_2^2\cos^2\theta)}{g} = \frac{1}{2}\frac{v_2^2 \sin^2\theta}{g} \qquad \text{(ii)}$$

From Torricelli's theorem, $v_2^2 = 2gh$

That, $y = h\sin^2\theta$

$y = h$ when $\sin\theta = 1$, i.e., $\theta = 90°$, therefore y can never exceed h.

6. For horizontal pipe, gravitational head is same at two points, so using Bernoulli's theorem,

$$h_1 + \frac{P_1}{\rho g} + \frac{1}{2}v_1^2 = h_2 + \frac{P_2}{\rho g} + \frac{1}{2}\frac{v_2^2}{g}$$

(P_1 and P_2: pressures at two points, V_1 and V_2: velocities of water at two points)

$$\frac{P_1}{\rho} + \frac{1}{2}V_1^2 = \frac{P_2}{\rho} + \frac{1}{2}V_2^2$$

$\Rightarrow$ $P_1 = 10$ cm of Hg $= h\rho g$

$= 10 \times 13.6 \times 980$ CGS units

$V_1 = 40$ cm/s; $V_2 = 50$ cm/s and $\rho = 1$ g/cc

$$\Rightarrow \frac{10\times13.6\times980}{1} + \frac{1}{2}(40)^2 = \frac{P_2}{1} + \frac{1}{2}(50)^2$$

$P_2 = 1.328\times10^5$ dynes/cm^2

$P_2 = 9.97$ cm of Hg

7. Diameter of tube at $A = 60$ cm

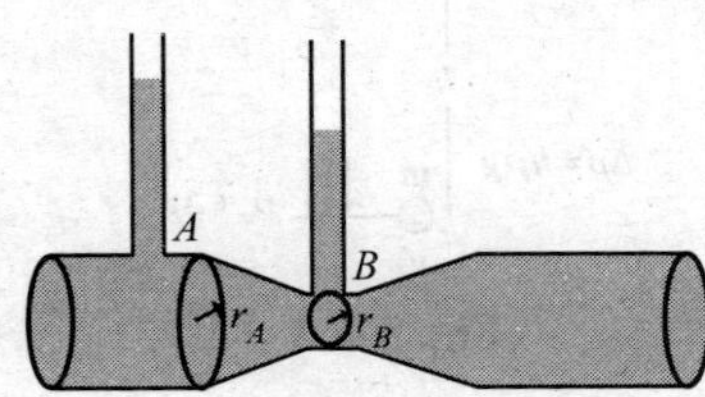

Fig. S-3.16

Radius at $A = r_A = 30$ cm, radius at $B = r_B = 10$ cm,

area of cross section at $A = \pi(30)^2$ cm^2 $= A_1$,

area of cross section at $B = \pi(10)^2$ cm^2 $= A_2$

$P_A - P_B = 100$ cm of water $= 100 \times 1 \times 980$ dynes/cm^2

Using the formula $Q = A_1A_2\sqrt{\dfrac{2(P_A - P_B)}{(A_1^2 - A_2^2)d}}$

substitute the values

$$Q = (900\pi)(100\pi)\sqrt{\frac{2\times(100\times1\times980)}{(900^2\pi^2 - 100^2\pi^2)}}$$

$\Rightarrow$ $Q = 1.4\times10^5$ cm^3/s $= 140$ l/s

8. In Pitot tube, velocity of liquids is given by

$$V = \sqrt{2\times1000\times20} = 200 \text{ m/s}$$

$$Q = Av = \pi r^2 v$$

$$= \frac{22}{7}\times7^2\times200 = 30800 \text{ cm}^3/\text{s} = 30.8 \text{ l/s}$$

9. For gases, velocity is given by

$$v = \sqrt{\frac{2h\rho g}{d}} = \sqrt{\frac{2\times0.1\times1000\times9.8}{1.3}} = 38.83 \text{ m/s}$$

10.

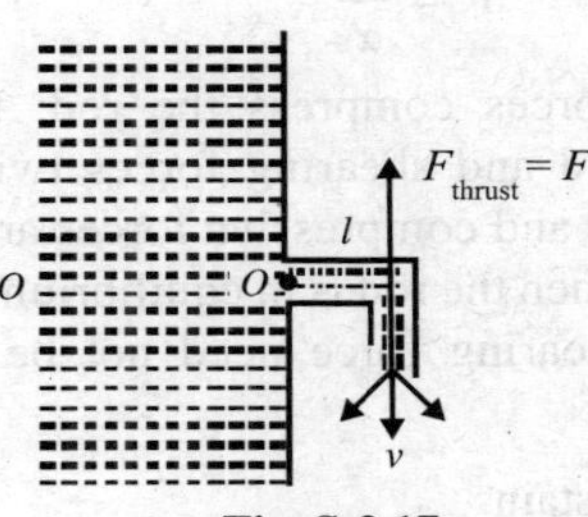

Fig. S-3.17

Velocity of flow $= v = \dfrac{Q}{\pi r^2}$

Reaction force $F = \rho A v^2$

$$\Rightarrow F = \rho\pi r^2\left(\frac{Q}{\pi r^2}\right)^2 = \frac{Q^2}{\pi r^2}\rho$$

Torque about $O = Fl = \dfrac{Q^2\rho l}{\pi r^2}$

Chapter 4

Exercise 4.1

1. Pressure is the external force per unit area, while stress is the internal force that comes into play from within a strained body acting transversely per unit area of the body.

2. (a) A, because the Young's modulus is given by the slope of the stress-strain plot. The slope is greater for (a)

(b) A, because ductility is given by the extent of plastic deformation. This is greater for (a)

(c) B, because for a brittle material the plastic region is small.

(d) A, because stress required for fracture is greater for (a).

3. (a) False. Elastic forces are not always conservative. Elastic forces are conservative as long as the loading and deloading curves are coincident, even if the curves are not linear. For example, rubber.

(b) False, because when Hook's law is obeyed, elastic forces are certainly conservative, but even when Hook's law is not obeyed but loading and unloading curves are coincident, then too, elastic forces are conservative.

(c) False, because a part of the work done is stored as permanent energy in producing permanent set.

4. Since the shaded segment is in translational equilibrium at rest; $\sum F = |F - F'| = 0$ in both the figures.

This gives $F' = F$.

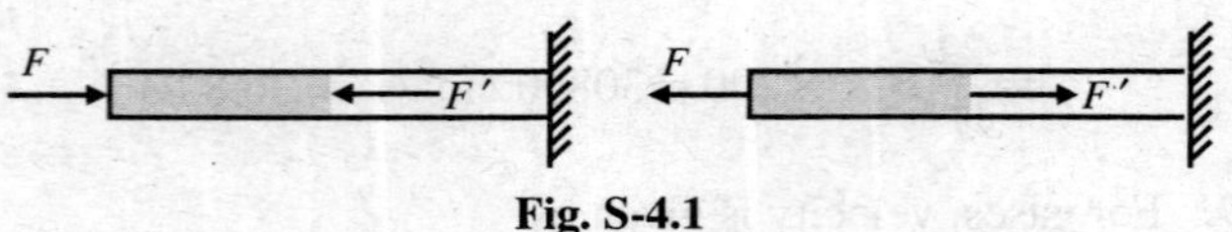

Fig. S-4.1

Since the rod is in rotational equilibrium,

$$\sum \tau = F'x - Fl = 0$$

This gives, $\quad F' = \dfrac{Fl}{x}$

Compressive forces compress the rod; tensile forces elongate the rod and shearing forces twist (bend) the rod. The tensile and compressive forces are equal to the applied force when the rod is in equilibrium horizontally, whereas the shearing force need not be equal to the applied force.

5. The volumetric stain

$$\frac{\delta V}{V} = -\frac{P}{B}$$

where $\quad P = \rho g h$

then $\quad -\dfrac{\delta V}{V} = \dfrac{\rho g h}{B} \qquad$ (i)

Since, the volume of the sphere is $V = 4/3\pi r^3$, we have

$$\frac{\delta V}{V} = \frac{3\delta r}{r} \qquad \text{(ii)}$$

Using Eqs. (i) and (ii), we have

$$\frac{\delta r}{r} = \frac{\rho g h}{3B}$$

6. Longitudinal or normal stress

$$\sigma_1 = \frac{100 \sin 30°}{5 \times 2} = 5 \text{ N/m}^2$$

Tangential stress

$$\sigma_t = \frac{100 \cos 30°}{5 \times 2} = 5\sqrt{3} \text{ N/m}^2$$

7.

$$p_{\text{gas}} = \frac{mg}{A} + p_a = \frac{50 \times 10}{50 \times 10^{-4}} + 1 \times 10^5$$

Fig. S-4.2

$$= 2 \times 10^5 \text{ N/m}^2$$

Bulk stress $= p_{\text{gas}} = 2 \times 10^5 \text{ N/m}^2$

8.

$$\Delta L_B = \Delta L_{AB} = \frac{FL}{AY} = \frac{MgL}{AY}$$

$$= \frac{10 \times 10 \times 0.1}{10^{-7} \times 2.5 \times 10^{10}} = 4 \times 10^{-3} \text{ m} = 4 \text{ mm}$$

$$\Delta L_C = \Delta L_B + \Delta L_{BC}$$

$$= 4 \times 10^{-3} + \frac{100 \times 0.2}{10^{-7} \times 4 \times 10^{10}}$$

$$= 4 \times 10^{-3} + 5 \times 10^{-3} = 9 \text{ mm}$$

$$\Delta L_D = \Delta L_C + \Delta L_{CD}$$

$$= 9 \times 10^{-3} + \frac{100 \times 0.15}{10^{-7} \times 1 \times 10^{10}}$$

$$= 9 \times 10^{-3} + 15 \times 10^{-3} = 24 \text{ mm}$$

9.

$$B = \frac{\Delta p}{-\dfrac{\Delta V}{V}} \Rightarrow \frac{\Delta V}{V} = -\frac{\Delta p}{B}$$

We know $\quad p = p_{\text{atm}} + h\rho g$

or $\quad m = \rho V = \text{const.}$

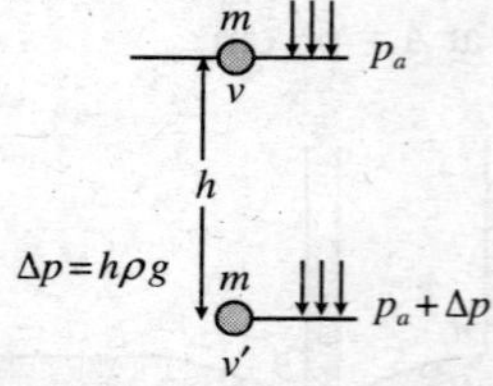

Fig. S-4.3

$$d\rho\, v + dv\rho = 0$$

$$d\rho\, V + dV\rho = 0$$

$$\Rightarrow \quad \frac{d\rho}{\rho} = -\frac{dV}{V}$$

i.e., $\quad \dfrac{\Delta\rho}{\rho} = \dfrac{\Delta p}{B}$

$$\Rightarrow \quad \frac{\Delta\rho}{\rho} = \frac{1}{100}$$

$$\frac{1}{100} = \frac{h\rho g}{B} \qquad \text{[assuming } \rho = \text{const.]}$$

$$h\rho g = \frac{B}{100} = \frac{1}{100\,k}$$

$$h\rho g = \frac{1 \times 1 \times 10^5}{100 \times 50 \times 10^{-6}}$$

$$h = \frac{10^5}{5000 \times 10^{-6} \times 1000 \times 10}$$

$$= \frac{100\times10^3}{50} = 2\,\text{km}$$

10. $L = 5\times10^{-2}\,\text{m} \Rightarrow \frac{F}{A} = \eta\frac{x}{L}$

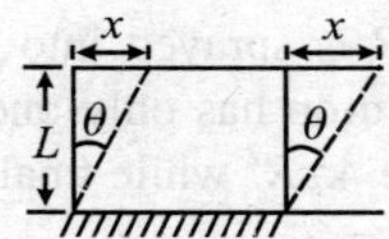

Fig. S-4.4

Strain θ

$$\theta = \frac{F}{A\eta} = \frac{1800}{25\times10^{-4}\times2.4\times10^6} = \frac{180}{25\times24} = \frac{3}{10}$$

$= 0.3$ radian

$$\frac{x}{L} = 0.3$$

$$\Rightarrow x = 0.3\times5\times10^{-2}\,\text{m} = 1.5\times10^{-2}\,\text{m} = 1.5\,\text{mm}$$

11. $k_{eq} = \frac{k_1k_2}{k_1+k_2} \Rightarrow T = 2\pi\sqrt{\frac{m}{k_{eq}}} = 2\pi\sqrt{\frac{m(k_1+k_2)}{k_1k_2}}$

where $k_1 = \frac{A_1y_1}{l_1}$ and $k_2 = \frac{A_2y_2}{l_2}$

12. Applying energy conservation

$$mg(h+x) = \frac{1}{2}\frac{k_1k_2}{k_1+k_2}x^2$$

where $k_1 = \frac{A_1y_1}{l_1}$ $k_2 = \frac{A_2y_2}{l_2}$

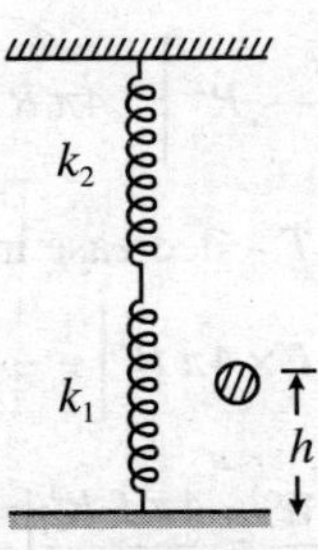

Fig. S-4.5

and $K_{eq} = \frac{A_1A_2y_1y_2}{A_1y_1l_2 + A_2y_2l_1}$

$$k_{eq}x^2 - 2mgx - 2mgh = 0$$

$$x = \frac{2mg \pm \sqrt{4m^2g^2 + 8mghk_{eq}}}{2k_{eq}}$$

$$x_{max} = \frac{mg}{k_{eq}} + \sqrt{\frac{m^2g^2}{k_{eq}^2} + \frac{2mgh}{k_{eq}}}$$

Exercise 4.2

1. $F - F_T = ma_{res}$

As boat moves with constant velocity $a_{res} = 0$

$$F = F_T$$

We know $F_T = \eta A\frac{dv}{dy}$, but $\frac{dv}{dy} = \frac{v_0 - 0}{D} = \frac{v_0}{D}$

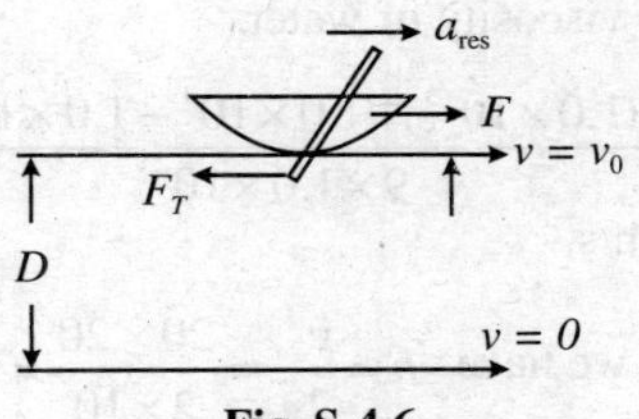

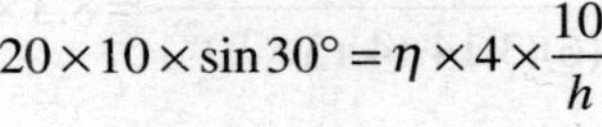

Fig. S-4.6

then $F = F_T = \frac{\eta A v_0}{D}$

2.

$$F = F' = \eta A\frac{dv}{dy} = mg\sin\theta \quad \left[\frac{dv}{dy} = \frac{v}{h}\right]$$

$$20\times10\times\sin30° = \eta\times4\times\frac{10}{h}$$

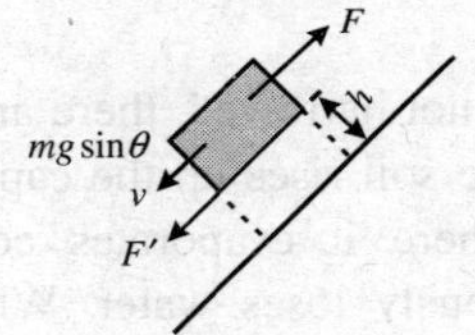

Fig. S-4.7

$$h = \frac{40\times10^{-2}}{100}\eta \quad [\eta = 10^{-1}\text{ poise} = 10^{-2}\text{ Ns/m}^2]$$

$= 4\times10^{-3}$ m $= 4$ mm

3. By Stoke's law, the terminal velocity of a water drop of radius r is given by

$$v = \frac{2}{9}\frac{r^2(\rho-\sigma)g}{\eta}$$

where ρ is the density of water, σ is the density of air and h the coefficient of viscosity of air. Here σ is negligible and $r = 0.0015$ mm $= 1.5\times10^{-3}$ mm $= 1.5\times10^{-6}$ m. Substituting the values:

$$v = \frac{2}{9}\times\frac{(1.5\times10^{-6})^2\times(1.0\times10^3)\times9.8}{1.8\times10^{-5}}$$

$= 2.72\times10^{-4}$ m/s

4. The velocity attained by the sphere in falling freely from a height h is

$$v = \sqrt{2gh} \qquad \text{(i)}$$

This is the terminal velocity of the sphere in water. Hence by Stoke's law, we have

$$v = \frac{2}{9}\frac{r^2(\rho - \sigma)g}{\eta}$$

where r is the radius of the sphere, ρ is the density of the material of the sphere.
σ (= 1.0×10^3 kg/m^3) is the density of water and η is coefficient of viscosity of water.

$$\therefore \quad v = \frac{2\times(1.0\times10^{-3})^2(1.0\times10^4 - 1.0\times10^3)\times10}{9\times1.0\times10^{-3}}$$

$$= 20 \text{ m/s}$$

From Eq. (i), we have $h = \frac{v^2}{2g} = \frac{20\times20}{2\times10} = 20\,\text{m}$

5. $F = \eta A\frac{dy}{dt} = \eta A\frac{v}{h}$

$$= 15\times2\times10^4\times\frac{10}{0.25} = 1.2\times10^7 \text{ dyne}$$

6. $F = 6\pi\eta r v$

$$\eta = \frac{F}{6\pi r v} = \frac{3.14}{(6)(3.14)(4)(5\times10^2)} = 8.3\times10^{-5} \text{ poise}$$

Exercise 4.3

1. When the soil is not harrowed, there are large capillaries in it. Water in the soil rises up the capillary holes to the surface from where it evaporates continuously. Thus the soil continuously loses water. When harrowed, all these capillaries are destroyed, and so capillary suction stops altogether. Thus, water is retained longer in the soil.
2. No, the water will not flow out. When the tube is taken out, a convex meniscus is formed and a force due to surface tension comes into play in the upward direction which keeps the water in the tube. Thus the length of water column remaining in the tube will be $l + h$.

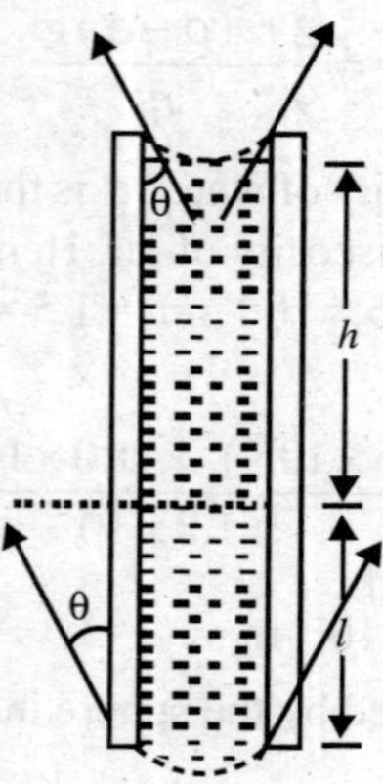

Fig. S-4.8

3. The excess pressure inside a soap bubble is inversely proportional to its radius. Hence, the pressure inside A is greater than the pressure inside (b) Air will flow from A to (b) therefore, A will become smaller and B will become larger.
4. If drop of radius R is sprayed into n droplets of equal radius r, then as a drop has only one surface, the initial surface area will be $4\pi R^2$ while final area is $n(4\pi r^2)$. So the increase in area

$$\Delta S = n(4\pi r^2) - 4\pi R^2$$

So energy expended in the process,

$$W = T\Delta S = 4\pi T\,[nr^2 - R^2] \quad \text{(i)}$$

Now since the total volume of n droplets is same as that of initial drop, i.e.,

$$\frac{4}{3}\pi R^3 = n[(4/3)\pi r^3]$$

or $$r = R/n^{1/3} \quad \text{(ii)}$$

Putting the value of r from Eq. (ii) in (i)

$$W = 4\pi R^2 T[(n)^{1/3} - 1].$$

5. Let n be the number of little droplets.
Since volume will remain constant, hence volume of n little droplets = volume of single drop

$$\therefore n\times\frac{4}{3}\pi r^3 = \frac{4}{3}\pi R^3 \quad \text{or} \quad nr^3 = R^3$$

Decrease in surface area = $n \times 4\pi r^2 - r\pi R^2$

or $$\Delta A = 4\pi\,[nr^2 - R^2] = 4\pi\left[\frac{nr^3}{r} - R^2\right]$$

$$= 4\pi\left[\frac{R^3}{r} - R^2\right] = 4\pi R^3\left[\frac{1}{r} - \frac{1}{R}\right]$$

Energy evolved $W = T \times$ decrease in surface area

$$= T\times4\pi R^3\left[\frac{1}{r} - \frac{1}{R}\right]$$

Heat produced, $Q = \frac{W}{J} = \frac{4\pi T R^3}{J}\left[\frac{1}{r} - \frac{1}{R}\right]$

But $$Q = ms\,d\theta$$

where m is the mass of big drop, s is the specific heat of water and $d\theta$ is the rise in temperature.

$$\therefore \frac{4\pi T R^3}{J}\left[\frac{1}{r} - \frac{1}{R}\right] = \text{volume of big drop} \times \text{density of water} \times \text{specific heat of water} \times d\theta$$

or, $$\frac{4}{3}\pi R^3 \times 1 \times 1 \times d\theta = \frac{4\pi T R^3}{J}\left(\frac{1}{r} - \frac{1}{R}\right)$$

or, $$d\theta = \frac{3T}{J}\left[\frac{1}{r} - \frac{1}{R}\right]$$

6. The surface tension of the liquid is

$$T = \frac{rh\rho g}{2}$$

$$= \frac{(0.025\,\text{cm})\,(3.0\,\text{cm})\,(1.5\,\text{gm/cm}^3)\,(980\,\text{cm/sec}^2)}{2}$$

$= 55$ dyne/cm

Hence, excess pressure inside a spherical bubble

$$\Delta p = \frac{4T}{R} = \frac{4 \times 55\,\text{dyne/cm}}{(0.5\,\text{cm})} = 440\ \text{dyne/cm}^2$$

7. Pressure inside the film is less than outside by an amount

$$P = T\left[\frac{1}{r_1} + \frac{1}{r_2}\right],$$

where r_1 and r_2 are the radii of curvature of the meniscus. Here $r_1 = t/2$ and $r_2 = \infty$, then the force required to separate the two glass plates, between which a liquid film is enclosed, is

$$F = P \times A = \frac{2AT}{t},$$

where t is the thickness of the film, A = area of film.

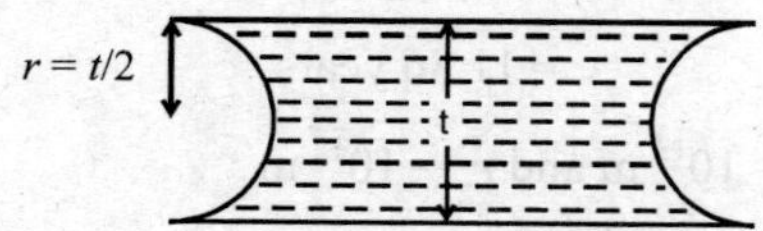

Fig. S-4.9

$$F = \frac{2A^2T}{At} = \frac{2A^2T}{V} = \frac{2 \times (40 \times 10^{-4})^2 \times (70 \times 10^{-3})}{0.05 \times 10^{-6}}$$

$= 45$ N

8. Let l be the length of the tube inside water. The forces acting on the tube are:

(i) Upthrust of water acting upward

$= \pi\rho^2 l \times 1 \times 980$

$$= \frac{22}{7} \times (0.14)^2\, l \times 980 = 60.368\ l\ \text{dyne}.$$

(ii) Weight of the system acting downwards

$= mg = 0.2 \times 980 = 196$ dyne

(iii) Force of surface tension acting downward

$= 2\pi\rho T$

$= 64.24$ dyne

Since the tube is in equilibrium, the upward force is balanced by the downward forces. That is,

$$60.368\ l = 196 + 64.24$$

$$= 260.24$$

$$l = 4.31\ \text{cm}$$

9. Given that surface tension of water,

$$T = 75\ \text{dyne/cm}$$

Radius $r = 0.1/2\ \text{mm} = 0.05\ \text{mm} = 0.005$ cm,

density $\rho = 1$ gm/cm³, angle of contact, $\theta = 0°$.

Let h be the height to which water rise in the capillary tube. Then

$$h = \frac{2T\cos\theta}{r\rho g} = \frac{2 \times 75 \times \cos 0°}{0.005 \times 1 \times 981}\ \text{cm} = 30.58\ \text{cm}.$$

But length of capillary tube, $h' = 5$ cm

(i) Because $h > h'/2$ therefore the first possibility does not exist.

(ii) Because the tube is of insufficient length, therefore the water will rise up to the upper end of the tube.

(iii) The water will not overflow out of the upper end of the capillary. It will rise only up to the upper end of the capillary.

The liquid meniscus will adjust its radius of curvature R' in such a way that

$$R'h' = Rh \qquad \left[hR = \frac{2T}{\rho g} = \text{constant}\right]$$

where R is the radius of curvature that the liquid meniscus would possess if the capillary tube were of sufficient length

$$R' = \frac{Rh}{h} = \frac{rh}{h'}$$

$$\left[R = \frac{r}{\cos\theta} = \frac{r}{\cos 0°} = r\right]$$

$$= \frac{0.005 \times 30.58}{5} = 0.0306\ \text{cm}$$

Chapter 5

Exercise 5.1

1. Gravitational force plays an important role for initiating the birth of stars, for controlling the entire structure of the universe and evolution of the universe. It helps to explain many natural phenomena.

2. It is so because the gravitational attraction of earth provides the necessary centripetal force to the moon for its orbital motion around the earth, due to which the moon is revolving around the earth.

3. We know that (time period)$^2 \propto$ (orbital radius)^3. Therefore, time period is larger in case of a satellite revolving away from the surface of earth.

4. If gravity suddenly disappears,
   (i) all bodies will lose their weights;
   (ii) we shall be thrown away from the surface of earth due to centrifugal force;
   (iii) the motion of planets around the sun will cease because centripetal force shall not be provided;
   (iv) motion of the satellite around earth will also be not possible as no centripetal force will be provided.

5. Let M be the mass of each copper sphere; and ρ be the density of copper. Then

$$M = \frac{4}{3}\pi R^3 \rho$$

Gravitational attraction between the two spheres is

$$F = \frac{GMM}{(2R)^2} = \frac{GM^2}{4R^2}$$

$$= \frac{G}{4R^2} \times (\tfrac{4}{3}\rho R^3 \rho)^2 = \tfrac{4}{9}\pi^2 G\rho^2 R^4$$

Hence, $F \propto R^4$ if ρ is constant.

6. Let $m_1 = m$, then $m_2 = M - m$. Gravitational force of attraction between the two masses when placed distance r apart will be

$$F = \frac{Gm\,(M-m)}{r^2}$$

Differentiating it w.r.t. m, we get

$$\frac{dF}{dm} = \frac{G}{r^2}\left[m\frac{d}{dm}(M-m) + (M-m)\frac{dm}{dm}\right]$$

$$\frac{G}{r^2}[m(-1) + M - m] = \frac{G}{r^2}(M - 2m)$$

If F is maximum, then $\frac{dF}{dm} = 0$

$$\Rightarrow \frac{G}{r^2}(M - 2m) = 0$$

or $M = 2m$ or $m = \frac{M}{2}$

7. Here, $F = K R^{-5/2} = mR\left(\frac{2\pi}{T}\right)$

or $$T^2 = \frac{4\pi^2 m R}{K R^{-5/2}} = \frac{4\pi^2 m}{K} R^{7/2}$$

Hence, $T^2 \propto R^{7/2}$

8. Here, semimajor axis of the elliptical orbit of the planet around the sun is

$$r = \frac{(r_1 + r_2)}{2}$$

Therefore, $T^2 \propto \left(\frac{r_1 + r_2}{2}\right)^3$

or $T \propto \left(\frac{r_1 + r_2}{2}\right)^{3/2}$

or $T \propto (r_1 + r_2)^{3/2}$

9. $\frac{GM\,m}{R^n} = mR\left(\frac{\pi}{T_2}\right)^2$

or $T^2 = \frac{R \times 4\pi^2 \times R^n}{GM} = \frac{4\pi^2 R^{(n+1)}}{GM}$

or $T = \frac{2\pi}{\sqrt{GM}} R^{(n+1)/2}$

or $T \propto R^{(n+1)/2}$

10. Here, $r_j = 5.2\, r_e$

$T_e = 1$ year

$T_j = ?$

Now, $\frac{T_j^2}{T_e^2} = \frac{r_j^3}{r_e^3}$

or $T_j = T_e\left(\frac{r_j}{r_e}\right)^{3/2} = 1 \times \left(\frac{5.2\, r_e}{r_e}\right)^{3/2}$

$= 11.86$ years

11. Here, $r_1 = 10^{13}$ m and $r_2 = 10^{12}$ m

$\frac{T_1}{T_2} = ?$

Now, $\frac{T_1^2}{T_2^2} = \frac{r_1^3}{r_2^3}$

$\frac{T_1}{T_2} = \left(\frac{r_1}{r_2}\right)^{3/2} = \left(\frac{10^{13}}{10^{3/2}}\right)^{3/2} = 10^{3/2}$

Now, $v = \frac{2\pi r}{T}$

Hence $\frac{v_1}{v_2} = \frac{r_1/r_2}{T_1/T_2} = \frac{10}{10^{3/2}} = \frac{1}{\sqrt{10}}$

12. Given $k = 10^{-13}\ \text{s}^2\ \text{m}^{-3}$

$$= 10^{-3}\left[\frac{1}{(24 \times 60 \times 60)^2} d^2\right]\left[\frac{1}{(1/1000)^3\ \text{km}^3}\right]$$

$$= 1.33 \times 10^{-14} d^2 \text{km}^{-3}$$

$T^2 = KR^3$ or $T = (KR^3)^{1/2}$

$\therefore$ $T = [1.33 \times 10^{-14} \times (3.84 \times 10^5)^3]^{1/2}$

$= 27.3d$

13. Here, $m_1 = 40$ kg, $m_2 = 60$ kg

$F = 4\, mgf = 4 \times 10^{-6} \times 10$

$= 4 \times 10^{-5}$ N

$G = 6 \times 10^{-11} \text{Nm}^2 \text{kg}^{-2}, r = ?$

Now, $F = G\dfrac{m_1 m_2}{r^2} \Rightarrow r = \sqrt{\dfrac{Gm_1 m_2}{F}}$

$= \sqrt{\dfrac{6 \times 10^{-11} \times 40 \times 60}{4 \times 10^{-5}}} = 0.06 \text{ m} = 6 \text{ cm}$

Exercise 5.2

1. As $g = \dfrac{GM}{R^2} = \dfrac{G}{R^2} \times \dfrac{4}{3}\pi R^3 \rho = \dfrac{4}{3}\pi G R \rho$, so $g \propto R\rho$

$\therefore \dfrac{g_1}{g_2} = \dfrac{R_1 \rho_1}{R_2 \rho_2}$

2. (a) The variation of g with height h is related by $g \propto a/r^2$ where $r = R + h$, i.e., part AB of the graph in the figure.

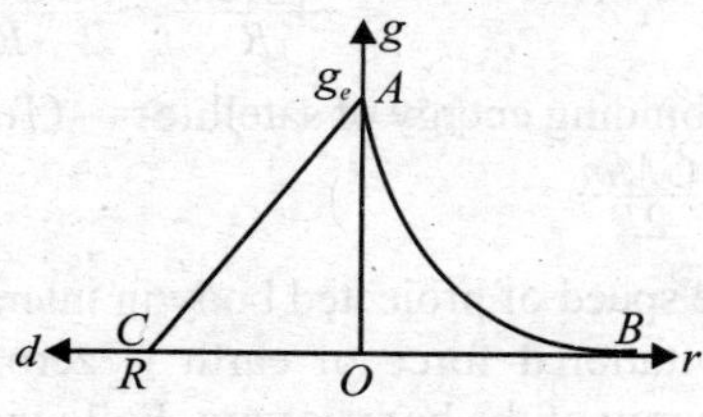

Fig. S-5.1

(b) The variation of g with depth is related by $g' = g(1 - d/R)$, i.e., $g \propto (R - d)$. Thus the variation of g and d is a straight line, i.e., part AC of the graph in the figure.

3. Since $g = GM/R^2$, if mass M and radius R are doubled, then g will reduce to half. Since time period $t = 2\pi\sqrt{l/g}$ so $t \propto \sqrt{g}$. Therefore, time period of a pendulum will be $\sqrt{2}$ times that on the earth. Thus the time period of seconds pendulum of the earth will be $2\sqrt{2}$ on the planet.

4. The value of g is larger at the poles than at the equator. If sugar is weighed in a physical balance, then there will be no difference. If it is weighed by a spring balance, calibrated at the equator, then 1 kg of sugar will have a lesser amount at poles.

5. When $r_1 > R$, the point lies outside the sphere. Then sphere can be considered to be a point mass body whose whole mass can be supposed to be concentrated at its centre. The gravitational intensity at a point distance r_1 from the centre of sphere will be

$$I_1 = GM/r_1^2 \quad \text{(i)}$$

When $r_2 < R$, the point P lies inside the sphere. The unit mass body placed at P will experience gravitational pull due to sphere of radius r_2, whose mass is

$$M = \frac{M \times \frac{4}{3}\pi r_2^3}{\frac{4}{3}\pi R^3} = \frac{M r_2^3}{R^3}$$

Therefore, the gravitational intensity at P will be

$$I_2 = \frac{GM r_2^3}{R^3} \times \frac{1}{r_2^2} = \frac{GM r_2}{R^3} \quad \text{(ii)}$$

So, $\dfrac{I_1}{I_2} = \dfrac{GM}{r_1^2} \times \dfrac{R^3}{GM r_2} = \dfrac{R^3}{r_1^2 r_2}$

6. When a person is sitting in an artificial satellite of earth, the gravitational pull on the person due to earth, i.e., the weight of the person due to earth is counterbalanced by the centrifugal force acting on the person. Because of this the net force acting on the person in satellite is zero and the person feels weightlessness. But when the person is standing on the moon, the gravitational pull on the person due to moon (which is left unbalanced) is responsible for his weight on the moon.

7. The weight of the body is the force with which the body is attracted by the earth towards its centre. Quantitatively, the weight of a body of mass m is equal to mg, where g is the acceleration due to gravity. At the centre of earth, $g = 0$, so weight of a body is zero at the centre of the earth.

8. No. The acceleration due to gravity, that is, gravitation as intensity at points outside the surface may be calculated by assuming the mass to be concentrated at the centre of the sphere. Hence, the distribution of mass over the earth does not affect the variation of g with height.

9. The tide-raising power of a body (sun or moon) is determined by the difference in the gravitational attraction between the water on the far and near sides, not by the total pull. The tide-producing force varies inversely as the cube of the distance. Thus, the relatively short distance separating the earth and the moon makes the moon's tide-raising force more than double that of the distant sun.

10. Mass effect dominates over the distance effect. We know that

$$F = \frac{GMm}{R^2}, \; M \propto R^3$$

Hence, $\quad F \propto R$

Hence, mass effect predominents.

11. Due to the rotation of the earth from west to east the stone will deviate slightly to the east.

12. $g = g - R\omega^2 \cos^2 \lambda$

At equator $\quad \lambda = 0, \cos\lambda = 1$

$g = g - R\omega^2$

$\Delta g = g - g = R\omega^2$

$= 6.67 \times 10^6 \times (7.27 \times 10^{-5})^2$

$= 3.36 \times 10^{-2} \text{ m/s}^2 = 3.36 \text{ cm/s}^2$

13. $g = \dfrac{g}{\left(1 + \dfrac{h}{R}\right)^2}$

Given, $\quad g = \dfrac{10}{100} g = \dfrac{g}{10}$

$$\frac{g}{10} = \frac{g}{\left(1 + \dfrac{h}{R}\right)^2}$$

$$1 + \frac{h}{R} = \sqrt{10} = 3.16$$

$$\Rightarrow \quad \frac{h}{R} = 2.16$$

$$\Rightarrow \quad h = 2.16\,R = 2.16 \times 6.37 \times 10^6 = 1.37 \times 10^7 \text{ m}$$

14. (a) (i) At height h, $g = \dfrac{g}{\left(1+\dfrac{h}{R}\right)^2}$

$$\Rightarrow \quad \frac{h}{R} = \frac{1600 \text{ km}}{6400 \text{ km}} = \frac{1}{4}$$

$$\therefore \quad g = \frac{g}{\left(\dfrac{1+1}{4}\right)^2} = \frac{16}{25}g = 9.8 = 6.27 \text{ m/s}^2$$

(ii) At depth x,

$$g_x = g\left(\frac{1-x}{R}\right) = g\left(1 - \frac{1600 \text{ km}}{6400 \text{ km}}\right)$$

$$= g \times \frac{3}{4} = 7.35 \text{ m/s}^2$$

(b) (i) At distance r from centre of earth above the earth's surface,

$$g = \frac{GM}{r^2}$$

$$\Rightarrow \quad \frac{dg}{dr} = -\frac{2}{r^3}GM$$

As $r = R + h$, $dr = dh$, $\dfrac{dg}{dh} = -\dfrac{2GM}{(R+h)^3}$

(ii) At distance x below earth's surface,

$$g = g\left(\frac{1-x}{R}\right)$$

$$\frac{dg}{dx} = -\frac{g}{R} = -\left(\frac{GM}{R^2}\right)\frac{1}{R} = -\frac{GM}{R^3}$$

$$\frac{G\frac{4}{3}\pi R^3\rho}{R^3} = -\frac{4}{3}\pi G\rho$$

15. We know that $\omega = \dfrac{2\pi}{24h} = \dfrac{2\pi}{24 \times 60 \times 60}\text{ s}^{-1}$

$$W = W - mR\omega^2 \text{ (at equator)}$$

When bodies are weightless $W' = 0$ and $W = mg$.

Therefore, $\quad 0 = mg - mR\omega^2$

$$\Rightarrow \quad \omega = \sqrt{\frac{g}{R}}$$

$$\therefore \quad \frac{\omega'}{\omega} = \frac{(9.78 \;\; 6.37 \times 10^6)^{1/2}}{(2\pi \;\; 24 \times 60 \times 60)} \approx 17$$

Exercise 5.3

1. As escape speed is given by $\sqrt{GM/R}$, therefore its values are different for different planets because of their different masses and different size.

2. The ability of a body (planet) to hold the atmosphere depends on: (i) acceleration due to gravity and (ii) surface temperature.

3. Moon has no atmosphere because the value of acceleration due to gravity 'g' on surface of the moon is small. Therefore, the value of escape speed on the surface of the moon is small (only 2.5 km s^{-1}). The molecules of the atmospheric gases on the surface of the moon have thermal speeds greater than the escape speed. That is why all the molecules of gases have escaped and there is no atmosphere on the moon.

4. The minimum energy required to free a satellite from the gravitational attraction is called binding energy. Binding energy is the negative value of total energy of satellite. Let a satellite of mass m be revolving closely around earth of mass M and radius R.

$$\text{Total energy of satellite} = \text{PE} + \text{KE} = \frac{-GMm}{R} + \frac{1}{2}mv^2$$

$$= \frac{-GMm}{R} = \frac{M}{2}\frac{GM}{R} = \frac{-GMm}{2R}$$

Therefore, binding energy of satellite = – (Total energy of satellite) $= \dfrac{GMm}{2R}$

5. Let v be the speed of projected body in interstellar space, where gravitational force of earth is zero. Hence, the potential energy of the body is zero. Following the law of conservation of energy, we have

$$\frac{1}{2}mv^2 + \left(\frac{-GMm}{R}\right) = \frac{1}{2}m(v')^2 + 0$$

$$(v')^2 = v^2 - \frac{2GM}{R} = v^2 - v_e^2 \; (\because v_e = \sqrt{2GM/R})$$

$$v = \sqrt{v^2 - v_e^2}$$

6. $v_e \sqrt{\dfrac{2GM}{R}} = \sqrt{\dfrac{2G}{M} \times \dfrac{4}{3}\pi R^3\rho} = R\sqrt{\dfrac{8\pi}{3}GK\rho}$

$$\frac{v_p}{v_e} = \frac{R_p}{R_e} = \frac{2R_e}{R_e} = 2 \quad \text{or } v_p = 2v_e$$

7. We know that the value of escape velocity of a particle from the surface of earth is given by

$$v_e = \sqrt{2GM/R}$$

So, the minimum kinetic energy needed for the particle to escape is given by

$$\frac{1}{2}mv_e^2 = \frac{1}{2}m \times \frac{2GM}{R} = \frac{GMm}{R}$$

As per question, the kinetic energy given to the particle is $(GMm/2R)$.

If h is the height attained by the particle from the surface of the earth, then its potential energy is $-GMm/(R+h)$.

According to the law of conservation of total energy,

KE + PE on the surface of earth = PE at height h.

$$\frac{1}{2}\frac{GMm}{R}+\left(\frac{-GMm}{R}\right)=\frac{GMm}{R+h}\Rightarrow h=R$$

or $$\frac{GMm}{2R}=-\frac{GMm}{R+h}\Rightarrow h=R$$

8. Magnitude of the potential energy per unit mass of the object at the surface of earth is

$$\frac{GM}{R}=E\text{ (given)}$$

Escape velocity, $v=\sqrt{\frac{2GM}{R}}=\sqrt{2E}$

9. Yes. The gravitational force is conservative. When a system is under the action of conservative forces, its total energy (kinetic + potential) is constant and hence, the energy of the solar system is constant. The gravitational forces on planets of the solar system are central forces directed towards the sun and hence, the moment of forces about the sun is zero. Thus, no torque acts on the planets and the angular momentum is constant.

10. The meaning of the total energy being negative is that the system (sun + earth) is a closed one, the earth is always bound to the attracting solar centre and never escapes from it.

11. E (total energy) = $\frac{1}{2}mv^2+\left(-\frac{GMm}{r}\right)$

where m is the mass of the satellite, M = mass of the earth and r = radius of the orbit. By the dynamics of circular motion,

$$\frac{GMm}{r^2}=\frac{mv^2}{r}$$

or $$v^2=\frac{GM}{r}$$

Hence, $$E=\frac{-1}{2}\frac{GM}{r}m=-\frac{1}{2}mv^2$$

$$=-k\text{ (kinetic energy)}$$

$$E_i\text{ (initial energy)}=-\frac{1}{2}\frac{GM}{r_0}$$

$$E_f\text{ (final energy)}=-\frac{1}{2}\frac{GM}{r}$$

By the principle of conservation of energy,

$$E_i=E_f+W\text{ (work done by frictional force)}$$

$$\Rightarrow\quad -\frac{1}{2}\frac{GM}{r_0}=-\frac{1}{2}\frac{GM}{r}+W$$

or $$\frac{1}{2}GM\left(\frac{1}{r}-\frac{1}{r_0}\right)=W\text{ (a positive quantity)}$$

$\therefore$ $$\frac{1}{r}>\frac{1}{r_0}\quad\text{or}\quad r<r_0$$

Also, $$K_0=-K+W$$

or $$K-K_0=W\text{ (a positive quantity)}$$

$\therefore$ $$K<K_0$$

12. No, the velocity of escape arises from the energy required to take it from the present condition to infinity. This is equal to the difference between energy on the surface of the earth and infinity. For a particle at rest,

$$\frac{1}{2}mv_e^2=\frac{GM}{R}\text{ (only potential energy)}$$

$\therefore$ $$v_e=\sqrt{\frac{2GM}{R}}$$

For a particle in circular motion,

$$E\text{ (energy)}=-\frac{GM}{R}+\frac{1}{2}mv^2$$

But $$\frac{mv_c^2}{R}=-\frac{GMm}{R^2}$$

(from dynamics of circular motion)

$\therefore$ $$E=-\frac{1}{2}\frac{GMm}{R}$$

$\therefore$ $$\frac{1}{2}mv_e^2=\frac{1}{2}\frac{GMm}{R}$$

or $$v_e=\sqrt{\frac{GM}{R}}$$

13. Conservation of energy implies

$$0+0=\frac{1}{2}mv^2-\frac{GMm}{R}$$

$\Rightarrow$ $$v=\sqrt{\frac{2GM}{R}}=\sqrt{\frac{2(R^2g)}{R}}$$

$$=\sqrt{2Rg}\approx 11.2\times 10^3\text{ m/s}$$

14. $\frac{1}{2}mv^2-\frac{GMm}{R}=0-\frac{GMm}{R+h}$

If we put $GM=R^2g$, we get

$$h=\frac{v^2R}{2gR-v^2}=2.5\times 10^4\text{ km}$$

15. The potential energy of satellite is given by

$$U=-\frac{GMm}{r}\propto-\frac{1}{r}$$

This is negative. It means that larger the radius, greater is the potential energy.

$$\frac{U_1}{U_2}=\frac{-r_2}{-r_1}=\frac{-(R+h_2)}{-(R+h_1)}$$

$$=\frac{-(6400+19200)}{-(6400+6400)}=\frac{-2}{-1}$$

Kinetic energy of satellite

$$k = \frac{GMm}{r^2}$$

Hence $\frac{k_1}{k_2} = \frac{r_2}{r_1} = \frac{2}{1}$

Total energy $E = \text{KE} + \text{PE} = -\frac{GMm}{2r}$

$\therefore$ $\frac{E_1}{E_2} = \frac{-r_2}{-r_1} = \frac{-2}{-1}$

As $-1 > -2$, satellite B has greater value of total energy.

16. $v_e = \sqrt{\frac{2GM}{R}} = \sqrt{\frac{2G\left(\frac{4}{3}\pi R^3\rho\right)}{R}}$

$$R = v_e\sqrt{\frac{3}{8\pi G\rho}} = 30894 \text{ m} = 31 \text{ km}$$

Exercise 5.4

1. Following are the basic conditions:
   (i) The rocket must take the satellite to a suitable height above surface of earth.
   (ii) From the desired height, the satellite must be projected with a suitable speed, called the orbital speed.
   (iii) In the orbital path of satellite, the air resistance should be negligible so that its speed does not decrease and it does not burn due to the heat produced.
2. The relation for orbital speed is $v = \sqrt{GM/(R+h)}$, where h is the height of the satellite above the earth's surface. Clearly, the smaller is the value of h, the greater is the value of v and vice versa. Hence, satellite revolving close to earth has larger speed.
3. Yes, it is so because the orbital speed of a satellite is independent of the mass of a satellite. Therefore, the speeds of the artificial satellites of different masses but of the same orbital radius will be the same.
4. No, this is because inside the satellite, there is no gravity efficiency, i.e., $g = 0$. As $t = 2\pi\sqrt{l/g}$, hence $t = \infty$ for $g = 0$. Thus the pendulum will not vibrate.
5. If a satellite of mass m is revolving in a circular orbit of radius r, with speed v around the earth of mass M, the centripetal force is provided by the gravitational pull, i.e.,

$$\frac{mv^2}{r} = \frac{GM\,m}{r^2}$$

or $mv^2r = GM\,m$

or $(mvr)v = \text{GM}\,m = \text{constant}$

or $Lv = \text{constant}$

where $mvr = L$ is the angular momentum of the satellite. As the air friction causes the retarding torque, it will decrease the angular momentum L of the satellite. Because of this the velocity of the satellite v increases in order to keep Lv = a constant.

6. We can learn more about the shape of the earth by studying the motion of an artificial satellite than by studying the motion of the moon because the motion of the satellite is governed by the value of g at each of its positions. Since the moon is very far away, the variation of g near the surface of the earth is averaged out at the position of the moon.
7. No, they are not in 'orbit' like an artificial satellite is. Because the whole of the gravitational pull on the artificial satellite provides the necessary centripetal force, but for objects at rest on the earth's surface, the whole of the gravitational pull (i.e., their weights) does not provide the necessary centripetal force; only a small fraction provides the necessary centripetal force. In short, bodies in 'orbit' are weightless but bodies on the earth at rest are not weightless.
8. No. For stability of rotational motion there should not be any torque produced by the force acting on it about the centre of the orbit. When the orbit does not lie in a plane passing through the centre, the gravitational pull will have a moment (torque) about the centre of the orbit.
9. The westward satellite will appear to have a shorter period than the eastward satellite, owing to the rotation of the earth itself from west to east.
10.

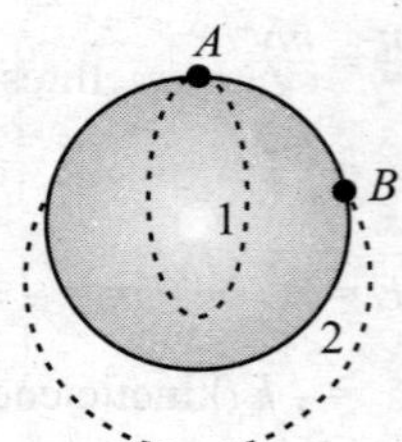

Fig. S-5.2

If the container is thrown in the direction opposite to the motion of the satellite, it will begin to move along an ellipse (say 1) lying inside the circular orbit of the satellites. The period of revolution will be slightly less than that of the satellites. Therefore, they can meet at the point of contact of the orbits (circular orbit of satellite and the elliptical orbit of the container) only after a great number of revolution. On the other hand, if the container is thrown in the direction of the satellite, the container begins to move along an ellipse like 2 in the figure. The velocity u should be such that during one revolution of the container, satellite B also makes one revolution and in addition covers the distance AB. So the container should be thrown in the direction of the first satellite.

11. The feeling of weight arises from the reaction of the ground on a man. In an orbiting satellite, the astronaut and the floor of the satellite both have the same acceleration (the centripetal acceleration towards the earth). Hence, the floor of the satellite offers no reaction to the astronaut and he feels weightless.

For circular motion of man,

$$\frac{GM \times m}{r^2} - N = \frac{mv^2}{r}$$

For circular motion of 'satellite + man' (mass m),

$$\frac{GMm}{r^2} = \frac{m\,v^2}{r} \Rightarrow \frac{GM}{r^2} = \frac{v^2}{r}$$

$$\therefore \quad \frac{mv^2}{r} - N = \frac{mv^2}{r} \Rightarrow N = 0$$

12. $T = 2\pi\sqrt{\frac{(R_e + h)^3}{GM_e}} = 2\pi\sqrt{\left\{\frac{\left(R_e + \frac{R_e}{2}\right)^3}{R_e^2 g}\right\}} = 2\pi\sqrt{\frac{27R_e}{8g}}$

$$= 2 \times 3.14\sqrt{\frac{27 \times 6.38 \times 10^8}{8 \times 9.8}} \approx 9308 \text{ s}$$

13. $T = 2\pi\sqrt{\frac{R^3}{GM}} = 2\pi\sqrt{\left\{\frac{R^3}{\left(G\frac{4}{3}\pi R^3\rho\right)}\right\}} = \sqrt{\frac{3\pi}{G\rho}}$

$$= \sqrt{\frac{3 \times 3.14}{6.67 \times 10^{-11} \times 8000}} \approx 4202 \text{ s}$$

14. Initial total energy of each satellite $= -\frac{GMm}{2r}$

Hence, total energy of both satellites

$$2 \times \left(-\frac{GMm}{2r}\right) = \frac{GMm}{r}$$

When satellites travelling in opposite directions collide, the final velocity v is given by

$$(2mv) = mv - mv = 0$$

$$\Rightarrow \quad v = 0$$

Therefore, the wreckage comes to rest and it will fall freely under gravity.

After collision, total energy of satellite is

$$2\left(-\frac{GMm}{r}\right) = \left(-\frac{2GMm}{r}\right)$$

15. (i) $\rho = \frac{M}{\frac{4}{3}\pi R^3}$

$$\frac{\rho_1}{\rho_2} = \frac{M_1}{M_2}\left(\frac{R_2}{R_1}\right)^2 = 0.605$$

(ii) $fg = -\frac{GM}{R^2}$

$$\frac{f_1}{f_2} = \frac{M_1}{M_2}\left(\frac{R_2}{R_1}\right)^2 = \frac{1}{11} \times \left(\frac{79}{42}\right)^2$$

(iii) $v_e = \sqrt{\frac{2GM}{R}}$

$$\frac{v_1}{v_2} = \sqrt{\frac{M_1 R_2}{M_2 R_1}} = \sqrt{\frac{1}{11} \times \frac{79}{42}} = 0.4135$$

(iv) Period near surface of planet is

$$T = 2\pi\sqrt{\frac{R^3}{GM}}$$

$$\frac{T_1}{T_2} = \sqrt{\left(\frac{R_1}{R_2}\right)^3 \frac{M_2}{M_1}} = \sqrt{\left(\frac{42}{79}\right)^3} \times \frac{11}{1} = 1.286$$

16. In elliptical orbit, total energy

$$E = -\frac{GMm}{2a} = \text{constant}$$

At position r, if v is orbital speed of satellite, then

$$\text{KE} = \frac{1}{2}mv^2 \text{ and PE} = \frac{GMm}{r}$$

$$\therefore \quad \frac{1}{2}mv^2 - \frac{GMm}{r} = -\frac{GMm}{2a} \Rightarrow v^2 = GM\left(\frac{2}{r} - \frac{1}{a}\right)$$